2009 International Conference on Electronic Packaging Technology & High Density Packaging

(ICEPT-HDP 2009)

Beijing, China
10 – 13 August 2009

Pages 1-636

Editors:

Keyun Bi
Jian Cai

IEEE Catalog Number: CFP09553-PRT
ISBN: 978-1-4244-4658-2

Copyright © 2009 by the Institute of Electrical and Electronic Engineers, Inc
All Rights Reserved

Copyright and Reprint Permissions: Abstracting is permitted with credit to the source. Libraries are permitted to photocopy beyond the limit of U.S. copyright law for private use of patrons those articles in this volume that carry a code at the bottom of the first page, provided the per-copy fee indicated in the code is paid through Copyright Clearance Center, 222 Rosewood Drive, Danvers, MA 01923.

For other copying, reprint or republication permission, write to IEEE Copyrights Manager, IEEE Service Center, 445 Hoes Lane, Piscataway, NJ 08854. All rights reserved.

This publication is a representation of what appears in the IEEE Digital Libraries. Some format issues inherent in the e-media version may also appear in this print version.

IEEE Catalog Number: CFP09553-PRT
ISBN 13:   978-1-4244-4658-2
Library of Congress No.: 2009904680

Additional Copies of This Publication Are Available From:

Curran Associates, Inc
57 Morehouse Lane
Red Hook, NY 12571 USA
Phone: (845) 758-0400
Fax: (845) 758-2633
E-mail: curran@proceedings.com

TABLE OF CONTENTS

A Novel MEMS Package with Three-Dimensional Stacked Modules 1
G. Xu, Q. Huang, W. Ning, Z. Ruan, L. Luo

Fabrication and Characterization of a Novel Wafer-level Chip Scale Package for MEMS Devices 5
Y. Cao, L. Luo

Study on a 3D Packaging Structure with Benzocyclobutene as a Dielectric Layer for Radio Frequency Application 9
F. Gang, X. Ding, L. Luo

Through-Silicon Via Filling Process Using Pulse Reversal Plating 15
X. Yang, H. Ling, D. Ding, M. Li, X. Yu, D. Mao

Low Temperature Cu-Sn Bonding by Isothermal Solidification Technology 20
Y. Rong, J. Cai, S. Wang, S. Jia

A PCB-based Package Structure 23
N. Wang, J. Cai, W. Yang, X. Dou

Collaborative Effect between Additives and Current in TSV Via Filling Process 27
K. Zou, H. Ling, Q. Li, H. Cao, X. Yu, M. Li, D. Mao

Comparison Study of Effective Power Delivery in Advanced Substrate Technologies for High Speed Networking Applications 31
J. Priest, R. Pomerleau, J. Savic, P. Aria, J. Xue

The Design and Fabrication of Multilayer Chip LC Filter with a Modified Structure by LTCC Technology 36
Y. Liu, H. Zhang, Y. Li

The Influence of SO_2 Environments on Immersion Silver Finished PCBs by Mixed Flow Gas Testing 40
S. Zhang, A. Shrivastava, M. Osterman, M. Pechr, R. Kang

Study on Reliability of Embedded Passives in Organic Substrate 47
W. Qiu, K. Pan, C. Yuan, J. Wang

Investigation on PCB Related Failures in High-Density Electronic Assemblies 52
L.N. Lu, H.Z. Huang, X.X. Su, B.Y. Wu, M. Cai

Nanoscale Mechanical Properties and Microstructure of 3D LTCC Substrate 57
Y. Zhang, J. Chen, S. Bai, M. Miao, J. Zhang, F. Mu, Z. Wang, S. Xia, Y. Jin

Process Research of LTCC Substrate with 3D Micro-channel Embedded 61
F. Mu, Z. Wang, Y. Zhang

Study of the Fine Line Process and Signal Integrity for Packaging Substrate 65
Y. Ma, J. Jiang, Z. Yang, M. Lu

Effects of Nitrogen on Wettability and Reliability of Lead-free Solder in Reflow Soldering 71
M. Dong, Y. Wang, J. Cai, T. Feng, Y. Pu

Solderability of Flash Gold Surface Finish 76
Y. Wang, M. Dong, J. Cai

Thermal Analysis and Testing of Multi Layer Ceramic-Metal Packaged LED 81
P. Jin, Q. Zhou, N. Wu, Q. Zhong

Multiscale Delamination Modeling of an Anisotropic Conductive Adhesive Interconnect Based on Micropolar Theory and Cohesive Zone Model 84
Y. Zhang, J. Fan, J. Liu

Finite Element Thermal Analysis for High Power Multi-chip Light Emitting Diode 88
L. Zhou, H. Chen, B. An, F. Wu, Y. Wu

A Design of an Optical Transceiver SiP 93
W. Gao, L. Wan, B. Li, J. Song, X. Zhang

Power Supply Analysis in Package and SiP Design 99
W. Dai

Effect of Fluid Dynamics and Device Mechanism on Biofluid Behaviour in Microchannel Systems: Modelling Biofluids in a Microchannel Biochip Separator 103
X. Xue, M.K. Patel, M. Kersaudy-Kerhoas, C. Bailey, M.P.Y. Desmulliez, D. Topham

Modeling of Manufacturing Processes of High Power Light Emitting Diodes: Wire Bonding Process 111
Z. Chen, Y. Liu, S. Liu

Advanced Package Design for Electronic and MEMS Applications Supported by FE Analyses and Deformation Measurements 117
J.P. Sommer, B. Michel, A. Kugler, H. Rank

A Novel Wafer Level Package Strategy for RF MEMS 122
    Z. Wang, Z. Liu

Design and Simulation of a Package Solution for Millimeter Wave MEMS Switch 125
    W. Yu, X. Liao

Packaging of Light Emitting Diodes for Display Backlights: Need for Multi-Disciplinary Design Tools 129
    Y.B. Lee, N. Strusevich, C. Bailey, C. Yin

A Simplified Computational Model for Solder Joints under Drop Impact Loadings 133
    T. An, F. Qin

Design of Miniaturized Bandpass Filters for GSM and GPS Applications Using Embedded Capacitor Material 139
    Y. Wang, L. Li, L. Wan

Moiré Method for Nanoprecision Wafer-to-Wafer Alignment: Theory, Simulation and Application 143
    C. Wang, T. Suga

Localized Induction Heating for Wafer Level Packaging 149
    M. Chen, W. Liu, Y. Xi, C. Lin, S. Liu

Molecular Dynamics of Nanofluids with Time-Dependent Thermal Conductivity in Heat Sink Dissipation 154
    X. Wang, H. Liu, W. Zhang, Z. Li, L. Chen

Hydrophobic Self-Assembly Monolayer Structure for Reduction of Interfacial Moisture Diffusion 158
    H.B. Fan, C.K.Y. Wong, M.M.F. Yuen

Thermal Numerical Simulation for Advanced Package Development 162
    G. Gao, H. Wang, G. Yang, H. Zhu

FE Simulation of Size Effects on Interface Fracture Characteristics of Microscale Lead-Free Solder Interconnects 166
    B. Li, L. Yin, Y. Yang, X. Zhang

Thermal Design of Power Module to Minimize Peak Transient Temperature 172
    X. Cao, K.D.T. Ngo, G. Lu

A Study of the Heat Transfer Characteristics of the Micro-channel Heat Sink 179
    S. Wang, Y. Zhang, Y. Fu, J. Liu, X. Wang, Z. Cheng

Fracture Simulation of Solder Joint Interface by Cohesive Zone Model 184
    T. An, F. Qin

Optimization Design for Packaging Device QFN Using a Prediction Model of the Neural-Genetic Algorithm 191
    M. Cai, D. Yang, L. Niu, M. Zhao, W. Chen

Lifetime Prediction of an IGBT Power Electronics Module under Cyclic Temperature Loading Conditions 198
    H. Lu, C. Bailey

An In-Depth Numerical Investigation into Packaging Design of Multi-Finger GaInP/GaAs Collector-Up HBTs 204
    H.C. Tseng, J.Y. Chen

A Study of Power Delivery Network in a GPS SiP 207
    J. Li, L. Wan, C. Liao

Modeling Thermal Fatigue in Anisotropic Sn-Ag-Cu/Cu Solder Joints 212
    S. Yang, Y. Tian, C. Wang, T. Huang

Numerical Analysis of Response of Indium Micro-Joint to Low-Temperature Cycling 214
    X. Cheng, C. Liu, V.V. Silberschmidt

A Thermal Model for Calculating Thermal Resistance of Eccentric Heat Source on Rectangular Plate with Convective Cooling Existing at Upper and Lower Surfaces 218
    X. Luo, Z. Mao, S. Liu

Modeling of Heat Transfer Performance for an Array of Micro Jets Impinging upon Dimpled Surface 223
    T. Geng, J. Fan, Y. Zhang, S. Wang

Full Thermal Parametric Model for Power WL-CSP Design 228
    Z. Yuan, Y. Liu, S. Martin, L. England, B. Lee

An Improved Substructure Method for Prediction of Solder Joint Reliability in Thermal Cycle 235
    F. Liu, L. Liang, Y. Liu

Warpage Prediction of Fine Pitch BGA by Finite Element Analysis and Shadow Moiré Technique 241
    K. Xue, J. Wu, H. Chen, J. Gai, A. Lam

Prediction of Die Failure in Copper-Low-K Flip Chip Package with Consideration of Packaging Process-Induced Stresses 246
    M. Zhao, D. Yang, L. Niu

Simulation of Fluid Flow and Heat Transfer in Microchannel Cooling for LTCC Electronic Packages 251
    J. Zhang, Y.F. Zhang, M. Miao, Y.F. Jin, S.L. Bai, J.Q. Chen

Research of Methodologies to Enlarge the Isolation in 3D Interconnection 255
L. Zhao, L. Yang, X. Sun, Y. Jin

Study on EBG Structure Combined with Decoupling Capacitor for Suppressing Ground Bounce Noise 259
W. Zhu, W. Chen, L. Li

Parametric Study of Warpage in Package-on-Package Manufacturing 263
C. Ren, F. Qin

Computational Modeling and Optimization for Wire Bonding Process on Cu/Low-K Wafers 268
W. Huang

The Principal Component Analysis of Cu Stud Bump Shaping Process Parameters 277
W. Mu, Z. Wu, C. Huang

Effects of Shape Parameters on Tensile Strength of Cu Bump 282
C. Huang, Y. Liang, T. Li

A New Embedded Helix-Type Inductor Using LTCC Technology for High Frequency Applications 286
D. Liu, Y. Liu, Y. Li

Modeling Solder Joint Reliability of VFBGA Packages under Board Level Drop Test Based on Dynamic Constitutive Relation with Thermal Effect 289
X. Niu, T. Chen, Z. Li, X. Shu

Effects of PCB Dynamic Responses on Solder Joint Stress 294
T. An, F. Qin

Influence of Strain Rate Effect on Behavior of Solder Joints under Drop Impact Loadings 299
T. An, F. Qin, J. Li

Local Plastic Zones of Wirebond Profiles Inspection Base on Curvature Estimation 303
H. Zhou, L. Han

First-Principles Calculations of Structural, Thermodynamic and Electronic Properties of Intermetallic Compounds in Solder 308
Y. Yang, H. Lu, C. Yu, J. Chen

A Core-Shell Structure Viscoelastic Model of Particulate-Filled Electronic Packaging Polymers 312
D. Gui, J. Liu, B. Chen, X. Miao, G. Zeng, D. Tian

Numerical Simulation on Heat Pipe for High Power LED Multi-Chip Module Packaging 317
D. Li, G.Q. Zhang, K. Pan, X. Ma, L. Liu, J. Cao

Board Level Drop Impact Reliability Analysis for Compliant Wafer Level Package through Modeling Approaches 322
C. Yuan, K.L. Pan, W. Qiu, J. Liu

FEM Study on the Effects of Flip Chip Packaging Induced Stress on MEMS 327
S. Wei, J. Tang, J. Song

Modeling and Simulation of SSN on FPGA Products 331
L. Ke, P. Zhou, L. Li

Coupled Simulation of Anisotropic Conductive Adhesive Bonding Process and Reliability Analysis of the Packaging 335
Z. Wang, Z. Yin, B. Tao, Y. Xiong

Miniaturized Printed Wire Antenna in Package for 2.4GHz Wireless Communications 341
X. Liu, Y. Wang, L. Li, L. Wan

Modeling of Copper Wire Bonding Ball Transient Temperature Behavior 345
T. Wang, Q. Zheng, R. Yu, Z. Li

Power Integrity Simulation for SiP Using GTLE 349
Y. Zhou, L. Wan, J. Li

First-Principles Based Modeling for Influence of Epitaxy and Packaging Induced Strains on Emission Properties of III-nitrides LED Chips 353
H. Yan, Z. Gan, X. Song, Z. Chen, J. Xu, S. Liu

Numerical Study on Thermal Management of LED Packaging by Using Thermoelectric Cooling 357
N. Wang, C. Wang, J. Lei, D. Zhu

Modeling of Thermal Phenomena in a High Power Diode Laser Package 362
E. Suhir, J. Wang, Z. Yuan, X. Chen, X. Liu

Freeform Lens for Application-Specific LED Packaging 367
F. Chen, K. Wang, Z. Liu, X. Luo, S. Liu

High Bright White LED Lens Formation by Vacuum Printing Encapsulation Systems (VPES) and its Packaging Resin 372
A. Okuno, O. Tanaka

250W QCW Conduction Cooled High Power Semiconductor Laser 375
J. Wang, Z. Yuan, Y. Zhang, E. Zhang, D. Wu, X. Liu

High Density Indium Bumping through Pulse Plating Used for Pixel X-Ray Detectors..................380
  Y. Tian, D.A. Hutt, C. Liu, B. Stevens

**Emerging Lead-free, High-temperature Die-attach Technology Enabled by Low-temperature
Sintering of Nanoscale Silver Pastes**..................385
  G. Lu, M. Zhao, G. Lei, J.N. Calata, X. Chen, S. Luo

A Novel Micro Pirani Gauge with Mono-wire Sensing Unit for Microsystem Application..................391
  Y. Qiu, L. Zhao, Y. Jin

On-Package Magnetic Materials for Embedded Inductor Applications..................395
  L. Li, D.W. Lee, K. Hwang, Y. Min, S.X. Wang

**Low-temperature Bonding of Laser Diode Chips on Si Substrates with Oxygen and Hydrogen
Atmospheric-pressure Plasma Activation**..................399
  R. Takigawa, E. Higurashi, T. Suga, R. Sawada

Integration of GaN Thin Film and Dissimilar Substrate Material by Au-Sn Wafer Bonding and CMP..................402
  S. Zhou, Z. Chen, B. Cao, S. Liu

Modeling of Nanostructured Polymer-Metal Composite for Thermal Interface Material Applications..................405
  Z. Hu, B. Carlberg, C. Yue, X. Guo, J. Liu

Characterization of a Photosensitive Dry Adhesive Film for Wafer Level MEMS Packaging..................409
  K. Zhao, C. Wang

**Optimization Design on Polybrominated Biphenyls (PBBs) Extraction from Plastics for RoHS
Directive**..................414
  L. Hua, X.P. Guo, J.K. Yang, H.N. Hou

The Design of a Terahertz Metamaterial Absorber Basing on LTCC Technology..................418
  Y. Xie, Y. Li, Y. Liu, H. Zhang, Q. Wen

**Direct-write Techniques for Maskless Production of Microelectronics: A Review of Current State-of-
the-art Technologies**..................421
  Y. Zhang, C. Liu, D. Whalley

Integrated Inductors on Silicon and Planarized Ceramic Substrates..................428
  J. Wang, J. Cai, X. Dou, S. Wang

Nano Resonator Simulation Fabrication and Packaging Consideration..................432
  W. Zhang, Z. Liu, Z. Wang

Optimization of Silver Paste Printed passive UHF RFID Tags..................436
  B. Gao, M.M.F. Yuen

Low-temperature Wafer Bonding Using Gold Layers..................440
  Y. Wang, J. Lu, T. Suga

Novel Pore-sealing of Ordered, Porous Silica, SBA-15 for Low-k Underfill Materials..................444
  K. Hsu, K. Chen, C. Cheng, J. Lee, J. Leu

**Study of Polyimide as Sacrificial Layer with O_2 Plasma Releasing for Its Application in MEMS
Capacitive FPA Fabrication**..................450
  S. Ma, Y. Li, X. Sun, X. Yu, Y. Jin

Fabrication and Hydrogen Sensing Properties of Titania Nanotubes..................454
  S. Bai, D. Ding, C. Ning, R. Qin, Y. Li, C. Chang, M. Li, D. Mao

Development of Lead-Free Solders with Superior Drop Test Reliability Performance..................458
  Y.W. Wang, C.R. Kao

Enhancing the Properties of a Lead-free Solder with the Addition of Ni-coated Carbon Nanotubes..................464
  S.M.L. Nai, Y.D. Han, H.Y. Jing, K.L. Zhang, C.M. Tan, J. Wei

**A Method to Produce Printed Circuit Boards with Embedded Semiconductors Using Stress Buffer
Layers**..................468
  W. Seo, Y. Koo, S. Park, N. Kang, G. Kim

Study of Tungsten Metallization Surface States for Multilayer Ceramic..................472
  W. Zhang, H. Jin

**Solder Cracking Mechanism Correlation to Alloy Composition under Thermal Cycling Stress (Part
1)***..................476
  S. Chang, R. Wang, Y. Xiang, G. Chang

**Morphological and Microstructural Evolution of Sn-patch in SnAgCu Solder with Ni(V)/Cu under
Bump Metallization**..................482
  K. Wang, J. Duh, S. Tsai

**Localized Recrystallization and Cracking Behavior of Lead-free Solder Interconnections under
Thermal Cycling**..................486
  H.T. Chen, T. Mattila, J. Li, W. Liu, M.Y. Li, J.K. Kivilahti

**Growth Kinetics and Microstructural Evolution of Cu-Sn Intermetallic Compounds on Different Cu
Substrates During Thermal Aging**..................493
  Z.Q. Liu, P.J. Shang, D.X. Li, J.K. Shang

Assessment of LF Solder Joint Reliability by Four Point Cyclic Bending 496
J. Wang, Y. Ye, J. Zhao, S. Liu, Y. Tu, S. Li, Z. Song

A Warpage of Wafer Level Bonding for CIS (CMOS Image Sensor) Device Using Polymer Adhesive 501
J. Park, J. Lee, M. Cho, J. Kim, G. Kim

Improved Thermal Performance of High-Power LED by Using Low-temperature Sintered Chip Attachment 505
T. Wang, G. Lei, X. Chen, L. Guido, K. Ngo, G.Q. Lu

Fundamental Studies on Whisker Growth in Sn-based Solders 509
M. Zhao, H. Hao, G. Xu, J. Sun, Y. Shi, F. Guo

Micro-structural and Interfacial Effects on the Dielectric Properties of High-k Aluminum/Epoxy Composites for Embedded Capacitors 513
C. Chen, S. Yu, R. Sun, S. Luo, L. Weng, R. Du

Comparative Study of Interfacial Reactions of High-Sn Lead-free Solders on Single Crystal Cu and on Polycrystalline Cu 517
Y. Cui, M. Huang

Phase Identification of Intermetallic Compounds Formed during In-48Sn/Cu Soldering Reactions 521
P.J. Shang, Z.Q. Liu, D.X. Li, J.K. Shang

Microstructural Characterization of Electroplating Sn on Lead-frame Alloys 525
Y. Wang, D. Ding, K. Galuschki, Y. Hu, A. Gong, S. Bai, M. Li, D. Mao

The Fabrication of Composite Solder by Addition of Copper Nano Powder into Sn-3.5Ag Solder 531
A. Nadia

Depiction of the Elastic Anisotropy of AuSn$_4$ and AuSn$_2$ from First-principles Calculations 535
R. An, C. Wang, Y. Tian

Microstructural Evolution of Sn-3.5Ag Solder with Lanthanum Addition 541
H. Lee, Y. Chen, T. Hong, K. Shih, C. Hsu

Optimal Packing Research of Spherical Silica Fillers Used in Epoxy Molding Compound 547
H. Mei, X. Du, L. Li, W. Tan

iNEMI HFR-free Program Report 551
H. Fu, S. Tisdale, M. Rausch, J. Davignon, S.H. Hall, R.C. Pfahl

Interfacial Reaction of Reballed BGAs under Isothermal Aging Conditions 557
L. Nie, M. Dong, J. Cai, M. Osterman, M. Pecht

Effect of Cu Addition in Sn-containing Solder Joints on Interfacial Reactions with Au Foils 566
W. Liu, C. Wang, L. Sun, Y. Tian, Y. Chen

Evolution of Ag$_3$Sn Compounds in Solidification of Eutectic Sn-3.5Ag Solder 570
H. Lee, Y. Chen, T. Hong, K. Shih

Indium Bump Fabricated with Electroplating Method 574
Q. Huang, G. Xu, L. Luo

Effect of Aging Time on Interfacial Microstructure of Sn-3.8Ag-0.7Cu Solder Reinforced with Co Nanoparticles 579
S.L. Tay, A.S.M.A. Haseeb, M.R. Johan

Behaviors of Palladium in Palladium Coated Copper Wire Bonding Process 586
B. Zhang, K. Qian, T. Wang, Y. Cong, M. Zhao, X. Fan, J. Wang

Effect of Electromigration on Intermetallic Compound Formation in Cu/Sn/Cu Interconnect 590
L.D. Chen, M. Huang

Solderability of Tinned FeNi under Bump Material 594
C. Chen, L. Zhang, J.K. Shang

Characterize the Microstructure and Reliability of Ultra Fine Pitch BGA Joints 598
L. Wang, Z. Zhao, Q. Wang, J. Lee

Reliability of Wafer Level Thin Film MEMS Packages during Wafer Backgrinding 603
J.J.M. Zaal, W.D. Van Driel, G.J.A.M. Verheijden, G.Q. Zhang

Super-hydrophobic Nickel Films with Micro-nano Hierarchical Structure Prepared by Electrodeposition for Appliance Industry 608
T. Hang, M. Li, A. Hu, D. Mao

Liquid-state Interfacial Reactions between Sn-Ag-Cu-Fe Composite Solders and Cu Substrate 611
X. Liu, Y. Zhao, M. Huang, C.M.L. Wu, L. Wang

Cure Kinetics Analysis and Simulation of Silicone Adhesives 615
Q. Zhang, B. Song, S. Wang, L. Xu, S. Liu

Growth and Recrystallization of Electroplated Copper Columns 619
J. Liu, C. Liu, P.P. Conway, J. Zeng, C. Wang

Electrochemical Corrosion Behaviour of Sn-8Zn-3Bi-XCr Solder in 3.5% NaCl 625
J. Hu, A. Hu, M. Li, D. Mao

The Influence of Silicon Content on the Thermal Conductivity of Al-Si/ Diamond Composites 632
Y. Zhang, X. Wang, J. Wu

Study on Microstructure and Properties of Al/SiCp Electronic Packaging Materials Embedded Metal Components 637
Z. Zhang, H. Zhang

Creep Characterization of Lead Free 80Au-20Sn Solder 642
F. Su, H. Pan

The Warpage Control Method in Epoxy Molding Compound 646
W. Tan, F. Zhou, X. Cheng, D. Ding, J. Wu

On Reliability of Chips bonded on Flex Substrates Using Thermosonic Flip-Chip Bonding Process with Nonconductive Paste 649
C. Chuang, J. Aoh, W. Chen

Dynamic Constitutive Relation of EMC over a Broad Range of Temperatures and Strain Rates 657
T. Chen, X. Niu, X. Shu, J. Zhang

Low K CMOS90 2N Gold Wire Bonding Process Development 661
M. Han, B. Yan

Preliminary Study of a New Process to Improve the Strength of Thermo-sonic Ball Solder Joint 666
Z. Yu, Z. Wang, J. Cheng, H. Tian

Improving the Toughness and Thermal Properties of Epoxy Resin Using for Electronic Packaging by Interpenetrating Polymer Network 673
B. Chen, D. Gui, J. Liu

Effects of Service Parameters on Thermomechanical Fatigue Behaviors of New Nano Composite Solder Joints 677
F. Tai, F. Guo, B. Liu, Z.D. Xia, Y.W. Shi

Failure Analysis of Halide of Epoxy Molding Compound Used for Electronic Packing 683
J. Ye, S. Bao, L. Ma, D. Lv

Characterization, Modelling, and Parameter Sensitivity Study on Electronic Packaging Polymers 687
L. Niu, D. Yang, G.Q. Zhang

Controlled Synthesis of Ortho-Substitution Ortho-cresol Novolac Resins 692
X. Tian, Z. Yang

Electrodeposition of Sn-Cu Solder Alloy for Electronics Interconnection 696
Y. Qin, A. Wassay, C. Liu, G.D. Wilcox, K. Zhao, C. Wang

Study on the Degradation of Sealed Organic Light-emitting Diodes under Constant Current 702
X. Xu, W. Zhu, Q. Wang, Z. Zhang, X. Jiang

Preparation and Performance of White LED Packaged YAG Phosphor 706
R. Guan, W. Zhao, S. Li

Nanoindentation Investigation of Copper Bonding Wire and Ball 710
X. Fan, K. Qian, T. Wang, Y. Cong, M. Zhao, B. Zhang, J. Wang

Wetting Behaviour of Lead Free Solder on Electroplated Ni and Ni-W Alloy Barrier Film 715
C.S. Chew, A.S.M.A. Haseeb, M.R. Johan

Parameter Design of Solder Die Bonding Based on DOE 719
J. Hu, X. Xie

Outgassing of Materials Used for Thin Film Vacuum Packages 722
Q. Li, J.F.L. Goosen, J.T.M. Van Beek, F. Van Keulen, G.Q. Zhang

Effects of Surface Finishes on the Intermetallic Growth and Micro-structure Evolution of the Sn3.5Ag0.7Cu Lead-free Solder Joints 727
G. Li, C. Tang, X. Yan, X. Xie

The Effect of Tape Casting Slurry System for Process Performance of Green Ceramic Tape Used for Electronic Packaging 732
P. Shi, C. Ren, H. Jin, L. Gao

Studies on Microstructure and Mechanical Properties of Sn-Zn-Bi-Cr Lead-free Solder 735
T. Luo, A. Hu, M. Li, D. Mao

Solder Joints Reliability with Different Cu Plating Current Density in Wafer Level Chip Scale Packaging (WLCSP) 739
K. Cao, K.H. Tan, C.M. Lai, L. Zhang

A Novel Research on Micro-ball Placement Machine Used for Wafer Level Package 744
J. Liu

Developing on IC Flip Chip Bonder Machine & Process 747
J. Guo

Effect of Ni Addition on the Sn-0.3Ag-0.7Cu Solder Joints 750
L. Wang, F. Sun, Y. Liu, L. Wang

Electrically Conductive Adhesives with Sintered Silver Nanowires 754
Z. Zhang, X. Chen, H. Yang, H. Fu, F. Xiao

Al/Ni Multilayer Used as a Local Heat Source for Mounting Microelectronic Components 758
J. Zhang, F. Wu, J. Zou, B. An, H. Liu

Dramatic Morphological Change of Interfacial Prism-type Cu_6Sn_5 in the Sn3.5Ag/Cu Joints Reflowed
by Induction Heating 763
L. Wang, H. Xu, M. Yang, M. Li, Y. Fu

Wetting Behavior of Electrolyte in Fine Pitch Cu/Sn Bumping Process by Electroplating 767
J. Jiang, J. Bi, Z. Chen, M. Li, D. Mao, T. Suga

Effects of the Matrix Shrinkage and Filler Hardness on the Thermal Conductivity of TCA 771
C. Yue, Y. Zhang, Z. Hu, J. Liu, Z. Cheng

Synthesis and Characterization of Nano $BaTiO_3$/Epoxy Composites for Embedded Capacitors 776
S. Luo, R. Sun, J. Zhang, S. Yu, R. Du, Z. Zhang

Influence of Leveler Concentration on Copper Electrodeposition for Through Silicon Via Filling 780
H. Ling, H. Cao, Y. Guo, H. Yu, M. Li, D. Mao

Characterization of Ag Nanofilm Metallization on Copper Chip Interconnect and Its Ultrasonic
Bondability 783
Y. Tian, S. Zhao, C. Wang

Electroless Plating of Copper Nano-coned Array for High Reliability Packaging 787
Z. Pan, A. Hu, T. Hang, Y. Duan, M. Li, D. Mao

Research on Self-constrained Sintering Low-temperature Cofired Ceramic 792
Y. Hu, T. Liang, Y. Li, B. Yang, Y. Lu

Fine Pitch and High Density Sn Bump Fabrication 794
J. Bi, J. Jiang, A. Hu, M. Li, D. Mao, T. Suga

Effect of [Au]/[Na_2SO_3] Molar Ratio on Co-electroplating Au-Sn Alloys in Sulfite-based Solution 798
X. Qing, M. Huang, J. Pan

Sequential Non-cyanide Electroplating Au/Sn/Au Films for Flip Chip-LED Bumps 802
Y. Liu, M. Huang

Study of Interfacial Reactions between Sn3.5Ag0.5Cu Alloys and Cu Substrate 806
L.C. Tsao, S.Y. Chang, W.H. Sun, S.F. Yen

Study on the Microstructure and the Shear Strength of Sn-0.7Cu-xZn 810
Y. Gao, Z. Luo, J. Zhao, L. Wang

Investigation of the Fundamental Interactions among the Ingredients of Flux by the Group
Contribution Method 814
Y. Jin, J. Hu, D. Lu

Study of Stencil Printing Technology for Fine Pitch Flip Chip Bumping 820
J. Yang, J. Cai, S. Wang, S. Jia

Processing and Properties of Cu-base and Co-base Amorphous Wires 826
W.B. Liao, Y. Zhang

Absorption of Ag_3Sn on Cu_6Sn_5 Intermetallic Compounds at Sn-3.5Ag-xCu/Cu Interfaces 829
N. Zhao, L. Wang, L. Wan, L. Cao

Prediction Model for Wire Bonding Process through Adaptive Neuro-Fuzzy Inference System 833
J. Gao, C. Liu, X. Chen, D. Zheng, K. Li

From Thin Cores to Outer Layers: Filling through Holes and Blind Micro Vias with Copper by
Reverse Pulse Plating 838
S. Kenny, B. Roelfs

Investigation of Thin Small Outline Package (TSOP) Solder Joint Crack after Accelerated Thermal
Cycling Testing 843
L.N. Lu, H.Z. Huang, B.Y. Wu, Q. Zhou, X.X. Su, M. Cai

On Variable Frequency Microwave Processing of Heterogeneous Chip-on-Board Assemblies 847
T. Tilford, S. Pavuluri, C. Bailey, M.P.Y. Desmulliez

Advanced Chip to Wafer Bonding: A Flip Chip to Wafer Bonding Technology for High Volume 3DIC
Production Providing Lowest Cost of Ownership 852
A. Sigl, S. Pargfrieder, C. Pichler, C. Scheiring, P. Kettner

The Effect of Plasma Etching Process on Rigid Flex Substrate for Electronic Packaging Application ... 857
K.C. Yung, H.M. Liem, H.S. Choy, H.F. Zheng, T. Feng, T.M. Yue

A Study on the Characterization of Quasi-Three-Dimensional PN Junction Capacitor 861
H. Wang, Y. Lv, W. Gao, L. Wan

Effect of PIII on the Adhesion Behavior of Epoxy Molding Compound-Nickel Interface 865
L. Liu, Q. Lu, Y. Wang, W. Dai, X. Zhang, Y. Li, X. Wu

Improving Board Assembly Yield through PBGA Warpage Reduction 869
L. Li, K. Hubbard, J. Xue

A Jetting System for Chip on Glass Package..874
H. Jia, Z. Hua, M. Li, Jinsong Zhang, Jianhua Zhang

A Multilayer Low Pass Filter Fabricated by Ferrite and Ceramic Cofiring System Based on LTCC Technology 881
Y. Li, Y. Liu, H. Zhang, L. Han, Z. Yang

Research of SMT Product Assembly Quality Management System Based on J2EE 885
H. Zhao, D. Zhou, Z. Wu, X. Cheng

Test Scheme of SOC Test with Multi-constrained to Reduce Test Time 890
C. Xu, J. Zhang, M. Zhang

Design and Modeling of Jet Dispenser Based on Giant Magnetostrictive Material 894
Z. Ge, G. Deng

Simulation and Experimental Study on Temperature Field of Fluid Jet-dispenser.........................900
Z. Ding, H. Hu, G. Deng

Automatic Plating Technology for Ceramic Packaging...906
C. Lu, S. Liu, L. Zhang

Analysis and Compensation of Perpendicularity Error for High Speed and High Precise IC Assembly Equipment Base on Coordinate Transformation 909
H. Chen, J. Quan, B. Peng, Z. Yin

Inspection of Miniaturised Interconnections in IC Packages with Nanofocus X-Ray Tubes and NanoCT 916
Z. He, Q. Wen, X. Huang

Advanced Moisture Diffusion Model and Hygro-Thermo-Mechanical Design for Flip Chip BGA Package...922
M. Tsai, F. Hsu, M. Weng, H. Hsu

Analysis the Performance of the Micro-Channels Cooler with Different Inlet Position 929
X. Wang, W. Zhang, H. Liu, L. Chen, Z. Li

Sn Whisker Concern in IC Packaging for High Reliability Application..934
J. Chang, B. Lee

Investigation of Mechanism for Spontaneous Zinc Whisker Growth from an Electroplated Zinc Coating 939
A. Baated, K. Kim, K. Suganuma

Effects of Thermal Aging on the Electrical Resistance of Sn-3.5Ag Micro SOH Solder Joints 943
J. Peng, F. Wu, H. Liu, L. Zhou, Q. Pan

Instability and Failure Analysis of Film-substrate Structure under Electrical Loading 947
Q. Wang, H. Xie, J. Liu, X. Feng, F. Dai

Reliability Study of RFID Flip Chip Assembly by Isotropic Conductive Adhesive through Computer Simulation 950
E.K.L. Chan, B. Gao, M.M.F. Yuen

Effect of Thermomigration in Eutectic SnPb Solder Layer 954
Y. Tao, L. Ding, Y. Yao, B. An, F. Wu, Y. Wu

Adhesion Behavior between Epoxy Molding Compound and Different Leadframes in Plastic Packaging 959
L. Xu, X. Lu, J. Liu, X. Du, Y. Zhang, Z. Cheng

Effect of Zn Addition on Microstructure of Sn-Bi Joint...963
Q.S. Zhu, H.Y. Song, H.Y. Liu, Z.G. Wang, J.K. Shang

Deformation Characteristics of Sn-3Ag-0.5Cu/Cu/Ni-xCu/Ti Joints after Mechanical Test 967
C. Peng, J. Duh

Studies on Microstructure of Epoxy Molding Compound (EMC)-Leadframe Interface after Environmental Aging 971
X. Lu, L. Xu, H. Lai, X. Du, J. Liu, Z. Cheng

The Effects of Bonding Parameters on the Reliability Performance of Flexible RFID Tag Inlays Packaged by Anisotropic Conductive Adhesive 974
X. Cai, X. Chen, B. An, F. Wu, Y. Wu

Effects of Design, Structure and Material on Thermal-Mechanical Reliability of Large Array Wafer Level Packages..979
B. Varia, X. Fan, Q. Han

Development of High Speed Cold Ball Pull as a Quick Turn Monitor for Solder Joint Reliability...........................990
Y. Wang, L. Cao

Influencing Factors and Solutions for Ball Short during Wire Bonding 994
Z. Meng, Y. Feng, S. Lee

Modeling Electrochemical Migration through Plastic Microelectronics Encapsulations...................998
W. Van Soestbergen, A. Mavinkurve, R.T.H. Rongen, L.J. Ernst, G.Q. Zhang

Microstructure Changes and Compound Growth Dynamic at Lead-free/Cu Interface under Different Conditions 1002
L. Qi, J. Huang, J. Niu, L. Yang, Y. Feng, X. Zhao, H. Zhang

Research of Structural Factors Effects on Drop Reliability 1007
J. Wang, X. Fu, X. Xie, J. Zhou, Q. Wang, Z. Lee

Effect of Stand-off Height on the Reliability of Cu/Sn-4.8Bi-2Ag/Cu Solder Joint 1013
H. Liu, L. Zhou, J. Li, F. Wu, Y. Wu

Enhancement of TBGA Substrate in Packing Drop Test 1016
K. Pun, C.Q. Cui

Quality and Reliability Challenges for Ultra Mobile Computing and Communication Application Processor Packaging 1023
D. He, W. Kang

Electromigration Analysis and Electro-Thermo-Mechanical Design for Semiconductor Package 1032
H. Hsu, S. Ju, J. Lu, H. Chang, H. Wu

Comparison of Thermal Fatigue Reliability of SnPb and SAC Solders under Various Stress Range Conditions 1038
C. Yang, Y.S. Chan, S.W.R. Lee, Y. Ye, S. Liu

Board Level Reliability Assessments of Thru-Mold Via Package on Package (TMV™ PoP) 1043
T. Hwang, D. Park, Jin-Seong Kim, Jin-Young Kim, Jae-Dong Kim, C. Lee

Prediction of IMC Formation during Interfacial Reactions: Application of CALPHAD Approach to Electronic Package 1049
H. Liu, W. Zhu, Z. Jin

Establishing Mixed Mode Fracture Properties of EMC-Copper (-oxide) Interfaces at Various Temperatures 1057
A. Xiao, G. Schlottig, H. Pape, B. Wunderle, O. Van Der Sluis, K.M.B. Jansen, L.J. Ernst

Effect of Thermal Cycling on Interfacial IMCs Growth and Fracture Behavior of SnAgCu/Cu Joints 1063
X. Li, F. Li

Effect of Miniaturization on the Microstructure and Mechanical Property of Solder Joints 1068
B. Wang, F. Wu, J. Peng, H. Liu, Y. Wu, Y. Fang

Numerical Simulation on Variable Width Multi-Channels Heat Sinks with Non-uniform Heat Source 1074
X. Wang, W. Zhang, H. Liu, L. Chen, Z. Li

Characteristic Analysis of Transducer Drive Current in Ultrasonic Wire Bonding Process 1078
S. Liu, F. Wang

Corrosion Characterization of Sn37Pb Solders and With Cu Substrate Soldering Reaction in 3.5wt.% NaCl Solution 1083
L.C. Tsao

Shape and Fatigue Life Prediction of Chip Resistor Solder Joints 1086
G. Zheng, C. Wang

Effect of Bonding Temperature and Power Setting on Transducer Velocity Using Principal Components Analysis in Thermosonic Bonding 1090
Y. Zhang, L. Han

Study of Thermal Fatigue Lifetime of Fan-in Package on Package (FiPoP) by Finite Element Analysis 1095
X. Yan, G. Li

Thermal-Mechanical Fatigue Reliability of PbSnAg Solder Layer of Die Attachment for Power Electronic Devices 1100
X. Xie, X. Bi, G. Li

Effects of Strain Rate and Temperature on Mechanical Behavior of SACB Solder Alloy 1105
G. Yuan, X. Yang, X. Shu

Thermal Stress Analysis and Structural Optimization of Ultra-thin Chip Stacked Package Device 1109
L. Li, X. Ma, X. Zhou

Prediction of Bending Reliability of BGA Solder Joints on Flexible Printed Circuit (FPC) 1114
J. Huang, Q. Chen, L. Xu, G.Q. Zhang

Failure Evaluation of Flexible-Rigid PCBs by Thermo-Mechanical Simulation 1120
L. Arruda, Q. Chen, J. Quintero

Fast Qualification Using Thermal Shock Combined with Moisture Absorption 1125
X. Ma, G.Q. Zhang, K.M.B. Jansen, W.D. Van Driel, O. Van Der Sluis, L.J. Ernst, C. Regard, C. Gautier, H. Fremont

Controlling the Morphology and Orientation of Cu_6Sn_5 through Designing the Orientations of Cu Single Crystals 1131
H.F. Zou, H.J. Yang, Z.F. Zhang

Thermal-Mechanical Failure and Life Analysis on CBGA Package used for Great Scale FPGA Chip 1135
W. Li, X. Zhang

Finite Element Analysis of Sn-Ag-Cu Solder Joint Failure under Impact Test 1139
G. Tang, B. An, Y. Wu, F. Wu

Study on Moisture Behavior in Flip Chip BGA Packages and Bake Process Optimization 1144
W.Q. Dai, Z.K. Hua, J.H. Dai, E.W. Pang, L. Jiang, C.Y. Li, P. Liao, J.H. Zhang

Study on Thermo-mechanical Reliability of Embedded Chip during Thermal Cycle Loading 1148
L. Niu, D. Yang, M. Zhao

The Influence of Plastic-package on the Voltage Shift of Voltage Reference in Analog Circuit 1152
Y. Jiang, J. Ju

Electrical Analysis of Mechanical Stress Induced by Shallow Trench Isolation 1155
Y. Jiang, J. Ju

The Effect of Thermal Cycling on Nanoparticle Reinforced Composite Lead-free Solder 1159
S. Chen, Z. Cheng, J. Liu, Y. Gao, Q. Zhai

Study of Isothermal Bending Fatigue Test 1165
M. Lou, L. Wen, Z. Chen, J. Zhou, Q. Wang, J. Lee

Testing Failure of Solder-Joints by ESPI on Board-Level Surface Mount Devices 1175
Y. Gao, J. Wang

Study on Shear Strength and J_c of EMC/Cu Interface with Cu Oxidation and Moisture Absorption 1179
X. Fang, Q. Fang, J. Wang, H. Yu, X. Shao

XPS Study on Epoxy/Ni Interface 1183
L. Liu, W. Xv

Shock Performance Study of Solder Joints in Wafer Level Packages 1185
A.S. Ranouta, X. Fan, Q. Han

Fatigue Evaluation of Power Devices 1196
K. Shinohara, Q. Yu

Advanced High Density Interconnect Materials and Techniques 1203
J. Wei, S.M.L. Nai, X.F. Ang, K.P. Yung

The High Balance Symmetric Balun for WLAN and WiMAX Application Using the Integrated Passive Device (IPD) Technology 1211
S. Wu, W. Lin, K. Wang, C. Huang, W. Yeh

New Packaging Technology Enabling Integration of Magnetics and Semiconductors in One Component 1215
A. Pot, H. Roehm, R.V.D. Berg, T.P. Sidiki

Parametric Study of Electroplating-based Via-filling Process for TSV Applications 1220
K.Y.K. Tsui, S.K. Yau, V.C.K. Leung, P. Sun, D.X.Q. Shi

Packaging and Assembly of 12-Channel Parallel Optical Transceiver Module 1225
Z. Li, W. Gao, J. Song, B. Li, L. Wan

Recent Advances in Laser Assisted Polymer Intermediate Layer Bonding for MEMS Packaging 1228
C. Wang, J. Zeng, Y. Liu

Wafer-Scale Hermetically Packaged MEMS Switches with Liquid Gallium Contacts 1233
Q. Liu

Underfill Study for Large Dice Flip Chip Packages 1237
A. Lin, C.Y. Li, M. Shih, Y. Lai, B. Appelt, A. Tseng

Development of a Novel Cost-Effective Package-on-Package (PoP) Solution 1243
P. Sun, V.C.K. Leung, D. Yang, D.X.Q. Shi

EcoDesign Technical Committee of JIEP (Japan Institute of Electronics Packaging) and Its Activity 1249
H. Hayashi

Artificial Neural Network Application in Vertical Interconnection Modeling 1253
Y. Liang, L. Li

A New Method to Fabricate Sidewall Insulation of TSV Using a Parylene Protection Layer 1257
M. Ji, Y. Zhu, S. Ma, X. Sun, M. Miao, Y. Jin

The Electrical, Mechanical Properties of Through-Silicon-Via Insulation Layer for 3D ICs 1261
S. Seo, J. Park, M. Seo, G. Kim

Through Silicon Via Filling by Copper Electroplating in Acidic Cupric Methanesulfonate Bath 1265
Q. Li, H. Ling, H. Cao, Z. Bian, M. Li, D. Mao

Package Heat Dissipation with Integrated Carbon Nanotube Micro Heat Sink 1270
X. Wang, H. Liu, J. Wang, W. Zhang, Z. Li

Author Index

Directed by
Chinese Institute of Electronics, China
Department of Higher Education, Ministry of Education, China
Department of High and New Technology, Development and Industrialization, Ministry of Science and Technology, China
Department of Telecommunication, Ministry of Industry and Information Technology, China
Industry and Information Committee, Beijing Municipality, China
China International Culture Exchange Center

Sponsored by
China Electronics Packaging Society (CEPS) of the Chinese Institute of Electronics, China
Tsinghua University, China
IEEE Component, Packaging, &Manufacturing Technology Society (IEEE-CPMT)

Organized by
Tsinghua University, China
Beijing Faith Consulting Co., Ltd.

A Novel MEMS Package with Three-Dimensional Stacked Modules

Gaowei Xu, Qiuping Huang, Wenguo Ning, Zugang Ruan, Le Luo
Shanghai Institute of Microsystem and Information Technology, Chinese Academy of Sciences,
Shanghai, 200050, China
E-mail: xugw@mail.sim.ac.cn

Abstract

A 3D stacked modular packaging technology was developed so as to meet the general requirements of MEMS and wireless communication on packaging. The general requirements include high-density, low-cost and high-yield etc. According to the requirements of modular package, a kind of accelerator (MEMS) and its modem circuits (IC) were incorporated by three stacked modules and a 3D stacked modular package was realized. In this package, every module was assembled by using traditional surfaced mounting technology (SMT) based on FR4 substrate. MEMS and IC devices were assembled in the same module. The vertical interconnection between modules was realized by means of solder paste printer, mechanical alignment apparatus and optimized reflow curve. The mechanical alignment apparatus was specially prepared for alignment and positioning of stacked modular assembly. The alignment of stacked modules was completed by the alignment-pin/hole method. In order to increase the alignment efficiency, the module substrate was design as paste-up, for example 3×3. The alignment precision of vertical interconnection of the stacked modules reaches to 0.068mm typically. The volume of the entire stacked module was about 19mm×19mm×8mm. Finally, the influencing factors on the vertical interconnection were discussed and shear strength test of the module was presented.

1 Introduction

Modular package of system refers to dividing the system into a few of subsystems with single function, and assembling each subsystem into the module with uniform standard e.g. dimension, finally assembling all the packaged modules into so-called system package.

3D stacked packaging includes bare-chip stack, single packaged-device stack and MCM (multi-chip module) stack. The bare-chip stack technology has found wide applications in Flash-SRAM and DSP/ASIC-memory etc. It can make the Si efficiency reach up to 200%-300% and reduce cost for about 20-40% [1]. However, for bare-chip stack there exist a few of problems such as limitation to one kind of chip, low yield, high cost and heat dissipation problem etc. [2] For single packaged-device stack, it has wide applications, simple process and high yield etc. However, with the development of MEMS and wireless sensor technologies single packaged-device stack cannot meet the demand of SiP (system in package) for higher density. 3D stacked MCM packaging technology can integrate the electronic, fluid or optical devices into one module and achieve the incorporation of MEMS and ICs, digital/analog devices. Currently, 3D stacked modular package can be realized by FR4 substrate and standard packaging technologies such as SMT (Surfaced mounting technology), COB (Chip on board) or FCOB (Flip-chip on board) [3-4] etc. FR4 substrate overcomes the shortcoming of the Low Temperature Co-firing Ceramic (LTCC) substrate such as complicated processes and high cost [5] etc. The ball planting and reflow method is used for vertical interconnection between stacked module layers [6, 7]. This method seems easy to realize, however generally requires multiple reflow, which can decrease the welding performance of the substrate and may decrease the reliability of vertical interconnection. If the welding of all modules is realized by solder paste print and one time of reflow will overcome above shortcoming. Furthermore, this method is compatible with traditional SMT technology and can avoid the effect of multiple reflow on welding performance of modules. Now the problem is that the "solder paste print and one time of reflow" method requires corresponding alignment equipment and positioning apparatus as well as optimal reflow curve. Most of the high precision alignment equipments are very expensive, furthermore, can only align single module, therefore its alignment efficiency is lower. To solve this problem, the mechanical alignment can be adopted. The mechanical alignment method has simple process and good repeatability and can avoid the error due to the movement and vibration during the reflow process. In addition to this, mechanical alignment method has no limitation to vacuum sucker and field range like optical alignment, therefore the substrate can be design as paste-up one. In this way, n×m modules can be simultaneously aligned therefore the alignment efficiency is increased.

In this paper, a kind of MEMS accelerator and its assorted modem circuit (including IC and passive component) were incorporated by stacked modular structure. Each layer was assembled by using SMT (typically COB technology) based on FR4 substrate. Three layers were vertically interconnected and reflowed by specially designed mechanical alignment /positioning apparatus. Finally, a few of factors affecting the reliability of interconnection were discussed and shear strength test of the module was presented.

2 Structure design

The schematic drawing of 3D stacked modular package was shown in Fig.1. The package structure was designed into three layers of modules: top layer, frame layer and bottom layer. The top layer module includes a MEMS accelerator, a square-wave generator chip and a phase-inverter chip. The bottom layer module includes three operational amplifier and one multi-way switch chip. The frame layer module is the structural as well as functional interface, which performs the electrical interconnection between the top and bottom modules and makes room for in-between components. The layouts of the three substrates were shown in Fig.2. The electrical interconnection between the top and bottom layers was conducted by the peripheral bonding pads and through holes of three PCBs. The parameters of modules are shown in Table 1. The dimension of stacked modules is about 19mm×19mm×8mm.

978-1-4244-4658-2/09 $25.00 © 2009 IEEE

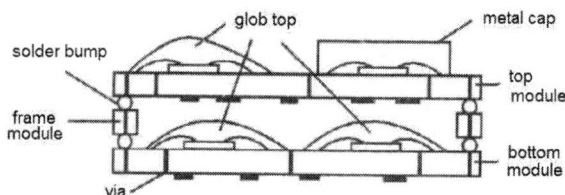

Fig.1 Cross-section of 3D stacked modular package

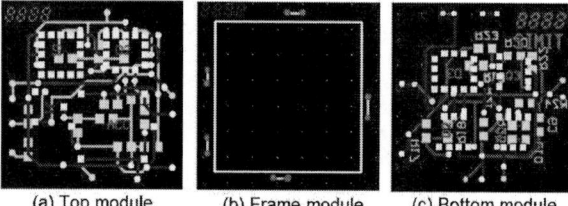

(a) Top module   (b) Frame module   (c) Bottom module

Fig.2 PCB layout of three modules

Table 1 Parameters of module

Item	Dimension(mm)
Pad	0.55
Line width	0.25
Diameter of Via	0.2
PCB	19×19×1.5
Single module	19×19×2

To improve the alignment efficiency and compatible with SMT, for each layer, the substrate was designed as a paste-up PCB, which consists of 9 (i.e. 3×3) same substrates. The layout of the bottom paste-up was shown in Fig.3. There are four alignment/positioning round-holes (with diameter 2mm) on each corner of the paste-up respectively.

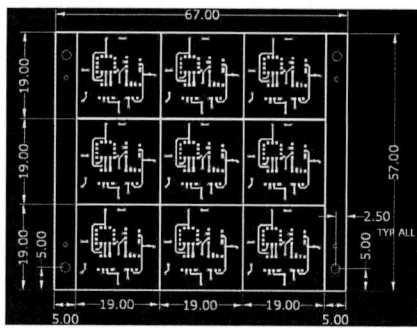

Fig.3 Typical paste-up of bottom module

3 Preparation and process

The process flow of the 3D stacked modular package was shown in Fig.4. The bare chips were adhered and wire-bonded to the top and bottom substrates (i.e. traditional COB technology) then were encapsulated by using Glob-top technology. The accelerator was protected by a special metal cap. After that, solder paste was stencil-printed and the necessary SMD (surface mounting device), generally passive components such as resistors and capacitors, were mounted on the top and bottom modules. After printing solder paste on the back side of the frame module, three modules (i.e. top, bottom and frame) were stacked in turn by tailor-made positioning apparatus and were reflowed through reflow batch. Finally, the 3×3 paste-up package was cut into nine 3D packages.

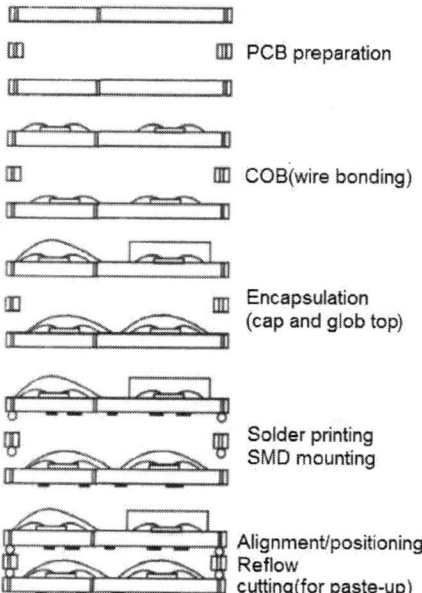

Fig.4 Process flow of 3D stacked modular package

4 Results and discussions

4.1 Alignment precision

The alignment/positioning plays an important role in interconnections of stacked modules and the alignment precision has a significant effect on the reliability of interconnections. Alignment includes mechanical alignment and optically assisted mechanical method. The latter adopts modern optical-mechanical system therefore it has high alignment precision. However the optical system used in commercial SMT equipment (for example flip-chip machine) can only perform alignment for single module and, as a result, has lower alignment efficiency. In addition, the sample dimension that the optical equipment is able to align is subject to the suction of its vacuum sucker and the visual field. In contrast, mechanical alignment apparatus generally positions by means of pin/hole and has no those limits, as a result, it has manageable process, good repetitiveness, low cost and wide applicability. The main attraction is that the pin/hole method has the advantage for paste-up, i.e. the module substrate can be designed as an "n×m" paste-up so that n×m modules can be aligned simultaneously, which will absolutely increase alignment efficiency.

A novel alignment/positioning apparatus that can simultaneously conduct stack alignment and positioning for 3×3 modules was developed as shown in Fig.5. This apparatus consists of positioning pin and positioning plate

(100mm×19mm×5mm). There are four round holes with diameter 2mm in the positioning plate. The positions of the four holes correspond to the four ones in the paste-up.

The alignment precision of the apparatus has significant effect on the shape and weld strength of solder bumps. In fact, alignment error coming from this apparatus is larger than that from optical method. Experiment showed that if the diameter of positioning pin is closer to that of positioning hole, the alignment precision will be higher and the profile the solder bumps will be more regular. However, if the alignment precision was extremely pursued it will be more difficult to conduct assembly. Measurement showed that the positioning error is about 0.068mm if the diameter of positioning pin is 1.90mm and that of hole is 2mm. On the other hand, the diameter of the pads in the module is 0.55mm, therefore the precision of the alignment apparatus satisfies the alignment requirement of the stacked modular package.

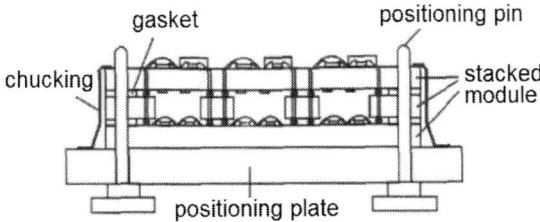

Fig.5 Schematic of alignment/positioning apparatus for 3D stacked modules

4.2 Optimization of reflow

Owing to the poor heat conducting performance of FR4 substrate the temperature distribution in the whole stacked modules is not uniform in the period of reflow, for example, there is a temperature difference 5-10°C between top and bottom layers of solder bumps. Furthermore, the temperature distribution in the same layer is also non-uniform, for example, for the paste-up PCB, the temperature along its periphery is higher than that at its center. In addition, if the soaking time is too short the scaling powder cannot thoroughly volatilize and may remain inside the solder ball and form the void therefore may affect the interconnection reliability, as shown in Fig.6. For that reason, the soaking period should be prolonged in the course of reflow. If the heating-up time of reflow is too short the bump weld may be insufficient.

Therefore, the reflow batch with 5 temperature zones was configured. The temperatures of every zone and the residence time of the sample in each zone were optimized. Finally the fitting curve of reflow was obtained, as shown in Fig.7. The longer soaking time (180s) and heating-up time (90s) were adopted and peak temperature was 215.6°C.

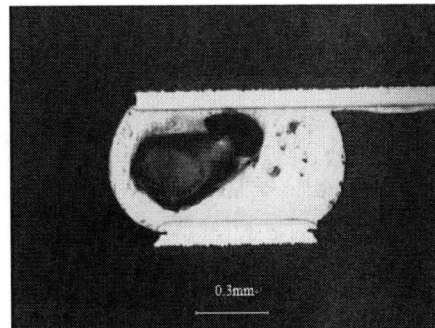

Fig.6 Cross-section of a bump after the reflow with shorter soaking time

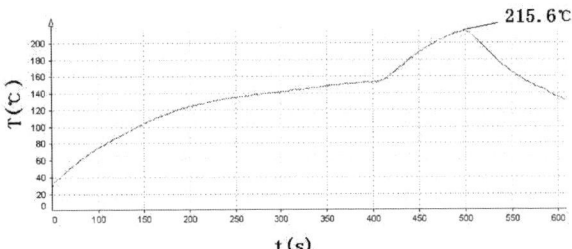

Fig.7 Fitting curve of reflow

Fig.8 shows the profile of the solder balls after reflow as per the curve in Fig.7. It can be seen that solder balls appear good profile and weld condition without void and crack. In addition, the height of bottom-layer balls is smaller than that of top layer. This phenomenon can be attributed the increasing pressure force on bottom balls coming from upper stacked modules.

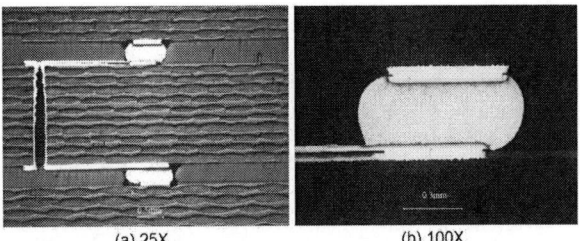

Fig.8 Cross-section of solder balls after optimal reflow

Table 2 Shear strength test result of solder balls

Item	1	2	3	4	5	6
Force(kg)	3.702	4.731	3.674	4.432	3.684	3.857
Pressure intensity (MPa)	30.56	39.05	30.32	36.58	30.4	31.84

According to corresponding strength test standard [8], the interconnection shear strength between module layers of sample was tested with result shown in Table 2. It can be seen that the average strength of single ball lies between 30MPa-40MPa, which is 5-7 times than the minimum strength that the standard provides.

Conclusions

A kind of accelerator (MEMS) and its modem circuits (IC) were incorporated as per three stacked modules and a 3D stacked modular package was realized. Every module was assembled by using traditional SMT (typically COB) technology on FR4 substrate. The vertical interconnection was realized by means of stencil print, specially prepared alignment/positioning apparatus and optimized reflow curve.

Newly developed alignment/positioning apparatus based on positioning pin/hole method can simultaneously align and position 3×3 module paste-up. If the diameter of positioning pin is closer to that of positioning hole, the alignment precision will be higher. However, if alignment precision was extremely pursued it will be more difficult to complete assembly. The positioning precision can reach up to 0.068mm.

Once reflow method was used to simultaneously complete the vertical interconnection and the welding of the surface mount device (SMD). The reflow curve was optimized. The longer soaking time (180s) and heating-up time (90s) were adopted. The peak reflow temperature was 215.6°C.

This package structure has good bonding strength, electrical performance and high reliability.

The traditional SMT and special mechanical alignment apparatus will made the assembly process simple, and lead to low cost. The paste-up method for alignment and positioning will be favorable to large scale manufacture.

References

1. Mark Klossner, Steve Babinetz, "Assembly Solutions for 3-D Stacked Devices", *SEMICON Singapore*, 2002, pp.A1-A10.
2. Yasuki Fukui, Yuji Yano, *et al*, "Triple-Chip Stacked CSP", *Electronic Components and Technology Conference*, 2000, pp.385-389.
3. Dipl-Ing, Kourosh Amiri Jam, *et al*, "Design, fabrication and test of electrical modules with several assembly layers", *MICRO-tec2003*, 2003, pp.13-15.
4. Kourosh Amiri Jam, Volker Groer, Klaus-Dieter Lang, *et al*, "Application of 3D-Stacking Technology for Sensor Integration and Miniaturization", *Micro System Technologies 2005*, 2005.
5. Rudolf Leutenbauer, Volker Grosser, Bernd Michel, *et al*, "The Development of a Top-Bottom-BGA (TB-BGA)", *Electronic Components and Technology Conference*, 1998, pp.1235-1240.
6. Pienimaa, S. K., Miettinen, J., Ristolainen, E., "Stacked Modular Package", *IEEE Transactions on Advanced Packaging*, Vol.27, No.3(2004), pp.461-466.
7. Yamazaki, T., Sogawa, Y., Yoshino, R., Kata, K., Hazeyama, I., Kitajo, S., "Real Chip Size Three-Dimensional Stackde Package", *IEEE Transactions on Advanced Packaging*, Vol.27, No.3 (2005), pp.397-403.
8. MIL-STD-883E, Method 1010. 7.

Fabrication and Characterization of a Novel Wafer-level Chip Scale Package for MEMS Devices

Yuhan Cao, Le Luo*

The State Key Laboratories of Transducer Technology

Shanghai Institute of Microsystem and Information Technology, Chinese Academy of Sciences (CAS)

865 Changning Road Shanghai, 200050 P.R.China

Tel: 86-021-62511070-5468; Mail: andycao@mail.sim.ac.cn

*Corresponding author: leluo@mail.sim.ac.cn

Abstract

This paper reports a hermetic MEMS package structure with silicon wafer as bonded cap at wafer-level scale. CMP followed by spraying chemical smoothing process is utilized to thin the N(100) silicon cap wafer to the thickness of 150μm after wafer-level Cu/Sn isothermal solidification bonding. Method for the thinning process and parameters for Cu/Sn isothermal solidification bonding process are researched and optimized. TMAH etching is used to open trenches penetrating through the cap wafer and then $Pb_{63}Sn_{37}$ bumps redistribution and reflow are utilized to achieve the final I/O interconnection. Such approaches have the most important features of simplified fabrication process compatibility with most MEMS processes and low-cost. Fine-leak tests as well as gross-leak tests have been done on the packaged samples in order to characterize the hermeticity of the bonded wafers which indicating an excellent leak rate of around 1.9×10^{-9} atm cc/s. The shear strength of the packaged samples has been measured and average shear strength of 19.5Mpa is achieved. Both results meet the demands of MIL-STD-883E.

1. Introduction

Nowadays with the trend of wider range of application of MEMS devices package has become the bottleneck for the batch production for the large part of manufacturing cost it represents. Wafer level chip scale packaging technology has been rapidly developed for its advantage of small size and low cost. Such technology mainly includes two methodologies, thin film approach and bonded cap approach [1-3]. The former involves the use of some sacrificial material to support deposition of an encapsulation layer, followed by the removal of the sacrifice layer. The latter involves the use of a cap wafer bonded to the device wafer. Thin film approach is widely adopted however limited by the complicated process and mechanical integrity of uncertainty for the thin thickness of the encapsulation layer as protection of the devices. Bonded cap approach also has two key process techniques to resolve, interconnection and bonding techniques. Vertical through silicon vias solution based on dry etching and sputtering or electroplating techniques is commonly adopted in wafer-level package structure for the advantages of small size and low parasitic loss [4]. However the fabrication of TSVs is costly which has limited the batch production. As to the bonding technique, there are multiple methods which can be roughly divided into three categories: (1) anodic bonding, (2) silicon fusion bonding, (3) intermediate film bonding. Anodic fusion bonding can be done directly by attaching silicon to sodium-rich glass while silicon fusion bonding by attaching silicon to silicon, however their usage is limited for the requirement of higher temperature during bonding process

and rigorously cleaner flat surface. Intermediate film bonding technique can be further divided according to the films as bonding media. Such films can either be low temperature sealing materials as organic materials (BCB, etc.) and solder, or high temperature sealing materials as glass frit and metal. By contrast, intermediate film bonding can be carried out at lower temperature, eliminating the requirement of flatness of bonding surface. However some disadvantages such as bad thermal stability and hermeticity of cured BCB film, wide bonding area for glass frit limit their applications [5].

In this paper a hermetic MEMS package structure with silicon wafer as bonded cap at wafer-level scale is proposed. Cu/Sn isothermal solidification bonding is firstly utilized to bond a N(100) orientation silicon wafer as cap to the device wafer followed by CMP process to thin the silicon cap wafer to 150μm thick. Photolithographic process and TMAH etching are then used to open trenches penetrating through the cap wafer, 300μm $Pb_{63}Sn_{37}$ bumps redistribution and reflow are subsequently utilized to accomplish the final I/O interconnection. Such approaches avoid the costly TSVs fabrication and satisfy the requirements for MEMS device package including low-cost, hermeticity, low parasitic loss, wafer-level chip scale and compatibility with most MEMS processes. Fine-leak tests as well as gross-leak tests have been done on the packaged samples in order to characterize the hermeticity of the bonded wafers which indicating an excellent leak rate of around 1.9×10^{-9} atm cc/s. The shear strength of the packaged samples has been measured using Dage series 4000B Bondtester and an average shear strength of 19.5Mpa is achieved. Both results meet the demands of MIL-STD-883E. A MEMS resonator with a resonant frequency of 1.3KHZ is integrated to monitor the long term stability of the vacuum in the chamber of the package structure.

2. Experiment

2.1 Packaging structure design

The schematic view of the packaging structure is shown in fig.1. Two key technique features are depicted by the picture. Firstly, $Sn_{63}Sn_{37}$ bumps redistribution technique is utilized to replace the TSV. Such a method contains two advantages: (1) Short length of the interconnection results in lower resistance and lower parasitic loss; (2) The replacement of TSV results in low cost. Secondly, Cu/Sn bonding technique is performed to bond the silicon cap to the die wafer as well as serving as space layer to alter the height between the two wafers for the move part of the MEMS device to be integrated in future.

978-1-4244-4658-2/09 $25.00 © 2009 IEEE

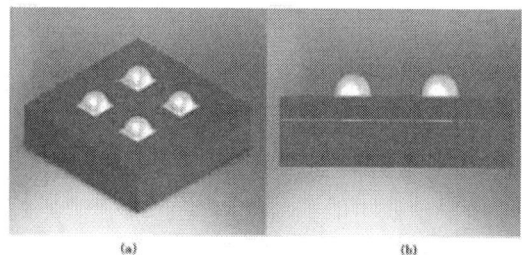

Fig.1 Schematic view of the packaging structure: (a) 3D, (b) 2D

2.2 Process flow

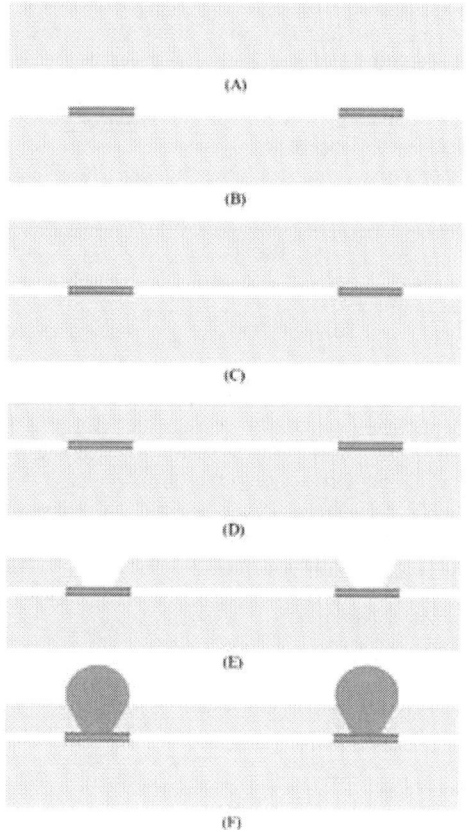

Fig.2 Schematic illustration of the process flow

Fabrication sequence for the packaging structure is illustrated in Fig.2. (A) A N(100) orientation silicon with a thickness of 450μm is used as die wafer. (B) TiW/Cu is sputtered on the wafer as seed layer and then 15μm thick photoresist is spin on the surface. Photolithographic process is performed to pattern the photoresist followed by 6μm Cu and 4μm Sn deposited as UBM layer and solder layer separately by electroplating to form the sealing ring structure. Addition step of Au sputtering on the Sn layer is performed to avoid the Sn oxidation at atmosphere. Photoresist is then removed by acetone and the remaining seed layer is removed by ion etching. (C) A 350μm, N(100) silicon wafer is used as cap

wafer and similar processes are performed on this wafer to form a 8μm thick Cu sealing ring structure and then the two wafers are bonded by Cu/Sn isothermal solidification technique under the control of optimized bonding parameters. (D) CMP process is used to thin the cap wafer to a thickness of 150μm. (E) 2μm thick oxide layer is deposited on the surface of the thinned cap wafer by PECVD followed by photolithographic process and TMAH etching to open trenches penetrating through the cap wafer exposing the upper surface of the Cu UBM layer. Another 2μm oxide is deposited to form a conformal isolation layer and another photolithographic process is then used to expose the upper surface the Cu UBM layer for electrical connection. (F) 300μm diameter $Pb_{63}Sn_{37}$ bumps are redistributed in the trenches. Thermal reflow is performed to accomplish the I/O interconnection.

3. Key features of the process
3.1 Wafer bonding process

Prior to bonding, surface treatment process is an important step for good bonding quality [6]. Because Sn is easy to be oxidized in open air even at room temperature, Au is sputtered on the surface. Besides plasma cleaning is also applied to remove the possible contaminations probably being the remaining photoresist on the wafer surface left by the previous photolithographic process. The power of plasma is controlled to be 500W and the cleaning time is set as 170s avoiding the Sn layer from melting.

After surface plasma cleaning, the wafers are firstly pre-aligned using KarlSuss MA6 which provides a critical resolution of ±1μm. Then, the aligned wafers are brought to SB6 bonding system to complete the bonding process. Static pressure of 200mbar is applied on the wafers throughout the whole bonding process. Its main utility is to break the hump of liquid Sn formed by conglomeration at the melting point to achieve good wetting condition. The temperature-time profile designed is as shown in Fig. 3 and the vacuum of the chamber is kept around 5×10^{-4} mbar during the whole process. All these parameters are optimized which has been reported in the previous work in details [7]. At the last step of the bonding process N_2 is filled up to cool the chamber. The temperature drop rate should be as fast as possible because the grain size at low temperature is always smaller than that at higher temperatures.

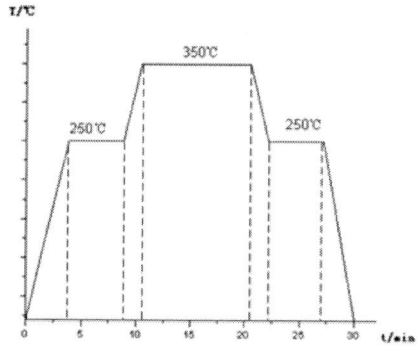

Fig.3 Temperature-time curve for Cu/Sn bonding

3.2 Wafer thinning process

Silicon cap wafer can also be reduce to a thickness of 150μm or below before bonding to the device wafer however it will be fragile during following processes and handling film to improve the mechanical integrity is usually demanded. In this work, the bonding process is performed first to improve the mechanical integrity to simplify the process. At the following step of thinning the cap wafer, three different wafer thinning processes are tried to get the optimized thinning method.

(1) Method 1: This process involves the use of common process of CMP.

(2) Method 2: This process is a wet etch process based on TMAH process which has the advantage of the possibility of batch etch in a tank system.

(3) Method 3: This process is a single wafer chemical spray etching process that is performed after CMP process to improve the smoothness of the silicon surface.

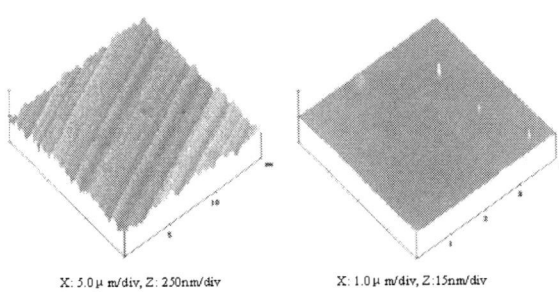

X: 5.0 μ m/div, Z: 250nm/div   X: 1.0 μ m/div, Z: 15nm/div

Fig.4 Three dimensional representation for method 1 (left) and method 3 (right)

The effects of the different thinning processes are tested. Method 2, thinning the wafer by TMAH wetting thinning provides a rough surface which needs to be smoothed by CMP before further utilization; however the roughness brings the uncertainty of mechanical integrity that the wafer is inclined to crack even in CMP flattening process. As a result, the feasibility of this method is firstly be negated. As to the other two methods, method 1 provides an average RMS roughness result at around 30nm, while method 3 produces the smoothest surface with an average RMS roughness at around 1nm. The AFM measuring results are shown in Fig. 4. These results reach the conclusion that method 3 is the optimized thinning process for this experiment.

3.3 Bump reflow process

The bump reflow process is performed in a five-zone oven with a peak temperature of 260°C held for 30s in N_2 ambient. The temperature of each zone is 100°C, 150°C, 220°C, 260°C, and 80°C respectively. The cooling rate is about 4°C/s.

4. Results and discussion

4.1 Composition of the bonding interlayer

Fig.5 shows the cross-sectioned view of the Cu/Sn interlayer after the bonding process. The IMC formed after bonding is composed of two components: a light grey colored layer and a dark grey colored layer. EDS analysis shows that the light grey colored layer is Cu_6Sn_5 and the darker grey colored layer is Cu_3Sn. The morphology of the Cu_6Sn_5 phase is not scallop-shaped but flat because during the isothermal aging process the scallops of Cu_6Sn_5 were flattened as a layer. The Cu_3Sn phase grows a lot and its thickness becomes comparable to that of the Cu_6Sn_5. It can be seen that the surface of the electroplated Cu layer is rough and not flat, but there is no obvious void formed at the interface between the Cu and IMC layer indicating good wetting during the bonding.

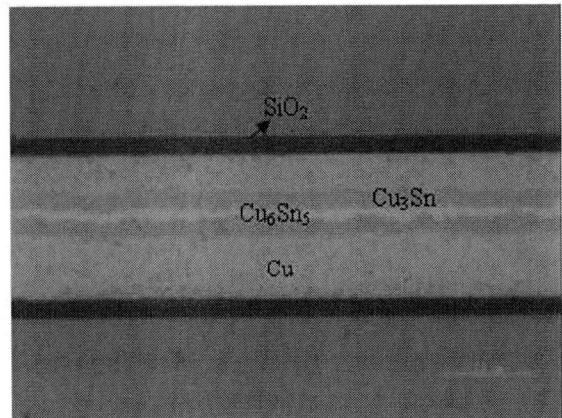

Fig.5 Cross-sectioned view of the Cu/Sn interlayer after bonding

4.2 Shear strength test

The shear strength of the samples has been measured using Dage series 4000B Bondtester.

As a reference, MIL-STD-883E, shear strength of 6MPa is needed. Measurement results of 20 samples show that the maximum shear strength is 25.3MPa, the minimum is 13.4MPa, and the average is 19.5MPa. The result demonstrates that the shear strength of the samples satisfies the MIL-STD-883E.

4.3 Hermeticity test

(1) Fine-leak test

Fine-leak tests as well as gross-leak tests have been done on the samples in order to characterize the hermeticity of the bonded wafers.

The cavity volume of the bonded sample is less than $0.001cm^3$ and according to MIL-STD-883E, the leakage limit is 5×10^{-8}atm cc/sec. In our test, the samples are first placed in a chamber, pressurized with helium soaked for 3h at 5bar, and transferred to the helium leak detector to be monitored. The reject limit is 1.9×10^{-9}atm cc/s which is well below the leakage limit.

(2) Gross leak test

The gross leaks are tested using fluorocarbon and are based on the "bubble method". The samples are placed under fluorocarbon liquid FC- 84 in a vacuum chamber full of N_2 with the pressure of 5 bar for 5h.The the samples are removed from the bath, dried in the air and immersed in fluorocarbon FC-40 which is maintained at 125 °C.If gross leak is present, any trapped helium will have ample time to escape from the cavity to be inspected. In our test, all 20 samples pass the gross-leak test successfully.

5. Conclusion

In this work, a hermetic MEMS package structure with silicon wafer as bonded cap at wafer-level scale is proposed. CMP process followed by spraying chemical smoothing process is utilized to thin the N(100) silicon cap wafer to the thickness of 150μm after wafer-level Cu/Sn isothermal solidification bonding. TMAH etching is used to open trenches penetrating through the cap wafer and then $Pb_{63}Sn_{37}$ bumps redistribution and reflow are utilized to achieve the final I/O interconnection. Method for the thinning process and parameters for Cu/Sn isothermal solidification bonding process are researched and optimized. The most important feature of this structure is the simplified fabrication process and low cost. Fine-leak tests as well as gross-leak tests have been done on the packaged samples in order to characterize the hermeticity of the bonded wafers which indicating an excellent leak rate of around 1.9×10^{-9} atm cc/s. The shear strength of the packaged samples has been measured using Dage series 4000B Bondtester and an average shear strength of 19.5Mpa is achieved. Both results meet the demands of MIL-STD-883E.

References

[1] Imed.Z, Michal Okoniewski,A Low-Temperature SU-8 Based Wafer-LevelHermetic Packaging for MEMS Devices, *IEEE TRANSACTIONS ON ADVANCED PACKAGING.*

[2] Rob N. Candler, Matthew A. Hopcroft, etc, Long-Term and Accelerated Life Testing of a Novel Single-Wafer Vacuum Encapsulation for MEMS Resonators, *JOURNAL OF MICROELECTROMECHANICAL SYSTEMS,* VOL. 15, NO. 6, December 2006.

[3] By Gilles Poupon, Nicolas Sillon, etc, System on Wafer: A New Silicon Concept in SiP, *Proceedings of the IEEE ,* Vol. 97, No. 1, January 2009.

[4] Junseok Chae, Joseph M. Giachino, Khalil Najafi, Fabrication and Characterization of a Wafer-Level MEMS Vacuum Package with Vertical Feedthrough, *Journal of Microelectromechanical Systems,* Vol 17, No. 1, Feb, 2008.

[5] Yi Tao, A.P.Malshe,W.D.Brown, Selective bonding and encapsulation for wafer-level vacuum packaging of MEMS and related micro systems, *Microelectronics Reliability,* 44 (2004), 251-258.

[6] Woonbae Kim, Qian Wang, Kyudong Jung, etc., Application of Au-Sn eutectic bonding in hermetric RF MEMS wafer level packaging, *9th Int'l Symposium on Advanced Oackaging Materials, IEEE,* pp. 215-219, 2004.

[7] Li Li, Jiwei Jiao, Le Luo, Yuelin Wang, Cu/Sn Isothermal Solidification Technology for Hermetic Packaging of MEMS, *Proceeding of the 1st IEEE International Conference of Nano/Micro Engineered and Molecular Systems,* January 18-21, 2006, Zhuhai, China.

Study on a 3D Packaging Structure with Benzocyclobutene as a Dielectric Layer for Radio Frequency Application

Fei Geng, Xiaoyun Ding, Le Luo

Shanghai Institute of Microsystem and Information Technology, Chinese Academy of Sciences,
Shanghai, China, Changning Road 865,
Email: gengfei@mail.sim.ac.cn, Tel: 021-62511070-5464

Abstract

A new wafer-level 3D packaging structure with benzocyclobutene (BCB) as interlayer dielectrics (ILDs) for multichip module fabrication is proposed for the application in radio frequency. The packaging structure consists of two layers of BCB films and three layers of metalized films, in which the monolithic microwave IC (MMIC), thin film resistors, strip lines and microstrip lines are integrated. Wet etched cavities fabricated on the silicon substrate are used for mounting the active and passive components. BCB layers cover on the components and serve as ILDs for interconnections. Gold bumps are used as electric interconnections between different layers, which eliminate the need of preparing vias by costly dry etching and deposition process. In order to get highly qualitied BCB films for the subsequent chemical mechanical planarization (CMP) and multilayer metallization processes, the BCB curing profile is optimized and the roughness of the BCB film after CMP process is controlled lower than 10nm. The thermal, mechanical and electrical properties of the packaging structure are investigated. The thermal resistance can be controlled below 2 °C/W. The average shear strength of the gold bumps on the BCB surface is around 70 N/mm². The performances of MMIC and interconnection structure in high frequency are optimized and tested. The S-parameters curves of the packaged MMIC shift slightly showing perfect transmission character. The insertion loss (S_{21}) change after the packaging process is less than 1db range at the operating frequency. And the return loss (S_{11}) is less than -8 dB from 10GHz to 15 GHz.

1. Introduction

Current electronic package trends to be denser, smaller and cheaper. Packaging technology has been developed from dual-in-line technology, surface mount technology (SMT), ball grid array (BGA) to 3D Multichip Modules (MCM) package. 3D MCM package holds promise for reducing interconnect delays by reducing interconnect length, which results in a reduction of interconnect associated parasitic capacitance and inductance. The vertical interconnection technology adopted in 3D package is smaller and less weight. These advantages meet with the needs in RF applications which must address special requirements, such as low loss and low parasitic interconnections. RF devices remain costly due to the high cost of fabrication and packaging process. The MMIC interconnection often requires high-precision machining due to the small wavelengths involved. Wire bonding is a common technology used for interconnection, but it has the disadvantage of high electrical signal loss. Thus, a low-cost compact reliable technique for integrating ICs with RF devices is of great interest.

As the application of MCM in high frequency range, the properties of interlayer dielectric material are important for the packaging performance. BCB is selected as dielectric material for its low dielectric constant (2.65), low dissipation factor (0.0008) and low curing temperature (250 °C). These characteristics are effective for high-frequency systems. The technology of utilizing BCB as ILD has already been realized in multilayer MCM package structure in some literatures[1-4]. In these works, a MCM packaging structure was achieved by using via dry etching in BCB to realize interconnection and due to the high cost of dry etching and the development of BCB material, photosensitive BCB was used in high frequency Ku-band wave multichip module extensively. However, in these cases the vertical interconnection between layers was all achieved by sputtering and electroplating. Such processes are simple but can not meet the requirement of high aspect ratio via (HARVi). As the aspect ratio of standard photosensitive BCB via is in the range of 1:4, large via is needed when the BCB thickness increases above 10 μm[5]. As a result, different techniques were proposed for HARVi in polymers. One approach is using plating gold bumps followed by BCB films coating. But this way can only be implemented on the pads of the substrate instead of the dies because of the electroplating process. Furthermore, very thin die (< 30μm) is needed to match with the height of the metal stud of HARVi[6,7]. Another approach is to prefabricate gold stud bumps on substrate by wire-bonding machine similar to our previous work[8]. However, this approach is not compatible with wafer-level package technology.

In this paper a novel wafer-level MCM packaging technology is proposed. Gold bumps and spun-on BCB films are used as HARVis and ILDs, which eliminate the need for dry etching. The fabrication process is as follows: Fisrt, the micro-machined cavities are formed on silicon substrate. Next, the MMIC's are embedded in these recessed cavities. After that, gold stud bumps are planted as multi-layer vertical interconnections and BCB films cover the multichip modules as ILDs. Because of the cavities, the BCB has to be thick enough to planarize the slot around the chips and at the same time reducing the capacitive coupling between the two layers sufficiently as well. The wafer-level MCM mentioned above proposes a new concept of integrating active and passive devices on BCB film with chips embedded in silicon substrate for radio frequency applications. This MCM packaging structure consists of two layers of BCB films and three layers of metalized films, in which the MMIC, thin film resistors, strip lines and microstrip lines are integrated. In the following sections, the performances of MMIC, transmission lines will be test and the influence to the circuit at high working frequency introduced by the package will be optimized.

978-1-4244-4658-2/09 $25.00 © 2009 IEEE

2. Packaging process

The fabrication processes of the wafer-level 3D MCM packaging structure are shown in Fig.1. The detailed processes are presented in the following sections.

Barrier layer (SiO₂: 2µm)

(a)

Etched cavities (110µm deep)

(b)

Ground plane (Metal layer 1)

(c)

Silver epoxy

(d)

Gold bumps

(e)

BCB

(f)

CMP

(g)

Metal layer 2 (Cr/Au: 200Å/3µm)

(h)

(i)

Fig. 1. Process flow for the packaging structure

Preparation of silicon substrate

A basic photolithographic process and anisotropic etching in 50 °C 40wt% potassium hydroxide (KOH) are used to form the cavities on the silicon substrate for the chip embedded. The silicon etching rate is approximately 10µm/hour. The etching depth of the silicon substrate is set as 110µm, as shown in Fig.2a. A TiW layer is sputtered on the silicon for improving adhesion. Then a gold layer is sputtered as seed layer. The contact pads and ground plane are patterned by another photolithographic process, as shown in Fig.2b.

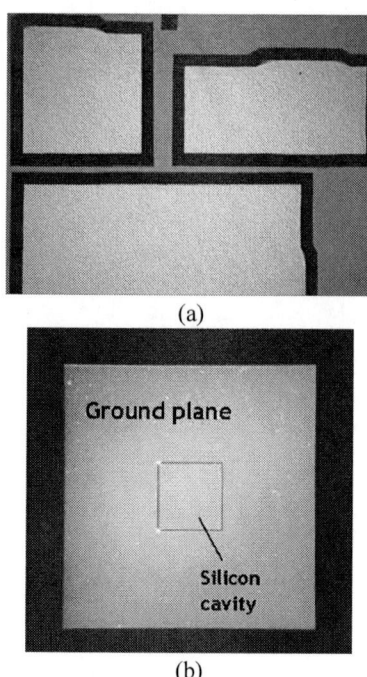

(a)

(b)

Fig. 2. Platform of Silicon substrate:
(a) after KOH etching; (b) with ground plane

Bump fabrication

The interlayer interconnection is through gold bumps followed by the dielectric film covering. The gold bumps are formed by wire-bonding machine on the contact pads. The tail filaments are pressed to column by a special capillary at the height of 25µm and 80µm in diameter, as shown in Fig.3.

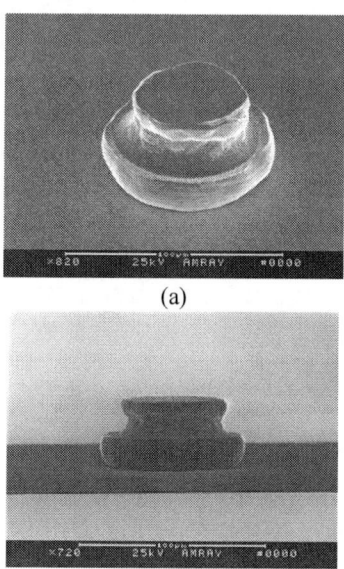

(a)

(b)

Fig. 3. Pressed gold bump: (a) top view; (b) lateral view

Coating and solidification of BCB film

The BCB coating process includes surface preparation, spin-coating, baking, and thermal curing. The thin film BCB layer (≤10μm) is easy to fabricate. But once the BCB thickness overpass 20μm, it is hard for BCB solidification and surface treatment, especially in multilayer BCB coating process. There are mainly two problems to confront. Firstly, if the cure temperature is too high, large volume of cracks will generate on the BCB surface(≥20μm), as shown in Fig.4a. Secondly, the rapid temperature change will lead to a large number of wrinkles on the BCB surface, as shown in Fig.4b.

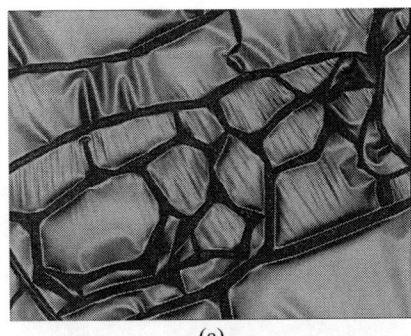

(a)

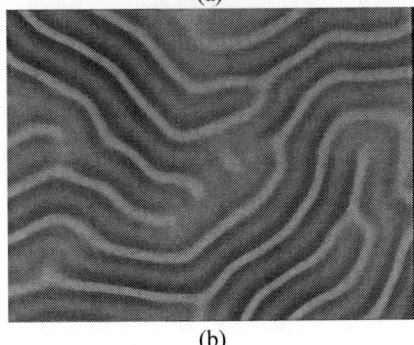

(b)

Fig. 4. 30μm thick BCB layer:
(a) crackled surface; (b) wrinkled surface

The first BCB layer is spun at 1000rpm rate for 20s and a thickness about 30μm BCB is achieved. The cure process is performed on an enclosed hot plate with flowing nitrogen to avoid oxidation. With the routine method, an 80% conversion is achieved. This partially cured BCB has enough unreacted bonds on the surface for enhanced metal-to-BCB adhesion. After the final metal deposition, the entire multilevel structure was full cured at 250 °C for 60 min. Due to the high coefficient of thermal expansion (CTE) of BCB, special care is necessary to achieve good adhesion to the gold coated silicon substrate.

CMP of BCB film

In order to remove the top coating of BCB film and achieve wafer level package (WLP), precise chemical mechanical planarization (CMP) technology is employed. Atomic force microscopy (AFM) and elipsometry of BCB films before and after CMP process show that higher down pressure and speed during polishing process will lead to a

higher removal rate at the expense of higher surface roughness, non-uniformity and scratch density.

The quality of the cured BCB film will influence the BCB removal rate and surface quality. The typical suggested temperature-time profile for BCB cure is 250 °C 1 hour in a box oven or furnace. The roughness of the BCB surface can reach 5.89nm measured by AFM, as shown in Fig.6a. In the subsequent experiment, it is found that the standard BCB cure process (250°C, 1hour) is not appropriate for the multilayer metallization and CMP process. So that the BCB cure condition should be optimized.

In order to promote the quality of BCB film, several ways are introduced to optimize the BCB cure profile for the following CMP step. The general solving method is to change the cure time and the cure temperature. On the one hand, if the cure time goes beyond needed, the BCB film will become too hard to remove which leads to a large volume of scratch after CMP process, as shown in Fig.5a. On the other hand, if the cure time is not long enough, the solvent in BCB will not be removed completely. And the BCB film will create defective surface after CMP process, as shown in Fig.5b.

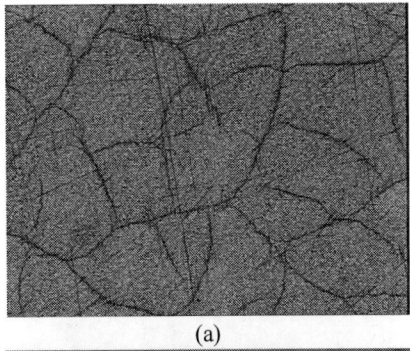

(a)

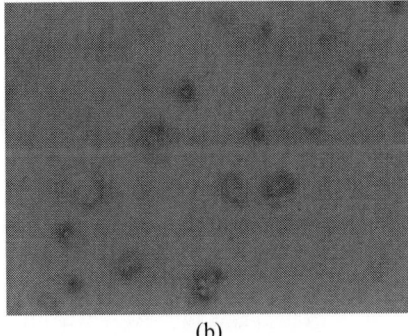

(b)

Fig. 5. The BCB surface after CMP process:
(a) Scratched surface; (b) Defective surface

Three kinds of cure profiles are tested to find out the optimized. As for the cure temperature (T_C), because the BCB film is less likely to generate cracks after cure process at 205°C, T_C is fixed at this temperature. As for the cure time, the three tested ones are 40min, 60min and 80min respectively. After CMP process, the results of BCB roughness measured by AFM are shown in Fig.6.

(a)

(b)

(c)

(d)

Fig. 6. The roughness of BCB surface:
(a) Standard cure without CMP Rz=5.89nm;
(b) T_C=205 °C, tC=40min, Rz=65.71nm;
(c) T_C=205 °C, tC=60min, Rz=12.39nm;
(d) T_C=205 °C, tC=80min, Rz=8.60nm

According to the AFM results, it can be seen that the longest cure time lead to the minimum roughness. So that the cure parameters is decided on T_C=205 °C, t_C=80min. Operating under this cure profile can achieve high quality BCB film after the CMP process. The roughness of BCB surface can reach 8.6nm, which is suitable for the subsequent metallization process.

Multilayer BCB coating

The multilayer coating process strongly relies on the planarizing property of BCB. A good planar surface determines the quality of subsequent interconnection steps. The structure of three metallization layers and two BCB layers has been fabricated, as shown in Fig.7. The yellow lines are gold metal line and dark areas in between are the BCB layers. A metalized via can be seen in the center region of the bottom BCB layer.

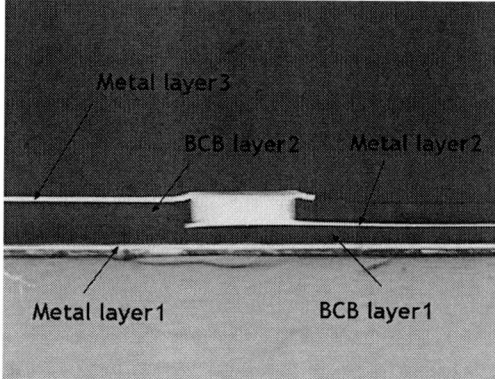

Fig. 7. Cross section of multilayer interconnection

Final interconnections and redistribution

One difficulty of this process is to improve the adhesion between BCB and metal layers. The most commonly used techniques for improving the adhesion is surface treatment of the BCB film, such as O_2 plasma. Because the adhesion between BCB and metal layer depends on their properties, one method to enhance adhesion is to optimize the thickness of the adhesion layer. Another method is to change the adhesion layer with different metals. Au layer is electroplated to achieve photo-defined interconnections and redistribution, as shown in figure 8.

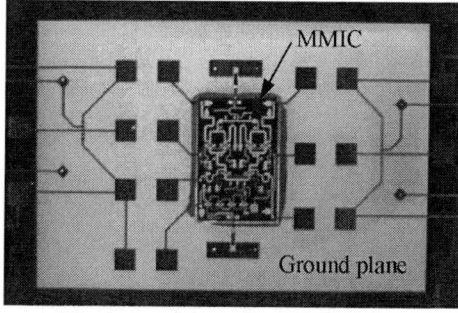

Fig. 8. Top metal line redistribution

3. Measurements

a. Shear strength Test

In order to investigate the mechanical reliability of the packaging structure, the shear strength of the gold bumps on the BCB surface is tested. The test load is 100g and the test speed is 100μm/s. The average shear strength of eighty samples is about 71.9 N/mm², as shown in Fig.9. The lower limit is 60N/mm² which meets the requirement of reliability of gold bumps.

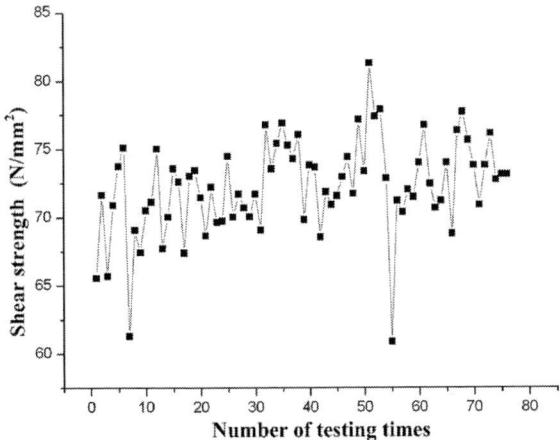

Fig. 9. Shear strength test of gold bumps

b. Thermal dissipation test

In order to evaluate the thermal dissipation behavior of this packaging structure, two types of thermal test chip are designed and fabricated in which thin resistance and diode are integrated for thermal resistance test. The thermal resistance of this packaging structure can be controlled within 2 °C/W. The temperature of the power module (1.1w) can be controlled below 80 °C with a heat sink. The detailed test structure was described in our previous work[9].

c. Measurements of transmission properties in RF range

In order to test the transmission properties of this multilayer interconnection structure, the packaging structure is applied to a low noise amplifier (LNA) chip to evaluate the effect of the package. The operating range frequency of the LNA is from 10GHz to 15GHz. Two layers of 25μm thick BCB is used and a section of CPW line is fabricated for the measurement of scattering parameters. The microstrip on the top layer is designed to be 50μm wide, as shown in Fig.11.

S-parameters are measured using Agilent 8722D network analyzer from DC to 20 GHz. The measured result after packaging is compared with the result of the bare chip. The measurement results of an embedded LNA chip are presented in Fig. 11, which shows that the insertion loss (S_{21}) change by the packaging structure is less than 1db range at the operating frequency. The return loss (S_{11}) is less than -8 dB from 10GHz to 15 GHz. It can be concluded that the S-parameters change slightly after packaging.

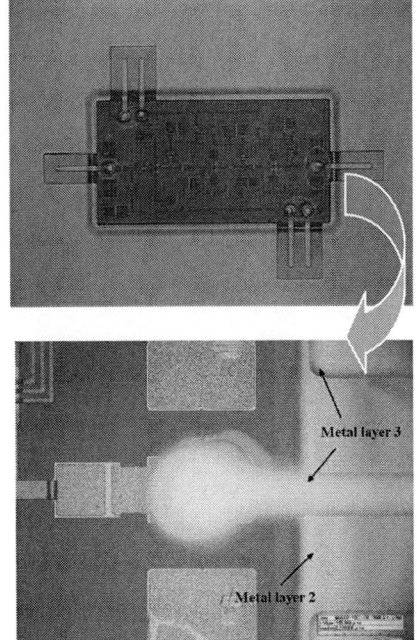

Fig. 10. Top view of the packaging structure

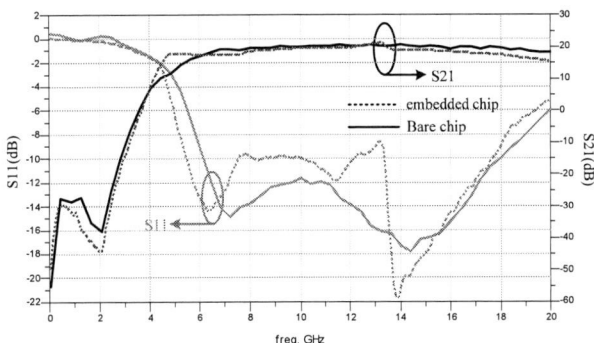

Fig. 11. S-parameters changes before and after being packaged

4. Conclusion

A wafer-level multi-layer packaging technology with chips and passive components embedded in silicon substrate for MCM applied in RF range is presented. BCB with low dielectric constant is selected as interlayer coating. The application of gold stud bumps achieves very short interconnections that limit cost and loss to the minimum. The processing parameters of BCB coating and curing are optimized to form steady BCB film. Furthermore, the influence of CMP process on BCB film is researched in detail and smooth BCB surface is achieved. The thermal, mechanical and electrical performances of this packaging structure are excellent. The thermal resistance can be controlled below 2 °C/W. Simultaneously, the temperature of the power modules (1.1w) is steady below 80 °C. The average shear strength of the gold bumps on the BCB surface is above 70 N/mm². The test results show good transmission properties of this packaging structure up to 20GHz. For a LNA chip, the insertion loss (S_{21}) change after packaging is less than 1db at the operating frequency and the return loss (S_{11}) is less than -8 dB from 10GHz to 15 GHz, both show slightly changes after packaging.

References

1. Percy B. Chinoy, Member, IEEE, and James Tajadod, "Processing and Microwave Characterization of Multilevel Interconnects Using Benzocyclobutene Dielectric", *IEEE Transactions on Components, Hybrids, and Manufacturing Technology,* 1993, 16 (7): pp. 714-719.

2. M.Topper, K. Buschick, and J.Wolf, "Embedding Technology-A Chip-First Approach using BCB", *International Symposium on Advanced Packaging Materials,* 1997,pp. 11-14.

3. H. Y. Li, M Tang, L. H .Guo and G. Q. Lo, "BCB (Benzocylcobutene) Process Integration for the RF Passive Device", *Electronics Packaging Technology Conference,* 2006, pp. 40-45.

4. Chul-Won Ju, Seong-Su Park, Seong-Jin Kim, Kyu-Ha Pack, Hee-Tae Lee and Min-Kyu Song, "Effects of O2/C2F6 Plasma Descum with RF cleaning on Via Formation in MCM-D Substrate using photosensitive BCB", *Electronic Components and Technology Conference,* 2001.

5. Geert J. Carchon, Walter De Raedt, and Eric Beyne, "Wafer-Level Packaging Technology for High-Q On-Chip Inductors and Transmission Lines", *IEEE Transactions on Microwave Theory and Techniques,* 2004, 52 (4), pp. 1244-1251.

6. K. Zoschke, J. Wolf, O. Ehrmann, M. Toepper, H. Reichl, "Copper/ Benzocyclobutene Multi Layer Wiring –A flexible base Technology for Wafer Level Integration of passive Components", *Electronics Packaging Technology Conference,* 2007, pp. 295-302.

7. F. Iker, D.S. Tezcan, R.C. Teixeira, P. Soussan, P. De Moor, E. Beyne and K. Baert, "3D Embedding and Interconnection of Ultra Thin (< 20 μm) Silicon Dies", *Electronics Packaging Technology Conference,* 2007, pp. 222-226.

8. Rodrigo Carrillo-Ramirez and Robert W. Jackson, "A Technique for Interconnecting Millimeter Wave Integrated Circuits Using BCB and Bump Bonds", *IEEE Microwave and Wireless Components Letters,* 2003, 13 (6), pp. 196-198.

9. Fei Geng, Jia-jie Tang, Le Luo, "Thermal Management and testing of MCM with embedded chip in Silicon Substrate" *International Conference on Electronic Packaging Technology & High Density Packaging,* 2008.

Through-Silicon Via Filling Process Using Pulse Reversal Plating

Xinxin Yang[1], Huiqin Ling[1], Dongyan Ding[1], Ming Li[1], Xianxian Yu[2], Dali Mao[1]

[1]School of Materials Science and Engineering, Shanghai Jiao Tong University, 800 Dongchuan Road, Shanghai, China
[2]Shanghai Sinyang Semiconductor Materials Co., Ltd, 1268 Wenhe Road, Shanghai, China
Email: dyding@sjtu.edu.cn

Abstract

Though silicon vias filled by pulse reversal current, and influences of frequency of pulse current and reverse current density are investigated. Chronopotentiometry was applied to analyze the principle of these effects. It was found that when frequency is too high, the reverse current was mainly consumed in process of charge-discharge of electric double layer; it cannot play the role of dissolving copper deposit on surface of via and orifice. For pulse reversal plating, low frequency is required. Generally frequency less than 40Hz is suitable in this system. High reverse current density could suppress copper growth on surface and at the orifice. It is beneficial for bottom-up via filling. However, if reverse current is too high, plating speed could be serious inhibited. In our study, 1:2 is the best ratio of forward and reverse current density.

Introduction

Through Silicon via (TSV) is a novel technology of 3-D chip interconnection [1-6]. By using this technology, size of stacked silicon chips is sharply reduced. TSV has the simplest structure and is expected to realize a high performance, high functionality and high-density LSI cube [7-10]. In recent years, TSV has been rapidly developed. However, many technological problems still exist, and the problem with via filling process is that it takes too long time.

The biggest obstacle of via filling process is the distribution of current density and Cu_2+ density in via [11-12]. Specifically speaking, on top of via, the density of current and Cu_2+ are relatively higher than those at the bottom. This situation will leads to higher plating speed on the top, leaving a void at the bottom of via. At first, people try to minimize this difference using low current density, prolonging via filling time. However, the obtained coating is still not satisfying. What's more, it is undesirable in industrial application for its low efficiency.

To solve the problem mentioned above, additives are added into plating solution to change the distribution of current density, and pulse current was used to accelerate the filling process. Jeng-Yu Lin filled the trenches of submicron level on wafer using pulse current [13], and investigated the effect of various frequencies of pulse current (PC) on the crystal orientation of copper deposit. It was found that when PC frequency was lower than 100 Hz, high (111)/ (200)ratio after annealing was achieved and the amount of void defects was reduced after chemical mechanical planarization CMP process. Chen Min-na[14] investigated the effect of organic additives on pulsed electroplating of Cu interconnect and find that proper concentration of additives can remarkably improve coverage, compactness and smoothness of the deposits. However, most people paid their attention on the effect of additives but less to effect of PC waveforms [15-17]. Recently pulse reversal plating was studied by P. Leisner [18]. It was

found that reverse current could dissolve copper deposit on surface and orifice preferentially which will increase the homogeneity of coating thickness in vias. It is applicable in via filling process.

In this article, vias are filled with different pulse reversal current waveforms, and influences of pulse frequency and density of reverse current are studied. Chronopotentiometry was applied to analyze the principle of these effects.

Experimental

Conical vias are filled in copper sulfate solution. The diameter of via is 110μm on the top while 40μm at the bottom, and the depth is 300μm. The electrolyte, supplied by Shanghai Sinyang Semiconductor Materials Co., Ltd., was composed of 160g/L $CuSO_4$, 60g/L H_2SO_4, 50ppm Cl-, 4ml/L 310A, 10ml/L 310S, 5ml/L 310L.

Two sets of experiments were carried out. First set is to find out the influence of cycle time by both chronopotentiometry and practical plating. Second set was to study the effect of various reverse current densities by electrochemical analyzer.

CHI600 electrochemistry workstation was used to test the potential response of different waveforms. The working electrode was Au film ($0.5cm^2$), counter electrode was Pt film and reference electrode was saturation mercury electrode (SCE).

Waveform used for plating in this article in shown in Fig. 1. To assure plating speed, t_f is set longer than t_r, with t_f: t_r equal to 4:1.

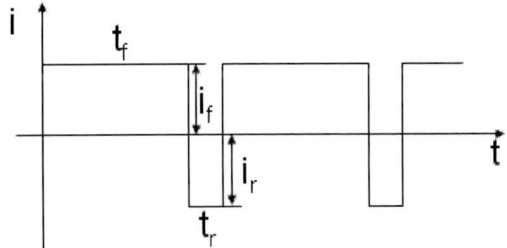

Fig. 1 Waveform used for plating,

In set 1, by changing periods of above-mentioned waveform, the relationship between potential response with time and period was researched. In set 2, by changing ratio of i_f and i_r, the impact of reverse current on hole-filling was investigated. And the forward current time was 20ms, the reverse current time was 5ms. The density of forward current is 0.4A/dm^2, and that of reverse current is changed from 0.4 A/dm^2 to 1.2 A/dm^2. Under the above-mentioned conditions, potential responses with time were tested and analyzed, and practical plating was performed to confirm the analysis.

Results and Discussion

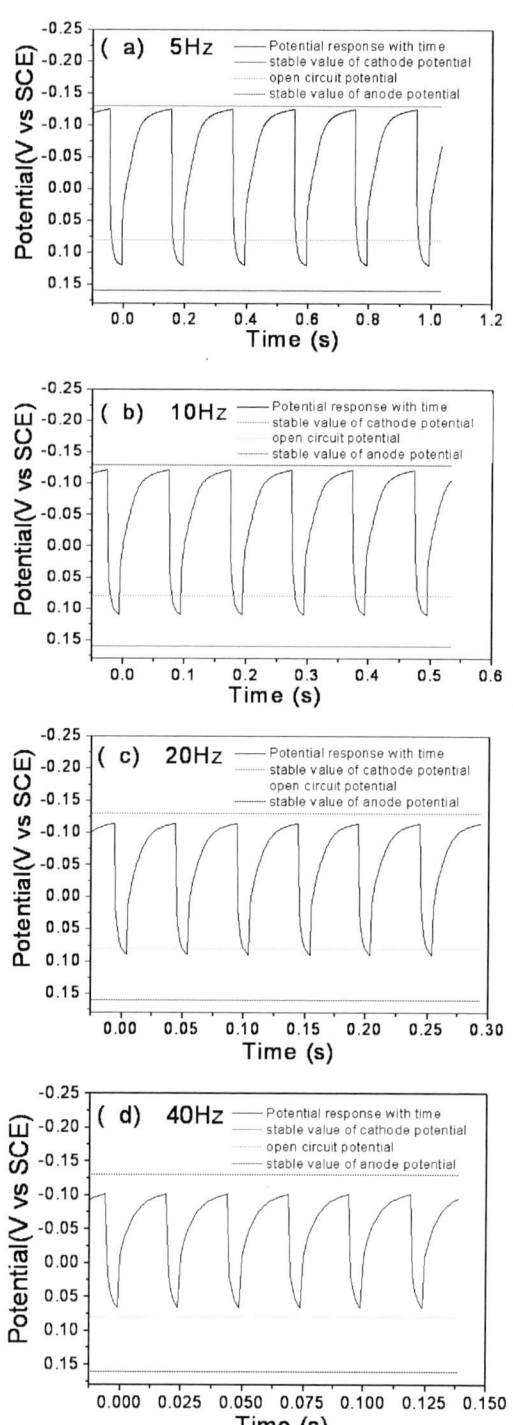

Fig. 2 Potential responses with time at various pulse frequencies, the applied forward/reverse current density is 0.4 /0.8 (A/dm²) (a) 5Hz, (b) 10Hz, (c) 20Hz, (d) 40Hz

It is widely known that capacitance effect exists in the process of plating. Because of this, when frequency of pulse

plating increases to a certain degree, the actual current is approximate to direct current.

In pulse reversal plating, the influence of frequency still exists. Potential responses with time at various pulse frequencies were shown in Fig. 2. Open circuit potential and stable values of cathode potential at 0.4 A/dm² and anode potential at 0.8A/dm².were also measured. From Fig. 2, we could clearly see that when frequency increases from 5 Hz to 10Hz, the maximum value of cathode potential decrease from -0.14V to -0.12V, and that of anode potential decrease from 0.12V to 0.11V. When frequency increases more, the maximum values of both anode and cathode potentials decrease more, too. So it is predicable that when the frequency increased to a certain degree, pulse reversal plating would be approximate to direct current plating, just like pulse plating.

In addition, when frequency is too high, as is shown in Fig. 2(d), the max value of anode potential is still lower than open circuit potential. It means that at high frequency, even though reverse current density is 0.8 A/dm², potential is still cathodic.

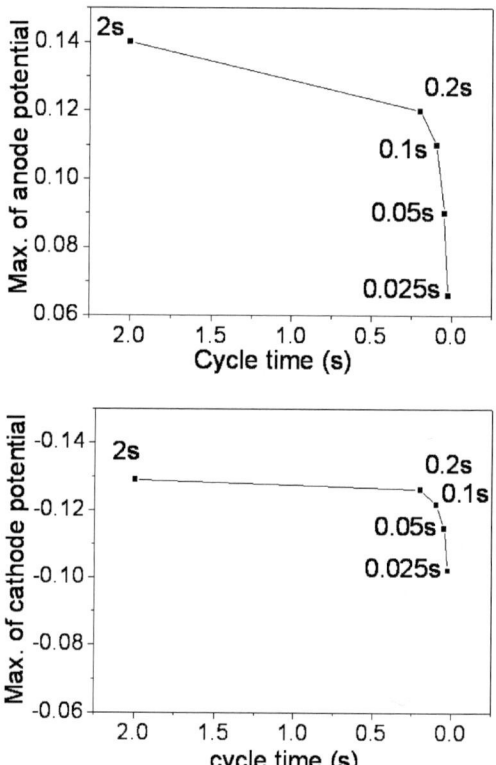

Fig. 3 Trend of maximum of anode and cathode potentials at various frequencies

In fact, the reverse current was mainly consumed in process of charge-discharge of electric double layer; it cannot play the role of dissolving copper deposit on surface of via and orifice. It means that for pulse reversal plating, low frequency is required.

Figure 3 is the trend of maximum values of cathode and anode potentials changing with frequency. From Fig. 3, we could see that when period is less than 0.2s, the maximum value of cathode potential decreases slightly when period shortened, and that of anode potential decreases sharply. When period increases to a certain degree, potential could reach stable value in single pulse time, and would change very little when frequency further decreased. According to the analysis in introduction, reverse current could dissolve copper deposit on surface and orifice preferentially, which is beneficial for bottom-up filling. Considering this, we should assure that the maximum value of anode potential is high. However, period could not be too long, either. This is because when period is at second-level, the plating process is not pulse reversal plating anymore. Long time of reverse current will cause some bad effect on the coating, for instance, oxidizing the copper coating.

Waveforms whose periods are 25ms and 100ms were chosen to perform practical via filling, and density of forward and reverse current are 0.4A/dm^2 and 0.8A/dm^2, section images were shown in Fig 4.

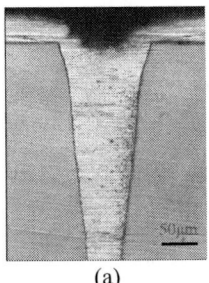 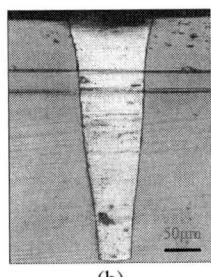

(a)   (b)

Fig. 4 Images of cross section of via with different periods. (a) 25ms, (b) 100ms

From these two images, we could see that on surface of wafer, copper growth is well-controlled when plating period is 100ms, when the period is 25ms, there is a copper coating of 20μm. This copper layer will bring some inconvenience in subsequent TSV process. By practical via filling, the effect of period discussed above was confirmed. In summary, in a certain scale, the longer period is, the higher maximum of anode potential will be, which will provide better inhabitation to the growth of copper on the surface and orifice. Therefore the filling result is preferred when period is 100ms.

Reverse current was known to dissolve copper deposit on surface and orifice preferentially and help to form the bottom-up filling in vias. However, whether the higher reverse current density is, the better filling result will be, is worth consideration. A set of experiments were designed to solve this problem. The parameters of these experiments were as follows: forward/reverse current time is 20ms/5ms, density of forward current is 0.4A/dm^2, and densities of reverse current are 0.4A/dm^2, 0.8A/dm^2, 1.2/dm^2, separately. Potential responses with time were measured; the results are shown in Fig. 5.

From Fig. 5, we could conclude that the increase of reverse current density not only increased the maximum value of anode potential, but also decreased that of cathode

potential. Because cathode potential directly determined plating speed, it is unbeneficial to via filling when reverse current is too high.

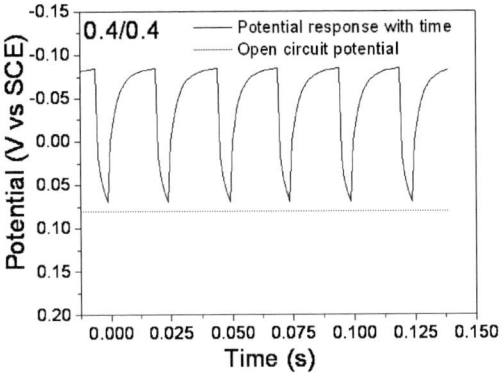

(a)

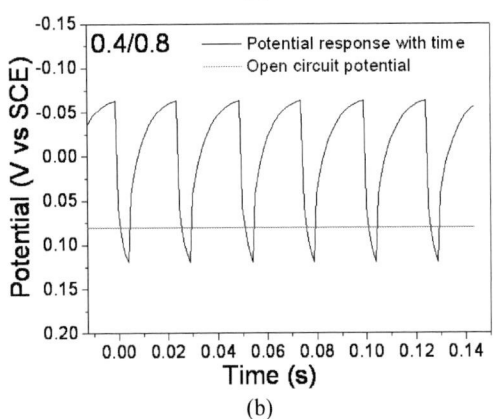

(b)

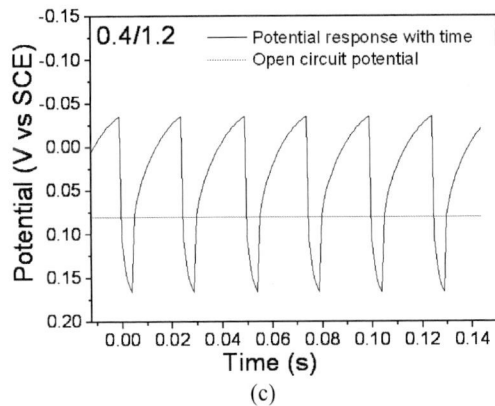

(c)

Fig. 5 Potential response with time at various reverse current densities (a) 0.4 A/dm^2, (b) 0.8 A/dm^2, (c) 1.2 A/dm^2

Practical plating was performed with the same parameters, and the results are displayed in Fig 6. When reverse current changes from 0.4 A/dm^2 to 0.8 A/dm^2, the maximum value of anode potential becomes higher than open circuit potential, which means reverse current begins to dissolve copper coating on surface and orifice. Practical plating results shows that the thickness of copper on surface is thinned from 30μm to 20μm.

978-1-4244-4658-2/09 $25.00 © 2009 IEEE

When reverse current continue to increase to 1.2 A/dm², little copper could be seen on surface. From these results, we could conclude that high reverse current density could suppress copper growth on surface and at the orifice. It is beneficial for bottom-up via filling. However, if reverse current is too high, plating speed could be serious inhibited, which is undesirable for industry application. Here 1:2 is the best ratio of if to ir.

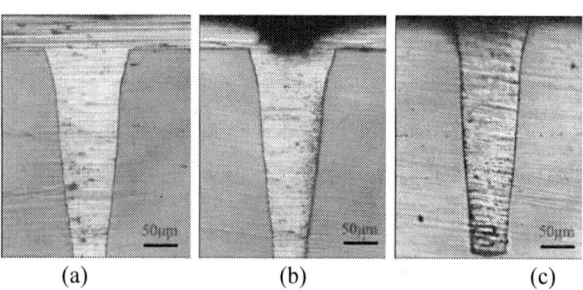

(a)   (b)   (c)

Fig. 6 Section image of via with different reverse current densities. (a) 0.4 A/dm², (b) 0.8 A/dm², (c) 1.2 A/dm²

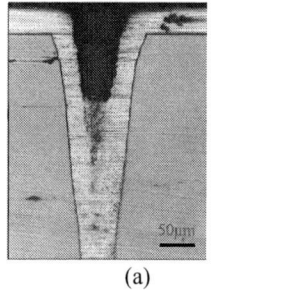 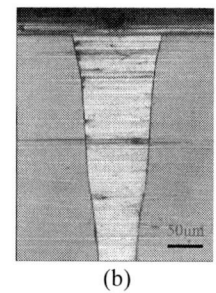

(a)   (b)

Fig. 7 Section image of via plated with direct current and pulse reversal current. (a). direct current, (b). pulse reversal current

Considering all the discussion above, an optimized waveform was selected for via filling, and the result was shown in fig. 7. Both of the two vias were filled at an average current of 0.2 A/dm² in 15 hours. Compared with via plated with direct current, via plated with pulse reversal current displayed its obvious superiority, and this result has proved that pulse reversal current could play a better role in TSV filling than direct current.

Conclusions

When frequency is too high, the reverse current was mainly consumed in process of charge-discharge of electric double layer; it cannot play the role of dissolving copper deposit on surface of via and orifice. For pulse reversal plating, low frequency is required. Generally frequency less than 40Hz is suitable in this system. High reverse current density could suppress copper growth on surface and at the orifice. It is beneficial for bottom-up via filling. However, if reverse current is too high, plating speed could be serious inhibited. 1:2 is the best ratio of forward and reverse current density. Pulse reversal plating is proved to be helpful in TSV filling process.

Acknowledgments

This work is sponsored by International Science and Technology Cooperation of China (No. 2008DFA51680), National Natural Science foundation of China (60876071), Shanghai nano technology promotion center (No. 0852nm06300). Wafer is supplied by Institute of Microelectronics, Singapore.

References

1. Tsui, Y.K. and S.W.R. Lee, "Design and fabrication of a flip-chip-on-chip 3-D packaging structure with a through-silicon via for underfill dispensing", *Ieee Transactions on Advanced Packaging*, Vol 28, No. 3(2005), pp. 413-420.
2. Kettner, P., et al., "New Technologies for advanced high density 3D packaging by using TSV process", *2008 International Conference on Electronic Packaging Technology & High Density Packaging*, Vols 1 and 2, (2008), pp. 43-45.
3. Jiang, T., S.J. Luo, "3D Integration-Present and Future", Eptc: *2008 10Th Electronics Packaging Technology Conference*, Vols 1-3, (2008), pp. 373-378.
4. Knickerbocker, J.U., et al. "3D silicon integration". *Proc 58Th Electronic Components & Technology Conference.* 2008, pp. 538-543.
5. Pozder, S., et al. "Progress of 3D integration technologies and 3D interconnects". *Proc of the IEEE 2007 International Interconnect Technology Conference.* 2007, pp. 213-215.
6. Spiesshoefer, S., Z. Rahman, et al. (2005). "Process integration for through-silicon vias." *Journal of Vacuum Science & Technology,* A 23(4): 824-829.
7. Spiesshoefer, S., et al. "Z-axis interconnects using fine pitch, nanoscale through-silicon vias: Process development". *Proc 54Th Electronic Components & Technology Conference*, Vols 1 and 2. 2004, pp. 466-471.
8. Spiesshoefer, S., et al., "Process integration for through-silicon vias", *Journal of Vacuum Science & Technology A*, Vol 23, No. 4(2005), pp. 824-829.
9. Tsui, Y.K. and S.W.R. Lee, "Design and fabrication of a flip-chip-on-chip 3-D packaging structure with a through-silicon via for underfill dispensing", *Ieee Transactions on Advanced Packaging*, Vol 28, No. 3(2005), pp. 413-420.
10. Kim, B., "Through-silicon-via copper deposition for vertical chip integration", *Enabling Technologies for 3-D Integration*, Vol 970, No. 2007), pp. 253-260.
11. Kim, B., et al. "Factors affecting copper filling process within high aspect ratio deep vias for 3D chip stacking". *Proc 56th Electronic Components & Technology Conference 2006*, Vol 1 and 2. 2006, pp. 838-843.
12. Beica, R., C. Sharbono, and T. Ritzdorf. "Through silicon via copper electrodeposition for 3D integration". *Proc 58Th Electronic Components & Technology Conference.* 2008, pp. 577-583.
13. Jeng-Yu Lin, et al. "Void Defect Reduction after Chemical Mechanical Planarization of Trenches Filled by Direct/Pulse Plating", *J. Electrochem. Soc.*, Vol 154, No 3(2007), pp. D139-D144.
14. Chen Min-na, et al. "Effect of Organic Additives on Pulsed Electroplating of Cu Interconnect", *Semiconductor Technology,* Vol 32, No 5(2007), pp. 387-390.

15. Zhang, Y., et al. "Fast copper plating process for TSV fill". *Proc 2007 International Microsystems, Packaging, Assembly and Circuits Technology Conference*. 2007, pp. 219-222.

16. Beica, R., C. Sharbono, and T. Ritzdorf. "Through silicon via copper electrodeposition for 3D integration". *Proc 58Th Electronic Components & Technology Conference*. 2008, pp. 577-583.

17. Beyne, E. "Solving Technical and Economical Barriers to the Adoption of Through-Si-Via 3D Integration Technologies". *Proc Eptc: 2008 10Th Electronics Packaging Technology Conference*, Vols 1-3. 2008, pp. 29-34.

18. P. Leisner, et al. "Recent progress in pulse reversal plating of copper for electronics applications", *Transactions of the Institute of Metal Finishing*, Vol 85, No 1(2007), pp. 40-45.

Low Temperature Cu-Sn Bonding by Isothermal Solidification Technology

Yibo Rong[1, 2], Jian Cai[1, 2], Shuidi Wang[1, 2], Songliang Jia[1, 2]

[1]Tsinghua National Laboratory for Information Science and Technology (TNList), Beijing 100084, CHN

[2]Institute of Microelectronics, Tsinghua University, Beijing 100084, CHN

Email: roberto_acmel@sina.com.cn, Tel: 86-10-62794957

Abstract

A low temperature wafer-to-wafer bonding technology for 3D packaging/integration based on Cu-Sn isothermal solidification (IS) technology is introduced in this paper. The fluxless bonding technique using Cu-Sn multilayer composites to produce higher re-melting temperature bonding layer is presented. The structure of the intermediate multi-layers and bonding patterns are designed, and the bonding process is optimized. The microstructure of bonding layer was investigated by SEM (Scanning Electronic Microscopy) and EDS (Energy Dispersive X-Ray Spectrometer). The compositions of the bonding layer show that there are intermetallic compounds (IMCs) with higher melting points. The bonding layers consist of Cu_6Sn_5 and Cu_3Sn phases. High strength of bonding layer has been detected, with average shear strength of 37.5MPa.

Introduction

With the development of microelectronic technology, 3D packaging/integration has been paid a lot of attention. As one of the important interconnect methods, Through Silicon Via (TSV) is getting more and more attention. The major technologies of TSV involve via formation (mostly by deep reaction ion etching), via filling by electroplating and inter-wafer bonding. Wafer-to-wafer bonding would be critical for TSV technology, especially when there are multi-layer interconnections. Bonding layer formed should not melt again in further bonding process. It leads to the requirement of higher re-melting temperature bonding layer.

As one of the crucial processes for 3D packaging, research and development of wafer-to-wafer bonding technology have become hot in both academia and industry. There are different wafer-to-wafer bonding methods for 3D packaging, including anode bonding, epoxy bonding, glass frit bonding and solder bonding, etc. All of these bonding technologies have their own advantages and disadvantages. For example, solder alloy bonding would have a lower bonding temperature and has been one of the promising bonding methods. However, the melting point of traditional solder joints/bonding layer would have a re-melting temperature no higher than the bonding temperature. This can not satisfied multiple layer bonding. Thus, the bonding temperature limited the post-processing temperature. This also restricts devices applications due to temperature constraint.

Another solution for wafer-to-wafer bonding is a liquid-solid inter-diffusion method, which is also called as isothermal solidification (IS) technology in some literature. Bonding layers by IS technology could achieve higher re-melting temperature structure.

IS Technology Mechanism and Bonding Couple Selection

The bonding structure of isothermal solidification (IS) technology includes a diffusion couple consist of two metals. One metal is with relatively lower melting point but the other is a higher melting point metal. At a constant temperature slightly higher than the melting point of the lower melting temperature metal, reaction or/and diffusion would take place between the liquid phase of lower melting temperature metal and the solid phase of the higher one. The reaction products would be intermetallic compounds (IMCs) with higher melting point. This means the bonding layer would have a higher re-melting temperature. So IS process gives a quantum jump to the post-processing temperature of the compound or solid solution fabricated. Additionally, as bonding would be occurred in lower temperature, this could induce lower stress for bonding system and devices on wafer.

For diffusion couple design, the low melting point metal is the key issue. Only a few of metals have lower melting temperature. Considering environmental issues, only Ga, In, Sn, Bi could be used for bonding. Table 1 shows the melting points of these elements. The melting temperature of Ga is too close to room temperature and the melting temperature of Bi is a little bit high. So In and Sn would be the candidates for diffusion couple. However, In is too easy to be oxidized. Then, Sn is the final decision for the lower melting point metal in diffusion couple.

Table 1 Low melting point non-contaminated metals

Element	Melting Point(°C)
Ga	29.8
In	156.6
Sn	232.0
Bi	271.3

For high melting point in diffusion couple, several metals would be selected. The common metals would be Au, Ag, and Cu. These metals could form different IMC with Sn and thus would achieve higher re-melting bonding structure. Compared with Au-Sn system, Cu-Sn bonding system would be lower costs. And the reaction rate is also higher. So Cu-Sn system was chosen as bonding media in this work.

Fundamental of Cu-Sn Bonding

To descript the fundamental of isothermal solidification, it is necessary to take a look on the Cu-Sn binary phase diagram. The Equilibrium phase diagram of Cu-Sn System is shown in Fig. 1 [1]. There are many equilibrium phases in this diagram. Among them, Cu_3Sn and Cu_6Sn_5 are considered to be two most important phases. Above 227°C, with tin composition of 99.3wt.%, reaction occurs between copper and tin [2]. And above 232°C, when tin becomes liquid, the

978-1-4244-4658-2/09 $25.00 © 2009 IEEE

reaction accelerates. For tin of 61 to 100wt.%, the alloy is a mixture of Sn and Cu_6Sn_5 with a melting point of 227°C. The alloy with tin composition of 38 to 61wt.% consists Cu_6Sn_5 and Cu_3Sn with a melting point of 415°C. And if tin composition reduces below 38wt.%, the melting point may even increase to 713°C [3].

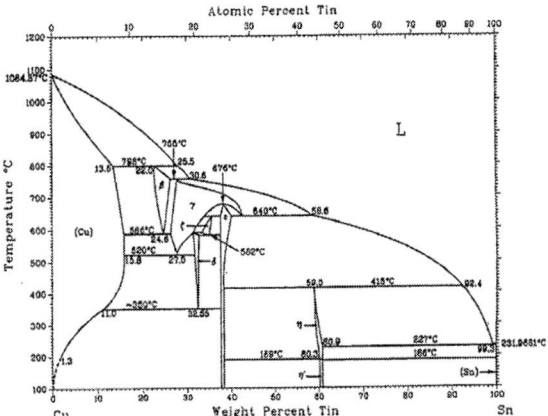

Fig. 1 Equilibrium phase diagram of Cu-Sn System

In order to realize a stable higher re-melting temperature of bonding structure, tin has been designed to be exhaust after reaction, while some copper would be left. The expected structure after bonding would be as those shown in Fig. 2, $Cu/Cu_3Sn/Cu_6Sn_5/Cu_3Sn/Cu$ or $Cu/Cu_3Sn/Cu$.

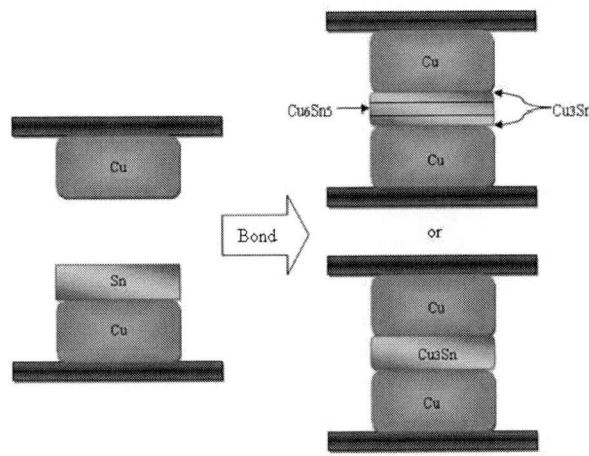

Fig. 2 Expected Cu-Sn Bonding Structures

If only Cu_3Sn left after reaction, then

$$\frac{S_{Cu} \cdot \delta_{Cu}}{W_{Cu}} = 3 \cdot \frac{S_{Sn} \cdot \delta_{Sn}}{W_{Sn}} \tag{1}$$

If only Cu_6Sn_5 left after reaction, then

$$5 \cdot \frac{S_{Cu} \cdot \delta_{Cu}}{W_{Cu}} = 6 \cdot \frac{S_{Sn} \cdot \delta_{Sn}}{W_{Sn}} \tag{2}$$

S: Area of Bonding
δ: Bump Height/Thickness
W: Mole volume of Cu or Sn
Hereby, this paper designed the thickness of Cu/Sn/Cu multilayer as 5μm/5μm/5μm before bonding.

Concerning the requirement of test, the figures of the top wafer and the bottom wafer are designed separately. Layout of the bonding pattern is shown in Fig. 3.

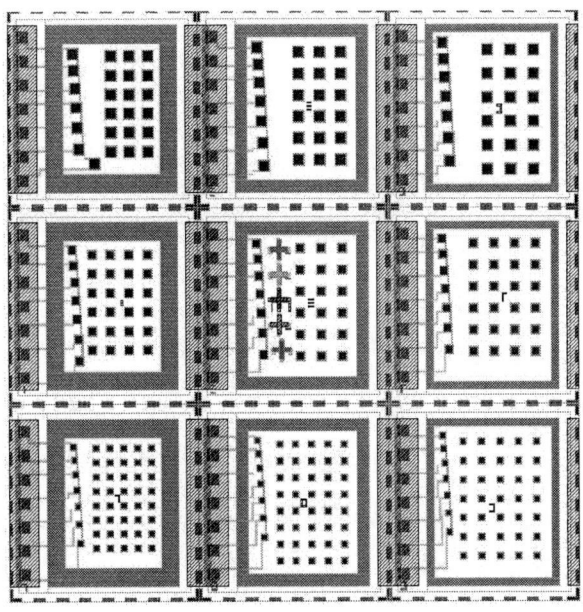

Fig. 3 Layout of bonding pattern

Experimental

Dummy wafers with Cu/Sn bumps were fabricated in-house. After carefully optimizing of fabricating process, excellent bump plating results are achieved. At a 260°C process temperature, 4 inches wafer-level Cu-Sn bonding, in which joints were almost void-free, was realized, including an interface with a melting point of 415°C. With further bonding and annealing steps, re-melting temperature of the joints could increase to 713°C. Details of the whole process are as following.

Firstly, this work prepared the wafers for bonding. TiW/Cu was sputtered as seed layer. And copper and tin bumps were electroplated as the diffusion couple for bonding.

When the wafers are ready, they are rinsed in plasma environment. By this way, impurities and oxides could be removed from the surfaces. Then, two wafers are aligned in SUSS MA6, and transferred into SUSS SB6 for bonding.

To avoid bubbles left between bonding patterns, this work select to process the whole bonding in vacuum, rather than in N_2 atmosphere. Heating until 160°C firstly and pre-bonding was preformed for 5 minutes at this temperature. Wafers would be heated until 260°C and bonded for 20 minutes at this temperature. The temperature of the bonding pair decreased to 200°C and inlet N_2 cutting down the cooling time. Besides, during the bonding process, pressure should be kept at 6kgf/cm². The thermal-pressure curve of bonding is shown in Fig. 4.

978-1-4244-4658-2/09 $25.00 © 2009 IEEE

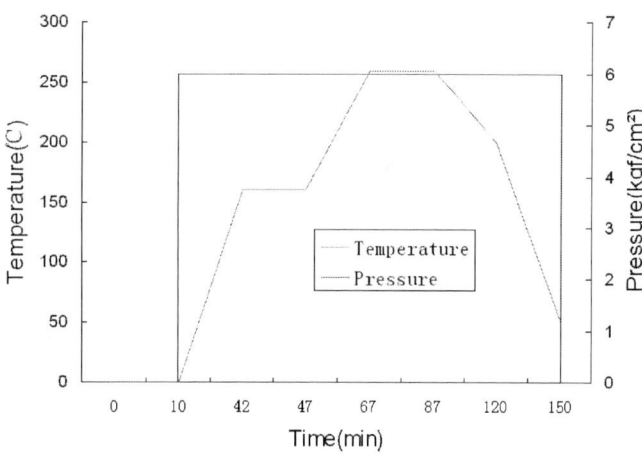

Fig. 4 Thermal-Pressure curve of bonding

Results and Discussion

To evaluate the microstructure of the bonding layer, SEM (Scanning Electronic Microscopy) is employed. Fig. 5 shows the SEM image of the cross-section of bonding intermetallic inter-layers. From this picture, there are three distinct phases existing in the joint between two copper phases. The EDS results of the three phases are listed in Tab. 2. The results show that the upper part of the bonding layer is Cu_3Sn, the middle one is Cu_6Sn_5, and the bottom one is Cu_3Sn.

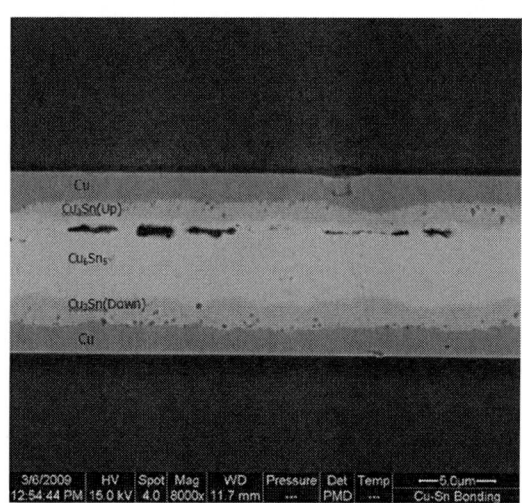

Fig. 5 Cross-section of bonding inter-metallic inter-layers

Table 2 EDS results of intermetallic inter-layers

IMC	Component	wt.%	at.%
Cu_3Sn(Up)	Cu	61.38	74.81
	Sn	38.62	25.19
Cu_6Sn_5	Cu	40.61	55.09
	Sn	59.39	44.91
Cu_3Sn(Down)	Cu	62.27	75.51
	Sn	37.73	24.49

Dage Series 4000 Shear Tester was utilized to measure the shearing strength of the bonding structure. The maximum shearing strength is as 47MPa and the minimum is as 33MPa. The average pressure is as 37.5MPa and the variance is 5.1MPa. Fig. 6 shows the surface of the bond after shearing

test. The image shows that peeling occurred between TiW and Si. This reveals the high reliability of the multi-layers of Cu-Sn bonding. Also from Fig. 6, we could figure out that the alignment accuracy during bonding is about 10μm.

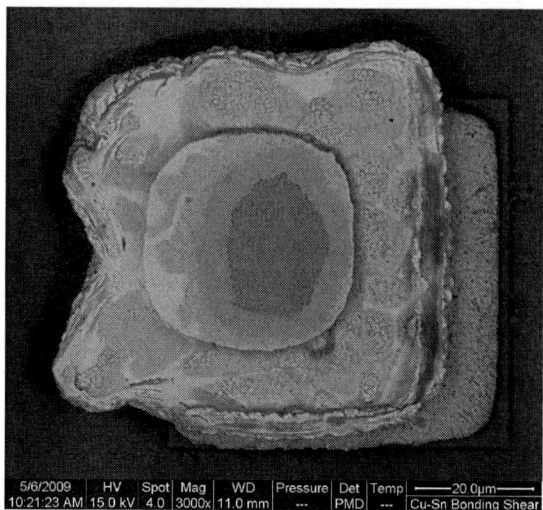

Fig. 6 Top-view of sheared bonding structure

Conclusion

Based on Cu-Sn isothermal solidification technology, bonding structures are designed and processes are optimized, Good bonding joints are achieved.

SEM results clearly show the microstructures of the joints consisting of Cu_6Sn_5 and Cu_3Sn inter-metallic phases.

In summary, the fluxless bonding method based on Cu-Sn isothermal solidification (IS) technology was demonstrated in the paper. At a 260°C process temperature, 4 inches wafer-level Cu-Sn bonding was realized, including an interface with a melting point of 415°C. Studies on the bonds reveal the basic bonding mechanism, and demonstrate the feasibility of this theory.

Acknowledgments

This work was supported by the National Basic Research Program of China, 2006CB302703.

The authors would appreciate Ms. Wenyan Yang, from the National Key Lab of Tribology, Tsinghua University, for her help on SEM analysis. The authors would also appreciate for the help from MEMSensing Microsystems Co., Ltd and Suzhou Institute of Nano-tech and Nano-bionics, CAS.

References

1. N. Saunders, A. P. Miodownik and etc., <u>Binary Alloy Phase Diagrams,</u> ASM International Metals Park, Ohio, USA, 1990: 1481.
2. S. Bader, W. Gust and H. Hieber, "Rapid formation of intermetallic compounds by interdiffusion in the Cu-Sn and Ni-Sn systems," *Acta metal*, 1995, 34, pp. 539-557.
3. Chin C. Lee, Yi-Dhia Chen, "High temperature tin-copper joints produced at low process temperature for stress reduction," *Thin Solid Films*, 1996, 286, pp. 213-218.

A PCB-based Package Structure

Na Wang[1,2], Jian Cai[1,2], Weikang Yang[1,2], Xinyu Dou[1,2]
[1]Tsinghua National Laboratory for Information Science and Technology (TNLIST)
[2]Tsinghua University, Beijing 100084, China
Email: wangna03@gmail.com

Abstract

A new package structure based on printed circuit board (PCB) as carrier substrate was proposed, aiming at providing an implementation of System-in Package (SiP) emphasizing small size and more importantly low cost. This structure is built upon a multilayer PCB, where open cavities are created in the top and bottom layers to house active dies, allowing package size reduction as well as protection against handling and impact. An application prototype was devised to integrate an RF IC and a baseband digital IC, typical of a wireless communication Rx system. Thermo-mechanical simulation models of the prototype are constructed to comparably assess certain aspects of its reliability and manufacturability. The results show comparable thermal performance compared with currently mainstream substrate material.

Thermal cycling test of the new structure to experimentally evaluate its reliability is planned with results to be reported in a separate report.

1. Introduction

IC packaging technologies that dominate the application space mostly use metal leadframe such as in Quad Flat Pack (QFP), or Bismaleimide Triazine (BT) as in ball grid array (BGA), as carrier substrate materials. Packages based on metal leadframe offer very low cost in volume production phase, as well as the low non-recurring engineering (NRE) and quick turn-around time in initial development phase. However, the limitations in pin count and package size/weight force IC designers to select BGA in many cases. BGA as a superior and mature technology, on the other hand, suffers in cost and cycle time. For many low-budget projects, designers are in a dilemma that one solution is out of reach while the other does not meet design requirements.

The solution proposed in this work aims at providing an alternative. This novel low cost packaging scheme puts cost and cycle time at priorities while trying to accommodate the needs for pin count and size/weight at the same time. It is a multilayer substrate-based structure similar to a BGA thus achieving the goals of high pin count and small size/weight. The substrate is, however, conventional low-cost PCB, thus achieving low cost and short cycle time. In extreme cases, this packaging method can even be utilized as a DIY approach (the fact is that many IC development efforts do not even command sufficient attentions from packaging houses).

It is not difficult to argue or even prove that current mainstream packaging technologies in the market place are not technologically superior. Proposing an alternative is equally as convenient. However, it is a vastly different game for a new technology to receive acceptance in the application markets. Reliability, size, weight, and cost are among the commonly mentioned factors that determine the outcome. A not-so-often discussed term is supply chain readiness, which sets timing for the new technology to take off. If timing is off, technological evolution force will push a given technology off its path with more advanced ones.

Built upon conventional multilayer PCB material, the proposed solution sacrifices packaging density (as compared to BGA) to gain benefits in low-cost and cycle time. With all its ingredients and manufacturing equipment readily available today, the new technology can be implemented to construct a SiP structure exhibiting features such as internal interconnects, IO array, accommodation of passive components, low cost (prototype and production), ease of testing, and short development cycle.

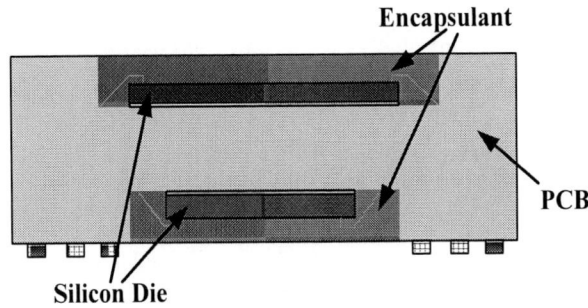

Fig.1 Schematic diagram of PCB-based Assembly

As demonstrated in Fig.1, the substrate is composed of PCB with multiple layers, which enables higher density, internal and external, interconnections than leadframe can provide. Cavities in the structure hold the active devices, allowing package size reduction. The cavities also provide protection against handling and impact. With this protection in place, the injection-molded top encapsulation as in BGA can be replaced with a variety of low cost processes such as dispensing liquid epoxy materials. The interconnection between bare die or dies and the PCB carrier is constructed by wirebonding process. Bond pads on the substrate can be either on the inner layers or on the top of PCB. The inner layer is preferred because it offers better protection and thinner overall package, as Fig.1 shows. The package's IO can be the pads on the bottom surface as in land grid array (LGA), or solder balls as in BGA. The finished package will have a BGA appearance.

In addition to the advantages in pin-count, cost, size/weight, and cycle time, this package also can be utilized with flexibility as a structure for IC or assembly level system integration (SiP for example). Multiple dies can be embedded in this structure since the PCB substrate allows interconnection within the package [1]. Passive components can be placed in the cavities or on the top. For finished packages, it is straightforward to further stack them up to form POP with SMT assembly equipment [2].

978-1-4244-4658-2/09 $25.00 © 2009 IEEE

Prototypes combining an RF IC and a digital IC are carried out as test vehicles for the proposed structure. Thermo-mechanical simulation of the structure is conducted to gain insights into certain aspects of its reliability and manufacturability [3] [4]. The simulation results presented here show that the low cost PCB-based packaging structure exhibits adequate performance and reliability for the application. Planned temperature cycle tests will be performed in the near future on the system to verify its reliability.

2. Design and Materials

Prototypes containing two ICs are carried out to demonstrate the design and processes of the proposed structure. The ICs in bare die form are embedded in the cavities of top and bottom layer respectively. The structure design, material selection, and process flow of the proposed package are discussed here.

2.1 PCB Substrate Preparation

Taking pad count, pad pitch and electric interconnection complexity into account, the number of layers of PCB is first to be determined. The PCB layer count is also limited by cost considerations and the capability of the vendor. For the prototype devised, the PCB carrier has six layers.

The size of cavities is determined by the die size plus space for pads on PCB. Redundancy needs to be spared for the cavity size according to the precision of bonders and wire loops. In our experiment the distances from the wirebonding pads to the edge of silicon and to the edge of the cavity are both set to 0.5mm.

Conductive traces are formed in each layer, providing connections between wirebonding sites and solder balls. Conductive traces should go around the solder balls at the bottom side which interfere with the routing of the conductive traces. Therefore, the exposed surface of plated-through vias at the bottom side should be insulated. Finally the pads are gold plated for wirebonding.

2.2 Die Attachment

The adhesive material is selected according to the thermal properties and mechanical performances. Then the bare dies are attached into the two cavities respectively. Adhesive is applied to achieve a 25 to 50μm wet bondline thickness, which dispenses with approximately 25 to 50% filleting on all sides of the die. Alternative dispense amounts may be used depending on the application requirements.

The depth of the cavity is dictated by the die thickness and bondwire loop height. Thinner die are desired to achieve a small overall package.

2.3 Wire Bond

Followed is the wire bonding operation. Bonding-wire loops are connected between wire-bonding pads formed on the IC die and wire-bonding sites on the top and bottom layer of the PCB substrate.

2.4 Encapsulant dispensation

After that, epoxy mold compound is underfilled into the cavities used as liquid encapsulation material. The material is injected to the cavities and molded. This is the interface that bonds dies and substrate, which is most vulnerable to crack. Therefore, the CTE, stress and strain properties of the material should be comprehensively considered for the choice of the encapsulant.

2.5 Solder Balls

Finally the bottom solder balls are formed into the pattern to comply with JEDEC Design Standard No.95, BGA package. The final package appears to be a BGA.

3. Finite Element Simulation

3.1 The Structure

Fig.2 schematically shows all the structure components of the finite element model, which consists of a PCB substrate, two silicon dies, and two encapsulation volumes. Considering the needs to simplify model construction and subsequent computation of solutions, certain parts of the structure such as bondwires, adhesives, and pads are excluded from the model. The exclusion of these details will not affect the overall conclusion since the structure's integrity is dominantly maintained by bulk materials. If they were included, the degrees of freedom of the model would have been dictated by the smallest feature in the structure, which easily leads to a model that pushes the limit of computing hardware.

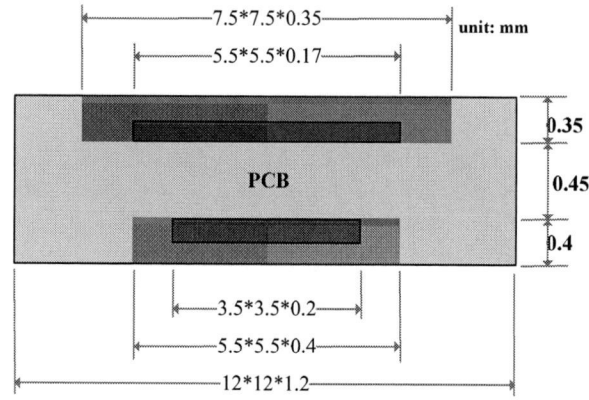

Fig.2 The simulation model

The dimensions of bulk materials are shown in Fig.2. Since vertical dimensions are of the same order of magnitude, a uniform mesh of good aspect ratio is expected as in Fig.3.

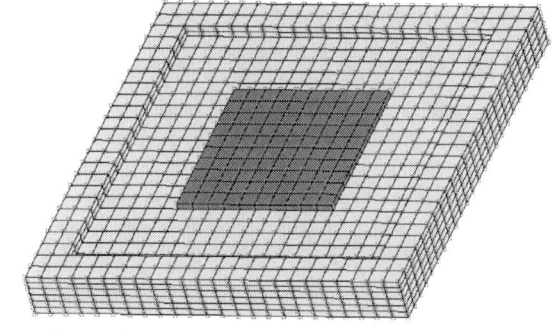

Fig.3 Three-dimensional finite element model

3.2 Material Properties

The materials of the chips and encapsulants are modeled with isotropic properties while the substrate is orthotropic. All

978-1-4244-4658-2/09 $25.00 © 2009 IEEE

the materials are assumed to be homogenous. Details of the material properties are listed in Table I.

Table 1 Material properties

Material		Young's Modulus (GPa)	Poisson's Ratio	Shear Modulus (GPa)	CTE (ppm/℃)
Silicon die		131	0.28	—	2.8
Encapsulant		14	0.37	—	18
FR4	In-Plane	19.7	0.18	3.7	16
	Out-Plane	9	0.39	2.9	55
BT	In-Plane	29	0.17	40	14
	Out-Plane	8.78	0.39	16.4	45

3.3 Boundary Conditions

The following assumptions and boundary conditions are used for the thermo-mechanical analysis. Due to symmetry at the x-z and y-z planes, the boundary condition of the model is: the displacements of the nodes on the symmetry x-z plane are constrained at the y direction. The displacements of the nodes on the symmetry y-z plane are constrained at the x direction. The displacement of the origin is $u=v=w=0$.

4. Results and Discussions

Since all of the materials are assumed linear and elastic, the stress and strain with 1□ of temperature elevation will yield all the information. Results for higher temperature elevation can be obtained proportionally. Reference state is selected to be stress-free with no deformation.

The purpose of the simulation is to compare the proposed package against the same structure with conventional BGA materials (BT in place of FR4, in particular).

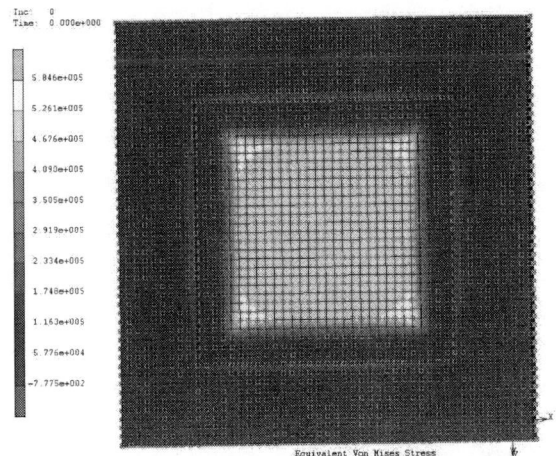

Fig.4 Equivalent Von Mises Stress of the structure (FR4)

Von Mises stress is convenient as a measure of load on material failure, which is displayed in all subsequent graphs. From Figs .4 and 5, high stresses occur along the edge of

silicon, the location of interface between carrier substrate, silicon, and encapsulation.

The maximum stress with FR4 substrate is 0.5846MPa, which shows a slight improvement over structure built upon BT substrate, 0.7132MPa. Although the CTE of FR4 material is slightly higher than that of BT, a significant drop in modulus leads to a less rigid structure, yielding lower stress. If the structure were to fail, it likely starts from interfacial locations. The lower stresses along the silicon edge as demonstrated by the simulation lead to less severe loading on the interfaces, thus a better reliability can be expected.

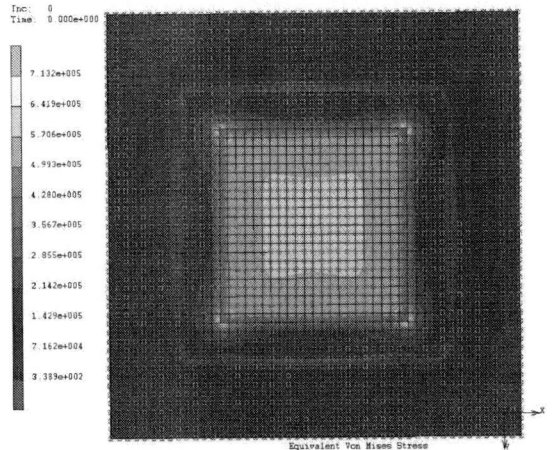

Fig.5 Equivalent Von Mises Stress of the structure (BT)

The crack strains of the two packages are also compared. From Figs. 6 and 7, the maximum crack strains also occur along the silicon edge. The maximum crack strain in the structure with FR4 material is 4.266×10^5, which is smaller than that of BT material, 5.093×10^5. Crack strain is a characteristics of the stress accumulation over multiple loading cycles. Consequently, it is expected that FR4 material with smaller stress will have lower probability of crack.

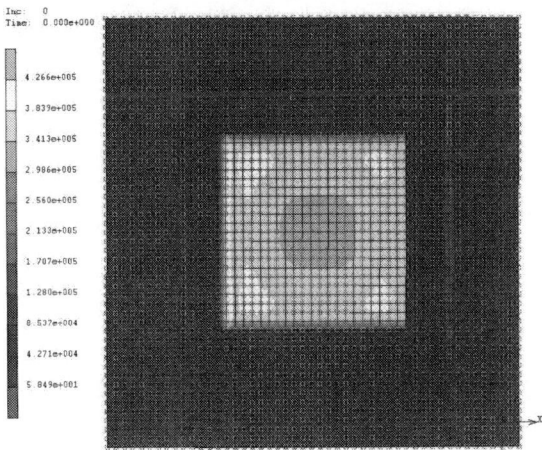

Fig.6 Equivalent crack strain of the structure (FR4)

The thermal strains of the two models are presented in Figs .8 and 9 respectively. It is observed that the distributions of the strain in two models are similar in pattern and magnitude (the maximum thermal strain with FR4 substrate is

4.856×10^{-5}, only slightly higher than that with BT substrate, 4.014×10^{-5}). The strain by itself is not related to material failure thus not a direct indicator of reliability. Its integral, however, indicates warpage of the package during subsequent assembly, which can be a measure of manufacturability. For a typical 100□ temperature elevation, the warpage predicted by the simulation models is .015mm and .013mm for FR4 and BT respectively.

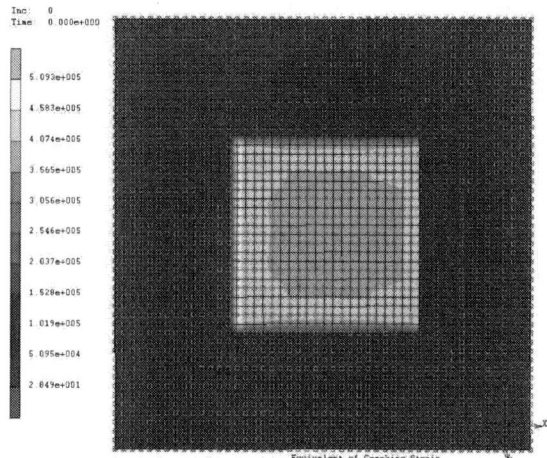

Fig.7 Equivalent crack strain of the structure (BT)

The results show that although the thermo-mechanical property of FR4 is interior to BT epoxy, its impact on reliability and manufacturability degradation is negligible.

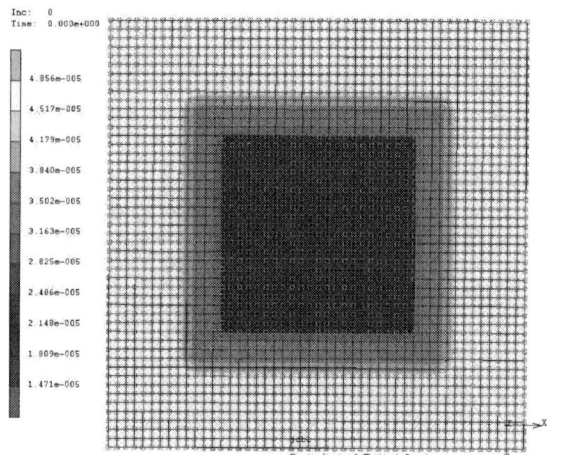

Fig.8 Equivalent thermal strain of the structure (FR4)

Conclusions

A new low-cost package structure based on PCB substrate carrier was proposed and prototyped. The core of this structure is a multilayer PCB, where cavities are made in the top and bottom layers to accommodate dies for the reduction of the package size. FEA simulation for the prototype that combines an RF IC and a digital IC showed that the proposed structure provides comparable reliability and performance with traditional substrate material.

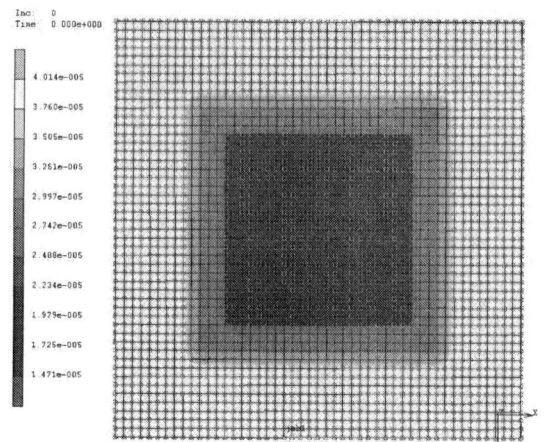

Fig.9 Equivalent thermal strain of the structure (BT)

References

1. C.T. Ko *et al*, "Embedded Active Device Packaging Technology for Next-Generation Chip-in-Substrate Package, CiSP", *Proceedings of IEEE Electronic Components and Technology Conference,* 2006.
2. K. Becker, E.Jung, A.Ostmann., T.Braun, A.Neumann., R.Aschenbrenner, H.Reichl, "Stackable System-On-Packages With Integrated Components", *IEEE Transactions on Advanced Packaging*, Vol. 27, No. 2(2004), pp. 268-277.
3. J.H.Lau, "Critical issues of wafer level chip scale package (WLCSP) with emphasis on cost analysis and solder joint reliability", *Electronic Manufacturing Technology Symposium [C], IEEE/CPMT/SEMI 29th International*, San Jose, USA , 2004.
4. S.M. Chang *et al*, "A Novel Design Structure for WLCSP With High Reliability, Low Cost, and Ease of Fabrication", *IEEE Transactions on Advanced Packaging*, Vol. 30, No. 3 (2007), pp. 377-383.

Collaborative Effect between Additives and Current in TSV Via Filling Process

Kaihe Zou[1], Huiqin Ling[1], Qi Li[1], Haiyong Cao[1], Xianxian Yu[2], Ming Li[1], Dali Mao[1]

[1]School of Materials Science and Engineering, Shanghai Jiao Tong University, 800 Dongchuan Road, Shanghai, China

[2]Shanghai Sinyang Semiconductor Materials Co., Ltd, 1268 Wenhe Road, Shanghai, China

Email: hqling@sjtu.edu.cn

Abstract

Deep via filling is one of key technologies of 3D packaging. Vias are commonly filled by electroplating. Since cupric transportation in vias is limited by diffusion, Current density is one of the most influential factors for copper plating in vias. We investigated new additive of accelerator, suppressor and leveler. Simulation of the competitive adsorption ability of accelerator and suppressor at different potential was studied. It was found that accelerator has more powerful adsorption ability than suppressor at high potential. Suppressor would form a passivating layer at the surface in the electroplating process, but the layer is easily disrupted by accelerator at high potential. We also investigated vias filling at different current density to prove our assumption. 0.4ASD was the best condition which got a fulfilled via without voids or seams. Conformal growth performance was attained at low current density and large current density would sealed the opening quickly, leaving seam at the bottom.

Introduction

In next generation IC packaging, 3D interconnect has been considered to be the solution not only for footprint shrinkage, but also for integration of different functional devices into one package. The heart of 3D silicon-based technology is through silicon via (TSV) which allows shortest chip-to-chip interconnections. [1] Via filling through copper electroplating is an extensively adopted approach in the interconnection metallization of integrated circuit(IC) chips. [2] During the electroplating process, some additives such as suppressor, accelerator or leveler should be selectively added to the plating electrolyte to realize the so-called superfilling. [2-8]

Superfilling, also called bottom-up filling, is a accepted filling performance to fulfill vias without voids or seams. [3-7] The bottom-up filling that the copper electrodeposition rate at the bottom of the via is dramatically greater than at the via opening may derives from a competition between electrolyte additives. [3, 5] Commonly, different additives are used to achieve the bottom-up growth. As shown in Fig.1 [8] Suppressors and levelers increase the potential at the opening of the via to control current crowding. Accelerators act as super catalysts activating the electrodeposition process at the via bottom, while chloride ions bridge the electron transfer of Cu^{2+}, anchor the suppressor, and promote the acceleration reaction.[6] Superconformal electrodeposition of copper is explained by the recently developed curvature-enhanced-accelerator coverage (CEAC) model, which is based on the assumptions that 1) the local growth velocity is proportional to the surface coverage of the accelerator, or catalyst, and 2) the catalyst remains segregated at the metal/electrolyte interface during copper deposition. [7] Accelerator formed Cu (I) (thiolate) inside the via and suppressor formed a layer at the sidewall and the opening of the via. [10, 11, 12]

Although the filling performance had been studied a lot, research work about the relationship between current density and additives was still lacking. The time of electroplating process can be reduced by enlarging current density which is quite important for its commercial application. The competitive of accelerator and suppressor determined the suitable current density. In our work, the adsorption effect of accelerator and suppressor at different potential is studied.

Experimental details

Linear voltammetry measurement was used to investigate the competitive adsorption effect of additives between accelerator and suppressors which were supplied by Shanghai Sinyang Semiconductor Materials Co., Ltd. Amperometric I-t curve was performed to investigate the competition between current and additives. The electrochemical tests utilized a conventional cell based on a columned flask with four ports which fixed the working electrode, saturated calomel reference electrode (SCE), and Pt counter electrode. The cell typically contained 250ml of electrolyte. The working electrode is exposed to the electrolyte with an area of 1 cm^2. For Amperometric I-t curve, the working electrode is firstly immersed in one solution for 10 minutes and then transfers to the cell. Considering the factor of response time, we do not record data in 10 seconds after beginning.

Microvias on the silicon wafer were 20μm in diameter and 100μm in depth. The silicon was then fixed on a copper sheet. Two phosphorus copper slices were used as anode and directly placed in the plating bath with a working volume of 1000ml. Agitation by continuous magnetic stirrer was sustained during electroplating to promote mass transfer. The composition of the copper electroplating solution contained: 40g/L Cu^{2+}, 0.05g/L Cl^-, 60g/L H_2SO_4, and 5 ml/L accelerator, 10ml/L suppressor, 5ml/L leveler. And these three additives were added to the solution with the sequence of accelerator, suppressor and leveler. The filling step was performed at room temperature. Only 20 cm^2 of copper sheet was exposed in the electrolyte. Three current densities including 0.2, 0.4 and 0.6ASD (A/dm^2) were performed to research the filling performance. These three samples were electroplated with same electric quantity. Cross-sectional images of samples were obtained by optical microscope.

Results and discussions

From our previous work, suppressor has high molecular weight, so it is hard to get into the bottom of the via. But accelerator usually adsorbs at the bottom of the via with its fast deposition. Accelerator inside the via has acceleration effect which makes a fast copper plating. And suppressor adsorbs on the surface and forms a passivating layer to inhibit the development of the limiting current density. [10] In real electroplating process, the adsorption relationship of additives has great meaning to get superfilling.

978-1-4244-4658-2/09 $25.00 © 2009 IEEE

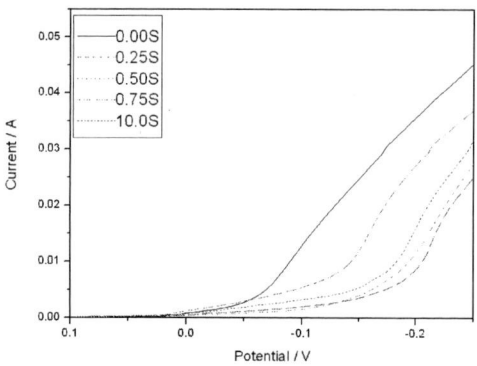

Figure 1. Linear voltammetry sweep curve with suppressor (S) varying from 0ml/L to 10 ml/L in a cupric sulfate electrolyte which also contained 10ml/L accelerator.

Figure 1 showed linear voltammetry sweep curve with the different quantities of suppressor in the electrolyte which also contained 10ml/L accelerator. The suppressor concentrations had been increased from 0 to 10 ml/L. Obviously the copper deposition was significantly inhibited along with suppressor increasing. The current was also inhibited to a quite low level under voltage of -0.2V. This was due to the polarizing effect which was considered as an inhibition effect by suppressor. All curves showed a fast rise of current when potential increased to a point. At the point, suppressor almost lost its suppression effect and current rose dramatically. This phenomenon may be caused by displacement of suppressor at high potential. The passivating layer may be disrupted by accelerator.

So we studied the displacement process at different potential to verify our assumption proposed above. As shown in Fig.2, at -0.05V, current had a little increase after 300 seconds. The increasing transition time was 100 seconds at -0.10V and current increased from beginning at -0.15V. When potential came to -0.20V and -0.25V, current had a dramatically increasing to a high steady level. Because the process was too fast, we recorded data from beginning. Current of these curves all started from more or less the zero. It was an evidence for that a suppressing layer had been formed on the surface of the electrode. Then accelerator in the electrolyte disrupted the layer. The replacement process became faster as potential increasing. It also showed that the adsorption ability of accelerator was enhanced greatly at high potential.

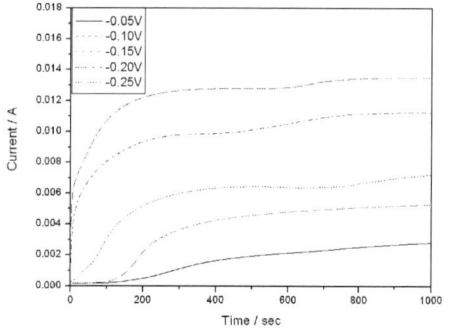

Figure 2. Amperometric I-t curve at different potential. The working electrode immersed in10 ml/L suppressor before being put in electrolyte contained 10 ml/L accelerator.

In contrast, we also studied the accelerator's displacement process by suppressor. As shown in Fig. 3, potentiostatic deposition from -0.05 to−0.25V in the solution contained 10ml/L suppressor yielded a declining current transient associated with progressive displacement of the suppressing layer formed by suppressor adsorption. The current had a quick decrease in about 100 seconds and then went to the same steady level nearly 0.05mA at -0.05 to -0.15 V. At -0.2V, it took 400 seconds before the current became steady. When the potential increased to -0.25V, current dropped at first and then had a little increasing peak at 50 second after beginning. The steady current was about 0.3 to 0.35mA. This may be caused by increasing active sites by accelerator's adsorption.

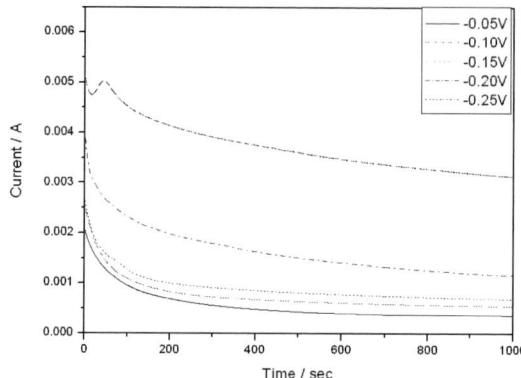

Figure 3. Amperometric I-t curve at different potentials. The working electrode immersed in 10ml/L accelerator before being put in electrolyte contained 10ml/L suppressor.

Accelerator in the electrolyte may have some effect to the displacement process. So we studied the displacement with accelerator in the electrolyte. As shown in Fig. 4, it showed a small descent at -0.05V to -0.15V compared to Fig. 3. When potential enhanced to -0.2 or -0.25V, the current obviously increased to a peak in about 50 seconds and then descended to 80% of the initial current. The reason of the peak in 50 seconds after beginning may also be caused by the widely increasing activation of active sites. At high potential, it cost longer time for suppressor to replace accelerator. It also showed the accelerator had more powerful adsorption ability at high potential.

Compared to Fig. 3, we could see that the accelerator in the electrolyte was an important factor which influenced the effect time of accelerator adsorbed on the electrode because the accelerator in electrolyte could be a supply source for consumption of accelerator on the electrode. The effect time had great meanings for bottom-up filling. Copper plating at the bottom of vias would be in a high speed for a compared longer time. But if the effect time was too long, the inhibition effect could not perform in the via filling process. So we still need more study for searching best quantity of accelerator to get the so-called superfilling.

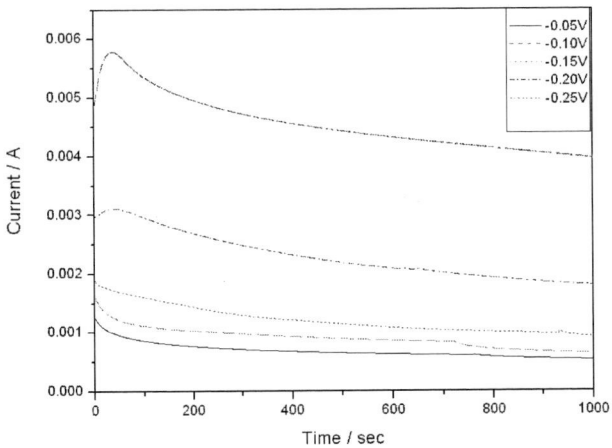

Figure 4. Amperometric I-t curve at different potential. The working electrode immersed in 10ml/L accelerator before being put in electrolyte contained 10ml/L suppressor and 10ml/L accelerator.

Because current density determined the potential in real via filling process, relationship between additive and potential could be used to search suitable current density. On the one hand, it was good to form the suppressing layer at the opening and the sidewall of the via. From this perspective, small current density would be better for the layer's keeping. On the other hand, high current density could activate accelerator at the bottom of the via. So three current densities including 0.2, 0.4 and 0.6ASD were investigated. According to optical micrograph, we could see 0.4ASD was the best condition which got a fulfilled via without voids or seams. Different via filling results was obtained with current density increasing.

(a) Current density was too small at 0.2 ASD that accelerator performed acceleration effect. Because of weak adsorption effect of accelerator at low potential, suppressor would adsorb instead of accelerator inside the via during electroplating process. Suppressor dominated the inside the via. The growth rates inside and outside the via were almost at the same level. Conformal growing performance was obtained. Diffusion of Cu^{2+} was not a key factor because of low growth rate. It may create seam in the middle of via. (b) When current density increased to 0.4 ASD, the adsorption ability of accelerator was enhanced. The molecular weight of accelerator was quite small, so it could have a fast deposition at the bottom. Growth rate at the bottom was very large. At the same time, the opening and sidewall were inhibited. The bottom was dominated by accelerator and the opening by suppressor. This was the so-called bottom-up. (c) The adsorption ability of accelerator became much powerful that the growth rate of opening was larger. Accelerator hindered the adsorption of suppressor. At this time, diffusion of Cu^{2+} was a key factor because it hard to get to the bottom when the growth rate inside the via was too fast. The Cu^{2+} was prior to plate at the opening. So it easily created seam at the bottom at this time.

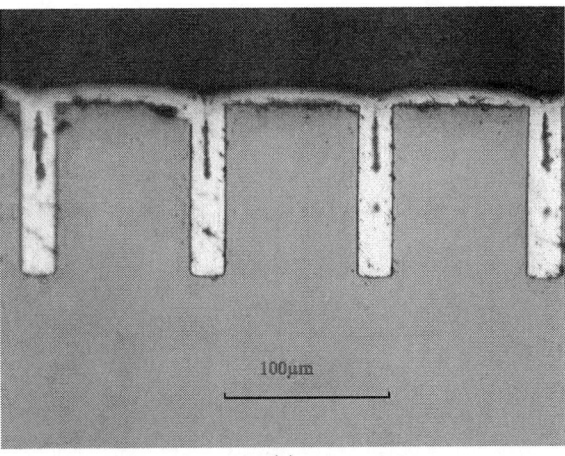

(a)

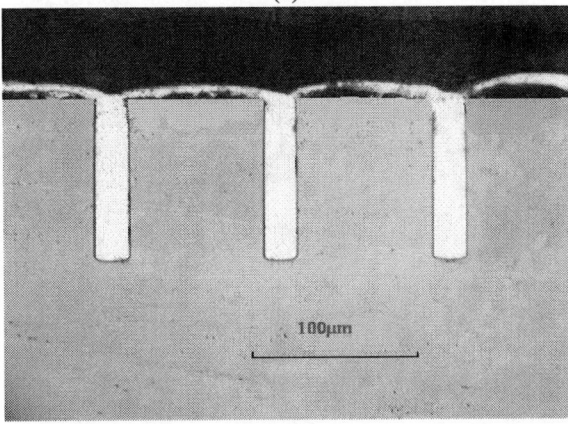

(b)

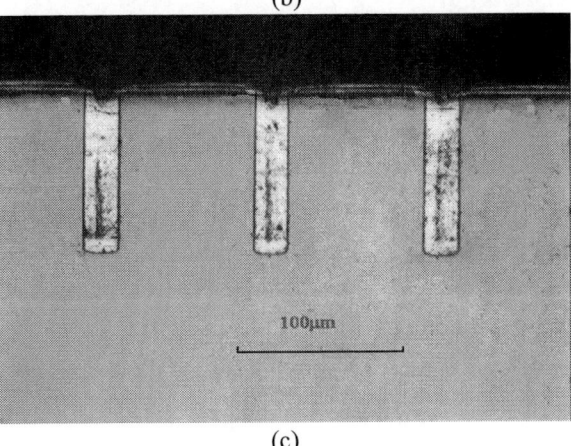

(c)

Figure 5. Vias (20μm in diameter and 100μm in depth) were filled with different current density: (a) 0.2 ASD; (b) 0.4ASD; (c) 0.6ASD.

These performances were caused by the changing adsorption ability of accelerator at different current density. The dominated additive at different location of the via changed as current increasing. Their illustration could be shown as Fig. 6.

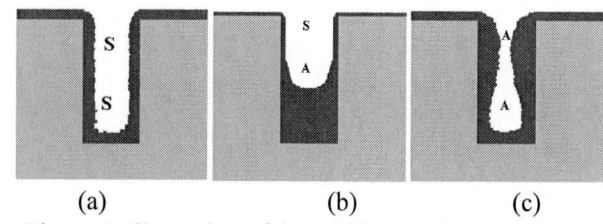

(a)   (b)   (c)

Figure 6. Illustration of three filling performances with variation of current densities with 0.2, 0.4 and 0.6ASD. The dominated additive inside via was shown in photos.

Conclusions

Adsorption ability of accelerator is enhanced along with potential increasing. The passivating layer formed by suppressor is easily disrupted by accelerator at high potential. When current density is small, also means potential is low, suppressor dominates inside the via, conformal filling was attained. At a suitable current density, The bottom was dominated by accelerator whereas the opening by suppressor. Thereby the so-called bottom-up filling was achieved. If current density is too large, accelerator dominates inside the via. Because the deposition rate is too high, Cu^{2+} is hard to reach the bottom and it is prior to plate at the opening which will seal the via quickly, leaving seam at the bottom.

Acknowledgments

This work is sponsored by International Science and Technology Cooperation of China (No. 2008DFA51680), National Natural Science foundation of China (60876071), Shanghai nano technology promotion center (No. 0852nm06300).

References

1. Yun Zhang, Thomas Richardson *et al*, Fast Copper Plating Process for TSV Fill, Enthone Inc., *Cookson Electronics*.

2. Dow, W.-P. and C.-W. Liu (2006). "Evaluating the Filling Performance of a Copper Plating Formula Using a Simple Galvanostat Method." *Journal of the Electrochemical Society* 153(3): C190-C194.

3. Dow, W.-P., H.-S. Huang, et al. "Interactions Between Brightener and Chloride Ions on Copper Electroplating for Laser-Drilled Via-Hole Filling." *Electrochemical and Solid-State Letters* 6(9): C134-C136.

4. Dow, W.-P, H.-S. Huang, et al. "Influence of Convection-Dependent Adsorption of Additives on Microvia Filling by Copper Electroplating." *Journal of the Electrochemical Society* 152(6): C425-C434.

5. Luhn, O, C. Van Hoof, et al. "Filling of microvia with an aspect ratio of 5 by copper electrodeposition." *Electrochimica Acta* 54(9): 2504-2508.

6. Moffat, T. P., D. Wheeler, et al. (2005). "Superconformal film growth: performance and quantification." *IBM J. Res. Dev.* 49(1): 19-36.

7. Beica, R., C. Sharbono, et al. (2008). Through silicon via copper electrodeposition for 3D integration. *Electronic Components and Technology Conference*, 2008. ECTC 2008. 58th.

8. Kondo, K., N. Yamakawa, et al. (2003). "Copper damascene electrodeposition and additives." *Journal of Electroanalytical Chemistry* 559: 137-142.

9. Hung, C.-C., Y.-L. Wang, et al. (2008). "Competitive Adsorption Between Bis(3-sodiumsulfopropyl disulfide) and Polyalkylene Glycols on Copper Electroplating." Journal of the Electrochemical Society 155(9): H669-H672.

10. Willey, M.J., J. Reid, and A.C. West, Adsorption Kinetics of Polyvinylpyrrolidone during Copper Electrodeposition. *Electrochemical and Solid-State Letters*, 2007. 10(4): p. D38-D41.

11. Willey, M.J. and A.C. West, SPS Adsorption and Desorption during Copper Electrodeposition and Its Impact on PEG Adsorption. *Journal of the Electrochemical Society*, 2007. 154(3): p. D156-D162.

12. Kim, S.-K., D. Josell, and T.P. Moffat, Electrodeposition of Cu in the PEI-PEG-Cl-SPS Additive System. *Journal of the Electrochemical Society*, 2006. 153(9): p. C616-C622.

Comparison Study of Effective Power Delivery in Advanced Substrate Technologies for High Speed Networking Applications

Judy Priest, Real Pomerleau, John Savic, Percy Aria, Jie Xue
Cisco Systems, Inc.
170 West Tasman Drive
San Jose, CA, 95135
Email: judyp@cisco.com, jixue@cisco.com

Abstract

High end networking and computing applications continue to drive silicon technologies for higher data rates and increased bandwidth. The push for silicon performance also drives a need for packaging performance to deliver clean and efficient power to the device. This paper compares the electrical performance of three new advanced packaging technologies designed to improve the DC and AC power delivery network in a network ASIC (Application Specific Integrated Circuit). A baseline component in conventional buildup with top-side capacitors was redesigned into three advanced package configurations: (1) a package utilizing a conventional buildup substrate, but with capacitors moved to the bottom ball-side of the package, (2) a package using a coreless substrate, and (3) a package using a coreless substrate, and with bottom ball-side capacitors. The package design, signal routing, device, and system and test environments are essentially unchanged, so the differences are attributed primarily to the substrate and package technology itself. Substrate, package, and system level electrical performance tests were performed and compared with the production baseline component design.

Motivation

Bandwidth across an interface, as defined as the rate of data throughput, is a key factor in determining network performance. As supply voltages drop, so does the amount of available noise margin. The ability to deliver and maintain stable voltage levels to devices on silicon through a power system from the supply to regulator to component, even during sustained switching events, is an area of analysis called "power integrity". Power integrity is analogous to, or even be considered a partial subset of signal integrity, although the correlation of the two is not obvious or direct.

One of the goals of a power delivery network design is to maintain as low an impedance as possible. Forming direct, electrically short, and multiple redundant contacts is one way to achieve this, but can perforate signal references and return paths if overused. Use of fully stacked microvias can also help, but this can be expensive to fabricate or introduce reliability risks.

In this particular example of a high speed networking device, a correlation between the peak-to-peak power rail noise and clock jitter has been observed through the use of a specialize purpose interposer board designed to decouple the power and noise from the ASIC and the main board [1]. It was reasoned that a similar result could be achievable using a single package solution and/or substrate technology that minimizes package inductance and provides adequate decoupling without the need of the interposer or additional changes to the motherboard.

The key motivation for this study is to quantify the relative improvement on overall power noise for various advanced substrate technologies, and then relate this back, in a quantifiable way, to the signal integrity and timing impact of the device in a system environment.

Substrate Comparison Study

A high volume production component was chosen as the test case for quantifying the impact that advanced substrates and new packaging concepts have on overall package power noise and timing integrity. In testing this ASIC, excessive clock jitter was tracked to supply noise found on the core and I/O power rails. The baseline component is a 40 mm x 40 mm BGA package with a 14 mm x 14 mm die utilizing a 4-4-4 "thin-core" substrate (200 μm core with 2 additional pre-preg layers; total core thickness is ~500μm) with top side Surface Mount Terminal (SMT) capacitors. The device was selected as the basis for comparison because the electrical performance was well understood, and the necessary diagnostic software and hardware test fixtures were readily available.

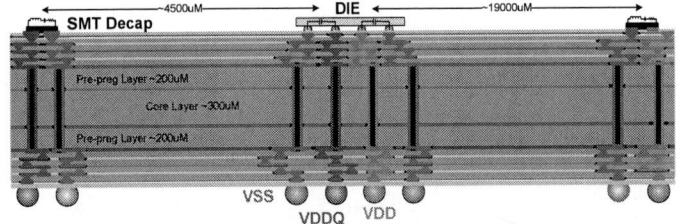

Figure 1 Thin core 4-4-4 baseline substrate case.

Baseline Case: This baseline package, shown in Figure 1, was redesigned into three different substrates and packages, intentionally preserving the same die bump, BGA ball out, and signal trace routing patterns as the baseline substrate wherever possible. The focus was to isolate the power delivery performance from other signal integrity improvements that could have been made with each substrate technology. Confining the design changes to the substrate power distribution system and capacitor assemblies made it easier to quantify the impact on the component timing and signal quality, and to rate its ability to deliver and maintain a stable voltage level under actual sustained switching events.

Figure 2 Bottom side of the BGA-side capacitor package. 0204 SMT decoupling caps mounted on the underside of the BGA between the BGA balls

Case 1: The first comparison case was using the same thin core 4-4-4 substrate and mounting capacitors on the bottom BGA side of the package under the die image, shown in Figure 2. Bottom side capacitor mounting on packages is achievable within today's supply chain and infrastructure using low profile (SMT) capacitors.

A total of 68 bottom-side 0204 SMT capacitors were added, 41 on the VDD rail and 27 on the I/O supply rail. As a result, 136 BGA balls were removed and traded off for capacitor pads.

Figure 3 Package with 10-layer coreless substrate

Case 2: The second comparison case was a BGA package utilizing a 9+1 ultra-thin coreless organic substrate, shown in Figure 3. Coreless substrate technology is still under development at many substrate manufacturers. There are currently has layer count limitations but has been demonstrated to be manufacturable for smaller body sizes with good assembly yield.

Because of the layer count reduction of the original 12-layer buildup baseline design to the 10-layer coreless

design, some minor layout changes were necessary to accommodate the design in a different technology [2].

Case 3: Finally, the third comparison case was an 8-layer coreless substrate package with fully stacked vias, with both top and bottom BGA side capacitors, shown in Figure 4. Bottom side capacitors are the same as Case #1. However the original top-side capacitors were replaced with higher performance 0306 Inter Digital Capacitors (IDC).

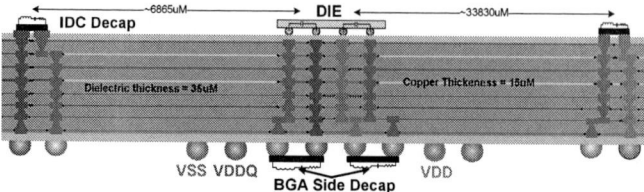

Figure 4 Package with 8-layer coreless substrate with top and bottom BGA side capacitors (similar to Figure 2)

Simulations and measurements were made at the substrate, package, and system level. Power plane impedance, core and I/O power noise were extracted using PowerSI [3] and then simulated in HSPICE [4] for all four cases.

All package styles were designed to be pin compatible for use on the same board. The characterization test board is shown in Figure 5.

Figure 5 Package test board

Special test points were designed on the substrates and custom lids were used to enable measurements of the power noise under different traffic patterns. Figure 6 shows a close up of the test socket used for the package measurements. Figure 7 shows a schematic diagram of the substrate measurement probe points and measurement methodology [5].

978-1-4244-4658-2/09 $25.00 © 2009 IEEE

Figure 6 Test socket

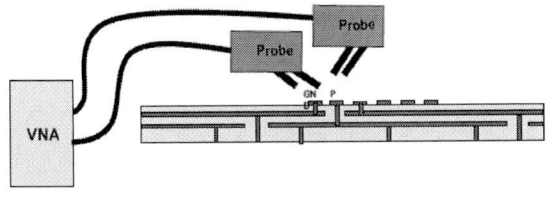

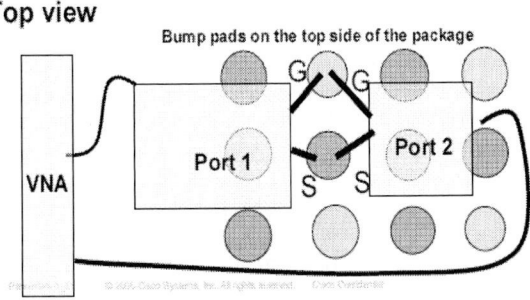

Figure 7 Side and top views of package test setup for impedance measurements on the substrate

Results

IR drop: The DC comparison for the various cases did not show significant differences between technologies in IR drop. This is because the ASIC test case used is a small die and IR drop is not a key factor.

Frequency response: The transfer impedance is shown in Figure 8 for the core voltage and in Figure 9 for the I/O voltage.

Ideally, the response would be a flat horizontal line across the frequency spectrum if all energy was completely transferred between the die, package, and board. In real physical systems, the frequency response shows a resistive "dip" at the intersection of the capacitive and inductive responses of the circuit. The higher in frequency and the lower the magnitude in the impedance profile that can be maintained, the more effective the system is in terms of

impedance transfer. That is of course, as long as anti-resonance peaks are not being created at the frequency of operation.

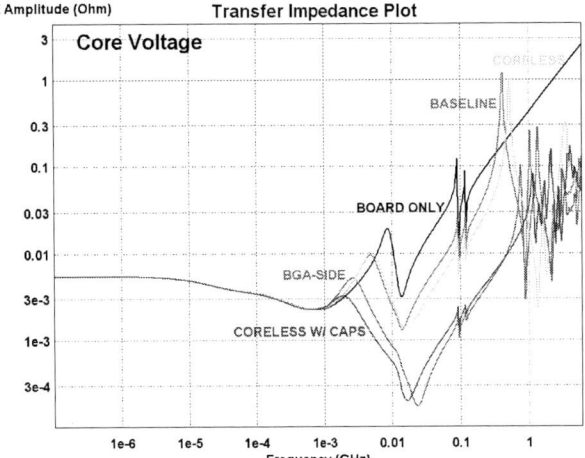

Figure 8 Core voltage transfer impedance plot

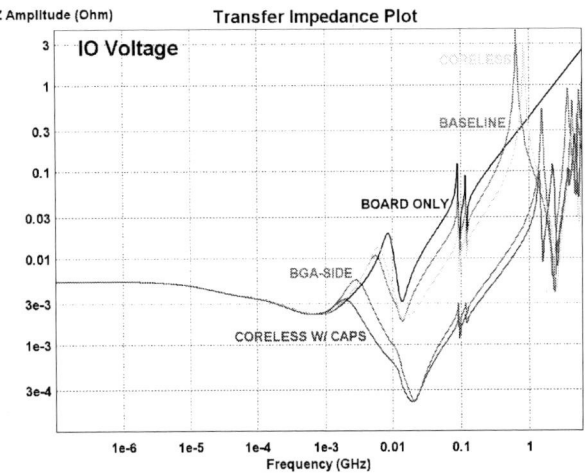

Figure 9 I/O voltage transfer impedance plot

The results show that coreless performs better than the baseline case. The coreless substrate would have yielded even better results if lower inductance top side capacitors were used. The BGA side capacitors and coreless with capacitors demonstrated the best performance in this comparison case.

Loop inductance: The loop inductance is the closed loop path as defined from the die through the package vias, distributed across substrate planes, to the decoupling capacitors, and back to the die. Loop inductance is an important metric in the design of overall power distribution system. In general, lower loop inductance improves the anticipated performance of the power delivery system. Table 1 shows a comparison of the computed path inductance of the original baseline system to the three other test substrates. All else being equal, physically shorter paths result in electrically shorter paths which will result in smaller loop inductance.

978-1-4244-4658-2/09 $25.00 © 2009 IEEE

Smaller effective loop inductance provides more stable power delivery to the core and I/O rails. If top-side capacitors are used, the overall loop inductance will not change. The most significant impact comes from the BGA bottom-side placement of capacitors, which provides very physically short connections and a substantial reduction of effective electrical loop inductance, on the order of 4.5X. The use of a coreless substrate eliminates the core layer and provides a more efficient path for the core supply, on the order of 3X loop inductance reduction. The improvement on the I/O supply of the coreless is partially attributed to the design change that was made in layer assignment to accommodate the redesign in this technology.

Table 1 Loop inductance table for package cases

Loop inductance	Vdd (core)	Vdd (I/O)
BASELINE design	32pH	91pH
#1: BGA side cap design	7pH	19pH
#2: Coreless design	10pH	11pH
#3: Coreless w/ caps	5pH	4pH

Capacitor ESL is not included as this is a vendor dependent parameter, and will affect results. Similarly, the capacitor placement location will also have a significant impact on results. In Case #3 (coreless with capacitors), the top side IDC capacitors were placed physically 75% further away than the SMD/T capacitors in the other cases, again, a change that was necessary as a result of the design remapping. However as with Case #1, the greatest impact came from the mounting of the bottom–side capacitors to the package.

Time domain noise: The true jitter is a circuit level phenomenon on the die was not practical to measure as this package comparison study was not conceived when the die was designed. However, the impact of loop inductance on maximum peak-to-peak core noise is a very useful metric and is measurable at the package level. The results are shown in Table 2.

Table 2 Summary of maximum peak-to-peak core noise measurement

	MIN (mV)	MAX (mV)	AVG (mV)	P2P (mV)	% IMPROVEMENT
BASELINE	943	1037	995	94	--
CORELESS	950	1042	1000	92	2
BGA-SIDE	954	1033	990	80	15
CORELESS W/ CAPS	969	1013	992	43	54

Peak-to-peak noise is not the only measure of jitter reduction. As time domain waveforms of the supply rail noise are viewed, it is important to examine the energy in the actual height and width of the pulses as well. In this respect, we also see a quantifiable improvement of all of the test cases when compared with the baseline case. The most significant is in the coreless with capacitor case.

Similar results are demonstrated for the maximum peak-to-peak I/O noise reduction, shown in Table 3.

All cases show an improvement in reducing the maximum peak-to-peak I/O rail noise, with the most significant improvement in the coreless with capacitor case with a 84% total noise reduction.

Table 3 Summary of maximum peak-to-peak I/O noise measurement

	MIN (mV)	MAX (mV)	AVG (mV)	P2P (mV)	% IMPROVEMENT
BASELINE	1238	1552	1397	314	--
BGA-SIDE	1339	1447	1397	107	66
CORELESS	1328	1449	1397	121	61
CORELESS W/ CAPS	1372	1422	1397	50	84

Impact on signal integrity, jitter, and timing: I/O jitter is easier to quantify as output clock jitter is readily measurable. Direct correlation between peak-to-peak I/O noise versus signal jitter contribution can be made. Results showed a 74% reduction for BGA side capacitors, a 66% reduction for coreless, and a 90% reduction for coreless with capacitors compared to the baseline case when measuring output clock jitter.

System level measurements were performed with 12 Gbps of traffic using four simultaneous interfaces. An Ixia pulse generator was used to generate packets. Peak-to-peak power rail noise and clock jitter were measured during both idle and active (traffic) modes. When compared to the baseline case, the BGA side capacitors showed 12% improvement in normalized clock jitter in idle, and 21% improvement with traffic flow.

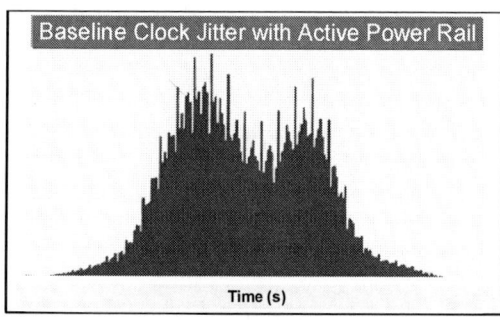

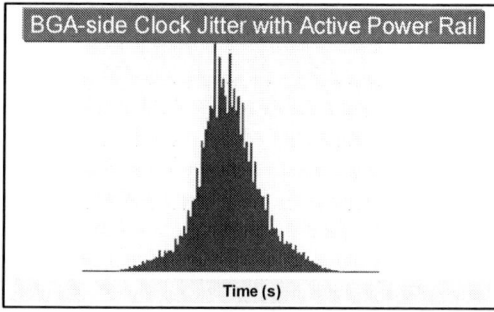

Figure 10 Clock jitter distribution for baseline case and BGA side capacitors case components

Core power noise was reduced 32% in active mode compared to the baseline case. The most interesting observation was that the core power noise in the BGA side capacitor case only increased by 7 mV for 12 Gbps traffic over the same package in idle mode, which is a 10X improvement over the baseline case between idle versus full traffic mode.

Figure 10 shows the jitter spectrum on the output clock. Reduction of peak-to-peak noise on the power rails of the chip is correlated to jitter reduction on the clock, illustrating the link between the integrity of the on-chip power delivery network to timing impact on the component and system level performance.

Conclusions

From a purely electrical performance standpoint, the comparison case showed, in order of increasing performance capability: baseline, coreless, BGA-side capacitors, coreless with BGA-side capacitors.

In terms of product implementation, the most viable alternative given development of manufacturing and assembly processes, is the BGA side capacitors, which offers a good solution for performance versus cost tradeoffs. Today, some percentage of the power and ground BGA balls will need to be removed and replaced by capacitors. An important factor in the effectiveness of this approach is a careful analysis of total number and locations of those balls to provide sufficient DC and AC power delivery to the component. This is highly dependent on the switching requirement of the die, and each device will have a unique solution. Once smaller form factor capacitors become commercially available, BGA balls will not need to be removed

Acknowledgements

The authors would like to thank Rick Brooks, Nicholas Dugbartey, Ken Hubbard, Matthias Kamm, Shu Li, Jane Lim, Subramanian Ramanathan, Paul Ruddy, Goutham Sabavat, Han Ta, and Tripp Worrell from Cisco Systems, Inc. for their support and assistance in the execution of this project. Additionally, the authors want to express gratitude to the individuals at our partner company's that helped design the advanced packages; including: Jinggoy Montenejo and Jon Aday, and all of our substrate partners.

References

1. J. Priest, *et al.*, "Performance Comparison of Advanced Substrate Technologies for High Speed Networking Applications", *Proceedings of SMTA Pan-Pacific Surface Mount Technology Association*, Kohala Coast, HI. February 10-12, 2009.
2. J. Savic, *et al.*, "Electrical Performance Assessment of Advanced Substrate Technologies for High Speed Networking Applications", *Proceedings of 2009 IEEE Electronic Component and Technology Conference*, San Diego, CA, May 26-29, 2009.
3. Sigrity PowerSI, v7.2, (2007), Sigrity, Inc.
4. Synopsis HSPICE, v2008.3, (2008), Synopsys Corporation.
5. I. Novak, "PicoHenries in Power Distribution Networks", *DesignCon 2000*.

The Design and Fabrication of Multilayer Chip LC Filter with a Modified Structure by LTCC Technology

Yingli Liu, Huaiwu Zhang, Yuanxun Li
State Key Laboratory of Electronic Thin Film and Integrated Devices
University of Electronic Science and Technology of China
Chengdu, 610054, China

Abstract

To solve the densification and thermal expansion mismatch problems, this paper proposes a novel method based on the simulations of the chip LC filter with the cut-off frequency at 10MHz with the aid of HFSS software. The difference of thermal expansion between ceramic and ferrite can be trailed off in some degree, because the ferrites layers were put in the middle of ceramic layers. The materials are adopted as ULF140 ceramic and LSF400 ferrites bought from Ferro Company. And the filter samples are successfully prepared by LTCC process. The interfacial conditions of the sample are systemically investigated by SEM and microcosmic results show that the heterogeneous layers are in good connection, with no evidence of cracks or delaminations. The electronic performances of the filter samples are carried on Agilent 8722ES equipment with the cut-off frequency is at 9.5MHz and the attenuation is beyond 25dB. All the measurements indicate that the method is very effective for the fabrication of multilayer chip LC filters.

1. Introduction

With the current explosive growth of communication technologies, it is generally accepted that low-temperature co-fired ceramics (LTCC) technology can fabricate integrated magnetic devices and passive components, such as resistors, circulators and capacitors, so that the technology contributes to the high integration of the microelectronic circuits [1-3].

As one of the basic passive components, low-pass filters (LPF) play an import role in RF circuits to realize the function of signal passing and are extensively used to construct a variety of RF components such as power amplifier modules, transceiver modules and voltage controlled modules [4-6]. And recently, multilayer chip LC filters with high attenuation have been developed as a promising electromagnetic interference device. They are made with a cofired and multilayer structure of ferrite, dielectric, and internal conductors. And the most important process in manufacturing chip LC devices is how to realize the cofiring match from the different materials. To solve the densification and thermal expansion mismatch problems, this paper proposes a novel method based on the simulations of the chip LC filter with the cut-off frequency at 10MHz.

2. TheStructure Design and Simulation for LBPs

A.. Circuit Model

The LPFs were constructed by traditional three-order Butterworth filter configuration based on the lumped-element L-C filter circuit. The typical circuit for LPFs was shown in Fig. 1 which is composed of inductors and capacitors by circuit simulation.

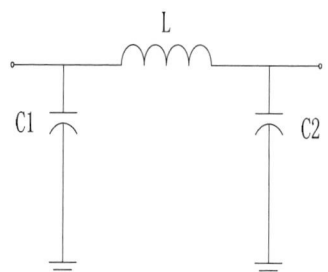

Fig. 1 The circuit model for low pass filters

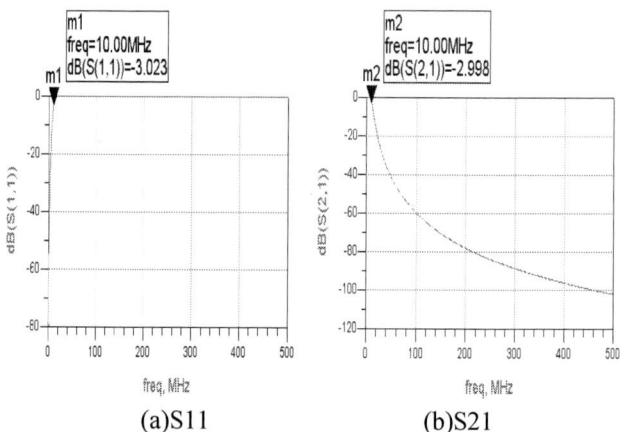

(a)S11   (b)S21

Fig. 2 The simulated results of two kind of band-pass filters by ADS software (a) for center frequency at 1.8GHz, (b) for center frequency at 1.3GHz

Fig. 2 depicts the simulated results. From Fig. 2, it can be seen that the attenuation over 25dB with the cut-off frequency at 10MHz could be achieved.

B. Structure Design for Inductors and Capacitors

The structures of the capacitors and the inductors embedded in the filters were adopted as shown in Fig. 3 and Fig. 4. The inductors were fabricated as Fig. 3 demonstrated with spiral structure which will benefit for improving the inductances. The capacitors were adopted as Fig. 4.

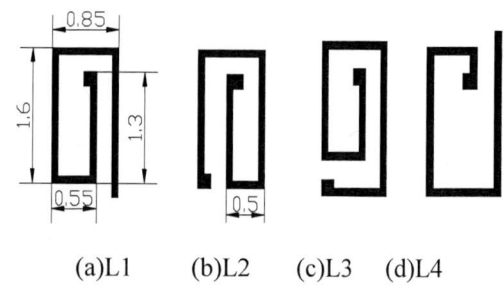

(a)L1   (b)L2   (c)L3   (d)L4

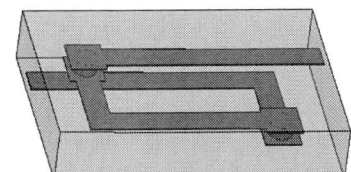

Fig. 3 The size of spiral inductors

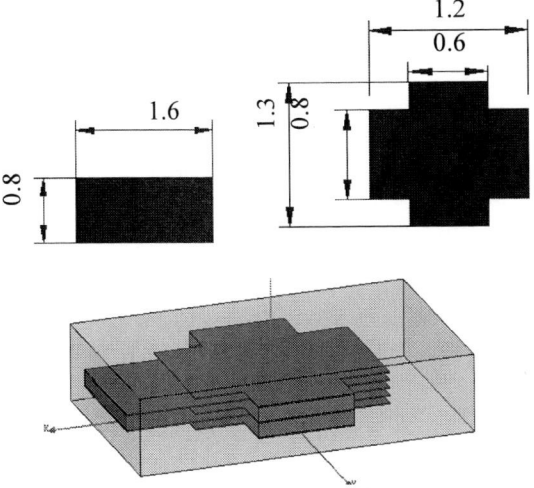

Fig. 4 The size of capacitors

C. Component Implementation

The structures of the capacitors and the inductors embedded in the filters were adopted as shown in Fig. 5. To adapt the size of packaging for LTCC chip electronic component, it demands compact routing structure, especially in the Z direction. Taking the different shrinkage of ceramic and ferrites materials into account, the design of the internal structure of multilayer chip LC filter is shown in Fig. 5. The ferrites layers were put in the middle of ceramic layers in order to trail off the difference of thermal expansion between ceramic and ferrite in some degree.

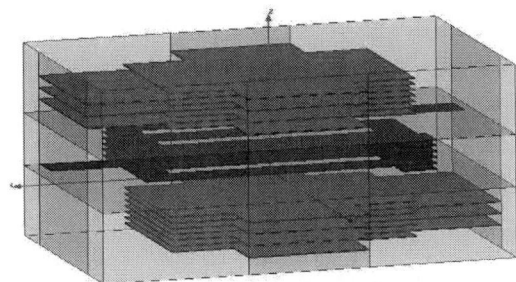

Fig. 5 The physical prototype for band-pass filters

The physical prototype for LPFs was shown in Fig. 5. The material was Ferro ULF101 ceramics with the dielectric constant of 100 and LSF400 ferrites with the permeability of 400. The thickness of each layer was 30μm. After optimized, the characteristics of the filters simulated can be achieved and presented in Fig. 6.

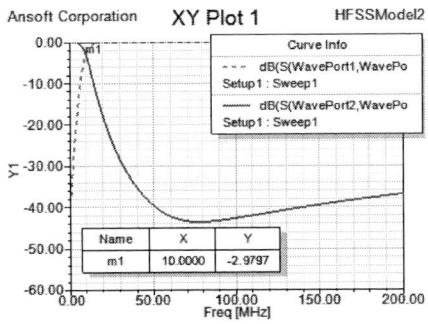

Fig. 6 The simulated results of low pass filters with the packaging size of 0805 by HFSS software

From the response of three dimension configurations, the cut-off frequency of the LPFs was 10MHz and the insertion losses were over 30dB. The difference from the circuit simulation by ADS software is mainly caused by that the equivalent circuit and an EM generated S-parameter database have been investigated with the collection of the electrical response by establishing the proper HFSS model for different available value, according to actual conditions for the fabrication of filters.

3. Experimental

For LTCC process technology, the compatibility of system materials with respect to shrinkage, thermal expansion coefficient and chemical compatibility must be considered, different shrinkage rate of the fired specimens during co-firing make the materials distort, which cause deviation of the designed component. While designing silk screen, the pre-cofiring tape must be larger than designed model. The graphics of silk screen and relative sizes for one unit which were shown in Fig. 3 and Fig. 4 will guarantee the good connection for each lays and formation of capacitors and inductors with the packaging size of 0805.

To investigate the cofiring characteristics of different materials, Fig. 7 shows that the sintering behaviors of the ferrites and ceramics. The shrinkage profiles of ferrites gives an outstanding resemblance with that of the ceramics, which demonstrates a good co-firing matching condition between these two materials.

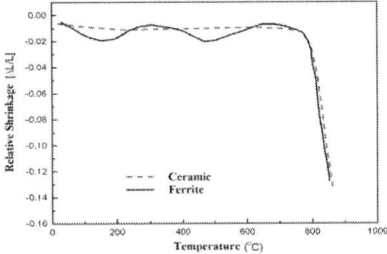

Fig. 7 The densification characteristics of the ferrite and ceramic

Fig. 8 gives back-scatter morphology near the interface between ferrites and ceramics in a multilayer sample prepared by tape casting and sintered at 920°C for 6h. Alternate heterogeneous layers are in good connection, with no evidence of cracks or delaminations. Both ferrite and ceramic layer show a dense microstructure and the corresponding grains grow normally.

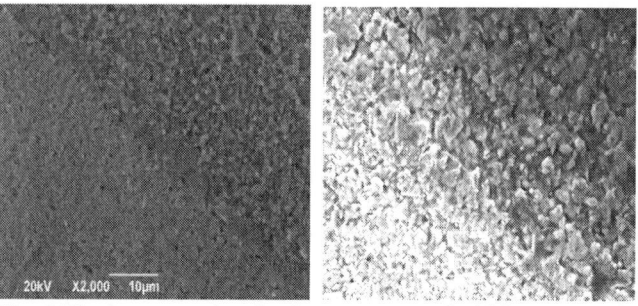

Fig. 8 The cross-sectional microstructure of a bi-layer sample

After the problems of cofiring for ferrites and ceramics, the conventional LTCC technology process was adopted to fabricate the laminated LPFs samples with the outline dimension of 2.0mm×1.2mm×0.9mm shown in Fig. 9.

Fig. 9 Experimental prototype for the samples

The measurements were carried on by Agilent 8722ES and the collected data was then calibrated to the desired reference plane by the thru-reflect line (TRL) technique through carefully designed calibration standards embedded in the same LTCC tie. The measured responses of the filters are shown Fig. 10. From Fig. 10, it can be clearly seen that the measured results agree with the simulated data basically. The cut-off frequency was at 9.5MHz and the attenuation was beyond 25dB. Thus, a good coincidence between simulated and measured data is observed. All the measurements indicate that the method is very effective for the fabrication of multilayer chip LC filters.

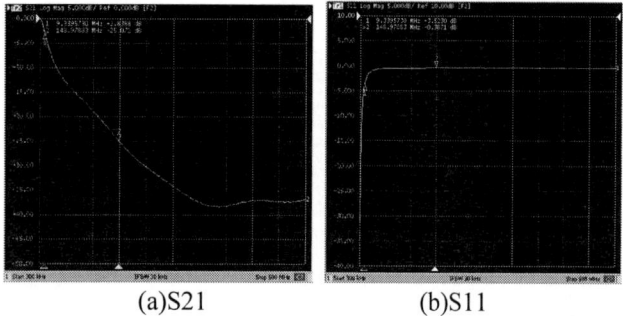

(a)S21   (b)S11

Fig. 10 The measurement results for the samples (a) S21, (b) S11

The differences between the measurement and the simulation are the cut-off frequency and the insertion losses are a little lower than the designed. It is believed that the deviation comes from the additional inductive and capacitive parasitic effects caused by the via. On the other hand, the process of LTCC technology also brings many errors which will introduce the manufacturing inaccuracy. Firstly, during the co-firing, the organic solvent in the silver conductor can not dispel all air bubbles, which cause the value of inductance and capacitance higher and make the cut-off frequency lower. Secondly, the effective value of the component can not be controlled accurately due to the difficulties of tiny manipulation problems including the printing and laminating.

Conclusions

The structure analysis of LTCC-based passive components is reported for the design of a small multilayer chip LPFs and the filter model was established according to the filter's circuit. The LPFs fabricated by LTCC process has small size (2.0mm×1.2mm×0.9mm) using dielectric material ULF101 and LSF400. The interfacial conditions of the sample are systemically investigated by SEM and microcosmic results show that the heterogeneous layers are in good connection, with no evidence of cracks or delaminations. And the filter samples are successfully prepared by LTCC process. The cut-off frequency for LPFs samples is 9.5MHz and the insertion loss is over 25dB respectively. The testing results are in a good agreement with the simulated data which will be helpful for the manufacture of passive components.

Acknowledgments

This work was supported by the Youth Fund of Sichuan Province under Grant No.08ZQ026-013, the Foundation for Innovative Research Groups of the NSFC under Grant No. 60721001, the Youth Fund of University of Electronic Science and Technology of China under Grant No. L08010301JX0725.

References

1. Brzezina, Greg, Roy, Langis, MacEachern, Leonard, "Design enhancement of miniature lumped-element LTCC bandpass filters," *IEEE Transactions on Microwave Theory and Techniques*, Vol. 57, No. 4(2009), pp. 815-823.
2. Lim Michele, Hui Fern; Vvan Wyk, *et al.*, "Internal geometry variation of LTCC inductors to improve light-load efficiency of DC-DC converters," *IEEE Transactions on Components and Packaging Technologies*, Vol. 32, No. 1(2009), pp. 3-11.
3. Mis, Edward, Dziedzic, Andrzej, *et al.*, "Microvaristors in thick-film and LTCC circuits," *Microelectronics Reliability*, Vol. 49, No. 6(2009), pp. 607-613.
4. Tomohiro Seki, Kenjiro Nishikawa, Yasuo Suzuki, *et al.*, "60GHz Monolithic LTCC Module for Wireless Communication Systems," *Proceedings of the 9th European Conference on Wireless Technology*, September, 2006, Manchester, UK, pp. 376-379.
5. Zhenhai Shao and Masayuki Fujise, "60 GHz Narrow Bandpass Filter Based on Circle Patch and LTCC," *IEEE, 6th International Conference on ITS Telecommunications Proceedings*, 2006, pp. 1173-1174.
6. Tao Yang, Bo Yan, Shuyi Wang, *et al.*, "A compact bandpass filter with two finite transmission zeros using

LTCC technology," *IEEE 2007 International Symposium on Microwave, Antenna, Propagation, and EMC Technologies For Wireless Communications*, pp. 293-296.

The Influence of SO₂ Environments on Immersion Silver Finished PCBs by Mixed Flow Gas Testing

Shunong Zhang[1], Anshul Shrivastava[2], Michael Osterman[2], Michael Pecht[2,3], Rui Kang[1]

[1]Dept. of System Engineering of Engineering Technology
Beijing University of Aeronautics and Astronautics
Beijing, 100191, P. R. China

[2]CALCE Center for Advanced Life Cycle Engineering
University of Maryland
College Park, MD-20742, U.S.A

[3]Dept. of Electronic Engineering
City University of Hong Kong

Abstract

This study focuses on the corrosion of immersion silver (ImAg) finished copper land patterns on printed circuit boards (PCBs) due to SO_2 exposure in a mixed flow gas chamber. Six test conditions were examined with varying concentrations, temperatures, relative humidity, and exposure times. The results indicated that there are two mechanisms of corrosion on ImAg-finished PCBs in an SO_2 gas environment: direct chemical corrosion and electrode reaction. No evidence shows that Ag_2S and Cu_2S or CuS were produced. In high humidity, chemical and electrode reaction both existed, and the corrosion products could included Ag_2O, $AgCl$, Ag_2SO_3, CuO, $CuCl_2$ and $CuCl$, In low humidity, the chemical corrosion was predominant, and the corrosion products could include Ag_2O, CuO. Passive films were formed on ImAg finished surface under long exposure time. The temperature from 30°C to 40°C did not have an obvious influence on the ImAg-finished PCBs.

1. Introduction

With the advent of the Restriction of Hazardous Substances (RoHS) legislation in Europe, which forbids the use of lead in electronics products, the electronic industry has moved to lead-free surface finishes for printed circuit boards (PCBs). The widely used SnPb hot air solder level (HASL) process, has been replaced with organic solderability preservative (OSP), immersion silver (ImAg), electroless nickel immersion gold (ENIG), and immersion tin (ImSn) processes. Lead-free HASL is beginning to see an increased use. [1-7]

Among the HASL-free finishes, ImAg and OSP are the preferred finishes for many applications, while ImSn and ENIG are used for niche applications [4]. Due to inherent processing difficulties with OSP boards [3], and also due to the many desirable characteristics of ImAg such as remaining solderable for up to 12 months prior to assembly, having little effect on signal loss due to the good conductivity and thinness of silver, and manufacturing costs of plating the ImAg is half the price in comparison to ENIG and comparable with ImSn finishes [6-7], ImAg boards are quickly becoming a popular PC board finish in the electronics industry. However, ImAg has been shown to have issues in high sulfur environments [3].

Researchers have conducted studies on the corrosion phenomena and mechanisms of ImAg PCBs [1-6]. Some studies have also compared ImAg-finished PCBs with other kinds of lead-free PCBs in the same batch experiments [1-2, 4-5]. Some researchers have used MFG tests to drive corrosion to conduct studies for the corrosion mechanisms of lead-free PCBs or evaluate its corrosion resistance.

On Cullen's study [1], Creep corrosion was reproduced by using a condensing vapor test, which used H_2S and added 1% hydrochloric acid to 0.1g/l sodium bisulfide to form a "sulfur chamber." All finishes exhibited creeping corrosion within 24 hours under this environment. Cullen also conducted Class III MFG testing and found that creep corrosion did not occur until the humidity increased to over 93% to create condensation.

Veale [2] used the MFG test (100ppb H_2S; 200ppb $NO2$; 200ppb SO_2; 20ppb Cl_2; temperature=28-29°C; 75%RH; test duration=20 days or 480 hours) to study the corrosion resistance of OSP, ImAg, ENIG, and ImSn. The progress of the test was monitored by copper reactivity, and the accumulated corrosion per day was 3500Angstroms/day, which is equivalent an ISA G1 (or Battelle class III). The results were that none of these coatings could be considered immune from failure in a Battelle class III environment, and ImSn and OSP could be expected to survive in a Battelle class II environment [4].

Xu *et al.* [4] conducted MFG testing on PCBs with OSP, ImAg, ImSn and ENIG finishes under a more severe environment (40°C; 69%RH; 1700ppb of H_2S; 200ppb of NO_2; 20 ppb of Cl_2; and 200ppb SO_2), which represented Battelle class IV and ISA class G2 conditions. The results after two days showed that all of the ImAg finishes were covered with grayish corrosion products, mostly Cu_2S. Four out of seven of the ImAg finishes showed fiber-assisted electrochemical migration after five days of MFG exposure. One sample was observed to have creep corrosion along the fiber (fiber-assisted creep corrosion) after 10 days. From the test results, it was found that the ImAg-finished boards were more susceptible to corrosion than the other three types of boards. A second issue associated with ImAg is the blistering or peeling of the conductive corrosion products from the surfaces. After 10 days of the MFG test most of the samples only showed minor blistering. Peeling and flaking were only observed after 40 days of MFG exposure.

In no previous studies has a test been conducted in a corrosive environment containing a single corrosive gas, such as H_2S or SO_2, during the analysis and qualification of ImAg as a PCB finish. Our study in reference [8] focuses on the corrosion of ImAg finished copper land patterns on PCBs under H_2S exposure. Twelve test conditions were examined with varying levels of H_2S, temperature, relative humidity and exposure times. The results indicated both direct chemical reaction corrosion and electrode reaction corrosion, especially galvanic corrosion. Temperature shows significant influences on ImAg finished surface PCBs. Tests found extensive corrosion on ImAg finished PCBs at $40^{\circ}C$ even in very low humidity. On ImAg finish surfaces, the corrosion is non uniform for the early period exposure, and corrosion modes mainly show as pitting, open mouth, particles, pits, the corrosion products mainly include Ag_2O and Cu_2O, as time goes on, the corrosion products mainly include CuS, Cu_2S, CuO, Ag_2S and form a passive film on the surface. ImAg finished PCBs are vulnerable to H_2S gas, non uniform severe corrosion was found to occur at defects on the ImAg surfaces. MFG test can produce creep corrosion on ImAg finished PCBs by only using H_2S gas. Dendrite corrosion products growing from the edge with solder mask usually is longer than from the edge without solder mask. The corrosion products of dendrites mainly included CuS or Cu_2S.

The current study focuses on the influence of SO_2 on ImAg-finished PCBs in MFG testing. Six case studies are presented, and the test results are compared.

2. Process of Samples Preparation

For this study, segments of unassembled boards with ImAg finish were used for test samples. The thickness of the ImAg finish is $0.23\sim0.36\mu m$ detected at 5 different locations by X-Ray Fluorescence Spectroscopy. An ImAg-finished PCB were segmented into many pieces using a Dremel tool, then were rinsed under running tap water, washed in an ultrasonic cleaner with de-ionized water, dried with ambient air and additional two hours in a temperature chamber ($110^{\circ}C$). Before MFG testing, three pieces were selected for inspection using environmental scanning electronic microscope (ESEM Model: QUANTA 200) and Energy Dispersive X-ray Spectroscopy (EDS). Each PCB piece was approximately 50mm×50mm with through holes and pads. Bare copper samples were also prepared in order to monitor the corrosion rate. The copper samples were prepared as per section 7.6.1 of ASTM B810-01a [11].

Fig. 1(a) shows the typical ImAg finished PCB segment sample before tests. Fig. 1(b) shows a typical ImAg surface under high magnification.

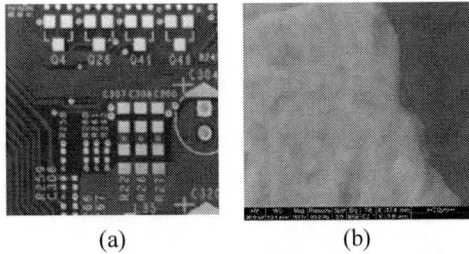

(a)   (b)

Fig. 1 The typical ImAg finished surface before tests

The elements on ImAg surface include Ag, Cu by a point EDS analysis under 30kv, and the elements on solder mask include C, O, Si, S, Ba, Ca. These findings were reconfirmed

in a separate SEM(S-530) & EDS (Link ISIS) system under high vacuum and 10kV, Fig. 2 (a) and Fig. 3 show a point analysis on ImAg finished surface, and Fig. 2 (b) and Fig. 4 show a point analysis on solder mask surface.

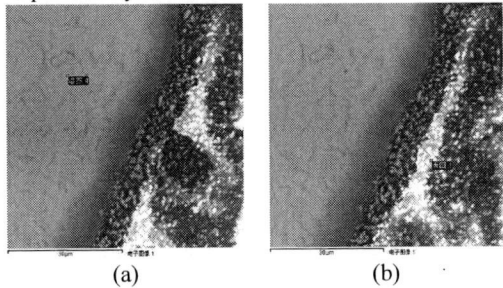

(a)   (b)

Fig. 2 Points analysis on ImAg finished surface and solder mask surface before tests

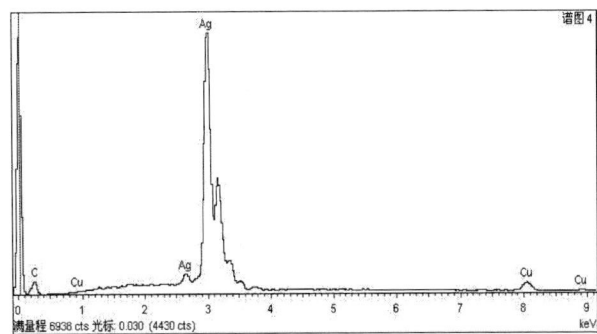

Fig. 3 A point analysis on ImAg finished surface by EDS (10kV)

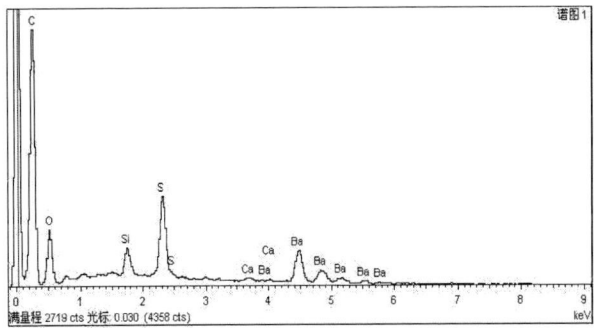

Fig. 4 A point analysis on solder mask surface by EDS (10kV)

3. Experiments and Results

Six test conditions, outlined in Table 1, were examined in this study. For each test, the PCB pieces and copper coupons were suspended by insulated wires in the MFG chamber. During the exposure, visual observations were conducted. ImAg samples were taken out on the exposure times listed in Table 1. Upon completing the set exposure time, test specimens were examined by ESEM (QUANTA 200) and their compositions were detected by EDS in order to study their characteristics. Some samples were also examined by ESEM (Model: QUANTA 600) & EDS, and by SEM (S-530) & EDS (link ISIS) again when needed.

Table 1 Test Matrix

①	②	③	④	⑤	⑥	⑦
1	30±1	75±1	70±30	5, 24, 48, 72, 120, 144, 168, 192, 216, 240	30	8
2	32±1	>95	50±20	72, 120, 168	15	5
3	32±1	<5	200, 400	48, 96	8	2
4	40±1	<5	400	48, 72	6	2
5	40±1	30	100±10	48	4	2
6	30±1	75±1	500±100	4.5	1	1

①Test; ②T (°C); ③%RH; ④Gas concentration (ppb); ⑤Exposure time (hours);
⑥Number of ImAg samples; ⑦Number of Cu samples.

The SO_2 concentrations of Test 1~Test 5 in Table 1 were obtained by calculation according to flow rate of SO_2 and air, because the instrument calibrating the concentration of SO_2 was failure at that time. The SO_2 concentration of Test 6 was obtained by new equipment: the Honeywell Single Point Monitor. The theory of calculating the SO_2 concentration in an MFG chamber is shown in Fig. 5, and the formula is shown in Equation (1).

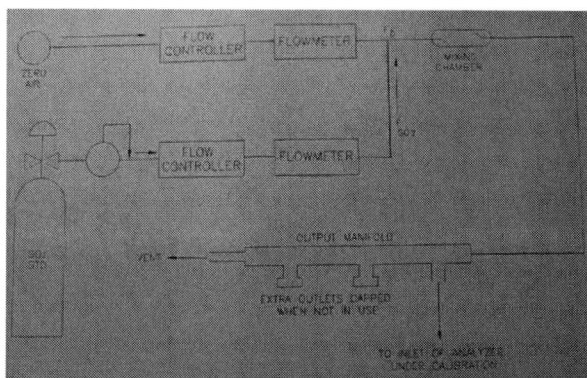

Fig. 5 The theory of calculating SO_2 concentration in MFG chamber

$$[SO_2]_{chamber} = \frac{[SO_2]_{STD} \bullet F_{SO_2}}{F_D + F_{SO_2}} \quad (1)$$

$[SO_2]_{chamber}$: the estimated SO_2 concentration in the MFG chamber, ppm;

$[SO_2]_{STD}$: concentration of the cylinder SO_2 standard, ppm, here was 1196 ppm;

F_{SO_2} : flow rate of the SO_2, slpm;

F_D : flow rate of the air, which include dry air and wet air, slpm.

Fig. 6 shows the samples of last exposure time in test 1~test 6. The samples did not show obvious change but looked a little bit faded in long exposure time (Fig. 6(a)~(e))

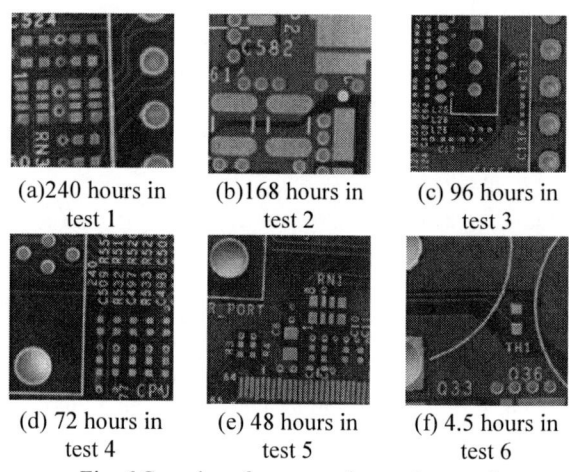

(a)240 hours in test 1   (b)168 hours in test 2   (c) 96 hours in test 3

(d) 72 hours in test 4   (e) 48 hours in test 5   (f) 4.5 hours in test 6

Fig. 6 Samples of exposure in test 1 ~ test 6

(1)Test 1

The first test condition included a temperature of 30±1°C, a relative humidity of 75±1%RH, and an estimated concentration of SO_2 of 70±30ppb. The exposure duration was 10 days. After 5, 24, 48, 72, 120, 144, 168, 192, 216 and 240 hours, at least one sample was taken out to check on these exposure times. The surfaces of samples after 240 hours exhibited many dark and bright points under ESEM (QUANTA 200). A point analysis by EDS on a dark point showed that the composition included Ag, Cu, and Cl (Fig. 7 (a) and Fig. 8). Another point analysis on a bright point showed that the composition included Ag, Cu, Cl, O, and C (Fig. 7 (b) and Fig. 9).

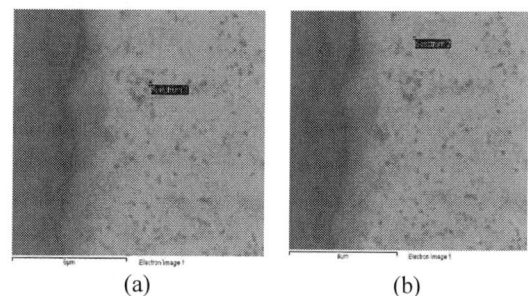

(a)   (b)

Fig. 7 The surface of sample after 10 days exposure time in test1

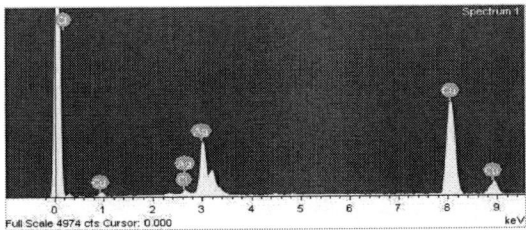

Fig. 8 A point analysis on a dark point after 10 days in test 1
(30kV)

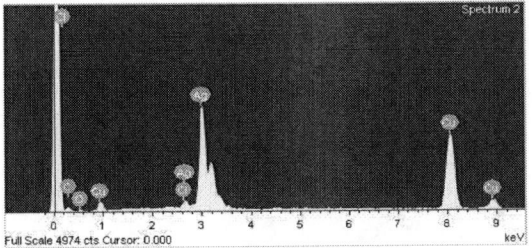

Fig. 9 A point analysis on a bright point after 10 days in test 1
(30kV)

(2)Test 2

The second test condition included a temperature of 32 ± 1°C, a relative humidity of > 95%RH, and an estimated concentration of SO_2 of 50 ± 20ppb. The exposure duration was 7 days. One sample was removed for inspection at 72, 120, and 168 hours. The samples after 168 hours also exhibited many dark and bright points, and corrosion products were found at the edges by ESEM (QUANTA 200) (Fig. 10 (a)). A 120 hours sample was checked by ESEM&EDS(Fig. 10 (b)), a point analysis by EDS on a bright point showed that the composition included Ag, Cu, Cl, S,O,C (Fig. 11).

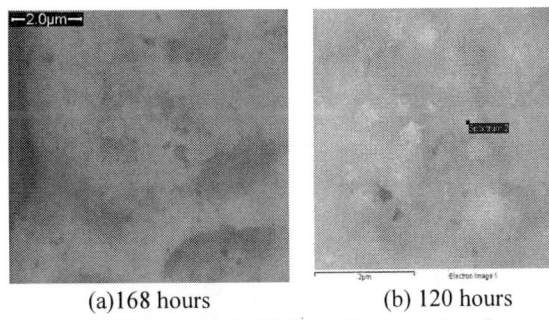

(a)168 hours   (b) 120 hours
Fig. 10 168 hours and 120 hours ImAg surface in test 2

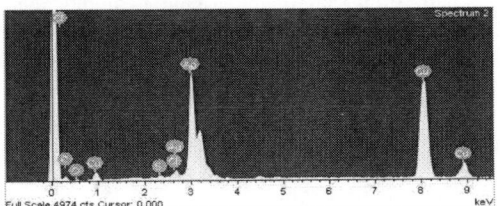

Fig. 11 A point analysis on a bright 'cotton' point after 120
hours in test 2 (30kV)

(3)Test 3

The third test condition included a temperature of 32 ± 1°C, a relative humidity of < 5%RH, and an estimated concentration of SO_2 of 200-400ppb(at first the concentration is 200ppb, after 48 hours it was changed to 400 ppb). The exposure duration was 4 days. One sample was checked after 48 and 96 hours. The surface of sample after 48 hours of exposure in test 3 showed almost no changes under ESEM (Fig. 12(a)), and the surface of the sample after 96 hours of exposure time shows a small number of dark points (Fig. 12(b)). A point analysis by EDS on a dark region showed that the composition included Ag, Cu, O, and C (Fig. 13).

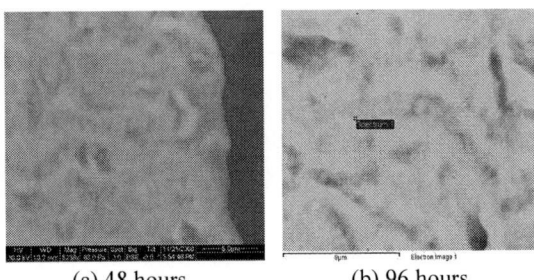

(a) 48 hours   (b) 96 hours
Fig. 12 The surfaces of samples after 48 and 96 hours
exposure time in Test 3

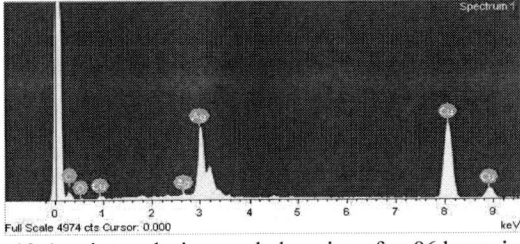

Fig. 13 A point analysis on a dark region after 96 hours in test
3 (30kV)

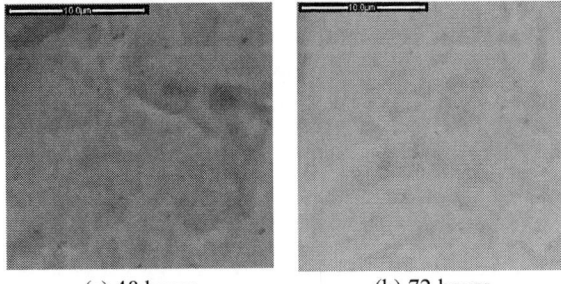

(a) 48 hours   (b) 72 hours
Fig. 14 The ImAg surfaces after 48 and 72 hours of exposure
time in test 4

(4)Test 4

The fourth test condition included a temperature of 40±1°C, a relative humidity of < 5%RH, and an estimated concentration of SO_2 of 400ppb, the exposure duration was 3 days. One sample was removed for inspection after 48 and 72 hours. Checking these samples with ESEM (QUANTA 200), the surface of the samples after 48 hours of exposure time showed some dark points (Fig. 14 (a)). The surface of the sample after 72 hours of exposure showed more dark points (Fig. 14 (b)).

(5) Test 5

The fifth test condition included a temperature of $40\pm1°C$, a relative humidity of 30%RH, and an estimated SO_2 concentration of 100ppb, the exposure duration was 2 days.

Checking the sample under ESEM (QUANTA 600) & EDS under high vacuum status, Sulfur was difficult to found on the surface (e.g. Fig. 15 (a) and Fig.16). A point analysis on the surface by SEM(S-530) & EDS (link ISIS) also showed that the surface was clear, and the composition only included Ag and Cu (Fig. 15 (b) and Fig. 17).

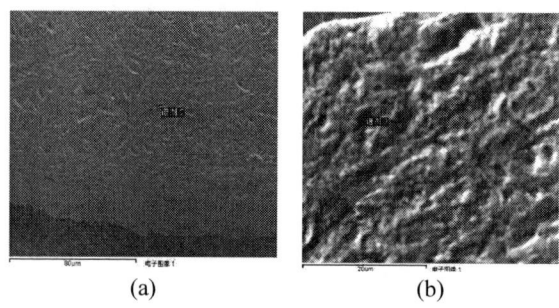

(a)   (b)

Fig. 15 ImAg surface under ESEM (QUANTA 600) and SEM (S-530) in test 5

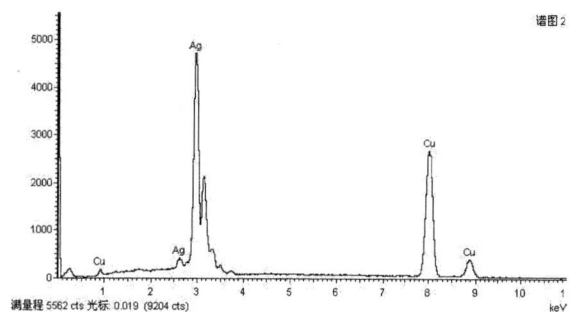

Fig. 16 A point analysis under high vacuum status by ESEM (Model: QUANTA 600) and EDS in test 5 (15kV)

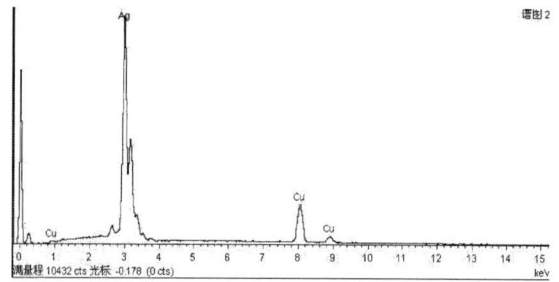

Fig. 17 A point analysis under high vacuum status by SEM (S-530) and EDS (link ISIS) in test 5 (20kV)

(6) Test6

The sixth test condition included $30\pm1°C$, 75%RH and 500 ± 100ppb SO_2. The exposure duration was 4.5 hours. A point analysis on the surface edge by ESEM (QUANTA 200) & EDS showed that the composition included Ag, Cu, Si, S, O, C, and Ba (Fig. 18(a) and Fig. 19), Si, C, Ba and some of

O, S should come from solder mask (Fig. 4). However, S was difficult to found on the ImAg surface by SEM (S-530) & EDS (link ISIS) (e.g. Fig. 18 (b) and Fig. 20).

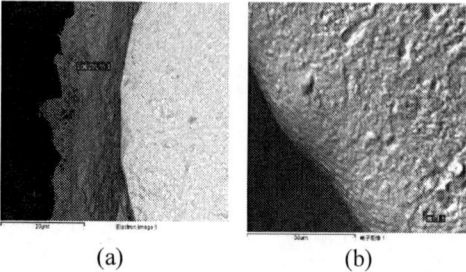

(a)   (b)

Fig. 18 ImAg surface under ESEM (QUANTA 200) and SEM (S-530) in test 6

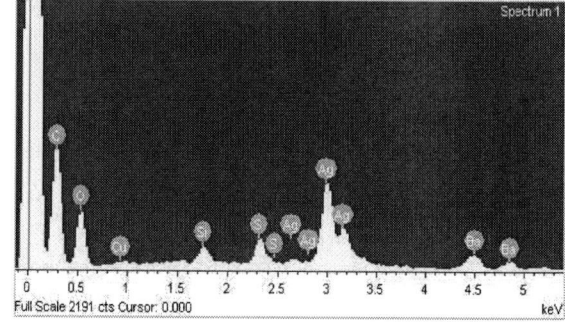

Fig. 19 A point analysis by ESEM (QUANTA 200) & EDS in test 6 (15kV)

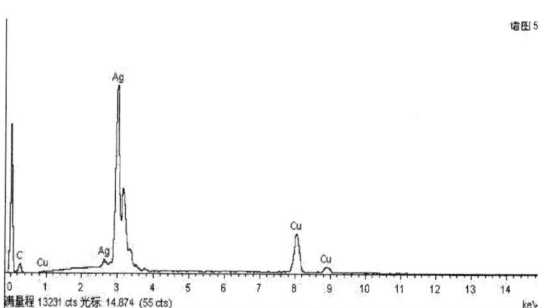

Fig. 20 A point analysis by SEM(S-530) & EDS (link ISIS) in test 6 (20kV)

4. Discussion

In these case studies, Chlorine was found in the corrosion products (test 1 and test 2). The Cl_2 cylinder was not connected the tube going to the flow meter when these experiments were conducted. Where did the Chlorine come from? Judging from the gas and air path, it should come from the water that was in the heater barrel for the purpose of increasing the humidity. However, the water was distilled water, which came from a distilled water instrument. But anyway, Chlorine should not be much more if it came from the water. So, there are SO_2, NO_2, O_2 and a small amount of Cl_2 in the MFG chamber. SO_2 is easy to dissolve in water, in the presence of high humidity, equations below exist [21]:

$$SO_2\,(g) = SO_2\,(aq) \quad\quad\quad (2)$$
$$H_2O + SO_2\,(aq) = H_2SO_3\,(aq) \quad\quad\quad (3)$$
$$H_2SO_3\,(aq) = H^+ + HSO_3^- \quad\quad\quad (4)$$

$$HSO_3^- = H^+ + SO_3^{2-} \qquad (5)$$

Other species also were generated, the ImAg surface in the MFG chamber constituted a rather complex electrode reactions systems, equations below exist [21-22]:

$$NO_2\,(g) = NO_2\,(aq) \qquad (6)$$
$$2NO_2\,(aq) + H_2O = HNO_2\,(aq) + H^+ + NO_3^- \qquad (7)$$
$$Cl_2\,(g) = Cl_2\,(aq) \qquad (8)$$
$$Cl_2\,(aq) + H_2O = HOCl\,(aq) + H^+ + Cl^- \qquad (9)$$
$$3HOCl\,(aq) = 2Cl^- + ClO_3^- + 3H^+ \qquad (10)$$
$$4H^+ + NO_3 + 3Cl^- = NOCl\,(aq) + Cl_2\,(aq) + 2H_2O \qquad (11)$$
$$Ag = Ag^+ + e \qquad (12)$$
$$O_2 + H_2O + 4e = 4OH^- \qquad (13)$$
$$2Ag^+ + 2OH^- = 2AgOH = Ag_2O + H_2O \qquad (14)$$
$$Ag^+ + SO_3^{2-} = Ag_2SO_3\,(s) \qquad (15)$$
$$Ag^+ + Cl^- = AgCl\,(s) \qquad (16)$$
$$Ag + Cl^- = AgCl\,(s) + e \qquad (17)$$

Because of the porosity of the ImAg finish, the following equations could exist:

$$Cu = Cu^+ + e \qquad (18)$$
$$Cu = Cu^{2+} + 2e \qquad (19)$$
$$Cu^+ = Cu^{2+} + e \qquad (20)$$
$$2Cu + H_2O = Cu_2O(s) + 2H^+ + 2e \qquad (21)$$
$$Cu_2O(s) + H_2O = CuO(s) + 2H^+ + 2e \qquad (22)$$
$$Cu_2O(s) + H_2O + 2OH^- = 2Cu\,(OH)_2(s) + 2e \qquad (23)$$
$$2Cu\,(OH)_2\,(s) = CuO\,(s) + H_2O(l) \qquad (24)$$
$$Cu^{2+} + 2Cl^- = CuCl_2\,(s) \qquad (25)$$
$$Cu + Cl^- = CuCl\,(s) + e \qquad (26)$$
$$2CuCl_2\,(aq) + SO_2 = 2CuCl(s) + 2HCl(aq) + H_2SO_4(aq) \qquad (27)$$

Some chemical reactions may also exist:

$$Ag + O_2 = Ag_2O \qquad (28)$$
$$Cu + O_2 = CuO \qquad (29)$$
$$Cu + O_2 = Cu_2O \qquad (30)$$
$$Cu + CuCl_2 = 2CuCl \qquad (31)$$
$$6CuCl + 3/2O_2 + 3H_2O = 2Cu_3Cl_2(OH)_4 + CuCl_2 \qquad (32)$$
$$Ag + SO_2 = Ag_2S + O_2 \qquad (33)$$
$$Cu + SO_2 = CuS + O_2 \qquad (34)$$

However, Fig. 13 in test 3 shows that Equations (33) and (34) were not likely to occur, the corrosion products if existed were mainly Ag_2O or CuO, Cu_2O rather than Ag_2S or CuS.

In high humidity, chemical and electrode reaction should have been present and the corrosion products mainly should have been Ag_2O, CuO, Cu_2O, $Cu\,(OH)_2$, Ag_2SO_3. The corrosion products of $AgCl$, $CuCl_2$, $CuCl$, etc. should also existed because Cl^- existed in water, even its amount is lower.

$AgCl$ and Ag_2SO_3 both are white in color [23-24]; Ag_2O is a fine black or dark brown powder [25]; $CuCl_2$ is a yellow-brown solid [29]; $CuCl$ is a colorless solid [30]; and CuO are black-colored powder [26]; Cu_2O is a brick red color[27]; Cu $(OH)_2$ is a pale blue, gelatinous solid [28], and none of them dissolves easily in water. Fig. 6 did not show what kind of corrosion products are much more than others, but Cu_2O and $Cu\,(OH)_2$ should not be much.

Long exposure time did not show severe corrosion on the ImAg surface, it should because passive films of these corrosion products above were formed on corrosion points or corrosion surface, and they will prevent gas in MFG chamber from producing corrosion again.

5. Conclusions

1) There should be two mechanisms of the corrosion of ImAg-finished PCBs in an SO_2 gas environment: direct chemical reaction and electrode reaction.

2) No evidence shows that Ag_2S and Cu_2S or CuS existed.

3) In high humidity, chemical and electrode reaction both existed, and the corrosion products could included Ag_2O, $AgCl$, Ag_2SO_3, CuO, $CuCl_2$, $CuCl$, etc. In low humidity, the chemical corrosion was predominant, and the corrosion products could include Ag_2O, CuO.

4) Passive films were formed on ImAg finished surface under long exposure time.

5) The temperature from 30°C to 40°C did not have an obvious influence on the ImAg-finished PCBs.

References

1. Cullen D., "Surface Tarnish and Creeping Corrosion on Pb-free Circuit Board Surface Finishes," *IPC Works*, 2005.
2. Veale, R., "Reliability of PCB alternate Surface Finishes in a Harsh Industrial Environment," *SMTA*, 2005
3. Mazurkiewicz, P., "Accereated Corrosion of Printed Circuit Boards Due to High Levels of Reduced Sulfur Gasses in Industrial Environments," *Proceedings of the 32nd International Symposium for Testing and Failure Analysis* Nov.12-16, Renaissance Austin Hotel, Austin, Texas, USA pp. 469-473, 2006.
4. Xu C., Flemming D., Demerkin K., "Corrosion resistance of PCB Surface Finishes," Alcatel -Lucent, Apex, 2007.
5. Schueller R., "Creep Corrosion on Lead-free Printed Circuit Boards in High Sulfur Environments," pp. 643-654, 2007.
6. Zhou, Y.L., Pecht, M., "Assessment of Immersion Silver Finished Circuit Board Assemblies Using Clay Tests," *Proceedings of ICRMS2009: International Conference on Reliability Maintainability and Safety*, Chengdu, China, July 21-25, 2009.
7. Wang, W.Q., Choubey, A., Azarian, M, Pecht, M., "An Assessment of Immersion Silver Surface Finish for Lead-free Electronics," *Journal of Electronic Materials*, 2009.
8. Zhang, S.N., Osterman, M., Shrivastava, A., Pecht, M., Kang, R., "The Influence of H_2S Environments on Immersion Silver Finished PCBs by Mixed Flow Gas Testing," *IEEE Transaction on Device and Material Reliability*, submitted.
9. ASTM Designation: B845-97: Standard Guide for Mixed Flowing Gas (MFG) Tests for Electrical Contacts.
10. ASTM Designation: B827-05: Standard Practice for Conducting Mixed Flowing Gas (MFG) Environment Tests.
11. ASTM Designation: B810-01a: Standard Test Method for Calibration of Atmospheric Corrosion Test Chambers by Change in Mass of Copper Samples.
12. Abbott, W.H., "The Development and Performance Characteristics of Mixed Flowing Gas Environment," *IEEE Trans. Components*, Hybrids, Manufacturing Technol., Vol. 11, No.1, March 1988, pp. 22-35.

13. ISA-S71.04-1985, "Environmental Conditions for Process Measurement and Control System: Airborne Contaminants," *Instrument Society of America*, 1985.

14. Telcordia GR-63-CORE Issue 2, Section 5.5, "Airborne Contaminants Test Methods," Nov. 2000.

15. CALCE Standard Operating Procedures, Mixed Flow Gas Chamber-Rev.A, September 1999.

16. CALCE Standard Operating Procedures for ESEM, [Model:Quanta 200F(26A6)].

17. Precision Calibration System Instruction Manual: Model 491 Interim. Kin-Tek Laboratories, Inc.

18. Operating Instructions: Trace Source ULED Permeation Sources. Kin-Tek Laboratories, Inc.

19. OptiSonde General Easten Chilled Mirror Hygrometer User's Manual.

20. 8270_74_80_84 Mass-Flow Control Boxes Instruction Manual, Matheson Gas Products.

21. Olof A. Svedung, Lars-Gunnar Johansson, and Nils-Gosta Vannerberg, "The Influence of NO2 and Cl2 at Low Concentrations in Humid Atomospheres on the Corrosion of Gold-Coated Contact Material," *IEEE Transactions on Components, Hybrids, and Manufacturing Technology*, Vol. CHMT-9, No.3, Sept. 1986.

22. Cao, C.N., Principles of Electrochemistry of Corrosion, Chemistry Industry press, Beijing, Feb. 2008.

23. Silver chloride, http://en.wikipedia.org/wiki/AgCl

24. silver sulfite, http://www.chemyq.com/xz/xz7/60855ennbm.htm

25. Silver oxide, http://en.wikipedia.org/wiki/Ag_2O

26. Copper(II) oxide, http://en.wikipedia.org/wiki/CuO

27. Copper(I) oxide, http://en.wikipedia.org/wiki/Cu_2O

28. Copper(II) hydroxide, http://en.wikipedia.org/wiki/$Cu(OH)_2$

29. Copper(II) chloride, http://en.wikipedia.org/wiki/$CuCl_2$

30. Copper(I) chloride, http://en.wikipedia.org/wiki/Cucl

Study on Reliability of Embedded Passives in Organic Substrate

Weiyang Qiu, Kailin Pan, Chaoping Yuan, Jiaopin Wang
School of Mechanical and Electrical Engineering, Guilin University of Electronic Technology
No.1 Jinji Road, Guilin, China, 541004
Email: qiuweiyang@163.com

Abstract

Embedded passives represent a promising solution regarding the reduction of size and assembly costs of system in package (SiP). In addition, the benefits also include improvements in electrical performance and reliability. Although embedded passives are more reliable by elimination solder joint interconnects, they also introduce other concerns such as deformation and component instability. More layers may be needed to accommodate the embedded passives, and various materials within the substrate may cause significant thermo-mechanical stress due to coefficient of thermal expansion (CTE) mismatch. Due to the embedded substrate include viscoelastic material, the Maxwell model was applied to describe the viscoelastic behavior of embedded capacitor in the FR4 substrate series, and the constitutive equations of the Maxwell model were obtained. The three dimensional finite element model of embedded capacitor is developed. And the thermo-mechanical deformation of embedded capacitor is investigated after accelerated thermal cycling (-55°C to 125°C) based on the American military standard (MIL). The result shows that the maximum displacement takes place at the both sides of the middle edges. And the deformation leads to the distance of capacitor electrodes increase, which results in decrease of capacitance. The biggest reduction of capacitance is approximate 10% after thermal cycling, but embedded substrate is baked before thermal cycling can prevent the capacitance from decreasing and the change of capacitor is less than 5%. From 100MHz to 1GHz, the signal integrity of micro-via is investigated and the model is developed. The signal micro-via including one ground-via have obvious resonance peaks around the 1GHz. Compare with one ground-via, the resonant peak of micro-via including four ground-via that array symmetrically can be inhibited. In addition, the insertion loss and return loss of the micro-via including four ground-vias are less than that including only one ground-via.

Introduction

According to ITRS 2007-2008, the growing demands of miniaturization, increased functionality, better performance and low cost for microelectronic products and packaging have been the driving force for new and unique solutions in system integration, such as SiP. [1-2] By removing these discrete passive components from the PCB surface and embedding them into the inner layers of substrate (Fig. 1), embedded passives can not only reduce the size and weight of the passives, but can also have many other benefits such as increased reliability, improved electrical performance and reduced cost. Specifically, due to their simplified structure and lack of solder joints,

embedded passives tend to have considerably less parasitic inductance and potential higher reliability than their surface-mount counterparts, so the MTBF (Mean Time between Failures) of embedded passives is significantly increased.

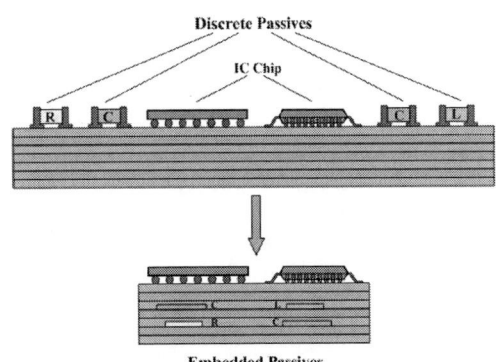

Fig. 1 Discrete passives and embedded passives

The Passives is a crucial part of modern electronic technology with a worldwide market of 25 billion US dollars, but now less than 2% of components are embedded. [3] There are four industry-developed standards about embedded passives. IPC-4821 was published in the year 2006, which specified embedded passive device capacitor materials for rigid and multilayer printed boards. IPC-2316 was published in the year 2007, which demonstrated design guide for embedded passive device printed boards. IPC-4811 was published in the year 2008, which specified embedded passive device resistor materials for rigid and multilayer printed boards. In addition, IPC-6017 is coming soon, which described qualification and performance specification for embedded passive printed boards. [4] For commercial application and increasing market share, the embedded passives technology has been studied in many foreign institutes and corporations. Han Seo Cho *et al.* had developed the processes for embedding components into organic substrate, and the discrete passives are embedded in cavities formed in the core layer of the PCB and the external electrodes of the discrete components are connected by a Cu platting and etching method without any additional laser via machining or soldering process. [5] Kai Zoschke *et al.* had studied the wafer level thin film fabrication of integrated passives devices (WL-IPD). [6] W. Jillek and W.K.C. Yung reported that the embedded capacitors are mostly based on barium-titanite with a dielectric constant of approximately 20, but the capacitance density is limited to a few nF/in^2. Ferroelectric material with a dielectric constant up to 2000 for embedded capacitors has been investigated but is not

978-1-4244-4658-2/09 $25.00 © 2009 IEEE

yet established. [7] Jiming Zhou *et al.* summarized the design, fabrication, and testing of an automotive engine controller, which was used to demonstrate the feasibility of embedded passives technology. [8] Yan Wei *et al.* had studied 3D integrated microwave modules based on LTCC. [9] Xu Gaowei had developed a new type of 3D multi-chips module based on embedded substrate. [10]

After embedded substrates is made, usually there are a number of devices will be assembled by SMT (Surface Mount Technology). Due to high temperature of the process, the thermo-mechanical reliability of embedded capacitors must be concerned. Embedded capacitor dielectric is viscoelastic material which depends on temperature and time, so it is important to develop a suitable material model to study the thermo-mechanical reliability of embedded capacitors. Due to there are micro-vias in the embedded substrate, the signal integrity analysis of micro-vias is also important.

Generalized Maxwell Model

The viscoelastic behavior can be represented by using various combinations of spring and dashpot elements, such as Maxwell model, Kelvin-voigt model and so on. The time-dependent behaviors include creep and stress relaxation. The temperature-dependent behaviors are shown in glass transition, which behaviors of polymer change from glass state to high elastic state. Compare with other material model, the generalized Maxwell model is usually applied to study viscoelastic behaviors of polymer (Fig. 2). Constitutive equations for a generalized Maxwell model are shown as follows. [11]

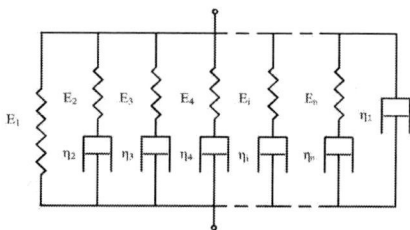

Fig. 2 Generalized Maxwell model

$$\sigma = \sum_{i=1}^{n} \sigma_i \tag{1}$$

$$\sigma_1 = E_1 \varepsilon + \eta_1 \dot{\varepsilon} \tag{2}$$

$$\sigma_i + \frac{\eta_i}{E_i} \dot{\sigma}_i = \eta_i \dot{\varepsilon} \qquad i = 2, 3, \cdots, n \tag{3}$$

Where: n = number of Maxwell elements.

Maxwell Model of Embedded Capacitors

The capacitor dielectric and solder mask are viscoelastic material, so the Maxwell model is used to study the viscoelastic behavior. Their property depend on temperature and time, the time temperature superposition principle is applied to model the viscoelastic behavior and the WLF (Williams-Landau-Ferry) equation (Equation 4)

is applied to unify the analysis of temperature and time, the affect of temperature is converted to variable of time.

$$\lg \alpha_T = -\frac{C_1(\theta - \theta_g)}{C_2 + (\theta - \theta_g)} \tag{4}$$

Where: $\theta < \theta < \theta_g + 100°C$, α_T = shift factor, C_1 = first material constant, C_2 = second material constant, θ_g = glass change temperature.

The dynamic mechanic analyzer (DMA) can be used to study the viscoelastic behavior. The master curve obtained from the experiment can be expressed as a function consisting of discrete number of Maxwell elements. The kernel functions are represented in terms of Prony series, which express as follows:

$$G = G_{\infty} + \sum_{i=1}^{n_G} G_i \exp\left(-\frac{t}{\tau_i^G}\right) \tag{5}$$

$$K = K_{\infty} + \sum_{i=1}^{n_G} K_i \exp\left(-\frac{t}{\tau_i^K}\right) \tag{6}$$

Where:

t = reduced time;

i = designation for each Maxwell element

G_{∞}, G_i = shear elastic moduli;

K_{∞}, K_i = bulk elastic moduli;

τ_i^G, τ_i^K = relaxation times for each element.

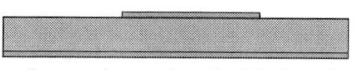

a. Prepare the core board with electrode

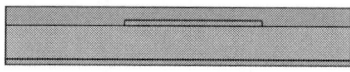

b. Spin coat capacitor dielectric

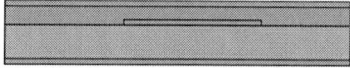

c. Laminate Cu foil

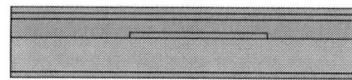

d. Laminate photoresist

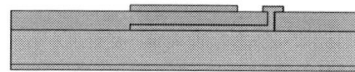

e. Pattern top electrode and micro-via

f. Deposit solder mask

Fig. 3 The process of embedded capacitor

Fabrication of Embedded Passives

The embedded capacitor process is based on a sequential build-up technology. First, the core board with bottom Cu electrode is prepared. Second, after patterning of the electrode, the capacitor dielectric can be spin coated. Third, Cu foil is laminated. Fourth, the photoresist is laminated on the Cu foil. Fifth, photoresist is exposed and pattern is developed. Then the top Cu electrode is made by etching Cu and micro-via is made for connecting bottom electrode. Finally, the solder mask is deposited. The process of embedded capacitor is shown in Fig. 3.

Finite Element Modeling of Embedded Passive

The finite element model of embedded capacitor is developed based on the above process. A 10mm×10mm FR4 core board is chosen, and the size of capacitor is 2.54mm×2.54mm. The thickness of capacitor dielectric layer is 45μm. The thickness of Cu foil is 15μm, and the thickness of solder mask is 30μm. Due to the dimension of micro-via (radius=0.05mm) is far less than the size of substrate, the micro-via is neglected and the structure is simplified as Fig.4.a. The capacitor dielectric and solder mask exhibit viscoelastic behavior, the generalized Maxwell models are used, and their Prony series are applied to analyze the embedded capacitor [12-13].

In order to represent the different layers of the embedded substrate, a shell 181 element is used to mesh layers except the FR4 layer, which is suitable to model laminated composite shells or sandwich construction. Due to FR4 layer is thick, the solid 45 element is applied to mesh that. The 1/2 finite element model is shown in Fig. 4.b.

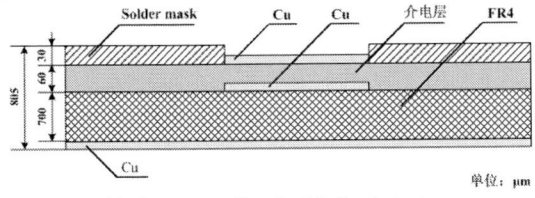

(a) Structure of embedded substrate

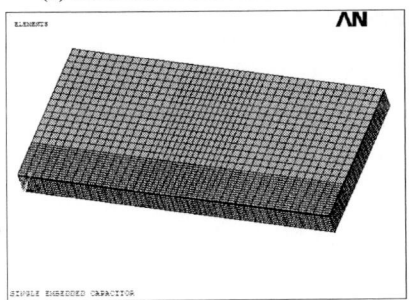

(b) 1/2 finite element model of embedded substrate
Fig. 4 The model of embedded substrate

Effect Caused by Thermo-Mechanical Deformation

The symmetry boundary condition is applied and the nodes on the bottom area are constrained in the Z direction. The embedded capacitor is cycled between -55°C and 125°C based on the American military standard (MIL). Due to the thermo-mechanical deformation of embedded capacitor is a important reason which leads to the changes of capacitance, the displacement of embedded capacitor is

important to study the change of capacitance. [14] After finishing 5 accelerated thermal cycling, the distribution map of displacement is obtained (Fig. 5). The result shows that the maximum displacement takes place at the both sides of the middle edges. The deformation leads to the distance of capacitor electrodes increase, which results in decrease of capacitance. In K J lee's investigate, the capacitance of embedded capacitor approximately decreases by 14% after thermal cycling test (-55 to 125°C 1000 cycles). [15] In some cases, the capacitors have also experienced large changes about 10% after thermal cycles (-55 to 125°C, 1000 cycles). The biggest change occurs in the initial 200-300 cycles, but the change is minimal after 300 cycles. After 1000 thermal cycles for baked samples (baked at 140°C/4hrs), the capacitance change lower than 5%. The change in capacitance with baking shows copper annealing that can possibly stabilized capacitors prior to thermal cycling. [16]

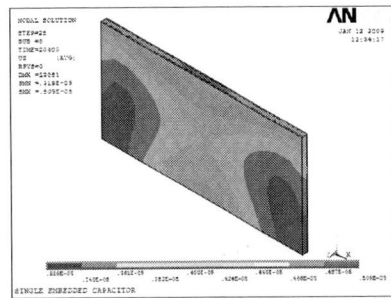

Fig. 5 The displacement (Z direction) of embedded capacitor

Signal Integrity Analysis of Micro-Vias in Embedded Substrate.

Due to the micro-via seriously impact on signal integrity, it must be considered in the signal integrity analysis, which is neglected as finite element analyzed above. The effect of the physical layout and distribution in multiple-via is studied in Lydia Lap Wai Leung's study. [17] In this work, the one signal via is investigated in the substrate. The simplified HFSS models are developed as shown in Fig. 6.a and Fig. 7.a. From 100MHz to 1GHz, the effect on signal integrity is predicted, and the simulation results of S parameters are obtained. The signal micro-via including one ground-via have obvious resonance peaks around the 1GHz (Fig. 6.b). But the signal micro-via include four ground-vias which array symmetrically and doesn't have a resonance peak (Fig. 7.b). Compared with one ground-via, the insertion loss (S12) and return loss (S11) of signal micro-via with four ground-vias are less than that with only one ground-via. So using the ground-vias in the embedded substrate can significantly improve the signal integrity.

978-1-4244-4658-2/09 $25.00 © 2009 IEEE

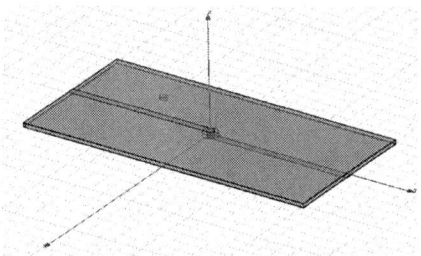

(a) Model of signal micro-via with one ground-via

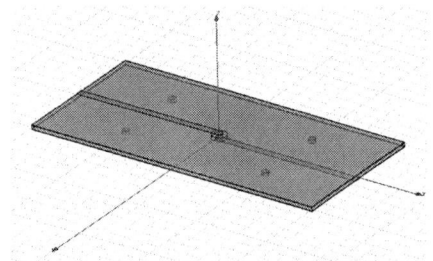

(b) S parameter simulation result

Fig. 6 The signal micro-via with one ground-via

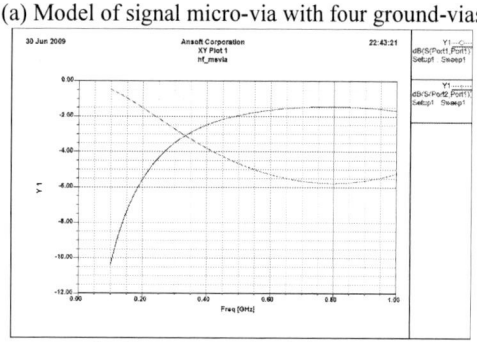

(a) Model of signal micro-via with four ground-vias

(b) S parameter simulation result

Fig. 7 The signal micro-via with four ground-vias

Conclusions

The Maxwell model is applied to describe the viscoelastic behavior of embedded capacitor in the FR4 substrate series, and the constitutive equations of Maxwell model are obtained. The 3D finite element model is developed to study the thermo-mechanic deformation, and the displacement of embedded capacitor is obtained. The result indicates that the maximum displacement takes place at the both sides of the middle edges, and the deformation leads to the change of capacitance after accelerated thermo

cycling. In addition, the range of capacitance change is approximately 10% after thermal cycling, but capacitors experienced less than 5% decrease after baked and thermal cycling. In the signal integrity analysis, the signal micro-via including four ground-vias which array symmetrically are used in the embedded substrate can inhibit the resonant peak and reduce signal loss.

Acknowledgements

This work is supported by the Key Laboratory of Manufacturing System and Advanced Manufacturing of Guang Xi Province Director Project Foundation (0842006_018_Z) and National Natural Science Foundation of China (600866002).

References

1. ITRS: International Technology Road Map for Semiconductors 2007 EDITION.
2. ITRS: International Technology Road Map for Semiconductors 2008 UPDATE.
3. http://www.inemi.org
4. http://www.ipc.org
5. Han Seo Cho, Sukhyeon Cho and Jihong Jo. *et al.*, "Highly reliable processes for embedding discrete passive components into organic substrates," *Microelectronics Reliability*, Vol. 48, No. 2 (2008), pp. 739-743.
6. Kai Zoschke, M. Jürgen Wolf and Michael Töpper. *et al.*, "Fabrication of Application Specific Integrated Passive Devices Using Wafer Level Packaging Technologies," *Proc IEEE 2007 Transactions on Advanced Packaging*, West Point, NY, USA, 2007, pp. 359-366.
7. W.Jillek, W.K.C. Yung, "Embedded components in Printed circuit boards: a processing technology review," *Adv Manuf Technol*, Vol. 25, No. 2 (2005), pp. 350-360.
8. Jiming Zhou, John D. Myers, "Embedded Passive Technology Application: Design and Fabrication of an Automotive Engine Controller," *Pro IEEE 2005 High Density Microsystem Design and Packaging and Component Failure Analysis*, Shanghai, China, 2005, pp. 1-7.
9. Yan Wei, Yu Shenglin, Fang xunlei, "Three Dimensional Integrated Microwave Modules Based on LTCC Technology," *ACTA ELECTRONICA SINICA*, Vol. 33, No. 11 (2005), pp. 2009-2012.
10. Xu Gaowei, Wu Yanhong and Zhou Jian. *et al.*, "Development of a Three-Dimensional Multichip Module Based on Embedded Substrate," *Semiconductors*, Vol. 29, No. 9 (2008), pp. 1837-1842.
11. Zhou Guangquan, Liu Xiaomin, <u>Viscoelastic Theory</u>, University of Science and Technology of China Press (He Fei, 1996), pp. 23-24.
12 Gnyaneshwar Ramakrishna, Fuhan Liu, and Suresh K. *et al.*, "Role of Dielectric Material and Geometry on the Thermo-Mechanical Reliability of Microvias," *Electronic Components and Technology Conf*, Vol. 52, No. 5 (2002), pp. 439-445.

13 Dunne, R.C., and Sitaraman. *et al.*, "Thermal and Mechanical Characterization of VialuxTM81: A novel Epoxy Photo-Dielectric Film (PDDF) for microvia Application," *Components and Packaging Technologies, IEEE Transactions*, Vol. 24, No. 3 (2000), pp. 436-444.

14 Kang J. Lee, Manoj Damani, Raghuram V. Pucha. *et al.*, "Reliability Modeling and Assessment of Embedded Capacitors in Organic Substrates," *IEEE Transactions on Components and Packaging Technologies*, Vol. 30, No. 1 (2007), pp. 152-161.

15 Kang Joon Lee, "Fabrication and Reliability Assessment of Embedded Passives in Organic Substrate," *Georgia Institute of Technology*, 2005, pp. 78.

16 Rabindra N. Das, Mark D. Poliks, and John M. Lauffer. *et al.*, "High Capacitance, Large Area, Thin Film, Nanocomposite Base Embedded Capacitor," *Endicott Interconnect Technologies, Inc.*

17 Lydia Lap Wai Leung, Kevin J. Chen, "Microwave Characterization and Modeling of High Aspect Ratio Through-Wafer Interconnect Vias in Silicon Substrates," *IEEE Transactions on Microwave Theory and Techniques*, Vol. 53, No. 8 (2005), pp. 2472-2478.

Investigation on PCB Related Failures in High-Density Electronic Assemblies

L. N. Lu[1], H. Z. Huang[1,*], X. X. Su[2], B. Y. Wu[2], M. Cai[3]

[1]School of Mechatronics Engineering, University of Electronic Science and Technology of China
Chengdu, Sichuan, 610054, P. R. China
[2]Flextronics Mobile & Consumer,
Zhuhai, P. R. China
[3]Guilin University of Electronic Technology, P. R. China
*Corresponding author: Email: hzhuang@uestc.edu.cn, Tel: 028-61830248, Fax: 028-61830229

Abstract

This paper presents an investigation on field returned open and short failures related to printed circuit board (PCB), including via hole crack, prepreg crack and insufficient circuit etching.

After an experimental study with cross section, time domain reflectometry (TDR), and finite element (FE) modelling, it was found that weak plating and corrosion induced via hole crack was a major root cause of interconnect open failures. Such a failure was always mistakenly treated as solder joint open. Except for via hole crack, PCB prepreg crack was another contributor for open failures. The prepreg material for some lot code in production was in poor quality in terms of long-term durability. Dye and pry and cross section investigation showed that for some field returned units, the pad-to-prepreg interface was the weakest adhesion part in the entire solder joint interconnects. Crack propagated across the whole support and hence pad cratering occurred upon overstress. For the circuit short problem, insufficient etching of copper traces and pads was found to be a dominant contributor in harsh environment for a long time. Particularly, corrosion induced dendrite growth brought a high risk of circuit short of insufficiently etched conductors for the fine-pitch lead-free application.

In summary, it is important to monitor and control the incoming quality of incoming PCBs so that the the risk of high field return rate and high cost for repair can be minimized.

Introduction

With the development of electronics towards low cost and high performance, electronic packaging engineers are integrating more and more chips and parts into the assembly, which has placed severe challenges on the long-term reliability of high density interconnection (HDI) [1]. Electronic manufacturing business units are suffering from a high field return rate of their products due to the degradation or damage of HDI within the product warranty period. In order to assess the exposure risk of those electronics in field and to reduce the cost for rework, a timely and thorough understanding of failure modes and mechanisms of field returned and repeated returned products is essentially critical to sustaining excellent engineering performance. By which effective corrective actions can be implemented immediately in terms of either design modification, or process improvement, or material control.

PCB Via Hole Crack

It is reasonable that electronic products are returned to the manufacturing factories for repair after the warranty period. However, many electronics are returned from field within a shorter time than the defect liability period. In extreme cases, 30 or 90 days return and repeated returns after shipment are encountered as infant mortality of the products. Of these failures, most are classified to be related to PCBA soldering open issues.

It frequently believed that assembly process has some problems, such as misprinting, contamination and improper reflow temperature, resulting in a poor soldering quality. Then, weak solder joints associated with these problems may crack upon overstress in delivery or service, causing a high rate of field return. Fig.1. shows a typical example that non-wetting between the BGA solder ball and PCB bad occurred and there was no intermetallic compound (IMC) layer at the interface. Right after assembly, this interconnect without metallurgical bond could escape from normal production functional tests as there was a direct contact of ball to the pad to carry electrical signal. However, deattachment occurred quickly in transportation or field usage.

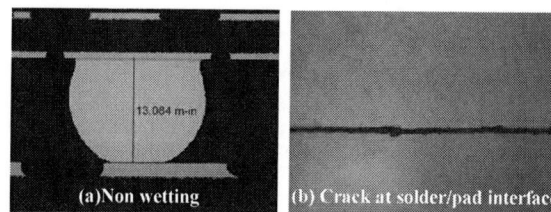

Fig. 1 Non-wetting induced solder joint open

Unfortunately, most of the time this is not the case as re-soldering or component replacement still could not heal the suspected open issue. If there is a solder joint open, after rework, new good solder joints were supposed to form so that failure would disappear. However, Fig. 2 shows an example of time domain reflectometry (TDR) curve of one suspected pin after component replacement. The impedance curve indicates a discontinuity in the electrical path of this pin. Cross section result in Fig. 3 shows that the solder joints were well formed and no crack was found. Therefore, the possibility of solder joint open is excluded and hence component replacement as a usual rework practice for suspected open units may overkill the original component. Always blaming the process induced solder joint open would be quite misleading in the real scenario. A further analysis of the interconnect by cross section study confirmed that the PCB via hole crack, as shown in Fig. 4(a), was the root cause of circuit open failure. Fig. 4(b) demonstrates another example of damage in copper barrel of via hole.

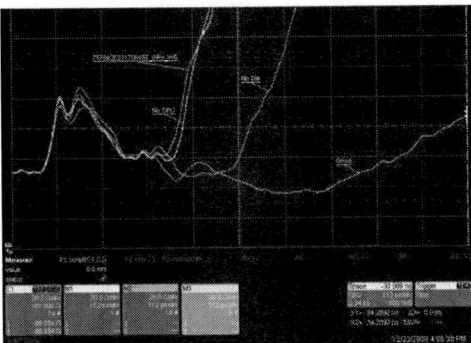

Fig. 2 TDR curve of suspected pin

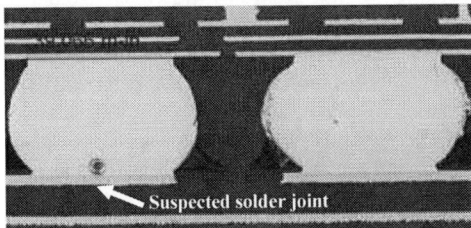

Fig. 3 Metallographic image of suspected solder joints

Microstructural analysis suggests that at least two dominant PCB fabrication defects should be responsible for this reliability failure from field service. One is that the drilling and plating quality is poor. The hole drilling process is poor, leaving very rough wall surface and hence a uniform copper plating cannot be achieved onto the via hole barrel. At some local places, the thickness of copper is not up to standard. Then, during thermal or mechanical experience, the weakest part may crack at first and then propagates across the entire interconnect, as shown in Fig. 4(a). The other one is that corrosion occurs at the copper barrel. After plating, some aggressive chemicals remain inside the via hole if the clean is insufficient. As shown in Fig. 4(b), these chemical residues may cause corrosion to etch away the copper in favorite humidity and temperature conditions in the field service. Actually, both failure modes suggest a reliability issue of PCB over time. At the early stage, there is a poor plating induced micro crack locally or corrosion induced weak point. As shown in Fig. 5, upon thermal or mechanical loadings, there is a significant stress concentration around this initial failure site. Therefore, we can image that crack will propagate and finally damage the whole interconnect.

It is important to minimize this via hole reliability risk of PCB by regular incoming inspection [2-3]. Although electrical measurement at time zero cannot tell the difference, cross section of representative samples from each lot can evaluate the plating quality in terms of via hole roughness and copper thickness. A benchmark for quality assessment is to check whether these parameters are up to IPC standard or not.

At the same time, ion chromatography examination can be applied to identify the level of ion contamination hence potential corrosion.

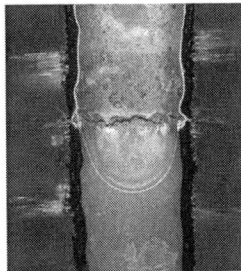 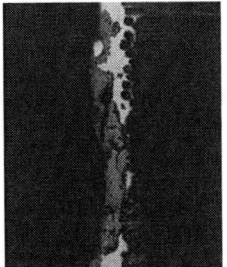

4(a) Weak plating induced   4 (b) Corrosion induced
Fig. 4 Examples of PCB via hole crack

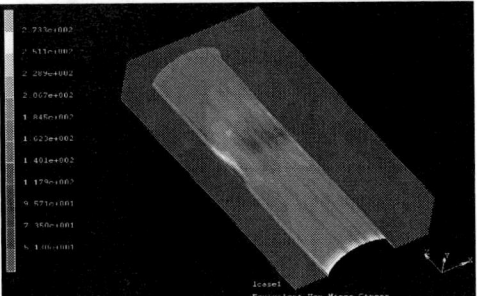

Fig. 5 Modelling of stress distribution in via hole

PCB Prepreg Crack

Except for the via hole crack, pad catering is another failure mode that rework station always find. It is generally encountered that after during component removal for rework, some BGA pads are lifted off [4-5]. This means that the pad has very poor adhesion to the PCB prepreg and it is broken and then taken away together with the solder, leaving a torn prepreg, as shown in Fig. 6.

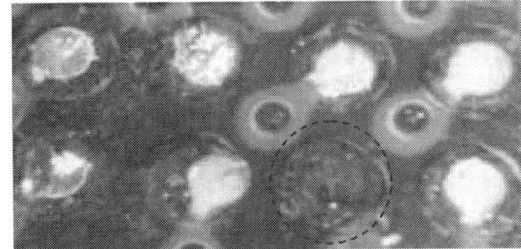

Fig. 6 Pad peel off after BGA removal

Cross section of another solder joint has found that serious prepreg crack occurred right below the pad and above the glass fiber, as shown in Fig. 7. In this case, the pad is prone to pad cratering upon slight stresses, causing electrical open failure in service or pad peel off in rework. It is understandable that such kind of cratering is more seriously associated with non soldermask defined (NSMD) pads. Dye and pry study of one BGA assembly of one PCBA from field return also confirms that most solder joint fractures were located at the pad-to-prepreg interface, as illustrated by mode Type 4 in Fig. 8. Then, we can say that the adhesion strength of this interface was smaller than that of those interfaces described by Type 1, 2 and 3. This weakest interface started to crack with the first priority in field service. This interface broke first due to poor adhesion strength and partial contact area. Therefore, dye penetrated into the crack in the analysis, as shown in Fig. 9. As shown in Fig. 10 and 11, the mapping of dye penetration distribution for the whole BGA shows that solder joints at outmost rows were more easy to crack and have higher crack percentage in individual solder joint than those of inner rows. It is not surprising to see such a dye penetration distribution since upon a uniform loading these locations would experience a higher stress than others.

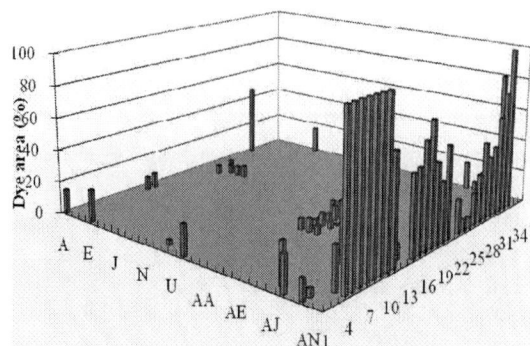

Fig. 11 Dye penetration percentage at different locations

It is noted that PCB prepreg crack sometimes occurs for a certain lot code of PCBs with those fresh samples, even at the new product introduction (NPI) phase. Also, early failures occur in the reliability testing. In order to assess the quality of these suspected PCB, a shearing study of the solder joints is performed after multiple reflow of a certain lot code of PCB. Solder joints were prepared with typical reflow process at different (1 to 4) cycles only by BGA balls and fresh bare PCB (without component). The ball and pad was 0.5mm in diameter. The shearing was conducted at 25.4mm/s with a height of 0.1mm. As shown in Fig. 12, ball shear test found that the percentage of crack at PCB to overall crack mode increased significantly with the number of reflow cycles. Meanwhile, the shearing force of PCB prepreg decreased substantially with the number of reflow cycles, as shown in Fig. 13. These suggest an obvious degradation of PCB prepreg with reflow cycles. Actually, the prepreg integrity of fresh PCB should meet the basic requirements of reflow cycles. Herein, one or 2 reflows was used to simulate the SMT assembly and 3 to 4 reflows for the rework process. Then, the shearing fracture is supposed to occur in the bulk solder, instead of prepreg. Unfortunately, this lot code of PCB failed to achieve a basic quality requirement.

Fig. 7 Cross section showing PCB prepreg crack

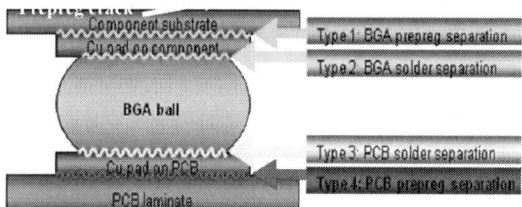

Fig. 8 Illustration of fracture mode in solder joint

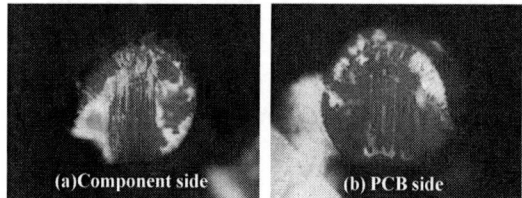

Fig. 9 Dye and pry images of solder joint

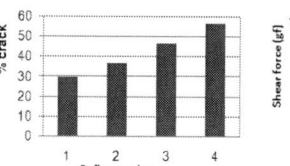 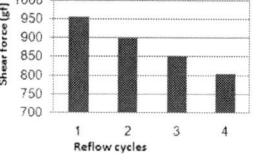

Fig. 12 Prepreg crack percentage vs reflow cycles   Fig. 13 Prepreg force vs reflow cycles

Prepreg crack is mainly a reliability issue of PCB since the crack occurs inside as a result of poor adhesion of copper to laminate. It should be pointed out that for the incoming PCB, the risk of even a slight prepreg crack cannot be underestimated. Although it cannot cause electrical failures immediately, prepreg crack is a precursor to pad cratering defect [6-7]. Once the prepreg has cracked through underneath the pad, there is no more support, and hence the trace may snap upon external stress. Therefore, we may suffer from pad cratering in field return boards after a period of time for PCB prepreg degradation. In addition, boards with pad

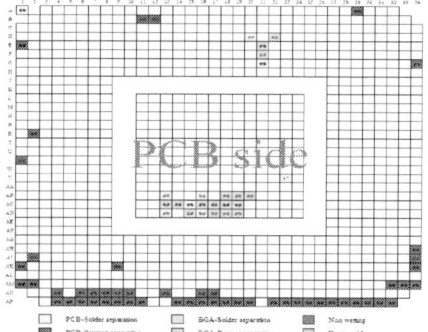

Fig. 10 Mapping of dye penetration locations for BGA pads

cratering are generally not suitable for repair as component removal may damage the whole pad and take it away. In this regard, a significant lost in cost can be imaged. Generally, any good-quality PCB should withstand several reflow cycles for assembly and rework without damage. Also, PCB is supposed to remain mechanical and electrical integrity upon thermal or mechanical stresses commonly encountered for all consumer electronics in service. Therefore, it is very important to have regular sampling for ball shearing study and cross section assessment so that the quality of prepreg can be monitored.

PCB Insufficient Etching

It is sometimes surprising to see a poor quality of incoming material of PCB by insufficient trace etching from it [8], as shown by the circles in Fig. 14. With optical examination, it was found that some copper burrs and residues remain around the traces and pads, which would cause a long term reliability concern of circuit short between adjacent conductors and pads. This is very dangerous for the long-term reliability of PCBA. With the aid of aggressive chemicals from flux residues, corrosion may occur on copper conductors and solders, causing electromigration to short the circuit.

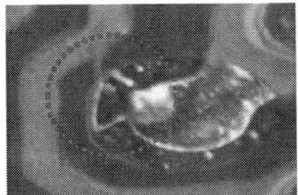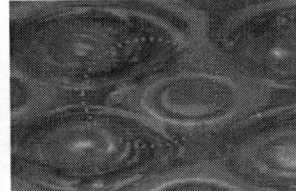

Fig. 14 Insufficient etching of copper on PCB

Generally, such an electromigration failure depends on the speed of dendrite growth. However, critical parameters are necessary to accelerate the failure. First, acid or halogen elements are needed to react with metal oxides so that a quick corrosion can take place. For most electronic assemblies, various remaining chemical species after PCB fabrication and flux residue after no-clean soldering process will provide such reactants. Second, moisture will act as a carrier and growth path for electromigration. Usually, electronic products will experience high humidity soaking during transportation and field service in harsh environment. Third, bias is a driving force for dendrite growth of metallic ions and species. No doubt that most circuit powered on may generate potential difference among adjacent conductors. Fourth, a fine-pitch application will be easier for potential circuit short. It is obvious that insufficient etching of copper will cause more short distance between conductors. With these four ideas in mind, we can expect that a soldering of boards with insufficient etching may fail due to dendrite growth when the PCBAs are subjected to a harsh environment (high temperature, high humidity, acid or salty atmosphere).

Actually, a temperature, humidity and bias test (55°C/85%RH and product powered on) has demonstrated such a potential risk by means of a fast growth of copper and tin dendrite within 500 hours, as shown in Fig. 15. The longest dendrite is about 200 micrometers. EDX analysis

result of the dendrite confirmed the composition to be mainly tin and copper, as shown in Fig. 16. An area scan mapping of element distribution has found that tin and copper migrate in parallel, as shown in Fig. 17. This suggested that both tin and copper oxides corroded quickly to accelerate the electromigration within 500 hours. Herein we recall that fine-pitch interconnects and insufficient etching will be more sensitive to this kind of dendrite growth induced failure.

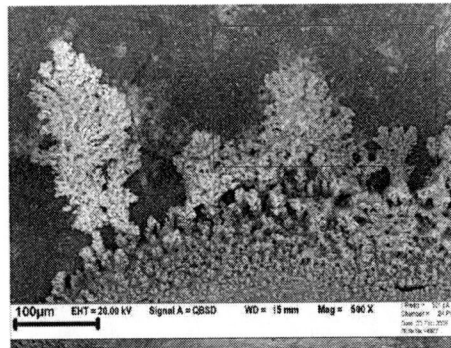

Fig. 15 Dendrite growth around solder joints

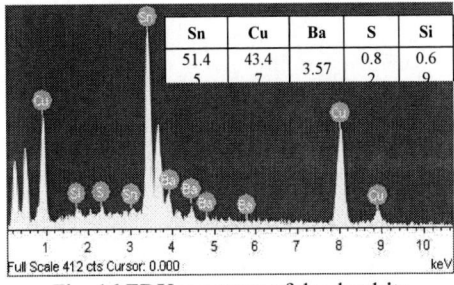

Sn	Cu	Ba	S	Si
51.45	43.47	3.57	0.82	0.69

Fig. 16 EDX spectrum of the dendrite

17(a) Distribution of Cu K_{a1}

17(b) Distribution of Sn L_{a1}

Fig. 17 Element mapping of the dendrite

Conclusions

In this paper, three commonly encountered PCB defects, via hole crack, prepreg crak and insufficient etching, are discussed with typical examples from field returned PCBAs and units after reliability testing.

It is found that due to a high apsect ratio of holes for the high density application, weak plating quality and copper corrosion are attributed to the via hole crack, resulting in a high field return rate. The copper plating thickness is not up to standard due to rough barrel surface after poor drilling process. The chemical residues inside via holes react with copper oxides and etch away the plating. Therefore, crack intiates at the weak plating locations or corroded points and then propagates across the whole via hole finally. PCB prepreg crack is found to another problem assocaited high rate of first return and repeated return. The organic prepreg material degrades to an acceptable level of poor quality, which cannot withstand normal reflow and rework cycles. Consequently pad cratering occurs in the field service. Insufficient of copper trace on PCB is also found during incoming quality control. This increases the risk of circuit short due to corrosion induced dendrite growth between adjacent conductors for fine-pitch lead-free application.

In summary, we must pay high attention to the incoming quality of incoming PCBs to minmize the risk of high field return rate and high cost.

Acknowledgement

The authors would like to thank the funding of National Natural Science Foundation of China under the contract number 60806029.

References

1. Mahulikar, D., Pasqualoni, A., Crane, J. and Braden, J., "Development of a cost effective high performance metal QFP packaging system," *Proc 43rd Electronic Components and Technology Conference*,. Jun. 1993, pp. 405-411.

2. Li-Na Ji and Zhen-Guo Yang, "Analysis on cracking blind vias of PCB for mobile phones," *Electronic Packaging Technology & High Density Packaging Conference*, Jul. 2008, pp. 1-6.

3. Tok, L., Xiong, Z. and Chua, K.H., "Plated through-hole via cracking in laminated substrate of heat slug PBGA packages," *Proc 5th Electronics Packaging Technology Conference*, Dec. 2003, pp. 589-594.

4. Dongji Xie, Wang, J., Him Yu, Lau, D., Dongkai Shangguan, "Impact Performance of Microvia and Buildup Layer Materials and Its Contribution to Drop Test Failures," *Proc 57th Electronic Components and Technology Conference*, Reno, NV May. 2007, pp. 391-399.

5. Sheng Liu, Yuhai Mei, Wu, T.Y., "Bimaterial interfacial crack growth as a function of mode-mixity," *IEEE Trans-Components, Packaging, and Manufacturing Technology*, Vol. 18, No. 3, Sep. 1995, pp. 618-626.

6. Dongji Xie, Chin, C., Kar Hwee Ang, Lau, D., Dongkai Shangguan, "A new method to evaluate BGA pad cratering in lead-free soldering," *Proc 58th Electronic Components and Technology Conference*, May. 2008, pp. 893-898.

7. Roggeman, B., Borgesen, P., Jing Li Godbole, G., Tumne, P., Srihari, K., Levo, T. and Pitarresi, J., "Assessment of PCB pad cratering resistance by joint level testing," *Proc 58th Electronic Components and Technology Conference*, May. 2008, pp. 884-892.

8. Sanka Ganesan and Michael Pecht, "Open trace defects in FR4 printed circuit boards," *Circuit World*, Vol. 32, No. 1, 2006, pp. 3-7.

Nanoscale Mechanical Properties and Microstructure of 3D LTCC Substrate

Yang-Fei Zhang[1], Jia-Qi Chen[1], Shu-Lin Bai[1]*, Min Miao[2,3], Jing Zhang[2], Fang-Qing Mu[4], Zheng-Yi Wang[4], Shao-Jun Xia[4], Yu-Feng Jin[2]*

[1]LTCS, Department of Advanced Materials and Nanotechnology,
College of Engineering, Peking University, Beijing, 100871, China
[2]National Key Laboratory on Micro/nano Fabrication Technology,
Peking University, Beijing, 100871, China
[3]Information Microsystem Research Institute,
Beijing Information Technology Institute, Beijing, 100101, China
[4]No.43 Research Institute of China Electronics Technology Corporation,
Hefei, 230022, China
*Corresponding author. Tel.: +86-10-62752536; fax: +86-10-62751789; E-mail address: jinyf@ime.pku.edu.cn
*Corresponding author. Tel.: +86-10-62753328; fax: +86-10-62757563; E-mail address: slbai@pku.edu.cn

Abstract

With the development of three-dimensional low temperature co-fired ceramic (LTCC) packaging substrate, the internal microscale structures have great effect on the strength, toughness and lifetime of LTCC microsystem, where the study of local or nanoscale mechanical properties and microstructure is critical, especial for the prediction of the local failure behaviors. In this study, the nanoscale mechanical properties and microstructure of 3D LTCC substrate have been investigated. The nanoscale mechanical properties were measured by nanoindentation method on both the top and lateral faces and discussed according to the profile of the residual pit. The Young's modulus and hardness measured by nanoindentation were compared for two faces. The microstructure was studied by using Scanning Electron Microscope and Scanning Probe Microscope. The results show that LTCC can be regarded as a macroscopically homogeneous isotropic material and a particle-reinforced composite at microscopic scales. The properties of the individual particle and matrix were successfully measured and discussed. Moreover, significant effect of the machining process of the embedded channels and cavities on the strength of the substrate and fluid flow in the microchannel are found and discussed according to the microstructures, including the defects found at the corner of the microchannel and impurities found inside the microchannel.

Introduction

Low temperature co-fired ceramic (LTCC) technique is newly developed to deal with the challenges of high density and multi-function in microelectronics packaging, multilayer MCM (Multi-Chip-Model) and SiP (System-in-Package), and provides a convenient medium for fabricating laminated three-dimensional (3D) structures [1-3]. The internal 3D structures such as channels and cavities can be served for heat dissipation and manufacturing or embedding various sensors into microsystem-level-packages including inertia measurement unit, fluid flow detector and micro-pump [4-7]. These microscale structures have great effect on the strength, toughness and lifetime of LTCC microsystem. The study of local or nanoscale mechanical properties and microstructure is critical, especial for the prediction of the local failure behaviors. LTCC was used to be considered as a single-phase isotropic material in most of previous works. This assumption is not suitable in some local failure situations such as the generation and propagation of the micro-cracks, where the heterogeneity of microstructures plays a key role. Nanoindentation has been used extensively to characterize hard materials such as metals and ceramic films [8]. It utilizes the load-penetration depth curves to calculate the hardness and Young's modulus. With the development of high resolution nanoindentation testing instruments, the individual characterization of each phase in LTCC becomes possible and accurate.

In our previous work, the chemical constituents, microstructure, some macro and micro mechanical behaviors of LTCC have been investigated [9-10]. In this work, the nanoscale mechanical properties on both of the top and lateral faces are investigated as well as the defects and impurities of the microstructure caused by the machining process of the embedded channels and cavities. The results are helpful to the reliability study of LTCC microsystems and optimization of 3D LTCC substrate machining process.

Experimental

In this study, solid substrate and substrate with a microchannel embedded were sintered at 850°C with 11 layers of DuPont 951 AT Green Tape with single layer height of $114 \pm 8\mu m$ [11]. The cross section of the microchannel is $1mm \times 0.1mm$. The top and lateral faces of the samples were polished with emery papers of 2000 grade finish and surface roughness after polishing descended from 210nm Ra to 36nm Ra, as measured by a Scanning Probe Microscopy (SPM) on a scan area of $60\mu m \times 60\mu m$.

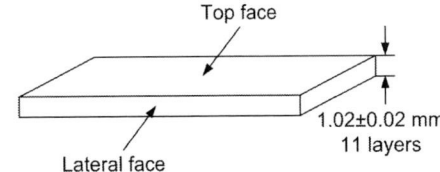

Figure 1 Schematic presentation of the top and lateral faces

Nanoindentation was performed on the top and lateral faces (indicated in Figure 1) of the solid substrate with maximum penetration load of 9 mN. The experiments were undertaken on TriboIndenter (Hysitron Co.) with a Berkovich diamond tip of a nominal radius of curvature equal to

978-1-4244-4658-2/09 $25.00 © 2009 IEEE

~100nm. The loading time, holding time and unloading time were fixed in 5 seconds. Twenty indents were made to obtain average values of the Young's modulus and hardness. During the experiments, the temperature was controlled at ~26°C and the relative humidity at ~43%. The microstructure was investigated by Scanning Electron Microscope (SEM) on Quanta 200FEG (FEI Co.) and Scanning Probe Microscopy (SPM) on TriboIndenter.

Results and Discussion

Both Young's modulus and hardness have the similar variation with the penetration depth and the close values for the top and lateral faces, which proves that the material is macroscopically homogeneous, as shown in Figure 2. The average modulus and hardness of the top face are 116.21 ± 6.64GPa and 8.85 ± 0.59GPa, respectively, while those of the lateral face are 115.00 ± 10.59GPa and 8.85 ± 1.14GPa. The values of modulus and hardness for lateral face are a little lower, which is considered to be caused by the laser cutting. The sintered samples were laser cut into samples for nanoindentation tests. Laser cutting is an optimum method for LTCC materials because the ceramics have a high absorption of the laser light and are easily decomposed under laser irradiation, by which residual stress, thermal deformation and breaking easily caused by other cutting methods for brittle, high melting point and hard materials can be avoided. It is well known that Al_2O_3 has a higher laser absorptance than SiO_2 [12]. Therefore, under the same laser cutting beam, Al_2O_3 in the matrix is more easily decomposed. As the modulus of Al_2O_3 is higher than that of SiO_2 [9], the decrease of Al_2O_3 content in matrix will lead to a smaller modulus, as well as the hardness.

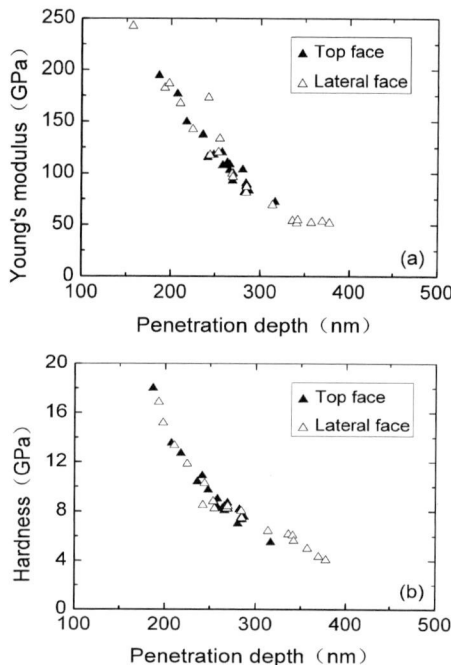

Figure 2 (a) Young's modulus and (b) hardness vs. penetration depth on the top and lateral faces

The modulus and hardness decrease with the increase of penetration depth for both the top and lateral faces. It is related to the microstructure of LTCC: indenting on the harder

location causes smaller penetration depth and larger values of modulus and hardness. In our previous work, LTCC is proved to be a material composed of homogeneously dispersed Al_2O_3 particles with an average radius of 0.71μm and the matrix of mainly Al_2O_3 and SiO_2. If the indenting is exactly acted on the single particle and pure matrix, we should have two distinct value groups, one with high value corresponding to particles, another with low value corresponding to matrix. In our study, except of large particles, many tiny particles are found to be embedded in the matrix. So for some measurements, the real response of the LTCC to the indentation force comes from the combination effect of both matrix and those tiny particles. This will result in the modulus and hardness values between that of particle and pure matrix. The dispersion and content of tiny particles may locally change place by place, which cause the continuous change of values measured.

The microstructures of two faces are similar, as shown in Figure 3, which again proves that LTCC is a particle-reinforced macroscopically isotropic but microscopically heterogeneous composite. Micromechanics have been greatly developed to predict the influence of phase morphology and distribution on effective elastic properties, non-linear behaviors, interface damage and interfacial failure damage for heterogeneous materials. The difficult of the local failure behaviors is the quantitative properties measurement of each phase.

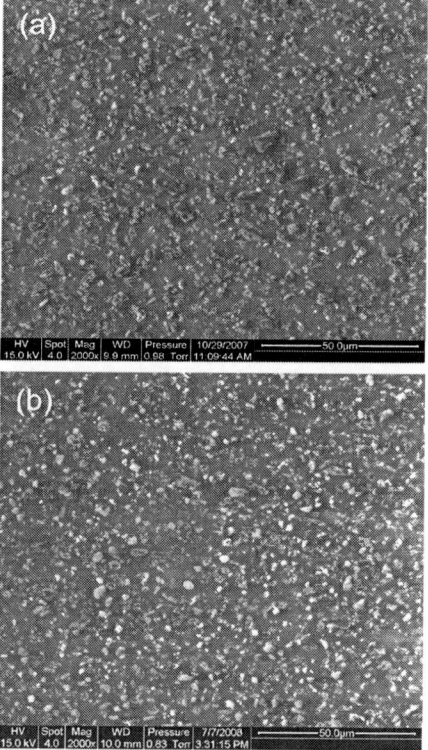

Figure 3 SEM photographs of the (a) top and (b) lateral faces after polishing

In order to investigate the properties of each phase, the nanoindentation tests were carefully carried out to make sure that the indenter was directly pressed on the individual particle and matrix for the top and lateral faces, as shown in Figure 4. The particle has larger modulus and hardness than the matrix and reinforces the substrate. For the top face, the moduli measured of the particle and matrix are 165.24GPa and 97.01GPa, while the hardnesses are 19.89GPa and 7.76GPa, respectively. For the lateral face, the moduli measured of the particle and matrix are 233.64GPa and 86.68GPa, while the hardnesses are 26.69GPa and 7.32GPa, respectively. The modulus and hardness of top face are a little smaller than those of the top face. This is because the indentation locations are manually chosen to make sure that the pits are on the individual particle and matrix, which limits the accuracy of the general properties measurement.

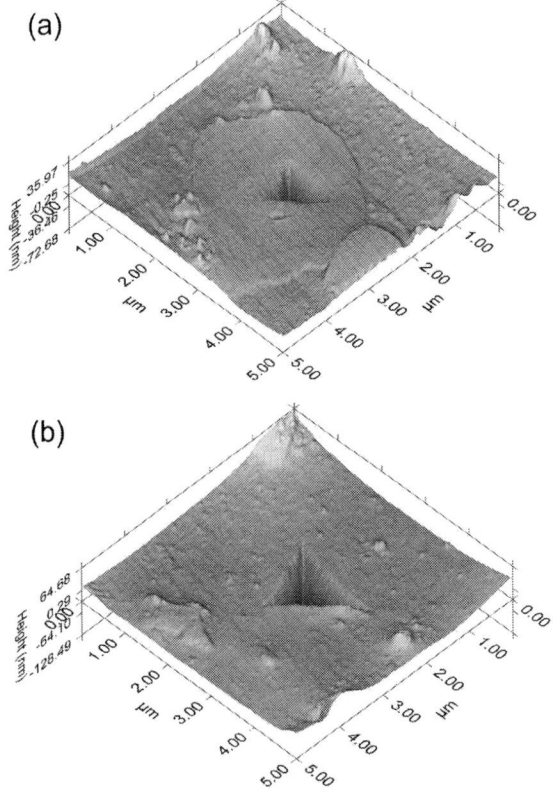

(a)

(b)

Figure 5 3D SPM photographs of pits on individual (a) particle and (b) matrix

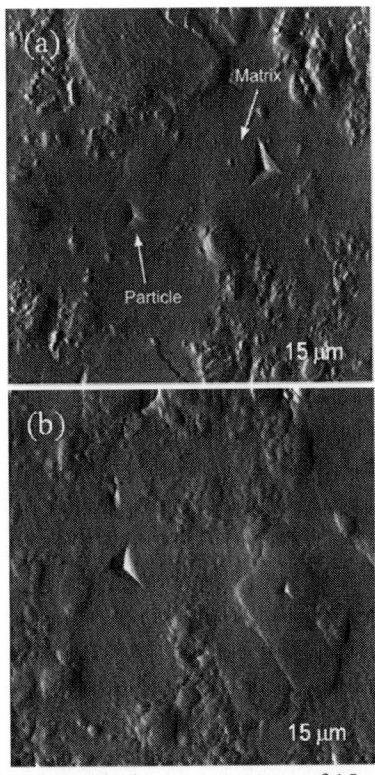

Figure 4 SPM photograph on a scan area of 15μm × 15μm on the (a) top and (b) lateral faces after nanoindentation tests

Figure 5 shows the 3D SPM photograph of pits after the nanoindentation tests on the individual particle and matrix. The pits show that the maximum load is not enough to cause micro-cracks or local damages. Moreover, no pile-up or sink-in is observed at the tip-sample contact boundary, which means the nanoindentation method is effective for LTCC.

In Figure 6, defects are found at the corner of the microchannel, while impurities are found inside the microchannel, which is due to the machining process of the embedded channels and cavities. These defects and impurities have great effect on the strength of the substrate and fluid flow in the microchannel and should be carefully controlled by optimization of the machining process.

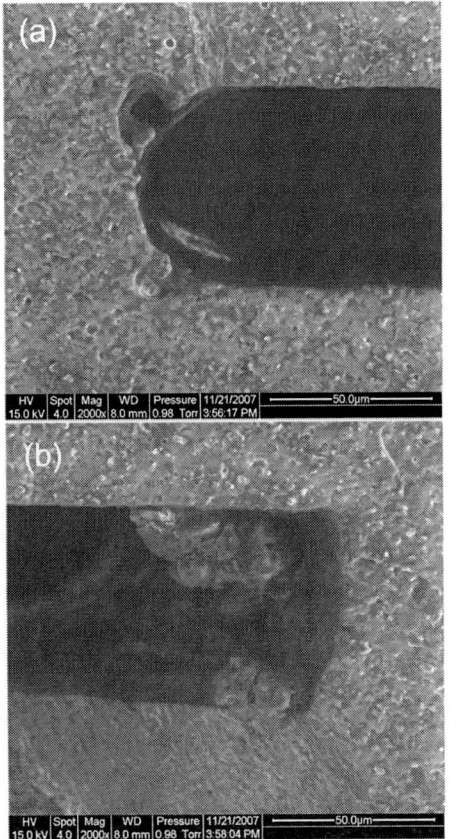

Figure 6 SEM photographs of fracture planes after three-point bending tests: (a) defects and (b) impurities

Conclusions

In conclusion, the LTCC substrate is proved to be a macroscopically homogeneous isotropic material and a particle-reinforced composite at microscopic scales. The modulus and hardness of individual particle and matrix are successfully measured and can be used to study the local failure behaviors. The machining process of 3D LTCC substrate should be improved to eliminate the defects and impurities in microchannels.

Acknowledgments

This work is supported by the National High Technology Research and Development program (863 Program) of China (No. 2007AA04Z352) and National Natural Science Foundation of China (No. 60501007). The authors also appreciate CETC No. 43 Research Institute, (Hefei, China) which kindly provided the specimens.

References

1. Imanaka, Y., <u>Multilayered low temperature cofired ceramics (LTCC) technology</u>, Springer (New York, 2005), pp. 1-18.
2. Bhedwar, H. C. *et al.*, "Low temperature co-fired ceramic tape system-an overview," *Proc 1ˢᵗ Int SAMPE Electronics Conf*, Santa Clara, CA, June. 1987, pp. 720-734.
2. Peterson, K. A. *et al.*, "Novel microsystem applications with new techniques in low-temperature co-fired ceramics," *Int J Appl Ceram Tec*, Vol. 2, No. 5 (2005), pp. 345-363.
3. Peterson, K. A. *et al.*, "LTCC in microelectronics, microsystems, and sensors," *Poc 15ᵗʰ Int Confer Mixed Design of Integrated Circuits and Systems*, 2008, 23-37.
4. Birol, H. *et al.*, "Fabrication of a millinewton force sensor using low temperature co-fired ceramic (LTCC) technology," *Sensor Actuat A-Phys*, Vol. 134, No. 2 (2007), pp. 334-338.
5. Gongora-Rubio, M. *et al.*, "The utilization of low temperature co-fired ceramics (LTCC-ML) technology for meso-scale EMS, a simple thermistor based flow sensor," *Sensor Actuat A-Phys*, Vol. 73, No. 3 (1999), pp. 215-221.
6. Neubert, H. *et al.*, "Thick film accelerometers in LTCC-technology-design optimization, fabrication, and characterization," *J Microelectron and Electron Packaging*, Vol. 5, No. 4 (2008), pp. 150-155.
7. Dannheim, H. *et al.*, "Lifetime prediction for mechanically stressed low temperature co-fired ceramics," *J Eur Ceram Soc,* Vol. 24, No. 8 (2004), pp. 2187-2192.
8. Oliver, W. C. *et al.*, "An improved technique for determining hardness and elastic modulus using load and displacement sensing indentation," *J Mater Res*, Vol. 7, No. 6 (1992), pp.1564-1583.
9. Zhang, Y. F. *et al.*, "Microstructure and mechanical properties of an alumina-glass low temperature co-fired ceramic," *J Eur Ceram Soc*, Vol. 29, No. 6 (2009), pp. 1077-1082.
10. Zhang, Y. F. *et al.*, "Microstructure and strength of low temperature co-fired ceramic substrates with channels and cavities," *Proc 2ⁿᵈ Integration and Commercialization of Micro and Nanosystems International Conference & Exhibition*, Kowloon, Hongkong, June. 2008, pp.243-248.
11. Eustice, A. L. *et al.*, "Low-temperature co-fired ceramics: A new approach to electronic packaging," *IEEE Proc 36th Electronic Components Conf*, Seattle, WA, 1986, pp. 37-47.
12. Zhang, J. H. *et al.*, "Investigation of the surface integrity of laser-cut ceramic," *J Mater Process Technol*, Vol. 57, No. 3-4 (1996), pp. 304-310.

Process Research of LTCC Substrate with 3D Micro-channel Embedded

Fang-Qing Mu[1], Zheng-Yi Wang[1], Yang-Fei Zhang[2]
[1]No. 43 Research Institute of China Electronics Technology Corporation
No.260 Jixi Road, Hefei Anhui, China, 230022
Email: mfq310@163.com, Tel: +86-0551-3667500
[2]LTCS, Department of Advanced Material and Nanotechnology, College of Engineering, Peking University, China

Abstract

Along with the development of the electrical technology, the power of the circuit rises rapidly, heat dissipation becomes a key problem in the design of the circuit. The 3D micro-channel coolers made with LTCC (low temperature co-fire ceramic) technology can absorb the heat of the chip and pass it to the external environment by the liquid circulation. In this paper, the critical manufacturing processes of LTCC multi-layer substrate with 3D micro-channel embedded: lamination and sintering are studied mainly. The special lamination sacrificial layer technique can prevent the 3D micro-channel from collapsing and deforming during the lamination. In addition, the sintering profile is optimized which helps avoid the crazing and delamination of the multi-layer substrate. In conclusion, the intact LTCC substrate with 3D micro-channel embedded can be fabricated using the optimized lamination and sintering process parameters, which makes the following heat dissipation experiment and design optimization more conveniently.

1 Introduction

The continuous high-speed development of integrated circuit chip technology, micro-system technology, and applications of electronic equipment has put forward a high demand for packaging technology. The electronic materials have taken place significant changes correspondingly in packaging material, structure, manufacture and packaging process. Above all, the packaging substrate which fixes and supports the chip, provides the interconnection and peripheral interface, is a platform of realizing the special function, and the corresponding manufacturing technology is a critical element of system package.

Along with the development of the electrical technology, the circuit power is rising rapidly, affecting the reliability and service life of the substrates and the chips severely The heat dissipation in the limited place becomes a key problem in the design of the circuit [1]. The micro-cooling technique which based on MEMS technology and fluid mechanics theory has gotten widespread concern of electronic industry for smaller volume, larger heat dissipation area, lower power consumption, lower cost of batch production and so on. It doesn't emphasize the size decrease of the devices, but puts emphasis on the constructing of micro-channel system to carry out the various microfluid functions. The micro-channel cooler can absorb the heat of the chip, and passes it to the external environment by the liquid circulation, achieving the purpose of heat dissipation [2-3].

The micro-channel is fabricated with LTCC material with the lamination sacrificial layer technique, but the related report of the manufacturing process of the LTCC substrate with 3micro-channel embedded isn't comprehensive enough.

The technical advantage of LTCC lies in its complicated liquid micro-channel, easy manufacturing, extreme thermal, chemical stability and hermetic sealing, resisting the chemical solvent and high temperature, and easy to clean. In this paper, the critical manufacturing processes of LTCC multi-layer substrate with 3D micro-channel embedded: lamination and sintering are studied mainly; the shape and the microstructure are detected by the X-ray and electron microscope.

2 Design and Fabrication of the 3D micro-channel

2.1 Design of the 3D micro-channel

LTCC 3D micro-channel is composed of channel layer, adjusting layer, inlet and outlet [4]. For the single LTCC green tape can be processed individually, the planar micro-channel can be realized easily by the punching process, and the integral 3D micro-channel can be formed after the stacking, laminating and sintering of all green tapes. The structure diagram of LTCC substrate with 3D micro-channel embedded is shown as Figure 1.

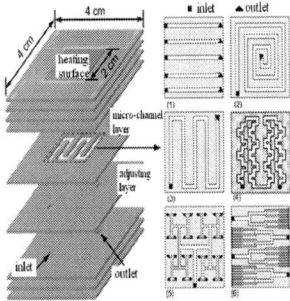

Figure 1 structure diagram of 3D micro-channel

In this study, the substrate with 3D micro-channel embedded is sintering by 15 layers of DuPont 951 AT green tape, and the dimension of the substrate is 40mm by 40mm. The dimension of the heating chip is 20mm by 20mm. The dimension of the cross section of the micro-channel is 0.2mm by 0.2mm. Heat source is on the top surface, and the heat is passed to the substrate steadily. The rectangle represents the inlet, triangle represents the outlet, and the minimum characteristic size of the micro-channel is 0.2mm.

2.2 Fabrication of the micro-channel

There are two principle methods of fabricating 3D micro-channel with LTCC material, as shown in Figure 2. One method is to laminate and sinter two parts and then connect them with adhesive independently (a). This method is very dependent on tightness and expansion of the adhesive material. It requires additional technological steps and restricts the workability of the substrate. The characteristic features of the second method (b) are the structuring in green

978-1-4244-4658-2/09 $25.00 © 2009 IEEE

condition, laminating of the complete structure and sintering in one step. This method does not disturb the normal LTCC process, but makes special demands on the lamination and sintering, which can prevent the 3D micro-channel from collapsing and deforming during the lamination and sintering process [5].

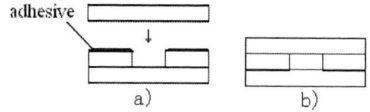

Figure 2 fabricating of micro-channel

The second method is chosen to fabricate the LTCC substrate with 3D micro-channel embedded after the comparison.

LTCC green tape is easy to process and can be used to fabricate the complicated micro-channel, but there are still some difficulties. During the low temperature sintering (< 900°C) phase, the glass viscosity will fall due to the lower softening point of the glass, so the glass element will deform. If there is no mechanical support, the micro-channel will collapse and jam [6]. The sacrificial layer material needs the micro-channel inserted during the fabrication of LTCC substrate with 3D micro-channel embedded, avoiding the collapse and jam of micro-channel during the lamination process [7].

3 Result and Discussion

3.1 The influence of lamination process on the formation of the 3D micro-channel

Lamination process is the general isostatic compaction technique used in the LTCC production line presently, which laminating the stack-up under the certain temperature, pressure and dwell time after the stacking and bag sealing of all green tapes. The lamination sacrificial layer technique is used to fabricate the LTCC substrate with 3D micro-channel embedded, using the sacrificial layer material to insert the 3D micro-channel.

3.1.1 The influence of lamination temperature on the formation of the 3D micro-channel

The lamination temperature must be determined when combining the 3D micro-channel with the sacrificial layer material. The rheological property of the green tape powder increases as the lamination temperature rising. This makes for the compactness between the green tapes; in the meantime the lamination temperature can't exceed the melting point of the sacrificial layer material. The lamination temperature is determined and optimized under the melting point of the sacrificial layer material, insuring the 3D micro-channel is inserted completely with the sacrificial layer material, and avoiding the collapse and jam of the micro-channel during the lamination process.

3.1.2 The influence of lamination pressure on the formation of the 3D micro-channel

The rheological property of the green tape power increases as the lamination pressure rises. In consequence, the adhesive acting and increasing of the combination force between the green tapes. The micro-channel will collapse or deform when the lamination pressure exceed greatly, and the multi-layer substrate will show delamination when the lamination pressure is too little. The lamination pressure is optimized when the lamination temperature and dwell time is determined, avoiding the substrate crazing or the 3D micro-channel collapsing.

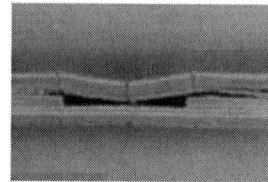

Figure 3 the collapsing of the 3D micro-channel

The collapsing of the 3D micro-channel is shown in Figure 3. It collapsed because of unsuitable lamination pressure.

After the optimization and comparison, the final lamination parameter is listed as:
·pressure: 3.0 kpsi
·dwell time: 600s
·temperature: 50°C

3.2 The influence of sintering process on the formation of the 3D micro-channel

There are two critical phases during the sintering process: the burn out of the polymer adhesive and the sintering of element itself.

The sintering of LTCC substrate with 3D micro-channel embedded is different from the sintering process of general LTCC substrate. The LTCC stack-up consists of more organic elements, so the burn out time should be prolonged and the heating speed should be decreased, making the sacrificial layer material oxidize and combust during the burn out process completely, leaving no residue. The burn out process is composed of two phases: the removal of the volatile organic element and the removal of the volatile product. The volatilization of the organic solvents and the removal of plasticizers and other small molecule organic elements take place in the first phase, and the generation, spreading, and volatilization of the volatile thermal decomposition take place in the second phase. If the organic element decomposes incompletely, the remnant small molecule organic element will decompose further during the hot temperature, leaving the carbon in the substrate. The CO and CO_2 will be generated when the carbon acts with the oxygen of the air, if they can't be discharged timely, the pore will form in the substrate, resulting in the defection of the substrate [7].

During the heating stage of the sintering process, the heating speed should be decreased, making the reaction completely and the removal of the pore in the substrate faster, so the ceramic is more compact after the sintering, avoiding the crazing of the LTCC substrate and the collapse of the micro-channel. The substrate shows clear crack formation as Figure 4.

978-1-4244-4658-2/09 $25.00 © 2009 IEEE

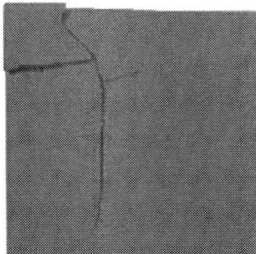

Figure 4 the crack formation of LTCC substrate with 3D micro-channel embedded

The sintering profile is optimized after the LTCC substrate with 3D micro-channel embedded is laminated. The final sintering profile is shown as Figure 5:

·dwell time: 2 hours at 480°C,

·dwell time: 15 minutes at the peak value of 850°C.

So the organic carrier can burn out completely, leaving the intact 3D micro-channel.

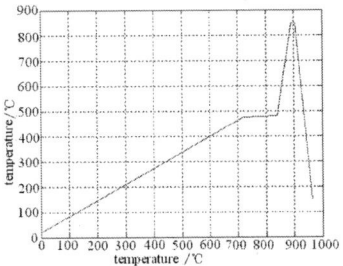

Figure 5 sintering profile of LTCC substrate with 3D micro-channel embedded

3.3 The shape and microstructure of 3D micro-channel

LTCC multi-layer substrates with 3D micro-channel embedded are fabricated by 15 layers of DuPont 951 AT green tape according to the optimized parameter, two kinds of them are shown as Figure 6.

Figure 6 two kinds of LTCC substrates with 3D micro-channel embedded

X-ray scanning pictures of two kinds of micro-channels are shown as Figure 7.

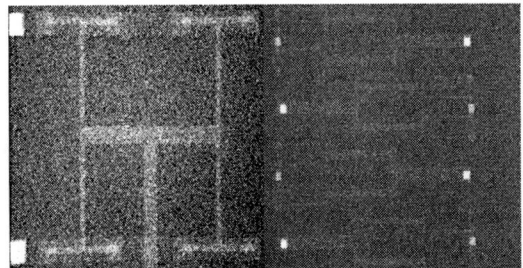

Figure 7 scanning picture of micro-channel with X ray

Sectional views of micro-channel are shown as Figure 8.

Figure 8 sectional view of 3D micro-channel

There is no element jammed in the micro-channel and the stack-up of two layers of micro-channel is precise. The micro-channel of sintering does not show clear jam and deformation which demonstrates the sacrificial layer material contributing to the mechanical support during the lamination. This process helps avoid the deformation of micro-channel, and prevent the collapsing and delamination of the multi-layer substrate during the burn out and sintering process, which illuminates the feasibility of fabricating complicated 3D micro-channel in LTCC substrates.

Conclusion

1) The 3D micro-channel is designed and consists a micro-channel layer, adjusting layer, inlet and outlet according to the material property of LTCC.

2) The critical processes of the LTCC substrate with 3D micro-channel embedded: lamination and sintering are studied mainly. The process of LTCC substrate with 3D micro-channel embedded is optimized and determined by the process experiment.

3) Large size LTCC substrate with complicated 3D micro-channel embedded is fabricated precisely using the optimized lamination and sintering process parameters without jam and collapse.

LTCC multi-layer substrate with 3D micro-channel embedded is a special substrate which supports high density interconnection, multi-chip, micro-system packaging, and has stronger practicability and good development outlook.. In this paper, the critical processes of the LTCC substrate with 3D micro-channel are studied mainly. The eligible substrate is developed and fabricated. The next work is to study the heat dissipation property of LTCC substrate with 3D micro-channel embedded, and optimize the design by the experiment result.

Reference

1. LIU Yi-cai, "Recent Development and Prospects of the Electronic Apparatus Cooling," *Chinese Journal of Electron Devices*, 3, 2006, pp. 296-300.

2. HE ye, LI Lei-min, YANG Tao, "Novel Microcooling Methods Based on MEMS Technology," *INSTRUMENT TECHNIQUE AND SENSOR*, 9, 2004, pp. 43-45.

3. HAN Zhen-yu, MA Ju-sheng, "Prograss of the research on LTCC technology for substrates," *ELECTRONIC COMPONENTS $ MATERIAL*, 19, 2000, pp. 31-33.

4. Keranen, K., Makinen, J.T.; Kautio, "Fiber Pigtailed Mutimode Laser Module Based on Passive Device Alignment on an LTCC Subsrate," *Advanced Packaging*, Vol. 29, No. 3, pp. 463-472.

5. R.Bauer, M.Luniak, L.Rebenklau, K.J.Wolter and W.Sauer, "Realization of LTCC-multilayer with special cavity applications," *International Symposium on Microelectronics*, pp. 659-664, 1997.

6. Petr Kosina, Martin Adamek, Josef Sandera, "MICRO-CHANNLE IN LTCC," *ELECTRONICS*, 9, 2008, pp. 109-114.

7. http://infoscience.epfl.ch/record/87959/files/2005%20-Birol%20couches%20sacrificielles%20carbone.pdf. "Fabrication of LTCC Micro-fluidic Devices Using Sacrificial Carbon Layers".

Study of the Fine Line Process and Signal Integrity for Packaging Substrate

Yi Ma, Jing Jiang, Zhiqin Yang, Minfei Lu
Shennan Circuits Co., Ltd.
East Gaoqiao Industrial Zone, Pingdi Street, Longgang District, Shenzhen, China
Email: mayi@scc.com.cn, Tel: +86 755 89300000-80914

Abstract

This paper presents the forming mechanism of fine line (less than 50μm) in manufacture of packaging substrate. Both subtractive and semi-additive lamination processes were applied for finer pitch substrate manufacture. In addition, signal integrity of the finer traces made on different types of material was tested. Based on these studies, design guidance of fabrication process, materials selection and layout design rule were established for high quality signal integrity applications. Control points of key processes in substrate fine traces manufacture were also described in details, such as choices of surface finish, image transfer, etch resist, foil, etc. Furthermore, the key significant technology in fine traces manufacture was analyzed particularly, and process control technique for improving the yield was obtained.

1 Introduction

At present, in the fast-growing field of electronic design, the electronic system consisting of integrated circuit chip is rapidly developing in the direction of large scale, small size and high-speed. With fast improvement of logic and clock frequency in electronic system, and increasing steepness of signal edge, how to deal with high speed signal becomes a key factor. Therefore, signal integrity problems, such as timing issues caused by interconnect delay, crosstalk and transmission line effects, etc. have to be confronted with in the design and manufacture process of high speed system.

Complying with the requirements of thinness, shortness and reliable performance for consuming electronic products, the line width of IC chips is getting increasingly finer. The required wiring density of package substrate, as the carrier of IC chip, is becoming higher and higher, so does the line width/space. It will be one of the major trends to increase package substrates wiring density. Currently, for the most advanced package substrate, the finest line width/space has reached 14μm/14μm standard, and it is expected to reach 10μm/10μm standard by 2012. [1] Thus the production of fine line will also be an enormous challenge. The more fine-line graphics, the higher of signal integrity demand, which bring on strict requirement of fine line graphic design in package substrate.

In order to meet these requirements, in this paper, we will firstly analyze the key technical points of fine line manufacture. Besides, signal integrity test is also carried out on the fine lines, which are produced by subtractive and semi-additive lamination processes for different materials, finally, the design principle of fabrication process, materials selection and layout design for high quality signal integrity applications, are obtained.

2 Process of fine line manufacture

There are mainly four fine line manufacture processes in electronic industry: subtractive, semi-additive lamination, thin film semi-additive and fully additive processes, as shown in Figure 1.

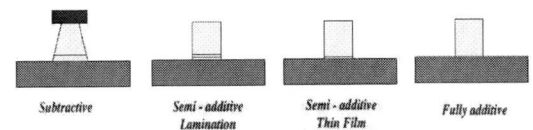

Figure 1 Main processes of fine line manufacture

Subtractive process is mainly used for line with the width/space of more than 35μm, and Semi-additive lamination process is applied for fabricating line with width/space from 25μm to 35μm. Regarding semi-additive thin film process, it's for the line less than 25μm width/space, while for fully additive, the corresponding line width/space is less than 10μm. In the following, subtractive and semi-additive lamination process will be mainly discussed.

3 Analysis of control points
3.1 Subtractive Process

The process flow of subtractive process is similar to normal PCB process, which consists of the following steps: core→lamination→exposure→developing→etching→stripping. Please refer to Figure 2 below for details.

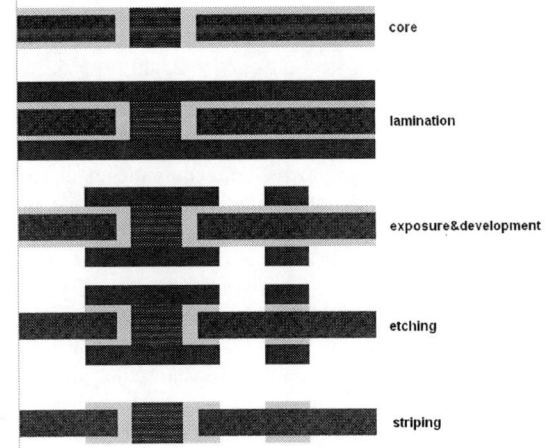

Figure 2 Flowsheet of subtractive process

Subtractive process is simple and its control parameters are mature. However, in etching process, the side etching is inevitable. Therefore, when the subtractive process was applied to fabricate fine traces, the principal factors, including copper foil selection, trace thickness, and dry film selection, should be properly taken into consideration.

3.1.1 The selection of copper foil

Currently, there are two kinds of Copper Foil. One is electrolytic copper foil, which mainly used in rigid PCB. And the other is rolled copper foil, used in flexible PCB. Correct profile of electrolytic copper foil for rigid PCB is the key to fabricate fine traces by subtractive process. Comparing the ordinary PCB copper foil with VLP (Very-low profile) copper foil, it's found that, the VLP copper foil is much easier to be etched and the etching factor is much higher than normal copper foil, when the line width/space is less than 50μm. (Please refer to table 1 for details). The main difference between these two kinds of copper foil is the profile. The surface Rz of the VLP copper foil is only about 4μm (Figure 4), while the surface Rz of the low profile copper foil is up to 9μm (Figure 3). The copper tumor at ordinary copper surface is hard to be etched due to deeply embedded in the substrate. The experiment data indicated that, the copper foil with smaller Rz should be chosen when fabricating the fine traces with line width/line space below 50μm. However, the smaller Rz will cause the binding power to be reduced between the copper and the substrate. Therefore, the binding power and etching ability should be balanced. In current stage, the VLP copper foil can satisfy the current requirements.

Table 1 Etching Comparison between Low profile copper foil and VLP copper foil

Copper Type	Copper Thickness	Etching speed (m/min)	Etching factor of 40μm line	Etching factor of 50μm line
Low profile	H OZ	5.8	2.000	4.057
Low profile	H OZ	5.5	2.138	3.287
VLP	H OZ	5.8	3.251	4.925
VLP	H OZ	5.5	2.909	5.582

3.1.2 The thickness of traces

In terms of fabrication process and requirements, there are two sections, Inner and outer graphic process in fine trace graphic manufacture of substrate. In the inner graphic process, the copper foil thickness is uniform, and mostly 1/2 oz, so line shape can be easily controlled in etching process and can meet the requirements of fine traces with line width/ space less than 50μm. In outer graphic process, the total copper thickness should be controlled strictly after plating. It means that, hole Cu thickness should achieve the required standard, and surface Cu thickness should lower than the maximum thickness. If the total copper thickness after plating is higher than the upper bound, the copper should be thinned to the maximum thickness. The relationship between the line width/ line space and copper thickness is shown in figure 5. If etching factor greater than 3 is the eligible standard of line,

then the corresponding maximum thickness of Cu thickness can gain from the figure bellow.

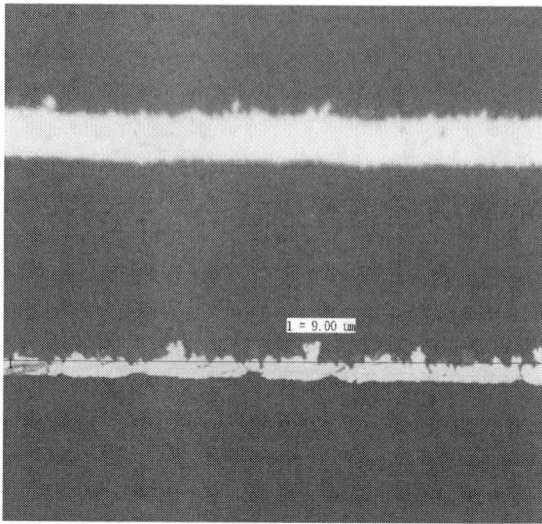

Figure 3 Cross-section of low-profile copper foil

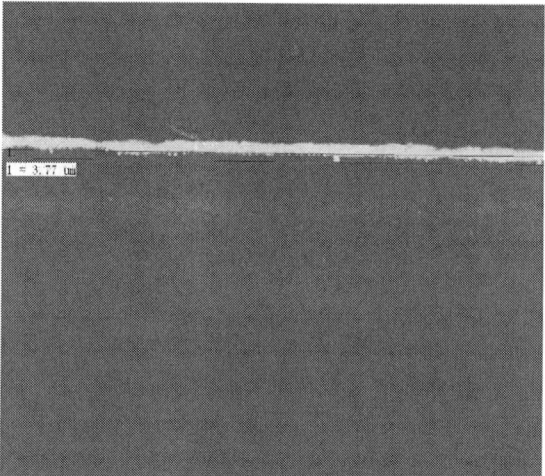

Figure 4 Cross-section of VLP copper foil

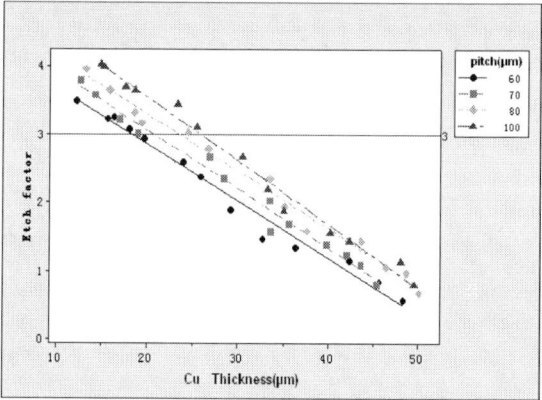

Figure 5 Etching factor at different line width/line space and Cu thickness

3.1.3 Selection of dry film

Excellent analysis ability of light resistance is necessary in graphic manufacture of fine traces. Dry film is widely used as light resistance in package substrate, so the selection of dry film is critical in graphic manufacture of fine traces. In etching of subtractive process, the etching solution not only fill the gap of the dry film, but also cover in the upper of the dry film, as shown in Figure 6.[2] The thinner dry film thickness, the better analysis ability of dry film. Dry film with 15-25μm thickness is widely used in industry nowadays.

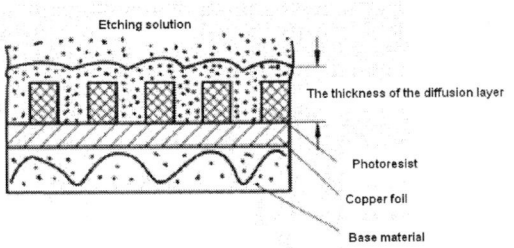

Figure 6 Etching mechanism of fine trace

3.2 Semi-additive lamination process

The process flow of semi-additive lamination process is composed of the following steps: core (3μm copper plus 2μm plating) →lamination →exposure →developing →plating →stripping →differential etching. Please refer to Figure 7 for details.

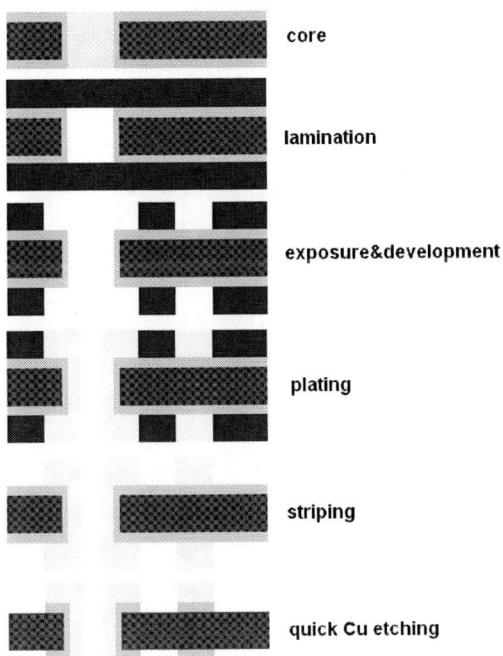

Figure7 Flowsheet of semi-additive lamination process

The line formed by differential etching after plating in semi-additive lamination process, thus the following factors:

the copper foil selection, dry film selection, and the etching solution selection should all be considered carefully.

3.2.1 Choice of copper foil

In semi-additive lamination process, copper-bottomed is etched by differential etching. Ultra thin copper foil with 3-5μm thickness is mainly applied in industry. Now, 2μm ultra thin copper has been put into production as well. Figure 8 shows the UTCF-ultra thin copper foil.

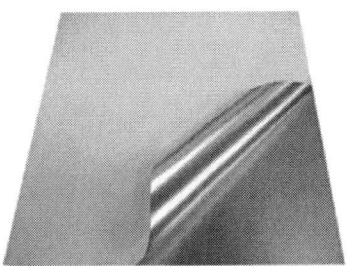

Figure 8 UTCF-Ultra thin copper foil

3.2.2 Choice of Dry Film

The thickness of dry film should be higher than the maximum value of the thickness of line in order to avoid clipping film in plating. The requirement of line thickness in substrates is mostly between 15 and 25μm. Considering the differential etching of the line thickness loss, the thickness of dry film should be 25μm and above. [3]

3.2.3 Selection of etching solution

In semi-additive lamination process, there is not protection at the top of the line due to the use of differential etching. From the experiments we found that, the top of the lines was etched to round edge by acid copper chloride etching solution (see Figure 9), However using sodium persulfate system etching solution there is not round edge at the top of the line, and the line appears to be rough (see Figure 10). Eventually, sulfuric acid hydrogen peroxide etching solution can get a better line shape (see Figure 11).

Figure 9 Line shape etched by acid copper chloride etching solution

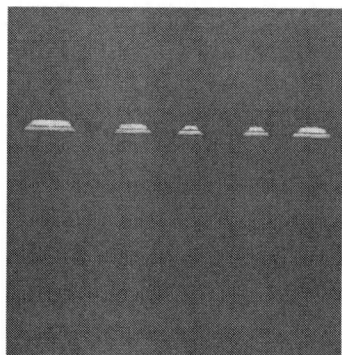

Figure 10 Line shape etched by sodium persulfate system etching solution

Figure 11 Line shape etched by sulfuric acid hydrogen peroxide etching solution

From the figure above, it can be seen that, due to fast etching of acid copper chloride solution, serious side etching is foreseen, and then the top of the lines is etched to round edge. Differently, for sodium persulfate and sulfuric acid hydrogen peroxide with slower etching speed, the round edge of the line is acceptable. Besides, this etching solution can make the copper surface be rough, improve the bonding ability with the resin. Consequently, sulfuric acid hydrogen peroxide etching solution should be chosen in semi-addictive process.

4 Signal integrity test

Signal integrity is a characteristic that the signal maintains its functional properties without any damage after transmission through lines, and can response with correct timing and voltage in the circuit. Generally speaking, the signal integrity can be mainly described as: transmission delay, reflection, crosstalk, simultaneous switching noise and electromagnetic compatibility. And the following aspects are closely related to the substrate manufacturing process: substrate process selection, substrate materials choice, and substrate layout design. Though designing a series of testing coupons, and using network analyzer to test the S21 characteristic of transmission line (S21 parameter is used to evaluate the attenuation of the transmission line, the smaller S21 absolute value, the better signal integrity), we summarize the rules of substrate design and manufacture process.

4.1 Selection of substrate processes

150mm straight line test coupons were produced by subtractive and semi-additive lamination process using BT material, respectively. Figure 12 shows the shape of the test coupon. S21 parameters of transmission lines were measured by network analyzer. The results are shown in Figure 13.

Figure12 150mm straight line test coupon

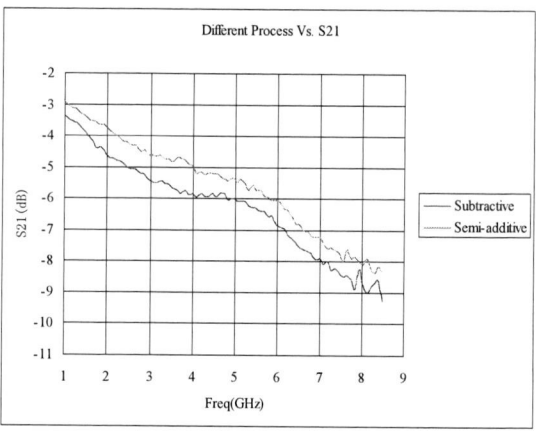

Figure 13 Test results by different processes

Comparing the curves with different processes in figure above, it can be concluded that, semi-additive lamination process with small side etching and roughness impacts less in signal integrity.

4.2 Selection of substrate material

150mm straight line test coupons were produced by three kinds of materials: FR4, BT, and ROGERS, respectively. S21 parameters of transmission lines were measured by network analyzer. The results are shown in Figure 14.

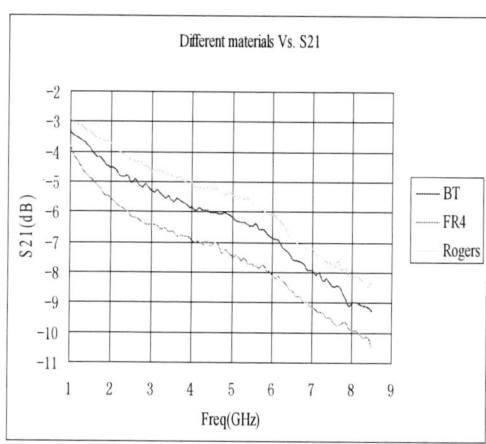

Figure 14 Test results of different materials

Figure 14 indicates that, electronic performance of the three kind of materials with the same subtractive process is ROGERS>BT>FR4. Considering the electronic performance and cost synthetically, BT material is preferred in substrate process.

4.3 Layout design of substrate

This paper presents design guidance of substrate layout in signal integrity from three aspects: signal line length, trace corner selection, and ground analysis.

4.3.1 Length of signal line

S21 parameters with different lengths (50mm, 100mm, and 150mm) of straight line test coupons are measured by network analyzer. Figure 15 shows the shape of the test coupons, and the test results show in Figure16.

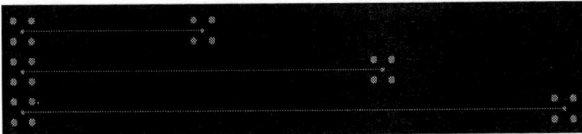

Figure 15 Different lengths of straight line test coupons

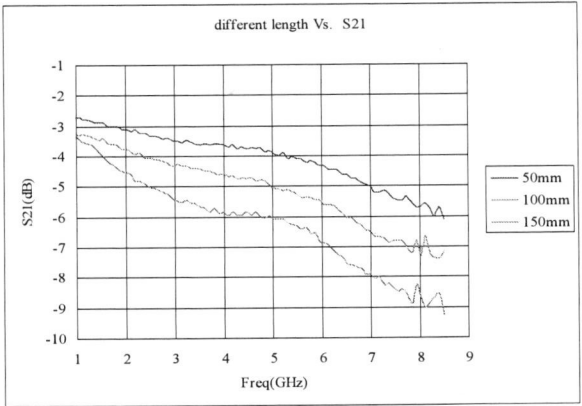

Figure 16 Test results of different length of straight line test coupons

It indicates from the figure above, the longer the signal line is, the bigger signal attenuation will be. Therefore, in wiring design, the signal line length should be as short as possible.

4.3.2 Selection of trace corner

S21 parameters with different trace corners (right angle, 45-degree angle, and fillet angle) of test coupons are measured by network analyzer. Figure 17 shows the shape of the test coupons, and the test results show in Figure18.

The test results show that, signal attenuation of right angle is the biggest, and the signal attenuation of 45-degree and fillet angle is almost the same. Corner point causes impedance discontinuity, therefore, corner should be avoided to the best, or 45-degress, or fillet angle is chosen to improve the signal integrity.

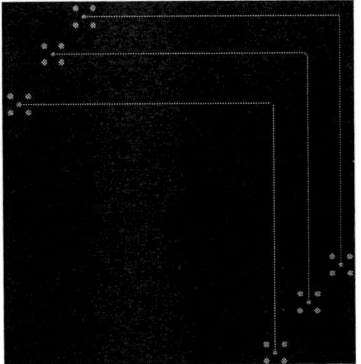

Figure 17 Different trace corners of test coupons

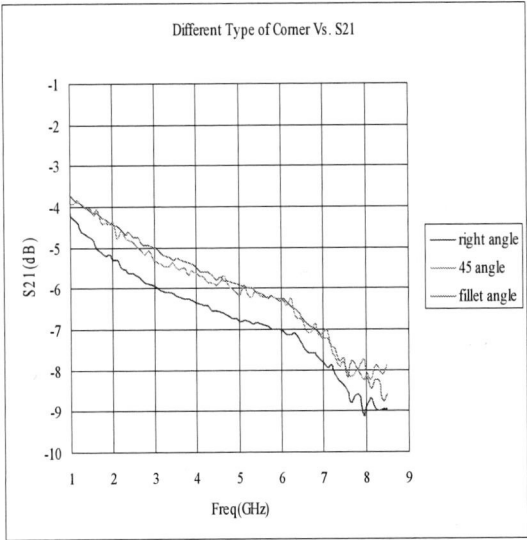

Figure 18 Test results of different trace corners of test coupons

4.3.3 Ground layer analysis

Three types of coupons are designed. First, there is not ground next to transmission lines. Second, the space of line and ground is 4mil. Third, the space of line and ground is 10mil. The graphics of different coupons are shown in Figure 19. S21of transmission lines are tested by network analyzer and the results show in Figure 20.

The results show that, ground design can improve signal integrity. The space between line and ground (4mil and 10mil) has little different effect on S21.

Figure 19 Different ground test coupons

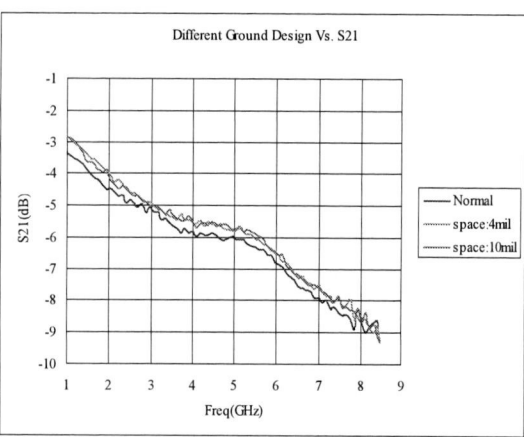

Figure 20 Test results of different ground test coupons

5 Conclusions

The main processes of fine line (less than 50μm) manufacture in packing substrate manufacture are subtractive and semi-addictive processes. Due to its simple process and easy control parameters, subtractive process is mainly used in inner graphic manufacture and outer graphic manufacture with line width/space larger than 35μm.

There are three key control points in subtractive process: Firstly, low profile copper can meet the requirements of signal transmission. Secondly, when line pitch is up to 60μm, the maximum of Cu thickness should less than 20μm. Thirdly, dry film thickness in area of 15-25μm is commended.

There are also three key control points in semi-addictive processes: Firstly, thin copper foil such as 2μm or 3μm thickness is suggested to choose. Secondly, dry film with 25μm thickness has better performance. Thirdly, sulfuric acid hydrogen peroxide etching solution can get a better line shape.

From signal integrity point of view, semi-additive lamination process is preferred in substrate process. BT material is the first selection in substrate materials. Regarding Layout design, the rules are as follows: minimize the length of signal lines, or if fold lines can not be avoided, using 45-degree angle or fillet angle. On top of it, adding ground design next to signal lines can further improve the signal integrity.

References

1. Dinghao Lin, "Technology Roadmap of Taiwan PCB Industry," *TPCA Magazine*, NO. 41 (2008), pp. 41.
2. Minbo Tian, *et al.*, "The high density packaging substrate," *TSINGHUA UNIVERSITY PRESS* (Beijing, 2003), pp. 556-563.
3. Dinghao Lin, "The countermeasure and challenge of the PCB Technology in the nowadays," *TPCA Magazine*, No.41(2008), pp. 43-63.

Effects of Nitrogen on Wettability and Reliability of Lead-free Solder in Reflow Soldering

Mingzhi Dong[1,2], Yuming Wang[3], Jian Cai[1,2], Tao Feng[4], Yuanyuan Pu[1,2]

[1]Tsinghua National Laboratory for Information Science and Technology
[2]Institute of Microelectronics, Tsinghua University, Beijing 100084, China
[3]Tsinghua-Flextronics SMT Lab, Beijing 100084, China
[4]The Linde Group, Linde Gases Division, Shanghai 201206, China

Abstract

Electronic assembly technology is in the transition from traditional Tin-lead to lead-free, which leading to challenge in many aspects. It is accepted that nitrogen can improve the wettability and reliability of lead-free solders; however, few systemic researches have been reported. In this work, the spreading ratio of Sn-Ag-Cu solder was measured through spreading tests under different atmosphere conditions. When oxygen concentration was 1000ppm and reflow temperature was 5°C lower than non-inerted reflow, the spreading ratio of Sn-Ag-Cu solder was comparable with the non-inerted one. PCBs and SMT components were assembled in nitrogen-inerted (oxygen concentration: 1000ppm) and non-inerted reflow. The nitrogen-inerted reflow temperature was 5°C lower. Pull and shear tests were employed as the evaluation of the soldering. Random vibration test and isothermal aging test were conducted as reliability evaluation. The morphology of the interfacial intermetallic compounds (IMCs) during wetting and reliability tests were studied as well.

Introduction

Lead-free soldering has become a global trend out of environmental protection and health concerns. However, comparing with Sn-Pb eutectic solder, high melting point and poor wettability of lead-free solders bring about great challenge to present electronic assembly technology [1]. It has been reported in recent years that nitrogen-inerted soldering can both widen the process window of reflow and improve the reliability of solder joints [2-5]. However, few systemic investigations have been performed on determining the specific atmosphere requirement for realizing the benefits and optimizing the nitrogen consumption. Also research data about effects of nitrogen on wettability and reliability of lead-free solders are far from enough.

This work aimed to investigate the effects of nitrogen on wetting behavior of Sn-Ag-Cu solders and reliability of solder joints after reflow. The wettability of solders was identified through spreading tests on copper plates in nitrogen and air. Reflow process was also conducted on PCBs assembled with different components. On purpose of cost reduction, the peak temperature of nitrogen-inerted reflow was set 5°C lower than non-inerted reflow. The results of reflow were evaluated and reliability tests were conducted, including vibration and isothermal aging. The morphology and composition of the interfacial IMCs during wetting and reliability tests were investigated by scanning electron microscope (SEM) and energy-dispersive spectrometer (EDS).

Experiment

In this study, Sn3.8Ag0.7Cu (SAC387) and Sn3.0Ag0.5Cu (SAC305) solder pastes were chosen for wetting spreading tests. Solder pastes were printed on pretreated copper plates and reflowed in different atmospheres. The oxygen concentrations were real-time monitored, ranging from 50ppm to 210,000ppm (Air), as listed in Table 1. The peak temperature of nitrogen-inerted reflow was set 5°C lower than non-inerted reflow. Fig. 1 shows the top view of as-reflow solders.

Table 1 Oxygen concentrations for wetting spreading tests

SAC387 / ppm	50	100	300	500	1000	1500	Air
SAC305 / ppm	-	100	300	500	1000	1500	Air

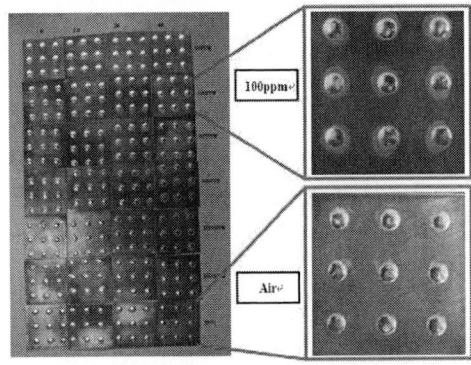

Fig. 1 The top view of as-reflow solders

Two kinds of PCB surface finishes have been chosen for this study. One is Electroless Nickel/Immersion Gold (ENIG) and the other is Organic Solderability Preservatives (OSP), respectively. Different components were assembled on PCBs, which include different pitches BGAs, QFPs, QFN and chip components. All the components were with lead-free solder balls or surface finishes. SAC305 solder paste was used for assembly. The boards were assembled by SMT convection reflow in air and nitrogen (oxygen concentration: 1000ppm). Fig. 2(a) and (b) show the temperature profiles measured on board and under BGA packages. The maximum temperatures under SBGA package is 232°C and 228°C for non-inerted reflow and nitrogen-inerted reflow, respectively. In this study, the boards were named as A-O (Air-OSP), A-E (Air-ENIG), N-O (nitrogen-OSP), and N-E (nitrogen-ENIG). The assembled testing vehicle is shown in Fig. 3. The reliability tests include pull test, shear test, vibration test and isothermal aging test, as well as cross-sectional microstructure analysis.

978-1-4244-4658-2/09 $25.00 © 2009 IEEE

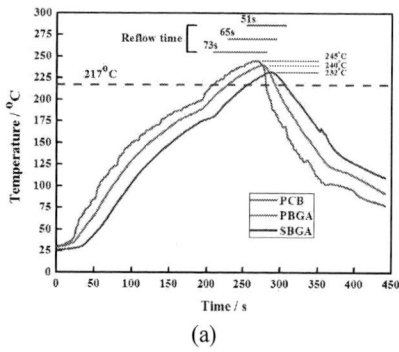 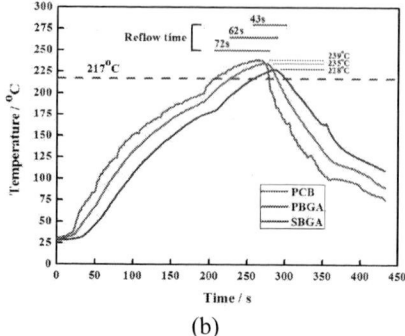

(a)   (b)

Fig. 2 Temperature profiles for (a) non-inerted reflow and (b) nitrogen-inerted reflow

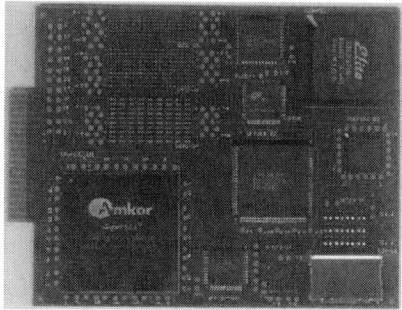

Fig. 3 Reflowed PCB (OSP)

Results and Discussions

I. Wetting Spreading Test

The spreading ratio of solder was calculated from Equation 1.

$$S_R = \frac{D-H}{D} \times 100\% \qquad (1)$$

where, S_R: spreading ratio;

H: height of the spread solder;

D: diameter of the solder, when it is assumed to be a sphere.

Fig. 4 shows the spreading ratios of two kinds of lead-free solders. It is seen that even with lower reflow temperature, the spreading ratios with nitrogen inerted would be comparative, in some cases, would be higher than the ones without inerted.

It is supposed that nitrogen can prevent solder and copper plate from oxidation, and then the wetting force of solder would increase, resulting in higher spreading ratios. It is also found that SAC387 solder has higher spreading ratios than SAC305.

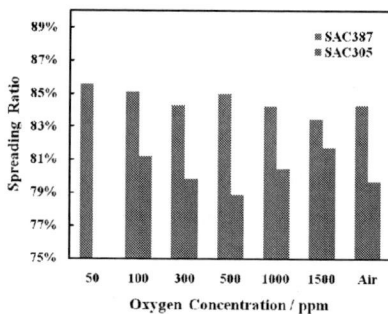

Fig. 4 Spreading ratios of SAC387 and SAC305 solders

Cross-section microstructures of SAC387 solder joints were characterized by SEM/EDS. Fig. 5(a) and (b) show the interfacial microstructures of the solder joints reflowed in nitrogen and air respectively. The interfacial IMC reflowed in air reveals more pores and dendritic grains, both of which are harmful to the reliability of solder joints. The thicknesses of the interfacial IMC layers were measured and shown in Fig. 6. The thickness of interfacial IMC layer increased with the concentration of oxygen, but was not quite remarkable. The EDAX results revealed the composition of the interfacial IMC was close to Cu_6Sn_5 for both in nitrogen and in air.

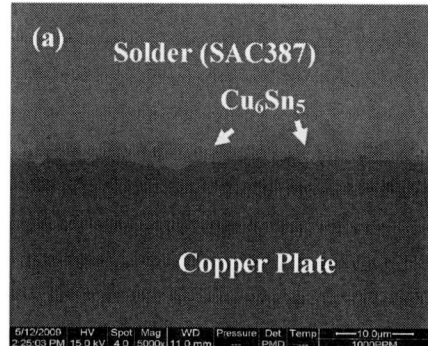 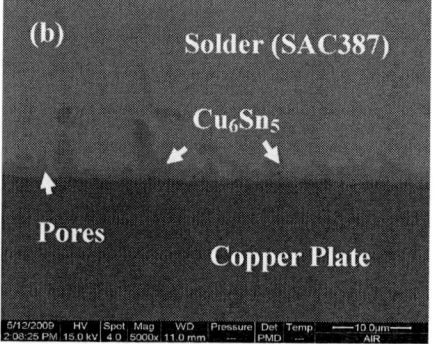

Fig. 5 Cross-section interfacial microstructure of solder joint reflowed in
(a) nitrogen (oxygen concentration: 1000ppm) and (b) air

978-1-4244-4658-2/09 $25.00 © 2009 IEEE

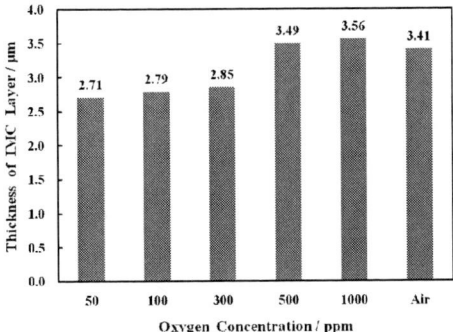

Fig. 6 Thickness of IMC layer (SAC387)

It is noticed that when oxygen concentration was 1000ppm, the spreading ratio was at least comparable or

higher than the one reflowed in air and the IMC layer had nearly the same thickness with the one in the air while showed more uniformity and less pores. Given these results, the reflow of PCBs was conducted with oxygen concentration at 1000ppm.

II. Reflow of PCBs and Reliability Tests

Leads of components were visually inspected by optical microscope. Fig. 7(a) and (b) show QFP lead reflowed in nitrogen (oxygen: 1000ppm) and air, respectively. Surface of solder joints reflowed in nitrogen was more uniform while the ones reflowed in air showed more micro-cracks. The reason could be explained as nitrogen can reduce the surface tension of liquid solder and prevent liquid solder from cracking during solidification.

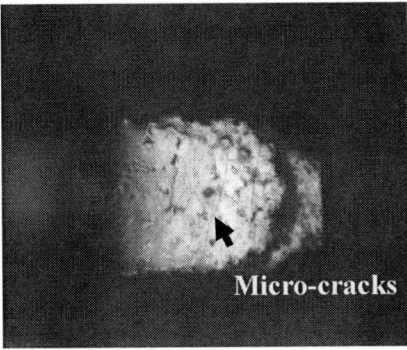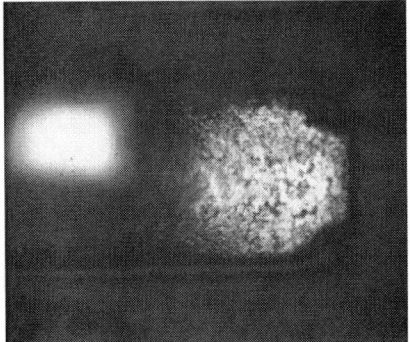

Fig. 7 QFP lead reflowed in (a) nitrogen (oxygen: 1000ppm) and (b) air

X-ray inspection indicated that voids existed in solder joints reflowed in both nitrogen and air, as shown in Fig. 8(a) and (b). The total volume percentage of voids in a solder ball is below 25% according to the X-ray inspection, which is acceptable for SMT.

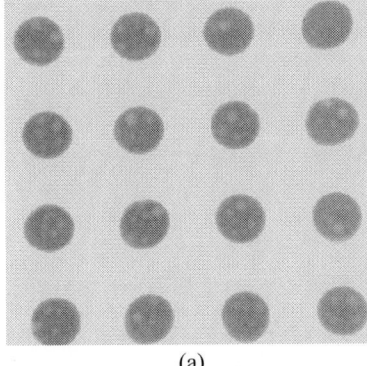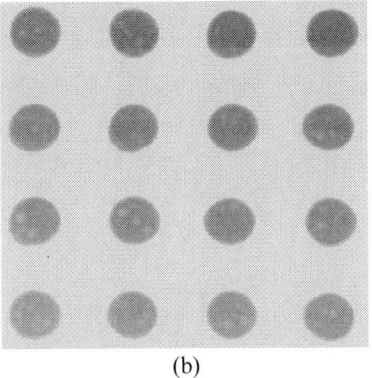

(a)   (b)

Fig. 8 X-ray inspection of 1.0mm pitch BGA reflowed in (a) nitrogen (oxygen: 1000ppm) and (b) air

Pull and shear tests were conducted to identify the mechanical properties of solder joints. The tensile strength of QFP leads reflowed in nitrogen was slightly lower than the one reflowed in air, as shown in Fig. 9. The shear forces of different components are shown in Fig. 10, indicating that

both nitrogen-inerted and non-inerted reflowed components reveal comparable shear strength.

Given that the temperature of nitrogen-inerted reflow was 5°C lower than non-inerted reflow, it is concluded that solder joints reflowed in nitrogen show comparable mechanical properties with those reflowed in air, even with lower reflow temperature.

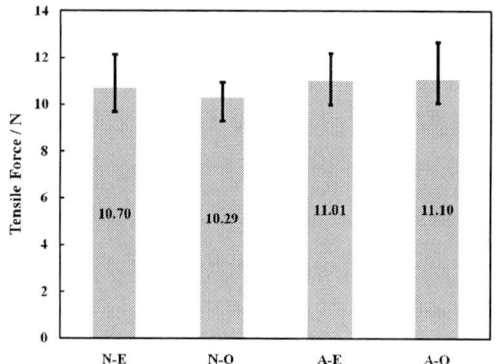

Fig. 9 Tensile forces of QFP leads reflowed under different conditions

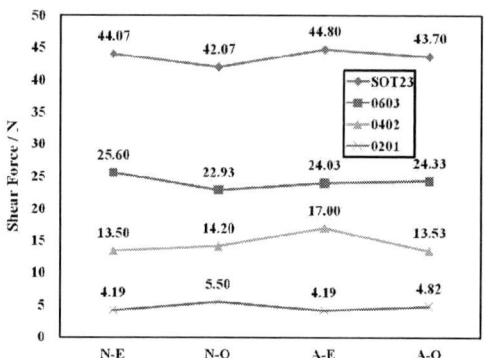

Fig. 10 Shear forces of components reflowed under different conditions

Every board passed the vibration and isothermal aging tests. The vibration applied on boards was in random mode, lasting 20mins for each board. The range of frequency was 20Hz to 2000Hz and the root mean square average (RMS) of the acceleration was 6.06G. The isothermal aging condition was 120°C for 120 hrs. Pull and shear tests were conducted

after vibration or isothermal aging. The tensile forces of QFP leads and shear forces of 0603 chip components are shown in Fig. 11. It is noticed that the mechanical strength of components dropped after either vibration or isothermal aging. Components reflowed in nitrogen showed slightly more reduction in mechanical strength.

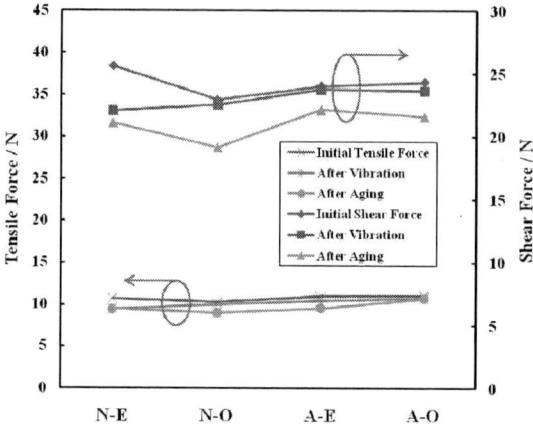

Fig. 11 Tensile forces of QFP leads and shear forces of 0603 chip components

The cross-section microstructures of solder joints after vibration or isothermal aging test were observed by SEM. Fig. 12(a) and (b) show the interfacial microstructures of SBGA reflowed in air and nitrogen, respectively. For the samples reflowed in air, some solder balls were found fractured from pads after vibration although the whole package passed the test. For the samples reflowed in nitrogen, no solder balls were found fractured from pads after vibration but some micro-cracks were detected in the interfacial IMC layer. This indicates that solder joints reflowed in nitrogen can withstand more high-intensity external force and nitrogen can strengthen the reliability of electronic components.

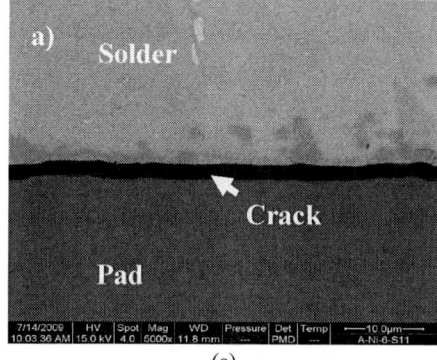

(a)

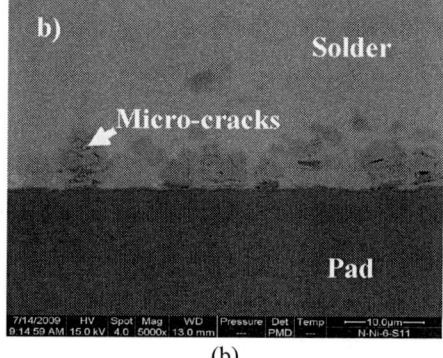

(b)

Fig. 12 Cross-section microstructures of solder joints after vibration, (a) reflowed in air and (b) reflowed in nitrogen(oxygen: 1000ppm)

The IMC layer grew thicker after isothermal aging (120°C for 120 hrs). As shown in Fig. 13(a), for QFP reflowed in nitrogen, the thicknesses of interfacial IMC layers were 3.8μm on lead-side and 2.5μm on PCB-side before isothermal aging. After aging the IMC layer became much thicker: 7.0μm on lead-side and 4.3μm on PCB-side, as shown in Fig. 13(b). The similar phenomenon were observed in components reflowed in air.

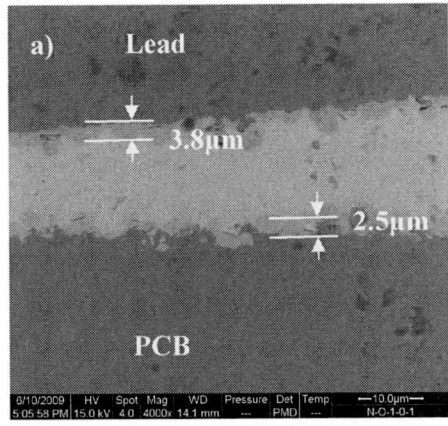 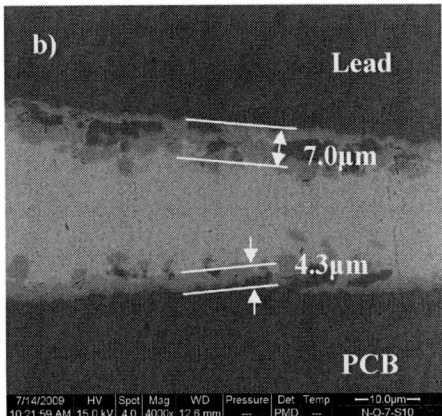

Fig. 13 Interfacial IMC layer of nitrogen-inerted reflowed QFP lead (a) before (b) after isothermal aging (120°C for 120 hrs)

Conclusion

The spreading ratio of SAC solder increased with decreasing oxygen concentration during nitrogen-inerted reflow, that is, nitrogen can enhance the expansibility of SAC solder, even at lower reflow temperature. The IMC layer thickness increased with oxygen concentration. When oxygen concentration was 1000ppm and reflow temperature was 5°C lower, both the spreading ratio of SAC solder and the IMC layer thickness were comparable with the non-inerted ones.

PCBs and SMT components were assembled in nitrogen-inerted (oxygen: 1000ppm) and non-inerted reflow. The nitrogen-inerted reflow temperature was 5°C lower. Visually inspection revealed that solder joints reflowed in nitrogen showed less surface micro-cracks. Components reflowed in nitrogen showed slightly lower tensile strength and comparable shear strength.

Reliability tests, including random vibration and isothermal aging, were conducted on the boards. All the boards reflowed both in nitrogen and in air, passed these tests. However, the mechanical properties degraded after vibration or isothermal aging. The tensile strength and shear strength of nitrogen-inerted reflowed samples dropped more. Observed by SEM, part of the BGA solder balls reflowed in air fractured from the pads, while nitrogen-inerted part showed no such cracks but micro-cracks in interfacial IMC. For both nitrogen-inerted and non-inerted reflow, interfacial IMC layers grew thicker after isothermal aging. For QFP reflowed in nitrogen, the thickness of interfacial IMC layer on lead-side was from 3.8μm to 7.0μm and on PCB-side was from 2.5μm to 4.3μm.

Acknowledgements

This work is supported by the National Engineering Laboratory on High Density IC Packaging Technology, China.

The authors would like to thank the staff from Tsinghua-Flextronics SMT Laboratory for performing the process of SMT assembling in this work. The authors gratrfully acknowledge Ms. Wenyan Yang for the help of SEM/EDS analysis.

References

1. Mulugeta Abtew, Guna Selvaduray, "Lead-free Solders in Microelectronics," *Materials Science and Engineering*, Vol. 27, 2000, pp. 95-141.
2. Adams S. M. , Stratton P. F. , "The effect of oxygen concentration on the comparative solderability of eutectic Sn/Ag/Cu and Sn/Pb," *Forth Pacific Rim International Conference on Advanced Materials and Processing*, Vols I And II, 2001, pp. 1035-1038.
3. Ling Hiew Pang, Stratton Paul, "Increasing Value in Lead-free Soldering with nitrogen," *7th International Conference on Electronics Packaging Technology, Proceedings*, 2006, pp. 428-430.
4. Belyakov Sergey A., "Nitrogen in Reflow Soldering of Lead-free Solders," *8th International Workshop and Tutorials on Electron Devices and Materials*, 2007, pp. 84-85.
5. Baated Alongheng *et al.*, "Sn-Ag-Cu Soldering Reliability Influenced by Process Atmosphere," *Proceedings of the 2007 International Symposium on High Density Packaging and Microsystem Integration*, 2007, pp. 110-114.

Solderability of Flash Gold Surface Finish

Yuming Wang[1], Mingzhi Dong[2,3], Jian Cai[2,3]
[1]Tsinghua-Flextronics SMT Laboratory, Beijing, 100084, China
[2]Tsinghua National Laboratory for Information Science and Technology
[3]Institute of Microelectronics, Tsinghua University, Beijing, 100084, China

Abstract

Flash gold with nickel and gold plating is a low-cost technique in PCB finishing, frequently used in the production of toys. The quality of plating has become one of the concerns for the performance and reliability of solderability. This study investigates the solderability of PCB produced by flash gold finish and traditional nickel and gold plating finish. For this purpose, solderability tests have been conducted. The test results with both techniques of plating layers are analyzed and compared. Their impact on solderability of PCB is discussed.

1 Introduction

Flash gold with nickel and gold plating is a low-cost technique, particularly when gold pre-plating is processed. With short time in plating, thinner concentration of gold can be achieved, only 1/10 of the normal thickness. This results a very thin layer of gold, only 0.05μm in thickness. The thin gold layer is weak in resisting the oxidation of nickel and then results in a series of quality problems, including soldering defects and poor reliability.

Little work has been done in the area of solderability of flash gold finish. There is a need to investigate the differences between flash gold finish and traditional nickel and gold plating finish, and to search a way for increasing the reliability of the solderability. The current paper is targeted at this area.

2 Experiment

For the flash gold PCB, nickel layer thickness was 7.0μm~7.5μm and gold layer thickness was 0.07μm~0.48μm. As the gold layer contains nickel, the thickness of the layer was not consistent. For the traditional nickel and gold plating finish PCB, nickel layer thickness was 5μm and gold layer thickness was 0.3μm. It can be seen that comparably, the nickel layer thickness by the flash gold PCB is slightly thicker than the traditional nickel and gold plating finish PCB, whereas the gold layer by the former is much thinner than the latter. This leads to the cost of the flash gold is much less than the nickel and gold plating.

Metronelec ST88 solderability tester was used for the tests. Two PCBs, with the flash gold and nickel and gold plating finishes, were tested, respectively. For each tested PCB, three mark points and a SOP were chosen for testing, as shown in Fig. 1.

The test was conducted according to IPC/EIA-J-STD-003B [1] with Sn3.0Ag0.5Cu lead-free solder. Test temperature was set to 265°C. The mounting temperature of the tested sample on the solderability tester is 45°. In the test, the test fixture was vertically moved down to allow the PCB to touch the liquid solder ball (Φ 4 mm) on the test device.

Wetting curve was recorded automatically. The tested fixture and sample are shown in Fig. 2.

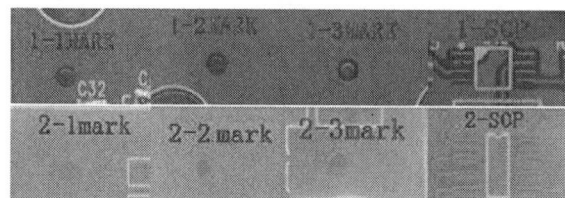

Fig. 1 Testing points on the PCBs

Fig. 3 shows the typical wetting curve. The wetting force goes down in the early test stage and then gradually increases to an equilibrium level. The buoyancy axis shown in the figure represents the datum line for wetting force. Table 1 shows the basic parameters, i.e. wetting time, immersing speed, exiting speed and immersing depth, used in the tests. Based on the specification IPC/EIA-J-STD-003B, the parameter determination and criteria are determined, shown in Table 2. The criteria set A in Table 2 was used in this study for assessing the fabrication quality.

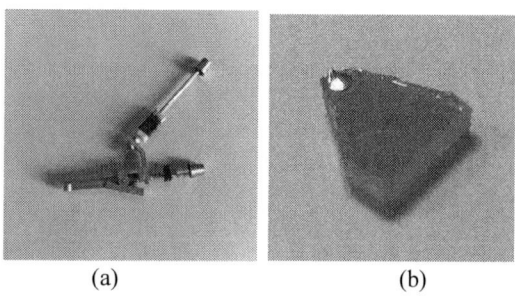

(a)   (b)

Fig. 2 (a) test fixture and (b) tested sample

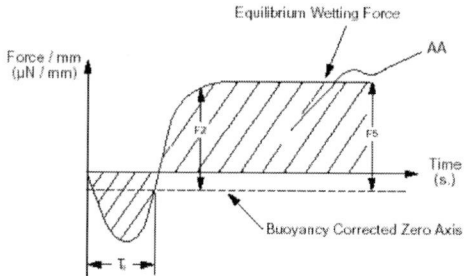

Fig. 3 Wetting curve [1]

Table 1 Test parameters

Soldering temperature/°C	265	Solder density/(mg·mm^{-3})		8.1	Surface tension/(mN·m^{-1})		390	
Wetting time/s	5	Immersing speed/(mm·s^{-1})		5	Exiting speed/(mm·s^{-1})		21	
Immersing depth/mm	1#PCB	1-1	0.5	1-2	0.5	1-3	0.1	1-SOP	0.5
	2#PCB	2-1	0.5	2-1	0.1	2-3	0.2	2-SOP	0.5
Note: 1-1 denotes the 1-1MARK. The rest can be deduced by analogy.								

Table 2 Wetting Balance Parameter and Suggested Criteria [1]

Parameter	Description	Suggested Criteria	
		Set A	Set B
T_0	Time to buoyancy corrected zero	≤ 1 second	≤ 2 second
F_2	Wetting force at two seconds from start of test	$\geq 50\%$ of maximum theoretical wetting force at or before two seconds	Positive value at or before two seconds
F_5	Wetting force at five seconds from start of test	At or above the value of F_2	At or above the value of F_2
AA	Integrated value of area of the wetting curve from start of test	$\geq$ area calculated using sample buoyancy and 50% maximum theoretical force	$>$ zero (0)

3 Results and Discussions

3.1 Wetting curves

The PCBs with two kinds of finishes were tested according to the test methods mentioned above. The wetting curves for the PCB with flash gold finish was plotted in Fig. 4 and the curves for nickel and gold plating finish was shown in Fig. 5.

1-1MARK

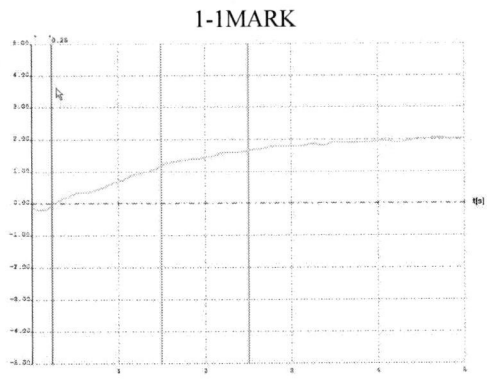

Note: Reach maximum slowly

1-2MARK

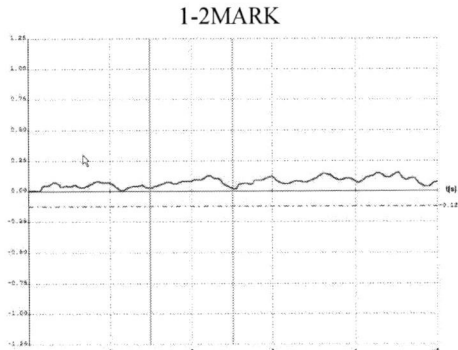

Note: Unqualified curve

1-3MARK

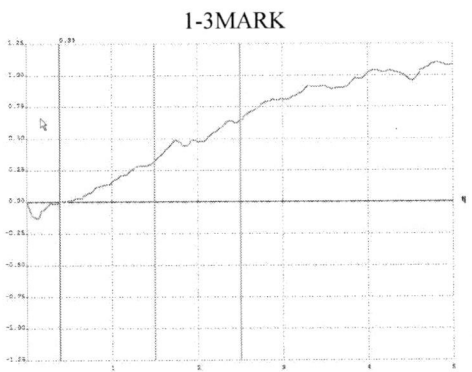

Note: Reach maximum slowly

1-SOP

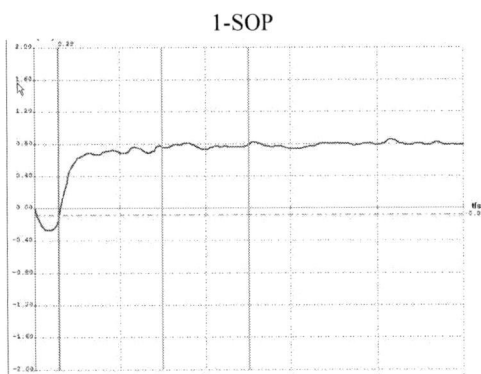

Note: Reach maximum quickly

Fig. 4 Wetting curves of flash gold PCB (PCB #1)

2-1MARK	2-2MARK

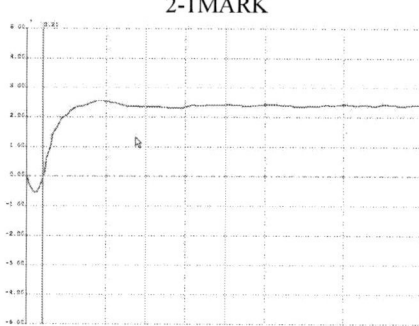 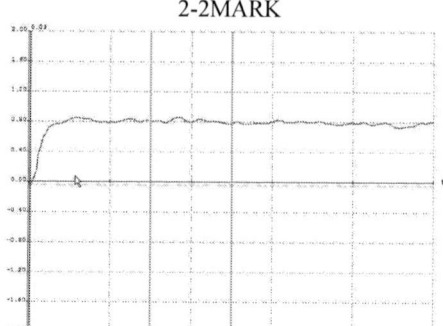

Note: Reach maximum quickly	Note: Reach maximum quickly
2-3MARK	2-SOP

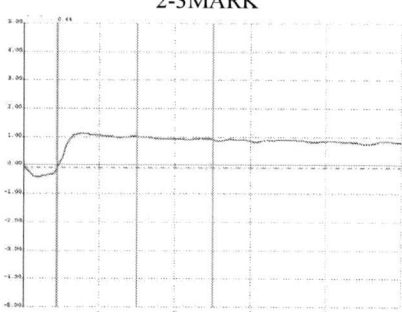 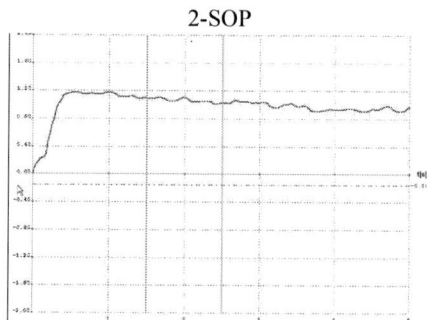

Note: Reach maximum quickly   Note: Reach maximum quickly

Fig. 5 Wetting curves of traditional nickel and gold plating PCB (PCB #2)

3.2 Calculation of maximum theoretical force and area under the wetting curve

From test principle, the difference between wetting and buoyancy forces denotes the tested force when a sample is immersing to solder, i.e.

$$F = F_m - F_a$$

where, F_m—Wetting force

F_a—Buoyancy force

The wetting force can be expressed as

$$F_m = t \times P \times \cos\theta$$

where, t—Surface tension of solder

P—The periphery of the specimen in millimeters

θ—Wetting angle of solder

Under optimal conditions, $\theta=0$, the wetting force reaches its maximum value. Therefore, here takes $\theta=0$ for the calculation of the maximum theoretical wetting force. This experimental sample touched liquid solder ball in 45° direction, thus,

$$F_a = \rho \times V \times g \times \cos 45°$$

where, ρ—Density of solder

V—The volume in cubic millimeters of the specimen that resides below the solder/board air interface as measured at the maximum depth of immersion

g—Gravitational constant

From above, the max theoretical force can be obtained, as

$$F_0 = t \times P - \rho \times V \times g \times \cos 45°$$

The area under wetting curve

$$AA_0 = 3 \times 0.5 \times F_0 - 2 \times F_a$$

The calculated data, based on the above equations, are shown in Table 3.

Table 3 Maximum theoretical force and area under wetting curve

Sample number	P/mm	V/mm³	F₀/mN	AA₀/mN·s
1-1Mark	3.45	0.40	1.33	1.94
1-2Mark	3.45	0.40	1.33	1.94
1-3Mark	3.45	0.008	1.35	2.02
1-SOP	3.40	0.40	1.30	1.91
2-1Mark	4.08	0.40	1.57	2.31
2-2Mark	4.08	0.008	1.59	2.39
2-3Mark	4.08	0.04	1.59	2.38
2-SOP	5.36	0.40	2.07	3.06

Table 4 Parameters of wetting curve

Test point	T_0/s	F_2/mN	F_5/mN	AA/mN·s	Maximum wetting force F_{max}/mN
1-1Mark	0.25	1.44	2.03	6.89	2.05
1-2Mark	0	0.09	0.08	0.35	0.14
1-3Mark	0.39	0.47	1.09	3.01	1.10
1-SOP	0.29	0.73	0.81	3.57	0.86
2-1Mark	0.21	2.34	2.41	11.13	2.57
2-2Mark	0.03	0.80	0.81	3.83	0.85
2-3Mark	0.44	0.93	0.82	4.11	1.14
2-SOP	0	1.11	0.97	4.93	1.18

Table 5 Mean values of wetting curve parameters

	T_0/s	F_2/mN	F_5/mN	AA/mN·s	Maximum wetting force F_{max}/mN
1#PCB	0.23	0.68	1.00	3.46	1.04
2#PCB	0.17	1.30	1.25	6.00	1.44

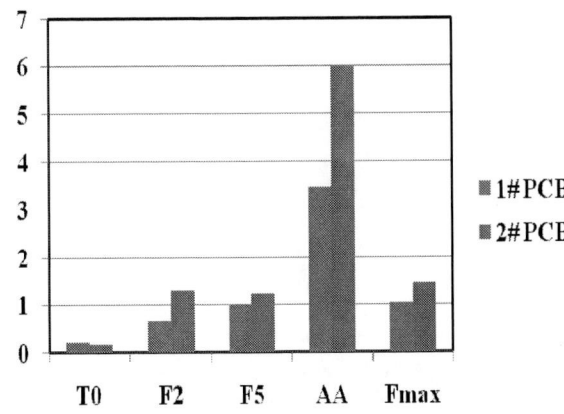

Fig. 6 Mean values of wetting curve parameters

3.3 Analysis

Table 4 shows the parameters of wetting curves. The mean values of the wetting parameters for each PCB are shown in Table 5 and also depicted in Fig. 6.

3.4 Conclusion of PCB solderability

Examining Table 4, Table 5 and Fig. 6, in terms of wetting force, the performance of PCB #1 is poorer than PCB #2. To judge whether each parameter is qualified or not against IPC standard, practical data are listed in Tables 6 through 9. It is found that PCB #1 has many unqualified items. Obvious difference between unqualified data and standard requirement is also observed. In contrast, PCB #2 has rare unqualified items and the difference is smaller between unqualified data and standard requirement. In summary, the solderability of PCB #1 is poorer than PCB #2.

Table 6 Judgment of qualified T_0

Sample number	T_0/s	$T_0 \leq 1s$
1-1Mark	0.25	qualified
1-2Mark	0	qualified
1-3Mark	0.39	qualified
1-SOP	0.29	qualified
2-1Mark	0.21	qualified
2-2Mark	0.03	qualified
2-3Mark	0.44	qualified
2-SOP	0	qualified

Table 7 Judgment of qualified F_2

Sample number	F_2/mN	$1/2F_0$/mN	$F_2 \geq 1/2F_0$
1-1Mark	1.44	0.66	qualified
1-2Mark	0.09	0.66	unqualified
1-3Mark	0.47	0.67	unqualified
1-SOP	0.73	0.65	qualified
2-1Mark	2.34	0.79	qualified
2-2Mark	0.80	0.80	qualified
2-3Mark	0.93	0.80	qualified
2-SOP	1.11	1.03	qualified

Table 8 Judgment of qualified F_5

Sample number	F_5/mN	F_2/mN	$F_5 \geq F_2$
1-1Mark	2.03	1.44	qualified
1-2Mark	0.08	0.09	unqualified
1-3Mark	1.09	0.47	qualified
1-SOP	0.81	0.73	qualified
2-1Mark	2.41	2.34	qualified
2-2Mark	0.81	0.80	qualified
2-3Mark	0.82	0.93	unqualified
2-SOP	0.97	1.11	unqualified

Table 9 Judgment of qualified AA

Sample number	AA/mN·s	AA_0	$AA \geq AA_0$
1-1Mark	6.89	1.94	qualified
1-2Mark	0.35	1.94	unqualified
1-3Mark	3.01	2.02	qualified
1-SOP	3.57	1.91	qualified
2-1Mark	11.13	2.31	qualified
2-2Mark	3.83	2.39	qualified
2-3Mark	4.11	2.38	qualified
2-SOP	4.93	3.06	qualified

3.5 Microstructure analysis of solder joint

Normal soldering process was made on flash gold PCB. The components were then removed. The cross-section microstructure was inspected under SEM. The result showed that the incomplete soldering was founded between solder and PCB. Dewetting phenomenon appeared in the interface, indicating that the PCB had problem of soldering, as shown in Fig. 7 (a). Black band between solder and interfacial intermetallic compound (IMC), an evidence of oxidation, was found, shown in Fig. 7 (b). From the view of the whole interface, the thickness of IMC layer was only 0.5μm, even after double heating treatment. This indicates that nickel diffusion is insufficient on the interface. The reason may be the poor solderability of flash gold PCB.

4 Conclusions

Compared with traditional nickel and gold plating finish, flash gold finish can produce much thinner gold layer on PCB. This leads to the increase of productivity and cost-efficiency of PCBs. However, some side effects are formed in this plating process, in terms of wetting force and solderability.

The solderability test shows that all three marked points for vetting force testing have indicated some weaknesses. In two marked points, i.e. Mark 1-1 and 1-3, the wetting force developed in a slow process and in the third point, e.g. Mark 1-2, the wetting force was very low. As a result, a low level of the integrated value of area of the wetting curve was produced. This leads to poor solderability behavior. The above results are coped with the theoretical analysis conducted by this paper.

It has been evident by the tests and SEM inspection that one of the major weaknesses of flash gold finish is its low resistance to the oxidation of nickel. This is an intrinsic weakness, as a result of the very thin gold layer and unevenly distributed material.

Acknowledgements

The authors would like to thank Telecommunication Science And Technology Meter Research Institute of Datang Telecom Group for supplying PCB samples for this work and acknowledge French manufacturer METRONELEC for providing solderability tester used in this work. The authors also wish to acknowledge Peian Jiang's support for sharing his experience on this subject.

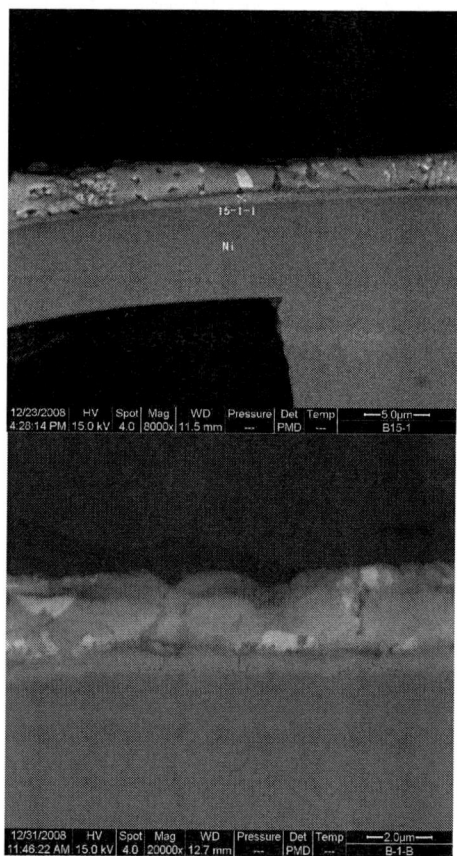

Fig. 7 Cross-section interfacial microstructure of solder joint on flash gold PCB

References

1. IPC/EIA J-STD-003B (2004) Solderability test for printed boards.

Thermal Analysis and Testing of Multi Layer Ceramic-Metal Packaged LED

Peng Jin, Qifeng Zhou, Na Wu, Qun Zhong

The Key Laboratory of Integrated Microsystems, Shenzhen Graduate School of Peking University,
Shenzhen university town, Xili, Shenzhen,   P.R. China,
Phone: 0755-26032269, Email: jinpeng@szpku.edu.cn

Abstract

We studied the thermal properties LED light unit composed of nine LED chips in a MLCMP (multi-layer ceramic-metal package) and compared various thermal dynamic models. We concluded that the direct measurement of forward voltage V_f is the most practical approach for in-situ testing. By calibrating the V_f to junction temperature T_j at various forward current in a temperature controlled oven, the linear relationship of V_f with respect to T_j was obtained and coefficient β was accurately extracted from experimental data.

Introduction

With the advancement of device structures, material systems, processing and packaging technologies, the luminous efficiencies of LEDs have improved significantly in the last decade from few lumens per watt to more than 150 lumens per watt in both the lab and mass production [1]. LED lighting is leading the third lighting revolution - replacing traditional light sources in office and household. Besides the optical, mechanical and electrical aspects, the thermal management is the most critical part of LED system. The junction to ambient thermal resistance R_{jc} controls the LED junction temperature (T_j) given the LED wattage, and the T_j is directly related with the LED luminous efficacy, degradation, life time, etc. Therefore, understanding the thermal dynamic LED lumiaire and monitoring the junction temperature to create feedback are essential for LED systems, such as LED roadside lighting.

Thermal dynamics

A typical LED thermal system consists of four elements: LED chip, chip bonding material, metal slug and heat sink. The thermal resistance and heat capacitance of each of element determine the system thermal response. Various techniques have been developed to study the thermal properties of the LED system. For example, the dynamical behavior can be studied as lumped and distributed networks similar to RC circuits as shown in Fig. 1 and Fig. 2, where the analogy is that resistance R_i corresponding to the thermal resistance and capacitance C_i to the thermal capacitance.

In Foster model, the distributed parameters such as τ_i are relatively easy to extract, and the stored energy is proportional to the temperature difference of the neighboring nodes, while in the actual system it is determined by the node temperature only. In this aspect, Cauer model is a better representation of the thermal characteristic, where capacitance is connected to the common ground and the node temperature determines the energy storage. Both Foster and Cauer model have been used in the transient thermal analysis, and after tedious calculation, the complete structure function of the heat flow path can be reconstructed [2, 3].

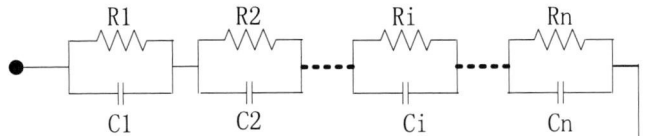

Fig. 1 Foster model

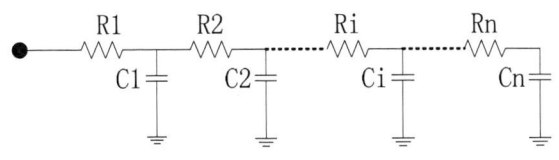

Fig. 2 Cauer model

In order to add a temperature feedback, Huang was able to fully characterized a LED system thermal dynamics to be 4th-order with three zeros following thermal step response tests [4, 5]. However, this approach needs to be adapted for each type of LED lumiaire and requires considerable amount of testing and analysis. Thermal transit tests require state of art instrument and upfront capital investment. Its inapplicability at in-situ field testing and limited resolution has prevented them from adopted as the industry standard. In thermal transit test, the output light energy needs to be subtracted from total input power in order to calculate the actual conducted energy, which also complicates the testing.

Forward voltage method

There are various ways to have non-contact measurement of junction temperature, such as Raman spectroscopy and emission peak shift, etc. However, the easiest and widely accepted way is through the forward voltage test.   The key for the forward voltage method is the determination of β, which is V_f to T_j coefficient.

A blue LED has complicated quantum-well hetero-junctions. However, its I-V relationship can be studied from the Shockley equation:

$$J_f = J_s(e^{qV_f/kT_j} - 1) \qquad (1)$$

where J_s is saturation current and can be expressed as

$$J_s = q \cdot n_i^2 \left(\frac{D_p}{N_D L_p} + \frac{D_n}{N_A L_n} \right) \qquad (2)$$

D_n, D_p is the diffusion constants of electron and hole respectively, and L_n and L_P are corresponding diffusion length, N_D and N_A are donor and acceptor concentration respectively, and n_i is the intrinsic carrier concentration which is expressed as

978-1-4244-4658-2/09 $25.00 © 2009 IEEE

$$n_i = \sqrt{N_C N_V} \exp(\frac{-E_g}{2kT_j}) \tag{3}$$

where N_C and N_V are effective density of state of conduction and valence band respectively. In the above equations, N_C, N_V, n_i, E_g are temperature dependent. In the case of non-degenerate doping, Xi [6, 7] has extracted V_f dependent on T_j to be

$$\frac{dV_f}{dT_j} = \frac{qf - E_g}{qT_j} + \frac{dE_g}{qdT_j} - \frac{3k}{q} \tag{4}$$

Further substitution of $E_g = E_0 - \dfrac{\alpha T^2}{\beta + T}$ get

$$\frac{dV_f}{dT_j} \approx \frac{k}{q} \ln(\frac{N_D N_A}{N_C N_V}) - \frac{\alpha T_j (T_j + 2\beta)}{q(T_j + \beta)^2} - \frac{3k}{q} \tag{5}$$

For GaN, the $\alpha = 0.77\text{meV/K}$, $\beta = 600\text{K}$, $n_i = 5.12 \times 10^{-10}$ cm^{-3}, we plotted dV_f/dT_j for various doping concentration, as shown in Fig.3. From Fig.3, we noticed a slight increase of coefficient for all doping levels. For a typical doping value of $N_D = N_A = 2 \times 10^{16}\text{cm}^{-3}$ the dV_f/dT_j is at -1.697mV/K at 300 K and -1.836 mV/K at 400 K, which is the functional junction temperature range for LED. The change rate is 0.082% per K.

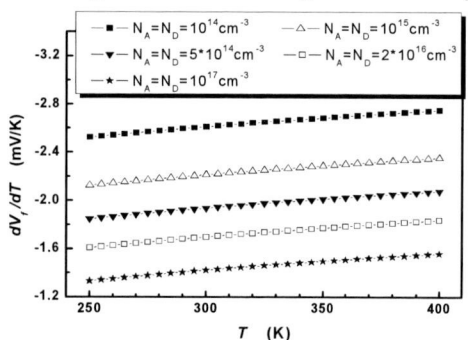

Fig.3 V_f temperature coefficient under various doping concentration

Experimental results

We studied the thermal characteristics of a multi-layer ceramic-metal packaged (MLCMP) LED array module, which offers advantage of compact size, SMT, moisture and UV resistance, etc. Nine 40 mil blue LED chips (BridgeLux) are directly mounted on the silver slug forming 3×3 serial-parallel connection, which can be drive up to 27 watt. The MLCMP is then soldered to a metal core PCB, as shown in Fig. 4.

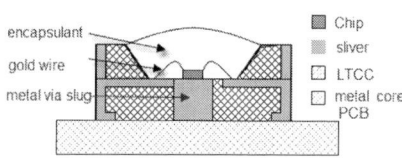

Fig. 4 The structure of MLCMP-LED unit

One significant feature of the MLCMP is the presence of a large heat sink slug just beneath the LED chip, see Fig. 5. The slug is made of 1.3 x 1.3 x 0.3 mm sintered silver, which offers nearly equivalent thermal conductivity as the bulk silver. The whole ceramic-metal package has superior thermal properties than that of the popular Luxeon LED package [8].

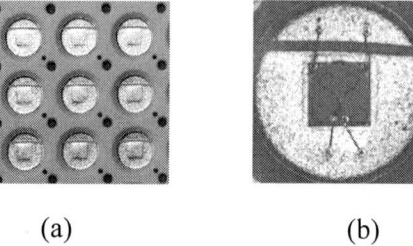

(a)   (b)

Fig.5 (a) photo of a 3 x 3 MLCMP, (b) 40 mil LED chip on top of the sliver slug

The experiment is carried out in a temperature controlled oven, as shown in Fig.6. A pulse current is applied after the unit is thermally stable, the duration of the pulse follows JEDEC51-1 procedures to avoid self heating [9].

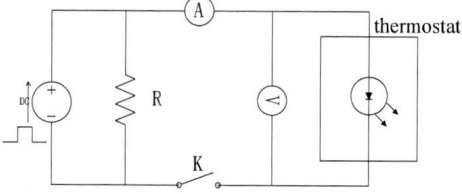

Fig. 6 Schematic of the testing system

The test pulse currents varied from 10 mA to 200 mA at various oven temperatures. The experimental results are plotted Fig. 7. In Fig.7, V_f and I_f are the total voltage drop, forward current of the whole 3 x 3 LED module respectively. In Fig. 8, β is the temperature coefficient of the LED module as well, which is three times the single LED chip temperature coefficient.

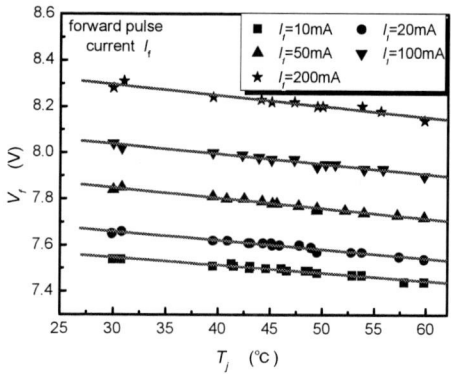

Fig.7 The relationship between forward voltage V_f and temperature T_j in various forward currents I_f

978-1-4244-4658-2/09 $25.00 © 2009 IEEE

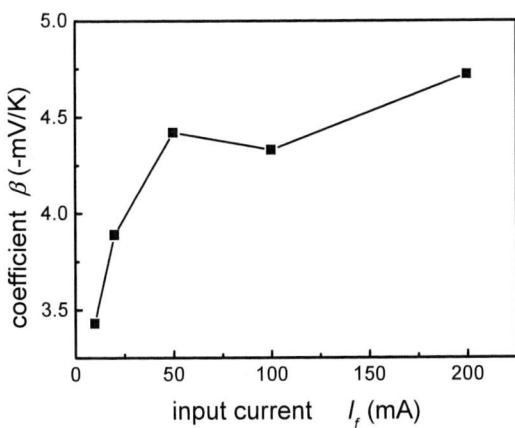

Fig. 8 Temperature coefficient β extracted from linear fitting

Conclusions

We studied various thermal models of a LED system and concluded that electrical measurement of V_f is the easiest way to obtain junction temperature. To further study the V_f and T_j relationship, we plotted theoretical temperature coefficient of V_f and noticed a slight shift as T_j increases.

Experimentally, the V_f results on a 3×3 power LED array confirm the linear relation of forward voltage toward junction temperature, and temperature coefficient β was extracted. However, β has shown forward current I_f dependency. A better understanding the coefficient β under high current is important for in-situ testing for LED lumiaire, where switching to lower current is impractical in a highly inductive circuit.

References

1. Nadarajah Narendran, "Improved Performance White LED", *Fifth International Conference on Solid State Lighting*, Proceedings of SPIE, Vol. 5941, pp. 45–50.
2. Vladimir Székely, Márta Rencz, "Thermal Dynamics and the Time Constant Domain", *IEEE Transactions on Components and Packaging Technologies*, Vol. 23, No. 3 (2000), pp. 587-594.
3. Paolo Emilio Bagnoli, Claudio Casarosa, Mario Ciampi, and Enrico Dallago, "Thermal Resistance Analysis by Induced Transient (TRAIT) Method for Power Electronic Devices Thermal Characterization—Part I: Fundamentals and Theory", *IEEE Transactions on Power Electronics,* Vol. 13, No. 6 (1998), pp. 1208-1219.
4. Bin-Juine Huang, Po-Chien Hsu, Min-Sheng Wu, Chun-Wen Tang, , "Study of system dynamics model and control of a high-power LEDlighting luminaire", Energy 32 (2007) 2187-2198
5. Bin-Juine Huang, Chun-Wen Tang, Min-Sheng Wu , "System dynamics model of high-power LED luminaire", *Applied Thermal Engineering*, Vol. 29, No. 4 (2009), pp. 609-616.
6. Y.Xi, E.F.Schubert, "Junction-temperature measurement in GaN ultraviolent light-emitting diodes using diode forward voltage method", *Applied Physics Letters*, Vol. 85, No. 12 (2004), pp. 2163-2165.
7. Yangang Xi, Thomas Gessmann, et al. "Junction Temperature in Ultraviolet Light-Emitting Diodes", *Japanese Journal of Applied Physicss*, Vol. 44, No. 10, 2005, pp. 7260–7266.
8. Jung Kyu Park, Hyun Dong Shin, Young Sam Park, et al, "A Suggestion for High Power LED Package Based on LTCC", *2006 Electronic Components and Technology Conference*, San Diego, California, June 2006, pp. 1070-1075.
9. EIA/JESD Standard No. 51-1, "Integrated Circuits Thermal Measurement Method – Electrical Test Method (Single Semiconductor Device)", pp. 1-29.

Multiscale Delamination Modeling of an Anisotropic Conductive Adhesive Interconnect Based on Micropolar Theory and Cohesive Zone Model

Yan Zhang[1], Jing-yu Fan[2]*, Johan Liu[1,3]

[1]Key Laboratory of Advanced Display and System Applications, Ministry of Education and SMIT Center
School of Mechatronics Engineering and Automation, Shanghai University, Shanghai, 200072, China
[2]Shanghai Institute of Applied Mathematics and Mechanics, Shanghai University, Shanghai, 200072, China
[3]SMIT Center & Bionano Systems Laboratory, Department of Microtechnology and Nanoscience
Chalmers University of Technology, SE-412 96 Gothenburg, Sweden
*Corresponding author, Email: jyfan@shu.edu.cn

Abstract

With the trend towards high performance, portability and low cost for electronic products, the packaging density in microsystem has become higher and the components have tended to be minimal. Flip-chip technology is superior because of its high packaging density and speed signal processing. There is an increasing interest in using electrically conductive adhesives as the interconnection material in flip-chip assembly due to its fine pitch I/Os, fewer number of processing steps, lower process temperature and environmental-friendliness. The interconnect behavior is a key issue in the long-term reliability of the system performance. In the present paper, a multiscale interface model was developed on the basis of micropolar theory, in which a cohesive zone model was also introduced to describe the possible delamination process that might occur when the interconnection was subjected to the thermal cycling loading. The delamination model was implemented using finite element method, and the numerical analysis for the reliability of an anisotropic conductive adhesive interconnect was made as the model application.

Multiscale Interface Formulation

Flip-chip technology is a significant development in surface mount technology (SMT) to improve cost and productivity in electronic packaging industry [1]. Flip-chip interconnect material usually involves solder or conductive adhesive to obtain the electrical and mechanical connection between the chip and the substrate. The anisotropic conductive adhesives (ACA) are widely used as the interconnection materials in microsystem packaging due to advantages like low processing temperature, fine pitch capability, compatibility with a wide range of substrates, flexible and simple processing and therefore low cost, environmental compatibility and so on. The applications can also be found in chip-on-glass, flat panel display areas [2,3].

The electrical performance and consequently the system reliability of the flip-chip assembly is highly dependent on the interconnect features. With the ever-increasing packaging density, the size difference between various components becomes more significant. Taking the anisotropic conductive adhesive for example, it is very suitable for ultra fine pitch assembly and has been widely adopted. Fig.1 shows the cross section of a flip chip bonding system, where Fig.1(a) is the photo indicating the chip, the ACA interconnect and part of the substrate, and Fig.1(b) provides a closer look at the thin-layered interconnection. For the complicated interconnect in microelectronics system, the relevant size effect shall be involved during the mechanical analysis [4,5]. There were also research works dealing with the composite with the size-dependent properties [6-8].

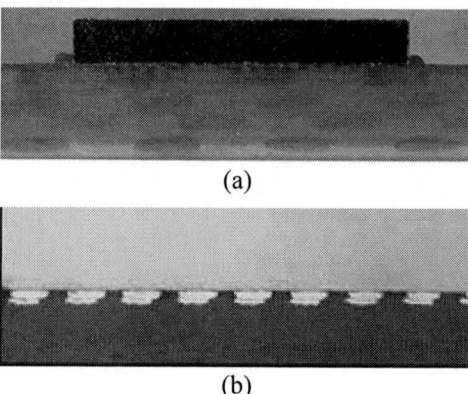

(a)

(b)

Fig.1 Photos of ACA flip-chip interconnect

In the interconnect modelling, the micropolar theory is used in order to describe the size effect of the interconnect in the bonding system. In the micropolar theory, a rotational degree of freedom is introduced in addition to the displacement, and this leads to a couple stress. Consequently, an additional parameter with the length dimension can be utilized to reflect the size effect of the considered material [9].

Following the idea in ref. [10], the strong discontinuity concept is adopted in order to reflect the influence of the interconnect interface on the material response under external loading, and the displacement vector $\mathbf{u}$ and rotation θ exhibit the discontinuities as

$$\mathbf{u} = \mathbf{u}_c(\mathbf{x},t) + Hs(\mathbf{x})\mathbf{d}_u(\mathbf{x},t) \qquad (1)$$

$$\theta = \theta_c(\mathbf{x},t) + Hs(\mathbf{x})\mathbf{d}_\theta(\mathbf{x},t) \qquad (2)$$

where the total displacement and rotation fields involve continuous portions $\mathbf{u}_c$ and θ_c, and discontinuous portions $\mathbf{d}_u = \mathbf{u} - \mathbf{u}_c$ and $d_\theta = \theta - \theta_c$, respectively. Hs is the Heaviside function by which the micropolar material can involve the variation due to the interconnect. The micropolar kinematical field at a planar interface is considered in the present paper.

After the interface is included, an insight into the detailed characteristics of the ACA interconnections is still expected. The responses of the constituents in the ACA interconnect structure at different scale levels are especially of interest in the present work. And the multi-scale behaviors of the interconnect have been considered.

978-1-4244-4658-2/09 $25.00 © 2009 IEEE

In order to obtain the micro-macro coupling of the interface response, a second order Taylor series expansion of the displacement is made as

$$\mathbf{u} = \bar{\mathbf{l}} \cdot \mathbf{x} + \frac{1}{2}\bar{\mathbf{g}} : (\mathbf{x} \otimes \mathbf{x}) + \mathbf{u}_f(\mathbf{x}) \tag{3}$$

$$\theta = \bar{\theta} + \theta_f(\mathbf{x}) \tag{4}$$

where the displacement gradient $\bar{\mathbf{l}}(\bar{\mathbf{x}}) = \bar{\mathbf{u}} \otimes \nabla$, the second-order displacement gradient $\bar{\mathbf{g}}(\bar{\mathbf{x}}) = \bar{\mathbf{l}}(\bar{\mathbf{x}}) \otimes \nabla$, and the fluctuation field $\mathbf{u}_f$ and θ_f represent the microscopic variation. The superimposed bar in the expressions denotes the corresponding macroscopic quantities. Here only the displacement is expressed as the second order series, which is assumed enough to describe the representative ACA structure responses on the macro and micro scale levels. More details concerning the higher order modelling of heterogeneous materials can be found in the refs. [11,12].

Interface Delamination Model

The increase in microsystem packaging density means that component sizes in the system are required to become smaller and smaller. Although its scale is relatively fine, the mechanical behavior of the interconnection layer is of great importance to the reliability of the entire packaging system. The conductive adhesive may fail in many ways, e.g. thermal stress failure caused by the mismatch between the substrate and the component, mismatch between the adhesive and adherent during the thermal cycling in its serving time, and so on. Fig.2 shows the delamination initiating between the conductive filler and the pad. The delamination initiation and propagation may induce open circuit and other failures in the microsystem.

Fig.2 Delamination in the interconnect

The thermal-mechanical fatigue behaviours of the interconnect are usually investigated macroscopically, while the microscopic mechanism of the fatigue crack growth is not quite clear by far. As a further development of the second order interface model, the delamination progress within the microstructure is studied.

The fluctuation $\mathbf{u}_f$ is considered in the same fashion as

$$\mathbf{u}_f = \mathbf{u}_{fc}(\mathbf{x},t) + H_s(\mathbf{x},t)\mathbf{d}_f(\mathbf{x},t) \tag{5}$$

In order to describe the process of delamination within the microstructure in the ACA interconnect, a cohesive zone model is advocated which utilizes a traction-separation law (or cohesive law) and is suitable for dealing with the interface fracture in the adhesive joint [13, 14].

The fracture process in the ACA joint is considered as a damage coupled to a plasticity process of the considered ACA interconnect structure. The traction vector $\mathbf{t}$ and the discontinuity $\mathbf{d}_f$ are the energy conjugated quantities, thus $\mathbf{t}$ is expressed as a function of $\mathbf{d}_f$ and the damage variable α as

$$\mathbf{t}(\mathbf{d}_f, \alpha) = (1 - \alpha)\hat{\mathbf{t}} \tag{6}$$

where $\hat{\mathbf{t}}$ is the effective traction vector specified with the recoverable portion of the discontinuity $\mathbf{d}_f$ as

$$\hat{\mathbf{t}} = \mathbf{K} \cdot \mathbf{d}_f^e \tag{7}$$

with $\mathbf{K}$ being the stiffness that governs the size of the elastic region, and the elastic portion of the fluctuation discontinuity is defined as

$$\mathbf{d}_f^e = \mathbf{d}_f - \mathbf{d}_f^p \tag{8}$$

To describe the propagation of the cohesive zone, the crack is determined based on the loading function F expressed in the components of the effective traction vector $\hat{\mathbf{t}}$. And the discontinuity evolution is expressed in the inelastic portion $\mathbf{d}_f^p$ as

$$\dot{\mathbf{d}}_f^p = \frac{\dot{\lambda}}{1-\alpha}\frac{\partial F}{\partial \hat{\mathbf{t}}} \tag{9}$$

in which $\dot{\lambda}$ is the plastic multiplier, and it has the relation with the evolution of the damage variable α as

$$\dot{\alpha} = \dot{\lambda}\frac{1}{S(1-\alpha)} \tag{10}$$

where S is a fracture parameter as

$$S = \frac{G_f^I}{\sigma_f} \tag{11}$$

In the above equation, G_f^I is the fracture energy of the considered material, and σ_f is the initial fracture stress in mode I, namely opening of the interface. The complete set of equations can be found in ref. [15].

Model-based FEM Simulation

After the delamination model for the interconnect is developed, it is implemented by the finite element method (FEM), and applied to predict the response of a typical ACA interconnect structure under the prescribed thermal-mechanical loading. The reliability of the flip-chip bonding structure can then be assessed.

The internal structures in the ACA interconnect, consisting of adhesive matrix, conductive fillers and component I/Os, is simplified as the geometry shown in Fig.3. The parameters in the interface delamination model, such as the fracture stress and the fracture energy can be obtained by the experimental method [16, 17].

Firstly, the involved microstructures with various sizes are numerically simulated to assess the ability of the interface model to reflect the size influence. Keeping the material parameters and loading conditions constant, the geometry of the considered structure is proportionally expanded. And the resulted stress distribution is then calculated.

978-1-4244-4658-2/09 $25.00 © 2009 IEEE

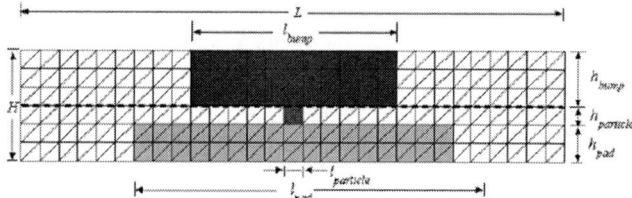

Fig.3 Geometry and mesh in the numerical simulation

Fig.4 shows the couple stress distributions of the different geometries. Comparing the stress profiles at different cases, it is shown that the couple stress is highly dependent on the length of the considered microstructure, which confirms the model's size-dependence.

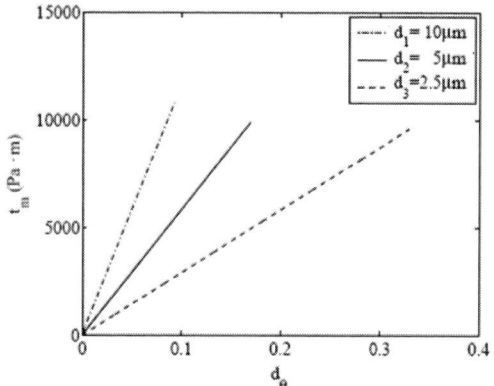

Fig.4 Couple stress distribution

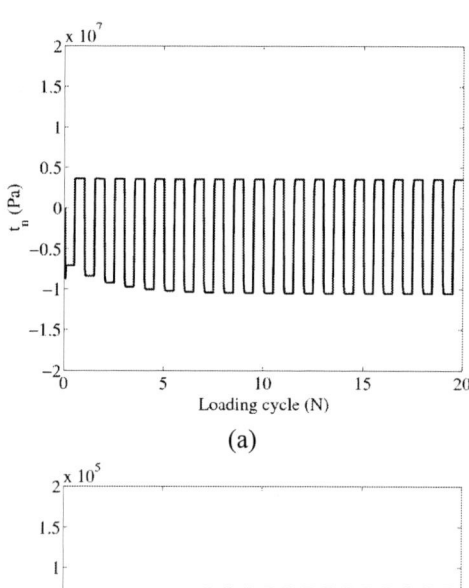

(a)

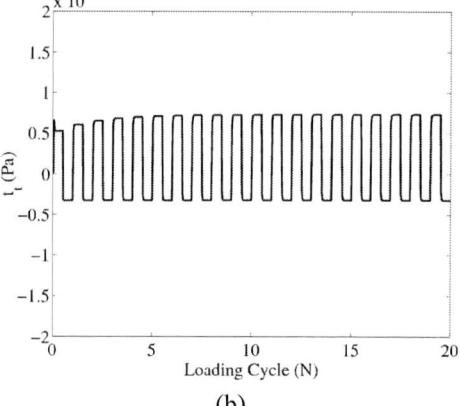

(b)

Fig.5 Stress curves during cyclic loading

Fig.5 shows the stress response during the thermal-mechanical cycling, where Fig.5(a) is the normal stress and Fig.5(b) is the tangential stress. It can be inferred that the stress profiles show a shake down behaviour with the increased cycle numbers, and the long-term magnitudes tend to be steady.

Conclusions

In the present paper, a micropolar delamination model has been established to describe the homogenized responses of the microstructure in the ACA interconnect. The second order Taylor series expansion and cohesive zone model have been introduced for modelling the failure progress. The micropolar delamination model has been implemented by the finite element method. The crack initiation and propagation of an anisotropic conductive adhesive microstructure under the thermal cycling loading has been simulated as the numerical example, and the reliability analysis has been made.

Acknowledgments

This work is supported by the National Natural Science Foundation of China (Grant No. 10702037). The supports from Shanghai Pujiang Program (08PJ14054) and the Innovation Program of Shanghai Municipal Education Commission (09YZ01) are also greatly appreciated.

This work is also supported by the FP7 IP Nanopack program, from the Swedish National Science Foundation under the project "Nanointerconnect" (621-2007-4660), from the Swedish Foundation for Strategic Research under the ProViking Program (PV08.08).

References

1. Tummala, R R. Fundamentals of Microsystem Packaging, (New York: McGraw-Hill, 2001).
2. Rizvi M J, Lu H, Bailey C, Chan Y C, Lee M Y, Pang C H. "Role of bonding time and temperature on the physical properties of coupled anisotropic conductive-nonconductive adhesive film for flip chip on glass technology," *Microelectronic Engineering*, Vol. 85, (2008), pp. 238-244.
3. Yim M J, Paik K W. "Recent advances on anisotropic conductive adhesives (ACAs) for flat panel displays and semiconductor packaging applications", *International Journal of Adhesion and Adhesives*, Vol. 26, (2006), pp. 304-313.
4. Yu S W. "Some Problems of Solid Mechanics for Micro-Electro-Mechanical Systems", *Journal of Mechanical Strength*, Vol.23, No.4, (2001), pp. 380-384.
5. Xun F, Hu G, Huang Z. "Size-dependence of overall in-plane plasticity for fiber composites", *International Journal of Solids and Structures*, Vol. 41, (2004), pp. 4713-4730.
6. Garikipati K. "Variational Multiscale Methods to Embed the Macromechanical Continuum Formulation with Fine-Scale Strain Gradient Theories", *Int J Numer Methods Eng*, Vol. 57, (2003), pp. 1283-98.
7. Dai C, Mühlhaus H B, Meek J, Duncan Fama M E. "Modelling of Blocky Rock Masses Using the Cosserat Method", *International Journal of Rock Mechanics and Mining Sciences*, Vol. 33, No. 4, (1996), pp. 425-432.

8. Ge X R, Huang M, Liu J. "Analysis on Layered Rock Mass Excavating with Spatial Elastic Couple-Stress Theory", *Chinese Jounal of Rock Mechanics and Engineering*, Vol. 19, No. 3, (2000), pp. 276-280.

9. Eringen A C. Microcontinuum Field Theory, (Springer Press, 1999).

10. Zhang Y, Larsson R, Fan JY, Liu J. "Interface Modelling of Microsystem Interconnections Using Micropolar Theory and Discontinuous Approximation", *Computers & Structures*, Vol. 85, (2007), pp. 1500-1513.

11. Kouznetsova V, Geers M G D, Brekelmans W A M. "Multi-scale constitutive modeling of heterogeneous materials with a gradient-enhanced computational homogenization scheme", *Int. J. Numer. Meth. Engrg.*, Vol. 54, (2002), pp. 1235-1260.

12. Larsson R, Diebels S. "A second order homogenization procedure for multi-scale analysis based on micropolar kinematics", *Int. J. Numer. Meth. Engng.*, Vol. 69, No. 12, (2007), pp. 2485-2512.

13. Feraren P, Jensen H M. "Cohesive zone modelling of interface fracture near flaws in adhesive joints", *Engineering Fracture Mechanics*, Vol. 17, (2004), pp. 2125-2142.

14. Elices M, Guinea G V, Gomez J, Planas J. "The cohesive zonel model: advantages, limitations and challenges", *Engineering Fracture Mechanics*, Vol. 69, (2002), pp. 137-163.

15. Zhang Y, Larsson R. "Homogenization of Delamination Growth in an ACA Flip-chip Joint Based on Micropolar Theory", *European Journal of Mechanics A/Solids*, Vol. 28, No. 3, (2009), pp. 433-444.

16. Cao L, Lai Z, Liu J. "Interfacial adhesion of anisotropic conductive adhesive on polyimide substrate", *Journal of. Electronic Packaging*, Vol. 127, (2005), pp. 43-46.

17. Cao L, Li S, Liu J. "Formulation and characterization of anisotropic conductive adhesive paste for microelectronics packaging applications", *Journal of Electronic Materials*, Vol. 34, No. 11, (2005), pp. 1420-1427.

Finite Element Thermal Analysis for High Power Multi-chip Light Emitting Diode

Longzao Zhou, Hongtao Chen, Bing An, Fengshun Wu, Yiping Wu
Wuhan National Laboratory for Optoelectronics
State Key Laboratory of Material Processing and Die & Mould Technology
Huazhong University of Science and Technology, Wuhan, 430074, China
Email: ypwu@mail.hust.edu.cn; Phone: 86-27-87543677

Abstract

In this paper, finite element method (FEM) numerical thermal simulation is performed on the given packaging configuration, and temperature as well as heat flux distribution are obtained. An optimal packaging configuration is obtained by simulating different materials and structures of the heat sink, submount, die attach, different die arrangement styles and die spacing parameters. A reasonable simplification for the high power multi-chip light emitting diode packaging body is performed and a precise FE model is established, and then a high-quality meshing is carried out. A comprehensive simulation is performed to get the temperature distribution of the chip for different materials and structure parameters of certain components; the mechanism of the change of the temperature is analyzed.

Simulation results show that an ideal temperature for chip junction under standard input power on the normal working conditions can be obtained through appropriate selection of heat sink structure and material, sub-mount material and structure parameter, die attach material and structure parameter, die arrangement style and spacing parameter.

1. Introduction

The increasing number of light emitting diode (LED) chips in one packaging module and the increasing input power of a single LED chip lead to the increasing LED junction temperature, which is challenging the packaging of the high power white multiple chip LED severely. LED junction temperature exceeding a certain limit will lead to a decrease of the quantum efficiency of radiation luminescence in the crystal of LED chip, the offset of optical wavelength and the reduction of LED chip's working life. Improving the packaging configuration and selecting appropriate packaging materials to increase the dissipation efficiency and decrease the temperature of the chip junction are the critical points to promote the application of the LED light source in the general illumination field, for which comes out the LED thermal management. It is through packaging structure and material optimization in chip, package and system level. LED chip junction temperature is maintained at a reasonable due to the enhanced thermal diffusing efficiency.

Two main methods for LED thermal management are numerical simulation and experiment. LED chip junction temperature is hard to measure experimentally due to its structural characteristics. Through thermal analysis in package and system level with numerical simulation, a great deal of experimental cost can be avoided and the numerical simulation results can be used in the optimization of packaging structure and material. Ivan Moreno studied the

impacts of different LED chip arrangement and different number of LED chip in the packaging module on the luminescence uniformity, and working performance for different chip array layout is obtained through radiation analysis [1]. Adam Christensen et al. studied the impact of chip array layout density, different LED input power density and different air convection heat transfer coefficient on the temperature distribution of the package and system level. Thermal stress of the chip is also obtained, and their results demonstrated that there is compressive stress in the active layer, which will change crystal's band gap and finally impact LED chip's thermal and optical properties [2]. Mehmet Arik et al. studied the impact of flip-chip solder bump distribution and morphology on the diffusing efficiency and the impact of defects in the die attach on the temperature distribution on the active layer, numerical simulation analysis gives the association between these defects and hot spots on the chip [3]. M. Shatalov et al. simulated a flip-chip packaging configuration: flip-chip was attached to the AlN substrate with Au80Sn20 alloy, which then mounted on the heat sink with fins. Simulation result shows that the thermal resistance between die and ambient is only 33°C/W [4]. Some researchers of Leader University in Taiwan compared the different configuration of GaN-based chip on the standard FR4 PCB, metal core PCB and thick Cu metal core PCB submount through numerical simulation, and their results show that chip junction temperature decreases with the increase of effect contact area of the chip [5]. Lianqiao Yang et al. studied the temperature distribution with GaN die in three different ceramic packaging configurations. Their results demonstrate that the larger effective contact area in the inside of the packaging module, the lower LED chip junction temperature [6].

In the above references, some authors focus on the basic research of packaging materials, some authors focus on the research of system-level heat sink, while others are focused on chip-scale packaging solutions. Few authors can give a more comprehensive set of chip-level, package level and system-level material and structural optimization of the overall solution. In this paper, FEM thermal simulation is performed on a given high power multi-chip modules LED light source (MCM-LED). Temperature and flux distribution on packaging body are obtained. An optimal packaging configuration is obtained by simulating different heat sink material and structure, different submount material and structure parameter, different die attach material and structure parameter, different die arrangement style and die spacing parameter. FEM thermal simulation results can be used for the packaging structure design and material selecting, which will highly reduce high power MCM-LED development cycle time and give enterprises priority in the market.

2. Packaging structure of the High Power MCM-LED

Packaging structure of the High power white MCM-LED is shown in Fig. 1. LED chip is Cree's product XB900, 8 of which are bonded on Si submount with Au80Sn20 alloy solder through reflow. Si submount, with silver-plated circuit, is pasted on the LTCC substrate. There is a yellow phosphor resin layer on the chip, on the top of the phosphor resin that is a silicone layer to paste resin lens.

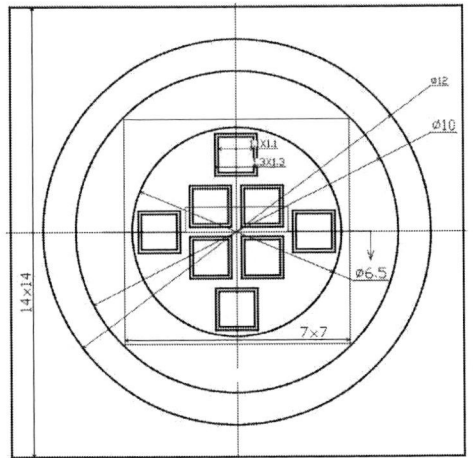

Fig. 1 MCM-LED light body packaging structure schematic

MCM-LED with heat sink is shown in Fig. 2. There are 24 fins in the heat sink, with a small curvature bending downward for the fins to increase the effective convection contact area. Fins are made of Al, which has a good heat conduction and convection heat transferring property.

Material properties of the MCM-LED packaging components are shown in Table 1. Heat pipe's density and specific heat capacity are presumed same as that of Al, whose thermal conductivity is 20000W/ (m · K). LED chip is made up of InGaN active layer and SiC substrate layer, the latter is a great deal thicker compared with the former, and so LED chip is presumed made of SiC, which is considered as a body heat source.

Fig. 2 MCM-LED with heat sink schematic

3. FEM simulation for MCM-LED

3.1 Modeling and meshing

Considering the complexity of the real MCM-LED packaging structure, some simplifications are carried out in the simulation. Firstly, the complex reflective cup is neglected and Si submount is simplified as flat. Secondly, only 1/4 of the packaging structure is molded considering its symmetry as shown in Fig. 3(a). The element type is SOLID 70 and about 700,000 elements are produced. MCM-LED model after meshing is shown in Fig. 3(b).

Table 1 Properties of the MCM-LED packaging components

Component	Material	Specific heat capacity C_V/[J/(Kg · K)]	Density ρ/[Kg/m^2]	Thermal conductivity λ/[W/(m · K)]
Die	SiC	1.0	3000	170
Die attach	Au80Sn20	15	15000	57
Submount	Si	712	2330	150
Fin	Al	910	2702	237
Heat pipe	*———*	910	2702	20000
Silicone	Silicone	1.0	1500	0.25
Lens	Polymide	1.3	1400	0.11

——— Heat pipe is not made of a certain material

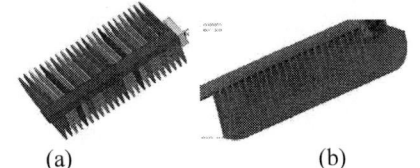

      (a)   (b)

Fig.3 Modeling and meshing: (a) 1/4 model; (b) 1/4 model after meshing.

3.2 Heat load and boundary conditions

The whole chip is looked as a heat source, with 1W input power on each chip, whose volume is 10^{-10} m^3 and heat generating rate per m^3 is 1010 W/m^3. The whole packaging model has initial temperature of 300K. Passive air convection heat transfer boundary condition is applied, with 300K's temperature for air fluid and 5W/ (m^2·K) for air convection heat transfer coefficient. Heat flux on the symmetrical section

978-1-4244-4658-2/09 $25.00 © 2009 IEEE

is 0 and radiation heat transfer is neglected due to the low temperature of the packaging model (the maximum temperature is less than 70°C).

3.3 Simulated results

Temperature and heat flux density distribution on the packaging model is shown in Fig.4. The maximum temperature is 54.344°C and the minimum temperature is 47.348°C. A good temperature gradient is obtained around chip, which means a good heat transfer property for the MCM-LED packaging model. Thermal conductivity of silicone and lens is smaller compared with that of die attach and Si submount, which means that the heat generated in the chip gets out of packaging model mainly through die attach, Si submount and finally transfers into ambient through heat sink. As shown in Fig.4 (b), heat flux mainly gathered in the area below the chip.

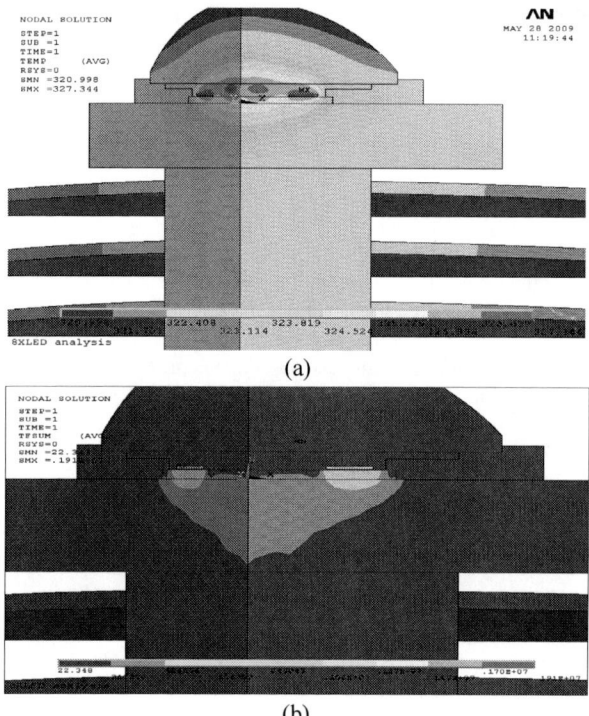

(a)

(b)

Fig 4 FEM simulation results: (a) Temperature distribution on the packaging model; (b) Flux distribution on the packaging model.

3.4 Experimental results and model validity

The temperature of the corner of the LTCC substrate has a certain relationship with that of the chip junction, so does the temperature of the chip junction and the input power. The temperature of the corner of the LTCC substrate is measured at different input powers thus chip junction temperature can be estimated in a simple experiment. The devices for measuring chip junction temperature are shown in Fig. 5. Wire was bonded on the MCM-LED lighting body electrodes, and electrodes were fixed on heat sink with screw and cooling agent, thermocouples and power, et al. Voltage values were put down and thermometer was read in steady state when

constant current was regulated at 350mA. The experimental results at three times are 55.775, 56.593 and 55.484°C, the average value of which is 55.951°C. FEM simulation result in Fig. 4(a) is 54.434°C, which is less than the average experimental results by 1.571°C, with an error of 2.7%. It indicates that FEM simulated result accords well with experimental measurement.

Fig. 5 Devices for measuring chip junction temperature

4. Optimization for MCM-LED packaging materials and structure parameters

4.1 Optimization of heat sink structure and chip layout

Chip junction temperature varies as thermal conductivity of the heat pipe changes, which is simulated when thermal conductivity is 1000, 5000, 10000 and 20000 W/ (m · K). As shown in Fig. 6, chip junction temperature decreases as thermal conductivity of the heat pipe increases. Heat sink's heat transfer property is determined by heat pipe's thermal conductivity and fins' thermal conductivity, once heat pipe's thermal conductivity is much larger than that of the fins, there is no use to increase heat pipe's thermal conductivity, and fins' thermal conductivity is the bottle neck to lower chip junction temperature.

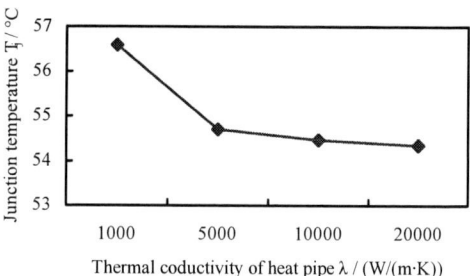

Fig. 6 Chip junction temperature varied with thermal conductivity of the heat pipe changes

Two different chip distribution forms are shown in Fig.7, chip junction accordingly are 54.286°C and 54.469 °C. As shown in Fig. 7, there is only a slight different between the two chip layouts for chip junction temperature, which are about 54°C. That is an ideal temperature for MCM-LED. Obviously circular chip layout is more compact in the same space, which is clearly conductive to improve optical output efficiency.

978-1-4244-4658-2/09 $25.00 © 2009 IEEE

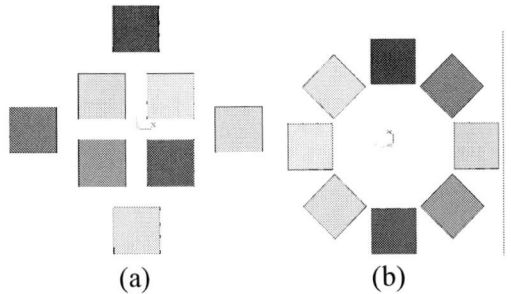

(a)   (b)

Fig. 7 Two different chip layouts: (a) Square chip layout; (b) Circular chip layout.

4.2 Optimization of heat sink materials and fin thickness parameters

Chip junction temperature is 56.67 and 60.017 °C when heat sink material is Cu and Al. Actually, Cu has larger thermal conductivity but costing much more. Al is a proper choice. The junction temperature changes with the fin's thickness changes, when the sum of fin's thickness and the space between two fins at 4 mm. As shown in Fig. 8, chip junction decrease as fin's thickness increases, yet there is a presumption that convection heat transfer coefficient is a constant value. Once the space between two fins is small enough that fins shade each other from contacting with air fluid, effective convention heat transfer coefficient will decrease sharply and so does the heat transfer property of the heat sink.

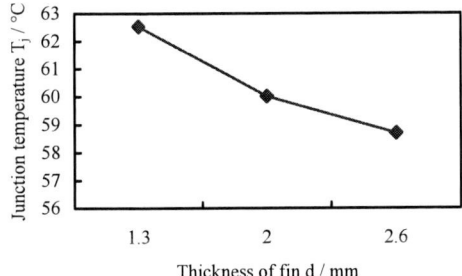

Fig. 8 Chip junction temperature varies as fin's thickness changes

4.3 Optimization of submount materials and structural parameters

The primary function of the submount is to absorb the heat generated in chip and then conduct it on heat sink. Considering the great difference of coefficient of thermal expansion (CTE), submount also serves as a buffer layer. Chip junction temperature is 60.017, 59.266 and 56.672°C when Si, Al and Cu are used. Obviously Al and Cu have a good heat conducting property, yet an insulating layer is necessary when multiple chip are fixed on submount, which will decrease their heat transfer property greatly. However, Si is insulated by itself, which is an ideal choice here. In addition, Si has a slightest difference with InGaN on CTE. Chip junction temperature varies as submount thickness changes, as shown

in Fig. 9. We simulated sub-mount thickness of 0.17, 0.27 and 0.37mm. As shown in Fig. 9, chip junction temperature increases slightly with submount thickness increases, which results from a slightly increasing thermal resistance. Considering electrical conducting and bonding reliability, too thin a submount is not preferred.

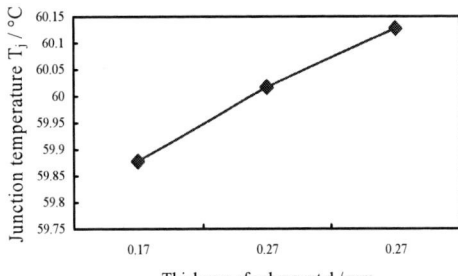

Fig. 9 Chip junction temperature varies as submount thickness changes

4.4 Optimization of Die attach materials and structural parameters

Die attach materials must have a large thermal conductivity to conduct heat generated by chip onto the submount effectively and timely to ensure chip junction temperature at a proper level. FEM simulation results are: chip junction temperature is 60.017,59.975 and 67.372°C when die attach material is Au80Sn20 alloy, solder paste and conductive adhesive accordingly. Although conductive adhesive serves well for electrical connections, it has a small thermal conductivity due to its organic matrix. Chip junction temperature varies as die attach thickness changes, as shown in Fig. 10, in which three points when die attach thickness is 0.01, 0.02 and 0.03mm is simulated. As shown in Fig. 10, chip junction temperature increases slightly when die attach thickness increases, which results from a slightly increasing thermal resistance. Considering its easiness to give rise to hollow and bonding defects, too small a die attach thickness is not preferred.

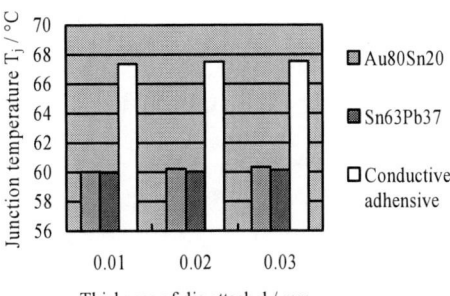

Fig. 10 chip junction temperature varies as die attach thickness changes

4.5 Optimization of Chip spacing parameters

Chip layout schematic is shown in Fig. 11. A,B,C and D

are the four chips in 1/4 MCM-LED packaging model, which are centro-symmetric on the origin and L can be used to represent the spacing between chips. Chip junction temperature varies as L changes, as shown in Fig. 12, in which 1.5, 1.7 and 1.9mm for L are simulated.

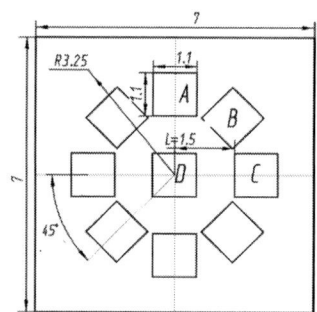

Fig. 11 Chip layout schematic (mm)

As shown in Fig. 12, chip junction temperature decreases as L increases, which primary should blame for the heat coupling between chips. Beside transfers downward, the heat generated in the chip will transfer horizontally in a certain distance, during which once another chip appears, t will be heat coupling between. As the spacing between chips increases, heat coupling decreases sharply. Considering packaging density and light output efficiency, too large a value for L is not preferred.

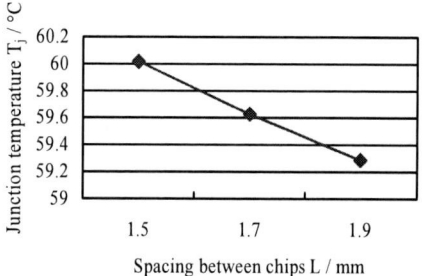

Fig 12 Chip junction temperature varies as L value changes

5. Conclusions

Based on FEM simulation, the optimization for MCM-LED packaging material and structure are carried out. Considering cost and effect, a too large thermal conductivity for heat pipe is not preferred, and 5000 W/ (m · K) is preferred in this paper. On the condition that the effective convection heat transfer coefficient is a constant, chip junction temperature decreases as fins' thickness increase. So far as InGaN dies, Si is the best submount choice when insulating between chips and CTE between chips and submount are taken into account. There is thermal coupling between chips, which weakens sharply as spacing between chips increases. Considering packaging density and the limit of the area of submount, too large a space is not preferred.

References

I. Ivan Moreno, "Configurations of LED arrays for uniform illumination," 5th Iberoamerican Meeting on Optics and 8th Latin American Meeting on Optics, Lasers, and Their Applications, Proc. of SPIE Vol. 5622 (2004), pp. 713-718.

2. Adam Christensen, Minseok Ha, Samuel Graham, "Thermal Management Methods for Compact High Power LED Arrays," Seventh International Conference on Solid State Lighting, Proc. of SPIE Vol. 6669 (2007), pp. 6690Z-1-19.

3. Mehmet Arik, Anant Setlur, Stanton Weaver, "Chip to System Levels Thermal Needs and Alternative Thermal Technologies for High Brightness LEDS," Journal of Electronic Packaging, Vol. 129 (2007), pp. 328-338.

4. M. Shatalov, P. Yadav, V. Adivarahan, et al., "Thermal analysis of flip-chip packaged 280 nm nitride-based deep ultraviolet light-emitting diodes," Appl. Phys. Lett., 86 (2005), pp. 201109-1–201109-3.

5. Chun-Jen Weng, "Advanced thermal enhancement and management of LED packages," International Communications in Heat and Mass Transfer, 36 (2009), pp. 245-248.

6. Lianqiao Yang, Sunho Jang, Woongjoon Hwang, et al., "Thermal analysis of high power GaN-based LEDs with ceramic package," Thermochimica Acta, 455 (2007), pp. 95-99.

A Design of an Optical Transceiver SiP

Wei Gao, Lixi Wan, Baoxia Li, Jian Song, Xu Zhang
Institute of Microelectronics, Chinese Academy of Sciences
B503,3#,BEITUCHENGWest, CHAOYANG District, Beijing, 100029, China
Telephone: 86-010-82995675-8008 86-010-82995591 Fax: 86-010-62021601
Email: galaxyvenus@126.com lixi.wan@gmail.com libaoxia@ime.ac.cn sj1203@163.com

Abstract

This paper presents a design of an optical transceiver SiP with an embedded capacitor which replace conventional filter network and suppress switching noise with a large bandwidth in multi-GHz PCBs. A SI design method was proposed based on Electromagnetic Analysis Method and Circuit Analysis Method. To study SI, a SPICE equivalent circuit model of the high speed I/O link was built based on Transmission Line Theory and Finite Elements Method by combination of field solver and circuit simulator. Following the design flow, a SI simulation was made by using the model. Furthermore, a thermal analysis and design were made by ANSYS. At last, an optical transceiver is fabricated and tested.

Introduction

Optical transceivers have been widely used in carrying data between racks, boxes and boards. The large bandwidth peripheral equipments and high-capacity data center require the optical transceiver to have more channels, higher data rate, smaller area, lower cost and better performance. To meet these demands, many researchers work on the design and optimization of optical transceiver structure. Previously, our member Baoxia Li [1] has shown the Optical Coupling structure of our optical transceiver. Tomoyuki Hino and others have also done a lot of works on optical transceiver in [2]-[5]. Until now, most works focus on the structure and optical design, few ones work on SiP design and optimization.

System-in-Package (SiP) is an advanced technology which integrates chips and passives into a single package. With a 3D-multiple-layer structure of heterogeneous materials, SiP faces a lot of challenges which involve electrical, materials, and mechanical technologies. The electrical problems issues are related to Power Integrity (PI) and Signal Integrity (SI), including low impedance power feed, large bandwidth filtering, controlled impedance, cross talk, etc. SiP also requires proper selection of materials to form the system-level package hierarchy, considering electrical performance, heat transfer and reliability. Particularly, as operating frequency in systems goes higher, the SI becomes more important. The old design methodology, which consists of design, building prototypes, test and find out problem, then repeat above steps until the package works, is no longer cost-effective when time to market is as important as performance and cost. Therefore, the design of SiP should take into account all these factors from the very beginning. Based on our design experience and research, some SI design rules were proposed, which could improve the performance of prototypes. Using optical transceiver as an example, we proposed farther an SI design flow for SiP.

The paper is organized as following: section 2 describes the PI design; section 3 shows SI design which contains SI design methodology, a SPICE model of the high speed I/O link and the simulation results; in section4, a thermal analysis result is shown. In section5, an optical transceiver is fabricated and tested; the conclusions are made in the last section.

PI Design

In order to design and fabricate the high density and high speed optical transceiver SiP with low cost, some new approaches were used. A novel embedded capacitor filter [6] was used in the optical transceiver PI design.

An embedded capacitor with two connections, which could be a filter with two ports: one serves as an input, while another as a output – just as the conventional low pass filter network composed of passive components does. As seen in Figure 1, the low pass filter can be called as Embedded Capacitor Filter (ECF).

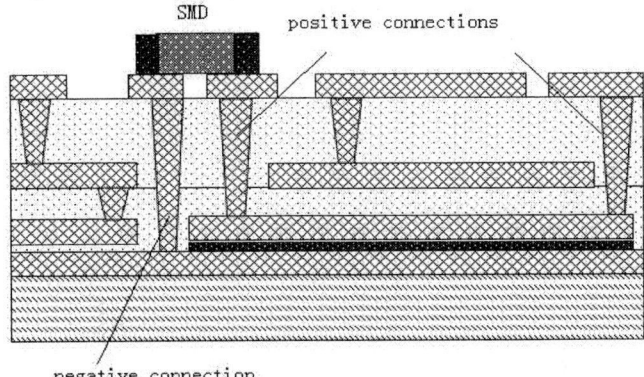

Fig. 1 A cross-section of an ECF in a package/PCB system

The ECF characterizes that the further the output point stays away from central input point, the better the performance will be in high frequency, which make it possible to achieve high density packaging with smaller dimensions and low cost. In the optical transceiver PI design, a power supply filter circuit of a commercial optical Transimpedance Amplifier Array [16] is needed, which takes up 1/3 area of the PCB. We replaced it with an embedded capacitor and a 100uF SMD capacitor and get a better filter performance. In Fig.2, a comparison of area between the optical transceiver with a conventional filter network and the one with an ECF is made. The size of the optical transceiver with a conventional filter network is 3cm*3cm, while the other is 1.6cm*2.5cm. Obviously, the area of optical transceiver is effectively reduced. Another advantage of the PI design using an ECF is that only one SMD was used in the filter, which saved 8 SMDs from the original design. It saves the cost of production.

978-1-4244-4658-2/09 $25.00 © 2009 IEEE

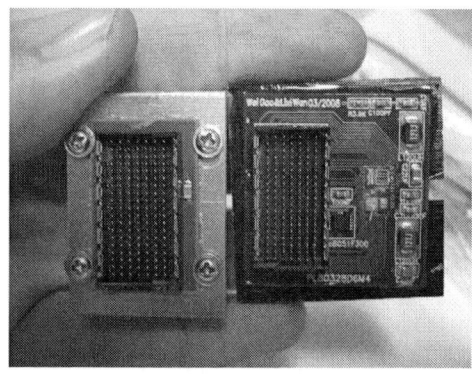

Fig.2 Size: Version1 (right) ($9cm^2$) down to Version 2 (left) ($4cm^2$)

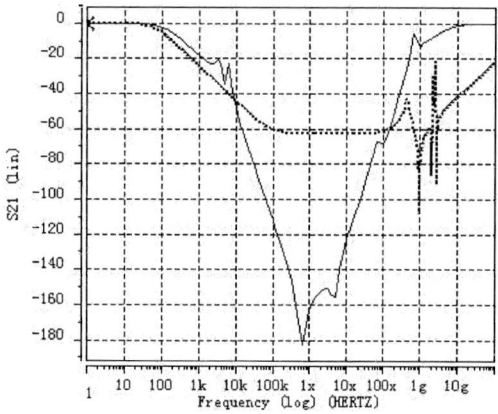

Fig.3 S21 comparison between conventional filter network (solid line) and ECF (dot line)

In Fig 3, the filtering performance of conventional filter network and ECF is also compared. Smooth response on wideband up to several GHz from ECF was observed. It's obvious that ECF has much higher filtering performance with less area, lower cost and a large bandwidth range from 50Hz to Multi-GHz.

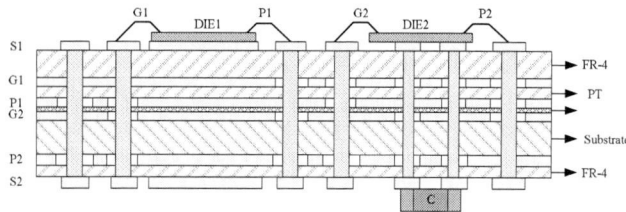

Fig. 4 A cross-section of the optical transceiver PCB Power system

The cross section of the optical transceiver PI design is shown in Fig 4. As both the driver of Vertical Cavity Surface Emitting Laser (VCSEL) and TIA of Planar PIN photo detectors (PD) are high speed chips, two pairs of Power and Ground layers are used to reduce the cross talk between them. A high K layer is used to form the ECF. To save cost,

through VIAs instead of Blind VIAs are adapted in the design and fabrication.

SI Design

The SI design of optical transceiver consists of four steps: impedance design, layout design, impedance simulation and layout simulation.

The first step (shown in Fig.5) is to calculate the impedance of high speed link, considering single line, differential pairs with or without ground line and coupling between two differential pairs. For the selected ply materials, by calculating the impedance of high speed lines with different widths and spaces, the most proper parameters of high speed line can be found.

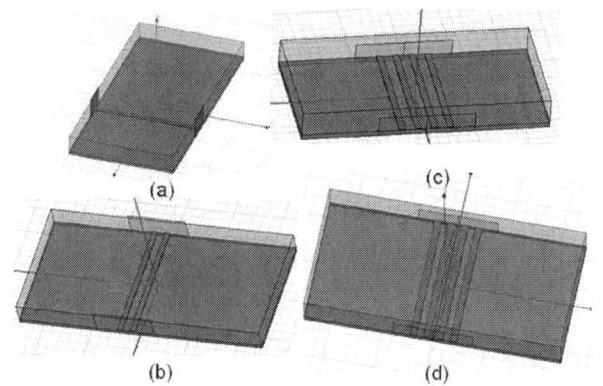

Fig.5 Design Models of High speed line with HFSS
(a) single line; (b) differential pairs; (c) coupling differential pairs without ground line; (d) coupling differential pairs with ground line

In Fig.6 the simulation results of differential pairs, coupling differential pairs with and without ground line are shown. The lines show that there are different values of characteristic impedance in these three cases with the same parameters. For differential impedance 100Ohm, there are different parameters for each situation.

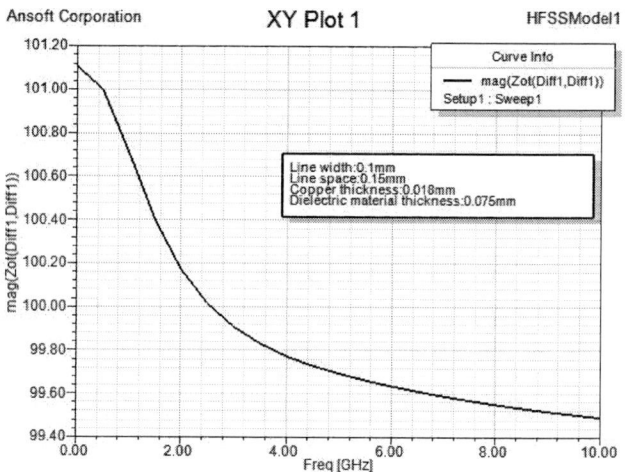

(a) Characteristic impedance of differential pairs

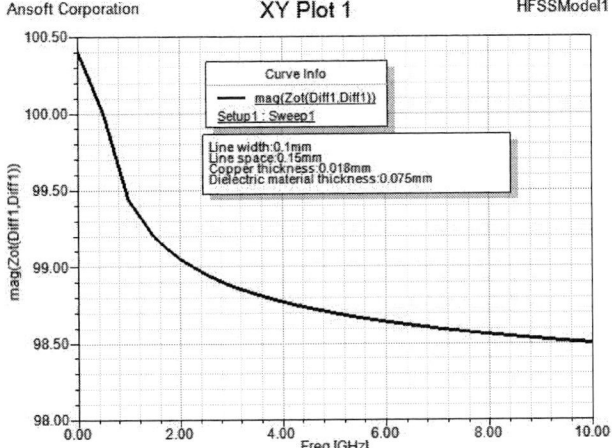

(b) Characteristic impedance of coupling differential pairs without ground line

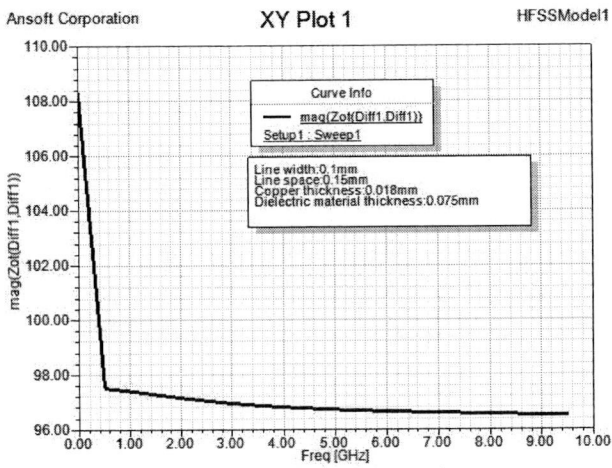

(c) Characteristic impedance of coupling differential pairs with ground line

Fig.6 The simulation results of differential pairs, coupling differential pairs with and without ground line

The second step is to do some detail design, including placing components and routing. The major principle for components placement is to make the routing easy and avoid interferences between different chips. Generally, the digital chips should be placed far away from analog chips; the I/O link should far away from other links; the high frequency chips should far away from low frequency chips; the high levels chips should far away from low levels chips; the sensitive circuits should far away from other circuits. The routing design should consider both transmit delay and insertion loss. As we known, the transmission line has distribution capacitors can be presented by $C_0 = \dfrac{t_d}{Z_0}$, where C_0 is unit distribution capacitor, t_d is unit transmit delay time and Z_0 is characteristic impedance. Due to the charge and discharge to distribution capacitors, the transmit delay occurs. When the signal return time beyond signal rise time,

impedance match is required. This will increase the design load and reduce signal quality. Hence, the length of routing should be controlled in the scope of impedance mismatch. According to empirical equations, the max length of routing can be calculated by $l_m \leq \dfrac{t_r}{2t_d{}'}$, where l_m is the max length of routing, t_r is signal rise time, $t_d{}'$ is unit time delay. When the transmission line has a capacitive load $t_d{}' = t_d\sqrt{1 + \dfrac{C_d}{C_0}}$, where C_d is the unit sum of capacitive load. For a Micro-strip line,

$$Z_0 = (\frac{87}{\sqrt{\varepsilon_r + 1.41}})\ln(\frac{5.98H}{0.8W + T})$$

$$t_d = 1.017\sqrt{0.475\varepsilon_r + 0.67}\ (ns/ft);$$

For a Strip-line,

$$Z_0 = \frac{60}{\sqrt{\varepsilon_r}}\ln(\frac{4H}{0.67\pi W(0.8 + \frac{T}{W})})$$

$$t_d = 1.0117\sqrt{\varepsilon_r}\ (ns/ft).$$

In another way, the transmission line routing on PCB is lossy. The insertion loss of the high speed lines should be controlled above -3db. This is another constraint to routing length. In Fig, 7, the insertion losses of different length lines are shown.

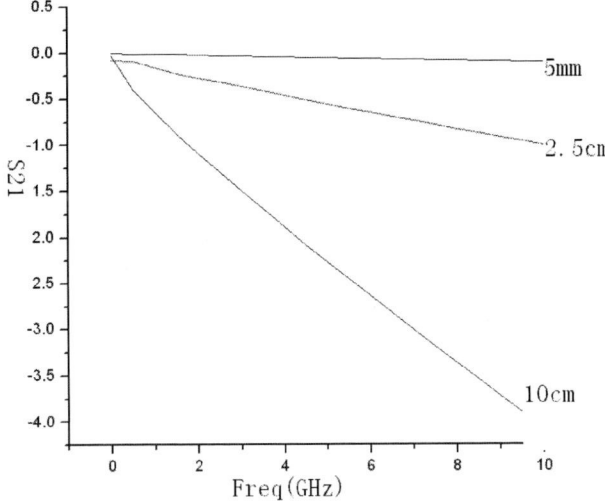

Fig.7 Insertion losses of different length lines df=0.02

Sometimes, a design of routing shape is made. In Fig. 8, the insertion losses of different turning shapes are compared. The Fig.8 shows that the C Type is better than B Type from 2GHz up to 10GHz and C Type takes less area than B Type. C Type is a better choice for high speed and high density design.

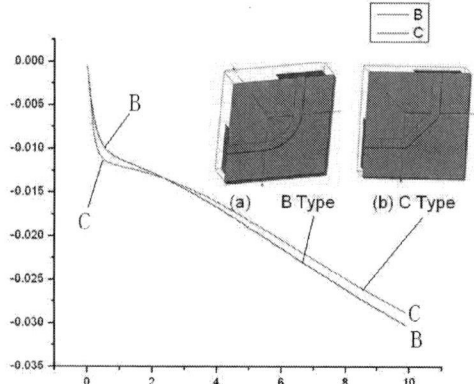

Fig.8 Insertion losses of different turning shapes

The third step is the impedance discontinuity simulation. Physically, a high speed link might be consist of a combination of I/O pads at TX/RX, package traces/pins, sockets, connectors, FR4 traces, and a backplane. The impedance discontinuities occur at the interfaces. A impedance discontinuity of the entire channel should be simulated to help select connectors, package type and ply materials which determined the size of I/O pad and dimensions of traces. Taking the optical transceiver as an instance, this step is shown in Fig.9 and 10.

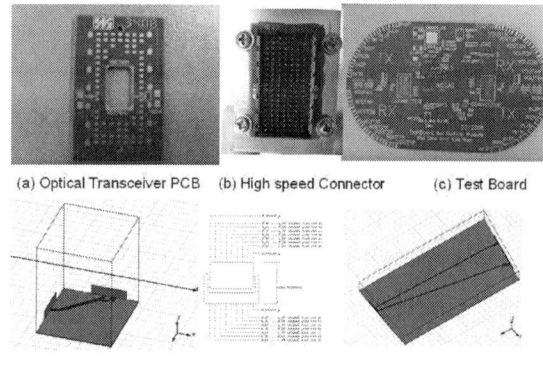

Fig.9 Impedance discontinuity Model in HFSS

In Fig.9 (a), (b), (c), a physical channel of optical transceiver is shown, which consists of a pair of wire bond lines, a pair of wire bond pads, a differential pairs on FR4, a socket high speed connector (female and male), connector pads, a two traces on BT and SMA pads. To obtain the SPICE Model of this channel, the physical model of every part is built in HFSS. By using the field solver HFSS, an Electromagnetic Analysis is made for every part except high speed connector. The SPICE Model of high speed connector was received from software provider. Based on Transmission Line Theory and Finite Elements Method, the entire channel can be treated as a transmission line with different distribution parameters which can be extracted from the results of Electromagnetic Analysis. An equivalent circuit model of the channel is built by using the distribution parameters in HSPICE. In Fig.10, impedance discontinuity simulation

results are shown. The results show that the insertion loss S21of this channel above -3db from 0~3GHz. It means that the channel can afford 3 Gb/s data rate.

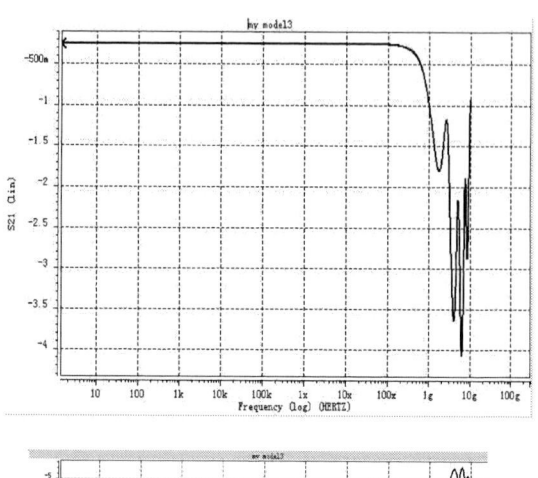

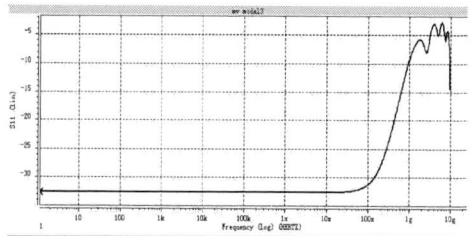

Fig.10 Impedance discontinuity simulation results in HSPICE

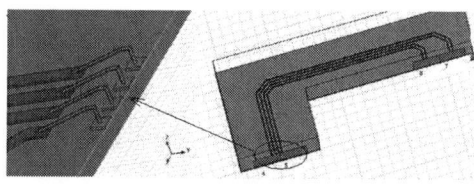

Fig.11 Model of two optical transceiver channels in HFSS

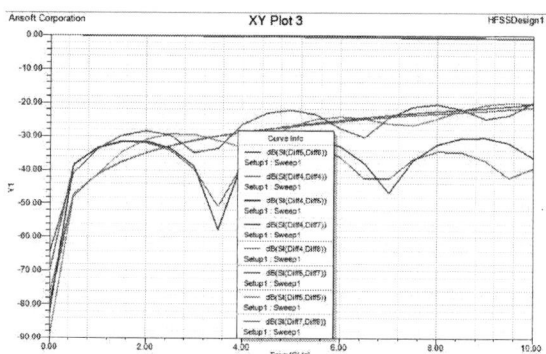

Fig.12 Simulation results of crosstalk below -20db from 0~10GHz

The last step of SI design is layout simulation. After layout, high speed graphics was imported into field solver, to build physical model and validate the crosstalk between high speed links to meet the performance requirements. Fig.11 shows the import model of two optical transceiver channels in HFSS. Fig. 12 shows the crosstalk between them. The S_{45}, S_{78} are crosstalk of proximal end below-30db, the S_{47}, S_{57} are crosstalk of distant end below -20db.

Thermal Analysis

The total power of the optical transceiver is about 1.6w, the area of heat source is $6*8mm^2$. That means a high heat density, so a thermal analysis is needed. A thermal analysis has been conducted to optimize the size of thermal fans by ANSYS. In Fig.13, the thermal results are shown.

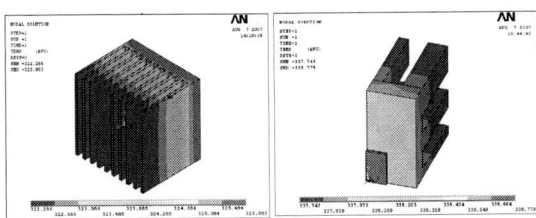

Fig.13 The thermal analysis results of optical transceiver.
(Convection coefficient: 30W/m2-oC;
Thermal fin size: 2x1x1cm3, Cu thickness: 2mm,
Work temperature: 66 □)

Fabrication and Test

According to design flow described, an optical transceiver has been fabricated and tested. In the case, a 2^{31}-1 Pseudorandom Bit Stream (PRBS) through coaxial-cable to SMA on test board edge, transmits from traces on BT, high speed connector, connector pads, traces on FR-4, wire bond pads, wire bond lines to VCSEL driver input pads, drives the VCSELs to produce optical signals. Through the optical coupling structure and the multimode fiber and optical coupling structure again, optical signals was converted back to electrical signals at PDs within the same transceiver module. The back-to-back eye-diagram at a data rate of 2.7 Gb/s for 2^{31}-1 NRZ- pseudorandom bit stream (PRBS) of one transmitter-to-receiver link is shown in Fig.14.

Conclusions

In this paper, a design of an optical transceiver is presented. A novel embedded capacitor filter is used in the PI design, which can effectively reduce the size of optical transceiver and yield a better high frequency filtering performance than conventional lowpass. The power/ground planes of both transmitter and receiver are separated to reduce interference. With proposed SI design method, a simulation of three types of differential pairs shows that the impedance of line is related to routing circumstance. Some SI design rules of layout are shown, which contains placing components, calculating transmit delay time and controlling length of line. The method to simulate the impedance discontinuity is studied by an example. A thermal analysis and design are made by ANSYS. The minimum size of thermal fan is $2*1*1$ cm^3.With the design method of an optical transceiver, an optical transceiver is fabricated and tested. The back-to-back eye-

diagram at a data rate of 2.7 Gb/s for 2^{31}-1 NRZ- PRBS of one transmitter-to-receiver link is shown

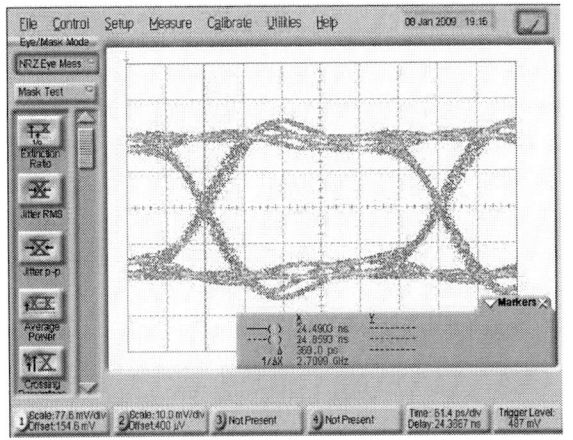

(a) The eye diagram of PRBS generator to oscillograph

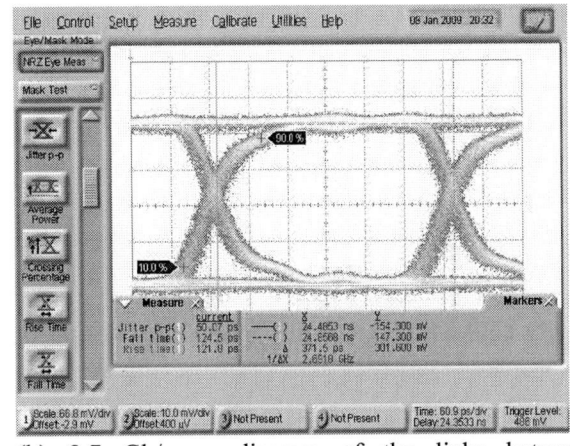

(b) 2.7 Gb/s eye-diagram of the links between 1 transmitter channel and 1 receiver channel in a single

Fig.14 Eye-diagram

Acknowledgments

This work was supported by Hi-tech Research and Development Program of China (863 Program) No. 2006AA01Z236, 2007AA01Z200.

References

1. Baoxia Li, "Low-cost High-efficiency 4 Channel Pluggable Parallel Optical Transceiver Using Optoelectronic MCM Packaging Technologies", *IEEE ICEPT,2008.*

2. Tomoyuki Hino, "A 10 Gbps x 12 channel Pluggable Optical Transceiver for High-speed Interconnections", *2008 Electronic Components and Technology Conference.*

3. Wood-Hi Cheng, "Low-Cost and Low-Electromagnetic-Interference Packaging of Optical Transceiver Modules", *JOURNAL OF LIGHTWAVE TECHNOLOGY, VOL. 22, NO. 9, SEPTEMBER 2004.*

4. S.M.Mitani, "Design and Fabrication of Power-Efficient VeSEL-Based Optical Transceiver", *2008 International Conference on Advanced Technologies for Communications.*

5. J.Ahadian, "A Quad 2.7 Gb/s Parallel Optical Transceiver", *2004 IEEE Radio Frequency Integrated Circuits Symposium.*

6. Wei Gao, Lixi Wan, etc. "A Novel Lowpass Filter with An Embedded Capacitor for Wideband Noise Suppression in Multi-GHz PCBs", *ICEPT, 2008..*

7. Paolo Pulici,etc. "Signal Integrity Flow for System-in-Package and Package-on-Package Device", *IEEE*, 2009.

8. Frank Y. Yuan, "Electromagnetic Modeling and Signal Integrity Simulation of Power/Grolund Networks in High Speed Digital Packages and Printed Circuit Boards", *IEEE.*

9. Q. J. Zhang, "Signal Integrity Optimization of High-speed VLSI Packages and Interconnects", *1998 Electronic Components and Technology Conference.*

10. GOH Ban Hok, "A Study of High Speed implementation for System on Chip on 2 layers Printed Circuit Board", *IEEE, ICIS2007 conference.*

11. Jianjian Song, "Effectiveness of PCB Simulation in Teaching High-Speed Digital Design", *IEEE, 2007.*

12. Frank B.J. Leferink, "Power and Signal Integrity and Electromagnetic Emission; the balancing act of decoupling, planes and tracks", *IEEE, 2007.*

13. Kuang Shenqing, "Air Isolated High-Speed PCB Connector's Signal Integrity Design", *IEEE, 2008.*

14. A. Ciccomancini Scogna, "Signal Integrity Analysis of a 26 Layers Board with Emphasis on the Effect of Non-Functional Pads", *IEEE, 2008.*

15. Huang Jimmy Huat Since, "Signal Integrity Analysis and Validation of GHz I/O Interface on Wire Bond Ball Grid Array Technology Package", *IEEE, 2006.*

16. Vitesse Semiconductor Corporation:"VSC7651 Datasheet-4X3.125Gps Optical Trans-impedance Amplifier Array", August, 2005.

Power Supply Analysis in Package and SiP Design

Wenliang Dai
Cadence Design Systems, Inc.
Floor 5, Building 8, No.690, Bibo Road, Pudong, Shanghai, 201203, P.R.China
Email: wldai@cadence.com

Abstract

This paper introduces a process that allows customers to do package power integrity (PI) analysis on the package side. The chip information such as the die circuits and current profiles as well as power delivery network circuit can be used at the board level to perform PI analysis. The die current profiles are used to obtain the target impedance. For the complicated package geometry structure, the 3D electromagnetic field solver is used to extract the package power supply model. In order to meet the target impedance, the required decoupling capacitors and location can be analyzed and placed according to the transfer impedance in frequency domain. The user can use the voltage ripples in time domain on power and ground nets for a direct verification process.

Introduction

Power Delivery Networks (PDNs) play a very important role during the analysis of Signal Integrity (SI) and Power Integrity (PI) in IC, packages and board design systems. The metallization of power distribution networks (PDN) in packages and PCBs is very complicated and is difficult to model. Systems operating at high speed often run into signal integrity or electromagnetic coupling issues that need special handling by way of added circuitry and rules of thumb. In such high speed designs, one often finds bypass capacitors that have been inserted to provide instantaneous current demands (deltI noise) when the IC is switching. This is typically for a digital IC that has considerable switching transient current.

Demands for large currents and the resulting lower voltage supply can occur when the device outputs changes state. More and more chip failures are being reported industry-wide, due to I/O cell simultaneous switching output. As the number of pins increase, the possibility of large supply noise on the chip and in the package due to simultaneous switching outputs increases as well. Compared to the core cells, output drivers consume more power, since they drive large off-chip loads. The simultaneous switching of many such outputs can create large current surges, resulting in detrimental effects on signal quality: Voltage collapse and ground bounce caused by the current surge can affect the signal quality of the output drivers as well as that of the drivers in the vicinity. Meanwhile, the supply noise and current surge in the on-chip and the package power network can be coupled into the signal, especially when the signal lines are referenced to the supply planes in a form of microstrip or striplines. In this case, the voltages at the power and ground pins of a device will "jump" when the device changes state. This phenomenon is often referred to as ground

or power bounce. It causes the most severe problems when multiple pins of the device switch in the same direction (all high or all low) at the same instant. By developing a design and analysis strategy that looks at power distribution system behavior in the frequency domain, designers can ensure that their power distribution system implementations will be both reliable and cost-efficient. The problem of providing very low impedance paths has become more severe with each succeeding generation of computer technology. As operating voltages continue to drop and currents increase, the allowable ripple on the supply planes also drops. Therefore, the power integrity analysis of whole system requires the simulation of board, package, and on-chip circuitry in the same simulation. Therefore, a package PI solution must support model formats used for the board and the package as well as on-chip circuitry.

Chip-Package-Board Design Brief

Today's high-performance ICs exhibit upwards of 2,000 IO pins requiring packages that sometimes exceeds 100 layers, which, in turn, go onto boards with more than 50 layers. This unprecedented complexity requires new design methodology, which allows designers to leverage knowledge of the design and manufacturing parameters that dictate the design of related system components. The EDA solutions should enable engineers to work concurrently with teams working across the three system domains-IC, package, and board. There are three flows each driven from a different perspective-board driven flows for complex high-speed signal topologies, IC driven for complex SoC designs utilizing custom packaging and package driven flows for cost constrained designs, standard packaging and SiP design as shown in Figure 1.

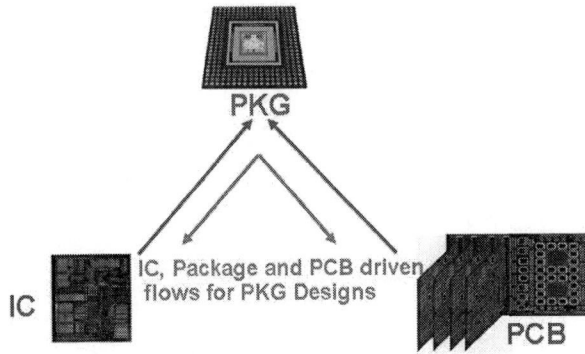

Figure 1. Silicon-Package-Board Scenario

All three scenarios must be supported by the design solution and the system integrator is the arbiter of constraints and tradeoffs. From design for high-speed, high-performance products to commodity markets, it is easy to integrate with the existing technology, allowing users to incrementally enhance their existing design flow with updates and best-of-breed technology to support all market sectors. Currently, above co-design flow can be supported by the Cadence Encounter and Virtuoso platform. The Allegro SiP methodology enables effective design chain collaboration. Allegro allows layers from both the IC design and the Package design to exist in the same design environment as shown in Figure 2.

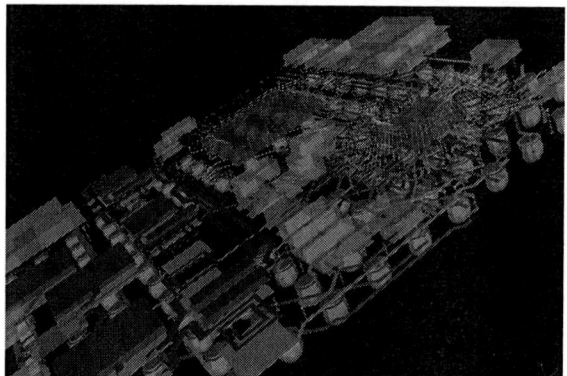

Figure 2. Cadence Allegro SiP layout structure view

Using Cadence Allegro Package/SiP design methodology, engineers can quickly optimize the power delivery network across ICs, packages, and PCBs-eliminating hardware re-spins and reducing both hardware costs and design cycles.

Power Integrity Analysis Focused on Board Design

The impedance versus frequency profiles of the power distribution system components including the voltage regulator module, bulk decoupling capacitors and high frequency ceramic capacitors are defined [1] and reduced to SPICE models shown as Figure 3. A sufficient number of capacitors are placed in parallel to meet the target impedance through typical power delivery network [2].

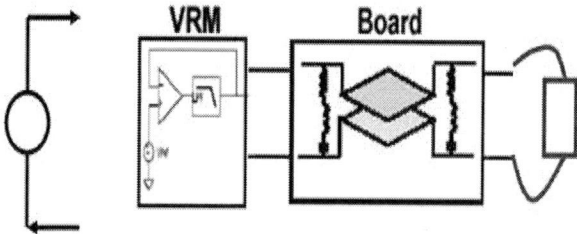

Figure 3. Typical power delivery network

An effective global PI solution needs to provide both the accuracy and capacity to handle the full-chip RLC power/ground (P/G) extracted network, and package models in S-parameter, transmission line, or RLCK format, and the board models. It must be able to simulate all these different models concurrently, with reasonable runtime and Spice-level accuracy. Board models are typically represented with a small number of lumped elements (RLCG); however, they could also include transmission line or additional S-parameter

models. A typical schematic of the circuit representation for board focused PI analysis is shown in Figure 4. Plane capacitance uses the power and ground planes of the circuit board to provide a distributed capacitor. Currently, the power/ground planes can be divided into unit cells with a lumped element model for each cell, as described in [2]. Each cell consists of an equivalent circuit with R, L, C and G components for a rectangular structure. Each unit cell can be represented using either a T or Π model [3-4]. The primary difference between those two models is an offset of half a unit cell. Both models however lead to similar results.

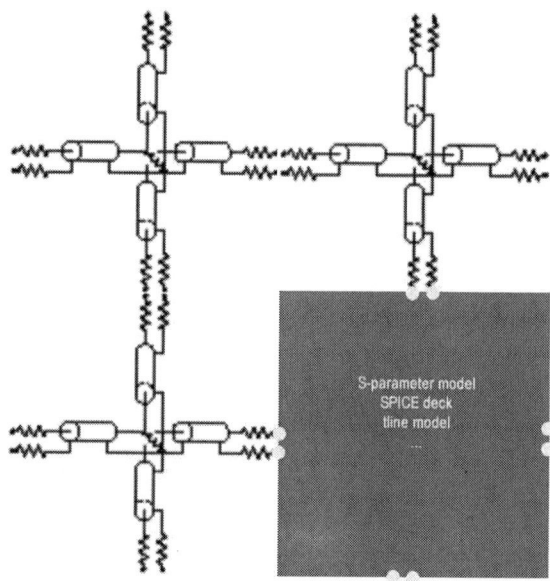

Figure 4. Board level power supply model

The SPICE models in Figure 4 are then analyzed in the frequency domain to find the impedance response. The power distribution systems require wide-band low impedance on the supply rails. The VRM only works at low frequency range.

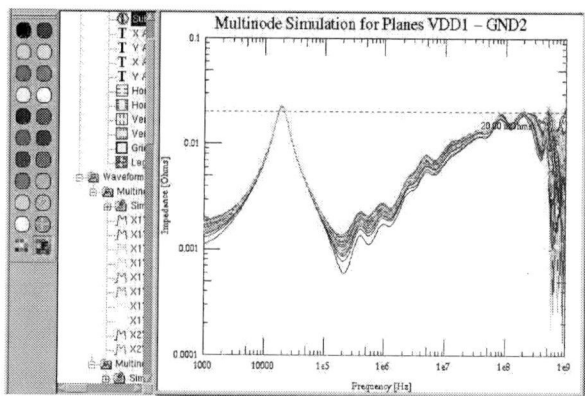

Figure 5. Board level impedance with de-capacitors

Even though decoupling capacitors are nominally a pure capacitance of a given value, in practice they include parasitic resistance and inductance, and perform best at their resonant frequency. At any other frequency, the effectiveness of the capacitor is reduced. Discrete capacitors work well at lower frequencies smaller than 1GHz because of their intrinsic and mounted inductance. Plane capacitance is rapidly becoming the solution of choice for high speed digital circuits.

Power Integrity Analysis Focused on Chip Design

On-chip power rail modeling requires extraction of power distribution environment including the RLC circuit of supply networks as well as the intrinsic and intentional decoupling capacitance like Figure 6.

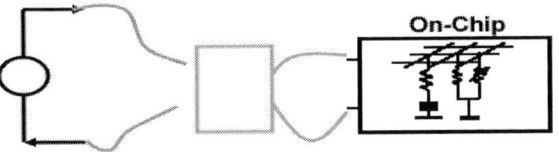

Figure 6. Chip level power integrity

To correctly predict the current profile caused by logic circuit within the core and IO power supply by transistor level spice circuit with VCD file. Using VoltageStorm tool in Cadence First Encounter platform, engineers can quickly view the dynamic IRDrop for on-chip power delivery network as shown in Figure 7.

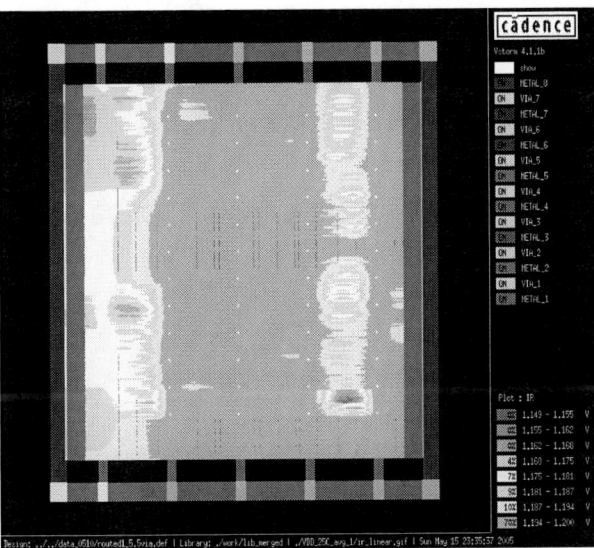

Figure 7. Chip level dynamic IRDrop results

Power Integrity Focused on Package/SiP Design

The package model is a key element in system PI analysis and a major contributor to rail collapse and ground bounce as well as coupling-induced signal distortion. A package model for power integrity analysis must have an accurate and complete frequency response of the supply network, the package signal routing, and any coupling between the two over a broad frequency range. The traditional lumped RLC model defines the lower frequency response of the package. However, it is becoming increasingly important for circuit simulators to support S-parameters, since a large number of

lumped RLCK circuit elements (depending on S-parameter waveform) or passivity enforced macro-model may be needed to achieve good approximation for time domain simulation. Different package model formats are possible in different stages of the design process. During package design phase, package models are often created with lumped elements or transmission line arrays. Later, when the package is available, package/SiP EM model can be derived from field solvers or measurements and reported as S-parameters, which accurately describe the frequency response of the circuit.

The board power supply circuit can be obtained from board level PI circuit netlist or measurement. The on-die power net subcircuit can be obtained through power analysis EDA tools such as Cadence Assura RC extraction while the current profile can be provided by VoltageStorm -- another Cadence IC product. Traditionally, the PI analysis uses rough switching current information of ICs to determine target impedance at board level. In this paper, the actual switching current profiles were adopted when determining target impedance, to specify more realistic target impedance like Z_{PDN} shown in Figure 8.

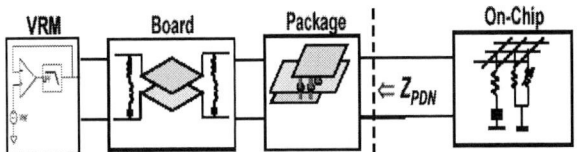

Figure 8. Whole system design for power integrity

With above VRM, board and package model and on-die circuit and current profile, the users can get the impedance curve at bump/die nodes as well as the pin nodes at the package level, obtain the target impedance by IC current profiles and view the practical voltage ripple resulted from the imported IC current profile in time domain [6].

The circuit netlist established based on Figure 8 can be simulated by spice-like simulators such as tlism. The resulting frequency dependant impedance will be used to select the type of decoupling capacitors and related locations. And the voltage ripples (for verification purpose) at die side pad are illustrated in Figure 9, which is resulted from the circuit with and without on-package decoupling capacitors (decaps). It can be seen that the voltage ripples on die pad without decaps is much larger than that with selected decaps.

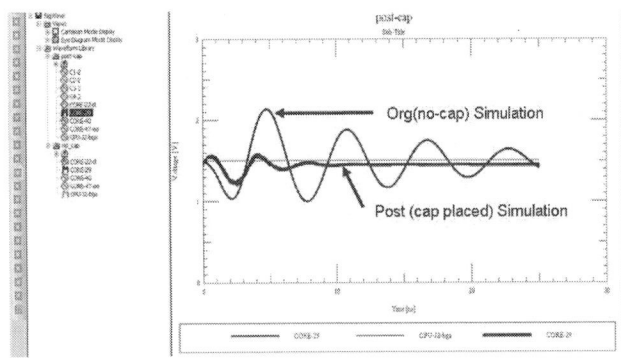

Figure 9. Voltage ripples on die pad w/o on-pkg decaps

Conclusions

Package/SiP power integrity analysis should consider the on-die information, the board power supply network and an accurate package EM model. Commonly, there is very small area left for decaps and often only 3~5 decaps can be used. Therefore, their selection and locations become important to the package designer, a problem which can be solved by package/SiP PI in the frequency domain. The power delivery network design can be further verified by voltage ripples in time domain. From the simulation results, the voltage ripples without on-package decoupling capacitors is much larger than that with on-package decoupling capacitors. It can be concluded that the package power delivery network can be upgraded by three methods: the improving package design, efficiently placing decoupling capacitors and optimizing on-chip & board supply network.

Acknowledgments

This is really a team project. I'd like to thank Cadence publication process team for their technical review and legal support, and the team members of Cadence High Speed Technology Center (HSTC) for their hard work and the Cadence Allegro High-speed R&D team for their valuable suggestions.

References

1   L. D. Smith, R. E. Anderson, D. W. Forehand, T. J. Pelc, and T. Roy, "Power distribution system design methodology and capacitor selection for modem CMOS technology," *IEEE Trans. Adv. Packag.*, vol. 22, no. 3, pp. 284-291, Aug. 1999.

2   W.L.Dai, "Power Integrity in IC-Package-Board Codesign Flow", *9th International IEEE CPMT Symposium on High Density Design, Packaging and Micros Integration (HDP'2007)*

3.   J K. Lee and A. Barber, "Modeling and analysis of multichip module. power supply planes," *IEEE Trans. Comp., Packag., Manufact. Technol. B*, vol. 18, pp. 628–639, Nov. 1995

4.   D. A. Al-Mukhtar and J. E. Sitch, "Transmission-line matrix method with irregularly graded space," *Proc. Inst. Elect. Eng. H*, vol. 128, pp. 299–305, Dec. 1981.

5.   Novak, "Reducing simultaneous switching noise and EMI on. ground/power planes by dissipative edge termination," in *Proc. 7th. Topical Meeting Elect. Perform. Electron. Packag.*, Oct. 1998, pp. 181-184.

6.   Cadence Allegro Package/SiP Power Integrity User Manual.

Effect of Fluid Dynamics and Device Mechanism on Biofluid Behaviour in Microchannel Systems: Modelling Biofluids in a Microchannel Biochip Separator

Xiangdong Xue[1*], Mayur K Patel[1], Maïwenn Kersaudy-Kerhoas[2], Chris Bailey[1], Marc P.Y. Desmulliez[2], David Topham[3]

[1]School of Computing and Mathematical Sciences, University of Greenwich, London SE10 9LS, UK
[2]MIcroSystems Engineering Centre (MISEC), School of Engineering & Physical Sciences
Heriot-Watt University, Edinburgh, EH14 4AS, UK
[3]School of Engineering and Design, Brunel University, Uxbridge, Middlesex, UB8 3PH, UK
[*]X.Xue@gre.ac.uk

Abstract

Biofluid behaviour in microchannel systems is investigated in this paper through the modelling of a microfluidic biochip developed for the separation of blood plasma. Based on particular assumptions, the effects of some mechanical features of the microchannels on behaviour of the biofluid are explored. These include microchannel, constriction, bending channel, bifurcation as well as channel length ratio between the main and side channels. The key characteristics and effects of the microfluidic dynamics are discussed in terms of separation efficiency of the red blood cells with respect to the rest of the medium. The effects include the Fahraeus and Fahraeus-Lindqvist effects, the Zweifach-Fung bifurcation law, the cell-free layer phenomenon. The characteristics of the microfluid dynamics include the properties of the laminar flow as well as particle lateral or spinning trajectories. In this paper the fluid is modelled as a single-phase flow assuming either Newtonian or Non-Newtonian behaviours to investigate the effect of the viscosity on flow and separation efficiency. It is found that, for a flow rate controlled Newtonian flow system, viscosity and outlet pressure have little effect on velocity distribution. When the fluid is assumed to be Non-Newtonian more fluid is separated than observed in the Newtonian case, leading to reduction of the flow rate ratio between the main and side channels as well as the system pressure as a whole.

1 Introduction

Microfluidic packaging has been gaining increased attention in Microsystems. A typical application is in the healthcare field, where microfluidic devices containing microchannels have shown an increasing number of applications in the biological and clinical areas. Computational Fluid Dynamics (CFD) can provide valuable insights at the early stage of the design on the performance such devices as well as the impact that the packaging might have on operational performance.

An important use of microfluidic devices in biological applications is in blood separation. Due to the difference in density between blood cells and plasma, human blood is usually separated by centrifugation where inertial forces tend to play a dominant role [1]. This approach, not so effective in microsystem devices, has led to an interest in adopting new microscopic separation methods, in which viscous forces, shear strain rate, surface tension and the geometrical effects of microchannels play important roles.

Separation methods using fluid mechanical effects may be classified into two categories: micro filter devices [2-3] and microchannel devices [4-5]. Among the latter, channel constriction [6-7], bending channel [8-9] and bifurcated channels [10-11] have been explored. In this paper, a T-shaped microchannel design [12-13] is investigated using CFD where the aim of the design is to separate the biofluid (i.e. blood) via multiple bifurcations. The resulting device consists of a main channel and a series of perpendicularly positioned branch/side channels. When the flow rate ratio between the main and side channels reaches a critical level, plasma is separated into the side channels.

The main difference between micro- and macrofluids comes from the effect that particles/cells have on the flow fields. Containing a vast number of particles, a macro system is generally considered as a single phase continuum flow, whereas with channel size comparable to the particles, a microsystem has to take into account particle behaviour on the bulk flow of the fluid. Consequently, for fluids such as blood, the suspension of cells in plasma, and their interactions, will affect the overall behaviour of the microfluidic device.

There are generally two modelling approaches in studying particle (or cell) performance in a bulk flow. One is the discrete method, in which the individual cells or a collection (PSI cell) are modelled by the immersed boundary method (IBM) or the immersed finite element method (IFEM) and the interactions between cells and the bulk flow are explicitly represented [14-15]. The other approach is the representation of the bulk flow field as a single-phase fluid with bulk properties to represent the affects of the cells [5, 12]. Of the two approaches, the former is mainly used for investigating detailed behaviour of particles and the flow in local regions. The latter is used for describing the global performance of the mixture (particles and fluid) of the flow in the system or device. In this paper, effort is concentrated on the latter approach.

The important phenomena for biofluid behaviour in microchannels has been explained by several laws or effects, including Fahraeus effect and Fahraeus-Lindqvist effect [16], Zweifach-Fung bifurcation effect [17-18] and the so-called cell-free or liquid-skimming layer effect.

The Fahraeus effect and Fahraeus-Lindqvist effect are strongly correlated, relating to the effect of changing channel cross-sections, such as a constriction, on flow hematocrit and its viscosity. It was observed [19] that, in a channel of 20μm diameter (this is the channel size in this study), the viscosity is about 50% of its normal value. The existence of a cell-free layer close to the channel walls can be from both effects indicated above. For laminar flow, the cell-free layer is always located in the layers close to the wall where the liquid can be extracted by appropriately designed apertures or branches on the wall. The Zweifach-Fung effect concerns cell behaviour at these bifurcations, inferring that particles have a tendency to travel through the channels with higher flow rate ratio.

In microfluidic channel systems, mechanical (i.e. geometrical) details also play an important role on the behaviour of the flow field. For the separation of blood plasma specifically, two geometrical effects of importance are channel constriction and channel bending. When blood passes through a channel constriction, the cells tend to travel faster and concentrate at the centre of the channel which results in a higher concentration of plasma near the wall, where it can be collected or skimmed from. A bending channel makes use of the centrifugal force induced on the particles when passing through the curved section, thus ensuring a higher concentration of plasma in the central region of the channel, where it can be collected or skimmered from.

Diluted and high sheared biofluids generally behave like Newtonian flows, but biofluids with higher cell concentrations show non-Newtonian behaviour. The difference is addressed in the current paper through the comparison of modelling results obtained by both Newtonian and non-Newtonian flows for a blood separator design. The biochip is introduced and analysed. This is followed by the modelled phase using CFD to investigate the flow field as a function of separation performance. Finally, comparisons of Newtonian and non-Newtonian flows are presented and discussed.

2 The Biochip and Fluid Dynamics

2.1 Device

Figure 1 shows the first prototype of the whole system separation device, together with system level controls at the micro level. The components at the system level include a syringe (1~4) for pumping the biofluid sample (e.g. blood) into the inlet (5), collection equipment (11,12) of the separated fluid (e.g. plasma) at the outlets (6,7) and concentrated fluid containing cells at the outlet (10), a counting equipment for examining cell concentration, and tubes and couplers between the micro and macro levels. The biochip (8), Figure 1(b), is the core part of the separator. Blood is pumped through the inlet at the top and flows downwards. After passing a constriction, the blood enters the main channel which contains a number of side branches where bifurcation occurs. Through these bifurcations, blood is separated into high concentrations of blood cells (main channel) and plasma (side channels).

In this example the size of the biochip is 7mm×10mm. The width of the main channel is 100 m and the constriction and branch side channels are 25μm and 20μm, respectively. The depth of all channels is 20μm. Due to the size of the biochip and associated micro-channel dimensions a high aspect ratio of length to cross-sectional area is evident.

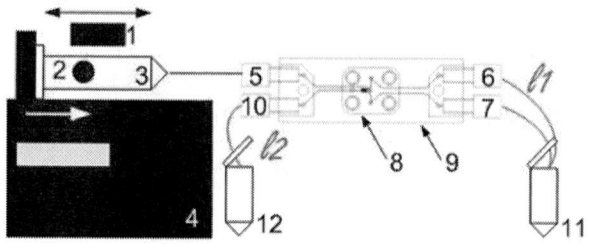

(a) the experimental setup

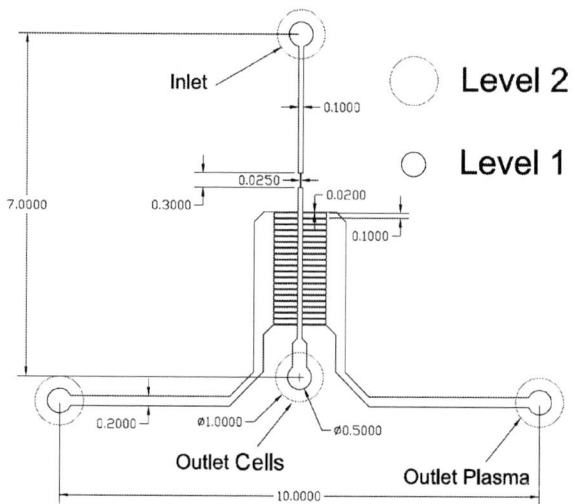

(b) the biochip in micro level, unit: mm

Figure 1 Separation device [20]

2.2 Device features and microfluidic dynamics

2.2.1 Microchannel

Microchannels have high surface-to-volume ratio. Assuming a no-slip condition for the fluid at the channel walls, the flow velocity cannot be developed to a high level in this micro space. High surface-to-volume ratio also means the flow in microchannels suffers relative high resistive force at its boundary surface on the wall. As a result, a laminar flow with low velocity, high pressure, high shear stain rate and low Reynolds number is common in microchannels, in which viscous forces play an important role on the behaviour of the flow field. When the channel size is reduced to 300μm or less [16], the Fahraeus effect and Fahraeus-Lindqvist effects are also important. A cell-free layer is formed close to the wall, which provides a distinctive feature for separating plasma from the bulk blood flow.

Spinning is an important mode for moving a particle. Under parabolic flow distribution, the flow velocity at the channel centre and near wall regions are substantially different. When a cell/particle is located to the left/right of the centre line of the channel, it experiences different flow velocities and shear stresses on the two opposite sides parallel to the bulk flow. This results in different forces acting on the two sides which, in turn, induce spinning. This difference in velocity results in a pressure drop across the particle and drives the cell towards the high velocity region, i.e. the channel centre. Here the relevant lift forces are known as the Magnus force [21] and Saffman force [22]. Note that these lift forces are produced when a particle is located to the left/right of the centre line of the channel. When the particle crosses the channel central line, this effect is substantially weakened.

2.2.2 Constriction

A constriction has three functions in terms of blood plasma separation: focusing the blood flow from the central region of the connected channel section, accelerating the flow and increasing the velocity difference between cells and plasma. The purpose of the first function is to concentrate the cells towards the channel centre. With the size effect, more plasma is being diffused to channel periphery region than blood cells.

The velocity changes when a flow passes a constriction can be explained via the mass conservation law, i.e. the continuity equation, as follow:

$$\frac{\partial \rho}{\partial t} + div(\rho v) = 0 \qquad (1)$$

where ρ is the density of the flow and v the velocity. For a well-developed incompressible flow, $\partial \rho / \partial t = 0$. Ignoring the small density change when the flow passes the constriction and expressing Eq. (1) in integral form, we get

$$\int_{chan} V_{chan} dA_{chan} = \int_{cons} V_{cons} dA_{cons} \qquad (2)$$

where V denotes flow velocity, A the area of channel cross-section; the suffixes *chan* and *cons* represent the main channel and constriction. Expressed with average forms, Eq. (2) becomes

$$\frac{V_{cons}}{V_{chan}} = \frac{A_{chan}}{A_{cons}} \qquad (3)$$

As the cross section of the constriction is 1/4 of that of the main channel, the velocity in the constriction increases by a factor of four where the average velocity can be around 200mm/s.

The change in relative velocities between blood cells and plasma with the constriction can be explained by the Fahraeus effect. The Fahraeus effect states that when blood of a given hematocrit (i.e. cell concentration), flows from a large reservoir into a tube of small diameter, the hematocrit in the tube decreases as the tube diameter decreases [16]. As the hematocrit at both sides of the constriction is the same and the thickness of the cell-free layer is relatively constant, a lower hematocrit level in the constriction means that the blood cells move faster than the plasma. As a result, blood cells are more likely to stay at the channel central region and the hydrodynamic effect becomes more effective.

2.2.3 Bifurcation

Zweifach-Fung bifurcation law is the most important effect for the current device. Based on this empirical law, the separation efficiency is a function of flow rate ratio between the main and side channels. It was observed [12] that, for a cell to channel diameter ratio of the order of 1, when the flow rate ratio reaches 6:1-8:1, nearly all cells will travel through the channel with the higher flow rate, leaving almost no cells travelling into the slower flow rate channel.

The effect of bending channels is also effective at bifurcations. As the side channels are perpendicularly connected with the main channel and the fluid flows into a side channel from the main channel, the fluid suffers a centrifugal force. As the density of blood cells is higher than plasma, under the inertia effect, blood cells have a tendency to move away from the bending flow to return to the flow in the main channel. The centrifugal force is directly proportional to the square of the flow velocity.

2.2.4 Channel size and outlet locations

The cross-section of the main channel is larger than the side channels. The bifurcation region is much closer to the outlet of the main channel than the outlets of side channels. The purpose of the above design is to produce a high flow rate ratio between the main and side channels for separation purpose.

The relationship of channel resistance to channel length, cross area and viscosity is expressed by Poiseuille's law, which states that, for the laminar incompressible fluid, the resistance of a channel is directly proportional to channel length and flow viscosity and inversely proportional to the fourth power of diameter (i.e. the square of the area) of a circular channel cross-section, expressed as follow.

$$R = C \cdot \frac{L\mu}{A^2} \qquad (4)$$

where R denotes channel resistance, L the channel length, A channel cross-section area, μ flow viscosity and C a coefficient. From Eq. (4), the ratio of channel resistance between the main and side channels can be expressed as

$$\frac{R_m}{R_s} = C' \cdot \frac{L_m \mu_m}{L_s \mu_s} (\frac{A_s}{A_m})^2 \qquad (5)$$

in which suffixes m and s denote the main and side channels, respectively.

Eq. (5) shows that the ratios of channel lengths and cross-sections between the main and side channels determine the channel resistance ratio, i.e. the flow rate ratio. It also shows that viscosity variation of the fluid has a strong effect on flow field. Hence, non-Newtonian behaviour of the fluid should be taken into account in modelling biofluid flows.

3 Modelling

3.1 Newtonian and Non-Newtonian Flows

In a Newtonian flow, the rate of shear stress and strain, i.e. viscosity, is constant. When human blood is considered as a Newtonian fluid, the bulk flow is modelled with a constant viscosity of 0.0035Pa.s and a density of 1060kg/m³. To investigate the effect of viscosity on flow field, horse blood is also modelled. For horse blood (Newtonian), the flow is modelled with a viscosity of 0.0047Pa.s [23] and a density of 1060kg/m³ [24].

When human blood is considered Non-Newtonian, two fluid models are implemented to the main channel and side channels, respectively. The fluid in the side channels and subsequent outflow region is modelled as plasma with the constant viscosity of 0.0015Pa.s and a density of 1025kg/m³. The flow in the main channel is modelled with a shear rate dependent non-Newtonian flow. The Carreau-Yasuda model is used for modelling the shear-thinning behaviour of the fluid, shown as follow:

$$\mu(\dot{\gamma}) = \mu_\infty + \frac{\mu_0 - \mu_\infty}{(1 + (\lambda\dot{\gamma})^b)^a} \quad (6)$$

where $\dot{\gamma}$ is the shear rate, μ_∞ and μ_0 are the infinite shear viscosity and the zero shear viscosity, respectively; λ, a and b are constants. Parameter values are as follows [25, 26]: μ_∞ = 0.0035Pa.s, μ_0 = 0.16Pa.s, λ =8.2s, a = 1.23 and b = 0.64. The fluid density is taken as ρ = 1060kg/m³. The Carreau-Yasuda model shows that non-Newtonian blood flow is a shear-thin flow.

3.2 Computational model

The three-dimensional Computational Fluid Dynamics (CFD) finite-volume package Ansys-CFX (CFX5) is used to perform the analysis. A thin-layer 3D model, in effect 2D, to save computer resources, is developed. Figure 2 shows the computational model together with the mesh at the first upstream bifurcation and the inlet side of the constriction. A three-layer fine mesh is constructed in the region close to the wall to represent the flow performance in the boundary layer. In total 605,243 mesh elements and 654,954 nodes are contained in the model. The coordinate origin is set at the left hand end of the last downstream side channel, as shown in the graphs below.

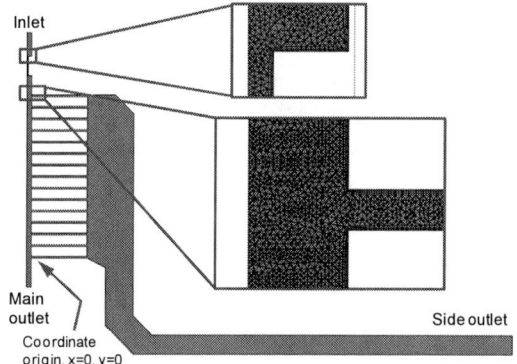

Figure 2 Computational model and the mesh in the first upstream bifurcation

Due to the symmetric nature of the device, and assuming that the flow is symmetric, only half of the device needs to be modelled. This consists of 15 side channels and one main channel as illustrated in Figure 2. For the boundary conditions, the inlet is a constant flow rate of 360μL/h (50mm/s). Both boundary conditions at the main and side outlets are set to 0Pa pressure. For a well-developed incompressible laminar flow, the flow field in the biochip is only determined theoretically by the relative pressure between the inlet and outlets. Thus, the result obtained from 0Pa outlet pressure is also suitable to other pressure levels.

4 Results and Comparison

4.1 Modelling biofluids using Newtonian flow

Figure 3 shows the Reynolds number in the main and side channels at the first upstream and the last downstream bifurcations, respectively. With lower flow rate, the Reynolds number in the side channels is one order of amplitude lower than the main channel. Laminar flow character is evident by the low Reynolds numbers.

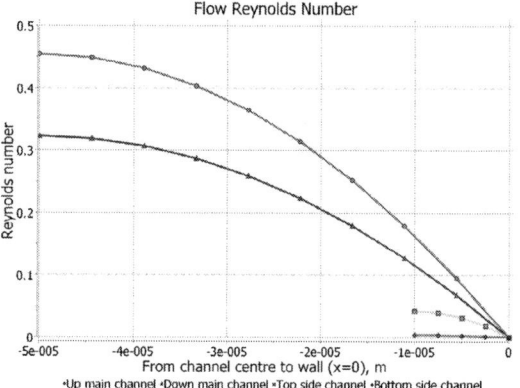

Figure 3 Reynolds number of channels

Figure 4 shows a close view of the 'creeping' flow pattern in the first upstream channel. The reduction of the fluid velocity from the centre of the main channel to the side channels is clearly visible. Only the fluid streamlines close to the wall of the main channel, where the cell-free layer is located, enters the side channels promoting separation.

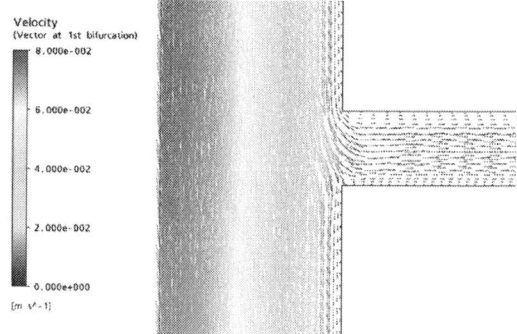

Figure 4 Velocity vector contour in the first upstream channel

Figures 5 through 7 show the velocity profile development through the microchannel biochip. Figure 5 shows the flow velocity profile within the constriction. The maximum velocity inside the constriction is about four times of that in the main channel. This difference in velocity is attributed to Eq. (3).

Figure 6 shows the velocity profile in the main channel. The parabolic profile of the velocity is closely related to a cross channel motion in the main channel, as discussed in section 2.2.1. The velocity distribution also indicates the difference of shear rate across the channel. The maximum

978-1-4244-4658-2/09 $25.00 © 2009 IEEE   106

shear strain rate appears on the boundary layer close to the wall.

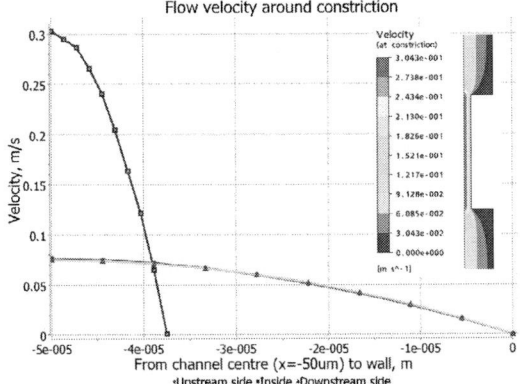

Figure 5 Flow velocity profile around constriction with velocity contour inserted

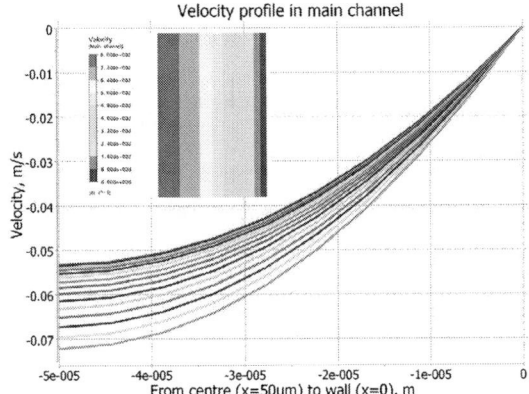

Figure 6 Flow velocity profile in the main channel with contour inserted

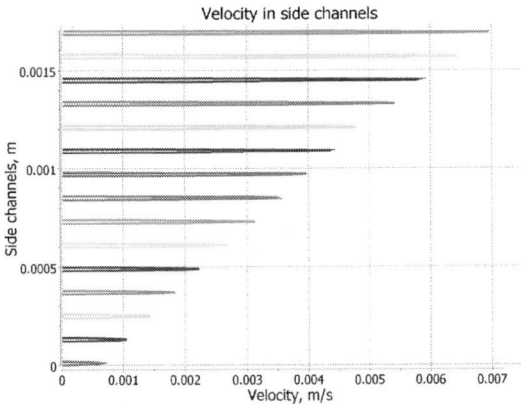

Figure 7 Flow velocity of side channels

Figure 7 shows the velocity of side channels. The velocities vary substantially upstream and downstream side of the channels, showing a linear reduction from upstream to downstream channels. This is because, in order to form high flow rate ratios between the main and side channels, the outlet

of the main channel is designed to be closer to the bifurcation region as opposed to the outlet of side channels (see Figure 1).

Horse blood is modelled with slightly higher viscosity than human blood fluid. The modelling result shows that the velocity distribution of horse blood flow is nearly identical as the human blood flow case, i.e. Figures 3 to 7. This implies that, for the flow rate controlled cases, the viscosity has little effect on the velocity distribution pattern. Combining the analysis in section 3.2 about the effect of the outlet pressure, we can conclude that, for a Newtonian flow, if the flow rate at the inlet is kept constant, the velocity distribution is independent of the viscosity and the outlet pressure.

Unlike velocity, the pressure distribution is affected by viscosity. Figures 8 and 9 show the pressure development at the two ends of side channels. The pressure drops for horse blood within the side channels and the pressure in the main channel are higher than human blood. This shows that the pressure drop increases with increase in fluid viscosity. Note that the increase in pressure is similar to the increase in viscosity (4.7/3.5), as expressed by Eq (5).

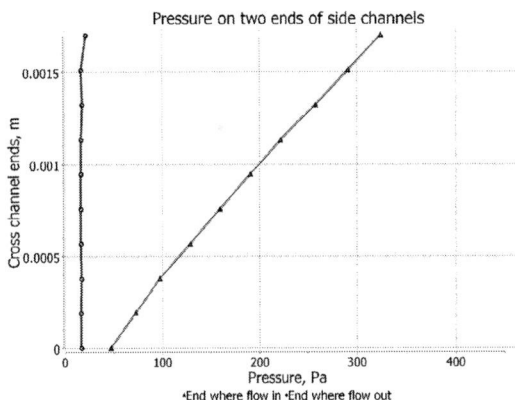

Figure 8 Flow pressure development at two ends of side channels, human blood flow

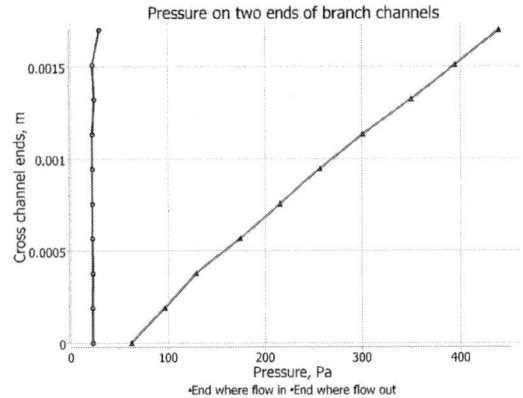

Figure 9 Pressure development at two ends of side channels, horse blood flow

The flow rate ratios at the bifurcations are shown in Table 1. All flow rate ratios are higher than the threshold value of

978-1-4244-4658-2/09 $25.00 © 2009 IEEE   107

6:1 to 8:1 [12] required for pure plasma separation from the bulk flow. Thus, numerically, this biochip can theoretically separate plasma by nearly 100% from the bulk blood flow. For easy of comparison, the flow rate ratio modelled by non-Newtonian blood flow is also presented in Table 1.

Table 1 Flow rate ratios at bifurcations for both cases
(No 1 refers the most upstream bifurcation)

Bifurcation number	Newtonian flow	Non-Newtonian flow
	Human/horse	Human blood
1	25.89	14.02
2	27.06	14.44
3	28.95	15.19
4	30.02	15.44
5	33.12	16.75
6	35.12	17.55
7	37.65	18.49
8	41.62	20.16
9	45.73	21.81
10	52.63	24.84
11	62.51	29.05
12	74.94	34.63
13	94.47	43.52
14	128.7	59.54
15	202.2	93.74

4.2 Modelling biofluids using non-Newtonian flow

4.2.1 Comparison of non-Newtonian and Newtonian flows

Figures 10 and 11 present velocity profiles of the main and side channels for non-Newtonian flow. Comparing these to the results obtained for Newtonian fluids (Figures 5 and 6), the velocities within the side channels are largely increased and the velocity of the main channel is gradually reduced, as the fluid goes downstream. As a result, the flow rate ratio of the bifurcations has reduced (see Table 1). All the flow rate ratios at the bifurcations are higher than the separation threshold, indicating a successful separation.

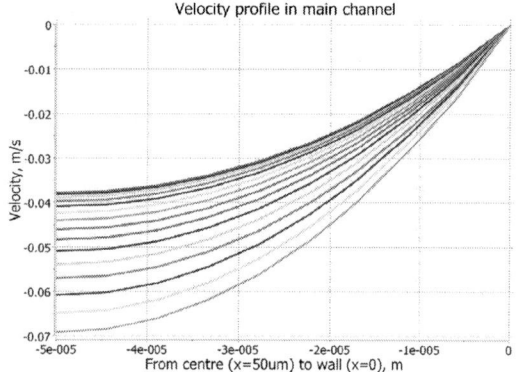

Figure 10 Flow velocity profile in the main channel

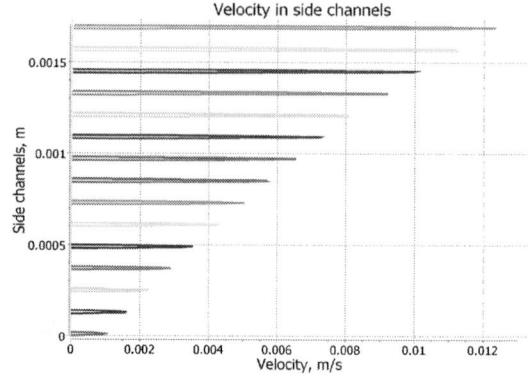

Figure 11 Flow velocity of side channels

Figure 12 shows the pressure development at the two ends of the side channels. The pressure expressed by the blue line (left) represents the pressure level at the outlet side of the side channels and the difference of two lines denotes the pressure drop over side channels. Both are largely decreased, as a result of the reduction of the fluid viscosity of the separated plasma from the bulk flow.

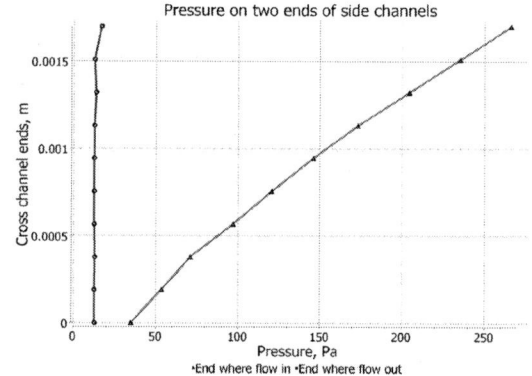

Figure 12 Pressure development at two ends of side channels, horse blood flow

In summary, comparing Newtonian and Non-Newtonian assumptions the flow rate of the side channels has largely increased and the flow rate of the main channel has gradually reduced as the fluid goes downstream. The pressures at the two sides of the side channels have reduced. Their difference, representing the pressure drop over the side channels, has also reduced. These lead to a microchannel biochip with low overall system pressure and fast and efficient plasma separation.

4.2.2 Effect of flow rate/ velocity on flow field

To investigate the effect of the flow velocity on flow field, three cases with different velocities (flow rates), i.e. 10mm/s (72µl/h), 50mm/s (360µl/h) and 200mm/s (1.44ml/h), are simulated. Figure 13 shows the shear strain rate in the cases of 10mm/s and 200mm/s of input velocities. The distribution patterns are similar between the two cases. The rate of change in shear strain is similar as the input velocities. The 200mm/s case experiences some 20 times of shear strain rate as the

10mm/s case in the region close to the wall. As blood has shear-thinning material behaviour, viscosity in the 200mm/s is lower than the 10mm/s case (see Eq. (6)).

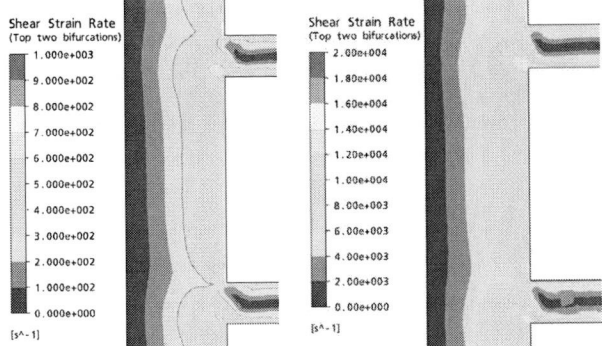

Figure 13 Shear strain rate at the top two bifurcations in the cases of 10mm/s (left) and 200mm/s (right) of input velocities

In microchannels, the shear strain rate is relevant to flow velocity. Flow velocity thus has an effect on the viscosity and this further affects the flow field. Table 2 shows maximum velocities at odd numbered bifurcations in the main and side channels. By defining the 10mm/s case as the reference case, the rates of increase in the maximum flow velocity by the other two cases can be obtained and are show in Table 3.

Table 2 Maximum velocities of three input velocity cases at odd numbered bifurcations in the main and side channels

No	Main channel (mm/s)			Side channels (mm/s)		
	10mm/s	50mm/s	200mm/s	10mm/s	50mm/s	200mm/s
1	13.64	69.17	277.9	2.595	12.34	47.98
3	11.87	60.85	245.5	2.161	10.01	38.99
5	10.38	53.91	218.4	1.752	8.047	31.28
7	9.203	48.34	196.5	1.386	6.536	25.76
9	8.288	44.04	179.4	1.059	5.049	19.99
11	7.619	40.90	166.9	0.757	3.520	14.04
13	7.169	38.82	158.5	0.478	2.230	9.006
15	6.952	37.83	154.3	0.209	1.009	4.143

Table 3 Change rate of velocity based on 10mm/s case at odd numbered bifurcations in the main and side channels

Bifurcation No	Main channel		Side channels	
	50mm/s	200mm/s	50mm/s	200mm/s
1	5.071	20.37	4.755	18.49
3	5.126	20.68	4.632	18.04
5	5.194	21.04	4.593	17.85
7	5.253	21.35	4.716	18.59
9	5.314	21.65	4.768	18.88
11	5.368	21.91	4.650	18.55
13	5.415	22.11	4.665	18.84
15	5.442	22.20	4.828	19.82
Average	5.273	21.41	4.701	18.63

From Table 3, the velocity distribution patterns are similar for the three cases, illustrating similar rates of change of velocities at different bifurcations in both the main and side channels. There are minor difference between the rate of change of the channel velocity and the input velocity, reflecting a limited effect of inlet velocity on the flow rate ratio. The average rates of bifurcation and input velocities from the 10mm/s case to the 50mm/s case are 5.3:5 for the main channel and 4.7:5 for the side channels. These values are much larger than those from the 50mm/s to 200mm/s case, i.e. 4:4 for the main and side channels. This shows that the effect of shear strain rate on viscosity and the flow field is nonlinear in behaviour. When velocity is high enough, the effect of the increased velocity becomes very limited.

The directions of the flow velocity change are different between the main and side channels. Velocity increase rates in the main channel are slightly higher than the difference of the inlet velocity values. By contrast, the rate of increase in the side channels is slightly lower than the rate of incease in the inlet velocities. This is because the viscosity in the main channel decreases with an increase in velocity, leading to more distribution of the volumetric flow through the main channel. In the side channels, due to the constant value of viscosity of the separated plasma, the increase in flow velocity cannot affect the flow rate through viscosity. As the total flow rate is fixed, the increase in velocity in the main channel leads to a decrease of velocity in the side channels. As a result, increasing input velocity for a non-Newtonian flow can result in a slight increase of the flow rate ratio between the main and side channels.

5. Conclusions

(1) The effect of mechanical (geometrical) features of the microchannel biochip on biofluid behaviour has been investigated. These features, i.e. microchannel, constriction, bifurcation, bending channel (or perpendicularly cross channels) and channel length ratio between the main and side channels, have shown strong effects on the biofluid hydrodynamics, bulk flow field and the separation efficiency. Some system improvement is therefore possible for the control of the flow field and the enhancement of the separation efficiency.

(2) Biofluid behaviour in the biochip is modelled with two Newtonian and one non-Newtonian flows cases. The comparison of two Newtonian flows cases shows that modelling has resulted in identifying that the velocity distribution in a flow rate controlled system does not depend on viscosity values. The comparison of non-Newtonian and Newtonian flow cases shows that modelling with the non-Newtonian representation; the results show an increase in plasma separation and an overall reduction of system pressure.

(3) For a non-Newtonian flow, an increase in the input flow rate can lead to an increase in the shear strain rate within the microchannel system and therefore, a decrease in the viscosity. Due to the different fluid features between the main and side channels, i.e. high cell concentration flow in the main channel and high plasma concentration flow in the side channels, the decrease in the viscosity affects more on the flow in the main channel than the side channels. More fluid volume will travel through the main channel than the side

channels. This leads to a slight increase in the flow rate ratio between the main and side channels.

(4) In terms of plasma/blood separation, numerically this biochip can separate plasma from the bulk blood flow. Consider the common behaviour among biofluids; this design may also be suitable to other separation purposes.

Acknowledgments

The authors gratefully acknowledge the financial support by the UK Engineering and Physical Sciences Research Council (EPSRC) in funding the Grand Challenge Project '3D-Mintegration' (www.3D-mintegration.com), referenced EP/C534212/1. The authors would also like to acknowledge the financial support of the Scottish Funding Council through the SRDG programme SCIMPS (Scottish Consortium in Integrated MicroPhotonics Systems).

References

1. Hester, J. P., Kellogg, R. M., Mulzet, A. P., Kruger, V. R., McCredie, K. B., Freireich, E. J., "Principles of blood separation and component extraction in a disposable continuous-flow single-stage channel," *Blood,* Vol. 54 (1979), pp. 254-68.
2. Carlson, R., Gabel, C., Chan, S., Austin, R., "Self-sorting of white blood cells in a lattice," *Phys. Rev. Lett.* Vol. 79 (1997), pp. 2149-2152.
3. Wilding, P., Kricka, L. J., Cheng, J., Hvichia, G., Shoffner, M. A., Fortina, P., "Integrated cell isolation and polymerase chain reaction analysis using silicon microfilter chambers," *Anal. Biochem.* Vol. 257 (1998), pp. 95-100.
4. Mohamed, H., Turner, J. N., Caggana, M., "Biochip for separating fetal cells from maternal circulation," *Journal of Chromatography A,* Vol. 1162 (2007), pp. 187-192.
5. Rainer, J., Roger, S., Carlo, E., "Microfluidic depletion of red blood cells from whole blood in high-aspect-ratio microchannels," *Microfluid Nanofluid,* Vol. 3 (2007), pp. 47-53.
6. Layek, G. C., Midya, C., "Effect of constriction height on flow separation in a two-dimensional channel," *Commun Nonlinear Sci Numer Simul,* Vol.12 (2007), pp.745-759.
7. Golia, C. and Evans, N. A., "Flow separation through annular constrictions in tubes," Experimental Mechanics, Vol 13, No. 4 (1973), pp. 157-162.
8. Blattert, C., Jurischka, R., Tahhan, I., Schoth, A., Kerth, P., Menz, W., "Microfluidic blood/plasma separation unit based on microchannel bend structures," *Proc of microtechnology in medicine and biology, 3rd IEEE/EMBS Special Topic Conf,* Honolulu, Hawaii, 12-15 May 2005, pp. 38-41.
9. Carlo, D. D., Edd, J. F., Irlmla, D., Tompkins, R. G., Toner, M., "Equilibrium separation and filtration of particles using differential inertial focusing," *Anal. Chem.* Vol. 80 (2008), pp. 2204-2211.
10. Pries, A. R., Secomb, T. W., Gaehtgens, P., Gross, J. F., "Blood flow in microvascular networks, experiments and simulation," *Circ Res.* Vol. 67 (1990), pp. 826-834.
11. Jafari, A., Mousavi, S. M., Kolari, P., "Numerical investigation of blood flow, Part I: in microvessel bifurcations," *Commun Nonlinear Sci Numer Simul,* Vol. 13 (2008), pp. 1615-1626.
12. Yang, S., Undar, A., Zahn, J. D., "A microfluidic device for continuous, real time blood plasma separation," *Lab Chip,* Vol. 6 (2006), pp. 871-880.
13. Kersaudy-Kerhoas, M., Kavanagh, D, Xue X., Patel M.K., Dhariwal R., Bailey C., Desmulliez M.P.Y., "Integrated biomedical device for flood preparation," *Proc of ESTC conf,* London, 1-4 September 2008, pp. 447-452.
14. Dzwinel, W., Boryczko, K. and Yuen, D. A., "Discrete-particle model of blood dynamics in capillary vessels," *Journal of Colloid and Interface Science,* 258 (2003), pp. 163-173.
15. Korin, N., Bransky, A. and Dinnar, U., "Theoretical model and experimental study of red blood cell (RBC) deformation in microchannels," *Journal of Biomechanics,* Vol. 40 (2007), pp. 2088-2095.
16. Brown, B. H. *et al.,* Medical Physics and Biomedical Engineering, Institute of Physics Publishing (Bristol, 1999)
17. Svanes, K., Zweifach, B. W., "Variations in small blood vessel hematocrit produced in hypothermic rats by micro-occlusion," *Microvasc Res,* Vol. 1 (1968), pp. 210-220.
18. Fung, Y. C. "Stochastic flow in capillary blood vessels," *Microvasc. Res.* Vol. 5 (1973), pp. 34-48.
19. Woodcock, J. P., "Physical properties of blood and their influence on blood-flow measurement," *Rep. Prog. Phys.* Vol. 39 (1976), pp. 65-127.
20. Kersaudy-Kerhoas, M., Dhariwal, R., Desmulliez, M.P.Y. and Jouvet, L., "Hydrodynamic blood plasma separation in microfluidic channels," *Microfluid Nanofluid,* DOI 10.1007/s10404-009-0450-5.
21. Dandy, D.S. and Dwyer, H. A., "A sphere in shear flow at finite Reynolds number: effect of shear on particle lift, drag, and heat transfer," *J Fluid Mech,* Vol. 216 (1990), pp. 381-410.
22. Saffman, P. G., "The lift on a small sphere in a slow shear flow," *J Fluid Mech,* Vol. 22 (1965), pp 385-400.
23. Dintenfass, L., Liao, F., "Plasma and blood viscosities, and aggregation of red cells in racehorses," *Clin. Phys. Physiol. Meas.,* Vol.3 (1982), pp.293-301.
24. Waite, L., Fine, J. F., Applied Biofluid Mechanics. 1st ed. McGraw-Hill Professional, (New York 2007), pp. 30.
25. Leuprecht, A., Perktold, K., "Computer simulation of non-newtonian effects on blood flow in large arteries," *Comput. Methods Biomech. Biomed. Eng.,* Vol. 4 (2001), pp. 149-163.
26. Abraham, F., Behr, M., Heinkenschloss, M., "Shape optimization in steady blood flow: A numerical study of non-Newtonian effects," *Comput. Methods Biomech. Biomed. Eng.,* Vol. 8, No. 2 (2005), pp. 127-137.

Modeling of Manufacturing Processes of High Power Light Emitting Diodes: Wire Bonding Process

Zhaohui Chen[1], Yong Liu[2,4], Sheng Liu[1,2,3*]

[1]Research Institute of Micro/Nano Science and Technology, Shanghai Jiao Tong University, Shanghai, P. R. China, 200240
[2]Wuhan National Laboratory for Optoelectronics
Huazhong University of Science & Technology, Wuhan, P. R. China, 430074
[3]Institutes of Microsystems, Huazhong University of Science & Technology, Wuhan 430074, P. R. China
[4]Fairchild Semiconductor Corp. 82 Running Hill Road, Mail Stop 35-2E, South Portland, ME 04106
*Corresponding author: Sheng Liu, Telephone: 86-13871251668, Email: victor_liu63@126.com

Abstract

In this paper, the wire bonding process of high power light emitting diodes (LED), which was simplified to consist of impact and vibration stages, was investigated by using a non-linear finite element method. Parametric studies were carried out to examine the effects of the ultrasonic vibration amplitude, the friction coefficient between the free air ball (FAB) and bond pad, the incline of LED chip on stress and strain distribution in the bond pad and ohmic contact layer of both p-type and n-type electrode structure. This numerical simulation work may provide guidelines for the wire bonding process virtual window development of high power lighting emitting diodes package.

Introduction

More and more research studies have been devoted to GaN based LEDs due to their significant impacts on illumination industry. The eventual goal of LED lighting is possible to replace the incandescent and halogen lamps because the latest white LED has significantly high efficient than the conventional bulb in terms of lumens per watt. Currently, high brightness LED has found applications in automotive front lighting, backlighting for large LCD displays, outdoor illumination, city improvement engineering and they are strong and promising candidates for the next generation general illumination applications [1].

The reliability of LED packaging possibly limits the LED from being applied to lighting and other industries. Defects in terms of voids, cracks, and delaminations are often generated in LED devices and modules during various manufacturing processes, accelerated testing, inappropriate material handling and field applications. Defects are mostly frequently induced in the early stage of process development. The effects of the defects have been proven to be critical to the reliability for the IC devices and systems [2, 3].

Wire bonding is one of the main processes of LED package. Improper bonding parametric may lead to the reliability problem of the electric interconnect of LED, such as bond pad cratering, peeling and cracking below the bond pad.

Many researchers have contributed to the studies of the wire bonding process. Failure estimation of a silicon chip and a GaAs chip during a gold wire bonding process was presented by Ikeda,T et al. [4]. A transient non-linear dynamic finite element framework which integrates the wire bonding process and the silicon devices under bond pad is developed by Yong Liu et al. [5]. The modeling and experiment study of ball stitch on bump (BSOB) wire

bonding process above a laminate substrate was also done by Yong Liu et al. [6]. Chang-Lin Yeh et al. [7, 8] conduct transient analysis to investigate response of the Cu/Low-K structure during the impact and ultrasonic vibration stage of wire bonding.

In this paper, numerical simulation was carried out by a nonlinear finite element method (FEM) to investigate the stress and strain distribution on the electrode structure of LED devices during the impact and ultrasonic vibration stage of wire bonding process. Parametric studies are also carried out to examine the effects of the amplitude of the ultrasonic vibration, the friction between the free air ball and bond pad and the incline of bond pad on the stress level and on the potential of structural defects in the ohmic contact layers of wire bonding interconnection.

Physical Model of Wire Bonding

In this work, the wire bonding process of high power LED is studied with a non-linear finite element framework. Typically the whole wire bonding process consists of four stages: the wire tip heated to generate a FAB, the z-motion of capillary and FAB, the impact of the FAB with bond pad and the input of ultrasonic wave energy. In our model, the whole wire bonding process on the bond pad of LED was simplified to consist of the impact stage and ultrasonic vibration stages, as shown in Fig.1. The model involves the capillary, the gold ball, the heat affect zone (HAZ), GaN layer and the sapphire substrate. The active layer of LED is not considered. The thicknesses of GaN layer and sapphire substrate are 3um and 50um respectively. The length of the structure is 100um.

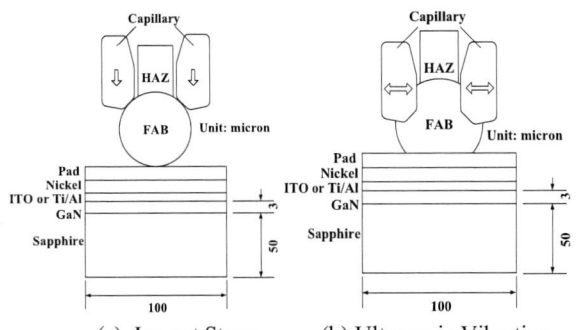

(a) Impact Stage   (b) Ultrasonic Vibration

Fig.1. Schematic diagram of wire bonding on LED chip: (a) Impact stage, (b) Horizontal ultrasonic vibration.

The p-type electrode structure of LED chip consists of the ITO layer as the ohmic contact layer, which is covered by the

978-1-4244-4658-2/09 $25.00 © 2009 IEEE

nickel layer and the gold bond pad. The titanium and aluminum layers are used as the ohmic contact layer in the n-type electrode structure. The thickness of ITO layer is assumed to be 0.1um in p-type electrode structure. In the n-type electrode structure of LED chip, the thicknesses of the titanium and aluminum layers are assumed to be 0.1um and 0.2um respectively. For both types of electrode structure, the thicknesses of the nickel layer and gold bond pad are chosen as 0.2um and 1um respectively.

The geometry of capillary is shown in Fig. 2. A specific set of parameters, d1=30um, d2=40um, d3=90, α=3°, β=90°, R1=20 um R2=R3=2um, are applied. The diameter of gold wire used in the package of high power LED is 25um. Generally the diameter of the FAB must be 1.5-4 times bigger than that of gold wire. The diameter of the FAB is chosen to be 50 um in present work.

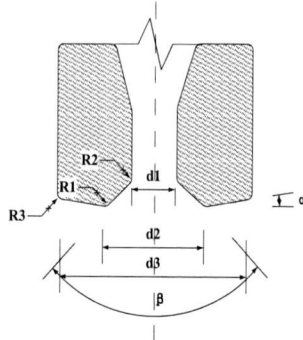

Fig.2. Geometry of capillary.

Method of Simulation

The actual wire bonding process is rather complex and modeling can not solve every detail of the bonding process. In the present study, in order to conduct an effective numerical simulation, the following assumptions are made:

(1) The thermo-stress induced by the difference of thermal expansion coefficient between the different metal layers of the electrode is not considered;

(2) The heat and temperature induced by plastic deformation and friction between FAB and bond pad is not included in this study; and

(3) The capillary is assumed as a rigid body due to much higher Young's modulus and hardness and the inertia force is not considered [5].

Two dimensional plane strain analyses were carried out using a general finite element code. The finite-element model is shown in Fig. 3. Since the bonder capillary is considered as a rigid body due to high hardness, this leads to the rigid and elastic-plastic contact pair between capillary and FAB. While the contact surfaces between FAB and gold bond pad are a non-linear contact pair with consideration of the dynamic friction. The bottom of sapphire is fixed and two sides are constrained in horizontal direction.

Fig.4 shows that the vertical and horizon position of the capillary over the wire bonding process. Two phases are defined in Fig.4, the first phase includes the contact impact with strain hardening; and the second phase deals with

horizontal ultrasonic vibration. In the impact stage, the capillary is supposed to move along a distance of 19um within duration of 2us. In the ultrasonic vibration stage, the amplitude of horizontal movement cycle of the capillary is assumed as 1um while the frequency is set to be 100 kHz. During the ultrasonic vibration stage, the displacement of capillary is supposed to be 1um to exert a bond force to the gold ball within duration of 20 us (similar to [8]).

The material parameters are listed in Table.1. FAB, bond pad, nickel, titanium and aluminum layers are nonlinear (bi-linear) materials, all of the rest of the materials are considered to be linear elastic [5-9].

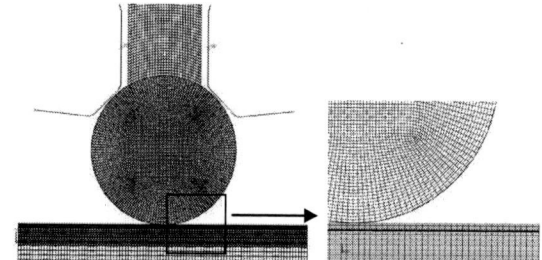

Fig.3. The finite element model.

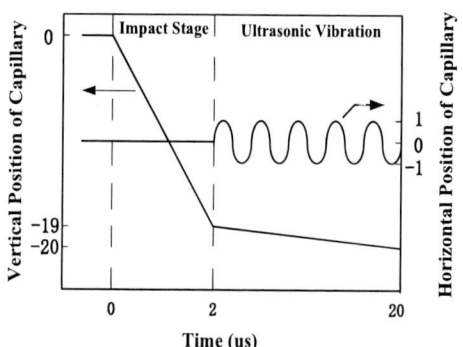

Fig.4. The position of capillary during wire bonding process

Table 1 Materials parameter.

Material	Modulus (GPa)	Poisson Ratio	CTE (ppm/℃)	Yield Stress (MPa)
ITO	160	0.335	8.6	-
Titanium	40	0.36	8.6	-
Aluminum	70	0.35	23	400
Nickel	219	0.31	13.4	620
PAD	30	0.44	14.2	110
HAZ	35	0.44	14.2	135
FAB	30	0.44	14.2	110
Sapphire	400	0.22	7.9	-
GaN	295	0.31	5.6	-

Modeling Results and Discussions

1. Deformation and Stress Distribution of Wire Bonding Process

In this study, the wire bonding process consists of impact and ultrasonic vibration stages. The deformed shapes and the

978-1-4244-4658-2/09 $25.00 © 2009 IEEE   112

von Mises stress contour at different times during the impact and ultrasonic vibration stage on the p-type electrode structure of LED are shown in Fig. 5. The amplitude and frequency of the ultrasonic vibration is chosen as 1um and 100 kHz respectively.

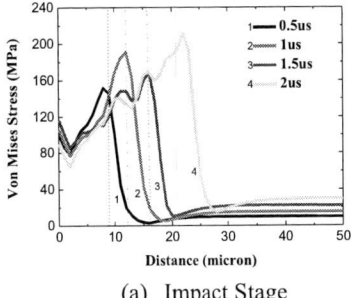

0.5us   1us

2us   4.5us

9.5us   22us

Fig. 5. Deformed shapes and von Mises stresses contour at different times of wire bonding process.

The distributions of von Mises stress along the middle of ohmic contact layer from center to right edge at different times are shown in Fig. 6 and Fig.7. The dots lines in the Fig.6 (a) and Fig.7 (a) represent the contact edge between the FAB and bond pad.

At the impact stage, the von Mises stress is increased with time and the maximum stress occurs close to the contact edge. The maximum stresses 211MPa and 182MPa in the ITO and Ti layer respectively occur at the end of the impact stage. While at the ultrasonic vibration stage, the von Mises stress appears periodically. The level of von Mises stresses is much higher than those in the impact stage. It can be concluded that

the defects such as peeling and cracks may be more possibly induced at the ultrasonic vibration stage during wire bonding process.

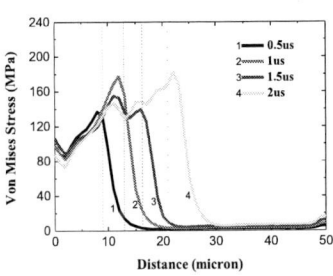

(a) Impact Stage

(b) Ultrasonic Vibration Stage

Fig. 6. The von Mises stress along the middle of ITO layer from center to right edge at different times in p-type electrode structure.

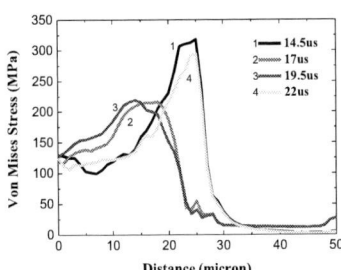

(a) Impact Stage

(b) Ultrasonic Vibration Stage

Fig. 7. The von Mises stress along the middle of Ti layer from center to right edge at different times in n-type electrode structure.

2. Effects of Amplitude of Ultrasonic Vibration

The effects of the ultrasonic vibration amplitude on the stress distribution in bond pad and ohmic contact layers during the wire bonding process were studied in this section. The results are listed in Fig.8 to Fig.11.

Fig.8 and Fig.10 show that the stresses in the bond pad increase greatly when the amplitude vibration changes from 0.5um to 1um. In the p-type electrode structure, the maximum von Mises stress increases from 185MPa to 266MPa and shear stress increases from 94MPa to 138MPa. While in the n-type electrode structure, the maximum von Mises stress increases from 198MPa to 258MPa and shear stress increases from 104MPa to 136MPa.

Fig. 9 and Fig.11 show that the stresses on the ohmic contact layer also increase with the greater ultrasonic wave amplitude. In the p-type electrode, the maximum von Mises stress increase from 283MPa to 324MPa and shear stress increase from 149MPa to 174MPa. While in the n-type electrode, the maximum von Mises stress increase from 283MPa to 294MPa and shear stress increase from 146MPa to 161MPa.

The above results disclose that the amplitude of ultrasonic wave has significant influence on the stress in bond pad and ohmic contact layers under bond pad. It is essential to control the ultrasonic wave energy input for avoiding the defects which may be induced by too high energy input during wire bonding process.

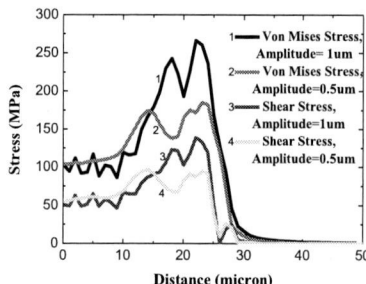

Fig.8 Distribution of von Mises and shear stresses in the bond pad from center to right edge with different amplitudes at the end of ultrasonic vibration stage in the p-type electrode structure.

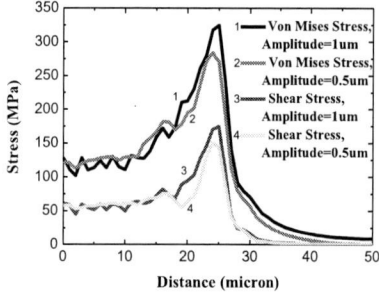

Fig.9. Distribution of von Mises and shear stresses in the ITO layer from center to right edge with different amplitudes at the end of ultrasonic vibration stage in the p-type electrode structure.

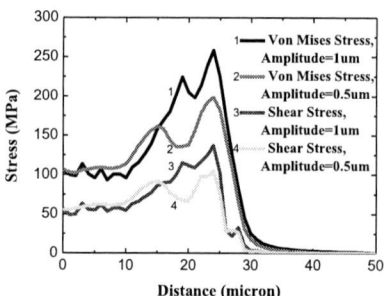

Fig.10. Distribution of von Mises and shear stresses in the bond pad from center to right edge with different amplitudes at the end of ultrasonic vibration stage in the n-type electrode structure.

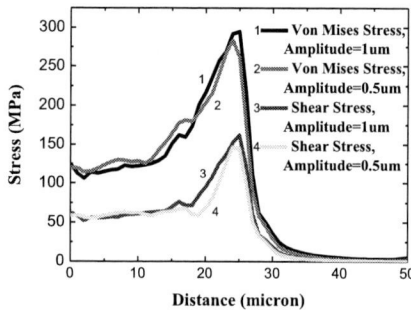

Fig.11. Distribution of von Mises and shear stresses in the Ti layer in the n-type electrode structure with different amplitudes at the end of ultrasonic vibration stage.

3. Effects of Friction between FAB and Bond Pad

Friction at interface between FAB and bond pad is a complicated multi-physics process, bonding occurs when the enough energy is available to overcome the active energy of barrier and surface oxidation [5, 6]. In this study, the effects of the friction coefficient on the stress distribution were studied by using two different friction coefficients 0.38 and 0.8 in the numerical analysis. The modeling results are listed in Fig.10 to Fig.13.

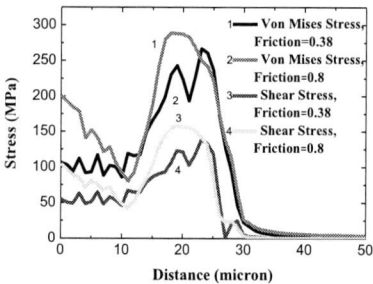

Fig.12. Distribution of von Mises and shear stresses in the bond pad in the p-type electrode structure with different friction coefficient at the end of ultrasonic vibration stage.

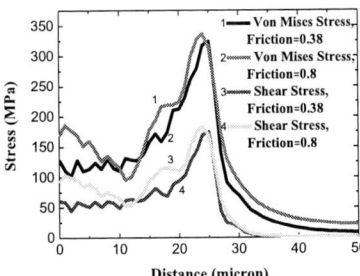

Fig.13. Distribution of von Mises and shear stresses in the ITO layer in the p-type electrode structure with different friction coefficient at the end of ultrasonic vibration stage.

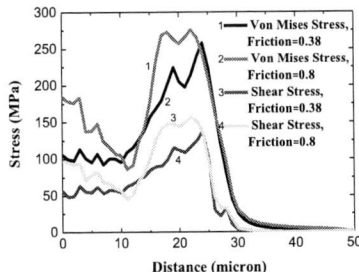

Fig.14. Distribution of von Mises and shear stresses in the bond pad in the n-type electrode structure with different friction coefficient at the end of ultrasonic vibration stage.

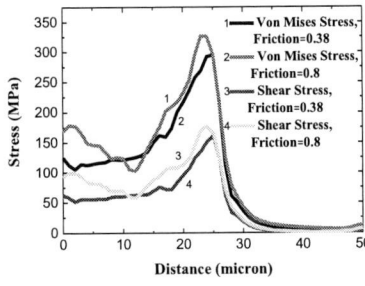

Fig.15. Distribution of von Mises and shear stresses on the Ti layer in the n-type electrode structure with different friction coefficient at the end of ultrasonic vibration stage.

From Fig.12 and 14, we can see that in the p-type electrode structure, the maximum von Mises stress in the bond pad of high friction case is 288MPa, greater than the value of lower friction case which is 266MPa. The maximum shear stresses in the bond pad of high and lower frication cases are 138MPa and 157MPa respectively. In the n-type electrode structure, the maximum von Mises stress in the bond pad of high friction case is 276Mpa, while in low friction case the value is 258MPa. The maximum shear stresses in the bond pad of high and lower frication cases are 136MPa and 151MPa respectively. In both type electrode structures, the high stress area on the bond pad of high friction case is larger than that of the low friction case.

Higher friction makes greater stress and lager area of higher stress level on the bond pad which may be good for wire bonding process. However, this will result in higher stresses in the ohmic contact layer under bond pad and cracks and delaminations may be induced there. From Fig.13 and Fig.15, we can see that the maximum von Mises stress and shear stress in the ITO layer increase from 324MPa to 336MPa and from 174MPa to 183MPa respectively. While in the Ti layer the maximum von Mises stress and shear stress increase from 294MPa to 325MPa and from 161MPa to 176MPa respectively.

4. Effects of Bond Pad Incline

The incline of the bond pad due to the surface mounted process of LED chip. The schematic diagram of the incline of bond pad and LED chip is shown in Fig.16. The incline angle of LED chip and the bond pad is set to be 1° in this analysis. The effects of the incline of bond pad on the stress distribution in the different metal layers were studied. The maximum von Mises stresses in different layers of both types electrode structures at both impact and ultrasonic stages of wire bonding are listed in the Table 1 and Table 2. The effects of bond pad incline are dramatic at the ultrasonic vibration stage, while at impact stage the influences of bond pad incline is less significance. At the end of ultrasonic vibration stage, the maximum von Mises stresses at right part of different metal layers of incline case increase compared to the flat case, while the values at left part of metal layer decrease. The incline of bond pad induces more rigorous condition for the metal layer under bond pad during wire bonding process. So in order to release the stress concentration in the ohmic contact layer, the incline of pad must be avoided through using more suitable process parameters in the LED chip surface mounted process.

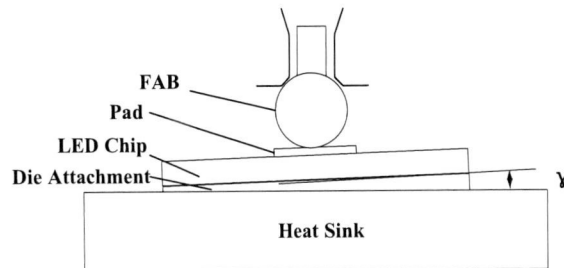

Fig.16. The schematic diagram of LED chip incline.

Table 2 The maximum von Mises stress on p-type electrode structure (MPa).

Time	Location	Flat		Incline (1°)		Change	
		L	R	L	R	L	R
2us	Pad	110	110	110	110	0	0
	Ni	232	227	247	238	+15	+11
	ITO	222	220	226	236	+4	+16
22us	Pad	288	286	237	411	-51	+125
	Ni	387	426	284	532	-103	+106
	ITO	266	344	279	396	+13	+52

Table 3 The maximum von Mises stress on n-type electrode structure (MPa)

Time	Location	Flat		Incline (1°)		Change	
		L	R	L	R	L	R
2us	Pad	111	110	110	111	-1	+1
	Ni	280	289	277	298	-3	+17
	Al	198	200	201	225	+3	+25
	Ti	188	191	188	209	0	+18
22us	Pad	297	291	244	335	-53	+44
	Ni	347	360	336	490	-11	+130
	Al	227	305	219	360	-8	+55
	Ti	220	310	215	358	-5	+48

Conclusions

Wire bonding process which is simplified to consist of impact and ultrasonic vibration stages on the p-type and n-type electrode structure of light emitting diodes has been studied by using a nonlinear finite element method in this paper. Summary of modeling results are listed as follows:

(1) The von Mises stresses in the ohmic contact layer increase with time and the maximum stress occurs close to the contact edge during the impact stage, while it appears periodically during the ultrasonic vibration stage of wire bonding process;

(2) The ultrasonic energy input during the wire bonding process of LED package should be controlled carefully since the ultrasonic amplitude has significant effects on the stress level in the ohmic contact layers under bond pad;

(3) Higher friction is good for the wire bonding process to form welding connection between FAB and bond pad because it makes greater stress and lager area of higher stress level on the bond pad. But it will also result in higher stress in the ohmic contact layer under bond pad;

(4) The incline of bond pad which is induced by wrong handing during the surface mounted process of LED chip do harm to the wire bonding process. The incline of bond pad induces much more stress concentration in the incline up side of ohmic contact layer; and

(5) This modeling work may provide guidelines for the parameters optimization of wire bonding process in the high power lighting emitting diodes package.

Acknowledgments

The authors wish to appreciate the support from NSFC Key Project with Contract Number 50835005..

References

1. Jin-Woo Park, Young-Bok Yoon, Sang-Hyun Shin, et al., "Joint structure in high brightness light emitting diode (HB LED) packages", *Materials Science and Engineering A 441* (2006), pp.357-361.

2. Jianjun Wang, Minfu Lu, Daqing Zou and Sheng Liu, "Investigation of interfacial fracture behavior of a flip-chip package under a constant concentrated load", *IEEE Transactions Electronics Packaging Manufacture Technology*, Vol. 21, No.1, (1998), pp79-87.

3. Jianjun Wang, Minfu Lu, Wei Ren, Daqing Zou, and Sheng Liu, "A study of the mixed-mode interfacial fracture toughness of adhesive joints using a multiaxial fatigue tester", *IEEE Transactions Electronics Packaging Manufacture Technology*,Vol. 22, No. 2, (1999), pp166-184.

4. Ikeda,T., Miyazaki, N., Kudo, K. et al, " Failure estimation of semiconductor chip during wire bonding process," *ASME J. of Electronic Packaging*, Vol.121, (1999), pp85-91

5. Yong Liu, Scott Irving, Timwah Luk, "Thermosonic wire bonding process simulation and bond pad over active stress analysis", *Proceedings of 54th Electronic Components and Technology Conference*, 2004, pp.383-391.

6. Yong Liu, Howard Allen, Timwah Luk and Scott Irving, "Simulation and experimental analysis for a ball stitch on bump wire bonding process above a laminate substrate", *Proceedings of 56th Electronic Components and Technology Conference*, 2006, pp.1918-1923.

7. Chang-Lin Yeh and Yi-Shao Lai, "Comprehensive dynamic analysis of wirebonding on cu/low-k wafers", *IEEE Transactions on Advanced Package* ,Vol. 29 No.2, (2006), pp. 264-270.

8. Chang-Lin Yeh, Yi-Shao Lai, "Transient simulation of wire pull test on cu/low-k wafers", *IEEE Transactions on Advanced Package* Vol. 29 No.3, (2006), pp631-638.

9. E. Medvedovski , N. Alvarez, O. Yankov, M.K. Olsson, "Advanced indium-tin oxide ceramics for sputtering targets", *Ceramics International*, 2008, pp1173-1182.

Advanced Package Design for Electronic and MEMS Applications Supported by FE Analyses and Deformation Measurements

Sommer, J.-P.[1], Michel, B.[1], Kugler, A.[2], Rank, H.[2]

[1]Fraunhofer Research Institution for Electronic Nano Systems (ENAS), Chemnitz

Micro Materials Center Berlin/Chemnitz

D-09117 Chemnitz, Germany

Otto-Schmerbach-Str. 19

Email: johann-peter.sommer@enas.fraunhofer.de

[2]Robert Bosch GmbH, Stuttgart

Abstract

Sensor packages have often to be adapted to current customer requirements. Dimensions, interior structures, or materials may change. In general, this influences strongly the thermally induced warpage during the wafer level manufacturing process and of the separated single packages which has to be controlled with respect to reliability issues. Therefore, the package manufacturer needs for a proved methodology: Numerical studies by means of finite element (FE) analyses combined with experimental warpage determination are proposed which enable reliability assessments before real parts are available.

Parametric studies by means of FEM are applied to the manufacturing process and its variations, taking into account carefully determined temperature-dependent material data, including cure shrinkage and visco-elastic behaviour of mold compounds. Several modelling strategies (batch module, one single module only, and an intermediate approach) are compared with respect to the computed bending for a typical set of parameters and with the real warpage of a thermally loaded test structure, measured by means of Thermo-Moiré.

Reliability and Design Process

One of the most important targets during the development of new advanced electronic products is to come to market as soon and economical as possible, yet with innovative properties and high reliability. This requires a reduced number of redesign loops and prototyping and can be promoted by FE analyses combined with experiments at suitable test structures.

For this purpose, a sufficiently short response time to any design idea and an evaluation after each FE run is necessary. Loading conditions from manufacturing, testing, storing, and operating should be taken into account. Prototypes are not required up to that time.

In addition, the deformation behaviour of test structures can be measured by means of microDAC, Thermo-Moiré, or a chromatic sensor in order to investigate the manufacturing steps or to evaluate the influence of unavoidable simplifications in the FE models.

This helps to make the right decision very early in the development process and to save time consuming and expensive meanders.

Advanced Packaging Concepts

Economical manufacturing often means smaller and denser packaging. Two of the most promising concepts are embedding chips into an organic board and the so-called system-in-package (SiP) concept.

The technology of chip embedding was developed within the European project Hiding Dies [1] and the German one called KRAFAS [2]. Within KRAFAS, a new generation of an adaptive cruise control (ADC) unit was developed. RF circuits in the 77 GHz band were built in a PCB which resulted in a smaller, 30% less expensive, less energy consuming, and highly reliable device.

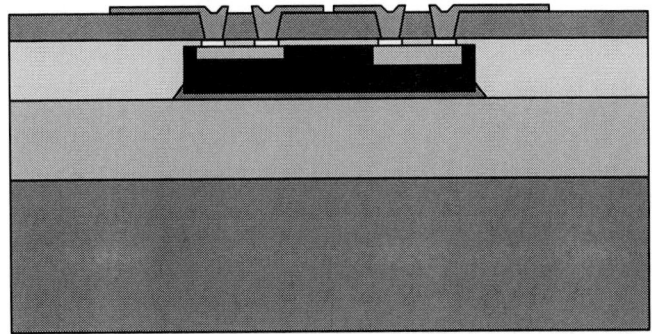

Fig. 1 Interconnection principle of chip embedding into sub-module and into device.

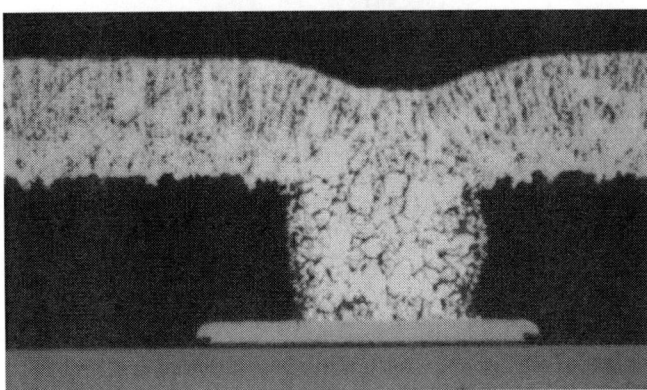

Fig. 2 Cross section of a via between the die and the Cu layer above.

The Si dies were molded directly into so-called submodules which were embedded later into the final systems, taking into account special demands for RF-suitable dielectric layers and interconnect (Cu vias). The latter were realised by laser drilling and galvanic metallisation. Fig. 1 points out the principle, and Fig. 2 represents an intersection through a typical via. The grain structure is very well visible. The detailed process flow for chip embedding, and a discussion of technical issues have been described in [3].

In many cases, a SiP consists of a sensor chip and a signal processing ASIC. In order to make mass production of SiP cheaper, chip area is reduced if possible, and batch module concepts are applied. Doing this, the Bosch company as an important manufacturer of such components presses ahead the project SmartSense [4]. Many components are molded simultaneously in an array and separated later by sawing. Unfortunately, warpage cannot avoided completely due to the different thermal expansion properties of the die and mold compound (MC). Therefore, some important targets of the project are to determine the temperature-dependent thermo-mechanical behaviour of MC, to measure the deformation of the mold modules after curing and during thermal cycling (TC), and to obtain agreement of precalculated and measured warpages.

Submodules in Cu-Si-MC Technology

In order to ensure precise positioning of the RF chips in KRAFAS, a submodule technology was developed combining Cu, Si, and MC, even though it was possible later to do it without the Cu leadframe. Supporting the package design, FE analyses and deformation measurements were combined.

The submodules were made as a 2 by 8 mold array and separated later. The different thermal expansion properties of the applied materials cause unavoidable warpage which was first studied numerically. Symmetry allowed for modelling a quarter, Fig. 3.

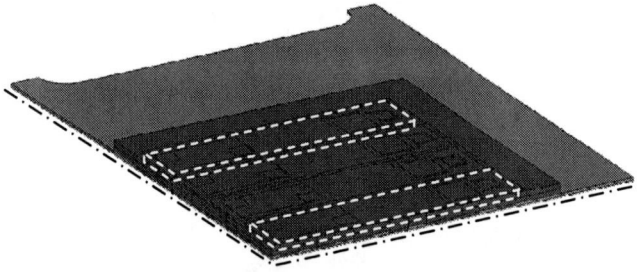

Fig. 3   Quarter of mold module. Innermost and outermost submodules marked.

Besides geometry, boundary conditions, and load description, an FE analysis requires a complete set of material parameters. Si and Cu can be regarded as linear-elastic, and MC is a visco-elastic material. Its mechanical properties have to be determined carefully, temperature- as well as time-dependent, see next chapter.

Cooling down from a stress-free assumed temperature (close to mold temperature) to room temperature (RT) is the first step of the calculation, Fig. 4, and separating the submodules the second one, Fig. 5, which consists of removing the finite elements in the saw trench. Rigid body motions must be suppressed in order to avoid numerical problems. Scaling is uniform in Fig. 4 and Fig. 5, respectively.

Warpage is reduced after sawing due to less constraint at the separated parts. Transforming the computed out-of-plane-displacements along the arrows in Fig. 5 to zero at the ends allows to compare them with corresponding warpage

measurements which have been carried out at RT applying a chromatic sensor [5].

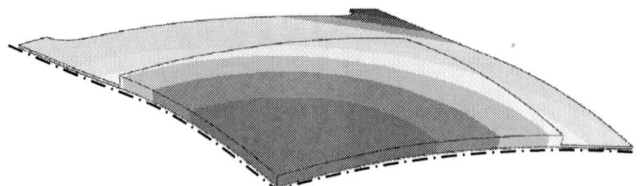

Fig. 4   Quarter of mold module after cooling to RT, deformation 10x.

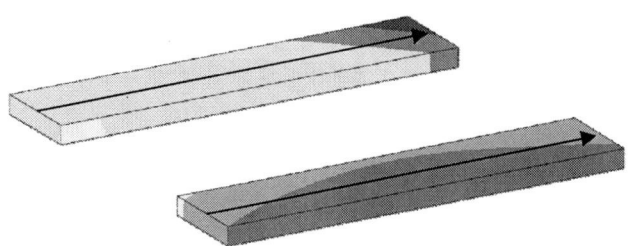

Fig. 5   Separated submodules: out-of-plane displacement.

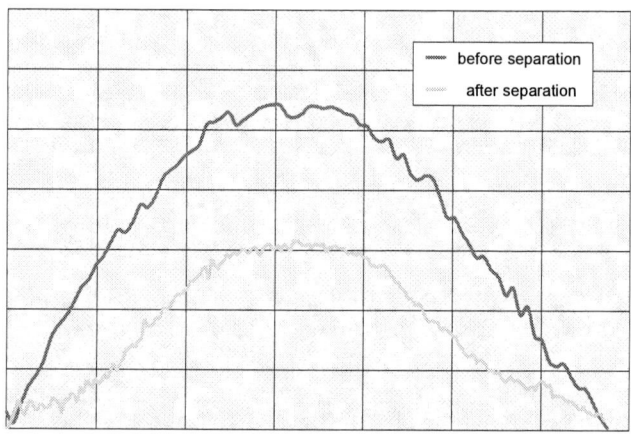

Fig. 6: Measured warpage of submodules

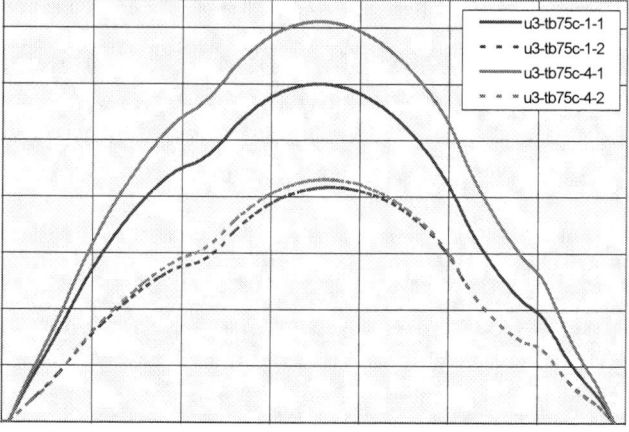

Fig. 7   Computed deformation of submodules. Solid curves before, dashed ones after separation. Red lines belong to inner parts, blue curves to outermost ones.
Fig. 6 and Fig. 7 are scaled uniformly.

No satisfying agreement could be achieved at the beginning, in spite of precise MC characterisation. But taking into account additional shrinkage after curing and adapting the coefficient of thermal expansion (CTE) to the experiment, the numerical and measured results were consistent, Fig. 6 and Fig. 7, [6, 7].

Within the applied FE tool ABAQUS, material parameters may be changed during one and the same calculation: It is possible to apply higher CTE values for the first cooling after mold curing, and to replace them later by that measured at completely cured specimens. This way, manufacturing steps and thermal cycling can be analysed in a single computation.

Constitutive Parameters

Reliable material data are necessary for reproducible numerical results. The temperature dependent material properties of pure elastic materials like copper or silicon in general can be derived from data sheets within an acceptable accuracy. In the case of visco-elastic mold materials, data sheets usually do not provide constitutive information in a manner well-adapted to the FE requirements [8].

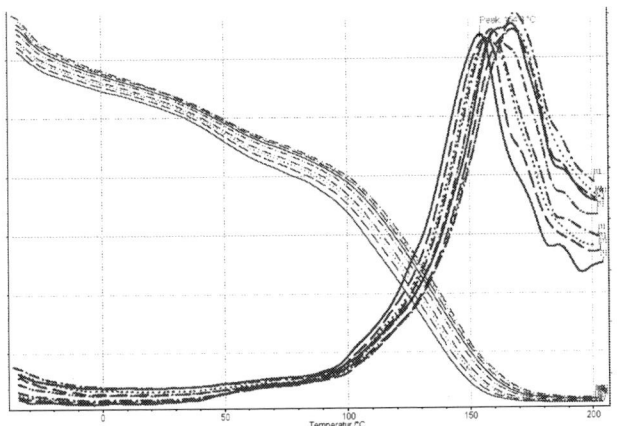

Fig. 8 Storage modulus (dropping with temperature) and tan δ, depending on frequency of loading.

Within the projects mentioned above, the coefficients of thermal expansion (CTE) were measured temperature-dependently by Thermo-Mechanical Analysis (TMA). For the visco-elastic characterisation of MC, Dynamic Mechanical Analysis (DMA) was used, a multi-frequency analysis with sinusoidal harmonic excitation. It provides the storage and loss moduli as well as tan δ, dependent on temperature. Typical frequency- and temperature-dependent curves are shown in Fig. 8. The modulus values were transferred into the time domain and fitted to a master curve via the well known WLF-time-shift function approach.

In case of the MC EME-G790S, this resulted in

$$\log a_T = c_1 (T-T_0) / (c_2 + (T-T_0)),$$
$$c_1 = 393, \quad c_2 = 3.3e-15 \text{ K}, \quad T_0 = 3.9e-16 \text{ K},$$

and the relaxation curves outlined in Fig. 9.

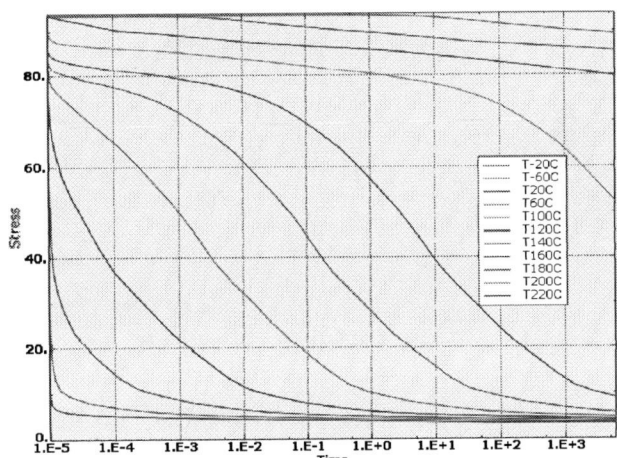

Fig. 9 Temperature-dependent Relaxation of EME-G790S.

Deformation Analysis for Systems-in-Package

Profitable mass production of SiP demands for taking into account new components, advanced packaging materials, and interconnection concepts as soon as possible. Besides the electrical functionality, the thermo-mechanical interaction of the components is often changed which influences directly reliability aspects. This can be investigated efficiently by means of FEA.

The R. Bosch Company is strongly engaged in the current SmartSense project. One of the tasks is the development of a generalised batch module concept for SiP applications. This includes controlling the unavoidable warpage from manufacturing process.

Fig. 10   Warpage of 1/4 molded batch module after curing.

Even if symmetry can be exploited, large FE models will be obtained because of many repeating structural details have to be taken into account, Fig. 10. Due to the visco-elastic properties of MC, superelements cannot be applied, and small computational increments are necessary.

How to reduce computational effort essentially without remarkable loss of accuracy?

As a matter of principle, a SiP is assumed to consist of a sensor chip and an ASIC, encapsulated with MC, and single symmetry. The interconnects are neglected here, Fig. 11.

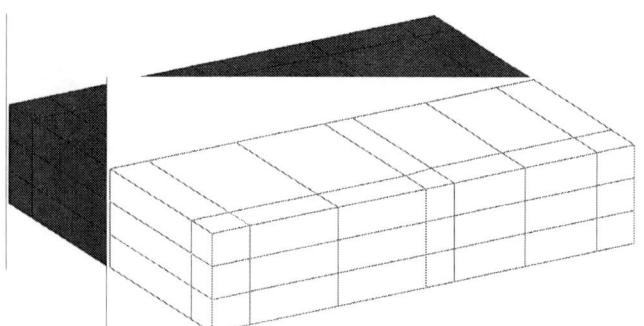

Fig. 11 SiP with ASIC (left) and sensor (right).

Three side faces (left, back, and right) correspond to the saw lines which are assumed to be plane but allowed to be inclined during thermal deformation. For the left surface, this can be realised numerically by forcing the x-displacement of each node P dependent on the nodes P_1 and P_2 as follows:

$$ux = (z_2-z)/(z_2-z_1) \, ux_1 + (z-z_1)/(z_2-z_1) \, ux_2$$

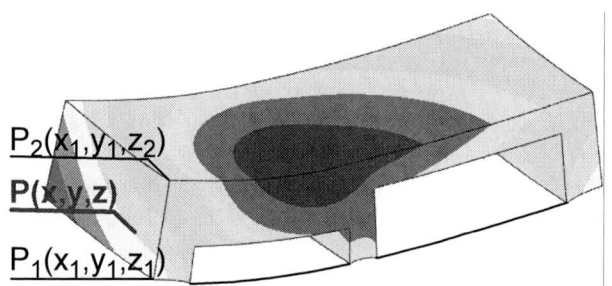

Fig. 12 Deformed SiP (100x, ASIC and sensor removed).

For the right and back surfaces corresponding constraints were applied. Within the FE system ABAQUS, this kind of multi-point-constraint (MPC) needs for a USER-routine, because of the MPC coefficients depend on the current nodal coordinates.

This approach reflects the boundary conditions at a centered SiP. It results in an essentially simpler FE model and enables a much denser discretisation. In order to check its permissibility, the model showed in Fig. 11 was extended by the neighbouring packages, Fig. 13.

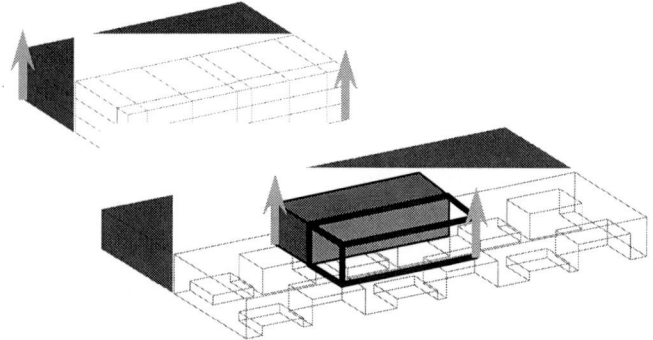

Fig. 13 Single SiP model and extended one.

The angle between the red arrows is equivalent to the bending radius of the component and should correspond to the bending radius at the centre of the mold module, Fig. 14. In order to confirm this, the computed bending radius has been compared with the diagonally measured curvature lines at real devices. As can be seen from

Table 1, an accuracy in the range of 4 % could achieved for the test setup with MC EME G 790 S. This confirms the validity of the proposed methodology which was applied to parametric studies, including chip and package dimensions.

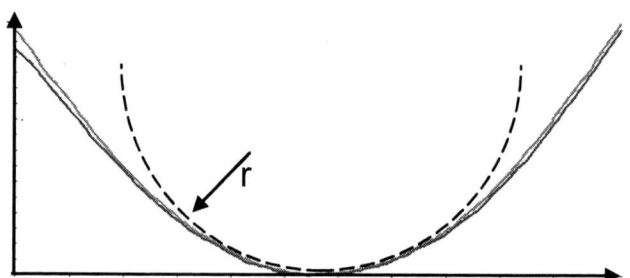

Fig. 14 Measured SiP warpage after curing and inscribed circle.

Table 1 Bending radii of SiP determined with different methods.

Model/Experiment	Bending Radius [mm]
FEA single SiP	960
FEA extended, 6 SiP	920
Themo-Moiré measurement	967

Doing this, a parametric ABAQUS-CAE model was set up, including the framework for the constraint formulations. The resulting input data sets had to be completed with the non-linear material data set, the USER MPC, and output definition. The resulting corner node displacements and the SiP dimensions defined the red vectors in Fig. 13, which were converted within an EXCEL file to the bending radii and allowed to represent the trends graphically we looked for.

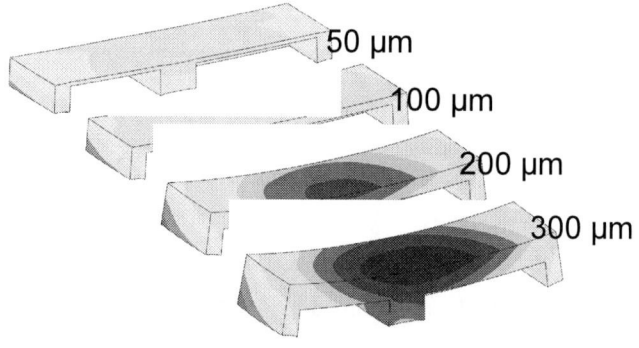

Fig. 15 SiP deformation after cooling from 180 °C to RT.

As an example, the mold coverage above the chips was varied. Best flat results were obtained for thin packages, but thermo-mechanical reliability is not only influenced by flatness, Fig. 15 and Fig. 16.

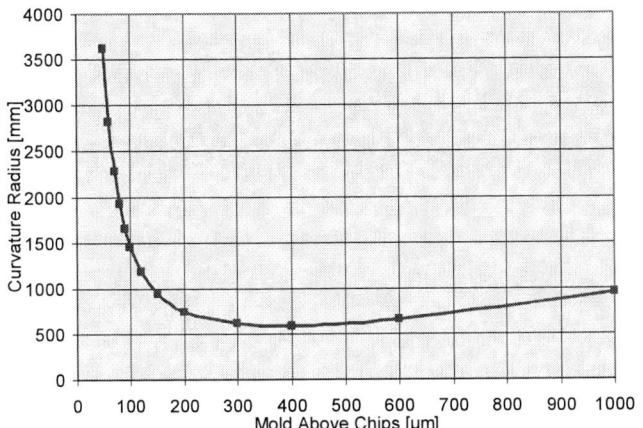

Fig. 16   Curvature of centered SiP packages vs. MC above chips.

Current work is concerned with the quantification of cure shrinkage for advanced MC, that means with the evaluation of differences in warpage between first cooling after curing and the behaviour during TC. Furthermore, the applicability of the optimisation tool OPTIMUS will be checked. This is not trivial because of the complicate interaction between ABAQUS-CAE, an EDITOR, ABAQUS-STANDARD, and EXCEL.

Summary – Conclusions

The FE analysis was confirmed to be a numerical tool well qualified to analyse and optimise new advanced electronic and sensoric products before any real parts are available.

In case of packages with organic encapsulation materials like mold compounds, special care has to be taken to the material data, usually determined for completely and dry MC. In order to obtain reproducible numerical results, shrinkage after curing has to be taken into account. One way to meet this requirement is to adapt the CTE for the first cooling after curing to the real deformation behaviour of suitable chosen test specimens or of first available components. Recommendable experimental methods are Thermo-Moiré as well as the application of a chromatic sensor. This way, the methodology outlined above combines the advantages of numerical and experimental analyses, minimises the risk of a one-side-interpretation, and leads to a better understanding of the deformation behaviour of the devices under construction which is an important aspect of thermo-mechanical reliability. It can be generalised and is free to be adapted to the development of new advanced technologies for electronic and sensoric products.

Acknowledgements

The authors would like to thank the European Commission for the financial support for the HidingDies project (IST-2002-507759). KRAFAS and SmartSense are projects funded by the Federal Ministry of Education and Research of Germany (BMBF, FKZ 16SV2178 and 16SV3675, respectively). Funding has been gratefully appreciated.

References

1. HIDING DIES website: http://www.hidingdies.net
2. KRAFAS basic information at website http://www.mstonline.de/foerderung/projektliste
3. Ostmann, A., De Baets, J., Kriechbaum, A., Kostner, H., Neumann, A.: Technology for Embedding Active Dies. European Microelectronics Conference 2005, Brugge, Belgium, June 12 - 15, 2005.
4. SmartSense website: http://www.mstonline.de/foerderung/projektliste/printable_pdf?vb_nr=V3022
5. Sommer, J.-P., Michel, B., Noack, E., Seiler, B.: Sensor Design Support for Automotive Applications by Means of FE Analyses and Micro Deformation Measurements. Proc. 7th International IEEE Conference on Polymers and Adhesives in Microelectronics and Photonics, Garmisch-Partenkirchen, August 17-20, 2008.
6. Sommer, J.-P., Michel, B., Noack, E., Seiler, B., Uhlig, P.: Finite Element Simulation and Micro Deformation Measurements - Contributions to the Development of Advanced Packages with Hidden Dies. 3rd Smart Systems Integration (SSI), Brussels, Belgium, March 10./11., 2009.
7. Sommer, J.-P., Michel, B., Noack, E., Seiler, B.: Thermo-Mechanical Pre-Optimisation of Radar Sensor Design by Means of FEA and microDAC Measurements. Proc. 2nd Electronics System-Integration Technology Conference, Greenwich, London, UK, September 1-4, 2008, 359-363.
8. Dudek, R., Walter, H., Auersperg, J., Michel, B.: Numerical Analysis for Thermo-Mechanical Reliability of Polymers in Electronic Packaging", Proc. of Polytronics 2007, Jan. 2007, Tokyo, Japan, 220-228.

A Novel Wafer Level Package Strategy for RF MEMS

Zheng Wang, Zewen Liu*
Institute of Microelectronics, Tsinghua University
Institute of Microelectronics, Tsinghua University, Beijing, 100084, China
*Email: liuzw@mail.tsinghua.edu.cn

Abstract

A novel wafer level package strategy for RF MEMS is presented in this paper, considering the important issues such as the seal material, RF feedthrough and interconnects method. BCB is chosen for the seal material due to its excellent RF performance. Analytical solutions are presented and applied for RF feedthrough design and optimization. Side-Wall CPW interconnect technology, which is a novel interconnect concept, is introduced in RF MEMS packaging instead of conventional Via-Through Method. The RF performance of the overall package is verified by 3D EM simulation software HFSS. The packaging structure keeps good RF performance up to 40GHz.

1. INTRODUCTION

With the rapid development of MEMS technology, the prominent RF performance of RF MEMS devices such as varactors [1], switches [2], and resonators has been attained. Considering that the movable friable parts of RF MEMS devices must be protected and packaged in a clean and invariable environment. These packages are required to exhibit minimum insertion loss as well as excellent match and should be manufacturable at a reasonable cost since packaging is generally the most costly step in the fabrication process. High-performance wafer-level packaging has been demonstrated as an almost best choice according to many research groups [3][4][5].

This paper presents a novel wafer level package strategy for RF MEMS with very low loss up to 40GHz. Some important issues such as the seal material, RF feedthrough and interconnect method are discussed in the packaging design. The structure of the packaging is shown in Fig.1.

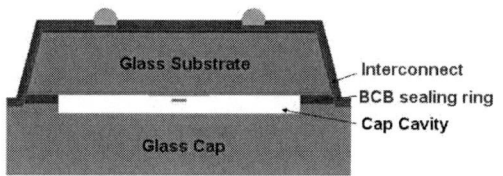

Fig.1 The structure of wafer level package strategy for RF MEMS

2. Packaging Design

In the Wafer Level Package strategy, there are some important issues we have to confront with. Seal material, RF feedthrough and interconnect method will be discussed as follows:

A) Seal material

BCB is the seal material in this WLP strategy. Compared to gold-to-gold thermo-compression bonding or glass frit bonding, using BCB for the bonding and the sealing material defines a relatively simple process and lower temperature which is rather significant to the reliability of RF MEMS.

In particular the liquid-like behavior of BCB observed during curing makes the sealing of cavities with protruding signal feedthrough rather straightforward. Moreover, BCB displays minimal outgasing, low moisture uptake, high chemical resistance, high bond strength, low processing temperature (<250°C), and low residual stress levels. Furthermore, its high resistivity (10^{19} Ωcm), low loss tangent (0.0008-0.002 in the range 1MHz-10GHz) and low permittivity (2.65) make BCB a very good candidate for high frequency (RF-MEMS) applications.[6]

B) RF Feedthrough—CPW Optimization

The coplanar waveguide (CPW) line, as the most common RF feedthrough method, is easy to design as 50Ω (18um/120um/18um=$G_0/S_0/G_0$). However, the performance of CPW under the cap wafer will detune due to the close proximity of the cap to the MEMS surface. Besides, the CPW line under BCB ring will also detune as the characteristic impedance will decrease. These challenges can be solved using different solutions as shown in Fig.2.

For CPW under the cap wafer, providing an etched cavity in the cap wafer will diminish the detuning effect without modify the dimensions of CPW ($G_2/S_2/G_2=G_0/S_0/G_0$). The quantitative method for detuning effect has been derived by applying the conformal mapping method based on the extension of the partial capacitance technique. A paper discussing the details of this approach have been submitted for publication elsewhere [7]. The equivalent relative permittivity can be expressed as

$$\varepsilon_{eff} = 1 + q_1(\varepsilon_{r1}-1) + q_2(\varepsilon_{r2}-\varepsilon_{r3}) + q_3(\varepsilon_{r3}-1) \quad (1)$$

where ε_{r1}, ε_{r2} and ε_{r3} are relative permittivities of substrate, air and cap wafer respectively and

$$q_1 = \frac{1}{2}\frac{K(k_1)}{K(k_1')}\frac{K(k_0')}{K(k_0)} \quad (2)$$

$$q_2 = \frac{1}{2}\frac{K(k_2)}{K(k_2')}\frac{K(k_0')}{K(k_0)} \quad (3)$$

$$q_3 = \frac{1}{2}\frac{K(k_3)}{K(k_3')}\frac{K(k_0')}{K(k_0)} \quad (4)$$

Then we can obtain the characteristic impedance of the packaged CPW:

$$Z_0 = \frac{1}{cC_{air}\sqrt{\varepsilon_{eff}}} = \frac{c\mu_0}{4\sqrt{\varepsilon_{eff}}}\frac{K(k_0')}{K(k_0)} \quad (5)$$

The detuning effect of CPW with cap wafer is realized by comparing the characteristic impedance of packaged CPW and that of unpackaged CPW using formulas in [7]. The cavity depth of 40μm is adequate to reduce the detuning of the

MEMS circuit to an acceptable level for the 50Ω CPW on the glass substrate for the glass cap wafer [7].

For CPW under the BCB ring, the dimension of CPW needs to be redesigned to achieve 50Ω again. The formulas in [7] can be used for this optimization too. For BCB with 10um thickness, the dimensions of CPW would be redesigned as 30um/96un/30um ($G_1/S_1/G_1$).

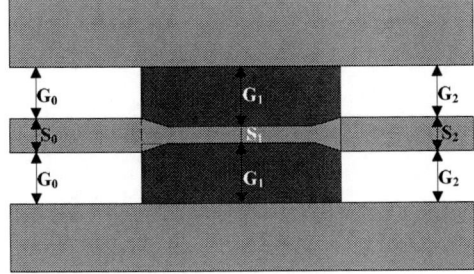

Fig. 2 Schematic top view and cross-section of RF Feedthrough

C) Interconnect Method

A novel interconnect concept for RF MEMS is presented in this paper as shown in Fig.3. This work introduces side-wall CPW interconnect technology without any via compared with the Via-Through Method [8]. So the protruding line outside the BCB ring will be connected with the CPW on the side-wall which can be leaded to the solder ball.

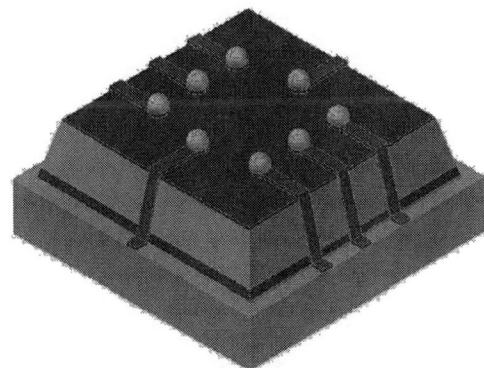

Fig. 3 the schematic view of side-wall interconnect method

3. Overall EM Simulation

To evaluate the RF effect of package, a 2mm long 50Ω CPW line without RF MEMS device is simulated only for convenience. The simulation structure of the CPW line without and with package is shown in Fig.4 (a) and Fig.4 (b).

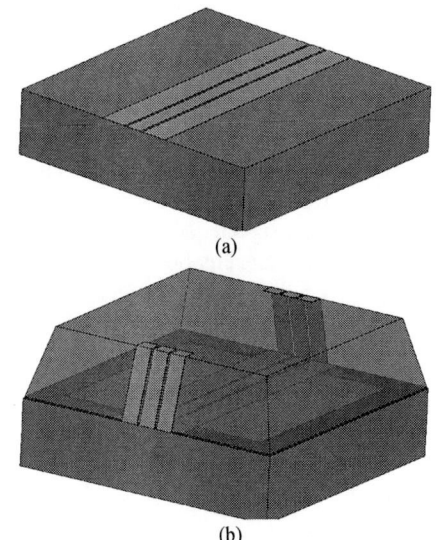

Fig.4 HFSS simulation structure without (a) and with (b) package

The overall EM simulation is accomplished by HFSS. The simulated S-parameters of the CPW line without and with package are presented in Fig.5 (a) and Fig.5 (b), respectively. The results clearly demonstrate the RF performance of the package is good. The extra insertion loss cost due to the package is less than 0.15dB according to the simulation results. Intriguingly, the return loss of the CPW with package drops at round 30GHz, which may be a certain harmonic point owing to the complex packaging structure. It opens an interesting method to further improve the RF performance at Ka band with the proposed packaging strategy.

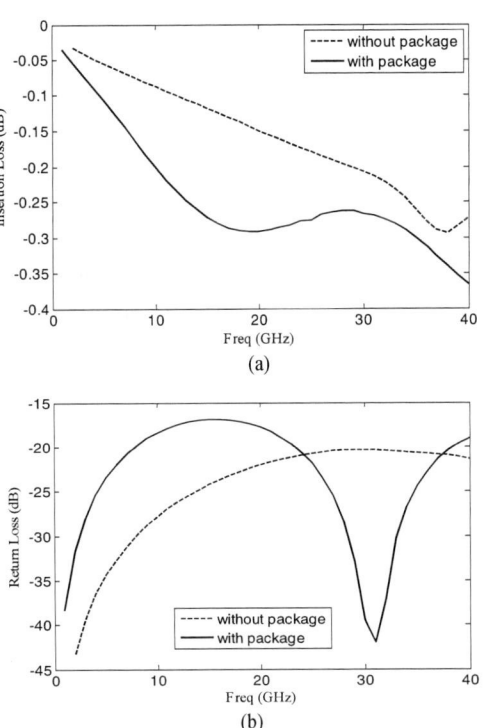

Fig.5 Simulated insertion loss (a) and return loss (b) without and with package

978-1-4244-4658-2/09 $25.00 © 2009 IEEE

4. Conclusions

In this paper a novel wafer level package strategy for RF MEMS is presented. BCB is chosen for the seal material as its outstanding RF performance and good hermeticity. New interconnect concept for RF MEMS is also presented using side-wall interconnects without via. Besides, analytic formulas are derived for CPW optimization under packaging. The cavity depth of 40μm is adequate to reduce the detuning of the MEMS circuit to an acceptable level for the 50Ω CPW on the glass substrate for the glass cap wafer.

Acknowledgments

This work is supported by the NSFC (National Sciences Foundation of China) with contract No. 60576048 and National 973 Project with contract No. 2007CB321504.

References

1. Lakshminarayanan B., Rebeiz, G., "High-Power High-Reliability Sub-Microsecond RF MEMS Switched Capacitors", *IEEE MTT-S International Microwave Symposium Digest*, 2007, pp.1801-1804

2. Hou Z., Liu Z. and Li Z., "Al/Au Composite Membrane Bridge DC-Contact Series RF MEMS Switch", in *Proceedings of the 2008 9th Conference on Solid-State and Integrated Circuit Technology*, 2008, pp. 2488-2491

3. Min B., Entesari K., Rebeiz G. M., "DC-50 GHz low-loss wafer-scale package for RF MEMS", in *Proceedings of the 34th European Microwave Conference*, 2004, v3, pp.1289-1291.

4. Schobel J., Buck T., Reimann M., Ulm M., and Schneider M., "W-band RF-MEMS subsystems for smart antennas in automotive radar sensors," in *Proceedings of European Microwave Conference*, Amsterdam, Netherlands, Oct. 2004, pp. 1305-1308

5. Jourdain A., Vaesen K., Scheer J.M., Weekamp J.W., van Beek J.T.M., and Tilmans H.A.C., "From zero to second level packaging of RF-MEMS devices," in *Proceedings of International Conference on MEMS*, Miami, FL, Jan. 2005, pp. 36-39.

6. Jourdain A., De Moor P., Pamidighantam S., Tilmans H.A.C., "Investigation of the hermeticity of BCB-sealed cavities for housing (RF-)MEMS devices", in *Proceedings of the IEEE Micro Electro Mechanical Systems (MEMS)*, 2002, pp. 677-680.

7. Wang Z., Liu Z., "An Analytical Method for Optimization of RF MEMS Wafer Level Packaging with CPW Detuning Consideration", in *Proceedings of* PIERS 2009 MOSCOW, (accepted).

8. Entesari K., Rebeiz G. M., "A Low-Loss Microstrip Surface-Mount K-Band Package", in *Proceedings of the 1st European Microwave Integrated Circuits Conference, EuMIC 2006,* 2007, pp. 537-540.

Design and Simulation of a Package Solution for Millimeter Wave MEMS Switch

Wencai Yu, Xiaoping Liao*
Key Laboratory of MEMS of Ministry of Education, Southeast University,
Nanjing, 210096, China
Email: xpliao@seu.edu.cn

Abstract

In the RF area MEMS devices are mainly used as switches that utilize mechanical movement to achieve a short or an open circuit in a RF transmission line. RF MEMS switches have shown great potential for the development of a wide range of low loss, low power and low cost systems. It is widely recognized that the advantages offered by the MEMS switch technology cannot be achieved without an appropriate package that will shield and protect the switch, as well as an interconnect technology that will allow for easy RF and DC signal transition in and out of the package. Low loss, wide bandwidth hermetic packages are therefore essential for practical high performance RF MEMS devices. The interconnection to such packaging techniques presents the largest challenge to RF performance. For an interconnection to function largely independent of frequency, the impedance of the interconnection must be maintained throughout the bonding interface. Although it is used very common in packaging ICs for its characteristics such as simplicity and low cost, wire bonding can not fulfil the requirement of RF MEMS packaging because of the defection existing in the package [1].

In this paper, we present the design, simulation and manufacturing of a package solution with lightweight, small size and short electric path length for a millimeter wave MEMS switch structure. To achieve this goal, we used the ultra thin silicon substrate as a packaging substrate. The via holes for vertical feed-through were fabricated on the thin silicon wafer by wet chemical processing. Then, via holes were filled and micro-bumps were fabricated by electroplating. To make up hermetic sealing, metal bonding was used in the sealing line. The switch before and after packaging were simulated by using Ansoft HFSS respectively, and S-parameters simulated show that the package structure has a small impact on the performance of switch, so that we can package the MEMS devices without loss and interference by using the vertical feed-through. Especially, by using the ultra thin silicon wafer we can realize a device package with low-cost, lightweight and small size. In addition, we can extend a 3-D packaging structure by stacking assembled thin packages.

Introduction

MEMS switches have been demonstrated with outstanding RF performance and great potential for the development of many low loss and low cost systems. MEMS technology has a wide range of prospects in wireless communication, and many military systems will be based on it. At present, a series of MEMS devices with high performance such as switches, varactors, and resonators have been developed, but the commercialization process of MEMS devices and systems has severly constrained because of the technical difficulties existing in the process of packaging. So, packaging has become the main factor that affects the cost of MEMS devices and systems in the market. Many researchers and experts in Micro-System industry view packaging as the only urgent need for successful commercialization, and the research on new packaging concept and dedicated design tools is ongoing.

The hermeticity and electrical connection of the packaging will impact greatly on the packaging performance. With the development of integrated circuits, advanced packaging technical develop continuously in order to adapt to the challenges posed by a variety of semiconductor processes and new materials.

The connection between chip and external pin is the critical part of the packaging. Wire bonding has been used very common as the electrical connection method for its simplicity, low cost, and broad applicability. However, wire bonding will significantly impact on the devices when the frequency increased, and circuits or devices are very sensitive to the parasitic effects introduced by wire bonding. So, wire bonding can be applied when the operating frequency is relatively low, and for high-frequency applications, it will impact greatly on the performance of MEMS devices.

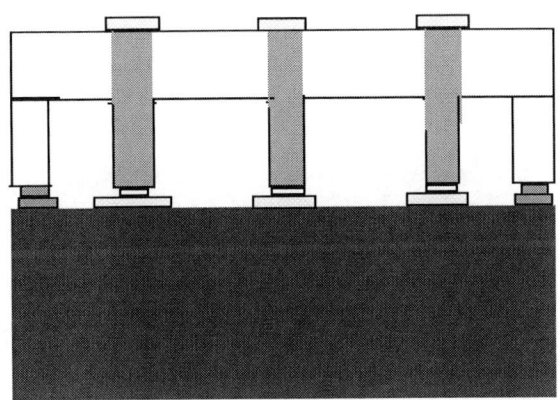

Fig. 1 The structure diagram of the package

In this paper, we designed ultra thin packaging that has thin silicon substrate as packaging substrate and vertical feed-through for low parasitic capacity for a millimeter wave MEMS switch. The package using vertical feed-through dose not need the wire bonding, which occurs the large parasitic capacity, to achieve signal propagation, so this method can achieve low loss and better performance at high frequencies. Also wafer level bumps can be fabricated using electroplating method to increase integration of micro bumps and reduce the cost [2].

Package design and manufacturing process

978-1-4244-4658-2/09 $25.00 © 2009 IEEE

By using vertical feed-through method, signal loss in the packaging interface will be significantly reduced, and the welding method can not be restricted since the signal lines do not have to contact with the packaging interface. Accordingly, the performance of packaging, such as hermeticity, mechanical strength will be significantly improved. In this paper, the connection between signal lines on the different layers is achieved by using vertical feed through method. Fig. 1 shows the structure diagram of the package.

In addition, in order to adapt to the needs of high-frequency, the length of feedings should be as short as possible. On the other hand, a certain distance should be maintained in order to minimize the influence introduced by cap wafer. For the reasons mentioned above, the cap wafer should be made appropriately thin.

Fig. 2 shows the manufacturing process of ultra thin cap wafer and via hole [3]. The process steps are as follows:

(1) Bonding between high-resistivity silicon cap wafer (1000 μm×1000 μm×150 μm) and the bearing silicon wafer.

(2) Deposition of silicon dioxide on the cap wafer, and the thickness is 5000 Å.

(3) Cavity formation. The size of cavity is 800 μm×800 μm.

(4) Deposition of sacrificial layer (5 μm).

(5) Through-hole etching. The radius of through-hole is 50 μm, and length is 155 μm.

(6) Copper plating in the through-hole.

(7) Remove sacrificial layer.

(8) Deposition of welding material.

(9) Deposition of Ti/Ni/Au layer on the switch chip.

(10) Bonding between cap wafer and chip wafer.

(11) Melt the wax layer and remove the bearing wafer.

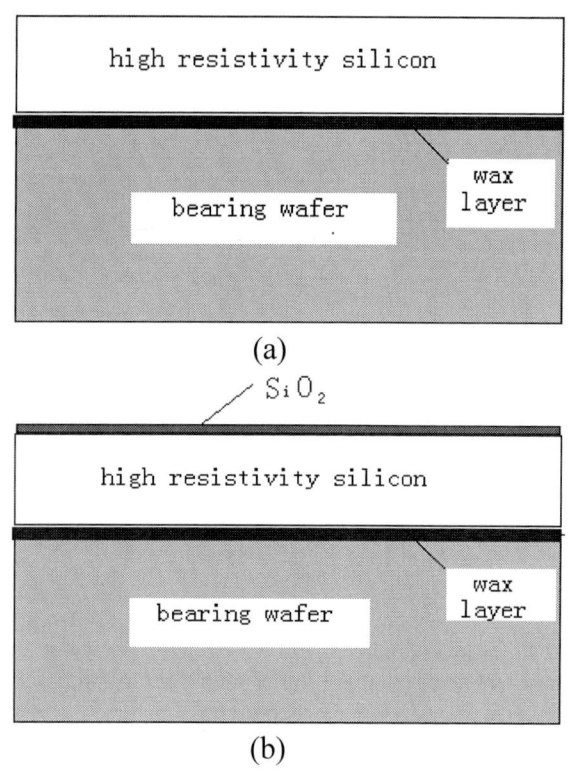

(a)

(b)

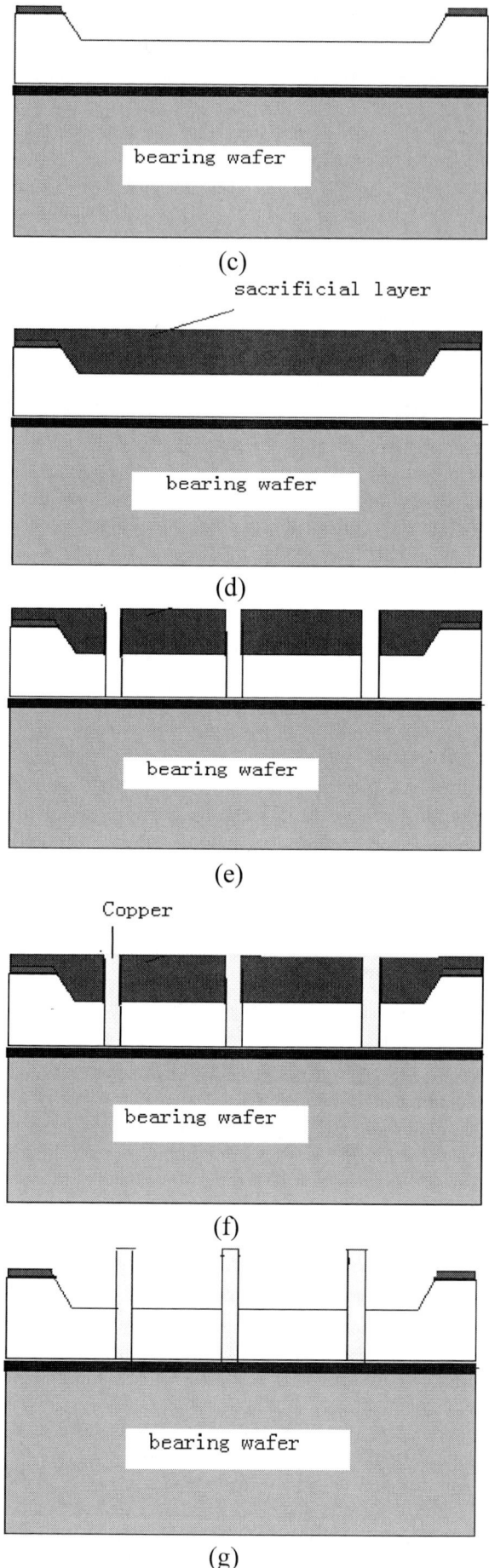

(c)

(d)

(e)

(f)

(g)

978-1-4244-4658-2/09 $25.00 © 2009 IEEE   126

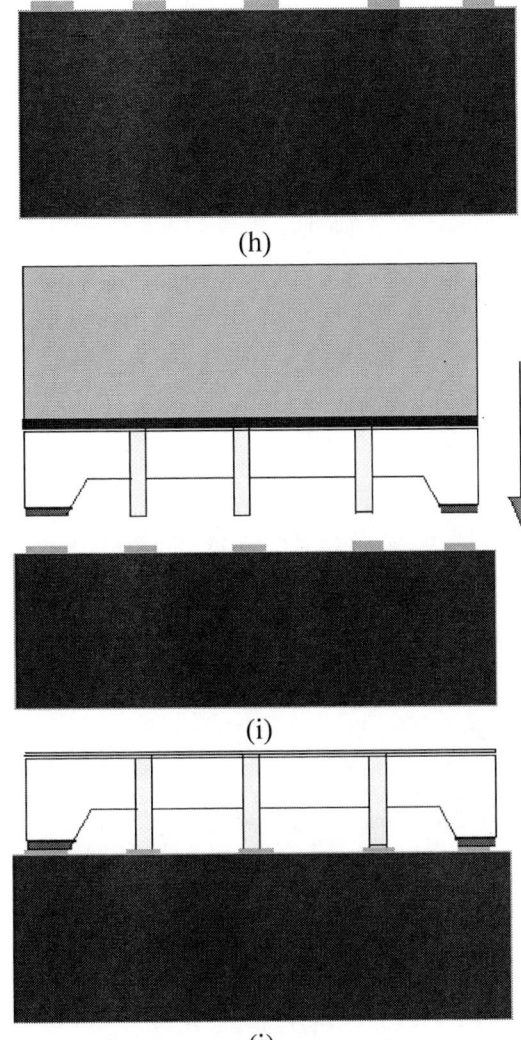

(h)

(i)

(j)

Fig. 2 Manufacturing process of cap wafer and via hole

Structure of Millimeter wave MEMS switch

The cross section of a metal membrane capacitive switch is shown in Fig. 3. The switches are built on high-resistivity silicon substrates (>10 kΩ·cm), with a 1-μm -thick layer of is fabricated on top of the silicon dioxide using 4.0-μm -thick aluminum coplanar waveguide transmission lines. The thick aluminum metallization system is compatible with CMOS circuitry and exhibits low losses at high frequencies. The bottom electrode of the switches is built using 0.4 μm of refractory metal. This film provides good conductivity for low loss and has a smooth surface finish. This finish is important for achieving good contact between the membrane and the lower electrode, minimizing any air gap. On top of the lower electrode is a thin film of silicon nitride. This film blocks the dc control signal from shorting out during switch activation, yet allows RF signals to capacitively couple from the upper membrane to the lower electrode. The metallic switch membrane consists of a thin aluminum less than 0.5 μm thick [4].

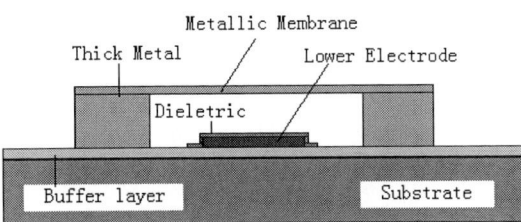

Fig. 3 Cross section of switch

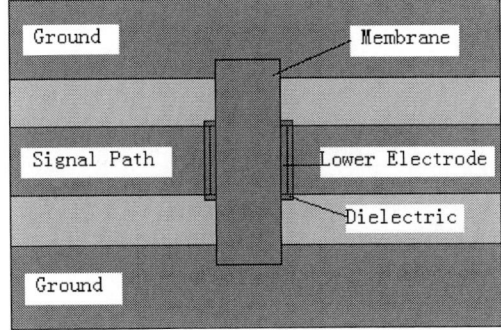

Fig. 4 Top view of switch

This membrane has high conductivity for low RF resistance and good mechanical properties. A top view of the RF MEMS switch element is shown in the photograph of Fig. 4. The thick transmission line metal connects to the lower electrode and dielectric materials to form the through path of a shunt switch. These coplanar waveguide lines have a width of 120 μm and a gap of 80 μm.

Simulation and results

The S-parameters of the switch before and after packaged can be simulated by Ansoft HFSS. The HFSS schematic of a millimeter wave MEMS switch and simulation results are shown in Fig.5 and Fig.6, respectively. Fig. 7 and Fig. 8 give the HFSS model of the packaged switch and its simulation results [5].

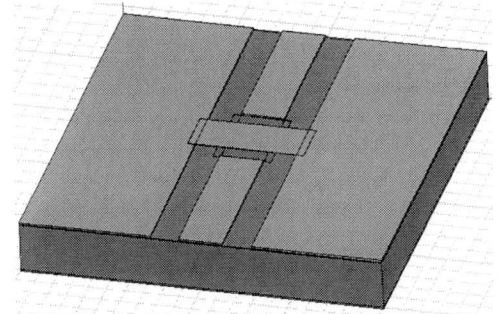

Fig. 5. HFSS model of a millimeter wave MEMS switch

It can be seen from Fig. 6 and Fig. 8 that the S21 parameter for an uncapped and capped switch is within -1.90 dB up to 40 GHz, the S11 parameter is between -12 dB and -15 dB for the uncapped switch, and -7.50 dB and -7.90 dB for the capped switch, at the frequency range of 30 dB to 40 dB.

978-1-4244-4658-2/09 $25.00 © 2009 IEEE

Accordingly, the package impacts more significantly on S11 than S21.

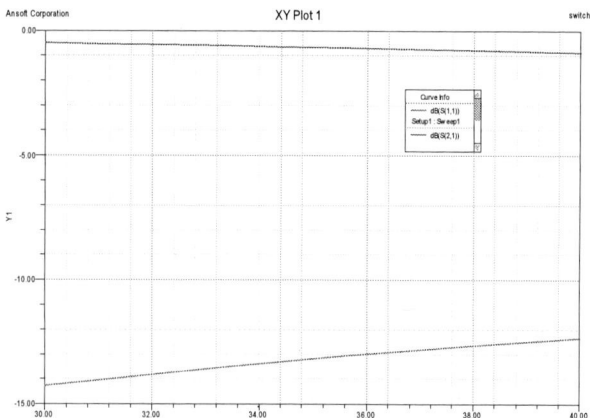

Fig. 6 Simulation results of switch without silicon cap

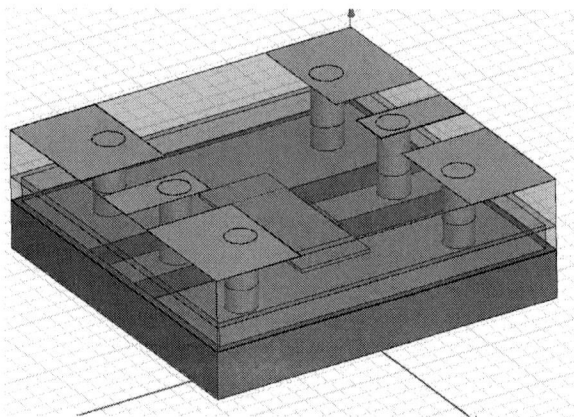

Fig. 7 HFSS model of the packaged switch

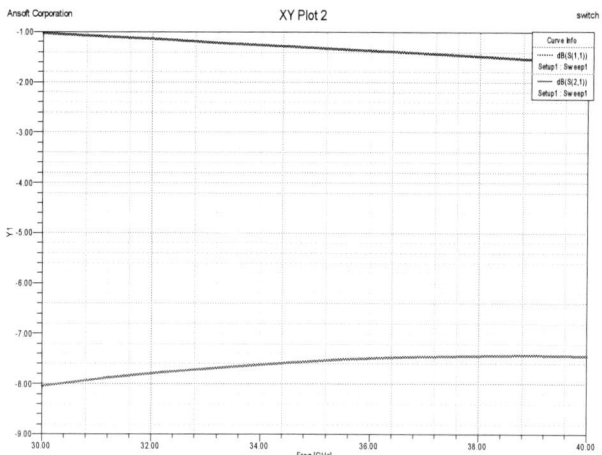

Fig. 8 Simulation results of the packaged switch

Conclusions

In this study, we present the design, simulation and manufacturing of a package solution with lightweight, small size and short electric path length for a millimeter wave MEMS switch. Ultra thin silicon as packaging substrate is used to reduce the weight and the thickness of the substrate in this packaging.

The switch before and after packaged was simulated by using Ansoft HFSS respectively, and the simulation results of the package show that the package structure has a relatively small impact on the performance of switch.

Acknowledgments

This work is supported by National Natural Science Foundation of China (60676043) and The National High Technology Research and Development Program of China (863 Program, 2007AA04Z328).

References

1. Reichl H, Grosser V, "Overview and development trends in the field of MEMS packaging," 14th IEEE International Conference on Micro Electro Mechanical Systems, January. 2001, pp. 1-5.
2. Yun-Kwon Park, Yong-Kook Kim, et al, "Innovation Ultra Thin Packaging for RF-MEMS Devices," 12th International Conference on Transducers, June. 2003, pp. 903-906.
3. Hanqin Wu, Xiaoping Liao, "RF Characteristic and Modeling of The RF MEMS Switch Packaging," 7th International Conference on Electronics Packaging Technology, Shanghai, China, August. 2006, pp. 1-4.
4. Goldsmith, L, Zhimin Yao, Eshelman, S, Denniston, D, "Performance of low-loss RF MEMS capacitive switches," IEEE Microwave and Guided Wave Letters, August. 1998, pp. 269-271.
5. Lannacci, J, Bartek, M, Tian, J, et al, "Electromagnetic optimization of an RF-MEMS wafer-level package," 20th Eurosensors Conference, September. 2006, pp. 434-441.

Packaging of Light Emitting Diodes for Display Backlights:
Need for Multi-Disciplinary Design Tools

Yek Bing Lee, Nadia Strusevich, Chris Bailey, Chun-Yan Yin
School of Computing and Mathematical Sciences, University Of Greenwich, Park Row,
Greenwich, London, SE10 9LS, United Kingdom
Colin Cartwright
University of Abertay, Dundee, UK
E-Mail: C.Bailey@gre.ac.uk

Abstract

This paper discusses the modelling tools used to predict the performance of a liquid crystal display using an array of light emitting diodes (LEDs) as the backlight. LEDs can provide a number of benefits such as adaptive light control, reduced form factor and other novel techniques to ultimately improve the performance of the display. Some of the design challenges in adopting LEDs are to fully understand their effect on the thermal management, reliability, optical and viewability performance of the display. For the display designer this requires integrated modelling and analysis tools.

The paper describes the current modelling efforts underway in this multi-disciplinary UK funded project that aims at identifying the intricate relationships between the optical, thermal and mechanical behaviour when developing a ruggedized electronic display using an array of LEDs as the backlight.

Introduction

Liquid Crystal Displays (LCDs) are used in a diverse range of environments: aircraft, boats, ATMs, etc., which are exposed to different environmental conditions including extremes of temperature and vibration and different ambient lighting conditions.

LCDs are moving towards the use of Light Emitting Diodes (LEDs) for their backlight technology [1-2]. The form factor for a comparable light output is significantly reduced from that required from Cold Cathode Fluorescent Lighting (CCFL). Fig. 1 illustrates a particular backlight design discussed in this paper.

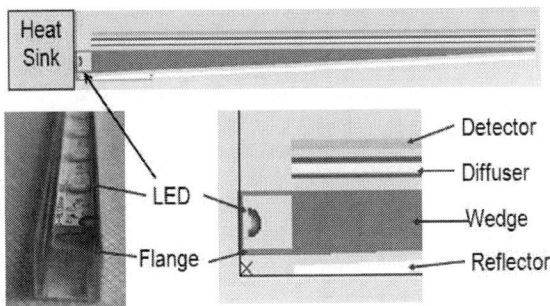

Fig. 1 Backlight design using LED's

Increasing the light output via LEDs is not without its obstacles. Regular LEDs are efficient in converting energy into light, which helps keep temperatures low. However, high powered LEDs are slightly less efficient, with more energy being converted to heat. This leads to a demanding thermal management solution, which in turn affects the performance and reliability of LED and subsequently the display.

LEDs themselves are designed with a thermal overload protection and an optimal operating temperature. The lifetime of the LED is intrinsically linked with the junction temperature, thus reducing this will result in a longer lifetime and higher reliability [3-4].

Towards a Multi-Disciplinary Design Environment

A core part of this project's effort is the development of a multidisciplinary design methodology that will enable the display design engineer to investigate the impact of different packaging solutions on thermal management, reliability, optical behavior and display viewability performance. Fig. 2 illustrates this environment.

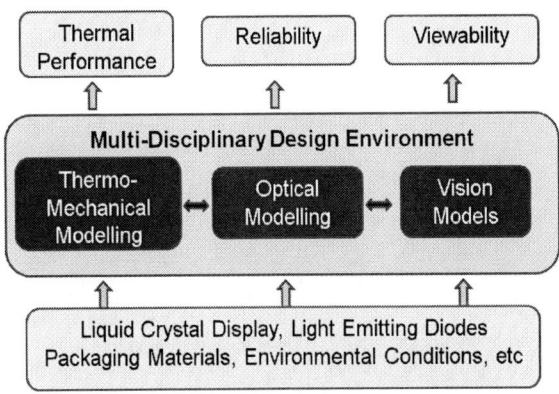

Fig. 2 Multi-Disciplinary Environment for Ruggedised Display Design

Thermal and Thermo-mechanical calculations can be undertaken by commercial computational fluid dynamics and finite element tools. These tools have the ability to predict important design parameters such as LED junction temperature, thermal resistances and stresses across the backlights and the whole display.

Optical modeling would normally be undertaken using raytracing techniques and a number of tools are available for this task. These tools can provide predictions for the luminance and other optical related parameters for a display design. In addition to raytracing vision models can be used to predict how good a display is when viewed by the human eye. In this study in addition to contrast ratio we are using

978-1-4244-4658-2/09 $25.00 © 2009 IEEE

experience from the field of human factors where and metric known as JND (just noticeable difference) is used [5-6]. This assumes that the amount of light that is sensitive to the eye does not follow a linear path, as shown in Fig. 3.

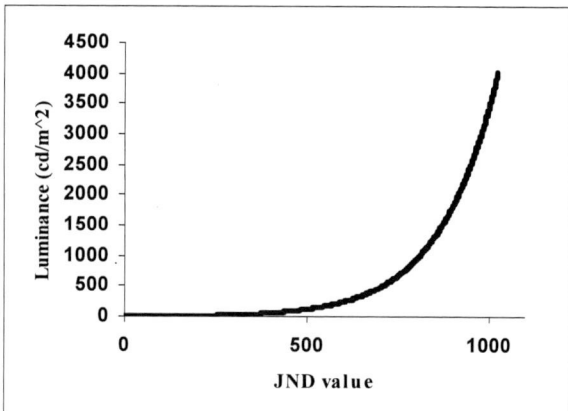

Fig. 3 JND Vs Luminance

A brighter display does not necessarily produce a display that is easier to read. Understanding how much light or JNDs are required to carry out certain tasks (i.e. interpreting static or moving characters) is a major activity in this project. Results from optical modeling that may be influenced by the thermo-mechanical behavior of the LED's and the whole display can be used as inputs to the JND calculation.

Example Display and Backlight Design

A 10.4" display stack is investigated for optical, mechanical and thermal analysis where the CCFL backlight is replaced with surface mounted LEDs. The LED backlight is a direct replacement and must perform as good as or even better than the CCFL, in terms of luminance, colour, beam angle and efficiency.

Using LEDs requires more intricate thermal management. A typical set up of the display components is shown in Fig. 4.

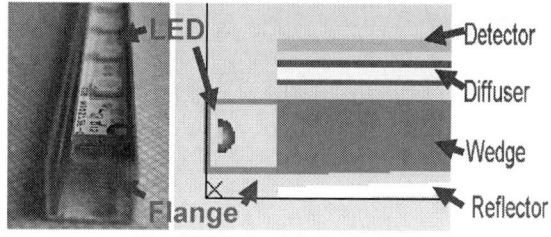

Fig. 4 LED Backlight design

With regards the layout of the LEDs there are two possibilities: (1) Portrait alignment and (2) Landscape alignment. Optically the dynamic range must supercede the CCFL, for the replacement to be beneficial. Fig. 5 illustrates the two possible LED alignments.

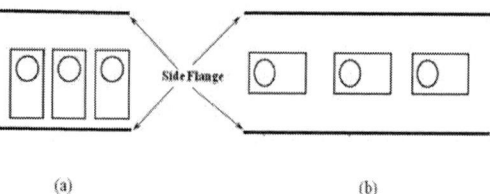

Fig. 5 (a) Portrait alignment, (b) Landscape Alignment LED strip

Optical Analysis of LEDs

A common problem with high brightness LED strips is the edge effects of the light not being distributed across the wedge, this can be broken down into two parts, initial insertion into wedge and across the wedge uniformity. CCFL produces a uniform light along its length, and the flange and optical stack are optimised for this. Typically a CCFL produces on average 350 nits, on a 10.4" display.

The LED strip does not inherently produce uniform light along itself, it is very dependent on the amount of light output by each LED and the distance between them. Ray tracing can aid the design engineer identify what layout design will provide light uniformity both into the wedge and then across the display. Fig. 6 illustrates a ray tracing calculation from an LED towards onto the wedge and then into the display.

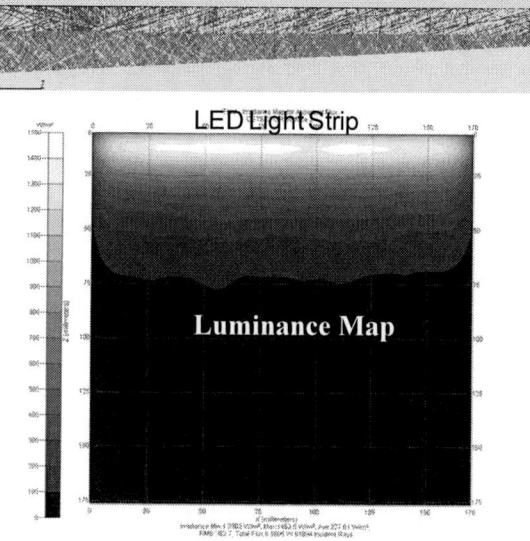

Fig. 6 Raytracing results for backlight

In this analysis different wedge materials can be investigated (e.e. glass or pmma). In the above analyis 10 W of light was used as the light source with 48% efficiency (i.e 48% of the light transmitted).

Thermal Analysis of LEDs

Fig. 7 illustrates the package design for an LED. Here the LED is mounted on a board lined with copper tracks and fixed on an aluminium frame.

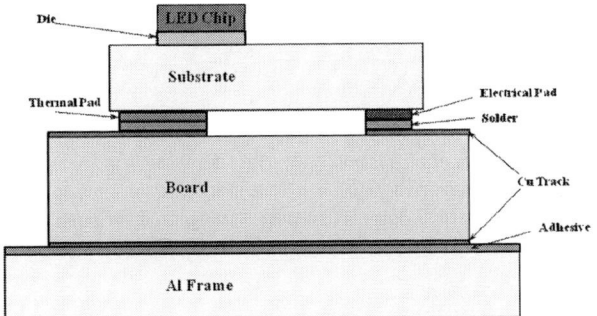

Fig. 7 The structure of the mounted LED package

The useful life time of an LED is highly dependent on its junction temperature. Here "junction" refers to the p-n junction within the semiconductor die. This is the region of the chip where photons, and subsequently heat, are generated.

Heat transfers within the package and the circuit board due to heat conduction, and from the free surfaces to the environment by means of natural convection and radiation.

A computational model of the above package has been analysed using the multi-physics software package PHYSICA [7]. Both portrait and landscape layouts have been investigated and the junction temperatures for each design predicted. In all calculations we assume that the ambient temperature is 50°C and that 50% of the applied power to the LED is generated as heat.

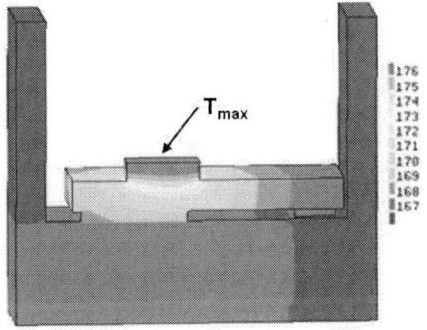

Fig. 8 Temperature Distribution for Portrait Design

The portrait and landscape designs correspond to how the LED packages are placed on the aluminium frame accompanied with two side aluminium flanges. These packages are placed in one line, next to each other. In the portrait design, the shorter sides of the package face the flanges, while in the landscape design the longer sides of the package face the flanges; see Fig. 5. In our study we assume the natural convection heat trasfer coefficient HTC = 32W/m²/K, and the applied power for each LED is $P=1.1$W.

Fig. 8 illustrates the temperature profile for the portrait design. Clearly we can see that the temperatures in the die are very high (more than 170°C). This is due to the very close proximity of the packages.

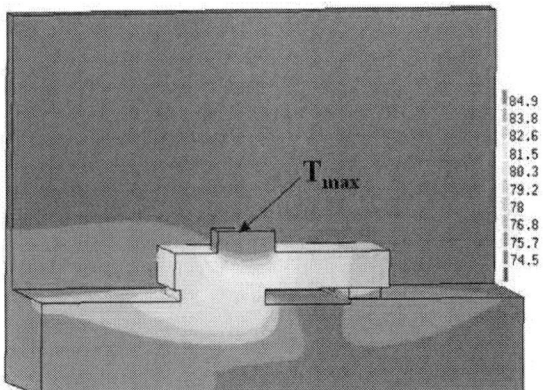

Fig. 9 Temperature Distribution for Landscape Design

Fig. 9 illustrates the temperature profile for the landscape design. Clearly we can see that the temperatures in the die are at much lower values (around 85°C). This is due to the closer coupling between the package and the surrounding flange and the total number of LED's in the strip in this case.

Table 1 reports the obtained results, where the junction-solder thermal resistance R_{j-s} is computed as the ratio of the difference between the corresponding temperatures to the heat power.

Table 1 Results for different LED layouts

	Junction Temperature (°C)	Solder Temperature (°C)	Frame Temperature (°C)	R_{j-s} (°C/W)
Portrait	177	170	167	12.8
Landscape	86.1	79	74	12.8

The next set of simulations study above LED package when mounted on an aluminium frame with the side flanges using the portrait design. In this study we have assumed the package to air heat transfer coefficient HTC = 30W/m²/K.

Fig. 10 illustrates a close up of the interconnection region between the LED package and the Aluminium frame.

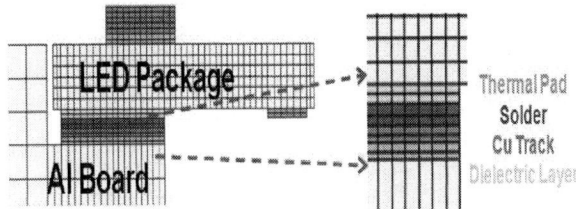

Fig. 10 Detailed view of Interconnection region

The simulations have been conducted for three different power regimes P_1=0.07W, P_2=0.18W and P_3=0.26W. Two values of the copper track thickness have been considered, 35μm and 70μm. The two materials for the dielectric layer and their properties are presented in Table 2.

978-1-4244-4658-2/09 $25.00 © 2009 IEEE

Table 2 The materials for the dielectric layer

	Thickness (μm)	Thermal Conductivity (W/m/K)	Density (kg/m³)	Specific heat (J/kg/K)
Material 1	75	2.2	1300	880
Material 2	35	6.0	2700	900

The results of our simulations are presented in Table 3. We report the values of the junction temperature and those of thermal resistence (junction-solder, junction-board and juanction ambient). In the last four columns, the cells with the grey background are related to the material 1, while the cells with transparent background correspond to material 2.

Table 3 The results for the device with Material 1 & 2 bonding layers

Cu Track Thickness (μm)	Power Regime (W)	Junction Temperature (°C)	$R_{j\text{-}s}$ (°C/W)	$R_{j\text{-}b}$ (°C/W)	$R_{j\text{-}a}$ (°C/W)
70	0.07	59.97	13.2	19.43	284.9
		59.85	13.43	16.00	281.4
70	0.26	87.003	13.19	19.26	284.6
		86.57	13.46	16.0	281.3
35	0.07	60.009	13.11	19.37	285.9
		59.89	13.43	16.10	282.6
35	0.18	75.795	13.13	19.40	286.4
		75.49	13.47	16.10	283.0
35	0.26	87.145	13.10	19.35	285.7
		86.71	13.45	16.06	282.4

To summarize, the first set of simulations show that the landscape design of the LED package produces a lower junction temperature than that achieved using the portrait design. Notice that the high values of the junction temperature in both cases are due to the choice of applied power (1.1 W) for each LED package and for the portrait design more LED's will be present.

Since the portrait design may be more attractive from the point of view of the optical performance of the backlight, we have conducted additional experiments with that design. We have observed that the junction temperature heavily depends on the power regime and is less affected by the copper track thickness or the dielectric bonding layer material.

An obvious challenge when using the portrait layout which will have more LED's than the landscape layout would be thermal management.

Conclusions

Development of a multi-disciplinary design environment for ruggedized displays that incorporate LED's as the backlight is a major component of the UK project ENDVIEW (http://endview.cms.gre.ac.uk). This project aims to deliver enhanced viewability of dynamic images through the application of high-performance LED arrays and novel optics systems.

In this paper we have discussed the vision behind the integrated design environment and provided some illustrations of both optical and thermal analysis capabilities. Further work will include closer coupling between thermal and mechanical behaviour of the LED array to provide predictions for its reliability. Similar analysis will also be undertaken for the whole display to assess its reliability. The affect of electro-thermal behaviour on the optical performance of the LED array will also be investigated through modelling and relevant optical parameters will be predicted for use in the JND prediction.

Acknowledgement

The authors acknowledge the financial support from the Technology Strategy Board for the ENDVIEW project (ProjectTP11/LLD/6/I/AFOO5K).

We also acknowledge the technical contributions and support from industrial partners which include GE Aviation. NCR, Design LED, Raymarine, National Physical Laboratory and Thin Film Solutions.

References

1. Lan Kim, Jong Hwa Choi, Sun Ho Jan, Moo Whan shin, "Thermal Analysis of LED Array System with Heat Pipe," *Joural of Thermochimica Acta*, volume: 455, pp. 21-25. 2007.
2. Yan Lai, Nicolas Cordero, "Thermal Management of Bright LEDs for Automotice Applications," *Proceedings of Thermal, Mechanical and Multi-Physics Simulation and Experimetns in Micro-Electronics and Micro-Systems (EUROSIME)*, Milano, Italy. April 2006. pp. 390-394.
3. Oon Siang Ling, "Thermal Mangment for application of High Power LED," white paper, "Thermal Mangment of OSTAR Projection Light Source-Application mote," *Opto Semiconductors*, OSRAM, 2006, pp. 1-16.
4. Digital Imaging and Communications in Medicine (DICOM) Part 14 Grayscale Standard Display Function, National Electrical Manufacturers Association
5. Barten. Peter, G, J., Contrast sensitivity of the human eye and its effects on image quality. SPIE Optical Engineering Press, 1999.
6. PHYSICA (1996-2007). Physica Ltd, 3 Rowan Drive, Witney, Oxon, United Kingdom, http://www.physica.co.uk.

A Simplified Computational Model for Solder Joints under Drop Impact Loadings

Tong AN, Fei QIN*
College of Mechanical Engineering and Applied Electronics Technology,
Beijing University of Technology, Beijing, 100124, China
Tel: +86-10-67392760; Fax: +86-10-67391617
*Email: qfei@bjut.edu.cn

Abstract

The effects of the moment, axial force and shear force induced during drop impact on the peeling stress of the solder joints were investigated by a 2-D beam model and a 3-D solid model of board level electronic package. It shows that the peeling stress is dominated by the bending stress and the maximum occurs at the PCB end. Results of the two models indicated that in the solder joint array only a few solder joints closed to the end of the package are stressed and the most solder joints inside the array are almost stress-free. Based on this observation, an approach was proposed to reduce the computation scale. By the approach, only 3 or 4 solder joints are necessary to be included in the computational model.

1 Introduction

Solder joints are functioned as mechanical support, thermal and electrical interconnections between electronic packages and the printed circuit board (PCB) in portable electronic products. Failure of the solder joint leads to malfunction of the products completely. Therefore, the mechanical behavior of solder joints during drop impact has attracted great attention.

The numerical models applied to the assembly of PCB-solder joints-component include beam model [1-3], shell-plate model [4, 5], and 3-D finite element model [6-9]. In Wong's work [2], the PCB was treated as a beam or a plate, and then the analytical solution of dynamic response of the PCB was predicted. However, solder joints were not taken into account in the research. In 1988, Suhir [10] used the elastic plate and springs to model the PCB-solder joints-component assembly and the axial stress solution of solder joints under static bending moment was given. Since the spring-like solder joints only transferred axial force in Suhir's model, the effect of bending moment and shearing force, which is regarded as the primary stress to the failure of solder joints, was not evaluated.

In board level packages, components are mounted to PCB through hundreds of solder joints. The diameter and height of solder ball are 0.35 mm and 0.3 mm in a BGA respectively. At the same time, the length of a PCB is usually about 100 mm. Hence there is a 10^4 order difference in the geometrical dimension in the assembly structure, and this difference leads to a large amount of elements in a 3-D finite element model. As a result, the computational cost, especially in a dynamic response analysis, is dramatically increased. Some researches were carried out to deal with this difficulty. Ren et al. [4, 5] used shell elements to model the PCB and component, beam elements for the solder joints trying to reduce the computational scale, but the detailed stress distribution across the solder joints can not be evaluated by this model. Zhu [6]

has investigated the solder joints stresses of a BGA package subjected to static loading by using the submodeling or the global-local approach. However, submodeling method is not efficient for path-dependent problems such as dynamic response analysis [11]. Tsai et al. [12] modeled a solder ball array by a uniform solder layer to simplify the computation. The model of solder joints have little influence on the deflection of PCB, however it is critical to the stresses of solder joints. More efforts are needed to balance the computational cost and the accuracy of predicted stress in solder joints.

In this paper, a 2-D beam model and 3-D solid model of a board level electronic package were used to investigate stress of solder joints under drop impact loadings. Based on the analysis, an approach was proposed to reduce the computational scale of the problem, and its feasibility was discussed.

2 Stress Analysis of the Solder Joint

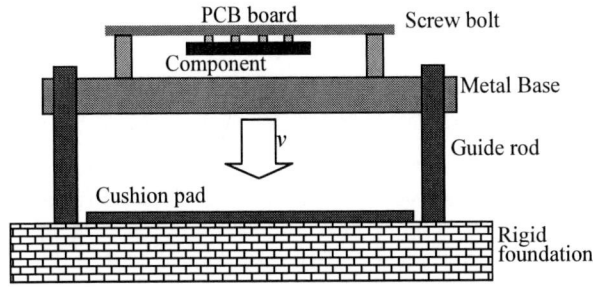

Fig. 1 Test setup of board level drop/impact

A typical board level drop impact test setup is depicted in Figure 1 as recommended by JEDEC [13]. The PCB and component assembly is mounted onto a metal base via screw bolts. In the test, the entire assembly is subjected to free fall along guide rods from a prescribed height, and the metal base impacts onto a rigid foundation then an impact loading is produced. A prescribed half-sine acceleration impulse can be achieved by manipulating the drop height and the dimension or material of cushion pad. Since the stiffness of metal base is exceedingly greater than that of the PCB, the half-sine acceleration impulse resulting from impact predominantly transmits to the PCB via the metal base and screw bolts with little distortion. Therefore the analysis of a board level drop impact test can be simplified by a model in which the PCB is alone subjected to the half-sine acceleration impulse at its points of mounting to the screw bolts, and the model is shown in Figure 2(a), in which $G(t)$ is the acceleration impulse. This

978-1-4244-4658-2/09 $25.00 © 2009 IEEE

approach was proposed by Tee et al. [9], and it is called the Input-G method.

The PCB and component bend under the act of $G(t)$. However, the difference of the bending stiffness between the PCB and component make the solder joints deforming under tensile/compressive stresses during PCB bends upward/downward. It has been indicated that the peeling stress, which is the normal stress vertical to the PCB, is the dominant stress component leading to failure of solder joints [14]. The bending deformation of the PCB and the component has great influence on the magnitude of the peeling stress [1].

Since the PCB has much larger warpage in its length direction than in the width direction, the PCB and component can be modeled as two strips of plate, as shown in Figure 2(b),and solder balls are modeled as cylindrical beams. Furthermore, due to the symmetry only one half of the model is analyzed, as shown in Figure 2(c). In this paper, the PCB strip and the component strip are modeled by Timoshenko beams and a bending moment M is applied to the right end of the PCB.

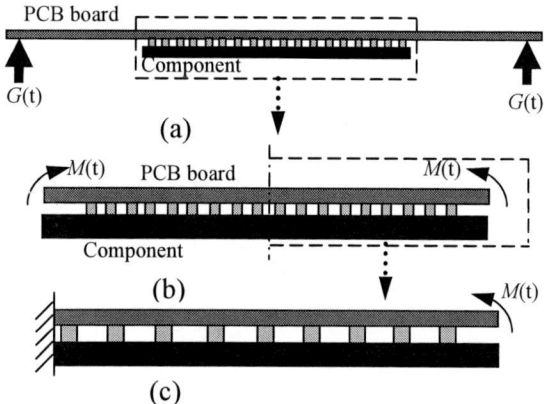

(a)

(b)

(c)

Fig. 2 Mechanical model for the board level packaging

Table 1 Model Parameters

	L (mm)	b,d (mm)	h (mm)	E (GPa)	v	ρ (g/cm^3)
PCB	10	1	1	24	0.28	2
Component	10	1	1	100	0.39	2.5
Solder joint	-	0.5	0.5	25	0.36	10

A finite element model was built according to Figure 2(c), wherein the PCB, package and solder joints are modeled with Timoshenko beam elements, and the model was analyzed by ABAQUS. The model parameters such as dimensions and materials are presented in Table 1, in which b and h are the width and thickness of the PCB and the component respectively, d is the diameter of solder column, E and v are the Young's modulus and Poission's ratio of materials, and ρ is the density. In the analysis, the load was a constant bending moment of 15 N·mm.

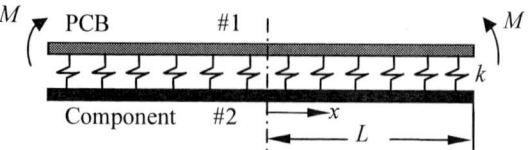

Fig. 3 PCB board-solder joints-component model of Suhir

In order to verify the model the axial stresses in solder joints obtained from this simulation were compared with Suhir's equation [7]. Suhir' model is shown in Figure 3, and it treated the solder joints as spring links with stiffness k, which only axial force $p(x)$ can be transferred. The axial force is:

$$p(x) = \frac{kML^2}{2D_1u^2}\left(\frac{u^2\chi(u)}{3}V_0(\alpha x) - \phi(u)V_2(\alpha x)\right) \quad (1)$$

where $\alpha = \sqrt[4]{k\dfrac{D}{4D_1D_2}}$, $u = \alpha L$, $D = D_1 + D_2$, D_1 and D_2 are the bending stiffness of the PCB and package respectively:

$$D_1 = \frac{E_1h_1^3}{12(1-v_1^2)}, \quad D_2 = \frac{E_2h_2^3}{12(1-v_2^2)} \quad (2)$$

The other terms in Equation (1) are

$$V_0(\alpha x) = \cosh\alpha x\cos\alpha x ,$$
$$V_2(\alpha x) = \sinh\alpha x\sin\alpha x$$
$$\chi(u) = \frac{6}{u^2}\frac{\cosh u\sin u - \sinh u\cos u}{\sinh 2u + \sin 2u} \quad (3)$$
$$\phi(u) = \frac{\cosh u\sin u + \sinh u\cos u}{\sinh 2u + \sin 2u}$$

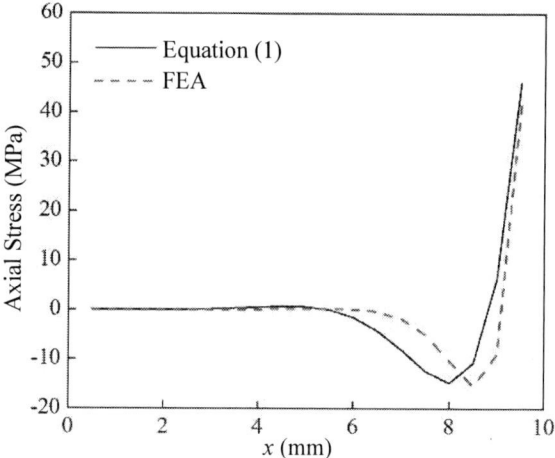

Fig. 4 Axial stress in solder joints

The axial stresses in solder joints calculated by Equation (1) and from the static finite element analysis (FEA) are shown in Figure 4. It shows that the FEA results agree quite well with Suhir's, both in the distribution and the maximum value of the stresses. The maximum value calculated by Equation (1) is 46 MPa, while that from the FEA is 41.5 MPa.

The stresses of solder joints are induced by bending moment, axial force, and shear force when the PCB subjected

978-1-4244-4658-2/09 $25.00 © 2009 IEEE

to bending moment, which are presented in Figure 5. The maximum stress occurs at the PCB end of the outer-most solder joint. It shows that the maximums of bending stress, axial stress and shear stress at the PCB end are 130.5 MPa,

41.5 MPa and 23.6 MPa respectively, which suggests that the bending moment produces much greater stress than the axial or shear force. At the component end, although the axial stress is the greatest among the all at the outer-most solder joint, it is in a relatively lower level when compared with the bending stress at the PCB end. Therefore the bending stress plays a dominant role in solder joints.

It is the peeling stress that dominates the failure of solder joints. The distribution of the peeling stresses in each solder joint are presented in Figure 6. It shows that the majority of solder joints are under low stress level, and there are only a few solder joints at the outer positions undergoing large peeling stress. The maximum peeling stress from FEA analysis occurs at the PCB end, and it is 172.1 MPa.

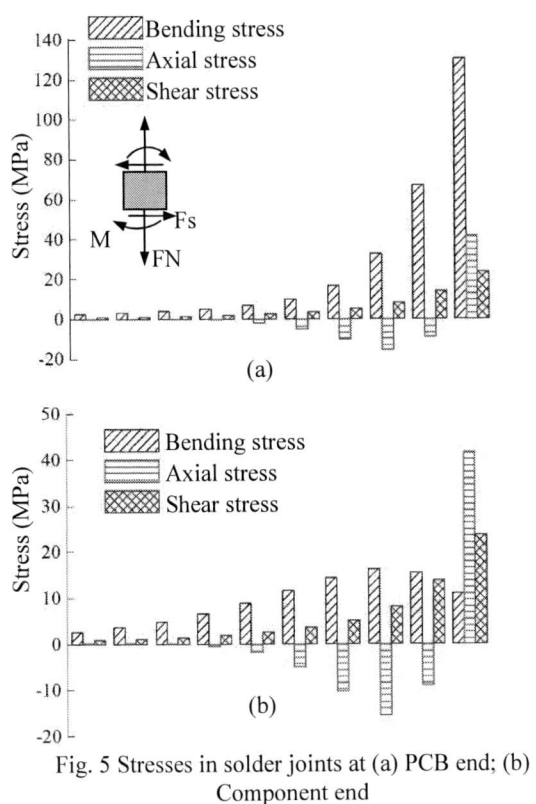

Fig. 5 Stresses in solder joints at (a) PCB end; (b) Component end

3 Simplified Model for Stress Computation of Solder Joints

As discussed in Section 2, only a few solder joints close to the edge of the component experience significant peeling stresses while the other majority are at quite low stress level that can be neglected. Based on this observation, one approach was proposed in this paper to reduce the computational model. By the approach, only a few outer solder joints are modeled and the other inner solder joints can be neglected from the model.

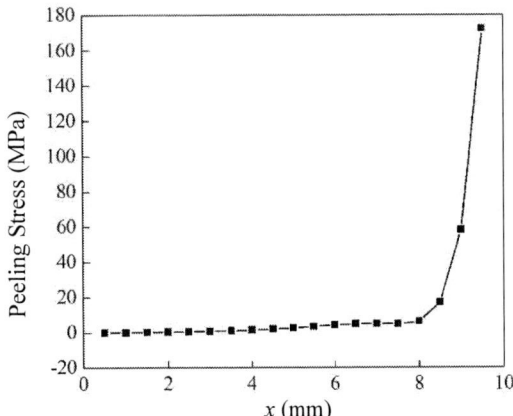

Fig. 6 Peeling stress in solder joints

In order to validate the approach, models containing 10, 5, and 3 solder joints were computed, respectively. The peeling stress and the PCB deflection of those models were compared carefully with a full model containing 19 solder joints. The comparison of stress distribution in each solder joint was presented in Figure 7, and there is no significant difference among these models. The difference of the peeling stress in percentage between the reduced and the full model is shown in Figure 8. The peeling stress obtained from the model with 10 solder joints is 0.06% greater than that from the full model, and model containing 5 solder joints predict 1.5% greater peeling stresses than that from the full model. Obviously, only 5 solder joints being included in the model is enough to predict satisfied peeling stress. The deflection curves of the PCB derived from various models are shown in Figure 9, which also suggests that it has no significant effect on the deflection of the PCB to neglect most of the inner solder joints.

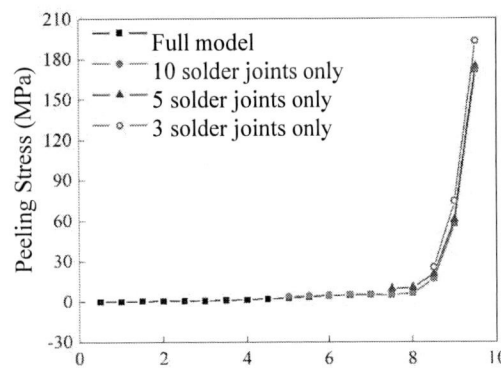

Fig. 7 Peeling stress in solder joints for different solder number

It indicates that most of inner solder joints can be neglected from the model when the maximum peeling stress or the PCB deflection are computed. The reduced model can not only guarantee the computational accuracy but also greatly reduce computational cost, especially when a 3-D model is used.

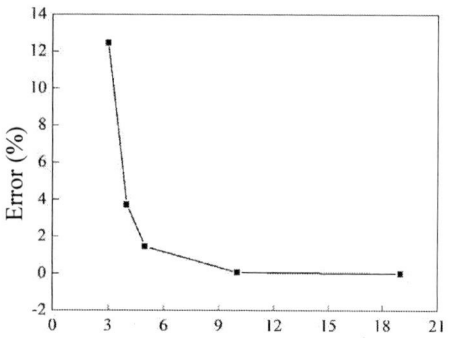

Fig. 8 Maximum peeling stress error in solder joints for different solder number

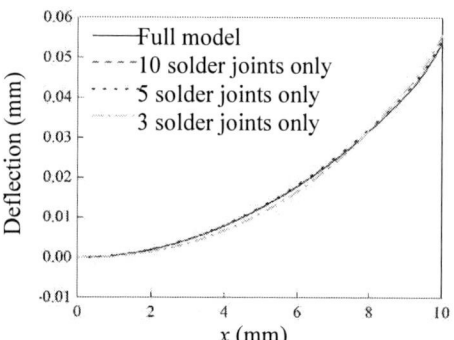

Fig. 9 Deflection of PCB board for different solder number

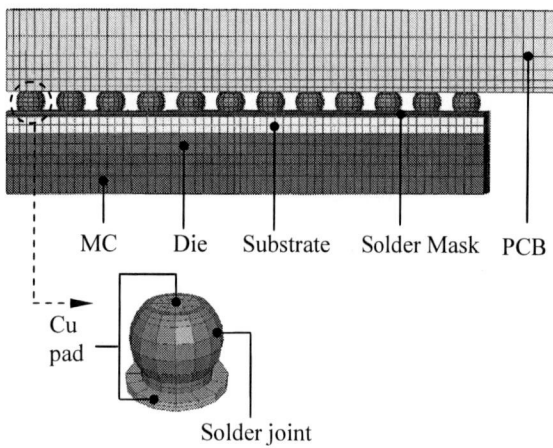

Fig. 10 3-D finite element model

A 3-D finite element model of board level electronic package was established, and used to validate the approach. Since the PCB has much larger warpage in the length direction than in the width direction and symmetry of the geometry and the loading condition in drop impact test, drop impact of a board level package was modeled as two cantilever beam system. The Input-G method [9] was used and the input acceleration $G(t)$ was applied at the far end of the PCB. This model can also be regarded as a sliced 3D

model. The mesh used is shown in Figure 10. In this analysis, a $6\times0.5\times1.02$ mm^3 package interconnects to a $50\times0.5\times1$ mm^3 PCB through Sn3.0Ag0.5Cu solder joints. The package contains a bare $3\times0.5\times0.26$ mm^3 silicon die, the solder mask, substrate and Cu pads are taken into account also. Diameter and standoff of a solder joint are 0.35 and 0.28 mm, respectively. The pitch between adjacent solder joints is 0.5 mm. The pad design is SMD on component side and NSMD on PCB side.

The element type is C3D8R defined by ABAQUS finite element software, and there are 22052 elements and 26258 nodes totally. The Johnson-Cook constitutive model of the Sn3.0Ag0.5Cu solder alloy was obtained from Ref. [15]. Other parameters used in the model are presented in Table 2. The impact acceleration is the test condition H defined by the JEDEC standard [16], which is featured as an half-sine impact acceleration pulse with a peak of 2900 G and duration of 0.3 ms.

Table 2 Parameters used in the finite element model

Materials	E (GPa)	v	ρ(kg/mm^3)
Die	131	0.23	2.33
Mold component	28	0.35	1.97
Cu pad	117	0.34	8.94
PCB	20	0.11	1.91
Solder mask	5	0.30	1.15
Sn3.0Ag0.5Cu solder	54	0.363	7.384
Substrate	26	0.11	2.00

Two simplified models which contain 4 and 3 solder joints were computed respectively. The peeling stresses of those models were compared carefully with a full model containing 12 solder joints. The histories of the peeling stresses of critical solder joint obtained from full model and simplified model were presented in Figure 11. The peeling stress in each solder joint obtained from different models was presented in Figure 12, and there is no significant difference of peeling stress distribution among these models. The maximum peeling stress obtained from the model with 4 and 3 solder joints are 3.8% and 14.3% greater than that from the full model. Obviously, only 4 solder joints being included in the model is enough to predict satisfied peeling stress.

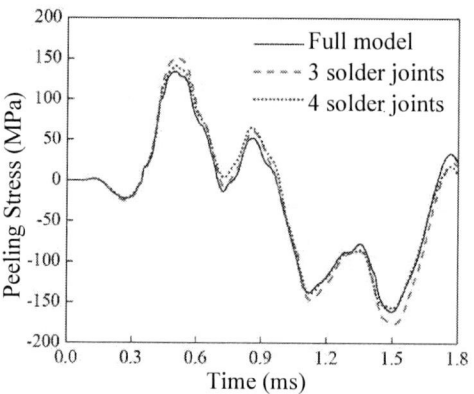

Fig.11 Peeling stress histories for different number of solder joints

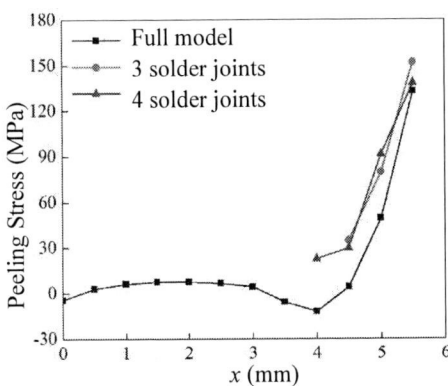

Fig.12 Peeling stress in solder joints for different solder number

In Table 3 the CPU time required to finish the entire calculation fro different model and the element number are compared. Obviously, the simplified mode proposed in this study can greatly benefit the computational efficiency.

Table 3 CPU time for different models

Solder joints number	CPU time	Elements number
12	14:30:16	22052
4	05:50:08	10380

Conclusions

The stress distribution in solder joint array was analyzed by the finite element method and the proposed beam model of a board level assembly under drop impact loading conditions. The maximum peeling stress occurs at the PCB end of the outer solder joint, and the stress induced by bending moment is dominant. The numerical results show that only a few solder joints close to the edge of the component experience greater peeling stresses while the other majority is under quite low stress level.

An approach to improve the computational efficiency was proposed and by the approach, only 4 outer solder joints are necessary to be modeled and the other majority solder joints can be neglected from the model.

Acknowledgments

The authors would like to thank financial support from the National Natural Science Foundation of China (NSFC) under the Grant No.10572010, and the Funding Project for Academic Human Resources Development in Institutions of Higher Learning Under the Jurisdiction of Beijing Municipality (PHR - IHLB).

References

1 E. H. Wong, K. M. Lim, N. Lee, S. Seah, C. Hoe and J. Wang. "Drop Impact Test-Mechanics and Physics of Failure," *Proceedings 4th Electronics Packaging Technology Conference*, Singapore, 2002, pp. 327-333.

2 E. H. Wong. "Dynamics of Board Level Drop Impact," *Transactions of the ASME. Journal of Electronic Packaging*, Vol. 127, No. 3, (2005), pp. 200-207.

3 E. H. Wong, Y. W. Mai, S. K. W. Seah. "Board Level Drop Impact-Fundamental and Parametric Analysis," *Trans. ASME, Transactions of the ASME. Journal of Electronic Packaging*, Vol. 127, No. 4, (2005), pp. 496-502.

4 W. Ren, J. J. Wang. "Shell-Based Simplified Electronic Package Model Development and its Application for Reliability," Analysis. Proceedings of the 5th Electronics Packaging T*echnology Conference*, Singapore, 2003, pp. 217-222

5 W. Ren, J. J. Wang, T Reinikainen. "Application of ABAQUS/Explicit Submodeling Technique in Drop Simulation of System Assembly," *Proceedings of the 6th Electronics Packaging Technology Conference*, Singapore, 2004, pp. 541-546.

6 L. P. Zhu. "Submodeling Technique for BGA Reliability Analysis of CSP Packaging Subjected to an Impact Loading," *Advances in Electronic Packaging. Pacific Rim/International , Intersociety Electronic Packaging Technical/Business Conference and Exhibition*, Kauai, Hi, United states, 2001, pp. 1401-1409.

7 T. Zhang, S. Rahman, K. K. Choi, K. Cho, P. Baker, M. Shakil, D. Heitkamp. "A Global-Local Approach for Mechanical Deformation and Fatigue Durability of Microelectronic Packaging Systems," *Transactions of the ASME. Journal of Electronic Packaging*, Vol. 129, No. 2, (2007), pp. 179-189

8 T. Y. Tee, H. S. Ng, C. T. Lim, E. Pek, Z. W. Zhong. "Application of Drop Test Simulation in Electronic Packaging," *4th ASEAN ANSYS Conference*, Singapore, 2002

9 T. Y. Tee, J. E. Luan, E. Pek, C. T. Lim, Z. W. Zhong. "Novel Numerical and Experimental Analysis of Dynamic Responses under Board Level Drop Test," *Proc. 5th Int. Conf. Therm. Mech. Simul. Exp. Microelectron. Microsyst. EuroSimE 2004*, Brussels, Belgium, 2004, pp. 133-140.

10 E. Suhir. "On a Paradoxical Phenomenon Related to Beams on Elastic Foundation: Could External Compliant Leads Reduce the Strength of a Surface-Mounted Device?.," Journal of Applied Mechanics, Transactions ASME, Vol. 55, (1988), pp. 818.

11 Y. S. Lai, T. H. Wang. "Verification of Submodeling Technique in Thermomechanical Reliability Assessment of Flip-Chip Package Assembly," *Microelectronics Reliability*, Vol. 45, No. 3-4, (2005), pp. 575-582.

12 T. Y. Tsai, C. L. Yeh, Y. S. Lai and R. S. Chen. "Transient Submodeling Analysis for Board-Level Drop Tests of Electronic Packages," *IEEE Transactions on Electronics Packaging Manufacturing*, Vol. 30, No. 1, (2007), pp. 54-62.

13 JESD22-B111. Board Level Drop Test Method of Components for Handheld Electronic Products, JEDEC Solid State Technology Association, Arlington, VA, 2003.

14 T. Y. Tee, J. E. Luan, E. Pek, C. T. Lim, Z. W. Zhong. "Advanced experimental and simulation techniques for analysis of dynamic responses during drop impact," *2004 Proceedings. 54th Electronic Components and Technology Conference*, Las Vegas, NV, USA, 2004, pp. 1088-1094.

15 Fei Qin, Tong An, Na Chen. "Strain Rate Effects and Rate-dependent Constitutive Models of Lead-based and Lead-free Solders," *Tran. ASME, Journal of Applied Mechanics*, (will be published in November 2009)

16 JESD22-B104C. Mechanical Shock, JEDEC Solid State Technology Association, Arlington, VA, 2004.

Design of Miniaturized Bandpass Filters for GSM and GPS Applications Using Embedded Capacitor Material

*Yunfeng Wang[1,2], Lei Li[1], Lixi Wan[2]
[1]Shenzhen Institute of Advanced Integration Technology
Chinese Academy of Sciences/The Chinese University of Hong Kong
1068 Xueyuan Avenue, Shenzhen University Town, Nanshan District, Shenzhen, 518055, China
[2]Institute of Microelectronics of Chinese Academy of Sciences
3#, Beitucheng West, Chaoyang District, Beijing, 100029, China
Email: yf.wang@siat.ac.cn

Abstract

Miniaturized bandpass filters in system in package were investigated for GSM and GPS applications using embedded capacitor material. Filters with novel LC resonator structure were proposed, and had small sizes of 3.5mm by 1.7mm and 3.1mm by 1.1mm, respectively. The filter for GPS had a 3dB bandwidth of 140MHz (1.52-1.66GHz), and another one for GSM was also 140 MHz (0.83GHz-0.97GHz). The loss mechanism of the novel structure was also investigated. The results showed that conductor loss had more influence on frequency properties than dielectric loss for the GSM and GPS applications. That is to say, if the frequency beyond multi-MHz, the conductor loss of this miniaturized resonator structure would dominate over the loss mechanism.

Introduction

The wireless communication systems are developing towards the direction of small size, low cost, and high performance. SOP/SIP (system-on-package/system-in-package), especially the embedded passives technology, is considered to be one of the most challenges and exciting technology to realize the advanced micro-systems, because the passive components take up a large real estate of a PCB, and the electrical performance and reliability are reduced by the longer interconnect and more solder joints. The embedded passives have drawn attractions to the RF front-end circuits and modules due to its low parasitic parameters, small size and high performance and reliability. Filters, baluns, Bluetooth modules, and power amplifier modules on organic substrate have been studied widely [1-5]. Among these modules, the RF filters play a most significant role. Various design and integration approaches for filters have been reported in many publications. But the conventional methods using SMDs can not meet the requirements for better performance and lower cost and size. The embedded passives are promising solutions due to its low parasitic parameters and small size, so embedded filters have been widely studied and also reported in several papers. Base on mature filter theory, better performance needs more passive components, and could not be achieved by reducing circuit orders. For example, band pass filter could be obtained by many orders of combinations of low pass and high pass filters or by many orders of LC resonators. Most researchers focused on designing different topologies and configurations of layout to reach the electrical specifications [6-8].

In this paper, the authors proposed miniaturized filters which should be realized with embedded capacitors [9]. The filters were implemented with plane spiral inductor geometries on the top electrode of embedded capacitor material, and one end of the filter was connected to the bottom electrode by a micro-via. Filters for GSM and GPS applications were designed with the size of 3.5mm by 1.7mm and 3.1mm by 1.1mm, respectively. 3dB bandwidth of GSM filter was 140MHz, and the one of GPS filter was 140MHz (1.52GHz-1.66GHz). The physical models were simulated by HFSS, and the loss mechanism influencing on the characteristics of the filter was also discussed, such as conductor loss and dielectric loss, etc. By modeling the filters with a perfect conductor or perfect dielectric, the influence of conductor loss and dielectric loss could be separated from each other.

Design and Discussion

a) Resonator

The filter was also a resonator which was implemented by spiral inductor geometry on the top electrode, and one end of the inductor was connected to the bottom electrode through a micro-via. Fig.1 (a) showed the typical resonator layout, (b) showed the cross section of the embedded capacitor, Fig.2 showed equivalent circuit modal of the filter, and the material specification were listed in Table 1.

Table1. Specifications of embedded capacitor

Structure	Dimensions	Material
Layer1	35μm	Copper
Dielectic	14μm	Dk=16 Df=0.005
Layer2	35μm	Copper

As shown from the Fig.1, w was the inductor width, s was the space between the winding, and d was the inner diameter of the spiral inductor. These planar spiral inductors were designed with rectangular shapes for finding out optimal geometry. Width and core area of the inductors varied for finding out the most optimal performance. The inductance and the capacitance between the spiral inductor and the bottom electrode comprised of the inductance and capacitance of the resonator, and the mutual inductance and parasitic capacitance among the inductor turns could be ignored because of weak electromagnetic coupling.

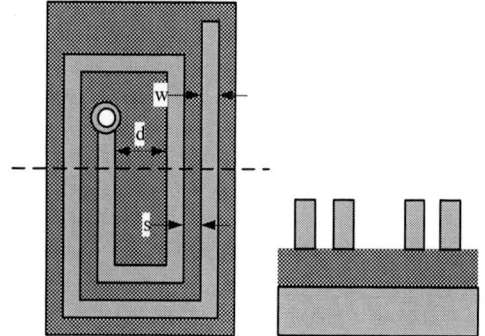

Fig. 1 (a) layout of resonator (b) cross section

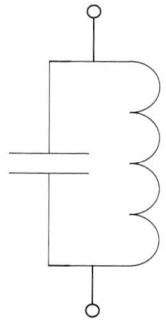

Fig. 2 equivalent circuit modal

b) Filter for GSM

A filter for GSM had been designed using the resonator structure mentioned above. As shown in Fig.3 (a), its dimension was 3.5mm by 1.7mm, and had 2.75 turns. Both w and s were 150μm, and the d was 200μm. The electrical characteristics of resonator or filter were simulated by Ansoft 3-D full wave electromagnetic software HFSS. The insertion loss and the return loss of these filters were specified in Fig. 4.

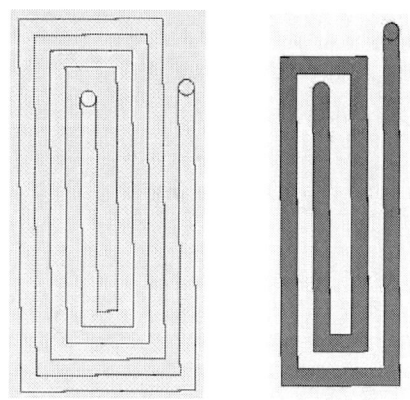

Fig.3 (a) filter for GSM   (b)filter for GPS

c) Filter for GPS

A filter for GPS also had been designed using the resonator structure mentioned above. As shown in Fig.3 (b), its dimension was 3.1mm by 1.1mm, and had 1.75 turns. Both w and s were 150μm, and the d was 200μm. The electrical characteristics of resonator or filter were simulated by Ansoft 3-D full wave electromagnetic

software HFSS. The insertion loss and the return loss of these filters are specified in Fig.7.

Result and discussion

Fig. 4 shows the insertion loss and return loss of the filter for GSM. The resonator exhibited bandpass feature, and the central frequency was close to 0.9GHz. The insertion loss and return loss at the central frequency were -2.57dB and -11.82dB. The bandwidth was 140MHz which from 0.83 to 0.97 GHz.

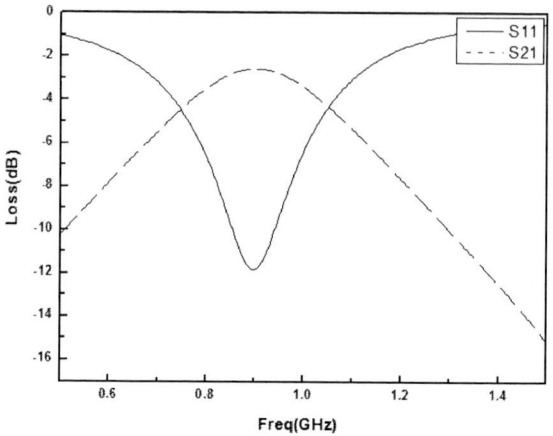

Fig.4 insertion and return loss of filter for GSM

To investigate the loss mechanism of this filter, the conductor and dielectric were modeled as perfect, respectively. Fig. 5 showed the insertion and return loss of the filter, when the metal was perfect electric conductor, and the dielectric loss tangent was 0.005. It showed that the insertion and return loss was -0.8dB and -20.82dB, respectively. Compared with Fig.4, the insertion loss increased 1.77dB and the return loss increase 9dB, in virtue of metal loss. It also could be found that the central frequency nearly shifted to 0.96 GHz. That was because the resistance of the filter was changed, and the impedance was not matching with 50ohm at 0.9GHz.

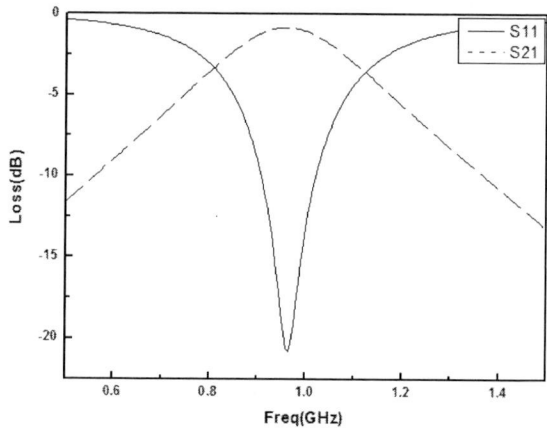

Fig.5 insertion and return loss of filter with perfect conductor for GSM

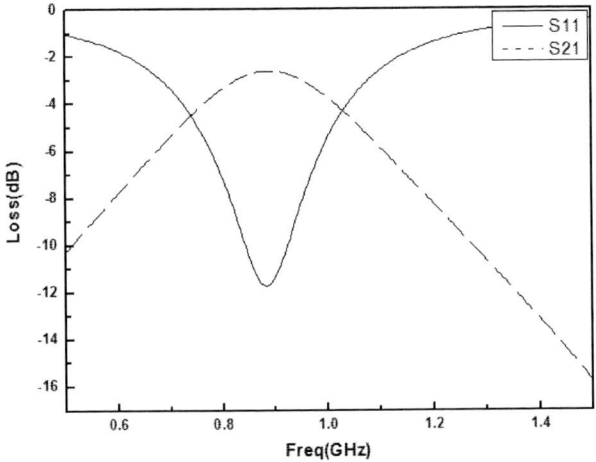

Fig. 6 insertion and return loss of filter with perfect dielectric for GSM

When the dielectric was perfect, the loss properties were shown in Fig.6. It showed that the insertion and return loss was almost the same as Fig.4. It showed that the influence of dielectric loss tangent on loss mechanism could be ignored, and the conductor metal loss was in the highest flight.

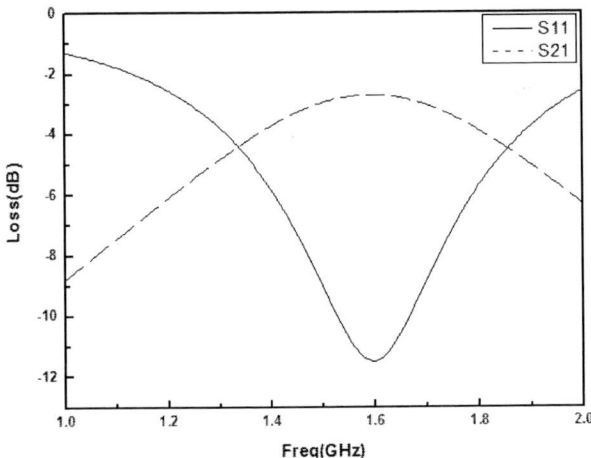

Fig.7 insertion and return loss of filter for GPS

Fig.7 showed the insertion and return loss of filter for GPS. The resonator exhibited bandpass feature, and it showed that the central frequency was close to 1.58 GHz. The insertion loss and return loss at the central frequency were -2.8dB and -13.5dB. The bandwidth was 140 MHz which from 1.52 to 1.66 GHz.

Fig.8 showed the insertion and return loss of the filter, when the metal was perfect electric conductor, and the dielectric loss tangent was 0.005. It could be seen that the insertion and return loss was -0.06dB and-41.5dB, respectively. Compared with Fig7, the insertion loss increased -2.8dB and the return loss increase -28dB, in virtue of metal loss. It also could be found that the central frequency nearly shifted to 1.7 GHz. The reason for that was the same as the shift of GSM filter.

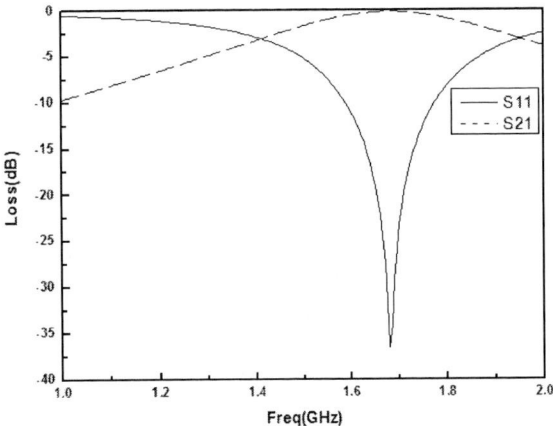

Fig. 8 insertion and return loss of filter with perfect conductor for GPS

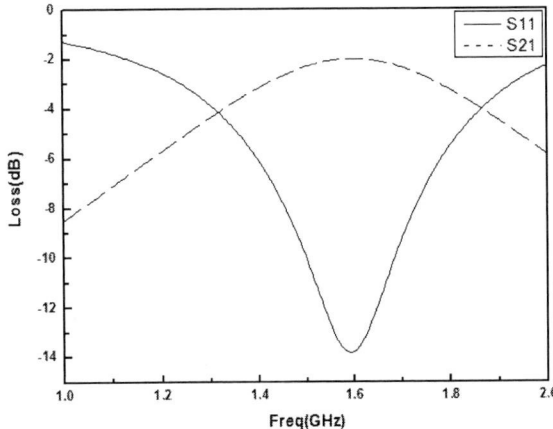

Fig.9 insertion and return loss of filter with perfect dielectric for GPS

When the dielectric was perfect, the loss properties were shown in Fig.9. Compared with Fig.7, the insertion loss decreased -0.82dB and the return loss increase -0.32dB, due to dielectric loss tangent.

Conclusion

The miniaturized bandpass filters using embedded capacitor had been proposed, which were designed and simulated by HFSS. A miniaturized filter for GSM with the size of 3.5mm by 1.7mm, and another one for GSM with the size of 3.1mm by 1.1mm, had been proposed. The results showed that the frequency characteristics could meet the wireless applications. Moreover, the loss mechanism of the miniaturized filters had also been studied, and the results showed that conductor loss had more influence on frequency properties than dielectric loss for the GSM and GPS applications. That is to say, if the frequency beyond multi-MHz, the conductor loss of this miniaturized resonator structure would dominate over the loss mechanism.

Acknowledgments

This work was supported by National High-tech Project (863) (Optical Interconnect Technology for High Speed Chips and Verification Platform NO. 2007AA01Z2a6.).

The authors would like to thank the SIAT faculty, research staff and students.

References

1. Chang-Sheng Chen., "Embedded Capacitors Technology in 2.4 GHz Power Amplifier with Multi-layer Printed Wiring Board (PWB) Process", int'l symposium on Electronic Material and Packaging, 2002.

2. Ching-Liang Weng., "Embedded Passive Technology for Bluetooth Application in Multi-layer Printed Wiring Board (PWB)", Electronic Components and Technology Conference, 2004.

3. Mekita F. Davis., "Integrated RF Architectures in Fully-Organic SOP Technology", IEEE Transacton on Advanced Packaging. 4. Dalmia,S., "Design of inductors in organic substrate for 1-3GHz wireless applications" IEEE MTT-S International,2002

5. Seung J.Lee., " Fully embedded High Q Passives and Band Pass Filters for Low Cost Organic RF SOP Applications". Electronic Components and Technology Conference. 2007.

6. Greg Brzezina., "A Miniature LTCC Bandpass Filter Using Novel Resonators for GPS Applications", Proceedings of the 37th European Microwave Conference, 2007.

7. Mohamadou Baba., "An Efficient Methodology for Design and Implementation of Embedded Bandpass filters for RF/Wireless Applications", Electronics Packaging Technology Conference, 2007.

8. Joong Keun Lee., "Design of Bandpass Filter for 900MHz ZigBee Application Using LTCC High Q Inductor ", APMC, 2005.

9. Yunfeng Wang., "A Study of RF Front-End Filters with Embedded Capacitor Technology", International Conference on Electronic Packaging Technology & High Density Packaging, 2008

Moiré Method for Nanoprecision Wafer-to-Wafer Alignment: Theory, Simulation and Application

Chenxi Wang, Tadatomo Suga
School of Engineering, the University of Tokyo
7-3-1 Hongo, Bunkyo-ku, Tokyo, 113-8656, Japan
Phone: + 81-3-5841-6495 E-mail: wang.chenxi@su.t.u-tokyo.ac.jp

Abstract

The two dimensional (2D) moiré centrosymmetric grating is developed to assist realization of high-precision wafer-to-wafer alignment and non-destructive measurement of misalignments for wafer bonding. Using these moiré patterns the misalignments in the order of ± 64 nm in X-Y axis can be resolved by a simple IR microscopy (5× objective) images. This value can be further improved to sub-10 nm range if the sub-pixel estimation is performed. For current status, the limit of the minimum resolved misalignment using moiré gratings is ~0.2 nm according to theoretical analysis. It can be applied for not only the future 3D integration of wafer-scale, but also the fabrication of 3D nanostructures and advanced lithography techniques.

1. Introduction

Aligned wafer bonding plays an important role for realization of 3D integration and nanostructure devices with accurate layer-to-layer interconnects [1-3]. Especially compared to microelectromechical system (MEMS) fabrication, 3D integration requires the alignment accuracy has to be improved 5~10 times [4]. For realization of nanostructure, alignment accuracy should be limited in the range of sub-100 nm or even smaller. A variety of alignment methods have been developed over the last decades. For most of alignment methods, the alignment process is generally performed by means of "cross-and-box" or "cross-on-cross" alignment marks, alternatively known as "alignment keys". Optical or infrared (IR) microscope is used to detect the positions of alignment marks [5]. Consequently, the alignment accuracy is mainly determined by three factors: the misalignment detection, the fabrication tolerance, and the resolution of position adjustment stage. It is known that the resolution of microscope (Re) is limited by diffraction effects according to Rayleigh's equation as follow [6]

$$Re = 0.61 \frac{\lambda}{NA} \quad (1)$$

where λ is the wavelength of the light and NA is the numerical aperture. Generally the wafers are inacceptable to be immersed in oil, thus, it reveals that the minimum resolution of the optical microscope is $Re \geq (0.61/1.0) \lambda \approx 0.6$ λ, for visible light $Re \geq 0.33$ μm and for IR light $Re \geq 0.67$ μm. The misalignment detection is limited by the resolution of microscope. In last decade, due to sub-pixel peak estimation, the minimum misalignment detection of 50 nm has been declared thus far using live IR image [7]. Furthermore, high NA (≥0.55) of microscopes implies a short depth of focus (±1 μm). It could be a problem since the gap between the top and bottom wafers during alignment is typically more than 10 μm. Hence, current available wafer-to-wafer alignment accuracies are in the range of 0.5~1 μm in laboratory-environments and in the range of 1~3 μm for commercial industry tools [5, 8].

Vernier patterns are extensively used to calibrate wafer-to-wafer alignment due to its higher accuracy more than just optics [9]. But that the verniers line-up is still difficult to be confirmed using optical microscope directly especially in nanometer scale. Moiré pattern is a promising candidate to resolve this problem as well as convenient for observation [10-11]. The moiré fringes produced by the superimposed gratings enable the manifestation of very small displacements in large moiré fringe movements [12]. In latest years, we developed a novel moiré fringe assisted alignment method using centrosymmetric square gratings, which have several advantages over previous moiré gratings [13-15]. For example, the moire fringes produced by two superimposed centrosymmetric gratings are highly sensitive to sub-pitch 2D misalignments, no relying on any external reference, and occupy relative smaller area.

In this article, we investigate the moiré fringes by simulation and estimate the minimum resolved misalignment using these moiré gratings, and subsequently test the actual moiré fringes created by these gratings in experiments.

2. Simulation of Moiré Fringes

We developed a 2D pattern of a centrosy-mmetric square grating, as illustrated in Fig. 1. It is composed of four sets of "L" shaped gratings as shown in Fig. 2. The second and fourth quadrants of the gratings have the pitch p1, whereas the first and third quadrants have the pitch $p2$. $p2$ is slightly different with p1. A discrepancy term Δp is given by $\Delta p = p2 - p1 > 0$. In this study, $p1 = 19$μm and $p2 = 20$μm, were chosen as an example. The square gratings on the top and bottom wafers are identical. Since the top wafer is turned over before alignment, consequently, the gratings on the two wafers with different pitches of four quadrants superimpose each other, and moiré fringes are achieved. The moiré fringe has a period of $P = p1 \times p2 / \Delta p = 380$ μm, which is much larger than p1 and $p2$. The designations Δx and ΔX hereafter refer to an actual misalignment and a moiré fringe displacement in the X-axis, Δy and ΔY refer to those in the Y-axis respectively. The ratio of moiré fringe displacement to actual misalignment (Z) is about 2×pavg/Δp = 39 ≈ 40, where pavg refers to the average of p1 and p2. The factor of two doubles the measurement sensitivity because of the counter motions of fringes along the X- or Y-axis in various quadrants, whose principle is similar to that of flipped line gratings. The size of the grating is set by 2×2 mm^2.

978-1-4244-4658-2/09 $25.00 © 2009 IEEE

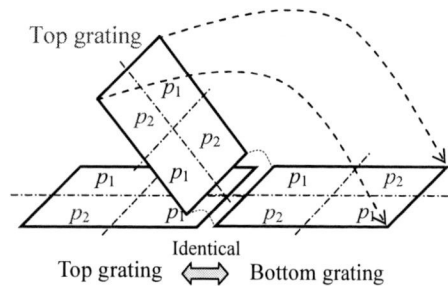

Fig. 1 Schematic illustration of identical top and bottom centrosymmetric square gratings.

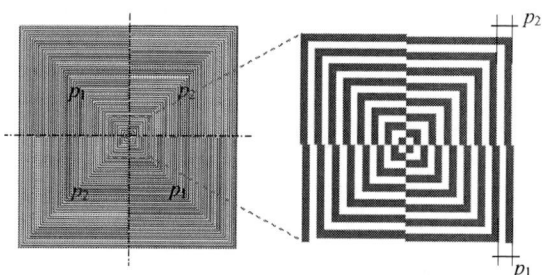

Fig. 2 Layout of centrosymmetric square grating

Simulation was carried out to investigate the relationships between misalignments and moiré fringes. The top grating is turned over and superimposed to the bottom grating by computer imaging. In Fig. 3(a), two square gratings are perfectly matched and the moiré fringe period P is 380 μm. When the top wafer is misaligned by 1 μm in the X-axis with respect to the bottom wafer, i.e., $\Delta x = 1$ μm, one vertical moiré fringe evolves into two mismatched moiré fringes, between which the distance is 40 μm. The difference between perfectly aligned and mismatched moiré fringes (corresponding to very small sub-pitch misalignments) can be judged without requiring any external reference. Thus, moiré fringes produced by two superimposed centrosymmetric gratings have higher sensitivity with small sub-pitch misalignment than conventional moiré gratings. With increasing the misalignments, the distance between two mismatched moiré fringes increase continuously [see Figs. 3(b)~(d)]. When the misalignment increase to $\Delta x = 10$ μm, the moiré fringe will match with each other again [Fig. 3(e)]. It means the mismatched moiré fringes repeat themselves when the two wafers are misaligned by a base period, which is the superior limit of measurement using the moiré gratings, namely Δx_{max}. It is obtained as

$$\Delta x_{max} = \frac{P}{Z} = \frac{p_1 p_2}{p_1 + p_2} \approx \frac{p_2}{2} = 10 \ \mu m \quad (2)$$

where P is the moiré fringe period and Z is the ratio of moiré fringe displacement to actual misalignment. If two wafers are misaligned beyond the base period x_{max}, such large

misalignments should be measured using alignment marks directly instead of using moiré pattern.

Moreover, whichever planar misalignment without rotations could be projected along two orthogonal directions, for instance, the misalignments of 1 μm and 2 μm in X- and Y-axis result in moiré fringes mismatched by 40 μm and 80 μm in X and Y axis respectively, i.e., $\Delta X = 40$ μm $\Delta Y = 80$ μm [Fig. 3(f)].

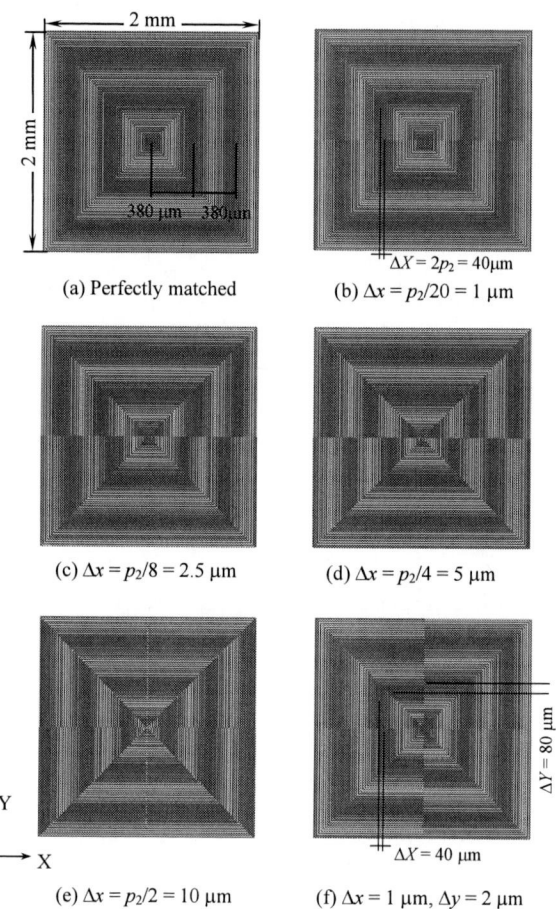

(a) Perfectly matched   (b) $\Delta x = p_2/20 = 1$ μm

(c) $\Delta x = p_2/8 = 2.5$ μm   (d) $\Delta x = p_2/4 = 5$ μm

(e) $\Delta x = p_2/2 = 10$ μm   (f) $\Delta x = 1$ μm, $\Delta y = 2$ μm

Fig. 3 Computer-generated moiré fringes produced by superimposing the two centrosymmetric square gratings with various misalignments.

3. Minimum Resolved Misalignment Using Moiré Pattern

We establish a mathematical model to describe moiré fringes theoretically and explore inferior limit of measurement, i.e., $\Delta x min$, the minimum resolved misalignment using the moiré grating. The interferometric approach is used to investigate the moiré fringe. For convenience, the top and bottom gratings are designated L_1 and L_2, respectively and they are of the same length, but L_1 has line pitch of p_1 whereas L_2 has pitch of p_2. Let the transmittance of grating L_1 be assumed sinusoidal as

$$T_1 = \frac{1}{2}\left(1 + \cos\frac{2\pi x_1}{p_1}\right) \quad (3)$$

978-1-4244-4658-2/09 $25.00 © 2009 IEEE   144

where x_1 represents the x coordinate of L_1. Similarly, the transmittance of grating L_2 may also be written as

$$T_2 = \frac{1}{2}\left(1 + \cos\frac{2\pi x_2}{p_2}\right) \qquad (4)$$

where x_2 represents the x coordinate of L_2. The misalignment Δx is given by the difference between x_1 and x_2, i.e. $\Delta x = x1 - x2$. When the light beam (the intensity of light is I_0) transmits through the superimposed gratings, then the overall transparent intensity distribution is determined by

$$I(x) = I_0 \cdot T_1 \cdot T_2 \qquad (5)$$

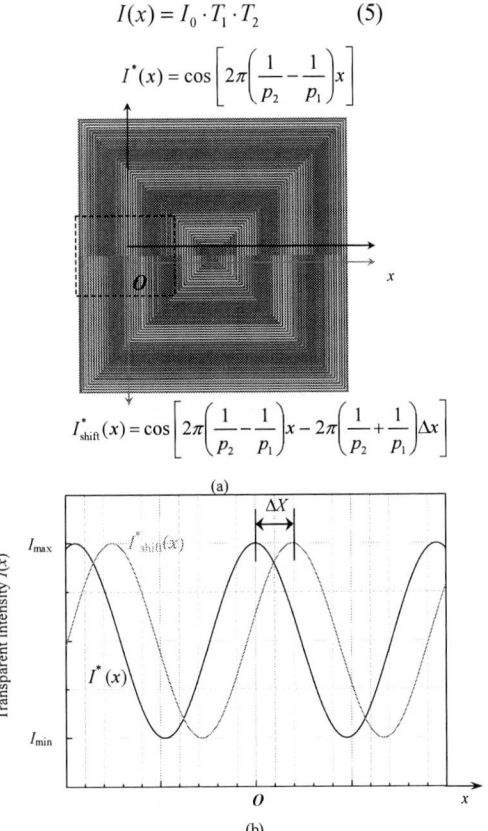

Fig. 4 Analysis of moiré fringes produced by two centrosyemmtric square moiré gratings with misalignment.

Moiré fringes are in fact modulated to this low spatial frequency component. Thus, the transparent intensity can be simplified and expressed approximately by

$$I^*(x) = I_0 \cos\left[2\pi\left(\frac{1}{p_2} - \frac{1}{p_1}\right)x_2 - \frac{2\pi \cdot \Delta x}{p_2}\right] \qquad (6)$$

For our centrosymmetric moiré gratings, similar to dual-line moiré gratings, two sets of moiré fringes move on opposite directions during the movement of the top grating to the bottom one, which doubles sensitivity of measurement. Thus, the transparent intensity can be modified by

$$I^*(x) = I_0 \cos\left[2\pi\left(\frac{1}{p_2} - \frac{1}{p_1}\right)x - 2\pi\left(\frac{1}{p_2} + \frac{1}{p_1}\right)\Delta x\right] \qquad (7)$$

$|\Delta x| > 0$ means there are some misalignments between the two gratings. Thus, one moiré fringes will be evolve into to two mismatched moiré fringes, which can be expressed as two sinusoidal curves, namely I*(x) and I*shift (x) [see Fig. 4]. And at least two sets of moiré fringes should be produced in a pair of superimposed centrosymmetric square gratings. So the relationship between the size of moiré square grating $L \times L$ and the moiré fringe period can be known as

$$L \geq 2P = 2\frac{p_1 \cdot p_2}{p_2 - p_1} = 2\frac{p_1 \cdot p_2}{\Delta p} \approx 2\frac{p_1^2}{\Delta p} \qquad (8)$$

ΔX represents a distance between the two peaks of sinusoidal curves, which is also equal to the distance between the mismatched moiré fringes in the X-axis. The distance ΔX should be larger than resolution of the microscope; otherwise the mismatched fringes will be unresolved. Thus it may be depicted as

$$\Delta X = Z \cdot \Delta x = \frac{p_1 + p_2}{p_2 - p_1} \cdot \Delta x \approx \frac{2p_1}{\Delta p} \cdot \Delta x \geq \text{Re} \quad (9)$$

where Re is the resolution of the microscope, defined as the shortest distance between two points that can still be distinguished by the microscopy image system. It is limited by Rayleigh's equation as mentioned previously. Δx and ΔX hereafter refer to an actual misalignment and a moiré fringe displacement in the X-axis, Z is the ratio of moiré fringe displacement to actual misalignment. For digital image, generally 1 pixel corresponds to the resolution of microscope. Recently the sub-pixel estimation technique using the computer-aided image processing was used to improve the resolution of microscope. It means displacement less than 1 pixel can be recognized, for example 1/10 pixel. For the moiré fringes, the value of

$$\text{Re}^* = \frac{\text{Re}}{C} \qquad (10)$$

where Re* is recognition resolution using image processing and C (≥ 1) is recognition coefficient of image processing. Considering the limitation of resolution of microscope and corresponding sub-pixel imaging processing technology, the resolved misalignment using the moiré grating may be described as

$$\Delta x \geq \frac{\text{Re}}{C} \cdot \frac{\Delta p}{2p_1} \qquad (11)$$

At least two moiré fringes should be captured in one image, thus the visual field with M×N pixel array (M≥N) is restricted by

$$N \cdot \text{Re} \geq 2P = 2\frac{p_1 p_2}{\Delta p} \approx 2\frac{p_1^2}{\Delta p} \qquad (12)$$

Based on the eqs.(11) and (12),

$$\frac{\Delta x \cdot 2p_1}{\Delta p} \geq \cdot \frac{\text{Re}}{C} \geq \frac{2}{N}\frac{p_1^2}{\Delta p} \qquad (13)$$

Here we may obtain

$$\Delta x \geq \frac{p_1}{C \cdot N} \qquad (14)$$

Based on the limitations of the recognition resolution using image processing eq. (11) as well as the visual field of image eq.(14), the resolved misalignment using the moiré grating may be described as

$$\Delta x \geq \Delta x_{min} = \max \left\{ \frac{Re}{C} \cdot \frac{\Delta p}{2p_1}, \frac{p_1}{C \cdot N} \right\} \quad (15)$$

where Δx_{min} is the minimum resolved misalignment of the centrosymmetric square moiré grating.

According to eq. (15), it is easy to know the discrepancy term Δp (= p2-p1 >0) is a critical factor to affect the minimum resolved misalignment. If Δp is chosen in nanometer scale to improve misalignment detectivity, the value of Δp can not regard as a constant value, but a random variable with standard deviation in fabrication process. On the basis the theory of probabilities, the fabrication tolerance of the discrepancy term Δp is given by

$$e_{\Delta p} = \sqrt{2} \cdot e \quad (16)$$

where e and $e_{\Delta p}$ refer to the fabrication tolerance of $p1$ (or $p2$) and Δp respectively. To ensure the moiré fringe must be produced by superimposed $p1$ and $p2$ in practice

$$\Delta p > \Delta e_{\Delta p} = \sqrt{2} \cdot e \quad (17)$$

In addition, to avoid diffraction effect, p1 should be limited by the wavelength of the light beam (λ)

$$p_1 \geq 2\lambda \quad (18)$$

Table 1 Parameters of moiré grating for misalignment detection.

Parameters of moiré grating design	
p_1, p_2	Pitch of moiré grating ($p_2 > p_1$), $p_1 \geq 2\lambda$ (Diffraction effect)
Δp	Discrepancy term, $\Delta p = p_2 - p_1 > \sqrt{2} \cdot e$ (Fabrication tolerance)
Parameters of optical microscope and Image processing	
Re	Resolution of microscope, 1 pixel = Re = 0.61λ/NA
Re*	Recognition resolution using image processing, Re* = Re/C (C = 10 for normal image processing)
N·Re	Size of visual field, N is the number of pixels in the capture image, e.g. N = 600
Misalignment detection using moiré grating	
Δx_{min}	Minimum resolved misalignment $\Delta x_{min} = \max \left\{ \frac{Re}{C} \cdot \frac{\Delta p}{2p_1}, \frac{p_1}{C \cdot N} \right\}$
L	Size of moiré grating, $L \geq 2P \approx 2\frac{p_1^2}{\Delta p}$

Therefore, all limitations for the parameters of moiré grating are demonstrated as above. The details are summarized in Table 1. For current status, p1 ~ 2 . 2.2 µm (IR wavelength ≥1.1 µm can pass through the silicon wafers), e ~ 10 nm for fabrication tolerance of e-beam lithography, and Re ~ 0.67 µm and Re* ~ 67 nm (C = 10) for the resolution of IR microscope and recognition resolution using image processing respectively. The pixel array is 1024 × 1024 (i.e. N ~ 1024). Therefore, the limit of the minimum resolved misalignment using moiré gratings is Δxmin ~ 0.2 nm. And the size of moiré gratings (L) should be $L \geq 2 \times 325$ µm = 650 µm.

Fig. 5 shows the minimum resolved misalignment using the moiré grating Δx_{min} as a function of p_2, Δp, the resolution of microscope (Re), and the recognition coefficient of image processing (C). Because Re is usually a constant for a microscopy imaging system, Fig. 5 provides us a guidance on how to choose the parameters of the moiré gratings for achieving a desirable misalignment detection.

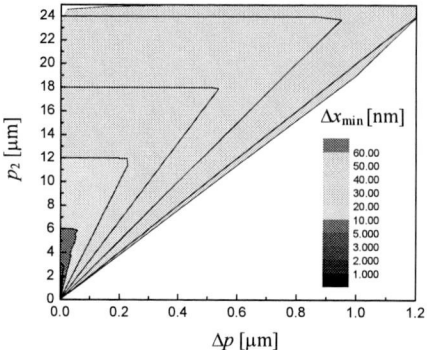

(a) Without sub-pixel image processing (C = 1)

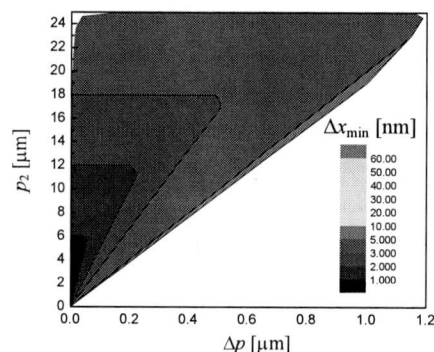

(b) With sub-pixel image processing (C = 10)

Fig. 5 Minimum resolved misalignment using moiré pattern (Δx_{min}) is as a function of p_2 and Δp (Resolution of microscope Re = 2 µm, Size of visual field with pixel array 800×600, i.e. N = 600), where $p_2 \geq 10\Delta p$.

4. Experimental Test

Topographical gratings were fabricated on double-side-polished silicon wafers. We tried to bond two wafers with high-precision alignment. And the misalignments in sub micronmeter or nanometer range will be measured using the square moiré pattern. To reduce the fabrication errors, the centrosymmetric square gratings were directly drawn by e-beam lithography without fabricating mask. In our study, the DRIE Bosch process was used to etch the grating grooves into silicon 6 µm deep. After etching, the photoresist was removed. Fig. 6 shows the process flow for fabrication of topographical moiré patterns on Si wafers.

A high-alignment wafer bonding tool was used to perform high-precision alignment and bonding of wafers. The alignment and bonding chamber is the main part of the wafer bonding tool. Fig. 7 shows the configuration of the alignment and bonding chamber, which consists of electrostatic chucks

978-1-4244-4658-2/09 $25.00 © 2009 IEEE

holding the top and bottom wafers, a piezoelectric walking table serving as the alignment table, IR camera/table for detection of alignment marks, and a bonding head to load the pressure. Three piezo elements installed into the bonding head and three Z axis units attached to the piezoelectric walking table are utilized to adjust the parallelism of the two wafers. The alignment used highly accurate image processing to recognize the position of the alignment marks on the top and bottom wafers and aligned the bottom wafer by means of the piezoelectric walking table to the top wafer. When the alignment process was completed, the two wafers were bonded with a pressure of 500 N.

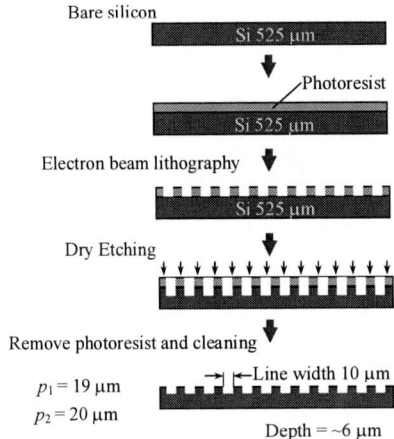

Fig. 6. Schematic process flow for the fabrication of the centrosymmetric square gratings on silicon wafers by DRIE etching.

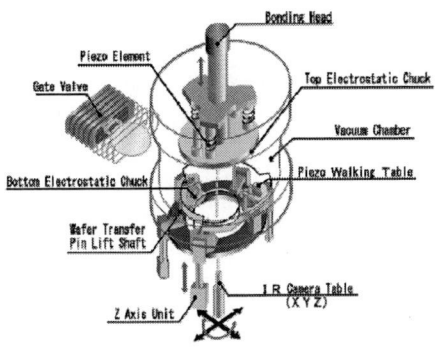

Fig. 7 Schematic of the wafer alignment and bonding chamber.

Here, we show an example how to detect the actual misalignments between the two wafers using moiré gratings. The CCD camera has a pixel array of 634×475 and the pixel size is 2.5 μm $\times 2.5$ μm. The moiré fringe is captured by a computer with 256 gray levels. The gray levels along dual lines, i.e. line 1 and 2, can be easily extracted from the digital image. For this IR image, a mismatched fringe of 1 pixel, which can be well detected, corresponds in our case to a ~2.5

μm mismatched moiré fringe distance. According to eq. (15), the minimum resolved misalignment Δxmin is ~64 nm without using sub-pixel estimation. If the image processing of sub-pixel estimation is carried out, for example recognition coefficient $C = 10$, the minimum resolved misalignment Δxmin can be improved to be ~6.4 nm.

Fig. 8(a) is the IR image for the moiré fringes produced by two superimposed topographical Si patterns. We detected the peak of transparent intensity for each moiré fringe and the distance between the mismatched moiré fringes (ΔX) was measured, as shown in Fig. 8(b). The ΔX is the average of ΔX_1, ΔX_2, ΔX_3 and ΔX_4, which is the mean value of 7 pixels and standard deviation of 1 pixel (i.e. 7 ± 1 pixels). The magnification factor of this moiré grating (Z) is 40 times. Consequently, the actual misalignment is $\Delta x = 0.448 \pm 0.064$ μm.

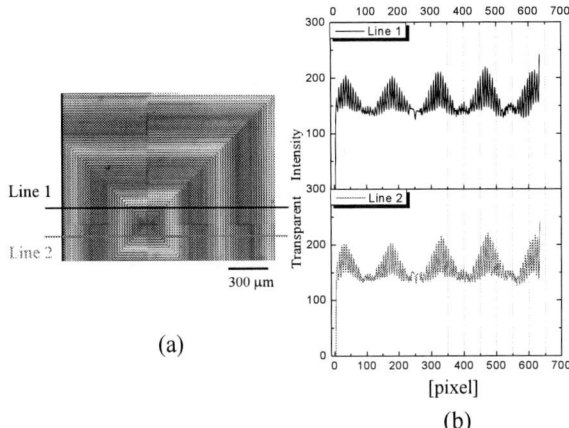

(a)

(b)

Fig. 8. Image processing for moiré fringes generated as IR image with $5 \times$ objective of Si/Si

Generally the accuracy of wafer-to-wafer alignment is determined by many factors, such as misalignment detection, wafer warpage and distortion, accuracy of nanopositioning stage, parallelism of the wafers and so on. For misalignment detection itself, by optimizing parameters of the square gratings, small misalignments in range of nanometers can be measured and feed back, thus high accuracy should be achieved. Three parameters need to be designed, i.e. p_1, p_2 and Δp, only two of which are independent due to the relationship $\Delta p = p_2-p_1$. We may expect to optimize the parameters of this square grating, e.g., Δp and p_2. Up to date, the most advanced lithography technologies have already accomplished gratings with fabrication tolerance e less than 10 nm (3 sigma). For example, we may choose $\Delta p = 0.1$ μm and $p_2 = 5$ μm for the moiré square grating. For our IR microscope ($20 \times$ objective) and image-processing system, Re $= 2$ μm for 1 pixel. According to eqs. (2) and (15), even without using the sub-pixel estimation ($C = 1$) the measurement range for planar misalignments is

$$20 \text{ nm} \approx \Delta x_{\min} \leq \Delta x \leq \Delta x_{\max} \approx \frac{p_2}{2} = 2.5 \text{ μm} \quad (19)$$

978-1-4244-4658-2/09 $25.00 © 2009 IEEE   147

If C = 10 for the sub-pixel estimation, $\Delta x_{min} \approx 2$ nm. Consequently, this moiré assist alignment method is possible applied for X-ray, e-beam and nanoimprint lithography techniques.

5. Conclusions

We developed practical and effective approach using 2D centrosymmetric square moiré patterns to measure the misalignment and assist high-precision alignment for wafer bonding. Moiré fringes produced by the two superimposed centrosymmetric square gratings are highly sensitive to very small misalignments and misaligned directions. The results of the simulation and experiment show that the mismatched moiré fringes can be identified easily, and the distance between the mismatched fringes are also measured using an IR microscope without requiring any external reference. Two pairs of the moiré square gratings are fabricated on the top and bottom wafers prior to bonding, thus, the alignment accuracy of bonded wafers are determined on wafer scale. In our experiment, the misalignments in the order of $\pm$ 64 nm in X-Y axis can be resolved by a simple IR microscopy (5× objective) images, as an example. This value can be further improved to sub-10 nm range if the sub-pixel estimation is performed. For current status, the limit of the minimum resolved misalignment using moiré gratings is ~0.2 nm according to theoretical analysis.

Acknowledgments

The authors would like to acknowledge Bondtech Co., Ltd. for their assistance on wafer bonding. Institute for Advanced Micro-system Integration (IMSI) consortium is appreciated for the financial support of this project.

References

1. M. Esashi, "Wafer Level Packaging of MEMS", *J. Micromech. Microeng.,* Vol. 18, No. 073001, pp. 1-13, 2008.
2. I. Radu, I. Szafraniak, R. Scholz, M. Alexe, and U. Gösele, "GaAs on Si Heterostructures Obtained by He and/or H Implantation and Direct Wafer Bonding", *J. Appl. Phys.,* Vol. 94, pp. 7820-7825, 2002.
3. A. W. Pool et. al., "Three-dimensional Integrated Circuits", *IBM J. Res. and Devel.,* Vol. 50, No. 4/5, pp. 491-506, 2006.
4. C. S. Tan, R. J. Gutmann, and L. R. Reif, Eds, Wafer Level 3-D ICs Process Technology, *New York: Springer,* 2008, pp. 59-63.
5. F. Niklaus, G. Stemme, J.-Q. Lu, and R. J. Gutmann, "Adhesive wafer bonding", *J. Appl. Phys.,* Vol. 99, No. 031101, pp. 1-28, 2006.
6. K. Okamoto, "Importance of wafer bonding for the future hyper-miniaturized CMOS devices", *ECS Trans.,* Vol. 16, No. 8, pp. 15-29, 2008.
7. M. M. R. Howlader, H. Okada, T. H. Kim, T. Itoh, and T. Suga, "Wafer level surface activated bonding tool for MEMS packaging", *J. Electrochem. Soc.,* Vol. 151, No. 7, pp. G461-467, 2004.

8. V. Dragoi, "Low temperature aligned wafer bonding", *Proc. of 1st International IEEE Low Temperature Bonding for Integration*, Tokyo, Japan, Nov.1-2, 2007, pp. 199-212.
9. S. Kawashima, M. Imada, K. Ishizaki and S. Noda, "High-Precision Alignment and Bonding System for the Fabrication of 3-D Nanostructures", *J. Micro electromech. Syst.,* Vol. 16, No. 5, 1140-1144, 2007.
10. M. Meinhold, J.-W. Jung and D. A. Antoniadis, "Sensitive Train Measurements of Bonded SOI Films Using Moiré", *IEEE Trans. Semicond. Manuf.,* Vol. 17, No. 1, 35-41, 2004.
11. A. Moel, E. E. Moon, R. D. Frankel, and H. I. Smith, "Novel On-axis Interferometric Alignment Method with Sub-10 nm Precision", *J. Vac. Sci. Technol. B,* Vol.11, No.6, pp. 2191-2194, 1993.
12. Y. Nishijima, "Moire Patterns: Their Application to Refractive Index and Refractive Index Gradient Measurements", *J. Opt. Soc. Am.,* Vol. 54, 1-8, 1964.
13. C. Wang and T. Suga, "Measurement of Alignment Accuracy for Wafer Bonding by Moiré Method", *Jpn. J. Appl. Phys.,* Vol. 46, No. 4B, 2007, pp. 1989-1993.
14. C. Wang, S. Taniyama, Y-H. Wang, and T. Suga, "High-Precision Alignment for Low Temperature Wafer Bonding", *J. Electrochem. Soc.,* Vol. 156, No. 3, 2009, pp. H197-201.
15. C. Wang and T. Suga, "A Novel Moiré Fringe Assisted Method for Nanoprecision Alignment in Wafer Bonding", *Proc. 59th Electronic Components and Technology Conf.*, San Diego, California, USA, May 26-29, 2009, pp. 872-878.

Localized Induction Heating for Wafer Level Packaging

Mingxiang Chen [1,2], Wenming Liu [1,2,*], Yanyan Xi [1,2], Changyong Lin [1,2], Sheng Liu [1,2]

[1]Division of MOEMS, Wuhan National Laboratory for Optoelectronics, Wuhan, China, 430074

[2]Institute of Microsystems, School of Mechanical Engineering, HUST, Wuhan, China, 430074

*Email: kevinliuwm@gmail.com, Phone: 86-27-87542604, Fax: 86-27-87557074

Abstract

Localized induction heating for wafer level packaging is discussed. This paper is to investigate the relationships between the geometry of solder loop and temperature distribution in induction heating. Using finite element method (FEM) and IR thermal imager, temperature distribution and variation are explored, which shows that the temperature on the solder loops is a function of the area and edge width in the induction heating. The temperature difference between the solder loop and its center on PCB is about 180°C, which shows an obvious localized heating effect. Thermal images of circle and square solder array indicate uniform temperature distribution on the PCB in the induction heating. Good agreement between measured and simulated temperature variation have been achieved.

1. Introduction

Wafer level packaging (WLP) is one of the promising packaging technologies for zero-level MEMS packaging owing to its fewer processing steps and lower cost, and it has increasingly become a key technology for materials integration in various areas of MEMS, microelectronics and optoelectronics [1-2]. It is noted that most MEMS devices are packaged after the release of their dynamic elements or the completion of the integrated circuitry, which both are temperature-sensitive. Hence, temperatures to which wafers are exposed during the packaging process and the exposure duration should be carefully controlled.

Some low temperature packaging techniques [3], such as low temperature solder packaging, adhesive packaging and surface activation packaging, are developed to avoid thermally-induced problems and damages by high packaging temperatures. Unfortunately, most low-temperature-packaging processes do not offer high quality bonds because packaging strength is also a function of temperature. Accordingly, there is an urgent need to develop a low temperature WLP technique to address this concern of meeting the packaging quality requirement. Localized heating and packaging technology, therefore, is presented [4-7]. That is, only the packaging or packaging area is heated, the other features on the wafer are still kept at a low temperature during the packaging process. In fact, many techniques have been used for localized heating and packaging, such as resistance heating [4], laser heating [6], microwave heating [7], and induction heating [8-9], et al.. Induction heating, due to its reliable, localized, accurate and energy-efficient heating in short time, provides many advantages over other localized heating methods and is commonly used for packaging applications. In the previous study [10], wafer level packaging using localized induction heating has been demonstrated for micro-device packaging. However, the temperature non-

uniformity among the micro solder loops on the wafer led to voids or packaging failure because of over-heating or under-heating. In addition, the temperature difference between packaging areas and those located inside micro-devices has not yet been considered.

In this paper, the principle of induction heating is firstly described, and the concept of localized heating for WLP is introduced. Then induction heating of metallic pattern array on the Printed Circuit Board (PCB) is simulated using finite element method, and induction heating experiments of different metallic pattern array on PCB are carried out. Finally, the results of the modeling and measured temperature distribution are compared and discussed.

2. Localized Induction Heating Principles and Modeling

2.1 Principles of Induction Heating

Induction heating is a complex combination of electromagnetic induction and heat transfer. As shown in Figure 1, when an alternative current passes through the induction coil (inductor), a magnetic field with the same frequency as the current is produced. According to Faraday law, eddy current is produced in the heating sample, then the sample is heated because of Joule effect.

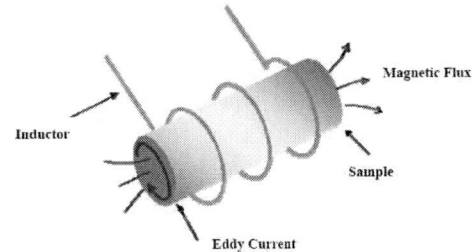

Fig. 1 Sketch of the principle of induction heating

Skin effect plays an important part in induction heating. That is, when induced current flow through the sample, the current focuses on the surface of the samples and then permeates into the sample. This permeating depth is called the skin depth, defined as,

$$\delta = \sqrt{\frac{\rho}{\pi \mu f}}$$

where ρ and μ are the resistivity and magnetic permeability of the material, respectively, f the frequency of the high-frequency source.

For wafer level packaging using induction heating, when the wafer is placed into the inductor, solder loops, made of metallic materials, can be inductively heated and melted for packaging, while the substrate, usually as an insulator or semiconductor, is still maintained at a low temperature during packaging, as shown in Figure 2. By optimizing of frequency

978-1-4244-4658-2/09 $25.00 © 2009 IEEE

of power supply, material, structure and geometry of solder loop, wafer level induction heating and packaging is promised.

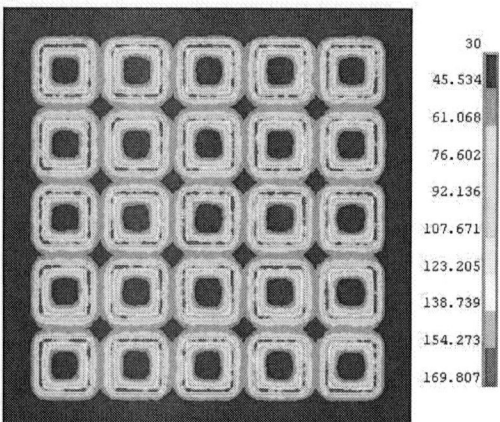

Fig. 2 Simulation of WLP by induction heating

2. 2. Modeling of Induction Heating for Packaging

The modeling of induction heating processes is a complex coupling of electromagnetic field analysis and thermal analysis. One of the major features of induction heating computation is that both the electromagnetic and thermal phenomena are tightly coupled because of the interrelated nature of the material properties. The temperature dependence of material characteristics such as the electrical conductivity, thermal conductivity and so on, is considered in the induction heating analysis. Obviously, these variations of physical properties make the induction heating nonlinear. This interrelated feature dictates the necessity of developing a special computational framework that is able to deal with these coupled effects. An indirect coupling framework [10], in this paper, is implemented to solve induction heating problems.

3. Experiment and Discussions

3.1. Experiment Setup

The induction heating system, as shown in Figure 3, consists of three parts, radio-frequency (RF) power supply, control system and packaging chamber. RF power has an output from 0 to 700 watt with f=13.56 MHz. so the same frequency harmonic magnetic field, which is vertical to the wafer and solder loops, is produced in the coil inside the chamber. During experiments, an IR thermal imager is used to monitor the temperature distribution of solder loops and PCB board.

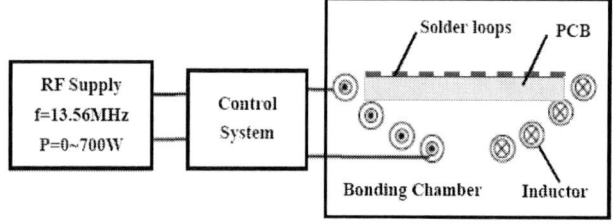

Fig. 3 Sketch of induction heating system

3.2. Solder Loops with Different Edge Width

Induction heating experiment is firstly performed on solder loops with different edge width on the PCB. The RF power is 300W. Figure 4 (b) shows the thermal image of solder loops in induction heating. At the 0.05 seconds, the temperature of the loop with 0.5 mm edge width is the highest, and with the edge width decreasing, the temperature of the loop also reduces. One reason for this could be that resistance of the solder loop increases due to the narrower loop edge width, and the eddy current induced reduces.

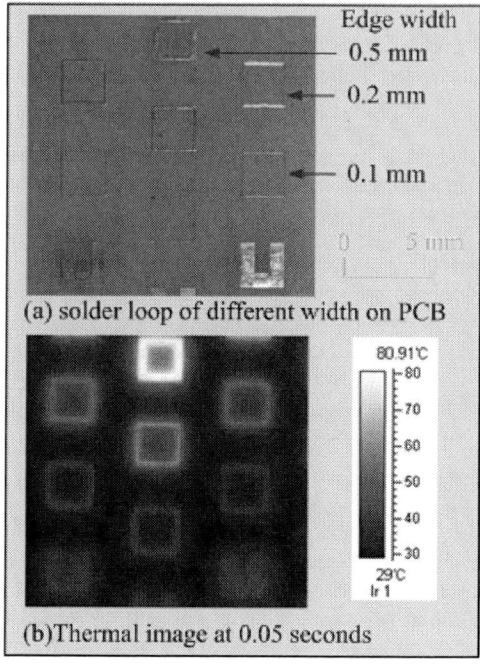

(a) solder loop of different width on PCB

(b)Thermal image at 0.05 seconds

Fig. 4 Thermal image of solder loops with different edge width

The temperature in the solder loops is recorded by the thermal imager as shown in Figure 5. The solder loop with 0.5 mm edge width is fast heated to about 120°C, and then the temperature increases at a lower rate. Other three kinds of loops have the similar heating tendency but with low heating speed. The possible reason might be the temperature-dependant properties of the solder and PCB board, such as the electrical conductivity, thermal conductivity, and heat capacity.

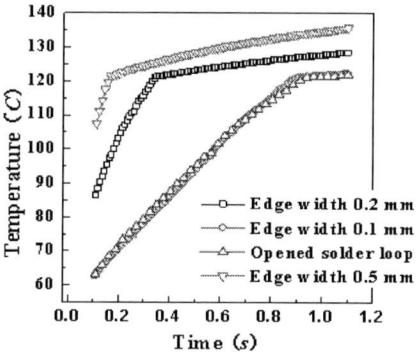

Fig. 5 Measured temperature variation in solder loop

Figure 6 shows the temperature comparison between simulation and experiment of solder loops on the PCB, seen also in Fig.4 (a). A relatively good agreement of between measurement and simulation is observed.

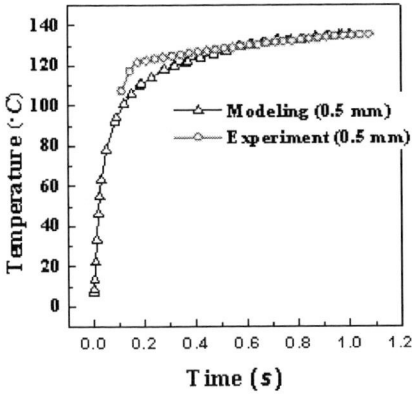

Fig. 6 Temperature variation of the solder loop
(Edge width is 0.5 mm, RF power is 300 W)

3.3. Solder Loops with Different Edge Length

Figure 7 (a) shows the solder loop array with different edge length on PCB. Each loop has the same edge width of 0.1 mm but with different edge length. The thermal image in Figure 7 (b) indicates that the larger the solder loop is, the higher the temperature is. In addition, these solder loop with the same geometry nearly have the same temperature. These findings accord with the relationship that the power absorbed by the sample in induction heating is the function of its area enclosed along external edges. These also suggest the reason that the open solder loop with 1 mm edge width is heated slowly as shown in Figure 6.

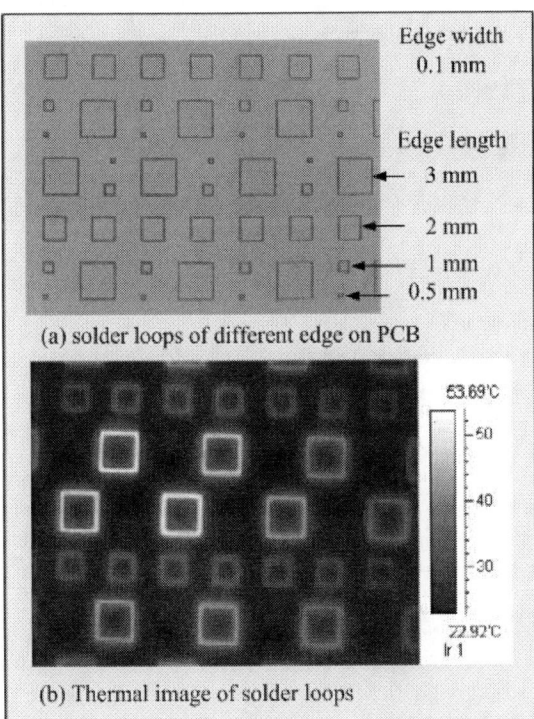

Fig. 7 Thermal image of solder loops with different edge length

Seen from the temperature variation curves in Figure 8, it also shows that the larger the edge length of the solder loop is, the quicker the temperature in the solder loop increase initially. Moreover, the temperature variation curve of the solder loop with 3 mm edge length changes the heating-up rate at about 120°C, which is almost the same as those in Figure 6. The inflexion of the temperature curve happens at a lower temperature in Figure 8, when the edge length of the solder loop decreases. When the solder array shown in Figure 7(a) is heated at 500W, the temperature on 3 mm- edge-length solder loop increased quickly to 360°C, as shown in Figure9, and then the loop opened. Before opening of the loop, simulation and measurement are basically consistent.

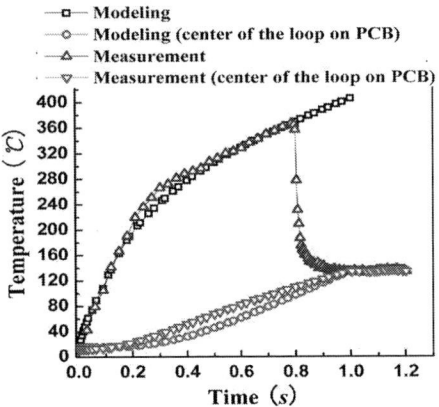

Fig. 8 Experimental temperature variation of the solder loops with different edge length (RF power is 300 W)

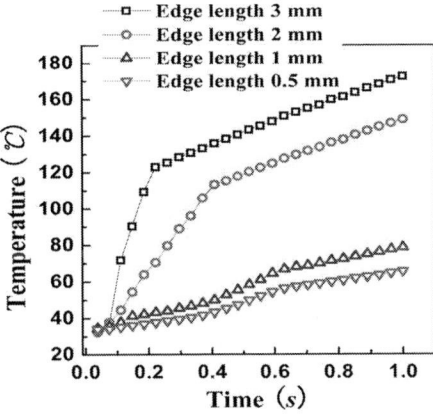

Fig. 9 Temperature variation of the solder loop
(edge width is 0.5 mm, RF power is 500 W)

When the RF power is increased to 600 W, the induction heating of the solder loops with different edge length is also investigated. Solder loops with 3 mm edge length are heated up more quickly and melted because of overheating, as shown in Figure 10. At 4 seconds, the temperature on PCB board is increased to about 480°C due to thermal conduction. The temperature is too high to see the solder loops with 0.5 mm and 1 mm edge length from the thermal imager. Figure 10(b) shows the temperature variation along L01 line. As can be seen, the temperature at the center of loops is below 300°C, while that on the loops is about 480°C, local heating effect is very obvious. Furthermore, the temperature on the loops with the same edge length of 3 mm is basically the same, which shows an even heating on the PCB substrate.

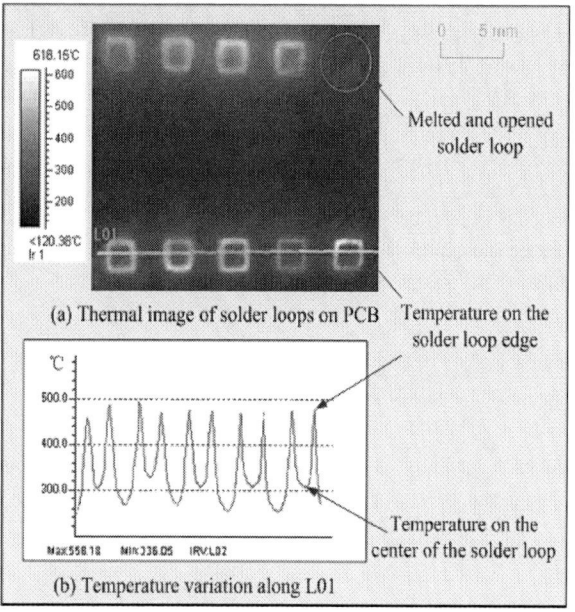

(a) Thermal image of solder loops on PCB

(b) Temperature variation along L01

Fig. 10 Thermal image of solder loop (RF power is 600W)

3.4. Same Solder Loops Array

Wafer level temperature uniformity by localized induction heating is also an important issue. In order to address this problem, induction heating experiments of the same solder loop array on PCB are performed.

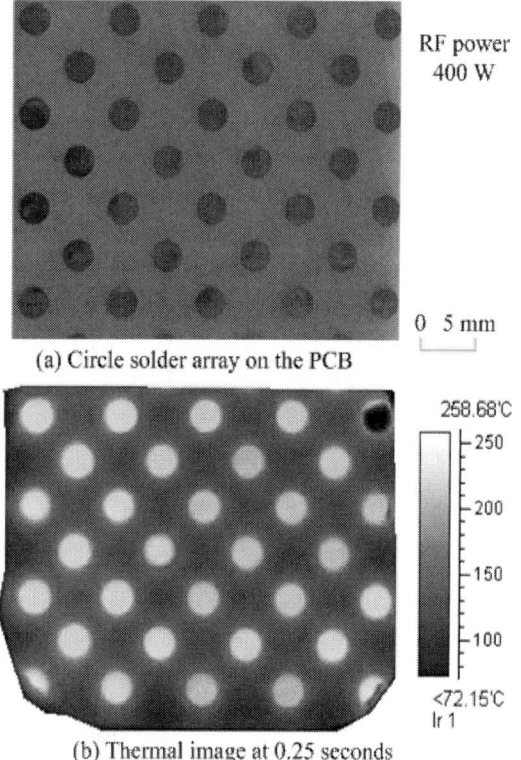

(a) Circle solder array on the PCB

(b) Thermal image at 0.25 seconds

Fig. 11 Thermal image of circle solder loop array

For the uniform magnetic field induced by a specially-designed inductor, the same solder loop array is heated very quickly. The thermal images, shown in Figure 11 and Figure 12, indicate a great consistency among the solder array. This suggests that the temperature uniformity for WLP by localized induction heating can be achieved in a well-designed magnetic field, provided that the shape and size of the solder loop are the same.

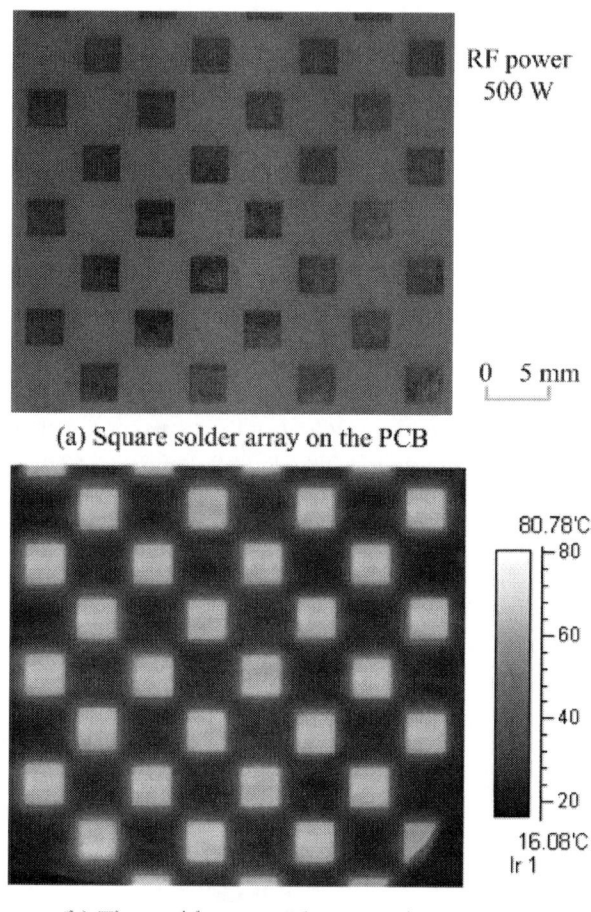

(a) Square solder array on the PCB

(b) Thermal image at 0.1 seconds

Fig. 12 Thermal image of square solder loop array

4. Conclusions

Localized induction heating for wafer level packaging is presented, and the relationship between the geometry of solder loop and the temperature distribution by induction heating is investigated. Firstly, indirect coupling method is introduced to model the induction heating process. Simulations and experiments show that, larger solder loops can be induced much more heat and heated more quickly, the heating rate is increased with the edge width of solder loops (range from 0.1 to 0.5mm). The temperature difference between the solder loops and their center on the PCB is about 180°C, which shows an obvious localized heating effect. For a uniform magnetic field, the temperature uniformity of same solder loop by induction heating is also promised. Good agreement between measured and simulated temperature variation is realized.

978-1-4244-4658-2/09 $25.00 © 2009 IEEE

Acknowledgements

This work was supported by High Tech Project of Ministry of Science and Technology in P.R. China with granted number of 2006AA04Z328 and Nature Science Foundation of China (NSFC) under grant number 50875012.

References

1. H. Ko. Wen, "Packaging of microfabricated devices and systems," Materials chemistry and physics, Vol. 42, No. 3 (1995), pp. 169-175.

2. Viorel Dragoi, "From magic to technology: materials integration by wafer bonding," *Proc. of SPIE*, Vol. 6123, (2006), pp. 1~15.

3. X. X. Zhang and Jean-Pierre Raskin, "Low-Temperature wafer Bonding: A study of void formation and influence on bonding strength," J. MEMS, Vol. 14, No. 2, (2005), pp. 368-382.

4. L. W. Lin, "MEMS post-packaging by localized heating and bonding," IEEE Transaction on adv. Packaging, Vol. 23, No. 4 (2000), pp. 608~616.

5. J. R. Mabesa, A. J. Scott1, X. Wu, "Localized heating/bonding techniques in MEMS packaging," *Proceedings of SPIE*, Vol. 5804, (2004), pp. 700~705.

6. M.J. Wild, A. Gillner, R. Poprawe, "Locally selective bonding of silicon and glass with laser," Sensors and Actuators, Vol. 93, No.1 (2001), pp. 63~69.

7. Jason Clendenin, Steven tung, Nasser Budra, et al., "Microwave bonding of silicon dies with thin metal films for MEMS applications," *Proc. 53rd Electronic Components and Technology Conference*, May. 2003, pp. 18~23.

8. Hsueh-An Yang, Weileun fang, "Localized induction heating solder bonding for wafer level MEMS packaging," J. Micromech. Microeng., Vol. 15, (2005), pp. 394~399.

9. M.X Chen, S. Liu. and Z.Y. Gan, "Selective induction heating for microsystem packaging", *Proc 7th ICEPT*, Shanghai, Aug., 2006, pp. 607-610.

10. W. M Liu,, M.X. Chen, Y. Xi, C. Lin and S. Liu. "Thermo-mechanical analysis of a wafer level packaging by induction heating," *Proc. ICEPT-HDP*, Shanghai, July, 2008, pp. 1-5.

Molecular Dynamics of Nanofluids with Time-Dependent Thermal Conductivity in Heat Sink Dissipation

Xiaojing Wang[*], Hongjun Liu, Wen Zhang, Zongshuo Li, Ling Chen
Shanghai University
Shanghai University, 224mail box, 149 Yan Chang RD, Shanghai, 20072, China
xjwang@mail.shu.edu.cn, 86-21-66136117

Abstract

Since the pioneering work by Tuckerman & Pease, lots of publications about heat sink have been researched in the last decade. Many enhancements are suggested in order to increase the critical heat current of heat sink including nanofluids which are solid-liquid mixtures composed of nanoparticles and basic liquid. The values of the thermal properties of nanofluids are enhanced to a large degree, even when the concentration of the nanoparticles is non-ignorably small. Molecular dynamics (MD) method can be used to generate and accurately trace the trajectories of the simulated particles and the interaction of the nanoparticles with the base fluid. In this study, MD is used to simulate the thermal conductivity of nanofluids obtained from the non-equilibrium MD (NEMD) approach under different temperatures. And simulations are based on the commercial software package FLUENT and treating the nanofluids as a two-phase mixture. Thermal resistance of four different coolants with different inlet velocities and heating powers are computed. Results show temperature-dependent thermal conductivity can't be neglected while simulating especially when the inlet velocity is large and the heating power is low.

Introduction

Micro channel heat sink (MCHS) has been studied over the past two decades for its capability to remove more heat from the very-large-scale integrated (VLSI) circuits. It is reported that a silicon micro-channel can remove 790W/cm^2 heat with a temperature rise of 71°C between the substrate and the coolant. The width and height of the silicon channel are 50μm and 302μm separately [1]. However as the technical development of the VLSI circuits, the traditional silicon MCHS can't meet the heat dissipation requirement of the advanced electronic devices, although lots of geometry optimizations have been reported by previous investigators (Goldberg [2], Phillips [3], and Knight et al [4]). In order to further enhance micro heat-sink performance, nanofluids is proposed as the cooling liquid. Nanofluids, first reported by Choi [5] in 1995, in which metallic or carbon-based nanoparticles with an averaged size of about 1-50 nm are suspended exhibit higher thermal conductivity values [5, 6]. Many theoretical models for the effective thermal conductivities of nanofluids are suggested by researchers because the traditional Maxwell [7] model can't explain the nanofluids' higher thermal conductivity than basic fluid. Xuan and Li [8] summarized the effective thermal conductivity of nanofluids as a function of both thermal conductivities of the carrier fluid and nanoparticle, in terms of particle volume fraction, surface area, and shape. Brownian motion of particles is considered by Jang and Choi [9], however, it has been proven that Brownian motion have smaller influence.

The present work attempts to estimate the thermal enhancement of water due to dispersed copper nanoparticles by Molecular Dynamics. The methodology, details of the computation, and results variation with temperature of the base fluid and the volume fraction of the nanoparticles, are explained in the follow section.

The micro channel flow and nanofluids are combined in the present study. Simulations are based on the commercial software package FLUENT & LAMMPS, and one-typical three-dimensional micro channel is modeled using copper-water nanofluids as cooling liquid considering that the thermal conductivity of nanofluids are temperature-dependent.

Heat Sink Performance

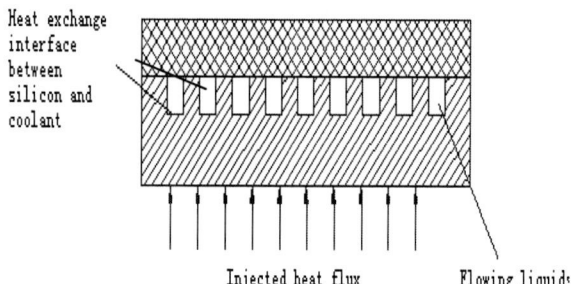

Fig. 1 Work demonstration of the heat sink

The heat sink is composed of three parts, the silicon substrate, micro channels and the glass plate. Originally, the substrate is attached closely with the chip. In order to simulate the heat generated by the chip, a constant heat flux is injected beneath the substrate. Flowing liquids are pumped into the micro channels with certain speed and exchanged energy with the heated interface, as shown in Fig. 1. The heat is taken away with the temperature rise of the flowing liquids. Because of the large amounts of channels in the heat sink, it is difficult to mesh the domain grids and takes extremely long time to obtain the results. For simplicity, one typical channel in Fig. 2 is chosen in the simulation due to symmetry.

The performance of the heat sink is measured by its thermal resistance R_{th}. Knight [4] expressed it as

$$R_{th} = \frac{T_{w,max} - T_{f,in}}{q''} \qquad (1)$$

Where $T_{w,max}$ and $T_{f,in}$ stand for the maximum wall temperature and the temperature of inlet flowing liquid and q'' represents the heat flux.

978-1-4244-4658-2/09 $25.00 © 2009 IEEE   154

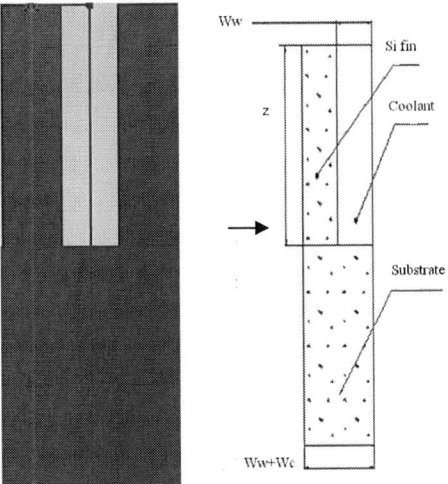

Fig. 2 One simple channel domain used for simulation

Theoretical analysis

Various approaches have been proposed to calculate the thermal conductivity of liquids at finite temperature with MD. In the equilibrium MD (EMD) approach, thermal conductivity k is calculated via Green–Kubo formulation [10].

In non-equilibrium MD (NEMD) approach, the MD model was divided into M regions along one direction of the nanofluids. The two regions at the two ends were defined as a hot and a cold region respectively. Based on the NEMD algorithm proposed by Ikeshoji and hafskjold [11], a constant energy was added to the hot region and the same amount of energy was removed from the cold region. The heat flux J in the system was defined as follows:

$$J = \frac{\Delta E}{\Delta t A_{cross}} \tag{2}$$

Where A_{cross} is the cross-sectional area. According to the temperature gradient along the nanofluids and the heat flux, the thermal conductivity, k, is obtained by

$$k = \frac{J}{\nabla T} \tag{3}$$

The Cu-O interaction [12] is taken proportional to the Al-O interaction with coefficient γ described as follows:

$$E_{cu-o} = \gamma f_0 \left(b_{Al} + b_o \right) \exp\left(\frac{a_{Al} + a_o - r}{b_{Al} + b_o} \right) \tag{4}$$

Where $\gamma = 0.4$

To model the interaction between the copper and hydrogen, the Lennard-Jones 12-6 potential is used and calculated by:

$$E = 4\varepsilon_{CuH} \left[\left(\frac{\sigma_{CuH}}{r} \right)^{12} - \left(\frac{\sigma_{CuH}}{r} \right)^6 \right], r \leq r_c \tag{5}$$

For copper and hydrogen [13],

$$\varepsilon_{CuH} = 1.6805 kcal/mol, \sigma_{CuH} = 0.258$$

The Morse Potential potential is used to account the interaction between the copper atoms in the nanoparticles. This makes sure that the copper atoms remain tightly connect together in the cluster. The Morse potential is given by:

$$E_{cu-cu} = De\left[\exp\left\{ -2\beta(r - R_e) \right\} - 2\exp\left\{ -\beta(r - R_e) \right\} \right] \tag{6}$$

where De, Re and β are the dissociation energy, equilibrium bond length and a constant with the dimension of the reciprocal of the distance respectively. For copper, the values for these are given below [14]:

De=0.3429eV; Re=2.8660Angstrom; β=1.3588 per Angstrom.

The TIP3P water model is used for the simulated water molecule. The two O−H bonds and the H−O−H angle in the molecule are held to be fixed and non-bonded interactions between the water molecules are comprised of a Lennard-Jones term between the oxygen atoms and a Coulomb potential as follows:

$$E = 4\varepsilon \left[\left(\frac{\sigma}{r} \right)^{12} - \left(\frac{\sigma}{r} \right)^6 \right] + \frac{1}{4\pi\varepsilon_0} \frac{q_i q_j}{r}, r \leq r_c \tag{7}$$

where ε and σ are equal to 0.1553kcal/mol and 3.166Angstrom respectively for Oxygen.

Results and Discussion

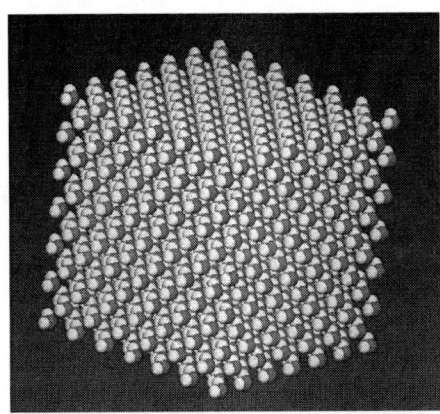

(a)

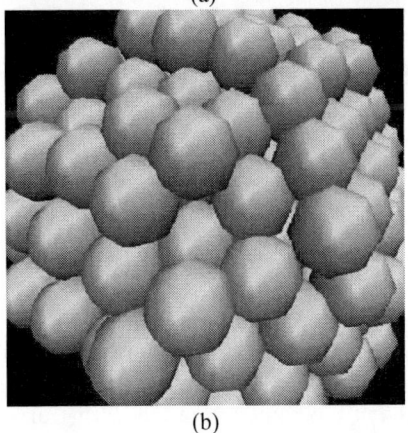

(b)

Fig. 3 Simulated water molecules and copper atoms. (a) Initialized water molecules in FCC structure. (b) Sphere copper cluster

In the present study, the reverse non-equilibrium method is applied, and water based copper nanofluid with 1% volume fraction under different temperatures are simulated. The sphere copper cluster has a diameter of 1 nm. So in the

978-1-4244-4658-2/09 $25.00 © 2009 IEEE   155

system, there are about 1005 water molecules, as seen in Fig. 3(a), containing 38 copper atoms, as seen in Fig. 3(b), in the center. At first the water molecules and copper atoms were arranged in the FCC structure, then the whole system is initially equilibrated for 40ps at desired temperature (NVT) controlled via a Nose'–Hoover thermostat and the integration time step was set to 1.0fs. A constant energy was then imposed on the system. NEMD simulations were run on the algorithm proposed by Ikeshoji and Hafskjold [11] using the ensemble of the constant number of particles, volume and energy (NVE) until a steady state temperature distribution achieved along the nanofluids. The temperature gradient was obtained in the next 10ps.

Thermal conductivity of 1% copper nanofluid under different temperatures can be seen in Fig. 4. There is a gradual increase in thermal conductivity of 1% copper nanofluids ranging from 0.680 to 0.851 as the rise of temperature. The blue star symbol depicted in Fig. 4 represents simulated values of thermal conductivity of 1% copper nanofluid at different temperatures and the red straight line is the fitting of a straight line adopted in the computation.

We consider the heat sink has a square surface (L*W), 1cm×1cm. The width of the channel and fin are both 50μm and the height of the fin is 400μm. The inlet coolant temperature is set at room temperature which is 293K. To validate the method, the thermal resistance is compared with the experimental data [1] in Table 1. w_c, w_w and z depicted in Table 1 are the channel width, fin width and channel height

of the heat sink respectively. P represents pressure drop, f stands for volume flow rate and $\dot{Q}$ is heating power. θ_{max} is maximum thermal resistance. Results show the data agree well with the experiment.

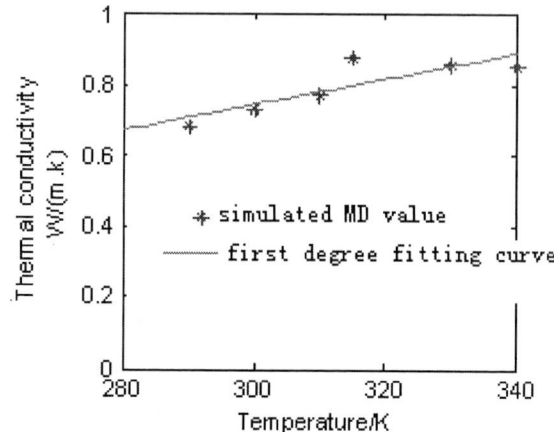

Fig. 4 Thermal conductivity of 1% volume Cu-Water nanofluid of different temperatures varying from 280K to 340K

Table 1 Comparison between the thermal resistance of Ref [1] and that of simulated results

$w_c(\mu m)$	$w_w(\mu m)$	$z(\mu m)$	$P(psi)$	$f(cm^3/s)$	$\dot{Q}(W/cm^2)$	$\theta_{max}(^{\circ}C/W)$	
						Ref[1]	Simulation results
56	44	320	15	4.7	181	0.110	0.107
55	45	287	17	6.5	277	0.113	0.104
50	50	302	31	8.6	790	0.090	0.091

In order to distinguish the difference between time-dependent nanofluids and non time-dependent nanofluids, there are four different coolants are compared which are pure water, 1% copper nanofluid with constant k which is 0.680 at the lowest temperature of 290K obtain via MD simulation results, 1% copper nanofluid with constant k which is 0.851 at the highest temperature of 340K and 1% copper nanofluid with time-dependent k obtained from the fitting of a straight line in Fig. 4.

Fig. 5(a) depicts thermal resistance of the MCHS using the four coolants mentioned above with different inlet velocities. From the figure, nanofluids do enhance the performance than water for the suspended copper cluster. Larger velocity means smaller thermal resistance but more slowly decrease. And constant k at higher temperature is more accurate than that at lower temperature especially when the inlet velocity is small.

The thermal resistance of the third and fourth coolant is almost the same when the inlet velocity is 3m/s but the difference among them becomes obvious when the inlet velocity reaches 10m/s.

Fig. 5(b) displays thermal resistance of the MCHS using four coolants which is also previous mentioned with different heating powers. As seen, the thermal resistance of the coolants is not dependent to the power except the copper nanofluid whose k is temperature-dependent. And when the power reaches the 500W, the thermal resistance is the lowest. It demonstrates that the effect of variable thermal conductivity can't be neglected.

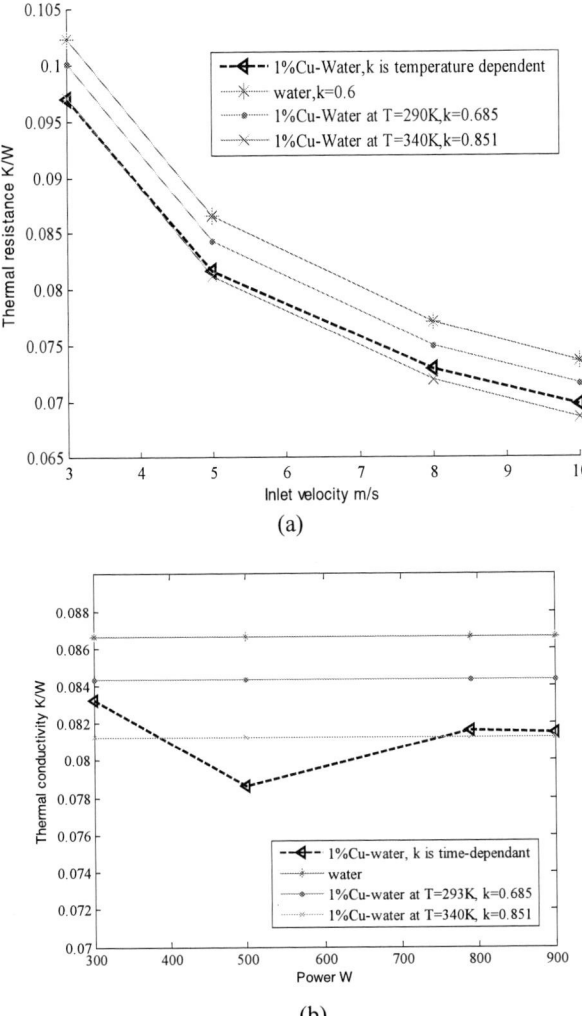

(a)

(b)

Fig. 5 Thermal resistance of the MCHS using four different coolants. (a) Different inlet velocities. (b) Different powers.

Conclusions

1) Thermal conductivity of 1% copper nanofluid is obtained via non-equilibrium MD (NEMD) approach. From simulated results, it is larger than pure water and varies significantly as the rise of temperature. Higher temperature means larger thermal conductivity.

2) The effect of temperature-dependent thermal conductivity is not significant in MCHS when the value of k is set at higher temperature with small volume flow rate. But the effect is obvious neither the value of k is set at lower temperature nor the volume flow rate is high.

3) When considering the temperature-dependent thermal conductivity of coolants, the thermal resistance is not non-dependent to the power any more.

Acknowledgments

The authors acknowledge the financial support from 863 program (No.2008AA04Z301), NSFC project (No.50876057) and Shanghai Municipal Education Commission (No.08YZ15).

References

1. D.B. Tuckerman, R.F. Pease, "High-performance heat sinking for VLSI," *IEEE Electronic Devices Letters*, EDL 2 (1981), pp.126–129.
2. N. Goldberg, "Narrow channel forced air heat sink IEEE Trans," *Comp. Hybrids Manuf. Technol.*, vol. CHMT-7 (1984), pp. 154-159.
3. R. J. Phillips, "Microchannel Heat Sinks," *Advances in Thermal Modeling of Electronic Components and Systems*, Vol. 2 (1990), Chapter 3, pp. 109-122, 165-171.
4. R.W. Knight, J.S. Goodling, D.J. Hall, "Optimal thermal design of forced convection heat sinks-Analytical," *ASME J. Electron Package*, 113 (1991), pp. 313–321.
5. S.U.S. Choi, "Enhancing thermal conductivity of fluids with nanoparticles," *ASME FED,* 231(1995), pp.99-103.
6. J. Koo, C.Kleinstreuer, "A new thermal conductivity model for nanofluids," *J. Nanoparticle,* Res. 6 (2004).
7. Maxwell JC, Electricity and magnetism. Clarendon Press, (Oxford, UK, 1873).
8. Y. Xuan, Q. Li, "Heat transfer enhancement of nanofluids", *Int. J. Heat Fluid Flow,* 21 (2000), pp. 58–64.
9. S.P. Jang, S.U.S. Choi, "The role of Brownian motion in the enhanced thermal conductivity of nanofluids," *Appl. Phys. Lett.*, 84 (2004), pp. 4316–4318.
10. K.V. Tretiakov, S. Scandolo, "Thermal conductivity of solid argon from molecular dynamic simulations," *Journal of Chemical Physics,* 120 (8) (2004), pp. 3765–3769.
11. T. Ikeshoji and B. Hafskjold, "Non-equilibrium molecular dynamics calculation of heat conduction in liquid and through liquid-gas interface," *Molecular Physics*, Vol. 81 (1994), pp. 251-261.
12. S.V. Dmitriev, N. Yoshikawa and Y. Kagawa 2004, "Misfit accommodation at the $Cu(111)/\alpha$-Al_2O_3 (0001) interface studied by atomistic simulation," *Comput. Mater. Sci.*, 29 95.
13. E. Duffour, P. Malfreyt, "MD simulation of the collision between a copper ion and an polyethylene surface: An application to the plasma-insulating material interaction," *Polymer*, Vol. 45, No. 13 (2004), p. 4565.
14. M. Han, Y.C. Gong, G.H. Wang, "Structure and dynamics of copper clusters," *J. Nanjing Univ.*, 30, 238 (1994).

Hydrophobic Self-Assembly Monolayer Structure for Reduction of Interfacial Moisture Diffusion

H. B. FAN, Cell K.Y. Wong, and Matthew M.F. Yuen
Department of Mechanical Engineering,
Hong Kong University of Science and Technology
Clear Water Bay, Kowloon, Hong Kong SAR, China

Abstract

Interfacial delamination is one of the primary concerns in electronic package design. Pop-corning during the solder reflow of plastic-encapsulated IC packages is a frequently occurred defect due to moisture penetration into the packages. Moisture absorption has a detrimental effect on the EMC/Cu interfacial adhesion and drastically reduces the reliability of plastic packages. To improve package reliability and to prevent interfacial delamination, it is important to design the EMC/Cu interface for high hydrophobicity and good adhesion.

The object of this paper is an investigation of both adhesion and moisture absorption at the EMC/Cu interface using MD simulations. Three kinds of models containing SAM1, SAM2 and a mixture of SAM1 and SAM2, have been used to evaluate the bonding energy and moisture absorption between EMC and SAM coated Cu substrate in this study. In each model, SAM1 or SAM2 or mixture of SAM1 and SAM2 chains were aligned on the copper substrate. MD simulations were performed at a given temperature using the constant-volume and temperature ensemble (NVT). Non-bond interactions cut-off distance of 1.25 nm with a smooth switching function was used in all simulations. The simulations were performed with an interval of 1 femto second (fs) in each MD simulation step. Moisture distribution and binding energy were calculated from simulation for each model. MD simulation results showed that the SAM1 has the higher bonding energy, while SAM2 has the higher hydrophobicity. It was also found that a mixture of SAM1 and SAM2 has both a higher bonding energy and a higher hydrophobicity which can be used as an interface promoter for adhesion and moisture inhibitor in electronic packages. This study shows that MD simulation can be an efficient tool for optimization of SAM to create a hydrophobic interface, which can provide useful pointers of the selection of the SAM structure.

Introduction

Moisture induced reliability concerns have been extensively studied in a package design. Pop-corning in plastic-encapsulated IC packages is a defect frequently occurring during solder reflow due to moisture penetration into the packages. Moisture absorption has a detrimental effect on the EMC/Cu interfacial adhesion and drastically reduces the reliability of the encapsulated package. Factors governing the interfacial delamination are mainly the moisture content and adhesion strength of the epoxy/copper interface at the target temperature. The loss of interfacial adhesion due to moisture is governed by the moisture diffusion rate combined with vapor pressure generated at the interface. Understanding interfacial adhesion subjected to different levels of moisture content is of significant interest to the electronic packaging industry.

Yee et al [1] have recommended the hydrophobic component in an epoxy system can hinder moisture uptake of the bulk epoxy. It implies that a hydrophobic interface may reduce the moisture content at an interface. Kinloch [2] suggested that the long alkyl chain of an vinyl silane with 20 carbons might impede water and improve bond durability. The argument is that the middle alkyl chains which are highly hydrophobic can obstruct water penetration and can thus improve the interface. Although these studies showed that a hydrophobic surface can reduce moisture absorption for interfacial joint, design of a hydrophobic interface with higher bonding energy for electronic devices still cannot be completely achieved. Moreover, as material properties at the interface are different from those of bulk materials, such as, the moisture diffusion along the EMC/Cu interface, it is important to better understand moisture diffusion at a molecular level.

Molecular modeling represents molecular structures numerically and simulates their behavior with the equations of quantum and classical physics and it is one of the fastest growing fields in science. The molecular dynamics (MD) method was first introduced by Alder and Wainwright in the late 1950's to study the interactions of hard spheres [3]-[4]. Mitsuhiro Fukuda and Satoru Kuwajima [5] used MD simulations to estimate diffusion of water cluster in amorphous polyethylene, which showed good agreement with the experimental values. The diffusion of water in mixture of water and poly was investigated using MD simulation by Muller- Plathe [6]. He concluded that water diffusion undergoes a pronounced change with the concentration and temperature when pure water mixed with high polymer concentration. Yarovsky, et al. [7] used MD simulations to investigate the diffusion constants for water molecules in the crosslinked polymer network composed of the epoxy resin of polymerization and different curing agents. Fan et al. [8] conducted MD simulations to investigate moisture diffusion into the epoxy molding compound (EMC) and the EMC/Cu interface. The MD results showed that the seepage along the EMC/Cu interface was more prevalent than moisture diffusion in the bulk EMC, and thus rendering it a dominant mechanism causing moisture induced interfacial delamination in plastic packages. However, not much research effort has been dedicated to the investigation on design of hydrophobic interface with higher interfacial bonding energy.

In this study, MD simulations were conducted to investigate wettability and the interfacial moisture diffusion at the EMC/Cu interface. Three kinds of models containing SAM1, SAM2 and a mixture of SAM1 and SAM2, were built. The MD simulation results show that SAM1 has the higher binding energy, while SAM2 has the higher hydrophobicity. The MD results also revealed that optimized SAM on Cu substrate for higher interfacial energy and more

978-1-4244-4658-2/09 $25.00 © 2009 IEEE   158

hydrophobicity can be achieved by the mixture of SAM1 and SAM2.

Molecular Dynamics Simulation

Wettability of SAM coated Cu substrate

In order to design a hydrophobic interface for reduction interfacial moisture diffusion, contact angle of water droplet on the SAM coated cooper surfaces was investigated using MD simulation. MD simulations are carried out for Cu surface coated with SAM1, SAM 2 and mixture of SAM1 and SAM2. Two SAM candidates are involved in this study and their chemical structures are shown in Figure 1.

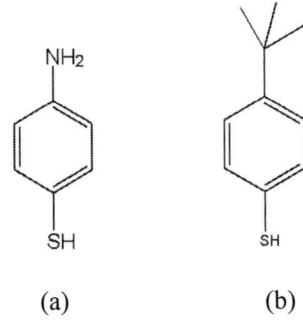

(a)   (b)

Figure 1: Chemical structures of (a) SAM 1 (b) SAM 2

In this simulation, the copper surfaces were cleaved from a crystal copper structure, corresponding to the (001) plane. As the non-bond cut off distance in the force field setting is 9.5A, the depth of copper surface used in the simulation was about 10A. A layer builder was used to build a sandwich and a large vacuum spacer was positioned at the top of the copper surface in order to avoid interaction across the mirror image in the z direction in the calculations. The unit cell was extended in the x and y directions to create a rectangular simulation box (50.22 x 50.22Å^2), periodic in the plane parallel to the surface. The SAM chains were initially placed on the copper substrate. Normally each sulfur atom in the chains should be covalently bonded to a copper atom. All simulations were carried out at the temperature of 298K, using the ensembles of the constant number of particles, constant-volume and constant temperature (NVT). MD simulations were performed to obtain optimized structure on the Cu substrate.

After the optimization of SAM on the copper substrate, a freely relaxed water dropt consisting of 670 molecules was centered on top of the SAM surfaces. The interactions among water molecules and between water molecules and SAM surface consist of van der Waals force and Coulombic force. The parameters used in the simulations are from the polymer consistent force files (PCFF) Accelrys Inc. San Diego CA, USA). A cutoff distance of 12.5Å was used for these nonbonding interaction forces. All the simulations were conducted at 25°C using NVT for about 500ps with an interval of 1 femto second (fs) in each MD simulation step.

Figure 2 shows the final configurations of water droplets on the two SAM-coated substrates. Water droplets on the two substrates were spread finally with different spheical shape of droplet cap. It is obviously seen that water contact angle for SAM2 coated Cu substrate is larger than that for SAM1

coated substrate. The stronger wettability of SAM1 coated substrate results from hydrophilic group N_2 in SAM1. While the hydrophobic group in SAM2 results in less interaction between water molecules and SAM 2, making the larger contact angle.

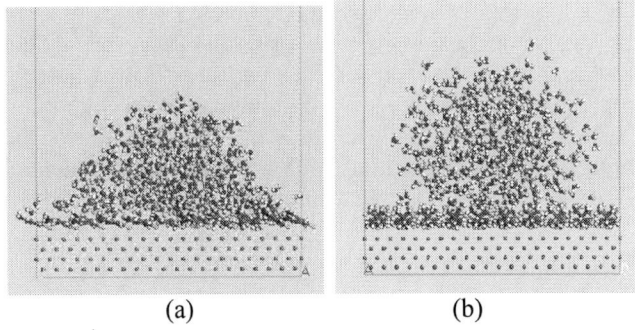

(a)   (b)

Figure 2: Final configuration of water droplets on (a) SAM1 coated Cu substrate (b) SAM2 coated Cu substrate

Interfacial moisture diffusion and bonding energy

Due to different wettability of SAM, optimized SAM structures on the Cu substrate should be designed to achieve a higher adhesion and lower interfacial moisture diffusion for the EMC-Cu interface. Considering the wettability o SAM and its ability to react with epoxy, three kinds of models containing SAM1, SAM2 and a mixture of SAM1 and SAM2, have been used to evaluate the moisture absorption and bonding energy between the EMC and SAM-Cu in this study.

In this study, the system consists of SAM, a fragment of Epoxy Molding Compound (EMC) and a Cu substrate. SAM solvent develops a thin monolayer on the Cu surface that allows direct linkage between Cu and epoxy resin. The fully cured epoxy network is composed of diglycidyl ether of bisphenol-A (DGEBA) epoxy and methylene diamine dianilene (MDA) curing agent and the model is same as that presented by Fan et al. [8]. Wong et al. [9] estimated interfacial bonding energy between the EMC and the SAM modified Cu substrate using MD simulations. They presented a method on how SAM was adsorbed on the Cu substrate and connected the EMC and the Cu substrate. Based on the same method, three models were built. For the model with SAM1, it was connected to the EMC by the covalent bonds formed between oxraine rings in the EMC chains and nitrogen atoms in SAM1 chains. For the model with SAM2, there is no any bonding between the EMC and SAM2. While for the third model, there is a mixture of 50% SAM1 and 50% SAM2, in which EMC was connected to SAM1 by the chemical bonds formed between them.

All MD models were built with a rectangular simulation box 2.54 x 2.54 nm^2 in the x and y directions, periodic in the plane perpendicular to the EMC-Cu interface. A large vacuum space was positioned at the top of the epoxy chains in order to avoid interaction across the mirror image in the z direction in the calculations. All the copper atoms were held rigid, while all the EMC chains and SAM molecules were allowed to move freely in all simulations. Water molecules were inserted in to the EMC/Cu interface. The mass ratio of water

978-1-4244-4658-2/09 $25.00 © 2009 IEEE   159

molecules to the EMC in both MD models is around 1.7%. Given the humidity condition, all the simulations were then carried out at a temperature of 85°C with a presumed moisture concentration value, using the using the ensemble of the constant-pressure and temperature (NPT) under 1 bar. All the system was equilibrated for about 100ps under 1 bar at 85°C with an interval of 1 femto second (fs) in each MD simulation. MD calculations were performed using the Discover module in Materials Studio software to find the thermal stable morphology and achieve a conformation with minimum potential energy for the whole system. Figure 3 shows the morphological configuration with the minimum potential energy for different SAM-coated substrates.

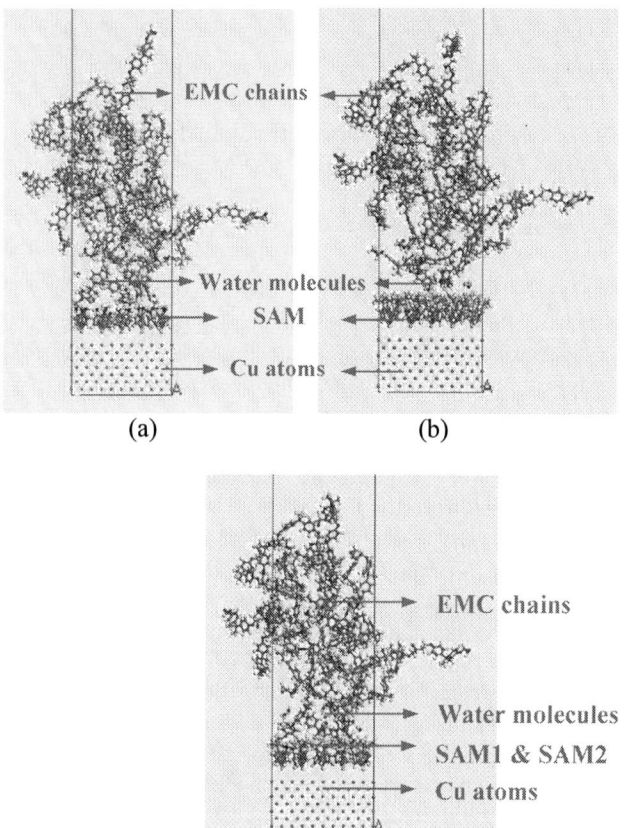

(a)   (b)

(c)

Figure 3: Morphological configuration of (a) SAM1 coated Cu substrate (b) SAM2 coated Cu substrate (c) Mixture of SAM1 and SAM2 coated Cu substrate.

Constants of moisture diffusion can be obtained in principle from diffusion trajectories $r(t)$ of water molecules determined during a MD simulation of a polymer packing model. The diffusion coefficients for the water molecules can then be calculated from the mean squared displacement, $s(t) = <|r(t) - r(0)|^2>$, of the water molecules averaged over time as follows:

$$D = \frac{1}{6N} \underset{t \to \infty}{Lim} \frac{d}{dt} \sum_{i=1}^{N} (r_i(t) - r_i(0))^2 \qquad (1)$$

where D is the moisture diffusion coefficient, $r_i(t)$ is the

coordinate of the center of the mass of the ith water molecule and N is the number of water molecules in the system.

Figure 4 shows the mean square displacement of water molecules against time calculated from MD simulations. The graphs were fitted using linear regression, $y = ax + b$, and the slop of the line, a, can be obtained (labeled in red in the figure). Since the value of the mean square displacement is already averaged over the number of atoms, N, the equation (1) can be simplified to:

$$D = a/6 \qquad (2)$$

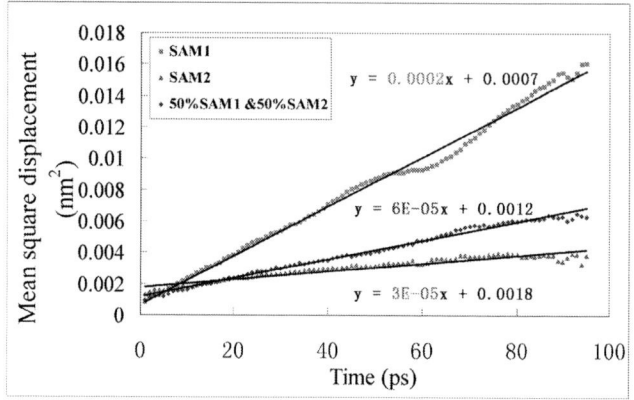

Figure 4: Mean square displacement of water molecules against time and the fitted line.

Normally, the interfacial energy is evaluated from the energy difference, ΔE, between the total energy of the whole system and the sum of the energies of individual parts as follows:

$$\Delta E = E_{tot} - (E_{EMC} + E_{Cu+SAM}) \qquad (3)$$

where E_{tot} is the total energy of the whole system, E_{EMC} is the energy of the EMC without the substrate, E_{Cu+SAM} is the energy of the SAM coated substrate without the EMC. The interfacial bonding energy, γ, is evaluated using the interfacial energy, ΔE, and the contact area, A, between the EMC and cuprous oxide substrate:

$$\gamma = \Delta E / 2A \qquad (4)$$

Results and Discussion

In this simulation, mean square displacement was used to analyze the moisture diffusion coefficient. Based on the slope of the linear regression fitting line for different mass ratio of water molecules to the EMC, the moisture coefficients for different cases are obtained by equation (2), and are listed in Table 1. From the above MD simulations, the interfacial moisture diffusion in three kinds of SAM coated Cu substrates are predicted. It is anticipated that moisture diffusion decreases when the substrate changes from SAM1, mixture of SAM1 and SAM2 to SAM2, which matches their wettability of these SAM-coated substrates. The reason is that the non-polar group in SAM can hinder interfacial moisture diffusion and water residence at the interface.

We also experimentally measured water contact angle on SAM1 and SAM2 treated Cu substrate. The contact angle for the SAM1 treated substrates was $50\pm1^{\circ}$, while the contact angle for SAM 2 is around $112\pm4^{\circ}$. These experimental results show the consistence with MD simulation results.

Based on equations (4), the interfacial bonding energy between the EMC and SAM coated substrate was calculated and the results were also listed in table 1. It shows that the interfacial bonding energy for the SAM 1 is the largest and that for SAM 2 is the smallest. The highest energy for SAM1 results from the covalent bonds formed between the SAM1 and EMC. The mixture of SAM1 and SAM2 still can provide a higher interfacial energy comparing with SAM2.

Table 1: Interfacial moisture diffusion coefficients and bonding energies for different SAM coated Cu substrates

	SAM1	SAM1 & SAM2	SAM2
Moisture diffusion coefficient at the EMC/Cu interface(mm^2/s)	3.33e-5	1.0e-5	5.0e-6
Interfacial bonding energy (J/m^2)	0.58	0.49	0.17

We also used tapered double cantilever beam (TDCB) test to evaluate fracture toughness the tensile adhesion between the EMC and SAM coated copper substrate under tensile mode. Experimental results showed that interfacial fracture toughness (G_{IC}) for SAM1, mixture of SAM1 and SAM2 and SAM2 only are 159 Jm^{-2}, 112 Jm^{-2} and 5 Jm^{-2} respectively. These experimental results are all consistent with the results from MD simulations in this study confirming that optimized SAM on the Cu substrate can improve adhesion between EMC and Cu substrate.

Both experimental and simulation results shows that mixture of SAM1 and SAM2 can not only improve adhesion by covalent bonds formed between EMC and SAM1 but hinder interfacial moisture diffusion by non-polar group in SAM2. These studies suggest a hydrophobic copper-epoxy interface can improve the long term reliability of an interfacial joint in electronic packages.

The results from MD simulations demonstrated moisture diffusion and bonding energy between the EMC and SAM coated substrate. Optimization of SAM on the Cu substrate can not only give higher adhesion but a hydrophobic copper-epoxy interface to reduce moisture absorption. Although MD simulations could not be directly compared to the experimental measurements, MD simulations are capable of generating in depth insight into the local molecular interactions andcan be used as a guide for further researches.

Conclusions

The paper is focused on wettability, interfacial moisture diffusion and bonding energy for different SAM coated Cu substrates using molecular dynamics simulations. MD simulation results show that the SAM1 has the higher binding energy, while SAM2 has the higher hydrophobicity. A mixture of SAM1 and SAM2 has both a higher bonding energy and a higher hydrophobicity which can be used as an interface promoter for adhesion and moisture inhibitor in electronic packages.

Acknowledgments

The project was supported by the Research Grant Council project (GRF 621907). The author would also like to thank Prof. Ping Gao for providing the software for the computer analysis.

References

[1] C. L. Soles, F. T. Chang, B. A. Bolan, H. A. Hristov, D. W. Gidley, and A. F. Yee, "Contributions of the nanovoid structure to the moisture absorption properties of epoxy resins," Journal of Polymer Science, Part B: Polymer Physics, vol. 36, pp. 3035-3048, 1998.

[2] [12] A. J. Kinloch, K. T. Tan, and J. F. Watts, "Novel self-assembling silane for abhesive and adhesive applications," Journal of Adhesion, vol. 82, pp. 1117-1132, 2006.

[3] B. J. Alder, T. E. Wainwright, "Phase Transition for a Hard Sphere System, Journal of Chemical Physics, Vol. 27, pp. 1208-1211, 1957.

[4] B. J. Alder, T. E. Wainwright, "Studies in Molecular Dynamics. I. General Method," Journal of Chemical Physics, Vol. 31, pp. 459-466, 1959.

[5] Mitsuhiro Fukuda, and Satoru Kuwajima, "Molecular-Dynamics Simulation of Moisture Diffusion in Polyethylene Beyond 10 ns Duration," J. Chem. Phys. 107 pp. 2149-2159, 1997..

[6] F. Muller-Plathe, Diffusion of Water in Swollen Poly (Vinyl alcohol) membranes Studied by Molecular Dynamic Simulation," J.Mmembr. Sci. 141, pp.147-154, 1998.

[7] Yarovsky I. and Evans E., "Computer Simulation of Structure and Properties of Crosslinked Polymers: Application to Epoxy Resin," Polymer, 43, pp.963-969, 2002.

[8] H. B. Fan, Edward, K. L. Chan, Cell, K.Y. Wong and M. F. F. Yuen, "Investigation of Moisture Diffusion in Electronic Packaging by Molecular Dynamic Simulation," J. Adhesion Sci. Technol. 20:1937-1947, 2006.

[9] Cell, K.Y. Wong, H. B. Fan, and M. F. F. Yuen, "Interfacial adhesion study for SAM induced covalent bonded Copper-EMC interface by Molecular Dynamics Simulation," IEEE Transactions on Components and Packaging Technologies, Vol. 31, pp.297-308, 2008.

Thermal Numerical Simulation for Advanced Package Development

Guohua Gao, HongHui Wang, GuoJi Yang, HaiQing Zhu

Nantong Fujitsu Microelectronics Co., Ltd.

No. 288, Chongchuan Road, Chongchuan, Economic Development Area, Nantong, Jiangsu, 226006, P. R. China

Email: gao.gh@fujitsu-nt.com

Abstract

In recent years, NFME has been developing advanced packages such as QFN, BGA, LGA, and SiP. In the new product development stage, thermal design needs to be taken into consideration to assess its thermal reliability. So, the accurate junction temperature prediction for the package is critical to electronic packaging design. In the industry, numerical models are often utilized to predict package thermal performance. In this paper, detailed HLQFP and BGA models were created, and a series of computational studies were conducted to obtain junction temperature and thermal resistance under JEDEC standards for natural and forced air convection conditions. In order to reduce computation time, model simplification was made from package geometry to the attached PCB test board. The simulated junction temperature values can also be used to estimate the maximum power dissipation allowed for the specific package. Thermal analysis helps determine whether the package will dissipate a sufficient amount of heat, or if it is still necessary to apply an external heat sink. Validated with experimental measurements, effective simulation can greatly help optimize advanced package structure design and material selection, and minimize the working temperature of packages to meet application requirements.

1. Introduction

With the evolution of IC advanced packaging technology, traditional package types will eventually be replaced with Flip Chip, Wafer Level Package and high-density SiP package. Along this technology development path, critical obstacles include heat, thermal stress and strain resulted from overheating. With the quantity of transistors increasing in one chip, more and more heat is generated. However, with a slight increase in chip size, the device heat density will increase disproportionally. Thus, overheating becomes the bottleneck in the development of microelectronics technology.

To assess component overheating, an effective thermal model of the developed package would be needed to predict the potential functional and reliability failures. This approach can offer a guideline for the feasible solutions as early as possible in the IC package design cycle. Ultimately, the goal is to provide optimal designs that meet or exceed IC component requirements within the given project schedule. JEDEC standards, achieved in the field of thermal characterization and chip packages modeling, provide the foundation in the thermal identification model.

Solutions to thermal reliability problems are sought at package-level, board-level, and system-level. Package-level thermal design seems to be the most important of the three, because it can produce the most significant effect with the least added cost.

With respect to the package-level thermal consideration, we will introduce our HLQFP thermal simulation and carry out studies on BGA and SiP packages. HLQFP has a different leadframe structural design (as shown in Fig.1). It adds an e-pad on the bottom surface of package, which leads to a more effective heat transfer and a much higher power density chip endurance than common QFP packages. We will do comparisons between common QFP and HLQFP by using the FEA method, and evaluate its IC power dissipation. And in this paper, methods for characterizing the transient thermal behavior of packages have been proposed and carried out, which is an area that needs urgent attention in future.

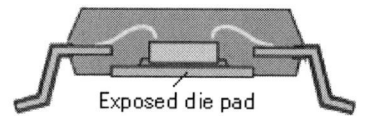

Fig.1. Cross-section graph of HLQFP

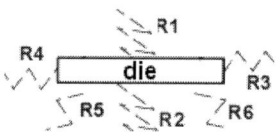

Fig.2. Thermal Dissipation path in a Device

From past studies, thermal resistance is a better parameter to measure the thermal performance of a package. Thermal resistance is analogous to electrical resistance in steady state. Fig.2 shows a simplified diagram of heat flow paths from the die surface to the package surface. Because of the specific package outline, the e-pad (shown in Fig.1) provides a better heat flow path (R5, R6, and R2). The standard definition of thermal resistance is $Rth=(T-Tref)/Q$, in which, T is the temperature of interest, such as junction or case, $Tref$ is the reference temperature, and Q is the steady-state power dissipation.

In an advanced SiP package, higher thermal resistance leads to poor heat transfer capability, which likely causes overheating problems. Lower thermal resistance will maximize the intrinsic performance and extend the life of semiconductor devices and help meet customers' needs for higher performance and lower cost. Understanding the role of thermal resistance is very important for thermal design and analysis for packages' heat transfer characteristics.

2. Simulation Assumption

As shown in Fig.3 (a) and (b), we present two kinds of LQFP256 models. In the HLQFP, the die pad is divided into two parts, connected by the tie bar, and the median downgoing die pad forms an e-pad, which is soldered directly to JEDEC PCB. When we created the BGA detailed model described in Fig.3 (c), we found that it is a complicated model because of its copper traces in the substrate.

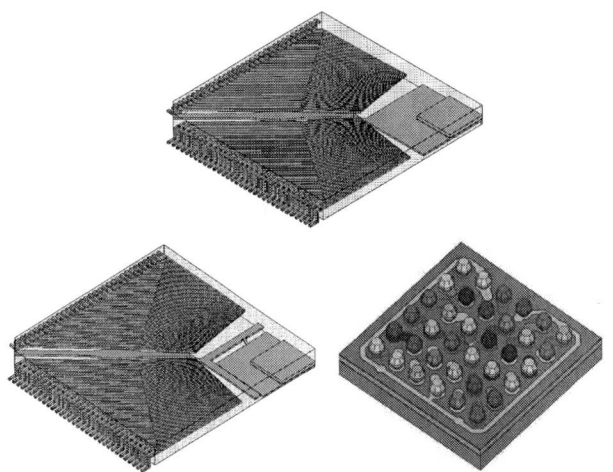

Fig.3. LQFP256 (a)HLQFP256 (b)BGA (c)models

Due to the symmetry, we only needed to build one quarter of the detailed three dimensional package models. Numerical simulation conditions for the above packages are summarized as following: the input power is 1 watt for the real assembled package, the selected PCBs are 1s0p single-layer and 2s2p 4-layer JEDEC standard test board with dimensions of 76.2mmx114.3mmx1.6mm. The control condition is ambient still air at 25°C temperature.

The material properties and test boards used in this study are described in Tab.1. The materials are commercially available in the each datasheet.

Tab.1. material properties

	Thermal conductivity(W/m°C)
Die Attach	2.5
Leadframe	170
Die pad	170
Au wire	317.9
Encapsulant	1
Solder paste	50.9
PCB dielectric	0.3
PCBtrace/plane	390

In the thermal simulation, the maximum junction temperature of each chip represents the thermal reliability level of an IC package. It is a complicated function of package structure, material property, input power, environment variables, and PCB layers.

In this paper, finite element analysis method was used to create detailed thermal models as shown in Fig.3. All the materials are assumed isotropic with constant physical property and all materials' interfaces are assumed perfectly bonded. The heat energy is applied on the chip to simulate heating power, and the heat transfer coefficient is calculated automatically in the models.

3. Result Analysis

3.1 HLQFP modeling

In order to compare two leadframes' heat transfer results, we did thermal simulations under general conditions indicated in **Part 2**. The temperature contour plots of HLQFP with the e-pad are shown in Fig. 4 and the thermal resistances are listed in Tab. 2. HLQFP package has an exposed pad that removes heat from the bottom surface of the package, which becomes the dominant thermal path for LQFP instead of the leads. It shows better thermal performance than LQFP package. In such instance, it is necessary to define top R_{jc} and bottom R_{jc} to avoid confusion over which surface is being referenced. In this way, the top surface is the surface of the package facing away from the PCB while the bottom surface is the surface of the package facing towards the PCB.

Tab.2. thermal resistance comparison

	R_{ja}	top R_{jc}	bottom R_{jc}
e-pad L/F	19.62	9.37	0.45
Ordinary L/F	37.41	13.76	11.49

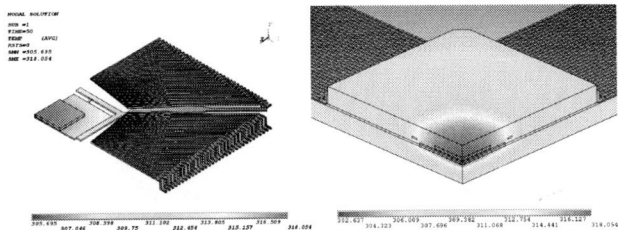

Fig.4. Temperature contour plots of HLQFP256

Within the JEDEC specification, single-layer and 4-layer configurations test boards are allowed for the thermal characterization. Because of its high copper trace density board design with buried power and ground planes, 2s2p 4-layer PCB gives a best thermal performance. Fig.5 shows thermal resistance values of R_{ja}, top R_{jc} and bottom R_{jc} for HLQFP with these two test boards. Note that all the materials and package geometries were held constant for these models.

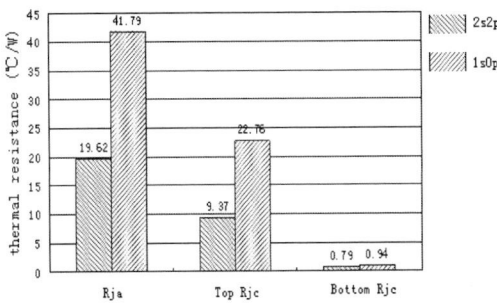

Fig.5. thermal resistance differences for two PCB

According to the model boundary setting, the surface temperature of the heat source drives both convection and radiation energy loss from the package. So, the hotter package surface will get more efficient convection and radiation heat loss to the ambient environment. Therefore, as shown in Fig.6, it is not surprising to note that R_{ja} decreases with increased power applied on the die. However, this effect is minimal as R_{ja} improved by only 2% when a package's power is doubled.

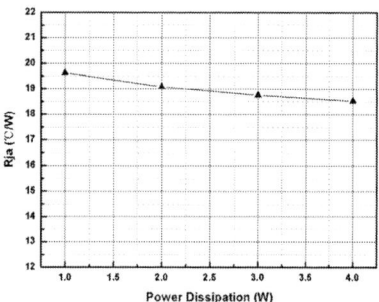

Fig.6. R_{ja} decreases with increased power

When a package reaches the thermal equilibrium state, the electrical power delivered is equal to the thermal heat dissipated, which is transferred to the surroundings. The maximum allowable power dissipation at a given ambient temperature is calculated using equation shown in **Part 1**. Over the maximum temperature, heat will destroy semiconductors. Therefore, we specify a maximum junction temperature as 125°C. If the junction temperature goes above this value, irreversible damage occurs. Typically, when we know about the ambient temperature of the operating environment, the thermal resistance of the IC package, and a specified maximum junction temperature, we can define the maximum power dissipation that can be applied to the particular package under the specified test conditions from Fig.7. Among the test board types, it is observed that the power dissipation with 2s2p PCB in the room temperature is optimal, reaching 5.41W.

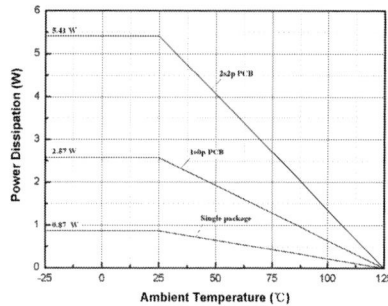

Fig.7. Power dissipation at different ambient temperature

The airflow is another important environment variable. As shown in Fig.8, by varying airflow velocity at the ambient temperature, thermal resistance R_{ja} changes greatly, and R_{jc} barely changes. Although we used the different test boards, all the variation trends are similar. So, R_{jc} is not changing with simulation environment, but has something to do with package structure, material property, input power, and PCB layers. R_{ja} is influenced by the different airflow velocity, but the change of R_{ja} will tend to reduce with the continuous increase of airflow velocity.

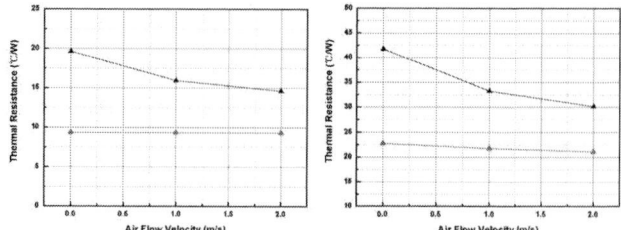

Fig.8. The relationship between thermal resistance and air flow velocity

3.2 BGA modeling

As discussed in the previous section, the air gap under the package to the board is secondary for most plastic packages when the package is mounted on the PCB in a normal way. This is due to the fact that the thermal resistance of the air gap is much larger than the junction to lead thermal resistance. However, if the e-pad is directly soldered to the board, the air gap will be replaced by the solder which will significantly reduce the resistance of the conduction thermal path to the board.

When we consider the BGA packages, they have no leads but solder balls directly soldered to the board with a large number of thermal via holes, about 80% of heat transfers from this path without other cooling devices. Evidently, superior thermal performance of BGA can be achieved by this way. Fig.9 plots the thermal analysis of a BGA, and the graph shows almost symmetrical temperature distribution in the package even though there were complicated copper traces in the substrate. Solder balls now become the dominant thermal path, and the position distant from die shows lower temperatures.

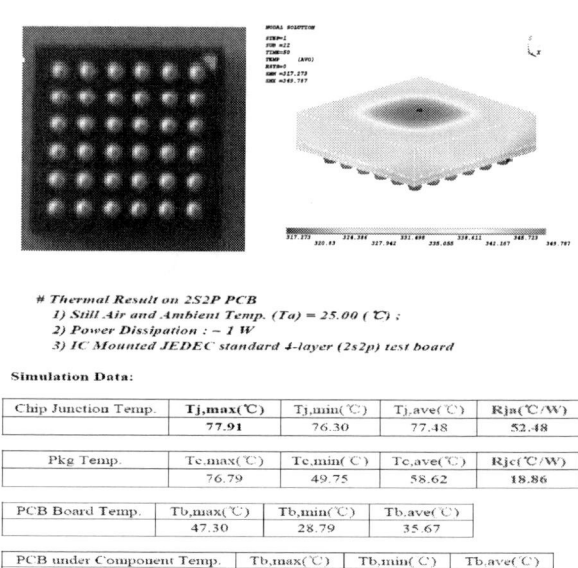

Thermal Result on 2S2P PCB
1) Still Air and Ambient Temp. (Ta) = 25.00 (°C) ;
2) Power Dissipation : ~ 1 W
3) IC Mounted JEDEC standard 4-layer (2s2p) test board

Simulation Data:

Chip Junction Temp.	Tj,max(°C)	Tj,min(°C)	Tj,ave(°C)	Rja(°C/W)
	77.91	76.30	77.48	52.48

Pkg Temp.	Tc,max(°C)	Tc,min(°C)	Tc,ave(°C)	Rjc(°C/W)
	76.79	49.75	58.62	18.86

PCB Board Temp.	Tb,max(°C)	Tb,min(°C)	Tb,ave(°C)
	47.30	28.79	35.67

PCB under Component Temp.	Tb,max(°C)	Tb,min(°C)	Tb,ave(°C)
	47.30	41.65	44.92

Fig.9. BGA 36 and its Temperature contour plots

By thermal simulation analysis, thermal resistance of BGA can be lowered by employing the next three aspects: material, package structure, and substrate design. Similar to QFP, the thermal resistance of BGA can be decreased by using materials with higher thermal conductivity. For package structure and substrate design, the use of thermal balls seems to be a low-cost solution. These balls play the thermal dissipation role by conducting heat to the printed wiring board; the use of two inner layers of substrate can lower thermal resistance, and heat from the die is transmitted to the inner layers via the die pad's through-holes, resulting in a effective heat dissipation path.

With the same method, three dimensional thermal modeling has been performed to determine the junction-to-air thermal resistance (R_{ja}) and junction-to-board thermal resistance (R_{jb}) for Flip chip and WLCSP package with standard 2s2p 4-layer boards (JESD51-9). Due to the wafer bumping technology, both packages can transfer heat out of dies effectively. Compared with other packages, R_{ja}, R_{jb} and R_{jc} all show lower values, thus better thermal performance.

3.3 SiP modeling and simulation

NFME has developed a large number of SiP packages of MCM and BGA forms. In thermal simulation for SiP, two important structure characteristic, such as 3D stacked dies and MCM package, have the obvious thermal cooling effect. The SiP package's cross-section and model graph are shown in Fig.10. Similar to BGA, most heat transfers from the chip down to the substrate and to the solder balls, while the upper path to air is secondary. The thermal analysis results demonstrate that in a stacked-die structure, the lower the die is, the lower the temperature. Dies with the same outline dimension in the SiP show similar temperature contour.

For SiP package, in order to spread the heat from the inside dies, we must shorten the heat transfer paths to reduce the thermal resistance in the paths. This can be realized by changing the layout design and the package structure, and by using material with a higher thermal conductivity, or by assembling a heat sink to reduce the heat concentration. With numerical simulation, we can adopt a more optimized thermal design, and decrease the impact of heat transfer problems.

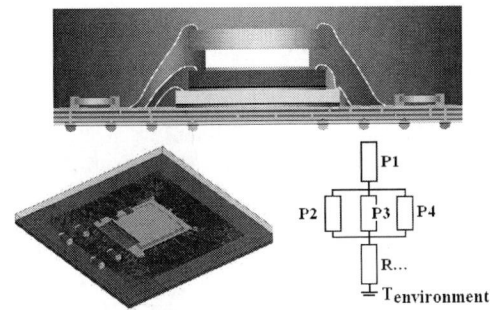

Fig.10. cross-section and thermal resistance net of SiP

4. The Comparison between Simulation Data and Measurement Data

As we all know, measuring the junction temperature of a package is very important. However, measuring the temperature of the chip after it has been installed in a system may be considered pointless, because the package has already been selected out of design stage. Nevertheless, such measurement data are useful for estimating the power dissipation of packages to be developed in future. Moreover, the simulation data can be validated with measurement data to validate and improve the models,.

In this study, the HLQFP256 has been selected to demonstrate the difference between simulated data and measured data conducted by thermal test chip technique, to confirm the thermal simulation approach. From the above comparison, the differences between the two sets of data are within 10%.

5. Conclusions

The following conclusions were drawn from this study: 1) thermal resistance of HLQFP256 has been calculated for varied parameters such as package structure, ambient temperature, power dissipation and the airflow velocity. It was shown that LQFP package with e-pad supplies a dominant heat transfer path to JEDEC test boards. And with heavy airflow, 4-layer PCB was helpful to the decrease of thermal resistance. 2) In the course of advanced package development, the effective thermal simulation method was used to analyze the thermal distribution of BGA and SiP detailed package model. We can use these packages' models in FEA analysis tools to make predictions about Rja and power dissipation which take both real environment and board conditions into consideration. 3) Changing the dies layout design and the package structure of SiP, and using materials with higher thermal conductivity can reduce the heat concentration. 4) we made the comparison between simulation data and measurement data, and found the difference to be below 10%.

Acknowledgments

The authors would like to acknowledge my colleagues in R&D center (Shen Haijun, Shi Jiangen) for their support in conducting experiments.

References

1. JESD51, http://www.jedec.org/.
2. Jim Benson, "Thermal Characterization of Packaged Semiconductor Devices," Intersil, 2002.
3. Harvey I. Rosten, Clemens J. M. Lasance, Jhon D. Parry, "The World of Thermal Characterization According to DELPHI," *IEEE Trans-CPMT-A*, Vol. 20, No. 4 (1997), pp. 384-391.
4. Zhongfa Yuan, Yong Liu, Scott Irving, Timwah Luk, Jiangyuan Zhang, "Thermal Numerical Simulation and Correlation for a Power Package," *ICEPT proceedings*, 2007.
5. Krishnamoorthi S., Desmond Y.R. Chong, Anthony Y.S. Sun, "Thermal Management and Characterization of Flip Chip BGA Packages," *Electronics Packaging Technology Conference*, (2004), pp. 53-59
6. Jingsong Li, Finite Element Methods, Beijing Post and Telecommunication University Publishing Company, 1999.

FE Simulation of Size Effects on Interface Fracture Characteristics of Microscale Lead-Free Solder Interconnects

Bin Li, Limeng Yin, Yan Yang, Xinping Zhang
School of Materials Science and Engineering, South China University of Technology,
Guangzhou, 510640, China
Email: mexzhang@scut.edu.cn; Tel: +86-20-22236396

Abstract

Understanding of interface fracture behavior of the solder joints has long been significant in reliability evaluation of electronic components and packages. The experimental and finite element methods were employed to characterize the fracture performance of "Cu wire/solder/Cu wire" sandwich structured butt microscale solder joints with different sizes (75 to 425μm in thickness and 200 to 300μm in diameter). In particular, linear elastic and elastic-plastic fracture mechanics approaches were used to quantitatively characterize the fracture performance of the predefined crack at the solder/IMC interface of both Pb-free (Sn-3.0Ag-0.5Cu) and Pb-contained (Sn-37Pb) solder joints. The simulation results show that the crack tip stress intensity factors (SIFs) for the crack at the solder/IMC interface, both K_{II} and K_I, decrease with decreasing thickness of the solder joint and increasing the loading rate; and this is coincident with the experimental results. Also, it has been seen that K_{II} is greater than K_I probably owing to the effect of Poisson contraction of the solder metal near the interfaces. It has also been shown that with increasing thickness of the solder joint, the orientation evolution of the high energy release rate area may result in the change in fracture position from the solder/IMC interface to the middle part of the joints.

1. Introduction

The miniaturization of electronic systems and products require high density packaging and thus has brought about increasingly fine pitch interconnects, and accordingly the size of solder interconnects is being continuously scaled down. As is well-known, the integrity and properties of the solder joints have significant influence on the reliability of packaging systems. Recently, a number of investigations have been carried out to characterize the size effect of mechanical behavior of microscale solder interconnects and the results reveal that the microscale solder interconnects show different mechanical behavior, for example exhibiting higher strength with decreasing the joint volume, from those with larger sizes [1-2]. However, little is known about the strengthening mechanism in the microscale interconnects, in particular for the interaction between interface intermetallic compounds (IMC) and mechanical constraint in the solder joints [3].

In addition, interfacial cracking has long been recognized as one of the predominant reasons for failure of the solder joints. It is commonly accepted that the crack tip of the dissimilar materials has the oscillation region which makes the stress distributions essentially different compared with homogeneous one. Currently, more and more researchers have attempted to extend the fracture mechanics approach to interfacial cracks to create a more reliable model for electronic packages [4-5]. As is well known, this is an extremely complicated problem. For example, the presence of friction in interfaces [6], the competition between the penetration and debond [7] and the path-dependent of J-integral [5] are challenging issues in the study of interfacial fracture behavior of the solder joints using fracture mechanics method.

This study aims to characterize the interfacial fracture behavior of the microscale joints of both Pb-free (Sn-3.0Ag-0.5Cu) and Pb-contained (Sn-37Pb) solders by finite element simulation and analysis with a comparison to the experimental investigation.

2. Experimental and FE model

The strength and fracture behavior of Sn-3.0Ag-0.5Cu and Sn-37Pb solder joints of different sizes (200−575 μm in diameter and 75−525 μm in thickness) were studied using a dynamic mechanical analyzer (DMA Q800, TA-Instruments) under a uniaxial micro-tension loading mode at ambient temperature [2]. A typical "copper/Sn-3.0Ag-0.5Cu/copper" sandwich structured butt joint with a thickness of 300 μm and a diameter of 300 μm is shown in Fig.1(a).

Two-dimensional FE model was established using ANSYS software (Version 10.0) based on the butt solder joint shown in Fig.1(a). As is well-known, the solder/IMC interfaces are usually regarded as the most dangerous position for migration in current loading [8] and also have been found as the a potential fracture initiation location in mechanical loading owing to the interfacial interaction [2]. In this study, a micro linear crack, with a length of 40μm, was set up to be parallel to the interface of the solder joints with different sizes, as shown in Fig.1(b), and the fracture behavior of the joints were simulated by FE analysis. Submodeling was used in the vicinity of the solder/IMC interface to obtain excessively fine meshes, see Fig.1(c).

For FE simulation in this study, the material properties of the solders (both tin-lead based and Sn-Ag-Cu based solders) and copper as well as Anand's constants were obtained from previous studies [2, 3, 9-14]. It should be indicated that for the convenience of the simulation and analysis, Sn-37Pb and Sn-40Pb solders are regarded identical since they have very similar physical and mechanical properties, as well as Sn-3.0Ag-0.5Cu and Sn-3.8Ag-0.5Cu solders.

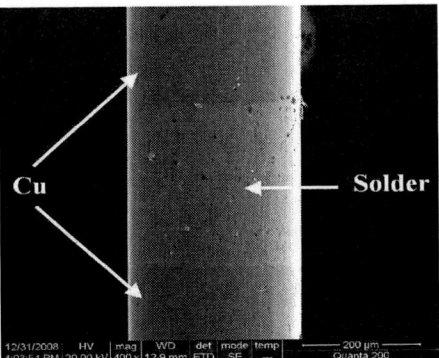

(a) SEM image of the solder joint

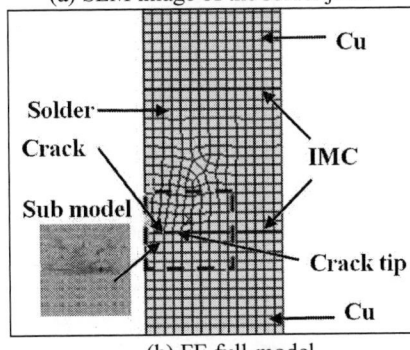

(b) FE full-model

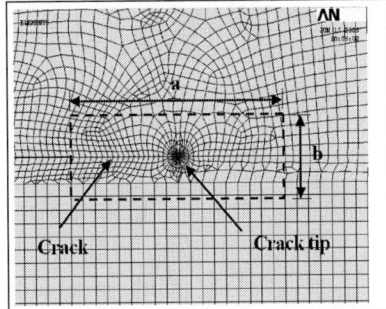

(c) Sub-model of a linear crack at the solder/IMC interface

Fig.1 Copper-wire/solder/copper-wire sandwich structured microscale solder joint (with both the diameter and thickness of 300μm) and FE model

3. Results and discussion

3.1 Crack driving force and fracture behavior of the solder joints with different sizes

(1) The variation of K_I and K_{II} values in the solder joints of different sizes

In our experimental results [2], the yielding stress of the microscale solder joints ranges from 50~70MPa. For FE simulation in this study, the uniaxial tensile load in the range of 0~40MPa was applied to one end of the "copper/solder/copper" butt joint and in the other end of the joint all degree of freedoms (DOFs) were constrained; this aims to meet the condition that the plastic deformation around the crack tip is small enough and linear-elastic fracture mechanics can still be used in the simulation. The loading is

far away from the crack tip thus the whole model can be regarded as plane strain. The quasi-static loading rate is 1N/min.

Fig.2(a) shows the distribution of Von Mises stress in the joint with a diameter of 200μm and a thickness of 175μm as well as a 40μm long linear crack at the interface. The region of maximum von Mises stress is in the vicinity of the crack tip, and a close look at the stress distribution indicates that actually the maximum von Mises stress region locates in the solder side of the joint interface, as shown in Fig.2(b).

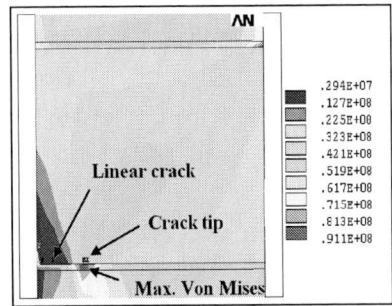

(a) In whole solder joint (full-model)

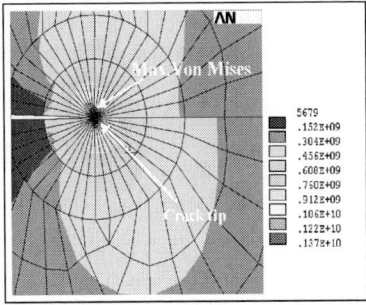

(b) At the crack tip (sub-model)

Fig.2 Von Mises stress distribution in the joint with a diameter of 200μm and a thickness of 175μm as well as a 40μm long crack at the interface

Based on the stress distribution and other parameters calculated, stress intensity factor (SIF) at the crack tip of the joint can be obtained. In FE simulation using ANSYS, a full-crack model was chosen and ANSYS automatically generated singular elements around the crack tip, thus the singularity of stress and strain could be simulated easily. Fig.3 shows the dependence of SIFs, both K_I and K_{II}, on diameter and thickness of the solder joints.

Clearly, SIF value decreases obviously with decreasing thickness of the solder joint from 450μm to 75μm (with a constant joint diameter of 200μm), as shown in Fig.3(a), and increases apparently with decreasing diameter of the solder joint from 600μm to 200μm (with a constant joint thickness of 200μm), as shown in Fig.3(b). It is worth noticing that at the same condition the crack driving force in Sn-3.0Ag-0.5Cu solder joints is obviously lower than that in Sn-37Pb solder joints. This means that the fracture properties of Sn-3.8Ag-0.5Cu solder joints are superior to that of Sn-37Pb solder joints.

978-1-4244-4658-2/09 $25.00 © 2009 IEEE   167

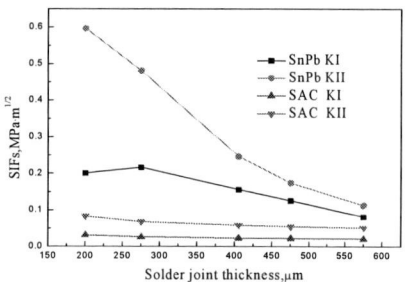

(a) SIFs vs thickness (at a constant diameter of 200μm)

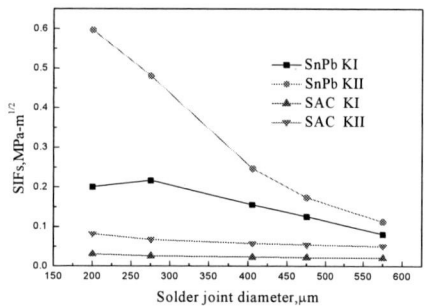

(b) SIFs vs diameter (at a constant thickness of 200μm)

Fig.3 The variation of K_I and K_{II} values in Sn-3.0Ag- 0.5Cu and Sn-37Pb solder joints with different sizes

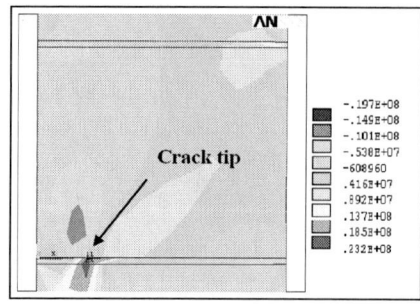

(a) In whole solder joint (full- model)

(b) At the crack tip (sub-model)

Fig.4 Von Mises stress distributions in the solder joint with a diameter of 200μm and a thickness of 175μm as well as a 40μm long crack at the interface

Moreover, it can be seen clearly in Fig.3 that for both Sn-3.0Ag-0.5Cu and Sn-37Pb solder joints, K_{II} value is obviously higher than K_I value. This would make the crack propagation more unstable [3]. The distributions of τ_{xy}, as shown in Fig.4, reveal that there are strong transversal interaction forces (i.e., $-\tau_{xy}$ and $+\tau_{xy}$) in the surrounding area of the crack tip, which are mainly caused by the effect of Poisson contraction [15] due to the visco-plastic deformation of the solder alloy. Accordingly, transversal fracture along the interface of the solder joints could be the main reason for the joint failure under tensile loading.

(2) Energy release rate and the fracture position/path in the solder joints

In our previous research [2], it had been shown that for the solder joint with a small diameter, the final fracture of the joint exhibited a typical necking fracture mode and fracture position located in the middle part of the joint. While for the joint with a relatively large diameter, the crack initiation was usually at the solder/IMC interfaces, but the final fracture took place at the interfaces or across the solder with an angle of about 45°. According to the results of micro-tension experiments by DMA in this study, the change in the joint thickness shows the similar trend in fracture modes of the solder joints. Fig.5 shows the morphology of the solder joints, with a diameter of 300μm and two different thicknesses of 100μm and 200μm respectively, after final fracture in quasi-static micro-tension loading. It can be seen that for the solder joint with a small thickness (i.e., 100μm), the rupture initiated at the interface near the outside edge of the joint and the final fracture occurred in the middle part of the joint, as shown in Fig.5(a). For the joint with a relatively large thickness, the fracture position of the joints mostly located in the middle, as shown in Fig.5(b). Table 1 shows the probability of fracture positions in Sn-3.0Ag-0.5Cu solder joints with different sizes. Clearly, the joint's thickness-to-diameter ratio (d/t), which is commonly regarded as the dominant factor for mechanical constraint in the joint, plays a key role in influencing the fracture position and mode of the joint.

However, it remains unclear why some cracks initiated at the interface near the outside edge of the joint and the final fracture occurred in the middle part as shown in Fig.5(a). To clarify this issue, a high level uniaxial tensile load up to 60MPa was applied to the joint, which is close to the ultimate tensile strength of the solder joints [2]. Besides, much more refined elements were used at the solder/IMC interface and in the solder joint to improve the numerical simulation accuracy. In addition, a parameter, the energy release rate (G), is used to determine the crack growth direction. As is commonly accepted, the direct calculation of G is the most reliable method for predicting the propagation direction of the crack. The crack tip propagates along the direction where the energy release rate has the maximum value in accordance with the principle of lowest energy. The value of G can be calculated by virtual crack extension method as the following equation [16]:

978-1-4244-4658-2/09 $25.00 © 2009 IEEE

$$G = -\frac{U_{a+\Delta a} - U_a}{B\Delta a} \qquad (1)$$

Where B is the thickness of the sample and set as B=1 in two-dimensional simulation; U is the total strain energy; a is the crack length. Extending the crack length for the second analysis by selecting all nodes in the vicinity of the crack tip and scaling them in the interface direction by the factor Δa, then the calculated G value fields for the solder joints with a constant diameter of 300μm and thicknesses of 100, 200 and 300μm are shown in Fig.6.

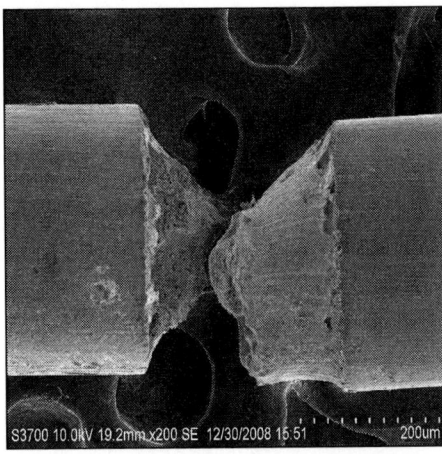

(a) With a thickness of 100μm

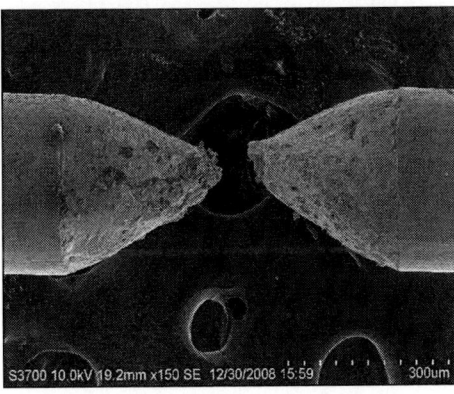

(b) With a thickness of 200μm

Fig.5 Morphology of the fractured joints with a diameter of 300μm and different thicknesses

In Fig.6, for the joint with a small thickness (e.g., 100μm), a strip with the negative energy release rate value exists from the crack tip to the upper right edge of the solder/IMC interface, see Fig.6 (a). Clearly, the crack tip would never propagate towards this strip. The area with high energy release rate appears between the strip and solder/IMC interface of the crack, and this means that the crack tip would be most likely to grow towards the interface. With increasing the joint thickness, it can be seen that the high energy release rate area rotates anticlockwise around the crack tip and

gradually passes the strip with the negative energy release rate value, then moves to another side as shown in Fig.6(b) to (c), and consequently the cracks propagate towards the middle part of the solder joint and the final fracture occurs there. A schematic illustration of the above-mentioned process for showing the change in G value field and for predicting the crack growth direction and fracture position is shown in Fig.6(d).

Table 1 Probability of fracture positions of Sn-3.0Ag-0.5Cu solder joints with different sizes

Diameter (μm)	Thickness(μm)	d/t ratio	Fracture position and probability
200	75	2.7	middle part: 45%; interfacial part: 55%
	125	1.6	middle part: 100%
	175	1.1	middle part: 100%
	225	0.9	middle part: 100%
275	125	2.2	middle part: 27%; interfacial part: 73%
	225	1.2	Interfacial part: 100%
	325	0.8	middle part: 100%
400	125	3.2	interfacial part: 100%
	225	1.8	middle part: 50%;interfacial part: 50%
	325	1.2	middle part: 100%
	425	0.9	middle part: 100%
475	125	3.8	interfacial part: 100%
	225	2.1	middle part: 33%;interfacial part: 67%
	325	1.5	middle part: 50%;interfacial part: 50%
	425	1.1	middle part: 100%
	525	0.9	middle part: 100%

From the results shown in Table 1, it is clear that the thicker of the solder joints, the more possible for them to fracture in the middle of the joint. Combining the results in Table 1 and Fig.6, therefore, it can be concluded that with decreasing d/t ratio, the orientation evolution of the high energy release rate area makes the fracture angle increasing and results in the crack propagation towards the middle part of the joint. The FE simulation results are in good coincidence with the experiments. By the way, it is worth indicating that our earlier study [2] has shown that the interfacial stresses and mechanical constrain effect in the joint have close relationships with the tensile strength of the solder joints.

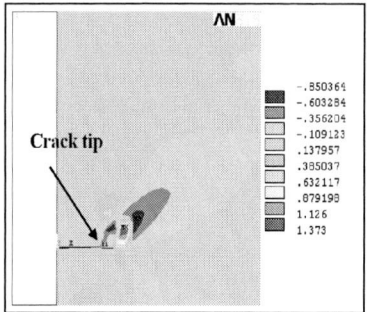

(a) With a thickness of 100μm

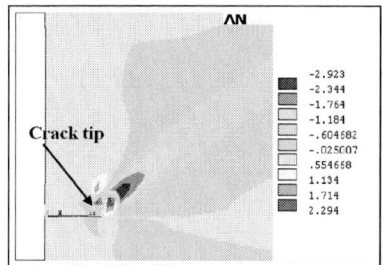

(b) With a thickness of 200μm

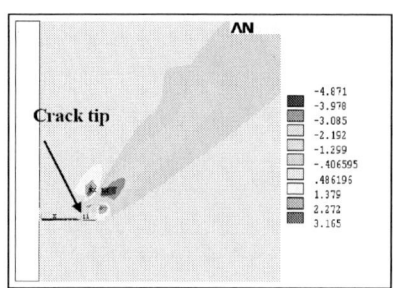

(c) With a thickness of 300μm

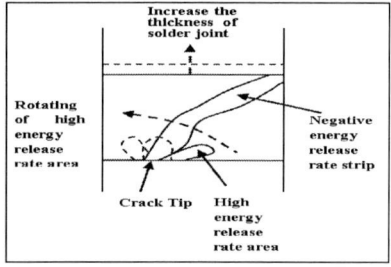

(d) Revolution of high energy release rate area

Fig.6 Energy release rate distribution in the joints with a constant diameter of 300μm and different sizes, and revolution of the high energy release rate area.

(3) J-integral value at the crack tip field

It is worth noticing that the interface cracks exhibit inherently different compared to those in homogenous materials and J-integral of the interface crack tip is also path-dependent. Previous studies have indicated that the energy release rate in dissimilar anisotropic media is proved to be nearly the same as J-integral with very flat path [5, 17-18].

In this study, the long side (b) of the rectangular path was parallel to the crack and the center of the path was located at the crack tip, see Fig.1(c). Setting a=40μm and b/a in the range of 0.1~0.5, then the calculation results show that J-integral is path-dependent in dissimilar materials, but the fluctuation range is limited even for very small b/a values, as shown in Fig.7. Clearly, the J-integral increases gradually with increasing thickness of the solder joints. This is also in good coincidence with the trend of SIFs.

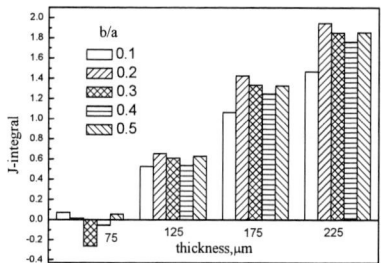

Fig.7 J-integral at different b/a values in the solder joints with a diameter of 200μm and thickness in the range of 75 to 225μm

3.2 The influence of the loading rate on fracture properties

To investigate the influence of the loading rate on fracture properties of the solder joints, the loading rate under tensile loading mode and the load level up to 40MPa is increased from 1N/min to 20N/min, and the dependency of SIFs on loading rate for the visco-plastic solder joints are shown in Fig.8. Clearly, for Sn-3.0Ag-0.5Cu solder joints, SIFs decrease sharply with the loading rate. This suggests that the fracture properties of the solder joints deteriorate with decreasing the loading rate. Similar to the results in the above Section 3.1, K_{II} value is greater than K_I in both Sn-3.0Ag-0.5Cu and Sn-37Pb solder joints.

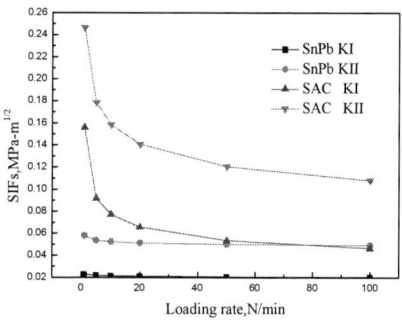

Fig.8 Dependency of SIFs on loading rate for the joint with a diameter of 225μm and a thickness of 405μm

978-1-4244-4658-2/09 $25.00 © 2009 IEEE

Conclusions

(1) The crack tip stress intensity factors (SIFs), both K_{II} and K_I for the crack at the solder/IMC interface, decrease with decreasing thickness of the solder joint and increasing the loading rate.

(2) For both Sn-3.0Ag-0.5Cu and Sn-37Pb solder joints, K_{II} is greater than K_I owing to effect of Poisson contraction of the solder metal near the interfaces of the joint.

(3) With increasing thickness of the solder joint, the orientation evolution of the high energy release rate area result in the change in fracture position from the solder/IMC interface to the middle part of the solder joints.

References

1. Zimprich, P., Saeed, U., Betzwar-Kotas, A., Weiss, B., Ipser, H., "Mechanical size effects in miniaturized lead-free solder joints," Journal of Electronic Materials, Vol.37, No.1 (2008), pp. 102-109.
2. Yin, L.M., Zhang, X. P., Lu, C., "Size and volume effects on the strength of microscale lead-free solder joints," Journal of Electronic Materials (in press, online available on 25 June 2009)
3. Alam, M.O., Lu, H., Bailey, C., Chan, Y.C., "Fracture mechanics analysis of cracks in solder joint intermetallic compounds," Proceedings of the 2nd Electronics System-Integration Technology Conference, ESTC, (2008), pp. 757-762.
4. Ozkan, U., Nied, H.F., "Finite element based three dimensional crack propagation simulation on interfaces in electronic packages," Electronic Components and Technology Conference, (2008), pp. 1606-1613.
5. Xu, B.L., Cai, X., Huang, W.D., Chen, Z.N., "Research of underfill delamination in flip chip by J-integral method," Journal of Electronic Packaging, Vol.126, No.1 (2004), pp. 94-99.
6. Sun, C.T., Qian, W., "A treatment of interfacial cracks in the presence of friction," International Journal of Fracture, Vol.94, No.4 (1998), pp. 371-382
7. Zhang, Z., Suo, Z.G., "Split singularities and the competition between crack penetration and debond at a bimaterial interface," International Journal of Solids and Structures 2007, Vol.44, No.13 (2007), pp. 4559-4573.
8. Liang, S.W., Chang, Y.W., Chen, C., "Effect of Al-trace dimension on joule heating and current crowding in flip-chip solder joints under accelerated electrimgiration," Applied Physics Letters, Vol.88, No.17 (2006), pp. 172108.
9. Chromik, P.R., Vinci, R.P., Allen, S.L., Notis, M.R., "Nanoindentation measurements on Cu-Sn and Ag-Sn intermetallics formed in Pb-free solder joints," Journal of Materials Research, Vol.18, No.9 (2003), pp. 2251-2261.
10. Wang, F.J., "Study on mechanical properties and size effects of lead-free BGA solder," Harbin Institute of Technology, PhD Dissertation (2006) (in Chinese)
11. Ma, H.T., Suhling, J.C., "A review of mechanical properties of lead-free solders for electronic packaging," Journal of Materials Science, Vol.45, No1.5 (2009), pp. 1141-1158.
12. Shangguan, D.K. et al., Lead-free solder interconnect reliability. Beijing, Publishing House of Electronics Industry (Beijing, 2008). (in Chinese)
13. Wang, G.Z., Cheng, Z.N., Becker, K., Wilde, J., "Applying Anand model to represent the visco-plastic deformation behavior of solder alloys," Journal of Electronic Packaging, Vol.123, No1.3 (2001), pp. 247-253.
14. Wang, Q., Zhang, Y.X., Liang, L.H., "Anand Parameter Test for Pb-Free Material Sn-Ag-Cu and life prediction for a CSP,". Proceedings of the 8th International Conference on Electronic Packaging Technology, ICEPT. 2007, pp. 4441437.
15. Chiang, Y.C., "The influence of Poisson contraction on matrix cracking stress in fiber reinforced ceramics," Journal of Materials Science, Vol.33, No1.13 (2001), pp. 3239-3246.
16. Swanson, J., "Release 10.0 Documentation for ANSYS," ANSYS Company (Pennsylvanian, 2006).
17. Suo, Z.G., "Singularities, interfaces and cracks in dissimilar anisotropic media," Proceedings of the Royal Society of London, Vol.427, No.1873 (1990), pp. 331-358
18. Shi, W. C., Kuang, Z. B., "J-Integral of Dissimilar Anisotropic Media," International Journal of Fracture, Vol.96, No1.4 (1999), pp. 37-42.

Thermal Design of Power Module to Minimize Peak Transient Temperature

Xiao Cao, Khai D. T. Ngo, Guo-Quan Lu
The Bradley Department of Electrical and Computer Engineering
Virginia Polytechnic Institute and State University
Blacksburg, VA 24060, USA

Abstract

This paper investigates the impact of the geometry of the heat spreader and the heat transfer coefficient of the heat exchanger on the steady-state and transient thermal performances of power semiconductor modules. Results show that the steady-state thermal resistances along the thermal flow path change with heat transfer coefficients owing to limited heat spreading effect. The transient thermal performance is mainly affected by the thermal capacitance in the power module. To minimize the transient thermal impedance without sacrificing weight, cost, and so on, the thickness of the heat-spreader should be selected to match the thermal time constant to the transient duration. Based on these results, a methodology is proposed to conservatively select the thickness of the heat spreader to maintain the silicon junction at a required peak transient temperature. The methodology is exemplified and verified by a thermal design for a medium-voltage power module.

I. Introduction

In recent years, there is a growing demand to increase the power density in power module. As a result, the heat flux generated by power semiconductor devices keeps on increasing. In some motor drive applications, the heat flux can be higher than $100W/cm^2$ in steady-state, and the peak heat flux can be as high as $300W/cm^2$ under transient conditions 0-2. In spite of the more stringent transient requirements, most publications in power electronics tend to focus on steady-state thermal management 3-4. The thermal resistances along the thermal flow path are often taken as constants independent the heat transfer coefficient of the heat exchanger.

In this paper, attention is paid to thermal design based on peak transient temperature, which is related to both steady-state and transient thermal conditions. The design parameters include the geometry of the heat spreader and the heat transfer coefficient of the heat exchanger. The dependence of thermal resistances on the heat transfer coefficient of the heat exchanger will be explored.

Fig. 1 shows the stack structure for thermal management of a typical power module. The IGBT is attached to a direct-bonded-copper (DBC) substrate via solder. The heat spreader underneath the DBC substrate can reduce the thermal resistance by spreading the heat over a larger cross-section. The whole power module is attached to the heat exchanger via thermal interface materials (TIM), such as thermally conductive epoxy, solder, or thermal grease. As shown in Fig. 2, the heat transfer coefficient can vary from 10 $W/m^2/°C$ to 10^5 $W/m^2/°C$, depending on the convection mechanisms 5.

During the operation of power module, the heat generated in the IGBT has to travel through the DBC substrate and the heat spreader before being dissipated in the heat exchanger.

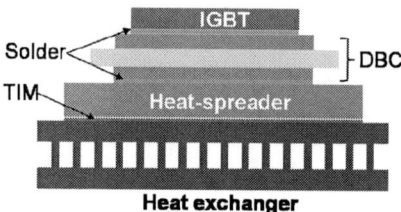

Fig. 1. Structure for conventional power module with heat exchanger.

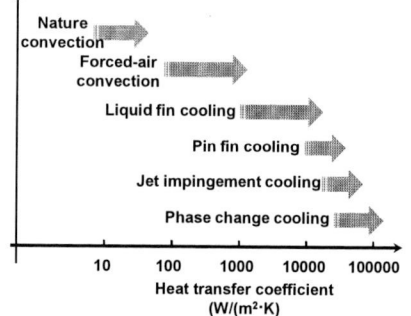

Fig. 2. Heat transfer coefficients for different cooling mechanisms.

The heat spreader can help spread heat to a larger area so that the thermal resistance can be reduced. In general, the thicker heat spreader can offer better spreading effect to achieve better thermal performance. However, for applications with high-performance heat exchanger, the spreading effect could be limited due to high heat transfer coefficient. In addition, the use of heat spreader increases the number of layers across which heat has to be conducted. This can compromise the thermal performance of the module.

During thermal transient, the heat spreader works as a thermal capacitance to store thermal energy, leading to lower peak temperature in the power module. However, a heat spreader with larger size does not necessarily result in performance lower junction temperature. The relation between heat propagation speed and the duration of the transient also has significant impact on the peak junction temperature.

In this paper, the impact of heat transfer coefficient and heat spreader geometry on heat removal is discussed, considering both steady state and transient. The design guideline for the heat spreader and heat transfer coefficient is provided based on finite-element simulation under steady-state and transient conditions.

In Section II of this paper, the steady-state and transient thermal performances of power module are analyzed based on finite-element simulation. In Section III, a methodology to design the heat spreader thickness and the heat transfer coefficient of the heat exchanger to meet both steady-state and transient thermal requirements for the power module is discussed. The methodology is verified experimentally in Section IV.

II. Steady-state and Transient Thermal Performance of a Power Module

A. Set-up for thermal simulation

For complicated multilayer structures, it is very difficult to obtain accurate thermal distribution analytically. Thus, finite-element ePhysics from Ansoft 6 is used herein to study how the heat propagates from the junction to the heat exchanger under steady state and transient conditions. EPhysics' ability to refine the mesh adaptively allows trade-off between numerical accuracy and computation time.

Before a simulation, a three-dimensional (3D) model of the power module is built in ePhysics as shown in Fig. 3. To avoid hot spots, sufficient distance should be allocated among neighboring dice to minimize thermal interaction. Thus, it suffices to consider a module with one die herein. The design methodology is still applicable to a module with multiple chips.

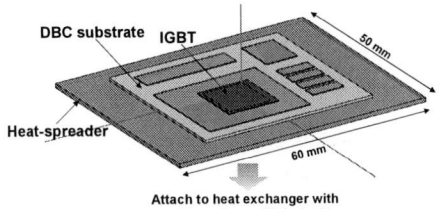

Fig. 3. Power module built in ePhysics.

The DBC substrate used in Fig. 3 has an alumina layer with a thickness of 0.625 mm sandwiched between copper layers with a thickness of 0.23 mm. The solder layer has a thickness of 0.1 mm. The IGBT area is larger than 1 cm^2 and is modeled with the thermal properties of silicon. The material properties used in the simulation are listed in TABLE I.

TABLE I
MATERIAL PROPERTIES USED IN THE SIMULATION

	Thermal Conductivity k	Specific Heat Capacity c	Mass Density ρ
Copper	400 $W/(m{\cdot}C)$	385 $J/(kg{\cdot}C)$	8933 kg/m^3
Alumina in DBC	35 $W/(m{\cdot}C)$	850 $J/(kg{\cdot}C)$	3960 kg/m^3
IGBT	148 $W/(m{\cdot}C)$	712 $J/(kg{\cdot}C)$	2330 kg/m^3
Solder	48 $W/(m{\cdot}C)$	167 $J/(kg{\cdot}C)$	8000 kg/m^3

For the medium-voltage power module considered herein, heat is convected away primarily by a heat exchanger mounted to the bottom of the power module. Therefore, the major part of the heat generated in the IGBT is transported via conduction to the bottom of the power module, and is then dissipated with a heat transfer coefficient h. The top and lateral surfaces of the power module can be considered adiabatic.

B. Steady-state thermal response

The most common parameter used to characterize thermal performance of electronic packages is thermal resistance, R_{th}. The resistance is defined by the following equation:

$$R_{th} = \frac{T_j - T_a}{P} \qquad (1)$$

where T_j is the junction temperature; T_a is the ambient temperature; and P is the power dissipation. Under steady state, both R_{th} and T_j are time-independent. In the simulation, the thickness of the copper heat spreader and the heat transfer coefficient of the heat exchanger were varied. As shown in Fig. 4, the junction-to-ambient thermal resistance of the power module can be reduced by increasing the heat transfer coefficient. However, once the heat transfer coefficient exceeds 20000 W/m^2/°C, the thermal resistance tends to saturate. This can be explained using the model shown in Fig. 5. Under steady state, each layer in the power module can be treated as a thermal resistor. Thus, the equivalent circuit of the power module is as drawn in Fig. 5. The heat exchanger with heat transfer coefficient of h is also considered as a resistor R in Fig. 5. R can be calculated based on the definition of heat transfer coefficient:

$$R = \frac{1}{h \cdot A} \qquad (2)$$

where A is the effective area for heat dissipation of the heat exchanger. The junction-to-ambient thermal resistance, R_{th}, is the summation of all resistances from the IGBT to the heat exchanger in Fig. 5. From (2), it can be seen that higher h leads to smaller R as well as smaller R_{th}. However, when h is higher than certain value, the resistance of R becomes negligible compared to those of other resistances, and further increase of h could not significantly reduce the thermal resistance of the package. The saturated thermal resistance of the power module can be calculated by the following equation:

$$R_{sat} = \sum_{i=1}^{6} R_i = \sum_{i=1}^{6} \frac{d_i}{k_i \cdot A_i} \qquad (3)$$

where R_i and d_i are the thermal resistance and the thickness of the ith layer, respectively; A_i is the effective heat conduction area, including heat spreading effect.

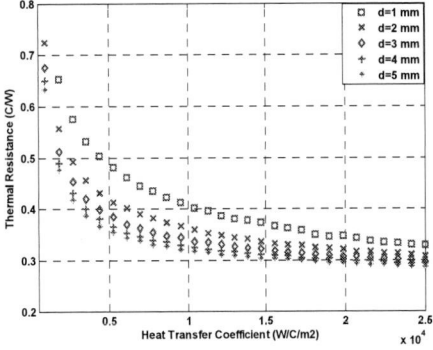

Fig. 4. Thermal resistance versus heat transfer coefficient, parametric with thickness of the heat spreader.

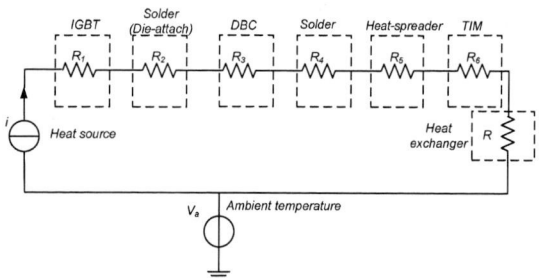

Fig. 5. Steady-state thermal equivalent circuit for the power module attached to the heat exchanger.

Fig. 6 shows the relation between the thermal resistance and the thickness of the heat spreader, parametric with the heat transfer coefficient. A thicker heat spreader spreads the heat more effectively, leading to lower thermal resistance. However, the increase of the thickness of the heat spreader also adds more thermal resistance along the direction of the heat flow. As a result, when the thickness of the heat spreader is larger than certain value, the thermal resistance goes to a saturation value determined by the thermal resistance of the other layers and the heat transfer coefficient. Fig. 6 also shows that with higher heat transfer coefficient, the thermal resistance tends to saturate even with a thinner heat spreader. For example, with h=1000 W/m^2/°C, the thermal resistance is reduced by 0.20 °C/W when the thickness of the heat spreader increases from 2 mm to 5 mm. However, with h=20000 W/m^2/°C, the thermal resistance is reduced by only 0.015 °C/W for the same change in the thickness of the heat spreader. This is caused by the reduction in heat spreading brought by high h value. Fig. 7 shows the heat flux distribution in the power module with different h values. With h=20000 W/m^2/°C (shown in Fig. 7(a)), the heat flux concentrates underneath the heat source. Thus, even with a thick heat spreader attached, the heat dissipation area would not dramatically increase. For the smaller h in Fig. 7(b), the heat flux is distributed over a larger area at the bottom of the power module, and increasing the thickness of the heat spreader can greatly help reducing the thermal resistance.

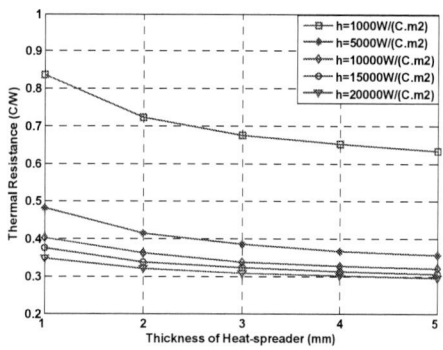

Fig. 6. Thermal resistance versus thickness of the heat-spreader, parametric with heat transfer coefficient.

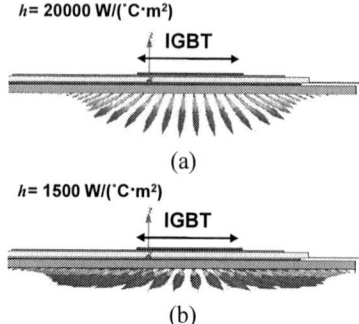

(a)

(b)

Fig. 7. Heat flux distribution at the bottom of the power module: (a) h = 20000 W/(°C·m^2); (b) h = 1500 W/(°C·m^2).

C. Transient thermal response

For a given pulse width of the heat flux, the transient thermal response can be characterized by a thermal impedance that is defined by (1) with R_{th} and T_j replaced by time-dependent $Z_{th}(t)$ and $T_j(t)$, respectively.

The simulated thermal impedance of the power module in Fig. 3 is described in Fig. 8 and Fig. 9. Different curves in Fig. 8 represent different heat transfer coefficients. When the transient duration is less 50 ms, the thermal impedance does not change for the range of heat transfer coefficient shown. For transient duration longer than 50 ms, a drastic change in heat transfer coefficient (from 100 W/m^2/°C to 25000 W/m^2/°C) only has slight impact on thermal impedance. This implies that the transient thermal response is a highly localized, passive phenomenon. The thermal masses provided by the DBC substrate and the heat spreader can absorb the transient thermal energy.

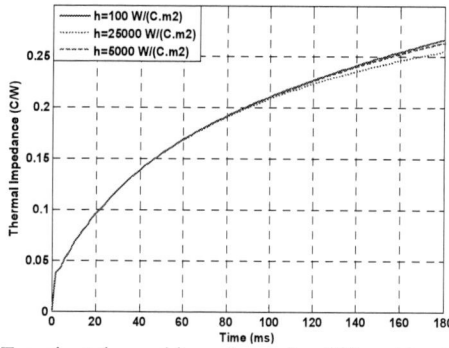

Fig. 8. Transient thermal impedance for different heat transfer coefficients and 1 mm heat-spreader thickness.

Fig. 9 shows the transient thermal performance for different thickness of the heat exchanger. In Fig. 9(a) and (b), the heat transfer coefficient is set to be 1000 W/m^2/°C and 25000 W/m^2/°C, respectively. Thanks to a larger thermal mass, thicker heat spreader offers lower thermal impedance. However, when the transient duration is less than 50 ms, the thermal impedance does not change with the thickness of the heat spreader. Under transient thermal condition, each layer in the power module can be considered as a capacitor in parallel with a resistor. Thus, the time constant of the ith layer can be calculated by the following equation:

$$\tau_i = R_i C_i = \frac{d_i}{k_i A_i} \cdot \rho_i c_i d_i A_i = \frac{d_i^2 \rho_i c_i}{k_i} \quad (4)$$

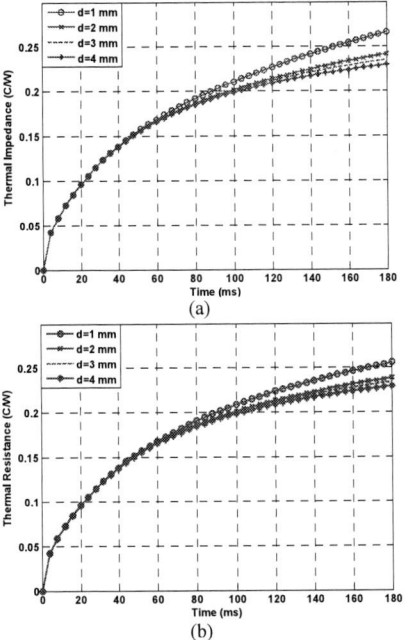

Fig. 9. Transient thermal impedance response for different thickness of the heat-spreader and h = 1000 W/ m²/°C; h = 25000 W/m²/°C.

The calculated time constant for each layer in Fig. 3 is listed in TABLE II. The time constant describes the heat propagation rate in each layer. It takes longer time for heat to pass through a layer with higher time constant. From TABLE II, the total time constant of the IGBT, the solder layer, and the DBC is 40 ms, which means for transient less than 40 ms, the thermal responses is dominantly determined by those three components. For the copper heat spreader, the time constant is proportional to the square of the thickness. For 1 mm thickness, the time constant for the assembly is around 49 ms. This explains why the thermal impedance does not change under transient shorter than 50 ms when the thickness of the heat spreader changes from 1 mm to 4 mm (shown in Fig. 9). Thus, to achieve an efficient thermal design based on transient specification, the thickness of the heat spreader should be selected to match the overall thermal time constant to the transient duration.

TABLE II
CALCULATED TIME CONSTANT FOR EACH LAYER IN THE POWER MODULE

	IGBT	Solder	DBC	Heat-spreader
Time Constant	0.5 ms	0.28 ms	39.4 ms	$d^2 \cdot 8.6$ ms

*d is the thickness of the heat-spreader in mm

III. Thermal Design of Power Module and Heat Spreader

In Section II, the impact of the heat spreader and the heat transfer coefficient on both steady-state and transient thermal performance has been studied. This Section describes the thermal design procedure for a power module based on the results obtained.

The flowchart for the thermal design of a power module in conjunction with the heat spreader is shown in

Fig. 10. The specifications for the design problem are illustrated in Fig. 11 and exemplified in TABLE III.

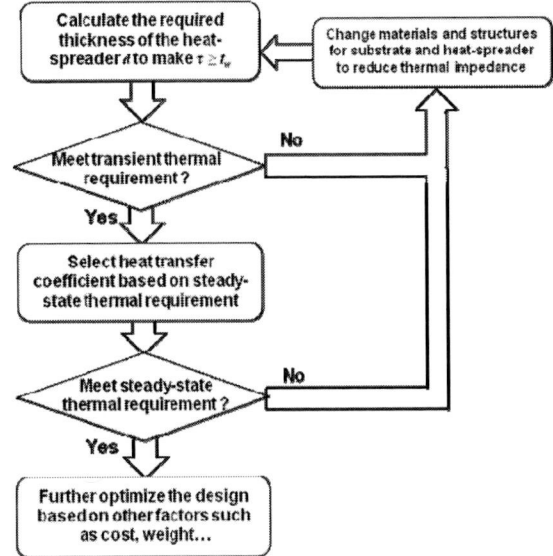

Fig. 10. Flowchart for thermal design of power module and heat spreader.

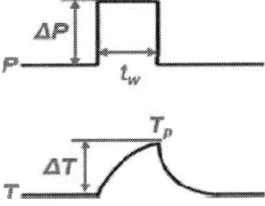

Fig. 11. Load profile and junction temperature response for IGBT.

For a given thermal load as shown in Fig. 11, P is the steady-state power dissipation in the IGBT and T is the temperature limit under steady state. ΔP and ΔT are the step change in power dissipation in the IGBT and the allowed junction temperature rise during transient, respectively. The peak transient temperature can be described as:

$$T_p = T + \Delta T = T_a + R_{th\_steady}P + Z_{th\_trans}(t_w)\Delta P \quad (5)$$

where R_{th_steady} is the steady-state thermal resistance and Z_{th_trans} is the transient thermal impedance of the power module and heat spreader.

As discussed in Section II, the transient thermal performance is mainly determined by the layers in the power module, and the heat spreader. Assume that τ is the sum of time constants for all layers in the power module and the heat spreader:

$$\tau = \sum \tau_i \qquad (6)$$

In order to achieve the lowest $\varDelta T$, τ has to be larger than the pulsewidth t_w. From the desired τ value, the thickness of the heat spreader can be calculated based on TABLE II. Fig. 12 shows the relationship between transient duration and the resulting thickness of the heat spreader. For transient duration of 120 ms, a heat spreader with the thickness of 3 mm is preferred. From Fig. 9, it can be observed that increasing the thickness of the heat spreader beyond 3 mm cannot further

TABLE III
THERMAL REQUIREMENTS FOR POWER MODULE AND HEAT SPREADER

Steady-state Power Dissipation (P)	35 W
Step Change of Power Dissipation ($\varDelta P$)	70 W
Ambient Temperature (T_a)	24°C
Peak Transient Temperature (T_p)	67°C
Pulsewidth for Transient (t_w)	120 ms
Steady-state Thermal Resistance ($R_{th\ steady}$)	0.7 °C/W
Transient Thermal Impedance ($Z_{th\ trans}$)	0.25 °C/W

reduce the transient thermal impedance. From Fig. 9, the transient thermal impedance also can be found to be 0.21°C/W, meeting the transient thermal requirement. However, if the transient requirement cannot be satisfied, the changes of materials and mechanical design in the stack have to be made to reduce the transient thermal impedance.

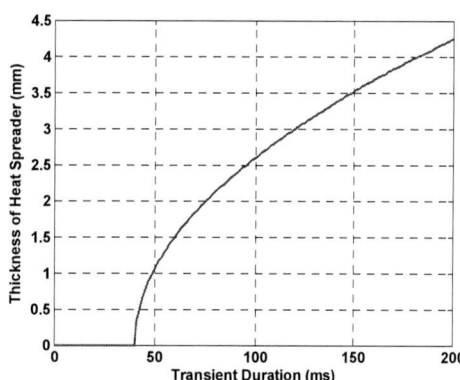

Fig. 12. Transient duration vs thickness of heat spreader.

With the designed thickness of the heat spreader, the steady-state thermal resistance can be plotted versus the heat transfer coefficient h as shown in Fig. 13. With the help of Fig. 13, the heat transfer coefficient of the heat exchanger can be found to meet the steady-state thermal requirement $R_{th\ steady}$. For example, to make the thermal resistance less than 0.7°C/W with a heat spreader with 3 mm thickness, a heat exchanger with h of 1000 W/m²/°C can be used. If the required thermal resistance is smaller than R_{sat} in (3), the increase of h cannot meet the steady-state requirement. Thus, materials and mechanical design in the stack have to be improved to reduce the thermal resistance.

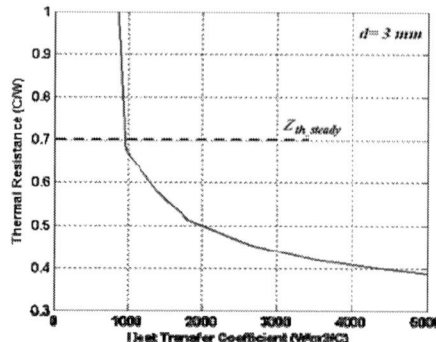

Fig. 13. Heat transfer coefficient vs steady-state thermal resistance when $d = 3$ mm.

It should be noticed that in the design procedure described in

Fig. 10, the thermal performance is the only aspect considered. Other factors could impose additional constraints on the geometry of the heat spreader or heat transfer coefficient of the heat exchanger. For instance, the thicker heat spreader induces larger thermo-mechanical stress in the bonding layer due to mismatch in coefficients of thermal expansion (CTE), leading to poor reliability 8. Thus, the heat spreader may not be allowed to be as thick as the design methodology suggests. To meet the transient thermal specifications, the heat transfer coefficient of the heat exchanger has to be increased. However, as shown in Fig. 8, the increase of h can only offer marginal improvement. Alternate materials and package structures need to be investigated in the future to increase the thermal capacitance near the IGBT.

IV. Experimental Results and Discussion

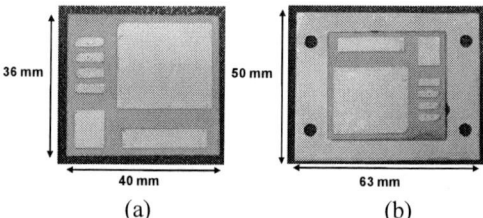

Fig. 14. Components used in experiment: (a) DBC substrate; (b) DBC substrate attached to the copper heat spreader.

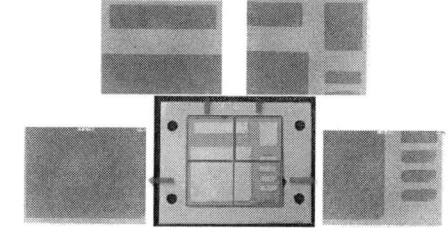

Fig. 15. X-ray images of a DBC substrate attached to a heat spreader by nano-silver paste.

Two power modules like the one shown in Fig. 3 were built to verify the analysis and design presented in the previous section. Recall that to meet the design requirements in TABLE III, a copper heat spreader with 3 mm thickness and a heat exchanger with h of 1000 $W/m^2/°C$ are desired. For comparison purpose, the thicknesses of the heat spreaders attached to these two modules are 1 mm and 3 mm. DBC is used as the substrate of the power module as shown in Fig. 14(a). Since nano-silver paste has higher thermal conductivity (80 $W/m/°C$), higher bonding strength, and lower void content for large bonding area compared to conventional solders 9, it was used to attach the substrate to the heat spreader (shown in Fig. 14(b)). Also, voids in the bonding layer can reduce the heat transfer area, resulting in higher thermal impedance of the power module 10. To evaluate the void percentage in the bonding layer, X-ray scan was performed after attachment of the heat spreader. As shown in Fig. 15, X-ray scan did not reveal voids.

After attachment of the substrate to the heat spreader, the IGBT was soldered to the substrate using SAC305 solder preform. The completed power module is shown in Fig. 16(a). To match the simulation conditions, the IGBT was covered by an insulation layer to prevent heat loss from the top surface of the power module. The X-ray image of the die-attach layer in Fig. 16(b) shows that the void percentage is less than 1%, causing only minor impact on the thermal performance of the power module. The whole module is mounted on a forced-air-cooled heatsink. The heat transfer coefficient of the heatsink is around 1000 $W/m^2/°C$.

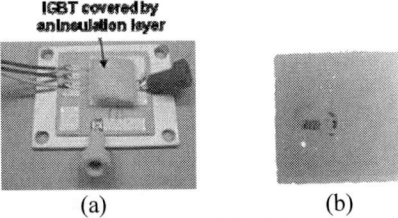

(a)   (b)

Fig. 16. Completed power module: (a) power module with heat-spreader; (b) X-ray image of die-attachment.

Because IGBT characteristics are highly dependent on junction temperature, it is difficult to generate the power dissipation waveform shown in Fig. 11 using a normal power supply. The heating control in the thermal impedance measurement system regulates the gate voltage of the IGBT to maintain constant power dissipation in the IGBT's active region. In this test set-up, a voltage source is connected across the IGBT. A closed-loop current regulator maintains the emitter current to be 16 A under test. By regulating both current and voltage in the IGBT, the power dissipated can be adjusted to any level. Fig. 17 shows the current, voltage, and power dissipation waveforms in the IGBT. With the designed power supply, the current has a step change with the amplitude of 16 A from 0 ms to 160 ms. Due to the output impedance of the power supply, the voltage across the IGBT is dropped from 5.2 V to 4.4 V. However, the power dissipation (shown in Fig. 17(c)) in the IGBT is still a pulse waveform with the amplitude of 70 W. The junction temperature of the IGBT can be measured by monitoring a temperature sensitive parameter within the power module 11. In this test, the forward voltage of a temperature sensor diode embedded in the IGBT is monitored.

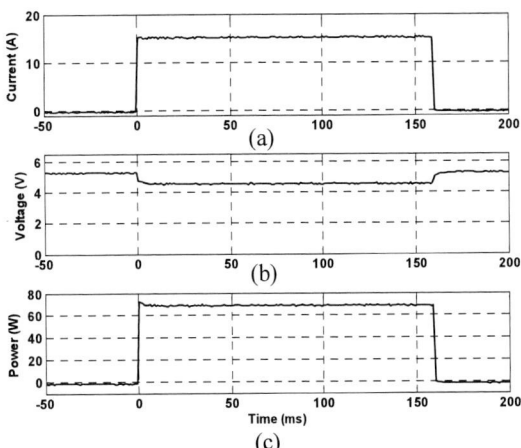

Fig. 17. Waveforms of IGBT; (a) current; (b) voltage; (c) power.

The steady-state thermal resistance of these two modules was measured. The power dissipation in power module was set to be 15 W and 35 W. The measurement results are shown in TABLE IV. It can be seen that for both modules, consistent measurement results can be achieved at different power levels. The measured results are around 10% higher than the simulation results. For the module with 3 mm thick heat spreader, the steady-state thermal resistance is 0.73°C/W, which is 5% higher than the designed value. This could be caused by the interfacial thermal resistance in the power module, which is not modeled in the simulations.

To measure the transient thermal impedance of the power module, the power dissipation as shown in Fig. 17(c) is generated in the IGBT. The transient temperature responses of these two modules are plotted and compared with the simulation results in Fig. 18. It can be seen that the measured and the simulation results match within noise uncertainty for both cases. The maximum transient thermal impedance for the power module with 3 mm thick heat spreader is 0.21°C/W at 120 ms, which agrees with the designed value. The noise in the measured waveforms is induced by the voltage measurement of the sensor diode using oscilloscope.

TABLE IV
EXPERIMENTAL RESULTS FOR STEADY-STATE THERMAL
RESISTANCE OF POWER MODULES

	Power Module with 1 mm heat-spreader	Power Module with 3 mm heat-spreader
Measured at P=15 W	0.89 °C/W	0.74 °C/W
Measured at P=35 W	0.91 °C/W	0.72 °C/W
Simulated value	0.84 °C/W	0.68 °C/W

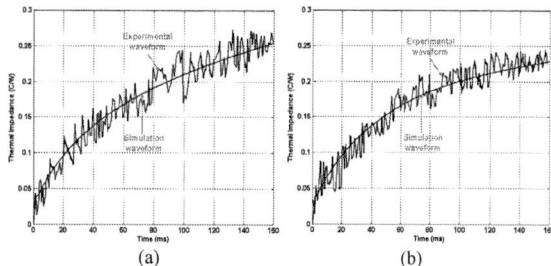

(a)   (b)

Fig. 18. Experimental results for transient response of the power module: (a) power module with 1 mm heat spreader; (b) power module with 3 mm heat spreader (heat transfer coefficient: 1000 W/m^2/°C).

V. Conclusion

In this paper, the impacts of the geometry of the heat spreader and the heat transfer coefficient of the heat exchanger on both steady-state and transient thermal performance of power module are investigated. It is observed that under steady-state conditions, the thicker heat spreader and the higher heat transfer coefficient result in lower thermal resistance. However, the heat spreading effect will be limited with the increase of the heat transfer coefficient. The transient performance is mainly dependent on the thermal capacitance in the power module. The heat transfer coefficient only has minor impact on transient thermal performance. To meet a specified peak transient temperature, the thickness of the heat spreader should be selected to result in a thermal time constant comparable with the transient duration.

Based on the results obtained, a methodology for thermal design based on peak transient temperature is proposed. Using the proposed methodology, the thickness of the heat spreader and the heat transfer coefficient were designed for a power module to achieve 0.70 °C/W thermal resistance under steady-state conditions and 0.21 °C/W thermal impedance under 120 ms transient. Experiment was performed to verify the design methodology. Results show that the design meet both steady-state and transient thermal requirements.

Acknowledgements

The authors would like to thank Mr. Tao Wang from Department of Materials Science Engineering and Mr. Pritish R. Parida from Department of Mechanical Engineering, Virginia Polytechnic Institute and State University, for their discussion and support in the production of this manuscript.

References

1. M.C Shaw, J.R Waldrop, S. Chandrasekaran, B. Kagalwala, X. Jing, E.R Brown, V.J Dhir, and M. Fabbeo, "Enhanced Thermal Management by Direct Water Spray of High-Voltage, High Power Devices in a Three-Phase, 18-hp AC Motor Drive Demonstration", Three-Dimensional Packaging for Power Semiconductor Devices and Modules", *in proc. Thermal and Thermomechanical Phenomena in Electronic Systems*, 2002, pp. 1007-1014.

2. Luc Meysenc, M. Jylhäkallio, and Peter Barbosa, "Power Electronics Cooling Effectiveness Versus Thermal Inertia", *IEEE Trans on Power Electronics*, vol. 20, no. 3, pp. 687-693, May 2005.

3. A.B. Lostetter, F. Barlow, and A. Elshabini, "An Overview to Integrated Power Module Design for High Power Electronics Packaging", *Microelectronics Reliability*, vol. 40, pp. 365-379, Mar 2000.

4. J. W Vandersande and J. P Fleurial, "Thermal Management of Power Electronics using Thermoelectricoolers", *in Proc. International Conference on Thermoelectrics*, Mar 1996, pp. 252-255.

5. Anandan, S. S. and Ramalingam, V, " Thermal Management of Electronics: A Review of Literature", *Thermal Science*, vol. 12, no. 2, pp. 5-22, Mar. 2008.

6. Ansoft, "User's Guide – ePhysics, Rev 3.0", www. Ansoft.com, Jul. 2006.

7. Bjorn Vermeersch and Gilbert De Mey, "A Fixed-Angle Heat Spreading Model for Dynamic Thermal Characterization of Rear-Cooled Substrates", *in Proc. Semiconductor Thermal Measurement and Management Symposium*, Mar 2007, pp.95-101.

8. Xiao Cao, Tao Wang, K. D. T Ngo, Guo-Quan Lu, "Height Optimization for a Medium-Voltage Planar Package", *in Proc. Applied Power Electronics Conference and Exposition*, Feb. 2009, pp.479-484.

9. Guo-Quan Lu, J.N. Calata, Zhiye Zhang, and John G. Bai, "A lead-free, low-temperature sintering die-attach technique for high-performance and high-temperature packaging", *in Proc. High Density Microsystem Design and Packaging and Component Failure Analysis Conference*, Jul. 2004, pp. 42-46.

10. Dimosthenis C. Katsis and J. D. van Wyk, " Void-Induced Thermal Impedance in Power Semiconductor modules: Some Transient Temperature Effects", *IEEE Trans on Industry Applications*, vol. 39, no. 5, pp. 1239-1246, Sep. 2003.

11. J. W. Sofia, "Analysis of Thermal Transient Data with Synthesized Dynamic Models for Semiconductor Devices", *IEEE Comp., Hybrids, Manufact. Technol.*, vol. 18, pp. 39-47, Mar. 1995.

A Study of the Heat Transfer Characteristics of the Micro-channel Heat Sink

Shun Wang[1], Yan Zhang[1], Yifeng Fu[2], Johan Liu[1, 2], Xiaojing Wang[3], Zhaonian Cheng[1]

[1] Key Laboratory of Advanced Display and System Applications, Ministry of Education
SMIT Center, School of Mechatronics Engineering and Automation
Shanghai University, P.O.B. 282, Shanghai 200072, China
[2]SMIT Center & Bionano Systems Laboratory, Department of Microtechnology and Nanoscience
Chalmers University of Technology, SE-412 96 Gothenburg, Sweden
[3]School of Mechatronics Engineering and Automation, Shanghai University, P.O.B. 282, Shanghai 200072, China
Email: jliu@chalmers.se

Abstract

Micro-channels are generally regarded as an effective method for the heat transfer in electronic products, and much effort has been put into improving their capacities. At the same time, carbon nanotubes have shown great potential in the field of heat transfer. This paper focuses on the micro-channel heat sinks combined with carbon nanotubes. A series of 3D models had been created, and the heat transfer characteristics were studied. Two kinds of numerical model were carried out by using Fluent. One model contained the whole micro-channel cooler structure, and the other was a simplified model with only one channel taken into simulation. A comparison of results by the two models indicates a 19.5% variation in the maximum temperature on the fins. Furthermore, five micro-channel cooler structures were constructed with different fin widths. The simulation results showed that with the decrease in the fin width, heat transfer ability was improved. This is attributed to the anisotropic thermal conductivity of the fin arrays.

1. Introduction

There are various methods for heat transfer management in high power integrated circuits, and micro-channel heat sinks are a prominent option because of their advantages including increased heat transfer area and low thermal diffusion length. At present, micro-channels are mainly made of silicon or copper. Even though these materials have similar coefficients of thermal expansion and good thermal conductivity, they fall short when compared with carbon nanotube (CNT). Assuming the same rules for transport in both thermal and electrical conductivity, the theoretical maximum thermal conductivity of CNT was calculated to be 6600 W/mK at 300K. Experiment indicated that the thermal conductivity measurements on mats of nanotube strands ranged from 1750 W/mK to 5850 W/mK[1]. The excellent physical properties of CNT have raised great interest in both thermal and electrical applications.

Although single-walled CNTs show great thermal properties, the CNTs are dependent on their orientation when bundled into groups. Figure 1 shows an overview of the array and displays numerous parallel CNTs gathering together to form a fin. This causes the thermal conductivity in the horizontal direction to be far lower compared with that in the vertical direction. Previous research revealed that the interface between multi-walled CNTs and their growth substrate and the interface established at the free ends of multi-walled CNTs could be significant thermal bottlenecks. Therefore, to promote a good thermal contact on the interfaces between the multi-walled CNTs and their growth substrate, adhesion layers consisting titanium, molybdenum, or chromium were often deposited onto the substrates before fabrication [2-4]. Other studies showed that the multi-walled CNT free ends' interface had significant higher resistance compared to that at the multi-walled CNT growth substrate interface; and this problem could be solved by using a thin layer of indium to weld the multi-walled CNT ends to opposing substrate [5]. There were also researches on the using of an impinging jet to enhance heat transfer [6]. Carbon nanotube grows at high temperatures of about 750°C. In order to prevent the chip from being damaged, the nanotubes could be grown on silicon substrates and thereafter transferred onto the chip[7]. The bond between the CNTs and the chip is crucial.

Figure 1:SEM imagine of the 25μm wide CNT fins

A single CNT can grow several nanometers in diameter and possess good array distribution. In addition, due to the small coefficient of thermal expansion of CNT, micro-scale channels can be constructed. It is of great potential to develop CNT-based micro-channel cooler to obtain good heat management for high power components. In this paper, numerical simulations were carried out to investigate the heat transfer characteristics of the micro-channel cooler. And various geometrical sizes of fin arrays and thermal conductivities have been taken into consideration.

2. CFD simulation models

2.1 Governing equations

In the simulation, the computational fluid dynamics model was set up for the studied case.

Continuity equation

$$\frac{\partial u}{\partial x} + \frac{\partial v}{\partial y} + \frac{\partial w}{\partial z} = 0 \tag{1}$$

Momentum equations

978-1-4244-4658-2/09 $25.00 © 2009 IEEE

$$u\frac{\partial u}{\partial x}+v\frac{\partial u}{\partial y}+w\frac{\partial u}{\partial z}=-\frac{1}{\rho}\frac{\partial p}{\partial x}+v(\frac{\partial^2 u}{\partial x^2}+\frac{\partial^2 u}{\partial y^2}+\frac{\partial^2 u}{\partial z^2}) \quad (2)$$

$$u\frac{\partial \upsilon}{\partial x}+v\frac{\partial \upsilon}{\partial y}+w\frac{\partial \upsilon}{\partial z}=-\frac{1}{\rho}\frac{\partial p}{\partial y}+v(\frac{\partial^2 \upsilon}{\partial x^2}+\frac{\partial^2 \upsilon}{\partial y^2}+\frac{\partial^2 \upsilon}{\partial z^2}) \quad (3)$$

$$u\frac{\partial w}{\partial x}+v\frac{\partial w}{\partial y}+w\frac{\partial w}{\partial z}=-\frac{1}{\rho}\frac{\partial p}{\partial z}+v(\frac{\partial^2 w}{\partial x^2}+\frac{\partial^2 w}{\partial y^2}+\frac{\partial^2 w}{\partial z^2}) \quad (4)$$

Energy equation

$$u\frac{\partial T}{\partial x}+\upsilon\frac{\partial T}{\partial y}+w\frac{\partial T}{\partial z}=\alpha(\frac{\partial^2 T}{\partial x^2}+\frac{\partial^2 T}{\partial y^2}+\frac{\partial^2 T}{\partial z^2}) \quad (5)$$

where u, v, w are the velocity components in the x-, y- and z-directions, p is the pressure, ρ is the fluid density, v is the dynamic viscosity, T is the temperature and α is the thermal diffusivity, respectively.

First of all, the mesh-dependence test was carried out. Three kinds of mesh numbers had been compared, and the maximum and minimum temperatures on the fin in various cases were calculated. Results were shown in table 1, and not much difference was found. Therefore the influence of amount of mesh can be neglected.

Table 1 Mesh-dependence test

mesh number	$T_{max,fin}$	$T_{min,fin}$
22080	38.43 °C	24.6 °C
1140	37.841 °C	24.512 °C
120	37.70 °C	24.99 °C

Then the simulation was carried out with the mesh number chosen according to the computation efficiency.

2.2 Simulation models

During the simulation, two types of models were taken into consideration. One model was a single channel, which can be seen as a local part of the micro-channel cooler. The other model was the whole micro-channel cooler, with the channel dimensions shown in Figure 2. The CNT array was grown on a 10mm × 10mm area. The thickness of the underneath silicon substrate was 50μm, and the heat flux was applied from the silicon bottom.

The following values were set in the simulation with the two geometries: two kinds of the fin widths were computed, namely 200μm and 100μm, where the heat flux was set to be 15W/cm^2. The initial temperature of the ambient water was set to be room temperature 20°C, and the inlet velocity of water was 0.1m/s. In the micro cooler simulation, the inlet and exit were lengthened in order to obtain a fully developed fluid field.

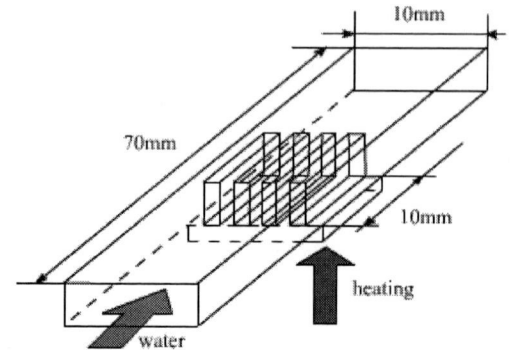

Figure 2 Simulation model of the micro-channel cooler

2.3 The temperature

Figure 3 show the temperature distribution of the single channel model. The results are shown in Table 2, with comparison to those of the micro-channel cooler model.

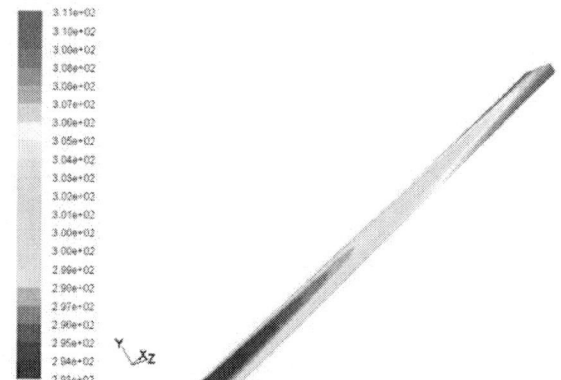

Figure 3 Temperature distribution in the single channel

Data from Table 2 shows the maximum variation can reach up to 19.5%. The reason may be that the inlet part plays an important role in the heat dissipation.

The velocity distribution of inlet part is parabola and the boundary condition near the wall is non-slip in the micro cooler model. While in the single channel model, the inlet velocity was prescribed as a constant.

Table 2 Comparison between the complete model and the simplified model

		$T_{max,bottom}$	$T_{min,bottom}$	$T_{max,fin}$	$T_{min,fin}$
200μm	Single channel	37.99 °C	26.00 °C	37.76 °C	25.86 °C
	Micro- cooler	32.10 °C	25.01 °C	31.60 °C	24.87 °C
100μm	Single channel	35.13 °C	22.81 °C	35.05 °C	21.58 °C
	Micro- cooler	29.36 °C	21.72 °C	29.18 °C	20 °C

2.3 The pressure drop

The pressure change in the channel was also calculated. Figure 4 shows the pressure contours of the single channel. The pressure drop between the channel inlet and outlet of the single model was compared with that of the micro cooler model, and the results are listed in Table 3.

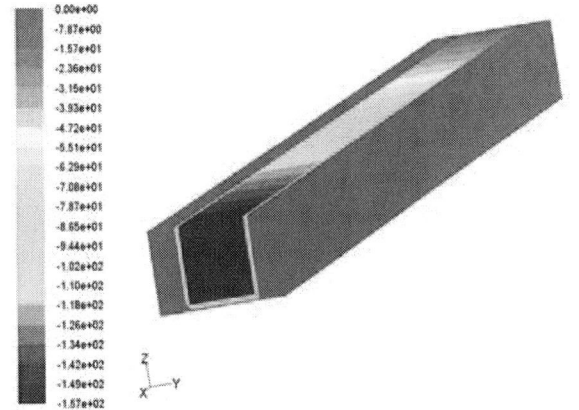

Figure 4 Pressure distribution of the single channel model

The difference in the pressure drop of the two models was obvious, which means the pressure results from the single channel could not be deduced to predict the flow field and the pressure distribution in the whole micro channel. The reason is that the streams around the fins had interacted with each other at the entrance and exit areas of the fin array. In addition, in the whole cooler model the front zone of the fins provided resistance force before the flow entered the micro-channel.

Table 3 Comparison between the micro cooler and single channel models

	500 μm	100 μm	20 μm
Single channel	157.3Pa	488Pa	10283.7Pa
Micro- cooler	439.9Pa	1249.3Pa	20457.4Pa

To obtain further analysis of the cooler, more simulations were carried out for the whole micro-channel cooler including the fin arrays.

2.4 Micro-channel cooler simulation

Simulations on five different widths of the fins were carried out in order to study the influence on the heat transfer performance. The heat power was generated through the silicon substrate from underneath. The thermal conductivity of the CNT fins in this model was assumed to be anisotropy with $300 W/mK$ in the vertical direction and $30 W/mK$ in the horizontal direction. The coolant material was set as water.

The micro cooler with different component power was calculated, as shown in figure 5. The inlet velocity of the water was $0.1 m/s$. The lines with square symbols were at a condition when the effect of the heater was set at 70W, the lines with triangle symbols represented 15W conditions. At lower power, the temperature gradient on the fin was relatively small. When the heat power rose to 70W, there was a substantial gap between the minimum and maximum fin temperature.

When studying the width of the fin by changing it from 500μm to 200μm, the cooler width was keep constant and the channels were of the same width as that of the fin. The temperature decreased as the fin width decreased, as shown in Figure 5. The lowest temperature was located at the beginning part of the micro-channel. As the flowing water absorbed the heat from the fins, the fluid temperature increased when passing through the channel. When the temperature of the fin was 55.05°C, which came close to the water's temperature of 53.71°C, the heat dissipation ability became limited.

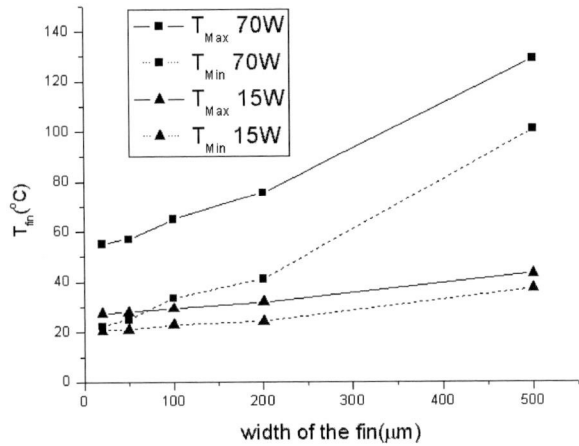

Figure 5 Temperature distribution on the fin.

Comparison of micro cooler with two inlet velocities was also made, as shown in figure 6. The lowest temperatures at the inlet tended to approach the temperatures of the water. The difference caused by velocity became inconspicuous as the width of fin decreased to 20μm.

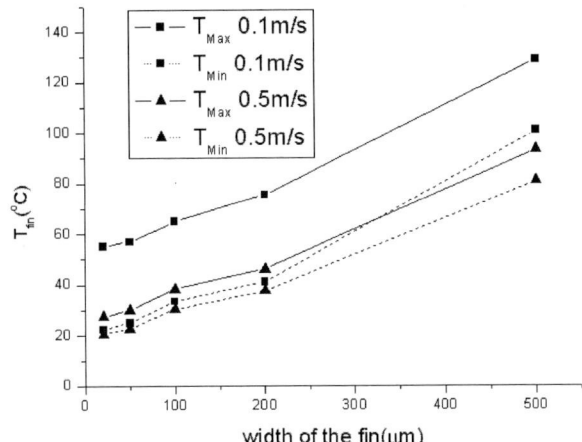

Figure 6 Temperature on the fin.

Although heat dissipation could be enhanced by reducing the width of the fin, it was at the cost of bringing a great pressure drop. This was particularly apparent as the width decreased from 50μm to 20μm, as shown in figure 7. Due to the adhesion strength between the CNT fins and the cooler bottom, low pressure is preferred. So a balance should be considered with respect to CNT-based micro cooler design.

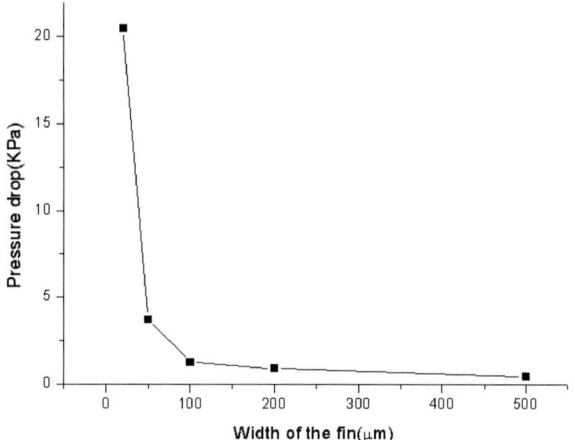

Figure 7 Pressure drop in the cooler

3.1 The effect of the thermal conductivity

As it can be inferred from numerous literature, the CNT bundles exhibit anisotropic conductive features. CNT array consists of numerous single CNTs so the thermal conductivity in the horizontal direction is not as good as the thermal conductivity in the vertical direction. In the present paper, various horizontal conductive values were applied to the fin to study the heat transfer efficiency of the micro-channel cooler. A heat flux of $100W/cm^2$ was used in the boundary condition. The thermal conductivity of CNT fin in the vertical direction was kept a 300W/mK in the studied cases.

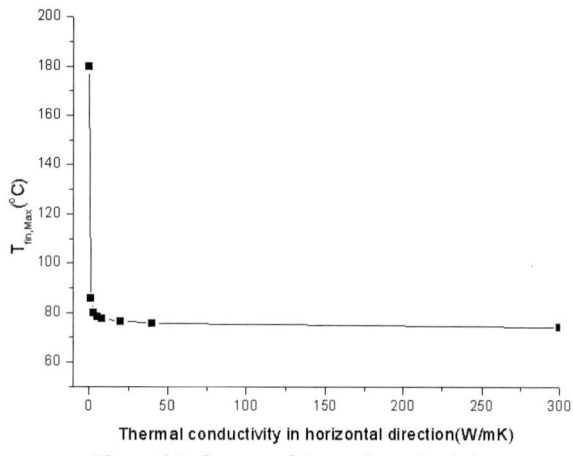

Figure 8 Influence of thermal conductivity

The calculation results are plotted in Figure 8. As the thermal conductivity in horizontal direction changed from zero to non-zero, the max temperature of the fin showed an abrupt drop, which means the heat transfer ability of the miro-channel decreased remarkably. Afterwards the slope of the temperature decreased smoothly.

Figure 9 provides a closer look at the initial stage of the temperature curve in Figure 8. It can be inferred that after $20W/mK$ in the horizontal conductivity, the maximum fin temperature tends to be steady.

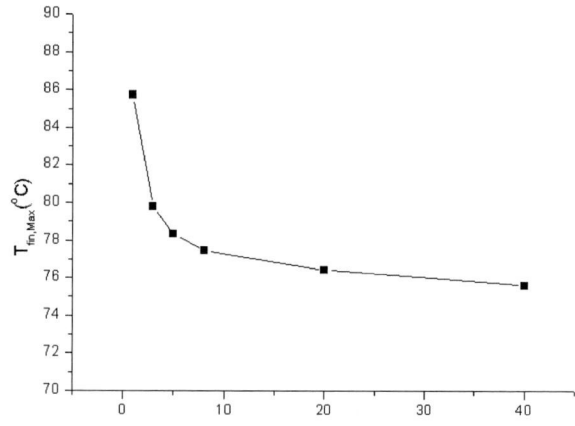

Figure 9 Influence of thermal conductivity

3.2 The effect of the width of the fin

Keeping the heat flux as $15W/cm^2$ in the boundary condition, inlet velocity of the water as 0.1m/s, and the temperature of initial water temperature as 20°C, the influence of fin width on heat transfer of the micro-channel cooler was studied.

To obtain an obvious trend, the thermal conductivity of the fin in the vertical direction was set at 800W/mK and 100W/mK in the horizontal direction. The results are listed in Table 4.

The temperature of the water at the end of the cooler stayed almost constant. This can be explained by the conservation of energy. The energy equation of the water in the model is expressed as

$$\frac{\int_0^{L_W} \rho_f c_{p,f} V(y,z)[T_{out}(y,z) - T_{in}]dydz}{L_W L_Z} = W_{chip} \qquad (6)$$

where $V(y,z)$ is the volume flow rate, $c_{p,f}$ is the specific heat of the fluid under constant pressure, $T_{out}(y,z)$ is the water temperature at the cooler exit, T_{in} is the water temperature at the cooler inlet, and L_Z and L_W are the height and width of the micro-channel cooler in the model, respectively.

In the numerical simulation, the adiabatic assumption was used for the boundaries of the cooler. So the energy that the chip/heater generates should be equal to the energy that water took away.

Comparing the fin temperature in the studied case, the difference in the maximum values $\Delta T = 11.4°C$ between the cases of fin width 500μm and 200μm, and decreases to $\Delta T = 2.7°C$ between those of fin width 200μm and 100μm.

978-1-4244-4658-2/09 $25.00 © 2009 IEEE   182

Temperature difference was not notable when going to the finer pitch. So, there is an optimized value of the fin pitch.

Table 4 Temperature change with different width of the fin

Fin width (μm)	500	200	100	50	20
Fin height (μm)			500		
$T_{chip,max}$ (°C)	43.53	32.10	29.38	27.91	27.38
$T_{chip,min}$ (°C)	38.11	25.01	22.81	21.11	20.48
$T_{fin,ave}$ (°C)	40.82	28.26	26.27	23.85	22.36
$T_{water,ave}$ (°C)	27.26	27.51	27.31	27.22	27.19

4 Conclusions

This simulation work presents a comparison between complete cooler models and simplified models. The results indicate a 19% variation of temperature. The stream field in simplified models could not fully develop like in complete cooler model. The heat transfer capability could be enhanced by reducing width of the channel or increasing the velocity of the inlet water, but would result in major pressure drop. A weak bond between CNT and substrate might ruin the microchannels. The results show that the temperature varies greatly when the horizontal thermal conductivity changes from 0 to 40. This simulation could provide reference resources for the future experiments.

Acknowledgments

This work was supported by the National Natural Science Foundation of China (Grant No. 10702037). The supports by Shanghai Pujiang Program (08PJ14054) and the Innovation Program of Shanghai Municipal Education Commission (09YZ01) are also greatly appreciated.

The authors acknowledge the financial support from the 863 program (No.2008AA04Z301), the Shanghai Science and Technology Commission with the contract no: 075007004 and the NSFC project (50876057).

This work is also supported by the FP7 IP Nanopack program, from the Swedish National Science Foundation under the project "Nanointerconnect" (621-2007-4660), from the Swedish Foundation for Strategic Research under the ProViking Program (PV08.08).

Reference

1. Mohanmed A., Srivastava O. D., "Temperature dependence of the thermal conductivity of single-wall carbon nanotubes", Nanotechnology, vol.12 (2001), pp. 21-24.
2. Wang X., Zhong Z. and Xu J., "Noncontact Thermal Characterization of Multiwall CarbonNanotubes", Journal of Applied Physics, vol. 97 (2005), pp. 064302.
3. Xu Y., Zhang Y., Suhir E. and Wang X., "Thermal Properties of Carbon Nanotube Array Used for Integrated Circuit Cooling", Journal of Applied Physics, vol. 100 (2006), pp. 074302.
4. Cola B. A., Xu J., Cheng C., Hu H., Xu X., Fisher T.S., "Photoacoustic Characterization of Carbon Nanotube Array Thermal Interfaces", Journal of Applied Physics, vol. 101 (2007), pp. 054313.

5. Tong T., Zhao Y., Delzeit L., Kashani A., Meyyappan M. and Majumdar A., "Dense Vertically Aligned Multiwalled Carbon Nanotube Arrays as Thermal Interface Materials", IEEE Transactions on Components and Packaging Technologies, vol. 30 (2007), pp. 92-99.
6. Jang S. P., Kim S. J., Paik K. W., "Experimental investigation of thermal characteristics for a microchannel heat sink subject to an impinging jet using a micro-thermal sensor array", Sensors and Actuators A, vol. 105 (2003), pp. 211-224.
7. Wang T., Jonsson M., Nystrom E., Mo Z., Campbell E.B. and Liu J., "Development and Characterization of Microcoolers using Carbon Nanotubes", Proceedings of 1st Electronics Systemintegration Technology Conference, Dresden, Germany, September 2006, pp. 881-885.

978-1-4244-4658-2/09 $25.00 © 2009 IEEE

Fracture Simulation of Solder Joint Interface by Cohesive Zone Model

Tong AN, Fei QIN*

College of Mechanical Engineering and Applied Electronics Technology,
Beijing University of Technology, Beijing, 100124, China
Tel: +86-10-67392760; Fax: +86-10-67391617
*Email: qfei@bjut.edu.cn

Abstract

In this paper, the damage fracture of solder joints in board level electronic package subjected to drop impact loadings was numerically simulated by the finite element method and the cohesive zone model. The solder-Cu pad interface was modeled by cohesive zone elements. The results show that fracture initiates at the edge of the PCB side and the damage of solder joint is affected greatly by the used material constitutive model in the simulation.

1 Introduction

Solder joints provide electricity connections and mechanical supports between the electronic component and the printed circuit board (PCB) in portable electronic products such as mobile phones and PDAs. Brittle fracture failure of the solder interconnection occurring at the pad/solder joint interface is the most common failure mechanism in microsystem packages. Therefore, the failure driving force and the critical factor that affects the failure of solder joints under drop impact have become great concern in the reliability design of the products.

Although the great efforts have been made to study the fracture of plastically-deforming solder joint in the past few years, the problem is not fully understood. Currently there are some methods used to model crack initiation and propagation in solder joints. One of the most commonly used approaches, which is an inherently empirical model, is the Coffin-Manson fatigue law [1, 2]. In empirical life models, the life time data which are obtained experimentally are related to measurable physical quantities, such as the inelastic strain range or the cyclic inelastic dissipation, and generally the damage mechanisms are unknown. The Coffin-Manson relationship is usually utilized for evaluating the low cycle fatigue life of solder joints under the thermal-mechanical loading [3, 4], as it is applicable to small-scale yield condition. Obviously, the parameters of the empirical model are not universally valid, since the physical quantity will not correlate with fatigue life if the load is beyond the tested condition. Another method which is a physics-of-failure based model is proposed by Paris et al. [5]. The Paris equation describes a relationship between stress intensity factor and crack growth rate which is applicable in linear elastic fracture mechanics. For the elastic-plastic materials, the J-integral can be applied in the function [6]. This method has been used to predict fatigue life of solder joints [7, 8]. However, since the plastic deformation theory of J-integral is based on small cracks and small strains, it may not be appropriate to utilize it for solder joints which experience large-scale yielding and great crack proceeding. The continuum damage mechanics (CDM) method [9, 10] can be used to account for material damage at the micro-scale

level. Some successful applications of this approach for the damage of solder joints have been reported [11-15]. Although there are some advantages of CDM such as that it is capable to predict the time to failure and the crack growth path during material degradation, its inherent computational complexity and a large number of material constants make it impractical to be utilized in solder joints damage analysis.

All the fracture mechanics approaches described above use one single parameter to govern the failure process, which can not describe the interaction between the fracture process and the plastic deformation. Another method, the cohesive zone modeling (CZM), which has the advantage of analyzing the interface fracture occurring large-scale plastic deformation [16], is an alternative to account to fracture process. Recently, some researches using the CZM approach in the solder joints failure have been demonstrated in the literatures. Yang [17] used a cohesive zone model to predict the low cycle fatigue life of solder joints under cyclic loading. Towashiraporn et al. [18] presented a hybrid model to simulate gradual fatigue crack propagation in solder joints under both isothermal and anisothermal conditions. Abdul-Baqi et al. [19] incorporated cohesive zones at phase boundaries in eutectic Tin-lead solder joint to predict accumulated fatigue damage under cyclic loading. Jing et al. [20] investigated fracture process of a single solder joint under constant velocity loading by CZM.

In this paper, dynamic fracture of solder joints under drop impact loading is numerically simulated by the finite element method. The solder-Cu pad interface is modeled as cohesive zone elements. The simulated results show that fracture emerges at the edge of the PCB side. It also be found that the damage of solder joint is effected greatly by the material constitute model using in the simulation.

2 Failure Model of Solder Joint

According to experimental results, there are three different kinds of possible failure modes which may appear in a solder joint [21, 22]. As shown in Figure 1, the failure modes were categorized as: Mode 1 stands for ductile failure in the bulk solder; Mode 2 represents interfacial delamination at the solder/IMC (intermetallic compound) interface; and Mode 3 means IMC cleavage. A mixture of the failure modes can be observed for both lead-containing and lead-free solders, while the portion of each failure mode within the solders is different. The failure of eutectic lead-containing solder (Sn37Pb) is always dominated by the Mode 1 during the drop impact process. For lead-free solders, such as SnAgCu, an increase in IMC thickness leads to the fracture through IMC layer rather than the bulk solder. With the increasing of the thickness of IMC layer, the amount of Mode 1 failure decreased greatly, and the failure mode was predominated by Mode 2 gradually, and then the amount of Mode 3 failure evidently increase as the progressive growth of IMC thickness.

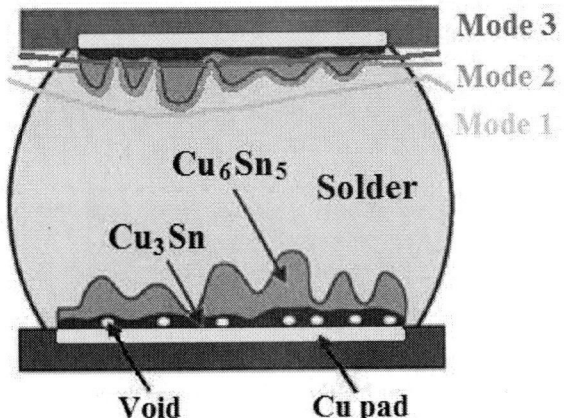

Fig.1 Schematic of failure modes

3 Cohesive Zone Model Approach

The cohesive zone modeling (CZM) approach has been considered as a simple and feasible analytical tool for non-linear fracture processes such as crack initiation, large-scale yielding condition, or irreversible unloading, since it provides the ability to account for the problems which are quite difficult to deal with by the linear elastic fracture mechanics. Barenblatt [23] and Dugdale [24] were the pioneers who proposed the CZM approach to model the fracture behavior of perfect brittle materials and developed it to perfect plastic materials by postulating the existence of a process zone at the crack tip. Needleman et al. [25-27] provided a basic CZM law and utilized it for investigating the void nucleation at a particle and the dynamic fracture growth process of the interface. A trapezoidal type of CZM was proposed by Tvergaard and Hutchinson [28-30] and used for a systematic study of crack growth resistance in elastic-plastic solids. Some attempts to use 2D cohesive zone models for monotonic impact loading have been reported [31, 32]. Camacho and Ortiz [31] adopted a linear type of CZM to simulate the multiple crack propagation along arbitrary paths under impact damage in brittle materials. In Geubelle and Baylor's research [32] the CZM was developed to simulate the spontaneous initiation and propagation of delamination process in thin composite plates under low-velocity impact. In order to make CZM suitable for cyclic loading, there are some studies working on extend CZM laws [33-35]. A 3D irreversible CZM model was developed by de-Andrés et al. [33], which enabled both the tracking of 3D fatigue crack fronts and the calculation of the fatigue life curves. Yang et al. [34] proposed a CZM for fatigue crack initiation and growth in quasi-brittle materials and artificially created a hysteresis loop between unloading and reloading paths. In Roe and Siegmund's work [35], fatigue crack growth was studied by using the irreversible CZM which considered the cyclic evolution of model parameters. A bilinear CZM was adopted by Zavattieri et al. [36, 37] to investigate the fracture of alumina-ceramic microstructures subjected to multi-axial dynamic loading.

The CZMs are approximate a prescribed relationship between surface traction and corresponding crack opening displacement, which is specific for each material and is independent of the geometry or global loading conditions. The traction-separation relations for most of the CZMs are such that, as the interfacial separates, the interface traction initially increases until achieving a maximum value, and then decreases and eventually goes to zero when complete separation occurs. The shape of traction–separation relationship may take various forms, but at least two material parameters are required to define the CZM, which are typically maximum cohesive traction and characteristic opening displacement. Two CZM formulations have been provided. One was proposed by Needleman [25]. In Needleman's model, a nominal traction field **T** which, in general, has both normal and shearing components, is supported an interface. A and B represents two material points which are initially positioned on opposite sides of the interface. The interfacial traction is taken to depend only on the displacement difference across the interface $\Delta \bar{u}_{AB}$. At each point of the interface, we define

$$u_n = \vec{n} \cdot \Delta \bar{u}_{AB}, \quad u_t = \vec{t} \cdot \Delta \bar{u}_{AB}, \quad u_b = \vec{b} \cdot \Delta \bar{u}_{AB} \quad (1)$$

$$T_n = \vec{n} \cdot \vec{T}, \quad T_t = \vec{t} \cdot \vec{T}, \quad T_b = \vec{b} \cdot \vec{T} \quad (2)$$

In equation (1) and (2), n, t, b constitute a coordinate system therefore the positive and negative u_n represents the interfacial separation increase and decrease respectively.

A constitutive relationship that gives the dependence of the tractions T_n, T_t and T_b on the displacements u_n u_t and u_b is used to describe the mechanical response of the interface. Here it is specified in terms of a potential $\phi(u_n, u_t, u_b)$, where

$$\phi(u_n, u_t, u_b) = -\int_0^u [T_n \mathrm{d}u_n + T_t \mathrm{d}u_t + T_b \mathrm{d}u_b] \quad (3)$$

With the separation of the interface, the magnitude of the tractions increases, reaches a maximum, and finally falls to zero which represents a complete decohesion. The magnitude of the tractions is taken to increase monotonically for negative u_n. Relative shearing between the interface leads to the progress of shear tractions, but the dependence of the shear tractions T_t and T_b on u_t and u_b is taken to be linear. The specific potential function used is

$$\phi(u_n, u_t, u_b) = \frac{27}{4} \sigma_{max} \delta \left\{ \begin{matrix} \frac{1}{2}\left(\frac{u_n}{\delta}\right)^2\left[1 - \frac{4}{3}\left(\frac{u_n}{\delta}\right) + \frac{1}{2}\left(\frac{u_n}{\delta}\right)^2\right] \\ + \frac{1}{2}\alpha\left(\frac{u_t}{\delta}\right)^2\left[1 - 2\left(\frac{u_n}{\delta}\right) + \left(\frac{u_n}{\delta}\right)^2\right] \\ + \frac{1}{2}\alpha\left(\frac{u_b}{\delta}\right)^2\left[1 - 2\left(\frac{u_n}{\delta}\right) + \left(\frac{u_n}{\delta}\right)^2\right] \end{matrix} \right\} \quad (4)$$

For $u_n \le \delta$, where σ_{max} is the maximum traction carried by the interface suffering a purely normal separation $(u_n \equiv u_t \equiv u_b \equiv 0)$, δ is a characteristic length and α specifies the ratio of shear to normal stiffness of the interface.

The interface tractions are obtained by differentiating equation (4):

978-1-4244-4658-2/09 $25.00 © 2009 IEEE

$$T_n = -\frac{27}{4}\sigma_{\max}\left\{ \begin{array}{l} \left(\dfrac{u_n}{\delta}\right)\left[1-2\left(\dfrac{u_n}{\delta}\right)+\left(\dfrac{u_n}{\delta}\right)^2\right]+\alpha\left(\dfrac{u_t}{\delta}\right)^2 \\ \left[\left(\dfrac{u_n}{\delta}\right)-1\right]+\alpha\left(\dfrac{u_b}{\delta}\right)^2\left[\left(\dfrac{u_n}{\delta}\right)-1\right] \end{array}\right\} \quad (5)$$

$$T_t = -\frac{27}{4}\sigma_{\max}\left\{\alpha\left(\frac{u_t}{\delta}\right)\left[1-2\left(\frac{u_n}{\delta}\right)+\left(\frac{u_n}{\delta}\right)^2\right]\right\} \quad (6)$$

$$T_b = -\frac{27}{4}\sigma_{\max}\left\{\alpha\left(\frac{u_b}{\delta}\right)\left[1-2\left(\frac{u_n}{\delta}\right)+\left(\frac{u_n}{\delta}\right)^2\right]\right\} \quad (7)$$

For $u_n \leq \delta$ and $T_n \equiv T_t \equiv T_b \equiv 0$ when $u_n > \delta$.

Another commonly used formulation of CZM is an idealized trapezoidal traction-separation law proposed by Tvergaard and Hutchinson [28]. δ_n, δ_t and δ_b denote the normal and tangential components of the relative displacement of the crack faces across the interface in the crack zone. When δ_n^c, δ_t^c and δ_b^c are critical values of displacement components and one non-dimensional separation measure is defined as $\lambda = \left[(\delta_n/\delta_n^c)^2 + (\delta_t/\delta_t^c)^2 + (\delta_b/\delta_b^c)^2\right]^{1/2}$, and the tractions drop to zero when $\lambda = 1$. The potential function used is

$$\phi(\delta_n,\delta_t,\delta_b) = \delta_n^c \int_0^\lambda \sigma(\lambda')d\lambda' \quad (8)$$

The interface tractions are given by

$$T_n = \frac{\partial\phi}{\partial\delta_n} = \frac{\sigma(\lambda)}{\lambda}\frac{\delta_n}{\delta_n^c} \quad (9)$$

$$T_t = \frac{\partial\phi}{\partial\delta_t} = \frac{\sigma(\lambda)}{\lambda}\frac{\delta_t}{\delta_t^c}\frac{\delta_n^c}{\delta_t^c} \quad (10)$$

$$T_b = \frac{\partial\phi}{\partial\delta_b} = \frac{\sigma(\lambda)}{\lambda}\frac{\delta_b}{\delta_b^c}\frac{\delta_n^c}{\delta_b^c} \quad (11)$$

Several types of CZMs have been developed and frequently associated with the finite element method to investigate fracture failure in various materials. In this study, we have used one of such models incorporated in the ABAQUS software [38]. The cohesive zone elements are implemented in a finite element model with zero volume placed between the regular finite elements at the interfaces of interest. The interfacial cohesive elements have their own behaviors which are different from the surrounding elements. The piecewise linear form of the cohesive laws is presented in Figure 2, which is determined by the initial stiffness $\mathbf{K}_0$, the peak value of the stress T_n^0 and the effective displacement at complete failure δ_n^f. It is presumed that once certain criterion has been met within the cohesive elements, the damage will begin. Progressive damage of the interface is described by a damage evolution law. A brief discussion of the cohesive mode used is given below.

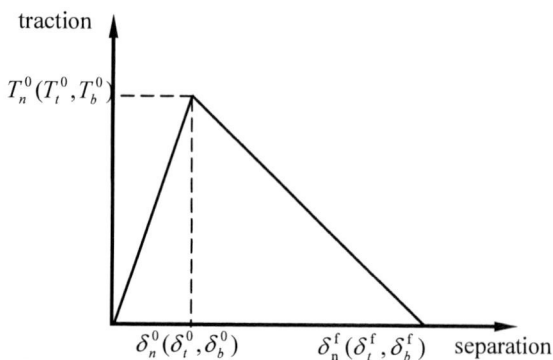

Fig. 2 Typical traction-separation response

The nominal traction stress vector $\mathbf{T}$, consists of three components T_n, T_t and T_b, which represent one normal and two shear tractions respectively. The corresponding separations are denoted by δ_n, δ_t and δ_b, and h_0 is defined as the original thickness of the cohesive element, the nominal strains can be defined as

$$\varepsilon_n = \frac{\delta_n}{h_0}, \ \varepsilon_t = \frac{\delta_t}{h_0}, \ \varepsilon_b = \frac{\delta_b}{h_0}, \quad (12)$$

The elastic behavior can then be written as

$$\mathbf{T} = \begin{Bmatrix} T_n \\ T_t \\ T_b \end{Bmatrix} = \begin{bmatrix} K_{nn} & K_{nt} & K_{nb} \\ K_{nt} & K_{tt} & K_{tb} \\ K_{nb} & K_{tb} & K_{bb} \end{bmatrix} \begin{Bmatrix} \varepsilon_n \\ \varepsilon_t \\ \varepsilon_b \end{Bmatrix} \quad (13)$$

Damage initiation. Damage initiation refers to the beginning of degradation of the response of a material point. The process of degradation begins when the stresses and/or strains satisfy certain damage initiation criteria. Here a quadratic nominal stress criterion was used and damage is assumed to initiate when a quadratic interaction function involving the nominal stress ratios (as defined in the expression below) reaches a value of one. This criterion can be represented as

$$\left\{\frac{\langle T_n\rangle}{T_n^o}\right\}^2 + \left\{\frac{T_t}{T_t^0}\right\}^2 + \left\{\frac{T_b}{T_b^0}\right\}^2 = 1 \quad (14)$$

In the discussion, T_n^0, T_t^0 and T_b^0 represent the peak values of the nominal stress when the deformation is either purely normal to the interface or purely in the first or the second shear direction, respectively. Before damage initiation, the cohesive zone is assumed to be elastic and its constitutive behavior is described by the traction-separation relationship shown in Figure 2, where T_n^0, T_t^0 and T_b^0 are properties of the interface representing the strength of the interface when it is subject to T_n, T_t and T_b respectively. The interface is isotropic in its own plane here, then $T_t^0 = T_b^0$. The symbol $\langle\ \rangle$ used in the discussion below represents the Macaulay bracket which signifies that a pure compressive deformation or stress state does not initiate damage.

Damage evolution. The damage evolution law describes the rate at which the material stiffness is degraded once the corresponding initiation criterion is reached. A scalar damage variable, D, represents the overall damage in the material and captures the combined effects of all the active mechanisms. It initially has a value of 0. If damage evolution is modeled, D monotonically evolves from 0 to 1 upon further loading after the initiation of damage. The stress components of the traction-separation model are affected by the damage according to

$$T_n = \begin{cases} (1-D)\overline{T}_n & , \quad \overline{T}_n \geq 0 \\ \overline{T}_n & , \quad \text{otherwise} \end{cases}$$
$$T_t = (1-D)\overline{T}_t \qquad (15)$$
$$T_b = (1-D)\overline{T}_b$$

where $\overline{T}_n$, $\overline{T}_t$ and $\overline{T}_b$ are the stress components predicted by the elastic traction-separation behavior for the current strains without damage.

To describe the evolution of damage under a combination of normal and shear deformation across the interface, it is useful to introduce an effective displacement defined as:

$$\delta_m = \sqrt{\langle \delta_n \rangle^2 + \delta_t^2 + \delta_b^2} \qquad (16)$$

For linear softening ABAQUS uses an evolution of the damage variable, D, namely:

$$D = \frac{\delta_m^f \left(\delta_m^{max} - \delta_m^0 \right)}{\delta_m^{max} \left(\delta_m^f - \delta_m^0 \right)} \qquad (17)$$

In the preceding expression and in all later references, δ_m^{max} refers to the maximum value of the effective displacement attained during the loading history. The assumption of a constant mode mix at a material point between initiation of damage and final failure is customary for problems involving monotonic damage (or monotonic fracture).

4 Drop Impact Simulation of a Board Level Package

A three-dimensional finite element model is established for a board level electronic package using hexahedron elements, see Figure 3(a). The model consists of a component in a size of 5.25×0.5×1.02 mm³ containing a bare silicon die in a size of 2.8×0.5×0.375 mm³ was interconnected to a 52.5×0.5×1 mm³ printed circuit board (PCB) through seven Sn3.0Ag0.5Cu solder joints, which the diameter, standoff and pitch were 0.35 mm, 0.28 mm and 0.7 mm, respectively. A solder mask, substrate and Cu pads were taken into account. Pad designs are solder mask-define on the component side and non-solder mask-define on the PCB side. A layer of intermetallic (IMC) placed between the Cu pad and the solder alloy is explicitly modeled, and the details of the Cu pad-IMC-solder ball are shown in Figure 3(b).

The linear elastic, elastic-plastic and strain rate-dependent Johnson-cook models are applied in the solder, and the material model was obtained from Reference 39. All other materials are modeled as linear elastic. Material parameters of the finite element model are listed in Table 1. Two layers of cohesive zone elements placed at both PCB and component side for the critical solder joint are used to model the solder-Cu pad interfaces to simulate interfacial fracture.

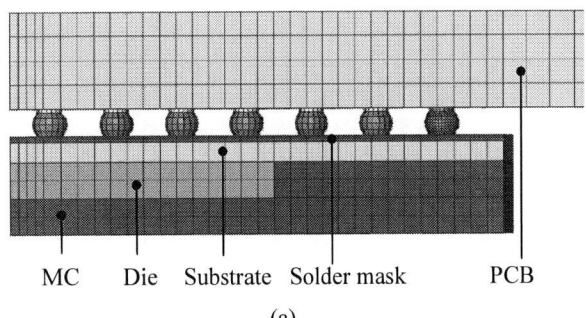

MC   Die   Substrate   Solder mask   PCB

(a)

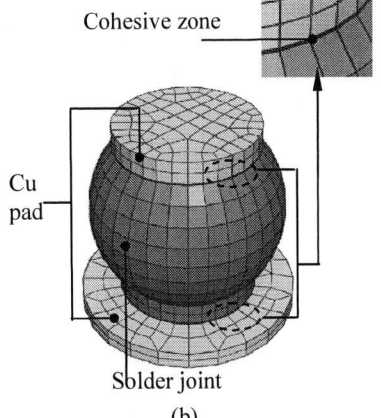

Cohesive zone

Cu pad

Solder joint

(b)

Fig. 3 Finite element model for test specimen meshed fully with Hex-shape elements (a) FEA mesh of test specimen and (b) FEA mesh of solder joint with cohesive zone element

Table 1 Material parameters used in the finite element model

Materials	E (GPa)	v	ρ (kg/mm³)
Die	131	0.23	2.33
Mold component (MC)	28	0.35	1.97
Cu pad	117	0.34	8.94
PCB	20	0.11	1.91
Solder mask	5	0.30	1.15
Sn3.0Ag0.5Cu solder	54	0.363	7.384
Substrate	26	0.11	2.00

The element type of IMC layer was COH3D8, an 8-node three-dimensional cohesive element and all the others were C3D8R which is an 8-node linear brick, reduced integration element in the ABAQUS [38]. There were 10,799 elements and 13,090 nodes totally in the model. The drop impact test condition B defined by the JEDEC [40] was used, which is characterized by an half-sine impact acceleration pulse with a peak of 1,500 G and a time duration of 0.5 ms.

5 Results and Discussion

Figure 4 shows the quadratic nominal stress damage initiation criteria variable which is evaluated as

$$\left\{ \frac{\langle T_n \rangle}{T_n^o} \right\}^2 + \left\{ \frac{T_t}{T_t^0} \right\}^2 + \left\{ \frac{T_b}{T_b^0} \right\}^2$$. It indicates whether the damage initiation criterion has been satisfied at a material point. The damage evolution process of solder joint is presented in Figure 5.

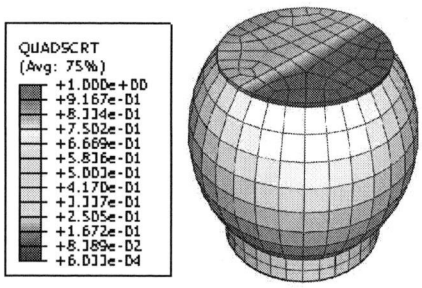

Fig. 4 The quadratic traction damage initiation criterion at integration points predicted by Johnson-Cook model

The simulation predicts that damage mainly develops at the edge of solder joint-Cu pad interface at the PCB side. The location of failure was captured initiated from the outer edge of the solder joint-Cu pad interface and propagated through the inner interface.

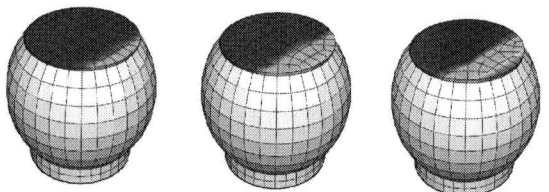

Fig.5 The evolution of the damage in the solder joint predicted by Johnson-Cook model

In order to show the influence of material models on behaviors of solder joints, elastic, elastic-plastic and strain rate-dependent Johnson-cook models were implemented in this analysis. The elastic-plastic model and the Johnson-Cook constitutive model of the Sn3.0Ag0.5Cu solder defined in Reference were used.

The overall value of the damage variable, D, indicates the scalar stiffness degradation of the cohesive elements. Histories of the damage computed by the three material models at critical point in the critical solder joint are plotted in Figure 6. It indicates that among the three material models, the elastic-plastic model predicts the greatest damage whilst the elastic model predicts the least damage. The Johnson-Cook model predicts medial results. This is reasonable because the strain rate effect is included in the Johnson-Cook model.

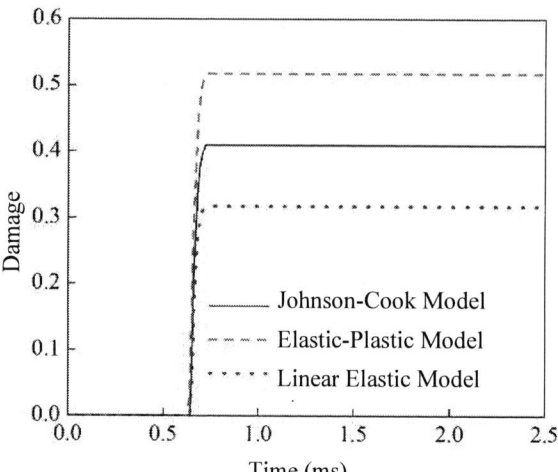

Fig. 6 The damage in the cohesive zone element predicted by different constitutive models

Conclusions

In this paper, the cohesive zone model was applied to predict solder joint failure in board-level drop impact test using the finite element method. The solder-Cu pad interfaces are modeled as cohesive zones. The simulated results show that failure occurs at the PCB side interface rather than the component side interface. Damage starts typically at the edge of the solder-Cu pad interface, and then progresses towards the center.

To close, it is necessary to mention that determining the material parameters of the cohesive zone is a rather difficult issue. Nevertheless, the results here demonstrate that the finite element method with a cohesive zone model is capable of simulating dynamic fracture in solder-Cu pad interfaces.

Acknowledgments

The authors would like to thank financial support from the National Natural Science Foundation of China (NSFC) under the Grant No.10572010, and the Funding Project for Academic Human Resources Development in Institutions of Higher Learning Under the Jurisdiction of Beijing Municipality (PHR - IHLB).

References

1 L. F. Coffin Jr. Transactions, American Society of Mechanical Engineers, TASMA, Vol. 76, (1954), pp. 931.

2 S. S. Manson. "Behavior of Materials Under Conditions of Thermal Stress," *NACA Techn. Note*, (1954), pp. 2933

3 T. D. Solomon. "Fatigue of Sn60Pb40 solder," *IEEE Trans. Component, Hybrid, and Manufacturing Technology*, CHMT29, (1986), pp. 423-433.

4 N. Nir, T.D. Dudderar, C.C. Wong, A.R. Storm. "Fatigue properties of microelectronic solder joints," *Transactions of the ASME, Journal of Electronic Packaging*, Vol. 113, No. 2, (1991), pp. 92-101.

5 P. Paris and F. Erdogan. "A critical analysis of crack propagation laws," *ASME Trans. J. of Basic Engineering*, Vol. 85, (1963), pp. 528-534.

6 N. E. Dowling and J. A. Begley. "Fatigue Crack Growth During Gross Plasticity and The J-Integral," *Mechanics of Crack Growth, Cracks and Fracture*, ASTM STP 590, (1976), pp. 82-103.

7 Y. H. Pao. "A Fracture Mechanics Approach to Thermal Fatigue Life Prediction of Solder Joints," *IEEE Trans. Components, Hybrids, and Manufacturing Technology*, Vol. 15, No. 4, (1992), pp. 559-570.

8 S. B. Lee, J. K. Kim. "A Mechanistic Model for Fatigue Life Prediction of Solder Joints for Electronic Packages," *International Journal of Fatigue*, Vol. 19, No. 1, (1997), pp. 85-91.

9 L. M. Kachanov. Introduction to Continuum Damage Mechanics, (Kluwer Academic Publishers, Dordrecht, 1986).

10 J. Lemaitre. A Course on Damage Mechanics, (second ed., Springer-Verlag, Berlin, Heidelberg, New York, 1996).

11 Y. Wei, C. L. Chao, H. E. Fang, M. K. Neilsen. "Characteristics of creep damage for 60Sn–40Pb solder materials," *Transactions of the ASME, Journal of Electronic Packaging*, Vol. 123, (2001), pp. 278-283.

12 H. E. Fang, C. L. Chow, F. Yang. "A method of damge mechanics analysis for solder material," *Key Eng. Mater.*, Vol. 145-149, (1998), pp. 367-374.

13 S. H. Ju, B. I. Sandor, M. E. Plesha. "Life prediction of solder joints by damage and fracture mechanics," *Transactions of the ASME, Journal of Electronic Packaging*, Vol. 118, (2001), pp. 193-200.

14 X. Zhang, S. W. R. Lee, Y. H. Pao. "A damage evolution model for thermal fatigue analysis of solder joints," *Transactions of the ASME, Journal of Electronic Packaging*, Vol. 122, (2001), pp. 200-206.

15 C. Basaran, C. Y. Yan. "A thermodynamic framework for damage mechanics of solder joints," *Transactions of the ASME, Journal of Electronic Packaging*, Vol. 120, (1998), pp. 379-384.

16 J. W. Hutchinson and A. G. Evans. "Mechanics of materials: Top-down approaches to fracture," Harvard University, Division of Engineering & Applied Sciences, Report MECH-351, (1999).

17 Q. D. Yang, D. J. Shim, S. M. Spearing. "A Cohesive Zone Model for Low Cycle Fatigue Life Prediction of Solder Joints," *Microelectronic Engineering*, Vol. 75, (2004), pp. 85-95.

18 P. Towashiraporn, G. Subbarayan, C. S. Desai. "A Hybrid Model for Computationally Efficient Fatigue Fracture Simulations at Microelectronic Assembly Interfaces. International," *Journal of Solids and Structures*, Vol. 42, (2005), pp. 4468-4483.

19 A. Abdul-Baqi, P. J. G. Schreurs, M. G. D. Geers. "Fatigue Damage Modeling in Solder Interconnects Using a Cohesive Zone Approach," *International Journal of Solids and Structures*, Vol. 42, No. 3-4, (2005), pp. 927-942.

20 J. P. Jing, F. Gao, J. Johnson, F. Z. Liang, R. L. Williams and J. Qu. "Simulation of Dynamic Fracture Along Solder-Pad Interfaces Using A Cohesive Zone Model. Engineering Failure Analysis," Vol. 16, (2009), pp. 1579-1586.

21 Y. S. Lai, P. F. Yang, C. L. Yeh. "Experimental Studies of Board-level Reliability of Chip-scale Packages Subjected to JEDEC Drop Test Condition," *Microelectronics Reliability*, Vol. 46, (2006), pp. 645-650.

22 S. M. Hayes, N. Chawla, D. R. Frear. "Interfacial fracture toughness of Pb-free solders," *Microelectronics Reliability*, Vol. 49, (2009), pp. 269-287.

23 G I Barenblatt. "Mathematical Theory of Equilibrium Cracks," *Advances in Applied Mechanics*, Vol. 7, (1962), pp. 56-129.

24 D. S. Dugdale. "Yielding of Steel Sheets Containing Slits," *Journal of the Mechanics and Physics of Solids*, Vol. 8, (1960), pp. 100-104.

25 A. Needleman. "A Continuum Model for Void Nucleation by Inclusion Debonding," *Journal of Applied Mechanics*, Vol. 54, (1987), pp. 525-531.

26 X. P. Xu, A. Needleman. "Void Nucleation by Inclusion Debonding in A Crystal Matrix," *Modelling and Simulation in Materials Science and Engineering*, Vol. 1, No. 2, (1993), pp. 111-132.

27 X. P. Xu, A. Needleman. "Numerical Simulations of Fast Crack Growth in Brittle Solids," *Journal of the Mechanics and Physics of Solids*, Vol, 42, No. 9, (1994), pp. 1397-1434.

28 V. Tvergaard and J. W. Hutchinson. "The relation between crack growth resistance and fracture process parameters in elastic-plastic solids," *Journal of the Mechanics and Physics of Solids*, Vol. 40, (1992), pp. 1377-1397.

29 V. Tvergaard and J. W. Hutchinson. "Effect of T-stress on mode I crack growth resistance in a ductile solid," *International Journal of Solids and Structures*, Vol. 31, (1994), pp. 823-833.

30 V. Tvergaard and J. W. Hutchinson. "Effect of strain-dependent cohesive model on predictions of crack growth resistance," *International Journal of Solids and Structures*, Vol. 33, (1996), pp. 3297-3308.

31 G. T. Camacho, M. Ortiz. "Computational Modeling of Impact Damage in Brittle Materials," *International Journal of Solids and Structures*, Vol. 33, No. 20-22, (1996), pp. 2899-2938.

32 P. H. Geubelle, J. Baylor. "Impact-induced Delamination of Laminated Composites: a 2D Simulation," *Composites PartB: Engineering*, Vol. 29, No. 5, (1998), pp. 589-602.

33 A. de-Andrés, J. L. Pérez, M. Ortiz. "Elastoplastic Finite Element Analysis Of Three-Dimensional Fatigue Crack Growth in Aluminum Shafts Subjected to Axial Loading. International," *Journal of Solids and Structures*, Vol. 36, No. 15, (1999), pp. 2231-2258.

34 B. Yang, S. Mall, K. Ravi-Chandar. "A Cohesive Zone Model for Fatigue Crack Growth in Quasibrittle

Materials. International," *Journal of Solids and Structures*, Vol. 38, No. 22-23, (2001), pp. 3927-3944.

35 K. L. Roe, T. Siegmund. "An Irreversible Cohesive Zone Model for Interface Fatigue Crack Growth Simulation," *Engineering Fracture Mechanics*, Vol. 70, (2003), pp. 209-232.

36 P. D. Zavattieri, P. V. Raghuram, H. D. Espinosa. "A Computational Model of Ceramic Microstructures Subjectred to Multi-axial Dynamic Loading," *Journal of the Mechanics and Physics of Solids*, Vol. 49, (2001), pp. 27-68.

37 P. D. Zavattieri, H. D. Espinosa. "AnExamination of The Competition between Bulk Behavior and Interfacial Behavior of Ceramics subjected to Dynamic Pressure-shear Loading," *Journal of the Mechanics and Physics of Solids*, Vol. 51, (2003), pp. 607-635.

38 ABAQUS 6.5 User's Manual, Hibbitt, Karlsson & Sorensen, Inc., 2004.

39 Fei Qin, Tong An, Na Chen. "Strain Rate Effects and Rate-dependent Constitutive Models of Lead-based and Lead-free Solders," *Tran. ASME, Journal of Applied Mechanics*, (will be published in November 2009).

40 JEDEC Solid State Technology Association, "Mechanical Shock," JESD22-B104C, 2004, Arlington, VA.

Optimization Design for Packaging Device QFN Using a Prediction Model of the Neural-Genetic Algorithm

Miao Cai, Daoguo Yang, Ligang Niu, Mingjun Zhao, Wenbin Chen
Guangxi Key Laboratory of Manufacturing System and Advanced Manufacturing Technology
Guilin University of Electronic Technology
Guilin, 541004, China
E-mail: caimiao105@gmail.com

Abstract

This study presents an optimal method to select the material and dimension parameters for designing microelectronics packaging devices loading hygro-thermal and vapor pressure. The failure mechanism for delamination of actual packaging devices is often a complex nonlinear function, which is a shortcoming of traditional methods. The proposed approach is a combination of Error back-propagation neural network (BPNN), principal component analysis (PCA) and genetic algorithms (GAs). First of all, PCA is employed to reduce the dimension and de-noise for the learning matrix of BPNN model. And then GAs is combined with the BPNN model to find the most appropriate linking weight with its global search feature. Secondly, the well-trained network model, which included a nonlinear function between the input parameters and corresponding outputs, is seen as a prediction tool to select optimal parameter size in order to reduce the J-integral value of interface cracking in the packaging device. Finally, optimal parameter groups can be achieved for the device after verification. The optimization results show the well-trained PCA-GA-BPNN model used the proposed approach, can be used well in the optimizing design of the microelectronics packaging device loading hygro-thermal and vapor pressure. Meanwhile, the model is available to reduce the fracture reliability problems, and is of much practical value.

1. Introduction

The delamination fracture is one of the main failure modes in microelectronics packaging devices in serviced high temperature conditions [1, 2]. The fracture is due to not only the effects of the environment factors, but also be influenced by the material and dimension parameters [3]. Recent years, more studies focus on the interface strength, but few concentrate on the optimal selection of the material and dimension according to interface fracture of electronics packaging devices. However, the purpose to study the reliability problem is to optimize the material and dimension according to fracture characteristic being studied. Therefore, it is necessary to develop an appropriate method to optimal design for the packaging devices.

Currently, most of the traditional optimization methods are based on the linear theory (or assumed to be linear systems). However, in reality the implied relationship with regarding to the failure mechanism for delamination of actual packaging devices is often a complex nonlinear function. This is the shortcoming of traditional methods.

Therefore, there is an urgent need for a forecasting and optimal design method on the nonlinear problem.

It has been reported that a three-layer feed-forward network can achieve arbitrary precision mapping of the continuous function. Error back-propagation neural network is a multilayer fully-connected feed-forward neural network. At present, the BPNN has been widely used [4-12]. But there are still many aspects that can be improved in using with the BPNN model. Firstly, the key problem is generalization (forecast) capacity for the study of BPNN forecast model [13], and the key issue of the generalization capacity is whether the generalization results is stable. If generalization results are instable, the forecasting model will be no practical value. In the generalization study of BPNN model, most researches are to look for the smallest network structure under the condition by giving the learning samples [7,13]. However, it is just a "passive" approach to improve the structure. Its essence is to find out the best generalization performance by finding the least number of hidden nodes. There are also many learning algorithms available to obtain appropriate network architecture [13]. However, these learning algorithms have their restrictions when they are used to improve generalization ability for a prediction model. The generalization performance of network model can be improved only when following requirements are meted: 1). A truly learning matrix reflecting the relation between input and output matrix is built. 2). The input dimension of learning matrix is lower. 3). The size of the network structure is smaller. Secondly, we can explore the sensitivity between the input and output factors using the implicit nonlinear relationship and generalization performance of a well-trained BPNN model, so as to realize a purpose of optimal design to the object variables. A preliminary extension to this method has been made recently by [9, 11]. The approach is a brand-new method for optimization design and has large potential for application.

In this paper, some methods, such as PCA, GAs and BPNN, are employed to look for an optimal way for packaging devices QFN loading hygro-thermal and vapor pressure. The J-integral value in the interface between bi-material is regarded as the optimal objective in different location, and the material and dimension parameters are selected for an optimal group of the device after this study. The rest of this paper is organized as follows. Section 2 describes the basic theories. Section 3 provides the details of building the optimal model with the proposed procedure of optimization for QFN. Section 4 discusses the results during the procedure. Concluding remarks are finally drawn in section 5.

2. Optimization approach

2.1 Theory basis

2.1.1 Back-propagation neural network

Error Back-propagation neural network (BPNN) has become the most extensive application of neural networks [14]. It has been reported that a three-layer feed-forward network, showed as Fig.1, can achieve arbitrary precision mapping of the continuous function. It is a good way to use the BPNN to fit the high nonlinear function from input data to corresponding output.

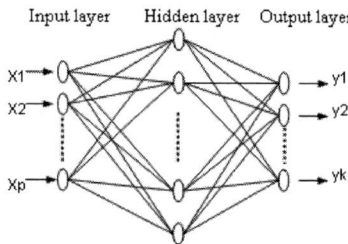

Fig. 1 BPNN topology

At present, for the BPNN model some shortcomings exist, such as poor stability between fitting and prediction, being difficult for determining network architecture, et al [7, 13, 15]. Therefore, it is necessary to take some improving method into the BPNN model.

2.1.2 Principal component analysis [16]

The essence of Principal Component Analysis (PCA) is an exploratory statistical analysis method, which scatters a group of variables in the information on to a certain number of integrated indicators (Principal components). By doing so, principal components (PC) are to use data to describe the internal structure of the data. In fact, it plays a role of reduced-dimension to data.

In this paper, in order to improve the structure of BPNN, The learning matrix is dealt with by employing PCA for the BPNN model as such to reduce the dimension and to de-noise. It constructs an initiative approach of low-dimensional learning matrix that reflects fully the relationship between prediction factors X_p and predictor Y_k. It also tries to improve generalization performance of BPNN to make it suitable for the need of predicting. Meanwhile, actually the new data f_q is to input into the BPNN model instead of X_p.

2.1.3 Genetic algorithms

Genetic algorithm (GA) simulates Darwin's natural selection, genetic selection and the process of biological evolution model. It is a search algorithm for global probability of the biological mechanisms based on a genetic variation and a natural selection. It is particularly applicable to solve the nonlinear problems that are complex and difficult for the traditional method [17].

The basic idea for optimizing BPNN with GAs is: it is to find the most appropriate linking weight and network structure using the global search feature of GAs. These years, many scholars over the world have established prediction models using optimizing neural network with the global search capability of GAs [8, 9].

2.2 The proposed optimization approach

According to the delamination of packaging device, here is an optimization method with the mentioned basic theories. Fig. 2 shows the proposed procedure of the method.

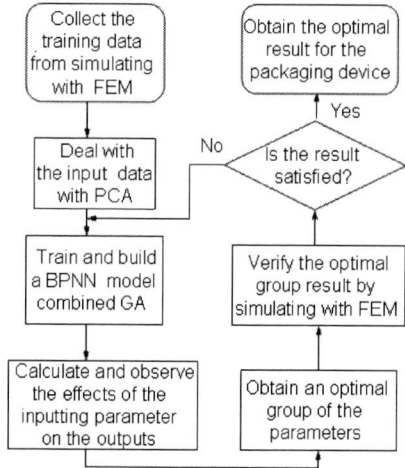

Fig. 2 The proposed procedure of optimization

Actually the proposed procedure of optimization uses the Φ function given as formula (2) which fitted by training GA-BPNN model.

$$Y = \min[\Phi(f_1, f_2, f_3, \cdots f_{q-1}, f_q)] \tag{2}$$

It is trying to find out the most excellent parameter combination f_q (corresponded to a group of X_p) so as to minimize Φ function.

3 Building the optimal model

3.1 Setting the parameters

The delamination along bi-material is the most important reason for the fracture of packaging devices. It is very important to reduce the fracture problems through improving the parameters of the material and dimension in order to optimize the interface reliability. In this paper, a new device QFN is selected to do optimal design. Its structure and dimension are showed in Fig. 3. The J-integral value of crack tips, which showed A~K in Fig. 3, is regarded as the optimal objective Y after getting the main crack of each interface. At the same time, the parameters of the material and dimension are seen as optimal variables X_p.

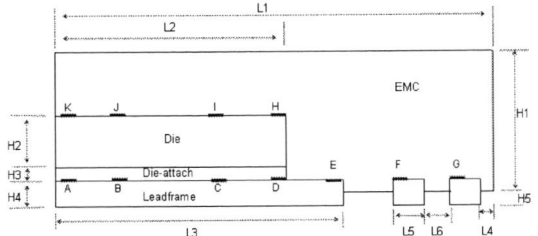

Fig.3 Structure and dimension of half QFN device

There are twenty variables showed as table 1, where DP-R is the ratio of die-size to die/pad-size. All variables have six levels among respective design space. Thinking about the visco-elasticity property of EMC or DA material in QFN, The selection mechanism [18, 19] of the Young's modulus E for the visco-elasticity material, shows as Fig. 4, is used to set the E levels of EMC-Ei and DA-Ei. Both selected levels have been given in table 2, where the glass transition temperature (Tg) of EMC is among 95~155°C, and 55~160°C for DA. In addition, other unchanged

parameters during the analyzing process are showed in table 3, where γ is the Poisson ratio, α is the thermal expansion coefficient and Lcr, Hcr is the length and the height of every crack respectively.

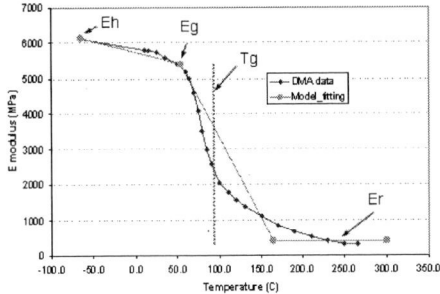

Fig.4 Selection mechanism of the Young's modulus E for the visco-elasticity material

Table 1 Parameter levels(where the E levels of EMC and DA get from table 2)

NO.	EMC		DA		COPPER		DIMENSION					
	E	α	E	α	E	α	L2	DP-R	H1	H2	H3	H4
	MPa	ppm/°C	MPa	ppm/°C	MPa	ppm/°C						
Factor	x1	x2	x3	x4	x5	x6	x7	x8	x9	x10	x11	x12
level1	EMC-E1	5	DA-E1	10	110000	5	1	0.7	0.7	0.2	0.01	0.1
level2	EMC-E2	6.4	DA-E2	18	118000	10	1.3	0.75	0.76	0.23	0.014	0.12
level3	EMC-E3	7.8	DA-E3	26	126000	17.6	1.6	0.8	0.82	0.26	0.018	0.14
level4	EMC-E4	9.2	DA-E4	34	134000	22	1.9	0.85	0.88	0.29	0.022	0.16
level5	EMC-E5	10.6	DA-E5	42	142000	27.7	2.2	0.9	0.94	0.34	0.026	0.18
level6	EMC-E6	12	DA-E6	50	150000	30	2.5	0.95	1	0.38	0.03	0.2

Table 2 E levels of EMC and DA

Level	EMC					DA			
	Tg(°C)	Eh(MPa)	Eg(MPa)	Er(MPa)		Tg(°C)	Eh(MPa)	Eg(MPa)	Er(MPa)
E1	98	12000	10500	500		56	1500	915	200
E2	107	16000	14400	700		81	3200	2710.4	320
E3	116	20000	18000	900		91	5100	4530	440
E4	125	24000	21600	1100		127	7400	6350	560
E5	138	28000	24900	1300		140	9200	8092.4	680
E6	152	32000	29440	1500		159	11000	9674	800

Table 3 Other parameters during the analyzing process

EMC	DA	COPPER	DIE			Dimension(mm)							
γ	γ	γ	E (MPa)	α (ppm/°C)	γ	L1	L3	L4	L5	L6	H5	Lcr	Hcr
0.25	0.33	0.33	169000	3	0.23	4.3	L2/(DP-R)	0.02	0.25	0.04	0.03	0.03	0.01

According to the need to train the BPNN model, the learning data should be comprehensive and well-distributed. Therefore, the uniform design table $U_{54}(6^{12})$ is used to simulate during FEM. Fig. 5 shows the loading

scheme, including the thermal, moisture, and vapour pressure loading that is used in the FE model.

3.2 Training the BPNN model

These models within different parameter group are simulated with the parametric modeling method by FEM.

Fig.6 shows the J-integral of different crack tip at interface DA/COPPER at different temperature. It shows that the positions C and D should be regarded as dangerous locations during the design process for this interface. With the same method, the positions E and H are seen as the main locations to optimal design for the interface EMC/COPPER and EMC/DIE respectively. Therefore, the locations C, D, E, H are the dangerous positions, and their J-integral values such as JC, JD, JE and JH, are regarded as the optimal objectives. The work of collecting data is finished at this time.

Before training the BPNN model, the inputting matrix is dealt with by employing PCA as such to reduce the dimension and to de-noise. The detail procedure and necessity combining PCA are analyzed by CAI Miao et al [20, 21]. Table 5 shows the principal component scores corresponding to the various factors Xi, where x13 is the level of temperature within the loading curve. There are 288(54*6) combinations data for training BPNN model when x13 has six levels such as -65°C, 25°C, 85°C, 200°C, 240°C, 260°C depended on the features of the result curves in Fig.7. Eight PCs f1, f2, f3, f4, f5, f6, f7 and f8 can be obtained through using the formula (1) based on table 4. Meanwhile, eight input nodes and four output nodes for BPNN model are fixed.

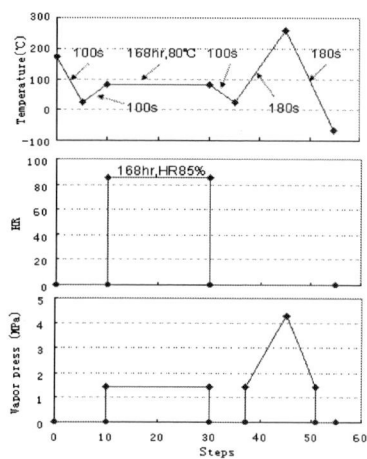

Fig. 5 The loading scheme during analyzing

GAs is also employed to optimize linking weights ω_{ji} and threshold θ_j for training BPNN in order to improve the convergence rate and avoid the weakness of the local flow minimum. During the training procedure, the hidden nodes number from 3 to 10 is used to analyze. After comparison, the network structure 8-6-4, which of the prediction and real results are showed in Fig.8, is selected according to its convergence rate and stability. Its learning mean square error (MSE) is 0.0011.

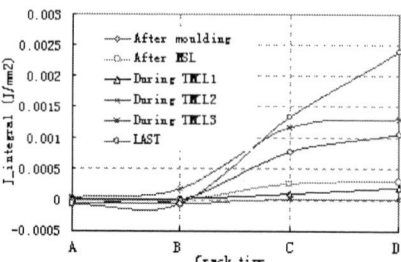

Fig.6 J-value of different crack at interfaceDA/COPPER

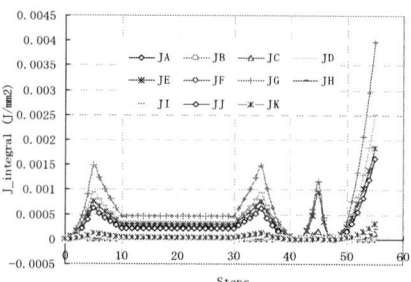

Fig. 7 J-value distribution during simulating procedure

Table 4 Principal component scores

PCs	f1	f2	f3	f4	f5	f6	f7	f8
1	0.01	0.2	-0.06	-0.24	0.38	0.58	0.21	0
x2	-0.49	0.38	0.13	-0.32	-0.12	-0.2	-0.11	0
x3	0.13	0.45	0.3	0.04	-0.47	0.16	0.13	0
x4	0.25	-0.3	-0.4	-0.28	-0.3	0.26	0.13	0
x5	-0.3	0.01	0.03	0.24	0.11	-0.3	0.76	0
x6	0.13	0.32	-0.24	0.44	0.23	-0.11	-0.43	0
x7	0.4	0.33	0.32	-0.17	0.17	0.14	0.19	0
x8	0.05	-0.38	0.39	0.51	-0.02	0.26	0.08	0
x9	0.13	-0.19	0.51	-0.21	0.46	-0.15	-0.18	0
x10	0.45	0.03	0.2	0.01	-0.38	-0.32	0	0
x11	0.09	0.36	-0.19	0.4	0.08	0.19	0.08	0
x12	0.43	0.06	-0.28	-0.11	0.27	-0.43	0.25	0
x13	0	0	0	0	0	0	0	1

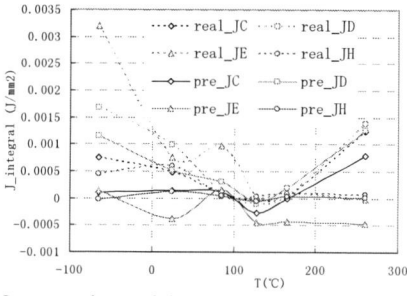

Fig.8 comparison of the prediction and real results

In addition, Fig.8 also shows that the well-trained BPNN model can not get accurate evaluation through predicting. Maybe, the relationship between the input factors and the output factors is too complex to obtain the accurate evaluation on the fracture problem. However, more attention should be paid on the same changing directions between the prediction results and the real results from FEM, but no on the orders of magnitude. Therefore, the well-trained BPNN model has a capacity to explore the variation characteristics of the interface energy. And it is the premise of the optimal design using the well-trained prediction model.

4. Discussions and results

4.1 Selecting optimal parameters

By using the well-trained GA-BPNN, the change process of J-integral can be observed following the changing size of input factors (such as the CTE α, the Young's modulus E, the thickness of DIE H2 and the thickness of DA H3 and so on) within the change scope of input factors in simulation data (showed as Fig.10).

Fig.9 shows that all J-integral decrease with the increasing of temperature, especially for crack C, D of interface DA/COPPER, and crack E of interface EMC/COPPER. Secondly, the J-value reduces to minimum at top temperature; it means that it is most easily to occur failure phenomenon at that moment. Therefore, the optimal parameters for these dangerous locations should be chose prior at top temperature 260 °C.

There are some negative J- integral values in Fig.10. It is a result of the status that both faces of crack close and squeeze mutually, as well as boost the growth of crack once the extrusion force is large enough [3]. So it is the ideal condition for the interface reliability only when the J-value of crack tip is close to zero.

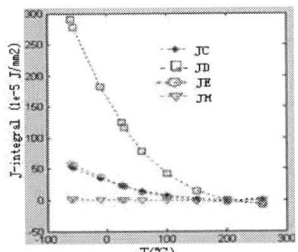

Fig.9 The relationship between T (°C) and J_ integral

Fig.10 (a) shows that with E level of EMC increasing crack C and D are in the opening condition and the opening degree of crack C and D is increasing. However, crack E is a transition from closure status to open, and crack H keeps in balance. It is known that the fracture occurs at interface DA/COPPER [1], and the effect of E level on crack D is bigger than C by combining Fig. 6. It is better to select a low E level EMC-E1 for reliability design. By the same analyzed and selected method, all parameters are selected for crack D prior. The selected results show as the group1 in table 5.

What more attention should be paid on is that the optimal combination from Fig.10 only is obtained by changing each parameter size base on the same parameter combination. Another optimal combination will be obtained using other basic combination. It is the reason of the existing collocation effects among parameters for device. So it has not only one optimal parameter combination, This conclusion is good message for design, because it gives a chance to select a suitable parameter combination for existing material and modern process technology. Additionally, in order to obtain best parameter combination by further comparing among some optimal combinations, the qualitative analysis capacity of the well-trained BPNN model can be used. The author explores the optimal combination through using the 54 groups in Table 4, and former four groups are showed in table 5 after comparison, where the comparative method show as formula (3).

$$J_{total} = \min(|J_{Ci}| + |J_{Di}| + |J_{Ei}| + |J_{Hi}|) \quad (3)$$

where i is ith optimal combination; J is the J-value; C, D, E, H is four main crack position in device respectively.

4.2 Verifying the optimal results

The FEM is employed to verify the optimal results again. Fig.11 shows that the J-integral results of the optimal combinations are improved greatly comparing with ten history data records at all locations. It is sure that the proposed optimal method can obtain effectively the optimal parameter combinations.

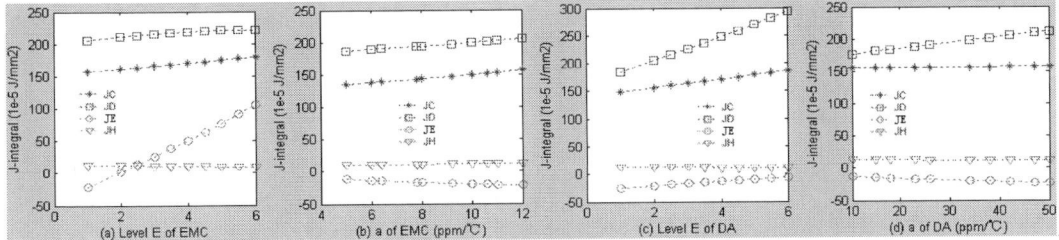

Fig.10 The relationship between each parameter and J- value of each crack tip

Table 5 The optimal group of the parameters

NO.	EMC		DA		COPPER		DIMENSION(mm)					
	E level	α ppm/°C	E level	α ppm/°C	E(MPa)	α ppm/°C	L2	DP-R	H1	H2	H3	H4
1	1	5	1	10	110000	8	2.5	0.7	1	0.38	0.01	0.2
2	6	12	1	10	150000	5	1	0.95	1	0.2	0.01	0.1
3	1	5	1	50	150000	5	2.5	0.7	1	0.38	0.01	0.2
4	6	12	4	10	110000	5	1	0.95	1	0.2	0.01	0.1

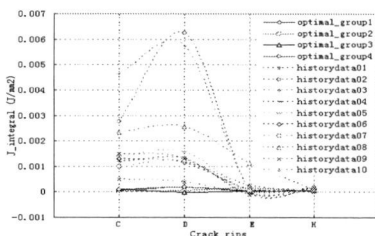

Fig. 11 Verifying the optimal results

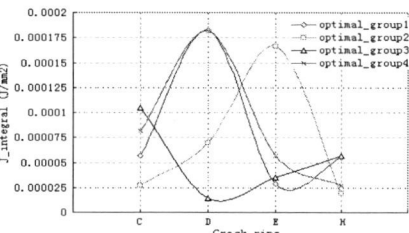

Fig.12 Comparison among the results of optimal groups

In order to meet the requirement for optimizing design, a suitable parameter combination can be selected from these optimal combinations based on existing materials and modern process technology. Fig. 12 shows the comparison results of four optimal groups. The third parameter group can be regarded as the best optimal combination if these materials exist and the processes allow.

5. Conclusions

This study proposes an optimal method using neural network, principal component analysis and genetic algorithms to select the material and dimension parameters for designing microelectronics packaging devices. As an example, it is taken to select the material and dimension parameters for designing microelectronics packaging device QFN loading hygro-thermal and vapor pressure. During analysis procedure, the fitness function is regarded as a optimal tool, which is utilized to explore the effects of each input parameter on the corresponding output results.

More complex relationship is observed among parameters in the system with hygro-thermal and vapor pressure than the system loaded only thermal which was showed by previous investigation [21]. The proposed approach can achieve efficiently the optimization for complex design in the packaging devices loading complication condition.

Acknowledgments

The research work in this paper is financially supported by the National Natural Science Foundation of China (NSFC) (Grant No. 60666002) and the Innovation Project of Guangxi Graduate Education (IPGGE) (Grant No.2008105950802M402)

References

1. W.D.van Driel, M.A.J van Gils, G.Q.Zhang, "Prediction of delamination in microelectronic packages," *IEEE.*, Vol. 6 (2005), pp. 676~681.

2. XIE Weidong, Suresh K. Sitaraman, "Investigation of interfacial delamination of a copper-epoxy interface under monotonic and cyclic loading: experimental characterization," *IEEE.*, Vol. 26, No. 4 (2003), pp. 447~452.

3. Cheng Ying-jun, Thermal dissiPation and thermal mechanical reliability study for MCM—L [D]. 2006, China.

4. HUANG Ruiyi, YANG Shaohua, LI KunLan, et al. "Application of BP neural network on the prediction of non-operation reliability of electronic," *Electronic Product Reliability and Environmental Testing*, Vol. 5 (2005), pp. 7-9.

5. Yousef Al-Assaf, Hany El Kadi, "Fatigue Life Prediction of composite materials using polynomial classifiers and recurrent neural networks," *Composite Structures*, Vol. 77 (2007), pp. 561–569.

6. J.A. Lee1, D.P. Almond, B. Harris. "The use of neural networks for the prediction of fatigue lives of composite materials," *Composites: Part A*, Vol. 30 (1999), pp. 1159–1169.

7. Jin Long, Kuang Xueyuan, Wang Haihong, et al. "Study on the over fitting of the artificial neural network forecasting model," *Acta Meteorologica Sinica*, Vol. 62, No. 1 (2004), pp. 63-70.

8. S.H. Mousavi Anijdan, H.R. Madaah-Hosseini. A. Bahrami, "Flow stress optimization for 304 stainless steel under cold and warm compression by artificial neural network and genetic algorithm," *Materials and Design*, Vol. 28 (2007), pp. 609-615.

9 T.S. Li, C.T. Su, T.L.Chiangc, "Applying robust multi-response quality engineering for parameter selection using a novel neural–genetic algorithm," *Computers in Industry*, Vol. 50 (2003), pp. 113-122.

10. Xu Liujie, Xing Jiandong, Wei Shizhong, et al. "Optimization of chemical composition of high speed steel with high vanadium content for abrasive wear using an artificial neural network," *Materials and Design*, Vol. 28 (2007), pp. 1031-1037.

11. ZHANG Junhong, XIE Anguo, SHEN Fengman, "Multi-objective optimization and analysis model of sintering process based on BP neural network," *Journal of Iron and Steel Research International*, Vol. 14, No. 2 (2007), pp. 1-5.

12. Yaomin Lin, Member, IEEE and Frank G. Shi, "Senior Member. Package design and materials selection optimization for overmolded flip chip packaging," *IEEE*, 2006, pp. 525-532.

13. WU Yan, ZHANG Liming, "A survey of research work on neural network generalization and structure optimization algorithms," *Application Research Computers*, Vol. 19, No. 6 (2002), pp. 21-25.

14. Shuang Cong, "MATLAB toolbox for the theory and application of the neural network," *HeFei: University of Science and Technology Press China*, 2003, pp. 45-60.

15. Hany El Kadi, "Modeling the mechanical behavior of fiber-reinforced polymeric composite materials using artificial neural networks," *Composite Structures*, Vol. 73 (2006), pp. 1-23.

16. Xiulin Yu, REN xuesong, Multivariate statistical analysis [M] Beijing:China Statistics Press, 1999.8pp. 154.

17. Xiaoping Wang, Li-Ming Cao, "Genetic algorithm theory, application and software," *XI'AN: Xi'an Jiaotong University Press*, China, 2002.1, pp. 18-65.

18. K.M.B.Jansen, L, Wang, D.G. Yang, "Constitutive modeling of moulding compounds," *IEEE*, 2004, pp. 890-894.

19. J.G.J.Beijer, J.H.J.Janssen, H.J.L.Bressers, "Warpage minimization of the HVQFN map mould," *IEEE*, 2005, pp. 168-174.

20. CAI Miao, YANG Dao-guo, ZHONG Li-jun, et al. "Prediction of fatigue life of packaging EMC material based on BP neural networks," *Electronic components and materials*, Vol. 27, No. 3 (2008), pp. 64-67.

21. CAI Miao, YANG Daoguo, LI Quan-yong, et al. "An optimum design method for nonlinear system using an improved neural network," IEEE, Wuhan of China: international conference on computer science and software engineering (CSSE 2008), 2008, Vol. 12, No. 1, pp. 94-97.

Lifetime Prediction of an IGBT Power Electronics Module under Cyclic Temperature Loading Conditions

Hua Lu[*], Chris Bailey
University of Greenwich
30 Park Row, London SE10 9LS
[*]Email: h.lu@gre.ac.uk

Abstract

The lifetime of an IGBT power electronics module under cyclic temperature loading conditions has been analyzed using Finite Element Analysis method. The failure mechanisms that have been taken into account are the fatigue of the chip-mount-down solder joint, the substrate attach solder joint, the busbar solder joint and the Aluminum wirebond. The results show that the lifetime of the module is about 1000 cycles under the -40 to 125C cyclic temperature loading condition. The critical failure location has been found to be the busbar solder joint and the lack of compliance of the busbar design is the cause of the problem. The objective of this paper is to demonstrate the methodology of using physics of failure approach for the reliability analysis of power electronics modules and highlights the important design parameters that affect the thermal-mechanical fatigue failures of these components.

Power Module Failure Mechanisms

Power electronic modules (PEM) are self-contained power electronics components that are widely used in aerospace, automotive and alternative energy generation and distribution applications. They play an important role in the conversion, control and delivery of electrical power [1]. PEMs have highly inhomogeneous structures. They are made of semiconductor, ceramic, copper, aluminum, polymer, and sometimes composite materials. These materials are assembled together in the packaging manufacturing process using soldering, direct bond copper (DBC), wirebond, and pressure contact interconnect techniques. In the service and accelerated qualification test conditions, the temperature of the power module changes with time. This gives rise to fluctuating stress and strain due to the mismatch of coefficient of thermal expansion in PEMs. The stress and strain will lead to the degradation and ultimately failure of the interconnection and the modules and this is one of the most important PEM failure mechanisms. As the current trends in the power electronics applications leads to ever greater currents passing through the components and ever smaller spatial profile, this failure is bound to become more and more critical to the PEM reliability.

In order to predict the reliability of PEMs effectively, it is crucial that the failure mechanisms are well understood and the lifetime are predicted accurately. In this paper, the methods of using Finite Element computer simulation combined with experimental results to predict the lifetime of PEMs (Figure 1) is described. This approach to reliability analysis is called the Physics of Failure (PoF) method. PoF method takes into account the root causes of failures and the lifetimes are predicted using computer simulations

appropriate for the underlying physical processes that have caused the failure.

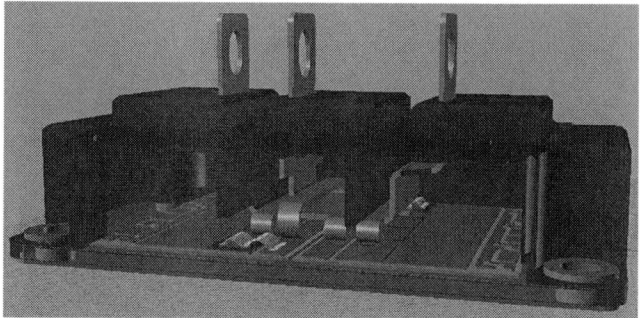

Figure 1. An IGBT power electronics module design.

Under cyclic electric or passive thermal-mechanical loading, the failures of PEMs may be caused by physical phenomena such as solder joints fatigue, wirebond, cracking fatigue, isolation substrate delamination [2-6]. In this work, four failure mechanisms are included in the analysis: the chip mount-down solder joint, the substrate mount-down solder joint, the busbar solder joint and the aluminum wire bond. In this paper, the design parameters that affect these failures are discussed and the lifetime of an ad hoc PEM design has been predicted.

Lifetime Prediction Methods

Busbar Solder joints

PEM busbars are made of copper. They serve as electric terminals and they are connected to the substrate using soldering technique. In Figure 2, a PEM with four busbars is shown.

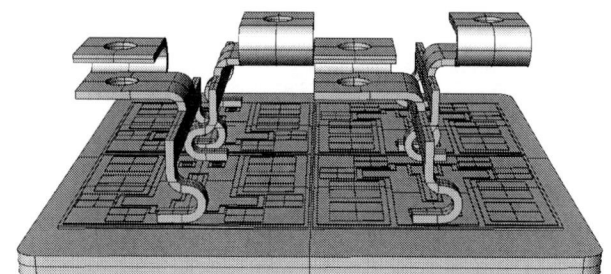

Figure 2. PEM with four busbars.

Busbars may fail because of the solder joint fatigue or delimination at the interfaces around the joint. For the solder joint fatigue mechanism, the lifetime can be predicted using similar prediction method that has been used in

978-1-4244-4658-2/09 $25.00 © 2009 IEEE

microelectronics solder joint reliability analysis [7]. In this type of analysis, the response of the solder joint under a prescribed cyclic loading condition is obtained using Finite Element Analysis method, and the damage indicators such as the accumulated plastic/creep strain per cycle $\Delta\varepsilon_p$ or plastic work density per cycle ΔW_p in solder joints can be calculated in the post-processing stage of the analysis. In general, damage indicator distribution is not uniform and as the crack grows, the geometry of the structure changes and this results in changes stress and damage indicator distribution. As an approximation, however, these are not taken into account and the maximum values or the average values in the solder volume where crack propagate are used for reliability prediction of busbars.

The mechanical properties of solder is highly nonlinear. Even at room temperature, creep is an important deformation mechanism. In this work it is assumed that the solder joints are made of Sn3.5Ag and the visco-plastic/creep constitutive equation for SnAg alloy is,

$$\varepsilon_{cr} = A \times \sinh^n(\alpha\sigma_e)\exp(\frac{-Q}{RT}) \quad (1)$$

where R is the gas constant, T is the temperature in Kelvin, σ_e is the von Mises equivalent stress, A, n, α, Q are material constants and their values are listed in Table I [8].

Table I: Creep parameters for solder materials.

	A(s)	n	α(1/MPa)	Q/R
SnAg	9.00E+05	5.5	0.06527	8690

To calculate the Sn3.5Ag solder joint lifetime after the damage indicator has been obtained, Equation 2 is used.

$$L/N = 0.00562\left(\Delta\varepsilon_p\right)^{1.023} \quad (2)$$

where L is the crack length and N is the number of load cycles that have caused this crack. The unit of L is in mm. The crack length is predefined by failure criteria. In this analysis, it is defined as Large solder joints.

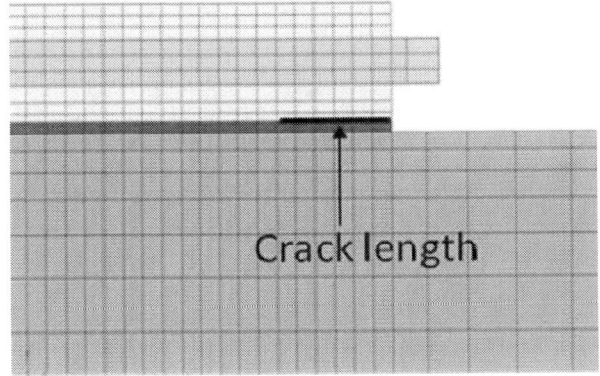

Figure 3. A 2D substrate solder joint model.

In PEMs, chip-mount and substrate solder joints are thin and much larger than microelectronics solder joints. For example, for the power module shown in Figure1, the width of the substrate is 58 mm and the solder thickness is 0.1mm and the aspect ratio is 580. In contrast, BGA and flip chip solder joints have ball shapes and the aspect ratio is in the order of magnitude of 1. In order to take into account of the effect of the stress distribution and changes caused by crack propagation, a method that is based on the Miner's linear damage accumulation is used in this paper for the lifetime prediction of these solder joints [9].

In illustrate how this method works, let's assume the 2D solder joint structure shown in Figure 3 is subject to cyclic loading and crack is expected to propagate along the solder-substrate interface where strain has the highest value. The crack path can be conceptually divided into a number of segments. Each segment can be defined as one or a number of Finite Element mesh elements in the solder joint are. These segments can be regarded as small solder joints and the damage indicator is a constant within each solder segment. For these small "solder joints", Equation 2 can be used to calculate the lifetime in terms of number of cycles to failure [8].

Damage indicators are not uniform in the solder joint. This means that the solder segments will not fail at the same time. At the edge the solder joint, the solder joint segment fails first and then the one next to it will fail and so on so forth, resulting in a crack propagation process that starts from the edge towards the centre of the solder joint. The lifetime of the first segment, i.e. the crack tip, can be calculated using Equation 2. The lifetime of the next segment has to be calculated differently because when the first segment is not cracked, the load on the second segment is not as severe as the load on the first segment, but after the failure of the first section, this second segment becomes the crack tip and load increases. In effect, this second segment will have experienced two load levels before it fails. Similarly, the third sections experiences three levels of load before its failure.

As it shown in [9], the shape of the damage indicator distribution remains very much the same and this means that the damage indicator value is a function of the distance from the crack tip only. By exploiting this characteristics of the large solder joint and by using the Miner's linear damage accumulation rule, the lifetime of each solder segment, NF_i, can be calculated using Equation 3.

$$\begin{pmatrix} \alpha_1 & 0 & \dots & 0 \\ \alpha_2 & \alpha_1 & \dots & 0 \\ \dots & \dots & \dots & \dots \\ \alpha_k & \alpha_{k-1} & \dots & \alpha_1 \end{pmatrix}\begin{pmatrix} NF_1 \\ NF_2 \\ \dots \\ NF_k \end{pmatrix} = N_1\begin{pmatrix} 1 \\ 2 \\ \dots \\ k \end{pmatrix} \quad (3)$$

where $\alpha_i = N_1/N_i$. N_i are the number of cycles to failure under the i_{th} load level and they are calculated using Equation 2.

978-1-4244-4658-2/09 $25.00 © 2009 IEEE

Wirebond lifetime

In PEMs, wirebonds carry strong electric currents and they are often made from heavy Aluminum wires. In thermal mechanical cyclic test environment, the difference between the thermal expansion coefficients of the wire and the substrate material causes mechanical fatigue at the wire-substrate interface (bond lifting) or at the bond heel (heel cracking) as shown in Figure 4. For wirebonds with no glob-top, the dominant failure will be bond-lifting under thermal-mechanical loading conditions. Under power cycling condition, however, this may not be the case because wire flexing will cause more damage to the heels. In this work, thermal-mechanical loading is applied and only the fatigue at the bond interface is assumed to be the failure mechanism.

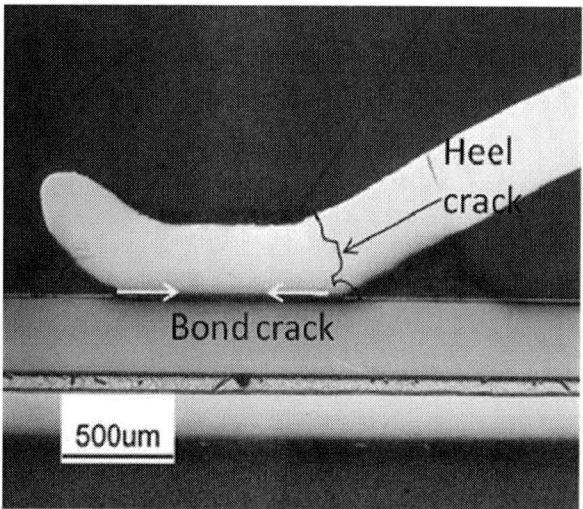

Figure 4. Possible wirebond failure sites.

In this work, the accumulated plastic strain per cycle is used as the damage indicator for the analysis of the wirebond reliability. For the failure at the bond interface, lifetime can be calculated from the following lifetime model.

$$N = 1.18\varepsilon_p^{-3.492} \qquad (4)$$

The failure criteria on which this lifetime model is based is a 90% reduction in shear strength.

Effects of Design Parameters on the Reliability of Power Modules

Busbars

The reliability of the busbars is to a great extent determined by their mechanical compliance. The compliance can be controlled by changing the thickness and the shape of the busbars. Figure 5 shows a computer model of typical busbar design. The bends in the busbar are used to increase the compliance.

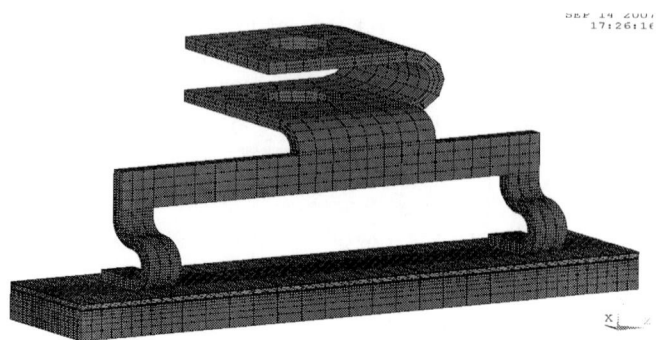

Figure 5. Finite Element model of a busbar.

The compliance can also be enhanced by reducing the thickness of the copper. As shown in Figure 6, the plastic work density in busbar solder joint decreases as copper thickness is reduced from 2.5 to 1 mm. In this analysis, the top- part of the busbar is constrained in the vertical direction.

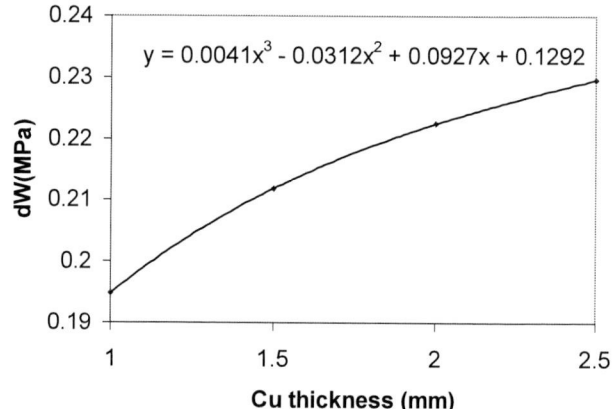

Figure 6. Plastic work density in busbar solder joint as a function of copper thickness.

Other parameters such as the copper's Young's Modulus and the yield stress have also been found to have little impact on the busbar solder joint.

Large solder joints

Table II shows the accumulated plastic strain values for 15 substrate designs. It can be seen that the reliability of large solder joint is mainly affected by its thickness. The size of the substrate will also has impact on substrate solder joint reliability because it takes longer for the cracks to propagate in a larger solder joint than in a small one. In contrast, in a non-underfilled flip chip or ball grid array, chip size greatly affect the reliability because the solder joints are not strong enough to stop the relative movement of the materials on the two sides of the joints and the stress in solder joints increase as chip size increases.

Table II: Effect of design parameters on reliability of substrate solder joint.

Solder thickness(mm)	Substrate thickness(mm)	Baseplate thickness(mm)	$\Delta\varepsilon_p$
0.1	60	9	0.0158
0.1	60	5	0.0144
0.1	54	9	0.0158
0.1	54	5	0.0144
0.1	57	7	0.0154
0.2	60	7	0.0092
0.2	54	7	0.0092
0.2	57	9	0.0094
0.2	57	5	0.0087
0.2	57	7	0.0092
0.3	60	9	0.0070
0.3	60	5	0.0066
0.3	54	9	0.0070
0.3	54	5	0.0066
0.3	57	7	0.0069
0.1	60	9	0.0158

Wirebonds

The reliability of wirebonds has been found to be affected by wire diameter, loop height, purity and manufacturing process. In Figure 7 and Figure 8 the maximum accumulated plastic strains at the end of three thermal-mechanical cycles are shown as functions of wire diameter and the loop height. In this case, the wire diameter has more important effect on the reliability than the loop height. However, by reducing the wire diameter, the current carrying capability is reduced and therefore may not always be an design option.

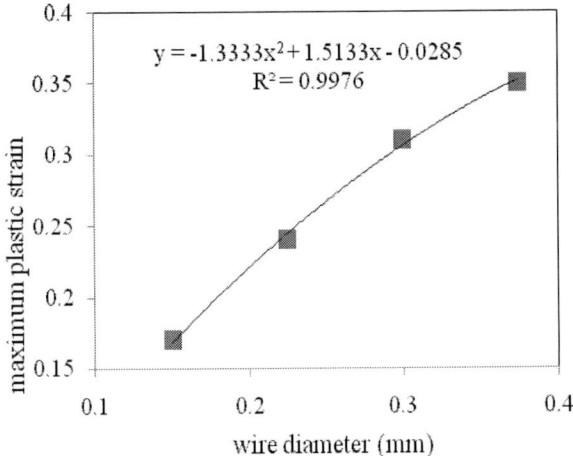

Figure 7. Plastic strain as a function of the wire diameter in Al wire.

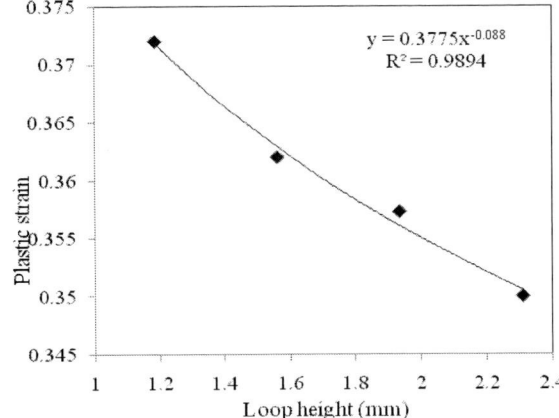

Figuire 8. Plastic strain strain as a function of the loop height.

Lifetime Prediction of The IGBT Module

In order to predict the lifetime of the IGBT module shown in Figure 1, three models have been used. In the first model, there are the three busbars mounted on the substrate using SnAg solder (Figure 9). There are two different busbar solder joint sizes in this PEM. They are 4x2x0.1mm and 4x2.5x 0.1mm respectively. The busbar in the middle has the larger solder joints.

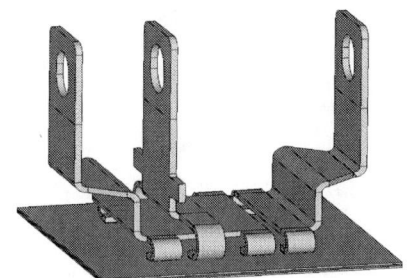

Figure 9. Busbar Model of the IGBT model.

Figure 10. Deformation of the PEM under a fixed thermal load of 100°C.

As busbars, baseplate and much of the substrate are made of copper, CTE mismatch is not the most important factor in causing damage in busbar solder joint. Instead, the boundary constraints may have great impact on the solder joint reliability. PEMs are usually filled with silicone gel and epoxy resin and as temperature changes they expand or contract to push the top of the casing up and down. This movement causes cyclic deformation the damage in the busbar solder joints. To include this effect, the expansion and contraction of the casing has been modeled and the the time dependent displacement values are used as mechanical loading conditions for the busbar simulation. Figure 10 shows the deformed shape of the PEM casing.

The chip mount-down and substrate solder joints are contained in one single model. Figure 11 shows half of the cross section of the model. The solder joints are both 100 microns in thickness. Figure 12 shows the stress and strain in the structure. While the stress is the highest in the silicon the strain concentration is in the solder joints.

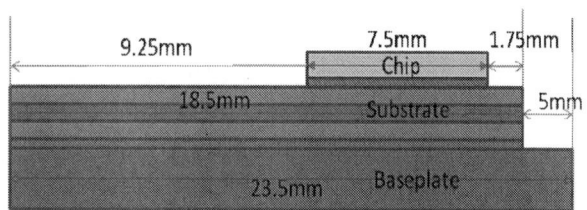

Figure 11. Substrate and chip mount-down solder joint model

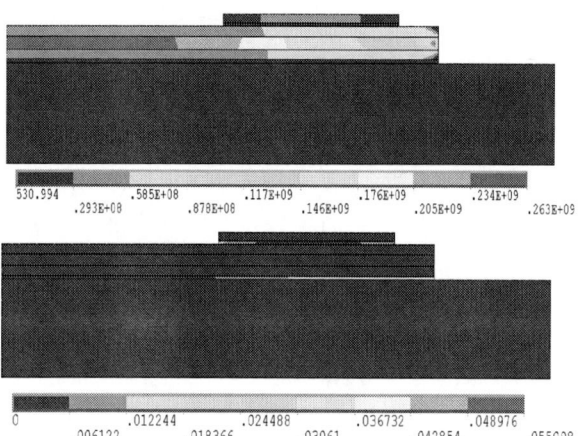

Figure 12. Stress (top) and strain (bottom) distribution in the solder joints

Part of the wirebond model is shown in Figure 13. The Al wire has a diameter of 375μm, the length of the bond foot is 1 mm, the loop length is 2.85 mm and the loop height is about 1.2mm. Figure 14 shows that plastic strain concentrates at the bod interface.

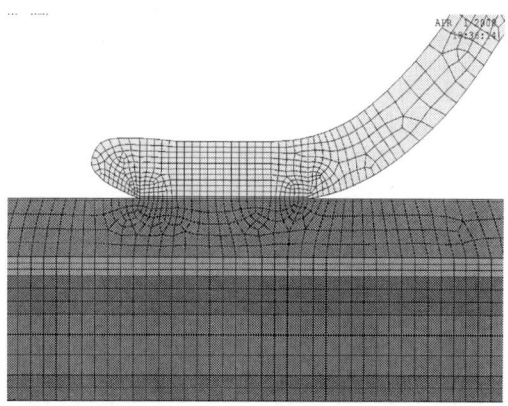

Figure 13. The wirebond model.

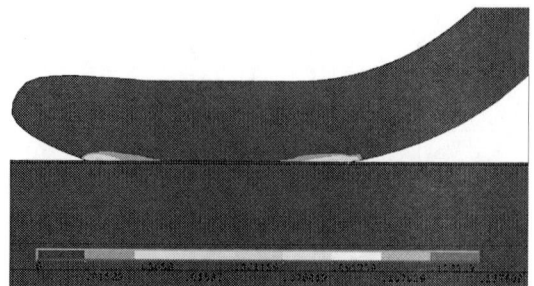

Figure 14. The accumulated plastic strain distribution after three thermal cycles.

Cyclic temperature loading conditions are applied to all models. The ramp time and dwell time are 15 minutes and the minimum and maximum temperatures are -40°C and 125°C respectively.

In order to define the lifetime, it is necessary to define the failure criteria. For the busbar, failure is defined as having 50% of area delamination while for the chip and the substrate solder failure is defined as the 20% area delamination. The failure criteria for the wirebonds is 90% reduction in shear strength.

The lifetimes of the components in the PEM are listed in Table III. For this particular design, the busbar solder joint is the first to fail and its lifetime is the lifetime of the whole PEM. The results also show that chip solder would fail before substrate solder although the damage indicator $\Delta\varepsilon_p$ has a lower maximum value in the chip solder. This is because the size of the chip solder is much smaller than the substrate solder.

Table III: Lifetime predictions

Component	Lifetime (number of cycles)
busbar	946
wirebond	10489
Chip solder	14889
Substrate solder	30000

Conclusions

Physics of Failure methods have been used to analyze the reliability of power electronics module designs. The lifetime of a power module has been predicted. Because of a lack of compliance, a busbar has been found to be the first to fail.

Acknowledgments

The authors wish to acknowledge the support of the Innovative electronics Manufacturing Research Centre (IeMRC) and the United Kingdom Technology Strategy Board for the project 'Modelling of Power Modules for Lifetime, Accelerated Testing, Reliability and Risk'. The authors would like to thank Semelab Ltd and Dynex Semiconductor Ltd. for their great contribution to the work and unreserved support. The author would also like to acknowledge the support from Goodrich Engine Control, SR Drives Ltd., Areva T&D Ltd and Rolls Royce Plc.

References

1. Sheng, W.W. and Colino, R.P., Power Electronic Modules, CRC Press (2005)

2. Shammas, N.Y.A., "Present problems of power module packaging technology", *Microelectronics Reliability*, Vol. 43, Issue 4 (2003), pp. 519-527

3. Pooch, M.-H., Dittmer, K.J., Gabisch, D., "Investigations on the damage mechanism of aluminum wire bonds used for high power applications", *Proc. EUPAC 96*, (1996) pp. 128-131

4. Günther, M., Wolter, K, Rittner, M, Nüthter, "Failure Mechanisms of Direct Copper Bonding Substrates", *Proceedings of Electronics Systemintegration Technology Conference* (ESTC), Dresden, Germany, 2006, pp.714-718

5. Dupont, L., Khatir, Z., Lefebvre, S., and Bontemps, S., "Effects of metallization thickness of ceramic substrates on the reliability of power assemblies under high temperature cycling", *Microelectronics Reliability*, vol. 46 (2006) pp.1766–1771

6. Yoshiyuki Nagatomo and Toshiyuki Nagase, "The study of the power Modules with High Reliability for EV Use", *Proceedings of the 17th International Electric Vehicle Symposium* (2000).

7. Syed, A., "Accumulated creep strain and energy density based thermal fatigue life prediction models for SnAgCu solder joints", *Proceedings of the 54th Electronic Components and Technology Conference*, pp.737-746, (2004)

8. Lau, J.H. (editor), Ball Grid Array Technology, McGraw-Hill (1995), p.396

9 Lu, H. Tilford, T., Bailey, C. and Newcombe D.R., "Lifetime Prediction for Power Electronics Module Substrate Mount-down Solder Interconnect", *Proceedings of The 2007 International Symposium on High Density Packaging and Microsystem Integration*, 2007, pp.40-45

An In-Depth Numerical Investigation into Packaging Design of Multi-Finger GaInP/GaAs Collector-Up HBTs

H. C. Tseng, J. Y. Chen

Department of Electronic Engineering, Kun Shan University
949, Da Wan Rd., Tainan, 71003, Taiwan
Email:hctseng3@ceg.com.

Abstract

A novel finite-element modeling approach is developed to design thermal-via packaging configurations of collector-up heterojunction bipolar transistors (C-up HBTs) in high-power amplifiers (HPAs). The thermal interaction between HBT fingers has been examined based on the temperature distribution phenomena in multifinger C-up HBTs. The results reveal that the overall compactness of thermal-via configuration can be further improved more than 33%.

1. Introduction

The GaInP heterojunction bipolar transistor (HBT), which can operate at low-power supply voltages and is capable of handling high power, is an attractive candidate for use as the active component in the dual-band system of high-power-amplifiers (HPA) applications [1]. Recently, the tendency is toward higher levels of system integration on a chip in a package, especially for cellular-phone communication systems. Inevitably, the miniaturization of HPAs is a must to meet the requirement of the system-on-chip (SOC) or system-in-package (SIP) development. The miniaturization of HPAs depends mainly on thermal management of HBTs. Furthermore, as systematically discussed in Kroemer's classical paper [2], an inverted transistor design with a smaller collector on top and a larger emitter at the bottom has speed advantages over the conventional emitter-up (E-up) HBT structure. Therefore, we investigate the collector-up (C-up) HBT that has a thermal-via packaging configuration in the backside, for the purpose of obtaining good heat radiation so that the thermal coupling between the fingers can be alleviated. For a modern cellular-phone communication system, it is crucial to reduce the temperature within HBTs during high-power operation for the improvement of power density, current gain, power-added efficiency, and reliability [3]. As regards the C-up HBT, a small HPA requires the HBT with a multifinger design to generate high-power signals and lower currents for each finger to reduce the collector temperature. In reality, the thermal interfering effect often influences the miniaturization of HPAs. In this work, through the multifinger GaInP C-up HBT, we demonstrated the thermal performance compared to the C-up tunneling-collector HBT with the original configuration, which is a large thermal-via structure [4], and we calculated the size effect of the thermal-via configuration on the temperature distribution using the finite-element modeling (FEM). For this advanced packaging-technology analysis, the FEM methodology was used to calculate temperature distribution around the transistor by adjusting the thicknesses of plated heat sink (PHS) layer and GaAs substrate. Simultaneously, the maximum operation temperature in the three-finger C-up HBT with different finger pitches was examined.

2. Finite-Element Modeling Approach

A three-finger GaInP-GaAs C-up HBT with a large thermal-via is demonstrated in Fig. 1. The thermal via is placed directly at the backside of the substrate. The thickness is normalized by the thickness t of the GaAs layer located on the etch-stop layer. The thickness of collector, base, emitter, PHS, and substrate are 0.8, 0.1, 0.2, 12.0, and 30.0, respectively.

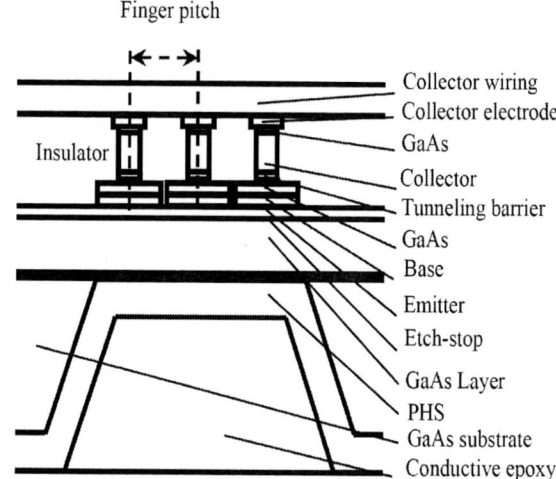

Fig. 1.   Schematic diagram of the three-finger collector-up HBT with a thermal via.

The overall dimension of the C-up HBT is 4.5t (width) by 40.0t (length). Five different finger-pitch configurations (15t, 18t, 21t, 24t, and 27t) were explored. The pitch of 15t is treated as a reference for comparison. Under this modeling strategy, the thermal resistance between interfaces of each layer is assumed to be negligible, and the thermal as well as mechanical properties of each layer are considered to be constant. The heat source is located at the collector layer. As to the boundary conditions, the natural convection with room temperature at the bottom surface of the thermal via and the adiabatic conditions at the rest of the surfaces are adopted.

3. Thermal Performance Analysis

In Fig. 2(a), a cross-sectional view of the three-finger C-up HBT analysis model is illustrated, and the temperature distribution in the C-up HBT with 15t finger pitch is displayed. It can be observed that heat transfers from the collector layer, through the emitter, GaAs and PHS layers, to the bottom. As seen in Fig. 2, most of the heat was spread from the PHS layer of the thermal via. The result obtained is comparable to those reported in [4], but with a much thinner PHS configuration.

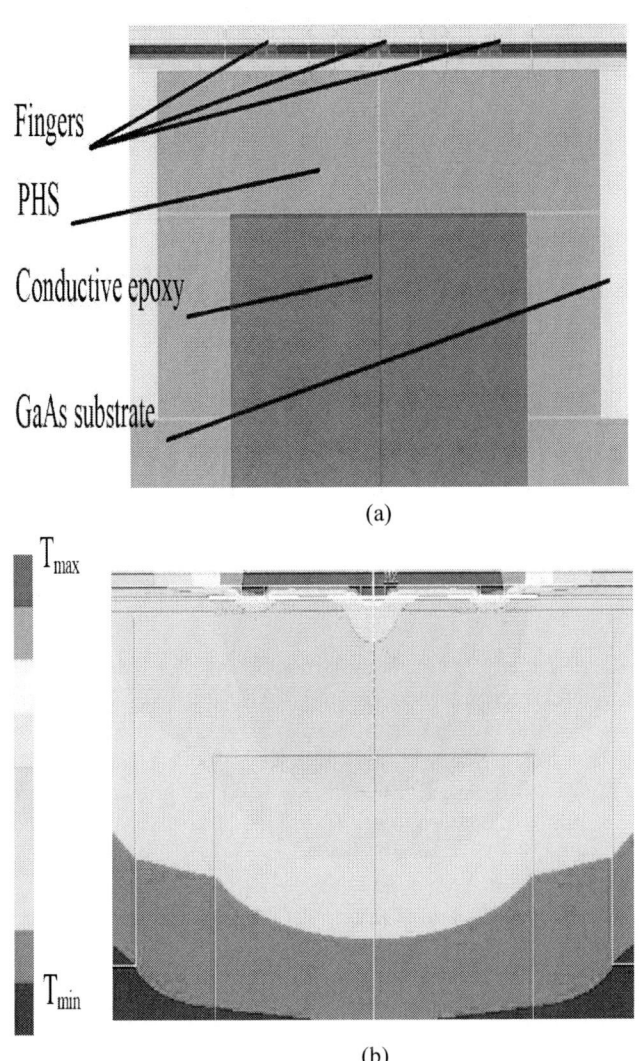

(a)

(b)

Fig. 2. Cross-sectional view of the three-finger collector-up HBT structure with 15t finger pitch. (a) Simulation model. (b) Temperature distribution within the HBT.

According to the aforementioned analysis, it is obvious that the PHS layer and the conductive epoxy constitute the major bulky portion of the HBT structure in the previous report [4]. Considering the key advantage of using C-up configuration is the shrinkage of device dimensions, it is very important that the thickness of bulk part must be reduced without sacrificing the thermal performance. From this point of view, the thickness of the thermal via underneath the HBT finger has to be thinned.

In Fig. 3, it is demonstrated that the proposed thermal-via packaging configuration can achieve the same thermal performance by decreasing 35% thickness of the substrate as well as 37.5% thickness of the PHS layer.

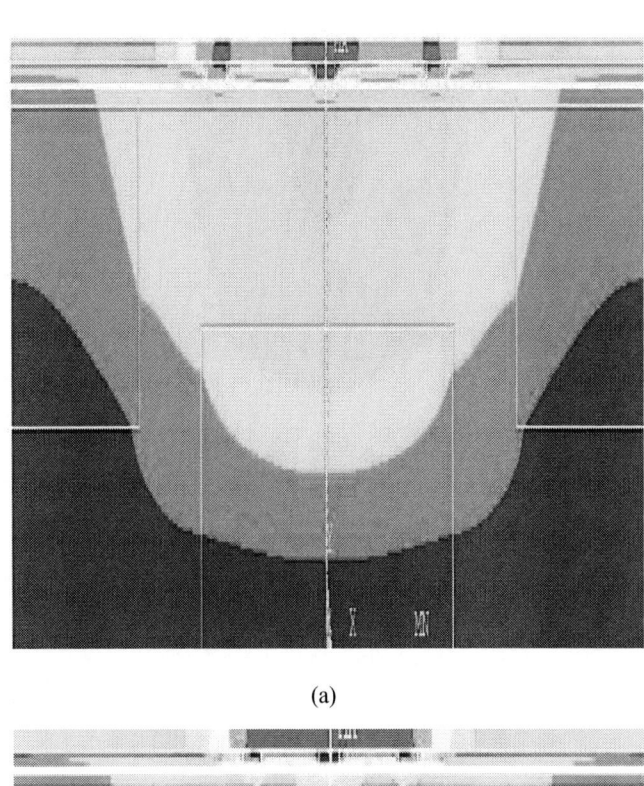

(a)

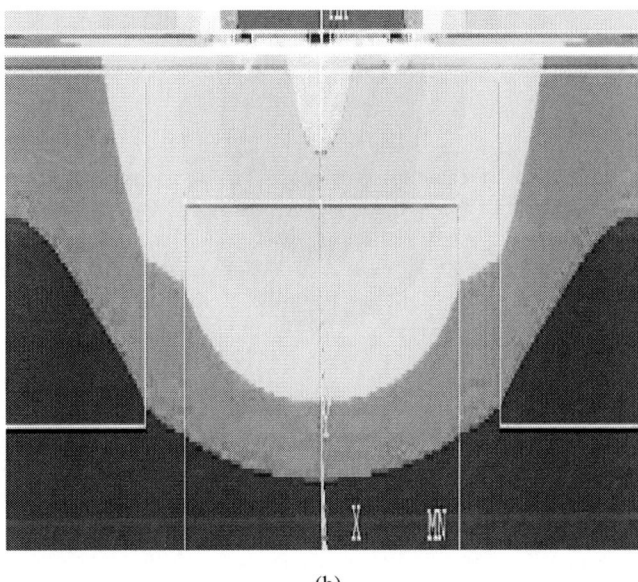

(b)

Fig. 3. Temperature distributions in the C-up HBT with different thicknesses of the thermal-via packaging configuration (finger pitch = 15t). (a) A thicker thermal-via structure. (b) A thinner thermal-via structure.

In contrast to the thicker thermal-via structure, for which thermal resistance was reduced by doubling the thickness of PHS layer, we can further reduce the thermal resistance of the packaging design, while decreasing both the thicknesses of PHS layer and the GaAs substrate simultaneously, so as to achieve the goal of miniaturizing HBT dimensions. For the present configuration, the overall thickness can be reduced by 33% and 42% for finger pitches of 15t and 21t, respectively.

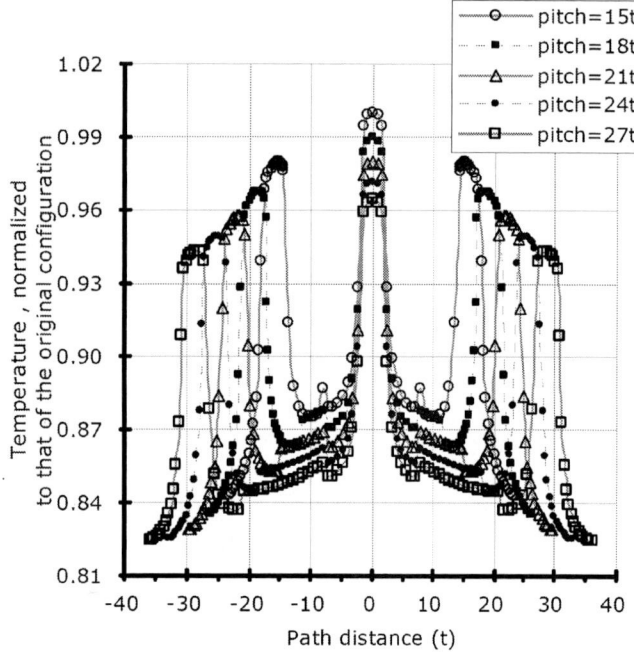

Fig. 4. Temperature distribution with different pitches at the top location of the collector.

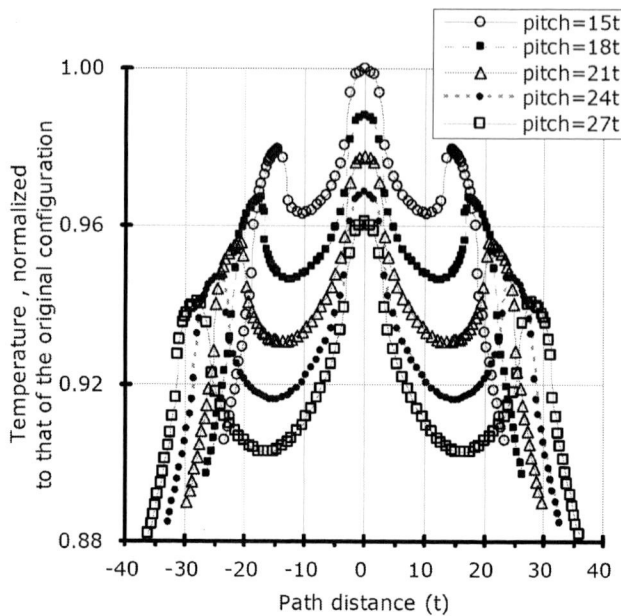

Fig. 5. Temperature distribution with different pitches at the bottom location of the collector.

Furthermore, temperature distributions at different locations of the collector region have been investigated to improve the HPA thermal performance. Figs. 4 and 5 show temperature distributions at the top and bottom locations of the collector. It is clear that the temperature of the center finger is the highest irrespective of pitch and location differences since it is affected by the heat from both sides.

Conclusions

The thermal performance of the multifinger GaInP C-up HBT with a thermal-via packaging configuration has been analyzed using the FEM approach, modified from our former paper [5]. The results reveal that the thickness of thermal-via structure can be further reduced by more than 33% through proper compact design for both PHS and substrate thicknesses, and the achieved thermal performance will not be deteriorated. On the other hand, the maximum operation temperature of the transistor can be reduced by the optimization of finger pitches. From this analysis, we conclude that thinning the thermal via together with a thinner GaAs substrate should be an effective access to the miniaturization of HPAs in future cellular-phone communication systems.

Acknowledgments

This work was supported in part by the NSC under Contract No. NSC 97-2221-E-168-040.

References

1. Liu, W., Kim, T., Ikalainen, P., and Khatibzadeh, A., "High linearity power X-band GaInP/GaAs heterojunction bipolar transistor," *IEEE Electron Device Lett.*, Vol. 15, No. 6 (1994), pp. 191-192.

2. Kroemer, H., "Heterostructure bipolar transistors and integrated circuits," *Proc. IEEE*, Vol. 70, No. 1 (1982), pp. 13-25.

3. Stenzel, R., Würfl, J., Richter, E., Pigorsch, C., and Klix, W., "Simulation of influence of heat removal on power gains of heterojunction bipolar transistor," *Proc. 22nd WOCSDIC*, Zeuthen, 1998, pp. 53-54.

4. Tanaka, K., Mochizuki, K., Takubo, C., Matsumoto, H., Tanoue, T., and Ohbu, I., "Investigation of Thermal stability in multifinger GaInP/GaAs collector-up tunneling-collector HBTs with subtransistor via-hole structure," *IEEE Trans. Electron Devices*, Vol. 53, No. 8 (2006), pp. 1759-1767.

5. Tseng, H. C., Lee, P. H., and Chou, J. H., "Improved design of thermal-via structures and circuit parameters for advanced collector-up HBTs as miniature high-power amplifiers," *IEICE Trans. Electron.*, Vol. E90-C, No. 2 (2007), pp. 539-542.

A Study of Power Delivery Network in a GPS SiP

Jun Li[1,2], Lixi Wan[2], Cheng Liao[1]

[1]Institute of Electromagnetics, Southwest Jiaotong University, Chengdu Sichuan 610031, China;
[2]Institute of Microelectronics, Chinese Academy of Sciences, Beijing, 100029, China
Email: zhiying19811123@hotmail.com

Abstract

The Power Delivery Network (PDN) in a Global Positioning System (GPS) SiP was studied in this paper. Each of power/ground planes was replaced by a novel embedded low-pass filter for multi-level power design. The equivalent circuit model, full-wave simulation and experiment were used to analyze the low impedance performance of this novel low-pass filter. A simple equivalent circuit model of normal signal theoretically showed that impedance matching, low parasitic parameters of transition via and the continuity of return path were the main factors for signal integrity (SI). To improve the signal quality, a "coaxial like" structure around one signal line was designed to shield the simultaneous switching noise in the RF chip and to protect signal energy radiation in the digital chip. The optimal insertion loss of digital and RF signal were less than -0.5dB. At last, some guidelines were summarized for mixed signal system especially high frequency and high speed system design.

Introduction

In recent years, there has been increasing interest in the development of System-in-Package (SiP) or System-on-Package (SoP) technology to integrate multi-chips into one package. SiP combines all electronic requirements of a functional system or a subsystem with high-performance digital Large Scale Integration (LSIs), Radio Frequency (RF), analog circuits, sensors, and so on. Global Positioning System (GPS) receiver is becoming commonplace in vehicles and hand-held electronics [1-2]. A GPS receiver contains low loss radio frequency (RF) front-end chip for receiving multi-mode signal and a baseband chip for managing the received data. Based on Complementary Metal Oxide Semiconductor (CMOS) technology, the RF chip integrates the low noise amplifier (LNA), mixer, Intermediate Frequency (IF) amplifier, Poly-phase Filter (PPF), Automatic Gain Control (AGC), analog-to-digital converter (ADC) etc. In the GPS receiver product, two chips were placed in one package side by side. The Simultaneous Switching Noise (SSN) or Ground Bounce Noise (GBN) produced by the baseband chip can cause the voltage level fluctuation. The noise not only affects digital circuit itself, but also disturbs analog and RF components in such small package [3-4].

In addition, the SSN will degrade signal quality of the RF chip as well [5-6]. The Signal Integrity (SI) can be affected by many other factors, including cross-talk, impedance mismatch, continuity return path and so on [7].

This paper mainly focuses on 1) designing a novel low-pass filter for low impedance; 2) optimizing continuity of return path for improving the signal quality and 3) analyzing the influence of RF signal coupling SSN through transition via. Some proposed guidelines can be applied to mixed signal system design, especially high frequency and high speed system.

The novel low-pass filter for multi-level power

A: Background of the novel low-pass filter

By combining of a planar embedded capacitor and power plane segmentation, a novel low-pass filter in replacement of power/ground planes in mixed system was proposed in the author's previous paper [3-4]. The test board was 10mm by 10mm and the thickness between power and ground was only 14μm with high DK=16. The "wide" and "gap" are equal to 0.5mm as shown in Fig. 1(a). Fig. 1(b) shows the equivalent circuit model of this novel low-pass filter. The equivalent capacitors between the power/ground planes, the equivalent inductor of the bridge, the effects of port connection, and the parasitic parameters of planes are the main factors to affect filter performance. Fig. 1(c) displays the insertion loss of three methods: equivalent circuit model, full wave electromagnetic simulator (HFSS) and experiment. The curves are matched very well, excepting for the resonance frequencies which slightly lower than simulation and circuit model due to parasitic inductance in the measurement. The 2-port scattering parameter measurement is carried out by Vector Network Analyzer (VNA) and plus on wafer probing system as shown in Fig. 1(d). The simulation technique has been validated by measurement technique, so we will use simulation method for further analysis.

B: Low-pass filter for GPS PI design

In GPS module, the RF and baseband chips are placed on the substrate side by side, and they have independent power supply as shown in Fig. 1(a). There are two different power levels, that is, 1.8V and 3.3V. Considering the low noise of margin of 1.8V, we put the power for 1.8V closer to chip side as shown in Fig. 2 (a). The design structure of Power/Ground is G/1.8V/3.3V/G, and then two signal layers are on the top and bottom respectively. Ground planes are continuous planes for shortest signal return path. The size of substrate is 12mm by 12mm. For wire bond BGA (WBBGA) design, the RF and baseband chips are connected to the substrate by wire bonding, and the package is connected to PCB (print circuit board) by BGA (ball grid array). As we know, the RF chip in GPS receiver is very sensitive to noise, so system must provide "pure" power. The maximum voltage fluctuation in the RF chip, ΔV, can be calculated by formula (1).

$$\Delta V = Z_{11}i_1 + Z_{12}i_2 + Z_{13}i_3 \quad (1)$$

Where Z_{11} is the input impedance of the RF chip (core voltage is 1.8V); Z_{12} is the transfer impedance between the digital chip (1.8V) and RF chip; Z_{31} is the transfer impedance between the digital chip (3.3V) and RF chip; i_1 is

978-1-4244-4658-2/09 $25.00 © 2009 IEEE

the maximum flow through the RF chip power_1.8V pin; i_2 is the maximum flow through the digital chip power_1.8V pin; i_3 is the maximum flow through the digital chip power_3.3V pin. The substrate is a passive system, so $Z_{12} = Z_{21}$ and $Z_{13} = Z_{31}$. Generally, the insertion loss should be as low as possible. Fig. 2(b) shows the simulation data of insertion loss, the S_{21} and S_{31} are below -40dB for a wide band. The resonant mode of Power/Ground cavity, $f_{TM_{10}/TM_{01}} = 3.125 GHz$, appears in the observation band due to bigger substrate dimension (12mm×12mm) compared to previous design. The insertion loss of this resonant mode is as low as -50dB in our design, so it will not degrade the signal quality severely.

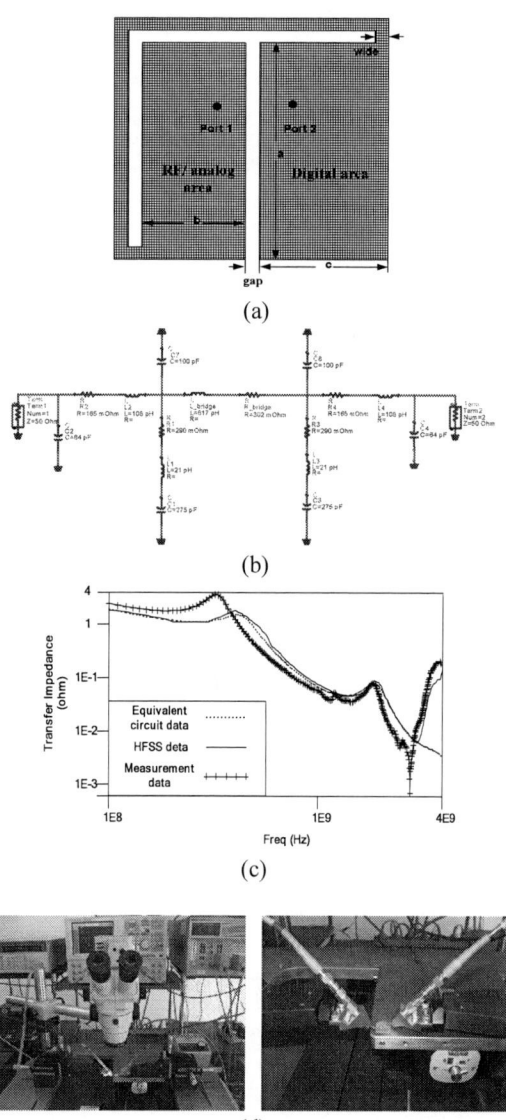

(a)

(b)

(c)

(d)

Fig. 1 The novel low-pass filter: (a) the top view of structure;
(b) the equivalent circuit model; (c) the transfer impedance;
(d) experimental device.

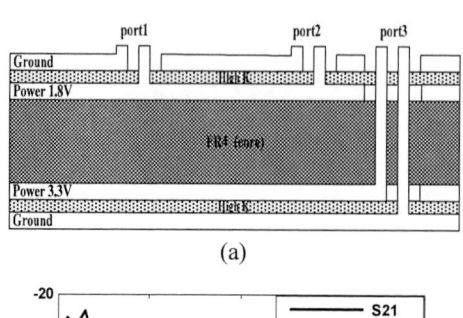

(a)

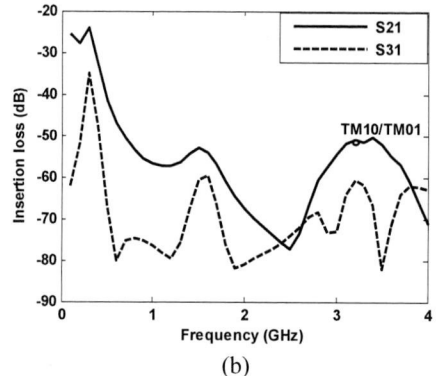

(b)

Fig. 2 (a) The cross section of multi-level power layers; (b) the insertion loss between ports.

PI design optimization to improve SI in the digital part

The signal return current actually flow through the inner impedance of power in DC and the return path is directly underneath the signal trace (lowest impedance) in high-frequency currents. Due to the "skin effect", high-frequency currents in conductors are surface currents. In GPS module, the chips are located in top of the substrate and all I/O pins have to be connected to the PCB by BGA. Therefore, the signal lines of both RF chip and baseband chip have to change reference plane through transition vias as shown in Fig. 3. The signal current will induce return current on the nearest plane and displacement current in dielectric material. The non-ideal return path causes additional inductance to degrade signal quality. Chia-yu Jin *et al.* [7] proposed displacement current would drive some resonant modes and the energy of signal would be absorbed by PDS (Power Distribution System), signal quality would degrade under resonant frequencies. Therefore, we designed two reference ground vias nearby the signal transition via to provide return path.

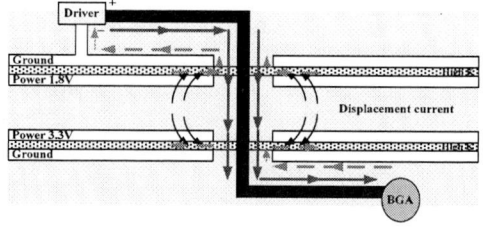

Fig. 3 The return path of signal changes reference planes.

Fig. 4 shows the cross section of signal changing reference plane in the digital part. When a driver switches to high level, a current is sent out one of signal pins and the current reenter the driver chip through its power pins to generate a close loop. When a driver switches to low level, sending current out one of signal pins, this current must eventually return the driver chip through ground pins. In GPS project, the length of trace is limited by the size and the thickness of substrate and the digital clock frequency is very low (16.368MHz), so the SI issue is not very serious. Considering the manufacture technology and the size of substrate, the normal signal line dimension is 100μm wide and 5461.5μm long (the total thickness of substrate is 461.5μm which result in low parasitic inductance), and the transition via is located in the middle of the trace. In the normal via design, the via and antipad are 100μm and 200μm in diameter. The simple equivalent circuit of the normal signal line was proposed as shown in Fig. 5. The terminal impedances of two ports are 50ohm, R_1, L_1, C_1 and R_2, L_2, C_2 represent line performance; L_{via}, C_{via1} and C_{via2} represent the performance of transition via; L_{g1} and L_{g2} represent the reference ground vias; although there are two reference ground vias nearby the transition via, part of energy will radiate to the substrate as usual and introduce additional inductance L_{add1} and L_{add2}.

Considering driver switches to low level, Fig. 6 shows the insertion loss of four cases: a) normal line and via; b) only change the width of signal line to 80μm (50ohm); c) only change the antipad diameter to 300μm (reduce the parasitic capacitors of transition via); and d) trace width is 80μm and antipad diameter is 300μm. 50ohm line design eliminates the reflection due to impedance mismatch. Reducing parasitic capacitors of transition via is more effective than impedance matching due to small electric size line. Both methods reduce the magnitude of insertion loss. However, resonance mode TM_{12}/TM_{21} and f_1 (induced by reference ground via inductance and the capacitor between line and reference plane) will not disappear completely.

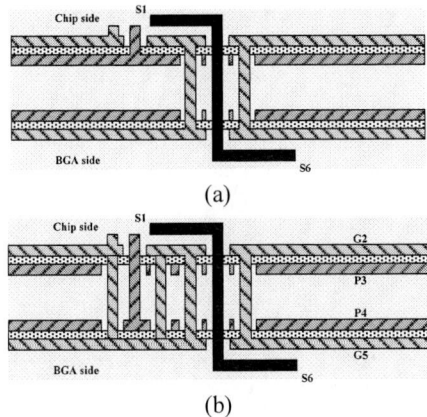

(a)

(b)

Fig. 4 The cross section for signal changes reference plane of six-layer WBBGA: (a) a driver switches to low level; (b) a driver switches to high level.

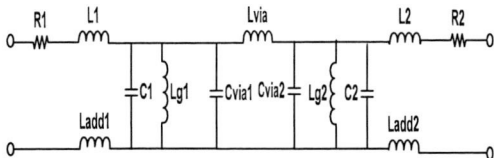

Fig. 5 The simple equivalent circuit of the normal signal line.

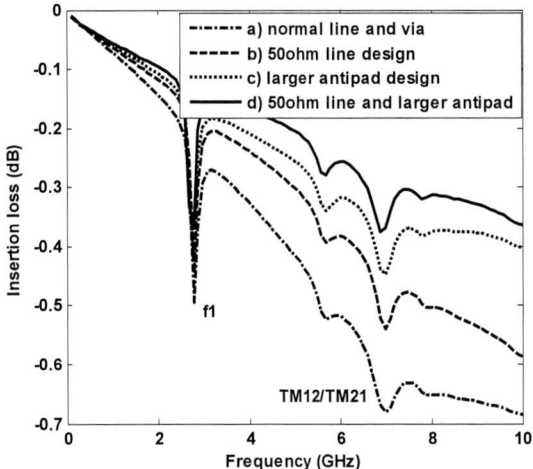

Fig. 6 The insertion loss of different cases.

When driver switches to high level, the current eventually return to power pins. There is discontinuity return path between power plane and ground plane, so signal quality degrades at some frequencies. Besides two reference ground vias nearby transition via, two more reference ground vias were designed near power via which connects a power pin of chip as shown in Fig. 4(b). Fig. 7 shows that such design can improve signal quality especially power_3.3V return path. For the power_1.8V return, the improvement is not very obvious, because the dielectric thickness is only 14μm and the capacitor between ground and power_1.8V is large enough for return path. The resonance frequency f_2 is induced by power connection via in simulator (when driver switches to low level, ground connection via was not included during simulation for the sake of simplicity).

In order to reduce the energy radiation, a "coaxial like" structure is adopted in this study, four reference ground vias around the tradition via and power connection via symmetrically. The combination of 50ohm line, low parasitic capacitors and "coaxial like" structure yield good performance as shown in Fig. 8. The curves are much smoother than previous design and the insertion loss is less than -0.5dB. In practice, the observation band for GPS is only below 1GHz. Our study can be used for other digital or mixed system design.

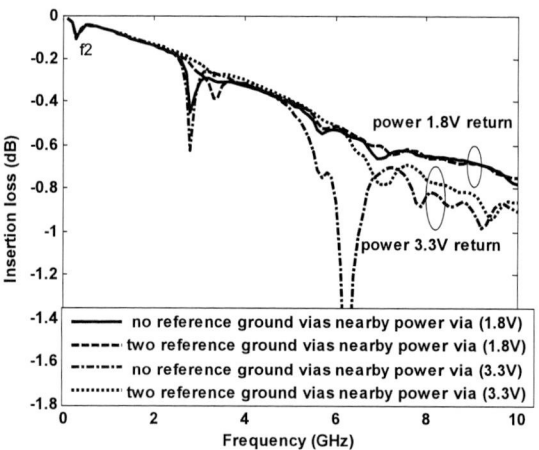

Fig. 7 The insertion loss of power pin return with and without reference ground vias.

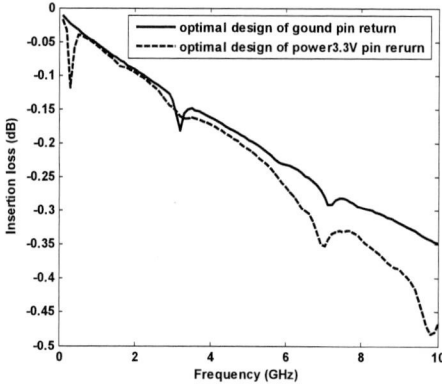

Fig. 8 The performance of optimal design.

Analysis of RF/ analog signal coupling SSN in the RF part

In previous study, the SSN was induced by transition via, while the SSN could disturb signal when signal changes reference planes. Replacing the signal line by differential pairs could result in better signal quality. For one port input mode, we have also done one single line design. As shown in Fig. 9, a model was set up in simulator to analyze RF signal coupling the SSN through transition via. Since the RF is 1.575GHz, we adopted 50ohm signal line, low parasitic parameters of transition via and two reference ground vias or "coaxial like" structure for return path. Fig. 10 is the comparison the performance between continuous plane and the novel low-pass filter. Both power and ground sides of the continuous plane remain continuous, with 50μm thick dielectric material FR4 in the middle. The SSN coupling of the novel low-pass filter is 20dB lower than that of continuous plane at 1.575GHz. There is little influence of coupling SSN through transition via and the signal quality is improved by the novel low-pass filter. The "coaxial like" structure was proven to have low radiation in previous work, so it has better shielding capability than two reference ground vias. SI design and the SSN coupling analysis can be used for

higher frequency or high speed design, such as Optical Transceiver, Ultra-wideband (UWB) Transceiver module.

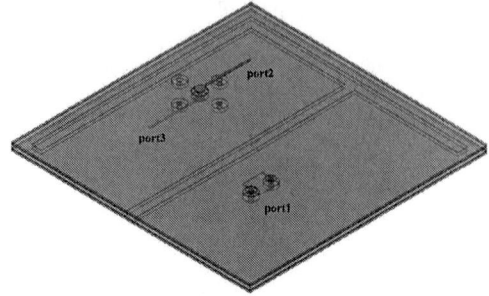

Fig. 9 The 3D view of the SSN coupling model.

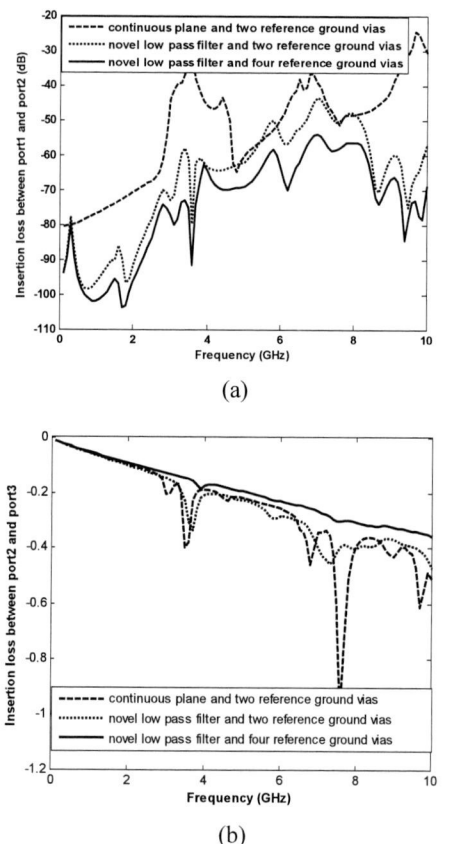

(a)

(b)

Fig. 10 The insertion loss (a) and signal quality (b) of continuous plane and the novel low-pass filter.

Conclusions

A Power Delivery Network in a GPS SiP was investigated in this work. The study covered the novel low-pass filter for multi-level power design, continuity of return path optimization and RF signal coupling SSN analysis. In the GPS module, the SSN produced by digital chip can cause voltage fluctuation in PDN and effect signal quality of the RF signal line. The novel low-pass filter for multi-level power supplies low impedance and protects the RF chip from the

SSN. Using "coaxial like" structure, impedance matching and low parasitic parameters of vias, the insertion loss of signal is less than -0.5dB. The influence of the RF coupling SSN through transition via can be ignored. The equivalent circuit models of the novel low-pass filter and normal signal line has been proposed to analyze the performance theoretically. The design guidelines for mixed signal system especially high frequency and high speed system can be summarized as following:

1. Thin dielectric layers are helpful for low impedance PDN and low parasitic inductance of vias for signal quality;

2. Small electric size line should be designed for SI consideration;

3. Impedance matching and low parasitic capacitor design should be used not only for the RF design but also the high speed digital design;

4. Differential pairs have better signal quality for high frequency and high speed design.

5. For mixed signal system, RF/analog and digital area should be distributed as far as possible; the return path of important RF and digital signal lines should not intersect.

6. The signal return path should be kept continuous and as short as possible, because discontinuous path can bring in the SSN and loss energy of signal.

7. Reference ground vias nearby signal line supply the nearest return path for changing reference planes situation, but introduce other resonant frequency in observation band. The "coaxial like" structure around single trace is able to shield noise and protect signal energy radiation effectively.

Acknowledgments

This work was supported by Hi-tech Research and Development Program of China (863 Program) No. 2006AA01Z236, 2007AA01Z200.

References

1. Patrick J. Zabinski, Barry K. Gillbert, *et al.*, "Example of a Mixed-Signal Global Positioning System (GPS) Receiver Using MCM-L Packaging," *IEEE Trans. Comp. Pack. and Manu. Tech.*, Feb. 1995, Vol.18, No.1, pp. 13-17.

2. Min Xu, David K. Su, *et al.*, "Measuring and Modeling the Effects of Substrate Noise on the LNA for a CMOS GPS Receiver," *IEEE 2000 Custom Integrated Circuits Conf.*, 2002, pp. 353-356.

3. Jun Li, Lixi Wan, *et al.*, "Improvement of power integrity with novel segmented power bus structures in RF/digital SOP," *2008 International Conference on Electronic Packaging Technology & High Density Packaging*, China, 2008, pp. 72.

4. Jun Li, Lixi Wan, *et al.*, "A novel embedded power filter structure research in System-in-Package," *Chinese Journal of Radio Science*, 2009, 24 (3). (in Chinese).

5. Jongbae Park, Hyungsoo Kim, *et al.*, "Modeling and Measurement of simultaneous Switching Noise Coupling Through Signal Via Transition," *IEEE Trans. Adv. Pack.*, 2006, Vol. 29, No. 3, pp. 548-559.

6. Xiaoxiao Wang, Donglin Su, "The Influence of Power/Ground Resonance to Via's SSN Noise Coupling in Multilayer Package and Three Mitigating Ways," *2006 Electronic Materials and Packaging*, Dec. 2006, pp. 1-5.

7. Chia-yu Jin, Chia-hsing Chou, *et al.*, "Improving signal integrity by optimal design of power/ground plane stack-up structure," *Proc. IEEE Electronics packaging Technology Conf.*, 2006, pp.853-859.

Modeling Thermal Fatigue in Anisotropic Sn-Ag-Cu/Cu Solder Joints

Shihua Yang[1], Yanhong Tian[2], Chunqing Wang[2], Tengfei Huang[1]
[1]Department of Electronics Packaging Technology, School of Materials and Engineering
[2]State Key Lab of Advanced Welding Production Technology
Harbin Institute of Technology, P.O. Box 436,
92, Xidazhi Street, Nangang, Harbin, P.R. China, 150001
Email: yangshihua@hit.edu.cn

Abstract

Properties of body-centered tetragonal β-Sn are highly anisotropic. The crystal orientation of the β-Sn has a marked effect on the stress state in the SnAgCu solder joints. A finite element analysis has been performed of the thermal cycling of the SnAgCu solder and solder joints. Significant amounts of stresses at the solder/Cu interface and in the plane of the grain boundary was found. The best Sn grain orientation is perpendicular to the solder/Cu interface.

Introduction

The near eutectic Sn-Ag-Cu solder joints are more than 95 atomic percent Sn and comprised of very limited β-Sn grains [1-3]. The β-Sn exhibits significant anisotropy in its elastic and thermal-expansion characteristics. Therefore, the crystal orientation of Sn grains of the Sn-Ag-Cu solder joints has a significant effect on the mechanical behavior, and it has been reported that the anisotropy of the Pb-free Sn-based solder joints can lead to anomalous early failure [4-6]. The main aim of this paper is to evaluate the stress state that arises at the solder/Cu interface and tin grain boundaries during thermal cycling, assuming that the tin grains remain elastic.

Simulation details

The reflow geometry of the solder joint was predicted using the Surface Evolver program (version 2.3). The Surface Evolver is an interactive program for the study of surfaces shaped by surface tension and energies. A surface is implemented as a simplicial complex, that is, a union of triangles. The Evolver evolves the surface toward minimal energy by a gradient descent method. Fig. 1 showed the general cross-sectional view of a reflowed BGA solder joint 300 μm in diameter and the finite element model predicted by the Surface Evolver. And Table 1 compared the solder joint shape with the predicted joint dimensions. It revealed a close agreement between the prediction and experimental values. The solder joints obtain from predictions were imported into the ANSYS software.

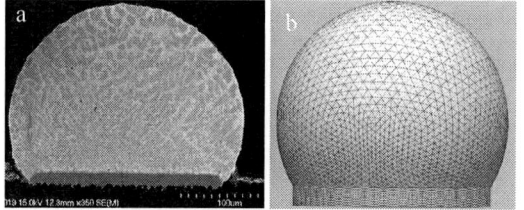

Fig. 1 Geometric profile comparison between (a) actual and (b) predicted solder joints

In the elastic analysis, the whole model was heated up from -25°C to 125°C. Properties at 25°C were considered as a reference state. The materials in the model were considered to be elastic. The coefficient of thermal expansion changed with temperature quoted from the ASM handbook and the elastic constants was introduced from reference [5]. The elements used were 3D solid elements with 20 nodes. One node on the bottom plane of Cu pad far away from grain boundary for the solder joints and on the underside of solder for the solder alloy were completely fixed. All the nodes on the bottom of the Cu pad for the solder joints and on the solder undersurface for the solder alloy were constrained not to move in the Z direction which was perpendicular to the pad surface. The Cu pad would be taken no account of when the misorientation between neighboring tin grains was studied.

Table 1 Comparsion of solder joint shape

	Experimental data	Surface evolver data
Solder ball maximum diameter	305 μm	317 μm
Solder ball height	213 μm	238 μm
Contact angle (degree)	55	58

The tin crystal orientation was defined by Bunge Euler angles φ_1-Φ-φ_2: 0-0-0, which has the c-axis of the crystal perpendicular to Cu pad (Z direction) and a-axis and b-axis coincide with the X direction and Y direction, respectively; 90-90-0, a, b, c axes coincide with Y, Z, X direction, respectively.

Results and Discussion

Fig. 2 shows the Von Mises stress distribution of the solder joint with Euler angles 0-0-0 at the maximum temperature. The greatest stress, 13.2 MPa, was small due to the small mismatch between the a-axis and b-axis of Sn and Cu pad. The stress distribution at minimum temperature was similar to that at highest temperature and the maximum stress was 3.9 MPa. The greatest stress of the 90-90-0 solder joint was up to 196.4 MPa at the maximum temperature and 57.7 MPa at the lowest temperature in the Cu pad and 126.9 MPa and 39.7 MPa in the bulk solder. The thermal mismatch was large in the 90-90-0 solder joint and this resulted in the very large stress at the interface. The stress can not be up to such large value in real joints because the plastic deformation will happen and relieve the stress. But it revealed that the crystal orientation has a significant effect on the stress state in the solder joints. The thermal mismatch between c-axis and Cu has the largest value and a, b-axis and Cu has the minimum value. Therefore, the best Sn orientation is 0-0-0, in which case the c-axis is perpendicular to the solder/Cu interface.

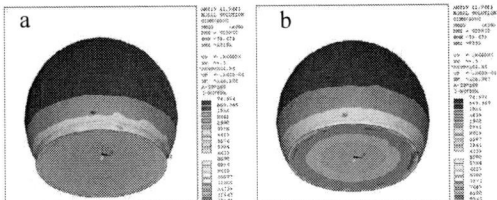

Fig. 2 Von Mises stress distribution at the maximum temperature (a) of the joint; (b) of the bulk solder

The misorientation between neighboring Sn grains also has an important effect on the stress state. Fig. 3 shows the schematic of a solder joint comprised of two Sn grains. The grain boundary was Y=0. The Euler angle of the left crystal was 0-0-0 and the right one 0-90-0. The elastic constants were obtained by coordinate transformation. The greatest stress was 124.5 MPa at the maximum temperature and 41MPa at the minimum temperature as shown in Fig. 4. It can be seen that the thermal mismatch arise from anisotropy of Sn will give rise to the large stress at the grain boundary.

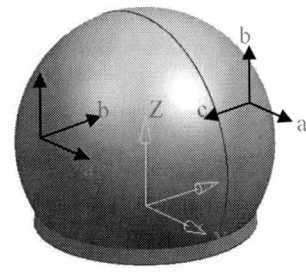

Fig. 3 schematic of a bi-crystal solder joint

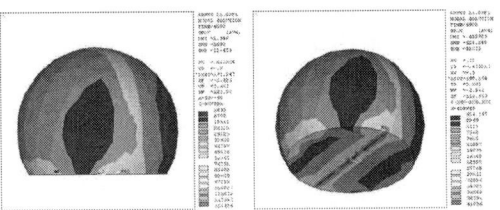

Fig. 4 Von Mises stress distribution (a) at the maximum temperature; (b) at the minimum temperature

It was also found that even a small Sn grains can lead to a large stress as shown in Fig. 5. The stress reached 145 MPa. It can be expected that the crack initiation will occur in this area. It means that the small Sn grains must be considered in the solder joints.

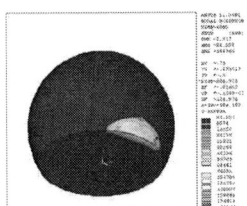

Fig. 5 Effect of small Sn grains on the Von Mises stress

Conclusions

The Sn crystal orientation in the SnAgCu solder joints has a significant effect on the stress. During the thermal cycling, mismatch between solder and Cu pad and between neighboring Sn grains will give rise to the large stress. The best Sn orientation is normal to the solder/Cu interface. The small Sn grains can lead to the large stress as well and must be considered during stress distribution analysis.

Acknowledgments

This work is financially supported by the National Natural Science Foundation of China under grant No. 50675047/ E052105 and Joint Project between Samsung Electronics Co. Ltd. (Korea) and Harbin Institute of Technology (HIT).

References

1. A. LaLonde, D. Emelander, J. Jeannette, C. Larson, W. Rietz, D. Swenson, and D.W. Henderson, "Quantitative Metallography of beta-Sn Dendrites in Sn-3.8Ag-0.7Cu Ball Grid Array Solder Balls via Electron Backscatter Diffraction and Polarized Light Microscopy," *J. Electron. Mater.*, Vol. 33, No. 12 (2004), pp. 1545-1549.

2. D.W. Henderson, J.J. Woods, T.A. Gosselin, J. Bartelo, D.E. King, T.M. Korhonen, M.A. Korhonen, L.P. Lehman, S.K. Kang, P. Lauro, D.Y. Shih, C. Goldsmith, and K.J. Puttlitz, "The Microstructure of Sn in Near-eutectic Sn-Ag-Cu Alloy Solder Joints and Its Role in Thermomechanical Fatigue," *J. Mater. Res.*, Vol. 19, No. 6 (2004), pp. 1608-1612.

3. B. Arfaei, Y. Xing, J. Woods, J. Wolcott, P. Tumne, P. Borgesen, and E. Cotts, "The Effect of Sn Grain Number and Orientation on the Shear Fatigue Life of SnAgCu Solder Joints," *Proc 58th Electronic Components and Technology Conf*, Lake Buena Vista, FL, 2008, pp. 459-465.

4. M.A. Matin, E.W.C. Coenen, W.P. Vellinga, and M.G.D. Geers, "Correlation between thermal fatigue and thermal anisotropy in a Pb-free solder alloy," *Scripta Mater.* Vol. 53, No. 8 (2005) pp. 927-932.

5. K.N. Subramanian, and J.G. Lee, "Effect of Anisotropy of Tin on Thermomechanical Behavior of Solder Joints," *J Mater Sci: Mater Electron*, Vol. 15, No. 4 (2004), pp. 235-240.

6. T.R. Bieler, H.R. Jiang, L. P. Lehman, K. Tim, E. Cotts, and B. Nandagopal, "Influence of Sn Grain Size and Orientation on the Thermomechanical Response and Reliability of Pb-free Solder Joints," *Trans. Compon. Packag. Technol.* Vol. 31, No. 2 (2008), pp. 370-381.

Numerical Analysis of Response of Indium Micro-Joint to Low-Temperature Cycling

X. Cheng, C. Liu, V.V. Silberschmidt
Wolfson School of Mechanical and Manufacturing Engineering, Loughborough University,
Loughborough, Leicestershire, LE11 3TU, United Kingdom
E-mail: X.cheng@lboro.ac.uk, Phone No.: +44(0)1509 227639

Abstract

In this study, the finite element analysis is chosen to investigate the thermo-mechanical properties of an indium joint exposed to low-temperature cycling. Based on the experimental data, the computational model is built, and the material model of indium is proposed for different joint thickness. The obtained results demonstrate that the outmost corner of the interface between the small indium joint and copper substrate is the weak site, while the large indium joint is characterized by a perfectly elastic core surrounded by uniformly plastic deformation during low-temperature cycling. With the proposed material model, the package-indium bump bonding with sensor and readout, used in the photon counting pixel detector Medipix 3, was studied under conditions of low-temperature cycling. The study provides an insight into the response of joints to thermal fatigue in indium joints during low-temperature cycling.

Keywords: Indium; the liquid nitrogen temperature; thermo-mechanical behavior; finite element analysis.

1. Introduction

Reliability of microelectronic packages under changing thermal conditions is one of the main concerns in their design due to the mismatch in coefficients of thermal expansion between the different layers in the packages. This mismatch can induce residual stresses resulting in thermal fatigue when the package is subjected to thermal loading due to power cycling of the equipment or service condition. For a hybrid pixel detector system that is used in the high energy physics experiments (e.g. CERN), reliability of solder bump bonds during service is required to provide excellent mechanical, electrical and thermal connections. And in terms of the system requirement and application, the normal service condition of solder bump bonds is in the cryogenic range, which is around and below the temperature of liquid nitrogen (77 K). Therefore, the thermo-mechanical behavior of solder bump bonds in cryogenic range is critical in determining the reliability of the connection.

In the present paper, we investigate the distribution and evolution of thermal stresses in pure indium bump bonds caused by temperature excursions down to the cryogenic region based on numerical analysis.

The hybrid pixel detector system comprises a crystalline semiconductor that is often bump-bonded to a pixellated application specific integrated circuit (ASIC) readout chips, for instance, the photon counting pixel detector Medipix 3. Indium has been widely used in the assembly of hybrid pixel detector system, because indium bump bonding can provide high I/Os, fine-pitch, small bumps (less than 20 μm), a low-temperature assembly process and greater ductility at cryogenic temperature. The research work by Hashimoto *et*

al. [1] predicted a longer fatigue life that can be obtained for pure indium than for Sn-Pb solder and indium alloys. It also demonstrated a good ductility without failure even after 300 cycles from the room temperature to the liquid nitrogen temperature (77 K). However, indium typically tends to relax under stress, and is rate-dependent material, which means the difference in the material's mechanical history changes the failure mechanism of the solder joint, and induces erroneous predictions. Based on our experimental work [2], the effect of low-temperature cycling from the room temperature (300 K) to the liquid nitrogen temperature (77 K) was simulated for the fundamental understanding of the thermo-mechanical behavior of pure indium bulk joint with different thickness. According to the proposed model, the initial thermal stress analysis of the indium joint, employed in the Medipix 3, was implemented with low-temperature cycling. The finite-element software ABAQUS, which is usually used to model the thermo-mechanical behavior of materials, was employed to simulate the thermo-mechanical behavior of indium joints under low-temperature cycling.

2. Numerical analysis on thermo-mechanical behavior of indium joints under low-temperature cycling

2.1 Numerical analysis of indium bulk joint with copper

In our experimental validation, two types of indium bulk joints were investigated: (i) large joints, for which the mechanical data can be obtained from the supplier (Indium® Corporation), and (ii) small joints that were studied with the use of specially designed tests to derive the relevant mechanical data at micro scale. The tension tests after different low-temperature cycles were carried out to evaluate the strain-stress curve and tensile strength as well as the failure mode for different thermal histories. The basic parameters and properties of indium and copper are given in Table 1 based on our experimental data [2]. The low-temperature cycling profile applied to indium joint is shown in Figure 1.

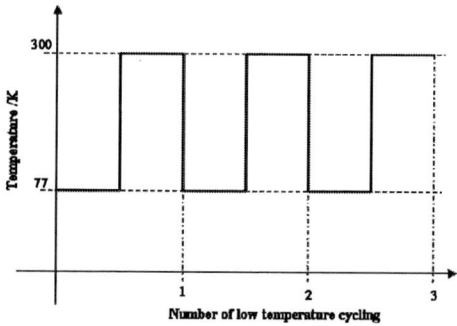

Figure 1. Low-temperature cycling profile of indium joints used in numerical analysis

978-1-4244-4658-2/09 $25.00 © 2009 IEEE

Table 1 Material properties and parameters used in simulations of In/Cu

Indium (large & small-joint)

Thickness / mm	2.0 (large) 0.3 (small)	CTE / μm/m K	33.0 at 300 K
			30.0 at 77 K
Mechanical properties /10^6 Pa		Large and small joints: Elasto-plastic with hardening based on [2]	

Copper

Thickness / mm	30.0	CTE / μm/m K	16.4 at 300 K
			8.21 at 77 K
Young's modulus / GPa	110	Mechanical properties	Linear elasticity
Poisson's ratio	0.343	Element type	C3D8R

One octant of the specimens with brick-shaped indium joints of different thickness, bonding two Cu plates, used in tensile testing in [2], was simulated due to its symmetry. This model offers a general insight into the characteristics of stress distribution in the indium joint when the specimen is exposed to a temperature decrease from the room temperature to that of liquid nitrogen. The thermal strain distributions for both joints denote that indium bulk joints with different thickness demonstrate different deformation behaviors. As seen in Figure 2, for small bulk joint, steady-state conditions in all direction with a constant strain increment per cycle are achieved after two cycles, which means sufficient levels of the temperature excursion amplitude can lead to irreversible macroscopic deformation called thermal fatigue (ratcheting) [3]. However, the deformation of large bulk joint increases linearly with increasing the temperature cycles without any transient behavior, especially in x- and y-direction. Regarding to the principal plastic strain distribution of both bulk joint in Figure 3, a core area inside large bulk joint (blue in Figure 3(b)) exhibits perfectly elastic behavior even after six cycles.

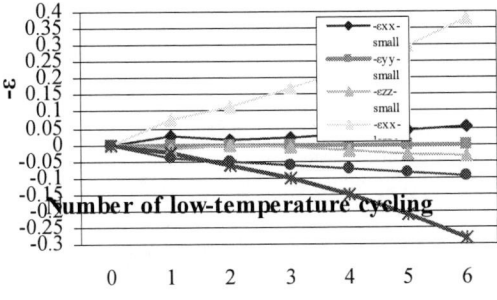

Figure 2. Evolution of axial thermal strains for both large and small indium bulk joints with low-temperature cycling

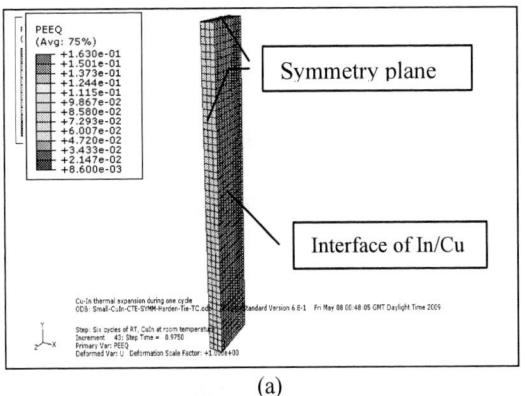

(a)

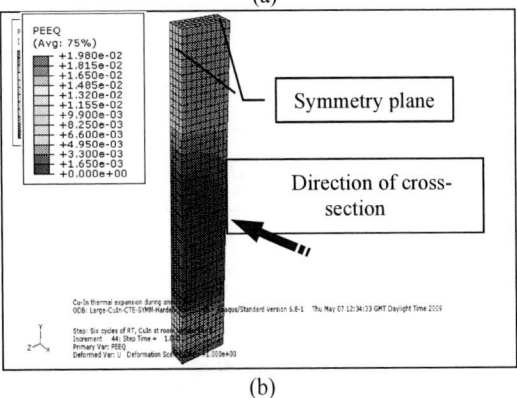

(b)

Figure 3. Distribution of plastic strain in indium bulk joints after six temperature cycles: (a) small joint; (b) cross-section of large joint after cut along x-direction.

The proposed model allows predictions of possible failure in both indium joints with low-temperature cycling. As shown in Figure 4, the results demonstrate that plastic deformations localize in the peripheral area of the small bulk joint, and in the peripheral corner of the interface between In/Cu during temperature excursion. As for the large bulk joint, it demonstrates an elastic core surrounded with plastic periphery, characterized by uniform plastic deformation during temperature excursions.

2.2 Numerical analysis of indium joint used in the Medipix 3

Indium bump bonds, which are employed in the Medipix 3 are 55 μm×55 μm pitches with around 22 μm bump diameter. The shape of the stress distribution at each pair of a sensor chip (silicon) and a readout chip (silicon) is related to the local in-plane thermal expansion mismatch between the indium and the readout chip and the sensor chip; and local in-plane expansion mismatch between readout chip and sensor chip as depicted in Figure 5. Shear strain is initiated due to the readout chip/sensor chip mismatch, while tensile strain is induced by the indium/chips mismatch. The difference in the linear coefficient of thermal expansion of indium and silicon with temperature is shown in Figure 6.

978-1-4244-4658-2/09 $25.00 © 2009 IEEE

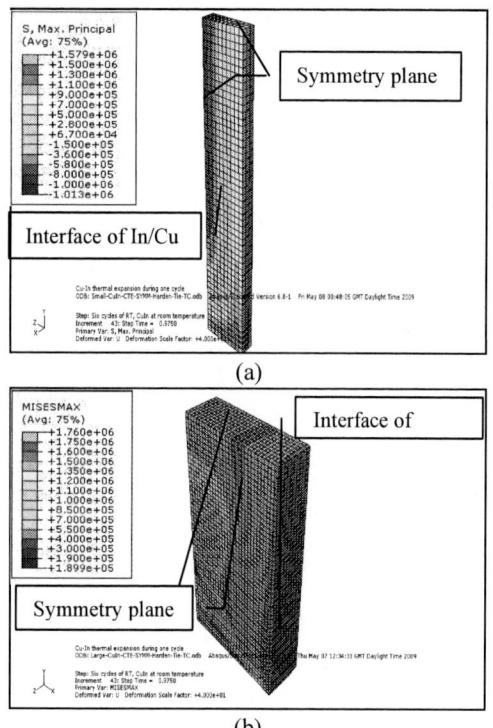

(a)

(b)

Figure 4. Von-Mises stress distribution in indium bulk joint after six temperature cycles:(a) small joint; (b) large joint.

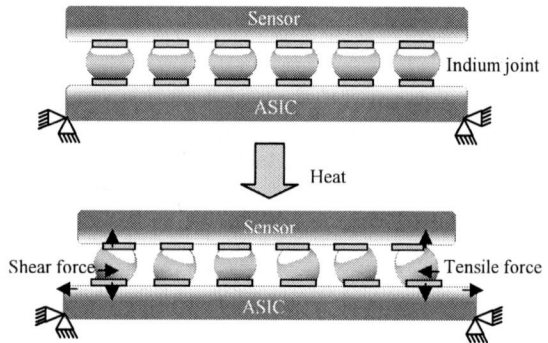

Figure 5. Solder joint subjected to thermal stress during thermal cycling due to thermal expansion mismatch between materials of join and readout and sensor chips

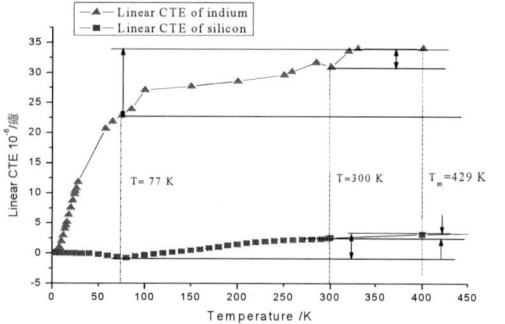

Figure 6. Llinear coefficient of thermal expansion of indium and silicon with temperature [4-9]

Based on the geometry and structure of the Medipix 3, only one quarter region of 6×6 array was built and simulated. The proposed material model of indium was adopted in this simulation. Other parameters used in simulations are shown in Table 2. The low-temperature cycling profile is the same as in pervious analysis (see Figure 1).

Table 2 Material properties and parameters used in simulations of indium joint used in Medipix 3

Silicon			
Young's modulus /GPa	168	CTE / μm/m K	2.57 at 300 K
			-0.77 at 77 K
Poisson's ratio	0.28	Mechanical properties	Linear elasticity

The evolution of Von-Mises stress with temperature cycling in the indium joint at the outmost chip site, corresponding to the maximum stress in the array, is plotted in Figure 7. The distribution illustrates that the stress in the indium joint exceeds the yield point immediately due to large temperature excursion, and residual stresses in indium joint rise with the increasing number of temperature cycles. The stress contours for the indium joint array are shown in Figure 8, indicating that the interface between indium and the sensor chip is the main weak site at low-temperature cycling, compared with the interface between indium and the readout chip.

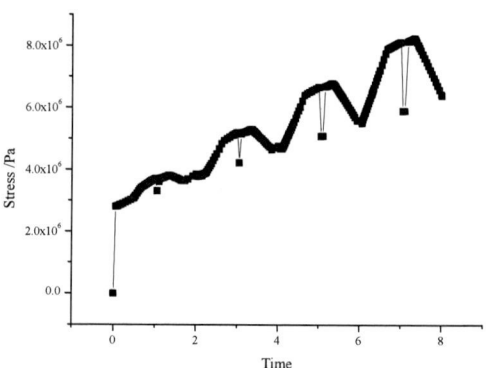

Figure 7. Evolution of Von-Mises stress in indium joint, the maximum stress in the array, with low-temperature cycles

The distributions of plastic strains in the indium joint in all direction were analyzed as well. The axial plastic strain distribution for the indium joint, which is near to the centerline of the chips, is plotted in Figure 9. Plastic strains in the x- and z-directions are larger than the y-directional one. Moreover, the thermo-mechanical behavior of indium with temperature cycling exhibits thermal ratcheting (thermal fatigue), i.e. the constant strain increment is achieved after each cycle.

978-1-4244-4658-2/09 $25.00 © 2009 IEEE

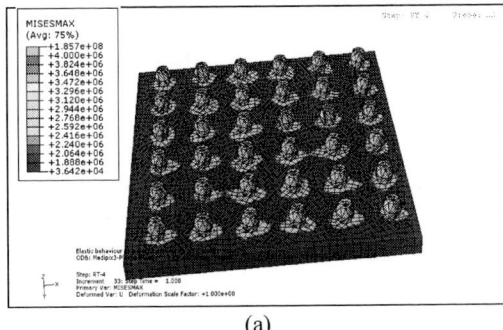

(a)

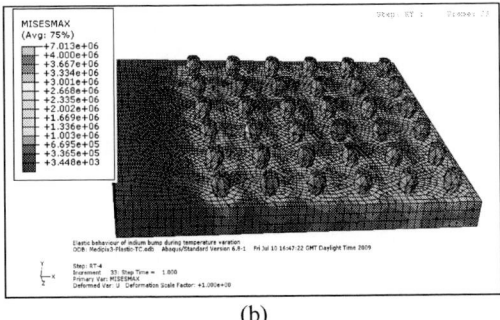

(b)

Figure 8. Distributions of Von-Mises stresses for indium joints after four temperature cycles: (a) indium joints with sensor chip after removing readout chip; (b) indium joints with readout chip after removing sensor chip.

Conclusions

A material model was proposed to analyze the thermo-mechanical behavior of indium with low-temperature cycling based on the understanding on physical performance and deformation behavior of indium at low temperatures. The finite element analysis demonstrates that indium has different mechanical behaviors in joints of different thickness, and thermal ratcheting (thermal fatigue) is the main characteristics of indium joint exposed to low-temperature cycling.

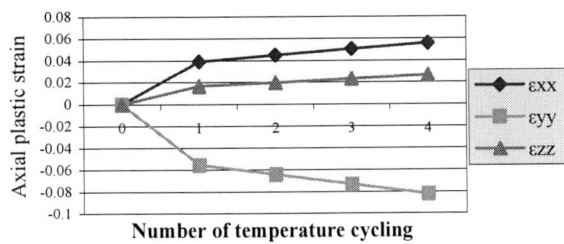

Figure 9. Evolution of axial plastic strain for indium joint with low-temperature cycles

Acknowledge

The technical support on the detailed information of Medipix 3 by Bob Stevens at Rutherford Appleton Laboratory (RAL) is gratefully acknowledged.

References

1. Hashimoto, K., "Flip-chip interconnection technology for packaging of VISL operated in liquid nitrogen", *IEICE Transactions*, Vol. 74, No. 8 (1991), pp. 2362-2368.

2. Cheng, X., Liu, C., and Silberschmidt, V. V., "Mechanical response of indium micro-joints to low temperature cycling", *Proc. 59th Electronic Components and Technology Conference*, San Diego, USA, May. 2009, pp. 1792-1795.

3. Silberschmidt, V.V., Rammerstorfer, F.G., Werner, E.A., Fischer, F.D. and Uggowitzer, P.J., "On material immanent ratchetting of two-phase materials under cyclic purely thermal loading", *Archive of Applied Mechanics*, Vol. 69, Issue. 9-10 (1999), pp. 727-750.

4. Collins, J.G., Cowan, J.A. and White, G.K., "Thermal expansion at low temperatures of anisotropic metals", *Indium. Cryogenics*, Vol. 7, Issue. 1-4 (1967), pp. 219-224.

5. Prakash, O., Ashok, R. and Dheer, P.N., "An apparatus for the measurement of thermal expansion of solids at low temperatures", *Journal of Physics*, Vol. 39, No. 6 (1992), pp. 655-660.

6. Smith, J.F. and Schneider, V.L., "Anisotropic thermal expansion of indium", *Journal of the less-common metals*, Vol. 7 (1964), pp. 17-22.

7. Sparks, P.W. and Swenson, C.A., "Thermal expansion from 2 to 40 K of Ge, Si, and Four III-V Compounds", *Physical Review*, Vol. 163, No. 3 (1967), pp. 779-790.

8. White, G.K., "Thermal expansion of reference materials: copper, silica and silicon". *Journal of Physics D: Applied Physics*, Vol. 6 (1973), pp. 2070-2078.

9. Yim, W.M. and Raff, R.J., "Thermal expansion of AlN, sapphire, and silicon", *Journal of Applied Physics*, Vol. 45, No. 3 (1974), pp. 1456-1457.

A Thermal Model for Calculating Thermal Resistance of Eccentric Heat Source on Rectangular Plate with Convective Cooling Existing at Upper and Lower Surfaces

Xiaobing Luo[1,2,*], Zhangming Mao[1], Sheng Liu[2]

[1] School of Energy and Power Engineering, Huazhong University of Science & Technology, Wuhan, China, 430074
[2] Wuhan National Lab for Optoelectronics, Huazhong University of Science & Technology, Wuhan, China, 430074
*Corresponding author: Telephone: 86-13971460283, Fax number: 86-27-87557074, Email: Luoxb@mail.hust.edu.cn

Abstract

Heat sources on rectangular plate with cooling at both upper and lower surfaces are common in electronic devices. A thermal model for calculating thermal resistance of eccentric heat source on plate with convective cooling at upper and lower surface is presented in this paper. The model was applied for calculating thermal resistance of a plate at three groups of boundary conditions. Simulations for calculating thermal resistance at the same boundary conditions were also done by software COMSOL. The comparison between the thermal resistances calculated by the presented model and the ones obtained by simulations showed that the proposed model is suitable for thermal resistance calculation of rectangular plate with cooling at both surfaces. The presented model can calculate the plate thermal resistance at following boundary conditions with high accuracy : I) Free convection at both surfaces; II) Free convection at upper surface and forced convection at lower surface.

Nomenclature

A_m, A_n, A_{mn}	Fourier coefficients
a	length of plate, [m]
b	width of plate, [m]
c	length of heat source, [m]
d	width of heat source, [m]
h_{equ}	equivalent heat transfer coefficient, [W/(m²·K)]
h_t	heat transfer coefficient at upper surface of plate, [W/(m²·K)]
h_b	heat transfer coefficient at lower surface of plate, [W/(m²·K)]
Q	heat rate generated by heat source, [W/(m²·K)]
R	thermal resistance of plate, [°C/ W]
R_s	thermal resistance of plate obtained by simulation, [°C/ W]
R_t	thermal resistance of upper plate, [°C/ W]
R_b	thermal resistance of lower plate, [°C/ W]
T_a	ambient temperature. [K]
t	plate thickness, [m]
t1	upper plate thickness, [m]
t2	lower plate thickness, [m]
$\overline{T}$	mean temperature of both surfaces, [K]
$\overline{T}_t$	mean temperature of upper surface, [K]
$\overline{T}_b$	mean temperature of lower surface, [K]

Greek symbols

$\overline{\theta}$	mean temperature excess of heat source area, [K]
λ_m	eigenvalues, $m\pi/a$
δ_n	eigenvalues, $n\pi/b$
β_{mn}	eigenvalues, $\equiv \sqrt{\lambda_m^2 + \delta_n^2}$
ϕ	spreading function
ζ	dummy variable

Subscripts

a	ambient
equ	equivalent
t	top
b	bottom

Superscripts

$\overline{(\bullet)}$	mean value

Introduction

Generally, there are many chip packages bonding on a broad mother board in electronic devices. Heat generated by dies conducts through packages and then transfers onto mother board. As dies and packages usually are much smaller than the substrate or mother board where they are located on, the heat dissipation processes can be treated as heat flux from a portion of surface conducting into a plate. The thermal resistances of mother board or heat spreader on which die is bonded commonly take significant impact on thermal characterization of electronic device or package. Therefore, it is important to find methods to calculate thermal resistance of heat sources on plate with various boundary conditions.

Thermal spreading resistance takes the majority part of total thermal resistance when heat dissipates from a planar heat source onto a larger cross-sectional plate. Kennedy [1] conducted research on the thermal spreading resistance of uniform heat-flux source on a finite cylinder. He obtained analytical solutions for a wide range of geometrical parameters, with the boundary conditions: 1) adiabatic sides and isothermal bottom surface, 2) isothermal sides and adiabatic bottom surface, and 3) isothermal side and bottom surfaces. Kadambi and Abuaf [2] presented analytic solutions for the three-dimensional as well as the axisymmetric steady-state and transient conduction equation with a uniform heat flux on the top surface and a convective heat transfer boundary condition at the bottom of the copper spreader. John and Krane [3] obtained the temperature field in the same rectangular geometries as in reference [2] and calculated

978-1-4244-4658-2/09 $25.00 © 2009 IEEE

thermal resistance using the temperature field, but here the boundary condition was constant and uniform temperature boundary. Yovanovich [4] and Negus et al [5] found the solution for the spreading thermal resistance of a single centered heat source on cylinder disk with a heat transfer coefficient at the lower surface. Lee [6] and Song et al. [7] analyzed the constriction and spreading resistance in a plate with a uniform heat flux region on one surface and a thermal boundary condition. They obtained the approximate equation of the constriction and spreading thermal resistance based on the analytical solutions of Yovanovich.

Geometric configurations have significant effect on the spreading thermal resistance. Yovanovich et al did series of studies on spreading thermal resistance for various geometrics [8-10]. Muzychka et al [11-13] presented analytical solutions of spreading thermal resistance for rectangular flux channels and discussed the influence of geometric and edge cooling. Yovanovich's spreading resistance formulaes are based on mean source temperature. Ellison [14] put forward the solution for the maximum spreading thermal resistance in rectangular geometry with one convective cooling surface. Moreover, Ellison provided extensive graphical results which are easy used for engineers.

In most of the studies, an adiabatic surface plane has been assumed. It is reasonable for free convection existing at the surface where heat sources located and heat sink at the other surface. However, there are many cases that both free convection exists at upper and lower surfaces of plate or heat sinks cooling exists at both surface planes in electronic devices. In these cases, if one surface is still treated as adiabatic condition as mentioned in the above references, it will bring unacceptable error for calculating thermal resistance. To solve the above problems, Kabir and Ortega [15] proposed an analytical solution to calculate temperature distribution at a case with both surfaces cooling. Thermal resistance was calculated based on the solution. They established a compact thermal model for BGA packaging by applying the solution.

In this paper, a thermal resistance model is presented for calculating thermal resistance of eccentric heat source on plate with convective cooling existing at both upper and lower surfaces. Simulation results are also obtained to compare with the analytical results. The comparison showed that the proposed model can calculate thermal resistance accurately, especially at the case of which free convection exists at top surface, no matter what the cooling condition of the bottom surface is.

Model development

Muzychka [16] et al analyzed thermal spreading resistance of eccentric heat sources on rectangular flux channels with cooling only at lower surface. However, heat sources on rectangular plates with both upper and lower surfaces cooling are common in electronic devices. Therefore, it is significant to find a method to calculate thermal resistance of plate with both surfaces cooling.

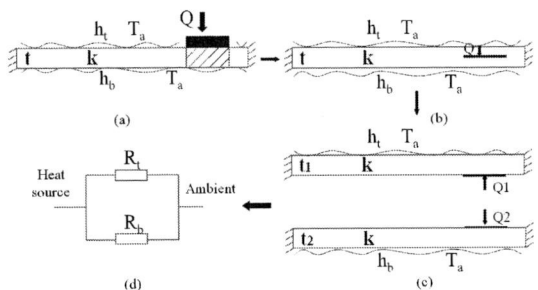

Fig.1. Analysis chart of thermal resistance calculation for eccentric heat source on isotropic plate with both surfaces cooling.

Fig. 1 (a) shows the rectangular plate with cooling at both surfaces. The sides of the plate are adiabatic. As the plate is usually very thin and its length and width commonly are one or two order larger than its thickness, the block under the heat source as the shaped part showed in Fig 1 (a) can be considered as an uniform temperature zone. As a result, the heat source can be assumed to locate in the plate as shown in Fig 1 (b). Heat generated by heat source dissipates into ambient in two paths: (1) a part of heat flows trough the upper part of the plate and transfers to ambient at upper surface; (2) the rest one conducts trough the lower part of the plate and transfers to ambient at lower surface. Therefore, the plate can be separated into two plates with one surface cooling as shown in Fig 1 (c). The total thermal resistance of the plate with both surfaces cooling is a parallel value of the two separated plates resistance as shown in Fig 1 (d) and is given by

$$R = \frac{1}{\dfrac{1}{R_t} + \dfrac{1}{R_b}} \quad (1)$$

where R_t is thermal resistance of upper plate and R_b is lower plate thermal resistance.

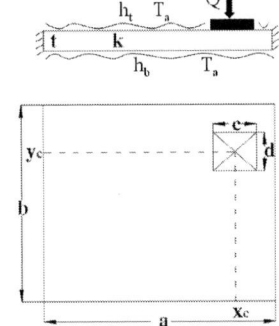

Fig 2. Eccentric heat source on isotropic plate with both surfaces cooling.

To calculate R_t and R_b, thickness t1 and t2 should be determined. The detailed methods will be discussed in the following two situations.

978-1-4244-4658-2/09 $25.00 © 2009 IEEE

Equal heat transfer coefficient at both surfaces

Since heat transfer coefficients at upper and lower surface are equal, the heat dissipation path of upper and lower plates are the same. As a result, t1 equals t2 and its value is 0.5t.

Unequal heat transfer coefficient at two surfaces

When h_t is unequal to h_b, thickness t1 and t2 can not be determined directly. As shown in the aforementioned part, for the situation of same heat transfer coefficients at two surfaces, t1 and t2 will be equal. This provides a hint that if an equivalent and same heat transfer coefficient at both surfaces can be found, the thickness t1 and t2 will be easily calculated.

For the case that an equivalent and same heat transfer coefficient exists at both surfaces, the total heat dissipation is

$$Q = h_{equ}(2ab - cd)(\overline{T} - T_a) \qquad (2)$$

where a,b and c,d are dimensions of heat source and plate as shown in Fig 2. $\overline{T}$ is mean temperature of both surfaces. Actually, the total heat dissipation is

$$Q = h_t(ab - cd)(\overline{T}_t - T_a) + h_b cd(\overline{T}_b - T_a) \qquad (3)$$

where $\overline{T}_t$ and $\overline{T}_b$ are mean temperatures of upper and lower surfaces, respectively.

As the plate is thin and its length and width are much larger than its thickness, the temperature distributions on upper and lower surface are nearly the same. Therefore, $\overline{T}_t$, $\overline{T}_b$ and $\overline{T}$ can be treated as the same parameters. Eliminating Q in Eqs (2) and (3), h_{equ} can be obtained as

$$h_{equ} = h_t \frac{ab - cd}{2ab - cd} + h_b \frac{ab}{2ab - cd} \qquad (4)$$

As analyzed before, in the case that t1 equals to t2, the values of both t1 and t2 are 0.5t. Heat transfer coefficients at both surfaces are substituted by an equivalent one calculated through Eq. (4). After these treatments, R_t equals to R_b. The total thermal resistance of the plate is

$$R = \frac{R_t}{2} \qquad (5)$$

R_t is thermal resistance of upper plate as shown in Fig 3. Muzychka [16] et al obtained an analytical expression of mean temperature excess for eccentric heat source on rectangular plate with cooling at one surface. The expression is exactly suitable for calculating R_t.

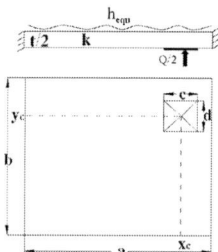

Fig 3. Upper plate with upper surface cooling.

Mean temperature excess of heat source obtained from the following general expression shows the its explicit relationships with the geometric and the thermal parameters of the system according to the notation in Fig 3,

$$\overline{\theta} = \frac{Q/2}{ab}(\frac{t/2}{k} + \frac{1}{h_{equ}}) + 2\sum_{m=1}^{\infty} A_m \frac{\cos(\lambda_m X_c)\sin(\frac{1}{2}\lambda_m c)}{\lambda_m c} +$$
$$2\sum_{n=1}^{\infty} A_n \frac{\cos(\delta_n Y_c)\sin(\frac{1}{2}\delta_n d)}{\delta_n d}$$
$$+ 4\sum_{m=1}^{\infty}\sum_{n=1}^{\infty} A_{mn} \frac{\cos(\delta_n Y_c)\sin(\frac{1}{2}\delta_n d)\cos(\lambda_m X_c)\sin(\lambda_m c)}{\lambda_m c \delta_n d} \qquad (6)$$

where

$$A_m = \frac{2Q\cos(\lambda_m X_c)\sin(\frac{1}{2}\lambda_m c)}{abck\lambda_m^2 \phi(\lambda_m)} \qquad (7)$$

$$A_n = \frac{2Q\cos(\delta_n Y_c)\sin(\frac{1}{2}\delta_n d)}{abdk\delta_n^2 \phi(\delta_n)} \qquad (8)$$

$$A_{mn} = \frac{8Q\cos(\lambda_m X_c)\sin(\frac{1}{2}\lambda_m c)\cos(\delta_n Y_c)\sin(\frac{1}{2}\delta_n d)}{abcdk\beta_{m,n}\lambda_m\delta_n\phi(\beta_{m,n})} \qquad (9)$$

where

$$\phi(\zeta) = \frac{\zeta\sinh(\zeta t/2) + h_{equ}/k\cosh(\zeta t/2)}{\zeta\cosh(\zeta t/2) + h_{equ}/k\sinh(\zeta t/2)} \qquad (10)$$

$$\lambda_m = \frac{m\pi}{a}, \delta_n = \frac{n\pi}{b} \qquad (11)$$

$$\beta_{m,n} = \sqrt{\lambda_m^2 + \delta_n^2} \qquad (12)$$

Thermal resistance R_t is defined as

$$R_t = \frac{\overline{\theta}}{Q/2} \qquad (13)$$

Therefore, R_t can be obtained by combining Eqs (6) to (13) and it is a function of dimensions and thermal parameters indicated in Fig 3, it does not have relation with Q. Since R_t have been calculated, R can be obtained by Eq (5).

Analysis and discussions

The presented model was used to calculate thermal resistance of a plate with three groups of boundary conditions. The three groups of boundary conditions are as following, I) equal heat transfer coefficient at both surfaces of the plate; II) the upper surface is free convective cooling and heat transfer coefficient is 5 W/(m²·K) while the lower surface is forced convection and heat transfer coefficient is variable; III) the upper surface is forced convection and heat transfer coefficient changes, the lower surface is free convective cooling and heat transfer is 5 W/(m²·K). Meanwhile, simulations for obtaining thermal resistance of the plate at the same three groups of boundary conditions were done by

978-1-4244-4658-2/09 $25.00 © 2009 IEEE

multiphysical analysis software COMSOL. Thermal resistance was obtained by

$$R = \frac{\bar{\theta}}{Q} \quad (14)$$

where $\bar{\theta}$ is temperature excess of heat source area obtained by simulation, Q is heat rate input in the simulation. Comparison between the results calculated by the presented model and the data obtained by simulations was used to demonstrate the feasibility of the model.

Dimensions of the plate and heat source are presented in Table 1. The plate thermal conductivity is 5 W/(m·K), the coordinate (x_c, y_c) is (20, 80) (mm). Input heat of the heat source is 3W.

Table 1. Dimensions of the plate and heat source

Component	Parameter	symbol	dimension
Heat source	Length, mm	c	4
	Width, mm	d	4
Plate	Length, mm	a	100
	Width, mm	b	100
	Thickness, mm	t	1

MATLAB was employed for the calculation of the plate thermal resistance at three groups of boundary conditions. 100 terms were used in each single summation and every double summation consists of 10000 terms.

The results obtained by the model and simulation at three groups of boundary conditions are shown in Tables 2 to 4. The relative errors between the results obtained by the model and simulation are also shown. Definition of the relative error is given by

$$error = \frac{R - R_s}{R_s} \times 100\% \quad (15)$$

where R_s is the plate thermal resistance obtained by simulation.

Table 2. Results and relative error obtained at boundary condition group I.

Heat transfer coefficient at both surfaces, W/(m²·K)	R, °C/W	Rs, °C/W	Relative error, %
1	165.686	162.605	1.89
2	129.342	126.415	2.32
3	113.335	110.525	2.54
4	103.548	100.833	2.69
5	96.702	94.071	2.80
6	91.543	88.985	2.88
7	87.462	84.970	2.93
8	84.120	81.690	2.97
9	81.311	78.940	3.00
10	78.904	76.588	3.02
20	65.103	63.243	2.94
50	50.343	49.382	1.94
100	40.487	40.430	0.14
200	31.479	32.441	-2.96
500	21.219	23.488	-9.66
1000	15.040	18.075	-16.79

Table 3. Results and relative error obtained at boundary condition group II.

Heat transfer coefficient at lower surface, W/(m²·K)	R, °C/W	Rs, °C/W	Relative error, %
10	85.707	83.018	3.24
20	74.093	71.410	3.76
50	59.645	57.144	4.38
100	49.610	47.444	4.56
200	40.141	38.536	4.16
500	28.661	28.190	1.67
1000	21.161	21.802	-2.94

Table 4. Results and relative error obtained at boundary condition group III.

Heat transfer coefficient at upper surface, W/(m²·K)	R, °C/W	Rs, °C/W	Relative error, %
10	85.720	83.494	2.67
20	74.113	72.448	2.30
50	59.667	59.097	0.96
100	49.631	50.198	-1.13
200	40.161	42.136	-4.69
500	28.680	32.677	-12.23
1000	21.177	26.542	-20.21

It is noted from Tables 2 to 4 that When the freee convection exists on the top surface, the relative error in most cases is less than ±5%. This application conditions are also suitable for most of the real cases. It also can be seen that in group II and III, the relative error is more than ±5% when the heat transfer coefficient at upper surface is more than 200 W/(m²·K). The possible reason for these larger relative errors is that heat generated by heat source flows into the plate and then part of heat flows back to the upper surface because of good cooling at this surface. The heat reflow makes the actual thermal resistance greater than the one calculated by the presented model, which is lack of consideration of heat reflow.

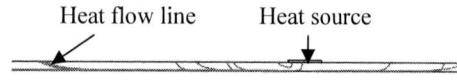

Fig 4. Heat flow line in the plate (h_t, 1000 W/(m²·K); h_b, 5 W/(m²·K))

It is clearly seen that the presented model is suitable for calculating thermal resistance of eccentric heat source on rectangular plate with both surface convective cooling, especially for the situations of low heat transfer coefficient at upper surface, which reflects most of the real applications.

Summary and conclusions

A thermal model for calculating thermal resistance of eccentric heat source on rectangular plate with both surface convective cooling was presented and it was applied for calculating thermal resistance of a plate with both surface cooling. Comparison between the results obtained by the model and simulations demonstrated that the presented model is suitable for calculating thermal resistance of the plate. The presented model can calculate the plate thermal resistance with good accuracy in the following two kinds of boundary conditions commonly happened in electronic devices: I) free convection at both surfaces; II) free convection at upper surface and forced convection at lower surface.

Acknowledgments

The authors would like to acknowledge the financial support from 973 Project of The Ministry of Science and Technology of China (2009CB320303).

References

1. D. P. Kennedy, "Spreading Resistance in Cylindrical Semiconductor Devices," *Journal of Applied Physics*, Vol. 31 , 1960, pp. 1490-1497.
2. V. Kadambi, N. Abuaf, "An Analysis of the Thermal Response of Power Chip Packages," *IEEE Transactions on Electron Devices*, Vol. ED32, No. 6, June 1985.
3. M. John, M. Krane, "Constriction Resistance in Rectangular Bodies," *Transactions of the ASME*, Vol. 113, December 1991, pp. 392-396.
4. M. M. Yovanovich, "General Solution of Constriction Resistance within a Compound Disk," *Progress in Astronautics and Aeronautics: Heat Transfer, Thermal Control, and Heat Pipes,* 1980, pp. 47-62, MIT Press, Cambridge, MA.
5. K. J. Nugus, M. M. Yovanovich, "Constriction Resistance of Circular Flux Tubes with Mixed Boundary Conditions by Linear Superposition of Neumann Solutions," *Proc ASME 22nd Heat Transfer Conf*, Niagara Falls, NY, Aug. 6-8, 1984.
6. S. Lee, S. Song, V. Au, and K. P. Moran, "Constriction/Spreading Resistance Model for Electronic Packaging," *Proc 4th ASME/JSME Thermal Engineering Joint Conf*, Vol. 4, 1995, pp. 199-206.
7. S. Song, S. Lee, and V. Au, "Closed Form Equation for Thermal Constriction/Spreading Resistances with Variable Resistance Boundary Condition," *Proc 1994 IEPS Conf*, 1994, pp. 111-121.
8. M. M. Yovanovich, "General Thermal Constriction Parameter for Annular Contacts on Circular Flux Tubes," *AIAA Journal*, Vol. 14, No. 6, 1976, pp. 822-824.
9. M. M. Yovanovich, Y. S. Muzychka, and J. R. Culham, "Spreading Resistance of Isoflux Rectangles and Strips on Compound Flux Channels," *Journal of Thermophysics and Heat Transfer*, Vol. 13, No. 4, 1999, pp. 495-500.
10. M. M. Yovanovich, J. R. Culham, and P. M. Teertstra, "Analytical Modeling of Spreading Resistance in Flux Tubes, Half Spaces, and Compound Disks," *IEEE Transactions on Components, Packaging, and Manufacturing Technology -Part A*, Vol. 21, 1998, pp. 168-176.
11. Muzychka Y. S., Culham J. R., Yovanovich M. M., "Thermal Spreading Resistances in Rectangular Flux Channels Part I-Geometric Equivalences," *Proc 36th AIAA Thermophysics conf*, June 23-26 Orlando, Florida AIAA 2003-4187.
12. Muzychka Y. S., Culham J. R., Yovanovich M. M., "Thermal Spreading Resistances in Rectangular Flux Channels Part II-Edge Cooling," Proc 36th AIAA Thermophysics conf, June 23-26 Orlando, Florida AIAA 2003-4188.
13. Muzychka Y. S., Yovanovich M. M., Culham J. R., "Influence of Geometry and Edge Cooling on Thermal Spreading Resistance," *Journal of Thermophysics and Heat Transfer*, vol. 20, No. 2, April-June 2006, pp 247-255.
14. G. N. Ellison, "Maximum Thermal Spreading Resistance for Rectangular Sources and Plates with Nonunity Aspect Ratios," *IEEE Transactions on Components and Packaging Technologies*, Vol. 26, No. 2, June 2003.
15. H. Kabir, A. Ortega, "A New Model for Substrate Heat Spreading to Two Convective Heat Sinks: Application to The BGA Package," *Proc 14th. IEEE SEMI-THERM™ Symposium*, San Diego, CA, 1998, pp. 24–30.
16. Y. S. Muzychka, J. R. Culham, and M. M. Yovanovich, "Spreading Thermal Resistances of Eccentric Heat Sources on Rectangular Flux Channels," *Journal of Electronic Packaging*, Vol. 125, 2003, pp. 178-185.

Modeling of Heat Transfer Performance for an Array of Micro Jets Impinging upon Dimpled Surface

Tian-yi Geng[1], Jing-yu Fan[1]*, Yan Zhang[2], Shun Wang[2]

[1]Shanghai Institute of Applied Mathematics and Mechanics, Shanghai University, Shanghai 200072, China
[2]Key Laboratory of Advanced Display and System Applications, Ministry of Education & SMIT Center
School of Mechatronics Engineering and Automation, Shanghai University, Shanghai 200072, China
*Corresponding author, Email: jyfan@shu.edu.cn

Abstract

The present paper paid particular attention to the flow and heat transfer performance for an array of micro jets impinging upon the dimpled surface by means of CFD modeling. The modeling of a three-by-four jet array impinging upon the smooth, concave- and convex-dimpled surfaces was carried out based on the solution of the steady, incompressible and three-dimensional RANS and energy equations by employing the RNG k-ε turbulence model. It has been demonstrated from the present study that the heat transfer performance for an array of the micro jets impinging onto the concave- or convex-dimpled surfaces tends to be somewhat attenuated or augmented in the impingement region, respectively, as compared with that onto the smooth one. And the variational trend with respect to the Nusselt number distribution in the inter-jet region appears to be inconsistent with that in the impingement region. In addition, the flow pattern and primary mechanism by which the smooth, concave- and convex-dimpled surfaces significantly affect the flow field and heat transfer are analyzed and discussed.

Introduction

In high power density electronic packaging applications, the design of the thermal control system should take two main requirements into consideration, namely the heat flux to be dissipated and the allowable temperature rise above the local ambient conditions[1]. With the ever-increasing heat flux level due to higher performance and further microminiaturization for a broad range of microelectronic components, more efficient and compact chip cooling techniques are becoming indispensable to the associated high heat flux removal since those conventional cooling techniques have been shown to be insufficient[2,3]. Numerous enhanced heat transfer solutions for the high heat flux removal in the chip and module level of microsystem packaging have been proposed in the past decades, among which including the jet impingement cooling and the surface microstructure heat sink[4].

A majority of previous experimental and numerical studies concerning the above-mentioned heat transfer enhancement techniques concentrated mainly on either a single round or slot jet impinging upon a flat and smooth target surface, where the highly localized cooling performance is desirable, or the surface microstructure heat sinks subjected to the parallel flows instead of the impinging jets. However, there are some inherent disadvantages for both cooling schemes. For the former, although the maximum level of the heat transfer occurs in the stagnation zone directly beneath the impinging jet, the abrupt reduction in the cooling effectiveness away from the impingement zone can yield large temperature variations along the surface of the heat-

dissipating device[5]. While for the latter, take the microchannel as an example, the relatively high temperature rise along the microchannel is an undesirable issue since the large temperature gradient will result in the thermal stresses that undermine the electrical reliability of the components, and besides, the excessive pressure drop is also an obstacle to practical applications since the microchannel heat sinks are typically used with miniature pumps with limited pumping power capability[6].

On the other hand, from the physical point of view, the heated chip with microstructure can be recognized as a somewhat roughened surface, and it has been proved that the roughened surface subjected to the jet impingement cooling is, to some extent, capable of the higher local heat transfer than the smooth one. Nevertheless, the excessively-protruded heat sink, such as conventional pin-fin, will inevitably induce a large pressure loss in the case of the common roughened surfaces subjected to jet impingement.

To obtain more uniform and efficient heat transfer rate and especially decrease the pressure drop as much as possible, the work in the present paper is an attempt to complement the micro jet array impingement cooling with the distinctive surface microstructure in order to improve the heat transfer performance in high power density electronic cooling. A 3×4 array of the micro jets impinging upon an alternative type of the roughened surfaces, namely a sort of the dimpled surface, is taken into consideration in the present paper due to its potential in heat transfer performance, such as light weight, low pressure penalty and low maintenance.

In this regard, relatively few experimental or numerical studies concerning the micro jet array impingement on the dimpled surface are available at present[7], and the different regions in the dimpled surface where the impinging jet or the peripheral interacting wall jet is dominant will show various heat transfer rates depending upon numerous parameters, which have not yet well-apprehended and even contradictory in the literature due to the high complexity of the local flow and heat transport processes induced by the dimpled surface configurations.

Therefore, the present paper pays particular attention to the combined effects of the micro jet parameters and the dimple geometry, in the form of the concave-dimpled and convex-dimpled surface configurations, on the flow field and heat transfer characteristics by means of CFD modeling.

Governing Equations and Numerical Methods

The schematic diagram of the geometry and the coordinate system was shown in Fig.1, and the flow field and thermal performance for an array of the micro jets (3×4 array from circular nozzles with the diameter D of 3mm and the spacing

H of 12mm) impinging upon the smooth, concave- and convex-dimpled heated surfaces, in which the pitch between the dimple-centres was 12mm, and the separation distance S between the nozzle and the heated surface (the flat-edge in the case of the dimpled surfaces) was selected as $S/D=3$, were numerically simulated based on the solution of the Reynolds averaging Navier-Stokes (RANS) and energy equations. In the present paper, only the case that the micro jets were in line with the dimple locations in the heated surfaces was taken into consideration for simplification, and as a result, each jet could impinge completely within the subjacent dimple.

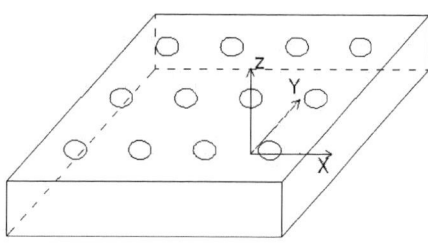

(a) Smooth surface

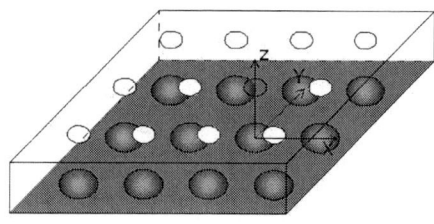

(b) Dimpled surface

Fig.1 The schematic diagram of the geometry and coordinates

The governing equations of mass, momentum, turbulent kinetic energy, turbulent energy dissipation rate and energy in the steady, incompressible and three-dimensional turbulent flow using the RNG k-ε turbulence model were numerically solved by means of the control-volume-based finite difference method, and the SIMPLE algorithm was adopted to solve the pressure-velocity coupling. The total grid number was over 600000 for the computational domain.

The corresponding boundary conditions specified over the overall computational domain included: the jet array inlet (fully developed and constant temperature boundary, and the specific jet Reynolds number Re=1.5×10⁴), the lateral spent air outlet (open pressure boundary), the uniformly heated bottom wall (no-slip and constant heat flux boundary, and the heat flux q=18860W/m²), and the upper surface (confined and adiabatic boundary). At all other boundary conditions being identical, the effects of the smooth, concave- and convex-dimpled surface configurations (with the dimple diameter D_c of 10.5mm and depth Δ of 3.1mm) on the flow structure and heat transfer performance were taken into consideration in the present paper.

Results and Discussion

The numerical simulation has been validated by comparing the computational results with the available experimental data[8] in terms of the local Nusselt number (Nu=hD/k, where h is the local convection heat transfer coefficient and k is the thermal conductivity of the fluid) profiles along x-direction at y=0, as shown in Fig.2. It has been found that the predicted Nu profile do not agree with the measured data only near the stagnation point adjacent to the lateral outlet. This discrepancy may be explained as due to the effect of the open pressure boundary. However, the simulated Nu profile in the central zone reveals that an acceptable agreement between the numerical results and the measured data is achieved.

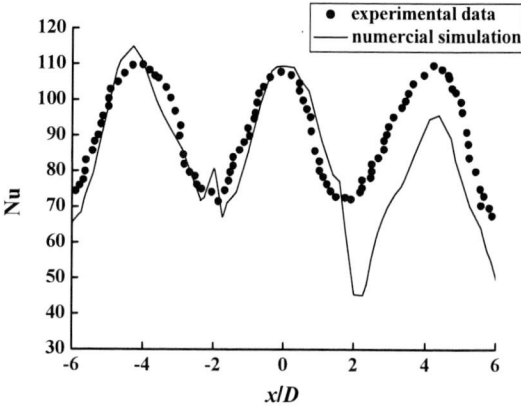

Fig.2 Comparison of the numerical simulation with the experimental data of the Nu profiles over the concave-dimpled surface along x-direction at y=0

The simulated Nu profiles over the smooth, concave-, and convex-dimpled surfaces along x-direction at y=0 are plotted in Fig.3, and the detailed Nu contour distributions over the smooth, concave- and convex-dimpled surfaces are plotted in Fig.4.

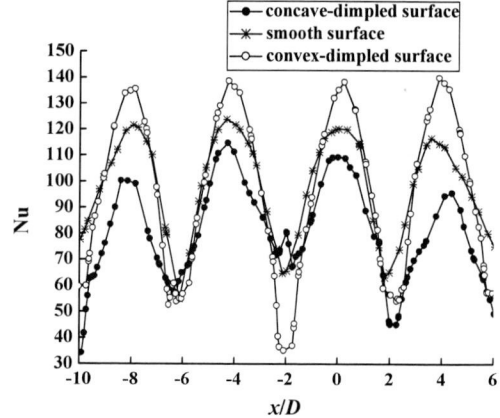

Fig.3 The simulated Nu profiles over the smooth, concave- and convex-dimpled surfaces along x-direction at y=0

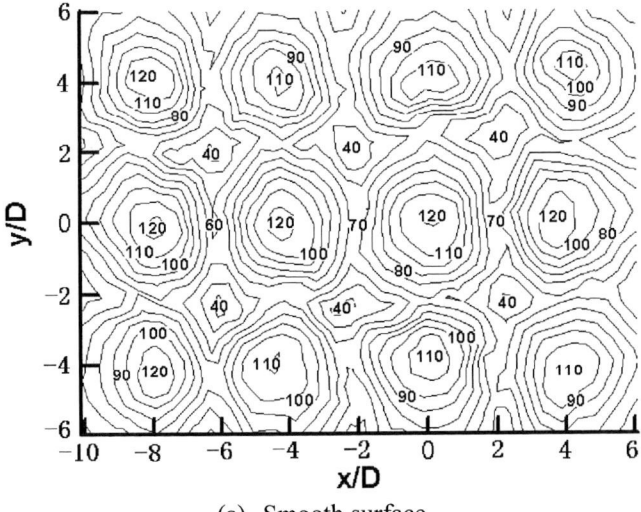

(a) Smooth surface

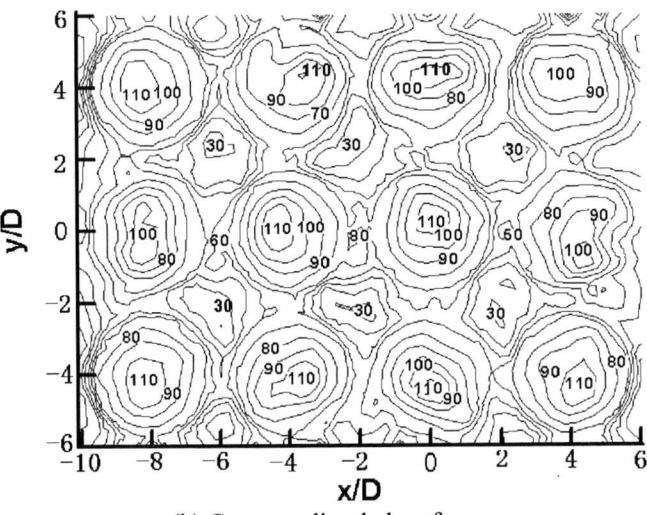

(b) Concave-dimpled surface

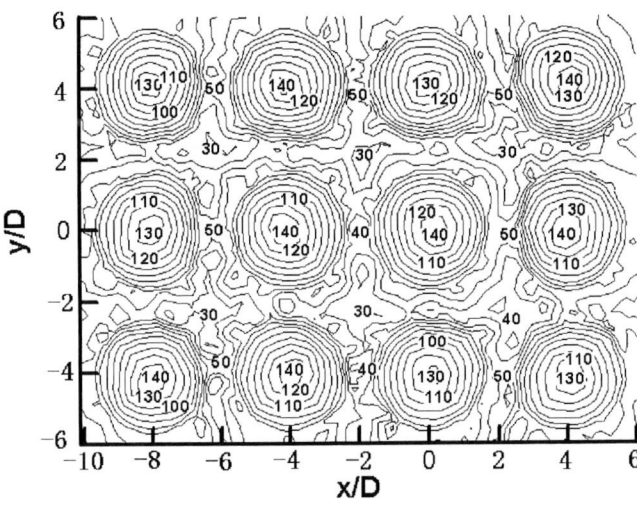

(c) Convex-dimpled surface

Fig.4 The simulated Nu distributions over the smooth, concave- and convex-dimpled surfaces

It can be seen from Fig.3 that the wavy Nu profiles along x-direction at $y=0$ show the same trend for different surface

configurations, maximal at the stagnation points underneath the jets and minimal at the locations between the jets, and the difference of the Nu distributions between the smooth and dimpled surfaces largely consists in their Nu values in amplitude. In the impingement region, a remarkable feature of the Nu profiles reveals that the convex-dimpled surface leads to the highest heat transfer rate, followed by the smooth surface, and the concave-dimpled surface exhibits the lowest heat transfer rate. This illustrates that in the present specified geometric parameter range, the heat transfer performance for an array of the micro jets impinging onto the concave- or convex-dimpled surfaces tends to be somewhat attenuated or augmented in the impingement region, respectively, as compared with that onto the smooth one. While in the inter-jet region, it is found that the variational trend with respect to the Nu distributions along x-direction appears to be inconsistent with that in the impingement region, and no determinate augmentation or attenuation of the heat transfer in the case of the dimpled surfaces can be achieved as compared with the case of the smooth surface, in some degree due to the effect of either the induced crossflow or the spent air outlet.

It can be seen from Fig.4 that the overall Nu contour distributing patterns for different surface configurations remain an appreciable similarity. There are twelve Nu contour humps centred at the jet stagnation points, indicating the local maximum level of the heat transfer. And there are also six Nu contour valleys located at the intersection region of four surrounding jets, indicating the local minimum level of the heat transfer. It can be also observed that the Nu contours over the convex-dimpled surface are much dense than those over the smooth surface, and the Nu contours over the concave-dimpled surfaces appear to be slightly sparser than those over the smooth surface.

It can be inferred that the local heat transfer in the vicinity of the stagnation point for the case of the convex-dimpled surface is higher than that for the case of the concave-dimpled surface, and this is attributed to the shorter relative distance between the jet nozzle and the stagnation point for the former. In both cases, there is the expanded heat exchange area exposed to the jet impingement as compared with the smooth surface, especially by which the average heat transfer within the concave-dimples is likely to be partly compensated.

As to the flow structure and heat transfer for an array jet impingement onto the dimpled surface, few experimental or numerical studies are available in the literature. To manifest the flow pattern modified by the concave- and convex-dimpled surface configurations, the velocity vectors in the x-z plane at $y=0$ are plotted in Fig.5. For the case of the smooth surface, each jet impinges vertically on the surface, resulting in the local maximum Nu value at the stagnation point, and then deflects towards the direction parallel to the surface forming the wall jet, as shown in Fig.5(a). In the inter-jet region, two adjacent wall jets in an opposite direction meet and give rise to the flow separation, and thus a pair of the recirculation vortices occurs in counter rotation, resulting in the local minimum Nu value at the separation point (or termed as the secondary stagnation point in some literature[3]).

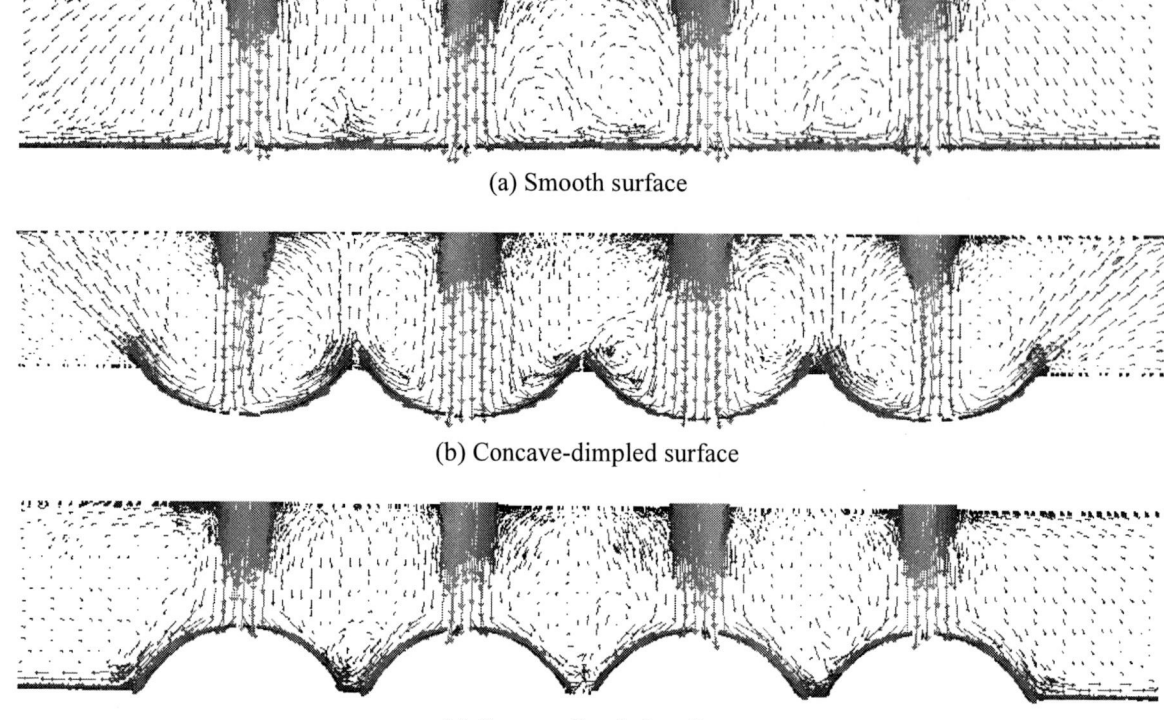

(a) Smooth surface

(b) Concave-dimpled surface

(c) Convex-dimpled surface

Fig.5 The velocity vectors in the x-z plane at y=0

It can be further inferred from Fig.5(b) and Fig.5(c) that the concaved- and convex-dimpled surface configurations (with the dimple geometric parameter $\Delta/D_c=0.3$) essentially influence the flow dynamics and have a consequent impact on the heat transfer in both the impingement region and the inter-jet region. In the case of the concave-dimpled surface, each jet impinges normally onto the subjacent concave dimple, with the considerable expansion of the impingement region and subsequently shrinkage of the wall jet region. The radially spreading wall jet inside the concave dimple develops along the peripheral border of the dimple, and is then detached from the exit edge of the dimple, forming a pronounced toroidal vortex appreciably interacting with the initial jet, as shown in Fig.5(b). While in the case of the convex-dimpled surface, each jet impinges directly onto the apex of the convex dimple, and then makes a less than 90° turn, generating the flow separation along the curved surface of the convex-dimple and forming a considerably weakening recirculation vortex away from the stagnation point, as shown in Fig.5(c). In both cases, the dimpled surface is likely to not only modify the gross flow field and relevant vortical structure, but also significantly affect the heat transfer in the impingement region and the inter-jet region in varying degrees.

It should be pointed out that the associated geometric and kinetic parameters in the present paper are not optimal for achieving the enhancement and uniformity of the heat transfer performance in the case of the jet array impingement, for example, each jet may initially impinge between the dimples (namely the staggered dimple mode according to [7]) or eccentrically impinge onto the dimples[8], which is still worthy of further investigation.

Conclusions

In this paper, the flow and heat transfer performance for an array of micro jets impinging upon dimpled surface were numerically investigated by means of CFD modeling. The comparison of the numerical simulation with the experimental data showed an acceptable agreement. And the effects of the different impingement surface configurations, including the smooth, concave- and convex-dimpled surfaces, on the flow and heat transfer were taken into consideration. The numerical results indicate that the heat transfer performance for an array of the micro jets impinging onto the concave- or convex-dimpled surfaces tends to be somewhat attenuated or augmented in the impingement region, respectively, as compared with that onto the smooth one. And the variational trend with respect to the Nusselt number distribution in the inter-jet region appears to be inconsistent with that in the impingement region. In addition, the flow pattern and primary mechanism by which the smooth, concave- and convex-dimpled surfaces significantly affect the flow field and heat transfer are analyzed and discussed.

Acknowledgments

This work was supported by the National Natural Science Foundation of China (10702037). The supports from Shanghai Pujiang Program (08PJ14054) and the Innovation Program of Shanghai Municipal Education Commission (09YZ01) were also greatly appreciated.

References

1. Beitelmal A H, Saad M A and Patel C D. "The Effect of Inclination on the Heat Transfer between a Flat Surface and an Impinging Two-Dimensional Air Jet", *International Journal of Heat and Fluid Flow*, Vol. 21, (2000), pp. 156-163.

2. Zhang Y, Fan JY and Liu J. "Numerical Investigation Based on CFD for Air Impingement Heat Transfer in Electronics Cooling", *Proceedings of the Seventh IEEE CPMT Conference on High Density Microsystem Design, Packaging and Component Failure Analysis*, Shanghai, 2005, pp. 360-364.

3. Etemoglu A B. "A Brief Survey and Economical Analysis of Air Cooling for Electronic Equipments", *International Communications in Heat and Mass Transfer*, Vol. 34, (2007), pp. 103-113.

4. Fan JY, Zhang Y and Liu J. "On the Heat Transfer Enhancement Based on Micro-Scale Impinging Jets with Microstructure Heat Sink in Electronics Cooling", *Proceedings of 2006 Conference on High Density Microsystem Design and Packaging and Component Failure Analysis*, Shanghai, 2006, pp. 171-175.

5. Sung M K, Mudawar I. "Experimental and Numerical Investigation of Single-Phase Heat Transfer Using a Hybrid Jet-Impingement/Micro-Channel Cooling Scheme", *International Journal of Heat and Mass Transfer*, Vol. 49, (2006), pp. 682-694.

6. Qu W, Mudawar I. "Measurement and Prediction of Pressure Drop in Two-Phase Micro-Channel Heat Sinks", *International Journal of Heat and Mass Transfer*, Vol. 46, (2003), pp. 2737-2753.

7. Ekkad S V, Kontrovitz D. "Jet Impingement Heat Transfer on Dimpled Target Surfaces", *International Journal of Heat and Fluid Flow*, Vol. 23, (2002), pp. 22-28.

8. Chang S W, Chiou S F and Chang S F. "Heat Transfer of Impinging Jet Array over Concave-Dimpled Surface with Applications to Cooling of Electronic Chipsets", *Experimental Thermal and Fluid Science*, Vol. 31, (2007), pp. 625-640.

Full Thermal Parametric Model for Power WL-CSP Design

Zhongfa Yuan[1], Yong Liu[2], Steve Martin[2], Luke England[2], Byoungok Lee[3]
[1]Fairchild Semiconductor, Suzhou, China
[2]Fairchild Semiconductor Corp., S.Portland, USA
[3]Fairchild Semiconductor, Bucheon, Korea
Email: yliu@fairchildsemi.com, Tel: 207-761-3155, Fax: 207-761-6339

Abstract

A new type of full thermal parametric model for power wafer level chip scale package (WL-CSP) is developed in this paper, which includes parametric WL-CSP and its adaptive parametric JEDEC thermal test board. By employment of the parametric model, package geometry parameters and the trace layout for PCB can easily be changed to meet the requirement of design, so that the influence of all geometry parameters to thermal performance can be investigated fast for the whole series of WL-CSP packages. The entire thermal simulation, including meshing, loading/boundary condition, solving, and post processing, is automated with ANSYS® parametric design language (APDL) coding. This paper introduces the construction of the parametric model for WL-CSP design, and both JEDEC low effective thermal board and high effective thermal conductivity boards with and without thermal vias are included in the model. To study impact of solder ball number, die size, terminal pitch on thermal resistances or parameters, extensive modeling tasks are run and related results are systemically investigated. As verification, a WL-CSP with 6 balls is actually tested finally, and results show that it is a good match between actual measurement and simulation results.

1. Introduction

WL-CSP becomes more and more popular in semiconductor industry due to its small profile and excellent electrical performance. However, the shrinkage of profile also results in a higher heat concentration, which brings wide concern throughout the industry. As a result of this, to make thermal prediction and its trend in the early design phase is extremely important for a robust design. Therefore, the objective of this paper is to develop a full parametric model using commercial software ANSYS®, which targets to solve above problems.

The parametric model herein follows JEDEC standards[1,2]. The thermal resistances θ_{JA}, ψ_{JB} and ψ_{JC} are considered in this paper to characterize thermal performance of WL-CSP. For traditional thermal analysis, when the package modeling engineers perform thermal simulation, solid models of both package and thermal board have to be built. This work takes most of the modeling time, therefore it makes the project cycle be longer. In addition, although every engineer follows JEDEC standard, different individual will generate different thermal board due to the tolerance of specified parameters which JEDEC provides (This seems unavoidable for actual manufacturing of thermal board, but can be avoidable for simulation.) and personal understanding of the general CAD drawing specification. Comparing with traditional thermal analysis, the thermal parametric model has two outstanding merits.

1) It improves the efficiency of modeling engineers significantly. By employment of the parametric model, modeling engineers don't have to spend much time to build model and meshing. What they need to do is just to set parameters based on requests from customers, and then the entire thermal simulation, including meshing, loading/boundary condition, solving, and post processing, will be dealt with automatically by APDL codes.

2) The parametric model can avoid any variation due to unclear or un-detailed declaration in JEDEC thermal test board especially for trace layout. This will eliminate the extra efforts for the investigator to track the variations and allow him to focus on the factors which are of primary concern.

This paper introduces the methodology for full parametric model for WL-CSP design. The worst and the best layouts for internal trace of PCB are included as extreme cases to evaluate the influence of internal trace layout to thermal resistances. Both low (1S0P) and high (2S2P, 2S2P with thermal vias) effective thermal conductivity JEDEC boards are included in the model. Then experiment validated empirical heat convection coefficients are applied to the parametric model. Extensive parameter based modeling and simulation are carried out to study the impact of solder ball number, die size, terminal pitch on thermal resistances.

Finally, the thermal resistance of a WL-CSP with 6 balls will be actually tested. Correlation between the modeling and test results will be presented and discussed.

2. Construction of the parametric model

As mentioned above, the parametric model consists of a WL-CSP package and a thermal test board. There are three types of thermal boards according to JEDEC standards which includes 1S0P, 2S2P and 2S2P with vias. The 1S0P is a simple case, so we will begin with that case, and take a 49-ball WL-CSP as an example to introduce how to construct the parametric model.

Figure 1 shows the structure of 49-ball WL-CSP which is mounted on a JEDEC 1S0P thermal test board. As shown in figure 1, the model consists of silicon die (grey color), solder balls (purple color), copper pads and traces (red color) and FR4 board (green color). The silicon die in the right bottom is set transparent so that the ball array can be shown clearly. All the parameters for the 1S0P parametric model are listed in table1.

978-1-4244-4658-2/09 $25.00 © 2009 IEEE

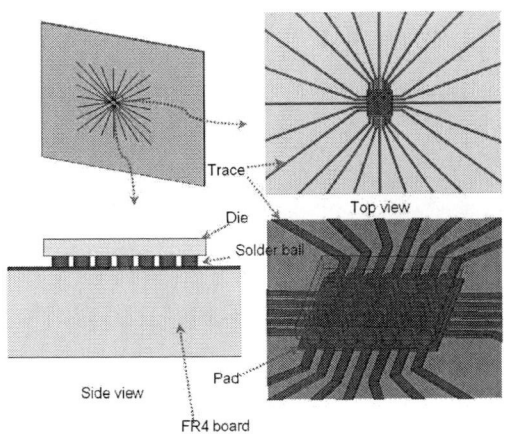

Figure 1. 49-ball WL-CSP mounted on JEDEC 1S0P thermal test board.

Table1. Specified parameters and its description

Parameter	Description	Note
l_si	length of silicon	
w_si	width of silicon	
h_si	height of silicon	
h_ball	height of solder ball	
d_ball	diameter of solder ball	
n1	lead number of long edge orientation	need input (n1≥2)
n2	lead number of short edge orientation	need input (n1≥2)
p	Pitch	
l_board	length of board	114.5mm [PKG≤40 mm] 139.5mm [40<PKG≤65 mm] 165.0mm [65<PKG≤90 mm]
w_board	width of board	101.5mm [PKG≤40 mm] 127.0mm [40<PKG≤65 mm] 152.5mm [65<PKG≤90 mm]
h_board	height of board	1.6-h_trace
w_trace	width of trace	40% of p for p>0.5 mm 50% of p for p≤ 0.5 mm
h_trace	height of trace	70 um for p>0.5 mm 50 um for p≤ 0.5 mm
lt	minimum length of trace	25 mm
s_inc	step increase of trace	can be adjusted according to ball number and pitch
l_i	initial shrinkage length of trace	can be adjusted according to ball number and pitch

Most parameters in table 1 are easily understood, so they will not be explained further. But for some trace-layout-related parameters which may cause confusion to readers, is illustrated in Figure2-3.

According to JESD51-9, traces to outer ball row should be flared to perimeter 25 mm from package body, as shown in Figure2-3. The distance from package body to perimeter is parameterized as "lt".

Parameter "l_i" indicates the length of maximum step of traces (see figure3): in the case of more balls with finer pitch, the parameter "l_i" should be specified as a bigger value so that the traces will not touch each other to have a better spacing. Parameter "s_inc" stands for the length difference between neighbor trace steps, for the same reason, "s_inc" also should be define properly.

One point should be noted that both "l_i" and "s_inc" are not specified in JEDEC standards. One advantage of the parametric model herein is: we can define values or rules for specifying above two parameters by ourselves. This allows us to eliminate the variation of models due to the uncontrolled factors.

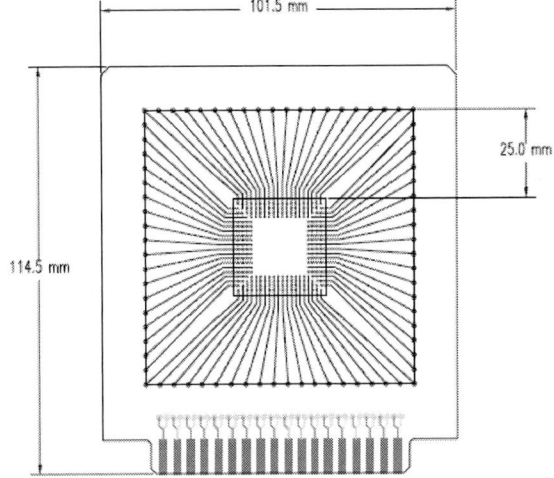

Figure 2. A typical trace layout for JEDEC 1S0P thermal test board

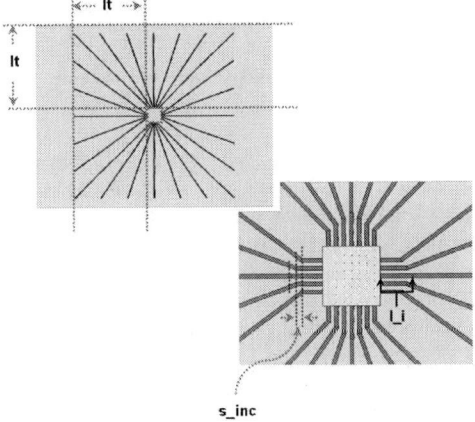

Figure 3. Illustration of trace layout related parameters

978-1-4244-4658-2/09 $25.00 © 2009 IEEE

For the parametric model, the major challenge herein is how to parameterize copper trace of thermal test board. For whatever side of board, the trace number should be either odd or even. For odd case, use a "*do…" commend to generate half of right side of traces except central one firstly. Then reflects traces along a horizontal symmetry axis following by the central trace generation. By this time, all the right side of traces have already been generated. Finally reflect all the right side of traces along a vertical symmetry axis. As the results of this, all traces along long side orientation are built. While for even case, it is similar to odd case. The difference is that the central trace should not be a special concern. (See figure4)

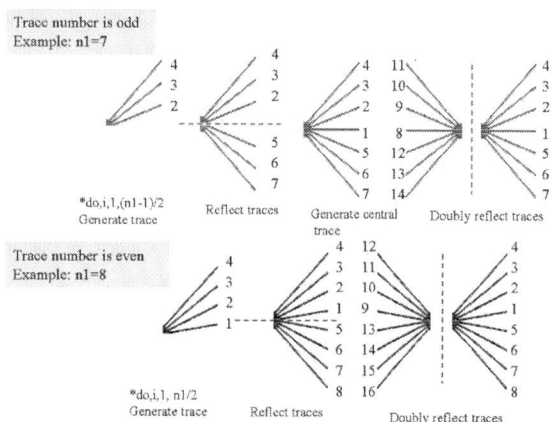

Figure 4. Trace generation flow chart

After generation of the traces along length of board orientation, hide all current traces and use same rule to build traces along short side orientation. One point should be noted that, the boundary traces (in figure 4, the trace number is 4,7,11,14 for odd case; and 4,8,12,16 for even case) belong to both neighbor sides. So when we build short side orientation traces, should use "*do,i,1,(n2-1)/2-1" instead of "*do,i,1,(n2-1)/2" for odd case, and use "*do,i,1,n2/2-1" instead of "*do,i,1,n2/2" for even case to avoid trace repeat in the same boundary (See figure 4).

According to JEDEC, internal pads should be connected to external traces. To have a simplified model to improve the efficiency, two extreme cases for internal trace layout are designed, which are also used to give the evaluation of internal trace layout to thermal resistances of packages. One is the best case, which uses a thin block (same thickness with trace, same area with die) to connect all pads, refer to figure 5 a). The other is the worst case: which only connects outer pads to traces and leaves internal pads that are isolated, see figure 5 b). Therefore, the actual trace connection under a WL-CSP will be between the two extreme cases.

Other parts including silicon die, solder ball and FR4 board is easily parameterized due to its simple geometry. So this part of work will not be given in details in this paper.

JEDEC 2S2P thermal board adds bottom signal trace layer and two buried layers to 1S0P thermal board. While for 2S2P with vias, one thermal via beneath each ball's pad is designed. These models employ the same trace generation methodology and the same internal trace design rule. So far all three types

of thermal boards (1S0P, 2S2P and 2S2P with vias) have been constructed and set up.

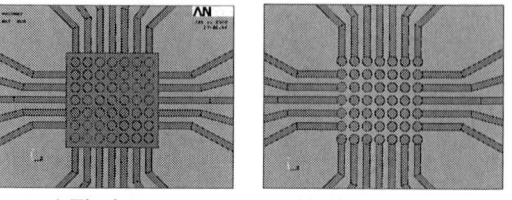

a) The best one   b) The worst one

Figure 5. Extreme internal trace layouts

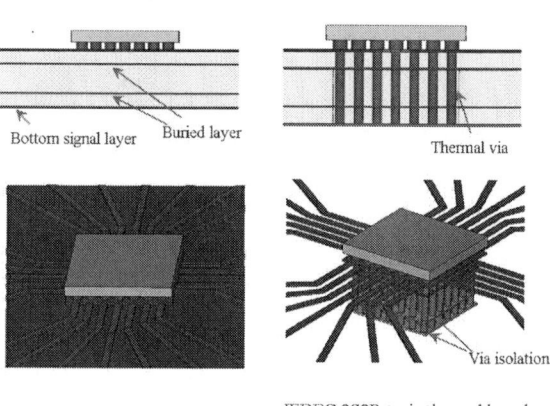

JEDEC 2S2P thermal board   JEDEC 2S2P + via thermal board

Figure 6. JEDEC high effective board

3. Thermal simulation automation procedure by using APDL

The thermal simulation automation procedure is enabled by using ANSYS APDL, the main procedure includes: 1) Construction of full parametric model, which is already introduced in previous section. 2) Selection of element type and definition of material properties. 3) Set material ID to volumes by use of entity selection function. 4) Mesh automation by combination of mesh command and entity selection function. 5) Define area components to apply film convection coefficient "H" and define volume components to apply heat generation rate. 6) Define arrays to store component names, and its values. 7) Solution setup and use "*do…" commend to apply loads or boundary conditions. 8) Results handling and calculation.

The thermal conductivities for all WL-CSP materials are listed in table 2.

Table 2 Material properties

Material	Thermal conductivity (W/m·C)
Silicon die	145
Solder ball	33
Copper trace & solder	386
FR4	0.4

4. Application of the parametric model

978-1-4244-4658-2/09 $25.00 © 2009 IEEE

The thermal performance of WL-CSP under natural convection condition is simulated by use of the parametric model developed here. A common method of characterizing a packaged device's thermal performance uses a definition "Thermal Resistance", denoted by the Greek letter "theta" or θ. For a semiconductor device, thermal resistance indicates the steady state temperature rise of the die junction above a given reference for each watt of power (heat) dissipated at the die surface. Its units are C/W. In additional to these theta thermal resistances, the thermal characterization parameters, denoted by "psi" can be useful. For a device powered up on an application board, these psi's provide a correlation between junction temperature and a certain reference temperature. The term "psi" is used to distinguish these from "theta" thermal resistances since not all heat is actually flowing between the points of temperature measurement with the psi's. They're not true thermal resistances for this reason [3]. In this paper, thermal resistance θ_{JA} and thermal parameters ψ_{JC} and ψ_{JB} are all used to character package performance, they are defined by the following equations:

$$\theta_{JA} =(T_J-T_A)/P$$
$$\psi_{JC} =(T_J-T_C)/P$$
$$\psi_{JB} =(T_J-T_B)/P$$

Where:

θ_{JA} = Thermal resistance junction to ambient, C/W

ψ_{JC} = Thermal parameter junction to case, C/W

ψ_{JB} = Thermal parameter junction to board, C/W

T_J = Die junction temperature, C

T_A =Ambient temperature, C

T_C = Case temperature, C

T_B = Board temperature, C (See figure7)

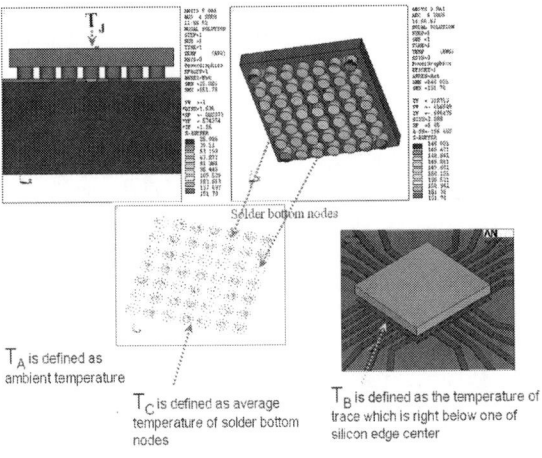

Figure7. Definition of different temperatures

For θ_{JA} evaluation, all three types of thermal boards are used , while for another two thermal parameters ψ_{JC} and ψ_{JB}, 1S0P and 2S2P thermal boards are employed separately. Impact of ball number, die size and pitch on these thermal resistance and parameters are studied systemically.

5. Results and discussions

5.1 Impact of solder ball number

In order to investigate the impact of solder ball number to thermal performance of WL-CSP, 3mm x 3mm die size and 0.4mm pitch are selected. The power is set as 1 watt, which is applied on the silicon die. The design ball arrays are as 2x2, 3x3, 4x4, 5x5, 6x6, 7x7 separately. Input these values to the parametric models includes 1S0P, 2S2P, and 2S2P+via thermal test boards. Figure 8 gives the temperature distribution of all the WL-CSP packages with the best internal trace design for 1S0P thermal board.

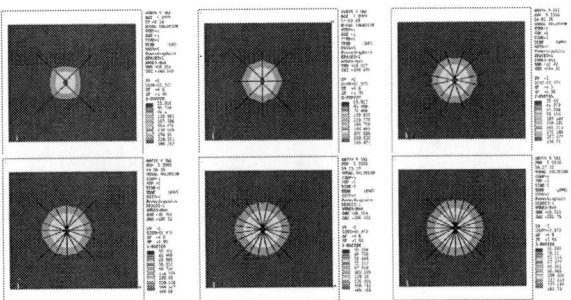

Figure 8. Temperature distribution with different ball array (Best internal trace design, 1S0P)

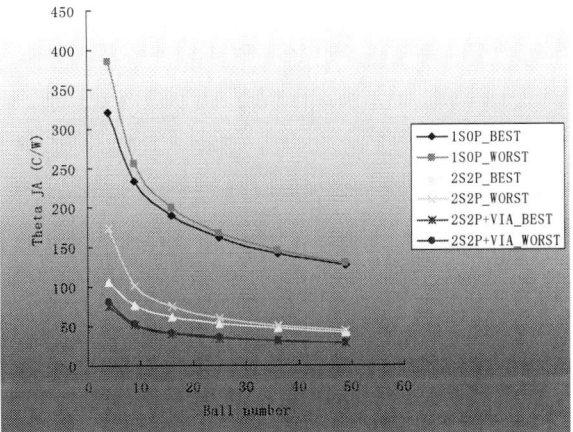

Figure 9. θ_{JA} according to ball number

All the results with two types of extreme cases for 1S0P, 2S2P, and 2S2P+via are summarized in figure9. It is clearly shown that:

1. Solder ball number is critical for thermal dissipation. Packages will dissipate heat with higher efficiency through more balls for all three types of boards, therefore more balls will result in a smaller θ_{JA}. So if viewed from thermal aspect, it is very important for a certain package to design enough balls to dissipate heat and secure die to work under safe temperature.

2. θ_{JA} curves for 1S0P is far above the ones for 2S2P and 2S2P+via, which means θ_{JA} is really board-dependant, obviously high effective boards dissipate heat from packages to ambient environment fast with higher efficiency.

3. The gap between two curves of extreme cases for a certain board reflects the role of internal trace layout design as the function of dissipating heat. Comparison between cases with and without vias tells whether internal trace layout design is invalid or valid for improving heat dissipation. This depends on the case if there is design for vertical heat transfer for thermal test board. In the case of thermal vias existing, heat transfers vertically from package to thermal board. In this case, internal trace layout will not work effectively due to its horizontal orientation. So for customer's application, we should suggest them focus on internal trace layout for the PCB without effective vertical heat transfer design, especially for the packages with lower ball density.

5.2 Impact of die size

For obtaining impact of die size to thermal performance of WL-CSP, two groups of ball arrays are selected including 2x2 and 4x4 cases. For 2x2(4 balls) case, designed die size includes 0.85x0.85, 1.25x1.25, 1.65x1.65, 2.05x2.05, 2.4x2.4 and 3x3; while for 4x4(16 balls) case, designed die size includes 1.65x1.65, 2.05x2.05, 2.4x2.4 and 3x3. For all these cases, pitch and power are kept unchanged: 0.4mm pitch and 1 watt power. Figure10 gives the temperature distribution of all the cases with the best internal trace design for 2S2P thermal board.

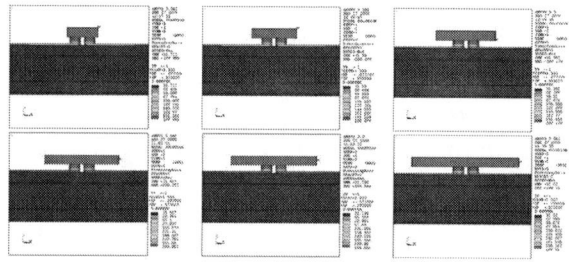

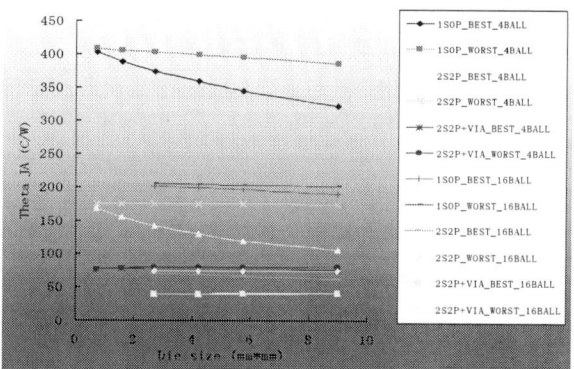

Figure 10. Temperature distribution with different die size (Best internal trace design, 2S2P)

the curves with 4 points are for 16-ball cases. The conclusion drawn from figure 11 is listed below:

1. For all the worst internal trace layout designs, the only change is die size, from the curve, we can see that these groups of curves appear flat. This means that increased surface due to the increased die size does not have significant impact on the dissipate heat, so θ_{JA} is not sensitive to die size for a given ball count case.

2. For the best internal trace layout designs except the cases with vias, θ_{JA} will decrease with increasing die size. Why does this happen? Why it does not have the same trend as the worst internal trace design? If we look back to the construction of the model, we can find that with die size increasing, the thermal test board is also changed. Since the copper block right below silicon die has same area as die, with the increasing die size, the copper block will also increase accordingly which enhances heat dissipation. This explains why in the best case, the θ_{JA} varies in decreasing trend as the die size increases.

3. The roles of internal trace layout as the function of dissipating heat, are summarized in "5.1 portion" which can also be shown in figure 11: internal trace layout has very limited impact on θ_{JA} for the thermal test board with thermal vias, and it should be a concern for WL-CSP with bigger size and lower ball density when they are mounted on the board without effective vertical heat transfer path.

5.3 Impact of pitch

To evaluate impact of pitch to thermal performance of the WL-CSP, five pitch designs including 0.35mm, 0.4mm, 0.5mm, 0.6mm and 0.7mm, with 3 types of thermal boards are prepared. For each numerical experiment, fix die size as 3mm x 3mm, ball number as 16(4x4), and input power as 1watt. Th curves of θ_{JA} vs pitch are illustrated in figure 12.

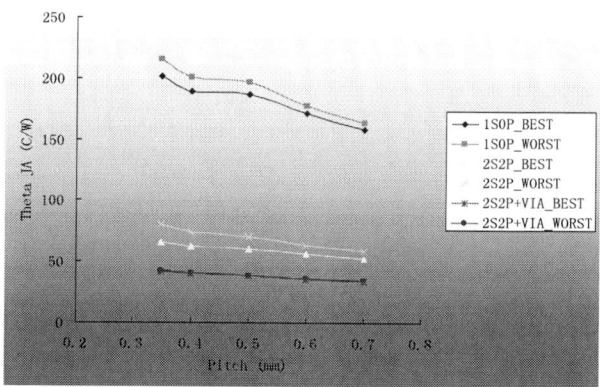

Figure 12. θ_{JA} according to pitch

Results show that the larger pitch has a smaller θ_{JA}, this might be due to the following two reasons:

Figure 11. θ_{JA} according to die size

The results for all the cases above are summarized in figure 10. The curves with 6 points are for 4-ball cases, and

978-1-4244-4658-2/09 $25.00 © 2009 IEEE   232

1. According to JEDEC standard, larger pitch will also need a wider trace, so this makes PCB dissipate heat with higher efficiency.
2. For all five designed pitches, larger pitch makes solder ball space more equally. Therefore it avoids heat crowds and results in a better heat dissipation.

5.4 Thermal parameter ψ_{JC} and ψ_{JB}

Thermal parameters ψ_{JC} and ψ_{JB} provide a correlation between junction temperature and case temperature or between junction temperature and board temperature. When we do correlations, it will be helpful. However, because they are not true thermal resistances, the trends of these parametric according to investigated factors are totally different with θ_{JA} sometimes, so it might be meaningless.. Figure 13-14 shows different trends with θ_{JA} for the impact of die size (refer Fig.11).

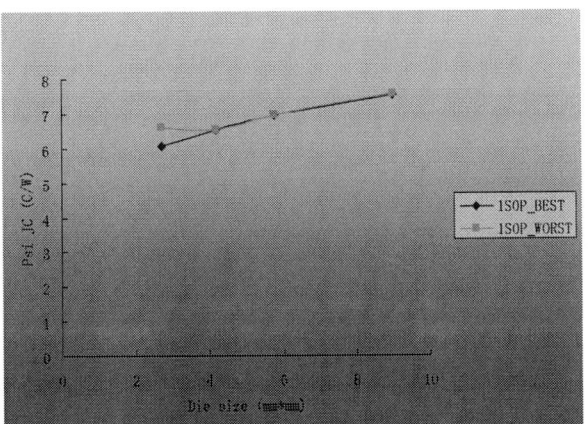

Figure 13. ψ_{JC} according to die size
(16balls, 0.4pitch, 1watt, 1S0P)

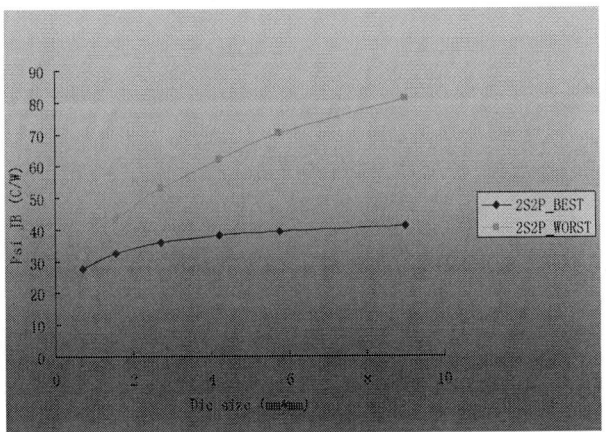

Figure 14. ψ_{JB} according to die size
(4balls, 0.4pitch, 1watt, 2S2P)

5.5 Thermal resistance measurement for a 6-ball WL-CSP

To correlate simulation result with the actual measurement, the thermal resistance test is performed for a 6-ball WL-CSP. The detailed package, thermal test board and test apparatus are listed in figure 15. This 6-ball WL-CSP has 0.65mm pitch and 1.92mm x 1.44mm die size. Three samples are prepared and tested. According to the test results in Table 3, the average θ_{JA} falls into the scope from 289 C/W to 309 C/W.

The model parameters were adjusted to make the actual test board. The parametric techniques of this paper were applied to generate the model for simulation. and calculating θ_{JA}. The simulated result shows that the θ_{JA} is 290C/W for the worst internal trace design and 283C/W for the best one. Table 4 gives the comparison of the simulated values and the actual test result. As shown in table 4, variation from simulation to the measurement is about 3.41%, which indicates very good correlation between the measurement and the simulation method in this paper.

Figure15. Actual θ_{JA} measurement set-up

Table 3. Thermal resistance measurement of 6-ball WL-CSP

Sample 1	Sample 2	Sample 3	Average. θ_{JA}
291.18	309.48	289.02	296.56

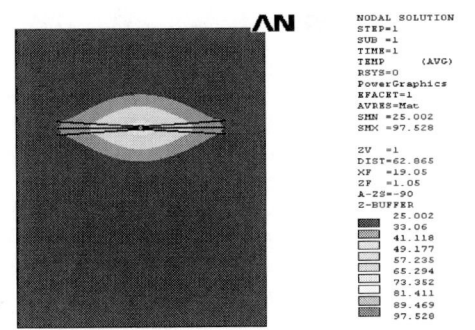

θ_{JA}, 290 C/W

Figure16. Simulated temperature distribution of 6-ball WL-CSP under 0.25watt power input
(1S0P, the worst internal trace layout)

978-1-4244-4658-2/09 $25.00 © 2009 IEEE

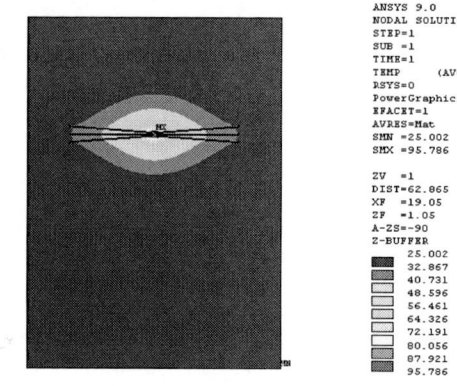

```
ANSYS 9.0
NODAL SOLUTION
STEP=1
SUB =1
TIME=1
TEMP        (AVG)
RSYS=0
PowerGraphics
EFACET=1
AVRES=Mat
SMN =25.002
SMX =95.786

ZV =1
DIST=62.865
XF  =19.05
ZF  =1.05
A-ZS=-90
Z-BUFFER
     25.002
     32.867
     40.731
     48.596
     56.461
     64.326
     72.191
     80.056
     87.921
     95.786
```

θ_{JA} , 283C/W

Figure17. Simulated temperature distribution of 6-ball WL-CSP under 0.25watt power input
(1S0P, the best internal trace layout)

Table 4. Simulatedθ_{JA} vs. Measured θ_{JA} for 6 –ball WL-CSP
(Unit: C/W)

	Simulation	Measurement
Min.	283 (best case)	289
Max.	290 (worst case)	309
Median or Average	286.5 [M]	296.6 [A]
Variation	3.41% off from measurement	

Conclusions

This paper develops a thermal simulation method for full parametric analysis for power WL-CSP, which includes parametric WL-CSP and its adaptive parametric JEDEC thermal test board. By employment of the parametric model, package geometry parameters can be easily set. The trace layout for PCB may also be accordingly changed to meet the requirement of package design, so that the influence of all geometry parameters to thermal performance can be easily investigated for whole series of WL-CSP packages. The entire thermal simulation, including meshing, loading/boundary condition, solving, and post processing, is automated with APDL coding.

Investigation of impact of ball number, die size and pitch on the thermal performance of WL-CSP shows:

1) Solder ball number is very critical to thermal performance of WL-CSP, more balls result in a smaller thermal resistance, therefore it is importance to design proper balls to secure die work under safe temperature.

2) Internal trace layout has significant impact on the thermal resistance for boards without effective vertical heat transfer design such as vias, so it should be a concern for designing a proper internal trace layout to achieve a better thermal performance, and fully makes of area right below the silicon die. Especially for the WL-CSP with bigger die size and lower ball density.

3) Equal distribution of balls on silicon die can avoid heat crowds and achieve a better performance for WL-CSP.

4) Correlation of a 6-ball WL-CSP between the simulation and the measurement results show that the method developed in this paper is reasonable and it can be used for WL-CSP thermal analysis and design.

Acknowledgments

The authors wish to thank the support from Package Development and Automation Department, Fairchild Semiconductor Corp.

References

1. Intergrated Circuits thermal Test Method Environment Conditions-Natual Convection (Still Air), EIA/JEDEC standard-JESD51-2.
2. Test Boards for Area Surface Mount Package Thermal Measurements, EIA/JEDEC standard-JESD51-9
3. Jim Benson., "Thermal Characterization of Packaged Semiconductor Devices", *Technical Brief*, December 2002.

An Improved Substructure Method for Prediction of Solder Joint Reliability in Thermal Cycle

Fei Liu[1], Lihua Liang[1], Yong Liu[1,2]

[1]Fairchild-ZJUT Microelectronic Packaging Joint Lab, Zhejiang University of Technology, Hangzhou, 310032, China
[2]Fairchild Semiconductor Corp., S.Portland, USA
Email: lianglihua@zjut.edu.cn, Tel: 0571-88320294

Abstract

With the current trend of less expensive, faster, and better electronic products, it has become increasingly important to evaluate the IC package and system performance early in the design stage using simulation tools. For the solder joint subjected to cyclic stresses generated during the thermal cycling, its reliability depends on its resistance to creep and fatigue. The approach for simulation in this paper includes advanced finite element modeling with sub-structuring method. To simplify the analysis the temperature cycle is divided into different portions, and correspondingly the material properties are specified at that temperature range. The material properties are assumed to be linear and the equivalent elastic modulus corresponding to a specified temperature is got based on similar critical solder joint deflection. All non-critical solder joints with equivalent elastic modulus are condensed into a super-element with substructure analysis. At different temperatures, non-critical solder joints with specified equivalent elastic modulus are condensed in a serial of super-elements. To use different super-elements, windows batch command and ANSYS batch mode are used for automatic super-element generation and analysis.

Introduction

Solder joint reliability and its fatigue life prediction are important in IC package development. A large number of solder joint life prediction models have been developed to predict the thermal fatigue life of solder joints during the past decades [1]. Many of them used finite element analysis approach for solder joint temperature cycle prediction which had the larger difference as compared with actual life. Of course the low accuracy life prediction is doubtful and can not be used to replace experimental tests. Therefore, it is necessary to develop an advanced finite element analysis method and to correlate well with the actual test results.

Assumptions should be made to simplify the model. Syed proposed the assumptions that have secondary influence on the accuracy of life prediction with ranging from structure details to mesh densities [2]. These assumptions for solders joints life prediction are adopted in this paper as well. Equivalent elastic modulus of nonlinear solder joints is achieved at different temperatures by comparing the deflection of critical solder joint so as to get linear property for non-critical solder joint and to use substructure method. Substructure method is introduced to condense nonlinear solder joints at different temperature with equivalent elastic modulus. Windows based batch command and super-elements are introduced in the paper to perform thermal cycles using substructures.

Model Description of a Stack Die Package

A package model with 72 solder joints is investigated in this paper. A 3D 1/4 symmetric model for a stack die package with bumps is created for solder life prediction. The finite element model and mesh are shown in Figure 1.

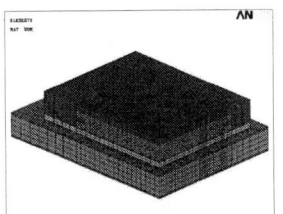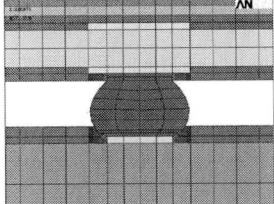

Figure 1 Model and Mesh of a Stack Die Package

The model contains parts with both linear and nonlinear materials. PCB core (FR4), Chip (silicon), Non-Conductive Attach (Proprietary) and solder joint are nonlinear materials while conductor(copper), mask(dry film), substrate core(BT) and mold cap(mold) are linear materials while others are nonlinear. Detailed material properties are listed in table 1.

The SnAgCu solder joint with time and temperature dependent deformation behavior usually fits with creep constitutive equation. This paper uses Hyperbolic Sine creep equation to describe solder joint deformation behavior [3]. The creep model was implemented in ANSYS® using TBOPT = 8 option for implicit creep equations.

$$\dot{\varepsilon} = A_1 [\sinh(a\sigma)]^n \exp\left(\frac{-H_1}{kT}\right) \qquad (1)$$

where, A_1=277984s^{-1}, a=0.02447MPa^{-1}, n=6.41, H_1/k=6500.

The temperature loads applied to the model with 2 temperature cycles, it found that error of the life prediction for solder joint is less than 1.5% [4]. This simulation is performed with 2 temperature cycles and each cycle is divided into 4 parts as shown in Figure 3. 125°C is given as reference temperature for model analysis with zero-stress state. The whole model experienced 165°C temperature ramp down from the beginning of 125°C to -40°C in 15 minutes and dwell at the low temperature of 125°C for another 15 minutes, then the temperature ramp back to 125°C and dwell at this temperature. Dwell time and ramp time are also 15 minutes.

Summary of Substructure Simulation Approach

Finite element method is taken to analyse the package reliability. But many parts of package with material nonlinear property make the simulation low efficiency. Therefore, researchers want to simplify the analysis structure by either eliminating detail of model or simplify the material properties. Of course, structure with linear materials is the simplest case because its analyses need no iteration. In a nonlinear analysis by substructure method, the researcher can substructure the linear portion of the model so that the element matrices for that portion need not be recalculated at every equilibrium iteration, thus it is a good candidate for the pakage analysis.

In this paper the package model contains many linear parts that can be condensed by substructure method. But substructure analysis with these linear parts is not the key point of this paper. The parts that we most care about are the solder joints, because the reliability of solder joints decides the life of package. To use substructure method at solder joints, the joints have to be simplified as linear property.

Three steps are taken to simplify the analyses. In the first step, we apply a unit temperature on the whole model so as to find out the location of critical solder joint which is most likely to fatigue. In the second step, we divide the temperature cycle into some parts then find equivalent linear property of elastic modulus for non-critical solder joints at different divided temperatures. In the third step we use equivalent elastic modulus at different temperatures to generate substructures which called super-element in ANSYS. Finally, run substructure analyses using the generated super-elements.

Table 1 Materials Parameters

Materials/Properties	Elastic Modulus /MPa	CTE /K^{-1}		Poisson's ratio
Solder Ball	61251-58.5*T	2E-5		0.36
Conductor(Copper)	128932	1.72E-5		0.34
Mask(Dry Film)	4137	3E-5		0.4
Substrate Core(BT)	24132	1.6E-5(XZ) 3.5E-5(Y)		0.3
Mold Cap(Mold)	15513	1.5E-5		0.25

PCB core (FR4)			
Elastic Modulus /MPa	Shear Modulus /MPa	CTE /K^{-1}	Poisson's ratio
27924-37*T(XZ)	12600-16.7*T(XZ)	1.6E-5(XZ)	0.39(XZ)
12204-16*T(Y)	5500-7.3*T(YZ&XY)	8.4E-5(Y)	0.11(Y)

Non-Conductive Attach (Proprietary)				
Temperature /K	233	298	423	523
Elastic Modulus /MPa	3844	866	125	119

Non-Conductive Attach (Proprietary)							
Temperature /K	223	268	278	283	288	298	473
CTE /K^{-1}	1.09 E-4	1.10 E-4	1.30 E-4	1.40 E-4	1.50 E-4	1.69 E-4	1.70 E-4

Chip (silicon)		
Elastic Modulus /MPa	CTE /K^{-1}	Poisson's ratio
162716	-5.88E-06+6.28E-08*T-1.60E-09*T^2+1.50E-13*T^3	0.28

The summary of substructure simulation approach is listed as bellows:

Step 1: Find the critical solder joint
- Build the 3D 1/4 symmetric model in x-z plane and extrude in y-direction for simulation.
- Symmetry constraints and coupled constraints are applied in x and z directions.
- Apply a unit temperature load and perform analysis for a load step.
- Find out the location of critical solder joint.

Step 2: Get equivalent elastic modulus
- Divide a temperature ramp of 165K into 5 parts. Thus 6 temperatures at 233K, 266K, 299K, 332K, 365K, and 398K are defined in a ramp stage for simulation. Therefore 6 equivalent elastic modulus need to be found at these temperatures accordingly.
- At a specified temperature, find an equivalent elastic modulus for non-critical solder joints which has less than 1% difference in deflection of critical joint.
- Specify all other temperatures and find there equivalent elastic modulus for non-critical solder joints.

Step 3: Substructure method

- At a specified temperature, modify the material parameters of non-critical solder joints with the equivalent elastic modulus at this temperature. Generate a super-element with the modified material property of non-critical solder joints.
- Specify all other temperatures and repeat modifying material property and generating super-elements with non-critical solder joints.
- Define the temperature load on generated super-elements and other part of the structure according to thermal cycle load conditions.
- Use different super-elements in thermal cycle according to temperature load for simulation of critical solder joint creep result. Batch method in simulation can save time and work amount.

Find the Critical Solder Joint

For a thermal cycle of a package the behavior of solder joint is the critical. The solder joint with maximum stress and strain is called critical joint. The behavior of the critical solder joint decides the whole package life prediction. So it is always focus on the critical solder joint in the simulation. A unit temperature is applied on the package and the critical solder joint can be found with the maximum stress and strain. The

978-1-4244-4658-2/09 $25.00 © 2009 IEEE

location of critical solder joint of the stack die package as shown in Figure 1 is shown in Figure 2.

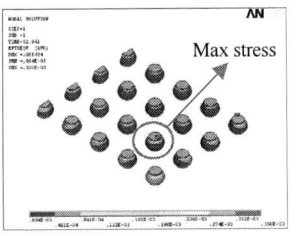 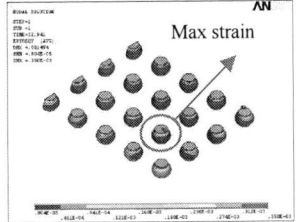

Figure 2 Location of Critical Solder Joint

Get Elastic Modulus

According to analyses conducted in reference [5], while mesh density is further increased the difference in accumulated creep strain and energy density is insignificant reduced, but simulation time has been considerably increased. Therefore, the critical solder joint was meshed in 2 elements in radial direction and 7 elements across bulk. Also, 2 elements across interface layer of the critical solder joint and 20 nodes solid185 element with enhanced strain formulation is attached for analysis. Mesh density of the critical solder joint can be seen in Figure 4.

Divide a temperature ramp of 165K into 5 parts. That is, 6 temperatures as 233K, 266K, 299K, 332K, 365K and 398K are defined in a ramp stage for simulation. Thus 6 equivalent elastic modulus need to be found at the corresponding temperature.

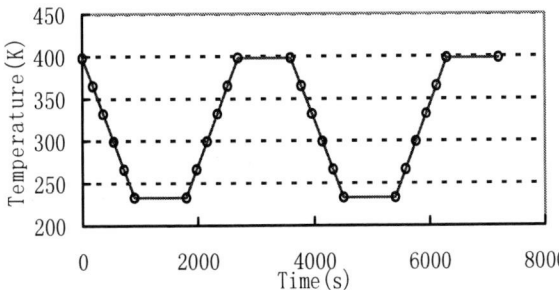

Figure 3 Two Thermal Cycles Divided

Use linear material property for non-critical solder joint by comparing the impact of deflection on solder joint. This process is carried out by finding out a proper elastic modulus for these non-critical solder joint which causes less than 1% difference at the influence on critical solder joint deflection.

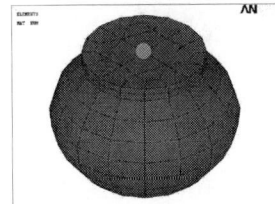 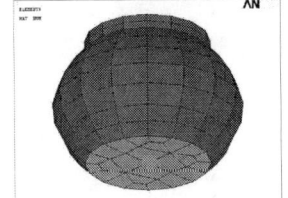

Figure 4 Location of Selected Nodes for Deflection Comparison

Find out a serial of equivalent elastic modulus corresponding to critical solder joint deflection in each

defined temperature. Use curve fitting to find the relationship between elastic modulus and temperature. A data fitting equation can be used to describe the equivalent elastic modulus varies with temperature:

$$EM = 0.0011T^2 - 0.7043T + 140.42 \ (MPa)$$

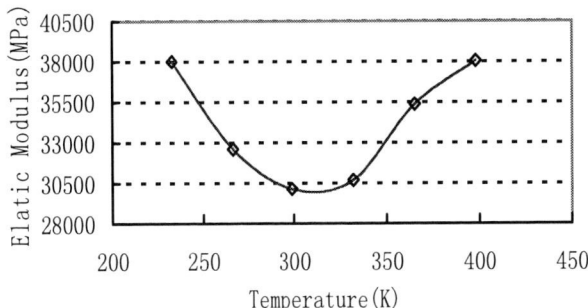

Figure 5 Equivalent Elastic Modulus at Different Temperatures

Substructure Method

Substructure method reduces computer time and allows solution of very large problems with limited computer resources. Nonlinear analyses with linear parts and analyses of structures including repeated geometrical patterns are typical candidates for employing the substructure. The elements that have constant stiffness, damping, and mass matrixes, and material properties do not change with time and temperature can be condensed into a super-element by performing substructure analysis. In a nonlinear analysis, one can substructure the linear portion of the model so that the element matrices for that portion need not be recalculated at every equilibrium iteration, thereby saving a significant amount of computer time [6].

Divide the temperature ramp into different parts and get a series of equivalent elastic modulus for non-critical solder joints according to elastic modulus curve. At a specified temperature, modify the material property of non-critical solder joints with the equivalent elastic modulus at this temperature. Then perform substructure analysis to condense these elements and generate a super-element with the modified material property of non-critical solder joints. Then specify all other temperatures and repeat modifying material property and generating super-elements with non-critical solder joints.

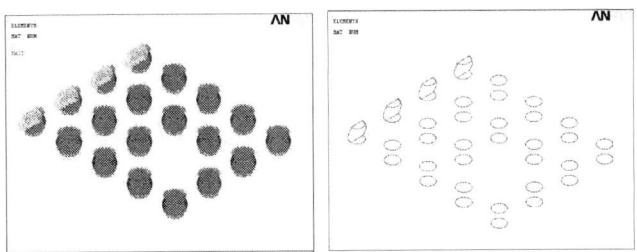

Figure 6 Generate Super-element with Non-critical Solder Joints

Solution with super-elements generated at different temperature can not be run with ANSYS directly. Once a step

978-1-4244-4658-2/09 $25.00 © 2009 IEEE   237

of solution finished, we must exit ANSYS and substitute the current super-element with the super-element which has been generated at the next specified temperature. However, this can not be done in ANSYS solution process, because ANSYS will not reread the sub file once it reads the file at the first load step and imports the file in process to create a super-element. But when ANSYS exited and restarts, the process will reread sub file so as to work with super-element. That means we can substitute the sub file when ANSYS exited. Then restarts ANSYS it will reread the super-element which we substituted before. This is a way of substructure method that can be adopted in some other structures which material properties can be represented by a set of linear properties.

In this example, we first define the temperature load on the generated super-elements and other part of the structure according to thermal cycle load condition. If temperature at this stage is 365K and at next stage is 332K, after ANSYS solution has been done at 365K and exit, the sub file which generated at 365K using equivalent elastic modulus at 365K should be substituted with the one generated at 332K. Of course the substitution of substructure file should base on the thermal cycle load condition. This analysis is performed with windows batch command and ANSYS batch mode.

Windows batch command has been used to generate super-elements and run ANSYS solution. Commonly batch command used to run ANSYS are as follows:

```
set ansys100_PRODUCT=ANE3FL
set ANS_CONSEC=YES
"C:\Program Files\Ansys Inc\V100\ANSYS\bin\intel\ansys100" -b -i file.inp -o file.out
```

where, file.inp is the input file with APDL and file.out is the output file with result output.

The flow chart of generation and solution process are shown in Figures 8 and 9.

Results Analysis

Both substructure analysis and full model solution were carried out at the same load and boundary condition. Substructure model with the simplified material property of non-critical solder joint and the full model with nonlinear property of creep in all solder joints are used in simulation for comparison. The finite element solution results of critical solder joint are listed in Figs 9 and 10 as bellows.

Figures 9 and 10 show the creep strain of critical solder joint at the end of first thermal cycle and at the end of second cycle. It can be seen from the chart that less than 25% difference in von Mises creep strain and 12% difference in creep strain intensity between the full nonlinear approach and the substructure method.

The life prediction models are given by Schuber when hyperbolic sine equation is used to simulate the creep behavior of SnAgCu solder [3]. This life prediction model parameters would depend on the constitutive equation used to simulate solder behavior. Exact parameters are given by Syed through actual test data and simulation work with a fitting curve [7-8]. The mean life of SnAgCu solder based on constitutive equation (1) can be predicted by using the following equations.

$$N_f = (0.175\varepsilon_{acc})^{-1} \tag{2}$$

$$N_f = (0.0069w_{acc})^{-1} \tag{3}$$

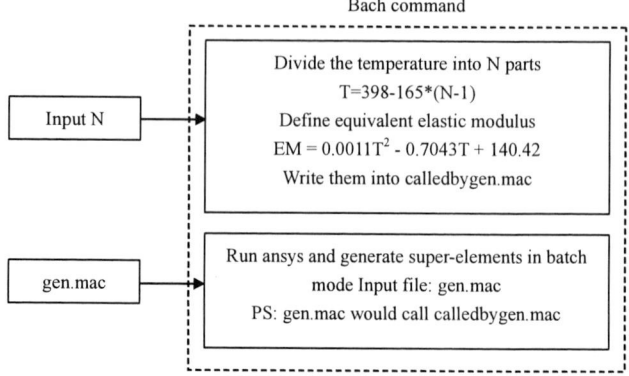

Figure 7 Generate Super-Elements

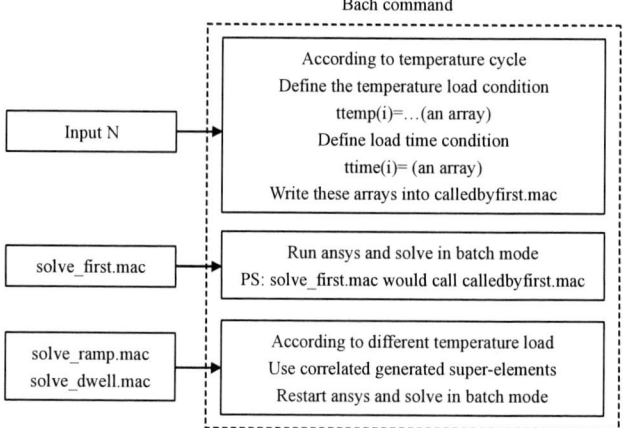

Figure 8 Batch Solution with Super-Elements

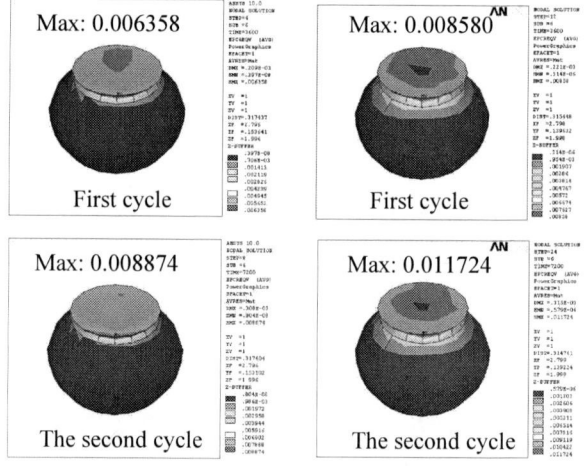

Full Nonlinear Approach   Substructure Method

Figure 9 Von Mises Creep Strain of Critical Solder Joint

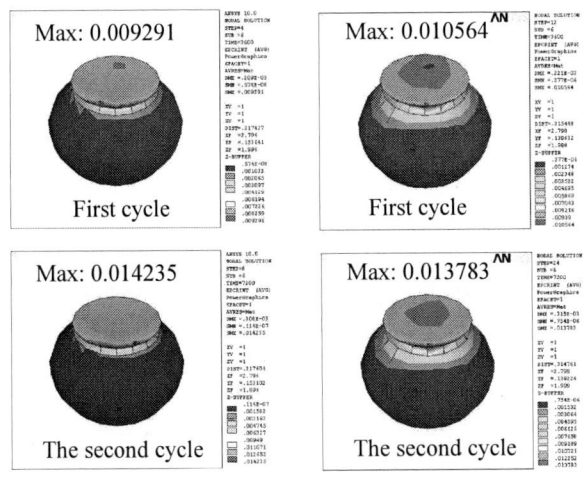

Max: 0.009291   Max: 0.010564

First cycle   First cycle

Max: 0.014235   Max: 0.013783

The second cycle   The second cycle

Full Nonlinear Approach   Substructure Method

Figure 10 Creep Strain Intensity of Critical Solder Joint

Different life prediction model used for predicting solder joint life are list in Table 2. The first life prediction model is based on the accumulated creep strain with equation (2) and the second one base on creep energy density with equation (3). Two types of life prediction results are in the range of SnAgCu solder experimental test results [3]. It is obvious in Table 2 that substructure method uses 11% less elements but saved 35% time, because it condenses the non-critical solder joints in a super-element. It is quite worth to mention that if all the elements in the model with linear property are condensed into super-elements with substructure method, we can save more time. That means computer time can be saved if substructure method is adopted to condense other linear elements.

Table 2 Different Life Prediction Methodologies

Life Prediction Results	Full Nonlinear Approach	Substructure Method
Number of Elements	39410	34786
CPU Time	14297	9384
Equation (2)	5462	5400
Equation (3)	1661	1000

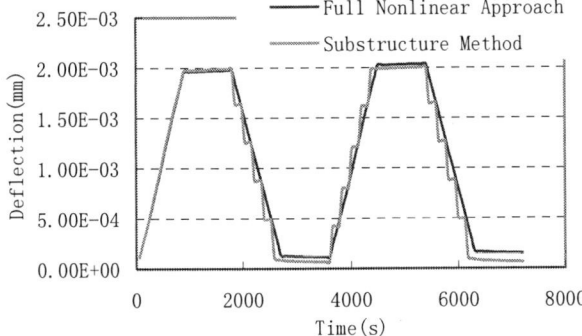

Figure 11 Comparison of critical solder joint deflection

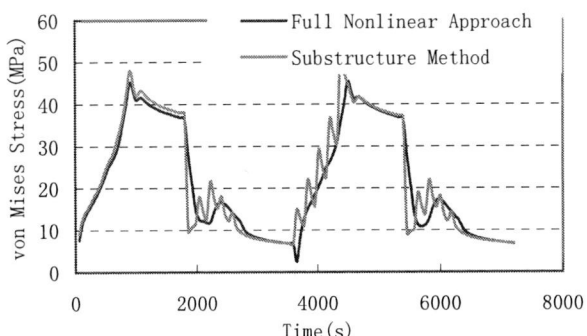

Figure 12 Maximum von-Mises Stress Comparison

Deflection of critical solder joint is compared so as to verify if the equivalent elastic modulus will have the significantly influence on critical solder joint deflection. The result on Figure 11 shows that the trend of deflections for both full non-linear method and the substructure method agrees very well. Thus it is reasonable to get equivalent elastic modulus by comparing deflection of critical solder joint.

As shown in Figures 9 and 10, the maximum von-Mises stress is at the same node. Further investigation find that the maximum von-Mises stress for both full nonlinear approach and substructure method is always at the same node. The maximum von-Mises stress with time is shown in Figure 12. We can see that the result by the full non-linear approach fits the substructure method well.

Summary and Conclusions

Thermal fatigue life prediction models are presented using Hyperbolic Sine constitutive equation for SnAgCu solder joints. An equivalent elastic modulus for non-critical solder joint is used to replace the nonlinear property of these solder joint. ANSYS substructure technique and batch method are used to simplify the model and simulation. The comparison of simulation results with the full non-linear method and the substructure shows the both methods agree very well. However, the substructure method with simplified procedures can significantly save the CPU time with less model element, and solving equations.

Acknowledgments

The authors gratefully acknowledge support from Zhejiang Science Foundation (ZSF) No. Y107365. The authors also thank the supports from Zhejiang University of Technology and Fairchild Semiconductor.

References

1. Lee, W. W., L. T. Nguyen, et al, "Solder joint fatigue models: review and applicability to chip scale packages." *Microelectronics Reliability* Vol. 40, No. 2 (2000), pp. 231-244.
2. Syed, Ahmer, "Predicting Solder Joint Reliability for Thermal, Power, & Bend Cycle within 25% Accuracy," *Proc 51st Electronic Components and Technology Conf*, 2001, pp. 255-263.
3. Schubert, A. "Fatigue life models for SnAgCu and SnPb solder joints evaluated by experiments and simulation,"

Proceedings 53rd Electronic Components and Technology Conference, 2003, pp. 603-610.

4. Xu Yangjian, Liu Yong, et al. "Finite element based solder joint fatigue life predictions for a same die stacked chip scale ball grid array package." *Journal of Zhejiang University of Technology*, Vol. 32, No. 6 (2004), pp. 668-673.

5. Syed, Ahmer, Kim, SeokBong; Lin, Wei, "Building accuracies in finite element models for life prediction of solder joints," *Proceedings of the 9th Electronics Packaging Technology Conference*, 2007, pp. 184-191.

6. ANSYS manual book, 2008.

7. Syed, Ahmer, "Accumulated creep strain and energy density based thermal fatigue life prediction models for SnAgCu solder joints," *Proceedings 54th Electronic Components and Technology Conference*, 2004(1), pp. 737-746.

8. Syed, Ahmer, "Finite Element Simulation and Life Prediction for Solder Joint Reliability," *Electronic Components and Technology Conference*, 2007 Short Course.

978-1-4244-4658-2/09 $25.00 © 2009 IEEE

Warpage Prediction of Fine Pitch BGA by Finite Element Analysis and Shadow Moiré Technique

Ke Xue[1,2], Jingshen Wu[1,2], Haibin Chen[1,2], Jingbo Gai[3], Angus Lam[4]
[1]Department of Mechanical Engineering, The Hong Kong University of Science and Technology
Clear Water Bay, Kowloon, Hong Kong
[2]Center for Engineering Materials and Reliability, Fok Ying Tung Graduate School
The Hong Kong University of Science and Technology, Clear Water Bay, Kowloon, Hong Kong
[3]Department of Aeronautics and Astronautics, Harbin Engineering University, Harbin, Heilongjiang Province, China
[4]ASAT Holdings Limited, Dongguan, China

Abstract

Warpage is one of the major concerns in manufacturing BGA, CSP, POP or QFN based array packages because a reasonably flat package is critical to successful singulation and board level assembly processes. Warpage of a package is a result of curing shrinkage of encapsulated mold compounds (EMC) and CTE mismatch between various packaging materials. In a completed package, all components are bound together by the crosslinked polymers, i.e. EMC and die-attaching adhesive (D/A). No component can expand or shrink freely. In a typical array package, warpage is in the form of either a 'crying' face (corners facing downward) or a 'smiling' face (corners facing upward), depending on the correlations of the packaging materials' mechanical, physical and chemical properties.

Fine pitch ball grid array (fpBGA) is a chip scale package offering a competitive solution for mobile applications. Package thickness ranges from 1.4mm down to 0.6mm with ball pitches as small as 0.4mm. Since the package is very thin, warpage control is a big challenge. In this study, an accurate finite element model, incorporated appropriate material properties was developed to predict the warpage of fpBGA under reflow condition. The experimental measurements of the warpage behavior of the fpBGA under solder-reflow condition were conducted using an Akrometrix TherMoire PS200. The experimental results were compared with the results of FE analysis, which provides feedbacks for modeling optimization. Effects of material properties and geometric parameters on thermal warpage were then studied using the optimized models.

Introduction

Reduction in size of portable products such as cellular phones and camcorders has led to the miniaturization of integrated circuit packages. Fine-pitch BGA (fpBGA) packages has been gaining its popularity due to compact in size and relatively low costing. The fpBGA is essentially a smaller version of the BGA package in which the pitch, bond-pads and balls are reduced in size. With further down-sizing in package height, reliability issues like die cracking and warpage could be big challenges during production engineering design process. [1]

The thermo-mechanical warpage of fpBGAs mainly results from the temperature drop to ambient after post-mold cure [2]. After the molding compound is heated to approximately 175 °C, it is injected into a cavity containing the silicon dies already die attached to the substrate. After solidifying, the fpBGA is held at a temperature of approximately 170 °C for several hours. In this state, it is assumed that the internal stresses are negligible, and the fpBGA is flat. Upon cooling down to an ambient temperature of approximately 25°C, the coefficient of thermal expansion (CTE) mismatch between adjacent materials causes the thermomechanical warpage. During the reflow process, warpage is much bigger issue since fpBGA package suffering very severe temperature ramps and drops. The temperature could reach as high as 260°C for a lead free solder. It would induce high warpage level which could damage the second level interconnection. The warpage of fpBGAs is being defined as the normal distance between two closest parallel planes that are able to encompass the fpBGA.. Unless the fpBGA is twisting due to an unusual 3D effect, the warpage is generally the vertical deflection from the horizontal seating plane. Reducing this warpage is important for improving the singulation process, reducing package bending and shearing stresses [3], minimizing coplanarity problems [4], and increasing solder joint reliability. For solder joint reliability in particular, the out-of-plane alignment of solder balls (coplanarity problem) due to thermo-mechanical warpage can result in unsoldered or mechanically weakened joints [5], shown in Figure 1.

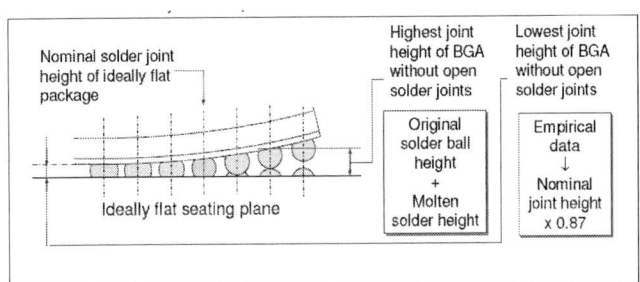

Fig. 1 Calculation of the Maximum Relative Displacement Immune from Open Solder Joints [6].

Finite Element Model

The package named 12 x 12mm fpBGA has been studied; its mold body size is 12mm x 12mm x 0.55mm. The substrate used here is bismaleimide triazine (BT), with a thickness of 0.20mm. The die size is 5.725mm x 6.22mm its thickness is 0.1524mm. The structure of fpBGA is shown in Figure 2. A 3-dimensional (3D) FEA model of the FBGA package was created to simulate the cooling condition from post mold cure temperature to room temperature. A half symmetry model of the fpBGA was being modeled using ANSYS 3D element type (Figure 3).This model takes advantage of geometry symmetry of fpBGA here to save calculation time. Half model

978-1-4244-4658-2/09 $25.00 © 2009 IEEE

is used instead of quarter model because in some cases the die does not sit exactly in the center.

To obtain as much modeling accuracy as possible, this 3D half model includes a lot geometry details which are usually neglected by other researchers investigating warpage behavior of packaging using 3D FEM modeling. For example, die attach fillet, thin solder mask and copper layers in substrate are also built in this model, as Figure 4 shown.

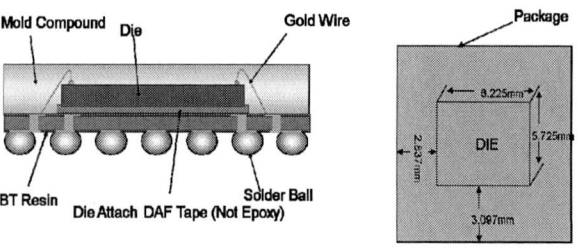

Fig.2 Schematic View of 12 x 12mm fpBGA

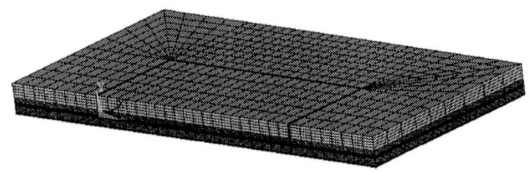

Fig.3 3D Half Model of 12 x 12mm fpBGA

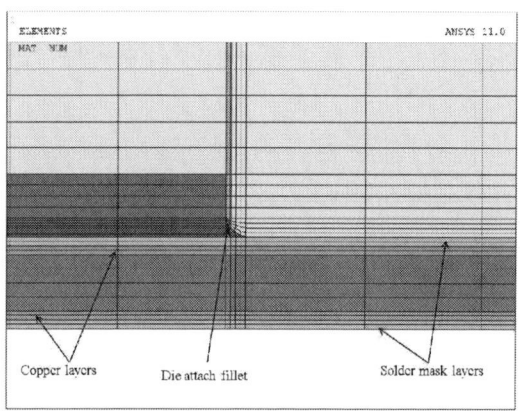

Fig.4 Geometry Details in 3D Model

Material properties of different constitutes are also carefully handled here. Especially for molding compound because it takes up most part of fpBGA .Studies have shown that the use of viscoelastic model representation for mold compound has generated close predictions in the package warpage measurements [7]. Here use thermo-elastic model to stand for the property of molding compound. Materials data comes from DMA tests results. The storage modulus as a function of temperature and frequency is shown in Figure 5. Die, copper and solder mask are modeled with linear elastic property while die attach is temperature dependent linear elastic property since it is very thin compared to the whole package structure. Linear orthotropic material properties were used for

the printed wireboard and the bismaleimide triazine (BT) substrate core. (See Table 1)

Table 1 Material Properties Used for fpBGA model

	Material model	Modulus (MPa)	Poisson ratio	CTE (ppm)
EMC	Thermo-elastic	22000	0.3	12, below Tg ; 46, above Tg ;
Die attach	Temperature dependent elastic	1870, at 25ºC ; 1460, at 50ºC	0.4	77, below Tg ; 260, above Tg ;
Silicon	Linear elastic	161000	0.21	2.6
Copper	Linear elastic	117000	0.34	17
Substrate	Orthotropic	16892 (x , y) 7377 (z)	0.39 (xz , yz) 0.11 (xy)	15 (x, y) 57 (z)
Solder mask	Linear elastic	2000	0.33	70

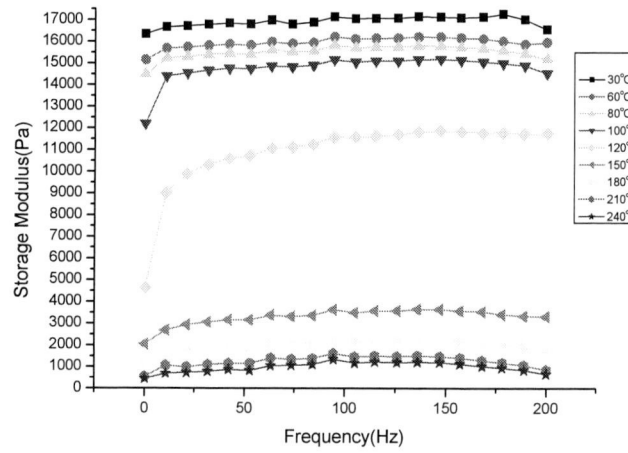

Fig.5 Storage Modulus of Molding Compound

The concept of effective coefficient of thermal expansion (CTE) is used here instead of actual CTE values for polymeric materials

$$\alpha_{ef} = \frac{\alpha_1 \left(T_g - T_{\min} \right) + \alpha_2 \left(T_{\max} - T_g \right)}{T_{\max} - T_{\min}}$$

where, T_g is the glass transition temperature, α_1 is the coefficient of thermal expansion (CTE) below T_g, α_2 is the CTE above T_g.

Introduction to Shadow Moiré Technique

Warpage can be measured by many different techniques such as the gauge indicator shim method, profilometry, interferometry and moiré methods. The gauge indicator shim method is fast, but ad-hoc in nature. The technique involves placing feeler gauges of different thicknesses under a package to determine package warpage. The gauge indicator shim method has poor accuracy, low resolution and cannot be used online. There are two types of profilometry: contact profilometry and laser profilometry. Contact profilometry measures the surface profile of the sample by using a stylus that contacts the sample surface, and determines depth deflection of the stylus at each point on the sample with respect to a reference point. The advantage of contact profilometry is that it provides high resolution results. The disadvantages are that the stylus has to contact the sample surface and extensive two-dimensional scanning is required to understand the full-field topology which makes it unsuitable for online use. Laser profilometry measures the surface profile of the sample by using a laser that reflects from the sample surface, and determines out-of-plane displacement based on the angle of the laser's reflection at each point on the sample with respect to a reference point. The advantages of laser profilometry are that it provides high resolution warpage results and it is a noncontact method. The disadvantage is that extensive two-dimensional scanning is required to understand the full-field topology which makes it unsuitable for online use [8].

Using interferometry, sample warpage can be determined by splitting a continuous laser beam. One split beam acts as a reference. The other beam is reflected back from a PWB surface and is combined with the reference beam creating interference. The interference is due to the phase shift between the two beams. The interference pattern is used to determine the sample warpage.

The shadow moiré technique is a non-contact, full field measurement method that generates a moiré fringe pattern. A moiré fringe pattern is a visual pattern that results from geometric interference between two periodic images. An example of a moiré fringe pattern is shown in Figure 6.

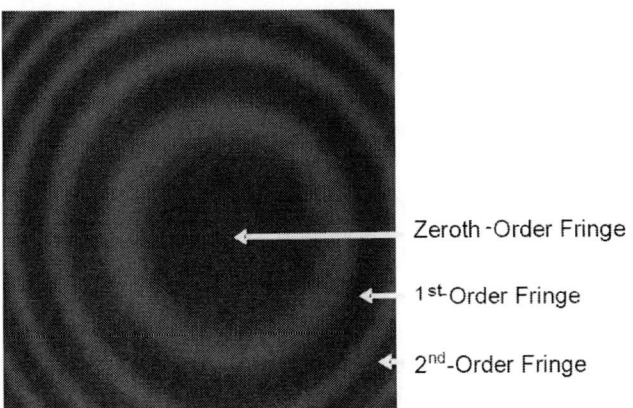

Fig.6 Moiré Fringe Pattern Showing Zeroth, First and Second Order Fringes

In a shadow moiré optical setup, as shown in Figure 7, one of the periodic images come from a glass grating, and the other is the shadow of the grating lines which is generated by the white light source on a surface being measured. Small variations between the measured surface and the reference glass grating are magnified by moiré fringes and give a quantitative measurement of surface topology.

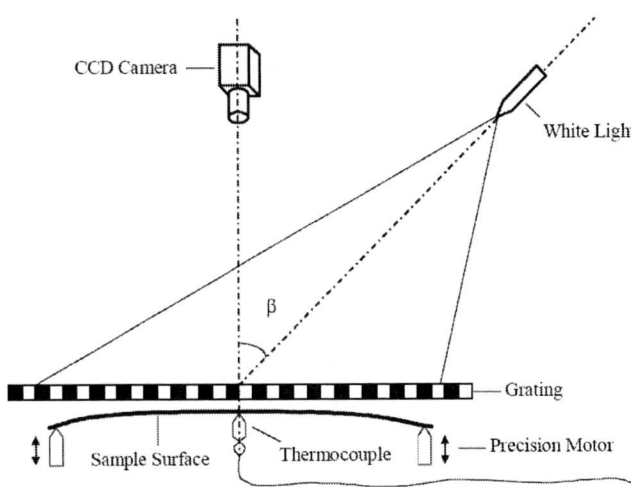

Fig.7 Shadow Moiré Optical Setup

The CCD camera can capture the moiré images and send them to a computer to calculate out-of-plane displacement with respect to the reference grating. Precision motors connected to the sample holder can move the surface up and down and give a high resolution for phase stepping. Thermocouples are used to measure the temperatures of different locations on the sample in real-time thermal processes. The resolution of the shadow moiré technique is equal to the pitch of the glass grating lines when $\beta = 45°$. The phase stepping technique can be used to increase the resolution by about 100 times [9].

In this part of work, a commercial thermal moiré equipment utilizing shadow moiré technique, TherMoiré PS200 (Figure 8), is used to measure the fpBGA samples warpage after molding to room temperature.

Samples and Experiments

Six 12 x 12 mm fpBGA samples were fabricated. After cooling down from mold temperature (175°C) to room temperature these samples warped in the form of convex and the warpage values are recorded.

Then a temperature increase was applied to these samples to simulate a real reflow process. The warpage behaviors of fpBGA samples at different temperatures were captured and recorded by TherMoiré PS200. The experiment results were used to compare with simulation results.

Results and Discussion

A thermal loading from 175°C to 25°C was prescribed to the finite element model, with the stress free state taken to be at 175°C. Perfect adhesion between constituent parts was also assumed in the modeling.

978-1-4244-4658-2/09 $25.00 © 2009 IEEE

The final warpage simulation result is shown in Figure 9. The package has a convex warpage when cooling down to room temperature and four corners have the maximum warpage values of 43.7μm.

The out-of-plane warpage of fpBGA at room temperature was measured and the 3D surface plot drawn by TherMoiré PS200 is shown in Figure 10 and Figure 11. The measured maximum warpage values are listed in Table 2.

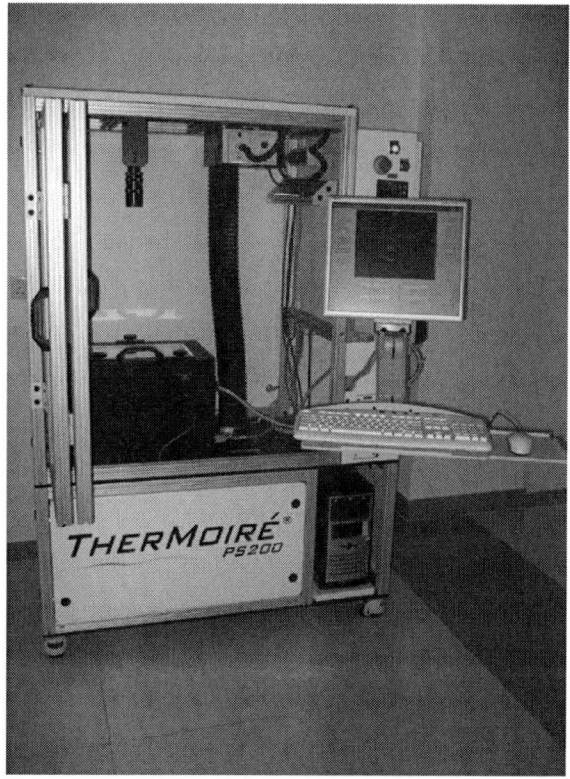

Fig. 8 TherMoiré PS200

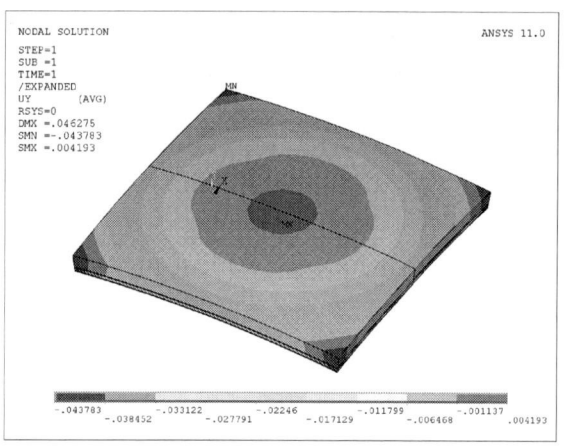

Fig.9 Warpage Plot in ANSYS

Table 2 Maximum Warpage Values at Room Temperature

Sample Number	Warpage value (μm)
#1	41
#2	35
#3	47
#4	58
#5	36
#6	30
Average	41
Standard Deviation	10

The simulation result is derived previously; the value is 43.7μm, with less than 7% error compared to the average value of measurement results.

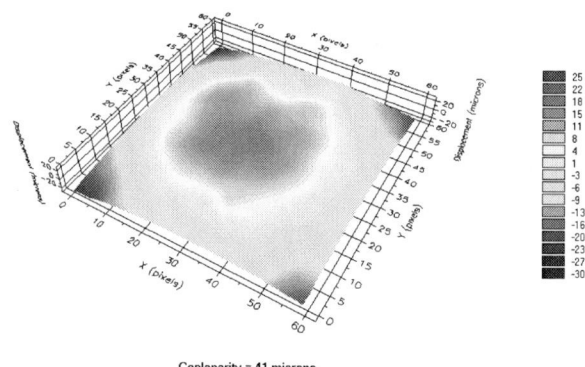

Coplanarity = 41 microns

Fig. 10 3D Surface Plot of fpBGA Sample

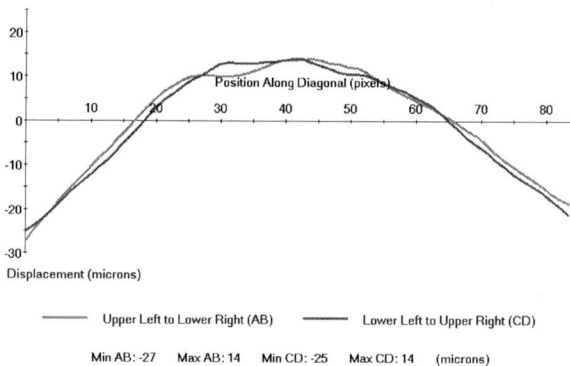

——— Upper Left to Lower Right (AB)   ——— Lower Left to Upper Right (CD)

Min AB: -27   Max AB: 14   Min CD: -25   Max CD: 14   (microns)

Fig. 11 Out-of-Plane Displacement along Diagonal Directions

To make a clear comparison between results coming finite element analysis and experiment work, the reference planes of modeling and measurement are set unified. The results are plotted in same coordinates. The two curves matched very well, show in Figure 12.

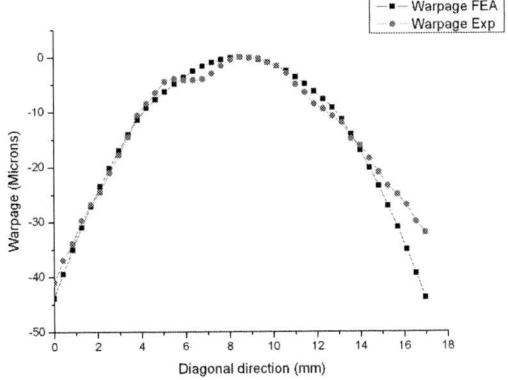

Fig.12 Comparison between FEA and Experiment Results

From the above comparison it can be said the elaborate construed finite element model could be used for the following computational.

The warpage behaviors of fpBGA during reflow was also measured and recorded, as shown in Figure 13 and Figure 14.

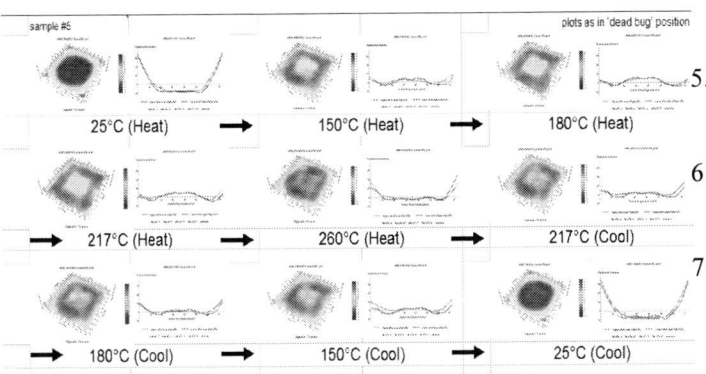

Fig.13 3D Warpage of fpBGA during Reflow Process

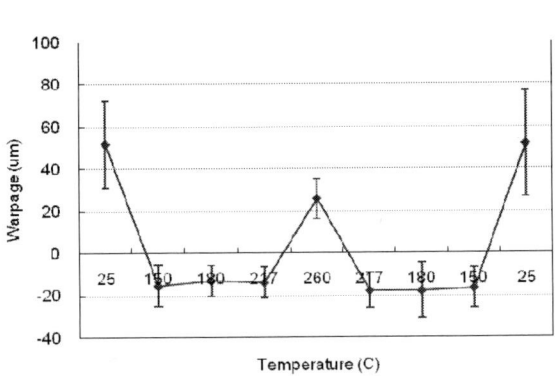

Fig.14 Maximum Warpage Values of fpBGA during Reflow Process

Conclusions

This paper focuses on the prediction and verification of thermally induced deformations during packaging and testing processes. The predicted FE deformations are compared with those obtained from shadow moiré measurements performed.

The results presented by this paper show that the shadow moiré technique can be used for out-of-plane deformation measurements of IC packages, with quantitatively reliable accuracy and high efficiency. Thermal deformations obtained from the non-linear FEM models match well with measured deformations for the real electronic packages. which provides feedbacks for modeling optimization. So the effects of material properties and geometric parameters on thermal warpage could be further studied using the optimized models.

References

1. Lau, J. H., Ball Grid Array Technology, McGraw Hill Publications, 1995.
2. Kelly, G. et al, "Importance of Molding Compound Chemical Shrinkage in the Stress and Warpage Analysis of PQFP's," *IEEE Trans-CPMT-E*, Vol. 19, No, 2, pp. 296-300, 1996.
3. Mirman, I., "Effects of Peeling Stresses in Bimaterial Assembly," *ASME J. of Electronic Packaging*, Vol. 1 13, pp. 43 1-433, 1991.
4. Miremadi, J., "Impact of PBGA-Ball-Coplanarity on Formation of Solder Joints," *Proc. 45th Electronic Components and Technology Conf,* Las Vegas, NV, pp. 1039- 1050, 1995.
5. Implementation of Ball Grid Array and Other High Density Technology, "*Joint Industry Standard, J-STD-013 by the Surface Mount Council*, pp. 32, July, 1996.
6. JEITA, "Measurement Methods of Package Warpage at Elevated Temperature and the Maximum Permissible Warpage", JEDITA 2003.
7. W.D. van Driel et al, "Prediction and Verification of Process Induced Warpage of Electronic Packages", *Journal of Microelectronics & Reliability*, 43 (5), 2003, pp. 765 - 774.
8. Tee, T.Y., Sivakumar, K., and L. Haurent, "Warpage Analysis and Viscoelastic Modeling of Block BGA," *Proc. Pacific Rim/ASME International Electronic Packaging Technical Conference and Exhibition*, 505-511, 2001.
9. Ding, H., Powell, R.E., Hanna, C.R., and I.C. Ume, "Warpage Measurement Comparison Using Shadow Moiré and Projection Moiré Methods", *IEEE Transactions on Components and Packaging Technologies*, 25(4), pp. 714-721, 2002.

978-1-4244-4658-2/09 $25.00 © 2009 IEEE   245

Prediction of Die Failure in Copper-Low-K Flip Chip Package with Consideration of Packaging Process-Induced Stresses

Mingjun Zhao, D. Yang, Ligang Niu

School of Mechanical & Electrical Engineering, Guilin University of Electronic Technology

Guilin, 541004, Guangxi, China

Email: zmj31415@163.com

Abstract

In Flip Chip package, the curing process of the underfill polymer will induce extra residual stress and strain fields. For simplicity reasons, in thermo-mechanical analyses, the curing induced stress state was usually neglected by assuming a so-called "stress-free" temperature. However, such simplification is not verified, in particular for advanced IC chips such as copper-low-k interconnects, which is very sensitive to the stress level it undergoes.

An investigation on the die failure issues in copper-low-k Flip Chip Package with consideration of packaging process-induced stresses was presented in this paper. Firstly, a cure-dependent viscoelastic model was applied to describe the properties of the underfill resin during the curing process and subsequent thermal cycling. Secondly, prediction of die fracture failure probability was conducted. Weibull statistics model was used to describe the probability distribution for the die strength test. Model parameters were obtained by fitting to the test results. Fracture failure probability of the die backside was calculated based on the Weibull statistics model and the stress states induced in the curing processes and test condition. Thirdly, the stress state on the copper-low-k layer was investigated. The results show that maximum stress occurs at top interface of the low-k layer structure. The cure-induced stresses play a significant role on the total stress level. The effect of the packaging process-induced stress cannot be simply neglected.

1. Introduction

The trend of electronic products is moving toward further miniaturization, better performance, high reliability and low cost. New packaging technologies have being developed and improved to achieve such a goal. Flip chip technology is a packaging technology that the IC chip is mounted on a substrate with the chip's active surface (active area) facing to the substrate. Die cracking is a major reliability concerns for high density flip chips due to mismatch thermal expansion between the silicon die and the substrate. Underfilling the flip chip can help to constrain the CTE mismatch locally and to couple the die and substrate mechanically, and thus significantly alleviates the stress on solder joints and thus extends the fatigue life of the flip chip. However, the curing process of the underfill polymer will induce residual stress and strain fields, which significantly impact packaging reliability. Meanwhile, with shrinking of CMOS device, cu/low-k interconnection has been adopted in the IC process. The Al/oxide process is replaced by cu damascene structures with oxide and low-k dielectrics. However, the low-k materials have lower moduli and poorer adhesion compared to other dielectric materials. The low-k

films generally have moduli less than 10 GPa, compared with 70 GPa for SiO_2. The process limitations and reliability issues of introducing low-k material to devices are of great concerns [1]. Therefore, it is vital to predict the die failure in copper-low-k flip chip package with consideration of packaging process-induced stresses.

So far, many works have reported concerns on the mechanical integrity of cu/low-k chip and its susceptibility to die failure in reliability test. Interfacial adhesion for copper/SiLK interconnects under flip-chip packaging conditions was measured and compared with the crack driving force derived from 2-D FEA models [2]. The porosity effect on film strength and adhesion of organosiloxane-based polymer films has been investigated. An attempt to understand the feasibility of packaging of porous ultra low- materials was reported. A characterization method for brittle low-k films has been reported and a finite element model has been developed to evaluate the effect of pore aggregation on elastic modules of thin films [3].

In order to investigate exactly the die failure in copper-low-k flip chip package with consideration of curing process of the underfill polymer, this paper firstly apply a cure-dependent viscoelastic model to describe the properties of the underfill resin during the curing process and subsequent thermal cycling. Finite element modeling was carried out to study the cure-induced stress fields and predict the die stress at the die backside and the interface stress at die bottom side (active side). Moreover, Weibull statistics model was employed to depict the probability distribution for the die strength test. Model parameters were obtained by four-point bending test. The failure probability of the die backside was calculated based on the Weibull statistics model. In the end, the stress state on the copper-low-k layer was investigated. The results show that maximum interface stress occurs at top interface of the Low-k stack layer.

2. Characterization of cure-dependent properties of underfill resin

The underfill resin for the Flip Chip package is a low viscosity, epoxy-based material and is filled with fused silica spheres. A previously developed cure-dependent viscoelastic model [4] is applied here to describe the material behaviour of this underfill resin during cure and the subsequent cooling and thermal cycling.

Assuming isotropy, the cure-dependent relaxation modulus functions can be decomposed into two independent parts, governed by the shear relaxation moduli and bulk relaxation moduli, respectively. The stress-strain relation is expressed as:

$$\sigma_{ij}(t) = \int_{-\infty}^{t} \left[2G[\alpha(\xi),T(\xi),(t-\xi)] \cdot \frac{d\varepsilon_{ij}^{d}}{d\xi} + K[\alpha(\xi),T(\xi),(t-\xi)] \cdot \frac{d\varepsilon_{v}^{eff}}{d\xi} \right] d\xi \quad (1)$$

where ε_{ij}^{d} represents the deviatoric strains, and ε_{v}^{eff} is the effective volumetric strain. Taking the shear relaxation modulus as an example, it can be considered as a sum of cure-dependent plateau (rubbery) and the transient part, described as:

$$G[\alpha,T,(t-\xi)] = G_{r}(\alpha,T) + [G_{g} - G_{r}(\alpha,T)] \cdot \sum_{n=1}^{N} g_{n} \cdot \exp\left[-\frac{t-\xi}{\tau_{n}(\alpha,T)} \right] \quad (2)$$

in which G_{g} and G_{r} are the glassy value and the rubbery modulus, respectively. $g_{n}(\alpha)$ are the weighting factors, with $\sum_{n=1}^{N} g_{n}(\alpha) = 1$.

DSC analysis is applied measure the cure kinetics parameters. The fitting parameters are obtained for the cure kinetics model.

Multi-frequency DMA is used to characterize the viscoelastic properties of the resin. Here only the experimental results are presented. Fig. 1 shows the shear storage moduli as functions of applied frequencies measured from DMA under different temperatures. Fig. 2 shows the storage modulus versus reduced frequency after being shifted along the frequency axis with reference temperature of 130°C. It can be seen that the frequency-temperature superposition principle is applicable to this resin.

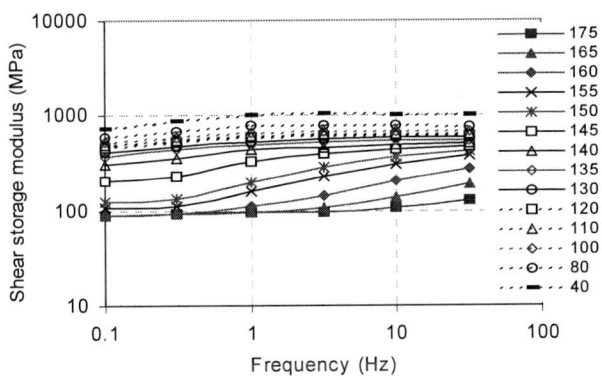

Fig. 1 Storage shear moduli as functions of frequencies of fully cured resin under different temperatures

Based on the DMA measurements, the model parameters can be acquired. The shear rubbery modulus is determined using low frequency DMA measurement at 180°C well above the T_{g}). The glassy modulus is determined as the glassy plateau of the storage modulus. Both are listed in Table 1. For the transient part, fourteen Prony terms are used. Table 2 presents the relaxation times and the corresponding best-fit normalized Prony coefficients.

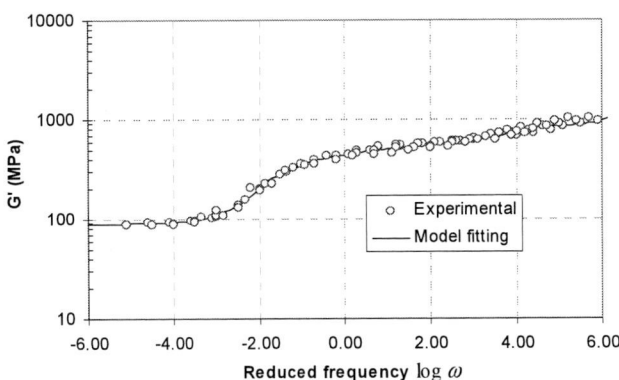

Fig. 2 Shifted storage shear moduli versus reduced frequencies and the master curve from the best-fitting, reference temperature =130°C.

Fig. 3 shows the shift factors that account for the temperature effect. It also shows different dependency trends at lower and higher temperatures. The two parts of the curve can be well fitted by using the WLF equation and Arrhenius equation, respectively. The two equations are rewritten below:

$$\log \alpha_{T} = -\frac{C_{1}(T - T_{ref})}{C_{2} + T - T_{ref}}, T \geq T_{s} \quad (3)$$

and

$$\log \alpha_{T} = \frac{\Delta H_{T}}{2.303R}(\frac{1}{T} - \frac{1}{T_{0}}), T < T_{s} \quad (4)$$

where C_{1} and C_{2} are constants, ΔH_{T} is the activation energy, T_{ref} and T_{0} are the reference temperature for WLF and Arrhenius equation, respectively; T_{s} is the switching temperature. The fitting values for these parameters are also listed in table 1.

Table 1: The parameters for viscoelastic model

G_{g} [MPa]	G_{rf} [MPa]	C_{1} [-]	C_{2} [K]	T_{ref} [K]	ΔH_{T} [KJ/mol]	T_{0} [K]
1081.6	83.8	10.6	44.4	406.4	102.9	427.7

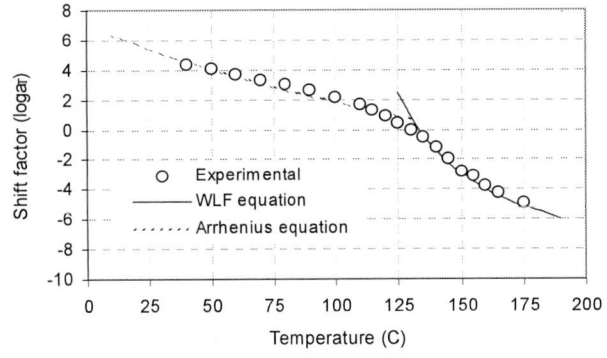

Fig. 3 Shift factor as a function of temperature

Table 2: Relaxation times and the normalized Prony coefficients

No.	Relaxation times τ_n [s]	Prony coefficients g_n
1	1.00E-08	0.220
2	1.00E-06	0.213
3	1.00E-05	0.0575
4	1.00E-04	0.0561
5	1.00E-03	0.0596
6	1.00E-02	0.0687
7	1.00E-01	0.126
8	1	0.157
9	10	0.0331
10	100	0.00310
11	1000	0.00087
12	10000	0.00309
13	1.00E+05	0.00252
14	1.00E+06	0.00032

3. Die strength and Weibull statistics model

From mechanical point of view, Silicon is a brittle material. As a consequence, its response to overstress is to fracture. Experimental measurement of the die strength is needed to conduct the simulation-based prediction of the die failure.

The strength σ_i was determined from four-point bending tests were done to determine the die strength. The test setup is schematically shown in Fig. 4. The following equation can be used to calculate the maximum stress "σ_i" that represents chip strength. ,

$$\sigma_i = 3FL/4\omega t^2 \tag{5}$$

where L is the support distance, F is the force, ω is the chip width, and t is the chip thickness.

By cumulative summation of the probability data P_i, the cumulative distribution of strength can be constructed. If it is normalized with respect to the total number of specimens in the batch, N, the resulting plot presents a measure for the probability of failure, P_i, where i refers to the specimen of strength σ_i.

$$P_i = \frac{i-3}{N+4}, \quad y_i = \ln\ln\frac{1}{1-P_i}, \quad x_i = \ln\sigma_i \tag{6}$$

It is known that for brittle materials $P_i(\sigma_i)$ can be fitted reasonably well using a two-parameter Weibull distribution:

$$P_i(\sigma_i) = 1 - \exp[-(\frac{\sigma_i}{\sigma_0})^m] \tag{7}$$

Variation of strength within a particular batch of specimens is characterized by a Weibull modulus or "m-value." σ_0 is known as the "scale parameter" and is roughly equal to the strength at the 63rd percent of probability of failure [5].

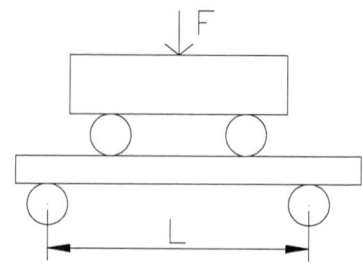

Fig. 4   A schematic diagram of four-point bending test

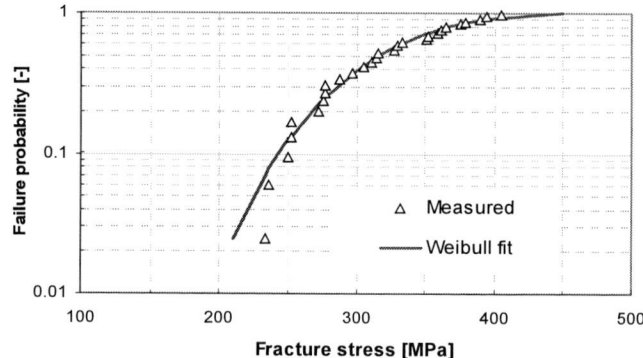

Fig. 5 Distribution of strength data for the chip

The resulting strength data was processed using the maximum likelihood procedure, and σ_0 and m are obtained, and here σ_0 =342.17MPa, m=6.009. Distribution of strength data for the chip was shown in Fig. 5. $P_i(\sigma_i)$ can be got if σ_i has been known after the upper two parameters were introduced to equation (7).

4. FE modeling of the Copper-Low-k Flip Chip package

4.1 Geometry and FEM mesh

A Finite Element thermo-mechanical modelling of a copper-low-k Flip Chip was carried out. In the package, 0.3mm thick silicon die, 4.25µm thick low-k, 15µm thick polyimide, 95µm thick underfill, 60µm high solder bumps, 35µm thick copper pads and 0.6mm thick substrate were considered.

2-D plane strain model was used. Because of the symmetry, only one half of the structure was simulated. The FEM meshes were presented in Fig. 6. Following boundary conditions were used: the nodes along the symmetry axis were fixed in x direction (u=0), and the node at the left bottom corner was fixed in x and y direction (u=v=0).

4.2 Thermal loading

The package was subjected to a thermal loading including the curing process and thermal cycles. First, the underfill was dispensed and cured for 60 min (a–b), then the package was cooled down to room temperature at a ramp rate of 10°C/min and then being hold for 30 min and subsequently three temperature cycles were applied. Fig. 7 shows the temperature profile with three thermal cycles used in the simulation, the high temperature is 125°C and the low temperature is -55°C in this process.

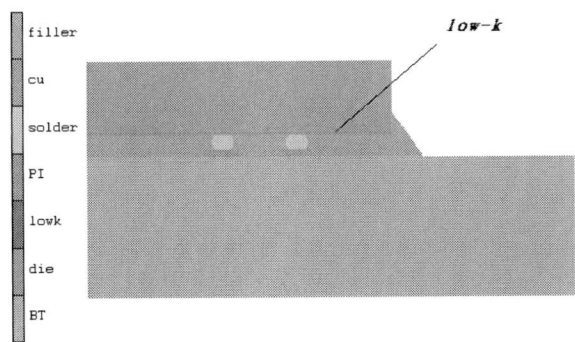

Fig. 6 A schematic illustration of the Flip Chip package

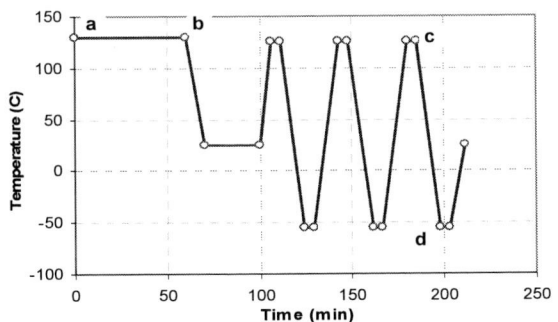

Fig. 7 Temperature profile

4.3 Material properties

The constitutive relation for curing resin as described before was implemented into the FEM code and applied here for the simulation of the curing of the epoxy underfill. The FR-4 substrate was considered to be orthotropic, other materials were assumed to be isotropic. The material properties used in the simulation are shown in Table 3.

Table 3 Material properties

Materials		E (GPa)	v	CTE (ppm/C)	G (GPa)
Silicon die		169	0.26	2.3	--
Copper pad		82.7	0.30	16.7	--
PI		3.5	0.3	23	--
Low-k		7.76	0.35	3.5	--
FR-4	In plane	20	0.11	13	17.6
	Out of plane	8.78	0.39	57	54.2

5. Results and discussion

5.1 Prediction of curing-induced die stress

Fig. 8 shows the horizontal stress component σ_{xx} on the die at the end of curing $t=60$min. It is found that the maximum value of σ_{xx} is 31.32MPa. The left bottom corner of the die for the stress is lager than other parts of stress. In other respects, Fig. 9 shows the distribution of the die stress at the die backside, and the maximum die backside stress is

2.68MPa at the distance to center 0.95mm. Fig. 10 shows the distribution of the die stress at the die bottom side (active side), the maximum stress is 19.74MPa at the lower left corner of the die, and the right side stress become gradually smaller. Based on the information after the FEM simulation, it is found that the curing induced residual stress is large at the die backside and the die bottom side (active side). Therefore, the residual stress should not be neglected and should be considered in the thermo-mechanical analysis.

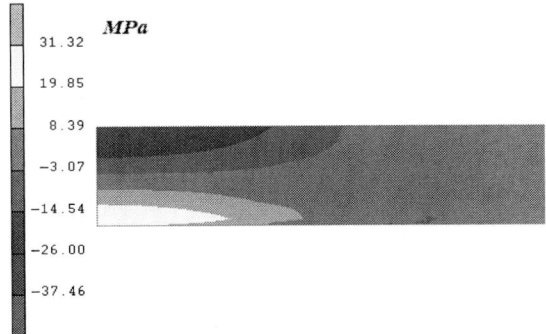

Fig. 8 Stress σ_{xx} at the end of curing (point B)

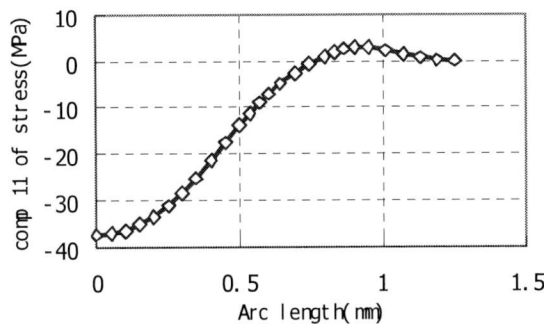

Fig. 9 The distribution of stress σ_{xx} at the die backside

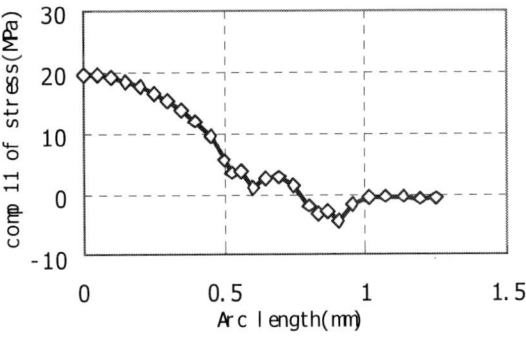

Fig. 10 The distribution of stress σ_{xx} at the die active side

5.2 Prediction of die failure

Fig. 11 shows the state of the stress at the low temperature of the third thermal cycle (see Fig. 7 point D), the maximum value of the horizontal stress is 81.10MPa at the die backside. It is significant to predict the crucial location that the die vertical crack appears. The die fracture failure probably was calculated by two-parameter Weibull statics model (see equation 7), and the result of the die failure probability is 0.0175%.

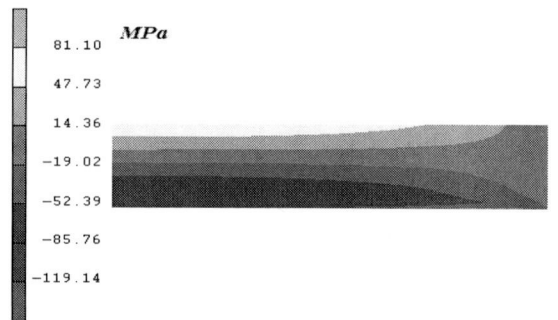

Fig. 11 Stress σ_{xx} at the low temperature of the third thermal cycle (point D)

5.3 Low-k layer stress analysis

Fig. 12 shows the maximum equivalent stress at the low temperature of the third thermal cycle (see Fig. 7 point D), it was known that the maximum equivalent stress was found at top interface of the low-k layer. That is because mismatch thermal expansion coefficient of low-k and the die, in addition, the low-k materials have lower moduli. So it is prone to crack at the upper of the low-k layer.

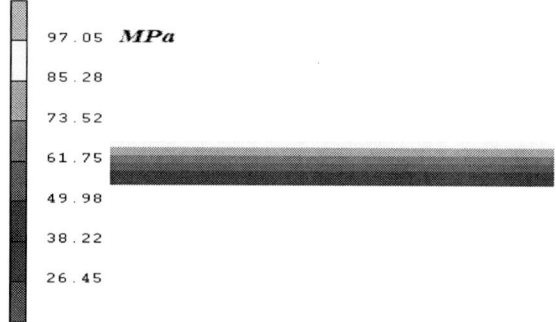

Fig. 12 The maximum equivalent stress at the low temperature of the third thermal cycle (point D)

6. Conclusions

The objective of this study is to predict the die failure in copper-low-k flip chip package with consideration of packaging process-induced stresses. Cured-induced stress was firstly studied at the end of curing. The maximum horizontal stress is separately 2.68MPa and 19.74MPa at the backside and the bottom side of the die. Moreover, the die failure probability was conducted by two-parameter Weibull statics model and 0.0175% die failure probability was attained at the low temperature of the third thermal cycle. Thirdly, the maximum equivalent stress of low-k layer was received, and the maximum stress locations were found at top interface of low-k layer and the stress value is 97.05MPa. Based on the paper's analysis, it is found that cure-induced stress should not be neglected and the effect on low-k is obvious from curing process and subsequent thermal cycling.

Acknowledgments

The research work in this paper is financially supported by the National Natural Science Foundation of China (NSFC) (Grant No. 60666002).

References

1. Mercado, L. L., Kuo, S.-M., et al, "Impact of Flip-Chip Packaging on Copper/Low-k Structures," *IEEE Transactions on Advanced Packaging*, VOL. 26, NO. 4(2003), pp. 433-440.
2. Ho, P. S. and Miller, M. R., "Interfacial Adhesion Study for Cu/SiLK Interconnects in Flip-Chip Packages," *Proc 51th Electronic Components and Technology Conf*, 2001, pp. 965-970.
3. Liang, Q. Z., Moon, K.-S., et al, "Low Stress and High Thermal Conductive Underfill for Cu/Low-k Application," Proc 58th *Electronic Components and Technology Conf*, 2008, pp. 1958-1962.
4. Yang, D.G., "Cure-Dependent Viscoelastic Behavior of Electronic Packaging Polymers: Modeling, Characterization, Implementation and Applications", Delft University of Technology, Delft, the Netherlands, 2007, ISBN 978-90-9022-180-9.
5. Bohm, C., Hauck, T., et al, "Probability of Silicon Fracture in Molded Packages," *Proc. EuroSimE 2004*, Brussels, May. 2004, pp. 75-82.

Simulation of Fluid Flow and Heat Transfer in Microchannel Cooling for LTCC Electronic Packages

J.Zhang[1,2], Y.F.Zhang[3], M.Miao[2,4], Y.F.Jin[2], S.L.Bai[3], J.Q. Chen[3]

[1]SHENZHEN Graduate School Of PeKing University

[2]National Key Laboratory on Micro/Nano Fabrication Technology, Peking University, Beijing, CHINA

[3]Department of Advanced Materials and Nanotechnology, Peking University, Beijing, CHINA

[4]Information Microsystem Research Institute, Beijing Science and Information Technology Institute, Beijing, CHINA

E-mail: jinyf@ime.pku.edu.cn; Tel.: +86-10-62752536; Fax: +86-10-62751789

Abstract

The hydrodynamic and thermal characteristics of microchannel networks are investigated by finite element analysis with commercial software Fluent. The simulation model is based on fabricated thick film LTCC substrate with 3D cooling microchannels. A comparison of the cooling performance among fractal-shaped microchannel, parallel microchannel, serpentine microchannel, spiral microchannel is also conducted numerically based on the same heat flux and the same mass flow rate. It is found that fractal-shaped microchannel facilitates the lowest fluid pressure drop and the most uniform temperature distribution over the substrate, and that the spiral microchannel network enables the smallest temperature rise.

Introduction

Low temperature co-fired ceramic (LTCC) is a relatively new technical option for high-end microelectronics packaging, Multi-Chip-Model (MCM) and System-in-Package (SiP) applications, where the heat dissipation and high frequency property are serious concerns. LTCC technology is now well established for avionics, automotive, computer, communications and military, thanks to its advantages of high conductivity, low dielectric constant, high quality factor (Q value), low sintering temperature, low surface roughness, high layer count, and small temperature coefficient of expansion, as well as the possibility of making three dimensional (3D) microstructures. Holger Neubert developed a way of manufacturing LTCC integrated leaf springs with an uniaxial piezoresistive accelerometer as an example. Thick-film accelerometers made of LTCC promise a higher operational temperature range and lower costs in small-volume production [1].

The development of the very large-scale integration (VLSI) technology has increased the circuit density and operating speed of microelectronic devices. Next generation of microprocessors and microelectronic components will have to dissipate heat flux in excess of 1000 W/cm2. New cooling techniques must be developed to dissipate such a high heat flux [3, 4]. Recently, many studies have been devoted to new technologies capable for high-efficiency cooling of integrated microsystems. LTCC technology enables development and manufacture of various microsystem applications, especially microfluidic systems, such as flow sensors, micropumps, microvalves, micromixers, microreactors, and polymerase chain reaction (PCR) devices [5-10]. Thelemann et.al have proved that the integrated microchannel cooling system in LTCC substrate can decrease the additional temperature more than 80% [11]. The two-layer structure can cool far more efficiently than single-layer microchannel with lower pressure drop and smaller temperature rise [12]. It is also found that the performance of the modified fractal-shaped cooling microchannel network can be greatly improved by local modifications of the channel size to eliminate hotspots on the bottom wall [3].

This paper presents numerical simulations based on fabricated thick film LTCC substrate with 3D cooling microchannels. Several cooling conditions are investigated. A comparison of the cooling performance among fractal-shaped microchannel, parallel microchannel, serpentine microchannel, spiral microchannel is also conducted.

Experimental and Simulation Procedure

The material used is a commercially available green tape DuPont 951 AT with a layer height of 114±8 m. The samples were laminated with 15 layers and sintered at 850 ℃ in vacuum. Six kinds of microchannel networks are supplied, including parallel, serpentine, spiral and another three fractal-shaped microchannel networks, as shown in Figure 1. The microchannel network was fabricated in the 4th and 5th layers of substrates with a cross section of 200 m × 200 m. A thick film resistor of 2.0 cm x 2.0 cm as the heating source was offered in the centre of the substrate surface. In this study, the fluid flow (water, and a temperatur 300 K) was controlled by a high resolution injection pump, while the temperature was measured by an infrared thermometer.

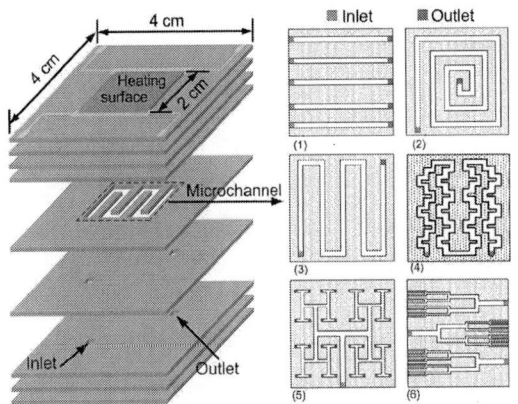

Figure 1. Scheme of LTCC substrate with microchannel

In this study, Finite volume method (FVM) was used to calculate the temperature, fluid pressure and flow velocity fields with commercial software Fluent. Mixture of mesh was selected and refined on the sidewalls of the channels, including tetrahedral, hexahedron, pyramid-shaped and wedge-shaped meshes. Convergence criterion for mass, velocity and energy are 10^{-4}, 10^{-4}, 10^{-7}. Thermal conditions of the solid-fluid conjugate wall is coupled.

Results and Discussion

Fig. 2 shows the temperature distribution at the center of the front wall of the heat sink, embedded with the fractal-shaped microchannel. For analysis microchannel network pattern is also projected to the surface. It can be seen that the temperature distribution on the area near the first branch is much lower. This kind of temperature distribution results from the fact that the coolant temperature is lower and the cooling capability is larger at the first branch. The channel has one inlet port and three outlet ports, shown in Fig. 3.

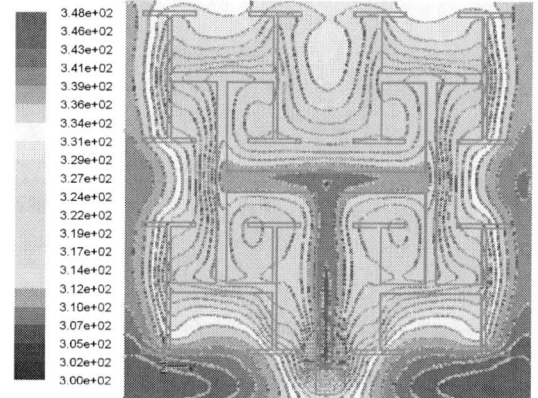

Figure 2 Temperature distribution at the center of the heated surface for fractal-shaped microchannel and microchannel network pattern is also projected to the surface

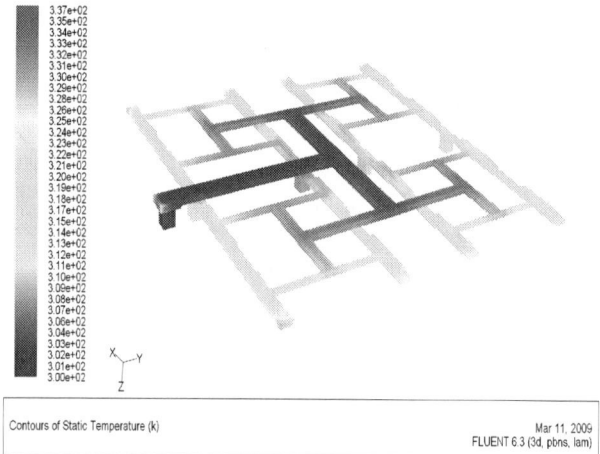

Figure 3 Temperature field of liquid in fractal-shaped microchannel network

The effects of microchannel network, mass flow rate at the inlet (V_{inlet}) and heat flux by the resistor are discussed according to additional temperature ($\triangle T$), pressure drop ($\triangle P$) and maximum flow velocity (V_{max}), as presented in Table 1. The temperature fields of parallel, spiral, and fractal-shaped microchannel networks based on the same heat flux (heat flux=2W/cm^2) and the same mass flow rate (V_{inlet}=0.108 kg/s) are shown in Figures 4-6. It is observed that the fractal-shaped microchannel network has the lowest pressure drop (4.68 KPa) and the more uniform temperature distribution, indicating that the fractal-shaped microchannel structure is the best channel for the fluid flow distribution. Controlling the pressure drop is important for the relatively small pressure provided by a micro-pump and electronic cooling application. The spiral microchannel network has the smallest additional temperature (21 K) due to the reasons that the sum of the length of spiral microchannel is the longest and the heat transfer area was increased by lengthening the channel. It can be seen that the temperature distribution on the areas near the outlet of the spiral microchannel network is much higher where hotspots of the heat sink are formed. Considering the thermal expansion coefficients between materials are different, temperature non-uniformity can produce potentially destructive thermal stresses and reliability concerns to the packages and devices.

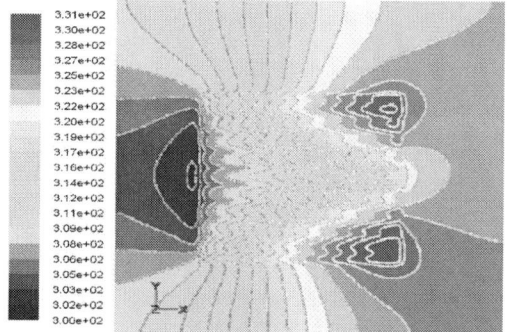

Figure 4 Temperature distribution at the heated surface of LTCC substrate with straight microchannel network with the mass flow rate of 0.108kg/s and the heat flux of 2w/cm^2

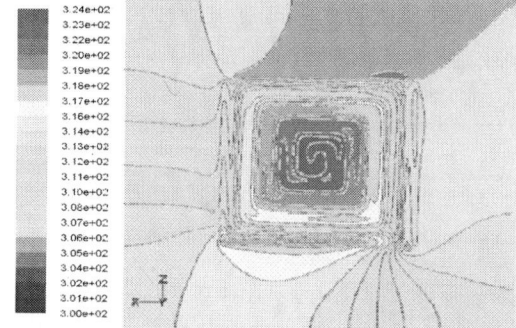

Figure 5 Temperature distribution at the heated surface of LTCC substrate with spiral microchannel network with the mass flow rate of 0.108kg/s and the heat flux of 2 w/cm^2

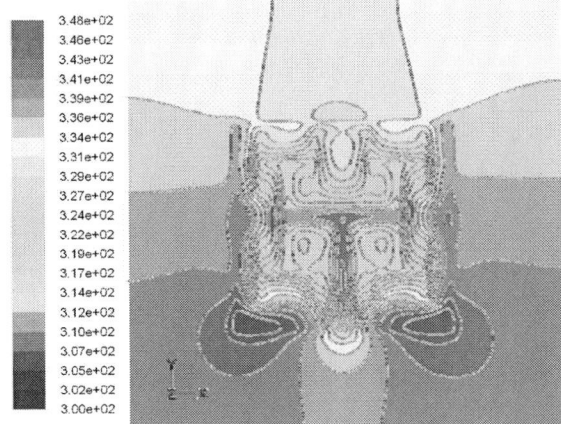

Figure 6 Temperature distribution at the heated surface of LTCC substrate with fractal-shaped microchannel network with the mass flow rate of 0.108kg/s and the heat flux of 2 w/cm^2

Table 1. Simulation results for straight, spiral and fractal-shaped microchannel networks

Sample	V_{inlet} (Kg/s)	Heat Flux (W/cm^2)	ΔT (K)	ΔP (KPa)	V_{max} (m/s)
fractal-shaped	0.18	4	73.53	8.99	1.30
fractal-shaped	0.18	3	55.14	8.99	1.30
fractal-shaped	0.18	2	36.76	8.99	1.30
fractal-shaped	0.108	2	48.30	4.68	0.81
fractal-shaped	0.036	2	101.03	1.25	0.29
straight	0.108	2	31.22	11.31	1.24
spiral	0.108	2	23.92	84.49	4.76

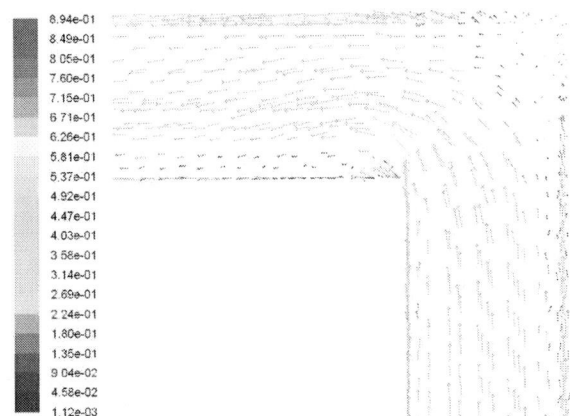

Figure 7 Flow velocity field at the first right-angle of the spiral microchannel

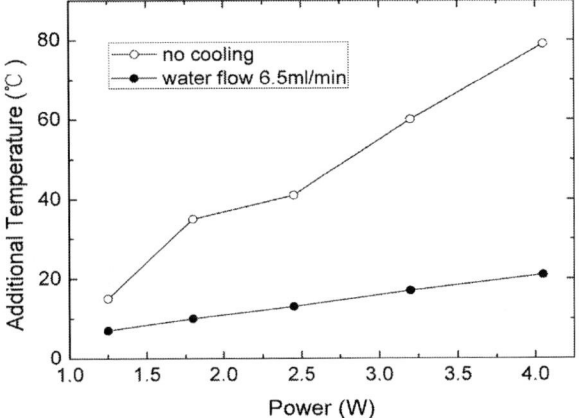

Figure 8 Temperature rise under cooling and no cooling condition with heat flux on the bottom wall of the spiral microchannel networks

It is also observed in Table 1 that the larger flow velocity enhances the cooling capability and requires a higher power pump. When the mass flow rate is larger than 0.18 Kg/s, turbulent flow appears and the simulation should be revised. Increasing the mass flow rate of the coolant can reduce the temperature rise and require higher pressure drop. The maximum temperature grows fast with the heat flux from 2 to 4 w/cm^2 and the water may be boiling. For higher level of heat flux, water should be replaced by other fluid such as metal liquid. The small flow velocity of the blue color location at the right angle of the microchannel (Figure 7) restrict the convective heat dissipation and suggests using arc angle instead and as far as possible with straight channels.

In this study, The experiment is being conducted and the additional temperature under cooling and no cooling condition with heat flux on the bottom wall of the spiral microchannel networks are presented in Fig. 8. It can be seen that the experiments prove the simulated anticipation. We can modify our work and design advanced liquid cooled LTCC substrates for high power applications in future.

Conclusions

In this paper, a 3D numerical simulation was conducted to study the hydrodynamic and thermal characteristics of LTCC microchannel networks. Based on the results presented above, conclusion as follows can be reached:

The spiral microchannel network has the smallest additional temperature but hotspots appear on the areas near the outlet.

Increasing the mass flow rate of the coolant can reduce the temperature rise and require higher pressure drop.

Straight channels restrict the convective heat dissipation.

Due to the structural limitation of the microchannel networks, the fractal-shaped microchannel has the lowest pressure drop but highest additional temperature. This kind of microchannel needs some modification on fabrication process.

Acknowledgments

The authors appreciate CETC No. 43 Research Institute,(Hefei, China) which kindly provided the specimens.

References

1. Holger Neubert, U. P. *et al*, "Thick Film Accelerometers in LTCC-Technology –Design Optimization, Fabrication, and Characterization," Actuator, (2008), PP.1-6.

2. X. Q. Wang, A. S. Mujumdar, C. Yap, "Characteristics of Tree-Shaped Microchannel Nets for Cooling of A Rectangular Heat Sink," Int. J. Therm. Sci., Vol. 45, (2006), pp.1103-1112.

3. F. J. Hong, P. Cheng, H. Ge, G. T. Joo, "Conjugate Heat Transfer in Fractal-shaped Microchannel Network Heat Sink for Integrated Microelectronic Cooling Application," Int. J. Heat Mass Transf., Vol. 50,(2007), pp. 4986-4998.

4. J. Darabi, M.M.O, "An electrohydrodynamic polarization micropump for electronic cooling", J.Microelectromech. Syst.(10) (2001), pp. 98–106.

5. Karol Malecha, Leszek J. Golonk,"Microchannel Fabrication Process in LTCC Ceramics", Faculty of Microsystem Electronics and Photonics, Microelectronics Reliability 48 (2008) 866–871.

6. J. Zhong, M. Yi and H.H. Bau, "Magneto hydrodynamic (MHD) pump fabricated with ceramic tapes", Sensors Actuators A 96 (2002), pp. 59–66.

7. M.R. Gongora-Rubio, L.M. Sola-Laguna, P.J. Moffett and J.J. Santiago-Aviles, "The utilization of low temperature co-fired ceramics (LTCC–ML) technology for meso-scale EMS", a simple thermistor flow sensor, Sensors Actuators A 73 (1999), pp. 215–221.

8. M. Yi and H.H. Bau, "The kinematics of bend-induced in micro-conduits", Int J Heat Fluid Flow 24 (2003), pp. 645–656

9. K. Malecha, D. Pijanowska, L.J. Golonka and W. Torbicz, "LTCC enzymatic microreactor", J Microelectron Electron Packaging 4 (2) (2007), pp. 51–56.

10. D. Sadler, R. Changrani, P. Roberts, C. Chou and F. Zenhausern, "Thermal management of BioMEMS: temperature control for ceramic-based PCR and DNA detection devices", IEEE Trans Compon Packaging Technol 26 (2003), pp. 309–316

11. T. Thelemann, H. Thust, G. Bischoff, T. Kirchner, "Liquid Cooled LTCC-Substrates for High Power Applications", Int. J. Microcircuits Electron. Packag., Vol. 23, (2000), pp. 209-214.

12. K. Vafai, L. Zhu, Int. J. Heat Mass Transfer 42 (1999) 2287–2297

13. Marko Hrovae, Darko Belavic, HanaUrsic, Jaroslaw Kita, Janez Holcl,Silvo Drnovsek lena Ci, "An Investigation of Thick-film Materials for Temperature and Pressure Sensors on Self-constrained LTCC Substrates", Electronics System-Integration Technology Conference (ESTC), Greenwich, Sept. 2008.2nd, pp.339-346.

14. K. A. Peterson, K. D. Patel, C. K. Ho, S. B. Rohde, C. D. Nordquist, C. A. Walker, B. D. Wroblewski, M. Okandan; "Novel Microsystem Applications with New Techniques in Low-Temperature Co-Fired Ceramics", International Journal of Applied Ceramic Technology, 2005, pp.19:345-363.

Research of Methodologies to Enlarge the Isolation in 3D Interconnection

Liwei Zhao [1], Long Yang [2], Xin Sun [1], Yufeng Jin[1*]

[1]National Key Laboratory of Micro/Nano Fabrication Technology, Peking University, Beijing, 100871, P. R. China
[2]Ansoft, 22F/04C Office Tower, Bund Center, 222 Yan An Road (East) Shanghai, 200002
*Tel: +86-10-62752536-22; Email: jinyf@ime.pku.edu.cn

Abstract

As the speed of digital systems continues to increase, digital design has entered a new realm that requires high integration and high complexity with limited dimension. Different from the parameter analysis in low frequency which mainly focuses on resistances and their effects, the coactions of resistances, inductances and capacitances are all needed to be taken into account in high frequency. This paper uses finite element method (FEM) to analyze the coupling and isolation of three types of structures, and presents a measure for the purpose of improving the performance of TSV.

Introduction

High-speed design will never omit the impact from interconnection, such physical IC chip, package, PCB, connector, cable, say a few. As for IC interconnection, there are two kinds of wires which are being considered into application, one is the lines extending along one single layer of IC wafers and represented by straight lines in the model; the another TSV, which is represented by vertical cylinders. So there can be three combinations in the scenario consisting of two layers of interconnects: a) double straight lines, b) one straight line and one TSV, c) double TSVs. In this paper, the research is developed by modeling these three forms of interconnections.

Simulation and Results

The basic models to be analyzed first are shown as Fig-1. The colored lines by amaranth are copper traces. The thick semitransparent gray cuboids are silicon wafers and the thin ones which lay between silicon wafers are air. There is a thick layer of air covers the highest silicon layer and this air layer is hidden. Fig. 2 shows the simulated S-parameters of the three models. NEXT (Near End Cross Talk) which equals to S13 and FEXT (Far End Cross Talk) which equals to S14 are the objects of major interests in this paper because they illustrate the electrical isolation of two traces in each combination. In the analyzed models, the setting of ports which supplying powers is imitating actual electric current. Port 1 is defined as the source and port 2 is the sink of the upper trace respectively, while port 3 is the source and port 4 is the sink of the lower line.

From simulation shown in Fig-2, we can draw a conclusion that while the NEXT are almost the same and the FEXT of double TSVs has a largest value which means the isolation of double TSVs is the best. The reason is that application of 3D-TSV increases the effective distance between signal wires, consequently gain improved isolation. This result shows that resistance is a main influencing factor to FEXT and NEXT because that longer distance of ports can brings larger resistance. [1]

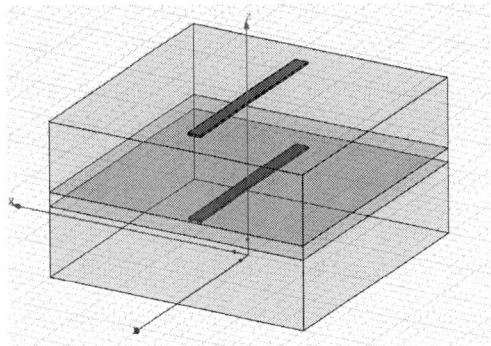

(a) double straight lines

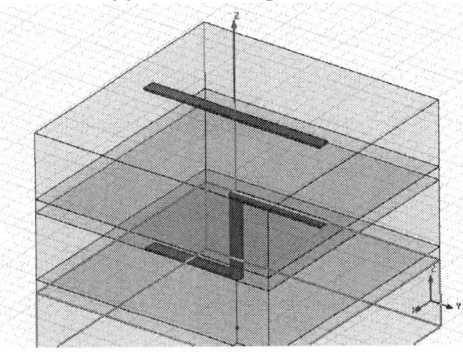

(b) one straight line and one TSV

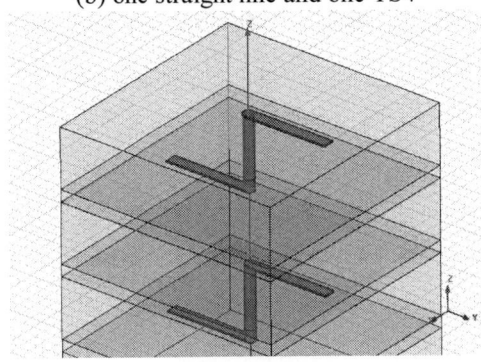

(c) Double TSVs

Figure-1 three Models (Ansoft/Q3D)

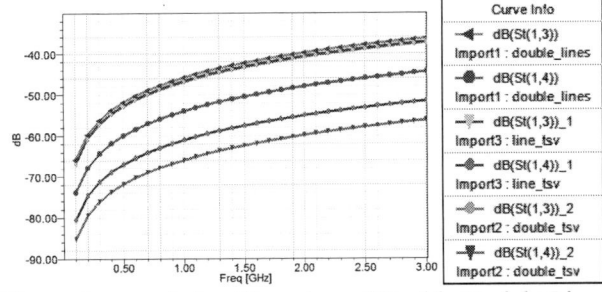

Figure-2 transmission Characters of Double Straight Lines, one straight line and one TSV, and Double TSVs

1. Adding isolation layers

978-1-4244-4658-2/09 $25.00 © 2009 IEEE   255

There are some proposed methods to improve the isolation of traditional condition. Ansoft/Q3D is used to verify the validity of these methods. The first one is adding layers with larger resistivity than silicon between wafers just like polyamide or SiO2 layers. The latter can be achieved by using SOI wafers. The transmission characters of models for comparison are shown in Fig-3, 4 and 5.

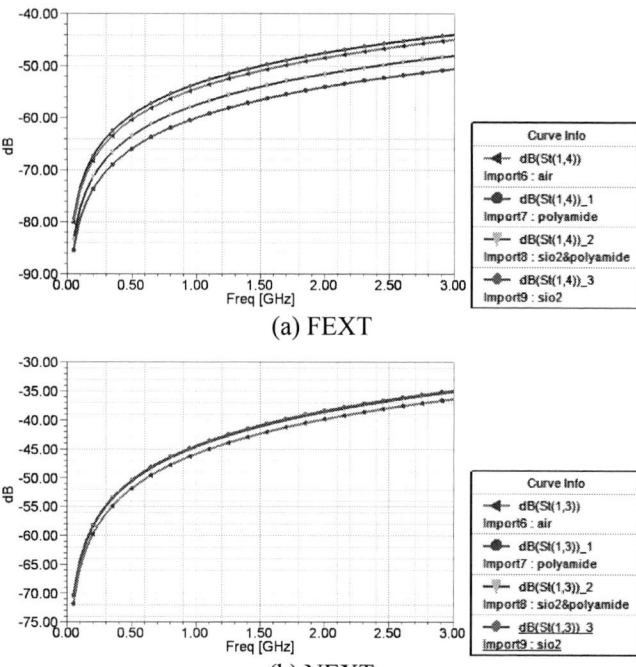

(a) FEXT

(b) NEXT

Figure-3 transmission Characters of Interconnection with Double Straight Lines: Without Isolating Layer, Insertion of Polymer Layer, Insertion of SiO₂ Layer, and both Insertion

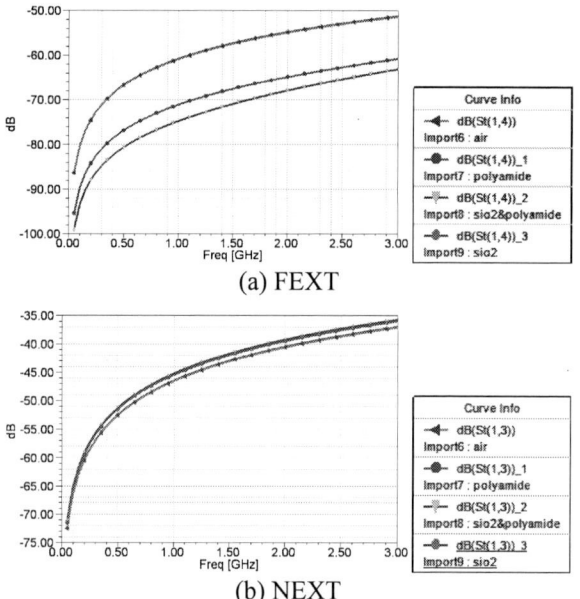

(a) FEXT

(b) NEXT

Figure-4 transmission characters of interconnection with one line and one TSV: basic model without isolating layers, isolating layer of polymer, isolating layer of SiO₂, and adding both

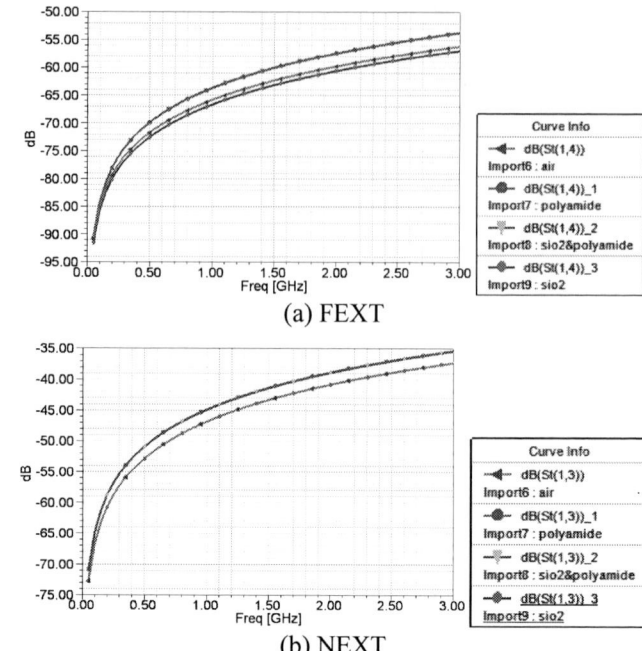

(a) FEXT

(b) NEXT

Figure-5 transmission Characters of Interconnection with Double Straight Lines: Without Isolating Layer, Insertion of Polymer Layer, Insertion of SiO₂ Layer, and both Insertion

From the complex changes shown in Fig-3, 4 and 5, it is concluded that crosstalk doesn't equal to traditional isolating characters in low frequency. In DC and low frequency, only resistance plays the main role in the mutual impact between two nets while at high frequency, mutual resistance, mutual capacitance and inductor impact the major impact between nets. Even the resistance is enlarged by adding polyamide or SiO₂, crosstalk doesn't induce as expected and disturbed by the admixture of mutual resistance, mutual capacitance and inductance. [2] When there is TSVs through layers and hence changing the electromagnetic field around SiO₂ or polyamide, the effect is bigger than that of only straight lines. The field of capacitance of one straight line and one TSV is shown in Fig-6. Not only the amount of capacitances but also the distributions are changed.

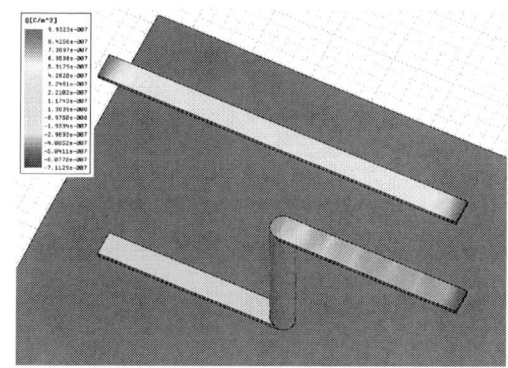

(a) air layer between wafers

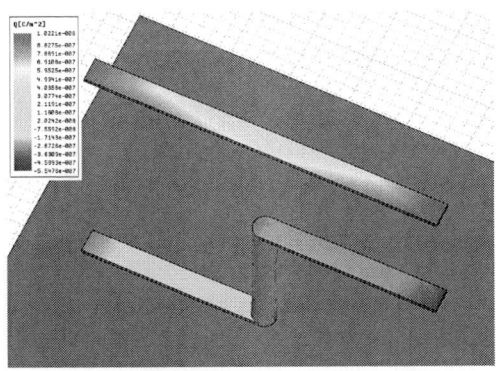

(b) polyamide layers between wafers

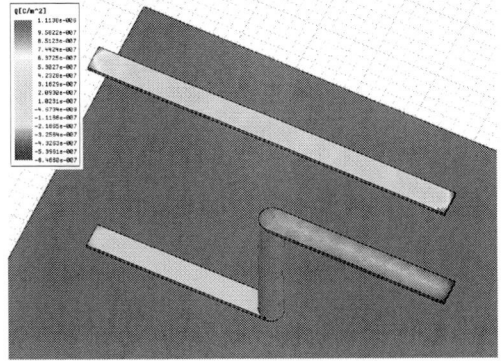

(c) SiO2 layer between wafers
Figure-6 distributed capacitance along traces.

FEXT and NEXT present incongruous currents which are affected by different transmission models based on relative positions. For example, the simplified distributed parameter model shown in Fig-1(a) can be concluding as circuits shown in Fig-7. Rd stands for distributed resistance, Rp for parasitic resistance, and Cp for parasitic capacitance. Rn designates to the resistance of port n. As shown below, the connection between port 1 and port 4 is cascade while the one between port 1 and port 3 is parallel. Different topology will impact on the FEXT and NEXT differently.

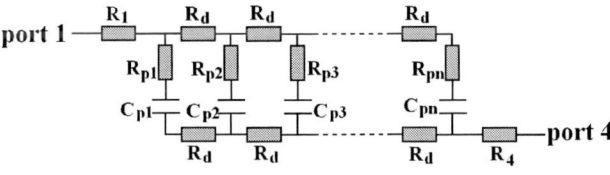

(a) cascade connection of FEXT

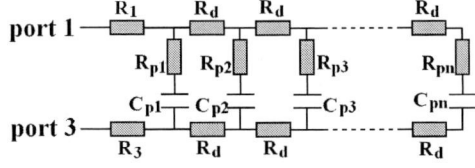

(b) Parallel connection of NEXT
Figure-7 predigest circuits of connections

So in 3D connection, adding isolating layers is not an effective way to reduce crosstalk in high frequency especially in the condition when the effect of parasitic capacitance is

larger enough to counteract the outcome of bigger resistivity. While adding interlayers of isolation is necessary in some cases of 3D connection, there is current flow that enhances the thickness of organics, bringing lower crosstalk and the curves are shown in Fig-8. Adding materials with small dielectric constant which can reduce the parasitic capacitance and large resistivity can be another solution.

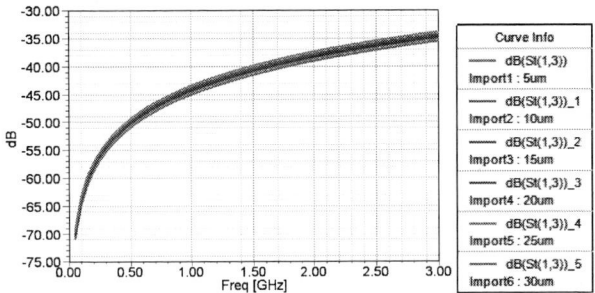

Figure-8 change of NEXT with different thickness of isolation layers

2. Changing relative position

The second method is changing the relative position of signal lines at different layers. For example, when the configuration of straight lines is changed to orthogonal from parallel shown as Fig-1 (a), the common area between the two signal lines is decreased, and the crosstalk (NEXT and FEXT) can be improved. When we make the same changes to the models in Fig-1 (b) and (c), the influence of TSV gives a neutral result. The transmission characters are shown and compared in Fig- 9. So we can get the conclusion once more that if there is a TSV, the distance between two ports plays an important role in crosstalk. Setting the traces on contiguous layers into orthogonal is already a universal regulation in designing PCB and it's also can be used in 3D connection with wafers because a balanceable crosstalk can be realized.

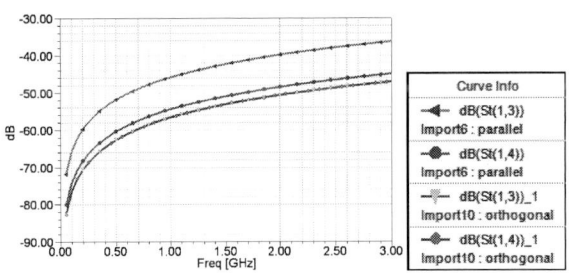

(a) NEXT and FEXT of straight lines and their orthogonal condition

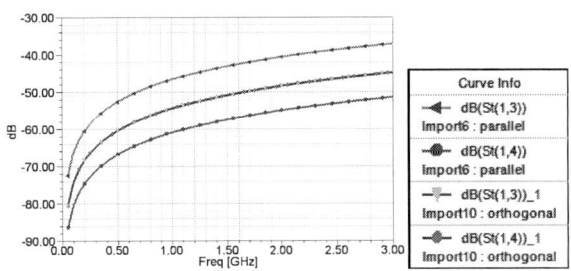

(b) NEXT and FEXT of combination of line and TSV and their orthogonal condition

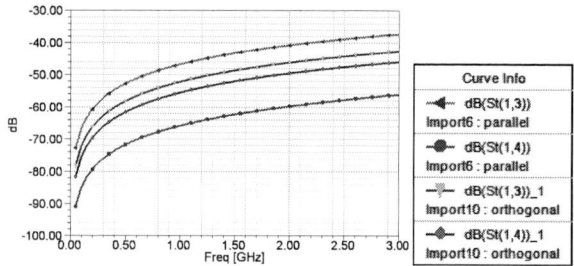

(c) NEXT and FEXT of TSVs and their orthogonal
condition

Figure-9 transmission Characters

3. Restricting the minimum thickness of wafers

There is a sweep of the model in Fig-1 (a) with different thickness of silicon wafer and the transmission characters are shown in Fig-10. We can get the conclusion that the isolation of 3D interconnection is improved when the thickness of wafer becomes larger. On one hand, bigger thickness of wafers can bring larger resistance and reduce crosstalk in high frequency. On the other one, thick wafer can enhance the punch through voltage in DC condition.

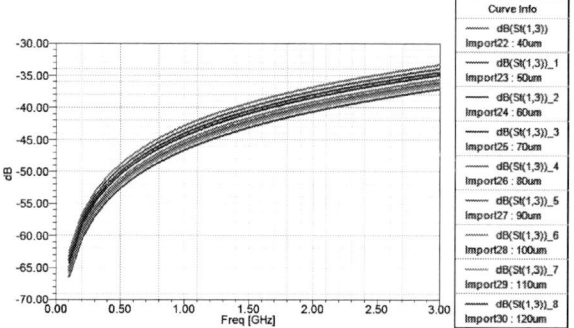

(a) NEXT of models with different thickness of wafer

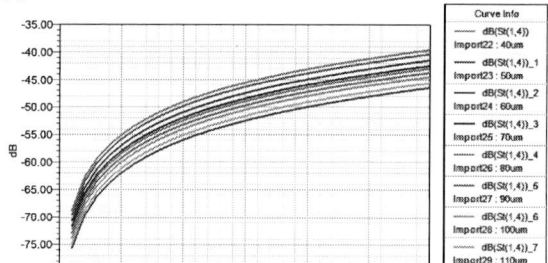

(b) FEXT of models with different thickness of wafer

Figure-10 Transmission Characters of Different Thickness
of Silicon Wafer

4. Reducing parasitic inductance

The effect of parasitic inductance is forming low-pass filters with parasitic capacitance and resistance. This filter can reduce the transmission of higher frequency and decrease the character of S12 and S34. Parasitic inductance in 3D connection is mainly distributed along the copper traces. The main solutions to reduce the parasitic inductance are enhancing the degree of homogeneity, adopting the optimal path of traces and ensure reliable electrical contact.

Conclusions

In transmission of high frequency, we can reduce the crosstalk by designing materials and structures of 3D TSV to change the parasitic parameters. Firstly, we can enhance the resistance between ports by adding isolation layers of high resistivity and lengthen the distance between ports of different traces by increasing the thickness of wafer or adopting orthogonal arrangements of traces. Second, we should reduce parasitic capacitance. When adding isolation layers, the characters of materials should be compromised by resistance and parasitic capacitance. The third parameter is parasitic inductance and can be reduced by enhance characters of copper traces.

Acknowledgments

The authors would like to express gratitude to Professor Min Miao for technique support in analyzing and improving design.

References

[1] S. J. Ok, C. Kim, and D. F. Baldwin, "High Density, High Aspect Ratio Through-Wafer Electrical Interconnect Vias for MEMS Packaging," *IEEE transactions on advanced packing*, vol. 26, no. 3, pp. 302-309, Aug. 2003.

[2] L. L. W. Leung, K. J. Chen, "Microwave Characterization and Modeling of High Aspect Ratio Through-Wafer Interconnect Vias in Silicon Substrates," *IEEE transactions on microwave theory and techniques*, vol. 53, no. 8, pp. 2472-2480, Aug. 2005.

[3] N.Ranganathan et al, "Through-Wafer Interconnection by Deep Damascene Process for MEMS and 3D Wafer Level Packaging," *Electronics packaging technology conference*, pp. 238-242, 2005.

[4] Jia-Sheng Hong, M. J. Lancaster, "Microstrip Filters for RF Microwave Applications," *John Wiley & Sons*, Inc., pp. 1-234, 2001.

Study on EBG Structure Combined with Decoupling Capacitor for Suppressing Ground Bounce Noise

Wenlong Zhu[1,2], Wenxiang Chen[1], Lei Li[2]

[1]Dept. of Electronic Engineering, Xiamen University, Xiamen, 361005, China

[2]Shenzhen institute of advance technology, Chinese Academy of Sciences/The Chinese University of Hong Kong, Shenzhen, 518067, China

Email: wl.zhu@sub.siat.ac.cn or zwlforever@163.com

Abstract

Ground bounce noise (GBN) is becoming one of the major challenges for designing high speed system and packages. It causes serious signal integrity (SI) problems in the high speed system. Traditional techniques center on adding more decoupling capacitors to create a low impedance path. Electromagnetic band gap (EBG) structure is used to suppress GBN combined with decoupling capacitors in this study. Different values and positions of capacitors are added to the design with EBG structure. The results show that the EBG structure functions like a band stop filter and prevents unwanted electromagnetic energy traveling through power/ground pair in stop band of the structure. The noise thus can't propagate within the power/ground pair and excite resonance. The combination of meander line bridged EBG structure and decoupling capacitors is able to achieve good GBN suppression performance.

1. Introduction

With the trends of fast edge rate, high clock frequencies and low voltage levels for high-speed digital systems, GBN or simultaneous switching noise (SSN) on the power/ground plane is becoming one of the major challenges for designing high-speed circuits and packages. The resonant modes among the power and ground plane pair excited by the GBN causes serious SI or PI problems for the system [1] [2]. A decoupling capacitor acts as a reservoir of charge, which is released when the power supply voltage at a particular current load drops below some tolerable level. Traditional techniques center on adding more decoupling capacitors to create a low impedance path at necessary frequency band. But design with decoupling capacitors can't provide elimination of GBN up to a few hundreds MHz due to the unavoidable lead inductance.

Recently, a concept of using EBG structure to suppress GBN between power/ground plane pair in high-speed printed circuit boards and packages was introduced. An EBG structure exhibits electromagnetic properties that have led to a wide range of applications to filter and antenna [3] [4] [5]. It functions like a band stop filter and prevents electromagnetic energy from traveling through the plane pair. The noise thus can't propagate within the plane pair and excite resonance. As a result, the GBN is suppressed and the characteristic impedance of the plane pair is lowered. The design with EBG structure is able to achieve high GBN suppression performance for frequency up to a few GHz ranges [6].But it's low frequency performance is lower.

In section 2, the principal of GBN suppression with EBG structure is discussed. Then the equivalent circuit of an EBG structure cell and the characteristics of meander line bridged EBG structure are studied. The performance of a power delivery system with meander line bridged EBG structure is compared with the design with decoupling capacitors only. The excellent GBN suppression performance of the proposed structure is validated numerically. In section 3, different values and positions of capacitors are added to the design with EBG structure. Their performance and characteristic are shown. Some conclusions are drawn in the last section. It is found that the power/ground plane pair designed with meander line bridged EBG structure combined with decoupling capacitors is able to achieve high GBN suppression performance for wider frequency range than designed with decoupling capacitors only.

2. Design of meander line bridged EBG structure for GBN suppression

In recent years, some exceptional schemes to suppress GBN in high speed system were achieved by using EBG structures to provide a high impedance surface on the power/ground planes by via, which were increased the cost [7]. Many pattern forms of EBG structures had been proposed. But they need an additional layer or extra array of via holes and then increase the manufacturing cost relative to power or ground planes with a periodic pattern only. Therefore, the plane with EBG structures is researched and it comprises two-layer power/ground planes. One layer consists of EBG structures and the other is a continuous metal plane.

The equivalent circuit of an EBG structure cell on power plane is shown in Fig.1. All inductors and capacitors except labeled L and C are the inductive or capacitive coupling factors between power plane and ground plane. L is derived from the inductive factor of meander line and C is derived from the gap coupling between two adjacent EBG cells [8]. The L and C values have a great impact on the stop bandwidth. While maintaining the same resonant frequency, if C is decreased, or L is increased, the stop band bandwidth is wider and performance improves. Therefore, to achieve a wider noise stop bandwidth, the inductive factor of two adjacent EBG cells should be larger, and the gap-coupling capacitance should be smaller.

978-1-4244-4658-2/09 $25.00 © 2009 IEEE

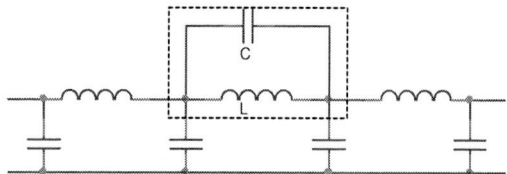

Fig.1. Equivalent circuit of an EBG structure cell

In this study, a meander line is used to connect two adjacent EBG cells which are shown in Fig .2 (a). The EBG structures have rectangular cells. These structures are characterized by periodic square metallic patches connected to each other by four meander lines. The critical parameters of the EBG patches are illustrated in Fig.2 (b).These include the patch size W, line width E and folded length F. The geometrical features of the patches directly relate to the GBN suppression performance of the structure. These structures have an inherent stop band where noise cannot propagate due to parallel inductance-capacitance resonance.

It's difficult to draw a closed form formulas for describing property of the whole periodic EBG structure using lumped elements model when structure is small enough comparing to wave length. The Equivalent circuit of an EBG structure cell which acts like a band stop filter is shown in Fig.1. There is no exact or even reasonably accurate closed formula that relates the geometrical parameters of an EBG structure cell to the frequency response of stop band filter. In this study, the effective band width of the EBG structure is generated directly using numerical tool.

The relative dielectric constant and thickness of the test board's substrate FR4 is 4.4 and 0.1 mm respectively. The dimension of entire test board is 47mm x 47mm. There are 9 EBG structure cells and three ports in the board as illustrated in Fig.2 (a). The dimensions W=15.8mm, E=3.8mm and F=4.2mm. All S-parameter measurements in this study are performed with a system reference impedance of 50 Ω. The simulation results for the reference board without EBG structures and with EBG structures are shown in Fig.3. From the S-parameters in the figure, we know that the performance of GBN suppression of the board with EBG structures is more effective than without EBG structures. The figure Illustrates maximum difference of 30dB in coupling reduction could be gained in the entire suppression band .These differences are noticeable in GBN suppression. But in low frequency range, the performance is not so high.

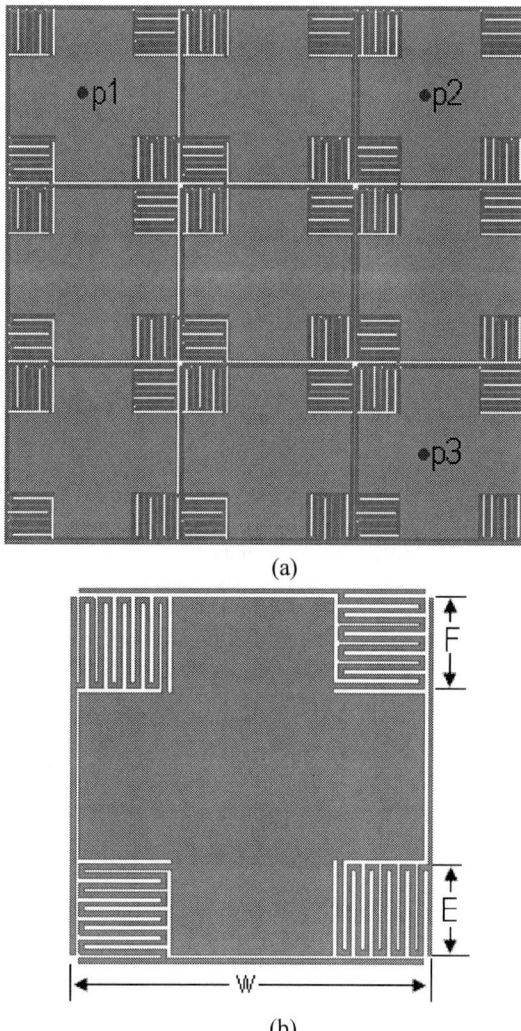

Fig.2. Meander line bridged EBG structure (a) Top view of the EBG structure (b) One cell of the EBG structure

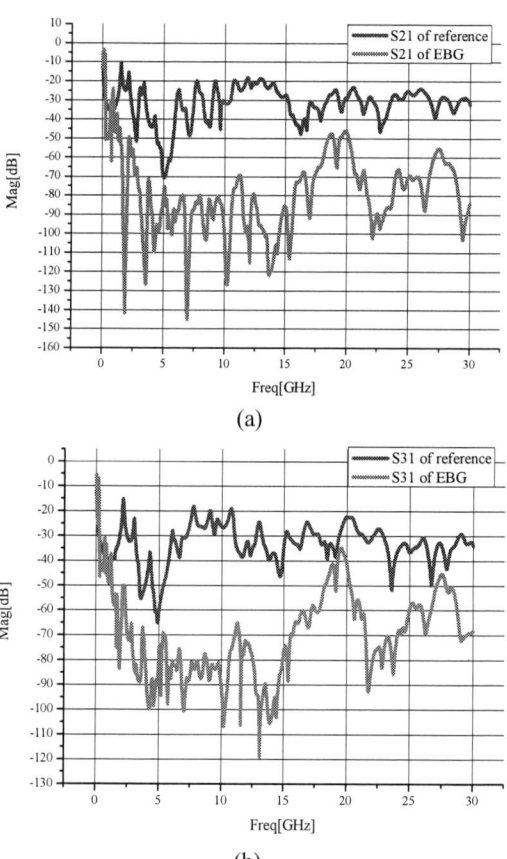

Fig.3. GBN suppression performance of meander line bridged EBG structure and reference board

3. EBG structure combined with decoupling capacitor

In the low frequency range, GBN or resonant modes excited by GBN can be suppressed by adding decoupling capacitors between power/ground plane pair. Fig.4 shows the GBN suppression performance of the reference board without decoupling capacitors and board with four decoupling capacitors near the three ports. The decoupling capacitor's value is 47pF, ESL is 10nH and ESR is 0.1Ω. Fig.4 shows that though the decoupling capacitors are applied to lower the PDN impedance and then to reduce the GBN voltage effectively. However, this technique has a limitation of frequency bandwidth by the self-resonance effect associated with the ESL of a capacitor. Typical self-resonance frequency usually is less than a few hundred megahertz and the GBN above several GHz becomes server in most modern digital boards. Different values of decoupling capacitors only affect the performance of low frequency range. The high frequency performance is only dependent on the values of ESL which is difficult to reduce due to capacitor process. The EBG structure's low frequency performance is degraded below GHz as shown in Fig.3. We combined those two techniques to improve the whole performance.

In order to study the performance of combination of those two techniques, two schemes are simulated by numerical tools. Four capacitors are placed uniformly near the three ports between the power/ground plane pair in the first case. Two decoupling capacitors are placed near the boundary of each EBG cell in the second case. The decoupling capacitor's value is 47pF, ESL is 10nH and ESR is 0.1Ω. Fig.5 shows the different impact of decoupling capacitor locations on GBN suppression performance. The second case has higher performance than the first case in the low frequency, but in some high frequency band, the first case has higher performance than the second case. Both have higher performance than reference board without EBG structure or with decoupling capacitors. The performance in the low frequency is dependent on ESL of decoupling capacitor and its location. By lowering the ESL, the self-resonance effect shift to higher frequency which can be eliminated by EBG structure.

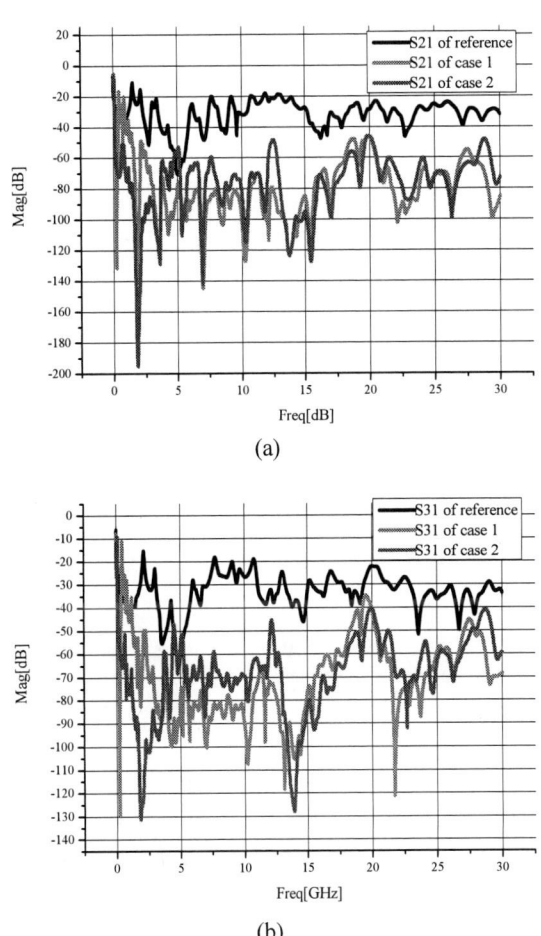

(a)

(b)

Fig.5. Different performances of different capacitor locations

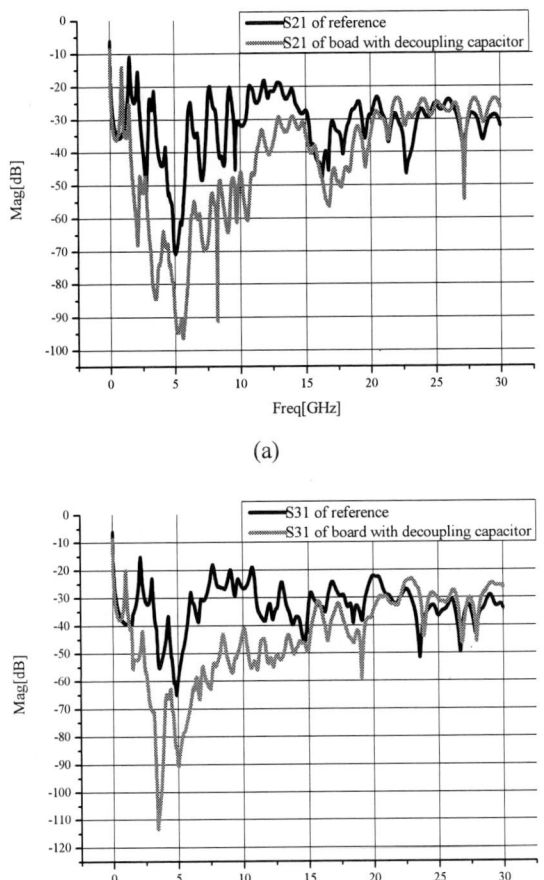

(a)

(b)

Fig.4.Reference and capacitor only board

Fig.6 shows the impact of decoupling capacitor's value on GBN suppression performance. Two decoupling capacitors are placed near the boundary of each EBG cell in both cases. The decoupling capacitor's value is 47pF in the first case, 470pF in the second case, ESL is 10nH and ESR is

0.1Ω.Diffent values of decoupling capacitors have different anti-resonant frequency which result in the peak of S parameters in low frequency range shown in Fig.6. Two cases have different performance in low frequency range but both improve the performance compared to the board with EBG structure only. Because their ESL and ESR are the same, their high frequency performances are the same as the board with EBG structure only. In order to improve the low frequency performance, the ESL and ESR should be reduced.

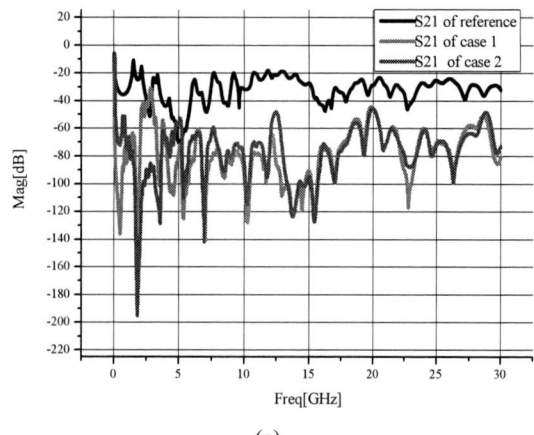

(a)

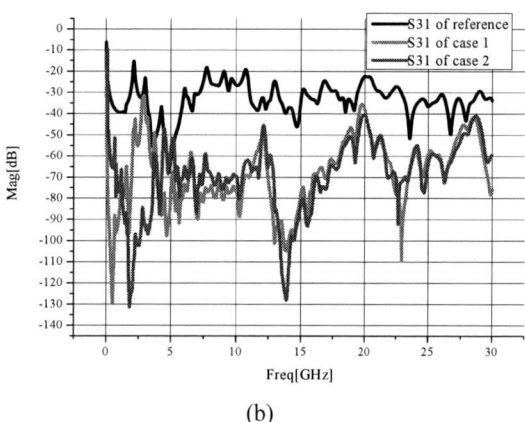

(b)

Fig.6.Different performances of different capacitor values

3. Discussion

The GBN behavior of a complete power delivery system, including package and PCB, is significantly different from that of considering only the package or PCB. Different techniques had been studied to suppress GBN and improve the performance of power delivery system. An EBG structure is designed and validated by numerical tool. The results show that the EBG structure can effectively improve the performance compare to reference board. The designed EBG cells connect to each other by meander lines which can increase the inductive factor. From the equivalent circuit of an EBG structure cell, the stop bandwidth is wider and the suppression performance improves.

Because the EBG structure has degraded performance and decoupling capacitor has good performance in low

frequency range. Those two merits are combined by adding decoupling capacitors to board with EBG structure. Different locations and different values of decoupling capacitors affect the performance of GBN suppression in low frequency. But their high frequency performance is the same as board with EBG structure only.

Conclusions

A meander line bridged EBG structure is designed for GBN suppression. The equivalent circuit of an EBG cell is used to analyze the suppression performance which is validated by numerical tool. The EBG structure has a good performance in high frequency. The decoupling capacitor shows low performance in high frequency and good performance in low frequency. Two merits are achieved by combining of these two techniques. The good performance of suppressing GBN can be achieved in wider frequency range.

Acknowledgments

This work was supported by China International Science and Technology Cooperation program (NO.2008DFA11010).

References

1. C.-S. Chang and M.-P. Houng, "Simultaneous switching noise mitigation capability with low parasitic effect using a periodic high-impedance surface structure," *Progress In Electromagnetics Research Letters*, Vol. 4 (1998), pp. 149-158.

2. Kamgaing, T. and O. M. Ramahi, "A novel power plane with integrated simultaneous switching noise mitigation capability using high impedance surface," *IEEE Microw. Wireless Compon.Lett.*, Vol. 13, No. 1, (January, 2003), pp. 21-23.

3. V. Radisic, Y. Qian, R. Coccioli, and T. Itoh, "Novel 2D photonic band gap structure for microstrip lines," *IEEE Microwave Guided Wave Lett*, Vol. 8 (1998), pp. 69-71.

4. A.R. Weily et al, "Linear array of woodpile EBG sectoral horn antennas," *IEEE Transactions on Antennas and Propagation*, Vol. 54, Issue 8 (2006), pp. 2263-2274.

5. B.-Q. Lin, J. Liang, Y.-S. Zeng, andH.-M. Zhang, "A novel compact and wide-band uni-planar EBG structure," *Progress in Electromagnetics Research C*, Vol. 1 (2008), pp.37-43.

6. Kuo-Chiang Hung1 et al, "Novel Fractal Electromagnetic Bandgap Structures to Suppress Simultaneous Switching Noise in High Speed Circuits," *PIERS Proceedings*, Cambridge, USA, July 2-6, 2008.

7. J.Park et al, "Double-stacked EBG structure for wideband suppression of simultaneous switching noise in LTCC-basd Sip application," *IEEE Microwave and Wireless Components Letters*, Vol. 16 (2006), pp.481-483.

8. Bobae Kim, Dong-Wook Kim,"Bandwidth Enhancement for SSN Suppression Using a Spiral-Shaped Power Island and a Modified EBG Structure for a λ/4 Open Stub," *ETRI Journal*, Vol. 31 (2009), pp. 201-208.

Parametric Study of Warpage in Package-on-Package Manufacturing

Chao REN, Fei QIN*

College of Mechanical Engineering and Applied Electronics Technology,
Beijing University of Technology, Beijing 100124, China
Tel: +86-10-67392173; Fax: +86-10-67391617
*Email: qfei@bjut.edu.cn

Abstract

In Package-on-Package (PoP) manufacturing, warpages on both top and bottom packages are concerned. Excess warpage causes solder joint opening, and results in the electrical connection failure of the assembled module. Many parameters of materials, geometry, and process contribute to the warpage of the package. The objective of this paper is to investigate effects of these parameters on the warpage. The dimensions of dies, package and substrate, materials of molding compound and substrates are investigated by the finite element method (FEM) in the processes of molding and reflow. After running simulation, analysis of variance is employed to identify the influential parameters. The results show that the coefficient of thermal expansion (CTE) of molding compound and substrate as well as die sizes have significant influence on the warpage. Based on this study, the influential parameters can be optimized within practical range to achieve the lowest warpage value.

Key words: package on package, warpage, finite element method

1 Introduction

Electronics industries expect to offer products that are smaller, lower cost, larger storage space, higher reliability and higher performance. Under these circumstances, miniaturization and high density 3D packaging technologies that can integrate much smaller form factor and multifunctional devices into portable products emerge and grow rapidly.

Package-on-package (PoP), which is an economical packaging solution for combining the logic and memory devices together to achieve system size reduction, is one of the major 3D packaging solutions. The top package of the PoP, which is a standard Fine-pitch Ball Grid Array (FBGA) package, contains staked high capacity memory dies. The bottom package, which in a general way is a Plastic Ball Grid Array (PBGA) package, has a high-density digital logic die in it. The top package and bottom package are integrated by solder balls. The solder balls provide clearance room for the mold cap of the bottom PBGA package [1]. Hence a PoP combines the FBGA and PBGA packages into together by the surface mount technology.

By adopting PoP packaging technology, the packaging mode can achieve a lower cost and faster turn benefits, when these two components are sourced from different IC suppliers and stacked on the printed circuit board (PCB). Prior to assembly, it allows separately testing to keep high yield [2]. Thus, a PoP has been rapidly used in portable products especially in 3G handsets due to its flexibility and testability.

However, package warpage due to mismatch of materials properties is a great challenge, especially for stacked package. Excess warpage may affect the downstream assembly processes due to the non-coplanarity of the package, and could result in the solder joint opening [3]. Hence it's important to control the PoP warpage during the assembly processes. Tzeng [4] and Sun [5-6] studied the effects of geometries and materials on the package warpage under temperature load cycling. Lai [7-8] used Taguchi method [9] to optimize thermo-mechanical reliability. However, few works were carried out to investigate evolution of the warpage during seamless manufacturing processes which reserve deformation history [10].

In this work, effects of the material properties, process conditions and geometry parameters on the warpage during the process of molding and reflow are investigated. The finite element simulation plan is that when one parameter is varied, the others are fixed to study the effects of various parameters on the package. Based on the parametric study, the identified material, geometry and process parameters can then be optimized to achieve the most reliable product.

2 Structures of Package on Package and Finite Element Models

2.1 Structures of PoP

Figure 1 is the schematic structure of a PoP [11], in which there are two dies in the top package while one die in the bottom package.

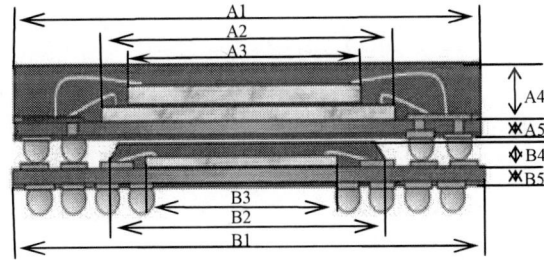

Fig.1 Schematic structure of a PoP

In the top package, A1 is size of the mold cap, and equals to the substrate size. A2 and A3 are sizes of the 1st and the 2nd dies, respectively. A4 is the thickness of the mold cap. A5 is the substrate thickness. In the bottom package, B1 is substrate size, B2 is the mold cap size, B3 is the die size, B4 is the mold cap thickness, and B5 is the substrate thickness. Detailed geometric data are listed in Table 1.

2.2 Finite element models

3D quarter symmetric finite element models of the top package and bottom package were constructed respectively.

Table 1 Geometries of all materials of the referenced PoP

	Materials	Length (mm)	Width (mm)	Thickness (mm)
Top	substrate	12	12	0.1
	1st die-attach film	8	8	0.025
	1st die	8	8	0.05
	2nd die-attach film	6	6	0.025
	2nd die	6	6	0.05
	mold cap	12	12	0.6
Bottom	substrate	12	12	0.2
	die-attach film	5	5	0.025
	die	5	5	0.05
	mold cap	7	7	0.25

In the models, the temperature loading was specially designed cooling down from 175°C to 25°C for the molding process. The materials of these models were provided with temperature dependent property and the reference temperature is 175°C. These materials were assumed to homogeneous and isotropic except the substrate.

The detailed materials properties in molding process are listed in Table2.

Table 2 Temperature-dependent material properties of molding process

Materials	Elastic Modulus (GPa)	Poisson's Ratio	CTE (ppm/°C)
substrate	17@25°C 15@ 125°C	0.25	20(xy)80(z)
molding compound	21 @ 25°C 3 @ 125°C 2 @ 174°C 0.01@175°C	0.25	17 @ 25°C 25 @ 125°C 35 @ 144°C 35 @ 175°C
die	130 @ 25°C 129@ 125°C	0.3	3 @ 25°C 3.5 @ 152°C
die-attach film	4.8 @ 25°C 0.01@100°C 0.005@200°C	0.4	245@ 25°C 300@ 100°C 300@ 125°C

After molding process, temperature ascends from 25°C to 260°C for the reflow process. In order to connect all the sequential procedure, the stress and warpage arose from the molding process are reserved to the reflow process. The temperature dependent materials properties in reflow process are listed in Table 3.

In this paper, a negative value represents concave shape while a positive value takes on convex shape, as shown in Figure 2. Figure 3 shows relationship between the warpage and the temperature. After transferred from the molding process, at 25°C, the simulated warpage pattern of the top package on the substrate side is concave and the warpage value is around -38um. By applying a reflow loading on the top package, the warpage pattern turns to be convex and becomes around 97um at the reflow peak temperature, 260°C. For the bottom package, its warpage pattern also changed through the processes. The warpage achieve the maximum positive value at 25°C while achieve the maximum negative value at 260°C. Since the package at the room temperature of 25°C is the final state of molding process, as well as the initial state of reflow process; the package at the peak temperature of 260°C endures the most severe thermal load, the two states of the package at 25°C and 260°C were intentionally investigated.

The abovementioned models were established as referenced benchmarks, and the same modeling method was adopted in the following material and geometrical optimization.

Table 3 Temperature-dependent material properties of reflow process

Materials	Elastic Modulus (GPa)	Poisson's Ratio	CTE (ppm/°C)
substrate	17@25°C 15@ 125°C	0.25	20(xy)80(z)
molding compound	21 @ 25°C 12@ 100°C 12 @ 125°C 12 @ 150°C 0.5 @ 183°C 0.5@ 200°C 0.5@ 260°C	0.25	17@ 25°C 25 @ 125°C 35 @ 144°C 35 @ 150°C 35 @ 200°C 50 @ 260°C
die	130 @ 25°C 129@ 125°C	0.3	3 @ 25°C 3.5 @ 152°C
die-attach film	4.8 @ 25°C 0.01@100°C 0.005@200°C	0.4	245@ 25°C 300@ 100°C 300@ 125°C

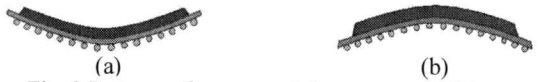

(a)   (b)

Fig. 2 Patterns of warpage: (a) concave and (b) convex

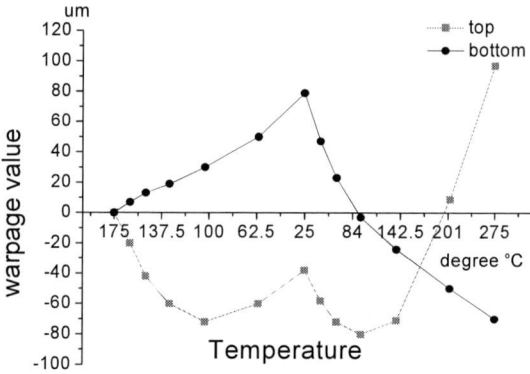

Fig. 3 Variation of package warpage with different temperatures and processes

978-1-4244-4658-2/09 $25.00 © 2009 IEEE

3 Effects of Materials

3.1 Effects of substrate materials

Firstly, the influence of CTE is focused. The CTE of X, Y orientation was changed from 23 to 11 ppm/°C, while Z orientation was from 98 to 28 ppm/°C simultaneously. Figure 4 show that a higher CTE of substrate can remarkably reduce the warpage both at 25°C and 260°C in the top package. This can be explained by that the combined CTE of the upper portion of the package (die and molding compound) approach to the lower portion of the package (substrate).

The case is different in the bottom package, where MC and dies are smaller than that in the top package, thus the substrate plays a more major role in the warpage. This suggests that a lower CTE is a suitable choice.

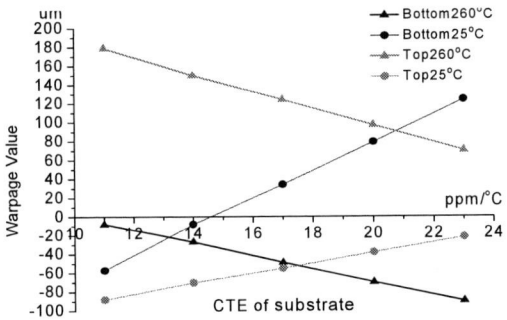

Fig. 4 Effects of the CTE of substrates

The effect of elastic modulus of the substrate was also investigated. The elastic modulus was changed from 17 GPa to 25 GPa at 25°C while from 15 GPa to 20 GPa at 125°C. The results show that the elastic modulus has insignificant influence on the warpgae, compared with the CTE, whether the top or the bottom package.

3.2 Effects of the molding compound

The material of molding compound is another crucial factor to warpage performance of the package. The CTE of MC was assigned values between 10 and 17 ppm/°C at room temperature. As shown in Figure 5, the variety of the CTE has a significant impact in the top package. Not only the warpage value but also the warpage pattern changes. The curves present distinct trends at different temperature.

Figure 5 also presents the relationship between warpage and the CTE of molding compound in bottom package that a higher molding compound CTE could reduce warpage value in some extent.

The effect of elastic modulus of the molding compound was also studied. The warpage has no obvious change when elastic modulus varies from 19GPa to 29GPa.

4 Effects of the Geometry

4.1 Effects of the mold cap thickness

Method of reducing thickness of mold cap is often used to meet the need of smaller packages. Effect of mold cap thickness on the warpage was investigated in this section. Figure 6 shows the relationship of the warpage and the thickness of mold cap in the top package. As the top package

contains two stacked dies, die thickness addition is 100um, the mold cap thickness is from 250 um to 900 um. As the ratio is less than 5, the warpage changes fast with the increasing of the ratio, but it tends to be steady when the ratio is greater than 5, no matter at 25°C or 260°C. There exists a minimum warpage nearby the ratio of 3.5.

For the bottom package, the die thickness fixes at 50 um, the mold cap thickness varies from 200 um to 325 um. As shown in Figure 7, the package always presents convex shape at 25°C while keep concave shape at 260°C. Warpage value decreases with the thickness increasing.

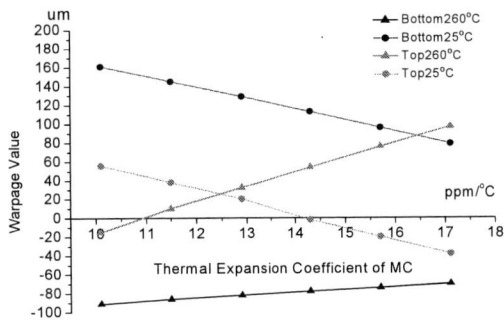

Fig. 5 Effects of the CTE of MC

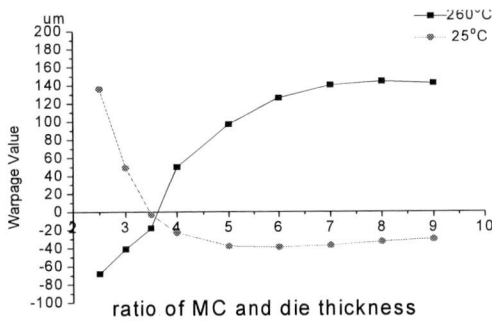

Fig. 6 Effects of the mold cap thickness: the top package

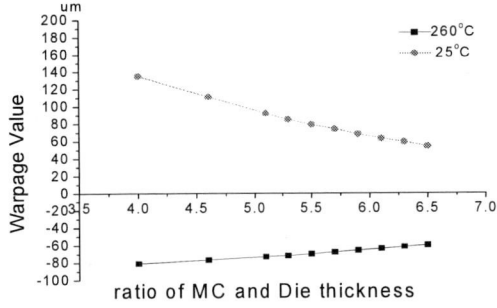

Fig. 7 Effects of the mold cap thickness: the bottom package

4.2 Effects of the substrate thickness

Ratio of substrate and die thickness also reacts on warpage. In the top package, when the ratio value ranges from 1 to 3.5, the curves of relationship are illustrated in Figure 8.

With the substrate thickness increasing, the warpage changes from negative to positive at 25°C. There is the slightest warpage when the ratio between 1.5 and 2. However, the warpage at 260°C shows a different trend. As a whole, the warpage keeps at a stable value after the ratio of 2. In the bottom package, Figure 9 shows the ratio has a small degree of impact for room temperature. The curve keeps steady decline at 260°C, thus a thicker substrate should be selected under meeting premise structure requirements.

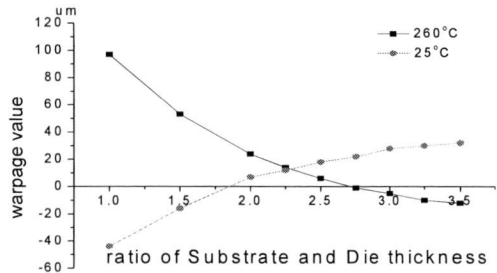

Fig. 8 Effects of the substrate thickness: the top package

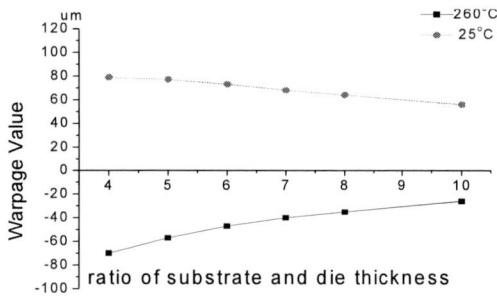

Fig. 9 Effects of the substrate thickness: the bottom package

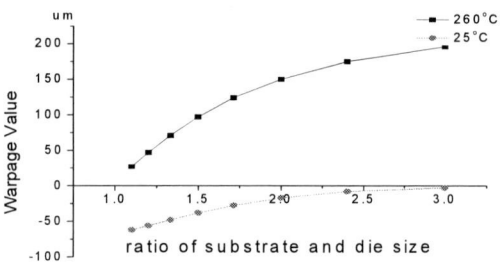

Fig. 10 Effects of the die size: the top

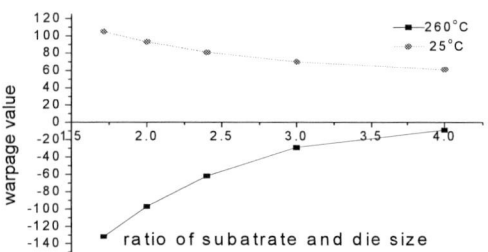

Fig. 11 Effects of the die size: the bottom package

4.3 Effects of the die size

As the dies have the lowest CTE value while the die-attach film has the highest one, variety of the die size can seriously change the deviation degree of combined CTE. In Figure 10, abscissa value is the ratio of substrate size and die size. The ratio values are obtained through changing die size while keep substrate size in a fixed value. According to Figure 10, warpage of the top package displays opposite trends when it at room temperature and peak temperature which attribute to the materials of molding compound is temperature dependent.

In the bottom package, increasing die size will indirectly reduce the warpage at room and peak temperature, especially at temperature of 260°C, as shown in Figure 11. The different performance between the top and the bottom package depends on their dissimilar structures, which will induce different CTE averaged.

Conclusions

A parametric investigation at molding and reflow process in package on package is conducted in this paper. A new finite element model and particular temperature dependent materials are adopted. The size of dies, thickness and material properties of substrate as well as molding compound are detailed studied. The effects of above mentioned parameters are summarized as below:

(1) For top package, higher substrate CTE, lower MC CTE can reduce the warpage effectively. When the thickness ratio of MC and dies is around 2, the ratio of substrate and dies about 2.5, the warpage exist the lowest value.

(2) For bottom package, thicker substrate and MC, smaller die, lower substrate CTE and higher MC CTE all can help to reduce the degree of the warpage.

(3) The die size as well as CTE of MC and substrate have significant influence on the warpage, while Elastic Modulus of MC and substrate almost no impact on the warpage.

Acknowledgments

The authors would like to thank financial support from the National Natural Science Foundation of China (NSFC) under the Grant No.10572010, and the Funding Project for Academic Human Resources Development in Institutions of Higher Learning Under the Jurisdiction of Beijing Municipality (PHR - IHLB).

References

1. Peng SUN, Vincent Chi-Kuen LEUNG, Bin XIE, Vivian Wei MA, Daniel Xun-Qing SHI,"Warpage Reduction of Package-on- Package (PoP) Module by Material Selection & Process Optimization", *2008 International Conference on Electronic Packaging Technology & High Density Packaging*, Pudong, Shanghai, July. 2008, pp.1-6.

2. Akito Yoshida, Jun Taniguchi, Katsumasa Murata, Morihiro Kada, Yusuke Yamamoto,"A Study on Package Stacking Process for Package -on- Package (PoP)", *IEEE 56th Electronic Components and Technology Conference*, San Diego, CA, May. 2006, pp. 825 – 830.

3. Moody Dreiza, Akito Yoshida, Kazuo Ishibashi, Tadashi Maeda, "High Density PoP (Package-on-Package) and Package Stacking Development", *IEEE 57th Electronic*

Components and Technology Conference, Reno, NV, May. 2007, pp. 1397-1402.

4. Y. L. Tzeng, N. Kao, E. Chen, Jeng Yuan Lai, Yu Po Wang and C.S. Hsiao,"Warpage and Stress Characteristic Analyses on Package -on -Package (PoP) Structure", *9th Electronics Packaging Technology Conference*, Singapore, December. 2007, pp. 482-487.

5. Wei Sun, W.H. Zhu, C.K. Wang, Anthony Y.S. Sun and H.B. Tan, "Warpage Simulation and DOE Analysis with Application in Package-on-Package Development", *9th. Int. Conf. on Thermal, Mechanical and Multiphysics Simulation and Experiments in Micro-Electronics and Micro-Systems, EuroSimE*, Freiburg im Breisgau, April . 2008, pp. 1-8.

6. Wei Sun, W.H. Zhu, Kriangsak Sae Le and H.B. Tan, "Simulation Study on the Warpage Behavior and Board-level Temperature Cycling Reliability of PoP Potentially for High-speed Memory Packaging", *2008 International Conference on Electronic Packaging Technology & High Density Packaging*, Pudong, Shanghai, July. 2008, pp. 1-8.

7. Yi-Shao Lai, Tong Hong Wang, and Ching-Chun Wang, "Optimization of Thermomechanical Reliability of Board-level Package-on- Package Stacking Assembly", *IEEE Trans. Comp. Package. Technol,* vol 29, no. 4, December (2006), pp. 864-868.

8. T. H. Wang and Y.-S. Lai, "Robust design in enhancing thermomechanical reliability of board-level flip-chip packages implemented with organic or silicon substrates," in *Proc. ANSYS Users Conf.*, Hualian, Taiwan, R.O.C., 2005, pp. 1.7–1.11.

9. G. Taguchi, "Quality engineering (Taguchi methods) for the development of electronic circuit technology," *IEEE Trans. Reliability*, vol. 44, no. 2, Jun. (1995), pp. 225–229.

10. Bin Xie, Peng Sun and Daniel Shi, "Design Advisor for Package-on-Package (PoP) Manufacturing", *2008 International Conference on Electronic Packaging Technology & High Density Packaging*, Pudong, Shanghai, July. 2008, pp. 1-7.

11. Hao Tang, Jonathan Nguyen, Jack Zhang, Irving Chien, "Warpage study of Package on Package configuation", *International Symposium on High Density Packaging and Microsystem Integration*, Shanghai, June. 2007, pp. 1-5.

Computational Modeling and Optimization for Wire Bonding Process on Cu/Low-K Wafers

Weidong Huang
Freescale Semiconductor (China) Ltd.
Xinhua Avenue 15, Xiqin Economic Development Zone, Tianjin, 300385, China
Email: Weidong.Huang@freescale.com

Abstract

A methodology is developed to use the explicit dynamic analysis results to reflect the real impact responses of wire bonding under different CV (constant velocity) settings. The optimal ranges of the process parameter settings for wire bonding on Cu/low-K wafers are determined by this way. The KNS Maxum bonder is selected as the vehicle for mapping the process settings into the simulation. The approximate mathematical model of the capillary motion during the real bonding impact is established. The loading forces linearly ascending with time on the capillary is proposed into the explicit dynamic analysis, and the analytical equation under this loading condition is deduced. Two assumptions are put forward for linking the real bonding responses and the simulation results, and specifying the impact time for the explicit dynamic analysis. A series of impact simulations under various force loading profiles are performed for the current wire bonding case on Cu/low-K wafers. The impact time is fixed on and the simulation results perfectly match the deduced analytical expression unless the time is larger than the impact time. By investigating the regression relations both in the impact simulation and the real bonding process, the simulated bond ball shape responses are mapped to the CV settings. Through defining the reasonable range of deformation contribution from impact, the CV and bond force ranges as the optimal process settings are deduced out to achieve the target bond ball shape after ultrasonic. Finally, a comparison is made to evaluate the process settings obtained both from numerical analysis and DOE (design of experiment).

1 Introduction

The increasing demands for high speed signal propagation have led to significant advances in IC (integrated circuit) fabrication. As the feature sizes of ICs are continually scaled down, it has become more critical that metal conductors that form the interconnections between devices as well as between circuits in the chip have low resistivities to achieve high speed signal propagation. In addition, the replacement of silicon dioxide with new materials, having a lower permittivity, will decrease the capacitive delay and reduce the power consumption. Therefore the use of diminutive copper interconnects encapsulated by low permittivity intermetal dielectric (IMD) material has become the major trend in chip designs. However, applying Cu/low-k technology into wafer fabrication brings more challenges to the manufacturing. Low-K materials are fragile and more prone to failures during assembly process and testing. For example, the bond ball together with the bond pad metal may peel away from the die surface right after ball bonding. Understanding the wire bonding characterizations of Cu/low-K structures accordingly becomes one of the important concerns at present.

Wire bonding is a quite complicated process due to real-time adjusting on bonding operation conducted by the bonding control program. Over the years, many researchers have published their contributions to the studies of the wire bonding process [1-15]. A few of these studies were focused on applying the numerical analysis on this subject including the wire bonding on Cu/low-K wafers [11-15]. The complicated nature of this process makes the numerical analysis difficult to be handled. Up to now, no numerical studies were reported with involving in determining the optimal bonding parameter settings for resulting in the expected responses. Actually, the key bonding parameter settings for 1st bonding point include the CV (constant velocity), the bond force, the USG (ultrasonic soldering generator) power and the bond time as addressed by a KNS Maxum bonder. Wire bond engineers always face the situations that need to adjust these parameters in bonding control programs for different wire bonding processes. These parameters can be determined by running DOE (design of experiment), but it will be time-consuming. Hereby, providing the valuable information and guidelines by numerical analysis for determining these parameters is significant, especially for the wire bonding on Cu/low-K wafers.

Two stages are usually figured to describe the whole wire bonding process: impact stage and ultrasonic vibration stage. In impact stage, the capillary leads the FAB (free air ball) to be smashed on the pad vertically with the initial constant velocity (CV). In subsequent ultrasonic vibration stage, bond force and ultrasonic power are synchronously applied on the capillary and last as bond time setting. The timescale of the impact stage is a critical concept to model this process. A KNS Maxum bonder specifies the CV range as 0.05~3mil/ms, from which it can be calculated that the timescale of impact stage is at millisecond level. It implies that the essential of the impact stage of wire bonding should be a quasi-static punching process. Therefore, modeling this stage by transient dynamic analysis with ANSYS/Mechanical or by implicit dynamic analysis with ANSYS/LS-DYNA, will consume huge computer CPU time to implement the analysis. Due to this reason, almost all research papers related with the numerical analysis on wire bonding modeled the impact stage by explicit dynamic analysis in which the impact stage was specified at microsecond level. That means a brief history at microsecond level has been utilized to describe an event at millisecond level. Feasibility of this replacement in timescale has never been studied.

This study develops a methodology that applies the explicit dynamic analysis results to reflect the real impact responses of wire bonding under different CV settings. The

---

978-1-4244-4658-2/09 $25.00 © 2009 IEEE

KNS Maxum bonder is selected as the vehicle for mapping the process parameter settings into the numerical analysis. The optimal ranges of CV and bond force as the process settings are come out for the current wire bonding case on Cu/low-K wafers to achieve the target bond ball shape after ultrasonic. These data are compared with the DOE results to confirm the feasibility of this methodology.

2 Approximate Mathematical Models
2.1 Equation for the Real Impact Stage

The impact stage of wire bonding process can be modeled as Fig. 1. Assuming capillary moves along x-axis, A is the position where the free air ball just touches the bond pad surface, B is the position where impact stage completes. Spring C and stepping motor D provide holding force to the capillary during reset stage and driving force during bonding stage. Spring C concentrates all elastic effects of the whole structure, so E and F are considered as rigid frameworks.

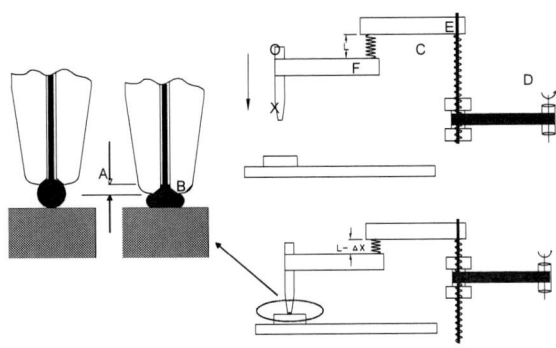

Fig. 1 Schematic wire bonding model

Assuming the framework E has the constant velocity CV during the whole impact stage, the displacement of E equals to $CV \cdot t$ where t represents time. Defining $x(t)$ as the displacement of capillary in regard to time, the change of the spring length can be calculated out as $CV \cdot t - x(t)$. Here the capillary is loaded by two forces. One is the force $F1(t)$ imposed by the framework F and another is the bounced force $F2(t)$ of the gold ball caused by the ball deformation. These two forces are in opposite direction. The motion equation of the capillary is written as

$$F1(t) - F2(t) = m\ddot{x}(t) \tag{1}$$

Here $F1(t)$ is rationally defined as being proportional to the change of the spring length,

$$F1(t) = k_1[cv \cdot t - x(t)] \tag{2}$$

Approximately, assuming $F2(t)$ is proportional to the ball deformation in term of the ball height change which equals to the displacement of the capillary,

$$F2(t) = k_2 x(t) \tag{3}$$

Then an approximate differential equation can be set up to describe the capillary motion during impact stage of wire bonding,

$$k_1[cv \cdot t - x(t)] - k_2 x(t) = m\ddot{x}(t) \tag{4}$$

Where k_1 and k_2 are the elastic coefficients and m the mass of the capillary.

The equation (4) can be further transformed for ease of solution,

$$\ddot{x}(t) + \frac{k_1 + k_2}{m} x(t) = \frac{k_1}{m} \cdot cv \cdot t \tag{5}$$

The general solution of equation (5) is:

$$x(t) = c_1 \cdot \cos\left(\sqrt{\frac{k_1 + k_2}{m}}t\right) + c_2 \cdot \sin\left(\sqrt{\frac{k_1 + k_2}{m}}t\right) + \frac{k_1}{k_1 + k_2} \cdot cv \cdot t \tag{6}$$

where c1 and c2 are the constants.

Considering the initial conditions $x(t)\big|_{t=0} = 0$ and $\dot{x}(t)\big|_{t=0} = cv$, the final solution of the equation (5) can be obtained:

$$x(t) = \left[\frac{k_2}{k} \cdot \sqrt{\frac{m}{k}} \cdot \sin\left(\sqrt{\frac{k}{m}}t\right) + \frac{k_1}{k} \cdot t\right] \cdot cv \tag{7}$$

where $k = k_1 + k_2$.

The 1st and 2nd derivatives of equation (7) are as follow:

$$\dot{x}(t) = \left[\frac{k_2}{k}\cos\left(\sqrt{\frac{k}{m}}t\right) + \frac{k_1}{k}\right] \cdot cv \tag{8}$$

$$\ddot{x}(t) = -\frac{k_2}{k} \cdot \sqrt{\frac{k}{m}} \cdot \sin\left(\sqrt{\frac{k}{m}}t\right) \cdot cv \tag{9}$$

Substitute (7) into (2), hence

$$F1(t) = \frac{k_1 k_2}{k}\left[t - \sqrt{\frac{m}{k}} \cdot \sin\left(\sqrt{\frac{k}{m}}t\right)\right] \cdot cv = \psi(t) \cdot cv \tag{10}$$

The impact stage of wire bonding shall be stopped when the bonder detects the contact between the bond head and the silicon die. The total impact time is determined by the contact detect mode which is a method of detecting contact surface. Generally Vmode is used for detecting contact for a KNS Maxum bonder. Vmode refers to velocity mode and contact surface is detected by a function of z-velocity. The contact threshold (%) controls the sensitivity of the bond head in detecting contact. For Vmode, it is mathematically expressed in percentage drop of CV. The contact is declared when the bend head velocity is equal to the relationship: (1- threshold) CV. Normally the contact threshold is set to 70%, so the servo controller will declare contact at the point where the velocity of z-axis has slowed to 0.3CV.

Substitute $\dot{x}(t) = 0.3 \cdot cv$ into equation (8), then getting

$$0.3 = \left[\frac{k_2}{k}\cos\left(\sqrt{\frac{k}{m}}t\right) + \frac{k_1}{k}\right] \tag{11}$$

From equation (11), it can be seen that the time that capillary touches down the bond pad surface has no relationship with CV value, it is only related with the threshold setting to validate the touchdown. In other words, it takes the same time for completing the impact stages with different CV settings. Defining this identical contact time as tc, then the maximum displacement of the capillary during impact stage can be expressed as

$$x(t_c) = \left[\frac{k_2}{k} \cdot \sqrt{\frac{m}{k}} \cdot \sin\left(\sqrt{\frac{k}{m}} t_c \right) + \frac{k_1}{k} \cdot t_c \right] \cdot cv = \alpha \cdot cv$$

(12)

Here α is a constant which is only related with tc.

During the real impact stage of wire bonding, F1(t) shall grow up as time increases. Based on equation (10), the maximum F1(t) can be expressed as

$$F1(\max\_real) = \frac{k_1 k_2}{k} \left[t_c - \sqrt{\frac{m}{k}} \cdot \sin\left(\sqrt{\frac{k}{m}} t_c \right) \right] \cdot cv = \psi(t_c) \cdot cv$$

(13)

2.2 Equation for Explicit Dynamic Analysis

For an explicit dynamic modeling, it's impossible to build all the models illustrated in Fig. 1 which including the holding frameworks and the motor. The Finite Element model relating to the impact stage could be reasonably simplified to only consist of the capillary, the gold ball and the die base. A puzzle then comes: how to apply the loads on the capillary to specify its motion characterization? It's the first nodus should be overcome in the analysis. The second nodus is to specify when the impact stage completes for the simulation time setting. Solving the two noduses is related with using the explicit analysis results to figure the real bonding responses.

For the first nodus, this study chooses applying force on the capillary as the loading method. A proper $F1(t)\sim t$ relation can be applied on the capillary to simulate the response of the bond ball. The explicit dynamic analysis always involves a rapid impact event at microsecond level. During this brief history, $F1(t)$ can be defined as increasing very fast. As an available attempt, this study defines F1(t) linearly ascending with time and being proportional to CV mapped from equation (10), i.e.

$$F1(t) = \gamma \cdot cv \cdot t = B \cdot t$$

(14)

where γ and B represent the proportional coefficients.

According to equation (1), the motion equation of the capillary can be also established for the explicit dynamic transient process, with still assuming $F2(t)$ equals to $k_2 x(t)$,

$$F1(t) - k_2 x(t) = m\ddot{x}(t)$$

(15)

Substitute equation (14) into equation (15), hence

$$\gamma \cdot cv \cdot t - k_2 x(t) = m\ddot{x}(t)$$

(16)

The general solution of equation (16) is

$$x(t) = c_3 \cdot \cos\left(\sqrt{\frac{k_2}{m}} t \right) + c_4 \cdot \sin\left(\sqrt{\frac{k_2}{m}} t \right) + \frac{\gamma \cdot cv}{k_2} t$$

(17)

where c_3 and c_4 are the constants.

It's necessary to consider the effect of the initial velocity on the explicit analysis results. Assuming the initial velocity CV has an upper limit value of 50mm/s which meeting the current 0.8mil gold wire bonding process, the impact time in the explicit analysis is assumed to 10us which is the uppermost assessed value for the current case. It is supposed

that the capillary only has the initial velocity CV to touch down on the gold ball without external forces loaded on it. Under this condition, the displacement of the capillary should never be more than 0.5um after impact. But actually the displacement of the capillary during bonding impact can be achieved up to 10um or more, so the effect of initial velocity on the results of the explicit dynamic analysis can be ignored. Thus the initial velocity of the capillary is set to zero for all simulations, however the CV setting still definitively affects the explicit analysis results through the force loading definition $F1(t) = \gamma \cdot cv \cdot t = B \cdot t$.

Therefore, considering the initial conditions $x(t)|_{t=0} = 0$ and $\dot{x}(t)|_{t=0} = 0$, the final solution of the equation (16) can be obtained,

$$x(t) = \frac{1}{k_2} \left[-\sqrt{\frac{m}{k_2}} \cdot \sin\left(\sqrt{\frac{k_2}{m}} t \right) + t \right] \cdot \gamma \cdot cv = \frac{1}{k_2} \left[-\sqrt{\frac{m}{k_2}} \cdot \sin\left(\sqrt{\frac{k_2}{m}} t \right) + t \right] \cdot B$$

(18)

Equation (18) is the capillary's motion equation which applying the linear force regarding to time into the explicit dynamic analysis, it specifies the motion characterization of the capillary if the assumption $F2(t) = k_2 x(t)$ is reasonable. The maximum $F1(t)$ during this analysis is

$$F1(\max\_explicit) = \gamma \cdot cv \cdot t_{impact} = B \cdot t_{impact}$$

(19)

where t_{impact} is the impact time when the impact stage stops.

2.3 Two Assumptions

For solving the second nodus mentioned in 2.2, two assumptions are introduced into this study:

(1) The real bonding impact and the explicit dynamic analysis will produce the same bond ball shape response only when an identical value of the maximum $F1(t)$ is reached both in the real process and the analysis.

(2) The impact time t_{impact} for the explicit dynamic analysis is specified by determining the moment when the bounced force $F2(t)$ presents a obvious abnormity away from its normal status.

Based on assumption (1), the link between the real process and the analysis can be set up

$$F1(\max\_real) = F1(\max\_explicit)$$

(20)

Substitute equation (13) and (19) into equation (20), the expression of impact time is obtained,

$$t_{impact} = \frac{\psi(t_c)}{\gamma}$$

(21)

In equation (21), t_c is the identical contact time in spite of various CV settings for the real processes, therefore t_{impact} is also a constant that being no relationship with CV settings in the impact simulation. Determining t_{impact} is the primal step for the impact simulation.

3 Finite Element Model

The finite element model is built with commercial software ANSYS LS-DYNA. Element PLANE162 is used with the axisymmetric option. The element is defined by four nodes having six degrees of freedom at each node:

978-1-4244-4658-2/09 $25.00 © 2009 IEEE

translations, velocities, and accelerations in the nodal x and y directions. The ½ FE model is illustrated in Fig. 2, where FAB (Free air ball), HAZ (heat affected Zone), capillary and Cu/low-k structure are included. The FAB diameter is 30.4um, the gold wire diameter is 0.8mil, and the capillary dimensions follow Table 1. The low K structure consists of various layers including Aluminium pad, passivation, TEOS dielectric, low K material, copper interconnections and the contact material bonded to silicon bulk. The silicon bulk hasn't been considered into the model because of assuming it as the fixed boundary condition during bonding. There is an assumption that no friction exits between each impact part. Elastic properties of the constituents in the model are listed in Table 2. In this study, the FAB and the copper interconnections are assumed bilinear plastic. The yield stress and the tangent modulus are 0.117 and 39GPa respectively for the FAB, and 0.33 and 121GPa respectively for the copper interconnections.

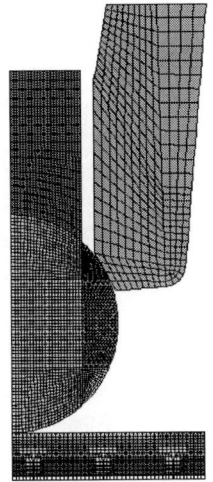

Fig. 2 FE model of wire bonding

Table 1 Capillary dimensional information

Capillary	Dimension
Hole size	24um
Cap CD (Chamfer Diameter)	29um
Cap T(Tip)	55um
Cap OR (Outer Radius)	3um
Face angle	8 degree
ICA (Inner Chamfer Angle)	60 deg

The capillary is defined rigid, so loading the body forces on the capillary is available. The linearly increasing body forces regarding to time will be loaded on the capillary according to equation (14). It's indispensable to decide the correct quantitative order of impact timescale which can smooth the computer calculating procedure. For this purpose, the trial simulations are performed at three different time orders 10^{-1}us, 10^{0}us and 10^{1}us with assigning some reasonable B values. It is found that timescale at 10^{-1}us level would leads to a smooth and rapidly completed analysis by computer. Therefore 10^{-1}us is chosen as the impact timescale for the explicit dynamic analysis of the current case study.

Table 2 Elastic properties of the constituents in the FE model

Material	E(GPa)	ν	ρ(g/cm3)
Copper interconnections	121	0.38	8.91
Passivation	32	0.24	1.31
TEOS dielectric	80	0.23	2.00
Low K	11	0.30	2.00
Contact material	80	0.23	2.00
Al Pad	69	0.33	2.71
Gold wire	39	0.43	19.3
FAB	39	0.43	19.3

4 Impact Simulations for Bond Ball Shape Responses
4.1 Impact Time Determination

Fig. 3 Force loaded on the capillary as the function of time

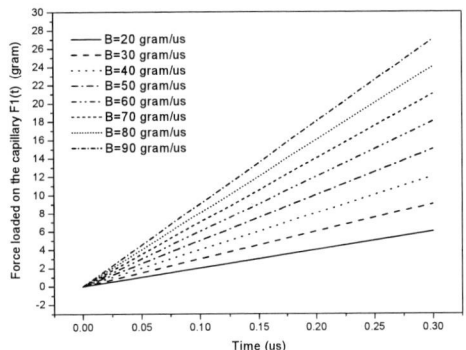

Fig. 4 Capillary displacement as the function of B value

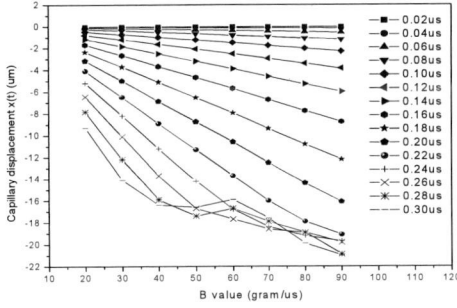

Defining the contact by Vmode is impossible because the initial capillary velocity is set to zero and the capillary velocity increases rapidly during the impact process. The test simulation runs are performed to determine the impact time. The B values are probingly set to 20gram/us, 30gram/us, 40gram/us, 50gram/us, 60gram/us, 70gram/us, 80gram/us and 90gram/us. Fig. 3 shows the forces loaded on the capillary under various B values according to $F1(t) = B \cdot t$. Fig. 4 illustrates the simulated results as the relations between the displacement x(t) of the capillary and the B value at various

978-1-4244-4658-2/09 $25.00 © 2009 IEEE   271

time points. It can be observed that the displacement of the capillary has a better linear and proportional relation regarding to the B value when the time is less than 0.2us. Recalling the equation (18),

$$x(t) = \frac{1}{k_2}\left[-\sqrt{\frac{m}{k_2}}\cdot\sin\left(\sqrt{\frac{k_2}{m}}t\right)+t\right]\cdot B$$

the simulation results of x(t)~B shown in Fig. 4 perfectly match the linear relation in analytical equation (18) unless the time is larger than 0.2us. Without question, the variation in x(t)~B relation at time larger than 0.2us must be caused by the abnormity of the bounced force F2(t), because there are only the stable forces $F1(t) = B\cdot t$ loaded on the capillary in the simulations. That means F2(t) deviates from its normal increase to a unstable status when time is larger than 0.2us. According to assumption (1) and (2) proposed in 2.3, 0.2us is specified as the identical impact time for all simulations under various B values.

4.2 Impact Responses under Various B Values

With assigning 0.2us as the impact time, now begins to investigate the wire bonding responses in impact stage under various B values. The B value is still respectively set to 20gram/us, 30gram/us, 40gram/us, 50gram/us, 60gram/us, 70gram/us, 80gram/us and 90gram/us. Fig. 5 shows the force profiles loaded on the capillary as the function of time $F1(t) = B\cdot t$, the maximum time is 0.2us and the maximum F1(t) can be calculated at 0.2us. For each B value, the maximum F1(t) respectively equals to 4gram, 6gram, 8gram, 10gram, 12gram, 14gram, 16gram and 18gram. This study assumes that F1(t) is the expression of impact force during the bonding impact. For a normal wire bonding process, the bond force setting during ultrasonic should be less than the maximum impact force to realize a stable bonding.

Fig. 5 Force loaded on the capillary with the impact time equaling to 0.2us

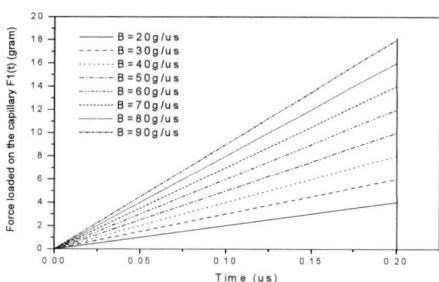

Fig. 6 illustrates the Von Mises stress distribution at 0.2us under various B values. It also reveals the bonding responses under various CV settings because of the relation $B = \gamma\cdot cv$, just γ is still unknown so far. Fig. 7 gives the simulated bond ball shape responses at 0.2us as the function of B. Fig. 8 gives the measured bond ball shape responses after ultrasonic produced by a KNS Maxum bonder under various CV settings. In Fig. 7 and Fig. 8, (a), (b) and (c) present the bond ball shape responses respectively in bond ball size (diameter), BBH (bond ball height) and BBR (bond ball height to bond ball size ratio). For the simulated results, BBH is assumed as

the height of the bottom of the capillary, it can be calculated by subtracting the capillary displacement from the initial height H0 of the bottom of the capillary (as illustrated in Fig. 9). Each data point in Fig. 8 represents the mean value of a group of the measured data with same bonding conditions. By comparing the relations in Fig. 7(a) and Fig. 8(a), it can be seen that the two curves have the similar tendence with increasing B or CV. But there is still obvious difference between them, because the measured ball responses include the effect of the ultrasonic stage while only impact stage is involved in the simulation.

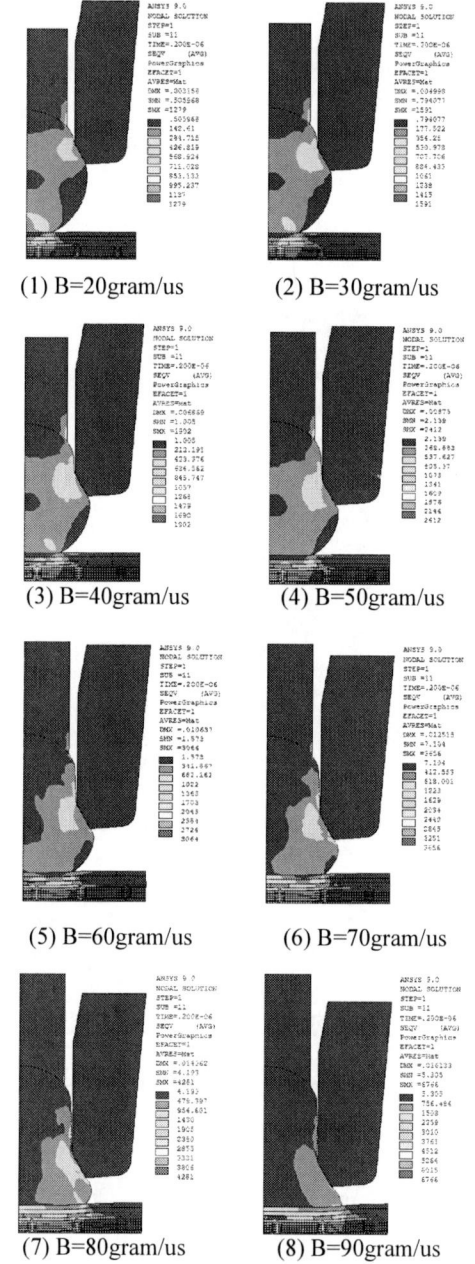

Fig. 6 Von Mises stress distribution at 0.2us under different B values (unit: MPa)

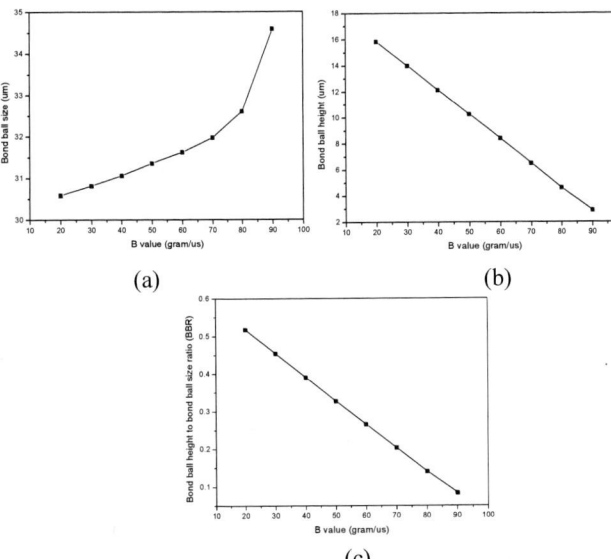

(a)

(b)

(c)

Fig. 7 Simulated bond ball shape responses as the function of B value for impact stage

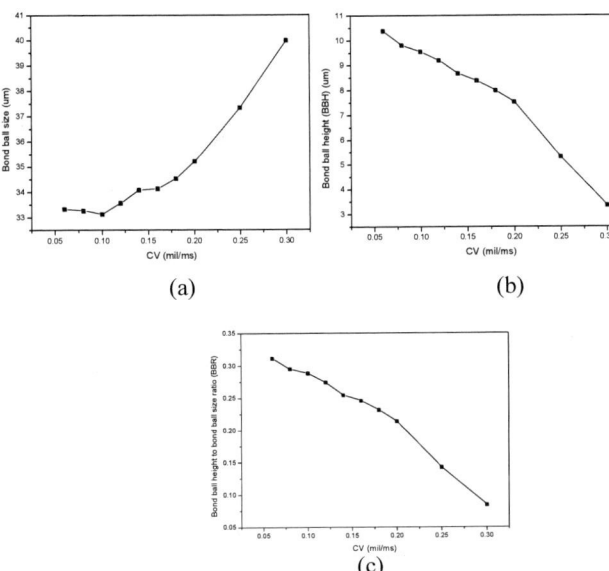

(a)

(b)

(c)

Fig. 8 Measured bond ball shape responses created by KNS Maxum bonder after ultrasonic stage

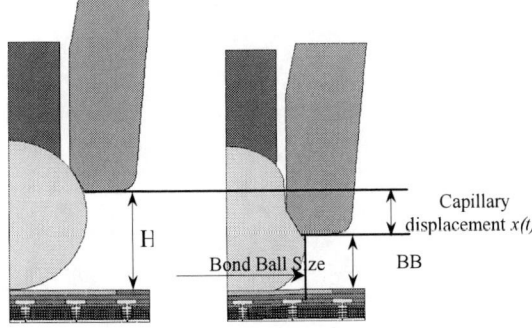

Fig. 9 Illustrated relation between BBH and capillary displacement

5 Parameter Setting Optimization for Bonding Process

A notable phenomenon is found when benchmarking the simulated data and the measured data. BBH from the impact simulation linearly decreases with increasing B as shown in Fig. 7(b), while the actual wire bonding process produces a bilinear relation between BBH and CV as shown in Fig. 8(b). This may give a hint to disclose the relation between B and CV. It is believed that the bilinear behavior in real bonding comes from the different affecting degrees of impact stage on the bonding responses due to different CV ranges. Both impact and ultrasonic play the important roles on bond ball shape formation when CV is a low value, while impact uniquely determines the final bond ball shape when CV is a high value. With a high CV, the impact could result in an enough short ball height and the ball has enough hardness to resist to be deformed by the ultrasonic. That's why the bilinear behavior exists in the measured data. According to Fig. 8(b), the impact can be regarded as the unique contribution to the ball shape formation when CV is higher than 0.2mil/ms. Therefore, the relation between BBH and CV without adding ultrasonic can be deduced by investigating the curve in Fig. 8(b) as long as CV is higher than 0.2mil/ms.

Recalling the equation (18) and substituting t=0.2us into it, then getting the capillary displacement right after impact for explicit analysis

$$x(0.2us) = \frac{1}{k_2}\left[-\sqrt{\frac{m}{k_2}} \cdot \sin\left(\sqrt{\frac{k_2}{m}}0.2us\right) + 0.2us\right] \cdot B = \beta \cdot B \tag{22}$$

where $\beta = \frac{1}{k_2}\left[-\sqrt{\frac{m}{k_2}} \cdot \sin\left(\sqrt{\frac{k_2}{m}}0.2us\right) + 0.2us\right]$

As mentioned before, BBH can be calculated by subtracting the capillary displacement from the initial height H0 of the bottom of the capillary, i.e.

$$BBH = H0 - x(0.2us) = H0 - \beta \cdot B \tag{23}$$

Recalling equation (12), the corresponding relation for the real bonding impact can be obtained

$$BBH = H0 - x(tc) = H0 - \alpha \cdot CV \tag{24}$$

Equation (23) and (24) theoretically reveal that BBH has a linear relationship with B and CV respectively for the explicit dynamic process and the real wire bonding if merely considering the impact stage. Equation (23) represents the relation in Fig. 7(b), and equation (24) represents the relation in Fig. 8(b) only when CV exceeds 0.2mil/ms.

It is obvious that $H0 - \beta \cdot B = H0 - \alpha \cdot CV$, hence

$$\gamma = \frac{B}{cv} = \frac{\alpha}{\beta} \tag{25}$$

At this moment, this study starts to fit the linear equations for BBH~B and BBH~CV to get α and β to calculate the γ value. At B=0 (namely CV=0), BBH equals to the initial height H0 which means no wire bonding operation happens. The height H0 at B(CV)=0 should be considered into the linear regressions.

978-1-4244-4658-2/09 $25.00 © 2009 IEEE
273

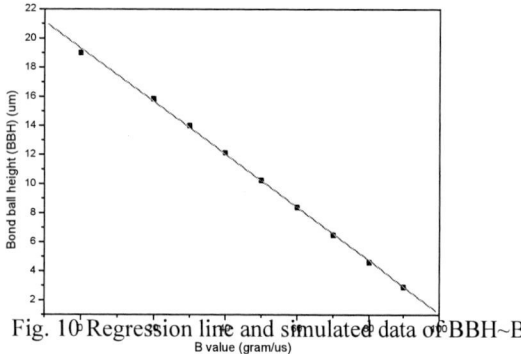

Fig. 10 Regression line and simulated data of BBH~B

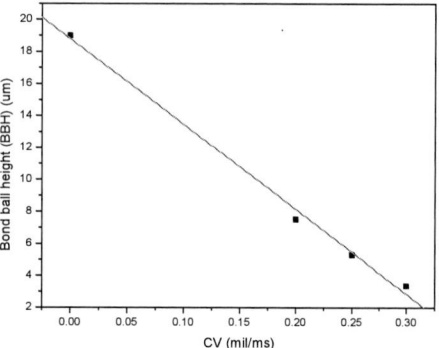

Fig. 11 Regression line and measured data of BBH~CV

Through linear regression on the data in Fig. 7(b) and Fig. 8(b) combining with H0 at B=0 (CV=0), the regressed lines have been gotten and shown in Fig. 10 and Fig. 11. It should be emphasized that the measured date point with CV<0.2mil/ms are not allowed to take part in the regression for BBH~CV. β and α are gotten as the slopes of the regressed lines and expressed as β = 0.188 um·us/gram and α = 53.18 um·ms/mil. Then γ can be calculated out as 53.18/0.188 = 282.9 ms·gram/(mil·us). The relation between B and CV is written as

$$B (g/us) = 282.9 \cdot CV (mil/ms) \qquad (26)$$

The regression relation between the simulated BBR and B value is also needed to set up for process optimization. Different from the regressions on BBH~B and BBH~CV, there is no theoretical equation to guide the regression on BBR~B. However, by observation on the data in Fig. 7(c), the linear regression is applied again on BBR~B. Fig. 12 shows the regressed line with the simulated data for BBR~B. The regression equation (further expand to BBR~CV) is expressed as

$$BBR = 0.637 - 0.0062 \cdot B (gram/us) = 0.637 - 1.754 \cdot CV (mil/ms) \qquad (27)$$

Based on the massive data analysis on the wire pull test after thermal aging, the reasonable BBR range after wire bonding with the best wire pull strength is found to be 20%~30% for 4N gold wire. If target BBR after ultrasonic stage is set to 25%, the target BBR just after impact is needed for the parameter setting optimization. Many wire bonding studies suggest that the optimal amount of deformation in impact stage should be slightly less than the total amount

(i.e.75-90% from impact, with the remaining deformation coming from steady-state force and ultrasonic energy). The initial BBR at CV=0 is 63.7% as calculated from equation (27). For deformation contribution 75%, the target BBR after impact is 25%+(63.7%-25%)•(1-75%)=34.7%. For deformation contribution 90%, the target BBR after impact is 25%+(63.7%-25%)•(1-90%)=28.9%. Accordingly for deformation contribution 75%~90%, the target BBR after impact is 28.9%~34.7%. The following inequations by considering the equation (27) can be established

28.9% <0.637 − 0.0062 • B (gram/us) < 34.7%

28.9% <0.637 − 1.754 • CV (mil/ms) < 34.7%

The optimal ranges for B and CV can be calculated out as

46.77< B (gram/us) <56.13

0.165<CV (mil/ms) <0.198   (28)

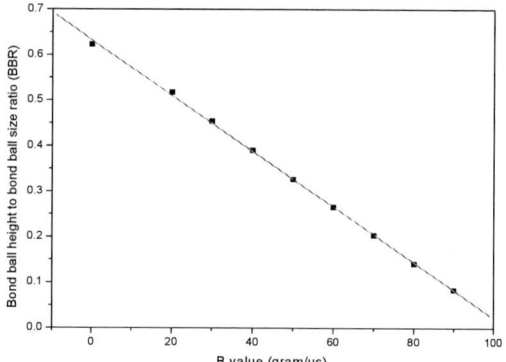

Fig. 12 Regression curve and simulated data for BBR~B

Further using the optimized B value range, the reasonable bond force range can be predicted for ultrasonic stage of this case. In an actual wire bonder control, the impact force should be control in a safe range. According to the technical requirement from the bonder vender, a stable wire bonding process requires the bond force range as

0.5 • Maximum impact force < Bond Force < Maximum impact force   (29)

Since here optimized B is from 46.77 to 56.13 gram/us, the range for maximum impact force can be calculated by multiplying B with impact time 0.2us. The maximum impact force range therefore is 9.35~11.2 grams. With considering the relation in (29), the bond force for this low K wire bonding case can be assigned in the range of 5~11 grams.

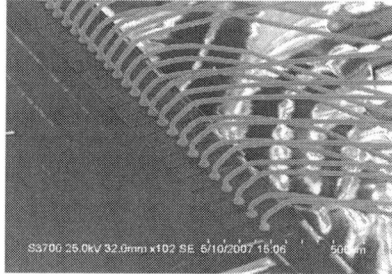

Fig. 13 Gold wire pattern bonding with the DOE optimized process parameters

Table 3 Optimized process parameter settings from numerical analysis and DOE

	USG power(mA)	Bond force (gram)	CV(mil/ms)
Numerical analysis	NA	5~11	0.165~0.198
DOE	75	8	0.18

DOE (design of experiment) on the KNS Maxum bonder has been also conducted to optimize the process parameter settings for this wire bonding case on Cu/low-K wafers. It experimentally determined the optimized process parameters for target ball size and maximum ball shear value. Fig. 13 shows the gold wire pattern bonding with the optimized process parameters resulted from DOE. Table 3 shows the optimized process parameter settings obtained both from numerical analysis and DOE. It is noted that the bond force and CV settings resulted from DOE are almost at the medial location of the optimal ranges resulted from the numerical analysis. Evaluated from this point of view, this study provides a novel and effective optimization methodology for process parameter settings of wire bonding on Cu/low-K wafers using numerical analysis based on the KNS Maxum bonder operation.

6 Conclusions

A methodology is developed to deal with using the explicit dynamic analysis results to reflect the real impact responses of wire bonding under different CV (constant velocity) settings. The optimal ranges of the process parameter settings for wire bonding on Cu/low-K wafers are determined by this way. The KNS Maxum bonder is selected as the vehicle for mapping the process settings into the simulation. The important results are summarized in the following.

1 The approximate mathematical model of the capillary motion during the real bonding impact is established. Also the analytical equation is deduced for loading the forces linearly ascending with time on the capillary. Two assumptions are put forward for linking the real bonding responses and the simulation results, and specifying the impact time for the explicit dynamic analysis.

2 A series of impact simulations under various force loading profiles are performed for the current wire bonding case on Cu/low-K wafers. The simulation results perfectly match the deduced analytical expression unless the time is larger than 0.2us. According to the two assumptions proposed in this study, 0.2us is fixed on as the impact time.

3 By investigating the regression relations both in the impact simulation and the real bonding process, the simulated bond ball shape responses are mapped to the CV settings. Through defining the reasonable range of deformation contribution from impact as 75-90%, the CV and bond force ranges as the optimal process settings are deduced out to achieve the target bond ball shape after ultrasonic. The optimal ranges for CV and bond force can be concluded as CV=0.165~0.198mil/ms and bond force=5~11grams.

4 A comparison is made to evaluate the process settings obtained both from numerical analysis and DOE (design of experiment). It confirms that this study has provided a novel and effective optimization methodology for process parameter settings of wire bonding on Cu/low-K wafers.

Acknowledgments

The author would like to thank Sonder Wang and C. L. Zhang for their supporting in wire bonding technology and experimental optimization.

References

1. R. G .McKenna and R. L. Mahle, 1989, "High impact bonding to improve reliability of VLSI die in plastic packages," in Proc. 39th Electron. Compon. Technol. Conf., Houston, TX, pp. 424-427.

2. B. Gonzalez, S. Knecht, H. Handy, and J. Ramirez, 1996, "The effect of ultrasonic frequency on fine pitch aluminum wedge wire bond," in Proc. 46th Electron. Compon. Technol. Conf., Orlando, FL, pp. 1078-1087.

3. Y. Tamura, Y. Miyahara, and H. Suzuki, 1998, "Analysis and application of vibration behavior for wirebonding capillary by transmission laser vibrometer," in Proc. SEMICON Int. Electron. Manuf. Technol. Symp., Austin, TX, pp. 72-75.

4. Y. Takahashi, M. Inoue, and K. Inoue, 1999, "Numerical analysis of fine lead bonding-effect of pad thickness on interfacial deformation," IEEE Trans. Compon. Packag. Technol., vol. 22, no. 2, pp. 291-298.

5. T.A. Tran, L. Yong, B. Williams, S. Chen, and A. Chen, 2000, "Fine pitch probing and wire bonding and reliability of aluminum capped copper bond pads," in Proc. 50th Electron. Compon. Technol. Conf., Las Vegas, NV, pp. 1674-1680.

6. C.V.Pham and K. Huth, 2001, "A new approach to the robust wire bonding," in Proc. Int. Symp. Adv. Packag. Mater.: Processes, Properties, and Interfaces, Braselton, GA, pp. 379-385.

7. V. Kripesh, M. Sivakumar, L. A. Lim, R. Kumar, and M. K. Iyer, 2002, "Wire bonding process impact on low-K dielectric material in damascene copper integrated circuits," in Proc. 52nd Electron. Compon. Technol. Conf., San Diego, CA, pp. 873-880.

8. J.W. Brunner, F. Keller, and T. Pan, 2003, "Optimization of wire bonding over Cu-low K pad stack," in Proc. 36th Int. Symp. Microelectron., Boston, MA, pp. 136-140.

9. M. Ikeda, H. Kudo, R. Shinohara, F. Shimpuku, M. Yamada, and Y. Furumura, 1998, "Integration of organic low-K material with Cu-damascene employing novel process," in Proc. IEEE Int. Interconnect Technol. Conf., San Francisco, CA, pp. 131-133.

10. Z.W.Zhong, K.S.Goh, 2000, "Analysis and experiments of ball deformation for ultra-fine-pitch wire bonding," Journal of Electronics Manufacturing, vol.10, no. 4, pp. 211-217.

11. T.C. Huang, M.S. Liang, T. T Chao, et al, 2002, "Wire bonding failure mechanisms and simulations of Cu low-K IMD chip packaging," in Proc. Adv. Metallization Conf., pp. 67-73.

12. D.Degryse, B. Vandevelde, and E. Beyne, 2004, "Mechanical FEM simulation of bonding process on Cu low K wafers," IEEE Trans. Compon. Packag. Technol., vol.27, no. 4, pp. 643-650.

13. G.K. Viswanath, Wang Fang, 2005, "Numerical analysis by 3D finite element wire bond simulation on Cu/low-K structures," Electronic Packaging Technology Conference, 2005. EPTC 2005. Proceedings of 7th, Volume 1, Issue, 7-9, pp. 215-220.

14. C.L. Yeh and Y.S. Lai, 2005, "Transient analysis of the impact stage of wire bonding on Cu/low-K wafers," Microelectron. Rel., vol. 45, no. 2, pp. 371-378.

15. C.L. Yeh and Y.S. Lai, 2006, "Comprehensive dynamic analysis of wirebonding on Cu/low-K wafers," IEEE Transactions on advanced packaging, vol. 29, no. 2, pp. 264-269.

The Principal Component Analysis of Cu Stud Bump Shaping Process Parameters

Wei Mu, Zhaohua Wu, Chunyue Huang

School of Machinetronic Engineering, Guilin University of Electronic Technology, Guilin 541004, China

Email: muwei@mails.guet.edu.cn

Abstract

As copper wire has excellent thermal conductivity, mechanical properties, low-cost advantages and copper stud bump process is similar the traditional wire bonding process, it makes copper stud bump technology better for medium and low I/O devices, and even in the future, it will expand into midrange I/O packages because it offers a large potential savings in packaging costs with the improved performance. However, the reliability of copper stud bump is crucial problem, and it is still hard to determine the critical process parameters which affect the reliability of copper stud bump. In this paper, by finite element method, the die chip having forty pads was used to build a model, and considering that: the first bonding joint process doesn't affect the second boding joint, so by using of software ANSYS/LS-DYNA, only one bonding joint model was built to decrease mass Computation, and then combine different bonding processing parameters to simulate the copper stud bump shaping processing. Based on this, collect the max ball shear stress during bonding as data samples, and then bonding processing parameters are analyzed by the method of principal component analysis through the software SAS software, take the ninety accumulative percentage as a boundary, extract the main characteristic factors from the bonding processing parameters to decrease the bonding processing parameters dimensions and eliminate the correlation among these bonding processing parameters. Then for the new defining main characteristic factors and original bonding processing parameters, the relative analyses are carried out to generate the main characteristic factors dates. This provides the basis with which the copper bonding reliability after bonding processing is predicted by the method of BP network.

Introduction

For flip-chip packaging applications, a bump process on LSI wafers is required due to increased chip circuit density, operating speed and performance [1]. Golden wire is widely used as a kind stud bump material; however, golden wire has a expensive price, material cost is very high especially in high density semiconductor packaging, moreover the harmful intermatellic compound between golden wire and aluminium pad can lead to generate void in bonding point, so the resistance increases rapidly, which affects the bonding joint performance severely. As copper wire has excellent thermal conductivity, mechanical properties, low-cost advantages and copper stud bump process is similar the traditional wire bonding process, it makes copper stud bump technology better for medium and low I/O devices, and even in the future, it will expand into midrange I/O packages because it offers a large potential savings in packaging costs with the improved performance and reliability.

How to effectively analyze the bonding processing parameters and forecast the boding reliability has always been people's concern. Jeon et al. [2] studied failure mechanisms of bond pad metal peeling by Experimental investigation and Numerical analysis. Guzmann and Mahaney [3] and Hu et al. [4] found that the ultrasonic power is a dominant variable in ball formation and has a critical effect on the bond shear strength. Liang et al [5] introduced a concept of reduced bonding parameter' and they were able to relate the bonding parameters directly to bond ability and ball bond reliability. Chu et al [6] utilized DOE (design of experiment) to optimize separately FAB parameters such as fire current, fire time and $0.9N_2/0.1H_2$ gas flow rate and bonding process parameters such as ultrasonic power, bonding force, bonding temperature and bonding time to obtain the feasible and stabile bonding process. Shu [7] made a lot of efforts to optimize the bonding parameters in order to improve the bond-ability characterized by the ball shear stress and the ball bond reliability determined by in ball bond degradation measurement.

Several numerical analyses of wire bonding process have been carried out for understanding the bonding mechanism. Among these works, Lin et al [8] performed Comprehensive dynamic analysis to simulate wire bonding on Cu/low-K wafers. Liu et al [9] discloses the stress and deformation impacts to both wire bonding and pad below device with strain rate, different ultrasonic amplitudes and frequencies, different friction coefficients, as well as different bond pad thickness and device layout under pad. Moreover, Takahasi et al [10] simulated wire deformation processes during thermo-compression bonding without ultrasonic vibration using finite element technique.

These researches studied below just are relevant to the some process parameters, and the whole processes are not considered. However, as for the bonding processing parameters, due to multifarious factors (structure of wedge tool, pad dimension, copper wire diameter, free air ball diameter, ultrasonic energy, bonding force and bonding temperature), as well as correlations among the processing parameters, so the key bonding processing parameters are hard to determine, therefore it is difficult to establish the corresponding mathematical model which is used to determine the relation between the max ball shear stress after bonding processing and the bonding processing parameters, therefore the reliability of copper stud bump is hardly forecasted. To solve this problem, the die chip having forty pads was used to build a model, and considering that: the first bonding joint process doesn't affect the second boding joint, so by using of software ANSYS/LS-DYNA, only one bonding joint model was built to decrease mass Computation, and then combine different bonding processing parameters to simulate the copper stud bump shaping processing. Based on this, collect the max ball shear stress during bonding as data samples, and then bonding processing parameters are analyzed by the method of principal

978-1-4244-4658-2/09 $25.00 © 2009 IEEE

component analysis through the software SAS software, take the ninety accumulative percentage as a boundary, extract the main characteristic factors from the bonding processing parameters to decrease the bonding processing parameters dimensions and eliminate the correlation among these bonding processing parameters. Then for the new defining main characteristic factors and original bonding processing parameters, the relative analyses are carried out to generate the main characteristic factors dates.

Figure 1 shows the bonding processing parameters.

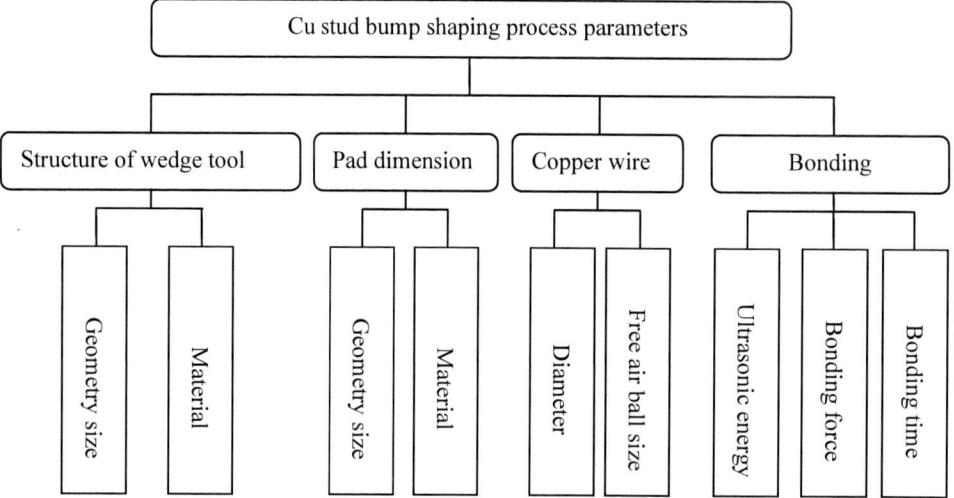

Fig 1 The bonding processing parameters

1 The simulation of stud bump bonding shaping process

1.1 The process and principle of wire bonding

Wire bonding technology is a process in which metal wire is used to connect the electrode leader of integrated circuit to the electrode leader of exterior base frame. The major methods of technology are hot-press, thermal-ultrasonic and ultrasonic.

The principle of wire bonding is that: by the applied bonding force the metal ball generates great plastic deformation; the sliding line makes the pad surface a step appearance and the thin film a corresponding concave convex groove; the surface oxide film is destroyed and activated and the connection is completed by the metal mutual diffusion. Figure 2 shows the schematic diagram of stud bump fabrication.

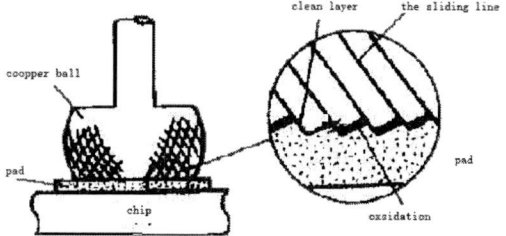

Fig 2 The schematic diagram of stud bump

1.2 The structure of capillary

Depending on different productions circuit performance damage and film layer shaping character, choosing the suitable capillary is a key process which can guarantee the quality and reliability of production. The material of most common capillary used is standard ceramic containing 99.9% Al_2O_3. Figure 3 shows the structure of capillary fabrication.

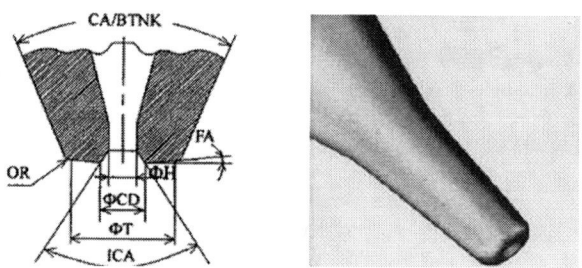

Fig 3 The structures of capillary

The structure of capillary is very precise and complicate and each part of the structure has a higher requirement and specific function. Experiments demonstrate that hole size, the end diameter, cone angel, the inner inclined plane angel and the face angel and so on are the crucial factors of capillary which play a determinative rule on stud bumping appearance.

The structures of capillary are chosen according to the copper wire diameter and then make a fine value adjustment to attain the continuous.

1.3 Key bonding process parameter

The factors affecting the copper stud bumping shaping are bonding time, bonding force, bonding energy and bonding temperature (the temperature here isn't considered temporarily). The standard parameters of copper stud bump shaping are showed as Table 1.

In this paper, the loading value is according to the standard parameters as showed table1: the loading force on capillary is 70gf; the loading ultrasonic energy is 100mA; the loading frequency is 60 KHz, and then the ultrasonic amplitude is 3.4μm. To save computing time, the ultrasonic action time is assumed to be 16.7μs.

Table 1 The standard parameters

Ultrasonic energy(mA)	Bonding force (gf)	Bonding time(ms)
100	70	20

1.4 Pad dimension

The pad thickness has a significant effect on bonding process. Generally think, we can enhance the protection of chip through the thickness of the pad. In order to verify the viewpoint validity, under the condition of the free air ball diameter 66μm and the aluminum pad length 132μm, two models are established with aluminum pad thickness 1 and 2 μm separately to check the chip stress field after loading the ultrasonic energy, it show as figure 4 . From the figure 4, we can see that it can effactually decrees the inner stress of chip by increasing the aluminum pad thickness. From the cloud graph of Figure 4 (a) and (b), it can be found that when the thickness reaches 2μm, the stress-focus phenomenon discovered under aluminum pad thickness 1μm can be alleviated. This demonstrates that it's meaningful to protect the chip by increasing the aluminum pad thickness.

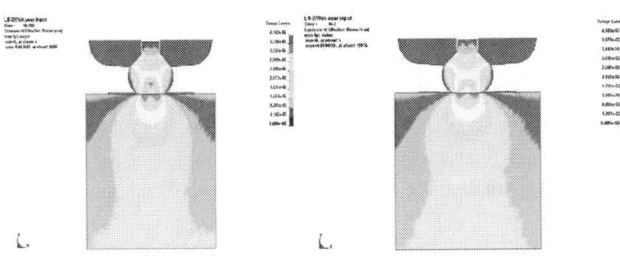

(a) The pad thickness 1μm   (b) The pad thickness 2μm
Figure 4 the stress distribution of different pad thickness

1.5 The simulation of bonding process

In this paper, the die chip used has forty pads, considering that: the first bonding joint process doesn't affect the second boding joint, so by using of software ANSYS/LS-DYNA, only one bonding joint model was built to decrease mass computation. Figure 5 shows the schematic diagram of chip.

In order to make model simple, here the pad oxide is not considered. The established model includes capillary, FAB

(Free Air Ball), aluminum pad and silicon chip. The local amplified graph of the finite model is showed as figure 6.

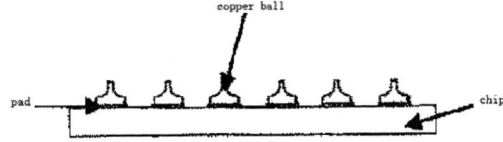

Fig 5 The schematic diagram of chip

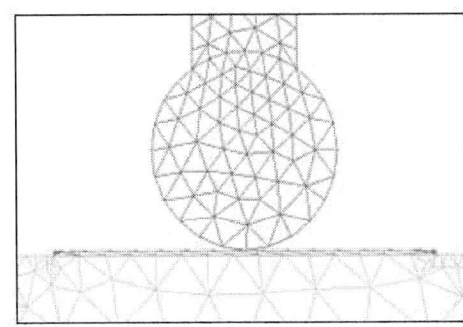

Fig 6 The local amplified graph of the finite model

Because the bonding is a symmetric process, the required information can be derived from the central plane. So in simulation process, the plane model is developed to analyze the problem quickly and clean up the whole thought, and then it puts a basis for the later further study.

And then define the boundary condition, define the contact and develop the loading, finally solve it. The process of the bonding can be observed by the tool of lsprepost. The max ball shear stress during bonding is viewed a objective factor. The figure 7 shows the distribution stress of bonding process.

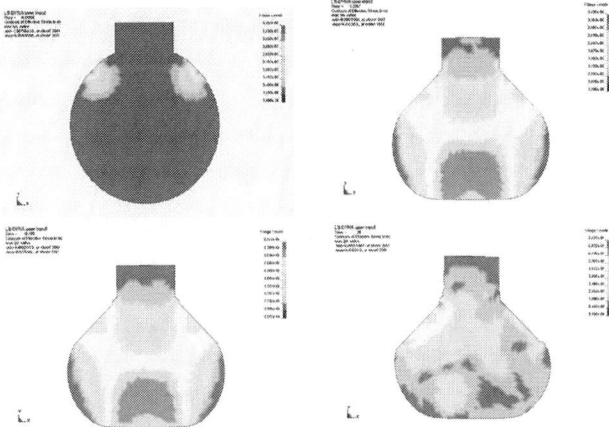

Fig 7 Stress nephogram of bonding processing

The principal component analysis

Principal Components Analysis (PCA) is a way of identifying patterns in data, and expressing the data in such a way as to highlight their similarities and differences [11]. Since patterns in data can be hard to find in data of high dimension, where the luxury of graphical representation is not available, PCA is a powerful tool for analyzing data. The other main advantage of PCA is that once you have found these patterns in the data, and you compress the data, i.e. by reducing the number of dimensions, without much loss of information.

Given a set of points characterized by N variables X_1, $X_2...X_p$, find new variables Z_1, $Z_2...Z_m$:

$$\begin{cases} z_1 = l_{11}x_1 + l_{12}x_2 + \cdots + l_{1p}x_p \\ z_2 = l_{21}x_1 + l_{22}x_2 + \cdots + l_{2p}x_p \\ \cdots\cdots\cdots\cdots\cdots\cdots\cdots\cdots\cdots\cdots\cdots\cdots \\ z_m = l_{m1}x_1 + l_{m2}x_2 + \cdots + l_{mp}x_p \end{cases}$$

In the equations above, coefficient l_{ij} is determined by the principles below:

(1) Z_i is irrelevant to Z_j ($i \neq j$; $i,j=1,2,...m$);

(2) Z_1 is the maximum among the variances of every linear combination; Z_2 is the maximum among the variances of linear combination of X_1, $X_2 ...X_p$ which are irrelevant to Z_1; Z_m is the maximum among the variances of linear combination of X_1, $X_2... X_p$ which are irrelevant to Z_1, $Z_2 ...Z_m$; After that, the new defined variables Z_1, $Z_2...Z_m$ are called the 1st, 2nd, mth major component of original variables X_1, $X_2...X_p$.

Z_1 takes up the max proportion among the total variances, and the variances of Z_2, $...Z_m$ are in a descending order. In the analysis to practical problems, the first maximum major components are usually picked in order to decrease the number of variable, grasp the principal contradiction and the simple the relationship among the variables.

From the analysis above, the process of determining the major components is finding the coefficients l_{ij}($i=1,2...m$; $j=1,2...p$) of the each major component loading by original variable x_i($i=1,2...m$;), in the view of math, we can know they are the Eigen values of correlation matrix of X_1, $X_2...X_p$.

In the paper, the realizing of the PCA is through software SAS. The PRINCOMP executes principal component analysis. Table 2 shows the each major Eigen value, each major Eigen percentage and cumulative percentage.

Table 2 the bonding processing parameters PCA analysis

No	Eigen value	Percentage	Cumulative percentage
1	4.79402185	0.4358	0.4358
2	1.61416342	0.1467	0.5826
3	1.16676280	0.1061	0.6886
4	1.15250264	0.1048	0.7934
5	0.75551798	0.0687	0.8621
6	0.55260454	0.0502	0.9123
7	0.41913115	0.0381	0.9504
8	0.34288357	0.0312	0.9816
9	0.12161370	0.0111	0.9927
10	0.06234058	0.0057	0.9983
11	0.01845777	0.0017	1.0000
12	0.00000000	0.0000	1.0000

Commonly, cumulative percentage is taken as 85-95%, so here the first 6 major components can satisfy the requirement.

Based on the principal component analysis, the data of the first 6 major components can be got by Correlation Analysis.

Conclusions

The finite element method by software ANSYS/LS-DYNA can simulate the copper stud bump shaping process effectively. The max stress can be seen during the bonding process and be collected as sample data. By Principal Components Analysis method, extracting the main characteristic factors from the bonding processing parameters can decrease the bonding processing parameters dimensions and eliminate the correlation among these bonding processing parameters. This provides the basis with which the copper bonding reliability after bonding processing is predicted by the method of BP network.

Acknowledgments

This research is sponsored by the Science Foundation of Guangxi Zhuang Autonomous Region Government (Grant No. 0832083) and Director Project Found of GuangXi Key Laboratory of Manufacturing System & Advanced Manufacturing Technology (Grant No.07109008_011_Z_).

References

1. R.Kiumi, J. Yoshioka, F. Kuriyama, N. Saito, M. Shimoyama, "Process development of electroplate bumping for ULSI flip chip technology", *Proceedings of Electronic Components and Technology Conference, IEEE*, pp. 711-716, 2002.

2. Jeon I, "The study on failure mechanisms of bond pad metal peeling, B: Numerical analysis," *Microelectronics Reliability*, Vol. 43, No. 12 (2003), pp. 2055-2064.

3. Mckenna RG, "High impact bonding to improve reliability of VLSI die in plastic packages," In: Proceedings of the 39th Electronic Component Conference, 1989, pp. 424-7.

4. Hu SJ, Lim GE, Foong KP, "Study of temperature parameters on the thermosonic gold wire bonding of high-speed CMOS," *IEEE Trans Compon Hybrids Manu Technol*, Vol. 14, No. 4 (1991), pp. 855-8.

5. Liang ZN, Kuper FG, Chen MS, "A concept to relate wire bonding parameters to bondability and ball bond eliability," *Microelectron Reliab*, Vol. 38 (1998), pp. 1287-91

6. Shu WK, "Fine pitch wire bonding development using statistical design of experiment," In IEEE 1995 Proceedings, 45th Electronic Components and Technology Conference, 1995, pp. 91-101.

7. Shu WK, "Fine pitch wire bonding development using statistical design of experiment," In IEEE 1995 Proceedings, 45th Electronic Components and Technology Conference, 1995, pp. 91-101.

8. C.-L. Yeh and Y.-S. Lai, "Comprehensive dynamic analysis of wirebonding on Cu/low-K wafers," IEEE Transactions on Advanced Packaging, Vol. 29, No. 2 (2006), pp. 264-270.

978-1-4244-4658-2/09 $25.00 © 2009 IEEE

9. Liu Yong. Irving S. Luk T, "Thermosonic wire bonding process simulation and bond pad over active stress analysis," 2004.

10. Takahasi Y, Shibamoto S, Inoue K, "Numerical analysis of the interfacial contact process in wire thermocompression bonding," *IEEE Trans Compon Pack Manuf Technol Part A*, Vol. 19 (1996), pp. 213-23.

11. Rao, C.R., "The Use and Interpretation of Principal Component Analysis in Applied Research," *Sankhya A*, Vol. 26 (1964), 329 -358.

Effects of Shape Parameters on Tensile Strength of Cu Bump

Chunyue Huang[1], Ying Liang[2], Tianming Li[3]

[1]School of Electro-Mechanical Engineering, Guilin University of Electronic Technology,
Guilin, 541004, China
[2]Chengdu Aeronautic Vocational and Technical College, Chengdu, 610021, China
[3]Dept. of Science and Technology, Guilin College of Aerospace Technology, Guilin, 541004, China

Abstract

Three parameters, Copper pad material, Pad thickness, and Bump Patterns were chosen as three control factors. By using an L9(34)orthogonal array the copper stud bump solder joints which have 9 different combinations of parameters were designed. The finite element analysis models of 9 copper stud bump solder joints were established by using ANSYS/LS-DYNA, and analyzed the finite element tensile simulation. Tensile strain data of copper stud bump solder joint were obtained under 9 different parameters combinations; it was analyzed through range analysis and variance analysis. The result shows that of the three parameters, the descending order of affecting the copper stud bump solder joint tensile strain is Copper pad material, Bump Patterns, and Pad thickness. With 95% confidence, Copper pad material has a significant effect on the copper stud bump solder joint tensile strain whereas bump patterns and pad thickness has little.

1 Introduction

The demand for electronic integrated circuit (IC) chips has been put forward with the development of the electronic products for the light, thin, short, small and multi-function. Flip-chip technology has been increasingly used for the demand. The flip-chip technology compared to traditional and tape bonding has obvious advantages: the highest packaging density, good electrical and thermal performance, good reliability and low cost. Therefore, flip-chip is a kind of adapt to the future electronic packaging development requirements technologies. The bump formation is the key process in flip-chip intergraded. [1] Golden bump flip-chip technology is one of flip-chip integrate technologies. The bump manufacture process is easy because it needn't redistribute the pad in the golden bump process. Besides, it's easy to produce the golden bump, the density of bump is very high, and the IC installation is very much (Eutectic welding, thermal-acoustic or heat-pressure bonding, bonding etc.), therefore, this technology has been widely used in the electronic integrate. [2]But, the golden bump defections will become increasingly prominent with the development of higher density integrate, at the same time, micro-electronics industry in order to reduce costs and improve reliability, have to find good process performance, low-cost metal materials to replace the high cost gold. Copper is the best replacer, the copper bump have some advantages in cost, electrical performance, thermal performance and mechanical performance [3] ,which can be used in FC technology. The copper stud bump technology may provide the lowest cost flip-chip integrate for low I/O density devices. The cost of producing copper stud bump is very low if used the high-speed automatic wire bonder aided the protection atmosphere system applied in producing spherical copper bump, this method can meet the cost requirement of flip-chip application. Due to the using is existing equipment and infrastructure, which not like sputtering or electroplating process increased more equipment investment, it's quite attractive that integrated the wire bonding technology into flip-chip bump production.

The copper stud bump bonded on the IC pad may be suffered a variety of sudden load, which can affect the bump reliability even to failure. Copper stud bump has better mechanical performances, which can enhance the copper bump resistance capacity of varieties sudden loads even the copper bump applying reliability. The copper bump form has directly impact on it's the mechanical properties included the shear strength. So, it is necessary to research the relation between the copper bump form and the shear strength. We can confirm the key parameter impacted on the tensile strength of copper stud bump through the research.

In this paper, the dynamic simulation of copper bump has been researched based on finite element simulation method. Three parameters, the copper bump form, pad materials, pad thickness were chose as control factors impacted the copper bump tensile strength. By using a $L_9(3^4)$ orthogonal array, 9 different tensile strength simulation models of copper stud bump under different parameters combinations were designed and nonlinear finite element dynamic simulated. Obtained 9 different groups strain of copper stud bump in shear process. Based on the strain results, the rang analysis and the variance analysis were carried out in order to analyze affection order of the copper bump form, pad materials and pad thickness to the strength of copper stud bump and their impacting significant.

2 Orthogonal design of impacting copper stud bump tensile strength parameters combinations

By using the orthogonal design [4], it only needs to arrange the parameters combinations of impacting the copper stud bump tensile strength, which can reduce the number of the simulation models of copper stud bump. These factors are: copper pad materials, pad thickness, bump form. Each critical factor has three levels, which are shown in Table 1. Since there are 3 three-level factors, a $L_9(3^4)$ orthogonal array with the first 3 columns can be used to arrange the combinations of the factors, which are shown in Table 2.

Table 1 Factors and their levels

	Factors	Levels		
		1	2	3
A	Pad material	Aluminum	Copper	Gold
B	Pad thickness	1.0(um)	1.5(um)	2.0(um)
C	Bump size(diameter ×height ×bottom contact length, um)	23.30 ×19.71 ×27.50	22.96 ×19.86 ×27.84	23.26 ×18.59 ×27.32

Table 2 Factors assignment to an $L_9(3^4)$ orthogonal array of copper stud bump

Trial No.	Pad material (A)	Pad thickness (B)	Bump Size(C)	strain
1	1	1	3	0.497876
2	2	1	1	0.313957
3	3	1	2	0.390137
4	1	2	2	0.496766
5	2	2	3	0.275149
6	3	2	1	0.381843
7	1	3	1	0.491588
8	2	3	2	0.305144
9	3	3	3	0.347178

3 Finite simulation of copper stud bump tensile strength

3.1 Finite simulation model of copper stud bump tensile strength

The finite simulation model of the copper stud bump had been build in the ANSYS/LS_DYNA, using plane quadrilateral element Plane 162 and mapped grid. Follow-plastic (related with strain-rate) material model has been chosen to copper stud bump and pad, which is anisotropy, kinematic hardening or isotropy and kinematic combing model related with the strain-rate. By adjusting the parameter β in 0 (kinematic hardening only)to 1(isotropic hardening only) to choose kinematic hardening or isotropic hardening. The strain-rate can be considered by Cowper-Symonds Model [5] using factors related with strain-rate to indicate the yield stress as following:

$$\sigma_y = \left[1 + \left(\frac{\varepsilon}{C} \right)^{\frac{1}{P}} \right] \left(\sigma_0 + \beta E_P \varepsilon_P^{eff} \right) \quad (1)$$

σ_0 -Initial yield stress, ε —Strain rate, C, P — Cowper-Symonds strain rate parameters, ε_P^{eff} —Effective plastic strain, E_P —Plastic hardening modulus, which can be given by equation 2 as following:

$$E_P = \frac{E_{tan} E}{E - E_{tan}} \quad (2)$$

The parameters include: Elastic Modulus (EX), Density (DENS), Poisson's ratio (NUXY), Yield strength and tangent modulus when defined this material model. The material model used in silicon chip is rigid body model, which can defined the rigid part of finite model to reduce the analysis time, greatly. The finite model of copper stud bumps as shown in Figure 1. The material parameters were shown in Table 3. The restriction of the model was constrained the freedom of all nodes on the bottom of silicon chip and exerted force on nodes which were on the top of the copper stud bump.

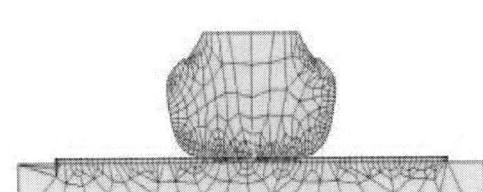

Fig. 1 The copper stud bump finite model

Table 3 material parameters

Meaning	Aluminum	Copper	Gold	silicon
density(kg/m^3)	2720	8900	19300	2329
Elastic modulus(Pa)	71e9	120.658e9	78.5e9	131e9
Shear modulus(Pa)	2.39e9	13e9	7.85e9	/
Yield stress (Pa)	145e6	307e6	205e6	/
Poisson's ratio	0.33	0.345	0.42	0.3

3.2 Tensile simulation result of copper stud bump

The tensile fracture process of copper stud bump was shown from Figure 2 to Figure 7.

The initial deformation of copper stud bump under the load was shown in Figure 2. As can be seen: under the load, the deformation appeared in the pad and bump contact position, and the greatest stress appeared in the right and lift sides of the bump corner.

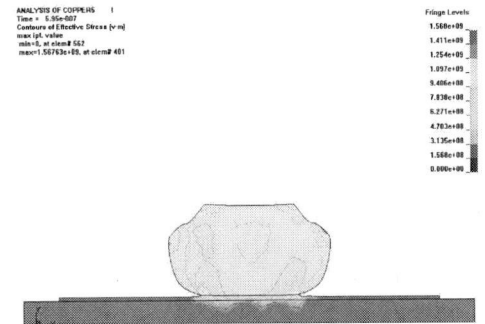

Fig. 2 Deformation of the model when t=5.95e-007s

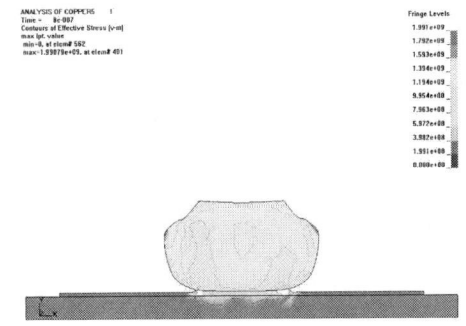

Fig.3 Deformation of the model when t=8e-007s

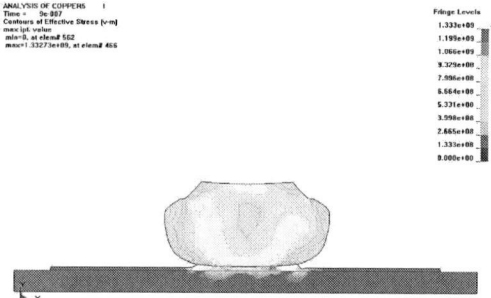

Fig.4 Deformation of the model when t=9e-007s

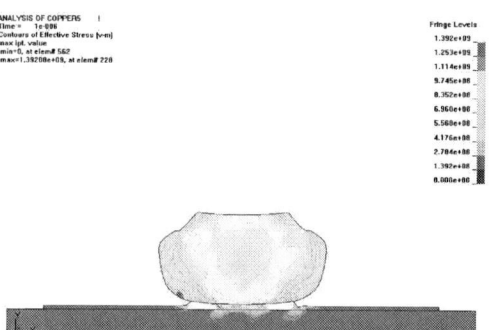

Fig.5 Deformation of the model when t=1e-006s

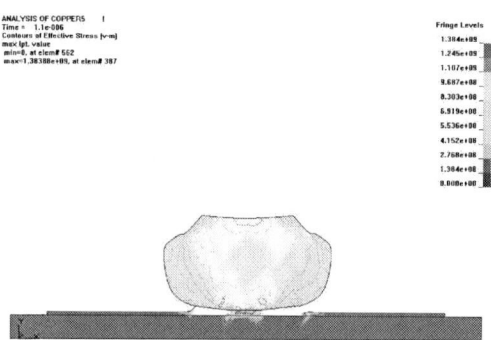

Fig.6 Deformation of the model when t=1.1e-006s

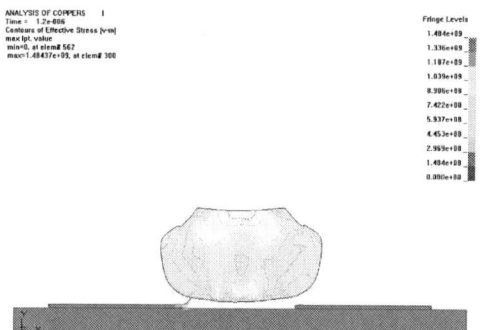

Fig.7 Deformation of the model when t=1.2e-006s

The fracture has been appeared in the corner between pad and bump, as well as signs of further fracture expansion as shown in Figure 3. As can be seen: the position of the greatest stress and deformation is still appeared the right and lift sides of the bump corner. As shown in Figure 4, the fracture expansion of the connection position increased. Figure 5 and Figure 6 compared with Figure 4, the pad position to reach failure strain with the process of the tensile. So, the cracks continued to increase until completely disconnected at last under the load, which was shown in Figure 7. From the whole simulation process, it can be concluded that the location of the maximum stress is always in the contact corner on both sides of the bump from the load beginning to the bump and pad becoming completely disconnected. And the fracture crack increasing is gradually expanded along with the tensile process.

4 Tensile strain data analysis of Cu stud bump

4.1 Range analysis

After finite element forming simulation, some geometric parameters can be obtained which contain height and diameter of bump. Based different forming process, the height and diameter of different morphology bumps are showed in table 2.

Table 4 Results of range analysis on
stress of Cu stud bump height

Trial No.	Time (A)	Press(B)	Energy(C)
$\overline{K}_{i1}$	0.49541	0.400657	0.395796
$\overline{K}_{i2}$	0.298083	0.384586	0.397349
$\overline{K}_{i3}$	0.373053	0.381303	0.373401
Range	0.197327	0.019354	0.023948
Range Order	1	3	2

The range analysis of every colonna can be obtained by orthogonal experiment; range analysis can be used to find out the best levels combination and to determine primary and secondary sequence of factors. The range of a factor is defined as the difference between the maximum level and the minimum level of the factor. The larger the range is. The more significant effect the factor has [6]. The results of range analysis are shown in Table 4.

The range data of Cu stud bump tensile strain is shown in Table 3, it can be conclude the order of the factor is $R_A > R_C > R_B$, there are 3 factors which contain the material of copper pad, the thickness of pad and the form of stud bump. The order of the factors impacting the tensile strain of Cu stud bump is the material of copper pad, the form of stud bump, and the thickness of pad.

4.2 Variance analysis

Range analysis method can only get the relative impaction of the experiment factors but can't confirm the specific impaction of each facto, and the impaction is significance or not, while, the variance analysis as a way of solving the experiment data can solve the basic problem by analyzing the experiment data, to research the factors impaction is significance or not. Therefore, the significance of those three factors, copper material, pad thickness and bump form to the bump tensile strain can be studied through the variance analysis of the copper stud bump tensile strain data.

The deviation square sum, degrees of freedom, variance and ratio of F calculated according to the above orthogonal method variance theory and the strain data of the copper stud bump in the Table 2, were shown in Table 5 as following.

From the data in Table 5, the ratio of F of the copper pad material, 354.485928, is bigger than the critical value $F_{0.01}(2,2)$, so with 95% confidence the copper pad material has a significant effect on the tensile strain of the copper stud bump. The ratio of F of the pad thickness and bump form are all less than the critical value $F_{0.05}(2,2)$, so with 95% confidence the pad thickness and bump form have little effect on the tensile strain of the copper stud bump.

Table 5 Results of analysis of variance on copper stud bump tensile strain

Variance resource	Sum of square	degrees of freedom	Variance	Ratio of F	Significance
Pad material	0.059529531	2	0.029764765	354.4859288	Highly Significance
Pad thickness	0.000643594	2	0.000321797	3.832467904	No Significance
Bump form	0.001077455	2	0.000538727	6.416013625	No Significance
Error	0.000167932	2	0.000083966	/	/
Sum	0.061418506	8	$F_{0.05}(2,2)= 19.0$ $F_{0.01}(2,2)= 99.0$		

5 Conclusions

(1)The result of the range analysis is that the descending affection order of copper stud bump in those factors, copper pad material, pad thickness and bump form is copper pad material, bump form and pad thickness.

(2) The result of the variance analysis is that the affection of copper pad material to the copper stud bump tensile strain is significant, wile the affection of pad thickness and bump form is no significant.

Acknowledgments

This research is sponsored by the Science Foundation of Guangxi Zhuang Autonomous Region Government (Grant No. 0832083) and Director Project Found of GuangXi Key Laboratory of Manufacturing System & Advanced Manufacturing Technology (Grant No.07109008_011_Z_).

References

1. Li Fuquan, WANG Chunqing, ZHNG Xiaodong, "Bump fabrication methods for flip chip," *Electronics Process Technology*. Vol. 24, No. 2 (2003), pp. 62-66.

2. GUO Hui, GUO Daqi, "Technology of many gold studs for flip-chip," *Electronics & Packaging*, Vol. 7, No. 6 (2007), pp. 18-20.

3. XU Hui, HANG Chun-jing, WANG Chun-qing, TIAN Yan-hong, "Contrast research on the reliability of gold and copper wire/ball bonding," *Equipment for Electronic Produacts Manufacturing*, Vol. 5 (2006), 23-28.

4. Yang De. Design and analysis of experiments. Beijing: Agriculture Press of China, 2002, pp. 171-172 (In Chinese).

5. LIU Yang, YAO Jiang-tao, LI Guo-lin. "Numerical Simulation of Warhead Penetrating into Multi-Layer Spaced Target," *Journal of Naval Aeronautical and Astronautical University*, Vol. 24, No. 2 (2009), pp. 144-148.

6. HE Shaohua, WEN Zuqing, LOU Tao, Design of experiments and data processing, Changsha: National University of Defense Technology Press, 2002, pp. 67-68.

A New Embedded Helix-Type Inductor Using LTCC Technology for High Frequency Applications

Dong Liu, Yingli Liu, Yuanxun Li

State Key Laboratory of Electronic Thin Film and Integrated Devices

University of Electronic Science and Technology of China, Chengdu, 610054, China

Email: Liudong525@gmail.com, Tel: 13981724080

Abstract

A chip inductor based on the CoTi-substituted hexa ferrite $BaFe_{12}O_{19}$ was designed by LTCC technology to realize the 0805 size of standard packaging and high frequency applications. Testing results show that the value of inductor is 2.45 µH and cut-off frequency is 508 MHz. And the measured results agree with the simulated data basically, which indicates that the samples can satisfy the need of high frequency inductors. The influences of each factor on the performances of inductor were also analyzed. It is an effective method to resolve the consistency problems between simulation and manufacture.

Introduction

Because of the rapid progress in the developments of wireless and mobile technology, further miniaturized inductor operating on higher frequencies are in great demand to fabricate high-density integrated devices. It is generally accepted that low-temperature Co-fired Ceramics and Ferrites (LTCC&LTCF) technology can satisfy such requirements. In addition, it also can be applied to manufacture other passive components , for example, resistors and capacitors. Thus, Nowadays LTCC is becoming the major technology for components integration[1-2].

Recently, M-type barium ferrite ($BaFe_{12}O_{19}$) has been found extensive applications in the fields of permanent magnets, wave-adsorbing, high-density magnetic recording and microwave devices. As the sintering temperature for the material is always higher than 1000°C and could not be satisfied with chip components manufacturing technology, low-temperature co-fired ceramics and ferrites process, the problems of how to realize the low-temperature firing and multifunction of barium ferrite are becoming the key technologies for the adaption to the developments of electronic components to the directions of sub-miniaturization, greater multifunctionality and excellent reliability[3-5].

In this paper, an embedded helix-type LTCC inductor prepared by Co-Ti substituted M-type Barium ferrite for high frequencies applications is described in details. Firstly, The influences of the Co-Ti substitution on the magnetic and electronic performances of barium ferrites are included, in order to find the optimal parameters for LTCC inductor fabrication. Secondly, according to the stimulation results of the inductor used in high frequency by the aid of HFSS software, the structure and the related size of the inductor are confirmed. Thirdly, a helix-type LTCC inductor with the 0805 size was successfully fabricated. Testing results show that the value of inductor is 2.45uH and cut-off frequency is 508MHz. The measured results agree with the simulated data basically, which indicates that the samples can satisfy the need of high frequency inductors.

Materials

The samples of $BaFe_{12-X} (Co\ Ti)_X O_{19}$ ferrites with X ranging from 0 to 1.5 in steps of 0.3 were prepared using traditional ceramic method. It is found that X=1.2 was the optimal parameters for LTCC inductor fabrication after compared the influences of different Co-Ti substitution on the magnetic and electronic performances[6].

The raw materials, $BaCO_3$, TiO_2, CoO and Fe_2O_3 with high purity of 99.9%, were employed as starting materials. They were mixed in a desired proportion to form $BaFe_{9.6} (Co\ Ti)_{1.2}O_{19}$ for 6 h ball-milling. The mixed powders were pre-sintered at 850°C for 2h in air. Then the pre-sintered powders and 3wt% (fusing assistant) BBSZ were mixed in a ball-milling pot again for 12h to improve the homogeneity of the heat-treated powder and to increase the surface area of the particles, and pressed into a disk under a pressure of 60MPa with 7% polyvinyl alcohol as lubricant. The size of pressed samples is Ø16mm×Ø8mm ×2.5mm. The disks were sintered at 920°C for 3 hours in air, then cooled with a rate of 1°C per minute.

Fig.1 shows the permeability varied with frequency within the range of 100 M to 1800M . From the results of Fig.1, the initial permeability is about 14.6 and the imaginary part of the relative permeability is about 0.007, which indicated over 100M to 500M that the barium ferrites pocesses excellent performances with high-frequency characteristics.

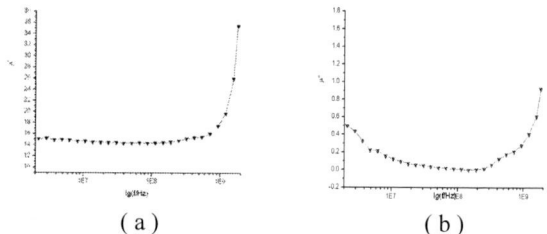

(a)   (b)

Fig.1(a) real part of relative permeability over frequency at 920°C Fig.1(b) imaginary component of relative permeability over frequency at 920°C

Fig.2 gives back-scatter morphology for barium ferrites sintered at 920°C for 2h. Fig.2(a) and Fig.2(b) indicate the samples with BBSZ and without BBSZ respectively. With the comparisons of the results, it can be seen that   the morphology of the samples without BBSZ is with some of cracks or delaminations. But under the function of BBSZ, the crystal grains of the samples with BBSZ grown up to 5um and a high dense microstructure can be achieved. It is explained that the additions of BBSZ through liquid sintering principal is an effective method to improve the sintering properties of barium ferrites.

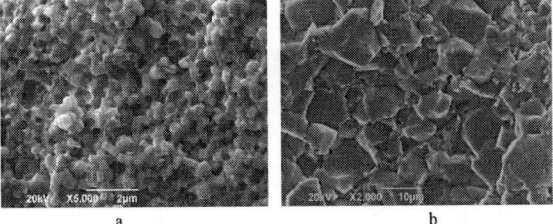

a   b

Fig.2(a)SEM of $BaFe_{9.6}(Co\ Ti)_{1.2}O_{19}$ without BBSZ at 920°C
(b) SEM of $BaFe_{9.6}(Co\ Ti)_{1.2}O_{19}$ adulterate with 3wt%BBSZ at 920°C

Inductor Structure and Fabrication

The physical prototype for inductor was shown in Fig.3. After optimized, the characteristics of the inductor simulated can be achieved and presented in Fig.5.

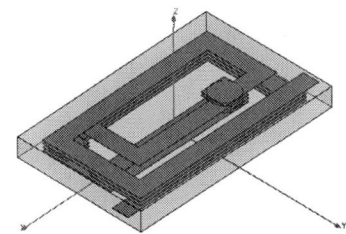

Fig.3 The physical proto type for inductor

The inductors were fabricated as Fig.4 demonstrated with special structure which will benefit for the improve the inductances.

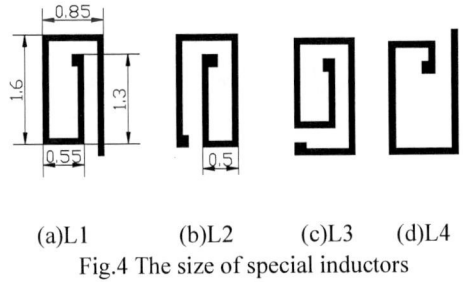

(a)L1   (b)L2   (c)L3   (d)L4
Fig.4 The size of special inductors

From the response of three dimension configurations in Fig.5., a cut off frequency at 560MHz, more than 2.52uH of the value of inductor can be achieved, which will satisfy the performances of inductor for high frequency applications.

Based on the material of $BaFe_{9.6}(Co\ Ti)_{1.2}O_{19}$, the helix inductor samples were fabricated by the conventional LTCC technology process with the outline dimension of 2.0mm×1.2mm×0.9mm shown in Fig.7.

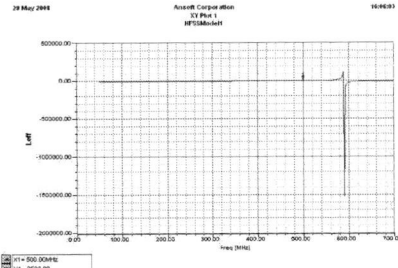

Fig.5 The stimulated results of inductor by HFSS software with the cut-off frequency at 560MHz

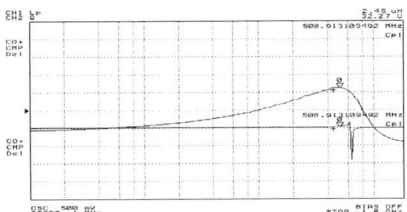

Fig.6 The testing results of inductor with cut-off frequency at 508MHz

The measurements were carried on by Agilent 4291B and the collected data was then calibrated to the desired reference plane by the thru-reflect line (TRL) technique through carefully designed calibration standards embedded in the same LTCC tie. The measured responses of the inductors are shown Fig.6. From Fig.6, it can be clearly seen that the measured results agree with the simulated data basically. The cut-off frequency was at 508 MHz and the inductance is 2.45 uH. Thus, a good coincidence between simulated and measured data is observed. All the measurements indicate that the method is very effective for the fabrication of multilayer chip inductors.

Fig .7 Experimental prototype for the samples

Conclusions

The structure analysis of LTCC-based passive components is reported for the design of a small multilayer chip helix-type inductor. The inductor fabricated by LTCC process has small size (2.0mm×1.2mm×0.9mm) using material $BaFe_{9.6}(Co\ Ti)_{1.2}O_{19}$. The cut-off frequencies for inductor samples are 508 MHz, and the value of inductor is 2.45uH respectively. The testing results are in a good agreement with the simulated data which will be helpful for the manufacture of passive components.

978-1-4244-4658-2/09 $25.00 © 2009 IEEE   287

Acknowledgments

This work was supported by the Youth Fund of Sichuan Province under Grant No 08ZQ026-013, the Foundation for Innovative Research Groups of the NSFC under Grant No. 60721001, the Youth Fund of University of Electronic Science and Technology of China under Grant No. L08010301JX0725.

References

1. Downey, D. F. *et al*, Ion Implantation Technology, Prentice-Hall (New York, 1993), pp. 65-67. [A book reference ...]

2. Wasserman, Y, "Integrated Single-Wafer RP Solutions for 0.25-micron Technologies," *IEEE Trans-CPMT-A*, Vol. 17, No. 3 (1995), pp. 346-351. [A reference to a journal article ...]

3. Shu, William K., "PBGA Wire Bonding Development," *Proc 46th Electronic Components and Technology Conf*, Orlando, FL, May. 1996, pp. 219-225. [A reference to a presentation at a Conference...]

4. Zhang. H. W Zhong. H Liu,B.Y. et al. "Electromagnetic properties of a new ferrite-ceramic low-temperature cocalcined (LTCC) composite materials," *IEEE Transactions on Magnetics*, Vol 41, No 10 (2005), pp 3454-3456.

5. Tomohiro Seki, Kenjiro Nishikawa, Yasuo Suzuki, et al. 60 GHz Monolithic LTCC Module for Wireless Communication Systems. Proceedings of the 9th European Conference on Wireless Technology, September, 2006, Manchester, UK, 376-379

6. G.B.Teh, D.A, Jefferson,J. Solid State Chem. 167 (1) (2002) 254-257.

Modeling Solder Joint Reliability of VFBGA Packages under Board Level Drop Test Based on Dynamic Constitutive Relation with Thermal Effect

Xiaoyan Niu[1,2], Tong Chen[1], Zhigang Li[1], Xuefeng Shu[1,*]

[1]Institute of Applied Mechanics & Biomedical Engineering, Taiyuan University of Technology
No.79 West Yingze Street, Taiyuan, 030024, Shanxi, China
[2]College of Civil Engineering and Architecture, Hebei University
No.180 Wusi East Road, Baoding, 071002, Hebei, China
*Email: shuxf@tyut.edu.cn, niu-xiaoyan2002@163.com, Tel: +86-351-6014455

Abstract

In this paper, according to the condition B in the drop test standard of JEDEC, create 3D finite element model of VFBGA (very-thin-profile fine-pitch BGA) packages in drop and impact, the material model of solder joints can be indicated by Cowper-Symonds model obtained from experiment data, taking into the consideration of strain rate and temperature effect. Comparing with traditional elastic for solder balls, this material model has better correlation with experimental measurement of dynamic strain, PCB center deflection. For conducting reliable analysis of VFBGA packaging in board-level drop condition, impact acceleration curve is applied to numerical simulation in input-G method.

1 Introduction

Board level solder joint reliability during drop impact is a great concern to semiconductor and electronic product manufacturers, especially for handheld or portable telecommunication devices such as mobile phone and PDA.[1] The mechanical shock resulted from mishandling during transportation or customer usage, may cause solder joint failure, which leads to malfunction of product. Under drop/impact loadings, solder joints in board level packages of the products, which are key parts functioned as electric connections and mechanical supports, Experience deformation in strain rate of $10^3 s^{-1}$. [2] However, there is not enough understanding to behavior of the solder joints under so high strain rates currently.

Board level drop test is convenient to characterize the solder joints performance, because it is more controllable than product level drop test. A few JEDEC standards [3-5] were released to provide standardized methods to conduct board level drop test. The actual drop test, however, is very expensive and time-consuming, and requires much manpower in measurement and failure analysis. In addition, very limited experimental results can be collected, so it is far from enough to observe the full dynamic responses of whole package during drop test. Especially for solder joint, it is almost impossible to measure its stress or strain response which in turn affects the solder joint reliability.

Numerical modeling is proved a cheap, fast and efficient approach in IC packaging for the purpose of design analysis and optimization. A validated drop impact model can depict the dynamic responses from outside to inside of package and thus enable researchers to understand the physics-of-failure. [6] However, the solder balls were always assumed to be elastic or simple elastoplastic in previous studies, this is not sufficient to capture the true mechanical behavior of solder balls during drop impact. Accurate material models are necessary to obtain meaningful results in computational simulation.

In this paper, according to JEDEC Standard JESD22-B111 by Joint Electron Device Engineering Council, we have established 3D finite element model of VFBGA packages in drop and impact, in which the material model of solder joints can be indicated by Cowper-Symonds model obtained from experiment data to be considered the effect of strain rate and temperature. Then the drop reliability of a VFBGA package is simulated using validated input acceleration (Input-G) method.

2 Numerical modeling for drop test

The test apparatus for the pulse-controlled board-level drop test is schematically shown in Fig.1. The drop table is released at a certain height to hit on the strike surface. As shown in Fig.2, the board-level test vehicle is mounted on the base plate with the corners fixed on the standoffs while the base plate is welded to the drop table. Ideally, the standoffs should move synchronously during the drop impact process. Defined by JEDEC [3, 5], an accelerometer is attached to the base plate at a specific location to monitor peak acceleration and duration of the half-sine impact pulse. The JEDEC drop test conditions are defined chiefly according to these two factors.

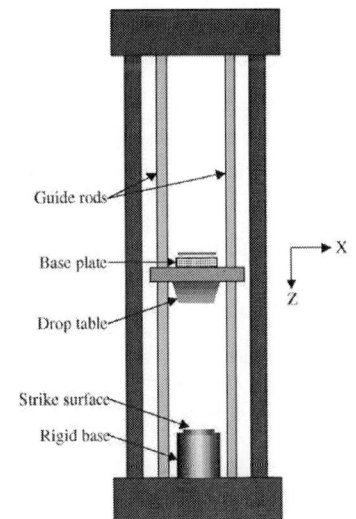

Fig.1. Schematic of drop test apparatus [3]

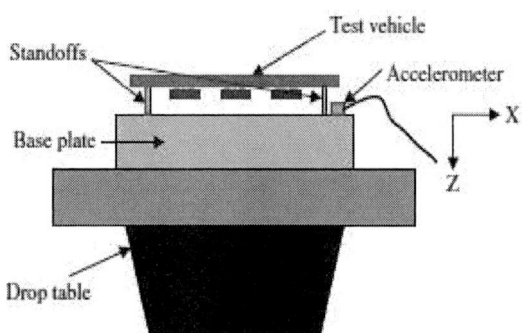

Fig.2. Schematic for board-level test vehicle mounted on drop table [3]

Finite element method is often used to calculate the dynamic response of the drop test board and to correlate the drop impact performance. It is difficult to include all involved parts in the finite element model. The size of the finite element model will be very large if all details of drop table, PCB, and all components on PCB are included. In this case the numerical difficulties will be encountered to deal with the contact simulation between the felt and drop table block. In order to reduce the computational time, the input-G method was developed by Tee and coworkers. [7, 8] The impact acceleration pulses, which can be measured from the actual test apparatus, are imposed on the supports of the board-level test vehicle as boundary conditions. Thus, the complex variations in friction, contact condition, or other unknown tester parameters, can be considered easily and accurately, because the effects of velocity before impact and contact surface/material are already included in the impact pulse curve. Due to the simplicity of input-G method, its solution time is three times faster than the free-drop model, using the same element mesh size of PCB and package.

In this paper, only one package is mounted on the center of test board. A very-thin-profile fine-pitch BGA (VFBGA) package with 181 solder balls is selected as a case study. Quarter 3D model is established due to symmetry (horizontal drop, component faced down). We consider a $10 \times 10 \times 0.8$ mm^3 VFBGA interconnected to a $132 \times 77 \times 1$ mm^3 standard 8-layer JEDEC drop test board with Sn3.0Ag0.5Cu solder joints. The package contains a $5.19 \times 5.19 \times 0.25$ mm^3 silicon die and a 0.26mm thick substrate. Solder balls of a diameter of 0.3mm are used. After reflow, the width and standoff of a solder joint become 0.35 mm and 0.24 mm, respectively. Openings of a solder joint on the package side and the test board side are 0.26 mm and 0.28 mm, respectively. The pitch between adjacent solder joints is 0.5 mm

Figure 3 shows the quarter symmetry finite element model with one package located at PCB center. The finite element model contains 233,831 nodes and 161,859 linear hexahedral solid elements. Detailed package geometry, solder balls, solder mask and pad design are included in the model. The pad design is SMD on component side and NSMD on PCB side. The analysis was carried out using ANAYA/LS-DYNA.

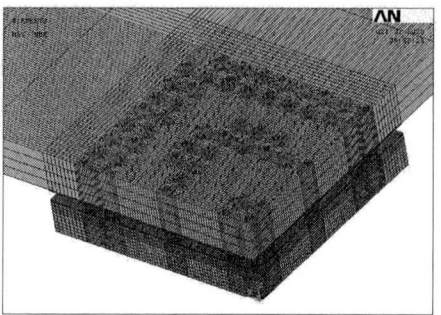

a)

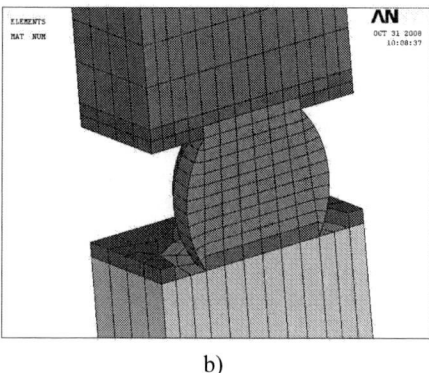

b)

Fig. 3 a) Quarter symmetry finite element model around the package; b) Detailed structure of solder joint

3 Material parameters

In this study, all constitutive components except for the solder alloy are assumed linearly elastic with the elastic properties presented in Table 1. Identical transversely isotropic material properties are assigned for the test board and the substrate, Table 2. In these tables, E is the Young' modulus, v the Poisson's ratio, ρ the mass density, and μ the shear modulus. Note that z denotes the out-of-plane direction while x and y in-plane directions.

Table 1. Elastic properties of components [9]

Component	E(GPa)	v	ρ(Kg/m^3)
Substate/test boars	Transversely isotropic	Transversely isotropic	1910
Die	131	0.23	2330
Compound	28	0.35	1890
Cu pad	117	0.38	8960
Solder mask	2.41	0.47	1100

Constitutive laws are very important in the numerical simulation of mechanical behavior of material. The cowper-Symonds material model is used to estimate deformation response and strain rate sensitivity of solders at high strain rate conditions. [10] The equation is as follows

$$\frac{\sigma}{\sigma_0} = 1 + \left(\frac{\dot{\varepsilon}}{C}\right)^{1/P} \tag{1}$$

where σ and σ_0 are dynamic and quasi-static stress, respectively; $\dot{\varepsilon}$ is strain rate; C (unit: s^{-1}) and P are Cowper-Symonds coefficients.

By logarithmic transformation of the Eqs. (1), we obtain

$$\ln\dot{\varepsilon} = P\ln(\sigma/\sigma_0 - 1) + \ln C \qquad (2)$$

Apparently, it shows a linear relationship between $\ln\dot{\varepsilon}$ and $\ln(\sigma/\sigma_0 - 1)$. From the experimental data [11] the values for C and P are presented in Table 3 for Sn3.0Ag0.5Cu. In addition, taking into account the electronic packaging components during normal working hours, the temperature is normally around 60°C. So we obtained the values of C and P at 60°C, it is used to express temperature effects on reliability of solder joints under board-level drop test conditions.

Table 2. Transversely isotropic properties of substrate and test board [9]

Substrate			
E_x, E_y (GPa)	16.8	E_z (GPa)	7.40
G_{xz}, G_{yz} (GPa)	7.59	G_{xy} (GPa)	3.31
v_{xz}, v_{yz}	0.39	v_{xy}	0.11
Test board			
E_x, E_y (GPa)	17.7	E_z (GPa)	7.80
G_{xz}, G_{yz} (GPa)	7.99	G_{xy} (GPa)	3.49
v_{xz}, v_{yz}	0.39	v_{xy}	0.11

Table 3. The parameters of the Cowper-Symonds equation for Sn3.0Ag0.5Cu solder

Material	ρ /Kg/m^3	E /GPa	T /°C	Cowper-Symonds parameters	
				C/s^{-1}	P
SnAgCu	7384	33.06	20	645.225	1.521
	7384	31.52	60	339.170	1.380

We investigate board-level drop reliability of VFBGA packages subjected to JEDEC drop test condition B, which features an impact pulse profile with a peak acceleration of 1500G and a pulse duration of 0.5 ms. This acceleration is defined as boundary condition to four corner screws of PCB, which is constrained in in-plane directions. For the two symmetrical planes, corresponding symmetrical constraint is applied.

4 Simulation results and discussion

4.1 Displacement of PCB

It is observed that the PCB cyclically vibrates, i.e. bends up and down during test due to impact inertial. Therefore, the PCB curvature is very important for evaluation of solder joint performance. Figure 4 shows the warpage distribution of PCB during the maximum downward bending. The PCB has much larger warpage difference in the length direction than in the width direction. It implies that the outer row of solder balls in the PCB length direction warps more and has higher bending stress level.

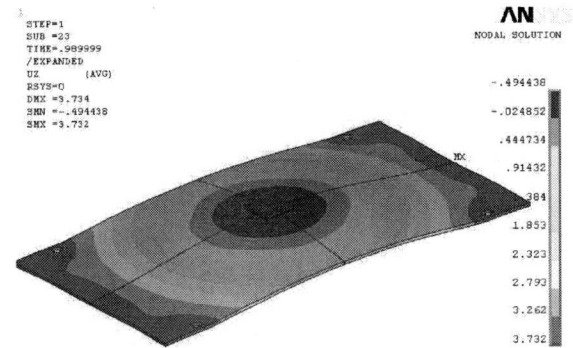

Fig. 4 PCB maximum warpage distribution

The relative displacement of any part of PCB against time to fixed screw can be obtained from modeling. Figure 5 shows the deflection of PCB center varying with time. The positive value means PCB bends up while the negative value denotes bending down of PCB.

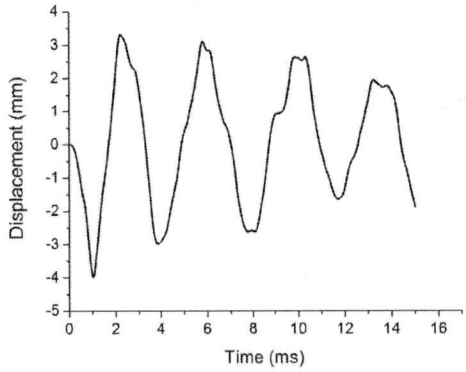

Fig. 5 Relative displacement of PCB center

4.2 Stress and strain behavior of critical solder joint

The reliability of solder joints during the drop impact is the main concern, as it affects the functionality of product. Figure 6 shows the variation of stresses in the critical solder joint with impact time, including normal stress in PCB length direction (S_x), first principal stress (S_1), vertical normal stress or peeling stress (S_z), shear stress (S_{xz}), and Von Mises stress (S_{eqv}). All the stresses vary cyclically under PCB vibration, corresponding to the measured dynamic strain of PCB. Among them, the peeling stress has close amplitudes in positive and negative directions. Comparing the five stresses at the same point, the first principal stress follows the same pattern as the peeling stress, especially during the periods of positive peaks, indicating that the peeling stress is the dominant stress component. Therefore, the solder joint peeling stress is critical during drop impact and can be used as failure criteria for the purpose of design optimization. Since the peeling stress is induced mainly by PCB bending or vibration, it can also be concluded that the PCB bending is the

978-1-4244-4658-2/09 $25.00 © 2009 IEEE

major failure mechanism of PCB subassembly under drop impact, especially for component mounted at PCB center with 4- screw fixation. Comparing Figure 6 b) with Figure 6 a), when the temperature increased from 20°C to 60°C, all the stresses are significantly lower. It indicates that the temperature effect in the reliability of solder joints during the drop impact is not negligible.

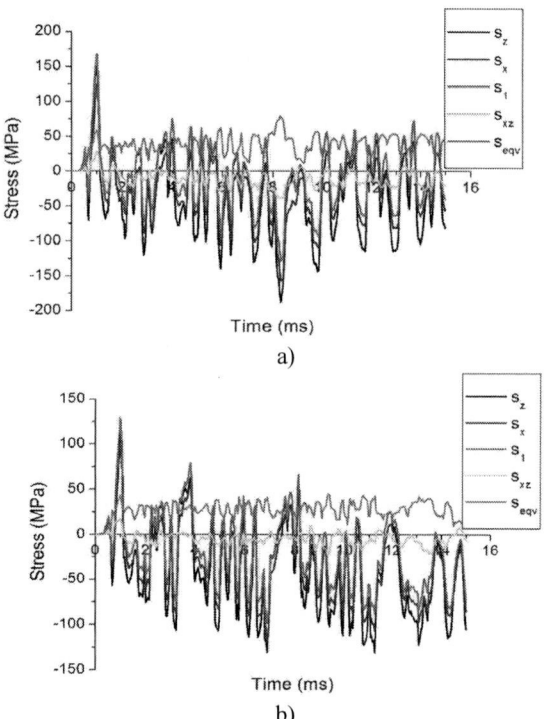

a)

b)

Fig. 6 Dynamic stresses during drop impact in the critical solder joint: a) at room temperature; b) at 60°C

Figure 7 shows the contour of normal stress on SnAgCu solder joints at room temperature and at 60°C under condition B at package side. From the figure, it is clear that corner solder joints incur the greatest stresses undergoing drop impact loads and therefore, they are more critical than solder joints at other locations. Stresses concentrate at the inner corner on the package side. Furthermore, it is found that the failure mode and critical solder ball location predicted by this modeling correlate well with test. Similarly, when the temperature increased from 20°C to 60°C, the peeling stress decreased from 148.032MPa to 116.525MPa.

a) at room temperature; b) at 60°C(unit: GPa)

In order to study the effects of strain rate, Figure 8 shows the effect of Cowper-Symonds model and linear elasticity on stress/strain at room temperature for SnAgCu, the maximum peeling stress exists on the package side. Known from the figure that it is obvious difference to get the maximum peeling stress of solder joint by using Cowper-Symonds model and linear elasticity, the effects of strain rate is very important during the drop impact. The stress of solder joint decreased to 148.03MPa from 324.23MPa when considering strain rate effect. Therefore, strain rate can not be ignored

during simulation. If according to elastic model, solder joint will be too hard to over-estimate the value of stress.

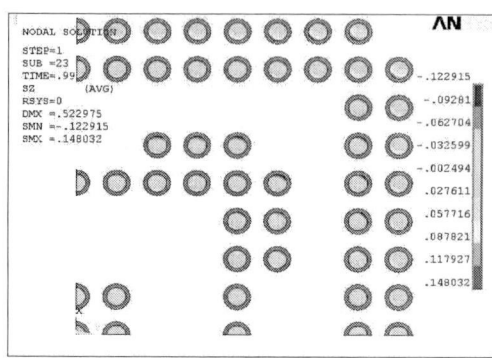

a)

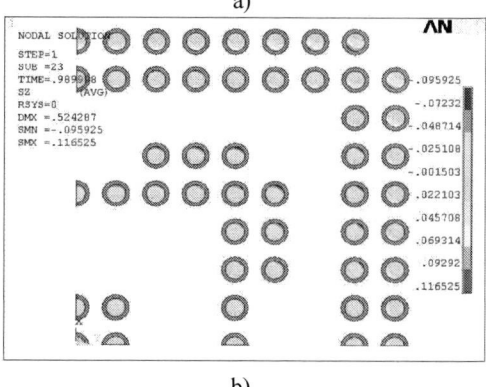

b)

Fig. 7 The contour of normal stress on SnAgCu solder joints under condition B at package side:

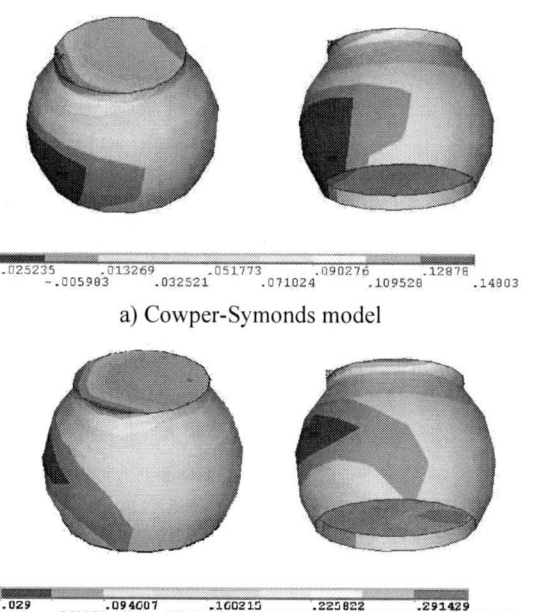

a) Cowper-Symonds model

b) Linear elasticity model

Fig. 8 The effect of Cowper-Symonds model and linear elasticity on stress/strain at room temperature for SnAgCu

5 Conclusions

In this paper, we investigate board-level drop reliability of VFBGA chip-scale packages subjected to JEDEC drop test condition B, which features an impact pulse profile with a peak acceleration of 1500G and a pulse duration of 0.5ms.

The numerical results indicate that the mechanical impact and bending of PCB caused by mechanical shock are the main reasons which induce the drop failure of solder joint in the drop impact conditions; it is not neglect to think over the strain rate and thermal effect of solders in numerical simulation; the peeling stress of solder joints which is located in the components/solder joint side is maximum at the outermost corner of BGA package, it indicates that the solder joint closed to components is easy to failure. Validated drop test model is a valuable design analysis tool to optimize the drop test performance of packages, saving coat, time, and manpower in performing the actual drop tests.

Acknowledgments

This project was sponsored by the National Natural Science Foundation of People's Republic of China (10672113), the Special PH.D Research Fund for Universities (20050112009), the Natural Science Foundation of Shanxi Province (2007011010) and the study abroad information of Shanxi Scholarship Council of China. (2008.30)

References

1. Tee, T. Y. *et al.*, "Impact life prediction modeling of TFBGA packages under board level drop test," *Microelectronics Reliability*, Vol. 44, No. 7 (2004), pp. 1131-1142.

2. S K W Seah, *et al.*, in the *4th Electronics Packaging Technology Conference*. (Singapore, 2002), pp. 120-125.

3. JEDEC Standard JESD22-B111, *Board Level Drop Test Method of Components for Handheld Electtronic Products*, 2003.

4. JEDEC Standard JESD22-B104-B, *Mechanical Shock*, 2001.

5. JEDEC Standard JESD22-B110, *Subassembly Mechanical Shock*, 2001.

6. Long Wen, *et al.*, "Dynamic Properties Testing of Solders and Modeling of Electronic Packages Subjected to Drop Impact", *2008 International Conference on Electronic Packaging Technology & High Density Packaging* (ICEPT-HDP 2008)

7. Tong Yan Tee, *et al.*, "Advanced experimental and simulation techniques for analysis of dynamic responses during drop impact[A]," *54th Electronic Components and Technology Conference*[C], 2004, pp. 1088-1094.

8. Tong Yan Tee, *et al.*, "Novel numerical and experimental analysis of dynamic responses under board level drop test[A]," *5th. Int. Conf on Thermal and Mechanical Simulation and Experiments in Micro-electronics and Micro- Systems*, 2004, pp.134-140.

9. Yi-Shao Lai, *et al.*, "Effects of different drop test conditions on board-level reliability of chip-scale packages," Microelectronics Reliability, Vol. 48 (2008), pp. 274-281.

10. Cowper GR, *et al.*, "Strain hardening and strain-rate effect on the impact loading of cantilever beams," *Division of Applied Mathematics Report 28*, Brown University; September 1957.

11. Niu Xiaoyan, "EXPERIMENTAL RESEARCH AND RELIABILITY ANALYSIS ON LEAD-FREE SOLDER JOINTS IN MICROELECTRONIC PACKAGING," Doctor thesis, TYUT, 2009.

Effects of PCB Dynamic Responses on Solder Joint Stress

Tong AN, Fei QIN*

College of Mechanical Engineering and Applied Electronics Technology,
Beijing University of Technology, Beijing, 100124, China
Tel: +86-10-67392760; Fax: +86-10-67391617
*Email: qfei@bjut.edu.cn

Abstract

The reliability of board level electronic package subjected to drop impact loadings is one of the most concerned issues. In this paper, a standard board level drop impact test was modeled as double cantilever beam model. The deflection and curvature of the Printed Circuit Board (PCB) and the component were compared with that derived from static analysis in order to understand the influence of dynamic response of the PCB on the solder joints. The results show that the difference of flexibility between the PCB and the package is the main driver for solder joint stress; the dynamic response of the PCB observably affects the stress in solder joints.

1 Introduction

Solder joints, which serve as mechanical, thermal and electrical interconnections between the electronic packages and PCB, are the most vulnerable area in a portable product when the product is subjected to accidental drop impact. As a result, a comprehensive study of mechanics behavior of solder joints is indispensable to understand the reliability of portable electronic products.

Since board level drop impact test is a crucial test for portable electronic products, it has attracted many researchers' attention. Some works on drop impact test simulation and experiment were reported over the past few years [1-6]. Tee et al. [7] found experimentally that it is the PCB bending that induces the solder joint failure in during the drop impact. The solder joint peeling stress is critical during drop impact and can be used as a failure criterion for the purpose of design optimization [8]. Some researchers suggests that the static test can substitute for drop impact test bases on the theory that peeling stress is induced mainly by PCB deflection [9], which means that the same magnitude stress can be reached so long as the PCB deflection derived from static test is equal to that from drop impact test. However, the dynamic response of PCB has great influence on solder joints due to higher order vibrating modes.

In this paper, a sliced 3D model of the PCB-solder joint-component system and the input acceleration (Input-G) method [10] were used to investigate the influence of PCB dynamic behavior on the solder joint stresses.

2 Drop Impact Simulation of a Board Level Package

Figure 1 shows the finite element model of a board level package. In this analysis, a component in a size of $6\times0.5\times1.02$ mm^3 was interconnected to a $50\times0.5\times1$ mm^3 PCB through Sn3.0Ag0.5Cu solder joints. The component contains a bare silicon die in a size of $3\times0.5\times0.26$ mm^3. A solder mask, substrate and Cu pads were taken into account. The diameter,

standoff and pitch of the solder joint were 0.35 mm, 0.28 mm and 0.5 mm, respectively. Pad designs are SMD on the component side and NSMD on the PCB side. The Young's modulus E, Poisson's ratio v and density ρ of various materials used in the finite element model are listed in Table 1.

The element type was C3D8R in the ABAQUS [11], and there were 22,052 elements and 26,258 nodes totally. The drop impact test condition H defined by the JEDEC [12] was used, which is characterized by an half-sine impact acceleration pulse with a peak of 2,900 G and a time duration of 0.3 ms.

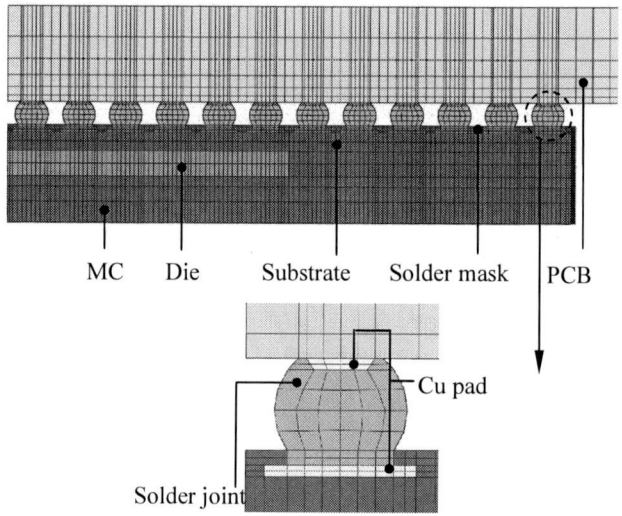

Fig. 1 Finite element model

Table 1 Parameters used in the finite element model

Materials	E (GPa)	v	ρ (kg/mm^3)
Die	131	0.23	2.33
Mold component (MC)	28	0.35	1.97
Cu pad	117	0.34	8.94
PCB	20	0.11	1.91
Solder mask	5	0.30	1.15
Sn3.0Ag0.5Cu solder	54	0.363	7.384
Substrate	26	0.11	2.00

3 Stress in Solder Joints under Drop Impact

3.1 Main driver for solder joint stress

Figure 2 shows the major mechanism of solder joint stress under drop impact [7]. Since the bending stiffness is different between the PCB and the component, the outermost solder joints experience tensile stress as the PCB bends downwards. On the other hand, when the PCB bends upwards, the outermost corner solder joints are under compressive stress. During the drop impact, the PCB bending induces repeated tensile and compressive stresses in the solder joints, and finally leads to the failure of the solder joints.

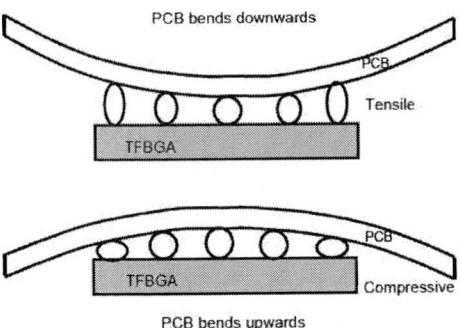

Fig. 2 Stress in solder joint during PCB bending under drop impact (Reprinted from ref. [7])

Although it is clear that the different flexing between the PCB and the package is the main failure driver for solder joints, an in-depth understanding of that how the PCB deflection influence the stresses in solder joints has not been discussed in great detail.

Based on the numerical model in Section 2, the history of maximum peeling stress in the critical solder joint was computed and is shown in Figure 3. Four time points have been chosen (1: 0.27 ms; 2: 0.51 ms; 3: 0.72 ms; 4: 0.855 ms) and the flexure curves of PCB and component and the peeling stresses in each solder joint are presented in Figures 4~7, which are adopted to discuss the influence of the deflection of PCB and component on the solder joints stresses.

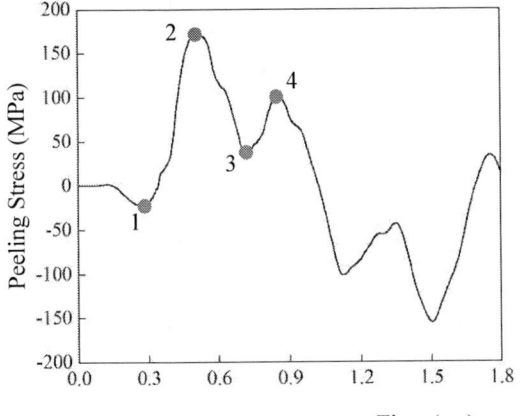

Fig. 3 The maximum peeling stress in the critical solder joint

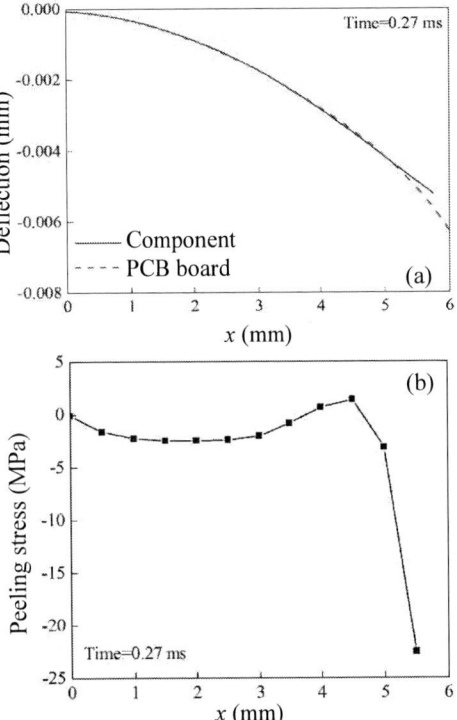

Fig. 4 (a) Deflection of PCB board and component; (b) Peeling stress of solders at 0.27 ms

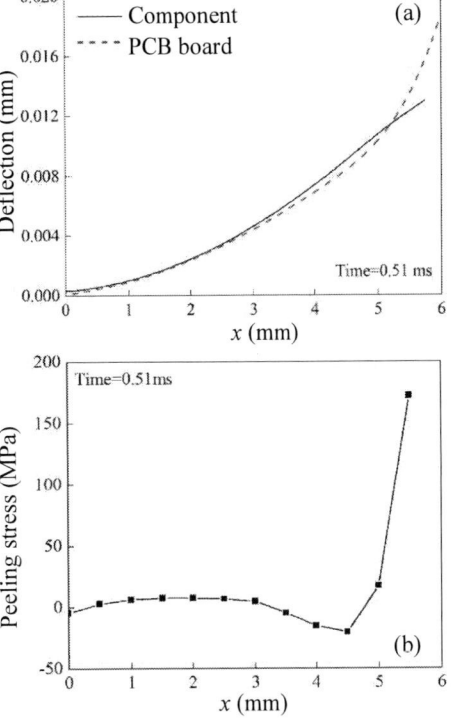

Fig. 5 (a) Deflection of PCB board and component; (b) Peeling stress of solders at 0.51 ms

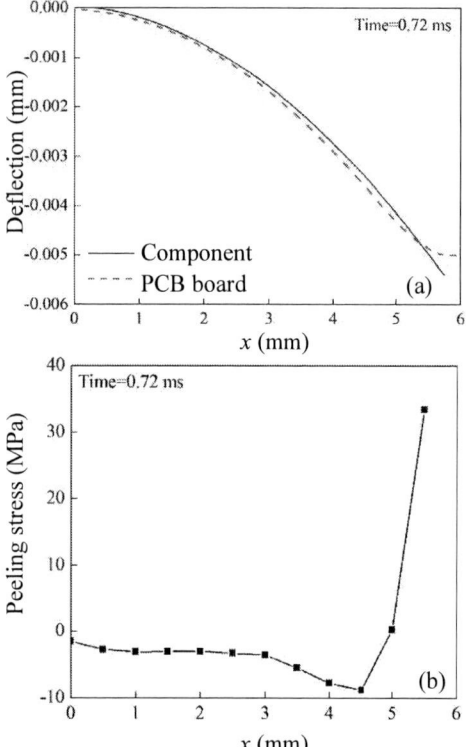

Fig. 6 (a) Deflection of PCB board and component; (b) Peeling stress of solders at 0.72 ms

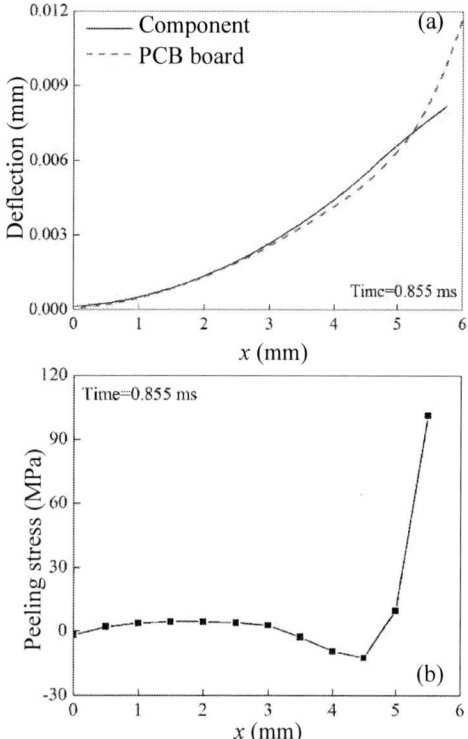

Fig. 7 (a) Deflection of PCB board and component; (b) Peeling stress of solders at 0.855

At the time of 0.27 ms, both PCB and component are upward, while the PCB bends much more badly due to its relative small bending stiffness (see Figure 4(a)), therefore the outermost solder joint suffers compressive stress at this moment, which is shown in Figure 4(b). A quite different phenomenon can be observed at the time of 0.51 ms. The PCB and component are downward at this moment, and the PCB bends more severity than the component does (see Figure 5(a)). Obviously, the critical solder joint experiences tensile stress as shown in Figure 5(b). It can be seen that the critical solder joint is under tensile stress at both 0.51 ms and 0.855 ms, while the peeling stresses of the solder joints obtained from time 0.51 ms are greater than that from time 0.855 ms, since the PCB deflection is larger at the anteriority moment. It indicates that the flexures of PCB and component have significant influence on the stress level of solder joints.

3.2 Dynamic vs. static analysis

The drop impact is physically a dynamic problem and the solution time of dynamic analysis is highly dependent on model and element size. As the complex electronic package model, the solution time can take up to days with the latest computation capability due to the small element size. Therefore, some researchers proposed static method for drop impact analysis for quick first-order estimation [9]. Since the solder joint failure is mainly due to PCB bending after drop impact, equivalent forces are applied to the entire PCB subassembly uniformly such that the maximum PCB deflection and strain are the same as experimental measurements in their research.

In order to provide a thorough understanding of the dynamics of PCB flexure and its influence on solder joints stresses, both dynamic and static analysis were performed. In the static analysis the maximum deflection at PCB obtained from dynamic analysis was applied to induce a bending load in the model, and the maximum peeling stress in the critical solder joint obtained from both analyses were compared and discussed carefully.

Figure 8 shows the flexure curve of PCB from both dynamic and static analysis with the same maximum deflection and the maximum peeling stresses of critical solder joints are given. At 0.27 ms, although the maximum deflection is the same, the peeling stresses obtain from dynamic and static analyzes are quite different from each other. In the dynamic analysis the transient motion of PCB is dominated by high level mode shape, therefore the flexure curves of PCB obtain from two analyzes are significantly different, which lead to the solder joint is suffered tensile stress in static while that is compressive stress in dynamic. At the time of 0.51 ms, although the solder joint experiences tensile stress in both analyses, the value of peeling stress in dynamic is 172.1 MPa which is nearly three times higher than 67.5 MPa that from static, which can be explained by the obviously great curvature of PCB in the dynamic analysis.

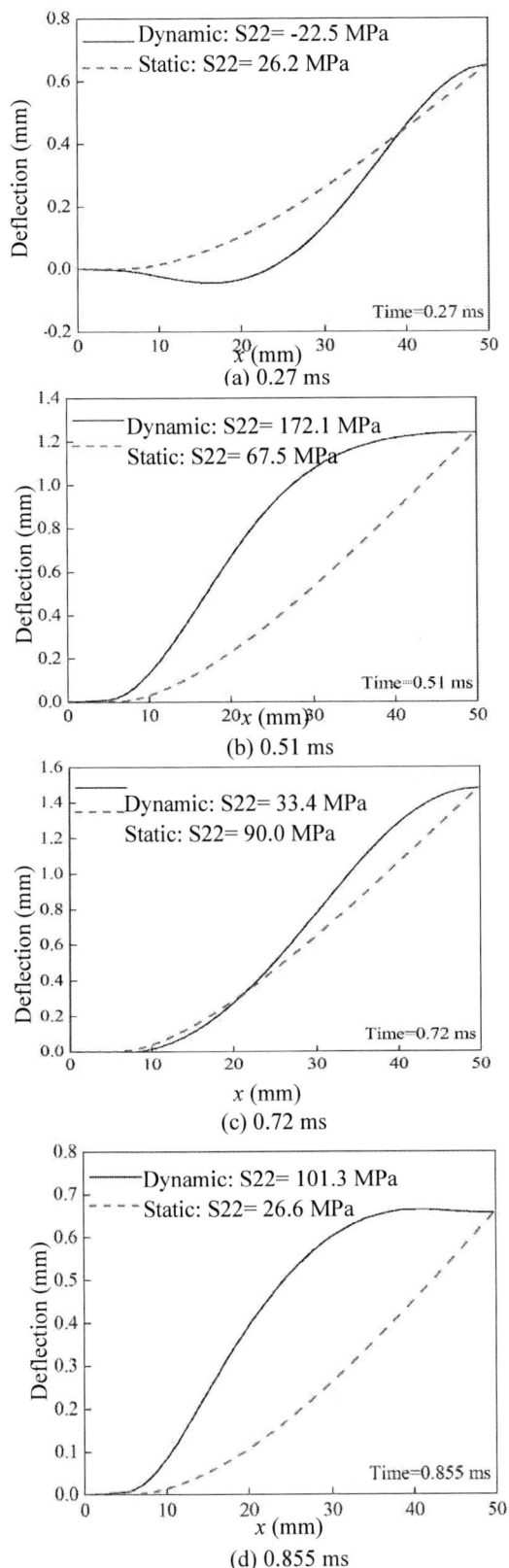

Fig. 8 PCB deflection and the maximum peeling stress comparison between dynamic and static analysis at different time

Conclusions

(1) The differential flexing between the PCB and the package was the main driver for solder joint stress. The stress and strain of solder joint on board side are greater than those on package side due to PCB deflection and inertial effect.

(2) The dynamic response of the PCB has great influence on the stress in solder joints. However, whatever the dynamic or static, the deflection and curvature of PCB play key roles to the peeling stress in solder joints.

Acknowledgments

The authors would like to thank financial support from the National Natural Science Foundation of China (NSFC) under the Grant No.10572010, and the Funding Project for Academic Human Resources Development in Institutions of Higher Learning Under the Jurisdiction of Beijing Municipality (PHR - IHLB).

References

1 Wong E H. "Dynamics of Board Level Drop Impact," *Trans. ASME, J. Electron. Packag.*, Vol. 127, (2005), pp. 200-207.

2 Wong E H, Mai Y W, Seah S K W. "Board level drop impact-fundamental and parametric analysis," *Trans. ASME, J. Electron. Packag.*, Vol.129, (2005), pp. 496-502.

3 Ren Wei, Wang Jianjun. "Shell-Based Simplified Electronic Package Model Development and its Application for Reliability," *Analysis. Proc.5th Electronics Packaging Technology Conference*, Piscataway, NJ, 2003, pp. 217-222.

4 Ren Wei, Wang Jianjun, Reinikainen T. "Application of ABAQUS/Explicit Submodeling Technique in Drop Simulation of System Assembly," *Proc. 6th Electronics Packaging Technology Conference*, United States, 2004, pp. 541~546

5 Zhu Liping. "Submodeling Technique for BGA Reliability Analysis of CSP Packaging Subjected to an Impact Loading," *Proc. Advances in Electronic Packaging*, New York, 2001, pp. 1401-1409.

6 Tee Tongyan, Luan Jing-en, Pek E, Lim C T, Zhong Zhaowei. "Novel Numerical and Experimental Analysis of Dynamic Responses under Board Level Drop Test," *Proc. 5th Int. Conf. Therm. Mech. Simul. Exp. Microelectron. Microsyst.* New York, 2004, pp. 133-140.

7 J. E. Luan, T. Y. Tee, E. Pek, C. T. Lim, Z. W. Zhong. "Modal Analysis and Dynamic Responses of Board Level Drop Test," *Proceedings of the 5th Electronics Packaging Technology Conference*, Singapore, 2003, pp. 233-243.

8 T. Y. Tee, J. E. Luan, E. Pek, C. T. Lim, Z. W. Zhong. "Advanced experimental and simulation techniques for analysis of dynamic responses during drop impact," *2004 Proceedings. 54th Electronic Components and Technology Conference*, Las Vegas, NV, USA, 2004, pp. 1088-1094.

9 T. Y. Tee, J. E. Luan, H. S. Ng. "Development and Application of Innovational Drop Impact Modeling Techniques," *55th Electronic Components and*

Technology Conference, Lake Buena Vista, FL, United states, 2005, pp. 504-512.

10 T. Y. Tee, J. E. Luan, E. Pek, C. T. Lim, Z. W. Zhong. "Novel Numerical and Experimental Analysis of Dynamic Responses under Board Level Drop Test," *Proc. 5th Int. Conf. Therm. Mech. Simul. Exp. Microelectron. Microsyst. EuroSimE 2004*, Brussels, Belgium, 2004, pp. 133-140.

11 ABAQUS 6.5 User's Manual, Hibbitt, Karlsson & Sorensen, Inc., 2004

12 JESD22-B104C. Mechanical Shock, JEDEC Solid State Technology Association, Arlington, VA, 2004.

Influence of Strain Rate Effect on Behavior of Solder Joints under Drop Impact Loadings

Tong AN, Fei QIN*, Jiangang Li
College of Mechanical Engineering and Applied Electronics Technology,
Beijing University of Technology, Beijing, 100124, China
Tel: +86-10-67392760; Fax: +86-10-67391617
*Email: qfei@bjut.edu.cn

Abstract

The strain rate dependent Johnson-Cook material model and the rate independent elastic-plastic model of lead-free solders were used to investigate influence of strain rate effect on the mechanical behavior of solder joints under drop impact loadings. Failure of the solder joints was predicted and the results were compared with the experimental observations. The strain rate effect of lead-free solders has no influence on the deflection of the PCB during the drop impact but has significant influence on the stress and strain in solder joints. The rate independent elastic-plastic solder material model always underestimates the stress and overestimates the strain of the solder joint. The material model that takes the strain rate effect into account can predict more realistic behavior of the solder joints.

1 Introduction

Currently, as a result of the harmfulness of the lead element to the environment, it is an inevitable development trend to use lead-free solder alloys, such as SnAg (tin and silver) and near-eutectic SnAgCu (tin, silver and copper) alloy, to replace lead-based solders. However, according to plenty of experiment results, the drop impact reliability of lead-free solder joints, which function as electric signal channels, thermal conductors and mechanical supports in microelectronic packages, exhibit orders of magnitude poorer performance than the eutectic Sn37Pb (63% tin and 37% lead in weight)[1]. Therefore the mechanics behavior of lead-free solder joint under drop impact has become one crucial problem for the reliability of portable electronic products.

As the mechanical properties of lead-free solders are more sensitive to strain rate than lead-based solders do[2,3], therefore it is necessary to take the strain rate effect into account when behavior of lead-free solder joint under drop impact loading are investigated. However, the rate-independent linear elastic or elastic-plastic material models are widely used for solder joints in the drop impact simulations [4-7], and leads to incorrect results.

In this paper, the strain rate dependent Johnson-Cook material model and the rate independent elastic-plastic model of lead-free solders were used to investigate influence of strain rate effect on the mechanical behavior of solder joints under four-point dynamic bending. Failure of the solder joints was predicted and compared with the experimental observations.

2 Four-Point Dynamic Bending Test

2.1 Experimental Procedures

The four-point dynamic bending test equipment and method used in this paper is exactly the same as that was applied in Ref. 8. In the test, a component of 13mm×11mm×1.4mm was mounted on the PCB with size of 75mm×40mm×1.0mm by 165 SnAgCu solder joints, and the installation angel is 45°. Diameter and standoff of the solder joint were 0.38 mm and 0.3 mm, respectively. Pitch of the solder joint in two directions is 0.580 mm and 0.570, respectively. The PCB was placed on two rolling supporters without any constrains, and a rigid trestle table was attached to the PCB, which allows the dynamic loading induced by a free fall steel ball being transmitted to the PCB. A cushion pad with diameter and height of 10 mm and 3.5 mm, respectively, was placed on the rigid trestle table. In order to measure deflection of the PCB, speckles were sprayed onto the side surface of the PCB, and a high speed camera was used to capture the deformation images of the speckle area. The digital image correlation method was used to measure the deflection of the PCB [8, 9]. The experiments were performed at room temperature.

2.2 Experimental Results and Discussion

Total five groups and six specimens for each group were tested. The five groups were labeled by the falling heights, 68 mm, 110 mm, 163 mm, 228 mm and 304 mm. The speckles were sprayed onto the side surface of PCB, and a high speed camera was used to record the deformation process of the speckle area. After the testing, the failure analysis was carried out and the digital image correlation software DICA1.0 developed by us was applied to calculate the PCB deflection. All the solder joints were not failure under the falling height of 68 mm and 110 mm. There was only one solder joint broken in one of all the six specimens when the falling height is 163 mm. This can be interpreted as no failure happened at this falling height. At height of 228 mm, 8 solder joints in five specimens failed, i.e., each specimen had 1.3 solder joints broken averagely. At the falling height of 304 mm, 23 broken solder joints were observed in five specimens and about 4 solder joints failed in each specimen averagely.

The experimental results indicate that when the falling height is less than 228 mm, the solder joints do not fail at all; when the height is more than 228 mm, the solder joints begin to fail and the number of broken solder joints increases with the increasing falling height. Therefore, the falling height of 228 mm is the critical height to solder joint failure in this research. The failure was observed at the interfaces between the PCB and the solder joints that in the outmost corner of the BGA array.

3 Numerical Model for the Four-Point Dynamic Bending

3.1 Finite Element Model

There were 165 solder joints used in the BGA array to connect the component and the PCB. If all the solder joints are modeled by fine solid finite elements, there is a great number of elements, plus the material nonlinearity and transient dynamic nature of the problem, the computational cost is huge. Hence certain simplification methodologies are needed. In this paper, only some critical solder joints which located at the diagonal outermost of the BGA array were treated as solder ball and were modeled by fine solid elements, and the others were modeled by a solder layer. This called solder layer-ball hybrid model is presented in Fig. 1(a). Tsai et al.[6] has shown that the solder layer model has no significant influence on the deflection of the PCB, and is capable to predict enough accurate deflection of the PCB. In Fig.1(b), 192 solid elements were used for each solder ball.

Since the rolling supporters, the trestle table and the steel ball have much greater stiffness than the PCB, they are assigned as rigid bodies. Contact surfaces between the rolling supporters, the PCB and the rigid trestle table were defined as hard contacts without friction, which implies that the contact is activated if and only if a pressure is built up between the two defined surfaces. The falling steel ball was modeled by element type C3D4 in the ABAQUS/Explicit[11], and contains 3289 elements; others were modeled by element type C3D8R, and there were 96243 elements totally.

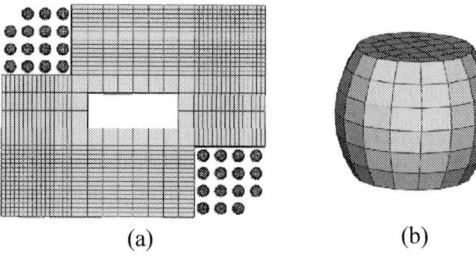

(a)   (b)

Fig. 1 Finite element models (a) solder layer-ball hybrid model; (b) solder ball model

The steel ball falls freely at height h. At the moment the ball contacts the cushion pad, its velocity is $v = \sqrt{2gh}$. In order to simplify the simulation, the free falling of the steel ball is neglected and instead of that, a near-contact velocity is applied as an initial condition. The cushion pad is treated as incompressible rubber-like materials, which are modeled by Mooney-Rivlin model; the PCB is modeled as orthotropic elastic material, and other parts including Cu pad, substrate, die and mold compound apply the elastic properties which are listed in Ref. [8].

3.2 Solder Joint Material Model

The rate-dependent Johnson-Cook model and rate-independent elastic-plastic material model presented in Ref. [3] were applied to the solder ball and solder layer. The Johnson-Cook model proposed by G. R. Johnson and W. H. Cook[10] in 1983 has been widely used to simulate large strain and high strain rate deformation process of metals. A general form of the Johnson-Cook model can be expressed as

$$\sigma = \left[A + B\left(\varepsilon^{\mathrm{p}}\right)^{n} \right]\left(1 + C \ln \dot{\varepsilon}^{*}\right)\left(1 - T^{*^{m}}\right) \qquad (1)$$

where σ, ε^{p} and $\dot{\varepsilon}^{*}$ are the von Mises flow stress, the equivalent plastic strain and the dimensionless strain rate, respectively. $\dot{\varepsilon}^{*}$ is defined as $\dot{\varepsilon}^{*} = \dot{\varepsilon}/\dot{\varepsilon}_{0}$, in which $\dot{\varepsilon}_{0}$ is a reference strain rate and a value of 0.001 s^{-1} was assigned to it in this research. T^{*} is the homologous temperature defined as $T^{*} = (T - T_{\mathrm{r}})/(T_{\mathrm{m}} - T_{\mathrm{r}})$, in which T_{r} and T_{m} are the reference temperature and melting temperature of the materials, respectively. The reference temperature was assigned to be room temperature here. There are five material constants, A, B, C, m and n, in Equation (1). Here A is the yield stress defined by the quasi-static compressive strain-stress data, B and n represent the effects of strain hardening, C is used to describe the strain rate effect, and m describes the effect of thermal softening. The Johnson-Cook model can be directly implanted into ABAQUS/Explicit [11]. The elastic-plastic model and the Johnson-Cook constitutive model of the Sn3.0Ag0.5Cu solder defined by Ref. [12] were used.

4 Strain Rate Effect

4.1 Strain Rate Effect on Solder Joint Stress

In order to show the influence of material models on behaviors of solder joints and the maximum PCB deflection, elastic, elastic-plastic and strain rate-dependent Johnson-Cook models were used in the analysis, respectively.

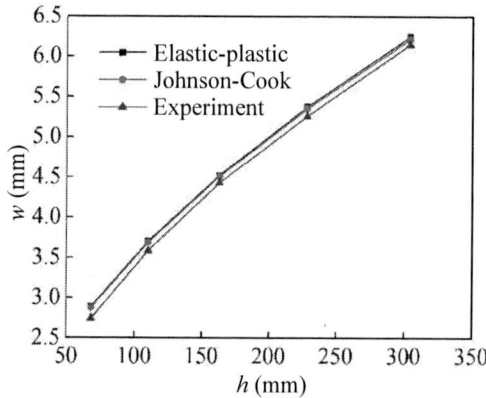

Fig.2 Maximum deflection of PCB center

Fig. 2 shows the maximum deflection at the PCB center. Deflections obtained by the experiment and the different material models agree quite well with each other, and it suggests that material models of solder joints have inappreciably influence on PCB deflection.

Deflection history of the PCB center, and von Mises stress in the critical solder joint at 163 mm falling height are presented in Fig. 3. The maximum von Mises stress in the critical solder joint occurs at the interface of PCB side, and it increases with the PCB deflection increasing. There is no obvious difference of the von Mises stress from two material models at early elastic stage of the deformation. After 1.2 ms, the solder joint enters into plastic deformation, however, the von Mises stress by Johnson-Cook model is clearly higher than that by elastic-plastic model due to the strain rate effect.

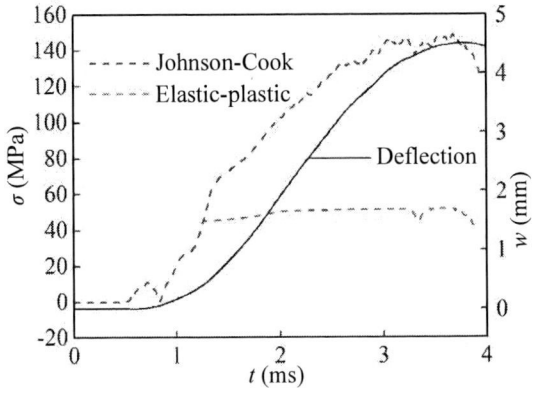

Fig. 3 Time histories of von Mises stress and deflection of PCB

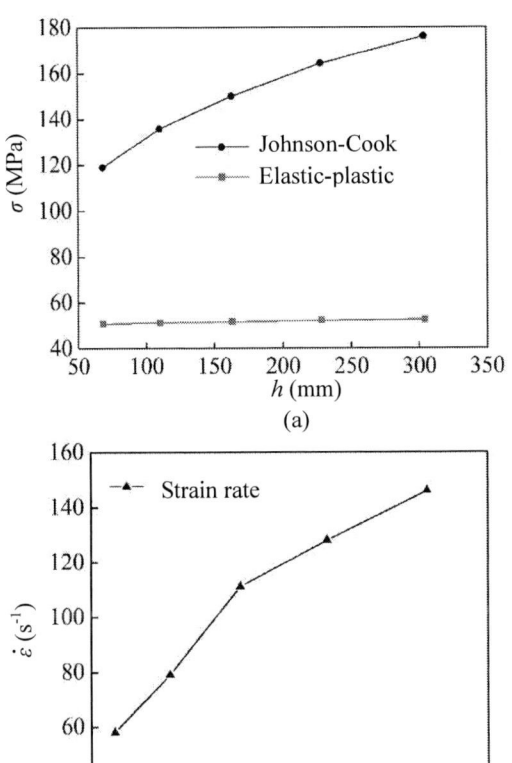

Fig.4 von Mises stress and strain rate in solder joints at different falling heights

The maximum von Mises stresses of the critical solder joint under various falling heights are shown in Figure 4(a). In the case of the Johnson-Cook model being applied to solder joints, the maximum von Mises stress increases with the increasing falling height because of the strain rate effect. However, for the elastic-plastic material model, the maximum von Mises stress has no significant increasing with the falling height increases. It suggests that by the rate-independent

elastic-plastic model, it is difficult to predict the solder joint failure according to von Mises stress. Figure 4(b) shows the strain rates under various falling heights by the Johnson-Cook model. The strain rate increases with the increasing falling height, and it reaches a peak value, 140 s^{-1}, at the height of 304 mm.

4.2 Strain Rate Effect on the Equivalent Plastic Strain

When the falling height was 163 mm, equivalent plastic strains computed by the different material models in the critical solder joint are plotted in Figure 5(a). At 1.2 ms, plastic deformation starts in the solder joint, and the equivalent plastic strain increases gradually. The equivalent plastic strain by the Johnson-Cook model is lower than that by the elastic-plastic material model, because the strain rate hardening is modeled by the Johnson-Cook model. Figure 5(b) shows the influence of the falling height on the equivalent plastic strain. It indicates that the equivalent plastic strain rises as the falling height increases.

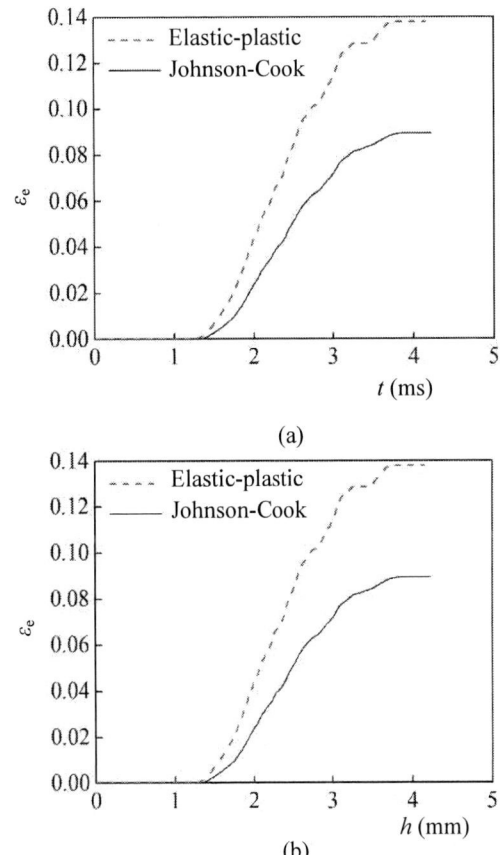

Fig. 5 Equivalent plastic strains in solder joints computed by different material models

4.3 Strain Rate Effect on the Solder Joint Failure Prediction

Taking the falling height of 228 mm as a critical height the solder joint begin to fracture, the maximum von Mises stress and the equivalent plastic strain of the critical solder

978-1-4244-4658-2/09 $25.00 © 2009 IEEE   301

joint computed by the two material models are plotted and compared with each other in Figure 6. At this height, the maximum von Mises stress is 164 MPa and the equivalent plastic strain is 0.11 by the Johnson-Cook model. This stress level reaches the tested strength of the solder [3]. At the same time, the maximum von Mises stress and equivalent plastic strain obtained from elastic-plastic material model is 52.2 MPa and 0.179 respectively. This suggests that, if stress is used as a failure or damage index, the stress by the elastic-plastic model is not able to predict the solder joint failure but the Johnson-Cook model can.

At the falling height of 304 mm, the maximum von Mises stress by the Johnson-Cook model is 176 MPa. According to the stress index, the two outer most solder joints fails, while their nearby inner solder joints, labeled by circles in Figure 6, do not break. This prediction agrees well with the experimental observation. If the equivalent plastic strain is considered as the failure index, the same result can be predicted by the Johnson-Cook model. However, the elastic-plastic material model predicts 3 more solder joint failure than the Johnson-Cook model.

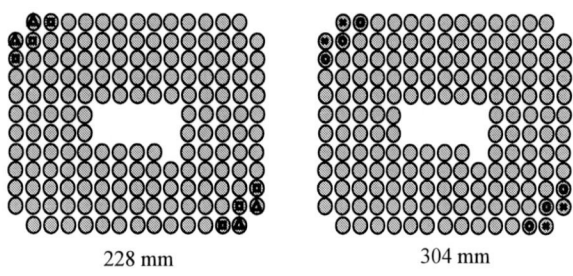

228 mm		304 mm	
Johnson-Cook	Elastic-plastic	Johnson-Cook	Elastic-plastic
⬢ 134MPa/0.072	52.0MPa/0.170	◉ 145MPa/0.082	52.4MPa/0.201
▲ 164MPa/0.110	52.2MPa/0.179	✖ 176MPa/0.130	52.5MPa/0.210

Fig.6 Failure prediction of solder joints

Conclusions

(1) The strain rate effect of solder material has little influence on PCB deflection, but has significant influence on the stress and strain of solder joint. The elastic-plastic material model predicts greater equivalent plastic strain while underestimates the solder joint stress.

(2) The rate-dependent Johnson-Cook material model can predict more realistic solder joint failure under dynamic bending whatever stress or strain is used as the failure index.

Acknowledgments

The authors would like to thank financial support from the National Natural Science Foundation of China (NSFC) under the Grant No.10572010, and the Funding Project for Academic Human Resources Development in Institutions of Higher Learning Under the Jurisdiction of Beijing Municipality (PHR - IHLB).

References

1. Zeng, K., and Tu, K. N. "Six Cases of Reliability Study of Pb-free Solder Joints in Electronic Packaging Technology," *Materials Science and Engineering R: Reports*, Vol. 38, No. 2, (2002), pp. 55-105.

2. Plumbridge, W. J. and Gagg, C. R. "Effects of strain rate and temperature on the stress-strain response of solder alloys," J*ournal of Materials Science: Materials in Electronics*, Vol. 10, No. 5, (1999), pp. 461-468.

3. Qin Fei, An Tong, Chen Na. "Strain rate effect and Johnson-Cook models of lead-free solder alloys," *2008 International Conference on Electronic Packaging Technology and High Density Packaging*. Shanghai, China: IEEE, 2008, pp. 734-739.

4. Tee, T. Y., Ng, H. S., Lim, C. T., et al. "Impact Life Prediction Modeling of TFBGA Packages under Board Level Drop Test," *Microelectronics Reliability*, Vol. 44, No. 7, (2004), pp. 1131-1142.

5. Desmond, Y. R. C., F. X. Che, Pang, J. H. L., et al. "Drop Impact Reliability Testing for Lead-free and Lead-based Solder IC Packages," *Microelectronics Reliability*, Vol. 46, No. 7, (2006), pp. 1160-1171.

6. Tsai, T. Y., Yeh, C. L., Lai, Y. S., and Chen, R. S. "Transient Submodeling Analysis for Board-Level Drop Tests of Electronic Packages," *IEEE Transactions on Electronics Packaging Manufacturing*, Vol. 30, No. 1, (2007), pp. 54-62.

7. Qin Fei, Bai Jie, An Tong. "Drop/Impact Stress Analysis of Solder Joints in Board Level Electronics Package," *Journal of Beijing University of Technology*, Vol. 33, No. 10, (2007), pp. 1038-1043. (in Chinese)

8. Qin Fei, Jin Ling, Yngve Wang. "Experimental and Numerical Study of Board Level Electronics Package with Dynamic Bending," J*ournal of Beijing University of Technology*, Vol. 34(Supp.), (2008), pp. 58-62 (in Chinese)

9. Qin Fei, Wei Jianyou. "Smoothing Algorithm for Strain Measurement in Digital Image Correlation Method," *Journal of Beijing University of Technology*, Vol. 34, No. 8, (2008), pp. 815-819 (in Chinese)

10. Johnson, G. R., Cook, W. H. "A Constitutive Model and Data for Metals Subjected to Large Strains, High Strain Rates and High Temperatures," *Proceedings of the 7th Int. Symp. on Ballistics*, The Hague, Netherlands, 1983, pp. 541-547.

11. ABAQUS 6.5 User's Manual, Hibbitt, Karlsson & Sorensen, Inc., 2004

12. Fei Qin, Tong An, Na Chen. "Strain Rate Effects and Rate-dependent Constitutive Models of Lead-based and Lead-free Solders," *Tran. ASME, Journal of Applied Mechanics* (will be published in November 2009).

Local Plastic Zones of Wirebond Profiles Inspection Base on Curvature Estimation

Hongjun Zhou, Lei Han

Key Laboratory of Modern Complex Equipment Design and Extreme manufacturing, Ministry of Education
Central South University, Changsha, 410083, China
E-mail: csu527@yahoo.com.cn

Abstract

In ultrasonic wire bonding, the local plastic zones play important roles in determining the wire profile. A way of detecting the local plastic zones using image processing was introduced in this paper. First we divided the video of wire bonding into continuous image sequences, after adaptive threshold segmentation, curve tracking, morphological opening and closing operator, finally we got the thinned curve. The curve was then smoothed with a small width Gaussian kernel in order to remove quantization noise and trivial details. The estimated curvatures distributions were calculated by means of point-to-chord distance accumulation. The local plastic zones of wire were detected based on the estimated curvature distributions by which the position was obtained. The results show that the wire local plastic zones can be detected accurately using this proposed method, and this method has strong robustness.

1. Introduction

Packaging interconnection is the most critical and most difficult part of the backend IC manufacturing. It influenced the cost and reliability of IC greatly. With the IC package to the development of the system-level integration, thinning and miniaturization, requests to the wire loop formation are getting higher and higher in the wire bonding [1]. Especially Wire Bond Looping in Stack Die Packages, as die thickness decreases, the space between the different wire looping tiers decreases accordingly. The wire bond loop height of the lower tiers needs to decrease to avoid wire shorts between the different layers of the loop. The top layer of the loop also needs to stay low to avoid exposed wire outside the molding compound [2]. In wire profile looping, the local plastic zones have an important impact on wire profiles. So detecting the local plastic zones of the wire is important to research the wire looping formation. And we can see the local plastic zones must have larger curvature. So we can detect the local plastic zones base on curvature estimation.

Calculating the curvature of the discrete planar curve, the main method is curvature scale-space (CSS) [3]. But the CSS curvature estimation technique is highly sensitive to the local variation and noise on the curve. In addition, the curvature estimation involves higher order derivatives of curve point-locations up to second order which cause errors and instability in results. And, the CSS technique requires appropriate Gaussian smoothing-scale selection which is a difficult task.

The purpose of this paper is to propose a new curvature estimation technique which overcomes the aforementioned problems associated with the existing CSS technique to detect the local plastic zones of wire bond profiles (curvature extreme points). We present a complete chord-to- point distance accumulation (CPDA) method for the discrete

curvature estimation which is used to inspect the local plastic zones. It can be used in researching the dynamic characteristics of the capillary and wire in the wire bonding looping formation.

2. Image Processing

The flow charts of image processing are shown as in Fig.1. After the image processing, then we can do the curvature estimation.

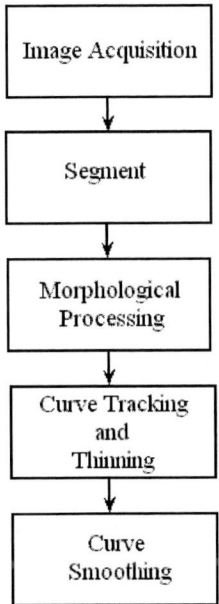

Fig.1. The flow charts of image processing

2.1 Image Acquisition

The wire bonding video which is obtained by the machine vision system is decomposed into continuous image sequences. In this paper, the video of Kulicke&Soffa company automatic wire bonder is used. The diameter of capillary is about 1.6mm. The diameter of gold wire is about 25.4μm. The time interval of each image sequence is one video frame which is about 0.12~0.24ms. Decomposing the video obtained a total of 413 images. Each image size is 320 × 240 pixels. The 256th and 257th images in sequence were shown as in Fig. 2.

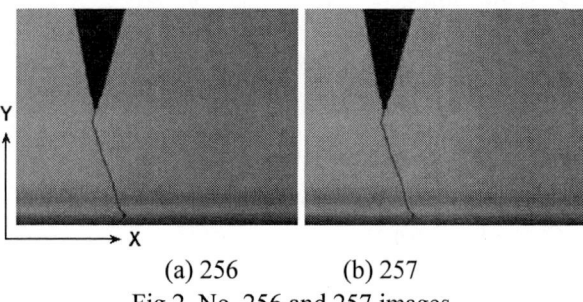

(a) 256   (b) 257

Fig.2. No. 256 and 257 images

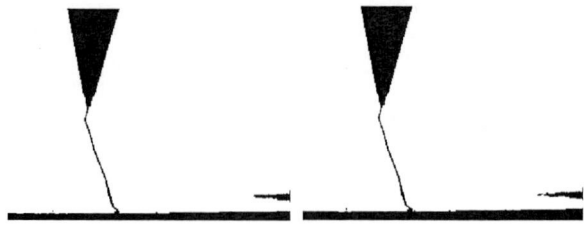

Fig.3. Segment results

2.2 Segment

There are many image segmentation methods. The classics are threshold method, region growing method, relaxation method, edge detection method, division combination method and so on. The modern methods are NN method, fuzzy clustering method and so on. The different method adapts to the different application fields. The method fitting some kinds of images maybe not fit the other kinds of images. There is no segmentation method fitting all images. So selecting the suitable segmentation method is very important.

In this paper, the capillary and the gold wire are moving unceasingly in the bonding process and the complex background various continually, so the contrast is not a constant. Thus choosing a fixed threshold value to segment the wires of all image sequences is not very good. Based on these characteristics, we choose the local dynamic threshold segment method, which is proposed by Kejun Wang and his partners [4]. The main idea of the method is calculating the mean $\mu(x,y)$ and variance $\sigma(x,y)$ of the points in $r \times r$ neighbor- hood of every points, then do segment using formula(1).

$$T(x,y) = \mu(x,y) + k \times \sigma(x,y) \qquad (1)$$

Where T(x, y) is the threshold, k is the coefficient of correction. If a pixel value is lower than the threshold, we consider it as wire domain. $\mu(x,y)$ and $\sigma(x,y)$ are calculated as follows[5].

$$\mu(x,y) = \frac{1}{r^2} \sum_{i=x-r/2}^{x+r/2} \sum_{j=y-r/2}^{y+r/2} f(i,j) \qquad (2)$$

$$\sigma(x,y) = \sqrt{\frac{1}{r^2} \sum_{i=x-r/2}^{x+r/2} \sum_{j=y-r/2}^{y+r/2} (f(i,j) - m(x,y))^2} \qquad (3)$$

The advantage of this method is that it can dispose aiming at every pixel well. Its disadvantage is that the speed is low, and it doesn't consider the boundary.

We use this method to segment the sequence image.Fig.3 shows the results after segmenting Fig.2

2.3 Obtain the Structure of the wire

There are three steps to obtain the structure of the wire after segmenting, morphological processing curve tracking and thinning. Morphological processing smooth the curve, fuses narrow breaks and long thin gulfs, eliminates small holes. Curve tracking remove the capillary and others from the image except gold wire part left.

Finally thinning process abstract the structure of wire pattern. The Image thinning is extracting a single pixel wide framework in the case of keeping the topological structure of the original image. In this paper we thin the wire image using the combination method of general conditional thinning and templates. Get rid of the special un-single pixel point after the general conditional thinning. The thinning result is as follows. But some of the wire skeletons are not perfect yet, we must prune them [6]. Fig.4 shows the results of the wire structure.

Fig.4. Results of the wire structure extraction

2.4 Curve Smoothing

The purpose of curve smoothing before curvature estimation is to reduce the effect of noise which may affect the curvature estimation severely. Noise may also be introduced by the edge detector.

In this paper, we smoothed the curve with a small width Gaussian kernel. Then do Gaussian smooth using formula (4), (5).

$$P_i = \sum_{j=-w}^{w} P_i \oplus_j G_j \qquad (4)$$

$$G_j = \frac{e - i^2 / (2\sigma^2)}{\sum_{j=-w}^{w} e - j^2 / (2\sigma^2)} \qquad (5)$$

Where P_i is the i th point; $\oplus$ denote model sum, namely $i \oplus j$ is $(i+j) \bmod n$; G_j is the Gaussian kernel; w is the filtering bandwidth; σ is the filter standard deviation.

978-1-4244-4658-2/09 $25.00 © 2009 IEEE

In general, short edges should be smoothed with small σ and long edges should be smoothed with large σ, because smoothing a short edge with a large σ may smooth out important features and smoothing a long edge with a small σ may leave a lot of noises on it. However, σ value depends on the amount of noise which is unknown. Therefore, determining appropriate σ for a given curve of any length is a difficult task.

Since large scale smoothing may smooth out many important features, we use small scale Gaussian smoothing to keep the effect of localization problem minimum. Because the curve length is between 100-200. For curves of length $n \leq 100$ we select scale $\sigma = 1$; for $100 < n \leq 200$ we select $\sigma = 2$. Fig.5 shows that using such small scale smoothing functions does not change the wire much. And we can see there are no any noises left in the image.

After smoothing, we extend the both ends of the smoothed curve. We extend each end by taking into account of the points of the same end. This curve extension prevents missing any local plastic zones near to the ends of the curve. The reason is, on points near to the ends, the number of chord movements becomes very low (zero at the ends) during the CPDA discrete curvature calculation. These phenomena results in low curvature estimation for a prominent corner (curvature extreme points) near to any end of the curve. The above curve extension reduces this effect. Noting that in the following section we do not calculate curvatures on the points of the extended parts, but we use the points on the extended parts to calculate the discrete curvature values on points near to the ends of the original smoothed curve.

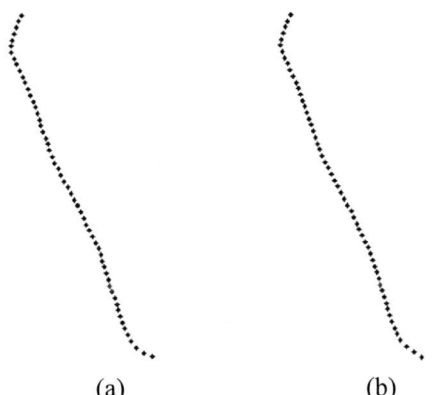

(a)   (b)

Fig.5. Results of Gaussian smoothing curves

3. Discrete Curvature Estimation

The formal definition of the chord-to-point distance accumulation (CPDA) [7] discrete curvature estimation technique is as follows. Let $P_1, P_2, P_3, \ldots$ be the n points of the parameterized curve $\Gamma(t) = (x(t), y(t))$, $1 \leq t \leq n$, as shown in Fig.6.

To measure the curvature $h_L(k)$ at a point P_k using a chord C_L of length L, we move the chord on each side of P_k at most L points while keeping P_k as an interior point. Note that according to [8], the chord-length L denotes the arc-length of the interior curve-segment as shown in Fig.6

Starting the movement from the point P_{k-L} when the two ends of the chord are at P_{k-L} and P_k respectively, we measure the perpendicular distance $d_{k,k-L}$ from P_k to the chord.

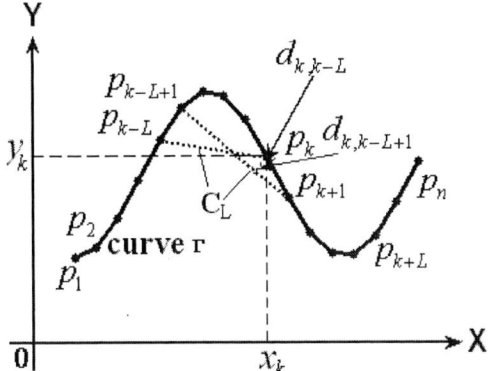

Fig.6. Schematic of distance from point to chord

Then we move the chord one point ahead when its two ends are at P_{k-L+1} and P_{k+1} respectively and measure the perpendicular distance $d_{k,k-L+1}$. The procedure continues by moving the chord one point at a time and stops when the both ends of the chord move to the points P_k and P_{k+L} respectively. Then we accumulate all distances to calculate the CPDA discrete curvature at the point P_k as

$$h_L(k) = \sum_{j=k-L}^{k} d_{k,j} \qquad (6)$$

Since $d_{k,k-L} = d_{k,k} = 0$, the above formula is simplified to

$$h_L(k) = \sum_{j=k-L+1}^{k-1} d_{k,j} \qquad (7)$$

Where, the distance $d_{k,j}$ is calculated as follows.

$$d_{k,j} = \frac{\left| (y_{k+L} - y_{k-L})x + (x_{k-L} - x_{k+L})y + (x_{k+L}y_{k-L} - x_{k-L}y_{k+L}) \right|}{\sqrt{(y_{k+L} - y_{k-L})^2 + (x_{k-L} - x_{k+L})^2}} \qquad (8)$$

And we can judge sign symbol of the distance by the determinant value [9], [10]. The determinant value is calculated by the matrix which is composed by P_{k-L}, P_k, P_{k+L} coordinate. Let the matrix $M_{k,L}(i)$ as follow.

$$M_{k,L}(i) = \begin{bmatrix} x_{i+L} - x_{i-L} & y_{i+L} - y_{i-L} \\ x_k - x_{i-L} & y_k - y_{i-L} \end{bmatrix} \qquad (9)$$

Then,

$$\left| M_{k,L}(i) \right| \begin{cases} > 0 & \text{sgn}(i,k,L) = 1 \\ = 0 & \text{sgn}(i,k,L) = 0 \\ < 0 & \text{sgn}(i,k,L) = -1 \end{cases} \qquad (10)$$

Where $\text{sgn}(i,j,L)$ represent the sign symbol of the distance from the point P_k to chord. In Fig.6 the distance $d_{k,k-L+1}$ is positive.

So the CPDA discrete curvature at the point P_k is expressed as follow.

$$h_L(k) = \sum_{j=k-L+1}^{k-1} \mathrm{sgn}(i,k,L)d_{k,j} \qquad (11)$$

The reference [10] proved the relation between the point-to-chord distance accumulation curvature and the curve curvature, and there was constant C meeting the follow formula.

$$\lim_{L \to 0} \frac{S_L(P_k)}{cL^3} = k(P_k) \qquad (12)$$

Where $k(P_k)$ is the curvature at point P_k.

4. Experiments and Analysis

In this section, we calculated the discrete curvature of the smoothed curve to inspect the local plastic zones (curvature extreme points). Note that the wire lengths of the image sequence are 0-200. So we choose three chords of medium lengths $L_1 = 10$, $L_2 = 20$, $L_3 = 30$ to calculate the CPDA distance curvature using(7).

Let $h_1(k)$, $h_2(k)$ and $h_3(k)$, where $1 \le k \le n$, be three curvature function using three chords of lengths L_1, L_2 and L_3 respectively. The absolute values of $h_j(k)$, where $1 \le j \le 3$, may range from zero to a long integer number depending on the types of corner on the curve. Moreover, the estimated curvature value of the same type of corner (for example, a corner with 50° angle) using any of the three chords may be different on different curves. Consequently, a single curvature threshold setting to detect the corners from all types of curves becomes problematic. To overcome this problem, we normalize the function using (9), so that the discrete curvature values are in the range [0, 1].

$$h'_j(k) = \frac{h_j(k)}{\max(h_j)}, \text{ for } 1 \le k \le n \text{ and } 1 \le j \le 3 \qquad (13)$$

Fig.7 shows the curvature for "wire 256" for different chord-lengths.

We can discover that the different L value was very small regarding the curvature extreme point's position and has little influence to the curvature result from Fig.7(b),(c),(d). When L value increased, the curvature distribution curve became flat gradually. We knew that L should take the small value by the formula (12). Moreover tests indicated that regarding the discrete curve, L value is too small, and then causes the result not to be unstable. In this paper, since the wire lengths were between 0 and 200, we chose L=5, 10 and15 to be possible to obtain satisfaction result. The curvature distribution could describe the curve bending strain.

A direct impact of wire's curve smoothing before the curvature estimating was also shown in Fig.7 (e). We see that though the average curvature function may contain a lot of local small peaks without smoothing [see the solid curve in Fig.7(e)], the effect of noise is reduced if the curve is smoothed prior to the curvature estimation [see dashed line in Fig.7(e)]. This will help distinguishing the curvature extreme points corresponding to the plastic zones unambiguously.

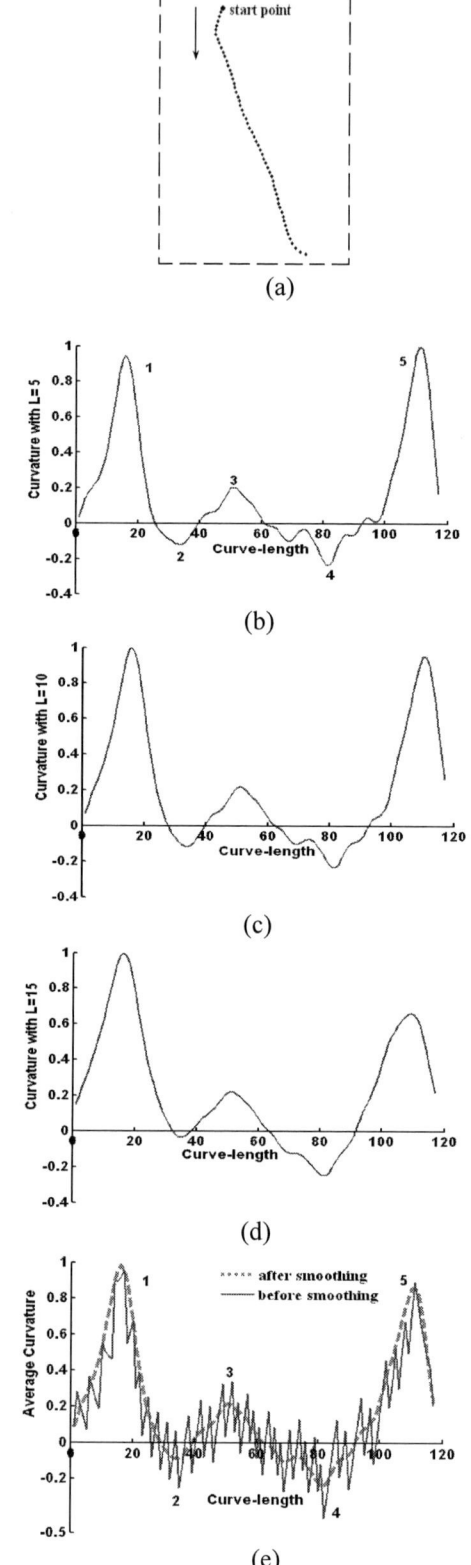

Fig.7. Normalized CPDA discrete curvature functions for "wire 256" in image sequence (see Fig.7(a)) with different chord-lengths: (b)L=5; (c)L=10;(d)L=15; and (e) the average curvature function with L=5,10 and 15,where "after smoothing" and "before smoothing" depict H(k) with and without Gaussian smoothing(σ =2.5), respectively.

And Fig.7 (e) indicated that there were five curvature extreme points. The curvature values were very high at the point 1 and 5. It denotes the bending deflections of wire were rather large in these regions. And the values were lower at the point 2,3 and 4. It denote there were small deflection in these zones. However, we attracted the critical attention to the strong local plastic zones. So we set the curvature-threshold Th=±0.4. In Fig.8, we can see only the curvature maxima points 1 and 5 were out of the curvature-threshold . The local plastic zones were marked as '□' as shown in Fig.9.

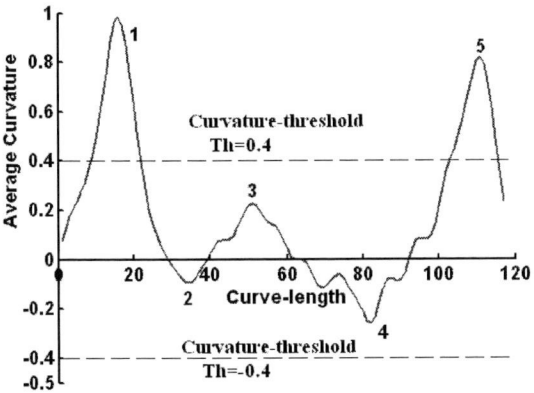

Fig.8. Set curvature-threshold

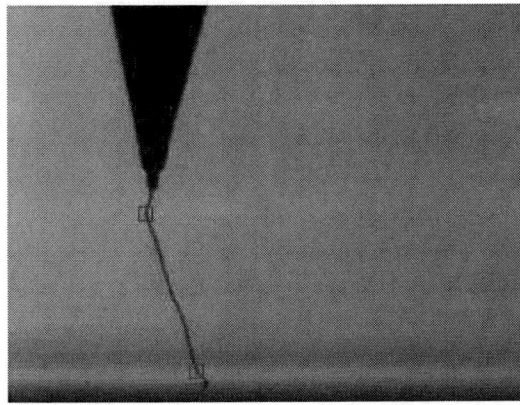

Fig.9.Test results of local plastic zones

Conclusions

The image sequences of wire bonding looping formation are analyzed. Thinned wires are obtained using image processing. The estimated curvature distributions of wire are calculated by means of point-to-chord distance accumulation. The positions of the wire local plastic zones are obtained by finding the extreme value points of curvature which are out of the curvature-threshold. This method has good accuracy and strong robustness. It can be used in researching the dynamic characteristics of the capillary and wire in the wire bonding looping formation.

Acknowledgments

This work has been supported by the China Department of Science & Technology Program 973 (No.: 2009CB 724203).

References

1. Lin TY. Moldability and reliability assessment on stacked100FSBGA. *Agere internal report*, July, 2002.
2. Jon Brunner, Ivy Wei Qin and Bob Chylak, "AdvancedWire Bond Looping Technology for Emerging Packages," *IEEE/SEMI Int'l Electronics Manufacturing Technology Symposium*.2004
3. A.Rattarangsi and R. T. Chin, "Scale-Based Detection of Comers of Planar Curves," *IEEE Transactions of Pattern Analysis and Machine Intell.*,vol.14,no.4pp.430-449,Apr.1992.
4. Kejun Wang, Yan Zhang, Zhi Yuan and Dayan Zhuang, "Hand Vein Recognition Based on Multi Supplemental Features of Multi-Classifier Fusion Decision," *IEEE International Conference on Mechatronics and Automation* June 25 - 28, 2006
5. Shi Zhao, Yiding Wang and Yunhong Wang, "Extracting Hand Vein Patterns from Low-Quality Images: A New Biometric Technique Using Low-Cost Devices," *Fourth International Conference on Image and Graphics*
6. Liu Zhi-cheng, Guan Li, Liu Xiang-bin, "Segmentation and smooth thinning of dorsal hand vein pattern," [J] *Computer Engineering and Applications*,2008.44 (13):182-184.
7. Mohammad Awrangeb. and Guojun Lu, "Robust Image Corner Detection Based on the Chord-to-Point Distance Accumulation Technique," *IEEE transactions on multimedia,* 10(6):1059-1072, 2008.
8. J.H.Han and T.T.Poston, "Chord-to-point distance accumulation and planar curvature: a new approach to discrete curvature,"*Pattern Recognition. Lett.*, Vol. 22, pp. 1133-1144,2001
9. Yang Shui-shan, He Yong-hui, Zhao Wan-sheng, Wang Zhen-long, " Steel strip edge split inspection based on curvature estimation ,"[J] *Infrared and Laser Engineering*, 2008, 37(4):634-638.
10. Peng Tie gen and Wu Ti-hua, "Study on parts recognitionand checking based on negative curvature minima,"*Journal of System Simulation*,2006,18(11): 3058-3062.(in Chinese)

First-Principles Calculations of Structural, Thermodynamic and Electronic Properties of Intermetallic Compounds in Solder

Yang Yang, Hao Lu*, Chun Yu, Junmei Chen

School of Materials Science and Engineering, Shanghai Jiao Tong University, Shanghai, 200240, PR China

*E-mail: shweld@sjtu.edu.cn

*Tel. and Fax: +86-21-34202548

Abstract

First principles calculations have been performed to investigate the structural, thermodynamic and electronic properties of four common intermetallic compounds (IMCs) formed at the solder joints of electronic packages, namely, Cu_6Sn_5, Cu_3Sn, Ni_3Sn_4 and Ag_3Sn. The theoretical heat of formation of Cu_6Sn_5 is close to that of Cu_3Sn, both of them are overestimated relative to the experimental results. In addition, Ni_3Sn_4 has the lowest heat of formation among these IMCs. The curves of total DOS near the Fermi level for the IMCs are mainly dominated by the Sn-*sp* hybridization states, M (M=Cu, Ni, Ag)-*sp* hybridization states and M-*d* states. The complicated bonding states for Ni_3Sn_4 may caused by the position of the main peak of Ni-*d* states.

Introduction

In electronic packaging, solder joints not only play an important role in the mechanical jointing, but also serve as the current path. On the other hand, IMCs widely exist at the interface of solder joints or inside the solders, and they have great effects on the reliability of solders joints. A thin, continuous and uniform IMC layer can improve the joint strength, while too thick IMC layer may become a reliability issue due to its brittleness. Cu_6Sn_5, Cu_3Sn, Ni_3Sn_4 are the typical IMCs at the interface. Meantime in the solder joints, a fine dispersion of Ag_3Sn can significantly improve its mechanical performance, whereas the formation of bulk platelike Ag_3Sn might seriously influence its mechanical performance [1]. Recent years, theoretical studies mostly focus on the mechanical properties of the IMCs in solder joints [2-4], and there are few reports concerning on the other properties of these IMCs [5, 6]. Thus it is necessary to investigate these properties, like structural and electronic properties.

The crystal structures of the common IMCs, namely, Cu_6Sn_5, Cu_3Sn, Ni_3Sn_4 and Ag_3Sn, have been extensively investigated using experimental methods. The crystal structure of Cu_6Sn_5 was reported as early as 1928 [7, 8]. Though there were many subsequent studies [9, 10], the detailed description of the crystal structure of Cu_6Sn_5 was given by Larsson et al. [11, 12]. Cu_6Sn_5 undergoes a phase transformation from high temperature phase (η) to low temperature phase (η') at approximately 460 K [10]. Similar to Cu_6Sn_5, the crystal structure of Cu_3Sn have also been studied for a long time [7, 13-18]. Recently, Cu_3Sn with the $D0_{19}$ structure was observed by using transmission electron microscopy [19]. For the structures of Ni_3Sn_4 and Ag_3Sn, the relevant references were mentioned by Lee et al. [3].

First-principle density functional theory (DFT) has been proved to be an effective theoretical means to study the properties of IMCs [5]. In this paper, based on the DFT theory, the structural, thermodynamic and electronic properties of these IMCs were investigated using the CASTEP plane-wave code [20] in the scheme of generalized gradient approximation (GGA-PBE) [21].

Structural and thermodynamic properties

The formation energy per atom of binary IMCs is can be expressed by the following equation:

$$\Delta E_f(A_m B_n) = \frac{1}{m+n} E_{A_m B_n} - \left(\frac{m}{m+n} E_A + \frac{n}{m+n} E_B\right) \quad (1)$$

where E_{AmBn} is the total energy of the IMCs mentioned above, E_A is the total energy per atom of A crystal, and E_B is the total energy per atom of A crystal. Here, the structure for Cu, Ni, and Ag crystals is $Fm\bar{3}m$ (225), and for Sn crystal, the structure is $I4_1/amd$ (141).

Our calculated lattice constants and heats of formation of the IMCs, comparing with the values from other theoretical and experimental works, are shown in Table 1 and 2. For Cu_6Sn_5, only the low temperature structure (η') was used as the initial configuration. Our calculated lattice constants are in good agreement with the experimental values. The values of heats of formation for IMCs in this work seem much greater than experimental values, but they have the same trendency. η'-Cu_6Sn_5 usually generated before ε-Cu_3Sn at the interface of joints, which is related to the content of Sn and is in accordance with the Cu-Sn binary phase diagram. Actually, the heat of formation of ε-Cu_3Sn is lower than that of η'-Cu_6Sn_5, as listed in Table 2. And Ni_3Sn_4 has the lowest heat of formation.

In addition, thermodynamic calculations also can be applied to analyze the interfacial reactions. The three interfaces, Solder/Cu_6Sn_5, Cu_6Sn_5/ Cu_3Sn and Cu_3Sn/Cu, are very common at the interface between solder and pads, but there still have some unsolved problems. Zeng et al. [27] studied the growth mechanisms of Cu_3Sn at the SnPb/Cu interface and held that the decomposition of Cu_6Sn_5 was the main mechanisms for the growth of Cu_3Sn. It is known that there are various possibilities for the growth mechanisms of Cu_3Sn,

(1) Cu_3Sn is formed by the reaction of Cu atoms from Cu pad and Sn atoms diffused through the grain boundaries of IMCs:

$3Cu + Sn \rightarrow Cu_3Sn$ $\quad \Delta E$=-47.2 kJ/mol

where ΔE is the reaction heat.

(2) Cu_3Sn is formed directly by the reaction of Cu_6Sn_5 and Cu atoms from Cu Pad:

$Cu_6Sn_5 + 9Cu \rightarrow 5 Cu_3Sn$ $\quad \Delta E$= -24.08 kJ/mol

(3) Cu_3Sn is formed by the decomposition of Cu_6Sn_5:

$$Cu_6Sn_5 \rightarrow 2Cu_3Sn+3Sn \quad \Delta E=20.23 \text{ kJ/mol}$$

According to values of the reaction heat for IMCs, we can find that the former two reactions are exothermic, while the third one is endothermic, which indicates that the decomposition of Cu_6Sn_5 needs external energy. And the result for the third reaction may be inconsistent with Zeng et al.'s deduction. Therefore, the growth mechanisms for Cu_3Sn are very complicated, and more works are needed to solve this problem.

Table 1 Crystallographic data of the intermetallic compounds

Phase	Lattice constants (Expt., Å)	Lattice constants (Present work, Å)
η'-Cu_6Sn_5	a=11.022 b=7.282 c=9.827 β=98.84°[11]	a=10.837 b=7.133 c=9.642 β=99.37°
ε-Cu_3Sn	a=5.49 b=4.32 c=4.74[14]	a=5.417 b=4.282 c=4.647
Ni_3Sn_4	a=12.214 b=4.06 c=5.219 β=105°[22]	a=11.956 b=3.976 c=5.143 β=104.97°
Ag_3Sn	a=5.968 b=4.78 c=5.184[23]	a=5.923 b=4.742 c=5.126

Table 2 Heats of formation of intermetallic compounds

Phase	This work (kJ/mol)	Works by others (kJ/mol)
η'-Cu_6Sn_5	-10.438	-3.205[5], -7.037, 298.15 K[10]
ε-Cu_3Sn	-11.789	−8.2, 298[24] −7.82, 298K[24]
Ni_3Sn_4	-42.834	−24.0, 298[24] −28.52, 298K[25] −30.30, 298K[25]
Ag_3Sn	-9.288	−4.2, 298[24] −4.5, 723K[26]

Electronic properties

The overlap population provides an effective method for assessing the covalent or ionic nature of a bond. A value of zero for the bond population indicates a perfectly ionic bond, while values greater than zero indicate increasing levels of covalency. And the negative value indicates the anti-bonding interaction in bonds. As listed in Table 3, the overlap population analysis indicates the covalent nature of various degrees for the bonds in Cu_3Sn, Cu_6Sn_5 and Ag_3Sn. For Ni_3Sn_4, part of Ni-Sn bonds are covalent bonds, and some bonds with negative value indicate their anti-bonding nature.

Table 3 Average bond length for bonds in intermetallic compounds and corresponding average overlap population (OP).

Phase	Bond	Bond length (Å)	OP
η'-Cu_6Sn_5	Cu-Cu	2.5756	0.153
	Cu-Sn	2.6807	0.321
ε-Cu_3Sn	Cu-Cu	2.6389	0.252
	Cu-Sn	2.6726	0.205
Ni_3Sn_4	Ni-Ni1	2.6142	0.01
	Ni-Ni2	2.6991	-0.04
	Ni-Sn1	2.5986	0.49
	Ni-Sn2	2.6085	-0.26
	Sn-Sn	2.8581	-0.71
Ag_3Sn	Ag-Ag	2.9395	0.247
	Ag-Sn	2.9274	0.202

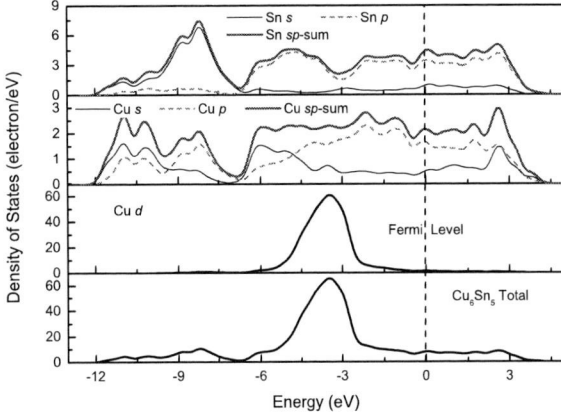

Fig. 1. Total and partial density of state (DOS) for η'-Cu_6Sn_5.

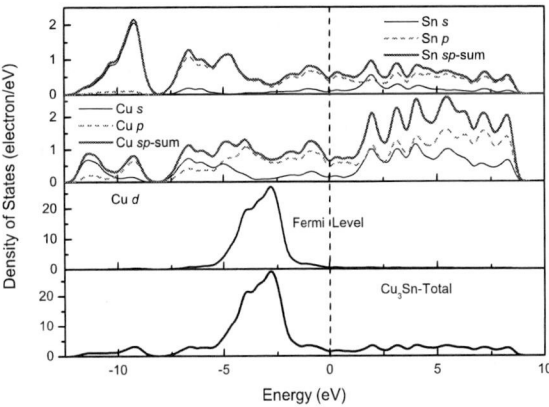

Fig. 2. Total and partial density of state (DOS) for ε-Cu_3Sn.

We also calculated the partial and total density of states (DOS) of these IMCs, as illustrated in Fig. 1, Fig. 2, Fig. 3 and Fig. 4. The DOS patterns of these IMCs are very similar. They all have a characteristic main peak which mainly dominated by M-d states. At the Fermi level, the total DOS consist of the Sn-sp hybridization states, M-sp hybridization states and M-d states. The main peak for Ni-d states in the DOS of Ni_3Sn_4 is more close to the Fermi level than that for Cu-d states in the DOS of the other three IMCs, that is, Ni-d states contribute more to the DOS at Fermi level. This may be one of the causes for the different bonding states in Table 3.

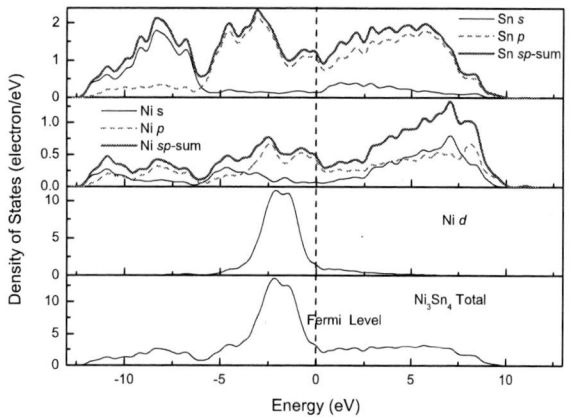

Fig. 3. Total and partial density of state (DOS) for Ni_3Sn_4.

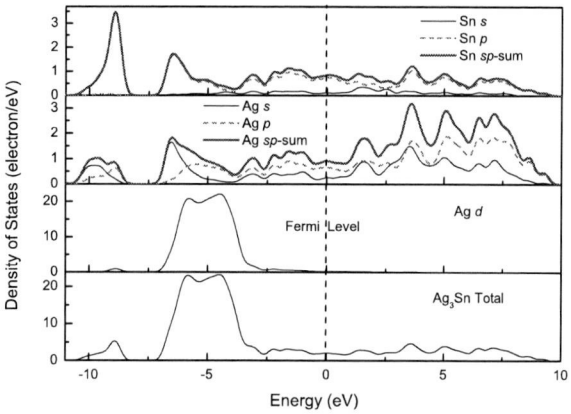

Fig. 4. Total and partial density of state (DOS) for Ag_3Sn.

Conclusions

In conclusion, our calculated lattice constants for IMCs agree well with the experimental values. The theoretical heats of formation are larger than the experimental values, but they have the same trendency. The overlap population analysis shows that Ni_3Sn_4 has the different bonding states with the other three IMCs. This may caused by the position of the main peak for Ni-d states.

References

1. S. K. Kang, D. Y. Shih, Sung K. Kang, Da-Yuan Shih, D. Leonard, D. W. Henderson, T. Gosselin, S. Cho, J. Yu, W. K. Choi, "Controlling Ag_3Sn plate formation, in near-ternary-eutectic Sn-Ag-Cu solder by minor Zn alloying, " *Jom*, Vol. 56, No. 6 (2004), pp. 34-38.
2. N. T. S. Lee, V. B. C. Tan, K. M. Lim, "First-principles calculations of structural and mechanical properties of

Cu_6Sn_5," *Applied Physics Letters*, Vol. 88, No. 3 (2006), 031913.
3. N. T. S. Lee, V. B. C. Tan, K. M. Lim, "Structural and mechanical properties of Sn-based intermetallics from ab initio calculations," *Applied Physics Letters*, Vol. 89, No. 14 (2006), 141908.
4. J. Chen, Y. S. Lai, C. Y. Ren, D. J. Huang, "First-principles calculations of elastic properties of Cu_3Sn superstructure," *Applied Physics Letters*, Vol. 92, No. 8 (2008), 081901.
5. G. Ghosh, M. Asta, "Phase stability, phase transformations, and elastic properties of Cu_6Sn_5: Ab initio calculations and experimental results," *J. Mater. Res.*, Vol. 20, No.11 (2005), pp. 3102-3117.
6. C. Yu, J. Liu, H. Lu, P. Li, J. Chen, "First-principles investigation of the structural and electronic properties of $Cu_{6-x}Ni_xSn_5$ (x=0, 1, 2) intermetallic compounds," *Intermetallics*, Vol. 15, No. 11, (2007), pp. 1471-1478.
7. J. Bernal, "The Complex Structure of the Copper-Tin Intermetallic Compounds," *Nature*, Vol. 122, No. 3063 (1928), pp. 54.
8. A. Westgren and G. Phragmén, "X-Ray Analysis of Copper-Tin Alloys," *Z. Anorg. Allg. Chem.*, Vol. 175, No. 1 (1928), pp. 80-89.
9. O. Carlsson and G. Hägg, "On the knowledge of crystal structures of some copper-tin phases," *Z. Kristallogr.*, Vol. 83, (1932), pp. 308-317.
10. A. Gangulee, G. Das and M. Bever, "An x-ray diffraction and calorimetric investigation of the compound Cu_6Sn_5," *Metall. Trans.*, Vol. 4, No. 9 (1973), pp. 2063-2066.
11. A. K. Larsson, L. Stenberg, and S. Lidin, "The superstructure of domain-twinned η'-Cu_6Sn_5," *Acta Crystallogr., Sect. B: Struct. Sci.*, Vol. 50, (1994), pp. 636-643.
12. A. K. Larsson, L. Stenberg, and S. Lidin, "Crystal structure modulation in η–Cu_5Sn_4," *Z. Kristallogr.* Vol. 210, No. 11 (1995), pp. 832-837.
13. K. Schubert, B. Kiefer, M. Wilkens, and R. Haufler, "Über einige metallische Ordnungsphasen mit grosser Periode," *Z. Metallkd.* Vol. 46, (1955), pp. 692-715.
14. W. Burkhardt and K. Schubert, "On Brass Like Phases with A3 Related Structure," *Z. Metallkd.* Vol. 50, (1959), pp. 442-452.
15. P. L. Brooks, E. Gillam, "The ε-Cu_3Sn phase in CuSn system," *Acta Metall.*, Vol. 18, No. 11 (1970), pp. 1181–1185.
16. M. Van Sande, R. De Ridder, G. Van Tendeloo, J. Van Landuyt, S. Amelinckx, "High resolution study of one-dimensional long period superstructures in Cu_3Sn with additions of zinc and nickel," *Phys. Status Solidi A*, Vol. 48, No. 2 (1978), pp. 383–394.
17. Y. Watanabe, Y. Fujinaga, H. Iwasaki, "Lattice modulation in the long-period superstructure of Cu_3Sn," *Acta Cryst. B*, Vol. 39, No. 3 (1983), pp. 306-311.
18. P. Villars, L.D. Calvet, Pearson's Handbook of Crystallographic Data for Intermetallic Phases (Ohio, 1991), pp. 3007.
19. Xiahan Sang, Kui Du, Hengqiang Ye, "An ordered structure of Cu_3Sn in Cu–Sn alloy investigated by

transmission electron microscopy," *Journal of Alloys and Compounds*, Vol. 469, No. 1-2 (2009), pp. 129-136.

20. S. J. Clark, M. D. Segall, C. J. Pickard, P. J. Hasnip, M. I. J. Probert, K. Refson, M. C. Payne, "First principles methods using CASTEP," *Z. Kristallogr.*, Vol. 220, No. 5-6 (2005), pp. 567-570.

21. J. P. Perdew, K. Burke, M. Ernzerhof, "Generalized Gradient Approximation Made Simple," *Phys. Rev. Lett.*, Vol. 77, No. 18 (1996), pp. 3865-3868.

22. G. Ghosh, "Interfacial microstructure and the kinetics of interfacial reaction in diffusion couples between Sn-Pb solder and Cu/Ni/Pd metallization," *Acta mater.*, Vol. 48, (2000), pp. 3719-3738.

23. C. W. Fairhurst, J. Cohen, "The crystal structures of two compounds found in dental amalgam: Ag_2Hg_3 and Ag_3Sn", *Acta Crystallogr. Sect. B: Struct. Crystallogr. Cryst. Chem.*, Vol. 28, (1972), pp. 371-378.

24. H. Flandorfer, U. Saeed, C. Luef, A. Sabbar, H. Ipser, "Interfaces in lead-free solder alloys: Enthalpy of formation of binary Ag–Sn, Cu–Sn and Ni–Sn intermetallic compounds," *Thermochimica Acta*, Vol. 459, No. 1-2 (2007), pp. 34-39.

25. A. N. Torgersen, H. Bros, R. Castanet, A. Kjekshus, "Enthalpy of formation for CoGe, CoSn, $Ni_{3.14}Sn_4$, $Ni_{3.50}Sn_4$, $AuCo_{1.66}Sn_4$, $AuNi_2Sn_4$ and $Au_{1.17}Pt_{1.82}Sn_4$," *J. Alloys Comp.*, Vol. 307, No. 1-2 (2000), pp. 167-173.

26. O. J. Kleppa, "A calorimetric investigation of the system silver-tin at 450°C," *Acta Metall.*, Vol. 3, No. 3 (1955), pp. 255-259.

27. K. Zeng, R. Stierman, T. –C. Chiu, D. Edwards, K. Ano, K. N. Tu, "Kirkendall void formation in eutectic SnPb solder joints on bare Cu and its effect on joint reliability," *Journal of Applied Physics*, Vol. 97, No. 2 (2005), 024508.

A Core-Shell Structure Viscoelastic Model of Particulate-Filled Electronic Packaging Polymers

Dayong Gui, Jianhong Liu, Bo Chen, Xin Miao, Guangfu Zeng, Deyu Tian
School of Chemistry and Chemical Engineering, Shenzhen University, Shenzhen, 518060, P.R.China
Email: dygui@szu.edu.cn, Telephone: +86 755 26558041

Abstract

On the basis of structural and mechanical characterization of particulate-filled epoxy materials, a new core-shell structure viscoelastic constitutive model is established for predicting mechanical performances of particulate-filled electronic packaging materials in this paper. The three-phases(layers) of core-shell particle and matrix constitute the core-shell structure model and the modified Kelvin viscoelastic model was combined with the structure model, which improve the mechanical property's calculation of electronic packaging polymers. With this model, the relationship between mechanical properties of the electronic packaging materials and modulus of matrix, solid particles content, particle size, gradation and modulus of the intermediate shell layer were constructed mathematically and analytical modeling of tensile strength and elongation of particle-filled electronic packaging polymers was conducted. The results show that the predictions of this model is in good agreement with experimental measurements for both tensile strength and elongation of particulate-filled electronic packaging materials.

Introduction

Electronic packaging polymers with excellent performances such as low stresses, heat resistance and moisture resistance have excessively been required in recent years. The core-shell structure particle combines both inorganic and organic good properties and has arose great interests in the field of composites. [1~3] Therefore particle with core-shell structure would be good filler of electronic packaging polymer for underfills.

If the required mechanical properties fail to maintain fine performance, the electronic packages will distort dramatically and even break, which will result in reliability problem of electronic packages. It's necessary to predict the mechanical properties of particulate-filled polymer for the development of advanced electronic packaging materials and its use in electronic packaging with high reliability.

In recent years, many efforts have been made to calculate and/or predict mechanical performances of particle filled composite.

In [4], the impact of processing conditions on warpage prediction of a plastic quad flat package (PQFP) was investigated. It was suggested that low temperature and longer molding time or high temperature and shorter molding time would result in less warpage. They also showed that the viscoelastic model predicted the warpage more accurately than the thermoelastic model.

Ernst *et al.* proposed a cure-dependent model for mechanical modeling of stress evolution induced during cure in a particle-filled electronic packaging polymer. [5] Yang, *et al.* carried out finite element modeling for cure-induced

warpage of plastic IC packages based on the above-mentioned cure-dependent viscoelastic model and indicated that warpage induced during the curing process has significant contribution on the total warpage of the map. [6]

Li *et al.* proposed a two-layer built-in model based on Christensen and Lo's three-layer sphere model and carried out analytical modeling of tensile strength of particulate-filled composites. [7]

Most studies were focused on the finite element modeling of the stress and strain of products induced during the process of packaging materials. The accurate prediction of both the tensile strength and the elongation of particulate-filled composite is still a very complicated problem because of the many variables that play definite roles in it.

In this study, a core-shell structure viscoelastic model is proposed and analytical modeling is performed for the mechanical properties of the particle-filled electronic packaging polymers.

Core-Shell Structure Viscoelastic Model

In particulate-filled polymer composites, solid filler is particle phase, and polymer is matrix phase, and the fillers adhere to matrix layer by means of absorption or bonding action. It is supposed that spherical particle with radius of a coated with an equivalent layer (shell) is embedded into matrix phase as shown in Fig.1. $b-a$ is the thickness of the intermediate layer. This is a three-phase model containing particle phase, intermediate phase, and matrix phase.

Fig.1 Three-phase model

According to the two-layer built-in model proposed by Li, [7] the average stress in the particle is obtained as follow:

$$\sigma_{af} = \frac{\dfrac{E_f - E_m}{E_m} + \left(3 - \dfrac{E_f}{E_m}\right)\dfrac{2\left(\arctan e^{ka} - \dfrac{\pi}{4}\right)}{k^2 a^2}\sinh ka}{1 + \dfrac{E_m}{E_f}\dfrac{2\left(\arctan e^{ka} - \dfrac{\pi}{4}\right)}{k^2 a^2}\sinh ka}\sigma_{am} \quad (1)$$

where σ_{am} is the average axial stress in the matrix, a is the radius of the particle, E_f, E_m are the elastic Young's modulus of the particle and matrix, respectively, v_m is Poisson's ratio of the matrix and k is a parameter depending on the radius of the particle and other mechanical properties as follows:

$$k = \frac{1}{a}\int_{\pi^{-0.5}}^{a} K dr = \frac{\ln a + 0.5\ln \pi}{a}\sqrt{\frac{E_m}{(E_f - E_m)(1+v_m)\ln 2}} \quad (2)$$

The tensile strength (ultimate tensile strength or yielding strength) of the composite, σ_{sc}, is obtained using the rule-of-mixtures method as follows:

$$\sigma_{sc} = f\sigma_{sf} + (1-f)\sigma_{sm} \quad (3)$$

where σ_{sm} is ultimate tensile strength of matrix, σ_{sf} is equal to σ_{af} when σ_{am} is replaced by σ_{sm} in $Eq. 1$. and f is the volume fraction of particles in the composite.

In the light of composition, structure and viscoelasticities of particulate-filled electronic packaging polymers, both particle phase and intermediate phase are supposed to meet Hooke's law so that they can be modeled with spring. The matrix is a viscoelastic polymer which can be described by Voigt(Kelvin) model. So an improved Kelvin viscoelastic model is constructed by connecting a spring element with a Kelvin model, corresponding to above-mentioned three-phase structure model, as shown in Fig. 2. The viscoelastic constitutive relation and three phase constitute model are combined, which improve the mechanical property's calculation of particulate-filled electronic packaging polymers.

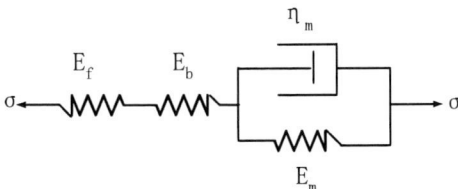

Fig.2 Improved Kelvin viscoelastic model

The fundamental differential equation of improved Kelvin model is:

$$\frac{d\sigma}{dt} + \frac{E_1 + E_2}{\eta_1}\sigma = E_1\frac{d\varepsilon}{dt} + \frac{E_1 E_2}{\eta_1}\varepsilon \quad (4)$$

where σ is the stress of materials, ε is the strain of materials, E_1 is the complex Young's modulus of filler (E_f) and intermediate phase (E_b), E_2 and η_1 are the Young's modulus and the viscosity (η_m) of matrix respectively.

Because $\varepsilon_1 = \varepsilon_f + \varepsilon_b$, $\quad \sigma = E\varepsilon$, it is obtained that

$$E_1 = \frac{E_f E_b}{E_f + E_b} \quad (5)$$

At initial time, E_1 is the glass modulus of composites(E_g):

$$E_b = \frac{E_f E_g}{E_f - E_g} \quad (6)$$

Tensile test is carried out at constant rate, i.e. displacement $u = \dot{u}t$ ($\dot{u}$ =constant):

$$\varepsilon = \frac{\Delta L}{L}, \quad \varepsilon = c_1 t \quad (7)$$

where $c_1 = \dot{u}/L$. The stress can be expressed as function of time by solving the differential equation (4):

$$\sigma = \frac{B}{A}t + (\frac{B}{A^2} - \frac{C}{A})e^{-At} - (\frac{B}{A^2} - \frac{C}{A}) \quad (8)$$

in which $A = \frac{E_1 + E_2}{\eta_1}$, $B = \frac{E_1 E_2}{\eta_1}c_1$, $C = E_1 c_1$

Thus, the ultimate tensile stress, i.e. tensile strength (σ_{sc}), can be calculated from equation (3) for particulate-filled electronic packaging polymers fastly. Then the corresponding t_{sc} can be obtained from equation (8). At last, equation (7) gives the ultimate tensile strain, i.e. elongation (ε).

However, Eq (3) is not the final expression predicting the tensile strength of the composite. Several factors should be considered before Eq (3) can be used, including (a) the degree of adhesion between the matrix and the particles as well as the degradation of the matrix properties due to the presence of the particles, (b) particles size distributions; and (c) particle clustering.

(a) Effect of the Degree of Particle Adhesion and Matrix Degradation

In particulate-filled composites, interfacial bonding plays a determinant role in the composite strength because the load transferred to the particles is through the interfacial bonding. The interfacial bonding depends on a number of factors, including the volume fraction of particles, compatibility between the particle and matrix, particle surface treatment, fabrication process, etc. Owing to these factors, only a portion of the particles are well bonded to the matrix. In addition, the matrix will also degrade because of the presence of particles and complications developed during the preparation of the composite. Papanicolaou and Bakos used the following modified rule-of-mixtures method to consider the effect of the degree of particle adhesion and matrix degradation on the composite tensile strength: [7]

$$\sigma_{sc} = mf^n\sigma_{sf} + (1-f^n)\sigma_{sm} \quad (9)$$

$$m \in [0,1], \quad n \in [0,1]$$

According to the meaning of m and n, $m = n = 1$ can be used in this ideal case. In the second case, perfect bond between the particles and the matrix is also used, but the matrix degrades linearly with the inclusion of particles. Therefore, $m = 1$ and $n = 1-f$.

(b) Effect of Particle Size Distributions

If the particle size distribution is known, the formula (9) of the tensile strength of particulate-filled electronic packing polymers can be modified as follow:

$$\sigma_{sc} = m\int_{a_{min}}^{a_{max}}\sigma_{sf}(a)fdP(a) + \sigma_{sm} - \sigma_{sm}\int_{a_{min}}^{a_{max}}f^n d[P(a)]^n \quad (10)$$

where a_{min} and a_{max} are the minimum and maximum radius of particles, respectively. $P(a)$ is the percent passing by volume of particles, i.e., $P(a)$ is the volume fraction of particles with sizes less than a. It describes the particle size distribution.

So far, the description of the core-shell structural viscoelastic model is completed. The composition, interface and cure parameters are attached to the model by the investigation of thermo-mechanical characterization and curing process of the materials. With this model, the relationship between mechanical properties of the electronic packaging materials and modulus of matrix, solid particles content, particle size, gradation and modulus of the interface shell layer is constructed mathematically. The elongation and tensile strength of packaging materials can be modeled with this relationship and its program code.

Experimental

1) Materials used and sample preparation

Diglycidyl ether of bisphenol A (DGEBA, epoxy equivalent weight EEW=196) used as matrix material was purchased from Huntsman, Switzerland. 4,4'-Diaminodiphenylmethane (DDM) was used as a curing agent and purchased from Shanghai Aladdin Chemical Reagent Co. Ltd. Silica particle was supplied kindly by Zhejiang Tongda Weipeng Electric Co., Ltd. and Taian Dacheng Powder Technical Co., Ltd. Silane coupling agent KH550, industrial grade, purchased from Shanghai Yaohua chemical group Co., Ltd. All solvents were commercial products and used without further purification.

The silica particles are all pretreated by coupling agent (KH550) in order to form good bonding action with epoxy resin. The content of silica particles in the composites were 0, 13, 30, and 40 wt %, respectively. The epoxy and the hardener were mixed in a stoichiometric ratio of 1:1.34. Epoxy resin and silica were well mixed at 130□ by mechanical stirring and then cooled down to 92□, and the hardener was added by manual stirring for 5 minutes. Afterwards, the mixture was degassed at room temperature for 30 minutes in a vacuum oven. The mixture then was poured into a polytetrafluoroethylene mold to be cured in an oven. The curing condition was 70□ for 2 hours as pre-curing, 120□ for 2 hours and 150□ for 2 hours as post-curing. After that, samples were cooled down to room temperature naturally. The formulation of the composites is listed in Table 1.

Table 1 Formulation of particulate-filled polymers

No	Epoxy/DDM	Silica	
	Content /wt%	Content /wt%	Diameter /μm
1	100	0	/
2	70	30	10
3	70	30	2, 6, 20
4	60	40	10

2) Characterization

The mechanical properties of the underfill and filled epoxy polymer were measured with CMT4304 universal testing machine (U.S. Waters Industrial Systems (China) Co., Ltd.) at room temperature according to the Chinese standard GB/T1040-92.

DMA was used to get creep strain curves of the underfill epoxy resin and to determine the viscosity coefficient as an input parameter for calculation. Bending creep tests were performed by using a TA-Instrument DMA Q800 at ambient temperature.

Results and Discussion

1) Experimental results

The tensile strength of the cured pure EP and silica-modified EP with various silica contents of EP are shown in Fig. 3(a). The tensile strength of the cured silica-modified EP improves with the increase in silica content. The tensile strength of the cured silica-modified EP reaches a maximum at 40wt% silica content of EP where the tensile strength improved 34.47% than that of the cured pure ER.

Fig.3 (b) depicts the elongation at the break for the cured pure EP and silica-modified EP with various silica contents of EP. There is an appreciable decrease of silica-modified EP with various silica contents of EP. However, when the silica content is added more, the elongation at the break of the cured silica-modified EP also reaches a maximum at 40wt% silica of EP.

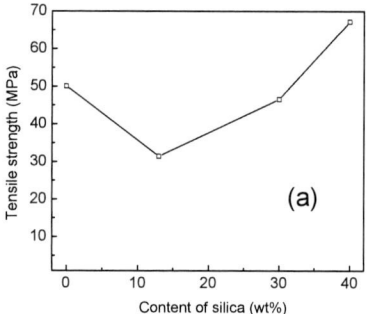

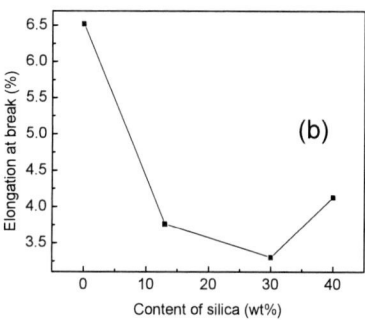

Fig.3 Effect of various silica contents on the mechanical properties of epoxy resin: the silica contents of EP are 0wt%(pure EP, Sample A); 13wt%(Sample B); 30wt%(Sample C); 40wt%(Sample D)

The measuring results of creep for underfill epoxy resin is shown in Fig. 4.

According to Kelvin viscoelastic model,

$$\sigma = E\varepsilon + \eta \frac{d\varepsilon}{dt} \tag{11}$$

The viscosity coefficient of the underfill epoxy resin is obtained from the curve of strain verse time shown in Fig.4 and Eq(11):

$\eta = 3.9216 \times 10^{10}$ Pa.s

978-1-4244-4658-2/09 $25.00 © 2009 IEEE

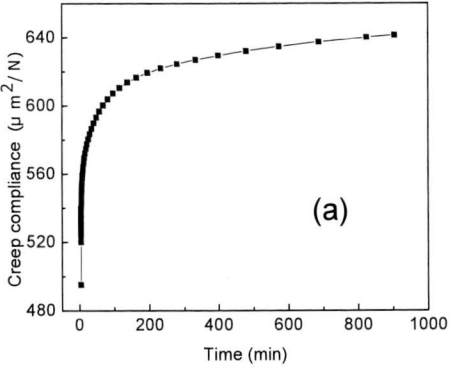

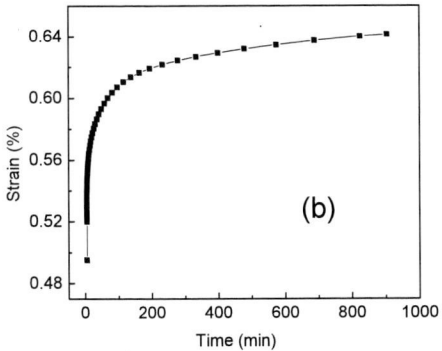

Fig.4 Creep-Compliance (a) and strain (b) curves of underfill epoxy resin at 30°C

2) Calculating results

The simulation calculation for experimental packaging materials formulas listed in Table 1 was carried out by utilizing the core-shell structural viscoelastic model and its program code. The comparison of calculated results with the test results are shown in Table 2.

Table 2 Comparison of calculated mechanical properties of composites with the test results

No	Tensile strength		Elongation	
	σ_{mexp}/MPa	σ_{mcal}/MPa	ε_{mexp}/%	ε_{mcal}/%
1	50.02	50	6.52	5.45
2	52.36	55.38	4.20	6.04
3	56.48	55.33	4.85	6.04
4	67.26	56.95	4.13	6.21

The results show that the predictions of this model is in good agreement with experimental measurements for both tensile strength and elongation of particulate-filled electronic packaging materials.

The effects of the content and size of particle on the mechanical properties of the particulate-filled epoxy resin are shown in Fig.5 and Fig.6 respectively.

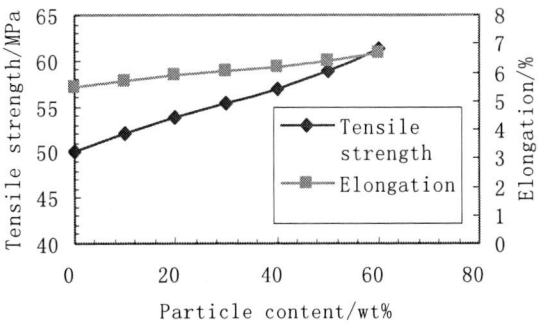

Fig.5 Effect of particle content on tensile strength and elongation of particulate-filled epoxy resin (particle diameter:10.36μm)

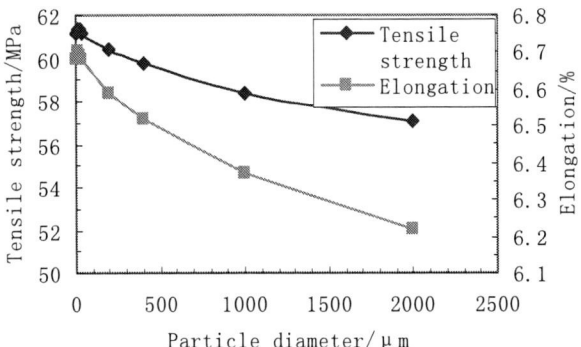

Fig.6 Effect of particle diameter on tensile strength and elongation of particulate-filled epoxy resin (content of particle:60wt%)

From the calculated results of Fig.5, both tensile strength and elongation increase with content of particle fillers. This is understandable because silica and epoxy resin bond more closely with the increase of mass fraction of silica. The trend shown in Fig.6 is also reasonable because the small particles have greater specific surface area and more bonding with matrix materials for same content of filler.

Conclusions

A core-shell structure viscoelastic model for predicting both tensile strength and elongation of particulate-filled electronic packaging polymers is proposed. The formulation of mechanical properties of the electronic packaging materials as a function of modulus of matrix, solid particles content, particle size, gradation and modulus of the intermediate layer are developed.

The analytical modeling results with this model and its program code are in good agreements with experimental measurements for both tensile strength and elongation of particulate-filled electronic packaging matcrials.

References

1. Stöber W., Fink A., "Controlled Growth of Monodisperse Silica Spheres in the Micron Size Range", *J. Colloid Interface Sci.*, No.26 (1968), pp. 62-69.
2. Caris C.H.M., Van Elven L.P.M., Van Herk A.M., German A.L., "Polymerization of MMA at the Surface

of Inorganic Submicron Particles", *British Polym. J.*, Vol.21, No.2 (1989), pp. 133-140.

3. Mandal T.K., Fleming M.S., Walt D.R., "Production of Hollow Polymeric Microspheres by Surface-Confined Living Radical Polymerization on Silica Templates", *Chem. Mater.*, No.12 (2000), pp. 3481-3487.

4. Yeung, D.T.S. *et al.*, "Warpage of Plastic IC Packages as a Function of Processing Conditions", *Journal of Electronic Packaging*, Vol.123, No.3 (2001), pp. 265-272.

5. Ernst L.J., van't Hof C., Yang D.G. *et al.*, "Mechanical Modeling and Characterization of the Curing Process of Underfill Materials", *ASME Journal of Electronic Packaging*, Vol.124, No.2 (2002), pp. 97-105.

6. Yang Daoguo, Jansen K.M.B., Wang L.G., Ernst L.J., Zhang G.Q., *et al.*, "Micromechanical Modeling of Stress Evolution Induced During Cure in a Particle-Filled Electronic Packaging Polymer", *IEEE Transactions on Components and Packaging Technologies*, Vol.27, No.4 (2004), pp. 676-683.

7. Li Guoqiang, Helms Jack E. and Pang Su-seng, "Analytical Modeling of Tensile Strength of Particulate-Filled Composites", *Polymer Composites*, Vol.22, No.5 (2001), pp. 593-603.

Numerical Simulation on Heat Pipe for High Power LED Multi-Chip Module Packaging

Dongmei Li[1], G.Q Zhang[1], Kailin Pan[1], Xiaosong Ma[1], Lei Liu[1], Jinxue Cao[2]

[1]School of Mechanical & Electronical Engineering, Guilin University of Electronic Technology, Guilin, China, 541004
[2]The Factory of Changxin Electronic, Zhejiang, China, 313100
Email: Lidongmei5611@163.com

Abstract

Light emitting diode (LED) as the new light source has the advantages of power saving, environment-friendly, long lifetime and no pollution compared with fluorescent and incandescent lights. But the disadvantage of LED is low light lumen that only 10%~20% input power transform into the light, and 80%~90% into the heat. The junction temperature of LED is so high as to induce the lifetime declining rapidly, luminous decay and reliability decreasing. Therefore, the effective thermal management is very important for the LED light system. In this work, a new packaging architecture the system in package (SiP) configuration is used in the high power LED packaging. The light system consists of nine chips that each chip is 1.2W. Copper/water miniature heat pipe (mHP) is chosen to dissipate heat based on the LED packaging structure and the input power of the system. The principles of the heat pipe are investigated to design and select the structure and size of the heat pipe. Capillary limit and boiling limit of the heat pipe are calculated to determine the maximum heat transfer and verify the design of the heat pipe. The heat pipe is seen as the thermal superconductor in axial, which take the place of the process of the phase exchange in the pipe. The axial thermal resistance of mHP estimated by the net of the thermal resistance is 0.15°C/W approximately. The system level heat and temperature distribution are investigated using numerical heat flow models. In this analysis, 3D finite volume model is developed to predict the system temperature with Icepak which is the professional software to analyze the temperature field of electronics. The result shows that the junction temperature of the source is under 70°C at the natural convection which is satisfied with the requirement of the LED working at under 120°C. It shows that the heat pipe is the effective solution for the LED light application dissipation. For the lower junction temperature, three factors including the height, the thickness and the fin numbers of the heat sink, respectively, are considered to be optimized by DOE (design of experiment). With the simulation results of Icepak, the optimal scheme that the lower junction temperature is 56.7°C obtained by the combination of optimization levels.

Introduction

As the fourth generation of lighting sources, LED are power saving, environmental-friendly, long lifetime and no pollution compared with fluorescent and incandescent lights. But the LED is sensitive to the temperature. Higher temperature induces the brightness attenuation, shorter lifetime, emission wavelength shifts, declination of the reliability and catastrophic chip failure. Especially, compact high power LED array is required to product high luminous output without inducing the high junction temperature. According to the ITRS reported that power density and junction to ambient thermal resistance for high performance chips are more than 100W/cm^2 and less than 0.2°C/W, respectively. New liquid and phase change active heat sinks are recommended to cool the high power device. [1]

Jung Kyu Park proposed a Multi Layer Ceramic-Metal Package (MLCMP) based on a LTCC for high power LED device, the thermal resistance of the MLCMP is less than 10K/W. [2] Adam Christensen applied heat sink to dissipate heat for LED array consisting of 25 LED devices. [3] His study suggested active cooling scheme that fan must be used to produce considerable convection to dissipate heat. Lan Kim analyzed the LED array including 6 LED devices with heat pipe, the analysis shown that the junction temperature of LED array with heat pipe at the air velocity of 7m/s is 63.3°C. [4] Liulin Yuan and Sheng Liu from Huazhong University of Science and Technology explored the thermal analysis of high power LED array (125W) packaging with a micro channel cooler, the study show that this cooling scheme with staggered fins achieves good thermal performance for multi-chip LED module. [5] J.P.Calame from Naval Research Laboratory in Washington investigated a variety of micro channel materials and configuration, the study shown that silicon and AIN micro channel coolers exhibit good thermal performance. Polycrystalline chemical vapor deposited (CVD) SiC micro channel coolers was the best to high power densities. [6]

In this study, the multi-chip packaging model consists of nine LED chips, which has the advantages of small package volume, high lumens, low cost. But the multi-chip model with hot spot density increase could limit the performance of the system. Based on the empirical values and equations on the heat pipe design, this study is to design and simulate the performance of miniature sintered heat pipe at the condition of natural convection. Miniature heat pipe is used to solve the problem of hot spot for the light application made of multi chips. The software Icepak is applied to analyze the temperature field distribution of the light system. It is a powerful tool for the fluid dynamics and thermal analysis.

Thermal model of LED light application

As to the lower of the LED light emitting efficiency, that 10%~20% of the input power transform into the light, the high luminance must be created by LED array, increasing the chip size and multi-chip packaging. In this paper, the structure of single blue chip covered with phosphor is shown in Fig. 1. The multi-chip packaging LED device consists of 9 LED chips that the whole power amount to 10W, as shown in Fig. 2. [7] Blue chips with vertical structure are mounted on the substrate. It has the same structure as the single chip with the different number chips.

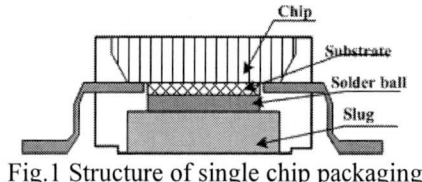

Fig.1 Structure of single chip packaging

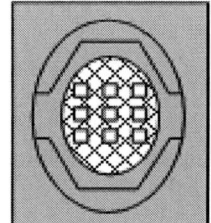

Fig.2 Schematic structure of multi-chip packaging

The whole structure of the heat transfer equipment is composed of the heat pipe, the copper block and the fins. The LED device is located on the copper block which enables the heat flux to be transmitted to the evaporator of the heat pipe. The structure of the light application is shown in Fig.3. So the design and select of the heat pipe is so important which determine the performance of the LED device. The capacity heat transmission should be calculated and the heat transfer performance should be evaluated based on the input power and the structure of the dissipation equipment.

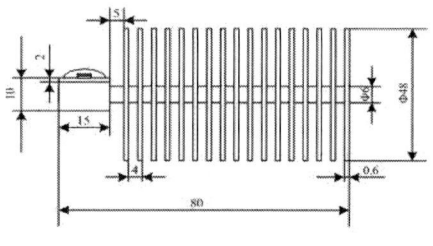

Units: mm

Fig. 3 Structure of the whole light application

The Calculation of the Heat Pipe [8-11]

The heat pipes using of the principle of phase transition to transmit heat flux, as shown in Fig.4, have the advantages of high conductive coefficient, low thermal resistance and small volume. It is wildly used to dissipate heat in electronics. The internal pressure of the heat pipe is negative 1.3×10^{-4}~1.3×10^{-1} and it is filled with the liquid appropriately. The heat pipe consists of the evaporator, the condensation and the adiabatic, but the adiabatic is not necessary. The heat flux of the source flows into the evaporator which enables the liquid evaporation. The heat is moved to the condensation and distributed to the air by convection, and the vapor was changed into the liquid flowing back to the evaporator by the wick structure which provides capillary pumping to circulation. Capillary pumping $\Delta P_{c\max}$ shown in Eq.1 is the driving force for the liquid circulation, $\Delta P_{c\max}$ is to overcome the viscous press drop ΔP_l in liquid and ΔP_v in vapor, and hydrostatic press drop ΔP_g. In this paper, the copper/water heat pipe has dimensions, outer diameter d_o =6 mm, vapor channel diameter d_v =4.8 mm, copper particles sinter wall

thickness δ =1mm, operating temperature 60°C, pipe length l = 80 mm, evaporation l_e = 15 mm, condensation 60mm, these parameters are selected and designed for heat transfer and thermal resistance of mHP.

$$\Delta P_{c\max} \geq \Delta P_l + \Delta P_v + \Delta P_g \qquad (1)$$

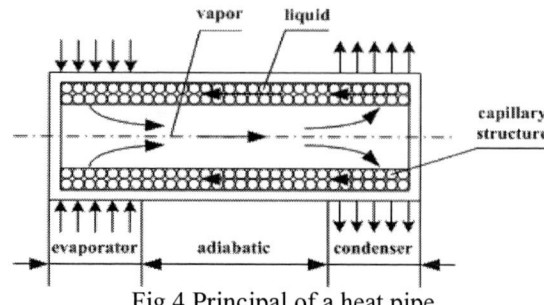

Fig.4 Principal of a heat pipe

In this work, sintered metal powder wick (75 μm—150 μm diameter) structure is the capillary wick structure that provides the capillary pressure for the fluid circulation. [10] Compared with wire and groove structure, the properties of sintered metal wick are lower thermal resistance of the wick, more connective areas with the inner surface of mHP that provides the lower thermal resistance and the small capillary radius which can provide the large capillary pressure. The sintered heat pipe can meet the requirement of the large heat flux with lower porosity. The maximize heat transport capability of the heat pipe is governed by several limited factors which should be consider when design a heat pipe, which are five primary heat pipe transfer limitations including viscous, sonic, capillary pumping, entrainment or flooding and boiling. Based on a large number of experiments, [10-11] for the low temperature heat pipe, the capillary and boiling limits are the most important restrictions of the maximum heat transport capability of the mHP. To find the maximum heat transfer, we need to analyze the capillary limit and boiling limit which are shown as Eq.2 and Eq.3, respectively.

$$Q_{c,\max} = \frac{2\sigma\cos\theta/r_c - \rho_l g l \sin\phi}{(8u_v/\pi r_4 \rho_v h_{fg} + u_l/KA_w \rho_l h_{fg})l_{eff}} \qquad (2)$$

$$Q_{b,\max} = \frac{2\pi l_e \lambda_e T_v}{h_{fg}\rho_v \ln(r_i/r_v)}(2\sigma/r_b - \Delta p_c) \qquad (3)$$

where, $Q_{c,\max}$ is capillary limit, σ is surface tension coefficient, r_c is effectively capillary radius, ρ_l is liquid density, d_v is inner diameter, u_v is gas viscosity, h_{fg} is vaporization latent of gas, A_w is wick cross section area, K is permeability of the wick, u_l is liquid viscosity, l_{eff} is effective length of the pipe, ϕ is angle of mHP inclination, l_e is evaporator length, T_v is the vapor temperature.

According to the conductive process of heat through the heat pipe, to get a rough estimate of the heat transfer and temperature drop that occurs in the heat pipe and it is common to represent it by the net of the thermal resistance shown as Fig.5.

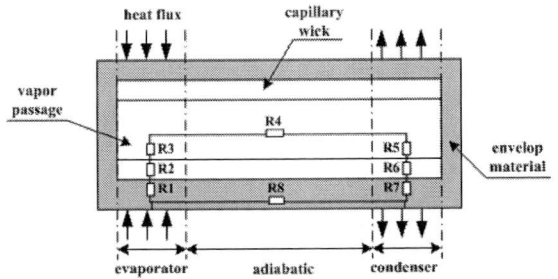

Fig.5 Equivalent thermal resistance network representative of a heat pipe

The total thermal resistance of a heat pipe is the sum of resistance due to conduction through the wall (R_1) and the wick (R_2), evaporation (R_3), axial vapor flow (R_4), condensation (R_5) and conduction losses through the wick (R_6) in the condenser and heat pipe wall (R_7). Thermal resistance R_3, R_4 and R_5 are generally neglected.

$$R_1 = \frac{\ln(d_o / d_i)}{2\pi l_e \lambda_s} \qquad (4)$$

$$R_2 = \frac{\ln(d_i / d_v)}{2\pi l_e \lambda_e} \qquad (5)$$

$$R_6 = \frac{\ln(d_i / d_v)}{2\pi l_c \lambda_e} \qquad (6)$$

$$R_7 = \frac{\ln(d_o / d_i)}{2\pi l_c \lambda_s} \qquad (7)$$

$$R_{HP} = R_1 + R_2 + R_3 + R_4 \qquad (8)$$

$$\lambda_e = \frac{\lambda_s [2\lambda_s + \lambda_l - 2\varepsilon(\lambda_s - \lambda_l)]}{2\lambda_s + \lambda_l + \varepsilon(\lambda_s - \lambda_l)} \qquad (9)$$

where, l_c is condensation length, R_{HP} is the resistance of mHP, λ_s is the thermal conductivity of the pipe shell, λ_l is the thermal conductivity of the liquid, λ_e is the thermal conductivity of the wick. [10] In generally, it is 50W/m· k. [13] The value of R_{HP} is approximately 0.15°C/W.

Simulation

The whole model is developed by Icepak, The most important objectives are to improve in the thermal model. These include mHP, fins, silicone grease, slug, solder ball, chip and substrate. These material properties are shown in table.1.

In Icepak, the typical parameters used to construct the model, are:

(1) Calculation domain is defined for the natural convection.

(2) The flow was treated as steady, incompressible and laminar in the external heat pip.

(3) Heat sink parts representing fin stacks on the heat pipe, auxiliary condenser and heater, and aluminum is the material for all the heat sink parts.

(4) The global ambient temperature defined as 20°C in this case.

Table.1 Material parameters of LED system

Material	Value(W/m· K)
copper/water heat pipe	386
chip	150
slug	386
solder ball	36
substrate	200
fin	178
cooper black	386

The model detail level was chosen by trying to obtain compromise between model complexity and accuracy. For example, chips, substrate and solder ball are modeled as blocks. Others are modeled as the real model. The whole model is meshed using unstructured mesh and non-conformal meshing between background and assembly. Element number is 193112 and node number is 205517. Due to there are almost no software to simulate the phase transition, the heat pipe is seen as the thermal superconductor with the very high thermal conductivity in axial of mHP, so the thermal conductivity of the heat pipe is the important parameter that comes from the Eq. $K_{eff} = L / AR$, L is the length, A is the effective cross section area, R is the thermal resistance, separately, of mHP. In generally, the range of the thermal resistance of the heat pipe is from 0.1 to 0.2. The range of the thermal conductivity is from 20000W/K to 50000W/K. In this simulation, the coefficient thermal conductivity K_{eff} is about 25163W/K in axial and 386W/K in radial. The simulation result, as shown in Fig.6, is that the junction temperature is approximately 62.5°C which is satisfied with the requirement of the LED working at under 120°C, and luminous decay, color offset are controlled effectively. So the result proves that the selection and the design of the heat pipe are reasonable.

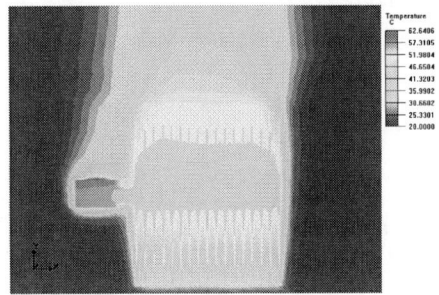

Fig.6 The cut plane of the simulation result

Result and Optimization

The simulation results provide temperature distribution within the field and the solid. The steady-state temperature of the source is less then 70°C. It is satisfied with the requirement of the LED working at under 120°C. The thermal resistance is very important parameter to measure the performance of the dissipation equipment. It is defined by:

$$R = \frac{T_j - T_{amb}}{Q} \qquad (10)$$

Where, T_j is the temperature of junction, T_{amb} is the temperature of ambient, Q is the input power. The lower the thermal resistance is, the higher the performance of dissipation equipment is. Resistance value of the whole model is calculated for model using the simulation data. It is approximately 3.67K/W when the temperature of the source is 56.7°C. The mHP with fins can provide the lower thermal resistance compared other dissipation structure. The fin pitch is too small to enable the air to remain cooler near the fin base. Since the surface area and fluid bulk temperature mostly remain the same throughout, then the fin geometry is the only parameter studied. One of the interesting features exhibited was the occurrence of a temperature overshoot, wherein the maximum module temperature, it can be explained by the increased downstream natural convection.

In this study, the optimal scheme of dissipation is selected by the DOE (design of experiment), which estimated the optimization levels of the influential factors to obtain the optimal scheme. [14] The factors include the number, the thickness, and the height of fins, respectively. Each factor has three levels, shown in table.2. The orthogonal table.3.L₉ (3*4) is used that each level of per factor is matched uniformly. The temperature is as evaluation index with the simulation of the software Icepak. Based on the range analysis, the important order of the factor is analyzed. The optimal scheme is the height 24mm, the thickness 0.6mm, the number 15 for the lower temperature, as shown in Fig. 7. The simulation result verified the above conclusion.

Table.2 Factors and Levels

Factors of fin	Level 1	Level 2	Level 3
number	15	17	20
Height	22	24	25
thickness	0.35	0.45	0.6

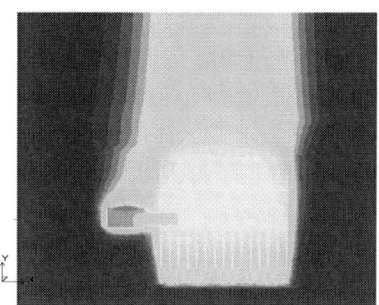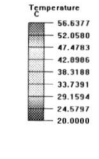

Fig.7 the cut plane of the optical scheme

Conclusion

High power multi-chip LED packaging meets the requirement of high luminous and low cost. The dispassion structure using miniature heat pipe with heat sink is a perfect solution which improves the thermal spreading capability to address the effect of hot spots problem. Compared with other

dissipation equipments, such as heat sink, heat sink with fan, mHP can decrease the resistance.

It is noted that the LED source has been supposed a whole block source, in further study, sub-model can be used to analysis the chip temperature more accurately; On the other hand, some parameters of the heat pipe are the empirical values, we should confirm these parameters through the experiment. But applying software icepack to simulate miniature heat pipe with fins to dissipate heat for high power LED are considered and have some reference value.

Table.3 Simulation results and range analysis results

	number	Height (mm)	Thickness (mm)	T(°C)
Case1	1(15)	1(22)	1(0.35)	74.9529
Case2	1(15)	2(24)	2(0.45)	64.0886
Case3	1(15)	3(25)	3(0.6)	64.4001
Case4	2(17)	1(22)	2(0.45)	69.4224
Case5	2(17)	2(24)	3(0.6)	62.4735
Case6	2(17)	3(25)	1(0.35)	75.3216
Case7	3(20)	1(22)	3(0.6)	70.5386
Case8	3(20)	2(24)	1(0.35)	74.2724
Case9	3(20)	3(25)	2(0.45)	68.7465
K1	203.4416	214.9319	204.5469	
K2	207.2175	200.8345	202.2575	
K3	213.5575	208.4662	197.4122	
k1=K1/3	67.8139	71.644	68.1823	
k2=K2/3	69.0725	66.9448	67.4192	
k3=K3/3	71.1858	69.4887	65.8047	
Range	3.3719	4.6992	2.3776	
Optimal scheme	15	24	0.6	56.6692

Acknowledge

This work is partly supported by the factory of Changxing electronics of ZheJiang province in China. The authors would like thank this factory for value discussions.

Reference

1. ITRS: international Technology Road Map for Semiconductors 2007 EDITION.
2. Jung Kyu Park, Ki Pyo Hong. *et al*, "Formation of Large Scale Via Slug for High Power LED Package ", *Ceramic Processing Research,* Vol.9.No.3 (2008), pp.262-266.
3. Adam Christen, Samuel Graham, "Thermal Effects in Packaging High Power Light Emitting Diode Arrays", *Applied thermal engineering*, Vol.29. (2009), pp.364-373.
4. Lan Kim, Jong Hwa Choi, "Thermal Analysis of LED Array System with Heat Pipe", *Thermochimica Acta*, Vol.499. (2007). pp.21-25.
5. Liulin Yuan, Sheng Liu, *et al*, "Thermal Analysis of High Power LED Array Packaging with Microchannel Cooler", *IEEE.2006 7th International Conference on Electronics Packaging Technology*, 2006,pp.1-5.

6. J.P.Calame, R.E.Myers, *et al*, "Experimental Investigation of Microchannel Coolers for the High Heat Flux Thermal Management of GaN-on-SiC Semiconductor Devices", *International Journal of Heat and Mass Transfer,* Vol.50.(2007), pp. 4767-4779.

7. Lan Kim, Woong Joon Hwang, "Thermal Resistance Analysis of High Power LED with Multi-chip Package", *IEEE. 2006 Electronic Components and Technology Conference,* pp. 1076-1081.

8. Zhuang jun, Zhang hong, Heat Pipe Technology and Engineering Application, Chemical Industry Press, (China, 2000), pp.15-66

9. Yu Jianzhu, Thermal Design and Analysis Techniques of Electronic Equipment, Higher Education Press, (China, 12001), pp.193-219.

10. Shijie Zhuo, "A study of the micro heat pipe with copper powder wick structure", National Taibei Science and Technology College. Master's thesis.

11. Zhang Hong, Zhuang Jun, "The analysis of heat transfer performance on micro heat pipe", *The 4th National conference on heat pipe,*(2000), pp.73-79.

12. L.L. Vasiliev, "Micro and Miniature Heat Pipe-Electronic Component Coolers", *Applied Thermal Engineering,* Vol.28 (2008), pp.266-273.

13. Y. Avenas, M. Ivanova, et al, "Thermal Analysis of Thermal Spreads Used in Power Electronics Cooling", *IEEE. 2002 37th IAS Annual. Conference Record of the Industry application conference,* Vol.1(2002), pp.216-221.

14. Yang de, Design and Analysis of Experiments, Chinese Agriculture Press, (China, 2000), pp.171-199.

Board Level Drop Impact Reliability Analysis for Compliant Wafer Level Package through Modeling Approaches

Chaoping Yuan, K.L.Pan, Weiyang Qiu, Jing Liu

School of Mechanical & Electronical Engineering, Guilin University of Electronic Technology

Guilin, China, 541004

Email: yuanchaoping@tom.com

Abstract

Board level solder joint reliability performance during drop test is a critical concern to semiconductor and electronic product manufacturers. In this paper, a new compliant Wafer Level Package technology is proposed which can accommodate the CTE mismatch between the chip and PCB substrate and consequently should be more reliable without the application of underfill. The purpose of this study is to explore the solder joint reliability of the new compliant WLP during drop impact and optimize the design of compliant lead. The input acceleration (Input-G) method is applied to simulate the exact drop test process subjected to JEDEC board level drop test conditions.

Several types of compliant lead shape with different sizes are studied by comparing and analyzing the dynamic responses of CWLP solder joints under board level drop test, the copper trace reliability of these types CWLP are compared, further more, the optimal parameters of the compliant lead are confirmed by design of experiment (DOE).

The board level drop test simulation illustrated the ability of the compliant Wafer Level Package to reduce the stress in the solder interconnects. It is observed that the highest stress appears in the copper trace, while copper trace can flex and effectively absorb the stress between the chip and the bump pad. This in turn will lead to an increase in the reliability of the assembly.

1. Introduction

Wafer level packages (WLP) are widely used for portable electronic packages due to their small package size, high frequency performance and low manufacturing cost. In the past, the traditional lead-type packages such as thin small outline package (TSOP) were used for dynamic random access memory (DRAM), but the increased performance requirements of DDR2 DRAM could not be met with TSOP, while WLP could meet these demands. However, a mismatch of the coefficient of thermal expansion (CTE) between the die and the PCB could cause stress in the package because of the mounting of the die onto the PCB, these will cause the solder joint to crack and fracture. The solder joint fatigue due to stress generated by the CTE mismatch will limit the adoption of WLP for large die sizes, so the current reliable die size limit for a non-underfilled WLP is about 5mm×5mm. [1,3,8]

To overcome the problem, it can be achieved by means of constraining the solder joint with the application of underfill between chip and substrate. To accommodate the thermal stress the whole system will bend. This in turn reduces the load on the solder balls and thus the accumulated inelastic strain within them, which increases the reliability of the

system. [1, 3] However, underfilling is an additional process during the assembly which will increase the assembly time, repair time and cost.

Therefore, several compliant interconnect technologies for WLP have been studied to eliminate the need for the application of underfill in recent years. One way to increase the reliability of the solder joints is to form a compliant polymer bump underneath the solder ball, a protective layer constituting a compliant polymer material is placed over the compliant bump to provide a gentle slope for the copper redistribution layers (RDL) trace that forms the primary electrical connection between the bond pad on the die and the metal pad on top of the compliant bump. [3] This compliant WLP have been proven to be highly reliable with conventional WLP. [3] Besides, Georgia Institute of Technology invented the so called "Sea-of-Leads (SoL)". One type of SoL, the "slippery type" SoL has a good compliancy in all directions, but is not easy to handle during board assembly. Another type, the SoL with embedded air gaps, is specially designed for out-of-plane (Z-direction) compliancy. [4, 5, 12, 13]

In this paper, a new type compliant wafer level package is investigated, to achieve a high flexibility in x-, y- and z-direction, the solder bump is located on a flexible lead, the flexible lead consists of a copper redistribution (RDL) embedded in a polymer-bridge which is located over an air-gap. [8] In the previous studies, FEM simulations of the proposed package in thermal cycling test have been studied, and it already represented brilliant performance. [13] Now, investigations of this type package during drop impact are carried out, with ANSYS/LS-DYNA simulations performed. The input acceleration (Input-G) method is applied to simulate the exact drop test process subjected to JESD22-B111. The reliability of CWLP in an accelerated test environment is analyzed, and three types of flexible lead shape are studied by comparing and analyzing the dynamic responses of CWLP. Furthermore, the optimal parameters of the compliant lead are confirmed by DOE.

2. Finite Element Modeling

The prototype is a WLCSP DRAM (Dynamic Random Access Memory) with 54 I/O, and is 1/2 symmetric distribution (Fig. 1). It is considered that $5.3 \times 2.5 \times 0.3 mm^3$ CWLP mounted on a $132 \times 77 \times 1 mm^3$ standard JEDEC drop test board using 95.5Sn-4.0Ag-0.5Cu solder. Solder ball radius is 0.0625mm. The layout of packages mounted on the test board following JESD22-B111 is shown in Fig. 2. Since the drop performance is a function of component location on the board, testing with components mounted on all 15 locations will provide useful information of the test board.

With board supported at 4 corners the worst case board curvature is at U8 location, so in this study, the option mounting just 1 component on the board is at U8 location. [2] JESD22-B111 drop test condition B was adopted. The peak acceleration, pulse duration and drop height are 1500G, 0.5ms and 112cm, respectively. The board is horizontal in orientation with components facing in downward direction during the test.

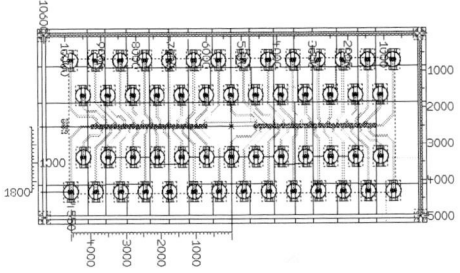

Fig.1 WLCSP DRAM layout and solder distribution pattern

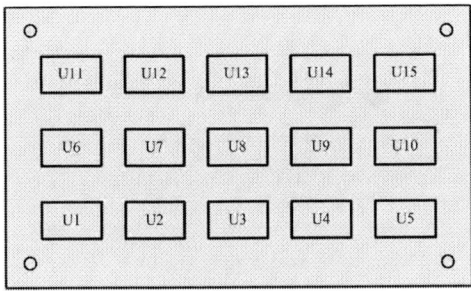

Fig. 2 Layout of test board subjected to JESD22-B111

In this paper, a 1/2 symmetric model is developed by using Input-G method. According to the Input-G method, the impact pulse can be considered as a PCB boundary condition in modeling and be input to PCB subassembly directly. [6, 14, 15] The drop table, fixture, contact surface, and friction of guiding rods are not simulated, but their complex effects are considered indirectly by using the same impact pulse as experiment (Fig. 3).

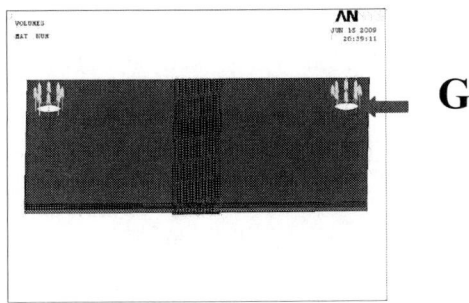

Fig.3 Input-G methed

The material properties are listed in table 1. [7, 8, 9, 11] In the previous studies, temperature-dependent elastoplastic model and bi-linear elastoplastic model are used for 95.5Sn-4.0Ag-0.5Cu solder joints and copper trace during the thermal cycling test simulation, respectively. However, during drop impact, the rate-dependent bi-linear elastoplastic model should be used, because the strain rate is much higher than that of normal material testing during the drop test. Unfortunately, currently there is no published material data with confidence for solder joint and copper trace at high strain rate. Nevertheless, rate-dependent model becomes harder under high strain rate, which is close to linear elastic model. [6, 7] In this study, all materials are assumed to be linear elastic except for PCB, which is considered as orthotropic material.

Table 1 Model material properties of package components

Material	Elastic modulus (KPa)	Poisson's ratio	Density (kg/mm³)
SnAg	4.85e7	0.36	7.44e-6
PCB	Ex:25e6 Ey:25e6 Ez:11e6	xy:0.11 yz:0.39 xz:0.39	1.82e-6
Chip	1.68e8	0.3	2.33e-6
Copper	9.7e7	0.35	8.96e-6
Polyimide	2.5e6	0.3	1.35e-6
Nickel	2.275e8	0.32	8.88e-6
Mask	5e6	0.3	1.15e-6

3. Result and Discussion

In this study, drop test for flexible lead CWLP is simulated to investigate the mechanical behavior of board and packages under drop impact. The reliability of solder joints during the drop impact is the main concern, as it affects the functionality of product. In this study the induced von Mises stress is used as the indicator of the reliability of the solder joints and copper trace. Fig. 4 and Fig. 5 show the von Mises stresses of solder joints and straight copper trace, respectively. The results show that the critical solder joint is at the outermost package corner, with stress concentration along the solder/PCB pad interface. The reason is that a combination of mechanical shock and PCB bending induce the solder ball interfacial failure. The bending stress is crucial to solder joint reliability. As shown in Fig. 5, it is observed that the highest stress appears in the copper trace, while for the conventional WLP the highest stress appears in the bump. [3,10] Copper trace can flex and effectively absorb the stress between the chip and the bump pad. So the stress of copper trace becomes the main issue in this analysis.

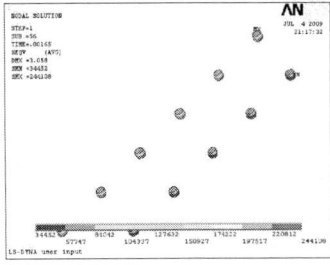

Fig. 4 Von Mises stress distribution of solder joints

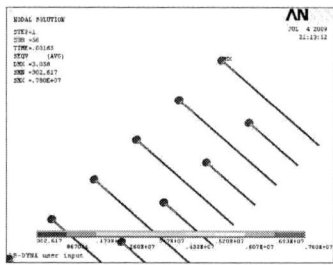

Fig. 5 Von Mises stress distribution of copper

To study the different layouts of copper trace, two trace layouts, the horizontal trace design and vertical trace design, are investigated. The horizontal trace design means that the trace display along the long axis of PCB, similarly, the vertical trace design means that the trace display along the short axis of PCB. The dynamic responses of copper trace for these two layouts are shown in Fig. 6. It reveals that the induced von Mises stress of the horizontal trace design is lower than that of the vertical trace design. The possible reason is that the loadings on horizontal trace layout are stretching loads, while the loadings on vertical trace layout are shearing loads. The rigidity of horizontal trace is much larger against stretching loads than the rigidity of vertical traces against shearing loads. [10] In this study, all trace layouts apply the horizontal trace design.

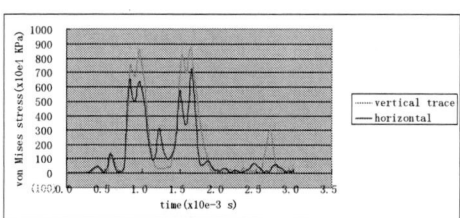

Fig. 6 Time history of von Mises
stress for different trace layouts

Moreover, three types of flexible trace CWLP are simulated in this study. As shown in Fig. 7, they are straight copper trace, banana shaped copper trace and zigzag copper trace, respectively. It can be found that stresses concentrated at the neck area between pad and trace. By comparing these types of copper trace, the induced von Mises stress for the straight copper trace is lower than that for the banana shaped copper trace and zigzag copper trace. To further study the dynamic behavior of copper trace, the time histories of von Mises stress are tracked as shown in Fig. 8. It could be concluded that the highest von Mises stress appears at 1.65ms, it lags behind 0.5ms which peak acceleration value appears at. The reason is that, when peak acceleration appears, PCB remains bending down trend until the maximum deformation arrives, which is close to the von Mises stress of copper trace. Besides, it is obvious that the induced von Mises stress for the straight copper trace is lower than that for the banana shaped copper trace and zigzag copper trace during the drop impact. The result during drop test simulation is

different from that during the thermal cycling test simulation which the induced von Mises stress for the straight copper trace is higher than that for the banana shaped copper trace and zigzag copper trace. [13] By expansion and contraction of flexible trace, it could absorb the stress generated during the thermal cycling. Obviously, banana shaped copper trace and zigzag copper trace are more flexible than straight copper trace.

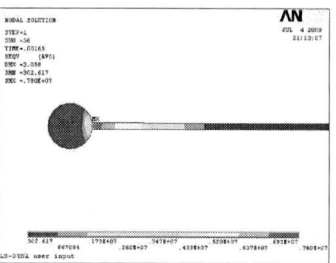

a. Straight copper trace

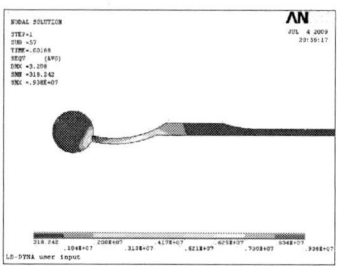

b. Banana shaped copper trace

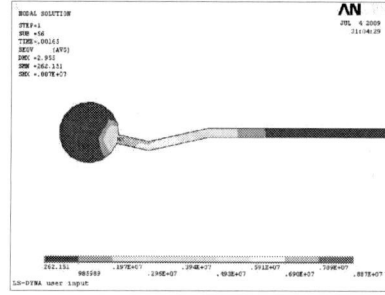

c. Zigzag copper trace

Fig. 7 Von Mises stress distribution of
different copper trace

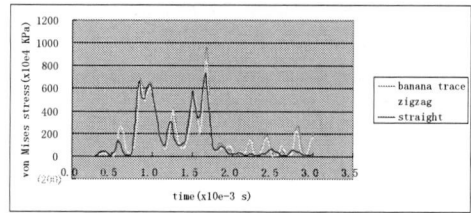

Fig. 8 Time history of von Mises stress
for different shapes of copper trace

978-1-4244-4658-2/09 $25.00 © 2009 IEEE

4. DOE and Range Analysis

In order to investigate the main parameters leading to excessive stress in the copper trace, DOE is performed. In this analysis, the shape of copper trace, the thickness and the width of copper trace, the height of air-gap are chosen as the four most important factors to carry out the analysis. All these factors have three levels, as shown in table2. An orthogonal experimental design is applied, and the orthogonal table L_9 (3^4) is adopted (Table3).

Range analysis is then used to deal with the DOE simulation results by using von Mises stress as evaluation index, as shown in Table 4. Range analysis is to calculate the range value of each column of the orthogonal table by the method of mathematical statistics. The range value is defined as the difference between maximum and minimum mean values. The bigger of the range value, the more important of the factor.

As shown in table 4, based on the value of the range (R), the sequences of the factors which affect the maximum von Mises stress of copper stress could be ranked as follows: the thickness of copper trace>the shape of copper trace>the width of copper trace>the height of air-gap. The factor trend of orthogonal test is shown as Fig. 9. It could be concluded that the main factors can be ordered as follows: B>A>C>D, which is same as the range analysis before. The optimal scheme is $B_3A_2C_3D_3$ by comprehensively considering with every factor's effect, i.e. the optimal parameters for CWLP are the thickness of copper trace 0.010mm, the shape of copper

straight shaped, the width of copper trace 0.020mm, the height of air-gap 0.015mm, respectively.

Table 2 Factors and levels

Levels / Factors	Level-1	Level-2	Level-3
Copper Trace Shape	Banana	Straight	Zigzag
Copper Trace Thickness(mm)	0.005	0.008	0.010
Copper Trace Width(mm)	0.015	0.018	0.020
Air-gap Height(mm)	0.010	0.012	0.015

Table 3 Orthogonal table L_9 (3^4)

Factors No.	1	2	3	4
No.1	1	1	1	1
No.2	1	2	2	2
No.3	1	3	3	3
No.4	2	1	2	3
No.5	2	2	3	1
No.6	2	3	1	2
No.7	3	1	3	2
No.8	3	2	1	3
No.9	3	3	2	1

Table 4 Simulation results of DOE and range analysis results

Run order	Trace shape (A)	Trace thickness (mm)(B)	Trace width (mm)(C)	Air-gap height (mm)(D)	The maximum von Mises stress (KPa)
1	banana	0.005	0.015	0.010	0.92796e7
2	banana	0.008	0.018	0.012	0.72942e7
3	banana	0.010	0.020	0.015	0.58860e7
4	straight	0.005	0.018	0.015	0.64793e7
5	straight	0.008	0.020	0.010	0.51347e7
6	straight	0.010	0.015	0.012	0.60810e7
7	zigzag	0.005	0.020	0.012	0.78778e7
8	zigzag	0.008	0.015	0.015	0.72074e7
9	zigzag	0.010	0.018	0.010	0.56235e7
Y_{1j}	0.74866e7	0.78789e7	0.75227e7	0.66793e7	
Y_{2j}	0.58983e7	0.65454e7	0.64657e7	0.70843e7	
Y_{3j}	0.69029e7	0.58635e7	0.62995e7	0.65242e7	
R_j	0.15883e7	0.20154e7	0.12232e7	0.05601e7	
Order	2	1	3	4	

5. Further Study

It is noted that all materials are assumed to be linear elastic except for PCB, which is considered as orthotropic material. Actually, materials should have rate-dependent behavior during drop impact. Further study should focus on the material properties of solder joint and copper trace. Besides, the result of DOE simulation is an approximation about geometric parameters for CWLP, more exact parameters should be obtained by other optimization method in the next step.

6. Conclusions

In this study, the board level drop test simulation for compliant wafer level package has been performed. It is noted that the highest stress appears in the copper trace, while for the conventional WLP the highest stress appears in the bump. The reason is that copper trace can flex and effectively absorb the stress between the chip and the bump pad. Besides, two trace layouts, the horizontal trace design and vertical trace design, were investigated. Results indicated that the horizontal trace design can bear more impact loading because its rigidity against impact load is higher. Moreover, comparing three types of flexible trace CWLP, the induced von Mises stress for the straight copper trace is lower than that for the banana shaped copper trace and zigzag copper trace which is different from the result during the thermal cycling test simulation.

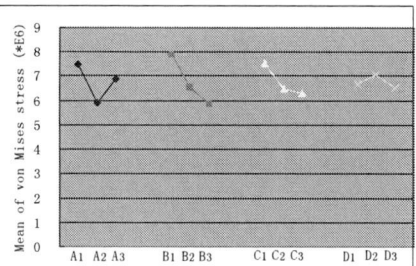

Fig. 9 The factor trend of orthogonal

From the DOE simulation and range analysis, the factors which affect the maximum von Mises stress of copper stress could be ranked as follows: the thickness of copper trace>the shape of copper trace>the width of copper trace >the height of air-gap. The optimal scheme is $B_3A_2C_3D_3$ by comprehensively considering with every factor's effect, i.e. the optimal parameters for CWLP are the thickness of copper trace 0.010mm, the shape of copper trace straight shaped, the width of copper trace 0.020mm, the height of air-gap 0.015mm, respectively.

Acknowledgments

The author gratefully acknowledges the financial support provided by National Natural Science Foundation of China (600866002) and the Innovation Project of Guangxi Graduate Education-Study on Compliant Bump Structure in SiP (Project No. 2008105950802M406).

References

1. ITRS: International Technology Road Map for Semiconductors 2005 EDITION.
2. JEDEC Solid State Technology Association, JESD22-B111: Board Level Drop Test Method of Components for Handheld Electronic Products; 2003.
3. Guilian Gao, Bel Haba, Vage Oganesian, Ken Honer, "Compliant Wafer Level Package for Enhanced Reliability," Proceedings of HDP'07, IEEE.
4. Chirag Suryakant Patel, "Compliant Wafer Level Package (CWLP)," School of Electrical and Computer Engineering Theses and Dissertations Georgia Tech Theses and Dissertations. May-2001.
5. Lunyu Ma, "Design and Development of Stress-engineered Compliant Interconnect in Microelectronic Packaging," School of Mechanical Engineering Theses and Dissertations Georgia Tech Theses and Dissertations. Aug-2003.
6. Jing-en Luan, Tong Yan Tee, "Solder Joint Failure Modes, Mechanisms, and Life Prediction Models of IC Packages under Board Level Drop Impact.," The Sixth

International Conference on Electronic Packaging Technology, 2005, pp. 382-388.
7. S. Wiese, S. Rzepka. "Time-independent Elastic–plastic Behaviour of Solder Materials," EuroSimE International Conference on Thermal and Mechanical Simulation and Experiments in Microelectronics and Microsystems, 2004, pp. 1893-1900
8. I.Eidner, K. Buschick, L. Dietrich, K.L.Pan, "Bump on Flexible Lead for Wafer Level Packaging," 41st Annual International Symposium on Microelectronics (IMAPS), providence, Rhode Island (USA), 2-6 November 2008.
9. I.Eidner, B. Wunderle, K.L.Pan, "Design Study of the Bump on Flexible Lead by FEA for Wafer Level Packaging, Thermal, Mechanical and Multi-physics Simulation and Experiments in Microelectronics and Microsystems," 2009. April 2009, pp. 1-7.
10. Chan-yen Chou, "Solder Joint and Trace Line Failure Simulation and Experimental Validation of Fan-out Type Wafer Level Packaging Subjected to Drop Impact," 19th European Symposium on Reliability of Electron Devices, Failure Physics and Analysis (ESREF 2008), August 2008, pp. 1149-1154.
11. Chang-lin Yeh, Yi-shao Lai, "Support Excitation Scheme for Transient Analysis of JEDEC Board-level Drop Test," Microelectronics and Reliability, Volume 46, April 2006, pp. 626-636.
12. Bakir, M.S.; Reed, H.A.; Mule, A.V.; Kohl, P.A.; Martin, K.P.; Meindl, J.D, "Sea of leads (SoL) Characterization and Design for Compatibility with Board-level Optical Waveguide Interconnection," Proceedings of the IEEE 2002, Volume , Issue , 2002 Page(s): 491-494.
13. Li Peng, "Research on Structural Design of Compliant Bump with Embedded MEMS Air-gap and Its Thermal Fatigue Reliability,"Guilin University of Electronic Technology School, 2009.
14. Tong Yan Tee, Jing-en Luan, "Novel Numerical and Experimental Analysis of Dynamic Responses under Board Level Drop Test," Thermal and Mechanical Simulation and Experiments in Microelectronics and Microsystems, 2004, EuroSimE 2004, January 2004..
15. Tong Yan Tee, Jing-en Luan, "Development and Application of Innovational Drop Impact Modeling Techniques," 7th EPTC. Conf, 2005, pp.524-512.

FEM Study on the Effects of Flip Chip Packaging
Induced Stress on MEMS

Songsheng Wei, Jieying Tang, Jing Song

Key Laboratory of MEMS of Ministry of Education, Southeast University, Nanjing, 210096, China

Abstract

Packaging induced stress may have considerable effect on performance of MEMS devices. This paper is focused on the performance change of MEMS devices after flip chip package. The warpage and stress distribution on die surface of flip chip package with different bumps layout are investigated using finite element method(FEM). Package of 4×4 all-array pattern is first studied. Then a comparison study between packages with bumps of 2×2,4×4 and 6×6 all-array patterns and one-circle 4×4, one-circle 6×6 and two-circle 6×6 peripheral patterns are presented. According to the FEM results, the packaging effect on the natural frequency of a flip-chipped microcbridge is case studied finally.

Introduction

Technology of micro-electro mechanical systems(MEMS) is a young and fast-developing field. Packaging is critical to the success of MEMS as they reach commercialization. With the increasing need for higher performance, low costs ,and higher reliability, flip chip packaging is becoming a promising technology for advanced MEMS packaging. [1] Since MEMS are inherently sensitive to stress, the performances of MEMS devices may be considerably changed after packaging due to packaging induced stress. Previous study have intensively validated that thermal mismatch between the materials may cause deformation and stress in the flip-chip structure. [2-4] However, little work mentions the detailed effect of the flip-chip packaging on the performance of MEMS devices, which is a key concern of a MEMS designer. In this paper, we focus on the effects of flip chip packaging induced stress on the performance of MEMS devices.

In Alander's and Chang's papers[5-6], bumps layout of flip chip packaging has been taken as a reliability factor of solder bumps, and some optimized layouts were developed. Besides, the disadvantages of uderfill have been persuasively discussed in Chang's paper. Inspired by their works, non-underfilled flipped chips with bumps of all-array patterns and peripheral patterns are studied through finite element analysis, and packaging induced warpage and stress distributions on die surface is derived. The all-array patterns include bumps array of 2×2,4×4,6×6 bumps. The peripheral-array patterns include the one-circle 4×4,one-circle 6×6 and two-circle 6×6 peripheral patterns. First, the 4×4 all-array pattern is investigated to find out the general rules. Then a comparison study between all-array patterns is present. Effect of stress distribution of 4×4 all-array pattern package on the natural frequency of a microbridge device is investigated to illustrate the package effect on MEMS devices. Finally, a conclusion is presented.

3D FEM models

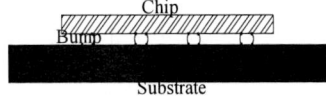

Fig .1 Model of flip chip packaging

Table 1 Parameters for the models used to describe different materials

Material	Young's modulus (MPa)	Poisson's ratio	CTE (ppm/k)
Die	131000	0.30	2.8
Solder	30000	0.35	21
FR4	20000	0.28	18

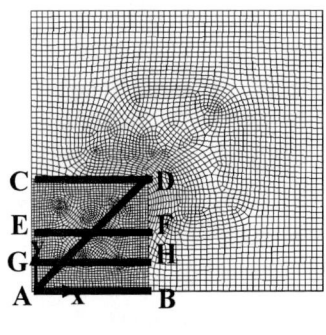

(a)

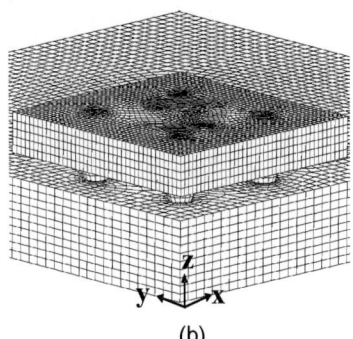

(b)

Fig. 2 Finite element model of flip chip package with bumps of 4×4 all-array pattern: (a) a top view; (b) a side view

As Fig.1 shows, a typical non-underfilled flip chip package consists of three constituents: silicon die, solder bumps, and print circuit substrate. [7] 3D finite element models of packages with bumps of all-array patterns and peripheral patterns are built in ANSYS respectively. The finite element model of 4×4 all-array pattern can be seen in Fig.2, symmetry was taken advantage of so only a quarter of segment is modeled. The length and width of the chip and the substrate are 8mm and 12mm respectively, and the height are 0.5mm and 1mm respectively. The radius of the solder bump is 0.25mm, and the Sn-Pb(63%-37%) solder is modeled. Element Solid45 is used and total 48041 elements are meshed. The material properties are given in Table.1. And a temperature load ranged from 25℃ to 125℃, according to general operational region of a MEMS sensor device, is used in the simulation.

General rules of packaging induced stress

Induced stress and warpage of a package with bumps of 4×4 all-array pattern is given in the Fig. 3. Distribution of plane stress components of Sx, Sy, and Sxy along line A-B, C-D, E-F and G-H(as Fig. 2 shows) on die surface are investigated.

As we can see from Fig.3(a), a remarkable change of the value of the three plane stress components along line A-B, C-D and E-F, decades of MPa at central positions to nearly zero on the edge of the die, occurs in X direction; while in Y direction, stress components of Sx are basically identical along any of these lines. Likewise, the stress components of Sy along line A-B, C-D, and E-F are basically identical in X direction, while drastically changed along Y direction (as Fig.3(b) shows). As Fig.3(c) shows, stress concentration occurs at places where a solder bump stands, and stress value in these regions may reach about two times of that outside. Stress component of Sxy is comparatively much small(less than 1MPa) on the whole die surface. The maximum warpage on the die surface is nearly 8μm and the maximum curvature has come to 1.0m-1 (as Fig.3(d))shows). The results of FEM can give some suggestions to practical MEMS designs: (1) As the performance of MEMS devices may have a remarkable change due to the packaging induced stress, the position of the device on the die surface has to be taken into account. Owing to an almost identical stress level in either direction, the direction of a MEMS structure such as a microbridge doesn't matter much when placed in the region of the die center. Since the stress level on the die edge is much lower, the package effect may be reduced in these regions. Besides, MEMS structures are not suggested to appear in the region of about 300-400μm around a solder due to the effect of the stress concentration. If the density of solder bumps exceeds a certain level, we should not place the devices between the bumps. (2) Effect of Sxy on the performance of MEMS devices are negligible when it comes to a practical design. (3) Packaging induced warpage of die surface has to be seriously taken into account. The warpage may leads to decrease of the separation between a MEMS structure and its substrate. For example, with a 400μm's long bridge device placed along line A-B, the maximum change of the separation can be determined as $kx^2/2$ (where k is curvature, and x is length of microbridge) and a result of 0.08μm, which is non-negligible for a common separation of 1-2μm in MEMS, can be calculated.

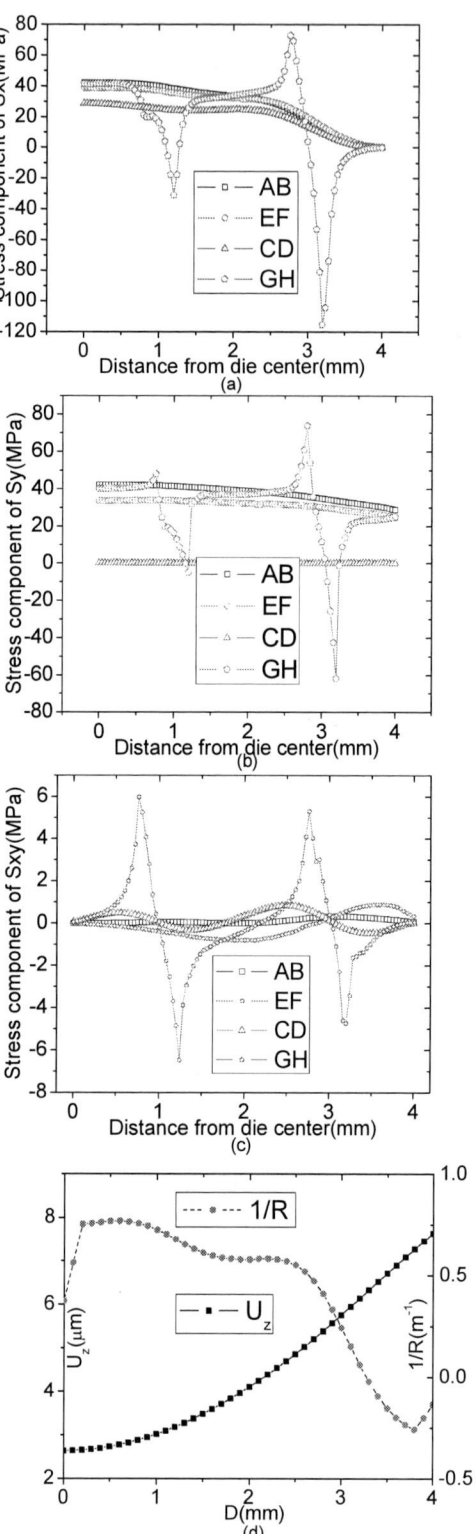

Fig. 3 Distribution of stress components and the warpage on die surface: (a) Sx along line A-B, C-D, E-F and G-H;(b) Sy along line A-B, C-D, E-F and G-H;(c) Sxy along line A-B, C-D, E-F and G-H;(d) Displacement of Z direction Uz and change of the curvature along line A-B

978-1-4244-4658-2/09 $25.00 © 2009 IEEE

Stress distribution of different bumps layout

The geometries of flip chip packages with bumps of 2×2,4×4 and 6×6 all-array patterns and one-circle 4×4, one-circle 6×6 and two-circle 6×6 peripheral patterns are given in Fig. 4. Stress components of Sx and Sy along line A-B on die surface of different patterns are derived with FEM so as to make a comparison(as Fig. 5 shows).

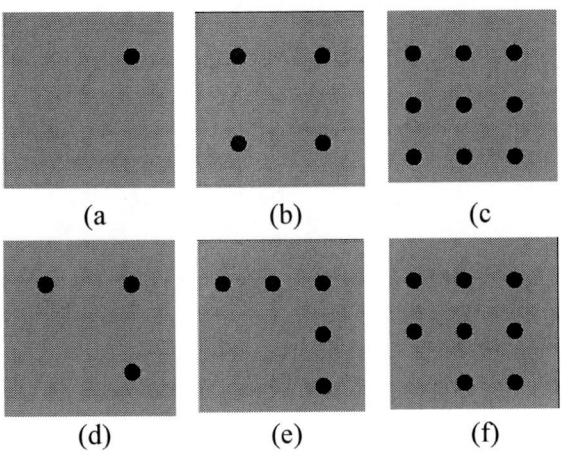

Fig. 4 Geometries of bumps-layouts(a quarter of segment): (a)2×2 all-array; (b)4×4 all-array; (c)6×6 all-array; (d) one-circle 4×4 peripheral pattern; (e)one-circle 6×6 peripheral pattern; (f)two-circle 6×6 peripheral pattern

The effect of Sxy is not concerned about as aforementioned. With increase of the solder bump, the warpage and values of Sx and Sy are all enhanced as we can see from Fig. 5.

Effect of packaging induced stress on MEMS performance

Natural frequency is a key parameter of design and test of MEMS devices. As a case study, a flip-chipped microbridge device with bumps of 4×4 all-array pattern on die surface is investigated. The length of the bridge is 400μm, and the height and separation from the substrate are both 2μm.

The natural resonant frequency of a microbridge is determined as[8]:

$$f = 1.028 \frac{t}{l^2} \sqrt{\frac{E}{\rho}[1 + 0.295 \frac{l^2}{t^2} \varepsilon]} \qquad (1)$$

where l is beam length,t is beam height, and ρ is beam density; ε is initial strain of beam, and E is Young's modulus. As Fig. 6 shows, the maximum relative change of the micro bridge's natural frequency at different places along line A-B is about 110%.

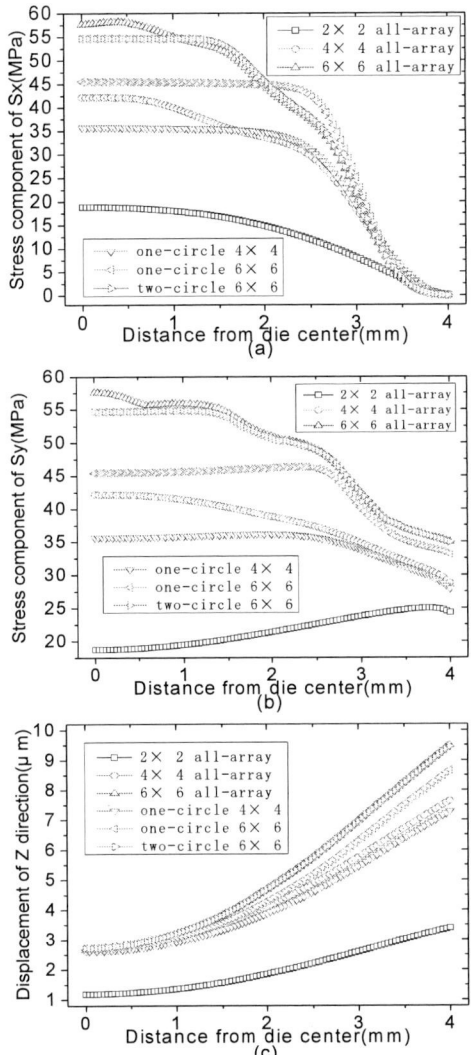

Fig. 5 Stress distribution and warpage of flip chip packaging with bumps of 2×2,4×4 and 6×6 all-array patterns and one-circle 4×4, one-circle 6×6 and two-circle 6×6 peripheral patterns

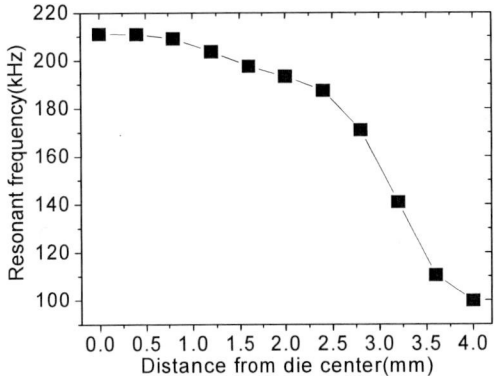

Fig. 6 Variation of natural frequency of a microbridge device according to different places along line A-B of 4×4 all-array pattern

Conclusions

In this paper, the distribution of the packaging induced stress of flip chip package with different bumps layout on die surface is investigate through FEM. The conclusions are as followings:

(1) In the region of a die center, the stress level is almost identical in all directions, and it is not important to place a MEMS structure in any direction. Placing the devices near to the edge of a die may decrease the packaging effect. Effect of the plane stress component of Sxy can be negligible.

(2) There is a significant stress concentration by the areas of solder bumps. The affecting scope is about 300-400μm in general. So if the density of solder bumps exceeds a certain level, we should not place the devices between the bumps.

(3) The warpage of a die surface may lead to a great change of the separation from substrate of a MEMS device, and thus affect its performances.

(4) With increase of the bumps density on die surface, packaging induced stresses are enhanced. Decreasing solder bumps can obtain better performances.

(5) Packaging induced stress will change the performance of MEMS devices under thermal loads. In this paper, it can lead to a 110% maximum relative change of the natural resonant frequency of a microbridge device, when the temperature changes from 25℃ to 125℃.

References

1. Rao R. T, Fundamentals of microsystems packaging. McGraw-Hill (New York, 2001), pp. 244-272.

2. Jing S, Ming L, An Q H, Ying J T, "Package level simulation and verification of microsystems," Sensors. 2007, pp. 99-102.

3. M. Kaysar Rahim, Jeffrey C S, "Die stress characterization in flip chip on laminate assemblies," Transaction on Components and Packaging Technologies, Vol. 28, No. 3(2005), pp. 415-429.

4. Mei Y C, Qing H, "Technologies of MEMS packaging," Chinese Journal of Tanceducer Technology, Vol. 24, No. 3(2005), pp.7-12.

5. Tapani A Pekka, H Eero R, "Solder bump reliability—issues on bump layout," Transaction on Advanced Packaging, Vol. 23, No. 4(2000), pp.715-720.

6. Chang, M, Liu, K. N. C, "Solder bumps layout design and reliability enhancement of wafer level packaging," Electronic Packaging Technology Proceedings, 2003, pp. 56-64.

7. Viraj A Daniel, L Peter, B. K. S, "Reliability issues on direct chip attach assemblies using reflow or no-flow underfill," Electronics Manufacturing Technology Symposium, November. 2002 , pp.73-77.

8. Kurte P, "Dynamic micromechanics on silicon techniques and devices," Transactiom on Electron Devices, Vol. 25, No. 10(1978), pp. 1241-1249.

Modeling and Simulation of SSN on FPGA Products

Lingzhi Ke[1,2], Peng Zhou[1], Lei Li[2]

[1]Dept. of Communication and Information System, Wuhan University of Technology, Wuhan, 430070, China
[2]Shenzhen institute of advance technology, Chinese Academy of Sciences/The Chinese University of Hong Kong,
Shenzhen, 518067, China
Email: Lz.ke@sub.siat.ac.cn, van_adams@163.com

Abstract

Simultaneous switching noise (SSN) and its behavior have recently become more and more critical in IC and other high-speed system designs [1][2]. This is attributed to ever-increasing speed, frequency, density, and power, as well as decreasing circuit dimensions and logic levels. The difference in a few mill volts may cause the system to fail. Therefore, it is very important to understand the characteristics of the SSN glitch of an active device for correct system level performance. In this paper, some methods to provide a complete picture of limitation characteristic behavior and its relationship to cause scheme, due to SSN, is demonstrated with FPGA system. Furthermore, model simulation confirms our postulations made on examination of experimental data and validates the methodology practical to SSN assessment in FPGA applications.

Introduction

One of the major concerns is SSN on the power/ground buses during the design cycle of high-speed digital communication systems. This is due to faster edge rates, lower voltage levels and higher integrations. The switching power dissipation becomes higher, and the average supply current is increased significantly, even if the power supply voltage is reduced by advances in semiconductor device technology. Combined with higher clock frequencies, the increased switching current changes (di/dt) produce considerable power supply switching noise that ultimately degrades the eye patterns and timing margins on critical clock and signal paths. This results in a limitation of the attainable clock frequencies in digital devices or systems. Moreover, this power supply's SSN can be a source of electromagnetic noise coupling and interference to nearby interconnections, circuits, and other devices that are difficult to circumvent. Hence, the switching noise must be under the designer's control to ensure reliable circuit, device, and system operation[2][6].

As we all know, SSN in an FPGA system may attribute to two primary factors: the mutual inductive coupling amongst switching I/Os and the impedance profile of a power distribution network (PDN) which includes die, package and printed circuit board (PCB). In essence, reducing SSN is a design cost issue. The former involves the designers' use of traditional methods. These include increasing the ratio of FPGA ground pins (or return-current pins) to I/O pins and/or minimizing the mutual inductive coupling which would sacrifice I/O densities. The latter involves increasing on-die capacitance and the addition of on-package decoupling capacitors which would improve PDN performance. However, this approach results in drastically increased costs. As FPGAs are programmable, they fit into a wide variety of user applications; therefore it is useful for designers to find a method to get the balance between costs and their own SSN budget.

This paper focuses on the SSN caused by FPGA output buffers [5]. It is organized as follows: Section II describes the causes of gate-level SSN, Section III give the details of the SSN model and provides a graphical description representing the results of improved mutual inductance and PDN, and Section V presents the conclusion.

Gate level SSN in FPGA

FPGA devices have unique attributes of structure on the behavioral level, register transfer level and gate level. Each level's analysis presents a different problem. In this section, we start with an overview of SSN in gate level systems.

The main structure of the gate level SSN system consists of two parts: the PDN circuit and the mimicking gate circuit. A hierarchical PDN consists of chip, package, and PCB level PDNs, as well as various structures such as via, ball, and wire bond interconnections, which connect the PDNs at different levels. In addition, the mimicking circuit has three elements: the driver, transmission line and receiver.

Figure 1 illustrates a simple representation of the FPGA gate level system model.

Figure 1 describes the particular model structure and scenario of quiet noise. This consists of a closed loop as a quiet pin is most frequently held at logic low or high with its return current path. This is mimicked by the adjacent I/Os, i.e., aggressor pins which form closed loops with their return current paths. However, when multiple I/Os switch simultaneously, they share a common return path such that their magnetic field is coupled amongst loops formed by the I/O and said path. In this way, SSN appears.

In brief, the relation between delta-I noise and the inductive coupling result in SSN can be expressed as follows:

$$\triangle v = L \, di/dt$$

Which L is the inductance of PDN. Delta-I, notated as di/dt is a source-drain current transition where input to a gate is changed from logic low to logic high or vice versa. Taking the aggregate current into consideration, the sum of all toggling I/Os hanging on the power rail can be estimated with individual drive strength. Δv is to referred as voltage dropped over the PDN such that the I/O buffer's input voltage is Δv less than supply voltage. It is illustrated by a piece-wise waveform close to PDN and represents power sag caused by the delta-I mechanism. A non-toggling I/O, often referred to as the "quiet" or "victim" pin, is pulled logic high or low to act as a voltage probe passing power rail noise through I/O trace to the outside of the package where it can be measured[2].

978-1-4244-4658-2/09 $25.00 © 2009 IEEE

When more aggressors toggle, the noise at the quiet pin is best described by:

$$\Delta v = \sum_{k=1}^{N} L_k \, di_k / dt$$

where N is the number of aggressors and the corresponding L_k is the mutual inductance between the quiet pin and aggressor pin. i_k is the current flowing through pin[5].

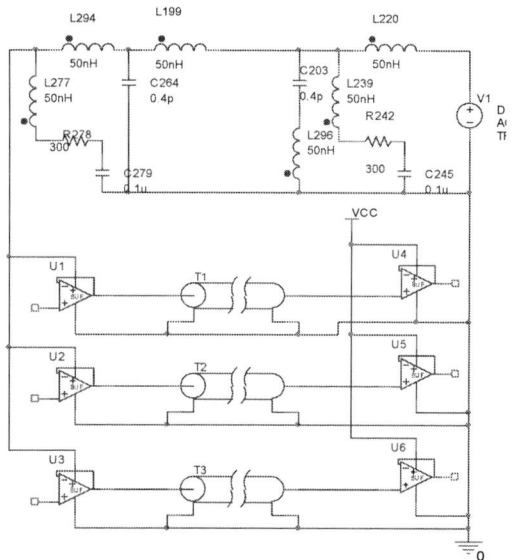

Figure 1. Schematic diagram of gate level test circuit

As described above, SSN is influenced by the PDN and mimicking gate circuit. However, validation on a FPGA devices and systems still troubled us, as well as detailed behaviors and their relationship to cause mechanisms. As a result, a model will be established to analyze it as followed.

Modeling and Simulation Analysis

With the growing possibility of increasing I/O counts, accurate SSN modeling is becoming more and more prominent. The SSN can be simulated using the SPICE or IBIS methods. The former is elaborate, but slow and highly computing intensive. The latter, which models the behavioral patterns of I-O buffers is fast but redundant for other interfaces. For this reason, SSN is analyzed using the most effective combination of SPICE and IBIS model drivers.

As mentioned above, significant amounts of SSN may be induced by a sudden switching of current at the chip, package, and PCB-level PDNs. The SSN is one of major sources of timing jitter and skew problems at high-speed serial I/O channels, clock distribution networks, electromagnetic interference (EMI) and noise coupling problems in high-density mixed signal integrated systems. When estimating the extent of SSN emission and evaluating the PDN designs in high-speed digital system designs, an efficient criterion is PDN impedance calculations at locations in the multilevel PDN. Precise estimations of the system's PDN impedance at locations is a critical prerequisite for secure and dependable designs of semiconductor systems[3].

In this example, the PDN of the FPGA gate level system is formed by a combination of multiple level networks creating a hierarchical interconnection structure[3], as illustrated in Fig. 2.

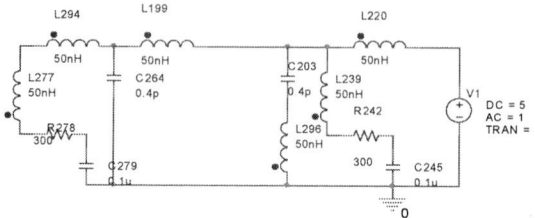

Figure 2. PDN Section in an FPGA System, With Factors of Chip, Package and PCB

This particular example consists of chip, package and PCB-level PDNs (as shown in Fig. 2), as well as various interconnection structures (via, ball, and wire bond) that connect the different levels to form a complete PDN. The model is necessary to determine the value of RLC circuit on each level of the PDN and to add component SPICE models of the interconnection structures to connect the PDNs when estimating the hierarchical PDN impedance. In other words, the different level PDNs' high-frequency electromagnetic coupling must be considered to guarantee accurate estimation of the hierarchical PDN impedance level and profile. This includes but is not limited to position and shift of resonance frequencies. The internal electromagnetic coupling occurs through the combination of conductive paths (of via, ball, wire bond) and parasitic electromagnetic coupling paths between metal plates at different levels PDNs. Figure 3 shows the system PDN impedance curves from the chip side, the same perspective that the chip would see during actual operations.

The impedance is obviously controlled in the specific event. The proper frequency is read at the approximately range of 10MHz to 1GHz [3]. The implication of the impedance feature is salient when the supply voltage will experience the most feature when transient current switches at this frequency spectrum. In Figure 4, a typical SSN measurement setup is given for mimicking gate circuit. In this example, the driver and the receiver are IBIS model, and the trace are coupling transmission line model [3].

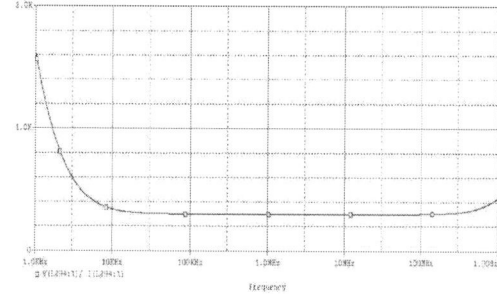

Figure 3. System PDN Impedance as Modeled and Simulation

978-1-4244-4658-2/09 $25.00 © 2009 IEEE   332

An IBIS file formatted utilizing ASCII contains the data required to behaviorally model the devices input, output and I/O buffers. Specifically, the data in an IBIS file is used to construct a model useful for performing SI simulations and timing analysis of PCB. The fundamental information needed to perform these simulations are generated by buffers I/V and switching (output voltage vs. time) characteristics. Note that the IBIS specification does not define an executable simulation model – it is a standard for the formatting and transfer of data. As such, the specification defines what the information included in an IBIS file represents and how it is gathered, but does not specify what the analog simulation application does with the data.

IBIS models are component centric. That is, a particular IBIS file would allow one to model an entire component, not just a particular input, output or I/O buffer. Therefore, in addition to the electrical characteristics of device buffers, an IBIS file includes a device's pin-to-buffer mapping, and the electrical parameters of the device's package.

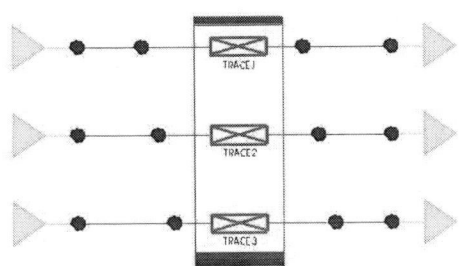

Figure4. Experiment Setup for SSN Characterization in mimicking gate circuit

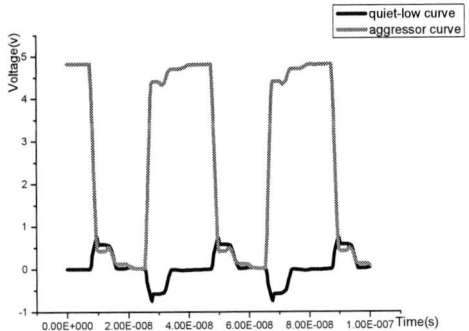

Figure 5. Simulation quiet-low noise waveform at a victim pin

A varying number of I/Os act as aggressors and toggle simultaneously in repetitive format. In the particular example of figure 4, 3 aggressors depicted. A single victim I/O net is kept either static high or static low, states which are realized by shorting its output to power rail or ground by FPGA programming. Noise is captured on the victim net, which refers to a local ground at the far end of the system with a high impedance and speed oscilloscope probe. When multiple

I/Os switch simultaneously, the transient currents within them create time-varying magnetic fields, which penetrate into the victim loop and induce a voltage noise. A typical quiet-low noise waveform is shown in Figure 5.

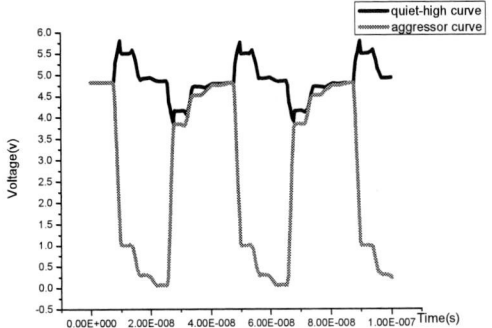

Figure 6. Simulation quiet-high noise waveform at a victim pin

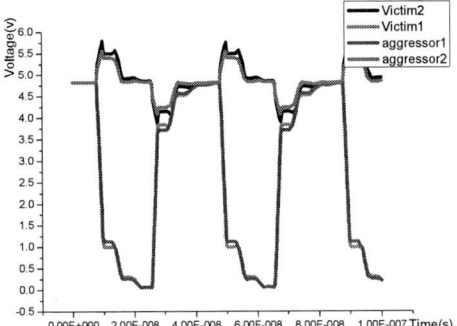

Figure7. Quiet-high noise compare with different PDN

A quiet-high noise scenario is illustrated in Figure 6. A quiet pin is held at logic high whilst a number of output buffers switch at the same time. The substantial number of transient currents not only generate magnetic fields coupled to the victim pin, but also causes a delta-I voltage drop when it flows through the inductive circuit components of the PDN. Compared quiet-high noise curves with different PDN of which is described in Figure 7, the mutual inductive coupling is still dominant for quiet-high noise. However, with an increased switching I/O number, the PDN effect may contribute more than the said inductive coupling [5].

Conclusions

From the systematic study performed the phenomenon of SSN in FPGA devises, two distinct SSN mechanisms were present: the mutual inductive coupling between I/Os and the PDN effect. Both contributions can be decomposed and quantified individually. In the case of the quiet-low noise illustrated from Figure 5, the mutual inductive coupling was always the primary factor. In the case of quiet-high noise, illustrated by Figure 6, with an increased switching I/O number the said noise may contribute to PDN effect[2][5].

978-1-4244-4658-2/09 $25.00 © 2009 IEEE   333

Acknowledgments

This work was supported by National High-tech Project (863) (Optical Interconnect Technology for High Speed Chips and Verification Platform NO. 2007AA01Z2a6.).

References

1. Varma, A.; Lipa, S.; Glaser, A.; Steer, M.; Franzon, P.; Electromagnetic Compatibility, 2004. EMC 2004. 2004 International Symposium on Volume 3, 9-13 Aug. 2004 Page(s):1000 - 1004 vol.3

2. Hong Shi; Geping Liu; Alan Liu; Pannikkat, A.; Kok Siang Ng; Yee Huan Yew; Electronic Components and Technology Conference, 2006. Proceedings. 56th 0-0 0 Page(s):8pp.   Digital   Object   Identifier 10.1109/ECTC.2006.1645652

3. Kim, Jaemin (Terahertz Interconnection and Package Laboratory, Korea Advanced Institute of Science and Technology (KAIST), Daejeon 305-701, Korea, Republic of); Jeong, Youchu; Kim, Jingook; Lee, Junho; Ryu, Chunghyun; Shim, Jongjoo; Shin, Minchul; Kim, Joungho Source: IEEE Transactions on Advanced Packaging, v 31, n 3, p 544-557, 2008

4. Yuan, F.Y.; Electrical Performance of Electronic Packaging, 1996., IEEE 5th Topical Meeting 28-30 Oct. 1996 Page(s):132 – 134. Digital Object Identifier 10.1109/EPEP.1996.5648073.

5. Geping Liu, Zhuyuan Liu, San Wong, Kaiyu Ren, Nafira Daud, "A Fast Algorithm to Instantly Predict FPGA SSN for Various IO Pin Assignments", DesignCon 2008, February 2008.

6. Wei Wei Lo   Smith, L.   Geping Liu   Man On Wong Daud, N; Applied Electromagnetics, 2007. APACE 2007. Asia-Pacific Conference on, DesignCon 2008, February 2008. Publication Date: 4-6 Dec. 2007, On page(s): 1-5.

Coupled Simulation of Anisotropic Conductive Adhesive Bonding Process and Reliability Analysis of the Packaging

ZhengJia Wang, ZhouPing Yin, Bo Tao, YouLun Xiong
State Key Laboratory of Digital Manufacturing Equipment and Technology
Huazhong University of Science and Technology
Wuhan, 430074, P. R. China
Email: wangzhengjia@gmail.com, yinzhp@mail.hust.edu.cn, taobo@mail.hust.edu.cn, famt@ mail.hust.edu.cn

Abstract

Anisotropic conductive adhesive (ACA) flip chip interconnection offer many distinct advantages over the traditional solder flip chip interconnection. But the reliability of the ACA is still a concerned issue. There are many studies on this topic, and most of them studies only one physics field was include in analysis each time. In fact, ACA bonding is a complicated process involving multiple physics and they usually occurred simultaneously and coupled with each other. In this study a numerical method is developed to solve coupled physics fields problems. The effects of all fields are taken into consideration by solving the equation of each physics field simultaneously. A parametric nonlinear finite element model was built for an ACA bonding process considering the effects of temperature and pressure. The deformation of particles and the stress development in packaging was investigated. Finally the optimal values were obtained from the results of analysis. The numerical simulation results match the experiments basically. The approach developed in the paper can be used to provide guideline for electronic packaging design and process parameter selection when using ACA as interconnection material.

1 Introduction

Using anisotropic conductive adhesive (ACA) as interconnection media in flip chip bonding is becoming more and more popular in electrical packaging. The ACA is cured and pressed under extern load and elevated temperature during ACA bonding process. The cured ACA matrix provides mechanical interconnection between die and substrate. And the deformed particles between the die and the substrate establish electrical path. In contrast to solder interconnection, the electrical path in ACA interconnection is constructed by the deformed particles between the chip and substrates, and the integrity of the packaging is maintained by the adhesion strength of cured matrix of ACA. Although ACA interconnection offers many distinct advantages over the solder interconnection, the reliability of ACA interconnection is still a concern issue. In order to improve the reliability of ACA interconnection the failure mechanism of ACA joints should be understood more clearly.

ACA bonding is a complicated process which involves many physics phenomena such as heat transfer, fluid flow and cure reaction, etc. Over the last few years, many researchers have studied the reliability of ACA joint. Most of those studies are based on experiments. For example, the curing temperature[1], the curing pressure[2-4], the bump height[5-7], and the material properties' effect on the reliability[8]

were studied. By comparing a series of experiments' results the optimum parameters were achieved in those researches which give guides for designers in designing similar applications.

Finite element analysis (FEA) has been widely used in analyzing the failure mechanism of the electronic packages to explore the directions of design improvement. There are many researches employing FEA to simulate stress/strain development and study the reliability. For example, the reflow process on the reliability of ACA bonding packaging[9], and the moisture induced failure in ACA interconnection were studied[10] respectively. Though physics phenomena involved in ACA bonding were occurred simultaneously and might couple with each other, those researches only studied one physics field each time[9, 10] or studied several physics fields by employing decomposition method supposing the different fields bring stains independently[11].

In this study a method is developed to solve coupled physical problems in a single model. First, each physics phenomena's government equation, which is usually partial differential equation, is constructed. And multiple physics field's effects are taken into consideration together by solving the equation simultaneously. Then the method is used to build a parametric nonlinear finite element model for a typical ACA bonding application which consider heating up the package, applying the assembly force, removal of assembly force, and cooling down to room temperature. The transient development of the temperature field during the bonding process was studied, and the deformation of particles and the stress development in packaging were investigated. Finally the optimal values were obtained from the finite elements analysis results.

2 Coupled Simulation Method of ACA Bonding Process

For a coupled multi-physics finite element analysis, the effect in each physics field should be taken into consideration simultaneously. All the govern equation of different physics fields should be solved in a single model. For finite element analysis of ACA bonding, two main physics fields analysis involved: heat transfer analysis and mechanical analysis. The degrees of freedom (DOF) of each node include temperature and displacement. They were solved simultaneously.

(1)The thermal analysis

The govern equation of heat transfer for 2D problems is:

$$\rho C \frac{\partial T}{\partial t} - \frac{\partial}{\partial x}(k_{xx}\frac{\partial T}{\partial x}) - \frac{\partial}{\partial y}(k_{yy}\frac{\partial T}{\partial y}) = Q \qquad (1)$$

Where T is the temperature, ρ is the density, C is the heat capacity, k_{xx} is the x component of thermal conductivity and k_{yy} the y component, Q is a heat source.

978-1-4244-4658-2/09 $25.00 © 2009 IEEE

Using compact notation, the relationship can be written as:

$$\rho C \frac{\partial T}{\partial t} - \nabla \cdot (k \nabla T) = Q \qquad (2)$$

The temperature fields described by Equations (2) are subjected to the initial conditions (at $t = 0$)

$$T = T_0 \qquad (3)$$

And boundary conditions

$$n \cdot (k \nabla T) = h(T_{\inf} - T) \quad \text{on} \quad S_h \qquad (4)$$

$$T = T_p \quad \text{on} \quad S_p \qquad (5)$$

Where T_p is the prescribed temperature on surface S_p, n the unit vector outward normal, h is the heat transfer coefficient on surface S_h where the convection takes place, and $T_{\inf}$ is the ambient temperature.

(2) Mechanics analysis

The equilibrium equations for plain strain problem:

$$\frac{\partial \sigma_x}{\partial x} + \frac{\partial \tau_{xy}}{\partial y} = -F_x \qquad (6)$$

$$\frac{\partial \sigma_y}{\partial y} + \frac{\partial \tau_{xy}}{\partial x} = -F_y \qquad (7)$$

Where Fx and Fy are volume forces, σ_x, σ_y the stress in x, y direction respectively, τ_{xy} is the shear stress.

Under the assumption of small displacements, the normal strain components and the shear strain components are related to the deformation as follows:

$$\varepsilon_x = \frac{\partial u}{\partial x} \qquad (8)$$

$$\varepsilon_y = \frac{\partial v}{\partial y} \qquad (9)$$

$$\gamma_{xy} = \frac{\partial u}{\partial y} + \frac{\partial v}{\partial x} \qquad (10)$$

Where ε_x, ε_y, γ_{xy} is normal strain in the x, y direction and shear strain respectively, and u, v is displacement components in x, y direction respectively.

The strain-stress relation is written as:

$$\begin{pmatrix} \sigma_x \\ \sigma_y \\ \tau_{xy} \end{pmatrix} = \frac{E(1-v)}{(1+v)(1-2v)} \begin{bmatrix} 1 & \frac{v}{1-v} & 0 \\ \frac{v}{1-v} & 1 & 0 \\ 0 & 0 & \frac{1-2v}{2(1-v)} \end{bmatrix} \begin{bmatrix} \varepsilon_x \\ \varepsilon_y \\ \gamma_{xy} \end{bmatrix} \quad (11)$$

Where E is the elastic module, which is function of temperature when material properties is temperature dependency. v is the Poisson's ratio. The strain-stress relation may be written using compact notation as:

$$\sigma = D\varepsilon \qquad (11)$$

So the relation between displacements and force can be written using compact notation as:

$$-\nabla \cdot (c \nabla u) = F \qquad (12)$$

Where u is the displacement. It is shown in equation(13). F is the force described by equation(14):

$$u = \begin{bmatrix} u \\ v \end{bmatrix} \qquad (13)$$

$$F = \begin{bmatrix} F_x \\ F_y \end{bmatrix} \qquad (14)$$

And the coefficient c is a 2×2 matrix, the elements in the matrix are:

$$c_{11} = \begin{pmatrix} 2G+\mu & 0 \\ 0 & G \end{pmatrix} \qquad (15)$$

$$c_{12} = \begin{pmatrix} 0 & \mu \\ G & 0 \end{pmatrix} \qquad (16)$$

$$c_{21} = \begin{pmatrix} 0 & G \\ \mu & 0 \end{pmatrix} \qquad (17)$$

$$c_{22} = \begin{pmatrix} G & 0 \\ 0 & 2G+\mu \end{pmatrix} \qquad (18)$$

Where G and μ are functions of temperature for non-linear material and descript as followings:

$$G = \frac{E}{2(1+v)} \qquad (19)$$

$$\mu = 2G \frac{\upsilon}{1-2\upsilon} \qquad (20)$$

(3) Analysis with thermal and mechanical coupled.

Considering the physics field involved in the process, the partial differential equations of the domain are composed of (2) and (13). The equation can be re-written as equation(21). In order to solve the temperature and the strain with a single method and simultaneously, the degrees of freedom of each node include temperature T, u and v.

$$\begin{bmatrix} \rho C & 0 \\ 0 & 0 \end{bmatrix} \frac{\partial}{\partial t} \begin{bmatrix} T \\ u \end{bmatrix} - \nabla \cdot \begin{bmatrix} k & 0 \\ 0 & c \end{bmatrix} \nabla \begin{bmatrix} T \\ u \end{bmatrix} = \begin{bmatrix} Q \\ F \end{bmatrix} \qquad (21)$$

The equation (22) does not include the strain induced by temperature variance. Thermal strain ε_{th} depends on the present temperature T, the stress-free reference temperature $Tref$, and the thermal expansion vector, α_{vec}

$$\varepsilon_{th} = \alpha_{vec}(T - T_{ref}) \qquad (22)$$

When the thermal strain and the initial strain were taken into consideration, the total strain was calculated by equation:

$$\varepsilon = \varepsilon_{el} + \varepsilon_{th} + \varepsilon_0 \qquad (23)$$

Where ε_{el} is the elastic strain and can be figured out from equation (22), ε_0 is the initial strain, ε is the total strain.

And the total stress was calculated by:

$$\sigma = D\varepsilon_{el} + \sigma_0 = D(\varepsilon - \varepsilon_{th} - \varepsilon_0) + \sigma_0 \qquad (24)$$

Where σ_0 is the initial stress.

From above equations, the deformation of particles in bonding process can be figure out, and the stress development in the package can also be studied. The reliability of the packaging can be investigated using those results.

3 Process modeling of ACA bonding

3.1 Description of bonding process

A typical ACA bonding process was as following: First the ACA was placed on substrate, the IC chip was flipped and aligned to the pad on the substrate, and then the ACA was cured by heat and pressure. Under high heat and pressure the conducting particles deformed and contacted with the chip and the substrate. The deformation is maintained by tensile stress in the cured adhesive. If the deformation of particles is not enough, loss of electric contact can occur when the adhesive expands in Z-axis direction, but if the tensile stress is excessive, delamination and cracking can occur in the adhesive. Both cases lead to failure of the packaging. The process variables [12], such as bonding force [13] and temperature [14], and the process conditions [15] all have effects on the reliability of the packaging.

The ACA bonding process is divided into curing stage and cooling stage according to mentioned previously. During the curing process, the whole packaging is heated to the elevated temperature for curing reaction comments, and the prescribed load was applied on the packaging to supply the contact of die and substrate. At cooling stage the load was removed and the packaging was cooled to room temperature.

At the curing stage, the temperature on the top side of the die is kept constant by the heat source and the pressure is applied on the top side of the die. The curing reaction commences after curing temperature arrives. Before ACA cured, it is in gel state and can not bear the force applied. The load was supported by the particles. The stress in whole packages can be calculated from the stress induced by force and the stress induced by thermal. At the curing stage the reference temperature for mechanical analysis is the room temperature, the present temperature was the result calculated from the thermal analysis. The initial stress of adhesive can be assumed as zero because it was in gel status when the force applied. The material properties such as Young's module and coefficient of thermal expansion in mechanical analysis are temperature related and the temperature was the value from thermal analysis.

After the curing reaction of adhesive finished, the force removed, and the whole package was cooled to room temperature naturally. The ACA is in solid state. The initial stress was the result calculated from the curing stage and the ACA matrix is zero stress because it is assumed not active in curing stage. The top side and the bottom side were in convective conditions and the ambient temperature is the room temperature.

3.2. Finite Element Model of a Typical ACA Application

(1) Model Definition and Material Properties of a Typical ACA Application

Radio Frequency Identification is a typical application of ACA bonding. It was selected as a typical model in this study. The chip size was 750x750μm^2 with 2 effective bumps, the bump size was 100x100μm^2, and the pitch was 300μm. The thick of the chip is 150μm and the height of bump is 20μm.The substrate was composed of 50μm thick PET and a layer of 30μm thick Al which was etched as antenna. The width of Al layer beneath the bump is 60μm .The ACA used had particles with diameter 3.5μm. 2D model was used for simplicity. Due to symmetry, only half the packaging was built. The SEM of the packaging and the model with mapped mesh was shown in Fig. 1.

Table 1 gives material properties of the packaging[6]. The temperature dependency of the Young's modulus of the ACA is shown in Fig. 2. The silicon die, the gold bump and metal pad in substrates are assumed elastic. The adhesive is a snap

curable ACA. Its elastic modulus is considered to be temperature-dependent. The filled particles of ACA are metallic particles made of nickel.

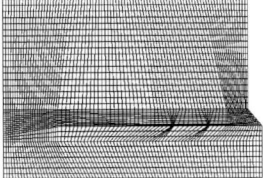

Fig. 1 The structure of an ACA interconnection packaging

Table 1 Material properties of packaging (units: SI)

	Density	Modulus	Spec heat	Conductivity	CTE(ppm)	Poisson's ratio
Silicon	2330	131e9	703	163	4.15	0.27
Nickel	8900	219e9	445	90.7	13.4	0.31
ACA	4000	See fig.5	800	150	57(<150℃) 133(>150℃)	0.45@ 30℃ 0.49@150 ℃
Copper	8700	110e9	385	400	17	0.35
Substrates	1530	2000e6-6000e6	1090	0.12	20	0.3

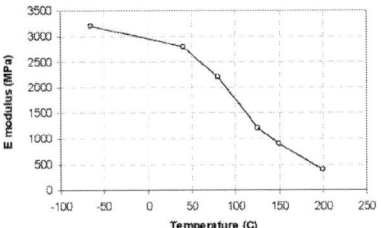

Fig. 2 Temperature-dependent elastic modulus of ACA

(2) Constraints and loads

At the curing stage, the ACA was in gel state and can not bear the force applied. For thermal analysis at this stage, and the temperature applied is T_{top} . The bottom side of the substrate is supported by a fixed plate and the temperature applied is T_{btm} .The left side of the model is the symmetry axis. The boundary condition for thermal analysis is: $n \cdot (k\nabla T) = 0$, Where n is the normal vector of the boundary, k is thermal conductivity. The right side is exposed to air, heat convection is the main method for heat exchange, so the boundary condition is given as: $n \cdot (k\nabla T) = h(T_{inf} - T)$, Where h is the heat transfer coefficient and T_{inf} is the ambient bulk temperature, and initial temperature is the room temperature T_0 . For mechanics analysis the force applied on the top side of die is F_y . The initial stress σ_0 and strain ε_0 are both zero, the reference temperature was the room temperature, the actual temperature T was the result calculated from the thermal analysis.

After the curing process finished, the force and the heat source was removed and the whole packaging was cooled

978-1-4244-4658-2/09 $25.00 © 2009 IEEE

down to room temperature naturally. All the components were activated for all physics fields. The initial stress was the result calculated from the curing stage. The reference temperature of all parts were the results at the end of the curing stage There was no load applied at this stage and both the top and bottom side of the packaging were in convective conditions.

4. Results and discussions

4.1 Thermal analysis of the bonding process.

Thermal analysis is very important for ACA bonding process. First the adhesive curing process comments only after it reaches the ACA curing temperature. The time of adhesive at the specified curing temperature decides the curing degree of the adhesive. The temperature history and temperature distribution is related to the material's thermal properties, the packaging geometry and the process condition. The ACA used here was DELO-MONOPOX-AC265 and the curing condition is 8 seconds at 180℃ and the curing reaction comments at 160℃. The time used here is 10 seconds to ensure the ACA be cured completely.

The die in electrical packaging has different sizes. Three sizes were selected in this study. They are $750\times750\times100\mu m^3$, $1500\times1500\times150\mu m^3$, $400\times400\times75\mu m^3$. The Teflon layer was used as normal, the top temperature was 180℃ and the bottom temperature was 150℃. The temperature history in adhesive was show in Fig. 3. It was found that the larger of the die size, the longer of the time needed for the adhesive to rise to specified temperature. But the time needed was short and differed slightly. This is because the packaging size used in this study is small, the die and the ACA can transfer heat efficiently due to their high thermal conductivity. The substrate with low thermal conductivity though, it is very thin in the study. So all the packaging can reach the setting temperature very quickly. On the other hand, the temperature of adhesive in equilibrium status was almost the same. So it's necessary to consider using different curing time for different packaging but using the same heat condition when the same adhesive was used.

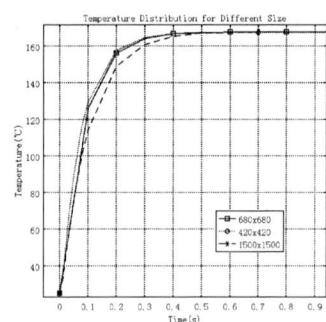

Fig. 3 Temperature history of different die size

4.2 Mechanic analysis of the bonding process

The bonding process is very critical to the ACA joint performance and reliability, since both mechanical integration and electrical interconnection are established in this process. The bonding pressure is applied to force the conductive particles to contact the electrodes. The electrical performance of the joint depends heavily on the deformation degree of particles. The deformation of the conductive particles induced by the used pressure is related to the contact resistance of the ACF joints. As the gap between Au bump and pad decreased with increasing bonding pressure, conductive ball deforms more and contact area between conductive fillers, bump and pad surface also become larger.

The residual stress due to the force applied during bonding and the coefficient of thermal expansion mismatch between components have great influences on the reliability of the packaging. The excessive bonding pressure might induce high compressive stress and internal stress in the epoxy adhesive. The stored elastic compression can be released and leads to a loss of the contact area during testing which results in the decrease of adhesive strength after the reliability test. Significant compressive stress is found to build up in the interface between the two contacts. This stress is believed to generate peel stress in the adhesive, which is probably the reason for catastrophic failure.

The different bonding process parameters not only change the contact resistance of ACF joints but also determine its adhesive strengths. Therefore, the electrical performance and adhesive strengths must both be considered to determine the optimum bonding parameters for reliability of interconnection.

In order to investigate bonding process parameters' effects on mechanical reliability and electrical performance, several cases were studied. The process parameters such as the force, the temperature applied on the top of the die and at the bottom of the substrate were used here. The von_mises and the deformation along the interface between the ACA and the substrate were illustrated in Fig. 4.

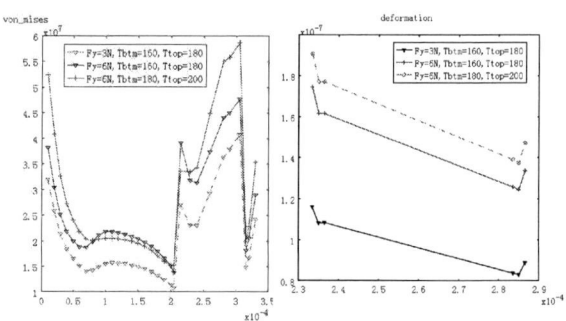

Fig. 4 von_mises stress and deformation under different conditions

The results showed that the larger pressure leads to bigger deformation of the particles, the higher temperature also help to keep the deformation of the particles. On the other hand, the large pressure and high temperature lead to high residual stress and impairs mechanical reliability of the packaging. The high temperature has slight effect on deformation of the particles but induce more residual stress than large pressure does. To consider the electrical performance and the mechanical reliability both simultaneously, the temperature is used as low as possible in case the productivity was satisfied.

However, the pressure should be used as large as possible provided that its stress is lower than the admissible stress.

A quality factor q_f was defined as equation (26) to indicate the performance of the packaging considering both the electrical and mechanical

$$q_f = \alpha \frac{deformation_{act} - deformation_{min}}{deformation_{max} - deformation_{min}} +$$

$$(1-\alpha)(1 - \frac{von\_mises_{act} - von\_mises_{min}}{von\_mises_{max} - von\_mises_{min}}) \qquad (25)$$

Here, α is the factor between 0 and 1 which indicates the importance of electrical performance in overall performance. If the deformation of the particles was taken into consideration only, α can be set to 1. $deformation_{act}$ is the actual deformation, $deformation_{max}$ is the allowed maximum deformation of the particles, $deformation_{min}$ is the minimum deformation needed to supply the steady electrical contact. von_mises_{act}, von_mises_{max} and von_mises_{max} is the actual von_mises stress minimum von_mises stress and allowed maximum von_mises stress respectively.

Using the quality factor as the evaluation index, the optimum process parameters can be obtained. The thick of Teflon was 100um in the study, the maximum value of top temperature was 200℃, the minimum value is 180℃, the maximum value of bottom temperature was 180℃, and the minimum value was 150℃. The force used was between 2N and 6N. And α was 0.5 in quality factor expression. The results showed that when the force applied was 6N and the temperature at the top/bottom side of the packaging were 180℃/150℃ the best parameters were obtained The results of the samples were tested by the reader illustrated as Fig. 5. The process parameters before optimized were: 3N,190℃/170℃. 89.8 percent of the samples pass the tests in that case. When the optimized process parameters were used, 96.4 percent of the samples pass the tests. The experiment results match the numerical simulation basically.

Fig. 5 Test equipment of the sample.

5. Conclusions

A method was developed for solving coupled fields problem. Using this method, a parametric multi-physics finite element model was built for a typical ACA bonding process. And a multi-physics analysis was carried out for ACA bonding process. It was found that the process temperature used not only related to the adhesive used, but also related to the process condition and geometry of the parts in the packaging. The results of mechanic analysis showed that the process force and process temperature both have effects on the reliability of the ACA bonding packaging, but the force has greater effects on the deformation of the metal particles in the ACA. The method can provide the best direction on how to improve the reliability within the design space. It is helpful for electronic packaging design and process parameter selection.

In this study a simple ACA interconnection packaging was used to state an example of multi-physics finite elements analysis. The same methodology can also be used to find the optimum designs of electronic packaging reliability. Only temperature and the applied force were considered in this study, and the material properties such as CTE and elastic module of ACA was not considered in the optimization process. Considering more factors' effects on the reliability of electronics packaging can be an interesting subject for further studies. And it is useful to build an integrated model on the bonding process, temperature cycling and external mechanical loading when the packaging was in-service.

Acknowledgments

This work is supported by the National Science Foundation of China under grant 50805060, 50625516, and the National Fundamental Research Program of China under Grant 2009CB724204.

References

1 Chen, X. Zhang, J. C. Jiao, L. Liu, Y. M. "Effects of different bonding parameters on the electrical performance and peeling strengths of ACF interconnection," *Microelectronics Reliability*, 446 (2006), pp.774-785.

2 Chan, Y.C. Luk, D.Y., "Effects of bonding parameters on the reliability performance of anisotropic conductive adhesive interconnects flip-chip-on-flex packages assembly ii. different bonding pressure," *Microelectronics Reliability*, 42 (2002), pp.1195-1204 .

3 Frisk, L. Seppala, A. Ristolanien, E. "Effect of bonding pressure on reliability of flip chip joints on flexible and rigid substrates," *Microelectronics Reliability*, 44 (2004), pp.1305-1310.

4 Wu, Y.P. Alam, M.O. Chan,Y.C. Wu, B.Y. "Dynamic strength of anisotropic conductive joints in flip chip on glass and flip chip on flex packages," *Microelectronics Reliability*, 44(2004), pp.295-302.

5. Pinardi, K. et al, "Effect of Bump Height on the Strain Variation During the Thermal Cycling Test of ACA Flip-Chip Joints," *IEEE Transactions on Components and Packaging Technologies*, 23 (2000), pp. 447-451.

6. Wu, C.M.L. Liu, J. Yeung, N.H. "The effects of bump height on the reliability of ACF in flip-chip," *Soldering & Surface Mount Technology*, Vol. 13, No. 1 (2001), pp. 25-30.

7. Chiang, W.K. Chan, Y.C. "Reliability of Anisotropic Conductive Film Joints Using Bumpless Chip—Influence of Reflow Soldering and Environmental Testing," *Journal of Electronic Packaging*, 127 (2005), pp.113-119.

8. Teo, M. et al., "Effects of Anisotropic Conductive Adhesive (ACA) Material Properties on Package Reliability Performance," *2003 Electronics Packaging Technology Conference*, 2003, pp. 718-725.

9. Yin, C.Y., et al., "Experimental and Modeling Analysis of the Reliability of the Anisotropic Conductive Films," *2003 Electronic Components and Technology Conference*, 2003, pp. 698-702.

10. Yin, C.Y.,et al., "Experimental and Modeling Analysis on Moisture Induced Failures in Flip Chip on Flex Interconnections with Anisotropic Conductive Film," *2005 International Conference on Asian Green Electronics*, 2005, pp. 172-177.

11. Mercado, L.L., et al., "Failure Mechanism Study of Anisotropic Conductive Film (ACF) Packages," *IEEE Transactions on Components and Packaging Technologies*, 26 (2003), pp.509-516.

12. M.J. Rizvi, C. Bailey, H. Lu, "Study of processing variables on the electrical resistivity of conductive adhesives," *International Journal of Adhesion and Adhesives*, Vol. 29, No. 5 (2009), pp. 488-494.

13. Yeung, N.H., Chan, Y.C., Tan, C.W., "Effect of Bonding Force on the Conducting Particle With Different Sizes," *Journal of Electronic Packaging, Transactions of the ASME*, 125 (2003), pp. 624-629.

14. Chan, Y. C. and Luk, Y. D. "Effects of bonding parameters on the reliability performance of anisotropic conductive adhesive interconnects flip-chip-on-flex packages assembly i. different bonding temperature," *Microelectron Reliab.*, 42 (2002), pp. 1185-1194.

Miniaturized Printed Wire Antenna in Package for 2.4GHz Wireless Communications

Xiaoli Liu[1],*Yunfeng Wang[1, 2], Lei Li[1], Lixi Wan[2]
[1]Shenzhen Institute of Advanced Integration Technology
Chinese Academy of Sciences/The Chinese University of Hong Kong
1068 Xueyuan Avenue, Shenzhen University Town, Nanshan District, Shenzhen, 518055, China
[2]Institute of Microelectronics of Chinese Academy of Sciences
3#, Beitucheng West, Chaoyang District, Beijing, 100029, China
Email: yf.wang@siat.ac.cn

Abstract

Miniaturized printed wire antennas in system in package were investigated for 2.4 GHz wireless communication systems. An inverted F antenna was proposed, and had a small area with 15mm by 15mm. The position of feed tap connected to the antenna arm greatly influenced the impedance of antenna, namely, it could adjust the position of feed tap to match the impedance to 50 ohm. The influences of chips and epoxy molding compounds on the impedance were also studied. Eventually, the inverted F antenna in package had a bandwidth and gain with almost 200MHz and 2.8dB, respectively, and could meet the demands of 2.4GHz wireless communication.

Introduction

Wireless communication system for local access network including Bluetooth and IEEE 802.11 has been widely employed in industrial, scientific and medical field. Market demands for more functionality, smaller size, and lower cost have driven the system to be more and more convergent. Upon that SIP (system in package) has become an encouraging solution meet the demands. SIP is a developing direction as important as SOC (system in chip) in semiconductor roadmap, and many active devices and passive devices have been implemented in SIP [1-3] to continuously minimize electronic system. As a necessary device in wireless system, Antenna plays an important role in transmitting and receiving. Traditional antennas can not be applied in mini wireless module or SIP directly due to their big size, so that their size limits the dimension decrease of whole wireless electronic system. For convergent wireless systems, miniaturized antennas have to be built into whole package as a necessary device. Many antenna structures for single, dual, or multiple bands have been proposed. As discussed in many papers [4-6], electronic current through feed point and the surfaces brings the near electromagnetic field of antenna. In fact, feed point, antenna shape and size could affect the antenna radiation performance very much. Different shapes and sizes of antenna feeds change the input impedance of antenna.

With the advantage of low cost, low profile and easy fabrication, printed wire antennas are suitable for SIP or module. In this paper, a 2.4 GHz inverted F antenna with smaller size and 100MHz bandwidth was proposed. It could be integrated in SIP for WLAN and Bluetooth applications. The antenna area was 15mm*15mm on a FR4

substrate (dielectric constant 4.4). The whole antenna length consisted of an inductive arm and a capacitive arm which related to an impedance matching problem [7]. The design and simulation results showed that this antenna had a small size and efficient bandwidth for in SIP for ISM application. Numerical results based on full wave simulation tool, Ansoft HFSS, characterized the antenna properties.

Design and Discussion

The inverted F antenna working at 2.4 GHz for WLAN or Bluetooth applications had been designed on a FR4 substrate with the size of 30mm by 30mm, and the metal on the opposite side was removed to improve the radiation. The width and thickness of antenna trace were 1mm and 0.035mm. The length of both line vertical and parallel to ground were 15mm.

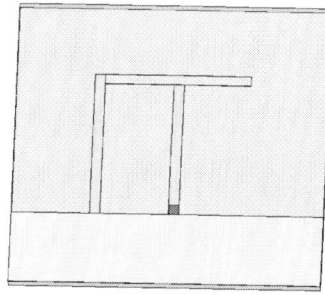

Fig.1 Inverted F antenna structure with a middle tap

As shown in Fig.1, the purpose of feed tap connecting to the parallel antenna arm was to adjust the impedance of the antenna to 50 ohm. The segments right and left to the feed tap were the capacitive and inductive arms respectively. If the position which the feed tap connected to changed, the impedance of the antenna would change too. It could be seen from Fig.3 and Fig.4. Fig.2 showed the impedance was 49-200j at 2.4GHz, when the tap was connected to the middle of the parallel arm. And Fig.3 showed the impedance was 60-20j at 2.4GHz, when the feed tap was connected to nearly the end of the parallel arm, just like Fig.2. It clearly could be seen that by changing the position of the connecting tap, the impedance nearly changed to 50 ohm. Therefore, if the feed position tapped right, the impedance of the antenna would match to 50 ohm. That also could be seen from the Fig. 5 which demonstrated that the return loss (S11) of the antenna was -23dB at 2.45GHz, and the bandwidth was more than

978-1-4244-4658-2/09 $25.00 © 2009 IEEE

100MHz. However, when the tap position at the middle of parallel arm, the central frequency shifted to 3.05GHz, and return loss was close to -2dB, just as Fig.6.

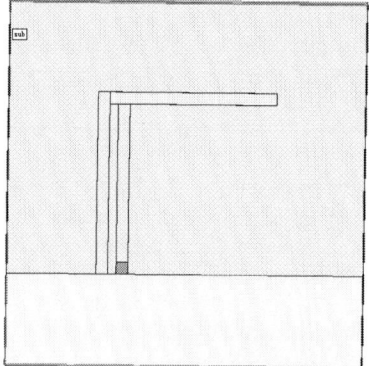

Fig.2 Inverted F antenna structure with a left tap

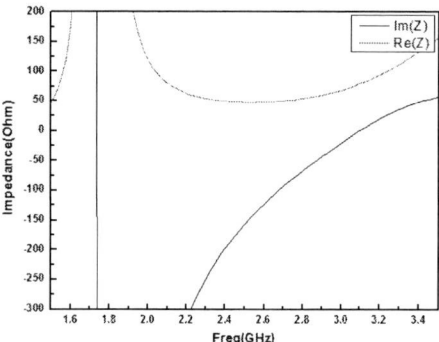

Fig.3 Impedance of structure with a middle tap

The antenna was used for the SIP, so both the chip and epoxy molding compound would affect its input impedance. In the process of antenna design, the influences should be considered.

Fig.7 was return loss of antenna near a chip, and the chip size was 5mm by 5mm. It could be seen that the chip has little influence to the impedance of antenna, and the return loss nearly had no change.

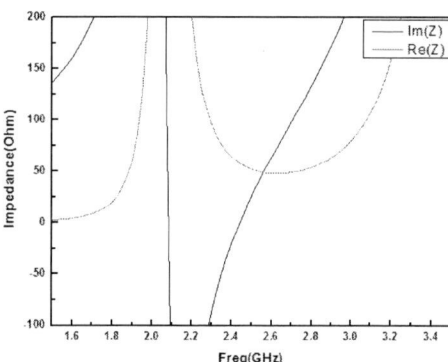

Fig.3 Impedance of structure with a left tap

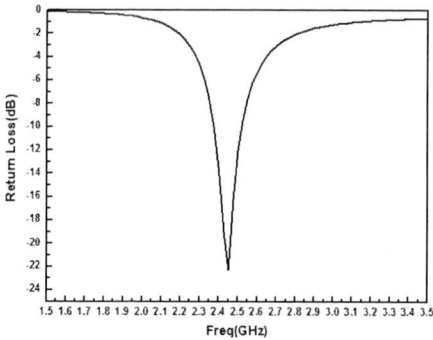

Fig.5 Return loss of inverted F antenna with a left tap

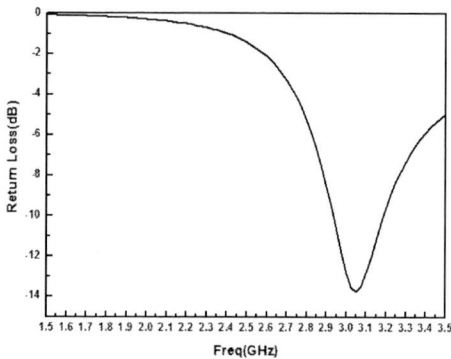

Fig.6 Return loss of inverted F antenna with a middle tap

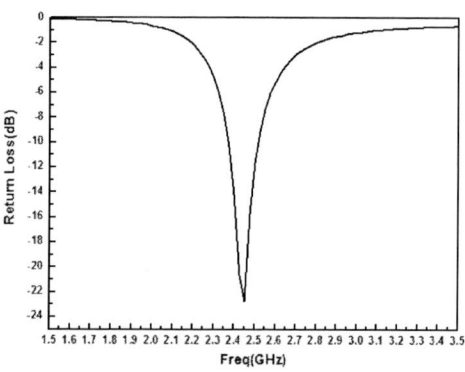

Fig.7 Return loss of the antenna near a chip

Fig.8 top view of a package

978-1-4244-4658-2/09 $25.00 © 2009 IEEE   342

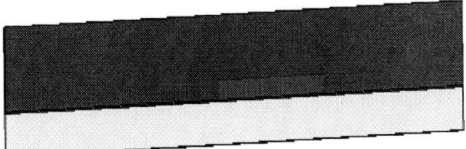

Fig.9 cross section of a package

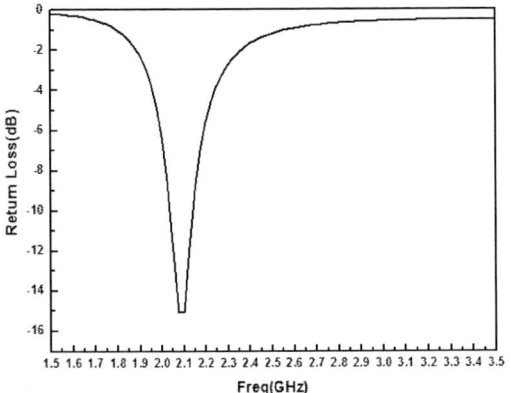

Fig.10 Return loss of antenna in package without tap adjusting

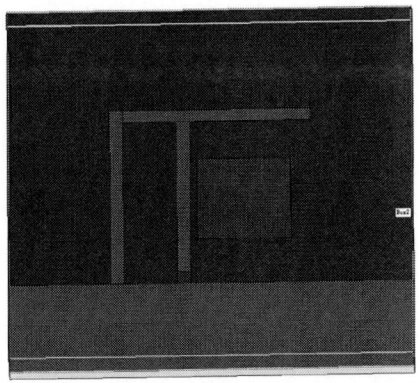

Fig.11 antenna in package with adjusted tap

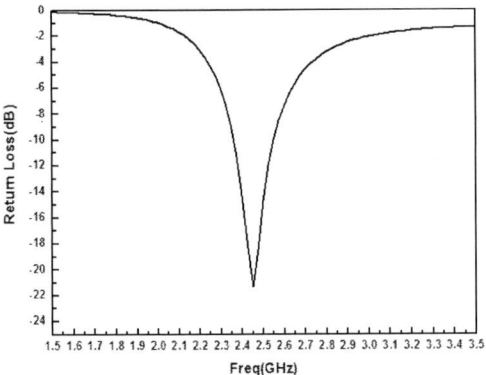

Fig.12 Return loss of antenna in package with adjusted tap

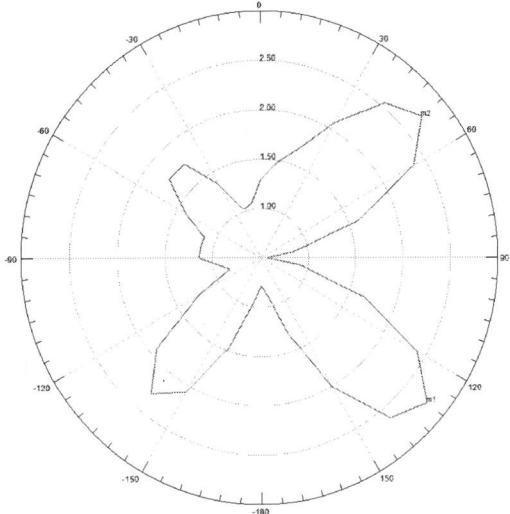

Fig.13 E-Plane radiation pattern of 2.4GHz antenna in package

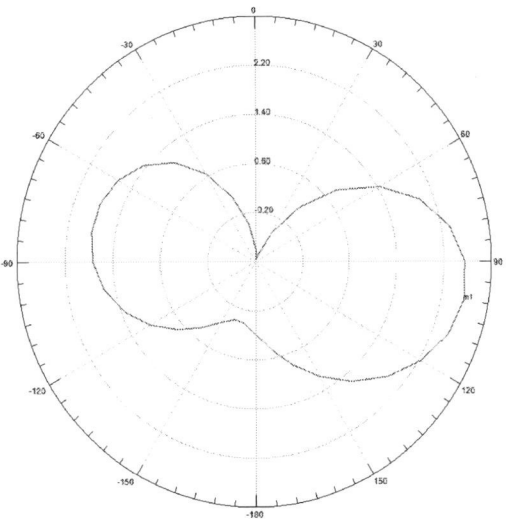

Fig.14 H-Plane radiation pattern of 2.4GHz antenna in package

As mentioned before, the impedance of inverted F antenna could be adjusted by changing the position of feed tap. In this module, the tap was shifted 3.3mm to the right, just as Fig.11, and Fig. 12 showed that return loss was -22dB at 2.45GHz, and the bandwidth was nearly 200MHz.

Fig.13 and Fig.14 showed the gain of E plane and H plane, respectively. It could be seen that at 2.4GHz the gains were 2.8 dB in E plane and 2.6dB in H plane.

Fig.8 showed a package model, which integrated antenna and chips, and Fig.9 was the cross section of package. Because the antenna was fully covered by the epoxy molding compound, the impedance had altered a lot. It showed from Fig. 10 that the central frequency shifted to 2.1GHz.

978-1-4244-4658-2/09 $25.00 © 2009 IEEE   343

Conclusions

In this paper, an inverted F antenna was designed for system in package. The size of antenna was 15mm by 15 mm, and the gain was nearly 3dB in package. It could meet the demands of 2.4 GHz wireless communication systems.

The package size in the simulation was 30mm by 30mm, but in the product application, the size could be reduced as small as the antenna, and the interconnection between the antenna and the chip could implement by wire bonding.

Acknowledgments

This work was supported by Shenzhen science and technology program (A study of embedded components in system-in-package).

The authors would like to thank the SIAT faculty, research staff and students.

References

1. Seung J.Lee. *et al*, " Fully embedded High Q Passives and Band Pass Filters for Low Cost Organic RF SOP Applications", Electronic Components and Technology Conference. 2007.

2. Greg Brzezina. *et al*, "A Miniature LTCC Bandpass Filter Using Novel Resonators for GPS Applications", Proceedings of the 37th European Microwave Conference, 2007.

3. Mohamadou Baba. *et al*, "An Efficient Methodology for Design and Implementation of Embedded Bandpass filters for RF/Wireless Applications", Electronics Packaging Technology Conference , 2007.

4. Wonbin Hong. *et al*, "Low-Profile, Multi-Element, Miniaturized Monopole Antenna", IEEE Transactions on Antennas and Propagation, 2009.

5. Xingping Lin. *et al* "A Low Profile L Shape Meander Circular Ultra Wideband antenna", Antennas and Propagation Society International Symposium, 2008.

6. Takemura, N. *et al* "Inverted-FL antenna with self-complementary structure", Antennas and Propagation Society International Symposium, 2008.

7. H.Y.David Yang, "Miniaturized Printed Wire Antenna for Wireless Communications", IEEE Antenna and Wireless Propagation Letter, 2005.

Modeling of Copper Wire Bonding Ball Transient Temperature Behavior

Techun Wang[1], QingYi Zheng[1], Rui Yu[1]* and Zhongyun Li[2]

[1]ASE Assembly & Test (Shanghai) Limited, No.669, Guoshoujing Road, Shanghai, 201203, China
[2]Mentor Graphics (Shanghai) Electronic Technology Co., Ltd., Rm.2901, 88 Shi Ji Da Dao, Shanghai 200120, China
*Tel: 86+21-50801060, Email: rui_yu@aseglobal.com

Abstract

Copper wire has been widely investigated in recent years as an alternative material for gold wire bonding of electronic packaging. Compared to the gold wire, copper has a higher electrical conductivity and thermal conductivity, better mechanical property and much slower IMC growth. However the bondability of copper wire is weaker than gold, and poor bonding can further degrade the reliability of the products.

Bonding material temperature is one of the key factors having a significant impact on the ball yielding deformation and the bondability. This report investigates the temperature evolution of the copper ball during the wire bonding process. Commercial software package FloEFD is used for the finite volume transient simulation. Melting and solidification of the copper ball is modeled with the equivalent specific heat method.

Results of the transient ball temperature simulations show the cooling profiles with rapid solidification finished only 1ms after the Electronic Flame Off (EFO) firing. The solidification is followed by a rapid temperature drop to below 200°C within only 10 ms.

Introduction

There are two major challenges for the copper wire to substitute the traditional gold wire in wire bonding electronic packaging. First is the oxidation of copper, and second problem appears to be the different mechanical property of copper during wire bonding. Copper is harder than gold, and damage can be resulted due to improper bonding control.

Before introducing the copper free air ball (FAB) formation and its subsequent behavior, we take a quick look at the wire bonding process. A typical time frame of the wire bonding process ball position can be outlined as Fig. 1. At a height of about 200 mils above the bonding surface, time interval T1 of arc firing less than 1 ms, an electronic flame off system provides an arc to melt the end of the wire and form a molten ball under the surface tension. After the arc turning off, the bonding capillary takes the ball under solidification to the bonding position in about 5 ms. After the first bonding of 15 ms, the capillary travel to loop the wire during a period of 75ms. Second bonding was set as 15ms, followed by an off lifting period of 10 ms. Then the capillary was again raised to the 200 mils height within 20 ms, waiting for next FAB firing.

For copper wire bonding, the forming gas of Nitrogen and Hydrogen gas have been used to provide a reducing atmosphere to avoid the oxidation of copper.

The bonding mechanical performance of metal wires and free air balls (FAB), are closely related to the temperature. [1] For example, higher temperature results in lower hardness and lower yield strength. Hardness of the bonding ball can have crucial impact on the bondability and reliability of the wire bonded packages. [2] [3] During the wire bonding electronic

flame off (EFO) process, evolution of the free air ball can vary according to process parameters, including the forming gas, EFO firing time and current. Geometry, mechanical property and microstructure of the ball are closely related to these controllable parameters. [4-8]

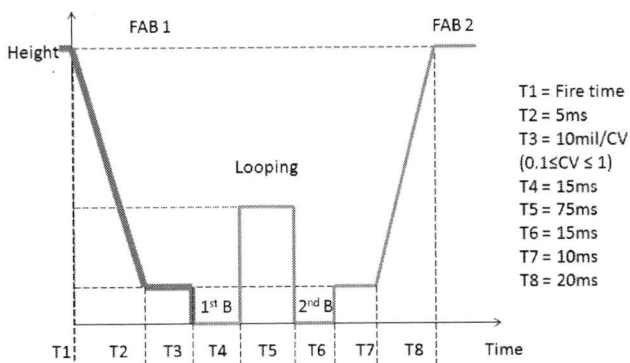

Fig. 1 Time history of a typical wire bonding process is shown by capillary tip height.

The free air ball formation can be considered as a melting-solidification process. [9-10]The shape and size can be predicted by a detail calculation on the heat transfer, molten metal roll-up and solidification process.

In this report, we try to use the software FloEFD of finite volume simulation to obtain the cooling profile of the copper free air ball. Through a simplified fixed ball shape, the heat transfer by conduction and melting-solidification has been modeled. Hopefully a practical cooling behavior can be extracted for the copper FAB for further reference on mechanical performance of subsequent wire bonding process.

Modeling

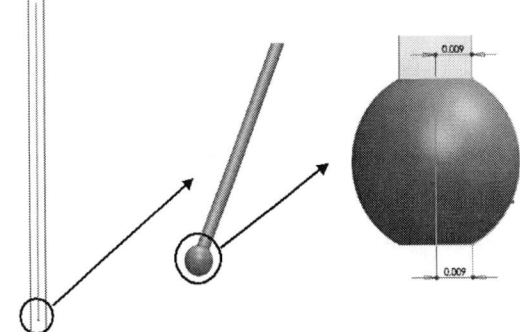

Fig.2 Model of copper wire and ball.

978-1-4244-4658-2/09 $25.00 © 2009 IEEE   345

The physical model is shown in Fig. 2. The total wire length of simulation is 8mm. A flat plane at the bottom of the copper ball is taken with the diameter as that of the wire. For simplicity and limitation of the software, the shape is fixed through out all the "melting and solidification" process. Only heat transfering of copper and solid-liquid phase change energy are considered. The thermal properties of the copper can be found in the software and is shown in Fig. 3.

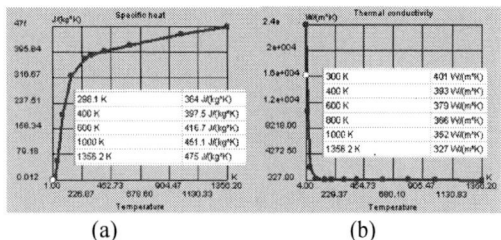

(a)   (b)

Fig. 3 Temperature dependent values of (a) copper specific heat and (b) thermal conductivity from FloEFD Engineering Database Library.

In this simulation, an equivalent specific heat method is used to include the fusion-solidification latent heat.

The equivalent specific heat CE =C + L0, where C is the true specific heat and L0 is the equivalent latent heat of fusion. And L0=L / (TL-TS), where L is the latent heat of fusion (204.8 J/g for copper)

TL is the temperature of liquid phase line and TS is the temperature solid phase line.

(TL–TS) can be considered as "latent heat affected temperature range", and has been taken 10°C, 1°C and 0.1°C in our simulation.

The melting was set to begin from the flat bottom plane, to generate the heat from arc plasma.

The forming gas condition is listed in table 1. Nitrogen/Hydrogen forming gas velocity 0.5m/s, corresponding to a flow of 0.8L/min for a 6mm diameter gas tube.

Table.1 Forming gas condition of FloEFD setting

Thermodynamic Parameters	
Parameters:	Pressure, temperature
Pressure	101325 Pa
Temperature	25 °C
Velocity Parameters	
Parameter:	Velocity
Velocity in X direction	0.5 m/s
Velocity in Y direction	0 m/s
Velocity in Z direction	0 m/s
Turbulence Parameters	
Concentration	
Type:	Volume fraction
Nitrogen	0.95
Hydrogen	0.05
Solid Parameters	
Initial solid temperature	25 °C

The computation time step of FloEFD is assigned as in Table 2, fine steps are needed for initial arc firing stage.

Table.2 Computation continuously increasing time step settings

Value t	Values f(t)
0 ms	0.001 ms
0.5 ms	0.005 ms
1 ms	0.01 ms
2 ms	0.02 ms
4 ms	0.04 ms
8 ms	0.08 ms
25 ms	0.25 ms
100 ms	1 ms

Three power levels with corresponding firing time setting are listed in table 3. All have the same total energy 0.5 mJ. Three latent heat affected temperature ranges are simulated for each power-time setting.

Table.3 Electronic flame off (EFO) arc heating power setting, corresponding firing time duration and latent heat affected temperature range settings of the equivalent latent heat model.

	Heat Power[W]	Heating time [ms]	Latent heat affected temperature range
1	2	0.25	10°C
2	2	0.25	1°C
3	2	0.25	0.1°C
4	1	0.5	10°C
5	1	0.5	1°C
6	1	0.5	0.1°C
7	0.5	1	10°C
8	0.5	1	1°C
9	0.5	1	0.1°C

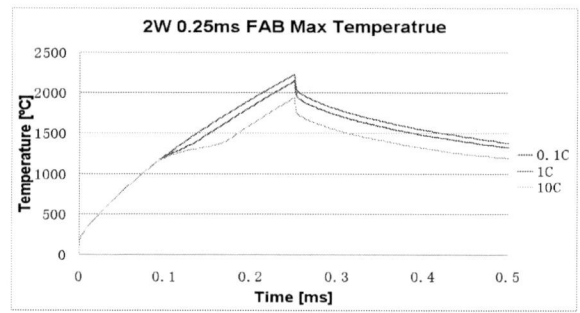

Fig. 4 The evolution of the maximum temperature in copper FAB during the EFO with 2W power arc firing within 0.25 ms. The narrower the latent heat affected temperature range assumed, the higher the FAB maximum temperature appeared to be. The highest temperature is 2225°C, still 370°C below the boiling point of copper, in the latent heat affected range 0.1°C setting.

Results and Discussion

For each combination of firing time and power, the maximum temperature of the ball has been monitored. It should be higher than the melting point of copper 1083°C and less than the boiling point 2595°C. For the highest power 2W with the shortest firing time 0.25 ms, the maximum temperatures during the EFO firing stage were simulated as 2225°C, 2144°C and 1940°C for latent heat affected temperature range of 0.1°C, 1°C and 10°C, respectively as shown in Fig. 4. The smallest latent heat affected temperature range results in a highest molten copper ball temperature.

The same trend was found for the 1W-0.5ms and 0.5W-1ms firing condition, as shown in Fig. 5 and Fig. 6 respectively.

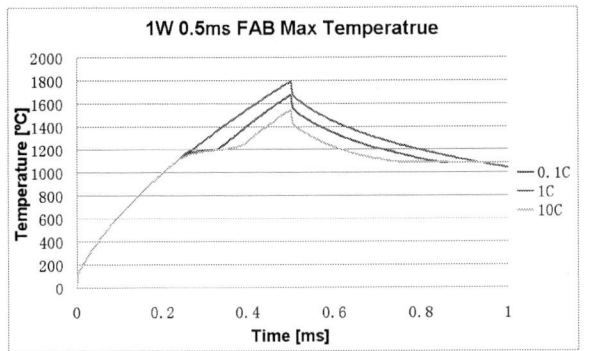

Fig. 5. The evolution of the maximum temperature in copper FAB during the EFO with 1W power arc firing within 0.5 ms.

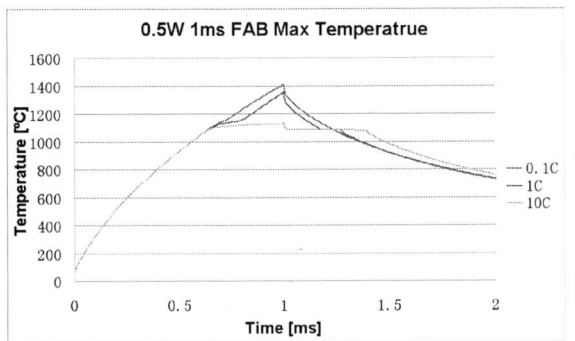

Fig. 6. The evolution of the maximum temperature in copper FAB during the EFO with 0.5W power arc firing within 1ms. The melting–solidification process appears to have the longest duration in the case of 10°C latent heat affected range setting, this can be due to the peak temperature of this setting is close to the melting point.

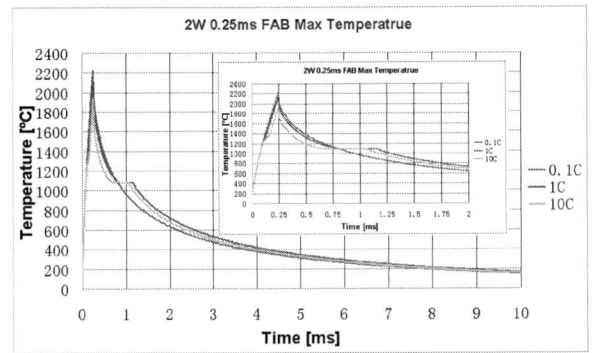

Fig. 7. The cooling profile of 2W power heated copper ball. The solidification can be observed to have a duration no more than 0.5ms followed by a temperature drop to below 200°C within 10 ms.

The simulation result of ball cooling for different latent heat affected temperature range can be obtained. Fig. 7. shows the cooling profile of the 2W heated ball.

For different power-time setting of total energy 2mJ, as the latent heat affected temperature range 1°C, the cooling profiles are shown in Fig. 8. The highest peak temperature has been achieved by the short time high power combination of 2W-0.25 ms, implying that approaching the copper boiling point may make a limiting condition for the firing power setting. Solidification finished for all heating processes almost within 1ms, and all cooled with a same rate to 200°C of bonding surface temperature within 10 ms. Compared to normal wire bonding process as shown in Fig. 1, the result suggests that the FAB temperature at first bond touchdown has been cooled to below the bonding surface temperature, as normally heated to about 200°C.

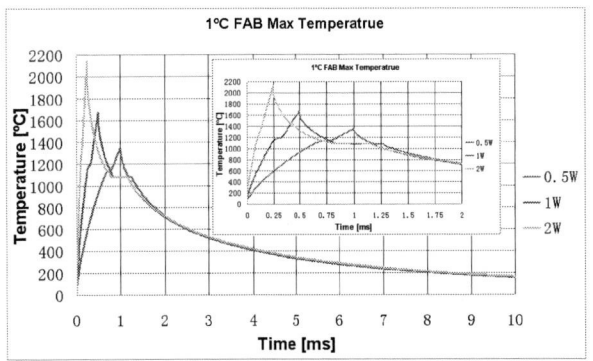

Fig. 8. FAB cooling curves for different power-firing time combinations. The latent heat affected temperature range is 1°C. All test balls have been cooled to below 200°C within 10 ms. Solidification durations are less than 0.5 ms and finished within 1.5 ms after EFO heating for all curves.

Conclusions

We used a commercial software package FloEFD to simulate the thermal behavior of the copper ball cooling during wire bonding EFO process. As we assume the equivalent latent heat method for a simplified fixed shape copper ball, the melting and solidification can be simulated with the transient thermal modeling. The latent heat affected range has been altered to evaluate its effect on the cooling profile of simulation. The results of various combinations of firing time and arc power show that a solidification of copper ball of ~40 um diameter occurred almost within 1ms after the start of EFO firing, followed by a temperature drop to 200°C within 10 ms. The time interval of the solidification may need further refinement by the experimental verification and the consideration of ball rolling-up deformation in the future.

Acknowledgments

We thank Calvin Lee of ASE Kaohsiung lab for important advices on this work.

References

1. J. Onuki, M. Koizumi, H. Suzuki, I. Araki and T. Lizuka, "Influence of Ball-Forming Conditions on the Hardness of Copper Balls", J. appl. Phys., Vol. 68, No. 11 (1990), pp. 5610-5614.
2. Frank W. Wulff, Christopher Breach, Dominik Stephan, Saraswati, Klaus Dittmer and Michel Garnier, "Further Characterization of Intermetallic Growth in Copper and

Gold Ball Bonds on Aluminum Metallisation", Semicon Singapore 2005.

3. Saraswati Ei Phyu Phyu Theint, D. Stephan, H. M. Goh, E. Pasamanero, D. R. M. Calpito, F. W. Wulff and C. D. Breach, "High Temperature Storage (HTS) Performance of Copper Ball Bonding Wires", Electronic Packaging Technology Conference, 2005. EPTC 2005. Proceedings of 7th, Vol. 2, pp.6.

4. C.J. Hang, C.Q. Wang, Y.H. Tian, M. Mayer, Y. Zhou, "Microstructural study of copper free air balls in thermosonic wire bonding", Microelectronic Engineering Vol. 85 (2008), pp. 1815–1819.

5. Hong Meng Ho, Jonathan Tan, Yee Chen Tan, Boon Hoe Toh and Pascal Xavier, "Modeling Energy Transfer to Copper Wire for Bonding in an Inert Environment", Electronic Packaging Technology Conference, 2005. EPTC 2005. Proceedings of 7th, Vol. 1, pp.6.

6. Jonathan Tan, Boon Hoe Toh, Hong Meng Ho, " Modeling of Free Air Ball for Copper Wire Bonding", Proc. 6th Electronic Packaging Technology Conference (2004), Singapore, pp. 711-717.

7. S. Murali, N. Srikanth and Charles J. Vath, "Grains, deformation substructures, and slip bands observed in thermosonic copper ball bonding", Materials Characterization Vol. 50, No. 1 (2003), pp. 39-50.

8. W. Qin, I.M. Cohen, P.S. Ayyaswamy, "Ball Size and HAZ as Functions of EFO Parameters for Gold Bonding Wire", EEP-Vol. 19, No. 1., Advances in Electronic Packaging – 1997, Vol. 1, ASME, pp. 391-398, 1997.

9. L. J. Huang, P. S. Ayyaswamy and I. M. Cohen, "Melting and solidification of thin wires: a class of phase change problems with a mobile interface-l. Analysis", Int. J Heat Mass Transfer. Vol. 38, No. 9 (1995), pp. 1637-1645.

10. I. M. Cohen, L. I. Huang and P. S. Ayyaswamy, "Melting and solidification of thin wires: a class of phase-change problems with a mobile interface-II. Experimental confirmation", Int. J. Heat Mass Transfer. Vol. 38. No. 9 (1995), pp. 1647-1659.

Power Integrity Simulation for SiP Using GTLE

Yunyan Zhou, Lixi Wan, Jun Li

Institute of Microelectronics, Chinese Academy of Sciences, Beijing, 100029, China

Email: zhouyunyan@gmail.com

Abstract

Power integrity (PI) simulation for system-in-package (SiP) is a bottleneck in SiP design flow. This paper presents a novel numerical algorithm for PI simulation in packaging structures. This algorithm is based on 2D Generalized Transmission Line Equation (GTLE), Finite Difference Frequency Domain (FDFD) and mesh division technique. The power distribution network is simulated using mesh division technique where the model of power distribution network is obtained by regarding each cell as a 2D transmission line. 2D GTLE is a group partial equation about voltage and current density distribution on a power/ground plane pair. After reduction, the voltage equation for 2D GTLE is obtained, which is a Helmholtz equation. One method to solve the Helmholtz equation is by the finite-difference scheme. The 2D Laplace operator can be approximated to solve the voltage equation. In this paper, the fringe effect is modeled by the addition of cells around edges which is efficient and easy to implement. Finally, the methodology described in prior sections has been implemented in a CAD tool. The results from our method were compared to those from a full-wave simulator to show efficiency in power integrity simulation.

Introduction

The density of transistors dramatically increases with the down-sizing in semiconductor technology. High-level integration with more functionality has been achieved through 3D packaging technique i.e. system-on-chip (SOC) and system-in-package (SiP). Novel electrical modeling technologies are highly demanded to tackle the challenges posed by the complexity of nano-scale integrated circuits(IC) as well as its package integration.

Usually, the whole package system can be divided into two networks: signal distribution network (SDN) and power distribution network (PDN), where the PDN is often designed as power-ground planes structure to reduce the ground's impedance. The global coupling effects such as the simultaneous switching noise (SSN) demand a system-level modeling methodology and co-simulation of signal and power integrities. Such a system level packaging EMC simulation is a great challenge to all available electromagnetic simulators [1], [2].

Based on the special features of the package structure, we propose to simulate the PDN by Generalized Transmission Line Equations (GTLE), Finite Difference Time Domain (FDTD) and Finite Difference Frequency Domain (FDFD) [3]. For the power/ground planes, the numerical methods, mesh division technique and finite difference method are used to solve these transmission lines equations and get their equivalent *RLCG* parameters.

The transmission line equation technique is well known for its capability in treating a wide range of electromagnetic problems with a great flexibility in terms of geometrical irregularity and material parameters. The equivalence between Maxwell's equation and circuit network allows this technique to solve complex problems in both time- and frequency-domains. It is known that the derivation of conventional transmission line equation (CTLE) is based on such an assumption of an infinite-length transmission line. Unfortunately, practical transmission lines are finite-length. When the CTLE are used in a finite-length unmatched uniform transmission line or arbitrary length nonuniform transmission line, the description of the CTLE for such line discontinuities needs further scrutiny [4], [5].

The reason is that when the nonuniform transmission line is generally treated as a cascading of many short uniform transmission lines, the discontinuities between any two neighboring segments are not only generate reflections, but also produce radiations. Furthermore, the equation is a one-dimensional approximation for a pair of conductor lines. This can't satisfy the simulation for power integrity (PI) in Sip. So it is necessary for extending the one-dimensional equation to 2- or 3-dimensions [6].

In this paper, based on the finite-length line concept and vector partial differential equation, we derive generalized transmission line equation (GTLE) by using circuit theory. However, the coefficients of the GTLE need to be determined by numerical methods, such as differential algebraic equations (DAE). With this equation, the power integrity in PCBs and SiP can be simulated easily and quickly. The GTLE can be solved in time domain or frequency domain. In this paper, a frequency domain solution is discussed with the Finite Difference Frequency Domain (FDFD) technique. Comparing with the methods based on Maxwell's equation, an obvious advantage of using GTLE is reduction of memory usage and saving processing time since the voltage and current density is only computed on the conductor and not in the entire volume space [7].

The rest of this paper is organized as follows: 1) A description of the GTLE for modeling 2-dimensionla plane is provided. 2) Discretization and numerical methods for GTLE are briefly described. 3) Modeling techniques for fringe effect and excitation is considered. 4) Results from our method are provided, and finally, conclusions are presents.

Generalized transmission line equation

In frequency domain, for an infinite-length nonuniform transmission line, the 1D CTLE equation can be expressed as

$$\begin{cases} \dfrac{\partial V(z)}{dz} = -\left(R + j\omega L\right)I(z) \\ \dfrac{\partial I(z)}{dz} = -\left(G + j\omega C\right)V(z) \end{cases} \quad (1)$$

where, V and I are the voltage and current for the infinitely short transmission line; R and L are the per-unit-length series resistor and inductance, respectively; C and G are per-unit-length shunt capacitance and conductance, respectively; ω is angular frequency.

Extending (1) to the GTLE:

$$\begin{cases} \nabla V = -\left(R_{uc} + j\omega L_{uc}\right) \cdot \vec{J}(z) \\ \nabla \cdot \vec{J} = -\left(G_{uc} + j\omega C_{uc}\right) V(z) \end{cases} \quad (2)$$

where, C_{uc}, G_{uc}, L_{uc}, R_{uc} are the parameters of capacitance, conductance, inductance and resistance per-unit-area for 2-dimensional problem or per-unit-volume for 3-dimensional problem. For a non-isotropic material in the structure, the L_{uc} and R_{uc} are tensors. Unlike CTLE, V and $\vec{J}$ in GTLEs are voltage and current density vector flowing through the domain. ∇V and $\nabla \cdot \vec{J}$ are the grads and divergence for voltage and current density vector, respectively. Equation (2) can also be deduced from electromagnetic field analysis, Maxwell equation. Same as CTLE, the GTLE describe the voltage and current density on a conductor with reference to a "ground" conductor. Therefore, the study will focus on the conductor instead of a pair of conductors.

By applying divergence to both sides of the first equation in equation (2), and replace $\nabla \cdot \vec{J}$ by the second equation, a new equation can be written as the following:

$$\begin{cases} \nabla^2 V + k^2 V = 0 \\ k^2 = -\left(j\omega L_{uc} + R_{uc}\right)\left(j\omega C_{uc} + G_{uc}\right) \end{cases} \quad (3)$$

Discretization and numerical methods for GTLE

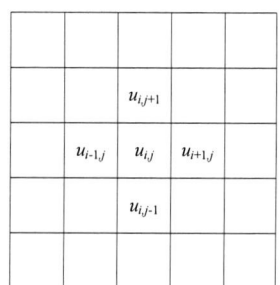

Figure 1 Mesh division of a PCB or substrate and discretization of voltage

Obviously, Equation (3) is a Helmholtz equation. In PI analysis, the ∇^2 is the transverse Laplace operator parallel to the planar structures. One method to solve the Helmholtz equation is by applying the finite-difference scheme. Using mesh division technique, the PCB or substrate in SiP is divided into many cells as shown in Figure 1. The 2-dimensional Laplace can be approximated as

$$\nabla^2 V = \left(u_{i-1,j} + u_{i,j-1} + u_{i+1,j} + u_{i,j+1} - 4u_{i,j}\right) / h^2 \quad (4)$$

where, h is the mesh length and u_{ij} is the voltage at node (i, j) for the cell-centered discretization as shown in Figure 1. Equation (4) is completed by assigning homogenous voltage and current density in one cell.

The discretization results in a well-known bedspring model for plane consisting of per-unit-cell resistance (R_{uc}) and

inductance (L_{uc}) between neighboring nodes, capacitance (C_{uc}) and conductance (G_{uc}) from each node to ground. For a parallel-plate transmission line of equal length and width (h), the resistance, inductance, capacitance and conductance per-unit-cell can be obtained as

$$\begin{aligned} R_{uc} &= 2\rho / d_R \\ C_{uc} &= \varepsilon_r \times 8.8542 \times 10^{-12} / d_C \\ L_{uc} &= \frac{\varepsilon_r \mu_r}{c^2 C_{uc}} \\ G_{uc} &= 2\pi f C_{uc} tg\alpha \end{aligned} \quad (5)$$

where, ρ is the resistivity of the conductor, ε_r, μ_r and $tg\alpha$ are the relative dielectric constant, relative permittivity and loss tangent of the material inside power plane and ground respectively, d_C is the distance between two planes, c is the light velocity in the material, d_R is the thickness of power plane, and f is the work frequency. The factor of "2" in resistance equation is because there are two conductor planes in this problem. The geometry and electrical model of a unit cell for a single plane pair is shown in figure 2.

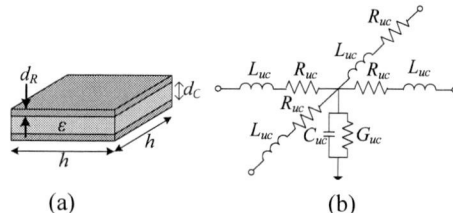

(a)   (b)

Figure 2 Geometry (a) and electrical model (b) of a unit cell for a single plane pair

The unit cell model shown in Figure 2(b) uses a common ground node. In a multilayered structure consisting of more than two planes, we use the nearest ground plane from the power plane as the reference common ground node.

Easily, boundary condition and connection condition between cells in GTLE can be confirmed by Kirchhoff laws. The boundary condition is current behavior on the boundary. The connection condition is that a voltage should keep in same at the connection region if two or more conductors connected together.

Fringe effect model

The model and solution discussed in the previous sections assumes that each unit cell plane-pairs of infinite extent along the lateral directions. However, fringing fields occur at edge discontinuities. This implies that both the per-unit-length inductance and capacitance will be different from that obtained from parallel plate formulae. This problem has been considered in [8], which proposes building a library that maps various geometries to model elements, and interpolating between these values. However, this technique requires the development of a large database that accounts for variations in dielectric height and permittivity, trace width and metal height, and can suffer from interpolation errors. The technique proposed in [9] modified the fringe fields by adding additional elements to edges, which is easy to implement.

In this work, we assume the space between power/ground planes is much less than the width and length of the power/ground plates. This allows the fringe effect of the conductors to be ignored. In order to solve the problem efficiently, we correct the field edge by adding additional cells to edges, as shown in figure 3. Then we use Kirchhoff law to compute the edge voltage and current density.

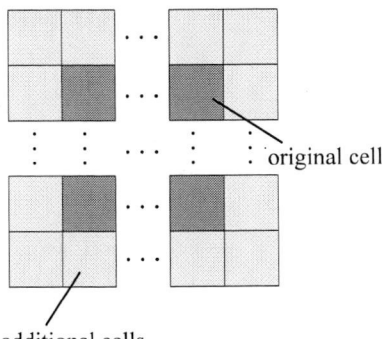

Figure 3 Sketch map of additional cells

Excitation models of GTLE

The problem discussed previously is based on normal cells which are not excitations; next we will present the modeling method for excitation port by GTLE. Obviously, for current excitation, I_e, the GTLE as equation (2) can be constructed as

$$\begin{cases} \nabla V = -\left(R_{uc} + j\omega L_{uc}\right) \cdot \vec{J}(z) \\ \nabla \cdot \vec{J} = -\left(G_{uc} + j\omega C_{uc}\right)V(z) + I_e \end{cases} \quad (6)$$

Simplification of equation (6) is

$$\begin{cases} \nabla^2 V + k^2 V = hI_e \\ k^2 = -\left(j\omega L_{uc} + R_{uc}\right)\left(j\omega C_{uc} + G_{uc}\right) \\ h = -\left(j\omega L_{uc} + R_{uc}\right) \end{cases} \quad (7)$$

The same as above, equation (7) can also be approximated by Laplace operator.

Results

The methodology described in prior sections has been implemented in a CAD tool. Simulations were performed to compare the methodology against full-wave simulations. Full-wave simulations were performed with finite-element-method based solver, HFSS. The above equations were applied to three test cases to illustrate the application of the GTLE method.

Test case 1: bared power/ground plane pair

At first, a bare power/ground plane pair of $1\times1cm^2$ square conductor pair with a dielectric of 20 μm thickness is tested. The dielectric was FR4 with ε_r=4.4. The current excitation and testing port are at (3.5, 5.7) (mm) and (6.5, 6.2) (mm), respectively. There is no component on the plane pair.

The insertion loss (S21) of the plane is shown in figure 4. In all cases, it can be seen that the accuracy of GTLE is comparable with HFSS. The reason of deviation between GTLE and HFSS is mainly about the modal difference between exciting and testing port. In HFSS, the excitation

port is one via resulting parasitic capacitance and inductance. Otherwise, in GTLE, the excitation is processed under equation (7).

This example was discretized using a cell size of 0.2 mm, resulting in about 2700 nodes, and required 34s of computation time for 100 frequencies simulated. But HFSS requires about 16 minutes of CPU time for 100 frequencies, resulting in a speed up of 28×.

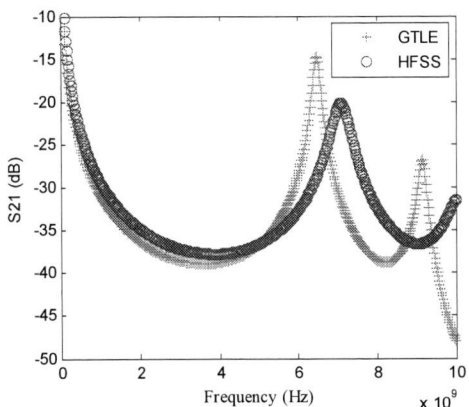

Figure 4 S21 for bared power/ground plane pair

Test case 2: low-pass filter

A novel low-pass filter of $1\times1cm^2$ square is shown in Figure 5(a). The thickness between power and ground was only 14um with high DK=16. The "wide" and "gap" are equal to 0.5mm as shown in Fig. 5(a). "a", "b", "c" figure 5(a) are equal to 9, 3.75, 4.75 respectively. The port 1 and 2 are at (3.5, 5.7) (mm) and (6.5, 6.2) (mm). The insertion loss (S21) of this low-pass filter is shown in Figure 5(b). In all cases, it can also be seen that the accuracy of GTLE is comparable with HFSS. And it is also resulting in a speed up of about 28×.

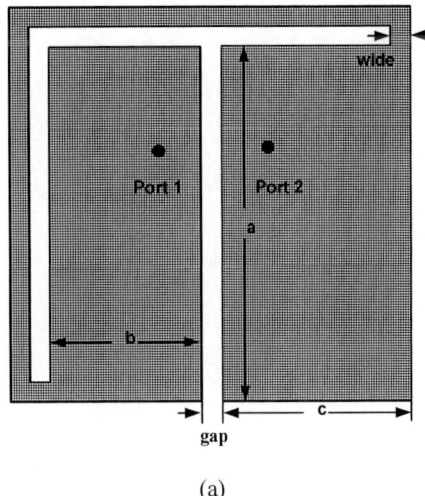

(a)

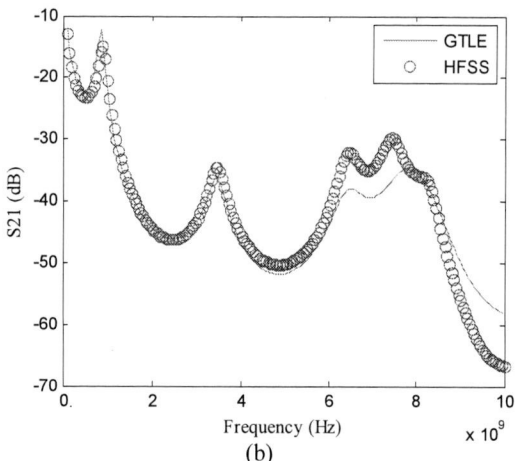

(b)

Figure 5 Simulation of the low-pass filter: (a) the top view of structure; (b) insertion loss (S21).

Test case 3: multilayered structure consisting of four planes

At last, a multilayered structure consisting of four planes is simulated, that is two power/ground pairs. The up two planes is 1×1 cm^2, and the down two planes is 2×2cm^2. In up power/ground pair, the dielectric is FR4 with ε_r=4.4 and 130 μm thickness. In down power/ground pair, the dielectric is ε_r=3.2 and 90 μm thickness. The distance between two power/ground pairs is 150 μm. The current excitation and testing port are at (3.5, 5.7) (mm) and (6.5, 6.2) (mm) in up power planes, respectively. The power and ground point connecting are at (2, 3) mm and (1, 6) mm.

The insertion loss (S21) of the plane is shown in Figure 6. From Figure 6, it can be seen the deviation of GTLE from HFSS. The reason of error is mainly about the stacking strategy to model a multilayered plane by short-circuit the cells between power and ground planes. In the future work, we will improve the GTLE in multilayered structure adapting the transmission line modal.

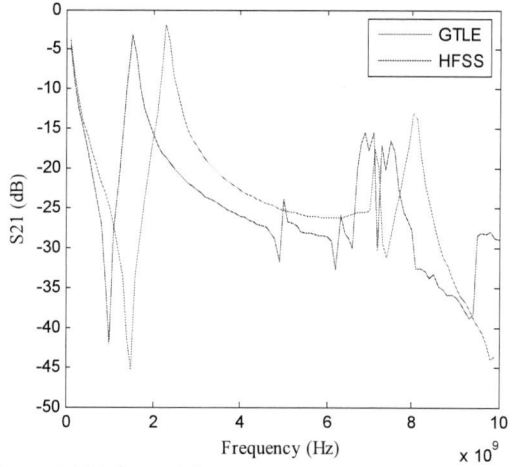

Figure 6 S21 for multilayered structure consisting of two power/ground pairs

Conclusions

Generalized Transmission Line Equation (GTLE) method was developed for simulation of power integrity in Sip structure. The fringe effect and excitation model are presented. Also the discretization and numerical methods for GTLE is presented in order to solve the problem efficiently. The methodology has been compared with full-wave simulators. The simulation results for three test cases matched well. Under similar condition, the GTLE was much faster than HFSS.

Acknowledgments

This work was supported by Hi-tech Research and Development Program of China (863 Program) No. 2006AA01Z236, 2007AA01Z200.

References

[1] S. W. Leung, et al. "A mathematical model for ground voltage fluctuation in PCBs", Asia-Pacific of Conference on Environmental Electromagnetic, IEEE, vol.1, 100-104, 2000.

[2] S. W. Leung, et al. "Modeling of the ground bounce effect on PCBs for high speed digital circuits", IEEE International Symposium on Electromagnetic Compatibility, on Vol. 1, 110–115, 2005.

[3] Wei-Da Guo, et al, "An Integrated Signal and Power Integrity Analysis for Signal Traces Through the Parallel Planes Using Hybrid Finite-Element and Finite-Difference Time-Domain Techniques", IEEE Transactions on Advanced Packaging, Vol. 30, Issue 3, pp.558-565, 2007.

[4] Lixi Wan, et al. "Design, simulation and measurement of embedded ecoupling capacitors for multi-GHz packages/PCBs", ICEPT2005 Proccedings, 108-112, 2005.

[5] Er-Ping Li, et al, "Advanced parallel algorithm for system-level EMC modeling of high-speed electronic package", EMC 2008, pp.1-5.

[6] Yuanqing Wang, et al. "Analysis of nonuniform coupled transmission lines using generalized transmission line equations", proceeding of APMC, IEEE, vol.3, 2005.

[7] Lixi Wan, et al. "Simulation of switching noise in multi-layer structures using generalized transmission line equation method", IEEE International Symposium on Electromagnetic Compatibility, Vol.2, 1026–1031, 2002.

[8] Ching-Chao Huang, et al, "Accurate Analysis of Multi-Layered Signal and Power Distributions Using the Fringe RLGC Models ", IEEE 13th Topical Meeting on Electrical Performance of Electronic Packaging, 103-106, 2004.

[9] Krishna Bharath, et al, "Efficient Simulation of Power/Ground Planes for SiP Applications", ECTC '07, 1199-1205.

First-Principles Based Modeling for Influence of Epitaxy and Packaging Induced Strains on Emission Properties of III-nitrides LED Chips

Han Yan[1,2], Zhiyin Gan[2,3], Xiaohui Song[2], Zhaohui Chen[2], Jingping Xu[1], Sheng Liu[1,2,3*]

[1]Department of Electronic of Science and Technology, Huazhong University of Science & Technology
[2]Wuhan National Laboratory for Optoelectronics
[3]School of Mechanical Science and Engineering, Huazhong University of Science and Technology
Luoyu Road, No.1037, Wuhan, China, Postal No. 430074, Tel./Fax: +86 27 87557074. Email: victor_liu63@126.com

Abstract

Undesired residual strain always exists in the epitaxial film inevitably as a result of lattice mismatch and thermal expansion mismatch between the substrate and the epitaxial film. The strain affects crystalline quality as well as optical and electrical properties of LED epitaxial film. In this paper, we report on the effects of strain on the emission properties of III-nitrides based LEDs grown on various substrates. As results, the band-gap energy of III-nitrides shows blue shift with increasing compressive strain and increases in tensile strain results in a decrease of band-gap energy. A linear relation of the band-gap of III-nitride with the biaxial strain is observed. In addition, III-nitrides materials with larger band-gap energy show a larger shift of emission energy, which means purple and ultraviolet LEDs will be more non-uniform than blue and red LEDs under the same strain states.

Introduction

Super-bright light emitting diodes (LEDs) have been intensely developed in recent years because of their many potential applications, such as automotive forward lighting, background lighting of liquid crystal displays (LCDs), flat panel displays, traffic signals, and chemical biological agent detection [1, 2, 3], to name a few. Besides, they are strong candidates for the next generation general illumination applications. Gallium nitride and related group-III-nitride compound semiconductors are characterized of wide and direct band-gaps, high temperature stability, high electron drift velocities, and high break-down electric fields [4], which make gallium nitride based white LEDs more suitable for general-purpose illumination light sources. Lacking of perfectly suited bulk substrates, wurtzite III-nitrides had to grow heteroepitaxially on various substrates currently [5], such as c-plane sapphire, 6H-SiC, and Si (111). Among these substrates, silicon is believed to be the most attractive for several advantages as low cost, large wafer size, high thermal conductivity and the potential integration of high-speed and high-power nitride devices with Si-based microelectronics [6].

In the past few years, metal organic chemical vapor deposition (MOCVD) has evolved as a leading technique for production of group III-V based optoelectronic and microelectronic devices, which has own an enormous achievement in gallium nitride based epitaxial layer [7]. However, residual strain always exists in the epitaxial film inevitably as a result of lattice mismatch and thermal expansion mismatch between the substrate and the epitaxial films induced by difference in lattice constants and thermal expansion coefficients of substrate and epitaxial film. The undesired strain affects crystalline quality as well as band-gap energies and effective masses of electrons and holes of LED epitaxial film [8]. The two important aspects that influence the strain of LED chips are the epitaxial growth process and the chip packaging. During the cooling process of post growth, group-III-nitrides film grown on sapphire and silicon were under compressive and tensile strain, respectively. Moreover, the temperature increase and decrease in packaging process also induced strains in LED chips.

To design and further improve the performance of gallium nitride based LED devices, a better knowledge of the effects of strains on optical properties of gallium nitride and its related ternary alloys is important. In one previous study of ours published recently, thermal stresses on LED chips were analyzed [9]. However, the influence of the undesired strains induced by the processes of epitaxial growth and packaging on the LED chips has not been clarified completely.

In this paper, we report on the effects of strains on the emission properties of III-nitrides based LEDs grown on various substrates. The emission properties of the III-nitrides were present theoretically by means of first principle calculations, and in particular the effects of strain have been discussed.

Modeling

The biaxial strain of III-nitrides epitaxial films grown hetero-substrates originates from thermal mismatch and lattice mismatch strains and can be expressed as:

$$\varepsilon_{xx} = \varepsilon_{thermal} + \varepsilon_{lattice}$$

where $\varepsilon_{thermal}$ and $\varepsilon_{lattice}$ are the strains induced by thermal mismatch and lattice mismatch, respectively.

The strain resulted from thermal mismatch can be calculated by following formula:

$$\varepsilon_{thermal} = \int_{growth}^{T_{work}} (\alpha_{Substrate} - \alpha_{Film}) dT$$

where $\alpha_{Substrate}$ are thermal expansion coefficients of the substrate and α_{Film} are thermal expansion coefficients of the corresponding epitaxial film deposited on substrate. The thermal expansion coefficient is of nonlinear dependence on temperature, however, sometimes it can be considered as a constant within the narrow range of temperature.

The lattice mismatch can be writing as follows:

$$\varepsilon_{lattice} = (L_{Film} - L_{Substrate}) / L_{Substrate}$$

where L_{Film} and $L_{Substrate}$ are the lattice constants of epitaxial layer and substrate, respectively.

Referring to the material thermal and mechanical parameters, it is not difficult to understand why an opposite strain states exist in gallium nitride epilayers grown on silicon versus on sapphire.

Our model considers the strain states in III-nitride epitaxial layers grown on various substrates. In this calculation, the diagonal elements of the strain tensor were set as follows: $\varepsilon_{zz} = 0$, and $\varepsilon_{xx} = \varepsilon_{yy}$.

In order to understand the effect of the strain on the emission properties, the band-gap energy of wurtzite III-nitrides under different compressive and tensile strains were calculated by using the CPMD (Car-Parrinello molecular dynamics) method [10] as implemented in the CPMD code. CPMD was a highly accurate first-principles calculation program, in which the inter-atomic forces were computed using density functional theory (DFT). The CPMD code was an implementation of DFT in the Kohn-Sham (KS) formulation and of the CP scheme. In the CP scheme, plane waves were used as the basis set to expand the Kohn-Sham orbital. The valence electrons were included in the calculation and the interactions between valence electrons and atomic cores being described by pseudo potentials. The CP method was distinguished from other first-principle implementations in that the atomic coordinates and electronic expansion coefficients evolve simultaneously.

In the present work, the interactions of the ion cores with the valence electrons were described by Trouiller-Martins norm-conserving pseudo potentials [11]. The electronic wave functions were expanded in a plane wave basis set with an energy cutoff of 100 Ry. The 3D electrons of gallium are treated as a part of the valence-band states. We used the local density approximation (LDA) to the density functional. The exchange functional was given by Becke (1988) and correlation energy expression by Lee-Yang-Parr (LYP), respectively. The CPMD code, version 3.11.1, was used. Periodic boundary conditions were applied along the crystallographic axial directions.

Under ambient conditions, III-nitrides crystallize in wurtzite structures. As for the lattice constants, the experimental values of a and c were listed in Table 1 [12]. The biaxial strains were applied on III-nitrides super-cell as shown in Fig.1.

Table 1. The lattice constants of III-nitrides.

	a_0 (nm)	c_0 (nm)
GaN	0.3189	0.5185
AlN	0.3112	0.4982
InN	0.354	0.570
$Al_{0.5}Ga_{0.5}N$	0.31505	0.50835
$In_{0.5}Ga_{0.5}N$	0.33645	0.54425

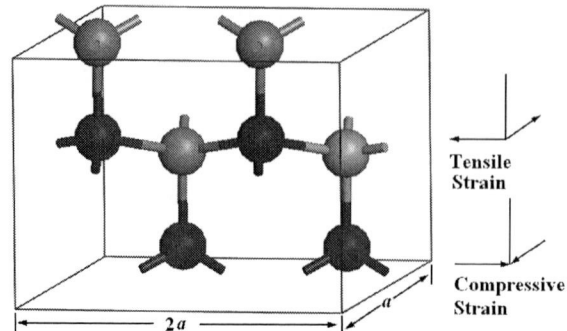

Fig.1 Biaxial tensile and compressive strain applied on III-nitride super cell.

Results and Discussions

It is found that a large tensile strain exists in gallium nitride film grown on silicon, while a small compressive strain is detected for epitaxial film grown on sapphire, due to the thermal expansion coefficients of III-nitrides are larger than that of silicon and lower than that of sapphire. In Fig. 2 and Fig. 3, the energies of band-gap are shown as a function of tensile and compressive strain.

For the case of III-nitride films grown on silicon substrate. Fig. 2 indicates that the band-gap of III-nitrides decreases with increasing in-plane tensile strain, implying that the band-gap decreases with increasing lattice constant of a axis induced by the tensile strain. In other words, with the increase of in-plane tensile strain, the emission wavelength shows red shift. The local density approximation result underestimates the band-gap energies of III-nitrides.

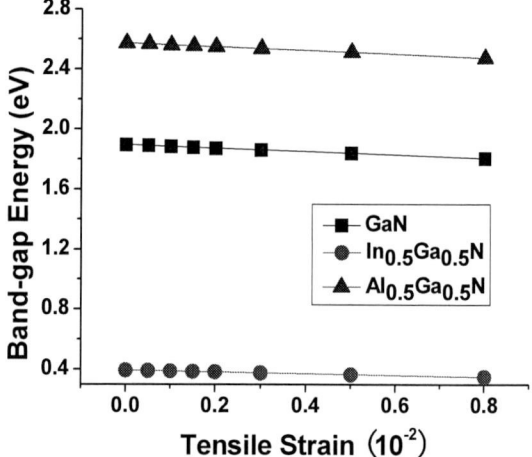

Fig.2 The relationship between the band-gap energy and tensile strain in III-nitride film.

For the case of III-nitride films grown on sapphire substrate as shown in Fig.3, the emission wavelength shows blue shift as the in-plane compressive strain increases.

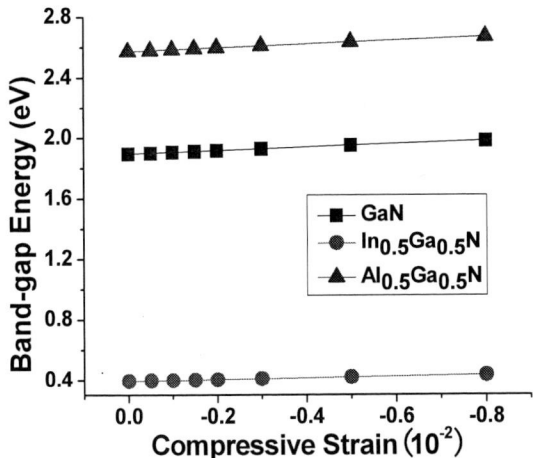

Fig.3 The relationship between the band-gap energy and compressive strain in III-nitrides film. Here the negative sign represents the compressive strain.

A linear relationship between the strain and emission energy bang-gap is observed. By linearly fitting of the calculation data, the biaxial strain dependence of the band-gap energy can be expressed as

$$E_g = E_{g0} + A\varepsilon_{xx}$$

where $\varepsilon_{xx} = (a - a_0)/a_0$, which represents the change in lattice constant along a axis. Here E_{g0} defines the emission energy of III-nitrides materials without strain. A is the linear coefficient characterizing the relationship between the band-gap energy and biaxial strain.

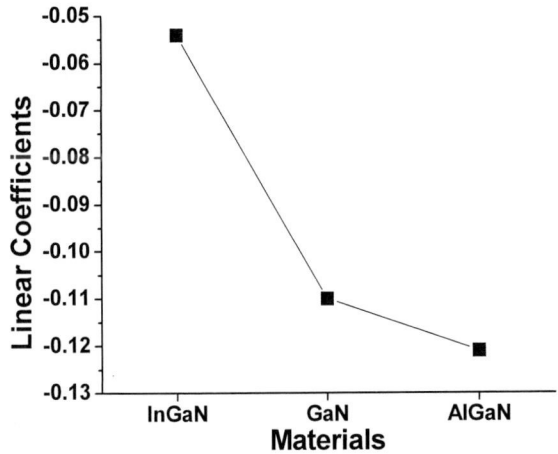

Fig. 4 The different linear coefficient A between III-nitrides.

As shown in Fig. 4, the absolute value of linear coefficient A of AlGaN is larger than that of GaN and InGaN. Indicates III-nitrides materials with larger band-gap energy show a larger absolute value of linear coefficient, which means purple and ultraviolet LEDs will be more non-uniform than blue and red LEDs under the same strain states.

In order to obtain a pure spectrum of LED chip, it is essential to obtain a uniform strain state in LED chip during the epitaxial and packaging process as well as chip operation, particularly for ultraviolet LEDs.

Conclusions

In summary, we have studied the strain effects on the emission properties of gallium nitride and its related ternary alloys through the calculation of band-gap energies. The emission properties of III-nitrides under tensile, compressive and no strain were analyzed and discussed. As a result, the band-gap energy of III-nitrides shows blue shift with increasing compressive strain and increases in tensile strain results in a decrease of band-gap energy. A linear relation of the band-gap of III-nitride with the biaxial strain was observed. In addition, strain-free III-nitrides materials with larger band-gap energy show a larger absolute value of linear coefficient.

Acknowledgments

The authors would appreciate the support of The National High Technology Research Program (863) of China (No. 2007AA04Z348) and Nature Science Foundation of China (NSFC) Key Project under Grant Number 50835005.

This work would appreciate the help from J. Hutter, M. Parrinello, D. Marx, *et al*, Computer code CPMD, version 3.11.1, IBM Zurich Research Laboratory, Copyright IBM Corp. and MPI-FKF Stuttgart 1990-2006, http://www.cpmd.org.

References

1. Park J.W., Yoon Y.B., Shih S.H., et, al. "Joint structure in high brightness light emitting diode (HB LED) packages", *Mater. Sci. Eng. A.* Vol. 441 (2006), pp.357-361.

2. Tadatomo K., Okagawa, H., Ohuchi Y., Tsunekawa T., Imada Y., Kato M. and Taguchi T., "High output power InGaN ultraviolet light-emitting diodes fabricated on patterned substrates using metalorgainc chemical vapor deposition", *Jpn, J. Appl. Phys.* Vol. 40 (2001) pp. 583-585.

3. Oder T.N., Kim K.H., Lin J.Y., et al., "III-nitride blue and ultraviolet photonic crystal light emitting diodes", *Appl. Phys. Lett.* Vol. 84, (2004) pp. 466-468

4. Harima H., "Properties of GaN and related compounds studied by means of Raman scattering" *J. Phys.: Condens. Matter* Vol. 14 (2002) pp. 967-993

5. Mo C.L., Fang W.Q., Pu Y., et. al. "Growth and characterization of InGaNblue LED structureon Si(1 1 1) by MOCVD", *J. Cryst. Growth* Vol. 285 (2005) pp. 312-317

6. Xiong C.B., Jiang F.Y., Fang W.Q., et al. , "The characteristics of GaN-based blue LED on Si substrate" *J. Lumin* Vol. 122-123 (2007) pp. 185-187

7. Nakamura S., Mukai T., Senoh M., Candela-class high-brightness InGaN/AIGaN double-heterostructure blue-light-emitting *Appl. Phys. Lett.* Vol. 64 (1994) pp. 1687.

8. Dems M., Nakwaski W., "Thermal and molecular stresses in multi-layered structures of nitride devices" *Semicond. Sci. Technol.* Vol. 18, (2003), pp. 733-737.

978-1-4244-4658-2/09 $25.00 © 2009 IEEE

9. Cheng T., Luo X.B., Huang S.Y., Liu S., "Thermal analysis and optimization of multi-chip LED packaging based on an analytical general solution", *59th Electronic Components and Technology Conf.*, San Diego, California, USA, May, 2009

10. Car R., Parrinello M., "Unified Approach for Molecular Dynamics and Density-Functional Theory", *Phys. Rev. Lett.* Vol. 55 (1985) pp. 2471-2474

11. Goedecker S., Teter M., Hutter J., "Separable Dual-space Gaussian Pseudopotentials", *Phys. Rev. B.* Vol. 54 (1996) pp. 1703-1710

12. Vurgaftman I.,. Meyer J. R.,. Ram-Mohan L. R, "Band parameters for III-V compound semiconductors and their alloys", *J. Appl. Phys.*, Vol. 89, (2001), pp. 5815-5875

Numerical Study on Thermal Management of LED Packaging by Using Thermoelectric Cooling

Nan Wang[1], Chang-hong Wang[2], Jun-xi Lei[1], Dong-sheng Zhu[1]
[1]The Key Lab of Enhanced Heat Transfer and Energy Conservation, Ministry of Education
South China University of Technology, Guangzhou, China
[2]Faculty of Materials and Energy, Guangdong University of Technology, Guangzhou, China
Email: wangnan65@163.com; chhong.wang@gmail.com

Abstract

The heat generated from LED (Light-emitting Diode) keeps rising with the rapid development of LED luminous flux and brightness in recent years. So it's urgent to find effective heat dissipation ways for thermal management of LED packaging. The finite element model of LED packaging heat dissipation, based on the thermal resistance model of flip-chip LED packaging and the principle of thermoelectric refrigeration, was founded by ANSYS in this paper. Performance parameters such as LED junction temperature (T_0), substrate temperature (T_{sub}), hot end and cold end temperature of Thermoelectric Cooler (TEC) (T_h, T_c), heat sink temperature (T_{sink}) and coefficient of performance (COP) were investigated systematically in the situation of different chip power and different TEC input current. The optimized parameters of packaging heat dissipation in different conditions were analyzed through comparing simulation results of the thermoelectric cooling method and the natural convection cooling method with and without heat sink.

1 Introduction

LED is widely used as light source because of its long life, small size, low energy consumption and fast response time. With the applications of high brightness LED, power supply is getting higher and higher. About 90% of the supply power is converted into heat dissipation. The high heat flux of LED chip surface causes low luminous efficiency, higher chip junction temperature [1]. It also changes the LED peak wavelength and significantly reduces operating lifetime and reliability. So it is urgent to find effective cooling methods for the thermal management of high-power LED packaging [2]. At present, as a new cooling solution, thermoelectric cooling is suitable for high-power LED thermal management because of its advantages.

Based on the thermal resistance model of flip-chip LED packaging and the principle of thermoelectric refrigeration, the finite element model of thermoelectric cooling LED packaging was founded in this paper. Performance parameters such as LED chip junction temperature, substrate temperature, hot end and cold end temperature of TEC, heat sink temperature and coefficient of performance are investigated systematically in the situation of different chip power and different TEC input current. The optimized parameters of packaging heat dissipation under different power conditions are analyzed by comparing the simulation results of the thermoelectric cooling method, natural convection cooling method with and without heat sink.

2 Analyses

2.1 Thermoelectric refrigeration principle

As a new cooling technology, thermoelectric refrigeration is rapidly developed since 1950s. It is Peltier effect that makes thermoelectric modules have a cooling effect. As Figure 1 shows that a thermoelectric element is coupled by a P-type semiconductor component and an N-type semiconductor component.

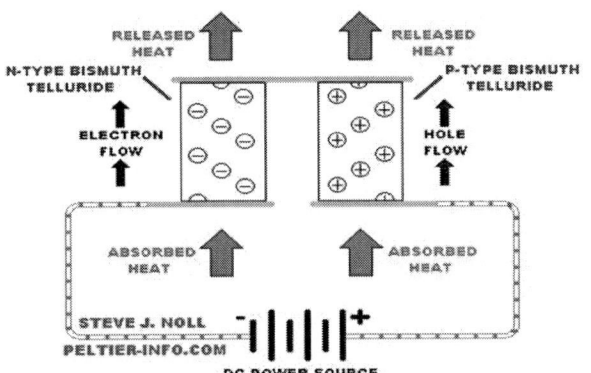

Figure1. Thermoelectric element

The temperature difference and heat transfer process will emerge in the joint after applying direct current. The holes in P-type semiconductor and the electrons in N-type semiconductor both migrate to the same orientation when the current direction is from P-type semiconductor to N-type semiconductor. The joint, which absorbs in the holes in P-type and electrons in N-type, generates heat in and becomes the hot end [3].

Assume that α, K and R is constant with the temperature, the cooling capacity Q_c and cooling efficiency COP of a TEC with N series can be expressed by:

$$Q_c = N[\alpha T_c I - \frac{1}{2}I^2 R - K(T_h - T_c)] \tag{1}$$

$$COP = \frac{Q_c}{W} = \frac{\alpha I T_c - \frac{1}{2}I^2 R - K \cdot \Delta T}{\alpha I \Delta T + I^2 R} \tag{2}$$

The equation (1) shows that thermoelectric refrigeration effect of TEC is consist of three effects. The Peltier effect $\alpha T_c I$ makes TEC generates heat in the hot end and absorbs heat in the cold end, which associated with the Joule effect $\frac{1}{2}I^2 R$ and the Fourier effect $K(T_h - T_c)$ caused by the temperature difference between the hot end and cold end.

978-1-4244-4658-2/09 $25.00 © 2009 IEEE

2.2 Thermal management of LED packaging

Thermal management of electronic equipment is a process to control temperature in the electronic equipment and ensuring the reliability of electronic system [4]. It applies suitable cooling technology and structural design in electronic component. A good LED packaging cooling system can lower LED junction temperature and extend LED life in the same power input. It can also increase power input and LED brightness in the same LED junction temperature range. Figure 2 is the LED heat dissipation packaging system using thermoelectric cooling technology. LED chip is arranged in 2×2 array. Heat generated from LED chip is conducted to TEC cold end through soaking block, thermal interface material (TIM) and substrate. The LED chip and substrate is cooled by the TEC cold end using the active refrigerating way. Heat generated from TEC hot end is conducted to ambient through lamp cap and heat sink. LED junction temperature is a significant index for LED heat dissipation performance. Because thermoelectric cooling is an active refrigerating technology, LED heat dissipation using TEC cooling can effectively lower LED junction temperature in comparison with traditional heat dissipation way such as air cooling.

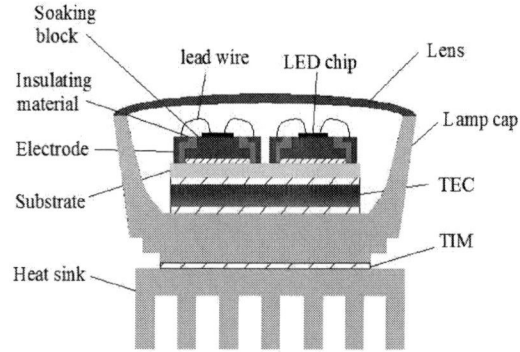

Figure2. Schematic of LED packaging
TEC cooling system

Thermal resistance has great effect on LED heat dissipation performance, which can be improved by lowering thermal resistance. Assuming that each LED chip has the same power P_i and the thermal resistance of TIM is neglected, the thermal resistance network of LED heat dissipation system using thermoelectric cooling can be express in Figure 3.

For single LED chip packaging TEC cooling system, the total thermal resistance can be expressed by $\theta_{total} = (T_0 - T_a)/P$. When the LED chip is in 2×2 array packaging, the total temperature difference and total thermal resistance can be expressed by [5]:

$$\Delta T_{total} = T_0 - T_a = (T_0 - T_{sub}) + (T_{sub} - T_a)$$
$$= P(\theta_{LED} + \theta_{soaking}) + 4P(\theta_{sub} + \theta_{TEC} + \theta_{cap} + \theta_{\sin k})$$

$$\tag{3}$$

$$\theta_{total} = \frac{\Delta T_{total}}{4P} = \frac{P(\theta_{LED} + \theta_{soaking}) + 4P(\theta_{sub} + \theta_{TEC} + \theta_{cap} + \theta_{\sin k})}{4P}$$
$$= \frac{\theta_{LED} + \theta_{soaking}}{4} + (\theta_{sub} + \theta_{TEC} + \theta_{cap} + \theta_{\sin k})$$

$$\tag{4}$$

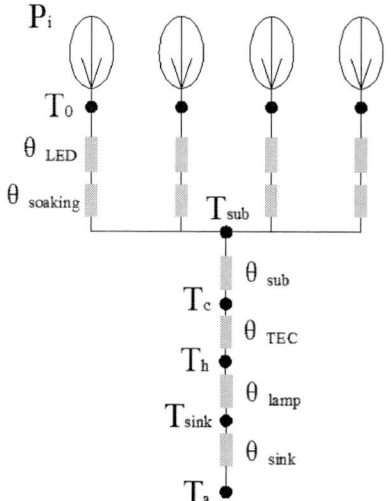

Figure3. Thermal resistance model of LED packaging TEC
cooling system

2.3 Modeling

Figure.2 shows that the heat dissipation method of LED packaging using thermoelectric cooling consists of three parts: thermal conduction among each packaging material, convective heat transfer between heat sink and ambient, convective heat transfer in the ambient. The heat transfer function of LED packaging system can be expressed by:

$$u\frac{\partial v}{\partial x} + v\frac{\partial v}{\partial y} = -\frac{1}{\rho_a}\frac{\partial p}{\partial x} + \eta(\frac{\partial^2 v}{\partial x^2} + \frac{\partial^2 v}{\partial y^2}) + g\rho\beta(T - T_a)$$

$$\tag{5}$$

$$u\frac{\partial T}{\partial x} + v\frac{\partial T}{\partial y} = a(\frac{\partial^2 T}{\partial x^2} + \frac{\partial^2 T}{\partial y^2}) + S$$

$$\tag{6}$$

Some hypotheses are made to simplify the model [6]: (1) LED chip is a uniform plane heat source; (2) Radiant heat is neglected; (3) Thermal expansion coefficients of the packaging material are neglected; (4) LED electrode, lead wire and insulated material are neglected; (5) Solder joints of the substrate are neglected.

Equation (5) and equation (6) may be solved by finite element method. The LED packaging TEC cooling system model is established by ANSYS 10.0, as shown in Fig.4:

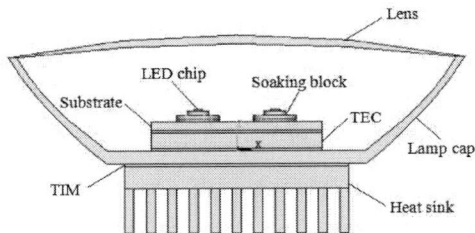

Figure4. Solid model of LED packaging TEC cooling system

The LED chips which in 2×2 array are uniformly arranged on Al$_2$O$_3$ substrate. The single chip size is 2×2×0.4mm, whose power is from 1W to 5W. Substrate and TEC with the size of 40×40mm are selected. The TEC type selected is TECI-12706M, whose the maximum current I_{max} is equal as 6.55A, maximum temperature difference ΔT_{max} is equal as 68.3 °C and maximum cooling capacity Q_{max} is equal as 63.73W. The heat sink size is 52×52×5mm, whose fin height is 10mm, fin thickness is 2mm and fin spacing is 3mm. The convective heat transfer coefficient of heat sink is 10W·m^{-2}·K^{-1}. The thermal conductivity of packaging material is shown in Table 1 [7-8].

Table1. Thermal conductivity of the LED packaging cooling system

	Thermal conductivity (W·m^{-1}·K^{-1})
LED chip	130
Soaking block	40
Substrate	200
TEC	1.5
TIM	5
Lens	0.2
Lamp cap	168
Heat sink	240

The power chip and TEC are mapped meshed because of large temperature gradient in this region; free mesh is made in other analysis regional. LED power is applied in the form of volume heat generation, while TEC cooling capacity and heat dissipating capacity are applied in the form of surface heat flux. Assuming that ambient temperature is constant at 25°C and the temperature inside lens is constant as 40°C, the finite element model of LED packaging is calculated by the Galerkin method in ANSYS.

3 Results and discussion

3.1 Temperature distribution of cooling system

When the LED power is equal as 4W and the performances of TEC is respectively: I=1.26A and Q_c=5W, the temperature distribution of LED cooling system is shown in Fig.5.

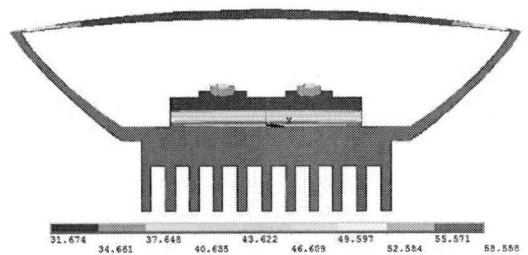

Figure5. Temperature distribution of the TEC cooling system

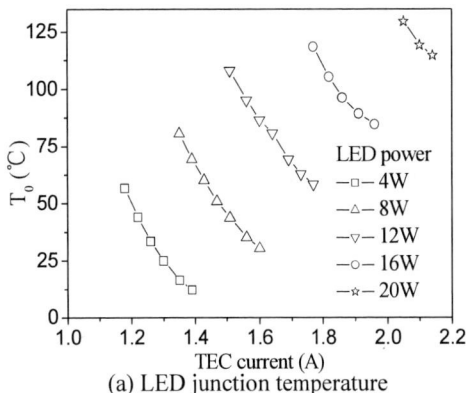

(a) LED junction temperature

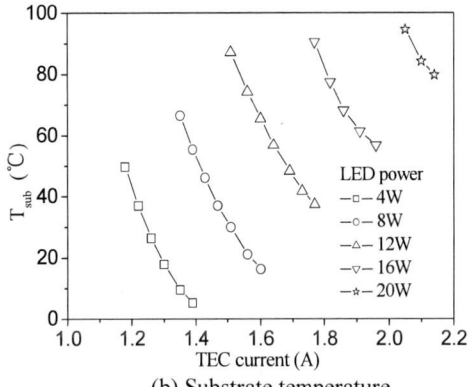

(b) Substrate temperature

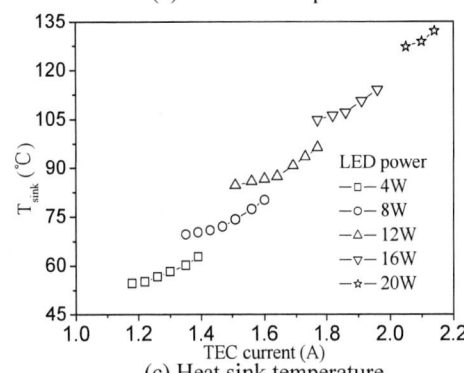

(c) Heat sink temperature

Figure 6 Temperature variation curve of LED packaging system with TEC current and chip power

The chip junction temperature, which usually regulated under 120°C, is a significant index for LED heat dissipation performance. Fig.6 (a) explains that the cooling capacity increases with TEC current and chip power increase. The heat absorbed from the substrate is increasing which lead to LED junction temperature drops. Moreover, the increasing extent of TEC cooling capacity versus TEC current becomes smaller. As a result, the decreasing extent of LED junction temperature versus TEC current also becomes smaller. On the other hand, with LED power increasing at the same TEC current, LED junction temperature is rising. Even LED power is 20W, LED junction temperature is still under 120°C while using TEC cooling.

As heat generated from LED chip is conducted to TEC cold end through soaking block and substrate, there is a heat dissipating path among LED chip, substrate and TEC cold end, the variation trend of substrate temperature versus TEC current in Fig.6 (b) is similar to LED junction temperature. Substrate temperature drops with TEC current increasing, and the decreasing extend of substrate temperature versus TEC current becomes smaller.

Heat from TEC hot side is conducted to heat sink and finally dissipates to ambient through natural convection. From Fig.6(c) we can conclude that with TEC current increasing at the same LED power, heat dissipating capacity of TEC is increasing, and heat sink temperature is also rising. But the heat sink size selected is small due to the limited LED size. As a result, heat sink temperature is rising rapidly with LED power increasing at the same TEC current. The heat sink temperature is all over 100°C when the LED power is over 16W.

3.2 Temperature difference between hot and cold side

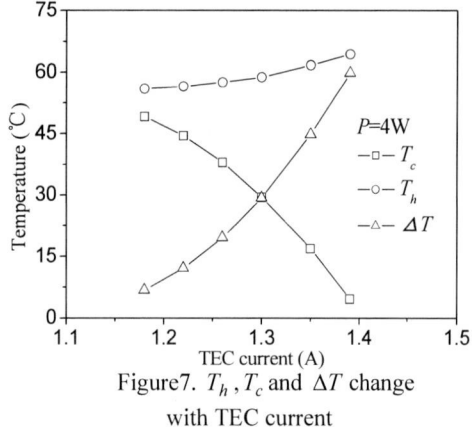

Figure7. T_h, T_c and ΔT change with TEC current

Figure7 gives that hot end temperature T_h, cold end temperature T_c and temperature difference between hot and end ΔT change with TEC current when the chip power is 4W. It shows that T_c drops when TEC current increases, but T_h and ΔT rise. TEC current and cooling capacity can not be increased any more when ΔT is closed to 65°C

because the maximum temperature difference of TEC is equal to 68.3 °C. There is a maximum TEC current input at certain LED power. On the other hand, chip junction temperature increases with TEC current decreasing. TEC current can not be decreased when the junction temperature is over 120 °C. So there is a minimum TEC current input at certain LED power.

3.3 COP and thermal resistance of cooling system

Figure8 shows that cooling efficiency COP and thermal resistance θ_{total} change with TEC current when the chip power is 12W.

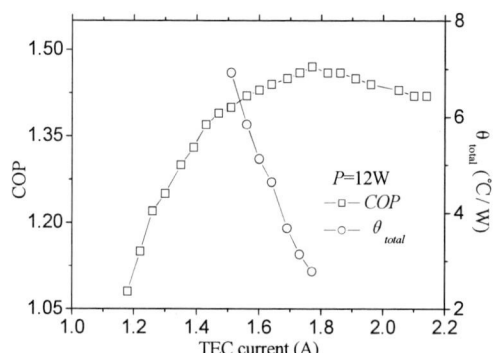

Figure8. COP and θ_{total} change with TEC current

When LED power is 12W, the thermal resistance of LED packaging cooling system decreases with the increase of TEC current. The COP firstly increases and then decreases with TEC current. There is a maximum COP versus TEC current. As mentioned in above that the LED current applied has a range at certain LED power, so COP and thermal resistance should be considered while applying TEC current. COP is the economic index for TEC performance, and the main principle for thermal management is to lower system thermal resistance as much as possible. As the COP decreasing extent is small versus TEC current, COP and thermal resistance tend to be better with TEC current increasing in most cases. So lowering system thermal resistance is mostly considered while applying TEC current at certain LED power. The optimal TEC current and the corresponding LED junction temperature at certain LED power are listed in Table 2.

Table2 TEC current and chip junction temperature

P (W)	I_{opt} (A)	T_0 (°C)
4	1.26	33.4
8	1.56	35.1
12	1.77	58.3
16	1.96	84.6
20	2.14	114.6

3.4 Comparison of different cooling method

The natural convection method with and without heat sink were selected in comparing with the TEC and heat sink method. The heat sink selected in the natural convection method is the same as the heat sink selected in the TEC and heat sink method. Properties of packaging material selected are also the same as Table 1. The optimal TEC current in Table 2 was selected as TEC input current in the TEC and heat sink method.

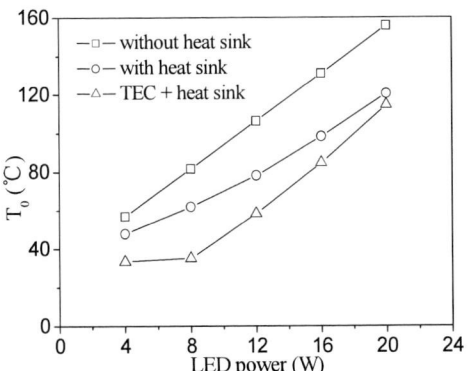

Figure9. Performance comparison of three cooling methods

Figure9 shows that LED junction temperature of the three heat dissipation methods all rise with increasing of LED power. The temperature rising extent of the heat dissipation method without heat sink is biggest, and the LED junction temperature is over 120 °C when LED power is over 14W. The method with heat sink is better, but the junction temperature is 120.5 °C when LED power is 20W. The method with TEC and heat sink is the best. The junction temperature can stay between 30°C to 40°C at low LED power, and the junction temperature is 114.6 °C when LED power is 20W. So only the TEC and heat sink method can satisfy the heat dissipation need for LED packaging system.

4 conclusions

(1) The finite element model of LED packaging heat dissipation was simulated based on the thermal resistance model of flip-chip LED packaging and the principle of thermoelectric refrigeration. The result shows that with TEC current increasing, LED junction temperature and substrate temperature drop and heat sink temperature rises.

(2) The result that LED current applied has a range at certain LED power was concluded after analyzing the relationship between TEC hot end, cold end temperature and TEC current. The COP and thermal resistance of LED packaging system were selected as the optimal parameters, and the optimal TEC current and LED junction temperature at certain LED power were concluded after optimizing.

(3) The result shows that the TEC + heat sink method is the best after comparing with the heat dissipation method with and without heat sink. The TEC + heat sink method can satisfy the heat dissipation need for LED packaging system when LED power is applied less 20W.

References

1. Arik M, Beckerb C, Weaverb S, et al. 2004, Thermal management of LEDs: Package to system. Proc SPIE, 5 (187), pp. 64-75.
2. Zhang C. J, and Wang C. Q. 2007, Thermal Analysis and Design of High – power White LED Package, Electronics Process Technology, 28(5), pp. 257-261.
3. Xu D. S., 1999, Thermoelectric refrigeration and applied technology, 2nd edition, Shanghai Jiaotong University Press, Shanghai, pp.7.
4. Yu J. Z., 2002, Thermal design and analysis technology for electronic device, High Education Press, Beijing, pp. 1.
5. Kim, L., Jong, H. C., Sun, H. J. et al. 2007, Thermal analysis of LED array system with heat pipe, Thermochimica Acta, 455, pp. 21-25.
6. Ma H. X., Qiang K. Y., Han Y. J. et al. 2007, Thermal Design of Ga N2based High2power LED Module, Semiconductor Optoelectronic, 28(5), pp. 627-630.
7. Qiang K. Y., Zheng D. S., Luo Y. 2006, Thermal Dispersion of GaN-based Power LEDs, Semiconductor Optoelectronic, 27(3), pp. 236-239.
8. Wang C. H. Research of enhanced heat transfer mechanism and characteristics optimization on the micro-electronic chip cooling system [D]. Guangzhou: South China University of Technology, 2009: 47-58.

Modeling of Thermal Phenomena in a High Power Diode Laser Package

Ephraim Suhir[1,2], Jingwei Wang[2], Zhenbang Yuan[3], Xu Chen[4], Xingsheng Liu[2,3]

[1]University of California, Santa Cruz, CA, University of Maryland
College Park, MD, and ERS Co., 727 Alvina Ct., Los Altos, CA 94024, USA
tel. 650-969-1530, cell. 408-410-0886, suhire@aol.com
[2]State Key Laboratory of Transient Optics and Photonics, Xi'an Institute of Optics and Precision Mechanics
Chinese Academy of Sciences
No. 17 Xinxi Road, New Industrial Park, Xi'an Hi-Tech Industrial Development Zone, Xi'an, Shaanxi, 710119, P.R. China
[3]Xi'an Focuslight Technologies Co., LTD
No. 60 Xibu Road, New Industrial Park, Xi'an Hi-Tech Industrial Development Zone, Xi'an, Shaanxi, 710119, P.R. China
[4]School of Chemical Engineering & Technology of Tianjin University
No. 92 Weijin Road, Nankai District, Tianjin, 300072, P. R. China

Abstract

High power solid state semiconductor laser arrays are widely used in today's industrial, military and bio-medical systems, as well as in various material processing technologies (welding, cutting, surface treatment, etc.). The current analysis has been motivated by the following major factors associated with the design and use of high power semiconductor laser packages: (1) adequate functional (optical) and mechanical (structural, "physical") reliability is certainly a must for the successful operation of the packages in question; (2) elevated temperatures, thermal stresses, strains and displacements are the major contributor to both functional and structural ("physical") laser package failures; examples are: output spectrum broadening, increase in the near-field nonlinearity-"smile effect", fatigue and brittle cohesive and adhesive cracking in the materials, and others; (3) predictive modeling, both numerical (finite-element-analysis (FEA) based) and analytical ("mathematical"), have proven themselves as an effective and cost-effective means to understand, analyze, predict and prevent thermal failures in packaging engineering in the most time- and cost-effective fashion.

Accordingly, the objective of our study is to demonstrate the use of predictive modeling in the analysis of the thermal phenomena in, and design-for-reliability of, high power laser packages. We employ FEA-based simulations to analyze the steady-state and the transient thermal behavior of a conduction cooled packaged semiconductor laser operated in a continuous wave (CW) mode. We show, as an example of the application of analytical modeling approach, how one could assess the size of the inelastic zones in a low-yield-stress solder at the interface between the semiconductor laser die and a copper heat sink ("sub-mount"). It is these zones that are responsible for the finite fatigue lifetime of the solder bond. It has been established that solder interfaces are the weakest link, as far as laser package reliability is concerned. This is due primarily to the low yield point of, and, as the consequence of that, high level of inelastic strains in, the solder materials. The solder material works therefore in a low-cycle-fatigue mode. At the same time, owing to its low yield stress, the interfacial solder material is able to provide a reasonably effective strain buffer between the low expansion semiconductor and the high expansion substrate. This "buffer" relieves the thermal stress in the semiconductor die, whose functional (optical) performance is absolutely crucial and cannot be compromised.

The developed FEA and analytical models enable one to analyze the thermal phenomena in high power laser devices and conduction-cooled packages, as well as to predict and prevent thermal failures.

Introduction

High power solid state semiconductor laser arrays are widely used in the today's industrial, military and bio-medical systems, as well as in various material processing technologies (welding, cutting, surface treatment, etc.). The ability to design and employ high-quality solder interfaces in high power laser device packages, to understand the physics of the behavior of these packages and interfaces, and to prevent possible functional (optical) and mechanical ("physical") failures is of obvious practical importance. Many failures in high power diode laser packages are directly related to the solder interfaces [1,2,3]. Thermal behaviors of, and thermal stresses in, solder interfaces are the major factors affecting the functional and structural performance of high power diode laser packages. If the accumulated heat cannot readily escape, the elevated temperatures and thermally induced stresses at the location of the p-n junction adversely affect the output power, the slope efficiency, the threshold current and the device lifetime. Elevated thermal stresses could cause spectral broadening and wavelength shift [4]. The emitting wavelengths vary, when the junction temperature of the emitters across the array is not controlled and/or is non-uniform. These circumstances make thermal management of high power devices a major challenge in the pump laser design, manufacturing and use.

Another reliability related challenge stems from the significant thermal-expansion (contraction) mismatch between the low expansion semiconductor material and the high-expansion material of the substrate (heat sink). This results in an elevated thermal stresses in the package structure. It has been determined [1] that the thermal stress in the epitaxial material of the laser diode could change the emitting wavelength by a factor on the order of $\sim 1*10\text{-}5$ eV/bar (~ 0.005nm/bar): thermally induced tensile strains cause red-shift and compressive strains cause blue-shift. Non-uniform thermal stress (thermal stress gradients) experienced by the emitters across the width of the array lead to uneven wavelengths, thereby broadening the output frequency spectrum.

Additional thermal-stress related reliability challenges stem from the employment of the low-yield-stress solder material. On one hand, a low yield stress is desirable, since it helps to reduce the stresses in the structurally vulnerable stress-sensitive semiconductor dies, whose optical behavior cannot be compromised. On the other hand, the use of low yield stress solders leads to a situation, when these solders operate, even at low thermal stress level, in a low-cycle-fatigue condition. This circumstance reduces dramatically the adhesive and the cohesive fatigue strength of the solder, and shortens considerably its life-time. It is important therefore, as long as a low yield stress solder has to be used, to be able to determine the sizes of the inelastic zones in the solder material: it is these zones that are responsible for the finite fatigue lifetime of the solder bond in the package structure. After the size of the inelastic zones is determined, the time-to-failure of the material could be assessed based on the existing more or less well developed and agreed upon methods (see, for instance, [5,6]).

Note that since the die attach solder interface can have a significant impact on the performance and reliability of a laser assembly, fluxless soldering processes are used now for the best quality possible, and voids should be controlled to the minimum. Large voids increase junction temperature and create hot spots. "Clusters" of small voids near the front facet are especially undesirable and could be detrimental to the device performance, as they can significantly raise the facet temperature and cause catastrophic optical mirror damage (COMD). It is particularly important that voids do not concentrate at the interfaces, especially at the ends of the bonding layer, where the thermally induced stresses are the highest: voids can add considerably to the stress concentration and initiate fatigue cracks.

In the study that follows we use, as a typical example of a high power laser package, a single-bar conduction-cooled-packaged 60W 808nm laser that operates in a continuous-wave state. We investigate the steady-state and the transient thermal behavior of the laser bars/arrays, discuss the strategies for optimizing the thermal design and thermal management, and develop a methodology for the evaluation of the size of the inelastic zones in the given bi-material die-substrate assembly in the packages of the type in question.

Steady-State Thermal Behavior

The single-bar conduction-cooled- packaged 60W 808nm semiconductor laser made by the Xi'an Focuslight Technologies Co., Ltd., is shown in Fig.1. A typical semiconductor laser consists of three major parts: the laser bar (chip/die), the cathode and the heat sink. The laser bar is schematically shown in Fig.2. The quantum well is the active region. It contains 19 emitters and is comprised of a cladding layer and a p-metal.

In our study the steady-state thermal behavior of a CS-Packaged semiconductor laser in Continuous Wave (CW) mode was simulated using FEA technique. The 19 emitters in the quantum well are the heat sources (heat producers) in the device. The bottom side of the device is kept at 25oC. As shown in Fig. 3, the distribution of the transverse temperature profile of quantum wells in steady state condition is basically parabola-shaped. At the emitter region, the heat flow is constrained in a small area. The temperature rise per unit length at the emitters' location is therefore the highest. The local temperature in the emitters located in the mid-portion of the emitting region is higher than at the peripheral portions. Peak temperature of the device reaches 51.8 oC. The thermal resistance R_{th} of a semiconductor laser can be determined as

$$R_{th} = \Delta T / \Delta Q \qquad (1)$$

where ΔT is the temperature rise in the quantum well, and ΔQ is the thermal power of the device. For a typical 808nm semiconductor laser, the photoelectric conversion efficiency is about 50%. This means that the thermal power, ΔQ, of a 60W (optical power) 808nm semiconductor laser is about 60W. According to the simulation data, the temperature rise, ΔT, in the quantum well of the device is 26.8 oC (51.8 oC minus 25 oC). Thus, the calculated thermal resistance in the device is about 0.472K/W.

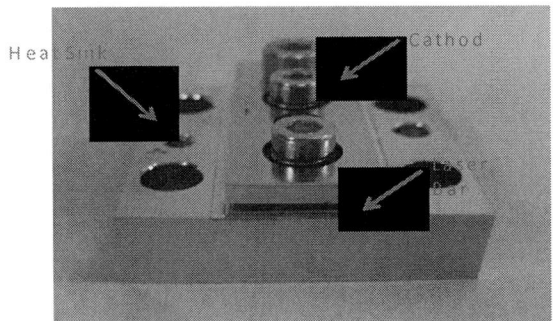

Fig. 1 A sample of single-bar 808nm 60W continuous-wave high power semiconductor laser

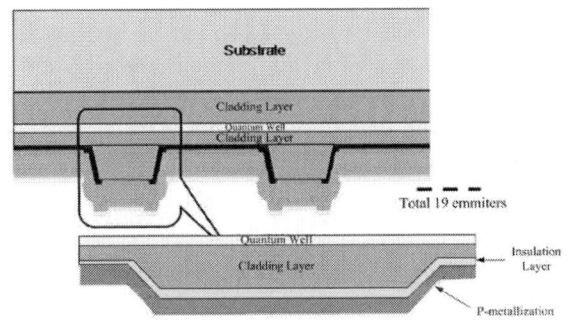

Fig. 2 Schematic of semiconductor laser bars

Transient Thermal Behavior

Transient thermal behavior is a crucial characteristic of the total thermal performance of a semiconductor laser. It reflects the physics of the heat propagation process inside the device. Particularly, by studying the transient thermal behavior of the laser and its package, one can understand better the contribution of each constituent material layer and the interfaces to total thermal resistance, as well as the possible failure modes and mechanisms in a semiconductor laser operated in a quasi-continuous (QCW) condition.

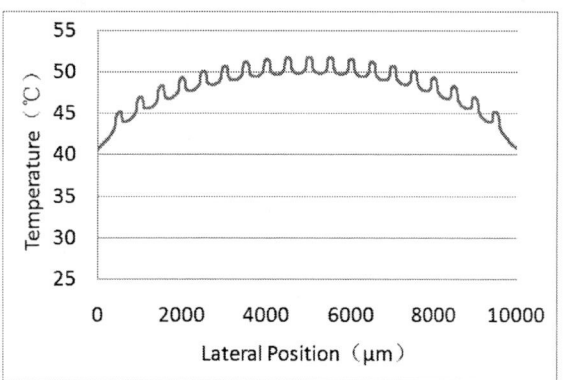

Fig. 3 Transverse temperature profile of quantum wells at steady state

The transient thermal behavior of the conduction-cooled-packaged semiconductor is shown as Fig. 4. The boundary conditions are assumed to be the same as in the steady-state thermal analysis. As one could see, it takes not less than 280ns for the temperature disturbance to be transmitted to the upper side. It takes 900ns for such a disturbance to reach the lower side of the solder layer. This means that the temperature difference between the upper side and the lower side of the solder could be quite large, if the device works in QCW mode whose pulse width is between 280ns and 900ns.

D. Schleuning et al. [7] have demonstrated that the reliability of a typical semiconductor laser could be only good for the controlled condition, e.g., for a short-pulse QCW operation or CW operation. When the device works during long-pulse QCW operation, the reliability of the device reduces rapidly. Our simulation gives a reasonable explanation to this phenomenon. When a semiconductor laser operates during a short pulse QCW mode (pulse length $\tau < 280$ns), heat generated in the previous pulse is able to dissipate completely before the next pulse takes place. There is simply not enough time for the heat to be transferred to the solder layer. When the device operates in a long pulse QCW mode (pulse length $\tau > 900$ns) or in a CW mode, temperature distribution in the solder layer has already reached the steady-state condition within a single pulse, and the temperature gradients in solder layer are relatively small. However, when the device operates in the QCW mode and the pulse width is between 280ns and 900ns, the reliability of the system might be rather low, because the heat generated by the previous pulse has been already transmitted to the upper side of the solder layer (i.e., prior to the next pulse), but not to the bottom side yet. This circumstance can cause significant temperature gradients in the solder layer, lead to intensive electromigration, and, consequently, to low considerably the reliability of the device. Thus, pulse widths between the durations of 280ns and 900ns should be avoided.

Let us choose one of the 19 emitters and draw at different time the horizontal and vertical temperature distribution curve. It can be seen that as the time progresses, the temperature distribution curve gradually stabilizes. At about 37μs, the temperature of each node increases with time. This means that the device has entered into the "Formal Status" of transient heat transfer [8]. The heat from the single emitter would not reach the center of the space (in Fig. 5, the temperature between 0μm and 500μm is Zero) and therefore

the heat from this single emitter would not influence the performance of the adjacent emitter. Because of that, when the pulse duration is under 37μs, this heat would not impact the neighboring emitters, even if one of the emitters does not operate, or if the heat generated by this emitter could not be removed. This will certainly contribute to the higher reliability of the device. In addition, as could be seen from Fig. 5, the difference between the two temperature curves is getting smaller from the initial moment within the same period, which indicates that the rise of the temperature is slowing down gradually.

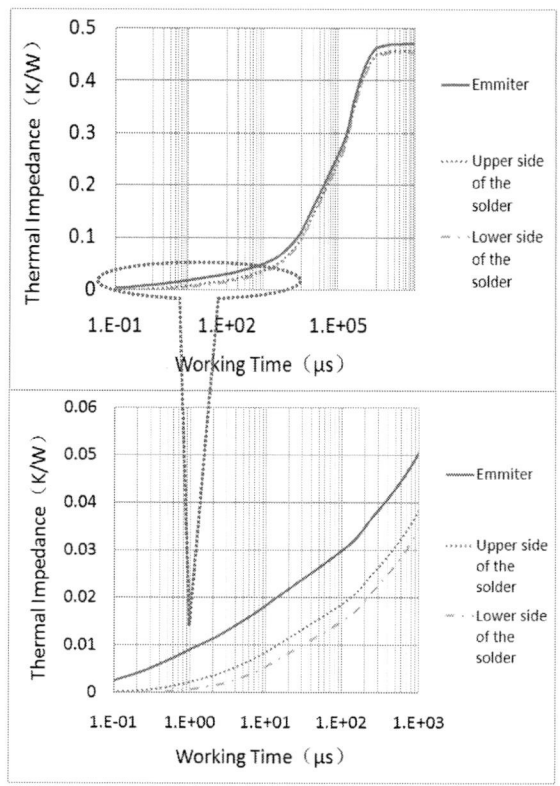

Fig. 4 Curves of thermal resistance versus transient time of a semiconductor laser structure without voids in solder layer

Fig. 6 indicates that the heat generated by the quantum well in the device would be conducted to the bottom of the heat sink 10ms later, which means that the TEC starts to work 10ms later than the device.

Size of the inelastic strain zones in a low-yield-stress solder bond

We developed an analytical ("mathematical") predictive model for the evaluation of the interfacial thermal stresses in a bi-material laser package assembly with a low yield stress solder bond (Fig. 7). The detailed of the developed model could be found in Ref. 2.

The solder is assumed in our analysis to be linearly elastic at a strain level below the yield point and ideally plastic at higher strains. Both the shearing and the peeling stresses in the solder material are considered, and their distribution along the bonded region is shown in Fig.8 for the carried out numerical example. The size of the inelastic zones at the peripheral portions of the bond was obtained based on the derived transcendental governing equation. This equation

978-1-4244-4658-2/09 $25.00 © 2009 IEEE

considers the redistribution of the interfacial stresses because of the plasticity effect. The numerical example is carried out for a 2.4mm long GaAs bar (chip) soldered using a soft silver-tin (96.5%Ag3.5%Sn) solder onto a copper substrate (submount). The estimated yield stress in shear for the solder material is 15MPa=1.530kg/mm^2=2176psi. The soldering temperature for the silver-tin solder alloy is about 158 °C, so that the change in temperature can be assumed to be 125 °C.

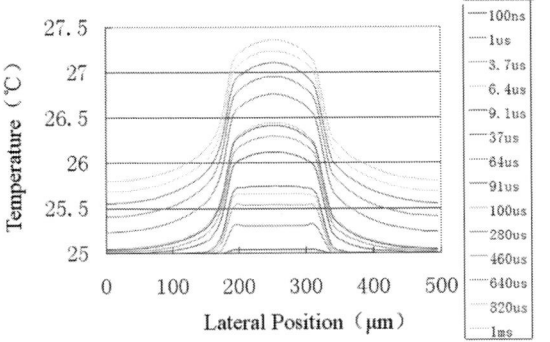

Fig. 5 Lateral temperature profiles of a single emitter (150μm in width)

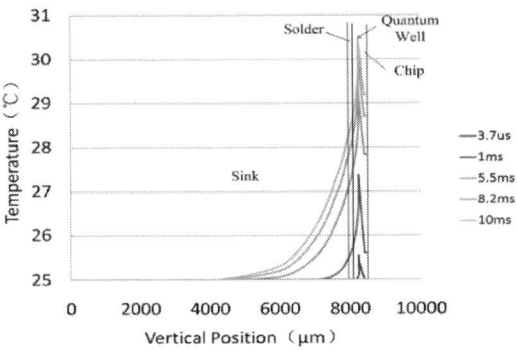

Fig. 6 Vertical temperature profiles of a single emitter during various heating time

We have shown that, in an approximate analysis, the shearing stress can be evaluated assuming that it is not coupled with the peeling stress. Then, after the shearing stress has been determined for the given structure and the given "external" thermal mismatch strain $\Delta\alpha\Delta t$ ($\Delta\alpha$ is the difference between the CTE's of the copper and the die, and Δt is the change in temperature from the soldering temperature to the temperature of interest), the peeling stress can be evaluated from the determined shearing stress. We have shown also that the interfacial peeling stress is proportional to the gradient of the shearing stress along the bond. Since we assumed in this analysis that the bonding material is ideally plastic above the yield point, the interfacial shearing stress becomes constant at this point, and the peeling stress becomes zero. It is noteworthy that a similar analysis for an elastic bond was carried out about 25 years ago (see Ref.9). Obviously, for solders that exhibit elasto-plastic behavior above the yield point, the stress distribution should be between the linearly elastic case addressed in Ref. 9 and linearly elastic – ideally plastic case addressed in this article and in Ref.2. The calculations show that the solutions for the idealized "linearly

elastic" and "linearly elastic – ideally plastic" situations form a rather narrow "bracket", so that obtaining a complicated solution for the general elasto-plastic case might not even be necessary for the assessment of the states of stress and strain in assemblies in question.

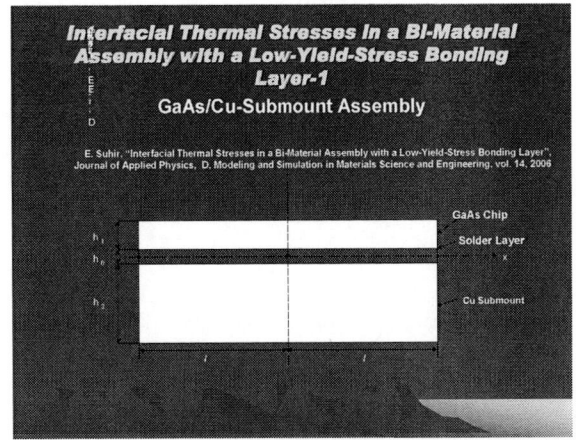

Fig.7. GaAs die bonded onto a cupper submount using soft and low-yield-stress silver-tin solder

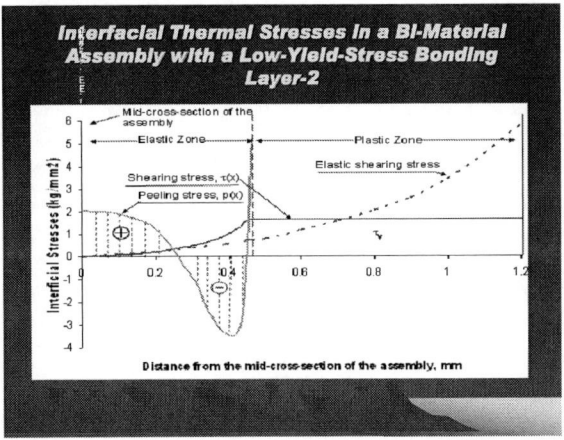

Fig.8. Predicted interfacial shearing and peeling stresses in the bonding layer

Conclusions

The following major conclusions could be drawn from the carries out analyses.

Simple and easy-to-use FEA and analytical predictive models have been developed for the evaluation of the thermal performance of high power semiconductor lasers and for the stress distribution in solder bonds in high power semiconductor laser array devices and packages. These models enable one to understand the physics of the addressed thermal phenomena and predict their thermal behavior and performance.

The simulation data indicated that the reliability of a conduction-cooled-packaged high power semiconductor laser array could be very high in an extremely short pulse ($\tau<280$ns) or in a short pulse (900ns$<\tau<37\mu$s) QCW mode, or in a CW mode. Pulse widths between 280ns and 900ns should be avoided; otherwise the device reliability could be compromised. If the pulse width exceeds 37μs, the working condition is the same as in the CW mode.

The methodology and simple analytical expressions developed in application to diode laser array packages can be used also well beyond this area of engineering for the analysis and design of bonded joints of the type in question, i.e. in a situation, when the peripheral zones of the bond exhibit inelastic deformations. The developed model can be used particularly for the development of design criteria (figures of merit) for assemblies bonded with solders.

Acknowledgments

This work is supported by the project of the "Open Research Fund of State Key Laboratory of Transient Optics and Photonics, Chinese Academy of Sciences".

References

1. X. Liu, R.W. Davis, L. C. Hughes, M. H. Rasmussen, and C.-E. Zah, "A Study on the Reliability of Indium Solder Die Bonding of High Power Semiconductor Lasers," *Journal of Applied Physics,* Vol. 100, Issue 1, 013104, 2006.
2. E. Suhir, "Interfacial Thermal Stresses in a Bi-Material Assembly with a Low-Yield-Stress Bonding Layer", *Modeling and Simulation in Materials Science and Engineering,* vol. 14, 2006, pp.1421-1432
3. E. Suhir, "Fiber Optics Structural Mechanics, and a New Generation of Nano-Technology Based Optical Fiber Cladding and Coating", *Invited talk at the Photonics West Conf., and SPIE Publication,* 2006
4. X. Liu, J. Wang, and P. Wei, "Study of the mechanisms of spectral broadening in high power semiconductor laser arrays", *Electronic Components and Technology Conference,* IEEE, 2008, pp1005-1010.
5. L. Anand, "Constitutive Equations for Hot-Working of Metals," *Int. J.Plast.,* 1, 2, 1985
6. S.B. Brown, K.H. Kim, and L. Anand, 1989, "An Internal Variable Constitutive Model for Hot Working of Metals," *Int. J. Plast.,* 5,2, 1989.
7. D. Schleuning et al, "Robust hard-solder packaging of conduction cooled laser diode bars," *the International Society for Optical Engineering, Proceedings of SPIE,* 2007: 645604(1)-645604(11)
8. S. Yang and W. Tao. "Heat Transfer", *3rd Edition. Higher Education Press,* 1998: 63-100. Beijing
9. E. Suhir, "Stresses in Bi-Metal Thermostats", *ASME Journal of Applied Mechanics,* vol. 53, No. 3, Sept. 1986.

Freeform Lens for Application-Specific LED Packaging

Fei Chen[1,2], Kai Wang[1,2], Zongyuan Liu[1,3], Xiaobing Luo[1,4], Sheng Liu[1,2,3*]

[1]Division of MOEMS, Wuhan National Laboratory for Optoelectronics, Wuhan 430074, China
[2]School of Optoelectronics Science & Engineering, Huazhong University of Science & Technology, Wuhan 430074, China
[3]Institute for Microsystems, School of Mechanical Science & Engineering
Huazhong University of Science & Technology, Wuhan 430074, China
[4]School of Energy & Power Engineering, Huazhong University of Science & Technology, Wuhan 430074, China
[*]Corresponding author: victor_liu63@126.com

Abstract

Traditional LED packaging always adopts hemisphere lens, although which can ensure high light output efficiency, its light beam is of circular symmetry and non-uniformity. Therefore, LED by traditional packaging cannot be applied in lighting directly and bulky secondary lenses have to be used, which will make the traditional lighting companies difficult to switch to LED lighting and the lamps are heavy. In this study, an effective freeform lens design method for extended light source was presented. With this method, we designed a freeform lens for street lighting, and integrated with this freeform lens, a novel application-specific LED packaging (ASLP), whose manufacturing process can be easily integrated into current LED packaging processes, was suggested. The results of both Monte Carlo ray tracing optical numerical simulation and the experiment show that the light beam is quite in agreement with the design target and the light output efficiency of this novel LED packaging is as high as traditional one. Moreover, comparing with traditional LED module with secondary optics, this novel ASLP module has advantages of higher system light output efficiency (~9%), compact size (~1/8) and convenience for end customers.

1. Introduction

As an emerging light source, high power white light emitting diodes (LEDs) have been increasingly used in various occasions due to their superior performance of low power consumption, high reliability, long lifetime and environmental protection [1-3]. Traditional LED packaging always adopts hemisphere lens, although which can ensure high light output efficiency, its light beam is of circular symmetry and non-uniformity, which is shown in Fig.1. Therefore, LED by traditional packaging cannot be applied in lighting directly, unless adding a secondary optical lens which will increase light loss, size as well as cost of the LED module to meet the requirements of different illumination applications (e.g. rectangular light patterns desirable in street lighting). However if we design a compact lens which can generate an application-specific light beam to replace the hemisphere lens in traditional LED packaging, the disadvantage mentioned above can all be avoided.

In this paper, we demonstrate a novel ASLP providing rectangular light pattern directly. Since the size of LED packaging lens can be compared with that of LED chip, the chip cannot be regarded as point light source approximately. An effective freeform lens design method for extended light source is suggested. Due to street lamps' significance in our LED community signaling the open arena of general lighting, with the method mentioned above, we design a freeform lens

for street lighting, and integrated with this freeform lens, a novel ASLP, whose manufacturing process can be easily integrated into current LED packaging processes, was suggested. The lens of polycarbonate (PC) material with refract index of 1.586 is designed to form a 32 meters long and 12 meters wide rectangular uniform illumination area at the height of 8 meters.

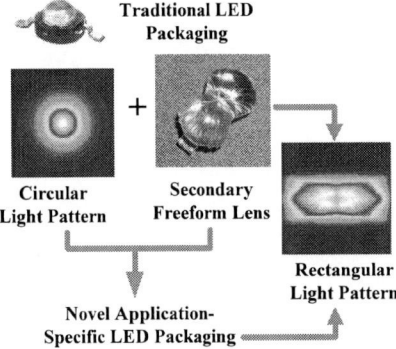

Fig. 1. Schematic of the concept of application-specific LED packaging.

2. Design method for the compact freeform lens

The design method approximately includes three main parts: establishing light energy mapping relationship [4] between the light source and the target, constructing lens and validating lens design by ray trace simulation.

2.1 Establishment of light energy mapping relationship

First of all, the one quarter of light source is divided into M×N grids with equal luminous flux. As shown in Fig. 2, the light source's intensity space distribution Ω is specified by coordinates (u, v), where u is the angle between light ray and X axis, and v is the angle between Z axis and the plane containing light ray and X axis. The volume of Ω represents the total luminous flux Φ_{total} of this one quarter light source and the luminous flux of unit object $d\Omega$ could be expressed as follows:

$$\phi(u,v) = \int_{v_j}^{v_{j+1}} \int_{u_i}^{u_{i+1}} I(u,v)\sin u \, du \, dv \qquad (1)$$

where I(u, v) is the light intensity distribution of light source. We divide the one quarter light source into M parts along u direction and N parts along v direction equally, and then each division point S(ui,vj) on the surface of Ω could be obtained through solving Eq. (2) and Eq. (3) as follows:

$$\int_0^{\pi/2}\int_0^{u_i} I(u,v)\sin u\,du\,dv = \frac{i\phi_{total}}{M}\quad (i=0,1,\ldots M)\tag{2}$$

$$\int_0^{v_j}\int_0^{\pi/2} I(u,v)\sin u\,du\,dv = \frac{j\phi_{total}}{N}\quad (j=0,1,\ldots N)\tag{3}$$

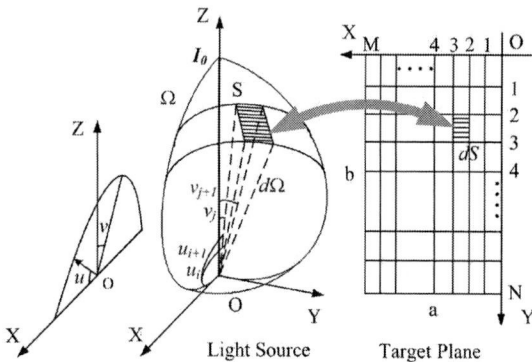

Fig. 2. Schematic of light energy mapping between the light source and target.

Secondly, to establish the mapping relationship with the light source, the one quarter rectangular target plane is also divided into M×N grids. The (x, y, z) Cartesian coordinates specify the points on the target plane. As shown in Fig. 2, the plane has been divided into M×N parts by rectangular grids dS. During LED packaging lens design, the size of LED chip and lens are comparable, thus the LED chip could not be regarded as a point light source assumption any more. Lights irradiated from edge area of the chip will deteriorate the illumination performance significantly. Rectangular grids with unequal area are adopted to overcome this problem. The width (W_{grid}) and length (L_{grid}) of each grid in the central area of target plane can be expressed as follows:

$$W_{grid,i} = \frac{C_{Wi}a}{M}\quad (i=0,1,\ldots M)\tag{4}$$

$$L_{grid,j} = \frac{C_{Lj}b}{N}\quad (j=0,1,\ldots N)\tag{5}$$

where C_{Wi} and C_{Lj} are optimization coefficients of grids which can optimize the light energy distributing on the target. We adjust these two coefficients to make the trend of illuminance distribution on the target plane is inverse of the trend of illumination performance deterioration caused by extended LED source, and then it will make up the deviation of illuminance distribution caused by the lights irradiate from the edge area of LED chip and obtain much better illumination performance.

According to edge ray principle [5], when the four end points of a grid on the light source are mapped correspondingly to the four end points of a target grid, the energy of the source grid is also transmitted to the target grid. Therefore if we desire to map the light energy in $d\Omega$ into the target grid dS, we should ensure that four rays, which construct the $d\Omega$ as boundary, irradiate at the four corresponding end points of the target grid dS after refracted by the freeform lens. Thus the light energy mapping

relationship between the light source and target plane have been established.

2.2. Construction of Freeform Lens

In this section we will work out the lens which can realize the mapping between the light source and the target plane. Since the compact freeform LED lens will package the LED chip, phosphor and gold wire, we focus on the construction of outside surface of the lens in this study. There are three main steps:

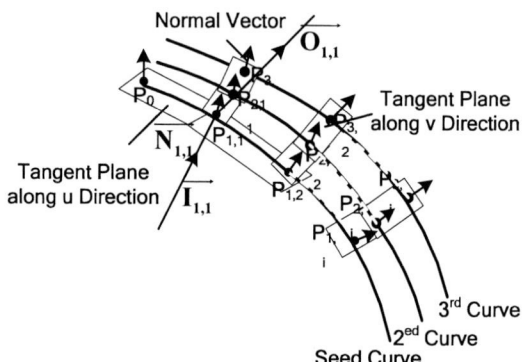

Fig. 3. Schematic of points generation on the outside surface of the lens.

Step 1. Construction of the seed curve. The seed curve is the first curve to generate other lens curves. As shown in Fig. 3, we fix a point P_0 as the vertex of the seed curve. The second point $P_{1,1}$ on the seed curve is calculated by the intersection of incident ray $I1,1$ and the tangent plane of the previous point. Then we can obtain the normal vector $N1,1$ of the second point according to the Snell's law expressed as follows:

$$\left[1+n^2-2n(\mathbf{O}\bullet\mathbf{I})\right]^{1/2}\mathbf{N}=\mathbf{O}-n\mathbf{I}\tag{6}$$

where I and O are the unit vectors of incident and refracted rays, N is the unit normal vector on the refracted point and n is the refraction index of the lens. Based on this algorithm we can obtain all other points and their normal vectors on the seed curve.

Step 2. Generation of other curves. First of all we calculate the second curve. As shown in Fig. 3, different from the seed curve algorithm, point i on the second longitude curve is calculated by the intersection of incident ray and the tangent plane of point i on the previous curve. Then the following curves, such as 3rd curve, 4th curve, et al., are easy to obtain based on this algorithm.

Step 3. Construction of surface. The lofting method is utilized to construct smooth surface between these curves.

2.3. Validation of Lens Design

Since it is costly to manufacture a real freeform lens, numerical simulation based on Mentor Carlo ray trace method is an efficient way to validate the lens design. According to the simulation results, optimize the coefficients of C_{Wi} and C_{Lj} until the illumination performance of LED packaging lens meets the requirements.

3. Novel ASLP Design

978-1-4244-4658-2/09 $25.00 © 2009 IEEE

In this study, to meet the requirements of street lighting, according to the design method mentioned above, compact PC freeform lens with refract indexs of 1.586 will be designed as examples for LED packagings to form a 32 meters long and 12 meters wide rectangular illumination area at the height of 8 meters (as shown in Fig. 4).

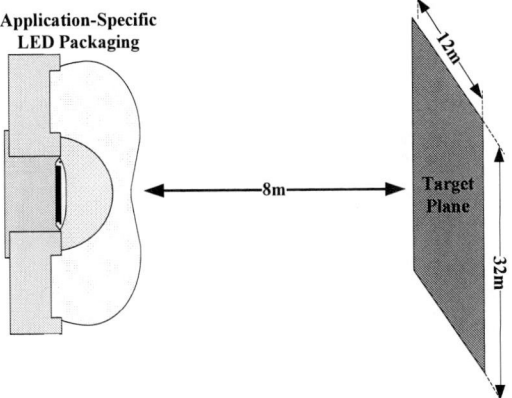

Fig. 4. Schematic of the design target.

3.1. Optical Modeling

Light intensity distribution curves (LIDCs) of LED chip and LED chip covered by phosphor layer are two key issues for accurate LED packaging optical modeling because most white LEDs are obtained either by integrating blue LED chips with yellow phosphor or by integrating RGB LED chips in one packaging. Vertical electrode LED chip with the size of 1.0 mm×1.0 mm×0.1 mm was adopted for lens design and the phosphor silicone layer with the thickness of about 70 μm was coated on the chip conformally. Figure 5 depicts that the measured LIDCs of the chip and the chip coated by phosphor are quite similar with the standard Lambert distribution. Therefore Lambert light source with the area of 1 mm×1 mm could be used as equivalent light source during optical design and simulation.

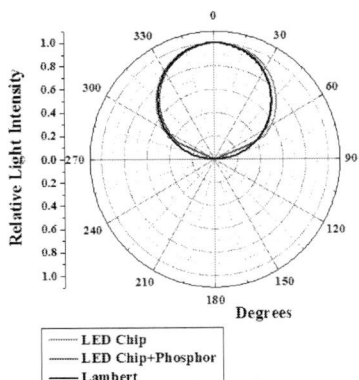

Fig. 5. Light intensity distribution curves of LED chip and LED chip coated by phosphor layer.

3.2. Design of A Novel Compact Freeform LED Packaigng Lens

A PC freeform LED packaging lens for street lighting was designed according to the method mentioned above. First of all, we used a surface light source with LIDC type of Lambert as the light source. Secondly, we divided the one quarter light source and target plane both into 100 (M)×100 (N)=10000 grids with the equal luminous flux and unequal area respectively and established light energy mapping relationship between these grids. Finally, we calculated the coordinates and normal vector of each point on the freeform surface and constructed the freeform lens utilizing these points. Figure 6 shows the designed lens. The volume and largest value of length, width and height of this lens are 45 mm³, 7.3 mm, 6.3 mm and 3.5 mm respectively, which are suitable for LED packaging with small volume.

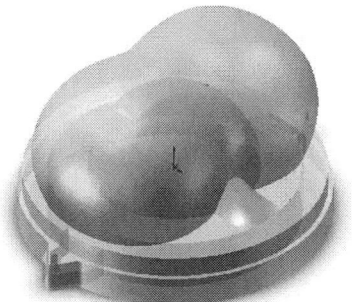

Fig. 6. A novel compact freeform PC lens for LED packaging.

3.3. A Novel Application-Specific LED Packaging

In this section we will integrate the compact freeform silicone lens with a traditional LED packaging to achieve a novel ASLP for street lighting. A novel ASLP design for street lighting was achieved by integrating this PC lens. As shown in Fig. 7, the optical structure of the ASLP is constructed by a 1.0 mm×1.0 mm×0.1 mm LED chip, phosphor, silicone and a freeform PC LED packaging lens (LPL), which is quite similar with traditional LED packaging except replacing the hemisphere LPL by the designed freeform PC LPL. This indicates that the manufacturing method of this ASLP is compatible with current LED packaging processes totally, which makes it easier for LED manufacturers to adopt this new technology with little change of existing processes.

Fig. 7. (a) The novel ASLP, (b) its detail optical structure.

978-1-4244-4658-2/09 $25.00 © 2009 IEEE   369

4. The Results of Monte Carlo Ray Tracing Optical Numerical Simulation and Experiments

We validate our design with both Monte Carlo ray tracing optical numerical simulation and the experiment method, the results show that the light beam is quite in agreement with the design target and the light output efficiency of this novel LED packaging is as high as traditional one.

4.1 Results of Monte Carlo Ray Tracing Optical Numerical Simulation

We simulated the optical performance of the ASLP and a LED module of traditional packaging numerically by the widely used Monte Carlo ray-tracing method and then the whole LED packaging was simulated by one million rays. The light output efficiency (LOE) of this novel LPL is 94.2% (considering Fresnel loss), which is slightly less than that of traditional hemisphere LPL of 95.0%. As shown in Fig. 8, more than 95% light energy of the ASLP uniformly distribute within an approximately rectangular light pattern with the length of 33m and width of 14m at the height of 8m, which is in agreement with the expected shape. Therefore the ASLP could be directly used for street lighting and no secondary optics are needed, which makes it convenient for LED fixtures designers and manufacturers to use and also will reduce the cost of LED fixtures further more.

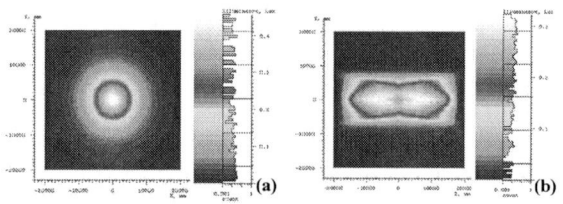

Fig.8 (a) The simulated light beam of traditional packaged LED, (b) The simulated light beam of ASLP.

4.2 Results of Experiments

As shown in Fig. 9, the freeform PC LPL was manufactured by injection molding method. The LOEs of traditional hemisphere PC LPL and the freeform PC LPL were measured by UV-VIS-near IR spectrum photo colorimeter measurement and integrating sphere. Results demonstrate that the average LOE of the freeform PC LPL reaches as high as 94.8%, which is slightly lower than that of traditional hemisphere PC LPL of 95.4%, and that is quite close to the simulation results. Moreover, considering the LOEs of secondary optics (e.g. freeform lenses) are always at a level of about 90% [6], the system LOE of LED fixture consists of ASLPs will be about 9% higher than that of traditional LED fixture.

Figure 10 depicts a white light ASLP for street lighting. An LED module for street lighting consists of an LED and a secondary optical element with the kind of freeform lens is shown in Fig. 11(c). From comparisons shown in Fig. 11, we can find that the height and volume of this novel ASLP are only about 1/2 and 1/8 of that of the LED module respectively, which provides an effective way for some size

compact LED illumination systems design and more design freedoms for new concept LED lighting fixtures.

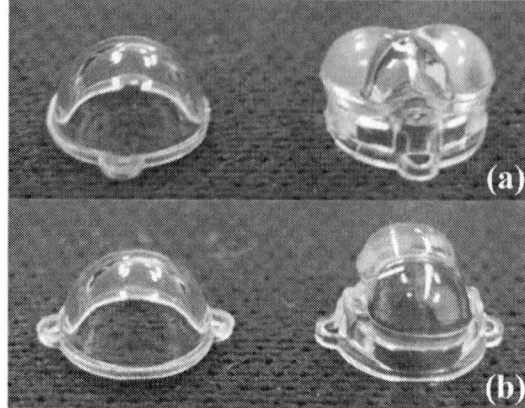

Fig. 9. (a) Front view of traditional hemisphere LPL (left) and the novel application specific freeform LPL (right) and (b) left view of these two LPLs.

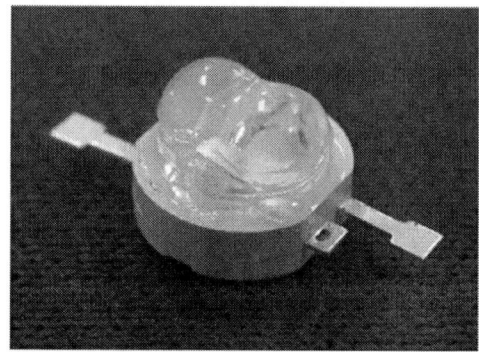

Fig. 10. White light ASLP.

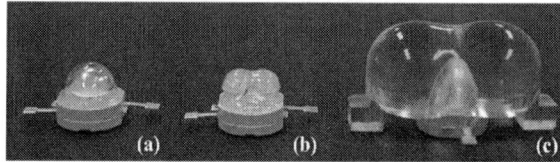

Fig. 11. (a) Traditional LED packaging, (b) ASLP and (c) traditional LED street lighting module consists of a traditional LED and a freeform secondary lens.

Figure 12(a) and Fig. 12(b) show the illumination performances of a traditional LED packaging and an ASLP respectively. The light pattern of the traditional LED packaging is circular with non-uniform illuminance distribution, while the ASLP redistributes LED's light energy distribution and forms a rectangular light pattern on the target plane, which is more uniform than the circular light pattern. The target plane is 58.5cm away from LED. We can find that most light energy of ASLP is distributed in a nearly rectangular area with length of 230cm and width of 83cm, and it will enlarge to 31.5m long and 11.4m wide at the height of 8m according to light rectilinear propagation principle, which is also quite in agreement with the expected performance.

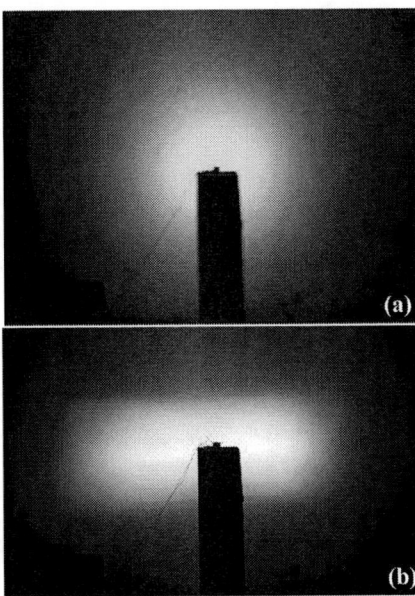

Fig. 12. Illumination performance of (a) traditional LED packaging and (b) ASLP.

5. Conclusions

In summary, an effective freeform lens design method for extended light source was suggested briefly and a compact freeform PC LPL was designed based on this method. Integrated with this new PC LPL, a novel ASLP was presented for street lighting. By comparing with the traditional LED illumination module, the novel ASLP has the advantages of low profile, small volume, high system light output efficiency, low cost and convenience for customers to use. Moreover, the ASLPs can also be designed to meet other LED lighting applications, such as backlighting for LCD display, automotive lighting, etc.. Therefore ASLPs will provide a more cost-effective solution to high performance LED lighting and probably become the trend of LED packaging.

Acknowledgments

This work was supported by Nature Science Foundation of China (NSFC) Key Project under grant number 50835005, NSFC Project under grant number 50876038, High Tech Project of Ministry of Science and Technology under grant number 2008AA03A184 and GuangDong Real Faith Optoelectronics Inc.

References

1. D. L. Evans, "High-luminance LEDs replace incandescent lamps in new applications," *Proc. SPIE 3002*, 142-153 (1997).
2. E. F. Schubert, *Light-Emitting Diodes* (Cambridge, 2006).
3. M. R. Krames, O. B. Shchekin, R. M. Mach, G. O. Mueller, L. Zhou, G. Harbers and M. G. Craford, "Status and future of high-power light-emitting diodes for solid-state lighting," *J. Display Technol.* 3, 160-175 (2007).
4. Parkyn, W. A, "Illumination lenses designed by extrinsic differential geometry", *Proceedings of SPIE*, Vol. 3428, 1998, pp. 389-396.

5. Harald, R., and Ari, R., 1994, "Edge-ray Principle of Nonimaging Optics," *J. Opt. Soc. Am. A*, 11(4), pp. 2627–2632.
6. K. Wang, S. Liu, F. Chen, Z. Y. Liu and X. B. Luo, "Effect of Manufacturing Defects on Optical Performance of Discontinuous Freeform Lenses," *Opt. Express 17*, 5457-5465 (2009).

High Bright White LED Lens Formation by Vacuum Printing Encapsulation Systems (VPES) and Its Packaging Resin

Atsushi Okuno (IEEE Senior Member), Osamu Tanaka
SANYU REC CO., LTD
3–5–1, 3–Chome, Doucho, Takatsuki-City, Osaka, JAPAN
TEL: 81–72–669–5231 FAX: 81–72–669–5230

ABSTRACT

White color LED has been investigated for lighting application with a big expectation of growth. And in terms of packaging of them in general, clear transparent epoxy resin has been mainly used in spite of problems such as coloring of the resin due to high temperature and UV light emitted from Chip itself. In order to solve this problem, we have investigated new materials such as silicon resin and specially modified epoxy resin etc. And then we would suggest that materials should be chosen depending upon LED specifications, packaging materials and also reliability test conditions etc. In addition we would like to suggest high density packaging technology using VPES, which meets very fine pitch, high density packaging designs, and meets cost effective mass production.

INTRODUCTION

In general, White LED composed of the Blue LED Chip as a source of light and an inorganic fluorescent substance which is so called YAG (Yttrium Aluminum Garnet) in a resin moiety. As this system generate yellow light, white color light is made as a blend of Blue and Yellow lights. Although there are other methods to make white color light, this is the present major system in the market [1-2].

Therefore development of Blue LED Chips of high brightness is dramatically progressed recently. Some examples of its applications are "Backlight sources of color liquid crystal display" such as mobile devices and as "Interior decorating illumination of car instrument panels".

Furthermore from ecological stand point LED has a big advantage because it does not use any hazardous materials like mercury in fluorescent lamps. So that LED is expected to have a big role of global environment protection. Therefore, development of brighter LED chips has been taken place eagerly for the sake of catching up with illumination of fluorescent lamps

And unfortunately the epoxy resin that encapsulates this Blue LED element changes its color by UV light and/or heat emitted by an LED element. Then it is said that this color change deteriorates the brightness of LED significantly. In order to solve this problem, several materials which improved UV and heat resistance have been proposed in the market, and we investigated them and developed some transparent resins which meet our VPES (Vacuum printing Encapsulation Systems) packaging process.

EXPERIMENTS

Development of Encapsulation Resin

With respect to general raw materials, there are transparent resins such as epoxy resin, silicone resin, and urea resin. Epoxy resin has been a main raw material from the consideration of cost, electric properties, and reliability, etc.

However, usual epoxy resin has limitation on heat and UV light tolerance because of its change, and then leads to brightness deterioration of LED lumps.

Instead of epoxy resin raw materials, it is necessary to select material that improves property of heat and UV resistance, such as silicone resin. However, because the adhesion property of silicone resin is weaker than that of epoxy resin, and its moisture permeability is high than that, it is not always the ideal material either.

Then, we studied both of these two materials (epoxy resin and silicone resin) and also a newly developed modified material.

Brightness deterioration

To minimize the brightness deterioration of Blue LED, we encapsulated Blue LED chips mounted on substrates with some resin samples. And we evaluated these.

Blue LED chips prepared were both Blue chip LED of 20mA type and the high-power chip of 350mA type. As to the mounted substrates, 20mA type chip was mounted on an organic substrate, and 350mA type was on a ceramic substrate with a heat sink attached respectively.

The work progress is as follows.

1. At first the silver paste (GA-6244M1: made by Sanyu Rec.Co., LTD) was applied on the each substrate. Then, Blue LED chip (20mA Blue LED chip: C460-MB290, 350mA Blue LED chip: C460-XB900: made by CREE.) was put on that. Afterwards, it was cured in 180 °C/1 hour.

2. They were wire bonded.

3. The dam was formed to enclose mounted LED chip by thixotropic resin, and then sample resin was poured inside of dam.

4. They cured at each curing conditions.

5. The brightness of Blue LED chip was measured with the spectrum brightness meter (CS-1100: made by Minolta).The measurement number was assumed to be 5 pieces. The electric current was a fixed current of 20mA or 350mA setting.

6. The sample had been kept in a high temperature and humid oven of 85°C/85%. The constant current of 20mA or 350mA was charged to the blue LED chips.

7. The brightness value after certain time was measured, and the relative brightness values based on the initial brightness were plotted.

Transparency change

The each sample sheet of 1mm thickness was made and measured the transparency at the initial stage. And they were kept in a 160°C oven for heat aging. After periodic certain time, transparency was measured and plotted on a same sheet. The coloring of the materials and transparency change was detected.

Sample preparation for VPES process

Samples were prepared by blending powder for thixotropic

material in resins by our conventional procedure. And we adjusted all samples to have controlled range of property such as viscosity and thixotropy, which could hold its droplet shapes.

RESULTS & DISCUSSION

The brightness test result of 20mA was shown in Fig.1, and that of 350mA was in Fig.2. In the case of 20mA, after 2000hr, Silicone resin and our specially modified resin had little decrease in brightness and were keeping about 99% of an initial brightness. Then relative brightness of both resin were keeping about 90% even after 4000hr. Therefore, both samples are anticipated to be good materials for this LED packaging. In addition to them, our developed epoxy resin was keeping 80% level after 2000hrs, also. It turns out in this examination that for this Blue LED chip of 20mA level this newly developed epoxy resin is worthwhile to use for the non sophisticated level devises. However, our newly developed epoxy resin showed big coloration and extreme brightness deterioration at Blue LED chip of 350mA level as shown in data.

When we looked at them carefully in a microscope, the color change of epoxy resin was observed significantly along the edge of LED Chip in brown color. But in silicon resin case, it was far from brightness deteriorating with time.

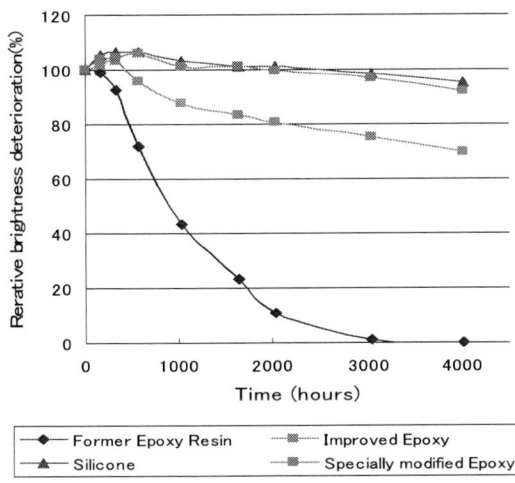

Fig. 1 Brightness deterioration of 20mA

On the other hand, the brightness deterioration happened on our specially modified resin. In the current of 350mA case, it showed 75% level after 2000hr. It seems that the deterioration by heat occurred to quite large extent at the current of 350mA. This was also confirmed when you looked at the heat edging examination result. The loss of transparency was so large in the case of both epoxy resin and specially modified resin at short wavelength area (From 350 nm to about 500nm). Of course epoxy resin showed the worst deterioration. Therefore we stopped the test at 500hrs. The detailed transparency data were shown in the following Fig. 3.

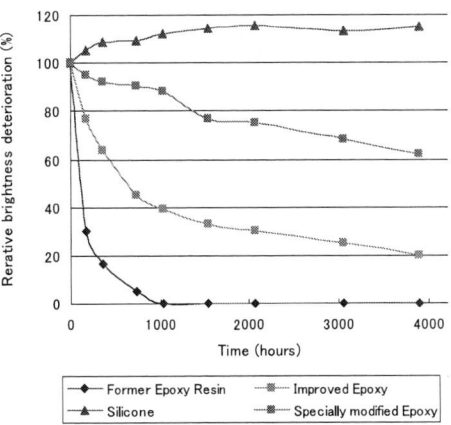

Fig. 2 Brightness deterioration of 350mA

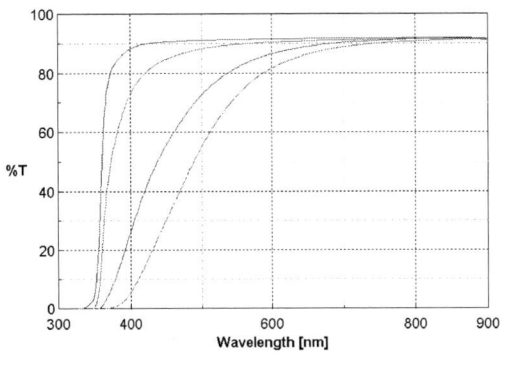

Epoxy resin

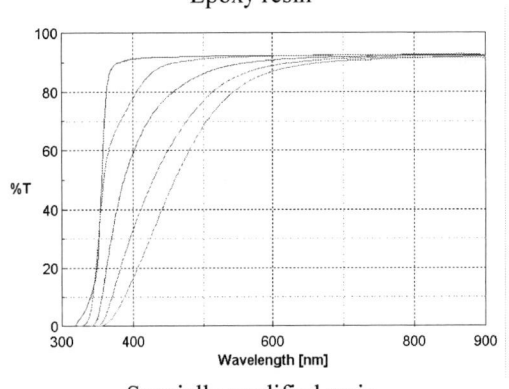

Specially modified resin

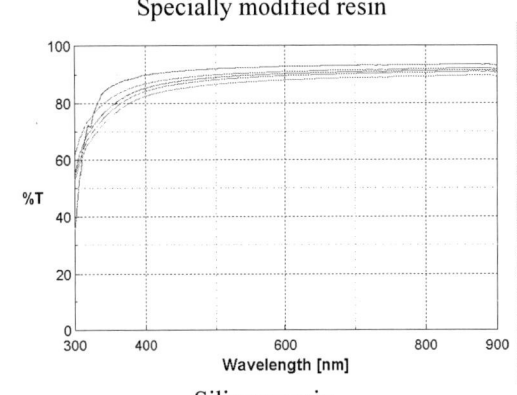

Silicone resin

———— : Initial ———┼———After 100hr ———┼———After 300hr
_____ : After 500hr _____ : After 1000hr

Fig. 3 The deterioration of transparency of each sample

From the above data, for high-power LED of 350mA class, we must select Silicone resin at present.

ENCAPSULATION METHOD

(VPES: Vacuum Printing Encapsulation Systems)

In general, in order to produce LED lens, transfer molding, casting molding and dispenser molding are common way of doing. However VPES (Vacuum Printing Encapsulation Systems) process should be another way of doing.

This is the method that builds lens height uniformly with a simple manner. Transfer molding method and Casting molding method must prepare the mold for lens molding. In other words, it takes an extra cost [3].

In case of dispenser molding, it may cause quite big unevenness of lens height [4]. The LED lens made by printing method keeps original shapes, and the field of vision square is expected to be very wide. The direction characteristic was shown in Fig.4. By controlling the thixotropy of the materials, we could develop suitable material samples of silicone and specially modified epoxy resin.

The transparency of the new encapsulating silicone sample decreased at the short wavelength area, because there is some gap of refractive index between the new silicone sample and powder thixotropic material (Fig. 5). But actually, it is not so dark as it was first thought, because of diffusion or reflect effect of light. Especially when fluorescent material like YAG is going to be blended, simple transparency is not the major item to consider.

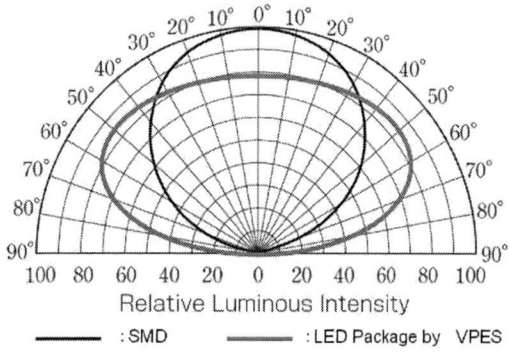

Fig. 4 Comparison of the direction characteristic

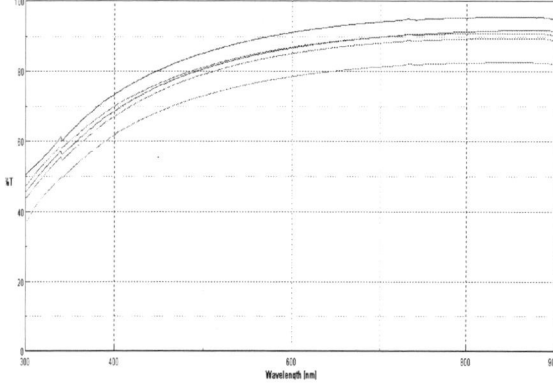

Fig. 5 The transparency of a new encapsulating sample

When we compare the package by VPES with that of SMD regarding the light extension to side direction, the LED package by the VPES method shows wider expansion of light than SMD package, even though strength of light of front side may get weaker a little bit (Fig. 4). Then in the case of VPES package the dot pitch can be designed wider than SMD package. In another word the mounting points can be reduced (Fig. 6).

LED Package by VPES

Fig. 6 Resonance of Light

In addition, VPES would enable to eliminate big air bubbles in packages, which may cause several potential problems, such as lens destruction, wire disconnection and unevenness of brightness.

CONCLUSION

We could develop new specially modified epoxy resin and Silicone resin materials which were suitable for Blue LED package using VPES process. They have good durability against heat and UV light and they are expected to show minimum rage of color change during the operation. We believe that the combination of VPES and our newly developed silicone resin offers quite large production capability easily for high brightness white LED in compact and high density dot matrix packages.

REFERENCE

1. K. Bando, *et al.*, "Development of high-bright and pure-white LED lamps", *J.Light & Vis.Environ.* Vol. 22, No. 1 (1998) pp. 2-5.
2. Japan patent (P3366586) S. Nakamura, other.
3. A.Okuno, "High reliability, High Density, Low Cost Packaging Systems for Matrix BGA &CSP Using VPES (Vacuum Printing Encapsulation Systems), *48th Electronic Components & Technology Conference*, pp. 109 (May, 1998).
4. Y. Miyawaki, N. Ooyama, A. Okuno, "Unique White LED Packaging Systems", *53rd Electronic Components & Technology Conference 2003*, pp. 1599-1601.

250W QCW Conduction Cooled High Power Semiconductor Laser

Jingwei Wang[1], Zhenbang Yuan[2], Yanxin Zhang[1], Entao Zhang[1], Di Wu[2], Xingsheng Liu[1,2]

[1]State Key Laboratory of Transient Optics and Photonics, Xi'an Institute of Optics and Precision Mechanics
Chinese Academy of Sciences
No. 17 Xinxi Road, New Industrial Park, Xi'an Hi-Tech Industrial Development Zone, Xi'an, Shaanxi, 710119, P.R. China ,
tel. 8629-88880786, fax.8629-88887075, wjw@opt.ac.cn
[2]Xi'an Focuslight Technologies Co., LTD
No. 60 Xibu Road, New Industrial Park, Xi'an Hi-Tech Industrial Development Zone, Xi'an, Shaanxi, 710119, P.R. China

Abstract

High power diode laser arrays (HPDLAs) have increased applications in pumping of solid state laser systems for industrial, military and medical applications as well as direct material processing applications. For quasi-continuous wave (QCW) conduction-cooled-packaged high power semiconductor lasers, the operational mode is commonly at lower duty cycle (DC), such as less 2% DC. However, operating high power laser diode arrays in long pulse regime of about 200μs, and high duty cycle (such as 8% DC), which greatly limits their useful lifetime, are demanded for some special application. Finite element numerical analysis based simulations to analyze the transient thermal behavior of a conduction-cooled-packaged semiconductor laser operated in QCW mode in this paper. This work describes the performance of laser diode arrays operating in long pulse and high duty cycle mode, including the characteristics of Power-Current-Voltage (LIV), thermal, near-field, and lifetime. This paper will then offer a viable approach for determining the optimum design and operational parameters leading to the maximum attainable lifetime. Based on the numerical simulation and analysis, a series of high power semiconductor lasers with good performances were produced.

Introduction

High power diode lasers (HPDLs) offer a variety of applications due to their higher electrical-optical conversion efficiencies, compact sizes and long life-times than the most prominent types of lasers by nearly an order of magnitude. High power semiconductor lasers, whether operated at continuous wave (CW) or QCW mode, including single emitters, arrays, stacks, and two dimension area array stack have found increased applications in pumping of solid state laser systems for industrial, science and technology research, military, antiterrorism, entertainment display and medical applications as well as direct material processing applications such as welding, cutting, and surface treatment [1,2,3].The optical-to-optical conversion efficiency of diode-pumped-solid-state-laser (DPSSL) is much higher than that of lamp-pumped-solid-state-laser, as the spectral width of the diode laser is very narrow. With continuing improvement of the power, electrical-optical conversion efficiency, reliability, and manufacturability of high power semiconductor lasers, and decreasing manufacturing cost, many new applications of high power semiconductor lasers are being enabled [4].The three key performance measures of high power semiconductor lasers are power, efficiency, and reliability. For QCW conduction-cooled-packaged high power semiconductor lasers, high reliability and long lifetime are required in practical application.

To obtain higher reliability and longer lifetime, packaging plays more important role in preparing these devices. For 808nm conduction-cooled-packaged high power semiconductor lasers, the semiconductor laser die attach solder interface can significantly impact the thermal performance and reliability of a laser assembly. Fluxless soldering processes are required and solder voids should be controlled to the minimum. Large solder voids would increase junction temperature and create hot spots. Solder voids near the front facet are especially detrimental as they can significantly raise the facet temperature and cause catastrophic optical mirror damage (COMD).

Consequently, it is significant to do some researches on the 808nm diode laser which operated at QCW mode, especially in long pulse and high duty cycle mode.

Thermal Performance and Analysis

Transient thermal behavior is a crucial characteristic of the total thermal performance of a semiconductor laser. It reflects the physics of the heat propagation process inside the device. Particularly, through studying the transient thermal behavior of a diode laser array and its package, one can understand better the contribution of each constituent material layer and the interfaces to total thermal resistance, as well as the possible failure modes and mechanisms in a semiconductor laser operated in a quasi-continuous (QCW) condition.

For HPDLA devices, although the thermal is a common problem in high power diode laser array products, the thermal behavior of 250W QCW conduction-cooled-packaged HPLDAs are rarely reported. Thermal behaviors of solder interfaces are the major factors affecting the functional and structural performance of high power diode laser packages. If the accumulated heat cannot readily escape, the elevated temperatures at the location of the p-n junction adversely affect the output power, the slope efficiency, the threshold current and the device lifetime. Elevated thermal could cause spectral broadening and wavelength shift [5]. The emitting wavelengths vary, when the junction temperature of the emitters across the array is not controlled and/or is non-uniform. These circumstances make thermal management of high power devices a major challenge in the pump laser design, manufacturing and use. It is more critical to investigate and develop thermal design and optimization.

The device of single-bar conduction-cooled-packaged 808nm 250W QCW semiconductor laser array to be simulated made by the Xi'an Focuslight Technologies Co., Ltd., is shown in Fig.1. A typical semiconductor laser consists of three major parts: the laser bar (chip/die), the cathode and the heat sink. The laser bar is schematically shown in Fig.2. The quantum well is the active region. It contains 62 emitters and is comprised of a cladding layer and a p-metal.

Figure 1 A sample of single-bar 808nm 250W QCW high power semiconductor laser array

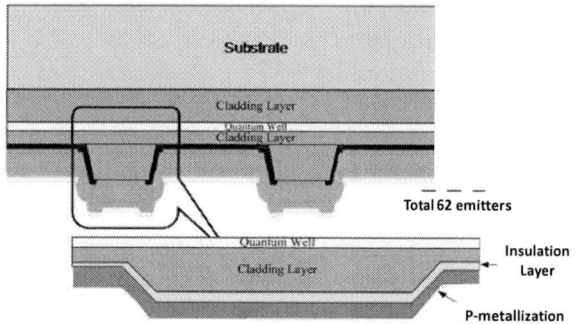

Figure 2 Schematic of semiconductor laser array

The 250W QCW high power semiconductor laser array is packaged in the conduction cooled structure, with the bottom of the heat sink fixed with a TEC (Thermal-Electric Cooler). The bottom TEC surface is kept at 25°C.

In our study the transient-state thermal behavior of a conduction-cooled-packaged semiconductor laser in QCW mode was simulated using FEA technique. Figure 3 shows the thermal behavior of 250W high power diode laser at pulse mode (400Hz, 200μs). The result shows the highest temperature in emitting region is 38.6°C when a single pulse 250A current is applied. The temperature in emitting region dropped down about 28°C within about 100μs if the drive current of 250Amp applied on 250W QCW conduction-cooled-packaged semiconductor laser is cut off.

Figure 4 (a) shows the curves of quantum well temperature versus transient time of a 250W QCW high power semiconductor laser array. During the initial 300μs, the zooming in curve of quantum well temperature versus transient time is illustrated in Figure 4 (b). From this curve, as we can see, the junction temperature rise is rapidly increased during the initial 300μs, illustrated in Figure 4 (b). The junction temperature rises from initial 25°C to 27.8°C. With the increment of operating pulse, from Figure 4 (a), the junction temperature rise is slowing down and trends flatness. The trend of temperature distribution curve via time in emitting region is like parabolic.

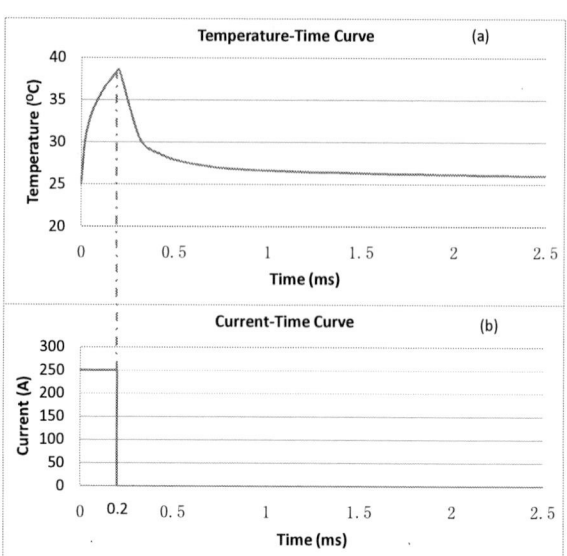

Figure 3 shows the thermal behavior of 250W QCW conduction cooled high power diode laser (a) the temperature versus time at pulse mode (400Hz, 200μs), (b) the operational current under the condition of a single pulse

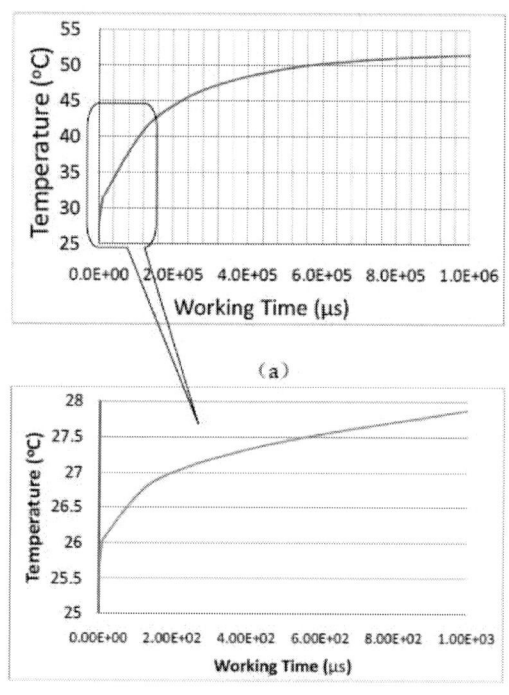

Figure 4 Curves of quantum well temperature versus transient time of a 250W QCW high power semiconductor laser.

Zhenbang Yuan's research paper indicated that if the pulse width is longer than 37μs, the working condition is the same as in CW mode. [6] It can be seen from Figure 4 that the increment of junction temperature rise is become smaller after a single-bar 250W QCW conduction-cooled-packaged high power semiconductor laser array operates 2500 pulse. That is to say, when the pulse width is longer than 37μs, the temperature distribution of a QCW high power semiconductor laser array comes to the steady state after it works for a couple of seconds.

978-1-4244-4658-2/09 $25.00 © 2009 IEEE   376

Consequently, for a single-bar 250W QCW conduction-cooled-packaged high power semiconductor laser array operating in high pulse width, the failure analysis could be considered from the aspect of continuous wavelength mode.

Based on the above analysis, one of the major challenges in single bar packaging is thermal management. Thermal management includes thermal design and process control to achieve voids "free" in bar bonding interface. Xingsheng Liu's work showed that voids in bar bonding interface can affect the performance of a laser bar including power and reliability significantly. Two process approaches are used to reduce solder voids in the bonding interface. One is to bond a laser bar using die bonders with controlled pressure, and protected environment under a designed temperature profile. The other approach is to use a vacuum solder reflow system. [7]

LIV testing

Based on the numerical simulation and analytical results, a 250W QCW conduction cooled high power semiconductor lasers using hard solder (Gold-Tin) bar bonding process were produced. Moreover, the performance parameters, including the optical and electrical parameters as well as thermal resistance, are tested and characterized by means of our home-made testing instruments.

A single-bar 250W QCW conduction-cooled-packaged high power semiconductor laser array is not collimated. The bottom of copper block heat sink is fixed on a cool end plane of TEC (Thermal-Electric Cooler). The TEC is set up to maintain at 25°C.

The LIV curves were obtained through the integration of integrating spheres, spectrometer and power meter. Measurement instruments are definitely calibrated before all of the performance parameters, including the optical and electrical parameters, are measured.

Figure 5 shows a practical example of the LIV testing of a single-bar 250W QCW conduction-cooled-packaged high power semiconductor laser array operating in high duty cycle (8%, 400Hz, 200μs).

Figure 5 Example of the LIV testing of a single-bar 250W QCW conduction-cooled-packaged high power semiconductor laser array operating in high duty cycle (8%, 400Hz, 200μs)

The results of a 250W QCW conduction cooled high power semiconductor lasers array with good performances were illustrated in the Figure 6. It can be seen that the higher power of 287W and slope efficiency of 1.3W/A at 250amps under the condition of pulse mode of 8% Duty Cycle were obtained. The full width of half maximum (FWHM) and full width of 90% energy (FW90%E) of spectrum is 3.2nm and 6.1nm, respectively. The measurement results indicate that the performance of 808nm 250W QCW conduction cooled high power semiconductor lasers array is pretty good.

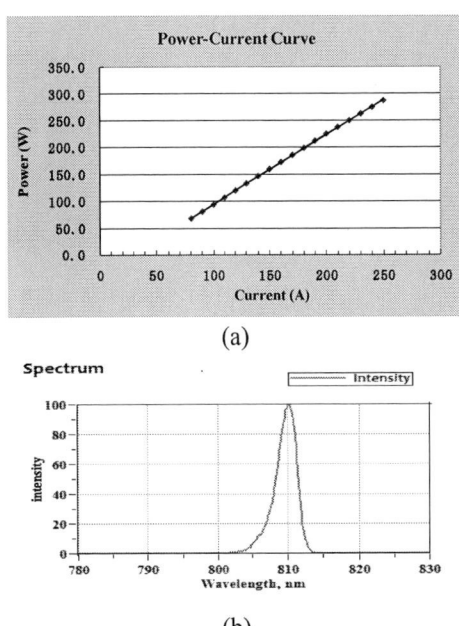

(a)

(b)

Figure 6 P-I curve and spectrum of a QCW (400Hz, 200μs) high power semiconductor laser

Near-field measurement

Our experimental setup schematic of near-field test system is illustrated in Figure 7. When measuring 250W QCW high power semiconductor laser array in the near-field test system, the output light of driven by a laser pulse diver went through the Optical Imaging System, making the intensity of output light imaged onto the photo-sensitive surface of the CCD camera. The signals from the photo-sensitive surface through the image grabbing card were converted to data signals and delivered to the analyzing software. Images are captured on a digital camera at a low drive current.

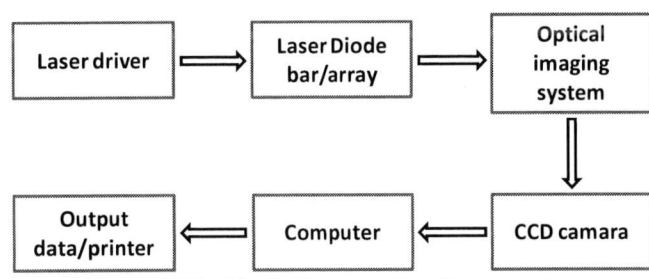

Figure 7 The block diagram of smile test system

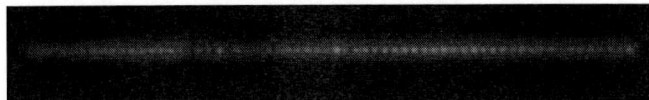

Figure 8 Near-field image of QCW 250W conduction- cooled-packaged high power semiconductor laser array

The measurement conditions of near field are 40Amp, degraded the intensity of the order of magnitude of 1E8. As it is seen from Figure 8, there are light and black laser spots in the near field imaging. The output light intensities of each emitter in a laser bar are not uniform. Possibly, this is caused by the inhomogeneous distribution of drive current on each emitter in a laser bar/array. Carriers in active region are not fully excited simultaneously. This leads to the output light inhomogeneity of high power semiconductor laser bar.

Our experiments result exhibits that the near-field performance of a 250W QCW conduction-cooled-packaged high power semiconductor laser is relatively good.

Lifetime testing

Compared with low-power CW counterparts, these HPDLAs suffer from shorter lifetimes and are more susceptible to degradation and premature failure. This is mainly due to the excessive localized heating and substantial pulse-to-pulse thermal cycling of the laser active regions. The thermally-induced stresses are even more dramatic when the required pump pulsewidth is increased from 200µs. [8]

The lifetime testing setup is automated using a single computer to set operational and environmental parameters, acquire and archive data, flag anomalous readings, and generate a number of warning and status alert messages when necessary, such as the over-temperature of laser bar and the shortage of cooling water. All the HPDLAs performance parameters are continuously monitored and recorded using a set of instruments for consistency and accurate comparative analysis and evaluation.

High power semiconductor laser arrays are tested under the general operational parameters chosen below:

- Drive current: 250 A
- Pulse duration: 200µs
- Rep. rate: 400Hz
- Heatsink temp. 25°C

An example of lifetime test data of single-bar 250W QCW conduction-cooled-packaged high power semiconductor laser array operating in 8% duty cycle (400Hz, 200µs) is illustrated in the Figure 9.

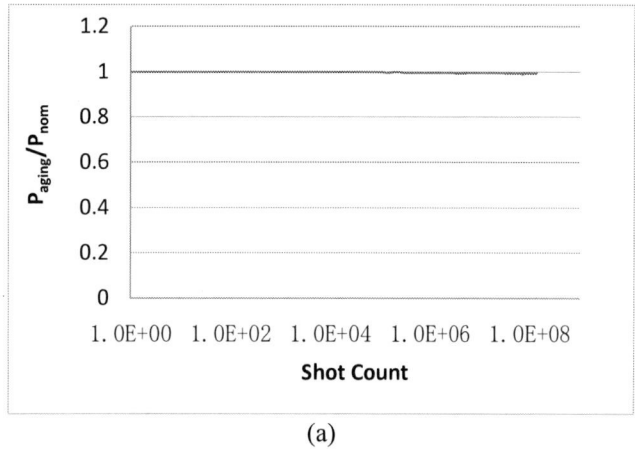

(a)

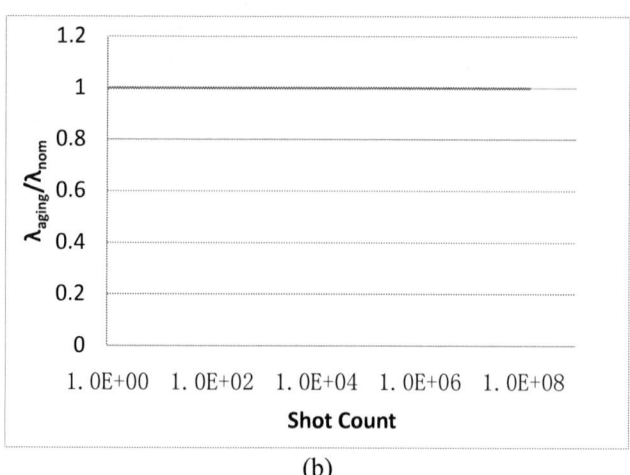

(b)

Figure 9 Example of lifetime test data of a single-bar 250W QCW conduction-cooled-packaged high power semiconductor laser array operating in high duty cycle (8%, 400Hz, 200µs)

As it is seen from Figure 9 (a) that the normalized ratio of output light power is almost constant after operating 1.2×10^8 pulse. The curve exhibited that the value of power reduction of the single-bar 250W QCW conduction-cooled-packaged high power semiconductor laser array is less than 2%. The wavelength shift is same to power; the shift difference is less 0.2nm. Results showed in Figure 9 indicate that the lifetime of single-bar 250W QCW conduction-cooled-packaged high power semiconductor laser array we prepared are acceptable. These devices have a high reliability.

Discussion

The current HPDLAs have an electrical to optical efficiency of about 50%. Therefore, when running a single bar 808nm 250W QCW high power semiconductor laser, about close to 300 W of peak power is generated in the form of heat, (24W average at 200µs pulse duration and 400 Hz prf). This excess energy primarily generated in the active area of the bars (light emitting surface), is quite substantial. Given that the total active area at the surface of each bar is on the order of 1.5 mm wide by 10 mm long (0.15cm2), yields a thermal density on the order of 160W/cm^2. If the accumulated heat cannot readily escape, the elevated temperatures at the location of the p-n junction adversely affect the output power, the slope efficiency, the threshold current and the device lifetime.

The level of impact of the long pulse operation may be roughly estimated by an Arrhenius relationship written as:

$$\text{Lifetime } (\tau) \propto (T_a - T_b)^{-N} \text{Exp}(E_a/kT_a)$$

Where lifetime (τ) is expressed as a function of junction temperatures T_a and T_b measured immediately after and before the generated pulse, the activation energy (E_a) and Boltzmann's constant (k). The leading term accounts for the thermal cycling fatigue due to mismatch of thermal expansion coefficients of different package materials and various layers of the laser bar. The power N in the expression above can have a value between 2 and 5 depending on the materials properties.

It is obvious from this Arrhenius equation that reducing the temperature difference before and after the pulse is the key for increasing the lifetime to an acceptable level. This may be achieved through careful selection of the HPLDA package type, specifications of the array considering the pumping requirements, and defining its operational parameters.

Many failures in high power diode laser packages are directly related to packaging technology, especially to the solder interfaces [9,10]. Thermal behavior in solder interfaces is a major factor affecting the functional performance of high power diode laser array packages. To provide a cost-effective and high performance HPDLAs, thermal management and optimization are needed. During practical manufacturing process, optimizing packaging design, thermal and thermal mechanics, thermal stress design and packaging process, such as the solder layer material, mounting substrate/heatsink material and thickness, and die bonding process, could improve the performance.

Conclusions

A 250W QCW conduction-cooled-packaged high power semiconductor laser array under the condition of 8% duty cycle (400Hz, 200μs) was presented in this paper. These devices were characterized using home-made measurement setup, and optical imaging techniques to study the performance of high power diode laser bar or array, including the LIV characteristics, near-field, and lifetime. The thermal behavior of the semiconductor laser was simulated by means of the finite element numerical analysis. Strategies in improving the performance of HPDLAs in terms of package architecture design and bar bonding process are proposed in the paper. Based on the numerical simulation and analytical results, 250W QCW conduction cooled high power semiconductor lasers using hard solder bar bonding process were produced.

Acknowledgments

This work is supported by the project of the "Hundred Talents Research Fund of Chinese Academy of Sciences".

References

1. Uwe Brauch, Peter Loosen, and Hans Opower, "High-Power Diode Lasers for Direct Applications", *Springer-Verlag (Berlin Heidelberg)* 2000, pp.1-2.
2. Brian Faircloth, "High-brightness high-power fiber coupled diode laser system for material processing and laser pumping", *Proceedings of SPIE* Vol. 4973, 2003, pp. 34-41.
3. E. Rugi, P. Mueller, P. Lambelet,"Scalable high brightness laser pump for aerospace applications", *IEEE*, 2003, pp81.
4. Michael A. Bolshov, Yuri A. Kuritsyn, "Laser Analytical Spectroscope", Ullmann's Encyclopedia of Introdustrial Chemistry, Six Edition, *Wiley (Weinheim, Germany)*, 2001, pp18-23.
5. X. Liu, J. Wang, and P. Wei,"Study of the mechanisms of spectral broadening in high power semiconductor laser arrays", *Electronic Components and Technology Conference,* IEEE, 2008, pp1005-1010.
6. Zhenbang Yuan, Jingwei Wang, Xu Chen, Di Wu, Xingsheng Liu, "Study of Steady and Transient Thermal Behavior of High Power Semiconductor Laser array," *Proceedings of 59th Electronic Components and Technology Conference (ECTC)*, 2009, accepted.

7. Xingsheng Liu and Wei Zhao, "Technology Trend and Challenges in High Power Semiconductor Laser Packaging", *Proceedings of 59th Electronic Components and Technology Conference (ECTC)*, 2009, accepted.
8. J. Yu, A. Braud, and M. Petros, "600mJ, Double pulsed 2 micron laser," Opt. *Lett.*, 28, 540-542, 2003.
9. X. Liu, R.W. Davis, L. C. Hughes, M. H. Rasmussen, and C.-E. Zah, "A Study on the Reliability of Indium Solder Die Bonding of High Power Semiconductor Lasers," *Journal of Applied Physics*, Vol. 100, Issue 1, 013104, 2006.
10. E. Suhir, "Interfacial Thermal Stresses in a Bi-Material Assembly with a Low-Yield-Stress Bonding Layer", *Modeling and Simulation in Materials Science and Engineering*, vol. 14, 2006, pp.1421-1432

High Density Indium Bumping through Pulse Plating Used for Pixel X-Ray Detectors

Yingtao Tian[1], David A. Hutt[1], Changqing Liu[1], Bob Stevens[2]

[1]Wolfson School of Mechanical and Manufacturing Engineering, Loughborough University
Loughborough, Leicestershire LE11 3TU, UK
[2]Central Microstructure Facility, Rutherford Appleton Laboratory
Chilton, Didcot, Oxfordshire OX11 0QX, UK
Tel: +44(0)1509 227678; Fax: +44(0)1509 227648; Email: Y.Tian2@lboro.ac.uk

Abstract

High density indium bump bonding is in high demand for devices which operate under cryogenic environments, such as pixellated X-ray detectors for high energy physics, due to the outstanding ductility of indium even at liquid helium temperatures. For these assembly applications, the connection pitch size is shifting to below 50 μm, such that the packaging density, i.e. I/Os, may exceed 40,000/cm^2. Electrodeposition is a promising approach to enable a low-cost and high yield bump bonding process, compared with conventional sputtering or evaporation which is currently utilized for small-scale production. Previous studies have shown the capability of electrodeposition to achieve high yield and high density indium bumps. The challenge exists to improve the bump height uniformity and consistency of electroplated indium bumps across the wafer at ultra-fine pitches with the highest yield. This paper is an initial investigation of the application of pulsed plating to the indium plating process and considers the influence of various current waveforms on the morphology and uniformity of the bumps. The results indicated that change in frequency and duty cycle did not have a significant influence on the indium bump morphology, but, together with the addition of a thief ring to the wafer design, pulse plating did have a noticeable impact on the bump height uniformity.

Introduction

The assembly of pixellated X-ray detectors used in high-energy physics, (e.g. the PILATUS detectors at the Diamond Light Source) requires the flip-chip interconnection of sensor and readout die, for which indium bump bonding has been successfully applied. In general, indium bump bonding is used as it does not interfere with the X-ray signal detection, offers ultra-fine pitch interconnections (<50 μm), is a low temperature process and can be carried out with a high yield (greater than 99%) [1]. Uniquely, indium can stay ductile under cryogenic environments even as low as liquid helium, at which these devices normally operate and feature high reliability. The current state-of-the-art indium bump bonding process developed by the Paul Scherrer Institute (PSI) employs photolithography, sputtering and evaporation steps to produce the indium bumps with a Ti/Ni/Au under bump metallization (UBM) [2]. The process uses two lift-off steps on the sensor die and one on the readout die and, briefly, consists of wafer cleaning and passivation after which a mask is applied and layers of titanium, nickel and gold are sputtered onto both of the sensor and readout wafers to coat the bondpads. Once the UBM is deposited, indium is evaporated onto the sensor die, and a thin indium layer is also deposited onto the readout chip to promote better adhesion during the

bonding step. After lift-off of the masks, the evaporated indium bumps on the sensor wafer are reflowed to form truncated indium spheres. Finally, the readout die is flip-chip bonded to the sensor using force to connect them.

While the PSI process has been successful at producing bonded devices, the many lithography and vacuum coating steps mean that it is time-consuming and relatively high cost. Alternative methods for larger scale, lower cost bumping is therefore sought. Electrodeposition has been investigated as a promising approach to enable low-cost, fine pitch and high yield bumping in a number of applications [3-7]. The process involves seed layer deposition across the whole wafer followed by photoresist patterning. After this, electroplating can take place into the resist apertures and, following resist stripping, the exposed seed layer is removed. Challenges of the electroplating process exist in the uniformity and consistency of plated bumps at the wafer scale with ultra-fine pitch and high yield. The main factors that can affect the electroplating process are current distribution and mass transport, which can be interdependent of each other [7, 8]. The wafer with the conductive seed layer can usually only be electrically connected through its edges to maximize the usage of the wafer surface for active devices. However, the resistance of the very thin seed layer will cause an Ohmic voltage drop from the edge to the centre of the wafer, known as the terminal effect, leading to uneven current distribution and non-uniformity of plating. Furthermore, the mass transport condition also varies across the wafer surface due to geometric effects, agitation conditions and bath temperature, etc. In order to generate indium bumps uniformly, the electrodeposition process including all possible variables must be precisely controlled.

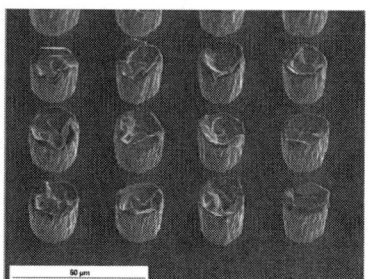

Figure 1 DC plated indium bumps: 15 μm diameter, 30 μm pitch size.

In our previous studies, preliminary experiments have been carried out on test wafers and the results have shown the capability of the electroplating method to generate indium

978-1-4244-4658-2/09 $25.00 © 2009 IEEE

bumps with ultra-fine pitch and very high yield (Figure 1) [9]. However, the results from DC plating indicated non-uniformity of the indium bump height across the wafer. Furthermore, the plated bumps had an uneven surface profile that implied a variation of current density and/or mass transport within the features, which was thought to be due to the current crowding near the aperture sidewalls.

Pulse plating has been widely used to assist in the electrodeposition of metals. With this technique, it is known that there is an additional pulsating diffusion layer adjacent to the cathode surface rather than only a single diffusion layer in the DC plating application. The use of a short current pulse allows a much higher peak current during the pulse-on period and during the pulse-off period the solution is able to recover the reactants. High frequency pulse plating has been shown to give significant changes in the electroplating bumping process, such as improved surface flatness, suppression of abnormal growth and reduced grain size [10]. To further investigate the application of pulse plating to indium electrodeposition, this paper presents the results of initial experiments carried out to examine the influence of current waveform on the bump forming process, using various duty cycles and frequencies with a constant average current density. Where possible, the experimental results will be compared with the results of DC electrodeposition which were conducted in the previous study.

Experimental Details

The electroplating bumping process flow is illustrated in Figure 2. Three inch diameter glass wafers were used for the experiments. Following the initial chemical cleaning assisted by ultrasonic agitation, a thin Ti/Cu seed layer was evaporated onto the test wafer. In this case, the seed layer provided a conducting path for electroplating, but also formed a part of the under bump metallization. The seed layer consisted of 100 nm Ti and 100 nm Cu for which the Ti was used to improve the adhesion of the copper deposit to the glass. Once the seed layer was deposited, a thick photoresist layer was spun, exposed and developed to form the required pattern. In this step, a positive photoresist AZ9260, was preferred as it was easier to remove after electroplating. The photoresist was around 20 μm thick. The photoresist pattern (Figure 3) consisted of many die sized areas distributed across the wafer within which, regions of different pad diameter and pitch sizes ranging from 15 μm to 25 μm were included as shown in the zoomed-in part of the diagram. To achieve more uniform current distribution across the wafer, a circular current "thief ring" (1 mm in width) was also designed surrounding the patterns – this was not used in the earlier study [9]. Indium bumps were electroplated onto the seed layer through the apertures in the pattern, after which the photoresist was removed in acetone. The top copper seed layer was then etched away using a chemical bath. The titanium layer was left at this stage as it can prevent indium from wetting the glass substrate during the following reflow. The indium bumps were then reflowed to form truncated spheres. For production, the bottom Ti seed layer would also need to be removed afterwards. However, as the main purpose

of this study was to evaluate the bump uniformity, the titanium layer was left on the wafer.

For the electroplating process, two wires were connected to the seed layer near the wafer edge, diametrically opposite to each other as shown in Figure 3. A sulfamate solution, supplied by Indium Corporation Inc. was used for plating. The solution comprises of $In(NH_2SO_3)_3$ 105.36 g/L, $NaNH_2SO_3$ 150g/L, HNH_2SO_3 26.4g/L, NaCl 45.84 g/L, dextrose and triethanolamine [11]. A 99.9% pure indium plate (8cm × 8cm) was utilized as the anode, having 100% anode efficiency. A computer controlled electroplating power system (PARSTAT® 2273, Potentiostat / Galvanostat, Princeton Applied Research, AMETEK) was employed to generate the three types of pulse plating waveforms and the parameters are listed in Table 1. Figure 4 shows the features of the pulsed current waveform. The duty cycle is defined as $[(t_{on}/(t_{on} + t_{off})) \times 100\%]$. The average current densities of the waveforms were all kept at 10 mA/cm^2, since a lower current density can make the deposit growth more controllable and this level matched the current density used for the previous DC plating trials. The electrodeposition rate was calculated as 0.3 μm/min according to Faraday's law, and all electroplating was conducted for 60 min at room temperature, thereby aiming to achieve 18 μm thick deposits.

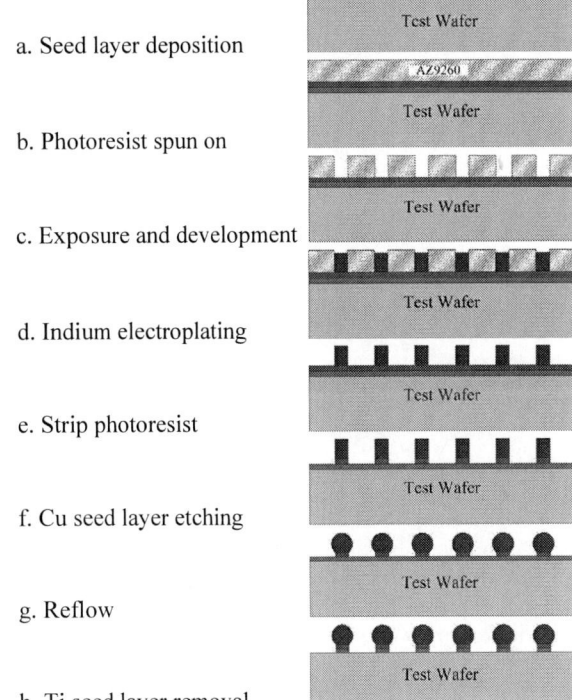

a. Seed layer deposition

b. Photoresist spun on

c. Exposure and development

d. Indium electroplating

e. Strip photoresist

f. Cu seed layer etching

g. Reflow

h. Ti seed layer removal

Figure 2 Schematic of the electroplating bumping process.

The characteristics of plated indium bumps were analyzed after reflow. The bump height was measured by using a Zygo NewView 5000 system and the uniformity over the whole wafer was assessed by measuring bumps selected from a number of areas. In addition, the bump morphologies and grain structures were also observed using Scanning Electron Microscopy (SEM) and Focused Ion Beam (FIB).

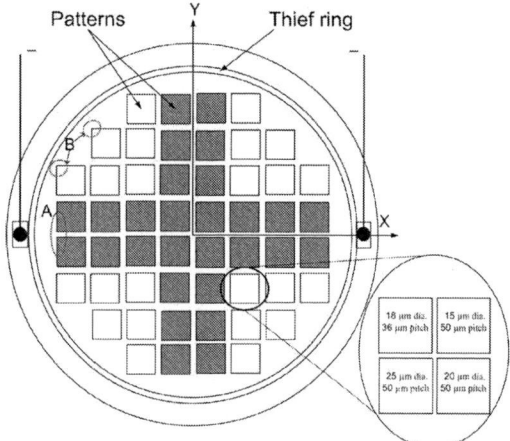

Figure 3 The layout of wafer patterns.

Table 1 Parameters of the pulse plating waveforms

Waveform Type	Frequency (Hz)	Duty Cycle	Peak Current Density (mA/cm^2)
1	100	20%	50
2	63	20%	50
3	100	10%	100

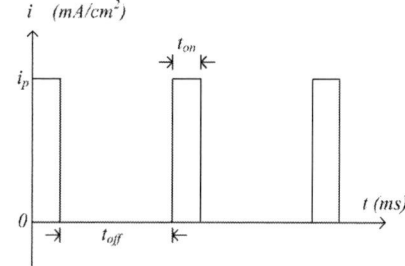

Figure 4 Features of the pulsed current waveform.

Results

1. As-plated bumps

Figure 5 shows SEM images of pulse plated bumps: the images were taken by tilting the sample by 52° and all the bumps were near the central area of the wafer. In terms of the morphology and profile of the as-plated bumps, there was little difference observed between the three types of waveforms used. As can be seen in Figure 5, the plated indium bumps had an uneven surface, similar to the DC plated bumps and the centre of the bumps tended to be lower than the edge of the bumps. Moreover, the 'thief ring' area was plated with indium as well.

As mentioned above, the photoresist on all the wafers was about 20 μm thick and based on the deposition rate and plating time, the bumps were expected to be around 18 μm high and appear with a column-like profile, as defined by the resist aperture. However, it was observed that bumps near the

edge of the wafer tended to have a "mushroom" profile, as they were plated out of the aperture and over the photoresist surface. Figure 6a was taken from the area A shown in Figure 3, while Figure 6b came from area B. This was despite the fact that the photoresist near the wafer edge is normally thicker than at the centre due to the physics of the spin coating process.

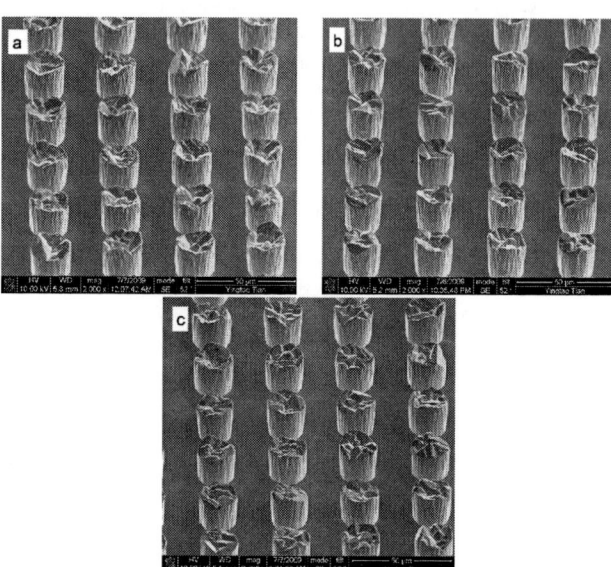

Figure 5 Overview of as-plated indium bumps with 18 μm diameter and 36 μm pitch size. a) waveform 1, b) waveform 2, c) waveform 3.

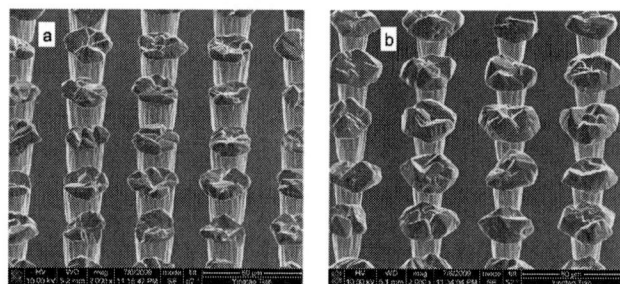

Figure 6 The mushroom features of plated bumps taken from a) area A, b) area B. (Plated using waveform 2).

2. Reflow

The as-plated bumps were difficult to characterize in terms of the volume of indium deposited, as defining the height of the uneven surfaces was problematic. By reflowing the bumps, the volume could be determined more readily, however before reflow it was necessary to remove the copper layer to prevent the indium spreading across the surface. The copper seed layer was etched away using a chemical bath that did not affect the Ti layer. The etchant slightly attacked the indium bumps while the very thin copper layer was removed, but did not noticeably reduce the volume of the deposit. A flux was then applied to the wafer and the bumps were reflowed using a conventional reflow oven. Figure 7 shows the reflowed bumps which were plated using current waveform 1. The yield of the reflowed bumps was very high and the truncated sphere shape made the measurement of the uniformity more accurate. Figure 8 presents the cross-section, prepared by FIB, of a reflowed bump. The arrow indicates the interface of the indium bump and copper substrate. Due to the very thin nature of the seed layer, it appeared that the copper had been largely consumed by the indium and it was noticed that there was a layer with a porous structure above the UBM. This is likely to be a serious problem for the future reliability and further investigation of the composition and microstructure of that interface and alternative UBM layers is still ongoing.

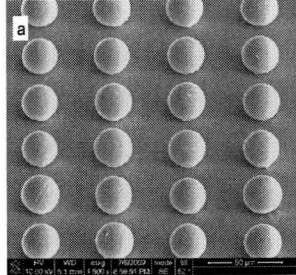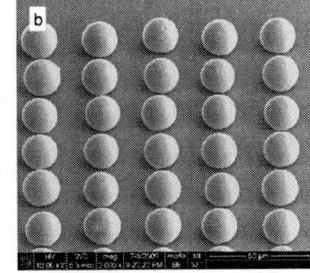

Figure 7 Overview of reflowed bumps (waveform 1). a) 20 μm diameter, 50 μm pitch, b) 18 μm diameter, 36 μm pitch.

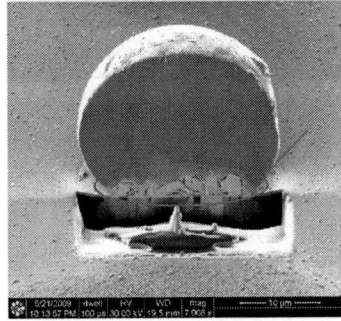

Figure 8 Cross-section of reflowed indium bump.

3. Uniformity Measurement

The bump height uniformity was measured after reflow by selecting points in different areas across the wafer. 20μm diameter bumps at 50μm pitch were chosen for comparison. Figure 9 shows the bump height measurement from the shaded die through the X and Y directions as indicated in Figure 3. At least one bump was measured within a single pattern and the bump height was averaged for the two die located at the same distance from the wafer centre. It is apparent that the bumps near the central area were lower than the bumps near the edge area of the wafer which reflected the terminal effect and there was little variation in the effect between the X and Y directions. The differences in bump height from centre to edge for waveforms 1 to 3 in the X direction were 3.64μm, 5.05μm and 5.05μm, respectively. As shown earlier in Figure 5 and Figure 6, near the edge area of the wafer, the bumps were over-plated and led to bumps with heights larger than the maxima shown in Figure 9. The highest bumps (at ~23μm) were found in the area B shown in

Figure 3, rather than area A which is physically closest to the electrical connection points.

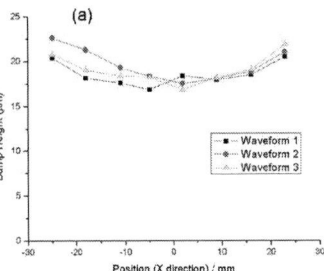

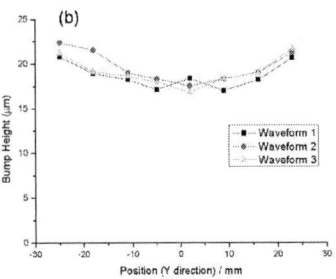

Figure 9 Bump height measurements after reflow: a) through X direction; b) through Y direction (defined by the shaded die in Figure 3).

Discussion

1. Effect of Pulsed Current on Bump Morphology

Within the plated features, due to the current crowding near the corner of the photoresist aperture, the ion concentration, i.e. mass transport condition, will be affected by the current distribution. This normally results in an uneven morphology of plated bumps within the feature scale. Comparing the bump morphology, it was found that the pulse plating resulted in a slightly more level as-plated deposit than the DC plating, as shown in Figure 5 compared with Figure 1. During pulse plating, there is a duplex diffusion layer developed, which consists of an outer stationary diffusion layer and inner pulsating layer [12]. The pulsating diffusion layer is much thinner and allows the consumed ions to be replenished during the pulse-off period. The current distribution and therefore mass transport condition within the feature scale is therefore expected to be more even than in the DC condition. Some researchers have demonstrated more even bump finish by using pulse plating [10]. However, under the condition of pulse plating used in this study, it was still observed that the centre of the bump was lower than the edge. Moreover, little impact of the change in frequency and duty cycle on the bump morphology and microstructure was observed. The reasons for this require further investigation, but could be because the deposition falls into the charge controlled region such that the mass transport does not play a significant role: the crowded current density near the corner of the feature could then lead to the uneven bump finish.

978-1-4244-4658-2/09 $25.00 © 2009 IEEE   383

2. Uniformity

Unsurprisingly, the terminal effect was still present for all the types of pulse plating because the electrical contact of the wafers was not physically changed. The bump uniformity is defined as [(maximum bump height – minimum bump height)/(2×average height) × 100%]. Thus, considering all die over the whole wafer, the overall bump uniformities corresponding to current waveforms 1 to 3 were 9.5%, 12.5% and 12.4%, respectively. The results indicated that the current waveform 1 gave the best bump height uniformity. It was also noted that the bumps in the central area of the wafer had a more uniform height. Considering the area within a 20 mm radius from the wafer centre, the uniformity corresponding to the current waveforms 1 to 3 was 4.2%, 4.8% and 6%, respectively. This could be a guide to designing the layouts of patterns where die requiring high uniformity are positioned near the centre of the wafer. In the previous DC plating work, the overall bump height uniformity was 18.5%, however it must be stressed that due to the differences in wafer design between these two studies, especially with the presence of the thief ring, it is not possible to draw conclusions regarding the influence of the pulsed current at this stage.

Purely considering the terminal effect, the closest plating area to the connection point should have the shortest electrical path, leading to the least voltage drop and ultimately the highest bump. However, in this work, little difference in the bump height distribution was seen between the X and Y directions despite the variation of difference distance to the connection points. This behavior is attributed to the presence of the thief ring. At the beginning of the plating process, the thickness of the seed layer on the ring was very thin and the current distribution across the wafer would be strongly affected by the terminal effect. However, once indium was deposited onto the ring, its thickness increased, together with its electrical conductivity, thereby providing a continuous electrical contact around the circumference of the wafer. This removed the directional effects caused by the two individual contact points such that the X and Y directions plated equally. Support for this interpretation is provided by the observation that the most over-plated bumps were found in the area B shown in Figure 3, which was not the closest area to the connection points, but was the closest area to the thief ring. The addition of a thief ring to the plating pattern could therefore be a straightforward method to provide a continuous electrical contact around the wafer rather than relying on multiple contact points as demonstrated in the literature [13].

Conclusion

Three types of pulse plating waveform for indium bump bonding were investigated in this study. The results showed that the change in the frequency and duty cycle used here did not have a significant impact on the bump morphology and microstructure. However, due to the terminal effect and other factors there was a variation in the bump height across the wafer which was influenced by the waveform used, with waveform 1 providing the best results. A thief ring was used within the wafer design and this helped to significantly reduce the terminal effect caused by single point electrical contacts.

Acknowledgement

The authors would like to acknowledge DTA funding from the UK Engineering and Physical Sciences Research Council (EPSRC) through the IeMRC.

Reference

1. M. Bigas, E. Cabruja, and M. Lozano, "Bonding techniques for hybrid active pixel sensors (HAPS)," Nuclear Instruments and Methods in Physics Research Section A: Accelerators, Spectrometers, Detectors and Associated Equipment, vol. 574, no. 2, pp. 392, 2007.
2. T. Rohe, C. Broennimann, F. Glaus et al., "Development of an Indium bump bond process for silicon pixel detectors at PSI," Nuclear Instruments & Methods in Physics Research, Section A (Accelerators, Spectrometers, Detectors and Associated Equipment), vol. 565, pp. 303.
3. T. Ritzdorf, and D. Fulton, Microelectronic Packaging, M. Datta, T. Osaka and J. W. Schultze, eds.: CRC Press, 2005.
4. P. Merken, J. John, L. Zimmermann et al., "Technology for very dense hybrid detector arrays using electroplated indium solderbumps," Advanced Packaging, IEEE Transactions on, vol. 26, no. 1, pp. 60-64, 2003.
5. J. Jiang, S. Tsao, T. O'Sullivan et al., "Fabrication of indium bumps for hybrid infrared focal plane array applications," Infrared Physics & Technology, vol. 45, no. 2, pp. 143-151, 2004.
6. M. Datta, "Electrochemical processing technologies in chip fabrication: challenges and opportunities," Electrochimica Acta, vol. 48, no. 20-22, pp. 2975-2985, 2003.
7. L. T. Romankiw, "A path: from electroplating through lithographic masks in electronics to LIGA in MEMS," Electrochimica Acta, vol. 42, no. 20-22, pp. 2985-3005, 1997.
8. M. Datta, and D. Landolt, "Fundamental aspects and applications of electrochemical microfabrication," Electrochimica Acta, vol. 45, no. 15-16, pp. 2535, 2000.
9. Y. Tian, C. Liu, D. A. Hutt et al., "Electrodeposition of indium for bump bonding," in Proc. 58th Electronic Components & Technology Conference, Florida, USA, 2008, pp. 2096-2100.
10. B. Kim, and T. Ritzdorf, "Electrical Waveform Mediated Through-Mask Deposition of Solder Bumps for Wafer Level Packaging," Journal of the Electrochemical Society, vol. 151, no. 5, pp. 342, 2004.
11. "Indium Sulfamate Plating Bath," Indium Coporation, Realised in 2007.
12. F. L. Jean-Claude Puippe, Theory and Practice of Pulse Plating,, Orlando, FL: American Electroplaters and Surface Finishers Society, 1986.
13. T. Homayoun, U. Cyprian, and B. B. M, Device providing electrical contact to the surface of a semiconductor workpiece during metal plating, U.S. Patent, Dec. 2002.

Emerging Lead-free, High-temperature Die-attach Technology Enabled by Low-temperature Sintering of Nanoscale Silver Pastes

Guo-Quan Lu[1], Meihua Zhao[1], Guangyin Lei[1], Jesus N. Calata[1], Xu Chen[2], and Susan Luo[3]

[1]Dept. of MSE and ECE, Virginia Tech, Blacksburg, VA 24061, USA
[2]School of Chemical Engineering, Tianjin University, Tianjin, 300072, China
[3]NBE Technologies, LLC, Blacksburg, VA 24060, USA

Abstract

A low-temperature joining technology based on silver sintering, which is being implemented in manufacturing of power electronics modules, holds the promise of achieving 5x higher temperature cycling capability, 3x better total module resistance, and chip junction temperature up to 175°C. Because of its RoHS compliance, the technology is also aggressively pursued by the makers of automotive power electronics components. However, a serious drawback of the technology is the use of high quasi-static pressure (> 40 MPa or 400 kg-force per cm² chip area) necessary to lower the sintering temperature of existing thick-film silver pastes to less than 300°C. In this paper, we describe the die-attach application of a nanoscale silver paste that can be sintered at low temperature without pressure. Mechanical, thermal, and thermo-mechanical properties of the sintered silver joints are reported.

I. Introduction

While many electronics manufacturers in the United States are still working to ensure that their products are compliant with European Directive 2002/95/EC on the Restriction of Hazardous Substances (RoHS), some European electronics manufacturers are moving beyond lead-free soldering technology for joining semiconductor chips. Pioneered by major power electronics companies in Germany, a low-temperature joining technology based on sintering of silver powder is emerging as a promising die-attach solution for high-performance, high-reliability, and high-temperature applications [1-3]. The technology has been shown to achieve 5x higher temperature cycling capability, 3x better total module resistance, and chip junction temperature up to 175°C. However, a serious drawback of the technology is the use of high quasi-static pressure (> 40 MPa or 400 kg-force per cm² chip area) that is necessary to lower the sintering temperature of existing thick-film silver pastes to less than 300°C. The need for such a large pressure has hampered quick adoption of the technology because it limits production throughput and places critical demands on substrate flatness and chip thickness.

In this paper, we describe the use of a nanoscale silver paste that can be sintered at low temperature without pressure, for attaching small (< 3 mm x 3 mm) and large (> 10 mm x 10 mm) semiconductor chips. The paste, upon heating, was characterized by thermogravimetric analysis (TGA) and shrinkage measurements. Mechanical, thermal, and temperature-cycling properties of the sintered joints were studied. Many of these properties have been reported elsewhere [4-9].

II. Nanosilver Paste Characterization

The nanoscale silver paste evaluated in this study was acquired from NBE Technologies, LLC (www.nbetech.com), Blacksburg, Virginia, which markets the material under the trade name of nanoTach®. To examine the weight-loss kinetics of the paste, we placed a small amount (tens of milligrams) of the material in a TA Instrument (Model Q500) and heated in air under a heating profile recommended by the supplier. The heating profile consists of four drying steps at 50°C, 75°C, 100°C, and 125°C and one sintering step at 275°C. The heating rate to each designated temperature was at 10°C per minute. Fig. 1 shows the TGA weight-loss curve along with the time-temperature heating profile. Some time-temperature points are labeled on the curve for clarity. The paste started to lose weight immediately upon heating because of evaporation of solvents in the paste formulation; and the weight loss ended at around 246°C signifying the completion of removal of all organics (solvents and binder molecules) in the paste.

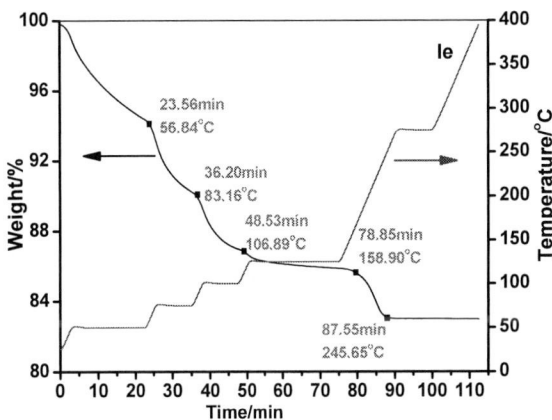

Fig. 1 TGA result for the nanoscale silver paste following the supplier's recommended heating profile.

To characterize the corresponding shrinkage profile of the paste arising from solvent and binder removal and densification of silver particles, we employed the optical setup (shown in Fig. 2) used by Choe et. al. [10] for their study of constrained-film sintering. We replaced the miniature high-temperature heater in their study with a large hot plate for better temperature uniformity and overall sample stability. Photos of the optical setup used to measure the thickness shrinkage profile of the paste films are shown in Fig. 3.

978-1-4244-4658-2/09 $25.00 © 2009 IEEE

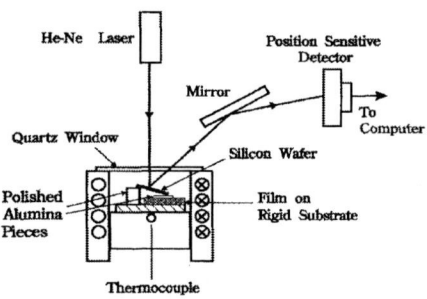

Fig. 2 Schematic of Choe *et. al.* [10] optical setup used for measuring shrinkage profiles of constrained-sintering films.

Fig. 3 Photos of the optical setup used in this study.

To obtain a thickness shrinkage profile of the paste, the material was stencil-printed on a ceramic substrate to cover about half of its surface. A small piece of thin (about 250-micron thick) alumina was placed on top of the film. A thick piece of alumina was placed on the uncovered area of the substrate. Then, a piece of silicon serving as mirror was carefully bridged across the thick and thin alumina. As one end of the silicon dropped in height due to film shrinkage, the silicon would tilt and its tilt angle was monitored by reflecting a laser beam off the silicon surface onto a position-sensitive detector. Fig. 4 is a plot of the thickness shrinkage profile of an initially 20-micron thick paste film.

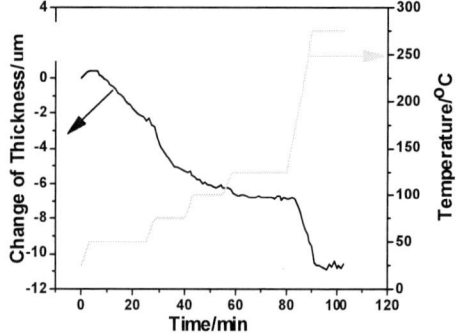

Fig. 4 Thickness shrinkage profile of a 20-μm thick paste film subjected to the same recommended heating profile.

The thickness of the paste film shrank by about 35% at the end of the final drying step and then shrank by another 20% at the end of sintering for a total shrinkage of 55%. Fig. 5 is a plot showing both the weight-loss and shrinkage profiles. The paste lost about 14% of its weight at the end of the final drying step and went on to lose only about 3.5% more at the end of sintering. This disparity in large film shrinkage versus small weight loss after drying suggests that a significant portion of the 20% film shrinkage was due to densification of the silver particles when the solvent and binder molecules were removed from the paste.

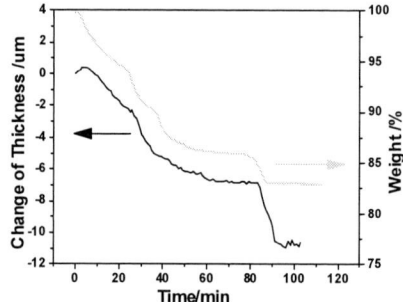

Fig. 5 Profile of weight-loss shown in Fig. 1 plotted together with thickness shrinkage profile in Fig. 4.

III. Die-attach Processing

We evaluated the die-attach application of the nanosilver paste by bonding power and optoelectronic semiconductor chips with dimensions ranging from 1 mm x 1mm to 12 mm x 12 mm. The bonding surfaces of the chips were metalized with either silver or gold. They were joined to various substrates, such as direct-bond copper (DBC), copper and Kovar, whose surfaces were plated with either silver or gold. Stencil-printing was used to apply a layer of the paste to a thickness of about 50 microns on the substrate. The small chips were bonded by directly mounting them on the wet paste and heating according to the recommended drying and sintering profile. For the large chips, the paste film was initially dried before the chip was mounted and sintered under a pressure of 1 to 5 MPa. Fig. 6 shows some examples of chips attached by the nanoscale silver paste.

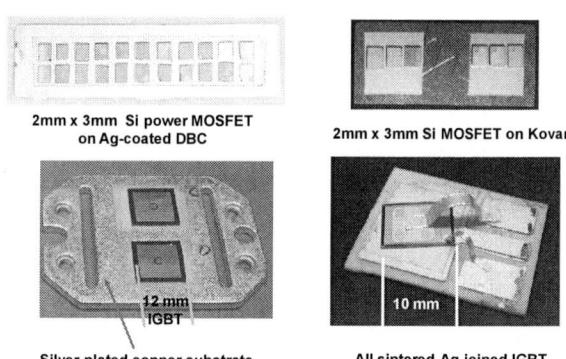

Fig. 6 Examples of chip attachment by low-temperature sintering of the nanoscale silver paste.

The reason for applying a low pressure during the attachment of large chips is three-fold: (1) the pressure ensures a full contact between the chip and the paste, which is considered to be problematic because of the surface roughness and curvature of the chip and the substrate, particularly when the chip is sufficiently large; (2) the pressure helps in securing the mounted chip against the outgassing, which comes from the organics burn-out during the sintering stage; and (3) the pressure helps to gain a more uniform and denser sintered microstructure. Fig. 7 illustrates the processing conditions for attaching the large chips and a simple setup for the pressure-assisted die-attach processing.

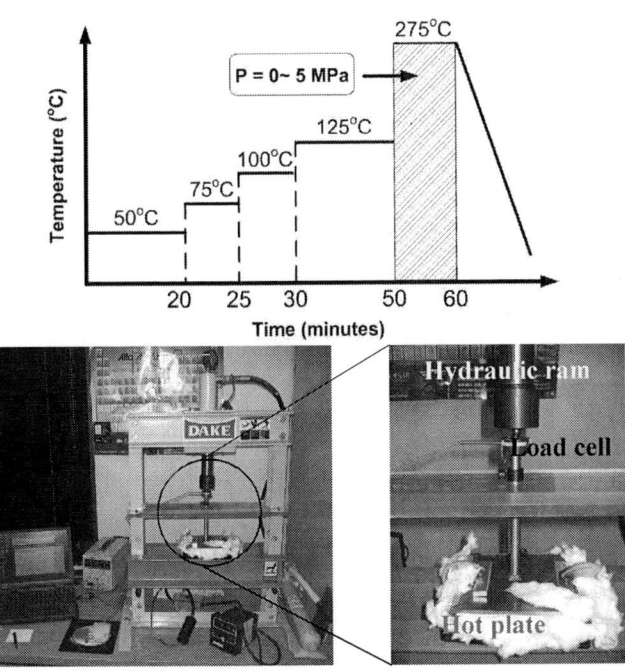

Fig. 7 Schematic of the heating profile and a custom setup for pressure-assisted attachment of large chips.

IV. Bonding Strength of Sintered Joint

Bonding strengths of the sintered chip attachments were determined by die-shear or lap-joint test. Measured strength values were generally in excess of 25 MPa. For the small-chip attachments, chips could be sheared off without cracking and the failure mainly occurred within the sintered layer. For large-chip attachments, it was impossible to determine bonding strength by die-shear test since chips were easy to crack. A qualitative evaluation of bonding strength of large chips on DBC substrate involves breaking the attached substrate over an edge. If chip cracks along with the DBC substrate without being delaminated, the bonding strength is said to be strong[2]. Fig. 8 shows a cracked large chip-to-DBC attachment demonstrating strong bonding strength of the sintered joint.

For quantitative measurements of the bonding strength of sintered joints in large-chip attachments, we fabricated lap-shear test samples made up of copper plates joined together by the nanoscale silver paste. All of the copper members were plated with nickel and then silver. The bonding area between two members is 10×10 mm^2. The joining process followed the same time-temperature heating profile. The shear strengths

of the lap-joint samples were measured using a ComTen 95 series tester (ComTen Industries, Pinellas Park, FL) at a rate of 8×10^{-5} m/s. Fig. 9 is a plot of the shear strengths of the Cu-to-Cu lap-joint samples bonded with the nanoscale silver paste at pressures up to 5 MPa. There were five samples made for each of the pressures. The average shear strengths are 7.7, 15.3, 21.3, and 31.6 MPa for applied pressure of 0, 1, 3, and 5 MPa, respectively.

Diode: 11 mm x 11 mm and 180- m thick

Fig. 8 Demonstration of strong bonding of a large chip sinter-joined on a DBC substrate.

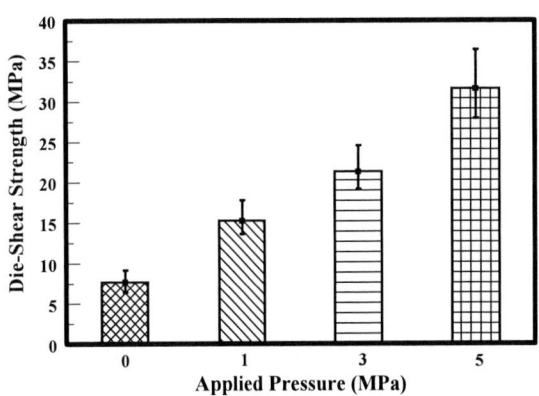

Fig. 9 The shear strengths of Cu-to-Cu lap-joint samples prepared at four different pressures; the error bars show maximum and minimum values obtained from five test samples at each pressure.

With increasing pressure from 0 to 5 MPa, the shear strength is significantly improved. This improvement is believed to come from: (1) better contact between the paste and chip/substrate during the binder burn-out and densification processes for diffusion of silver atoms across the two interfaces to develop stronger bonding; and (2) tighter contact between the nanoparticles throughout the binder burn-out process that results in denser microstructure after sintering. These observations are consistent with the cross-sectional SEM micrographs in Fig. 10 showing the microstructures of the large chip attachments sintered at pressures of 0 MPa, 1 MPa, and 5 MPa. The improved bonding strength, both at the interfaces and within the sintered microstructure, is expected to enhance the electrical and thermal properties of the chip attachment.

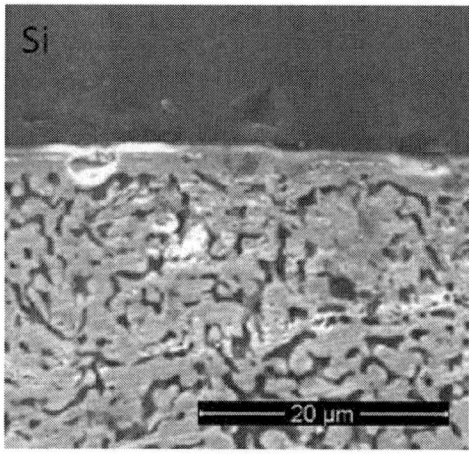

(a) 0 MPa

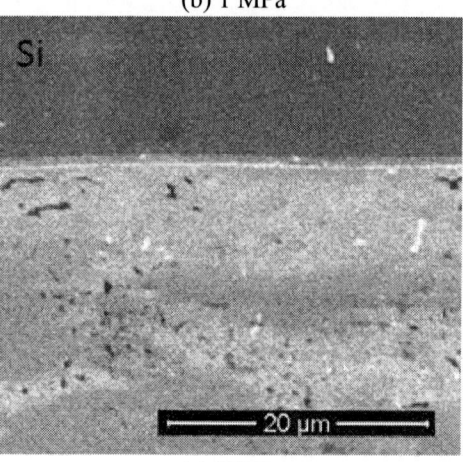

(b) 1 MPa

(c) 5 MPa

Fig. 10 Cross-sectional SEM images of the sintered joints at: (a) 0 MPa; (b) 1 MPa; and (c) 5 MPa.

V. Thermal Performance of Sintered Joint

Since the sintered joints are essentially pure silver, they are expected to have excellent thermal conductivity. Furthermore, since adhesion at the joint interfaces is established by Ag-Ag or Ag-Au atomic diffusion, one would expect the sintered joints to have low interfacial thermal resistance. However, it is not easy to characterize the thermal properties of a thin layer of joint sandwiched between two good thermal conductors. An indirect method was used to determine the thermal conductivity of the silver joint. It involved measuring the electrical resistance of a resistor pattern formed by sintering the nanosilver paste and then using the Wiedemann-Franz relationship between electrical and thermal conductivity of a simple metal to calculate the thermal conductivity. By this method, the thermal conductivity of the sintered silver was estimated to be about 200 W/mK.

The excellent thermal performance of sintered silver joint was demonstrated by Wang *et. al.* [11] in a study of the effect of die-attach thermal properties on the performance of light-emitting diodes (LEDs). Their findings are reported in the proceedings of this conference. They obtained an improvement of about 20% in light output from LED chips attached by the nanosilver paste over those attached by solder paste. Based on a recent study by Liu *et. al.* [12] on thermal solutions for chip-on-board packaging of high-power LED chips, we estimated that the thermal conductivity of the sintered silver attachment is at least 100 W/mK.

VI. Thermo-mechanical Reliability of Sintered Joint

One of the main reasons for the rising interest in the low-temperature chip joining technique by silver sintering is the improved thermo-mechanical reliability of sintered joints, especially in applications requiring higher chip junction temperatures. The single-metal sintered joint is free of brittle intermetallic phases that are the culprits of fatigue failure in soldered joints. Silver melts at 961°C, much higher than any of the soft and hard solders; thus the sintered joint operates at a much lower homologous temperature, translating to low inelastic deformation or accumulation of damage during temperature or power cycling. Chen et. al.[13] studied the tensile behavior of free-standing films of sintered nanosilver paste and found the elastic modulus to be about 10 GPa, drastically lower than that of bulk silver. The low modulus is a result of about 20% porosity in the sintered film. The porosity lowers thermal performance of the sintered joint (still better than solder), but its compliance is expected to improve thermo-mechanical reliability.

Plotted in Fig. 11 is the die-shear strength versus aging time obtained on sintered small-chip attachments after annealing at 300°C. The bonding strength is basically unchanged after a long period of aging. This can be attributed to the fact that there is no presence or growth of intermetalic phases in the joint. Fig. 12 shows the die-shear strength vs. temperature cycles from cycling small chip attachments between -40°C and 125°C. We also cycled sintered chip attachments between -55°C and 250°C to show the thermo-mechanical reliability over a large temperature range. To do this, we had to attach chips on direct-bond-aluminum (DBA) substrates since DBC or direct-bond-copper substrates failed too quickly (well under 100 cycles) under such large □T cycling. Fig. 13 shows the die-shear strength vs. number of cycles of small chips on DBA. We see that even under such extreme temperature cycling condition, the sintered joints lasted over 800 cycles. More temperature and power cycling studies of the sintered joint are underway in our laboratory.

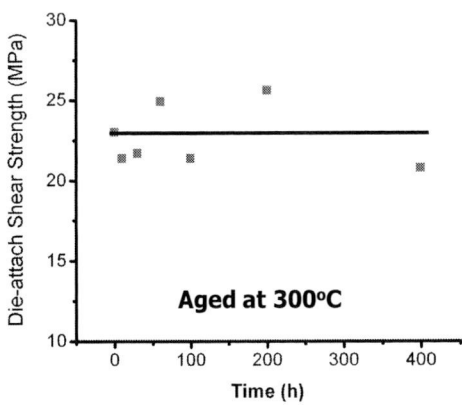

Fig. 11 Result of an aging test done at 300°C on sintered small chip attachments.

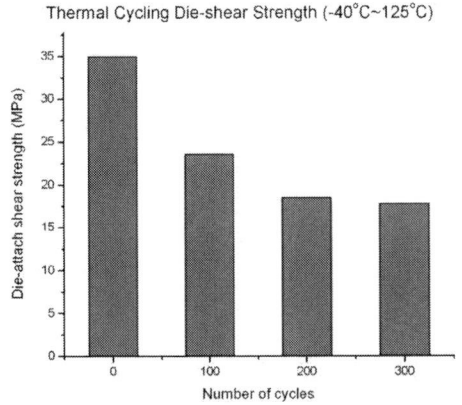

Fig. 12 Die-shear strength of sintered chip attachment versus number of temperature cycles between -40°C and 125°C.

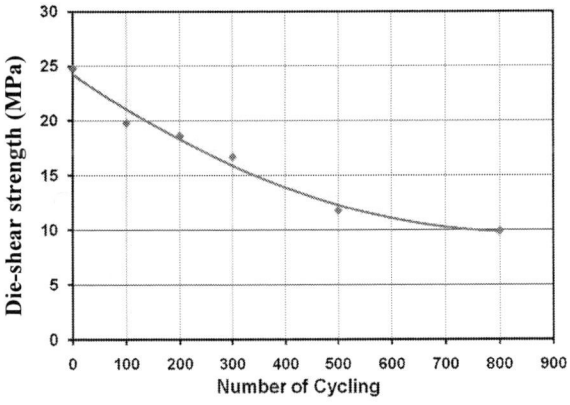

Fig. 13 Die-shear strength versus number of temperature cycles between -55°C and 250°C from sintered chip attachments on direct-bond-aluminum substrate.

VII. Summary and Conclusions

The European RoHS Directive is forcing many manufacturers of automotive power electronics components to search for alternative die-attach materials. The low-temperature joining technique based on silver sintering is emerging as a promising lead-free die-attach solution for high-performance, high-reliability, and high-temperature applications. A nanoscale silver paste, which can be sintered at low temperatures without or with minimal pressure, offers ease of chip attachment for superior mechanical, thermal, and thermo-mechanical properties. Without the need for significant retooling, the low-temperature sintering technology, enabled by nanoscale silver paste, could rapidly emerge as a competitive lead-free die-attach solution in the near future.

Acknowledgments

The authors are grateful for financial support from the US Office of Naval Research (#: N00014-09-1-0566), US National Science Foundation (SBIR-STTR Award No. 0740927), Chinese National Natural Science Foundation of China (#: 50528506), and the Program of Introducing Talents of Discipline to Universities (#: B06006).

References

1. Scheuermann, U., and P. Beckedahl, "The Road to the Next Generation Power Module - 100% Solder free Design," *in The 5th International Conference on Integrated Power Electronics Systems (CIPS 2008)*, 2008, Nuremberg, Germany.
2. Schulze, E., C. Mertens, and A. Lindemann, "Low Temperature Joining Technique – a Solution for Automotive Power Electronics," *in PCIM*, 2009, Nurnberg, Germany.
3. Mertens C., J.R., and R. Sittig, "Top-side Contacts with LTJT," *in the 35th Annual IEEE Power Electronics Specialists Conference*, 2004, Aachen, Germany.
4. Bai, J.G., Z. Z. Zhang, J. N. Calata, and G-Q. Lu, "Low-Temperature Sintered Nanoscale Silver as a Novel Semiconductor Device-Metallized Substrate Interconnect Material," *IEEE Transactions on Components and Packaging Technologies*, Vol. 29, No. 3 (2006), pp. 589 - 593.
5. Bai, J.G., and G-Q. Lu, "Thermomechanical Reliability of Low-Temperature Sintered Silver Die-Attached SiC Power Device Assembly," *IEEE Trans. on Device and Materials Reliability*, Vol. 6, No. 3 (2006), pp. 436 - 441.
6. Calata, J.N., T. G. Lei, and G-Q. Lu, "Sintered nanosilver paste for high-temperature power semiconductor device attachment," *Int. J. Materials and Product Technology*, Vol. 34, No. 1/2 (2009), pp. 95-110.
7. G.-Q. Lu, J.N. Calata, G. Lei, and X. Chen, "Low-temperature and Pressureless Sintering Technology for High-performance and High-temperature Interconnection of Semiconductor Devices," *in 8th International Conference on Thermal, Mechanical, and Multiphysics Simulation and Experiments in Micro-Electronics and Micro-Systems*, EuroSime 2007, 2007, London, UK, IEEE.
8. G.-Q. Lu, J.N. Calata, and T.G. Lei, "Low-temperature sintering of nanoscale silver paste for power chip attachment," *in the 5th International Conference on Integrated Power Electronics Systems (CIPS'08)*, 2008, Nuremberg, Germany.

9. Wang, T., X. Chen, G-Q. Lu, and G-Y. Lei, "Low-Temperature Sintering with Nano-Silver Paste in Die-Attached Interconnection," *Journal of Electronic Materials*, Vol. 36, No. 10 (2007), pp. 1333-1340.

10. Choe, J.W., J.N. Calata, and G-Q. Lu, "Constrained-Film Sintering of a Gold Circuit Ink," *J. Mater. Res.*, Vol. 10, No. 4 (1995), pp. 986 - 994.

11. Wang, T., G. Lei, G-Q. Lu, X. Chen, and L. Guido, "Improved Thermal Performance of High-Power LED by Using Low-temperature Sintered Chip Attachment," *in 2009 ICEP-HDP*, 2009, Beijing, China.

12. Liu, J.G., D. Xiong, and P. Panaccione, "Chip-on-Board Packaging and Thermal Solutions for 100W Single Large Area LED," *in 5th International Conference and Exhibition on Device Packaging*, 2009, Scottsdale/ Fountain Hills, AZ.

13. Chen, X., R. Li, K. Qi, and G-Q. Lu, "Tensile Behaviors and Ratcheting Effects of Partially Sintered Chip-Attachment Films of a Nanoscale Silver Paste," *J. of Electronic Materials*, Vol. 37, No. 10 (2008), pp. 1574-1579.

A Novel Micro Pirani Gauge with Mono-wire Sensing Unit for Microsystem Application

Yunsong Qiu[1, 2], Lei Zhao[3], Yufeng Jin[1, 2*]

[1]Shenzhen Graduate School of Peking University, Shenzhen 518055, China
[2]National Key Lab of Micro/ Nanometer Fabrication Technology, Institute of Microelectronics,
Peking University, Beijing 100871, China
[3]Nanjing Dajing Hi-tech, Inc., Floor7, No.6, Ji E Lane, Zhujiang Road, Nanjing 210018, China
*Tel: +86-10-62752536-22; Email:jinyf@ime.pku.edu.cn

Abstract

This paper reports a novel, simple, low-cost and high sensitivity micro Pirani gauge to monitor long-time pressure change and stability inside a vacuum package. The packaging structure is very simple with one wire as sensing part only. Two ends of a metal wire with a certain temperature coefficient of resistance are bonded to the pads of the ceramic structure with a cavity for microsystem application. The metal wire and the packaged part can be formed as a micro Pirani gauge. The dynamic range of this novel Pirani gauge in our experiment is from 1Pa to 100Pa.

Introduction

Many microsystem devices such as accelerometers, resonators and gyroscopes have to meet the requirements of low pressure environment in a sealed small cavity. The vacuum level in the cavity affects the performance and sensitivity of microsystem devices deeply. So real-time measuring the pressure inside a vacuum package is very important. Traditional methods for measuring leak rates in packages include Helium leak testing and Q factor extraction. However, Helium leak testing equipment is expensive and limited to a leak measurement resolution of 10-12cm3/s or worse, also it is impossible to do in-situ and real-time test. Q factor measurement is limited by sensor drift and requires complex readout electronics and long test time [1]. Micro Pirani gauges have been widely used for in-situ vacuum measurement as its low-cost and high sensitivity.

Most of reported micro Pirani gauges are integrated on devices using micromachining technology. Such as depositing a thin film on dielectric membrane as the heater [1], [2] and using a micro bridge as the heater [3].The former method may cause the membrane collapse to the substrate when the releasing process of the dielectric membrane is carried out. The residual stress of the membrane could not only be large enough to make the gap distance uneven across the whole membrane but also make the thermal-resistive property unstable. The latter method also uses bulk micromachining technology to form micro bridge structure, but the micro bridge will bend to the substrate if it is long and thin which is needed to enhance the sensitivity.

In this paper, a novel, simple, low-cost and high sensitivity micro Pirani gauge is presented. It does not need micromachining technology and will not affect the structure of the devices. A metal wire with a certain temperature coefficient of resistance is chosen and the metal wire is welded on the pads of the ceramic structure with a cavity using wire bonding technology. The metal wire as a heater and the packaged body can be formed as a micro Pirani gauge. Fig.1 shows the schematic design of the packaging structure.

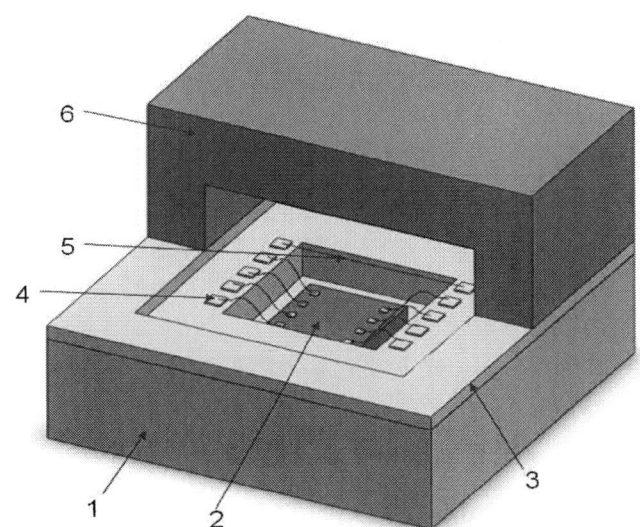

Fig.1 The schematic packaging structure
1-ceramic structure with a cavity 2-chip 3-solder 4-wire bonding 5-metal wire 6-cap

Theory and Design

The Pirani gauge was invented in 1906[4].It is based on the principle that the heat loss to ambient of a hot wire depends on the pressure of its surrounding gas. The pressure inside the vacuum package can be measured according to the

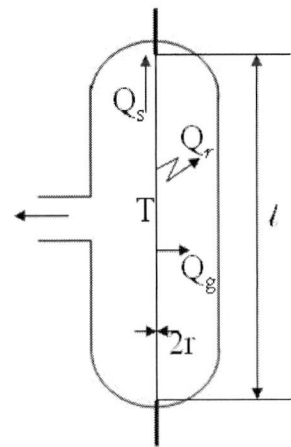

Fig.2 Heat dissipation of the metal wire

temperature change of the thermally sensitive wire [5]. In this paper, a metal wire which is a temperature-sensitive resistor as the heating and sensing element is used. Applied a certain current, the metal wire from which some heat transfer to its surroundings will be at a certain temperature when the thermal equilibrium is reached.

The total thermal conductance (GT) of the metal wire contains three major mechanisms: the solid conduction (Gs) along the supporting anchors, the radiation conduction (Gr) and the gas conduction (Gg) [6].It is simply modeled as shown in Fig.2.The thermal equilibrium equation is:

$$G_T = G_s + G_r + G_g \qquad (1)$$

According the Stefan-Boltzmann Law, the radiation conduction (Gr) is as following:

$$Q_r = \sigma(\varepsilon_2 T_2^4 - \varepsilon_1 T_1^4) 2\pi r l \qquad (2)$$

where б is the Stefan-Boltzmann constant, l is the radiation coefficient of the inner surface of the tube, □2 is the radiation coefficient of the hot wire, T1 is the temperature of the tube, T2, r and l are the average temperature, semidiameter and length of the hot wire.

The solid conduction (Gs) can be modeled as:

$$Q_s = 2 \times 0.239 K_\omega A_0 (dT/dl) \qquad (3)$$

where Kω is the thermal conductance of the hot wire,A0 is the cross-sectional area of the hot wire and dT/dl is the temperature gradient along the wire[4].

As shown in Fig.3, the radiation conduction can be ignored at low temperature. Solid conduction is independent of pressure. The gas conduction can be divided into two parts according to the Knudsen number ($K_n = \lambda / r$, □is mean free path of the gas, and r is the feature size of the domain). At high pressure, the gas is a continuum regime ($K_n \ll 1$) and its thermal conductance is nearly constant. At low pressure, the gas can be modeled as in a free molecule regime ($K_n \gg 1$) and its thermal conductance is linearly sensitive to the pressure in the package [5]. The gas thermal conduction can be expressed as following when the gas is in a molecule regime:

$$Q_g = k_f \alpha p (T_2 - T_1) 2\pi r l \qquad (4)$$

where kf is the thermal conductance of free molecule gas, is the accommodation coefficient of the surface of the hot wire and p is the pressure inside the package.

The Pirani gauge will be work effectively when the gas conduction (Gg) is dominant of the total thermal conductance. The lower limit of the dynamic range is determined by Gs/Gg, therefore, in order to obtain large dynamic range, a Pirani gauge should to be designed to minimum solid conduction and maximum gas conduction.

Fig. 1 shows the schematic graph of our design. A metal wire such as platinum, tungsten, aurum or silicon-aluminum is welded on the pads of the substrate using the wire bonding technology. Apparently the metal wire and the whole packaged body can be seemed a Pirani gauge which is separate from the devices in the package. In this paper, a silicon-aluminum wire is used with the diameter of 25□m.The length of the heating part is 8mm.

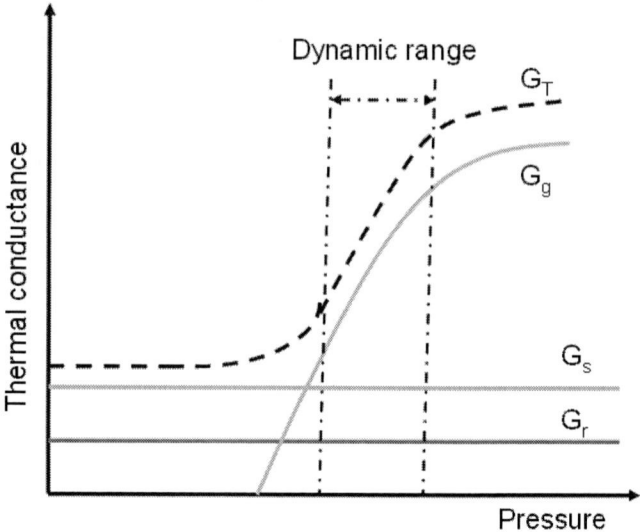

Fig.3 Thermal conductance vs. pressure

Simulation and Experimental Results

Both simulation and experiment are carried out to investigate this novel sensing method. In the aspect of simulation, the software of CFD-ACE+ was used to simulate the change trend of the voltage drop and temperature of the metal wire when the pressure inside the structure was changed.

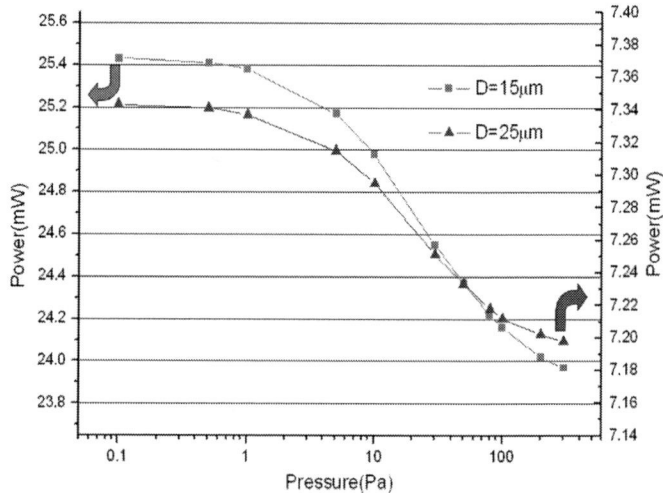

Fig.4 Simulation results for 25μm and 15μm diameter wires, I=100mA

Fig.4 shows the heating power versus ambient pressure for gauges with 25μm and 15μm diameter silicon-aluminum wires. They are both 8mm long. It can be seen that the Pirani gauge with a 15μm diameter wire shows higher sensitivity and larger dynamic range. It is consistent with the theory. As indicated in equation (3), the solid conduction (G_s) is proportional to the cross-sectional area of the metal wire. G_s is reduced while A_0 is smaller. The proportion of gas conduction to total thermal conductance will be larger. Also the resistance of the metal wire is increased as A_0 is smaller when the length is the same. So the heat generated under a same current will be more and the change quantity of equilibrium temperature is larger under various pressures, so as to the resistance and voltage drop.

To see the effect of the length, simulation for a 5mm long wire is carried out. The diameter is 25μm. The simulation results are showed in Fig.5. The results show that the gauge with a 5mm long wire has lower sensitivity. The change quantity of voltage drop is small, as in 10^{-5}V. As presented in equation (4), the gas conduction will be reduced when the length of the metal wire is shortened.

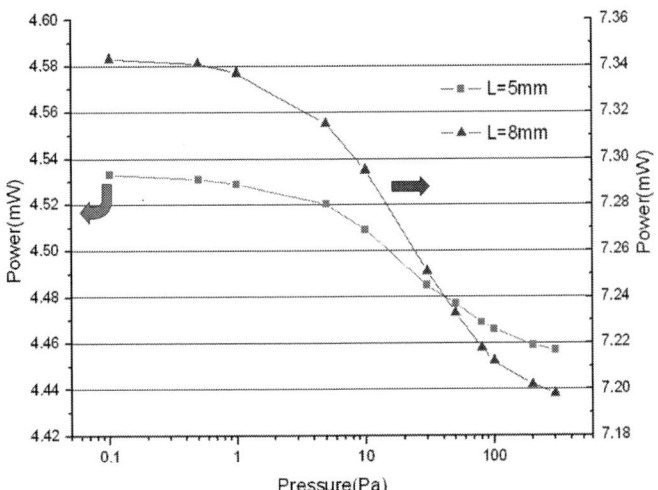

Fig.5 Simulation results for 5mm and 8mm long wires, I=100mA

Therefore the proportion of gas conduction to total thermal conductance becomes smaller and the sensitivity will be lower.

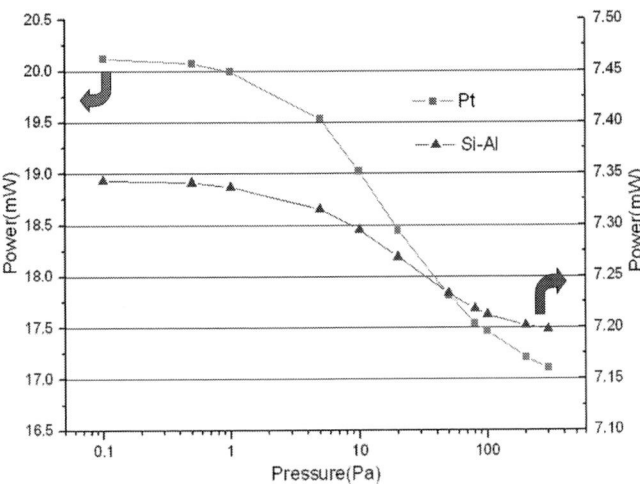

Fig.6 Simulation results for platinum and silicon-aluminum, I=100mA

Fig.6 is the simulation results for platinum and silicon-aluminum wires. They are both 8mm long and the diameter is 25□m. Compared to the silicon-aluminum wire, both dynamic range and sensitivity are improved by using platinum. The reason is like the analysis in Fig.4. The resistivity of platinum is larger than silicon-aluminum material and the heat generated under a same current will be more. The change

quantity of the resistance and voltage drop is larger under various pressures.

The curves which are corresponding to the right y-axis are the same. They are the simulation results for a 8mm long silicon-aluminum wire which the diameter is 25μm. From the figures above, it can be seen that length, diameter and material of the wire are the factors which are affecting the performance of Pirani gauge.

Based on the simulation results, the experimental study was carried out. A silicon-aluminum wire is used with the diameter of 25□m. The heating part is 8mm long. An "open" reference structure which did not contain a cap was realized to obtain calibration curves over pressure [7]. Through measuring the different characteristics of the metal wire under circumstance with various pressures, the pressure inside the package will be monitored. The sensing part was tested inside a small chamber in which the pressure can be regulated.

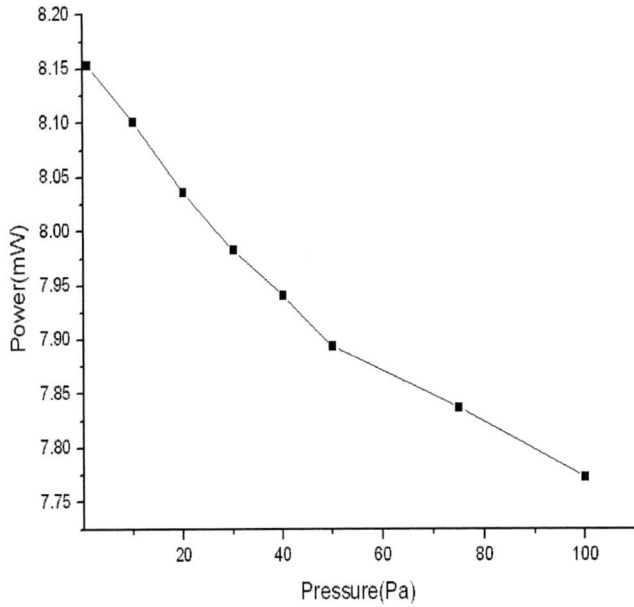

Fig.7 Heating power as a function of pressure, I=100mA

Traditionally, a four-point probe method is used to test the Pirani gauge. A constant current source applied a constant current to the metal wire, and the voltage drop was measured using a common multimeter. Under a certain current, as the pressure is changed, the heat conduction through gas will be different. And the equilibrium temperature and the resistance of the metal wire will be different. Many series of experiments are carried out under different currents and different pressures or using different samples. And curves of power versus pressure under certain current will be obtained. A typical result curve is presented in Fig.7. The dynamic range of the Pirani gauge is from 1Pa to 100Pa at least. It is can be widen if other metal materials such as platinum or tungsten are used.

Conclusions

A novel, simple micro Pirani gauge has been developed to do in-suit and real-time monitoring pressure change inside a micro vacuum package. The fabrication process of the Pirani gauge is very simple that a metal wire with a certain temperature coefficient of resistance is welded on the pads of the substrate using wire bonding technology. Based on the Pirani principle, the metal wire is the sensing and heating part. This method does not affect the structure of the packaged devices. And the using efficiency of the space of the package is improved. The dynamic range of the Pirani gauge is from 1Pa to 100Pa which can be extended using other metal materials with larger resistivity and temperature coefficient of resistance or thinner wire.

Acknowledgments

The authors would like to thank Prof. Fuming Mao from SouthEast University. This work is supported by National High-tech Research Development Plan (863 Program). (No. 2007AA04Z352)

References

1. Stark, B.H., Yuhai Mei,Chunbo Zhang, Najafi, K. "A doubly anchored surface micromachined Pirani gauge for vacuum package characterization", *Proc. MEMS 2003 Conference,* 19-23 Jan., 2003 Kyoto, Japan, pp. 506-509.

2. F.T Zhang, Z. Tang, J. yu, R.C Jin,"A micro-Pirani vacuum gauge based on micro-hotplate technology", *Sensors and actuators A,* 126 (2006),pp. 300-305.

3. J. Mitchell, G. R. Lahiji, and K. Najafi, " An Improved Performance Poly-Si Pirani Vacuum Gauge Using Heat-Distributing Structural Supports", *J. Microelectromech. Syst.,* Vol. 17, pp. 93-102, 2008.

4. Yuzhi Wang and Xu Chan,*Vacuum Technology,* 2007, pp. 326–332.

5. J. Chae, B. H. Stark, and K. Najafi, "A micromachined Pirani gauge with dual heat sinks," *Proc. 17th IEEE Int. Conf. MEMS: Maastricht MEMS Tech. Dig.,* Jan. 25–29, 2004, pp. 532–535.

6. J. Shie, B. C. S. Chou, and Y. Chen, "High performance Pirani vacuum gauge," *J. Vac. Sci. Technol. A, Vac. Surf. Films,* vol. 13, no. 6, pp. 2972 – 2979, Nov. 1995.

7. B. C. S. Chou and J. Shie, "An innovative Pirani pressure sensor", *Proc. Int. Solid State Sens. Actuators Conf. (Transducers),* Jun. 16–19,1997, pp. 1465–1468.

8. I. Simon, S. Billat, T. Link, P. Nommensen, J. Auber and Y. Manolil, "In-situ pressure measurements of encapsulted gyroscopes", *Transducers and Eurosensors '07, The 14th International Conference on Solid-State Sensors, Actuators and Microsystems,* Lyon, France, June 10-14, 2007

9. B. C. S. Chou, Y. Chen, M. Ou-Yang, and J. Shie, "A sensitive Pirani vacuum sensor and the electrothermal SPICE modeling," *Sens. Actuators A, Phys.,* vol. 53, no. 1, pp. 273–277, May 1996.

On-Package Magnetic Materials for Embedded Inductor Applications

Liangliang Li[1], Dok Won Lee[2], Kyu-Pyung Hwang[3], Yongki Min[4], Shan X. Wang[2, 5]

[1]Department of Materials Science and Engineering, Tsinghua University, Beijing, 100084, China
[2]Department of Materials Science and Engineering, Stanford University, 476 Lomita Mall, Stanford, CA 94305, USA
[3]Intel Corporation, 1900 Prairie City Road, Folsom, CA 95630, USA
[4]Intel Corporation, Chandler, AZ 85226, USA
[5]Department of Electrical Engineering, Stanford University, 476 Lomita Mall, Stanford, CA 94305, USA
liliangliang@mail.tsinghua.edu.cn

Abstract

Emerging applications in microwave communication, RFIC, and power delivery system are driving the need for the miniaturization, cost-reduction and quality optimization of inductors. In our prior work, high-quality-factor and low-resistance embedded inductors were fabricated on low-cost printed circuit board (PCB) for system-on-package applications. The inductor with a CoFeHfO magnetic core patterned by sand-blasting experimentally produced an inductance gain of ~12% over the air-core inductor. In order to further improve the inductance gain, we looked at alternative processes and materials to use in the fabrication of magnetic cores. In this paper, we investigated and compared the magnetic properties of CoFeHfO magnetic cores on the PCB substrate patterned by both wet-etching and sand-blasting. The magnetic domain images by Kerr microscopy, hysteresis loops and permeability spectra clearly showed that the CoFeHfO core patterned by wet-etching had a 3x larger permeability value and softer magnetic properties than the core patterned by sand-blasting. With the implementation of the CoFeHfO core by wet-etching, the theoretic calculation predicted an inductance gain of 32% for our embedded inductor. Besides CoFeHfO, we also explored the magnetic properties of CoTaZr film deposited on the same PCB substrate and measured its permeability value to be ~600, which theoretically indicated a gain of 128% for the inductor we designed.

Introduction

Recently, much research has been carried out on the integration of inductors on a packaging substrate, which provides plenty of usage area, has negligible substrate loss and is low-cost [1-4]. In order to improve the properties of the integrated inductors, some investigation has been done to implement magnetic materials into the inductors [5-6]. However, the behavior of the magnetic materials on a packaging substrate has not been studied thoroughly and so the enhancement of the performance of the inductors was limited. In this paper, we will study the relationship between the magnetic material and inductor performance.

In our prior work, we fabricated high-quality-factor (Q ~ 80) and low-resistance (RDC ~ 10 mΩ) embedded inductors on printed circuit board (PCB) for an application of the DC/DC buck converter [7]. Furthermore, we successfully integrated soft magnetic material CoFeHfO into the fabrication processes of the inductors. Figure 1(a) shows the photo of an embedded magnetic inductor with a ground ring fabricated on PCB and Fig. 1(b) is its schematic. In Fig. 1(b), two blue rectangular bars are the CoFeHfO magnetic cores, which were patterned by sand-blasting in the fabrication

processes [7]. For the air-core inductor with the same geometry, the magnetic cores are absent in Fig. 1(b).

To enhance the inductance gain, we fully explored the magnetic behavior, including the permeability spectra, magnetic domain images and hysteresis loops, of CoFeHfO bars patterned on PCB by both wet-etching and sand-blasting. From the experimental results, we concluded that the CoFeHfO magnetic core patterned by wet-etching has very good soft magnetic properties and high-frequency performance and is a very promising candidate for implementation into embedded inductors. For the inductor design in Fig. 1, the theoretic calculation predicted an inductance gain of 32% with a CoFeHfO core patterned by wet-etching. Besides CoFeHfO, we also performed the permeability test on a CoTaZr film deposited on the same packaging substrate. The real part of the permeability of the CoTaZr film was measured to be ~600, indicating a theoretic inductance gain of 128%.

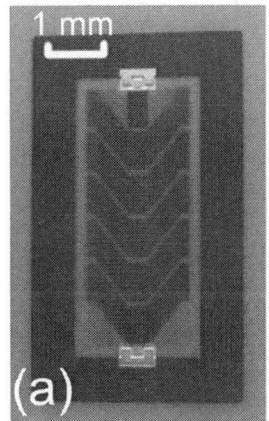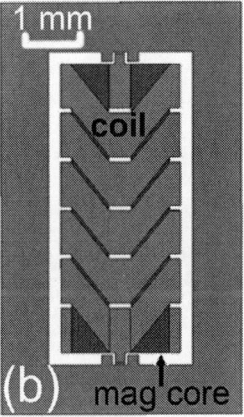

Fig. 1. (a) Photo of an embedded inductor with a ground ring fabricated on PCB. (b) Schematic of the embedded inductor. Two blue rectangular bars are the CoFeHfO magnetic cores.

Experimental Procedure

4 μm-thick CoFeHfO film was deposited on the organic dielectric layer (Ajinomoto Fine-Techno, Japan) of a PCB substrate by pulsed-dc reactive sputtering [7, 8]. The dielectric layer was an insulating layer to separate the magnetic cores from the copper conductor underneath. After the magnetic film deposition, wet-etching or sand-blasting was applied to pattern the blanket film into the rectangular bars. Figure 2

978-1-4244-4658-2/09 $25.00 © 2009 IEEE

shows the patterning procedures. For wet-etching (Fig. 2(a)), we mixed 1:6 BOE (buffered oxide etchant: 34% NH_4F, 7% HF, 59% H_2O) and DI water at a ratio of 1: 9 to form a unique etchant for CoFeHfO material. The etch-rate is ~0.2 μm/min. Shipley 3612 photoresist was exposed and developed to form a mask before wet-etching and stripped by acetone after etching. For sand-blasting, a photo-sensitive resin was used as a hard mask and the blanket film was etched away by high-speed particles. After sand-blasting, the mask was stripped by 3% NaOH solution at 45 °C.

The frequency dependence of the permeability of the magnetic samples was measured by a Ryowa PMF-3000 thin-film permeameter with an HP 8753ES network analyzer. The hysteresis loops of the samples were obtained from an SHB 109 BH loop tracer. The magnetic domain images were obtained with a custom-made Kerr imaging microscope.

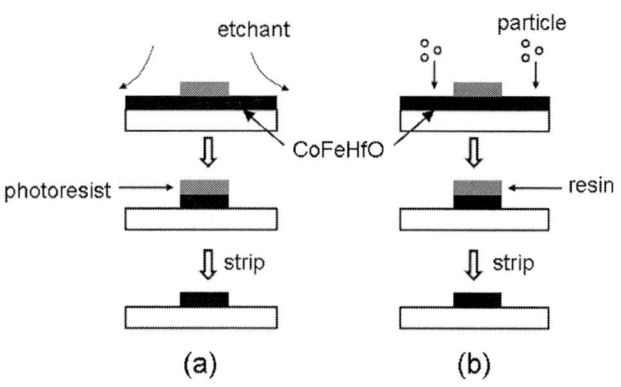

Fig. 2. Patterning procedures of CoFeHfO magnetic cores by wet-etching (a) and sand-blasting (b).

Results and Discussion

The width of the CoFeHfO magnetic cores patterned by wet-etching is 300 μm or 600 μm. The width of the cores patterned by sand-blasting is 300 μm or 700 μm. The length of all these cores is ~6000 μm. The easy axis of the CoFeHfO material is along the width of the bars [7, 9].

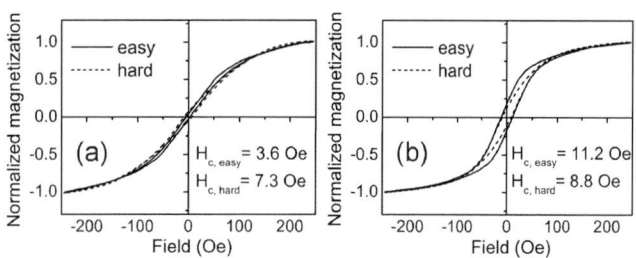

Fig. 3. Hysteresis loops of the CoFeHfO magnetic bars patterned by wet-etching (a) and sand-blasting (b).

First, we measured the hysteresis loops of the magnetic cores patterned by both methods, shown in Fig. 3. The values of the coercivity, H_c, of the samples by wet-etching (3.6 Oe along the easy axis and 7.3 Oe along the hard axis) are much smaller than those of the samples by sand-blasting (11.2 Oe

along the easy axis and 8.8 Oe along the hard axis), which indicates that the cores by wet-etching have a softer magnetic property.

Secondly, to verify the previous results, we performed Kerr microscopy on both kinds of samples to study the magnetic domain patterns. Figure 4(a) and (c) are the magnetic bars patterned by wet-etching with a width of 300 μm and 600 μm, respectively. Figure 4(b) and (d) are the bars by sand-blasting with a width of 300 μm and 700 μm, respectively. The stripe-like domains are not visible for the bars by sand-blasting, but can be clearly observed for the bars by wet-etching. These Kerr images also prove that the CoFeHfO cores by wet-etching have a good soft magnetic property.

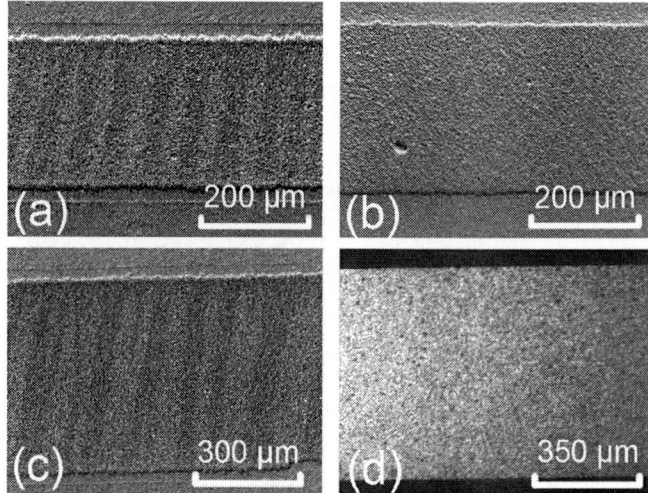

Fig. 4. Kerr images of the CoFeHfO bars on PCB. Magnetic domains are visible for (a) and (c). (a) 300 μm-wide bar patterned by wet-etching; (b) 300 μm-wide bar patterned by sand-blasting; (c) 600 μm-wide bar patterned by wet-etching; (d) 700 μm-wide bar patterned by sand-blasting.

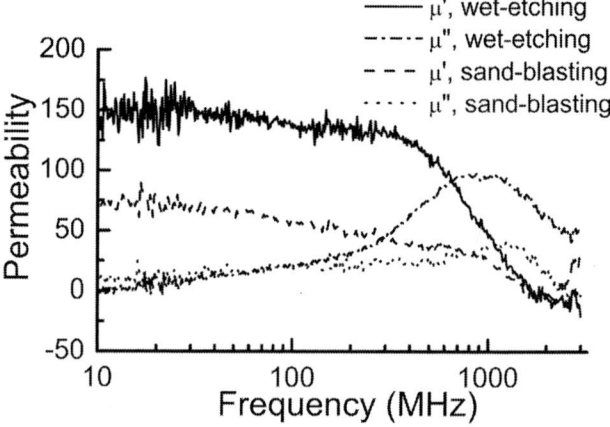

Fig. 5. Permeability spectra of the CoFeHfO magnetic bars patterned by wet-etching and sand-blasting.

Thirdly, we investigated the high-frequency performance of the magnetic cores, which affects embedded-inductor properties including inductance, resistance and quality factor. Figure 5 shows the permeability spectra of the samples. The solid line and dash-dot line represent the real part, μ', and imaginary part, μ'', of the permeability of the cores by wet-etching, respectively. The dashed line and dotted line represent μ' and μ'' of the cores by sand-blasting, respectively. The value of μ' of the cores patterned by wet-etching is ~150 at 10 MHz, which is 3× larger than that of the cores patterned by sand-blasting and close to the permeability value of the CoFeHfO film on Si wafer [9].

The experimental data above show that wet-etching can produce a CoFeHfO core with much better magnetic properties than sand-blasting. The reason why magnetic material degrades significantly after sand-blasting is that the CoFeHfO film is corroded in the NaOH solution during the stripping off of photo-sensitive resin. Figure 6 is the photo of the CoFeHfO film soaked in 3% NaOH solution at 45 °C for 30 minutes. The color of the magnetic bars changed from silver to brown after soaking, and the film peeled off at the edge of the sample. It is very likely that the patterning procedure of sand-blasting damaged the magnetic film and turned out deteriorated magnetic properties. In summary, the choice of patterning method is critical to the fabrication of magnetic cores in embedded magnetic inductors. For CoFeHfO film on PCB, wet-etching is a promising method.

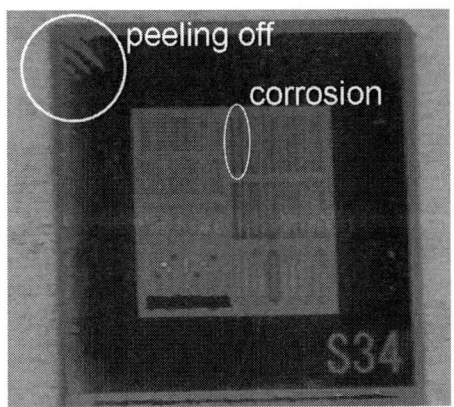

Fig. 6. Photo of CoFeHfO magnetic film soaked in 3% NaOH solution at 45 °C for 30 minutes. Some magnetic bars were corroded and their color changed from silver to dark. The film peeled off at the edge of the sample.

With the permeability data in Fig. 5, the inductance gain, g, of the inductor (Fig. 1) with a magnetic core can be theoretically calculated with the following equations:

$$g = \frac{L_{mag} - L_{coil}}{L_{coil}} = \frac{L_{core}}{L_{coil}} \quad (1)$$

$$L_{core} = \frac{1}{2}\frac{\mu_0 \mu_{eff} N^2 w_m t_m}{l_m} \quad (2)$$

$$\mu_{eff} = \frac{\mu_r}{1 + N_d(\mu_r + 1)} \quad (3)$$

where L_{mag} and L_{coil} represent the inductances of the magnetic and air-core inductors, respectively; L_{core} is the contribution of the magnetic core to the inductance, μ_o is the

permeability of vacuum, μ_{eff} is the effective permeability of magnetic core, μ_r is the relative permeability of magnetic material, N is the number of the turns of the copper coil, w_m is the width of magnetic core, t_m is the thickness of the core, l_m is the length of the core and N_d is the demagnetization factor whose value depends on the geometry of the magnetic core [10]. In Ref. [7, 11], L_{mag} of the inductor with a CoFeHfO core patterned by sand-blasting and L_{coil} were measured as 3.25 nH and 2.88 nH, respectively. Both the experimental data and theoretical calculation (by Equations 1-3) showed a gain of 12%. However, for the same inductor design with the CoFeHfO core patterned by wet-etching, the theoretic inductance gain could be as large as 32% due to the 3x increase of μ_r.

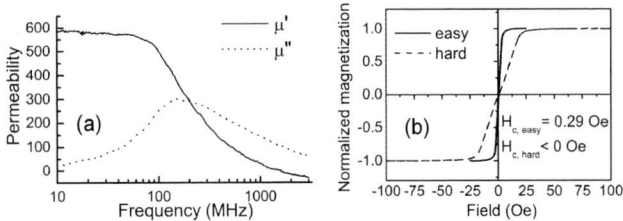

Fig. 7. (a) Permeability spectra of the CoTaZr film deposited on PCB. (b) Hysteresis loops of the CoTaZr film.

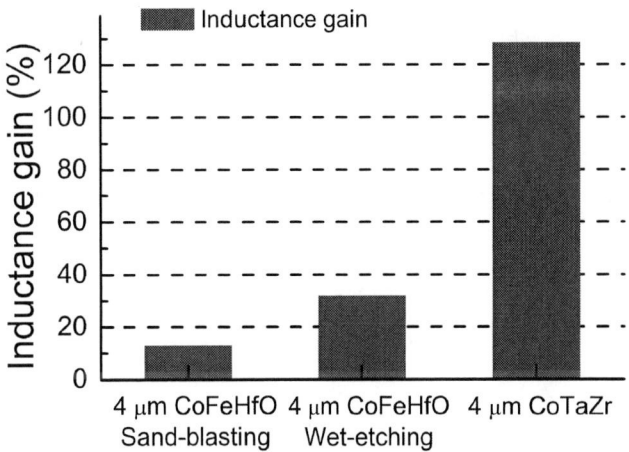

Fig. 8. Inductance gains with different magnetic cores. 4 μm-thick CoFeHfO core patterned by sand-blasting experimentally produced a gain of 12%. The theoretic calculation predicts the gains of 32% and 128% for 4 μm-thick CoFeHfO core patterned by wet-etching and 4 μm-thick CoTaZr film, respectively.

Besides CoFeHfO, we also investigated the magnetic properties of CoTaZr film on PCB. CoTaZr film was deposited on the dielectric layer of PCB by rf-bias sputtering [12]. Figure 7(a) shows the permeability spectra of 3 μm-thick CoTaZr blanket film. The solid line and dotted line represent μ' and μ'' of the CoTaZr film, respectively. The value of μ' of CoTaZr at 10 MHz is ~600, which is 4x larger than CoFeHfO. It also should be noted that μ' of the CoTaZr material decreased quickly at 100 MHz due to its low anisotropy value, but μ' of the CoFeHfO film can sustain to 500 MHz. Figure 7(b) shows the hysteresis loops of the CoTaZr film, indicating a very soft magnetic property (The coercivity is 0.29 Oe along the easy axis and too small to be measured along the hard axis). According to Equations 1-3, the inductance gain could be greatly increased to 128% for the inductor in Fig. 1 if the magnetic core is made of the CoTaZr film with a permeability value of 600. Figure 8 summarized the inductance gains of the embedded inductors with varied magnetic cores.

Conclusion

A method of wet-etching to pattern CoFeHfO blanket film into the core of the embedded inductors on PCB was successfully demonstrated. The CoFeHfO cores patterned by wet-etching showed much softer magnetic properties and better high-frequency performance than the cores we previously patterned by sand-blasting. According to the theoretic calculation, the inductance gain of the embedded inductor could be increased to 32% by implementing the CoFeHfO core by wet-etching. The core by sand-blasting was corroded by the NaOH solution during the stripping off of photo-sensitive resin, and as a result, it only gave a gain of 12%.

Besides those of CoFeHfO, the magnetic properties of CoTaZr material deposited on PCB were investigated. The relative permeability of the CoTaZr film is around 600 at low frequencies, which could increase the inductance gain to 128% theoretically. On the other hand, the permeability decreases quickly when the frequency is more than 100 MHz, which limits its applications in high frequencies.

In summary, a suitable magnetic material along with an appropriate patterning method on a packaging substrate can produce a magnetic core with superior magnetic properties, which in turn can greatly increase the inductance gain for embedded magnetic inductors.

Acknowledgment

This work is supported in part by Intel Corporation and National Science Foundation.

References

1. H. Lee and J. Y.Park, "Characterization of fully embedded RF inductors in organic SOP technology," *IEEE Trans. Adv. Packag.*, Vol. 32, No. 2 (2009), pp. 491-496.

2. W. Yun, V. Sundaram and M. Swaminathan, "High-Q embedded passives on large panel multilayer liquid crystalline polymer-based substrate," *IEEE Trans. Adv. Packag.*, Vol. 30, No. 3 (2007), pp. 580-591.

3. D M. F. Davis, A. Sutono, S.-W. Yoon, S. Mandal, N. Bushyager, C.-H. Lee, K. Lim, S. Pinel, M. Maeng, A. Obatoyinbo, S. Chakraborty, J. Laskar, E. M. Tentzeris, T. Nonaka, and R. R. Tummala, "Integrated RF architectures in fully-organic SOP technology," *IEEE Trans. Adv. Packag.*, Vol. 25, No. 2 (2002), pp. 136-142.

4. S. A. Chickamenahalli, H. Braunisch, S. Srinivasan, J. Q. He, U. Shrivastava, and B. Sankman, "RF packaging and passives: design, fabrication, measurement, and validation of package embedded inductors," *IEEE Trans. Adv. Packag.*, Vol. 28, No. 4 (2005), pp. 665-673.

5. M. Ludwig, M. Duffy, T. O'Donnell, P. McCloskey and S. C. Ó. Mathùna, "PCB integrated inductors for low power DC/DC converter," *IEEE Trans. Power Electron.*, Vol. 18, No. 4 (2003), pp. 937–945.

6. D. H. Bang and J. Y. Park, "Ni-Zn ferrite screen printed power inductors for compact DC-DC power converter applications," *IEEE Trans. Magn.*, Vol. 45, No. 6 (2009), pp. 2762-2765.

7. L. Li, D. K. Lee, K. Hwang, Y. Min, T. Hizume, M. Tanaka, M. Mao, T. Schneider, R. Bubber and S. X. Wang, "Small-resistance and high-quality-factor magnetic integrated inductors on PCB," *IEEE Trans. Adv. Packag.*, to be published.

8. L. Li, A. M. Crawford, S. X. Wang, A. F. Marshall, M. Mao, S. Thomas, and R. Bubber, "Soft magnetic granular material Co-Fe-Hf-O for micromagnetic device applications," *J. Appl. Phys.*, Vol. 97, No. 10 (2005), pp. 10F907-1-3.

9. L. Li, D. W. Lee, M. Mao, T. Schneider, R. Bubber, K. P. Hwang, Y. Min and S. X. Wang, "High-frequency responses of granular CoFeHfO and amorphous CoZrTa magnetic materials," *J. Appl. Phys.*, Vol. 101, No. 12 (2007), pp. 123912-1-4.

10. D.-X. Chen, E. Pardo, and A. Sanshez, "Demagnetizing factors for rectangular prisms," *IEEE Trans. Magn.* Vol. 41, No. 6 (2005), pp. 2077-2088.

11. D. W. Lee, L. Li and S. X. Wang, "Embedded inductors," in *Materials for Advanced Packaging*, Springer (Berlin, 2008), pp. 460-480.

12. D. W. Lee and S. X. Wang, "Multiple magnetic resonances in permeability spectra of thick CoTaZr films," *J. Appl. Phys.*, Vol. 99, No. 8 (2006), pp. 08F109-1-3.

Low-temperature Bonding of Laser Diode Chips on Si Substrates with Oxygen and Hydrogen Atmospheric-pressure Plasma Activation

Ryo Takigawa[1], Eiji Higurashi[1,2], Tadatomo Suga[1] and Renshi Sawada[3]

[1]Department of Precision Engineering, School of Engineering, The University of Tokyo, Tokyo 113-8656, Japan.
[2]Research Center for Advanced Science and Technology, The University of Tokyo, Tokyo 153-8904, Japan.
[3]Department of Mechanical Engineering, Kyushu University, Fukuoka 819-0395, Japan.
E-mail: takigawa.ryo@su.t.u-tokyo.ac.jp

Abstract

Surface activated bonding (SAB) method with atmospheric-pressure plasma treatment is an effective approach to develop low cost, low damage, and low temperature bonding technology. In this research, not only conventional low-pressure plasma treatment (Ar RF plasma) but also atmospheric-pressure plasma treatment (Ar+O_2, Ar+H_2) was investigated for low-temperature Au-Au surface-activated bonding (150°C). In the case of Au thin film to Au thin film bonding, enough bonding strength was not obtained with Ar+O_2 atmospheric-pressure plasma treatment due to Au_2O_3 formed on Au surface. However, by using Au microbump (diameter at the top: 5 µm, height: 2 µm, and pitch: 10 µm), strong bonding strength was obtained with all these plasmas. Semiconductor laser diodes chips were successfully bonded to Si substrates wiht Au microbumps at low temperature (150°C) in ambient air using Ar+H_2 atmospheric-pressure plasma treatment.

1. Introduction

Hybrid integration of laser diode chips on silicon (Si) optical bench is attractive approach for the realization of high functional and compact optical devices, such as microsensors and communication modules [1-2]. As an alternative to conventional AuSn (80 wt%Au, 20 wt%Sn) [3] solder bonding (process temperature is over 300°C), we have been developed Au-Au low-temperature direct bonding of laser diode chips with a surface-activation process using low-pressure argon (Ar) radio frequency (RF) plasma [4]. Room temperature bonding of vertical-cavity surface-emitting laser (VCSEL) chips on Si substrates with gold (Au) microbumps has also been developed by low-pressure Ar RF plasma [5]. Recently, atmospheric plasma treatment has been demonstrated to be effective for ultrasonic bonding [6]. However, in the case of ultrasonic bonding the enough positioning accuracy for passive alignment of optical components has not been obtained, because bonding tool is oscillated by the ultrasonic wave.

In this research, not only low-pressure Ar RF plasma but also atmospheric-pressure plasma treatment (Ar+O_2, Ar+H_2) was investigated for low-temperature Au-Au surface-activated bonding. In addition, low-temperature bonding of laser diode chips on Si substrates with Au microbumps was performed.

2. Experiment

2.1 Bonding Test Sample

In our experiments, we used Si chips (size: 400µm×450µm, thickness: 100µm) and Si substrates (size: 2.4mm×1.7mm, thickness: 375µm) to evaluate bonding strength. The Si chips with Ti (100nm) /Au (600nm) thin film deposited by electron beam (EB) evaporation on part (200µm×300µm) of the Si chips were used. The Ti layer insures adhesion of the Au to Si.

The Si substrates with Ti (100nm)/Au (500nm) thin film deposited by EB evaporation or with Au microbumps were used. For the fabrication of Au microbumps, 100 nm of Ti and 2.5µm of Au were deposited by EB evaporation. Then, the Au microbumps were made by standard photolithography and dry etching. The diameter at the top of Au microbumps and the bump pitch were approximately 5µm and 10µm, respectively. The height of the Au microbumps was 2µm. Figure 1 shows scanning electron microscope (SEM) photograph of Au microbumps on Si substrate.

Table 1 Ar RF plasma treatment conditions

Plasma Power	100 W
Gas	Ar: 10 sccm
Treatment time	240 s

Table 2 Atmospheric pressure plasma treatment conditions

Plasma Power	150 W
Gas	Oxygen: 27.0 mL/min Hydrogen: 0.03 L/min Ar: 2.14 L/min
Gap distance	6 mm
Treatment time	Oxygen: 30 s Hydrogen: 60 s

Table 3 Bonding conditions

Bonding temperature	150 °C
Contact load	600 MPa
Bonding time	30 s
ambience	ambient air

2.2 Bonding process

SAB is a direct bonding method that joins two clean surfaces using the adhesive force of surface atoms at room temperature or low temperature. After organic contaminants (carbon) on the surfaces of Au were removed by surface activation process using an Ar RF plasma [7] or atmospheric-pressure plasma treatment [8] (Ar+O_2, Ar+H_2), Au-Au bonding was carried out only by contact in ambient air with applied static pressure. It is required that the time of the exposure of the bonding surface in air after plasma cleaning is as short as possible. The time from the activation to the bonding is about 5 min. The conditions of surface activation by Ar RF plasma and atmospheric-pressure plasma are

978-1-4244-4658-2/09 $25.00 © 2009 IEEE

summarized in Table 1 and Table 2, respectively. Table 3 shows the bonding conditions in these experiments.

3. Results and Discussion

3.1 Bonding strength

Figure 2 shows the measured die shear strength of Si chips bonded on Au thin film (bonding temperature: 150 °C). In the case of Au thin film to Au thin film bonding, strong shear strength was obtained with Ar RF low-pressure plasma and Ar+H_2 atmospheric pressure plasma treatment. On the other hand, the enough bonding strength was not obtained with Ar+O_2 atmospheric-pressure plasma treatment. To evaluate surface activation effects, surface analysis by x-ray photoelectron spectroscopy (XPS) has been performed. Figure 3 shows XPS spectra of Au 4f and Au_2O_3 4f on the Au surface before and after plasma irradiation. The peak intensity of Au_2O_3 appeared in the case of only Ar+O_2 atmospheric-pressure plasma treatment. Figure 4 shows XPS spectra of Au 4f and Au_2O_3 4f on the Au surface after heating at 150 °C for 60 s (bonding conditions).

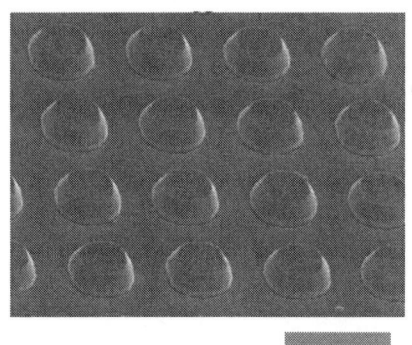

10 μm

Figure 1 Au microbumps on Si substrate

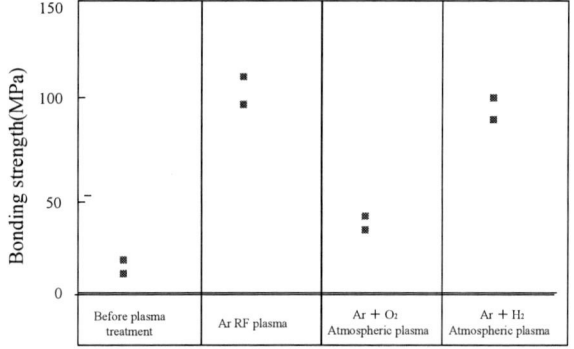

Figure 2 Measured die-shear strength on Si chips bonded on Au thin film treated with various plasmas (bonding temperature: 150 °C).

Although it is well known that Au_2O_3 can be spontaneously dissociated to initial Au in air at room temperature, Figure 4 means that Au_2O_3 exits in the bonding process. These results indicate that the bonding strength decreased due to Au_2O_3 formed by Ar+O_2 atmospheric-pressure plasma treatment.

Figure 5 shows the measured die shear strength of Si chips bonded on Au microbumps treated with various plasmas

(bonding temperature: 150 °C). In the case of Au thin film to Au microbump bonding, strong shear strength was obtained with all these plasmas. It is considered that Au microbumps are easier to deform than thin films and this deformation leads to new surface appearance such that interatomic attraction occurs. Au microbumps were more effective than Au thin film for improving the bondability.

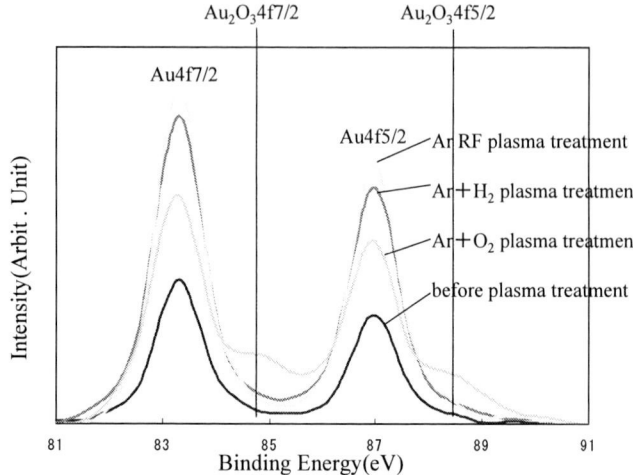

Figure 3 XPS spectra of Au 4f and Au_2O_3 4f on the Au surface before and after plasma irradiation

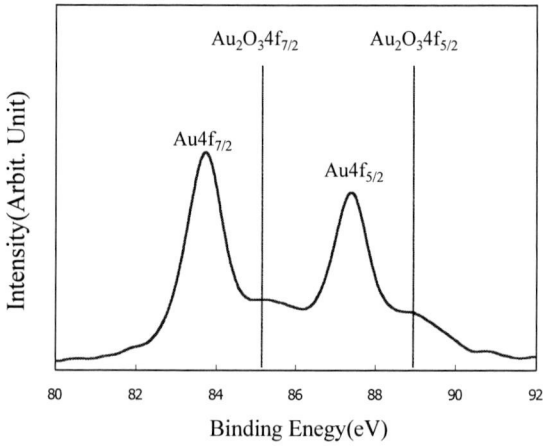

Figure 4 XPS spectra of Au 4f and Au_2O_3 4f on the Au surface with Ar+O_2 atmospheric-pressure plasma treatment after heating (150 °C, 60 s)

3.2 Laser diode chips bonding

GaAs-based VCSEL chips were bonded to Si substrates with Au microbumps at low temperature in ambient air with Ar+H_2 atmospheric-pressure plasma treatment (bonding temperature: 150 °C, contact load: 600MPa (900 gf), treatment time: 60 s). The optoelectronic characteristics of VCSEL chips were measured by light-current-voltage (L–I–V) inspection to detect laser degradation. Figure 6 shows the typical L–I–V characteristics before and after bonding. The measurement was performed by the continuous wave operation. The VCSEL chips functioned normally after bonding. The threshold current did not increase and output power did not decrease after bonding. The measured results of

978-1-4244-4658-2/09 $25.00 © 2009 IEEE   400

light-current-voltage (L–I–V) characteristics indicated no significant degradation of laser diode chips after bonding.

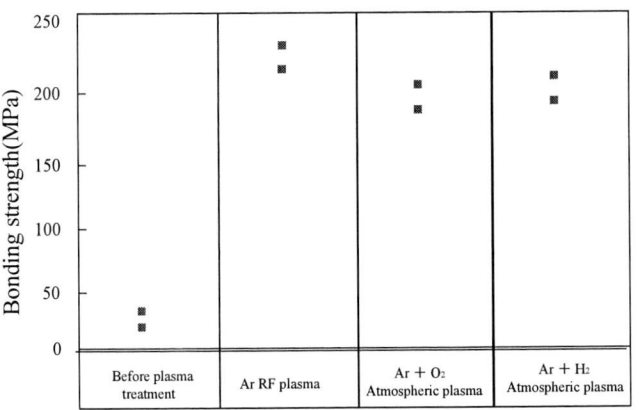

Figure 5 Measured die-shear strength on flip-chips bonded on Au microbumps treated with various plasmas (bonding temperature: 150 °C).

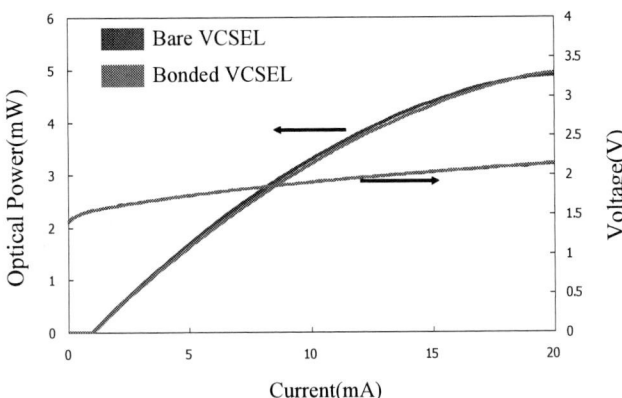

Figure 6 Comparison of the L-I-V characteristics of the VCSEL chip before and after bonding with Ar+H$_2$ atmospheric pressure plasma treatment

4. Conclusions

We investigated low-temperature Au-Au bonding (150 °C) with atmospheric-pressure plasma (Ar+O$_2$, Ar+H$_2$). In the case of Au thin film to Au thin film bonding, the bonding strength was not obtained with Ar+O$_2$ atmospheric-pressure plasma treatment due to Au$_2$O$_3$ formed on Au surface. In the case of Au thin film to Au microbump bonding, the strong bonding strength was obtained with all plasma treatment. Semiconductor laser diodes chips were bonded to Si substrates with Au microbumps at low temperature (150 °C) in ambient air with Ar+H$_2$ atmospheric-pressure plasma treatment. The measured results of L–I–V characteristics indicated no significant degradation of laser diode chips after bonding. It is expected that bonding technique using Au microbumps with atmospheric-pressure plasma activation is useful for low cost optical packaging because vacuum pumps and large vacuum chambers are not required.

Acknowledgments

This work was supported in part by Global COE Program (Global Center of Excellence for Mechanical System Innovation), Ministry of Education, Culture, Sports, Science and Technology (MEXT), Japan and the 2006 Industrial Technology Research Grant Program from New Energy and Industrial Technology Development Organization (NEDO) of Japan.

References

1. E. Higurashi *et al.*, "An integrated laser blood flowmeter," *J. Lightw. Technol.*, vol.21, no.3 (2003), pp.591-595.
2. E. Higurashi *et al.*, "Micro-encoder based on higher-order diffracted light interference," *J. Micromech. Microeng.*, vol.15, no.8 (2005), pp. 1459-1465.
3. H. Okamoto *et al.*, Eds., *Phase Diagram of Binary Gold Alloys*. Metals Park, OH:ASM, 1087, pp.278-289.
4. E. Higurashi *et al.*, "Low-Temperature Bonding of Laser Diode Chips on Silicon Substrates Using Plasma Activation of Au Films," *IEEE Photon.Technol.Lett.*, vol.19, no.24 (2007), pp. 1994-1996.
5. R. Takigawa *et al.*, "Room-Temperature Bonding of Vertical-Cavity Surface-Emitting Laser Chips on Si Substrates Using Au Microbumps in Ambient Air," *Appl. Phys. Exp.*, vol.1, no.11 (2008), pp.112201-1-112201-2.
6. J.M Koo *et al.*, "Effect of Atmospheric Pressure Plasma Treatment on Transverse Ultrasonic Bondingof Gold Flip-Chip Bump on Glass Substrate," *Jpn. J. Appl. Phys.,* vol.47, no.24 (2008), pp. 4309-4313.
7. T. Suga *et al.*, "Surface activated bonding for new flip chip and bumpless interconnect systems," in *proc. 52nd Electron. Components Technol. Conf.*, San Diego, CA, 2002, pp.105-111.
8. Y. Sawada, "Applications of Atmospheric-Pressure Glow Plasma," *J. Plasma Fusion Res.*, vol.79, no.10 (2003), pp. 1022-1028.

Integration of GaN Thin Film and Dissimilar Substrate Material by Au-Sn Wafer Bonding and CMP

Shengjun Zhou[1], Zhaohui Chen[1], Bin Cao[2], Sheng Liu[1, 2, 3] *

[1]Research Institute of Micro/Nano Science and Technology
Shanghai Jiao Tong University, Shanghai 200240, P. R. China
[2]Wuhan National Laboratory for Optoelectronics
Huazhong University of Science & Technology, Wuhan, 430074, P. R. China
[3]Institute of Microsystems, Huazhong University of Science & Technology, Wuhan, 430074, P.R.China
victor_liu63@126.com

Abstract

GaN thin film grown on sapphire substrate of 50mm*50mm in size are successfully bonded and transferred onto Si substrate using Au-Sn wafer bonding followed by grinding, chemical mechanical polishing (CMP) and dry etching. The GaN/sapphire structures are integrated to receptor Si substrate by thermal pressure bonding process. The bonding medium comprises Au-Sn multilayer composite deposited directly on the object to be bonded. Sapphire substrate is detached from GaN epi-layer by combining mechanical grinding, CMP with dry etching process. The dissimilar Si substrate provides support to the GaN-based LED epitaxial layer during the CMP process, and takes the role of conducting and high heat-dissipating substrate. The CMP can remove or reduce most of the scratches produced by mechanical grinding, recovering both the mechanical strength and wafer warpage to their original status and resulting in a smoother surface. The results have been presented.

1. Introduction

Recently, the Group-III nitride based semiconductors have emerged as the leading materials for realization of high performance light emitters from ultraviolet (UV) to the blue and green spectral regions[1,2,3]. Generally, GaN-based light-emitting diodes (LEDs) are grown on the sapphire (Al_2O_3) substrate. GaN epitaxial sturcture grown on sapphire substrate has poor crystal quality due to the large mismatch in lattice constants and thermal expansion coefficients between the nitride epi-layer and sapphire substrate. Besides, the poor thermal conductivity of sapphire substrate has been identified to the primary limitation for the application of high power GaN-based LED. Therefore, it is necessary to replace the sapphire substrate with a secondary substrate which has higher thermal conductivity.

Typically, GaN/sapphire structures are integrated onto dissimilar substrate material such as silicon through wafer bonding and laser lift-off (LLO) process [4, 5, and 6]. Wong successfully transferred the GaN epi-layer onto Si substrate by using transient-liquid-phase Pd-In wafer-bonding process followed by laser lift-off [7]. The advantage of Pd-In bonding system lies in that it can produce a high-melting point bonding layer under a lower bonding temperature. However, the LLO method has some drawbacks. During the LLO process, the temperature should be above 900^0C at the GaN/sapphire interface due to the absorption of radiation, which leads to the local decomposition of film at the GaN/sapphire interface, and the decomposition can cause a variety of deformations and create localized stresses that can produce defect and cracks.

Such cracks significantly decrease yields. Another drawback of this approach lies in that the bonding layer can be affected, because the bonding layer is only several microns away from GaN/sapphire interface [8]. In order to solve these problems, an alternative method to replace sapphire substrate with a high thermal conductivity substrate is combining the wafer bonding technique with mechanical grinding, chemical mechanical polishing and dry etching process.

In our study, the LED wafer was first flip bonded to the dissimilar substrate such as silicon by Au-Sn wafer bonding technique in order to detach sapphire substrate from GaN epitaxial layers. The dissimilar substrate provides support to the GaN-based LED epitaxial layer during the chemical mechanical polishing (CMP) process, and takes the role of conducting and high heat-dissipating substrate. The sapphire substrate was removed by integrating mechanical grinding with CMP and dry etching process.

2. Experiments

To investigate the Au-Sn wafer bonding and detach sapphire substrate from GaN epitaxial layer, LED wafer and Si wafer were used for experimental samples. The LED structure consists of a 375-um-thick sapphire substrate, a 2-um-thick unintentionally doped GaN, a 2-um-thick n-GaN layer, an active region with ten periods of InGaN/GaN multiple quantum well, and a 0.2-um-thick p-GaN. GaN epitaxial layer was grown on the sapphire substrate by Metal Organic Chemical Vapor Deposition (MOCVD). The GaN-based LED sample was clean with a standard solvent. A 390nm thick indium-tin-oxide (ITO) was deposited on the upper surface of p-GaN using an electron beam deposition system, followed by annealing at 550°C in flowing nitrogen for 10min to obtain a good ohmic contact to p-GaN. A 500nm thick Al reflective layer was deposited to prevent light from reaching the Si substrate. After the deposition of reflective Al layer, Sn was deposited on the top of Al layer in a thermal evaporator followed by the deposition of an outer gold layer. The outer gold layer forms Au-Sn compound that can reduce tin oxidation. Before the wafer bonding process, a heavily doped (111) oriented silicon was used for receptor substrate. The heavily doped silicon was cleaned by standard organic acid solution, and the Si substrate was deposited with chromium and gold where the chromium layer enhanced the adhesion. The ohmic contact layer and bonding layer configuration were shown in Fig.1.

978-1-4244-4658-2/09 $25.00 © 2009 IEEE

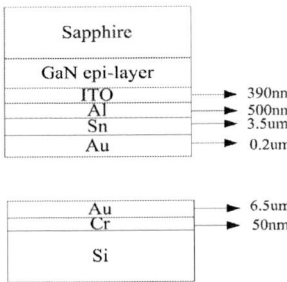

Fig.1. The ohmic contact layer, reflective metal layer and multilayer Au-Sn composite structure showing the bonding configuration.

During the bonding process, the Si substrate was laid on a graphite boat, and GaN/sapphire was placed on the Si substrate. To ensure good contact between the GaN/sapphire and Si substrate, the GaN/sapphire was held down with considerable pressure. The optimum bond force has to be researched to obtain void free bonding because voids existing along the bonding interface will cause high localized stress, consequently increase the probability of wafer cracking and raise the chip operating temperature [9]. Additionally, the optimization should take into account the possibility of spillage of melted solder shorting the active region of the LED and consequently form Schottky contact [10].

After wafer bonding process, mechanical grinding, chemical mechanical polishing (CMP) and dry etching process were integrated to remove sapphire substrate for fabricating thin-film LEDs. First, sapphire substrate was thinned from 375um to around 100um by mechanical grinding including coarse and fine grinding, then sapphire substrate was thinned from 100um to about 8um by CMP, finally dry etching was used to remove the remaining sapphire substrate. The etching was carried out in an ICP etcher (Oxford Plasma lab system 100). The process flow was shown in Fig. 2.

Fig. 2. Process flow for fabrication of thin-film LEDs. (a) GaN/Sapphire was flip bonded onto Si substrate using Au-Sn eutectic bonding. (b) Sapphire substrate was thinned from 375um to around 100um by mechanical grinding. (c) Sapphire substrate was thinned from 100um to about 8um by CMP. (d) The remaining sapphire substrate was removed by ICP dry etching.

In order to provide a smooth surface for the n-contact formation, the undoped GaN was etched away to expose the n-GaN layer by ICP etching. Then, the GaN epi-layer was cleaved into micro-units which suppressed the transverse propagation of light emanating from the active region and directed the light to the top surface of n-GaN through a short path. The Cr/Pt/Au (20nm/20nm/1200nm) films were deposited on the exposed n-GaN layer and back surface of the Si substrate to form ohmic contact.

3. Results and discussion

The scanning electron microscope (SEM) cross-sectional image of bonding interface was shown in Fig. 3. As shown in the figure, the continuously uniform Au-Sn bonding layer was formed at the interface.

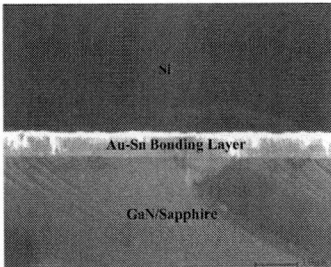

Fig.3. The SEM cross-sectional image of bonding interface

The thinning of sapphire substrate was done by fast mechanical grinding followed by CMP. Fig. 4 demonstrated the results of surface microscopy of grinded and polished sapphire substrate. As shown in Fig. 4, the sapphire surface appeared some scratches, which were produced by mechanical grinding. After polished by CMP, the scratches could be reduced or eliminated. It was worth noting that grinding generated wafer warpage because of damaged layer created during the grinding process. The CMP removed or reduced most of the damage produced by coarse and fine grinding, recovering both the mechanical strength and wafer warpage to their original status and resulting in a smoother surface [11].

Fig. 4. Surface microscopic images of sapphire substrate. (a) grinded sapphire substrate. (b) polished sapphire substrate.

After the sapphire substrate was thinned to about 8um by grinding and CMP, the BCl_3/Ar gas combinations were used to etch the remaining sapphire substrate. We used the 2000W /150W (ICP/RF power), 5mTorr, 90% BCl_3/10% Ar condition at a total flow rate of 60sccm to etch the remaining sapphire substrate. The etch rate was up to 1809 A/min. As the total etch time can take up about 45 minutes, the sapphire wafer temperature should be controlled and maintained by backside helium cooling during the etch process.

4. Conclusions

GaN thin-flims were successfully bonded onto Si using Au-Sn wafer bonding technique. Sapphire substrate was detached from GaN-based LED epi-layer by integrating mechanical grinding with CMP and dry etching process. The CMP can remove or reduce most of the scratches produced by mechanical grinding, recovering both the mechanical strength and wafer warpage to their original status and resulting in a smoother surface.

Acknowledgments

This work was supported by a key project of the National Natural Science Foundation of China under Project 50835005. The authors would like to acknowledge AquaLite Optoelectronics Co., Ltd. for the support in some experimental processes.

References

1. S. Nakamura, T. Mukai, M. Senoh, "Candela-class high-brightness InGaN/AlGaN double-heterostructure blue-light-emitting diodes ," *Appl. Phys. Lett. Vol. 64, No.13 (1994), pp.1687-1689.*

2. A. Bell, R. Liu, F. A. Ponce *et al*, "Light emission and microstructure of Mg-doped AlGaN grown on patterned sapphire," *Appl. Phys. Lett., Vol. 82, No.3 (2003), pp.349-351.*

3. F. Dwikusuma, D. Saulys, T.F. Kuech, "Study on sapphire surface preparation for III-Nitride heteroepitaxial growth by chemical treatments," *J. Electrochem. Soc., Vol.149, No.11 (2002), pp. G603- G608.*

4. W.S. Wong, T. Sands, N.W. Cheung, "Damage-free separation of GaN thin films from sapphire substrates," *Appl. Phys. Lett. Vol.72, No.5 (1998),pp. 599-601.*

5. L. X. Tan, J. Li, S. Liu, "Modeling and simulation of laser lift-off process for LED's substrates," *2007 8th International Conference on Electronic Packaging Technology.*

6. P. R. Tavernier, D. R. Clarke, "Mechanics of laser-assisted debonding films," *J. Appl. Phys., Vol. 89, No. 3 (2001), pp. 1527-1536.*

7. W. S. Wong, A. B. Wengrow, Y. Cho, *et al*, "Integration of GaN thin films with dissimilar substrate materials by Pd-In metal bonding and laser lift-off," *J. Electron. Mater., Vol. 28, No. 12 (1999), pp. 1409-1413.*

8. S. C. Hsu, C. Y. Liu, "Fabrication of thin-GaN LED structures by Au-Si wafer bonding," *Electrochem. Solid State Lett., Vol. 9, No. 5 (2006), pp.G171- G173.*

9. G. S. Matijasevic, C. Y. Wang, C. C. Lee, "Void free bonding of large silicon dice using gold-tin alloys," *IEEE Transactions on Components, Hybrids, and Manufacturing Technology, Vol. 13, No. 4 (1990), pp. 1128-1134.*

10. T. S. Thang, D. Sun, H. K. Koay *et al*, "Characterization of Au-Sn eutectic die attach process for optoelectronics Device," *2005 International Symposium on Electronics Materials and Packaging, Tokyo, Japan, December, 2005, pp. 118-124.*

11. R. R. Tummala, M. Swaminathan, Introduction to System-on-Package (SOP) , McGraw-Hill (New York, 2008), pp. 128-142.

Modeling of Nanostructured Polymer-Metal Composite for Thermal Interface Material Applications

Zhili Hu[1,2,3], Björn Carlberg[2], Cong Yue[1], Xingming Guo[3], Johan Liu[1,2]

[1]Key Laboratory of Advanced Display and System Applications, Ministry of Education and SMIT Center
School of Mechatronics Engineering and Automation, Shanghai University, Shanghai 200072, China
[2]SMIT Center and BioNano Systems Laboratory, Department of Microtechnology and Nanoscience
Chalmers University of Technology, SE-412 96, Göteborg, Sweden
[3]Shanghai Institute of Applied Mathematics and Mechanics, Shanghai University,
Yanchang Road 149, Shanghai 200072, P.R. China
Corresponding author Email: jliu@chalmers.se

Abstract

Previous studies have discovered a unique type of nanostructured polymer-metal composite for thermal interface material with effective thermal conductivity of 8 W/mK. It is a promising result but extensive efforts are still required to further enhance the thermal conductivity. Therefore, this paper will try to help the process with modeling and simulation. Calculations reveal the alignment of the fibers have insignificant influence. Therefore volume percentages of fiber together with mean interface temperature become the dominating parameters of effective thermal conductivities of thermal interface material. Based on this approximation, simulation was taken which showed good results in comparison with experimental data. However, the preferred volume percentage of fibers (33%) was slightly too large according to the surface image of thermal interface material.

Introduction

According to Moore's law, since the middle of 20th century the integration density of microelectronic systems has increased exponentially. Related with this, microprocessors have traditionally exhibited power densities doubling every 36 months [1, 2] and, despite the temporary halt due to the introduction of multi-core processors, the trend is expected to continue. Hence the need of efficient thermal management of microelectronic packages has emerged, with 2007 iNEMI Roadmap identifying thermal management as a research priority. This includes the need for "new interface materials" [3].

iNEMI's thermal management road map requires thermal conductivity of thermal interface materials (TIM) to be as high as 10 W/(mK). Despite the existence of various kinds of high thermal conductivity materials, e.g. metals and ceramics, iNEMI's aim is difficult to achieve. The reason arises from the complexity of electronic packaging. For example, if pure metals are used between die and heat spreader, the coefficient of thermal expansion mismatch between die and heat spreader and TIM will cause delamination. If ceramic pads are put between die and heat spreader directly, the interface resistance will nullify the thermal conductivity. Thus, novel ideas emerge [4]. One method is thermal conductive adhesive (TCA) which consists of highly conductive metals, e.g. silver flakes, mixed into low conductive polymers, e.g. epoxy. With considerable efforts, TCA's present thermal conductivity of roughly 9W/(mK) is approaching iNEMI's requirement [5]. In the meantime an inverse idea emerges which involves mixing low conductive polymers with high conductive metals. This idea has been successfully implemented by Carlberg et al., who embedded polyimide fibers into a eutectic In/Bi/Sn alloy [6]. This alloy has very low melting point which helps to diminish the interface resistance between adjacent components. The purpose of polyimide fiber is to prevent failures due to mismatch in the coefficient of thermal expansion. The total thermal resistances are as low as 8.5 Kmm^2/W at bond-line thicknesses of approximately 70 µm, corresponding to an effective thermal conductivity of 8 W/(mK). This is a promising result but further studies are required.

On the other hand, TIM shows great potential on the models and simulations. TCA, for example, has been extensively studied with modeling and simulation, and valuable results were acquired. Previous research discovered the filler effect on thermal conductivity of TCA [5]. Subsequent studies predicted the anisotropic behavior of TCA by taking the geometry of filler into account and their results matched aforementioned experimental data [7]. Based on the model, further improvement of TCA's performance was made possible. The shrinkage of epoxy was pointed out as the main factor for thermal conductivity of metal filler-based adhesives [8]. The success of modeling and simulation on TCA encourages modeling and simulation on nanostructured polymer-metal composite mentioned above.

Modeling and simulation

Geometry configurations of the TIM were given by Carlberg et. al. The total thickness was roughly 100 µm, the pore size was estimated by visual inspection to 5-15µm, the mean fiber diameter was 2.1µm, and the surface metallization was approximately 300 nm thick [6]. All these dimensions were much larger than the electron's mean free path in metal, thus the usage of finite element method with continuum theory without quantum effect was rendered possible. A type of software called COMSOL was used to implement the simulation. Most necessary material parameters were given by the manufacturers. Thermal conductivity k of the alloy was 19 W/(mK) at 40.1°C, and 0.52 W/(mK) for polyimide fibers. Electrical conductivity of alloy was 1.6428e+06 Ω-1cm-1, which was much lower than that of pure indium (1.1947e+07 Ω-1cm-1, at 20°C). This decrease was caused

978-1-4244-4658-2/09 $25.00 © 2009 IEEE

by the increase of crystal boundaries, which caused strong electron and phonon scattering.

The volume percentage of fiber, which was of outmost importance for the simulation work, was not given by Carlberg *et al.*. To solve this problem, a photo of the material's surface was processed with Matlab and its pixels were analyzed. By calculating the percentage of lighter or darker pixels, the area percentage of fiber could be estimated, which would approach the volume percentage of fiber if the fibers were dispersed evenly. The percentage was evaluated to 23-28%. The value could not be made more precise due to vague borders between fiber and metal in the photo. Moreover, only one photo was taken of TIM surface, which could hardly represent the overall structure of TIM. To compensate these inaccuracies, this paper will examine several different volume percentages of fiber, with emphasis on three percentages that approach the aforementioned predicted values; 23%, 28% and 33%.

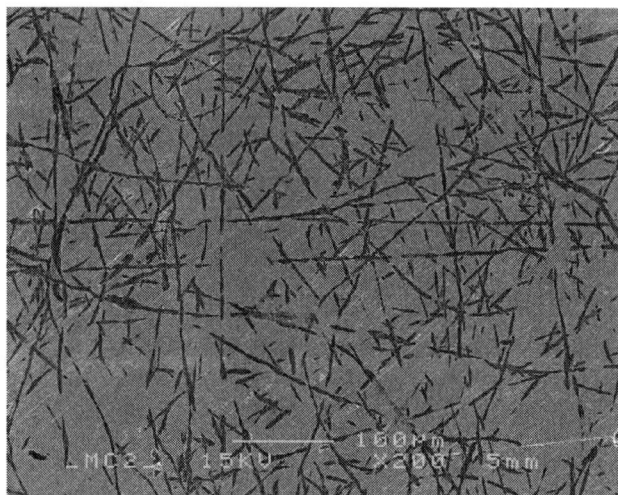

Fig. 1. Photo image of the TIM's surface. The dark lines consist of fibers and the lighter area represents the alloy.

Several cases have been considered to test the influence of different alignment of fibers. 3D or 2D models were used of fibers that were embedded inside the alloy. The fibers have diameters on micrometer scales; therefore their cross-sections are likely to be shaped as circles, and in the models they will be regarded as cylinder and circles for 3D and 2D respectively. In the TIM, the alloy occupies spaces surrounding the fibers. Likewise, in the geometry model, the alloy fills space around the cylindrical or circular fibers, forming a cube or square. The corresponding geometric models and theirs simulation results are listed below.

Case 1: A single fiber with cylinder shape inside cubic alloy cell. Volume percentage of fiber is 27%, thermal conductivity k of the TIM drops to 42.6% of that of pure alloy.

Case 2: Several holes parallel to bottom surface inside cubic alloy. The holes' positions in vertical cross-section plane of alloy cell are random. A volume percentage of fiber is 27%, and k drops to 44.5% of that of pure alloy. The holes are in fact distributed among 97.5% of the height of the cube

in order to avoid tangent contact between fiber and boundary of alloy cell, due to the difficulty to mesh near tangent area.

Case 3: A lot of random parallel holes inside the alloy. Volume percentage of fiber is 27%, and k drops to 43.7%. 2D was used in order to save computational cost.

(a) Case 1

(b) Case 2

(c) Case 3

Fig. 2. Several geometric models for polymer-metal composite TIM in COMSOL. In case 1 two adjacent cubes are shown, the lower one stands for the TIM, the upper one stands for pure alloy.

All these cases show insignificant difference of thermal conductivity. This facilitate our further work since only the volume percentages of fiber and not their alignment are to be considered in order to make a qualitative prediction. In what follows, the simple case 1 model will be used and this model will be simplified further to a 2D model.

The melting point of the alloy used by Carlberg et. al. was merely 60°C, which implies that it could be in liquid state during usage. As only the thermal conductivity of solid alloy is given by the manufacturer, it is necessary to make an estimation of thermal conductivity of liquid state alloy. To do that, the contribution of phonon to thermal conductivity is supposed to vanish when the alloy liquefies. Thus, the thermal conductivity of liquid alloy depends simply on electron transportation. The Wiedemann-Franz law predicts the contribution of electron transportation to the thermal conductivity as

$$\frac{K}{\sigma} = LT \tag{1}$$

and with the Lorenz number, is equal to

$$L = \frac{K}{\sigma T} = \frac{\pi^2}{3} \left(\frac{k_B}{e} \right)^2$$

$$= 2.44 \times 10^{-8} \, \mathrm{W\,\Omega\,K^{-2}} \tag{2}$$

As a result, contribution by electron conductance to the thermal conductivity of the alloy is estimated as 13.7 W/(mK) at 68°C, 13.3W/(mK)W/(mK) at 60°C and 13.0 W/(mK) at 52°C. In particular at 40.1°C, this contribution is 12.5 W/(mK). By making comparison with empirical thermal conductivity of the alloy of 19 W/(mK) at 40.1°C, the contribution by phonon transportation to the thermal conductivity is estimated as 6.5 W/(mK) at the given temperature. This value can be approximated as a constant unaffected by temperature, because the Debye temperature of metal elements composing the alloy is close to the experiment environment temperature [9].

In the work of Carlberg et. al. several data were taken at 60°C, which was the melting point of the alloy, thus implying a mixture of solid and liquid alloy. The volume percentage of liquid alloy was not given and we have to fit it according to the experimental data. The volume percentage is calculated to be 37.5%, corresponding to a bulk thermal conductivity of 17 W/(mK).

Results and Discussion

Combining the thermal conductivities of bulk solid and liquid alloy together with the decreased thermal conductivity caused by fiber, effective thermal conductivities as a function of mean interface temperature with different volume percentages of fiber (marked as 'v%') are shown in Figure 3. To verify these results, experimental data were compared. In Carlberg et. al.'s experiment, samples with different assembly pressures were tested with those with higher assembly pressure showing slightly higher thermal conductivity. A possible explanation would be that high assembly pressure can reduce contact thermal resistance. To minimize the influence of contact thermal resistance, only the experimental data with highest assembly pressure are shown in the figure.

The results of the simulation correspond with the experimental data. The test results with volume percentages of fiber of 33% coincide roughly with the experimental data at 52°C and 68°C. Furthermore, simulations with volume percentages of fiber of 23% or 28% show slightly higher thermal conductivity compared with the experimental data. This suggests an underestimation of volume percentage of fiber from surface image of TIM (23-28%). This underestimation is possibly caused by the inaccuracy during the experiment, since the only surface image taken by Carlberg et. al. was from the top of TIM and without any polishing. Therefore, despite the previous calculation,

volume percentage of fiber is suggested to be 33% by this paper.

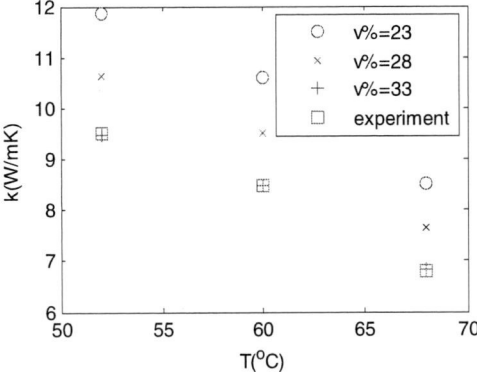

Fig. 3. Effective thermal conductivities as a function of mean interface temperature, according to simulation with different volume percentages of fiber (marked as 'v%'), experiment data and theoretical prediction.

A theoretical prediction of thermal conductivity as function of mean interface temperature can be made based on parameters obtained above. By combining phonon's contribution to the thermal conductivity of 6.5 W/(mK) together with electron's contribution to the thermal conductivity according to the Wiedemann-Franz law, the curve with volume percentages of fiber of 33% is plotted in Figure 5 as solid line. This curve may not be accurate when it approaches 60°C.

Some predictions cannot be verified based on the model above, e.g. effect of thermal conductivity of polyimide, or volume percentage of polyimide, on final thermal conductivity of TIM. These values are checked by simulations and are shown in Figure 5. It can be seen that the thermal conductivity (k) of polyimide has small influence on TIM's thermal conductivity. When k = 0.52 W/mK, the thermal conductivity of TIM is unaffected by the thermal conductivity of polyimide. This, on the other hand, shows quite good agreement with Figure 4, since heat flux's inside area corresponding to polyimide is almost 0. Our studies conclude that further attempts to increase the thermal conductivity of polymers are futile as it cannot in reality exceed 1 W/mK.

However the volume percentage of polyimide has great influence on TIM's thermal conductivity. Figure 6 shows how the thermal conductivity of TIM with 10% volume percentage of polyimide is more than twice of the conductivity value when+ volume percentage of polyimide is 50%. This indicates the benefit to decrease the volume percentage of polyimide during production.

The TIM in this paper shows much potential, since the currently-used In/Sn/Bi alloy has low thermal conductivity. If an alloy with much higher thermal conductivity and low melting point, e.g. pure In which has a thermal conductivity of 83W/mK and a melting point of 175°C. If such metal or alloy could be applied, with consistent volume percentage of polymer, the final thermal conductivity of TIM will be revolutionary.

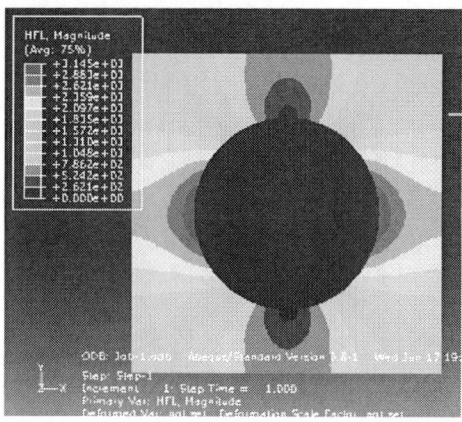

Fig. 4. Field of heat flow in 2D case 1 model. Blue color of the round polyimide displays the very low heat flow.

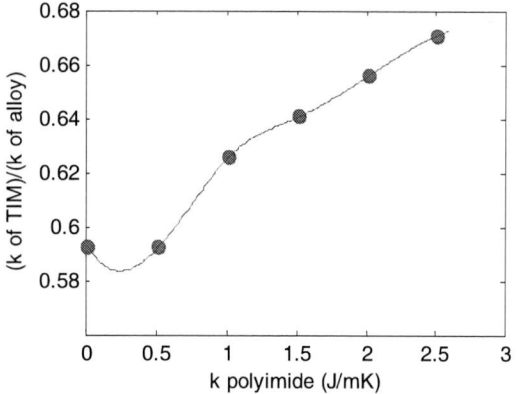

Fig. 5. Effect of thermal conductivity of polyimide on thermal conductivity of TIM. The circles represent simulation data and the curve is plotted by fitting technique. k refers to thermal conductivity.

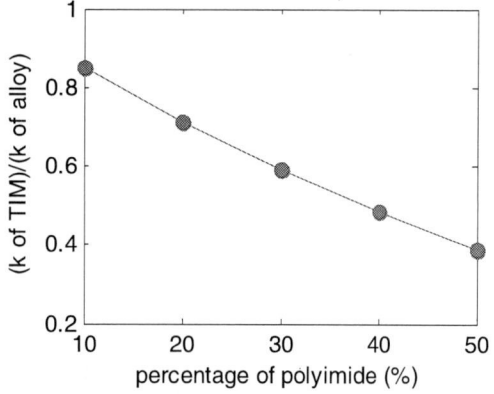

Fig. 6. Effect of volume percentage of polyimide on thermal conductivity of TIM. The Circles represent simulation data and the curve is plotted by fitting technique.

Conclusion

A unique type of nanostructured polymer-metal composite for thermal interface material has been discovered with effective thermal conductivity of 8 W/(mK). While the result shows potential, further studies are required to further

enhance the thermal conductivity. Calculations reveal the alignment of fibers have insignificant influence and conclude that the volume percentages of fiber together with the mean interface temperature are the dominating parameters that determine TIM's thermal conductivities. Based on this approximation, simulations were conducted and correspond well with experimental data. However, the preferred volume percentage of fiber (33%) is slightly too large in accordance to the calculation of surface image of TIM, which suggests the need of a better image.

Acknowledgments

The authors acknowledge the financial support from 863 program (No.2008AA04Z301), NSFC project (50876057). The authors at Chalmers University of Technology would like acknowledge the financial support by the Seventh Framework Programme of the European Union (Project name: Nano Packaging Technology for Interconnection and Heat Dissipation under the contract No: 216176).

This work is also supported by Shanghai University Graduate Innovation Fund, and "SEC E-Institute: Shanghai High Institutions Grid" project.

References

1. Mahajan, R., Nair, R., Wakharkar, V., Swan, J., Tang, J., and Vandentop, G., "Emerging Directions for Packaging Technologies," *Intel Technology Journal*, Vol. 6, No. 2 (2002), pp. 62-75.
2. Viswanath, R., Wakharkar, V., Watwe, A., and Lebonheur, V., "Thermal Performance Challenges from Silicon to Systems," *Intel Technology Journal*, Vol. 4, No. 3 (2000), pp. 1-16.
3. iNEMI, "iNEMI 2007 Research Priorities," *Herndon*, VA 2007.
4. Liu, J., Wang, T., Carlberg, B., Inoue, M., "Recent progress of thermal interface materials", *ESTC (2008)*, 2nd1-4 Sept. 2008 Page(s):351 - 358.
5. Yan Zhang, Cong Yue, Johan Liu, Zhaonian Cheng and Jing-yu Fan, "Study of the Filler Effect on the Effective Thermal Conductivity of Thermal Conductive Adhesive", *ICEP (2009)*, 638-642.
6. Carlberg, R., Wang, T., Fu, Y., Liu, J., Shangguan, D., "Nanostructured Polymer-Metal Composite for Thermal Interface Material Applications", *978-1-4244-2231 Electronic Components and Technology Conference (2008)*.
7. Cong Yue, et al, "Influences of Filler Geometry and Content on Effective Thermal Conductivity of Thermal Conductive Adhesive". *in Proceedings - Electronic Components and Technology Conference. 2009.* pp. 2055-2059.
8. Bin Su, Jianmin Qu, A Micromechanics Model for Electrical Conduction in Isotropically Conductive Adhesives during Curing. *9th Int'l Symposium on Advanced Packaging Materials.* pp. 145-151.
9. Kittel, Charles, Introduction to Solid State Physics, *Wiley*, (1990).

978-1-4244-4658-2/09 $25.00 © 2009 IEEE

Characterization of a Photosensitive Dry Adhesive Film for Wafer Level MEMS Packaging

Kun Zhao*, Changhai Wang

School of Engineering & Physical Sciences, Heriot-Watt University
Edinburgh EH14 4AS, UK
E-mail: kz24@hw.ac.uk

Abstract

In this paper we present the results of initial studies of a new dry film polymer, PerMX developed by Du Pont for wafer level packaging applications. PerMX is a high resolution permanent photosensitive dry film material with low temperature processing capability. Dry film is easy to deposit on large wafers compared to the liquid form materials. We have carried out initial chip scale characterization of the material for MEMS packaging applications. A PerMX film of thickness of 20 μm was used to create sealing rings on glass wafers. The outer dimensions of the polymer rings are 5mmx5mm with a track width of 100 μm or 400 μm. Glass substrates with a polymer ring were bonded to polished silicon chips for investigation of the polymer as a bonding material for MEMS packaging. A range of bonding parameters was studied in order to obtain high bond quality. Shear test and gross leak test were conducted to evaluate the bond strengthen and hermeticity. The results show that the PerMX polymer can produce a strong bond between the glass and silicon substrates. Bond strengths of ~7 kgf have been obtained for the samples with a bonding track width of 100 μm. The samples produced under the optimal bonding conditions passed the gross leak test. In addition the surface roughness was also measured and it is less than 60 nm. These results show that the PerMX is a potentially suitable material for MEMS packaging applications.

1. Introduction

MEMS devices are being used in an increasing number of applications in air bag accelerometers, pressure sensors, optical switches, inkjet heads, RF switches and recently energy scavenging devices. Polymer materials are attractive for MEMS packaging as polymers have high fracture strength, low Young's modulus and low cost. They can be deposited on a variety of substrates and the fabrication techniques can be simple and flexible. Furthermore, polymers can be biocompatible and therefore can be used for packaging of MEMS and sensors for biological applications [1].

Packaging is one of the most important manufacturing processes for MEMS as it is essential for protection and operation of the devices. Currently a critical challenge in MEMS manufacture and packaging is the development of low temperature bonding processes for encapsulation and packaging of the devices and systems. As the MEMS and NEMS based devices consist of micro and nanoscale mechanical structures, some devices are particularly sensitive to the processing temperature in device fabrication and packaging. Excessive temperature rise can cause failure of operation due to the thermally induced stress in the structures. Thus low temperature processes for device assembly and packaging are desirable. With the rapid development of new materials for MEMS applications, polymers will play a significant role in wafer level MEMS packaging, particularly for temperature sensitive devices. Polymers can create strong, chemically and thermally stable bonds. Unlike direct fusion and anodic bonding, polymer based joining has a high tolerance to non-uniform substrate surfaces [2, 3].

2. Properties of the PerMX film

The PerMX material is a multi-purpose permanent polymer film with high resolution, low curing temperature and good mechanical and electrical properties. Table 1 shows the properties of the PerMX material. The material can be used as a dielectric material for redistribution and also as bonding and structuring material.

Table.1 Properties of the PerMX polymer [4]

Tone	Negative
Aspect Ratio	4:1
Develop Solvent	PGMEA
Rinse Solvent	IPA
Cure C/min/atm	150/30/air
Tg	220 °C
Thermal Stability @ 5%wt Loss	300 °C
CTE	50 ppm
Tensile Strength	75 MPa
Young's Module	3.2 GPa
Elongation at break	5
Moisture Uptake	<1
Dielectric Constant	2.9 @ 50%
Dielectric Loss Tan	0.006@10MHz
Breakdown Voltage V/m	4.6×10^{7}
Acid Resistance	Very Good
Alkali Resistance	Very Good
Solvent Resistance	Very Good
Adhesion to Si and SiN	Good
Storage Condition, Life	RT, 12 month

As a dry film material, PerMX offers a number of benefits such as high productivity, good thickness control, and fewer process steps for packaging applications. It is easier and safer to handle than the solvent based polymers [5]. It is highly suitable for cavity based packaging application. The film has sufficient thickness for joining a planar cap to a MEMS die to produce a cavity to house the MEMS devices without the need to fabricate a recess in the cap as it is the case for silicon direct bonding and anodic bonding in MEMS packaging.

978-1-4244-4658-2/09 $25.00 © 2009 IEEE

3. Design and fabrication of bonding rings

The fabrication process of PerMX bonding rings is shown in Fig.1. Like other dry photoresist films in electronic manufacture, the PerMX film is deposited on a substrate wafer by lamination. The lamination step is to provide an intimate contact between the PerMX film and the substrate eliminating air entrapment and to ensure good adhesion. The hot roll lamination process can be carried out under the temperature of 65 - 85°C at a rolling speed of 0.3-1.0 m/min and at a pressure of 45-55 psi [4].

For bonding ring fabrication, 4-inch diameter glass wafers were laminated with the PerMX film. The film lamination process was carried out by IZM Berlin. Then the PerMX film was exposed to the UV light using a photomask for producing the sealing ring structures. Before exposure the polyester coversheet film for film protection was removed for improved side wall geometry. A broadband mask aligner (Tamarack) was used as the exposure source. For better resolution an I-line filter was used to produce exposure at 365 nm. After UV exposure the wafer was baked at 95°C on a hotplate for 4 minutes to ensure a consistent developing time. During immersion development, the unexposed PERMX was dissolved in PGMEA (Propylene glycol monomethyl ether acetate). The UV exposed areas in the PerMX film was left on the glass wafer defining the bonding ring patterns. The PerMX patterns on the wafer were rinsed in Isopropanol (IPA). In the subsequent processing step, the polymer structures were baked at 120 °C for 30 seconds to remove the solvent. Finally the glass wafer was diced into square chips each containing a PerMX bonding ring. The dimensions of the glass chips are 10mm × 10mm.

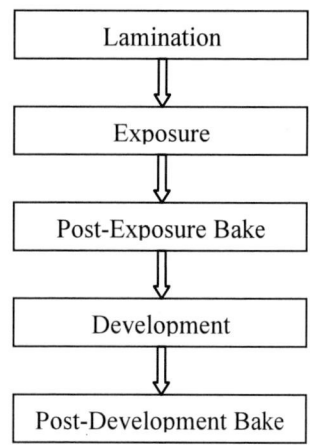

Fig.1. Illustration of process flow for fabrication of PerMX bonding ring structures.

Fig. 2 shows optical pictures of fabricated PerMX square rings on a glass substrate. The outer dimensions of the PerMX rings are 5mm × 5mm and the track widths are 100 μm and 400 μm respectively. The thickness of the PerMX ring was measured to be 20 μm on a Zygo white light interferometer Before bonding the surface roughness of the PerMX bonding rings was also measured using the Zygo white light interferometer. Fig. 3 shows the surface profile of the PerMX

rings after the fabrication process. The corresponding roughness of the polymer surface is 60 nm.

(a) 100 μm track width

(b) 400μm track width

Fig.2. Optical pictures of fabricated square PerMX rings on glass substrates.

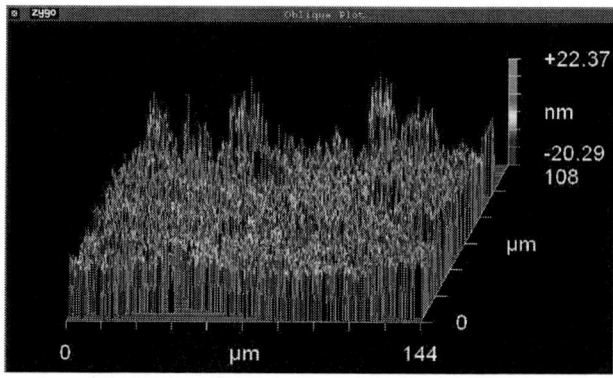

Fig. 3. Surface roughness of the PerMX polymer after fabrication.

4. Chip scale bonding

The glass chips with PerMX bonding rings were bonded onto planar silicon substrate surfaces using a hotplate to produce microcavities for investigation of PerMX as a bonding material for wafer level MEMS packaging. In the bonding process the silicon substrate was placed on the hotplate at room temperature and the glass chip with the PerMX ring was placed onto the silicon substrate. The substrates were then heated to the selected experimental temperature. A bonding force was applied to the substrate assembly using a metal load. After the bonding process is completed the hotplate was turned off to allow the bonded sample to cool to the room temperature.

In the initial experiments, the bonding temperature was fixed at 150°C while the bonding load was varied. This temperature is recommended by the material manufacturer for curing the PerMX polymer. In order to investigate the effect of bonding force on bond quality, a range of bonding loads was used ranging from 0.2 kg to 1.0 kg at a step size of 0.2 kg. The bonding time was fixed at 30 minutes. For the load of 0.2 kg, it was found only about half of the substrates could be bonded together. This results from the insufficient contact between the PerMX bond ring and the silicon surface. For the load of above 1.0 kg, the substrates could not be bonded together. The substrates became separated after cooling to room temperature. It is believed that the delamination is a result of the relaxation of the interfacial stress that was built up during the bonding and cooling process under high bonding pressure. Based on the analysis of the results for all of values of the bonding load, the best load for bonding PerMX rings of 100 μm of track width is 0.4 kg. The corresponding bonding pressure is 2 MPa.

To study the effect of bonding temperature on bond strength, further bonding experiments were carried out at three different temperatures: 120°C, 150°C and 180°C. The bonding load and time were 0.4 kg and 30 minutes. It was found that defect free bonding could not be produced consistently at 120°C and 180°C of bonding temperature. A typical sample with defects is shown in Fig. 4. On the other hand defect free samples can be produced consistently at the bonding temperature of 150°C and bonding load of 0.4 kg. Fig. 5 shows the optical pictures of a bonded glass-silicon substrate assembly. It can be seen that the quality of the PerMX bond is very good.

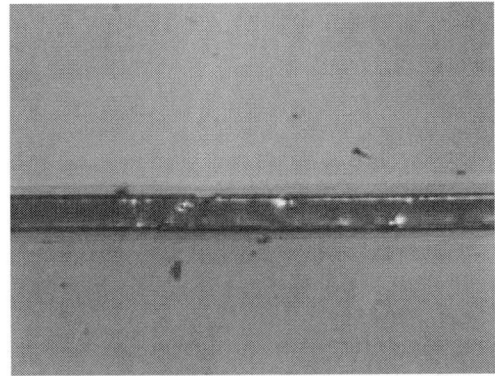

Fig. 4. Optical picture of a section of the PerMX ring between a bonded glass and silicon assembly produced at a bonding temperature of 120°C.

The bonding time was also varied in the bonding experiments. Bonding times of 10, 20 and 30 minutes were used for PerMX bonding under the temperature of 150°C and the load of 0.4 kg for PerMX rings of 100 μm of track width. On the basis of optical inspection, it was found that the best PerMX bond is obtained for the bonding time of 30 minutes.

In total more than 40 samples were produced using PerMX rings of track widths of 100 μm and 400 μm at 150°C for 30 minutes. Bonding loads of 0.4 kg and 1.6 kg were used for the

100 μm and 400 μm of track width respectively to maintain the same bonding pressure. Optical inspection shows a yield of about 90% for defect free bonding.

A full sealing ring of 100 μm track width

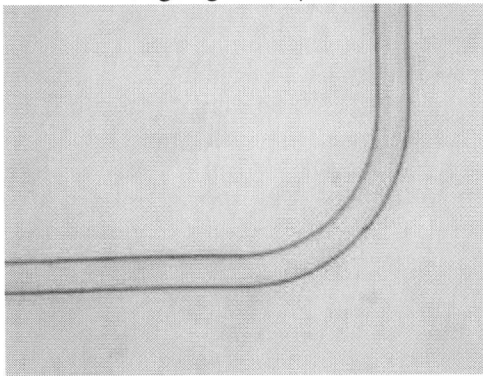

(b) Enlarged corner section of the ring

Fig. 5. Optical pictures of a PerMX ring between glass and silicon after bonding illustrating good bond quality.

5. Reliability studies

5.1 Gross leak test

Gross leak test is often carried out to investigate the hermeticity of MEMS packages. The gross leak test is based on the MIL-STD-883E standard, Method 1014.9 [6]. A hotplate and two Fluorinert liquids of different boiling temperatures are normally used. They are FC-84 with a boiling point of 80°C and FC-40 with a higher boiling point of 155°C, both produced by 3M. For gross leak test the samples were immersed in the FC-84 liquid for 24 hours. Then they were taken out and immersed into the FC-40 liquid at 110°C one by one for testing. Bubbles will emerge from the defect places in the polymer seal causing the leakage. The inspection period for each sample is about 30 seconds. Big leak is shown by a stream of bubbles, while more than two large bubbles originating from the same point of the sample also indicate a leak and the sample fails the gross leak test. A beaker containing FC-40 liquid was placed on the hotplate and maintains at 110°C during the gross leak tests.

Leak test was conducted for the samples bonded at 150°C for 30 minutes at 2 MPa of bonding pressure. All of the defect-free samples shown under optical inspection with 100μm wide PerMX track width passed the gross leak test. However, 90% of the samples with 400μm track width failed the gross leak test. Fig. 7 shows an example of the defect causing gross leak. Fine leak test will be conducted in the future to investigate the hermiticity of the samples which passed the gross leak test.

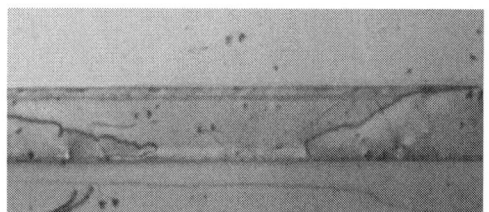

Fig. 6. Defect in PeMX ring caused by gross leak

5.2 Shear test

Shear test has been carried out on the samples bonded onto the silicon substrates using a shear tester to investigate the bond strength of the PerMX polymer. Table 3 shows the results of shear test, track height after shear test and the corresponding deformation of the track after bonding for the samples bonded at the load of 0.4 kg and 30 minutes of bonding time. The best shear strength is for the sample bonded at 150°C. Table 4 summarizes the similar results as in Table 3 but for different bonding loads. The best shear strength is for the load of 0.4 kg. The shear strength is also good for the load of 0.6 kg and 0.8 kg. The shear test confirms the best bonding conditions for the 100 μm wide polymer track are 150°C, 0.4 kg and 30 minutes for the bonding temperature, load and time respectively.

Table 3. Shear strength for samples bonded at different temperatures for the bonding load of 0.4 kg.

Temperature (°C)	Shear strength (kg)	Track height (μm)	Deformation (%)
120	2.3	17.5	12.3
150	6.9	18.4	7.8
180	3.0	17.0	14.9

Table 4. Shear strength for samples bonded at 150°C for different bonding loads .

Load (kg)	Shear strength (kg)	Track height (μm)	Deformation (%)
0.2	2.8	-	-
0.4	6.9	18.4	7.8
0.6	4.4	16.8	16.1
0.8	5.0	16.8	16.2

The surfaces of the substrates were inspected under an optical microscopy. In general most of the bonding rings are complete and are on the glass substrates. This shows that the adhesion of the laminated PerMX film to the glass wafer is stronger than the adhesion to the silicon substrate after bonding. It was observed that the failure induced damage to the PerMX rings occurs mostly at the corner location of the rings. Fig. 7 shows a typical picture of damaged corners of a ring during the shear test. Fig. 8 shows a 3D image of the track surface of the PerMX polymer ring after shear test. It shows that the surface become rougher than that before bonding

because of the drag force of the shear tool. More detailed study is under way to determine the relation between surface roughness after shear test and the shear strength.

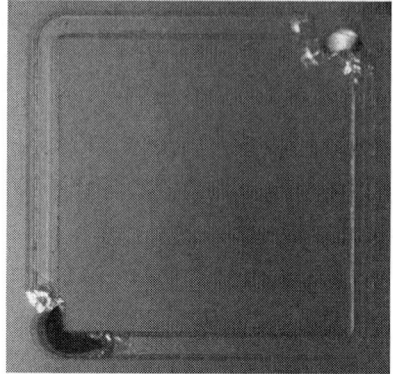

Fig. 7. PerMX track of 400μm after shear test

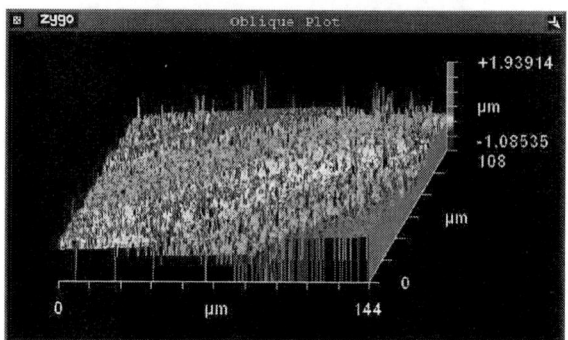

Fig. 8. Surface roughness of PerMX after shear test

6. Conclusions

Initial characterization of a photosensitive dry adhesive film has been carried out to evaluate its potential for wafer level MEMS packaging applications. The PerMX sealing rings were fabricated on glass substrates and bonded onto silicon substrates in order to investigate the properties of polymer bonding under different conditions. A high-yield PerMX bonding procedure was established for hotplate based bonding. Defect-free sealing has been demonstrated for 100 μm wide track of PerMX rings using the bonding conditions of bonding temperature of 150°C for 30 minutes at a pressure of 2 MPa. The shear test showed good bond strength of ~7 kgf. The samples bonded using 100 μm of track width of PerMX polymer passed gross leak test. But most of the samples with 400 μm of track width failed the gross leak test. The shear strength is also not higher than that of the 100 μm wide track samples despite of much larger bonding area. Fine leak test and more reliability tests will be undertaken. The low temperature bonding capability of the PerMX material at ~150°C with good bond strength is very attractive for packaging of temperature sensitive MEMS devices. It has been shown that the PerMX polymer has the potential to be used as a bonding material for encapsulation of MEMS and other microscale devices.

Acknowledgments

The authors are grateful to Mr Russ Crockett and Mr Werner Liebsch of Du Pont UK and Du Pont Germany respectively for provision and lamination of the PerMX film on wafers.

References

1. Mohamed Gad-el-Hak, MEMS Design and Fabrication, *CRC Press* (FL USA, 2006).

2. Niklaus, F. et al, "Low-temperature Full Wafer Adhesive Bonding," *Journal of Micromechanics and Microengineering*, Vol. 11, No. 2 (2001), pp. 100–107.

3. Wang, C. H. et al, "Chip Scale Studies of BCB Based Polymer Bonding for MEMS Packaging," *Proc 58th Electronic Components and Technology Conf, Lake Buena Vista*, FL, May 2008, pp. 1869-1873.

4. Preliminary Data Sheet & Processing Information for PerMX Series polymers, *Du Pont*.

5. Jorge P. et al, "Microlithographic Polymer Film – A Dry Film Solution for Advanced Packaging, Proc of IMAPS International Conference and Exhibition on Device Packaging," *Scottsdale, Arizona*, 18-19 March 2007.

7. Department of Defense of USA, Test Method Standard: MIL-STD-883E, 1997.

Optimization Design on Polybrominated Biphenyls (PBBs) Extraction from Plastics for RoHS Directive

L. Hua[1, 2, 3], X. P. Guo[1], J. K. Yang[2], H. N. Hou[3]

[1]School of Chemistry and Chemical Engineering, Huazhong University of Science and Technology, China, 430074
[2]School of Environmental Science and Engineering, Huazhong University of Science and Technology, China, 430074
[3]School of Chemistry and Life Science, Hubei University of Education, China, 430205
E-mail: txuehua@163.com, txuehua001@yahoo.cn, Tel: 086-027-87943696

Abstract

Extracting polybrominated biphenyls (PBB) from electronic and electrical equipment is of high concern due to RoHS directive. PBBs were so toxic for human and environment, they are applied largely in electronic and electrical equipment as flame retardant. In this thesis, a new method was developed to predict the optimal conditions of semivolatile organic polybrominated biphenyls (PBB) extraction from plastic for Enviromental protection. A feed forward type of artificial neural network (ANN) model design was used to investigate the effects of four independent variables, namely, the ratio of solvent, stirring speed (rpm), extraction temperature (°C), extraction time (h) on the response, the acquired ratio of PBB. The independent variables were coded at four levels and their actual values selected on the basis of results of single-factor experiment. The model was initially trained by the analytical data with function approximation principle in MATLAB environment to reveal the real engineering world of extract process. Then dimensions of the trained result were reduced from n-D to 2-D, In which, a visual contour plot and simulated curve were displayed, an optimal extract processing is achieved with 9.654% maximal acquired ratio of PBB. In order to validate the method, at the optimal point, the simulated result generated by proposed model was checked with the real experiment, it founded they kept a good agreement with each other. Thus proves that the mathematical model developed for resolving the PBB extraction from plastics is very effective and accurate. It is also a useful tool to reveal the real parameters effect on productivity.

Modern computational and experimental tools have matured to a stage where they can provide substantial insight into engineering processes involving extracting some contents from herbs, vegetable oil, rubber, plastics etc, and the successful application of optimization design can help to improve the process productivity. Extracting the flame retardant as polybrominated biphenyls (PBBs) from electronic and electrical equipment is of high concern due to RoHS directive [1]. PBBs were so toxic for human and environment, it is applied in electronic and electrical equipment as flame retardant. When the EE wastes disposed unsoundly at the end of life, some hazardous substances emit into air, water, soil which are a strict threaten to human. So to determine such substances for EE industries has a profound meaning. Many methods to determine such hazardous substance such as GC-MS, ICP-OES, AES, IC etc. are developed[2-4], however, there is a very high requirement for sample preparation. Soxhlet extraction is a good digest technology to prepare samples, in this thesis, a new method was developed for predicting the optimal digest conditions of semivolatile organic compounds PBB in plastic with soxhlet extrat technology. During experiment, Uniform design is accepted to array data due to the uniformation consideration. Conventional optimization design was to hold one but all other variables as constant while methodically changing one at a time which is called one variable at a time or OVAT [5]. But the major pitfall in this procedure is that it cannot quantitatively explore the interaction among all variables and does not describe the net effect of the various combined conditions on the response [6]. The rapid and continuous development in processing design will require a new proposal to meet goals for increased performance, robustness and visualization. To date, the majority of efforts in optimization of extraction process have relied on gradient-based search algorithms, polynomial-based response surface methodologies (RSM), local gradient-based method and so on [7]. However, the challenge of these approaches are that it renders to search insufficiently due to the objective functions discontinuous over the broad design space or too resource-intensive due to unrestricted "brute force" search schemes. So here, based on global optimization problems especially for multidisciplinary ones, a matrix method was developed to predict the extract process of PBB from plastics which are compositions of packaging, sealed, adiabatic, insulating materials in electronic and electrical equipment (EEE). By establishing a functional artificial neural network (ANN) which maps the relationship between the response and variables, using a global optimization line-up competing algorithm (LCA) to train the network [8], simulation curve and the optimal contour plots were produced in 2-D plane. The model would not only serve as a visual aid to have a clearer picture about the effect of different variables on the responses in the form of animation but also enable to locate the region where the properties are optimized. It is also possible to predict the combination of independent variables which will result in optimal acceptance. It helps to understanding how the result is achieved and increases the reliability to the result.

1. Experimental

1.1 Instrumental

With reference to IEC 62321,US-EPA 1613:1994, US-EPA Method 8270C: 1996, plastic samples were grinded by cooling and milling followed by soxhlet extraction. Samples were detected by Clarus 500 Gas Chromatograph/TurboMass Gold Mass Spectrometer (PerkinElmer Inc., USA) coupled with TurboMassTM software for operation. Experiments were designed and optimized by using visual programme developed in MATLAB v.6.5. Soxhlet apparatus with condenser (100ml of soxhlet extractor with 100ml round flask) was employed. Heating mental to fit the round flask was adopted at soxhlet apparatus. A temperature (100 - 300 °C) can kept constant at setting point. Rotary evaporator equipped with a variable temperature water bath and vacuum source was performed. Other glassware such as beaker, funnel, vials, and disposable pipettes were used.

1.2 Reagents

Calibration solutions (purchased from Advanced Technology & Industrial Co. LTD, Hong Kong); Sulfuric acid (AR grade); Toluene, Tetrahydrofuran (≥99.9%, HPLC grade); Acetone, hexane should be free of interferences. Dehydrated sodium sulfate, silica gel, acid silica gel: Mix 2.24g of silica gel with 1.76g of concentrated sulphuric acid.

1.3 Extraction

Flame retardant containing PBBs were often added in such industrial plastic matrices as ABS, PS, PE, PUR, Epoxy resin etc. Because they were non-soluble polymers, precipitation and liquid-liquid extraction were difficult, the soxhlet extraction was suitable. The samples were grinded to a size of 1mm3 by the combination of cutting and milling under cooling with liquid nitrogen. Then they were extracted by soxhlet system with magnetic stirring. For this case, toluene and tetrahydoffuran mixture were employed as extraction solvents. After finishing, the extracted was combined with filtrated solution and washed with H2SO4 until colorless and then with hexane-rinsed water to make them neutral. At the same time, 13C12-labeled PBB(s) was used as recovery/internal standards before starting the extraction. In order to purify the samples, the one end of a glass column (20mm × 350mm) with 250ml of reservoir was stopped with solvents rinsed glass wool. In sequence, 1g of dehydrated sodium sulfate, 1g of slica gel, 4g of acid modified silica gel, and 1g of dehydrated sodium sulfate were added with 30ml of hexane to rinse the column. Sample concentrated was added and the column was eluted with additional 130ml of hexane. The solution evaporated to a volume of about 1 - 2ml and transfer to a brown glass bottle using glass disposable pipette for GC-MS detection. These optimum parameters of extraction were determined by this optimization experiment.

1.4 Experimental design-Development of a feed forward artificial neural network model

In this work, the structure of a feed forward artificial neural network model has been developed to predict extract process as shown in Fig.1. For most of the engineering applications, in general, back propagation neural network (BPNN) in combination with gradient descent optimization method as learning algorithm which is a powerful tool for the least square problem. So in order to determine the suitable PBB extract conditions, this kind of neural network used as a mechanism for mapping the function. When the three input parameters extract temperature, extract time, the ratio of solvent are entered into the system, and then the system will predict the suitable acquired ration of PBB (output). Any neural network will not be able to predict the output parameters accurately unless the network is properly trained with accurate data, hence sufficient care must be taken in the preparation of training data. An optimization scheme requires

large amounts of data, which means that the training data must cover the wide variety of possible ranges. The various operating conditions that are used for training the network were given in Table 1. The total data set generated for training the network (4×4×4×4=256) and this forms the neural network's reasoning domain, actually experimental data was acquired by uniform design which is a better method for uniformly arranging experiment. After defining the domain, experimental results were generated from the ANN simulation model. The steps are the following: (1) Define some functions: Input parameters in a vector form (Xi), Xi = $[$ $xi1$ $xi2$ $xi3$ $...xim$ $]^T$, where m is the number of input parameter, so each of 256 points contains n-dimensions space information. i is the number of batches of data, if there are n batches of data, then $i=n$. Y is the actual experimental object function vector; Network output vector Yt = $[$ $y1$ $y2$ $y3$ $...y_q$ $]$, where q is the number of object functions; Mean value Y0 is employed for distinguishing whether such a network training point is an outlier or not, Y0 = Y$[$ ($y$max + $y$min) /2$]$, wherein, $ymax$, $y$min are maximum and minimum of object function (y), respectively; ξ is the error limit of training point,$\xi=\min\frac{1}{2}$ (Y - Yt)^2 , ξ<0.001. Vectors of weights are W, V. (2) Input parameters (Xi) enter the ANN network, neural networks simulate human functions such as learning from experience by adjusting the values of the weights (W$_1$, W$_2$) and biases, the results are mapped onto 2-D plane (Z$_1$, Z$_2$), where Z$_1$ = W$_1$ Xi , Z$_2$ = W$_2$ Xi , W$_1$ =$[w_{01}$ w_{11} $w_{21}...w_{n1}$ $]$, W$_2$ = $[w_{02}$ w_{12} w_{22} $...w_{n2}$ $]$. (3) Then with employment of another weight coefficient V, polynomial-based non-linear simulation process was achieved, where polynomial-based function vector F = $[1$ $f(z_1)$ $f(z_2)$ $f(z_1 z_2)$ $f(z^t_1)$ $f(z^t_2)]^T$, t = 2 ,3 ; $f(x)$ = 1 + e $^{(-x)}$, Weight vector: V = $[$ V$_1$ V$_2$ V$_3$ $...$V$_q$ $]^T q×f$, V$_k$ = $[$ v_{k1} v_{k2} v_{k3} $...v_{k(f+1)}$ $]$, k = 1 ,2 ,3 , ..., q. Training of a network requires repeated cycling through the data, each time adjusting the values of the weights and biases to improve performance. Each pass through the training data is called an epoch or an iterative and the BPNN learns through the overall change in weights accumulation over many epochs or iteratives. Training continues until the error target is met or until the maximum number of neurons is reached.

Table 1 The experimental data

Extraction parameters	Operating conditions			
	Level 1	Level 2	Level 3	Level 4
Extract temperature (°C)	150	200	250	300
Extract time (hours)	12	16	20	24
Stirring speed (rpm)	45	60	75	90
Ratio of solvents	40:60	50:50	70:30	80:20

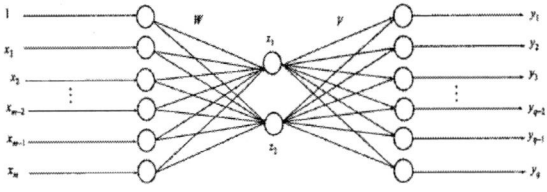

Fig.1 the artificial neural network (ANN) training model

2 Results

2.1 The error of iterative algorithm

The error of neural network training is shown in Fig.2, the final error changing of training with evolving 100 generations is minimal to 0.003. It is shown that neural network predictions are acceptable for time-dependent model. Convergence of the BPNN is investigated with respect to the complexity of the required function approximation, it is demonstrated that NN can give a very good approximations to non-linear global optimization problem and kept a high convergeuniformity and can commendably shield the noise contamination in the training data.

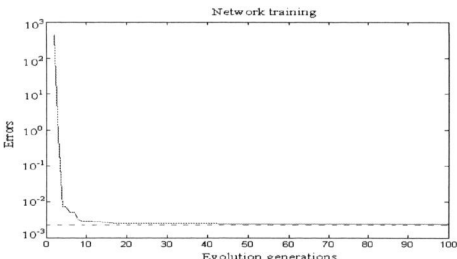

Fig.2 The error changing of network training with evolution generations

2.2 The result of function simulating

In Fig.3, the simulation value (marked with "x") very close to actual value (marked with "o") further demonstrated that the trained network predicted the extract coefficients have an acceptable accuracy defined to be the experimental error. LCA algorithm with two weights creates a powerful pruning scheme that can be used for tuning feed-forward connectionist models and provides a designer-specified degree of accuracy in mapping the functional relationship. In this work, a novel matrix method for multi-input-multi-output on MATLAB manipulated flat has superior interpolation ability when compared to conventional feed-forward network, so that the simulation calculation of functions is ideal.

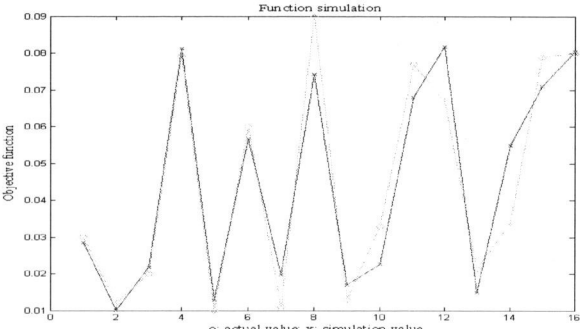

Fig.3 The curve of function simulating

2.3 Contour plotting

Objective contour plotting in Fig.4 gives a visual picture of the relationship between parameters and the acquired extract ratio. Therefore, on the basis of this figure, by adjusting inserted or extrapolated distance between the two known points (12, 15), the calculating result of the optimal conditions in computer is that the extract temperature is 200oC, extract time is 24h, stirring speed is 60rpm, the ratio

of solvent is 40:60, the mapping 2-D coordinate of Z1 axis is 1.012, the coordinate of Z2 is -0.0342, the point is marked with "*" as shown in Fig.4. The acquired optimal extract ratio is 9.654%. In order to check the validity of simulation, do experiment with reference to optimal parameters, the actual value of extract ratio is 9.652%, the error between the calculation and experiment is 0.002% which shows that the developed model to simulate the real process of extracting PBB from plastic has a good precision due to a good accordance between the predictive and the experimental data.

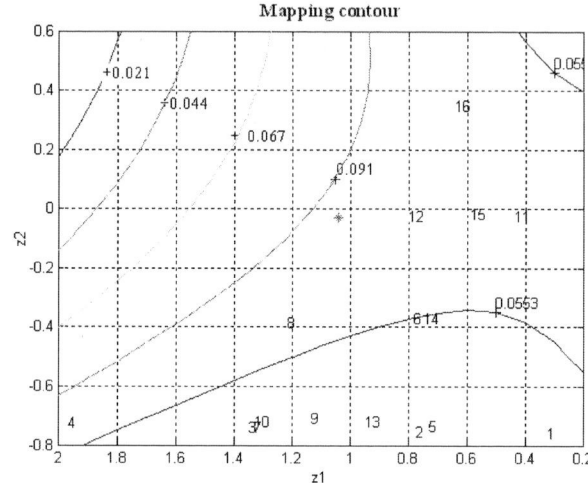

Fig.4 The mapping contour of simulation in computer

3 Conclusions

In this present work, attempts have been made to develop an optimal design model for PBB extraction from plastic, which was based on a back propagation artificial neural network system (BPNN). By using a LCA algorithm to reduce dimensions from n-D to 2-D and map, the space information between variables and objectives can be visualized. By adjusting inserted or extrapolated distance between the two known points, the optimal space point was searched and objective contour was plotted, the result is found as follows: the ratio of mixed solvent is 40:60, the extract time of 24h, and the extraction temperature of $200^{\circ}C$, the stirring speed is 60rpm. In the proposed model, it showed that the maximum acquired ration of PBB was 9.654% at this optimal point. The result shows a good agreement between the predictive and the experimental data. So it can be concluded that the trained BPNN has a predictive accuracy equal to or better than the accuracy of the experimental measurements using only 6.2% of the data acquired during the experimental test.

Acknowledgements

The authors would like to acknowledge the vital fund support from Hubei province education office on research on reuse of arsenic sulfide residue and high purity arsenic extraction technology from arsenic sulfide residue with zero pollution (D20093103).

References

1. Directive 2002/95/EC of the European Parliament and of the Council on the restriction of the use of certain hazardous substances in electrical and electronic equipment. *Official Journal of the European Union.* 2003, 1, 27. pp. 19-37.

2. V. Yusa, O. Pardo, A. Pastor, et al, "Optimization of a microwave-assisted extraction large-volume injection and gas chromatography–ion trap mass spectrometry procedure for the determination of polybrominated diphenyl ethers", polybrominated biphenyls and polychlorinated naphthalenes in sediments. *Analytical Chim.* Acta 557, 2006, pp. 304 - 313.

3. B. Gevao, M. Al-Bahloul, A.N. Al-Ghadban, et al, *Chemosphere 64* (2006) 603.

4. J. Tollb¨ack, C. Crescenzi, E. Dyremark, J. Chromatogr. A 1104 (2006) 106.

5. Kukreja, T. R., Kumar, D., Prasad, K., et al, "Optimisation of physical and mechanical properties of rubber compounds by response surface methodology-Two component modeling using vegetable oil and carbon black", *Journal of European polymer,* 2002, 38, pp. 1417 - 1422.

6. Chang, Y. N., Huang, J. C., Lee, C. C., et al, *Enzyme Microb Technol 2002,* 30, pp. 889.

7. Montgomery DC. In: Design and analysis of experiments. *New York: John Wiley;* 1991. pp. 229-341.

8. Yan L. X., Ma D. X., "A new algorithm for continuous variable global optimization line-up competing algorithm (LCA)", *Journal of Hubei Polytechnic University,* Vol. 14, No. 1,2, Jun. 1999.

The Design of a Terahertz Metamaterial Absorber Basing on LTCC Technology

Yunsong Xie[1*], Yuanxun Li[1], Yingli Liu[1], Huaiwu Zhang[1], Qiye Wen[1]

[1]State Key Laboratory of Electronic Thin Film and Integrated Devices,
University of Electronic Science and Technology of China
[1]No. 4, Section 2, North Jianshe Road, Chengdu, P. R. China
*Email: xieyunsong@gmail.com

Abstract

Using CST, a LTCC (Low Temperature Co-fired Ceramics) technology basing design of the metamaterial absorber in terahertz region (MM) is proposed. The metamaterial absorber is constructed by silver plane and a specially designed ERR structure. The investigation to the electric field, surface current and absorption power density distribution indicates that the metamaterial absorber we design is somewhat different from the ones have been proposed, the reason of which is thought to be the wide line width comparing to the working wavelength, however, the working mechanism is consist with the one we has reported. And we have also pointed out the EM characteristic and thickness of the isolation layer is crucial in obtaining the unit absorption.

1. Introduction

Terahertz electromagnetic (EM) waves, with frequency typically ranging from 0.1 THz to 100 THz, have been less explored than those in the contiguous spectral regimes. The numerous present and future applications for these waves, such as time-domain spectroscopy [1], biomedical [2] and security [3], and ultrahigh speed electronics, together with advances in their generation and detection have triggered the interest in THz signal routing systems [4]. Metamaterial, formed either from periodic or random arrays of scattering elements [5] is thought to be the ideal candidate for the terahertz devices.

Metamaterial absorber (MA), first reported in 2008, is a kind of three-layer metamaterial with thickness significantly smaller than the wavelength [7]. It can absorb the EM wave perfectly in a narrow frequency band, which makes it be the ideal candidates for bolometric pixel elements in whole EM spectrum. Lots of attention has been paid on this novel metamaterial [6-9]. However, most of the fabrications of terahertz metamaterial absorber are basing on the lithography and spin coating technologies, which contains at least three independent steps and naturally can not be applied for large-scale production.

In the other hand, Low-Temperature Co-fired Ceramic (LTCC) has become an attractive material system for innovative designs over the past years. This technology enable the devices can be produced in a certain pipeline, and the line width of which is suitable for fabrication of terahertz device. In this paper, we report the design of metamaterial absorber with nearly unit absorptance at 0.14THz basing on the technological limitation of the LTCC, and the investigations to the field distribution of MA are also proposed.

2. Design methodology

The ceramic with ε=7.8+0.011i and μ=1 is used to separate the gold plane and ERR structure from each other.

The metal in the simulation is silver with conductance of 6.17×107 S/m. The absorptance characteristic of the metamaterials highly relates to the parameters of the structures. It is assumed that the narrowest line width which the LTCC technology can fabricate is 50μm, and the thickness of ceramic layer is possible to be controlled. Fig. 1 depicts the ERR structure in MA which we design: A=300μm, T=50μm, D=25μm. In the simulations, the thicknesses of the metal and polyimide slice are 1μm and 12μm respectively. The system is constructed in z axis from negative direction to positive direction is EM port- vacuum- ERR- ceramic- silver plane. In all the simulations, the TEM waves are radiated by port propagating along z axis and the electric fields are parallel to x axis. In order to investigate the relationship between the ERR structure and the MA, we also simulate the metamaterial with only ERR structure, which is labeled as MERR.

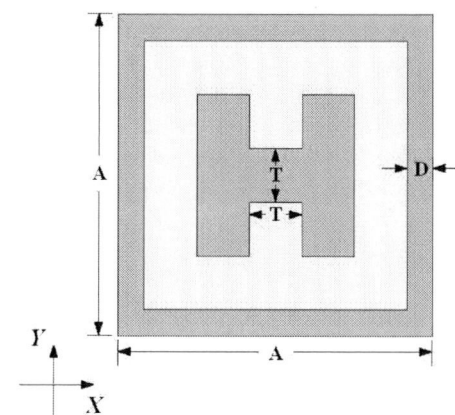

Fig. 1. The dimension of the ERR structure in the metamaterial absorber.

3. Results and discussion

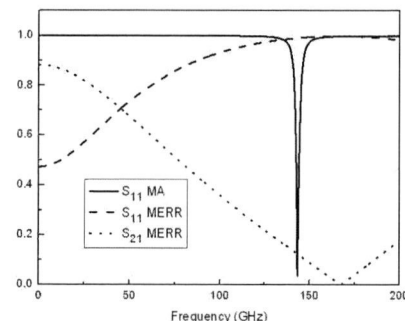

Fig. 2. The S-parameters of MA and MERR

Utilizing CST, the S-parameters of MA and MERR are obtained and shown in Fig. 2. The MA, just like those have been reported, displays a near unit absorption peak at 143

978-1-4244-4658-2/09 $25.00 © 2009 IEEE   418

GHz with bandwidth about 2 GHz, which is extremely narrow comparing to the work frequency. So the MA we design is capable to be well applied in bolometer. And the LC resonance of MERR locates at 167 GHz, with only 24 GHz higher than the absorption frequency of MA, which is familiar to the results we have reported [10]. It also can be found that the S-parameter shapes of the MERR are quite different from the general ones, since the bandwidth of the MERR is much wider than the counterparts. Understanding basing on the proposed transmission line model [11] suggests that, this phenomenon is produce by the rising of the effective induce because of the wide line width comparing to the dimension of the unit cell. To verify this idea and find the different between this MA and the general ones, we investigate the electric field of the MERR and MA.

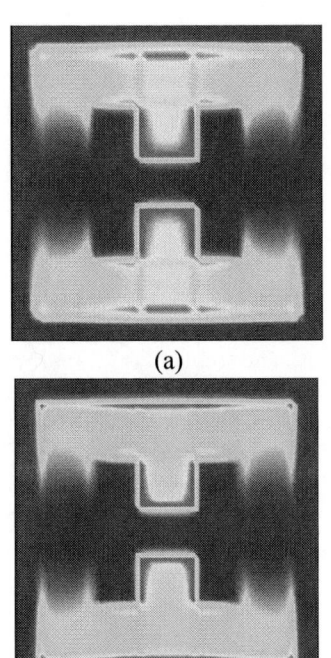

(b)

Fig. 3. The electric field amplitude of the MERR (a) and MA (b).

Fig. 3 demonstrates the electric field amplitude distribution of MA and MERR, we can find in Fig. 3 (a) that, in MERR, the strongest electric field locates at the top, bottom of the framework and the edge of the split in the ERR structure. Generally, the strongest electric field distributes in the split of the ERR structure as reported [9], if using the transmission line model to understand the MERR, the effective capacitance is provided by the coupling between the top and bottom of the split. However, the MERR we design does not satisfy this rule. Since the place with strong electric field is able to supply more effective capacitance, the effective capacitance of MERR is largely provided by the coupling between two vicinal ERR structure and self-coupling of the salient area. It is thought that this is because the split distance is rather large comparing to the wavelength, which cause the coupling between the top and bottom of the split is very weak. And it also can be pointed out that the addition of the silver plane does not lead to much change to the electric field

distribution, so the electric coupling between the silver plane and the ERR structure is rather weak.

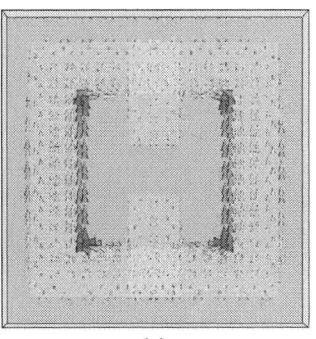

(a)

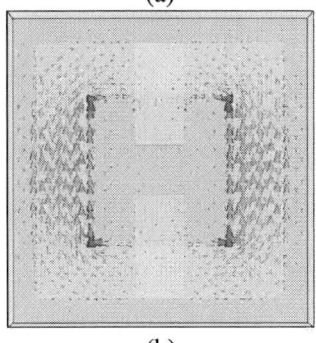

(b)

Fig. 4. The surface current distribution of the MERR (a) and MA (b).

Then we investigate the surface current distribution of the MERR and MA as shown in Fig. 4. It can be pointed out that the surface current distribution of both MERR and MA is very similar to the results that have been reported. In MERR, it can be noticed that the strongest surface current locates at the inner edge of the framework of the ERR structure, and there is also some surface current at the outer edge of the framework, which can be interpreted by the skin effect in the high frequency. The addition of the silver causes significant change to the current. Dose not satisfy the skin effect, the surface current in the framework of ERR structure almost uniformly distributes at the arm of the ERR structure, the reason of which is thought to be the too close distance between the silver plane and ERR structure. The close distance might lead to the magnetic coupling between these to metal structure and also produce some coupling inductance, which causes the deduction of the resonance frequency as we discussed above. In order to study the absorption mechanism of the metamaterial absorber, we investigate the absorption power distribution inside the metamaterial absorber.

Fig. 5 illustrates the average absorption power densities in x-y plane inside MA, which are calculated using the equation:

$$P(z) = \frac{\iint P(x,y,z)dxdy}{\iint dxdy}$$

Where P, simulated using CST Microwave Studio, is the absorption power densities distribution inside MA, and it is important to mention that the power of the EM wave is 1W in the simulation. It is can be notice in Fig. 5 that the absorptance peaks at the position near z=0mm, yielding the value 3×108 W/m3, where locates the ERR structure. The absorption power density is similar to the results that we has reported [10], except that the power loss density in the ceramic isolation layer is very strong among the space, which is thought to be caused by the narrow space between the two metal structures. But it is still thought the working mechanism is suitable for this metamaterial absorber.

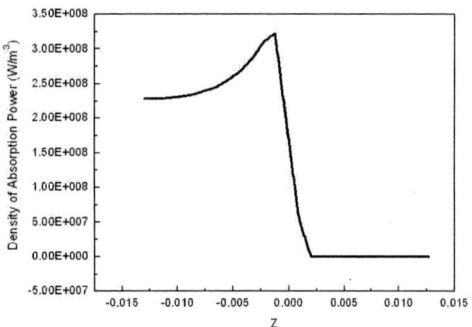

Fig. 5. The absorption power distribution of the MA.

The discussion above shows that the metamaterial absorber is able to produce the unit absorptance, but the wide line width and narrow isolation layer comparing to the working frequency leads to the field distribution of which is quite difference from the ones have been reported. It must be pointed out that the EM characteristic (ε and μ) and the thickness of the isolation layer in this metamaterial absorber are needed to be precisely controlled to obtain the unit absorptance.

Conclusions

The THz working metamaterial absorber basing on the technological limitation of the LTCC is proposed. Although the electric field, surface current and absorption power distribution is more or less different from the ones that have been proposed, the working mechanism of which is thought to be consist with the results we reported. And it is pointed out that the EM characteristic and thickness of the isolation layer is crucial in obtaining the unit absorption.

Acknowledgments

This work was supported by the National Basic Research Program (973) under Grant No 2007CB310407, the Youth Fund of Sichuan Province under Grant No.08ZQ026-013, and the Foundation for Innovative Research Groups of the NSFC under Grant No. 60721001, the Youth Fund of University of Electronic Science and Technology of China under Grant No. L08010301JX0725.

References

1. Zhang, J. and Grischkowsky, D., "Waveguide terahertz time-domain spectroscopy of nanometer water layers", *Optics letters,* Vol. 29, No. 1617, 2004.
2. Nagel, M.; Haring Bolivar, P.; Brucherseifer, M. et al, "Integrated THz technology for label-free genetic diagnostics", *Applied Physics Letters,* Vol. 80, No. 154, 2002.
3. Federici, J.F.; Schulkin, B.; Huang, F.; et al, "THz imaging and sensing for security application explosives, weapons and drugs", *Semiconductor Science and Technology,* Vol. 20, No. S266, 2005.
4. Ishikawa, A.; Zhang, S.; Genov, D. A.; et al, "Deep Subwavelength Terahertz Waveguides Using Gap Magnetic Plasmon", *Physical Review Letters,* Vol. 102, No. 043904, 2009.
5. Chen H.T.; Padilla, W. J.; Zide, J.M.O; et al. "Active terahertz metamaterial devices", *Nature,* Vol. 444, pp. 597-600, 2006.
6. Landy, N. I.; Sajuyigbe, S.; Mock, J. J., et al. "Perfect Metamaterial Absorber", *Physical Review Letters,* Vol. 100, No. 207402, 2008.
7. Tao, H.; Landy, N. I.; Bingham, C. M.; et al, "A metamaterial absorber for the terahertz regime: Design, fabrication and characterization", *Optics Express,* Vol. 16, pp. 7181-7188, 2008.
8. Hu, T, C.; Bingham, M.; Strikwerda, A. C.; et al, "Highly flexible wide angle of incidence terahertz metamaterial absorber: Design, fabrication, and characterization", *Physical Review B,* Vol. 78, No. 241103(R), 2008.
9. Avitzour, Y.; Y. A., Urzhumov; and Shvets, G., "Wide-angle infrared absorber based on a negative-index plasmonic metamaterial", *Physical Review B,* Vol. 79, No. 045131, 2009.
10. Li Y. X.; Xie Y. S.; Zhang H. W.; et al, "The strong non-reciprocity of metamaterial absorber: characteristic, interpretation and modelling", *Journal of Physics D: Applied Physics,* Vol. 42, No. 095408, 2009.
11. Azad, A. K.; Taylor, A. J.; Smirnova, E.; et al. "Characterization and analysis of terahertz metamaterials based on rectangular split-ring resonators", *Applied Physics Letters,* Vol. 92, No. 011119, 2008.

978-1-4244-4658-2/09 $25.00 © 2009 IEEE

Direct-write Techniques for Maskless Production of Microelectronics:
A Review of Current State-of-the-art Technologies

Yan Zhang, Changqing Liu, and David Whalley
Wolfson School of Mechanical and Manufacturing Engineering, Loughborough University
Wolfson School of Mechanical and Manufacturing Engineering, Loughborough University, Loughborough, Leicestershire,
United Kingdom, LE11 3RQ
Email: Y.Zhang6@lboro.ac.uk Telephone: 44(0)1509 227519

Abstract

Recently, there has been growing interest in direct-write methods for the manufacturing of microelectronic products, as the entire electronics industry sector is aiming towards low cost, rapid manufacturing and shorter time-to-market, as well as reduced environmental impacts. This paper will review the main direct-write techniques, most of which have been invented or seen significant development during the last decade. These techniques include droplet-based direct writing, such as inkjet printing, filament-based direct writing, such as the Micropen and nScrypt processes, tip based direct-write methods, and laser beam direct writing. For each category, only a few examples are presented, although there are a number of specific methods and variants within each of these categories.

Introduction

Direct-write technology has many advantages over traditional, often subtractive, processes where, for example, a film deposition process is followed by subsequent removal of much of this layer.

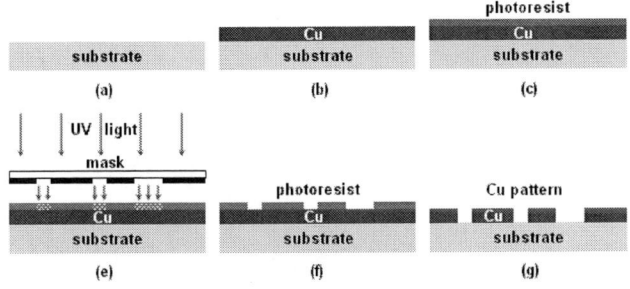

Figure-1 A traditional subtractive manufacturing process

Figure-1 is a schematic illustration of a conventional lithographic based subtractive manufacturing process. In this case a conductive layer is first deposited onto the substrate followed by photoresist deposition on top of it. Afterwards masks are used to transfer the required conductor pattern onto the photoresist layer. After development of the photoresist an etching process is used to remove unwanted areas of the conductive layer, leaving the required conductive pattern. As can be seen from the above, traditional manufacturing processes involve a lot of steps, and the temporary application and then removal of all or part of various layers of materials. Such processes therefore create large quantities of material wastes and are also often highly energy consuming such that a 2 gram microchip has been estimated to consume 1.7kg of non-renewable resources [1].

Direct writing is therefore emerging as a potential alternative approach for the manufacturing of microelectronic devices and other micro-engineered components such as MEMS and micro-fluidic devices. Existing lithography techniques also use masks to replicate the CAD created images that define the pattern of each layer, which usually involve additional tooling and manufacturing steps. This will increase the potential for human error and the overall manufacturing cost and time [2, 3]. The subtractive removal of the materials being patterned to form a part of the product, such as a copper conductor layer, and of temporary layers such as photoresist, which do not form a part of the finished product, also generates significant waste and creates environmental problems. Direct-write techniques promise an added value in that they can be incorporated directly with the output from CAD/CAM software to generate circuit patterns, which can be subsequently tailored for digital imaging. It is agile as any changes made within the CAD system, for example to include new design features or necessary modifications, can be immediately implemented without the delays associated with manufacture of revised photo-tools. Therefore, it enables the rapid prototyping of a new product from its initial design, in comparison with the costly and time-consuming mask generation and lithography processes. In addition to being more cost effective, flexible and providing a rapid turn-round, as a data-driven process, direct-write techniques can minimize the usage of materials and potentially reduce energy consumption [4], thereby reducing the large environmental burdens associated with micro-engineered product manufacture. In many direct writing processes, materials can be deposited only where needed and no further etching or removal is necessary. This simplifies the manufacturing process and reduces the requirement for processing and disposal of hazardous and toxic substances. Such data driven processes also provide the possibility of greater customisation of products to specific applications.

Definition and classification

Many different variants of a definition of direct writing have been offered in various papers by different researchers [5-7]. In this paper, direct writing is defined, by combining several of these previous definitions, as additive techniques enabling the deposition of electronic components and functional or structural patterns, out of different kinds of materials, directly following a preset layout in a data driven way without utilizing masks or subsequent etching processes. After material deposition onto the substrate is completed, a further heat treatment process, such as curing or sintering, is often needed in order for the deposited material to achieve its full performance.

978-1-4244-4658-2/09 $25.00 © 2009 IEEE

A variety of direct-write methods have been created and developed over recent years. However, no matter what devices are involved for implementation, or whatever the fundamental principle behind each technique is, most of the existing direct-write techniques can be classified into one of these four categories: (1) droplet-based direct writing; (2) filament-based direct writing, which is sometimes referred to as a continuous approach or flow-based technique; (3) tip based direct writing; and (4) laser based direct writing.

Droplet-based direct writing

Droplet based direct write methods rely on ejection of droplets of a liquid material from a single nozzle or multiple nozzles. The main sub-categories of droplet-based methods are those based on ink jet printing and the Aerosol Jet process. For these processes the jetted material must be a liquid at temperatures compatible with the process equipment.

Inkjet printing has demonstrated its remarkable power in areas such as home and office personal computer based printing and is also widely used in industrial and commercial applications, for example in high speed contactless marking of foodstuffs and packaging. The idea of applying inkjet printing to fields such as electronic manufacturing and solid free-form (SFF) fabrication developed during the 1980s [5], but only recently applications in printed electronics, MEMS, wireless communication, etc, have become the driving force behind the adaptation of inkjet printing [5].

There are two key types of inkjet type technologies, continuous jetting and drop on demand (DoD). For DoD systems there are two main types of actuation method; used in commercial inkjet printers, piezoelectric inkjet nozzles and thermal inkjet nozzles.

In a thermal inkjet nozzle, when printing of a dot is required, an electrical pulse is applied to a small ohmic heater in the nozzle, and the generated heat vaporizes a small quantity of the volatile ink so as to form a bubble. When this happens it creates a pressure difference between the interior and exterior of the nozzle, resulting in a droplet being ejected from the orifice and propelled towards the printing substrate. As the heater then cools the bubble collapses and a pressure difference between the interior and exterior of the ink cavity again occurs, but ink is now sucked into the cavity from the reservoir to replace the jetted material and restore the pressure balance.

In a piezo-type inkjet head, the actuation method relies on deformation of a piezo-electric material to create the pressure to eject a droplet from the nozzle. Piezo-electric materials are capable of transforming external mechanical inputs into electrical outputs and vice versa. So when a changing electrical voltage is applied to the piezo-material, it deforms in response to the signal input. If the voltage waveform applied to the piezo-material is transient, a deformation pulse corresponding to the voltage pulse should be expected. Material expansion due to this voltage propels an ink droplet from the orifice and ink is then drawn from the reservoir to refill the cavity in compensation for the pressure discrepancy when the voltage drop disappears and the piezo material returns to its previous steady state dimensions. The maximum temperature at which a material may be jetted is limited by the piezo-electric material, which will cease to function above some temperature threshold.

In a continuous mode inkjet system, a transducer typically made from piezoelectric material is used to generate pressure waves within a pressurised ink reservoir in response to the continuous application of an alternating current. As the transducer may be remote from the reservoir, continuous mode jetting does not have the same temperature limits as DoD jetting. Pressurisation of the reservoir ensures a continuous stream of the material being jetted passes through the nozzle and issues from the orifice, where it breaks up into individual uniform droplets in response to the waveform. The droplet stream subsequently passes through an electrostatic charging field to acquire an electrostatic charge so that use of a deflection field allows the droplet stream to be directed onto the required locations on the substrate. Those not deflected by the deflection field are collected into a gutter and may be recycled for later reuse.

Both types of system are available from a range of vendors as either a print head or a full printing system. There are several print-head assemblies to choose from depending on the specific application requirements. The nozzle orifice diameter controls the droplet volume and therefore feature resolution that can be achieved. The resolution also depends on several other factors such as wettability of the ''ink'' onto the substrate, ink curing speed, and so on. Microfab has successfully demonstrated its piezo based printing platforms' abilities in solder bumping, embedded passives, and for printing of other material types such as dielectric materials and adhesives, components of fuel cells, organic light-emitting diodes, waveguides, micro lenses and so on. Figure-2 is an illustration of a flip-chip with a perimeter array of jetted solder bumps with 60μm balls on 150μm centres. For continuous systems droplet sizes range from 20μm to 1mm, with 150μm being the typical feature size. For DoD systems, the achievable droplet diameter ranges from 15μm to 100μm [8].

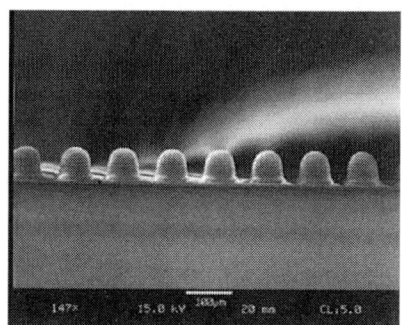

Figure-2 Jetted flip-chip solder bump [9]

Aerosol Jet is a registered trademark of Optomec®. The Aerosol Jet system is mainly composed of two key components, the atomizer and the deposition head, marked as ① and ② in Figure-3 respectively. The raw material to be deposited must be in a liquid form and is first placed into an either ultrasonic or pneumatic atomizer, which is utilized to generate a dense vapour of material droplets between 1-5 microns in size - a process known as 'mist generation'. Then the generated mist, or aerosol, is transferred into a tightly confined jet, within what is known as the deposition head, by a gas flow running through the atomizer and out into the deposition head. The aerosol stream brought into the deposition head is further focused by a second gas flow introduced into the jet. The aerosol stream and the newly introduced sheath gas flow interact with each other and form a co-axial annular flow, which then leaves the deposition head through a nozzle attached to it and lands on the substrate. Optomec® offers a laser module that can be integrated into the system to locally complete thermal post-processing of the deposited material, e.g. by sintering of a nano or micro particles, and for some materials such as gold and silver has been shown to achieve properties as good as the bulk materials, but without damaging substrates which have low heat endurance capability. The printing process of the Aerosol Jet system provides a non-contact way of printing, like inkjet printing, which makes it compatible with processes where contamination of the substrate must be avoided.

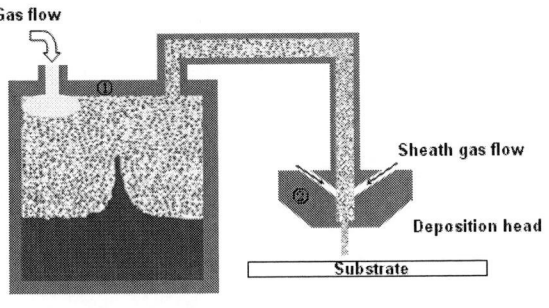

Figure-3 Illustration of the Optomec® Aerosol Jet operating principle

The Aerosol Jet system can create lines as fine as 10 microns with a minimum pitch of 20 microns, or as wide as 150 microns. The material handling ability in terms of viscosity falls into a wide range, i.e. between 0.001 and 2.5 Pascal second (Pa·s), which enables deposition of a wide range of materials [10].

Figure-4 An example of silver lines written using the Aerosol Jet system over a trench [10]

Figure-4 shows an example of silver lines written over a trench using Aerosol Jet printing, where the line width is 60μm and the trench depth is 500μm. Aerosol Jet printing is particularly suited to such 3D applications as its deposition head can be tilted at an angle to follow the contour of the substrate without contact. Compared with traditional inkjet

printing, Aerosol Jet has the following advantages: its stand-off distance can be adjustable between 1mm to 5mm instead of being fixed as in an inkjet printing system; and the aerosol is a continuous stream composed of high density micro-droplets which are tightly focused resulting in a fine feature definition ability and the nozzle is also more clog resistant [11].

Filament based direct writing

Filament based direct writing (FBDW) also requires a liquid material for dispensing, but differs from the inkjet direct writing processes in that the material flows continuously instead of being jetted as droplets and the material viscosity may therefore vary over a much greater range. Two commercially available examples of this process are discussed here: MicroPen and nScrypt.

The working principle of MicroPen writing can be described as follows:

Material to be deposited is first loaded into a syringe, which is connected with the writing head, referred to as a "block" by the system vendor. The material is squeezed out of the syringe into the writing head by compressing the plunger of the syringe using a pneumatic ram. The material transferred into the "block" is then pressurized up to almost 14MPa (2,000 psi), from where it flows into a micro-capillary writing tip, which deposits the material onto the substrate. A schematic configuration of the MicroPen writing system is presented in Figure-5.

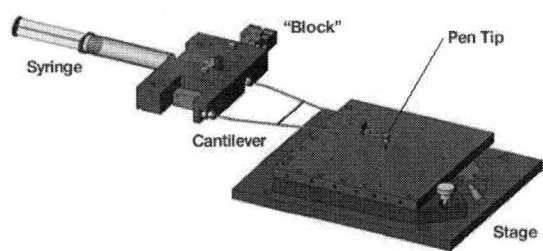

Figure-5 Schematic of the MicroPen system [12]

MicroPen direct writing can deposit potentially any liquid material ranging in viscosity from 0.005 to 500 Pa·s, which is quite a large range. MicroPen not only deals with a wide range of materials, but also possesses the capability of depositing materials onto a range of substrates, including flexible and non-planar substrates. The MicroPen technique is also a non-contact technology, with the micro-capillary writing tip not touching the surface.

Another filament based direct writing technique is the nScrypt dispensing technique, where a direct-print dispensing tool is integrated with nScrypt's novel pump called Smart Pump™, which is illustrated in Figure-6.

The integrated smart pump is able to dispense materials with a viscosity of up to 1,000 Pa·s with accurately controlled air pressure, valve opening and dispensing height [13]. When dispensing is initiated, a valve opens so as to allow the material to be dispensed to flow through the dispensing tip onto the substrate. Once dispensing stops, the valve closes to block material from leaking. One advantage of the smart pump is that it includes a sucking-back movement of materials into the dispensing nozzle when deposition is terminated. A negative pressure is maintained in the dispensing tip chamber to induce material sucking back when deposition is ceased. This feature allows the orifice to be left clean and clear without -any agglomeration of-material or the possibility of nozzle clogging, and delivers a consistent start every time dispensing commences. Consequently line printing can be precisely controlled and the width of the dispensed material can be maintained consistent without bulges at the ends.

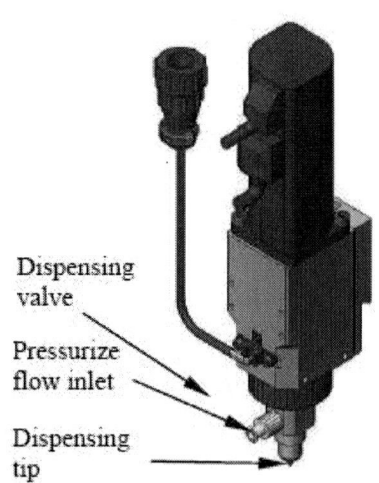

Figure-6 The nScrypt Smart Pump ™ [13]

Tip based direct writing

The dip-pen nanolithography (DPN) technique utilizes an atomic force microscope (AFM) tip to deliver molecules from the dispensing AFM tip onto substrates, which is especially useful in nanoscale applications.

Piner, *et al*, discovered this technique while investigating a problem that had baffled researchers in atomic force microscopy for a long time that condensed water can either be transported from the substrate to the AFM tip or vice versa depending on the relative humidity and substrate wetting properties [14]. Due to the capillary effect, the AFM tip is capable of dispensing ink adhering to it, from previous dipping, onto substrates which have an affinity for the ink, in a similar way to a dip pen writing on a piece of paper, as shown in Figure-7.

With traditional AFM cantilever tips, dip-pen nanolithography can create features as fine as 12nm in line width and 5nm in spatial resolution. This resolution can be further improved by using sharper cantilevers, but the finest resolution that this process can achieve is still uncertain [5].

Thermal dip pen nanolithography (tDPN) is a modified version of dip pen nanolithography utilising an atomic force microscopy tip which is compatible with heat. The pen tip is first coated with the material to be dispensed in a solid form. When heated, the coated material on the pen tip will melt due to heat transfer and will flow over the pen tip surface onto the substrate. It is claimed that using meltable inks has many advantages. Writing can be switched on and off readily and the ink flow rate can be varied easily by controlling the tip

temperature. Complex 3D structures can be generated through writing layer by layer with the previously deposited layer being solid. And it is compatible with other traditional semiconductor manufacturing methods, since thermal dip pen nanolithography can be performed in vacuum [15].

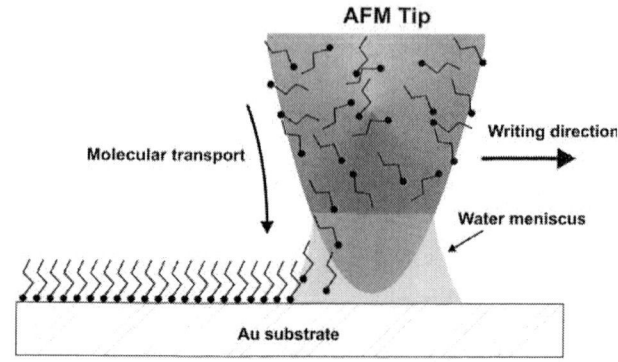

Figure-7 Schematic illustration of DPN [14]

The finest feature researchers have achieved with tDPN is close to 75 nm wide. Researchers has also successfully demonstrated writing indium metal onto a substrate and the potential operating temperature is claimed to be much higher than the melting point of indium, as high as 1000°C. One key feature of tDPN lies in that it enables ink to be deposited with such abundant thermal energy as to organize into monolayers prior to its solidification [15].

Polymer pen lithography is a high-throughput and low-cost tip based direct writing method, which is another variant of dip pen nanolithography. It does not use an actual AFM tip to dispense materials, but utilizes polymer tips instead. Its high-throughput feature compensates for DNP's weakness at the micro scale, which makes it a very promising method for large scale manufacturing. Thousands of, or even millions of, polymer pens can be made together into a polymer pen array which by dispensing materials onto the substrate simultaneously could rapidly create much larger features than individual pens [16]. When the tips of the polymer pen array contact the substrate, their ink coating can be delivered onto substrate.

Laser based direct writing

Laser based direct writing techniques rely on using a laser to process materials so that they can be deposited onto a substrate. The popularity of lasers in direct writing results mainly from a subtractive process called laser micro-machining, however there are many other types of laser based methods such as laser chemical vapour deposition (LCVD) [17], electro-less plating or electroplating with laser assistance [18,19], matrix assisted pulsed laser evaporation direct write (MAPLE-DW) [20], etc.. Some laser based patterning methods are subtractive methods, but are still sometimes considered direct writing, however this paper only introduces additive methods of depositing materials, in line with definition given at the beginning as their being an additive technique without etching.

978-1-4244-4658-2/09 $25.00 © 2009 IEEE

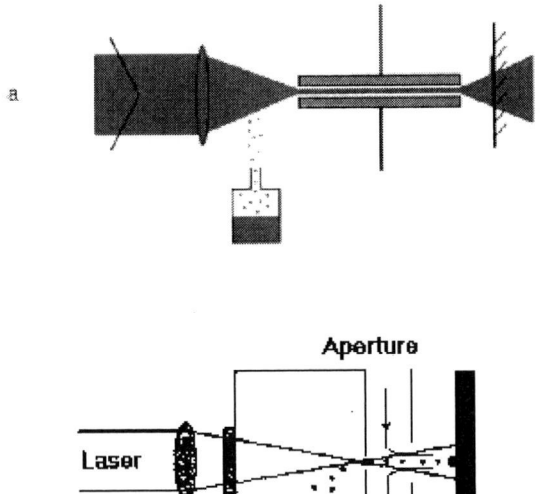

deposited on a laser-transparent support as a thin film. The support with the thin film on it is brought close to the substrate. The laser is focussed at the interface between the transparent support and the thin film layer where it induces some of the target material to vaporise and propel the rest of the thickness of the material onto the receiving substrate.

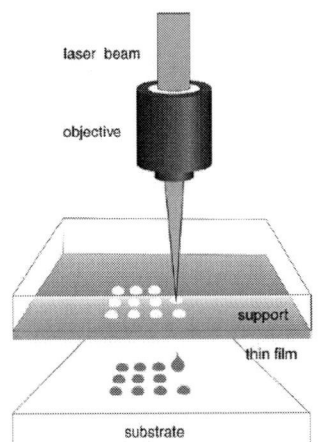

Figure-9 Laser induced forward transfer scheme [21]

Figure-8 Schematic of a (a) laser-guided direct-write (LGDW) system and (b) flow-guided direct write (FGDW) system [5]

Figure-8 a and b illustrate flow-guided and laser-guided direct writing techniques respectively. In these techniques a liquid precursor or colloid suspension is first transformed into an aerosol usually by ultrasonic atomization devices or pneumatic nebulisation depending on the particle size of the starting liquid precursor or colloid suspension, just like the mist generation in the Aerosol Jet system. This dense aerosol is then introduced into the deposition devices and focused by the optical and hydrodynamic forces occurring due to momentum transfer between the laser beam and aerosol particles into a narrow beam. For the laser-guided direct-write system, the laser and the generated mist are both transmitted through a hollow optical fibre to guide the particles towards the substrate. The deposition rate ranges from 1 to 10,000 particles per second. But for flow guided direct write, the deposition rate can be higher than 10,000 particles per second, and particle velocity can increase from 1m/s to 10m/s. The accuracies for LGDW and FGDW are 2μm and 25μm respectively [5]. In flow guided direct write, a sealed chamber is fed with the generated aerosol by a gas flow and these particles come out of an orifice with diameter of one millimetre. The particles are then joined by a second gas flow which forms a cylindrical sheath to protect the mist. These combined streams then pass through another small hole to be further focused again and deposited onto the substrate with the assistance of the laser beam.

Laser induced forward transfer (LIFT) is another laser based direct write technique. The general principle is illustrated in Figure-9. The material to be written is first

Conclusions

In this paper, each of the main types of direct writing technique were discussed and examples presented. The focus has been on the basic principles of those methods, together with some examples of applications. Table-1[5, 10, 14-16, 22] compares the main features of some of the methods discussed above. For some of these processes the deposition rate quoted is volumetric whilst for other it is a linear deposition rate.

Even though direct-write processes are beginning to demonstrate their potential impact on future microelectronic manufacturing, there are still some significant technical barriers and challenges to be overcome before any further breakthrough can be achieved in terms of widespread industrial application. For instance, in inkjet printing processes, not only the ink formulation poses a great challenge, in terms of rheological properties, particle suspension, wetting and adhesion behaviour, etc. [23,24], but also the interactions of the ink with the substrates are critical in determining the stability, integrity and resolution of such ink-jetted micro sized structures. Since features to be deposited are tailored with the functionality of the materials formulated in ink, it is vitally important to consider the evolution of the materials during processing. Furthermore, full integration through creation of multilayered structures or full 3D approaches require precise positioning and alignment accuracy, and the achievable feature resolution also demands some further developments in the fundamental understanding of these processes.

978-1-4244-4658-2/09 $25.00 © 2009 IEEE

Table-1 Comparison of direct-write method features

DW techniques			Resolution	Material viscosity range (Pa·s) or types		Writing speed	3D capability*
Droplet-based DW	Inkjet	Continuous	Droplet size 20μm-1mm typically 150 μm	<0.01		60mm3/s	••
		DOD	Droplet size 15 - 200μm	<0.04		0.3mm3/s	
	Aerosol Jet		Line width 10-150μm thickness 10nm-5μm	<2.5		0.25mm3/s (single nozzle)	•••
FBDW			50μm	<1,000		50mm/s typically	•••
Tip-based DW	DPN		12nm line width and 5nm spatial resolution	Thiol molecules, macromolecules, nanoparticles		0.2-5 μm/s	•
Laser beam DW	LGDW		2 μm	Non-absorbent droplets and solid particulates	Close to that of Aerosol Jet	10-4mm3/s	••
	FGDW		25μm	Atomizable fluids and colloids		0.25mm3/s	••
	LIFT		10-200μm	Solids		3 -50mm/s	••

* • little 3D capability, •• moderate 3D capability, ••• excellent 3D capability

References

1 E. D. Williams, R. U. Ayres, and M. Heller, "The 1.7 Kilogram Microchip: Energy and Material Use in the Production of Semiconductor Devices", *Environmental Science and Technology*, Vol 36, No. 24 (2002), pp. 5504-5510.

2 S. Gamerith and A. Klug, "Direct Ink-Jet Printing of Ag-Cu Nanoparticle and Ag-Precursor Based Electrodes for OFET Applications", *Advanced Functional Materials*, Vol 17 (2007), pp. 3111-3118.

3 S. Jeong, K. Woo, *et al*, "Controlling the Thickness of the Surface Oxide Layer on Cu Nanoparticles for the Fabrication of Conductive Structures by Ink-Jet Printing", *Advanced Functional Materials*, Vol 18 (2008), pp. 679-686.

4 M. Mäntysalo, P. Mansikkamäki, *et al*, "Evaluation of Inkjet Technology for Electronic Packaging and System Integration", *Proceedings of the 57th IEEE Electronic Components and Technology Conference*, Reno, Nevada, May. 2007, pp. 89-94.

5 A. Piqué and D. B. Chrisey, Direct-Write Technologies for Rapid Prototyping Applications: Sensors, Electronics, and integrated Power Sources, (Academic Press, 2002), pp. 1-551. San Diego.

6 J. A. Lewis and G. M. Gratson, "Direct Writing in Three Dimensions", *Materials Today*, Vol 7/8 (2004), pp. 32-39.

7 C. J. Robinson, B. Stucker, *et al*, "Integration of Direct-Write (DW) and Ultrasonic Consolidation (UC) Technologies to Create Advanced Structures with Embedded Electrical Circuitry", *Proceedings of the 17th Solid Freeform Fabrication Symposium*, Austin, TX, August. 2006, pp. 60-69.

8 D. Wallace, *et al*, "Ink-jet as a MEMS Manufacturing Tool", *Proceedings of the SMTA Pan Pacific Microelectronics Symposium*, Hawaii, January. 2006.

9 D. J. Hayes, *et al*, "Printing System for MEMS Packaging", Proceedings of *SPIE Micromachining and Microfabrication Conference*, San Francisco, CA, October. 2001, pp.206-214.

10 B. King and M. Renn, "Aerosol Jet Direct Write Printing for Mil-Aero Electronic Applications", http://www.optomec.com/

11 B. E. Kahn, "The M3D Aerosol Jet System, an Alternative to Inkjet Printing for Printed Electronics", *Organic and Printed Electronics*, Vol 1 (2007), pp.14-17.

12 http://www.micropen.com

13 B. Li, P.A. Clark and K.H. Church, "Robust Direct-Write Dispensing Tool and Solutions for Micro/Meso-Scale Manufacturing and Packaging", *Proceedings of the ASME International Manufacturing Science and Engineering Conference*, Atlanta, Georgia, October. 2007.

14 R. D. Piner, J. Zhu, *et al*, " 'Dip-Pen' Nanolithography", *Science*, Vol 283 (1999), pp. 661-663.

15 P.E Sheehan, W.P. King, *et al*, "Thermal Dip Pen Nanolithography", *NRL Review Chemical/Biochemical Research* (2006), pp.1-2.

16 F. Huo, Z. Zheng, *et al*, "Polymer Pen Lithography", *Science*, Vol 321 (2008), pp. 1658-1660.

17 K. William, J. Maxwell, *et al*, "Freeform Fabrication of Functional Microsolenoids, Electromagnets and Helical Springs Using High-pressure Laser Chemical Vapor

Deposition", *Proceedings of the 12th IEEE International Conference on Micro Electro Mechanical Systems*, Orlando, Florida, January. 1999 pp. 232-237.

18 R.J. Von Gutfeld, *et al*, "Laser-Enhanced Plating and Etching: Mechanisms and Applications", *IBM Journal of Research and Development,* Vol 26 (1982), pp. 136-144.

19 R.J. Von Gutfeld, *et al*, "Laser Enhanced Electroplating and Maskless Pattern Generation", *Applied Physics Letters,* Vol 35 (1979), pp. 651-653.

20 A. Pique, *et al*, "A novel Laser Transfer Process for Direct Writing of Electronic and Sensor Materials", *Applied Physics A,* Vol 69 (1999), pp. S279-S284.

21 M. Colina, P. Serra, *et al*, "DNA Deposition through Laser Induced Forward Transfer", *Biosensors and Bioelectronics,* Vol 20 (2005), pp.1638-1642.

22 K.K.B. Hon, *et al*, "Direct Writing Technology – Advances and Developments", *Manufacturing Technology*, Vol 57 (2008), pp.601-620.

23 D. Kim, S. Heong, *et al*, "Direct Writing of Silver Conductive Patterns: Improvement of Film Morphology and Conductance by Controlling Solvent Compositions" , *Applied Physics Letters*, Vol 89 (2006), pp. 264101-1~264101-3.

24 I. Shim, Y. Lee, *et al*, "An Organometallic Route to Highly Monodispersed Silver Nanoparticles and Their Application to Ink-Jet Printing", *Materials Chemistry and Physics*, Vol 110 (2008), pp. 316-321.

Integrated Inductors on Silicon and Planarized Ceramic Substrates

Jianwei Wang[1,2], Jian Cai[1,2], Xinyu Dou[1,3], Shuidi Wang[1,2]

[1]National Laboratory of Information Science and Technology in Tsinghua University, Beijing, China

[2]Institute of Microelectronics, Tsinghua University, Beijing, China

[3]Institute of Information Technology, Tsinghua University, Beijing, China

Jianwei.wng@gmail.com, 13269336548

Abstract

Passive integration is one of the important issues for system miniaturization in wireless applications on different substrate. Integrated inductors were designed and realized on both silicon and planarized ceramic substrate. Planarized ceramic substrate has the advantages such as lower cost than polished ceramic substrate and has other advantages such as lower dielectric constant, higher bulk resistance than silicon substrate. The L values of fabricated inductors on planarized ceramic substrate are about 1nH, the peak values of Q are about 30 and the corresponding frequency is about 10GHz. By contrast, for the Inductors fabricated on silicon substrate with the same process, the L values is more or less the same with those on ceramic substrate, the Q peak value is about 7 lower, the corresponding frequency is about 5GHz.

Introduction

As the fast development of miniaturization technologies of wireless communication, the requirements of data transfer speed and communication band width are more and more restricted and the leading trend moves to the higher carrier frequency. One of the difficulties for moving to the higher frequency realm is that the performance of RF passive components, especially the integrated inductors will be deteriorated largely. For now, there are many researches developed in the realm of integrated inductor with the aim to improve the performance, such as the ferromagnetic RF integrated inductors and MEMS inductors [1-4]. To some extent the performance improved, but still exist the difficulties such as high density integration, higher performance requirement, standard technical process.

One of the solutions is on-chip inductors, which are fabricated on silicon substrate. The inductors would be realized above dielectric layer using wafer-level packaging (WLP) techniques. Another choice is the inductors integrated based on ceramic substrate. Firstly ceramic substrate has much higher resistivity and lower dielectric constant than silicon substrate which leads to the decrease of substrate loss, so higher Q value performance can be achieved. Furthermore, as the requirements of portable and miniature equipment growing, the area of the IC becoming a precious resource, but in many IC especially the RFIC, the passive devices take up a large portion of the whole chip. So, the inductors integrated on ceramic substrate can be applied in the SiP, which saves much area resource.

Consideration of Ceramic Substrate

To integrate micro-inductors on ceramic substrate, the substrate's surface roughness should meet requirements. The surface roughness data of popular electronic ceramic substrates are shown in Table 1. However, the substrate with smoother surface is very expensive. In this paper, a planarized

96% alumina substrate was selected, with a total thickness of 1 mm. The planarized layer is about 20μm thick and with a surface roughness (Ra) of 52nm. Fig. 1 shows the cross-section of the planarized ceramic substrate.

Table 1 Surface roughness of ceramic substrate

Ceramic substrate	Roughness(Ra)
99.6% abrasive substrate	<10nm
99.6% thin film substrate	<125nm
96% thick film Substrate	200nm~800nm

Fig. 1 cross-section of planarized ceramic substrate

For WLP inductors on silicon substrate, 4 inch silicon wafers (low resistance, P type, 500μm in thick) are employed.

Model and Calculation

The popular circuit model for inductor on silicon substrate is shown in Fig. 2. L_S is the self-inductance of the spiral inductor. R_S is series-resistance of inductor itself. C_S is coupling capacitance between spiral and underpass pattern. C_{OX} is dielectric capacitance between spiral metal and Si substrate; C_{si} and R_{si} are Si substrate capacitance and resistance, respectively [5].

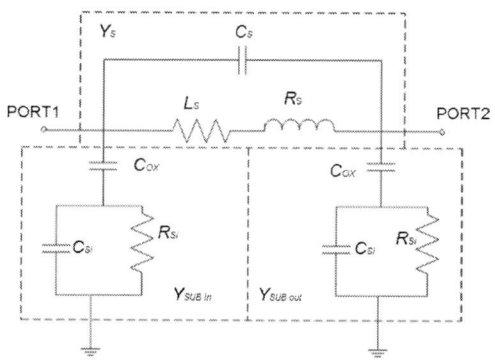

Fig. 2 Circuit model for inductor on Si substrate

For the inductors on ceramic substrate, C_{ox} does not exist, and the substrate resistance R_{si} is large enough to be equivalent to open circuit. So based on these thoughts we can get the modified circuit model for inductor on ceramic substrate which is shown in Fig. 3.

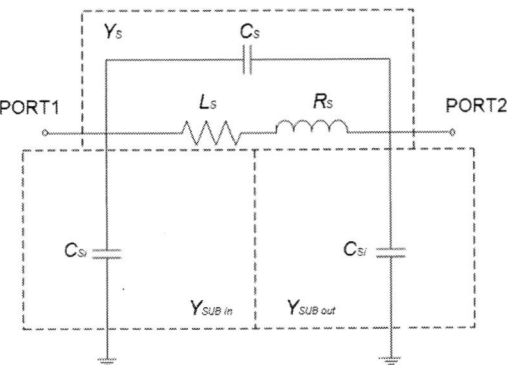

Fig. 3 Circuit model for inductor on ceramic substrate

The lumped-element equivalent circuit can be changed into two port network to calculate Y parameters from which the L and Q of the inductors can be drawn [6].

$$L = \frac{1}{2\pi f} \text{Im}\left\{\frac{1}{Y_{11}}\right\}$$

$$Q = \frac{\text{Im}\{1/Y_{11}\}}{\text{Re}\{1/Y_{11}\}} = -\frac{\text{Im}\{Y_{11}\}}{\text{Re}\{Y_{11}\}}$$

Inductors Structure and Design

11 different inductors were designed. They are different in radius, turns and width of spiral, the layout is shown in Fig. 4, the specific size of each inductor is showed in Table 2. There are three layers in the vertical structure, named metal1, dielectric layer and metal2 [7]. In this study, both Metal 1 and metal 2 are electroplated Cu. SU8, a kind of photoresist dielectric, was used as the dielectric layer. SU8 has a good insulated property and the dielectric constant of which is 3.2 [8]. In the layout, there is an open circuit cell for every inductor to eliminate the effect of output pad in test.

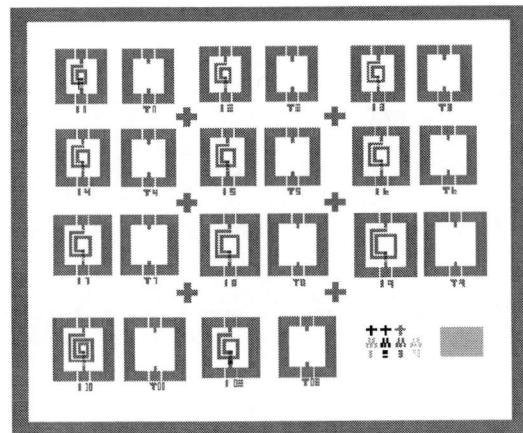

Fig. 4 layout of inductors

Table 2 illumination of inductors layout

	Turn/N	Radius/R (µm)	Width/W (µm)	Space/S (µm)
L1	1.5	30	25	20
L2	1.5	35	25	20
L3	1.5	40	25	20
L4	1.5	45	25	20
L5	1.5	50	25	20
L6	1.5	55	25	20
L7	1.5	60	25	20
L8	1.5	70	25	20
L9	1.5	80	25	20
L11	2.5	30	25	20
L12	1.5	50	30	20

Fabrication Process

The inductors were fabricated on silicon and ceramic substrate in-house. The fabrication processes are shown in Fig. 5, and the details are as follows:

(1) For ceramic substrate: rinsing the ceramic substrate with ethanol and acetone in the ultrasonic environment, sputtering TiW/Cu as the seed layer. The thickness of TiW adhesion layer is 1000Å which used to adhere to the substrate and prevent diffusion. The thickness of Cu layer is 2000Å which acts as the seed layer of plating Cu. For silicon substrate: cleaning the substrate with H_2SO_4 and H_2O_2, growth silicon dioxide layer of 1000 Å; the process of sputter TiW/Cu is same as that on the ceramic substrate.

(2) Spin coating the photoresist AZ-4620, with a thickness of 15µm. The spin rate is 3500rpm and time is 40s. The photoresist exposure process follows as exposing 40 second and developing 2 minute.

(3) Electroplating Cu to 7µm with the current density of 1 ASD for 45 minute.

(4) Removing photoresist with acetone, etching the seed layer of TiW/Cu.

(5) Spin coating SU8 as the dielectric layer. The thickness is 10µm with the spin speed of 3000rpm. After exposure and developing, the baking time for ceramic substrate should be longer than silicon substrate, as the thermal conductivity for ceramic substrate is lower than silicon substrate.

(6) Sputtering TiW/Cu as the seed layer, with the same parameter as step (1).

(7) Spin coating the photoresist and develop the spiral figure, with the same parameter as step (2).

(8) Electroplating Cu to 7µm.

(9)   Removing the photoresist and etching off the seed layer.
(10)   Finished inductors for testing.

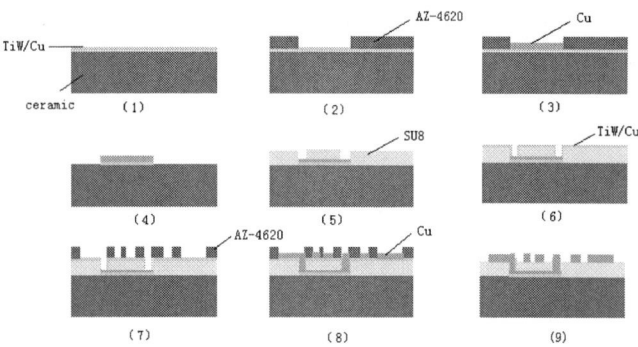

Fig. 5 Process flow of inductors fabrication on ceramic
substrate

Results and Discussion

Fig. 6 to Fig. 7 show the optical photo of fabricated
integrated inductors on planarized ceramic substrates. Table 3
shows the comparison of test value and design value of the
inductors, from which we can see that the fabrication is well
done. To test the inductors, firstly DC test was set up to make
sure the structure is well finished. In this test the resistance
should lower than 0.5Ω. And then higher frequency testing
was employed. The testing equipments for high frequency
testing include Cascade microprobe station and HP8722ES
network analyzer. The S parameters can be measured from the
network analyzer and transformed to the Y parameters.
Finally get the L and Q value can be calculated from the Y
parameters.

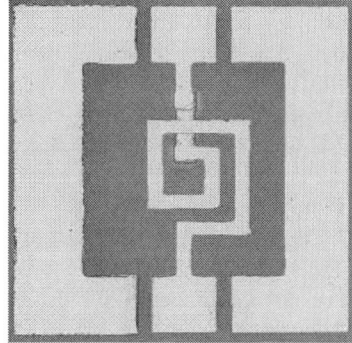

Fig. 6 Optical photo of fabricated inductors on ceramic

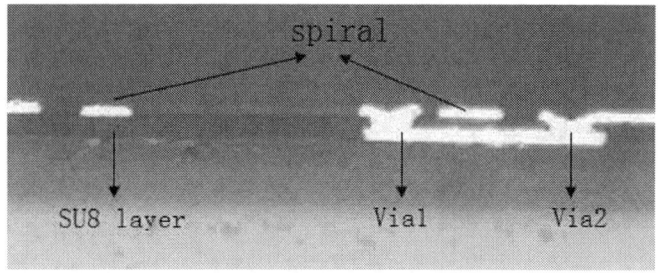

Fig. 7 Cross-section of inductor on ceramic substrate

Table 3 Comparison of test value and design value

Unit: μm	R		S		W	
	Design value	Test value	Design value	Test value	Design value	Test value
L1	30	30.5	20	19.72	25	26.1
L2	35	35.7	20	20.16	25	24.9
L3	40	40.5	20	19.73	25	25.8
L4	45	45.6	20	20.60	25	25.8
L5	50	50.8	20	20.17	25	26.3
L6	55	55.9	20	19.73	25	26.7
L7	60	60.9	20	19.72	25	26.3
L8	70	71.0	20	20.16	25	25.8
L9	80	81.7	20	19.72	25	26.3
L11	30	30.0	20	19.28	25	25.8
L12	50	50.8	20	19.73	30	30.6

Fig. 8 and Fig. 9 show the measured results of inductance
of fabricated inductors on ceramic and silicon substrate. From
L1 to L7, the inductor's radius varies from 30μm to 60μm,
increasing by the step of 5μm while from L7 to L9 the radius
varies from 60μm to 80μm, increasing by the step of 10μm.
L11 is a 2.5 round spiral inductor. It's obvious that the value
of inductance increases as the radius increases, and the same
as the rounds of spiral. Through the comparison of Fig. 8 and
Fig. 9, we can see that the L value is more or less the same
between silicon and ceramic substrate.

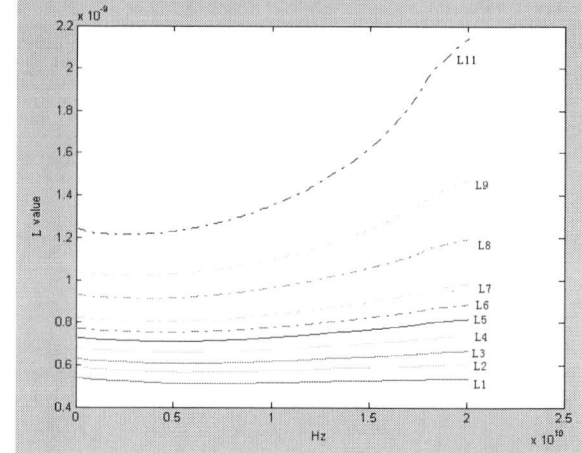

Fig. 8 Measured inductance of inductors on ceramic substrate

Fig. 10 shows the comparison of Q value of fabricated
inductors on ceramic and silicon substrate. We can see from
the figure that the peak Q value of inductors based on ceramic
substrate is about 30, which is 7 larger than those on silicon

substrate. And the frequency of peak is about 10GHz for inductors on ceramic, while it's 5GHz for the silicon one. So the performance in the higher frequency is largely improved.

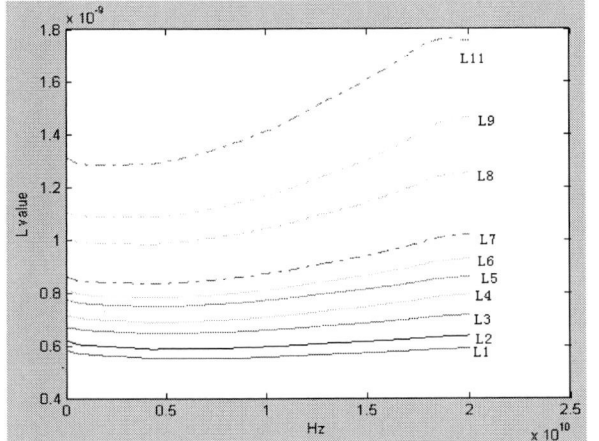

Fig. 9 Measured inductance of inductors on silicon substrate

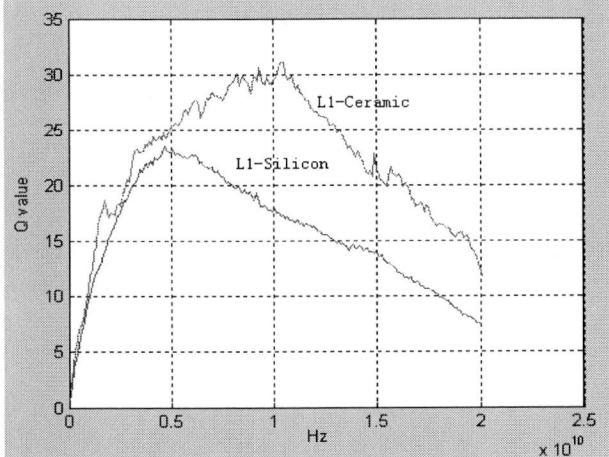

Fig. 10 Comparison of Q value of inductors on ceramic and silicon substrate

Conclusion

In conclusion, the integrated inductors based on ceramic substrate show better performance than the ones on silicon substrate. By planarized ceramic substrate, it can not only get higher peak value of Q, but also get great improvement in the realm of high frequency. So it can be applied in the RF integrated circuit and can also be integrated in the substrate in system package (SiP).

Acknowledgments

This work is supported by the National Engineering Laboratory on High Density IC Packaging Technology, China.

The authors would like to appreciate Mr. Yue Hou from the Nuclear Energy Academe of Tsinghua University for the providing of planarized ceramic substrates.

References

1. C. Patrck Yue, *et al.*, "On-Chip Spiral Inductors with Patterned Ground Shields for Si-Based RF IC's," *IEEE JSSC*, Vol. 33, No. 5 (1998), pp. 743-752.

2. Joachim N. Burghartz, *et al.*, "Multilevel-Spiral Inductors Using VLSI Interconnect Technology," *IEEE Electron Device Letters*, Vol. 17, No.9 (1996), pp. 428-430.

3. J.B.Yoon, *et al.*, "3-D Construction of Monolithic Passive Components for RF Micromachining Technology," *IEEE Trans. Microwave*

4. M. Yamaguchi, *et al.*, "Sandwich-type ferromagnetic RF integrated inductors". *IEEE Trans. Micro. Theory Tech.*, Dec. (2001), 49, pp. 2331-2335.

5. Yue. C. P, *et al.*, "Electron A physical model for planar spiral inductors on silicon". *Devices Meeting. 1996. International 8-11*, Dec. 1996, pp. 155-158.

6. A. M.Niknejad, *et al.*, "Analysis, Design and Optimization of Spiral Inductors and Transformers for Si RFIC's," *IEEE JSSC*, Vol.33, No.10(1998), pp. 1470-1481.

7. Tao Feng, "Study on High-Q RF Inductor using WLP Technology," (2008), Institute of Microelectronics, Tsinghua University, Beijing, China.

8. http://www.microchem.com/products/pdf/SU8_2-25.pdf 2009-4-14.

Nano Resonator Simulation Fabrication and Packaging Consideration

Wei Zhang, Zewen Liu*, Zheng Wang
Institute of Microelectronics, Tsinghua University
Beijing, 100084, China
*Email: liuzw@tsinghua.edu.cn

Abstract

A study on simulation, fabrication and packaging of metallic nano resonator is presented. The nano resonator is modeled with an equivalent circuit and simulated using HSPICE. CMOS compatible process is used to fabricate the nano resonator with a minimum modification The suspended part of the device is released using HF vapor-phase etching. Packaging of nanoscale device is conceived based on the features, fabrication process and the application of the nano resonator.

1. Introduction

With the development of microelectronics, a new generation of systems called Nanoelectromechanical systems (NEMS) emerged as an extension of wide researched Micro-electromechanical systems (MEMS). NEMS, in particular, nanomechanical resonators have generated great interest in both the scientific and engineering communities as they have the potential for ultrasensitive sensor applications such as biological sensors [1], high frequency signal processing [2], low power communication devices [3] and research of quantum phenomena[4][5], etc.

From the view point of the application, how to pick-up of the weak electric signal from the nano resonator is one of the most important tasks to be addressed. It need a fundamental understanding of the electro-mechanical coupling starting from the modeling of the nano resonator. Several works had been proposed in this research. When the nano resonator is actuated by electrostatic force, it can be modeled by an electrical circuit and hence the frequency response of nano resonator can be obtained.

Special difficulty appears if one intends to use the resonator as sensor devices, for example, the gas sensor. The tiny size nanoscale feature of the movable structure and the weak electric signal bring new challenges for NEMS devices packaging. The NEMS device is more sensitive to the stress compared with its MEMS counterpart so that special consideration should be taken into account to prevent the NEMS device from being damaged. The application of the NEMS device should also be considered. For instance, vacuum sealing is not needed if the NEMS device is intended for sensing the pressure, the temperature or moisture while vacuum packaging is indispensable for most NEMS devices.

Wafer Lever Packaging (WLP) will be preferred for the NEMS devices, but traditional (WLP) such as silicon wafer bonding [6][7] requires high process temperature. A lot of research has been conducted to develop low temperature wafer bonding packaging [8][9]. Thin film encapsulation [10][11] is possible solution for NEMS packaging applications. With careful choice of the suitable sacrificial materials, one can realize WLP for devices with the movable suspended structure, which advantages such as reducing wasted area and increasing integration complexity[12][13].

In this paper, we will describe the modeling, fabrication and packaging considerations of a nano resonator.

The total chip area of obtained resonator is smaller than $0.5 \times 0.5 mm^2$.

2. Design and simulation

As shown in Fig. 1, a doubly clamped nanomechanical resonator is mainly consisted of the beam, with dimension in nano scale. Its fundamental resonance frequency is determined with the structural and material parameters of the beam: $f_0 = \omega_0/2\pi = 1.03\sqrt{E/\rho}(t/l^2)$, where l denotes the length, w the width, t the thickness, E the Young's modulus and ρ the mass density. The beam material can be semiconductor, metal, or dielectrics with a metallic layer. In our design, the beam material is Gold and its E is 7×10^{10}Pa, for a nano resonator with length l=4.5μm, width w=2μm, thickness, t=200nm, the fundamental resonance frequency is 20.45MHz.

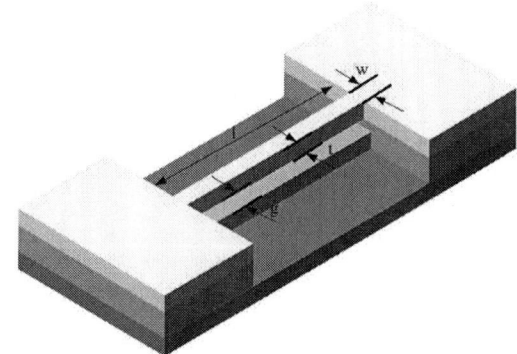

Fig. 1　Schematic drawing of the nano resonator

When applying an ac drive voltage V_{ac} and dc voltage V_{dc} to the drive electrode and the resulting motion can be detected by measuring the current flowing through the device (Fig. 2). The electrical property of a nano resonator can be modeled by an equivalent circuit [14][15][16][17] which is a series RLC circuit as shown in Fig. 3. C_p is the static capacitance when V_{dc} is applied. C_m is the modeled capacitance in the circuit，L_m is the modeled inductance in the circuit, R_m is the modeled resistance in the circuit。Modeling with HSPICE, a peak voltage is observed over a range of frequencies when a current source is applied to the circuit (Fig. 4), which indicates the electric resonant frequency and mechanical resonant frequency are quite close and can be equal to each other if the resonator structure parameters are optimized.

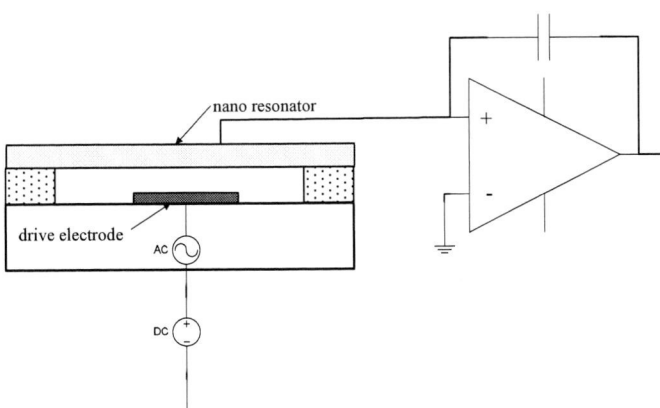

Fig. 2   Schematic view of the nano resonator and the
measurement system

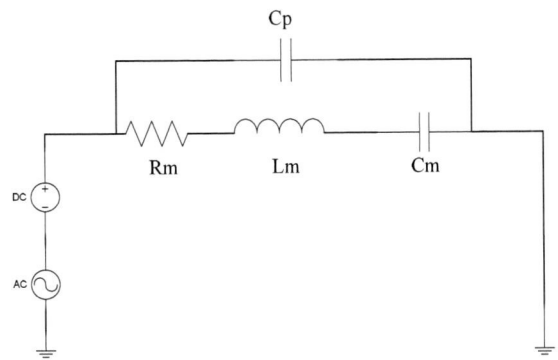

Fig. 3   Equivalent circuit model

The static capacitance of the system C_p can be calculated by

$$C_p = C_0(1 + \kappa V_{dc})$$

where κ is electromechanical coupling parameter, which can be determined by

$$\kappa = \frac{\varepsilon}{2k}\frac{lw}{g^3}V_{dc}^2$$

The value of the capacitance C_m in the model is

$$C_m = \frac{\eta\varepsilon}{k}\frac{lw}{g^3}V_{dc}^2 C_0$$

where η is the modeled constant whose value is 10^7 and ε is the dielectric constant in the vacuum. k is the spring constant which can be derived using the expression:

$$k = 32\,Ewt^3/l^3$$

When there is no dc voltage applied, the capacitance C_o is

$$C_0 = \varepsilon\frac{lw}{g}$$

The value for other component is calculated as follows:

$$L_m = \frac{1}{(2\pi f_0)^2 C_m}$$

$$R_m = \frac{1}{Q}\sqrt{\frac{L_m}{C_m}}$$

Assuming that Q is 10000, the equivalent circuit parameters of the fabricated resonator could be determined from those equations presented above. Modeling the nano resonator with HSPICE, a considerable change of voltage across the circuit is observed over a range of frequencies when a current source is applied to the circuit. That means if we choose a proper L_m, the electrical oscillating peak can match the mechanical resonating frequency very well. More work is on the way to test the response of the nano resonator experimentally.

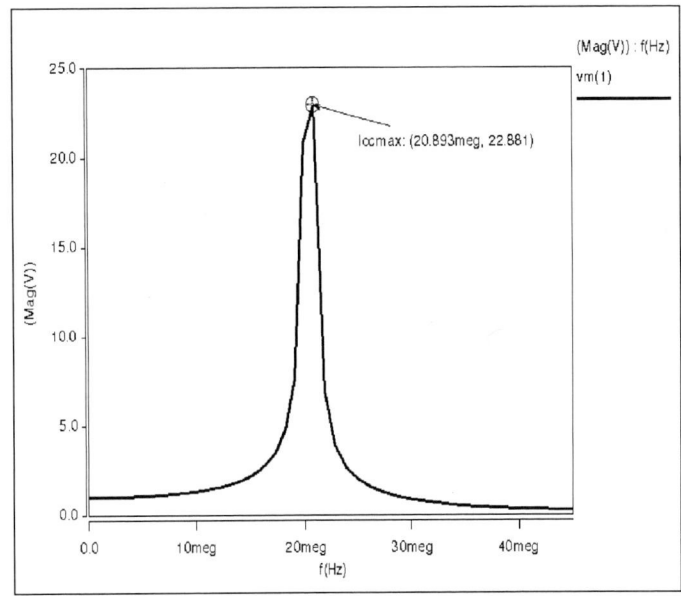

Fig. 4   Frequency response of electromechanical model of the nano resonator

3. Fabrication

The fabrication process is shown in Fig. 5. After the growth of a 1-μm-thick thermal oxide to 10Ωcm p-type Si(100) wafer, a protective layer of SiN was deposited using low pressure chemical vapor deposition (LPCVD). Pt was sputtered and patterned with ion beam etching (IBE) as down electrode. A sacrificial layer of SiO_2 was then deposited by a plasma-enhanced chemical vapor deposition (PECVD) process. After Au sputtering and patterning, the underlying oxide was released using HF vapor phase etching. Fig. 6 shows a scanning electron microscope of a released nano resonator. The size of the designed nano resonator is length l=4.5μm, width w=2μm, thickness, t=200nm.

978-1-4244-4658-2/09 $25.00 © 2009 IEEE   433

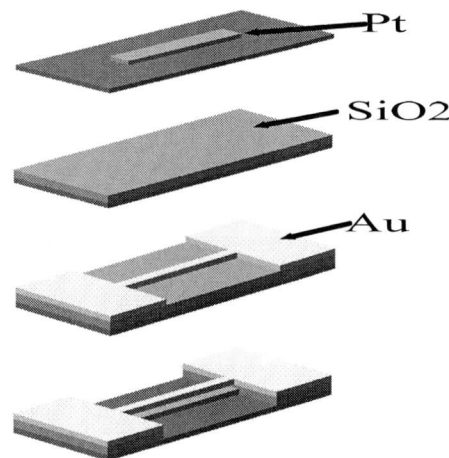

Fig. 5   The fabrication sequence for the nano resonator

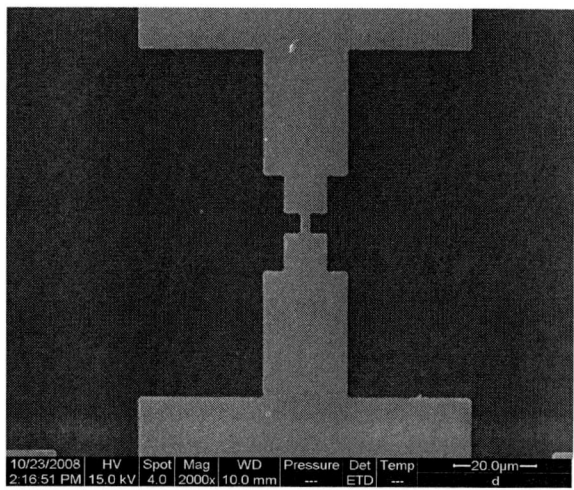

Fig. 6   SEM image of a released nano resonator

4. Packaging

The packaging process for nano resonator is shown in Fig. 7. First, a sacrificial layer of dioxide is deposited on the unreleased nano resonator using Plasma Enhanced Chemical Vapor Deposition (PECVD) and patterned. Then a layer of SiN is deposited and micro holes with the size of 5μm width are etched using Reactive Ion Etching (RIE). One big access port for quick releasing is also created. A polymer layer is deposited after releasing the device with HF vapor phase etching. The thickness of the polymer is different depending on the application of the nano resonator. A thin film of polymer will be grown if the nano resonator is used as a sensor such as temperature sensor. However, if hermetic vacuum packaging is required, the thickness of the polymer should be thick enough so that the following layer of metal such as Au can be sputtered to seal the package.

Conclusions

A metallic nano resonator is analyzed and modeled with an electrical circuit. The nano resonator is fabricated using traditional .process and released by HF vapor-phase etching. Packaging of nano resonator is a challenging task with the scaling down of feature size of devices. A similar approach for packaging of NEMS is presented based on previous research on wafer-level packaging of MEMS by so many researchers.

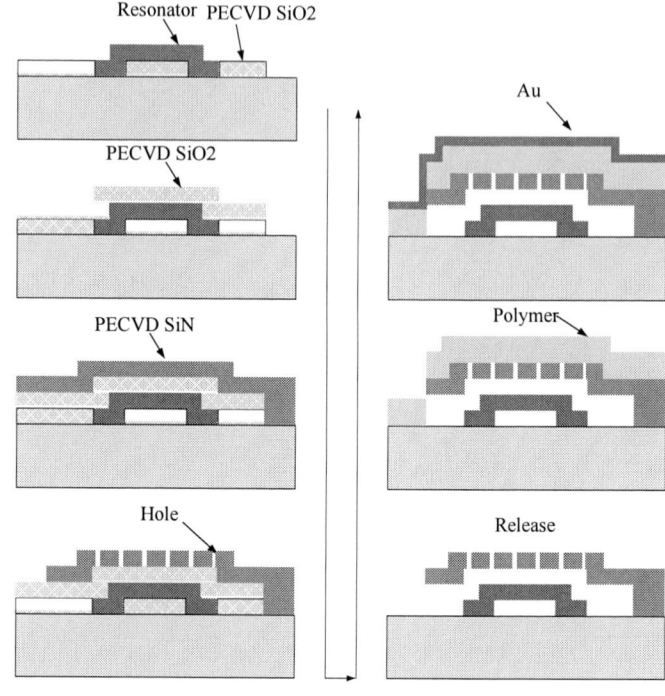

Fig. 7   Polymer-based packaging process of nano resonator

Acknowledgements

This work is supported by the NSFC (National Sciences Foundation of China) with project No. 60576048.

References

1. B. Ilic, Y. Yang, and H.G. Craighead. "Virus detection using nanoelectromechanical devices," *Appl. Phys. Lett.*, Vol. 85, No. 13(2004), pp. 2604-2606.
2. C.T.-C. Nguyen, A.-C. Wong, and D. Hao, "Tunable, switchable, high-Q VHF microelectromechanical bandpass filters," *Proc IEEE International Solid-State Circuits Conf.*, San Francisco, CA, Feb. 1999, pp. 78-79.
3. C.T.-C. Nguyen, "Frequency-selective MEMS for miniaturized low-power communication devices," *IEEE Trans. Microwave Theory Tech.*, Vol. 47, No. 8(1999), pp. 1486~1503.3.
4. A. Cho, "Physics—Researchers race to put the quantum into mechanics," *Science*, Vol. 299(2003), pp. 36–37.
5. M.D. LaHaye, O. Buu, B. Camarota, *et al*. "Approaching the quantum limit of a nanomechanical resonator," *Science*, Vol. 304(2004), pp. 74~77.
6. T. Tsuchiya, Y. Kageyama, H. Funabashi, *et al*. "Polysilicon vibrating gyroscope vacuum encapsulated on-chip micro chamber," *Sens Actuators A Phys*, Vol. 90, No. 1-2(2001), pp. 49-55.
7. B. Lee, S. Seok, and K. Chun, "A study on wafer level vacuum packaging for MEMS devices," *J. Micromechan. Microeng.*, Vol. 13, No. 5(2003), pp. 663–669.
8. T. Itoh, H. Okada, H. Takagi, *et al*. "Room temperature vacuum sealing using surface activated bonding method,"

Proc The 12th international conference on solid states sensors, actuators and Microsystems, Boston, MA, Jun., 2003, pp. 1828–1831.

9. Y.K. Kim, E.K. Kim, S. W. Kim, *et al*. "Low temperature epoxy bonding for wafer level MEMS packaging," *Sens Actuators A Phys*, Vol. 143, No. 2(2008), pp. 323-328.

10. L.C. Chomas, Y.N. Hsu, S. Friends, *et al*, "Low-cost manufacturing/packaging process for MEMS inertial sensors," *Proc 36th international symposium on microelectronics (IMAPS 2003)*, Boston, MA, Nov., 2003, pp. 389–401.

11. A. Hochst, R. Scheuerer, H. Stahl, *et al*, "Stable thin film encapsulation ofacceleration sensors using polycrystalline silicon as sacrificial and encapsulation layer," *Sens Actuators A Phys*, Vol. 114 , No. 2–3(2004), pp. 355–361.

12. M. Pejman, J.J. Paul, A. K. Paul, *et al*. "Characterization of a polymer-based MEMS packaging technique," *Proc IEEE 11th International Symposium and Exhibition on Advanced Packaging Materials Processes, Properties and Interfaces*, Atlanta, GA, Mar. 2006, pp. 139-144.

13. P.J. Joseph, P. Monajemi, F. Ayazi, *et al*. "Wafer-Level Packaging of Micromechanical Resonators," *IEEE Transactions on Advanced Packaging*, Vol. 30, No. 1 (2001), pp. 19-26.

14. M.W. Putty, S.C. Chang, R.T. Howe, *et al*. "One-port active polysilicon resonant microstructures," *Micro Electro Mechanical Systems, 1989, Proceedings, An Investigation of Micro Structures, Sensors, Actuators, Machines and Robots. IEEE*, Salt Lake City, UT, Feb. 1989, pp. 60-65.

15. C.T.-C. Nguyen and R.T. Howe, "An integrated CMOS micromechanical resonator high-Q oscillator," *IEEE Journal of Solid-State Circuits*, Vol. 34, No. 4 (1999), pp. 440-455.

16. T. Lamminmäki, K. Ruokonen, I. Tittonen, *et al*. "Electromechanical analysis of micromechanical SOI-fabricated RF resonators," *Proc 2000 International Conference on Modeling and Simulation of Microsystems - MSM 2000*, San Diego, CA, Mar., 2000, pp. 217-220.

17. G. Abadal, Z.J. Davis, B. Helbo, *et al*. "Electromechanical model of a resonating nano-cantilever-based sensor for high-resolution and high-sensitivity mass detection," *Nanotechnology*, Vol. 12, No. 2 (2001), pp. 100-104.

Optimization of Silver Paste Printed passive UHF RFID Tags

Bo Gao, Matthew M.F. Yuen
Department of Mechanical Engineering
Hong Kong University of Science and Technology
Clear Water Bay, Kowloon, Hong Kong
E-mail: megb@ust.hk

Abstract

Passive UHF (Ultra High Frequency) RFID (Radio Frequency Identification) is a promising technology for products tracking in logistics or routing packages in supply chain. Usually, the UHF RFID antenna is made by etched copper, or aluminum. These etching process introduce chemical wastes during the etching process. Recently, printing silver paste becomes popular due to its environmental friendly manufacturing process. However, there are two drawbacks on printed silver paste RFID antenna. One is low conductivity compared to metal and the other is the high material cost. In this paper, we exam the effects of printing thickness on the performance of UHF RFID tags and summarize the optimized thickness for different read range requirements. A commercial conductive silver paste is printed on PET film as RFID tag antennas. Both finite element simulations and experiments are conducted to evaluate the effects of printed thickness. The simulation results presents that the 10 um thick RFID antenna exhibits relatively good radiation efficiency. The results indicate that 10 um thick RFID tag antenna presents the minimum turn-on power of 18 dBm to 22 dBm which is enough for most applications. The simulation and measurement results present that thinner printed silver paste RFID tag antenna is a potential solution for low cost UHF RFID tags. The cost of these tags could be lower than current etched Al/Cu RFID tags because of material saving.

I. Introduction

The recent development of radio frequency identification (RFID) technology for item-level tracking has been accelerated due to pressing industry demand. RFID technology is an automatic identification method which can be used conveniently for product tracking. It has the advantage that the track process requires only minimal human input and thus further reduces the labor cost in logistics operation. RFID has many potential applications in different areas such as item identification and retail management. It makes it possible to locate the items at the right place. Passive UHF RFID tags are commercially used in cases such as pallet and container tracking because of its low cost.

A RFID system typically consists of transponders and transceivers such as tag and reader. The objectives of RFID system is enabling the tag be read by a RFID reader. Usually, there is no power source inside the passive RFID tag, so the voltage needed to power RFID COMS chip is obtained from reader via remote activation. The reader transmits a modulated signal to the tag and the tag backscatters a signal with identification data to reader at the same time. A passive UHF RFID tag usually includes an antenna and a RF chip which has a memory to store identification data.

Usually, the UHF RFID antenna is made by etched copper, or aluminum. These etching process introduce

chemical wastes during the etching process. Recently, printing silver paste becomes popular due to its environmental friendly manufacturing process. However, there are two drawbacks on printed silver paste RFID antenna. One is low conductivity compared to metal and the other is the high material cost. In order to maximize the read range, the thickness of printed RFID tag antenna should be higher than skin depth. Since the cost of silver paste is high, the thick RFID antenna material costs more than traditional RFID tag antennas.

Previous research papers indicated that the silver paste printed RFID antenna exhibited good antenna efficiency [1] [2]. But these papers only focus on the benchmarking of different materials; there are stills lots of outstanding issues on printed RFID antenna. In this paper, we exam the effects of printing thickness on the performance of UHF RFID tags and summarize the optimized thickness for different read range requirements. A commercial conductive silver paste is printed on PET film as RFID tag antennas. Two types of tag antennas, dipole based directional and quasi-isotropic antennas, are chosen to evaluate the effects of skin depth on RFID tag performance [3]. We validate these designs by conducting finite element analysis and experiments.

II. Evaluation method design and simulation

To evaluate the effects of silver paste thickness on RFID tag performance, we applied finite element simulations to design RFID tags. The designs of two types of modified dipole are shown in figure 1. Figure 1 (a) is a directional RFID tag antenna and figure 1 (b) is an isotropic RFID tag antenna. As shown in figure 2, it is found that the radiated power of isotropic design at any direction is almost the same. To match with Alien's RFID straps, we fine tuned the impedance of RFID tag antennas.

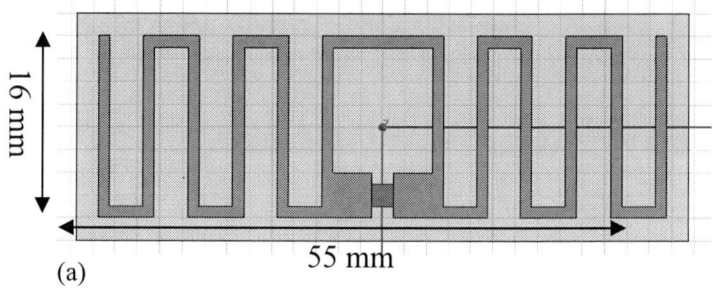

(a)

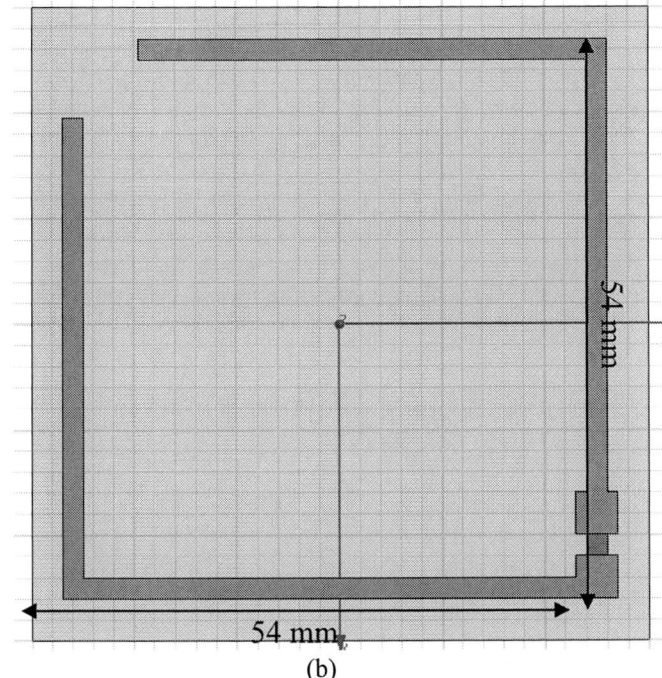

54 mm

54 mm

(b)

Figure 1. Design layout of directional RFID tag antenna and isotropic RFID tag antenna

The radiation efficiency is selected as the factor to evaluate the thickness effects. The radiation efficiency is defined as the ratio of the power delivered to the radiation to the power received. The loss of antenna mainly due to ohmic loss, especially for dipole based antennas. In simulations, we defined different antenna layer thickness and simulated the radiation efficiency.

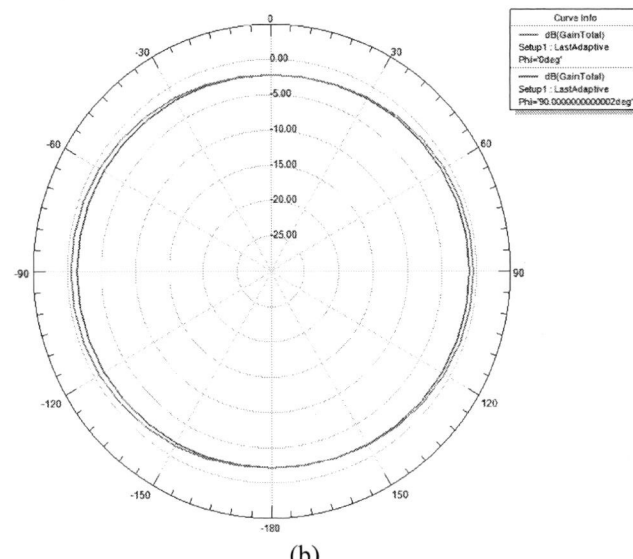

(b)

Figure 2. Radiation pattern of directional RFID tag antenna and isotropic RFID tag antenna

The thickness of 1 um, 3 um, 5 um, 10 um, and 20 um were constructred for each designs. These simulated radiation efficiency were shown in figure 3. It is found that the raidation efficiency increases with the thickness. However, the radiation efficiecy does not change too much when it is thicker than 10 um. Since the silver paste's resistivity is 3.4*10-7 ohm*m, the calculated skin depth from equation (1) at 1 GHz is around 10 um. When the silver paste thickness is much thicker than two skin depth, the current flow is only in the skin of material. From our results, we can find that 20 um smaples show best performance, which is consistant with skin depth theory.

$$\delta = \frac{1}{\sqrt{\pi\mu_o}}\sqrt{\frac{\rho}{\mu_r f}} \approx 503\sqrt{\frac{\rho}{\mu_r f}} \qquad (1)$$

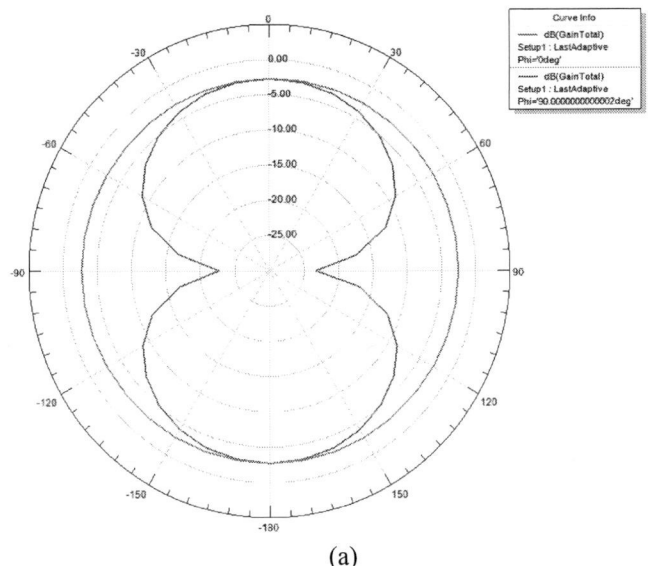

(a)

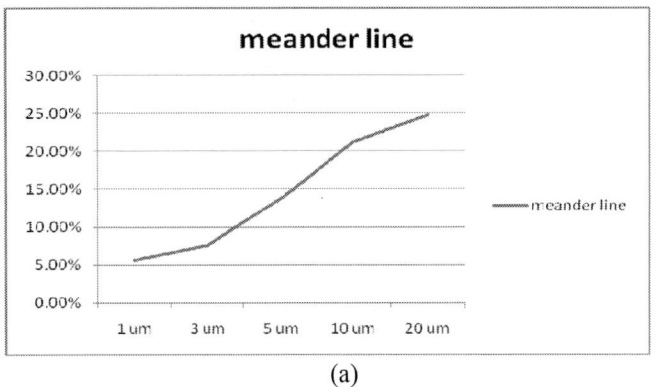

(a)

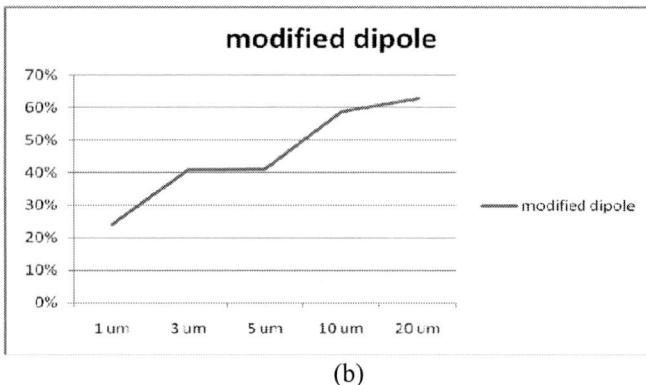

(b)

Figure 3. Radiation efficeny at different thickness (a) meander line designs (b) modified dipole designs

III. Experiment results

To validate the simulation results, we used screen printing technology to pattern RFID tag antennas on 50 um thick PET films. The silver paste with resistance of 15mΩ/25μm was printed by screen printers. Then the printed silver paste is cured at 110°C for 10 minutes. To evaluate the effects of thickness, we print the RFID tag for different times to form different layer thickness. In this paper, we prepared three different thickness, 3 um, 8 um, and 10 um as shown in figure 4.

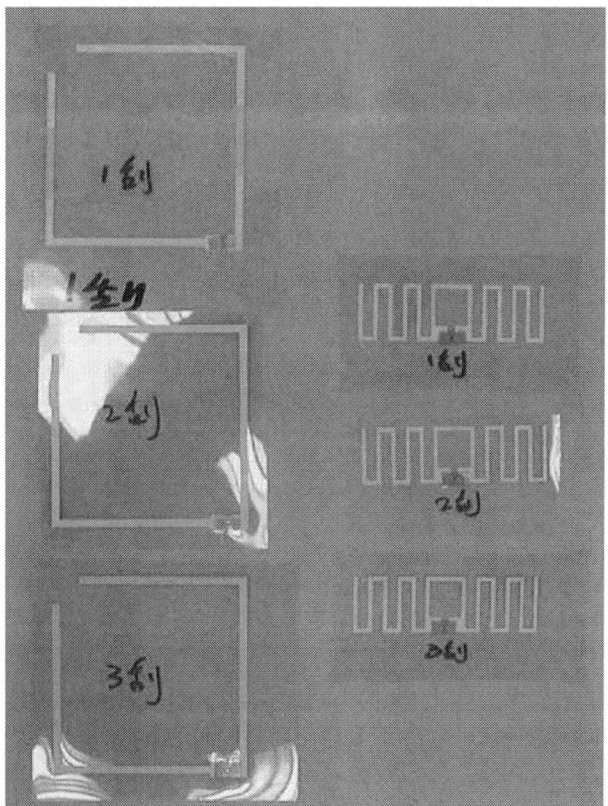

Figure 4 Screen printed RFID tag with different thickness

Then, we placed the RFID tag with bonded RFID straps at a fixed distance of 1 meter in anechoic chamber. The output power of RFID testing system was tuned to be minimal to read RFID tags. The turn-on power is recorded to indicate the performance of RFID tags.

It can be observed from figure 5 that the turn on power of 8 um thick RFID tags are the same with 10 um thick ones. This is likely due to the small differences in the radiation efficiency.

Thicker layer RFID tags are found to be more efficient than thinner one due to the low ohmic losses. It is also found that the effects of layer thickness are validated for both directional RFID tag antenna designs and isotropic RFID tag antenna designs.

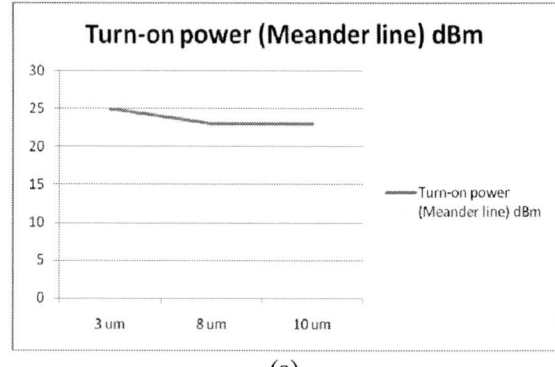

(a)

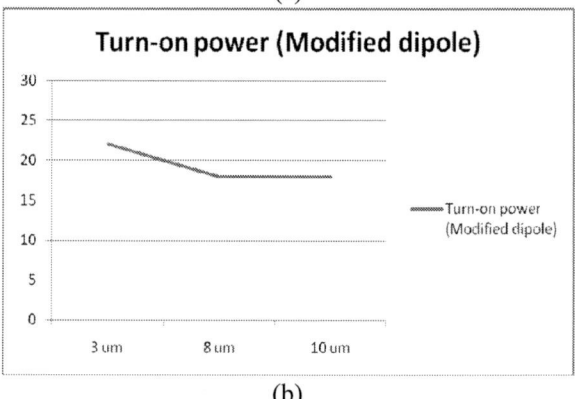

(b)

Figure 5 Measured turn-on power of (a) meander line design and (b) modified dipole.

Conclusions

In this paper, two types of tag antennas, dipole based directional and quasi-isotropic antennas, are chosen to evaluate the effects of skin depth on RFID tag performance. Both finite element simulations and experiments are conducted to evaluate the effects of printed thickness. The simulation results presents that the 10 um thick RFID antenna exhibits better radiation efficiency, compared to thinner designs. These dipole-based antennas are designed to match with Alien's UHF RFID strap. Prototypes are prepared by screen printing technology. The measurements are conduct in an anechoic chamber by RFID tester. The results indicate that 10 um thick RFID tag antenna presents the minmual turn-on power of 22 dBm and 18 dBm which is enough for most applications. The simulation and measurement results present that thinner printed silver paste RFID tag antenna is a potential solution for low cost UHF RFID tags. The cost of

978-1-4244-4658-2/09 $25.00 © 2009 IEEE

these tags could be lower than current etched Al/Cu RFID tags because of material saving.

References

1   Afzal Syed, Kenneth Demarest, Daniel D. Deavours, "Effects of Antenna Material on the Performance of UHF RFID Tags." *IEEE RFID 2007*. March 26--28, 2007, Grapevine, TX, pp. 57-62.

2   Nikitn, P. V; Lam, S; Rao, K.V.S; "Low Cost Silver Ink RFID Tag Antennas." *Transactions on Antennas and Propagation Society International Symposium,* July 2005, pp. 353–356.

3   Rao, K.V.S. Nikitin, P.V. Lam, S.F., "Antenna design for UHF RFID tags: a review and a practical application", *IEEE Transactions on Antennas and Propagation,* Dec. 2005, Volume: 53, Issue: 12 pp. 3870- 3876

Low-temperature Wafer Bonding Using Gold Layers

Ying-Hui Wang, Jian Lu, Tadatomo Suga

Department of Precision Engineering, School of Engineering, the University of Tokyo
Hongo 7-3-1, Bunkyo-ku, Tokyo 113-8656, Japan
Email address: Wang.Yinghui@su. t.u-tokyo.ac.jp, Tel: +81-(0)3-5841-6495

Abstract

The bonding possibility of gold layers was investigated at 25~200°C in wafer scale using a surface activated bonding (SAB) method. The interconnections of Si-to-Si and Si-to-PZT substrates were confirmed using different thickness of gold layers. The influence of surface roughness, vacuum condition, wafer temperature and rolling was studied. The bonded samples were observed using a scan acoustic microscope (SAM). The bonding energy was measured using razor blade test and the bonding strength was evaluated using tensile test. The microstructures on interfaces and the fractured surfaces after tensile test were observed using a scanning electron microscope (SEM) and an optical microscope. The interface of the bonded wafers was nearly void free. The gold layers were effective on metal diffusion and plastic deformation and therefore enlarge the bonded areas.

Introduction

Recently, wafer bonding technique has become a highly effective way for material integration in various areas such as microelectromechanical systems (MEMS), microelectronics, microfluidic devices, bio-MEMS, vacuum packaging and optoelectronics. Conventional bonding processes need high temperature heating. For example, subsequent annealing at 800~1100°C is necessary to bond Si-Si wafers in the conventional process to ensure the bond strength and eliminate voids, and more than 300°C heating is necessary to bond metal-film wafers using thermo-compression bonding method [1-3]. High temperature heating may cause thermal stress problem and induces defects that influence the reliability of devices [4].

Surface activated bonding (SAB) is well known as a useful method to acquire great bond strength at low temperature. The idea of the SAB method is based on the very strong metal or covalent bonding energy between two atomic clean surfaces. The clean surfaces can be obtained by dry process such as argon fast atom beam (Ar-FAB), ion beam, or Ar radio plasma pretreatment in a certain vacuum condition. In previous investigations, various materials have been bonded successfully with high bond strength, such as Al, Cu, Au, Ti, Ni and their alloys as well as the combinations to ceramics and semiconductors including SiC, Si_3N_4, Al_2O_3, AlN, diamond and silicon at room temperature in vacuum [5,6]. Au, Au/Sn and lead-free bumps with flip-chip structures have been successfully bonded not only at low temperatures in vacuum or N_2 under atmosphere pressure but also in ambient air by SAB [7-11]. However, the bonding between semiconductors and oxides need high vacuum condition, critical surface roughness, and also the Si-to-SiO_2 bonding by the SAB method is weak [12]. A wafer bonding technique by a functional multilayered thin metal film is believed a promising candidate to realize the bonding between different materials with less requirements on vacuum and surface roughness. Moreover, VLSI circuits need electrical interconnects using metal interlayer. Thin film metal layers are therefore expected to be used in wafer bonding for being effective to realize the electrical interconnects.

In this paper, the wafer bonding through gold layers was investigated at low temperatures by the SAB method. It started from Si-to-Si bonding and applied to the bonding of Si-to-PZT substrates using different thickness of gold layers and size. The influence of bonding environment and wafer temperature was studied.

Experimental Details

3-mm to 10-mm squared Si substrates, 4-inch (100) Si wafers, and 1-μm-thick Si wafers with PZT thin film (deposited by sol-gel process on Si wafers) were used in bonding experiments. Gold layers were deposited on Si substrates using electroplated process or electron beam evaporation. The thickness of electroplated gold films is 10 μm and the thickness of electron beam evaporation gold films is 100 nm or 300 nm. The schematic views of the bond pairs were shown in Fig. 1.

The bonding experiments were carried out using SAB bonders. The sample surfaces were activated by Ar-plasma or Ar-FAB sources. After activation, the samples were bonded under a certain load at 25~200°C in vacuum or ambient air. In the case of Ar-plasma pretreatment process, the pretreatment chamber is operated for activating sample surfaces under 5-10 Pa pressure with Ar plasma background for 30 s (RF-100W), and then the samples were transferred into the bonding chamber to be bonded together with certain bonding force for 30 seconds at a specified temperature. The bonded sample was then returned back to the pretreatment chamber. The whole process was carried out in air expect the step of Ar plasma pretreatment, and the exposure time of the bonding pair in air before assembly was 2-10 min. In the case of Ar-FAB irradiation, the bonding experiments were carried out using a SAB wafer bonder. There are three sections in the bonding chamber; a bonding head, a bonding stage, and a camera unit. The bonding head is used for holding the upper wafer and applying a specific pressure for a certain time. During the bonding experiment, when the vacuum condition reached the specified value (5×10^{-4} Pa ~ 2×10^{-6} Pa), the two wafer surfaces were etched by two Ar-FAB sources at the voltage of 1.5 kV and current of 50-60 mA simultaneously. The vacuum pressure is around 1×10^{-1} Pa during Ar-FAB irradiation. After the activation, the top wafer was moved down and then those two wafers were bonded together in vacuum or in ambient air.

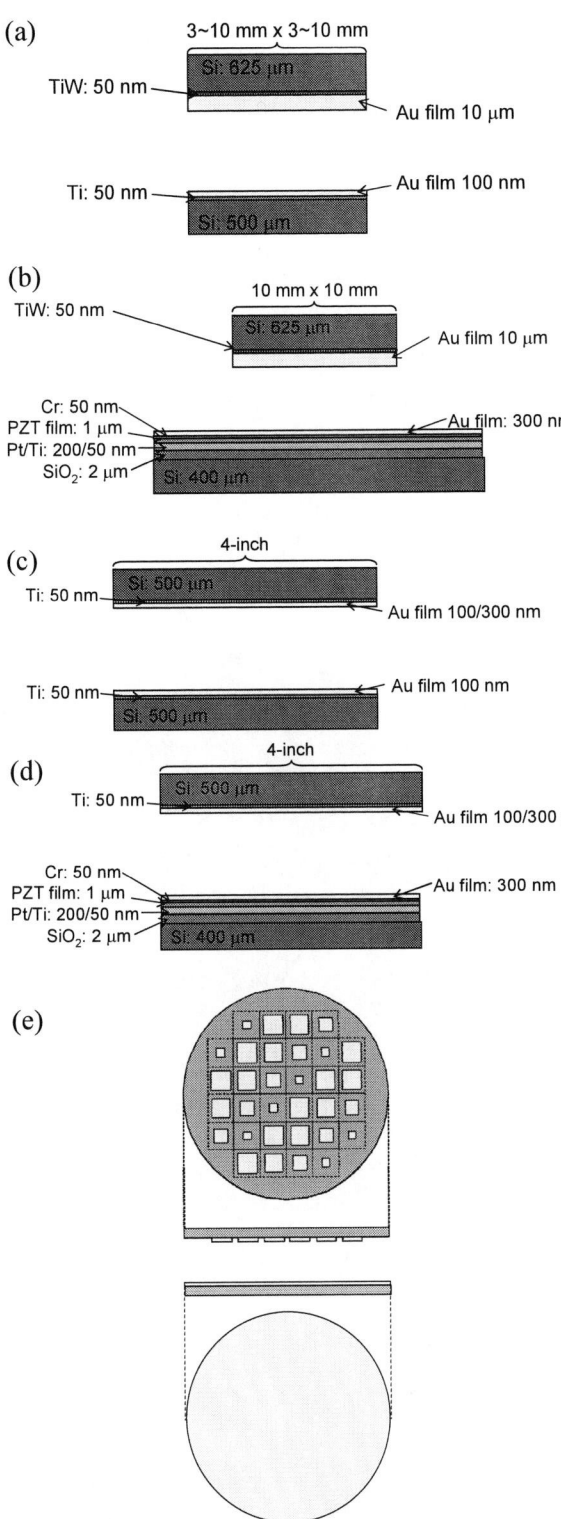

Fig.1. Schematic view of the bond pairs.

The bonded samples were observed using a scan acoustic microscope (SAM). The bonding energy was measured using razor blade test. The bonding strength was evaluated using tensile test and dicing test. The surface roughness of the Au films was measured by atomic force microscopy (AFM). The microstructures on the bond interfaces and the fractured surfaces after tensile test were observed using a scanning electron microscope (SEM) and an optical microscope.

Results and Discussion

Intermediate gold layers are assumed to be effective on plastic deformation which may compensate the gaps between bonding interface which caused by wafer warpage and surface roughness during interconnecting ductile materials [13]. The bonding of gold layers was studied from using 10-μm-thick Au film which was flattened by a planarization technique after electroplating [14]. The surface roughness was around 2 nm. Activated by Ar-plasma pretreatment, the bonding possibility of Si-to-Si and Si-to-PZT film using gold layers was confirmed at 25~200°C in ambient air by the SAB method. Figure 2 shows SEM cross-sectional images of Si-to-Si bonding which was performed at 100°C under 12.5 MPa. The influence of bonding temperature on tensile strength under the bonding pressure of 12.5 MPa was shown in Fig. 3. It was found that 100°C is enough under such condition. Although the bonding strength at room temperature is less than 1 MPa, strong bond was achieved at larger bonding pressure. Figure 4 shows an example of the fracture surfaces of the bonded Si-to-PZT film sample using gold layers at room temperature under the bonding pressure of 50 MPa. The fracture occurred not on bond interface but on adhesive TiW layers.

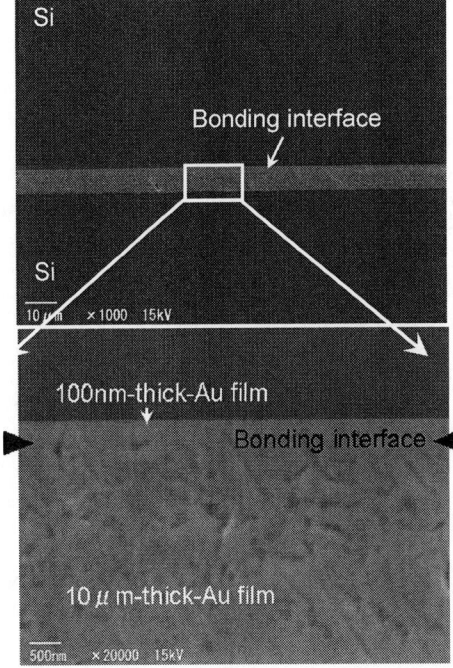

Fig.2. SEM cross-sectional images of the Si-to-Si bonding interface using gold layers (100°C, 12.5 MPa).

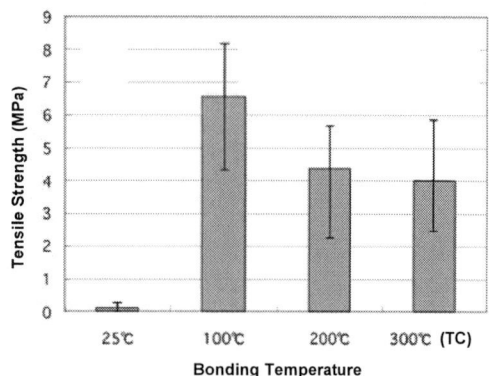

Fig.3. Influence of bonding temperature on tensile strength (ambient air, bonding pressure: 12.5 MPa).

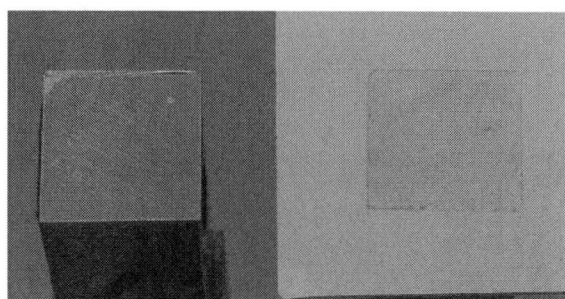

Fig.4. Fracture surfaces of the bonded Si-to-PZT film using gold layers (room temperature, 50 MPa).

Using thinner gold layers in larger wafer scale under lower bonding pressure is more promising for practical applications. The bonding feasibility of thin Au films in 4-inch wafer scale was therefore further investigated in this study.

The surface roughness of 100-nm-thick Au thin films after Ar-FAB irradiation for 200 s is around 7 nm. Both of them are 10 times larger than the critical value of surface roughness on Si bare wafer bonding [15]. Under the bonding pressure of 1.2 MPa, the Au-Au wafers were bonded under 1×10^{-5} Pa at room temperature. More than 50% bond areas were achieved as shown in Fig. 5(a). Adding metal films between Si wafers was effective to increase the ability of contact deformation. The bonding energy of Au-Au wafers was about 0.39 J/m^2 on average. The average value of tensile tests was about 12.4 MPa. Figure 6 shows the images of fracture surfaces after tensile test.

In order to reduce the voids and compare the influence of vacuum pressure, the influence of vacuum pressure and temperature on Au-Au bonding was investigated. Figure 5(b) and (d) compared the SAM images on the bond interface under 1.0×10^{-5} Pa and 5×10^{-4} Pa. The bonding energy was still larger than 1.0 J/m^2 under 5×10^{-4} Pa. Nearly all of the areas on the wafers were bonded and no obvious difference was detected between them. The wafers were heated to 150 °C or 200°C during bonding. The bonding energies were much increased when heating the wafers with Au thin films in the bonding. As shown in Fig.5 (a), (b) and (c), the bonded

areas performed at 150 and 200°C were larger than those of wafers bonded at 25°C. No obvious difference was detected between the interfaces bonded at 150°C and at 200°C. The Au grains may diffuse and recrystallize. They were assumed to be effective on diffusion and plastic deformation, which may help the bonding wafer to compensate the small gaps caused by rough surfaces and therefore enlarge the bonded areas.

In order to verify the function of 100 nm-thick Au thin films on plastic deformation, the bonded wafers were compared before and after rolling under 75 kgf. As shown in Fig.7, the small voids were disappeared on the bond interfaces. 100 nm-thick Au thin films can not only reduce the recrystallization temperature but also make the plastic deformation easier.

Since Au film bonding is not sensitive to vacuum pressure, the patterned wafers with Au thin film on Si substrates was tried to bond with Au thin film on Si or PZT films at 150°C in ambient air after Ar-FAB irradiation. As shown in Fig .8, nearly all the patterned parts were well bonded with few voids. The bonding using Au thin films can be applied in MEMS packaging and the interconnections between other brittle and easy oxidized materials in wafer scale at low temperatures by the SAB method.

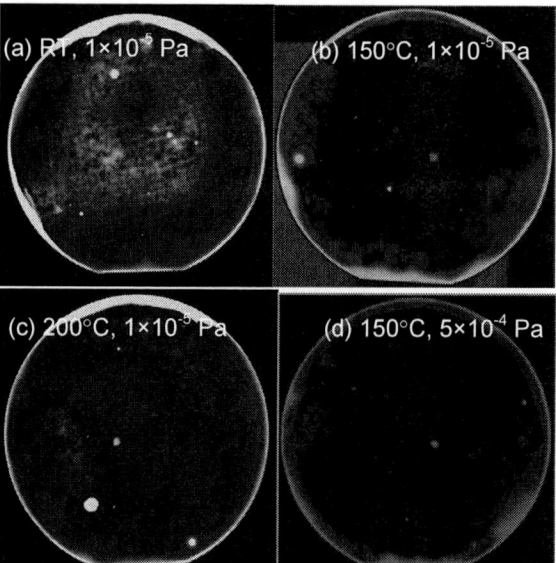

Fig.5. SAM images of the bonded wafers at (a) room temperature/1×10^{-5} Pa, (b) 150°C/ 1×10^{-5} Pa, (c) 200°C/1×10^{-5} Pa, and (d) 150°C/5×10^{-4} Pa.

Fig.6. Fracture surfaces of diced the Au-Au chip bonded at room temperature after tensile test.

978-1-4244-4658-2/09 $25.00 © 2009 IEEE

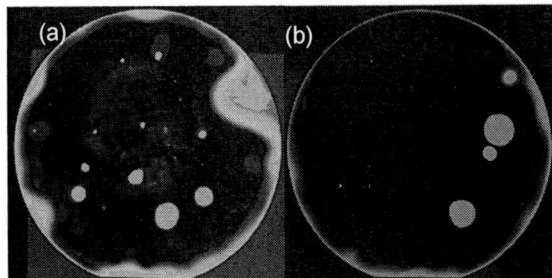

Fig.7. SAM images of the wafers bonded at room temperature (1×10-5 Pa): (a) before rolling and (b) after rolling.

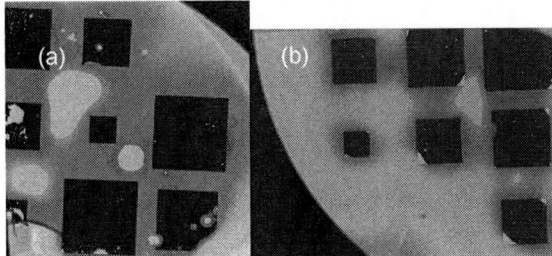

Fig.8. SAM images of the wafers (a) Si-to-Si and (b) Si-to-PZT bonded using gold thin films at 150˚C.

Conclusions

The wafer bonding through gold layers was confirmed at 25-150° by the SAB method. It was successfully applied on the interconnection between the bonding of Si-to-Si and Si-to-PZT film. Intermediate gold layer was found effective to reduce the requirements on surface roughness and vacuum conditions. The bonding interface was nearly void free with great bonding strength. It is promising to be applied in MEMS packaging and interconnections between various brittle and easy oxidized materials in wafer scale.

Acknowledgments

The authors would like to thank Mr. Taniyama, Mr. Hattori and Mr. Oshikawa for their assistance on experiments. Mr. Kamata in Japan Laser Ltd. Co., is specially appreciated for the SAM observation. This work is supported by Institute for Advanced Microsystem Integration (IMSI) and Japan Society for the Promotion of Science (JSPS).

References

1. Shimbo M., Furukawa K., Fukuda K., and Tanzawa K., "Silicon-to-silicon direct bonding method", Journal of Applied Physics, Vol.60 (1986) 2987–2989.

2. Kurman B.K., and Mita S.G., "Gold-Gold (Au-Au) Thermocompression (TC) Bonding of Very Large Arrays", Proc 42nd Electronic Components and Technology Conf, 1992, pp. 883-889.

3. Kim H. W., and Kim N. H., "Structural invesigations of gold-to-gold wafer bonding interfaces", Mater. Sci. and Eng. B, Vol.110, (2004), pp.64-47.

4. Han B., Verma K., Chopra M., Park S., and Li L., "Effect of substrate CTE on solder ball reliability of flip chip PBGA package assembly," Proc. Surface Mount Int./Adv. Electron. Mfg. Technol., 1997, pp. 43-52.

5. Suga T., Takahashi,Y. Takagi H., Gibbesch B., and Elssner G., "Structure of Al-Al and Al-Si3N4 Interfaces Bonded at Room-temperature by Means of the Surface Activation Method", Acta Metallurgica et Materialia, Vol. 40, (1992), S133–S137, Suppl. S.

6. Suga T., Otsuka K. "Bump-less Interconnect for Next Generation System Packaging", Proc 41st Electronic Components and Technology Conf, 2001, pp.1003-1008.

7. Matsuzawa Y., Itoh T., and Suga T., "Room-temperature Interconnection of Electroplated Au Micro-bump by Means of Surface Bonding Method", Proc 51st Electronic Components and Technology Conf, 2001, pp. 384-397.

8. Suga T., Itoh T., Xu Z., Tomita M., and Yamauchi A., "Surface Activated Bonding for New Flip Chip and Bumpless Interconnect Systems", Proc 52nd Electronic Components and Technology Conf, 2002, pp. 105-111.

9. Tomita M., Xu Z., Itoh T., and Suga T., "Low Temperature Flip Chip Bonding by SAB Method", Meeting Abstracts and Program of International Semiconductor Technology Conference, 2002, No.80.

10. Wang Y.H., Matiar R H., Nishida K., Kimura T., and Suga T., "Study on Sn-Ag Oxidation and Feasibility of Room Temperature Bonding of Sn-Ag-Cu Solder", Materials Transactions, Vol.46, (2005), pp. 2431-2436.

11. Wang Y.H., Nishida K., Hutter M., Kimura T., and Suga T., "Low-Temperature Process of Fine-Pitch Au-Sn Bump Bonding in Ambient Air", Japanese Journal of Applied Physics, Part1, Vol.46, No.4B (2007), pp.1961-1967.

12. Takagi H., Maeda R., Chung T. R., and Suga T., "Low-temperature direct bonding of silicon and silicon dioxide by the surface activation method", Sensors and Actuators A, Vol.70 (1998) 164-170.

13. Itoh T., and Suga T., Necessary load for room temperature vacuum sealing, Journal of Micromechanics and Microengineering, Vol. 15 (2005) S281-S285.

14. Mizukoshi M., "New Planarization Technique by High Precision Diamond Cutting for Packaging", SEMICON Japan, 2004.

15. Takagi H., Maeda R., Chung T. R., Hosoda N., and Suga T., "Effect of surface roughness on room-temperature wafer bonding", Japanese Journal of Applied Physics, Vol.37 (1998) 4197-4203.

Novel Pore-sealing of Ordered, Porous Silica, SBA-15 for Low-k Underfill Materials

Kuo-Yuan Hsu[1], Kuei-Yue Chen[1], Ching-Yuan Cheng[2], Jhih-Jhao Lee[2], Jihperng Leu[1]*

[1]Department of Materials Science and Engineering, Chiao-Tung University, Taiwan, China
1001 University Road, Hsinchu, Taiwan 30049
[2]Synchrotron Radiation Research Center
101 Hsin-Ann Raod, Hsinchu Science Park, Taiwan 30076
*E-Mail: jimleu@mail.nctu.edu.tw, Tel: +886-3-5131420

Abstract

Underfill materials, consisting of organic resin and inorganic filler, have been widely employed in flip-chip technology for preventing the failure of solder joints during packaging process. In order to meet the requirements of high frequency device applications, low-dielectric-constant underill is highly desired to alleviate the power consumption issue. However, it is disadvantageous to introduce low-k into organic resin because of its high cost and low volume fraction. Alternative approach is to use highly porous silica filler with appropriate pore sealing prior to mixing with epoxy. In this paper, a siloxane compound, polymerized vinyl-trimethoxysilane (poly-VSQ), was synthesized and applied to an ordered, porous silica structure, SBA-15 to study its effectiveness in pore sealing and interaction with SBA-15 matrix for low-k underfill applications. We have developed a convenient pore sealing treatment using poly-VSQ to achieve a 10.9% reduction (from 3.2 to 2.85) in dielectric constant of underfill material with 15% filler content by fully retaining the high porosity (61%) of SBA15. Moreover, excellent mechanical strength could be maintained as 3.0GPa without any interfacial delamination.

Introduction

As semiconductor device continues scaling down to 45 nm node and beyond, package demands smaller and high I/O. Flip-chip package, which utilizes area array of solder bumps to connect IC chip and substrate, has been widely adopted in the packaging of microprocessors, graphic chips, and DSP chips due to high I/O density and short interconnects. However, the most challenging problem in the reliability of flip-chip packaging is the solder joint fatigue which is mainly induced by thermal mechanical stress during temperature cycle due to coefficient of thermal expansion (CTE) mismatch between silicon chip and substrate such as organic substrate. However, the solder fatigue can be alleviated with an underfill material with mechanical property ranging from rigid to compliant depending on the requirements dominated by the solder type and the low-k dielectrics in the copper interconnect. [1-2]

In addition to thermal-mechanical properties, the electric requirements of underfill materials necessitates low dielectric constant to alleviate the RC delay and power consumption for high frequency, which can be illustrated by Eqs. (1)-(3) [3-5]:

$$C = \varepsilon \, L \, T \, / \, W \qquad (1)$$

$$RC = 2\,\rho\,\varepsilon\left(\frac{4L^2}{W^2} + \frac{L^2}{T^2}\right) \qquad (2)$$

$$P \propto 2\pi \cdot fV^2 \varepsilon \cdot \tan\delta \qquad (3)$$

where C is capacitance, ε is dielectric constant, L is line length, T is thickness, W is pitch, ρ is resistivity, and P is power consumption. As the development toward smaller T and W of solder bumps, but still in 60-100 µm range in the next 2-3 technology nodes, the increase of RC delay may not be an urgent issue. However, high operating frequency would cause an increase of power consumption and crosstalk noise if ε is fixed. [6] Therefore, underfill materials shall possess low dielectric constant to alleviate power loss, and excellent mechanical properties to support solder joints for high-frequency applications.

However, it is disadvantageous to introduce low-k into organic resin because of its high cost and low volume fraction in underfill materials. Alternative approach is to use highly porous silica filler with appropriate pore sealing prior to mixing with epoxy for reducing the dielectric constant. In order to accomplish high efficiency of pore sealing, major treatments involves additive and surface modification based on thin film processes. On the surface modification, the radiation treatment provides distinct polarity to prevent the moisture or impurity filling into the pore. [7-8]. In addition, the plasma deposition, like atom layer deposition (ALD) and chemical reaction are utilized as additive modification to achieve the pore-sealing and further enhance the strength to resist the mechanical or chemical damage. [9] A new approach of pore sealing treatment involves the use of sacrificial material such as D3 which is burned out during the cure step. [10] Unfortunately, sacrificial material may cause interfacial delamination in the underfill materials during the thermal curing step. [10] Moreover, chemical grafting reaction has been employed to modify the particle surface. Reaction initiator is attached on the silica filler surface first, and the oligomerization occurs. [11] Furthermore, the complexity process and monomer attach onto inside wall are the urgent issues.

In this study, a novel pore-sealing technology has been developed by retaining the high porosity ($k_{vacuum} = 1$) of SBA-15 to accomplish low dielectric constant (low-k) underfill materials. In this paper, a siloxane compound, polymerized vinyl-trimethoxysilane (poly-VSQ), was synthesized and applied to an ordered, porous silica structure, SBA-15 to study its effectiveness in pore sealing and interaction with SBA-15 matrix. The synthesis and the polymerization degree of VSQ were first investigated by nuclear magnetic resonance (^{1}H-NMR) and gel permeation chromatography (GPC). Then pore sealing treatment was characterized by small angle XRD, Brunauer-Emmett-Teller (BET), dielectric constant and dynamic mechanical analyzer (DMA). (This can be improved if I have more time.)

Experimental

978-1-4244-4658-2/09 $25.00 © 2009 IEEE

Synthesis of SBA-15 Materials

In this study, SBA-15 material was synthesized using Pluronic 123 triblock copolymer (EO20-PO70-EO20; Aldrich) as a template. First, Pluronic 123 (4 g) was dissolved in 2.0 M HCl (150 g) under slow stirring at room temperature. When solution became transparent, tetraethylorthosilicate (TEOS, Aldrich) (8.85 g) was added and heated to 35°C. The resultant solution was stirred at 35°C for 20 hrs, followed by aging at 35°C under static condition for 24 hrs. The solid product was recovered by filtration and dried at 100 °Cfor 2 hrs. Finally, the template was removed from the as-made porous material by calcination at 550°C for 6 hrs. [12]

In addition, the structure and pore volume of SBA-15 were characterized by small-angle x-ray diffraction (XRD) using beamline 17A at Synchrotron Research Center, Taiwan, and Brunauer-Emmett-Teller (BET, Quantachrome NOVA-1000A).

Polymerization of pore-sealing material and pore sealing treatment

Vinyl-trimethoxysilane (VSQ, Alfa Aesar) was selected as the pore sealing material because it possessed three methoxysilane groups which could react with the hydroxyl group on the SBA15 surface and its size can be designed to exceed the pore size of SBA15 for sealing the pore entrance effectively. Thus, the molecular weight (MW) of polymerized VSQ (poly-VSQ) based on free-radical polymerization was first investigated using initiator ranging from 1 wt% to 5 wt%. 10 g VSQ was stirred in a round-bottom flask at room temperature and degassed by N_2 for 1 hr, then initiator, dicumyl peroxide (DCP, Acros), was added into the flask and heated to 165°C for 10 hrs. The molecular structure of poly-VSQ was characterized by nuclear magnetic resonance (^{1}H-NMR, Varian unity 300MHz spectrometer), while the MW was measured by gel permeation chromatography (GPC, Water 1515).

The pore sealing of SBA-15 was carried out by dissolving poly-VSQ (10 g) with 4 different MWs in toluene (Aldrich, 10 ml), then adding SBA15 (1 g) into the solution with stirring at 50°C for 12 hrs. The solid products (P1, P2, P3, and P4), whose conditions were summarized in Table 1, was recovered by filtration and dried at 100°C for 2 hrs.

The pore volume, specific area, porosity and pore-size distribution of porous silica with or without pore sealing treatment were characterized by BET. Furthermore, SBA15 structure was examined by small angle XRD to study the integrity of the ordered, open pores with and without poly-VSQ treatment.

Underfill materials

In order to validate whether the porosity in SBA-15 has been successful retained by pore sealing treatment, the underfill materials consisting of 85% organic components and 15% filler based on SBA-15 was examined in terms of dielectric constant reduction and interfacial adhesion between the filler and resin. Specifically, the organic components contained epoxy resins, bisphenol-A based Epikote 828 (Shell) and a low viscosity bisphenol-F based epoxy resin Epikote 862 (Shell), a hardener, methyl hexahydrophthalic anhydride (MeHHPA, Aldrich), and a catalyst, 2-ethyl-4-methylimidazole (2E4MI, Acros). For the fillers, SBA-15

with or without pore sealing treatment (P1 through P4) were employed to prepare four low-k underfill materials (U1, U2, U3, and U4) for comparative study. The curing of underfill was carried out at 180°C for 2 hrs.

In addition, the dielectric constants and moduli of cured underfill materials were measured by RF impedance-material analyzer (HP 4291B) and dynamic mechanical analyzer (DMA, PerkinElmer DMA 7), to examine the efficiency of pore sealing treatment and any interfacial adhesion issue caused by poly-VSQ sealing materials.

Results and discussion

Characterization of polymerization of VSQ

The VSQ precursor and poly-VSQ prepared by free-radical polymerization were first examined by ^{1}H-NMR and GPC. As shown in Figure 1, two main peaks were observed at 3.5 ppm and 6.0 ppm, which could be attributed to the proton of carbon-hydrogen bond (C-H) and vinyl group in the VSQ precursor, respectively. Moreover, the integrated area of protons at 3.5 ppm and 6.0 ppm could be measured as 27 : 9, thus, the ratio of 3 indicated a pure VSQ consisting of three methoxylsilane groups and one vinyl group.

^{1}H-NMR spectrum of poly-VSQ illustrated in Figure 2 showed the disappearance of the vinyl group at 6.0 ppm, but new peaks of C-H bonding at 1.3 and 0.8 ppm because initiator, DCP attacked the vinyl group in the initiation step. Furthermore, the peaks at 3.5, 1.3 and 0.8 ppm became broader than those of the VSQ precursor presumably due to the broad MW distribution of polymerized VSQ. The new peaks indicated that the vinyl bond had opened and linked to another VSQ unit and the vinyl group was transferred to alkyl group according to the polymerization scheme illustrated in Figure 3.

Subsequently, the molecular weights of poly-VSQ were characterized by GPC and summarized in Table 2. The MW of poly-VSQ increased from 148.4 to 9380. Next, we estimated the size of poly-VSQ in comparison with the pore size of SBA-15 (5.87 nm). The Equation (4) described the relationship between molecular weight and sphere size of polymer. [13]

$$R = 1.33 \, \alpha \kappa \, M^{1/2} \qquad (4)$$

where R was radius of polymer, α equaled 1.2, κ equaled $670*10^{-3}$ Angstrom and M was the molecular weight of polymer. Thus, the sizes of poly-VSQ with various MWs including their PDI were summarized in Table 2. The poly-VSQ with MW of 1410 showed a pore size 4.0 nm which was less than the open hole size (5.87 nm) of SBA15. In contrast, the size of poly-VSQ was much larger than the open hole in SBA-15 if MW is 9380 (19 nm) or greater. The correlation between MW (or poly-VSQ size) and their effectiveness in pore sealing will be discussed and addressed in the subsequent section.

Characterization of SBA15 and pore sealing treatment

The structure of SBA15 based on synthesis method by Zhao et al [12] should possessed significantly regular pores and high surface area. Small-angle XRD was employed to examine the structure of SBA15 with and without pore sealing treatment as shown in Figure 4. For pure SBA15 (P1), significant peaks appeared at 2θequal to 1.0, 1.5, 1.7 and 2.2, indicating (100), (110), (200) and (210) reflections associated with *p6mm* hexagonal symmetry. The high intensity of (100) indicated the regular through holes in SBA15 structure. For SBA15 treated with poly-VSQ with MWs of 4125 or 9380 (P3 and P4), their significant diffraction peaks of (100), (110), (200) and (210) were similar to those of pure SBA15 (P1). This indicated that the inner structure of regular pores in SBA-15 was not destroyed due to the much larger sizes of poly-VSQ which presumably sealed the entrances of pores only. However, pore sealing treatment with poly-VSQ with MW of 1410 (P2) showed the disappearance of (200) peak, but enhanced intensity of (110). The alteration was attributed to pore sealing material filling into open pores and attaching on the inside wall of SBA15, thus destroying the regular structure and high porosity. Consequently, poly-VSQs with high molecular weight were effective in sealing only the pore entrance without filling into the open pore when their size were much larger than 5.87 nm.

Furthermore, the surface area and pore volume of SBA15 with various pore sealing treatment (P1-P4) obtained by Brunauer-Emmett-Teller (BET) were summarized in Table 3. Pure SBA15 (P1) possessed 450 m^2/g surface area and 0.72 cm^3/g pore volume with 61% porosity. However, the surface area and pore volume of P2-P4 were reduced to < 50% of original value (P1) because of the pore-sealing by poly-VSQ pretreatment. Moreover, P4 exhibited the lowest surface area 25 m^2/g, pore volume 0.038 cm^3/g and lowest porosity as 7% when poly-VSQ with the highest molecular weight (9380) was used. It implied that poly-VSQ material with high molecular weight could seal the pore more efficiently than low molecular weight poly-VSQ material due to effective blockage of the pore entrances. In contrast, the less reduction of surface area and pore volume in P2 indicated that poly-VSQ of small size (< 4 nm) may flow into the open pores and attach onto the sidewall of open pores in limited extent due to high viscosity and surface tension. Thus, there are still many open pores remained in P2 case.

Figure 5 showed the pore size distributions of porous silica after four different pretreatment (P1 to P4). For SBA15 (P1), the pore size curve was very sharp and high in intensity. It implied that pores of SBA15 were uniform and regular. Moreover, P4 exhibited a much reduction of pore size from 5.87 nm to 1.88 nm, in addition to a very low pore volume. This implied that the pores of P4 might have been sealed completely, except a few micro pores. For P2 and P3, their pore volumes were about the same and their pore size distribution were broader than P1 and P4. Thus, poly-VSQ material in P2 was believed to fill in the open pore and attach along the sidewall to reduce the pore size and the pore volume. Furthermore, P3 showed smaller pore size with a broader peak and less pore volume than P2. The high polydispersity index (PDI) associated with high MW will be further taken into consideration. Due to the broad molecular weight distribution, the polymer size also exhibited broad distribution ranging from smaller size, which could fill into open pores, to larger size, which could only attach to the entrance of pores and outer surface of SBA15. In addition, small angle XRD of P3 showed peaks as perfect as pure SBA15. Thus it implied that smaller size poly-VSQ of P3 might be obstructed by larger size poly-VSQ and remained

the regular structure, therefore the pore sealing situation of P3 was a mixed mode of P2 and P4.

Based on calculated polymer size, BET and small angle XRD analysis, a simple model of pore sealing treatment using various poly-VSQ was proposed and schematically illustrated in Figure 6. When molecular weight of poly-VSQ material was very small (below a lower threshold), the poly-VSQ material could fill in the open pore of SBA15 and attach onto the sidewall, which reduced the pore diameter, the pore volume and surface area to some degree, but rendered the SBA15 less useful as a low-k filler. Comparatively, if molecular weight of poly-VSQ was larger than the open pore size, the poly-VSQ could not enter the open pores and may effectively seal the pore entrance and outer surface of SBA15. Thus, a mixed model was applicable if broad molecular weight distribution was taken into consideration.

Low-k underfill materials using SBA15 with pore sealing

In conjunction with the pore sealing treatment of SBA15, 4 different underfill low-k materials were prepared to investigate the effectiveness of pore sealing in the reduction of dielectric constant and their impact on mechanical strength and interfacial adhesion. Table 4 summarized the dielectric constant, modulus and loss factor of underfill materials (U1-U4) with different poly-VSQ pore sealing treatment. U1 showed the highest dielectric constant (3.2) due to the resin flowed into the open pore during curing process. In contrast, U4 possessed lowest dielectric constant (2.85). It implied that high effectively pore sealing treatment of U4, whole porosity was retained by the pore sealing treatment. Although P4 showed the lowest porosity in BET measurement, it could form a barrier layer on the surface to avoid the resin flow in during curing process. In addition, U2 (3.0) and U3 (2.92) also decreased 6.2% and 8.75%, respectively. The dielectric constant decreased was due to the retaining porosity by reduced pore diameter or blocked pores. In composite material, its dielectric constant could be predicted by Equation (5) [14]:

$$Logk = \sum Vi \cdot Logki \qquad (5)$$

where k are the dielectric constant of composite materials, V_i and k_i are the volume fraction and dielectric constant of i-th component, respectively. There were two assumptions when applying this equation; namely: (1) The low-k underfill material was an uniform mixture of epoxy resins, silica and pores, and (2) The volume faction retained by pore sealing materials after curing and left as pores with k=1.

Based on Eq. (5), the theoretic dielectric constants of U1 to U4 were calculated to be 3.20, 3.06, 3.00 and 2.83, respectively. In particularly, the results of U1 and U4 were very close to their theoretic values. The result showed resin could flow into the open holes in SBA15 (U1) easily during the curing process. However, the dielectric constant in U4 could achieve its theoretic value because organic resin could not flow in the open holes, whose entrance had been sealed efficiency, during curing step. On the contrary, U2 and U3 deviated from the theoretic values to a large extent, due to partial pore sealing. Fortunately, the experimental data (U2 and U3) were still lower than their theoretic values because the pore sizes had been reduced to 5.3 and 2.8 nm, which were small enough to obstruct the resin flowing into the deep pore, and pore blockage to some degree.

In addition, the moduli of low-k underfill materials were characterized to study mechanical strength of underfill and the interfacial adhesion effect. From Table 4, U1 and U4 both possessed the same higher modulus of 3.0 GPa, while U2 and U3 showed modulus of 2.8 GPa and 2.9 GPa, which were 6.67% and 3.3% lower than U1 and U4. The higher modulus of U1 was due to resin flowed into open hole completely during curing process and formed more dense structure by reducing porosity. However, U4 also possessed higher modulus due to the efficacious pore sealing treatment. The pore sealing materials of P4 sealed on the surface of SBA15 and formed smooth surface with better adhesion with resin. In contrast, the degradation of modulus in U2 and U3 was due to imperfect pore sealing treatment of SBA 15. Thus incompatible interface could affect mechanical strength and induce crack and delamination. [10] In order to prove the hypothesis, we calculated the theory moduli of composite material by Eq. (6),

$$\frac{1}{E(L)} = \sum \frac{Vi}{Ei} \qquad (6)$$

where Vi and Ei are the volume fraction and modulus for i-th component assuming perfect adhesion at various interfaces. [15] Calculation by the Eq. (6), the theoretic moduli was 3.0 compared to each modluli, U1 and U4 achieved the theoretic value. The result could be proved the perfect interface. However, U2 and U3 degraded 3.3% and 6.67%, respectively. It is believed that higher surface roughness of filler in U2 and U3 caused by pore sealing treatment could result in more voids at filler/expoxy interface and in turn the degradation of the mechanical strength. The hypothesis was further confirmed by the loss factor, tanδ, obtained from DMA measurement. A high tanδ indicated high internal energy dissipation due to poor adhesion. From Table 4, the highest loss factor was U2, 0.040 and the lowest loss factor was U4 as 0.003. However, U1 and U3 indicated the same loss factor as 0.032. Comparing the modulus and loss factor, U4 possessed the highest modulus and loss factor. Therefore, underfill materials with higher loss factor possessed reduced mechanical strength primarily due to deteriorated adhesion at epoxy/filler (SBA15) interface.

Poly-VSQ played a role as coupling agent on the SBA15 surface and improved the adhesion better than U1 without pore sealing treatment. Furthermore, U2 and U3 also sealed by poly-VSQ, however the less efficiency of pore sealing leaded to more roughness surface of SBA 15 and induced defect interface between organic and inorganic with void during curing process. Consequently, the surface of silica filler could not be wetted by resin completely. The defects would induce cracking during thermal cycling process and destroyed the mechanical strength.

Conclusions

A novel pore sealing treatment on highly ordered, porous SBA15 filler for low-k underfill application has been successfully developed using a polymerized vinyl trimethoxysilane (poly-VSQ) as the pore sealing material. The

polymer size and distribution dictated by the MW and PDI from free-radical polymerization, were found to play critical role in the effectiveness of pore sealing of SBA15. For poly-VSQ with MW of 9389 (~19 nm particle size), the sealing materials could seal the pore more efficiently than low molecular weight poly-VSQ material due to its effective blockage of the pore entrances as validated by small-angle XRD and BET. Moreover, such novel pore sealing method maintained perfect interfaces such as epoxy/SBA15 and epoxy/poly-VSQ in the underfill composite as confirmed by loss factor and mechanical strength.

In summary, an easy and convenient pore sealing treatment using poly-VSQ has achieved a 10.9% reduction (from 3.2 to 2.85) in dielectric constant of underfill material with 15% filler content by fully retaining the high porosity (61%) of SBA15, and excellent mechanical strength, 3.0 GPa as that in pure SBA15 system. Moreover, the excellent adhesion between epoxy and poly-VSQ/SBA-15 porous silica in the underfill materials was attained to preserve its mechanical strength when pore sealing pretreatment was applied.

Acknowledgments

The authors wish to thank the instrumentation support by Industrial Technology Research Institute (ITRI), Taiwan.

References

1. Z. Zhang *et al,* "Recent Advances in Flip-Chip Underfill: Materials, Process, and Reliability," *IEEE T. ADV. Packaging*, Vol. 27, No. 3 (2004), pp. 515-524.
2. H. Y. Chen. *et al,* "Thermal Behavior Analysis of Lead-free Flip-Chip Ball Grid Array Packages with Different Underfill Material Properties," 8th *International Conference on Electronic Packaging Technology and High Density Packaging (ICEPT-HDP) Conf*, Shanghai, China, July. 2008, pp. 28-34.
3. P. S. Ho *et al,* Low Dielectric Constant Materials for IC Applications, Springer (Berlin, 2003), pp. 1-19.
4. S. P. Murarka *et al,* Interlayer Dielectrics for Semiconductor Technologies, Elsevier/Academic Press (Boston, 2003), pp. 1-6.
5. Z. J. Yang *et al,*, "Process for Cu matel and low dielectric materials," *NDL Communication*, Vol. 7, No. 4 (2001), pp. 40-46.
6. Z. Feng *et al,* "RF and mechanical characterization of flip-chip interconnects in CPW circuits with underfill," *IEEE T. Microw. Theory*, Vol. 46, No. 12 (1998), pp. 2269-2275.
7. A. Furuy *et al,* " Ta penetration into template-type porous low-k material during atomic layerdeposition of TaN ", *J. Appl. Phys., Vol.* 98, (2005), pp. 094902.
8. Y. Travaly *et al,*" A theoretical and experimental study of atomic-layer-deposited films onto porous dielectric substrates" *J. Appl. Phys.*, Vol. 98(2005), pp. 083515.
9. K. Maex *et al,* "Low dielectric constant materials for microelectronics", *J. Appl. Phys.*, vol. 93, (2003), pp. 8793
10. K. Y. Hsu *et al,* "Novel Pore-sealing Technology in the Preparation of Low-k Underfill Materials for RF Applications," 8th *International Conference on Electronic Packaging Technology and High Density Packaging*

(ICEPT-HDP) Conf, Shanghai, China, July. 2008, pp. 462-467.

11. J. Moreno *et al*," Well-Defined Mesostructured Organic-Inorganic Hybrid Materials via Atom Tran sfer Radical Grafting of Oligomethacrylates onto SBA-15 Pore Surfaces" *Chem. Mater*. Vol. 20 (2008), pp. 4468–4474.

12. D. Zhao *et al*, "Triblock Copolymer Syntheses of Mesoporous Silica with Periodic 50 to 300 Angstrom Pores", *Science*, vol. 279 (1998), pp. 548-552.

13. L. H. *et al*, Introduction to Physical Polymer Science, Springer (Berlin, 1998), pp. 92-102.

14. Y. Rao *et al*, " A precise numerical prediction of effective dielectric constant for polymer-ceramic composite based on effective-medium theory", *IEEE T. Compon. Pack.*, Vol. 23, No. 4 (2000), pp. 680-683.

15. T. Murayama *et al*, Dynamic Mechanical Analysis of Polymeric Material, Elsevier (New York, 1978), pp. 71-81.

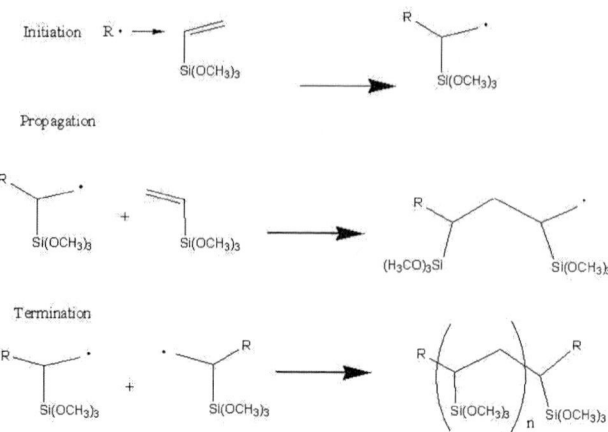

Figure 3 Polymerization scheme of VSQ

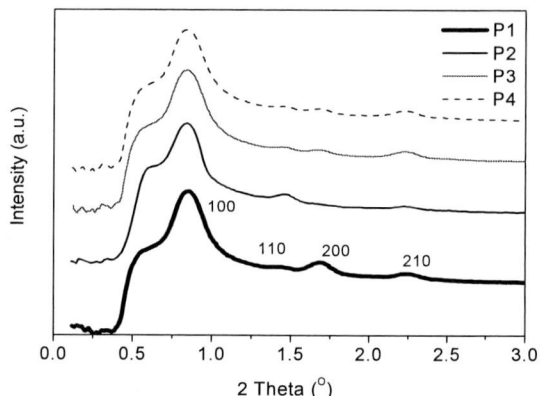

Figure 4 Comparison of pore sealing treatment by small-angle XRD

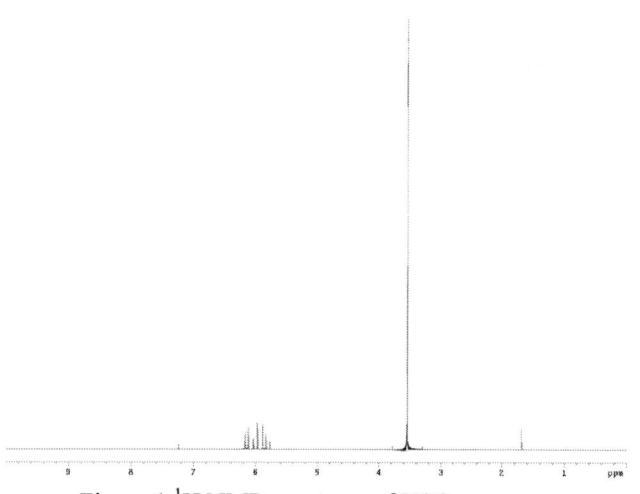

Figure 1 ^{1}H NMR spectrum of VSQ precursor

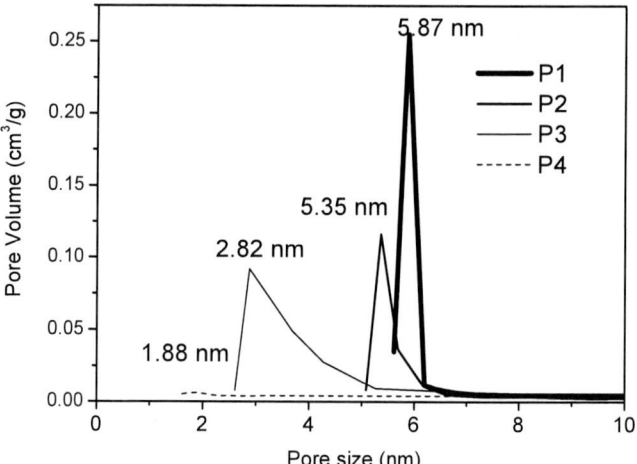

Figure 5 Pore size distribution of pore sealing treatment

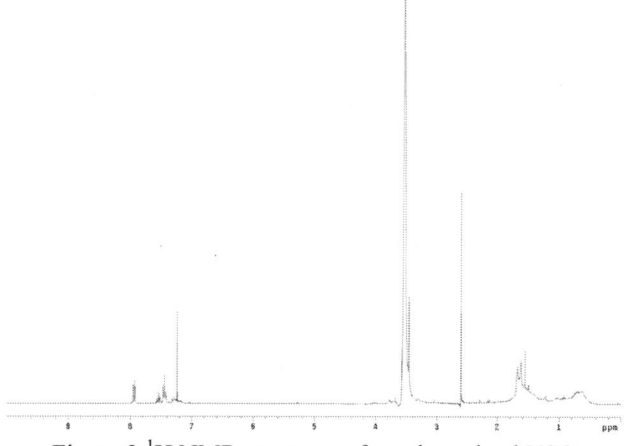

Figure 2 ^{1}H NMR spectrum of a polymerized VSQ

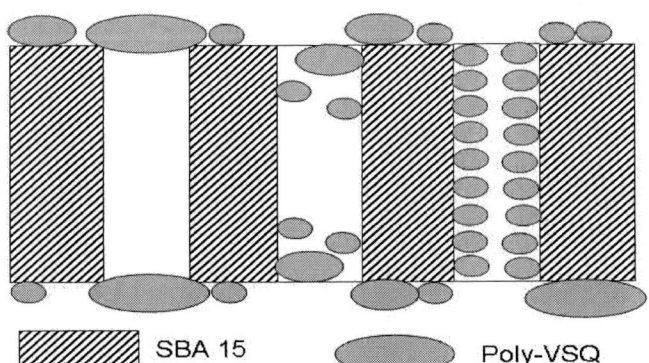

U1	3.20 (3.20)	3.0	0.032
U2	3.00 (3.06)	2.8	0.040
U3	2.92 (3.00)	2.9	0.032
U4	2.85 (2.83)	3.0	0.030

SBA 15   Poly-VSQ

Figure 6 Mechanism of pore sealing treatment

Table 1 Conditions of poly-VSQ and SBA-15 for pore sealing treatments

Cases	Porous Silica SBA15	Poly-VSQ	
		VSQ	DCP
P1	1 g	None	None
P2	1 g	10 g	0.10 g (1%)
P3	1 g	10 g	0.25 g (2.5%)
P4	1 g	10 g	0.50 g (5%)

Table 2 MW, PDI and size of polymerized VSQs

	Molecular weight	PDI	Particle size of VSQ (nm)
PureVSQ (Precursor)	148	None	1.0
Polymerized VSQ (1.0% initiator)	1410	1.96	4.0
Polymerized VSQ (2.5% initiator)	4125	3.66	12.0
Polymerized VSQ (5.0% initiator)	9380	5.98	19.0

Table 3 Surface area, pore volume and porosity of various pore sealing treatment by BET

	Surface area (m^2/g)	Pore volume (cm^3/g)	Porosity (%)
P1	450.0	0.72	61
P2	158.5	0.29	39
P3	85.0	0.18	28
P4	25.0	0.038	7

Table 4 Dielectric constant, modulus and loss factor of underfill materials with different poly-VSQ pore sealing treatments

	Dielectric constant (theoretic data)	Modulus (GPa)	tanδ (at R.T.)

Study of Polyimide as Sacrificial Layer with O₂ Plasma Releasing for Its Application in MEMS Capacitive FPA Fabrication

Shenglin Ma[1], Ying Li[1, 2], Xin Sun[1], Xiaomei Yu[1*], Yufeng Jin[1, 2]

[1]National Key Laboratory on Micro/Nano Fabrication technology, Peking University, Beijing 100871, P. R. China
[2]Peking University Shenzhen Graduate school, Shenzhen 518055, P. R. China
*Telephone: (86-10) 62752536-15, Email: yuxm@ime.pku.edu.cn.

Abstract

Polyimide (PI) was a good candidate as the sacrificial layer for its compatibility with CMOS technology. This paper first presented a new patterning method of PI film and then investigated the relationships among undercut rate, the undercut limit length and the releasing hole size in the releasing step, which was helpful and important for its popularity and its application in MEMS capacitive FPA (Focal plane array) fabrication. A new patterning approach of PI film was successfully developed in ICP chamber. The patterning approach selected a PECVD SiO₂ layer as patterning mask. The optimized ICP PI recipe was composed of an O₂ flow of 180sccm, an electrode power of 400W and bias plate power 200W. With the optimized ICP PI recipe, a vertical etching rate about 0.5-0.6μm/min with a lateral etching rate 0.13μm/min was realized. With 1μm PI sacrificial layer, a 0.22μm/min undercut rate was achieved in a normal barrel etcher. Based on our experimental facts, the optimized releasing hole size was 5μm×5μm and the distance between lateral releasing holes should be fewer than 17 μm for effectively and completely releasing.

Introduction

Polyimide (PI) has been widely used in IC fabrication due to its good thermal and electrical performance. Its novel application in fabrication of MEMS devices such as resonator, pressure sensor and RF switch demonstrated its high thermal and chemical stability, and good compatibility with CMOS technology [1-6, 8]. Different surface process modules based on PI sacrificial layer using different releasing ways have been reported [1-2, 4]. One kind of the surface process module use fully-cured PI as sacrificial layer and then pattern the Polyimide using hard mask, in the end the structure will be released through O₂ plasma isotropic etching of PI sacrificial layer. Another kind of the surface process module patterns the pre-cured PI sacrificial layer and then fully cure the patterned polyimide layer with following structure layer deposition and releasing through O₂ plasma [1-2, 4].

MEMS capacitive FPA is a rising technology for infrared (IR) imaging for its easy integration with its readout electronics and low NETD [5-7]. To enhance its integration and performance, post-CMOS surface process is preferred to fabricate the MEMS capacitive FPA [5-6]. While unlike these MEMS devices such as resonator, RF switch that have large releasing area [1-4], Capacitive FPA has to minimize its releasing area to improve the capacitive FPA integration and the sensing capacitor of every pixel of the capacitive FPA. Meantime, in order to improve the response of every pixel IR detector of the FPA, the distance between the up and bottom plate should be as small as possible, which means the sacrificial layer should be as thin as possible. Very small releasing areas and rather small releasing depth will have to bring in challenges for the surface process using PI sacrificial layer. Besides that, the critical dimension of the MEMS capacitive FPA calls for good patterning method of the polyimide sacrificial layer.

In this paper, we first presented our new patterning method of PI film and then investigated the relationships among the undercut rate, undercut limit length and releasing hole size at very small sacrificial layer thickness.

PI film deposition and patterning

Commercial polyimide is usually supplied as polyamic acid precursors dissolved in an N-methyl-2-pyrrolidone (NMP) based solvent carrier suitable for spin coating applications. After coated on the wafer, the polyamic acid precursors will be transformed into polyimide film after a series thermally curing process which is also called polymerization. The fully-cured polyimide is well known for thermal stability exceeding 400°C and mechanical toughness [9].

Commercial PI was utilized in this work. The whole thermally curing process mainly consisted of two periods of thermal curing that was illustrated in Fig. 1[9].The first period of curing staying at about 200°C aimed to evaporate the solvents. The second period keeping at about 300°C guaranteed the absolute evaporation of solvents and full polymerization and finally yielded a high TG temperature of the cured polyimide [1, 9].

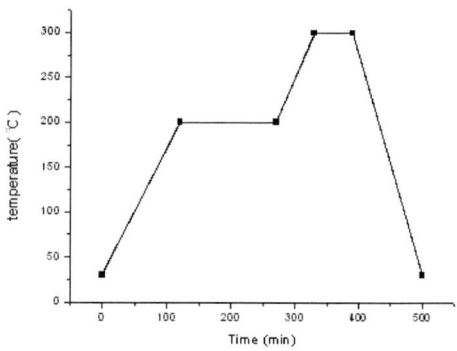

Fig.1 the thermal process curve

The PI film thickness before and after fully-cured versus the spin speed was studied and the results were plotted in Fig 2. It was easy to find out that fully-curved PI film would decrease about 30% in its thickness. This also gave the evidence that patterned pre-cured Polyimide film with fine

978-1-4244-4658-2/09 $25.00 © 2009 IEEE

lines would deteriorate after being fully-cured. Therefore, surface process module based on polyimide sacrificial layer with O_2 releasing was fit for the design that requested fine lines.

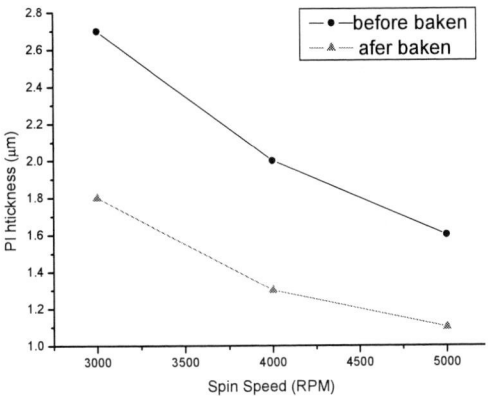

Fig.2 Polyimide thickness versus the spinning speed

PI film patterning involved choosing of mask materials and its relative anisotropic etching method. Usually, PECVD SiO_2, PECVD SiC and PECVD SiNx were proposed to do this job [1, 2]. We tried the PECVD SiO_2, SiNx and SiC as the mask materials. The experiment showed that PECVD SiO_2 was the best candidate for its low stress, good adhesion and easy removal. The optimized procedure was to pattern the PECVD SiO_2 with BHF and remove it after patterning of the PI film. No obvious etching of PI film was found in this procedure. While the other hard mask materials including PECVD SiNx, SiC confronted some problems, such as monitoring of the etching stop, rough surface of polyimide, hard to remove and so on.

For good patterning of PI film, A STS ICP system was elected to gain anisotropic etching. In our experiment, an O_2 flow of 180sccm and an electrode power of 400W were chosen. In order to improve the ratio of vertical to lateral etching rate, different bias plate power was tried. A batch of 4' Si wafers with about 2.7 μm polyimide was fabricated by the same coating and curing process and then a PECVD SiO_2 about 0.5μm was deposited on the polyimide layer as mask layer. The PECVD SiO_2 was patterned using BHF etching. The typical ICP polyimide recipes were illustrated in Tab.1.

Tab.1 ICP Polyimide recipes

	1#	2#	3#
Bias plate power(W)	120	160	200

Fig. 3 was a group of SEM photos illustrating the etching characteristics of PI film in the ICP chamber. The photos showed that lateral etching occurred although bias plate power was employed to collimate the O_2 plasma. That may be due to the quick chemical reaction between the O_2 plasma and polyimide. Contrast the Fig.3(b) and Fig.3(c), it could be concluded that the vertically speedup of the O_2 plasma due to the rise of the bias plate power effectively enhanced anisotropic etching of polyimide and steep sidewall was got.

(a)

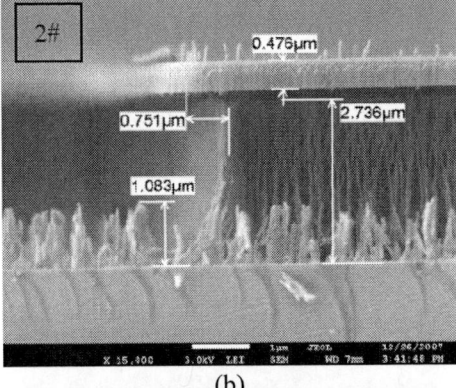

(b)

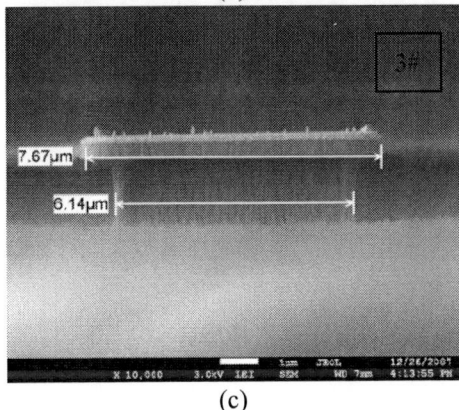

(c)

Fig.3 SEM photos of ICP PI under different bias plate power

Compare the three photos, thread like remaining was observed on the substrate of PI film which was etched under lower bias plate power. When the bias plate power was raised to 200W, no thread like remaining was observed. This may be ascribed to the chemical reaction between O_2 plasma and PI. With the help of O_2 plasma bombardment this chemical reaction may produce new products which hinder the chemical reaction. As the bombardment was not strong enough, the thread-like remaining was hard to be cleared off the surface. With the rise of the bias plate power, the thread like remaining was effectively to overcome. While the bias plate power should not be too high, because it would result in more and stronger reflection of the precipitated O_2 plasma which would strengthen lateral etching.

The experimental results showed that the ICP PI recipe should be composed of an electrode plate power of 400W, an

978-1-4244-4658-2/09 $25.00 © 2009 IEEE

O_2 flow rate of 180sccm and a bias plate power of 200W. Under this condition, a vertical etching rate of 0.5-0.6µm/min with a lateral etching rate about 0.13µm/min was successfully gained. The ratio of vertical etching rate to lateral etching rate reached about 4.

Experiment

To evaluate the capacity of the PI sacrificial layer, a simple experiment was designed and done to investigate its limit undercut length, under etch rate, under different releasing holes at very thin PI sacrificial layer thickness. This experiment was done in a normal barrel O_2 plasma etcher. The experiment had only one mask which was covered with a series of squares and circles whose size range from 5µm×5µm to 100µm×100µm. The Fig.3 showed the basic process flow.

First a 1µm PI film was coated and cured on the silicon wafer with 8000Å SiO_2 as etching stop layer and then a 1µm SiO_2 was deposited with PECVD as its mask (a). Then SiO_2 was patterned as the mask for PI releasing (b). Finally the silicon wafer was placed in a normal barrel O_2 plasma etcher to etch the polyimide(c). Fig.4 was a group of photos that illustrated the PI releasing process in the barrel O_2 plasma etcher.

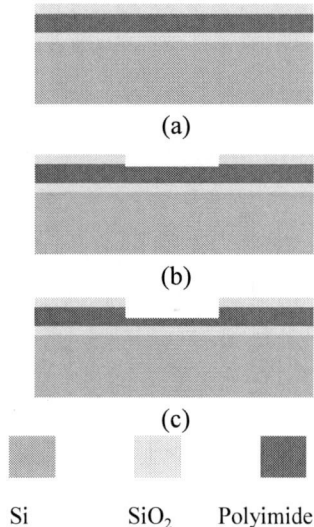

(a)

(b)

(c)

Si SiO_2 Polyimide

Fig.4 basic flow of the designed experiment for measuring the undercut rate and the limit undercut length

Throughout the releasing period, the undercut of every releasing hole kept pace with each other although they have different shapes and sizes [Fig.5]. From this point, we could make a conclusion that the shape and size of the releasing hole above 5µm×5µm entangle or a circle with its diameter 5µm had not much impact on the undercut rate and the optimized releasing hole was a circle with its diameter 5µm when the PI sacrificial layer was only about 1 µm.

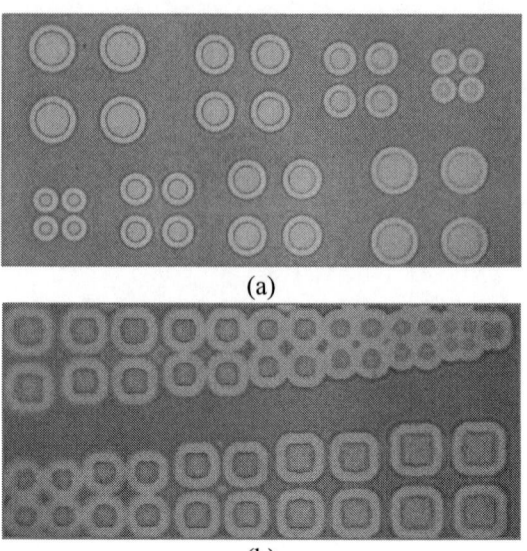

(a)

(b)

Fig.5 the photos of releasing hole in the releasing process

At the beginning of the releasing, the under etching rate kept about 0.22µm/min. When the undercut reached about 17µm, no obvious undercut was observed even adding more etching time. This may be due to the limit deliver capacity of the O_2 plasma. Therefore, based on this fact, the distance of the lateral releasing holes should be fewer than 17 µm for successfully releasing.

Conclusion

Surface process based on polyimide sacrificial layer was a promising technology for its compatibility with CMOS technology. For its application in MEMS capacitive FPA fabrication, a new patterning method of the PI film was developed for fine lines and anisotropic etching. According to our experiment results, the best procedure for Polyimide layer patterning was using PECVD SiO_2 as its mask, then patterning the mask using BHF solution and in the end ICP PI film with following removing the mask with BHF solutions. The optimized ICP PI film recipe consisted of an electrode plate power of 400W, an O_2 flow of 180sccm and a bias plate power of 200W. Under this condition, a vertical etching rate of 0.5-0.6µm/min with a lateral etching rate about 0.13µm/min was successfully achieved.

To minimize the releasing hole and therefore get more useful structure area, experiment was designed and done to evaluate the surface process module with PI sacrificial layer and optimize the design of releasing holes size and pitch. With 1µm polyimide sacrificial layer using O_2 plasma releasing, a undercut rate about 0.22µm/min, the optimized releasing hole size was about a circle with its diameter 5µm and the distance between the lateral releasing hole should be fewer than 17µm for successfully and effectively releasing.

A high density MEMS capacitive FPA based on this optimized surface process module now is being fabricated to test this optimized surface process module.

Acknowledgments

The authors would like to thank Xiaobao Geng and Xuejiao Fan for acquainting SEM photos and also thank the staff in the National Key Laboratory on Micro/Nano Fabrication technology.

References

1. A Bagolini, *et al*, "polyimide sacrificial layer and novel materials for post-processing surface micromachining," *Journal of Micromechanics and Microengineering*, 12 (2002), pp. 385 - 389.

2. H.T.M.Pham, *et al*, "polyimide sacrificial layer for an all-dry post-process surface micromachining module," *the 12th International Conference on Solid State Sensors, Actuators and Microsystems*, Boston, June 8-12, 2003, PP813 - 816.

3. L S Pakula, *et al*, "Fabrication of a CMOS compatible pressure sensor for harsh environments," *Journal of Micromechanics and Microengineering*, 14 (2004), pp. 1478 – 1483.

4. Meili Hu, *et al*, "Research for polyimide as a sacrificial layer in MEMS device", *Proc. of SPIE*, Vol. 5774 (2004), pp. 642-645

5. Weidong Wang, *et al*, "FEA simulation, design, and fabrication of an uncooled MEMS capacitive thermal detector for infrared FPA imaging," *Proc. of SPIE*, Vol. 6206 (2006).

6. S. Huang, *et al*, "Application of polyimide sacrificial layers for the manufacturing of uncooled double- cantilever microbolometers," *Materials Research Soc.*, Vol. 890 (2006), pp. 15.1 - 15.6.

7. Scott R. Hunter, *et al*, "High Sensitivity Uncooled Microcantiliver Infrared Imaging Arrays," *Proc. of SPIE*, Vol. 6206 (2006).

8. M.Fernandez-Bolanos, *et al*, "Polyimide sacrificial layer for SOI SG-MOSFET pressure sensor", *Microelectronic Engineering*, 83 (2006), pp. 1185 - 1188.

9. 2000 PI2610 Data Sheet (Detroit, MI: H D Microsystems).

Fabrication and Hydrogen Sensing Properties of Titania Nanotubes

Shuo Bai[1], Dongyan Ding[1], Congqin Ning[2], Rui Qin[1], Yan Li[1], Chengkang Chang[1], Ming Li[1], Dali Mao[1]

[1]Lab of Microelectronic Materials and Technology, State Key Laboratory of Metal Matrix Composites
School of Materials Science and Engineering, Shanghai Jiao Tong University, Shanghai 200240, China
[2]State Key Laboratory of High Performance Ceramics and Superfine Microstructure
Shanghai Institute of Ceramics, Chinese Academy of Sciences, Shanghai 200050, China
Email: dyding@sjtu.edu.cn, Tel: +86-21-34202741

Abstract

TiO2 nanotubes have wide applications in gas sensing applications such as oxygen sensors and hydrogen sensors. In this work, crystallized TiO2 nanotubes self-assembled on Ti substrate were fabricated through anodization and heat-treatment. The anodization parameters for fabricating optimized nanostructures were investigated and hydrogen sensing properties of the nanotubes were tested. It was found that the anodization temperature and voltage had a great impact on the formation of uniform nanotubes or nanostructures. And the nanotubes assembled on Cu-coated PCB demonstrated a good hydrogen sensing properties at room temperature.

Introduction

Titania nanostructure has aroused considerable scientific interests due to its wide application in heterojunction solar cells, water photolysis, fuel cells, molecular filtration, tissue engineering [1,2], photocatalytic devices [3], and lithium-ion batteries [4], exhibiting highly variable and tunable wetting behavior [5,6]. As a promising semiconductor, titania is sensitive to atmosphere. The resistance of titania nanotubes could increase significantly when an oxidative atmosphere deprives the carriers of titania. On the other hand, the resistance of titania nanotubes with a length of micrometer scale would reduce by about 8.7 orders of magnitude when exposed to alternating atmospheres of nitrogen containing 1000ppm hydrogen and air [7].

The anodization of pure Ti and its alloy to fabricate self-assembled titania nanotubes has been reported in many literatures. To date, however, some key factors which could affect the formation of titania nanotubes have not been well studied. In the present work, we investigate the effect of two important parameters (anodization temperature and voltage) on the formation of highly ordered array of titania nanotubes. A nanotube-based sensor was assembled on Cu-coated PCB for testing the hydrogen sensing properties of the crystallized titania nanotubes.

Experimental

Pure Ti sheets were used as a substrate for anodization. The sheets were ultrasonically cleaned in absolute alcohol before anodization. Electrochemical anodization was carried out with a DC Voltage Stabilizer and Pt negative electrode. The electrolyte used is 1M $(NH_4)_2SO_4$ containing 0.5wt% NH_4F. The sheets were anodized at 15V at different temperatures, and at 20°C with different voltages.

Some of the anodized samples were heat-treated at 450°C [8]. Circular Pt electrodes with a thickness of 100nm were deposited onto surfaces of the nanotube samples through sputtering (Fig. 1). Copper wires were connected to the Pt electrode with conductive paste. And the other ends of the copper wires were connected to Cu-coated PCB. The above assembled nanotube sensor was then put into glass chamber flowing with either compressed air or diluted H_2 gas. A Keithley 2700 multimeter was used to test the hydrogen sensing properties of the nanotube sensor assembled on the Cu-coated PCB. All of the nanotube samples were characterized with X-ray diffraction (XRD), scanning electron microscope (SEM, FEI SIRION 200) equipped with Energy Dipersive X-ray analysis (EDXA).

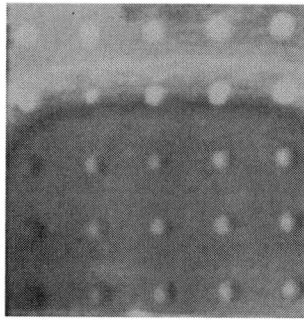

Fig. 1 Pt electrode deposited onto the nanotube sample.

Results and Discussion

The anodization temperature was found to affect the fabrication of TiO_2 nanotubes through determining both the dissolution on the top of the as-anodized nanotubes and the erosion at the bottom of the as-anodized nanotubes [9, 10]. Nanopores layer at the top of the anodized surface could not dissolve completely when the sample was anodized at 15V and 10°C (in Fig. 2(a)). But the nanopores layer dissolved completely when the anodization temperature was increased to 20°C (Fig. 3(a)). The average length of nanotubes shown in Fig. 2 is about 700nm with an average diameter of 50nm, and the average length of the nanotubes shown in Fig. 3 is about 1 μm an average diameter of 50nm. Obviously, Fig. 2(b) and Fig. 3(b) indicate that the average length of the as-anodized nanotubes was much longer when the sample was anodized at a higher temperature, which suggest that the erosion rate at the bottom of the TiO_2 nanotubes should be accelerated at higher anodization temperature.

Fig. 4 shows longer nanotubes fabricated at 15V and 30°C. The average length of nanotubes shown in Fig. 4 is about 1.2 μm. This also proves that the erosion speed at the bottom of the TiO_2 nanotubes could be accelerated by the higher anodization temperature.

978-1-4244-4658-2/09 $25.00 © 2009 IEEE

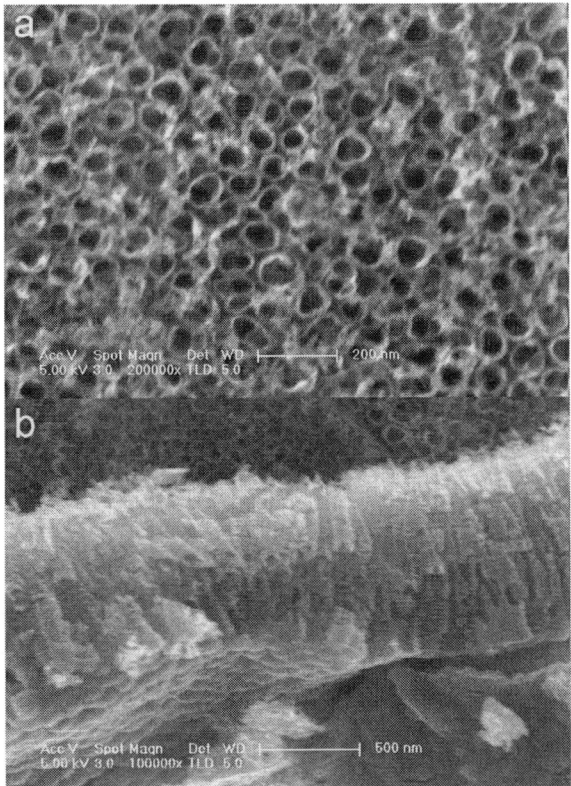

Fig. 2 SEM images of sample anodized at 15V and 10°C.
(a) top view (b) cross-section.

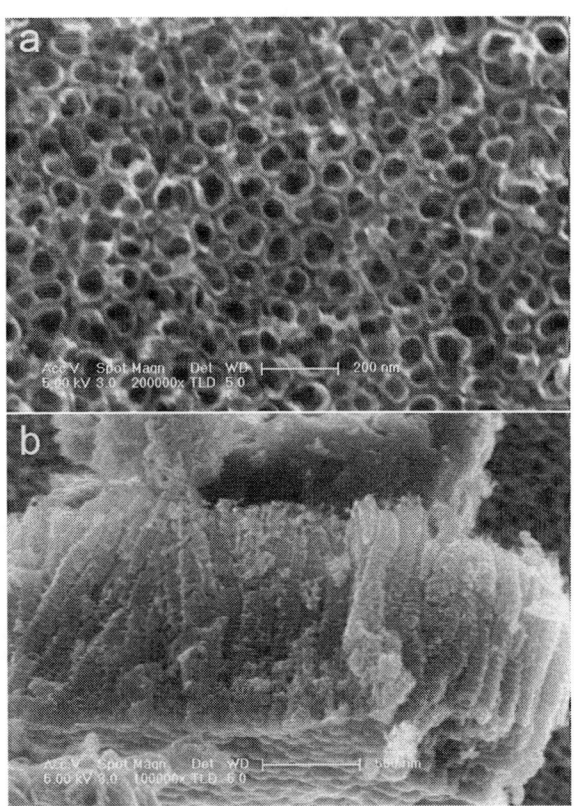

Fig. 4 SEM images of sample anodized at 15V and 30°C.
(a) top view (b) cross-section

Fig. 3 SEM images of sample anodized at 15V and 20°C.
(a) top view (b) cross-section.

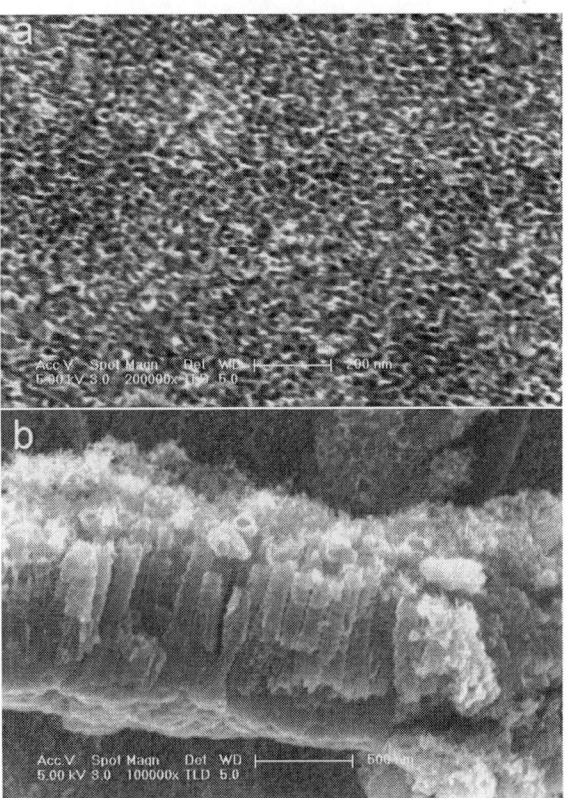

Fig. 5 SEM images of sample anodized at 5V and 20°C.
(a) top view (b) cross-section.

978-1-4244-4658-2/09 $25.00 © 2009 IEEE

The anodization voltage is another key factor during the anodization process. Figures 5 and 6 present the nanostructures obtained at different anodization voltage. At a lower anodization votage of 5V, there was a nanopores layer left on top of the nanotube surface (Fig. 5a), which indicates that the dissolution speed on top of the nanotubes layer was slow when the anodization was carried out at a lower voltage. And Fig. 5(b) shows shorter nanotubes than Fig. 2(b), which proves that higher voltage accelerates the erosion speed at the bottom of the nanotubes. But when the anodization voltage was increased to a high level like 30V, the wall of the nanotubes dissolved, which destroyed the nanotubular structures (Fig. 6).

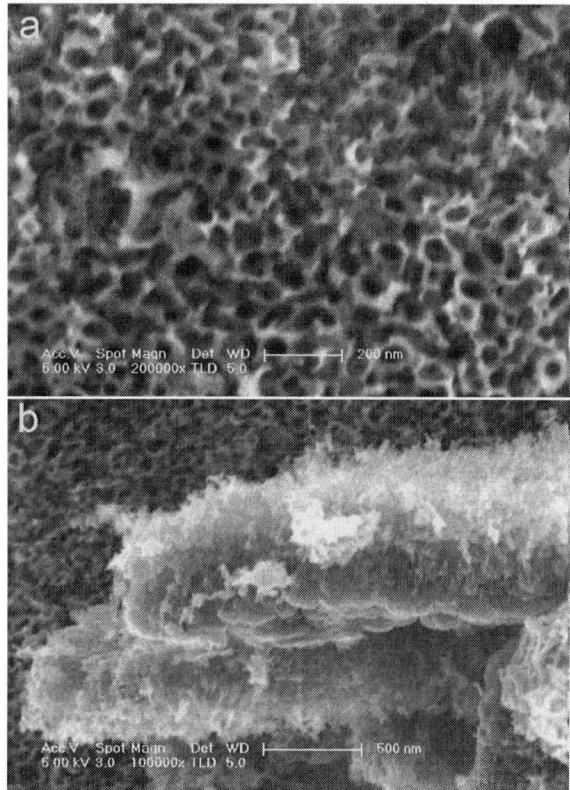

Fig. 6 SEM images of sample anodized at 25V and 20°C. (a) top view (b) cross-section.

From the above analysis, it could be found that anodization at 15V and 20°C should be a good choice to fabricate uniform and highly aligned nanotube arrays. Thus, in the following experiments, we directly heat-treated the as-anodized amorphous nanotubes (fabricated at 15V and 20°C) to obtain crystallized nanotubes. Our XRD analysis of the crystallized nanotubes proved that the nanotubes were anatase titania. As mentioned above, we fabricated the nanotube sensor with these crystallized titania nanotubes.

During testing the hydrogen sensing properties of the titania nanotubes, the resistance of the sample decreased when the sample exposed to H_2 containing atmosphere. Fig. 7 shows the response of TiO_2 nanotubes at different H_2 concentrations. The response increased with increase of the H_2 concentration. The response of the TiO_2 nanotube was 1% at 100ppm H_2 and up to 15% at 1000ppm H_2.

Fig. 8 shows the response curves of the nanotube sensor for different H_2 containing atmospheres. One cycle of the testing usually took 1000s. Good response repeatability at the same concentration (such as 600 ppm and 1000 ppm) could be found.

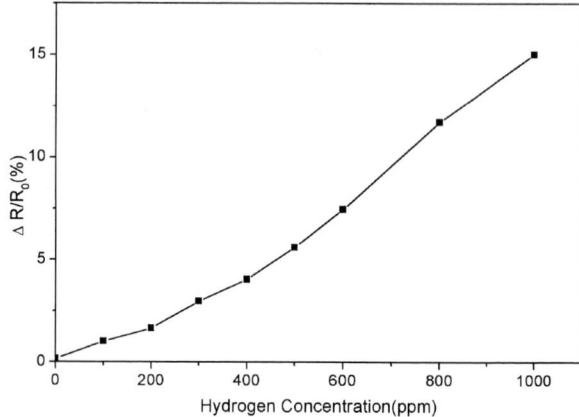

Fig. 7 Response of TiO_2 nanotube layer at different H_2 concentration.

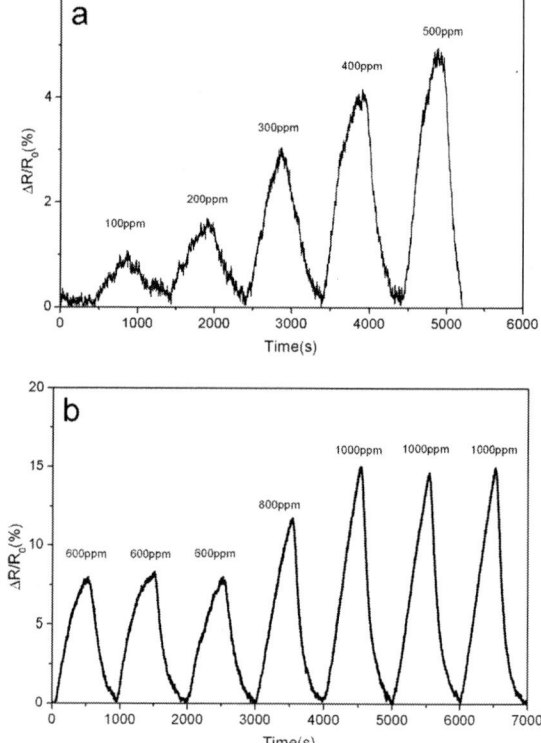

Fig. 8 Response of TiO_2 nanotube layer at dilute H_2 concentration. (a) 100-500ppm (b) 600-1000ppm.

Conclusions

Through affecting the dissolution speed on top of the nanotube layer and the erosion speed at the bottom of the nanotube, anodization temperature and voltage were shown to be key factors for fabrication of highly ordered array of

TiO_2 nanotubes. Crystallized TiO_2 nanotube array connected and assembled on Cu-coated PCB substrate exhibited different response in H_2 atmospheres at room-temperature.

Acknowledgments

This work was supported by Shanghai Pujiang Program (No. 07pj14047) and 863 Plan of China (No. 2006AA02A1). We thank the contribution from SEM lab at Instrumental Analysis Center of SJTU.

References

1. Hueso, L., Mathur, N., "Dreams of a hollow future," *Nature*, Vol. 427, (2004), pp. 301.

2. Tenne, R., Rao, C. N. R., "Inorganic nanotubes," *Philos. Trans. R. Soc. Ser. A*, Vol. 362, (2004), pp. 2099.

3. Fujishima, A., Honda, K., "Electrochemical photocatalysis of water at a semiconductor electrode," *Nature*, Vol. 238, (1972), pp. 37-38.

4. Kavan, L., Gratzel, M., Gilbert, S. E., Klemenz, C., Scheel, H. J., "Electrochemical and Photoelectrochemical Investigation of Single-Crystal Anatase," *J. Am. Chem. Soc.*, Vol. 118, (1996), pp. 6716-6723.

5. Wang, R., Hashimoto, K., Fujishima, A., Chikuni, M., Kojima, E., Kitamura, A., Shimoshigoshi, M., Watanabe, T., "Light induced amphiphilic surface," *Nature* ,Vol. 388, (1997), pp. 431-432.

6. Feng, X., Zhai, J., Jiang, L., "The fabrication and switchable superhydrophobicity of TiO_2 nanorod films," *Angew. Chem. Inc. Ed.*, Vol. 44, (2005), pp. 7463-7465.

7. Paulose, M., Varghese, O. K., Mor, G. K., Grimes, C. A., Ong, K. G., "Unprecedented ultra-high hydrogen gas sensitivity in undoped titania nanotubes," *Nanotechnology*, Vol.17, (2006), pp. 398-402.

8. Min-Hyun Seo, Masayoshi Yuasa, Tetsuya Kida, Jeung-Soo Huh, Kengo Shimanoe, and Noboru Yamazoe, "Gas sensing characteristics and porosity control of nanostructured films composed of TiO_2 nanotubes", *Sensors and Actuators B: Chemical*, Vol. 137, (2009), pp. 513-520.

9. Hiroaki Tsuchiya, Jan M. Macak, Andrei Ghicov, et al, "Nanotube oxide coating on Ti-29Nb-13Ta-4.6Zr alloy prepared by self-organizing anodization," *Electrochimica Acta*, 52, (2006), pp. 94-101.

10. Taveira L V, Macák J M, Tsuchiya H, Dick L F P, and Schmuki P, Initiation and growth of self-organized TiO_2 nanotubes anodically formed in $NH_4F/(NH_4)_2SO_4$ electrolytes. *J. Electrochem.*, Vol. 152 (10), (2005), pp. 405-410.

Development of Lead-Free Solders with Superior Drop Test Reliability Performance

Y. W. Wang, and C. R. Kao*
Department of Materials Science & Engineering
Taiwan University
Taipei City, Taiwan, China
*E-mail: crkao@ntu.edu.tw   Phone/Fax: +886-2-33663745

Abstract

The objective of this study is to investigate the alloying effects of Fe, Co, and Ni on the interfacial reactions between solder and Cu. Emphasis is placed on a systematic comparison study on the effect of Fe, Co, and Ni additions. Solders with simultaneous Fe and Ni addition as well as simultaneous Co and Ni addition are also prepared in order to investigate whether there is any interaction between the alloying elements.

The results of this study can be summarized as below:

(1) In multiple reflow study using the Sn2.5Ag0.8Cu, Sn2.5Ag0.8Cu0.03Fe, Sn2.5Ag0.8Cu0.03Co, Sn2.5Ag0.8Cu0.03Ni, Sn2.5Ag0.8Cu0.03Fe0.03Ni, and Sn2.5Ag0.8Cu0.03Co0.03Ni solder over Cu substrate, Cu_6Sn_5 was the only reaction product for all the different solders used.

(2) Reflows using the solder without doping produced a thin, dense layer of Cu_6Sn_5. The additions of Fe, Co, or Ni transformed the microstructure into a much thicker Cu_6Sn_5 with many small trapped solder regions between the grains.

(3) The amount of Cu_6Sn_5 formed at the interface increased with the number of reflows.

(4) In solid state aging study, both Cu_6Sn_5 and Cu_3Sn formed, but the additions of Fe, Co, or Ni produced a much thinner Cu_3Sn layer in all cases in this study. The growth of Cu_3Sn followed the parabolic kinetics.

(5) The simultaneous addition of 0.03 wt.% of Fe and Ni was the most effective in reducing the Cu_3Sn thickness. The Cu_3Sn thickness was only one-third that of Sn2.5Ag0.8Cu.

(6) The additions of Fe, Co, or Ni in an amount as small as 0.03 wt.% were effective in reducing the Cu_3Sn thickness at 160°C for at least 2000 hrs.

Introduction

In the past few years, the SnAgCu family of solders has obtained a wide acceptance as a replacement for the PbSn eutectic solder in electronic applications. The current main research thrust in lead-free solder alloy development is on the enhancing or fine-tuning the various properties of SnAgCu through the addition of minor alloying elements. For example, Ni was evaluated for its potential as a minor alloying element to Sn-based lead free solders, and so were Ge, Fe, Co, Zn, Bi, Mn, Ti, Si, and Cr. Those previous studies point to the fact that minor alloy additions indeed can enhance the solder performance in many different respects. One of the more note-worthy alloying elements is Ni. It was shown that Ni addition to Sn3.5Ag (3.5 wt.% Ag, balance Sn) in amounts as minute as 0.1 wt.% could substantially hinder the Cu_3Sn growth during soldering as well as during the subsequent solid-state aging. The Cu_3Sn growth had been

linked to the formation of micro voids, which in turn increased the potential for brittle interfacial fracture. Recently, it indeed was shown that drop test performance increased for solders joints with just a small amount of Ni addition (< 1 wt. %).

In the past few years, the SnAgCu family of solders has obtained a wide acceptance as a replacement for the PbSn eutectic solder in electronic applications. But there are still some problems in lead-free systems. Examples include package compatibility, creep, and Kirkendall voids. In the reactions between Sn-based solders and Cu substrate, the formation of Kirkendall voids within the Cu_3Sn layer had been reported by many research groups. The formation of Kirkendall voids accompanying the Cu_3Sn growth raises serious reliability concerns because excessive voids formation increases the potential for brittle interfacial fracture [1-4]. It is reasonable to assume that the amount of Kirkendall voids can be reduced by reducing the Cu_3Sn thickness. People try to add the minor elements to improve its properties. For example, Ni was evaluated for its potential as a minor alloying element to Sn-based lead-free solders [5, 6], and so were Ge [6], Zn [6], Mn [6], Ti [6], Si [6], Cr [6], Fe [7], and Co [7]. These previous studies point out minor alloy addition can strengthen the solder performance in many different respects. One of the more noteworthy thing is minor elements can hinder the Cu_3Sn growth during soldering as well as during the following solid-state aging. The objective of this study is to examine whether Fe and Co additions have a similar effect as Ni does.

Experimental

The solder used for this study are Sn2.5Ag0.8Cu0.03x (x=Fe, Co, Ni). Six different solders were prepared from 99.999% purity elements. The solder ball compositions were verified by using ICP-AES (Inductively Coupled Plasma Atomic Emission Spectroscopy), and it was estimated that the reported compositions had a maximum 0.002 wt% uncertainty. Each solder ball had a mass of 10 mg. The solder ball was placed on a Cu soldering pad with 600 □m diameter wettable region. The reactions during the reflow and during the solid state aging were studied separately. The reflow temperature profile had a peak temperature of 235°C and 90 s duration during which the solder was molten. The nominal ramp rate and cooling rate were both 1.5°C/s. The number of reflows was 1, 3, 5, or 10 times. Some of the samples were subjected to solid state aging at 160°C for 500, 1000 or 2000 hrs after being reflowed once. The samples were then mounted in epoxy, sectioned by using a low-speed diamond saw, and metallurgically polished in preparation for characterization. The reaction zone for each sample was examined using an optical microscope and a scanning

978-1-4244-4658-2/09 $25.00 © 2009 IEEE

electron microscope (SEM). The compositions of the reaction products were determined using an electron microprobe (EPMA), operated at 15 keV. In microprobe analysis, the concentration of each element was measured independently, and the total weight percentage of all elements was within $100\pm1\%$ in each case. For every data point, at least four measurements were made and the average value was reported.

Reactions during Reflow and Aging

Figure 1 shows the backscattered electron micrographs for the samples with different alloy additions that were reflowed for one time. The symbol SAC stands for the Sn2.5Ag0.8Cu solder without any minor element addition, and SAC0.03Fe stands for Sn2.5Ag0.8Cu with 0.03 wt.% Fe addition, etc. Comparing Fig. 1 (a) to Fig. 1 (b)-(f), one notes that the additions of Fe, Co, or Ni to SAC made the amount of Cu_6Sn_5 at the interface increase substantially. For the SAC solder, the Cu_6Sn_5 layer was thin, continuous, and void-free, consistent with what had been reported in the literature. With the additions of Fe, Co or Ni, Cu_6Sn_5 became thicker, and had many voids between the Cu_6Sn_5 grains.

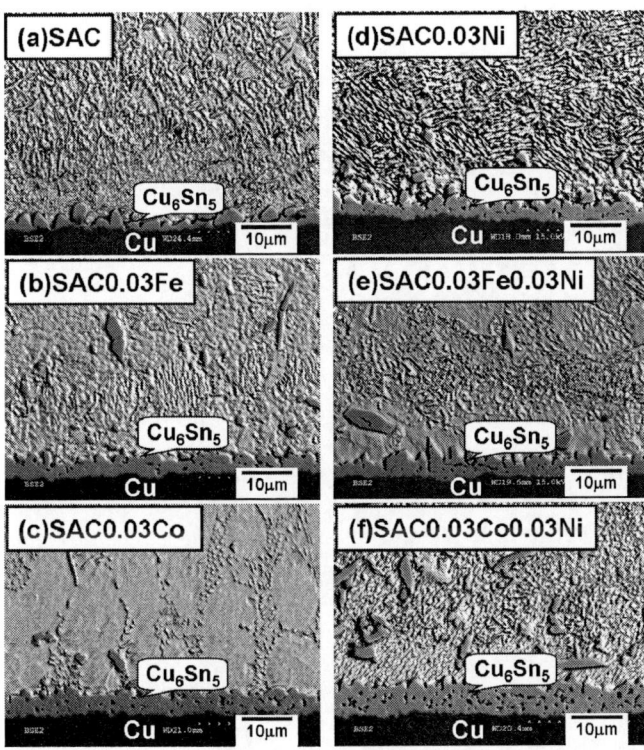

Fig.1 Backscatter electron micrographs of as-reflowed solder joints that had been reflowed at 235°C for 90 sec: (a)Sn2.5Ag0.8Cu,(b)Sn2.5Ag0.8Cu0.03Fe,(c)Sn2.5Ag0.8Cu 0.03Co,(d)Sn2.5Ag0.8Cu0.03Ni,(e)Sn2.5Ag0.8Cu0.03Fe0.0 3Ni,and(f)Sn2.5Ag0.8Cu0.03Co0.03Ni. The additions of minor elements to solder have a porous needle-shaped morphology IMC.

Figure 2 shows the evolution of the microstructures as the number of reflows increased 3, 5, and 10 times. Under all conditions, only Cu_6Sn_5 was observed as before. Nevertheless, in a previous study [5], it was reported Cu_3Sn

would eventually appear when the reaction time reached 9 hrs at (240°C). For all the solders used in this study, the amount of Cu_6Sn_5 at the interface increased with the number of reflows.

The alloy additions had different effects between Cu_6Sn_5 and Cu_3Sn. Any alloy addition to SAC, increased the Cu_6Sn_5 thickness substantially compared to SAC. Different additions had a different thickening effect on Cu_6Sn_5, and the thickness followed the following trend for different types of additions: SACCo>SACCoNi≅SACFe>SACFeNi>SACNi>SAC.

In addition to the reaction during reflow, the reaction between solid solders and Cu at 160°C is also studied. After aging at 160°C, Cu_3Sn appeared with a layered microstructure between Cu_6Sn_5 and Cu, as shown in Fig. 3. Their thicknesses were digitally measured by using the software "OPTIMAS". The average thickness of Cu_6Sn_5 or Cu_3Sn layer was calculated by dividing the measured area of Cu_6Sn_5 or Cu_3Sn by the length of the interface.

The additions of Fe, Co, or Ni substantially reduce the Cu_3Sn to Cu_6Sn_5 thickness ratio. Adding Fe, Co, or Ni increased the Cu_6Sn_5 thickness, the additions, however, substantially decreased the Cu_3Sn thickness as shown in Fig. 4. The Cu_3Sn thickness was only one-third that of SAC solder. Adding both Fe and Ni was the most effective, while adding Co was slightly less effective. On the other side, for all the minor elements used, the thicknesses of both Cu_6Sn_5 and Cu_3Sn increased with the aging time. The Cu_3Sn thicknesses for various conditions were plotted in Fig. 4. The thickness data are plotted in Fig. 4 (a) for Cu_6Sn_5, and in Fig. 4 (b) for Cu_3Sn. As shown in Fig. 4 (a), for all the cases, Cu_6Sn_5 grew thicker with the aging time, although the growth kinetics did not follow the parabolic kinetics very well. The growth kinetics for Cu_3Sn, shown in Fig. 4 (b), had a better fit with the parabolic kinetics. Even though the concentration of the addition elements was very low (0.03 wt. %), some of the addition elements did get incorporated into Cu_6Sn_5 and Cu_3Sn. The amounts of these incorporated elements are shown in Table 1 for solder joints that had been added at 160°C for 2000 hrs. As these concentrations were determined using an electron microprobe, which had an accuracy of about one percent, the data in Table 1 only serves to qualitatively illustrate the existence of these elements in Cu_6Sn_5 and Cu_3Sn. The Cu_6Sn_5 and Cu_3Sn contain a small amount of Ni.

The reason Fe, Co, and Ni are effective in reducing the Cu_3Sn thickness is unclear at this moment. Several theories have been proposed, including thermodynamic arguments [8] and kinetic arguments [9, 10, 11]. It is highly likely that adding Fe, Co, or Ni somehow reduced the interdiffusion coefficient of Cu_3Sn relative to Cu_6Sn_5. It is widely known a phase with a higher interdiffusion coefficient will grow faster at the expense of its neighboring phase that has a lower interdiffusion coefficient. The mechanism explaining how adding Fe, Co, or Ni can reduce the Cu_3Sn interdiffusion coefficient is still lacking. If we can find the mechanism, it is helpful to solder joint reliability. More studies are needed to clarify this point.

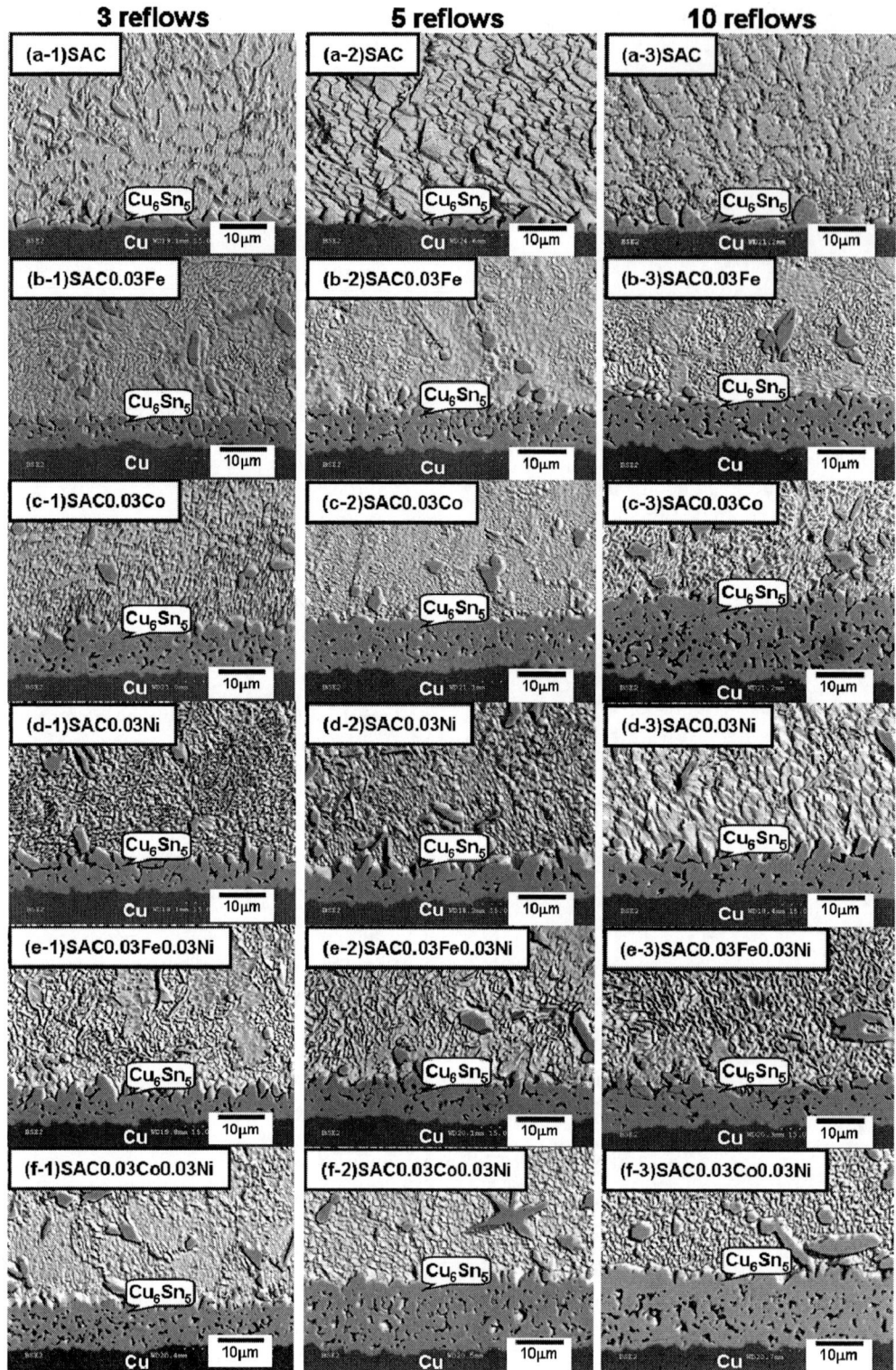

Fig.2 Evolution of the microstructures for different solders as the number of reflows increased from 3 to 5, and then to 10.

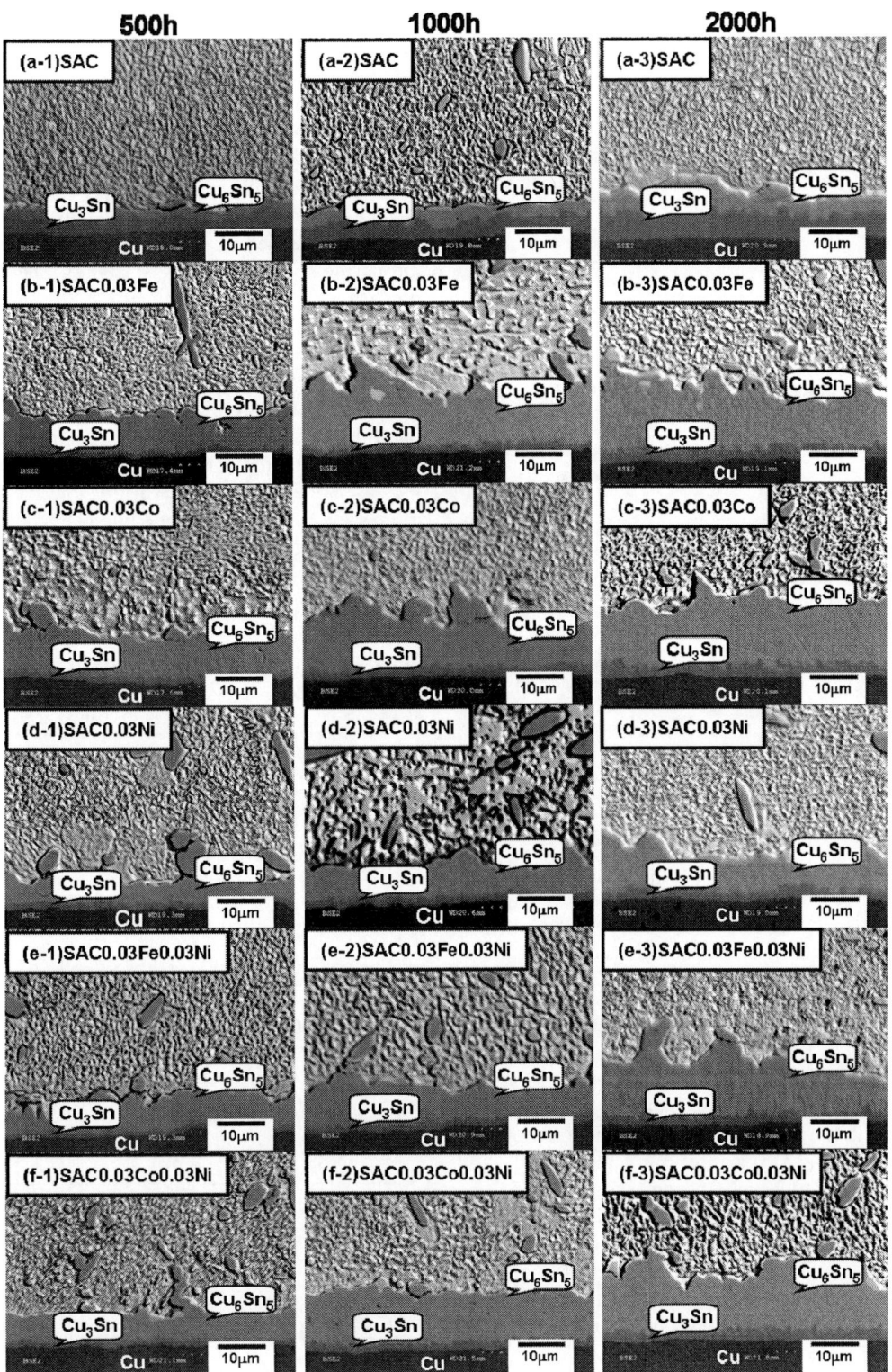

Fig.3 Backscatter electron micrographs of the interfacial IMC layers in the specimens after aging at 160 °C for 1000h: (a)Sn2.5Ag0.8Cu,(b)Sn2.5Ag0.8Cu0.03Fe,(c)Sn2.5Ag0.8Cu0.03Co(d)Sn2.5Ag0.8Cu0.03Ni,(e)Sn2.5Ag0.8Cu0.03Fe0.03Ni, and (f)Sn2.5Ag0.8Cu0.03Co0.03Ni. The addition of minor Fe, Co, and Ni to Sn2.5Ag0.8Cu are able to substantially hinder the Cu_3Sn growth in the reaction between solder and electroplated Cu substrate.

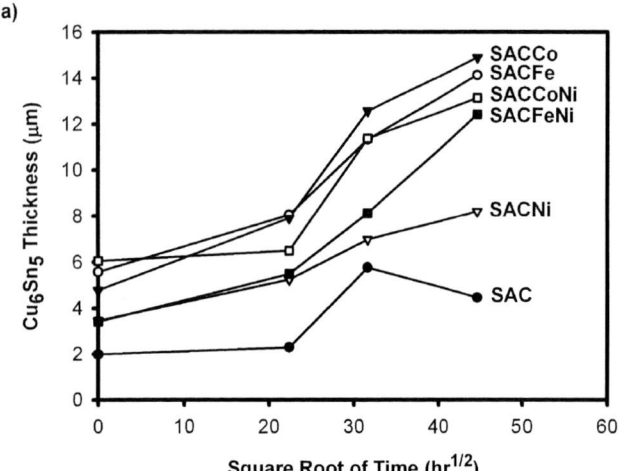

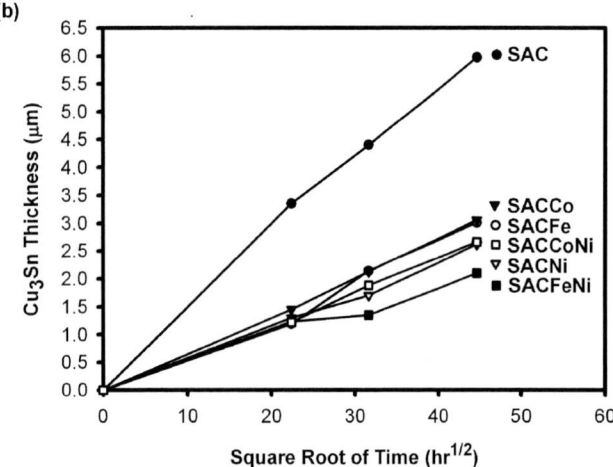

Fig.4 Layer thickness versus the aging time at 160°C for (a) Cu_6Sn_5 and (b) Cu_3Sn. The thickness of Cu_6Sn_5 increased with the minor elements addition. On the other hand, the Cu_3Sn decrease with the minor elements addition.

The mechanism explaining how minor elements can reduce Cu_3Sn

This study showed that doping the SAC with a small amount of Fe, Co, or Ni would not change the type of the reaction product with the Cu substrates. The reaction product was always Cu_6Sn_5 during a typical reflow. The Cu_3Sn phase only formed when the reflow time became very long [5] or during the solid-state aging. Although the addition of Fe, Co, or Ni did not change the type of the intermetallic compound, the amount of compound at the interface did increase substantially with Fe, Co, or Ni additions. The growth of Cu_6Sn_5 shown in Fig. 4 (a) did not follow the parabolic kinetics or the linear kinetics. This was because during reflow there were several concurrent processes. The first was the growth of the compound. The second was the dissolution of the compound into the molten solder. The third was the ripening of the Cu_6Sn_5 grains, pointed out by Gusak and Tu [12].

The key observation of this study was adding Fe, Co, or Ni at an amount as small as 0.03 wt.% was able to reduce the Cu_3Sn thickness substantially, as shown in Fig. 4 (b). The reason why minor element addition is effective in reducing

the Cu_3Sn thickness is unclear at this moment. Several theories have been proposed, including thermodynamic arguments [8] and kinetic arguments [13]. It was likely that the Ni addition somehow increased the ratio of interdiffusion flux through the Cu_6Sn_5 layer and the Cu_3Sn layer.

It is widely known that a phase with a higher interdiffusion coefficient will grew faster at the expense of its neighboring phase that has a lower interdiffusion coefficient [13]. The mechanism explaining how the Ni addition can change the ratio of interdiffusion flux is still unclear. More studies are needed to elucidate this point. Figure 4 (a) shows during the solid-state aging at 160°C. The parabolic kinetics did not describe the growth of Cu_6Sn_5 very well. However, the growth of Cu_3Sn did follow the parabolic kinetics closely, as shown in Fig. 4 (b). Figure 4 (b) shows simultaneously adding 0.03 wt.% of Fe and Ni was the most effective in reducing the Cu_3Sn thickness. The Cu_3Sn thickness was only one-third the SAC solder. Because the Cu_3Sn growth had been linked to the formation of micro voids, which in turn increased the potential for brittle interfacial fracture [2], a thinner Cu_3Sn might translate into better solder joint strength. More studies on the strength of solder joints with and without Fe, Co, or Ni additions are required to fully confirm this promising suggestion.

Table 1 Amounts of alloy elements in Cu_6Sn_5 and Cu_3Sn after aging at 160°C for 2000 hrs. The symbol SAC represents Sn2.5Ag0.8Cu. (unit: at.%)

Alloys	Cu_3Sn			Cu_6Sn_5		
	Fe	Co	Ni	Fe	Co	Ni
SAC0.03Fe	0			0.1		
SAC0.03Co		0.7			0.2	
SAC0.03Ni			1.2			0.9
SAC0.03Fe0.03Ni	0.1		0.8	0.1		0.4
SAC0.03Co0.03Ni		0.5	0.4		0.3	0.4

Conclusions

(1) Reflows using the solder without doping produced a thin, dense layer of Cu_6Sn_5. The additions of Fe, Co, or Ni transformed the microstructure into a much thicker Cu_6Sn_5 with many small trapped solder regions between the grains.

(2) The addition of 0.03 wt.% of Fe, Co or Ni was effective in reducing the Cu_3Sn thickness. The Cu_3Sn thickness was only one-third that of Sn2.5Ag0.8Cu.

(3) The Cu_3Sn growth had been linked to the formation of micro voids, which in turn increased the potential for brittle interfacial fracture, a thinner Cu_3Sn might translate into a better solder joint strength.

(4) Simultaneously adding 0.03 wt.% of Fe and Ni was the most effective in reducing the Cu_3Sn thickness.

Acknowledgments

This work was supported by the Science Council of Taiwan through grant NSC 95-2221-E-002-443-MY3 and Accurus Scientific Co., LTD. It is a great pleasure to acknowledgement Ms. S. Y. Tsai for EPMA analysis in National Tsing Hua University.

References

1. Ahat, S., Sheng, M., and Luo, L., "Microstructure and Shear Strength Evolution of SnAg/Cu Surface Mount Solder Joint during Aging," *Journal of Electronic Materials*, Vol. 30, No. 10 (2001), pp. 1317-1322.

2. Zeng, K., Stierman, R., Chiu, T. C., Edwards, D., Ano, K., and Tu, K. N., "Kirkendall Void Formation in Eutectic SnPb Solder Joints on Bare Cu and Its Effect on Joint Reliability," *Journal of Applied Physics*, Vol. 97 (2005), pp. 024508(1)-024508(8).

3. Chiu,T. C., Zeng, K., Stierman, R., Edwards, D., and Ano, K., "Effect of Thermal Aging on Board Level Drop Reliability for Pb-Free BGA Packages," Proc 54th Electronic Components and Technology Conf, 2004, pp. 1256-1262.

4. Date, M., Shoji, T., Fujiyoshi, M., Sato, K., and Tu, K. N., "Impact Reliability of Solder Joints," Proc 54th Electronic Components and Technology Conf, 2004, pp. 668-674.

5. Tsai, J. Y., Hu, Y. C., Tsai, C. M., and Kao, C. R., "A Study on the Reaction between Cu and Sn3.5Ag Solder Doped with Small Amounts of Ni," *Journal of Electronic Materials*, Vol. 32, No. 11 (2003), pp. 1203-1208.

6. Anderson, I. E., and Harringa, J. L., "Suppression of Void Coalescence in Thermal Aging of Tin-Silver-Copper-X Solder Joints," *J Electronic Materials*, Vol. 35 , No. 1 (2006), pp. 94-106.

7. Anderson, I, E., and Harringa, J. L., "Elevated Temperature Aging of Solder Joints Based on Sn-Ag-Cu: Effects on Joint Microstructure and Shear Strength," *Journal of Electronic Materials*, Vol. 33, No. 12 (2004), pp. 1485-1496.

8. Gao, F., Takemoto, T., and Nishikawa, H., "Effects of Co and Ni Addition on Reactive Diffusion between Sn-3.5Ag Solder and Cu during Soldering and Annealing," *Materials Science and Engineering*, Vol. A420, (2006), pp. 39-46.

9. Yu, H., Vuorinen, V., Kivilahti, J., "Effect of Ni on the Formation of Cu_6Sn_5 and Cu_3Sn Intermetallics," Proc 56th Electronic Components and Technology Conf, 2006, pp. 1204-1209.

10. Ho, C. E., Yang, S. C., and Kao, C. R., "Interfacial Reaction Issues for Lead-Free Electronic Solders," *Journal of Materials Science: Materials in Electronics*, Vol. 18 (2007), pp. 155-174.

11. Garner, L., Sane, S., Suh, D., Byrne, T., Dani, A., Martin, T., Mello, M., Patel, M., and Williams, R., "Finding Solutions to the Challenges in Package Interconnect Reliability," *Intel Technology Journal*, Vol. 9 (2005), pp. 297-308.

12. Gusak, A. M., and Tu, K. N., "Kinetic Theory of Flux-Driven Ripening," *Physical Review*, Vol. B66 (2002), pp. 115403(1)-115403(14).

13. Laurila, T., Vuorinen, V., and Kivilahti, J. K., "Interfacial Reactions between Lead-Free Solders and Common Base Materials," *Materials Science and Engineering*, Vol. R.49 (2005), pp. 1-60.

Enhancing the Properties of a Lead-free Solder with the Addition of Ni-coated Carbon Nanotubes

S.M.L. Nai[1*], Y.D. Han[2, 4], H.Y. Jing[2], K.L. Zhang[3], C.M. Tan[4], J. Wei[1, 3*]

[1]Singapore Institute of Manufacturing Technology, 71 Nanyang Drive, Singapore 638075
[2]School of Materials Science and Engineering, Tianjin University, Tianjin, P.R. China 300072
[3]School of Electronics Information Engineering, Tianjin University of Technology, Tianjin, P.R. China 300384
[4]School of Electrical and Electronic Engineering, Nanyang Technological University, Singapore 639798
*Email: mlnai@simtech.a-star.edu.sg; jwei@simtech.a-star.edu.sg

Abstract

In this study, 0.05wt.% of Ni-coated multi-walled carbon nanotubes (Ni-CNTs) were successfully introduced into the 95.8Sn-3.5Ag-0.7Cu solder using the powder metallurgy technique, to synthesize a new lead-free composite solder system. The samples were characterized in terms of their microstructural and mechanical properties. Microstructural analysis of the polished samples revealed uniformly distributed intermetallic phases throughout the solder matrix and EDS analysis identified the phases as Ag_3Sn and Cu_6Sn_5. Furthermore, with the addition of Ni-CNTs, the tensile results showed an improvement in 0.2% yield strength (~13%) and ultimate tensile strength (~15%), when compared with that of unreinforced SnAgCu solder. Nanoindentation tests were also conducted to assess the creep performance of the samples. The results showed that addition of Ni-CNTs improved the creep resistance of the composite solder. The results in this study convincingly established that nanotechnology coupled with composite technology in electronics solders can lead to the enhancement of mechanical properties. Thus, these advanced interconnect materials will benefit the microelectronics assembly and packaging industry. An attempt is made in this study to interrelate the mechanical properties of the resultant composite solder with the presence of Ni-CNTs.

Introduction

In recent years, the implementation of strict legislations to ban the use of lead-containing solder has led the move towards the development and use of lead-free solders. Among the commercially available lead-free solders, Sn-Ag-Cu alloys which have relatively low melting temperature and good compatibility with common commercial components, are one of the leading alternatives to replace the widely used Sn-Pb solders. However, with increasing functional requirements and shrinking of electronic components, conventional solder technology can no longer guarantee the reliability of IC-to-board joint. Thus, in order to fulfill the ever-stricter requirements, new interconnect solder materials equipped with a combination of good properties have to be developed [1]. It has been proven that one of the effective ways to improve the performance of a conventional solder, is to intentionally introduce foreign phases into the conventional solder matrix, forming a composite solder [2-12].

Since the discovery of carbon nanotubes (CNTs) in 1991 by Iijima [13], there has been increasing scientific and research interest in CNTs. CNT is well-known for its excellent mechanical, thermal and electrical properties and

in recent year, it is used as a filler material in composites as it yielded promising enhancement in conventional monolithic materials [14-20]. Kim *et al.* [18, 20] incorporated the CNTs in Cu matrix. It was observed that the ultimate tensile stress, hardness, yield stress and compressive stress of Cu reinforced with CNTs were much higher than that of the monolithic Cu. In another study, Sun *et al.* [19] reported that the presence of CNTs can effectively improve the tensile strength of Ni.

Accordingly in the present study, Ni-coated multi-walled CNTs were chosen as the reinforcement material. 0.05 wt. % of Ni-coated CNTs was incorporated into 95.8Sn-3.5Ag-0.7Cu solder using the powder metallurgy technique. Monolithic SnAgCu solder was also synthesized and characterized as a basis of comparison. The unreinforced and composite solders were characterized in terms of their microstructural properties and mechanical properties. The mechanical performance of the solders was assessed using the tensile tests and nanoindentation creep tests.

Experimental Procedures

Powder metallurgy route was used to synthesize the nanocomposite solders. 95.8Sn-3.5Ag-0.7Cu solder (particle size range: 5-15 μm) was used as the matrix material and Ni-coated multi-walled carbon nanotubes (Ni-CNTs) with outer diameter: 10-20 nm and length 30 μm, was used as the reinforcing material. Sn–Ag–Cu solder was reinforced with 0.05wt.% of Ni-CNTs. The processing procedures involved blending the pre-weighed solder powder and Ni-CNTs, at a speed of 50 rpm for 10 hours. The mixture was then compacted and sintered for 2 hours at 175°C. Finally, the sintered billet was extruded at room temperature, using an extrusion ratio of 20:1.

Microstructural characterization studies were conducted on metallographically polished extruded samples to investigate: (i) the presence and identity of the intermetallic phases in the solder samples, (ii) presence and distribution of Ni-CNTs in the solder matrix, (iii) interfacial integrity between the matrix and Ni-CNTs, and (iv) the planar surface indentations on the solder sample. The EVO 50 XVP Scanning Electron Microscope (SEM) equipped with Energy Dispersive Spectroscopy (EDS) and the JSM 6340F Field-Emission Scanning Electron Microscope (FE-SEM) were used.

The tensile properties of the extruded samples in the form of smooth bars were determined in accordance with ASTM test method E8M-96 using an Instron Microtester with a crosshead speed set at 0.254mm/min on round tension test specimens of 5 mm diameter and 25 mm gauge length.

Prior to nanoindentation tests, all the solder samples were grinded and metallurgically polished to produce a

978-1-4244-4658-2/09 $25.00 © 2009 IEEE

smooth surface. Nanoindentation tests were performed on a Nano Test nanoindentation testing platform (Micro Materials Ltd., Wrexham, UK), at room temperature, using a diamond Berkovich indenter. A load-control method was used during the nanoindentation test whereby the samples were subjected to a constant ramp rate of 0.05 s^{-1} and at a maximum load of 40 mN.

Following the load ramp, the sample was held for 300s at the maximum load so as to ensure that the sample undergoes sufficient creep and for the determination of creep response. For all the tests, there was a 60s holding period after 90% unloading to allow for thermal drift correction. In all cases, at least 12 indentations were made and the results were then averaged to reduce the noise of the load-depth curves. In this study, the surface effects could be neglected as the data were acquired at depths greater than 2000nm.

Results and Discussion

Figure 1 shows a representative SEM micrograph of the microstructure of the SnAgCu and composite solder samples. As observed in both composite and unreinforced solder systems, there was presence of intermetallic compounds uniformly distributed throughout the solder matrix. These intermetallic phases were identified using the EDS (see Figure 1(b)) and the phases present in the solder matrix are namely: β-Sn, Ag_3Sn and Cu_6Sn_5.

For solder materials synthesized using the powder metallurgy route, the Ag_3Sn and Cu_6Sn_5 intermetallic compounds exhibited granular morphology as shown in Figure 1(a).

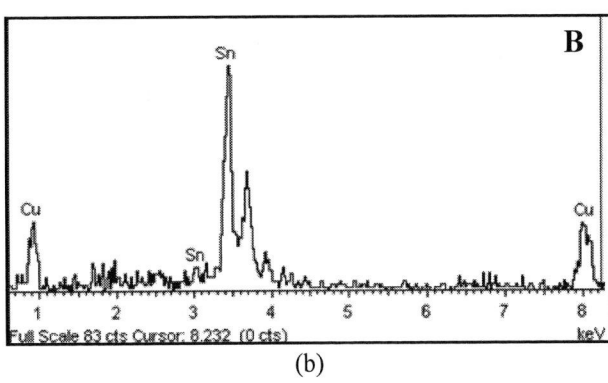

(b)

Figure 1. (a) Representative SEM micrograph and (b) EDS analysis showing the presence of different phases in the SnAgCu solder. (Presence of Au phase resulted from sample coating).

For the composite solder samples, a fair degree of uniformly distributed Ni-CNT clusters was observed throughout the solder matrix. The length of the Ni-CNTs used in the present study is ~ 30 μm. Their high aspect ratio and the strong van der Waals forces caused the Ni-CNTs to attract one another. Thus Ni-CNTs have a tendency to entangle together, resulting in Ni-CNT clustering rather than homogenous dispersion. Figure 2 shows the representative FE-SEM micrograph of the presence of Ni-CNTs on the fracture surface of the composite solder.

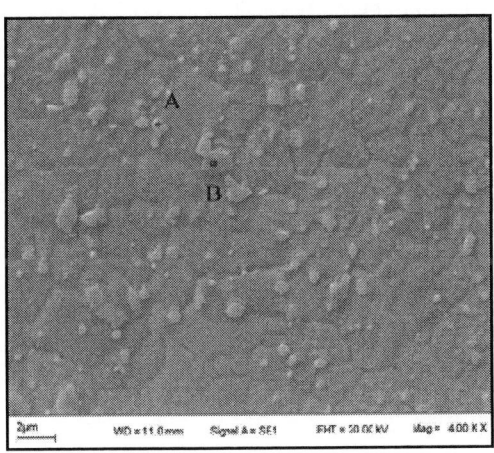

(a)

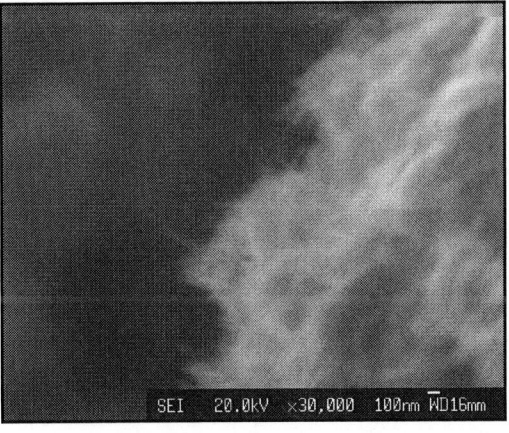

Figure 2. Representative FE-SEM micrograph showing the presence of Ni-CNTs on the fracture surface of the composite solder.

The results of ambient temperature tensile tests revealed that the average strength values improved with the addition of 0.05 wt.% of Ni-CNTs (see Figure 3). An improvement of ~13% for 0.2% yield strength (YS) and ~15% for ultimate tensile strength (UTS) was observed when compared to that of the unreinforced SnAgCu material.

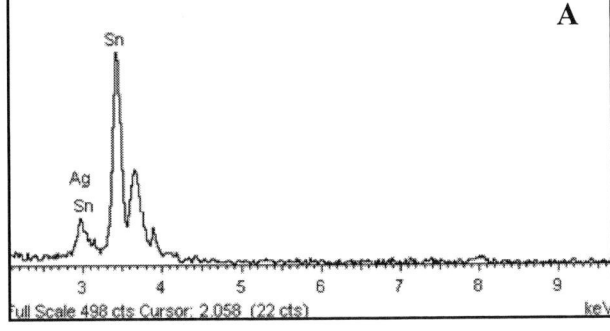

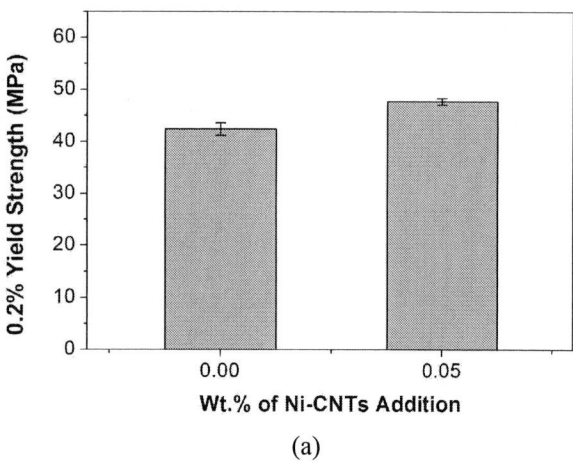

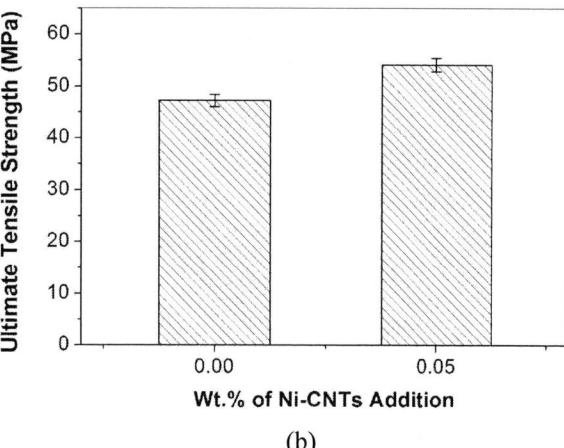

Figure 3. Effect of Ni-CNTs on the (a) 0.2% yield strength and (b) ultimate tensile strength of the solder samples.

The improvement in strength of the composite solders could be attributed to: (a) the progressive increase in dislocation density due to coefficient of thermal expansion (CTE) mismatch ($\Delta\sigma_{CTE}$) [21] between SnAgCu and Ni-CNT, and (b) the elastic modulus (EM) mismatch ($\Delta\sigma_{EM}$) between solder matrix and Ni-CNT. The strength of a reinforced matrix can be defined by [21]:

$$\sigma_{my} = \sigma_{mo} + \Delta\sigma \tag{1}$$

where σ_{my} and σ_{mo} are the yield strength of the reinforced and unreinforced matrix, respectively. $\Delta\sigma$ represents the total increment in yield stress of the reinforced SnAgCu matrix and is estimated by [22]:

$$\Delta\sigma = \sqrt{\left(\Delta\sigma_{CTE}\right)^2 + \left(\Delta\sigma_{EM}\right)^2} \tag{2}$$

The presence of Ni-CNTs as reinforcements (see Figure 2) acts as obstacles to hinder the initiation of dislocation motion in the solder matrix. Therefore, a higher initial stress is required.

Figure 4 shows the typical indentation load-depth curves of the SAC and SAC/0.05wt.%Ni-CNT solder samples. Both the SAC and SAC/0.05wt.%Ni-CNT solder samples exhibited the typical indentation behavior of significant permanent deformation and insignificant elastic recovery,

which can be observed from the almost vertical slope of unloading curve. The absence of steps and discontinuities on the curves indicated that no cracks and fractures occurred during the indentation [23, 24].

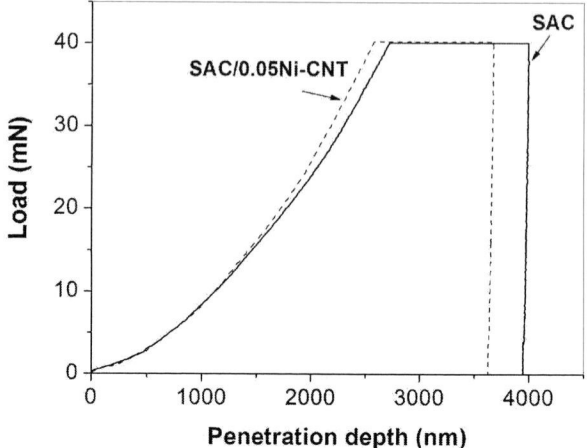

Figure 4. Typical indentation load-depth curves of SAC and SAC/0.05wt.%Ni-CNT solder samples.

As observed in Figure 4, there was a slight difference in the increase of penetration depths for the SAC and SAC/0.05wt.%Ni-CNT solders, as the load increased, just before subjecting to a holding duration at the maximum load. However, during the holding period, a more significant difference in the penetration depth was noted between the two materials. The creep process is operative during the holding time. A considerable amount of creep strain at the maximum load was found for all the samples. The holding segment at the maximum load is necessary to allow for the dissipation of creep displacement (which corresponds to the change in penetration depth under the maximum load for a holding time of 300s). The penetration depths of the SAC solder at the maximum load ranged from 2720 to 3990 nm (Figure 5). While, for the case of SAC/0.05wt.%Ni-CNT sample (see Figure 5), a smaller creep displacement was noted (penetration depths ranged from 2580 to 3670 nm). Hence, the composite sample reinforced with Ni-CNTs appeared to be more creep resistant than its unreinforced counterpart.. The rate of displacement (slope of the curve in Figure 5) during the holding period under constant load was observed to be high for the initial stage of holding segment but was observed to decrease with the increase in holding time.

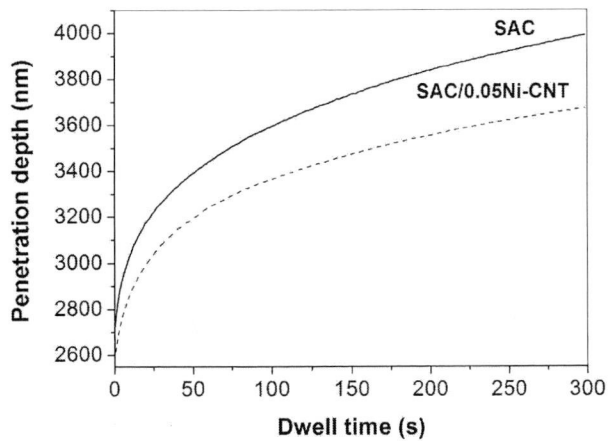

Figure 5. The change in penetration depth under holding time of 300s, at a maximum load of 40mN.

Conclusions

1. Sn-3.5Ag-0.7Cu based solder composite systems containing 0.05 wt. % of Ni-coated CNTs was successfully synthesized using the powder metallurgy technique.

2. Microstructural studies revealed uniformly distributed intermetallic phases throughout the solder matrix and EDS analysis identified the phases as Ag_3Sn and Cu_6Sn_5.

3. With the addition of Ni-CNTs, the tensile results revealed an improvement of ~13% in 0.2% yield strength and ~ 15% in ultimate tensile strength.

4. Nanoindentation creep results showed that addition of Ni-CNTs improved the creep resistance of the composite solder.

Acknowledgments

The authors acknowledge the support received for this research work from the Singapore Institute of Manufacturing Technology and NSFC (60806030). The authors would also like to thank Liu Yuchan from Singapore Institute of Manufacturing Technology for her help in the nanoindentation tests.

References

1. M. Abtew and G. Selvaduray, *Mater. Sci. Eng.: R: Reports*, vol. 27(2000), p. 95.

2. D. C. Lin, G. X. Wang, T. S. Srivatsan, Meslet Al-Hajri and M. Petraroli, *Mater. Lett.*, Vol. 57, No. 21(2003), p. 3193.

3. J. L. Marshall, J. Calderon, J. Sees, G. Lucey and J. S. Hwang, *IEEE Trans.* Components, Hybrids and Manufacturing Technology, Vol. 14, No. 4(1991), p. 698.

4. H. Mavoori and S. Jin, *JOM – Journal of the Minerals Metals & Materials Society*, Vol. 52, No. 6(2000), p. 30.

5. L. Wang, D. Q. Yu, S. Q. Han, H. T. Ma and H. P. Xie, *International Conference on the Business of Electronic Product Reliability and* Liability, Vol. 50(2004).

6. S. M. L. Nai, J. Wei and M. Gupta, *J. Electron. Mater.*, Vol. 35, No. 7(2006), pp. 1518-1522.

7. S. M. L. Nai, J. Wei and M. Gupta, *Proceedings of ASME IMECE 2006, International Mechanical Engineering Congress and Exposition*, Chicago, Illinois, USA, Nov. 2006.

8. S. M. L. Nai, J. Wei and M. Gupta, *Thin Solid* Films, Vol. 504, No. 1 – 2(2006), pp. 401-404.

9. S. M. L. Nai, J. Wei and M. Gupta, *Solid State Phenomena*, Vol. 111(2006), pp. 59-62.

10. P. Babaghorbani, S. M. L. Nai and M. Gupta, *J. Mater. Sci.: Mater. in Electronics*, Vol. 20(2009), pp. 571-576.

11. M. E. Alam, S. M. L. Nai and M. Gupta, *J. Alloys and Compounds*, Vol. 476, No. 1 – 2(2009), pp. 199-206.

12. S. M. L. Nai, J. Wei and M. Gupta, *Journal of Alloys and Compounds*, Vol. 473, No. 1 – 2(2009), pp. 100-106.

13. S. Iijima, *Nature*, Vol. 354(1991), p. 56.

14. L. Valentini, J. Biagiotti, J. M. Kenny and S. Santucci, *Compos. Sci. Technol.*, Vol. 63(2003), p. 1149.

15. C. Balazsi, Z. Shen, Z. Konya, Z. Kasztovszky, F. Weber, Z. Vertesy, L. P. Biro, I. Kiricsi and P. Arato, *Compos. Sci. Technol.*, Vol. 65, No. 5(2005), p. 727.

16. A. Mamedov, N. A. Kotov, M. Prato, D. M. Guldi, J. P. Wicksted and A. Hirsch, *Nat. Mater.*, Vol. 1(2002), p. 190.

17. S. I. Cha, K. T. Kim, S. N. Arshad, C. B. Mo and S. H. Hong, *Adv. Mater.*, Vol. 17(2005), p. 1377.

18. K.T. Kim, S.I. Cha, S.H. Hong and S.H. Hong, *Mater. Sci. Engin. A*, Vol. 430(2006), p. 27.

19. Y. Sun, J. Sun, M. Liu and Q. Chen, *Nanotechnology*, Vol. 18(2007), p. 1.

20. K. T. Kim, J. Eckert, S. B. Menzel, T. Gemming, S.H. Hong, *Applied Physics Letters*, Vol. 92(2008), p. 121901.

21. L. H. Dai, Z. Ling and Y. L. Bai, *Compos. Sci. Technol.*, Vol. 61(2002), p. 1057.

22. T. W. Clyne and P. J. Withers, An Introduction to Metal Matrix Composites, Cambridge University Press, 1993.

23. X. D. Li, H.S. Gao, W.A. Scrivens, D.L. Fei, X.Y. Xu, M.A. Sutten, A.P. Reynolds and M.L. Myrick, *Nanotechnology*, Vol. 15(2004), p. 1416.

24. B. Lim, C.J. Kim, B. Kim, U. Shim, S. Oh, B.H. Sung, J.H. Choi and S. Baik, *Nanotechnology*, "Components, Hybrids and Manufacturing Technology, Vol. 17(2006), p. 5759.

A Method to Produce Printed Circuit Boards with Embedded Semiconductors Using Stress Buffer Layers

Won Seo[1], Young-Mo Koo[2], Se-Hoon Park[3], Nam-Ki Kang[3], and Gu-Sung Kim[1]

[1]Kangnam University, Yongin 446-702, Korea
[2]EPWorks Co., Ltd., ESIP Lab., Gyeonggi 464-070, Korea
[3]Korea Electronics Technology Institute, Gyeonggi 463-816, Korea
E-mail : scanfs@naver.com, gkim@kangnam.ac.kr

Abstract

In the case of embedded chip type CSPs (Chip Scale Packages) made by embedding chips in PCBs (Printed Circuit Boards), problems occur due to differences in the CTE (Coefficient of Thermal Expansion) between the PCBs and the chips. This study tested a method to solve this problem by inserting thermal SBLs (Stress Absorbing Layers) into the embedded chips, thereby improving the reliability of the connection between the chip and the PCB. This study focused on a production method used to form a SBL on a Silicon Active Layer in order to make embedded PCBs and the material used as SBLs was absorbent in the polyimide family, which was composed of a material containing at least 30% polysiloxanes. For the experiment, two types of samples were used, including: 20mm × 20mm sized silicon active layers without any SBL formed on them; those with an SBL formed only on the top surface. The samples were produced and placed on the PCBs. To examine the effects of the physical damage, 3 point bending tests were conducted and the results were analyzed.

Introduction

Since the extremely small, lightweight electronic devices of today cannot use general forms of semiconductor packages, CSPs (Chip Scale Packages) in similar forms as those of semiconductor chips are made and used frequently. These CSPs include lead frame CSPs made using the copper or alloy 42 lead frames used as a main material of semiconductor packages. Generally used like as sealed with plastic or ceramic material as they are and other CSPs made using general substrates such as printed circuit board, both of which can be easily seen in the market now [1].

To satisfy continued demands for packages suitable to modern electronics, including mobile products, efforts to make the smallest and the most reliable semiconductor packages with an economically high performance are continuing unceasingly [2]. These demands are indispensable in the efforts to design parts with more input/out terminals and improved electric properties. The SOP (System-on-Package) was developed to meet those demands.

Basic structures used to make the SOP include semiconductor active element chips; passive elements that can drive the chips; and PCBs (Printed Circuit Boards) needed to make the SOP in the form of semiconductor packages [3]. Therefore, when seen from the viewpoint of semiconductor element chips, the chips are considered as being embedded in PCBs and thus, they are generally called embedded chips or embedded PCBs.

Fundamental problems occur in these embedded PCBs, because the CTE of the PCBs is quite different from that of the semiconductor chips. Thus, the reliability of the SOPs employing them becomes compromised due to the possibility of damages or burns resulting from the differences in the CTEs [4-5]. If the size of the semiconductor chips is reduced or the thickness of the semiconductor chips is increased, the sensitivity to damages will be remarkably reduced. However, in the case of multi-stack packages requiring semiconductor chip specifications of at least 7mm×7mm in size and less than 80 μm in thickness, the mechanical stress caused by differences between the CTE of the PCB and the chip affects the copper electric wiring connected between the external electrodes of the semiconductor chip and the PCB [6]. The added stress causes cracks or breakdowns to the wiring, thereby impairing the reliability of the joint part of the wiring. That is, if the chip is heated while being used, the chip and the board will expand. When the heat has been removed, the chip and the board will shrink. Since the chip and the board expand and shrink at different rates, they will impose stress on the joint part between the chip and the board, i.e., the copper wiring part. Moreover, during physical tests of produced and embedded PCBs, mechanical loads on the product may impose stress on the surface of the semiconductor chip that has become extremely thin, thereby causing cracks or disconnections on the chip. Finally, problems can arise from the lamination process, which is used to laminate, heat and pressurize insulating layers with a copper film layer on one or both sides of the PCB. After the process of inserting the embedded chip into the PCB during the lamination process, if the multiple attached embedded chips are not even or have local prominences and depressions, or if the local balance of the lamination roller is not accurate, the surfaces of the thin semiconductor chips will be damaged. Thus, the chips may be spoiled during the process. These hidden factors will be revealed during the period of reliability tests, thereby compromising the reliability of the product [7-8].

Various attempts have been made to solve these problems and in general, there were cases where stress buffer layers were composed in general package technologies. However, no case can be found where buffer layers were applied to chip embedding process technologies. Therefore, at this moment, embedded PCBs are made and applied without any absorbing layer when chips are small or entire chips are thick. In the case of semiconductor elements, in which the chips are becoming larger and thinner, the residual stress or deformation in the edges of the chips increases greatly compared to the central part. This result indicated that the prevention of disconnections between semiconductor chips

978-1-4244-4658-2/09 $25.00 © 2009 IEEE

and metal parts of the PCBs in existing structures was not sufficient [9-10]. This research study focused on a method to produce embedded chips and PCBs with new processes and improved connection reliability, especially between the chips and the boards, as a way to solve those problems. In this study, embedded chips' thermal SBLs were used to improve the reliability of the joint parts between the semiconductor chips and the PCBs (for example, reliability of the copper wiring part), and a method to produce them was tested. To that end, 20mm × 20mm sample chips with copper wiring as Daisy-Chains were produced.

As explained earlier, the PCB with embedded semiconductor chips that had thermal SBLs produced according to this study improves the reliability of the joint part considerably. Moreover, the SBL made according to this study will absorb or discharge various kinds of stresses in printed circuit production processes, such as thermal stress, when installing produced packages and when using them for a long time. Therefore, the lifetime of PCBs will be extended and the lifetime of electronic products, such as mobile phones, will also be extended.

Experimental Method

In order to examine the performance of PCBs with thermal SBLs, the chips that would be embedded were produced first. The produced chips had copper wiring as Daisy-chains in order to monitor chip damage, and the damaged copper wiring served the role of electric conduction when the chips were embedded in the PCBs. The chips to be embedded were made by forming SiO_2 films on silicon boards and sputtering on the films using Ti/Au. After that, the designed copper wiring pattern was affixed on the chips, and 4 μm-thick virtual Daisy chains were formed on the chips through Cu plating. The Ti/Au around the Daisy-chains was then removed through ion milling. After that, the back surfaces of the chips were ground thin to 150 μm, and were cut into 20mm × 20mm sized chips by sawing them into the PCBs. Figure 1 shows the production processes used to make the chips to be embedded in the PCBs. Fig. 2 shows the pattern formed on the chips, which was used to monitor phenomena occurring while the chips were being embedded in the PCBs.

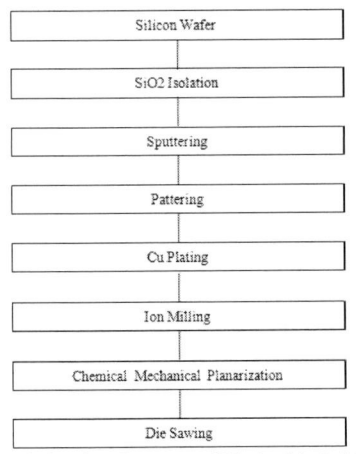

Fig. 1 Fabrication Process of Embedded chips

Before the produced chips are embedded in the PCBs, thermal SBLs will be formed on their front and back surfaces. The thermal SBL to be applied between the semiconductor active circuit layer and the PCB was composed of materials that can absorb 5 μm or more of thermal stress. The thermal SBL is to be applied to the embedded semiconductor chips and the outer area of the cavity of the PCB, under the part where the semiconductor chips have been embedded. The layer that can absorb thermal stress was composed of a material in the polyimide family containing at least 30% polysiloxane, which enables photo imaging and is easily processed by laser.

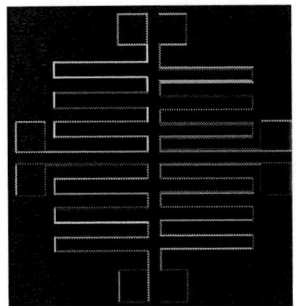

Fig. 2 Daisy-chains Pattern

Table 1 Material Properties of Thermal Stress buffer layer

Material	Silicone
Residual Stress(MPa)	6.4
Modulus(GPa)	0.37
Dielectric Constant(1Khz)	3.1
Dissipation Factor(1KHz)	0.002
Moisture Uptake(%)	<0.2

$$\kappa = \frac{6h_f(1-\nu_s)E_f}{h_s^2(1-\nu_f)E_s}(\alpha_f - \alpha_s)\Delta T \qquad (1)$$

Basically, the warpage occurring in the joint part between the chip and the PCB can be expressed as the modulus of the thermal SBL applied on the chip, multiplied by the CTE [11]. As shown in Equation (1), E_f represents the modulus and α_f the CTE. To minimize warpage, the thermal SBL used in this study was selected among materials with a very small modulus. Table 1 shows the properties of the material used as the thermal SBL [11].

In order to examine the effect of the thermal SBL on the chip that is actually embedded in a PCB, three samples were produced. Fig. 3 is a simulated photo of the two kinds of samples made differently by thermal SBLs. (a) is a photo of a chip in the same form as that of existing PCBS, without any thermal SBLs. (b) is a form of PCBs with a thermal SBL on the top surface with copper wiring.

978-1-4244-4658-2/09 $25.00 © 2009 IEEE   469

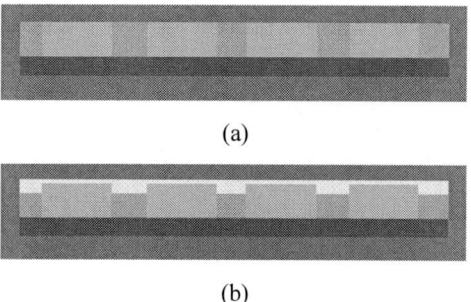

(a)

(b)

Fig. 3 Structure of SBLs:(a) PCBs without any thermal SBLs.
(b) PCBs with a thermal SBL on the top surface

The produced samples were embedded in the PCB through the lamination processes. During the lamination processes, the chip and the PCB were joined together under the temperature of 200~250°C. In this process, the chip may become damaged and voids may be made in the joint part. Since, in the course of embedding the chip in the PCB, a process to laminate and pressurize other material on both sides of the insulating layer is carried out, SAT(Scanning Acoustic Tomograph) to photos of the embedded chip installed in the PCB were taken to examine the effect of thermal SBLs on the embedded chip.

If an external physical impact is imposed on the embedded PCB, the impact will not be directly transmitted to the embedded chip, but the thermal SBLs will serve the role to mitigate the physical impact. To examine this process, an experiment was conducted, using 3 point bending tests, to investigate the degree that the thermal SBLS disperse and mitigate the impact. The results were analyzed.

Results

In the course of installing the embedded chip in the PCB, after going through the lamination process, the state of the joint part of the chip embedded in the PCB was identified using SAT photos. 10 SAT photos were taken of each type of 25mm × 25mm embedded PCBs, including those without any thermal SBL, those having a thermal SBL on the top surface. The photos were analyzed as shown. Consequently, it was evident that joint parts were more properly formed without having cracks or voids in the embedded chips with thermal SBLs than in the embedded chips without any thermal SBL. The 10 embedded PCBs, made for each of the different embedded chips, were checked with SAT photos and based on the result, eight pieces out of 10 were defective in the case of the embedded chips without any thermal SBL. In the case of those chips with a thermal SBL formed on the top surface, four pieces out of 10 were defective. In the case of the embedded chips without any thermal SBL and those having thermal SBLs formed on the top surface, a difference of 40% was shown.

Table 2 Failed rate of Embedded PCBs

	Any thermal SBL	SBL formed on the top surface
No. of Samples	10	10
Failed PCBs	8	4

The defects occurring between the embedded chips and the PCBs were in fact voids caused by an imperfect joining of the two layers and cracks on the chips. Those problems can be identified in Fig. 4. In the case of (a), it is evident that no void or crack on the chip occurred, and the embedded chip and the PCB were completely joined together. In the photo under (b), the part shown in black indicates the occurrence of voids. From this, it can be identified that, although the embedded chip was installed in the PCB, the chip and the PCB were not completely joined. In the case of (c), it is evident that cracks occurred on the chip due to the impacts imposed on it during the lamination process.

Through the analyses of the SAT photos, the problematic occurrence of voids, caused by improper joining of the embedded chips and the PCB while the embedded chips were being installed in the PCBs, and the problem of cracks occurring in the chips due to the impacts occurring during the lamination process, could be identified. It was evident that embedded chips with thermal SBLs were more completely joined by at least 40% than those without any thermal SBL.

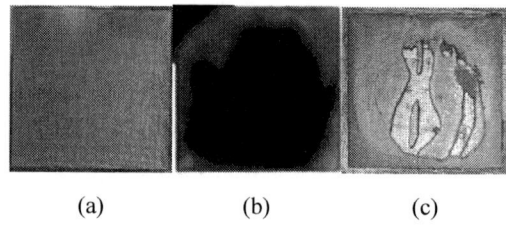

(a)   (b)   (c)

Fig. 4 Scanning Acoustic Tomograph of embedded PCBs:
(a) No error (b) Void (c) Crack

In the course of the above-mentioned experiment, 3 point bending tests were conducted to examine the degree to which the thermal SBL would absorb and mitigate exogenous, physical impacts imposed on a completely embedded chip in a PCB.

The 3 point bending tests were conducted on embedded PCBs in which 20mm × 20mm embedded chips without any thermal SBL, those with a thermal SBL on the top surface were all completely embedded. SAT photos were taken.

The results of the 3 point bending tests can be seen in Fig. 5. Fig. 5 shows the stress given to embedded chips, as forces are imposed in a red-line graph. Fig. 4(a) shows the result of a 3 point bending test conducted on an embedded chip without any thermal SBL. In the graph, the point at which the red line is the highest is the time point at which the embedded chip was completely destroyed. The values of the flexure stress and flexure strain at this time point can be accurately seen through Fig. 4(a). The time taken for the chip to be completely destroyed was 114sec and the load imposed was 25.8N. The flexure stress given to the chip was 290.6Mpa, flexure load was 2.6kgf. (b) is the result of the test conducted on an embedded chip with a thermal SBL formed on the top surface. The point at which the red line in the graph is the highest is the time point at which the embedded chip was completely destroyed. As shown in Fig. 4(b), the time taken for the chip to be completely destroyed was 129sec and the load imposed was 27.9N. The flexure stress given to the chip was 313.7Mpa and the flexure load was 2.8kgf.

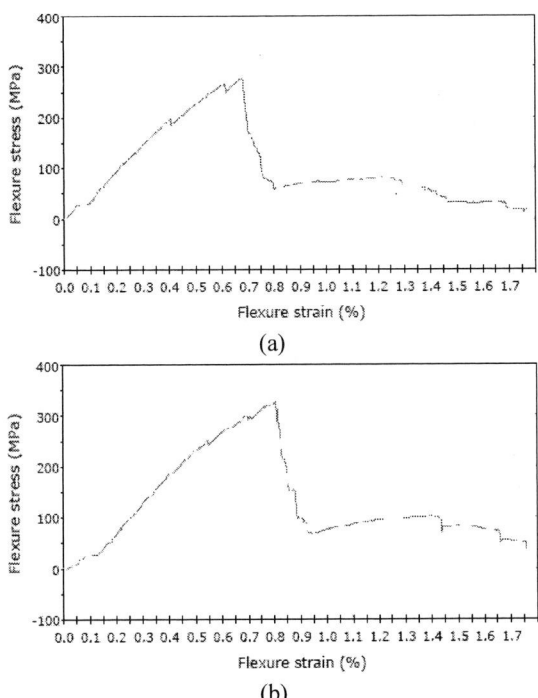

(a)

(b)

Fig. 5 3 point bending test graph: (a) an embedded chip without any thermal SBL (b) an embedded chip with a thermal SBL formed on the top surface

The 3 point bending tests showed that when external, physical impacts are imposed, embedded chips with thermal SBLs mitigate and absorb the external impacts better compared to embedded chips without any thermal SBL. The results show that an embedded chip with a thermal SBL only on the top surface can endure around 23Mpa higher for flexure stress and around 0.2kgf higher for flexure load than an embedded chip without any thermal SBL.

Conclusions

This study attempted to find solutions to minimize the physical impact imposed on embedded chips when installing the embedded chips in PCBs. The study also found a method to help minimize the voids and cracks in the joint area between each chip and each PCB caused by differences in the CTEs using thermal SBLs. Based on the results of the experiments conducted, improved joint areas between the chips and the PCBs could be identified through SAT photos of installed embedded chips. The embedded chips with thermal SBLs formed on the top surface showed 40% higher reliability compared to embedded chips without any thermal SBL. 3 point bending tests were conducted to examine the degree to which the thermal SBL would absorb and mitigate impacts when exogenous, physical impacts are imposed on an embedded chip. The result showed that an embedded chip with a thermal SBL only on the top surface would endure around 23Mpa higher for flexure stress and around 0.2kgf

higher for flexure load than an embedded chip without any thermal SBL.

References

1. Rao R. Tummala, Fundamentals of Microsystems Packaging, Mc-Graw-Hill (New York ,2001), pp. 24-26.
2. K. Takahashi, M. Sekiguchi, "Through Silicon Via and 3-D Wafer/Chip Stacking Technology," *VLSI Circuits 2006 Digest of Technical Papers*, pp.89-92.
3. William D. Callister, Jr., Materials Science and engineering:an introduction, Wiley (New York, 1994), pp. 646-655.
4. Bongtae Han, Yifan Guo, "Determination of an effective coefficient of thermal expansion ofelectronic packaging components: a whole-field approach," *Components, Packaging, and Manufacturing Technology, Part A, IEEE Transactions*, (1996), pp.240-247
5. Suhl, D., "Thermally induced IC package cracking," *Thermal Phenomena in Electronic Systems*, 1990. I-THERM II., *InterSociety Conference*, pp53-56.
6. Draney, N.R. Jun Jun Liu Jiang, T., "Experimental investigation of bare silicon wafer warp," *Microelectronics and Electron Devices*, 2004 IEEE Workshop, pp120-123.
7. Suga, T. Takahashi, A. Saijo, K. Oosawa, S., "New fabrication technology of polymer/metal lamination and itsapplication in electronic packaging," *Polymers and Adhesives in Microelectronics and Photonics*, 2001. *First International IEEE Conference*, pp.29-34.
8. Chang-Chun Lee Hsin-Chih Liu Ming-Chih Yew Kuo-Ning Chiang, "3D structure design and reliability analysis of wafer level package with bubble-like stress buffer layer," *Thermal and Thermomechanical Phenomena in Electronic Systems*, 2004. *ITHERM '04. The Ninth Intersociety Conference*, pp. 317-324.
9. L. Marton, K. Lark-Horovitz, Vivian A. Johnson, Solid State Physics, Academic Press (New York, 1959), pp. 307-321.
10. Rajsuman, R., "Testing a system-on-a-chip with embedded microprocessor," *Test Conference Proceedings. International*, 1999 , pp. 499-508.
11. Dow Corning, "Wafer Level Packaging Materials & Processes", SEMI KOREA Tutorial, (2009).

Study of Tungsten Metallization Surface States for Multilayer Ceramic

Wenjuan Zhang, Huajiang Jin
Hebei Semiconductor Research Institute
P. O. Box 179-41, Shijiazhuang Hebei, P. R. China
E-mail: chinapackage@163.com
Phone: 0311-87091842

Abstract

Ceramics are widely used in high reliable microelectronic packaging for their good electricity, thermal, mechanical characteristics and dimension stability [1]. Metallization of multilayer ceramic is one of the key techniques affecting the packaging quality and reliability. Metallization conductor should have appropriate resistivity, compact structure and can be co-fired with ceramic etc. There are many reports about Mo-Mn metallization technology, but rare about multilayer ceramic tungsten metallization especially about its surface states [2]. Surface states of metallization affect the adhesive strength, airtightness character and reliability of nickel-plating directly. The tungsten metallization surface state of three pastes was analyzed with SEM and EDS. Study the influence of viscosity, printing parameters, conductor thickness, sintering cycle etc. In order to analyze the effects of microstructure to the performance and surface quality of products, metallization surface states of every product was observed. The experiments indicated that two of the three pastes have alumina grains in the metallization surface after printing. According to the printing parameters, the printing process influences the surface state but doesn't have obvious effect on grains. The amounts of grains decrease with the increase of conductor's thickness. After fired, the ceramic grains in the conductor grow up apparently. The results indicated that adhesion of the metallization can be controlled by adjusting the sintering temperature and soaking time. For the ceramic grains exists in the nickel layer, the nickel layer has many impurities which affect the roughness of nickel-plating apparently. In a word, control metallization state and optimize process parameters are very important to the quality and reliability of products.

1. Introduction

Ceramics have good electricity property, thermal behavior, mechanical property and dimensional stability. For multilayer ceramics can meet the requirements of miniaturized, high density, multifunction, high reliability, high speed, high power et al, they have been widely used in high reliable packages. Alumina ceramic has high mechanical strength, good electric properties, for example low dielectric in high frequency, high volume resistance and insulation resistance et al, so they are widely used in ceramic packages and have yield good economic performance. At the same time, the metallization of alumina have attained quick development, but almost all the metals are Mo-Mn, W metallization is rare reported. The surface state and microstructure of tungsten metallization not only affect the bonding strength and airtightness but also the surface state and reliability of plating.

Alumina metallization samples were prepared with co-firing process in this paper, the influences of screen printing parameters and sintering conditions to the surface states were studied. The influence of metallization microstructure to the nickel plating was analyzed with SEM and Energy spectrum.

2. Experiments

2.1 Samples Preparation

Print the samples with three different pastes which have different viscosity. Every paste prints three thickness of metallization film then write down the printing parameters. After printing, the green sheets were fired at the condition of 1530~1630°C while N2:H2=1:4. After fired, the metallization surface was plated nickel for the metal brazing. The fabrication process is as follows.

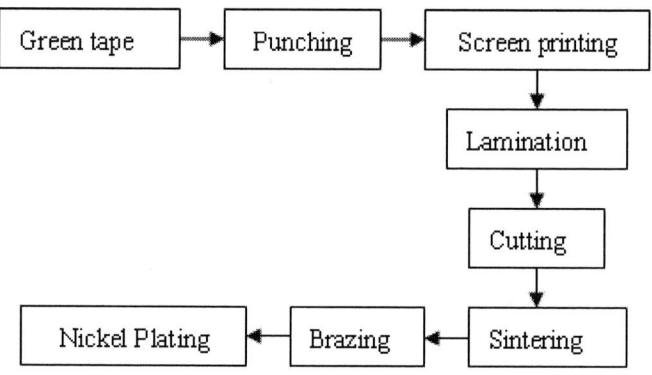

Fig. 1 The fabricating process of samples

2.2 Performance measurement

Measure the thickness of the nickel plating with X ray-thickness tester, monitor the dimensional precision and thickness of the metallization lines. Analyze the samples with SEM and Energy spectrum.

3. Results and Discussion

3.1 The influence of different printing parameters to the dimensional precision and roughness

Monitor the thickness of the metallization with laser-thickness tester. Change one parameter while fixed the others, the results are shown in Table 1.

The results show that the thickness of the metallization is increasing with the screen height in certain extent but the dimensional precision is reduced, because the lifting of the screen height has increased the distortion of the screen. At the condition of the screen height 2.0mm, comparing the influence of print speed and squeegee pressure to the roughness. From Table 2, we can see that increase the squeegee pressure will enhance the roughness of metallization, because increased pressure has deepened the impression of the screen [3-4].

978-1-4244-4658-2/09 $25.00 © 2009 IEEE

Table 1 The influence of the screen height to the dimensional precision

Projected Dimension Screen Height	Thickness and precision(μm)			Width and precision(μm)		
	10	15	20	0.1	0.3	0.5
2.0	±0.2	±0.3	±0.5	-0.004	-0.009	-0.012
2.5	±0.8	±0.9	±1.0	-0.007	-0.011	-0.014
3.0	±1.2	±1.2	±1.5	-0.009	-0.015	-0.019

Table 2 The influence of the printing parameter to the roughness

Squeegee pressure (MPa) Print speed(mm/s)	1.8	2.3	2.7	3.0
30-50	0.48	0.45	0.58	0.63
50-70	0.40	0.33	0.46	0.54
70-90	0.59	0.55	0.69	0.79

The samples in Fig. 2 are two samples with different metallization thickness, from the results we can see that with the increase of metallization thickness the impurity at surface are obviously reduced. Analyzed the impurity with energy spectrum the results indicated that the impurity are alumina which floated from the paste for low density.

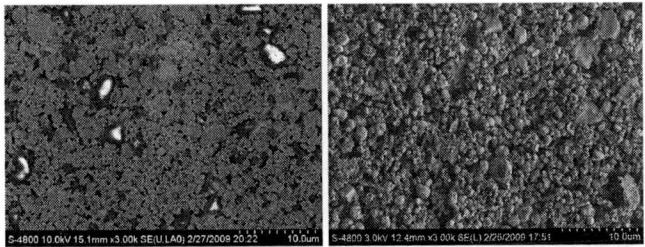

Fig. 2 The surface state of metallization with different thickness

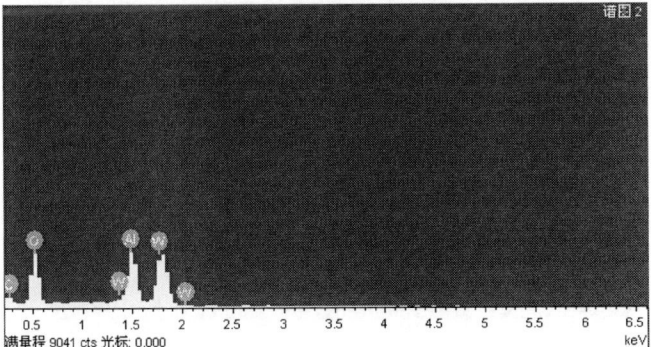

Fig. 3 The energy spectrum analysis of impurity in the metallization

3.2 The influence of firing condition to the microstructure and adhesion

Prepared the samples at the optimizing condition, then fired with the different sintering schedule. After fired the grains of alumina grow apparently. The sintering temperature is increased from number 1 to 4 samples. Test the adhesion of samples sintered at different temperature. The results are as follows:

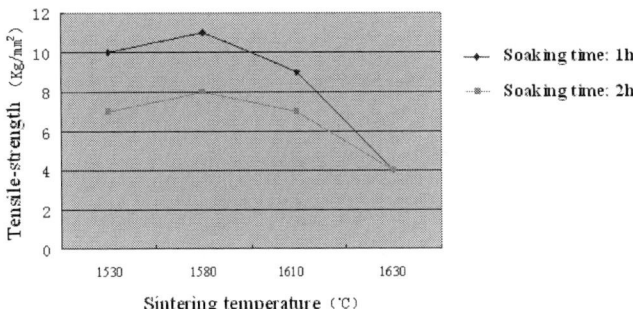

Fig. 4 The influence of sintering condition to the tensile-strength of metallization

Fig. 4 shows the variation of tensile-strength to sintering temperature and soaking time. At 1580°C the tensile-strength is best, but obviously decreased at 1630°C. The sintering process of metallization is a complex physicochemical process, during that process the tungsten powder formed a porous cavernous body structure, while the inorganic addition agent formed new glass phase filling the porous structure. The glass phase in the ceramic was infiltrated into the porous structure for the capillary action, then the metallization was combined hard with the ceramic for the glass phase solidified [5-6]. That's why the tensile-strength of co-firing ceramic metallization is better than after burning metallization. From the results, we can see the osmotic at 1530°C is not obvious, while the temperature reach 1630°C crystal grain of tungsten growth lead the porous cavernous structure disappeared. The adhesion was decreased sharply for the glass phase infiltrated the metallization was stayed in the interface between metallization and ceramic. Analyzed the micro-structure of samples number 2 and 4, results are as follows.

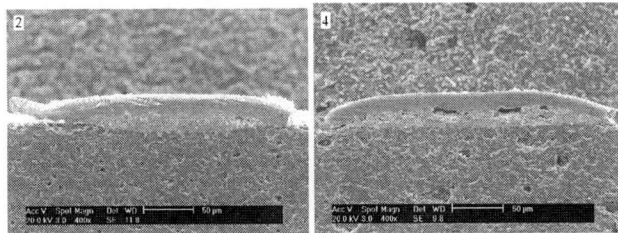

Fig. 5 The section plane of different samples

From the section plane, we can see the metallization states of number 2 are obviously better than number 4. The holes in the metallization of number 4 lead to the decrease of adhesion obviously. Although the samples of metallization is compact after printing but many holes generated after sintering for material transport. So the sintering condition is very important to the surface state and adhesion of the metallization.

3.3 The influence of metallization to the plating

The surface state of metallization directly determined the roughness, adhesion and reliability of plating. In this experiment we choose samples with different roughness, the results are as follows.

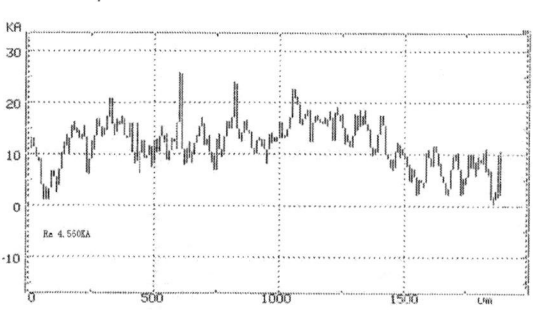

Fig. 6 The patterns of plating A

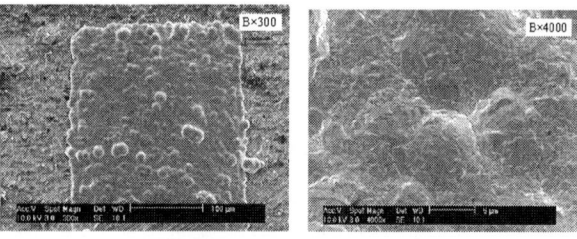

Fig. 7 The roughness of plating A

Fig. 6 shows the patterns of plating A, we can see that the surface state of plating is compact and homogeneous. The results show that the roughness of plating A is 0.4560KA.

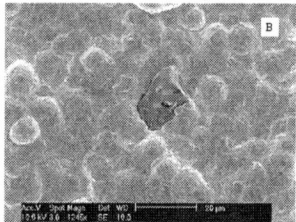

Fig. 8 The patterns of plating B

Fig. 8 shows the patterns of plating B, we can see that the surface state of plating is rough and loose; many small holes exist in the plating. From the patterns we can see the impurity which obviously influenced the dependable and roughness of plating. According to the results forward, the impurities are alumina grains. Analyzed the impurity with energy spectrum, results are as follows.

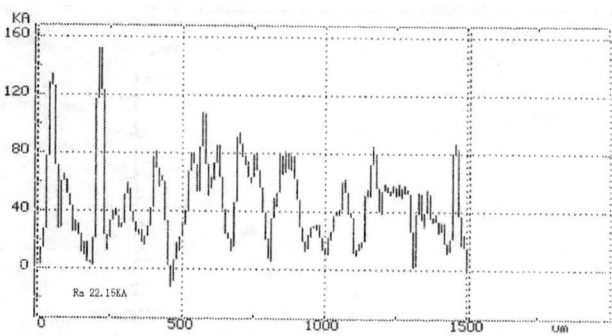

Fig. 9 The roughness of plating B

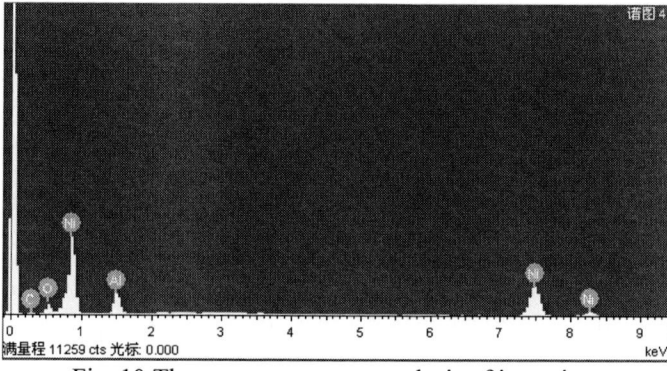

Fig. 10 The energy spectrum analysis of impurity

The results show that the roughness of plating B is 22.15KA, which is obviously worse than plating A. From the energy spectrum analysis results we can see the impurity is alumina, which exists in the tungsten paste. This is because the grain of alumina growth exceptional at the illegitimate sintering condition.

4. Conclusions

(1) The influence of printing parameters to the metallization surface state is obviously, regulating the parameters can reduce the roughness to 0.33μm. Increase the thickness can reduce the alumina grains in the surface obviously.

(2) The influence of firing condition to the patterns and adhesion of metallization is obviously, the results show that the variation of adhesion is affected by osmotic capacity of the glass phase from ceramic to metallization.

(3) At the same condition, the surface state and reliability of plating is obviously affected by the state of ceramic and metallization, the alumina grains in the metallization have obviously increase the roughness and reduce the reliability.

References

1. Juxian Zhang, "Microstructure of metallized aluminum ceramics applied in vacuum electronics components of domestic and oversea," *Vacuum Electronics*, Vol. 4, No. 24 (2003).

2. The 13th electronic components conference papers of Chinese electronic institute, 6, 2004, pp.39.

3. T.-X,W. Z. Sun,L.-D.Wang, Y. H. Wang, "Effect of Surface Energies on Screen Printing Resolution," *IEEE Transactions on Components, Packaging, and Manufacturing Technology-Part B*, Vol. 2, No. 19 (1996).

4. R. W. VEST, "Materials science of thick film technology," *Am. Ceram. Soc. Bull.*, No. 65 (1986), pp. 631-636.

5. Mattox, D. M, "Role of Manganese in the Metallization of High Alumina Ceramics," *Amer. Ceram. Soc. Bull.*, Vol. 64, No. 10 (1985).

6. Longqiao Gao, "Secondary metallization technology of ceramic-metalpackage," *Electronic Technology*, Vol. 4, No. 23 (2002).

Solder Cracking Mechanism Correlation to Alloy Composition under Thermal Cycling Stress (Part 1)*

Shining Chang, Rocky Wang, Yu Xiang, and Gibson Chang
Mitac Computer, ShunDe, China
No. 1, Shunda RD, Lunjiao, Shunde, Foshan, Guangdong

Abstract

Interconnect reliability of Micro BGA in enterprise and portable electronics were studied. Two Micro BGA with similar dimension but different alloy compositions were used in the life testing to observe crack propagation mechanism under thermal cycling stress. Standard metallurgical x-section, dye penetration and SEM were performed at time zero as well as after thermal cycling on 1st pass Micro BGA-solder-PCB interconnect system as well as on reworked.

Solder crack propagation path along substrate-solder and solder-PCB pad interface were observed on both alloy systems, SAC305 and SAC125Ni. The stochastic trend of failure interface was rationalized with CTE variation and phase recrystallization in microstructure. Observation of crack initiation at grain boundary or triple-grain junction and recrystallization induced grain growth and coarsening were referred to earlier study. The study indicate that the Interconnect life under thermal cycling stress and cracking propagation were highly correlated to alloy composition but less related to 1st pass SMT reflow or low cycle reworked process. The crack initiation mechanism and second phase strengthening correlation to phase dispersion are in-depth discussed.

Introduction

Study on eutectic Sn-Pb and lead-Free SAC305 based on thermal cycling testing has demonstrate great value on determining the life time prediction in enterprise system and game console systems [1, 2]. Correlation of residual stress to fatigue life using 4-point bending and accelerated thermal cycling mechanism also show its engineering logic stringency for low strain rate loading induced fracture/crack propagation [3]. However thermal cycling has been found inadequate to address solder-Cu interfacial fracture mode induce from high loading speed impact load [4] which is the major source of energy in portable electronic user environment.

With Micro BGA usage in both enterprise computer system and portable electronic, interconnect reliability of component with alloy composition variation on a test vehicle is studied. Both thermal cycling for enterprise server and drop/impact test for portable electronic will be conducted to determine life expectancy in scopes of user environment. In this study of Part 1, a test vehicle with Mico BGA on U_EMMC location as shown in Fig. 1 was used to emulate a server products that functions for tele-signal control.

U_EMMC adopts BGA from two vendors: Supply A and Supply B. There are slight dimensional and configuration difference as shown in Table 1. The solder ball pitch was 0.50mm and ball diameter 0.30mm. The solder ball count was 169 I/O for Supply A and 153 I/O for Supply B. Thermal cycling test is conducted on the Mico BGA-test vehicle interconnect system to determine the life time variation due to package type and alloy composition.

Fig. 1 Test vehicle for Micro BGA study

Table 1 Packaging data comparison between Supply A's BGA and Supply B's BGA

Vendor	Package Size	Ball Diameter	Ball Pitch
Supply A	12mm×16mm×1.2 mm	0.30mm	0.50mm
Supply B	11.5mm×13mm×1.2mm	0.30mm	0.50mm

Experiment Procedure

BGA substrate pad applies ENIG. PCB pad applies LF-HASL. Pad size was 0.25mm and pad pitch 0.50mm with solder mask defined. Micro BGA from Supply A had a ball composition of SAC305, melting at 217°C and Supply B had a ball composition of SAC125Ni, melting at 227°C. Stencil applies 0.10 thick, 0.28mm squared aperture with round corners (45°). Type 4 solder paste of SAC305 was printed on the test vehicle. The two packaging BGAs use the same SMT profile and rework profile. Material property of PCB pad, paste solder, solder ball, BGA pad was listed in Table 2.

Table 2 Material property of PCB pad, paste solder, solder ball, BGA pad

Item	Supply A	Supply B
PCB Pad surface finish	LF-HASL	LF-HASL
Paste Solder Alloy	SAC305 (mp:217°C)	SAC305 (mp:217°C)
Solder Ball Alloy	SAC305 (mp:217°C)	*SAC125Ni* (mp: 227°C)
BGA Pad surface finish	ENIG	ENIG

In this experiment the two main factors that affect solder reliability are reflow profile and solder ball alloy composition. The experimental process flow is illustrated in Fig. 2.

978-1-4244-4658-2/09 $25.00 © 2009 IEEE

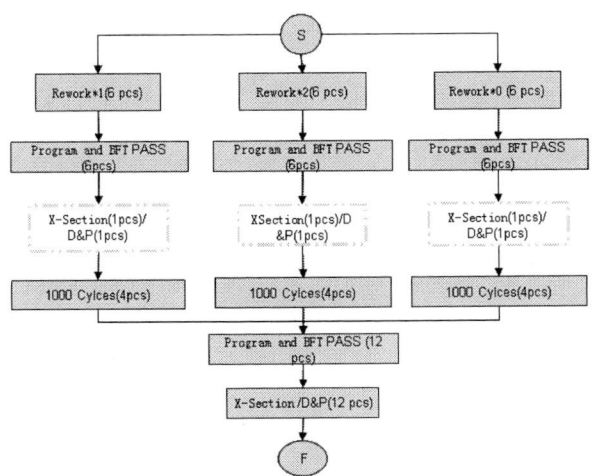

Fig. 2 Experiment Flow Chart

Table 3 Main Profile Parameters

Item	Peak Temperature	Time above 220°C	Time above 230°C
SMT Profile	244°C	60-66s	42-45s
Rework Profile	243°C	115s	62s

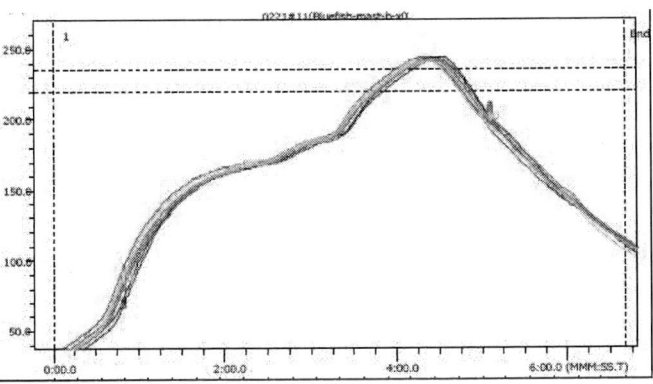

Fig. 3 SMT Profile

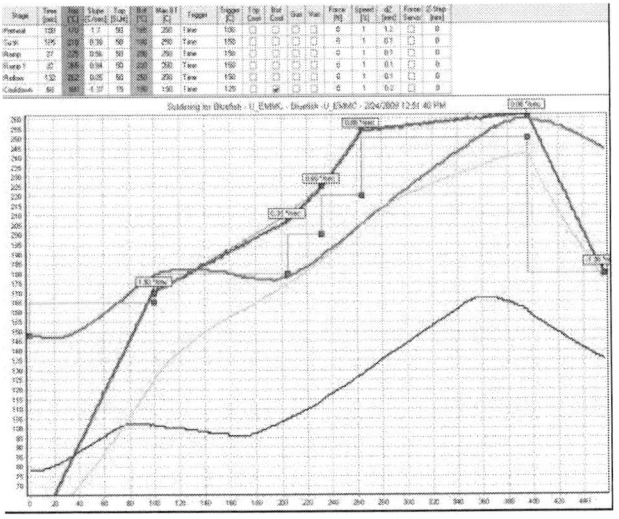

Fig. 4 Rework Profile

The reflow profile control variable, reflow vs. rework profiles are shown in Fig. 3 and 4. Table 3 lists the main parameters of these two profiles. To ensure good solder joint, test vehicles were verified by x-section and dye-pry before thermal cycling.

The following thermal cycling conditions were employed to evaluate the performance of the package. The 0~100°C temperature cycle with ramp rate 12°C/min and 10 min dwells, 1 cycle for 40min. Thermal cycling profile was shown in Fig. 5. Failure analysis method applies Dye-Pry and X-Section. To illustrate failure mechanism, SEM/EDS was also used.

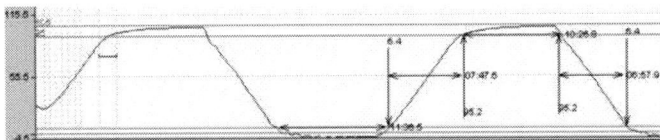

Fig. 5 Thermal Cycling Profile

Experiment Results

The time zero solders joint integrity before thermal cycling test is verified by x-section and dye penetration. X-Section results indicate that IMC thickness was less than 6 μm; copper thickness was between 20 μm and 30 μm; the solder joint height was from 206 μm to 230 μm as shown in Table 4. Dye and Pry didn't find any joint crack. Fig. 6 show X-Section photos for Supply A and Fig. 7 show X-Section photos for Supply B. Dye penetration photo for Supply A (Fig. 8) and Supply B (Fig. 9) indicate no obvious crack initiation and penetration at time zero.

Table 4 X-Section Data before Thermal Cycling

Status	PPID	Sample Name	PIN	IMC Thickness (um)			Copper Thickness (um)			Solder Joint Height (um)
				①	②	③	①	②	③	
Rework*0	000F	U_EMMC (Supply B)	A1	3.36	3.84	2.40	18.69	22.49	21.45	206.92
			A7	2.95	3.50	2.81	18.69	19.72	18.69	213.49
			A14	3.36	3.16	2.95	19.03	21.45	19.72	207.61
	001E	U_EMMC (Supply A)	A1	5.35	3.36	4.87	32.87	34.60	33.74	217.99
			A7	3.91	3.50	4.12	35.29	32.87	33.91	216.96
			A14	3.91	4.12	3.91	34.60	33.91	34.26	214.88
Rework*1	0004	U_EMMC (Supply B)	A1	3.16	3.16	2.81	18.69	20.76	20.76	220.07
			A7	2.95	3.36	2.81	16.96	19.38	16.61	225.61
			A14	3.57	3.50	3.29	18.69	20.76	17.65	219.72
	0019	U_EMMC (Supply A)	A1	5.69	5.35	5.21	28.03	31.83	29.07	212.46
			A7	5.42	5.14	5.42	29.07	30.80	30.80	210.38
			A14	5.76	4.60	5.01	28.03	29.76	29.76	210.38
Rework*2	000G	U_EMMC (Supply B)	A1	3.70	3.91	3.71	19.72	22.49	22.49	229.76
			A7	3.57	3.70	3.29	16.96	21.45	18.69	229.07
			A14	3.57	3.50	3.16	17.65	21.45	18.69	222.49
	001X	U_EMMC (Supply A)	A1	5.69	5.90	5.42	27.34	28.03	29.07	214.19
			A7	5.21	5.56	5.01	28.03	30.80	26.99	212.11
			A14	5.56	5.90	5.56	28.03	29.07	29.76	208.65

After 1000 ATC cycles, cracks were found on BGA from both Supply A and Supply B. There are two fracture modes at solder-BGA interface or solder-PCB interface. Supply B has both crack fracture modes and Supply A has the solder-BGA interface crack. By inspecting the crack size and crack counts, Micro BGA from Supply B has more severe crack propagation than Supply A as shown in Table 4. Both Fig. 10 show X-Section micrograph and Fig. 11 show Dye penetrant picture from Supply A Micro BGA indicate the crack initiation at corner of solder to Cu pad intersection then penetrate into the solder after thermal cycling repetitive stress. However both Fig. 12 X-section micrograph and Fig. 13 Dye & Pry picture indicate the interfacial fracture at solder to pad substrate and PCB Cu pad interfaces. After 1000 ATC cycles, the x-section results show the IMC thickness is less than 6 µm for Micro BGA from both Suppliers and cupper thicknesses are 20 µm and 30 µm respectively. Standoff height of Micro BGA are on the average 210 µm and 230 µm on average for Supply A and B as shown in Table 7 and 8.

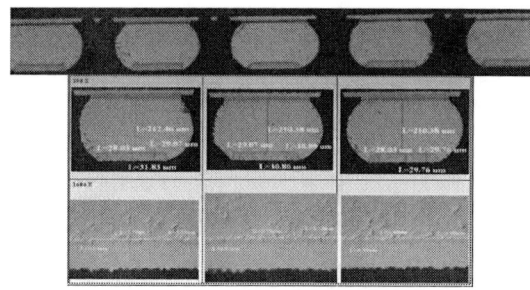

Fig. 6 X-Section photo from Supply A's BGA before thermal cycling (SN: 0019 –Rework*1)

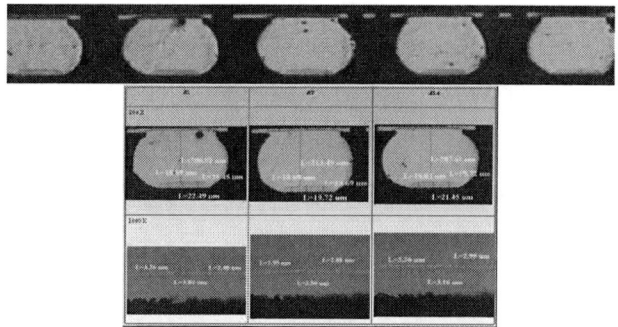

Fig.7 X-Section photo from Supply B's BGA before thermal cycling (SN: 000F –Rework*0)

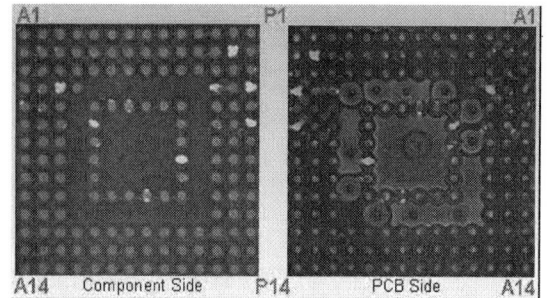

Fig. 8 Dye and Pry photo from Supply A's BGA before thermal cycling (SN: 000K –Rework*1)

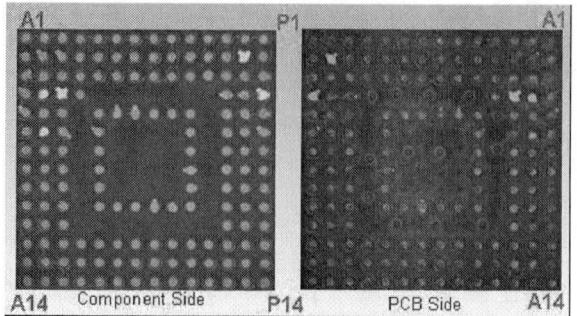

Fig. 9 Dye-Pry photo from Supply B's BGA before thermal cycling (SN: 0018 –Rework*2)

Table 5 Test result from Supply A Micro BGA after thermal cycling

SN	Test Stage	Rework	BFT	X-Section	Dye-Pry
001H	After T/C	0X	PASS	PASS	
001J		0X	PASS	1 crack 12% (less than 50%)	
001W		0X	PASS	\	PASS
001L		0X	PASS	\	PASS
001A		1X	PASS	PASS	
001M		1X	PASS	1 crack 17% (less than 50%)	
001C		1X	PASS	\	PASS
001Y		1X	PASS	\	PASS
001U		2X	PASS	PASS	
0015		2X	PASS	PASS	
0024		2X	PASS	\	1 crack 10%(less than 50%)
001G		2X	PASS	\	PASS

Table 6 Test result from Supply B Micro BGA after thermal cycling

SN	Test Stage	Rework	BFT	X-Section	Dye-Pry
000A	After T/C	0X	PASS	1 crack in 5 over 50%	\
000B		0X	PASS	4 cracks <50%	\
0008		0X	PASS	\	1 crack in 6 over 50%
000C		0X	PASS	\	6 cracks in 12 over 50%
0001		1X	PASS	4 cracks <50%	\
0006		1X	PASS	1 crack in 5 over 50%	\
000I		1X	PASS	\	1 crack in 2 over 50%
000E		1X	PASS	\	2 cracks in 9 over 50%
0005		2X	PASS	6 cracks<50%	\
000D		2X	PASS	4 cracks <50%	\
000J		2X	PASS	\	1 crack in 4 over 50%
0002		2X	PASS	\	1 crack in 5 over 50%

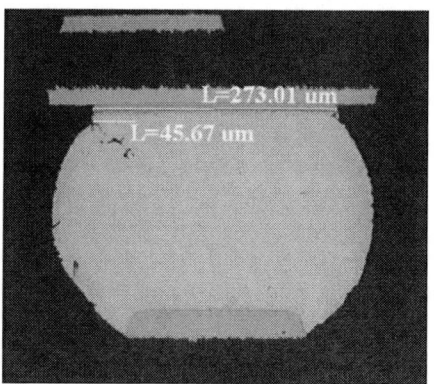

Fig. 10 X-Section photo from Supply A Micro BGA after thermal cycle

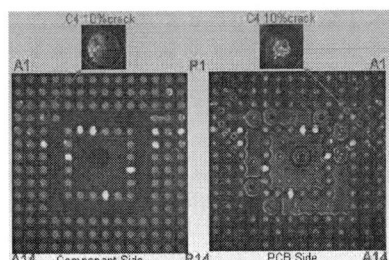

Fig. 11 Dye-Pry photo from Supply A Micro BGA after thermal cycle

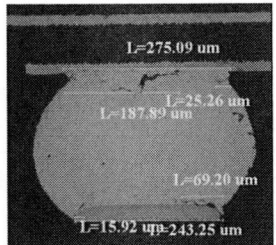

Fig. 12 X-Section photo from Supply B Micro BGA after thermal cycle

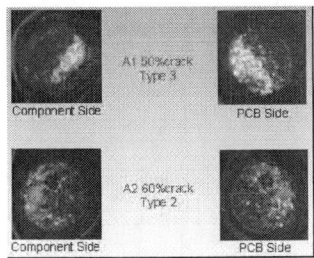

Fig. 13 Dye-Pry photo from Supply B Micro BGA after thermal cycle

Table 7 X-Section Data from Supply A Micro BGA after thermal cycle

Status	PPID	Sample Name	PIN	IMC Thickness (um)			Copper Thickness (um)			Solder Joint Height (um)
				①	②	③	①	②	③	
Normal	001H	U_EMMC (Supply A)	A1	3.57	4.05	2.95	31.53	30.56	29.33	213.65
			A7	4.12	3.70	2.81	32.60	31.60	30.60	212.73
			A14	3.95	4.05	3.50	31.83	32.87	31.23	218.24
	001J		A1	3.70	4.12	3.70	31.83	32.60	30.50	214.15
			A7	4.16	5.36	4.16	30.60	34.60	29.67	218.93
			A14	4.46	4.13	3.57	32.53	32.87	30.43	215.12
Rework* 1	001A	U_EMMC (Supply A)	A1	5.76	5.90	5.78	27.34	26.30	28.03	215.22
			A7	3.91	5.01	5.62	29.07	29.07	29.07	213.15
			A14	5.01	5.42	5.01	28.72	29.76	31.83	218.69
	001M		A1	5.42	5.90	5.01	26.99	27.34	24.22	216.96
			A7	5.56	5.56	5.76	29.07	27.34	29.03	216.96
			A14	5.21	5.36	5.01	26.30	26.30	25.26	219.72
Rework* 2	001U	U_EMMC (Supply A)	A1	5.42	5.56	5.21	28.05	25.95	29.07	211.42
			A7	5.35	5.35	5.90	26.30	28.03	29.07	213.15
			A14	5.97	5.21	5.42	26.90	28.03	29.07	219.03
	0015		A1	4.05	4.32	3.77	26.76	28.72	30.80	219.03
			A7	4.46	4.25	4.80	28.72	30.80	28.72	216.96
			A14	4.05	4.12	4.12	28.03	30.80	30.10	222.49

Table 8 X-Section Data from Supply B Micro BGA after thermal cycle

Status	PPID	Sample Name	PIN	IMC Thickness (um)			Copper Thickness (um)			Solder Joint Height (um)
				①	②	③	①	②	③	
Rework *0	000A	U_EMMC (Supply B)	A1	3.29	2.95	3.57	19.72	17.65	18.69	221.80
			A7	3.16	3.16	3.29	15.22	18.69	17.65	222.84
			A14	3.36	3.16	2.95	20.42	20.42	19.72	219.72
	000B		A1	3.57	3.36	3.50	19.72	18.62	20.76	220.76
			A7	3.16	3.16	3.36	16.61	16.61	18.69	225.26
			A14	3.50	3.50	3.50	20.42	20.42	19.72	222.84
Rework *1	0001	U_EMMC (Supply B)	A1	3.70	2.95	3.16	17.99	17.65	20.42	224.57
			A7	3.70	3.57	3.29	17.65	17.65	18.69	230.10
			A14	3.70	3.50	3.70	18.69	18.69	20.76	230.10
	0006		A1	3.50	3.50	3.16	22.49	17.65	21.45	217.65
			A7	4.25	3.91	3.50	18.69	20.76	21.80	220.76
			A14	4.12	3.84	3.91	20.76	18.69	22.49	221.80
Rework *2	0005	U_EMMC (Supply B)	A1	3.50	3.91	3.70	18.69	18.69	21.45	228.03
			A7	3.91	3.70	3.36	19.72	18.69	22.49	230.10
			A14	3.16	2.81	3.36	19.72	20.42	22.49	230.10
	000D		A1	3.57	3.16	3.91	18.69	19.72	20.76	225.26
			A7	3.50	3.50	3.36	20.42	17.65	16.96	231.14
			A14	3.70	3.91	3.57	18.69	19.38	20.42	230.10

Discussion

Metallurgical analysis on IMC thickness, Cu pad thickness and standoff height for both Micro BGA show that these after reflow and rework soldering material characteristics fulfill industrial specification.

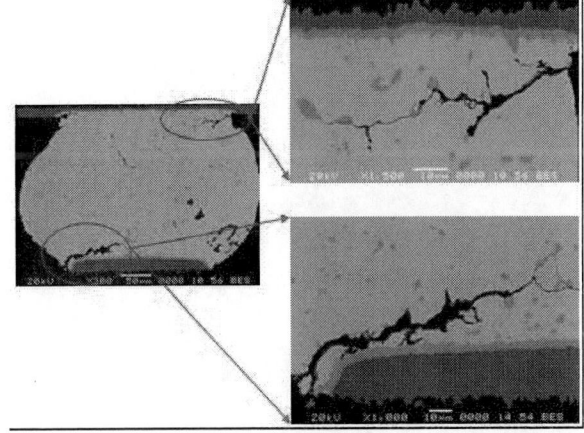

Fig. 14 Supply B Micro BGA solder crack configuration (SN: 0006)

The Fracture micrograph of Supply B Micro BGA show predominated interfacial fracture in nature in the beginning as well as in the crack propagation stage as shown in Fig. 14 as compare to inter-solder-fracture in Micro BGA from Supply A with higher Ag contain in SAC305. as shown in Fig. 15. The SEM dark vs. bright phase mapping images in Fig. 16 as well as EDX analysis indicate the Cu_6Sn_5 and Ag_3Sn two phases microstructure in the Sn-rich matrix. The crack initiation and second phase strengthening mechanism as hypothesized in our early 2000 study [6] as well as recent publication [7] related to recrystallization and grain growth also applied here.

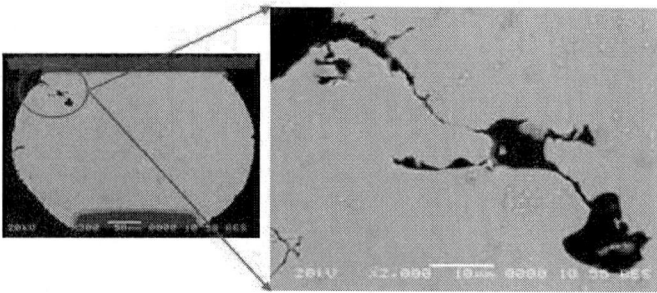

Fig. 15 Supply A Micro BGA solder crack configuration (SN: 001M)

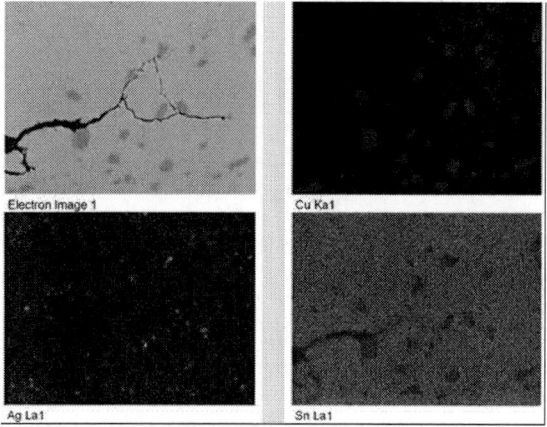

Fig. 16 SEM dark/bright mapping indicate crack propagation around IMC in the solder ball of Micro BGA from Supply B (SN: 0006)

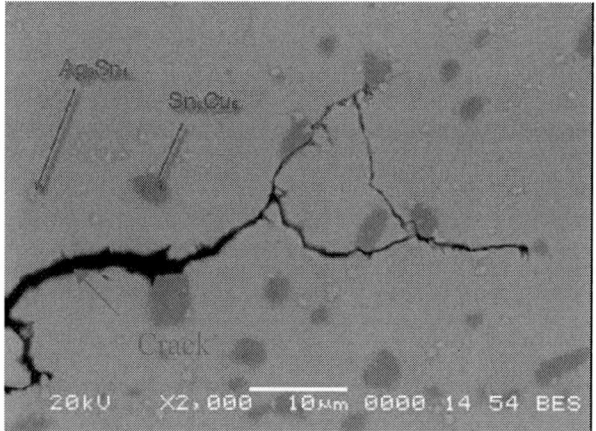

Fig. 17 IMC Phases in solder from Micro BGA Supply B

Crack initiation mechanism:

(1) The material property at corner of solder to Cu pad interface is the highest in stress concentration to serve as precursor for crack initiation. The CTE mis-match due to material property difference induced from thermal stress cause crack propagation as shown in Fig. 15 and 16.

(2) Second Phase strengthening:

The crack penetration path is depended on material strength distribution in the component and IMC interface. Crack tip penetration when near second phase can cause further stress concentration but the energy of concentration can be further released by recrystallization which induced grain growth and grain coarsening. As a result, crack initiated from grain boundary or triple-grain junction and then propagated. When the crack met Cu_6Sn_5 or Ag_3Sn phase, it steered clear of these IMC phase and propagated other ways as can be illustrated graphically by fracture in Fig. 17.

Figs. 18 and 19 represented the Ag_3Sn dispersion in the Micro BGA from Supply A with higher silver content at solder-substrate or solder-PCB pad interface as compare to Fig. 20 and 21 for Supply B. The dispersion density of second phase is about twice in Micro BGA from Supply A as compare to Supply B. Even through physical meaning of trace of Ni content in the alloy composition in Micro BGA from Supply B is not totally clear, the fatigue life and crack propagation under thermal cycling show that the joint strength in Sn-Ag-Cu based material is positively increase with silver content. The higher is the Silver content the higher the density of Ag_3Sn dispersed in the Sn-matrix so as let to higher reliability attribute to second phase strengthening under low deformation rate stress.

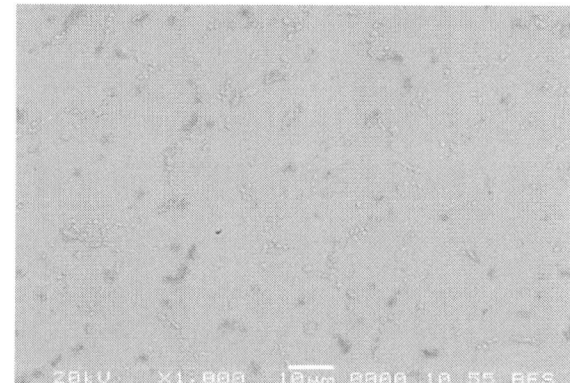

Fig. 18 Location near solder-BGA interface in Supply A Micro BGA solder ball (SN: 001M)

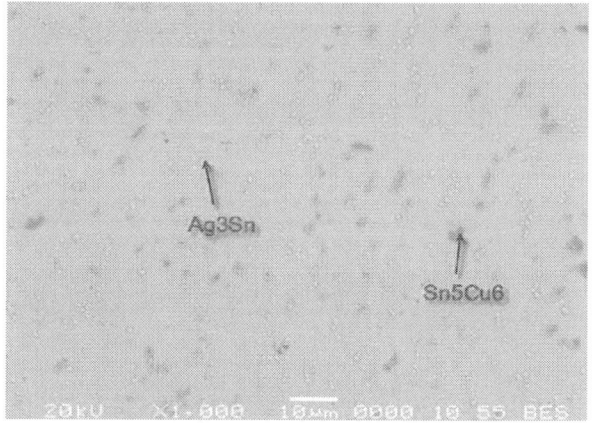

Fig. 19 Location near solder-PCB interface in Supply A Micro BGA solder ball (SN: 001M)

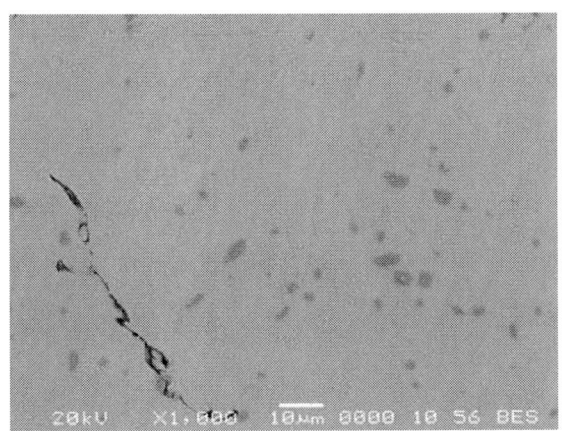

Fig. 20 Location near solder-BGA interface in Supply B Micro BGA solder ball (SN:0006)

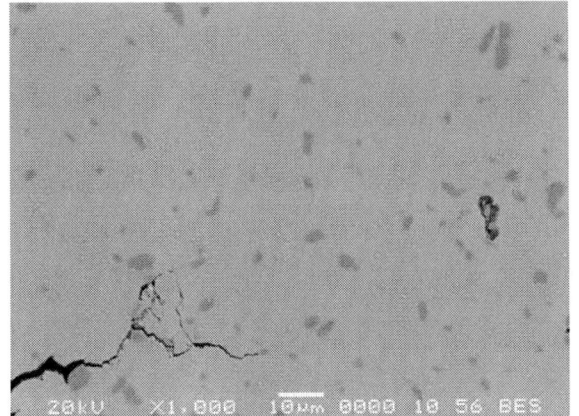

Fig. 21 Location near solder-PCB interface in Supply B Micro BGA solder ball (SN: 0006)

Conclusions

For low strain-rate loading thermal stress, the interconnect reliability of Micro BGA with higher silver content performed better. Even through the physical meaning of trace Ni content and package size variables need further study, the microstructure show that the dispersion density of second phase Ag_3Sn can be correlated to crack propagation and thermal fatigue life.

Based on the physical property and interconnect configuration, it is logic to hypothesize the crack initiation is induced from interface stress concentration. With drastic second phase dispersion density difference crack path deviation due to present of second phase and positive relationship to fatigue life, it is reasonable to attribute higher fatigue life to second phase strengthening.

Future Study

On going Thermal cycling and Cycle to Failure date are in statistic analysis. The Ni content and package size are variables under consideration to decouple from correction of second phase strengthening to fatigue life. Further study on higher strain-rate loading impact stress is in planning to apply to portable electronic user environment as Part II of the study.

Acknowledgments

The authors wish to gratefully acknowledge our Sr. Management C.J. Lin and DF for their leadership and Dr. Paul Wang technical in-depth editing during the course of the project reporting. The team also would like to express our heartfelt gratitude to our families and friends during the time of this study.

*This study is a work of collaboration from JDM/ODM and Package Suppliers for knowledge sharing and facilitation of technology evolution. No product endorsement, business implication, product and process reliability, and working preference in any forms are associated with this publication.

References

1. Paul P.E. Wang, "65nm FCBGA Reliability for Next Generation Gaming Device-Part I: A Methodology for Streamline Product Design Cycle and 2nd and System Level Qualification", *Proceeding of 2007 Pan Pacific Microelectronics Symposium*, 01, 2006 and *The Journal for Surface Mount and Electronics Assembly*, August Issue, 2007.

2. Paul P.E. Wang, Ken Kochi, Livia Hu, Heather McCormick, Dave Ellison, Sameer Ahmed, Hiroshi Tabuchi, and Vincent Hool, "Reliability Study of FCBGA Packages", *SMTA International Conference*, Chicago,IL 2001 and *Proceeding of APEX Technical Conference*, January 2002, pp. 17-2-1,

3. Paul Wang and Damian Hujic, "Residual Stress Correlation to ATC Reliability Scale in the u-PGA-solder-PCB Pad System-An Empirical methodology for Life Prediction under Residual Stress", *Pan Pacific Microelectronic Symposium*, 02, 2005, Apex 2005 Best Conference Paper.

4. Zhao, X. J, "Improvement of Mechanical Impact Resistance of BGA packages with Pb-free solder bumps", *Proceeding of the 8th Electronic packaging Technology Conference*, Singapore 2006, pp. 174-178.

5. JESD- A104C

6. Dr. Paul P.E. Wang and Dr. Shelgon Yee , "World Wide Deployment of CSP Assembly Process - CSP Stress/Strain Energy Distribution Modeling and Fatigue Resistance Study", , *Journal of Surface Mount Technology*, July 2000, Vol. 13, Issue 3, pp. 17-26 and *proceeding of Pan Pacific Microelectronics Symposium*, January, (2000).

7. Ahmer Syed, TaoSeong Kim, and Se Woong Cha, "Impact of Package Design and Materials on Reliability for Temperature Cycling, Bend, and Drop Loading Conditions", *SMTA Journal*, Vol. 21, Issue 3, 2008.

Morphological and Microstructural Evolution of Sn-patch in SnAgCu Solder with Ni(V)/Cu under Bump Metallization

Kai-Jheng Wang[1], Jenq-Gong Duh[1,*], Su-Yueh Tsai[2]

[1]Department of Materials Science and Engineering, Tsing Hua University, Hsinchu, Taiwan
[2]Precision Instrument Center, Tsing Hua University, Hsinchu, Taiwan
*corresponding author: E-mail: jgd@mx.nthu.edu.tw Fax: 886-3-5712686

Abstract

In flip chip technology, the Ni(V)/Cu multi-metallic thin-films is a widely used under bump metallization (UBM), for doping 7 wt.% V into the Ni target can eliminate the magnetism of Ni during sputtering [1,2]. It was noted that V in the Ni(V) layer did not react with solders and intermetallic compounds (IMC) during reflow and aging process, yet a Sn-rich phase, as the so-called "Sn-patch", would form in the Ni(V) layer [3]. The possible reason of Sn-patch formation may be the fast Sn diffusion from the solder matrix to the Ni(V) layer. However, the formation mechanism of Sn-patch and the detailed composition variation and structure evolution in Sn-patch were not fully discussed yet. In this study, Sn-patch would be analyzed by a field emission electron probe X-ray microanalyzer (FE-EPMA) and a transmission electron microscope (TEM) to elucidate the composition redistribution and the microstructure evolution, respectively. There existed concentration redistribution of the constituent element in Sn-patch, and its microstructure also varied with the aging time. On the basic of the detailed characterization by FE-EPMA and TEM, it was revealed that Sn-patch was consisted of crystalline Ni and amorphous Sn-rich phase after reflow, while V_2Sn_3 formed with amorphous Sn-rich phase during aging. A possible formation mechanism of Sn-patch was proposed, which could be employed to explain the corresponding composition variation and structure evolution associated with the Sn-patch formation.

Experimental procedure

The Ti/Ni(V)/Cu UBM was sputtered 0.1/0.5/0.3 μm on a Si wafer. Afterward Sn3.0Ag0.5Cu solder was jointed with the Ti/Ni(V)/Cu UBM at 250°C for 90 sec. The ball height of the solder joints was 120 μm, and the diameter of pads was 80 μm. Finally, the specimens were treated at 125 and 200°C, respectively, for 500, 1000 and 2000 h.

The cross-section samples were mounted with the G2 resin, and then prepared by a cross-section polisher (CP; SM-09011, JEOL, Japan) to avoid the damages from the mechanical polishing process. The interfacial morphologies were observed with a field emission scanning electron microscope (FE-SEM; JSM-7600F, JEOL, Japan). The compositions of IMCs and Sn-patch were measured by an FE-EPMA (JXA-8500F, JEOL, Japan). The structure of Sn-patch was further identified with a TEM (2010F, JEOL, Japan) at an accelerating voltage of 200 kV, and the compositional line-scan profile was also analyzed by a TEM-EDX. The TEM samples were prepared by focus ion beam (FIB; Dual beam 835, FEI). The operation voltage for FIB was lowered to 5 or 2 kV. The sample was tilted about 52° with respect to the ion beam and both sides were polished to remove the surface damage resulted from the ion beam.

Discussions

Fig. 1(a) is the cross-section image of the SAC solder jointed with Ti/Ni(V)/Cu UBM. From the magnified image (Fig. 1(b)), the Cu layer on UBM was consumed by the solder during reflow, and then Cu_6Sn_5 was formed at the interface between SAC solder and Ni(V) layer. In addition, some bright gray areas, identified as Sn-patch, existed in the Ni(V) layer near the grain boundary of Cu_6Sn_5 after reflow. The composition of Cu_6Sn_5 was measured by a FE-EPMA, as 50.5 ± 0.9 at.%Cu-3.6 ± 1.2 at.%Ni-45.9 ± 0.3 at.%Sn, and Ni atoms substituted for Cu atoms in Cu_6Sn_5. Therefore, the binary Cu_6Sn_5 phase would be marked as a ternary $(Cu,Ni)_6Sn_5$ phase.

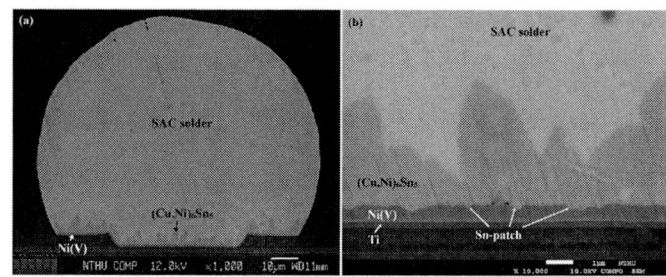

Fig. 1: (a) The interfacial morphology of Sn3.0Ag0.5Cu solder joint with Ti/Ni(V)/Cu UBM after reflow, (b) the magnified image of interface.

The cross-section images of SAC solder joints under various heat treatments are shown in Fig. 2. The average thickness of $(Cu,Ni)_6Sn_5$ grew to 2.1 μm at 125°C for 2000h. The amount of Sn-patch was increased slowly during aging, as shown in Figs. 2(a)-(c). The interfacial morphologies of solder joints after heat treatment at 200°C are shown in Figs. 2(d)-(f). The thickness of $(Cu,Ni)_6Sn_5$ ranged from 2.1 to 3.3 μm, and the amount of Sn-patch was increased after aging for 500 h. Another phase, Ni_3Sn_4, formed under $(Cu,Ni)_6Sn_5$ after aging at 200°C for 1000 h, was identified with the composition 8.6 ± 1.5 at.%Cu-34.4 ± 1.7 at.%Ni-57.0 ± 0.3 at.%Sn. The binary Ni_3Sn_4 phase was marked as a ternary $(Ni,Cu)_3Sn_4$ phase due to the Cu incorporation. The phase transformation was due to the fact that V in the Ni(V) layer did not react with solder during reflow and aging process, and only Ni in the Ni(V) layer diffused to solder and $(Cu,Ni)_6Sn_5$. As a result, $(Ni,Cu)_3Sn_4$ would be formed when the Ni content was close to the saturated concentration in $(Cu,Ni)_6Sn_5$. The detailed mechanism about phase transformation between $(Cu,Ni)_6Sn_5$ and $(Ni,Cu)_3Sn_4$ has been reported [4,5].

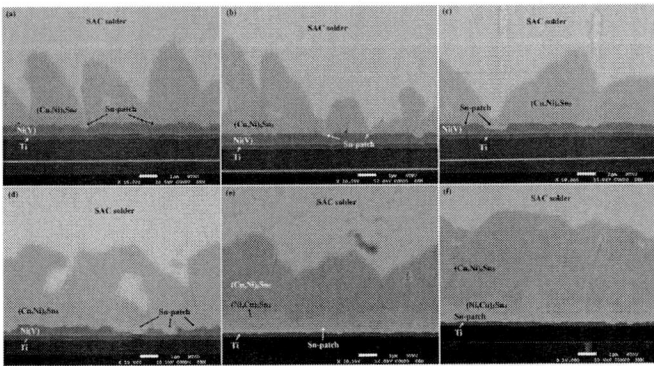

Fig. 2: The interfacial morphologies of solder joints after aging at 125°C for (a) 500 h, (b) 1000 h, (c) 2000 h, and at 200°C for (d) 500 h, (e) 1000 h, (f) 2000 h.

Besides, the Ni(V) layer was replaced by Sn-patch after aging at 200°C for 1000h, as shown in Fig. 2(e). The composition of Sn-patch was measured by FE-EPMA and TEM-EDX shown in Table 1. The Ni and Cu concentration in Sn-patch decreased from 1000 to 2000 h due to the fact that Ni and Cu atoms would be continuously diffused from Sn-patch to the solder side to make IMC grow after the Ni(V) layer was replaced by Sn-patch.

Table 1: The compositions (at.%) of Sn-patch after aging at 200°C for various periods.

Aged time	Cu (at.%)	Ni (at.%)	Sn (at.%)	V (at.%)
1000h	7.0±1.3	5.4±0.9	68.0±1.2	19.6±1.0
	-	4.9±0.4	71.5±1.0	23.6±0.6[*]
2000h	6.0±0.5	3.6±1.0	71.5±0.4	18.9±0.1
	-	0.4±0.4	71.4±2.1	28.2±1.8[*]

*: The composition was measured by TEM-EDX.

In order to observe the microstructure and to measure the detailed composition of Sn-patch, a TEM-EDX was used. In Fig. 3(a), Sn-patch was observed near the grain boundary of IMC, since the diffusion rates of Sn and Ni at the grain boundary were much faster than those in the grain. Fig. 3(b) is the diffraction pattern of Sn-patch, composed of two different patterns. The lighter spots were diffraction pattern of Ni, and the darker ones were V_2Sn_3. The Ni, V and Sn compositional line-scan profiles were analyzed by TEM-EDX (Fig. 3(c)), and the arrow in the Fig. 3(a) indicated the line-scan direction. The Ni content of Sn-patch near the $(Cu,Ni)_6Sn_5$ side was lower than that near the Ni(V) side. The composition of Sn-patch was also measured by TEM-EDX, which was 49.3 at.%Ni-17.0 at.%V-33.7 at.%Sn and 75.4 at.%Ni-7.5 at.%V-17.1 at.%Sn near the $(Cu,Ni)_6Sn_5$ side and Ni(V) side, respectively.

It should be pointed out that the composition of Sn-patch was not uniform, and it was the reason that the measured compositions in this study were different from those in literatures [3,6,7]. The discrepancy may be attributed to the sample type as well as methodology used in the quantitative analysis of the electron microanalysis. In literatures, the

composition in Sn-patch was usually done randomly for two or three points, and the average of composition was reported [3,6,7]. In this way, it was difficult to exhibit the exact composition variation for the non-uniform sample. In contrast, this study revealed more precise and detailed composition variation of Sn-patch by a series of systematic analysis using TEM-EDX after reflow and aging processes, in addition to the FE-EPMA quantitative analysis.

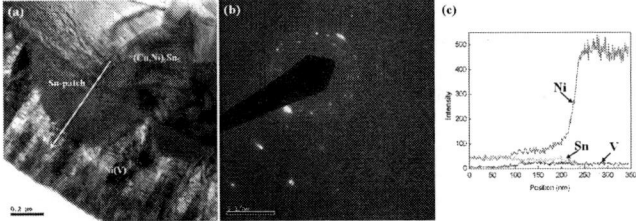

Fig. 3: (a) TEM image, (b) diffraction pattern, and (c) compositional line-scan profile of Sn-patch after reflow.

The TEM image and compositional line-scan profiles of solder joint aged at 125°C for 1000 h are shown in Fig. 4. Sn-patch was also observed near the grain boundary of $(Cu,Ni)_6Sn_5$ (Fig. 4(a)). The Ni content varied with the position of Sn-patch where the direction of compositional line-scan was marked in Fig. 4(a). The compositions of Sn-patch were analyzed by a TEM-EDX, and they were 15.3 at.%Ni-25.5 at.%V-59.2 at.%Sn and 46.0 at.%Ni-12.9 at.%V-41.1 at.%Sn near the $(Cu,Ni)_6Sn_5$ side and Ni(V) side, respectively. The composition variation in Sn-patch at 125°C was similar to that after reflow.

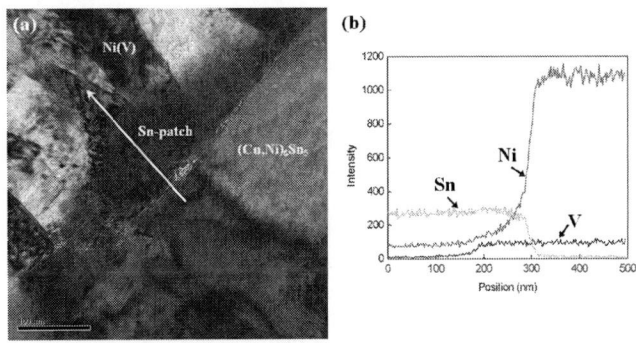

Fig. 4: (a) TEM image and (b) compositional line-scan profile of Sn-patch after aged at 125°C for 1000 h.

The structure of Sn-patch was polycrystal with amorphous phase after aging at 200°C for 2000 h, as shown in Fig. 5(a). Fig. 5(b) was the diffraction pattern of Sn-patch. An amorphous ring and diffraction rings, contributed by polycrystal V_2Sn_3, were revealed. The major elements of Sn-patch were identified as Sn and V (Table 1). However, the ratio of V to Sn was not 2:3, as V_2Sn_3. It may be due to the fact that an amorphous region was formed at the interface between IMC and the Ni(V) layer, when the Sn atom diffused to the Ni(V) layer fast. This phenomenon was called "solid-state amorphization", and the formation mechanism was widely discussed in literatures [8,9]. In this study, the amorphous phase in Sn-patch would be a Sn-rich phase caused by the fast diffusion of the Sn atom. As Sn-patch was consisted of the crystal V_2Sn_3 and the amorphous Sn-rich phase, the Sn content in Sn-patch would exceed 60 at.%.

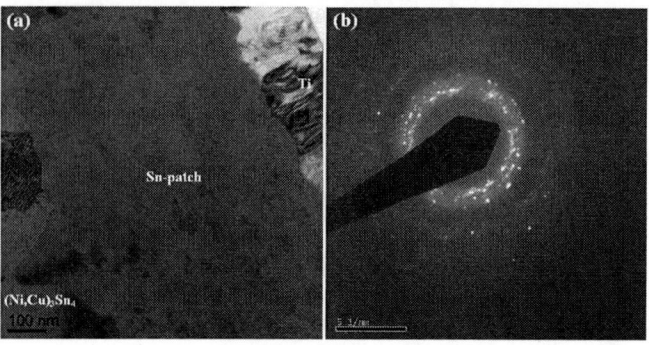

Fig. 5: (a) TEM image and (b) diffraction pattern of Sn-patch after aged at 200°C for 2000 h.

Fig. 6 illustrates the possible mechanism of Sn-patch formation.

(i) After reflow, $(Cu,Ni)_6Sn_5$ formed at the interface between SAC solder and Ni(V) layer. During thermal aging, the Ni atom in Ni(V) layer diffused to solder to form IMC, while the Sn atom in solder matrix moved toward Ni(V) layer to form Sn-patch with V and un-reacted Ni in Ni(V) layer near the grain boundary of $(Cu,Ni)_6Sn_5$.

(ii) Due to the fast diffusion of Sn atom to Ni(V) layer, an amorphous phase was produced in Sn-patch (Fig. 6(a)). In addition, the Sn atom was easily aggregated near the $(Cu,Ni)_6Sn_5$ side, as shown in Fig. 3(c).

(iii) When the Sn concentration in Sn-patch exceeded the stoichiometry of V_2Sn_3, V_2Sn_3 would begin to form in the Sn-patch (Fig. 6(b)). The V_2Sn_3 pattern was shown in Fig. 3(b), and the ratio of composition V to Sn in Sn-patch near $(Cu,Ni)_6Sn_5$ was close to 2:3 after reflow, which was another evidence for the presence of V_2Sn_3 phase. Sn would diffuse to Sn-patch, grew continuously during aging. Therefore, more and more V_2Sn_3 was formed at the interface between $(Cu,Ni)_6Sn_5$ and Sn-patch (Fig. 6(c)).

(iv) In TEM-EDX results, the Sn content in Sn-patch was evidently increased from 17 at.% to 59 at.% at 125°C for 1000 h. Finally, the Sn-patch replaced the Ni(V) layer (Figs. 2 and 5(a)). The amorphous Sn and crystalline V_2Sn_3 were revealed (Fig. 5(b)).

(v) It should be noted that even the Ni(V) layer was replaced by Sn-patch, the Ni atom in amorphous Sn phase of Sn-patch would continuously diffuse toward solder, leading to the decrease of Ni content in the Sn-patch, as shown in Table 1.

Conclusions

The interfacial reactions and Sn-patch formation of Sn3.0Ag0.5Cu solder joint with Ti/Ni(V)/Cu UBM were studied. Only $(Cu, Ni)_6Sn_5$ was observed after aging at 125°C. $(Cu, Ni)_6Sn_5$ grew gradually slowly because of the slow diffusion rate of Ni at 125°C. As the Ni content in $(Cu, Ni)_6Sn_5$ was close to the saturated concentration, $(Cu,Ni)_6Sn_5$ was partially transformed to $(Ni, Cu)_3Sn_4$. Sn-patch was observed in Ni(V) layer near the grain boundary of $(Cu, Ni)_6Sn_5$ in the solder joint. The composition and microstructure in Sn-patch varied with aging duration. The Sn

content increased in Sn-patch with the aging time, while Ni content was decreased. The structure of Sn-patch was consisted of Ni crystal and amorphous Sn-rich phase after reflow. The Sn-patch was later transformed to V_2Sn_3 crystal along with an amorphous Sn-rich phase during aging. The possible formation mechanism of Sn-patch was proposed, which could be employed to explain the corresponding composition variation and structure evolution associated with the Sn-patch formation.

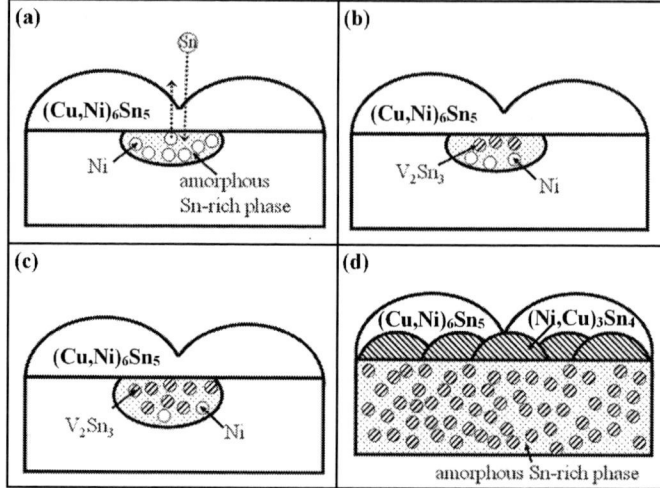

Fig. 6: Schematic plots of Sn-patch formation at interface of SAC solder and Ni(V)/Cu UBM.

Acknowledgments

Financial support from the contract No. NSC-97-2221-E-007-021-MY3 is acknowledged.

References

1. C. Y. Liu, K.N. Tu, T. T. Sheng, C.H. Tung, D.R. Frear and P. Elenius, "Electron microscopy study of interfacial reaction between eutectic SnPb and Cu/Ni(V)/Al thin film metallization", *J. Appl. Phys.* 87, 750 (2000).

2. G. Y. Jang and J. G. Duh, "Elemental redistribution and interfacial reaction mechanism for the flip chip Sn-3.0Ag-(0.5 or 1.5)Cu solder bump with Al/Ni(V)/Cu under-bump metallization during aging", *J. Electron. Mater.* 35, 2061 (2006).

3. C. H. Tung, P. S. Teo and C. Lee, "Interface microstructure evolution of lead-free solder on Ni-based under bump metallizations during reflow and high temperature storage", *IEEE Trans. Device Mater. Reliab.* 5, 212 (2005).

4. C. Y. Li, G. J. Chiou, and J. G. Duh, "Phase distribution and phase analysis in Cu_6Sn_5, Ni_3Sn_4, and the Sn-rich corner in the ternary Sn-Cu-Ni isotherm at 240 degrees", *C. J. Electron. Mater.* 35, 343 (2006).

5. C. E. Ho, Y. W. Lin, S. C. Yang, C.R. Kao, and D.S. Jiang, "Effects of limited Cu supply on soldering reactions between SnAgCu and Ni," *J. Electron. Mater.* 35, 1017 (2006).

6. S. W. Chen and C. C. Chen, "Interfacial reactions in Sn-0.7wt.%Cu/Ni-V couples at 250 degrees", *C. J. Electron. Mater.* 36, 1121 (2007).

7. S. W. Chen, C. C. Chen and C.H. Chang, "Interfacial reactions in Sn/Ni-7 wt.%V couple", *Scr. Mater.* 56, 453 (2007).

8. Z. J. Zhang and B. X. Liu, "Solid-state Reaction to synthesize Ni-Mo metastable alloys", *J. Appl. Phys.* 76, 3351 (1994).

9. K. W. Kwon, H.J. Lee, and R. Sinclair, "Solid-state amorphization at tetragonal-Ta/Cu interfaces", *Appl. Phys. Lett.* 75, 935 (1999).

Localized Recrystallization and Cracking Behavior of Lead-free Solder Interconnections under Thermal Cycling

H. T. Chen[1, *], T. Mattila[2], J. Li[2], X. W. Liu[3], M. Y. Li[1], J. K. Kivilahti[2]

[1]Shenzhen Graduate School, Harbin Institute of Technology, Shenzhen, 518055, China
[2]Laboratory of Electronics Production Technology, Helsinki University of Technology, P. O. Box 3000, 02015, Finland.
[3]Laboratory of materials science, Helsinki University of Technology, P.O. Box 6200, 02015, Finland
*E-mail: chenht@hit.edu.cn

Abstract

The failure mechanism of lead-free solder interconnections under thermal cycling has been studied by cross-polarized light microscopy, scanning electronic microscopy (SEM), and nanoindentation test. From the results of finite element modeling (FEM), it was found that the critical solder interconnection was located at the chip corner, and the stress was concentrated at the outer neck region beneath the ball grid arrays (BGA) component. The FEM results were in good agreement with the experimental observation. Two failure modes of the interconnections were identified: one is the intergranular or transgranular cracking through many small equiaxed recrystallized grains and the other is the transgranular cracking in few large irregularly shaped recrystallized grains. The results show that the localized recrystallization makes the Ag$_3$Sn intermetallic compounds (IMC) coalesce and distribute sparsely, which leads to the degradation of the recrystallized microstructure and easy propagation of the cracks.

Introduction

Solder interconnections have been extensively used in microelectronics packaging industry as electrical circuit paths and mechanical connections. Solder interconnection reliability receives increasing attention in recent years due to the miniaturization trend and the adoption of the lead-free solder alloys [1-4]. Many researches have been done to study the reliability issue, and coefficient of thermal expansion (CTE) mismatch of different packaging materials in electronic products due to the environment change and power on/off has been established as the main cause for the solder interconnection failure [5]. The IMCs at the interfaces have been assumed to play an important role in dictating the overall performance of solder interconnections, and excessive IMC growth will degrade the reliability of solder joints due to its inherent brittle nature [6]. However, it was found that the cracks were initiated and propagated in the solder bulk instead of in the interfacial IMC of the solder interconnections under thermal cycling [7-11]. Therefore, the mechanical behavior of the solder bulk needs to be further investigated to have a better understanding of the failure mechanism of the solder interconnections.

From the microscopic view, the damage of the solder materials is the accumulative result of the dislocation movements, such as the climb of the edge dislocations and the cross-slip of the screw dislocations. Tin and its alloys have high homologous temperature (generally above 0.5 T/Tm at room temperature), which makes the dislocations have enough activity to overcome the pining effect, and then the recovery and recrystallization may occur during the normal service.

However, it has been well established that recrystallization is generally favored in metals with low stacking fault energy, however, tin is a high stacking fault energy material, in which, recovery should be the dominant process and consume most of the stored energy required for recrystallization [12]. Therefore, recrystallization should be not easy to occur in tin and its alloys. Nevertheless, the recrystallization behaviors in tin and its alloys with the refined grains in the local deformation zone have been reported in recent years [7-11]. It has been assumed that the recrystallization was driven by the stored energy from the localized inelastic deformation and through the process of boundary migration. The recrystallization in solder interconnections is often accompanied by cracks because the fine recrystallized grains assist the cracks in propagating along the grain boundaries [8-9].

Experimental procedure and finite element modeling (FEM)

It was shown in the Fig.1, the component used in this

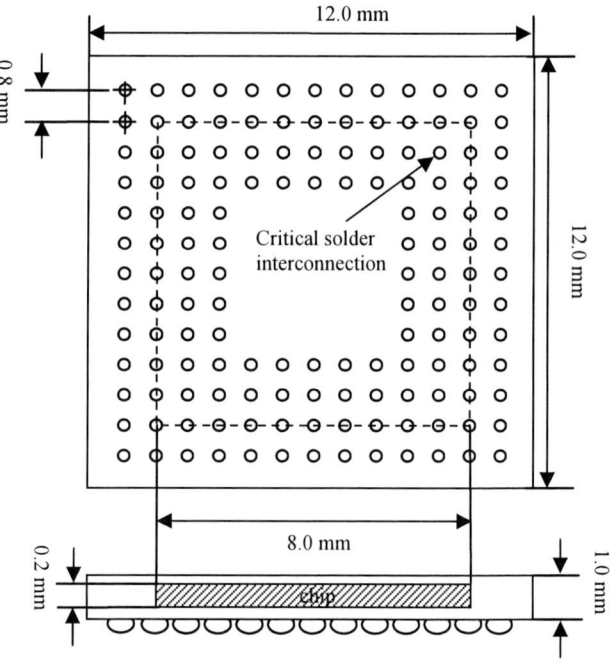

Fig.1 Dimensional drawing of the component

experiment was a 12 mm x 12 mm chip scale packaged (CSP) BGA, which has 144 bumps with a pitch of 0.8 mm and a ball diameter of 0.5 mm. No under bump metallization was applied on the copper pads of the component. Ni with immersion Au (Ni|Au) or Cu pads with organic solderability

978-1-4244-4658-2/09 $25.00 © 2009 IEEE

preservative (Cu|OSP) was used on the pads of the printed wiring board (PWB).

Solder interconnections with three after-reflow compositions, Sn3.1Ag0.5Cu(SAC), Sn3.1Ag0.52Cu0.24Bi (SACB), and Sn-1.1Ag-0.52Cu-0.1Ni(SACN), were formed by using different combinations of commercial solder pastes and solder bumps. The component boards were thermally cycled (ESPEC ENX12-7.5CWL) between -40°C and + 125°C with a cycle time of 42 minutes (21 minutes for each ramp without dwells at the extremes). Samples of each interconnection composition were removed from the oven every 10 days to observe the microstructural evolution.

The thermal cycled samples were cut along the diagonal of the component with the low speed diamond saw. To reveal the microstructure of the critical solder interconnections, the cross-section of the samples were prepared by standard metallographic methods. Tin is a birefringent material and grains with different orientations appear in different colors under cross-polarized light, therefore, optical microscopy (Olympus BX60) with cross-polarized light was used to discern different grains in solder interconnections. The cross-sectioned samples must be carefully polished to remove the artifacts introduced by the grinding or polishing. In order to have a detailed view of the microstructure, the cross-sections of the solder interconnections were also characterized by SEM (JEOL6330F) under backscattered electron (BSE) mode.

Nanoindentation was performed on the well-polished cross-sections of the solder interconnection using a TriboIndenter® (Hysitron Inc, USA) nanomechanical testing instrument. With an ultra-high resolution in load/displacement control, the device allows exploration of property fluctuation across different phases/regions at sub-micro scales. It was expected that information from mechanical characterization could provide insights for better understanding the cracking behavior of the solder connections. In each indentation cycle, a 15-second holding was performed before unloading for minimizing the viscoplastic effect on the measured properties. A Berkovich indenter with a 130 nm tip curvature was used in the measurement.

The finite element modeling was carried out by the commercial finite element software ANSYS v. 11.0. Due to symmetry, only one-fourth of the component board was modeled. The whole model was composed of a global model with coarse meshes and a local model with fine meshes. The constraint equations were employed in order to tie together the meshes in the global and local models, and the displacements were transferred along the boundary between the two models. Each solder interconnection was roughly meshed with 96 elements in the global model while the finer meshes with 3060 elements were applied for the diagonal solder interconnections in the local model. The local model had realistic shape of the solder interconnection and detailed pad design (solder-masked-defined pad on the component side and non-solder-masked-defined pad on the PWB side).

The elastic material properties for thermomechanical simulation are listed in Table I [13],[14]. The solder material was modeled as viscoplastic using Anand's constitutive equations with parameters provided by Reinikainen [15],[16]. The Anand's constitutive model is composed of a flow equation and three evolution equations that describe strain

hardening or softening during the primary and secondary creep stages [15]. All the other materials were assumed as isotropic except the PWB and the substrate, which were orthotropic. The 3-D element type SOLID45 was assigned to all the materials except the solder interconnections. The solder interconnections were modeled with the element type VISCO107, which is a viscoplastic solid element for Anand's constitutive model and is designed to solve rate-dependent large strain plasticity problem [17]. The symmetry boundary conditions were applied as mechanical constraints and the center node of the bottom of PWB was fixed in order to prevent free rigid body motion.

Table I [13], [14] Elastic Material Properties for Thermal-Mechanical Simulation

Materials		Young's Modulus (GPa)	CTE (ppm/°C)	Poisson's Ratio
Silicon		131	2.7	0.3
FR-4	In-plane	17.7	16	0.28
	Out-of-plane	7.72	84	
Substrate	In-plane	18.8	15	0.39
	Out-of-plane	8.27	57	
Mold Compound		15.8	15.5	0.3
SnAgCu		49-0.07T	21.3+0.017T	0.35
Copper		117	17	0.38

FEM results and analysis

The simulation results of the global model show that the critical solder interconnection is located at the chip corner due to the large CTE mismatch between the chip and the PWB, which is in good agreement with the observed experimental results (see Fig. 2). As compared to the component-corner solder interconnection, the critical solder interconnection has higher normal stress (out-of-plane direction), von Mises stress and viscoplastic strain energy density. The maximum normal stress, von Mises stress and viscoplastic strain energy density of the critical solder interconnection are 41.73 MPa, 45.13 MPa and 0.415 MPa, respectively. The corresponding calculated stresses and viscoplastic strain energy density of the component-corner solder interconnection are 17.95 MPa, 20.49MPa and 0.034MPa, respectively. In Fig. 2, the contour plots of the normal stress and the viscoplastic strain energy density were extracted from the diagonal cross-section of the critical interconnection at the the lowest testing temperature. The von Mises stress has the similar distribution as the viscoplastic strain energy density. As shown in Fig. 2(a) and (b), the maximum normal tensile peeling stress and the maximum viscoplastic strain energy density are both located at the outer corner near the component side of the solder interconnection. The contour plot of the viscoplastic strain energy density shows that the outer corner region near the component side of the interconnection is under the most severe distortion. The normal stress distribution shows that the outer corner region is in tensile state while the inner corner in compression state. It is well known that a crack initiated at the corner will propagate more easily with the help of the high tensile stress as compared to the case when it is in compression. Therefore, the outer corner of the critical interconnection is relatively more susceptible to failure as compared with the inner corner. Furthermore, the contour plot of the viscoplastic strain energy density again indicates that the outer corner region near the component side has the most plastic work stored in the material, which provides the driving force for the microstructural changes.

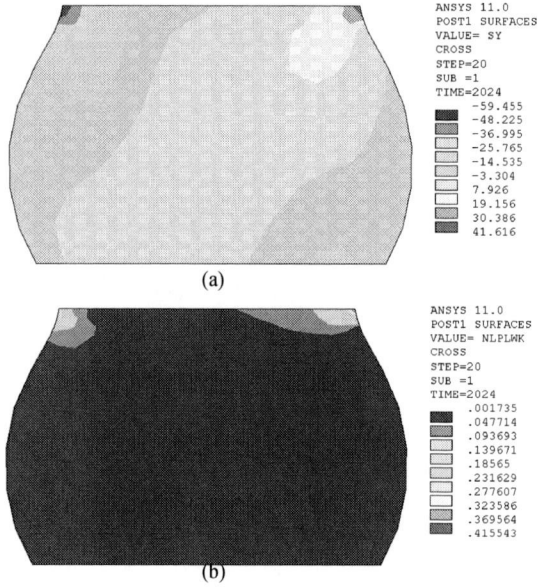

(a)

(b)

Fig.2 (a) Normal stress distribution in y direction in the local model (b) Viscoplastic strain energy density distribution in the local model

Localized cracking and recrystallization behaviors in lead-free solder interconnections

No significant difference was found between the as-solidified microstructures with the NiAu and OSP surface finishes in this experiment (see Fig.3). It can be observed clearly under cross-polarized light microscopy that few grains, typically only one, were found in the SAC and SACB interconnections, as shown in Fig.3 (a)-(d). Detailed examination revealed that β-Sn cells were surrounded by eutectic structures, which were typically composed of Cu_6Sn_5, Ag_3Sn and β-Sn matrix, in the microstructure of the solder interconnection. However, no obvious β-Sn cells with thick eutectic structure around was observed in the microstructure of SACN solder interconnections due to the low silver content (1.1 wt.%) in SACN solder alloy, as shown in Fig.3 (e) and (f). It is interesting to note that more grains were found in SACN solder interconnections compared with the interconnections with other compositions.

The recrystallization and cracking behaviors were localized on the outer neck region of the solder interconnections, which was the inelastic strain concentration area, as shown in Fig.2 (b). The colonies in the as-solidified solder interconnections were replaced by refined grains in the outer neck region after the recrystallization. Two recrystallization and cracking behaviors were identified for the solder interconnections under the thermal cycling test. One is the cracking through many small equiaxed

recrystallized grains intergranularly or transgranularly. In the interconnections with the intergranular cracking, the zigzag cracks were located a little bit far from the interface compared with the straight crack in the interconnection with transgranular cracking, and all the recrystallized grains were localized around the cracks. The other is transgranular cracking in few large irregularly shaped recrystallized grains. The crack was located quite near the interface with straight path almost parallel to the interfacial IMC. Large deformation zone can extend far away from the crack-tip.

1. Cracking through many small equiaxed recrystallized grains intergranularly and transgranularly

Recrystallization could occur more readily in SACN solder interconnections, and the cracks prefer to propagate in the recrystallized region, as shown in Fig.4. A detailed examination of the recrystallized microstructure found that the densely distributed small Ag_3Sn particles in the as-solidified SACN/OSP solder microstructure coalesced into larger ones during the thermal cycling, and then Ag_3Sn IMC became sparsely distributed gradually in the recrystallized region. The sparse distribution of the IMC particles leads to the weakening of the recrystallized microstructure. According to the classical grain boundary strengthening theory, more grain boundaries, which can be obstacles for dislocation movements, contribute to the strengthening of the solder bulk [18]. However, more grains in the recrystallized region of low-silver solder interconnections do not have that strengthening effect, and cracks propagated more readily in recrystallized microstructure, leading to the earlier failure in the thermal cycling test. Solder alloy strength itself, rather than the number of the grain boundaries, dictates the solder interconnection reliability. It has been reported that the low silver concentration in SACN solder alloy leads to less Ag_3Sn IMC dispersion and then the less strengthened interconnections [11].

The mechanical properties of the recrystallized and un-recrystallized microstructures in SACN/OSP solder interconnection were studied by nanoindentation test, as shown in Fig.5. The cross-polarized images show clearly how the indents were placed at the desired regions within the recrystallization region. While, the topographic features, grains boundaries and fine precipitates at each vary locations are revealed by the high resolution SPM (scanning probe microscopy) images. The nanoindentation hardness and elastic modulus values were shown in Fig.6, in which the averages from 12 to 18 indents are reported. The elastic modulus of microstructure does not change significantly after recrystallization. However, the hardness of the recrystallized microstructure is much lower than that of un-recrystallized one due to the sparsely distributed IMCs after recrystallization. It should be noted that the hardness of the recrystallized microstructure has a very slight variation, and it correlates well with the characteristics of IMC, such as the size and distribution, than the grain orientations. As a consequence, the combination of the degraded recrystallized microstructure and the stress concentration determines that the recrystallized neck region is the weakest link in the solder interconnection, and then it is not surprising to see that the cracks were often observed to propagate in the recrystallized region. The local inelastic deformation caused by CTE

978-1-4244-4658-2/09 $25.00 © 2009 IEEE

mismatch during thermal cycling offers the driving force for recrystallization and corresponding microstructural change. As shown in Fig.7, a detailed examination of the microstructure revealed that transgranular cracking instead of intergranular cracking frequently occurred in SACN/OSP solder interconnections. Due to the low strength of the recrystallized microstructure without original densely distributed IMCs, the cracks tend to propagate transgranuarly especially when the stress state, orientation and geometry of the grains are not in favour of cracking along the grain boundaries. Occasionally, the cracks were found propagating intergranularly when the recrystallized grain size is small.

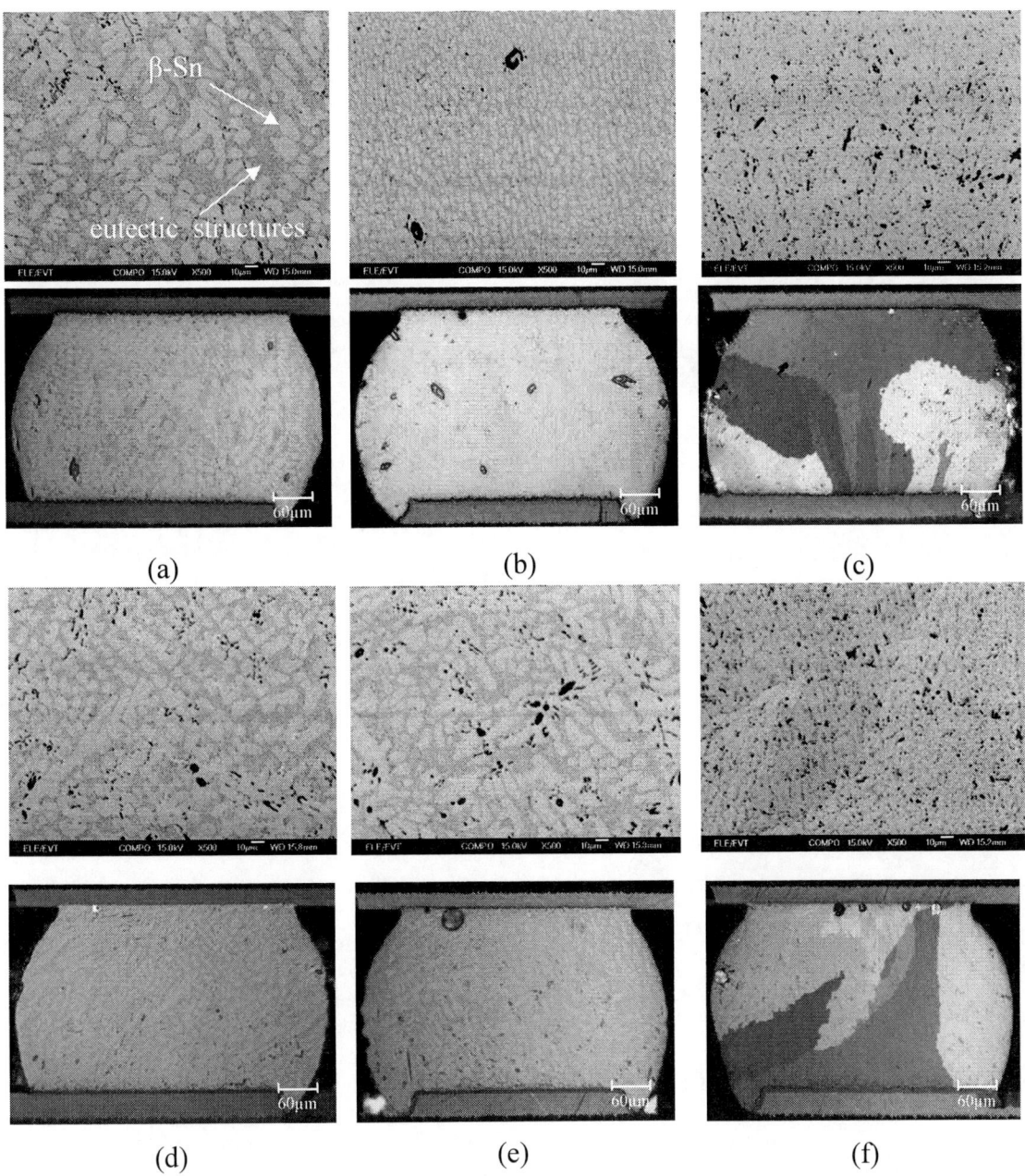

Fig.3 SEM and polarized images of as-solidified solder interconnections with different surface finishes on PWB (a) SACB/OSP (b) SAC/OSP (c) SACN/OSP (d) SACB/NiAu (e) SAC/NiAu (f) SACN/NiAu

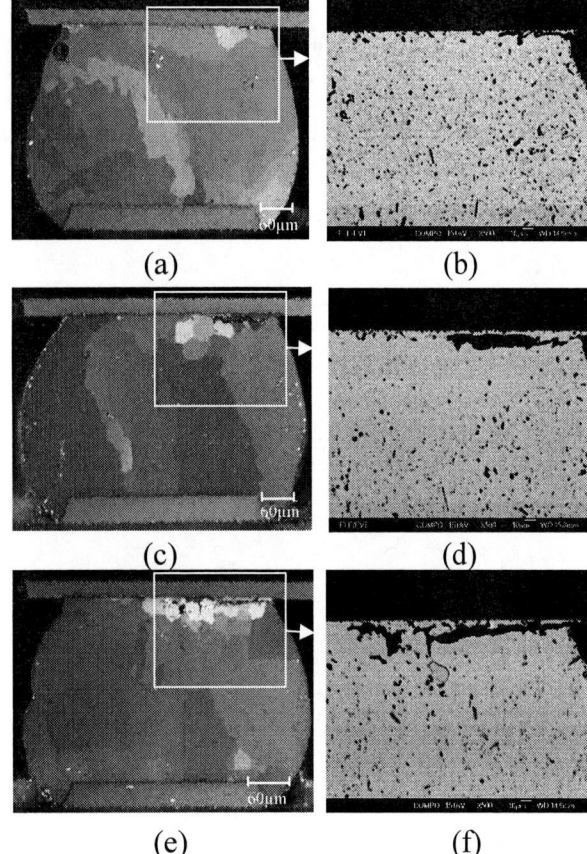

(a)   (b)

(c)   (d)

(e)   (f)

Fig.4 Microstructures of SACN/OSP solder interconnections after thermal cycling for 60 days (a) Cross-polarized image (b) SEM images; 80 days (c) Cross-polarized image (d) SEM images; 90 days (e) Cross-polarized image (f) SEM images

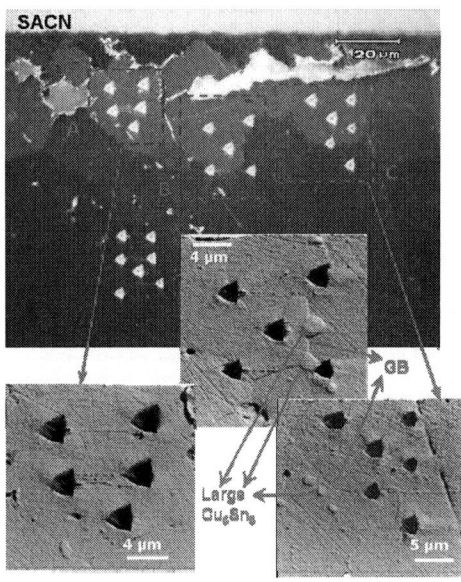

Fig.5 Image under optical microscope with cross-polarized light after indentation test

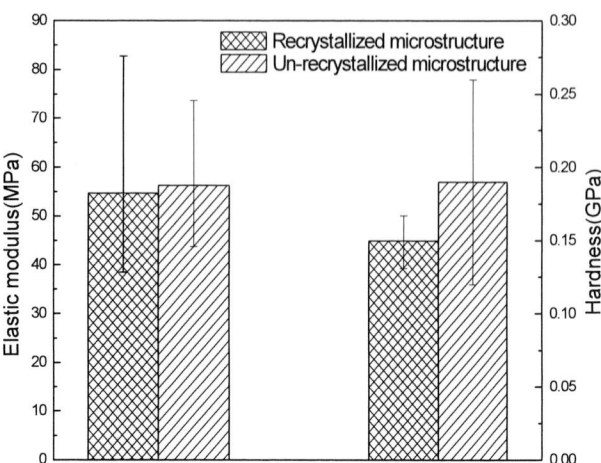

Fig.6 Mechanical properties of recrystallized and un-recrystallized microstructures

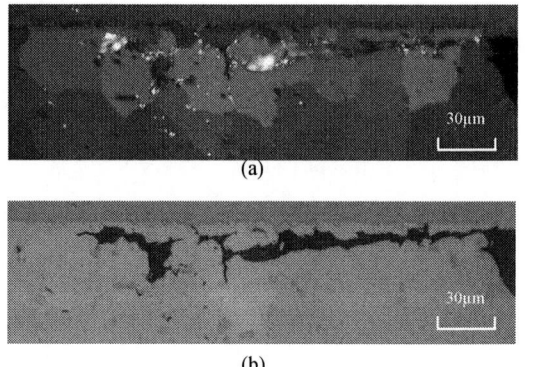

(a)

(b)

Fig.7 High magnification cross-polarized (a) and bright field (b) images of recrystallized region in Fig.4. (e)

Recrystallization assisted cracking, was also observed in SACN/NiAu solder interconnections, as shown in Fig.8. Intergranular cracking occurred along the grain boundaries of the refined recrystallized grains. The grain boundary sliding is the main mechanism for cracking along the grain boundaries, and the decohesions of the grain boundaries were frequently observed, as shown in Figs.8(a) and (b). It has been assumed that the cracking along the recrystallized grain boundaries needs less energy than the transgranular cracking [9]. In fact, the cracking behavior is closely related to the orientation and size of the recrystallized grains and their corresponding stress status. If the cracks have a zigzag type path, these grains may be under shear stress condition and the cracks are highly likely to follow the refined grain boundaries. If the cracks have the tendency to propagate along the interfacial IMC instead of towards the solder bulk under the high tensile stress, the transgranular cracking behavior is highly likely to occur.

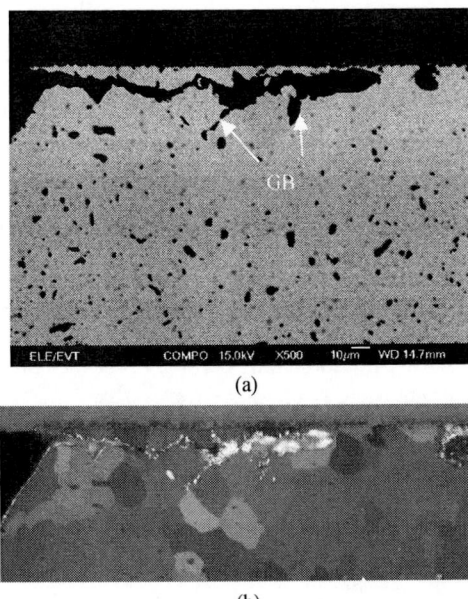

(a)

(b)

Fig.8 SEM (a) and cross-polarized (b) images of SACN/NiAu solder interconnections after thermal cycling for 90 days

2. Transgranular cracking in few large irregularly shaped recrystallized grains

As shown in Fig.9, another typical failure mode is the transgranular cracking in few irregularly shaped large recrystallized grains, and the straight crack is quite near the IMC layer. This failure mode is frequently observed in SACB solder interconnections, which have better performance in resisting the creep and fatigue in the thermal cycling. It has been reported that Bi addition can strengthen the solder matrix due to solid solution strengthening and precipitating strengthening [19]. It is interesting to note that there is a big difference in the microstructure between the recrystallized and unrecrystallized regions. No typical β-Sn cells with surrounding eutectic structure as in the unrecrystallized region was observed in the recrystallized microstructure of SACB/NiAu solder interconnections. Similar to what has been found in SACN solder interconnections, Ag_3Sn IMC particles became sparsely distributed in the recrystallized area and the small Ag_3Sn IMC particles coalesced into larger ones. It is assumed that the inelastic deformation provides the energy needed for the IMC agglomeration. Compared with the zigzag type cracking in SACN/NiAu solder interconnections, the cracking behavior of the strengthened SACB eutectic alloy tends to be more brittle, and the crack seems to propagate more abruptly through the large recrystallized grains. Compared with the inelastic deformation zone in SACN solder interconnections, the deformation zone in Bi-containing solder interconnection can extend to a large distance away from the crack-tip because the stress can not be released as readily as in the ductile low-silver solder interconnections. Furthermore, no grain boundary decohesion as in SACN solder interconnections was observed for the irregularly shaped recrystallized grains. It means that the intergranular cracking is not always an energy-saving mode for tin alloys, and the cracking path is determined jointly by many factors, such as grain orientation and size, stress status, and IMC distribution. The combination of the stress

concentration and sparsely distributed IMCs makes the crack initiate and propagate in the recrystallized region at the outer corner of the interconnections.

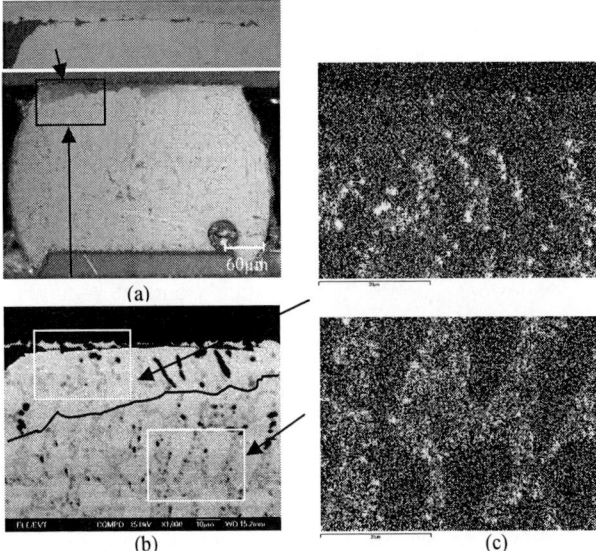

(a)

(b)   (c)

Fig.9 Cross-polarized and SEM images of SACB/NiAu solder interconnections after thermal cycling for 90 days (a) Cross-polarized image. (b) SEM image. (c) Ag mapping of Fig.(b).

Both type of recrystallization and cracking behaviors were observed in the SAC solder interconnections with two surface finishes. As shown in Fig.10, it is interesting to note that large recrystallization grains were found ahead of the crack-tip after the crack was first initiated at the corner of SAC/OSP solder interconnections. That is to say, the recrystallization behavior is a relatively independent process, which seems only dictated by the strain, and the cracks can be initiated before as well as after the recrystallization. It is assumed that the severe recrystallization behaviour occurred ahead of the crack-tip, where the deformation exceeds a threshold value [20]. The zigzag intergranular cracking through small recrystallized grains was found at the outer neck region of the SAC/NiAu solder interconnections as well, and the recrystallized grains were just localized around the crack.

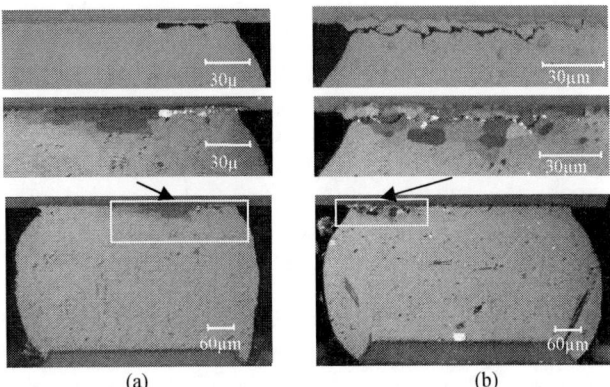

(a)   (b)

Fig.10 Cross-polarized images of solder interconnections after thermal cycling for 90 days (a) SAC/NiAu (b) SAC/OSP

Conclusions

978-1-4244-4658-2/09 $25.00 © 2009 IEEE

1. The critical solder interconnection was under the chip corner, and the stress concentration area was at the outer neck region of the interconnection beneath the chip.

2. No significant differences were found out between the microstructures with NiAu and OSP surface finishes.

3. More grains were found in the as-solidified microstructure of SACN solder interconnections under cross-polarized light microscopy compared to that of the SAC and SACB solder interconnections, which typically have only one grain.

4. Many equiaxed refined grains were observed in the recrystallized region of SACN solder interconnections and cracks can propagate both transgranularly and intergranularly. The path of cracks was determined mainly by the grain size and orientation, stress status, and IMC distribution.

5. Transgranular cracking through few large irregular shaped recrystallized grains were found in SACB solder interconnections, and the inelastic deformation zone can extend to a large distance away from the crack-tip.

6. Nanoindentation test shows that the weakening has occurred in the recrystallized microstructure due to the sparse distribution of IMC after recrystallization.

7. The local inelastic deformation caused by CTE mismatch during the thermal cycling offers the driving force for the recrystallization, leading to the IMC coarsening in the recrystallized area.

Acknowledgments

The author would like to thank Dr. V. Vuorinen for his valuable discussion and help in the SEM studies.

References

1. K. Zeng and K.N. Tu, "Six cases of reliability study of Pb-free solder joints in electronic packaging technology," *Materials science and engineering*, R 38, 2002, pp. 55-105.

2. J.W. Kim, D.G. Kim, W.S. Hong and S.B. Jung, "Evaluation of solder joint reliability in flip-chip packages during accelerated testing," *Journal of electronic materials*, Vol. 34, No. 12 (2005), pp. 1550-1557.

3. M. Ering, P. J. G. Schreurs, G. Q. Zhang, and M. G. D. Geers, "Microstructural damage analysis of SnAgCu solder joints and an assessment on indentation procedures," *Journal of materials science: Materials in electronics*, Vol. 16 (2005), pp. 693-700.

4. H.W. Chiang, J.Y. Chen, M. C. Chen, J. C.B. Lee, and G. Shiau, "Reliability testing of WLCSP lead-free solder joints," *Journal of electronic materials*, Vol. 35, No. 5 (2006), pp. 1032-1040.

5. W. W. Lee, L. T. Nguyen, and G. S. Selvaduray, "Solder joint fatigue models: review and applicability to chip scale packages," *Microelectronics Reliability*, Vol. 40, Issue 2 (2000), pp. 231-244.

6. T. Laurila, V. Vuorinen, J.K. Kivilahti, "Interfacial reactions between lead-free solders and common base materials," *Materials science and engineering*, R. 49 (2005), pp. 1-60.

7. S. Terashima, K. Takahama, M. Nozaki, and M. Tanaka, "Recrystallization of Sn grains due to thermal strain in Sn-1.2Ag-0.5Cu-0.05Ni solder," *Materials Transactions*, Vol. 45, No. 4 (2004), pp. 1383-1390.

8. D.W. Henderson, J.J. Woods, T.A. Gosselin, J. Bartelo, D.E. King, T.M. Korhonen, M.A. Korhonen, L.P. Lehman, E.J. Cotts, S.K. Kang, P. Lauro, D.Y. Shih, C. Goldsmith, and K.J. Puttlitz, "The microstructure of Sn in near-eutectic Sn-Ag-Cu alloy solder joints and its role in thermomechanical fatigue", *Journal of Materials Research*, 19, 6, (2004), pp. 1608-1612.

9. T.T. Mattila, V. Vuorinen, and J.K. Kivilahti, "Impact of printed wiring board coatings on the reliability of lead-free chip-scale package interconnections," *Journal of materials research*, Vol. 19, No. 11 (2004), pp. 3214-3223.

10. A.U. Telang, T.R. Bieler, A. Zamiri and F. Pourboghrat, "Incremental recrystallization/grain growth driven by elastic strain energy release in a thermomechanically fatigued lead-free solder joint," *Acta Materialia*, Vol. 55 (2007), pp. 2265-2277.

11. J.J. Sundelin, S.T. Nurimi, T.K. Lepistö, "Recrystallization behavior of SnAgCu solder joints," *Materials Science and Engineering*, A 474 (2008), pp. 201-207.

12. D. Hardwick, C. M. Sellars, and W. J. McG. Tegart, "The occurrence of recrystallization during high-temperature creep", *Journal of the Institute of Metals*, Vol. 90 (1961), pp. 21-22.

13. C. Kanchanomai, Y. Miyashita, Y. Mutoh, and S. L. Mannan, "Influence of frequency on low cycle fatigue behavior of Pb-free solder 96.5Sn–3.5Ag," *Materials Science and Engineering A*, Vol. 345, Issues 1-2 (2003), pp. 90-98.

14. J. Lau and W. Dauksher, "Effects of ramp-time on the thermal-fatigue life of SnAgCu lead-free solder interconnections", in Proc. 55th Electron. Comp. Technol. Conf., Lake Buena Vista, FL, May 31-June 3, 2005, pp. 1292-1298.

15. L. Anand, "Constitutive equations for rate-dependent deformation of metals at elevated temperatures," *J. Eng. Mater. Technol. ASME*, Vol. 104, No. 1 (1982), pp. 12-17

16. T. O. Reinikainen, P. Marjamäki, and J. K. Kivilahti, "Deformation characteristics and microstructural evolution of SnAgCu solder interconnections," in Proc. 6th EuroSim Conf., Berlin, Germany, April 18-20, 2005, pp. 91-98.

17. Release 11.0 documentation for ANSYS.

18. J.D. Verhoeven, "Fundamentals of physical metallurgy," *John Wiley&Sons Inc.* (1975), pp.515-519.

19. J. Zhao, L. Qi, X. Wang and L. Wang, "Influence of Bi on microstructures evolution and mechanical properties in Sn-Ag-Cu lead-free solder," *Journal of Alloys and Compounds*, Vol. 375, Issues 1-2 (2004), 196-201.

20. S. Canumalla, S. Mathew and S. K. Saha, "Damage Evolution in Sn62Pb36Ag2 Solder of a Chip Scale Package under a Monotonic Shear Stress", Electronic Components and Technology conference, New. Orleans, Louisiana USA. May 27-30, 2003, pp. 611-619.

Growth Kinetics and Microstructural Evolution of Cu-Sn Intermetallic Compounds on Different Cu Substrates during Thermal Aging

Z.Q. Liu[1]*, P.J. Shang[1], D.X. Li[1], J. K. Shang[1, 2]

[1]Shenyang National Laboratory for Materials Science, Institute of Metal Research, Chinese Academy of Sciences
Shenyang 110016, China, Email: zqliu@imr.ac.cn
[2]Department of Materials Science and Engineering
University of Illinois at Urbana-Champaign, Urbana, IL 61801, USA

Abstract

The microstructure of the eutectic SnBi/Cu interface was investigated by transmission electron microscopy (TEM). In-situ and ex-situ thermal aging were conducted to study the nucleation and growth of Cu_3Sn and Cu_6Sn_5 on both polycrystalline and single crystalline Cu substrates. The growth kinetic analysis showed that although the growth of total IMCs (Cu_3Sn + Cu_6Sn_5) was similar on single and polycrystalline Cu substrates, the Cu_3Sn grew faster on polycrystalline Cu substrate than that on single crystal Cu substrate, while Cu_6Sn_5 grew slowly on polycrystalline Cu during thermal aging.

Introduction

In SnBi/Cu joint system, Bi does not react with Cu or Sn to form any intermetallic compound (IMC). Thus Cu_6Sn_5 forms first at the interface during reflow, while Cu_3Sn forms later between Cu and Cu_6Sn_5 by solid-state reaction to satisfy the requirements of local equilibrium [1-4]. Formation of these IMC layers at the interface is an indication of good bonding between solder and the metal substrate. However, due to the stress concentration at the interface and the brittle nature of the IMCs, defects can develop easily in the IMC layer, which may take the form of Kirkendall voids or microcracks. It has been reported that Cu_3Sn layer is prone to voiding during solid-state aging. Therefore, the evolution of these interfacial IMCs during thermal aging plays a great role in solder joint reliability [5-8].

In this work, the growth of Cu_3Sn and Cu_6Sn_5 at the interface of SnBi/Cu solder joint during in-situ and ex-situ thermal aging was investigated by TEM. Different growth behavior and mechanisms were observed on polycrystalline and single crystalline Cu substrates.

Experimental procedure

The oxygen-free high-conductivity polycrystalline Cu and (100), (111) single crystal Cu plates were cut by electrodischarge machining to the size of $10 \times 2.5 \times 2$ mm^3. Copper surfaces were ground and carefully polished by 0.5 μm diamond paste, and then rinsed in acetone, methanol alcohol and distilled water in an ultrasonic bath. The two copper sheets were covered with a commercial eutectic SnBi solder paste, aligned, clamped, and heated to the reflow temperature to form a solder joint before the specimens were air-cooled to room temperature. The reflowing process was held at 443 K for about 3~5 seconds. The as-reflowed samples were then cut into 500μm-thick thin slices before they were aged in silicone oil at 393 K for up to 360 hours.

Both scanning electron microscope (SEM) and TEM samples were prepared to measure the thickness of different IMC layers for kinetic studies. A JSM-6301F field emission SEM equipped with an Oxford Link ISIS EDS system was used for large-scale microstructural analysis, while JEM-2010 and FEI Tecnai F30 electron microscopes were used to carry out TEM observations. To prepare TEM foils, the slices were mechanically grounded to a final thickness of 40 μm, and ion-milled (Gatan model 691 PIPS) under 5.0 keV and 5 □A at a low milling angle (less than 6°).

Results and discussion

During reflow and following thermal aging, two layers of IMC - Cu_6Sn_5 and Cu_3Sn, formed and grew at the solder interface. To study the growth kinetics, the individual and total thickness of the IMC layer was measured on each sample at different aging time. For the thickness of Cu_3Sn, it was mainly measured from the TEM images, since the Cu_3Sn layer was too thin to be detected by SEM after reflow or at the initial aging stage. On the other hand, after long-term solid-state aging, the interfacial Cu_3Sn or Cu_6Sn_5 layer grew so thick that the TEM specimen could not cover all of the interfacial IMC layers. In such a case, only SEM images were used to measure their thickness as shown in Fig.1, which are the cross-sectional images of the interfaces formed between solder and Cu after solid-state aging for 360hrs at 393K.

The growth kinetics of individual Cu_3Sn and Cu_6Sn_5 as well as the total IMC ($Cu_3Sn+Cu_6Sn_5$) layer on different Cu substrates were expressed by plotting the square of measured thickness as a function of the aging time, which was shown in Fig.2. It can be clearly seen that the growth rates of total IMC layer on polycrystalline Cu and single crystal Cu are similar. The three growth kinetic lines are almost parallel although the average thickness of IMC layer on polycrystalline Cu is slight thicker than that on (100) and (111) single crystal Cu. However, the individual growth behaviors of Cu_3Sn and Cu_6Sn_5 were different on polycrystalline and single-crystal Cu substrates. According to Fig.2, after reflow the Cu_6Sn_5 layer is thicker on polycrystalline Cu than on single-crystal Cu. However, during thermal aging the Cu_6Sn_5 phase on single crystal Cu was thickened quickly than that on polycrystalline Cu. That is to say, the growth rate of Cu_6Sn_5 layer on polycrystalline Cu is slower than that on (100) and (111) single crystal substrates. By contraries, with the prolonging of solid state aging, the growth rate of Cu_3Sn layer on polycrystalline Cu is faster than that on single crystal Cu (also see Fig.1). Between single crystal Cu (100) and (111) substrates, there is less difference on the growth rate of Cu_3Sn or Cu_6Sn_5.

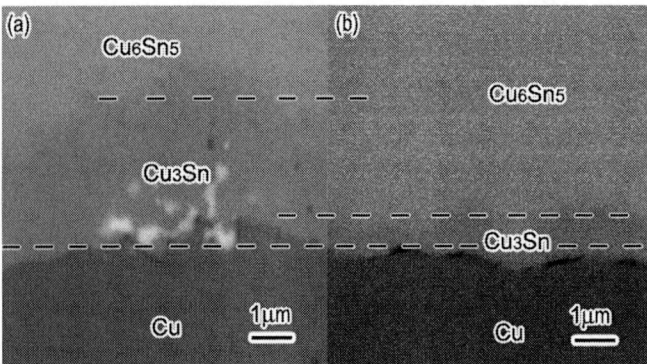

Fig.1 SEM images showing the interfacial microstructures on (a) polycrystalline Cu, (b) single-crystal (100) Cu after the solders were solid-state aged for 360 hrs at 393K. The average thickness of Cu_3Sn layer was indicated by dashed lines.

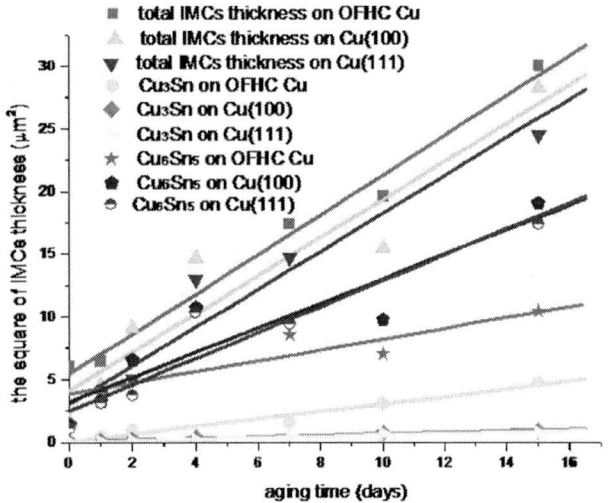

Fig.2 Square of the thickness of individual and total IMC ($Cu_3Sn + Cu_6Sn_5$) layers as a function of aging time at 393 K on different Cu substrate.

The different growth behaviors of IMC on polycrystalline and single-crystal Cu substrates could be understood concerning the reactions at the solder interface. At the Cu/Cu_3Sn interface, in-situ thermal aging in TEM revealed that Cu_3Sn grew faster into Cu substrate than into solder side, which agrees with the marker experiments done by other researchers [9, 10]. After reflow, polygonal Cu_3Sn grains formed on polycrystalline substrate, while that on single crystal substrate has a colomnar morphology. During solid-state aging, on polycrystalline Cu new Cu_3Sn grains nucleated and grew at both the Cu/Cu_3Sn and Cu_3Sn/Cu_6Sn_5 interfaces, which increased the thickness of Cu_3Sn layer. On single crystal Cu new Cu_3Sn grains only nucleated at the triple junction sites at the Cu/Cu_3Sn interface and form a second equiaxial Cu_3Sn layer beneath the initial columnar Cu_3Sn layer [11], while on the Cu_3Sn/Cu_6Sn_5 interface the columnar Cu_3Sn grains keep growing directly and drives Cu_3Sn/Cu_6Sn_5 interface shift to Cu_6Sn_5 side. Fig. 3 shows the morphology of Cu_3Sn/Cu_6Sn_5 interface. On single-crystal substrate the columnar Cu_3Sn grain grew into Cu_6Sn_5, while on

polycrystalline substrate new Cu_3Sn grains nucleated at the Cu_3Sn/Cu_6Sn_5 interface. The growth processes of IMC layer on polycrystalline and single-crystal substrate were depicted in Fig.4 and Fig.5, respectively. The small grain size and the fast grain boundary diffusion, contributed to the quick growth of Cu_3Sn on polycrystalline substrate.

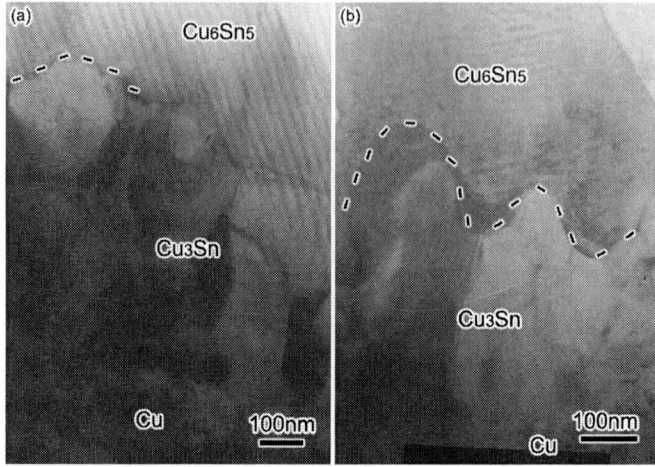

Fig.3 The morphology of Cu_3Sn/Cu_6Sn_5 interface on (a) polycrystalline and (b) single-crystal Cu substrates.

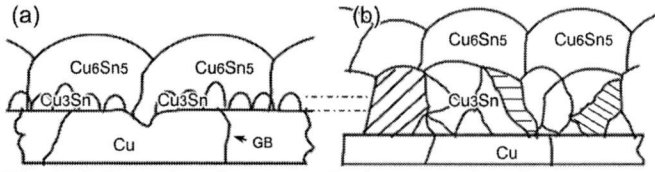

Fig.4 Schematic diagram of the growth of Cu_3Sn on polycrystalline Cu after (a) reflow and (b) solid-state aging, showing the inhomogeneous nucleation on Cu and the heterogeneous growth at both Cu/Cu_3Sn and Cu_3Sn/Cu_6Sn_5 interfaces.

After reflow the total IMC thickness is determined by the thickness of Cu_6Sn_5, which is the main IMC formed during reflowing and has the highest content of Sn among Cu-Sn IMCs. In the following solid-state aging, thickening of IMC layer depends on the extension of Cu/Cu_3Sn and Cu_6Sn_5/solder interfaces into Cu and solder, respectively. At the solder/Cu_6Sn_5 interface, Cu diffused through volume or grain boundary to the solder/Cu_6Sn_5 interface and reacted with Sn to form Cu_6Sn_5, which drove the solder/Cu_6Sn_5 interface moving to solder side. However, during wetting reaction a substantial amount of Bi was left behind at the Cu_6Sn_5/solder interface due to the consumption of Sn from SnBi solder to form Cu_6Sn_5. This Bi-rich layer would act as a barrier layer for Sn diffusion and decrease the growth rate of Cu_6Sn_5 at solder/Cu_6Sn_5 interface. On the other hand, at the Cu_6Sn_5/Cu_3Sn interface the Cu_6Sn_5 was consumed simultaneously by the growth of Cu_3Sn. Therefore, the thickening of Cu_6Sn_5 layer was slow during solid state aging, especially on polycrystalline substrate where the growth of Cu_3Sn was faster than that on single-crystal Cu substrate.

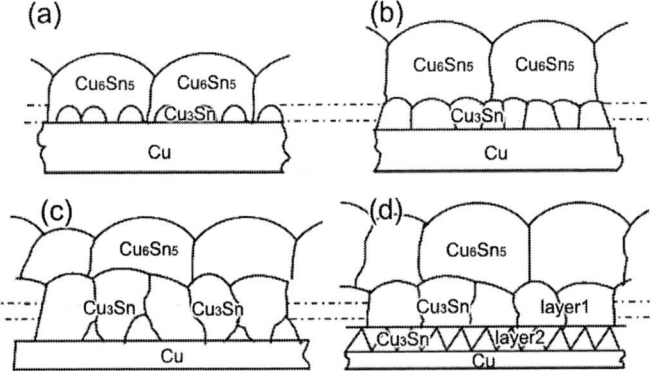

Fig.5 Schematic diagram of the growth of Cu_3Sn on single-crystalline Cu during reflow and solid-state aging: (a) nucleation of directional Cu_3Sn grain during reflow; (b) formation of columnar layer; (c) nucleation of new Cu_3Sn grain at triple junction site of grain boundaries during solid-state aging; (d) formation of equiaxed Cu_3Sn layer (layer 2) between columnar Cu_3Sn layer (layer 1) and Cu during solid-state aging.

Conclusions

TEM investigations were carried out to study the growth behavior of Cu-Sn IMC in the reactions of eutectic SnBi solder on both polycrystalline and single crystal Cu substrates. The growth kinetics analysis showed that Cu_3Sn grew faster on polycrystalline Cu substrate than that on single crystal Cu substrate. This was attributed to more nucleation sites and smaller grain size of Cu_3Sn formed on the polycrystalline Cu, and the larger effects of grain boundary diffusion on the polycrystalline Cu than on single-crystalline Cu. Nevertheless, Cu_6Sn_5 grew slowly on polycrystalline Cu during solid state aging due to the consumption by the growth of Cu_3Sn at Cu_3Sn/Cu_6Sn_5 interface and the decrease of growth rate of Cu_6Sn_5 at solder/Cu_6Sn_5 interface. The growth of total IMCs ($Cu_3Sn + Cu_6Sn_5$) was similar on single and polycrystalline Cu substrates. The different growth kinetics of Cu_3Sn and Cu_6Sn_5 layer on polycrystalline and single-crystal substrates, could be understood concerning the microstructural evolution during thermal aging, which has been discussed and described.

Acknowledgments

The authors gratefully acknowledge the financial support from the National Basic Research Program of China under Grant No. 2004CB619306 and the Hundred Talents Program of the Chinese Academy of Sciences.

References

1. Tu, K.N., "Interdiffusion and reaction in bimetallic Cu-Sn thin flims," *Acta Metall.*, Vol. 21 (1973), pp. 347-354.
2. Li, J.F., Mannan, S.H., Clode, M.P., Whalley, D.C., Hutt, D.A., Wasserman, Y., "Interfacial reactions between molten Sn-Bi-X solders and Cu substrates for liquid solder interconnects," *Acta Mater.*, Vol. 54 (2006), pp. 2907-2922.
3. Vianco, P.T., Erickson, K.L., Hopkins, P.L., "Solid-state intermetallic compound growth between copper and high-temperature, Tin-rich solders.1. experimental-analysis," *J. Electron Mater.*, Vol. 23 (1994), pp. 721-727.
4. Choi, S., Bieler, T.R., Lucas, J.P., Subramanian, K.N., " Characterization of the growth of intermetallic interfacial layers of Sn-Ag and Sn-Pb eutectic solders and their composite solders on Cu substrate during isothermal long-term aging," *J. Electron Mater.*, Vol. 28 (1999), pp. 1209-1215.
5. Laurila, T., Vuorinen, V., Kivilahti, J. K., "Interfacial reactions between lead-free solders and common base materials," *Mater. Sci. Eng. R.*, Vol. 49 (2005), pp.1-60.
6. Tu, K.N., Zeng, K., "Tin-lead (SnPb) solder reaction in flip chip technology," *Mater. Sci. Eng. R.*, Vol. 34 (2001), pp. 1-58.
7. Liu, P.L., Shang, J.K., "Segregant-induced cavitation of Sn/Cu reactive interface," *Scripta. Mater.*, Vol. 53 (2005), pp. 631-634.
8. Chen, C. M., Chen, C. H., "Interfacial reactions between eutectic SnZn solder and bulk or thin-film Cu substrate," *J. Electron Mater.*, Vol. 36 (2007), pp. 1363-1371.
9. Tu, K.N., Thompson, R.D., "Kinetics of interfacial reaction in bimetallic Cu-Sn thin films," *Acta Metall.*, Vol. 30 (1982), pp. 947-952.
10. Paul, A., Kodentsov, A.A., Van Loo, F.J.J., "Intermetallic growth and Kirkendall effect manifestations in Cu/Sn and Au/Sn diffusion couples," *Z. Metallkd.*, Vol. 95 (2004), pp. 913-920.
11. Shang, P.J., Liu, Z.Q., Li, D.X., Shang, J.K., "Directional growth of Cu_3Sn at the reactive interface between eutectic SnBi solder and (100) single crystal Cu," *Scripta Mater.*, Vol. 59 (2008), pp. 317-320.

978-1-4244-4658-2/09 $25.00 © 2009 IEEE

Assessment of LF Solder Joint Reliability by Four Point Cyclic Bending

Jifan Wang, Yuming Ye, Junying Zhao, Sang Liu, Yunhua Tu, Song Li, Zhiwei Song
Huawei Technologies Co., LTD.
Shenzhen, Guangdong, P.R.China
wangjifan@huawei.com

Abstract

Quick and effective reliability assessment method is one of aspired targets for product development and test. Base on specific design board, this paper study the effect of four-point cyclic bend test parameters, such as deflection and temperature to lead-free solder joint fatigue life and failure mode, compare the performance of lead-free joints in four-point cyclic bend and accelerated temperature cycling (ATC), explore the application of four-point cyclic bend test method in lead-free solder joint reliability assessment.

Introduction

In filed of electronic packaging & Interconnection, solder joint reliability is always the major concern [1]. To assess the board-level reliability, specially the surface mount solder joint reliability, people develop kinds of test method, along which accelerated temperature cycling(ATC) test get wide acknowledged and applied for base on it can properly predict solder joint thermal fatigue life in filed. But ATC test need very long time and industry are developing faster and effective substitute method, such as Thermal Twist Cycling (TTC), Mechanical Shear Fatigue, Four-point Cycle Bend, et al. [2-4].

For current lead-free transition, faster and effective assess method can reduce product R&D test period, get reliability information in time and help enterprise better master market opportunity. In research such faster and effective test methods; the obvious premise is failure mechanism must be same as filed performance. With crucial test parameter design (e.g. DOE), people can make clear the influence of such parameters to test result, compare with known ATC data then can verify the validity and suitable application circumstance.

There are several famous research institutes report the achievement and progress in faster and effective test methods for board-level reliability, these may make deepen influence on reliability assessment methods as the research advance and relationship between experiment and field specified. The new method may used as qualitatively and/or quantitatively in joint reliability assessment.

Experimental

Sample preparation

The four-point cyclic bend test samples include three different sizes of commercial dummy BGA packages, which have special daisy chain. The test board design conforms to that prescribed in JEDEC Standard JESD22-B113 [5], and the solder joints/paste is SAC305, the surface finishes of BGA and PCB are OSP. A standard lead-free surface mount temperature profile is used for soldering reflow. The samples are thoroughly examined through X-ray, daisy chain electrical on/off, and cross-section to verify assembly quality, make sure the O.K.'s enter the further reliability test.

Figure 1 Assembly Samples Passed Pre-examine

Design of Experiment (DOE)

Base on industry progress in four-pint cyclic bend experiment, we choose deflection and temperature as main factors to investigate. Design two levels for every factor and eight components per test condition, test group as show in Table 1. Additionally we select 24 components to make ATC test for there BGAs respectively.

Table 1 DOE for Four-point Cyclic Bend test

Std	Run	Component Type	Test Temperature (°C)	Deflection Amplitude (mm)
12	1	BGA228-0.5-12	25	1
5	2	BGA256-1.0-17	100	1.5
8	3	BGA256-1.0-17	25	1
11	4	BGA676-0.8-23	25	1
6	5	BGA228-0.5-12	100	1
4	6	BGA676-0.8-23	25	1.5
7	7	BGA676-0.8-23	100	1.5
9	8	BGA256-1.0-17	100	1
1	9	BGA228-0.5-12	100	1.5
10	10	BGA676-0.8-23	100	1
2	11	BGA228-0.5-12	25	1.5
3	12	BGA256-1.0-17	25	1.5

Test vehicle and method

The test vehicle and method list in Table 2 and the high temperature chamber is designed specially for four-point cyclic bend test at elevated temperature. All samples in four-point cyclic bend test with displacement-controlled mode, and in-situ monitor component's resistance in four-point cyclic bend and ATC test [6].

Table 2 Test Equipment and Method

Expe item	Four-point Cyclic Bend	ATC
Test Equipments	Instron5080, Agilent34970, High Temp Chamber	ATC Chamber, Event Detector
Parameters	Load and Support Span: 110mm/75mm, 1Hz; Sinusoidal	0~100°C, DT =RT=15min, 1Hour/cycle
Reference	JESD22-B113	IPC-9701
Failure Criteria	Monitor Dummy Component Daisy Chain Resistance > 300Ω	

Result and Discussion

Four-point cyclic bend test platform is shown in Figure 2, which comprise of several parts, three different lead-free BGAs are test according to DOE one this vehicle.

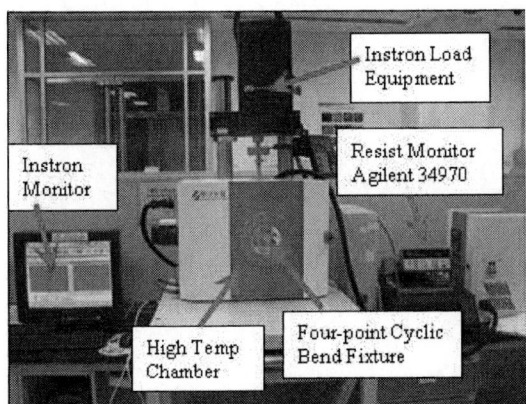

Figure 2 Schematic of Four-point Cyclic Bend test platform

Factors' effect to lead-free cyclic bend fatigue life

Firstly analyze the factor's effect to test result with Minitab [7], as Figure 3 illustrate. Apparently, the deflection amplitude and test temperature are predominant to lead-free solder joint cyclic bend fatigue life, while the inter-effect of three BGAs is not notable.

Then make Weibull distributions of the experiment fatigue life. From the twelve test conditions, it demonstrates that the deflection has different degree of influence to three lead-free BGAs, as shown in Figure 4:

1) Within the two test temperature, the lead-free samples' characteristic fatigue lives in 1.0mm deflection amplitude are obvious higher than 1.5mm's, climb up 2.7 times averagely;

2) The cyclic bend fatigue lives of lead-free samples reduce as the BGA package size grows at the same test temperature [8-9].

The Weibull slops of all test conditions are basically analogous, especially for component BGA228 and BGA256; these suggest the failure mode of lead-free samples in four-point cyclic bend is similar.

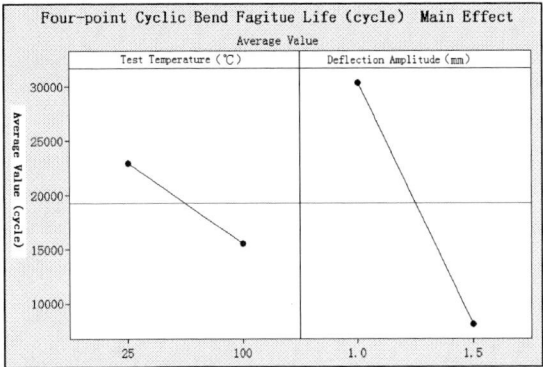

Figure 3 Factor's Main Effect in Four-point Cyclic Bend

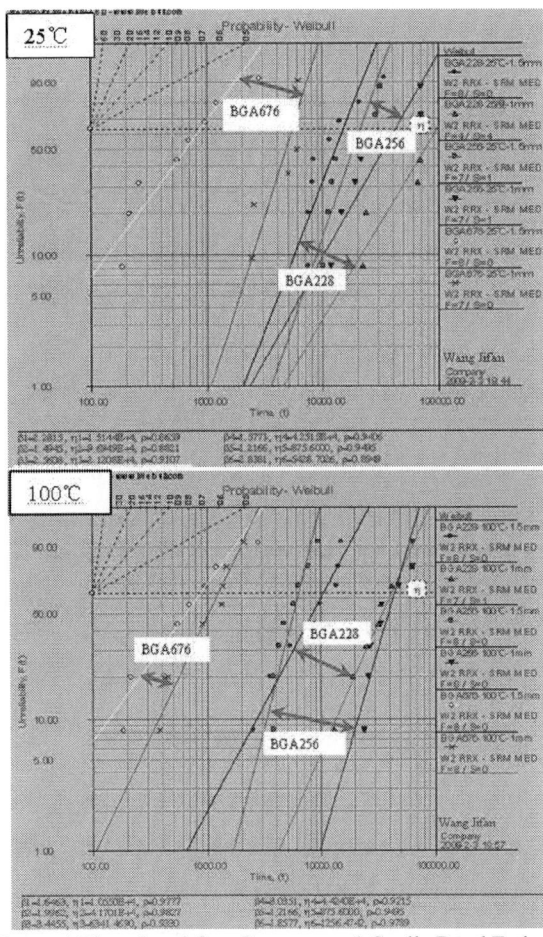

Figure 4 Three Lead-free Components' Cyclic Bend Fatigue Lives Vs Deflection Amplitude at 25°C and 100°C

Figure 5 displays the lead-free samples' bend fatigue lives decline with test temperature rise, this coherent with industry study [10]. Compare with 100°C test cases, lead-free samples'

978-1-4244-4658-2/09 $25.00 © 2009 IEEE   497

bend fatigue lives are two times at 25°C, this may attribute to the higher temperature reduce the fatigue resistance of lead-free solder.

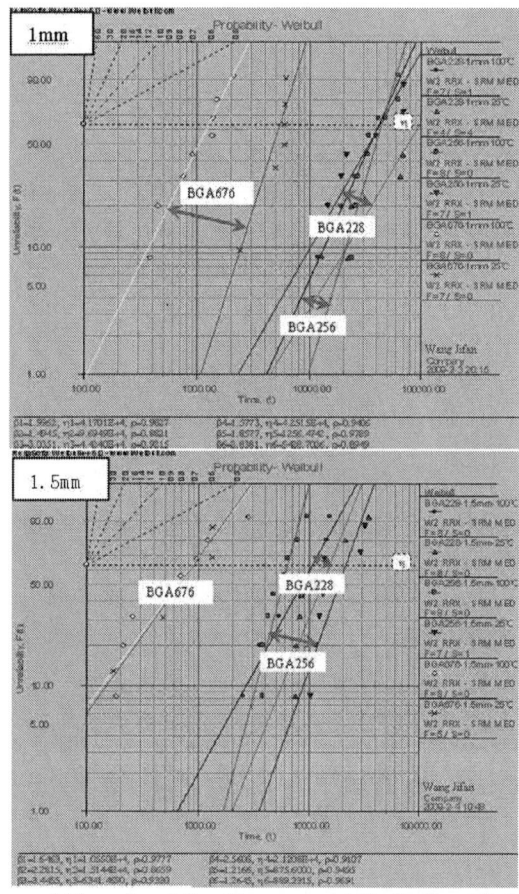

Figure 5 Three Lead-free Components' Cyclic Bend Fatigue Lives Vs Test Temperature at 1mm and 1.5mm

Lead-free joints' failure analysis in four-point cyclic bend

Primarily analyze the failure joints location and distribution in four-point cyclic bend with dye & pry, it indicates that the failed joints mainly distribute on the out line and corners of inner line that parallel to support bar (Figure 6), because such locations suffer larger stress in the same test condition for package enhances bending rigidity[11-12]. For large package and thick body, this effect will be more severe to samples.

Then cross-section three BGAs' solder joint in out two lines, which is vertical to each other. The result exhibits four types of fracture mode are found: ①fracture at component side; ②fracture at PCB side; ③ pad crater and ④trace open.

Statistic failure modes of the three BGAs in twelve test conditions, which shows the majority is fracture at component side, especially for sample BGA228 and BGA256, this is correspond with Weibull slopes shown in Figure 4, and also implies the solder joint fatigue stress in four-point cyclic bend is moderate; while for component BGA676, cross-section shows the failure modes are complex, all the four types can be found, and the failure modes as shown in Figure 7(③,④)

compose of about 10% respectively. This phenomenon maybe related to the larger component package size (23mm×23mm) and thicker body (2.45mm, while the BGA228 and BGA256 are both 0.8mm), thicker package enhances the rigidity of component, and then the lead-free joints suffer higher bend stress in same test condition [13].

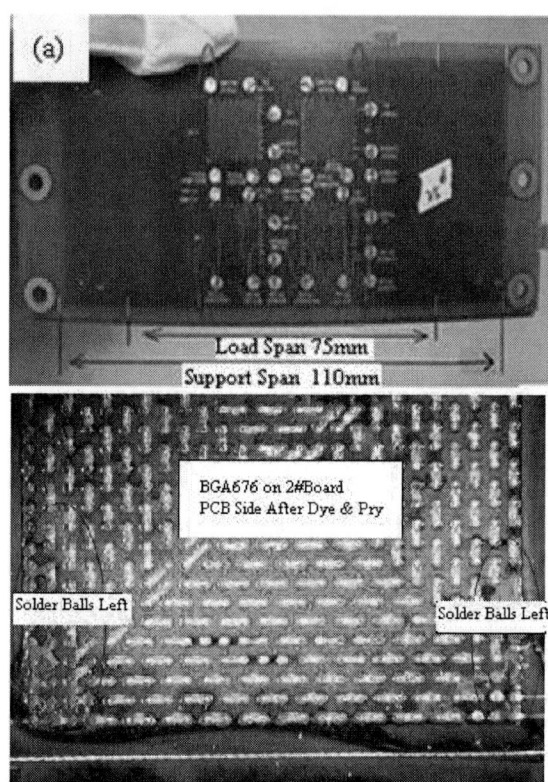

Figure 6 Failure Joints Distribution in Four-point Cyclic Bend (a) Sample Panorama (b) PCB side

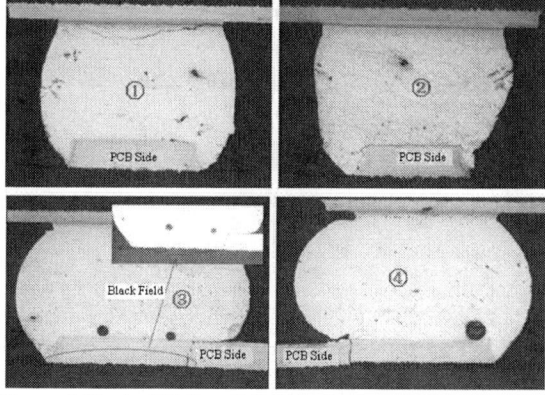

Figure 7 Typical Failure Mode in Four-point Cyclic Bend

Characteristic failure of lead-free solder joint in four-point cyclic bend is found, which indicate as Figure 8: crack initiate at interface of solder and IMC, propagate horizontally for a distance then swerved to Cu pad and induce trace open. This phenomenon only happens at PCB side, and possible reason is

PCB pad is NSMD design, the solder contacts one part of leaked trace in assembly process. In cyclic bend fatigue test, when crack propagates along the interface of solder/IMC to the contiguity of Cu trace and PCB pad, for the structure integrity transforms notably, the strength of Cu trace is weaker than solder/IMC combination then propagates direction change. On the same time, IMC fracture characteristics can be observed at PCB side as show in Figure 8 (b), this may be in cyclic bend loading, the separate IMC from solder continue bend to crack vertically as main crack propagate.

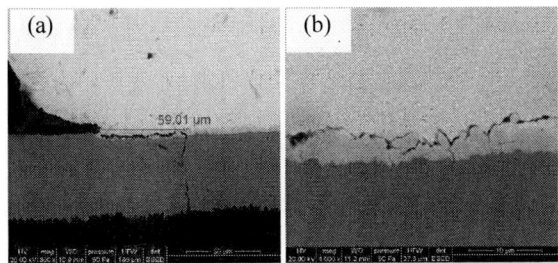

Figure 8 Characteristics Failure of Lead-free Solder Joint in Four-point Cyclic Bend (a) Trace Open (b) IMC Crack

Lead-free joints' failure analysis in ATC

Thermal fatigue is the main stress of surface mount joint suffered in field and ATC test, there are two kinds of joint failure mode observed in thermal fatigue condition (Figure 9): ①joint failed at component side, crack started at interface of solder/IMC, propagate horizontally till joint fracture; ②crack initiate at interface of solder and component/PCB side, run through solder body at 45°C till fracture, as show below. This proves the SAC joint failure mode is similar to SnPb in creep fatigue circumstance.

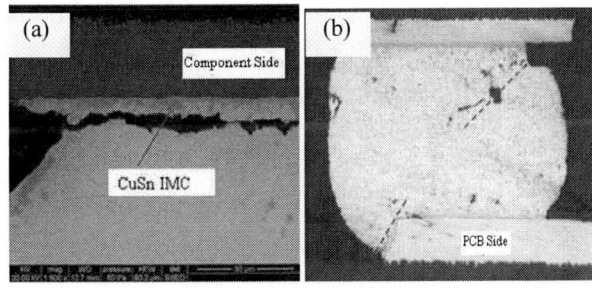

Figure 9 Characteristic Failure Mode in Thermal Fatigue Stress (a) Horizontal Crack (b) 45°C Propagate Crack

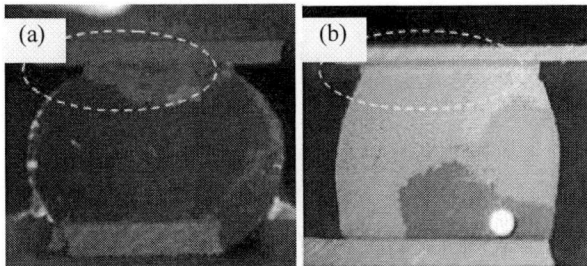

Figure 10 Metallographic structure difference under polarized light (a) ATC (b) Four-point Cyclic Bend at 1.0mm/ 100°C

From Figures 8-9, it shows the SAC joint image in optical microscope and SEM is similar between four-point cyclic bend and ATC, while under polarized light, metallographic structure difference of lead-free joint can be found between such two load circumstances, as shown in Figure 10. It can be found that in ATC test, the crack appears different color, which indicates a clearly recrystallization zone. While in four-point cyclic bend test, joint recrystallization is not observed at crack initiation area, the metallographic structure remains the same as pre-test. This maybe attributes to higher temperature and longer dwell time, for atom's diffusion ability is higher at ATC, and can recrystallize easier.

Characteristic life: four-point cyclic bend Vs. ATC

Compare the characteristic life of lead-free joints in four-point cyclic bend and ATC; presume upon the relationship between such two methods in assessment lead-free joint reliability [14]. From characteristic life trend analysis in Figure 11, it seems that in four-point cyclic test condition of 1.5mm/25°C and 1.0mm/100°C, lead-free joints' characteristic lives between four-point cyclic bend fatigue test and ATC have the same variation trend.

Combine with failure mode analysis of test samples, it indicates that in four-point test condition of 1.5mm/25°C, BGA256 and BGA676 samples reveal some(more than 10%) non-component side failure mode as show in Figure 12, these are clearly different to ATC samples'. While in test condition of 1.0mm/ 100°C, three BGAs' major failure mode is fracture at component side as shown in ATC. So we think in some condition (e.g. known some type of component's ATC life, while assembly process has a change, like rework), through four-point cyclic bend fatigue test can qualitative estimate lead-free joint reliability, such method can quickly get reliability information with the same variation trend as ATC. For the BGAs' in this experiment, we recommend use four-point cyclic bend with test condition of 1.0mm/100°C.

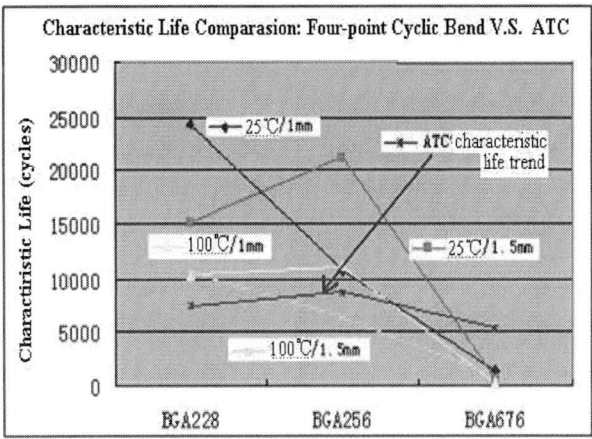

Figure 11 Characteristic lives variation trend: Four-point Cyclic Bend Vs. ATC

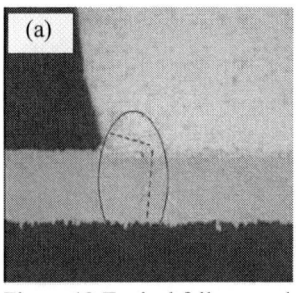 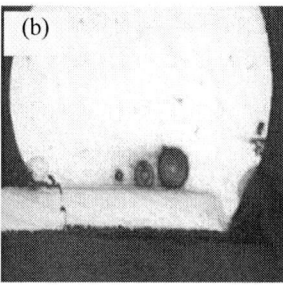

Figure 12 Typical failure mode in test condition 1.5mm/ 25°C (a) BGA256 (b) BGA676

Conclusions

Base on specific design board, the effect of four-point cyclic bend test parameters, such as deflection, temperature to lead-free solder joint fatigue life and failure mode are studied, at the same time compare the life performance of lead-free joints in four-point cyclic bend test and ATC, get following information:

1) Temperature and deflection have great effect to lead-free joint characteristic life and failure mode in four-point cyclic bend test. When test temperature from 100°C decline to 25°C or deflection amplitude from 1.5mm down to 1.0mm, three BGAs' bend fatigue characteristic lives increase averagely 2 times or 2.7 times respectively;

2) There are four typical failure modes found in four-point cyclic bend. For test samples with recommend condition, the major failure mode and characteristic life variation trend can be consistence with ATC;

3) The crack propagates at PCB side along interface of solder/IMC then change direction induce trace open, and IMC crack vertically as main crack propagate are characteristic failure found in cyclic bend test;

4) Under polarized light, lead-free joints' metallographic structure in four-point cyclic bend test conditions is different from ATC's, there is no distinct joint recrystallization zone at crack initiation an propagation area;

5) Develop faster and effective reliability assessment method is meaningful to industry, and this field needs more attention to explore under the condition of failure mechanism/mode and characteristic life variation trend coherent to ATC.

Acknowledgments

The author would like to express their appreciation to Zhou Xin, Sun Fujiang and Luo Meichun for their support on project development. Thanks Guangdong-Ministry of Education Strategic Alliance of Lead-free Electronic Manufacturing for supply experiment resources.

References

1. IPC-SM-785: "Guidelines for Accelerated Reliability Testing of Surface Mount Solder Attachments", Nov. 1992.

2. Robert Darveaux and Ahmer Syed, "Reliability of Area Array Solder Joint in Bending", *SMTA*, 2000, pp. 313-324.

3. Ahmer Syed, "Predicting Solder Joint Reliability for Thermal, Power, & Bend Cycle within 25% Accuracy," *Proceedings of 51st Electronic Components and Technology Conference*, 2001, S08p1.

4. J.D.Wu, *et al.*, "An experimental Study of Failure and Fatigue Life of a Stacked CSP Subjected to Cyclic Bending," *Proceedings of 51st Electronic Components and Technology Conference*, 2001,S30p6.

5. JEDEC Solid State Technology Association, JESD22-B113: "Board Level Cyclic Bend Test Method for Interconnect Reliability Characterization of Components for Handheld Electronic Products," 2006.

6. IPC-SM-785.Guidelines for Accelerated Reliability Testing of Surface Mount Solder Attachments.1992.

7. Statistical software, http://www.minitab.com.

8. Yi-Shao Lai, *et al.*, "Cyclic bending reliability of wafer-level chip-scale packages," *Microelectronics Reliability*, 47 (2007), pp. 111-117.

9. Lei L. Mercado, Betty Phillips, *et al.*, "Use-Condition-Based Cyclic Bend Test Development for Handheld Components," *Proceedings of 54st Electronic Components and Technology Conference*, 2004, pp. 1279-1287.

10. F.X. Che, John H.L. Pang, *et al.*, "Modeling Board-Level Four-Point Bend Fatigue and Impact Drop Tests," *Proceedings of 56st Electronic Components and Technology Conference*, 2006, pp. 443-448.

11. Jennifer Nguyen, David Geiger, *et al.*, "Solder Joint Characteristics and Reliability of Lead-Free Area Array Packages Assembled Under Various Tin-Lead Soldering Process Conditions," *Proceedings of 57st Electronic Components and Technology Conference*, 2007, pp. 1340-1349.

12. Brian Roggeman, Michael Meilunas, "Investigation of Solder Joint Reliability Under Various Bending Loads," 2007, Unovis.

13. Pardeep K. Bhatti1, Min Pei, *et al.*, "Reliability Analysis of SnPb and SnAgCu Solder Joints in FC-BGA Packages with Thermal Enabling Preload," *Proceedings of 57st Electronic Components and Technology Conference*, 2006, pp. 601-606.

14. Ilho Kim and Soon-Bok Lee, "Fatigue Life Evaluation of Lead-free Solder under Thermal and Mechanical Loads," *Proceedings of 57st Electronic Components and Technology Conference*, 2007, pp. 95-104.

A Warpage of Wafer Level Bonding for CIS (CMOS Image Sensor) Device Using Polymer Adhesive

Jae-Hyun Park[1a], Ji-Young Lee[1], Min-Kyo Cho[2], Jae-June Kim[2], and Gu-Sung Kim[1*]

[1]Kangnam University, Yongin 446-702, Korea

[a]Present Address : EPWorks Co.,Ltd., ESIP Lab. Gyeonggi 464-070, Korea

[2]EPWorks Co.,Ltd., ESIP Lab. Gyeonggi 464-070, Korea

E-mail : windmaster@naver.com, *gkim@kangnam.ac.kr

Abstract

The polymer adhesive bonding technology using wafer-level technology was investigated to adhere silicon to glass wafer and it analyzed warpage caused in cemented wafer and the degree of intensity. We executed the wafer adhesion depending on temperature (130°C, 190°C), the pressure (5000N, 8000N), the height of the adhesive layer (10μm, 20μm) and the adhesive time (process time, the time for temperature rising) of each of the silicon and glass wafer. The warpage was measured using three-dimensional measuring equipment and the results were caused by the differences of CTE and the physical stress. It was also confirmed that the more the temperature of Si wafer, adhesive pressure and adhesive layer was lowered in order to improve the warpage results, the more warpage decreased, and that the adhesive time and temperature differences of glass wafer were relatively insufficient factors. To judge the degree of wafer adhesion, the shear intensity was tested and it showed that the higher the adhesive temperature of glass wafer was, the more degree of shear intensity it showed, and that the other conditions showed little effects. Also, in the center of the adhesive wafer where the warpage occurred showed that the more it was getting to the edge, the more shear intensity decreased, and that the stress related with the occurrence of warpage also had effects on the state of adhesion.

Introduction

Wafer level packaging technology has received a lot of attention as mass production that aims at realizing light, thin, short, and small features in semiconductor production technology. Recent electronic packaging technology is used in CIS, Micro Electro Mechanical Systems (MEMS), etc., where highly integrated and highly reliable products are developed and produced using wafer level packaging technology. These advancements are focused on core technologies such as wafer bonding and Through Silicon Via (TSV) [1-2]. Wafer bonding technology is also realized in highly integrated devices such as memory and System In Package (SIP). Following this trend, the size of the applied wafers also increases, demanding advanced technologies for 8-inch and 12-inch wafer [3]. The increased wafer size must have an important technological basis of reliability. Hence, demand is higher than ever for research regarding warpage phenomena of wafers following changes in thermal characteristics, uniform bonding conditions, and bonding reliability.

In general, wafer bonding is divided into the following three methods: direct, anodic, and intermediate layer. In the direct bonding method, the initial bonding is done by the hydrogen bonding force of the surface of both substrates and a covalent bond is formed through a thermal treatment process at a high temperature (over 800°C). It is often used for bonding identical substrates but has problems related to high-temperature process treatment [4-6]. The anodic bonding binds the glass and silicon substrates at 300°C~400°C and applies high voltage to form SiO_2 covalent bonding. Even though this is a relatively low-temperature process, it has shortcomings related to high electrical reliability and thermal stress that result from coefficient of thermal expansion (CTE) differences that limit the selectivity of substrates [7-9]. These two methods have large limitations in applied fields and actual implementation. On the contrary, the intermediate layer bonding method uses diverse bonding layers for bonding, such as metal and polymer, which enables the pattern of bonding. Moreover, the selectivity of bonding temperature is relatively easy compared with the other methods. As a result, a number of research projects are in progress that fit specific usages, such as wafer stacking, sensors, MEMS package, etc [10-12].

Chip scale packages (CSPs) for CMOS Image Sensor (CIS) must have cover glass to protect the sensor from the outside environment and a low-temperature bonding process. This process should be performed within a temperature range that does not pair the sensor, which is extremely sensitive to temperature. However, the intermediate layer bonding method has significantly weaker bonding strength, compared with the direct and anodic bonding methods. Hence, continuous efforts are necessary to develop materials that can enhance performance and improve the process. Also when using the intermediate layer bonding method, a hermetic sealing reliability test is required.

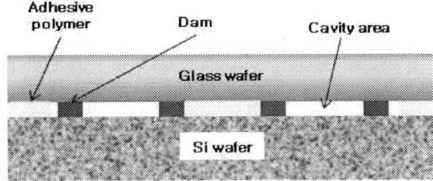

Fig. 1 Schematic of intermediate layer wafer bonding using polymer for CIS

In this study, an 8-inch, low-temperature intermediate layer bonding process was performed to package CIS devices, using a cover glass to protect the image sensor and a type of Si wafer bonding. As shown in Fig. 1, a dam structure was developed for a bonding that forms a cavity for the Si and glass wafer and maintains a uniform height. Next, wafer bonding was performed by patterning the bonding polymer. The wafer warpage was measured according to the bonding process, and the impact of each variable (bonding temperature and pressure, the intermediate bonding layer's height, bonding and heat-up times) on the warpage phenomenon was analyzed.

Experiments

An 8-inch bare Si wafer and borosilicate glass wafer were used for image sensor packaging. A borosilicate glass wafer has higher heat and acid resistance, compared with other glass wafers, and has a low CTE resulting in less difference between the Si wafer in its CTE. Moreover, this kind of wafer is often used in image sensors and display areas due to its light weight and high transmission rate. After evaluating the bonding characteristics, glass transition temperatures, and curing temperatures of the bonding polymers, one was selected that is appropriate for low-temperature bonding.

Table 1 Material properties of test samples

Type	CTE (ppm/°C)	Young's Modulus (Gpa)	Shear modulus (Gpa)	Height (μm)
Si	2.8	190		750
Glass	3.18	70.9	28.9	500
SiO$_2$	0.6	69	30~100	0.1

Table 2 Properties of adhesive polymer

Type	Tg(°C)		CTE(-65~150°C) (ppm)	Tensile Strength (MPa)	Elastic Modulus (GPa)	Water Absorption (%)
	DMA	TMA				
Polymer	215	201	56	90	22	1.5

After evaluating the bonding characteristics, glass transition temperatures, and curing temperatures of the bonding polymers, one (Table 2) was selected that is appropriate for low-temperature bonding.

The process flow for creating the sample is depicted in Fig. 2. An 1000Å oxide film was deposited on a Si wafer by plasma enhanced chemical vapor deposition (PECVD) for uniform bonding between the polymer and the oxide film. On the glass wafer, a dam pattern was formed with thicknesses of 10μm and 20μm using a bonding polymer, with the purpose of maintaining uniform height of bonding. The dam height was set at a minimum and a maximum that allow the use of the polymer without damaging the micro lens of the image sensor. The dam pattern was designed so that the image sensor and pad part could form a cavity space. After curing was completed, its role as a physical supporter was realized. After forming the dam structure, the wafer bonding layer was patterned by placing bonding polymer about 3μm (±0.5μm) higher than the dam pattern. The polymer bonding layer pattern showed dispersion due to the temperature and pressure that occurs during the wafer bonding. The bonding was

performed so that the dispersion did not affect any other patterns.

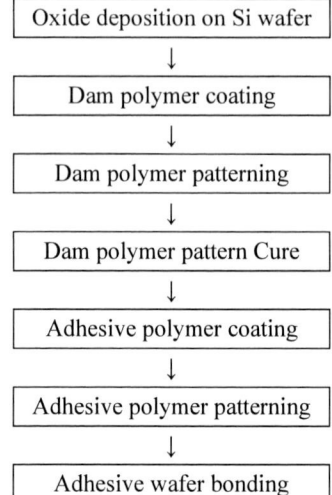

Fig. 2 Wafer level bonding fabrication process flow

The bonding process was conducted in vacuum conditions (1.53×10^{-4}mbar) using Electronic Visions Group's (EVG) 520 bonder after the unit processing of the Si and glass wafer. The wafer bonding device used in this experiment is a double-sided heating system with a heater that controls the upper and lower wafers separately. In case of bonding different materials with heterogeneous features, such as thermal expansion coefficient, the bending phenomenon of a substrate can be minimized by increasing or decreasing the temperature. Considering that the bonding polymer has a Tg point around 200°C, the temperature was set below 190°C to satisfies the curing condition. The bonding appearances of the Si and glass wafer were observed following the changes in temperature. The overall bonding by the polymer was observed from the minimum bonding pressure of 5000N, and the effects of various pressures were examined up to 8000N, due to a limit of the device used in this experiment. The process time for this experiment was conducted from 60 minutes to 90 minutes, including the polymer's curing time. The heat-up times were set at 5°C/minutes and 20°C/minutes for the experiment.

In the presence of stress induced from physical pressure, the stacking wafer that bonded at the wafer level has various deformations due to the different physical characteristics of the materials, which results in warpage. Such warpage causes alignment misses or mechanical damage to rear process, such as CMP and bump processes, and dicing, depending on its level. It can also cause troubles in durability test in a reliability test at final product. For the warpage (Fig. 3) induced by Si, SiO$_2$, glass layer after the wafer bonding process using polymer, the residual stress was checked by curvature measurements [13].

$$\sigma = \frac{E_s}{1-v_s} \frac{t_s^2}{6t_f}\left(\frac{1}{R} - \frac{1}{R_0}\right) \qquad (1)$$

Where E_s is Young's modulus, v_s is a Poisson's ratio of the silicon wafer, and t_s and t_f are thicknesses of the silicon and the polymer, respectively. R_0 and R are the silicon wafer radius of curvature before and after the bonding, respectively. To measure the warpage, the experimental wafer was laid on a level table and its virtual criteria points were formed to measure heights at each point using 3-D non-contact measuring equipment.

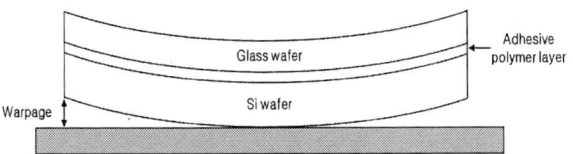

Fig. 3 Schematic diagram of warped wafer

Results and Discussion

Fig. 4 presents cross-sectional images of the wafer sample where the wafer is bonded, forming a height of 10μm and 20μm. The figure also shows cross-sections of the Si, glass, and polymer bonding layer and the formation of the cavity space. From the figure, it can be confirmed that the polymer bonding layer used as intermediate layers maintains a height of 10μm and 20μm with the dam pattern. It turned out that the dam pattern maintains the bonding layer height for the cavity space pattern within the process factors considered in this study of the low-temperature bonding process, such as temperature, pressure, and time.

Fig. 5 presents the warpage response to the wafer bonding condition. Warpage response to the Si wafer temperature changes from 130~190°C was 224.8~251.8μm,

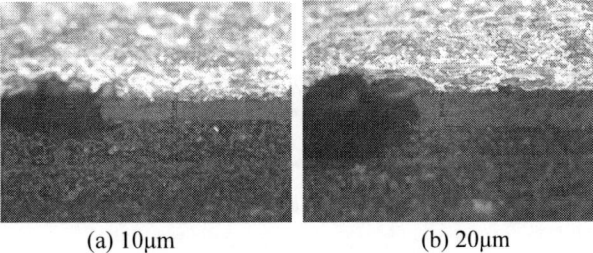

(a) 10μm   (b) 20μm
Fig. 4 Cross-sectional image of wafer bonding

showing a wider changing range, compared with the glass wafer temperature, which was 235.4-241.2μm. Meanwhile, the thermal mismatch between the low-CTE SiO_2 film deposited on the Si wafer and the Si wafer causes an edge load phenomenon, resulting in warpage. This result shows the widest range compared with the results from other process elements. The multi-layered structure had the most severe warpage in response to the thermal mismatch. The reason of less warpage response in the case of a glass wafer can be explained as follows. The warpage effects are larger for the SiO_2 thin layer deposited on Si and Si substrate due to the CTE gap. The intermediate bonding layer, which has a low Young's modulus and its cavity space serves as a buffer layer, lightly influences the wafer's overall warpage results. As for

the bonding pressure, the warpage was 247μm at 8000N and decreased to 229.6μm when the pressure was lowered to 5000N. In a multi-layered structure, interior stress is induced in proportion to the bonding pressure, and this increases the warpage along with the stress from the thermal mismatch. The lower the bonding pressure, the smaller the interior stress, resulting in less warpage. The warpage of the bonding wafer with a 20μm-high intermediate bonding layer was 244.6μm, while the warpage dropped to 232μm when the height was 10μm, indicating a lower height implies less severe warpage. The intermediate bonding layer used in this study has a larger difference of Young's modulus compared with Si and glass wafers with a small value of 22GPa. The range of deformation due to substrate stress is narrower with a lower bonding layer that absorbs the bonding stress from thermal and physical stress occurring in both sides of wafer during the bonding process, and a smaller cavity space, which results in decreased warpage. The warpage was 238.8μm and 237.9μm at a heat-up time of 20°C/min and 5°C/min, respectively, showing that the warpage lessens under the condition of longer heat-up time, i.e., the one with 5°C/minutes. The bonding time of 60 minutes yielded a warpage phenomenon of 241.5μm, which decreased to 235.1μm when the time was changed to 90 minutes. Warpage became severer with a shorter process time and diminished with a longer bonding process time. The thermal and physical deformations during the bonding process occur at the polymer bonding layer, while inducing tensile residual stress at the same time. Due to the general characteristics of the polymer, the residual stress increases as the deformation takes place in a shorter time. It was confirmed that such tensile residual stress was combined with stress that occurs from a difference in CTE and eventually increases the warpage.

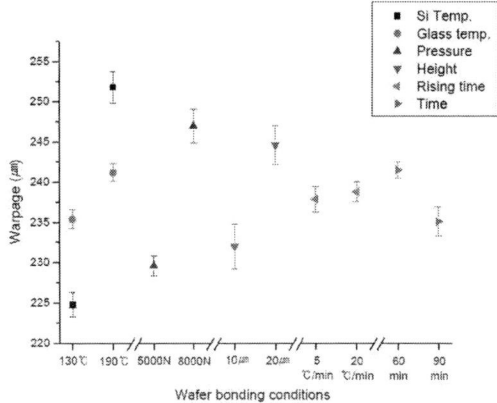

Fig. 5 Warpage test results with bonding temperature, bonding pressure, height of intermediate layer, rising time and bonding time

Conclusion

This study analyzed the polymer bonding method and its optimization regarding wafer level packaging for semiconductor devices that require low-temperature processing, such as image sensors and MEMS and intermediate layers. Specifically, uniform maintenance of intermediate layers and cavity space, heterogeneous substrates, and the polymer intermediate layers' warpage phenomenon were examined by controlling each process variable. Moreover, after dicing a sample of wafer level bonding, the bonding strength was measured at the chip level. The conclusions from each experiment and the results and interpretation are summarized below.

This study involved an experiment to examine whether the intermediate layers and cavity space uniformly maintain their heights when process conditions change, such as bonding temperature, bonding pressure, bonding time, intermediate layer height, heat-up time, etc. The experimental results confirmed a range of $\pm 1\mu m$ when $10\mu m$ and $20\mu m$ were chosen as target levels for the intermediate layers and cavity space, respectively.

When Si and glass wafer polymer are used for wafer bonding, warpage occurs in both cases, where the results change most sensitively to the changes in temperature due to different CTEs. Within the control range of this study, the warpage results decreased with lower bonding temperature, bonding pressure, and height of polymer bonding layer and higher processing time. Meanwhile, it turned out that the heat-up time was irrelevant to the results.

References

1. W. H. Ko, J. T. Suminto, and G. J. Yeh, "Bonding techniques for microsensors," *Micromachining and Micropackaging for Transducers*, Amsterdam, The Netherlands: Elsevier (1985).

2. M. A. Schmidt, "Wafer-to-wafer bonding for microstructure formation," *Proc. of the IEEE*, Vol. 86, Aug. 1998, pp. 1575-1585.

3. Yole Developpement, "Market trends for 3D stacking," EMC-3D, Yole Roadmap (2007).

4. P. W. Barth, "Silicon fusion bonding for fabrication of sensors, actuators and microstructures," *Sens. Actuators A*, Vol. 23 (1990), pp. 919-926.

5. C. Harendt, H. -G. Graf, B. Hofflinger, and J. E. Penteker, "Silicon fusion bonding and its characterization," *J. Micromech. Microeng.*, Vol. 2 (1992), pp. 113-116.

6. Q. Y. Tong, G. Cha, R. Gafiteanu, and U. Gosele, "Low temperature wafer direct bonding," *J. Microelctromech. Syst.*, Vol. 3 (1994), pp. 29-35.

7. T. Rogers and J. Kowal, "Selection of glass, anodic bonding conditions and material compatibility for silicon-glass capacitive sensors", *Sens Actuators A*, Vol. 46-47 (1995), pp. 113-120.

8. H. Henmi, S. Shoji, Y. Shoji, K. Yoshimi, and M. Esashi, "Vacuum packaging for microsensors by glasssilicon anodic bonding," *Sens Actuators A*, Vol. 43 (1994), pp. 243-248.

9. Y. T. Cheng, L. Lin, and K. Najafi, "A hermetic glasssilicon package formed using localized aluminum/silicon-glass bonding," *J. Microelectromech. Syst.*, Vol. 10 (2001), pp. 392-399.

10. M. Chiao and L. Lin, "Hermetic wafer bonding based on rapid thermal processing," *Sens Actuators A*, Vol. 91 (2001), pp. 398-402.

11. F. Niklaus, P. Enoksson, P. Griss, E. Kalvesten, and G. Stemme, "Low-temperature wafer-level transfer bonding," *J. Microelectromech. Syst.*, Vol. 12 (2001), pp. 525-531.

12. Maik Wiemer, Chenping Jia, Michael Toepper, and Karin Hauck, "Wafer bonding with BCB and SU-8 for MEMS Packaging," *Electronics system integration technology conference*, 2006, pp. 1401-1405.

13. G. G. Stoney, *Proc. R. Soc.* London, Ser. A, 82, 172 (1909).

Improved Thermal Performance of High-Power LED by Using Low-temperature Sintered Chip Attachment

T. Wang[1,2], G. Lei[2], X. Chen[1], L. Guido[2,3], Khai Ngo, G-Q. Lu[2,3,4]

[1]School of Chemical Engineering, Tianjin University, Tianjin, P.R. China;
[2]Dept. of Materials Science and Engineering, Virginia Tech, USA;
[3]Dept. of Electrical and Computer Engineering, Virginia Tech, USA;
[4]School of Material Science and Engineering, Tianjin University, Tianjin, P.R. China
213 Holden Hall (0237), Virginia Tech, Blacksburg, VA 24061
Phone: 540-231-8423; Fax: 540-231-8919; Email: gqlu@vt.edu

Abstract

Chip attachment is an important step in packaging light-emitting diode (LED) chips. The attachment material provides mechanical support and heat dissipation. As high-power LED chips are targeted at general illumination applications, the thermal property of the die-attach material is critical to the light output and degradation of the lighting device. Silver epoxy, lead-free solder paste, and nanosilver paste, were evaluated as the attachment material for LED chips. The interconnected chips were tested at current-density levels up to $3 \times 10^2 A/cm2$. We found that at high levels of current density, the chips that attached using the nanosilver paste had a significantly higher light output than the lead-free soldered chips, which in turn had a higher light output than the epoxy-glued chips. This can be explained by the higher thermal conductivity of the sintered silver than that of solder and epoxy. Higher thermal conductivity of the attachment material results in a lower junction temperature at a given level of current density, thus causing a higher light output. Our results suggest that LED performance can be significantly improved with a die-attached nanosilver paste.

Introduction

The promise of enormous energy savings by replacing incandescent light bulbs with energy-efficient LED or solid-state lights has led to rapid surge of investments and expectation in high-power LED technologies. However, significant technical challenges have to be overcome before requirements for performance (such as brightness or lumens per lamp, and efficiency or luminous efficacy), reliability or lifetime, and cost of LEDs can be met for their widespread applications in general illumination. One of the key technical advancements needed to ensure development of reliable and high-power LED lights is removal of high heat flux generated by LEDs [1-3]. Recently, the technical community has paid increasing attention to thermal management of HB-LEDs by developing highly conductive heat spreader substrates and high-conductivity chip-attach materials [4-6]. Chip-attach material is important for packaging of high-power LED chips because it has to reliably bond the chip to package and effectively dissipate heat to keep chip-junction temperature low. Liu *et. al.* [4] discussed various thermal solutions for chip-on-board packaging of high-power LED chips and showed that the chip-junction temperature could be drastically reduced by using a die-attach material with thermal conductivity greater than 100 W/mK.

In this work, we used the temperature dependence of LED light output [7] to evaluate relative thermal performance of high-power LED chips bonded by three different types of die-attach materials in their paste form: silver epoxy, lead-free solder, and nanosilver powder. The nanosilver paste has been widely studied [8-14] as a lead-free solder replacement for attaching power semiconductor chips. It has been shown to provide high-performance, high-reliability, and high-temperature packaging of SiC power devices. From their datasheets, the three attachment materials have thermal conductivities covering a wide range, and they rely on uniquely different processes to form the joints. Thus, they are good representations of materials to study the effects of die-attach thermal properties and processing on LED performance and reliability.

Experiments

High-power, blue GaN-based LED chips of 1 mm × 1 mm in size with gold back metallization were acquired from a commercial vendor. The three types of die-attach materials studied were: (1) silver-filled epoxy from Emerson & Cuming, Inc.; (2) lead-free solder paste (95.8Sn-3.5Ag-0.7Cu); and (3) nanoscale silver paste or nanoTach® from NBE Technologies, LLC (www.nbetech.com). Substrates used for attaching the LED chips were made from a direct-bond-copper (DBC) alumina board. Electrical circuit patterns were etched on the substrates and then electroless-plated with nickel and gold. Stencil-printing was used to apply a layer of the die-attach material on substrate followed by chip placement.

Chip attachments were formed by using heating profiles recommended by the respective suppliers of the materials: with the silver epoxy, die-attach joints were formed through cross-linking of long-chain polymer molecules or commonly known as polymer curing at peak temperature of 150°C; with the lead-free solder paste, joints were formed through melting at peak temperature of 250°C and then solidification of the solder alloy; and with the nanosilver paste, joints were formed through densification of nanoparticles of silver at sintering temperature of 300°C. Thermal conductivities of the three different joints claimed by their suppliers are: 10 W/m-K for the epoxy, 50 W/m-K for the solder, and 200 W/m-K for the sintered silver. X-ray inspection of the joints shown in Fig. 1 reveals good structural integrity, i.e. no voids or cracks, in all three.

After chips were attached, they were bonded with six 25-micron gold wires on each electrode. Use of multiple gold wires was necessary to power the chips at high current density of 3×10^2A/cm^2. Relative optical performance of the attached LED chips was evaluated one at a time under a microscope outfitted with a silicon detector at its camera port. Relative light output as a function of electrical current was measured. To characterize the thermal performance of each chip attachment, a fine thermocouple was fixed on the substrate near the edge of the LED chip to monitor temperature versus electrical current.

Fig. 2 shows the experimental setup and sample configuration.

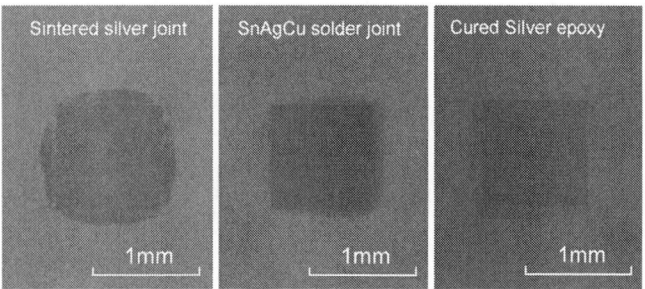

Fig. 1 X-ray images of LED samples bonded by the three different die-attachment materials

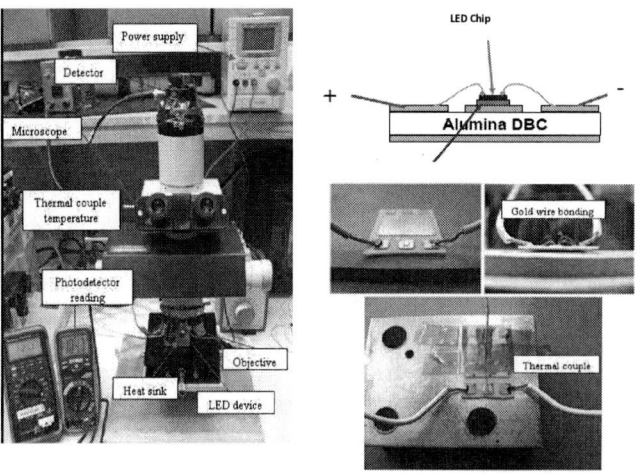

Fig. 2 An optical setup for measuring relative performance of attached LED chips and configuration of a sample under test

Results and Discussion

Fig. 3 shows the temperature dependence of LED light output. LED chips attached by the three different die-attach materials were heated, one at a time, on a hotplate from room temperature to 80°C. At every increment of five degrees, the chip temperature measured on the substrate near the chip edge was allow to reach steady state, after which a low electrical current of 140 mA was used to drive the chip and a reading of light-output intensity was taken. The figure shows nearly identical dependence of light output on temperature for the three die-attach materials.

Plotted in Fig. 4 are the traces of relative light-output intensity versus low electrical current obtained on LED chips attached by the three die-attach materials: silver epoxy, lead-free solder, and nanosilver paste. The electrical current was increased manually from 0 to 350 mA at 50 mA per step. At

each current level, enough time was given to allow the device to reach steady state, and then readings of light output and temperature were taken. Plots of the substrate temperature at the edge of the LED chip versus electrical current are shown in Fig. 5. The ambient temperature was at 200°C for all the tests.

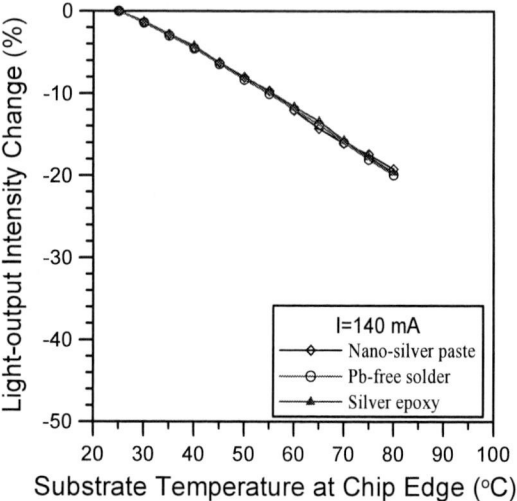

Fig. 3 Nearly identical temperature dependence of light output of LED chips attached by the three different die-attach materials

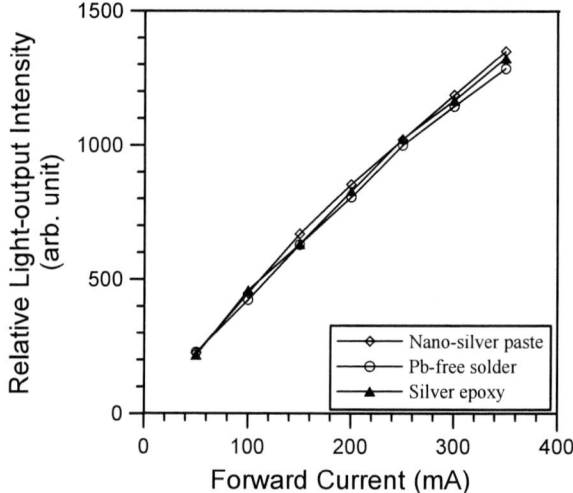

Fig. 4 Plots of steady-state relative light output intensity versus low electrical current of LED chips attached on DBC substrate using the three different die-attach materials.

At the low levels of electrical current, the differences in light-output intensity of the three types of LEDs are slight. This is because the heat power generated by the chip is low, thus the temperature difference between the chip junction and DBC substrate, which is a product of heat power and thermal resistance of the die-attach layer, is small. As seen in Fig. 5, the largest substrate temperature difference amongst the three types of LED devices is only about 2°C at 350mA even though the die-attach materials have very different values of thermal conductivity. The substrate temperature in the LED attached by the nanosilver paste was slightly higher than that in the soldered LED, which was slightly higher than that in the glued LED. Higher substrate temperature means smaller temperature difference between the chip junction and substrate because of higher thermal conduction of the die-attach layer.

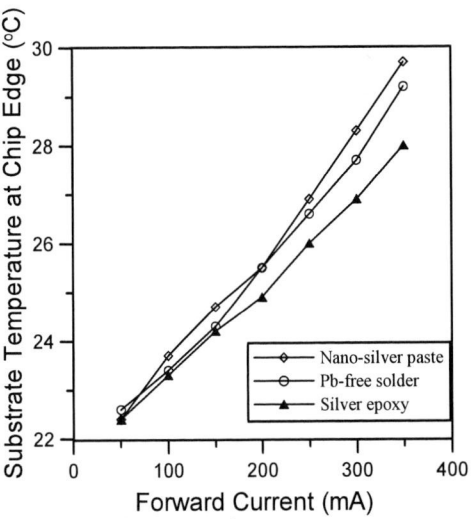

Fig. 5 Plots of steady-state substrate temperature versus low electrical current of the three LED samples in Fig. 4

As the electrical current was pushed to much higher levels, self-heating of the chip became so profound that there can be a large temperature difference between the chip junction and the DBC substrate if thermal resistance of the die-attach layer is large. Fig. 6 shows the substrate temperature versus current measured on the three LEDs. The current was increased continuously at a rate of 6 mA/s from 0 to 3 A, and the substrate temperature was read at every 150 mA intervals. The temperature differences amongst the three LEDs grew wider with increasing current. The largest difference between the nanosilver-sintered LED and silver-epoxied LED reached more than 10°C at 2.5 A. Beyond 2.5 A, the epoxied LED was irreversibly burned.

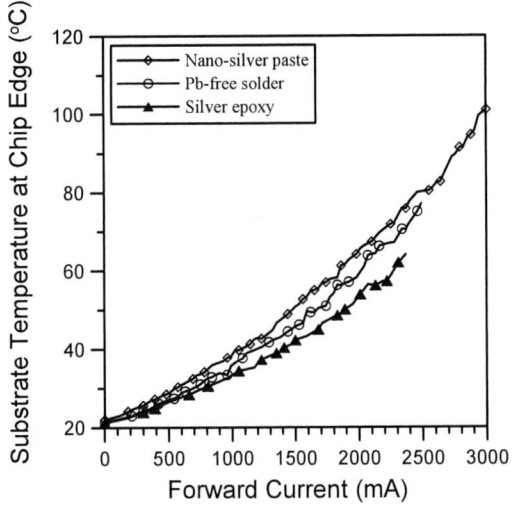

Fig. 6. Plots of substrate temperature vs. current ramped up at 6mA/s on LED chips attached by the three types of materials.

The low substrate temperature near the epoxied LED chip means that the chip junction temperature was high. One expects that the light-output intensity in the epoxied LED would be lower than those of the other two. This was confirmed from the plots of relative light-output intensity versus current shown in Fig. 7 obtained under the same current ramping condition. It is seen that the nanosilver-sintered LED chip had over 15% more light output than the soldered and epoxied LED chips. This can be attributed to better thermal performance offered by the sintered nanosilver joint than the other two die-attach materials. Given its high reliability and high-temperature capability demonstrated in power semiconductor device packaging, the nanosilver paste also seems to be a superior die-attach material for high-power LED packaging.

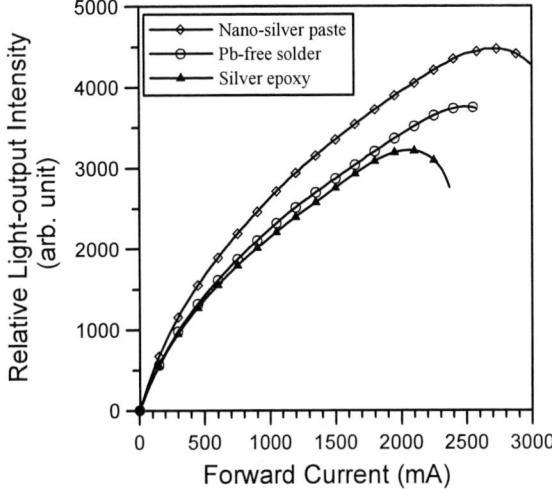

Fig. 7 Relative light-output intensity vs. electrical current measured on the three LED devices under the same test condition as in Fig. 6

Summary and Conclusion

Using the temperature dependence of LED light-output intensity, the effect of die-attach thermal property on thermal performance of high-power GaN blue light-emitting devices was characterized. An emerging die-attach material, nanoscale silver paste processed by low-temperature sintering, was evaluated along with two commonly used die-attach materials, silver epoxy and lead-free solder paste. We found that chip-to-substrate joints formed by the sintered nanosilver paste gave significantly improved thermal performance, thus more light output than the joints formed by the two conventional materials.

Acknowledgments

The authors are grateful for financial support from the US National Science Foundation under Award Number EEC-9731677, US National Science Foundation (SBIR-STTR Award No. 0740927), Chinese National Natural Science Foundation of China (#: 50528506), and the Program of Introducing Talents of Discipline to Universities (#: B06006). We are also grateful to Dr. Jesus N. Calata's many helpful suggestions on the experiment.

References

1. Arika M., C.B., S. Weaverb, and J. Petroskic, "Thermal Management of LEDs: Package to System," *Proceeding of the Third International Conference on Solid State Lighting, SPIE.*, San Diego, CA., 2004, pp. 64-75.

2. Haque, S., D. Steigerwald, S. Rudaz, B. Steward, J. Bhat, D. Collins, F. Wall, S. Subramanya, C. Elpedes, P. Elizondo, P. S. Martin, "Packaging Challenges of High-Power LEDs for Solid State Lighting," *Society of Photo-Optical Instrumentation Engineers (SPIE)*, 2003, pp. 881-886.

3. Zweben, C., "New Material Options for Light-Emitting Diode Packaging," *Light-Emitting Diodes: Research, Manufacturing, and Applications VIII, Proceedings of SPIE.*, San Jose, CA., 2004, pp. 173-182.

4. Liu, J.G., D. Xiong, and P. Panaccione, "Chip-on-Board Packaging and Thermal Solutions for 100W Single Large Area LED," *5th International Conference and Exhibition on Device Packaging*, Scottsdale/Fountain Hills, AZ., 2009.

5. Welsh, A. J., L. Fu, and W. So, "Die-Attach Epoxy Reliability of InGaN LEDs," *Solid State Lighting II, Proceeding of SPIE.*, Seattle, WA., 2002, pp. 68-73.

6. Zhang, K., and M. F. Yuen, "Heat Spreader with Aligned CNTs Designed for Thermal Management of HB-LED Packaging and Microelectronic Packaging," *7th International Conference on Electronics Packaging Technology, ICEPT '06*, 2006, pp. 1-4.

7. Schubert, E. F., "Light-Emitting Diodes," United Kingdom /Cambridge: The Edinburgh Building, 2003.

8. Bai, J. G., Z. Z. Zhang, J. N. Calata, and G-Q. Lu, "Low-Temperature Sintered Nanoscale Silver as a Novel Semiconductor Device-Metallized Substrate Interconnect Material," *IEEE Transactions on Components and Packaging Technologies*, 29(3) (2006), pp. 589 - 593.

9. Bai, J.G., and G-Q. Lu, "Thermomechanical Reliability of Low-Temperature Sintered Silver Die-Attached SiC Power Device Assembly," *IEEE Trans. on Device and Materials Reliability*, 6(3), 2006, pp. 436 - 441.

10. Bai, J. G., T. G. Lei, J. N. Calata, and G-Q. Lu, "Control of nanosilver sintering attained through organic binder burnout," *Journal of Materials Research*, 22(12), 2007, p. 7.

11. Calata, J. N., T. G. Lei, and G-Q. Lu, "Sintered nanosilver paste for high-temperature power semiconductor device attachment," *Int. J. Materials and Product Technology*, 34(1/2), 2009, pp. 95-110.

12. Chen, X., R. Li, K. Qi, and G-Q. Lu, "Tensile Behaviors and Ratcheting Effects of Partially Sintered Chip-Attachment Films of a Nanoscale Silver Paste," *J. of Electronic Materials*, 37(10), 2008, pp. 1574-1579.

13. Lu, G.-Q., J. N. Calata, and T. G. Lei, "Low-temperature sintering of nanoscale silver paste for power chip attachment," *the 5th International Conference on Integrated Power Electronics Systems (CIPS'08)*, Nuremberg, Germany, 2008.

14. Wang, T., X. Chen, G-Q. Lu, and G-Y. Lei, "Low-Temperature Sintering with Nano-Silver Paste in Die-Attached Interconnection," *Journal of Electronic Materials*, 36(10), 2007, pp. 1333-1340.

Fundamental Studies on Whisker Growth in Sn-based Solders

Mengke Zhao, Hu Hao, Guangchen Xu, Jia Sun, Yaowu Shi, Fu Guo
College of Materials Science and Engineering, Beijing University of Technology
Beijing, 100124, P. R. China, E-mail: zhaomengke@emails.bjut.edu.cn

Abstract

Sn whisker growth is considered as a crucial reliability issue in the electronic packaging industry, especially with the massive application of Pb-free solder alloys. In this paper, we report our fundamental studies on Sn whisker growth through accelerated tests. Excessive rare earth element addition is one way we employed to investigate the kinetics of whisker growth. Several aspects of the morphological features were characterized and modelled. The growth mechanism was clarified through observing the diffusion path of oxygen atoms to form rare earth oxides. Our attempt to clarify the growth mechanism of the whiskers observed during electromigration test was be presented based on the experimental efforts on the one-dimensional joint specimens. Such mechanism was investigated from a series of comparative experiments to understand whether mass movement from electrical current or internal stresses from Joule heating induced diffusion plays the crucial role to form such whiskers.

Introduction

Metal whiskers are hair-like crystal structures commonly grown from the surfaces finished with pure metals or alloys of the enriched component. The dimensions of whiskers are usually of some microns in the diameter but up to a length of hundreds of microns [1]. However, the reliability issue caused by the growth of metal whisker could not be neglected. The key problem is the short circuit of electronic products. Whiskers can grow between adjacent pins and cause either a transient short as the whisker is burned open, or a permanent short.

So far, five explanations on the growth theories of tin whisker are as follows: the dislocation movement mechanism [2], recrystallization mechanism [3], oxide layer rupture mechanism [4], IMCs oxidation and decomposition mechanism [5], hydrogen-induced whisker growth mechanism [6]. Although the driving force for whisker growth is still under discussion, it is commonly believed that the compressive stress inside the solder plays an important role in whisker formation and growth.

It is widely known that by adding rare earth elements, the internal compressive stress is induced through the chemical reaction. Hence the whisker growth can be accelerated. Meanwhile, the electromigration of the metal atoms can also build up a compressive stress field on the anode side.

Therefore in this study, we employed this two different accelerated tests to investigate the mechanism for whisker growth.

Experimental procedure

Accelerated tin whisker growth test was conducted by the following two ways. One is excessive rare earth elements addition, the other is electromigration.

In the excessive RE elements addition test, the pure Sn, Ag and Cu metals were used as raw materials. The pure metals were mixed and then melted in an Al_2O_3 ceramic crucible at 550°C-600°C for 30 mins. With the help of a stainless steel bell with holes on the sidewall, the RE was pushed into the liquid alloy. After the RE melted, the melted alloy was held for 2 hours. To homogenize the solder alloy, mechanical stirring was performed every 10 minutes using a stainless steel rod. During the melting, a eutectic salt of KCl and LiCl, the weight ratio of which is 1.3:1, was used over the surface of the liquid solder to prevent oxidation. The melted solder was chill cast into a rod, and then cold-rolled into solder sheet with thickness of 0.1 and 0.2mm.

In the electromigration test, the eutectic SnBi solder balls, the diameter of which are 500μm were obtained from Accurus Scientific Co., Ltd.. A solder ball was placed in the U-shaped groove on the Al board, and then stuck by two Cu wires, the diameter of which is also 500μm. The temperature of the set-up was raised to 150°C in few minutes before cooling in the air in order to obtain the fine eutectic microstructure.

In order to facilitate the following test procedures, the joints were cold mounted in the epoxy resin. In order to get the current density of 104A/cm2, the sample should experience further grind and polish process to reach the maximum cross-section area. The schematic illustration of the solder joint is shown in Fig. 1.

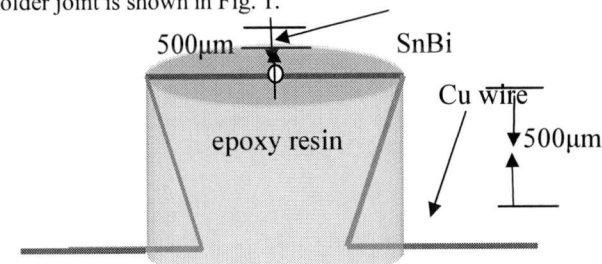

Fig. 1 The schematic illustration of the solder joint

Results and Discussion

3.1 Accelerated whisker growth through excessive RE elements addition

3.1.1 The morphological features of Sn whisker

In our experiments, different morphology of tin whiskers were observed. They are formed due to the compressive stress induced by chemical reactions. As mentioned above, five explanations on the growth theories of tin whisker are generally accepted. In the oxide layer rupture mechanism, it can be said that the strain energy is the driving force, whereas the surface energy is the resistance force for tin whisker growth [4]. Therefore, most of tin whiskers tend to exhibit cylinder-shaped, which could reduce the increment of surface energy during their growing. However, apart from cylinder-shaped, some long tin whiskers can also reduce surface energy. Fig. 2 shows the morphology of long tin whisker. As can be seen, the root of tin whisker tends to grow along their tip and form common surface when they are long enough.

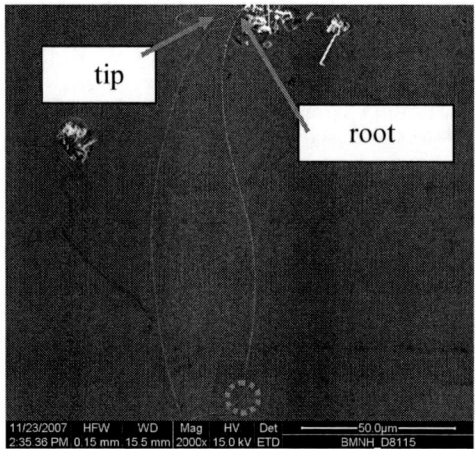

Fig. 2 Contact phenomenon for long tin whiskers

By balancing the strain energy against the surface energy in a unit length of tin whisker,

$$\pi R^2 \varepsilon = 2\pi R\gamma \qquad (1)$$

where R is radius of the whisker, γ is surface energy per unit area, ε is strain energy per unit volume. Rearranging Equation (1), we find that

$$R = \frac{2\gamma}{\varepsilon} \qquad (2)$$

Since γ and ε are constant, it's generally believed that tin whiskers should exhibit invariant cross section during their growth. However, it is different in our study results. Sn whiskers with non-constant cross section were observed in the experiment, as shown in Fig. 3.

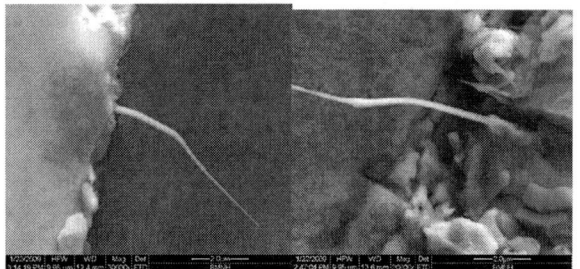

Figure 3 Tin whiskers with non-constant cross section

In the previous study, researches believed that tin whiskers keep constant cross section during growth since the strain energy per unit volume released by tin whisker growth is constant [4]. This is due to the slow, homogeneous

diffusion of Cu into Sn layer, which lead to a homogeneous compressive stress. However, in our work the chemical reaction between RESn3 and oxygen produces a fluctuant driving force for rapid growth of tin whiskers. Since the oxidization of RESn3 phase is unbalanced, and the driving force will vary from time to time, the strain energy per unit volume released by tin whisker growth is no longer invariable. A new problem appears, according to Equation (1). When ε varies from time to time, then the diameter of the tin whisker will vary from time to time. Obviously, most of tin whiskers observed in our work present constant cross section. So this energy model should be modified. Considering that the growth rate of tin whiskers in our work is very fast, we could not ignore the energy of motion of tin whisker, thus we get new formulas as follows:

$$\pi R^2 \varepsilon = 2\pi R\gamma + E \qquad (3)$$

$$E = \frac{1}{2}\left(\pi R^2 L \rho\right)V^2 \qquad (4)$$

$$R = \frac{4\gamma}{2\varepsilon - L\rho \cdot V^2} \qquad (5)$$

where E is energy of motion of the whisker, L is the length of the whisker, ρ is the density of the whisker, and V is the growth rate of the whisker. In the new energy model, the growth rate V plays a very important role in maintaining the R constant. When ε varied in small range, the growth rate V will adjust thereupon in order to maintain R constant. That is V will increase with the increasing of ε, and will decrease with the decreasing of ε, and R will keep constant all the time. However, when ε varies in great range, the adjustment of V will lose ability to maintain R constant again. That is because not only driving force but also Sn atoms are needed for tin whisker growth. For example, when ε increases greatly, tin whisker will have a tendency to grow in a very rapid rate. However, if now the diffusion of tin atoms to the root of tin whisker is too slow to satisfy the rapid growth of tin whisker, then necking phenomenon will happen. That is the diameter of tin whisker will decrease to adapt the insufficient supply of tin atoms and then another balance is reached [7].

3.1.2 The mechanism of whisker growth

Rapid tin whisker growth was found on the surface of the oxidized RESn3 (RE including La, Ce, Y and Er) precipitates during aging in air. In order to clarify the mechanism for the growth of tin whiskers in our work, specimens that had been stored at room temperature for 1152 hrs were etched by FIB across the RESn3 phase. The cross section reveals that the RESn3 phase react predominantly with oxygen to form oxide and simultaneously release tin atoms due to the high activity of rare earth element: RESn3 + O2 → Sn + RExOy. The diffusion of oxygen into RESn3 lattice will induce volume expansion, and the expansion will result in a compressive stress due to the constraint of the solder matrix. The results will sequeeze the released tin atoms out of the surface of the RESn3 phase to form tin whiskers.

978-1-4244-4658-2/09 $25.00 © 2009 IEEE   510

3.2 Accelerated whishker growth through electromigration

In our work, the sample experienced current stressing and isothermal aging process in order to accelerate the growth of tin whisker.

In order to clarify the influence of Joule heating, the temperature variation of the solder joint surface was monitored during the electromigration test. The result shows that the Joule heating effect could not be neglected since it contributes approximately 50°C~60°C to the surface of the solder joint. When the ambient temperature is 80°C, it can finally reach 121°C under a current density of $10^4 A/cm^2$, whereas when the ambient temperature is 90°C, it can finally reach 143°C under the same current density. Therefore it can be concluded that the solder joint experienced a sharp increase in temperature under Joule heat effect, and then remains constant. The temperature variation are illustrated in Fig. 4.

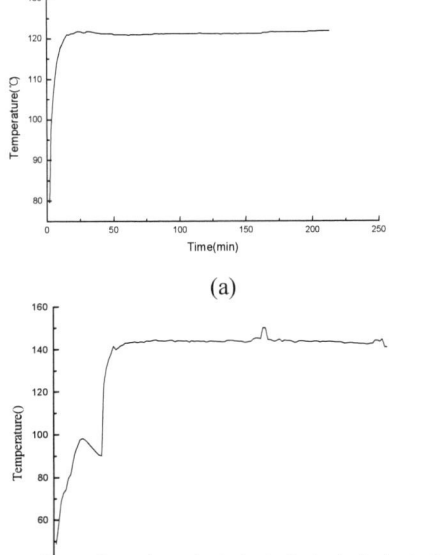

(a)

(b)

Fig. 4 The temperature variation under Joule heating effect plus ambient temperature
(a) ambient temperature: 80°C
(b) ambient temperature: 90°C

In the electromigration test, it was found that solder overflow film plays an important role in the formation and growth of tin whisker. Actually, it is a thin film of solder alloy which is overflowed on Cu substrate. Compared with bulk solder, it has relatively large interface with Cu substrate, which made it easier for the chemical reaction between Sn and Cu atoms. Thus, the compressive stress induced by the formation and growth of Cu6Sn5 intermetallic compounds (IMCs) is consequently higher than the bulk solder. This is the reason why whiskers are more likely to be formed on the overflowed solder film. In this accelerated test, the conditions for the formation of such overflowed solder film were investigated. Results showed that it could be formed after 12h

under the Joule heating effect plus 80°C of ambient temperature, while under the Joule heating effect plus 90°C of ambient temperature, the formation of such overflowed solder film only need 3hrs. Besides, the overflowed region became larger with the increase of Joule heating time.

After the overflowed solder film is formed, the samples are then placed in the oven with constant ambient temperature for isothermal aging. In this period, tin whiskers are formed and kept growing, which could release the compressive stress formed before. Fig. 5 shows the enlarged SEM image of the whiskers. The striations on the surface of these whiskers proved that they are extruded from the substrate.

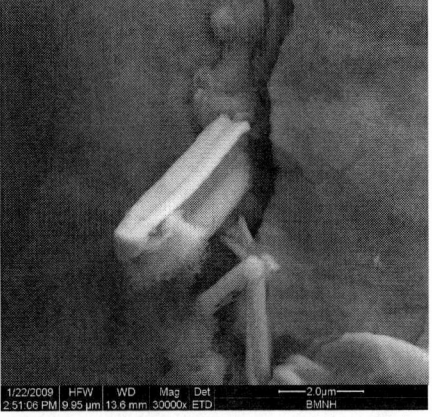

Fig. 5 The enlarged SEM image of the whiskers

In order to reveal the growth mechanism, it is necessary to take the second grinding and polishing from the side of the samples in order to investigate the interior microstructure.

In the overflowed solder film, massive Cu_6Sn_5 intermetallic compounds (IMCs) were formed under the Joule heating effect plus 80°C or 90°C of ambient temperature. Fig. 6 shows the enlarged SEM image of the side microstructure.

Fig. 6 The enlarged SEM image after second grinding and polishing

As can be seen in the figure, Bi-rich phase (white contrast) and Sn-rich phase (light grey contrast) both exist in the overflowed solder film and the whisker, that is, the whisker exhibits a multi-phase composition. Growth of massive Cu_6Sn_5 IMCs could provide the compressive stress to the upper layer. Different from traditional scallop-shaped or layer-shaped IMCs, the morphology of Cu_6Sn_5 IMCs at the bottom was more inclined to the reticular structure. It's believed that this interlaced Cu_6Sn_5 IMCs could prevent the continuous diffusion of Cu atoms from the substrate to the overflowed solder film, which therefore prohibit the further reaction between Sn and Cu atoms, and thus stop the growth of whiskers.

Conclusions

Several study was carried out to reveal the growth mechanism of tin whiskers under the condition of electromigration and isothermal aging, as well as the addition of rare earth elements. Both experiments reached the results that the formation and growth of tin whiskers is driven by a compressive stress. In EM test, the compressive stress derives from the growth of Cu_6Sn_5 IMCs, whereas in RE addition test, it comes from the volume expansion induced by the chemical reaction between RE and oxygen. Both compressive stress could be released through the growth of whiskers.

Acknowledgments

The authors greatly appreciate the financial support of this work from Beijing Natural Science Foundation Program and Scientific Research Key Program of Beijing Municipal Commission of Education (KZ200910005004), the Funding Project PHR (IHLB), and New Century Excellent Talents Program by China Ministry of Education (NECT-04-0202).

References

1. C.C.Hu, Y.D.Tsai, C.C.Lin, G.L.Lee, S.W.Chen, T.C.Lee, T.C.Wen, "Anomalous growth of whisker-like bismuth–tin extrusions from tin-enriched tin–Bi deposits," *Journal of Alloys and Compounds*, 472 (2009), pp. 121-126.
2. Peach M O., *J Appied Physics*, 23 (1952), pp. 1401-1403.
3. Ellis W C, Gibbons D F, et al, "Growth and Perfection of Crystals," 1958, pp. 102-120.
4. Tu K. N., Chen C., Wu A. T, "Stress analysis of spontaneous Sn whisker growth," *J Mater Sci: Mater Electron*, 18 (2007), pp. 269-281.
5. Jiang B., Xian A. P., "Spontaneous Growth of tin whisker on tin-rare earth alloys," *Philosophical Magazine Letters*, 87 (2007), pp. 657-662.
6. Jiang B., Xian A. P., "Whisker growth on tin finishes of different electrolytes," *Microelectron Reliab*, 48 (2008), pp. 105-110.
7. Hao Hu, Shi Yaowu, et al, "Cross section changing phenomenon of tin whisker in Sn3.8Ag0.7Cu1.0Er solder," *Acta Metallurgica Sinica*.

Micro-structural and Interfacial Effects on the Dielectric Properties of High-*k* Aluminum/Epoxy Composites for Embedded Capacitors

Chong Chen[1,2], Shuhui Yu[1]*, Rong Sun[1], Suibin Luo[1], Lvqian Weng[2], Ruxu Du[1]

[1]Shenzhen Institute of Advanced Technology, Shenzhen, Guangdong, China, 518055

[2]Harbin Institute of Technology Shenzhen Graduate School, Shenzhen, Guangdong, China, 518055

Corresponding author: yuushu@gmail.com, Tel: +86-755-26803560, Fax: +86-755-26803589

Abstract

Electrically percolative composites have attracted much attention because they exhibit a high dielectric constant at a critical concentration of the conductive fillers. Aluminum (Al) is well known as a fast self-passivation and low-density metal. The thin and dense passivation layer forms an insulating boundary layer outside the metallic spheres (core-shell) which lowers the dielectric loss and realizes high filler loading levels. In this article, the influences of the content of Al, the size of Al core, the thickness of Al_2O_3 shell, and the two interfaces of polymer-Al_2O_3 and Al_2O_3-Al on the dielectric behaviors of the resulting percolative composites have been systematically analyzed and interpreted. For the composite containing 70 vol.% of 1.0 μm aluminum, a dielectric constant of 95.49 and a low dielectric loss of about 0.05 over a frequency range of 100Hz~10MHz was achieved. The broad peak over a wide volume fraction range of the percolation threshold concentration makes the material reproducible for use.

Introduction

Recently, with the rapid development of electronic industry, multi-function and further miniaturization trend of the integrated circuits (ICs) calls for replacing discrete passive components (capacitors, resistors, and inductors) with embedded passives. Among the embedded passive components, special interest has focused on capacitors because of their wide applications and the capability to improve the electric performance [0]. High dielectric constant (high-*k*) film is the key material for the development of embedded capacitors. Electrically percolative composites have attracted much attention because they possess a high-*k* at the critical metallic concentration in the polymer matrix.

Many investigations have been conducted on the high-*k* percolative materials with different conductive fillers. Rao et al. [0] reported a high-*k* of 2000 in Ag-flake/epoxy percolative materials. Dang et al. [0, 0] reported a high-*k* value of 400 in a Ni-poly (vinylidene fluoride) composite. Similar enhancements of high-*k* in other metal-polymer systems were observed by Fan et al [0]. However, the benefits of these high-*k* values in the metal/polymer composites were counteracted by the high dielectric losses, e.g., values of dielectric loss of 0.24 [0], 0.18 [0], 0.5 [0], and > 2 [0]. It is also risky to prepare percolative dielectrics with a threshold composition due to the abrupt variation in the vicinity of the threshold. According to Rao et al. [0], the dielectric constant increased from 200 to 2000 with the filler content increasing from 11.03 vol.% to 11.43 vol.% and decreased to nearly zero at 11.52 vol.%.

In order to achieve a reproducible low-loss percolative composite, a self-passivation core-shell metal aluminum was used to fabricate percolative composites [0]. Xu et al. reported a high-*k* of 109 and a low dissipation factor of about 0.02 in the Al/epoxy composites [0]. However, the influence of the interface on the dielectric behaviors remains unclear. In the present study, we have investigated the aluminum particles surface and interface effect on the dielectric properties. The influences of the content of Al, the size of Al core, the thickness of Al_2O_3 shell, and the interface of Al_2O_3-Al on the dielectric behaviors of the resulting percolative composites have also been systematically analyzed and interpreted.

Experiment

The bisphenol-A epoxy resin (E51: $C_{21}H_{24}O_4$), from Jitian Chemical Company, China, was used as a polymer matrix. The hardener was 4, 4'-diaminodiphenyl sulphone (DDS: $C_{12}H_{12}N_2O_2S$, from Kebang Chemical Company, China). The onset curing temperature and the peak curing temperature of the E51/DDS system was determined using differential scanning calorimetry (DSC Q20, from TA Instruments) and was found to be 160°C and 212°C, respectively. Aluminum particles (Hongwu Nanomaterial Technologies Corporation, China) with particle sizes of 100 nm, 1.0 μm, and 5.0 μm, respectively, were used as fillers. The particles were treated at 640°C for 5 min and 30 min, respectively, in order to oxidize the shell with varied thickness. The silane coupling agent (CA) γ-aminopropyltrimethoxysilane (KH550: $NH_2(CH_2)_3Si(OC_2H_5)_3$) was used to modify the particle surface.

Prior to use, the raw aluminum particles were modified in 5.0 g distilled water and 95.0 g absolute ethanol media with 10.0 wt.% Al and 0.5 wt.% KH550. Ammonia was used to adjust the pH value to about 9 in order to catalyse the condensation reaction. Then, the aluminum particles were rinsed with ethanol three times in order to remove the unreacted silane coupling agent and the adsorbed water. After filtration, the modified aluminum particles were dried at 80°C for 24 h.

The modified aluminum powder was mixed with solvents followed by the addition of the epoxy resin and curing agent under agitating condition. A composite paste was obtained after vaporizing most of the solvents. The paste was coated on Cu foils with bar coating method. After baking and solvent evaporation, two pieces of the Al/epoxy composites coated Cu foils were laminated at 214°C under a pressure of 10 MPa for 1 hour. The thickness of the dielectric film in the laminated capacitor was about 25 μm.

To measure the dielectric properties of the capacitor components, the laminates were patterned with photolithography method. The thickness of the capacitor components was measured using a XP-2 stylus profiler (from Ambios Technology, United States). Capacitance and dielectric loss were tested with an impedance analyzer (Aglient, Model 4294). Dielectric constant was calculated from the measured thickness and capacitance. The surface chemistry of the aluminum particles was characterized with an FTIR (Nicolet380, United States). Thermogravimetric analysis was conducted with TGA (Model Q600, from TA Instruments.) under nitrogen atmosphere, with a heating rate of 10°C/min. A TecnaiF20 TEM was used to analyze the particle size of aluminum powder and the thickness of its oxide layer. Morphologies of the Al composites were observed with a SEM (JSM-5910LV, from Japan).

Results and Discussion

Fig. 1 presents the TGA results of the 1.0 μm aluminum particles untreated and treated with silane coupling agent . As shown in the figure, there is more weight loss for the modified powder from room temperature up to 400 which suggests the existence of KH550 organic on the aluminum particles surface. For the untreated aluminum particles, the weight dramatically increases at the temperature around 660 °C, because aluminum particles get enough energy to react with O_2 and form thicker aluminum oxide, which corresponds to the endothermic peak in the heat flow curve. In contrast, there is no abrupt change in both the weight and heat flow curves below 700°C for the treated aluminum particles which could degrade at an elevated temperature. The reason is that the decomposition of the silane coupling agent results in silica on the aluminum particles surface, which could significantly reduce the oxidation rate of aluminum at temperatures above 600°C.

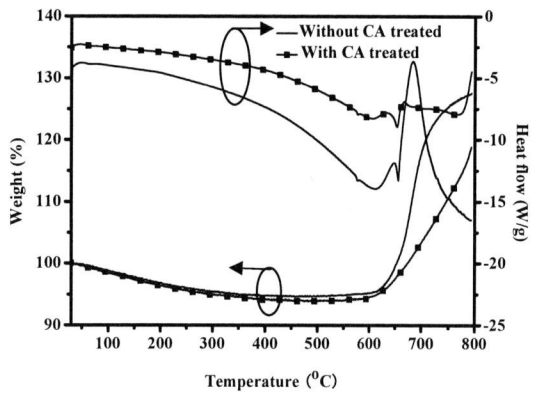

Fig. 1 TGA curves of the untreated and CA treated aluminum particles

Fig. 2 presents the SEM images of the Al/epoxy composites containing the Al particles untreated and treated with silane coupling agent, respectively. Compared with the composite containing the untreated Al particles, the composite with treated Al shows less voids or pores and the Al particles are dispersed homogenously in the epoxy resin matrix.

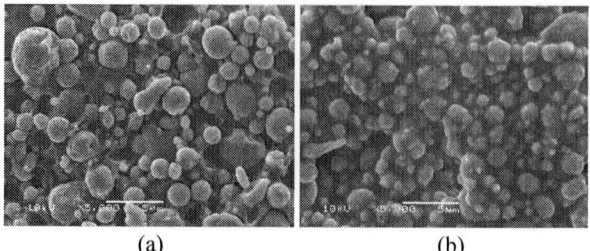

(a)   (b)

Fig. 2 SEM images of the Al/epoxy composites with Al (a) untreated and (b) treated with silane coupling agent

Fig. 3 presents the frequency dependence of the dielectric constant and loss of the Al/epoxy composites containing the untreated Al particles and that treated with silane coupling agent. The content of the Al particles is 70 vol.% and the average diameter is 1.0 μm. Compared with the composite containing the untreated Al, the dielectric constant of the composites with treated Al increases by about 51%, from 63.42 to 95.49 at 1 MHz. The improvement of dielectric constant is attributed to the better dispersion of the aluminum particles in the polymer matrix. The dielectric loss for both composites is about 0.05 over the frequency range of 100Hz~10MHz.

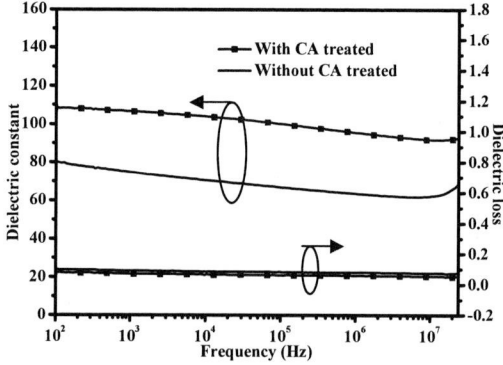

Fig. 3 Frequency dependence of the dielectric constant and loss of the Al/epoxy composites untreated and treated with silane coupling agent at 70 vol.% of the 1.0μm Al loading level

Fig. 4 shows the dielectric constant and loss of the epoxy composites filled with 100 nm, 1.0 μm and 5.0 μm Al particles, respectively, as a function of the Al volume fraction, at a frequency of 1 MHz and room temperature. The 1.0 μm Al particles and 5.0 μm Al particles can be loaded up to 80 vol.%. Because of the self-passivated insulating oxide layer on the aluminum particle surface, the aluminum/epoxy composites remained to be insulated even at high filler loading levels. However, for the 100 nm Al filler, it is difficult to disperse the Al particles uniformly when the filler loading is higher than 50 vol.%. Generally speaking, the dielectric constant of the composites increases with the filler loading level. The composite filled with 1.0 μm Al shows a higher dielectric constant than that filled with 5.0 μm Al, and the maximum dielectric constants are 95.49 and 86.34, respectively, at a frequency of 1 MHz and room temperature. At the same filler loading level (30~60 vol.%), the composite containing 100 nm Al exhibits higher dielectric constant than the other two. The dielectric constant is 38.43 at 50 vol.% loading for the composites filled with 100 nm Al.

A low dielectric loss of less than 0.08 is observed for all the Al composites as shown in the figure, which can be attributed to the thin and dense passivation layer on the metal surface. In general, the dielectric loss of the material is partly resulted from interfacial loss and conduction loss. The interfacial loss originates from the excessive polarized interface induced by the fillers. In the Al/epoxy composites, there are two interfaces which are epoxy-Al_2O_3 and Al_2O_3-Al. The epoxy-Al_2O_3 interface can be assimilated to ceramic/epoxy as Al_2O_3 usually is recognized ceramic, thus the dielectric loss of epoxy-Al_2O_3 is usually low, amounted to that ceramic/epoxy. Although the Al cores are conducting, the insulating Al_2O_3 layer confines the electrons within an aluminum particle. Therefore, the dielectric loss of Al/epoxy composites is mainly determined by the interfacial loss in the interface epoxy-Al_2O_3.

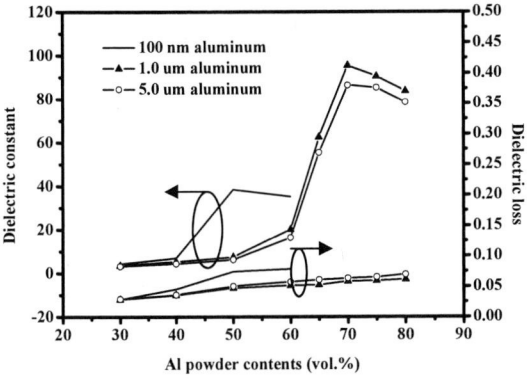

Fig. 4 Dielectric constant of aluminum filled composites as a function of Al powder contents (@ 1 MHz)

For both the 1.0 μm and 5.0 μm Al, a moderate increase in the dielectric constant is observed when the Al content is below 50 vol.%. The increase of the dielectric constant becomes significant when the Al content is above 60 vol.%. When the Al volume fraction is below 60 vol.%, the variation of the dielectric constant shows a good agreement with the typical power law [0].

$$\overline{\varepsilon} / \varepsilon_D = \left| (f - f_c) / f_c \right|^{-q} \tag{1}$$

where ε_D, f_c, f and q are the relative dielectric constant of the matrix, the percolation threshold, Al volume fraction, and a critical exponent, respectively. As shown in Fig. 4, a broad peak forms over a wide volume fraction range near 70 vol.%. There are a couple of reasons accounting for this phenomenon. On the one hand, according to equation (1), metal composites with high percolation threshold have a wider volume fraction range of the percolation threshold

concentration, compared to the composites with low percolation threshold, as shown in the Fig. 5. On the other hand, the porosity of the sample with 80 vol.% of Al could significantly higher than those of the other samples with less Al [0]. Therefore, the observed maximum dielectric constant point should not be a real percolation threshold. It is only a point where the influence of porosity on the dielectric constant overwhelms the contribution from adding Al. Nevertheless, for practical applications, this is definitely an advantage. It is that the broad peak over a wide volume fraction range of the percolation threshold concentration makes the material reproducible for use.

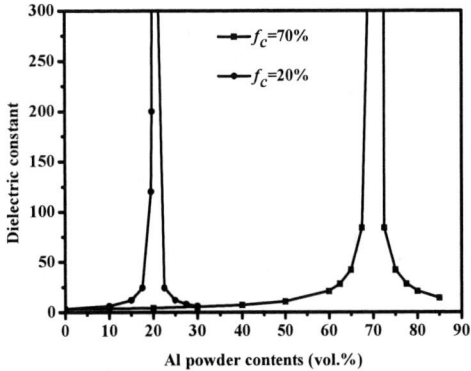

Fig. 5 Variation of dielectric constant data simulated according to Equation (1) f_c=20% and f_c=70%. (q=1)

Fig. 6 illustrates the dielectric constant versus filler loading of the composites filled with 1.0 μm Al@15nmAl_2O_3, 1.0 μm Al@40nmAl_2O_3 and 1.0 μm Al_2O_3 particles. The maximum dielectric constant of the composites containing 1.0 μm Al@40nmAl_2O_3 particles is about 71, which is lower than 95 of 1.0 μm Al@15nmAl_2O_3 particles. Therefore, the thickness of Al_2O_3 layer has significant influence on the dielectric behavior of the Al composites. The 1.0 μm Al_2O_3 composites show a very low dielectric constant, which indicates the importance of the metallic core for Al composites.

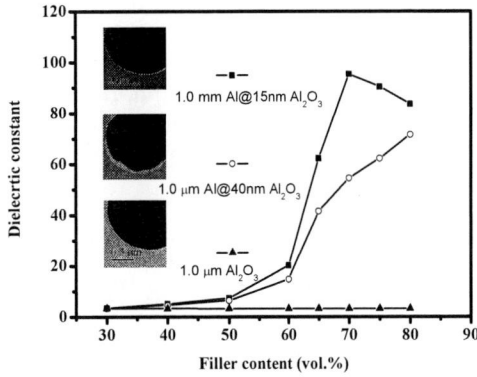

Fig. 6 Variations of dielectric constant with the filler contents of the Al@Al_2O_3 particles(@ 1 MHz)

Conclusions

A high-k percolative composite material was developed by using self-passivated and low density Al as the filler. The surface of the Al particles were treated with silane coupling agent. The DSC/TGA analysis confirmed that the agent was successfully grafted on the aluminum particle surface, which significantly enhanced the dielectric properties and compatibility of Al-epoxy interface of the resulting composites. With the insulating aluminum oxide layer, a high loading level of aluminum can be obtained while the composite materials still shows good mechanical properties. For the composite containing 70 vol.% 1.0 μm aluminum, a dielectric constant of 95.45 and a low dielectric loss of about 0.05 at 1 MHz were achieved. Formation of the broad peak over a wide volume fraction range of the percolation threshold concentration is due to the high percolation threshold and the porosity of the composites. It makes the material reproducible for practical applications. Therefore, the easy processing, flexibility and good dielectric behavior of the Al/epoxy percolative composites potentially make the composites particularly attractive for practical applications.

Acknowledgments

The authors acknowledge the financial support from National Natural Science Foundation (50807038).

References

1. Dang Z. M., Lin Y. Q., Xu H. P., et al, "Fabrication and Dielectric Characterization of Advanced BaTiO$_3$/ Polymide Nanocomposite Films with High Thermal Stability, " *Adv. Funct. Mater.*, Vol. 18 (2008), pp. 1-9.

2. Rao Y., Wong C. P., *IEEE Proc. Electron. Technol. Conf.*, 2002, pp. 920.

3. Dang Z. M., Lin Y. H., Nan C. W., "Novel ferroelectric polymer composites with high dielectric constant," *Adv. Mater.*, Vol. 15 (2003), pp. 1625.

4. Dang Z. M., Shen Y., Nan C. W., "Dielectric behavior of three-phase percolative Ni-BaTiO$_3$/Polyvinylidene fluoride composites," *Appl. Phys. Lett.*, Vol. 81 (2002), pp. 4814.

5. Fan L. Kumashiro Y., Wong C. P., *IEEE Proc. Electron. Compon. Technol. Conf.*, 2003, pp. 167.

6. Xu J. W., Moon K. S., Tison C. et al, "A Novel Aluminum-Filled Composites Dielectric for Embedded Passive Applications," *IEEE Trans. on Adv. Pac.*, Vol. 29, No. 2 (2006), pp. 295.

7. Efros A. L., Shklovskii B. I., "Critical behaviour of conductivity and dielectric constant near the metal-non-metal transition threshold," *Phys. Status. Solidi.*, Vol. 76, No. 2 (1976), pp. 475.

8. Qi L., Lee B. I., Chen S., Samuels W. D., Exarhos G. J., "High-dielectric-constant silver-epoxy composites as embedded dielectrics," *Adv. Mater.*, Vol. 17 (2005), pp. 1777-1781.

Comparative Study of Interfacial Reactions of High-Sn Lead-free Solders on Single Crystal Cu and on Polycrystalline Cu

Yipeng Cui, Mingliang Huang*

Electronic Packaging Materials Laboratory, School of Materials Science & Engineering, Dalian University of Technology
Dalian, 116024, China
*Corresponding author, E-mail: huang@dlut.edu.cn, Tel: 86-411-84706595

Abstract

This work compared the interfacial reactions of single crystal Cu and polycrystalline Cu with the high-Sn solders. The morphology aspects of Cu_6Sn_5 intermetallic compound (IMC) grains formed between (001) single crystal Cu and high-Sn solder bath (pure Sn, Sn-0.7Cu, Sn-3Cu) at 250°C and 300°C for different durations ranging from 10 s to 10 min were investigated. The Cu_6Sn_5 grains formed in all solder baths at 250°C displayed common scallop-type, while those formed in pure Sn and Sn-0.7Cu baths at 300°C displayed rooftop-type in early stage of the interfacial reaction (less than 10 min). Both the top-view morphology and cross-section morphology of the interfacial Cu_6Sn_5 IMC formed on single crystal Cu at 300°C are different from those formed on polycrystalline Cu. From the top-view, the rooftop-type Cu_6Sn_5 grains formed on (001) single crystal Cu are elongated along two perpendicular directions; from the cross-section view, the IMC layers present a relatively planar intermetallic structure. Furthermore, Cu_6Sn_5 grains formed between Sn-0.7Cu/(001)Cu partly changed from rooftop-type to scallop-type when the reaction time is up to 10 min; Cu_6Sn_5 grains formed between pure Sn/(001)Cu remained rooftop-type even after the reaction for up to 10 min. It is considered that reaction temperature and Cu concentration in the solder bath have an effect on the formation of rooftop-type Cu_6Sn_5 grains. The dissolution rate of single crystal Cu in the high-Sn solder is higher than polycrystalline Cu in solder bath at 300°C, which indicated that the morphology of Cu_6Sn_5 affects the dissolution of substrate Cu.

1. Introduction

During the solid/liquid interfacial reaction, the formation of the intermetallic compound (IMC) and the dissolution of substrate are integrated processes of the same overall mechanism of soldering reaction. Recently, due to the environmental concern of the traditional Pb-bearing solders, the research of lead-free solder alloys has been widely concerned around the world. However, all the lead-free solders are high tin-containing alloys which lead to more substrate metal dissolution and thicker IMC layer formation than those of Sn-Pb alloys. The formation of the IMC is desirable for interconnects but excessive metal dissolution and uncontrolled growth of the IMC is detrimental to the reliability of the joint. Many papers about the above-mentioned issues have been reported. Huang et al [1] investigated the role of Cu content in the dissolution kinetics of Cu substrate in high Sn solders during the solid/liquid reaction. It was shown that small additions of Cu (0.7wt.%, 1.5wt.%) in high-Sn solders dramatically decrease the dissolution rate of Cu at low temperature. During reflow soldering process, the IMC exhibits scalloped morphology.

However, a recent study by Suh et al. [2] revealed that elongated and rooftop-type Cu_6Sn_5 grains formed on (001) Cu single crystal substrate. The elongations align along two perpendicular directions.

The present paper focused on the formation of IMC on single crystal Cu substrate and the dissolution of single crystal Cu substrate, and then the results were compared with those of polycrystalline Cu substrate. Because single crystal has no grain boundary, all the atoms grow along one direction, single crystal Cu was chosen as substrate in this study, which will provide insight into understanding the growth mechanism of IMC and the dissolution of substrate.

2. Experiments

The high-Sn solders used in this study were pure Sn, Sn-0.7Cu, and Sn-3Cu (all in wt.%). Sn-0.7Cu and Sn-3Cu solder were prepared from pure Sn (99.99%) and Cu (99.9%) in a solder-pot by heating up to 350°C for 2h. Then, the single crystal Cu sheets (Φ=8mm) were spark-cut from the Cu single crystal rod, ensuring that the wetting surfaces were all parallel to the (001) planes. Both sides of the thin single crystal Cu sheets were carefully ground and polished. The thickness of each Cu sheet was measured by a screw micrometer.

The solder baths containing 250g solder were maintained at 250±2°C, 300±2°C, respectively. Polished Cu sheets were coated with RMA flux to improve the wettability, then they were vertically dipped into the solder baths at different temperatures, and the duration time ranged from 10 s to 10 min. To inhibit the growth of IMC in cooling procedure, all the samples were quenched in water. Then all the samples were deeply etched with nitric acid to remove the Sn matrix so that the interfacial IMC grains were fully exposed. The top view and cross-section microstructure of the IMCs were observed using scanning electron microscope (SEM) and the remaining thickness of Cu sheets were examined with an optical microscope.

3. Results

3.1 Interfacial reactions between single crystal Cu and pure Sn bath (300°C)

First of all, pure Sn was used as a solder. Fig. 1 showed the morphology of IMC formed on the Sn/(001) Cu at 300°C for different reaction times. It is obvious that the Cu_6Sn_5 grains displayed a rooftop-type rather than a scallop-type structure, which is quite similar to previous observations on Cu substrates [2, 5]. Furthermore, the elongations of rooftop-type Cu_6Sn_5 grains go along two perpendicular directions, as shown in Fig. 1. The size of the Cu_6Sn_5 grains increased with the increasing dipping time. Further increasing the dipping time up to 150s, 300s and 450s, there are no obvious change in the morphology of the Cu_6Sn_5 grains as shown in Fig. 1(b), Fig. 1(c) and Fig. 1(d).

978-1-4244-4658-2/09 $25.00 © 2009 IEEE

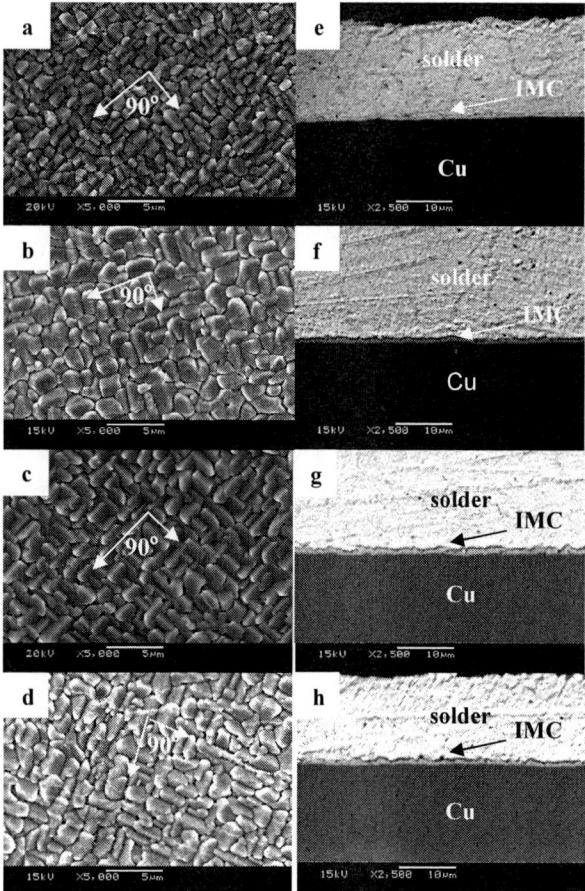

Fig. 1 Micrographs showing the morphology of Cu_6Sn_5 grains formed on Sn/(001)Cu at 300°C for (a)10 s, (b)150s, (c)300s, (d)450s. The left column is top view, and the right column is cross-section view, respectively.

The right column showed the cross-section morphology of IMC formed on Sn/(001)Cu. Fig. 1(e) showed the backscattered electron micrographs(BEM) of the IMC layers formed on Sn/(001)Cu at 300°C for 10 s. The Cu_6Sn_5 layer is very thin. When the reaction time increased, the Cu_6Sn_5 layer becomes thicker. Obviously, the IMC layer presented a relatively planar intermetallic structure rather than a scallop structure, the cross-section structure is consistent with the top-view structure. The average IMC thickness recorded in the experiment was low (when the reaction time is up to 450s, the thickness of IMC is only a little more than 1μm).

Because of the large solder volume compared with the Cu sheet, the concentration of Cu in the pure Sn solder was far lower than the solubility of Cu in the solder. Not only the substrate metal but also the IMC dissolved into the molten solder. A fast diffusion of copper from the IMC/solder interface would not allow the IMC to grow beyond a certain thickness. Consequently, a thin intermetallic layer make more copper diffuse across the IMC/solder interface and eventually into the molten solder.

Based on the above-mentioned phenomena, it is reasonable to suppose that the nucleation, growth and ripening

behaviors of Cu_6Sn_5 on Cu single crystal substrate may be quite different from the conventional case of wetting on a polycrystalline Cu surface.

3.2 Interfacial reactions between single crystal Cu and pure Sn bath (250°C)

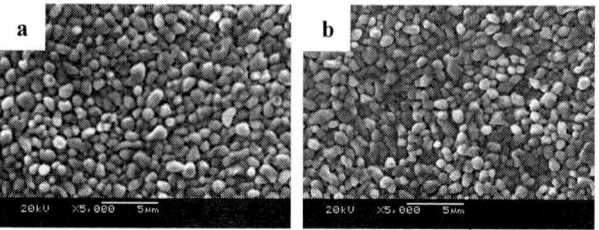

Fig. 2 Micrographs of scallop-type Cu_6Sn_5 grains formed on Sn/(001) Cu at 250°C for (a)10 s, and (b)600 s

Fig. 2 showed the morphology of Cu_6Sn_5 grains formed on the Sn/(001) Cu at 250°C for different dipping times. Unlike the rooftop-type structure of Cu_6Sn_5 grains formed at 300°C, the Cu_6Sn_5 grains formed at 250°C displayed the common scallop-type structure for different dipping times ranging from 10 s to 10 min.

3.3 Interfacial reactions between single crystal Cu and Sn-0.7Cu bath (250°C)

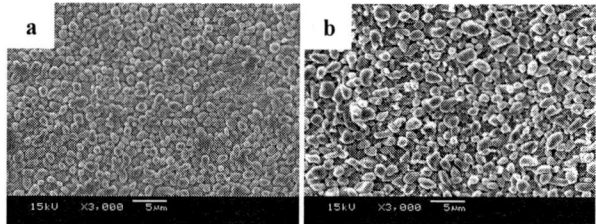

Fig. 3 Micrographs of scallope-type Cu_6Sn_5 formed on Sn-0.7Cu/(001)Cu at 250°C for (a) 10 s, and (b) 600 s.

Fig. 3 showed the SEM micrograph of Cu_6Sn_5 formed on Sn-0.7Cu/(001) Cu at 250°C. Similar to Cu_6Sn_5 formed on Sn/(001) Cu at 250°C, the Cu_6Sn_5 grains displayed the common scallop-type structure for different dipping times ranging from 10 s to 10 min. There was no rooftop-type Cu_6Sn_5 grain observed in the samples.

3.4 Interfacial reactions between single crystal Cu and Sn-0.7Cu bath (300°C)

Fig. 4 showed the morphology of Cu_6Sn_5 grains formed on Sn-0.7Cu/(001) Cu at 300°C for different dipping times. Fig. 4(a) showed the top view of Cu_6Sn_5 grains on the (001) Cu/Sn-0.7Cu at 300°C for 10 s. It is clear that the morphology of Cu_6Sn_5 grains was similar to that formed on the pure Sn/(001) Cu at 300°C. The cross-section morphology of Cu_6Sn_5 grains appeared a flat structure, as shown in Fig. 4(b). The elongations of Cu_6Sn_5 grains also exhibited along two perpendicular directions. Further increasing the dipping time to 150 s, as shown in Fig. 4(c), there was no obvious change in the morphology of the Cu_6Sn_5 grains. When the dipping time increased up to 10 min, a change in the morphology of the Cu_6Sn_5 grains occurred, as shown in Fig. 4(d). The Cu_6Sn_5 grains in the areas marked A appeared rooftop-type, but the Cu_6Sn_5 grains in the area marked B appeared common scallop-type. Fig. 4(e) and Fig. 4(f) is the enlarged view of the two areas. This indicated that some rooftop-type Cu_6Sn_5 grains change into scallop-type Cu_6Sn_5 grains after dipping for 10 min. From the cross-section microstructure shown in Fig. 4(g), the IMC layer also appeared scallop-type, corresponding to Fig. 4(f).

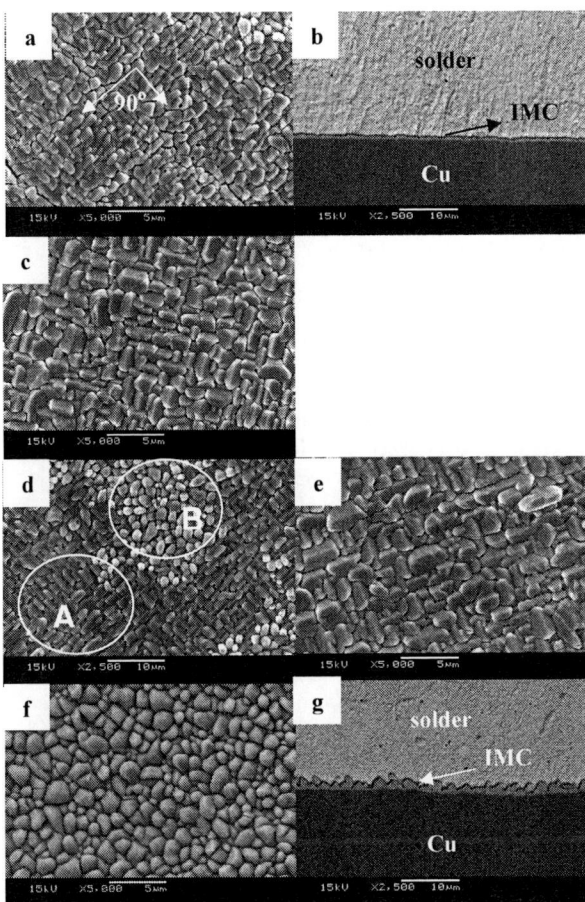

Fig. 4 Micrographs showing the morphology of Cu_6Sn_5 grains formed on Sn-0.7Cu/(001)Cu at 300°C for (a) 10s, top-view; (b) 10s, cross-section; (c)150 s, top-view; (d),(e),(g) 600s, top-view; (f) 600 s, cross-section.

3.5 Interfacial reactions between single crystal Cu and Sn-3Cu bath (250°C and 300°C)

Since the solubility of Cu in Sn is about 2.65 wt.% at 300°C and 1.6 wt.% at 250°C [1]. Sn-3Cu has supersaturated Cu at both 250°C and 300°C. For the reaction in the Sn-3Cu bath, the overall amount of decrease of the Cu sheet can be accounted for the conversion of Cu into the intermetallics phase. There is no driving force for Cu to dissolve into the bath.

Fig. 5 is the morphology of Cu_6Sn_5 grains formed on Sn-3Cu/single crystal Cu at 250°C and Sn-3Cu/polycrystalline Cu at 250°C and 300°C for different dipping times. Similar to

the morphology of Cu_6Sn_5 grains formed on Sn-3Cu/polycrystalline Cu at 250°C for 30 s, as shown in Fig. 5(c), the morphology of Cu_6Sn_5 grains formed on Sn-3Cu/(001) Cu at 250°C for 30 s are scallop-type, as shown in Fig. 5(a), As we can see from Fig. 5(d), 5(f), with the reaction time increasing to 10 min, Cu_6Sn_5 formed on polycrystalline Cu and single crystal Cu are still scallop-type. With reaction temperature increasing to 300°C, there is no changing in the morphology of Cu_6Sn_5. No rooftop-type Cu_6Sn_5 grains were observed both at 250°C and 300°C.

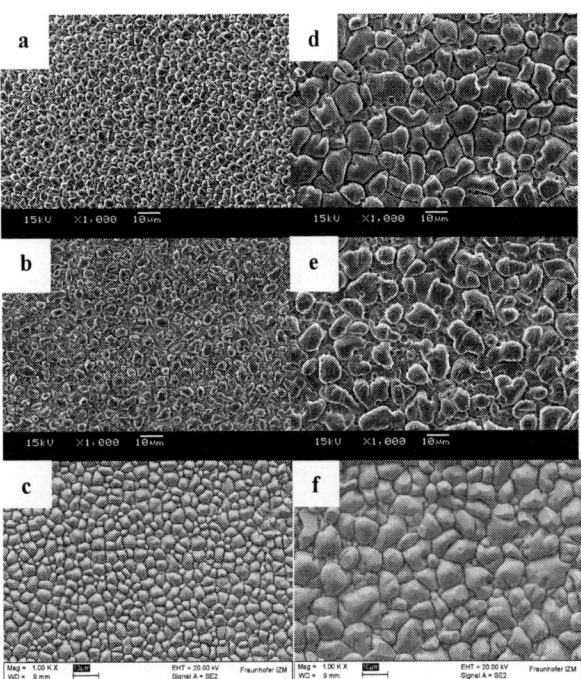

Fig. 5 Micrographs showing the morphologies of Cu_6Sn_5 formed on (a) Sn-3Cu/(001)Cu at 250°C (b) Sn-3Cu/(001)Cu at 300°C (c) Sn-3Cu/ polycrystalline Cu at 250°C. The left column is 30 s and the right column is 10 min, respectively.

3.6 the dissolution of single crystal Cu in the high-Sn solder(pure Sn, Sn-0.7Cu)

The dissolution of Cu in the molten solders involves the consuming of Cu by interfacial IMC formation of $\eta(Cu_6Sn_5)$ and thin layer $\varepsilon(Cu_3Sn)$, and the direct dissolution of Cu into the liquid solders.

Fig. 6 showed the dissolved single crystal Cu sheet thicknesses in pure Sn and Sn-0.7Cu. It exhibited an approximately linear increase with increasing reaction time. As shown in Fig. 6, the slope was the Cu dissolution rate in the molten solders. At 300°C, the dissolution rate of single crystal Cu in pure Sn and Sn-0.7Cu is 29 μm/min and 18.2 μm/min, respectively, in contrast, the dissolution rate of polycrystalline Cu is 22 μm/min and 9.6 μm/min, respectively. This indicated that the morphology of Cu_6Sn_5 affected the dissolution of the Cu substrate.

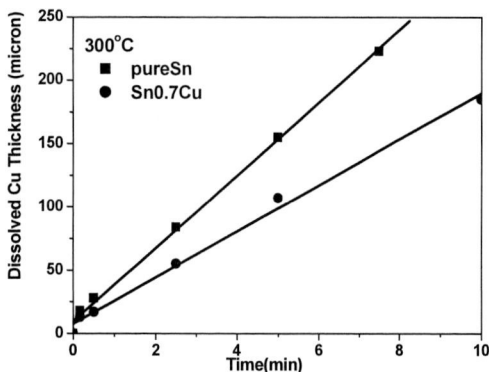

Fig. 6 Dissolved single crystal Cu sheet thicknesses in pure Sn and Sn-0.7Cu at 300°C.

4. Discussion

N.Zhao el at. [9] proposed that in molten Sn-0.7Cu, there is only short-range order (SRO) and its liquid structure is similar to that of pure Sn. While in Sn-2Cu solder, there is not only SRO but medium-range order (MRO). The Cu_6Sn_5-phase-like clusters will form in the liquid Sn-2Cu solder. Only the temperature reaches 400°C, MRO structure can disappear. In our experiment, the Cu concentration of Sn-3Cu is higher than that of Sn-2Cu, so there is also MRO in the Sn-3Cu. The MRO with Cu_6Sn_5-phase-like clusters in molten solder in liquid solder will quicken the formation of interfacial IMC in the very early stage of soldering process. Because of isotropic interface in liquid solder, the formation of Cu_6Sn_5-phase-like clusters is non-directional. Once the crystal nucleus of Cu_6Sn_5 formed in the molten solder, the single crystal Cu substrate has no impact. So the Cu_6Sn_5 grains formed in the Sn-3Cu bath at 250°C and 300°C do not present rooftop-type.

Why the morphology of Cu_6Sn_5 grains was transformed from rooftop-type into scallop-type. H.F.Zou el at. [5] point out that with increasing reflow time, a Cu_3Sn layer would form between the Cu_6Sn_5 layer and Cu substrate. The scallop-type Cu_6Sn_5 grains were firstly formed during the short reflow time (below 600 s). However, the Cu atoms coming from the Cu substrate would react with these rooftop-type Cu_6Sn_5 grains to form the Cu_3Sn layer at longer reflow times (above 600 s). In this case, the nucleation of rooftop -type Cu_6Sn_5 grains would decrease with increasing reflow time, and the scallop-type Cu_6Sn_5 grains would begin to form.

Conclusions

(1) When the samples were dipped into pure Sn bath, Sn-0.7Cu bath at 250°C, the Cu_6Sn_5 grains present common scallop-type.

(2) Rooftop-type Cu_6Sn_5 grains aligning along two perpendicular preferential appeared on the Sn/(001)Cu single crystal and Sn-0.7Cu/(001) Cu at 300°C (less than 10 min). It indicated that the formation of rooftop-type Cu_6Sn_5 grains depended not only on the substrate but also on the interface reaction temperature.

(3) Cu_6Sn_5 grains formed on Sn-0.7Cu/(001)Cu at 300°C change from rooftop-type to scallop-type with increasing reaction time to 10 min.

(4) When the (001) single crystal Cu samples were dipping into the Sn-3Cu bath (supersaturated Cu), not rooftop-type Cu_6Sn_5 grains but scallop-type Cu_6Sn_5 grains formed at both 250°C and 300°C.

(5) At 300°C, the dissolution rate of single crystal Cu in pure Sn and Sn-0.7Cu is 29 μm/min and 18.2 μm/min, respectively, in contrast, the dissolution rate of polycrystalline Cu in both solder bath is 22 μm/min and 9.6 μm/min, respectively. This indicates that the morphology of Cu_6Sn_5 affects the dissolution of substrate Cu.

Acknowledgments

This work is supported by National Key Technologies R&D Program (2006BAE03B02-2), NSFC key program (U0734006), and Key Laboratory program in Liaoning Province (20060133).

References

1. M.L. Huang, T. Loeher, A. Ostmann, H. Reichl, "Role of Cu in dissolution kinetics of Cu metallization in molten Sn-based solders," *Appl Phys Lett*, 86 (2005), 181908.
2. J. O. Suh, K. N. Tu, N. Tamura, "Dramatic morphological change of scallop-type Cu_6Sn_5 formed on (001) single crystal copper in reaction between molten Sn-Pb solder and Cu," *Appl Phys Lett*, 91 (2007), 051907.
3. Suh JO, K.N.Tu, N Tamura, "Preferred orientation relationship between Cu_6Sn_5 scallop-type grains and Cu substrate in reactions between molten Sn-based solders and Cu," *J. Appl Phys*, 102 (2007), 063511.
4. Suh JO, "Orientation distribution, morphology, and size distribution of Cu_6Sn_5 intermetallic compound in reactions between molten solder and Cu in flip chip solder joints," Ph.D thesis 2006.
5. H. F. Zou, H. J. Yang, Z. F. Zhang, "Morphologies, orientation relationships and evolution of Cu_6Sn_5 grains formed between molten Sn and Cu single crystals," *Acta Materialia*, 56 (2008), pp. 2649-2662.
6. A. Hayashi, C. R. Kao and Y. A. Chang, "Reactions of solid copper with pure liquid tin saturated with copper," *Scripta Materialia*, Vol.37 (1997), pp. 393-398.
7. K. N. Tu, A. M. Gusak, M. Li, "Physics and materials challenges for lead-free solders," *J Appl Phys.*, Vol.93, No. 3 (2003), pp. 1335-1353.
8. T.Laurila, V.Vuorinen, J.K.Kivilahti, "Interfacial reactions between lead-free solders and common base materials," *Mater Sci Eng R.*, 49 (2005), pp. 1-60.
9. Ning zhao, Xuemin Pan et al, "The liquid structure of Sn-based lead-free solders and the correlative effect in liquid-solid interfacial reaction," *13th International Conf on Liquid and Amorphous Metals Conference Series 98*, 2008, 012029.
10. M. Schaefer, R. A. Fournelle, and J. Liang, "Theory for intermetallic phase growth between Cu and liquid Sn-Pb solder based on grain boundary diffusion control," *J Electron Mater.*, Vol.27, No. 11 (1998), pp. 1167-1176.

978-1-4244-4658-2/09 $25.00 © 2009 IEEE

Phase Identification of Intermetallic Compounds Formed during In-48Sn/Cu Soldering Reactions

P. J. Shang[1], Z. Q. Liu[1]*, D. X. Li[1], J. K. Shang[1, 2]*

[1]Shenyang National Laboratory for Materials Science, Institute of Metal Research
Chinese Academy of Sciences, Shenyang, 110016, China
[2]Department of Materials Science and Engineering
University of Illinois at Urbana-Champaign
Urbana, IL 61801, USA
Email: zqliu@imr.ac.cn, jkshang@uiuc.edu

Abstract

Transmission electron microscope observations and precise diffraction analyses were performed on the interfacial reaction between In-48Sn and Cu at the temperature range from 160°C to 250°C for up to 90 minutes. The results indicated that two different morphologies formed between In-48Sn and Cu at the temperature below 200°C: small-grain $Cu_2(In,Sn)$ at the Cu side and large-grain $Cu_2(In,Sn)$ at the solder side. If the soldering temperature is above 200°C, the $Cu_6(In,Sn)_5$ phase is the first phase formed at the solder/Cu interface. However, the subsequent solid-state diffusion of In and Sn atoms through the initial $Cu_6(In,Sn)_5$ layer drove the formation of $Cu_9(In,Sn)_4$ phase between $Cu_6(In,Sn)_5$ and Cu substrate.

Introduction

Due to the toxicity to human health and environment, Pb usage was limited in microelectronics industry in recent years. The search for suitable Pb-free alloys has become one of the key research activities in microelectronics [1, 2]. Several Pb-free alloy systems have been developed in the last decade. Among the Pb-free alloy families, the In-48Sn eutectic alloy has attracted much attention because of its low melting temperature, higher ductility, better wettability, and longer fatigue life [3]. Moreover, on account of the low melting point, this eutectic solder alloy has also been used for diffusion solder process and as thermal switch and so on [4,5].

During soldering, intermetallic compounds (IMCs) formed between In-48Sn alloy and Cu substrate. Several studies have focused on the interfacial IMCs, especially on the phase identification [3-11]. Roming et al [12] dipped Cu based metal into the molten solder twice, and then the samples were placed in alumina boats and aged in air at the temperature ranging from 60°C-110°C, two kinds of IMCs, $Cu_2(In,Sn)$ and Cu_2In_3Sn, were found at the interface. Moreover, based on their investigations, the IMCs layer growth was dominated by the growth of Cu_2In_3Sn, the $Cu_2(In,Sn)$ layer formed during dipping but did not undergo subsequent growth. A subsequent investigation was carried out by Vianco et al [13] who hot dipped Cu in molten 50In-50Sn solder and then solid-state aged at 75, 80 and 100°C for up to 200 days. They found that the IMCs layer contained a thin $Cu_{17}Sn_9In_{24}$ layer near the solder matrix and a thicker $Cu_{26}Sn_{13}In_8$ layer on the side of the Cu substrate. A latest work performed by Sommadossi et al [4] reported two IMCs, $Cu_2(In,Sn)$ and Cu_2In_3Sn, at the interface after experienced a diffusion soldering process from 180-400°C. Another recent work by Chuang et al. [6] reported that ε-$Cu_3(In,Sn)$ and η-$Cu_6(In,Sn)_5$ were the possible phases at the interface after soldering the In-49Sn/Cu system at 60-

110°C. As shown above, for the phase identification between InSn solder and Cu during soldering, the results are not entirely in agreement with each other. Moreover, it must be noticed that all of the above results were obtained through electron probe microanalyzer (EPMA) or Energy Dispersive X-Ray Spectrometer (EDS) methods. The detailed proof for the phase identification, especially the crystallographic structure of the IMCs, was absent in earlier reports. In this work, we present the characterization of Cu/In-48Sn/Cu reaction resulting from the diffusion soldering process, which joins the components of a high melting point using a thin interlayer of solder material of a low melting point. The assembly is held at a temperature above the melting point of the interlayer of solder material, until all the low melting point interlayer has transformed into solid intermetallic phases. The resulting joint allows applications at service temperatures much higher than the low manufacture temperature [4]. Transmission electron microscopy (TEM) which is good at phase identification was used to investigate the interfacial microstructure formed between In-48Sn solder and Cu substrate.

Experimental Procedure

The oxygen-free high-conductivity copper (OFHC) polycrystalline Cu were cut by electro-discharge machining to the size of $10\times2.5\times2$ mm^3. The In-48Sn solder used in this study was prepared by melting high purity In and Sn into an ingot, cold rolled into a 1mm thick foil, and then cut into the size of 10×2.5mm^2. Copper surfaces were grinded and carefully polished by 0.5 μm diamond paste, and then rinsed in acetone, methanol alcohol and distilled water in an ultrasonic bath. Two copper sheets were soldered together with the eutectic SnIn alloy to form a copper-solder sandwich. Before soldering, several brass wires with diameter of 50μm were placed in-between two copper sheets for control of the solder thickness. The soldering process was performed at the temperature of 160°C, 180°C, 200°C and 250°C for 3 seconds to 90 minutes.

The soldered sample was cut across the cross section, and then grinded and carefully polished by BUEHLER polishing suspension with 0.05μm Al_2O_3 powder. Scanning electron microscopy (SEM) morphology observations were carried out on FEI Quanta 600 SEM equipped with an Oxford Link ISIS EDS system. To prepare TEM foils, the slices were mechanically grinded to a final thickness of 40 μm, and ion-milled (Gatan model 691 PIPS) under 5.0 keV and 5μA at a low milling angle (less than 6°). A JEM-2010 and an FEI Tecnai F30 electron microscope were used to carry out TEM observations.

978-1-4244-4658-2/09 $25.00 © 2009 IEEE

Results and Discussion

Fig. 1 shows SEM cross-sectional morphologies of eutectic SnIn solder and Cu joints after soldering at the temperature of 160°C, 180°C, 200°C and 250°C for 90 minutes. It can be seen that the microstructure of eutectic SnIn solder is composed of Sn-rich γ phase and In-rich β phase before soldering, as shown in Fig. 1(b), which was also proved by TEM observations. The existence of γ phase and β phase were identified by the SAEDP analysis, as shown in Fig. 2(b) and 2(e). It seems that the grain size of γ-InSn₄ phase (Grain A in Fig. 2(a)) is larger than that of β-In₃Sn phase (Grain C in Fig 2(d)). The EDS analyses for grain B in Fig. 2(a) shows that it consists of 63.56 at.% Cu, 15.30 at.% In and 21.14 at.% Sn.

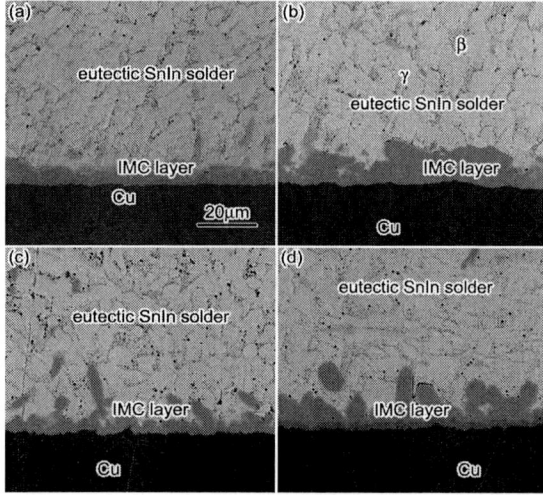

Fig. 1 SEM images showing the cross-sectional morphologies of eutectic SnIn solder and Cu solder joints after soldering for 90 minutes at different temperatures: (a) 160°C, (b) 180°C, (c) 200°C and (d) 250°C

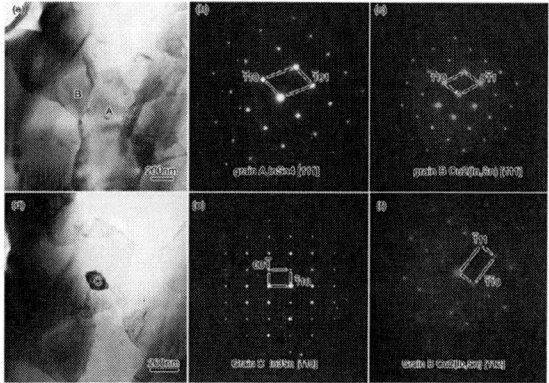

Fig. 2 Microstructure of the solder and the corresponding selected area electron diffraction patterns (SAEDPs): (a)(d) TEM bright images; (b) SAEDP of grain A along [111] zone axis; (c)(f) SAEDPs of grain B along [111] and [112] zone axis; (e) SAEDP of grain C along [110] zone axis

EDPs by tilting the grain B in TEM are shown in Fig. 2(c) and 2(f). The hexagonal Cu₂In lattice with dimensions of a=b=0.4292nm, c=0.5232nm, α=β=90°, γ=120° [14] was

adopted to index the EDPs along [111] and [112] zone axis in Fig. 2(c) and 2(f), respectively. The rotation angle from [111] zone axis to [112] zone axis measured in the experiment was 16.64° which was close to the theoretical value of 17.07°. Therefore, the crystal structure can be identified as hexagonal Cu₂In lattice dissolved with some Sn atoms, which corresponds to the Cu₂(In,Sn). It can be seen that during soldering, Cu atoms can diffuse through IMC layer into solder, and react with Sn and In to form IMC.

When the soldering temperature was below 200°C (including the 200°C itself) for 90 minutes, only one IMC phase was observed at the solder/Cu interface. Moreover, the IMC layer was not uniform. With the increment of soldering temperature, the thickness of IMC layer increased. At the temperatures above 180°C for 90 minutes, some IMC was separated from the IMC layer at the solder/IMC interface and moved gradually into the solder. However, the IMC/Cu interface kept intact. At the temperature of 250°C for 90 minutes, there are two kinds of IMC formed at the eutectic SnIn solder/Cu interface, as shown in Fig. 1(d).

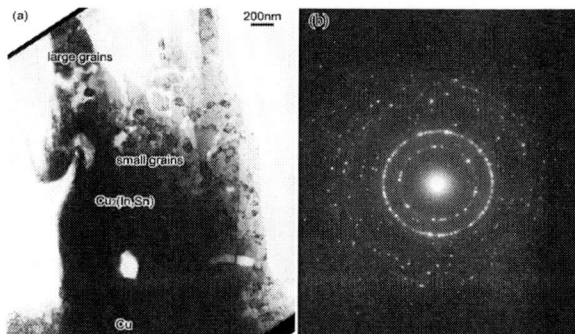

Fig. 3 TEM BF image (a) of eutectic SnIn solder/Cu joint after soldering at 160°C for 3 seconds and corresponding SAEDPs (b) of a large area of the small grain phase at the Cu side

Although only one kind of IMC phase below 200°C was detected in SEM, the microstructure and crystal structure in IMC layer was not clear based on the SEM observations. Therefore, TEM analyses were carried out when the sample was soldered at different conditions. Fig. 3 shows the interfacial microstructure of In-48Sn solder /Cu joint and corresponding selected area electron diffraction pattern of a large area of the small grain phase at the Cu side after soldering at 160°C for 3 seconds. It indicated that two different morphological IMCs formed at the solder/Cu interface. A fine-grain IMC layer, the grain size of which is about 50nm, developed at the Cu side, which was verified in the SAEDP by the polycrystalline ring, as shown in Fig. 3(b).

At the solder side, the grain size of IMC is large, which is about 500nm. This is the same as the results of Sommadossi et al [4]. The EDPs results indicated that the crystal structure of both of IMC layers were Cu₂In-type, but some Sn atoms were dissolved into the Cu₂In lattice and took up the substitutional sites of In atoms. Therefore, the IMC at the interface corresponds to the Cu₂(In,Sn) phase. During soldering process, the large-grain Cu₂(In,Sn) phase firstly nucleated and grew at the solder/Cu interface. Once the large-grain IMC

978-1-4244-4658-2/09 $25.00 © 2009 IEEE

layer formed at the solder/Cu interface, Sn and In atoms had to diffuse through the large-grain $Cu_2(In,Sn)$ layer to react with Cu at the large-grain IMC/Cu interface to form the small-grain $Cu_2(In,Sn)$ phase by a solid-solid reaction process.

With the increment increase of soldering temperature, the interfacial IMC phase changed. Fig. 4 shows the interfacial microstructure of the In-48Sn solder and Cu interface and the corresponding SAEDPs results after soldering at 200°C for 10 minutes. The SAEDPs, as shown in Fig. 4(b) and 4(c), were indexed with the crystallographic structure of Cu_6Sn_5 which has the monoclinic lattice with dimensions of a=1.103nm, b=0.7294nm, c=0.983nm, β=98.82°. The rotation angle from $[\bar{2}51]$ zone axis to $[\bar{3}54]$ zone axis measured in experimental process was 26.2° which was consistent with the theoretical value of 26.63°. Moreover, the EDS analysis shows that some In atoms were dissolved into the Cu_6Sn_5 crystal lattice to form $Cu_6(In,Sn)_5$ IMC layer at the solder/Cu interface.

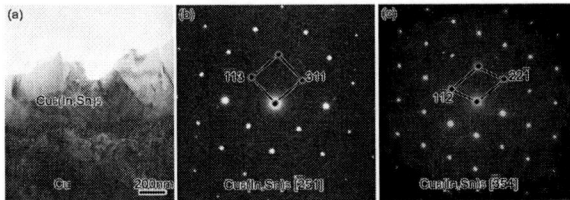

Fig. 4 TEM bright field image (a) and corresponding selected area electron diffraction patterns (b) and (c) of IMC at the In-48Sn solder/Cu interface after liquid-state aging for 10 min at 200°C

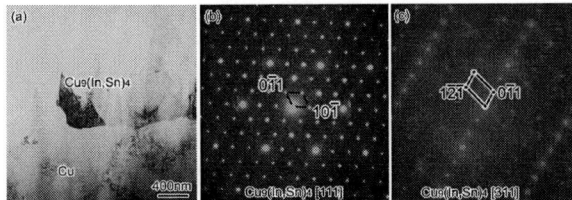

Fig. 5 TEM bright field image showing the interfacial microstructure between eutectic SnIn and Cu (a) and SAEDPs (b) and (c) of interfacial IMC beside the Cu substrate after liquid-state aging for 30 min at 250°C

Fig. 6 TEM bright field image showing the interfacial microstructure between eutectic SnIn and Cu (a) and SAEDPs (b) -(c) of interfacial IMC beside the solder after liquid-state aging for 30 min at 250°C

According to the above results, it can be seen that the crystal structures of interfacial IMC are affected by the Fig. 1(d) shows that with increasing soldering temperature to 250°C, two types of IMCs formed at the In-48Sn solder/Cu

interface. Fig. 5 shows the microstructure and the corresponding SAEDPs of IMC at the Cu side. The EDPs in Fig. 5(b) and 5(c) were indexed as [111] and [311] of Cu_9In_4 which has a cubic lattice with dimensions of a=b=c=0.9097nm, α=β=γ=90° [14]. The rotation angle from [111] to [311] zone axis measured in experimental process was 28.78°, which was near the theoretical value of 29.50° in Cu_9In_4 structure. Moreover, TEM EDS analysis indicated that these IMCs have concentration of 73.89 at.% Cu, 12.17 at.% In and 13.94 at.%Sn. Therefore, the IMC formed at the Cu side between In-48Sn and Cu could be identified as $Cu_9(In,Sn)_4$ which has a Cu_9In_4 crystal structure. In the same way, the IMC at the solder side between In-48Sn and Cu could be identified as $Cu_6(In,Sn)_5$ which has a Cu_6Sn_5 crystal structure, as shown in Fig. 6.

soldering temperature and the diffusive supply of reactive elements. At lower soldering temperatures, the In-48Sn solder/Cu interfacial reaction was dominated by the formation of Cu_2In-type IMC through the reaction of the low melting point In atoms with Cu . However, the higher soldering temperature caused the higher melting point Sn to become the predominated reactive species; therefore, Cu_6Sn_5-type IMC formed firstly at the solder/Cu interface. However, it seems that during solid-state diffusion process, the Cu-In IMC is easier to form. Therefore, with the increase of aging time, Sn and In atoms diffused through the initial solid IMC layer to react with Cu to form Cu-In type IMC at the interface. Moreover, since the sizes of Sn and In atoms are very similar, Sn atoms could easily take the substitutional sites of In or Sn in Cu_2In, Cu_9In_4 crystal lattices. In the same way, the In atoms also can take the substitutional sites of Sn in Cu_6Sn_5 crystal lattices. These substitutions lead to the formation of Cu-In-Sn IMCs at the In-48Sn/Cu interface.

Conclusions

The reactions between In-48Sn and Cu were studied by using TEM BF image and precise diffraction analyses at the temperature range from 160°C to 250°C for up to 90 minutes. The major findings of this investigation are summarized as follows:

1. Below 200°C, only one type of IMC was observed at the In-48Sn/Cu interface. Diffraction analyses indicated that both IMC layers have Cu_2In-type structure. However, the IMC showed two different morphologies: large-grain IMC at the solder side and small-grain IMC at the Cu side.

2. When the soldering temperature was increased to 200°C, $Cu_6(In,Sn)_5$ was the only IMC layer formed at the In-48Sn/Cu interface. Sn is the dominant element during the process of liquid InSn solder reacting with Cu to form the $Cu_6(In,Sn)_5$ phase.

3. At 250°C, two kinds of IMCs formed at the In-48Sn/Cu interface, $Cu_6(In,Sn)_5$ at the solder side and $Cu_9(In,Sn)_4$ at the Cu side. The $Cu_6(In,Sn)_5$ phase developed through a solid-liquid reaction while the $Cu_9(In,Sn)_4$ phase came from a solid-solid reaction.

Acknowledgments

The authors gratefully acknowledge the financial support from the National Basic Research Program of China under Grant No. 2004CB619306, and the Hundred Talents Program of the Chinese Academy of Sciences.

References

1. Li D, Liu C, Conway PP, "Characteristics of Intermetallics and Microstructural Properties During Thermal Ageing of Sn-Ag-Cu Flip-Chip Solder Interconnects," *Mater Sci Eng A*, Vol. 391 (2005), pp. 95-103.

2. Wang FJ, Yu ZS, Qi K, "Intermetallic Conpound Formation at Sn-3.0Ag-0.5Cu-1.0Zn Lead-Free Solder Alloy/Cu Interface During As-Soldered and As-Aged Conditions," *J Alloys Comp*, Vol. 438 (2007), pp. 110-115.

3. Kim DG, Jung SB, "Interfacial Reactions and Growth Kinetics for Intermetallic Compound Layer Between In-48Sn Solder ang Bare Cu Substrate," *J Alloys Comp*, Vol. 386 (2005), pp. 151-156.

4. Sommadossi S, Gust W, Mittemeijer EJ, "Characterization of the Reaction Process in Diffusion-Soldered Cu/In-48at.%Sn/Cu Joints," *Mater Chem Phys*, Vol. 77 (2002), pp. 924-929.

5. Susan DF, Rejent JA, Hlava PF, Vianco PT, "Very Long-Term Aging of 52In-48Sn (at.%) Solder Joints on Cu-Plated Stainless Steel Substrates," *J Mater Sci*, Vol. 44 (2009), pp. 545-555.

6. Chuang TH, Chang SY, Tsao LC, Weng WP, Wu HM, "Intermetallic Compounds Formed During the Reflow of In-49Sn Solder Ball-Grid Array Packages," *J Electron Mater*, Vol. 32, No. 3 (2003), pp. 195-200.

7. Wu HF, Chiang MJ, Chuang TH, "Selective Formation of Intermetallic Compounds in Sn-2oIn-0.8Cu Ball Grid Array Solder Joints with Au/Ni Surface Finished," *J Electron Mater*, Vol. 33, No. 9 (2004), pp. 940-947.

8. Sommadossi S, Huici J, Khanna PK, Gust W, Mittemeijer EJ, "Mechanical Properties of Cu/In-48Sn/Cu Diffusion-Soldered Joints," *Z Metallkd*, Vol. 93 (2002), pp. 496-501.

9. Sommadossi S, Guillermet AF, "Interfacial Reaction Systematics in the Cu/In-48Sn/Cu System Bonded by Diffusion Soldering," *Intermetallics*, Vol. 15 (2007), pp. 912-917.

10. Roy R, Sen SK, Sen S, "The Formation of Intermetallic in Cu/In Thin Films," *J Mater Res*, Vol. 7, No. 6 (1992), pp. 1376-1386.

11. Roy R. Pradhan SK, De M, Sen SK, "Structural Characterization of the CuIn Intermetallic Phase Produced by Interfacial Reactions in Cu/In Bimetallic Films," *Thin Solid Films*, Vol. 229 (1993), pp. 140-142.

12. Romig Jr AD, Yost FG, Hlava PF, "Intermetallic Layer Growth In Cu/Sn-In Solder Joints," *Proc Conf Microbeam Analysis* (San Francisco, CA: San Francisco Press, 1984), pp. 87-92.

13. Vianco PT, Hlava PF, Kilgo AC, "Intermetallic Compound Layer Formation Between Copper and Hot-Dipped 100In, 50In-50Sn, 100Sn, and 63Sn-37Pb Coatings," *J Electron Mater*, Vol. 23, No. 7 (1994), pp. 583-594.

14. Che GC, Ellner M, "Powder Crystal Data for the High-Temperature Phases Cu_4In, $Cu_9In_4(h)$ and $Cu_2In(h)$," *Powder Diffraction*, Vol. 7, No. 2 (1992), pp. 107-108.

Microstructural Characterization of Electroplating Sn on Lead-frame Alloys

Yiqing Wang[1], Dongyan Ding[1], Klaus-Peter Galuschki[2], Yu Hu[2], Angela Gong[2], Shuo Bai[1], Ming Li[1], Dali Mao[1]

[1]Lab of Microelectronic Materials & Technology, School of Materials Science and Engineering
Shanghai Jiao Tong University, Shanghai 200240, China
[2]SMT Technology & Material, Corporate Technology, Siemens Ltd., China
Email: dyding@sjtu.edu.cn, Tel: +86-21-34202741

Abstract

Electroplating Sn plays an important role in the lead-free age because of its excellent solderability and many other advantages. In this work, bright Sn was electroplated onto both C194 and FeNi42 alloys. The plating parameters, substrate effect, barrier layer effect and IMC formation were investigated through detailed microstructural characterization of the Sn films, barrier layers and cross-sections. It was found that the current density, plating time, substrate type and barrier layer play an important role in determining the fine structures of the Sn deposition. Interfacial reactions to form intermetallic compounds (IMC) was also observed. Replacing the traditional Ni barrier with Ni nanocone barrier was found to result in a quite different IMC formation/distribution.

1. Introduction

Due to environmental considerations and especially the RoHS policy, the IT/IC industry has come to an age of lead-free electronics [1-2]. As a good alternative to the widely used Sn-Pb solderable coatings, electroplating Sn has received much attention in recent years [3,4]. Unfortunately pure Sn and its alloys have to face the tin whisker problems, which have not been fully understood and controlled for more than five decades [5-7]. To date, various kinds of literatures have reported on the formation mechanisms of tin whiskers. As tin whiskers could cause great damage in high-reliability electronic products such as lead-frames, connectors and capacitors, numerous efforts have been done to find right mitigation methods or recommendations.

In view of the fact that microstructure of the electroplating Sn film plays an important role in determining the formation and growth of tin whiskers during a storage or product application, a systematic study of the relationship (among the electroplating process, plated microstructures and the whisker formation) is highly needed for better understanding the whisker formation mechanism and further developing feasible mitigation methods. Thus, in the present work, we use a commercial plating solution to fabricate bright Sn films on lead-frame alloys. Fabrication parameters and microstructures of the Sn depositions were investigated through both scanning electron microscope and metallography.

2. Experiments

Strip foils of two kinds of lead-frame alloys (C194 and FeNi42) were used as the original substrates. Typical chemical compositions of the C194 alloy and FeNi42 alloy are Cu-2.2Fe-0.1Zn-0.03P and Fe-42Ni-1.0Co-0.8Mn-0.3Si-0.1Al (wt%), respectively. Fig. 1 shows surface morphologies of the original foil substrate. Uneven texture-like surfaces induced by a rolling process could be found on the lead-frame alloy surfaces. The FeNi42 substrate presents reduced defects than the C914 substrate does.

The lead-frame foils were cut into desired samples, washed with deionized water and conducted with electrolytic degreasing before a microetching of the alloy substrate.

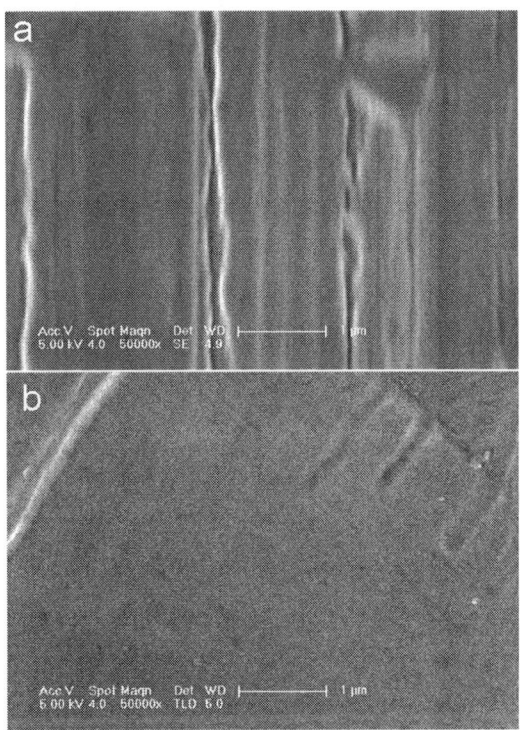

Fig. 1 SEM morphologies of as-received lead-frame foil substrates.
(a) C194, (b) FeNi42

After the above mentioned surface treatment of microetching, a surface activation with methylsulfonic acid was also carried out before electroplating Sn. Fig. 2 presents the surface morphologies of the as-activated alloy substrates. Compared with the original surfaces of the lead-frame substrate (Fig. 1), the as-activated substrates present a quite different surface morphology. Many small holes could be seen on the surface of the FeNi42 substrate. It suggests that the microetching and activation play an important role in determining the surface roughness or nucleation/growth of electroplating Ni or Sn depostion.

The plating solution for bright Sn was supplied by Shanghai Sinyang Semiconductor Materials Co., Ltd. The current density of the DC plating ranged from 0.5A/dm^2 to 4A/dm^2. Bright Sn films were deposited onto the surface of both the C194 copper alloy and the FeNi42 alloy. The plating

978-1-4244-4658-2/09 $25.00 © 2009 IEEE

parameters especially the plating time were controlled to obtain different coating thicknesses ranging from 2 μm to 10 μm. After electroplating the Sn-coated samples were washed with deionized water, dried and stored for further evaluation.

To fabricate electroplating Sn on Ni-coated lead-frame alloys, the above mentioned microetched foil substrates were directly electroplated with two kinds of Ni barrier layers, i.e., traditional Ni film and Ni nanocone film developed by Ming Li et al. from Shanghai Jiao Tong University, China. The plating solutions for the traditional Ni barrier mainly consist of nickel chloride and boric acid. And the other solution for the Ni nanocone barrier also includes some crystallization additives. The thickness of the Ni barrier layer used in this work varied from 2 μm to 4 μm. The Ni-coated substrate were then activated with methylsulfonic acid and electroplated with bright Sn to obtain the Sn/Ni composite coatings on the lead-frame foils.

The surface morphologies of the original foil substrate, activated samples, as-plateded samples were investigated with scanning electron microscope (SEM) equipped with Energy dispersive X-ray analysis (EDXA). Cross-sectional optical microscope (OM) analyses of grinded and polished samples were conducted to investigate the coating thickness and interfacial reactions. X-ray fluorescence analysis of the coating thickness as well as the composition was also conducted with a CMI900 from Oxford Instruments.

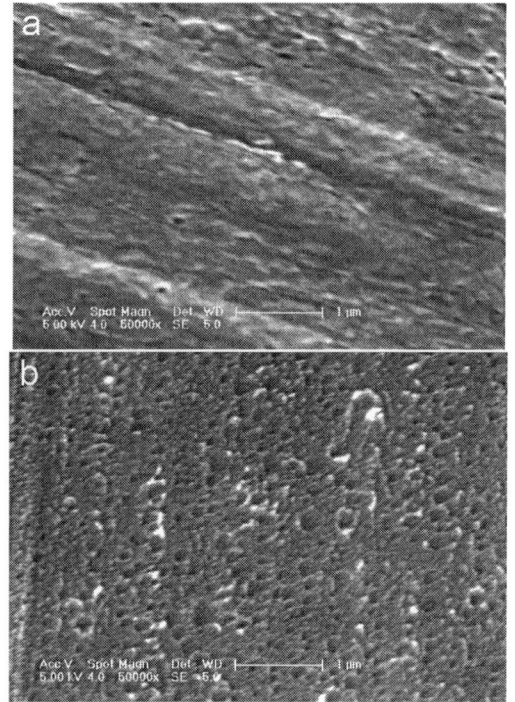

Fig. 2 SEM morphologies of as-activated substrates. (a) C194, (b) FeNi42.

3. Results and Discussion

3.1 Electroplating Sn on lead-frame alloys without Ni barrier layer

As mentioned above, bright Sn could be directly deposited onto the C194 substrate. Figs. 3 and 4 present the effect of current density on the thickness and the surface morphologies of the Sn films, respectively. It can be found that, with increase of the current density, the thickness of the Sn film increased from 2.6 μm (at 0.5A/dm^2) to 15.4 μm (at 2A/dm^2). However, a further increase of the current density to 4A/dm^2 resulted in a decrease of the film thickness to only 9.6 μm. And a variation of the current density could lead to quite different surface morphologies or surface grain size of the Sn deposition. The average grain size of the as-plated bright Sn could be higher than 1.5 μm under a current density of 0.5A/dm^2 (Fig. 4(a)). And it could be decreased rapidly to less than 100 nm under a higher current density of 4A/dm^2 (Fig. 4(e)). Our experiments also revealed that under a high current density of 4A/dm^2, large quantity of hydrogen bubbles could appear in the bath during the electroplating process. The wave-like surfaces shown in Fig. 4(e) also prove that a relatively higher current density should not be used for fabrication of uniform Sn films on the lead-frame alloy. According to the above analyses, in the following work, we will only discuss the Sn films fabricated at the current density of 1A/dm^2.

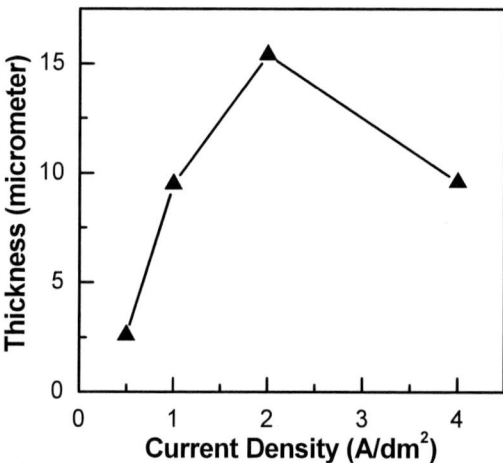

Fig. 3 Effect of current density on the thickness of Sn deposition on C194 substrate. The deposition time is 15 minutes.

Fig. 5 and Fig. 6 present SEM morphologies of 2 μm (in thickness) Sn deposited onto the C194 substrate and FeNi42 substrate, respectively. It can be found that the surface of the Sn/FeNi42 system is flatter than that of the Sn/C194 system. The surface grains of the Sn/FeNi42 system tend to exist with a tetragonal shape (Fig. 4(b)), and the surface grains of the Sn/C194 system present irregular form (Fig. 4(a)). Obviously, when we fabricate bright Sn on lead-frame alloys, the type of the substrate could affect the morphology or grain size of the bright Sn deposition.

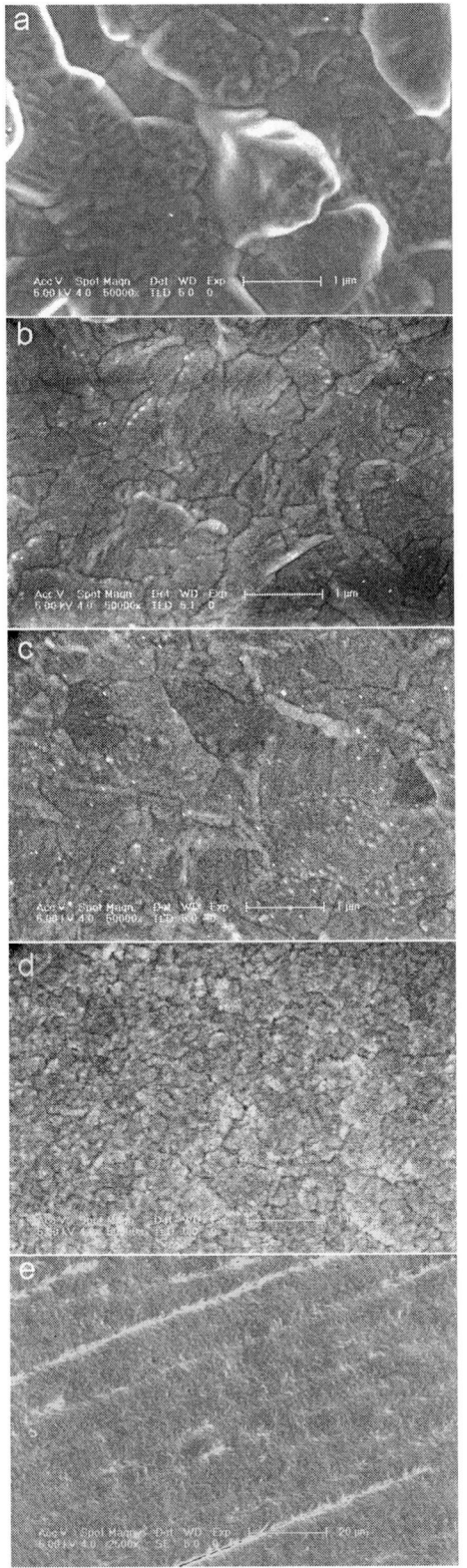

Fig. 4 SEM morphologies of Sn deposition obtained at different current density. (a) 0.5A/dm^2, (b) 1A/dm^2, (c) 2A/dm^2, (d) and (e) 4A/dm^2

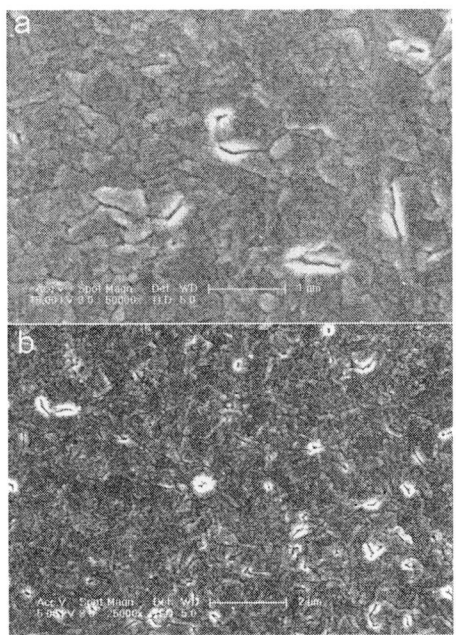

Fig. 5 SEM morphologies of 2μm Sn deposition on C194 substrate. (a) high magnification image, (b) low magnification image.

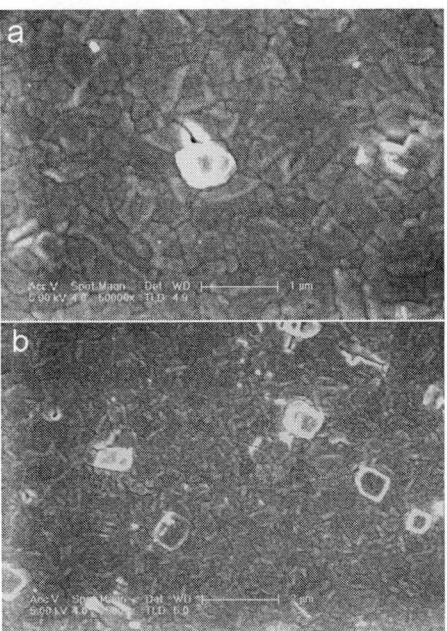

Fig. 6 SEM morphologies of 2 μm Sn deposition on FeNi42 substrate. (a) high magnification image, (b) low magnification image

However, this substrate effect became weaker as the thickness of the Sn film increased to 5 and 8 μm. As shown in Fig. 7, thicker bright Sn deposited on both the C194 and FeNi42 substrates could yield much more flat surface. And tetragonal Sn crystals no longer exist as they present in the thinner film system. Compared with the thinner films shown in Figs. 5 and 6, the thicker Sn films shown in Fig. 7 present a much more dense structure, which is preferred for further tin whisker evaluation. Obviously, the substrate effect could be almost ignored with increase of the film thickness to a certain degree.

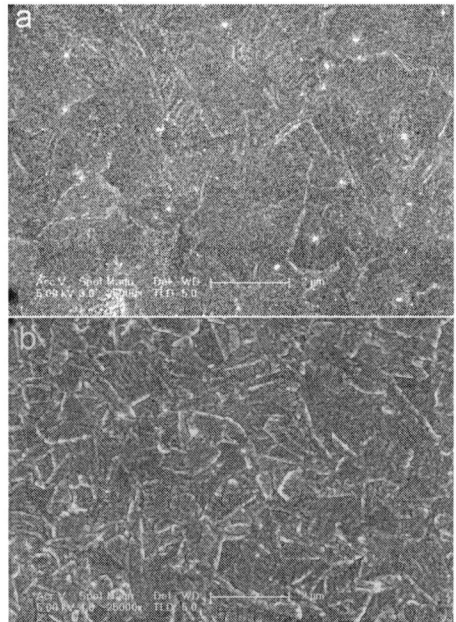

Fig. 7 SEM morphologies of 8 μm Sn on (a) C194 substrate, (b) FeNi42 substrate

3.2 Electroplating Sn on Ni-coated lead-frame alloys

Electroplating Ni has been reported to be a good candidate as barrier layer to prevent or retard the formation of tin whiskers. In our experiments, we explored two kinds of Ni

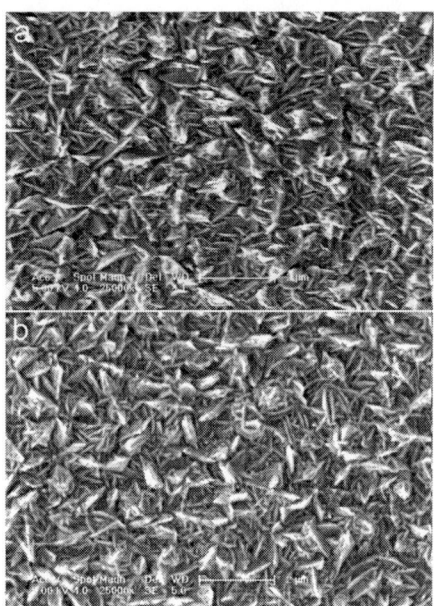

Fig. 8 Traditional Ni films deposited onto the lead-frame alloy substrates. (a) Ni on C194, (b) Ni on FeNi42.

barrier layer. One was the traditional Ni film deposited onto lead-frame alloys (Fig. 8). The other was a novel barrier of Ni nanocone array (Fig. 9). As shown in Fig. 8, the surface of traditional Ni barrier fabricated here consist of many slender Ni crystals that stacks together on both the C194 and FeNi42.substrates. While in Fig. 9, the novel Ni barrier mainly consists of large quantity of Ni nanocones self-assembled on top of a thin layer of dense Ni film.

Fig. 10 and Fig. 11 present SEM morphologies of bright Sn films deposited onto the traditional Ni-coated substrate and the nanocone-coated substrate, respectively. Similar to the previous findings with the Sn/C194 system, an increase of the Sn film could lead to flat surface, densified deposition and reduced substrate effect on the surface microstructure of the Sn films. A close examination of the surface morphologies of the two Sn/Ni systems revealed that the barrier type could also lead to different surface morphology of the Sn deposition. That is to say, we could adjust the microstructure of the bright Sn films by using various kinds of barrier structures to change the fine structures of electroplating Sn, and thus further control the tin whisker formation.

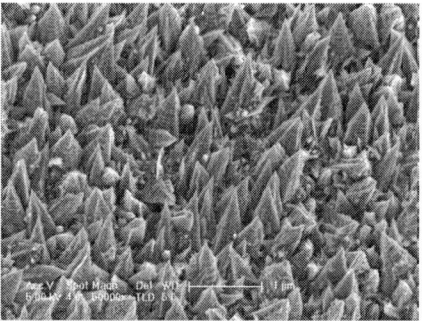

Fig. 9 Ni nanocone barrier deposited onto C194 substrate.

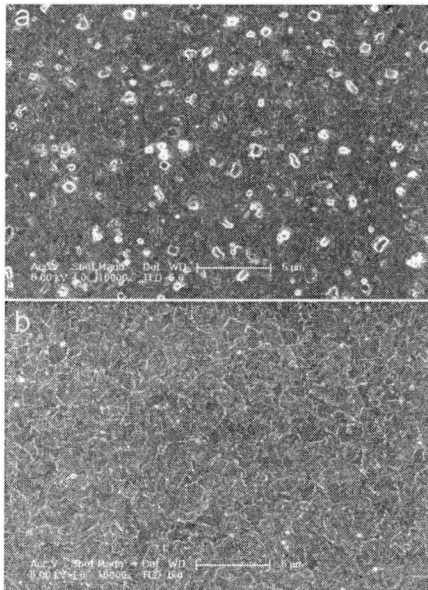

Fig. 10 Sn deposited onto traditional Ni-coated C194 substrate. (a) 2 μm Sn, (b) 5 μm Sn

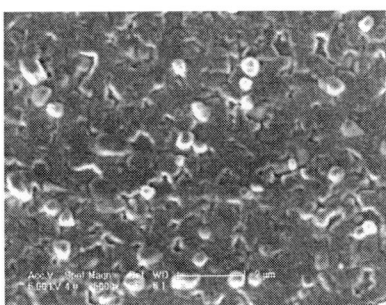

Fig. 11 2 μm Sn deposited onto Ni nanocone-coated C194 substrate.

3.3 IMC formation

Some of the electroplated samples (with or without Ni barriers) have been grinded and polished for a cross-section analysis through optical microscope (OM) observations. Figs. 12 and 13 are cross-sectional images of the bright Sn deposited onto C194 substrate and FeNi42 substrate, respectively.

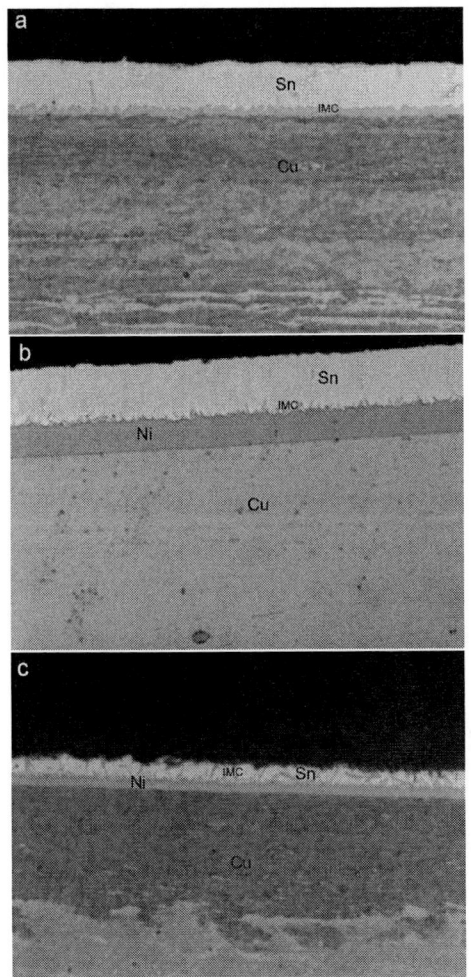

Fig. 12 Cross-sectional images of (a) Sn on C194 substrate, (b) Sn on traditional Ni-coated C194 substrate, (c) Sn on Ni nanocone-coated C194.

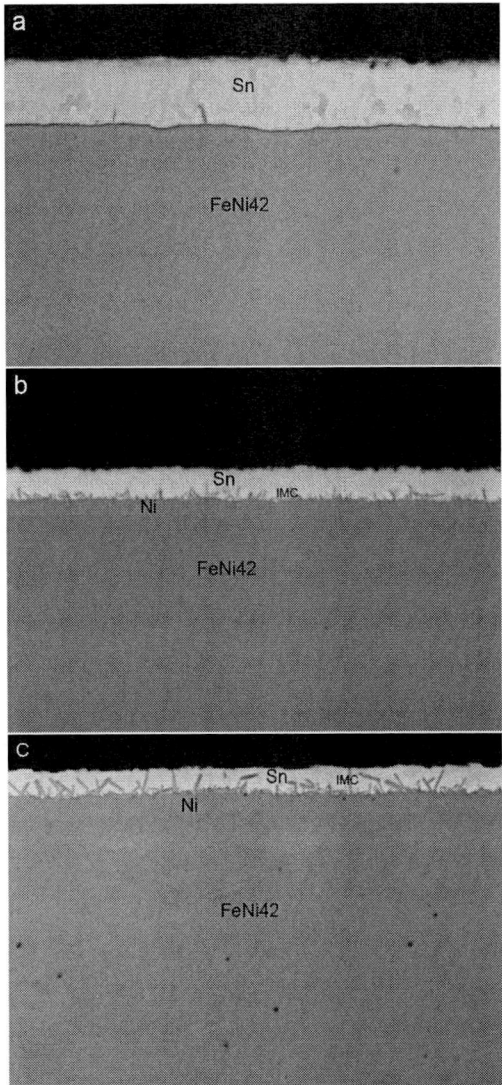

Fig. 13 Cross-sectional images of (a) Sn on FeNi42 substrate, (b) Sn on traditional Ni-coated FeNi42 substrate, (c) Sn on Ni nanocone-coated FeNi42 substrate

Intermetallic compounds (IMC) resulting from interfacial reaction/diffusion during electroplating or short time aging at room temperature were found for both substrate systems. Without a Ni barrier layer, a rough IMC layer (usually Cu_5Sn_6) could be found at the Sn/C194 interface (Fig. 12(a)). Whereas, rare IMC could form at the Sn/FeNi42 interface (Fig. 13(a)). Such a difference in IMC formation should be mainly attributed to the difference in substrate type.

When the two kinds of Ni barrier layers were introduced to the Sn/C194 or Sn/FeNi42 interface. A quite different characteristic of IMC formation could be found. As shown in Figs. 12(b) and 13(b), the introduction of traditional Ni barrier could lead to a formation of large quantity of short needle-like IMC (Ni_3Sn_4) at the Sn/Ni interface. However, this was not the case with the novel Ni-nanocone barrier. As shown in Figs. 12(c) and 13(c), the introduction of Ni-nanocone barrier could result in the formation of very large needle-like IMC inside the Sn film rather than at the Sn/Ni interface. Obviously, all of the barrier layers enhanced the formation of needle-like IMC at the either the Sn/Ni interface or inside the Sn film. Thus, when the Ni barrier layer exists, the substrate effect seems to be not so evident.

At present, we have no ideas of how the tin whiskers will form on the above Sn/Ni/LF systems especially the novel Ni-nanocone system. But one thing is certain that we have opened up a new way to control the IMC formation and later the stressing status in the electroplating Sn film by using novel barrier structures. We believe that, through further optimization of the microstructures or IMC reaction of electroplating Ni nanocones and electroplating Sn, we could finally minimize or even control the formation of tin whiskers.

4. Conclusions

In summary, bright Sn was electroplated onto lead-frame alloy substrates with or without Ni barrier layers. The plating parameters, substrate effect and barrier effect were evaluated through detailed microstructural characterization. As the lead-frame alloys (C194 and FeNi42) have different chemical composition and surface topographies, there was a substrate effect during the deposition of bright Sn especially relatively thinner films. With increase of the plating time, both the thickness and average surface grain size of the tin films increased, and the densification of the Sn films also increased. The effect of Ni barrier on the Sn depositions and the IMC formation was so significant that considerable efforts should be given to optimize the fine microstructures of both the Ni barrier (especially the novel nancone barrier) and the electroplating Sn.

Acknowledgments

We thank the financial support from Corporate Technology, Siemens Ltd., China and Shanghai Pujiang Program (No. 07pj14047). And we also thank for the support from Shanghai Sinyang Semiconductor Materials Co., Ltd.

References

1. Tu, K. N., Gusak, A. M., "Physics and materials challenges for lead-free solders," *J. Appl. Phys.*, Vol. 93 (2003), pp. 1135-11153.
2. Lee, B., Lee, D., "Spontaneous growth mechanism of tin whiskers," *Acta Metall.*, Vol. 46 (1998), pp. 3701-3714.
3. Dittes, M., Oberndorff, P., Petit, L., "Tin whisker formation-results, test methods, and countermeasures," *Proc. IEEE Elec. Comp. Conf.*, 2003, pp. 822-826.
4. Choi, W. J., Lee, T. Y., Tu, K. N., "Sturctures and kinetics of Sn whisker growth on Pb-free solder finish," *Proc. IEEE Elec. Comp. & Tech. Conf.*, 2002, pp. 628-633.
5. Franks, J., "Metal whiskers," *Nature*, Vol. 177 (1956), pp. 984.
6. Dunn, B. D., "Whisker formation on electronic materials," *Circuit World*, Vol. 2 (1976), pp. 32-40.
7. Tu, K. N., Zeng, K., "Reliability issues of Pb-free solder joints in electronic packaging technology," *Proc. IEEE Elec. Comp. & Tech. Conf.*, 2002, pp. 1194-1199.

The Fabrication of Composite Solder by Addition of Copper Nano Powder into Sn-3.5Ag Solder

Aemi Nadia, University Malaya
Mech Engineering, University Malaya, 50603 KL, Malaysia.
aeminadia@um.edu.my

Abstract

Copper nano powder was incorporated into Sn-3.5Ag solder at 3wt%, by mechanically alloyed (MA) by ball milling in a planetary ball mill at room temperature. The rotation speed was kept constant at 150 rpm with different milling time, 26 hours, 78 hours and 120 hours. The milled solder was mixed with flux and reflowed at 240°C for 60 seconds. XRD result confirmed the formation of Cu_3Sn after 120 hours of milling time. The melting temperature of the samples decreased as a function of milling time. The wetting angle was improved by 48% for the composite solder Sn-3.5Ag-nano-3.0Cu for 120 hours ball milled, compared to the Sn3.5Ag solder. Increased of hardness was obtained as a result of incorporation of nanoparticle.

Introduction

For years, Sn-Pb solders have been used extensively as interconnect materials. However, a recent environmental concerns, and the increasing awareness of health risk associated with lead (Pb) containing solder alloys has pushed the electronics industry towards lead-free and there are environmental concerns over the amount ending up in landfill. [l-2] In Europe, the waste electrical and electronic equipment (WEEE) directive by EU has banned the use of Pb in consumer, and together with the restriction of hazardous substances (ROHS) compliance has claimed, Pb is the most common material that must be eliminated.

So an effort to develop high performance and environmentally friendly solders, this research was done to investigate processing method for the production of nanoparticle reinforced composite based lead free solder. Sn-Ag-Cu (SAC) alloy has immense potential because of its good wettability, higher strength, and superior resistance to creep and thermal fatigue, when compared to eutectic Pb-Sn solder. In fact, few organizations, such as NEMI and SMTA recommend a solder alloy type SAC396, whilst JEIDA recommends SAC 305 as the perfect candidates to replace Sn-Pb solder. In this study, a composite SAC solder is being fabricated. Previous researches were done as reported by Guo F.J in Composite lead-free electronic solders, nanoparticle additions to solder materials have been demonstrated to result in increased reliability of solder joints. [3] Hardening the solder with the addition of nanoparticles and stabilizing the grain structure should result in increased reliability. The reinforcement particles should not be prone to significant coarsening during reflow and service. Due to thermo-dynamic considerations, fine particles tend to coalesce to reduce the high interfacial energy associated with small particles. A number of methods are currently being investigated for the production of composite solder, such as mechanical mixing, in situ method and mechanical alloying - ball milling. In this research, the mechanical alloying is chosen as the method to

produced SAC composite solder. Mechanical alloying (MA) is a nonequilibrium processing technique for producing composite metal particles with submicron homogeneity by the repeated fracture of powder particles cold welding. This method is believed to offer good processablity, precise control over the solder composition, and produce more homogeneous mixture. [4-5] In this method, Cu nano powder was incorporated to Sn 3.5Ag solder prior to the milling process.

The present work focuses on the study of the morphology and thermal properties of ball milled Sn-Ag-nanoCu solder. The wetting characteristics and mechanical properties of composite solder form also were also studied as a function of milling period.

Experiment

To fabricate the Sn-Ag-Cu composite solder, Cu nano powder (purchased from Sigma Aldrich, 40nm, 99% purity) was incorporated into Sn 3.5Ag solder powder (AIM Solder, 30-40 μm, purity 99.9%), at 3.0wt%. The mechanical alloying- ball milled then was performed with high energy planetary ball mill (Retsch, PM400) at constant speed 150 rpm at room temperature, with different milling time, 26 hours, 78 hours and 120 hours. The zirconia milling balls with diameter of 1cm were used with ball to weigh ratio maintained at 2:1. The morphology of ball milled powder was studied using scanning electron microscopic (Philips XL40). The resultant composites solder were analyzed using an x-ray diffractometer (Philips Xpert MPD) using a wavelength of Cu Kα (λ:1.5405 Å). The melting temperature of composite solder is measured using differential scanning calorimetry tester (DSC 820, Mettler Toledo). The flux (RMA type, Kester Solder Corp) then, was added to make the samples into the paste form. It was found, with addition of Cu nano powder, HCl acid was needed as activator to aid the reflow process. The ratio of flux to HCl was 5:1. The samples with averagely 0.2g and and opening of 6.5mm were deposited onto Cu substrates (30mm×30mm). The samples were then reflowed at 240°C, for 60 seconds. The inductive couple plasma test was carried out to measure the percentage of copper retained in the fabricated solder. The wetting characteristics were studied using optical microscope. The hardness of samples was measured using microhardness tester (MVK-H2, Mitutoyo with load of 10gf).

Result and Discussion

Figure 1 shows the SEM image of as received Sn-3.5Ag, Sn-3.5Ag-3nano-Cu after ball milling for 26 hours, 78. The size of the initial Sn-3.5Ag powder is seen 30-40μm in diameter in Fig (1a). As received Cu nano powder reportedly has <50 nm dimension. During the ball milling, the powders were flattened, cold-welded, fractured, and rewelded. After ball milling for 26 hours, plastic deformation refined the particles and increased the grain-boundary area. This

978-1-4244-4658-2/09 $25.00 © 2009 IEEE

explaining the morphology of the powders, which were flattened into pieces in Fig 1(b). With a longer milled for 78 hours, the pieces started to be welded and joined each other, as observed in Fig 1 (c). After 120 hours of milling time, the welded powders were observed with irregular shape, and estimated dimension of 50-70μm along the long axis. The raw materials, Sn, Ag, and Cu powders, were ductile material. When the ductile to ductile reaction involved, the particles were flattened to platelets shapes and cold welded to form a composite structure. [6]

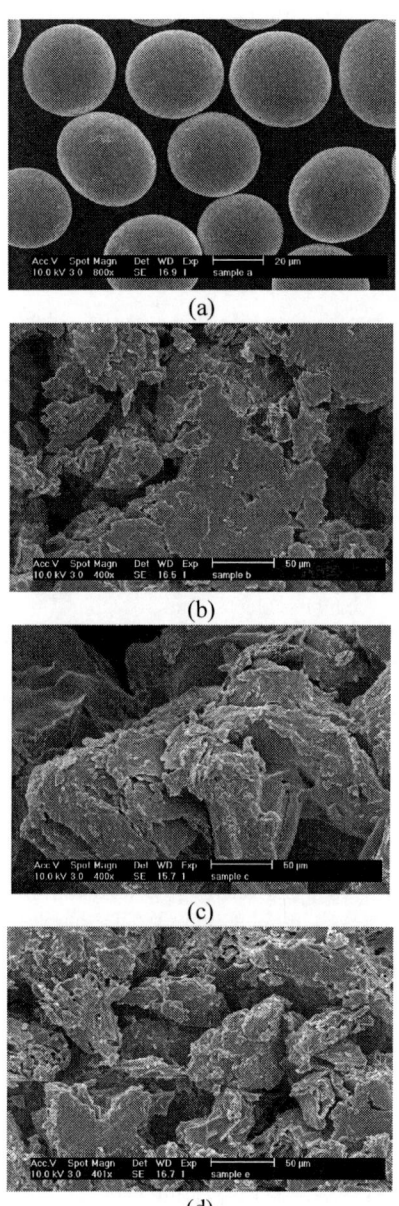

(a)

(b)

(c)

(d)

Figure 1 The SEM image of powder of a) as-received Sn-3.5 Ag b) Sn-3.5Ag-3nanoCu after 26h ball milling c) Sn-3.5Ag-3nanoCu after 78 H ball milling d) Sn-3.5Ag-3nanoCu after 120 H ball milling.

To confirm on the successful alloying of the milled powder, X-ray diffraction was done. Figure 2 shows the x-ray diffraction pattern of as-received powder Sn-3.5Ag and Sn-3.5Ag-3nanoCu powder milled for 120 hours. The x-ray diffraction peaks of Sn (211) and (311), Ag_3Sn (211) were detected on as received Sn-3.5Ag. After 120 hours milling of Sn-3.5Ag-3.0Cu, an extra peak of Cu_3Sn (002) appeared, together with the initial peaks of Sn and Ag_3Sn. This suggested that the kinetic energy initiate the chemical reaction between Sn-3.5Ag powder and nano Cu powder, and forming intermetallic compound of Cu_3Sn without any external heating.

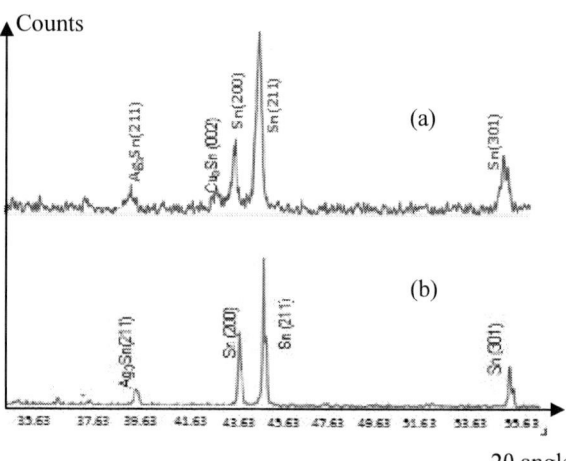

Figure 2 The XRD result of a) Sn-3.5Ag-3-nanoCu after 120 H ball milled b) As-received Sn-3.5Ag

Table 1 shows the melting temperature of the samples, measured by DSC. Melting temperature is determined as onset values of endothermic peaks in heat flow curves. The melting point of as-received Sn 3.5 Ag solder was 219.10°C. After ball milling for 26 Hours, the melting temperature decreased to 217.40°C. With increased of milling period to 78 hours and 120 hours, the melting temperature was dropped to 214.70° and 213.60° respectively. This suggested that a longer milling period resulted a higher excess energy in the system, which reduced the total enthalpy of melting. Reddy reported there was a depression of melting temperature of 5°C for SAC 305 with nanocrystalline structure produced by milling. [7] It was found in that study; the reduction of grain size can lead to the reduction of melting point as the excess energy stored in the grain boundaries as well as the capillarity reduces the total enthalpy of melting.

However, another factor also contributed to the depression of melting temperature, which was the Cu nano powder with higher surface energy added into the solder alloy. The solidification temperature was affected by the addition of Cu nano powder because of a reaction of the nano powder with tin. It is suggested that part of the Cu nano-particles dissolved in the molten solder, and changed the alloy composition of the solder. The copper nano powders were completely dissolved inside the solder, and produced a Sn-Ag-Cu ternary alloy with co-existence of the Sn- Ag binary system. [8]

Table 1 The melting point of solder as measured by DSC

Samples	Melting temperature (°C), ±1
As received Sn 3.5 Ag	219.10
Sn 3.5 Ag 3Cu, 26H Milling Time	217.40
Sn 3.5 Ag 3Cu, 78 Milling Time	214.70
Sn 3.5 Ag 3Cu, 120 Milling Time	213.60

The Cu concentration in the Sn-Ag-Cu solder alloy produced by MA is shown in Table 2. The test was carried out on the sample after the reflow process. It was found that, when 3wt% Cu added, the solder the same percentage of Cu, 2.95wt% was retained inside the composite solder. The ICP-OES derived concentration of the as-fabricated solder powders by the MA process was fairly close to the nominal one, which indicating that, the Cu added was almost completely retained inside the composite solder. This result also derived that mechanical milling process is a good processability method, which can produce a homogeneous composite solder.

Table 2 The Cu concentrations analyzed by ICP-OES from the SnAgCu solder alloy produced by MA process.

Nominal Cu Concentration in Synthesis Concentration (wt.%)	ICP Measured Cu, Concentration (wt.%)
3.0	2.95

Wetting is crucial for soldering, as it plays an essential role in ensuring good bonding between the solder material and the substrate. In this study, the wettability of samples is evaluated by wetting angle. The wetting angle of samples is measured as shown in Fig. 3. The wetting angle of Sn 3.5 Ag solder was measured at 30.14°. It was found, a longer ball milling period, would result in a better contact angle, as observed on the sample after 26 hours of ball milling, whereas the wetting angle was drastically dropped to 22.02°. After 78 hours and 120 hours of milling, the wetting angle continued to decrease to 18.94° and 15.14°. It is well known there are many factors contributing to the wetting characteristics, such as the types of substrate, solder composition, impurities, soldering atmosphere, flux, pre tinning type and thickness, storage conditions and time as well as on reflow conditions. [9-10]

Figure 4 shows the hardness value for the samples, which were taken after reflow process. The hardness of Sn-3.5Ag solder was measured at 15.7 HV. With the addition of nano powder into Sn-3.5Ag by MA, the hardness value was increased. It was observed, the sample with 3wt% nano Cu and milled for 26 hours, the hardness measured was 22.11 HV. However, the milling period did not appreciably affect the hardness value, as for the sample with 3wt% nano Cu added milled for 78 hours and 120 hours. The increased of

hardness value for the samples is linked to the presence of nano Cu powder. It has been observed that reinforcement particles of size ~1μm or smaller tend to stabilize a very fine grained microstructure. [11] The addition of Cu nano powder is believed to improve the mechanical properties of resulting composites, with a dispersion-strengthening effect. The Cu particles dispersed through the composite microstructure act as potential sites for pinning the movement of dislocations and the grain boundary and thus facilitates a slowdown in grain growth rate.

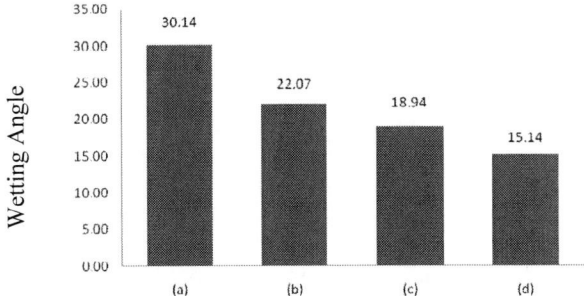

Figure 3 The wetting angle measured on a) Sn-3.5 Ag solder, b) Sn-3.5Ag-3nanoCu after 26 hour ball milling c) Sn-3.5Ag-3nanoCu after 78 hour ball milling and d) Sn-3.5Ag-3nanoCu after 120 hour ball milling

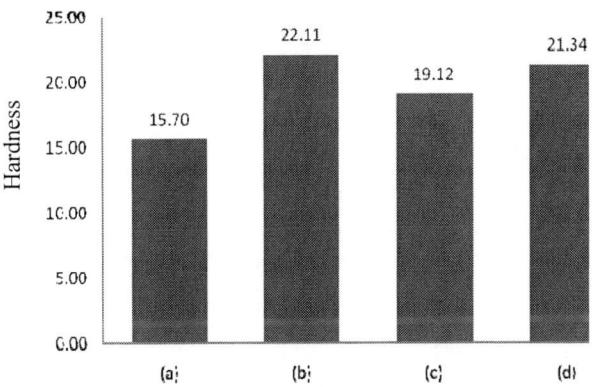

Figure 4 Hardness Value measured on sample a) Sn-3.5 Ag solder, b) Sn-3.5Ag-3nanoCu after 26 hour ball milling c) Sn-3.5Ag-3nanoCu after 78 hour ball milling and d) Sn-3.5Ag-3nanoCu after 120 hour ball milling

Conclusions

Composites solder Sn-Ag-Cu was successfully synthesized with the incorporation of Cu nano powder into Sn-3.5Ag solder powder by mechanical alloying. The high kinetic energy during fracture mechanism changed the initial well formed powders to the irregular shapes. The mechanical alloying processing caused chemical reaction between tin and copper. It also has been found that, a longer milling period lowered melting point, and reduced contact angle. However, there is no significant effect of milling period on the hardness of composite solder.

References

1. K. Zeng, *et al.*, "Six cases of reliability study of Pb free solder joints in electronic packaging electronic," *Material Science and Engineering*, Vol. 38 (2002), pp. 55-105.

2. K. N. Tu, *et al.*, "Physics and materials challenges for lead-free solders," *Journal of Applied Physics*, Vol. 93 (2003), pp. 1335-1353.

3. Jianbiao Pan, *et al.*, "Effect of Reflow Profile on SnPb and SnAgCu Solder Joint Shear Force," *IPC Printed Circuits Expo, APEX and the Designers Summit*, California, Feb. 2006.

4. William J, "Second Generation Lead Free Solder Alloys – A Challenge to Thermodynamics," *Plumbridge. Monatshefte f€ur Chemie*, Vol. 136 (2005), pp. 1811-1821.

5. Won Kyoung Choi, *et al.*, "Interfacial microstructure and joint strength of Sn-3.5Ag-X (X = Cu, In, Ni) solder joint", *J. Mater. Res.*, Vol. 17, No. 1 (2002).

6. Szu-Tsung Kao, *et al.*, "Controlling Intermetallic Compound Growth in SnAgCu/Ni-P Solder Joints by Nanosized Cu6Sn5 Addition," *Journal of Electronics Materials*, Vol. 35 (2006), pp. 486-493.

7. Bhupal Reddy, *et al.*, "The effect of ball milling on the melting behavior of Sn-Cu-Ag eutectic alloy," *Journal of Materials Science*, Vol. 44 (2009), pp. 257-2263.

8. D.C. Lin, *et al.*, "Microstructural development in a rapidly cooled eutectic Sn-3.5% Ag solder reinforced with copper powder," *Powder Technology*, Vol. 166 (2006), pp. 38-46.

9. S.M.L. Nai *et al.*, "Influence of ceramic reinforcements on the wettability and mechanical properties of novel lead-free solder composites," *Thin Solid Films*, Vol. 504 (2006), pp. 401-404.

10. Richards, B. P., *et al.*, "An Analysis of the Current Status of Lead-Free Soldering", *Journal of Electronic Material*, Vol. 31 (2002), pp. 1122-1128.

11. Fu Guo, "Composite lead-free electronic solders", *Journal Material Science: Mater Electron.*, Vol. 18 (2007), pp. 129-145.

12. Abtewa. M, *et al.*, "Lead free Solder in Microelectronics," *Materials Science and Engineering*, Vol. 27 (2000), pp. 95-141.

13. C.M.L. Wua, *et al.*, "Properties of lead-free solder alloys with rare earth element additions," *Materials Science and Engineering*, Vol. 44 (2004), pp. 1-44.

14. Szu Tsung Kao, *et al.*, "Effect of Cu Concentration on Morphology of Sn-Ag-Cu Solders by Mechanical Alloying," *Journal of Electronics Materials*, Vol. 33 (2004), pp. 1445-145.

15. Feng Gao, *et al.*, "Co-Sn intermetallic compounds in Sn-3.0Ag-0.5Cu-0.5Co lead-free solder alloy," *Materials Letters*, Vol. 62 (2008), pp. 2257-2259.

16. C.M. Liu, *et al.*, "Reflow Soldering and Isothermal Solid-State Aging of Sn-Ag Eutectic Solder on Au/Ni Surface Finish," *Journal of electronic Materials,* Vol. 30 (2001), pp. 1152-1156.

17. Kim YS, *et al.*, "Effect of Composition and Cooling rate on Microstructure and Tensile Properties of Sn-Zn-Bi," *J. Alloys and Comp.*, Vol. 352 (2003), pp. 237-245.

18. Mayappan R, *et al.*, "Wetting Properties of Sn-Pb, Sn-Zn and Sn-Zn-Bi Lead Free Solders," *Journal Teknologi.*, Vol. 46 (2007), pp. 1-14.

19. Feng Gao *et al.*, "Characterization of Co-Sn intermetallic compounds in Sn-3.0Ag-0.5Cu-0.5Co lead-free solder alloy," *Materials Letters*, Vol. 62 (2008), pp. 2257-2259.

20. Masazumi Amagai, "A study of nanoparticles in Sn-Ag based lead free solders," *Microelectronics Reliability*, Vol. 48 (2008), pp. 1-16.

Depiction of the Elastic Anisotropy of AuSn₄ and AuSn₂ from First-principles Calculations

Rong An[1,2]*, Chunqing Wang[2], Yanhong Tian[2]
[1]School of Chemical Engineering and Technology
Harbin Institute of Technology, Harbin 150001, China
anyieng@gmail.com, +86-451-86418359
[2]Department of Electronics Packaging Technology, School of Materials Science and Engineering
Harbin Institute of Technology, Harbin 150001, China
wangcq@hit.edu.cn, +86-451-86418725

Abstract

As the solder joints become increasingly small and contain only a few grains, their mechanical properties cannot be determined from conventional mechanical tests as with bulk samples, and there may be a considerable variation in mechanical behavior from joint to joint because of the anisotropy of mechanical properties. In this paper, elastic constants of single-crystal AuSn₄ and AuSn₂ were preliminarily determined through first-principles calculations to characterize their polycrystalline elastic behavior and elastic anisotropy. The ideal bulk, shear, and Young's moduli, as well as Poisson's ratios were determined using the Voigt-Reuss-Hill method. These compounds exhibited distinct anisotropy in Young's modulus, which may be partially responsible for the discrepancy in the experimental results.

1 Introduction

AuSn₄ and AuSn₂ are well known to many researchers in the field of electronic packaging. Generally, the two intermetallic form at the interface between most tin-based solders and the pad with gold finish in solder joints in electronics during reflow, after aging or during service [1-2]. The formation of these compounds is necessary to obtain high-performance bonds, and is unavoidable in most situations owing to its metallurgical characteristics. As an average intermetallic, these compounds are more brittle than solder, so they have a significant effect on the reliability of solder joints, just like the compound Cu₆Sn₅ [3], Ni₃Sn₄ [4] growing at the solder-pad interface. As solder joints are made smaller, the intermetallic occupy a greater proportion of the whole joint and may severely compromise the future reliability of the joint. The properties of these intermetallic and their effects on overall joint reliability are not yet well understood. In order to model and predict the behavior of a solder joint accurately, it is essential to have a good knowledge of their properties. However, few studies of the mechanical properties of these Au-Sn compounds themselves exist in the literature [5-6]. This is attributable mainly to the difficulty in preparing the single-phase samples and the limitations of the available experimental techniques.

The elasticity of the Au-Sn compounds is the most fundamental mechanical property for modeling and prediction of the behavior of the solder joint, because the elastic stiffness is important to understand how these intermetallic respond to applied mechanical perturbations. Nevertheless, there is no good agreement in the experimentally obtained Young's modulus values [5-6], as shown in Table I; for example, the modulus of AuSn₄ measured using the ultrasonic technique

with cast samples after annealing is 71GPa [5], whereas the value determined by microindentation testing on interfacial layer is up to 39GPa [6]. Moreover, the miniaturization of microelectronic packages requires the use of finer solder joints; as the solder joints become increasingly small (less than 100μm in diameter) and contain only a few grains, their mechanical properties, however, cannot be determined from conventional mechanical tests as with bulk samples, and there may be a considerable variation in mechanical behavior from joint to joint because of the anisotropy of mechanical properties. Thus, in addition to characterizing the morphology and crystallographic orientation of the grains in joints, there is a strong need to determine the elastic constants of these intermetallic to represent their elastic anisotropy. To the best of our knowledge, however, no research on this particular aspect has been found in the literature.

First-principles calculation is being widely utilized for the prediction of material properties by virtue of the increase of computing power and the development of the density functional theory (DFT). Lee et al. [7] provided evidence on the applicability of DFT calculations with the pseudopotential method for Sn-based intermetallic compounds. It can be seen from these two examples that the experimental data on singlecrystal elastic constants can be enriched using the Voigt-Reuss-Hill method, and bounds can be placed on polycrystalline elastic moduli, so that they can be compared with existing polycrystalline data. In our study, we computed the full set of elastic constants by performing the first-principles pseudopotential total energy calculations on both AuSn₄ and AuSn₂. Our aims are (i) to determine the ideal elastic properties of AuSn₄ and AuSn₂ and (ii) to investigate their elastic anisotropy.

2 Methodology

AuSn₄ crystallizes in an orthorhombic structure (*Aba2*) crystal belonging to the space group 41 (Fig. 1a), which can be described by three cell parameters *a*, *b*, and *c* [8]. The Au atom occupies the 4a site (0, 0, 0), and there are two Sn sites: Sn1 located at 8b positions (0.1639, 0.3395, 0.1205), and Sn2 resides at 8b positions (0.3312, 0.1642, 0.8591). The crystal structure of AuSn₂ is depicted in Fig. 1b. The compound crystallizes in a hexagonal lattice with the *Pbca* space group (No. 61) [9]. It has 24 atoms per unit cell; the Au atom occupies the 8c site (0.010, 0.895, 0.110), and there are two Sn sites: Sn1 located at 8c positions (0.845, 0.250, 0.082), and Sn2 resides at 8b positions (0.130, 0.545, 0.190). We used their conventional cells for all the *ab initio* calculations.

978-1-4244-4658-2/09 $25.00 © 2009 IEEE

The CASTEP code [10, 11] was used in the present calculations, wherein the Vanderbilt-type ultrasoft pseudopotential [12], the PW91 form [13] of the generalized gradient approximation (GGA), and a plane-wave basis set were employed to describe electron–ion interactions, to take into account exchange–correlation effects, and to represent electronic wavefunctions, respectively. The cut-off energy of plane-wave basis sets was 360 eV for all the calculations on either $AuSn_4$ or $AuSn_2$. For these compounds, a 6×6×4 Monkhorst-Pack [14] k-point mesh was employed to approximate the definite integral over the Brillouin zone. Increasing the plane-wave cut-off energy to 540eV and the k-point mesh to 8×8×6 changed the total energy by less than 0.01eV/atom and lattice constants by less than 0.2%. Therefore, the present computations were precise enough to reproduce the ground-state properties.

Table I Experimental Young's Moduli of $AuSn_4$ and $AuSn_2$

Material	Specimen/technique	ν	E (GPa)
$AuSn_4$	Bulk/Resonance [5]	0.312	71.1
	Interfacial layer/Nanoindentation [6]	-	39
$AuSn_2$	Interfacial layer/Nanoindentation [6]	-	103

Table II Calculated Elastic Stiffness (C_{ij}) and Compliance (S_{ij}) Constants of Single-Crystal $AuSn_4$ and $AuSn_2$.

			11*	22*	33*	44*	55*	66*	12*	13*	23*
$AuSn_4$	360eV, 6×6×4 k points	C_{ij} (GPa)	126	107	129	33	22	45	31	44	46
		S_{ij} (10^{-3}/GPa)	9.3	11	9.8	31	46	22	-1.6	-2.6	-3.5
$AuSn_2$	360eV, 6×6×4 k points	C_{ij} (GPa)	170	154	64	33	20	43	52	57	35
		S_{ij} (10^{-3}/GPa)	8.6	7.7	23	30	50	23	-1.3	-6.9	-3.1

*11, 22, etc. are the tensor subscripts.

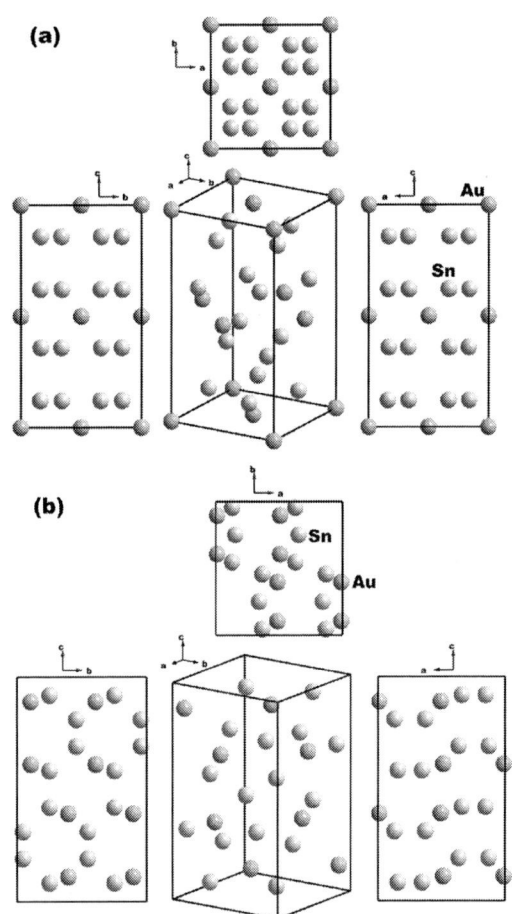

Fig. 1 The crystal structures of (a) $AuSn_4$ and (b) $AuSn_2$

Table III Bounds on the Elastic Properties of Polycrystalline $AuSn_4$ and $AuSn_2$

	Bound	K (GPa)	G (GPa)	E (GPa)	ν
$AuSn_4$	Voigt	67.09	35.96	91.52	0.2727
	HS upper	66.69	34.97	89.30	0.2768
	VRH	66.64	34.70	88.71	0.2781
	HS lower	66.56	34.59	88.45	0.2785
	Reuss	66.18	33.45	85.87	0.2837
$AuSn_2$	Voigt	75.22	35.38	91.75	0.2967
	HS upper	70.13	33.07	85.73	0.2962
	VRH	66.82	32.24	83.32	0.2922
	HS lower	65.17	31.26	80.85	0.2932
	Reuss	58.41	29.10	74.87	0.2864

The unit cell was fully optimized in order to obtain the equilibrium crystal structure. Lattice parameters and internal atomic coordinates were independently modified to minimize the total energy and interatomic forces. The Broyden-Fletcher-Goldfarb-Shanno (BFGS) minimization scheme [15] was used for geometry optimization. The tolerances for geometry optimization were selected as follows: difference in total energy within $5×10^{-6}$eV/atom, maximum ionic Hellmann–Feynman force within 0.01eV/Å, maximum ionic displacement within $5×10^{-4}$Å, and maximum stress within 0.02GPa.

We set the strain of the optimized equilibrium cells to a finite value by applying a given homogeneous deformation to optimize the internal atomic coordinates and calculate the resulting stress; each of the second-order elastic constants was determined by means of a least-squares linear fit of stress against strain. Four strain patterns (the first with nonzero xx and yz components, the second with nonzero zz and xy

components, the third with nonzero yy component, and the fourth with nonzero xz component) were employed to generate the stresses related to all nine independent elastic constants of the orthorhombic system. Three positive and three negative amplitudes were used for each strain component with a maximum strain of 0.4%. This technique has been successfully used for many compounds, such as Ag_3Sn [16], Cu_3Sn, [17] and AuSn [18].

The compliance tensor S was calculated as the inverse of the stiffness tensor, $S = C^{-1}$. Polycrystalline elastic parameters, such as bulk modulus and Young's modulus, were estimated from the compliance tensor components using the Voigt method and the Reuss method [19], and the Hashin-Shtrikman (HS) bounds [20] were also used to place tighter bounds within the Voigt and Reuss bounds. In addition, the Voigt-Reuss-Hill (VRH) average [19] was employed to determine the theoretical polycrystalline elastic property.

3 Results and discussions

The theoretical lattice parameters of $AuSn_4$ and $AuSn_2$ are listed in Table II, together with the experimental data for comparison. The computed lattice constants a, b and c are consistent with the reported experimental data, that is, for $AuSn_4$, $\Delta a/a < 4.6\%$, $\Delta b/b < 0.06\%$, $\Delta c/c < 0.3\%$; for $AuSn_2$, $\Delta a/a < 1.3\%$, $\Delta b/b < 1.5\%$, $\Delta c/c < 0.6\%$. Therefore, the present first-principles computation is sufficiently reliable to reproduce the equilibrium crystal structures of these two intermetallic.

The elastic stiffness determines the response of a crystal to an imposed strain (or stress) and provides information about bonding characteristics near the equilibrium state. Investigating the elastic stiffness is essentially the first step to understanding the mechanical properties of a solid. Table III includes the full set of theoretical second-order elastic constants of $AuSn_4$ and $AuSn_2$. Since their second-order elastic constants have not been reported in the literature, it is impossible to compare the present theoretical elastic constants with others. Fortunately, the Voigt method and the Reuss method [19] can be used to estimate the polycrystalline modulus, and the Hashin-Shtrikman (HS) bounds [20] can also be used to place tighter bounds within the Voigt and Reuss bounds. In addition, the Voigt-Reuss-Hill (VRH) average [19] can be employed to determine the theoretical polycrystalline elastic properties.

Table IV lists the isotropic bulk (K) and shear (G) moduli calculated from the corresponding singlecrystal data using the VRH approximation. The Young's modulus E and Poisson's ratio v were calculated from K and G using the interrelationship of these four elastic parameters based on the isotropic elasticity of the materials, and the results are also listed in the table. The VRH Young's modulus of $AuSn_4$ is 89GPa, which is larger than that (71GPa) obtained using resonance [5] and that (39GPa) determined using nanoindentation [6] (Table I). The VRH average on the Poisson's ratio of $AuSn_4$ (0.278) is smaller than the reported experimental value (0.312). Similarly, the VRH average of $AuSn_2$ on Young's modulus (83 GPa) is much different from the value determined through nanoindentation (103 GPa). [6] These unusual negative deviations from the experimental

values can be derived not from the calculation error but from the elastic anisotropy, which will be discussed later in this paper.

Pugh [21] introduced the quotient of bulk modulus to shear modulus, K/G, as an indication of the extent of fracture range in metals. A high value of K/G is associated with ductility and a low value with brittleness. There are several examples that involve some representation based on the quotient K/G; these include the evaluation of brittleness for trialuminide alloys [22], the determination of a transition from plasticity to brittleness in a bulk amorphous steel [23], the discovery of the ceramics with unusual plastic behavior [24], and the explanation for the ductility of a ternary Y-Si-O silicate [25]. Theoretical K/c_{44} and K/G values are 2.02 and 1.92, for $AuSn_4$; the K/c_{44} and K/G values of $AuSn_2$ are 2.02 and 2.07. Although the attribution of the extent of fracture range solely to the K/G (or K/c_{44}) ratio is highly simplistic, it nevertheless indicates the tendency of brittleness for the two Au-Sn intermetallic compounds.

Both $AuSn_4$ and $AuSn_2$ are noncubic crystals; their elastic anisotropy can be measured using three dimensionless quantities A_G, A_K, and A_E, defined as $A_G = (G_V - G_R)/(G_V + G_R)$, $A_K = (K_V - K_R)/(K_V + K_R)$, and $A_E = (E_V - E_R)/(E_V + E_R)$, [26-27] respectively. The subscripts V and R here designate the Voigt and the Reuss averaging schemes and they represent the maximum and minimum limits of the true polycrystalline elastic moduli. A gives the relative magnitude of the elastic anisotropy present in crystals. It is always positive and is zero for crystals which are elastically isotropic. For $AuSn_4$, the anisotropy factors, viz., A_G, A_K, and A_E (in percent), are 3.62%, 0.683%, and 3.18%; and for $AuSn_2$, these factors are up to 9.74%, 10.1%, and 12.6%, respectively. The results indicate that $AuSn_2$ is much more anisotropic than $AuSn_4$, which are also in accordance with the following data on the orientation-dependent Young's moduli in these crystals.

A three-dimensional surface representation of elastic anisotropy is an illustrative way of showing the variation of elastic modulus with crystallographic direction. The directional dependence of the Young s modulus for orthorhombic crystals can be given by [28]

$$E = \begin{bmatrix} l_1^4 s_{11} + l_2^4 s_{22} + l_3^4 s_{33} + 2l_1^2 l_2^2 s_{12} \\ + 2l_1^2 l_2^2 s_{12} + 2l_1^2 l_3^2 s_{13} + 2l_2^2 l_3^2 s_{23} \\ + l_1^2 l_2^2 s_{66} + l_1^2 l_3^2 s_{55} + l_2^2 l_3^2 s_{44} \end{bmatrix}^{-1} \quad (1)$$

where s_{ij} are the single-crystal elastic compliance constants in the two suffix notation, and l1, l2, and l3 are the direction cosines to the a, b, and c axes, respectively. In this representation, an isotropic system would have a spherical shape, and so the degree of deviation of the geometry from a sphere indicates the degree of anisotropy in a specific property of system. The representations on the directional dependence of the Young's moduli of both $AuSn_4$ and $AuSn_2$ show clear deviations from a spherical shape as shown in Fig. 2 and Fig. 3, but the more deviation occurs in $AuSn_2$. Therefore, the two intermetallic have a high degree of anisotropy in Young's modulus, but the degree in $AuSn_2$ is much higher than that in $AuSn_4$.

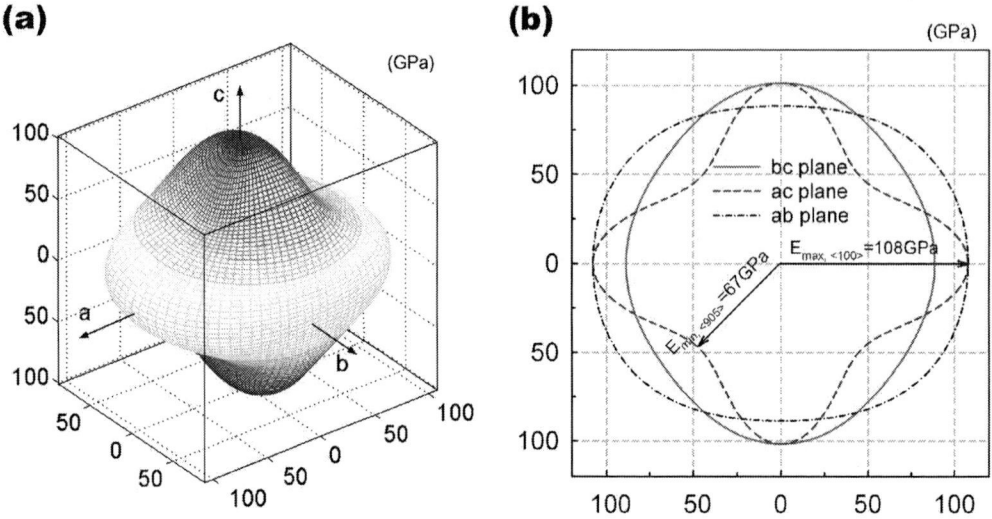

Fig. 2 (a) Directional dependence of Young's modulus in AuSn$_4$ (b) Plane projections of the directional dependence of Young's modulus on (b) the (100), (010), and (001) planes

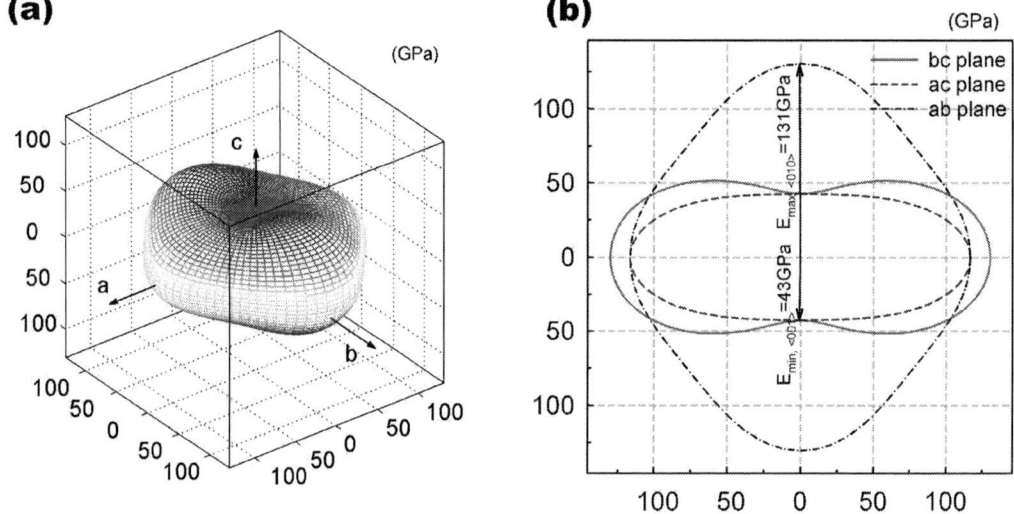

Fig. 3 (a) Directional dependence of Young's modulus in AuSn$_2$ (b) Plane projections of the directional dependence of Young's modulus on the (100), (010), and (001) planes

The anisotropic Young's moduli of both the compounds in the (100), (010), and (001) planes are depicted in Fig. 2b and Fig. 3b. For either AuSn$_4$ or AuSn$_2$, significant in-plane elastic anisotropy appears in the (100), (010) and (001) planes, which is consistent with the atomic arrangements in itself. The maximum Young's modulus of AuSn$_4$ is 108GPa, which is in the [100] direction; the minimum is 67GPa, which is along the [905] direction (Fig. 2b). For AuSn$_2$, the maximum modulus of 131GPa is reached in the [010] crystallographic orientation; the minimum of 43GPa is obtained along the [001] direction, as shown in Fig. 3b. The difference between the maximum and minimum moduli of AuSn4 is 41GPa, which is 46% of the VRH Young's modulus. This large difference implies that the distinct disagreement in its experimentally obtained Young's modulus

can be attributed to the strong elastic anisotropy plus the differences induced during sample preparation and the restriction imposed by limited experimental techniques available. This explanation is partially supported by making the comparisons between the maximum-minimum difference of Young's modulus and the bound of the reported experimental results. Examples include the cases of Cu$_6$Sn$_5$ and Ag$_3$Sn; these two Sn-based intermetallic fall within orthorhombic crystal structure. The difference between the maximum and minimum moduli of Cu$_6$Sn$_5$ (calculated using the elastic constants in Ref. 7) is more than 20GPa, and the experimental Young's moduli summarized in Ref. 7 show a wide range from 85GPa to 125GPa. Here again, the calculated difference of Ag$_3$Sn (using the data in Ref. 16) is 34GPa, and the reported experimental values are scattered within the large range of 70GPa to 94GPa. For $AuSn_2$, the difference of Young's modulus between the maximum and minimum is up to an unusual large value of 88GPa, which is larger than itself VRH Young's modulus. This is graphically illustrated in the surface representation of its elastic anisotropy. If the test samples primarily have the [010] texture, it is obvious that the experimentally obtained Young's moduli will be much larger than the VRH Young's modulus. So far it has been impossible to quantitatively compare the modulus values obtained with samples that possess different textures because the textures of samples are not readily accessible. However, because of such distinct anisotropy of elasticity, it is clear that, the fewer grains there are in the intermetallic compounds of the increasingly small solder joints, the greater the difference between the Young's modulus value of the bulk samples and that measured with the intermetallic in the solder joints.

Conclusions

The elastic constants of single-crystal $AuSn_4$ and $AuSn_2$, have been preliminarily determined from first-principles calculations using the pseudopotential plane-wave method. The ideal bulk, shear, and Young's moduli, as well as Poisson's ratios, of these two polycrystalline intermetallic compounds were calculated using the VRH and HS method. $AuSn_4$ exhibits quite distinct anisotropy in Young's modulus, with a difference of 41GPa between its maximum and minimum modulus; $AuSn_2$ shows a much higher degree of anisotropy than $AuSn_4$, that is, the maximum-minimum difference of the Young's moduli of $AuSn_2$ is 88GPa. In addition to the metallurgical imperfections of test samples and the restrictions imposed by the limited experimental techniques available, high elastic anisotropy is one of the causes of the discrepancy in experimentally and theoretically obtained elastic modulus values. It is believed that with time the elastic anisotropy of these intermetallic will become a more important consideration in the accurate characterization of the mechanical behavior of the small solder joints.

Acknowledgments

This work is financially supported by the National Natural Science Foundation of China under grant No. 50675047/ E052105, the National High-tech R&D Program (863 Program) of China under grant No. 2007AA04Z314, and Joint Project between Samsung Electronics Co. Ltd. (Korea) and Harbin Institute of Technology (HIT). The authors thank J. Cheng and J.C. Zhu, Harbin Institute of Technology, for their valuable help with the execution of CASTEP code, and the Shanghai Supercomputer Center (SSC) for supercomputing resources provided.

References

1. P.G. Kim, K.N. Tu, Morphology of wetting reaction of eutectic SnPb solder on Au foils, J. Appl. Phys. 80, (1996) 3822–3827.
2. L.C. Shiau, C.E. Ho, C.R. Kao, "Reactions between SnAgCu lead-free solders and the Au/Ni surface finish in advanced electronic packages," *Solder. Surf. Mount Technol.* 14, (2002) pp. 25-29.
3. A.C.K. So, Y.C. Chan, "Reliability studies of surface mount solder joints- effect of Cu-Sn intermetallic compounds," *IEEE Trans. Compon. Packag. Manuf. Technol. Part B Adv. Packag.* 19 (1996) pp. 661-668.
4. Y.C. Chan, P.L. Tu, C.W. Tang, K.C. Hung, J.K.L. Lai, "Reliability studies of mBGA solder joints-effect of Ni-Sn intermetallic compound," *IEEE Trans. Adv. Packag.* 24 (2001) pp. 25-32.
5. G. Ghosh, "Elastic properties, hardness, and indentation fracture toughness of intermetallics relevant to electronic packaging," *J. Mater. Res.* 19 (2004) pp. 1439-1454.
6. R.R. Chromik, D.-N. Wang, A. Shugar, L. Limata, M.R. Notis, R.P. Vinci, "Mechanical properties of intermetallic compounds in the Au-Sn system," *J. Mater. Res.* 20 (2005) pp. 2161-2172.
7. N.T.S. Lee, V.B.C. Tan, K.M. Lim, "First-principles calculations of structural and mechanical properties of Cu_6Sn_5," *Appl. Phys. Lett.* 88 (2006) 031913.
8. R. Kubiak, M. Wolcyrz, "Refinement of the crystal structures of $AuSn_4$ and $PdSn_4$," *Journal of the less-common metals.* 97 (1984) pp. 265-269.
9. K. Schubert, H. Breimer, R. Gohle, "On the constitution of the gold - indium, gold - tin,. gold - indium - tin and gold - tin - antimony systems," *Z. Metallkd.* 50 (1959) pp. 146-153.
10. M.C. Payne, M.P. Teter, D.C. Allan, T.A. Arias, J.D. Joannopoulos, "Iterative minimization techniques for ab initio total-energy calculations: molecular dynamics and conjugate gradients," *Rev. Mod. Phys.* 64 (1992), pp. 1045-1097.
11. M.D. Segall, P.J.D. Lindan, M.J. Probert, C.J. Pickard, P.J. Hasnip, S.J. Clark, M.C. Payne, "First-principles simulation: ideas, illustrations and the CASTEP code," *J. Phys.: Condens. Matter* 14 (2002) pp. 2717-2744.
12. D. Vanderbilt, "Soft self-consistent pseudopotentials in a generalized eigenvalue formalism," *Phys. Rev. B* 41 (1990) pp. 7892-7895.
13. J.P. Perdew, Y. Wang, "Accurate and simple analytic representation of the electron-gas correlation energy," *Phys. Rev. B* 45 (1992) pp. 13244-13249.
14. H.J. Monkhorst, J.D. Pack, "Special points for Brillouin-zone integrations," *Phys. Rev. B* 13 (1976) pp. 5188-5192.
15. T.H. Fischer, J. Almlöf, "General methods for geometry and wave function optimization," *J. Phys. Chem.* 96 (1992) pp. 9768-9774.
16. N.T.S. Lee, V.B.C. Tan, K.M. Lim, "Structural and mechanical properties of Sn-based intermetallics from ab initio calculations," *Appl. Phys. Lett.* 89 (2006) 141908.
17. R. An, C. Wang, Y. Tian, H. Wu, "Determination of the elastic properties of Cu_3Sn through first-principles calculations," *J. Electron. Mater.* 37 (2008) pp. 477-482.
18. R. An, C. Wang, Y. Tian, "Determination of the elastic properties of Au_5Sn and $AuSn$ from ab initio calculations," *J. Electron. Mater.* 37 (2008) pp. 968-974.
19. R. Hill, "The elastic behaviour of a crystalline aggregate," *Proc. Phys. Soc. A* 65 (1952) pp. 349-354.
20. J.P. Watt, "Hashin-Shtrikman bounds on the effective elastic moduli of polycrystals with orthorhombic symmetry," *J. Appl. Phys.* 50 (1979) pp. 6290-6295.

21. S.F. Pugh, "Relations between the elastic moduli and the plastic properties of polycrystalline pure metals," *Philos. Mag.* 45 (1954) pp. 823-843.

22. C.L. Fu, "Electronic, elastic, and fracture properties of trialuminide alloys: Al_3Sc and Al_3Ti," *J. Mater. Res.* 5 (1990) pp. 971-979.

23. X.J. Gu, A.G. McDermott, S.J. Poon, and G.J. Shiflet, "Critical Poisson's ratio for plasticity in Fe-Mo-C-B-Ln bulk amorphous steel," *Appl. Phys. Lett.* 88 (2006) 211905.

24. D. Music and J.M. Schneider, "Elastic properties of MFeN (M= Ni, Pd, Pt) studied by ab initio calculations," *Appl. Phys. Lett.* 88 (2006) 031914.

25. J.Y. Wang, Y.C. Zhou, and Z.J. Lin, "Mechanical properties and atomistic deformation mechanism of γ-$Y_2Si_2O_7$ from first-principles investigations," *Acta Mater.* 55 (2007) pp. 6019-6026.

26. D.H. Chung and W.R. Buessem, "The elastic anisotropy of crystals," *J. Appl. Phys.* 38 (1967) pp. 2010-2012.

27. D.H. Chung and W.R. Buessem, "The Voigt-Reuss-Hill approximation and elastic moduli of polycrystalline MgO, CaF_2, β-ZnS, ZnSe, and CdTe," *J. Appl. Phys.* 39 (1968) pp. 2777-2782.

28. J.F. Nye, <u>Physical Properties of Crystals</u>, Oxford: Oxford University Press, 1985.

Microstructural Evolution of Sn-3.5Ag Solder with Lanthanum Addition

Hwa-Teng Lee[1], Yin-Fa Chen[1], Ting-Fu Hong[2], Ku-Ta Shih[1], Che-wei Hsu[1]
[1]Department of Mechanical Engineering, Cheng Kung University,
1, University Road, Tainan 70101, Taiwan
[2]Department of Materials Engineering, Pingtung University of Science and Technology,
1, Hseuhfu Road, Neipu, Pingtung 91201, Taiwan
E-mail: htlee@mail.ncku.edu.tw

Abstract

The goal of this research is to evaluate the effects of La addition on the microstructure and microhardness of Sn-Ag based solders. Sn-3.5Ag-xLa ternary alloy solders were prepared by adding 0-1.0wt% La into a Sn-3.5Ag alloy. Copper substrates were then dipped into these solders and aged at 150°C for up to 625 hours. The microstructure and microhardness of the as solidified solder and the aged solder/copper couples were investigated. Experimental results show that Sn-3.5Ag-xLa solders are composed of β-Sn, Ag_3Sn, and $LaSn_3$ phases, and that their microstructure is refined with La addition. After isothermal storage, the Ag_3Sn and IMC layer can be effectively depressed by adding a small amount of La element, and the size and amount of $LaSn_3$ compounds does not change perceptibly with storage time. For as-cast solder, the addition of La increased the microhardness of the Sn-Ag solder due to the refining effect of Ag_3Sn particles in the eutectic zone and increased formation of $LaSn_3$ compound. After isothermal storage, the microhardness of solders was decreased with the increasing coarsening of Ag_3Sn compounds as aging time was increased. However, coarsening of Ag_3Sn compounds was retarded by La addition. Therefore, La addition helped to improve the microhardness and thermal resistance of the solder joints.

1. Introduction

Sn-3.5Ag eutectic solders are recognized as having the greatest potential due to their favorable mechanical and electrical properties, high-temperature stability, and good wettability on a variety of surfaces [1, 2]. Sn-3.5Ag eutectic solder contains a uniform dispersion of fine Ag_3Sn compounds, which greatly enhance its mechanical strength. However, these compounds experience a coarsening effect when the solder is exposed to high temperature for extended periods of time, and thus the solder loses its mechanical strength [3]. Therefore, the feasibility of improving the mechanical properties of eutectic Sn-3.5Ag solder by adding small quantities of appropriate alloying elements has attracted significant interest in recent years.

Recent studies have reported that the addition of small amounts of RE elements to Sn-Ag solder systems has a number of beneficial effects. For example, Wang et al. [4] showed that the addition of 0.25~0.5wt% lanthanum (La) and Cerium (Ce) yields a significant improvement in the wettability of Sn-3.5Ag solder alloys. Many other researchers have shown that the addition of RE metals improves the mechanical properties of Pb-free solder systems by causing a grain refinement effect and more uniform distribution of the intermetallic compounds (IMCs) within the solder matrix [5-7]. In addition, it has been shown that RE dopants reduce the thickness of the IMC interfacial layer in solder joints [8, 9]

and yield notable improvements in the hardness, tensile strength, shear strength, thermal fatigue, and creep resistance of a wide variety of Pb-free solders [7, 10-12].

Although recent work has highlighted quite a few advantages to adding RE elements into solders, the theoretical background of such work remains unclear. In order to investigate the formation mechanisms and relationships of Ag_3Sn and $LaSn_3$ in the Sn-Ag-based solder, different amounts of La were added to the eutectic Sn-3.5Ag solder to form a ternary solder alloy. The present research investigates how La additions influence the microstructure and microhardness of Sn-Ag-La solder mounted on Cu substrate after various amounts of aging at elevated temperature.

2. Experiment

In general, RE elements have high activity and readily oxidize when exposed to air. Due to the high melting point and the susceptibility of La to oxidization, the current solders were prepared in a vacuum environment created within a sealed quartz tube. A total of four different solders were prepared, namely a conventional Sn-3.5Ag solder and three Sn-3.5Ag-xLa alloy solders with 0.05, 0.1, 0.5, and 1.0 wt% La additions, respectively. The Sn, Ag and La used to fabricate the various solders were all high purity metals (i.e. a wt% greater than 99.9 wt% in every case). In the fabrication process, the solder mixtures were maintained at a temperature of 1000°C for 5 h and shaken every 30 min to ensure homogeneous mixing of the solder components. Finally, the tube was quenched in water and the solid solder removed.

Two different specimens were used in the current study, namely the bulk solder specimen (15 × 10 × 5 mm) and the solder joint specimen (15 × 10 × 5 mm). The bulk solder specimen was formed by pouring the molten solder at 300°C into a stainless steel case, and then cooled in air. The solder thus produced was used to fabricate the solder joint specimen with a high purity (>99.95%) copper slice. The solder joints were isothermally stored at 150 °C for up to 625 hours to investigate their microstructure and thermal resistance.

The microstructures of the samples were observed with an optical microscope (OM, Leica Metallux 3) and an environmental scanning electronic microscope (E-SEM, FEI Quanta 400F). The chemical compositions of the structures were analyzed by Energy Dispersive X-ray Spectroscopy (EDS) performed in the E-SEM. The mechanical properties of a material are frequently quantified by its hardness. Accordingly, the present study used a Vickers microhardness tester (AKASHI MVK-H1) with an indentation load of 20 g and a dwell time of 10 s to measure the hardness of the various Sn-Ag-La solders. To ensure accurate results, twenty different indents were performed on the polished surfaces of

978-1-4244-4658-2/09 $25.00 © 2009 IEEE

each sample. The average of the individual measurements was then taken as the Vickers hardness number.

3. Results and discussion

3.1. Effect of La additions on microstructure of Sn-3.5Ag solder

For observing the microstructure evaluation with La addition, specimens were characterized by microscopy. Fig. 1 shows a series of Sn-3.5Ag-xLa solder OM images. Fig. 1(a) reveals the microstructure of eutectic Sn-3.5Ag solder. Due to the non-equilibrium solidified processes, the Sn-3.5Ag eutectic solder is composed of primary β-Sn grains with Ag_3Sn compounds. It forms into a network structure, where the β-Sn grains have an approximately spherical shape and an average size of 30-60μm. It is noticed that the 0.05-1.0 wt% La addition resembles the original Sn-3.5Ag microstructure, as shown in Figs. 1(b)-1(e). Compared to the original Sn-3.5Ag solder, the addition of La tends to refine the β-Sn grains. The β-Sn grains are reduced to an average size of 15-30 μm, and change significantly as La addition is increased.

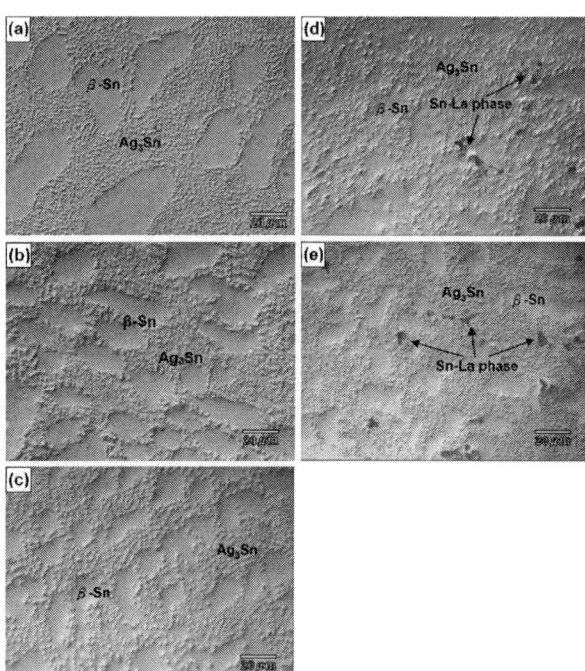

Fig. 1 OM micrographs showing as-cast microstructures of: (a) Sn-3.5Ag, (b) Sn-3.5Ag-0.05La, (c) Sn-3.5Ag-0.1La, (d) Sn-3.5Ag-0.5La, and (e) Sn-3.5Ag-1.0La solders.

Figs. 2(a) to 2(c) present SEM micrographs of the as-cast Sn-3.5Ag-xLa solders with 0.1, 0.5, and 1.0 wt% La additions, respectively. The SEM images in Fig. 2 show that the addition of La to the present Sn-3.5Ag solder system resulted in the formation of $LaSn_3$ compounds. Note that the morphologies of $LaSn_3$ compounds observed with OM (Fig. 1) and with SEM (Fig. 2) are different. It can be seen in higher magnification images that peeling the Sn phase off the $LaSn_3$ compounds exposes the $LaSn_3$ compounds. This phenomenon also implies a weak adhesion or bonding force between the Sn matrix and $LaSn_3$ compounds. Since Sn can

easily be removed from $LaSn_3$ compounds by polishing, the peeled-off surface revealed the $LaSn_3$ compound. The peeled-off Sn structure is consistent with the findings presented by Chen et al. in their publication of ref. [15]. They found that the RE compound can be snowflake shaped, clump shaped, spherical, or quadrangular. As discussed above, both the size and the number of the $LaSn_3$ compounds increase with increases in the amount of La. Inspection revealed $LaSn_3$ particles in the Sn-3.5Ag-0.1La solder with an average size of around 2-5 μm. However, with a La addition of 0.5 wt% or more, the average $LaSn_3$ particle size increases to approximately 5-40 μm.

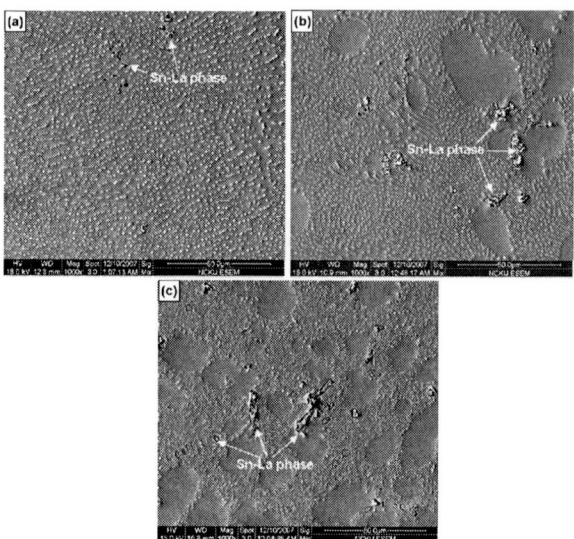

Fig. 2 As-cast microstructure of SEM micrographs of Sn-3.5Ag-xLa solders: (a) Sn-3.5Ag-0.1La (b) Sn-3.5Ag-0.5La, and (c) Sn-3.5Ag-1.0La

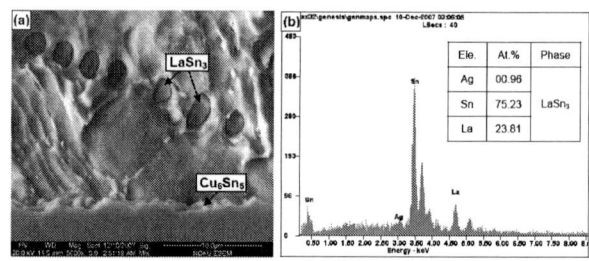

Fig. 3 Microstructure of as-soldered Sn-3.5Ag-1.0La solder joint and compositional analysis of the oval-like particles: (a) SEM image, and (b) EDS analysis results.

Fig. 3(a) presents an SEM image of the Sn-3.5Ag-1.0La solder joint microstructure as etched. As shown in this figure, the etching process removed the β-Sn phases from the solder matrix and exposed the IMC components, which have an oval-like shape and an average size of around 2-4 μm. Fig. 3(b) presents EDS analysis results showing that those IMCs contain mainly Sn and La, with a Sn:La ratio of 75.23:23.81. This confirms that the particles are $LaSn_3$ phase.

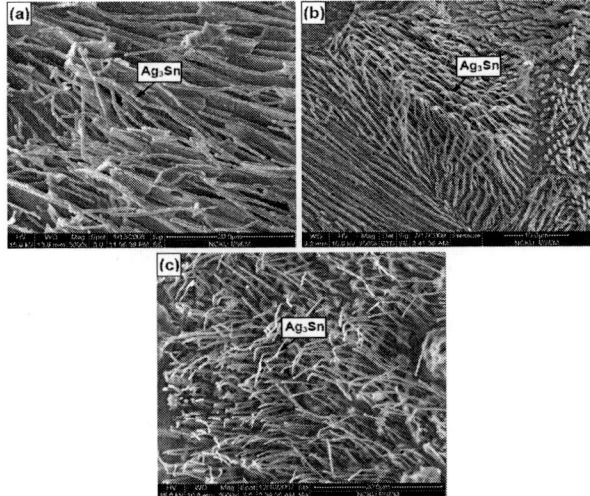

Fig. 4 Morphology of Ag₃Sn phases in: (a) as-cast Sn-3.5Ag solder, (b) as-cast Sn-3.5Ag-0.05La solder, and (c) as-cast Sn-3.5Ag-1.0La.

The SEM images in Fig. 4 show more clearly that the La addition not only affects the morphology of the β-Sn matrix but also helps to depress the coarsening of Ag₃Sn particles. Figs. 4(a) to (c) show the morphology of Ag₃Sn in the deep-etched Sn-3.5Ag, Sn-3.5Ag-0.05La, and Sn-3.5Ag-1.0La solders, respectively. It can be found that a finer and more uniform distribution of Ag₃Sn can be obtained after La addition. According to Chen's research [13], the RE atoms tend to aggregate at boundaries of primary dendrites. In this study, La has low solubility in β-Sn and Ag, and therefore it only exists at the boundaries between β-Sn and Ag₃Sn. The growths of β-Sn grains and Ag₃Sn are consequently restrained by these La precipitates; therefore, this indicates another contributor to the decrease in the size of solder grains and IMCs [4].

3.2. The effect of thermal aging on microstructure and IMC layer

Fig. 5 presents optical micrographs of the Sn-3.5Ag-xLa solder following aging at 150°C for 625 hours. It can be seen that the coarsening level of Ag₃Sn compounds in the Sn-3.5Ag is more significant than those in the La-containing solder. The phenomenon indicates evidently that adding La element can depress the growth of Ag₃Sn compounds during aging process.

It should be noted that aging has only a slight influence on LaSn₃ compounds. LaSn₃ compounds are found in the matrix when the amount of La exceeds 0.05 wt%. It can be seen in Figs. 6(a)~(c) that although both the size and number of LaSn₃ compounds increase with increasing La content in the as-cast samples, those of LaSn₃ compounds do not have noticeable change after aging at 150°C for 625 hours (Figs. 6(d)~(e)). This indicates that most of the La has been consumed in the formation of LaSn₃ compounds during solidification, and not retained in the solder matrix, due to its low solubility with Sn and Ag.

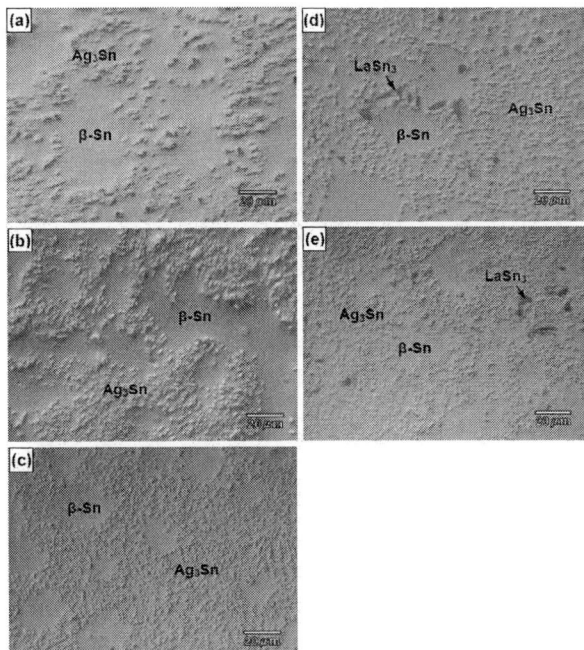

Fig. 5 Sn-3.5Ag-xLa solders aged at 150°C for 625 hours. (a) Sn-3.5Ag, (b) Sn-3.5Ag-0.05La, (c) Sn-3.5Ag-0.1La, (d) Sn-3.5Ag-0.5La, and (e) Sn-3.5Ag-1.0La

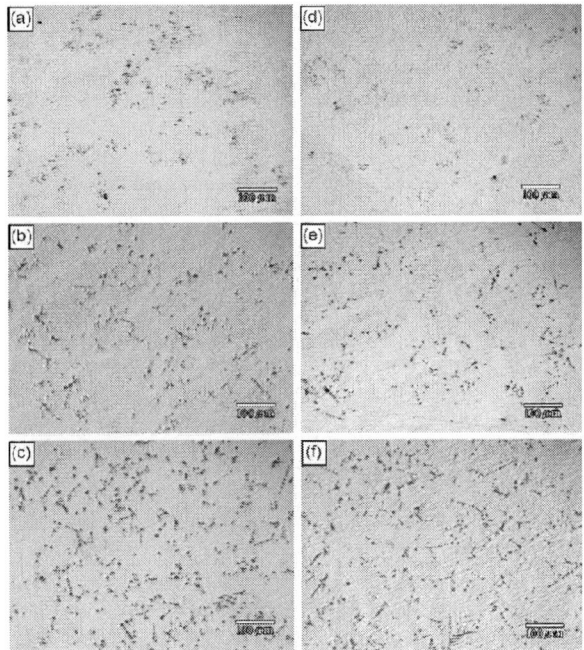

Fig. 6 OM micrographs of as-cast and aged Sn-3.5Ag-xLa solders: (a) Sn-3.5Ag-0.1La (as-cast), (b) Sn-3.5Ag-0.5La (as-cast), (c) Sn-3.5Ag-1.0La (as-cast), (d) Sn-3.5Ag-0.1La (150°C×625 h), (e) Sn-3.5Ag-0.5La (150°C×625 h), and (f) Sn-3.5Ag-1.0La, (150°C×625 h).

Fig. 7 presents SEM images of the solder joints. After soldering, a continuous layer of intermetallic compounds (IMCs) which exhibit a scallop like morphology is formed at the interface of the solder/Cu substrate. EDS analysis identified the compound at the interface as Cu_6Sn_5 phase. From Figs. 7(a) and (b), it can be seen that La helped to depress the coarsening of IMCs. This effect can be observed more clearly from the samples after aging at 150°C for 625 hours (Figs. 8(a) and (b)), where the IMC layer of Sn-3.5Ag-0.5La solder appeared to be thicker than that of the Sn-3.5Ag-1.0La solder. Since La has a higher affinity to Sn, La addition will reduce the driving force for Cu-Sn IMC layer formation. Lowering the activity of Sn will depress the growth of the Cu_6Sn_5 IMC layer [8]. After aging, the IMC layer grew into two different layers; next to the Cu substrate was the Cu_3Sn, which was slightly darker than the Cu. Above the Cu_3Sn was the Cu_6Sn_5 seen in the aged solder joint. The IMC layer grew thicker with increasing aging time. The mean thickness of the interfacial IMC layer of Sn-3.5Ag-xLa solder is shown in Fig. 9. By measurement, IMC thickness was proportional to the square root of time in Sn-3.5Ag-La solder, as illustrated in Fig. 9. The growth trend of the IMC layer was the same as that with Sn-3.5Ag solder, but the IMC growth rate was slower than with Sn-3.5Ag, especially Sn-3.5Ag-1.0La. Hence, the tin diffusion rate was decreased by the La atoms. The main factor of thickness and roughness of the IMC layer is the length of time for which the solder remains in the liquid state. According to Dudek et al[14], $LaSn_3$ compounds formed first during the solidification process. A larger La addition causes a higher undercooling level, which prompts solder to nucleate $LaSn_3$ compounds, and also a faster solidification process. The La compounds provide heterogeneous nucleation sites not only for the solder itself, but also for IMCs in the soldering and aging processes. A large number of heterogeneous nucleation sites can decrease the size of Sn dendrites and IMCs, and hence help to refine the microstructure of solder and its interface layer with Cu substrate. Furthermore, a lower IMC thickness forms due to a short time for reaction with Cu substrate in liquid state.

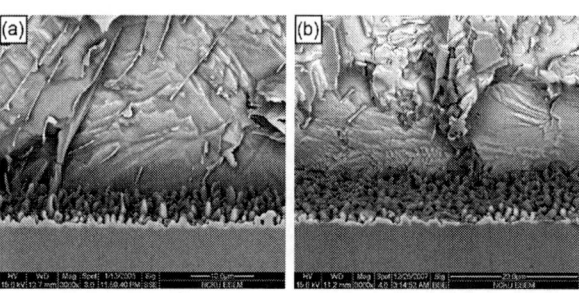

Fig. 7 SEM micrographs showing intermetallic layer in solder / Cu joints: (a) Sn-3.5Ag (as-soldered), (b) Sn-3.5Ag-0.5La (as-soldered).

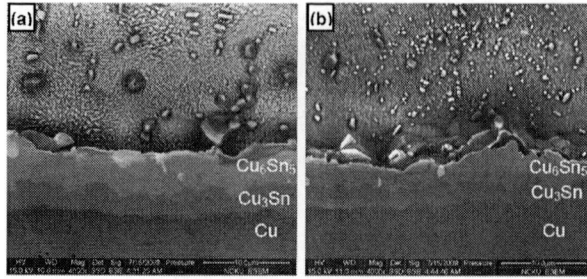

Fig. 8 SEM micrographs showing intermetallic layer in solder / Cu joints: (a) Sn-3.5Ag-0.5La (150°C×625h), (b) Sn-3.5Ag-1.0La (150°C×625h).

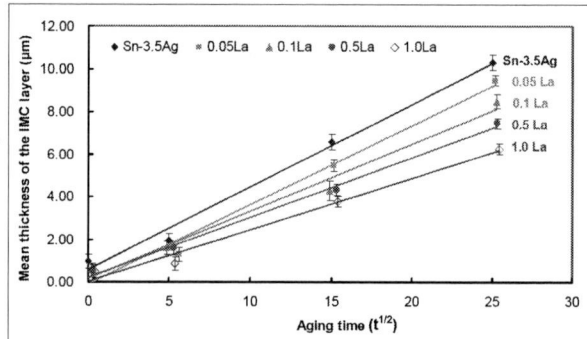

Fig. 9 The IMC growth behavior of Sn-3.5Ag-xLa solder joint

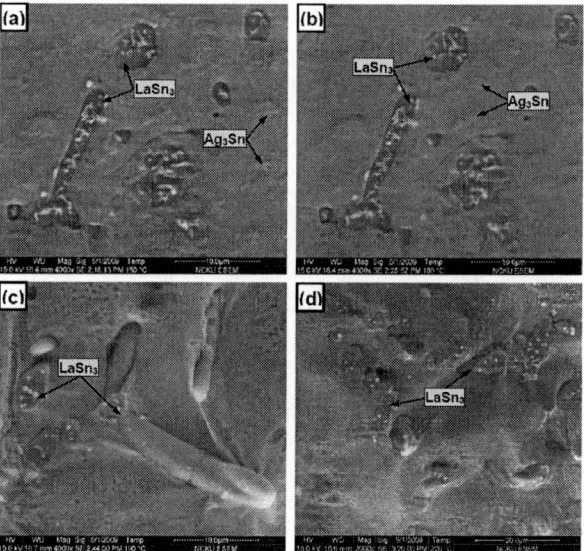

Fig. 10 The observation of $LaSn_3$ during 150-220°C carried out by ESEM with heating stage.

In order to understand the formation of $LaSn_3$ compounds during the solidification process, the environment mode of ESEM with heating stage was used in this study. The Sn-3.5Ag-1.0La was heated in SEM from room temperature to 220°C, as shown in Fig. 10. In Fig. 10(a-b), the Ag_3Sn still can be seen in the matrix during the 150-180°C stage. When the Sn-3.5Ag-1.0La was heated to 220°C (Fig. 10(c-d)), the

Ag_3Sn was melted into the solder matrix. Below 221°C, partially melted phases are present as primary phases, which are β-Sn+Ag_3Sn. However, $LaSn_3$ compounds remained in the melting solder. This result proves that the $LaSn_3$ compounds provide heterogeneous nucleation sites for solder and IMCs in the soldering and solidification processes.

3.3. The effect of adding La on microhardness of Sn-Ag solder

Because La cannot be highly dissolved in the solder matrix, there was no significant solid solution hardening effect found in the Sn-Ag-La solders. Therefore, the factors which affect the microhardness of Sn-3.5Ag-xLa solders are the degree of refinement and the amount of $LaSn_3$ compounds. Fig. 11 presents the variation of solder microhardness with different La contents and aging period of time. In the as-cast samples, the addition of La increases the microhardness of the solder material. As mentioned before, adding La led to refinement of the Ag_3Sn particles in the eutectic zone and increased the formation of $LaSn_3$ compounds. This also enhanced the strength of the La-containing solder. Furthermore, the microhardness of solders decreased with the aging time, due to the coarsening of Ag_3Sn compounds.

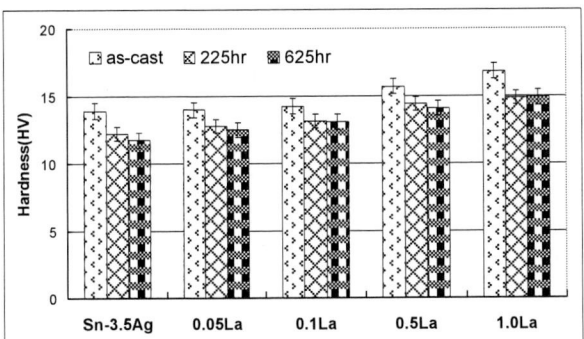

Fig. 11 Variation of solder microhardness with La content and aging time at 150°C

After aging, coarsening of the Ag_3Sn compounds decreased the dispersion strengthening effect in the solder matrix. However, it is apparent that the microhardness of the aged Sn-3.5Ag-xLa samples was significantly higher than that of the Sn-3.5Ag binary solder. As discussed earlier, La has a refining effect on both the β-Sn grains and the Ag_3Sn compounds (see Fig. 1). We can thus conclude that La suppressed the coarsening effect of the Ag_3Sn compounds during thermal aging, yielding corresponding improvements in the microhardness and thermal resistance of the solder as a result. However, with La additions in excess of 0.5 wt%, most of the La reacted with Sn, forming $LaSn_3$ compounds. Only a small portion of the La atoms were gathered at the boundaries. Therefore, in samples with La addition in excess of 0.5 wt%, there was no obvious suppression of the coarsening of the Ag_3Sn compounds during thermal aging. Based on microhardness results, it is recommended that no greater than 0.5 wt% La be added to the Sn-3.5Ag solder.

4. Conclusions

The relation of cooling rate to the real three-dimensional morphology of Ag_3Sn compound of Sn-3.5Ag solder was investigated in this study by two kinds of cooling experiment combining with deep etching. The results are summarized as follows.

1. The microstructure of Sn-3.5Ag-xLa solder is similar to that of Sn-3.5Ag, which exhibits a network structure. The size of β-Sn grains and Ag_3Sn in the solder decreases significantly with La addition. $LaSn_3$ phases are found in the matrix when the level of La addition exceeds 0.1 wt%. Ag_3Sn particles can be effectively depressed by adding a small amount of La element after aging.

2. $LaSn_3$ compounds are found in the matrix when the amount of La exceeds 0.05 wt%. The size and number of $LaSn_3$ compounds increases with increasing La content. When the level of La addition exceeds 0.5 wt%, a large number of $LaSn_3$ compounds are formed. However, the size and amount of $LaSn_3$ compounds did not change perceptibly after aging.

3. The thickness and roughness of scallop-shaped IMC layer can be effectively depressed by adding a trace amount of La element after soldering and isothermal aging.

4. When the Ag_3Sn was melted into the solder matrix, the $LaSn_3$ compounds remained in the melted solder. Therefore, the $LaSn_3$ compounds can provide heterogeneous nucleation sites for solder and Ag_3Sn in the soldering and solidification processes.

5. The refined microstructure and the large number of $LaSn_3$ compounds in the Sn-3.5Ag-xLa solder give rise to a hardening effect. The volume fraction of these particles increases with increasing La addition. La addition helps to improve the microhardness and thermal resistance of the solder joints.

Acknowledgments

The authors would like to thank the Science Council of Taiwan for the final support of this work under contract No. NSC 97-2221-E-006-021-MY3.

References

1. ESPEC Technology Report, 1(2002).
2. M. Abtew and G. Selvaduray, "Lead-Free Solders in Microelectronics," *Materials Science and Engineering*, Vol. 27 (2000), pp.95-141.
3. D.R. Flanders, E.G. Jacobs, and R.F. Pinizzotto, "Activation energies of intermetallic growth of Sn-Ag eutectic solder on Copper substrates," *J. Electron. Mater.*, Vol. 26 (1997), pp.883-887.
4. L. Wang, D.Q. Yu, J. Zhao, and M.L. Huang, "Improvement of Wettability and Tensile Property in Sn-Ag-RE Lead-Free Solder Alloy," *Materials Letters*, Vol. 56 (2002), pp. 1039-1042.
5. Z. Xia, Z. Chen, Y. Shi, N. Mu, and N. Sun, "Effect of Rare Earth Element Additions on the Microstructure and Mechanical Properties of Tin-Silver-Bismuth Solder," *Journal of Electronic Materials*, Vol. 31 (2002), pp.564-567.
6. C.M.T. Law and C.M.L. Wu, "Microstructure Evolution and Shear Strength of Sn-3.5Ag-RE Lead-Free BGA Solder Balls," *Proceeding of the Sixth IEEE CPMT*

Conference on High Density Microsystem Design and Packaging and Component Failure Analysis, Shanghai, Jul. 2004, pp. 60-65.

7. C.M.L. Wu, D.Q. Yu, C.M.T. Law, and L. Wang, "Microstructure and Mechanical Properties of New Lead-Free Sn-Cu-RE Solder Alloys," *Journal of Electronic Materials*, Vol. 31 (2002), pp. 928-932.

8. X. Ma, Y. Qian, and F. Yoshida, "Effect of La on the Cu-Sn Intermetallic Compound (IMC) Growth and Solder Joint Reliability," *Journal of Alloys and Compounds*, Vol. 334 (2002), pp. 224-227.

9. C.M.T. Law, C.M.L. Wu, D.Q. Yu, L. Wang, and J.K.L. Lai, "Microstructure, Solderability, and Growth of Intermetallic Compounds of Sn-Ag-Cu-RE Lead-Free Solder Alloys," *Journal of Electronic Materials*, Vol. 35 (2006), pp. 89-93.

10. X. Mal, Y.Y. Qian, F. Liu, and F. Yoshida, "High Reliable Solder Joints Using Sn-Pb-La Solder Alloy," *8th Failure Analysis of Integrated Circuits (IPFA 2001)*, Singapore, 2001, pp. 63-66.

11. L. Liang, Q. Wang, and Z. Zhao, "Effect of Cerium Addition on Board Level Reliability of Sn-Ag-Cu Solder Joint," *8th International Conference on Electronic Packaging Technology(ICEPT 2007)*, 2007.

12. H. Hao, J. Tian, Y.W. Shi, Y.P. Lei,and Z.D. Xia,"Properties of Sn3.8Ag0.7Cu Solder Alloy with Trace Rare Earth Element Y Additions", *Journal of Electronic Materials*, Vol. 36 (2007), pp.766-774.

13. Z.G. Chen, Y.W. Shi, Z.D. Xia, and Y.F. Yan, "Study on the Microstructure of a Novel Lead-Free Solder Alloy SnAgCu-RE and Its Soldered Joints," *Journal of Electronic Materials*, Vol. 31 (2002), pp. 1122-1128.

14. M. Dudek, R. Sidhu, N. Chawla, and M. Renavikar, "Microstructure and Mechanical Behavior of Novel Rare Earth-Containing Pb-Free Solders," *Journal of Electronic Materials*, Vol.35 (2006), pp. 2088-2097.

Optimal Packing Research of Spherical Silica Fillers Used in Epoxy Molding Compound

Hujie Mei, Xinyu Du, Lanxia Li, Wei Tan

Henkel Huawei Electronics Co. Ltd. (HHE)

Songtiao Industrial Park, Lianyungang, Jiangsu 222006, China

jammy.mei@cn.henkel.com, 0518-85155323

Abstract

Optimal packing of spherical silica fillers used in epoxy molding compound (EMC) field was studied through founding theoretical and experimental models based on Dinger-Funk Equation. The former model kept the superposition contentment of calculated particle size distribution (PSD) curve with Dinger-Funk Equation distribution curve while the later model ignored the superfine particle distribution area. 8 batches of samples with the same formulation except for filler compositions were prepared including 2 controls and 6 calculated ones with 3 different distribution modulus (0.34, 0.37&0.40) around the theoretical optimal one 0.37. Finished materials were characterized base on standard epoxy molding compound test matrix. Compared with 2 controls, flowability of materials through calculation was increased 30%~50% while keeping steady gel time and the same level of Flash&Bleed. In the mean time, viscosity of materials decreased significantly from 175poise to 25poise. It was suggested with applying the calculation method based on the algorithm Dinger-Funk Equation, the flowability of EMC could be obviously improved. The experimental model seemed slightly inferior to the theoretical one. Besides known benefiting physical mechanical properties, such as dielectric property, lower coefficient of thermal expansion and higher coefficient of thermal conductivity, the optimization of filler PSD can help to increase the filler content of epoxy molding compound and reduce viscosity, hence to improve workability and reliability of future packaging of advanced high density packaging for 3D integration.

Introduction

With the rapid development of information technology and semiconductor industry since the integrated circuits (IC) firstly came out in 1958, advanced high density packaging for 3D integration have become the necessarily developing trends [1-4]. It requires bringing people electronic products with enough multifunction and excellent performance but relatively lower cost. Plastic packing occupying 90% semiconductor devices can be widely used in commercial fields with good productivity and high efficiency and low cost, in which EMC exceeds 90% market occupation [5-6].

In the EMC component, silica occupies the most weight content varying from 60% to 92% for its excellent properties, such as dielectric property, low coefficient of thermal expansion (CTE), and high coefficient of thermal conductivity (CTC) [7-11]. One of the typical superiority is its low cost and good adhesion with EMC resin system following with excellent physical mechanical properties. Contrasted with angular silica, the advantage of spherical silica fillers is offering great flowability and lower viscosity and specific gravity. Besides shape type, degree of sphericity

and specific surface area (SSA) are the critical factors for contribution to optimal density packing with lower porosity degree, especially for particle size distribution [12-14]. Hence optimal packing of spherical silica can help to increase filling content highly reaching or exceeding 90% in EMCs and popularly follow the IC development trends more than ever.

The latest particle continuously distribution theory, Dinger-Funk Equation (1), was chosen as the research theoretical model for typical particle distribution theory and earlier packing theory were modified in it [15-17]. Another experimental model was founded based on the theory too. The difference was that the later one ignored the fine distribution area. 3 distribution factors (n = 0.34, 0.37&0.40) were determined around the theoretical optimal value (n = 0.37)[18] for considering practical errors in mix and preparation process. Besides 2 controls with cut 75 microns, other 4 types of spherical silica with different PSD and filler cut were chosen as the research objects. After computation according to the 2 models, 8 batches finished sample were characterized based on standard test methods of EMC. Then the optimal model and distribution factor and silica assemble matrix were determined to give the research application potential.

$$F(D) = \frac{CPFT}{100} = \frac{D^n - D_s^n}{D_L^n - D_s^n} \tag{1}$$

Experimental Part

Materials

6 types of spherical silica (HS-0~5) were chosen for which were widely used presently or have great application potential in HHE. HS-0 was the research target for its known excellent performance while directly used in EMC, and HS-1 was the major research object for it is usually mix-used though its cut is the same as HS-0. The other 4 smaller than HS-0 and HS-1 were chosen as the adjustive additive. Their code, filler cut and median diameter were listed in Table 1.

Table 1. The data of different kinds of spherical silica.

Code	Filler cut (μm)	Median diameter (μm)
HS-0 (target)	*75*	*15*
HS-1	75	18
HS-2	18	4.5
HS-3	36	5
HS-4	7	1.5
HS-5	1.5	0.5

Henkel standard molding compound with 85% ash was used as the basic formulation in the research. 2 controls were 100% filled with HS-0 and HS-1, and the difference of the other 6 samples was the silica components which were calculated according to the 2 models. Some of the fixed composition, including hardener, catalyst, wax, flame retardant and ion getter were premixed through ball milling.

Characterization

Gel Time (GT) was measured on the hot plate at 175°C. Spiral Flow (SF) and Flash&Bleed (F&B) were measured by using HB-121-75T heat press at 175°C, respectively in Spiral Flow mould and Flash&Bleed Mould. Ash was tested in baking oven at 700°C for 8h. Viscosity was tested with SHIMADZU CFT-500D Capillary Rheometer under 20kgf at 150°C. All tests were performed trice parallelly.

Results and Discussion

Experimental model was derived from comparison the particle size distribution curve of HS-0 with HS-1 and Dinger-Funk Equation. Compared with HS-1 (Figure 1), HS-0 owned less large particle and more fine particle, hence smaller median diameter. But the most important thing for HS-0 is its excellent performance, such as ultra high flowability and high load and controllable F&B without adding other silica. Cumulative distribution curves of HS-0 and Dinger-Funk Equation with different distribution factors (n = 0.27, 0.34, 0.37, 0.40 and 0.45) were compared too, which was listed in Figure 2. Dinger-Funk Equation curves sank as the increase of n value. Surprisingly, curve of HS-0 kept good superposition contentment with the calculated curve (n = 0.37) in area that particle size is larger than 6μm. Furthermore, less particle distribution were detected in fine particle area (1~6μm) of HS-0, especially nearly zero distribution in superfine area (0~1μm). Thus, a new experimental model was set up for trials which obeyed the Dinger-Funk Equation but ignored the fine particle area. In the same time, it also can save cost of raw material of EMC for known superfine silica is more expensive than large ones.

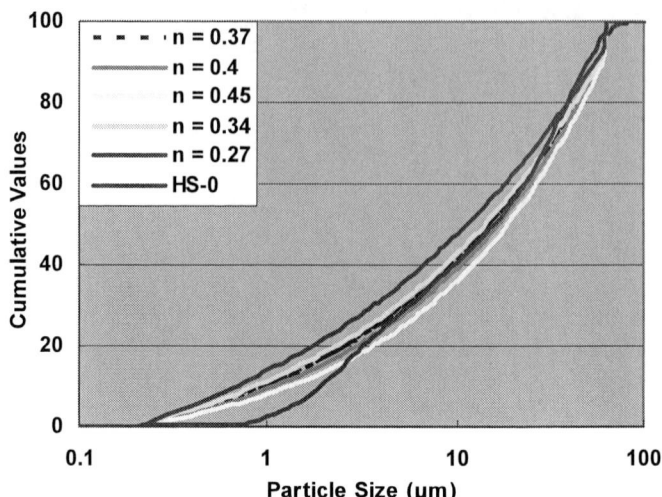

Figure 2. Cumulative distribution curves of HS-0 and Dinger-Funk Equation.

All PSD data were from suppliers and adjusted uniformly to the same distribution range from 0.04μm to 500μm with 100 data points. For computation, cumulative distribution curves of Dinger-Funk distributions with 3 kinds of distribution factor were drawn as the standard curves firstly. Then, a series composition of the 5 research objects were calculated out by superposition calculated curves with those standard curves according to the 2 models. The results listed in Table 2, showed HS-3 and HS-4 had no any contribution to both models. The reason is perhaps that other 3 silica fillers have cover their distribution area. Furthermore, more fine particles (HS-5) were shown in the results of theoretical model than experimental one while occupation of large particle HS-1 kept fixed. The difference of those results kept in line with the difference of the 2 models. Last, large particles increased with the increase of n value as the distribution trends of Dinger-Funk distributions.

Table 2. Calculation results of different models.

n	Computational models	HS-1	HS-2	HS-3	HS-4	HS-5
0.34	PSD-3401[a)]	76	14	0	0	10
	PSD-3402[b)]	76	17	0	0	7
0.37	PSD-3701	80	12.5	0	0	7.5
	PSD-3702	80	15	0	0	5
0.4	PSD-4001	83	10	0	0	7
	PSD-4002	83	14	0	0	3
a) 3401 means n is equal to 0.34 and theoretical model.						
b) 3402 means n is equal to 0.34 and experimental model.						

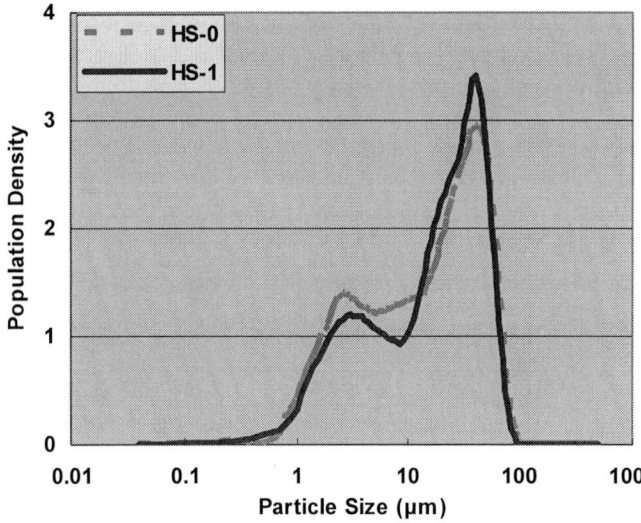

Figure 1. Population density curves of HS-0 & HS-1.

Table 3. Characterization results of 8 samples.

GR9810-1P		Control		n					
				0.34		0.37		0.4	
EMC Code		PSD-HS-0	PSD-HS-1	PSD-3401	PSD-3402	PSD-3701	PSD-3702	PSD-4001	PSD-4002
GT	s	29	28	27	27	29	28	27	29
SF	inch	48	43	61	58	57	58	64	57
F&B	0.25	4.1	4.9	3.5	3.9	4.3	4.4	4.9	5.8
	0.5	4	5.1	4.6	4.3	3.6	5.5	5.3	5.9
	1	2.5	5	5.2	4	3.8	4.8	2.8	5.7
	2	3	3.9	3.7	4.3	2.8	5.7	3.6	4.5
	3	50.9	30.2	95.6	63.6	79.2	97.5	96.7	5.2
Ash	%	85.27	85.29	85.19	85.23	85.17	85.19	85.14	85.22
Viscosity	Pa·s	10.5	17.5	14.6	3.6	2.5	16.9	3.8	10.7

The characterization results were averaging from 3 parallel tests and listed in Table 3. All ash results of 8 samples were around the theoretically calculated one 85.20%, and standard deviation of GT results were only 0.9. So it was proved that there was nearly very few chance of big experiment error and the test results were reliable. PSD-HS-0 showed better flowability and F&B performance than PSD-HS-1 as expectation. Generally speaking from the results, seldom difference was found in F&B performance for all samples, but surprisingly the flowability has been remarkably improved, the max one was PSD-4001 which increased nearly 50% compared with control PSD-HS-1 (Figure 3). Moreover, the viscosity has been obviously decreased compared with control PSD-HS-1. Viscosity of 3 samples, PSD3701 and PSD3402 and PSD4001, were even much lower than control PSD-HS-0 which has been proved with excellent performance. The viscosity of the lowest one PSD3701 was only 14% of control PSD-HS-1 (Figure 4). Comprehensively concluded, the theoretical model was slightly better than the experimental model. Results of both models showed improved silica filler matrixes based on HS-1 have lowered the viscosity and gotten much better flowability, even have exceeded the performance of the target silica HS-0.

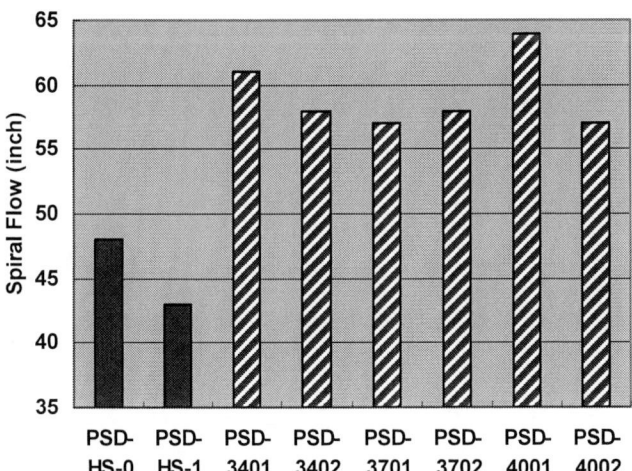

Figure 3. Spiral Flow column chart of 8 samples.

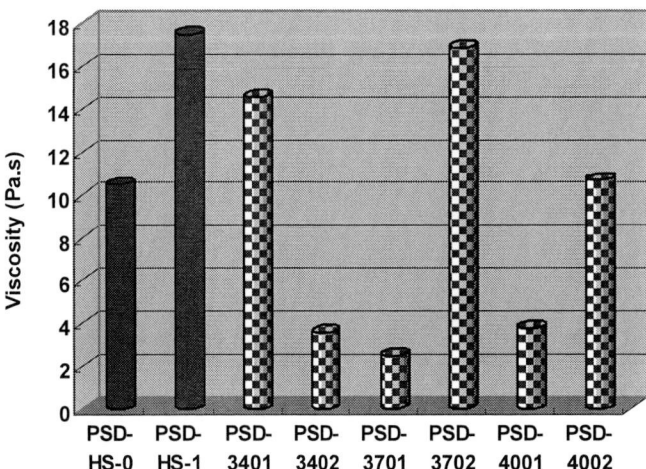

Figure 4. Viscosity cylinder chart of 8 samples.

Conclusions

Besides theoretical computational model, Dinger-Funk Equation, a novel experimental model was founded according to the PSD specialty of research target, which is known for its excellent performance. The difference was the original one required superposition of all distribution area and the later one ignored fine particle distribution area, which was proved in the computational results. Computational filler matrix also showed HS-3 and HS-4 had no any contribution to both models for others have covered those particle area. Furthermore, large particles increased with the increase of n value as the change trends of Dinger-Funk distribution. Characterization data showed both models benefited lowering viscosity and increasing flowability for the optimization of PSD, the experimental model was slightly inferior to the theoretical one. Moreover, improved silica filler matrixes based on HS-1 have maximally lowered the viscosity to 14% and gotten maximally 50% increase in flowability than HS-1 itself, even exceeded the performance of the target silica HS-0. The research has exhibited great guide significance to apply the optimal packing of spherical silica fillers in EMC field.

References

1. Garrou, P. *et al*, <u>Handbook of 3D Integration: Technology and Applications of 3D IC</u>, John Wiley & Sons Inc (New York, 2008).

2. Kettner, P. *et al*, "New Technologies for advanced high density 3D packaging by using TSV process," *2008 ICEPT-HDP*, Shanghai, China, July. 2008, pp. 20.

3. Erik, H. M., "Future semiconductor detectors using advanced microelectronics with post-processing, hybridization and packaging technology," *Nucl. Instrum. Meth. A*, 2000, 541(1-2): 274.

4. Manabu, B. *et al*, "Trends and opportunities of system-in-a-package and three-dimensional integration," *Electron. Comm. Jpn. 2*, 2005, 88(10): 37.

5. Kuang Y.X., Zhao G.Y. "Rapid development of 3D electronic packaging technology," *Electronic & Computer Design World*, 1998, 75 (9): 47.

6. Yang F.W., Zhao Y.M. "Situation and Development of Epoxy Resins for Electronic Packaging," *Electronic Process Technology*, 2001, 22 (6): 238.

7. Yang, D.G., *et al.*, "Effect of filler concentration of rubbery shear and bulk modulus of molding compounds," *Microelectron. Reliab.* 2007, 47(2-3): 233.

8. Ryuji, M. et al. "Particle size distribution of spherical silica gel produced by sol-gel method," *J. Non-Cryst. Solids* 1990, 121(1-3): 389.

9. Barry, J. B. *et al.* "Epoxy/SiO2 interpenetrating polymer networks," *Polym. Advan. Technol.* 1996, 7(4): 333.

10. Steven, R. D. *et al.* "Formation of silica/epoxy hybrid network polymers," *J. Non-Cryst. Solids* 2003, 315(1-2): 197.

11. Sun, Z. X. Electronics Chemistry, Chemical Industry Press (Bejing, 2002), pp. 212.

12. Olhero, S. M. *et al.* "Influence of particle size distribution on rheology and particle packing of silica-based suspensions," *Powder Technol.* 2004, 139(1): 69.

13. Ryuji, M. *et al.* "Particle size distribution of spherical silica gel produced by sol-gel method," *J. Non-Cryst. Solids* 1990, 121(1-3): 389.

14. Han, M. F. et al. Preparation and Process of Nonmetallic Mineral Materials, Chemical Industry Press (Bejing, 2004), pp.166.

15. Funk J E, Dinger D R. *Am. Ceram. Soc. Bull.* 1988, 67(5): 890.

16. Dinger D R. "One Dimensional Packing of Spheres (Part I): Monodisperse and Bimodal Distributions," *Am. Ceram. Soc. Bull.* 2000, 79(2): 71.

17. Dinger D R. "One Dimensional Packing of Spheres (Part II): Continual Distribution," *Am. Ceram. Soc. Bull.* 2000, 79(4): 83.

18. Liu, H. B. "Particle Size Distribution and Packing Theories," *Journal of the Chinese Ceramic Society*, 1991, 19(2): 164.

iNEMI HFR-free Program Report

Haley Fu[1], Stephen Tisdale[2], Martin Rausch[2], John Davignon[2], Stephen H. Hall[2], Robert C. Pfahl[3]
[1]iNEMI, China,
[2]Intel Corporation, USA,
[3]iNEMI, USA,
Email: haley.fu@inemi.org, Stephen.Tisdale@Intel.com, martin.rausch@intel.com,
John.J.Davignon@Intel.com, Stephen.H.Hall@Intel.com, bob.pfahl@inemi.org

Abstract

The electronics industry is aggressively pursuing the removal of potentially toxic compounds from their products, including the halogenated flame retardants (HFRs) that were once widely used in electronics housings and cases and are still used extensively in printed circuit boards. Several leading electronics companies have publicly stated their intent to remove brominated and/or halogenated flame retardants from some or all of their products. The International Electronics Manufacturing Initiative (iNEMI), an industry-led consortium, is working with a number of its OEM and supply chain members to assess the feasibility of a broad conversion to HFR-free PCB materials. While IPC and JEDEC are developing halogen-free standard specifications and numerous companies have compliant materials, significant questions remain regarding overall readiness to make a transition to these materials. This paper will discuss results & conclusions from the completed iNEMI HFR-free PCB Materials Project, as well as outline current projects, which include the HFR-free High-Reliability PCB project, the HFR-free Signal Integrity Project and the HFR-free PCB Material Development Project.

Introduction

The European Union's Restriction on the use of certain Hazardous Substances (RoHS) Directive prohibits the use of polybrominated biphenyls (PBBs) and polybrominated diphenyl ethers (PBDEs) in nonexempt electronic equipment. These compounds can be used as flame-retardants and some of these substances have been shown to present unacceptable risks to human health and the environment.

A key requirement that is governed by Underwriters' Laboratory (UL) is the ability to meet the flammability standard of UL 94-V0. In general, thermosetting resins, alone or in combinations with other additives widely used in the electronic industry for PCB laminate applications, meet these requirements only because they contain approximately 30-40% brominated aromatic epoxy components, based on the resin, or approximately 17% to 30% bromine (based on the total resin weight). Although these brominated compounds have excellent flame-retardant properties, they also have some undesirable properties when incomplete burning occurs. The chemical decomposition of aromatic bromine compounds release free bromine radicals and hydrogen bromide, which are highly corrosive.

Non-halogenated alternative fire retardant material systems are being developed and introduced into products today. These systems typically use nitrogen compounds, phosphorus based compounds, or a combination of both.

Some of these may be incorporated into the backbone of the polymer as is done with TBBPA in epoxy. These flame retardant systems are currently available for some printed wire boards and engineered plastic applications. It is important to note that the reliability of many alternative flame-retardants has not been fully qualified at the assembly level. Product developers will need to address whether substitutes can meet the same technical and functionality requirements, whether they will decrease product safety or reliability, and what the tradeoffs may be.

Low-Halogen Electronics

The trend toward low-halogen materials in electronic products has created a need for supply chain alignment on the maximum levels of bromine (Br) and chlorine (Cl) allowed in electronic materials and systems that are identified as low halogen (or "halogen-free" and/or "BFR/CFR/PVC-free"). A common definition of maximum halogen levels for low-halogen components and materials will enable the development of compliant material sets. The following definition of Low Halogen ("BFR/CFR/PVC-Free") electronics is supported by iNEMI and its member companies (Dell, HP, Intel, Lenovo, Cisco, Sun, Tyco Electronics etc.).

A **component** [1] must meet all of the following requirements to be Low Halogen ("BFR/CFR/PVC-Free"):

1)   All printed board (PB) and substrate laminates shall meet Br and Cl requirements for low halogen as defined in IEC 61249-2-21 and IPC-4101B per 1a below (refer to IEC and IPC standards for actual requirements).

1a – Non-halogenated epoxide with a glass transition temperature of 120°C minimum. The maximum total halogens contained in the resin plus reinforcement matrix is 1500 ppm

---

[1] Other than those terms listed below, the definitions of terms used in this position statement, such as "component," are in accordance with IPC -T-50 and/or JESD88.

Plastic

Any of a group of synthetic or natural organic compounds produced by polymerization, optionally combined with additives (organic or inorganic fillers, modifiers, etc) into a homogeneous material capable of being molded, extruded, coated, printed or cast into various shapes and films.

PVC copolymer

Copolymers are polymers derived from two or more monomers. Highly chlorinated PVC copolymers, block polymers, and congeners are not considered acceptable alternatives to PVC for low-halogen components.

with maximum chlorine of 900 ppm and maximum bromine being 900 ppm.

2) For components other than printed board and substrate laminates:

Each plastic within the component contains < 1000 ppm (0.1%) of bromine [if the Br source is from BFRs] and < 1000 ppm (0.1%) of chlorine [if the Cl source is from CFRs or PVC or PVC copolymers].

iNEMI HFR-Free Project Portfolio

While alternative non-halogenated material systems are being developed and introduced into products, the reliability of many of these alternative flame retardants has not been fully qualified at the assembly level. Data is needed to address whether these substitutes can meet the same technical and functionality requirements as standard products, whether they will decrease product safety or reliability, and what the trade-offs may be.

iNEMI has launched several projects to identify technology readiness, supply chain capability, and reliability characteristics for HFR-Free alternatives to conventional printed circuit board materials and assemblies. The projects are:

- HFR-Free PCB Material Evaluation (completed project)
- HFR-Free High Reliability PCB (active project)
- HFR-Free Leadership Program (new program)
  - HFR-Free PCB Materials
  - HFR-Free Signal Integrity
- PVC Alternative Initiative (new initiative)

HFR-Free PCB Material Evaluation

Working together with the materials supplier base and volunteer printed wiring board manufacturers, this project evaluated the electrical, mechanical and reliability attributes of "halogen-reduced" materials using known designs from IBM and Intel Corporation.

The adoption of the halogen free alternatives requires that the laminates have minimal impact on the electrical, mechanical, electro-migration, chemical resistance, thermal, moisture absorption, and rheological properties. In addition, adhesion to copper, the oxide treatment, and to the laminate itself needs to be sufficient. Processing and assembly performance of the laminate products must meet design requirements.

Upon identification of suitable halogen free PWB laminate materials, a series of tests were performed to evaluate the electrical, thermal, and physical properties of the new commercially available materials. Materials passing the initial screen were then used to build test vehicles to evaluate other material properties. Three test vehicles were adopted in the project.

IBM SMASPP2z Electrical Test Vehicle

This test vehicle design is geared specifically toward the assessment of the dielectric material electrical properties and total loss using the Short Pulse Propagation (SPP) technique [1]. The test vehicle is an 8 layer design, MP1-V2-S3-V4-V5-S6-V7-MP8, with two stripline structure designs, one each on layers 3 and 6, which are used for the material analysis in this effort. See Fig. 1. There are also microstrip structures in the test vehicle design on layers 1 and 8 which can be used when those structures are of interest.

The launch structures are designed around a high performance bolt-on SMA connector, rated at 26GHz.

The SMASPP2z test vehicle is approximately 280 × 125 mm [11" × 5"], which allows 6-up on typical panel layout.

The two stripline structures on layers 3 and 6 allow for the assessment of a resin rich and a resin poor design point respectively. This effectively defines the range of dielectric constant and effective loss tangent for any given material.

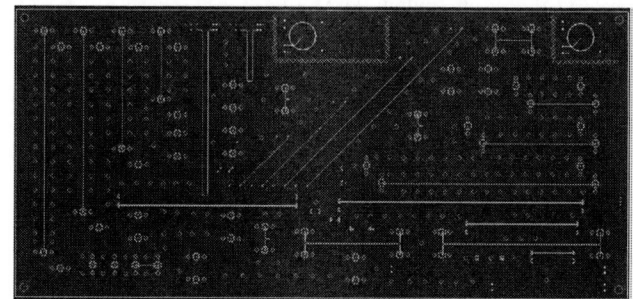

Fig. 1 SMASPP2z Test Vehicle Layout, Showing One Layer of Internal Wiring

HOP31B Test Vehicle

This test vehicle is specifically geared toward the assessment of the propensity of a given laminate material to exhibit quality issues such as cracks and/or delamination upon exposure to multiple cycles of higher assembly process reflow conditions associated with mixed solder assembly (MSA) and/or full lead free solder assembly processes.

The HOP31B test vehicle is approximately 140mm × 100 mm [5.5" × 4"].

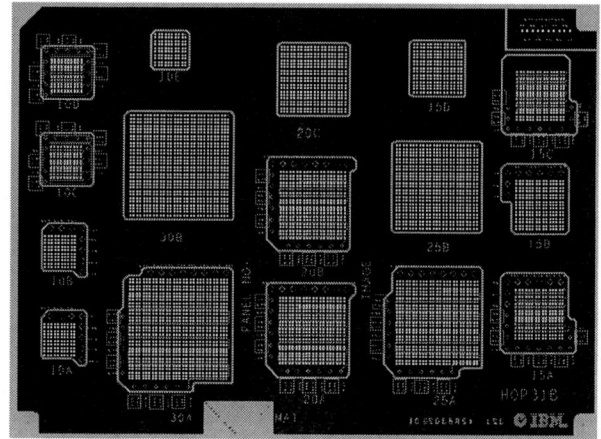

Fig. 2 HOP31B Test Vehicle Layout

The HOP31B test vehicle contains various size PTH arrays, all of which consist of 0.2mm [0.008"] PTHs on a 0.8mm [0.031"] pitch. See Fig. 2. These PTH arrays have been shown to be sensitive to the laminate properties for which the test vehicle was designed to screen.

The HOP31B test vehicle shares the same 10 layer cross section definitions with the MEBII test vehicle, which is described in a latter portion of this paper. As with the MEBII test vehicle, two varieties of the HOP31B test vehicle were constructed, targeting a 40 mil thickness and 80 mil thickness, respectively.

Process Simulation Conditions:

The test vehicles were exposed to a specific set of assembly process conditions as follows.

SMASPP2z: The SMASPP2z test vehicles were exposed to an overnight moisture removal bake at 125°C, along with 3×, 245°C IR reflow processes.

The bake was implemented to put all laminates on a more level playing field with regards to potential impact of moisture content on the test results. **NOTE:** The SMASPP2z test vehicles were inadvertently exposed to a 165°C overnight bake instead of the defined 125°C bake.

Immersion Ag was used as the surface finish whenever possible. This was done to reduce the impact (on probing) of Cu oxide formation during reflow process simulation.

HOP31B: The HOP31B test vehicles were exposed to a matrix of conditions, all of which were preceded by an overnight moisture removal bake at 125°C. **NOTE:** The 40 mil test vehicles were inadvertently exposed to a 165°C overnight bake.

The bake was implemented to put all laminates on a more level playing field with regards to potential impact of moisture content on the test results.

The 4 cell matrix of assembly process simulations consisted of the following conditions, emulating mixed solder assembly (MSA, 245°C) and full Pb-free (260°C) reflow process conditions.

- 3×, 245°C peak temperature
- 5×, 245°C peak temperature
- 3×, 260°C peak temperature
- 5×, 260°C peak temperature

MEB II Evaluation

The Material Evaluation Board II (MEB II) is Intel's 2nd generation multifunctional test vehicle which contains test structures for the electrical, thermal, and mechanical performance evaluation. The test vehicle is designed to evaluate performance across a full working panel used at the PCB fabrication facility. The design is modular and can be broken into 4 quadrants for ease of testing and handling after fabrication. The test structures contained on the MEB II include: 1) Registration coupons for soldermask, drill to innerlayer and drill to outerlayer copper structures in the 4 corners of the working panel; 2) All laminate and copper plane/laminate coupon structures for flexural modulus and thermo-mechanical testing (TMA, DMA, DSC); 3) Through via, buried via, and microvia IST coupons; 4) Through via, buried via, and microvia in-line daisy chain structures for tight pitch CAF testing; 5) Trace and space capability coupons for all copper layers; 6) Hi Pot/Capacitance coupons; 7) Performance network analyzer electrical structures with GSG micro-probe contact points; 8) Moisture

diffusivity coupons with SMA connection structures; and 9) BGA pad structures for Cold Ball Pull test evaluation.

The MEB II was constructed as a 10 layer board at 2 different thicknesses (40 and 80 mils or 1 and 2 mm). The construction was a 1-8-1+ double lamination with microvias 1 to 2, 2 to 3, 9 to 8, and 10 to 9; buried vias from 2 to 9; and through hole via structures 1 to 10.

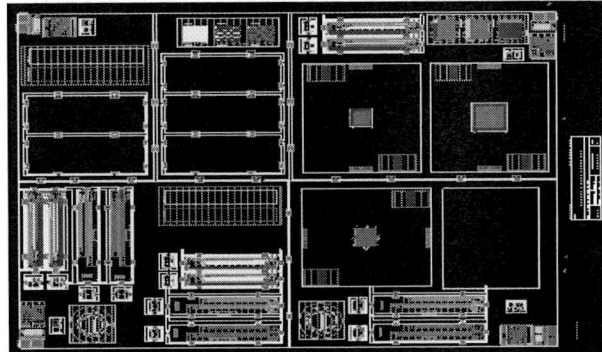

Fig. 3 MEB II Test Vehicle Layout

Overall, the team's investigation showed that not all halogen-free materials are equivalent, and none is equivalent to the FR4 baseline used (See Table 1). Findings include:

- The halogen-free materials generally had higher dielectric constant (Dk) values and lower loss tangent (Df)
- Most of the halogen-free laminate materials did not exhibit resin cracking/delamination (issues normally associated with an incompatibility with higher reflow temperatures)
- In time-to-delamination tests at 260°C, 288°C and 300°C, the bromine-free materials in general did not show problematic results and were in line with brominated materials used by the electronics industry
- Moisture absorption at room temperature at 24 hours showed relatively low values, with a number of materials being below 0.20%
- Exposure to hydrothermal conditions pointed out interlayer weakness in some materials
- Most of the materials exhibited relatively good adhesion to copper
- In-plane expansion is similar to brominated materials, in the range of 17-22 ppm/°C
- Out-of-plane data indicate relatively lower values with an average of around 45 ppm/°C below Tg
- The significance of the differences in the halogen-free material properties and performance will be dependent on the design and demands of the products in which they are incorporated. The iNEMI project team recommends individual testing of any material for its intended application prior to mass production.

Table 1 Material Evaluation Summary

Mat'l	Dk	Df	H$_2$O Absorb	Tg	CTE	Flex	Td	T260/Cu	T288/Cu	Peel Strength	IST	CAF	UL94V0	Shock	Vibe	Temp Cycle	Cold Ball Pull
A																	
B																	
C																	
D																	
E																	
F																	
G																	
H																	
I																	
J																	
K																	

Color Code
Equal to or better than FR4 (No issue)
Marginal vs FR4 (Issue not clear)
Worse than FR4 (Clear issue)
No Data

HFR-Free High Reliability PCB

This project is a follow-on to the HFR-Free PCB Material Evaluation Project. Its focus is to identify technology readiness, supply capability and reliability characteristics for "HFR-free" alternatives to conventional printed wiring board materials and assemblies, based on the requirements of the high-reliability market segment. Project goals are to:

- Identify commercially viable materials
- Benchmark past work and key in on critical knowledge gaps and technical issues
- Build on industry knowledge and capability, including the iNEMI HFR-Free PCB Material Evaluation Project
- Design test vehicles and test methodologies and, leveraging prior investigations, carry out the necessary testing to characterize viable materials

The team has identified seven HFR-free materials to be evaluated, and a typical halogenated material as a control. Test vehicle lay-ups have been completed for the material evaluation testing as well as the board-level reliability testing and test vehicles are currently being built utilizing the materials being supplied by each of the seven material suppliers, with testing to begin as soon as all materials / TVs have been received.

HFR-Free Leadership Program

Several major OEMs are evaluating the elimination of HFRs from their PCB materials. While mobile phone manufacturers are well along in this effort, the next area of impact will likely be driven by high volume client computer applications.

An industry-wide conversion to HFR-free materials faces numerous challenges:

- Reliability of materials with alternative flame retardants has not been fully qualified

- Complete "technology envelopes," or technical specifications, have not been established for various product applications
- Incomplete design knowledge in segments of the supply chain increases risk of conversion issues
- A rapid complete conversion of computer products will have a major impact on the supply chain and needs to be coordinated

iNEMI is working with a number of our OEM members to assess the feasibility of a broad conversion to HFR-free PCB materials. Although IPC and JEDEC are developing halogen-free standard specifications and numerous companies have compliant materials, significant questions remain regarding overall readiness to make a transition to these materials. For example:

- What electrical properties are needed to meet high speed signaling requirements?
- With many HFR-free materials showing higher stiffness, what mechanical properties are needed to ensure system reliability isn't degraded?
- Can design modifications reduce sensitivity to electrical and material properties?

As a result, two projects have been established: HFR-Free Signal Integrity and HFR-Free PCB Materials.

HFR-Free Signal Integrity

The iNEMI HFR-Free PCB Signal Integrity Work Group has been in formation for the past 3 months and presently has 13 companies actively involved. The work groups' goal is to identify the critical electrical parameters of HFR-Free dielectric materials, set limits on those parameters so that signal integrity will not be jeopardized for high speed buses, and communicate those electrical limits to the material suppliers so they can focus resources on producing products the industry requires, which in turn should help ensure adequate volume for product launches and reduced costs.

A critical goal of the work group is to develop a common measurement methodology that will ensure consistent and accurate evaluation of the electrical parameters of the dielectric and allow apples-apples comparison of the industries HFR-Free materials. The measurement methodology will also be used by the material suppliers as the "standard evaluation method" for reporting the electrical values of the critical parameters on the material data sheets.

The work group has identified 4 critical electrical parameters: 1) dielectric permittivity (Dk); 2) loss tangent (Df); 3) moisture absorption (specifically how it affects Dk and Df) and 4) breakdown voltage. Based on the high-speed signaling needs required by each company, the limits of each parameter will be identified. Samples of HFR-Free materials will then be evaluated (using the common measurement methodology) and the results will be mapped into the requirements providing a design data base that companies can use to select adequate materials for products.

HFR-Free PCB Materials

The iNEMI HFR-Free PCB Materials Work Group has been in formation for the past 3 months and presently has 17 companies actively involved in the PCB Materials Work Group. The work groups' goal is to identify material/technology limitations involved in transitioning to HFR-free PCB materials.

The work group (WG) has identified 24 "areas of concern" (See Table 2), including thermo-mechanical performance characteristics and is now determining if these are in the critical path for the halogen-free PCB material transition. The WG is beginning its' evaluation of metrologies and test methods which can quantify these characteristics. The

WG will build test boards to verify the sensitivity of the test methods/metrology ability to quantify the areas of concern. It is expected that the test methods will include standard metrology already used in the industry with modified limits or ranges required in some cases. New test methods/metrologies will be developed if no standard method can be applied. The test methods will become part of a test suite used to evaluate HF laminate materials and provide feedback from the OEM/ODM to the laminate suppliers. This project will identify the technical risks of a broad transition to HF PCB platforms/systems, initially focusing on Desktop and Mobile notebook including the assessment of manufacturing and supply chain capacity for high volume HF transition.

PVC Alternative Initiative

This new initiative is one of the projects proposed at the iNEMI Sustainability Summit (September 2008). This effort will investigate "PVC-free" alternative materials, focusing on:

- Environmental lifecycle assessment (LCA) comparing PVC compounds with PVC-free compounds for US-based detachable desktop power cord applications (cable, connectors and wire)
- Cradle-to-grave LCA, including end-of-life aspects (recycling, incineration, landfill, etc.)
- Comparison of equivalent functional units that meet UL requirements
- Performance testing to gain a better understanding of the electrical, mechanical and safety aspects of PVC free alternatives

Table 2 Areas of Concern by HFR-free PCB Materials Project

	Areas of Concern			
1	UL Fire ratings (V0-V1)	13	Fracture Toughness of Resin / Resin Cohesive Strenght
2	Glass transition temperature (Tg)	14	Stiffness/Flexural Strength
3	Decomposition temperature (Td)	15	Copper Pad Adhesion (CBP/Hot Pin Pull/ Shear or Tensile)
4	Coefficient of thermal expansion (z-axis and x-, y-axes)	16	Co-Planarity Warpage characteristics
5	Moisture absorption	17	Long term life prediction, (IST or thermal shock test)
6	UL CTI rating	18	CAF resistance
7	MOT Maximum Operating Temperature	19	Delamination characteristics under mechanical or thermal stress
8	Punchability/Scoring/Breakoff Performance	20	Resin system dependency/hardening/curing agents
9	PCB fabrication process, drill wear, lamination & desmear cycle	21	Affect of Fillers
10	Rework (Pad Peeling)	22	Plastic and elastic deformation characteristics
11	Micro and macro hardness	23	Shock & Vibe and Drop test data
12	Electrical Properties (Dk & Df)	24	Transient Bend

Conclusions

Our investigation has shown that not all halogen free materials are equivalent, and none are equivalent to our FR4 baseline. Compared to the baseline material, we see generally higher Dk values and lower Df values for halogen free materials. The significance of the halogen free material

property and performance differences will be dependent on the design and demands of the products in which they are incorporated.

While the dielectric constant was relatively unaffected by the assembly reflow processes, the effects on the effective loss tangent were more noticeable, albeit not consistent. This may be due to the varying moisture content of the laminate materials as manufactured at the laminate vendor and PCB fabricator, and the relative sensitivity of the loss tangent vs. the dielectric constant.

The thermal expansion coefficient measurements for the various bromine free laminates, both below and above the glass transition temperature, indicate that the in-plane expansion is similar to the brominated materials, in the range of 17-22 ppm/°C. The out of plane data indicate relatively lower values with an average of around 45 ppm/°C below Tg. The lower out of plane expansion is due to the constraining properties of the fillers used to impart V0 rating.

This project teams recommend the individual testing of any material for the specific application by the designer prior to mass production.

Acknowledgments

iNEMI HFR-Free PCB Materials Evaluation Project Team, Fabricators & Laminate Suppliers

> ➢ Albemarle, Cisco, Clariant, HP, IBM, ITEQ, Sun Microsystems, Vitronics-Soltec, Hitachi Chemical, Nan Ya, Shengyi, Isola, Panasonic, TUC, Multek, Chin Poon, GCE, Meadville, Nan Ya PCB, E&E, Sanmina, PWB Interconnects, IST

iNEMI HFR-Free Hi Reliability PCB Project Team, Fabricators & Laminate Suppliers

> ➢ Agilent, Albemarle, Celestica, Cisco, Dell, Doosan, Elite Material Co., HP, Intel, IST, ITEQ, Nan Ya, Rohm & Haas, Sun, LG Micron

iNEMI HFR-Free Leadership Teams - the Signal Integrity and PCB Materials Work Groups

> ➢ Albemarle, Celestica, Cisco, Dell, Delphi, Dossan, EMC, Flextronics, Foxconn, Shengyi, HP, Huawei, IBIDEN, IST, Intel, ITEQ, Lenovo, Quanta, Nan Ya, etc.

References

1. A. Deutsch, T.M. Winkel, G. Kopcsay, C. Surovic, B. Rubin, G. Katopis, B. Chamberlin, R. Krabbenhoft, "Extraction of Er(f) and tan d(f) for Printed Circuit Board Insulators Up to 30 GHz Using the Short Pulse Propagation Technique," *IEEE Transactions On Advanced Packaging*, Vol. 28, No. 1 (2005), pp. 4 - 12.

Interfacial Reaction of Reballed BGAs under Isothermal Aging Conditions

Lei Nie[1], Mingzhi Dong[2,3], Jian Cai[2,3], Michael Osterman[1], Michael Pecht[1]

[1]Center for Advanced Life Cycle Engineering, University of Maryland
[2]Tsinghua National Laboratory for Information Science and Technology
[3]Institute of Microelectronics, Tsinghua University, Beijing, 100084, China

Abstract

Reballing involves removing the original solder balls and attaching new solder balls on the pads of ball grid array components. In this study, 676 I/O BGAs with Sn3.0Ag0.5Cu (SAC305) balls were reballed with eutectic tin-lead solder. The original metallization of the pads on the BGAs was Cu/Ni with immersion Au. Two solder ball removal methods were applied: the low-temperature wave solder method and the solder wick method. The preform solder ball re-attachment method was used to attach new solder balls. The components were investigated after the solder ball removal and solder ball re-attachment procedures. In comparison with non-reballed BGAs, the components using low-temperature wave solder method had phase change in the interfacial IMC layer, while the components using solder wick method had no phase change. Isothermal aging was used to investigate IMC growth and IMC composition change. There was no phase change in non-reballed and reballed BGAs after isothermal aging.

Introduction

While the majority of electronic equipment manufacturers have converted to lead-free production, there are some electronic equipment manufacturers that are excluded or exempt from global regulations and continue to produce lead-based electronics [1-4]. For these manufacturers, reballing lead-free ball grid array (BGA) packages with tin-lead solder balls has become a common solution for the shortage of tin-lead BGA parts. In the reballing process, the original solder balls are removed and new solder balls are placed on the exposed pads of the package. The reballing process has two major steps: solder ball removal and solder ball re-attachment. Multiple methods are available for the removal and re-attachment steps.

In the reballing process, the reballed BGAs go through extra heating procedures, solder ball removal, and solder ball re-attachment. This process leads to thicker interfacial intermetallic compounds (IMCs). Since solder failure in the BGA packages often occurs at the component side interface, the interfacial IMCs resulting from the reballing process should be studied.

There is not much literature addressing reballed BGAs. The ball attachment strength of reballed BGAs is a topic that researchers focused on [5-6]. There is plenty of literature addressing the interfacial IMCs in mixed solder joints and lead-free solder joints [7-8]. Currently, there is little literature talking about interfacial IMCs on reballed BGAs. In the reballing process, if the solder ball composition changes, the interfacial IMC composition may change also. To investigate the interfacial IMCs in the reballed BGAs will help achieve a better understanding of reballed BGA durability.

The reballing studies we initiated at 2007 included a series of contents, the solder ball attachment strength evaluation of the reballed BGAs, the interfacial IMCs composition identification of the reballed BGAs, and the durability of the reballed assemblies under thermomechanical and mechanical loading condition. The interfacial IMCs in the reballed BGAs was investigated in this paper.

In this study, 676 I/O BGAs with Sn3.0Ag0.5Cu (SAC305) balls were reballed with eutectic tin-lead solder. The original metallization of the pads on the BGAs was Cu/Ni with immersion Au. Two reballing processes were examined. One group of BGAs was reballed using the solder wick removal method and the preform re-attachment method. The other group of BGAs was reballed using the low-temperature wave solder removal method and the preform re-attachment method. A set of non-reballed BGAs was used as a control. For purposes of comparison, both non-reballed (SAC) BGAs and reballed (SnPb) BGAs were exposed to two thermal aging conditions, 100°C for 24 hrs and 125°C for 350 hrs. The effect of the reballing method on the interfacial IMCs was evaluated in this study, and the effect of isothermal aging on the composition and growth of interfacial IMCs was studied as well.

Experiment

In this study, 676 I/O lead-free plastic overmold ball grid array (PBGA) components with Sn3.0Ag0.5Cu (SAC305) solder balls were selected to be reballed with eutectic tin-lead solder. The ball diameter of the 676 I/O PBGA was 630μm with 1.00mm pitch (Amkor). The original metallization of the pads on the PBGA was Cu/Ni with immersion Au. The experimental procedure carried out on these components was divided into two parts. Part I investigated interfacial changes due to two different solder ball removal procedures. In Part II the microstructural development of the interfacial IMCs of reballed components due to isothermal aging was examined. Figure 1 gives an overview of the complete test program.

The reballing process for the first part involved two major steps: solder ball removal and solder ball re-attachment. In this study, two solder ball removal methods and one solder ball re-attachment method were examined (see Table 1). The solder wick and low-temperature wave solder methods were examined for solder ball removal. The solder ball preform method was selected for solder ball re-attachment.

In the low-temperature wave solder process, the component was held by a vacuum nozzle and suspended into the solder wave for a sufficient time to remove the solder balls (see Figure 2). The solder in the wave was eutectic SnPb and the wave temperature was 220°C. The low-temperature wave solder ball removal process requires special equipment, but it can be developed as an automated process suitable for high-volume conversion. It also resolves

978-1-4244-4658-2/09 $25.00 © 2009 IEEE

the handling issue due to the ability to automate the process. This method is much more controllable, but also more expensive, in comparison to the solder wick method.

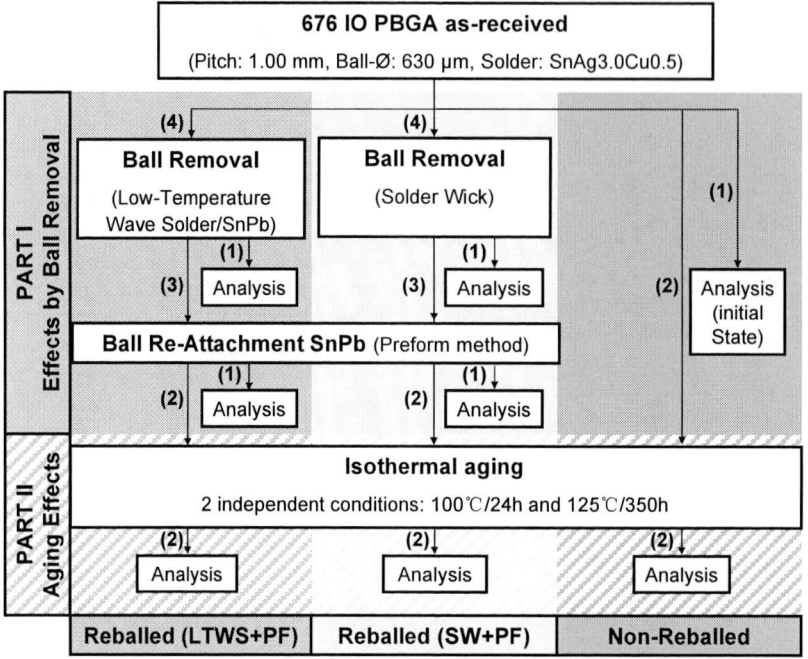

Figure 1 Overview of the test program carried out on interfacial IMC changes with different solder ball removal procedures (low-temperature wave solder "LTWS" and solder wick "SW": PART I) and isothermal aging (PART II). The numbers in parentheses indicate the number of samples for each step. After every step cross-sections were prepared for analysis

Table 1 Reballing process matrix

Part	Reballing Type	Solder Ball Removal Approach	Solder Ball Re-Attachment Approach
676 IO PBGA	LTWS+PF	Low-temperature wave solder	Preform method
676 IO PBGA	SW+PF	Solder wick	Preform method

In the solder wick process, the package is fixed with the solder balls facing upward. A soldering iron is used to heat a copper braid (wick), which is manually wiped (by tweezers) over the solder ball (see Figure 3). The braid wire melts the solder balls and picks up the molten solder. In this study, the solder iron temperature was set to 250°C. The solder wick method is inexpensive and good for low volume work.

During both solder ball removal processes, the temperature was measured by a K-type thermocouple. A hole was drilled at the plastic mold side to place a thermal couple above the die as shown in Figure 4. Figure 5 and Figure 6 show the thermal profiles measured for low-temperature wave solder and solder wick methods, respectively. The maximum package temperature measured at the die surface was 164.5°C for the low-temperature wave solder method, while the maximum package temperature was 180°C for the solder wick method. The low-temperature wave solder method has a shorter preheat and contact time than the solder wick method, which reduces the thermal shock and damage

to the components. The contact time for the low-temperature wave solder procedure was 18 seconds, while the contact time for the solder wick method was 42 seconds.

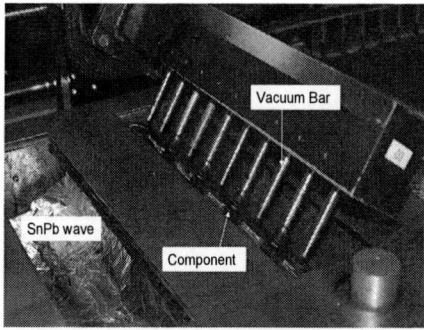

Figure 2 Solder ball removal by low-temperature wave solder

Figure 3 Solder ball removal by solder wick

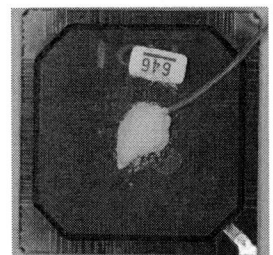

Figure 4 Thermocouple placement

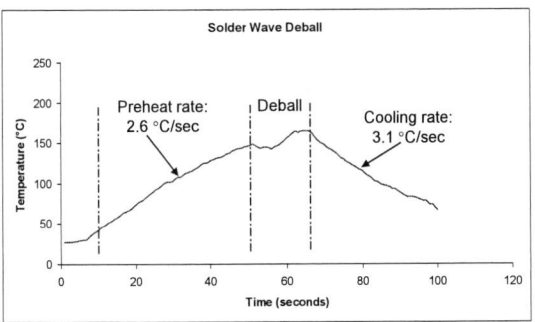

Figure 5 Thermal profile (above die) of low-temperature wave solder method

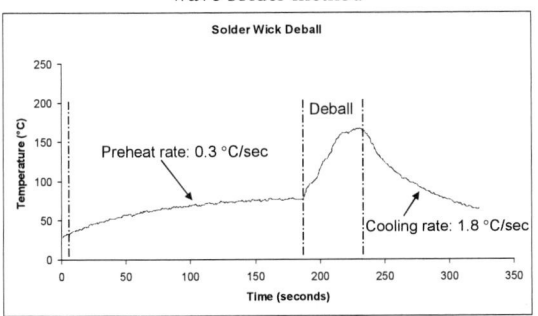

Figure 6 Thermal profile (above die) of solder wick method

The solder ball re-attachment procedure in the reballing process places new solder balls on the pads of the package. The preform method was selected in this study. The BGA preform technique, which was originally developed by Raychem, consists of an array of solder balls sandwiched between a lamination of carriers. Carriers control the pitch, progression, and alignment. Carriers are made of water-dispersible paper laminates that are easily removed after reflow. In the solder ball re-attachment procedure, the preform and component are matched by the fixture and reflowed together. Figure 7 shows the preform product. The

temperature profile above the die is also documented as shown in Figure 8. The maximum package temperature measured at the die surface was 213°C. The preheat rate and the cooling rate were 1.3°C/sec and 1.65°C/sec, respectively.

Figure 7 SolderQuik™ BGA Preform

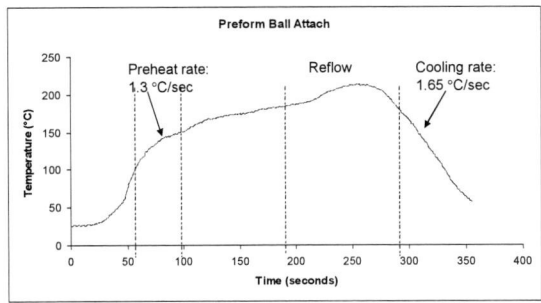

Figure 8 Thermal profile (above die) of preform method

To investigate the interfacial IMCs of reballed BGAs using different reballing methods, all the reballed BGA components were subjected to thermal aging. For purposes of comparison, non-reballed lead-free BGA components were also subjected to isothermal aging. Two isothermal aging conditions, 100°C for 24 hrs and 125°C for 350 hrs, were examined. Table 2 lists the sample matrix. The samples after solder removal and the samples after solder ball re-attachment, including the non-aged and aged ones, were cut, molded, and cross-sectioned. An environmental scanning electron microscope (FEI Quanta FEG ESEM 200) and an energy-dispersive X-ray spectroscopy (Oxford Instruments INCA 4.09) were used to investigate the composition of IMCs as well as the IMC growth.

Table 2 Sample matrix of non-reballed and reballed BGAs

676 I/O BGA	Non-Reballed			Reballed (LTWS+PF)			Reballed (SW+PF)		
	Non-aged	100°C 24 hrs	125°C 350 hrs	Non-aged	100°C 24 hrs	125°C 350 hrs	Non-aged	100°C 24 hrs	125°C 350 hrs

Results and Discussions

In order to investigate the interfacial IMC composition of reballed BGAs, it is necessary to investigate the interface after the solder ball removal procedure. Also, the investigation of the interfacial IMC in non-reballed SAC components helps to give a better understanding of the interface after solder ball removal. The interfacial IMC composition and IMC thickness of the reballed components were investigated and discussed in this study as well.

Interfacial IMC Analysis of Non-Reballed BGAs (Initial State)

As shown in a previous study [9]错误！未找到引用源。, Ni_3Sn_4 is the first phase formed and the only phase observed at the interface when Ni reacts with liquid Sn (<260°C). However, the IMC formation is different when the solder material contains Cu, i.e., SAC305 solder. Studies revealed that if the Cu concentration in the solder exceeded 0.3-0.4wt.%, the IMC formation at the interface was $(Cu,Ni)_6Sn_5$ [10]. The $(Cu,Ni)_6Sn_5$ IMC is based on the Cu_6Sn_5 lattice, and Ni dissolves into the lattice and replaces Cu atoms.

The non-reballed components were the BGAs with SAC305 solder balls. SAC305 solder reacted with the Ni pad during the soldering process. Since SAC305 solder contained 0.5wt.% Cu, the expected IMC formed at the interface after soldering was $(Cu,Ni)_6Sn_5$. There was only one layer of interfacial IMC that was found at the interface of non-reballed BGA, as shown in Figure 9. The energy-dispersive X-ray spectroscopy (EDS) results of the IMC formation at the interface is shown in Table 3. Based on the atom percentage of the different elements, the IMC was confirmed to be $(Cu,Ni)_6Sn_5$.

Table 3 EDS detection results of interfacial IMCs in non-reballed BGAs (non-aged)

Element	Ni	Cu	Sn
Atom%	33.00	20.49	46.50

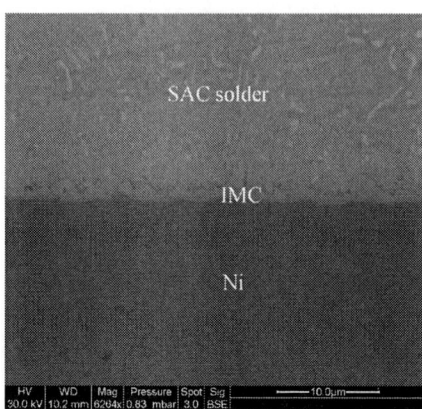

Figure 9 Non-reballed BGA interface (non-aged)

Interfacial Changes Due to Reballing Process

As mentioned in Section 2, the package surface may be damaged during the solder ball removal procedure. The surfaces of the packages were investigated after the solder removal procedure. Figure 10 shows an image of a pad using low-temperature wave solder without any damage. Damage to the solder mask induced by the solder wick method is shown in Figure 11.

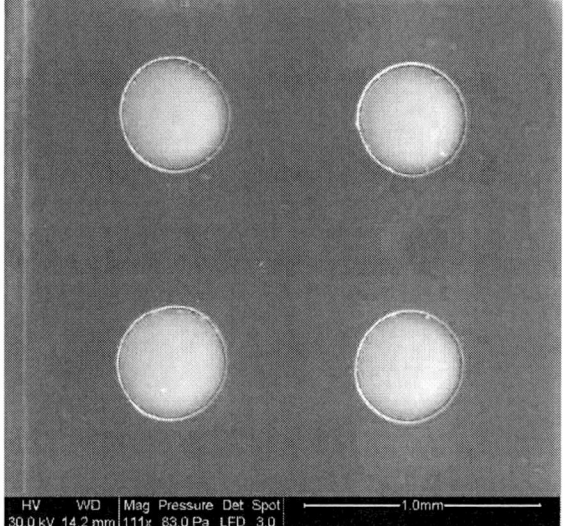

Figure 10 Good solder mask of a BGA after low temperature wave solder method

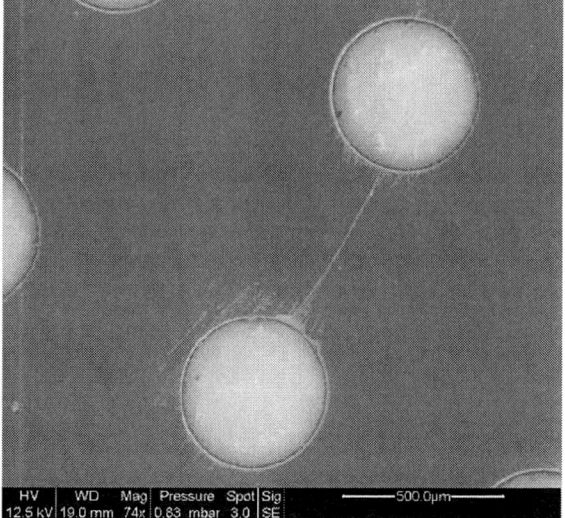

Figure 11 Solder mask damage on a BGA induced by the solder wick method

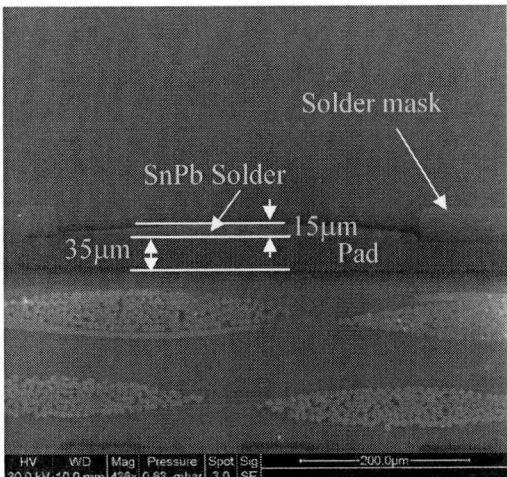

Figure 12 Cross-section of BGA pad area after solder ball removal using low-temperature wave solder method

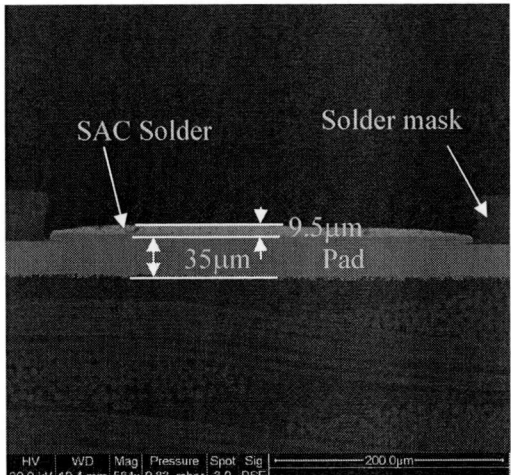

Figure 13 Cross-section of BGA pad area after solder ball removal using solder wick method

Two BGA components were cross-sectioned after the solder ball removal step to investigate the thickness and the composition of the residue solder. Figure 12 shows the cross-section of a component using the low-temperature wave solder method and Figure 13 shows the cross-section of a component using the solder wick method. The residue solder of a sample using the low-temperature wave solder method was SnPb with a thickness of 15 microns, and the residue solder of another sample using the solder wick method was SAC with a thickness of 9.5 microns. Since the parts went through a SnPb solder wave in the low-temperature wave solder removal procedure, the original SAC solder was removed and left a layer of SnPb solder on the package pad.

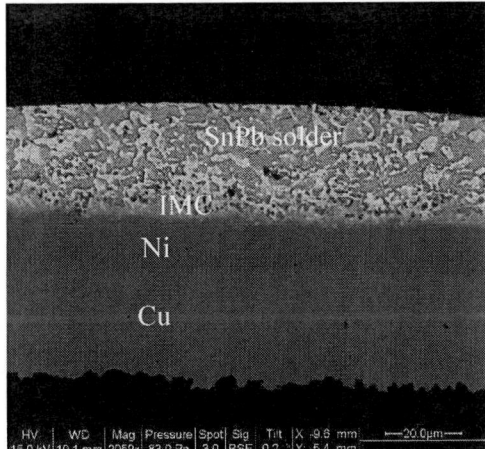

Figure 14 Interface of the BGA after low temperature solder wave method

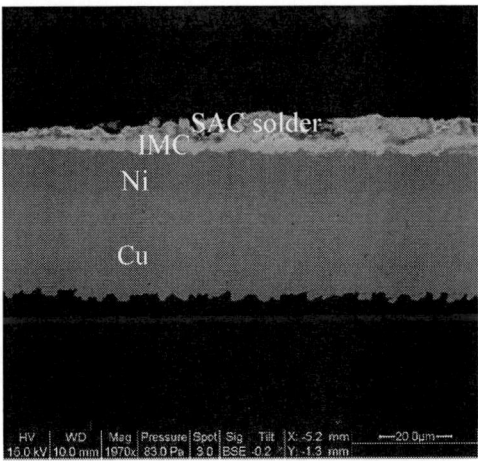

Figure 15 Interface of BGA after solder wick method

The BGA components after solder ball removal had different solder materials. Thus it was important to find out the interfacial IMCs composition to get a better understanding of reballed BGAs. Figure 14 and Figure 15 show the components after the solder removal procedure. Based on the EDS results shown in Table 4, the interfacial IMCs were ternary Cu-Ni-Sn IMCs, but the compositions were different.

Table 4 EDS detection results of interfacial IMC in the samples after solder ball removal procedure

Element (atom %)	Ni	Cu	Sn
Sample after low-temperature wave solder	26.84	24.23	48.94
Sample after solder wick	8.98	45.82	45.20

For the component using the low-temperature wave solder removal method, there was a layer of Cu-Ni-Sn IMC at the interface. During the attachment procedure, it was a four-layer structure reacted with each other at the interface for reballed (LTWS+PF) BGAs, SnPb solder ball/SnPb solder/Cu-Ni-Sn IMC/Ni pad. The composition of the interfacial IMCs in the reballed (LTWS+PF) BGAs (shown in Table 5) was close to $(Ni,Cu)_3Sn_4$. For the component using the solder wick removal method there was a layer of Cu-Ni-Sn IMC at the interface and a layer of SAC solder. During the attachment procedure, it was also a four-layer structure reacted with each other at the interface for reballed (SW+PF) BGAs, SnPb solder ball/SAC solder/Cu-Ni-Sn IMC/Ni pad. The composition of the interfacial IMCs in the reballed (SW+PF) BGAs was also shown in Table 5, which was close to $(Cu,Ni)_6Sn_5$.

Table 5 EDS detection results of interfacial IMCs of Reballed BGAs after solder ball re-attachment

Element (atom %)	Ni	Cu	Sn
Reballed (LTWS+PF) BGA	25.55	16.77	57.68
Reballed (SW+PF) BGA	24.35	29.76	45.89

In order to get a better understanding of the interfacial IMCs changes after the solder ball removal and solder ball re-attachment procedures, the EDS composition results of interfacial IMCs in the non-reballed components, the components after solder ball removal, and the components

after solder ball re-attachment were plotted, as shown in Figure 16.

There was a phase change in the component using the low-temperature wave solder method after the solder ball re-attachment procedure. The non-reballed SAC BGA (initial state) had an interface IMC composition close to $(Cu,Ni)_6Sn_5$. During the solder ball removal procedure, the SAC solder melted, and the interfacial IMC layer dissolved in liquid solder [11]. Liquid Sn reacted with Cu and Ni, which formed a new layer of interfacial IMC at the interface. After attaching the SnPb solder balls on the pads of the component, the Cu amount in the interfacial IMC layer dropped compared with the component after solder ball removal procedure, as shown in Figure 16. Because there was no Cu in the SnPb solder balls, there was a large driving force for the Cu dissolved into the bulk solder and formed Cu-Sn IMCs in the reflow process (see Figure 17). In the meantime, the interfacial IMC layer dissolved into the liquid solder, and Ni from the component pad kept diffusing into the interfacial IMC layer, which formed a new layer of interfacial IMC with a composition close to $(Ni,Cu)_3Sn_4$. The main reason for the phase change at the interface is the Cu amount: there was not enough Cu (0.3-0.4wt.%) to maintain $(Cu,Ni)_6Sn_5$. The interfacial IMCs changed to a more stable state, $(Ni,Cu)_3Sn_4$, which was based on the Ni_3Sn_4 lattice with Cu substituting for the Ni atoms.

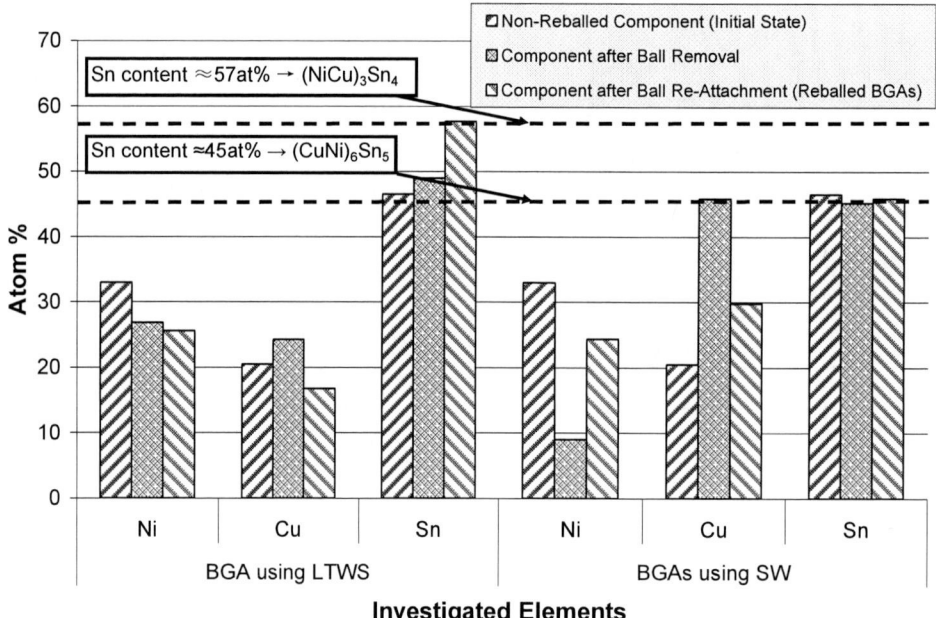

Figure 16 Element percentage of the interfacial IMCs in theBGAs after the solder ball removal and solder ball re-attachment procedures

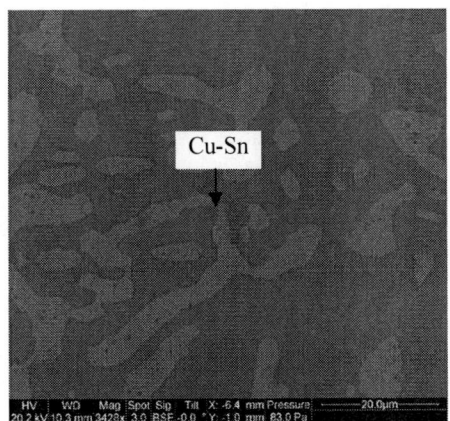

Figure 17 Bulk solder of Reballed (LTWS+PF) BGA

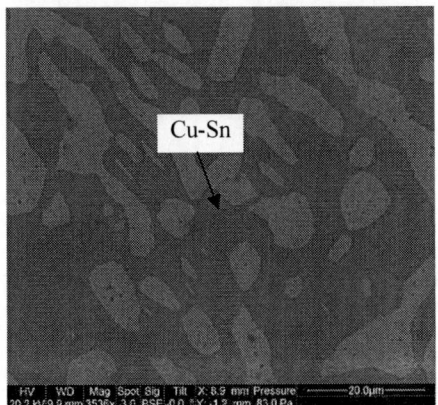

Figure 18 Bulk solder of Reballed (SW+PF) BGA

For the components using the solder wick method, the Cu amount was found to increase dramatically after the solder ball removal procedure. This was because a copper braid was used to remove the melted solder in the solder wick method. The Cu dissolved into the liquid Sn during the solder ball removal step, and the solder wick method gave enough contact time (42 seconds) for the Cu dissolution. Thus, the composition of the interfacial IMCs was close to $(Cu,Ni)_6Sn_5$ but with a high Cu concentration. In the solder ball re-attachment procedure, because SnPb solder balls contained no Cu, the Cu in the interfacial IMC

layer dissolved into the bulk solder to form Cu-Sn IMCs (see Figure 18). This also explained why the Cu amount in the interfacial IMC decreased after the solder ball re-attachment procedure. However, the interfacial phase did not change in the whole reballing process, even though the Cu and Ni concentration changed in the ball removal and ball re-attachment procedures.

Interfacial IMC Analysis under Isothermal Aging Condition

Both the reballed and non-reballed BGAs were subjected to isothermal aging conditions to evaluate the interfacial IMC growth. There were two isothermal aging conditions used in this study, 100°C for 24 hours and 125°C for 350 hours. As discussed in Section 3.2, for the reballed BGAs using different solder ball removal methods, the interfacial phase changed. EDS detection was used to document the interfacial IMC composition of reballed BGAs in order to study the effect of isothermal aging on the interfacial phase change. The interfacial IMC composition of non-reballed BGAs was also studied as a reference. Table 6 lists EDS results of interfacial IMC composition of non-reballed and reballed BGAs. In order to get a better understanding of the interfacial IMC changes.

For the non-aged reballed LTWS+PF BGAs, the interfacial IMC composition was close to $(Ni,Cu)_3Sn_4$. For the aged reballed LTWS+PF BGAs, there was only one layer of interfacial IMC with a composition close to $(Ni,Cu)_3Sn_4$. However, the composition of the interfacial IMC layer remained the same even after aging. A similar situation was found in reballed SW+PF BGAs, where the interfacial phase, which was close to $(Cu,Ni)_6Sn_5$, stayed the same even after aging. Only one layer of interfacial IMC was found in non-aged and aged samples.. But the Cu amount of the interfacial IMC decreased slightly as aging time increased. Figure 19 shows the interfacial IMCs in the non-aged reballed SW+PF BGA.

For the non-reballed BGAs, the interfacial IMC composition was close to $(Cu,Ni)_6Sn_5$, which was the only phase found even after aging for 350 hours at 125°C. However, the Cu amount in the interfacial IMC almost remained at the same level as the aging time increased, which was different than what occurred in the reballed BGAs. As aging time increased, the interfacial IMC layer grew and became thicker.

Table 6 EDS detection results of interfacial IMCs of reballed and non-reballed BGAs

Element (atom %)		Ni	Cu	Sn
Reballed (LTWS+PF)	Non-Aged	25.55	16.77	57.68
	100°C/24hrs	26.60	16.14	57.26
	125°C/350hrs	26.83	15.28	57.89
Reballed (SW+PF)	Non-Aged	24.35	29.76	45.89
	100°C/24hrs	26.14	27.93	45.93
	125°C/350hrs	32.71	24.76	43.53
Non-Reballed	Non-Aged	33.00	20.49	46.50
	100°C/24hrs	30.48	24.22	44.54
	125°C/350hrs	29.73	24.45	45.81

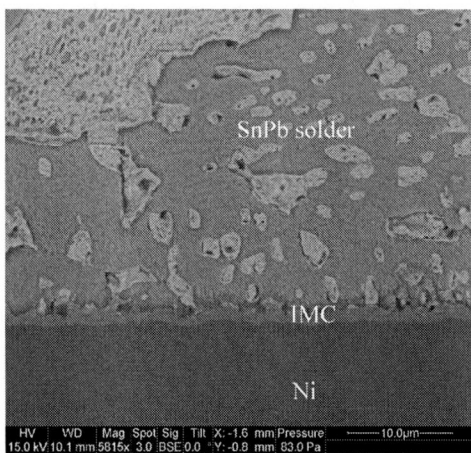

Figure 19 Reballed (SW+PF) BGA interface (non-aged)

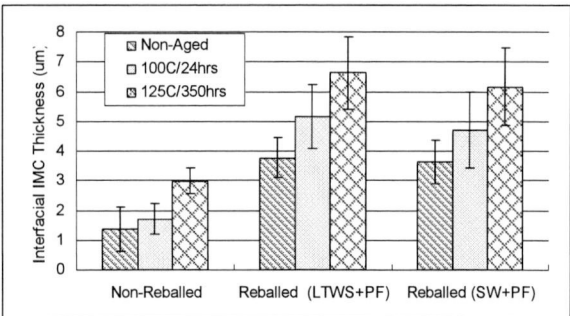

Figure 20 Interfacial IMC thickness of non-reballed and reballed BGAs

The IMC composition was investigated, and the interfacial IMC thickness was documented. The interfacial thicknesses of IMCs of non-reballed and reballed parts are plotted in Figure 20. For each sample, 27 measurements were calculated. The mean and standard deviations were plotted.

The non-reballed parts had an interfacial IMC thickness of $2.98\pm0.44\mu m$ after aging at $125°C$ for 350 hours. However, the reballed parts had a much thicker interfacial layer due to multiple thermal procedures in the reballing process. In the solder ball removal and solder ball attachment procedures, the solder was in a liquid state, which could react much faster than a solid state. For reballed LTWS+PF and reballed SW+PF components, although the IMC compositions were different, the interfacial IMC layer thicknesses were equivalent to each other.

Conclusion

In this paper, the interfacial IMC composition of the reballed BGAs using different reballing method was investigated. Here, 676 I/O BGAs with SAC305 solder balls were subjected to the reballing process. Two solder ball removal methods were used in this study, and the reballed BGAs had eutectic SnPb solder balls. The components using the solder wick solder removal method had a layer of SAC solder on the pad, while the components using the low-temperature wave solder method had a layer of SnPb solder on the pad. The interfacial IMCs of the parts were investigated after solder removal procedure, and the ternary Cu-Ni-Sn IMCs were found. However, the Ni concentration in the Cu-Ni-Sn IMC of the part using the solder wick method was lower than that of the part using the low-temperature wave solder method.

The components using the low-temperature wave solder method had phase change after the ball removal and ball re-attachment procedures. There was a layer of $(Ni,Cu)_3Sn_4$ at the interface in the reballed (LTWS+PF) components after reflow. For the components using the solder wick method, there was no phase change at the interface after the ball removal and re-attachment procedures. However, the Cu concentration in the interfacial IMCs changed dramatically after the solder ball removal procedure because of the copper dissolution from the copper braid used in the solder wick method. In both types of reballed BGAs, the Cu concentration decreased after the solder ball re-attachment procedure. The reason was that the Cu dissolved into bulk solder and formed Cu-Sn IMCs during the reflow procedure.

After isothermal aging, there was no phase change at the interface in the non-reballed and the reballed BGAs. The non-reballed BGA has a interfacial IMC layer with the composition as $(Cu,Ni)_6Sn_5$. The interfacial IMC composition in the reballed (LTWS+PF) and the reballed (SW+PF) was $(Ni,Cu)_3Sn_4$ and $(Cu,Ni)_6Sn_5$ respectively. The reballed BGAs had a much thicker layer of IMCs than the non-reballed BGAs. The interfacial IMCs of the non-reballed and reballed BGAs grew after being subjected to isothermal aging. For reballed (LTWS+PF) and reballed (SW+PF) components, although the IMC compositions were different, the interfacial IMC layer thicknesses were equivalent to each other.

Acknowledgements

The authors would like to thank Russ Winslow and Minerva Cruz from Six Sigma for performing the reballing operator on the lead-free ball grid arrays in this study and sharing their extensive knowledge on the subject. The authors would also like to thank the members of the CALCE Electronic Products and Systems Consortium for sponsoring the work presented in this paper. The authors appreciate Mark Zimmerman for reviewing and editing this paper.

References

1. S. Ganesan and M. Pecht, Lead-free Electronics, New York: John Wiley and Sons Inc., 2006.
2. M. Pecht, Y. Fukuda, and S. Rajagopal, "The Impact of Lead-free Legislation Exemptions on the Electronics Industry," *IEEE Transactions on Electronics Packaging Manufacturing*, Vol. 27, No. 4, pp. 221-232, October 2004.
3. R. Ciocci and M. Pecht, "Questions Concerning the Migration to Lead-free Solder," *Circuit World*, Vol. 30, No. 2, pp. 34-40, 2004.
4. L. Nie, M. Pecht, and R. Ciocci, "Regulations and Market Trends in Lead-free and Halogen-free Electronics," *Circuit World*, Vol. 33, No. 2, pp. 4-9, 2007.
5. L. Nie, M. Osterman, M. Pecht, F. Song, J. Lo and S. K. Lee, "Solder Ball Attachment Assessment of Reballed Plastic Ball Grid Array Packages," *Proceedings of IPC*

APEX EXPO 2008, Las Vegas, NV, March 30– April 3, 2008.

6. F. Song and S.W.R. Lee, "Reliability Assessment on the Re-Balling of PBGA from SnPb to Pb-free Solder Spheres," *Proceedings of the 1st Electronics System Integration Technology Conference*, Vol. 1, pp. 474-480, September 2006.

7. A. Choubey, M. Osterman, and M. Pecht, "Microstructure and Intermetallics Formation in SnAgCu BGA Components Attached with SnPb Solder under Isothermal Aging," *IEEE Transactions on Device and Materials Reliability*, Vol. 8, Issue 1, pp. 160-167, 2008.

8. A. Choubey, H. Yu, M. Osterman, M. Pecht, F. Yun, L. Youghong, and X. Ming, "Intermetallics Characterization of Lead-free Solder Joints under Isothermal Aging," *Journal of Electronic Materials*, Vol. 37, No. 8, pp. 1130-1138, August 2008.

9. T. Laurila, V. Vuorinen, and J. K. Kivilahti, "Interfacial Reaction between Lead-free Solders and Common Base Materials," *Materials Science and Engineering*, R 49, pp. 1-60, 2005.

10. C. E. Ho , R. Y. Tsai , Y. L. Lin , C. R. Kao, "Effect of Cu concentration on the reactions between Sn-Ag-Cu solders and Ni," *Journal of Electronic Materials*, Vol. 31 No. 6, pp. 584-590, June 2002

11. D.Q. Yu, C.M.L. Wu, C.M.T. Law, L. Wang and J.K.L. Lai, "Intermetallic compounds growth between Sn–3.5Ag lead-free solder and Cu substrate by dipping method," *Journal of Alloys and Compounds*, Vol. 392, pp. 192-199, 2005.

Effect of Cu Addition in Sn-containing Solder Joints on Interfacial Reactions with Au Foils

Wei Liu[1, 2*], Chunqing Wang[1], Lining Sun[2], Yanhong Tian[1], Yarong Chen[1]
[1]Department of Electronics Packaging Technology,
State Key Lab of Advanced Welding Production Technology,
[2]Mechanical Engineering Post-Doctoral Mobile Research Station,
436#, Harbin Institute of Technology,
92 West Dazhi Street, Harbin 150001, P.R. China

Abstract

Pure Sn, Sn-0.7Cu, Sn-1.5Cu and Sn-3.0Cu solder joints were fabricated by a laser reflow soldering method, and then they were aged at 125°C for various periods. After 600 hours treatment, we found that thickness of Intermetallic Compounds (IMCs) in the pure Sn solder joints was 54μm. However, in the solder joints with Cu addition, formation of $AuSn_4$ IMCs was inhibited, and other two kinds of Sn-Cu-Au ternary phases were found in stead. In the Sn-3.0Cu solder joints, the IMCs and phases were only 34μm in thickness, moreover, growth of the IMCs and phases did not follow the parabolic law.

1. Introduction

Au films have been commonly used as protective layers on pads due to their excellent antioxidant ability and wettability for Sn-bearing solder alloys [1-3]. In order to protect other metals beneath them from oxidation, the Au films must keep certain thickness. Nowadays, the size of electronical devices is getting smaller and smaller. As a result, the content of Au in the solder joints will be increased greatly. Moreover, in some thin film circuits found in MEMS and micro-sensors, entire pads and circuits are totally made of Au, of which the Au is used as a soldering layer, and thickness of the Au layers can reach about several microns.

Generally, Au on pads will react with Sn in solders rapidly to form Au-Sn IMCs during reflow or aging processes. If the content of the IMCs in solder joints exceeds a certain value, reliability of the solder joints will be degraded dramatically [1,4-5]. Therefore, it is critical to find ways to inhibit formation and growth of the Au-Sn IMCs.

Kim and Tu found that reaction speed between Au foils and liquid solders (95Pb-5Sn, 63Sn-37Pb, pure Sn, 96Sn-4Ag, 57Bi-43Sn and 77.2Sn-20In-2.8Ag) is quite different [6]. Mita et al. found that Ni addition in Sn-bearing solder alloys will accelerate growth of the IMCs at the interfaces of the alloys and Au foils [7]. Reaction kinetics between Au and Sn-xwt.%Si solder alloys has been reported by Furuto and Kajihara [8]. They found that Si addition in the alloys can restrain growth of the Au-Sn IMCs. Zn in the Sn-Zn or Sn-Zn-Bi alloys will react with Au on the pads to form Au-Zn IMCs has been reported widely [9-14]. Results of the studies described above remind us that formation and growth of Au-Sn IMCs may be restrained effectively by introducing certain alloying elements into the Sn-bearing solders.

Yen et al. reported that Cu addition in the Sn-bearing solder alloys can inhibit formation of the Au-Sn IMCs [15]. The maximum amount of Cu addition in their study was 1.0wt.%. They supposed that the mechanism of IMCs' growth can be described by the parabolic law. However, to authors' knowledge, effects of high concentration of Cu addition on formation and growth of Au-Sn IMCs in Sn-bearing solder joints have not been reported.

In this study, we investigated the effect of 0.7, 1.5 and 3.0wt.%Cu addition in Sn-bearing solders on interfacial reactions between the solders and Au foils during an isothermal aging process. Variation of the IMCs and phases' thickness at the interfaces of the solders and Au foils was also inspected. The results could be helpful to restrain formation and growth of Au-Sn IMCs, and improve reliability of solder joints with high Au content.

2. Experimental procedures

Cylinders (1.5mm in diameter, 0.5mm in thickness) of pure Sn and Sn-xwt.%Cu (x=0.7, 1.5 and 3.0) solders, and commercial Au foils with a dimension of 2mm × 2mm ×0.1mm were used in this study. The solder cylinders and Au foils were ultrasonically cleaned by acetone first, and then the solder cylinders were dipped into a soluble flux, afterward, they were placed onto the Au foils.

A continuous wave (CW) Nd: YAG laser, of which the beam is 0.38mm in diameter, was applied to fabricate the solder joints. In order to control the formation of void defects and IMCs within the specimens, laser reflow parameters for the pure Sn and Sn-xwt.%Cu solder joints were determined through experiments as 11W, 590ms and 11W, 580ms, respectively. The solder joints were then aged at 125°C for 96, 216, 384 and 600 hours.

The solder joints were sectioned and mounted in epoxy resin, then they were grounded, polished and etched by a mixed solution containing 0.3ml concentrated HCl, 2.9ml concentrated HNO_3, 84.5ml C_2H_5OH and 12.3ml H_2O. Scanning Electron Microscopy (SEM) was carried out to analyze the microstructure of the solder joints. Energy Dispersive X-ray Spectrometer (EDX) was utilized to identify element composition of the IMCs or phases within the specimens.

The following process was used to find out thickness of the IMCs or phases within the specimens. First, total areas of the IMCs or phases near the interface of the solder and Au foil were calculated by an image processing software, then

---

*Corresponding author. Tel.: 86-451-86418359; Fax: 86-451-86416186.
E-mail address: w_liu@hit.edu.cn, liuw@eptech.hit.edu.cn

divided by length of the interface, thus the value would be thickness of the IMCs or phases. Six data points from each aging condition were collected to calculate average thickness of the IMCs or phases.

3. Results and discussion

Fig. 1 shows cross-sections of the solder joints fabricated by the laser reflow soldering process. In the pure Sn solder joints, $AuSn_4$ and Au-rich IMCs layers were formed at the interface of the solder and Au foil. Apart from the $AuSn_4$ IMCs layer, platelet-like and nubbly $AuSn_4$ IMCs were found within the solder as well (Fig. 1(a)). In the Sn-0.7%Cu solder joints, $AuSn_4$ IMCs can also be found at the interface and within the solder. Little Cu containing IMCs or phases were observed within the solder joints (Fig. 1(b)). When the content of Cu addition was increased to 1.5wt.%, Sn-Cu-Au ternary phases in nubbly morphology were detected near the interface of the solder and foil. In addition, within the specimen, Cu_6Sn_5 IMCs which located about 40μm away from the interface were observed. The Cu_6Sn_5 IMCs were covered by a thin layer of Sn-Cu-Au ternary phases, which showed a smaller gray scale than that of the Cu_6Sn_5 IMCs (Fig.1 (c)). In the Sn-3.0Cu solder joints, Cu_6Sn_5 IMCs with larger size (approximate 5μm in diameter) as compared with those in the Sn-1.5Cu solder joints were formed and also covered with a layer of Sn-Cu-Au ternary phases (Fig.1 (d)).

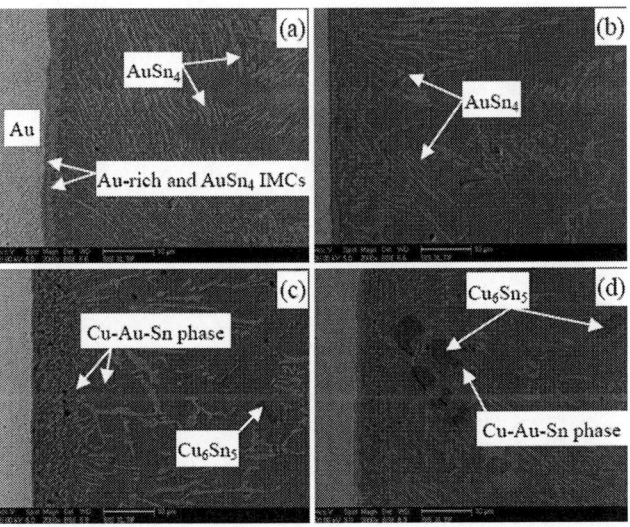

Fig. 1 Interface of solder joints fabricated by the laser reflow process: (a) pure Sn (b) Sn-0.7Cu (c) Sn-1.5Cu (d) Sn-3.0Cu

Formation and growth characteristics of Au-Sn IMCs in the pure Sn solder joints are provided in Fig.2. With 96hours treatment, $AuSn_4$, $AuSn_2$ and AuSn IMCs layers were found at the interface. Thickness of the IMCs layers was 28, 4 and 3μm, respectively. In addition, the shape of $AuSn_4$ IMCs away from the interface changed from platelet-like to nubbly (Fig.2 (a)). With the increment of aging time, thickness of the IMCs layers was also increased (Fig. 2 (b) and (c)). After 600hours aging treatment, thickness of the $AuSn_4$ IMCs and total thickness of the Au-Sn IMCs reached 40 and 54μm, respectively (Fig.2 (c)).

Formation and growth characteristics of IMCs and phases of the Sn-bearing solder joints with Cu addition were also studied and shown from Fig. 3 to Fig. 5.

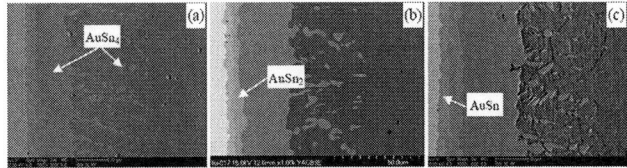

Fig. 2 Interface of the pure Sn solder and Au foil aged at 125°C for different time: (a) 96hours (b) 216hours (c) 600hours

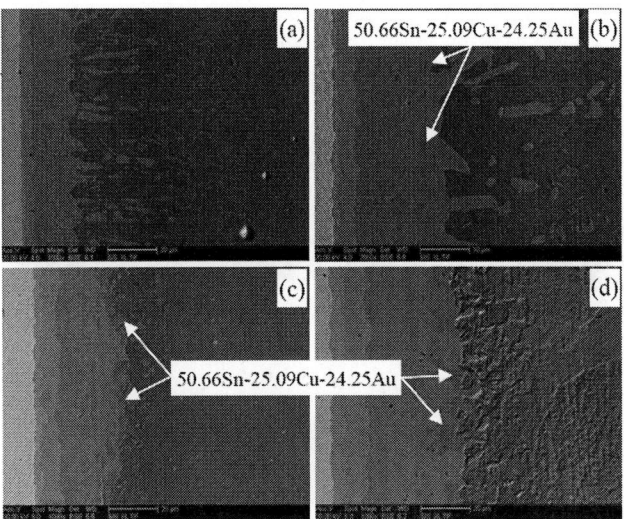

Fig. 3 Interface of the Sn-7.0Cu solder and Au foil aged at 125°C for different time: (a) 96hours (b) enlarged image of the solder joints aged for 96hours (c) 384hours (d) 600hours

The results from the Sn-0.7 solder joints were provided in Fig.3. Similarly with the pure Sn solder joints, $AuSn_4$, $AuSn_2$ and AuSn IMCs layers were also detected at the interface after 96hours treatment. Moreover, many 50.66Sn-25.09Cu-24.25Au (in atom percentage) ternary phases in nubbly morphology were observed within the $AuSn_4$ IMCs layer and solder (Fig. 3 (a) and (b)). Total thickness of the $AuSn_4$ IMCs and ternary phases was only about 21μm. With the increase of aging time, more 50.66Sn-25.09Cu-24.25Au ternary phases with larger sizes as compared with those in the solder joints aged for 96hours were congregated near the interface or within the $AuSn_4$ IMCs layer (Fig. 3 (c) and (d)). As the aging period was extended to 600hours, total thickness of the $AuSn_4$ IMCs and ternary phases was increased to 38μm (Fig. 3 (d)).

Fig.4 shows the results from the Sn-1.5Cu solder joints aged for different time. 50.66Sn-25.09Cu-24.25Au ternary phases were observed within the $AuSn_4$ IMCs layer after 96hours treatment (Fig.4 (a)). With the increase of aging time, parts of the $AuSn_4$ layer were replaced by the 50.66Sn-25.09Cu-24.25Au ternary phases. In addition, $AuSn_4$ IMCs which located away from the interface disappeared, and Sn-Cu-Au ternary phases with larger gray scale than those of the 50.66Sn-25.09Cu-24.25Au ternary phases formed instead. Composition of the ternary phases was confirmed to be 48.23Sn-38.77Cu-13.00Au. As the aging time was increased to 600hours, total thickness of the AuSn₄ IMCs and ternary phases was 26μm.

The formation and evolution process of IMCs and phases in the Sn-3.0Cu solder joints are shown in Fig.5. After 96hours aging treatment, the Cu_6Sn_5 IMCs were completed covered by a thin layer of Sn-Cu-Au ternary phases, which exhibit similar gray scale as those of the 48.23Sn-38.77Cu-13.00Au ternary phases. With the increment of aging time, both amount and sizes of the Cu_6Sn_5 IMCs decreased. On the contrary, sizes of the 48.23Sn-38.77Cu-13.00Au ternary phases were increased obviously. As the aging time was increased to 600hours, all Cu_6Sn_5 IMCs were consumed to form 48.23Sn-38.77Cu-13.00Au ternary phases. Moreover, parts of the 48.23Sn-38.77Cu-13.00Au ternary phases, which were close to the AuSn₄ IMCs, changed to 50.66Sn-25.09Cu-24.25Au ternary phases. Total thickness of the AuSn₄ IMCs and ternary phases was only 15μm.

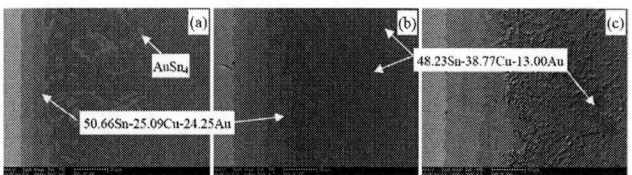

Fig. 4 Interface of the Sn-1.5Cu solder and Au foil aged at 125°C for different time: (a) 96hours
(b) 384hours (c) 600hours

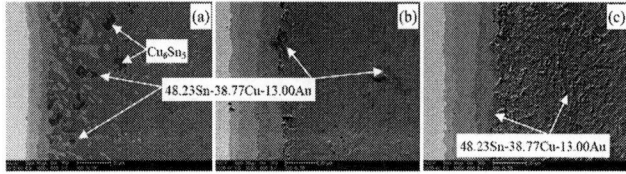

Fig. 5 Interface of the Sn-3.0Cu solder and Au foil aged at 125°C for different time: (a) 96hours
(b) 384hours (c) 600hours

A plot of total thickness for the IMCs and phases versus square root of aging time is demonstrated in Fig.6. The result indicated that total thickness of the IMCs and phases in the pure Sn, Sn-0.7Cu and Sn-1.5Cu solders fit linear relationship with square root of aging time. That means growth of the IMCs and phases could be described by the following parabolic law:

$$d = d_0 + \sqrt{Dt}$$

where t is aging time; d is total thickness of the IMCs and phases in the solder joints aged for the period of t; d_0 is thickness of the IMCs and phases in the laser reflowed solder joints; D is growth coefficient of the IMCs and phases. The coefficient D in the pure Sn, Sn-0.7Cu and Sn-1.5Cu solder joints was determined by the least-squares method. The values were 9.19E-16, 7.83E-16 and 6.79E-16m²/s, respectively. It can be seen that the D values in the Sn-0.7Cu and Sn-1.5Cu solder joints were decreased 14.8% and 26.1% as compared with that in the pure Sn solder joints. However, growth of the IMCs and phases in the Sn-3.0Cu solder joints did not follow the parabolic law. Total thickness of the IMCs

and phases decreased from 40μm in the condition of 384hours aging to 34μm in the condition of 600hours aging. The restraining effect of 3.0wt.%Cu addition in the Sn-bearing solder on the increase of the IMCs and phases' thickness was the most remarkable among the Sn-bearing solders.

The results shown above indicated that Sn in the solder alloys normally reacts with Au foils to form AuSn₄, AuSn₂ and AuSn IMCs during the aging process. However, when Cu is added into the Sn-bearing solders, Au which possibly comes from the decomposition of AuSn₄ IMCs will be involved into the interactions with Cu and Sn to form Sn-Cu-Au ternary phases. Because Cu_6Sn_5 IMCs can supply enough Cu during the reflow or at the beginning of the aging processes, the Au will react with Cu and Sn to form 48.23Sn-38.77Cu-13.00Au ternary phases. When the Cu_6Sn_5 IMCs are consumed, or supplement of Au is superfluous, the 48.23Sn-38.77Cu-13.00Au ternary phases will react with Au to form 50.66Sn-25.09Cu-24.25Au ternary phases. By comparing the results of the present and previous studies [15], it can be inferred that the Sn-Cu-Au ternary phases have a higher driving force for formation than that of AuSn₄ IMCs, and the phases are more stable than AuSn₄ IMCs in the reflow or aging conditions.

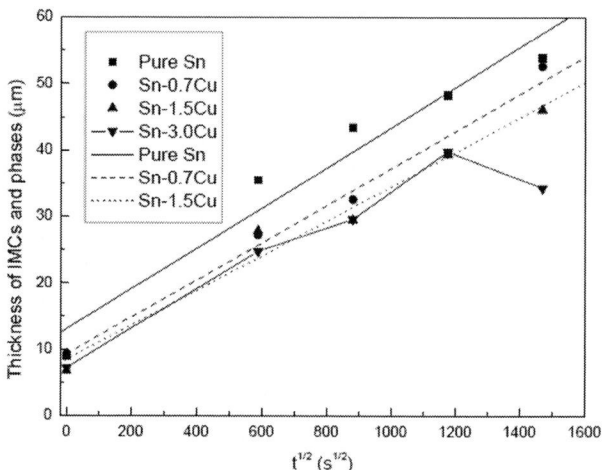

Fig. 6 a plot of total thickness for the IMCs and phases versus square root of the aging time

In the solder with 0.7wt.%Cu addition, because supplement of Au near the interface of the solder and AuSn₄ IMCs was ample, the small Cu_6Sn_5 IMCs would be consumed to form 50.66Sn-25.09Cu-24.25Au ternary phases during reflow or at the beginning of aging processes. With the increase of aging time, aggregation of 50.66Sn-25.09Cu-24.25Au ternary phases near the interface would occur, and then the phases would be surrounded by the growing Au-Sn IMCs. When the content of Cu addition was increased to 1.5wt.%, 50.66Sn-25.09Cu-24.25Au and 48.23Sn-38.77Cu-13.00Au ternary phases were formed near the interface of the solder and Au-Sn IMCs in the reflow process. With the increment of aging time, the 48.23Sn-38.77Cu-13.00Au ternary phases near the interface would react with Au to form 50.66Sn-25.09Cu-24.25Au phases. However, the 48.23Sn-

38.77Cu-13.00Au ternary phases located away from the interface and would not change to the 50.66Sn-25.09Cu-24.25Au ternary phases totally because of the deficient supplement of Au.

In the Sn-3.0Cu solder joints, many Cu_6Sn_5 IMCs were formed near the interface of the solder and $AuSn_4$ IMCs. Moreover, parts of them might even contact with the $AuSn_4$ IMCs. In the earlier aging process, Au would react with the Cu_6Sn_5 IMCs to form 48.23Sn-38.77Cu-13.00Au ternary phases, which would cover the surface of the Cu_6Sn_5 IMCs. With increasing of aging time, 48.23Sn-38.77Cu-13.00Au ternary phases would be formed continuously until the Cu_6Sn_5 IMCs were consumed. The 48.23Sn-38.77Cu-13.00Au ternary phases would react with Au to form 50.66Sn-25.09Cu-24.25Au ternary phases in succession. As a result, the large amount of Cu_6Sn_5 IMCs and 48.23Sn-38.77Cu-13.00Au ternary phases near the interface would consume Au continuously during the aging process. As for the $AuSn_4$ and Cu_6Sn_5 IMCs, 1.0mol Au and 1.0mol Cu in the IMCs will occupy 4.0mol and 0.83mol Sn, respectively. However, if Cu, Au and Sn evolve to the 50.66Sn-25.09Cu-24.25Au ternary phases, 1.0mol Au and 1.0mol Cu can only consume about 2.0mol Sn. Accordingly, amount of the consumed Sn will be reduced approximately 59% because of the 50.66Sn-25.09Cu-24.25Au ternary phases' formation. This may be the reason why total thickness of the Au-Sn IMCs and phases stopped increasing in the condition of 384 hours aging, even decreased in the condition of 600hours aging, and growth of the IMCs and phases did not follow the parabolic law.

Conclusions

The Sn-bearing solder joints with different content of Cu addition were fabricated by the laser reflow soldering method in the present study. After the following treatment at 125°C for various periods, 50.66Sn-25.09Cu-24.25Au and 48.23Sn-38.77Cu-13.00Au ternary phases were formed in the solder joints. Meanwhile, formation of $AuSn_4$ IMCs was inhibited greatly and its extent became obviously with increasing of the content of Cu addition. As for the Sn-3.0Cu solder joints concerned, increasing of the IMCs and phases' thickness did not follow the parabolic law. The possible reason is that the large amount of Cu_6Sn_5 IMCs near the interface of the solder and Au foil will consume Au continuously to produce Sn-Cu-Au ternary phases during the aging process. Moreover, the formation process of 50.66Sn-25.09Cu-24.25Au ternary phases will release much Sn. As a result, total thickness of the Au-Sn IMCs and phases would stop increasing during the extended aging periods.

Acknowledgments

This work is financially supported by the National High-tech R&D Program (863 Program) of China, Grant No. 2007AA04Z314 and the National Natural Science Foundation of China under grant No. 50675047/ E052105.

References

1. Jacobson D. M., jumpston G., "Gold coatings for fluxless soldering," *Gold. Bull*, Vol. 22, No.1 (1989), pp. 9-18.
2. Zeng K., Tu K. N., "Six cases of reliability study ofPb-free solder joints in electronic packaging technology," *Materials Science and Engineering*, R38 (2002), pp. 55-105.
3. Kim P. G., Tu K. N., "Morphology of wetting reaction of eutectic SnPb solder on Au foils," *J. Appl. Phys*, Vol. 80, No.7 (1989), pp. 3822-3827.
4. Mei Z., kaufmann M., Johnson P., "Brittle interfacial fracture of PBGA Packages soldered on electroless nickel/immersion," Proceedings of the 48th Electronic Components and Technology Conference, Seattle, Washington, USA, Piscataway, USA, IEEE, May. 1998, PP. 952-961.
5. Hung S. C., Zheng P. J., Ho S. H., Lee S. C., Chen H. N., Wu J. D., "Board level reliability of PBGA using flex substrate," *Microelectron. Reliab*, Vol. 41, (2001), pp. 677-687.
6. Kim P. G., Tu K. N., "Fast dissolution and soldering reactions on Au foils," *Mater. Chem. Phys*, Vol. 53, (1998), pp. 165-171.
7. Mita M., Miura K., Takenaka T., kajihara M., Kurokawa N., Sakamoto K., "Effect of Ni on reactive diffusion between Au and Sn at solid-state temperatures," *Mat. Sci. Eng. B-Solid*, Vol. 126, (2006), pp. 37-43.
8. Foruto A., Kajihara M., "Influence of Si on reactive diffusion between Au and Sn at solid-state temperatures," *Mat. Sci. Eng.*, A-Struct, Vol. 445-446, (2007), pp. 604-610.
9. Chang S. C., Lin S. C., Hsieh K. C., "Phase reaction in Sn–9Zn solder with Ni/Au surface finish bond-pad at 175 °C ageing." *J. Alloy. Compd*, Vol. 428, (2007), pp. 179-184.
10. Kim K. S., Yang J. M., Yu C. H., Jung I. O., Kim H. H., "Analysis on interfacial reactions between Sn–Zn solders and the Au/Ni electrolytic-plated Cu pad," *J. Alloy. Compd.*, Vol. 379, (2004), pp. 314-318.
11. Yu C. H., Kim K. S., Kim H. I., Jeon H. J., "Influence of interfacial reaction layer on reliability of chip-scale package joint from using Sn-37Pb and Sn-8Zn-3Bi solder," *Journal of ELECTRONIC MATERIALS*, Vol. 34, No.2 (2005), pp. 161-167.
12. Lee C. Y., Yoon J. W., Kim Y. J., Jun S. B., "Interfacial reactions and joint reliability of Sn–9Zn solder on Cu or electrolytic Au/Ni/Cu BGA substrate," *Microelectronic Engineering*, Vol. 82, (2005), pp. 561-568.
13. Sharif A., Chan Y. C., "Retardation of spalling by the addition of Ag in Sn–Zn–Bi solder with the Au/Ni metallization," *Materials Science and Engineering A*, Vol. 445-446, (2007), pp. 686-690.
14. Yoon J. W., Jun S. B., "Solder joint reliability evaluation of Sn–Zn/Au/Ni/Cu ball-grid-array package during aging," *Materials Science and Engineering A*, Vol. 452-453, (2007), pp. 46-54.
15. Yen Y. W., Jao C. C., Hsiao H. M., Lin G. Y., Lee C., "Investigation of the Phase Equilibria of Sn-Cu-Au Ternary and Ag-Sn-Cu-Au Quaternary Systems and Interfacial Reactions in Sn-Cu/Au Couples," *J. Electron. Mater*, Vol. 36, (2007), pp. 147-158.

Evolution of Ag₃Sn Compounds in Solidification of Eutectic Sn-3.5Ag Solder

Hwa-Teng Lee[1], Yin-Fa Chen[1], Ting-Fu Hong[2], Ku-Ta Shih[1]

[1]Department of Mechanical Engineering, Cheng Kung University,
1, University Road, Tainan 70101, Taiwan
[2]Department of Materials Engineering, Pingtung University of Science and Technology,
1, Hseuhfu Road, Neipu, Pingtung 91201, Taiwan
E-mail: htlee@mail.ncku.edu.tw

Abstract

The relation between cooling rate and morphology of Ag₃Sn IMCs was investigated in this study. The morphology of Ag₃Sn intermetallic compounds of eutectic Sn-3.5Ag solder was investigated after solidification at different cooling rates. As the nucleation time of Ag₃Sn depended on cooling rate, the morphology and size of Ag₃Sn compounds were affected by cooling rate. The three types of Ag₃Sn compound during different cooling rate solidification were found to be particle-like, needle-like, and plate-like in Sn-3.5Ag solder. The results show that as the cooling rate decreases, the morphology of Ag₃Sn formed in Sn-3.5Ag solder transforms progressively from particle-like to plate-like.

1. Introduction

Eutectic Sn-3.5Ag solder is one of the promising candidates to replace the conventional Sn-Pb solder, primarily because of its excellent mechanical properties [1]. Sn-3.5Ag solder has a eutectic point of 221°C and contains a uniform dispersion of fine Ag₃Sn compounds, which greatly enhance its mechanical strength. However, the microstructure of Sn-Ag solder alloy is easily affected by different cooling rates due to the relatively low melting point. It has been demonstrated that the size and morphology of Ag₃Sn compounds vary significantly with cooling rate. A survey of the literature shows that three types of morphology of Ag₃Sn, particle-like, needle-like, and plate-like (or bulk), can be found in Sn-Ag solder alloy joints. Different cooling rates and Ag content result in different types of Ag₃Sn. Bulk or plate-like Ag₃Sn has especially been widely studied and discussed in recent investigations. The large plate-like Ag₃Sn compounds that form in solder joints under high Ag content or slow cooling conditions can seriously degrade mechanical performance and reliability [2-4]. It has been reported that the β-Sn phase requires large undercooling relative to Ag₃Sn to begin nucleation. The difference in undercooling allows the Ag₃Sn plates to nucleate and grow relatively large before final solidification [5, 6].

The fine particle-like Ag₃Sn was formed in Sn matrix when the solder cooled in fast cooling rate (24 °C/s), and at slow cooling rate (0.1 °C/s) was formed the needle-like Ag₃Sn in Sn matrix [7, 8]. It also be reported that the particle-like Ag₃Sn changed to needle-like as cooling rate decreased [7, 9, 10]. However, defining the morphology of Ag₃Sn in two-dimensional micrographs provides inconsistent results. Some studies have defined the morphology of Ag₃Sn compounds with two-dimensional imagery. This approach, however, cannot determine the actual shape. Sidhu *et al.* [10] further used the three-dimensional microstructure visualization confirmed the consisting of needle-like Ag₃Sn compounds in the Sn matrix. However, the relation of the three-dimensional

morphology of Ag₃Sn compounds to the cooling rate remains unclear. In order to understand the evolution of three-dimensional Ag₃Sn compound morphology at various cooling rates, continuously cooled solder (resembling a Jominy end-quench test) and solder cooled in water, air, and a furnace were examined in this study. Observations of deeply etched samples were also used to define the morphology of three-dimensional Ag₃Sn compounds.

2. Experiment

In order to determine the evolution of Ag₃Sn intermetallic compounds in Sn-3.5Ag solder through solidification at different cooling rates, two types of experiments were used. The first experiment employed the continuous cooling method, which is similar to the Jominy end-quench test and can obtain a continuous temperature gradient, as shown schematically in Fig. 1. The cooling rates varied with distance from the bottom (the cooling source). The molten solder (250°C) was cast into a square tube made of 1 mm thick stainless steel with dimensions of 10 × 10 × 40 mm. The surface of the square tube was surrounded by an adiabatic band to ensure heat transfer from the bottom to the bulk aluminum and cool water. The bulk aluminum had the dimensions of 300 × 100 × 20 mm. The cool water was maintained at 7°C. The cooling rates at various points were measured using K-type thermocouples with a diameter of 0.32 mm (American Wire Gage, AWG No. 28). The 4 measuring points (P1-P4) in the molten solder were evenly spaced in 10 mm increments. The solidified solder was removed from the square tube when the molten solder had cooled to room temperature.

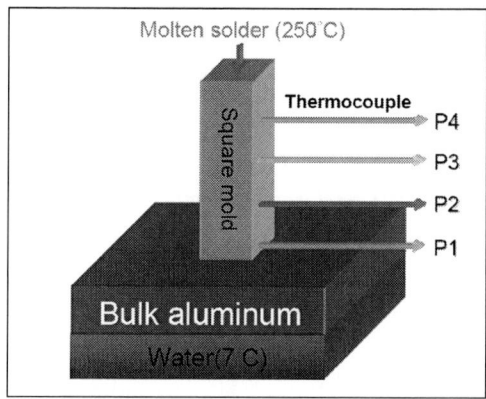

Fig. 1 Schematic diagram of continuous cooling experiment and temperature measurement

In order to observe the evolution of the microstructure, the solidified solder was cut along the center line in the longitudinal direction. These specimen were etched with a solution of HNO_3 and C_2H_5OH in a ratio of l:10, respectively. After deep etching, the specimens were then cleaned using an ultrasonic vibrator in order to allow observation of the morphology of the Ag_3Sn intermetallic compounds via scanning electron microscopy (SEM). Metallographical microstructure and morphology of the solders were investigated by optical microscopy (OM) using a Leica Metallux 3 and by environmental scanning electronic microscopy using an FEI Quanta 400F ESEM.

3. Results and Discussion

The relation between cooling rate and morphology of Ag_3Sn IMCs was investigated. The continuous cooling experiment was used to observe the continuous conversion of Ag_3Sn IMCs in Sn-3.5Ag solder. Fig. 2 presents the cooling curves of the various locations. The temperature history of solder solidification was recorded from 250°C (molten) to solidification. There are four steps in the solidification process. The first step is the molten solder cooling to undercooling. The second step, in which the nucleation of the eutectic phase begins, is recalescence, which is the abrupt temperature rise of a liquid caused by the release of latent heat from crystal growth initiated by nucleation in a supercooled liquid. After recalescence, the solidification continues to the end of nucleation. This is the third step, which is nucleation of the eutectic phase. The solidification is completed when the cooling temperature falls below the eutectic point of Sn-3.5Ag solder. In the fourth step, the solid solder cools to room temperature. The phenomenon of recalescence and the stage of eutectic solidification can be seen clearly in the cooling curves P2-P4. According to Gibbs' phase rule (F=C-P+1=2-3+1=0), no degrees of freedom are left and temperature does not change in this region. Therefore, the eutectic stage appears in the cooling curve. Only P1 exhibits no clear recalescence, and the eutectic stage shown in its cooling curve. Because P1 was located in the rapid cooling zone, most of the latent heat was carried away rapidly by the bulk aluminum. Hence, the phenomenon of recalescence and eutectic stage cannot be seen in the cooling curve of P1.

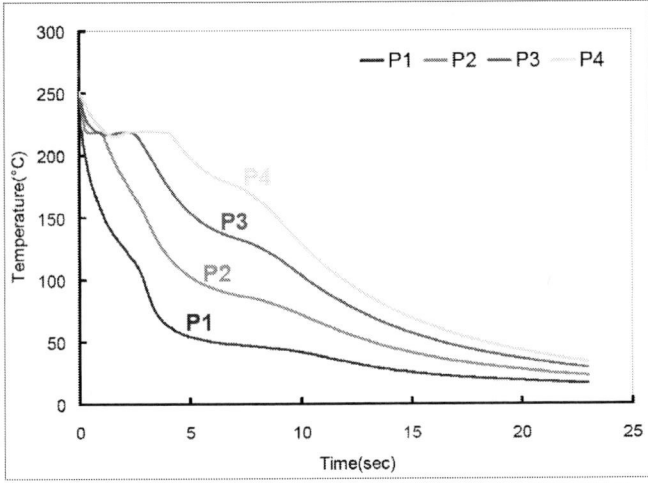

Fig. 2 The cooling curves of the various locations (P1-P4)

The analysis of the cooling curve can be divided into three sections, as shown in Fig. 3. The first section is from the 250°C molten temperature to the undercooling point. This section can be defined as the primary cooling rate (CR_1). The second section is from the undercooling point to the end of nucleation, defined as the nucleation time. The third section is from the end of nucleation to 150°C. The cooling rate of this section was measured from the end of nucleation to 150°C because below this temperature, the microstructure and IMCs undergo no significant change [11]. This section can be defined as the secondary cooling rate (designated by CR_2). The primary and secondary cooling rates calculated from the results of the cooling curve are presented in Table 1. The nucleation time also can be determined from the results of the cooling curve; these results presented in Table 2.

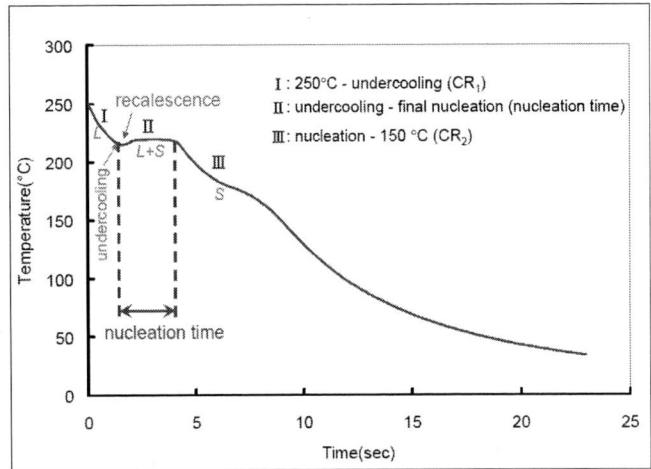

Fig. 3 The analysis of the cooling curves

Table 1 The primary and secondary cooling rate of different point (P1-P4)in continuous cooling experiment

Position	P1	P2	P3	P4
CR_1 (°C/s)	263.6	116	43.9	27.1
CR_2 (°C/s)	69.6	35.7	24.7	14.6

Table 2 The nucleation time of different point (P1-P4) in continuous cooling experiment

Position	P1	P2	P3	P4
Nucleation time (s)	0.01	0.78	1.75	3.01

It can be seen that each cooling rate is higher than the typical cooling rate in the reflow oven. Because the changing temperature of solder was measured directly by thermocouples, this measurement differs from measurements of the reflow process in a reflow oven. In the reflow process, the thermocouples do not contact or measure the solder directly, so the actual temperature change of the solder cannot be obtained. Therefore, the cooling rate of this study is higher than the typical cooling rate.

Fig. 4 presents the microstructure images of the various locations (P1-P4). It can be seen that the microstructure was obviously affected by cooling rate. P1 was located at the top of the bulk aluminum. Therefore, when the molten solder contacted the bulk aluminum, the molten solder near the cooling source was chilled immediately. The cooling curve of P1 clearly shows the extreme cooling rate and short solidification time; these factors caused the microstructure of P1 to be finer than that of others. The β-Sn grains of P1 are approximately spherical in shape and have an average size of 5 μm, as shown in Fig.4 (a). The cooling rate decreased as the distance increased from the measuring points to the bottom. The evolution of the microstructure depended on the cooling rate. Since the cooling rate of P2 was lower than that of P1, the microstructure of P2 became coarser than that of P1. The β-Sn grains of P2 have an average size of 10-20 μm, as shown in Fig.4 (b). The β-Sn grains of P3 have an average size of 20-40 μm, as shown in Fig.4 (c). The cooling rate of P4 is the slowest, resulting in a coarser microstructure. The β-Sn grains of P4 have an average size of 30-60 μm, as shown in Fig.4 (d). From the optical micrographs of the continuously cooled solder, it is apparent that the size of β-Sn grains increased as cooling rate decreased. However, the relation between the cooling rate and volume fraction of β-Sn phase was not obvious. In Snugovsky's research, it was found that in Ag and Cu concentrations of less than Sn-Ag-Cu eutectic, the volume fraction of Sn dendrites tended to increase with cooling rate, while for Ag and Cu concentrations greater than Sn-Ag-Cu eutectic, they tended to decrease [12].

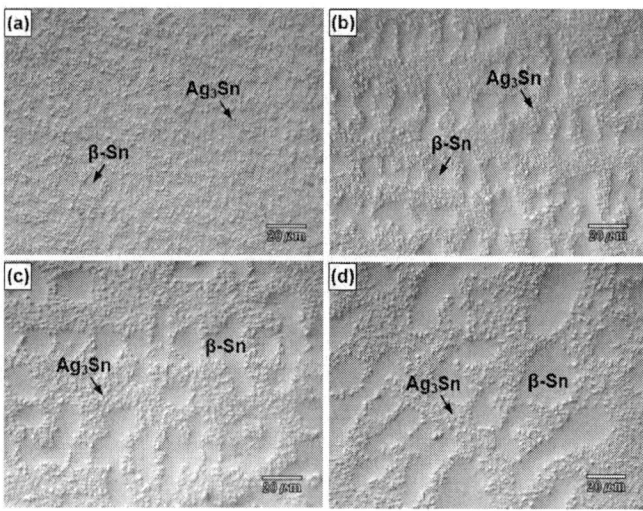

Fig. 4 The microstructure of the various locations (P1-P4) in continuous cooling experiment.(a)P1, (b)P2, (c)P3, and (d)P4.

Fig. 5 presents the three-dimensional (3-D) morphology of Ag₃Sn compounds formed at various cooling rates (P1-P4). As can be seen, the etching process resulted in the removal of the β-Sn phases from the solder matrix, leaving the Ag₃Sn compounds exposed. The morphology of Ag₃Sn can thus be clearly defined because the three dimensional Ag₃Sn compounds are exposed completely after etching process. Fig. 5 reveals that the morphology and size of Ag₃Sn compounds changed with the cooling rate.

P1 was the fastest cooling rate in these four locations in the continuous cooling experiment. The primary cooling rate of P1 was about 263.6°C/s (CR₁). The undercooling increased with the cooling rate, and the nucleation rate was sensitive to undercooling. Large undercooling resulted in a high nucleation rate. Therefore, the finest microstructure was found in the P1 rapid cooling zone. The high cooling rate means that the latent heat was carried away very quickly, which reduced the solidification time. The nucleation time of P1 was about 0.01s from initial nucleation to final nucleation. Many nano

plate-like Ag₃Sn compounds were found in the P1 cooling condition, as shown in Fig. 5 (a). The average size of nano plate-like Ag₃Sn compounds was approximately 80-900 nm.

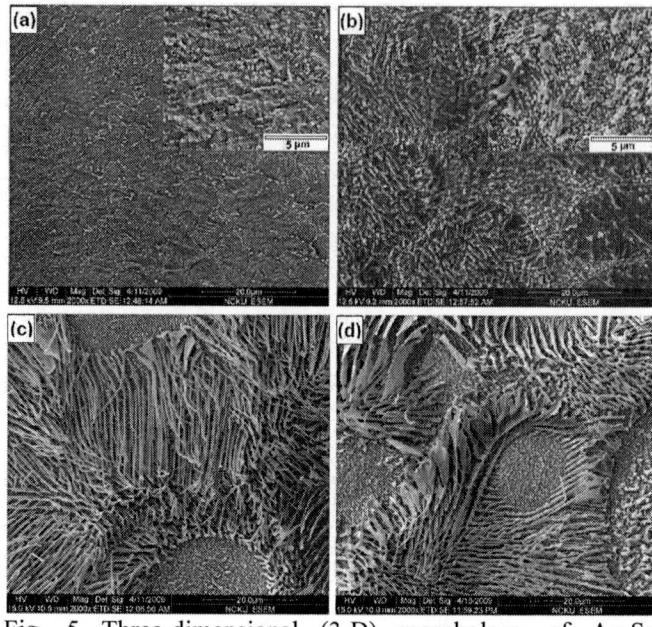

Fig. 5 Three-dimensional (3-D) morphology of Ag₃Sn compounds at various cooling rate (P1-P4), (a) P1, (b) P2 (c) P3, and (d) P4.

The primary cooling rate of P1 was about 263.6°C/s (CR₁) in the first section of the P1 cooling curve. The secondary cooling rate of P1 was about 69.6°C/s (CR₂). After nucleation finished, the number of Ag₃Sn compounds was determined. Grain growth occurred only as the temperature dropped from final nucleation to 150°C. This period was still very short when the secondary cooling rate of P1 was fast, about 69.6°C/s (CR₂). With such a short time to grow, the Ag₃Sn compounds did not grow significantly. In this sense, the cooling rate from the 250°C molten temperature to the eutectic temperature is most important in the growth of Ag₃Sn compounds during solder solidification. Therefore, the microstructure of the solder was affected significantly by the primary cooling rate (CR₁).

The primary cooling rate was about 116°C/s (CR₁) in the P2 cooling condition. Although the CR₁ of P2 was lower than the CR₁ of P1, the CR₁ of P2 was still fast for solder. In this condition, the nucleation time of P2 was about 0.78s from initial nucleation to the end of nucleation. With the extended time in the nucleation zone, longer than in the P1 condition, the Ag₃Sn compounds grew larger. The Ag₃Sn compounds evolved from only nano particle-like to a combination of nano particle-like and micro needle-like, as shown in Fig. 5(b). The size of the nano particle-like Ag₃Sn compounds was approximately 200-900 nm. The size of micro needle-like Ag₃Sn compounds was approximately 0.4-1.2 μm.

The primary cooling rate was about 43.9°C/s (CR₁) when the cooling condition was at P3. The CR₁ of P3 was smaller than the CR₁ of P2. The nucleation time of P3 was about 1.75 s from initial nucleation to the end of nucleation in this condition. As a result, the Ag₃Sn compounds were in the nucleation and growth zone longer than in the P2 cooling condition, with the result that the Ag_3Sn compounds grew larger than those in the P2 condition. The Ag_3Sn compounds evolved from the nano particle-like and micro needle-like to only micro needle-like, as shown in Fig. 5(c). The size of the needle-like Ag_3Sn compounds increased to approximately 10-30 μm. In addition, some the tails of the needle-like Ag_3Sn compounds began to develop a plate-like morphology.

The primary cooling rate was about 27.1°C/s (CR_1) in the cooling condition P4. The CR_1 of P4 was the slowest of these four various locations. Since the nucleation time of P4 was around 3.01s from initial nucleation to final nucleation in this condition, the longest of the conditions, the nucleation and growth zone was the longest of the cooling conditions during the solidification process, with the expected increase in size. As with the P3 condition, the morphology of Ag_3Sn compounds were still micro needle-like, and the tails of the needle-like Ag_3Sn compounds grew into small plate-like compounds, as shown in Fig. 5(d). However, the needle-like compounds increased in size and coarsened, growing to approximately 20-40 μm. Comparison of these SEM micrographs with the OM micrographs in Fig. 4 reveals that the etching process exposed the real 3-D morphology of Ag_3Sn IMC. However, the 3-D morphology of Ag_3Sn IMC appears to differ from the 2-D morphology of OM micrographs in the same sample, as in the 2-D OM micrograph, the morphology of the Ag_3Sn compounds appears to be particle-like. However, it should be noted that the OM micrograph is not a complete micrograph of Ag_3Sn IMC. Therefore, using the complete 3-D morphology to define the micrograph of Ag_3Sn provided a more accurate representation. From results of the continuous cooling experiment, the evolution of Ag_3Sn compounds as cooling rate changes from P1 to P4 can be summarized as follows: nano particle-like → nano needle-like → micro needle-like → micro coarse needle-like with small plate-like compounds.

4. Conclusions

The three types of Ag_3Sn compounds that formed in Sn-3.5Ag solder during solidification at different cooling rates were found to be particle-like, needle-like, and plate-like. During the solidification process, cooling rate determines the nucleation time, which in turn affects the size and morphology of Ag_3Sn compounds. The Ag_3Sn compounds coarsen with increases in nucleation time. The evolution of Ag_3Sn compounds as cooling rate decreases (and nucleation time thus increases) can be summarized as follows: nano particle-like → nano needle-like → micro needle-like → micro coarse needle-like with small plate-like tails.

Acknowledgments

The authors would like to thank Science Council of Taiwan for the final support of this work under the contract of No. NSC 97-2221-E-006-021-MY3.

References

1. ESPEC Technology Report, 1 (2002).

2. D.W. Henderson, T. Gosselin, and A. Sarkhel, "Ag_3Sn Plate Formation in the Solidification of Near Ternary Eutectic Sn-Ag-Cu alloys," *Journal of Material Research*, Vol. 17 (2002), pp. 2775-2778.

3. K.S. Kim, S.H. Huh, and K. Suganuma, "Effects of Intermetallic Compounds on Properties of Sn–Ag–Cu Lead-free Soldered Joints," *Journal of Alloys and Compounds*, Vol. 352 (2003), pp. 226-236.

4. K.S. Kim, S.H. Huh and K. Suganuma, "Effect of Cooling Speed on Microstructure and Tensile Properties of Sn-Ag-Cu alloys," *Materials Science and Engineering*, Vol. A333 (2002), pp. 106-114,.

5. I. Ohnuma, M. Miyashita1, K. Anzai, X. J. Liu1, H. Ohtani, R. Kainuma, and K. Ishida, "Phase Equilibria and the Related Properties of Sn-Ag-Cu Based Pb-free Solder alloys," *Journal of Electronic Materials*, Vol. 29 (2000), pp. 1137-1144.

6. K.W. Moon, W. J. Boettinger, U. R. Kattner, F. S. Biancaniello, and C.A. Handwerker, "Experimental and Thermodynamic Assessment of Sn-Ag-Cu Solder Alloys," *Journal of Electronic Materials*, Vol. 29 (2000), pp. 1122-1136.

7. F. Ochoa, J.J. Williams, and N. Chawla, "Effects of Cooling Rate on the Microstructure and Tensile Behavior of a Sn-3.5wt.%Ag Solder," *Journal of Electronic Materials*, Vol. 32 (2003), pp. 1414-1420.

8. F.Ochoa, X.Deng, and N.Chawla, "Effects of Cooling Rate on Creep Behavior of a Sn-3.5Ag Alloy," *Journal of Electronic Materials*, Vol. 33 (2004), pp. 1596-1607.

9. J.M. Song, J.J. Lin, C.F. Huang, and H.Y. Chuang, "Crystallization, Morphology and Distribution of Ag_3Sn in Sn–Ag–Cu Alloys and Their Influence on the Vibration Fracture Properties," *Materials Science and Engineering*, Vol. A466 (2007), pp. 9-17.

10. R.S. Sidhu and N. Chawla, "Three-dimensional (3D) Visualization and Microstructu Re-based Modeling of Deformation in a Sn-rich Solder," *Scripta Materialia*, Vol. 54 (2006), pp. 1627-1631.

11. Q. Hu, Z.S. Lee, Z.L. Zhao, and D.L. Lee, "Study of Cooling Rate on Lead-free soldering microstructure of Sn-3.0Ag-0.5Cu solder," *International Conference on Asian Green Electronics*, Shanghai, Mar. 2005, pp. 156-160.

12. L. Snugovsky, P. Snugovsky, D. D. Perovic, and J. W. Rutter, "Effect of Cooling Rate on Microstructure of Ag-Cu-Sn Solder Alloys," *Materials Science and Technology*, Vol. 21 (2005), pp. 61-68.

Indium Bump Fabricated with Electroplating Method

Qiuping Huang, Gaowei Xu, Le Luo

ShangHai Institute of Microsystem and InformationTechnology, Chinese Academy of Sciences,
Shanghai, China,
No. 865 Changning Road, Shanghai, 200050
E-mail: huangqp@mail.sim.ac.cn

Abstract

Indium solderbumps are usually used in interconnection between focal plane arrays (FPAs) and Si read out integrated circuits (ROICs) by flip-chip bonding. The fabrication of indium bump array is a critical technology in this process. In this paper, the 16×16 indium bump array was fabricated by electroplating method. The indium bump is 100μm in pitch and 40μm in diameter. Lift-off method and IBE process were adopted to try to remove the seed layer. Ti/Pt/Au(200 Å/300 Å/800Å) by sputtering method and Ti/Pt/Au/ep Au(200 Å/300 Å/800Å /3-4μm) by electroplating after sputtering were investigated as UBM (under bump metallization) of indium bump. The reliability of indium bumps with different UBM was evaluated by cross-section analysis and shear test.

1.Introduction

Indium bumps are always used in the flip-chip integrated Hybrid infrared focal plane arrays and the Si read-out integrated circuit [1,2,3]. The reasons for adopting the indium as interconnecting material lie in the fact that indium has many good properties and is suitable for this kind of applications. Indium has excellent plasticity which can overcome the mismatch between the detector substrate (GaAs, InSb HgCdTe) and the substrate even at very low temperature. The good performance in electrical conductivity and solderability of indium are necessary for signal transduction and flip-chip process. Its relative ease of making very small solder bumps is another critical advantage for indium using in this application [4,5,6].

Currently two primary methods are employed to fabricate indium bump, one is evaporation method and the other is electroplating technique (also called UV-LIGA) [7,8]. In the evaporation technique, indium is firstly deposited onto the substrate and then a lift-off process is performed. In the electroplating method, indium is directly electrically deposited onto the UBM (seed layer). Compared with the evaporation technique, the electroplating method is a cost effective, flexible and reliable indium bump fabrication method. In this paper, electroplating method was used to fabricate indium bump.

As we know, seed layer is necessary in the electroplating process. But in indium bump fabrication process, how to remove the seed layer is a difficulty because Indium is a so active and tender metal compared with Ti/Pt/Au. In this experiment, two methods are tried to remove the seed layer. One is the lift-off process, the other is IBE method. The chemical etching can't remove Ti/Pt/Au seed layer without destroying indium bump.

UBM is other important factor for indium bump fabrication due to it directly relates to bump reliability.

Usually Ti/Pt/Au, Ti/Ni/Au, Cr/Au are used as UBM in indium bump fabrication. In the experiment of Patrick Merken [6], they suggest that the thickness of Au in UBM should be limited to 1000 Å, or the contact between the indium and substrate becomes brittle. So Ti/Pt/Au(200Å/300Å/800Å) UBM layer is used in experiment. In the experiment of Lili Ma, indium bump fell off from the substrate because the Au was exhausted by indium during doing the temperature cycling experiment and vibration experiment [9]. Increasing the thickness of the Au in the UBM is a probable solution to prolong the life of indium bump. GEC Hirst Research Centre carried out some works focus on the reaction occuring between molten indium and gold metallization. They found that the intermetallic compound $AuIn_2$ formed as a continuous interfacial layer between indium and gold above the indium melting point of 157°C. Once the continuous interfacial established, the $AuIn_2$ can act as an effective barrier between the indium and gold. In order to prevent exhausting the gold and have enough Au reaction with Indium to form continuous $AuIn_2$ barrier layer in UBM, Ti/Pt/Au/ep Au (200Å/300Å /800Å /3-4μm) is also designed as UBM in the experiment [10]. Increasing the thickness of UBM to prolong the life of bump is a widely adopted method in bump fabrication process [11]. Through the experiment, we hope to find out which UBM is suitable for indium bump.

2. Experiment

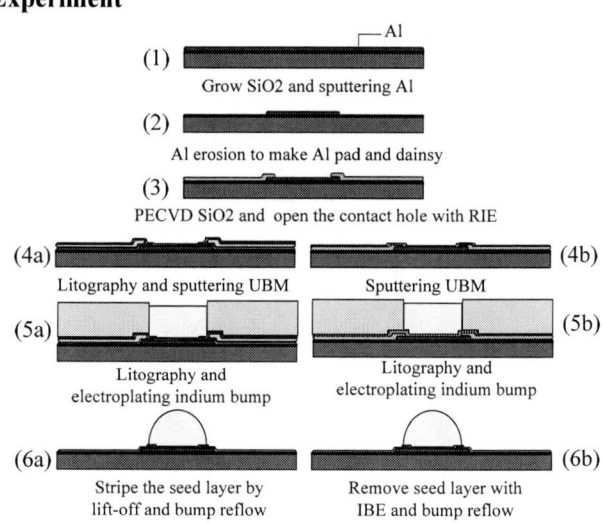

Fig.1.Schematic of electroplated indium bump fabrication process

In this experiment, indium bump are fabricated on 4-inch silicon wafer and the details are described as follows:

978-1-4244-4658-2/09 $25.00 © 2009 IEEE

(1). SiO_2 is grown On the Si wafer by thermal oxidation and then Al is sputtered on it with magnetron sputtering, both thickness are 5000Å;

(2). Al pad and daisy chain are fabricated with phosphoric acid erosion followed lithography. Photoresist S1912 with thickness of 1.7μm is used.

(3). SiO_2 passivation layer is deposited on the wafer with PECVD (Plasma enhanced chemical vapor deposition) and then lithography and RIE (Reactive Ion Etch) erosion with CF_4 by step are performed to make contact holes;

(4a). Lithography followed by deposition of Ti/Pt/Au (200Å /300Å /800Å) as UBM. In these three metal layers, titanium acts as an adhesive layer, platinum is a barrier layer and gold is a wettable layer.

(5a). Lithography of the AZ9260 with the thickness of 30μm. To obtain the photoresist with thickness of 30μm, AZ9260 is spun at 600rmp for 30s.

(6a). Indium electroplating is performed in an indium sulfamate plating bath at room temperature using GW instek DC power supplier. The solution comprises of $In(NH_2SO_3)_3$ 105.36g/L, $NaNH_2SO_3$ 150g/L, HNH_2SO_3 26.4g/L, NaCl 45.84g/L dextrose 8g and triethanolamine 2.29g, PH 1.5-2. This kind of indium electroplating bath has nearly 100% electric current efficiency. A pure indium plate (99.9999% indium) was used as anode. Silicon wafer was submerged in plating bath and contacted as cathode. DC plating current density was controlled in the range of 5mA-15mA. Smaller current density can make the indium deposit growth more controllable. This kind of electroplating condition can obtain indium bump array with excellent uniformity. After plating indium on the wafer, lift-off process is done by putting the wafer into hot acetone. After seed layer was striped, indium bump was reflowed and became indium ball.

The main differences between the two fabrication methods lies in the step 4 and step 6, which are showed in Fig.1.4(b)-Fig.1.6(b). In the second fabrication method, Ti/Pt/Au (200Å/300Å/800Å) as UBM is deposited directly onto wafer without lithography in step 4 and IBE is used to remove the seed layer but not lift-off process. Large beam current and long time etch can destroy the bump in IBE process. So in the IBE process, beam current and etching time must be strictly controlled at a low level. Beam current at 150mA and time for 1min for each time is preferred. To obtain the Ti/Pt/Au (200Å/300Å/3-4μm), Au plating process was performed before indium electronic deposition.

After bump electroplating and removing the seed layer, the samples with different UBM were reflowed in a 5-zone falcon 8500 reflow oven with 5RMA flux of America indium corporation. The reflow temperature was 80°C, 110°C, 140°C, 170°C, 180°C respectively and the time was 30s for each zone. After the first reflow, the samples were divided into two groups and then experienced different thermal process. The samples in one group were reflowed for another 2, 6, 10, 12 times under the same condition of the first reflow. The samples from another group would experience aging process at the temperature of 140°C for 24h.

The cross-sections of the samples were made for observation and analysis of intermetallic compound and its growth with SEM (scan electric microscope) and EDX (Energy Dispersive X-ray Analysis). Shear test was used to evaluate the strength of the indium bumps. Shear speed was 100um/sec and the shear height was 10 microns. At least 50 bumps were measured for every sample.

3. Results and discussions

3.1 Indium bump fabrication

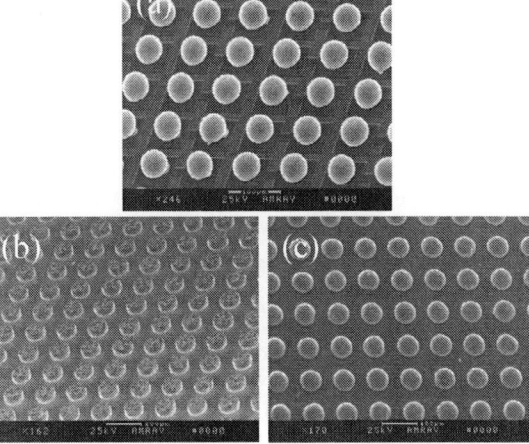

Fig.2.Indium bump fabricated with lift-off process (a); indium bump after experience IBE process (b) and after reflow (c)

Fig.2(a) shows the results of indium bump fabrication by electroplating in which the seed layer was removed by lift-off method after reflow at the temperature of 170°C for 30s. Fig.2(b) shows the indium bump after just removing the UBM with IBE and the reflowed indium bump is showed in Fig.2(c). Fig. 2(b)-(c) show that the IBE process almost has no negative effect on indium bumps.

Uniformity is a critical index to evaluate the quality of bump array. To investigate the uniformity of the indium bump in the experiment, the height of more than 40 indium bumps were measured in six arrays of different area on wafer with Olympus STM6. The average height of the indium bumps shown in the Fig.2(a) is 43.4μm and that of 35.7μm in Fig.2(c), the uniformity is about 1.23% and 0.92% respectively. The uniformity can be calculated followed the formula: Uniformity=$(H_{max}-H_{min})/(H_{max}+H_{min})$. Indium bumps with good uniformity can be fabricated with electroplating.

3.2 Interface evolution of indium bump and UBM during reflow and aging processes

Fig.3 shows the cross-sections of samples with Ti/Pt/Au/Ep Au UBM after 1, 5, 9, 13 times reflows under the condition described above. It shows that the continuous wavy type and the floating-island type of intermetallic compounds exist at the interface of indium and gold where they contact. Both types of intermetallic compounds grew with the increase of reaction time. As we know, Au has little solubility in solid indium but at the temperature of 170°C, Au atom dissolve rapidly into liquid indium, which is the cause to form a sunken curvature at the interface of indium and gold and irregular structure of IMC. Fig.3(a)-(d) show that the granular IMC grew in size while decreasing in number. These phenomena implied that a ripening effect occurred with the growth of intermetallic compounds, which is consistent with paper [12].

Spot EDX and line scan experiment was carried out on 13 times reflows sample and the results show in the Fig.4 and Fig.5. At the spot showed in Fig.4(a), the atom proportion of Indium element is 65.54% and that of gold is 34.46%, which is very close to 2:1 . This indicates that the main IMC at this area is $AuIn_2$. But at the spot showed in the Fig.4(d) EDX result indicates that the indium and Au is in the radio 33.61:66.39,which is close to 1:2. This result shows that several phases coexist in this area because Au_2In is not a stable phase in Au-In IMC. Line scan in Fig.5(a) is carried out from bottom to top. Obviously the intensity of gold atom (Fig. 5(b)) becomes weaker and weaker and that of indium (Fig.5(c)) become stronger and stronger. But the intensity of indium changes faster than that of Au. This indicates that the Au is the main diffused element in this thermal process and indium almost can't dissolve into Au substrate. Fig.4(a) and Fig.5(a) show that many voids exist at the interface of the indium and gold. This kind of voids caused mainly by kirkendall effect in which the different solubility and interdiffusion lead to the appearance of voids. Between the two spots showed in Fig.4(a) and Fig.4(d), another kind of voids appeared. EDX results of the two spots indicated phase change occurred during the thermal process. Phase change process may accompany with volume shrinkage of phase grain, which lead to the appearance of this kind of voids. These phenomena also always occur in Cu/Sn, Ni/Sn systems afrer experiencing solid-liquid reaction process [13, 14].

Fig.6 shows the metallographic image of indium bump with Ti/Pt/Au (200Å/300Å/800Å) UBM after 1 times and 13 times reflow respectively. Experiment result shows that interface of indium and UBM has no obvious change. It indicates that Pt is a good barrier layer for indium bump.

Cross-sections of two samples which experienced aging process at 140°C for 24h are shown in Fig.7. Ti/Pt/Au/Ep Au and Ti/Pt/Au were used as UBM in these two samples respectively. Fig.7(a) shows that Au in Ti/Pt/Au/Ep Au were almost exhausted by indium and a thick IMC layer appeared between indium bump and the substrate after this thermal process. The IMC make the soft indium bump become brittle, which may harm the reliability of packaging. So the Ti/Pt/Au(200Å/300Å/800Å) UBM by sputtering is suitable in the fabrication of indium bump, just as showed in Fig.7(b).

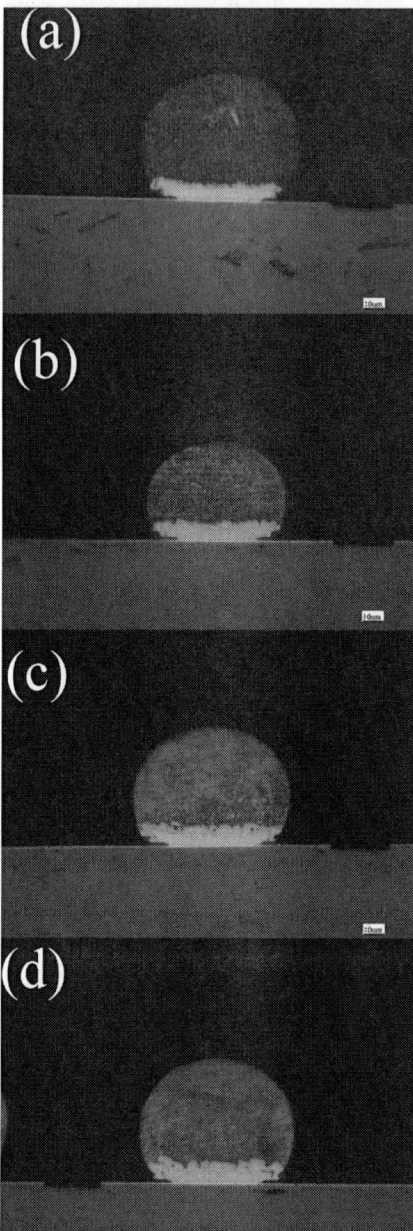

Fig.3.Metallographic images of indium bump with Ti/Pt/Au/ep Au UBM after different reflow times: (a)1 times; (b) 5 times; (c) 9 times; (d) 13 times

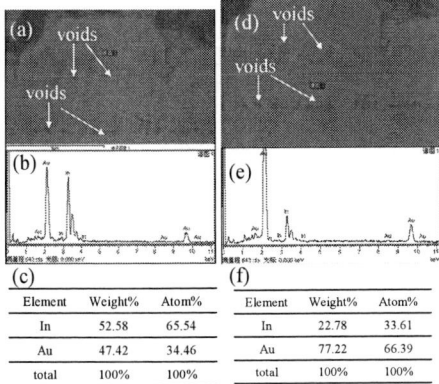

Element	Weight%	Atom%
In	52.58	65.54
Au	47.42	34.46
total	100%	100%

Element	Weight%	Atom%
In	22.78	33.61
Au	77.22	66.39
total	100%	100%

Fig.4.EDX spot (a) and(d) and results ((b), (c),(e),(f)) for13 reflow times samples with Ti/Pt/Au/Ep Au UBM

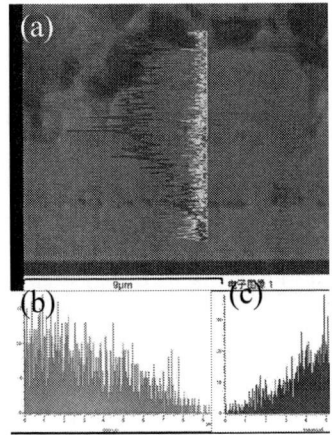

Fig.5.Line scan results of 13 reflow times indium bump with Ti/Pt/Au/Ep Au UBM

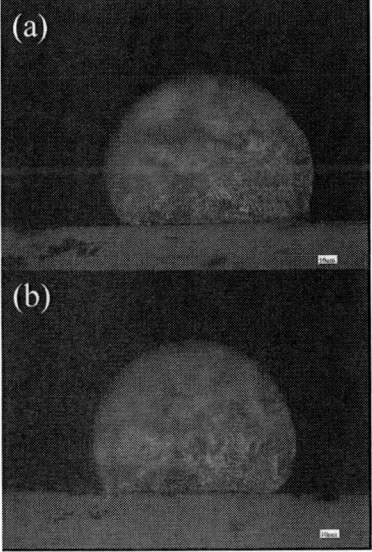

Fig.6. Metallographic image of indium bump with Ti/Pt/Au (200Å/300Å/800Å) UBM with different reflow times:(a) 1 times; (b) 13 times

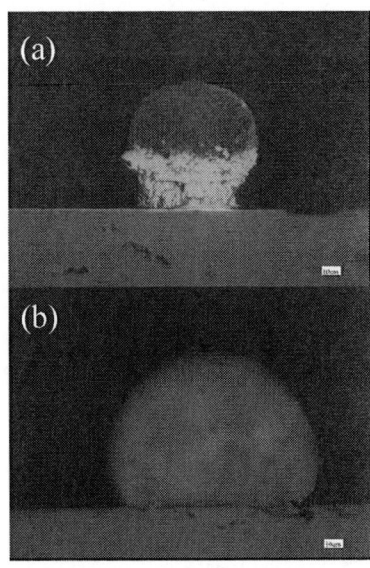

Fig.7.Cross-section of samples with Ti/Pt/Au/Ep Au UBM (a) and Ti/Pt/Au UBM (b) after aging 24h at 140°C

3.3 Shear strength test

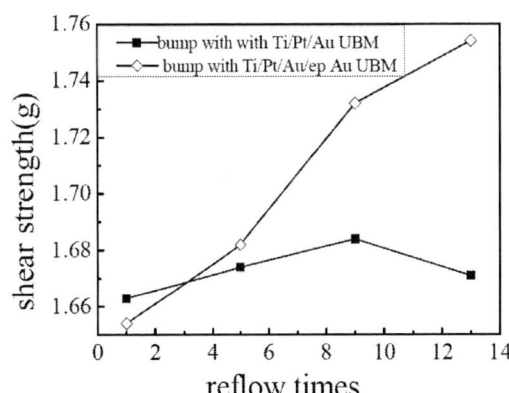

Fig.8. Shear strength of indium bumps with Ti/Pt/Au UBM and Ti/Pt/Au/Ep Au UBM after multi reflows

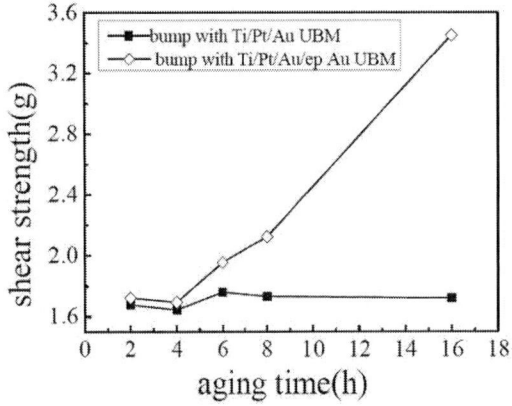

Fig.9. Shear strength of indium bumps with Ti/Pt/Au UBM and Ti/Pt/Au/Ep Au UBM experiencing different aging time at 140°C

Fig.8 and Fig.9 show the shear strength of indium solder

bumps with different UBM layer and different thermal process. Shear strengths of samples with Ti/Pt/Au (200Å/300Å/800Å) UBM change little both in reflow and aging process. This result indicates that the Ti/Pt/Au (200Å/300Å/800Å) UBM won't change the plasticity of indium bump. But for the samples with Ti/Pt/Au/Ep Au UBM, shear strengths have significant change. Shear strengths increase with increasing the reflow times and aging time. These increasing IMC cause shear strengths increase. Longer reflowing and aging time lead to the formation of thicker and more irregular IMC layer. Au/In IMC is very brittle. These brittle IMC diffused into soft indium bump during the thermal process and change indium plasticity. The results of shear test are consistent with the cross-section analysis.

4. Conclusions

(1) Indium solder bumps of 16×16 full-area array with excellent uniformity were prepared with electroplating. Lift-off process and IBE are two effective methods to remove the seed layer.

(2) The continuous type of Au/In IMC layer may act as an effective barrier, but thick Au layer in Ti/Pt/Au/ep Au UBM can make indium bump brittle. Pt is a good barrier for indium bumps. Au with thickness less than 1000Å is suitable in Ti/Pt/Au UBM for indium bump fabrication.

(3) Increasing the reflow times and aging time, the thickness of Au-In IMC increase for Ti/Pt/Au/ep Au UBM. The voids at the interface of Au/In are caused by kirkendall effect. Voids in IMC results from the volume shrinkage in phase change process.

(4) With the increasing of reflow times and aging time, shear strength nearly have no change for indium bump with Ti/Pt/Au (200Å/300Å/800Å) UBM but have a significant change for indium bump with Ti/Pt/Au/ep Au UBM. Au in the Ti/Pt/Au/ep Au UBM diffuses into indium solder to form Au-In IMC, which changes the plasticity of indium.

References

1. D Scribner *et al*, "A Retinal Prosthesis Technology Based on CMOS Microelectronics and Microwire Glass Electrodes", *IEEE Transactions on Biomedical Circuits and Systems*, Vol.1.No.1 (2007), pp. 73-84.

2. Joachim John *et al*, "High-density hybrid interconnect methodologies", *Nuclear Inst. and Methods in Physics Research* ,A531(2004), pp. 202-208.

3. Han Seo cho *et al*,"Compact packaging of optical and electronic components for on-board optical interconnects," *IEEE Transactions on Advanced Packaging,* Vol.28, No.1, pp. 114-120.

4. Jutao Jiang *et al*, "Fabrication of indium bumps for hybrid infrared focal plane array applications," *Infrared physics and Technology,* 45(2004),pp. 143-151.

5. Ch.Broennimann *et al*, "Development of an indium bump bond process for silicon pixel detectors at PSI," *Nuclear Instrument and Methods in Physics Research*, A,565(2006), pp. 303-308.

6. P Merken *et al*, "Technology for very dense hybrid detector arrays using electroplated indium solderbumps," *IEEE Transactions on Advanced Packaging,* Vol.26, No.1, Feb. 2003, pp. 60-64.

7. J.Breibach *et al*, "Development of a bump bonding interconnect technology for GaAs pixel detectors," *Nuclear Instrument and Methods in Physics Research,* A 470 (2001), pp. 576-582.

8. Y Tian *et al*, "Electrodeposition of indium bump bonding," *Electronic Components and Technology Conference, 2008.*

9. L Ma, S Bao, J Peng, Z Du, "Failure Analysis of In/Au solder Joints," *Electronic Packaging Technology, 2006. ICEPT06, 7th International Conference.*

10. DM Jacobson, G Humpston, Gold coating for fluxless soldering, Gold Bull,1989,22(1), pp. 9-18.

11. Xiaoqin Lin, Le Luo, "Lead Free SnAg Solder Bumping with Size sub 100 Microns Electronic Packaging Technology," *ICEPT06, 7th International Conference.*

12. Y.M.Liu,T.H.Chuang, "Interfacial Reactions between Liquid Indium and Au-Deposition Substrates," *Journal of ElECTRONIC MATERIALS,* Vol.29, pp. 405-410.

13. Y.C.Chan, P.L.Tu, C.W.Tang, *et al*, "Reliability Studies of micro BGA Solder Joints - Effect of Ni-SnIntermetallic Compound," *IEEE Trans Adv Pack.*, 24(2001), pp. 25-32.

14. Y.Watanabe, Y.Fujinaga, H.Iwasaki, "Lattice Modulation in the Long-Period Superstructure of Cu3Sn," Acta. Cryst.B39 (1983), pp. 306-311.

Effect of Aging Time on Interfacial Microstructure of Sn-3.8Ag-0.7Cu Solder Reinforced with Co Nanoparticles

S.L. Tay, A.S.M.A. Haseeb, Mohd. Rafie Johan
Department of Mechanical Engineering, University of Malaya
50603 Kuala Lumpur, Malaysia
E-mail:clengtay@yahoo.com

Abstract

This paper investigates the growth of intermetallics in Sn-3.8Ag-0.7Cu (SAC) solder with and without Co nanoparticle reinforcement. Co reinforced composite was prepared by mechanically mixing 0.5 wt% and 1.5 wt% Co nanoparticles respectively into Sn-3.8Ag-0.7Cu solder paste. The formation and growth of the intermetallic compounds (IMC) at the solder joint interface were evaluated after thermal aging at 150°C for up to 1008 hours. In the solder joint, Cu_6Sn_5 intermetallic was observed on Cu substrate, followed by Cu_3Sn intermetallic formation between Cu_6Sn_5 and Cu after prolonged aging. The thickness of both IMC increased linearly as a function of square root of aging time. The Co nanoparticle reinforcement tended to suppress the growth of Cu_3Sn intermetallic layer. However, the total of IMC thickeness was almost the same for the specimen with or without Co nanoparticle reinforcement.

Introduction

An intermetallic compound is a solid state phase that can form during the metallurgical reaction between solder and substrate material [1]. η-phase (Cu_6Sn_5) and ε-phase (Cu_3Sn) are the typical intermetallic compounds that form in the Sn-Cu system. The IMCs usually occur in scallop-shape and planar-shape, respectively at Sn-based lead-free solder interface [2,3].

The formation of a thin IMC layer is desirable as it enhances a strong bonding at the interface. However, excessive IMC layers promote a brittle failure mode or degrade the mechanical properties of the solder joint [3-10]. Thus, the formations of IMC, IMC layer growth and microstructure of the bulk solder have a crucial effect on the solder joint reliability in microelectronic packaging.

IMC layer grows thicker under high temperature work condition or during isothermal aging for a long time [11]. One of the methods to obtain thinner IMC layer is the composite method by adding suitable particles into solder to stabilize the solder microstructure [4]. Several of researchers found that the addition of Ni and Co into the Sn-Ag solder could prevent the increase of intermetallic thickness after thermal annealing [12-13].

In this study, the composite solder was synthesized by mixing Co nanoparticle into SAC387 solder paste. The effect of addition Co nanoparticles on the IMC layer after high temperature storage is presented.

Experimental Procedures

Nanocomposite solder was prepared by mixing Sn-3.8Ag-0.7Cu solder paste (Indium Corporation of America, Singapore) with nominal 0.5 and 1.5 wt% of cobalt nanoparticles (~28 nm, Accumet Materials, Co., USA). Solder paste was blended with cobalt nanoparticles by manual blending for approximately 30 minutes to ensure uniform distribution of reinforcing particles. Copper sheets (30 mm x 30 mm x 3 mm) were used as substrates for solder preparation. Before reflow, the copper sheets were cleaned with soap detergent to remove oil, cleaned with detergent to remove greasy material, dipped in a solution of 50% HNO_3 to remove oxide and then rinsed thoroughly in distilled water followed by drying with ethanol. After that, 0.2 g of as-prepared nanocomposite was placed on copper sheet through a mask with an opening of 6.5 mm and the height of 1.24 mm. The copper sheet was then placed over a hotplate maintained at 250°C for 45 seconds. After reflow, the flux residue on the top of the solder was cleaned with acetone. A chemical analysis of solder and flux-residue was done by inductively coupled plasma mass spectrometry to measure the actual amount of Co presented. The high temperature storage (HTS) was performed in an oven at 150°C for time period ranging from 24 hours to 1008 hours. All the specimens were then cut by diamond saw and prepared for cross-sectional metallographic preparation. The mounted specimens were then ground with various grades of abrasive papers (grade 100-1200), followed by polishing with polycrystalline diamond suspension with a particle size of 6 μm and 1 μm. These were then finally polished with colloidal silica suspension with a particle size of 0.02 μm. The morphology of the IMCs was observed under a scanning electron microscopy (SEM). SEM images were obtained under back scatter electron (BSE) imaging. The chemical composition of the IMCs was revealed by using energy dispersive spectroscopy (EDS). An operating voltage of 15 kV was used in the examination.

Results and Discussions

Fig.1 (a)-(i) reveals the interfacial microstructure between Sn-3.8Ag-0.7Cu solder and Cu substrate after reflow and after high temperature storage for 24 to 1008 hours. A scallop shaped intermetallic layer (lighter layer) was formed after reflow. After prolonged aging another intermetallic layer (darker layer) planer in shape was formed in between the first intermetallic layer and Cu substrate.

978-1-4244-4658-2/09 $25.00 © 2009 IEEE

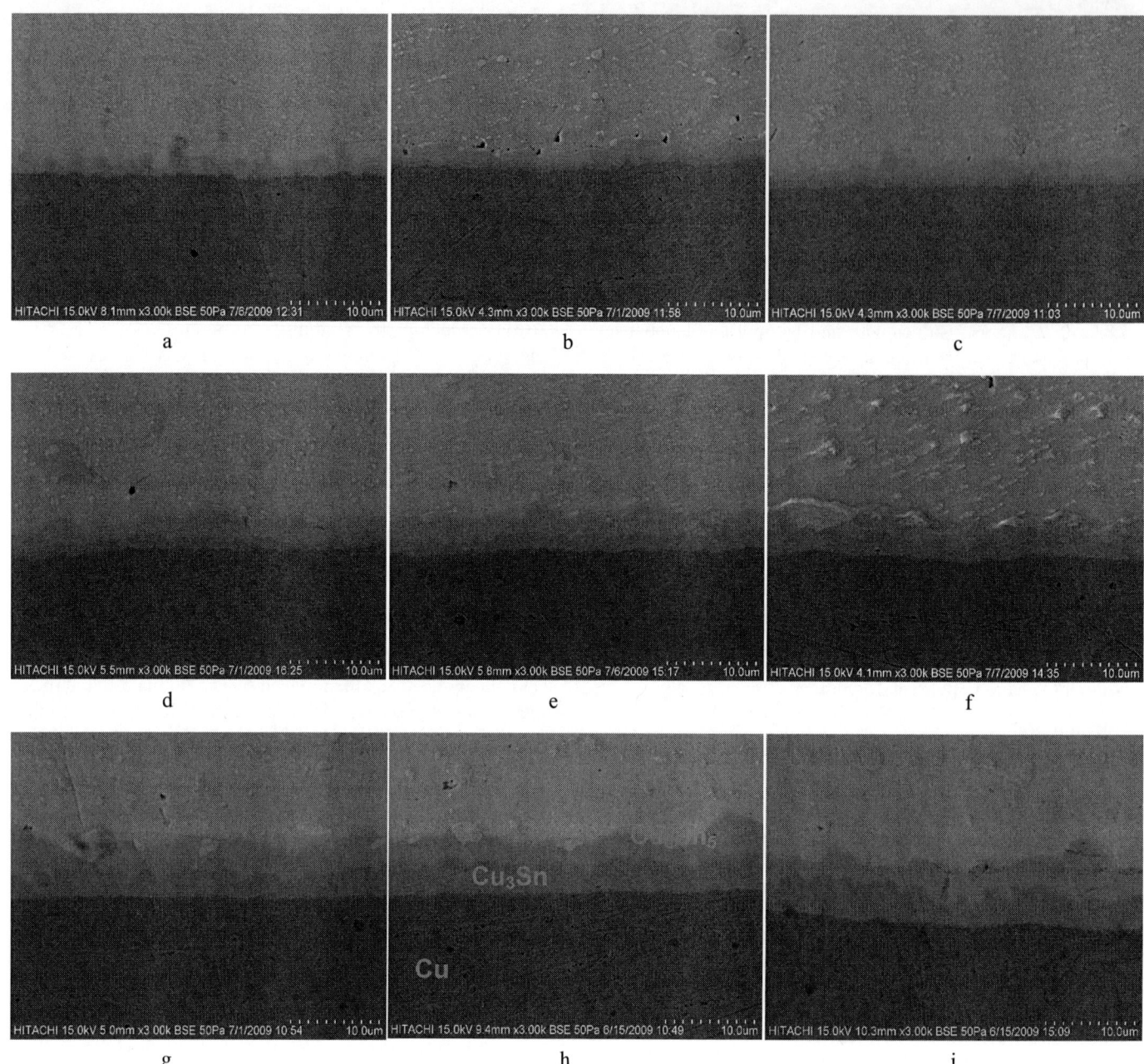

Figure 1 Cross-sectional SEM micrographs of Sn-3.8Ag-0.7Cu solder interfaces after aging at 150°C for (a) 0h, (b) 24 h, (c) 48h, (d) 96h, (e) 168h, (f) 336h, (g) 504h, (h) 840h and (i) 1008h.

Energy dispersive spectroscopy was used to determine the composition of each layer. Cu_6Sn_5 and Cu_3Sn are the two main reaction products that result from interfacial reactions between SAC and Cu substrate. Cu_6Sn_5 is the first compound formed, whereas Cu_3Sn formed later compound during prolonged aging.

In the microstructure Cu_6Sn_5 appears lighter in color while Cu_3Sn appears darker. In the present case, one reaction product, Cu_6Sn_5 was formed in as-reflowed SAC solder. When the specimen was aged for 24 hours or longer at 150°C, another reaction product, Cu_3Sn was formed. After prolonged aging, Cu_3Sn layer became more evident and thicker. During isothermal aging, Cu_6Sn_5 was formed by interdiffusion of Cu and Sn, while Cu_3Sn was formed by the reaction between the

Cu substrate and Cu_6Sn_5 layer, according to the reaction $Cu_6Sn_5 + Cu \rightarrow Cu_3Sn$ [1, 14].

Since the diffusion rate of tin is faster than that of Cu in the Sn-Cu system, tin rich IMC Cu_6Sn_5 layer is easier to form rather than Cu_3Sn is more favorable at this interface. The composition of tin in IMC near to the Cu substrate was decreased, thus the formation of Cu_3Sn more favorite [3].

The interfacial reaction layer, Cu_6Sn_5 exhibited the typical scallop shape in the as-reflowed SAC solder. When the aging time extended to 24 hours, the Cu_6Sn_5 layer became smoothened. This is consistent with the Choi et al. [15] who reported that during aging IMC changed into a layer type when soldering is done for less than 120s. The change in morphology to a layer type during solid state aging has been

978-1-4244-4658-2/09 $25.00 © 2009 IEEE

explained by the tendency to reduce interfacial energy between IMC and solder.

Fig.2 shows that the IMC layer increases as a function of square root of aging time. The IMC growth thus follows the classical diffusion theory or Fick's Law [18]. The total IMC layer increased from 1.87 ± 0.45 μm to 6.29 ± 0.87 μm for the specimen during aging for 1008 hours.

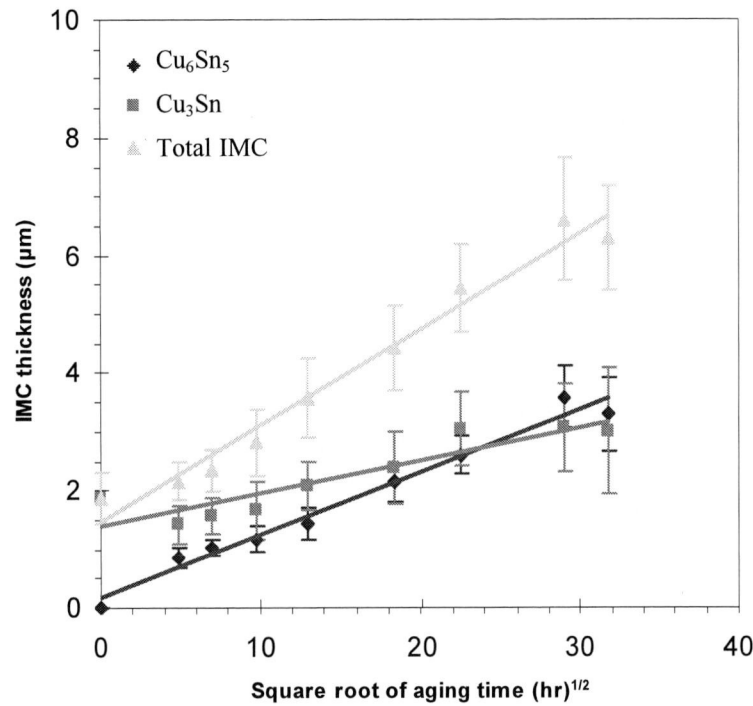

Figure 2 Growth of interface IMC compound with respect to square root of aging time for SAC solder.

The thickness ratio of two IMC layers (Cu_6Sn_5 and Cu_3Sn) became approximately 1:1 after aging for 168 hours. Chan *et al.* [15] reported that during the interaction between Sn-3.5Ag-0.75Cu and electroless tin plated Cu pads in FlexBGA packages, the ratio of Cu_6Sn_5 and Cu_3Sn layer thickness became approximately 1:1 after aging for 96 hours and the Cu_3Sn intermetallic layer achieved maximum thickness after aged for 168 hours. Ag_3Sn intermetallics were found dispersed in the bulk solder. Ag_3Sn precipitate coarsened as a function of aging time. Others researchers also reported similar observation [17].

Fig. 3 (a-i) and Fig. 4 (a-i) show the intermetallic morphology development during aging at 150°C for SAC specimens doped with 0.18 wt% and 0.75 wt% cobalt nanoparticles, respectively. Only one reaction product, Cu_6Sn_5 was found in the as-reflowed SAC samples. The second reaction product, Cu_3Sn was observed after aging for 48 hours, as compared to 24 hours for the pure SAC solder.

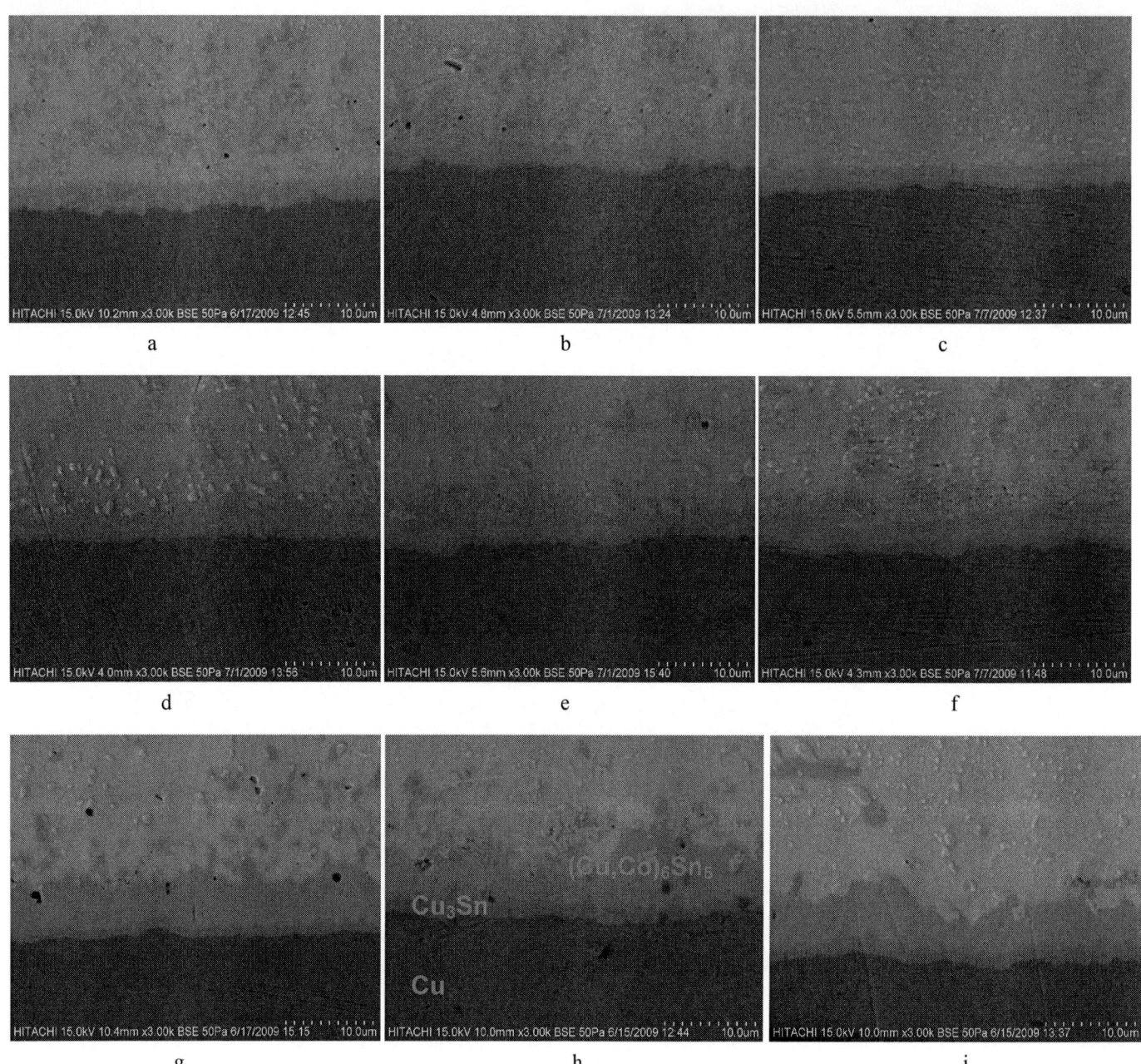

Figure 3 Cross-sectional SEM micrographs of Sn-3.8Ag-0.7Cu-0.18 nano-Co solder interfaces after aging at 150°C for (a) 0h, (b) 24 h, (c) 48h, (d) 96h, (e) 168h, (f) 336h, (g) 504h, (h) 840h and (i) 1008h.

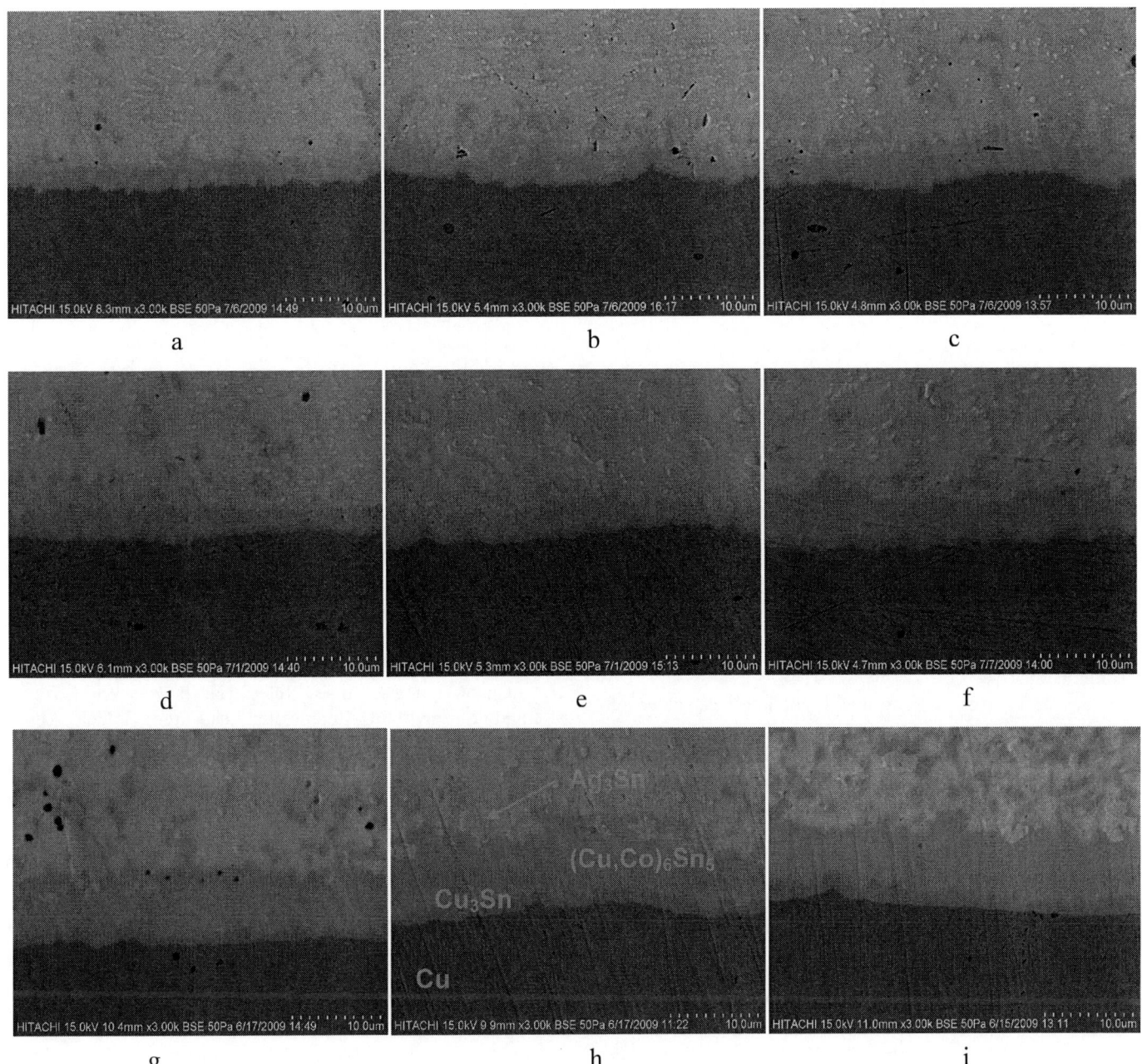

Figure 4 Cross-sectional SEM micrographs of Sn-3.8Ag-0.7Cu-0.75 nano-Co solder interfaces after aging at 150°C for (a) 0h, (b) 24h, (c) 48h, (d) 96h, (e) 168h, (f) 336h, (g) 504h, (h) 840h and (i) 1008h.

For the nano Co-doped solders, a small amount of Co is seen to be present in the Cu_6Sn_5 phase, as indicated by the energy dispersive spectroscopy analysis. Apparently, there was no Co incorporation into the Cu_3Sn IMC. This finding is similar to that obtained by Manauis *et al.* [2] who reported that nickel only incorporated into the $(Cu, Ni)_6Sn_5$ IMC, and there was no Ni present in Cu_3Sn IMC for the Sn-1.2Ag-0.5Cu solder ball alloy doped with Ni.

Cu_3Sn layer much thinner in the specimens doped with cobalt nanoparticles seemed at all aging time as compared to pure SAC (Compare Fig.1 with Fig.3 and Fig.4). A number of researchers [10, 12, 14] observed that the addition of Ni or Co into Sn-based solder tended to suppress the formation of Cu_3Sn by raising the activation energy of diffusion or altered the interdifussion of Sn/Cu. However, the total IMC layer for all the cases were almost the same. It was believed that by suppressing the growth of Cu_3Sn, the reliability of the solder joint could be improved. It is because Cu_3Sn is harder and more brittle than Cu_6Sn_5. [17]. Further investigations on this study have to clarify.

Fig. 5 and Fig. 6 show the growth of different IMC layers as a function of square root of aging time for the specimen with nano Co reinforcement. Results show that both of Cu_6Sn_5 and Cu_3Sn IMC layer increases as a function of square root of aging time which indicates diffusion as the main mechanism.

978-1-4244-4658-2/09 $25.00 © 2009 IEEE

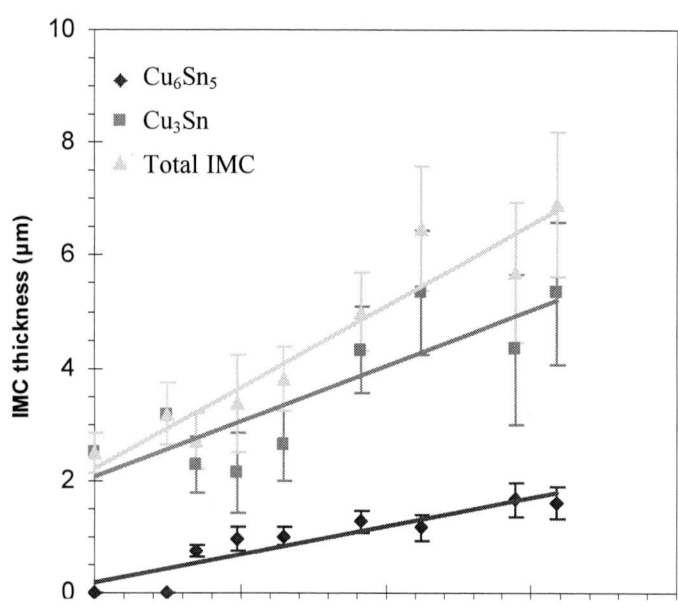

Figure 5   Growth of interface IMC compound with respect to square root of aging time for SAC-0.18 nano-Co.

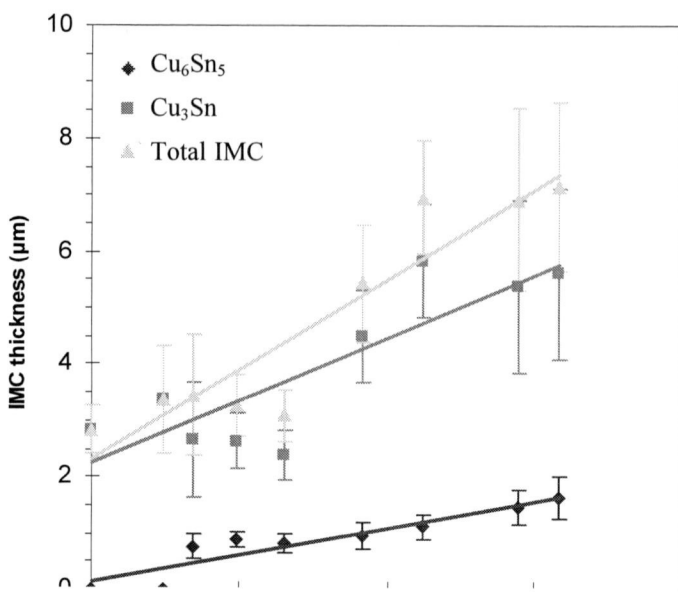

Figure 6   Growth of interface IMC compound with respect to square root of aging time for SAC-0.75 nano-Co.

Conclusion

Two IMC layers were formed in the interface solder joint, which were Cu_6Sn_5 and Cu_3Sn. Cu_3Sn only formed after aging for 24 hours for the specimen without Co-reinforcement and after 48 hours for the specimen reinforced with cobalt nanoparticles. Ag_3Sn intermetallic formed in the bulk solder tended to coarsen with the increased of aging time. The IMC layer increased linearly as a function of square root of aging time. With the nano Co reinforcement, growth of Cu_3Sn tended to be suppressed.

Acknowledgments

The authors would like to acknowledge the financial support from Ministry of Science, Technology and Innovation (MOSTI, with the project no. 13-02-03-3072) and Institute of Research Management and Consultancy (IPPP), University of Malaya with the project no.PS071-2008C.

References

1. M. Amagai, M. Watanabe, M. Omiya, K. Kishimoto, and T. Shibuya, "Mechanical Characterization of Sn-Ag-based Lead-free Solders," *Microelectroni. Reliab.* Vol. 42, (2002), pp. 951-966.

2. M. B. Manauis and A.L. Amorsolo, "Effects of Thermal Aging on Intermetallic Compound Formation in Sn-1.2%Ag-0.5%Cu Solder Alloy with Nickel Dopant on Copper Organic Solderability (Cu OSP) Substrate Surface Finish," *Proc 4th Int. Conf. on Recent Advances in Materials Advances in Materials Environment and 2nd Asian Symposium on Materials and Processing,* Penang, Malaysia, June. 2009.

3. H.T. Lee, and M.H. Chen, " Influence of Intermetallic Compounds on the Adhesive Strength of Solder Joints'" *Mater. Sci. Eng. A* Vol. 333, (2002), pp. 24-34.

4. V. Sivasubramaniam, N.S. Bosco, J. Janczak-Rusch, J.Cugnoni, and J. Botsis, "Interfacial Intermetallic Growth and Strength of Composite Lead-Free Solder Alloy through Isothermal Aging," *J. Electron. Mater.* Vol. 37, No. 10, (2008), pp.1598-1604.

5. D.R. Frear, and P.T. Vianco, "Intermetallic Growth and Mechanical Behavior of Low and High Melting Temperature Solder Alloys," *Metal. Mater. Trans. A-Phys. Metall. Mater. Sci.* Vol. 25, No.7, (1994), pp. 1509-1523.

6. J. Liang, N. Dariavach, and D. Shangguan, <u>Metallurgy, processing and reliability of lead-free solder joint interconnections</u>, ed. E. Suhir, Y.C. Lee, and C.P. Wong, Springer (USA, 2007), pp.351-409.

7. I. Ahmad, A. Jalar, B.Y. Majlis, and R. Wagiran, " Reliability of SAC405 and SAC387 as Lead-free Solder Ball Material for Ball Grid Array Packages," *Int. J. Eng. Tech.* Vol. 4, No.1 (2007), pp.123-133.

8. Z. H. Huang, P. P. Conway, and R.C. Thomson, "Microstructural considerations for Ultrafine Lead Free Solder Joints," *Microelectroni. Reliab.* Vol. 47 (2007), pp. 1997-2006.

9. D.Z. Li, C. Q. Liu, P. P. Conway, " Characteristics of Intermetalllics and Micromechanical Properties during Thermal Ageing of Sn-Ag-Cu flip-chip Solder Interconnects," *Mater. Sci. Eng. A,* Vol. 391(2005), pp. 95-103.

10. K. Suganuma, "Advances in Lead-free Electronics Soldering," *Current Opinion in Solid State and Mater. Sci.* Vol.5 (2001), pp.55 -64.

11. H.F. Zou, Q.S. Zhu, and Z. F. Zhang, "Growth kinetics of Intermetallic Compounds and Tensile Properties of Sn-Ag-Cu/Ag Single Crystal Joint," *J. Alloys Compd.* Vol. 461(2008), pp. 410-417.

12. F. Gao, H. Nishikawa, and T. Takemoto, "Additive Effect of Kirkendall Void Formation in Sn-3.5Ag Solder Joints on Common Substrates'" *J. Electron. Mater.* Vol. 37 No.1 (2008), pp. 45-50.

13. P.L. Eu, M.D. Dr., T.L. Wong, N. Amin, I. Ahmad, Y.L. Mok, and A.S.M.A. Haseeb, " A Srudy of SnAgNiCo vs Sn3.8Ag0.7Cu C5 Lead Free Solder Alloy on Mechanical Strength of BGA Solder Joint," Proc *10th Electron. Packaging Tech. Conf.,* Singapore, Dec. 2008, pp.588-594.

14. Y. H. Lee, H. T. Lee, "Shear Strength and Interfacial Microstructure of Sn-Ag-xNi/Cu Single Shear Lap Solder Joints," *Mater. Sci. Eng. A*, Vol.444 (2007), pp. 75-83.

15. W. K. Choi, and H. M. Lee, "Effect of Soldering and Aging Time on Interfacial Microstructure and Growth of Intermetallic Compounds between Sn-3.5Ag Solder Alloy and Cu Substrate," *J. Electron. Mater.* Vol. 29, No.10 (2000), pp.1207-1213.

16. F. Guo, J. Lee, S. Choi, J.P. Lucas, T. R. Bieler, and K. N. Subramanian, "Processing and Aging Characteristics of Eutectic Sn-3.5Ag Solder Reinforced with Mechanically Incorporated Ni Particles," *J. Electron. Mater.* Vol. 30, No.9 (2001), pp.1073-1082.

17. C. F. Chan, S.K. Lahiri, P. Yuan, and J. B. H. How, "An Intermetallic Study of Solder Joints with Sn-Ag-Cu Lead-free Solder," *Proc 3rd Electron. Packaging Tech. Conf.,* Singapore, 2000, pp.72-80.

18. F. Song, and S.W. R. Lee, "Investigation of IMC Thickness Effect on the Lead-free Solder Ball Attachment Strength: Comparison between Ball Shear Test and Cold Bump Pull Test Results," *Proc 56th Proceedings in Electronic Components and Technology Conf.,* San Diego, California, 2006.

Behaviors of Palladium in Palladium Coated Copper Wire Bonding Process

Binhai Zhang[1], Kaiyou Qian[2], Ted Wang[2], Yuqi Cong[2], Mike Zhao[2], Xiangquan Fan[1], Jiaji Wang[1]
[1]Department of Material Science, Fudan University, No. 220, Handan Road, Shanghai, 200433, China
[2]ASE Assembly & Test (Shanghai) Limited, No. 669, Guoshoujing Road, Shanghai, 201203, China
Tel: 86+21-55664588, Email: 072030005@fudan.edu.cn

Abstract

Because of its high thermal conductivity, great electrical property and low cost, copper wire is considered to replace the conventional gold wire and becomes widely used in IC assembly processes recent years. However, copper wire bonding also has its limitations. Copper oxidation, its high hardness and yield strength are two main disadvantages that manufacturers concern most. The application of copper wire coated with Palladium is a solution to prevent copper oxidation during the bonding process. Nevertheless, Pd coated copper wire brings in new possible influences to bonding interface and its reliability. This paper gave a systematic study on behaviors of Palladium in Palladium coated copper wire during the bonding process. SEM and EDS were used to analyze Pd distributions in copper wire, FAB (free air ball), bonded ball and its interface. It was shown that Pd distribution changed with bonding process going on and the factors that might cause these changes were discussed.

Introduction

In IC assembly processes, gold wire is widely used to connect IC chip pads to the lead frame because of its great electrical property and stable chemical property. But in the past few years, as the continuous rise of gold price, manufacturers have been investigating ways to replace the conventional gold wire with various new materials. Among these, copper wire is most promising [1-3]. Compared to gold wire, copper wire has the following merits which semiconductor manufacturers have great interest in.

1) Copper wire has high pull strength allowing for a reduced wire diameter, which leads to reduced pad size and pad pitch. This contributes to higher packing density of IC chips [4-5].

2) Copper wire has high thermal conductivity, low electrical resistance, contributing to better device performance.

3) Copper wire is much cheaper than gold wire, which reduced the overall cost of IC assembly.

Despite these merits, copper wire bonding also has its drawbacks. Copper oxidation, high hardness and yield strength are two main disadvantages that manufacturers concern most. Especially due to surface oxidation promoted mainly by heat condition from EFO process, bond strength are much lower [6]. The application of copper wire coated with Palladium is a solution to prevent copper oxidation. However, coating Pd into copper wire brings in new possible influences to bonding process and reliability [7].

Here, we give a systematic study on behaviors of Palladium in copper wire bonding process with Palladium coated. Strength changes are tested through a series of mechanical experiments. SEM and EDS are used to analyze Pd distributions in copper wire, FAB (free air ball), bonded ball and its bonding interface.

Experiment

We use two different kinds of Copper wires for test result comparison. One is pure 4N Copper wire while the other is coated with Pd. The diameters of both two wires are 25μm and the thickness of Palladium coating layer, the oxidation-resistant metal, is approximately 100nm. In this paper, for comparison convenience, all the bonding parameters such as EFO current and EFO time for two kinds of wire are the same.

First, we designed a series of experiments to find out how Pd coating has an influence on Cu wire bonding process. For mechanical experiment, we use Dage Corporation's Dage 4000 electronic test system, which has two mechanical test modes: shear test mode and pull test mode. The experiment parameters are in the table below. The first bonds strengths were measured by a shear test mode, and the second bonds by a pull test mode.

Table 1 experiment parameters in shear & pull test

	Shear test mode	Pull test mode
Height	5μm	5μm
Speed	300μm/s	300μm/s

Then, in order to further study the mechanism that leads to mechanical properties change, SEM and EDS methods were used to analyze Pd distributions in copper wire, FAB (free air ball), bonded ball and its bonding interface.

Results & Discussion

A. Bond Strengths Comparison of two different wires

The shear stress comparison for first bonds of two different wires is shown in Fig. 1. Results in two sets of bonding parameters both verified that copper wire with Pd coated has lower shear bond strengths at first bonds than that of pure 4N Cu wire. As all the bonding parameters in each set are the same, we get to the conclusion that Pd distribution at bonding interface has an impact on bond strengths.

The tensile stress comparison for second bonds of two different wires is shown in Fig. 2. In this test, both samples were heated for 16 hours under 180°C to better study the oxidation-resistant metal Pd's behavior. Results showed that copper wire with Pd coated has higher tensile bond strengths at second bonds than that of pure Cu wire. We believe that Pd metal layer joined in the forming of complex intermetal compounds (IMC) under temperature aging, which made much higher second bonds strengths.

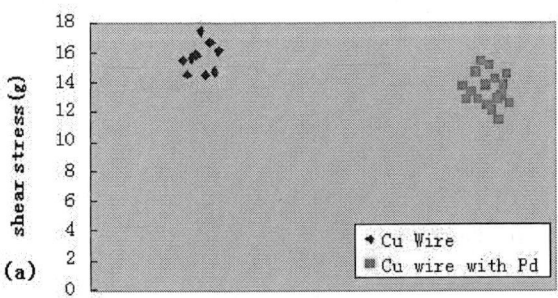

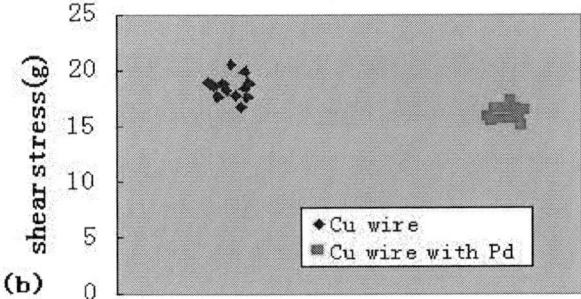

Fig.1. Shear stress comparison of copper wire and Cu wire with Pd coated(set a:EFO current is 80mA; set b: EFO current is 90mA)

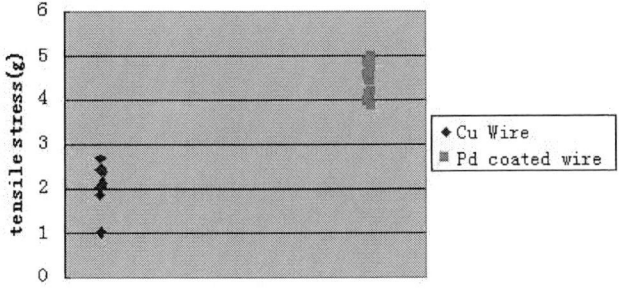

Fig. 2 tensile stress comparison of copper wire and Cu wire with Pd coated(both heated for 16 hours under 180°C)

B. Pd distributions on free air ball(FAB) of Pd coated copper wire

From the shear stress test (shown in Fig. 1), we found that Cu wire with and without Pd coating provide significant changes on mechanical properties at bonding interface. Therefore, original Pd distribution and its changes during the bonding process is what we intend to study below. In this paper, Pd's discussion will be focused on first bonds.

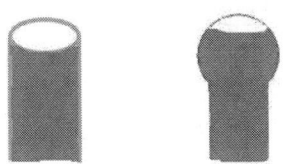

Fig. 3 flat headed wire tail (left) and Pd distribution after forming FAB (gray area is Pd coating)

As shown in Fig. 3, we first assume the section of Cu wire tail coated with Pd is horizontal flat-headed at the beginning , which means, after forming free air ball (FAB), Pd distribution is supposed to be spherical symmetric. And few Pd will be found on spherical crown of FAB. After bonding process, therefore, interface of bond and pad will be almost the same as that of pure 4N copper wires, containing few Pd. But this assumption doesn't accord with the results of shear stress test.

Using Energy-dispersive of X-ray (EDX), we analyzed actual Pd distribution on FABs. The EDX data are listed in Fig. 5 below, and SEM image of Fig. 4 shows locations of EDX detecting points.

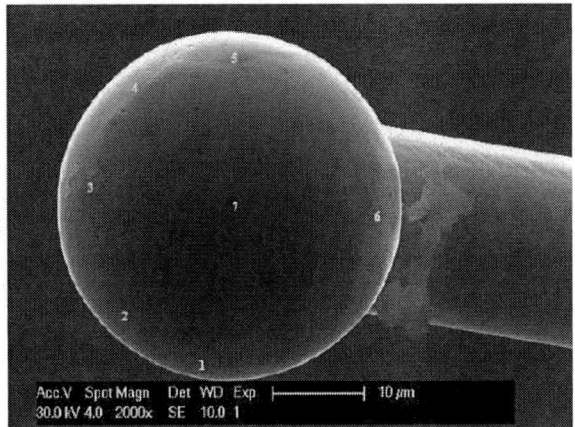

Fig. 4 Free air ball's SEM images of Cu wire coated with Pd (Number 1-7 shows locations of EDX detecting points)

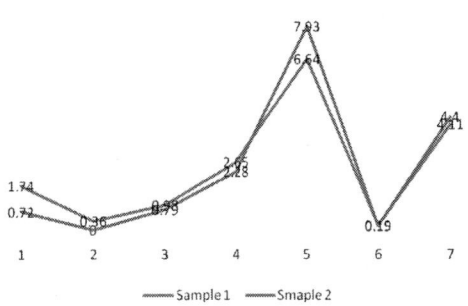

Fig.5 Pd content at different locations (No. 1-7) on FAB of Cu wire coated with Pd (two sets of data come from different FABs of the same wire)

EDX results show that Pd distribution is actually unsymmetrical and Pd signals were detected on the part of spherical crown of FAB. In order to clearly analyze Pd distribution on bonding area, more symmetrical points were detected on the spherical crown of FAB by EDX, shown in Fig. 6 and Table 2.

Bonding area's EDX results further verify Pd distribution is unsymmetry. The content of Pd in Orange area is much higher than gray area (shown in Fig. 7), and along the red arrow's direction, Pd's content reduces gradually. (4.87 to 4.62 to 1.15 to 0.34 to 0.07).

Through these experiments, we believe that Pd's unsymmetrical distribution leads to Pd's existence in part bonding area of FABs. In this situation, Pd partially joined in the Original Cu-Al intermetallic compounds (IMC) and changed mechanical properties of IMC, thus changes first bond's shear stress.

To explain the forming of unsymmetrical distribution, we propose an assumption that the tilt angle of wire tail's section is the main reason (Fig. 8). SEM photos confirm our assumption, showing that most of wire tail's sections have a sloping angle (Fig. 9).

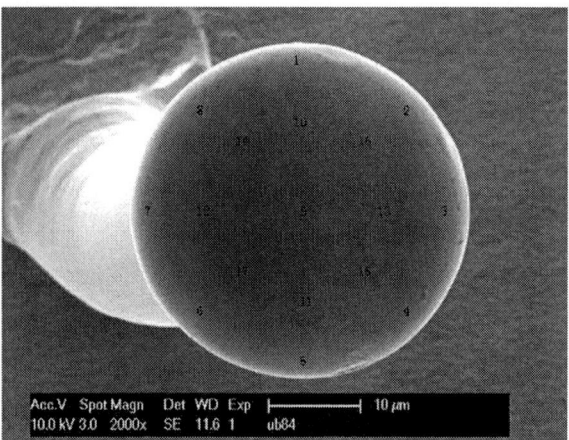

Fig. 6 bonding area's SEM images of Cu wire coated with Pd (Number 1-17 shows locations of EDX detecting points)

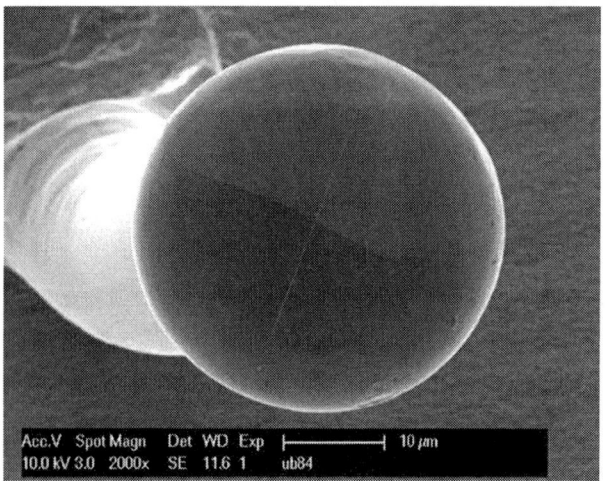

Fig.7 Pd content sketch image on FAB bonding area of Cu wire coated with Pd

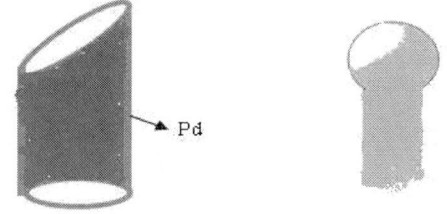

Fig. 8 imaginary actual section of wire tail (left) and Pd distribution after forming FAB (gray area is Pd)

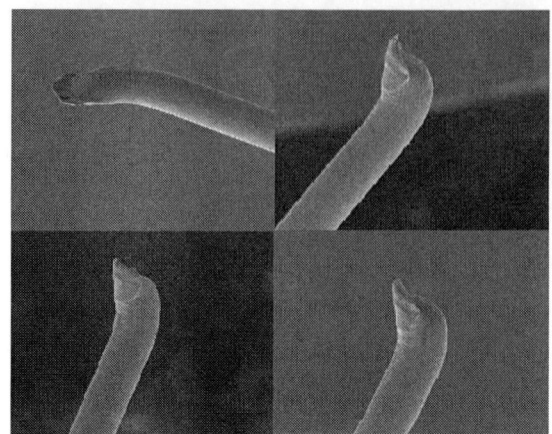

Fig. 9 SEM photo of wire tail showing tilt angle

Table 2 EDX results of number 1-17 points on Fig. 5

	Pd (At%)	Ag (At%)	Cu (At%)
01	2.53	1.11	96.36
02	4.87	38.35	56.78
03	4.63	40.20	55.17
04	0.85	10.72	88.43
05	0.89	9.80	89.31
06	0.07	8.11	91.82
07	0.08	3.54	96.38
08	3.45	0.74	95.81
09	1.15	1.25	97.60
10	3.26	0.07	96.67
11	0.38	1.96	97.66
12	1.56	0.71	97.78
13	2.56	1.53	95.91
14	6.14	0.25	93.11
15	0.88	2.98	96.14
16	4.42	0.43	95.15
17	0.34	2.17	97.49

C. Pd distributions on first bonds of copper wire with Pd coated

We are concerned about the changes of Pd distribution and Pd's possible joining into Cu-Al IMC layer. By packing with resin, rubbing and a series of polishing, a smooth cross section of bond was prepared. Similarly using EDX method above, we found that Pd distribution remained unsymmetrical (Fig. 10 and 11), Pd was only existed in the location of No.4-8 which was in accord with Chapter B's result.

Then we focus on Pd distribution on the possible IMC zone, which is directly related to first bond's shear stress. Though the exact thickness of the IMC layer could not be determined by EDX analysis due to its detecting limit, we use quasi-continuous EDX method to measure IMC thickness approximately (about 2-3μm, shown in Fig. 12). Afterwards, a few points in possible IMC area were detected from left to right by EDX, trying to find out Pd content distribution. The EDX results are shown in Fig. 13. We found that Pd distribution in IMC area is still unsymmetrical; form left to right, Pd content reduced gradually. No Pd signals were detected in IMC area right of point 6. We believe that Pd's existence in left part of IMC layer lead to the decrease of shear stress at first bonds. And the length of IMC contained Pd is related directly with the tilt angle of wire tail's section.

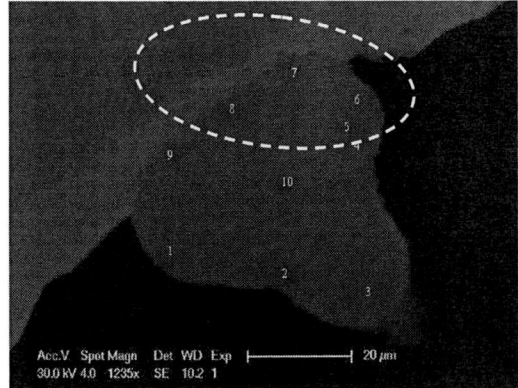

Fig.10 Bond's cross section's SEM images of Cu wire coated with Pd (No. 1-10 shows locations of EDX detecting points)

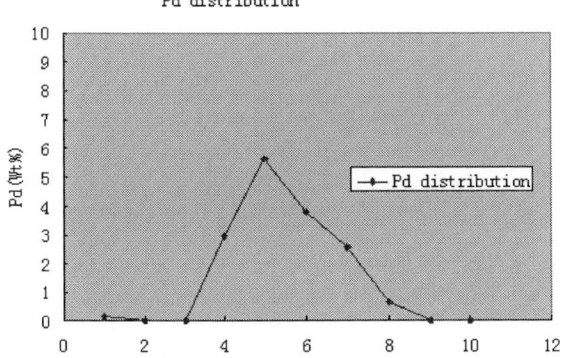

Fig.11 Pd distribution on cross section of bond coated with Pd (number 1-10 points showing in Fig.10)

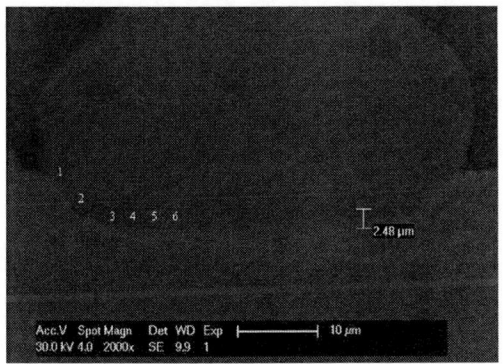

Fig.12 Bond's cross section's SEM images of Cu wire coated with Pd (Number 1-6 shows locations of EDX detecting points)

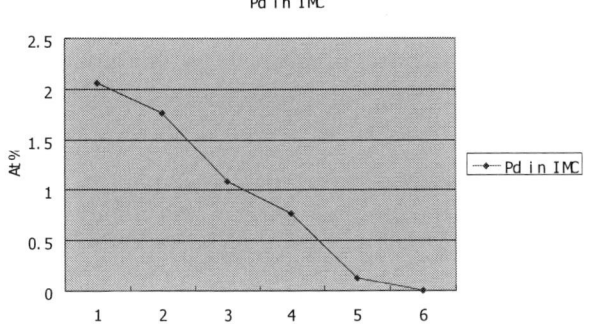

Fig.13 Pd distribution in possible IMC area (number 1-6 from left to right)

Conclusions

From a series of mechanical experiment and discussion, it is verified that coating copper wire with Pd prevents copper from oxidation, but meanwhile reducing the shear stress at first bonds.

We also find out that wire tail's shape decides Pd's distribution: flat-headed wire tail leaded to Pd's symmetrical distribution, while actual wedge-shaped wire tail caused Pd's unsymmetrical distribution on the surface of free air ball (FAB), so as to the bond ball and Cu-Al IMC interface, which were further confirmed by SEM and EDX analysis. The mechanism of Pd's behavior in IMC will be further studied in next research.

References

1. G. G.Harman, Wire Bonding in Microelectronics, 2nd ed., McGraw-Hill (New York, 1997), pp. 1-2.
2. T. W.Ellis et al., "Copper: Emerging material for wire bond assembly," *Solid State Technol.*, Vol. 43, No. 4 (2000), pp. 71-77.
3. Salim L. Khoury, David J. Burkhard, David P. Galloway, "A Comparison of Copper and Gold Wire Bonding on Integrated Circuit Devices," *IEEE Trans. on components, hybrids, and manufacturing technology*, Vol. 13, No. 4 (Dec. 1990).
4. R. R. Tummala, Fundamentals of Microsystems Packaging, 1st ed., McGraw-Hill Professional (New York, 2001), pp. 346-355.
5. S. Ouimet et al., "Overmold technology applied to cavity down ultrafine pitch PBGA package," *IEEE Trans. Adv. Packag.*, Vol. 22, No. 2 (May. 1999), pp.123-128.
6. S.Kaimori, T.Nanoka, and A.Mizoguchi, "The Development of Cu Bonding Wire with Oxidation-Resistant Metal Coating," *IEEE Trans. Adv. Packag.*, Vol. 29, No. 2 (May 2006), pp.227-231.
7. Sang-ah Gam, Hyoung-joon Kim, Jong-soo Cho, Yong-jin Park, "Effects of Cu and Pd Addition on Au Bonding Wire/Al Pad Interfacial Reactions and Bond Reliability," *Journal of ELECTRONIC MATERIALS*, Vol. 35, No. 11 (2006).

Effect of Electromigration on Intermetallic Compound Formation in Cu/Sn/Cu Interconnect

L. D. Chen, Mingliang Huang*

Electronic Packaging Materials Laboratory, School of Materials Science & Engineering, Dalian University of Technology,
Dalian, 116024, China
*Corresponding author, E-mail: huang@dlut.edu.cn, Tel: +86-411-84706595

Abstract

The effect of electromigration on the solid state interfacial reaction of pure Sn and Cu was investigated. Two electron current densities ($1.0 \times 10^4 A/cm^2$ and $5.0 \times 10^3 A/cm^2$) were applied to line-type Cu/Sn/Cu interconnect at $150^\circ C$. The same types of intermetallic compound (IMC), Cu_6Sn_5 and Cu_3Sn, formed at the Sn/Cu interfaces independent of electric current. A high current density caused a polarity effect where the IMC layer at the anode grew significantly thicker than that at the cathode. The growth of IMC was enhanced by electric current at the anode, comparing with that of the samples without applying current. The growth of IMC at the anode followed a parabolic growth rule. After current stressing for 100 hours with a current density of $1.0 \times 10^4 A/cm^2$, microcrack formed at the interface between IMC and solder at the cathode side, while there was no crack formed even after stressing for 200h with a current density of $5.0 \times 10^3 A/cm^2$.

Introduction

Electromigration is a mass transportation caused by the directional electron flow. It is basically a diffusion phenomenon under a driving force. Investigations of this phenomenon can be traced back about a century, with the first observation reported by Geradin (1861) in molten alloys of lead-tin and mercury-sodium [1].

There are two trends in microelectronic packing industry, the first is that due to health and environmental concerns, lead-free solders are used to replace Sn-Pb solder. Among all the candidates Sn-rich alloys, such as Sn-3Ag-0.5Cu, Sn-0.7Cu, Sn-9Zn and Sn-3.5Ag are leading candidates to replace SnPb. In these alloys, the properties of tin dominate the solder behavior because of the high content of tin (more than 90wt.%). Therefore it is important to understand the interfacial reaction between Sn and metallization.

The second is that the diameter of the flip-chip solder joints is shrinking to accommodate the continuing rise of the I/O density. Currently, the design rule requires that each flip-chip solder joints of 50 μm in diameter carries 0.2A, which means that the average current density in such a joint is about $10^4 A/cm^2$ [2] The International Technology Roadmap for Semiconductor (ITRS) projections indicated that electromigration is a near-term issue in high current density packages [3].

C.Y. Liu et al. [4] studied the electromigration behavior of eutectic SnPb, they found that at room temperature Sn is the dominant diffusing species while at $150^\circ C$ Pb is the dominant diffusing species. X. F. Zhang et al. [5] found an abnormal polarity effect after direct current was applied on Sn-Zn solder/Cu interconnect where the IMC at the cathode was thicker than that at the anode, this finding is in contrast to the Sn-Pb, Sn-Ag-Cu interface, where the interconnects have a

polarity effect that the IMC formed at the anode side is thicker than that at the cathode side [6-7]. Obviously the behavior of electromigration is very complex process in solder interconnects. Until now there is no literature report on the effect of current on the interfacial reaction of Sn and Cu. So it is interesting to study the interfacial reaction of pure tin interconnect under current stressing.

In an actual flip-chip joint, the cross section of a solder bump is two orders of magnitude larger than that of an interconnect wire, so at the entrance of a flip chip joint there must be a very significant current crowding occurs. Also due to the different electrical resistances and thermal capacities between solder bump and interconnect wire, it is possible that larger heat accumulates at the chip side than at the substrate which will lead to a considerable temperature gradient across the solder joint. This temperature gradient can cause additional thermomigration. So electromigration in a real flip-chip joint is very complex. Therefore, in order to understand the fundamentals of the electromigration mechanism in solder joints, an appropriate solder joint configuration should be designed to minimize other effects such as thermomigration and current crowding.

In this study we chose line-type Cu/Sn/Cu interconnect to investigate the effect of electromigration on interfacial reaction. Because of its symmetric design this type of interconnect avoids current crowding effect and theromigration.

Experimental Procedure

A Cu/Sn/Cu line-type interconnect was prepared. Before soldering, two small copper plates, 7mm wide, 5mm thick, and 12mm long, were cut from a big copper plate. Of each platelet one face (5mm×7mm) was carefully ground and polished using sandpaper and 1.5μm, 0.5μm diamond paste to make a smooth surface. The polished faces were finally rinsed with ethanol. Both of the cleaned substrates were immediately plated with Sn for 3 mins to avoid void formation during soldering. After plating with Sn, the two Cu sticks were aligned and fixed before the assembly was immersed into Sn bath at $260^\circ C$ for 10s. The space between the two Cu sticks was controlled by two stainless steel lines (D=500μm). After water cooling the soldered interconnects were cut into slender bars with a cross-section of roughly 600μm×600μm by electric discharge machining. Then, the samples were ground and polished down to 300μm×300μm.

Electromigration experiment was conducted by applying a direct current to both ends of the line-type interconnect by using coppering wires as the electrical leads. The solder joint temperature was monitored by a K-type thermocouple placed next to the solder interconnect. The current density in the interconnect were 5.0×10^3 A/cm^2 and $1.0 \times 10^4 A/cm^2$, the temperature of the samples was $150 \pm 4^\circ C$. The current stressing was applied for 20h, 50h, 100h and 200h. For comparison,

978-1-4244-4658-2/09 $25.00 © 2009 IEEE

samples aged at 150°C without a current passing through were also prepared. After the aging treatment, samples were removed from the oven and cooled in air to room temperature.

Fig. 1 shows the Cu/Sn/Cu line-type interconnect after polishing. As shown in Fig. 1, Sn and Cu were well metallurgy soldered, no microvoid was observed after soldering.

Fig. 1 Line-type Cu/Sn/Cu interconnect after poling

Fig. 2 shows schematically the experimental setup for electromigration experiment. The interconnect was placed on a sheet of glass. Both ends of the sample were fixed by a copper clamp. Two copper clamps, were connected to a constant current source. The entire fixture was then immersed in an oil bath held at a constant temperature of 423K.

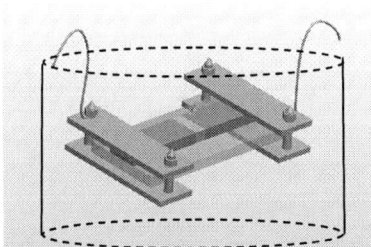

Fig. 2 Schematic of the interconnect samples with passage of electric current

After current stressing, the interconnects were polished and then the solder was etched by using acid (3vol.% HCl+5 vol.%HNO$_3$ + 92vol.% C$_2$H$_5$OH) for a few seconds to reveal the microstructure. The microstructure evolutions with different current densities were investigated by scanning electron microscopy (SEM). The chemical composition of the precipitates and phases between the solders and the substrates were analyzed with EDX.

Results and discussion

The as-soldered interface is shown in Fig. 3. A scallop Cu$_6$Sn$_5$ IMC layer of less than 1μm was formed at the solder/substrate interface during the soldering process. The formation of IMCs at the interface between Cu and Sn is very crucial to achieve a strong joint strength. However, excessive amount of compound formation will cause deterioration in the

mechanical strength of the solder joint, due to its brittle nature. No obvious Cu$_3$Sn layer was observed at the interface between Cu$_6$Sn$_5$ and Cu using SEM.

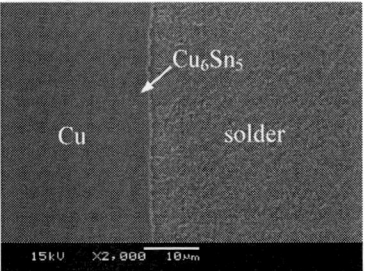

Fig. 3 SEM micrographs of the interface between Cu and Sn after soldering

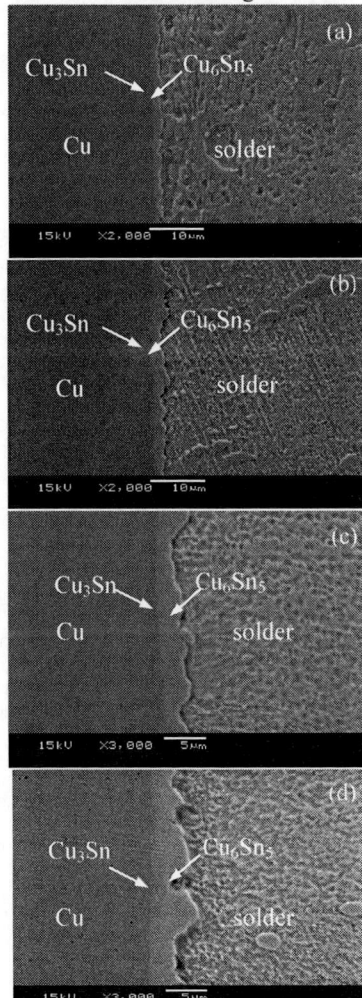

Fig. 4 SEM micrographs of the solder joints after aging at 150°C (without current stressing) for (a) 20 h, (b) 50 h, (c) 100 h, and (d) 200 h

Fig. 4 shows the interfacial reaction of samples without current loading, the scallop-type IMC transforms into layer-type IMC after aging.

As the aging time increasing the IMC became thicker and thicker. The IMC reached 3.6μm when the interconnect was aged at 150°C for 200h. Due to the electronic industry continuously shrinks the solder bump size and uses thick under bump metallization (UBM) structure to satisfy the trend of miniaturization, so this excessive IMC growth problem becomes more and more serious. The thickness of IMC at both side of the interconnect are the same.

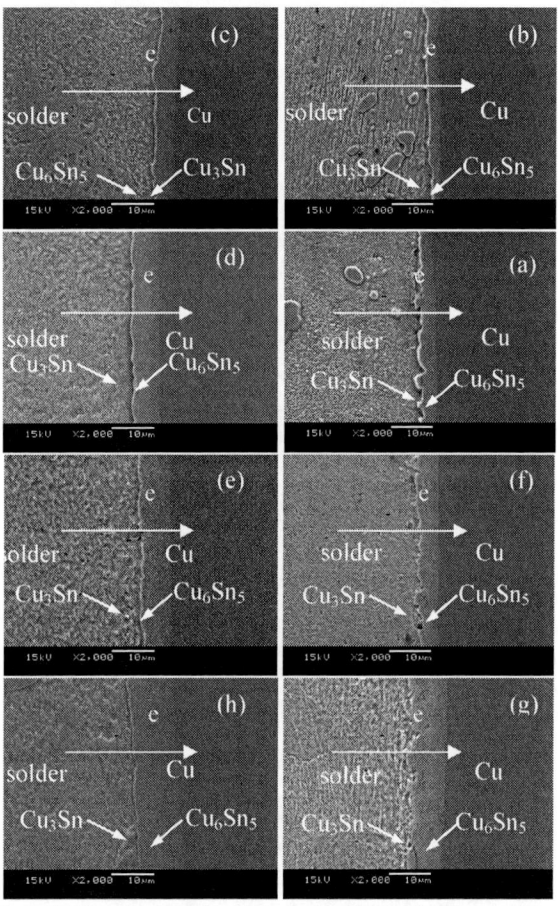

Fig. 5 SEM micrographs of the solder joints (at the anode) at 150°C after current stressing for (a) 20 h, (b) 50 h, (c) 100 h, (d) 200 h with current density of $1.0 \times 10^4 A/cm^2$; and (e) 20 h, (f) 50 h, (g) 100 h, (h) 200 h with current density of $5.0 \times 10^3 A/cm^2$

Fig. 5 gives a series of solder interconnect interfaces after being loaded with current density of $1 \times 10^4 A/cm^2$ and $5.0 \times 10^3 A/cm^2$ for different durations. The major difference between Fig. 4 and Fig. 5 is that the driving force for the case without current loading is chemical potential while the driving forces for the samples loaded with direct current include

chemical potential, electron wind force and electrostatic field force.

As shown in Fig. 5, the IMC at the anode side changed from scallop-type to layer-type and the thickness of the IMC became thicker and thicker as the loading time increasing. After the interconnect was applied for 200h with current density of $1.0 \times 10^4 A/cm^2$, the thickness of the IMC was 12.1μm which was 3.36 times thicker than that without current case. When the applied current density was $5.0 \times 10^3 A/cm^2$, the IMC at the anode side reached 7.8μm, which was thicker than the no-current case but thinner than the sample applied with $1.0 \times 10^4 A/cm^2$. Gan and Tu [5] also found that electromigration enhanced the growth of IMC at the anode side when compared with the no-current case. Their result confirmed our results. The thickness of IMC in the left column of Fig. 5 was thicker than that in the right column, which indicated that the IMC at the anode side grew more quickly when higher current density was applied.

To clearly show the growth kinetics of the IMC, we plotted the IMC thickness as a function of the square root of loading time. The measured thickness was the total thickness of the Cu_6Sn_5 and Cu_3Sn layers. The solid round symbols in the plot reflect the thickness of the anode side when a $1.0 \times 10^4 A/cm^2$ current was applied to the samples, the solid triangle symbols represent the IMC of the anode side when a $5.0 \times 10^3 A/cm^2$ current was applied to the samples and the solid square symbols represent the corresponding data for the no-current case.

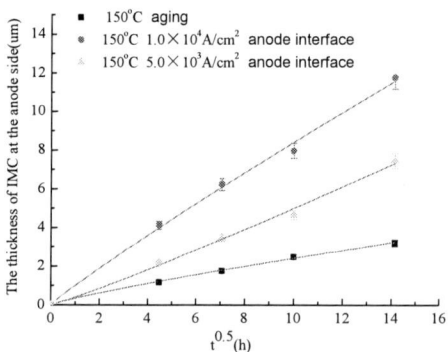

Fig. 6 Variation of the IMC thickness with the square root of time at 150°C

As can be seen from Fig. 6, the growth of IMC at the anode side had a parabolic dependence on time since the thickness of IMC increased linearly with square root of time. That means the interfacial reaction at the anode side was determined by the diffusion process. Electromigration enhanced the diffusion velocity. The growth of IMC depended upon the mass fluxes flowing in and out. When the inward atomic fluxes were larger than the outward fluxes the IMC grows; otherwise, the IMC shrinks.

The thickness data of the anode were higher than the no-current case. So, IMC grew more quickly at the anode than the

978-1-4244-4658-2/09 $25.00 © 2009 IEEE

no-current case, electric current enhanced the growth of IMC at the anode.

The growth of the IMC at the cathode was much more complicated due to the void formation. So in this study we focused on the IMC growth of anode, the no-current case were used as a reference. Fig. 7 shows microcrack formed at the cathode side after current stressing for 100 h with a current density of 1.0×10^4 A/cm^2, while there was no crack formed at the anode side. There was no crack observed at both anode and cathode sides when a current of 5.0×10^3A/cm^2 was applied.

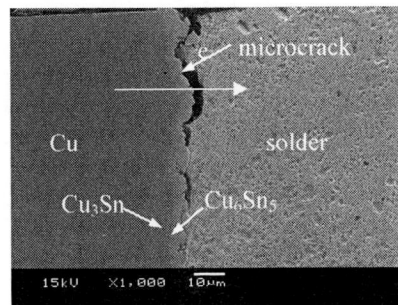

Fig. 7 Microcrack formed at the cathode side after current stressing for 100 h with current density of 1.0×10^4A/cm^2

As shown in Fig. 7, the microcrack was formed at the interface between IMC and solder. The reason of why microcracks were formed at the cathode side was that due to the applied current, the electron will deliver its momentum to the atoms. Atoms therefore moved forward the anode side and then generated a back stress which will draw voids to the cathode side, as the stressing time increased more voids will accumulated and then microcracks will be formed. If microcracks were formed in real flip chip solder joint it will deteriorate the reliability of the solder joint.

Conclusions

The electromigration behavior for the Cu/Sn/Cu interconnect was observed after current stressing at 150°C with two current density (1.0×10^4A/cm^2 and 5.0×10^3A/cm^2). The following conclusions can be drawn:

1. Electric current enhanced the growth of IMC at the anode, compared with those in the samples without current loading. The growth of IMCs at the anode had a parabolic dependence with time, which was same as the no-current case. However, the growth rate increased when current was applied and the growth rate became higher when larger current was applied to the sample.

2. The IMC had a scallop-type morphology when it formed after soldering, and it transformed into layer-type morphology after current stressing which was similar to the no-current case.

3. After stressing for 100h with current density of 1.0×10^4A/cm^2, mircocracks formed at the interface between IMC and solder at the cathode side, while there were no cracks

observed even after stressing for 200h with a current density of 5.0×10^3A/cm^2.

Acknowledgments

This work is supported by National Key Technologies R&D Program (2006BAE03B02-2), NSFC key program (No.U0734006), Key Laboratory program in Liaoning Province (20060133) and Key program in Dalian (2006A11GX005).

References

1. Paul S Ho and Thomas Kwok, "Electromigration in metals," *Rep. Prog. Phys.*, Vol. 52 (1989), pp. 301-348.
2. K.N. Tu, "Recent advances on electromigration in very-large-scaleintegration of interconnects," *J. Appl. Phys.*, Vol. 94 (2003), pp. 5451-5473.
3. Assembly and Packaging Section. International Technology Roadmap for Semiconductors (2005 Edition, ITRS: San Jose,CA), p. 2 Table 93a (http://www.itrs.net/Links/2005ITRS/Home2005.htm).
4. C.Y. Liu, C. Chen, C. N. Liao, and K. N. Tu, "Microstructure-electromigration correlation in a thin stripe of eutectic SnPb solder stressed between Cu electrodes," *Appl. Phys. Lett.*, Vol. 75(1999), pp. 58-60.
5. X. F. Zhang, J. D. Guo, and J. K. Shang, "Abnormal polarity effect of electromigration on intermetallic compound formation in Sn-9Zn solder interconnect," *Scripta Materialia*, Vol. 57(2007), pp. 513-516.
6. T.Y. Lee, K.N. Tu, S.M. Kuo, and D.R. Frear, "Electromigation of eutectic SnPb solder interconnects for flip chip technology," J. Appl. Phys. Vol. 89(2001), pp. 3189-3194.
7. Y.C. Hsu, C.K. Chou, P.C. Liu, C. Chen, D.J. YAO, T. Chou, and K.N. Tu, "Electromigration in Pb-free SnAg3.8Cu0.7 Solder Stripes," *J. Appl. Phys.*, Vol. 98(2005), pp. 033523-1-033523-6.
8. H. Gan and K. N. Tu, "Polarity effect of electromigration on kinetic of intermetallic compound formation in Pb-free solder V-groove samples," *J. App Phys.*, Vol. 97(2005), pp. 063514-1-063514-110.

Solderability of Tinned FeNi under Bump Material

Cai Chen[1], Lei Zhang[1], J. K. Shang[1,2]

[1]Shenyang National Laboratory for Materials Science
Institute of Metal Research, Chinese Academy of Sciences, Shenyang 110016, China
Email: lzhang@imr.ac.cn, Tel: +86-24-2397-1551
[2]Department of Materials Science and Engineering
University of Illinois at Urbana-Champaign, Urbana, IL 61801, USA

Abstract

Finding new under bump materials (UBM) is of significant importance with the shrinkage of pitch size in advanced high density packaging. FeNi alloy has been found to be a good UBM candidate for its slower interfacial reaction during reflow and aging. However, solderability of FeNi has to be improved. 58Fe42Ni and 92Fe8Ni alloy platings with electrodeposited tin finish were investigated by wetting balance method in a SnAgCu liquid bath. Results showed that tin finish can reduce the initial wetting time t_b of 58Fe42Ni and change the non solderable 92Fe8Ni to be the same wetting performance as 58Fe42Ni. The maximum wetting force of tin finished both FeNi alloys were larger than the bare alloy. However, after the maximum was reached, a slight decline of the wetting force was observed. The decline depended on the thickness of tin finish, with the thicker tin finish resulting in a higher steady plateau on the wetting force curve.

Introduction

In the microelectronic package industry, under Bump Metallization (UBM) has a significant effect on the reliability of solder joint. To gain a good solderability and reliable interconnection, intermetallic compound (IMC) is essential. Traditional UBM materials, such as copper and electroless nickel, can be quickly consumed by tin-rich solder during reflow and a thick brittle IMC layer is formed subsequently [1, 2, 3, 4]. Numerous research results show that thick IMC could easily become crack initiations which eventually lead to the failure of interconnection [5, 6, 7].

FeNi alloy has been found to exhibit a much slower IMC growth rate during reflow [8], and SnAgCu solder balls on electrodeposited FeNi layer shows that FeNi layer can be used as a reliable under bump metallization material. However, FeNi oxidizes easily in air which poisons the solderability of FeNi as the UBM layer [9]. Tin has been used as a surface finish to protect the substrate from oxidation for a long time [10, 11, 12]. The main purpose of our present work was to investigate whether tin can be used as a protective surface finish on electroplated FeNi layer.

Wetting balance method is one of the most import tools in the determination of quantitative wetting properties. During the test, a coupon immerses into a molten solder bath, and the force exerted on the plate is measured by the balance. The wetting curve reveals the dynamic relation between the time and force, as shown in Fig.1. It records the evolution of wetting force and typical time point such as wetting start time t_b when the wetting force begins to increase.

Metallurgical reaction between liquid solder and UBM materials at the interfaces is an essential process for bonding formation, and the compounds formed at the interface have obvious effects upon the wetting properties. Thus, the wetting performance should have a strong relationship with the intermetallics formed. Tin finished FeNi plates may exhibit specified interfacial metallurgy different from traditional UBM based on Copper or Nickel. From the wetting start time and the wetting force results from wetting balance test, the solderability of the new UBM was evaluated and the interfacial reaction between the molten solder and the substrates was discussed.

Experimental Details

Pure copper sheet (0.2mm thick) was cut into a dimension of 10 mm × 10 mm × 20 mm, polished with 0.5 μm diamond paste, ultrasonic cleaned in acetone for 30 minutes to remove the grease, and finally dried with pure nitrogen gas gun. Two kinds of FeNi alloy with different Fe contents (wt %) were selected, nominally 58Fe42Ni alloy and a high Fe content alloy 92Fe8Ni. The content of the Fe and Ni was controlled by the ratio of Fe and Ni ion in the electrolyte solution [10]. FeNi alloy electroplating was carried out under a current density of 1.5A/dm² at 318K for 30 minutes. The thickness of the FeNi plating was in a range of 3~4 μm. After flushing in distilled water, rinsing in alcohol for a few second and drying with dry nitrogen air, the sample was quickly transferred into tin electroplating bath. Tin electroplating was carried out under a current density of 1.5A/dm² at room temperature for different times 1 min, 2 min, 4 min, 6 min, and 10 min.

Solderability of electroplated specimens both as-electroplated and stored for a certain time was measured on an SKC-8H (DM Analytical Instruments) wetting balance with an accuracy for wetting force measurement of better than 0.02mN. Wetting balance test was performed at 445K in Sn3.8Ag0.7Cu liquid solder using RMA flux. The immersion depth was 4 mm and the immersion velocity was 20 mm/s.

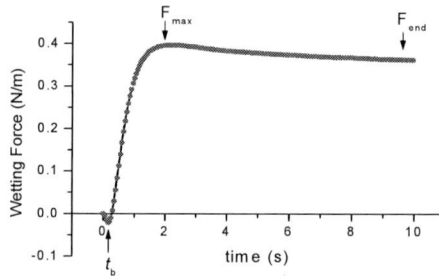

Fig. 1 A wetting curve of tin finished FeNi alloy plating

The typical data points, initial wetting time t_b, maximum wetting force F_{max} and final wetting force F_{end}, on the curve are marked.

Surface morphology and cross-section image of the samples after test were taken by using scanning electron microscopy (SEM). Composition of the platings and interfacial products were analyzed by the associated energy dispersive X-ray spectroscopy (EDS, Oxford link ISIS300) on the SEM. The surface of platings was analyzed in ESCALAB 250 UHV surface analysis system by X-ray photoelectron spectroscopy (XPS) with monochromatized Al Kα excitation line. By using an integrated Ar-ion gun attachment in the system for surface sputtering, the composition depth profiles were obtained.

Results and Discussion

The thickness of electroplated tin finish was calculated from the mass increment by using the following formula:

$$h = \frac{m}{\rho \cdot s}$$

where h represents the thickness of tin finish; m represents the weight of tin finish; ρ represents the density of tin finish, and s represents the electroplating area of the sample. Here we use the volume density about 7.28g/cm^3.

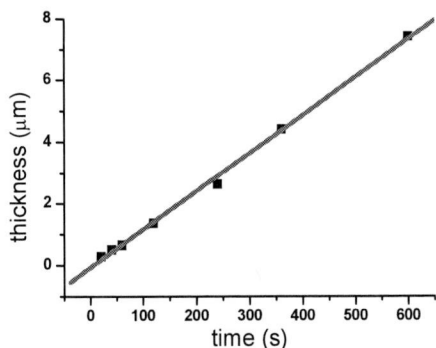

Fig. 2 The electroplating velocity of tin finish

The validation of this method to determine tin plating thickness has been proven by other researchers. According to the calculation results, the increment rate of tin thickness was around 0.74μm/min. It followed a linear relationship with time expansion, as shown in Fig. 2.

Fig. 3 (a) and (b) shows the surface morphologies of the original FeNi alloy platings, 58Fe42Ni and 92Fe8Ni, respectively. The 58Fe42Ni plating showed a fine crystal and smooth surface. In contrast, 92Fe8Ni exhibited a polygon crystal and much rougher surface. Such morphologies might be due to crystalline growth mechanism with difference lattice structure. According to the research of Guo etc. [10], 58Fe42Ni has a face centered cubic (fcc) structure while 92Fe8Ni has a body centered cubic (bcc) structure. As shown in Fig. 3 (c)-(d), the morphology of tin finish changed from granular to lamellar structure as the tin layer getting thicker; and the grain size grows as well.

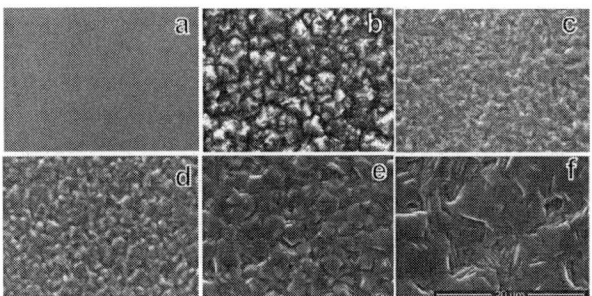

Fig. 3 Surface morphology of original 58Fe42Ni electroplating (a); original 92Fe8Ni electroplating (b); 58Fe42Ni electroplating with tin finish of (c) 1 min; (d) 2 min; (e) 4 min; (f) 10 min, respectively;

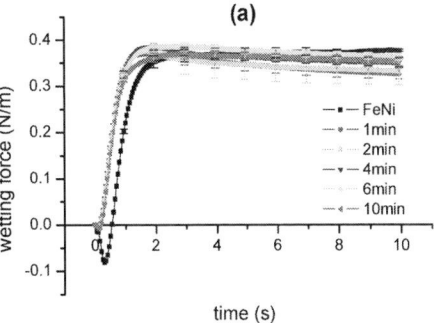

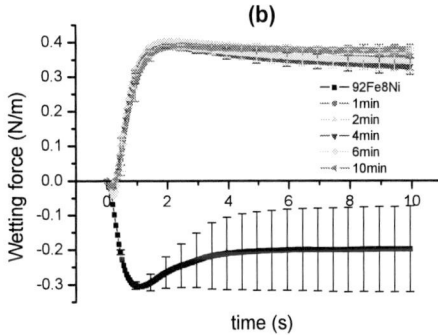

Fig. 4 Wetting curves of various tin-finished FeNi alloy: (a) as-plated 58Fe42Ni and tin finishing for 0min, 1min, 2min, 4min, 6min, 10min; (b) as-plated 92Fe8Ni and tin finished for 0min, 1min, 2min, 4min, 6min, 10min

The wetting curves of various tin plated FeNi specimens are illustrated in Fig. 4. The initial wetting time t_b, maximum wetting force F_{max} are the two key parameters to evaluate solderability. In Fig. 4 (a), the initial wetting times t_b is 0.54s for as-plated 58Fe42Ni and 0.15s after tin finish, which shows that tin can remarkably reduced t_b. It is obvious that tin finish can reduce the time for solder becomes wettable on the sample compared with the bare FeNi. The value of t_b does not change with the thickness of tin. The final wetting forces F_{end} were nearly the same for both bare 58Fe42Ni and tin-coated 58Fe42Ni. It should be noted that the wetting force may reach

978-1-4244-4658-2/09 $25.00 © 2009 IEEE

a climax force F_{max}. Before the end of the test, it declined 5% ~ 10% at most, which depend on the thickness of tin finish. Declination of wetting force became smaller as tin finish getting thicker.

Fig. 4(b) shows that 92Fe8Ni with no tin finish cannot be wetted by SnAgCu solder. Whereas, specimens with tin electroplating even for 1 min had a good solderability. The wetting force also appears a maximum and declination for this high iron baring alloy with tin finish. The wetting performance is quite similar to 52Fe48Ni with tin finish. It implies that the function of tin finish on wetting is independent of the FeNi alloy composition as the tin thickness increasing.

Fig. 5 show the solderability of 92Fe8Ni alloy electroplated 2min tin after stored in ambient atmosphere for a couple of times. The wetting force curves were repeatable, F_{max} and F_{end} has no obvious change as the storage time extended more than 30 min. The wetting initiation time t_b for originally tinned sample is 0.24s, and changed to 0.28s after 86 hours exposure in air. Although the difference is not that obvious, a small increment of t_b is aroused from the oxidation layer growth on top of tin finish, since the initiation of wetting is very sensitive to the state of the surface oxidation. However, the layer thickness growth of tin oxidation follows a parabolic curve. It means longer exposure time does not obviously increase the thickness of oxide layer. So preformed oxide layer can prevent tin from further oxidation. It was undoubted that FeNi alloy was away from oxidation by tin coverage. That is one of the keys for a good solderability.

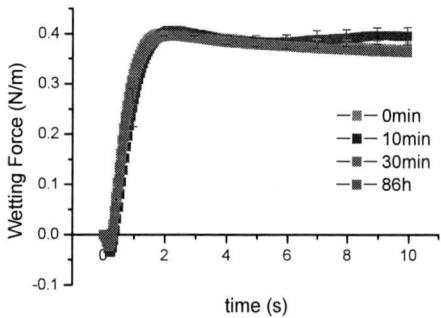

Fig. 5 Wetting curve of 92Fe8Ni alloy with 2min tin finishstored in air for 0min, 10min, 30min, and 86h.

Fig. 6(a) demonstrates the coupon specimen with the tin finish immersed in a solder bath. Due to the surface tension of liquid solder on the coupon, a meniscus with elevated liquid formed on both sides of the coupon. Around the tip of the liquid meniscus, three regions can be separated. In Region 1, the finished tin does not melt and has no effect on the wetting performance; Region 2 is a transitional region, where a part of tin finish closer to the liquid meniscus melted and take part into the wetting and Region 3 is the part fully underneath the liquid meniscus. Fig. 6(b) is a cross-section view of the transitional region 2 after the wetting balance test. Closer to the meniscus is on the left side, where the solder covered the coupon was re-solidified from the tin melt after the wetting balance test. One right side, the tin covered coupon is the original electroplated tin finish. In between, a part of the coupon is open and without any tin overlayer. It indicates that dewetting phenomenon happened during the wetting balance test.

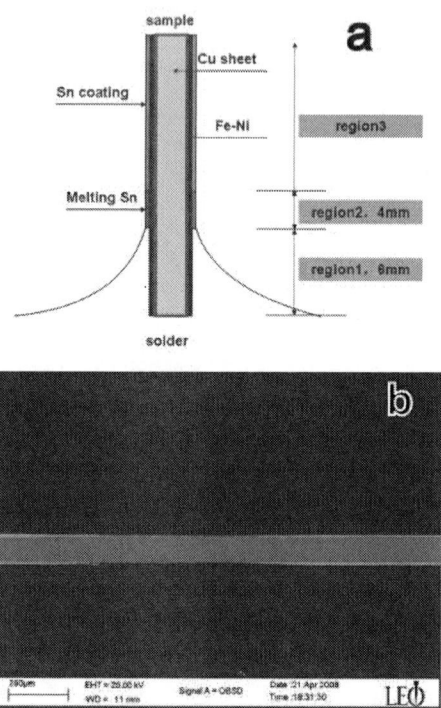

Fig. 6 (a) A schematic illustration of the steady state that the test specimen immersion in a liquid solder bath during wetting balance test; (b) SEM cross-section view of a part of Region 2, where liquid solder dewetting happened.

Dewetting cannot be attributed to the oxidation at the interface between FeNi and Tin since no oxygen was found at the interface, according to XPS depth profile analysis of chemical composition in Fig. 7. Oxygen only existed on the surface of the tin finish and formed tin oxide with a thickness of less than 30 nm. During the wetting balance test, the coupon dipped into the liquid solder, and a part of tin finish ahead of the wetting front melted. The instantaneous contact angle between molten tin and FeNi substrate was almost zero. This state was unstable. The thin tin liquid film was broken and receded from the FeNi substrate. Then it moved towards the formation of the intrinsically contact angle, which is the thermodynamic stable contact angle of the liquid tin solder on the FeNi alloys. It was also observed that the liquid solder can oxidize at a higher test temperature of 545K, at which the RMA flux tends to loss its activity rapidly. If there is excessive molten tin, dewetting phenomenon must be suppressed because there was no enough space for the liquid tin film broken and receding. That is why the decline of the wetting force was smaller when the tin finish was thicker.

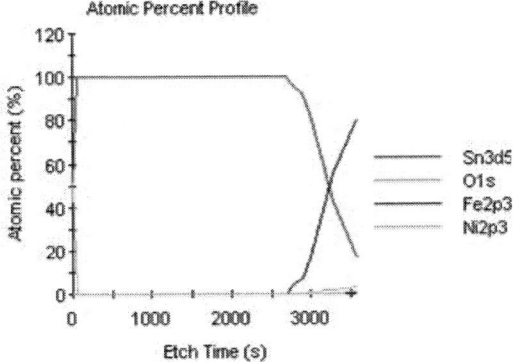

Fig .7 Depth profiles of various elements on 92Fe8Ni substrate with 2 min tin finish after stored in ambient atmosphere for 86 hours at room-temperature.

For the FeNi with tin finish to be used as the UBM layer, IMC between FeNi and tin finish should also be taken into consideration. There are two effects if IMC grows too thick. First, the interface changes from solder/FeNi to solder/IMC, and the liquid-solid surface energy changes correspondingly. However, the wettability of $FeSn_2$ and Ni_3Sn_4 has not been investigated by far. Second, there is no new IMC comes out when solder reflows on IMC layer, whether a reliable interconnection can be formed stays unknown. It is already known that Cu_6Sn_5 IMC comes out as soon as a layer of tin electroplated on Copper. However in our FeNi/Sn system, $FeSn_2$ and Ni_3Sn_4 could hardly be found under the resolution of SEM after electroplating and stored in ambient atmosphere for several days. So the FeNi with tin finish is a reasonable candidate to serve as the UBM for flip chip or other advanced packaging technology. Longer time aging under higher temperature is needed to examine the IMC growth between FeNi substrate and tin finish.

Conclusions

Solderability of FeNi with tin finish was examined by wetting balance measurement. Tin finish can remarkably reduce the initial wetting time t_b. The final wetting force Fmax showed little difference for bare 58Fe42Ni and the one with tin finish. Solderability did not decay after the sample was stored in ambient atmosphere for 86 hours, which indicates that tin layer can be a protective finish on FeNi. Thus FeNi with improved solderability by tin finish can be used as a potential UBM.

Acknowledgments

This research was supported by the Natural Science Foundation of China (Grant No. 50501022) and the National Basic Research Program of China (Grant No. 2004CB619306).

References

1. Gagliano, Robert A.; Fine and Morris E., "Thichening Kinetics of Interfacial Cu6Sn5 and Cu3Sn layers during reaction of liquid Tin with Solid Copper," *J. Electron. Mater.*, Vol. 32, No. 12 (2003), pp. 1441-1447.

2. D. Q. Yu , and L. Wang, "The growth and roughness evolution of intermetallic compounds of Sn-Ag-Cu/Cu interface during soldering reaction," *Journal of Alloys and Compounds*, Vol. 458, No. 1-2 (2008), pp. 542-547.

3. Z. Chen, M. He and G. J. Qi, "Morphology and Kinetic Study of the Interfacial Reaction between the Sn-3.5Ag Solder and Electroless Ni-P Metallization," *J. Electron. Mater.*, Vol. 38, No. 12 (2004), pp. 1465-1472.

4. J. Y. Tsai, Y. C. Hu, C. M. Tsai, and C. R. Kao, "A Study on the Reation between Cu and Sn3.5Ag Solder Doped with Small Amounts of Ni," *J. Electron. Mater.*, Vol. 32, No. 11 (2003), pp. 1203-1208.

5. Shawkret Ahat, et al, "Effect of Aging on the Microstructure and Shear Strength of SnPbAg/Ni-P/Cu and SnAg/Ni-P/Cu Solder Joints," *J. Electron. Mater.*, Vol. 29, No. 9 (2000), pp. 1105-1109.

6. Y. C. Chan, Alex C. K. So, J. K. L. Lai, "Growth Kinetics Studies of Cu-Sn Intermetallic Compound and its Effect on Shear Strength of LCCS SMT Solder Joints," *Mater. Sci. Eng. B.*, Vol. 55, No. 1-2, pp. 5-13.

7. Y. C. Chan, et al, "Effect of Intermetallic Compounds on the Shear Fatigue of Cu/63Sn-37Pb Solder Joints," *IEEE Trans CPMT-Part B*, Vol. 20, No. 4 (1997), pp. 463

8. N. Dariavach, P. Callahan, J. Liang and R. Fournelle, "Intermetallic Growth Kinetics for Sn-Ag, Sn-Cu, and Sn-Ag-Cu Lead-free Solders on Cu, Ni, and Fe-42Ni Substrates," *J. Electron. Mate.*, Vol. 35, No. 7 (2006), pp. 1581-1592.

9. J. J. Guo, L. Zhang, A. P. Xian and J. K. Shang, "Solderability of Electrodeposited Fe-Ni Alloys with Eutectic SnAgCu Solder," *J. Mater. Sci. Technol.*, Vol. 23, No. 6 (2007), pp. 811.

10. B. D. Barker, "Electroless deposition of Metals," *Surf. Technol.*, Vol.12 (1981), pp. 77-88.

11. R. Subramarian, M. Selvam and K. N. Srinivasan, "Electroless Plating," *Bullet Electrochem.*, Vol. 41 (1988), pp. 25-34.

12. J. Henry, "Electroless Plating," *Metal. Finishing* 1A, (1989), PP. 361-370.

Characterize the Microstructure and Reliability of Ultra Fine Pitch BGA Joints

Lei Wang, Zhenqing Zhao, Qian Wang, Jaisung Lee
Samsung Semiconductor China R&D Co., Ltd.
No. 15, Jin Ji Hu Road, Suzhou Industrial Park, Suzhou, China 215021
Tel: 86512-6288 8288-8827, E-mail: lei25.wang@samsung.com

Abstract

The solder ball pitch of BGA packages in mass production now is normally above 0.5mm. To fulfill the future demand of package miniaturization, the 0.4mm ultra fine pitch BGA solder joint was researched in this paper. The solder volume shrinkage along with the ball pitch decrease was found to affect the joint microstructure, which could be characterized in two aspects: the Sn dendrites in bulk solder became finer, and the so called "cross-interaction" in interfacial layer occurred. The joint reliability variation due to volume shrinkage was paid much attention to. Four different alloy content solder balls were evaluated through thermal cycling and drop reliability test, to find out the most proper soldering materials on both NiAu and OSP pad finishes. Our research result indicated that reliable ultra fine pitch joint could be gotten if proper materials and processes adopted. Sn1.0Ag0.5Cu0.02Ni was recommended as the solder alloy composition for the ultra fine pitch BGA usage.

1. Introduction

As the trend of electronic products towards miniaturization and multi-function integration continues, BGA packages with thin narrow configuration and high I/O terminals are becoming more preferred. The solder ball pitch and volume keeps shrinking, however, the demands on joint reliability turns to be more rigorous, for example, good drop impact reliability requested in portable products. Thus research on the property and reliability of small solder joints is needed.

The published research work on solder ball usually concentrated on the theoretical and laboratory studies on several aspects, such as solder alloy properties, interfacial reactions between solders and pads, dopant addition etc. The solder balls adopted in discussion are mostly normal size with diameter in the range of 0.45mm-1.27mm, for which pitch is more than 0.8mm [1-3]. Research on small solder joint is lacking. As a matter of fact, the dominant solder ball attach technology now in BGA mass production is pitch 0.5mm-0.8mm. In our research, 0.4mm ultra fine pitch BGA solder joint was investigated. The joint microstructure variation caused by solder volume shrinkage was studied. The effect of different solder alloy content on joint reliability was evaluated. Compared with normal size solder joints, the small joints tend to be weak in resisting thermal cycling fatigue, but there's some possible improved method proposed in this paper. The objective of our study is to characterize the microstructure and reliability properties of such small solder joints, find out the most suitable soldering materials to get reliable ultra fine pitch BGA joints.

2. Experiment procedure

The solder ball size adopted in this study is diameter 0.25mm. Four kinds of solder balls with different alloy contents were employed, and both NiAu and OSP (Cu) pad finishes were engaged. These materials made up eight groups of combination, as listed in Table 1, to be evaluated in reliability test.

Table 1 The material list of reliability test groups

Group	Solder composition	Pad finish
1N	Sn3.0Ag0.5Cu-0.0075Ge-0.003Ni	NiAu
1O	Sn3.0Ag0.5Cu-0.0075Ge-0.003Ni	OSP
2N	Sn2.0Ag0.5Cu-0.02Ni-0.08Ge	NiAu
2O	Sn2.0Ag0.5Cu-0.02Ni-0.08Ge	OSP
3N	Sn1.2Ag0.5Cu0.02Ni	NiAu
3O	Sn1.2Ag0.5Cu0.02Ni	OSP
4N	Sn1.0Ag0.5Cu-0.02Ni-0.08Ge	NiAu
4O	Sn1.0Ag0.5Cu-0.02Ni-0.08Ge	OSP

The normal solder ball attach process on NiAu pad is shown as follows:
Flux dotting → Ball placing → N_2 protected ball attach reflow → Flux cleaning.

For S/B/A on OSP pad, pre-cleaning process is particularly conducted ahead of the forenamed S/B/A process to get high yield. [4,5] The pre-cleaning process is as follows:
Flux dotting → N2 protected pre-cleaning reflow → Flux cleaning.

The devices were then mounted onto PCB boards through normal SMT process. The pad finish on PCB board is OSP(Cu).

The joint microstructure was analyzed by optical microscope SEM and EDX.

Drop test and TC test was conducted separately to evaluate the board level reliability of those 0.4mm pitch solder joints. The drop test parameters are correlated to condition B (1500Gs, 0.5millisecond duration, half sine-pulse) of JEDEC JESD22-B104B. The TC test condition is -65°C ~150°C of 29.5 minutes per cycle.

3. Result

3-1. Variation of joint microstructure due to solder volume decrease

To investigate the variation of joint microstructure caused by solder volume shrinkage, three different size solder balls with diameter 0.45mm, 0.3mm and 0.25mm were employed. The solder composition is Sn3.0Ag0.5Cu. The pad finish on substrate side is NiAu and on PCB side is OSP(Cu). Fig. 1

shows the metallographic images of the solder bulk microstructure. The solder bulk consists of primary Sn dendrites and an inter-dendritic Sn-Ag eutectic phase. The Sn dendrite size in 0.3mm solder joint shrinks almost 1/2 of that in 0.45mm joint (Note the different gauge in the images), and the dendrites become much finer in 0.25mm joint. It could conclude that as the solder diameter decreases, the microstructure in the joint becomes finer.

The most remarkable effect of solder volume shrinkage is on the interfacial reaction between solder and pads. For normal size solder joint like 0.45mm diameter, the interfacial reactions on substrate NiAu pad and PCB OSP pad proceed independently. The compound layer on the Au/Ni/Cu side is $(Cu,Ni)_6Sn_5$, and on the OSP(Cu) side is scallop Cu_6Sn_5 compound layer, as shown in Fig. 2(a). However, when the joint diameter decreased to 0.30mm, the so called "cross-interaction" happened, as shown in Fig. 2(b). Some Cu_6Sn_5 grains could be seen detached from the interfacial Cu_6Sn_5 layer on Cu pad and diffused into the bulk solder. Simultaneously some Cu_6Sn_5 grains could also be detected near the original $(Cu,Ni)_6Sn_5$ layer at the Ni interface [6-9]. According to prior published paper, the IMC grains scattered near the interface of Sn-Ag-Cu solder and Ni finish are usually Cu-Ni-Sn ternary compound. So the formation of Cu_6Sn_5 on the Ni side indicated that the interfacial reaction on the Ni side did not conduct independently, but was affected by the cross-interaction of the opposite Cu side. The cross-interaction became more seriously for the 0.25mm diameter solder joint (Fig. 2(c)). That's because the Cu diffusion from Cu side to Ni side turned to be easier when distance was shorten.

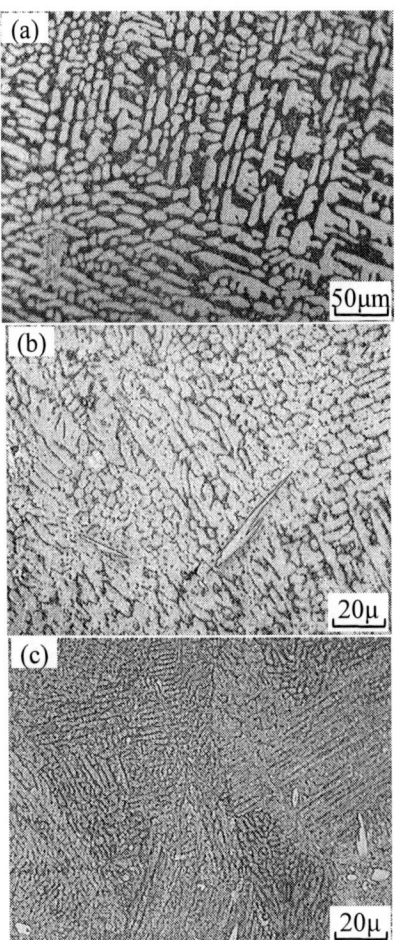

Fig. 1 The solder bulk microstructures of different diameter solder joints
(a) 0.45mm; (b) 0.3mm; (c) 0.25mm

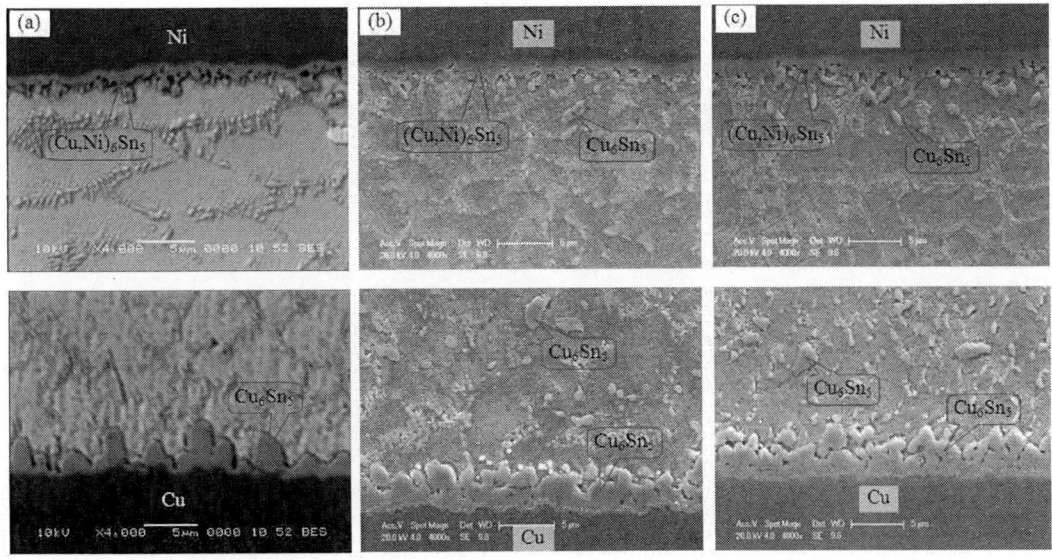

Fig.2 The SEM images of interfacial microstructure of different diameter solder joints
(a) 0.45mm; (b) 0.3mm; (c) 0.25mm

3-2. Drop reliability study

Firstly the drop reliability of the ultra fine pitch joint was compared with the normal pitch joint: 0.8mm ball pitch, 0.35mm ball diameter. The substrate pad finish is NiAu and solder composition is Sn3.0Ag0.5Cu. The comparison graphs are shown in Fig. 3.

It was forecasted that the drop performance of the ultra fine pitch joint would be worse than normal pitch joint [10, 11]. However, the experiment graphs and scale data (characteristic life when 63.2% failure happened) indicated that their drop reliabilities are quite similar. This should attribute to two changes: (1) The solder ball amount of the ultra fine pitch ball layout increased. For the specimen engaged in experiment, the ball amount for 0.8mm pitch ball layout is 64 ea, and for 0.4mm pitch ball layout is 172 ea. (2) Four dumb balls were added on four corners. It is well known that the most serious stress concentration lies in corner joints. These dump balls would resist high stress and help to protect functional balls. So it could conclude that the ultra fine pitch joint could also achieve demanded drop reliability if proper solder ball layout design was adopted.

Then the drop reliabilities of the eight groups' materials were compared with each other to find out the best material combination. The result is shown in Figure 4. According to the graphs and characteristic life data, the sequence of the eight material groups from the best drop reliability to the worst is 3O, 2O, 3N, 4O, 4N, 1N, 2N and 1O. The drop reliability of Sn2.0Ag0.5Cu0.02Ni on NiAu pad and Sn3.0Ag0.5Cu on OSP pad are obvious poor compared with other groups, thus the two combinations should be abstained. For both NiAu pad and OSP pad, Sn1.2Ag0.5Cu0.02Ni is the best solder ball choice to get high drop performance, and Sn1.0Ag0.5Cu0.02Ni is another recommendation.

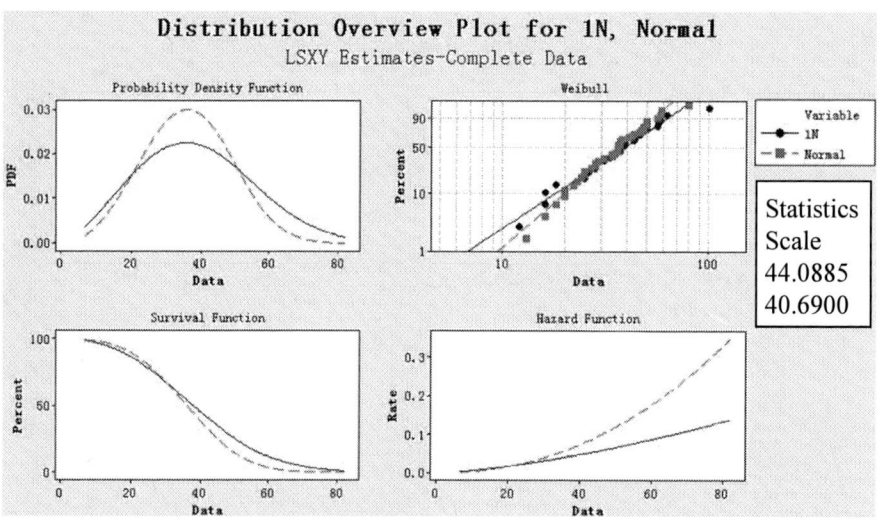

Fig. 3 Drop reliability comparison of ultra fine pitch joint with normal pitch joint

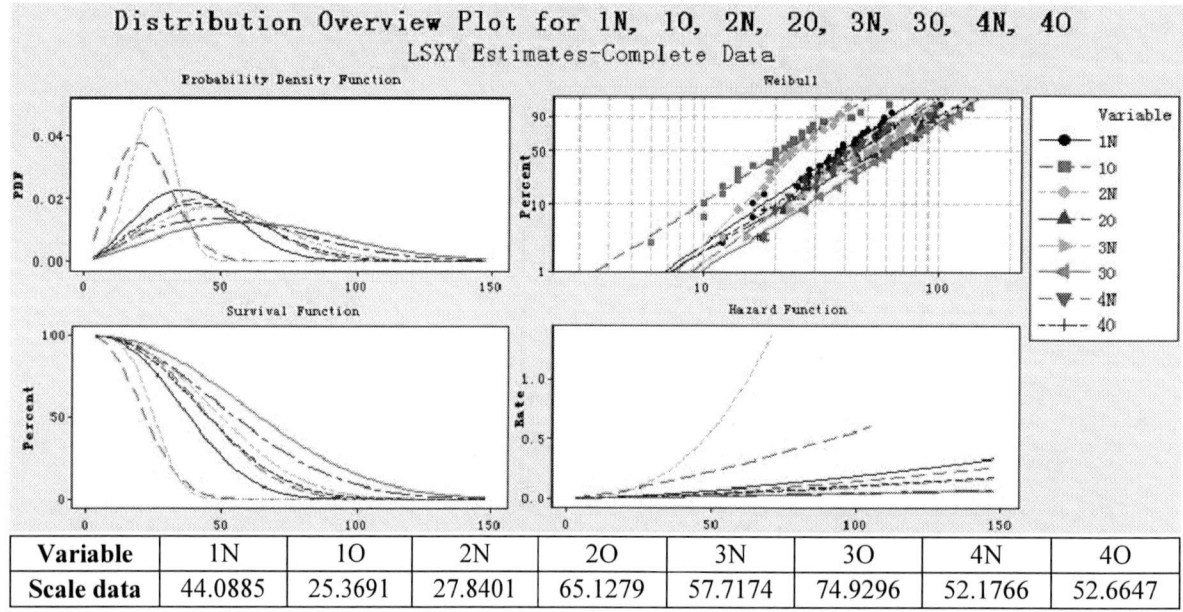

Variable	1N	1O	2N	2O	3N	3O	4N	4O
Scale data	44.0885	25.3691	27.8401	65.1279	57.7174	74.9296	52.1766	52.6647

Fig. 4 Drop reliability test graphs of different solder and pad combination

3-3 Thermal cycling reliability study

Thermal cycling test was conducted to evaluate the small joints' resistance to thermal fatigue. The test result of the eight groups' materials is shown in Fig. 5. According to the graphs in Fig. 5(1), the sequence of the eight material groups from the best TC reliability to the worst is 4N, 1N, 4O, 2N, 3O, 2O, 3N and 1O. The TC reliability of Sn3.0Ag0.5Cu on OSP pad is quite poor and all the joints failed in 500 cycles, so Sn3.0Ag0.5Cu solder should be abstained for the ultra fine pitch BGAs considering joints' thermal-mechanical reliability. For both NiAu pad and OSP pad, Sn1.0Ag0.5Cu0.02Ni is the best solder ball choice to get high performance in resisting thermal-mechanical fatigue.

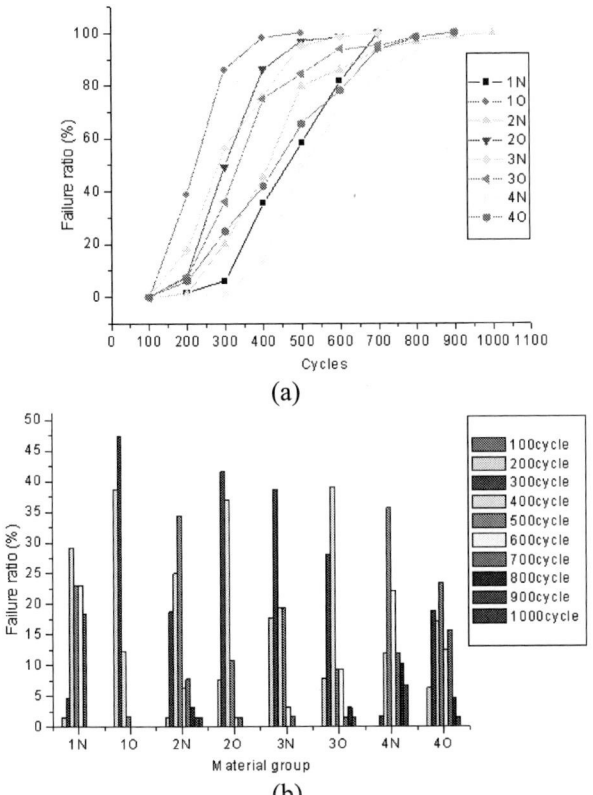

(a)

(b)

Fig. 5 TC reliability test graphs of different solder and pad combination

It should be noticed that majority of the TC lifetimes for these ultra fine pitch joints are in the region of 300 cycles ~ 500 cycles, as shown in Fig. 5(2). This TC lifetime is quite short compared with normal pitch solder joints: 0.5mm pitch, 0.3mm diameter solder ball. The comparison result is shown in Fig. 6. The solder ball compositions are Sn3.0Ag0.5Cu and Sn1.0Ag0.5Cu0.02Ni, the substrate pad finish is NiAu. The graphs in Fig. 6 show that the TC life seriously deteriorated as the solder joint volume decreased. This is due to the poor stress absorption capability of the small solder bulk [12]. The displacement caused by CTE mismatch of chip, mold body and PCB board under thermal cycling from -65°C to 150°C leads to thermal mechanical stress in solder joints. Same displacement would cause much higher stress as the joint volume decreased.

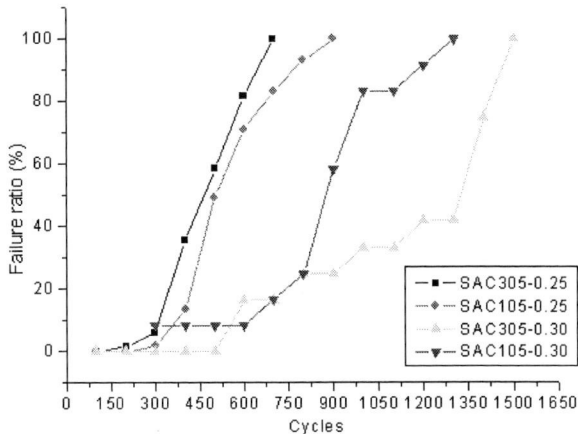

Fig. 6 TC reliability comparison of ultra fine pitch joint with normal pitch joint

This result indicates that the ultra fine pitch joint is sensitive to thermal cycling circumstance. To satisfy the long life usage in serious TC condition, it is recommended to improve the solder property (adjust composition, add dopants etc.) or modify the packaging process such as add under-filling.

4. Conclusions

The solder joint microstructure varied when the solder ball size decreased from 0.45mm to 0.25mm. The Sn dendrites in bulk solder became much finer as solder volume shrunk. Cross-interaction of two interfacial reactions tends to occur when solder ball diameter decrease to 0.3mm, and this interfacial behavior turns to be more serious in smaller joints.

According to the board level reliability study result, the ultra fine pitch joint could reach equal drop reliability as the normal pitch joint if proper solder ball layout adopted; but the small joint showed poor resistance in thermal cycling fatigue, and under-filling might be an improved method.

Based on the evaluation result of different solders on both NiAu pad and OSP pad, Sn1.2Ag0.5Cu0.02Ni is the best solder ball choice to get high drop performance, Sn1.0Ag0.5Cu0.02Ni shows good integrated performance of drop and TC reliability. Reliable ultra fine pitch joint could be gotten if proper materials adopted.

References

1. T. Laurila, V. Vuorinen, J. K. Kivilahti, "Interfacial reactions between lead-free solders and common base materials," *Materials Science and Engineering*, 49 (2005), pp. 1-60.
2. Hansen Garcia Sy, Jackson Hsu, "Miguel Jimarez, New Robust Process Improvement for BGA Solder Ball Attach First Pass Yield," *2007 9th Electronics Packaging Technology Conference*, pp. 797-803.
3. C.M.L. Wu, D.Q. Yu, C.M.T. Law, L. Wang, "Properties of lead-free solder alloys with rare earth element additions," *Materials Science and Engineering*, 44 (2004), pp. 1-44.
4. David Chang, Frank Bai, Y.P. Wang, C.S. Hsiao, "The Study of OSP as Reliable Surface Finish of BGA

Substrate," *2004 Electronics Packaging Technology Conference*, pp. 149-153.

5. Yoon-Chul Sohn, Young-Kun Jee, Jin Yu, "Comparison between Electroless Ni(P)/Au and Cu OSP as a Surface Finish Layer of Mobile Application," *2005 International Symposium on Microelectronics*, pp. 191-198.

6. C.E. Ho, Y.W. Lin, S.C. Yang, "Volume Effect on the Soldering Reaction between SnAgCu Solder and Ni," *IEEE*, (2005).

7. Lei Wang, Zhenqing zhao, "Cross-Interaction of Different Pad Finishes in SMT Joints and Its Effect on Board Level Reliability," *IMAPS*, (2007).

8. S.J. Wang and C.Y. Liu, "Study of interaction between Cu-Sn and Ni-Sn interfacial reactions by Ni-Sn3.5Ag-Cu sandwich structure," *Journal of Electronic Materials* 32 (11) (2003), pp. 1303-1309.

9. C.M. Tsai, W.C. Luo, C.W. Chang, "Cross-interaction of under-bump metallurgy and surface finish in flip-chip solder joints," *Electronic Materials*, 33 (2004), pp. 1424-1428.

10. E.H. Wong, K.M. Lim, Norman Lee, "Drop impact test – Mechanics & Physics of failure," *2002 Electronic Packaging Technology Conference*, pp. 327-333.

11. Christian Bizer, Bernd Rakow, Rainer Steiner, "Drop Test Reliability Improvement of Lead-free Fine Pitch BGA Using Different Solder Ball Composition," *2005 Electronics Packaging Technology Conference*, pp. 255-261.

12. John H.L. Pang, T.H. Low, B.S. Xiong, Xu Luhua, C.C. Neo, "Thermal cycling aging effects on Sn–Ag–Cu solder joint microstructure, IMC and strength," *Thin Solid Films* 462–463 (2004), pp. 370–375.

Reliability of Wafer Level Thin Film MEMS Packages during Wafer Backgrinding

J.J.M. Zaal[1], W.D. van Driel[1, 2], G.J.A.M. Verheijden[3], G.Q. Zhang[1, 4]

[1]Delft University of Technology department PME, Mekelweg 2, 2628 CD Delft, the Netherlands
[2]NXP Semiconductors, Nijmegen, the Netherlands
[3]NXP Semiconductors, Leuven, Belgium
[4]NXP Semiconductors, Eindhoven, the Netherlands
J.J.M.Zaal@TUDelft.nl

Abstract

With the ever-growing number of MEMS resonator applications, an research effort is required. One important step to consider in the design and evaluation phase is the packaging of such a resonator. Packaging has proven to be a problem and this research focuses on the backgrinding of wafers. Backgrinding introduces a tape application and removal step on the active side of the wafer. This paper shows the induced stress profiles on a Wafer Level Thin Film Package (WLTFP) during the tape application step, the stress profiles during removal and evaluates several options to lower the forces acting on the WLTFP in order to increase the survivability.

Introduction

An increasing number of semiconductor companies are researching MEMS resonators. This is initiated by the number of potential applications for MEMS resonators, which shows a dramatic increase. Common example applications are filters [1] and oscillators [2-3]. Less straightforward applications are the use of resonators to measure the presence of a specific gas or particle type [4], or to measure extremely low gas pressures [5]. Thermometers [6] and gyroscopes [7] are also among the possible applications for MEMS resonators.

MEMS devices in general are usually vulnerable to the external loads subjected to it and packaging and lifetime problems are a major roadblock [8]. In an earlier study assembly influences on MEMS structures have been investigated [9].

One method to obtain a hermetic cavity for a resonator is by means of a Wafer Level Thin Film Package (WLTFP). Such a construction consists of a thin film cap, which is released by etching away a sacrificial layer, and a sealing structure to close the etching holes needed for the transport of the etchant. The main advantage of this method towards others such as local assembly is that all caps can be made in one time. Another advantage is the fact that the packaging and sealing takes place in a controlled and clean environment minimizing the contamination. Fig. 1 depicts an example of a WLTFP package.

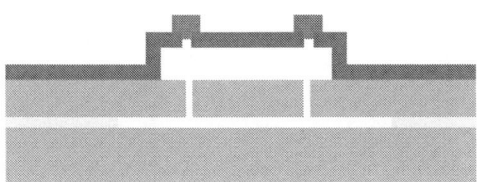

Fig. 1 An example of a WLTFP package on top of a MEMS

To fit the resonators in small IC packages the die needs to be very thin. This is achieved by grinding the wafer to the required thickness after processing. The grinding process can be divided in three steps: the grinding tape application on the active side, the actual grinding process and the removal of the grinding tape.

Grinding process and imposed loads

Tape is applied to the wafer several times during the common assembly process. During both grinding and dicing tape is used. The tape application process can damage the WLTFP due to the applied pressure needed to sufficiently adhere the tape to the wafer. When tape is applied to the WLTFP side of the wafer the tape will land on the part of the WLTFP that is sticking out above the surface. The application pressure will then force the tape to contact the remaining wafer surface. If tape is applied to the backside of the wafer one also needs to apply pressure to the wafer. Careful support to the WLTFP side of the wafer is needed since the soft tape is on the other side and cannot protect the WLTFP from rigid objects. All objects or surfaces used to support or pressurize the tape will land on the WLTFP's first, thereby creating high cap loads. Cap loading mechanisms during the tape application process are depicted in Fig. 2.

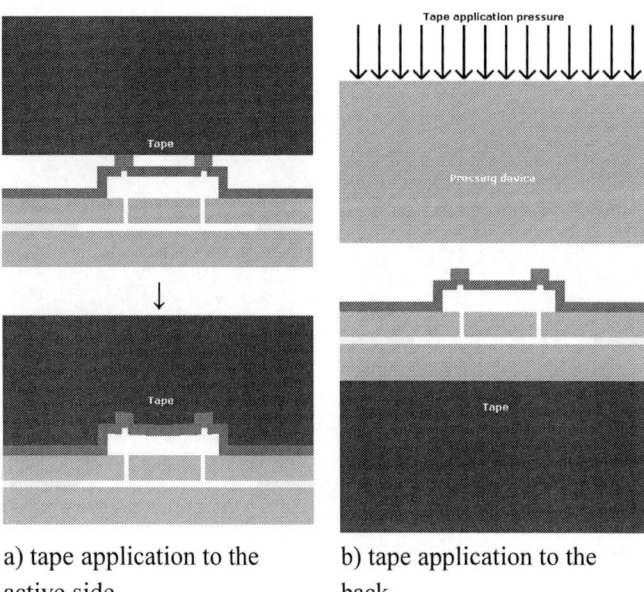

a) tape application to the active side

b) tape application to the back

Fig. 2 Tape application to the front and backside of a wafer

The tape removal process on the active side is hazardous since the adhesive forces will apply tensile forces on the WLTFP cap and plugs.

Experimental observations

Pictures showing damaged WLTFP's due to tape application and removal on the active side of a wafer are depicted in Fig. 3.

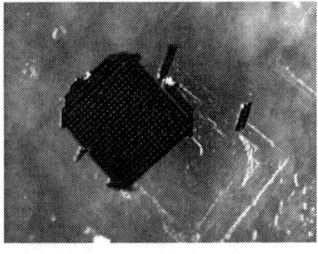
a) cap sticking on the tape

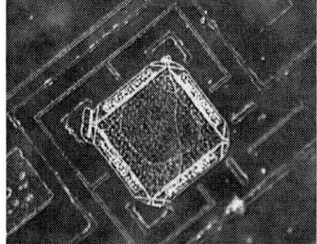
b) resonator sticking on tape

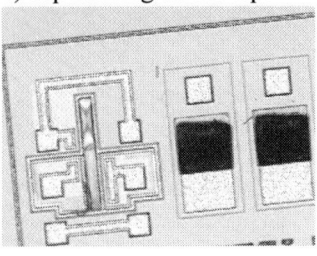
c) cavity missing a cap

d) cavities missing caps

Fig. 3 Taping process damage

In Fig. 3(a) & (b) one can see parts of a WLTFP sticking to the grinding tape after detaping a wafer. In Fig. 3(c) & (d) resonator structures are depicted where the caps are missing due to a tape / detape step. In all pictures it is very clear that the cracks almost everywhere exactly follow the edge of the cavity.

When the WLTFP cap fails it will stick to the grinding foil. A cross-sectioned example is depicted in Fig. 4.

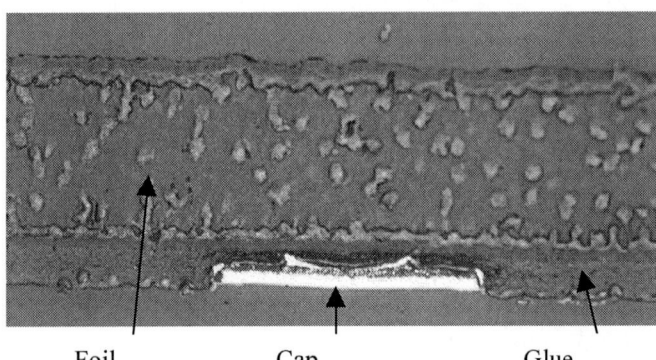

Foil   Cap   Glue

Fig. 4 Cap embedded into the grinding tape

The forces during detaping prove to be high enough to destroy a WLTFP cap. Special geometrical and/or design features are required for the WLTFP in order to survive this assembly step.

Tape application simulation

In order to improve the grinding process several steps of the process are simulated. First step is the application of the foil. A model representation is depicted in Fig. 5.

In this 2D plane strain simulation the wafer is made of ordinary silicon, the capping material is SiN and the tape

(Nitto-Denko BT-150EDL) is composed of a base PET film and an acrylic glue. The properties of the glue were estimated since it could not be measured. The foil properties, which are almost equal to the base film properties since the glue layer is very soft and thin are: 50 MPa Young's Modulus and a poisson ratio of 0.4. The Young's modulus is measured using a DMA. The complete model contains about 33000 elements and uses contact bodies to prevent the tape from penetrating the MEMS. In this simulation any stress induced by the present thermal mismatch [10] is not included.

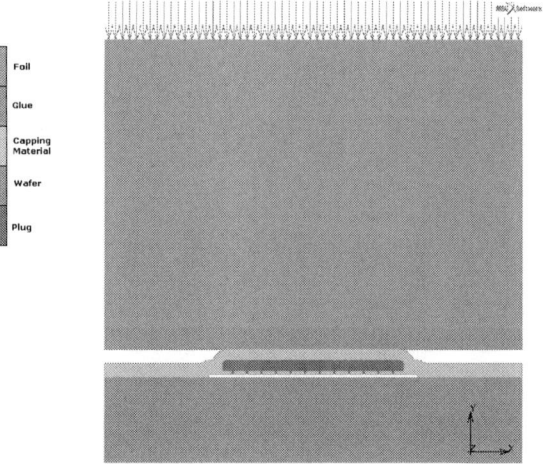

Fig. 5 Tape application simulation

After applying the pressure (depicted in Fig. 5) to the model the deformed tape is formed around the WLTFP. This is depicted in Fig. 6.

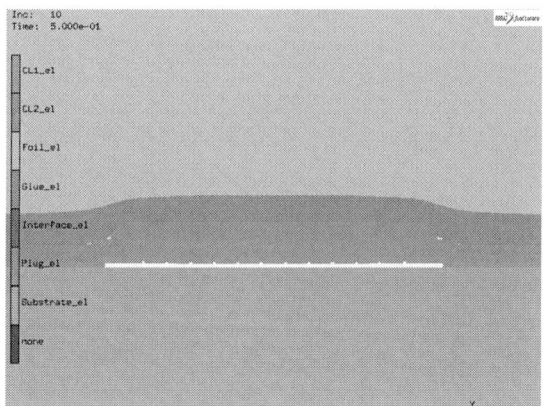

Fig. 6 Deformation of the tape around the cap

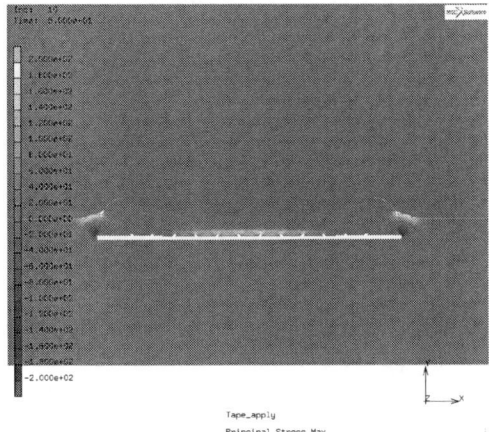

Fig. 7 Stress distribution caused by tape application pressure

The highest stresses (around 200 MPa) are found in the area of the step near the edge of the cavity as depicted in Fig. 7. This indicates that careful design of the step geometry (in- and out of plane) is required to prevent the WLTFP from failing during either tape application or grinding. Both processes induce a pressure load on the tape and can be investigated using one simulation.

Detaping simulation

The other taping step in the grinding process, the removal of the tape can also prove to be hazardous since the tape will exert a pulling force on the WLTFP's. The materials used for the WLTFP's (SiN) are not very resistant against tensile stress and already present cracks can easily be progressed.

The detaping simulation uses the technique of Cohesive Zones (CZ) [11]. The adhesion strength is set to 4Nm and the critical opening to 1μm. In a further study we will measure the exact adhesion strength values.

The WLTFP geometry is equal to the model as before but the glue layer is assumed to be fully around the cap, as seen in the experiment in Fig. 4. The model geometry is depicted in Fig. 8.

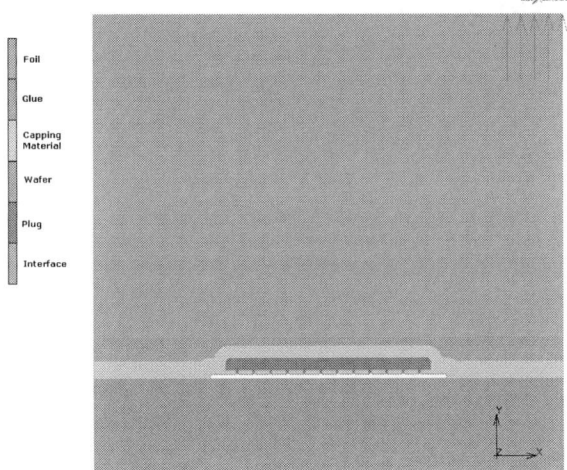

Fig. 8 Detape simulation model geometry

When one pulls the foil away from the wafer stress results as depicted in Fig. 9 are found.

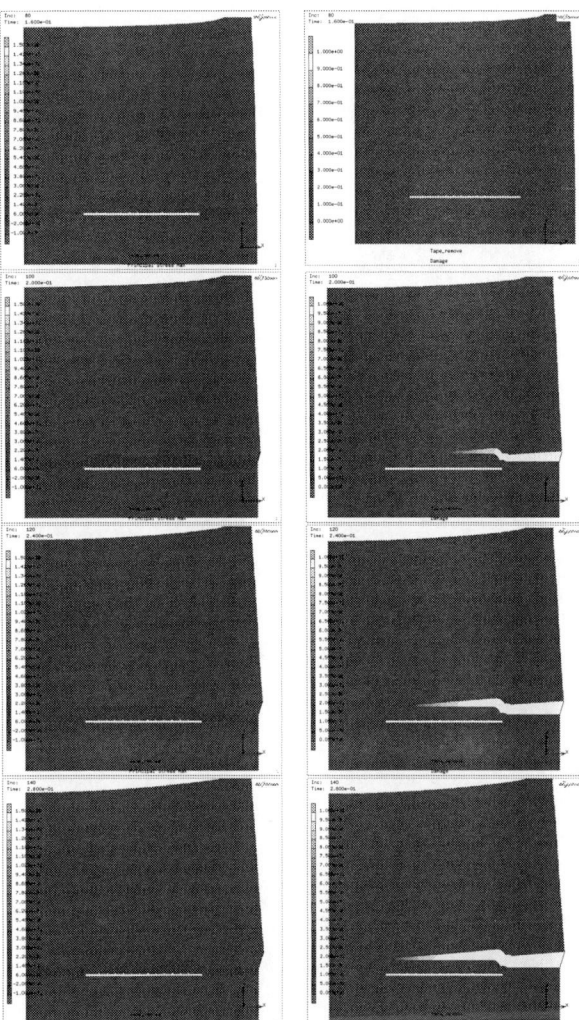

Fig. 9 Stress (left column) and damage (right column) during detaping

From Fig. 9 one can conclude that the stress on the cap is at its highest while the actual delaminating region is moving over the cap. From the simulation the point of the highest principal stress (around 200MPa) can found, this is depicted in Fig. 10.

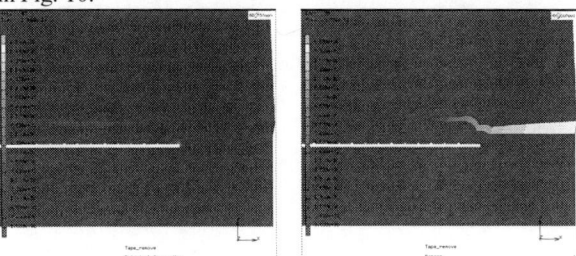

Fig. 10 Point of highest principal stress (left) and the corresponding damage (right)

From these simulations it is clear that the adhesion of the grinding tape is very well capable of imposing high loads and progressing cracks, this is confirmed by the earlier presented pictures of failed WLTFP's.

In order to lower the stress a simulation test is done using a tape only adhered to the cap. This tape could be stiffer and less compliant in the earlier stages. The model contains the same material as before and the geometry is changed to the shape depicted in Fig. 11.

Fig. 11 Modified detaping model

This model uses the same CZ properties so an easy comparison can be made. Fig. 12 depicts the stress levels at the moment of highest stress.

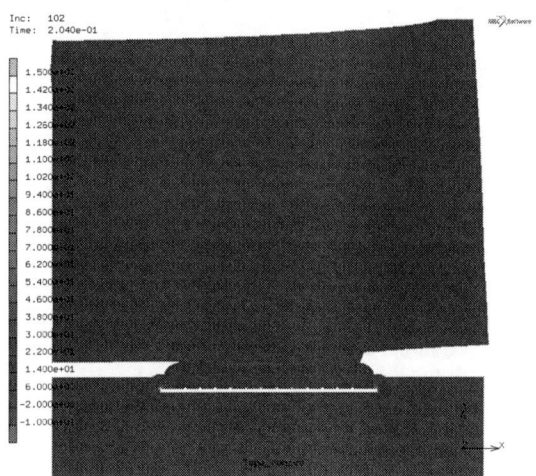

Fig. 12 Maximal principal stress

The damage levels corresponding to these stresses indicate that the stress is at its highest level while the first elements are almost fully damaged. This is depicted in Fig. 13.

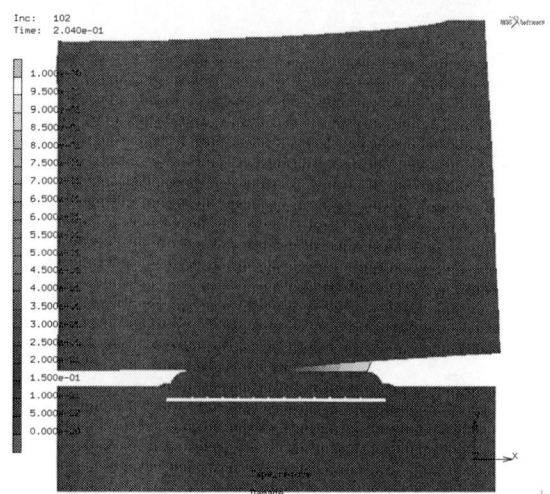

Fig. 13 Damage at the moment of the highest principal stress

The stress levels are reduced by 1/4 and such a modification to the grinding foil should improve the chance the WLTFP survives the grinding step. An overview of stresses during assembly is depicted in Fig. 14.

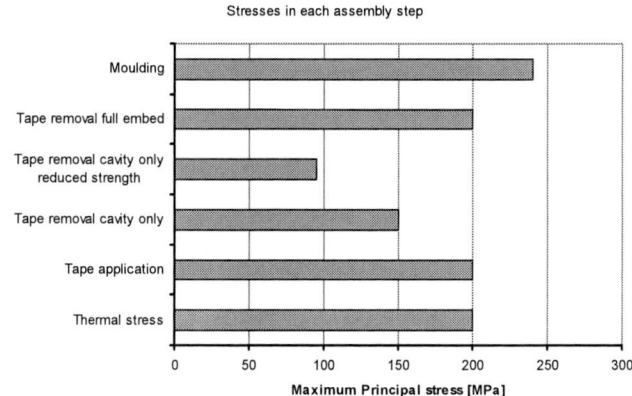

Fig. 14 Stresses during each assembly step

Another measure that could be taken to lower the grinding load is the choice of a different grinding foil that adheres not very well or is curable after the grinding. To investigate the effect of the adhesion several simulations are done using different cohesive energies.

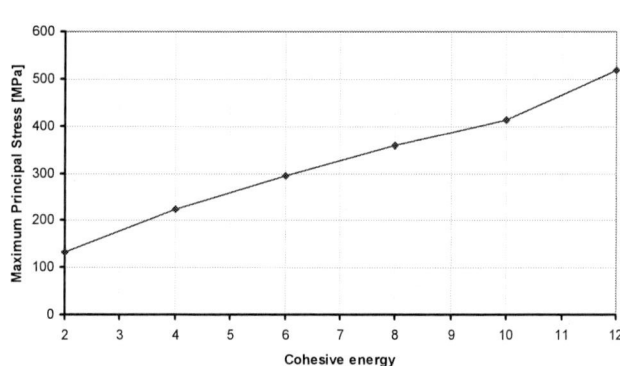

Fig. 15 Cavity stress due to grinding foil adhesion properties

From Fig. 15 one can conclude that the foil adhesion properties have a very strong influence on the results and a careful selection of a grinding foil with the lowest useable adhesive properties is advised.

Conclusions

From simulations and experiments it has been shown that grinding a wafer to its required thickness can cause significant problems concerning yield and survivability.

Experimental results show cracks throughout the WLTFP caps usually following the side of each cavity. It is also shown in an experiment that the WLTFP cap can be fully embedded into the glue layer that is present on the grinding foil.

A numerical model is set up to investigate loading levels due to grinding and the application of grinding tape. These loadings show to be significant and capable of initiating cracks when combined with initial (thermal) stress levels and/or small defects in the structure.

The detaping step is investigated by the evaluation of several numerical models. When the WLTFP cap is fully embedded in the glue the stress has proven to be 33% higher compared to the case where the grinding foil only adheres to the cap. The effects of the foil adhesion on the stress are significant and it is advised to carefully select the most suitable grinding foil.

Future work

Future work will cover reinforced cavity edge structures and the effects of planerisation on the cavity stresses.

Acknowledgments

The authors gratefully acknowledge the financial support by the Dutch Ministry of Economic Affairs in the framework of the Point One project MEMSLand. [http://www.memsland.nl]

References

1. Yan J., Seshia A.A., Phan K.L., Steeneken P.G., van Beek J.T.M., "Narrow Bandwidth Single-Resonator MEMS Tuning Fork Filter," *Proceedings of IEEE International Frequency Control Symposium joint with the 21st European Frequency and Time Forum*, 2007, pp. 1366-1369.

2. Hsu W.T., Pai M., "The New Heart Beat of Electronics - Silicon MEMS Oscillators," *Proceedings of the 57th Electronic Components and Technology Conference*, 2007, pp. 1895-1899.

3. Steeneken P.G., Ruigrok J.J.M., Kang S., van Beek J.T.M., Bontemps J., Koning J.J., "Parameter Extraction and Support-Loss in MEMS Resonators", *Comsol Conference 2007*, 2007

4. Voiculescu I., Zaghloul M.E., McGill R.A., Houser E.J., Fedder G.K., "Electrostatically Actuated Resonant Microcantilever Beam in CMOS Technology for the Detection of Chemical Weapons," *IEEE Sensors Journal*, Vol. 5, No. 4 (2005), pp. 641-647.

5. Suijlen M., 2007, "Ontwerp MEMS druksensor : rapport ontwerpproject NXP Nijmegen," *Final report design project NXP Nijmegen.*

6. Jha C.M., Bahl G., Melamud R., Chandorkar S. A., Hopcroft M. A., Kim B., Agarwal M., Salvia J., Mehta H., Kenny T. W., "Cmos-Compatible Dual-Resonator MEMS Temperature Sensor with Milli-Degree Accuracy,"

Proceedings of Solid-State Sensors, Actuators and Microsystems Conference, 2007, pp. 229-232.

7. Zaman M.F., Sharma A., Ayazi F., "High Performance Matched-Mode Tuning Fork Gyroscope," *Proceedings of the 19th IEEE International Conference on Micro Electro Mechanical Systems*, 2006, pp. 66-69.

8. Hsu T-R., "Reliability in MEMS Packaging," *Proceedings of Reliability Physics Symposium*, 2006.

9. Zaal J.J.M., van Driel W.D., Zhang G.Q., "Challenges for MEMS Packaging Reliability," *Smart System Integration and Reliability (MicroNanoReliability '07)*, 2009.

10. Zaal J.J.M., van Driel W.D., van Beek J.T.M., Zhang G.Q., "Assembly Induced Failures in Thin Film MEMS Packages," *Proceedings of the 10th EuroSimE conference*, 2009.

11. Zhang G.Q., van Driel W.D., Fan X.J., "Mechanics of Microelectronics," Series: Solid Mechanics and Its Applications, Vol. 141, ISBN: 1-4020-4934-X, (2006).

Super-hydrophobic Nickel Films with Micro-nano Hierarchical Structure Prepared by Electrodeposition for Appliance Industry

Tao Hang, Ming Li, Anmin Hu, Dali Mao

Lab of Microelectronic Materials & Technology, School of Materials Science and Engineering,
Shanghai Jiao Tong University, Shanghai 200240, China
Email: mingli90@sjtu.edu.cn

Abstract

Super-hydrophobic Nickel films were prepared by a simple and low cost electrodeposition method. The surface morphology of the films characterized by FE-SEM exhibit hierarchical structure with micro-nano cones array, which can be responsible for their super-hydrophobic characteristic (water contact angle is over 120 degree) without any chemical modification. The wettability of the film can be varied from super-hydrophobic (water contact angle 143 degree) to relatively hydrophilic (water contact angle 87 degree) by controlling the size of the micro-nano cones. The mechanism of the hydrophobic characteristic of nickel films with this unique structure was illustrated by several models. Such novel micro-nanostructure and its special wettability are expected to be applied in miniature aquatic devices, micro-fluid devices, satellite antenna, and conductors with self-cleaning surfaces etc.

Introduction

Super hydrophobic surfaces, with water contact angle larger than 150°, have great potentials not only in fundamental researches but also in so many practical applications of electrical industry, such as micro-fluidic devices, satellite antenna, and conductors with self-cleaning surfaces.[1-9] Conventionally, super hydrophobic surfaces have been produced mainly in two ways. One is to create a rough structure and the other is to modify by materials with low surface free energy. [10] Many methods have been developed to produce rough surfaces, including plasma etching [11], anodic oxidization [12], chemical vapor deposition [13], phase separation [14] and molding [8]. However, most of the synthetic surfaces reported so far are expensive, or can not be easily scaled-up to create large-area surfaces, which limit their applications. One example is the preparation of a super hydrophobic nano-structured carbon film [15]. However, the use of alumina oxide template and the high temperature (900°C) pyrolysis procedure limited the application of this method.

Here, we present a facile and economic method to fabricate large-area super hydrophobic nickel micro-nano hierarchical structure by electrodeposition with a crystallization modifying agent. Nickel, as an essential engineering material, has the properties of high hardness, good wear and corrosion resistance, and it can be electrodeposited to the surface of many different metals, thus it has potential applications in the fabrication of varied engineering materials. Since nickel is also a ferromagnetic material, anisotropic magnetic Ni nanostructures are expected to be used in magnetic sensor and memory devices. These large-area super hydrophobic materials can also be anticipated to serve as a microwave absorbing interfacial functional materials. [16]

Experiment

The Nickel films were electroplated onto commercial pure (99.5%) 4cm×2cm Cu plates. These Cu plates were first rinsed with pure water, and then, anodized at 0.25A/cm2 in a solution (10g/l KOH, 10g/l santomerse, 70g/l deoil powder) for 20sec. After anodizing, these plates were rinsed pure water. And then these Cu plates were insulated using insulating rubber (TECHBOND), left only 2 cm×2cm area in each one. After this, they were activated with 0.4M H_2SO_4 for 20sec, and finally rinsed with pure water. The electro deposition solution was composed of analytical pure $NiCl_2 \cdot 6H_2O$ (providing Ni ions) 1mol/L, H_3BO_3 (pH buffer) 0.5mol/L, and crystal modifier 1.5mol/L respectively dissolved in deionized water. A pure Ni plate (99.9%) was used as an anode while as-sputtered slides used as cathode for Ni deposition. The temperature of deposition solution was kept at 60 °C and pH value was 4.0. The deposition current density were varied from 2.0 A/cm^2 to 10 A/cm^2 .The deposition time was 30 seconds, 1 minute, 5 minutes and 20 minutes respectively to obtain different Nickel micro-nano hierarchical structure. And the four samples were marked as A, B, C and D.

The morphologies of the Nickel micro-nano hierarchical structure films were observed by FE-SEM (FEI SIRION200) and samples were rinsed for 10 minutes by ethanol, then dried under blower then dried on the hot plate at 80°C overnight for contact angles (CA) measurement on a Data physic OCA20 contact angle system at ambient temperature. Water droplets (about 4μL in volume) were carefully dropped onto the surface of the nickel films. The average value was obtained by measuring the CA value at five different positions of the same sample.

Result and discussion

Fig. 1 (a) ~ (d) show the SEM images of the Nickel deposited prepared in different conditions. The morphology of the Nickel film shown in Fig. 1(a), deposited in high current density ($10A/dm^2$) for very short period of time (30 seconds), is relatively smooth, only sphere-like nanoparticles can be observed. But in Fig. 1(b) and (c), one dimensional nanocones array was formed in lower current density and longer deposition time. The bottom diameters of individual nanocone grown for 1 minute at $5A/dm^2$, and 5 minutes at $2A/dm^2$ were estimated to be about 100 nm and 300 nm, and the heights were 200 nm and 700 nm, respectively. The density of cones was estimated to be about 5×100 cones/cm^2 and 1×100 cones/cm^2. As the deposition time extending, it can be seen from the fig. 1(d) that, small nanocones were generated on the big microcones, which resulted in the surface of Nickel film coarser and spinous.

978-1-4244-4658-2/09 $25.00 © 2009 IEEE

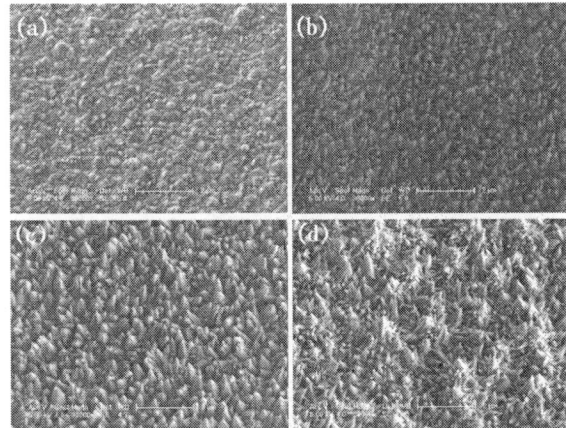

Fig. 1 SEM image of Nickel films electroplated for different current densities and times. (a) 10A/dm² 30second; (b) 5A/dm² 1minute; (c) 2A/dm² 5minutes; (d) 2A/dm² 20minutes.

The growth mechanism of the micro-nano cones can be explained as the directional growth of nuclei by electrodeposition.[17,18] At the beginning of electrodeposition, a great amount of nuclei form nearly simultaneously and then grow up outward directionally into micro-nano cones. As time went on, the original cones were growing bigger and bigger while some new nucleus formed on the surface of them were also growing into small cones. And finally, the so called micro-nano cones hierarchical structure has come into being.

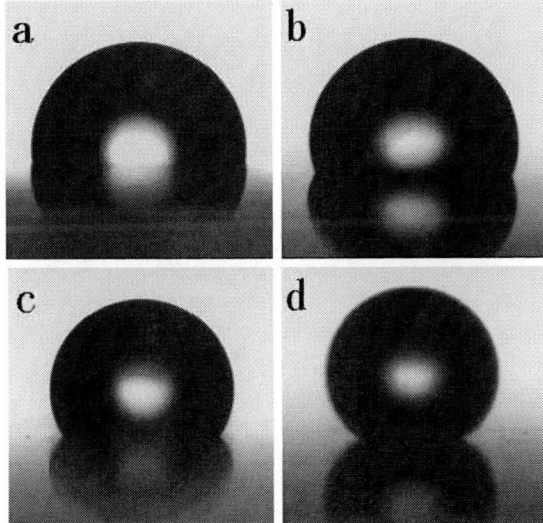

Fig.2 Optical graphs of a water droplet(3.5µL)on the nickel film of micro-nano hierarchical structure (a)-(d) correspond to SEM image of Fig.2 (a) ~ (d)

The wettability of the nickel films were traced by contact angle measurement shown in Fig. 2(a) ~ (d). It is easy to find that the water contact angle of film without any micro-nano cones structure is the lowest as 93°. Since the nanocones appeared and grew bigger into microcones, the contact angle

varied from 109° to 123°. For the micro-nano cones hierarchical structure, the contact angle of the nickel film without any chemical modifier was as high as 143°.

According to Cassie's law for surface wettability [19], such micro-nano structures can be regarded as heterogeneous surfaces composed of solid and air. The apparent contact angle θ' of the film is described by:

$$\cos\theta' = -1 + f(r\cos\theta + 1) \qquad (1)$$

Where f is the area fraction of micro-nano cones, 1-f is the area fraction of air on the film surface and θ is the contact angle of the smooth nickel. Using measured values of θ' and θ (87°), the air fraction between the film and the water surface can be deduced from the equation, listed in Table 1. Available air is trapped in spaces between the micro-nano cones to form a cushion at the film–water interface that prevents the film from being wetted. That is why sample B, C, and D have better hydrophobic characteristic than sample A.

Compare sample B with sample C. The nickel cones of sample C are bigger than those of sample B thus lower in distribution density. The air fraction is increased from sample B (42%) to sample C (61%). It is concluded that the increase of the cone size will result in a larger water contact angle.

Table 1 Water contact angle and air fraction of the Ni films deposited in different condition.

Sample	A	B	C	D
Current density (A/dm²)	10	5	2	2
Deposition time (min.)	0.5	1	5	20
Water contact angle (°)	93	109	123	143
Roughness factor	2.8	3.1	3.5	3.9
Air fraction	18%	42%	61%	83%

However, for sample D, the contact angle is larger than sample C, though the cone size of sample D is no bigger than sample C. This phenomenon is due to the hierarchical micro-nano structure. This unique structure of the nickel film surface (shown in Fig. 3a) is similar to the leg of water strider (shown in Fig. 3b) which is confirmed to be responsible for its water resistance. [20] The model of such a surface can be inferred as describe by Feng [10]. The surface roughness was considered on top of the cones. By introducing the fractal formula from triadic Koch curve model as a roughness factor, a mathematic model was evolved as follows to describe the contact angle on a rough surface and that on a smooth surface of the same solid:

$$\cos\theta' = -1 + f\,[1 + (L/l)^{D-2}\cos\theta] \qquad (2)$$

Here, $(L/l)^{D-2}$ is the surface roughness factor, where L and l are, respectively, the upper and lower limit scales of the fractal behavior of the surface, and D is the fractal dimension. f is still the area fraction occupied by micro-nano cones. For an engineering surface, it is very difficult to know these parameters precisely. Therefore, the calculations of the hierarchical micro-nano structure Nickel films by equation (2) will be investigated in the future.

The stability of the super-hydrophobicity of the obtained surface after storage in air for various time intervals was also investigated. The water contact angle on the hierarchical surface still remains larger than 140° after storage about 2 weeks in air, showing the long term stability of the fabricated surface. It is also noted that the fabricated nickel film shows super-hydrophobicity in the *p*H range from 1 to 14. There is no obvious effect of acid on the wettability of the surface. These results are very important for the application of Nickel as an engineering material with super-hydrophobicity in the wide *p*H range of corrosive liquids.

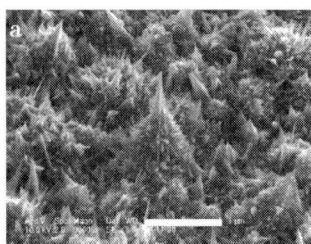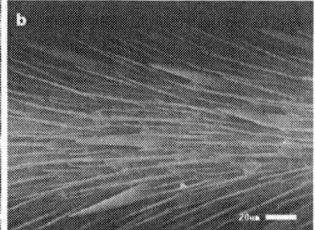

Fig. 3 SEM images of (a) Micro-nano hierarchical structure Nickel film; (b) Leg of water strider. [20]

Conclusions

A simple method of Ni micro-nano cones hierarchical structure fabricated by electrodeposition was studied in this work. FE-SEM shows that the surface morphology of the films exhibit hierarchical structure with micro-nano cones array. The wettability of the film is characterized by water contact angle test which can be varied from super-hydrophobic (water contact angle 143 degree) to relatively hydrophilic (water contact angle 87 degree) deeply depends on the surface morphology of the films. The mechanism of the hydrophobic characteristic of nickel films with this unique structure was illustrated by several simple models. Optimized parameters are expected to be obtained to fabricate Ni micro-nano structure with better hydrophobic performance in the future.

Acknowledgments

This work was sponsored by National Natural Science Foundation of China (No. 90406013), Shanghai Pujiang Program (No. 05PJ14065) and Shanghai Nanotechnology Promotion Center (No. 0452nm030). We thank Analytical and Testing Center of Shanghai Jiao Tong University for the help on SEM observation.

References

1. Erbil H Y, Demirel A L, Avci Y, et a.l Transformation of a simple plastic into a super hydrophobic surface. *Science,* 2003, 299(5611):1377-1380.
2. Aussillous P, QuéréD. "Liquid marbles". *Nature,* 2001, 411:924-927.
3. Wasserman, Y, "Integrated Single-Wafer RP Solutions for 0.25-micron Technologies," *IEEE Trans-CPMT-A*, Vol. 17, No. 3 (1995), pp. 346-351. [A reference to a journal article]
4. Shu, William K., "PBGA Wire Bonding Development," *Proc 46th Electronic Components and Technology Conf*, Orlando, FL, May. 1996, pp. 219-225. [A reference to a presentation at a Conference.]

5. Jiang L., Wang R., Yang B., et al. "Binary cooperative complementary nanoscale interfacial materials". *Pure. Appl. Chem.,* 2000, 72:73-81.
6. Lafuma A, Quéré D., "Super hydrophobic states". *Nature Mater.,* 2003, 2:457-460.
7. Genzer J., Efimenko K., "Creating long-lived super hydrophobic polymer surfaces through mechanically assembled monolayers". *Science,* 2000, 290:2130-2133.
8. Gu ZZ., Uetsuka H., Takahashi K., et al "Structural color and the lotus effect". *Angew. Chem. Int. Ed.,* 2003, 42(8):894-897.
9. Sun T., Wang G., Liu H., et al "Control over the wettability of an aligned carbon nanotube film". *J Am Chem. Soc.,* 2003, 125:14996-14997.
10. Bico J., Marzolin C., Quéré D., "Pearl drops". *Euro. Phys. Lett.* 1999, 47(2):220-226.
11. Cyranoski D. "Polar bears fuel row over Alaskan oil". *Nature,* 2001, 414:240.
12. Feng L., Li S., Li Y., et al "Super-Hydrophobic Surfaces: From Natural to Artificial". *Adv. Mater.,* 2002, 14, 24: 1857
13. Chen W., Fadeev A., Heich M., "Ultrahydrophobic and ultralyophobic surfaces: Some comments and examples". *Langmuir,* 1999, 15, 3395.
14. Tsujii K., Yamamoto T., Onda T., et al "Super oil-repellent surfaces". *Angew. Chem. Int. Ed. Engl.,* 1997, 36, 1011
15. Wu Y., Sugimura H., Inoue Y., et al "Thin films with nanotextures for transparent and ultra water-repellent coatings produced from trimethylmethoxysilane by microwave plasma CVD". *Chem. Vap. Deposition,* 2002, 8, 47
16. Nakajima A., Abe K., Hashimoto K., et al "Preparation of hard super-hydrophobic films with visible light transmission." *Thin Solid Films* 2000, 376, 140
17. Feng L., Yang Z., Zhai J., et al "Superhydrophobicity of nanostructured carbon films in a wide range of pH values". *Angew. Chem. Int. Ed.,* 2003, 42:4
18. Zhou X., Gui G., Zhi L., et al "Large-area helical carbon microcoils with superhydrophobicity over a wide range of pH values". *New Carbon Materials* 2007, 22, 1
19. Albalat R., Gomez E., Muller C. et al, "Electrochemical Nucleation Of Nickel On Vitreous Carbon Electrode" *Journal of Applied Electrochemistry,* Vol.21 (1991), pp. 709-715.
20. Abyaneh M.Y., Fleischmann M., "The Electro-crystallization of Nickel". *Journal of Electroanalytical Chemistry,* Vol.119 (1981), pp. 187-191.
21. Cassie A. B. D., Baxter S., "Wettability of porous surfaces". *Trans. Faraday Soc.* 1944, 40: 546-551.
22. Gao X., Jiang L. "Water-repellent legs of water striders". *Nature* 2004, 432: 36

Liquid-state Interfacial Reactions between Sn-Ag-Cu-Fe Composite Solders and Cu Substrate

Xiaoying Liu[1], Yanhui Zhao[1], Mingliang Huang[1], C.M.L.Wu[2], Lai Wang[1*]

[1]Department of Materials Engineering, Dalian University of Technology 116023, P. R. C

[2]Department of Physics and Materials Science, City University of Hong Kong, Hong Kong

*Correspondent author E-mail: wangl@dlut.edu.cn

Tel: 86 0411-84707636, Fax: 86 0411-84709284

Abstract

Growth kinetics and interfacial morphologies of the intermetallic compound (IMCs) between Sn-3Ag-0.5Cu-xFe(0, 0.5wt.%, 1wt.%) composite solders and Cu substrates were investigated by reflowing for different durations at 250°C. Fe particles were deposited quickly in the vicinity of IMCs of the as-reflow samples due to the higher density of Fe than that of Sn-Ag-Cu. They formed a region about 30μm wide where the volume percentage of Fe particles could reach 19%. The isothermal equation of chemical reaction and phase diagrams is used to explain the effect of Fe on the growth kinetics of IMCs under liquid-state conditions. It is found that Fe can effectively retard the growth of Cu_6Sn_5 and Cu_3Sn during liquid-state aging and reduce the size of Cu_6Sn_5 particles. Some local small cracks were observed in the Cu_6Sn_5 particles near interfaces of SAC solder alloys after reflowing for about five minutes. Such cracks were found in the other composite solders only thirty minutes reflow.

1. Introduction

Concerned about environment pollution and the biological hazardness of Pb in the Sn-Pb solder, alloys are raised during the past decades. Thus, there has been an increasing interest in developing alternative lead-free solders in recent years. One of the most favored lead-free soldering alloys is SnAgCu. Its reliability and mechanical properties have been widely studied during the past few years [1, 2]. The current main research thrust in lead-free solder alloy development is on enhancement or fine-tuning the various SnAgCu properties through addition of minor alloying elements [3, 4].

Composite solders are known to have good reliabilities because the dispersed particles can serve as obstacles to grain-boundary sliding, grain growth, crack growth, and furthermore, stress uniformly [5]. Dispersed particles introduced to the solder alloys, on the other hand, do not coarsen easily since the elements or compounds involved can be chosen to have low solubilities and diffusivities (e.g., Fe) or no reactivity (e.g., oxide particles) with the matrix solder alloy. In composite solders, the reinforcements can be intermetallics, metallic powders, carbon fibers or fine oxide particles. Besides that some researchers also used other reinforcing particles such as Ni [6], RE [7] and Carbon nanotubes [8]. The presence of large amounts of dispersoids, however, tends to cause other deteriorating side effects in the composite solders, such as porosity trapping and higher viscosity.

In Sn-rich solders, they are usually in contact with Cu lands. In soldering processes, intermetallic compounds (IMCs) are the major reactants at the interface between the liquid solder and the metal substrate. For instance, interfacial Cu_3Sn and Cu_6Sn_5 IMCs are formed between some Sn-based solder alloys and a Cu under-bump metallization (UBM) [9]. Many studies on the various aspects of reactions between solder and Cu have been carried out. Despite the large number of studies, very little is known about the effect of Fe particles segregation in composite lead-free solders, which is essential to understand the wetting behavior in the soldering process and the subsequent evolution of these IMCs.

The aim of this study is to examine the effect of Fe additions on IMC thickness, growth rate and grain size of the Cu_6Sn_5 intermetallic compound during the soldering reaction of the composite lead-free solders on Cu substrates.

2. Experimental Procedure

In this study, Cu was used as a substrate and Sn-3Ag-0.5Cu with added Fe particles was employed as solder pastes. Sn-3Ag-0.5Cu solder powders (purity 99.9%) with the size of 30~45μm, a commercial water soluble RMA flux and Fe particles (purity 99.99%) with the size of 1~3μm were blent adequately to form the composite solder pastes. Four solder systems were synthesized. They are Sn-Ag-Cu with 0wt.%, 0.5wt.% and 1wt.% of Fe particles. Just prior to soldering, the coupons were acid pickled to remove surface oxide and contaminates. The solder pastes, with a standard diameter of 5mm and a height of 1mm, were laid on the Cu coupon and put in an oven with a set temperature of 250±1°C, then removed from the oven and cooled in air. The solder bump were subsepuently cross-sectioned in epoxy resin. Each sample was mechanically polished and etched in a solution of 5%HNO$_3$-2%HCl-93% methanol for a few seconds to observe the microstructures of solder matrix and IMC layer. The whole soldering joints/Cu sheets were put into the etchant of 10% HNO_3 with ultrasonic wave cleaning for about 10 min. When the IMC grains were exposed and the solder ball was dissolved, the microstructures and compositions of IMC layer from top views were analyzed by EPMA and SEM.

In order to calculate the mean thickness of IMC layer, measurement of the average thickness of the total intermetallic layers was performed by quantitative image analysis. The area of the IMC layers was measured and then divided by the layer length to obtain the average value.

3. Results and Discussion

According to literature [9] on composite solders, methods to introduce the desired reinforcement particles into solder matrices may be classified as two: a mechanical mixing method and an in-suit method. Nevertheless, little literature has been reported about the effect of segregation on the interfacial reactions between Sn-based solder and Cu substrate.

The density of Fe (7.86g/cm^3) is higher than that of Sn-3Ag-0.5Cu (7.4g/cm^3), so Fe particles began to deposit to the vicinity of IMCs of the as-reflowed samples and they formed

978-1-4244-4658-2/09 $25.00 © 2009 IEEE

a iron-rich area about 30μm height ahead of IMCs. Fig.1 displays the cross-sections of composite solders/Cu interfaces. The volume percentage of Fe in this iron-rich area can reach up to 19% when only 1% Fe was added. As can be seen in Fig.2, the pure Fe particles are surrounded by $FeSn_2$ phase. This layer is only hundreds of nanometers thick after reflow for 1 min. With time increasing, the thickness of $FeSn_2$ gets to about 1μm.

Fig. 3 shows the backscattered electron micrographs for cross sections of the samples with different percentage of Fe additions that were reflowed at 250°C for different times. The typical scallop-type Cu_6Sn_5 phase is clearly visible in all cases. Adding Fe definitely refines the IMC thickness of composite solder joints.

Fig. 1 The micrograph of SAC-1Fe as reflowed for 2 min

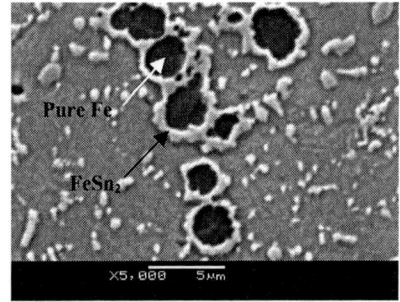

Fig. 2 The micrograph of pure Fe particles surrounded by $FeSn_2$ phase

The plot of IMC thickness as a function of time for SAC and SAC-Fe solders is shown in Fig. 4. It is known that if IMC growth is controlled by grain boundary diffusion, the n value is 1/3; while when the growth is controlled by volume diffusion, n is 0.5. According to the experiment measured IMC thickness of Sn-3Ag-0.5Cu/Cu joint, the value of n is 0.45, the same as former literature reported [10], while the n of composite solder is 0.37. This indicates that the interfacial reaction of SAC solder is still volume diffusion controlled during liquid state aging. Fig. 5 represents the top views of interfacial intermetallics of joints with different composite solders, followed by solder etched away. On the Cu substrate, the scallop and hexagonal-shape intermetallic can be seen in Fig. 5. These grains which are confirmed to be Cu_6Sn_5 decrease with the addition of Fe particles. The plot of Cu_6Sn_5 grain size after reflowed for different times is shown in Fig. 6. The average diameter of Cu_6Sn_5 particle of SAC/Cu joint as reflowed 30s at 250°C is about 3μm, while it is only 1.4μm of the SAC-1Fe solder. The retarding effect of Fe on the IMC formation and growth is obvious.

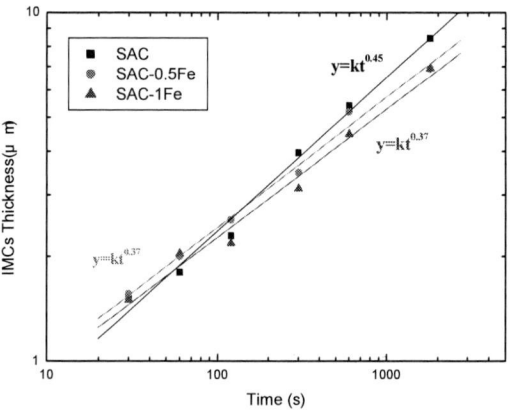

Fig. 4 The plot of IMC thickness as a function of time for three solders

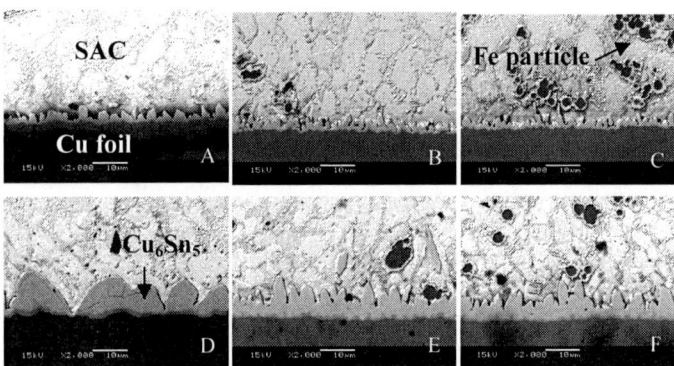

Fig. 3 Backscattered electron micrographs for samples as reflowed for different times (A) SAC 2min (B) SAC-0.5Fe 2min (C) SAC-1Fe 2min (D) SAC 30min (E) SAC-0.5Fe 30min (F) SAC-1Fe 30min

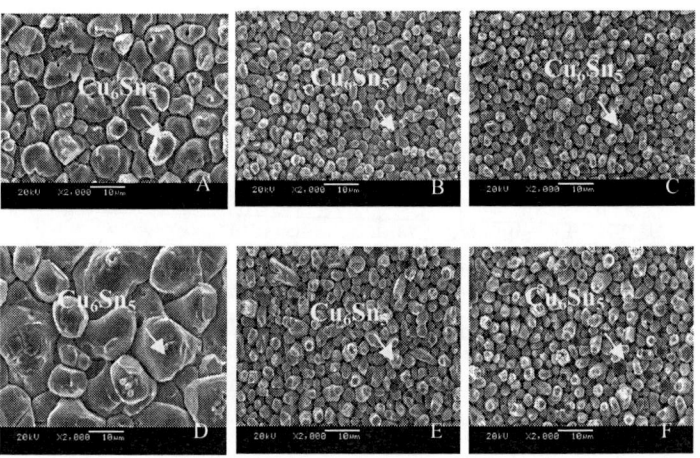

Fig. 5 Top views of IMC grains (A) SAC 10min (B) SAC-0.5Fe 10min (C) SAC-1Fe 10min (D) SAC 30min (B) SAC-0.5Fe 30min (C) SAC-1Fe 30min

As shown in Figs. 3 and 5, there are several differences in the morphology and thickness of the IMC layer for the three solder joints. For the as-reflowed solder joints, the IMC layer of the SAC/Cu joint is the thickest and the grain morphology

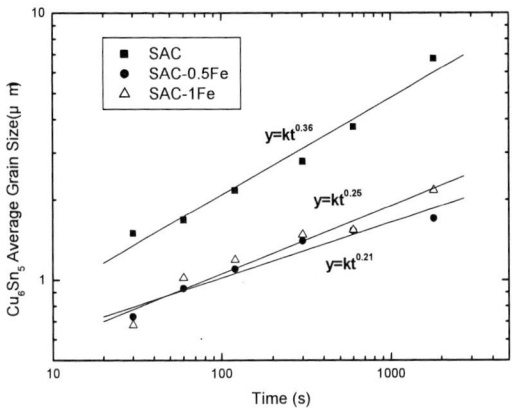

Fig. 6 Average grain sizes of Cu_6Sn_5 particles as-reflowed

is the coarsest. It is well-known that the interfacial reaction between liquid Sn-based solder and Cu substrate can be described as

$$6Cu(s) + 5Sn(l) \rightarrow Cu_6Sn_5(s) \quad (1)$$

According to isothermal equation of chemical reaction [11]:

$$\Delta G = \Delta G^0 + RT \ln \frac{\alpha_{Cu_6Sn_5}}{\alpha_{Cu}^6 \alpha_{Sn}^5} \quad (2)$$

As Cu and Cu_6Sn_5 are solid phases and can be obtained:

$$\alpha_{Cu_6Sn_5} = 1; \ \alpha_{Cu} = 1; \ \alpha_{Sn} \approx \gamma_{Sn}[Sn] \quad (3)$$

Therefore, we can get

$$\Delta G = \Delta G^0 + RT \ln \frac{1}{\gamma_{Sn}^5[Sn]} \quad (4)$$

Where ΔG and ΔG^0 are the Gibbs function of molar reaction and the standard Gibbs function of molar reaction, R is the gas constant, T is the absolute temperature, $\alpha_{Cu,Sn,}$, α_{Cu}, α_{Sn} are the activities of Cu_6Sn_5, Cu atom and Sn atom in molten solder, γ_{Sn} is the activity coefficient of Sn atom. $[Sn]$ is the concentration of Sn atoms. In the following discussion, it is assumed that the value of γ_{Sn} remained approximately constant in different liquid solders. With decreasing concentration of Sn atoms for the reaction of Sn and Fe, the ΔG will intensely increase, leading to more difficult reaction between Sn and Cu atoms. So Fe addition has an effect on IMC thickness and grain size.

Fig. 3(a) shows the microstructures of SAC/Cu joint as reflowed for 2 min. It can be seen there are microcracks in the IMC grains. For composite solders/Cu joints, however, it is clear to see that some local cracks started within the Cu_6Sn_5 particles or along the interface when reflowing time increases to 10min. Based on the observation above, there might be some explanations for the formation of local cracking within Cu_6Sn_5 particles or near the interfaces. Firstly, the difference

in the Young's modulus between SAC and Cu_6Sn_5 is greater [2]. The larger difference in their Young's modulus should lead to a higher stress concentration near the interface. Moreover, the Cu_6Sn_5 is a brittle system. It is much easier to fracture when large Cu_6Sn_5 particles are formed on SAC/Cu joints. So the larger Cu_6Sn_5 will lead to the decreasing of mechanical properties and service lifetime in the electronic assembly.

4. Conclusions

(1) With the addition of Fe particles in Sn-3Ag-0.5Cu solders, the growth kinetics of the interfacial IMC decreased. The dependence follows: $y=kt^n$. The constant n was calculated to be 0.45 for SAC, 0.37 for composite solders. IMC growth of SAC is controlled by volume diffusion while that of composite solder is controlled by grain boundary diffusion.

(2) During the liquid interfacial reaction, the size of Cu_6Sn_5 will increase with reflowing time and get smaller for the addition of Fe. The Cu concentration in solder joints which is reduced with the decreasing consumption of Cu substrate can lead to the quick dissolution of IMC.

Acknowledgments

This work is supported by National Key Technologies R&D Program (2006BAE03B02-2), NSFC key program (U0734006), Key Laboratory Program in Liaoning Province (20060133). The authors would like to thank Prof. Lawrence Wu at City University of Hongkong for his cooperation in this study.

References

1. Jeong-won Yoon, SEUNG-boo Jung, "Effect of isothermal aging on intermetallic compound layer growth at the interface between Sn-3.5Ag-0.75Cu solder and Cu substrate," *Journal of Materials Science*, 39 (2004), pp. 4211-4217.
2. H.F.Zou, Z.F.Zhang, "Solid-state and liquid interfacial reactions between Sn-based solders and single crystal Ag substrate," *Journal of Alloys and Compounds*, 469 (2009), pp. 207-214.
3. C.M.L.Wu, D.Q.Yu, et al., "Preperties of lead-free solder alloys with rare earth element additions," *Materials Science and Engineering R*, 44 (2004), pp. 1-44.
4. H.T.Chen, C.Q.Wang, et al., "Effect of Cu diffusion through Ni on the interfacial reactions of Sn3.5Ag0.75Cu and SnPb solders with Au/Ni/Cu substrate during aging," *Materials Letters*, 60 (2006), pp. 1669-1672.
5. Seong-Yong Hwang, Joo-Won Lee, and Zin-Hyoung Lee, "Microstructure of a Lead-Free Composite Solder Produced by an In-Situ Process," *Journal of Electronic Materials*, Vol.31, No.11 (2002).
6. Fangjie Cheng. Hiroshi Nishikawa. et al., "Microstructural and mechanical properties of Sn-Ag-Cu lead-free solders with minor addition of Ni and/or Co," *Journal of Materials Science*, 43 (2008), pp. 3643-3648.
7. Guangdong Li, Yaowu Shi, et al., "Effect of rare earth addition on shear strength of SnAgCu lead-free solder joints," *Journal of Materials Science: Materials in Electronics*, 20 (2009), pp. 186-192.
8. K.Mohan Kumar, V.Kripesh, et al., "Single-wall carbon nanotube (SWCNT) functionalized Sn-Ag-Cu lead-free

composite solders," *Journal of Alloys and Compounds*, 450 (2008), pp. 229-237.

9. J.Shen, Y.C.Chan, "Research advances in nano-composite solders," *Microelectronics Reliability*, 49 (2009), pp. 223-234.

10. D.Q.Yu, L.Wang, "The growth and roughness evolution of intermetallic compounds of Sn-Ag-Cu/Cu interface during soldering reaction," *Journal of Alloys and Compounds*, 458 (2008), pp. 542-547.

11. C.H.P.Lupis, Chemical Thermodynamics of Materials, Elsevier Science Publishing Co. Inc. New York, (1983).

Cure Kinetics Analysis and Simulation of Silicone Adhesives

Qin Zhang[1,2], Bin Song[2], Simin Wang[3], Ling Xu[3], Sheng Liu[1,2]*

[1]School of Mechanical Science and Engineering, Huazhong University of Science and Technology
Wuhan, Hubei, 430074, China
[2]Wuhan National Laboratory for Optoelectronics Division of MOEMS
Wuhan, Hubei, 430074, China
[3]School of Materials Science & Engineering, Huazhong University of Science and Technology
Wuhan, Hubei, 430074, China
E-mail: victor_liu63@126.com; Tel: 027-87542604; Fax: 027-87557074.

Abstract

A nonlinear transient heat transfer finite element model based on commercial finite element software, ABAQUS, is developed to simulate the curing process of silicone adhesive. The curing reaction process and specific heat of the silicone adhesive were investigated using a differential scanning calorimetry (DSC) to obtain material parameters for numerical simulations. Curing reaction kinetics was derived. Using the user subroutine HETVAL of ABAQUS, temperature distributions inside silicone can be evaluated by solving the heat conduction equations including internal heat generation produced by curing reactions. Good agreement between experimental data and numerical analysis by ABAQUS is obtained.

Introduction

Silicone materials are well established products used in many applications. For electronics applications, silicone adhesives have demonstrated good performance. They have a high degree of flexibility to significantly increase the reliability and life of the devices because of the stress-relieving properties. They can also maintain the stress-relieving nature over a wide temperature or humidity range.

The physical, chemical, and mechanical properties of the polymer are significantly affected by the curing behavior. The silicone adhesive requires the proper temperatures and time to achieve full cure and the best properties. Thus a thorough understanding of the vulcanization mechanism is desirable for better control of the curing process of silicone adhesives.

The silicone adhesive studied in this work is a thermally curable, addition-curing, two-part, LTV silicone rubber. A hydrosilylation reaction will happen when sufficient heat is applied. Assuming that each cross-linking can releases the same amount of energy, the heat released in curing reaction can be utilized to monitor the curing process. Differential scanning calorimetry (DSC) technology is very capable of measuring thermal characterization of many materials, especially polymers [1,2,3,4].

There have been many studies on the modeling or simulation of curing process of polymers [5,6,7,8]. In these works, the thermal properties in the cure circle were calculated mostly with one- or two-dimensional finite difference method. Finite element analysis was used just in recent years. The numerical software people used to study the curing process were the developed special-purpose or general-purpose software, just like ANASYS and ABAQUS. The internal heat generation produced by cure reactions can be accurately evaluated by the user subroutine HETVAL of ABAQUS. In this work, the curing process of the silicone adhesive was investigated by DSC. The cure reaction kinetics and the specific heat of the silicone adhesive were derived. A finite element model based on ABAQUS is developed to simulate the curing process of silicone adhesives.

Experiment

Differential scanning calorimetry

The measurements were carried out using a Diamond DSC (PerkinElmer Instruments). The curing process was studied in non-isothermal conditions. The program was as follows. The samples were heated from 50°C to 200°C. The heating rate was 5°C/min. The experiment was repeated three times with fresh sample to investigate the repeatability. All the scans were performed under nitrogen air. The heat flow-temperature plot is shown in Fig. 1.

Specific heat values of the silicone adhesive were also measured by DSC. Empty pan, silicone adhesives and a standard material, sapphire, were heated between the initial and final isothermal stages at 50°C and 200°C under the same conditions. The heating rate was 5°C/min and the scans were performed under nitrogen air. Comparing the heat flow obtained for the silicone sample above that given by an empty pan, with the heat flow obtained for sapphire, the specific heat of the silicone adhesive was calculated as shown in Fig. 2.

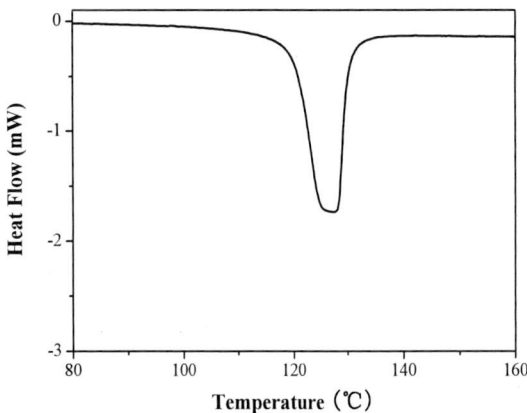

Fig. 1 Dynamic DSC curve at 5°C/min heating rate

978-1-4244-4658-2/09 $25.00 © 2009 IEEE

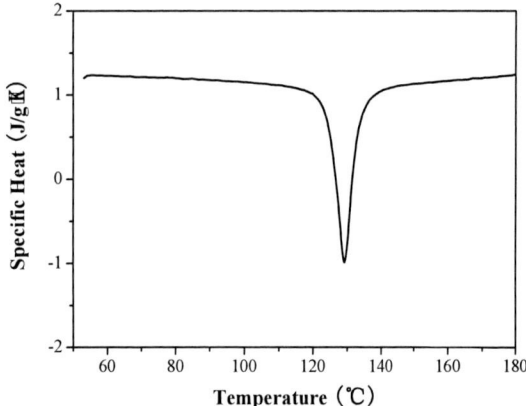

Fig. 2 Specific heat vs. temperature curve of the silicone adhesive

Thermogravimetric (TG) test

Thermogravimetric (TG) test scanning runs were also carried out to measure the any loss in weight in curing process. The mass loss is less than 1% at the temperature range from 50°C to 170°C, and it is less than 0.1% in the curing circle. The value of loss in weight is so small that we can draw the following conclusion -- there is no mass loss in curing reaction.

Cure kinetics model

Vulcanization reaction model is generally described by differential equation as follows:

$$\frac{d\alpha}{dt} = k(T) \cdot f(\alpha) \tag{1}$$

where t is time, $f(\alpha)$ is reaction mechanism function, and k is the temperature dependence rate constant which follows the Arrhenius law:

$$k = A \exp\left(-\frac{E}{RT}\right) \tag{2}$$

where A is pre-exponential factor, E is the activation energy, R is the gas constant (8.314J/mol·K) and T is the absolute temperature.

Mechanistic and phenomenological methods can both characterize a vulcanization reaction. Phenomenological method was used by a lot of researchers because it is simple and highly accurate, despite it cannot describe the mechanism of curing. Usually, in the numerous phenomenological kinetic models, the silicone vulcanization reaction is described by autocatalytic model, Kamal-Sourour model, or n th order reaction kinetics models. In this paper, kinetics model was analyzed using n th order reaction kinetics of a chemical reaction:

$$\frac{d\alpha}{dt} = k(T) \cdot (1 - \alpha)^n \tag{3}$$

where n is reaction order.

Table 1 Cure kinetics parameters of the silicone adhesive

Cure Kinetics Parameters		
Pre-exponential Factor	A (1/s)	9.29e70
Activation Energy	E (kJ/mol)	553.87
Reaction Order	n	1.28
Total Heat of Reaction	H (J/g)	15.64

Following the methods in ASTM E2041, the kinetics parameters of the crossing reaction were estimated without any constraints on them by multiple linear regression analysis. The parameters are listed in Table 1.

Numerical analysis

It is assumed that there is no silicone flow, no volume change, no mass loss, and isotropic in curing process of the silicone adhesive. Based on these assumptions, the one-dimensional time dependent energy balance is given as follows:

$$\rho C_P \frac{\partial T}{\partial t} = \frac{\partial}{\partial z}\left(K \frac{\partial T}{\partial z}\right) + \rho H \frac{d\alpha}{dt} \tag{4}$$

where $\rho H \dfrac{d\alpha}{dt}$ is present the heat flux of internal heat generation of the silicone adhesive. By Equation 3, we have

$$\frac{d\alpha}{dt} = A \exp\left(-\frac{E}{RT}\right) \cdot (1 - \alpha)^n \tag{5}$$

The Equation 5 can be discretized as

$$\begin{aligned}
\alpha\big|_t &= \alpha\big|_{t-\Delta t} + \Delta t \frac{d\alpha}{dt}\bigg|_{t-\Delta t} \\
&= \alpha\big|_{t-\Delta t} + \Delta t\left[A\exp\left(-\frac{E}{RT_{t-\Delta t}}\right) * \left(1 - \alpha\big|_{t-\Delta t}\right)^n\right]
\end{aligned} \tag{6}$$

The Equation 6 leads to a numerical solution of Equation 4 at each increment using the user subroutine HETVAL of ABAQUS.

A nonlinear transient heat transfer finite element model is developed to simulate the curing process of silicone adhesives based on ABAQUS. Let the DSC scanning with 10mg silicone adhesives as the numerical example, the finite element model is shown in Fig. 3. The silicone adhesive is filled in aluminium pan, and an aluminium disk on the top of the silicone adhesive. The material properties of aluminium and silicone adhesive can be found in Table 1 and Table 2. The inside diameter of the aluminium pan is 5mm, and the thickness is 0.1mm. Due to axisymmetric, a two-dimensional model is meshed. There are 155 elements which type is 4 node axisymmetric element. The model was subjected to a described temperature cycle in DSC scanning. Assume the surfaces which contact with the heating wire is insulate. The convection coefficient of the surfaces which contact with the nitrogen air is specified as 5W/m²·K.

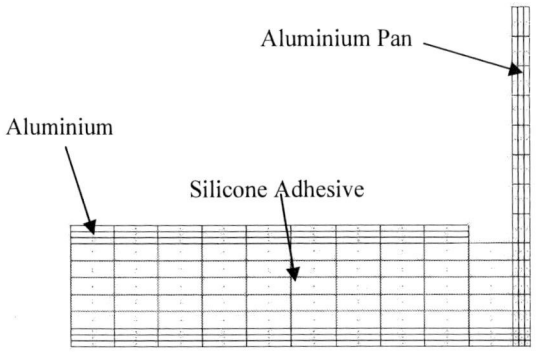

Fig. 3 A typical finite element model for DSC scanning

Table 2 Material properties of aluminium and silicone adhesives

	Aluminium	Silicone
Density ρ (g/mm^3)	2.7e-3	1.1e-3
Specific Heat c_p (J/g.K)	0.88	As shown in Fig. 2.
Thermal Conductivity K (J/s.m.K)	247	0.22

Results and Discussion

Using the user subroutine HETVAL, the degree of cure on each integration point can be obtained. Define the average of the all the degrees of cure on the integration points is the degree of cure of silicone sample, the curve for conversion of the silicone adhesive versus temperature was observed. The plots of the experimental data and the data obtained by numerical simulation using ABAQUS for the silicone adhesive are shown in Fig. 4. Similarly, the rate of conversion of the silicone adhesive was defined, and the rate of conversion of the silicone adhesive vs. temperature curve was observed, as shown in Fig. 5. As we can see, the results obtained from the numerical analysis were in good agreement with the experimental results. In detail, the experimental data and numerical analysis data are well in agreement at the initial stages of the cure reaction. But in the later stages, the conversion data based on numerical analysis are slightly over estimated until the curing process completed. The biggest value over estimated for the silicone adhesive is about 15%. The discrepancy may have come from errors in cure kinetics model. Too large increment size also may bring deviations.

Fig. 6 and Fig. 7 show isotherm contour plots at different times in curing process. From the figures, it is evident that the temperature gradients within the silicone adhesive increase with temperature increasing. But the temperature gradients are very small because of the small amount of the silicone adhesive.

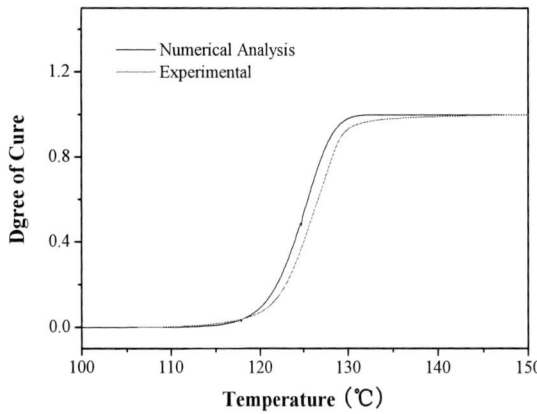

Fig. 4 Comparison of experimental data with numerical analysis data: conversion vs. temperature curves

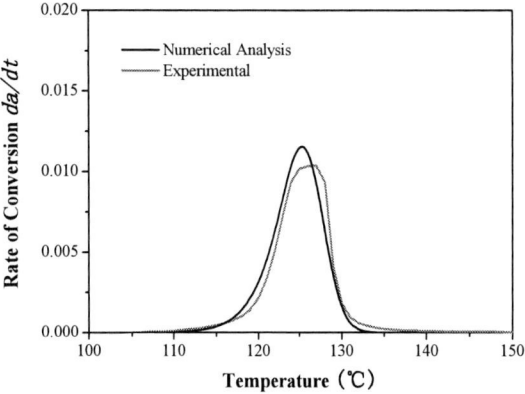

Fig. 5 Comparison of experimental data with numerical analysis data: rate of conversion vs. temperature curves

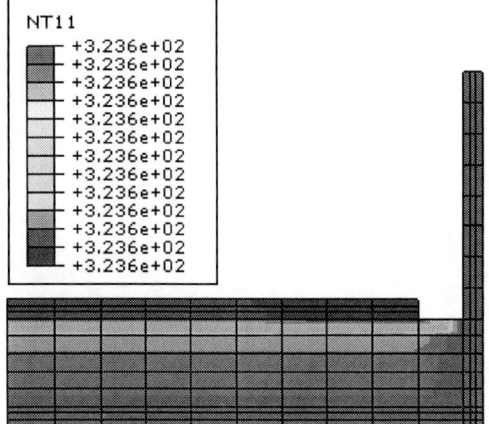

Fig. 6 Temperature distribution at t=0.01s

978-1-4244-4658-2/09 $25.00 © 2009 IEEE   617

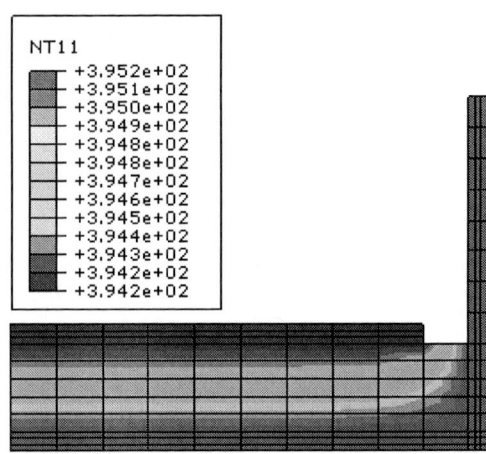

Fig. 7 Temperature distribution at t=858s

Conclusions

In the present work, the curing process of silicone adhesives is measured, evaluated and simulated. The DSC technology was used to measure the heat released in the curing process, and the measurement program was simulated by finite element method. The key in analysis of the curing process is cure kinetics model which can be used to develop heat generation. The developed code in user subroutine HETVAL has successfully simulated the exothermic behavior of curing process. The results obtained from the numerical analysis are in good agreement with the experimental results.

Acknowledgments

The authors wish to appreciate the support from NSFC Key Project with Contract Number 50835005 and National Project 973 with Contract Number 2009CB320203.

References

1. H.-J. Flammersheim, J. R. Opfermann, "Investigation of Epoxide Curing Reactions by Differential Scanning Calorimetry-formal Kinetic Evaluation," *Macromolecular Materials and Engineering*, Vol. 286, No. 3 (2001), pp. 143-150.
2. A. Yousefi, P. G. Lafleur, R. Gauvin, "Kinetic Studies of Thermoset Cure Reactions: A Review," *Polymer Composites*, Vol. 18, No. 2 (1997), pp. 157-168.
3. L. M. Lopez, A. B. Cosgrove, J. P. Hernandez-Ortiz, T. A. Osswald, "Modeling the Vulcanization Reaction of Silicone Rubber," *Polymer Engineering and Science*, Vol. 47, No. 5 (2007), pp. 675-683.
4. R. B. Prime, C. Michalski, C. M. Neag, "Kinetic Analysis of A Fast Reacting Thermoset System," *Thermochimica Acta*, Vol. 429, No. 2 (2005), pp. 213-217.
5. S. Yi, H. H. Hilton, M. F. Ahmad, "A Finite Element Approach for Cure Simulation of Thermosetting Matrix Composites," *Computers & Structures*, Vol. 64, No. 1-4 (1997), pp. 383-388.
6. Z.-S. Guo, S. Du, B. Zhang, "Temperature Field of Thick Thermoset Composite Laminates during Cure Process," *Composites Science and Technology*, Vol. 65, No. 3-4 (2005), pp. 517-523.
7. S. Y. Pusatcioglu, J. C. Hassler, A. L. Fricke, H. A. McGee Jr., "Effect of Temperature Gradients on Cure and Stress Gradients in Thick Thermoset Castings," *Journal of Applied Polymer Science*, Vol. 25, No. 3 (1979), pp. 381-393.
8. D. C. Blest, B. R. Duffy, S. Mckee, A. K. Zulkifle, "Curing Simulation of Thermoset Composites," *Composites: Part A*, Vol. 30, No. 11 (1999), pp. 1289-1309.

Growth and Recrystallization of Electroplated Copper Columns

Jun Liu[1], Changqing Liu[1], Paul P Conway[1], Jun Zeng[2], Changhai Wang[2]

[1]Wolfson School of Mechanical and Manufacturing Engineering, Loughborough University
Loughborough, Leicestershire, LE11 3TU, UK
[2]School of Engineering and Physical Sciences, Heriot-Watt University
Edinburgh, EH14 4AS, UK
c.liu@lboro.ac.uk

Abstract

The present paper addresses the formation and evolution of microstructure and crystal structure of electroplated copper columns. Copper columns of different size have been plated on Au seed layer for a series of periods of time from half an hour to three hours and characterized by Focused Ion Beam (FIB) and XRD. It was found that not only the growth process during plating but also accompanying recrystallization and post-plating spontaneous self-annealing account for the interesting microstructure with bi-modal or tri-modal grain size distribution. The evolution of crystal structure of the copper columns during plating and self-annealing after plating was also studied and discussed. The results indicate that the presence of organic additives is not essential for self-annealing to occur in copper columns, which is against the popular belief with respect to electroplated copper films.

Introduction

Electrodeposition has been widely used in electronic manufacturing industry for producing micro- or nano-components of electronic devices e.g. flip-chip bumps, copper interconnects replacing Al interconnects, micro-copper columns, through silicon vias, nanowires etc. Deposit of a wide variety of microstructure ranging from single crystal [1] to nano-crystal [2] can be obtained using different deposition parameters and/or with assisting techniques. Microstructure largely determines electrical and mechanical performance of deposit. For example, electroplated copper produced by pulse plating featuring a high density of nano-twins combines ultra-high strength (ten times higher tensile strength than normal-grain copper) and comparable electrical conductivity [3]. However, the understanding to mechanisms behind the various microstructures, especially the single crystal one and the nano-twinning, by electrodeposition is still too limited at present to facilitate microstructure tailoring for specific applications. Moreover, microstructure evolution of copper deposit due to recrystallization at room temperature (also termed self-annealing or abnormal grain growth), which has been observed in electroplated Cu films during the last decade and gaining increasing concerns [4-11], adds further complexity. It is also suggested that recrystallization may even take place along with the deposition [11-12]. In this paper, we will use the term self-annealing for post-plating recrystallization and the term recrystallization for during plating. Self-annealing of electroplated copper is often accompanied by four typical phenomena, namely decrease of resistance, stress relaxation, microstructure evolution and crystal structure evolution.

It typically causes up to 20% decrease of deposit resistance and considerable stress relaxation within hours or days depending on the thickness of the deposit and deposition parameters [5, 7, 13-14]. Xu et al. [12] performed in-situ measurement of the stress during high-frequency pulse plating of copper and found periodical stress changes during the pulse-on and pulse-off time, which indicates the stress relaxation can immediately occur during the pulse-off time.

A typical feature of microstructural evolution is secondary grain growth, that is, growth of large grains in the matrix of small grains which usually results in bi-modal grain size distribution in the case of incomplete transformation. Brongersma et al. [4] proposed a two-step grain growth mechanism based on the experimental observation on the FIB cross-sectional microstructure of plated copper. Two steps refer to a very rapid crystallization occurs from the top surface down just after deposition followed by a slower lateral recrystallization resulting in large secondary grains. Yin et al. [15], however, reported quite the opposite finding that the microstructure transformation starts and nucleates from the bottom of the deposit near the deposit/substrate interface and then propagate towards the free surfaces. Hau-Riege et al. [6] performed in situ TEM observation of the grain growth in free-standing electroplated copper films at room temperature, starting minutes after the plating process. The results seem to suggest that the observed twins resulted from recrystallization.

Crystal structure also evolves due to self-annealing. Detavernier et al. [13] reported that Cu(111) peak became much narrower and underwent a splitting into $CuK\alpha1$ and $CuK\alpha2$ components during storage, which was attributed to the decrease of defect density within the plated film. Pantleon et al. [9] found similar phenomenon of peak splitting in both Cu (111) and Cu (200) peaks. Additionally, they performed detailed and quantitative analysis of the crystal structure of self-annealed copper film of different thickness in terms of grain size, grain orientation and grain boundary character using in-situ XRD and EBSD [9, 16-17]. It was concluded that crystallographic texture of electrodeposits changes as a result of multiple twinning during self-annealing and more pronounced twinning occurred for thicker film.

What is the driving force behind these intriguing phenomena? Harper et al. [5] proposed that the driving force is the grain boundary energy density in the fine-grained as-deposited film and ascribed the observed incubation time for recrystallization to the Ostwald ripening of the pinning species along the grain boundaries. And subsequent unpinning of certain grain boundaries leads to rapid secondary grain growth. Detavernier et al. [13] studied the thermodynamics and kinetics of microstructural evolution in copper films by estimating the magnitude of possible driving forces including grain boundaries, stacking faults, dislocations, surface energy, elastic strain and Zenner pinning. They concluded that a high density of dislocations and/or stacking faults is the primary driving force for self-annealing while Zenner pinning by impurities from the plating bath is not crucial although it may strongly influence the kinetics (e.g. the incubation time). Hau-Riege et al. [6], however, proposed that impurities play important roles in self-annealing of electroplated copper films. First, the normal growth of grains in untransformed matrix is impaired by solute drag due to the impurities. The resulting stabilized small grain size contributes to a high driving force for self-annealing. Second, impurity rejection contributes to the energy which drives the transformation to a large-grained structure. Third, accumulation of rejected impurities at the perimeters of growing grains can slow the rate of transformation. Pantleon et al. [9, 16-17] believed that their EBSD results together with in-situ XRD analysis had confirmed that multiple twinning in copper electrodeposits as the mechanism for microstructure evolution at room temperature.

Kinetics of self-annealing process has been given much attention too. An incubation time of a few hours was observed before the onset of self-annealing [5, 15]. Once it begins, the rate of microstructural transformation remains nearly constant until the end of the process [15]. A number of factors have been found to influence the kinetics of self-annealing. Firstly, it has been widely accepted that deposit thickness strongly influences self-annealing kinetics [4-5, 7, 9, 13]. Generally, the thicker the deposit is, the faster the self-annealing; when the it is less than a critical value, e.g. 250 nm [13] no self-annealing can be observed at all. Harper et al. [5] proposed a simple model to describe the inverse dependence of transformation time on deposit thickness, which was also used to estimate the incubation time and critical minimum thickness. Yin et al. [15] described the transformed fraction as a function of self-annealing time using the Avrami equation and obtained the Avrami constants, from which it was inferred that 3D growth and nucleation saturation has occurred. Secondly, deposition parameters e.g. current density, have effects on the self-annealing kinetics. It was found [10] that the microstructures of self-annealed Cu deposits obtained using different current density look similar while the transformation rate for the high current density (32 mA/cm^2) sample was much higher than that of the low current density (8 mA/cm^2) sample. Thirdly, plating bath additives are believed to be critical to the self-annealing kinetics. Brongersma's group [4, 7] concluded that organic additives in plating bath are necessary for self-annealing to occur, but slows down the process for higher concentration. Vas'ko et al. reported that self-annealing occurs only if both the two organic additives the 3-N,N-dimethylaminodithiocarbamoyl-1-propanesulphonic acid (DPS) and polyethylene glycol (PEG) are present in the plating bath. It is worth noting that all above-reviewed work in which self-annealing phenomena were experimentally observed used commercial copper plating bath which normally contains organic additives or lab-made solutions with addition of organic additives. Finally, the kinetics of the self-annealing of electroplated Cu damascene trenches were studied and it was found that the topography or

geometry of the trench, e.g. trench width and spacing, has influences on the recrystallization [14].

To sum up, the driving force behind the self-annealing of electroplated copper still remains controversial. Whether it occurs during plating has not been confirmed. Additionally, most of above-reviewed work concerns electroplated copper films. Since the microstructural transformation is direction-selective, the shape of plated samples is anticipated to affect the self-annealing process. The growth and recrystallization with respect to cylindrical interconnects such as copper columns and through-wafer interconnects has not been reported. A firm understanding to the mechanism of growth during plating, possible accompanying recrystallization and post-plating self-annealing, which is of great significance in predicting and tailoring the microstructure, crystal structure and mechanical and electric performance of copper interconnects, has yet to be established. In this paper, we will address some issues regarding the growth, recrystallization and self-annealing of electroplated copper columns. Copper columns of different diameters were electroplated on Au seed layer for a series of period of plating time using same other deposition parameters. The samples plated for different time were used to represent different points of time over three hours of deposition for the ex-situ study of the growth and recrystallization process during plating. The microstructural evolution, texture development and the effects of pad size are studied and discussed. Finally, in-situ XRD was performed on one plated sample to study the post-plating self-annealing.

Experiment

Test samples were 3-inch Si wafers metallised with sputtered Ti adhesion layer of 150nm thick followed by Au seed layer of 100nm. The metallised wafers were then lithographically patterned to provide a photoresist template for deposition of copper columns of 10μm, 15μm, 20μm and 25μm in diameters and 50μm in pitch.

A two-electrode cell is step up with a test wafer held by a commercial-level six-contact-point wafer holder as the working electrode and a copper disk as the counter electrode. The area ratio of the copper disk to plating area is greater than 200:1. The electrochemical workstation, PARSTAT® 2273 of Princeton, has been used to supply constant current for galvanostatic plating. Lab-made acidic copper sulfate solution containing 166g/l $CuSO_4 \cdot 5H_2O$ and 200g/l H_2SO_4 and deionized water is used as the electrolyte. Galvanostatic electrodeposition experiments were carried out for a series of periods of time under current density of 4mA/cm^2 and at room temperature without agitation.

Results

Fig. 1 shows the surface morphology of copper columns plated for different time from half an hour to three hours. It is prominent that all these columns feature extraordinarily large and well-faceted grains protruding from the finer matrix. In the column plated for half an hour, the protrusions show morphology of rectangular pyramids with blunt tips and rough surfaces. They are basically in similar size and on the edge of the column. In the column plated for one hour, more large grains and impingement of some of them are observed as shown Fig. 1(b). The protrusions basically remain the

978-1-4244-4658-2/09 $25.00 © 2009 IEEE

pyramid shape while the surfaces are better formed. For 2h and 3h, however, the shape of the protrusions evolves into frustums of rectangular pyramids. Triangular surfaces also appear at the base of the pyramid as marked by arrows in Fig. 1(c) and Fig. 1(d). Meanwhile, they have grown larger. In the column plated for three hours, the large grains have nearly grown across the whole surface.

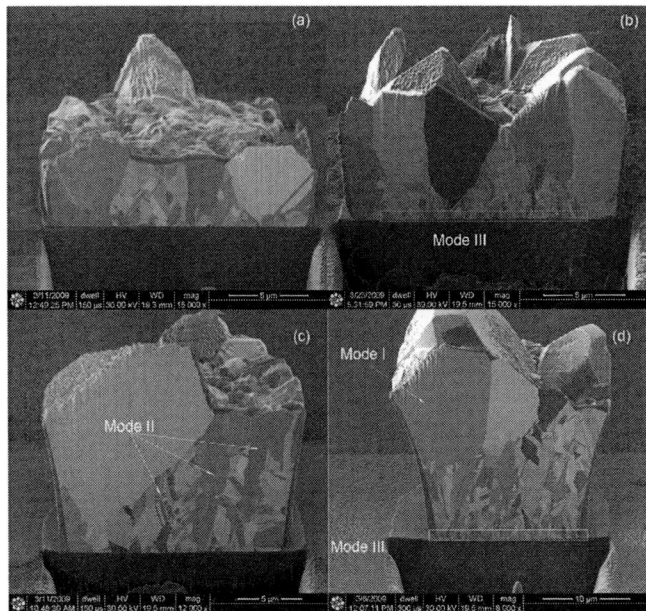

Fig. 2 Cross-sectional microstructure of copper columns plated for a) 0.5 h, b) 1 h, c) 2h, d) 3h using additive-free electrolyte

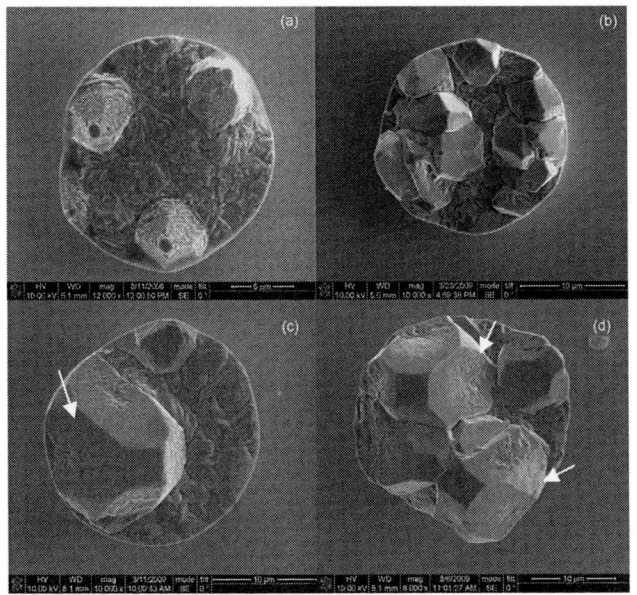

Fig. 1 Top view of copper columns plated for (a) 0.5 h, (b) 1 h, (c) 2 h and (d) 3h using additive-free electrolyte

Fig. 2 shows the FIB images of the cross-sectional microstructure corresponding to those shown in Fig. 1. All these columns show bi-modal or tri-modal distribution of grain size. Firstly, there are a few large grains on the top and/or the edge of columns. These large grains are defined as Mode I grains in this paper for succinct presentation. Secondly, near the deposit/substrate interface are a great many ultrafine grains (Mode III) as shown in Fig. 2(b) and Fig. 2(d). Thirdly, mixture of a number of basically columnar grains of median size and many twins (Mode II) are between large grains and ultrafine grains. The microstructure of the columns plated for short time (e.g. Fig. 2(a) and Fig. 2(b)) look considerably different from the corresponding bottom part of the columns plated for longer time (e.g. Fig. 2(d)). On the contrary, covering the bottom one third of the Fig. 2(d), we can see its upper part looks broadly similar to the Fig. 2(c). In the same way, Fig. 2(a) corresponds to the upper half of the Fig. 2(b). And so on. These characteristics of the observed surface morphology and microstructure basically correspond to the features of secondary grain growth as a result of recrystallization. Therefore, it is safe to come to the conclusion that it is the recrystallization during plating and/or self-annealing after plating that finally results in the final structures.

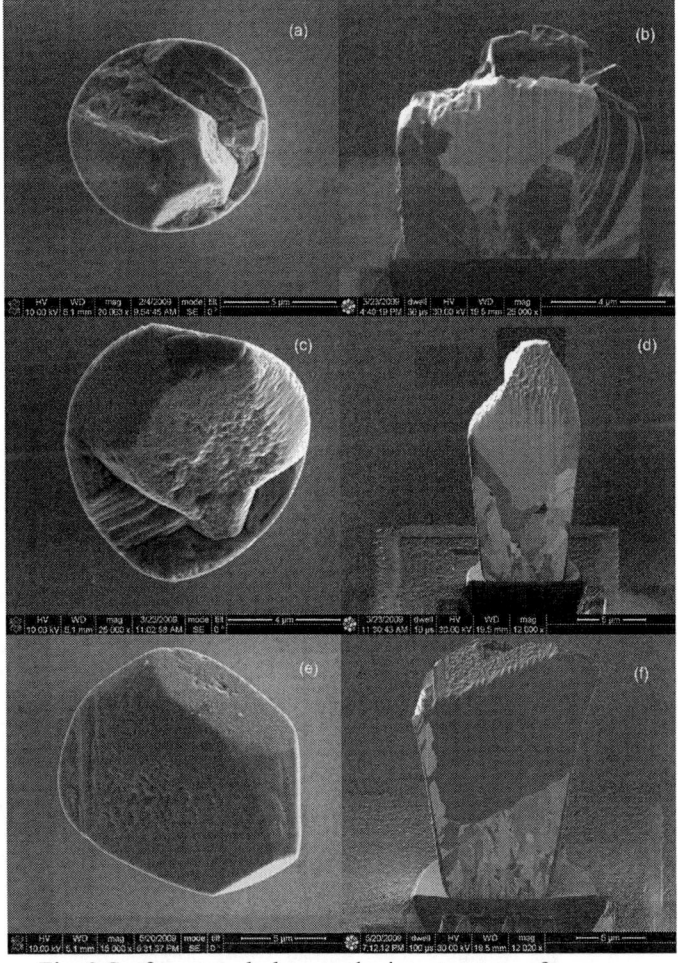

Fig. 3 Surface morphology and microstructure of copper columns of 10μm in nominal diameter plated for (a) (b) 1 hour, (c) (d) 2 hours and (e) (f) 3 hours

Fig. 3 shows the surface morphology and microstructure of smaller columns of nominal diameter of 10μm plated for different time. It can be seen that the size of the Mode I grains is comparable with the diameter of the column. Compare to Fig. 1, the size of the Mode I grains increases and their morphology has changed too. It shows the morphology of frustums of rectangular pyramid in the 10μm column plated for one hour while in the 15μm columns it is not observed until plated for two hours. The Mode I grains in the 10μm columns plated for two hours and three hours, however, show ridge-like morphology rather than the frustum of rectangular pyramid in the 15μm columns. Additionally, it is interesting that the upper half of the column plate for three hours is basically a single crystal. Such size effects on the microstructure may be attributed to change of influence of the recrystallization in two directions, i.e. from top surface down and laterally from outside inwards. Brongersma et al. [4] proposed in electroplated copper films self-annealing basically follows the two-step growth mechanism i.e. a primary rapid recrystallization from top surface down followed by a slower lateral one. It is probably owing to lower energy barrier for recrystallization with respect to the free surface that the primary recrystallization from top surface down takes the priority over the lateral one. For copper columns, however, the cylinder surface is free surface too. That is probably why the Mode I grains described above tend to grow on the edge of columns. It is hard to determine based on the present results which way is the primary one. Nevertheless, the resulting microstructure as shown in Fig. 3(b) and Fig. 3(d) seems to suggest that later recrystallization is promoted when the column diameter decreases.

Fig. 4 shows the overview of arrays of copper columns. It can be seen that not all the columns have Mode I grains on top. The distribution of Mode I grains is shown in Fig. 5. The number of Mode I grains in each column follow the normal distribution. And most of them are on the edge of the columns.

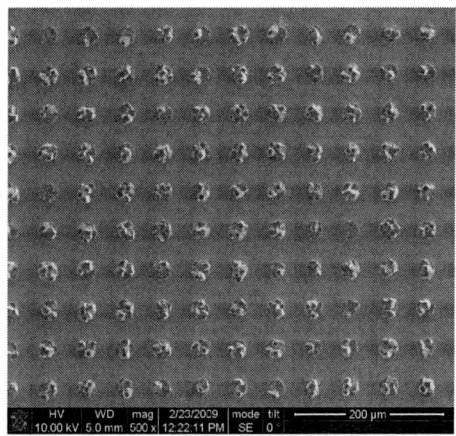

Fig. 4 Overview of several arrays of copper columns plated for three hours

Fig. 6 shows XRD profiles of the samples plated for different time as well as a bare Si wafer and an as-metallised

wafer before plating for comparison. Cu (111) and Cu (200) peaks are present in all the samples plated for different time. The samples plated for one hour or more have much stronger Cu (200) peaks. Cu (220) peak is only observed in the sample plated for three hours. Cu (311) peaks, although extremely weak, are only detected in the samples plated for two hours or more. It has been proposed that Cu (111) texture should develop for surface energy driven growth, (100) texture for strain energy driven growth in the elastic regime, and (110) texture for strain energy driven growth in the plastic regime [18]. Accordingly, the columns plate for one hour or longer showing preferred (100) orientation must have underwent growth or recrystallization driven by the strain energy in elastic regime. Only those plated for three hours with low-intensity (110) and (311) texture being developed have enough stress to lead to plastic strain which drives the growth or recrystallization of the (110) texture. It is very interesting to notice that new Au (311) peaks, which is absent in the as-sputtered samples, appear in the plated samples. It indicates that deposition has changed the crystallographic texture of the Au seed layer. It is probably that Cu diffuses into the Au substrate and changes its structure considering that Au atoms are much larger than Cu atoms and usually difficult to diffuse into Cu. The intensity of Au (311) peak increases with the deposition time, which may indicate that the changes are likely to be made mainly during plating.

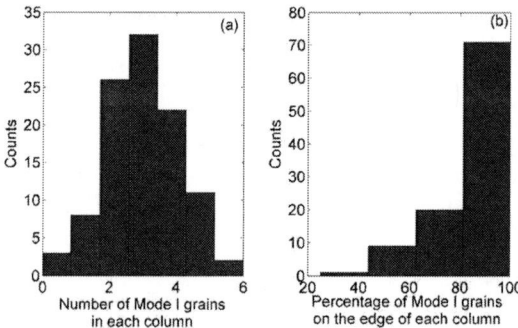

Fig. 5 Mode I grain distribution (a) frequency counts of Mode I grains in each column (b) Percentage of the Mode I grains on the edge of each column

Fig. 7 shows the selected XRD line profile for 111 and 200 peaks of one sample a series of periods of time from a few minute until 10 days after plating. The change of diffracting intensity of 111 planes was negligible until 10 days after deposition while for 200 planes changes were observed within 100 minutes but remained nearly unchanged after that. It indicates that the self-annealing may have preferences in orientations during different stages. Overall, however, the amplitude of the changes of the peak density is small and peak narrowing and splitting, which have been previously reported in literature [9, 13], are not observed at all. It can be seen that they have already had Kα1 and Kα2 components and been quite narrow as deposited. The measurement for the as-deposited profile was performed 15 minutes after plating and it took about 17 minutes to collect the data for these two peaks. The deposit may have already

978-1-4244-4658-2/09 $25.00 © 2009 IEEE

undergone major recrystallization within such short time after plating or even during plating. As explained earlier, it was believed that the presence of organic additives is necessary for self-annealing to occur [4, 19]. However, our findings concerning copper columns are against this popular belief based on observations on thin films. Organic additives, which can be partially co-deposited with metal atoms, may retard the recrystallization during plating and save the energy for subsequent spontaneous self-annealing. Meanwhile, it was reported that self-annealing is slowed down for higher concentration of organic additives. Without the action of organic additives, the self-annealing in electroplated copper film is probably too fast to be experimentally captured.

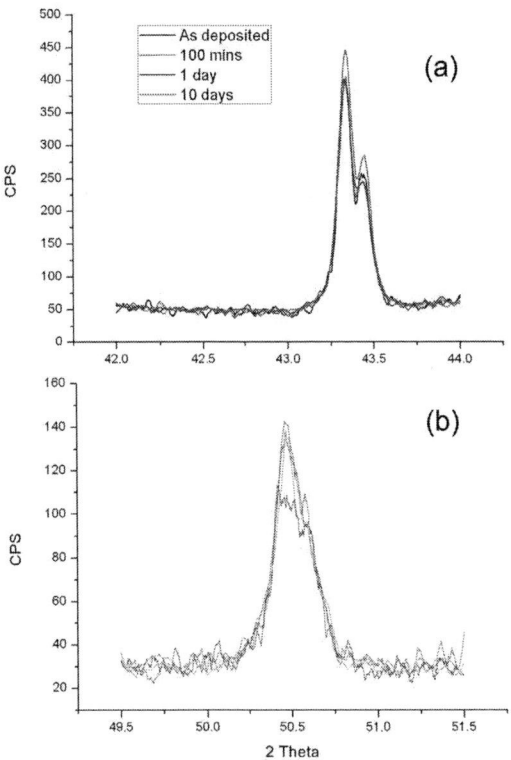

additives may account for the faster kinetics of recrystallization and self-annealing of electroplated copper columns.

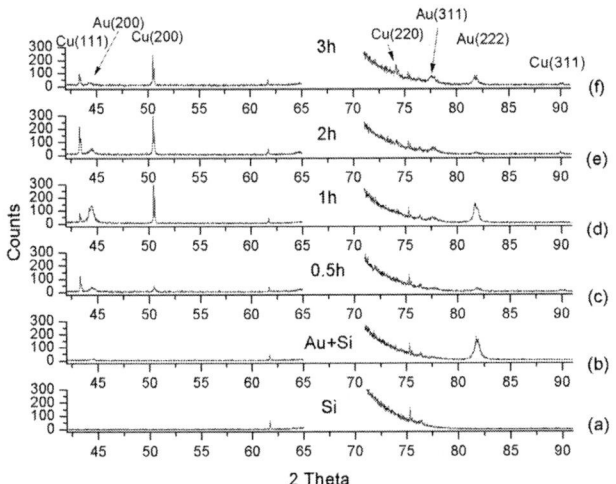

Fig. 6 XRD line profiles of (a) bare Si wafer, (b) Si with sputtered Au, (c) plated for 0.5 hour, (d) plated for 1 hour, (e) plated for 2 hours and (f) plated for 3 hours

Fig. 7 XRD line profiles for the 111 and 200 texture, (a) as-deposited, (b) 100 minutes, (c) 24 hours and (d) 10 days after deposition of a sample plated for one hour

Discussion

As a room-temperature process, electro-crystallization is well off the equilibrium. High density of defects is usually expected in copper electrodeposit. The passage of electric current through deposit may also influence atomic diffusion process during plating as suggested by the recent work [20]. For deposition of electronic interconnects e.g. copper columns and through silicon vias, shape and size effects should also be considered. These various factors make the formation process of electroplated copper column very complex. Complete interpretation of the final structure of copper deposit requires consideration of not only the growth process during plating, which are governed by deposition parameters and conditions, but also the simultaneous recrystallization accompanying the growth and spontaneous self-annealing after plating.

In the present case, the ultrafine Mode III grains are probably the original as-deposited grains; the Mode II and Mode I grains are more likely to be the combined action of original deposition, accompanying recrystallization as well as self-annealing after plating. With current results, it is hard to determine which process is dominant or mainly responsible for the final structure. Compared to thin films (<5μm thick) in literatures, the thicker deposit (about 5-25μm), the cylindrical shape of copper columns as well as the absence of organic

Conclusions

In conclusion, the electroplated copper columns show microstructure with bi-modal or tri-modal grain size distribution, i.e. large grains on the top and preferably the edge (Mode I), ultrafine grains near the deposit/seed layer interface (Mode III) and mixture of columnar grains and twins, having undergone growth, simultaneous recrystallization and spontaneous self-annealing after plating. The copper columns plated for longer time show preferred Cu (200) orientation. The low intensity peaks including Cu (311) and Cu (220) were only found in XRD line profiles of the columns plated for longer than two hours. New Au (311) peak, which is absent in as-sputtered Au, was found in plated samples and its intensity increase with deposition time. Slight changes of the crystal structure were observed by the in-situ XRD and it was found that the changes have preference in orientation in different stages of self-annealing. Finally, our results indicate the presence of organic additives is not essential for self-annealing of copper column to occur, which is against the popular belief for the case of electroplated films.

Acknowledgments

This research was funded by Innovative electronic Manufacturing Research Centre (IeMRC) in UK. The authors would like to thank Dr David Ross from Loughborough University for the XRD analysis and useful discussion.

References

1. Dobrev, D., Vetter, J., Angert, N., and Neumann, R., "Electrochemical growth of copper single crystals in pores of polymer ion-track membranes," *Appl. Phys. A-Mater.*, 69 (2), (1999), pp. 233-237.

2. Xu, L., Dixit, P., Miao, J., Pang, J.H.L., Zhang, X., Tu, K.N., and Preisser, R., "Through-wafer electroplated copper interconnect with ultrafine grains and high density of nanotwins," *Appl. Phys. Lett.*, 90, (3), (2007), pp. 033111-033113.

3. Lu, L., Shen, Y., Chen, X., Qian, L., and Lu, K., "Ultrahigh Strength and High Electrical Conductivity in Copper," *Science*, 304, (5669), (2009), pp. 422-426.

4. Brongersma, S.H., Richard, E., Vervoort, I., Bender, H., Vandervorst, W., Lagrange, S., Beyer, G., and Maex, K., "Two-step room temperature grain growth in electroplated copper," *J. Appl. Phys.*, 86, (7), (1999), pp. 3642-3645.

5. Harper, J.M.E., Cabral, J.C., Andricacos, P.C., Gignac, L., Noyan, I.C., Rodbell, K.P., and Hu, C.K., "Mechanisms for microstructure evolution in electroplated copper thin films near room temperature," *J. Appl. Phys.*, 86, (5), (1999), pp. 2516-2525.

6. Hau-Riege, S.P., and Thompson, C.V., "In situ transmission electron microscope studies of the kinetics of abnormal grain growth in electroplated copper films," *Appl. Phys. Lett.*, 76, (3), (2000), pp. 309-311.

7. Lagrange, S., Brongersma, S.H., Judelewicz, M., Saerens, A., Vervoort, I., Richard, E., Palmans, R., and Maex, K., "Self-annealing characterization of electroplated copper films," *Microelectron. Eng.*,50, (1-4), (2000), pp. 449-457.

8. Murakami, M., Moriyama, M., Tsukimoto, S., and Ito, K., "Grain Growth Mechanism of Cu Thin Films," *MATERIALS TRANSACTIONS*, 46, (7), (2005), pp. 1737-1740.

9. Pantleon, K., and Somers, M.A.J., "In situ investigation of the microstructure evolution in nanocrystalline copper electrodeposits at room temperature," *J. Appl. Phys.*, 100, (11), (2006), pp. 114319-114317.

10. Yin, K.B., Xia, Y.D., Zhang, W.Q., Wang, Q.J., Zhao, X.N., Li, A.D., Liu, Z.G., Hao, X.P., Wei, L., Chan, C.Y., Cheung, K.L., Bayes, M.W., and Yee, K.W., "Room-temperature microstructural evolution of electroplated Cu studied by focused ion beam and positron annihilation lifetime spectroscopy," *J. Appl. Phys.*, 103, (6), (2008), pp. 066103-066103.

11. Zhang, X., Tu, K.N., Chen, Z., Tan, Y.K., Wong, C.C., Mhaisalkar, S.G., Li, X.M., Tung, C.H., and Cheng, C.K., "Pulse Electroplating of Copper Film: A Study of Process and Microstructure," *Journal of Nanoscience and Nanotechnology*, 8, (2008), pp. 2568-2574.

12. Xu, D., Sriram, V., Ozolins, V., Yang, J.-M., Tu, K.N., Stafford, G.R., and Beauchamp, C., "In situ measurements of stress evolution for nanotwin formation during pulse electrodeposition of co per," *J. Appl. Phys.*, 105, (2), (2009), pp. 023521-023526.

13. Detavernier, C., Rossnagel, S., Noyan, C., Guha, S., Cabral, J.C., and Lavoie, C., "Thermodynamics and kinetics of room-temperature microstructural evolution in copper films," *J. Appl. Phys.*, 94, (5), (2003), pp. 2874-2881.

14. Lingk, C., and Gross, M.E. "Recrystallization kinetics of electroplated Cu in damascene trenches at room temperature," *J. Appl. Phys.*, 84, (10), (1998), pp. 5547-5553.

15. Yin, K.B., Xia, Y.D., Chan, C.Y., Zhang, W.Q., Wang, Q.J., Zhao, X.N., Li, A.D., Liu, Z.G., Bayes, M.W., and Yee, K.W., "The kinetics and mechanism of room-temperature microstructural evolution in electroplated copper foils," *Scr. Mater.*, 58, (1), (2008), pp. 65-68.

16. Pantleon, K., Gholinia, A., and Somers, M.A.J., "Quantitative microstructure characterization of self-annealed copper films with electron backscatter diffraction," *physica status solidi (a)*, 205, (2), (2008), pp. 275-281.

17. Pantleon, K., and Somers, M.A.J., "X-ray diffraction investigation of self-annealing in nanocrystalline copper electrodeposits," Scr. Mater., 55, (4), (2006), pp. 283-286.

18. Thompson, C.V., "Texture evolution during grain growth in polycrystalline films," *Scripta Metallurgica et Materialia*, 28, (2), (1993), pp. 167-172.

19. Vas'ko, V.A., Tabakovic, I., Riemer, S.C., and Kief, M.T., "Effect of organic additives on structure, resistivity, and room-temperature recrystallization of electrodeposited copper," *Microelectron. Eng.*, 75, (1), (2004), pp. 71-77.

20. Chen, K.-C., Wu, W.-W., Liao, C.-N., Chen, L.-J., and Tu, K.N., "Observation of Atomic Diffusion at Twin-Modified Grain Boundaries in Copper," *Science*, 321, (5892), (2008), pp. 1066-1069.

Electrochemical Corrosion Behaviour of Sn-8Zn-3Bi-XCr Solder in 3.5% NaCl

Jing Hu, Anmin Hu, Ming Li, Dali Mao
Shanghai Jiao Tong University
Dongchuan Road 800, Minhang, Shanghai, China
huluobo1234@sohu.com

Abstract

This paper has explored the using of Cr micro-alloying method to enhance the corrosion resistance of Sn-Zn-Bi alloy. In addition, the influence of long-term aging on solder corrosion resistance has been studied. The electrochemical corrosion behaviour of Sn-8Zn-3Bi-XCr solder in 3.5% NaCl solution was investigated by using potentiodynamic polarization methods, scanning electron microscopy (SEM), energy-dispersive X-ray spectroscopy (EDX) and X-ray an X-RAY diffractometer (XRD) analysis.

The results obtained from polarization study show that: adding a small amount of Cr can improve the corrosion resistance of Sn-Zn-Bi alloy. It shifted the corrosion potential towards more noble values. This change was also reflected in the current density of solder alloy in passivation, corrosion rate, respectively. The oxides of zinc were responsible for the formation of passive film. Presence of Cr atoms in the oxide layer also improved the passivation behaviour of solders to a certain extent and the Sn-9Zn-3Bi-0.5Cr has the best resistance property. We can see from the SEM that it has proportional passive film. Long-term aging decreases the corrosion resistance property and it became poorer with prolonging the aging time. However, it is also more difficult to form passive film.

1. Introduction

The eutectic Sn-37% Pb alloy has extensively been used in the microelectronic packaging industry for many years, mainly due to the unique combination of low melting point, suitable wettability, high strength and low cost. [1][2][3] Despite all of these, they are now being replaced with the Pb-free solder alloys because of their high toxicity and the environmental hazards. The non-toxic Pb-free solder alloys must fulfill a number of other requirements in both economic and physical/chemical points of view. Several investigations have been carried out on lead-free solder alloys such as Sn-Zn, Sn-Ag. In this context, the Sn-Zn eutectic alloy has emerged as a promising lead-free solder alloy. [4][5] It has attracted great attentions because of its low cost. More importantly, it has a melting point (198.5°C) very close to conventional Sn-37wt.% Pb alloy (183°C) which is much lower than that of other Sn-based alloys e.g. Sn-Cu (227°C), Sn-Ag (221°C) and Sn-Ag-Cu (217°C) [6].And its tensile stress and creep resistance are better than that of Sn-37 Pb alloys[7]. However Zn is active both chemically and metallurgically. The presence of Zn in the solder alloy results in wettability decrease and corrosion behaviour of the alloy.

Many investigations added a third element to improve the deficiencies in the Sn-9Zn binary alloys, such as Al [8], Bi, In [9]. It can be found that the addition of Bi can improve the wettability of the solder alloy by lowering the melting temperature [10-12]. Chih-ming et al found the Bi atoms tended to segregate around the Zn-rich phase. Kim et al. [13] have pointed out that the small addition of Bi in the Sn-Zn solder alloy was able to intensify the solder joint strength due to the precipitation of Zn-rich phase. The melting temperature of the solder alloy also decreases from 198.4 to 186.1°C.

Although the addition Bi can effectively increase the mechanical properties of the Sn-Zn solder joint, such as the solder joint strength, the wettability. The oxidation resistance of Sn-Zn serious solder alloys still exhibits insufficient reliability characteristics. Junxiang Jiang et al have found Bi added alloys were rather susceptible to oxidation and corrosion. [14] Previous studies different alloying elements, such as La, Cr, Ti, Al, on Sn-Zn-based solder, Cr was proved as the best choice. Sn-9Zn-Cr alloy has finer microstructure and better oxidation resistant and plasticity than Sn-9Zn alloy while maintaining the same level of thermal and mechanical properties [15]. When Cr addition reach to 0.1wt.%, oxidation resistance of Sn-9Zn alloy is the best 16].

However, there is very lack of information about the electrochemical corrosion behaviour of Sn-Zn-Bi-XCr solders in NaCl solution. In addition, a systematic investigation is necessary to understand the corrosion behaviour after aging progress. The present work thus investigates the corrosion behaviour of Sn-9Zn-3Bi-XCr solder with and without aging progress in 3.5% NaCl solution by potentiodynamic polarization methods. Scanning electron microscopy (SEM), energy-dispersive X-ray spectroscopy (EDX) and X-ray diffractometer (XRD) were used to obtain further information about the surface composition and chemical state of various elements formed due to corrosion reactions on the surface of the solder.

2. Experiment

A series of alloys of Sn, Zn, Bi, and Cr were prepared from pure metals (greater than 99.9% pure). The Cr content of the Sn-9Zn-3Bi-XCr solders investigated was 0-0.5 wt%. The solder constituents were then weighed and melted in a crucible followed by natural cooling in air. Samples 30mm×30 mm×2.5mm in size were sliced out of the alloys.

The solder formed after cooling was polished using 600 to 1200 SiC paper grade, respectively, rinsed with distilled water and then ethanol followed by cleaning in an ultrasonic cleaner. The specimen was then dried in hot air and placed in the test cell. Potentiodynamic polarization measurements were conducted in a rectangular cell containing 3.5% NaCl solution which was aerated with N2 with purity of 99.9% for 30 min before the test .A Pt wire and a KCl-saturated calomel electrode were used as the counter and reference electrodes, respectively. The potential was scanned in the anodic direction in the range of −2000 to +500mV at a scan rate of 1 mV/s. The surface morphology and elemental composition of the various elements in the solder were determined by analyzing the corrosion product formed on the specimen by SEM and EDX techniques.

The compositions of the Sn-9Zn-3Bi-XCr solders were measured by XRD using Si powder as an internal standard. The measurement was performed at 25°C using graphite-monochromated Cu Kαradiation at 50kV with a measurement range of 2θ=20°~70°,a scan speed of 4.5°/min and a sampling step of 0.02°.

3. Results and discussion

3.1. Electrochemical corrosion behaviour of the solders

Fig. 1 shows the polarization curves of the Sn-9Zn-3Bi-XCrsolder alloys with varying Cr content (0-5 wt%) and the results are listed in Table 1. Since all corrosion tests were conducted in deaerated NaCl solution, the only feasible cathodic reaction is evolution of hydrogen from water reduction (cathodic region AB).

$$2H_2O + 2e^- \leftrightarrow 2(OH)^- + H_2$$

However, the anodic polarization part consists of several regions, namely BC, CD, DE, and EF. At point B with increasing potential in the anodic direction, zinc dissolution occurs according to the following reactions. [17, 18]

$$Zn + 2OH^- = Zn(OH)_2 + 2e^- \quad Esce= -1208mV$$
$$Zn + 2OH^- = ZnO + H_2O + 2e^- \quad Esce= -1211mV$$

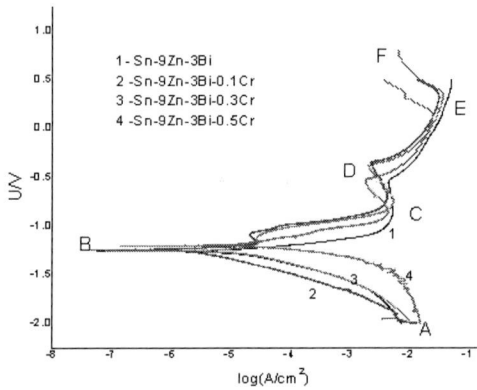

Fig. 1. The potentiodynamic polarization curves obtained from de-aerated 3.5% NaCl solution for the Sn-9 Zn-3 Bi-XCr.

For Sn-9Zn-3Bi, the active dissolution of zinc continues with increasing potential until the zincate concentration reaches a critical value and supersaturates the surface of the solder (point C). These insoluble zincate salts (ZnO or Zn(OH)₂) cover the surface of the corroded samples and create a plateau region (CD) where the current density is found to be independent of potential over an approximate range of about 350mV.This plateau appears only when we doesn't add Cr to the solder. Abayarathna et al. [19] have reported that the zincate film formed in the region CD is not protective and the film formation is a mass-transport limited process. Fig.2 shows an example of the surface morphologies of Sn-9Zn-3Bi-XCr alloys after polarize. Our results on SEM

(Fig.2a) also support that the Sn-9Zn-3Bi solder is not completely passive.

However adding a small amount of Cr to Sn-9Zn-3Bi solders, it is observed (Fig. 1) that the current does not remain constant in the region CD over a large range of potential. Onset of passivation for the solder starts at point C where the current density drops with increasing potential. Increasing the content of Cr shifts the current density towards lower values. Region CD represents the passive region of the solder. The passive region consists of high atomic concentration of Zn and O [20].We can get the same result from XRD analyses which will be introduced in the next section. Now it is generally accepted that zinc passivation begins with the precipitation of zincateion as Zn(OH)₂ or ZnO [21-27]. It is reported that there may be high concentration of Cr in hypo-surface. Cr is oxidated by metathesising Zn. The oxidation of Cr restrains the zincateion in the bulk of solder. The zincateion forms passive film but Zn dose not go on oxidizing. These oxidation of Cr block the active sites of the electrode and inhibit the oxygen evolution reaction which is partly responsible for the breakdown of the passive film. This can make the zincateion layer thiner, only overcast the surface. That could be useful for the corrosion resistance.

Comparing varying Cr content solders, (Fig. 2) it is obviously that with increasing the content of Cr, the passive film becomes more uniform and compact. The solder without Cr is not able to form complete passive film. The passive film of 0.1wt% Cr solder is not obvious. Many flat extrusions formed on the Sn-9Zn-3Bi-0.1Cr surface. But the tropism is different. For the Sn-9Zn-3Bi-0.3Cr solder, the passive film is fragmentary. It can be seen a uniform passive film for Sn-9Zn-3Bi-0.5Cr solder. Presence of Cr atoms in the oxide layer also improved the passivation behaviour of solders to a certain extent.

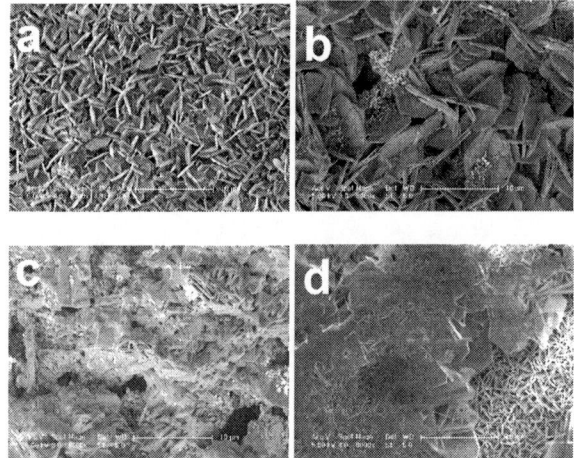

Fig. 2 SE micrographs of Sn-9 Zn-3Bi-XCr solder polarized (a) Sn-9Zn-3Bi (b) Sn-9Zn-3Bi-0.1Cr (c) Sn-9Zn-3Bi-0.5Cr (d) Sn-9Zn-3Bi-0.5Cr

A sharp increase in current density is observed at point D with the increase in potential .The passive film is broken. The breakdown of the passive film can be caused by the presence of Cl⁻ on the surface other than the oxygen evolution reaction described above. SEM shows a mixture of sheet-like structures with lots of pores and openings on the surface. Some people reported that it is Cl⁻ caused. We can also get the same results from EDX and XRD analysis. Metals are tending to be passivation in acidic or neutral solution. It is more difficult to form passive film in alkalescence solution. Further more, halogen especially Cl⁻, can prevent the passive film formation obviously. Pistorius and Burstein [28] have reported that chloride ions penetrate into the oxide film and form solid metal chloride which causes the mechanical breakdown of the film. Abd El Rehim et al. [29] have also found that halide ions stimulate the active dissolution of zinc and result in breakdown of the passive film. EDX analysis given following also shows the presence of oxygen on the surface of the solder. $SnCl_2$ may react with oxygen to form oxychlorides. Windholz et al. [30] have reported that $SnCl_2$ is able to absorb O_2 from air to form insoluble oxychloride.

On the other hand, for the Sn-9Zn-3Bi-0.5Cr solder the passive film may be divided into two different layers. These species are formed at different time. Fig. 3 presents the micrograph of the Sn-9Zn-3Bi-0.5Cr solder after polarize. The Sn-9Zn-3Bi-0.5Cr solder form the first passive film at region CD. As increasing the potential, the passive film becomes thicker, and then broken. At last, the film will break off. On the new surface thus forms new passive film. At point E the current density shifts lower values. It may be a new region for passivation.

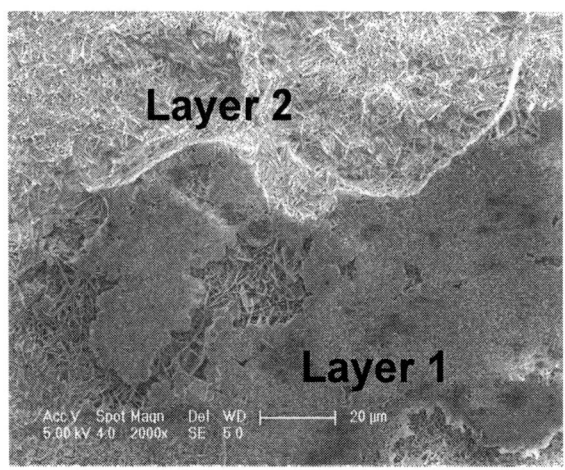

Fig.3 SE micrographs for two kinds of passive film

Table 1 Polarization results of the various solder alloys.

Solder alloys	Ec(V)	IC(A/cm²)	Ip(V)(A/cm²)	Emax (V)	Imax(A/cm²)
Sn-9Zn-3Bi	1.2509V	1.57×10^{-7}	4.25×10^{-3}	0.463	0.0296
Sn-9Zn-3Bi-0.1Cr	1.2569V	2.78×10^{-7}	2.03×10^{-3}	0.481	0.0297
Sn-9Zn-3Bi-0.3Cr	1.2386V	2.57×10^{-7}	2.38×10^{-3}	0.350	0.0285
Sn-9Zn-3Bi-0.5Cr	1.1997V	1.39×10^{-7}	1.66×10^{-3}	0.344	0.0208

Ec: corrosion potential of solder alloy; Ic:current density of solder alloy at Ec Ip: current density of solder alloy in passivation; Emax: potential for the maximum current density; Imax: maximum current density.

Table 1 shows the effect of varying Cr content on the parameter obtained from the polarization curves. The Sn-9Zn-3Bi solder alloy has an corrosion potential of −1.2509V Ec, which slightly increases to −1.1997 Ec with increasing the Cr content in solder alloy to 0.5wt%. However, the corrosion potential of the solder alloy decreases to −1.2569 VSCE as 0.1 wt% Cr is added to the Sn-9Zn-3Bi solder alloy because the Zn-rich phase is formed in the Sn matrix. This breaks off the continuity of the matrix and can't block the diffusion of Zn [31].

The 0.1wt% Cr addition slightly decreases the corrosion resistance of the Sn-9Zn-3Bi solder alloy .However, when Cr content reaches 0.3wt%, an opposite trend is observed. The solder containing 0.5 wt% Cr has the most positive corrosion potential. The anodic potential increases only when the content of Cr reaches a certain content. They must block the diffusion of Zn. It can also get that 0.5 wt% Cr has a very lower Ic value of 1.39×10^{-7}A/cm² than the other solders mentioned in Table 1.Its Ip is also the lowest. These results suggest an inhibition of the anodic dissolution reaction. Fig. 1 also presents a decrease in the corrosion rate and Ip of the Sn-9Zn-3Bi-XCr with an increase in the Cr content from 0 to 0.5 wt%. All above suggests that Sn-9Zn-3Bi-0.5Cr has the best corrosion resistance, the reliability of the solder is also expected to increase to a certain extent which might be useful to the electronic industry.

3.2 The influence of long-term aging

The aging progress may change the microstructure of the solder thus change the properties. Then we had to make sure the influence of long-term aging. Polished samples were exposed to a constant temperature of 150°C and to a constant humidity of 85%RH for 4 or 9 days in environmental testing equipment. Then we did the same electrochemistry experiment mentioned above. The Sn-9Zn-3Bi-0.3Cr and Sn-9Zn-3Bi solder alloy are set as an example. It can be seen from Table 2 that the Ecorr value shifts towards more negative potentials with increasing the aging time. The long-term aging decreases the corrosion resistance. However the influence is not obvious. Fig.4~Fig.6 show the surface of Sn-9Zn-3Bi-XCr solder alloys with and without aging progress after polarize .Our results on SEM also support this observation. We can see that even Sn-9Zn-3Bi-0.3Cr solder is not able to form intact passive film. The restrain to the anode dissolution became poorer.

978-1-4244-4658-2/09 $25.00 © 2009 IEEE   627

Table 2 Polarization results of the various solder alloys after long-term aging

Solder	Ec original	Ec 4 day aging	Ec 9 day aging
Sn-9Zn-3Bi	-1.2509V	-1.2535V	-1.2676V
Sn-9Zn-3Bi-0.3Cr	-1.2386V	-1.2391V	-1.2436V

Fig.4 SE micrographs for original Sn-8Zn-3Bi-XCr polarized
(a) Sn-8Zn-3Bi (b) Sn-8Zn-3Bi-0.3Cr

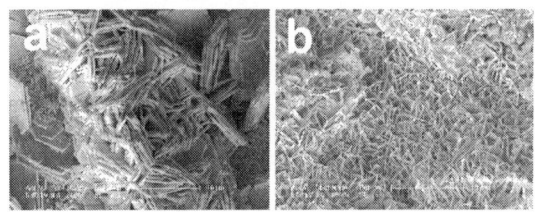

Fig.5 SE micrographs for Sn-8Zn-3Bi-XCr with 4 days aging
polarized (a) Sn-8Zn-3Bi (b) Sn-8Zn-3Bi-0.3Cr

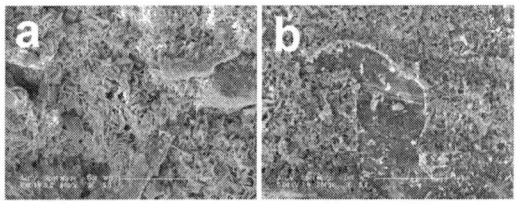

Fig.6 SE micrographs for Sn-8Zn-3Bi-XCr with 9 days aging
polarized (a) Sn-8Zn-3Bi (b) Sn-8Zn-3Bi-0.3Cr

3.3 Microstructure evolution of Sn-9Zn-3Bi solder alloys

The microstructure of the as-cast Sn-9Zn-3Bi-XCr alloys
are shown in Fig.7.Two types of Zn morphologies, primary
Zn and eutectic Zn, exist in the Sn matrix of all alloys,
although their compositions are very close to the eutectic
point according to the Sn-Zn equilibrium diagram.[32]

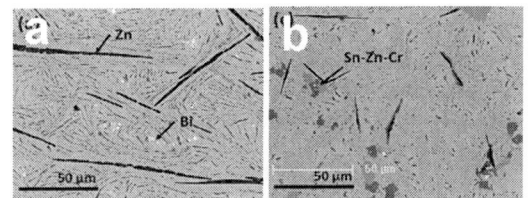

Fig.7 SE micrographs for Sn-8Zn-3Bi-XCr (a)Sn-8Zn-3Bi (b)
Sn-8Zn-3Bi-0.3Cr

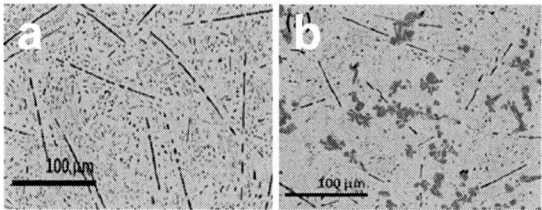

Fig.8 SE micrographs for Sn-8Zn-3Bi-XCr with 9 days
high temperature aging (a) Sn-8Zn-3Bi (b) Sn-8Zn-3Bi-0.3Cr

The primary and eutectic Zn becomes much small size as
adds 0.3% Cr into the Sn-9Zn-3Bi solder. Another new phase
Sn-Zn-Cr appears. Some people also regard it as Cr-rich
phase. The β-Sn phase content decrease, a part of it change
into Cr-rich phase and separate out. The alloy microstructure
becomes more compact. Because of the Zn phase becomes
scrappy, the anode dissolve speed decreases. Thus the
corrosion resistance ability of Sn-9Zn-3Bi-0.3Cr solder is
stronger than Sn-9Zn-3Bi solder. This result is in accordance
with that of electrochemistry experiment.

The solder are aging at high temperature for 150°C. Fig. 8
shows the microstructure of the solder after 9 days time aging.
After long-term aging, the unequilibrium solidified coarser
Zn-rich phase also becomes scrappy. The high aging
temperature results in the eutectic glomerates. In addition,
according to the Sn-Bi equilibrium diagram, the solubility of
Bi can reach to 20wt% in the Sn matrix at 150°C. That exceed
the Bi content in solder alloy, the Bi dissolves completely. We
can't see simple Bi phase in the solder after aging. Besides,
for the Sn-9Zn-3Bi-0.3Cr solder, the Cr-rich phase change
obviously. Before aging, there are two kinds of Cr-rich phase.
One mainly contains Zn and its colour is undertone. The other
mainly contains Sn and its colour is fuscous. After aging, the
mainly Sn content Cr-rich phase disappears. There may be
much Zn phase change into Cr-rich phase. So the zincate salts
content decreases. That is harmful for the formation of the
passive film.

3.4 Surface analyses by EDX and XRD

For the component study, the analyses results of different
kinds of solders are summarised in table 3. The Sn-9Zn-3Bi-
0.5Cr oxygen content is much lower than that of Sn-9Zn-
3Bi.For Sn-8Zn-3Bi, there are three point five oxygen atom
every one zinc atom. Zinc is the most active element in the
solder alloy, and it is the first oxidized composition.

Table 3 Chemical compositions of solder alloys polarized

Sn-8Zn-3Bi			Sn-8Zn-3Bi-0.5Cr		
Element	Weight%	Atomic%	Element	Weight%	Atomic%
O	33.23	70.70	O	11.87	39.72
Cl	3.11	2.98	Cl	4.88	7.37
Sn	21.14	19.39	Sn	31.46	37.04
Zn	37.23	6.06	Zn	45.22	14.19
Bi	5.29	0.86	Bi	6.57	1.68

The oxygen atom is much more than zinc atom confirms that besides Zinc there are other elements oxidized.

By contrast, for Sn-9Zn-3Bi-0.5Cr solder, the number of oxygen atom and the zinc atom is almost the same. This means only zinc is oxidized to form passive film but no element else is oxidized. From Sn content, the amount of Sn in Sn-8Zn-3Bi-0.5Cr is much more than that of Sn-8Zn-3Bi. It is found that Sn exists mainly in simple substance form. It isn't corroded. All the results prove that the Sn-9Zn-3Bi-0.5Cr solder is more difficult to be corroded. We have already got this conclusion above. Fig.9 shows the selected part of Sn-8Zn-3Bi-0.5Cr solder when we did EDX analyses. It is mentioned above that on the surface of Sn-8Zn-3Bi-0.5Cr solder there are passive film. As increasing the potential, the passive film becomes thicker, and then broken.

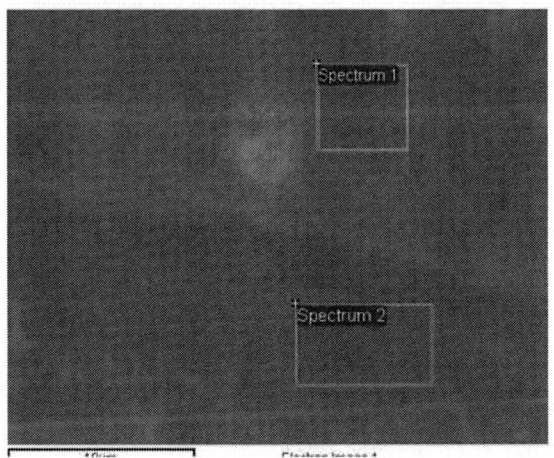

Fig.9 The SE micrographs for two kinds of passive film

At last, the film will break off. Table 3 shows the content of the surface after the film breaks off. In Fig.9, it is spectrum 2. The content of passive film is summarized in table 4. In Fig.9, it is regarded as spectrum 1. Comparing with Table.3 and Table.4, the content of the two part is obviously different.

Table 4 content of Sn-8Zn-3Bi-0.5Cr solder passive film

Element	Weight%	Atom %
O	26.53	57.50
Cl	11.27	11.03
Zn	55.78	29.59
Sn	6.42	1.88

It is found that the amount of chlorine element is much more than Table.3 shows and the amount of oxygen atom is more than that of zinc atom. These insoluble zincate salts absorbs oxygen in the air forming zincate oxychloride. This can make the passive film mechanical broken finally. From Fig.9, it can be seen spectrum 2 dislocated at the fringe of the passive film. It would break very soon. Oxygen and zinc begin to assemble largely is the symbol of this part.

The XRD results are shown in Fig.10. It analyses the component of the Sn-8Zn-3Bi-0.3Cr solders after different aging days.

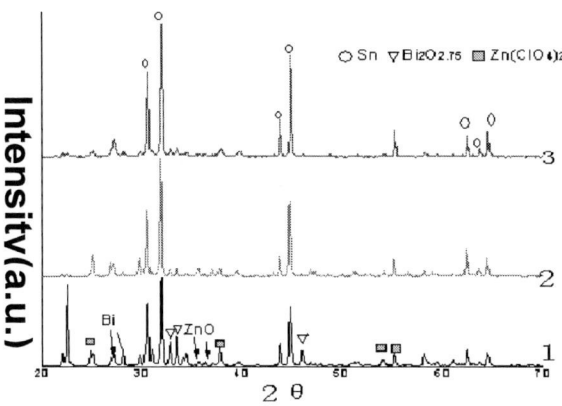

1. Original solder; 2. 4days time aging; 3. 9 days time aging
Fig.10 The XRD pattern at room temperature for the Sn-8Zn-3Bi-0.3Cr solders.

It is found that species decreases after aging progress and the longer aging time the less species. There are monophasic tin, bismuth oxid and zincate oxychloride in the solder.Sn exits mainly in simple substance. Without aging progress there is still come ZnO. However after aging progress, ZnO almost disappears. Thus it is more difficult to form passive film.

4. Conclusion

The following conclusions can be made from the study:

1. The corrosion current density (Icorr) decreases and the corrosion potential (Ecorr value) shifts towards more negative values with increase in the Cr content from 0.1 to 0.5 wt% in the solder.

2. Passivation behaviour is noted in the Sn-9Zn-3Bi-XCr solders having Cr content. Increasing the content of Cr, the

978-1-4244-4658-2/09 $25.00 © 2009 IEEE

passive film becomes more uniform and compact. The Sn-9Zn-3Bi-0.5Cr solder alloy has the best corrosion resistance.

3. The oxides and hydroxides of zinc are responsible for the formation of passive film.

4. The long-term aging decreases the corrosion resistance of Sn-9Zn-3Bi-XCr solder.

5. There are two kinds of Cr-rich phase in the Sn-9Zn-3Bi-XCr solder. That is useful for passivition.

Acknowledgments

This work is sponsored by Shanghai nano technology promotion center (Contact No. 05nm05043) and Shanghai pujiang program (Contact No.05PJ14065). We thank the Instrumental Analysis Center of Shanghai Jiaotong University, for the use ofthe AES, SEM and ICP equipment.

References

1. W.H. Zhong, Y.C. Chan, M.O. Alam, B.Y. Wu and J.F. Guan, "Joint Strength and Interfacial Microstructure Between Sn-Ag-Cu and Sn-Zn-Bi Solders and Cu Substrate", *J. Alloys Compd*, Vol. 414 (2006), pp. 123-130.

2. R.A. Islam, B.Y. Wu, M.O. Alam, Y.C. Chan and W. Jillek, "Effect of Cooling Rate on the Room Temperatuer Impression Creep of Lead-free Sn9Zn and Sn-9Zn-3Bi Solders", *J. Alloys Compd*, Vol. 392 (2005), pp. 149-158.

3. K.K. Mohan, V. Kripesh, L. Shen, K. Zeng and A.A.O. Tay, "Variable aeautectic Temperature Caused by Inhomogeneous Solute Distribution in Sn-Zn System", *Mater. Sci. Eng.*, A 423 (2006), pp. 57-63.

4,5. Chih-ming Chen, Yu-min Hunga and Ching-hsuan Lina "Effect of Substrate Metallization on Interfacial Reactions and Reliability of Sn-Zn-Bi Solder Joints", *Journal of Alloys and Compounds*, Vol. 475 (2009), pp. 238-244.

6. T. Ichitsubo, E. Matsubara, K. Fujiwara, M. Yamaguchi, H. Irie, S. Kumamoto and T. Anada, "Comparative Study of Wetting Behavior and Mechanical Properties(microhardness) of Sn-Zn and Sn-Pb Solders", *J. Alloys Compd.*, Vol. 392 (2005), pp.200-205.

7. K.I. Chen, S.C. Cheng, S. Wu and K.L. Lin, "Interfacial Reaction Between molten Sn-Bi-X Solders and Cu Substrate for Liquid Solder Interconnects", *J. Alloys Compds*, Vol. 416 (2006), pp. 98-105.

8. K.L. Lin, T.P. Liu, "Investigations of Interfacial Reactions of Sn-Zn Based and Sn-Ag-Cu Lead-free Solder Alloys as Replacement for Sn-Pb Solder", *Oxid. Met.*, Vol. 50 (1998), pp. 255.

9. M. McCormack, S. Jin, H.S. Chen, D.A. Machusak, "Microstructural Evolution of Sn-Zn-Bi Solder/Cu Joint During Long-time Aging at 170°C", *J. Electron. Mater.*, Vol. 23 (1994), pp. 687.

10. Y.S. Kim, K.S. Kim, C.W. Hwang and K. Suganuma, "Properties of Low Melting point Sn-Zn-Bi Solders", *J. Alloys Compd.*, Vol. 352 (2003), pp. 237.

11. S.W. Yoon, J.R. Soh, H.M. Lee and B.J. Lee, "Effect of Sample Perimeter and Temprature onSn-Zn Based Lead-free Solders", *Acta Mater.*, Vol. 45 (1997), pp. 951.

12. M. McCormack and S. Jin, "Compositional Effects on the Microstructure and Vibration Fracture Properties of Sn-Zn-Bi Alloys", *J. Electron. Mater.*, Vol. 23 (1994).

13. Y.S. Kim, K.S. Kim and C.W.H.K. Suganuma, "Effect of Composition and Cooling Rate on Microstructure and Tensile Properties of Sn-Zn-Bi Alloys", *J. Alloys Compds.*, Vol. 352 (2003), pp. 237-245.

14. Junxuang Jiang, Jae-Ean Lee, Kuun-Soo Kim, "The Effect of Crosshead Speed on the Joint Strength Between Sn-Zn-Bi Lead-free Solders and Cu Substrate", *Katsuaki Suganuma JALCOM-16751.*

15. X. Chen, M. Li, X.X. Ren, A.M. Hu and D.L. Mao, "Study on the Properties of Sn-9Zn-xCr Lead-free Solder", *J. Electron. Mater.*, Vol. 35 (2006), pp. 1734.

16. X. Chen, A.M. Hu, M. Li and D.L. Mao, "Effect of A Trace of Cr on Intermetallic Compound Layer for Tin-Zinc Lead-free Solder Joint During Aging", *J. Alloys Compd.*, Vol. 460 (2008), pp. 478-484.

17. M.W. Laitmer, "Retardation of Spalling by the Addition of Ag in Sn-Zn-Bi Solder with the Au/Ni Metallization", *J. Alloys*, pp. 39.

18. M. Pourbaix, "Prediction of Interface Reaction Products Between Cu and Various Solder Alloys by Thermodynamic Calculation", *J. Alloys*, pp. 404.

19. D. Abayarathna, E.B. Hale, T.J. O'Keefe, Y.M. Wang, D. Radovic, "Nanoindention Study of Zn-based Pb-free Solders Used in Fine Pitch Interconnect Applications", *Corros. Sci.*, Vol. 32, No.7 (1991), pp. 755.

20. Udit Surya Mohanty, Kwang-Lung Lin, "Investigations on Microhardness of Sn-Zn Based Lead-free Solder Alloys as Replacement of Sn-Pb Solder", *Materials Science and Engineering*, A 40 (2005), pp. 34-42.

21. JO.M. Bockris, Z. Nagy, A. Damjanovic, "Globular-to-needle Zn-rich Phase Transition During Transient Solidification of A Eutectic Sn-9Zn Solder Alloy", *J. Electrochem. Soc.*, Vol. 116 (1969), pp. 719.

22. M.C.H. Mc Kubre, D.D. Macdonald, "Interfacial Microstructures and Solder Joint Strengths of the Sn-8Zn-3Bi and Sn-9Zn-1Al Pb-free Solder Pastes on OSP finished printed circuit boards", *J. Electrochem. Soc.*, Vol. 128 (1981), pp. 524.

23. Y.C. Chang, G. Prentice, "Corrosion Mechanism of Tin-Zinc Alloys in Neutral Medium", *J. Electrochem. Soc.*, 131 (1984), pp. 1465.

24. Y.C. Chang, G. Prentice, "Evaluation of Corrosion Behavior of A New Class of Pb-free Solder Materials", *Electrochim. Acta*, Vol. 31 (1986), pp. 579

25. P.L. Cabot, M. Cortes, F.A. Centellas, J.A. Garrido, E. Perez, "Corrosion Measurements by Titration (CMT) : Copper, Sn-Zn Alloy and Aluminium", *J. Electroanal. Chem.*, Vol. 201 (1986), pp. 85.

26. X. Shan, D. Ren, P. Scholl, G. Prentice, "EIS Behavior of Anodized Zinc in Cloride Enviroment", *J. Electrochem. Soc.*, Vol. 136 (1989), pp. 3594.

27. A. Hugot Le Goff, S. Joiret, B. Saidani, R. Wiart, "Solderability of Sn-9Zn-0.5Ag-1In Lead-free Solder on Cu Substrate", *J. Electroanal.Chem.*, Vol. 263 (1989), pp. 127.

28. P.C. Pistorius, G.T. Burstein, "Electroplating and Corrosion Baviour of Tin-Zinc Alloy", *Corros. Sci.*, Vol. 38 (1994), pp. 1525.

29. S.S. Abd El Rehim, S.M. Abd El Wahab, E.E. Fouad, H.H. Hassan, "Electrochemical Behavior of A New Solder Material (Sn-In-Ag)", *Mater. Corros.*, Vol. 46 (1995), pp. 633.

30. M. Windholz, S. Budavali, R.F. Blumeti, E.S. Otterbein, "Electrochemical Corrosion Behaviour of Lead-free Sn-8.5Zn-XAg-0.1Al-0.5Ga Solder in 3.5% NaCl Solution", Merck Index, 10th ed., MercK WhiteHouse Station, NJ, 1983, pp. 1257.

31. Xi Chen , Anmin Hu, Ming Li and Dali Mao, "Study on the Properties of Sn-9Zn-xCr Lead-free Solder", *Journal of Alloys and Compounds*, Vol. 460 (2008), pp. 478-484.

32. Moser, J.Dutkiewicz, W.Gasior, J.Salawa, "Oxidation Behavior of Sn-Zn Solder Under High-temperature and High-humidity Conditions", Binary Alloy Phase Diagrams, ASM International, 1992, pp. 209-214.

The Influence of Silicon Content on the Thermal Conductivity of Al-Si/ Diamond Composites

Yang Zhang, Xitao Wang, Jianhua Wu

State Key Laboratory for Advanced Metals and Materials, University of Science and Technology Beijing

257# University of Science and Technology Beijing, Beijing, P.R. China, 100083

Email: xitao.wang@gmail.com, Tel: 86-010-62333751

Abstract

Diamond reinforced Al-based composites with excellent comprehensive properties (i.e., high thermal conductivity, flexible coefficient of thermal expansion) are showing great potential for high power density packaging.

In this study, effect of the Silicon addition on the thermal conductivity and interfacial characteristic of the Al-Si/diamond composites are investigated. Al-Si alloy containing 1.0, 2.0, 3.0, 4.5, 7.0, 10.0, 14.0, 20.0 wt. % Si are prepared as the matrix and Al-Si/diamond composites are fabricated by pressure infiltration under 800°C and 5Mpa. The analysis of the SEM and the element distribution maps detected by the EDX shows that, as the Si addition, the Si skeleton of the AlSi-eutectic phase segregates to the diamond particle surface and form a better bonding between the reinforcement phase and the matrix. The highest thermal conductivity of the composites gets at Al-1.0Si/diamond as 248 W/m K. The results, which are calculated using the differential effective medium scheme (DEM), indicate that the interfacial thermal conductance (h) shows first an increase and then a decrease with Silicon content. Overall, Si addition can raise the h.

1. Introduction

Over the last decade, along with the continuing miniaturization of the electronic devices, power density increases remarkably, resulting in a significant growth of the heat flux on the chips. For instances, chip heat flux for CMOS microprocessors have reached values close to 100 W/cm^2 and are planned to amplify above 200 W/cm^2 at mid-term [1]. In this condition, thermal aspects are becoming increasingly important for the reliability of the electronic devices. Therefore, effective thermal management is the key issue.

Thermal management material with high performance is primary method to improve the thermal management. The ideal materials working for thermal dissipating should have a suitable coefficient of thermal expansion (CTE) to match with chip materials and a high thermal conductivity (TC). The conventional thermal management materials, such as Cu-W, Cu-Mo and Al-SiC, are no longer meet the requirements, especially their thermal conductivity are getting insufficient [2]. Thus, a new kind of materials must be developed as the thermal management materials in order to dissipate the heat generated in these electronic devices effectively.

Diamond has the highest thermal conductivity of all natural materials (typically 600-2000 W/(m*K) at room temperature) and low CTE (0.8×10^{-6}/K at room temperature) [3]. On the other hand, by adding an appropriate reinforcement phase in the proper metal matrix, not only the thermal properties of the metal matrix composites (MMC) can be controlled and tailored, but also the attractive physical properties of reinforcement phase and matrix can transfer to the composites. Hence, the diamond particle reinforced MMC would be effective in improving the thermal management situation significantly. Especially the use of Aluminum as the matrix appears attractive in terms of the lower material cost and lower density.

In the composites, interface between the reinforcement phase and matrix plays a crucial role in determining the thermophysical properties. In order to improve the interfacial adhesion, the common alloying element Silicon is added to the Aluminum matrix. The Silicon content is varied in order to investigate its influence on the composites thermal transport properties.

In the present work, Al(Si)/diamond composites are fabricated by pressure infiltration with the different matrix materials of Al-xSi alloys (x=0, 1.0, 2.0, 3.0, 4.5 ,10.0 ,14.0, 20.0 wt.%). The microstructure and crystallography of the Al(Si)/diamond are investigated by scanning electron microscope (SEM). and X-ray diffraction. Thermal conductivity of each composite is obtained through thermal diffusivity measurements and interfacial thermal conductance is calculated. The influence of alloying element Silicon on the interfacial characteristics is studied.

2. Experimental procedures

2.1 Materials

Al and Al-xSi alloys with different Silicon content (x=1.0, 2.0, 3.0, 4.5, 10.0, 14.0 20.0 wt.%) are used as the matrix materials. The Al-Si alloys are prepared from the Al 99.99 and Al-20.0Si by the arc melting method. Synthetic diamond single crystals of MBD-4 grade are used in experiment. Average size of the diamond powder is 74μm.

2.2 Composites synthesis

Al-Si/diamond composites samples are prepared by liquid metal pressure infiltration. The diamond powders are tap-packed in a graphite cylinder mold under a pressure of 1 MPa. Enough block matrix materials are placed on top of diamond powder bed. The graphite mold with powder and metal blocks is heated to 800°C. The holding time at 800°C is about 10 min. Then, the diamond powders are infiltrated with the metal melt under a pressure of 5 MPa. The maximum applied pressure of 5 MPa is maintained for 2 min. The composites could be obtained after stripping from the graphite mold.

2.3 Characterization

A ZEISS SUPRA 55 scanning electron microscopy (SEM) is used to observe the microstructures and the fracture surfaces of the composite samples. In order to study the influence of the Silicon alloying element on the interface, some investigation using SEM and X-ray diffraction is done

978-1-4244-4658-2/09 $25.00 © 2009 IEEE

on diamonds released from the composites by simple chemical etching.

The thermal diffusivity measurements of the composite samples with a diameter of 10mm and a height of 3mm are performed at room temperature by using the laser flash technique. The thermal conductivity, λ (W/(m*K)), is calculated using the relation:

$$\lambda = K \cdot \rho \cdot C_p \tag{1}$$

where K (m^2/s) is thermal diffusivity, ρ (g/cm^3) is density and Cp (J/g·K) is the specific heat capacity. The densities of the composites are examined by Archimedes method.

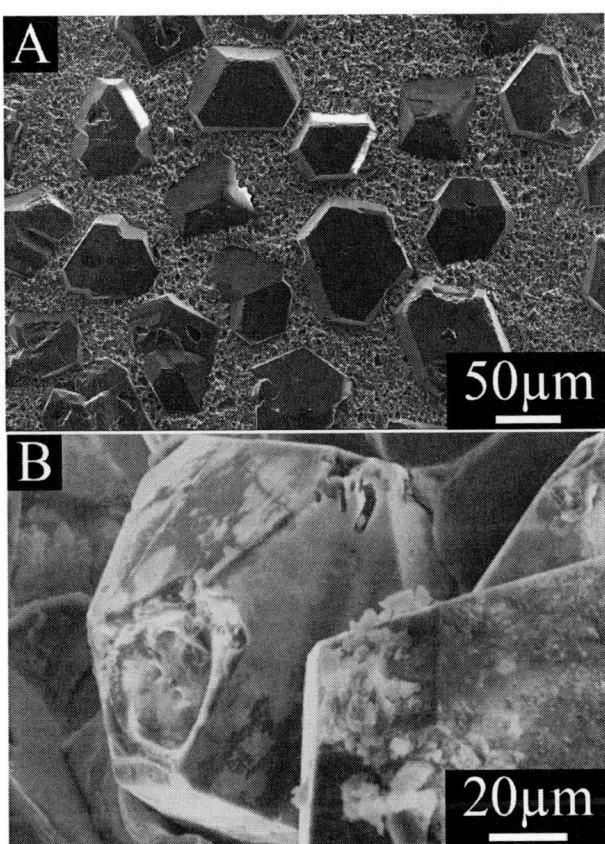

Fig. 1 SEM micrograph of the structure of Al-10.0Si/diamond composite (A) and diamond particle released from Al-4.5Si/diamond fragments by chemical etching (B)

3. Results and discussion

3.1 Microstructure and interfacial characterization

The observations by the SEM show that all Al(Si)/diamond composites with the different matrix have the approximately similar features of the microstructure. Now, Al-10.0Si/diamond composite is taken as the example and Fig.1(A) shows its microstructure. The block phase is diamond particles. The distribution of the diamond particles is uniform and the reinforcement phase matches well with the metal matrix. There are no obvious defects (such as holes, etc), with a high density. It demonstrates that the melting matrix metal is able to penetrate the pore space of the particles uniformly under a certain pressure.

The detailed studies on the interface between the diamonds and matrix display that, in the Si-containing composite samples, some phases like the skeleton remain on the diamond particles surface. Fig.2 shows the interfacial microstructure of the Al-4.5Si/diamond composites and the element distribution maps detected by the EDX in SEM. They illustrate that the phases like the skeleton on the diamond particle surface contain high level of Si. In accordance with the analysis of the Al-Si phase diagram (eutectic concentration is 12.6 wt.%), we can confirm that the rich Si skeleton is the AlSi-eutectic phase segregated in the diamond particle surface. This phenomenon is emphasized by the observations on the diamond particles released from the Al-4.5Si/diamond fragments, as shown in Fig.1(B). On the naked diamond particle surface, there is obvious Si skeleton adhesion.

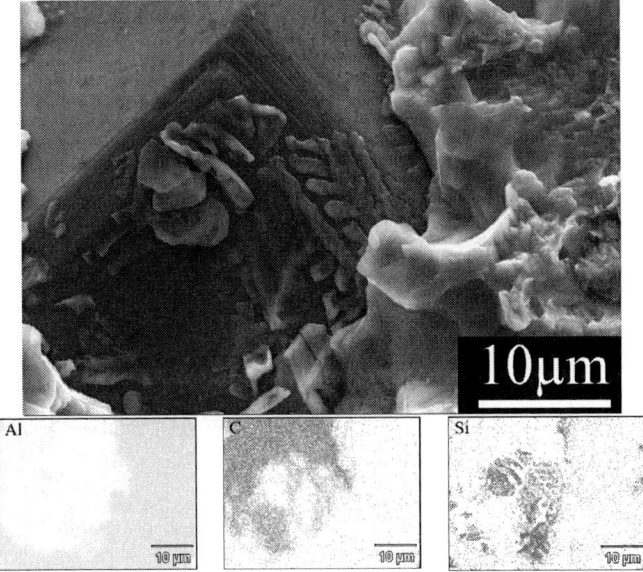

Fig. 2 Interfacial microstructure of Al-4.5Si/diamond composites and EDX map scanning of elements as Carbon, Aluminum and Silicon

Fig.3(A) and (B) show the fracture surface of the Al/diamond and Al-4.5Si/diamond composites, respectively. They can reflect the bonding between the particles and the matrix. In the Al/diamond composite sample, the fracture obviously occurs between the particles and matrix, baring the original diamond particle shape. However, the fracture surface of the Al-4.5Si/diamond composite is different. Here, the metallic matrix adhered on the surface of the diamond particles can be observed and the combination between the particles and the matrix is more compact.

Based on the above studies, as a result of Si addition, the Si skeleton of the AlSi-eutectic phase can precipitate in the diamond particle surface preferentially with the lower energy compared with the rich Al of the eutectic phase. It can reduce the surface energy between the diamond and matrix, simultaneously, promote the interfacial bonding in the Al(Si)/diamond composites. These improvements in the composites are the most desirable ones because of its

978-1-4244-4658-2/09 $25.00 © 2009 IEEE

consistent character and higher strength, and maybe bring the benefit to the thermal properties.

Fig. 3 Fracture surface of Al/diamond (A) and Al-4.5Si/diamond composites (B)

3.2 The influence of the Si addition on the thermal conductivity and the interfacial thermal conductance

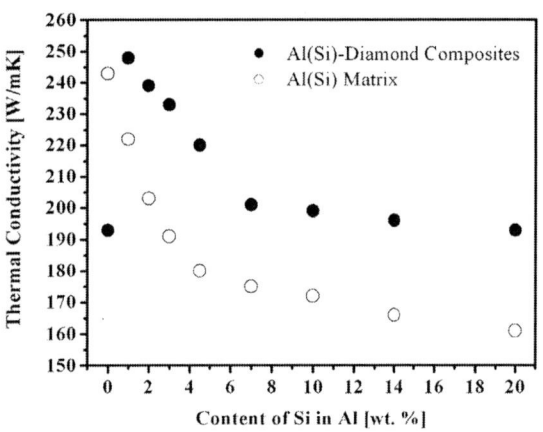

Fig. 4 Thermal conductivity of the Al-Si matrix materials and Al-Si/diamond composites with different Si content [4, 5, 6, 7]

The thermal conductivity of the matrix materials alloyed with different Si content and the Al-Si/diamond composites are listed in the Fig.4 respectively, with the composites'

thermal conductivity calculated by Equation (1). From Fig.4, it shows that the thermal conductivity of the Al-Si/diamond composites indicates the maximum value as 248 W/(m*K) at 1.0 wt.% of Si, much higher than the pure Al matrix composite as 193 W/(m*K). And then, the value decreases with increasing Si content up to 20 wt.%. Therefore, the thermal conductivity of the Al(Si) matrix reduces along with the increase of the Si. Reasons for this phenomenon are discussed as follows.

Interfaces in composites are crucial to the thermal conductivity. The interface acts as a thermal barrier, and it determines the contribution of high thermal conductive reinforcement to final properties. In keeping with usual practice in the theory of heat transfer, thermal resistance can be characterized by an interfacial heat transfer coefficient or thermal conductance, $h(W \cdot m^{-2} \cdot K^{-1})$, defined as the proportionality constant between the heat flux through the boundary and temperature drop across it [8, 9].

The Hasselman-Johnson model (H-J) and differential mean-field scheme (DS), based on the effective medium theory (EMT), [10, 11] are the key predictive schemes for the thermal conductivity of the particle reinforced composites with the interfacial thermal conductance. In has been showed that the DS has the superior ability in the case of high volume fraction of the reinforced particles and intermediate phase contrast between the thermal conductivity of matrix and particle.

DS processes the thermal conductivity at high volume fraction in the way of calculus, taking the surrounding field as the composites at each stage. The thermal conductivity at low volume fraction can be taken in the following from:

$$\frac{K_c}{K_m} = \frac{\left[2\left(\frac{K_p}{K_m} - \frac{K_p}{h \cdot a} - 1 \right)V_p + \frac{K_p}{K_m} + 2\frac{K_p}{h \cdot a} + 2 \right]}{\left[\left(1 - \frac{K_p}{K_m} + \frac{K_p}{h \cdot a} \right)V_p + \frac{K_p}{K_m} + 2\frac{K_p}{h \cdot a} + 2 \right]} \quad (2)$$

where K_i with $i = c$, m and p denote the thermal conductivity of the composite, the alloy matrix and the particle, V_p is the volume fraction of the particle, a is the radius of the particle idealized as sphere and h is the interfacial thermal conductance. In order to simplify Equation (1) and treat the particle with a non-ideal interface, K_p^{eff} is defined as Equation (3):

$$K_p^{eff} = \frac{K_p}{1 + \frac{K_p}{h \cdot a}} \quad (3)$$

which for $h \to 0$, $K_p^{eff} \to 0$, and $h \to \infty$, $K_p^{eff} \to K_p$.

In the case of small V_p and K_p^{eff} introduced, Equation (2) change to

978-1-4244-4658-2/09 $25.00 © 2009 IEEE

$$\frac{K_c}{K_m} = 1 + 3 \frac{K_p^{eff} - K_m}{K_p^{eff} + 2K_m} V_p \qquad (4)$$

Equation (4) transforms into the differential form

$$\frac{dK}{K} = 3 \frac{K_p^{eff} - K}{K_p^{eff} + 2K} \cdot \frac{dV}{1 - V} \qquad (5)$$

With the edge condition $K_c = K_m$ at $V_d = 0$, the DS approach can obtain on integrating Equation (5)

$$(1 - V_d)^3 = (\frac{K_c}{K_m})^{-1} (\frac{K_c - K_p^{eff}}{K_m - K_p^{eff}})^3 \qquad (6)$$

The interfacial thermal conductance of the Al-Si/diamond in this work can be back-calculated by the Equation (6), assuming a thermal conductivity of 1800 W/(m*K) [6] for the diamond particle as well as 74μm and the volume fraction of 0.6 respectively, to study the effect of the Si addition.

The results are illustrated in Fig.5. By adding the Si, the interfacial thermal conductance can be remarkably increased. The maximum value of h, about half of that of Al/diamond sample, is reveals between 2.0 and 3.0 wt. % Si-added composites. But, an excessive amount of Si reduces the interfacial thermal conductance. With increasing Si content, it approaches to 50% of its maximum value, but still higher than that of the Al/diamond.

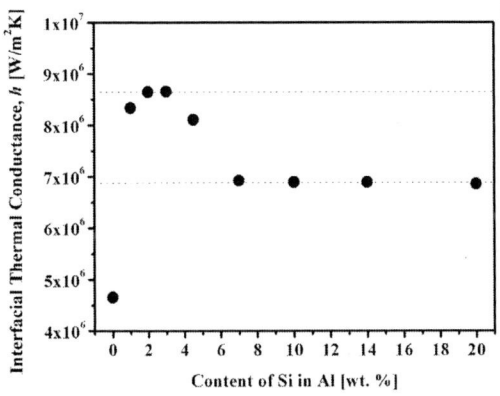

Fig. 5 The influence of the Si content on the interfacial thermal conductance (h) of the Al-Si/diamond composites

Because of the huge difference between the CTE of the alloy matrix and of the diamond, the interface suffers the thermal stress during the solidification of the matrix. When the interfacial bonding is weak, the interface will be broken, which causes the occurrence of cracks and flaws. This phenomenon can reduce the interfacial thermal conductance of the composites dramatically. The Si addition has the positive effect. The precipitations of the AlSi-eutectic phase on the diamond particle surface preferentially during the solidification can release the thermal stress. This modification

can lower the broken of the interface to improve the interfacial morphology. On the other hand, the eutectic Si phase provides a transitional layer between the diamond particle and the matrix. These improvements result in an increase in the h. However, as raise of the Si addition, the amount of the AlSi-eutectic phase of low thermal conductivity, which segregating to the particle surface, is increasing. This negative effect can lead to a decrease of the h.

In combination with the two influences, the interfacial thermal conductance increases firstly, and then obtains the optimal point. After that, the interfacial thermal conductance falls gradually and arrives at a steady value finally.

Conclusions

The influence of the different Si content on the thermal conductivity and the interfacial characteristics of the Al-Si/diamond composites fabricated by the pressure infiltration have been investigated, and it has been shown that:

(1). As the Si addition, the Si skeleton of the AlSi-eutectic phase segregates to the diamond particle surface and form a better bonding between the reinforcement phase and the matrix.

(2). Highest thermal conductivity of the composites indicates at Al-1.0Si/diamond as 248 W/m K. It is desired for the thermal management applications.

(3). There are positive and negative effects of the Si addition on the interfacial thermal conductance (h). In combination with the two influences, h obtains the optimal point at about 2.0 - 3.0 wt.% Si, then reduces gradually and arrives at a steady value finally.

Acknowledgments

This work is financially supported by Cultivation Fund of the Key Scientific and Technical Innovation Project, Ministry of Education of China (No. 707007) and Building Projects of Scientific Research and Scientific Research Base, Bejing Muicipal Commission of Education (No. 2008100071601Y).

References

1. Barcena, J. et al, "Innovative packaging solution for power and thermal management of wide-bandgap semiconductor devices in space application," *Acta Astronautica*, Vol. 62, No. 6-7 (2008), pp. 422-430.
2. Zweben, C, "Thermal materials solve power electronics challenges," *Power Electronics Technology*, Vol. 32, No. 2 (2006), pp. 40-47.
3. Field, J. E, The properties of natural and synthetic diamond (London, 1992), p. 681.
4. Woodcraft, A. L, "Predicting the thermal conductivity of aluminum alloys in the cryogenic to room temperature range," *Cryogenic*, Vol. 45, No. 6 (2005), pp. 421-431.
5. Molina, J. M. et al, "The effect of porosity on the thermal conductivity of Al-12 wt. % Si/SiC composites," *Scripta Materialia*, Vol. 60, No. 7 (2009), pp. 582-585.
6. Ruch, P. W. et al, "Selective interfacial bonding in Al(Si)-diamond composites and its effect on thermal conductivity," *Composites Science and Technology*, Vol. 66, No. 15 (2006), pp. 2677-2685.

7. Hasselman, D.P.H. *et al*, "Effective thermal conductivity of composites with interfacial thermal barrier resistance," *Journal of Composite Materials*, Vol. 21, No. 6 (1987), pp. 508-515.

8. Nunes, R. *et al*, <u>ASM handbook, Vol. 2: properties and selection: nonferrous alloys and special-puepose materials</u> (Ohio, 1990), pp. 41-243.

9. Clyne, T. W. *et al*, An introduction to metal matrix composites (Oxford, 1994).

10. Every, A. G. *et al*, "The effect of particle size on the thermal conductivity of ZnS/diamond composites," *Acta Metallurgical et Materialia*, Vol. 40, No. 1 (1992), pp. 123-129.

11. Tavangar, R. *et al*, "Assessing predictive schemes for thermal conductivity against diamond-reinforced silver matrix composites at intermediate phase contrast," *Scripta Materialia*, Vol. 56, No. 5 (2007), pp. 357-360.

AUTHOR INDEX

An, B. 88, 758, 954, 974, 1139
An, R. 535
An, T. 133, 184, 294, 299
Ang, X. F. 1203
Aoh, J. 649
Appelt, B. 1237
Aria, P. 31
Arruda, L. 1120
Baated, A. 939
Bai, S. 57, 454, 525
Bai, S. L. 251
Bailey, C. 103, 129, 198, 847
Bao, S. 683
Berg, R. V. D. 1215
Bi, J. 767, 794
Bi, X. 1100
Bian, Z. 1265
Cai, J. 20, 23, 71, 76, 428, 557, 820
Cai, M. 52, 191, 843
Cai, X. 974
Calata, J. N. 385
Cao, B. 402
Cao, H. 27, 780, 1265
Cao, J. 317
Cao, K. 739
Cao, L. 829, 990
Cao, X. 172
Cao, Y. 5
Carlberg, B. 405
Chan, E. K. L. 950
Chan, Y. S. 1038
Chang, C. 454
Chang, G. 476
Chang, H. 1032
Chang, J. 934
Chang, S. 476, 806
Chen, B. 312, 673
Chen, C. 513, 594
Chen, F. 367
Chen, H. 88, 241, 486, 909
Chen, J. 57, 204, 251, 308
Chen, K. 444
Chen, L. 154, 590, 929, 1074
Chen, M. 149
Chen, Q. 1114, 1120
Chen, S. 1159
Chen, T. 289, 657
Chen, W. 191, 259, 649
Chen, X. 362, 385, 505, 754, 833, 974
Chen, Y. 541, 566, 570
Chen, Z. 111, 353, 402, 767, 1165
Cheng, C. 444
Cheng, J. 666
Cheng, X. 214, 646, 885
Cheng, Z. 179, 771, 959, 971, 1159

Chew, C. S. 715
Cho, M. 501
Choy, H. S. 857
Chuang, C. 649
Cong, Y. 586, 710
Conway, P. P. 619
Cui, C. Q. 1016
Cui, Y. 517
Dai, F. 947
Dai, J. H. 1144
Dai, W. 99, 865, 1144
Davignon, J. 551
Deng, G. 894, 900
Desmulliez, M. P. Y. 103, 847
Ding, D. 15, 454, 525, 646
Ding, L. 954
Ding, X. 9
Ding, Z. 900
Dong, M. 71, 76, 557
Dou, X. 23, 428
Du, R. 513, 776
Du, X. 547, 959, 971
Duan, Y. 787
Duh, J. 482, 967
England, L. 228
Ernst, L. J. 998, 1057, 1125
Fan, H. B. 158
Fan, J. 84, 223
Fan, X. 586, 710, 979, 1185
Fang, Q. 1179
Fang, X. 1179
Fang, Y. 1068
Feng, T. 71, 857
Feng, X. 947
Feng, Y. 994, 1002
Fremont, H. 1125
Fu, H. 551, 754
Fu, X. 1007
Fu, Y. 179, 763
Gai, J. 241
Galuschki, K. 525
Gan, Z. 353
Gang, F. 9
Gao, B. 436, 950
Gao, G. 162
Gao, J. 833
Gao, L. 732
Gao, W. 93, 861, 1225
Gao, Y. 810, 1159, 1175
Gautier, C. 1125
Ge, Z. 894
Geng, T. 223
Gong, A. 525
Goosen, J. F. L. 722
Guan, R. 706

AUTHOR INDEX

Gui, D. 312, 673
Guido, L. 505
Guo, F. 509, 677
Guo, J. 747
Guo, X. 405, 414
Guo, Y. 780
Hall, S. H. 551
Han, L. 303, 881, 1090
Han, M. 661
Han, Q. 979, 1185
Han, Y. D. 464
Hang, T. 608, 787
Hao, H. 509
Haseeb, A. S. M. A. 579, 715
Hayashi, H. 1249
He, D. 1023
He, Z. 916
Higurashi, E. 399
Hong, T. 541, 570
Hou, H. N. 414
Hsu, C. 541
Hsu, F. 922
Hsu, H. 922, 1032
Hsu, K. 444
Hu, A. 608, 625, 735, 787, 794
Hu, H. 900
Hu, J. 625, 719, 814
Hu, Y. 525, 792
Hu, Z. 405, 771
Hua, L. 414
Hua, Z. 874
Hua, Z. K. 1144
Huang, C. 277, 282, 1211
Huang, H. Z. 52, 843
Huang, J. 1002, 1114
Huang, M. 517, 590, 611, 798, 802
Huang, Q. 1, 574
Huang, T. 212
Huang, W. 268
Huang, X. 916
Hubbard, K. 869
Hutt, D. A. 380
Hwang, K. 395
Hwang, T. 1043
Jansen, K. M. B. 1057, 1125
Ji, M. 1257
Jia, H. 874
Jia, S. 20, 820
Jiang, J. 65, 767, 794
Jiang, L. 1144
Jiang, X. 702
Jiang, Y. 1152, 1155
Jin, H. 472, 732
Jin, P. 81
Jin, Y. 57, 251, 255, 391, 450, 814, 1257

Jin, Z. 1049
Jing, H. Y. 464
Johan, M. R. 579, 715
Ju, J. 1152, 1155
Ju, S. 1032
Kang, N. 468
Kang, R. 40
Kang, W. 1023
Kao, C. R. 458
Ke, L. 331
Kenny, S. 838
Kersaudy-Kerhoas, M. 103
Kettner, P. 852
Kim, G. 468, 501, 1261
Kim, J. 501
Kim, Jae-Dong 1043
Kim, Jin-Seong 1043
Kim, Jin-Young 1043
Kim, K. 939
Kivilahti, J. K. 486
Koo, Y. 468
Kugler, A. 117
Lai, C. M. 739
Lai, H. 971
Lai, Y. 1237
Lam, A. 241
Lee, B. 228, 934
Lee, C. 1043
Lee, D. W. 395
Lee, H. 541, 570
Lee, J. 444, 501, 598, 1165
Lee, S. 994
Lee, S. W. R. 1038
Lee, Y. B. 129
Lee, Z. 1007
Lei, G. 385, 505
Lei, J. 357
Leu, J. 444
Leung, V. C. K. 1220, 1243
Li, B. 93, 166, 1225
Li, C. Y. 1144, 1237
Li, D. 317
Li, D. X. 493, 521
Li, F. 1063
Li, G. 727, 1095, 1100
Li, J. 207, 299, 349, 486, 1013
Li, K. 833
Li, L. 139, 259, 331, 341, 395, 547, 869, 1109, 1253
Li, M. 15, 27, 454, 525, 608, 625, 735, 763, 767, 780, 787, 794, 874, 1265, 486
Li, Q. 27, 722, 1265
Li, S. 496, 706
Li, T. 282
Li, W. 1135
Li, X. 1063

AUTHOR INDEX

Li, Y. 36, 286, 418, 450, 454, 792, 865, 881
Li, Z. 154, 289, 345, 929, 1074, 1225, 1270
Liang, L. 235
Liang, T. 792
Liang, Y. 282, 1253
Liao, C. 207
Liao, P. 1144
Liao, W. B. 826
Liao, X. 125
Liem, H. M. 857
Lin, A. 1237
Lin, C. 149
Lin, W. 1211
Ling, H. 15, 27, 780, 1265
Liu, B. 677
Liu, C. 214, 380, 421, 619, 696, 833
Liu, D. 286
Liu, F. 235
Liu, H. 154, 758, 929, 943, 963, 1013, 1049, 1068, 1074, 1270
Liu, J. 84, 179, 312, 322, 405, 619, 673, 744, 771, 947, 959, 971, 1159
Liu, L. 317, 865, 1183
Liu, Q. 1233
Liu, S. 111, 149, 218, 353, 367, 402, 496, 615, 906, 1038, 1078
Liu, W. 149, 486, 566
Liu, X. 341, 362, 375, 611
Liu, Y. ..36, 111, 228, 235, 286, 418, 750, 802, 881, 1228
Liu, Z. 122, 367, 432, 493, 521
Lou, M. 1165
Lu, C. 906
Lu, D. 814
Lu, G. 172, 385, 505
Lu, H. 198, 308
Lu, J. 440, 1032
Lu, L. N. 52, 843
Lu, M. 65
Lu, Q. 865
Lu, X. 959, 971
Lu, Y. 792
Luo, L. 1, 5, 9, 574
Luo, S. 385, 513, 776
Luo, T. 735
Luo, X. 218, 367
Luo, Z. 810
Lv, D. 683
Lv, Y. 861
Ma, L. 683
Ma, S. 450, 1257
Ma, X. 317, 1109, 1125
Ma, Y. 65
Mao, D. 15, 27, 454, 525, 608, 625, 735, 767, 780, 787, 794, 1265
Mao, Z. 218

Martin, S. 228
Mattila, T. 486
Mavinkurve, A. 998
Mei, H. 547
Meng, Z. 994
Miao, M. 57, 251, 1257
Miao, X. 312
Michel, B. 117
Min, Y. 395
Mu, F. 57, 61
Mu, W. 277
Nadia, A. 531
Nai, S. M. L. 464, 1203
Ngo, K. 505
Ngo, K. D. T. 172
Nie, L. 557
Ning, C. 454
Ning, W. 1
Niu, J. 1002
Niu, L. 191, 246, 687, 1148
Niu, X. 289, 657
Okuno, A. 372
Osterman, M. 40, 557
Pan, H. 642
Pan, J. 798
Pan, K. 47, 317, 322
Pan, Q. 943
Pan, Z. 787
Pang, E. W. 1144
Pape, H. 1057
Pargfrieder, S. 852
Park, D. 1043
Park, J. 501, 1261
Park, S. 468
Patel, M. K. 103
Pavuluri, S. 847
Pechr, M. 40
Pecht, M. 557
Peng, B. 909
Peng, C. 967
Peng, J. 943, 1068
Pfahl, R. C. 551
Pichler, C. 852
Pomerleau, R. 31
Pot, A. 1215
Priest, J. 31
Pu, Y. 71
Pun, K. 1016
Qi, L. 1002
Qian, K. 586, 710
Qin, F. 133, 184, 263, 294, 299
Qin, R. 454
Qin, Y. 696
Qing, X. 798
Qiu, W. 47, 322

AUTHOR INDEX

Qiu, Y. 391
Quan, J. 909
Quintero, J. 1120
Rank, H. 117
Ranouta, A. S. 1185
Rausch, M. 551
Regard, C. 1125
Ren, C. 263, 732
Roehm, H. 1215
Roelfs, B. 838
Rong, Y. 20
Rongen, R. T. H. 998
Ruan, Z. 1
Savic, J. 31
Sawada, R. 399
Scheiring, C. 852
Schlottig, G. 1057
Seo, M. 1261
Seo, S. 1261
Seo, W. 468
Shang, J. K. 493, 521, 594, 963
Shang, P. J. 493, 521
Shao, X. 1179
Shi, D. X. Q. 1220, 1243
Shi, P. 732
Shi, Y. 509, 677
Shih, K. 541, 570
Shih, M. 1237
Shinohara, K. 1196
Shrivastava, A. 40
Shu, X. 289, 657, 1105
Sidiki, T. P. 1215
Sigl, A. 852
Silberschmidt, V. V. 214
Sommer, J. P. 117
Song, B. 615
Song, H. Y. 963
Song, J. 93, 327, 1225
Song, X. 353
Song, Z. 496
Stevens, B. 380
Strusevich, N. 129
Su, F. 642
Su, X. X. 52, 843
Suga, T. 143, 399, 440, 767, 794
Suganuma, K. 939
Suhir, E. 362
Sun, F. 750
Sun, J. 509
Sun, L. 566
Sun, P. 1220, 1243
Sun, R. 513, 776
Sun, W. H. 806
Sun, X. 255, 450, 1257
Tai, F. 677

Takigawa, R. 399
Tan, C. M. 464
Tan, K. H. 739
Tan, W. 547, 646
Tanaka, O. 372
Tang, C. 727
Tang, G. 1139
Tang, J. 327
Tao, B. 335
Tao, Y. 954
Tay, S. L. 579
Tian, D. 312
Tian, H. 666
Tian, X. 692
Tian, Y. 212, 380, 535, 566, 783
Tilford, T. 847
Tisdale, S. 551
Topham, D. 103
Tsai, M. 922
Tsai, S. 482
Tsao, L. C. 806, 1083
Tseng, A. 1237
Tseng, H. C. 204
Tsui, K. Y. K. 1220
Tu, Y. 496
Van Beek, J. T. M. 722
Van Der Sluis, O. 1057, 1125
Van Driel, W. D. 603, 1125
Van Keulen, F. 722
Van Soestbergen, W. 998
Varia, B. 979
Verheijden, G. J. A. M. 603
Wan, L. 93, 139, 207, 341, 349, 829, 861, 1225
Wang, B. 1068
Wang, C. 143, 212, 357, 409, 535, 566, 619, 696, 783, 1086, 1228
Wang, F. 1078
Wang, H. 162, 861
Wang, J. 47, 362, 375, 428, 496, 586, 710, 1007, 1175, 1179, 1270
Wang, K. 367, 482, 1211
Wang, L. 598, 611, 750, 750, 763, 810, 829
Wang, N. 23, 357
Wang, Q. 598, 702, 947, 1007, 1165
Wang, R. 476
Wang, S. 20, 179, 223, 428, 615, 820, 395
Wang, T. 345, 505, 586, 710
Wang, X. 154, 179, 632, 929, 1074, 1270
Wang, Y. 71, 76, 139, 341, 440, 525, 865, 990, 458
Wang, Z. 57, 61, 122, 335, 432, 666, 963
Wassay, A. 696
Wei, J. 464, 1203
Wei, S. 327
Wen, L. 1165
Wen, Q. 418, 916

AUTHOR INDEX

Weng, L. 513
Weng, M. 922
Whalley, D. 421
Wilcox, G. D. 696
Wong, C. K. Y. 158
Wu, B. Y. 52, 843
Wu, C. M. L. 611
Wu, D. 375
Wu, F. 88, 758, 943, 954, 974, 1013, 1068, 1139
Wu, H. 1032
Wu, J. 241, 632, 646
Wu, N. 81
Wu, S. 1211
Wu, X. 865
Wu, Y. 88, 954, 974, 1013, 1068, 1139
Wu, Z. 277, 885
Wunderle, B. 1057
Xi, Y. 149
Xia, S. 57
Xia, Z. D. 677
Xiang, Y. 476
Xiao, A. 1057
Xiao, F. 754
Xie, H. 947
Xie, X. 719, 727, 1007, 1100
Xie, Y. 418
Xiong, Y. 335
Xu, C. 890
Xu, G. 1, 509, 574
Xu, H. 763
Xu, J. 353
Xu, L. 615, 959, 971, 1114
Xu, X. 702
Xue, J. 31, 869
Xue, K. 241
Xue, X. 103
Xv, W. 1183
Yan, B. 661
Yan, H. 353
Yan, X. 727, 1095
Yang, B. 792
Yang, C. 1038
Yang, D. 191, 246, 687, 1148, 1243
Yang, G. 162
Yang, H. 754, 1131
Yang, J. 414, 820
Yang, L. 255, 1002
Yang, M. 763
Yang, S. 212
Yang, W. 23
Yang, X. 15, 1105
Yang, Y. 166, 308
Yang, Z. 65, 692, 881
Yao, Y. 954
Yau, S. K. 1220

Ye, J. 683
Ye, Y. 496, 1038
Yeh, W. 1211
Yen, S. F. 806
Yin, C. 129
Yin, L. 166
Yin, Z. 335, 909
Yu, C. 308
Yu, H. 780, 1179
Yu, Q. 1196
Yu, R. 345
Yu, S. 513, 776
Yu, W. 125
Yu, X. 15, 27, 450
Yu, Z. 666
Yuan, C. 47, 322
Yuan, G. 1105
Yuan, Z. 228, 362, 375
Yue, C. 405, 771
Yue, T. M. 857
Yuen, M. M. F. 158, 436, 950
Yung, K. C. 857
Yung, K. P. 1203
Zaal, J. J. M. 603
Zeng, G. 312
Zeng, J. 619, 1228
Zhai, Q. 1159
Zhang, B. 586, 710
Zhang, E. 375
Zhang, G. Q. 317, 603, 687, 722, 998, 1114, 1125
Zhang, H. 36, 418, 637, 881, 1002
Zhang, J. 57, 251, 657, 758, 776, 890, 1144
Zhang, Jianhua 874
Zhang, Jinsong 874
Zhang, K. L. 464
Zhang, L. 594, 739, 906
Zhang, M. 890
Zhang, Q. 615
Zhang, S. 40
Zhang, W. 154, 432, 472, 929, 1074, 1270
Zhang, X. 93, 166, 865, 1135
Zhang, Y. 57, 61, 84, 179, 223, 251, 375, 421, 632, 771, 826, 959, 1090
Zhang, Z. 637, 702, 754, 776, 1131
Zhao, H. 885
Zhao, J. 496, 810
Zhao, K. 409, 696
Zhao, L. 255, 391
Zhao, M. 191, 246, 385, 509, 586, 710, 1148
Zhao, N. 829
Zhao, S. 783
Zhao, W. 706
Zhao, X. 1002
Zhao, Y. 611
Zhao, Z. 598

AUTHOR INDEX

Zheng, D. 833
Zheng, G. 1086
Zheng, H. F. 857
Zheng, Q. 345
Zhong, Q. 81
Zhou, D. 885
Zhou, F. 646
Zhou, H. 303
Zhou, J. 1007, 1165
Zhou, L. 88, 943, 1013
Zhou, P. 331
Zhou, Q. 81, 843
Zhou, S. 402
Zhou, X. 1109
Zhou, Y. 349
Zhu, D. 357
Zhu, H. 162
Zhu, Q. S. 963
Zhu, W. 259, 702, 1049
Zhu, Y. 1257
Zou, H. F. 1131
Zou, J. 758
Zou, K. 27

2009 International Conference on Electronic Packaging Technology & High Density Packaging

(ICEPT-HDP 2009)

Beijing, China
10 – 13 August 2009

Pages 637-1273

Editors:

Keyun Bi
Jian Cai

IEEE Catalog Number: CFP09553-PRT
ISBN: 978-1-4244-4658-2

**Copyright © 2009 by the Institute of Electrical and Electronic Engineers, Inc
All Rights Reserved**

Copyright and Reprint Permissions: Abstracting is permitted with credit to the source. Libraries are permitted to photocopy beyond the limit of U.S. copyright law for private use of patrons those articles in this volume that carry a code at the bottom of the first page, provided the per-copy fee indicated in the code is paid through Copyright Clearance Center, 222 Rosewood Drive, Danvers, MA 01923.

For other copying, reprint or republication permission, write to IEEE Copyrights Manager, IEEE Service Center, 445 Hoes Lane, Piscataway, NJ 08854. All rights reserved.

***This publication is a representation of what appears in the IEEE Digital Libraries. Some format issues inherent in the e-media version may also appear in this print version.**

IEEE Catalog Number: CFP09553-PRT
ISBN 13:   978-1-4244-4658-2
Library of Congress No.: 2009904680

Additional Copies of This Publication Are Available From:

Curran Associates, Inc
57 Morehouse Lane
Red Hook, NY 12571 USA
Phone: (845) 758-0400
Fax: (845) 758-2633
E-mail: curran@proceedings.com

TABLE OF CONTENTS

A Novel MEMS Package with Three-Dimensional Stacked Modules 1
G. Xu, Q. Huang, W. Ning, Z. Ruan, L. Luo

Fabrication and Characterization of a Novel Wafer-level Chip Scale Package for MEMS Devices 5
Y. Cao, L. Luo

Study on a 3D Packaging Structure with Benzocyclobutene as a Dielectric Layer for Radio Frequency
Application 9
F. Gang, X. Ding, L. Luo

Through-Silicon Via Filling Process Using Pulse Reversal Plating 15
X. Yang, H. Ling, D. Ding, M. Li, X. Yu, D. Mao

Low Temperature Cu-Sn Bonding by Isothermal Solidification Technology 20
Y. Rong, J. Cai, S. Wang, S. Jia

A PCB-based Package Structure 23
N. Wang, J. Cai, W. Yang, X. Dou

Collaborative Effect between Additives and Current in TSV Via Filling Process 27
K. Zou, H. Ling, Q. Li, H. Cao, X. Yu, M. Li, D. Mao

Comparison Study of Effective Power Delivery in Advanced Substrate Technologies for High Speed
Networking Applications 31
J. Priest, R. Pomerleau, J. Savic, P. Aria, J. Xue

The Design and Fabrication of Multilayer Chip LC Filter with a Modified Structure by LTCC
Technology 36
Y. Liu, H. Zhang, Y. Li

The Influence of SO_2 Environments on Immersion Silver Finished PCBs by Mixed Flow Gas Testing 40
S. Zhang, A. Shrivastava, M. Osterman, M. Pechr, R. Kang

Study on Reliability of Embedded Passives in Organic Substrate 47
W. Qiu, K. Pan, C. Yuan, J. Wang

Investigation on PCB Related Failures in High-Density Electronic Assemblies 52
L.N. Lu, H.Z. Huang, X.X. Su, B.Y. Wu, M. Cai

Nanoscale Mechanical Properties and Microstructure of 3D LTCC Substrate 57
Y. Zhang, J. Chen, S. Bai, M. Miao, J. Zhang, F. Mu, Z. Wang, S. Xia, Y. Jin

Process Research of LTCC Substrate with 3D Micro-channel Embedded 61
F. Mu, Z. Wang, Y. Zhang

Study of the Fine Line Process and Signal Integrity for Packaging Substrate 65
Y. Ma, J. Jiang, Z. Yang, M. Lu

Effects of Nitrogen on Wettability and Reliability of Lead-free Solder in Reflow Soldering 71
M. Dong, Y. Wang, J. Cai, T. Feng, Y. Pu

Solderability of Flash Gold Surface Finish 76
Y. Wang, M. Dong, J. Cai

Thermal Analysis and Testing of Multi Layer Ceramic-Metal Packaged LED 81
P. Jin, Q. Zhou, N. Wu, Q. Zhong

Multiscale Delamination Modeling of an Anisotropic Conductive Adhesive Interconnect Based on
Micropolar Theory and Cohesive Zone Model 84
Y. Zhang, J. Fan, J. Liu

Finite Element Thermal Analysis for High Power Multi-chip Light Emitting Diode 88
L. Zhou, H. Chen, B. An, F. Wu, Y. Wu

A Design of an Optical Transceiver SiP 93
W. Gao, L. Wan, B. Li, J. Song, X. Zhang

Power Supply Analysis in Package and SiP Design 99
W. Dai

Effect of Fluid Dynamics and Device Mechanism on Biofluid Behaviour in Microchannel Systems:
Modelling Biofluids in a Microchannel Biochip Separator 103
X. Xue, M.K. Patel, M. Kersaudy-Kerhoas, C. Bailey, M.P.Y. Desmulliez, D. Topham

Modeling of Manufacturing Processes of High Power Light Emitting Diodes: Wire Bonding Process 111
Z. Chen, Y. Liu, S. Liu

Advanced Package Design for Electronic and MEMS Applications Supported by FE Analyses and
Deformation Measurements 117
J.P. Sommer, B. Michel, A. Kugler, H. Rank

A Novel Wafer Level Package Strategy for RF MEMS 122
    Z. Wang, Z. Liu

Design and Simulation of a Package Solution for Millimeter Wave MEMS Switch 125
    W. Yu, X. Liao

Packaging of Light Emitting Diodes for Display Backlights: Need for Multi-Disciplinary Design Tools 129
    Y.B. Lee, N. Strusevich, C. Bailey, C. Yin

A Simplified Computational Model for Solder Joints under Drop Impact Loadings 133
    T. An, F. Qin

Design of Miniaturized Bandpass Filters for GSM and GPS Applications Using Embedded Capacitor Material 139
    Y. Wang, L. Li, L. Wan

Moiré Method for Nanoprecision Wafer-to-Wafer Alignment: Theory, Simulation and Application 143
    C. Wang, T. Suga

Localized Induction Heating for Wafer Level Packaging 149
    M. Chen, W. Liu, Y. Xi, C. Lin, S. Liu

Molecular Dynamics of Nanofluids with Time-Dependent Thermal Conductivity in Heat Sink Dissipation 154
    X. Wang, H. Liu, W. Zhang, Z. Li, L. Chen

Hydrophobic Self-Assembly Monolayer Structure for Reduction of Interfacial Moisture Diffusion 158
    H.B. Fan, C.K.Y. Wong, M.M.F. Yuen

Thermal Numerical Simulation for Advanced Package Development 162
    G. Gao, H. Wang, G. Yang, H. Zhu

FE Simulation of Size Effects on Interface Fracture Characteristics of Microscale Lead-Free Solder Interconnects 166
    B. Li, L. Yin, Y. Yang, X. Zhang

Thermal Design of Power Module to Minimize Peak Transient Temperature 172
    X. Cao, K.D.T. Ngo, G. Lu

A Study of the Heat Transfer Characteristics of the Micro-channel Heat Sink 179
    S. Wang, Y. Zhang, Y. Fu, J. Liu, X. Wang, Z. Cheng

Fracture Simulation of Solder Joint Interface by Cohesive Zone Model 184
    T. An, F. Qin

Optimization Design for Packaging Device QFN Using a Prediction Model of the Neural-Genetic Algorithm 191
    M. Cai, D. Yang, L. Niu, M. Zhao, W. Chen

Lifetime Prediction of an IGBT Power Electronics Module under Cyclic Temperature Loading Conditions 198
    H. Lu, C. Bailey

An In-Depth Numerical Investigation into Packaging Design of Multi-Finger GaInP/GaAs Collector-Up HBTs 204
    H.C. Tseng, J.Y. Chen

A Study of Power Delivery Network in a GPS SiP 207
    J. Li, L. Wan, C. Liao

Modeling Thermal Fatigue in Anisotropic Sn-Ag-Cu/Cu Solder Joints 212
    S. Yang, Y. Tian, C. Wang, T. Huang

Numerical Analysis of Response of Indium Micro-Joint to Low-Temperature Cycling 214
    X. Cheng, C. Liu, V.V. Silberschmidt

A Thermal Model for Calculating Thermal Resistance of Eccentric Heat Source on Rectangular Plate with Convective Cooling Existing at Upper and Lower Surfaces 218
    X. Luo, Z. Mao, S. Liu

Modeling of Heat Transfer Performance for an Array of Micro Jets Impinging upon Dimpled Surface 223
    T. Geng, J. Fan, Y. Zhang, S. Wang

Full Thermal Parametric Model for Power WL-CSP Design 228
    Z. Yuan, Y. Liu, S. Martin, L. England, B. Lee

An Improved Substructure Method for Prediction of Solder Joint Reliability in Thermal Cycle 235
    F. Liu, L. Liang, Y. Liu

Warpage Prediction of Fine Pitch BGA by Finite Element Analysis and Shadow Moiré Technique 241
    K. Xue, J. Wu, H. Chen, J. Gai, A. Lam

Prediction of Die Failure in Copper-Low-K Flip Chip Package with Consideration of Packaging Process-Induced Stresses 246
    M. Zhao, D. Yang, L. Niu

Simulation of Fluid Flow and Heat Transfer in Microchannel Cooling for LTCC Electronic Packages 251
    J. Zhang, Y.F. Zhang, M. Miao, Y.F. Jin, S.L. Bai, J.Q. Chen

Research of Methodologies to Enlarge the Isolation in 3D Interconnection 255
    L. Zhao, L. Yang, X. Sun, Y. Jin

Study on EBG Structure Combined with Decoupling Capacitor for Suppressing Ground Bounce Noise...259
    W. Zhu, W. Chen, L. Li

Parametric Study of Warpage in Package-on-Package Manufacturing 263
    C. Ren, F. Qin

Computational Modeling and Optimization for Wire Bonding Process on Cu/Low-K Wafers................268
    W. Huang

The Principal Component Analysis of Cu Stud Bump Shaping Process Parameters 277
    W. Mu, Z. Wu, C. Huang

Effects of Shape Parameters on Tensile Strength of Cu Bump..282
    C. Huang, Y. Liang, T. Li

A New Embedded Helix-Type Inductor Using LTCC Technology for High Frequency Applications.....286
    D. Liu, Y. Liu, Y. Li

Modeling Solder Joint Reliability of VFBGA Packages under Board Level Drop Test Based on Dynamic Constitutive Relation with Thermal Effect...289
    X. Niu, T. Chen, Z. Li, X. Shu

Effects of PCB Dynamic Responses on Solder Joint Stress..294
    T. An, F. Qin

Influence of Strain Rate Effect on Behavior of Solder Joints under Drop Impact Loadings 299
    T. An, F. Qin, J. Li

Local Plastic Zones of Wirebond Profiles Inspection Base on Curvature Estimation 303
    H. Zhou, L. Han

First-Principles Calculations of Structural, Thermodynamic and Electronic Properties of Intermetallic Compounds in Solder 308
    Y. Yang, H. Lu, C. Yu, J. Chen

A Core-Shell Structure Viscoelastic Model of Particulate-Filled Electronic Packaging Polymers..........312
    D. Gui, J. Liu, B. Chen, X. Miao, G. Zeng, D. Tian

Numerical Simulation on Heat Pipe for High Power LED Multi-Chip Module Packaging 317
    D. Li, G.Q. Zhang, K. Pan, X. Ma, L. Liu, J. Cao

Board Level Drop Impact Reliability Analysis for Compliant Wafer Level Package through Modeling Approaches..322
    C. Yuan, K.L. Pan, W. Qiu, J. Liu

FEM Study on the Effects of Flip Chip Packaging Induced Stress on MEMS...................................327
    S. Wei, J. Tang, J. Song

Modeling and Simulation of SSN on FPGA Products..331
    L. Ke, P. Zhou, L. Li

Coupled Simulation of Anisotropic Conductive Adhesive Bonding Process and Reliability Analysis of the Packaging..335
    Z. Wang, Z. Yin, B. Tao, Y. Xiong

Miniaturized Printed Wire Antenna in Package for 2.4GHz Wireless Communications......................341
    X. Liu, Y. Wang, L. Li, L. Wan

Modeling of Copper Wire Bonding Ball Transient Temperature Behavior 345
    T. Wang, Q. Zheng, R. Yu, Z. Li

Power Integrity Simulation for SiP Using GTLE 349
    Y. Zhou, L. Wan, J. Li

First-Principles Based Modeling for Influence of Epitaxy and Packaging Induced Strains on Emission Properties of III-nitrides LED Chips 353
    H. Yan, Z. Gan, X. Song, Z. Chen, J. Xu, S. Liu

Numerical Study on Thermal Management of LED Packaging by Using Thermoelectric Cooling...........357
    N. Wang, C. Wang, J. Lei, D. Zhu

Modeling of Thermal Phenomena in a High Power Diode Laser Package 362
    E. Suhir, J. Wang, Z. Yuan, X. Chen, X. Liu

Freeform Lens for Application-Specific LED Packaging...367
    F. Chen, K. Wang, Z. Liu, X. Luo, S. Liu

High Bright White LED Lens Formation by Vacuum Printing Encapsulation Systems (VPES) and its Packaging Resin...372
    A. Okuno, O. Tanaka

250W QCW Conduction Cooled High Power Semiconductor Laser 375
    J. Wang, Z. Yuan, Y. Zhang, E. Zhang, D. Wu, X. Liu

High Density Indium Bumping through Pulse Plating Used for Pixel X-Ray Detectors..380
Y. Tian, D.A. Hutt, C. Liu, B. Stevens

Emerging Lead-free, High-temperature Die-attach Technology Enabled by Low-temperature Sintering of Nanoscale Silver Pastes..385
G. Lu, M. Zhao, G. Lei, J.N. Calata, X. Chen, S. Luo

A Novel Micro Pirani Gauge with Mono-wire Sensing Unit for Microsystem Application......................391
Y. Qiu, L. Zhao, Y. Jin

On-Package Magnetic Materials for Embedded Inductor Applications..395
L. Li, D.W. Lee, K. Hwang, Y. Min, S.X. Wang

Low-temperature Bonding of Laser Diode Chips on Si Substrates with Oxygen and Hydrogen Atmospheric-pressure Plasma Activation..399
R. Takigawa, E. Higurashi, T. Suga, R. Sawada

Integration of GaN Thin Film and Dissimilar Substrate Material by Au-Sn Wafer Bonding and CMP..................402
S. Zhou, Z. Chen, B. Cao, S. Liu

Modeling of Nanostructured Polymer-Metal Composite for Thermal Interface Material Applications..................405
Z. Hu, B. Carlberg, C. Yue, X. Guo, J. Liu

Characterization of a Photosensitive Dry Adhesive Film for Wafer Level MEMS Packaging...................409
K. Zhao, C. Wang

Optimization Design on Polybrominated Biphenyls (PBBs) Extraction from Plastics for RoHS Directive...414
L. Hua, X.P. Guo, J.K. Yang, H.N. Hou

The Design of a Terahertz Metamaterial Absorber Basing on LTCC Technology................................418
Y. Xie, Y. Li, Y. Liu, H. Zhang, Q. Wen

Direct-write Techniques for Maskless Production of Microelectronics: A Review of Current State-of-the-art Technologies...421
Y. Zhang, C. Liu, D. Whalley

Integrated Inductors on Silicon and Planarized Ceramic Substrates...428
J. Wang, J. Cai, X. Dou, S. Wang

Nano Resonator Simulation Fabrication and Packaging Consideration..432
W. Zhang, Z. Liu, Z. Wang

Optimization of Silver Paste Printed passive UHF RFID Tags..436
B. Gao, M.M.F. Yuen

Low-temperature Wafer Bonding Using Gold Layers...440
Y. Wang, J. Lu, T. Suga

Novel Pore-sealing of Ordered, Porous Silica, SBA-15 for Low-k Underfill Materials.........................444
K. Hsu, K. Chen, C. Cheng, J. Lee, J. Leu

Study of Polyimide as Sacrificial Layer with O_2 Plasma Releasing for Its Application in MEMS Capacitive FPA Fabrication..450
S. Ma, Y. Li, X. Sun, X. Yu, Y. Jin

Fabrication and Hydrogen Sensing Properties of Titania Nanotubes...454
S. Bai, D. Ding, C. Ning, R. Qin, Y. Li, C. Chang, M. Li, D. Mao

Development of Lead-Free Solders with Superior Drop Test Reliability Performance.............................458
Y.W. Wang, C.R. Kao

Enhancing the Properties of a Lead-free Solder with the Addition of Ni-coated Carbon Nanotubes.................464
S.M.L. Nai, Y.D. Han, H.Y. Jing, K.L. Zhang, C.M. Tan, J. Wei

A Method to Produce Printed Circuit Boards with Embedded Semiconductors Using Stress Buffer Layers...468
W. Seo, Y. Koo, S. Park, N. Kang, G. Kim

Study of Tungsten Metallization Surface States for Multilayer Ceramic..472
W. Zhang, H. Jin

Solder Cracking Mechanism Correlation to Alloy Composition under Thermal Cycling Stress (Part 1)*...476
S. Chang, R. Wang, Y. Xiang, G. Chang

Morphological and Microstructural Evolution of Sn-patch in SnAgCu Solder with Ni(V)/Cu under Bump Metallization..482
K. Wang, J. Duh, S. Tsai

Localized Recrystallization and Cracking Behavior of Lead-free Solder Interconnections under Thermal Cycling..486
H.T. Chen, T. Mattila, J. Li, W. Liu, M.Y. Li, J.K. Kivilahti

Growth Kinetics and Microstructural Evolution of Cu-Sn Intermetallic Compounds on Different Cu Substrates During Thermal Aging..493
Z.Q. Liu, P.J. Shang, D.X. Li, J.K. Shang

Assessment of LF Solder Joint Reliability by Four Point Cyclic Bending 496
J. Wang, Y. Ye, J. Zhao, S. Liu, Y. Tu, S. Li, Z. Song

A Warpage of Wafer Level Bonding for CIS (CMOS Image Sensor) Device Using Polymer Adhesive 501
J. Park, J. Lee, M. Cho, J. Kim, G. Kim

Improved Thermal Performance of High-Power LED by Using Low-temperature Sintered Chip Attachment 505
T. Wang, G. Lei, X. Chen, L. Guido, K. Ngo, G.Q. Lu

Fundamental Studies on Whisker Growth in Sn-based Solders 509
M. Zhao, H. Hao, G. Xu, J. Sun, Y. Shi, F. Guo

Micro-structural and Interfacial Effects on the Dielectric Properties of High-k Aluminum/Epoxy Composites for Embedded Capacitors 513
C. Chen, S. Yu, R. Sun, S. Luo, L. Weng, R. Du

Comparative Study of Interfacial Reactions of High-Sn Lead-free Solders on Single Crystal Cu and on Polycrystalline Cu 517
Y. Cui, M. Huang

Phase Identification of Intermetallic Compounds Formed during In-48Sn/Cu Soldering Reactions 521
P.J. Shang, Z.Q. Liu, D.X. Li, J.K. Shang

Microstructural Characterization of Electroplating Sn on Lead-frame Alloys 525
Y. Wang, D. Ding, K. Galuschki, Y. Hu, A. Gong, S. Bai, M. Li, D. Mao

The Fabrication of Composite Solder by Addition of Copper Nano Powder into Sn-3.5Ag Solder 531
A. Nadia

Depiction of the Elastic Anisotropy of $AuSn_4$ and $AuSn_2$ from First-principles Calculations 535
R. An, C. Wang, Y. Tian

Microstructural Evolution of Sn-3.5Ag Solder with Lanthanum Addition 541
H. Lee, Y. Chen, T. Hong, K. Shih, C. Hsu

Optimal Packing Research of Spherical Silica Fillers Used in Epoxy Molding Compound 547
H. Mei, X. Du, L. Li, W. Tan

iNEMI HFR-free Program Report 551
H. Fu, S. Tisdale, M. Rausch, J. Davignon, S.H. Hall, R.C. Pfahl

Interfacial Reaction of Reballed BGAs under Isothermal Aging Conditions 557
L. Nie, M. Dong, J. Cai, M. Osterman, M. Pecht

Effect of Cu Addition in Sn-containing Solder Joints on Interfacial Reactions with Au Foils 566
W. Liu, C. Wang, L. Sun, Y. Tian, Y. Chen

Evolution of Ag_3Sn Compounds in Solidification of Eutectic Sn-3.5Ag Solder 570
H. Lee, Y. Chen, T. Hong, K. Shih

Indium Bump Fabricated with Electroplating Method 574
Q. Huang, G. Xu, L. Luo

Effect of Aging Time on Interfacial Microstructure of Sn-3.8Ag-0.7Cu Solder Reinforced with Co Nanoparticles 579
S.L. Tay, A.S.M.A. Haseeb, M.R. Johan

Behaviors of Palladium in Palladium Coated Copper Wire Bonding Process 586
B. Zhang, K. Qian, T. Wang, Y. Cong, M. Zhao, X. Fan, J. Wang

Effect of Electromigration on Intermetallic Compound Formation in Cu/Sn/Cu Interconnect 590
L.D. Chen, M. Huang

Solderability of Tinned FeNi under Bump Material 594
C. Chen, L. Zhang, J.K. Shang

Characterize the Microstructure and Reliability of Ultra Fine Pitch BGA Joints 598
L. Wang, Z. Zhao, Q. Wang, J. Lee

Reliability of Wafer Level Thin Film MEMS Packages during Wafer Backgrinding 603
J.J.M. Zaal, W.D. Van Driel, G.J.A.M. Verheijden, G.Q. Zhang

Super-hydrophobic Nickel Films with Micro-nano Hierarchical Structure Prepared by Electrodeposition for Appliance Industry 608
T. Hang, M. Li, A. Hu, D. Mao

Liquid-state Interfacial Reactions between Sn-Ag-Cu-Fe Composite Solders and Cu Substrate 611
X. Liu, Y. Zhao, M. Huang, C.M.L. Wu, L. Wang

Cure Kinetics Analysis and Simulation of Silicone Adhesives 615
Q. Zhang, B. Song, S. Wang, L. Xu, S. Liu

Growth and Recrystallization of Electroplated Copper Columns 619
J. Liu, C. Liu, P.P. Conway, J. Zeng, C. Wang

Electrochemical Corrosion Behaviour of Sn-8Zn-3Bi-XCr Solder in 3.5% NaCl 625
J. Hu, A. Hu, M. Li, D. Mao

The Influence of Silicon Content on the Thermal Conductivity of Al-Si/ Diamond Composites 632
Y. Zhang, X. Wang, J. Wu

Study on Microstructure and Properties of Al/SiCp Electronic Packaging Materials Embedded Metal Components 637
Z. Zhang, H. Zhang

Creep Characterization of Lead Free 80Au-20Sn Solder 642
F. Su, H. Pan

The Warpage Control Method in Epoxy Molding Compound 646
W. Tan, F. Zhou, X. Cheng, D. Ding, J. Wu

On Reliability of Chips bonded on Flex Substrates Using Thermosonic Flip-Chip Bonding Process with Nonconductive Paste 649
C. Chuang, J. Aoh, W. Chen

Dynamic Constitutive Relation of EMC over a Broad Range of Temperatures and Strain Rates 657
T. Chen, X. Niu, X. Shu, J. Zhang

Low K CMOS90 2N Gold Wire Bonding Process Development 661
M. Han, B. Yan

Preliminary Study of a New Process to Improve the Strength of Thermo-sonic Ball Solder Joint 666
Z. Yu, Z. Wang, J. Cheng, H. Tian

Improving the Toughness and Thermal Properties of Epoxy Resin Using for Electronic Packaging by Interpenetrating Polymer Network 673
B. Chen, D. Gui, J. Liu

Effects of Service Parameters on Thermomechanical Fatigue Behaviors of New Nano Composite Solder Joints 677
F. Tai, F. Guo, B. Liu, Z.D. Xia, Y.W. Shi

Failure Analysis of Halide of Epoxy Molding Compound Used for Electronic Packing 683
J. Ye, S. Bao, L. Ma, D. Lv

Characterization, Modelling, and Parameter Sensitivity Study on Electronic Packaging Polymers 687
L. Niu, D. Yang, G.Q. Zhang

Controlled Synthesis of Ortho-Substitution Ortho-cresol Novolac Resins 692
X. Tian, Z. Yang

Electrodeposition of Sn-Cu Solder Alloy for Electronics Interconnection 696
Y. Qin, A. Wassay, C. Liu, G.D. Wilcox, K. Zhao, C. Wang

Study on the Degradation of Sealed Organic Light-emitting Diodes under Constant Current 702
X. Xu, W. Zhu, Q. Wang, Z. Zhang, X. Jiang

Preparation and Performance of White LED Packaged YAG Phosphor 706
R. Guan, W. Zhao, S. Li

Nanoindentation Investigation of Copper Bonding Wire and Ball 710
X. Fan, K. Qian, T. Wang, Y. Cong, M. Zhao, B. Zhang, J. Wang

Wetting Behaviour of Lead Free Solder on Electroplated Ni and Ni-W Alloy Barrier Film 715
C.S. Chew, A.S.M.A. Haseeb, M.R. Johan

Parameter Design of Solder Die Bonding Based on DOE 719
J. Hu, X. Xie

Outgassing of Materials Used for Thin Film Vacuum Packages 722
Q. Li, J.F.L. Goosen, J.T.M. Van Beek, F. Van Keulen, G.Q. Zhang

Effects of Surface Finishes on the Intermetallic Growth and Micro-structure Evolution of the Sn3.5Ag0.7Cu Lead-free Solder Joints 727
G. Li, C. Tang, X. Yan, X. Xie

The Effect of Tape Casting Slurry System for Process Performance of Green Ceramic Tape Used for Electronic Packaging 732
P. Shi, C. Ren, H. Jin, L. Gao

Studies on Microstructure and Mechanical Properties of Sn-Zn-Bi-Cr Lead-free Solder 735
T. Luo, A. Hu, M. Li, D. Mao

Solder Joints Reliability with Different Cu Plating Current Density in Wafer Level Chip Scale Packaging (WLCSP) 739
K. Cao, K.H. Tan, C.M. Lai, L. Zhang

A Novel Research on Micro-ball Placement Machine Used for Wafer Level Package 744
J. Liu

Developing on IC Flip Chip Bonder Machine & Process 747
J. Guo

Effect of Ni Addition on the Sn-0.3Ag-0.7Cu Solder Joints 750
L. Wang, F. Sun, Y. Liu, L. Wang

Electrically Conductive Adhesives with Sintered Silver Nanowires 754
Z. Zhang, X. Chen, H. Yang, H. Fu, F. Xiao

Al/Ni Multilayer Used as a Local Heat Source for Mounting Microelectronic Components 758
J. Zhang, F. Wu, J. Zou, B. An, H. Liu

Dramatic Morphological Change of Interfacial Prism-type Cu_6Sn_5 in the Sn3.5Ag/Cu Joints Reflowed by Induction Heating 763
L. Wang, H. Xu, M. Yang, M. Li, Y. Fu

Wetting Behavior of Electrolyte in Fine Pitch Cu/Sn Bumping Process by Electroplating 767
J. Jiang, J. Bi, Z. Chen, M. Li, D. Mao, T. Suga

Effects of the Matrix Shrinkage and Filler Hardness on the Thermal Conductivity of TCA 771
C. Yue, Y. Zhang, Z. Hu, J. Liu, Z. Cheng

Synthesis and Characterization of Nano $BaTiO_3$/Epoxy Composites for Embedded Capacitors 776
S. Luo, R. Sun, J. Zhang, S. Yu, R. Du, Z. Zhang

Influence of Leveler Concentration on Copper Electrodeposition for Through Silicon Via Filling 780
H. Ling, H. Cao, Y. Guo, H. Yu, M. Li, D. Mao

Characterization of Ag Nanofilm Metallization on Copper Chip Interconnect and Its Ultrasonic Bondability 783
Y. Tian, S. Zhao, C. Wang

Electroless Plating of Copper Nano-coned Array for High Reliability Packaging 787
Z. Pan, A. Hu, T. Hang, Y. Duan, M. Li, D. Mao

Research on Self-constrained Sintering Low-temperature Cofired Ceramic 792
Y. Hu, T. Liang, Y. Li, B. Yang, Y. Lu

Fine Pitch and High Density Sn Bump Fabrication 794
J. Bi, J. Jiang, A. Hu, M. Li, D. Mao, T. Suga

Effect of [Au]/[Na_2SO_3] Molar Ratio on Co-electroplating Au-Sn Alloys in Sulfite-based Solution 798
X. Qing, M. Huang, J. Pan

Sequential Non-cyanide Electroplating Au/Sn/Au Films for Flip Chip-LED Bumps 802
Y. Liu, M. Huang

Study of Interfacial Reactions between Sn3.5Ag0.5Cu Alloys and Cu Substrate 806
L.C. Tsao, S.Y. Chang, W.H. Sun, S.F. Yen

Study on the Microstructure and the Shear Strength of Sn-0.7Cu-xZn 810
Y. Gao, Z. Luo, J. Zhao, L. Wang

Investigation of the Fundamental Interactions among the Ingredients of Flux by the Group Contribution Method 814
Y. Jin, J. Hu, D. Lu

Study of Stencil Printing Technology for Fine Pitch Flip Chip Bumping 820
J. Yang, J. Cai, S. Wang, S. Jia

Processing and Properties of Cu-base and Co-base Amorphous Wires 826
W.B. Liao, Y. Zhang

Absorption of Ag_3Sn on Cu_6Sn_5 Intermetallic Compounds at Sn-3.5Ag-xCu/Cu Interfaces 829
N. Zhao, L. Wang, L. Wan, L. Cao

Prediction Model for Wire Bonding Process through Adaptive Neuro-Fuzzy Inference System 833
J. Gao, C. Liu, X. Chen, D. Zheng, K. Li

From Thin Cores to Outer Layers: Filling through Holes and Blind Micro Vias with Copper by Reverse Pulse Plating 838
S. Kenny, B. Roelfs

Investigation of Thin Small Outline Package (TSOP) Solder Joint Crack after Accelerated Thermal Cycling Testing 843
L.N. Lu, H.Z. Huang, B.Y. Wu, Q. Zhou, X.X. Su, M. Cai

On Variable Frequency Microwave Processing of Heterogeneous Chip-on-Board Assemblies 847
T. Tilford, S. Pavuluri, C. Bailey, M.P.Y. Desmulliez

Advanced Chip to Wafer Bonding: A Flip Chip to Wafer Bonding Technology for High Volume 3DIC Production Providing Lowest Cost of Ownership 852
A. Sigl, S. Pargfrieder, C. Pichler, C. Scheiring, P. Kettner

The Effect of Plasma Etching Process on Rigid Flex Substrate for Electronic Packaging Application 857
K.C. Yung, H.M. Liem, H.S. Choy, H.F. Zheng, T. Feng, T.M. Yue

A Study on the Characterization of Quasi-Three-Dimensional PN Junction Capacitor 861
H. Wang, Y. Lv, W. Gao, L. Wan

Effect of PIII on the Adhesion Behavior of Epoxy Molding Compound-Nickel Interface 865
L. Liu, Q. Lu, Y. Wang, W. Dai, X. Zhang, Y. Li, X. Wu

Improving Board Assembly Yield through PBGA Warpage Reduction 869
L. Li, K. Hubbard, J. Xue

A Jetting System for Chip on Glass Package..874
    H. Jia, Z. Hua, M. Li, Jinsong Zhang, Jianhua Zhang

A Multilayer Low Pass Filter Fabricated by Ferrite and Ceramic Cofiring System Based on LTCC
Technology 881
    Y. Li, Y. Liu, H. Zhang, L. Han, Z. Yang

Research of SMT Product Assembly Quality Management System Based on J2EE 885
    H. Zhao, D. Zhou, Z. Wu, X. Cheng

Test Scheme of SOC Test with Multi-constrained to Reduce Test Time 890
    C. Xu, J. Zhang, M. Zhang

Design and Modeling of Jet Dispenser Based on Giant Magnetostrictive Material 894
    Z. Ge, G. Deng

Simulation and Experimental Study on Temperature Field of Fluid Jet-dispenser......................................900
    Z. Ding, H. Hu, G. Deng

Automatic Plating Technology for Ceramic Packaging...906
    C. Lu, S. Liu, L. Zhang

Analysis and Compensation of Perpendicularity Error for High Speed and High Precise IC Assembly
Equipment Base on Coordinate Transformation 909
    H. Chen, J. Quan, B. Peng, Z. Yin

Inspection of Miniaturised Interconnections in IC Packages with Nanofocus X-Ray Tubes and
NanoCT 916
    Z. He, Q. Wen, X. Huang

Advanced Moisture Diffusion Model and Hygro-Thermo-Mechanical Design for Flip Chip BGA
Package...922
    M. Tsai, F. Hsu, M. Weng, H. Hsu

Analysis the Performance of the Micro-Channels Cooler with Different Inlet Position................................929
    X. Wang, W. Zhang, H. Liu, L. Chen, Z. Li

Sn Whisker Concern in IC Packaging for High Reliability Application...934
    J. Chang, B. Lee

Investigation of Mechanism for Spontaneous Zinc Whisker Growth from an Electroplated Zinc
Coating 939
    A. Baated, K. Kim, K. Suganuma

Effects of Thermal Aging on the Electrical Resistance of Sn-3.5Ag Micro SOH Solder Joints 943
    J. Peng, F. Wu, H. Liu, L. Zhou, Q. Pan

Instability and Failure Analysis of Film-substrate Structure under Electrical Loading 947
    Q. Wang, H. Xie, J. Liu, X. Feng, F. Dai

Reliability Study of RFID Flip Chip Assembly by Isotropic Conductive Adhesive through Computer
Simulation 950
    E.K.L. Chan, B. Gao, M.M.F. Yuen

Effect of Thermomigration in Eutectic SnPb Solder Layer 954
    Y. Tao, L. Ding, Y. Yao, B. An, F. Wu, Y. Wu

Adhesion Behavior between Epoxy Molding Compound and Different Leadframes in Plastic
Packaging 959
    L. Xu, X. Lu, J. Liu, X. Du, Y. Zhang, Z. Cheng

Effect of Zn Addition on Microstructure of Sn-Bi Joint...963
    Q.S. Zhu, H.Y. Song, H.Y. Liu, Z.G. Wang, J.K. Shang

Deformation Characteristics of Sn-3Ag-0.5Cu/Cu/Ni-xCu/Ti Joints after Mechanical Test 967
    C. Peng, J. Duh

Studies on Microstructure of Epoxy Molding Compound (EMC)-Leadframe Interface after
Environmental Aging 971
    X. Lu, L. Xu, H. Lai, X. Du, J. Liu, Z. Cheng

The Effects of Bonding Parameters on the Reliability Performance of Flexible RFID Tag Inlays
Packaged by Anisotropic Conductive Adhesive 974
    X. Cai, X. Chen, B. An, F. Wu, Y. Wu

Effects of Design, Structure and Material on Thermal-Mechanical Reliability of Large Array Wafer
Level Packages...979
    B. Varia, X. Fan, Q. Han

Development of High Speed Cold Ball Pull as a Quick Turn Monitor for Solder Joint Reliability............990
    Y. Wang, L. Cao

Influencing Factors and Solutions for Ball Short during Wire Bonding 994
    Z. Meng, Y. Feng, S. Lee

Modeling Electrochemical Migration through Plastic Microelectronics Encapsulations...........................998
    W. Van Soestbergen, A. Mavinkurve, R.T.H. Rongen, L.J. Ernst, G.Q. Zhang

Microstructure Changes and Compound Growth Dynamic at Lead-free/Cu Interface under Different Conditions 1002
L. Qi, J. Huang, J. Niu, L. Yang, Y. Feng, X. Zhao, H. Zhang

Research of Structural Factors Effects on Drop Reliability 1007
J. Wang, X. Fu, X. Xie, J. Zhou, Q. Wang, Z. Lee

Effect of Stand-off Height on the Reliability of Cu/Sn-4.8Bi-2Ag/Cu Solder Joint 1013
H. Liu, L. Zhou, J. Li, F. Wu, Y. Wu

Enhancement of TBGA Substrate in Packing Drop Test 1016
K. Pun, C.Q. Cui

Quality and Reliability Challenges for Ultra Mobile Computing and Communication Application Processor Packaging 1023
D. He, W. Kang

Electromigration Analysis and Electro-Thermo-Mechanical Design for Semiconductor Package 1032
H. Hsu, S. Ju, J. Lu, H. Chang, H. Wu

Comparison of Thermal Fatigue Reliability of SnPb and SAC Solders under Various Stress Range Conditions 1038
C. Yang, Y.S. Chan, S.W.R. Lee, Y. Ye, S. Liu

Board Level Reliability Assessments of Thru-Mold Via Package on Package (TMV™ PoP) 1043
T. Hwang, D. Park, Jin-Seong Kim, Jin-Young Kim, Jae-Dong Kim, C. Lee

Prediction of IMC Formation during Interfacial Reactions: Application of CALPHAD Approach to Electronic Package 1049
H. Liu, W. Zhu, Z. Jin

Establishing Mixed Mode Fracture Properties of EMC-Copper (-oxide) Interfaces at Various Temperatures 1057
A. Xiao, G. Schlottig, H. Pape, B. Wunderle, O. Van Der Sluis, K.M.B. Jansen, L.J. Ernst

Effect of Thermal Cycling on Interfacial IMCs Growth and Fracture Behavior of SnAgCu/Cu Joints 1063
X. Li, F. Li

Effect of Miniaturization on the Microstructure and Mechanical Property of Solder Joints 1068
B. Wang, F. Wu, J. Peng, H. Liu, Y. Wu, Y. Fang

Numerical Simulation on Variable Width Multi-Channels Heat Sinks with Non-uniform Heat Source 1074
X. Wang, W. Zhang, H. Liu, L. Chen, Z. Li

Characteristic Analysis of Transducer Drive Current in Ultrasonic Wire Bonding Process 1078
S. Liu, F. Wang

Corrosion Characterization of Sn37Pb Solders and With Cu Substrate Soldering Reaction in 3.5wt.% NaCl Solution 1083
L.C. Tsao

Shape and Fatigue Life Prediction of Chip Resistor Solder Joints 1086
G. Zheng, C. Wang

Effect of Bonding Temperature and Power Setting on Transducer Velocity Using Principal Components Analysis in Thermosonic Bonding 1090
Y. Zhang, L. Han

Study of Thermal Fatigue Lifetime of Fan-in Package on Package (FiPoP) by Finite Element Analysis 1095
X. Yan, G. Li

Thermal-Mechanical Fatigue Reliability of PbSnAg Solder Layer of Die Attachment for Power Electronic Devices 1100
X. Xie, X. Bi, G. Li

Effects of Strain Rate and Temperature on Mechanical Behavior of SACB Solder Alloy 1105
G. Yuan, X. Yang, X. Shu

Thermal Stress Analysis and Structural Optimization of Ultra-thin Chip Stacked Package Device 1109
L. Li, X. Ma, X. Zhou

Prediction of Bending Reliability of BGA Solder Joints on Flexible Printed Circuit (FPC) 1114
J. Huang, Q. Chen, L. Xu, G.Q. Zhang

Failure Evaluation of Flexible-Rigid PCBs by Thermo-Mechanical Simulation 1120
L. Arruda, Q. Chen, J. Quintero

Fast Qualification Using Thermal Shock Combined with Moisture Absorption 1125
X. Ma, G.Q. Zhang, K.M.B. Jansen, W.D. Van Driel, O. Van Der Sluis, L.J. Ernst, C. Regard, C. Gautier, H. Fremont

Controlling the Morphology and Orientation of Cu₆Sn₅ through Designing the Orientations of Cu Single Crystals 1131
H.F. Zou, H.J. Yang, Z.F. Zhang

Thermal-Mechanical Failure and Life Analysis on CBGA Package used for Great Scale FPGA Chip 1135
W. Li, X. Zhang

Finite Element Analysis of Sn-Ag-Cu Solder Joint Failure under Impact Test 1139
G. Tang, B. An, Y. Wu, F. Wu

Study on Moisture Behavior in Flip Chip BGA Packages and Bake Process Optimization 1144
W.Q. Dai, Z.K. Hua, J.H. Dai, E.W. Pang, L. Jiang, C.Y. Li, P. Liao, J.H. Zhang

Study on Thermo-mechanical Reliability of Embedded Chip during Thermal Cycle Loading 1148
L. Niu, D. Yang, M. Zhao

The Influence of Plastic-package on the Voltage Shift of Voltage Reference in Analog Circuit 1152
Y. Jiang, J. Ju

Electrical Analysis of Mechanical Stress Induced by Shallow Trench Isolation 1155
Y. Jiang, J. Ju

The Effect of Thermal Cycling on Nanoparticle Reinforced Composite Lead-free Solder 1159
S. Chen, Z. Cheng, J. Liu, Y. Gao, Q. Zhai

Study of Isothermal Bending Fatigue Test 1165
M. Lou, L. Wen, Z. Chen, J. Zhou, Q. Wang, J. Lee

Testing Failure of Solder-Joints by ESPI on Board-Level Surface Mount Devices 1175
Y. Gao, J. Wang

Study on Shear Strength and J_c of EMC/Cu Interface with Cu Oxidation and Moisture Absorption 1179
X. Fang, Q. Fang, J. Wang, H. Yu, X. Shao

XPS Study on Epoxy/Ni Interface 1183
L. Liu, W. Xv

Shock Performance Study of Solder Joints in Wafer Level Packages 1185
A.S. Ranouta, X. Fan, Q. Han

Fatigue Evaluation of Power Devices 1196
K. Shinohara, Q. Yu

Advanced High Density Interconnect Materials and Techniques 1203
J. Wei, S.M.L. Nai, X.F. Ang, K.P. Yung

The High Balance Symmetric Balun for WLAN and WiMAX Application Using the Integrated Passive Device (IPD) Technology 1211
S. Wu, W. Lin, K. Wang, C. Huang, W. Yeh

New Packaging Technology Enabling Integration of Magnetics and Semiconductors in One Component 1215
A. Pot, H. Roehm, R.V.D. Berg, T.P. Sidiki

Parametric Study of Electroplating-based Via-filling Process for TSV Applications 1220
K.Y.K. Tsui, S.K. Yau, V.C.K. Leung, P. Sun, D.X.Q. Shi

Packaging and Assembly of 12-Channel Parallel Optical Transceiver Module 1225
Z. Li, W. Gao, J. Song, B. Li, L. Wan

Recent Advances in Laser Assisted Polymer Intermediate Layer Bonding for MEMS Packaging 1228
C. Wang, J. Zeng, Y. Liu

Wafer-Scale Hermetically Packaged MEMS Switches with Liquid Gallium Contacts 1233
Q. Liu

Underfill Study for Large Dice Flip Chip Packages 1237
A. Lin, C.Y. Li, M. Shih, Y. Lai, B. Appelt, A. Tseng

Development of a Novel Cost-Effective Package-on-Package (PoP) Solution 1243
P. Sun, V.C.K. Leung, D. Yang, D.X.Q. Shi

EcoDesign Technical Committee of JIEP (Japan Institute of Electronics Packaging) and Its Activity 1249
H. Hayashi

Artificial Neural Network Application in Vertical Interconnection Modeling 1253
Y. Liang, L. Li

A New Method to Fabricate Sidewall Insulation of TSV Using a Parylene Protection Layer 1257
M. Ji, Y. Zhu, S. Ma, X. Sun, M. Miao, Y. Jin

The Electrical, Mechanical Properties of Through-Silicon-Via Insulation Layer for 3D ICs 1261
S. Seo, J. Park, M. Seo, G. Kim

Through Silicon Via Filling by Copper Electroplating in Acidic Cupric Methanesulfonate Bath 1265
Q. Li, H. Ling, H. Cao, Z. Bian, M. Li, D. Mao

Package Heat Dissipation with Integrated Carbon Nanotube Micro Heat Sink 1270
X. Wang, H. Liu, J. Wang, W. Zhang, Z. Li

Author Index

Study on Microstructure and Properties of Al/SiCp Electronic Packaging Materials Embedded Metal Components

Zhiqing Zhang, Hongyu Zheng

The 13th Research of Institute of China Electronics Technology Group Corporation, Shijiazhuang, 050051, China

Abstract

With the development of microelectronics and semiconductor, the density of electronic packaging increases quickly, which result in the high demand to the material. Aluminum reinforced with silicon carbide particles composites is one of potential materials for electronic packaging with its excellent properties such as low density, high thermal conductivity, lower coefficient of thermal expansion etc. In order to obtain the final product, packaging component of Al/SiCp must to be bonded with different materials in practical application. It is significant practical to research the infiltration bonding of aluminum silicon carbide electronic packaging materials and metal component.

Al/SiCp electronic packaging materials embedded metal components are fabricated by gas pressure infiltration. The microstructure of interface is observed by EDS and XRD, Bend strength is in order to test the joint strength.

The result shows that during the fabrication process of Al/SiCp electronic packaging materials, the reliable joint between composites and solid metal (FeNi$_{50}$, Ti) can be realized. In the research of the infiltration bonding of Al/SiCp/FeNi$_{50}$, it is found that (Fe,Ni)$_2$(Al,Si)$_5$ layer and (Fe,Ni)$_4$(Al,Si)$_{13}$ layer all exist in the interface reactive layer of the joint. With the increment of the infiltration bonding temperature, thickness of the former approximately stay the same, and the relationship between the latter and temperature is near linear. As a result of the formation of intermetallic compound, bending strength of joint is low. Interface bending strength is up to 46% of that of Al/SiCp (467MPa), when the infiltration bonding temperature is 670°C.

On the study of the infiltration bonding of Al/SiCp/Ti, it is found that there wasn't any interface reactive layer in the joint. Mechanical and metallurgic combinations were all found in the interface at lower temperatures. When the infiltration bonding temperature is 710°C, atoms in the Al/Ti interface occurs interdiffusion and the interface, which has a 10μm thick diffusion layer, was bonded well. About the mechanical properties, it is found that, the bend strength of Al/SiCp/Ti is much more than that of Al/SiCp/FeNi$_{50}$ and with the increase of infiltration bonding temperature, the bend strength increased first and then decreased.

1. Introduction

Al/SiCp electronic packaging materials with its excellent physics and mechanical properties such as low density, high thermal conductivity, lower coefficient of thermal expansion, have widespread application prospect in aviation, astronautics and electronic packaging domain [1]. But in the practice application, Al/SiCp often must be assembled to make the final product with different materials, such as metal, ceramics [2]. The use of some traditional bonding processes is restricted because of the huge difference between the Al/SiCp

compound materials basal body and the reinforced body, as well as between composite material and the alloy in the physical chemistry and mechanical property. This is the key element which hinders the using of Al/SiCp. Although some advanced process which is friction stir welding and transient liquid phase diffusion bonding (TLP-DB) can partially solve above problems, but the cost is excessively high and the working procedure is complex.

At present, a new bonding method - Concurrent Integration technology is applied in the field of composite materials bonding. Concurrent Integration manufacturing technology, that is, directly to the components such as lead, sealing ring and substrate embedded in the molding of the preform, forming a direct bonding through infiltration. This method can connect metal components (lead, sealing ring) and the molten aluminum matrix state in situ. The interface is high reliability and the seal performance is very good. It was also reported that the Concurrent Integration manufacturing technology can save packaging costs by up to 50% [3]. Therefore, it is necessary to study the application of the Al/SiCp electronic packaging materials embedded metal components.

This paper explores the Al/SiCp electronic packaging materials embedded metal components which was made by pressure infiltration technology. The influence of the infiltration process parameters on joint strength and connectivity is also studied in the paper.

2. Experimental Procedures

2.1 Experiment materials and method

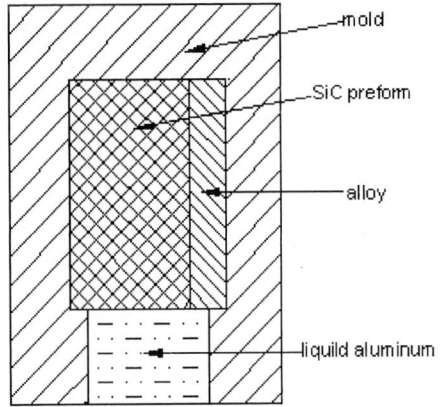

Fig. 1 Schematic diagram of specimens before infiltration

This paper chooses commercial α-SiC particles as the reinforcement, and the average size of the reinforcement is 5 μm and 50 μm; the main components (mass fraction, %) of self-made aluminum alloy are Si (8 ~ 10), Mg (2 ~ 3) and the remaining aluminum. In view of the common ground of the metal components, this article only studies FeNi alloys and titanium FeNi$_{50}$ alloy which can match well with soft glass and ceramics is chose.

The preform is made by dry-press process. Firstly, the proportional SiC particles is ball milled for 6 h; secondly, adding the additives such as water, phenolic resinafter and so on; thirdly, press molding in the 50 tons of press. After water removal, degreasing, sintering, the volume fraction of about 65% can be achieved. The dimensions of Iron-nickel alloy and titanium is 40 mm × 20 mm × 1 mm. Oxide film and dirt on the alloy piece are removed by 1200 # and 1800 # metallographical sand paper, and then followed by chemical cleaning , water cleaning and ultrasonic cleaning. Finally, metal components are stored in acetone solution. Figure 1 is the schematic diagram of specimens before infiltration. First, vacuumizing to 0.02 atm, preheating to 700°C; second, liquid aluminum alloy overheats to 850°C, pressurizing to 1.2 MPa; third, unloading pressure after holding pressure for 10 min. Al/SiCp composite materials can be obtained, at the same time the bonding between composite and alloy is completed. In this paper, the infiltration process is completed when the preheating temperature is at 670°C, 700°C, 730°C, 760°C respectively.

2.2 Performance test

In this paper, the microstructure of the specimen is analyzed by way of optical metalloscope, Olympus PM-T3. Spot scanning and line scanning have been made by scanning electron microscope (SEM) and energy spectrum analysis (EDS), and the standards is Q/AH0268-2003. X-ray diffraction test equipment for the test is PANalytical X'Pert PRO. Bending strength is obtained by domestic electronic universal testing machine, done three times, taking the arithmetic mean.

3. Results and Discussion

3.1 Microstructure observation

The microstructure of Al/SiCp embedded metal components is shown in Fig. 2. As can be seen, the obtained Al/SiCp composite is uniform and dense in microstructure and reliable in connection. The interface of the materials is bonded well, and there is no flaw such as crack, porosity existence in it. Fig. 2 is the microstructure of Al/SiCp embedded FeNi$_{50}$ alloy. From the figure we can see that the upper light gray region is FeNi$_{50}$ alloy and the central dark part is interfacial reactive layer, the thickness of which is 40 μm. Fig. 2(b) shows the interface microstructure of Al/SiCp embedded titanium. The interface layer is white aluminum alloy matrix; the upper part of the figure is pure metal titanium. There is a 10μm thick diffusion layer between them and there is no generation of intermetallic compounds. This is because of the low infiltration temperature. During the infiltration, aluminum and titanium only mutually diffuse in the short distance scope of interface. Away from the original interface, aluminum and titanium have no time to fully diffuse and stay the same composition.

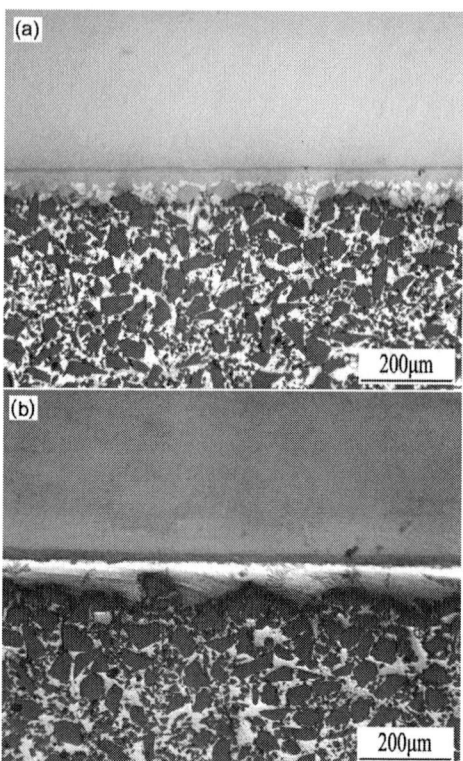

Fig. 2 Microstructure of Al/SiCp embedded alloy piece

3.2 Phase analysis

In order to study the diffusion of different elements, line scanning is used to analyse the joint. Fig. 3 gives the elements distribution of the interface region. The composite is made by Concurrent Integration manufacturing technology. The preheat temperature is 690°C and the holding time is 10min. Scanning position is shown in Fig. 3(a). The total length of elements line scanning is 182μm, both sides of the interface is about 60μm long. The left part of the interface is FeNi$_{50}$ alloy, the right part of the interface is Al/SiCp composites and interface layer is in the middle.

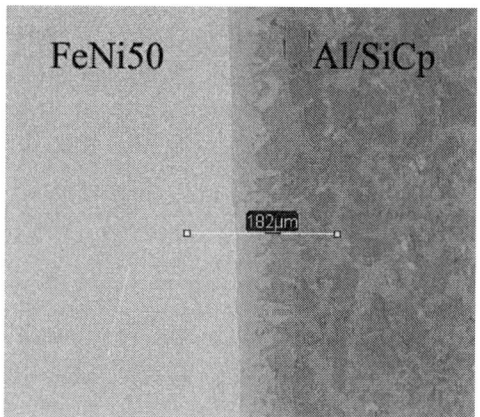

(a) Scanning position

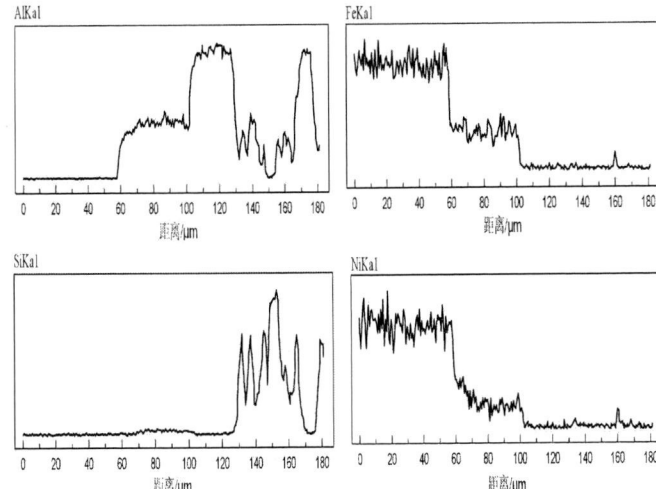

(b) The line scanning curve of elements

Fig. 3 Concentration profile of alloy elements in Al/SiCp/FeNi$_{50}$ interface

From Fig. 3(b) we can see, for FeNi$_{50}$ alloys, composite materials, as well as the interface layer, the distribution of the main elements such as Al, Fe, Ni and Si, is obviously different on both sides of the interface. The distribution of Silicon is nearly 0 in the side of FeNi$_{50}$ alloy, fluctuant in the Al/SiCp side, rich in the interfacial reactive layer. For iron and nickel, the distribution is very similar, concentration distribution on the left side is significantly higher than that on the right side, and they in the interface layer. However, when the distance is 60 ~ 70μm, iron decreases very quickly, unlike the slow declining of nickel. This is because of the different diffusion rate of them. In the first 60μm, the content of aluminum is nearly 0, in 60μm place increases suddenly, and then the distribution is balance. But in 100μm place it is up to maximum. In the Al/SiCp region, the line scanning curve of aluminum is fluctuant. In the interface region, aluminum, iron, nickel all have a transition zone of the same composition, and silicon is rich, indicating that intermetallic compounds generates among the four elements.

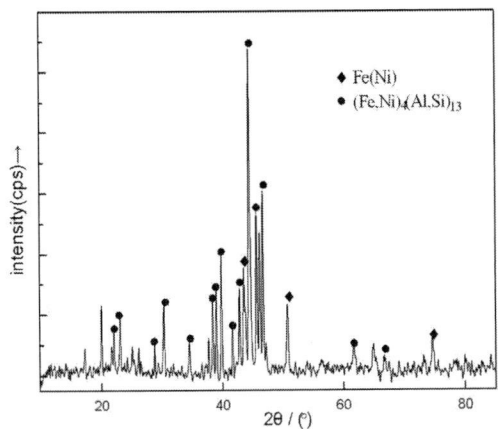

Fig. 4 XRD pattern of fracture surface on FeNi$_{50}$ side

Studies have shown that Fe$_2$A$_{l5}$ intermetallic compound phase often appears in hot-dip aluminizing of iron. The

process of gas pressure infiltration is similar to hot-dip aluminizing, so there must be (Fe, Ni)$_2$(Al, Si)$_5$ generation [4]. In order to further determine the phase composition of the interface layer, X-ray diffraction method is applied to analyze the fracture on FeNi$_{50}$ side. The results are shown in Fig. 4. Diffraction analysis results shows that the intermetallic compounds (Fe, Ni)$_4$(Al, Si)$_{13}$ and a small amount of iron-nickel solid solution exist in the fracture. This is because the fracture occurred in the interface reactive layer near the FeNi$_{50}$ side.

3.3 Effect of infiltration bonding temperature on reactive layer

When the infiltration pressure is 1.2MPa, holding time is 10min, interfacial structure of Al/SiCp/FeNi$_{50}$ joint with different temperature is shown in Fig. 5. As Can be seen that, the joint interface is formed by the layered material.

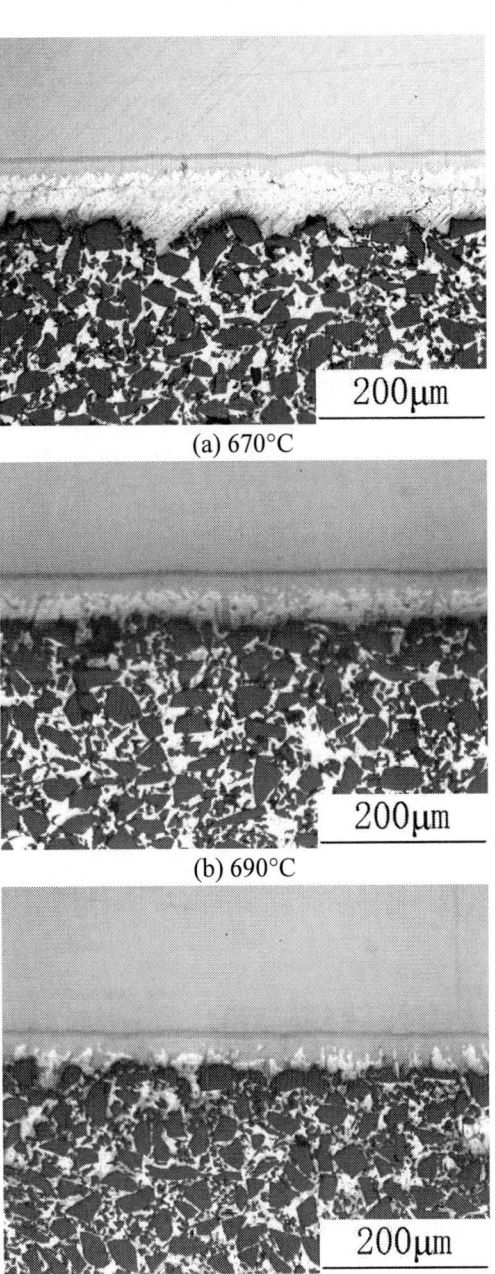

(a) 670°C

(b) 690°C

(c) 710°C

(d) 730°C

Fig. 5 Effect of infiltration bonding temperature on structure of Al/SiCp/FeNi$_{50}$ joint

(P=1.2MPa, t=10min)

The joint is consisted of FeNi$_{50}$, interfacial reactive layer, aluminum alloy matrix and Al/SiCp composites from top to bottom. Reactive layer on FeNi$_{50}$ side (dark part) and reactive layer on aluminum side (light part) constitute the interfacial reactive layer together.

Al/SiCp composite at the bottom of the figure is uniform and dense in microstructure and reliable connection with FeNi$_{50}$ alloy.

When the infiltration bonding temperature is below 730°C, there is no defects such as cracks, pores in the interface layer that combinates well. At 730°C, cracks are found in the reactive layer. With the increment of the infiltration bonding temperature, the reactive layer on FeNi$_{50}$ side grow slowly, its thickness approximately stays the same. The change of temperature influences less on it. The thickness of the reactive layer on aluminum side gradual increases with the increment of the temperature. The relationship between them is near linear, until the growth of aluminum matrix to the whole layer, as shown in Fig. 6. The formation of intermetallic compound has a direct impact on the joint performance.

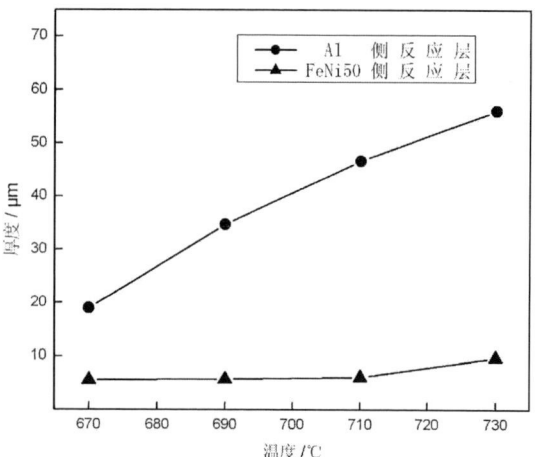

Fig. 6 The relation between reactive layer and infiltration jointing temperature

3.4 Effect of infiltration bonding temperature on the mechanical properties of joint

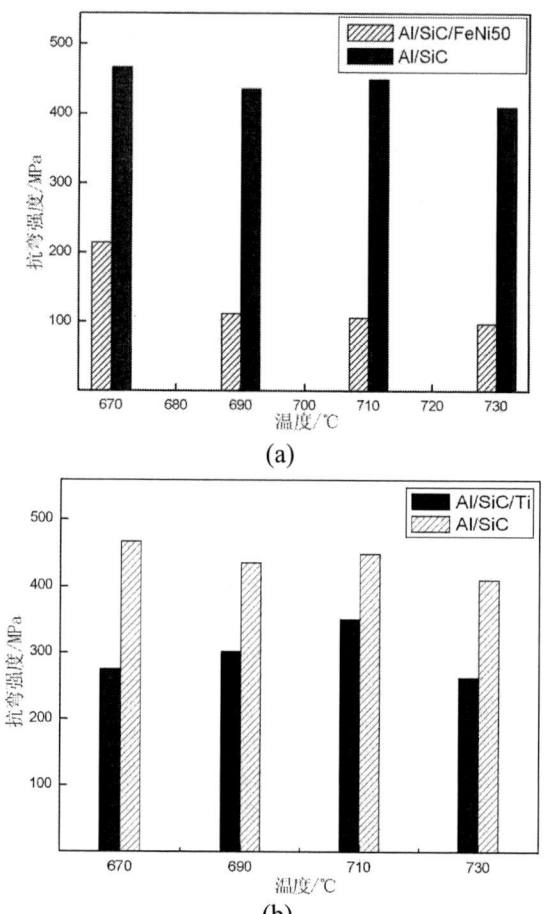

(a)

(b)

Fig. 7 Relation between bending strength and infiltration bonding temperature

The infiltration bonding temperature impacts remarkably on the joint strength of Al/SiCp/FeNi$_{50}$. When the temperature is 670°C, due to low infiltration bonding temperature, inferior diffusion ability of atoms, intermetallic compounds thickness is thinner (24.6μm) than that of higher temperature. The interface bending strength of joint is similar to that of aluminum alloy (223MPa), only 215Mpa. It is approximately up to 46% of that of Al/SiCp (467MPa) (Fig.7 (a)). Fig. 7(b) indicates that, during the range of 670°C~730°C, the joint bending strength of Al/SiCp/Ti increases gradually with the increment of temperature. When the temperature is 670°C, Al/SiCp/Ti's bending strength (275 MPa) is up to 59% of that of the composite. At 710°C, the value of the bending strength is 351 MPa, about 80% of the Al/SiCp (449 MPa). And the bonding effect is very well. However, when the temperature is excessively high, a reduction is to appear. This is due to aluminum-titanium interface elements have taken place in diffusion, formed a thin diffusion layer. The interface is metallurgical combination and bonded well. When the preheating temperature continues to rise, the coarse crystallization will cause a decline in the quality of joint. So there is an optimum value in the pressure infiltration, which is the same as Up Going's research [5].

Therefore, the mechanical property can be significant improved through the reasonable control of the process parameters.

Conclusions

(1) Al/SiCp electronic packaging materials embedded metal components are fabricated by gas pressure infiltration. During the fabrication process of Al/SiCp, the reliable joint between composites and solid metal ($FeNi_{50}$, Ti) can be realized, the interface layer is uniform and dense in microstructure and the bending performance is well.

(2) $Al/SiCp/FeNi_{50}$ interfacial reactive layers are common formed by $(Fe, Ni)_2(Al, Si)_5$ layer and $(Fe, Ni)_4(Al, Si)_{13}$ layer. With the increment of the infiltration bonding temperature, thickness of the former approximately stay the same, and the relationship between the latter and temperature is near linear.

(3) The bend strength of Al/SiCp/Ti is much more than that of $Al/SiCp/FeNi_{50}$ and with the increase of infiltration bonding temperature, the bend strength increases first and then decreases.

References

1. Molina J M, Saravanan R A, Arpon R, et al., "Pressure infiltration of liquid aluminium into packed SiC particulate with a bimodal size distribution," *Acta Materialia*, 50 (2002), pp. 247-248.

2. Yi Wang, Jie Liu, "Package technology of high performance Al/SiCp," *Electronic Mechanical Engineering*, 22(3) (2006), pp. 27-29.

3. Adams R W, Novich B E, Fennessy K Y, "Concurrent IntegrationTM of Al/SiC MMIC Packages," *ISHM'95 Proceeding*, 1995.

4. C.W.Su, J.W.Lee, C.S.Wang, C.G.Chao, T.F.Liu, "The effect of hot-dipped aluminum coatings on Fe-8Al-30Mn-0.8C alloy," *Surface And Coatings Technology*, 202 (2008), pp. 1850-1852.

5. Guoqing Xu, Gang Zeng, Jitai Niu, "The diffusion jointing process of aluminum/titanium," *Welding*, 3 (2000), pp. 21-23.

Creep Characterization of Lead Free 80Au-20Sn Solder

Fei Su, Haiyan Pan

Institute of Solid Mechanics, Beijing University of Aeronautics and Astronautics

37 Xueyuan Road, Beijing, P. R. China, 100083

sufei@buaa.edu.cn

Abstract

In this paper, steady-state creep behavior of lead free 80Au-20Sn solder was investigated at three different temperatures (i.e. 25°C, 75°C and 125°C) and a range of stress levels. Hyperbolic-Sine model was employed to characterize the creep properties of the lead free solder, constants in the model were determined through data fitting and were compared with that of other lead free solders.

1 Introduction

In electronic packages such as flip chip and BGA, solder joints act as electrical, mechanical and thermal interconnections, their failure usually means the failure of whole packages. In practice, with the power on and off and the coefficient of thermal expansion (CTE) mismatch between different packaging materials, these solder joints have to suffer cyclic inelastic deformations that lead to low cycle fatigue failures. Fatigue life of solder joints is a critical problem to electronic packaging industrials, and various models were developed and employed to predict the fatigue life of solder joint [1-2], which is more important in the stage of product design. In most of these models, inelastic strain or strain energy of solder joint accumulated in one thermal cycle is an important control factor. For reason of easy manufacturability, low melting point alloy e.g. 37Pb-63Sn, are often used as solder material. But these materials tend to creep at low stress level and their working temperature, as a result, creep-caused component dominated in the inelastic strain/strain energy of solder joint accumulated in thermal cycle and became a key role in the determination of fatigue life. As such, investigation on creep property of solder material is very important and necessary.

Due to the legislation on ban of lead (Pb) in electronic assemblies and products [3-4], lead free solder materials have to be developed and their mechanical properties must be characterized as well. Until now, some lead free solder materials such as Sn-Ag-Cu and Sn-Cu are widely used in industry, and their mechanical properties including creep characterizations were investigated [5-9]. Among these lead free solder materials, 80Au-20Sn is thought to be hard and highly creep resistant, so it is often used as soldering material in laser-diode packages or optical MEMS [10-12], where creep of soldering material may affect the optics alignment greatly and is tried to avoid. Besides, its applications as solder joint in flip chip are also reported [12-13]. However, its mechanical properties are not well studied and relevant experimental data is rarely reported. In this paper, we investigated the creep behavior of this material: its steady state strain rate at various stress levels and temperatures were tested. Then experimental data were fitted to Hyperbolic-Sine model, constants in this model are determined through data fitting.

2 Experimental Design

2.1 About Specimen

For reasons of cost, standard specimen was not used. We have two specimen designs for the creep test.

A. Lap-shear test specimen

In this design, a piece of 80Au-20Sn perform (2mm by 2mm by 0.2mm) was placed between two brass substrates. Contact surface of the brass substrate was chemically processed as that for solder pad. Then solder perform was bound to the two brass substrates with a reflow oven especially for lead-free solder material. For even-distribution of shear stress during the test, a triangular notch was fabricated next to the reflowed solder. Specimen prepared in this way is illustrated in Fig. 1.

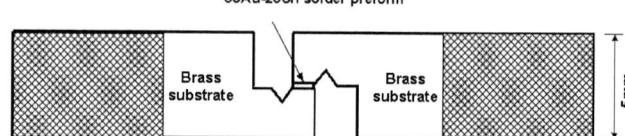

(a) Lap-shear specimen for creep test

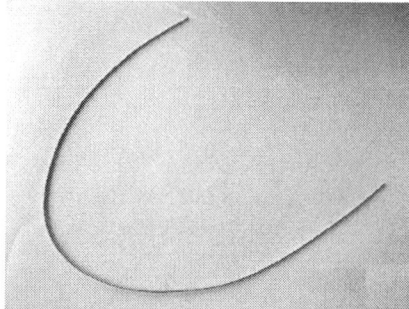

(b) Ribbon perform specimen for creep test

(c) Micro-tester for the creep test

Fig. 1 Specimen design and micro-tester used in the experimental investigation

Advantage of this specimen is that it can mimic the real case, some practical factors that may affect the creep test results can be included. For example, reflow process induced change to microstructure and thus mechanical properties of solder; the existence of intermetallic compound between

978-1-4244-4658-2/09 $25.00 © 2009 IEEE

solder and brass substrate, etc. However, void is usually trapped in the reflowed solder and this will affect the testing results more seriously.

B. Ribbon specimen

Ribbon perform specimen with width of 1.0mm and thickness of 0.4mm was also used for creep test. In each test, a section of ribbon (about 15mm) was set in a micro-tester, and a specially designed fixture was used to hold the specimen and prevent it from slipping. Gauge length of this specimen was set to 5.0mm.

In this paper, all experimental results of creep test were based on the ribbon specimen. Creep test with the lap-shear specimen is still on going for future comparison.

Microstructure of the 80Au-20Sn specimen was shown in Fig. 2(a). EDX analysis showed a high Sn content of 35-40 wt.% in the dark domain while the light domain has a smaller Sn content of 10-12wt.%. The dark and light domains can be confirmed to be δ and ζ phases with the phase diagram as shown in Fig. 2(b).

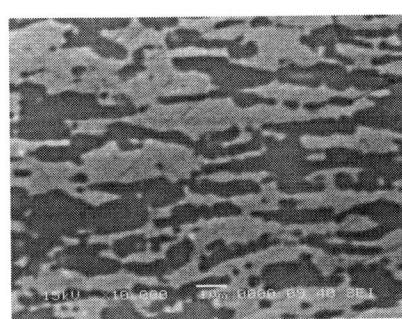

(a) Microstructure of 80Au-20Sn solder before reflow (ζ and denote different phases)

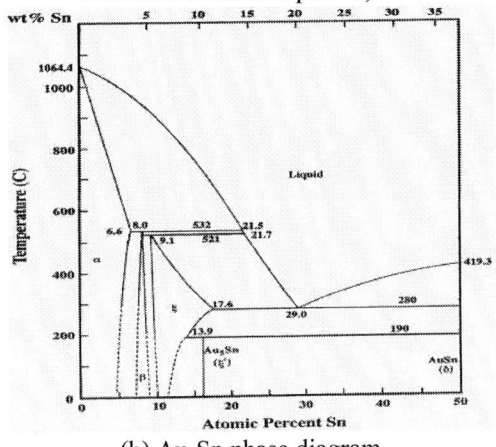

(b) Au-Sn phase diagram

Fig.2 Microstructure and binary phase diagram of the 80Au-20Sn solder

2.2 About the Micro-tester and Test Conditions

A micro-tester with thermal chamber was employed to perform the creep test, a sensitive extensometer was used to measure the specimen extension. Technical parameters of the testing system are listed as following:

Load cell: two load cells with full scale of 10N and 250N respectively.

Resolution: 0.1% full scale of the load cell (i.e. 10mN and 250mN respectively); 10nm for the extensometer.

Temperature : up to 250°C

With the micro-tester, creep tests were carried out for three different temperatures (i.e. 25°C, 75°C and 125°C) at different stress levels shown in Table 1.

3 Experimental Results

The experimental results of steady state creep strain rate versus applied normal stress on 80Au-20Sn solder specimen are given in Fig. 3(a). The steady state creep strain rate will increase with temperature and applied normal stress.

By comparing with the experimental results of 95.5Sn-3.8Ag-0.7Cu reported by Pang et. al [9], one can find that much higher stress level is needed for 80Au-20Sn to achieve the same strain rate at the same temperature. For example, about 5 times of stress level is needed at 75°C to achieve a strain rate of $1.0e^{-5}$. So, creep resistance of 80Au-20Sn is much higher than that of 95.5Sn-3.8Ag-0.7Cu, and furthermore, most of the reported lead free solders.

Table 1 Experimental conditions for creep test of 80Au-20Sn

Stress levels (MPa)	25°C	75°C	125°C
5			√
7.5			√
10			√
12.5			√
15			√
16.7			√
25		√	√
37.5		√	
50		√	
60		√	
75		√	
87.5		√	
100	√	√	
125	√		
150	√		
175	√		
200	√		
250	√		
300	√		

4 Creep Constitute Models

Based on various assumed creep mechanisms, some creep constitute models are developed to characterize the relationship between steady state creep strain rate and temperature and stress levels. Among these models, the Hyperbolic-Sine model as stated below is widely used.

$$\dot{\varepsilon} = C[\sinh(\alpha\sigma)]^n \exp(-\frac{Q}{kT}) \qquad (1)$$

where $\dot{\varepsilon}$ is steady state creep strain rate, C and α are material constants, σ is applied stress, n is the stress exponent, Q is creep activation energy, k is Boltzmann's constant, and T is absolute temperature.

The experimental data of creep test for 80Au-20Sn solders shown in Fig. 3(a) was fitted to Eq. (1). Through the curve fitting technique, the creep constitute model of 80Au-20Sn was determined as in Eq. (2)

$$\dot{\varepsilon} = 8.66 \times 10^{12} \left[\sinh(0.0062\sigma) \right]^{2.17} \exp\left(-\frac{14292.7}{T} \right) \quad (2)$$

All constants in Eq. (1) are thus determined as well. Comparison between the 'theoretical' and experimental results is shown in Fig. 3(b). In general, error between these two results lies within 15%.

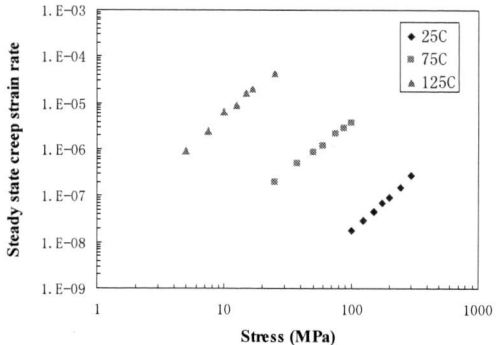

(a) Experimental results of 80Au-20Sn creep test

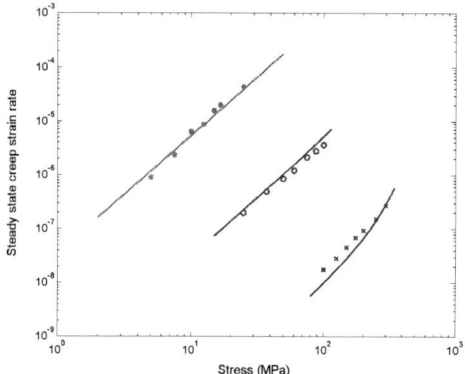

(b) Steady state creep strain rates by Hyperbolic-Sine model and its comparison with experimental results

Fig. 3 Creep behavior of 80Au-20Sn solder

Activation energy is an inherent characterization of material that describes the easiness of creep. Usually, it's high at low temperatures and low at high temperatures. In Eq. (1), the activation energy is defined over a wide range of temperatures and it is only meaningful in average sense. In Eq. (2), this value is determined as 118.7 KJ/mol for 80Au-20Sn over the temperature range of 25-125°C by curve fitting technique. To estimate the activation energy of 80Au-20Sn at 75°C which is average of the experimental temperatures and compare it with the one from curve fitting, a differential creep test was designed: creep test was performed at 75°C for some time first; then temperature was suddenly increased by dT=10°C. The experimental results are shown in Fig. 4, where

a steady state strain rate change of $d\dot{\varepsilon} = 9.1 \times 10^{-6}$ was found. With these data, activation energy of 80Au-20Sn at 75°C can be estimated by:

$$d\dot{\varepsilon} = C\left[\sinh(\alpha\sigma)\right]^n d\left[\exp\left(-\frac{Q}{kT}\right)\right]$$

$$= C\left[\sinh(\alpha\sigma)\right]^n \exp\left(-\frac{Q}{kT}\right) d\left(-\frac{Q}{kT}\right)$$

$$= \dot{\varepsilon} \times \frac{Q}{kT^2} dT$$

$$Q(T) = \frac{d\dot{\varepsilon}(T) \times kT^2}{\dot{\varepsilon}(T) \times dT}$$

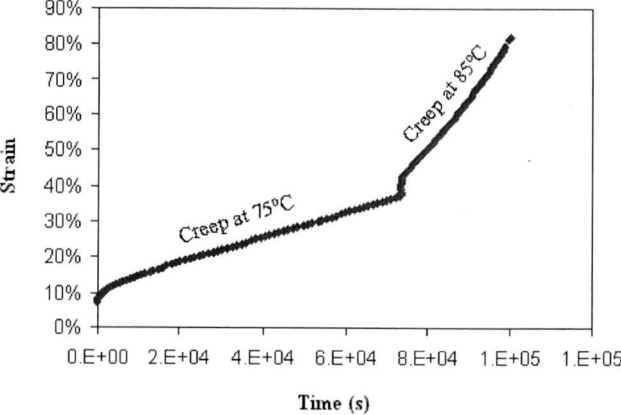

Fig. 4 Creep test with temperature changing from 75°C to 85°C Suddenly

With above equations, activation energy of 80Au-20Sn at 75°C is determined as 254.3 KJ/mol. Compared with the reported activation energy of Sn-Ag-Cu (45-54 KJ/mol) [9], the activation energy of 80Au-20Sn is much higher.

Conclusions

Characterization to creep behavior of 80Au-20Sn lead free solder was performed at different temperatures and stress levels. With the experimental data and curve fitting technique, creep constitute model of 80Au-20Sn in the form of Hyperbolic-Sine function was determined. Activation energy of 80Au-20Sn was experimentally studied further and compared with that of Sn-Ag-Cu lead free solders, the experimental results quantitatively revealed the high creep resistance of 80Au-20Sn solder.

Acknowledgments

The authors would like to express their appreciations to Dr. Xiong Bingshou from Nanyang Technological University (Singapore) and Dr. Dudek Rainer from IZM (Germany) for their helpful discussions and suggestions. Also, the support of "Fanzhou" founding (No. 2007507) was appreciated.

References

1. Kilinski T. J, Lesniak JR, Sandor BI, "Modern Approaches to Fatigue Life Prediction of SMT Solder Joints," Lau JH, editor, Solder Joint Reliability: Theory and Applications, (New York, 1991) pp.384-405.
2. Pao Y. H., "A Fracture Mechanics Approach to Thermal Fatigue Life Prediction of Solder Joints," *IEEE Transaction Components, Hybrids, and Manufacturing Technology,* Vol. 15, No. 4, (1992), pp.559-570.

3. Edwin Bradley *et. al.*, NEMI Pb-free Task Group Report, *2002 APEX Free Forum National Electronics Manufacturing Initiative*, INC. January 23, 2002.

4. Toyoda Y., "The Latest Trends in Lead-Free Soldering," *4th International Symposium on Electronic Packaging Technology*, Beijing, China, August. 2001

5. Igoshev V.I. and Kleiman J.I., "Creep Phenomena in Lead-Free Solders," *Journal of Electronic Material*, Vol. 29,No.2 (2000), pp.244-250.

6. Neu R.W., Scott D.T. and Woodmansee M.W., "Thermomechanical behavior of 96Sn-4Ag and Castin Alloy", *ASME Transactions, Journal of Electronic Packaging*, Vol. 123, No.3, (2001), pp. 238-246.

7. Plumbridge W. J., Gagg C. R. and Peters S., "The Creep of Lead-free Solders at Elevated Temperature", *Journal of Electronic Materials*, Vol. 30, No.9(2001).

8. Guo F., Lucas J. P. and Subramanian K. N., "Creep behavior in Cu and Ag particle-reinforced composite and eutectic Sn-3.5Ag and Sn-4.0Ag-0.5Cu non-composite joints," *Journal of Materials Science: Materials in Electronics,* Vol. 12(2001), pp. 27-35.

9. Pang H. L John., Xiong B. S. and Low T. H., "Creep and Fatigue Properties of Lead Free Sn-3.8Ag-0.7Cu Solder," *IEEE Proceedings of 2004 Electronic Components and Technology Conference*, June 1-4, 2004, pp.1333-1337.

10. Fujiwara K., Method of mounting a semiconductor laser device, in U.S. Patent. Nov 30, 1982.

11. Kim W., Wang Q., Jung K., Hwang J. and Moon C. "Application of Au-Sn eutectic bonding in hermetic RF MEMS wafer level packaging". *Proceedings of 9th Int'l Symposium on Advanced Packaging Materials. IEEE*, 2004. pp. 215-219.

12. Pittroff W., Barnikow J., Klein A., et al. "Flip chip mounting of laser diodes with Au/Sn solder bumps: bumping, self-alignment and laser behavior". *Proceedings of Electronic Components and Technology Conference, IEEE*, 1997. pp. 1235-1241.

13. Zakel E. and Reichl H., Flip chip assembly using the gold, gold-tin and nickel-gold metallurgy. ed. Lau. J., McGraw-Hill(New York, May 1995).

14. Lau J., Solder Joint Reliability: Theory and Application. Van Nostrand Reinhold, (New York, 1997).

The Warpage Control Method in Epoxy Molding Compound

Wei Tan, Fang Zhou, Xingming Cheng, Dong Ding, Juan Wu
Henkel Huawei Electronics Co., Ltd.
Songtiao Industry Park, Lianyungang, Jiangsu, China, 222006
Wei.Tan@cn.henkel.com

Abstract

This paper is to study the warpage impact of filler content and special additives on BGA packages. The special additives have significant impact on the warpage. The additive A has great influence on the warpage at different temperature; the warpage will flatten when more additive A is added. From the warpage study of single unit from room temperature to 260°C, all the warpage is in 50um when additive A is about 1%. The additive A can flatten the warpage of the package and keep the warpage at same level even with different silicon occupation in the package. The filler content shows great impact on the warpage by control the shrinkage after cured, the lower shrinkage; the package will move to more crying.

Introduction

In recent years, the plastic ball grid array (PBGA) has become one of the most popular packaging alternatives for high I/O devices in the industry due to its special features such as high input/output, low-profile, high heat dissipation, high electrical performance, and assembly self-alignment. However, coplanarity of the package which is called warpage is one of important issues related to the solder ball reliability. However, the coplanarity issue would become more severe, the package is getting larger and the temperature of solder reflow is increasing (due to the application of lead-free solder). [1]

There are many materials in the BGA that have effect on the BGA's warpage such as molding compound, die ,die attach and substrate properties.

Generally, the warpage is caused by the different coefficient of thermal expansion between molding compound and die, die attach in the BGA packages. From the molding compound side, there are many materials in the molding compound have effects on the warpage.

Experiment

The molding compounds were prepared by dry blending the ingredients, melt-mixing, extrusion into a sheet form, fine-grinding into powder, then palletizing into pre-forms.

Standard transfer molding techniques were used to fabricate various test specimens for DMA, shrinkage, and molded array strips for warpage. In mold cure time was 120sec at 175°C followed by a four-hour post mold cure at the same temperature for all test parts. Shrinkage was measured on 5x0.5x0.25 inch bar. Tg and modulus were measured by DMA, a TA instrument model, from 25°C to 280°C at 3°C per minute. Molded array strips were prepared by using a 4 cavity molds on 0.2mm thick BT laminate substrate with 6mm*6mm die on 12mm*12mm package which is shown is figure 1.

Warpage measurements were made by the highly sensitive Shadow Moire' method. The positive number is smile face as molding compound upside.

The single unit warpage is referred to warpage of after sawed from the block. The unit is heated to the different temperature on the shadow moiré to determine the warpage at different temperature.

Figure 1 The warpage test vehicle

Results and discussion

The effect of different filler content on the unit warpage at different temperature was shown in Figure 2.

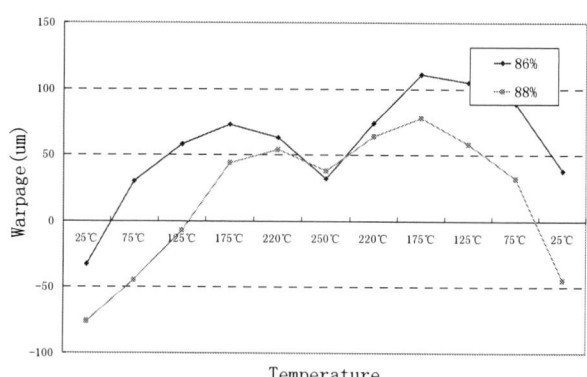

Figure 2 The effect of filler content on single unit warpage from 25°C to 260°C

It can be found that filler content shows great impact on the warpage at different temperature. The warpage will move to more crying when more filler is added to the molding compound. The 88% filler content has less warpage variation at different temperature than 86% filler content which is due to the big different shrinkage between 88% and 86% filler content which is 0.02 and 0.22 separately.

The effect of the four special additives on the unit warpage at different temperature is shown at Figure 3. The filler content of molding compound is 86%. The different additive has great impact on the warpage. The warpage will flatten when more additive A added. From the warpage study of single unit from room temperature to 260°C, all the warpage is in 50um when additive A is about 1%. The warpage when use additive B is much smiling , when add the additive C and D in the molding compound, the unit warpage at different temperature has great variation and much larger.

978-1-4244-4658-2/09 $25.00 © 2009 IEEE

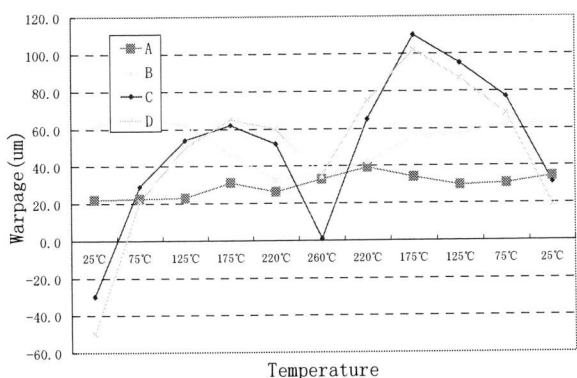

Figure 3 The effect of special additive on the single unit warpage from 25°C to 260°C

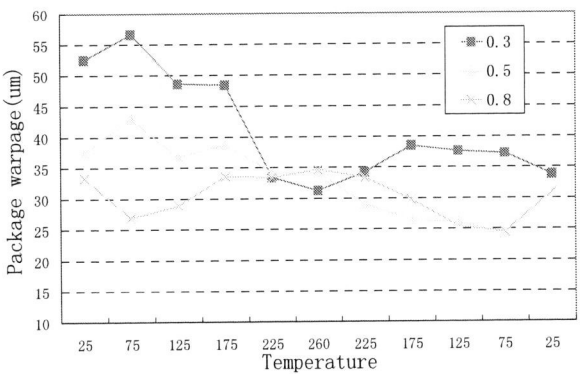

Figure 4 The effect of amount of additive A and additive B on the single unit warpage from 25°C to 260°C

The shrinkage, modulus and Tg of molding compound which use additive A,B,C and D are shown in Table 1.

Table 1 the basic properties of molding compound with different special additive

Elastomer	A	B	C	D
filler content	86	86	86	86
Shrinkage %	0.18	0.2	0.22	0.15
Storage Modulus at 260°C(MPa)	980	1000	890	920
Tg C (Tan d)	178	180	175	168

From the Table 1, it can be concluded that the four special additives and basic properties of molding compound have no significant difference. That's mean the different warpage performance is caused by the special additive.

The warpage of molding compound was to be tested at room temperature to 260°C with a different amount (0.8%, 0.5%, 0.3%) of special additive A was shown in Figure 4. From the Figure 4, it can be found that, with decrease of the special additive A's amount, the warpage move to more smiling, the variation of warpage between high temperature and low temperature also increased.

Table 2 The basic properties of molding compound with different amount of special additive A

Amount of additive A	0.3	0.5	0.8
filler content	86	86	86
Shrinkage %	0.21	0.20	0.18
Storage Modulus at 260°C(MPa)	1230	1150	980
Tg (tanδ) /°C	175	180	178

The basic properties of the molding compound are shown in Table 2. From the Table 2, it can be found that the higher content of special additive A, the lower storage modulus of molding compound at 260°C, the Tg of tanδ and shrinkage are similar. According above analysis, the good and stable warpage performance of molding compound is caused by the special additive A.

The effect of additive A with 0.3%, 0.5% and 0.8% amount in the molding compound on the block warpage which has large die (58% silicon occupation) and small die (28% silicon occupation) is shown in Figure 5 (after molding) and in Figure 6 (after PMC for 4 hours).

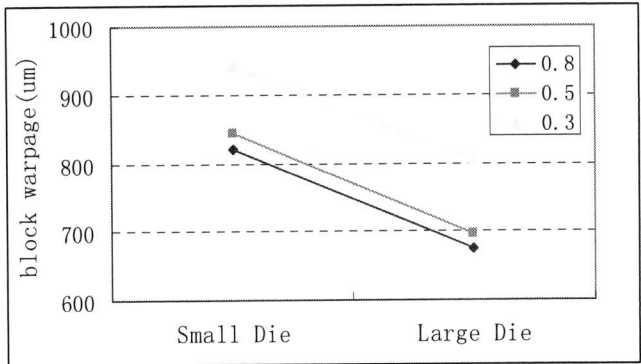

Figure 5 The effect of additive A on different silicon occupation in the package after molding

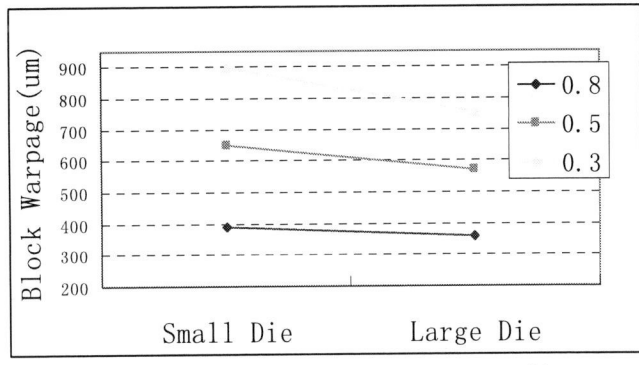

Figure 6 The effect of additive A on different silicon occupation in the package after PMC

From the Figure 5 and Figure 6, special additive A can achieve good balance between large and small silicon occupation. After PMC for 4 hours at 175°C, adding 0.8 percent special additive A can keep very flatten warpage even with large and small silicon occupations.

It is also can be found that more special additive A added, the more crying the warpage is.

Conclusions

1. Filler content has a significant effect on the warpage, the higher filler content, the warpage will move to crying (molding compound upside), and the variation of warpage at different temperature is lower.

2. Different special additives have different effects on the warpage. Among the four special additives used in this study, the additive A has best performance on warpage even at different temperature and different silicon occupation. All the warpage can be in 50um from room temperature to 260°C.

3. When adding 1% of special additive A, the warpage can keep at same level with different silicon occupation.

Future work

Key mechanism of the effect of special additive on warpage is required for further studies. How to balance the molding compound properties to suit for different package design with one compound should be studied.

References

1. H. Lau, C. P. Wong, N. C. Lee, and S. W. Ricky Lee, Electronics Manufacturing with Lead-Free Halogen-Free, and Conductive-Adhesive, Handbooks, McGraw-Hill, (2003).

On Reliability of Chips bonded on Flex Substrates Using Thermosonic Flip-Chip Bonding Process with Nonconductive Paste

Cheng-Li Chuang[1], Jong-Ning Aoh[2], Wei-How Chen[2]
[1]Department of Occupational Safety and Health, Chung Shan Medical University, Taiwan
[2]Department of Mechanical Engineering, Chung Cheng University, Taiwan
*E-mail address: luke@csmu.edu.tw

Abstract

The purpose of this study is to verify the reliability of chips bonded on flex substrates using thermosonic flip-chip bonding process with a non-conductive paste (NCP). High temperature storage (HTS) test, temperature cycling test (TCT), pressure cooker test (PCT) and high temperature/high humidity (HT/HH) test were conducted to investigate the reliability of chips bonded on flex substrates. The environmental parameters for reliability tests were complied with the JEDEC standards. After various reliability tests, the peeling test and microstructure observation on tested specimen were performed to evaluate the reliability.

The bonding strength increased with increasing the storage durations of HTS test. The HTS test provided sufficient thermal energy to promote atomic interdiffusion between Au bumps and Cu electrodes. A metallurgical bonding between Au bump and the Cu electrode was formed, and the bonding strength is thus improved. The bonding strength of chips and flex substrates assembly without applying ultrasonic in bonding process was decreased with increasing the storage durations of PCT. The typical failure for PCT was the interfacial delamination between NCP and flex substrates. Approximately 80% specimen exhibited fully separated after PCT at 336 h, implying the NCP cannot withstand the PCT test and lost its adhesion strength. The mean bonding strength of chips and flex substrates assembly with an ultrasonic power of 14.46 W in bonding process slightly varied with increasing storage durations of PCT, and standard deviation of bonding strength increased dramatically in the range of storage durations from 196 h to 336 h. Although the adhesion strength of NCP decreased, a part of Au bumps well bonded on Cu electrodes as the storage duration increased to 336 h. Applying the adequate ultrasonic power to bonding process was not only to improve the bonding strength, but also the bonding strength could be maintained in high level after PCT. There was no significantly change in bonding strength for chips bonded on flex substrates after TCT test. It shows that specimen has great reliability for TCT test. The bonding strength increased with increasing storage durations of HT/HH test. Neither cracks nor defects at boding interface are observed.

The reliability for chips bonded the flex substrate using thermosonic flip-chip bonding process with NCP meets the requirements stated in JEDEC specifications, exception of the adhesion strength of NCP for PCT should be improved.

Introduction

Adhesives have been widely used in microelectronics packaging for chips and substrates assembly [1-3]. The flip-chip bonding process with adhesives provides several advantages compared to conventional flip-chip bonding with solder bumps. It meets the environmental requirement, since the adhesive was lead free. There was no under-fill was required after flip-chip bonding with the adhesive, which is possible to reduce the manufacturing cost. The bonding strength of chips onto substrates was derived from the cured strength of adhesives. In our previous study [4], the thermosonic flip-chip bonding with NCP has been successfully applied to chips onto flex substrates. The ultrasonic power play an important role in scraping the NCP away from the surface of Cu electrodes, and then Au bumps directly bonded onto Cu electrodes to form a successful electrical path between chips and the substrates. With an appropriate value of ultrasonic power not only to scrape NCP away from the surface of Cu electrodes, but also provides a part of thermal energy to form the metallurgical bonding between Au bumps and Cu electrodes. Thus, this bonding technology is expected to offer several distinct advantages and enables achieving the low cost and excellent performance comparing with existed schemes.

Several studies investigated the reliability of chips onto substrates using adhesives [5-6]. Teh et al [5] verified the reliability of chips bonded on the rigid substrates using thermal compression bonding process with NCP. The experimental results indicated that NCP can easily pass the requirements of TCT and HTS test. However, most of specimen appeared interfacial delamination and opening between bumps and electrodes after reliability test of MST and PCT. Both MST and PCT failures were contributed to excessive stress generated due to moisture vaporization at high temperature. Similar experimental results were obtained for chips onto flex substrate using NCP after reliability tests [6]. The delamination appeared at interface between chips and the flex substrates after PCT. Up to now, no published literature was found to examine the reliability of chips and flex substrates assembly using thermosonic flip-chip bonding process with NCP.

The objective of this study is to verify the reliability of chips onto flex substrates using thermosonic flip-chip bonding with NCP. The HTS test, TCT, PCT and HT/HH test were conducted to investigate the reliability of chips bonded on flex substrates. The environmental parameters for reliability tests were complied with the JEDEC standards. After reliability tests, the peeling test and microstructure observation on tested specimen were performed to evaluate the reliability.

Experimental methods

A thermosetting type of commercial NCP was used in this study. The non-conductive paste was deposited on the surface of flex substrates, and a chip with eight Au bumps to be bonded onto Cu electrodes over the flex substrates using thermosonic flip-chip bonding process. The flip-chip bonding experiments were performed using an automatic thermosonic flip-chip bonder developed by the Industrial Technology Research Institute (ITRI). The deposited layers of bond pads and Cu electrodes were stated in our previous work [7]. To evaluate the effect of ultrasonic power on the reliability of chips bonded on flex substrates, two kinds of bonding process for chips and flex substrates assembly were performed in this study. The first bonding process is applied an ultrasonic power of 14.46 W to flip-chip bonding process with NCP for chips and flex substrates assembly. The other was no ultrasonic power applying in flip-chip bonding process to simulate the thermal compression bonding process with the NCP. Other bonding parameters are fixed at a bonding force of 10 N, a preheating temperature of 80°C, a curing temperature of 140°C and a cured time of 40 s.

After chips and flex substrates assembly, the specimen was subjected to various reliability tests, including HTS test, TCT, PCT and HT/HH test. The storage durations and tested parameters were complied with the JEDEC standards [8-11], as shown in Table 1. The peeling test was conducted to evaluate the bonding strength after the reliability tests.

Table 1 Lists of storage durations and environmental parameters for various reliability tests. [8-11]

	HTS	HT/HH	TCT	PCT
Test Conditions	+150 °C	+85°C/85% RH(no bias)	+125 °C /- 55 °C	+121 °C/ 100%RH 2atm
Read Point	200,400,60 0,800,1000 (hrs)	200,400,60 0,800,1000 (hrs)	100,300,50 0,800,1000 (cycles)	24,48,96,1 68,240,336 (hrs)
Test Duration	1000 hrs	1000 hours	1000 cycles	336 hs

Optical microscopy (OM) and scanning electron microscopy (SEM) were conducted to examine changes at bonding interface after various reliability tests. The fracture mode and the fracture mechanism after peeling test also verified using SEM and OM. Line scanning of the field-emission Auger electronic spectroscopy (FEAES) was used to determine the atomic inter-diffusion between Au bumps and Cu electrodes after reliability tests.

Results and discussion

The effects of ultrasonic power on the bonding quality

To identify the effects of the ultrasonic power on the bonding quality for chips and flex substrates assembly, the applied ultrasonic power to bonding process was set at 0 W and 14.46 W. And other bonding parameters were fixed, 10N in bonding force, 80°C in preheat temperature, 140°C in cured temperature and 40 s in cured time. Figure 1(a) shows an un-deformed Au stud bump and a layer of NCP existed in bonding interface between the Au bump and the Cu electrode when the applied ultrasonic power was zero. This is an inactive interconnect between Au bumps and Cu electrodes since the Au bump cannot be directly contact with the Cu

electrode to form an electrical path. In contrast to Au bumps bonded onto Cu electrodes without applying ultrasonic power, the Au bump directly bonded on the Cu electrode to form a conductive path between chips and the flex substrates when the ultrasonic power of 14.46 W was applied to bonding process, as shown in Fig. 1(b). This experimental result indicates that NCP was scraped away from the surface of Cu electrodes by ultrasonic power during thermosonic flip-chip bonding process, and Au bumps thus bonded on Cu electrodes. An electrical path between chips and flex substrates can be achieved. Applied an adequate value of the ultrasonic power is an effective way to increase yields of the successful interconnects between chips and flex substrates.

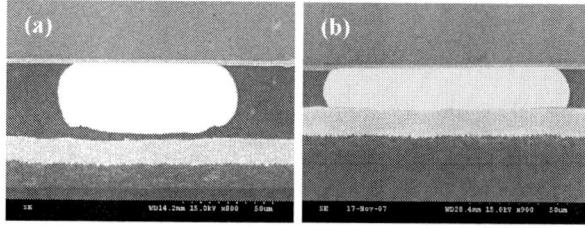

Fig. 1 SEM micrographs of (a) cross-section of chip bonded on the flex substrate without application of ultrasonic power, (b) cross-section of chip bonded on the flex substrate with an ultrasonic power of 14.46 W in flip-chip bonding process.

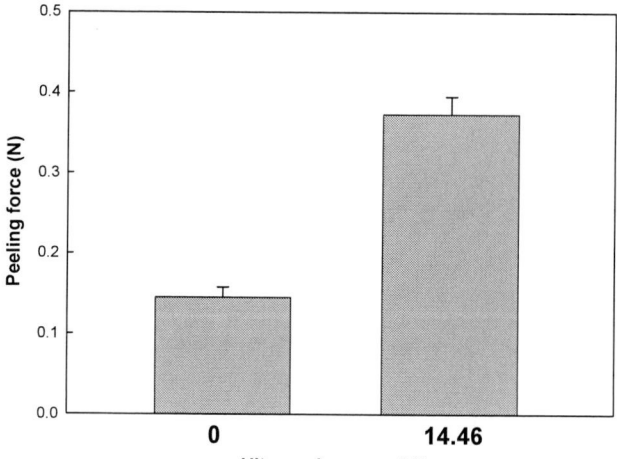

Fig. 2 The effect of ultrasonic power on the bonding strength of chips and flex substrates assembly.

The influence of the ultrasonic power on bonding strength of chips and flex substrates assembly was shown in Fig. 2. A higher bonding strength can be obtained when applied an ultrasonic power of 14.46 W to flip-chip bonding process with NCP than that flip-chip bonding without providing ultrasonic power. The increased bonding strength is assumed that was contributed to metallurgical bonding formation between Au bumps and Cu electrodes. After peeling test for chips and flex substrates assembly without applying ultrasonic power in bonding process, the fractured interface appeared a smooth surface without any scraping or any Au

978-1-4244-4658-2/09 $25.00 © 2009 IEEE

bump residual, indicating fractured trace were propagated along between the NCP and Cu electrodes. Thus, the specimen of chips and flex substrates assembly without applying ultrasonic power in bonding process was designed to simulate the thermal compression bonding process with NCP. The peeling test after reliability test for this specimen was used to verify the adhesion strength of NCP.

High temperature storage test

Figure 3 displays the effect of storage durations on the bonding strength of chips and flex substrates assembly. No significantly change in bonding strength was observed for chips and flex substrates assembly without applying ultrasonic power in bonding process. The cured level of NCP was approximately 100% and no aging was found, the NCP thus maintained a steady adhesion strength for HTS test at storage durations varied form 0 h to 1000 h, indicating the NCP is able to withstand the HTS test. For chips and flex substrates assembly with an ultrasonic power of 14.46W, the bonding strength increases with increasing storage durations, revealing a good reliability. After HTS test at storage duration of 200 h, the fractured morphology of chips and flex substrates was shown in Fig. 4. A layer stuck on Au bump over the chip and a concavity was formed on the surface of flex substrates, as shown in Fig. 4(a) and 4(b). The dispersive spectrometer (EDS) was used to verify the composition of the layer stuck on Au bump. Figure 4(c) and 4(d) show morphology of the layer stuck on Au bump and the EDS spectrum, respectively. Only Cu peak was observed in the EDS spectrum, indicating that layer stuck on Au bump is Cu electrode. This analytical result indicates that the Cu layer was transferred to Au bump after peeling test. The bonding strength of Au bumps and Cu electrode are even higher than adhesion strength between Cu electrodes and polyimide substrate. A similar fractured morphology of chips and flex substrates was found after HTS test at storage duration of 1000 h. A sheet of Cu layer peeled off from the interface between Cu electrodes and the polyimide substrate, and the Cu layer transferred to the surface of chips (Fig. 5(a)), a concavity was thus formed on the fractured surface of flex substrates, as shown in Fig. 5(b). The observation on fractured morphology after peeling test consists with the changes of bonding strength at various storage durations, as shown in Fig. 3.

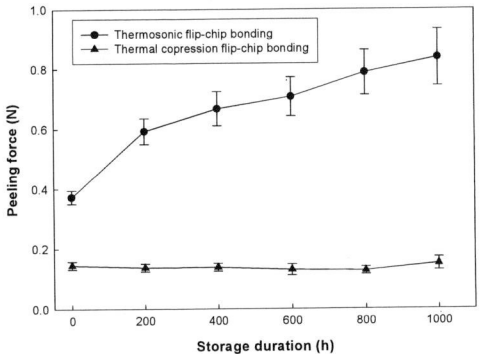

Fig.3 Relationships between bonding strength and storage durations of HTS for chips bonded on flex substrates heated at 150°C.

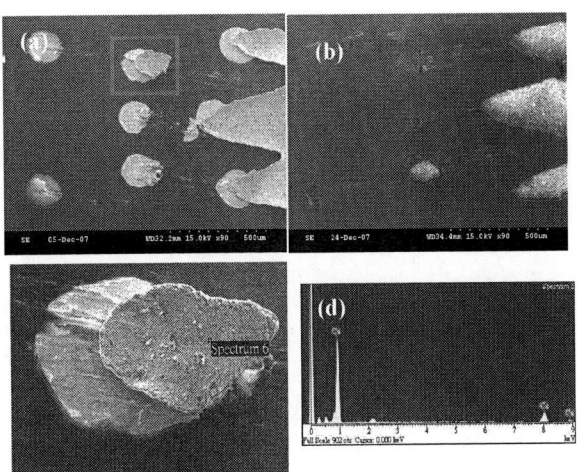

Fig. 4 SEM micrographs show (a) the fractured morphology of chips after peeling test, (b) the fractured morphology of flex substrates after peeling test, (c) a layer bonded on Au bump shown in (a) with a larger magnification, (d) the EDS analytical spectrum. The peeling test was performed after HTS test at storage

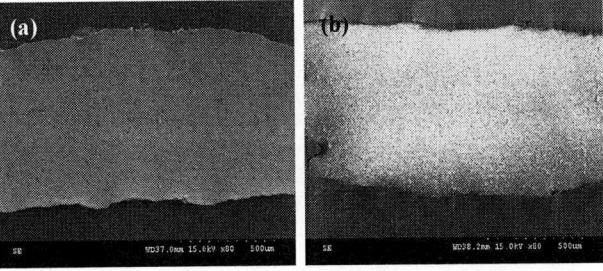

Fig. 5 SEM micrographs show (a) the fractured morphology of chips after peeling test, (b) the fractured morphology of flex substrates after peeling test. The peeling test was performed after HTS test at storage duration of 1000 h.

To investigate the possible change at bonding interface after HTS test for 200 h and 1000 h, the interface between Au bumps and Cu electrodes was examined using SEM. Neither delamination nor other defects were found at bonding interface as shown in Fig. 6(a) and Fig. 6 (b). A line scanning of FEAES was conducted to determine atomic interdiffusion between Au bump and the Cu electrode. A narrow atomic interdiffuion trace was observed at bonding interface for an as received specimen, as shown in Fig. 7. The atomic interdiffusion for an as received specimen could be attributed to thermal energy from cured process, 140°C for 5 s. In contrast to the narrow atomic interdiffusion trace for an as received specimen, the broadened atomic interdiffusion trace is observed after HTS test at 1000 h, as shown in Fig. 8. The atomic interdiffusion was promoted by the high temperature and storage durations, thus, resulted in increased bonding strength. This analytical result can be used to interpret the bonding strength increased with the storage durations of HTS test, as shown Fig. 3. Applying ultrasonic power to flip-chip bonding process with NCP for chips and flex substrates assembly, not only the bonding strength was improved, but also the reliability of HTS test can be guaranteed.

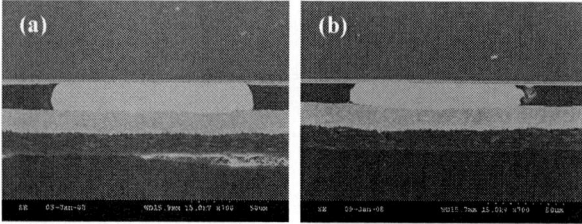

Fig. 6 SEM micrographs show (a) the cross section of Au bump bonded on the copper electrode after HTS test at storage duration of 200 h, (b) the cross section of Au bump bonded on the copper electrode after HTS test at storage duration of 1000

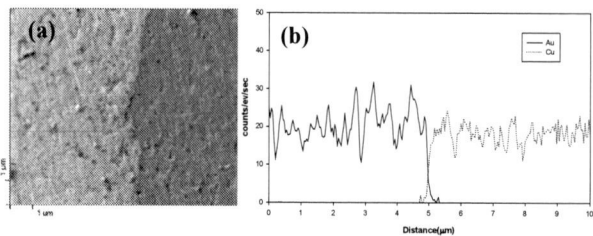

Fig. 7 Line scanning results on the cross section of Au bump and Cu electrode interface. (a) Cross section on interface between Au bump and Cu electrode, (b) atomic interdiffusion of Au bump and Cu electrode

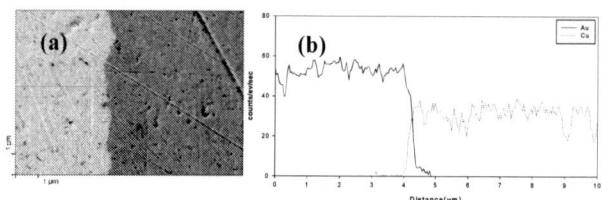

Fig. 8 Line scanning results on the cross section of Au bump and Cu electrode interface. (a) Cross section on interface between Au bump and Cu electrode after HTS test at storage duration of 1000 h, (b) atomic interdiffusion of Au bump and Cu electrode after HTS test at storage duration of 1000 h.

Temperature cycling test

The effect of TCT on the bonding strength of chips and flex substrates assembly was shown in Fig. 9. The variation of bonding strength for chips and flex substrates assembly without applying ultrasonic power in flip-chip bonding process after TCT at 1000 cycles is very small. The adhesion strength of NCP has a stable performance after TCT at 1000 cycles, which could be met the requirements stated in the JEDEC specifications. After TCT, the bonding strength of chip and flex substrates assembly with applying an ultrasonic of 14.46 W in flip-chip bonding process also varied in a narrow zone. To investigate the possible change in bonding

interface after TCT at 100 cycles and 1000 cycles, the cross-section of Au bumps bonded on Cu electrodes was examined using SEM. As shown in Fig. 10, the Au bump firmly bonded on Cu electrode and no defects was found at bonding interface. This observation on bonding interface could be used to explain why bonding strength has a good reliability for TCT at 100 cycles and 1000 cycles, as shown in Fig. 9.

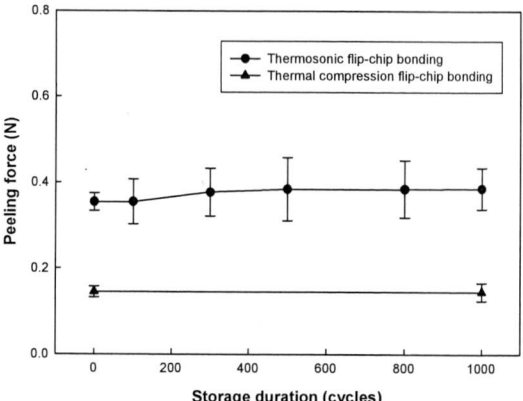

Fig. 9 Relationships between bonding strength and storage durations of TCT for chips bonded on flex substrates heated at 125°C/-55°C.

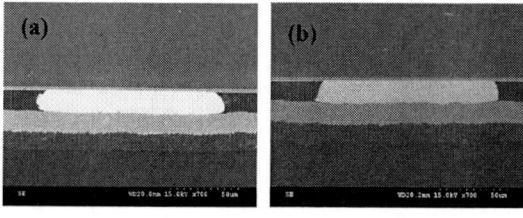

Fig. 10 SEM micrographs show (a) the cross section of Au bump bonded on the copper electrode after TCT at storage duration of 100 cycles, (b) the cross section of Au bump bonded on the copper electrode after TCT at storage duration of 1000 cycles.

Pressure cooker test

To investigate the reliability of chips and flex substrates assembly after PCT, the chips together with flex substrates were subjected to a highly humid (100% RH), a highly pressure (2 atm) and a highly temperature (150°C) for various storage durations. The influence of PCT on bonding strength of chips and flex substrates assembly is illustrated in Fig. 11. The bonding strength of chips and flex substrates assembly without applying ultrasonic in bonding process was decreased with increasing the storage durations. The typical failure for PCT was the interfacial delamination between NCP and flex substrates. Approximately 80% specimen exhibited fully separated after PCT at 336 h, as shown in Fig. 12. The adhesion strength of NCP was deteriorated after PCT. This phenomenon could be used to explain the bonding strength of chips and flex substrates assembly without applying ultrasonic power in bonding process decreased with

978-1-4244-4658-2/09 $25.00 © 2009 IEEE   652

increasing storage durations of PCT. This experimental result also indicates that adhesion strength of NCP for PCT should be improved. The mean bonding strength of chips and flex substrates assembly with an ultrasonic power of 14.46 W in bonding process slightly varied with increasing storage durations of PCT, and standard deviation of bonding strength increased dramatically in the range of storage durations from 168 h to 336 h, as shown in Fig. 11. To find out the changes at bonding interface between chips and flex substrates, the SEM was used to examine the cross section of Au bumps bonded on Cu electrodes. The Au bump well bonded on the Cu electrode, no delamination or others defects were found at bonding interface after PCT for 24 h and 96 h. as shown in Fig. 13. The slightly decreased bonding strength in the range of storage duration from 24 h to 96 h could be contributed by the lost adhesion of NCP. As storage duration extended to 168 h, a delamination was found at bonding interface between NCP and the chip, as shown in Figs. 14(a) and 14(b). However, a sound bonding interface can be found at interface between Au bump and the Cu electrode, as shown Fig. 14(c). The delamination also was found at chip edge and a crack also appeared at bonding interface between Au bump and the Cu electrode when the storage duration of PCT was increased to 240 h, as shown in Figs. 15(a) and 15(b), respectively. Similar defects of delamination were also found at bonding interface of NCP/flex substrate and NCP/chip when storage time increased to 336 h, as shown in Figs. 16(a) and 16(b). Although the NCP lost its adhesion and delamination was found at bonding interface after PCT at storage duration of 336 h, a part of Au bumps well bonded on the Cu electrodes, as shown in Fig. 16(c). The bonding strength of chips bonded on flex substrates using thermosonic flip-chip bonding process with NCP were attributed to cured strength of NCP and the bonding strength of metallurgical bonding between Au bumps and Cu electrodes. As mentioned in the observation on the cross section of chips bonded on flex substrates, the adhesion of NCP decreased and the delamination were found at bonding interface of NCP/flex substrate and NCP/chip when storage durations of PCT were increased from 168 h to 336 h. However, a part of Au bumps well bonded on Cu electrodes as the storage duration increased to 336 h. The bonding strength after PCT at various storage durations was attributed by the interaction of the NCP lost its adhesion strength and the increased bonding strength for metallurgical bonding formed between Au bump and Cu electrode. Theses observations on the changes of bonding interface after PCT at various storage durations could be used to interpret the standard deviation of bonding strength dramatically increased with increasing the storage duration from 168 h to 336h, as shown in Fig. 11. Figure 17 shows the fractured morphology of the chips after PCT at various storage durations. The fractured morphology of the chip has a good shine after PCT at storage duration 24 h, as shown Fig. 17(a). With increasing the storage durations of PCT, the color of chips gradually transformed into dark, as shown in Figs. 17(b) and 17(c). The phenomenon indicates that NCP cannot withstand high moisture under high-pressure condition after PCT for extending storage durations, and delaminations were found at bonding interface of NCP/chip and NCP/flex

substrate. The moisture penetrated to bonding interface between NCP and chip, which results in standard deviation increased after PCT at storage durations from 168 h to 336h, as shown in Fig. 11. These experimental results imply that applied adequate ultrasonic power to thermosonic flip-chip bonding process not only to improve the bonding strength of chips and flex substrates assembly, but also the bonding strength could be maintained in high level after PCT at various storage durations.

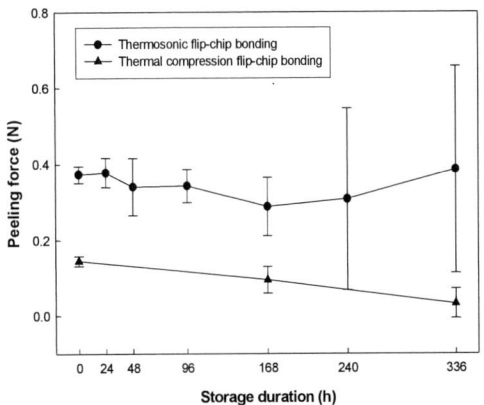

Fig. 11 Relationships between bonding strength and storage durations of PCT for chips bonded on flex substrates subjected at 121°C, 100% RH and 2 atm.

Fig. 12 The chips and flex substrates was fully separated after PCT at storage duration of 336 h. the chips and flex substrates assembly without applying ultrasonic power to flip-chip bonding process.

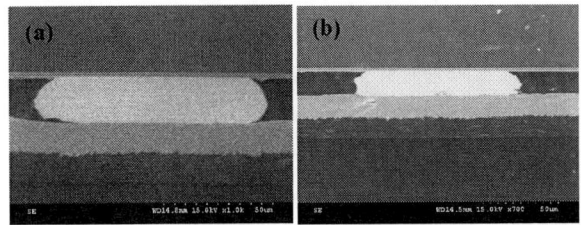

Fig. 13 SEM micrographs show (a) the cross section of Au bump bonded on the copper electrode after PCT at storage duration of 24 h, (b) the cross section of Au bump bonded on the copper electrode after PCT at storage duration of 96 h.

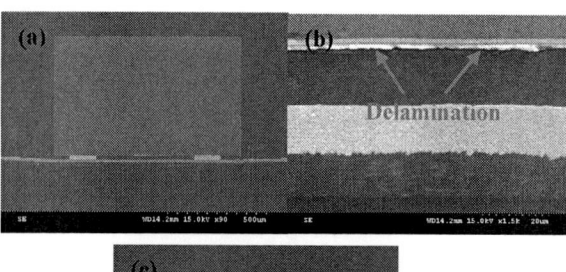

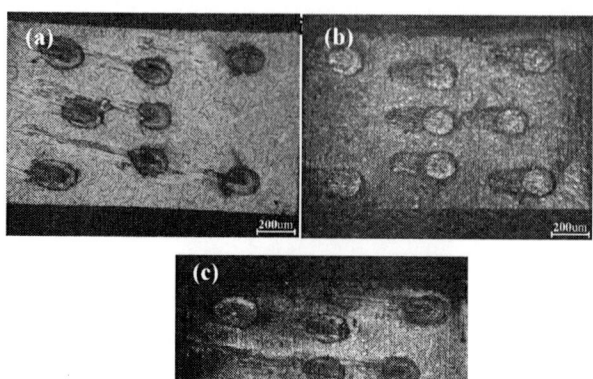

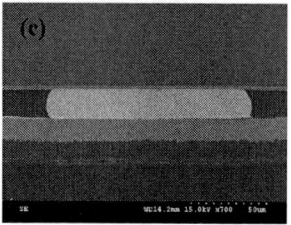

Fig. 14 SEM micrographs show (a) the cross section of chip bonded on flex substrate, (b) the delamination appeared at bonding interface between chip and NCP, (c) a Au bump well bonded on the Cu electrode without any defects

Fig. 17 Optical micrographs show the fractured morphology of chips after PCT at various storage durations, (a) 24 h, (b) 240 h, (c) 336 h.

High temperature/High humidity test

Figure 18 shows the relationship between bonding strength and storage durations of HH/HT test for chips bonded on flex substrates with and without applying ultrasonic power in flip-chip bonding process. For chips and flex substrates assembly without applying ultrasonic power, the bonding strength varied in a narrow range, indicates NCP with a good reliability after HH/HT test at various storage durations. As chips and flex substrates assembly with an ultrasonic power of 14.46 W, the bonding strength was increased with increasing storage durations of HH/HT test, as shown in Fig. 18. The cross-section of bonding interface between Au bumps and Cu electrodes were examined using SEM, no significant defects or cracks were found at bonding interface after HH/HT test at 200 h and 1000 h, as shown in Fig. 19. A clear interdiffusion trace between Au bump and Cu electrodes was obtained for chip bonded on flex substrates with an ultrasonic power of 14.46 W after HH/HT test at storage duration of 1000 h, as shown in Fig. 20. The bonding strength was improved by increasing the interdiffusion between Au bump and Cu electrodes. This experimental result can be used to explain that the higher bonding strength was obtained for chips and flex substrates assembly with the ultrasonic power after HH/HT test. In this study, the ultrasonic power is effective in improving bonding strength for chips bonded on flex substrates after HH/HT test. The reliability of HH/HT test for chips and flex substrates assembly using thermosonic flip-chip bonding process with NCP should not be a concern of issue.

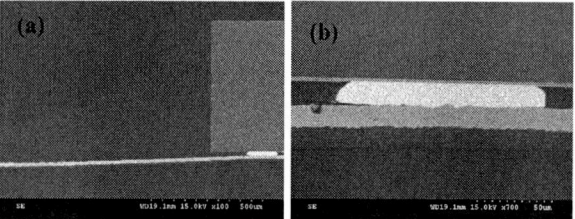

Fig. 15 SEM micrographs show (a) the delamination appeared at bonding interface between NCP and flex substrate, (b) the crack appeared at bonding interface between Au bump and Cu electrode. The specimen were subjected PCT at storage duration of 240 h.

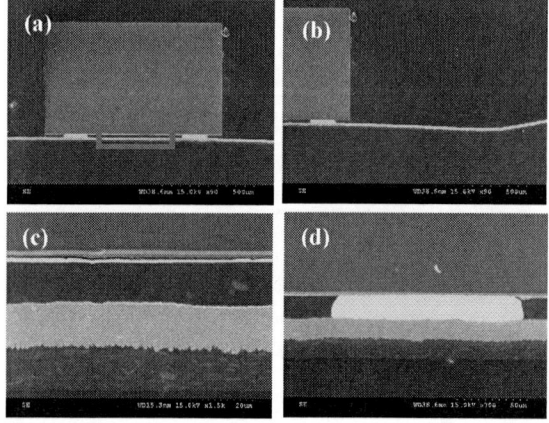

Fig. 16 SEM micrographs show (a) the cross section of chip bonded on flex substrate, (b) the delamination appeared at bonding interface between NCP and flex substrate, (c) the delamination appeared at bonding interface between NCP and chip, (d) a Au bump well bonded on Cu electrode without any defects. The specimen were subjected PCT at storage duration of 336 h.

978-1-4244-4658-2/09 $25.00 © 2009 IEEE

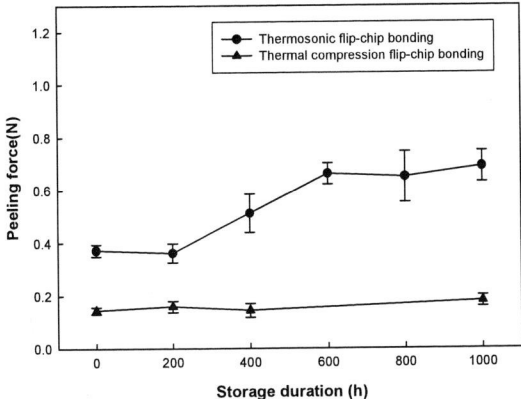

Fig. 18 Relationships between bonding strength and storage durations of HT/HH test for chips bonded on flex substrates subjected at 85°C, 85% RH.

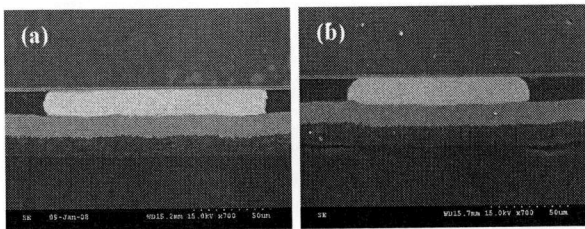

Fig. 19 SEM micrographs show (a) the cross section of Au bump bonded on the copper electrode after HH/HT test at storage duration of 200 h, (b) the cross section of Au bump bonded on the copper electrode after HH/HT test at storage duration of 1000h.

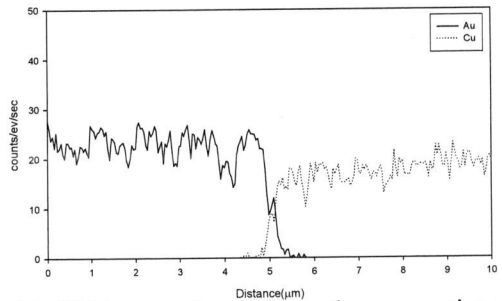

Fig. 20 Line scanning results on the cross section of Au bump and Cu electrode interface. The specimen was subjected to HH/HT test at storage time of 1000h.

Conclusions

The reliability of chips and flex substrates assembly using thermosonic flip-chip bonding process with the NCP has been verified in this work. For chips and flex substrates assembly without applying ultrasonic power, the NCP maintained a steady bonding strength after HTS test, TCT and HT/HH test, indicating that NCP has good reliability. Approximately 80% specimen, chips and flex substrates assembly without applying ultrasonic power in bonding process, exhibited fully separated after PCT at storage duration of 336 h. This observation implies that NCP cannot withstand the high pressure and high humidity testing condition, and lost its

adhesion strength. Therefore, the adhesion strength of NCP should be improved for PCT. The bonding strength increased with increasing the storage durations of HTS test for chips and flex substrates assembly with the ultrasonic power of 14.46 W in flip-chip bonding process with NCP. As the HTS test provides sufficient driving force to promote the atomic interdiffusion between Au bumps and Cu electrodes, a metallurgical bonding was formed, and the bonding strength was thus increased. The mean values of bonding strength maintained in high level after PCT at various storage durations for chips and flex substrates assembly applying an ultrasonic power in bonding process. The standard deviation of bonding strength increased when the storage durations was extended from 168 h to 336 h. Delaminations were found at bonding interface of chip/NCP and NCP/flex substrate after PCT at storage durations from 168 h to 336 h. However, a part of Au bumps firmly bonded on Cu electrode after PCT at storage duration of 336 h. The standard deviation of bonding strength increased after PCT was attributed to interaction between the NCP lost its adhesion strength and the increased bonding strength for Au bump well bonded on the Cu electrode. Applying an adequate ultrasonic power in flip-chip bonding process can provide a high level bonding strength after PCT. There was no significantly change in bonding strength for Au bumps bonded on the flex substrates after TCT test. It shows the specimen has great reliability for TCT test. The bonding strength increased with increasing tested durations of HT/HH test. Neither cracks nor defects at boding interface are observed. The reliability of HT/HH test for chips bonded on flex substrates using thermosonic flip-chip process with NCP meets the requirements stated in JEDEC standards.

The reliability for Au bumps bonded the flex substrate using the thermosonic bonding process with NCP can meet the requirements stated in JEDEC specifications; exception of the adhesion strength of NCP for PCT should be improved.

Acknowledgments

This study was granted by the Science Council of Taiwan, under grant number NSC-96-2212-E-040-006. The authors would like to express their appreciation to MRL of ITRI and its south branch for their assistances in providing experimental facilities.

References

1. D. Wojciechowski, J. Vaneteren, E. Reese, H.W. Hagedorn, "Electro-conductive Adhesives for High Density Package and Flip-Chip Interconnections", *Microelectronics Reliability*, Vol. 40 (2000), pp. 1215.
2. M. A. Uddin, M. O. Alam, Y. C. Chan, H. P. Chan, "Adhesion strength and contact resistance of flip chip on flex packages-effect of curing degree of anisotropic conductive film", *Microelectronics Reliability*, Vol. 44 (2004), pp. 505.
3. S. M. Chang, J. H. Jou, A. Hsieh, T. H Chen, C. Y. Chang, Y. H. Wang, C. M. Huang, "Characteristic Study of Anisotropic Conductive Film for Chip-on-Film Packaging", *Microelectronics Reliability*, Vol. 41 (2001), pp. 2001.
4. C.L. Chuang, Q. A. Liao, H. T. Li, S. J. Liao, G. S. Huang, "Increasing the bonding strength of chips on flex

substrates using thermosonic flip-chip bonding process with nonconductive paste", *Microelectron. Engineering. (revised).*

5. L. K. The, E. Anto, C. C Wong, S. G. Mhaisalkar, E. H. Wong, P. S. Teo, Z. Chen, "Development and reliability of nonconductive adhesive flip-chip packages", *Thin solid film*, Vol. 462 (2004), pp. 446.

6. W. K. Chiang, Y. C. Chan, B. Ralph, A. Holland, "Processability and reliability of nonconductive adhesives in fine pitch chip-on-flex application ", *Journal of electronic materials*, Vol. 35, No. 3 (2001), pp. 443.

7. C. L. Chuang, "Increasing of Bondability and Bonding Strength of Gold Stud Bumps onto Copper Pads with a Deposited Titanium Barrier Layer", *Microelectronic Engineering*, Vol. 84 (2007), pp. 551.

8. JEDEC standard, JESD22-A-103-B, "High temperature storage life", 2001.

9. JEDEC standard, JESD22-A-102-C, "Pressure cooker testing", 2000.

10. JEDEC standard, JESD22-A-101-B, "Temperature humidity test", 1997.

11. JEDEC standard, JESD22-A-104-B, "Temperature cycling test", 2000.

Dynamic Constitutive Relation of EMC over a Broad Range of Temperatures and Strain Rates

Tong Chen[1], Xiaoyan Niu[1,2], Xuefeng Shu[1,*], Jianwen Zhang[1]

[1]Institute of Applied Mechanics & Biomedical Engineering, Taiyuan University of Technology
No.79 West Yingze Street, Taiyuan, 030024, Shanxi, China
[2]College of Civil Engineering and Architecture, Hebei University
No.180 Wusi East Road, Baoding 071002, Hebei, China
*corresponding author: shuxf@tyut.edu.cn, tengyafan@sina.com, +86-351-6014455

Abstract

In this paper, dynamic properties of EMC were studied sat different temperatures and different strain rates. Firstly EMC was investigated by quasi-static tests. Secondly a series of dynamic compressive experiments of EMC were conducted using the Split Hopkinson Pressure Bar (SHPB) at high strain rates. Corresponding measurements were conducted at temperatures ranging from 20°C to 200°C. The results indicate that the yield strength and flow stress of EMC increase remarkably with the increase of strain rate. However, the yield strength of EMC is almost unchanged with the increase of temperature which is ranging from 20°C to 200°C.

1 Introduction

Epoxy molding compound (EMC), one of the largest three kinds of raw materials in semiconductor industry, performs the significant functions on the capabilities of discrete and IC [1]. Especially the plastics packages of micro-electronics, it has the character of lowly coat, simple craft as well as large-scale and roboticized production. Although the plastics packages doesn't belong to airtight seal, with the gradual improvement of package material and the continuous development of a new craft technology and mold, the reliability of plastics packages has been improved greatly. Nowadays, plastics packages has been the mainstream of all kinds of packages in the consumption field of general public electronics products [2].

There have been many papers published regarding moisture or thermo mechanical behavior. Nevertheless, the increase in the use of mobile telephones and computers that plastics packages are regularly loaded at strain rates higher than those previously considered. For example, even dropping a mobile phone on the floor can produce strain rates of the order of 1000 s-1. Therefore, in order to understand more fully the processes of these events, measurements of the EMC strength must now be extended to higher strain rates. However, there are very few published experimental researches on the high strain rate properties of EMC.

The purpose of this paper is to study the dynamic properties of EMC at different temperatures and different strain rates. The Split Hopkinson Pressure Bar (SHPB) of 8mm was used. A series of dynamic compressive experiments of EMC were conducted at high strain rates. Measurements were conducted at temperature ranging from 20°C to 200°C. The effects of strain rate and temperature were estimated by using of Cowper-Symonds constitutive model.

2. Experimental techniques

2.1 Theory

The theory of the SHPB procedure is based on one dimensional wave propagation in aligned bars as shown in Fig.1. When a striker bar impacts an incident bar, a compressive elastic wave travels through the incident bar. When this wave reaches the specimen, some portion of the wave is reflected at the specimen surface, while the remaining portion travels through the transmitter bar. The traveling waves can be measured by strain gages and signal conditioner. Three equations are required to calculate the true flow stress-true strain relationship of a test material from the experimental strain data [3].

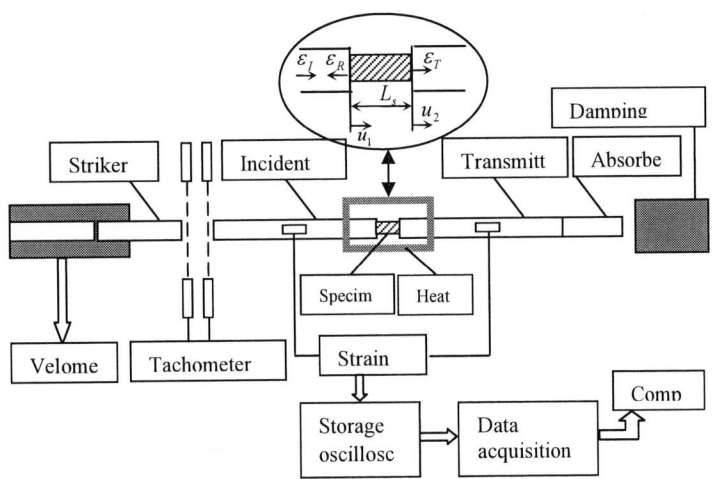

Fig. 1 Schematic of the split Hopkinson Pressure bar apparatus

$$\sigma(t) = \frac{A}{2A_0} E[\varepsilon_i + \varepsilon_r + \varepsilon_t] \qquad (1)$$

$$\dot{\varepsilon}(t) = \frac{C_0}{L_0}[\varepsilon_i - \varepsilon_r - \varepsilon_t] \qquad (2)$$

$$\varepsilon(t) = \int_0^t \frac{C_0}{L_0}[\varepsilon_i - \varepsilon_r - \varepsilon_t] dt \qquad (3)$$

where $C_0 = \sqrt{E/\rho}$ is the elastic stress wave speed in pressure bars, E and ρ are Young's modulus and the density of pressure bars, respectively, L_0 and A_0 are the original length and area of specimen and A is the cross-section area of pressure bars, ε_i, ε_r and ε_t are the recorded incident, reflected and transmitted strain pulses with time shifted from the strain gauge locations to the interfaces between pressure bars and specimen according to the elastic wave speed in pressure bars. Therefore, the flow stress-strain relationship of a material can be determined using the experimental values of ε_i, ε_r and ε_t, and Eqs. (1)-(3).

2.2 SHPB experimental system

The compressive SHPB system for high temperature tests consists of two pressure bars, an impact loading apparatus, a heater system, and measuring instruments [4]. A schematic diagram of the test system used in this work is shown in Fig. 1. The incident bar, transmitter bar, and striker bar were made of Inconel 718, and were 8 mm in diameter. Each pressure bar was 800 mm in length, and the striker bar was 250 mm in length. A launching device was used to accelerate the striker bar to produce a compressive impact were on the incident bar. Compressed nitrogen gas was employed to accelerate the striker bar. Several measurement devices were adopted to measure the strain voltages, velocity of the striker bar, and temperature of the specimen.

2.3 Specimens

EMC is selected for the specimens in this investigation shown in Table 1. Cylindrical specimens are injected from the EMC powder and have a 6mm nominal diameter and 3 mm height. The processes of making specimens are as follows. Firstly the EMC flour is poured into steel-cast mold. Then heat up the mold to 180°C and keep it five minutes in order to make the EMC flour dissolve completely. After that, the EMC flour is compressed tightly. Finaly, take out the sample when the mold cooling down naturally. All samples are injected at a temperature of 180°C. Then they are cooled down to room temperature. Over 5 experiments are conducted for each striker impact velocity. In addition, the quasi-static compression tests for solders are also performed on MTS material experiment machine for the purpose of comparison with the results of dynamic tests.

3. Results and discussion

The true flow stress-true strain relationships of the EMC between 20°C and 200°C, at a strain-rate of 600s^{-1}, is shown in Fig. 2. The flow stress almost unchanged as the temperature increased. This indicates that the work hardening effect will not change between 20°C and 200°C as the temperature increased. To determine the effect of the strain-rate, tests at different strain-rates and room temperatures were also performed. The results are shown in Fig. 3. The dependency of stress on the strain rate for EMC is clearly shown in the figure, which characterizes the strain rate sensibility. Increasing strain rate leads to higher yield stress and flow stress. Take Stress-strain curves of EMC at 300s^{-1} strain rate for example, the yield strength at high strain rate (about 25MPa) is exceeding about two times of that in quasi-static (about 14MPa) condition, and the flow stress at high strain rate (25~50MPa) is much higher than that of quasi-static condition (25MPa), shown as in the Fig. 4. The typical time-strain rate curve for EMC indicates that it is a kind of quasi brittle material, which is shown as in the Fig. 5.

The following equation was suggested by Cowper and Symonds to describe the material behaviour at different strain rates [5]:

$$\frac{\sigma'}{\sigma} = 1 + \left(\frac{\dot{e}}{D}\right)^{1/q} \qquad (4)$$

where σ' and σ are dynamic and quasi-static stress, respectively; $\dot{e}$ is strain rate; D and q are Cowper-Symonds coefficients. With the coefficients D=662.95 and q=0.3698, the Cowper–Symonds equation gives reasonable agreement with the experimental data for EMC assembled by Symonds [6] is shown in Fig. 6.

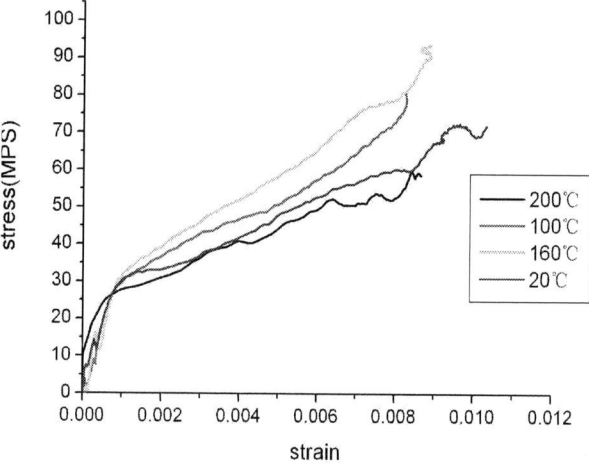

Fig. 2 Stress-strain curves for EMC at four different temperatures and a strain rate of 600 s^{-1}

Table.1 General physical properties of the EMC

EMC propert y	Silica content (%)	Density (g/cm3)	Flow length(cm)	Gel time (sec)
	77.4	1.99	95	27

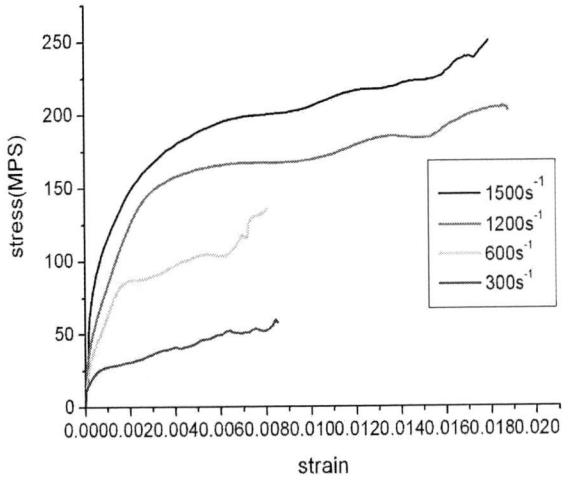

Fig. 3 Stress-strain curves for EMC at different strain rates

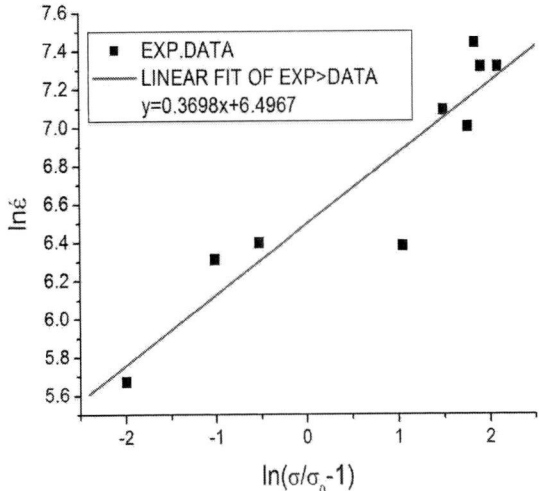

Fig. 6 Experimental data and Cowper-Symonds constitutive fitting for EMC

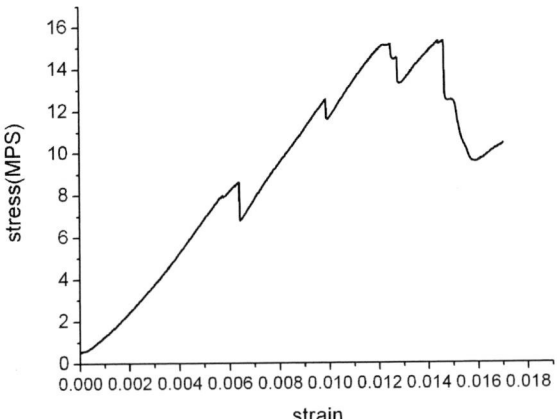

Fig. 4 Typical Stress-strain curve for EMC at the quasi-static compression tests

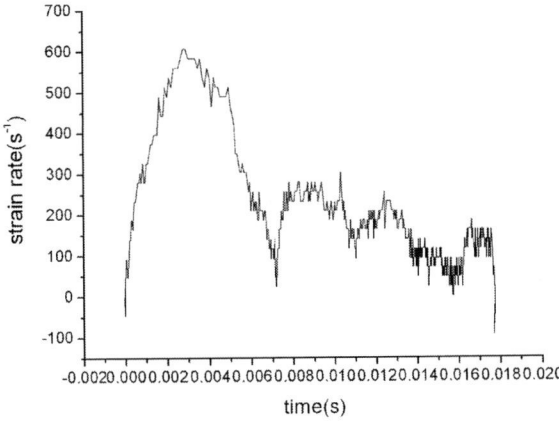

Fig. 5 Typical time-strain rate curve for EMC

4. Conclusions

In this paper, the thermal effect and strain rate dependent dynamic properties of EMC is further studied and measured by a Split Hopkinson Pressure Bar facility. The results indicate that the dynamic yield strength of the EMC is greater than their static yield strength and it increases remarkably with the increase of strain rate. When the ambient temperature is ranging from 20°C to 200°C, the EMC yield strength is almost unchanged. The material constants in cowper-Symonds model were determined from experiments. This paper gives dynamic constitutive relation including strain-rate effect and temperature effects for EMC, in order to further study the reliability of electronic products and to provide a research foundation.

Acknowledgments

This project was sponsored by the National Natural Science Foundation of People's Republic of China (10672113; 10772131), the Special PH.D Research Fund for Universities (20050112009), the Natural Science Foundation of Shanxi Province (2007011010) and the study abroad information of Shanxi Scholarship Council of China. (2008.30)

References

1. Xie Zhao-wen, "The Impact of temperature on the performance of epoxy molding compound," *Electronics&Packaging*, Vol. 7, No. 12(2007).
2. S.L. Liu*, G. Chen, M.S. Yong, "EMC characterization and process study for electronics packaging," *Thin Solid Films*, Vol. 462–463 (2004), pp.454–458.
3. Lindholm US, "Some experiments with the split Hopkinson pressure bar," *J. Mech Phys Solids*, Vol.12(1964), pp.317-335.
4. Yang H, Min O, Park K, "Temperature dependence of dynamic behavior of titanium by the high strain-rate compression test," *Proceeding of the Fourth International*

Symposium on Impact Engineering, Kumamoto, Japan, 16-18. July. 2001, vol. II, pp. 571-576.

5. Cowper GR, Symonds PS, "Strain hardening and strain rate effects in the impact loading of cantilever beams," Report No. 28, Brown University Division of Applied Mathematics, 1957.

6. Symonds PS, "Survey of methods of analysis for plastic deformation of structures under dynamic loading," Division engineering report BU/NSRDC/1–67, Brown University, 1967.

Low K CMOS90 2N Gold Wire Bonding Process Development

Ming-chuan Han, Bei-yue Yan

Freescale Semiconductor (China), Limited

No. 15, Xinghua Avenue, Xiqing D.A., Tianjin, China

Ming-chuan.Han@freescale.com, Bei-yue.Yan@freescale.com

Abstract

The electronics industry is mainly driven by the demand for smaller, faster, higher complexity, more reliable and cheaper device. Reduction in bond pad patch and low K wafer were introduced into mass production for miniaturation and signal transmission integrity. High safety requirement and longer working life demand high reliability performance especially in automotive and aerospace filed. Although trends in electronic packing have been focused on the development of flip chip, Wire bonding still cements its position in chip interconnection technology. In IC assembly, gold wire is the key material in wire bonding process to connect chip to substrate. 2N gold wire with specific dopant is applied in many types of package for stringent reliability requirement field. This study will focus on 2N gold wire application in wire bonding process.

Bonding wire, bonding capillary, and wire bonding parameters, was selected as critical factors in this study. This paper describes the key 1st bond parameter optimization on CMOS90 low K device. Gold wire and capillary design were studied. Tail scrubs function of KNS maxum plus wire bonder was also studied to find out how to produce robust 2nd bond. The Newly-designed capillary from SPT aimed to improve wire stitch pull strength was also studied to improve 2N gold wire 2nd bond stitch bond which have weak stitch bond strength result from its big hardness. Critical responses such as Ball size, Ball height bonded ball diameter ratio, bonded ball placement, wire pull strength, ball shear strength, and stitch pull strength were studied to understand the wire bonding effect of low k device. DOE (Design of Experiment) and RSM (response surface methodology) was used to optimize the wire bond process. Thermal aging test coupled with wire pull and ball shear test with recording failure mode were studied. The IMC (Inter Metallic Coverage) of the Au-Al and cratering were also tested.

The combination of the low K material and harder 2N wire introduction leaves a narrow process window for packaging assembly, and raises challenges to wire bonding process in terms of process manufacturability. The studies showed that 2N wire were found to be a good replacement of traditional 3N wires on improving bonding integrity of the low K device. Capillary geometry design is also critical to achieve a better IMC coverage, robust 2nd bond. The newly-designed capillary from SPT aimed to improve 2N wire stitch pull strength.

I. Introduction

Owing to the rapid development of technology and fast-changing customer's demand in recent years, the trends in

Microelectronics manufacturing are still size reduction of components, and increasing functionality coming along with decreasing production costs [1]. Under these conditions, it is a great challenge to semiconductor assembly and package industry regarding to its process capability and material selection.

In IC assembly, gold wire is the key material in wire bonding process to connect chip to substrate. Gold alloy wire is being used to substitute for traditional gold wire to get better reliability performance. The industry has successfully developed a solution to inhibit or retard IMC growth by adding certain elements (like Pd) to the bonding wires. Nevertheless, the gold purity drops from 4N (99.99%) to 2N (99%). The increase in hardness brings the problem in determining optimal parameters to wire bond with alloy gold wire in BGA package.

The drive towards device miniaturization and high level integration had lead to the development of low-k material /Cu structure. As 90nm and 65nm technology ramped into mass production in recent years, the need for copper metallization, combined with the use of low-k dielectrics materials, becomes a critical challenge in all phases of semiconductors manufacturing, but especially in final assembly plants. Bonding on such low modulus materials may be resulted in bond pad cupping, cratering and ILD delamination, lowering yield and reducing the device reliability. The introduction of harder 2N gold wire makes it more different to control 1st bond and 2nd bond process. This study aims to identify and optimize wire bonding process in order to get robust, reliable wire bond process. [2]

II. Challenges to assembly process

Wire bonding is the most widely used first-level interconnection technology between chip and package terminals. Today more than 90% of IC chips used in the word are with wire bonds because its speed, flexibility, higher productivity [3]. Wire bonding is capable of fine pitch bonding and is in principle applicable to all peripheral pad pitch. Higher quality bonds are vital to the performance of IC device. The parameter setting is very critical to successful wire bond process. The key parameters such as bond power, bond force and bond time should be optimized to achieve higher wire bond quality.

The characteristic of low K materials are as below: poor mechanical characteristics, low mechanical strength, and low adhesive strength, reduced thermal conductivity, higher coefficient of thermal expansions. Its softer and spongy characteristics are a challenge to the existing wire bond tools and process capabilities. Bonding on such low modulus materials has resulted in bond pad bond pad metal peeling or crack during wire bonding, poor ball shear

978-1-4244-4658-2/09 $25.00 © 2009 IEEE

strength, and NSOP (non stick on pad) issue. These issues lead to overall assembly yield and device reliability. Although 2N wire could address ball lift issue in ultra fine pitch process. The introduction of harder 2N gold wire makes it more different to resolve all those issues. Bonding over such low modulus materials has caused the top metallization layer to cup and deflect under the ball, and potential ILD layer delamination during assembly or reliability testing may occur [4, 5, 6].

The biggest problem of 2N wire combined with low-k device bonding is 1st bond and 2nd bond bondbility. The process for low-k device bonding so very especially for ultra fine pitch bonding .Relatively high rejected rates caused by Pad peeling and NSOP were observed. Harder FAB and harder wire result in NSOP, crating, NSOL and low stitch pull strength. These all leave a narrow process window for packaging assembly and raises challenges to wire bonding process in terms of process manufacturability and package reliability [7, 8]. The only solution to the problems was the parameter optimization. As bond pad pitch and bond pad opening shrinking at the same time, the targeted bonded ball size becomes smaller. The smaller bonded ball size is more sensitive to the low k bond pad structure, bond window and reliability performance [9].

III. Wire bonding process development

A56 µm fine wire pitch and CMOS90nm low K technology die was selected as the test vehicle to optimize the low k wire bond process and assess its reliability performance. The following are details of the die / package information:

a) Package description: Standard MAP (Mold Array Process) BGA 17*17*1.4mm, 256I/ O, 1 mm pitch)

b) Die size: 9.39x6.38 mm

c) Die thickness: 7mils

d) Bond pad passivation opening: 52µm

e) Bond pad pitch: 56 µm.

f) Bond pad Al composition and thickness: Al / 0.5%Cu, 10kA

g) Bonding wire: 23um gold wire (2N gold purity: 99%)

h) Low K type: BD1 (Black Diamond 1)

i) Wafer fabrication technology: CMOS90nm.

To achieve a robust wire bonding process on fine pitch, low k wafer devices, the following responses have been identified based on past fine pitch wire bonding evaluation and low K wafer evaluation:

Consistent free air ball

Consistent bonded ball diameter (BBD)

Consistent ball height / ball size ratio

Stable / sufficient inter-metallic coverage at different thermal aging stages

Wire pull with recording failure mode at different thermal aging stages

Ball shear and its failure mode at different thermal aging reading point

Stitch pull test with recording failure mode

Bond pad cratering

Package level reliability test
Experiment:

Based on the experience on CMOS90 consumer low K device, in the first part of the experiment, the 2nd bond parameter 2N wire was optimized. 2 types of capillary with different tip surface processing were evaluated to improve 2nd bond bondability. An experiment matrix was generated by MINITAB® software to screen out critical factors. Finally, the experiment was designed and optimized using a response surface with a quadratic function to establish 2nd bond parameter windows, stitch pull strength, 2nd bond crescent integrity and 2nd bond related alarm in wire bond process as response. The optimized bond parameters were chosen and a conformation run was conducted to check the validation and accuracy of the design of experiment.

The next experiment is FAB parameter optimization and 1st bond parameter. The objective of the experiment was to identify a set of baseline parameter that could be used for 2N gold wire bonding on the low k device. IMC check, creating test and high temperature storage test was conducted to test reliability.

The following there variables were regarded as key sub-systems of the wire bonding on fine pitch low k devices: Bonding wire, capillary, and wire bonding parameters. A thorough analysis and study were carried out on these there sub-systems and the results are discussed in the following sections.

Discussion:

A. Capillary selection

The selection of the capillary was constrained by two factors, fine pitch limitation and low k structure. Figure1 shows key dimensions of the capillary.

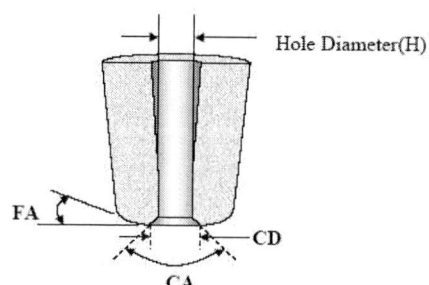

Figure 1 Key dimension of capillary

For fine pitch wire bonding (56µm wire pitch), the empirical data shows hole (H) of the capillary should be about 4 to 5 um larger than wire diameter to ensure good loop control and reduce damage to the wire during wire feeding / looping. Chamfer Diameter (CD) is over 5 um larger than the Hole to ensure sufficient ultrasonic transfer. The empirical value of T is 1.15xbond pad pitch. For a 52 um pad opening, the target BBD (bonded ball diameter) is about um 42µm, after considering the bonding accuracy of +/-2.5 um. Chamber angle is selected as 60 degrees to have a good FAB centering. The Face Angle (FA) was selected as 11 degrees to allow for good second bond quality.

In general, the initial capillary selection based on 47um fine pitch is as follows:

Hole: 28um, CD: 35 um, T: 63 um, Face angle: 11 degrees, CA: 60 degrees.

B. 2nd parameter optimization

Recently, the industry has successfully developed alloy gold wire to inhibit the formation of the IMC and improve mechanical prosperity by adding certain elements to the bonding wires. This leads to new gold wires having a different composition / purity – 2N 99% and 3N 99.9% of gold wire. Pd is main dopant in 2N wire. When a certain content Pd was added to gold bulk, it enhances the mechanical properties, but has negligible reduction in the electronic conductivity. Table1 shows the properties comparison between different wire types.

Table1. Gold wire properties comparison

Wire properties comparison				
Wire type		2N	3N	4N
General Properties				
Wire Diameter (mm)		25±0.5	25±0.5	25±0.5
Breaking Load (g)	Room Temp.	>11	>8	>8
	High Temp. 250^{0}C	>10	>7	>7
Elongation (%)	Room Temp.	12~17	2-7	2-7
Physical Property				
Hardness (HV, 10mN, 5s)	Free Air Ball (FAB diam.44 mm)	60-70	45 - 55	45 - 55
Elastic Modulus (GPa)		>85	>75	>75
Fusing Current (A, Length=5mm)		0.51	0.5	0.5
Resistivity (μOhm cm) @ 20^{0}C		3.3	2.3	2.3
Component Property				
Au Purity (%)		99	99.9	99.99

2N wire shows higher broken load, which makes mare Difficult in 2nd bond quality control and narrow 2nd bond parameters window, IN preliminary study low stitch issue was found. The wire bonder was stopped by EFO open or short tail alarms. A type of capillary with modified tip surface was evaluated to improve stitch pull strength and 2nd bond bondability. Figure 2 shows stitch pull strength comparison between capillary with polished tip surface and SI type capillary modified tip surface. Figure 2 shows top view of SI capillary.

tip surface could harder 2N wire strongly, we could achieve desirable 2nd bond bondability. SI type was selected as bonding tools.

With the selected bonding tool, we were able to focus our next efforts on the wire bond process parameters window study. The experiment was designed and optimized using a response surface with a quadratic function.

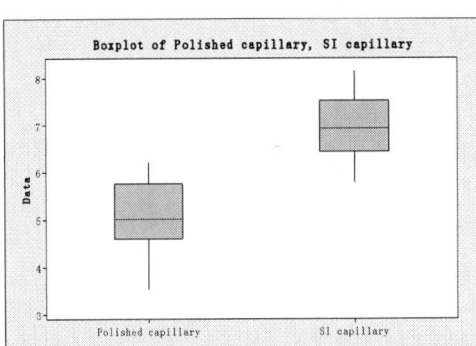

Figure 3 Box plot of stitch pull strength

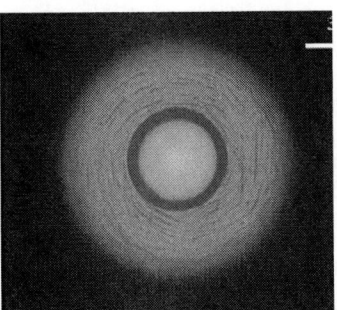

Figure 2 The top-view image of SI capillary tip
Courtesy of SPT

From the data showed in Fig.3, SI capillary could improve stitch pull strength. The SI capillary with roughen

The purpose of the wire bonding process optimization is 2nd bond quality and reliability, which is related with harder 2N gold wire.

978-1-4244-4658-2/09 $25.00 © 2009 IEEE

For the 2nd bond process optimization, there are many parameters including bonding time, 2nd USG current, 2nd bond force, and tail scrub parameters.

After parameter screening DOE, key factors were screened. During the RSM experiment, 2nd bond USG current and tail scrub cycle were selected as the critical input variables. The stitch strength, 2nd bond crescent integrity before and after stitch pull test and stitch pull failure mode are response. After optimization, all responses all fall into satisfactory ranges and meet specification.

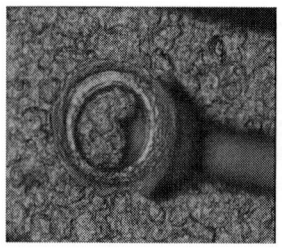 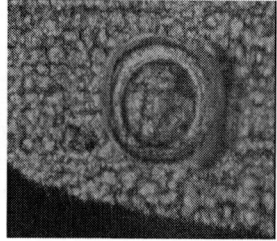

Before stitch pull test   After stitch pull test

Figure 4 2nd bond images

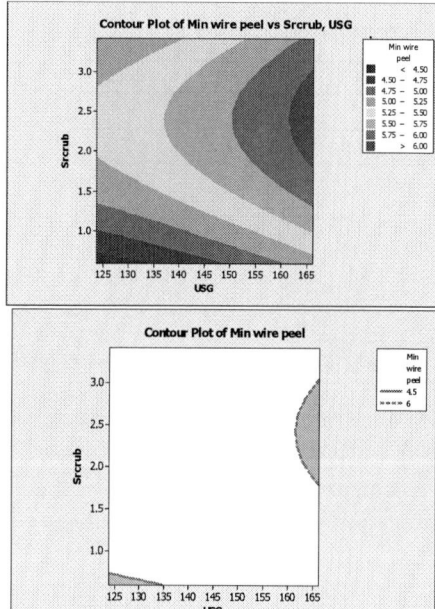

Figure 5 RSM experiment

C. FAB parameter optimization

During wire bond process, a free air ball is formed on a wire tip with an electronic discharge between the wire and the electrode wand. The FAB then descends on a capillary for compression onto bond pad to from the first bond. The consistent FAB size is critical to wire bonding process. First bond shape, size, and height are proportional to the volume of the FAB. Too large or too smaller FAB size would result in golf ball, NSOP and clogged capillary.

2N gold wire with 99% purity is much harder than 4N or 3N wire. If we exactly copy 3N or 4N wire EFO parameter, smashed or golf club ball would encountered. 2N wire is more sensitive to tail length or fire level. If we shorten tail length, Heat concentration is more even distributed over total wire tip surface. More symmetrical FAB will be gotten. Fire angle should fall in the range from 45° to 60°. EFO current and EFO also need to be optimized to get a desirable FAB shape and size.

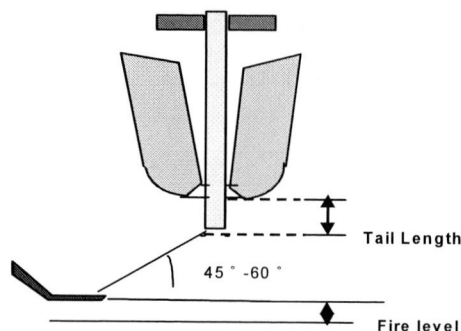

Figure 6 Fire level diagram

D. 1st bond parameter optimization

Bonding wire plays a key role in low k / fine pitch wire bonding. An appropriate selected wire type can provide a larger wire bond process window, bond ability and reliability. First of all, consistent FAB size is necessary. Secondly, NSOP and lifted metal during wire bonding are not acceptable. Thirdly, a smoothly and well formed inter-metallic layer between Au and Al at different reading point of heat treatments during reliability test is necessary.

A thermal aging storage (HTS) has been widely applied by the industry to evaluate gold wire reliability performance and the quality of the inter-metallic. There are two major failure modes during wire pull test at each thermal aging test read points: break at neck and break at IMC (Inter-Metallic Compound). Neck break is acceptable, the other failure mode, which is the low strength ball lift, and depicts the weakened Au and Al interface, is not acceptable.

The ball lift failure during wire pull test usually occur due to impurities in the bond interface, poor inter-metallic formation because of the inappropriate bonding parameters/ wires/ tools or the formation of the Kirkendall void.

The formation of Kirkendall void can't be evitable, but it can be inhibited. The industry has successfully developed 3n and 2N alloy gold wire to inhibit the formation of the IMC to suit to increasing demand for high reliability. The Pd is main inertia dopant in gold wire. When Pd was added into Au matrix, a Pd atom replaces an Au atom in the crystal structure. The diffusion speed is much faster than that of Au, When Au diffused into aluminum side, Pd atoms will formed a barrier to prevent Au atom diffusing into aluminum. This is why 2N alloy gold wire shows ultimate reliability. [10]

978-1-4244-4658-2/09 $25.00 © 2009 IEEE

Based on CMOS90 low K device 3N wire recipe.2N wire 1st bond parameter was optimized. It was found that 2N wire needs different parameter setting such as ultrasonic power, FAB size, and bond force as compared to 3n o 4N wire. These were due to the higher bulk material hardness of the wire and different dopant. . While the wire pull test post different thermal aging test shows no ball lifted was found up to 1008 hours high temperature storage. Figure 7 shows the ball lift % vs. thermal aging reading point during wire pull test.

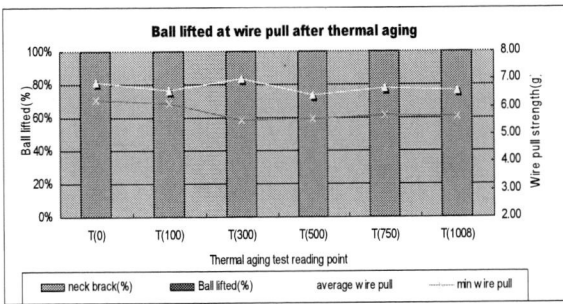

Figure 7 ball lift % vs. thermal aging reading point

After IMC and cratering check, IMC converge I above 60%, harder FAB couldn't cause and cratering issue.

With the selected bonding wire, bonding tool and bonding equipment, we were able to focus our next efforts on the wire bond process parameters window study.

For the first bond process optimization, there are many parameters including bonding time, bonding power, bonding force, tip, C/V, EFO current and FAB size.

After parameter screening DOE, key factors were screened. During the RSM experiment, contact power, bond force and bond power were selected as the critical input variables. The ball size, ball height, BB ratio all fall into satisfactory ranges, Thermal aging read point shows zero ball lift up to 1008 hours with the optimized bonding wire, bonding tool and parameters.

E. Reliability Test

The bonded samples with the optimized bonding wire, tool, and process parameter recipe were molded with green compound. The reliability tests were carried out using the test conditions as JEDEC standards. The packaged sample (BGA 17*17*1.4mm, 256I/ O, 1mm pitch) successfully passed MSL3@260C preconditioning followed by 500 T/C, High temperature storage 1008hours and 96hours autoclave with zero electrical test failure, and no package delamination or die crack assembly defects observed. No ball lift issue was found on decapped units after 500 T/C and 1008 hours high temperature storage stress test.

F. Conclusions

The combination of the low K ILD with the fine pitch leaves a narrow process window for packaging assembly, and raises challenges to wire bonding process in terms of process manufacturability and package reliability. By using Design of Experiment with MINITAB® software, the significant variables to the response for 56 bond pad pitch

wire bonding process were identified. Bond integrity tests thermal aging test are used to characterize the wire bonding process development. The studies showed that 2Nwires was found to be a good option on improving bonding integrity of the low K device. Capillary geometry and tip surface roughness are critical to achieve a better IMC coverage, good 1st bond and 2nd bond quality. Ultra fine pitch wire bonding process was established on low K device. The MAP BGA packages could successfully pass reliability test.

Acknowledgments

The authors wish to thank Freescale PSD for their great help and valuable discussions and suggestions throughout the low K wire bonding process development. The authors would like to thanks Colin. Zhang form SPT for their support in this project.

References

1. Jonathna Tan, et al., "Wre-bonding Porcess development for low-*K* Materials," *Microelectronic Engineering*, Vol. 81 (2005), pp. 75-82
2. T.A. Tran, L. Yong, B. Williams, S. Chen and A.Chen, "Fine-pitch probing and wire bonding and reliability of aluminum capped copper bond pads," *Proc. HDI conf., Denver, CO*, Apr. 2000, pp. 390-395
3. George Harman, Wire bonding in Microelectronics Materials, Processes, Reliability and Yield, McGraw-Hill (1997), pp. 2-10
4. Vaidyanthan Kripesh, Mohandass Sivakumar, "Wire bonding Process Impact on Low-k dielectronic Material in Damascene Copper Integrated Circuits", Electronics Components and Technology Conference, 2002
5. T. Scherban, B.Sun, J.Blaine, C.Block, B.Jin, E.Andideh., "Interfacial adhesion of copper-low k interconnects", in IEEE International Interconnect Technology Conference Proceeedings, 2001, pp. 257-259
6. P.Nunam, "The challenge of Low k, Issues and Considerations for Accelerated Performance", *Yield Management Solutions*, Spring 2000, pp. 17
7. Yeh, C.-L. et al, "Impact Analysis of Wire bonding on Cu/low-k structures," Proc IMAPS 2003, Boston, MA, Nov. 2003
8. Geoge G. Harman, "Wire Bonding to Advanced Copper, Low-K Integrated Circuits, the Metal/Dielectronic Stanks, and Materilas Considerations," *IEEE transaction on components and package technology*, Vol. 25 (2002), pp. 677-683
9. K.S. Goh, Z.W.Zhong, "A new bonding-tool solution to improve stitch bondability Wre-bonding Porcess"
10. Shankarara K.Prasad, Advanced Wirebond interconnection technology, Kluwer Academic (1997), pp. 65-70

Preliminary Study of a New Process to Improve the Strength of Thermo-sonic Ball Solder Joint

Zhai Yu[1, 2], Zhao Wang[1], Jun Cheng[1], Hw Tian[2]

[1]ShenZhen Institute of Advanced Integration Technology

Chinese Academy of Sciences and The Chinese University of Hong Kong

Shenzhen, 518055, China, Email: zhai.yu@sub.siat.ac.cn, zhao.wang@siat.ac.cn, jun.cheng@siat.ac.cn

[2]School of Mechanical Engineering, Southwest Jiao Tong University, Chengdu, 610031, China,

Email: hwtian@swjtu.edu.cn

Abstract

The separation of golden ball and die pad due to the poor bonding strength of 1st ball solder point is the main failure mode in thermo-sonic ball wire bonding. In this paper, nonlinear elastic and elasto-plastic quasi-static analysis of FAB (free air ball) in ultrasonic vibration process is carried out by finite element method with ANSYS. We study the evolution way of compressive stress distributions of FAB and sliding area in the contact interface under elastic and elasto-plastic deformation. According to the results of the analysis and the mechanism of micro-slip based on elastic theoretical assumption, we propose a new bonding process control that can improve the bonding strength of 1st ball solder point effectively.

1. Introduction

Thermo-sonic ball wire bonding has a large market share at the current chip-pins package because of its flexibility and cost effectiveness. The bonding strength of solder joints at both ends of wire is the important indicator to protect the connection of chip and external electronic device. As far as the state of the craft, the real bonding area of 1st ball solder point is usually a long elliptical shape [1]. There is no connection in the center district, which it is known as 'donut' phenomenon. If the unconnected area on craft could be reduced or eliminated, it could greatly improve the bonding strength of 1st ball solder point and improve the reliability and stability of entire chip package. To achieve this goal, it is need to clarify the mechanism of bonding.

There are many papers concerned the effects of process parameters, such as bonding force, bonding time, ultrasonic power, and so on [2,3]. However, we also have to pay more time to optimize the process in order to find the best parameter arrangement because there is still lack of a quantitative understanding of the bonding mechanisms. The reason for this lies in difficulty in interpreting the 'donut' phenomenon.

Bonds made at liquid-nitrogen temperature prove that thermo-sonic ball wire bonding is a solid-state process [4]. It is so far generally accepted that a relatively contaminant-free surface is one of the requirements to form metallurgical bonding. In bonding process, relative motion at interface plays a very important role. In-situ and in-process monitoring of bonding process has shown that stick-slip motion is necessary for high strength bonding [5] known as micro-slip model. Although the mechanism of micro-slip based on elastic theoretical assumption may explain the "donut" phenomenon [5,6,7], an significant plastic deformation of FAB (free air ball) is created by the ultrasonic energy and bonding pressure energy in bonding process. It is not easy to understand the effect of the plastic deformation and how bond pattern is dependent on process parameters in bonding process.

The dynamic process of thermo-sonic ball bonding involves a variety of nonlinear physical phenomena and multi-field, including contact dynamics, plastic deformation of metal materials, tribology and so on. And due to the high speed (vibration frequency =100 KHz), small displacement (few microns) and short time-consuming (tens of microseconds) in ball bonding process, it is not easy to study the dynamic process by experiment. With the rapid development of computer technology, numerical simulation has become a well research method for the bonding process. At present, some researchers have used Finite Element Analysis (FEA) to simulate the stress distributions of FAB in bonding process [8,9,10], but they all used the fixed bonding force, only consider pure mechanical bonding loads with static methods and there are few researches on the slip in the bonding contact interface in ultrasonic vibration process. In this paper, a 3-D model to simulate the thermo-sonic ball wire bonding is proposed and investigated. Based on the elastic micro-slip model, how the compressive stress distributions of the FAB affect the sliding area in the contact interface under elastic and elasto-plastic deformation and finally affect the bonding strength will be studied.

2. Finite element analysis

2.1. Assumption and FEA Model

The actual bonding process involves many factors. In order to get effective simulation, we need to take the following assumptions and simplifications:

a) Assumed that the FAB made by high-voltage discharge is a standard sphericity

b) When considering plastic, the FAB is a rate-independent bilinear isotropic plastic material

c) Without regard to chemical reaction in bonding process.

d) Assumed that the material of bonder capillary and pad is rigid.

e) Take no account of the friction heat, assumed that the temperature of the FAB is the same as substrate all the time.

f) Without regard to the influence of ultrasonic energy on yield strength of the FAB [11].

g) The simulation is quasi-static FEM analysis, not dynamic analysis.

Both the elastic and elasto-plastic models of thermo-sonic ball wire bonding are considered in the study. The model consisted of a bonder capillary, FAB, and bond pad, as show in Fig.1. (Due to the symmetry, only half of the structure was adopted). The diameter of the FAB is 70μm, the wire has a cylindrical shape of radius 12.5μm. The size of the bonder capillary is referred to the SBN-46165-745E-ZP34T that made by SPT corporation and the pad is a 200μm *100μm *10μm rectangle block. The bond pad and bonder capillary are rigid materials, which are much harder than the FAB and assumed not to be deformed. The material parameters of FAB are listed in Table 1. Since the bonder capillary is considered as a rigid body, this leads to the rigid and elastic nonlinear contact pair under elastic condition and the rigid and elasto-plastic nonlinear contact pair under elasto-plastic condition between FAB and bond pad. Mesh was created for the symmetric loading geometry using the FE code ANSYS, as shown in Fig.2. Supposed the friction coefficient of the contact interface is 0.4. The FEA model is composed of 7168 solid186 elements and 31265 nodes. This ensures sufficient resolution and accuracy of the results while maintaining a reasonable time needed for computation time. The elastic case and the elasto-plastic case have the same analysis condition except for the material of the FAB.

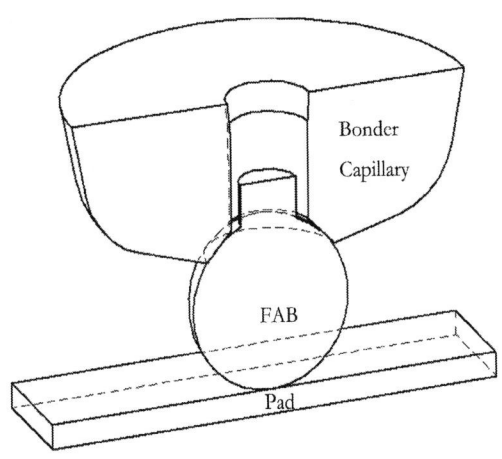

Fig. 1 The geometry of bonder capillary, FAB, and pad

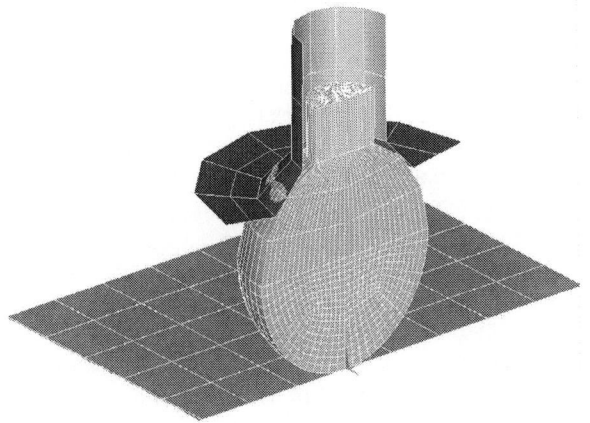

Fig. 2 The mesh of thermo-sonic ball bonding

Table 1 Materials Parameters of Au at 200°C

Material	EX (GPa)	PRXY	Yied stress (GPa)	Tensile modulus (GPa)
Au	68.6	0.44	0.0327	21.5

2.2. Analysis steps and solving

In general, the real process of 1st ball solder point of thermo-sonic ball wire bonding can be refined as follows: Firstly, the FAB is formed by using high voltage discharge. Secondly, it's pressed by the bonder capillary toward the bond pad. Finally, apply ultrasonic vibration to form a bond. We focus on the last two steps in finite element analysis. Control the compression displacement of FAB (bond shift) instead of pressure and control vibration amplitude instead of ultrasonic power. The process is divided into several micro-steps: First, move down a certain bond shift (H=5μm) on the FAB. Considering there is no time factor in static analysis, but we still use a time value to specify the end of each load points and time units can be arbitrary. Define the first loading time is 100s. Then apply a horizontal oscillation movement of the capillary with an amplitude of 1μm (vibration amplitude) on the FAB. Oscillating time is 200s. Repeat the procedure until the total load-steps are 25 and the total loading time is 2500s. The bond shift and the vibration amplitude are shown in Fig. 3 and Fig. 4 respectively.

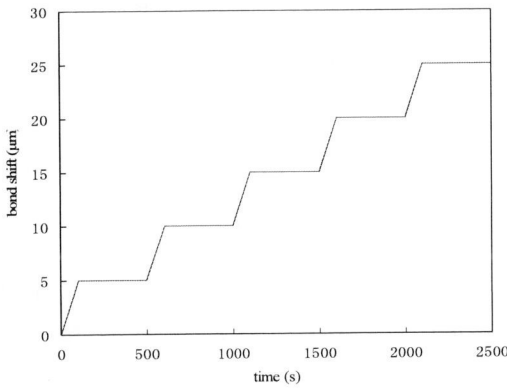

Fig. 3 The bond shift

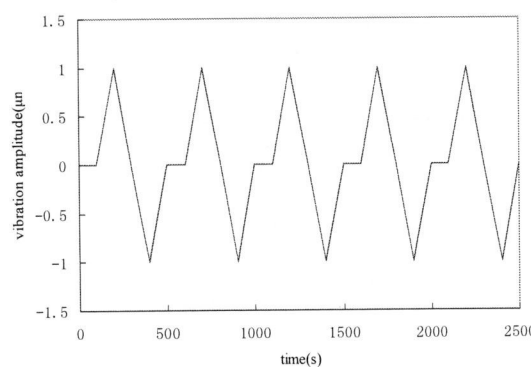

Fig. 4 The vibration amplitude

3. Results and Discussion

3.1 Stress distributions of the FAB

In elastic analysis, the bond shift was increased until it reached 25μm. The nodal result of compressive stress distributions and contact pressure distributions of the FAB for H (bond shift) =2μm, 5μm, 15μm and 25μm are shown in Fig.5- Fig.6.

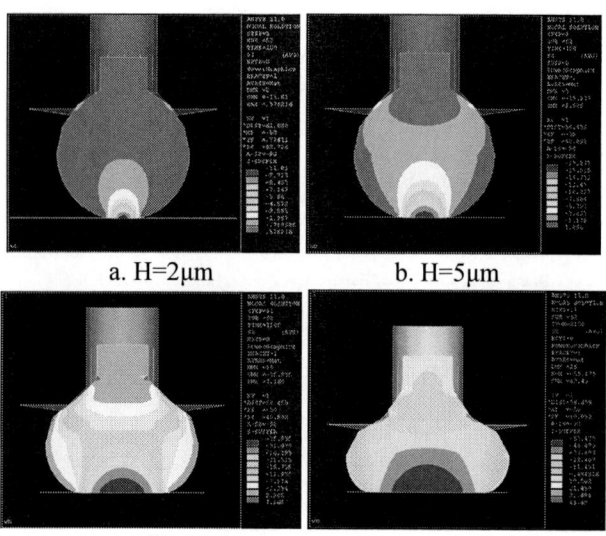

a. H=2μm   b. H=5μm

c. H=15μm   d. H=25μm

Fig. 5 Compressive stress distributions of the elastic FAB under pressure

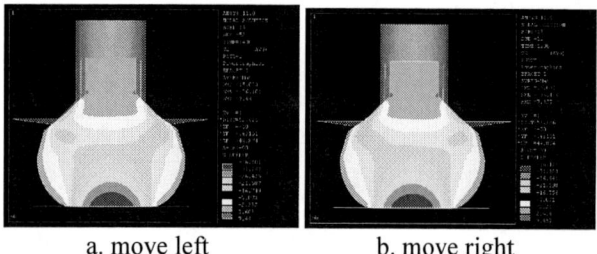

a. move left   b. move right

Fig. 6 Compressive stress distributions of the elastic FAB under vibration when H=15μm

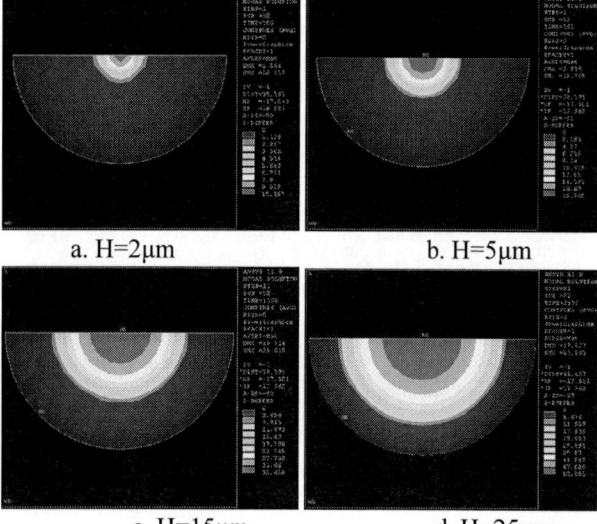

a. H=2μm   b. H=5μm

c. H=15μm   d. H=25μm

Fig. 7 Contact pressure distributions of the elastic FAB

Fig. 5 shows that there is a degressive stress ring at the section of the FAB. The maximum compression stress (Z-stress) occurred in the middle of the contact area and expand to the peripheral area gradually as the bond shift increase. When applying a vibration amplitude on the FAB for a sinusoidal period, the compression stress in bonding processing moves as bonder capillary moves left and right under pressure. Fig. 6 shows the typical result of compression stress under vibration when H=15μm.

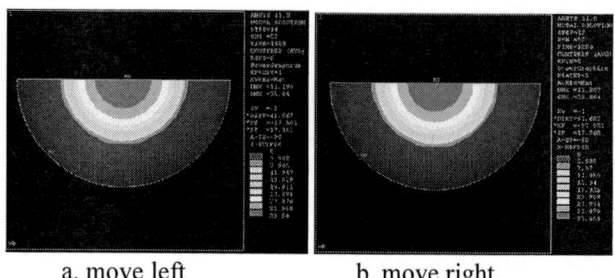

a. move left   b. move right

Fig. 8 Contact pressure distributions of the elastic FAB under vibration when H=15μm

The contact pressure distributions over the whole contact interface between the FAB and bond pad are studied in more details with reference to Fig.7. It's clearly seen that the maximum contact pressure occurred in the central of the contact interface. And there is a standard degressive stress ring in the contact interface. It's extending outside when the bond shift increase. There is little impact of ultrasonic vibration on the contact pressure distribution of the FAB when apply a vibration amplitude on the FAB under the bond shift H=15μm, as shown in Fig.8. Those results of elastic analysis meet the the Hertz contact theory under the elastic conditions.

In elasto-plastic analysis, some results are different from those of the elastic analysis. The bond shift is same as the elastic analysis. Fig.9 shows the deformation and compression stress distributions of the FAB for bond shift H=2μm, 5μm, 15μm and 25μm. At the low bond shift, the compression stress distribution of the FAB is the same as in elastic analysis. The FAB is still in condition of elastic deformation (Fig.9 a). When the bond shift increases, it's obvious that the maximum compression stress areas appear firstly at the periphery of the contact interface (Fig.9 c) and then gradually expand to both sides. The variation law of compression stress under the elasto-plastic deformation is the same as the evolution of plastic region [13]. When apply vibration amplitude on the FAB under the bond shift H=15μm, the compression stress in bonding processing moves the same as in elastic case, shown in Fig.10. By way of contrast the maximum compression stress circle, which is obvious different from elastic case.

Fig.11 shows the contact pressure distributions over the contact interface for bond shift H=2μm, H=5μm, 15μm and 25μm. We can see that the evolve way of contact pressure are the same as compression stress. The initial position of maximum stress may be connected to the geometry of the bonder capillary.

It can also be observed that impact of ultrasonic vibration on the contact pressure distribution of the FAB is quite clear when apply a vibration amplitude on the FAB under the bond shift H=15μm, as shown in Fig.12. When the bonder capillary moves in one direction, the contact pressure of the direction is higher than that of another direction. In conclusion, plasticity has a great impact on compressive stress distributions and contact pressure distributions of the FAB in comparison with elastic analysis.

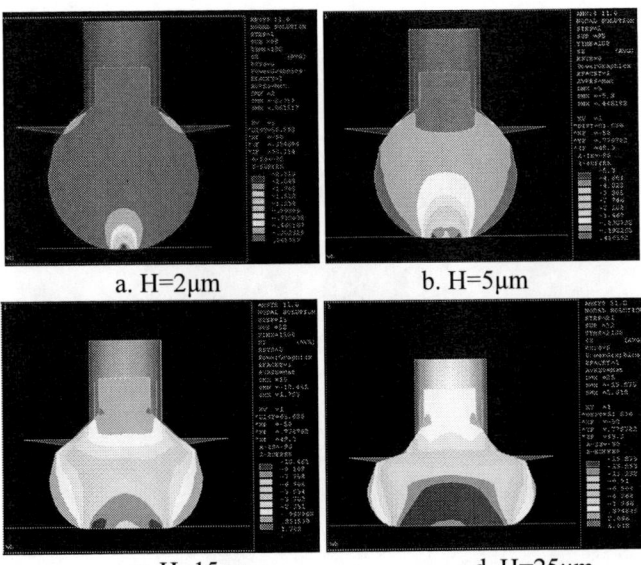

a. H=2μm   b. H=5μm

c. H=15μm   d. H=25μm

Fig. 9 Compressive stress distributions of the elasto-plastic FAB under pressure

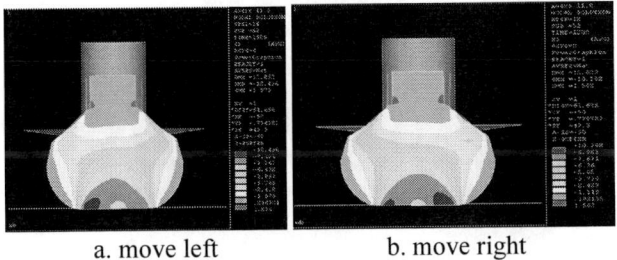

a. move left   b. move right

Fig. 10 Compressive stress distributions of the elasto-plastic FAB under vibration when H=15μm

3.2 Sliding result in the contact interface

According to the micro-slip theory, in bonding process, the relative motion at interface plays a very important role. So studying the slip law of the interface is significative. Since the model is symmetrical, the representative nodes with red dot in the contact area are study in detailed. The center of the contact area is marked with 'o'. Extend from the center to periphery, marked from '*1' to '*8'(*=A,B,C), as is shown in Fig.13. In order to express the slip law of the node in the process of vibration more clearly, the simulation is from 0s to 2000s. The curve of slip displacement versus time of the marked node in contact area under elastic deformation and elasto-plastic

deformation are shown in Fig.14. Fig.15 shows relationships of distributed node versus distributed areas of contact pressure under elasto-plastic deformation. We can simplify and summarize some points as follows:

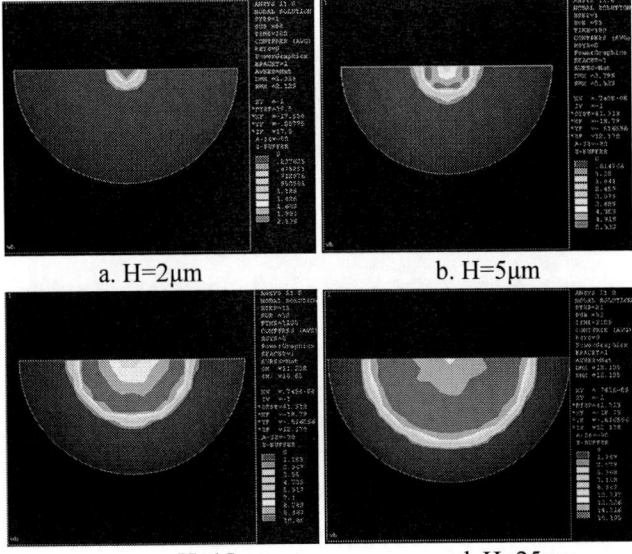

a. H=2μm   b. H=5μm

c. H=15μm   d. H=25μm

Fig. 11 Contact pressure distributions of the elasto-plastic FAB under pressure

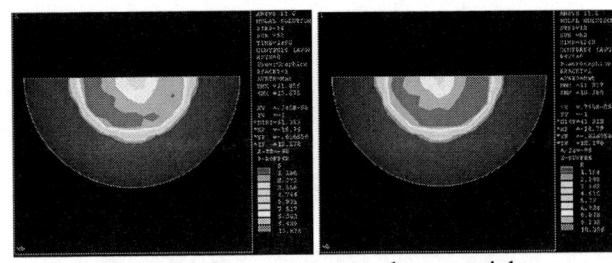

a. move left   b. move right

Fig. 12 Contact pressure distributions of the elasto-plastic FAB under vibration when H=15μm

a) The Fig.14-b(1-3) are typical elasto-plastic slip curves. Combining Fig.14-b(1-3) and Fig.15, we can see that when H=5μm, nodes A1-A3, nodes B1-B2 and nodes C1-C3 are in contact area, node A4, node B3, and node C4 are in perimeter zone, other nodes don't touch the pad. The slip curves of nodes A1-A4, nodes B1-B3 and nodes C1-C4 have a big jump because of tangential displacement when contacting. Then imposing supersonic vibration, the slip curves of nodes A1-A3, nodes B1-B2 and nodes C1-C3 are straight lines to explain that those nodes are nearly motionless. But slip curves of node A4, node B3 and node C4 appear concussion though node A4 is not clear, they shows nodes around perimeter zone slip. When H=10μm, nodes A1-A4, nodes B1-B3 and nodes C1-C4 are in contact area, node A5, node B4, and node C5 are in perimeter zone, other nodes don't touch the pad. The slip curves shown the same result as H=5μm, so do H=15μm and H=20μm. Those results are in accord with micro-slip theory

b) Fig.14—a(1-3) shows the similar discipline of node slip under elastic deform as Fig.14—b(1-3) under elasto-plastic deform. .The major difference of Fig.14—a(1-3) and Fig.14—b(1-3) is that the slip displacement IEEE has a difference of one order of magnitude. Elasto-plastic nodes slip displacement is larger than elastic nodes. When the vibration amplitude of capillary is 1μm, the slip amplitude of elastic node is about 0.002μm and the slip amplitude of elasto-plastic node is about 0.02μm. It shows that Micro-slip theory is suitable to elastic condition and elasto-plastic condition.

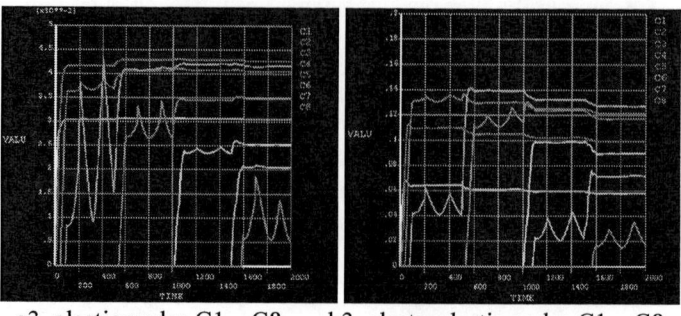

a3. elastic nodes C1---C8   b3. elasto-plastic nodes C1---C8

Fig. 14 The curve of slip displacement versus time of the marked node

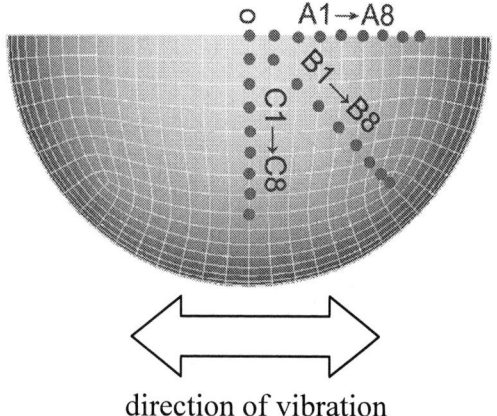

direction of vibration

Fig. 13 Nodes in contact area

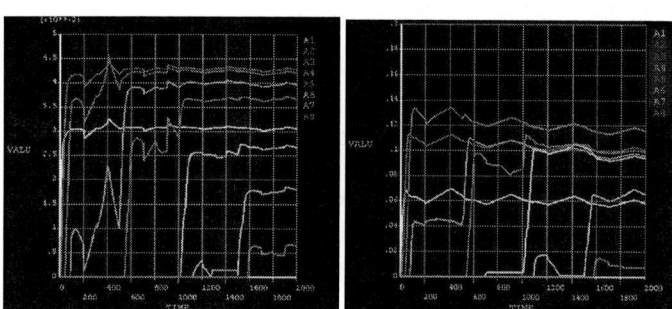

a1. elastic nodes A1---A8   b1. elasto-plastic nodes A1---A8

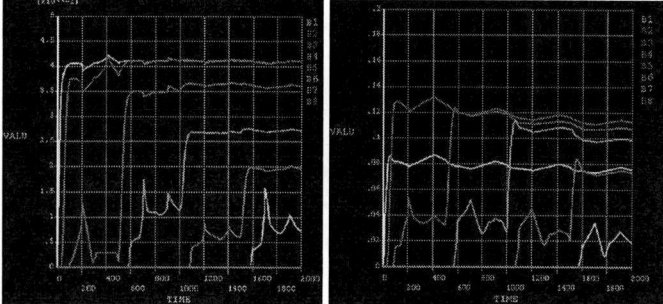

a2. elastic nodes B1---B8   b2. elasto-plastic nodes B1---B8

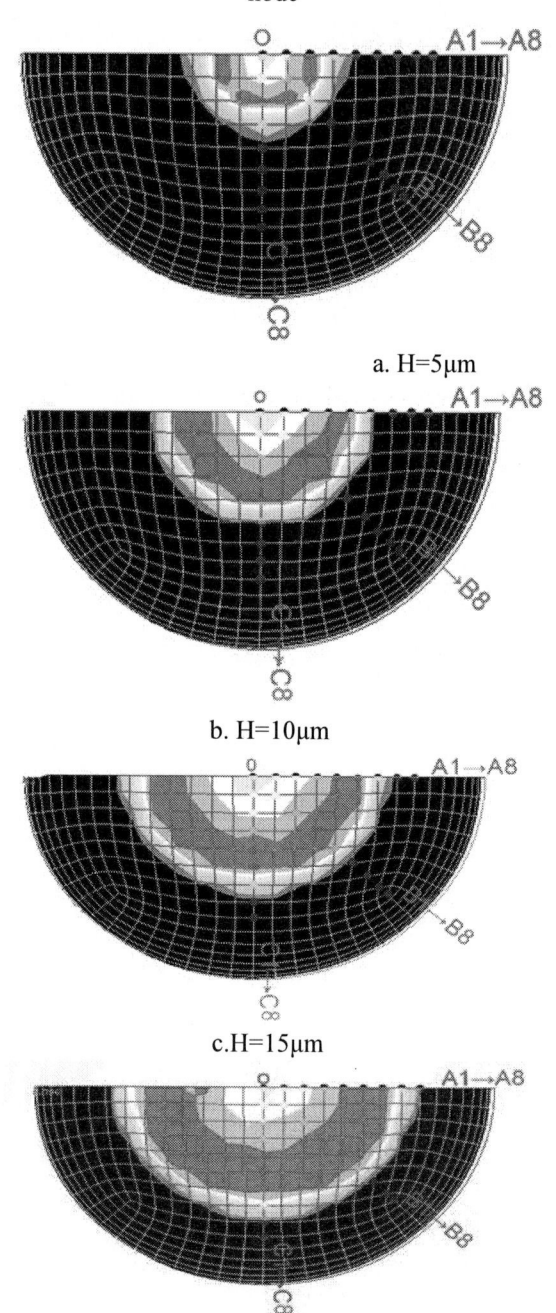

a. H=5μm

b. H=10μm

c. H=15μm

d. H=20μm

Fig. 15 Distributed node versus distributed areas of contact pressure under elasto-plastic deform

c) Fig.16 shows marked node movement under the loading of capillary that first move down bond shift of H=15μm one-stop, then apply a horizontal oscillation movement of the capillary with an amplitude of 1μm on the FAB six period. The results show that the slip amplitude won't change when the number of vibrations period increasing. Based on micro-slip theory the material in the vibration zone will come into connection. And the simulations do not include the material connection algorithm.

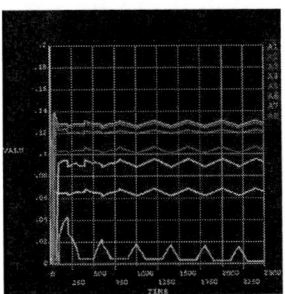 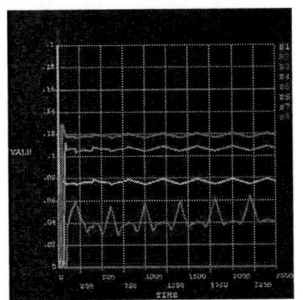

a. elasto-plastic nodes A1---A8   b. elasto-plastic nodes B1---B8

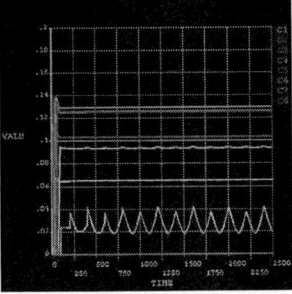

c. elasto-plastic nodes C1---C8

Fig.16 The curve of slip displacement versus time when the marked node vibrate many times under the condition of H=15μm

According to above results, we know that under both elastic condition and elasto-plastic condition, the slip law of contact surface is similar regardless of the difference of the stress distributions. They all indicate that only the perimeter area generates micro-slip. They also explain the "donut" phenomenon, but the slip displacement has a difference of one order of magnitude. This could be plastic lead to the molecule of the FAB generating some dislocation. It also reveals that through several micro-bond shifts, more contact area participates in the micro-slip vibration and connecting. It proves that a new process that will be mentioned is feasible.

4. Conclusions

Based on the results of FEA and the micro-slip theory, we propose a new bonding process replacing bond force control with bond shift control, which can improve the bonding strength of 1st ball solder point effectively. The process is as follows: By controlling bond shift to replace the bond force control. First, impose a suitable bond shift and ultrasonic power on the FAB to make the interface of FAB and pad in the status of stick-slip to form a donut bonding area that the small circle can be ignored [12], then adjust the ultrasonic power according to the increased bond shift to expand interface. Thus the new contact area will be in status of micro-slip in accordance with the micro-slip theory, while the previously formed bonding is still at rest because of the binding force and friction. In this way, the conjunction control of bond shift and ultrasonic power is able to form a circular bonding area to achieve higher strength requirements with least unconnected area. In this paper, through FEA to simulate the bonding process, it helps people understand the mechanism of 'donut' phenomenon from the standpoint of mechanics and tribology and put forward a new process to improve the strength of thermo-sonic ball bonding.

Further work includes the process need to be confirmed by experiment and studying the impact of displacement amplitude and friction on the slip displacement by FEA.

Acknowledgments

The work described in this paper is partially supported by the National Natural Science Foundation of China (NSFC 60806050).

References

1. HARMAN G G, "Experimental Model of the Microelectronic Ultrasonic Wire Bonding Mechanism," *Proceeding 10th Annual Reliability Physics Symposium*, Vol. 8, No. 4 (1972), pp. 49-56.
2. Sheaffer, M, Levine L, "How to optimize and control the wire bonding process," *Part II, Solid State Technol.*, 1991, pp .67-70.
3. Mayer M, Schwizer J, "Ultrasonic bonding: understanding how process parameters determine the strength of Au-Al Bonds," *Proc. Intl. Symposium on Microelectronics,* IMAPS, Denver, Co, USA, 2002, pp. 626-631.
4. Harman G G, Leedy K O, "An experimental model of the microelectronic ultrasonic wire bonding mechanism," *10th annual proceedings reliability physics symposium,* 1972, pp. 49-56.
5. Schwizer J, Mayer M, Bolliger D, et al., "Thermosonic ball bonding : friction model base on integrated microsensor measurements," *IEEE Intl. Electronic Manufacturing Technology Symposium IEMT'99*, Austin Texas, Oct. 18-19, pp. 108-144.
6. Chen G K C, "The role of micro - slip in ultrasonic bonding of microelectronic dimensions," *International Hybrid Microelectronic Symposium*, 1972, pp. 5A11-5A119.
7. Mindlin RD, "Compliance of elastic bodies in contact," *Trans. ASME, Series E, Journal of Applied Mechanics,* Vol. 16 (1949), pp. 259-268.
8. Ikeda T, Miyazaki, Kudo K, et al, "Failure estimation of semiconductor chip during wire bonding process," *ASME J of Electronic Packaging,* Vol. 121, No.2 (1999), pp. 85-91.
9. Zhong Z W, Gon K S , "Analysis and experiments of ball deformation for ultra-fine-pitch wire bonding," *J of Electronics Manufacturing,* Vol. 10, No.4 (2000), pp. 211-217.
10. Takahashi Y, Inoue M, "Numerical study of wire bonding analysis of interfacial deformation between wire and pad," *ASME J of Electronic Packaging,* Vol. 124, No. 1 (2002), pp. 27-36.
11. Langenecker B, "Effects of ultrasound on deformation characteristics of metals," *IEEE transactions Oil sonic and ultrasonic,* Vol. 13, No. 1 (1966), pp. 1-8.

12. Jun Qi, Ngar Chun Hung, Ming Li, Deming Liu, "Mechanism Analysis of Process Parameters Effects on Bondability in Ultrasonic Ball Bonding," *High Density Microsystem Design and Packaging and Component Failure Analysis,* June 2005, pp. 1-5.

13. Chang WR, Etsion I, Bogy DB, "Static friction coefficient model for metallic rough surfaces," *Journal of Tribology, Transactions of the ASME*, Vol. 110 (1988), pp. 57-63.

Improving the Toughness and Thermal Properties of Epoxy Resin Using for Electronic Packaging by Interpenetrating Polymer Network

Bo Chen, Dayong Gui, Jianhong Liu

School of Chemistry and Chemical Engineering, Shenzhen University, Shenzhen, 518060, China.
Bo Chen, Telephone: 13421370592, Email: chenbo533349@163.com;
Dayong Gui, Telephone: 15818546712, Email: dygui@szu.edu.cn;
Jianhong Liu, Telephone: 13902987715, Email: liujh@szu.edu.cn

Abstract

A series of Polydimethylsiloxane-Epoxy resin interpenetrating polymer networks (PDMS-ER IPNs) were synthesized with Hydroxyl-terminated Polydimethylsiloxane (PDMS), Diglycidyl ether of bisphenol A (Epoxy resin, ER), Toluenediisocyanate (TDI), 1,4- Butanediol and Dibutyltin dilaurate (DBTDL) as catalyst. The reaction completion was confirmed by the disappearance of the characteristic -NCO peak (at 2270cm^{-1}) in the IR spectrum. Cured of PDMS-ER IPN and unmodified ER were investigated by measuring its tensile strength, break elongation, thermo-gravimetric curve, water absorption and the morphology of fractured surfaces. The results show that the PDMS-ER IPN with correct ratios has excellent mechanical properties with better break elongation, tensile strength, bending strength and flexural modulus than the unmodified ER. The toughening effect of the IPN was demonstrated by the SEM micrographs of the fractured surface of the samples. Water absorption tests and thermogravimetric analysis showed that the PDMS-ER IPN exhibited better water resistance and thermal stability than the unmodified ER. The PDMS-ER IPN with high toughness and thermal resistance is suitable for the use of electronic packaging.

1. Itroduction

Epoxy resin (ER) is an important polymeric material with outstanding properties of high modulus, high electrical resistance, and excellent adhesive properties. It is widely used as composites, coatings, structural adhesives, and microelectronics. However, the main disadvantage associated with its application is related to brittleness, thus improving the fracture toughness of ER becomes an interesting topic [1-2]. Polydimethylsiloxane (PDMS) is remarkable for its low dielectric loss, low water absorption, high electrical resistance, good flexibility and thermal stability as well as chemical inertness, so PDMS may act as the desirable candidate for improving the toughness and thermal stability of ER. But PDMS is immiscible with ER, a simple blend can only lead to an obvious phase separation [3]. Interpenetrating polymer network (IPN) is polymer alloys consisting of at least two polymers, held together by permanent entanglements with covalent bonds between the chains of the two different types of polymer, and which exhibit better properties than the pure polymer. The Toluene diisocyanate (TDI) was also used as a coupling agent between the ER and hydroxyl-terminated polyether [4]. There are two kinds of approaches to prepare an IPN, namely sequential and simultaneous polymerization. Sequential IPN is generally prepared by swelling the first-formed network with the second monomer, which is then polymerized in situ. Simultaneous IPN results from both reactions proceed simultaneously [5].

In this study, the Polydimethylsiloxane-Epoxy resin IPN (PDMS-ER IPN) was synthesized from the hydroxyl-terminated polydimethylsiloxane and diglycidyl ether of bisphenol-A (E51, DGEBA) to modify the characteristics of the latter. The effects of modification for the cured materials were investigated by measuring their tensile strength, bending strength, break elongation, thermo-gravimetric curve, water absorption and the morphology of fractured surfaces.

2. Materials and sample preparation

2.1 Materials

Diglycidyl ether of bisphenol-A (DGEBA, E51, Huntsman, Switzerland), Polyether amine (D230, Huntsman, Switzerland), Isophorone diamine (IPDA, Huntsman, Switzerland), Hydroxyl-terminated Polydimethylsiloxane(PDMS, 42mP·s(25°C), Shenzhen Chuying Chemical Industry Co., China), Toluene diisocyanate (TDI, Tianjin Damao Chemical Reagent Co., China), Dibutyltin dilaurate (DBTDL, Shanghai Chemical Reagent Co., China), 1, 4-Dutanediol(BDO, Shanghai Jingchun Reagent Co., China).

2.2 Sample preparation

For the reaction, PDMS, TDI, ER and Dibutyltin dilaurate (DBTDL) as catalyst were added to a 100 mL three-necked round-bottomed flask under a nitrogen atmosphere. The mixtures were stirred using magnetic stirrer for about 1.5 hrs at 60°C. The product was then reacted with the 1, 4-butanediol for about 1.0 hr at the same temperature to form the PDMS-ER IPN. The reaction completion was confirmed by the disappearance of the characteristic -NCO peak (at 2270cm^{-1}) in the IR spectrum. Then, a stoichiometric amount of curing agent based on the epoxy groups was added. After the addition of the curing agent, the mixture was gently mixed, degassed in vacuum oven for 40 mins, and cast into molds with appropriate dimensions for tensile. The samples were cured at 80°C for 2 hrs, 120°C for 2 hrs and 150°C for 2 hrs.

2.3 Methods and measurements

The mechanical tests were measured with CMT4304 universal testing machine (U.S. Waters Industrial Systems (China) Co., Ltd.) at room temperature according to the Chinese standard GB/T9341-2000, GB/T1040-92. The results represented an average of five specimens.

Thermal stability was determined with a thermogravimetric analyzer (TGA) (STA409PC, NETZSCH Instruments Co., Germany) over a temperature range of 30-900°C at a heating rate of 10°C/ min.

Scanning electron microscope (SEM) (JSM-5910LV, JEOL Instruments Co., Japan) was employed to examine the morphology of the surfaces of fractured samples.

The water absorption tests were carried out with the conditioned specimens were entirely placed in a constant temperature and humidity box which maintained the temperature at 40±1°C, humidity at 85±3% for a definite period according to the Chinese standard GB/T2423.9-2001. The specimens were weighed immediately to obtain the weight, W_1. The percentage increase in the weight of the samples was calculated by using the formula $(W_1 - W_0)/W_0 \times 100\%$, W_0 is the initial weight of specimens. (LHP-60, Shanghai Hongdu Electronic Technology Co., China).

3. Results and discussion

3.1 Mechanical properties test

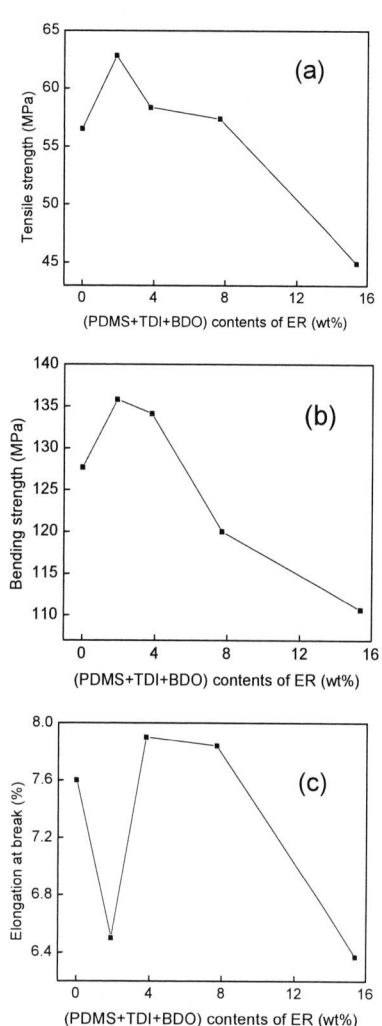

Fig.1 Effect of the (PDMS+TDI+BDO) content of ER on the mechanical properties of the cured PDMS-ER IPN: the (PDMS+TDI+BDO) contents of ER are 0wt% (neat ER, Sample A）; 1.9wt% (Sample B) ; 3.8wt% (Sample C) ; 7.7wt% (Sample D) ; 15.4wt% (Sample E）

The tensile strength and bending strength of the cured neat ER and PDMS-ER IPN with various (PDMS+TDI+BDO) contents of ER are shown in Fig. 1(a-b). Both the tensile strength and bending strength of the cured PDMS-ER IPN increased within the addition of 3.8wt% the (PDMS+TDI+BDO) contents of ER. When the (PDMS+TDI+BDO) contents are more than 3.8wt%, the tensile strength and bending strength decrease moderately. The tensile strength and bending strength of the cured PDMS-ER IPN reach a maximum at 3.8wt% the (PDMS+TDI+BDO) contents of ER where the tensile strength and bending strength of the cured PDMS-ER IPN improved 11.1% and 6.4% than that of the cured neat ER, because the crosslink density of the cured PDMS-ER IPN may reach a maximum. When the (PDMS+TDI+BDO) were added, the morphology and interpenetration degree of the PDMS/ER matrix may break up and result in bad mechanical properties.

Fig. 1(c) depicts the elongation at the break for the cured neat ER and PDMS-ER IPN with various (PDMS+TDI+BDO) contents of ER. There is an appreciable decrease at the (PDMS+TDI+BDO) contents of ER is 1.9wt% where maybe the crosslink density of the cured PDMS-ER IPN may reach a maximum and the tensile strength and bending strength of the cured PDMS-ER IPN reach the maximum, a higher crosslink density in the PU/ER matrix restricts the molecular mobility, micro-phase separation and the PDMS dispersion degree. When the (PDMS+TDI+BDO) contents are about 3.8wt%, the interpenetration degree of the PU/ER matrix is higher and the grafting reaction between PDMS and ER is perfect, which result in excellent elongation at the break. Moreover, when the (PDMS+TDI+BDO) contents are above 3.8wt%, the dispersion degree, morphology and interpenetration degree of the PDMS/ER matrix may deteriorate and cause brittleness.

3.2 SEM analysis

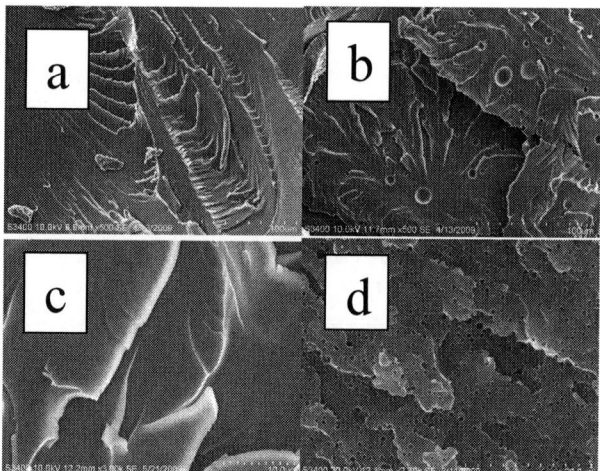

Fig.2 SEM micrographs of the cured neat ER and PDMS-ER IPN with 3.8 wt% (PDMS+TDI+BDO) contents of ER: (a) 0 wt% (neat ER ,Sample A×500; (b) 3.8 wt%(Sample C×500); (c) higher magnification of (a) ×3000; (d) higher magnification of (b)×2700

Fig. 2(a) shows SEM micrographs of fractured surface of the cured neat ER (Sample A). The micrograms for the neat epoxy system specimens exhibit smooth and homogeneous surface, stress orientation consistent, these are the characteristics of brittle break. The fractured surface of cured PDMS-ER IPN with 3.8 wt% the (PDMS+TDI+BDO) contents of ER (Sample B) in Fig. 2(b) is rough and stress orientation inconsistent, these are the features of fracture toughness. The microstructure of polymers is very important in determining their physical properties. The microstructure is consistent with the mechanical properties. Fig. 2(c-d) are higher magnification SEM micrographs of fractured surface of Sample A and Sample B, the fractured surface of Sample B reveals that PDMS particles of 8μm or less in size are dispersed in ER in a "sea-island" structure ("island" of PDMS dispersed in the "sea" of ER). The cavity formation and deformation seem to be the toughening mechanism. Nevertheless, Sample A even at higher magnification (Fig. 2(c)) than Sample B (Fig. 2(d)) the SEM micrographs of fractured surface is still smooth and homogeneous.

3.3 Thermal Stability

Table 1 Thermal Properties of the Cured PDMS-ER-IPNs and neat ER

(PDMS+TDI+BDO)/ER (wt%)	Temperature of on-set deflection (°C)	Temperature of weight loss 50% (°C)
0 (neat ER)	357.7	385.4
1.9	363.4	389.5
3.8	363.4	389.6
7.7	361.9	389.5
15.4	361.3	389.4

The thermal stability behavior of the cured PDMS-ER IPNs with various (PDMS+TDI+BDO) contents of ER was studied by TGA under a nitrogen atmosphere. Table 1 lists some characteristic transition temperatures for the various cured PDMS-ER IPNs and neat ER. The cured neat ER shows an on-set deflection temperature at 357.7°C, and a major weight-loss temperature (i.e., 50% weight loss) at 385.4°C. All the cured PDMS-ER IPNs exhibit better thermal stability than does the cured neat ER.

This result may be attributed to the thermal stability of the incorporated Si-O-Si structure of the PDMS.

3.4 Water absorption

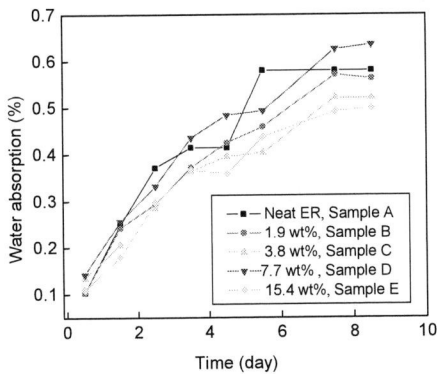

Fig.3. Water absorption of the cured neat ER and various PDMS-ER IPNs

Water absorption in the package was found not only to plasticize the epoxy resin, causing a lowering of the T_g and in turn affecting mechanical response, but also to cause package cracking [6]. This cracking is caused by evaporation and expansion of absorbed moisture in the package. So the applications of ER are limited because its otherwise excellent properties are greatly affected by water absorption. It is thus important to study the change in water adsorption before and after modification. The number of hydrophilic groups, free volume and morphology induced by phase separation of PDMS are significant in determining the ultimate water absorption. Water absorption measurements were carried out according to the Chinese standard GB/T2423.9-2001 and the results are shown in Fig. 3. The cured PDMS-ER IPNs show a lower water absorption rate and the balance of water absorption more slowly than the cured neat ER does except sample D. The cured neat ER has higher water absorption because of the existence of many hydrophilic groups. The lower water absorption of the cured PDMS-ER IPNs result from increased crosslink density and decreased the number of hydrophilic groups of the cured neat ER by the reaction between PDMS and ER. With the (PDMS + TDI + BDO) contents of ER increasing in sample D and sample E, the rate of water absorption and water absorption show two opposite trends in comparison with other components, because with the increase of the (PDMS + TDI + BDO) contents of ER, the phase separation in samples D expand and more easily lead to water infiltration in it; the phase separation of sample E may be greater separation than samples D do, but the excess PDMS with low surface tension will migrate to ER surface, PDMS with hydrophobic groups attached to ER surface lead to a reduction in the rate of water absorption and water absorption. A decrease in water absorption means that the material has more stable properties, which is interesting for practical applications [7].

4. Conclusions

Various PDMS-ER IPNs were synthesized by IPN with Polydimethylsiloxane, Toluene diisocyanate, 1, 4-Dutanediol and Epoxy resin. The results indicate that the cured PDMS-ER IPN with content of 3.8 wt% (PDMS+TDI+BDO) of ER shows better elongation at break, tensile strength, bending strength and lower water absorption than the cured neat ER, so it's suitable for electronic packaging. These characteristics will make it an ideal candidate for electronic applications.

References

1. Jansen B. J. P., *et al*, "Rubber-Modified Glassy Amorphous Polymers Prepared via Chemically Induced Phase Separation. 2. Mode of Microscopic Deformation Studied by in-Situ Small-Angle X-ray Scattering during Tensile Deformation," *Macromolecules*, Vol. 34, No. 12 (2001), pp. 4007-4018.

2. Rebizant V., *et al*, "Chemistry and Mechanical Properties of Epoxy-Based Thermosets Reinforced by Reactive and Nonreactive SBMX Block Copolymers," *Macromolecules*, Vol. 37, No. 21 (2004), pp. 8017-8027.

3. Lin R. H., *et al*, "In situ FT-IR and DSC investigation on the cure reaction of the dicyanate/diepoxide/diamine system," *Polymer International*, Vol. 50, No. 10 (2001), pp. 1073-1081.

4. Soares B. G., *et al*, "Toughening of an Epoxy Resin with an Isocyanate-Terminated Polyether," *Journal of Applied Polymer Science*, Vol. 108, No. 1 (2008), pp. 159-166.

5. Sperling L. H., *et al*, "The Current Status of Interpenetrating Polvrner Networks," *Polymers for Advanced Technologies*, Vol. 7, No. 4 (1996), pp. 197-208.

6. Ho T. H., *et al*, "Modification of epoxy resins with polysiloxane thermoplastic polyurethane for electronic encapsulation: 1," *Polymer*, Vol. 37, No. 13 (1996), pp. 2733-2742.

7. Jia Q. M., *et al*, "Synthesis and characterization of polyurethane/epoxy interpenetrating network nanocomposites with organoclays," *Polymer Bulletin*, Vol. 54, No. 1 (2005), pp. 65-73.

Effects of Service Parameters on Thermomechanical Fatigue Behaviors of New Nano Composite Solder Joints

F. Tai, F. Guo, B. Liu, Z.D. Xia, Y.W. Shi

College of Materials Science and Engineering, Beijing University of Technology

Beijing, 100124, China, E-mail: guofu@bjut.edu.cn

Abstract

Thermomechanical fatigue (TMF) resulted from the mismatch in the coefficient of thermal expansion (CTE) between solder and substrate would degrade the mechanical properties of solder joints during service. In this research, nano-structured polyhedral oligomeric silsesquioxane (POSS) reinforcing particles were incorporated into a promising lead-free solder, eutectic Sn-3.5Ag solder, by mechanically mixing to improve service reliability of the base solder matrix. Three different temperature profiles, with the same temperature extremes of -40°C to 125°C, different dwell times at temperature extremes and ramp rates, were applied in the tests. Microstructural characterization of surface damage and residual shear strength of these solder joints were carried out after 0, 100, 250, 500 and 1000 TMF cycles. Results obtained from this study were used to analyze the effect of service parameters on the TMF behaviors of the nano composite solder joints. Experimental results indicated that the nano composite solder joints exhibited better TMF performance than the eutectic Sn-Ag solder joint. It also proved that the nano composite solder joints that experienced longer dwell time and slower ramp rate exhibited less surface damage accumulation and less decrease in shear strength of solder joints.

Introduction

In surface mount technology (SMT), one of major cause for the failure of electronic components is due to failure of solder joints from Thermomechanical fatigue (TMF) caused by the mismatch in the coefficient of thermal expansion (CTE) between the solder and Cu substrate [1]. Hence, it is very significant to evaluate residual properties and surface damage accumulations of solder joints after TMF.

The eutectic Sn-3.5Ag solder is considered as a promising Pb-free solder substitute for its non-toxic, comparable wetting characteristics and better mechanical properties [2, 3]. TMF damage accumulation and resultant residual mechanical properties of Sn-Ag based solder joints have been documented in many publications [1, 4, 5]. Several approaches have been used to improve mechanical reliability of solder joints, especially TMF performance. The composite solder approach has been considered as a potential method to improve the TMF properties of solder joints. In the fabrication of traditional composite solders, reinforcing particles have tendency to coarsen. The new nano-structured materials of polyhedral oligomeric silsesquioxanes (POSS) incorporated into the solders can solve this problem through promoting bonding between nano-structured reinforcing particles and matrix and preventing the inert particles from reacting with solder matrix in service [6]. Besides, the service parameters in the TMF profile, such as temperature extreme, dwell time,

ramp rate and so on, are important to the TMF properties of solder joints [7]. The effects of service parameters on TMF damage behavior and residual shear strength were need to be comprehend systemically in order to enhance TMF resistance of lead-free solder joints.

Due to these considerations, surface damage evolution and residual mechanical behavior of the nano-structured POSS particles reinforced composite solder joints that experience different TMF service conditions have been investigated and compared with the eutectic Sn-Ag solder joints in this research. Besides, this study also investigate the effects of different ramp rates and dwell times at temperature extremes during TMF on the damage accumulation and residual shear strengths of nano composite solder joints.

Experimental Procedures

1. Polyhedral Oligomeric Silsesquioxanes (POSS)-TriSilanolCyclohexyl

The nano-structured materials technology of POSS, with appropriate organic groups, can produce suitable means to promote bonding between nano-structured reinforcing particles and matrix, and can prevent the inert particles from reacting with solder matrix in service. The nano-structured POSS reinforcing particles used in this research is the POSS-TriSilanolCyclohexyl. The anatomy of the POSS-TriSilanol molecule used in this study is shown in Fig. 1. In the soldering, the surface-active groups promote bonding between POSS molecules and matrix. This bonding is achieved, POSS molecules will no longer react with the matrix, and can not coarsen or agglomeration during service.

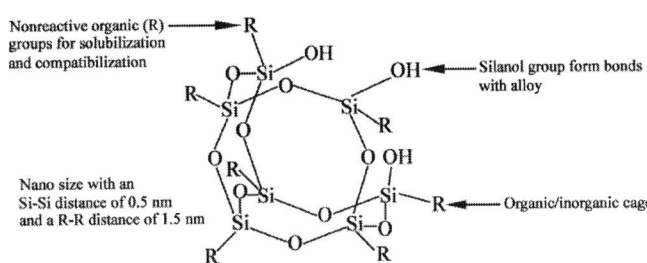

Fig. 1 The anatomy of the POSS-TriSilanol molecule used in this study

The nano composite solders were prepared by incorporating the nano-structured POSS-TriSilanolCyclohexyl reinforcing particles into the eutectic Sn-3.5Ag solders by mechanical-mixing methods. It has been proved that those nano composite solders can improve service performance, such as mechanical property, creep property and reliability of solder joint [8, 9]. According to the results of the nano-composite solder joints in our former researches, the best weight fraction of the nano-structured POSS-TriSilanolCyclohexyl reinforcing particles is 3 wt. % [9]. The morphology of the nano-structured POSS-TriSilanolCyclohexyl particulates observed using transmission electron microscopy (TEM) have been reported in our former research [9].

2. TMF Property Testing

The current study dealt with solder joints made with the eutectic Sn-Ag solder and its nano composite solder pastes to compare surface damage accumulation and residual shear strength properties due to TMF. The detailed shape and dimensions of the single shear-lap solder joint and soldering procedure used in this research have been reported earlier [9]. Cross sections of solder joints were metallurgically polished prior to subjecting them to the TMF testing. The procedures of metallurgical polishing have been documented [9].

The TMF temperature profiles with controllable ramp rate and dwell times were designed in this study. According to the former TMF studies on the eutectic Sn-3.5Ag solder joint, it indicated that the transition between shear banding and grain boundary sliding occurred at a temperature range in the neighborhood of 125°C [7]. So the thermal cycling between -40°C and 125°C was imposed on the solder joints in the TMF testing. The ramp rates of the TMF cycle were 8.25°C /min and 33.3°C /min. The dwell times at temperature extremes of TMF cycle were 10mins and 15mins. The three different TMF temperature profiles with different ramp rates and dwell times are given in Fig. 2 in this study.

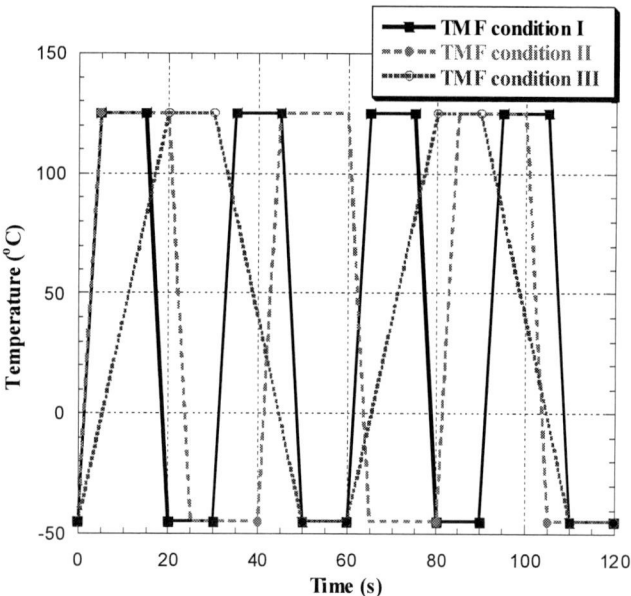

Fig. 2 TMF temperature profiles in this research

In this current study, surface damage accumulation at the same area was observed in eutectic Sn-Ag solder and nano composite solder joints after 0, 100, 250, 500 and 1000 TMF

cycles by scanning electron microscopy (SEM). Besides, the residual shear strength of solder joints result from TMF were assessed under a tensile testing machine at a shear speed of 0.01mm/s at room temperature. Moreover, the TMF properties of nano composite solder joints in different TMF conditions were compared in order to investigate the effects of service parameters on TMF performance of nano composite solder joints.

Results and Discussion

1. The Microstructure of As-fabricated Solder Joints

Representative microstructures of the as-fabricated eutectic Sn-3.5Ag solder and nano omposite solder joints were shown in Fig. 3. As can be seen in Fig. 3 (a), the eutectic Sn-3.5Ag solder matrix microstructure is characterized by eutectic Ag3Sn phase located around Sn grains. There are scallop shaped Cu6Sn5 intermetallic compound (IMC) layer presented at the interface between theCu substrate and the solder matrix. The microstructure characterization of the nano composite solder was similar to that of the eutectic Sn-3.5Ag solder, as shown in Fig. 3 (b).

2. Surface Damage Accumulation in Sn-Ag Based Solder Joints during TMF

The surface damage accumulation of the thermomechanically-fatigued eutectic Sn-3.5Ag solder joints under TMF condition I were shown in Fig. 4. TMF surface damages in the eutectic Sn-3.5Ag solder joint occurred at the interface between Cu substrate and the solder matrix, and extended into solder matrix with an angle of 45o, shown in Fig. 4 (a). Such damages and cracks indicated that the stress, which promoted the damage accumulation and crack extension in the eutectic Sn-3.5Ag solder joint, was shear stress. The surface damage at interfacial IMC layer became more serious with the increasing of TMF cycles. The surface damage accumulation and cracks of resulting from TMF in the eutectic Sn-3.5Ag solder joints was found to be the worst after 1000 TMF cycles, as shown in Fig. 4.

The TMF damage accumulations of the nano composite solder joint after various TMF cycles at the same TMF condition are illustrated in Fig. 5. From Fig. 5 (a), the TMF damages inhabited at the Cu substrate/solder interface, and did not extended into the solder matrix. This phenomenon was different from what was observed in the eutectic Sn-3.5Ag solder joint. Besides, the nano composite solder joints showed no such surface damage because there was not serious surface damage present in the interface of solder joints even after 1000 TMF cycles, as shown in Fig. 5 (d). As comparison of Fig. 4 and Fig. 5, it suggests that the nano composite solder joint exhibits less surface damage than the eutectic Sn-3.5Ag solder joint under the same TMF condition I. It provided the reason for better TMF performance of the nano composite solder joint as compared to the eutectic Sn-3.5Ag solder joint.

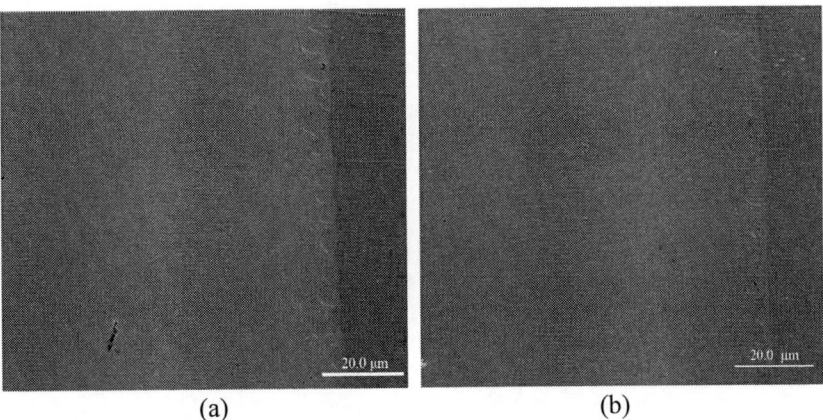

(a)   (b)

Fig. 3 The microstructures of the as-fabricated eutectic Sn-3.5Ag solder and its nano composite solder (a) eutectic Sn-3.5Ag solder, (b) nano composite solder

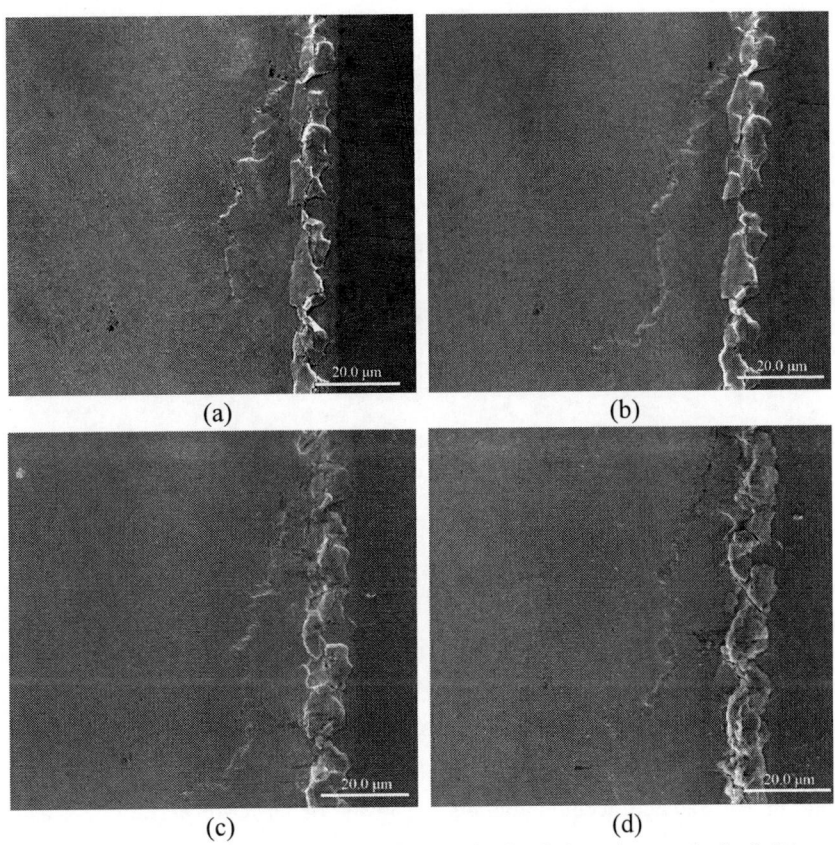

(a)   (b)

(c)   (d)

Fig. 4 The TMF surface damage accumulation of the thermomechanically-fatigued eutectic Sn-3.5Ag solder joints after various TMF cycles under TMF condition I (a) 100 cycles, (b) 250 cycles, (c) 500 cycles, (d) 1000 cycles

In this nano-structured POSS particles reinforced composite solders, Si-O cages that can be strongly bonded to Sn at the Sn grain boundaries are present. Active radicals that were present at the surface of such cages have used to strongly bond the Si-O cage to Sn, especially at the grain boundary. Such grain boundary reinforcements could improve the strength of solder joints, especially TMF properties. Besides, Cu_6Sn_5 IMC layers present in the eutectic Sn-3.5Ag solder matrix coarsened significantly during TMF, and developed large cracks. If cracks occurred at the boundaries, they would increase the damage accumulation during TMF and deteriorate the residual shear strength of the eutectic Sn-3.5Ag solder joint. There are no serious cracks resulting from

TMF surface damage accumulation in nano composite solder joint. So the nano composite solder joint posed better TMF performance as compared to the eutectic Sn-3.5Ag solder joint.

3. Comparison of Surface Damage Accumulation in Nano Composite Solder Joints with Different Ramp Rates and Dwell Times

Microstructures comparing surface damage accumulation after 1000 TMF cycles in the nano composite solder joints studied the different ramp rate and dwell times at temperature extremes are presented in Fig. 6. The amount of visible surface damage in the nano composite solder joints after 1000TMF cycles was similar for specimens experiencing different ramp rate and dwell time at temperature extremes as illustrated in and Fig. 6. As can be seen in Fig. 6 (a), only

small surface damages in the nano composite solder joint with slower ramp rate and constant dwell time, while TMF surface

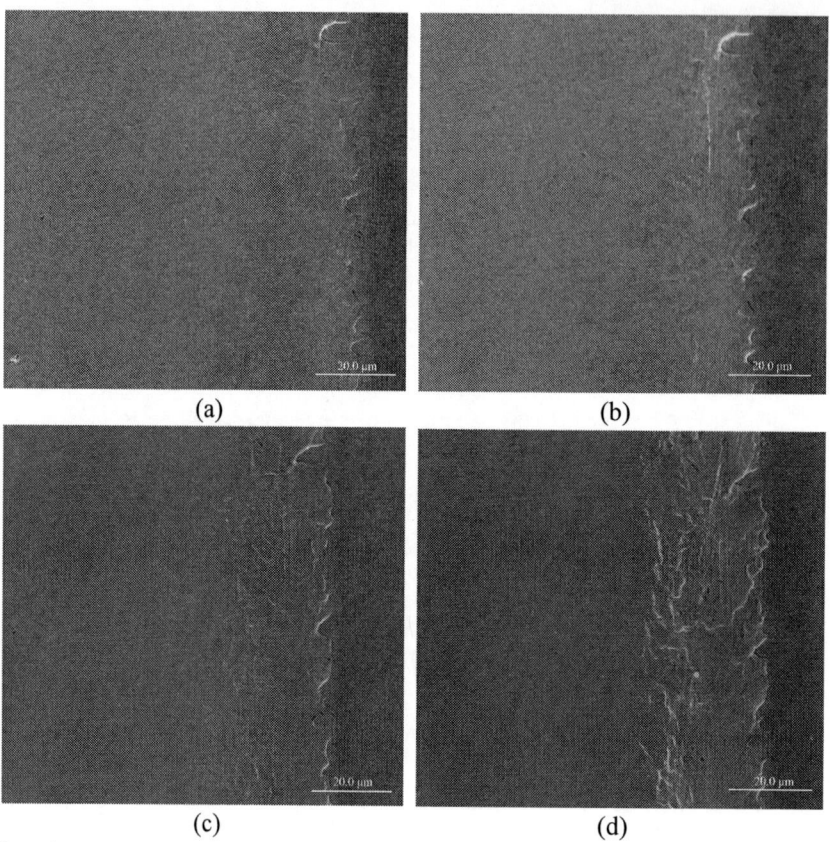

(a)   (b)

(c)   (d)

Fig. 5 The TMF surface damage accumulations of the nano composite solder joint after various TMF cycles under TMF condition I (a) 100 cycles, (b) 250 cycles, (c) 500 cycles, (d) 1000 cycles

damages were deeply located at the Cu substrate/solder interface in the nano composite solder joint with faster ramp rate (shown in Fig. 5 (d)). The reason is that the faster ramp rate might cause higher internal stresses due to the mismatch in coefficient of thermal expansion (CTE) between solder and substrate, which give rise to more serious damages of the interfacial joining. As shown in Fig. 6 (b), the nano composite solder joints exhibited no serious surface damage with longer dwell time at temperature extremes and constant ramp rate of 33°C/min. However, some surface damage accumulation is observed in the nano composite solder joint for shorter dwell time at temperature extremes, as can be seen in Fig. 5 (d). During the dwell time, stress relaxation will take place, resulting in TMF damage repaired by recovery / recrystallization / crack healing, or additional crack growth from residual stresses [7]. The cyclic straining at high temperature had a tendency to heal the damage encountered. This study revealed there was lesser surface damage when the solder joints experienced a longer dwell time at temperature extremes.

4. Residual Shear Strength of Sn-Ag Based Solder Joints during TMF

Compared to the residual shear strengths of the eutectic Sn-3.5Ag solder and nano composite solder joints under TMF condition I was provided in Fig. 7. As can be observed from Fig. 7, these is an initial obvious drop in residual shear strength as a consequence of 250 TMF cycles in the eutectic

Sn-Ag solder and nano composite solder joints. Then, the stabilized decrease of the residual shear strength was noted between 250 and 1000 TMF cycles. The reason is that surface damage generation and accumulation occurred from very early stages of TMF as above mentioned. Besides, the shear strength of the nano composite solder joint decreased with the increasing of TMF cycles, while was slower than the decreasing of shear strength of the eutectic Sn-3.5Ag solder joint. So it indicated that the nano composite solder joints showed better TMF performance as compared to the eutectic Sn-3.5Ag solder joint.

The residual shear strength of the nano composite solder joints that experienced different ramp rates and dwell times at temperatures were also compared in this section. The effects of different service parameters on the residual shear strength of nano-composite solder joint after various TMF cycles were illustrated in Fig. 8. The residual shear strength reduced sharply after 100 cycles and dropped when TMF cycles increased at both TMF conditions. As can be seen in Fig. 8, the shear strength of nano composite solder joint decreased from 55.09 MPa to 46.31 MPa after 1000 TMF cycles at the ramp rate of 33°C/min (TMF condition I), while reduced to 48.02 MPa at the ramp rate of 8.25°C /min (TMF condition III) with the same dwell time. It indicated that faster ramp rate had brought the solder joints less time for the healing of cracks. Therefore, the shear strengths of nano composite solder joint were decreased excessively after TMF cycles at a faster ramp rate. Besides, as can be seen in Fig. 8, the drop in the residual shear strength was approximately 13 % after 1000 TMF cycles for nano composite solder joints that experienced longer dwell time at temperature extremes.

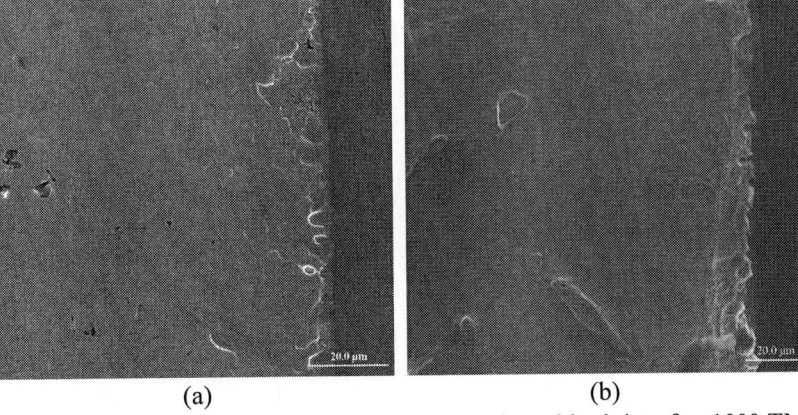

(a)   (b)

Fig. 6 The microstructure of thermomechanically-fatigued nano composite solder joint after 1000 TMF cycles under different TMF condition (a) slower ramp rate, (b) longer dwell time

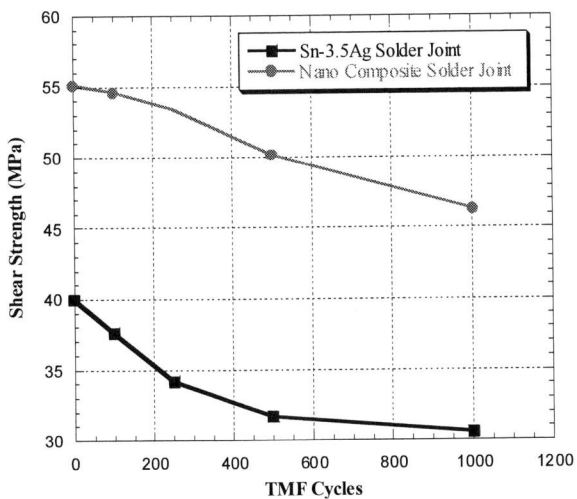

Fig. 7 Compared to the residual shear strengths in solder joints made with the eutectic Sn-3.5Ag solder and nano-composite solder under TMF condition I.

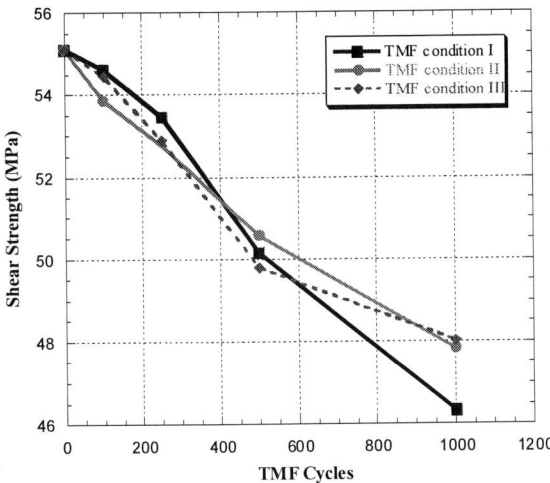

Fig. 8 The residual shear strength of nano composite solder joint after various TMF cycles under different TMF conditions

While, the decrease in residual shear strength is less for nano composite solder joints that experienced shorter dwell

time at temperature extremes (approximately 16 % after 1000 TMF cycles). Residual shear strength of the nano composite solder joints after TMF is consistent with the observation presented in the previous sections. Because the specimens that experience faster ramp rate and longer dwell time at temperature extremes of TMF exhibit less damage accumulation than those with slower ramp rate and shorter dwell time at temperature extreme.

Conclusions

The following conclusions can be made based on this present study.

1. The TMF properties of the nano composite solder joint was better than that of the eutectic Sn-3.5Ag solder joint.

2. More serious surface damage at the interface between Cu substrate and solder matrix in the nano composite solder joint at a faster ramp rate (33°C/min) were observed. No obvious surface damage accumulation can be noted in the nano composite solder joints that experience longer dwell time at temperature extremes even after 1000 TMF cycles.

3. A comparison of decrease in shear strength of the nano composite solder joints that experience different ramp rates and dwell times at temperature extremes at TMF indicated that the nano composite solder joints that experience longer dwell time at temperature extremes or slower ramp rate exhibited less decrease of residual shear strength.

4. In general, the POSS reinforcing particles to Sn-Ag solder to produce nano composite solder bring better TMF performance.

Acknowledgments

The authors acknowledge the support of this work from New Star Program of Beijing Science and Technology Commission (2004B03).

References

1. S. Choi, "Thermomechanical fatigue behavior of Sn-Ag solder joints," *Journal of Electronic Materials*, Vol. 29, No. 10 (2000), pp. 1249-1256.

2. F. Guo, "Evaluation of creep behavior of near-eutectic Sn-Ag solders containing small amount of alloy additions," *Materials Science and Engineering A*, Vol. 351 (2003), pp. 190-199.

3. F. Guo, "Creep behavior in Cu and Ag particle-reinforced composite and eutectic Sn-3.5Ag and Sn-4.0Ag-0.5Cu non-composite solder joints," *Journal of Materials Science: Materials in Electronics*, Vol. 12, No. 1 (2001), pp. 27-35.

4. J. G. Lee, "Residual-Mechanical Behavior of Thermomechanically Fatigued Sn-Ag Based Solder Joints," *Journal of Electronic Materials*, Vol. 31, No. 9 (2002), pp. 946-952.

5. P. Liu, "Effect of Ramp Rate on Microstructure and Properties of Thermomechanically-Fatigued Sn-3.5Ag Based Composite Solder Joints," *7th International Conference on Electronics Packaging Technology*, Shanghai, China, Aug. 2006, pp. 823-826.

6. H. P. Shawn, "Developments in Nanoscience: Polyhedral Oligomeric Silsesquioxane (POSS)-Polymers," *Solid State and Material Science*, Vol. 8, No. 1 (2004), pp. 21-29.

7. K. N. Subramanian, "Assessment of factors influencing thermomechanical fatigue behavior of Sn-based solder joints under severe service environments," *Journal of Materials Science: Material Electron*, Vol. 18, No. 1-3 (2007), pp. 237-246.

8. A. Lee, "Development of Nano-Composite Lead-Free Electronic Solders," *Journal of Electronic Materials*, Vol. 31, No. 11 (2005), pp. 1399-1407.

9. F. Tai, "Microstructure evolution and mechanical properties of Sn-Ag based composite solder joints during isothermal aging," *Acta Materiae Compositae Sinaca*, Vol. 25, No. 5 (2008), pp. 8-13.

Failure Analysis of Halide of Epoxy Molding Compound Used for Electronic Packing

Jianhai Ye, Shengxiang Bao, Lili Ma, Dechun Lv

State Key Laboratory of Electronic Thin Films and Integrated Devices

University of Electronic Science and Technology of China, Chengdu, 610054, China

E-mail: yjhyq@uestc.edu.cn Tel: +8613980637304

Abstract

As the continuously development of microelectronics industry, higher demand has been put forward toward the electronic packing technology. Being one of the three most important packaging materials (lead frame, spun gold, Epoxy Molding Compound) used for semiconductor, EMC has directly affected the performance of the semiconductor product. In this article, the transistor (model SOT32) was researched. Obtaining the packaging EMC of the transistor through the chemistry method, the elemental compositions of the EMC were analyzed by the scanning electron microscope (SEM) and the energy dispersive spectrum analysis (EDS). On discussing the differences in the microstructure and the elemental compositions between the failure and the qualified samples, the failure mechanism of the electronic components which was because of the halide of the EMC was lodged and the corresponding improve measures were discussed.

1 Introduction

The continuous development of the microelectronics manufacturing industry led to the development of electronic packing technologies, higher demand has been put forward to the material properties; packaging materials and packaging technologies are inseparable from each other. In the 1950s, with the rapid development of the semiconductor devices, integrated circuits, the ceramics, metal and glass packaging were difficult to cope with large-scale industrialization and the costs reduction. The first experimental study to substitute the above material with the plastic was started in the US and Japan. Through the unceasing selection of raw materials and the techniques of production, finally the EMC, which made the ortho-cresol epoxy molding compound for the main body material, as the main packing material was determined.

As one of the three important electronic packing materials, the EMC has become the mainstream of the current package in the large scale integrated circuit (VLSI) both at home and abroad; according to incomplete statistics, about 95% above micro electronic components are the plastic packaging components. However, while in general use in the electronic packing, the plastic materials also face a series of serious challenges from the green environmental protection aspect to no bromine, antimony-free at the same time. In the paper, the transistor (SOT32) was researched. Obtaining the packaging EMC of the transistor through the chemistry method, the elemental compositions of the EMC was analyzed by the SEM and EDS. On discussing the differences in the microstructure and the elemental compositions between the failure and the qualified samples, the failure mechanism of the electric components which was because of the halide of the EMC was lodged and the corresponding improve measures were discussed.

2 Experimental Details

2.1 The samples selected

The samples used were SOT23 (Small Outline Transistor 23, a packaging structure commonly used for transistor with three airfoil pins.), its typical structure is shown in Figure 1, mainly used in small power transistor package, also used for the package of field-effect transistor, diode and resistor network compound transistor.

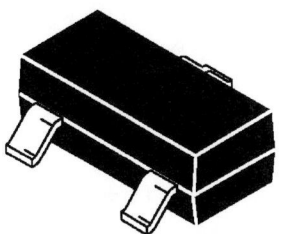

Fig. 1 imagine of SOT23

Select certain transistor samples separately from two different plastic materials used for package, divide them into two groups, number A# and B#.

2.2 Electric parameters test

Check out the appearance of the two groups selected samples firstly; and then analyze the interrelated electric parameters which were measured by the curve tracker and multiplexer tester one by one, to identify the abnormal failure of electric parameters.

The results of the electric parameters test show: the transistor packaging used A# plastic found no electrical parameters failure while the B# plastic had, there were 12 grains had electric parameters Vfp failure in the 84 grains transistors. The Vfp failure is that: when added test condition if, the transistor's direct impulse voltage value failure. Respectively take the qualified sample of A# package and the electric parameters failure of B# package transistors as the study objects to carry on the contrast research.

2.3 Chemical opening

The key to analyze the transistor electric parameters failure mechanism is to open it in a non-destructive way, and then analyze the internal microscopic structure and the external packaging plastic materials respectively. Chemical opening is a process to remove the plastic materials of the transistor surface by chemical means, it requires to remove the surface plastic materials well and cannot destroy the internal structure at the same time. In this study, the plastic materials used were a kind of epoxy resin.

Both the Au line and the Cu line are used as welding leads in the wire bonding, as a result of the different chemical property of gold and copper, there are two chemical solutions used for chemical opening, which are the 65% nitric acids and

98% strong sulfuric acid mixed solution. So the solution used to corrode the transistor's bonding leads for opening are prepared as follow.

Corrode Au thread product:
fuming HNO_3: strong H_2SO_4=5 : 1;
Corrode Cu thread product:
fuming HNO_3: strong H_2SO_4=3 : 1.

Preparation of the samples to be opened: weld the sample iron welding wire on the bonding jumper, pay attention to keep the the chip side facing up, place the sample as request.

Operation steps: firstly set the temperature of the warm table to HI, pour about 40mm opening solution into the glass beaker, and put it onto the warm table heating to the ebullition; put the prepared samples into the glass beaker and carry on the corrosion(corrosion time: gold wire sample 30□3 seconds, copper wire sample 25□3 seconds); after opening, take it out, wash twice with deionized water and then dehydrate with 98% alcohol, air dries.

The transistor after chemical opening is shown as Figure 2.

3 Samples analysis and results discussion

The internal micro-structure of the two groups samples (qualified and failure samples) were analyzed and compared by the SEM and EDS, the internal structure of the transistors after opening are both shown as Figure 2. Compared with the qualified sample there were no difference in the internal

micro-structure and the elemental compositions of the failure sample, so conclude initially that the failure of the transistor was not caused by the internal micro-structure.

The external plastic may also cause the transistor failure, in the electric parameters test there were failure transistors packaged with B# EMC but no with the A# EMC, therefore it was necessary to analyze the plastic elemental compositions both A# and B# EMC. The A# and B# EMC were both belong to a kind of epoxy resin. Use the SEM and EDS to analyze the microstructure and the elemental compositions of them, the selected regions for the analysis of the plastic materials were around the chip. The idiographic process of analysis were as follows:

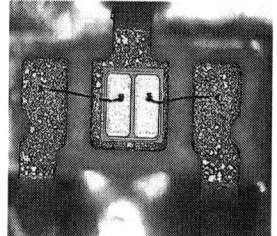

Fig. 2 the transistor after opening

 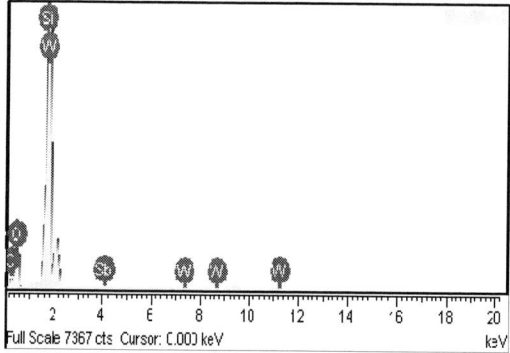

Fig. 3 SEM imagine and EDS element analysis of A$^{\#}$ EMC

 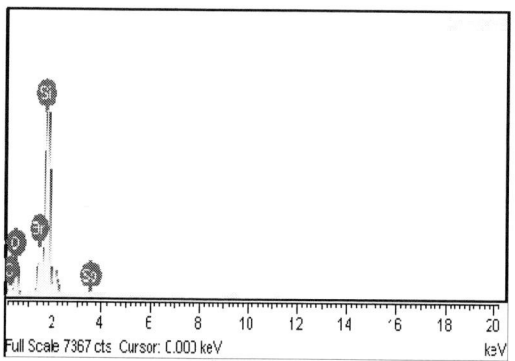

Fig. 4 SEM imagine and EDS element analysis of B$^{\#}$ EMC

Table 1 The elemental compositions of EMC A# and B#   Weight%

Element	CK	OK	SiK	BrL	SbL	WM	Total
A #	18.90	35.78	33.88	----	0.75	10.69	100.00
B #	22.57	35.73	33.71	5.26	2.73	----	100.00

From Table 1 we can know that the basic elements components of the A# EMC were Si-33.88%, O-35.78% and C-18.90%, there were also a small amount of W and Sb at the same time. The basic elements components of the B# EMC were Si, O and C, the same as the A# EMC, Br-5.26% held a certain proportion, Sb was 2.73% about 3 to 4 times compared with the A# EMC. In the elemental composition,the Br and Sb act as corrosive toward the eutectic layer.

When bonding with the gold thread, the eutectic layer of aluminum and gold contains a lot of Au4Al and Au5Al2, they can be oxidized and generate Al2O3 and Au, which would lead to failure of the welding spot. The existence of Br (HBr or CH3Br) intensified the occurrence of this process, Br (HBr or CH3Br) releases more Br through its own catalytic reaction to corrode once more. The chemical reaction equation are as follow:

$$Au_4Al + 3Br = 4Au + AlBr_3$$
$$2AlBr_3 + 6O = Al_2O_3 + 6Br$$

Flame retardant plastic material is the most basic materials in the EMC, it contains the trioxide of Sb-Sb_2O_3, which can lead to the Au-Al bonding failure, can intensify the process of Br denudation Au-Al bonded point, and cause the failure of the electronic components finally.

Usually, the broad use of EMC in the electronic packaging industry is fireproofing resin plastic material, which often contains bromide and releases Br ions in the high-temperature condition. Br or other halogen elements react with aluminum can form the unstable halogenide of aluminum, which can form inner cavities on the surface of the Au and Au-Al eutectic layer, lead to break off in welding spot or open circuit. Cl or Br is in existence in the form of catalyst, in the plastic components they can reduce the spread cycle of Au proliferates to the welds plate's of Al.

Fom the above analysis and comparison between the A# EMC and the B# EMC in the microstructure and the elemental compositions can infer that the causes for the failure of the transistor is because of the existence of Br element in the B# EMC. On the view of current practical application, the A# EMC has lower probability than B# EMC in causing the bonding failure of the micro electronic devices when used for packaging, so in the optimization of mechanical strength and electric parameters, the A# EMC had obviously much smaller impact of failure.

Conclusions

Form the SEM images, the surface morphology of A# EMC was with smaller particle size and distribution of uniform density; the B# EMC had larger particle size plastic material, and the distribution was sparse uneven.

From the EDS analysis of elemental composition can be seen that the main elements in A#, B# plastic materials are

similar, B# EMC contains a certain proportion of Br, while A# not.

The Br element is introduced from the flame retardant, the halogen elements are highly active and can form halide with other elements easily, which can form inner cavities on the surface of the Au and Au-Al eutectic layer, lead to break off in welding spot or open circuit, lead to the failure of the electronic components at last.

As one of today's mainstream electronic packaging materials in the electronic industry, EMC are being affected by environmental protection aspect lead-free, non-bromine, and antimony-free flame-retardant, and many other serious challenges, therefore abandon the original epoxy bromide and antimony containing flame-retardant system, the mission to develop new environmentally friendly fire retardant resin molecular structure or change the intrinsic flame retardant is no time to delay.

References

1. J. P. rusius, Che-Yu Li, "Research Summarization of Electronics Packaging," *Micro-electronics Technology*, Vol. 24, No. 5 (1996), pp. 41-46.
2. G. V. Clatterbraugh, J. A. Weiner, and H. K. Charles, Jr., "Gold-Aluminum intermetallics: Ball bond shear test and thin film reaction couples," *IEEE Trans. Compon. Packag. Technol.*, Vol. CHMT-7, No. 4 (1984), pp. 349–356.
3. L. T. Nguyen, "Reactive Flow Simulation in Transfer Molding of IC Packages," *1993 Proceedings. 43rd IEEE Electronic Components and Technology Conference*, 1993, pp. 375-390.
4. S. K. Kang, "Gold-to-Aluminum bonding for TAB application," *IEEE Trans. Compon. Packag. Technol.*, Vol. 15, No. 6 (1992), pp. 998–1004.
5. R. Gale, "Epoxy degradation induced Au–Al intermetallic void formation in plastic encapsulated MOS memories," in *Proc. IEEE Int. Rel. Phys. Symp.*, 1984, pp. 37–47.
6. P. McCluskey, K. Mensah, et al, "Reliability of commercial plastic encapsulated microelectronics at temperatures form 125 C to 300 C," in *Proc. IEEE Aerosp. Conf.*, 2000, pp. 445–450.
7. A. A. Gallo, "Effect of mold compound components on moisture degradation of gold-aluminum bonds in epoxy encapsulated devices," in *Proc. IEEE Int. Rel. Phys. Symp.*, 1990, pp. 244–251.
8. K. N. Ritz, W. T. Stacy, E. K. Broadbent, "The microstructure of ball bond corrosion of failure," in *Proc. IEEE Int. Rel. Phys. Symp.*, 1987, pp. 28–33.
9. H. Chang, K. Hsieh, T. Martens, A. Yang, "Wire-bond void formation during temperature aging," *IEEE Trans. Compon. Packag. Technol.*, Vol. 27, No. 1 (2000), pp. 155–160.

978-1-4244-4658-2/09 $25.00 © 2009 IEEE

10. K. Mitani, U. M. Gösele, "Wafer Bonding Technology for Silicon-On-Insulator Applications: A Review," *Journal of Electronic Materials*, Vol. 21, No. 7 (1992), pp.669-676.

Characterization, Modelling, and Parameter Sensitivity Study on Electronic Packaging Polymers

Ligang Niu[1], D. Yang[1], G.Q. Zhang[1,2]
[1]Guilin University of Electronic Technology, Guilin, China
[2]Delft University of Technology, Delft, the Netherlands
gstnlg@yahoo.com.cn

Abstract

Thermosetting polymers are widely used in electronic packaging. For instance, epoxy molding compound is extensively used as an encapsulant for electronic packages to protect the IC chips from mechanical and chemical hazards. It is well known that molding compounds show not only strong temperature dependent but also time dependent behavior. The thermo-mechanical behavior of these polymer constituents determines the performance, such as functionality and reliability, of the final products. In this paper, experimental characterization was carried out to investigate the time and temperature dependent properties of the selected molding compound. Thermal Mechanical Analysis (TMA) was applied to measure the CTE of the materials. Dynamical Mechanical Analysis (DMA) measurement was performed using temperature and frequency sweep modes. Finite element modeling was conducted on typical QFN (Quad Flat Non-lead) package device. Three material models, i.e., full viscoelastic model, temperature-dependent elastic model (1Hz DMA data) and constant elastic model, are used respectively to describe the behavior of the molding compound. The output responses of the simulations are von Mises stress distribution, package warpage and interlaminar stresses. The results show that the von Mises stress and package warpage are significantly different when considering the EMC as full viscoealstic, temperature-dependent elastic and constant elastic.

Introduction

According to the developing trends of microelectronics, the development and application of polymers becomes one of the bottlenecks for the microelectronic industry. With the introduction of new packaging materials there are many new requirements to improve devices reliability. Therefore, understanding the thermo-mechanical behavior of packaging polymers is critical for the development of packages. In order to be able to conduct process and structure optimization and to predict thermo-mechanical reliability more accurately and further to provide a base for virtual thermo-mechanical prototyping of electronic packaging, a fundamental understanding of the thermo-mechanical behavior of the packaging polymers and its influence on the reliability of the packages is imperative. Therefore, an appropriate material model should be established, and material characterization and numerical implementation should be carried out.

In thesis [1], the authors used DMA experiments and time-temperature superposition to verify WLF equation and the viscoelastic model were available to characterize the electronic packaging polymers.

Z. Xiong [2] point out that the epoxy molding compound is often modeled as an elastic material in the analysis of plastic-encapsulated IC package. However, many polymers show some viscoelastic behavior which causes its Young's modulus to be not only temperature-dependent but also time-dependent.

Yeung [3] studied the viscoelastic model property of an EMC and found that a viscoelastic model predicted the warpage of an IC package more accurately than an elastic model.

J. de Vreugd [4] described the viscoelastic properties of molding compound in detail using DMA and pointed out the CTE of the molding compound varying over temperature.

X. R. Zhang [5] pointed out that linear elastic and temperature-dependent elastic moduli models are simple to use, but they cannot model the stress relaxation behavior that a viscoelastic model can offer.

Although the material of EMC exhibits viscoelastic behavior, the material is commonly regarded as temperature-dependent elastic one or constant elastic one in the microelectronics reliability analysis. Numerous studies of microelectronics packaging reliability have been reported involving EMC; however, few considered the CTE is different below and above the glass transition temperature [6].

In the analysis of plastic-encapsulated IC packages, the epoxy molding compound is often modeled as an elastic material. However, many polymers show some viscoelastic behavior which causes its Young's modulus to be not only temperature-dependent but also time-dependent [7].

Frank Feustel [8] pointed out that finite element analyses (FEA) have established as effective method for reliability assessment of flip chip assemblies. The simulation results are significantly dependent on the selected material models.

The objective of this work was to characterize the EMC's viscoelastic property using Generalized Maxwell model after experiment. And QFN was chose as the research objective. Finite Element Method (FEM) was employed to investigate the von Mises stress and package warpage of QFN under thermal cycling.

Experiment

DMA experiment

Dynamic mechanical analysis (DMA) is a commonly used method for characterization of mechanical properties (modulus and damping) for viscoelastic materials over a spectrum of time (or frequency) and temperature. A Dynamic Mechanical Analyzer (TA Instruments, Q800) was used to determine the storage modulus of the samples. Figure 1 shows the samples with the form of strips having dimensions of about 50mm×10mm×2mm. A single cantilever mode under 16 different frequencies from 0.1Hz to 100Hz was employed for the tests. The specimen was mounted on the DMA and heated from room temperature to 250°C at a heating rate of 1°C/min.

Fig. 1 The samples of EMC

Figure 2 presents the storage modulus as functions of frequencies of EMC measured from DMA under different temperature levels. Application of time-temperature superposition (TTS) principle to the multiplexing frequency mode analysis of the epoxy molding compounds was illustrated. The procedure for performing TTS involves data extracted from the multiplexing frequency experiments followed by shifting the data with respect to a reference curve to generate a master curve. Reference curves for this epoxy molding compound were chosen to be at temperature 110°C. The result of shifting along the frequency axis is shown in Figure 3. It is noted that if compounds are subjected to the same oscillatory frequency, higher experimental temperature condition will result in lower modulus of the same brand of material.

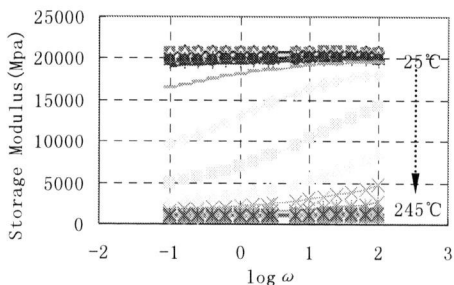

Fig. 2 Storage modulus as functions of frequencies of EMC under different temperatures

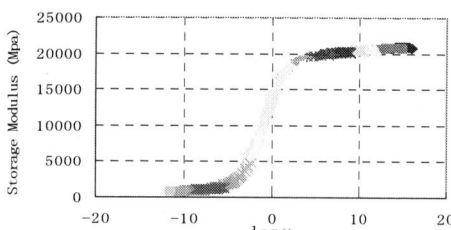

Fig. 3 The master curve for storage modulus

A viscoelastic material's properties represent both viscous and elastic characters at the same time. Mechanical models often been used to describe viscoelastic behavior of materials. The most common model for viscoelastic materials is the Generalized Maxwell model. The Generalized Maxwell model can be represented by viscous dampers and elastic springs connected in series. The mathematical form of the Generalized Maxwell model could be expressed as followed:

$$E(t) = E(\infty) + (E(0) - E(\infty))\sum_{i=1}^{N} C_i \, exp\left(-\frac{t}{\lambda_i}\right) \quad (1)$$

Where $E(t)$ is the relaxation modulus, t is time; N is the number of elements, C_i is the coefficient of the ith element,

λ_i is the relaxation time of the ith element, $E(0)$ is the elastic modulus at $t = 0$, and $E(\infty)$ is the elastic modulus at t approached ∞ [1].

In order to obtain the parameters to meet the FE software require, the relationships of elastic constants were used to get the shear modulus G' and bulk modulus K', respectively. Then, the follow equations were used to obtain the shear relaxation coefficients and bulk relaxation coefficients, which are listed in Table 1.

$$G'(\omega) = G_\infty + \sum_{n=1}^{N} G_n \frac{\omega^2 \tau_n^{\,2}}{1 + \omega^2 \tau_n^{\,2}} \quad (2)$$

$$K'(\omega) = K_\infty + \sum_{n=1}^{N} K_n \frac{\omega^2 \tau_n^{\,2}}{1 + \omega^2 \tau_n^{\,2}} \quad (3)$$

Where G_∞ and K_∞ are shear and bulk relaxation modulus at t approached ∞ respectively; G_n and K_n are the shear and bulk relaxation coefficients respectively; τ_n is the relaxation time.

Table 1 Relaxation times and coefficients

τ_n (s)	G_n (MPa)	K_n (MPa)
1.00E-04	470.9730223	755.1111094
1.00E-02	159.7143531	1437.786689
1.00E-01	246.6564779	1896.398461
1.00E+00	238.2582743	3077.800045
1.00E+01	258.2268206	1837.50515
1.00E+02	460.2942028	4265.965146
1.00E+03	379.7287416	863.462001
1.00E+04	43.84011256	781.0249913
1.00E+05	546.0840982	1045.854248

TMA experiment

CTE measurement of the EMC samples were performed on a Thermal Mechanical Analyzer (TA Instrument, TMA Q400) using an expansion probe. The size of the samples was 4mm×4mm×2mm. The samples were mounted on the TMA and heated to 250°C from room temperature at a heating rate of 10°C/min. The coefficients of thermal expansion were determined from the slope of the plot between dimension change and temperature.

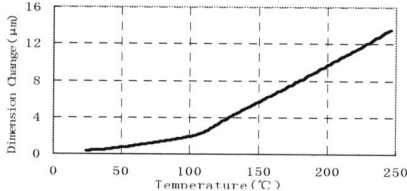

Fig. 4 The curve for dimension change of EMC at different temperature

Figure 4 depicts a TMA trace of the EMC and shows a glass transition temperature of 110°C. The coefficients of thermal expansion (CTE) below and above the glass transition temperature are 11.28ppm/°C and 39.31ppm/°C, respectively. The moderate glass transition temperature maintains good dimensional stability of the integrated circuit package and the low CTE of EMC matches that of the chip to minimize the thermo-mechanical stresses during temperature cycling [9].

FE modeling of QFN package

Geometry model

A QFN device was chosen as the research objective. The type of QFN is RFT6100 and dimension is 6mm×6mm×0.9mm with a total of 40 lead-in pins. The packaging physical structure of QFN device is shown in Figure 5 [10, 11].

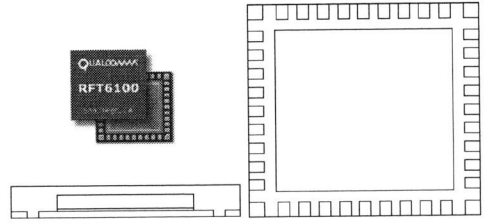

Fig. 5 The device and geometry configuration of QFN

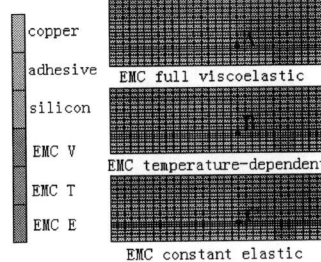

Fig. 6 Geometry of the finite element models

Figure 6 shows the geometry of the finite element models which are based on the real type of electronic packaging. The geometry of the assembly is represented by a 2D model. It contains only one half of the QFN.

Material properties

The silicon die and copper pads are considered elastic and isotropic. The die adhesive is simplified, because it is thin and not the main research objective in this paper. A constant poisson's ratio is assumed for the EMC. The material properties used in the simulation are shown in Table 2 and Table 3.

Table 2 Characteristic parameters of materials

Material	E (Mpa)	ν	CTE (ppm/°C)
Copper	82700	0.3	16.7
Adhesive	See Tab.3	0.4	48
Silicon	131000	0.3	2.8
EMC	viscoelastic	0.3	11.28, T<110°C 39.31, T>110°C°C

Table 3 Young's modulus of adhesive

E(MPa)	640	140	100
T(°C)	25	150	250

In the temperature-dependent elastic model, the temperature-dependent Young's modulus of the EMC was obtained by the 1 Hz response from the modulus curves in Figure 2 [2]. At the constant elastic model, the storage modulus at room temperature was chosen as the Young's modulus.

Load model

According to the temperature cycling standards JESD22-A104C [12], thermal cycling test condition G and soak mode condition 4 were chosen as thermal loading condition, which was assumed to be within the state-steady condition. The typical thermal loading history imposed on the QFN is shown in Figure 7. It can be seen that for each cycle the temperature was between -55°C and +125°C, with 10°C/minutes ramp, 10 minutes hold at hot and cold, respectively. Three full cycles were executed. The starting temperature was the reference temperature at which the QFN device was assumed stress-free. The total time of temperature cycles was 168min.

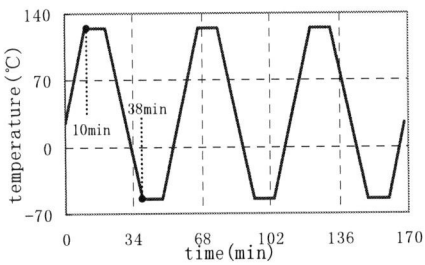

Fig. 7 Representative temperature profile

Results and discussion

The thermal stress in the microelectronics device is mainly induced by the mismatches of mechanical properties of the component materials [7]. Three different material properties were applied to EMC as described above: full viscoelastic model, temperature-dependent elastic model and constant elastic model. The von Mises stress distributions are significantly different as depicted in Figure 8. Due to mismatch of CTE of silicon die, die adhesive and EMC, the maximum von Mises stress appears at the connections of this three materials, and the connections A, B and C are listed in Figure 6. At 10 minute, 38 minute, the maximum von Mises stress was obtained in EMC constant elastic model and EMC full viscoelastic model, respectively. The von Mises stress at point A, B and C (see figure 6) based on three different EMC properties during the thermal loading are presents in Figure 9. It can be seen that the EMC constant elastic model will over-estimate the von Mises stress than the EMC full viscoelastic model when the temperature is above 45°C. If the temperature below 45°C, the von Mises stress when considered EMC constant elastic will be under-estimated than the other two models.

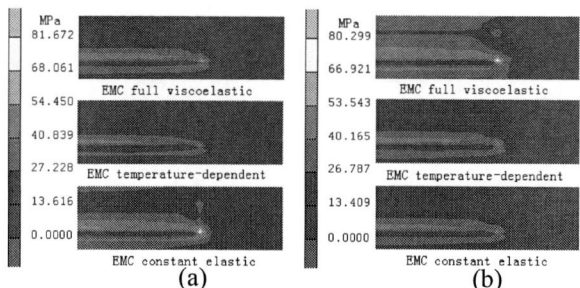

(a)   (b)

Fig. 8 Contour plots of von Mises stress at 10min (a), 38min (b)

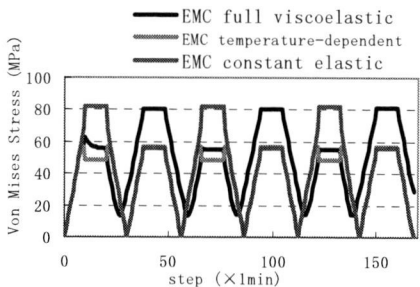

Fig. 9 History of the von Mises stress vs loading time at point A, B and C

Figure 10 presents the package warpage patterns for the QFN models based on the three different EMC properties at 10 minute and 38 minute of thermal loading. Here we take the displacement of left top corner node as indication of the strip warpage. It can be seen that the packages bend upwards at 10 minute, and the warpage of full viscoelastic model is most seriously. But the packages bend downwards at 38 minute, and the warpage of constant elastic model is most seriously.

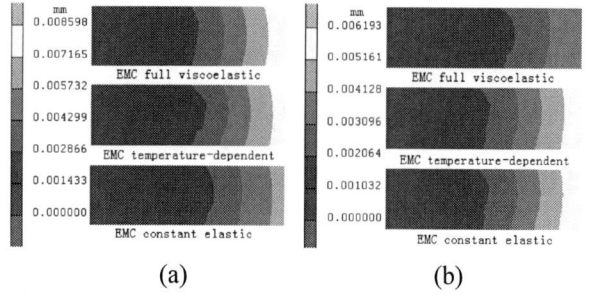

(a)   (b)

Fig. 10 Contour plots of warpage at 10min (a), 38min(b)

The interlaminar stresses, σ_{yy} and τ_{xy} are discussed in this paper. Generally, the interface between two different materials is a weak link due to imperfect adhesion and stress concentrations [13]. The normal stress σ_{yy} will control the fracture mode I, i.e., opening mode, while τ_{xy} will be responsible for the mode II. The stresses were concentrated at the point A, B and C (see Figure 6), and the delamination always begin at this point. Figure 11 presents the interlaminar stresses at point A, B and C during thermal cycle loading. It can be seen that the EMC based on different properties will cause different interlaminar stresses. EMC based on constant

elastic will over-estimate the normal stress σ_{yy} at high temperature, and under-estimate the shear stress τ_{xy} at low temperature.

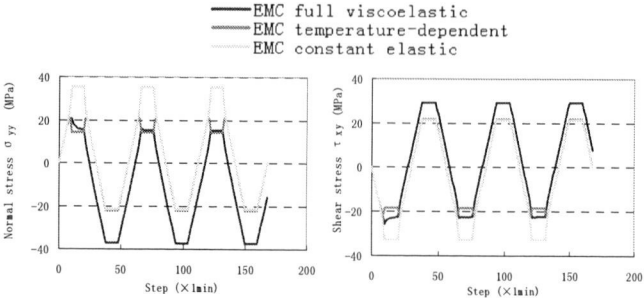

Fig. 11 Interlaminar stresses at point A, B and C during thermal cycle loading

Acknowledgments

The research work in this paper is financially supported by the National Natural Science Foundation of China (NSFC) (Grant No. 60666002).

Conclusions

In this study, experimental characterization was carried out to investigate the time and temperature dependent properties of the selected molding compound. QFN (Quad Flat Non-lead) package was simulated by FEA and von Mises stress was obtained through thermal cycle based on the thermal cycling standard JESD22-A104C.

The results show that the maximum von Mises stress present to the connection of silicon die, die adhesive and EMC, because of mismatch of the materials used in packaging. The von Mises stress and warpage are significantly different when considering the EMC as full viscoelastic, temperature-dependent elastic and constant elastic. When the temperature is above 45°C, the EMC constant elastic model will over-estimate the von Mises stress than the model considered EMC as full viscoelastic, and the von Mises stress will be under-estimated when considering the EMC as a constant elastic material or temperature-dependent elastic material. The maximum upward warpage is present to EMC full viscoelastic model at high temperature, while the maximum downward warpage is present to EMC constant elastic model at low temperature. EMC based on constant elastic will over-estimate the normal stress σ_{yy} at high temperature, and under-estimate the shear stress τ_{xy} at low temperature. The results indicate that the effects of viscoelasticity of the molding compound may not be neglected.

References

1. Chi-Hong Shue, et al., "Post-Mold Cure Process Simulation of IC Packaging," *IEEE*, Vol. 6 (2008), pp. 106-110.
2. Z. Xiong, A. A. O. Tay, "Modeling of Viscoelastic Effects on Interfacial Delamination in IC Packages," *Electronic Components and Technology Conference*, (2000), pp. 1326-1330.
3. T.S. Yeung, M.M.F. Yuen, "Viscoelastic analysis of IC package warpage, Sensing, modeling and simulation in

emerging electronic packaging," *ASME 1996*, EEP-Vol. 17, pp. 101-107.

4. J. de Vreugd, K.M.B.Jansen, A. Xiao, et al., "Advanced Viscoelastic Material Model for Predicting Warpage of a QFN Panel," *Electronic Components and Technology Conference*, (2008), pp. 1635-1639.

5. X. R. Zhang, John H. L. Pang, "On the Moduli of Viscoelastic Materials," *Electronics Packaging Technology Conference*, (2002), pp. 318-322.

6. Yeong K. Kim, "Viscoelastic Effect of FR-4 Material on Packaging Stress Development," *IEEE TRANSACTIONS ON ADVANCED PACKAGING*, Vol. 30, No. 3 (2007), pp. 411-419.

7. C. P. WONG, RAJA S. BOLLAMPALLY, "Thermal Conductivity, Elastic Modulus, and Coefficient of Thermal Expansion of Polymer Composites Filled with Ceramic Particles for Electronic Packaging," *Journal of Applied Polymer Science*, Vol. 74, (1999), pp. 3396-3403.

8. Frank Feustel, Steffen Wiese, Ekkehard Meusel, "Time-Dependent Material Modeling for Finite Element Analyses of Flip Chips," *Electronic Components and Technology Conference*, (2000), pp. 1326-1331.

9. Kalyan Ghosh, Mark McCabe, "Advances in Flip Chip Underfill Technology for Lead-free Component Packaging," *Electronic Component and Technology Conference*, (2005), pp. 1480-1485.

10. Rao R. Tummala, "Fundamentals of Micro-system Packaging," MeGraw-Hill (New York, 2001).

11. http://www.dzsc.com.

12. JEDEC Solid State Technology Association. JESD22-A104C, "Temperature Cycling," May, 2005.

13. A. Xiao, L. G. Wang, W. D. van Driel, et al., "Thin Film Interface Fracture Properties at Scales Relevant to Microelectronics," *ICEPT*, (2007), Shanghai, China, pp. 300-304.

Controlled Synthesis of Ortho-Substitution Ortho-cresol Novolac Resins

Xiaowei Tian, Zhenguo Yang[*]

Department of Materials Science, Fudan University, Shanghai, China

220 Handan Rd, Shanghai, 200433, China

Email: zgyang@fudan.edu.cn; Phone: 86-21-65642523

Jiang-Yan Sun, Zheng Ji, Chuang Jiang

Shanghai Sinyang Semiconductor Materials Co., Ltd.; Shanghai 201616; China

Abstract

The structure of ortho-cresol novolac resin is very important for the synthesis of ultra-high purity ortho-cresol novolac epoxy resin used as electronic encapsulating materials. But the relationship between the synthesis conditions and the structure of ortho-cresol novolac resin has not yet been completely clarified. This paper mainly studied the controlled synthesis of different proportion of ortho-substitution ortho-cresol novolac resins under different reaction conditions, such as molar ratio of ortho-cresol to formaldehyde, different catalyst types, reaction time. Then, the structure of different ortho-substitution ortho-cresol novolac resins were investigated by many measurements, such as FTIR, 1H-NMR, 13C-NMR, and the softening point. And the relationships between the structure and the reaction conditions were thus studied. The results indicated that, the ortho-cresol novolac resin with the needed proportion of ortho-substitution was synthesized through the adjustment of the reaction conditions.

1 Introduction

Novolac resins are the first thermosetting resins to be synthesized by the polycondensation between phenol and formaldehyde under acid catalysts. During the past decades, novolac resins have grown continuously in terms of volume and applications, such as molding compounds, electrical insulators, mechanical parts, and friction material binders [1]. However, the relationship between the synthesis conditions and the structure of novolac resins has not yet been completely clarified. One of the reasons for this is that the catalyst used in the polymerization.

Phenols react with aldehydes only at ring positions located ortho and para to the hydroxyl group. Thus, conventionally known novolac resins obtained by using a strong acid as a catalyst, are characterized by having a preponderance of para-para and para-ortho methylene bridges [2]. High ortho-substitution novolac resins prepared in the presence of salts of bivalent metal, which an ortho-directing effect occurs, are characterized by having linear chains consisting of high proportion of ortho-ortho methylene bridges [5]. Bender shows that a novolac resin with a high proportion of ortho-ortho methylene links could be cured by hexamethylenetetramine at a higher rate than novolac resin made with conventional acid catalysts [2]. And high-ortho novolac resins are being preferred for many applications, such as photoresists, molding materials.

Most researches taken on high-ortho novolac resins were based on trifunctional phenol compounds, which means there were no ortho- or para-substitute on phenol ring. Culbertson discloses a process for preparing high-ortho novolac resins

wherein phenol/formaldehyde reactants in molar ratio of 1.2 ~ 2: 1 are reacted in an organic solvent under anhydrous conditions, and the reaction is carried out at a temperature of 115°C ~ 145°C in the presence of divalent metal salts, and the proportion of ortho-ortho methylene bridges is about 80 ~ 95 % [6]. Based on this patent, Linden discloses a high ortho novolac resins with a softening point of about 80°C, which is used for hot melt adhesives [7]. Nanjo et al discloses a synthetic route of liquid high ortho phenolic novolac resin. The reaction employs two catalysts, and indicates that a high ortho resin will not be produced if the divalent metal catalyst is present at the start of the reaction [8]. Zou et al have studied the effects of 19 kinds of divalent metal catalysts. By adjust the use of divalent metal catalyst and reaction conditions, the proportion of ortho-ortho methylene bridges of novolac resins can be controlled from 40 ~ 90 % [3]. Casiraghi et al have developed a novel synthetic route to all- ortho novolac resins by reaction of metal phenolates with para-formaldehyde in aprotic poorly donating media. The unique combination of metal phenolate and a poorly donating solvent play a leading role in the ortho- specific methylene bridging of the phenolic nuclei [2].

In this paper, the different ortho-substitution ortho-cresol novolac resins (OCN) were synthesized by the polycondensation between ortho-cresol and para-formaldehyde under bivalent metal salt catalyst. And the relationship between the different proportion of ortho-substitution and the reaction conditions was also studied. The research on the synthesis of bifunctional phenolic compounds was useful for the investigation of novolac resin reaction mechanisms.

2 Experimental

2.1 Synthesis of the ortho- cresol novolac resin

Ortho-cresol (CP), para-formaldehyde (powder, > 95%), oxalic acid (powder, CP), zinc acetate (ZnAc2, CP), and calcium chloride (CaCl2, CP) were purchased from Sinopharm Chemical Reagent Co., Ltd., China. All solvents used were commercial products and used without further purification.

Into a three necked round-bottom flask with a thermometer, a stirrer and a water separator were added 1 mol of ortho-cresol, 0.85 mol of para-formaldehyde, 10 grams of toluene (10 wt % of ortho- cresol) as solvent, and 1.08 grams of zinc acetate as a catalyst. The solution was stirred and reacted at 125 ~ 135°C for six hours. There was an initial mild exotherm, which was easily moderated by regulating the source of heat. Water born during the reaction was then removed azeotropically with the toluene. After reaction, the appropriate amount of methyl isobutyl ketone (MIBK) was added into the flask to dissolve the reaction mixture. So the reaction mixture was washed by deionized water till that the water phase's pH was 7. After water washed, MIBK was desolventized under the condition of 200°C and minus pressure for a certain time. After removal of volatile matter by distillation, the solid ortho-cresol novolac resin (OCN) was obtained, which the softening point was about 100°C.

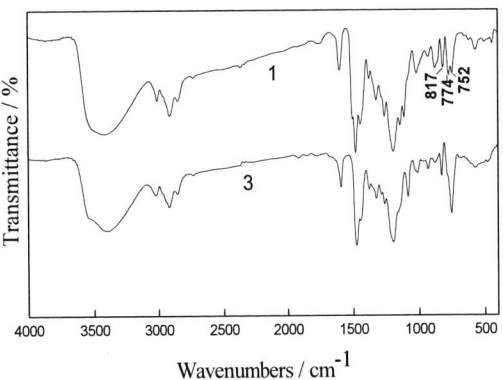

Fig. 2 FTIR spectra of different ortho- cresol novolac resins

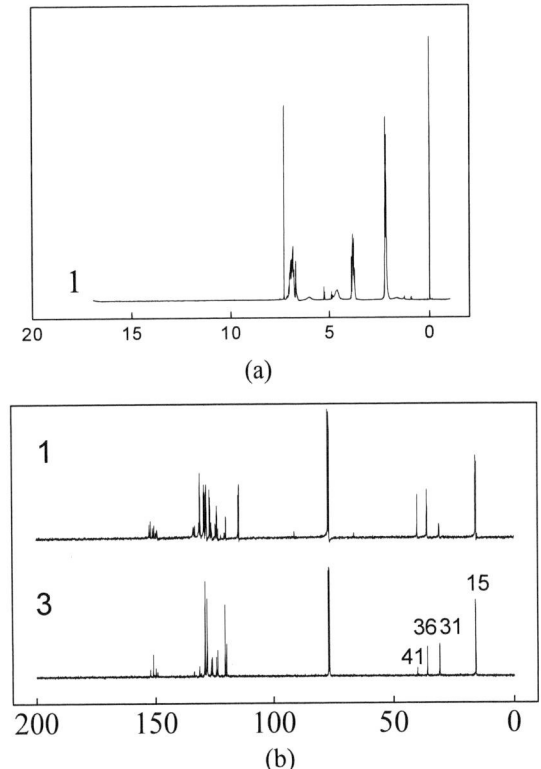

Fig. 1 Schematic diagram of the synthesis of ortho- cresol novolac resin

2.2 Characterizations

2.2.1 Nuclear magnetic resonance spectroscopy

^{1}H and ^{13}C nuclear magnetic resonance (^{1}H and ^{13}C NMR) characterizations were carried out by Bruker DMX500 NMR spectrometer using chloroform-d6 as the solvent and tetramethylsilane (TMS) as internal standard at 298 K.

2.2.2 Fourier transform infrared spectroscopy

FTIR spectra of the novolac resins were recorded on a Nicolet Nexus 470 FT-IR Spectrometer. The novolac resins were dissolved in acetone and spread onto KBr windows, and each spectrum was scanned 32 times at a resolution of 4 cm^{-1} from 400 cm^{-1} to 4000 cm^{-1}.

2.2.3 The softening point measurement

The softening point of ortho- cresol novolac resin was measured by Ball and Ring method according to GB 12007.6-89.

3 Results discussion

3.1 FTIR spectra studies

The chemical structures of different ortho- cresol novolac resins were confirmed with FTIR. FTIR spectra (Fig. 2) showed the characteristic absorptions of novolac resin at 3424 cm^{-1} (stretching of hydroxyl), 1605 cm^{-1} (stretching of aromatic ring), 3011, 2853 cm^{-1} (stretching of methyl). Special attention was paid to the stretching peaks of different diphenylmethane- type methylene bridges between aromatic rings, which were at 752 cm^{-1} (ortho, ortho-), 774 cm^{-1} (ortho, para-), and 817 cm^{-1} (para, para-), respectively. As shown in Fig.2, it was difficult to distinguish different ortho-cresol novolac resins with different ortho- substitution.

3.2 ^{1}H-NMR and ^{13}C-NMR spectra studies

Typical ^{1}H-NMR spectra of the different ortho-cresol novolac resin that we synthesized, taken in chloroform-d$_6$, are shown in Fig. 3(a). Under these experimental conditions, the signals of the aromatic ring protons (named as ArH) were at 8.6 ~ 6.9 ppm, the signals of the methylene protons in ortho, ortho- and ortho, para- diphenylmethane- type methylene bridges (named as o, o-, and o, p- RArCH$_2$ArR) were at 4.2 ~3.8 ppm, and the signals of the methylene protons in para, para- diphenylmethane- type methylene bridges (named as p, p- RArCH$_2$ArR) were at 3.7 ppm. The peak at 2.2 ppm originated from the residual ortho-cresol.

Fig. 3 (a) ^{1}H-NMR and (b) ^{13}C-NMR spectra of OCN

The different ortho-cresol novolac resins are also identified by ^{13}C-NMR in Fig. 3(b). And the spectral peak assignments are shown in Table 1, according to the method of Park et al. [5]. The sketch of ortho-cresol novolac resin is shown in Fig. 4. Particular attention was paid to the zone at 30 ~ 45 ppm of the methylene bridges between two aromatic rings. The peaks around 31, 36, and 41 ppm were attributed to the methylene carbons in o, o-, o, p-, and p, p-diphenylmethane- type methylene bridges. The peaks at 41, 36, and 31 ppm were different between different ortho- cresol novolac resins. The relative quantities of various methylene bridges were calculated based on the ^{13}C-NMR spectrum [4]. The results were shown in Table 2. As shown in Table 2, the proportion of o, o- methylene bridge in OCN was increased from 16.1 % to 55 %.

Table 1 The chemical shifts of different carbons in OCN

No.	Chemical shift (ppm)	Carbon atom
1	15	CH$_3$
2	31	Ar-CH$_2$-Ar (o, o-)
3	36	Ar-CH$_2$-Ar (o, p-)
4	41	Ar-CH$_2$-Ar (p, p-)
5	115~ 135	Ar C
6	149~ 153	Ar C-OH

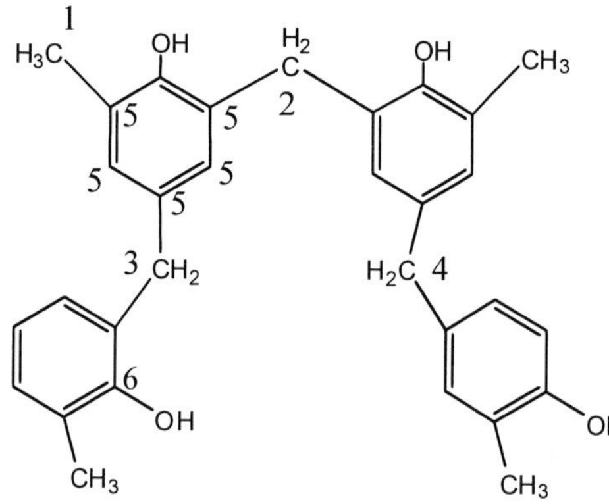

Fig. 4 The sketch of ortho- cresol novolac resin (OCN)

Table 2 Proportion of different methylene bridges in OCN

Polymer	Catalyst	Proportion of methylene bridges (%)		
		o, o-	o, p-	p, p-
1	Oxalic cid	16.1	51.6	32.3
2	CaCl$_2$	47.6	42.9	9.5
3	ZnAc$_2$	50.0	40.0	10.0
4	ZnAc$_2$	55.2	13.8	31.0

3.3 Synthesis of ortho- cresol novolac resins

In order to synthesize different ortho- substitution OCN, different bivalent metal salts were used under different reaction conditions, such as the molar ratio of ortho- cresol to formaldehyde, the metal-ion/ortho-cresol molar ratio, reaction temperature.

An excess of ortho-cresol is necessary for novolac resin. The ortho- cresol/ formaldehyde (OC/ F) molar ratio was altered from 1.05 to 1.25. Fig. 5 shows the proportion of o, o-methylene bridges of OCN with different OC/F ratio. When the OC/ F ratio was 1.18, the proportion was highest.

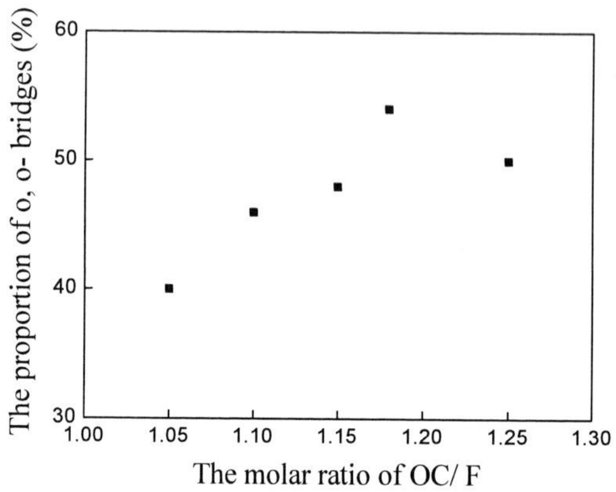

Fig. 5 Proportion of o, o- methylene bridges of the synthesized resins versus the OC/ F molar ratio. The polymers were synthesized under a molar ratio of ZnAc$_2$/ OC = 0.05 at 130°C for 6 h.

Fig. 6 shows a curve of the proportions of o, o- methylene bridges of different OCNs versus the metal-ion/ortho- cresol molar ratio. During the synthesis, the formation of chelate rings between the metal, formaldehyde, and ortho- cresol was an important step to determine the proportion of o, o-methylene bridges. When the catalyst was CaCl$_2$, the proportion of o, o- methylene bridges of OCNs was around 48 % under any metal-ion/ortho- cresol molar ratio, as shown in Table 2. When ZnAc$_2$ used as catalyst, the proportion was directly proportional to the metal-ion/ortho- cresol molar ratio. When the metal-ion/ortho- cresol molar ratio was 0.05, the proportion reached the peak. When the metal-ion/ OC ratio was greater than 0.05, the proportion was decreased to 51 %.

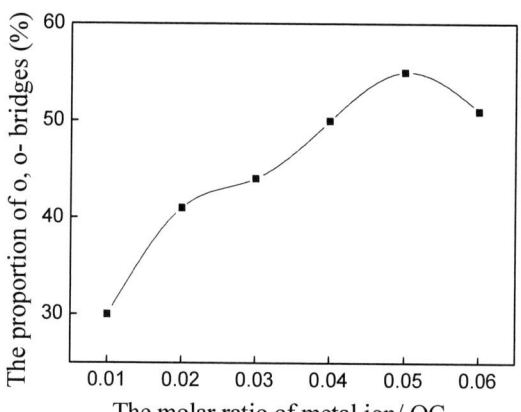

Fig. 6 Proportion of o, o- methylene bridges of the synthesized resins versus the metal-ion / OC molar ratio. The polymers were synthesized under a molar ratio of OC/ F = 1.18 at 130°C for 6 h.

The effect of the reaction time is shown in Fig. 7. When the reaction time was too long, the proportion was maintained at a certain level with little change. This phenomenon is different to the synthesis of high- ortho phenol novolac resin [3]. For an economic consideration, the reaction time had to be 6 h.

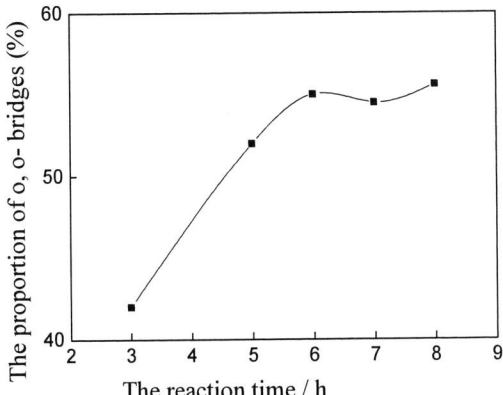

Fig. 7 Proportion of o, o- methylene bridges of the synthesized resins versus the reaction time. The polymers were synthesized under a molar ratio of OC/ F = 1.18 and ZnAc$_2$/ OC = 0.05 at 130 °C

4 Conclusions

This study was taken to investigate the effects of various conditions on the synthesis of different ortho-substitution ortho-cresol novolac resins.

Among the used bivalent metal salts, the proportions of o, o- methylene bridges were close, and the best catalyst was ZnAc$_2$. Using ZnAc$_2$ as catalyst, the OC/F molar ratio is 1.18, the metal ion/ ortho- cresol molar ratio is 0.05, and the reaction time is 6 h. Thus, we can obtain an orhto-cresol novolac resin whose proportion of o, o- methylene bridges is above 50 %.

With the changing of catalyst and reaction conditions, the desired ortho- cresol novolac resin with certain proportion of o, o- methylene bridges can be synthesized. Then, it will be useful for the synthesis of ortho-cresol novolac epoxy resins with different viscosity.

References

1. Knop A., Pilato L. A., <u>Chemistry and Application of Phenolic Resins</u>, Springer-Verlag (Berlin, 1979), pp. 28.
2. Giovanni CasiraghP, Giovanni Sartori, Franca Bigi, et al, "A Novel Synthetic Route to All-ortho Novolac Resins," *Makromol. Chem*, Vol. 182 (1981), pp. 2151-2159.
3. Jianying Huang, Miaoqing Xu, Qiang Ge, et al, "Controlled Synthesis of High-Ortho-Substitution Phenol–Formaldehyde Resins," *J. Appl. Polym. Sci.*, Vol. 97 (2005), pp. 652-658.
4. Pindiwe Sitetyana, <u>An Optimisation Study into the Synthesis of o-Cresol Novolacs</u>, Rand Afrikaans University (2004).
5. Park B. D., Riedl B., *J. Appl. Polym. Sci.*, Vol. 77 (2000), pp. 841.
6. Harry M. Culbertson, "Process for preparing high ortho novolac resins," U. S. 4097463, 1978-06-28.
7. Gary L. Linden, "Storage stable 100 % solids high ortho phenol formaldehyde hot melt adhesives," U. S. 5182357, 1993-01-26.
8. Motoyuki Nanjo, et al, "Process for producing quick-curing phenolic resin," U. S. 4299947.

Electrodeposition of Sn-Cu Solder Alloy for Electronics Interconnection

Yi Qin[1,2*], Abdul Wassay[2], Changqing Liu[2], G.D.Wilcox[1], Kun Zhao[3], Changhai Wang[3]

[1]Department of Materials
[2]Wolfson School of Mechanical and Manufacturing Engineering
Loughborough University, Leicestershire, LE11 3TU, UK
[3]School of Engineering and Physical Sciences
Heriot-Watt University, Edinburgh, EH14 4AS, UK
* Corresponding author. E-mail address: Y.Qin@lboro.ac.uk

Abstract

Eutectic Sn-Cu alloy is one of the lead-free solder candidates for electronics interconnection. The methanesulphonic acid based baths were investigated for the electrodeposition of Sn-Cu alloys. The effect of iso-octyl phenoxy polyethoxy ethanol (Triton X-100) as a surfactant was studied, its addition proved to facilitate achieving the near eutectic composition of Sn-Cu alloy with improved microstructure. Different morphologies of deposited films were analysed using scanning electron microscopy (SEM). X-ray diffraction (XRD) results indicated that a biphasic structure of β-Sn and Cu_6Sn_5 was present in the as-electroplated film. Fine pitch near eutectic Sn-Cu solder bumps were also produced on a test wafer.

1 Introduction

With the on-going pace of miniaturization of electronic devices, the demand for flip-chip technology has been growing continuously to satisfy such high density and fine pitch interconnections. One major step in the flip-chip interconnection process route involves the deposition of solder alloys on to the bondpads of the chips. Many different methods for this have been employed. Among popular bumping technologies, evaporation, screen printing and electroplating are mainstream in industry today. Evaporation was the first method used for solder bumping, and shows attractive features in high-reliability bumps, but it is relatively costly because of expensive infrastructure and inefficient material use. Flip-chip bumping by screen printing has also been extensively implemented for many years, and proven to be a viable low-cost solution for bumping at relatively large pitch sizes. However, limitations in print quality and solder paste volume, due to stencil manufacturing limitations and solder particle size, mean that existing printing methods are not likely to move significantly below a typical pitch of 150-200 μm, while even finer pitch sizes would be desirable for high density flip-chip packaging. Meanwhile, flip-chip bumping by electroplating shows its attractiveness where features of high volume production and fine pitch bumping are particularly addressed, and high yield at reasonable cost achieved [1, 2].

Currently, implementation in the electronic industry to replace lead-containing solder alloys with lead free solders has posed numerous challenges for the electrodeposition of lead-free alloys. Electroplating of alternative lead-free binary alloys, such as eutectic Sn-Cu [3-4], Sn-Ag [5-7] and Sn-Bi [8], as well as the ternary systems such as Sn-Ag-Cu [9], has been extensively pursued. Though codeposition of Pb and Sn via electrodeposition is relatively straight forward as both elements exhibit similar standard electrochemical reduction potentials (Sn^{2+}: -0.136V and Pb^{2+}: -0.125V, vs. standard hydrogen electrode) , the large difference between other metals in lead-free solder alloys, such as Cu/Cu^{2+} and Sn/Sn^{2+} (ΔE_{Cu-Sn}=0.573 V) [3], makes the electroplating of Sn-Cu alloys difficult. It is well known that precise control of alloy composition in such a multi-component system is critical to solder joint formation and reliability, yet the complexity of multi-component electroplating renders the control difficult. Moreover, there is a particular challenge to electrodeposit the eutectic Sn0.7wt.%Cu alloy, as only such a small amount of Cu is required as the alloy constituent.

The electrolyte stability is also a universal problem for the electrodeposition of tin and its alloys, as in aqueous solutions, tin (II) ions are readily oxidised to tin (IV) by atmospheric oxygen and anodic oxidation [10]. In this paper, the electroplating baths were based on methanesulphonic acid (MSA), which is reported to be a good reducing agent that helps in controlling the oxidation of stannous ions [11]. It also helps in achieving high deposition rates, and can be used over a wide range of current densities. Thiourea was used as a chelating agent for copper ions, thereby improving the bath stability and also bringing close the deposition potentials of tin and copper. Iso-octyl phenoxy polyethoxy ethanol (Triton X-100) was added as a surfactant for improvement in the microstructure of deposits [10]. As a result, Sn-Cu deposits films were electrodeposited on copper coupons and consequently characterised. Fine pitch near-eutectic Sn-Cu solder bumps were produced on test (glass) wafers with a patterned photoresist, proving the feasibility of Sn-Cu alloy solder bumping for flip-chip interconnection using current fabrication processes.

2 Experiment

Apparatus

All the electroplating experiments were carried out at room temperature (~ 20°C) in a 500 mL cell, with potential and current control via a potentiostat (PARSTAT 2273, Princeton Applied Research, AMETEK) and computer. Copper coupons, of 1cm² working area defined by chemically inert tape, were used as the working electrode (cathode). A Pt plate (4cm×4cm) was used as the counter electrode (anode). The distance between the cathode and anode was fixed at 3cm for all experiments. To measure the working electrode potentials, a saturated calomel electrode (SCE) was used as a reference electrode, against which all electrode potentials were measured.

Prior to experimentation, copper working electrodes were immersed in 20wt.% potassium hydroxide solution for 5 min, and then 30 s in a 50vol.% nitric acid (S.G. 1.51) solution. They were then rinsed with deionised water and quickly immersed into the electroplating bath.

Materials

The electroplating baths were based on methanesulphonic acid, and prepared by chemicals from Fisher Scientific Inc® and Sigma-Aldrich Co®, with details in Table 1. $Sn(CH_3SO_3)_2$ and $CuSO_4$ were used as sources providing Sn^{2+} and Cu^{2+} ions for deposition. A relatively large amount of CH_4SO_3 (100g/L) was included for the purpose of solution stability. $(NH_2)_2CS$ was added to chelate Cu^{2+} ions, and different concentrations of Triton X-100 were examined for its effect as a surfactant.

Table 1 Electroplating bath constituents for the production of Sn-Cu alloys

Chemicals/Electroplating parameters	Concentration/Electroplating conditions
Tin methane sulphonate	70 g/L
Copper sulphate	0.002~0.02 M
Methane sulphonic acid	100 g/L
Thiourea	0.2 M
Triton X-100	10~50 ml/L
pH	< 1
Temperature	Room temperature (~20°C)

Cathodic polarisation curves

The polarisation curves were recorded in the potential range between open-circuit potential to approximately -1.0 V versus SCE with a potentiostat (PARSAT 2273) under computer control. Three independent cathodic polarisation curves were recorded for each electrolyte-material combination, and one typical curve is reported in each case. All potentials are quoted versus SCE. The potential sweep rate was set at 1 mV/s.

Solder Bump formation

Sn-Cu solder bumps were produced on test wafers, which were made of glass for experimentation. The wafers were sequentially coated with 100nm Ti and then 100nm Cu as under bump metallisation (UBM) by evaporation from above the glass substrate. This provided a conductive path for the electroplating. Once the seed layer was deposited, the photoresist (AZ 9260, approximately 15μm thick) was spin-coated and developed to form the required pattern, which included a matrix of different diameter circles and pitch sizes ranging from 20μm to 60μm.

Sn-Cu alloy bumps were then electroplated onto the seed layer under potentiostatic conditions through the apertures in the pattern. The photoresist was removed in acetone to reveal the solder bumps on the wafer after electroplating. For flip chip application, the seed layer was then etched away to electrically isolate each microbump, and reflowed to form truncated spherical bumps which could be readily used for the flip chip assembly. This study will focus on the bumping processes up to the Sn-Cu electrodeposition stage.

Characterisation of the deposits

The surface morphology of the deposited films was characterised by SEM (Scanning Electron Microscopy). WDS (Wavelength Dispersive X-Ray Spectroscopy) was used to measure the compositional constituents of the deposits, and phase information was obtained by XRD (X-Ray Diffraction) with an X-ray source of Cu K_α radiation, scanning in a 2θ angle from 20° to 100° with a step of 0.02°/s.

3 Results and Discussion

Electrodeposition behaviour

The reduction of metallic species in methanesulphonic acid electrolytes and the cathodic behaviour of the baths have been analysed in Figs. 1, 2 and 3. The polarisation curves for the co-deposition of tin-copper alloys are initially given in Fig. 1. The electrodeposition baths used were made by adding different concentrations of $CuSO_4$ (0 M, 0.002M, and 0.02M respectively) into a 70 g/L $Sn(CH_3SO_3)_2$ and 100g/L CH_4SO_3 solution, in order to investigate the effect of $CuSO_4$ on the Sn-Cu alloy deposition. As the objective Sn-Cu alloy contains only 0.7wt.% of Cu, it is presumed that the proper $CuSO_4$ concentration in the electrolyte should be in the range between 0.002 M and 0.02 M, which were examined.

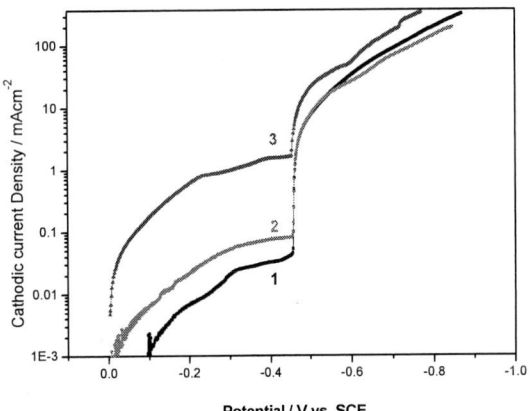

Fig. 1 Cathodic potentiodynamic polarisation curves for the deposition of tin-copper alloys from a solution containing 70 g/L $Sn(CH_3SO_3)_2$, 100g/L CH_4SO_3 plus different concentrations of $CuSO_4$: 0 M (curve 1), 0.002 M (curve 2) and 0.02 M (curve 3), respectively. Potential sweep rate 1 mV/s.

As can be observed, the deposition of tin commenced at a potential of -0.44 V in all the three curves with sharp increases in current densities, regardless of $CuSO_4$ concentrations. Before this potential, plateaux of current densities were shown in the curves corresponding to different $CuSO_4$ concentrations, which were attributed to the deposition of copper at limiting current densities (curves 2 and 3) and presented a typical mass-transport controlled mechanism. The higher limiting current density (approximately 1 mA/cm², curve 3) responded to a higher concentration of $CuSO_4$ (0.02 M), while the lower concentration (curve 2) showed a much lower value (less than 0.1 mA/cm^2). Meanwhile, the co-deposition of Sn-Cu was achieved from the commencement of tin deposition at the potential of -0.44 V, where copper was depositing at its limiting current density. In all the three curves, as the potential became more negative than -0.7 V, the current densities reached even higher values, greater than 200 mA/cm^2. This proved that a methanesulphonic acid based system was capable of deposition at a relatively high rate. However, it was noticed that hydrogen evolution could be a concurrent process as the potential approached -1.0 V.

The effect of thiourea on the co-deposition of the Sn-Cu alloy is illustrated in Fig. 2. The most noticeable difference with the presence of thiourea (curve 2 compared to curve 1) was the rest potential of the copper coupon in the solution. With the absence of thiourea, the rest potential of copper (curve 1) was close to 0 V, comparable to results shown in Fig. 1. Nonetheless, when 0.2 M thiourea was added in the solution (curve 2), the rest potential of the copper coupon decreased to -0.44 V, which was precisely the potential of the commencement for tin electrodeposition.

It was also observed from experiments that a greyish layer started to form on the copper coupon surface as soon as it was immersed into the solution. This was because in the presence of thiourea as a complexing agent for copper ions, the electrode potentials of Sn and Cu differed so remarkably from their standard electrode potentials (Sn^{2+}: -0.136V and Cu^{2+}: +0.337V), that even the order in the electrochemical series changed. Thus, the chemical coating of copper with tin was conducted without applying an external current by a replacement reaction [12]:

$$Sn^{2+} + Cu \rightarrow Sn + Cu^{2+}$$
Equation (1)

As a result, the surface of the copper coupon was covered with a Sn layer deposited by a displacement reaction, and the rest potential of cathode turned to be that of tin. By comparing these two curves, though thiourea showed a strong chelating effect with copper, its effect on the deposition of tin proved to be negligible.

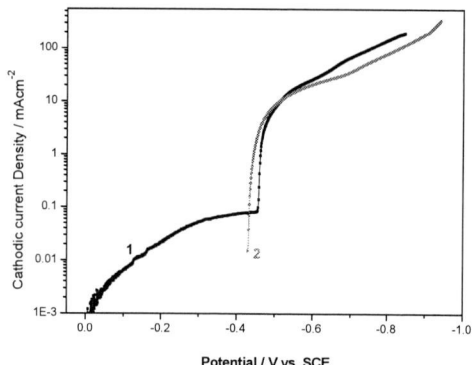

Fig. 2 Effect of thiourea on the cathodic potentiodynamic polarisation curves for the deposition of tin-copper alloys from a solution containing 70 g/L Sn(CH$_3$SO$_3$)$_2$, 100g/L CH$_4$SO$_3$ and 0.002 M CuSO$_4$: without thiourea (curve 1); with 0.2 M thiourea (curve 2). Potential sweep rate 1 mV/s.

Triton X-100 was added to the electrolyte as a non-ionic surfactant for improvement to the microstructure of the deposits. The cathodic polarisation curves illustrating the effect of Triton X-100 are presented in Fig. 3, with different concentrations of Triton X-100 (10 ml/L and 50 ml/L) added to the solutions consisting of 70 g/L Sn(CH$_3$SO$_3$)$_2$, 100g/L CH$_4$SO$_3$, 0.002 M CuSO$_4$ and 0.2 M thiourea. As Fig 3 shows, the previously discussed deposition of Sn on copper coupons was not affected by the addition of Triton X-100 in all the three curves (all sharing the same rest potential of about -0.44V). However, it did change the overall shape of the polarisation curves.

Compared to curve 1 (no addition of Triton X-100), with the presence of 10 ml/L Triton X-100 (curve 2), it showed a strong inhibitive effect on tin electrodeposition. This was even more the situation with a larger Triton X-100 concentration of 50 ml/L (curve 3), which implied that a stronger effect was apparent at a higher concentration. For example, without Triton X-100, the current density reached 200 mA/cm^2 at a potential of -0.9 V, while it was only 40 mA/cm^2 and 15 mA/cm^2 with the presence of 10 ml/L and 50 ml/L Triton X-100 respectively. This inhibitive effect on tin deposition also changed the surface morphology of deposits, which is discussed later.

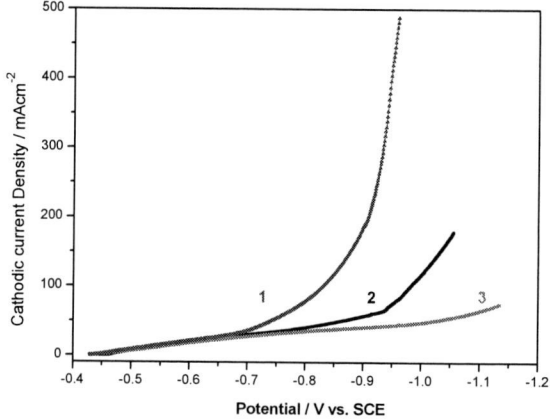

Fig. 3 Effect of Triton X-100 on the cathodic potentiodynamic polarisation curves for the deposition of tin-copper alloys from a solution containing 70 g/L Sn(CH$_3$SO$_3$)$_2$, 100g/L CH$_4$SO$_3$, 0.002 M CuSO$_4$ and 0.2 M thiourea: without Triton X-100 (curve 1); with 10 ml/L Triton X-100 (curve 2), with 50 ml/L Triton X-100 (curve 3). Potential sweep rate 1 mV/s.

Film composition

The composition of electroplated Sn-Cu films on copper coupons under galvanostatic conditions was analysed by WDS. Fig. 4 shows the copper contents in the electrodeposited films at current densities of 15, 20 and 25 mA/cm^2 from baths containing 70 g/L Sn(CH$_3$SO$_3$)$_2$, 100g/L CH$_4$SO$_3$, 0.2 M thiourea, 10 ml/L Triton X-100, and different concentrations of CuSO$_4$ (curves 1, 2 and 3 corresponding to 0.004 M, 0.005 M and 0.007 M respectively). The horizontal broken line in the figure shows the eutectic composition of the Sn-Cu alloy (Sn0.7wt.%Cu). It can be seen that in general, at one particular current density, the copper content in the deposits increased proportionally as the $CuSO_4$ concentration in the electrolytes increased. At a $CuSO_4$ concentration of 0.004 M, the copper content tended to be relatively unstable, and was the furthest away from the eutectic value (0.7wt.%). As the $CuSO_4$ concentration increased to 0.007 M, the copper contents were stabilised at the eutectic level, which was the aimed for composition of this Sn-Cu solder alloy.

Microstructure of electroplated films

The surface morphologies of electrodeposited Sn-Cu films at current densities of 15, 20 and 25 mA/cm^2 under galvanostatic conditions from a bath containing 70 g/L $Sn(CH_3SO_3)_2$, 100g/L CH_4SO_3, 0.007 M $CuSO_4$ and 0.2 M thiourea (without Triton X-100) were analysed by SEM as indicated in Fig. 5. During each electroplating trial, the total electrical charge was kept constant at 1500 C/dm^2. It is clearly shown that all these micrographs appear similar over thewhole current density range examined. As can be observed, the most characteristic feature of these images is that the deposits consist of closely-packed small nodules with sizes of less than 5 μm. All the samples showed a grey and dull surface appearance.

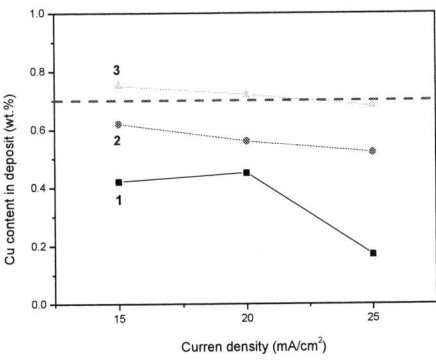

Fig. 4 Cu content (wt.%) in electrodeposited Sn-Cu films vs. galvanostatic current density from baths containing 70 g/L $Sn(CH_3SO_3)_2$, 100g/L CH_4SO_3, 0.2 M thiourea, 10 ml/L Triton X-100, and 0.004 M (curve 1), 0.005 M (curve 2) and 0.007 M (curve 3) $CuSO_4$ respectively. Measurements carried out by WDS. Broken line represents the objective copper content, 0.7wt.%.

Fig. 6 shows the surface morphology of samples produced at the same range of current densities (15, 20 and 25 mA/cm^2) from the same electrolyte except with the addition of 10 ml/L Triton X-100. It was clearly found that the most characteristic feature of all the images in this figure are the well defined crystalline surfaces, which give the samples a bright and shiny look, a clear comparison to those samples from electrolytes not containing Triton X-100. Similar to the images in Fig. 5, the change in current density (from 15 to 25 mA/cm^2) did not result in any clear distinction in the surface morphology of deposits over the current densities examined. This revealed

that Triton X-100 was acting in its role of surfactant over the whole range of current density investigated.

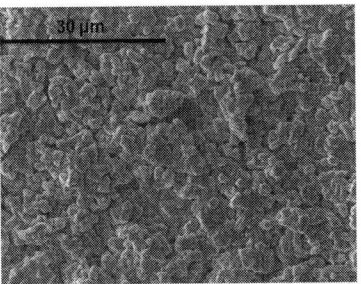

(a) current density=15 mA/cm^2

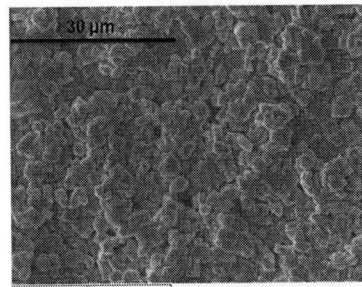

(b) current density=20 mA/cm^2

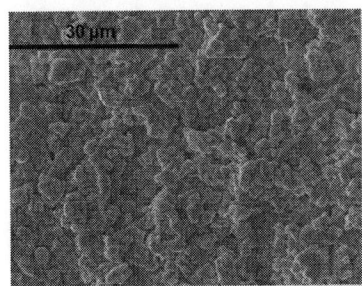

(c) current density=25 mA/cm^2

Fig. 5 SEM micrographs of Sn-Cu films electrodeposited under galvanostatic conditions from a bath containing 70 g/L $Sn(CH_3SO_3)_2$, 100g/L CH_4SO_3, 0.007 M $CuSO_4$ and 0.2 M thiourea, electrical charge passed= 1500 C/dm^2.

The XRD analysis of the Sn-Cu film electrodeposited from the bath containing 70 g/L $Sn(CH_3SO_3)_2$, 100g/L CH_4SO_3, 0.007 M $CuSO_4$, 0.2 M thiourea and 10 ml/L Triton X-100 at a current density of 20 mA/cm^2 is shown in figure 6. The result is analysed on the basis of a powder diffraction database from JCPDS-ICDD 2002$^©$. As shown in Fig. 7, all the diffraction peaks can be attributed to β-Sn and the Cu_6Sn_5 phase, which means that the deposited Sn-Cu alloy films from the methane sulphonic acid based bath have a β-Sn and Cu_6Sn_5 biphasic structure. This result also matches the indications from the Sn-Cu equilibrium phase diagram [13].

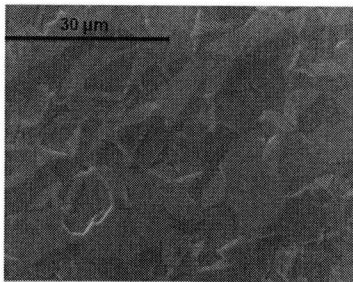

(a) current density=15 mA/cm^2

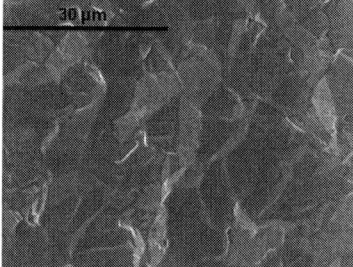

(b) current density=20 mA/cm^2

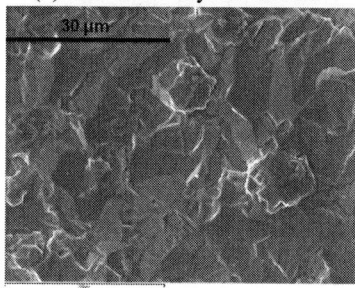

(c) current density=25 mA/cm^2

Fig. 6 SEM micrographs of Sn-Cu films electrodeposited under galvanostatic conditions from a bath containing 70 g/L Sn(CH$_3$SO$_3$)$_2$, 100g/L CH$_4$SO$_3$, 0.007 M CuSO$_4$, 0.2 M thiourea and 10 ml/L Triton X-100, electrical charge passed= 1500 C/dm^2.

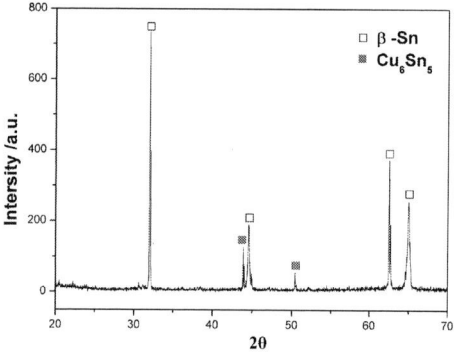

Fig. 7 XRD analysis of a Sn-Cu film electrodeposited under galvanostatic conditions from a bath containing 70 g/L Sn(CH$_3$SO$_3$)$_2$, 100g/L CH$_4$SO$_3$, 0.007 M CuSO$_4$, 0.2 M thiourea and 10 ml/L Triton X-100, electrical charge passed= 1500 C/dm^2, current density=20 mA/cm^2.

Bump fabrication

Sn-Cu solder bumps were produced on test wafers from a bath containing 70 g/L Sn(CH$_3$SO$_3$)$_2$, 100g/L CH$_4$SO$_3$, 0.007 M CuSO$_4$, 0.2 M thiourea and 10 ml/L Triton X-100 at a current density of 20 mA/cm^2. Fig. 8 shows the as-electroplated bumps with a pitchsize of 50 µm (25 µm diameter). As characterised before (Fig. 4), a near eutectic Sn-Cu alloy composition is achieved under such conditions. This has demonstrated the feasibility of producing real fine pitch solder bumps for miniaturised flip-chip application from the processes studied.

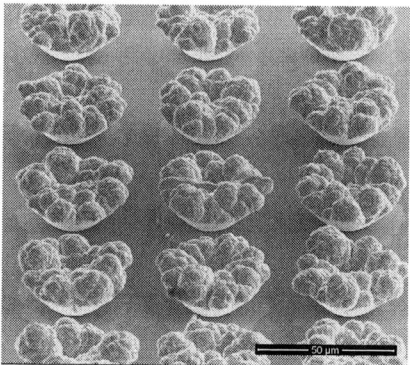

Fig. 8 SEM micrographs of as-electrodeposited Sn-Cu solder bumps formed under galvanostatic conditions from a bath containing 70 g/L Sn(CH$_3$SO$_3$)$_2$, 100g/L CH$_4$SO$_3$, 0.007 M CuSO$_4$, 0.2 M thiourea and 10 ml/L Triton X-100, electrical charge passed= 1500 C/dm^2, current density=20 mA/cm^2.

Conclusions

Methane sulphonic acid based baths for the electroplating of Sn-Cu alloys, as well as the effectiveness of Triton X-100 as a surfactant additive, were investigated. The morphology of the deposits consisted of small nodules with sizes of less than 5 µm with the absence of Triton X-100. Bright Sn-Cu films could be achieved when Triton X-100 was added to the electrolyte, which showed a strong inhibitive effect on the electrodeposition of Sn. Compositional analysis shows that near eutectic Sn-Cu alloys have been electrodeposited at a current density ranging from 15 to 25 mA/cm^2 with the addition of Triton X-100. XRD analysis revealed that the deposited Sn-Cu alloy films have a β-Sn and Cu$_6$Sn$_5$ biphasic structure. Fine pitch (50 µm), near eutectic Sn-Cu solder bumps were produced on a test wafer, which proved to be practical for producing Sn-Cu solder bumps through procedures examined in the current paper.

References

1. M. Juergen Wolf, G. Engelmann, L. Dietrich, and H. Reichl, "Flip chip bumping technology--Status and update," *Nuclear Instruments and Methods in Physics Research Section A: Accelerators, Spectrometers, Detectors and Associated Equipment*, Vol. 565 (2006), pp. 290-295.

2. P. A. Gruber, L. B. Langer, G. P. Brouillette, D. H. Danovitch, J. L. Landreville, D. T. Naugle, V. A. Oberson, D. Y. Shih, C. L. Tessler, and M. R. Turgeon,

"Low-cost wafer bumping," *IBM Journal of Research and Development*, Vol. 49 (2005), pp. 621-639.

3. S. Arai, Y. Funaoka, N. Kaneko, and N. Shinohara, "Electrodeposition of Sn-Cu Alloy from Pyrophosphate Bath," *Electrochemistry*, Vol. 69 (2001), pp. 319-323.

4. N. Kaneko, M. Seji, S. Arai, and N. Shinohara, "Sn-Cu Solder Bump Formation from Acid Sulfate Baths Using Electroplating Method," *Electrochemistry*, Vol. 71 (2003), pp. 791.

5. S. Arai, H. Akatsuka, and N. Kaneko, "Sn-Ag Solder Bump Formation for Flip-Chip Bonding by Electroplating," *Journal of The Electrochemical Society*, Vol. 150 (2003), pp. C730-C734.

6. S. Arai and T. Watanabe, "Electrodeposition of Sn-Ag Alloy with a Non-Cyanide Bath," *Electrochemistry*, Vol. 65 (1997), pp. 1097-1101.

7. Y. Qin, G. D. Wilcox, and C. Liu, "Electrodeposition of Sn-Ag solder alloy for electronics interconnection," *Electronics System-Integration Technology Conference, 2008. ESTC 2008. 2nd*, 2008, pp. 833-838.

8. M. Fukuda, K. Imayoshi, and Y. Matsumoto, "Effect of polyoxyethylenelaurylether on electrodeposition of Pb-free Sn-Bi alloy," *Electrochimica Acta*, Vol. 47 (2001), pp. 459-464.

9. K. Bioh and R. Tom, "Electrochemically Deposited Tin-Silver-Copper Ternary Solder Alloys," *Journal of The Electrochemical Society*, Vol. 150 (2001), pp. C53-C60.

10. S. Joseph and G. J. Phatak, "Effect of surfactant on the bath stability and electrodeposition of Sn-Ag-Cu films," *Surface and Coatings Technology*, Vol. 202 (2008), pp. 3023-3028.

11. N. M. Martyak and R. Seefeldt, "Additive-effects during plating in acid tin methanesulfonate electrolytes," *Electrochimica Acta*, vol. 49 (2004), pp. 4303-4311.

12. E. Huttunen-Saarivirta, "Observations on the uniformity of immersion tin coatings on copper," *Surface and Coatings Technology*, Vol. 160 (2002), pp. 288-294.

13. T. Massalski, "Binary Alloy Phase Diagrams," *ASM*, (1996).

Study on the Degradation of Sealed Organic Light-emitting Diodes under Constant Current

Xiao-Ming Xu*[1,2], Wen-Qing Zhu[1,2], Qiang Wang[1,2], Zhi-Lin Zhang[1,2], Xue-Yin Jiang[1,2]

[1]School of Material Science and Engineering, Shanghai University, Shanghai 20072, China,
[2]Key Laboratory of Advanced Display and System Application Ministry of Education, Shanghai University
No. 149 Yan Chang Road, Shanghai 200072, China
Email: samlxu@gmail.com

Abstract

Degradation characteristics of sealed OLED devices under four different constant currents were investigated by using OLEDs aging tester. We found that the higher the applied current density was, the faster luminance decayed. There was a relationship between luminance and lifetime when the current density was within 200mA/cm^2. In addition, spectrum characterization indicated that the peak wavelength didn't shift during the aging process, which implied that the carriers recombination occurred near the NPB/Alq$_3$ interface all along even when OLEDs luminance dropped down to half of the initial value, indicating that the degradation wouldn't lead to the deviation of the carriers recombination region. Through the metallurgical microscopy observation, the number of bubbles and dark spots were noticed to increase continuously and the non-emissive area kept growing around the bubbles. It showed that bubbles would badly accelerate the formation of dark spots. By investigating luminance degradation under constant current, we draw a conclusion that OLEDs luminance degradation is not entirely coulombic. After introducing an acceleration coefficient, the formular could be applicable until the current went beyond some degree.

1. Introduction

Since Tang and Van Slyke [1] reported the first vacuum-deposited double layer OLED, many studies have been conducted by academic and industrial researchers due to the high potentials for use in low-cost, mechanically flexible, lightweight display and lighting applications [2], OLEDs consist typically of an anode for hole injection, a hole transport layer (HTL), a recombination layer, an electron transport layer (ETL), and a cathode. Carriers are injected into the device by applying an electric field and are transported to the emissive region.

The researches on OLEDs at present lay stress on the luminance efficiency and tremendous progress has been achieved by leaps and bounds. Recently white organic light-emitting diodes with fluorescent tube efficiency have been proposed [3]. However, several inveterate problems are still in urgent need of solution. OLEDs degradation is one of the main issues to make this technology sufficiently reliable for mass production. The cause of the degradation of OLEDs is complicated, which has been attributed to various mechanisms. For example, weakening of the bonding strength between interfaces could probably lead to the structural deformation of OLEDs [4]. Bubble structures owing to polymer photoelectrochemical reaction may also be a cause for the device degradation [5]. H. Aziz et al. suggested that cathode delamination resulted from the moisture in the air

created nonemissive spots [6-8]. Other reasons were also considered [9-12].

In this paper, OLEDs were fabricated by thermal evaporation and encapsulated with epoxy seal and cover glass. The degradation characteristics of the OLED devices under constant current were studied with OLEDs aging tester. Its optical and electrical properties, as well as the surface morphology were particularly investigated.

2. Experiments

Here we used patterned indium-tin-oxide(ITO)-coated glass as the substrate and the anode, which had a sheet resistance of 30Ω/□. The thickness of the ITO film was 80nm. N,N'-diphenyl-N,N'-bis(3-methylphenyl)-1,1'-diphenyl-4,4'-diamine (NPB) acted as the HTL material and Tris-(8-hydroxyquinoline)-aluminum (Alq$_3$) as the ETL and EML material. LiF and Al were served as the electron-injection layer and the cathode respectively. The basic device structure and the chemical structures of the HTL and the ETL materials are shown in Fig. 1, the luminance area was 5×5mm^2.

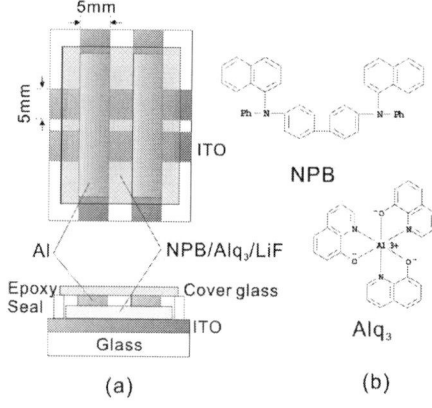

Fig. 1 (a)Basic device structure and (b) chemical structures of NPB and Alq$_3$

The fabrication of the OLED samples followed the standard procedures. Indium tin oxide (ITO) glass was ultrasonically cleaned with acetone and ethanol (for 15 minutes in each), and rinsed with de-ionised water, then followed by 10 minutes UV-ozone treatment. Films were thermally evaporated under the base pressure of 2×10^{-4}Pa. Film thicknesses were monitored with calibrated quartz crystal monitors. Samples were fabricated with structures of NPB(20nm)/ Alq3(50nm)/ LiF(0.7nm)/ Al(100nm). After removal from the vacuum system, the devices were encapsulated with epoxy seal and cover glass immediately in the clean room. This low-cost encapsulation method has been proved to protect the OLEDs from O_2 or moisture effectively in our experiment.

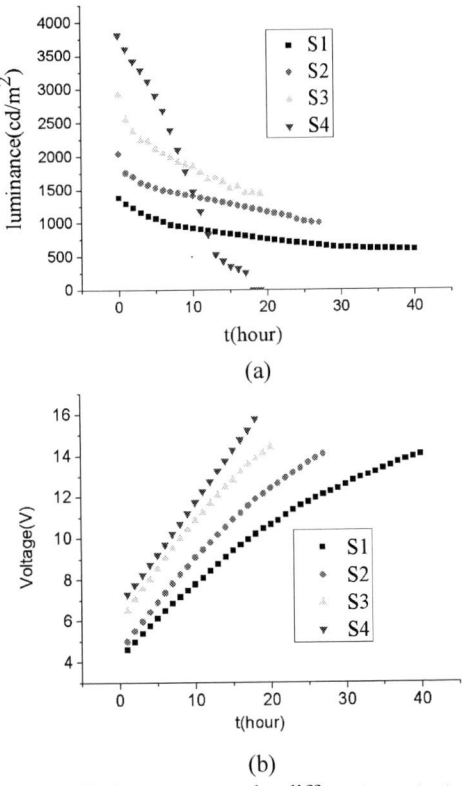

(a)

(b)

Fig. 2 Lifetime curves under different constant current densities. (a)luminance decay curve; (b)voltage augment curve

Table 1 half-life and initial luminance

Samples	$t_{1/2}$(h)	L_0(cd/m²)
S1	50	1381
S2	28	2045
S3	17	2925
S4	8	3812

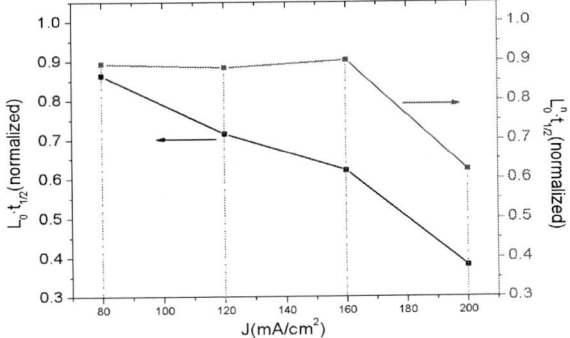

Fig. 3 $L_0 \cdot t_{1/2}$ and $L_0^n \cdot t_{1/2}$ versus current density

A homemade multi-channel OLEDs aging tester was employed to test the degradation behavior, which was capable of testing 16 samples simultaneously. The values of luminance and voltage were recorded automatically by computer once an hour. Spectrum characterization and electroluminescence (EL) intensity were measured by PR650 Spectra Scan Colorimeter. The current-voltage (I-V) characteristics were recorded by Keithley model 2400 programmable voltage-current source. All the measurements were carried out at room temperature under ambient condition (25°C, 45%).

3. Results and discussion

To investigate the influence of constant current on aging characteristics, we prepared four OLED samples which were fabricated and encapsulated with the same processes. Constant current densities of 80mA/cm², 120mA/cm², 160mA/cm², 200mA/cm² were applied on the four samples (henceforth, S1, S2, S3, S4, respectively). Each Luminance was monitored before it dropped to 50% of the initial value. Fig. 2 shows the curves of luminance and voltage versus degradation time. Table 1 shows the data of the lifetime under different constant currents.

Fig. 2 showed obviously that, during the degradation process, device luminance declined continuously, and driving voltage rose. Additionally, the higher the constant current we applied, the shorter the half-life became. C. W. Tang et al. thought OLEDs luminance degradation should be coulombic [13]. According to this opinion, half-life and initial luminance have the relationship as follows:

$$L_0 \cdot t_{1/2} = \text{Constant} \qquad (1)$$

Afterward, another formula was suggested, which introduced an acceleration coefficient [14]:

$$L_0^n \cdot t_{1/2} = \text{Constant} \qquad (2)$$

Here L_0 represents the initial luminance, $t_{1/2}$ means half-life, n is an acceleration coefficient, which has been estimated to be about 1.45 in our experiment.

In order to prove whether these two equations are applicable, we analyzed the values of L0 · t1/2 and L0n · t1/2 of the four samples, as shown in Fig. 3.

In the case of S1~S3, there were significant differences of L0 · t1/2, but the values of L0n · t1/2 were identical when the acceleration coefficient was involved. However, the degradation behavior of S4 was not the same as the others, no matter whether n was taken into account or not.

It indicated that Eq. (2) was more applicable than Eq. (1) in our experiment. Therefore, OLEDs luminance degradation is not entirely coulombic. Heat induced by current accelerated the device failure so that an acceleration coefficient is essential. Furthermore, S4 wasn't in accordance with Eq. (2) and it degraded much faster than the other three samples. This could be attributed to the reason that the operating temperature was so high that intense chemical reaction and serious crystallization in the organic solids occurred [6].

To further study the degradation behavior of the device during the aging process, the electroluminescence spectrums and corresponding chromaticity coordinates of the four samples were measured simultaneously. Fig. 4 showed the results of S2, which is similar with the others. Inset graph was the corresponding chromaticity coordinates (x,y) shifting curve.

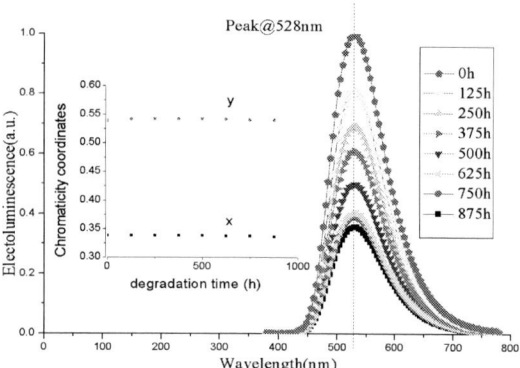

Fig. 4 EL spectrums and chromaticity coordinates (inset graph) at different time of the aging process

It was found that with the aging time passed, EL peak wavelength was invariable while EL intensity kept decreasing. Chromaticity coordinates (x,y) were almost unchangeable, too. This indicated that the carriers recombination region would not move during the aging process.

To observe the transformation surface morphology of the degraded OLEDs, metallurgical microscopy were employed. We took photos when the luminance dropped down to some value (75%, 50%, 10%) in the aging process of the samples. Fig. 5 showed the photos of S2.

Through the surface morphology observation, we found large amounts of bubbles appeared, which may be filled with some gas released from chemical reaction [9]. Then dark spots (non-emissive areas) formed around the bubbles and the area grew bigger. With the expandation of the non-emissive areas, luminance thereby reduced. Owing to the the separation of cathode from organic layers, resistance rose, resulting in the increase of driving voltage as the current was changeless. The bubbles would badly damage the current uniformity so that it could accelerate the formation of dark spots.

Conclusions

In conclusion, degradation properties of OLEDs under different constant currents were investigated with a multi-channel aging tester. The relationship between initial luminance and half-life indicated that OLEDs luminance degradation is not entirely coulombic. The equation with an acceleration coefficient is applicable within a certain limits. Excessive high operating temperature would lead to intense chemical reaction and serious crystallization in the organic solids, which resulted in much faster failure of OLED devices.

Acknowledgments

This work has been supported by National High Technology Research and Development Program of China(863)(No.2008AA03A336), Project of Science and Technology Commission of Shanghai Municipality

(08DZ1140702), Project of Shanghai Municipal Education Comission (2006AZ010).

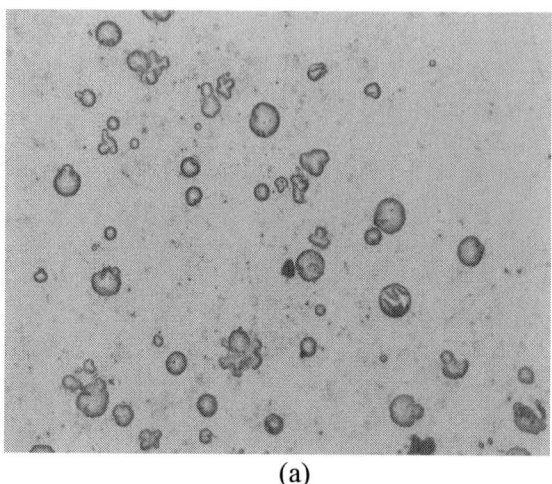

(a)

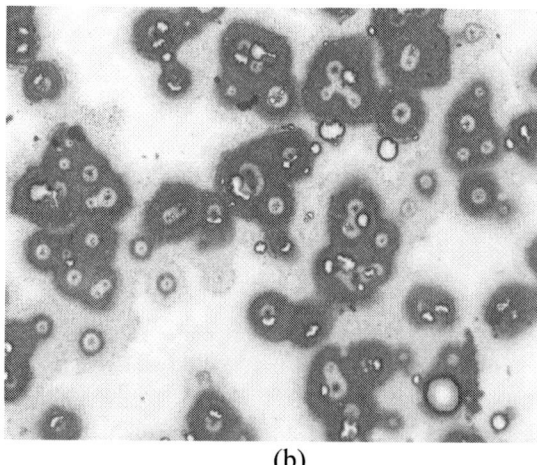

(b)

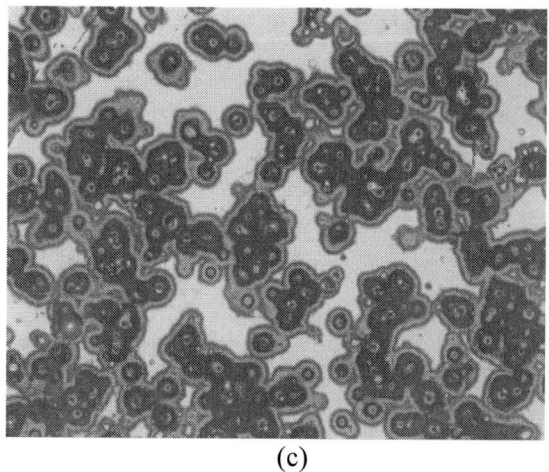

(c)

Fig. 5 Microscope images in the aging process of OLEDs, 75%(a), 50%(b) and 10%(c) of the initial luminance, respectively

References

1. C.W. Tang, S.A. Vanslyke, "Organic electroluminescent diodes," *Appl. Phys. Lett.*, Vol. 51, No. 12 (1987), pp. 913-915.

2. Stephen R. Forrest, "The path to ubiquitous and low-cost organic electronic appliances on plastic," *Nature*, Vol. 428 (2004), pp. 911-918.

3. Sebastian Reineke, Frank Lindner, Gregor Schwartz, et al, "White organic light-emitting diodes with fluorescent tube efficiency," *Nature*, Vol. 459 (2009), pp. 234-239.

4. H. J. Shin, M. C. Jung, J. Chung, et al, "Degradation mechanism of organic light-emitting device investigated by scanning photoelectron microscopy coupled with peel-off technique," *Appl. Phys. Lett.*, Vol. 89, No. 6 (2006), pp. 063503.

5. Lin Ke, Soo-Jin Chua, Keran Zhang, et al, "Bubble formation due to electrical stress in organic light emitting devices," *Appl. Phys. Lett.*, Vol. 80, No. 2 (2002), pp. 171-173.

6. H. Aziz, Z. Popovic, S. Xie, et al, "Humidity-induced crystallization of tris (8-hydroxyquinoline) aluminum layers in organic light-emitting devices," *Appl. Phys. Lett.*, Vol. 72, No. 7 (1998), pp. 756-758.

7. H. Aziz, Z. D. Popovic, N. Hu, et al, "Degradation Mechanism of Small Molecule-Based Organic Light-Emitting Devices," *Science*, Vol. 283 (1999), pp. 1900-1902.

8. Hany Aziz, and Zoran D. Popovic, "Degradation Phenomena in Small-Molecule Organic Light-Emitting Devices," *Chem. Mater.*, Vol. 16, No. 23 (2004), pp. 4522-4532.

9. M. Schaer, F. Niiesch, D. Berner, et al, "Water Vapor and Oxygen Degradation Mechanisms in Organic Light Emitting Diodes," *Adv. Funct. Mater.*, Vol. 11, No. 2 (2001), pp. 116-121.

10. J. McElvain, H. Antoniadis, M. R. Hueschen, et al, "Formation and growth of black spots in organic light-emitting diodes," *J. Appl. Phys.*, Vol. 80, No. 10 (1996), pp. 6002-6007.

11. R. Czerw, D. L. Carroll, H. S. Woo, et al, "Nanoscale observation of failures in organic light-emitting diodes," *J. Appl. Phys.*, Vol. 96, No. 1 (2004), pp. 641-644.

12. Yichun Luo, Hany Aziz, Zoran D. Popovic, et al, "Degradation mechanisms in organic light-emitting devices: Metal migration model versus unstable tris(8-hydroxyquinoline)aluminum cationic model," *J. Appl. Phys.*, Vol. 101, No. 3 (2007), pp. 034510.

13. S. A. Van Slyke, C. H. Chen, and C. W. Tang, "Organic electroluminescent devices with improved stability," *Appl. Phys. Lett.*, Vol. 69, No. 15 (1996), pp. 2160-2162.

14. Philipp Wellmann, Michael Hofmann, Olaf Zeika, et al, "High-efficiency p-i-n organic light-emitting diodes with long lifetime," *SID Int. Symp. Digest Tech. Papers*, Vol. 13, No. 5 (2005), pp. 393-397.

Preparation and Performance of White LED Packaged YAG Phosphor

Rongfeng Guan, Wenqing Zhao, Shuaimou Li

Institute of Materials Science and Engineering, Henan Polytechnic University

No. 2001, Century Avenue, Jiaozuo 454003, Henan Province, China

Email:Rongfengg@163.com, Phone: 86-15995740133

Abstract

In recent years, the preparation technology of the white light LED packaged YAG phosphor develops very quickly. The preparation methods include solid-phase method, sol-gel method, precipitation method, combustion method and so on. The traditional solid-phase method is still dominant in all of them, whereas this method also has shortcomings such as a high synthesis temperature, longer reaction time, bulky grains and larger hardness. So it requires a new approach to achieve the phosphor synthesis. Using metal nitrate as raw materials and citric acid as a complexing agent, we prepared $Y_3Al_5O_{12}:Ce^{3+}$ yellow phosphor by the sol-gel method (sol-gel). The X-ray powder diffraction (XRD) shows that it has been pure yttrium aluminum garnet phase (YAG) at 900°C. The infrared absorption spectra of the powders show that YAG phase has been formed when the calcination temperature is 900°C, because there are the characteristic vibrational peaks of metal and oxygen atoms, which are in the low-frequency area at $724cm^{-1}$, $792cm^{-1}$ peak. With scanning electron microscopy (SEM) analysis of its morphology, the result proves the grain diameter ranges from 1 to 3μm, averaged at 2μm. Their fluorescence spectra were studied using fluorescence spectrometer and the results show luminous intensity reach at the maximum when calcination temperature up to 1200°C.

Introduction

White LED [1] has the advantages of small power consumption, long life, environmental protection, and it will be expected to become the most promising high-tech field in the 21st century. According to the principles of optics and colorimetry, there are a variety of programs to achieve white-light LED. Today the most commonly used method of making white light LED is usually efficient InGaN/GaN-based blue LED. Its issued Blue-ray excites YAG:Ce rare-earth phosphor, then YAG has been stimulated to emit yellow light and the remaining blue light and the yellow light mix together to form white light. However, the fine luminescent and ultrafine YAG:Ce powder is one of keys to the technology.

Traditional yellow YAG phosphor commonly uses high-temperature solid-state reaction [2-3] to obtain. It has many disadvantages, such as high sintering temperature, long cycle of synthetic course, the hard and large granular powder after firing, the difficult dealing in the end and so on. This method to synthesize powder also has to subject to ball milling, which will undermine the crystal, so brightness and luminous efficiency of LED will reduce significantly. In order to achieve the lower sintering temperature, improve product performance and reduce costs, wet synthesis [4-5] is tried to use to produce YAG phosphor at home and abroad at present.

Sol-gel method [6-7] is a new method of wet-chemical synthesis in recent years. Compared with the traditional high

temperature solid state reaction method, It has many advantages, such as high reactivity of the starting material, the uniform composition, the low temperature synthesis, and the small size synthetical powder. For these reason, the method catches people's attention. In this paper, this method was adopted to synthesize $Y_3Al_5O_{12}:Ce^{3+}$ rare-earth phosphor.

Experimental materials

Raw materials are aluminum nitrate $((Al(NO_3)_3.9H_2O)$, AR, 99.99%); citric acid $(C_6H_8O_7.H_2O$, AR); nano-yttria $(Y_2O_3$, AR, 99.99%); cerium nitrate $(Ce(NO_3)_3.6H_2O$, AR, 99.99%,); nitric acid $(HNO_3$, AR).

Experimental process [8-11]

The quality of Y_2O_3 (purity ≥ 99.99%), $Al(NO_3)_3.9H_2O$ (purity ≥ 99%) and $Ce(NO_3)_3.6H_2O$ (purity ≥ 99.99%) are weighed with some chemical dosage. Y_2O_3 is dissolved in dilute nitric acid to become $Y(NO_3)_3$ solution. $Al(NO_3)_3.9H_2O$ is added to the mixing $Y(NO_3)_3$ solution, then the mixture adds into the dissolved $Ce(NO_3)_3$ solution. A certain amount of citric acid is added into the metal solution, wih the material ratio of citric acid to total metal ions as 1:1.The mixture stir in a water bath at 80~85°C until colloid is formed; the produced colloid then is dried in the vacuum ovens for 12h at the temperature of 120°C. In this process, water evaporates gradually; at the same time a large number of gas comes out. Finally the sample turns into a xerogel. Remove the sample from oven and make the sample subject to pulverisation in agate mortars. At last samples are sintered under the high temperature furnace at 800°C, 900°C, 1000°C respectively.

Infrared analysis

Fig. 1(a)~(c) are the infrared absorption spectra of sintering samples under 800°C to 1000°C respectively. In Fig.1(a), the infrared absorption spectra of sample can not be seen the characteristic vibrational peak of the metal and oxygen atoms of YAG crystal phase after sintering under 800°C. Fig. 1(b) can be seen that when the calcination temperature up to 900°C, the peak $724cm^{-1}$, $792cm^{-1}$ is located at low-frequency area, which is the characteristic vibrational peak of the metal and oxygen atoms of YAG crystal phase. Al-O stretching and bending frequencies depend on the number of ligand atoms and the formation of vibrating keys. The peak $792cm^{-1}$ attributed to the AlO_4's symmetric stretching vibration; the peak $724cm^{-1}$ represents the AlO_4's antisymmetric stretching vibration. All these peaks demonstrate the formation of YAG phase. The peaks $3857cm^{-1}$, $3744cm^{-1}$, $3614cm^{-1}$, $2369cm^{-1}$, $1743cm^{-1}$, $1696cm^{-1}$ are caused by the OH stretching and bending vibration of adsorbed water. According to standard map of citric acid, carbonyl peak of citric acid appears at the $1710cm^{-1}$, but the carbonyl peak in Fig. 1 appears at the $1600~1500cm^{-1}$ range, which indicate C=O double bonds have been moved to low-frequency. In other words the carbonyl ligand has coordinated with the metal atom. In order to make full integration of the reactant, excess citric acid is added.

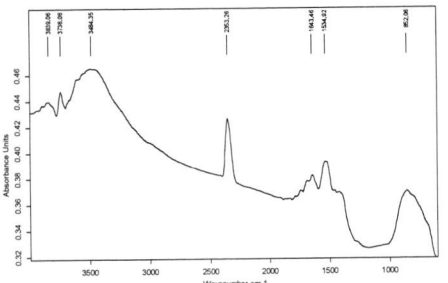

(a) Infrared absorption spectra of sintering samples under 800°C

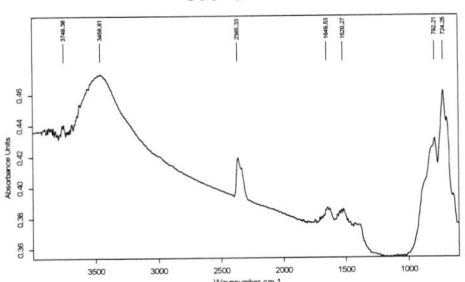

(b) Infrared absorption spectra of sintering samples under 900°C

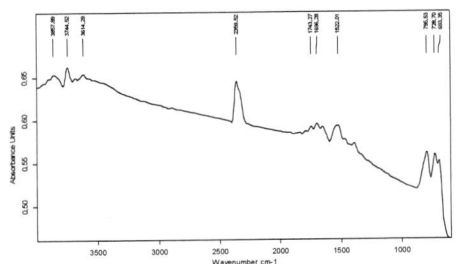

(c) Infrared absorption spectra of sintering samples under 1000°C

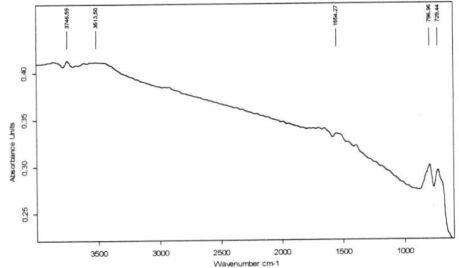

(d) Infrared absorption spectra of sintering samples under 900 °C at the restored atmosphere

Fig. 1 Infrared absorption spectra

From Fig. 1(c) can be seen, with the calcination temperature increasing, the intensity of these peaks decreased,

while the characteristic peaks of YAG become stronger. Fig. 1(d) is infrared absorption spectra of sintering samples under 900°C at the restored atmosphere. It can be seen that the characteristic peaks still exist, but more clearly, accompanied with less impurities in the sample.

X-ray diffraction

The crystal structure of samples were analyzed by X-ray diffraction (XRD) of DX-1000 type from Fangyuan Instrument Company of Dandong. (Cu, Kα, λ = 0.15418nm)

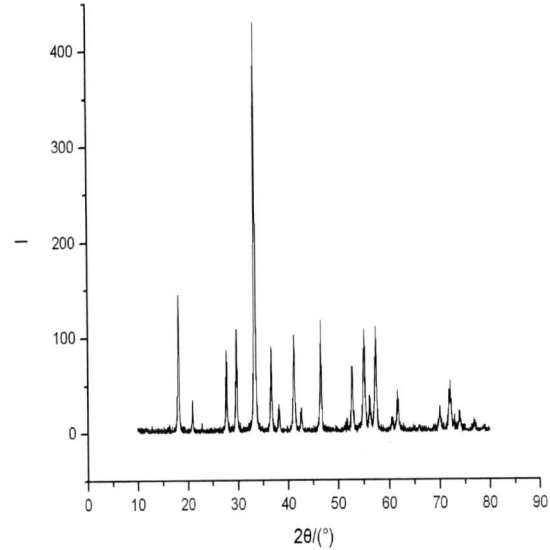

Fig. 2 X-ray diffraction pattern of sintering samples under 900°C

Fig. 2 is X-ray diffraction pattern of sintering samples under 900°C. Compared with standard spectra, the diffraction pattern of the sample and the PDF card are exactly the same with each other, without any other diffraction peaks of miscellaneous phases. It confirms the same conclusion with the test results of the infrared spectra before. In the sol-gel process of preparation of YAG phosphor, YAG phase can be synthesized at the temperature of 900°C. Compared with the traditional high temperature solid-phase method, the sintering temperature declines nearly 500°C to synthesize the sample powders.

Scanning Electron Microscope (SEM)

Morphology and grain size of samples were observed by scanning electron microscope (SEM) of JSM-6360 type from Japan's JEOL Company.

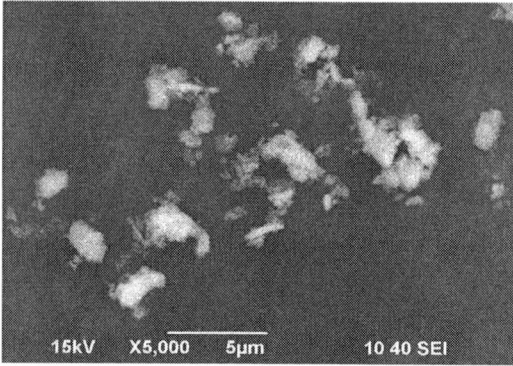

(a) SEM figure of sintered samples under 900°C

(b) SEM figure of commercial samples
Fig. 3 SEM figure

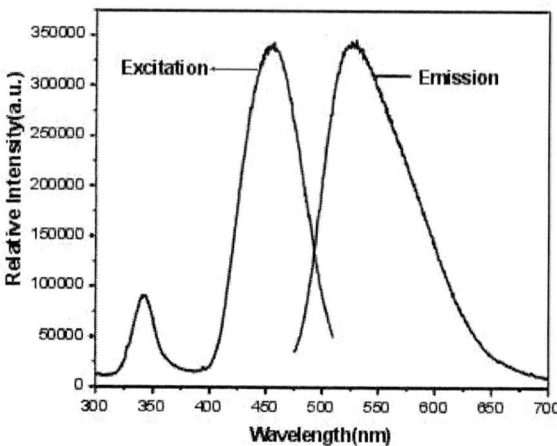

Fig. 4 Excitation spectrum and emission
spectrum of Ce:YAG phosphor

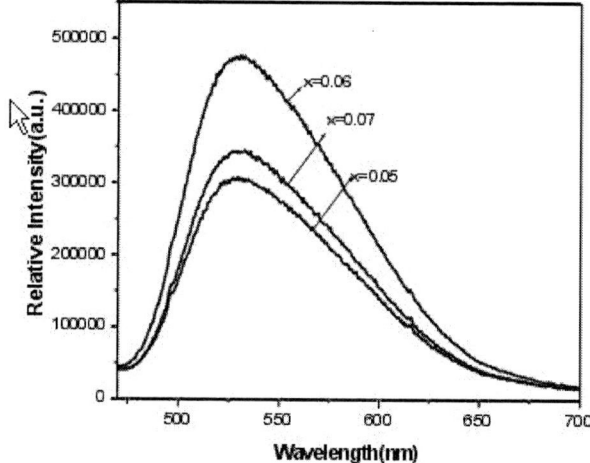

Fig. 5 Ce:YAG emission spectrum with doping
amount of X for different values

Fig. 3 is the SEM figures of the sintered sample under 900°C and commercial sample respectively. In previous reports, the sol-gel method usually made powder reunion. The formed gel network contains a large number of adsorbed water. In the capillary force of water molecules, the gel particles get together with hydrogen bonds. With the temperature increasing, the adsorbed water will be removed and hydrogen bonding will form between the non-bridging hydroxyl of the two particles. Heat continues to rise up, and non-bridging hydroxyl makes water molecules out, while the true chemical bond forms between particles. Such reunion caused by chemical bonds is known as a hard reunion. The hard reunion of powder has been hardly dispersed after formation, so the powder which is made by sol–gel method normally has poor dispersion. However, good dispersion can be seen from the sample in Fig. 3(a). But there is still a phenomenon of hard reunion in the sample. Fig. 3(b) is the SEM figure of commercial sample, the average particle size of the sample can be seen at 12μm; the particles show good in the overall uniformity. The sample size obtained in the experiment is roughly between 1 to 3μm, with an average particle size of 2μm; the particles show good uniformity, but there are still hard agglomerate phenomenon in the sample.

Spectral properties

After sample preparation, the emission spectra of samples were tested. The equipment is the fluorescence spectrometer of Jasco PF-6500 type from Japan (excitation wavelength at 200nm~800nm).

Fig. 4 is the excitation spectrum and emission spectrum [12] of Ce:YAG phosphor. As can be seen from the chart, excitation spectrum of Ce:YAG is double-peak structure; in UV region there is a excitation peak at 340nm;in the visible region there is a maximum excitation peak at 465nm. This is determined by 4f energy levels of Ce^{3+} dividing into two teams of $2F_{5/2}$ and $2F_{7/2}$ as a result of 4f spin coupling of Ce^{3+}. $2F_{5/2}$ is the base spectrum; the excitation peak at 340nm corresponds to the $2F_{5/2}$-5d transition; the excitation peak at 465nm corresponds to the $2F_{7/2}$-5d transition. The emission spectrum of phosphor is excited by visible light of 465nm; light intensity was measured by monitoring different wavelengths. As shown in Fig. 4, the emission spectra is a wide spectrum in visible region; the strongest emission peak is at 530nm and it belongs to the 5d-4f transition of Ce^{3+}. The peak shape and position of excitation and emission spectra of $Y_{3-x}Al_5O_{12}:Ce^{3+}_x$ phosphor do not change with the change of doping concentration, but the relative intensity increases with the increasing of Ce^{3+} content. When x=0.06, it reaches to the maximum, then reduces in the following, as shown in Fig. 5.

978-1-4244-4658-2/09 $25.00 © 2009 IEEE

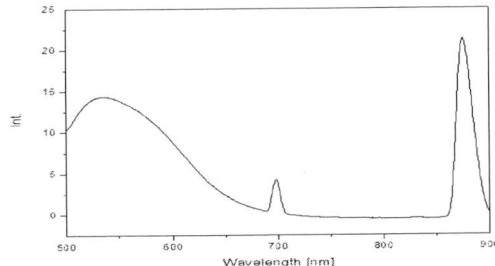

(a) Emission spectrum of the Ce:YAG under 800°C

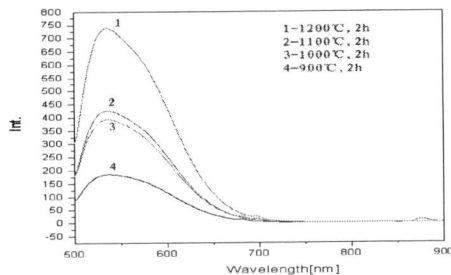

(b) Emission spectrum of the Ce:YAG under 900°C~1200°C

Fig. 6 Emission spectrum of the Ce:YAG

Fig. 6(a) ~ (b) show luminescence intensity of Ce:YAG phosphor under the different sintering temperature. Fig. 6(a) is emission spectrum of the Ce:YAG under 800°C; it is shown that other peaks exist in addition to the characteristic emission spectra of YAG phosphor, and the intensity of characteristic emission peak is very weak. This is because the sample has not yet formed the complete YAG phase at 800°C. From Fig. 6(b), it is known that the emission intensity of phosphor increases with the annealing temperature. This is because the grain of powder also increases with the increasing of annealing temperature; the sample has better crystalline state and it results to high luminous efficiency; whereas the greater the particle size is, the smaller the stimulated light scattering is , it will help stimulated light to be absorbed more effectively.

Conclusions

(1) Using citric acid as a complexing agent and the sol-gel method, the crystal structure of prepared phosphor sample is YAG structure; the crystallization temperature is 900°C, which is much lower than traditional high temperature solid-phase synthesis (above 1400°C).

(2) From the test of scanning electron microscope, very small particles can be seen, with an average particle size of 2μm; the synthesized phosphor sample is better than commercial sample; the particles of experimental samples also have good uniformity, but it still has hard agglomeration phenomenon in the sample.

(3) The excitation spectrum of prepared Ce:YAG is a double-peak structure; in UV region there is a excitation peak at 340nm;in the visible region there is a maximum excitation peak at 465nm. Its emission spectrum has a wide spectrum in the visible region, with the strongest emission peak at 530nm, which belongs to the 5d-4f characteristic transition of Ce^{3+}. With the temperature increasing, the intensity of emission spectra of the samples also increases.

Acknowledgments

The authors acknowledge the following projects for financial support: Key Scientific and Technological Research Projects of Henan Province (No.072102240027), Dr Fund of Henan Polytechnic University (No.648602).

References

1. Tang Y J, Wang Y G,Yang Z J et al, "Fabrication and Properties of White Luminescenc Conversion LEDs," *Chinese Journal of Luminescence*, Vol.22, No.10 (2001), pp. 91-94.
2. Zhang Qiwu, Fumio Satio, "Mechanochemical solid reaction of yttrium oxide with luminaleadin to synthesis of yttrium aluminum garnet," *Powser Technology*, Vol.129 (2003), pp. 86.
3. Ming-Shyong Tsai, Wen-Chuan Fu, Guang-Mau Liu, "Effect of pre-aging pH on the formation of yttrium aluminum garnet powder (YAG) via the solid state reaction method," *Journal of Alloys and Compounds*, Vol.440 (2007), pp. 309-314.
4. Shi Shi-Kao, Wang Ji-Ye, "Microstructure and Luminescent Properties of Tb Doped YAG Phosphor by Combustion Synthesis with Glycine," *Chinese Journal of Inorganic Chemistry*, Vol.18, No.4 (2002), pp. 431-434.
5. Guo Xianzhong, Devi P.Sujatha, Ravi et al, "Phase evolution of yttrium aluminium garnet(YAG)in a citrate-nitrate gel combustion process," *Journal of Materials Chemistry*, Vol.14, No.8 (2004), pp. 1288-1292.
6. Li Yong-Xiu, Min Yu-Lin, Zhou Xue-Zhen et al, "Coating and Stability of YAG:Ce^{3+} Phosphor Synthesized Using Inorganic-organic Hybrid Gel Method," *Chinese Journal of Inorganic Chemistry*, Vol.19, No.11 (2003), pp. 1169 – 1174.
7. Peter A.Tanner, Po-Tak Law, Lianshe Fu, "Preformed sol-gel synthesis and characterization of lanthanide ion-doped yttria-alumina material," *Physica status*, Vol.199, No.3 (2003), pp. 403-415.
8. Y. H. Zhou, J. Lin, S. B. Wang et al, "Preparation of $Y_3Al_5O_{12}$:Eu phosphors by citric-gel method and their luminescent properties," *Optical Materials*, Vol.20, No.1 (2002), pp.13-22.
9. Liu Chun-jia, Yu Rui-min, Xu Zhi-wei et al, "Crystallization, morphology and luminescent properties of YAG:Ce^{3+} phosphor powder prepared by polyacrylamide gel method," *Transactions of Nonferrous Metals Society of China*, No.17 (2007) , pp. 1093-1099.
10. Junying Zhang, Zilong Tang, Zhongtai Zhang et al, "Preparation of SrAl₂O₄:Du,Eu Phosphor by Sol-gel Processing and Its Luminescent Properties," *Key Engineering Materials*, (2002), pp. 224-226.
11. H.M.H. Fadlalla, C.C. Tang, E.M,Elssfah et al. "Synthesis and characterization of single crystalline YAG:Eu nano-sized powder by sol-gel method," *Materials Chemistry and Physics*, Vol.109, No. 2-3 (2008), pp.436-439.
12. Suya Feng, The synthesis and spectral properties of Ce:YAP crystal and Ce:YAG nanophosphor, Henan University (Kaifeng, 2007), pp. 54-55.

Nanoindentation Investigation of Copper Bonding Wire and Ball

Xiangquan Fan[1], Kaiyou Qian[2], Techun Wang[2], Yuqi Cong[2], Mike Zhao[2], Binhai Zhang[1], Jiaji Wang[1]

[1]Department of Material Science, Fudan University, No. 220, Handan Road, Shanghai, 200433, China
[2]ASE Assembly & Test (Shanghai) Limited, No. 669, Guoshoujing Road, Shanghai, 201203, China
Tel: 86+21-55664588, Email: 072030001@fudan.edu.cn

Abstract

Copper is the most cost-effective alternative to gold which is used in wire bonding. Hardness of bonding wire and FAB (free air ball) is double-edged swords, wire with higher hardness and strength can build more reliable loop, but damage the bounding pad on the chip more easily. Precisely measuring hardness is meaningful for improving the performance and reliability of the bonding process. Nanoindenter is a powerful experimental technique for detecting mechanical properties of materials. In this work, the mechanism of nanoindentation and the selection of hardness test parameters were studied. Hysitron's nanoindenter with Berkovich tip was used to measure hardness of 0.8mil (20µm) 4N(99.99%wt of copper) copper ball bonding wire, FAB and bonded ball. It is confirmed that copper wires get harder mainly in ball formation and ball bond form process.

Introduction

Wire bonding was the most important technology for electrical connecting the chip and lead frame [1-2]. Copper ball bonding may be a cost-effective alternative to traditional gold ball bonding [3]. Copper has greater electrical and thermal conductivity than gold, which can enhance the performance of packaged chip. Copper wire has higher tensile strength than gold wire, the copper wire pull strengths are about 50% higher than gold wire [4]. Generally speaking, hardness of metal is proportional to its strength and copper has higher hardness than gold which could cause some reliability issues in bonding process. For an example, silicon crater of copper wire bonding may results from pull test [5]. After optimizing the process condition, crater after pull test could be avoided. But other less serious problems, such as bond pad peeling (Fig 1(a)) after pull test, chip interconnection layer crack (Fig 1(b)) will occur. Therefore hardness of copper is a key parameter which should be well characterized and controlled in the bonding process.

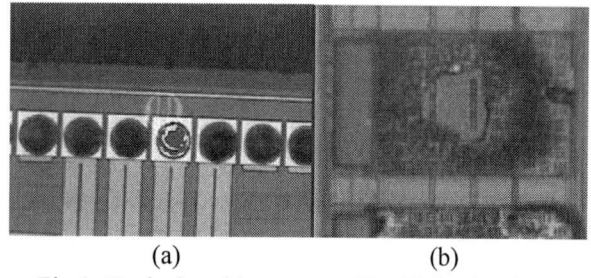

(a)   (b)

Fig 1. Typical problems caused by higher hardness of Copper (a) Al pad peeling   (b)Chip interconnection crack

For hardness characterization, there are many criteria and methods. The most common one for metal is static indentation hardness test [6]. Static indentation hardness which indent sphere, diamond pyramid or other pyramid to a sample, form indentation on the sample, is measured with dividing the total load by indentation area. There are various static indentation hardness tests (Fig 2), each of them has different load range. Each hardness test value can't exactly convert into others.

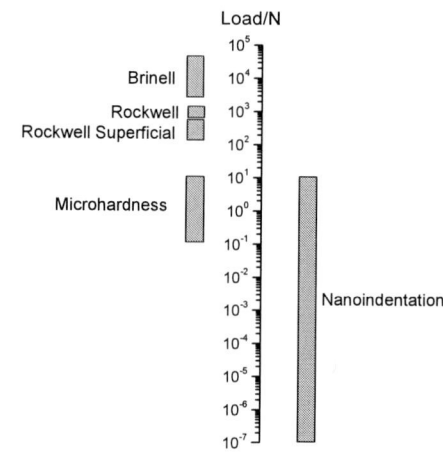

Fig 2. Load of various static indentation hardness

For copper material has less hardness than other metals bonding wire, the testing load should be controlled to lower level and its typical load is 10 milinewton [7]. That's why we can only test copper hardness by nanoindentation or an expanded microhardness tester [7-9]. Compared to conventional Microhardness test with Vickers tip, nanoindentation (also called depth sensing indentation and instrument indentation testing [8]), which allows determining mechanical properties at penetration depths as low as 20 nm, can detect hardness more precisely and nondestructively. The possibility to carry out tests in so small scales makes this technique one of the tools chosen to characterize mechanical properties of thin films, coatings and small scale wire. On the other hand, conventional microhardness test only consider plastic deformation in calculating hardness, nanoindentation can consider both plastic and elastic deformation. Because a perfect tip shape is difficult to achieve, Berkovich tip, which has no slight offset compared to the Vickers indenter, is used in the experiment commonly.

Theoretical Aspects

When tests were carried out with Berkovich indenters the registered data can be analyzed using the method proposed in 1992 by Oliver and Pharr [9]. Fig 3 is a load and unload curve, in this curve "a" is the load part, "b" is the unload part and "c" is the tangent line at maximum load. In Fig 3 there are many depths need to explain, h_{max} is the maximum indentation depth at F_{max}, h_r is point of intersection of the tangent "c" to curve "b" at F_{max} with the indentation depth-axis, h_p permanent indentation depth after removal of the test force.

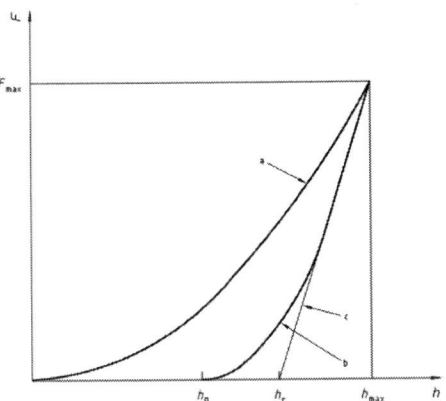

Fig 3. An load and unload curve

In order to calculate the hardness of a tested material, parts of the data from the upper portion (20%~95% F_{max}) of the unloading curve (Fig 3) are fitted by a power-law relation of the form:

$$F = B(h-h_p)^m \qquad (1)$$

Where F is the indenter load, m and B are empirical constants determined after unloading data fitting, h_p is the residual depth, and h is the elastic displacement.

The contact stiffness, S, is defined as the slope of tangent line of unload curve at F_{max}.

$$S = \left.\frac{dF}{dh}\right|_{h=h_{max}} = mB(h_{max}-h_p)^{m-1} \qquad (2)$$

As hardness is defined as:

$$\mathrm{H} = \frac{\mathrm{F}}{A} \qquad (3)$$

Where F is the load of the indentation, usually we choose the maximum load F_{max}, A is the contact area A_c. Fig 4 show the cross section of indentation, if the contact depth (h_c) and the type of indenter tip is known, A_c can be calculated.

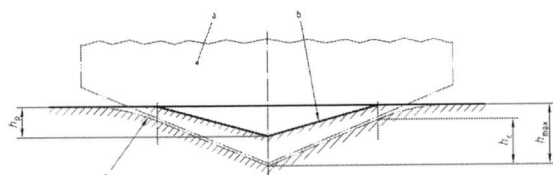

a Indenter; b surface of residual plastic indentation in test piece; c surface of test piece at max indentation depth and test force

Fig 4. Cross section of indentation

As the unloading from h_{max} to h_f is elastic, the relation between h_c h_{max} and dF/dh is: [9]

$$h_c = h_{max} - \varepsilon\frac{F_{max}}{dF/dh} \qquad (4)$$

Where h_c is contact depth, ε is about 0.75 for Vickers and Berkovich indenters. If the indenter tip is ideal triangular base pyramids, $A_c=24.5h_c^2$. The contact area function of a real Berkovich indenter tip which used in the experiment can be express as:

$$A_c = 24.5h_c^2 + \sum_{i=0}^{n} C_i h_c^{1/2i} \qquad (5)$$

Where n normally set as 3~7 for a Berkovich tip whose radius of curvature is less than 100nm, C_i can be determined by experiment of reference samples.

Microstructure of copper wire and ball

Hardness may strongly relate to the microstructure of materials. Especially for polycrystal of metal , as indentation is smaller than the grain size, indent at different orientation, location or boundary of polycrystal may get different hardness results.

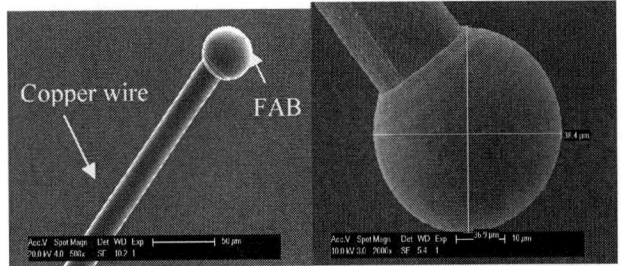

Fig 5. SEM image of FAB

First, we observed the morphology of copper FAB by SEM (Fig 5), a smooth and round FAB was got. The diameter of the FAB is about twice diameter of copper wire, about 36~38μm. This round FAB and wire couldn't perform nanoindentation directly. Mounting in epoxy or other resin is necessary.

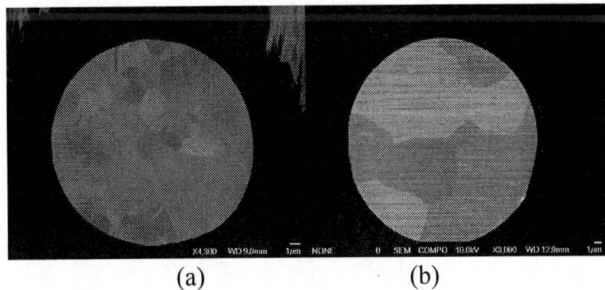

(a)   (b)

Fig 6. Back scattered electron image of copper wire (a) and FAB (b)

When copper wire and FAB were mounted in epoxy resin, cross-section polish was performed to get a strain free surface. SEM with back scattered electron image was used to observe its polycrystal microstructure. As shown in Fig 6, the grain size of the copper wire is about 1-2μm, the grain size of FAB is more than 5μm. FIB was also used to observe the surface microstructure (Fig 7). Based on these two images of grain we could assume FAB with columnar grain structure and wire with equiaxed grain structure.

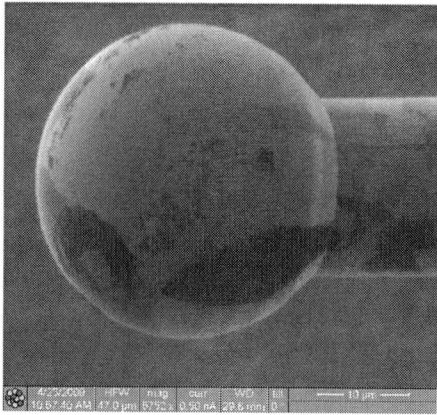

Fig 7. Scanning ion microscope (SIM) image of FAB

Hardness test

Three different kinds of sample were prepared for hardness test, original copper wire, FAB and bonded ball. Each sample was cleaned with acetone and ethanol, followed by blowing dry and mounting in epoxy resin. After grinding and polishing, surface roughness of the sample cross section was less than 10nm which was adequate in flatness for nanoindentation test.

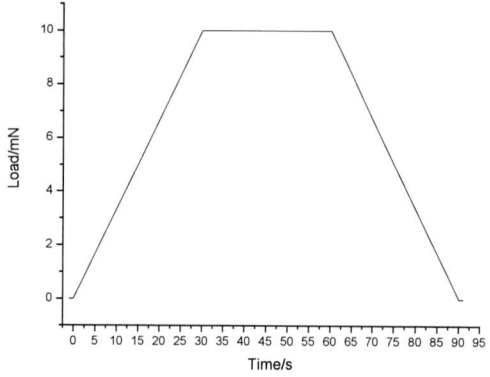

Fig 8. Load function of nanoindentation

Showing the load function of the experiment Fig 8, all tests below were taken under this load function. The spacing between different indentations should be larger than 10μm which can reduce the risk of interaction between these indentations. A typical load-displacement curve under the load function is shown in Fig 9. The horizontal line at the maximum load region means creep of the material. Hardness can be calculated using the method which explained in the previous section from the data of Fig 9 that includes elastic deformation, plastic deformation and creep of the material.

The Hysitron's system is capable of "SPM" style imaging. In this type of imaging, the Berkovich tip is moved in a raster scan pattern across a sample surface using the TriboScanner three-axis piezo positioner. A typical image of indentation is in Fig 10. The indentation is an equilateral triangle and the size of whose side is about 2μm. Testing indentation can be aligned in location within 1μm.

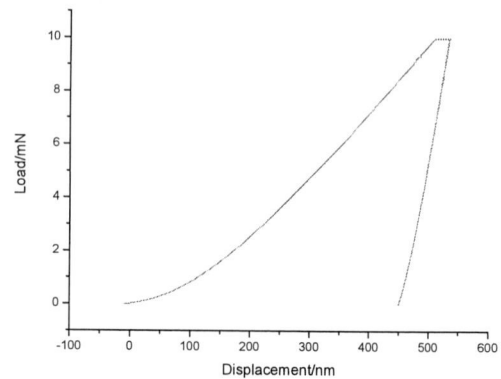

Fig 9. A typical curve of copper wire nanoindentation

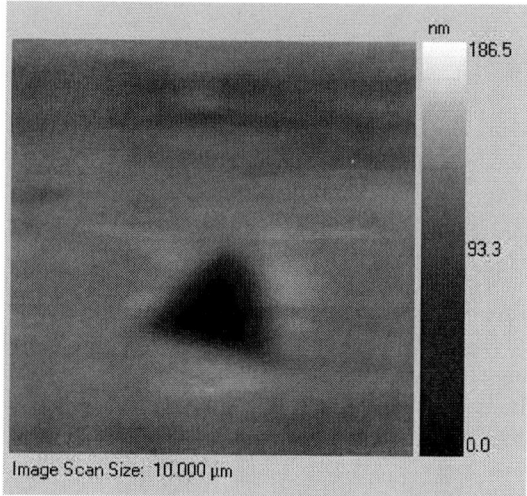

Fig 10. In-situ image of indentation on copper wire

Results & Discussion

In order to get the mean hardness of copper wire, each wire was indented several indentations, as shown in Fig 11. Several samples were tested. The average hardness of copper wire was 1.46GPa, the standard deviation (SD) of these hardness was 0.124GPa. The standard deviation was calculated according to formula (6).

$$\sigma = \sqrt{\frac{\sum_i (H_i - \overline{H})^2}{n-1}} \qquad (6)$$

Where σ: standard deviation, H_i :hardness, n: test times.

The free air ball (FAB) is formed in electronic flame off (EFO) process[1]. Because the diameter of the FAB is about twice the diameter of the wire, more indentations can indent in the FAB sample, as shown in Fig 12. The average hardness of FAB is 1.51GPa, the standard deviation calculated according to [6] is 0.145GPa. Both the hardness and standard deviation of FAB are higher than original copper wire.

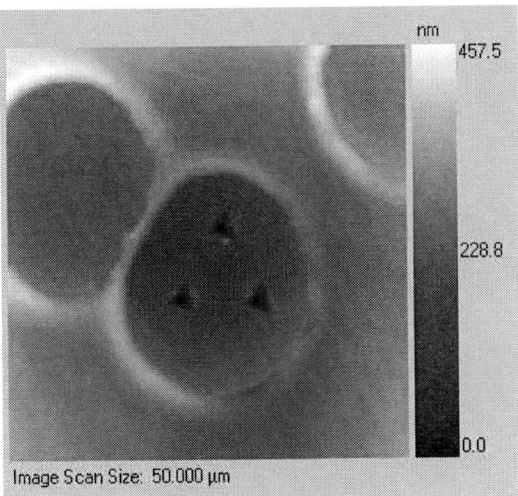

Fig 11. Indentation of copper wire

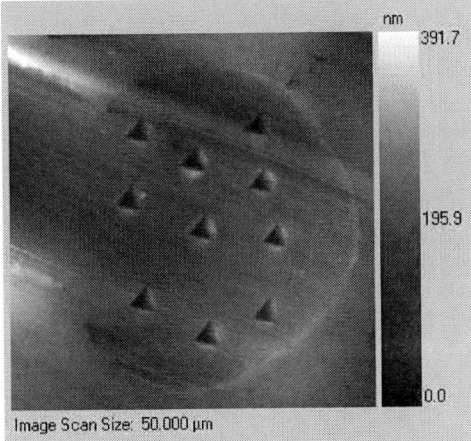

Fig 12. Indentation in FAB

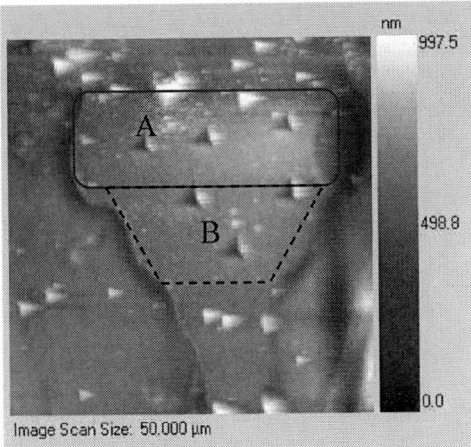

Fig 13. Indentation in two different areas of bonded ball

During the wire bonding process, Ultrasonic generator (USG) may further strain harden the copper FAB and the strain in bonded ball is uneven.[7] The bonded ball can be divided into two parts, part A and part B, as shown in Fig 13. Both these two parts are evolved from FAB. The average hardness of bonded ball (both area A and B are considered) is 1.65GPa, the standard deviation (SD) is 0.209GPa.

For the highest hardness and highest standard deviation in the bonded ball, we think the hardness distribution is not uniform in the bonded ball, not only the difference between area A and B, but also within area A and B. The hardness can be described as the function of the position. This may be a topic to investigate in the future. Comparison of the hardness of copper wire, FAB and Bounded ball is shown in Fig 14, from which we can see the gradually increasing hardness and their standard deviation.

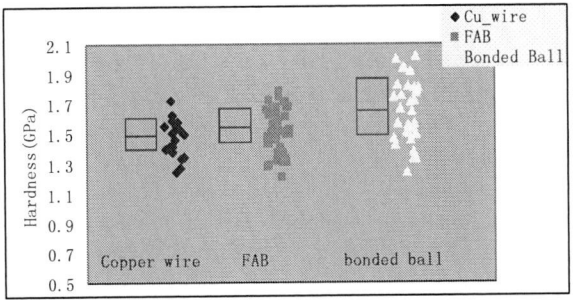

Fig 14. Hardness comparison of copper wire, FAB and bonded ball

As shown in the above, the grain size of FAB is greater than original copper wire, but the hardness of FAB is higher than that of copper wire. This appears not following the Hall-Petch relation. Hall-Petch formula reflects the relationship between macro mechanical property and the grain size. In this hardness tests, nanoindentation only reflect the mechanical property of small area around the indentation. As shown in Fig 9, the depth of indentation is about 500nm, for a theoretical Berkovich indenter, the side of indentation is about 2μm. According to the microstructure we studied, nanoindentation in copper wire may reflect the mean property of different grains, but in FAB this only reflect the property of a single grain. The hardness value of FAB is determined by substructure in a single grain. Because original copper wire is made under annealing state, and FAB experiences rapid heating and cooling process, which will make the hardness value of FAB be a little greater than copper wire.

The standard deviation (SD) of FAB hardness value mainly reflects the difference among grains, and the SD of bonded ball hardness mainly reflects uneven strain harden effect. The SD of FAB hardness is smaller than that of bonded ball and greater than that of copper wire, that's because uneven strain harden effect is greater than different hardness among grains.

Conclusions

From the result of experiment and discussion above, the following conclusions can be drawn.

1) Nanoindentation with in-situ image system is an advanced tool for testing the hardness of copper bonding wire, FAB and bonded ball when the testing parameters, such as the maximum load level, load velocity and spacing between indentations, etc. are optimized.

2) The average hardness of tested copper wire, FAB and bonded ball are 1.46GPa 1.51GPa and 1.65GPa respectively, showing as bonding process in progress, the material hardness increases.

3) The SD of hardness value of copper wire, FAB and bonded ball is 0.124GPa, 0.145GPa and 0.209GPa respectively, showing as bonding process in progress, the heterogeneity of material hardness is enlarged.

References

1. G. G.Harman, <u>Wire Bonding in Microelectronics</u>, 2nd ed. McGraw-Hill (New York, 1997), pp. 1-2.

2. R. R. Tummala, Fundamentals of Microsystems Packaging, 1st ed. McGraw-Hill Professional (New York, 2001), pp. 346-355.

3. L. Levine and M. Sheaffer, "copper ball bonding," *semiconductor international,*1986, pp. 126-129.

4. J. Kurtz, D. Cousens, and M. Dufour, "copper wire ball bonding," *34th Electronic Components Conference*, New Orleans, 1984, pp. 1-6.

5. K. Toyozawa, K. Fujita, S. Minamide, and T. Maeda, "Development of copper wire bonding application technology," *Electronic Components and Technology Conference, 1990. Proceedings., 40th*, 1990, vol.1, pp. 762-767.

6. Taihua Zhang and Yemin Yang, "Developments and applications of nano-hardness techniques," *Advances in mechanics,* No. 3 (2002), pp. 349-363.

7. Z. W. Zhong, H. M. Ho, Y. C. Tan, W. C. Tan, H. M. Goh, B. H. Toh, and J. Tan, "Study of factors affecting the hardness of ball bonds in copper wire bonding," *Microelectronic Engineering,* vol. 84, No. 2(2007), pp. 368-374.

8. J. A. Bailey, J. Barsom, and P. Blau, <u>Mechanical Testing and EvaluationASM international</u> (Ohio, 2002), pp. 495-515.

9. W. C. Oliver and G. M. Pharr, "An improved technique for determining hardness and elastic-modulus using load and displacement sensing indentation experiments," *Journal of Materials Research,* vol. 7, No. 6 (1992), pp. 1564-1583.

Wetting Behaviour of Lead Free Solder on Electroplated Ni and Ni-W Alloy Barrier Film

C. S. Chew[1]*, A. S. M. A. Haseeb[2], M. R. Johan[3]

University of Malaya

Department of Mechanical Engineering, 50603 Kuala Lumpur, Malaysia.

Tel: +603-7967 4492; Fax: +603-7967 5317

Email: [1]chewcheesean@hotmail.com, [2]haseeb@um.edu.my, [3]mrafiej@um.edu.my

Abstract

In this work, electrodeposited Ni-W alloys were studied as barrier films between Cu substrate and lead free Sn3.8Ag0.7Cu (SAC387) solder. Wetting angle and spreading rate of SAC387 on Ni-W alloys with different tungsten contents were investigated and compared with that on Ni film deposited from Watt's bath. Ni-W alloy films of about 1.5 to 2.5 μm were deposited from ammonia-citrate bath on copper substrate. Results show solder/Ni-W alloy system and solder/Ni system exhibited almost the same spreading rate and wetting angle. The wetting angle does not change significantly with tungsten content in the range of 11.3 to 18.0 at.%. Metallographic observation reveals that while intermetallic compound on Ni is continuous, that on Ni-W alloy film is discontinuous.

1. Introduction

At present, the electronic industry is actively shifting from lead solders to lead free solders due to environmental concerns. Sn based solders are the most suitable materials for the replacement of lead solders. Interfacial reaction between Ag based solder and Cu substrate is an important reliability issue. Sn reacts rapidly with copper to form Cu-Sn intermetallic compound (IMC) that weakens the solder joints [1]. Further reductions in solder bumps size aggravate the reliability concern. These developments have extended demands for the solder materials and the diffusion barrier (barrier film) that reduce the consumptions of solder materials.

The main purpose of barrier/thin film is to hinder the diffusion in between lead free materials and Cu substrate. A few studies have been reported on consumption of a barrier films, such as the Co-P alloy film, Co-W-P alloys film, Ni film and Ni-P alloy film during reflow [2-7]. Jin (2006) reported that a thin layer of nickel was effective in controlling the diffusion rate of copper [7]. Although Ni is currently used as a barrier layer between lead free solder and copper substrate, studies Sharif et al., 2007 revealed that Ni film may not be adequate for high temperature application because nickel film reduced the mechanical strength of the joints during high temperature long time liquid state annealing due to Ni_3Sn_4 intermetallic compound formation at the solder interface. He suggested that nickel phosphorous film can be as a good choice as barrier film in lead free solder technology [8]. But during prolonged reflow, new crystals of Ni_2P, Ni_5P_4 and NiP_2 are also formed and the intermetallic compounds will influence the reliability of solder joint [1].

For electronics industry soldering applications, a solder joint with satisfactory fillet formation is desired for minimum stress concentration. In order to achieve this, a solder spreading characteristic with low contact angle is needed. Spreading rate (Japanese Industrial Standard-JIS Z 3197-1896), SR of the solder is calculated by Equation (2) and the solder used for testing is considered as a ball shape.

$$D = 1.2407 \, V1/3 \qquad (1)$$
$$SR = (D - H)/D \times 100\% \qquad (2)$$

H – height of spread solder (mm)

V – mass/ specific density of the SAC387 solders

A low contact value can be explained with either chemistry or physics approaches. The physics approach includes manipulation of surface tension of materials involved in soldering process. The chemical reaction between solder and base metals often play a more important role in soldering process. In the reported case of Ni substrate, some researches explained the decreasing in contact angle is cause by the formation of interface IMC layer [9-10].

In this research, electrodeposited Ni-W alloy films were studied as barrier films between SAC387 solder and Cu substrate. Ni-W alloys are known to possess a good thermal stability and exhibit lower diffusion rate [10-11]. The aim of this work is therefore to investigate the effect of tungsten content on the performance of barrier film. For comparison purposes, Ni films were studied. Watt's bath and ammonia citrate bath were used for electrodeposition of Ni film and Ni-W alloy films, respectively on copper sheet [12]. The characteristics of Ni-W alloy deposit were investigated by optical microscopy and energy dispersive X-ray spectroscopy (EDX).

2. Experimental procedures

2.1 Coating preparation

In this work, ammonia citrate baths were used for electrodeposition. Table 1 shows the bath composition and parameters of Ni-W alloy deposition [12]. A 1L beaker was used as a container. The bath was covered by PMMA cover with three holes for placing thermometer, substrate and platinised titanium anode. At first, the 0.3 mm thickness as received Cu substrate was cut into 30 mm × 30 mm area. It was then washed with detergent and rinsed under taped water follow by deionized water. The oxide layer was removed by dipping in 10% sulphuric acid for a few seconds. It was finally washed in deionized water and dried with acetone. The cleaned Cu substrate was then placed in the heated ammonia citrate bath for Ni-W alloy deposition. The bath pH was adjusted by sulphuric acid or ammonia solution. Deposition time was set at 60 min. The W content of the films were found out by energy dispersive X-ray (EDX) .

978-1-4244-4658-2/09 $25.00 © 2009 IEEE

Table 1 Components concentrations and operating temperature for Ni-W alloy films deposition

Components	Concentration(mol/L)
Nickel sulphate, $NiSO_4.6H_2O$	0.06-0.22
Sodium citrate, $Na_3C_6H_5O_7.2H_2O$	0.50
Sodium Tungstate, $Na_2WO_4.2H_2O$	0.14-0.28
Ammonium Chloride, NH_4Cl	0.5
Sodium Bromide, NaBr	0.15

Operating Parameters	Tyipcal Range
pH	8.5 to 9.2
Bath Volume	500 ml
Bath temperature	$80 - 85°C$
Stirring Condition	Mild stirring with magnetic bath
DC current density	$10 \ mA/cm^2$
Substrate	30 mm thickness of Cu sheet

2.2 Solder preparation and reflow

After the coating, shinny surface was obtained which was rinsed with deionized water and dried with acetone. A SAC387 solder paste was then placed on electrodeposited substrates by using a jig with 6.5mm opening. The samples were reflowed on a hot plate at 250°C and melt it for about 45 seconds. Now, SAC387 solder was spread over electroplated substrates. After cooling, the residual flux was removed by acetone and the height of solder was calculated according the Equation (2). After calculating the spreading rate, cross section samples were prepared by standard metallographic methods to examine the microstructure of the solder and the formation of IMC at the interface. The microstructures were observed by optical microscopy.

3. Results and Discussion

3.1 EDX Analysis

The W contents of the three baths analysed by EDX are shown in Table 2. It is found that by decreasing of the bath sodium tungsten concentration and increasing of nickel sulphate concentration as shown in Table 2, a range of W contents from 11.3 to 18.0 at.% W is obtained.

Table 2 W content of Ni-W alloys electrodeposits obtained from various baths

$NiSO_4.6H_2O$ (mol/L)	$Na_2WO_4.2H_2O$ (mol/L)	Deposited W content (At. %)
0.22	0.07	11.3
0.10	0.14	16.0
0.06	0.14	18.0

3.2 Wetting angle and spreading rate

Figs. 1 (a) and (b) show the wetting angle and spreading rate on Ni and Ni-W film substrates, respectively. The angles are comparable in both cases. The wetting angle on pure Ni film is $13.15°\pm5.4$ (Fig. 2(a)). The graph in Fig. 2(a) shows some variation of wetting angle with W content. However, given the accuracy of those results, it can be suggested that incorporation of W does not significantly increases the wetting angle. For Ni-W alloy wetting angle obtained is within the range 12.93-14.30°. In the reported case of Ni substrate, some researches explained the decreasing in contact angle is cause by the formation of interface IMC layer, Ni_3Sn [10].

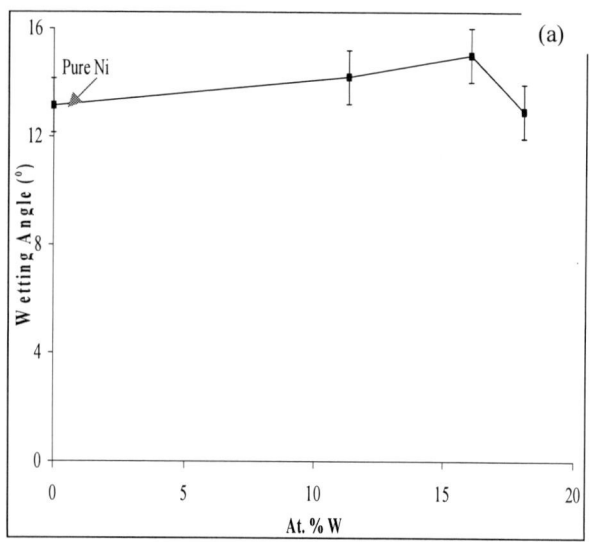

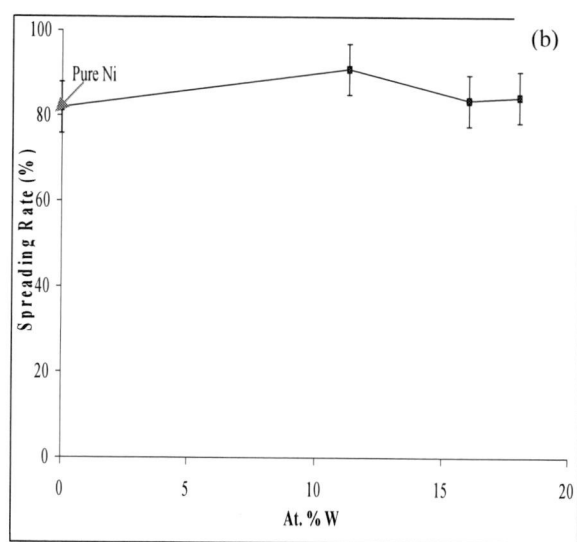

Fig. 1 SAC 387/Ni – X at. % W for (a) wetting angle and (b) spreading rate

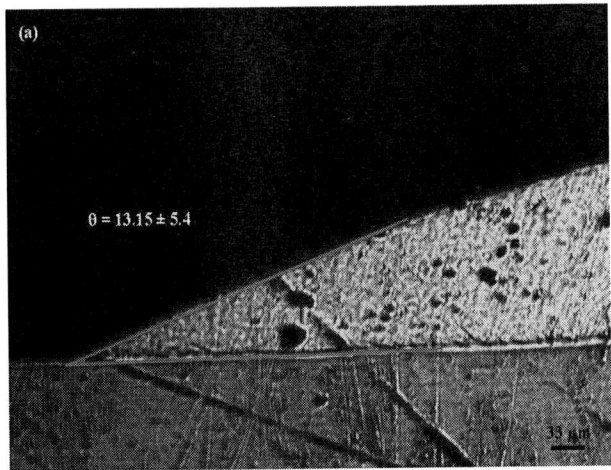
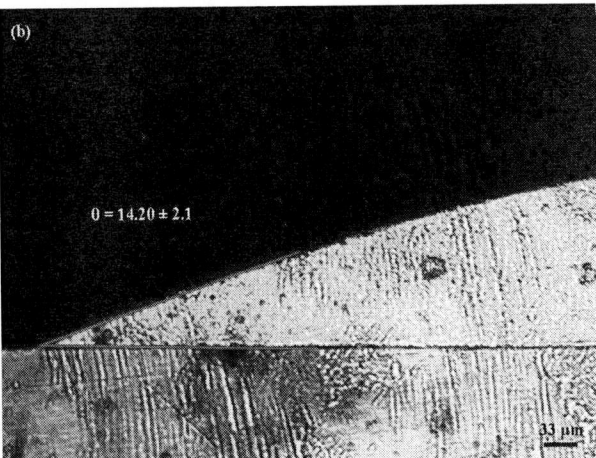

Fig. 2 Optical microscope images of a cross section of solder on different substrates: (a) SAC387 on Ni substrate, (b) SAC387 on Ni-18.3 at. % W alloy substrate after reflow

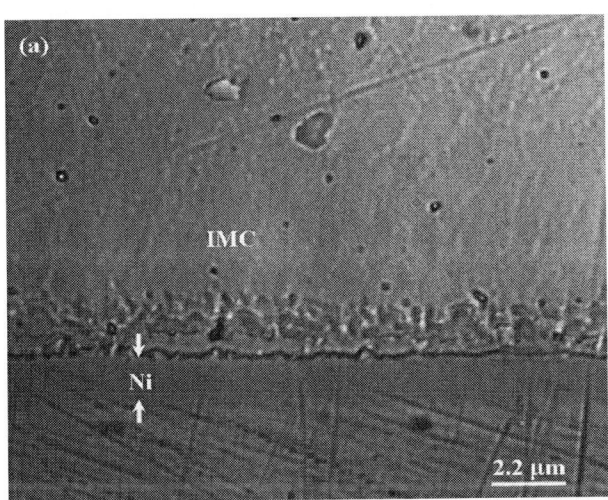
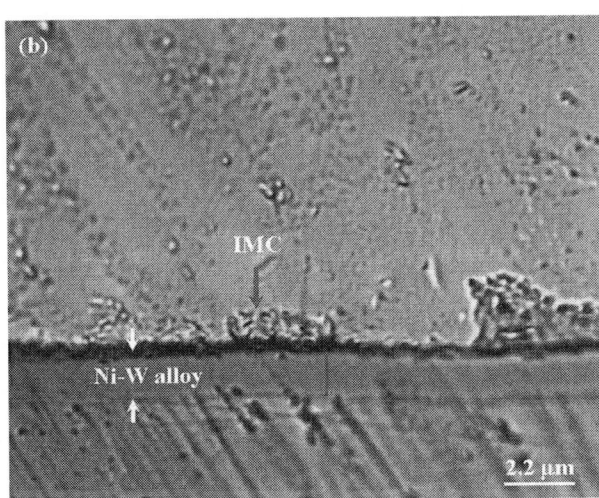

Fig. 3 Optical micrographs of interface IMC characterisation: (a) SAC387 on Ni substrate, (b) SAC387 on Ni-18.3at.% W alloy

The spreading rate of solder on pure Ni is calculated as 81.83%±1.0. The spreading rate shows no clean trend with respect to W content in Ni-W alloys. Considering the experiments accuracy, it is behaved that W content does not significantly influence the spreading behaviour. The observation is in line with that with respect to the wetting angle.

The differences of interface morphologies of solder on Ni and Ni-W films are shown in Fig. 3 (a) and (b), respectively. IMC layer in Ni seems to grow continuously, whilst IMC grow on Ni-W films discontinuously. The reason for the discontinuous growth of IMC on Ni-W film is not clean. However, further study is necessary to understand the performance of Ni-W alloy films as barrier.

Conclusions

The Ni-W alloys layer has been investigated with EDX and optical microscopy. The conclusions obtained are listed as below:

(1) The range of W contents electrodeposited in three baths is obtained in range of 11.3 to 18.0 at. % W.

(2) W content in Ni-W alloys does not significantly increase the wetting angle nor influence the spreading behaviour.

(3) IMC layer in Ni grow continuously, however IMC grow on Ni-W film discontinuously.

Acknowledgments

One of the authors (C. S. Chew) would like to acknowledge Institute of Research Management and Consultancy (IPPP) University of Malaya for finance support under project no. PS070/2008C.

978-1-4244-4658-2/09 $25.00 © 2009 IEEE

References

1. Alam, M. O., Chan, Y. C. and Hung, K., " Interfacial reaction of Pb-Sn solder and Sn-Ag solder with electroless Ni deposit during reflow," *Journal of Electronic Materials*, Vol. 31, No. 10 (2002), pp. 1117-1121.

2. Liang, M. W., Yen, H. T. and Hsieh T. E., "Investigation of electroless cobalt-phosphorous layer and its diffusion barrier properties of Pb-Sn solder," *Journal of Electronic Materials*, Vol. 35, No. 7 (2006), pp. 1593-1599.

3. Wu, W. C., Hsieh, T. E. and Pan, H. C., "Investigation of electroless Co(W,P) thin film as the diffusion barrier of underbump metallurgy," *Journal of Electronic Materials*, Vol. 155, No. 5 (2008), pp. D369-D376.

4. Anhock, S. *et al*, "Reliability of electroless nickel for high temperature application," *Proc Advanced Packaging Materials : Processes, Properties and Interfaces, International Conf*, Braselton, GA, USA, Mar. 1999, pp. 256-261.

5. Stepanova, L. I., Bodrykh, T. I. and Sviridov, V. V., "Electroless Ni-P and Ni-W-P films as a barrier for thermostimulated diffusion of gold into semiconductor," *Proc 3rd High Temperature Electronics European Conf*, Berlin, Germany, Apr. 1999, pp. 101-106.

6. Wang, S. J., Kao, H. J. and Liu, C. Y., "Correlation between interfacial reactions and mechanical strengths of Sn(Cu)/Ni(P) solder bumps," *Journal of Electronic Materials*, Vol. 33, No. 10 (2004), pp. 1130-1136.

7. Jin Y. G. *et al*, "Investigation of Sn/Cu/Ni ternary alloying in lead free solder bump application," *Proc 7th Electronics Packaging Technology International Conf*, Shanghai, China, Aug. 2006, pp. 1-4.

8. Sharif, A., Chan, Y. C., "Effect of substrate metallization on interfacial reactions and reliability of Sn-Zn-Bi solder joints," *Microelectronic Engineering*, Vol. 84. No. 2 (2007), pp. 328-335.

9. Amore, S., Ricci, E., Borzone, G. and Novakovic, R., "Wetting behaviour of lead-free Sn-based alloys on Cu and Ni substrates," *Materials Science and Engineering*, Vol. 495 (2008), pp. 108-112.

10. Lee, N. C., <u>Reflow soldering processes and troubleshooting: SMT, BGA, CSP, and flip chip technologies</u>, Newnes (London United Kingdom, 2002), pp. 21-22.

11. Wang, L. *et al*, "Effect of sputtered Cu film's diffusion barrier on the growth and field emission properties of carbon nanotubes by chemical vapor deposition," *Applied Physics A. Materials Science & Processing*, Vol. 90. (2008), pp. 701-704.

12. Haseeb, A.S.M.A. and Bade, K, "LIGA fabrication of nanocrystalline Ni-W alloy micro specimens from ammonia-citrate bath," Microsyst Tecournal, Vol. 14, No. 3 (2008), pp. 379-388.

Parameter Design of Solder Die Bonding Based on DOE

Jun Hu[1], Xinpeng Xie[2]

1. Guangdong Yuejing High-Tech Co, Ltd, No.10,Nan Xiang 2Rd. Science City, Guangzhou, China, 510663
2. South China University of Technology, Guangzhou Higher Education Mega centre, Guangzhou, China, 510006
Email 1:hjrainy@163.com, Phone 1:+86-20-82075328-8165
Email 2: 296172036@qq.com, Phone 1: +86-13760701554

Abstract

In this paper, DOE methodology was used to set parameters that may exert an influence on voids in the solder layer of MOSFET's products. Parameters of temperature and delay-times were set. With different parameters, voids rate in solder layer changes. Finally we can find the main factor affecting voids rate, thus to reduce the voids rate in solder layer and get high-powered semiconductor device.

1. Introduction

Solder joints are widely used to join two metallic materials. This provides a bond that has electrical conductivity, low thermal impedance, and mechanical durability. Depending on the particular application in which they are being utilized, solder joints may be relied upon for their ability to accommodate thermal expansion stresses, to form a mechanically sound joint that is stable across a range of temperatures, to resist moisture, and to provide low thermal impedance. In the semiconductor device applications, a solder joint can be used to join an active device such as a microprocessor semiconductor die to a heat spreader [1]. Used in this way, low thermal impedance and uniform heat dissipation are keys to satisfactory solder joint performance.

A conventional method for forming a solder joint between metal surfaces typically includes depositing flux on the metal surfaces, placing a solder material between the surfaces, and then heating the solder material to form the solder joint. In the conventional method, the flux, which outgases as the solder is heated, can disperse into the solder material and form "voids", which can reduce the mechanical strength of the solder joint. This is problematic for semiconductor device applications because a void in a solder joint can act as effective local insulator resulting in sharply increased thermal impedance around the void. The presence of insulating voids in a solder joint used to attach an active semiconductor device to a heat spreader, for example, may result in overheating, damage, and ultimately failure of the device. Thus, voids can undesirably reduce the effectiveness of solder joints [2].

2. Experiment

2.1 Materials and Equipment

In this experiment, we use TO220 silver plating frame, Hua Run Hua Jing's MCS5A6008A Power VDMOS chip (dimension: 3.96mm × 3.96mm × 0.29mm, P_{tot}: 125 W) as materials. Die bonding machine uses ASM SD890A. Voids measurement equipment is Shimadzu's SMX-1000.

2.2 DOE

For chip bonding, solder under liquid state condition infiltrates well with lead frame and chip back by intermetallic compound (IMC), thus formed good electrical connection and good mechanical connection.

It is easy to forms void in junction between solder and chip's back, and junction between solder and leadframe.

We used an orthogonal design to screen out the best parameters to get the lowest void rate.

We select 4 factors including the maximum temperature and important location's delay time. Each factor takes 5 levels, a total of 25 sets of tests.

Table 1 Factors and levels of DOE

Parameters/units	Level 1	Level 2	Level 3	Level 4	Level 5
Max. Temp(/°C)	340	350	360	370	380
Delay time 1 (/ ms)	20	50	100	200	300
Delay time 2 (/ ms)	100	200	300	400	500
Delay time 3 (/ ms)	20	50	100	200	300

3. Results and discussion

Using X-ray instrument Shimadzu's SMX-1000 to observe voids, we can see the images below.

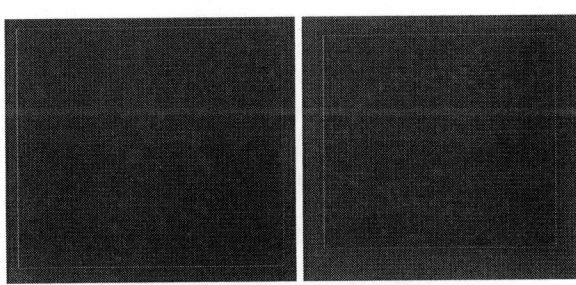

Fig.1-a X-ray Images showing voids

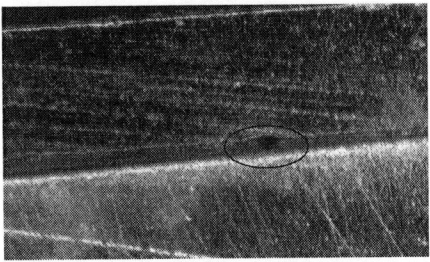

Fig.1-b Sectional view of sample
Fig.1 Solder void in die attach layer

Fig.1-a shows the typical void image observed by X-ray machine. The voids in the right side are obviously less than the voids in the left. In the red box, white spots are voids. Test results showed that voids' rate (void area in chip area) is between 0.5% ~ 5%.

Fig.1-b shows sectional view of sample. In the middle of Fig.1-b is solder between chip and frame. We can see voids really exist between chip and frame, thus affecting device performance [3].

Table 2 Experiments array and the data

	Temp /°C	Time-1 / mS	Time-2 / mS	Time-3 / mS	Voids' rate /%
1	340	20	100	20	1.10
2	340	50	300	200	1.45
3	340	100	500	50	1.18
4	340	200	200	300	1.08
5	340	300	400	100	0.96
6	350	20	500	200	0.72
7	350	50	200	50	0.49
8	350	100	400	300	0.32
9	350	200	100	100	1.26
10	350	300	300	20	0.75
11	360	20	400	50	1.38
12	360	50	100	300	0.65
13	360	100	300	100	0.83
14	360	200	500	20	0.87
15	360	300	200	200	0.93
16	370	20	300	300	0.64
17	370	50	500	100	0.57
18	370	100	200	20	0.91
19	370	200	400	200	0.67
20	370	300	100	50	1.03
21	380	20	200	100	1.38
22	380	50	400	20	1.45
23	380	100	100	200	0.77
24	380	200	300	50	2.34
25	380	300	500	300	1.30

Table 2 shows the result we get in our experiment. The 8th parameter can result lowest voids' rate. That is: Max Temp-350°C, Delay time 1-100ms, Delay time 2-400ms, Delay time 3-300ms.

Table 3 the experiments results of void

		Temp /°C	Time-1/ ms	Time-2/ ms	Time-3/ ms
Voids' Rate /%	K_{11}	1.154	1.044	0.962	1.016
	K_{21}	0.708	0.922	0.958	1.284
	K_{31}	0.932	0.802	1.202	1
	K_{41}	0.764	1.244	0.956	0.908
	K_{51}	1.448	0.994	0.928	0.798
	R^1	0.740	0.442	0.274	0.486

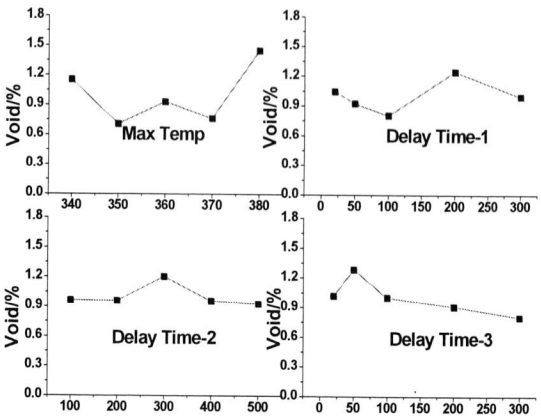

Fig.2 Each factor's affection to void's rate

We can see the affection relation obviously in Fig.2. Fig.2 shows the comprehensive effect of each factor's different levels to voids' rate. Factors' affecting sequence to voids' rate is: Max Temp>Delay time 3>Delay time 1>Delay time 2. Experimental results show that temperature effect most to voids' rate.

Considering average level of each factor, the best parameter is: Delay time 1-100ms, Delay time 2-500ms, Delay time 3-300ms.

Table 3 and Table 4 calculated F value. Compared F value with Fa in Standard manual, thus we can calculate the significance of each factor. The bigger F value, the more significant influence of the factor to voids' rate.

Table 4 Data results for variance analysis

	Temp /°C	Time-1 /ms	Time-2 /ms	Time-3 /ms		
T_1	0.77	0.22	-0.19	0.08	-0.61	
T_2	-1.46	-0.39	-0.21	1.42	0.58	
T_3	-0.34	-0.99	1.01	0	-0.45	
T_4	-1.18	1.22	-0.22	-0.46	0.46	
T_5	2.24	-0.03	-0.36	-1.01	0.05	$T=\sum Y_i$ =0.03
T_1^2	0.5929	0.0484	0.0361	0.0064	0.3721	$T^2=0.00$ 09
T_2^2	2.1316	0.1521	0.0441	2.0164	0.3364	$T^2/25=0.$ 000036
T_3^2	0.1156	0.9801	1.0201	0	0.2025	
T_4^2	1.3924	1.4884	0.0484	0.2116	0.2116	
T_5^2	5.0176	0.0009	0.1296	1.0201	0.0025	
$\sum T_i^2/5$	1.8500	0.5340	0.2557	0.6509	0.2250	
$\sum T_i^2/5-$ $T^2/25$	1.8500	0.5339	0.2556	0.6509	0.2250	

Normally, in condition factor's F value is higher than F0.01 is said to be highly significant. While F value is higher than F0.05 and smaller than F0.01 is called significantly. When F value is higher than F0.10 but smaller than F0.05, this factor has influence to results. With F value smaller than F0.10, it has no influence to results.

In this experiment, $F_{0.01}(4,4)=15.9770$, $F_{0.05}(4,4)=6.3882$, $F_{0.10}(4,4)=4.10725$. F value has already been calculated in Table 5. Thus concluded factors' affecting sequence to voids' rate is: Max Temp>Delay time 3>Delay time 1>Delay time 2, which is the same as shown in Fig. 2.

Table 5 ANOVA

	Sum of Squares	df	Mean Square	F
Temp	SA=1.85	4	0.4625	8.409090909
Time-1	SB=0.53	4	0.1325	2.409090909
Time-2	SC=0.26	4	0.065	1.181818182
Time-3	SD=0.65	4	0.1625	2.954545455
Within Groups	SE=0.22	4	0.055	1
Total	ST=3.51	20		

4. Conclusions

From this experiment, we can conclude the best parameter to get lowest voids' rate is: Max Temp as 350°C, Delay time 1 as 100ms, Delay time 2 as 500ms, Delay time 3 as 300ms.

Factors' affecting sequence to voids' rate is: Max Temp>Delay time 3>Delay time 1>Delay time 2. Temperature has the biggest influence to voids' rate.

Voids' rate should be less than 5%, single void rate should be less than 1%. Thus we can guarantee device's good electrical connection and good mechanical connection.

Acknowledgments

This work was financially supported by Guangdong Yuejing High-Tech Co, Ltd, technical supported by South China University of Technology and Guangdong Yuejing High-Tech Co, Ltd.

References

1. Tameerug, Phanit, US Patent, WO 2009/002536 A1.
2. JIANG Jialing, CHEN Ling, "Discussion of life prediction for fatigue-creep interaction", *Chinese journal of materials research*, Vol. 21, No. 5 (2007), pp. 537-541.
3. Laxmidbar Biswal, Arvind Krishna, "Effect of Solder Voids on Thermal Performance of a High Power Electronic Module", 2005 Electronics Packaging Technology Conference.

Outgassing of Materials Used for Thin Film Vacuum Packages

Q. Li[1,2], J.F.L. Goosen[2], J.T.M. van Beek[3], F. van Keulen[2], G.Q. Zhang[2]

[1]Materials innovation institute, Mekelweg 2, 2600GA, Delft, The Netherlands

[2]Department of Precision and Microsystems Engineering, Delft University of Technology,
Mekelweg 2, 2628CD, Delft, The Netherlands

[3]NXP Semiconductors, Eindhoven, The Netherlands

Email: q.li@m2i.nl; q.li@tudelft.nl, phone: +31-015-2783522

Abstract

In this paper, outgassing of the thin film vacuum package is analyzed by integrated MEMS (Microelectromechanical system) resonator and RBS (Rutherford backscattering spectroscopy). Outgassing behavior of different thin film packaging materials is studied by exodiffusion experiments. It is demonstrated that a high temperature pre anneal will greatly reduce the outgassing but can not eliminate it.

Thin film vacuum packaging

Thin film packaging technology is essential for miniaturization and high level integration of semiconductor components. It allows for the encapsulation of MEMS or ICs (Integrated Circuits) at the wafer level using micromachining technology before the standard IC packaging process [1].

The typical steps for thin film packaging are illustrated in Fig. 1. The packaging process normally starts after the MEMS is fabricated and can be easily integrated into the back end of the MEMS process. The critical steps include sacrificial layer deposition, first thin film deposition, etching holes on the thin film, releasing the sacrificial layer through the etched holes, and sealing the package by another thin film layer.

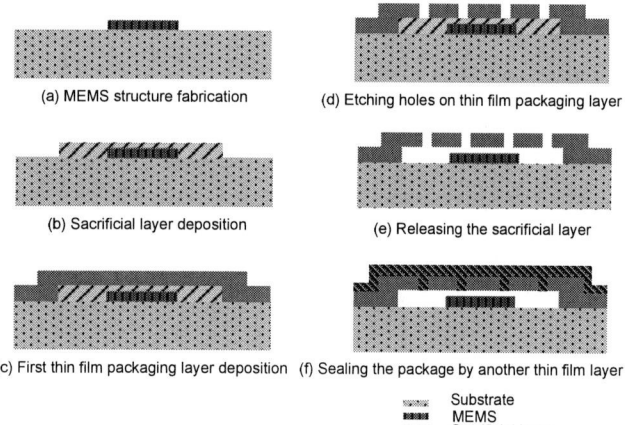

Fig. 1. A conceptual process flow of creating a thin film package around a MEMS device.

The thin film package protects the free standing micro structures on the wafer and isolates the contamination in the environment. Compared with another wafer level packaging technique, the bonded package, it greatly reduces the size and cost of the package, and simplifies the process. Moreover, some thin film packages not only play a role in protection of the device, but also are a functional part of the whole systems [1]. For example, thin film vacuum packages are crucial parts of many MEMS to reduce air damping [2]. Normally, the vacuum inside the package is realized by sealing the packaging in a vacuum environment, for instance, by vacuum thin film deposition.

The successful packaging of such MEMS by thin film technology relies on the long-term stability of the thin film material.

Outgassing of the thin film materials

Vapor deposition technologies like PECVD (Plasma Enhanced Chemical Vapor Deposition) and LPCVD (Low Pressure Chemical Vapor Deposition) are typical ways to fabricate thin films [3]. Outgassing is one of the biggest concerns for the thin film vacuum package, since some gas molecules, like Hydrogen, may be incorporated inside the thin films during the deposition process [3] and may outgas under certain conditions. When outgassing happens, the freed gas will enter the inside of the thin film package and spoil the vacuum. This may happen after a long period of time or due to environmental changes such as thermal loading.

Thermal loading is likely to happen in the MEMS process. For instance, a PECVD silicon nitride deposition with a thermal loading of about 300°C to 450°C may be used for the final anti scratch step for the wafers. Moreover, the temperature of standard IC packaging, which is often carried out after the thin film packaging, may reach several hundred degrees Celsius. The vacuum thin film packages should at least be able to withstand these thermal loadings.

In this paper, outgassing of the vacuum thin film package after an anneal is investigated by use of an integrated MEMS resonator and RBS analysis. Outgassing behavior of different thin film packaging materials is studied by vacuum annealing Exodiffusion experiments. A method to reduce the outgassing is proposed and evaluated.

Outgassing measurement by integrated MEMS and RBS

Vacuum thin film PECVD silicon nitride packages are fabricated and investigated. The advantage of this package is that it is a low temperature packaging process. Therefore it can be easily integrated at the back end of the MEMS process, even after the bond pad fabrication step, without damaging any structure on the wafer. The effect of the outgassing on the packages is studied using integrated MEMS resonator and RBS analysis before and after an anneal of 450 °C.

Inside the package, a MEMS resonator is present. This resonator (Fig. 2) functions as a pressure sensor inside the thin film package, since its Q (Quality) factor depends on the ambient pressure. The physics of the air damping of the resonator shows three different regimes when going from the high vacuum to air pressure. When the ambient pressure is very low, the Q factor of the resonator is not pressure dependent and will reach its maximum value. When the ambient pressure is higher, the Q factor is inversely proportional to the pressure. At a higher pressure regime, the Q factor will not purely be determined by the pressure [4, 5]. These regimes are called the intrinsic, molecular and viscous regimes, respectively.

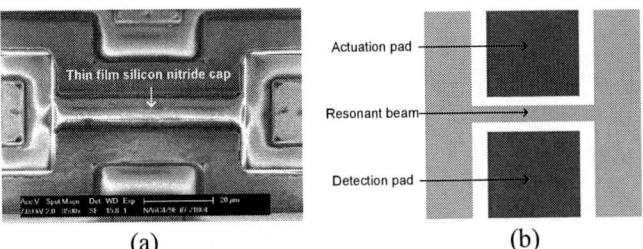

(a)   (b)

Fig. 2. (a) Top view of a thin film PECVD silicon nitride packaged MEMS resontor. (b) Schematic of a clamped-clamped MEMS resontor.

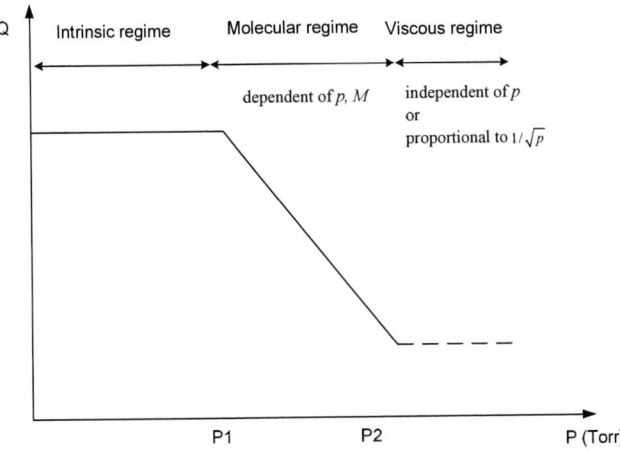

Fig. 3. Dependence of the Q factor on the pressure from high vacuum to air pressure, P1 and P2 are demarcation points between different pressure regimes.

The Q factor of our MEMS resonator could reach 6000 in high vacuum and will drop to about 200 in the viscous regime. After the thin film nitride package is sealed, the Q factor of the integrated resonator is 1800 which is equivalent with 50 Torr, assuming the gas inside is nitrogen. Then, we have carried out a 450°C anneal of the package. Afterward, the resonator is tested again and the Q factor drops to 718 (Fig. 4), which is equivalent with 200 Torr assuming the gas inside is nitrogen. This indicates a pressure increase inside the package due to the anneal. Two reasons could cause the pressure increase inside the package, namely the outgassing of the material and leaking of the package. As the Q factor did not change for a period of four months, the pressure was stable, which indicates that the package is hermetic. The reason of the pressure change inside the thin film silicon nitride package is found to be the outgassing of the inner thin film silicon nitride layer during the anneal.

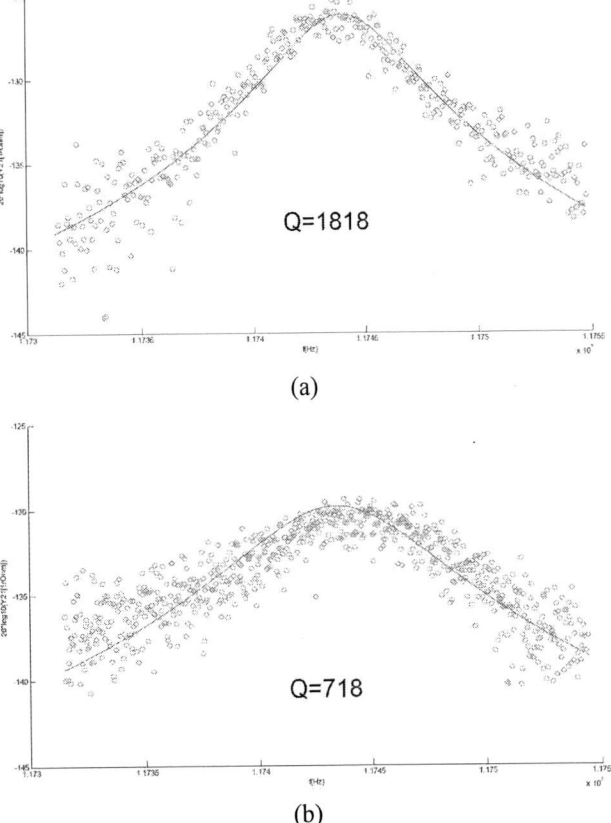

(a)

Q=1818

(b)

Q=718

Fig. 4. Q factor of the thin film packaged resontor before (a) and after a 450°C anneal (b).

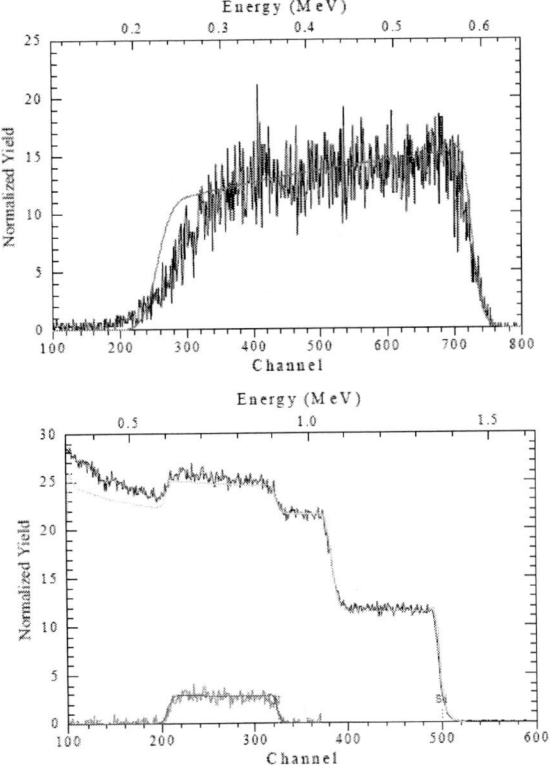

Fig. 5. Top: ERD (Elastic Recoil Detection) H spectrum of a reference sample (model in red); bottom: RBS Si (Silicon) and N (Nitrogen) spectrum (model in green and blue).

978-1-4244-4658-2/09 $25.00 © 2009 IEEE

To understand the microstructure change of the PECVD thin film silicon nitride due to the 450 °C anneal, RBS analysis has been carried out on the thin film package layer before and after the anneal. A composition change of the thin film silicon nitride is found. The thin film nitride has 2.5E18 atoms/cm² with Si: N: H = 1.00: 1.10: 0.50 before the anneal (Fig. 5) and has 2.4E18 atoms/cm² with Si: N: H = 1.00: 1.10: 0.44 after (Fig. 6). The reduction of the H atoms inside the thin film confirms the outgassing of Hydrogen from the material during the anneal. The H outgassing causes the pressure increase inside the thin film silicon nitride package.

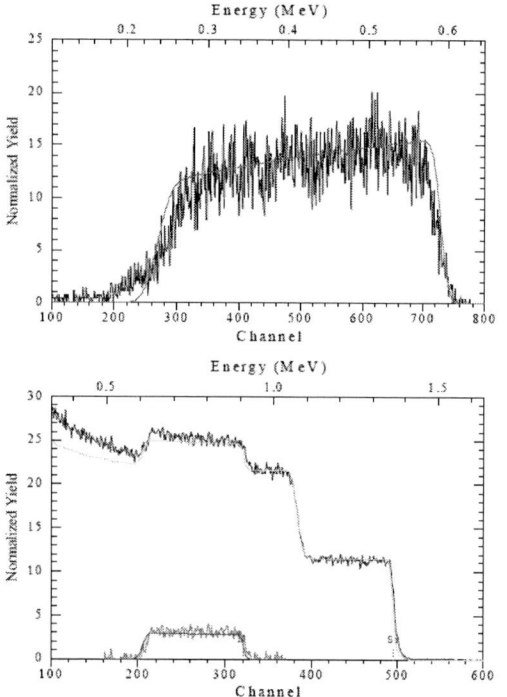

Fig. 6. Top: ERD H spectrum of the sample (model in red); bottom: RBS Si and N spectrum (model in green and blue) after the 450 °C anneal.

From the above experiments, we find that the PECVD thin film silicon nitride is very unstable as the outgassing already occurred under a thermal loading of only 450 °C.

Outgassing analysis by Exodiffusion

To understand the outgassing behavior and find a good material with less outgassing, several potential thin film packaging materials are investigated using an advanced vacuum annealing system (Exodiffusion) (Fig. 7). The Exodiffusion system composes of a tube for loading the sample, the gas analysis system, temperature control system and pumps.

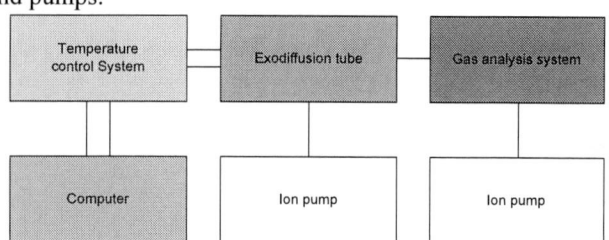

Fig. 7. Schematic of the setup for the Exodiffusion system.

The Exodiffusion system allows for the annealing to take place in vacuum, which eliminates the contamination from the environments. The system also enables the measurement of the amount of gas to come out from the thin film by recording the pressure increase in the Exodiffusion tube. Moreover, a gas analysis system is connected to the Exodiffusion tube, which could provide information about the species of the outgas.

Four types of potential thin film packaging materials are selected: Poly silicon, LPCVD silicon nitride, silicon dioxide and PECVD silicon nitride. All films have a thickness of 300nm and are subjected to an anneal of 450°C for 5 hours.

All four types of thin films show outgassing during the anneal. The data is fitted by the diffusion behavior based exponential equations (curving red lines in Fig. 8 to Fig. 11) and the maximum outgassing amount is also projected by the fitting (straight red lines in Fig. 8 to Fig. 11). The Poly silicon shows the lowest outgassing (Fig. 8) with a 0.8 mTorr in pressure. PECVD silicon nitride shows the highest outgassing (Fig. 9) with a 9.1 mTorr pressure increase in the Exodiffusion tube. The silicon dioxide (Fig. 10) and the LPCVD silicon nitride (Fig. 11) show pressure increase due to outgassing of 1.6 mTorr and 1.2 mTorr respectively.

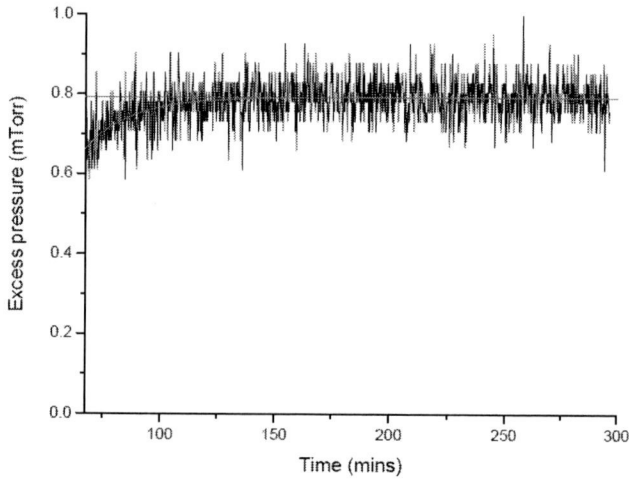

Fig. 8. Outgassing of the Poly silicon during a 450 °C anneal.

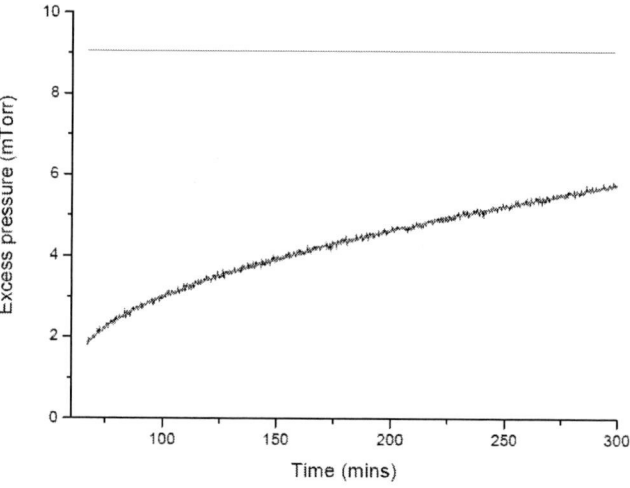

Fig. 9. Outgassing of the PECVD silicon nitride during a 450 °C anneal.

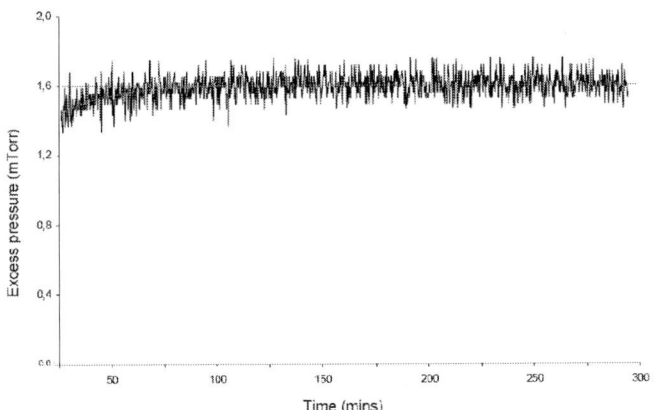

Fig. 10. Outgassing of the silicon dioxide during a 450°C anneal.

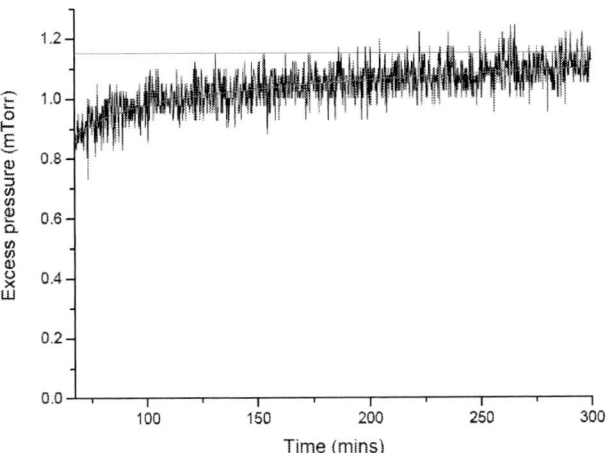

Fig. 11. Outgassing of the LPCVD silicon nitride during a 450°C anneal.

The above study shows that most thin film materials, even those fabricated by high temperate deposition, will still outgas under a low temperature thermal load, which remains a concern for the reliability of thin film vacuum packages.

Methods to reduce the outgassing of thin film package material

The measurements show that the amount of outgassing is different for the four thin film materials. The high temperature deposited thin films LPCVD silicon nitride and Poly silicon shows a better stability and less outgassing. However, for some applications these high temperature depositions could be incompatible with the MEMS process.

To reduce the outgassing of a low temperature deposited thin film, we propose to carry out an extra high temperature anneal before the sealing of the package. This high temperature anneal will accelerate outgassing of the thin film, which should greatly reduce the outgassing at a subsequent lower temperature anneal. To test the idea, a pre annealing at 700°C is carried out on 300nm thick PECVD silicon nitride. Then, the thin film is annealed at a lower temperature (450°C) to see whether the high temperature pre anneal would eliminate the outgassing at a lower temperature. Fig. 12 (with

same kind of fitting in Fig. 8 to Fig. 10) shows that the pre anneal will reduce the outgassing but could not eliminate it. The PECVD silicon nitride still outgas at a lower temperature but the outgassing amount is greatly reduced to about 0.2 mTorr pressure increase in the Exodiffusion tube compared to 9.1 mTorr pressure increase by the direct anneal at 450°C (Fig. 9).

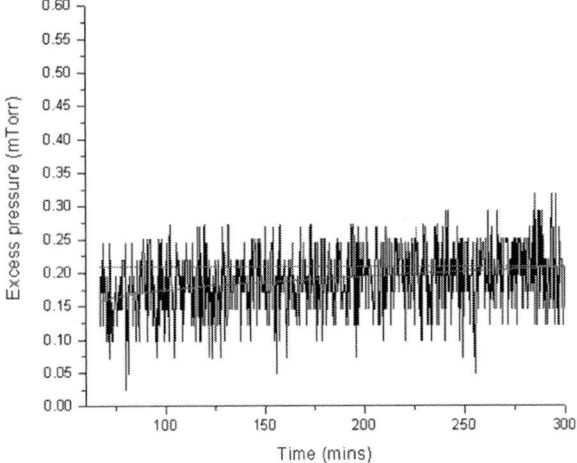

Fig. 12. Outgassing of the PECVD silicon nitride during a 450°C anneal after the anneal of 700°C for 4 hours.

Conclusions

The outgassing behavior of thin film packaging materials is studied in this work. The Q factor change of the thin film PECVD nitride packaged MEMS resonator after the 450°C anneal, indicates that the pressure increased inside the package. It is caused by the outgassing of the thin film layer towards the inside of the package. Hydrogen is found to be the most important outgassing product and is present in a very high concentration inside thin film PECVD silicon nitride.

From the Exodiffusion experiments, we find that all tested thin film materials (Poly silicon, LPCVD silicon nitride, silicon dioxide and PECVD silicon nitride) with the same thickness, outgas under a thermal loading of only 450°C. The comparison of the outgassing of different thin film materials shows that poly silicon had the lowest amount of outgassing. The pre annealed PECVD silicon nitride shows much less outgassing at a lower temperature (450°C) compared with the direct outgassing at 450°C. It indicates that pre annealing will greatly reduce the outgassing but can not eliminate it. This will be a critical concern when the high vacuum is needed inside the thin film package.

Acknowledgments

This research is carried out under project number MC3.05230 in the framework of the Research Program of the Materials innovation institute M2i (www.m2i.nl).

References

1. Hsu, T. R., MEMS Packaging, INSPEC, 2004.
2. Reuter, D., Bertz, A., Werner, T., Nowack, M., Gessner, T., "Thin Film encapsulation of microstructures using Sacrificial CF-Polymer", *Proc TRANSDUCERS & EUROSENSORS' 07*, Lyon, France, June 2007, pp. 343-346.

3. Smith, D. L., Thin-film deposition: principles and practice, McGraw-Hill, (New York, 1995).

4. Blom, F. R.; Bouwstra, S.; Elwenspoek, M.; Fluitman, J. H. J., "Dependence of the quality factor of micromachined silicon beam resonators on pressure and geometry", *Vacuum Sci Technol B*, Vol. 10, Issue 1 (1992), pp. 19-26.

5. Newell, W. E., "Miniaturization of Tuning Forks", *Science*, Vol. 161, No. 3848 (1968), pp. 1320-1326.

6. Lee, B.; Seok, S.; Chun, K., "A study on wafer level vacuum packaging for MEMS devices", *Micromech. Microeng*, Vol. 13 (2003), pp. 663-669.

Effects of Surface Finishes on the Intermetallic Growth and Micro-structure Evolution of the Sn3.5Ag0.7Cu Lead-free Solder Joints

Guoyuan Li[1], Chuan Tang[2], Xueyou Yan[1], Xinpeng Xie[1]

[1]School of Electronic and Information Engineering, South China University of Technology, Guangzhou China
[2]Department of Computer Science and Technology, Guangdong University of Finance, Guangzhou, China
Email: phgyli@scut.edu.cn

Abstract

The growth rate and kinetics of the intermetallic formation, intermetallic grain morphology, and micro-structural evolution for the solder with I-Ag, OSP and ENIG finish were investigated. Comparing the intermetallic thickness growth rate in Sn3.5Ag0.7Cu solder joint on PCB Cu metallizations with I-Ag, OSP and ENIG finish, the OSP had the highest intermetallic growth rate followed by I-Ag and ENIG. The growth rate for I-Ag was a little bit lower than OSP finish. ENIG served as a good barrier to block the IMC formation. These results suggested that the PCB finishes had a big impact on the intermetallic growth rate. The activation energy was determined to be 30.83kJ/mol and 32.67 kJ/mol for I-Ag and OSP finishes respectively. The growth mechanism of the intermetallic layer was compared in terms of IMC thickness and grain size evolution during isothermal ageing. Results reveal that the selection of solder alloys and PCB finishes played an important role in the morphology, microstructure evolution, growth of the intermetallics formed in the solder joint interface.

1. Introduction

With continuous miniaturization in electronic products, the size of electronic components and thus the solder joints are being continuously scaled down. This makes the brittle nature intermetallic compound (IMC) layer at the interface tending to occupy a large volume percentage of the solder joint, which is quite crucial to the failures of the portable devices such as cellular phones, digital cameras, and MP4 caused by brittle fracture or impact fatigue of the solder joints. Lead-free solder alloy Sn-Ag-Cu is widely used for Pb-free soldering in electronic industry because of its advantages in mechanical properties and solderability [1-4]. However, the intermetallic compound (IMC) growth in Sn-Ag-Cu solder joints is faster than that in eutectic Sn-Pb solder joints [5-6], which influence the long-term reliability of solder joints. Hence the prevention of excessive growth of IMC in solder joint becomes a challenging task for materials researchers. It is well know that the various materials in an interconnect joint will interact with each other and the joints microstructure will evolve during service. On the other hand, surface finishes are commonly used to protect the copper metallizations prior to reflow. Some surface finishes also function as a diffusion barrier, forming a stable bond to copper via the formation of intermetallic compounds with the solder. The selection of an appropriate finish and solder alloy combination therefore influences the properties and reliability of the joints [7-12]. For this reason, it is important to develop our understanding of the nature and dynamics of the interfacial interactions between lead-free solder and the associated metallizations and

finish materials. The purpose of this research is to investigate the effects of surface finishes on the intermetallic growth and micro-structure evolution of the solder joints.

2. Experimental

In this research, the most popular Sn-Ag-Cu lead-free solder was selected to investigate the interaction between lead-free solder and Cu substrate with different surface finishes. The finishes selected were Immersion Ag (I-Ag), Electroless Nickel Immersion Gold (ENIG), Organic Solderable Preservative (OSP) and Immersion Sn (I-Sn). The details of which are shown in Table 1. To prepare the test samples, the Sn3.5Ag0.7Cu solder pastes applied on the PCBs were all fine pitch formulations (flip-chip packaging grade). Photo-resist coating was used to prevent over-flow of solder during reflow. A consistent amount of solder thickness and volume was achieved with the use of an industrial stencil (ϕ, Diameter 0.8mm and thickness 5 mil). The solder joints were re-flowed through a forced industrial convection reflow oven (Heller 1800) at a peak temperature of 250°C for 60 seconds, and then aged in the Blinder M115 aging oven at different temperatures for different times respectively.

Table 1 Printed circuit board surface finishes used in this research

Finish	Ave. thickness (μm)	Supplier	Standard	Mfg. Process
I-Ag	0.038	Enthone	SMT*	Alpha Level immersion
ENIG	0.02 (Au) /6.8(Ni)	Enthone	SMT*	Electroless Ni Immersion Gold
OSP	0.4	Enthone	SMT*	Entek
I-Sn	1.38	Enthone	SMT*	Immersion

For metallographic observations, specimens were first mounted in Klarmount and cross-sectioned perpendicularly to the solder-Cu interface of the solder joint. They were then successively grounded down to 1000 grit on a silicon carbide paper under water cooling. Polishing was performed using 5 μm Al_2O_3 suspension followed by 0.25 μm diamond paste. The specimens were then etched in a dilute solution of 2% concentrated Hydrochloric Acid (HCl), 6% concentrated HNO_3 and 92% H_2O for about 30 seconds. This process would provide the contrast between the Cu_6Sn_5 and the solder matrix. To study the microstructural evolution of the intermetallic grains and confirm the IMC phases by X-ray diffractometer (XRD), the solder alloys on the top of the IMCs were dissolved out chemically by using 13% (by volume) nitric acid. Scanning electron microscopy (SEM) was used to study the microstructural morphology of IMCs.

978-1-4244-4658-2/09 $25.00 © 2009 IEEE

Energy dispersive x-ray analysis (EDX) and XRD were used to characterize the composition of the IMCs. The mean thickness measurement of the total IMCs was performed using SEM and image analysis. To determine the average intermetallic thickness, the total area of the intermetallic was acquired using image pro-analysis software, and then divided by the length of the image.

3. Results and Discussion
3.1 Microstructure evolution of solder joints

For comparison of the intermetallic growth in solder joints with different finishes, the intermetallic thickness was measured for the as-soldered solder joints and aged solder joints at temperature of 120°C for1000 hours. As shown in Fig.1, the intermetallic thickness in the solder joint with I-Sn finish was the thickest and all the others were the about the same after reflow. After ageing for 1000 hours, the IMC growth in solder joint with OSP finish was the largest followed by I-Sn, I-Ag and ENIG.

For the Sn3.5Ag0.7Cu lead-free solder reaction with Cu pad, different intermetallic compounds were found at the solder joint interface depending on the finishes. Copper-tin based (Cu-Sn) intermetallic layers was found in solders joints

when reflowed on the PCB coated with I-Ag, OSP and I-Sn finishes, while Nickel based (Ni-Sn) intermetallic layers were found in the solder joints with ENIG finish.

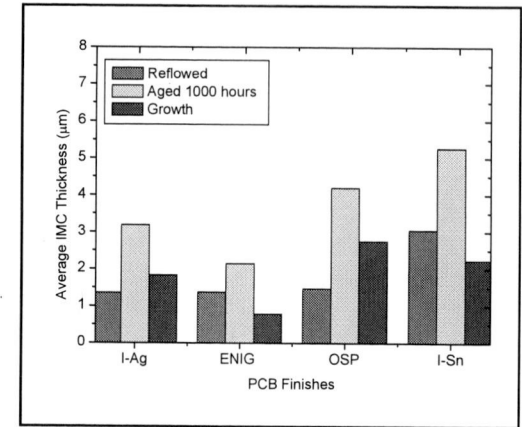

Fig. 1 Intermetallic Growth with different finishes for as-soldered and after ageing 1000 hours solder joints

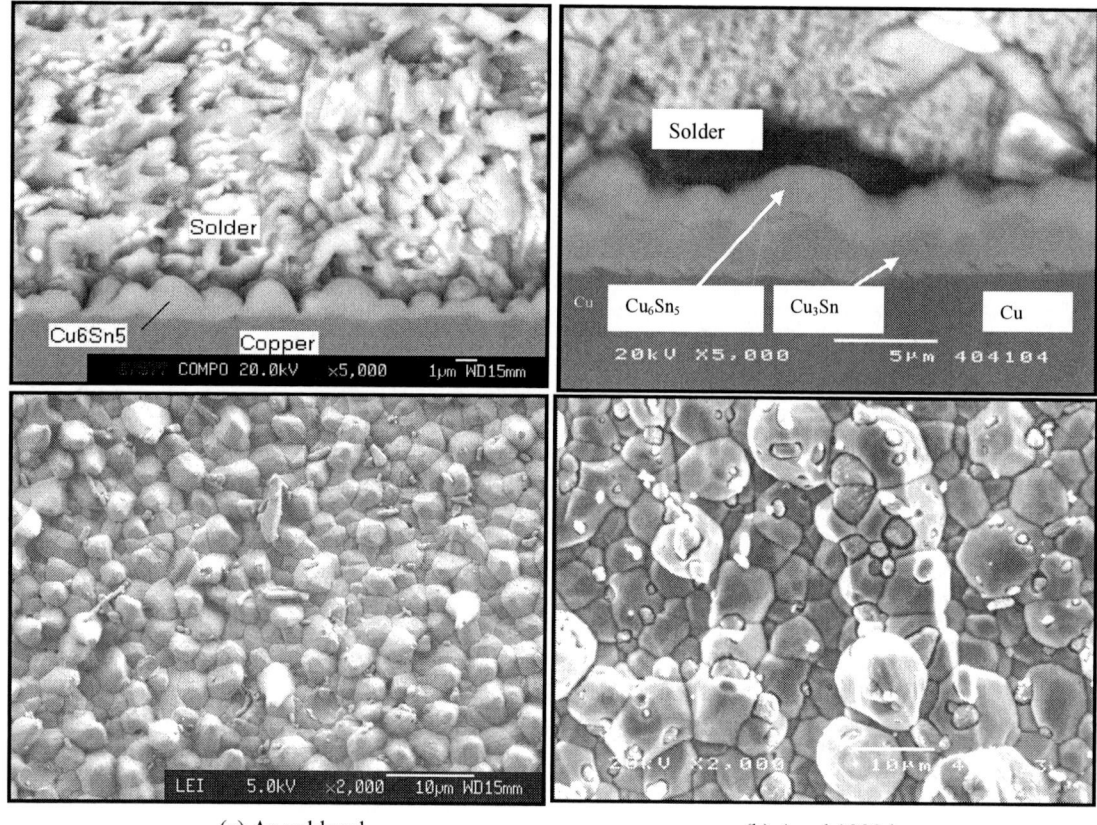

(a) As-soldered   (b) Aged 1000 hours

Fig. 2 Cu-Sn Intermetallic layer Cross-sectional and top view

The Cu-Sn Intermetallic layer Cross-sectional and top view of the solder joints were shown in Fig.2. The intermetallic layer formed in the as-soldered solder joint was found to be Cu_6Sn_5. The morphology micrograph revealed the scallop-like grains of the Cu_6Sn_5 intermetallic layer, Fig.2 (a). After an isothermal ageing of the solder joint at 120°C for 200 hours, a Cu_3Sn layer was found in between the Cu_6Sn_5 intermetallic layer and the PCB Cu metallizations. After ageing for 1000 hours at 120°C, the total thickness of the intermetallic layer increased and the morphology became more planar. The uniformity of the intermetallic grain size in the as-soldered condition had also decreased due to the effect of ageing.

Ni-Sn based intermetallic layer was found in solder joints with ENIG (Electroless Nickel Immersion Gold) finish, as shown in Fig.3. The EDX result confirmed that the intermetallics to be Ni_3Sn_4 and Cu-Ni-Sn phase. However, the IMC layer was not detected using the XRD acquisition mainly due to their thickness (~1.5 μm). After aging for 1000 hours at 120°C, the morphology of the Ni-Sn intermetallic appeared less rugged with rounded edges and the layer had more planar appearance. There was no significant increase in thickness but some intermetallic were found to spall away from the interface into the solder region as shown in the cross-sectional view SEM image in Fig.3 (b).

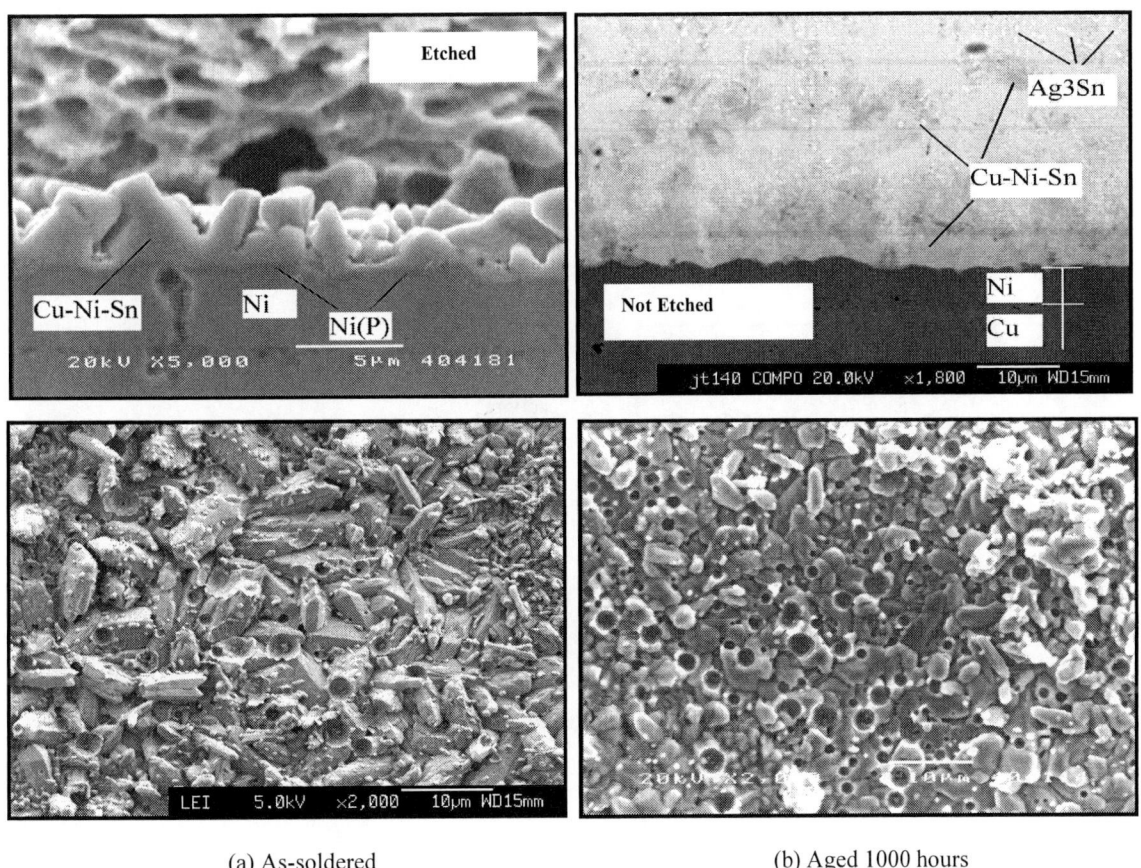

(a) As-soldered   (b) Aged 1000 hours

Fig. 3 Ni-Sn Intermetallic layer Cross-sectional and top view

To identify the element concentration profile across the intermetallic (IMC) layer, the atomic percentage concentration of different elements presented across the intermetallic (IMC) layer and the solder region near the interface was acquired using the EDX (Energy Dispersive X-ray) spectroscopy. In our experiment, 10 EDX points were located across these two regions as shown in Fig. 4. As discussed earlier the Cu-Sn intermetallic layer formed at the interface for the I-Ag, OSP and I-Sn finishes. The concentration profile across Cu-Sn IMC layer aged 1000 hours was measured as shown in Fig.5. After reflow, the

initial intermetallic layer formed was small (~1.5 to 2.8 μm) and only one EDX point can be located at the center of the IMC layer. After 1000 hours of ageing, the IMC layer thickness had increased significantly (average of 5-7 μm) and the thickness was sufficient to accommodate five EDX points. From the results, EDX points 2 lied at the region of the total IMC layer where Cu_3Sn compound formed after 1000 hours of ageing. The atomic percentage acquired was very close and agreed with the theoretical at. % values of Cu_3Sn compound. Similarly, EDX point 3 and 4 are located within the Cu_6Sn_5 region and at. % acquired were also very close and agreed

978-1-4244-4658-2/09 $25.00 © 2009 IEEE   729

with Cu_6Sn_5 intermetallic compound. The concentration level at each point of the IMC layer provided a good profile indicating the inter-diffusion at the interface between the Cu from the PCB metallizations and the Sn from the Solder.

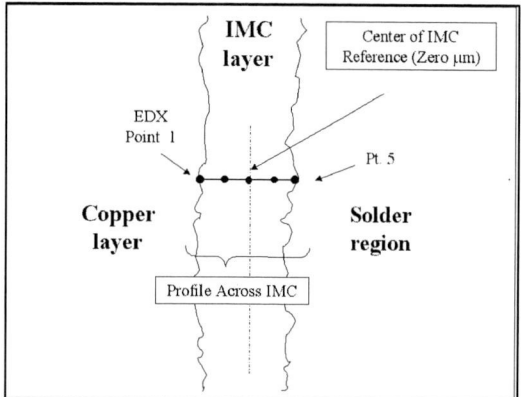

Fig. 4 EDX points acquisition across IMC layer

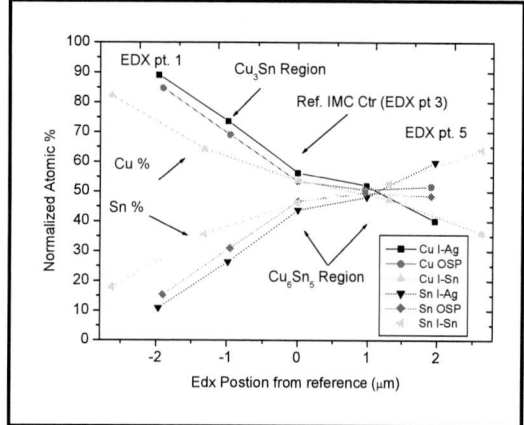

Fig.5 Concentration profile across Cu-Sn IMC layer aged 1000 hours

For ENEG finish, Ni-Sn intermetallic layer was observed Results show that the Nickel (Ni) layer remained intact and the intermetallic layer formed was very thin (~0.7 to 1.5 μm) for the as-soldered solder joint. After 1000 hours of ageing, no significant increase in thickness was found in the Ni-Sn intermetallic layer. Hence, with the same method used, only one EDX point can be located at the center of the intermetallic layer.

3.2 IMC Growth kinetics

The IMC thicknesses in solder joints aged at different aging temperatures for different time were measured in order to determine the growth rate and activation energy of the IMC formation. The relationship between the thickness of IMC layer and aging time is generally considered to follow Fick's law:

$$X = \sqrt{Dt} \qquad (1)$$

where X is the average IMC thickness of the layer, t the aging time and D the interdiffusion coefficient. Fig. 6 shows the total thickness of IMC formation aged at 120°C against the ageing time $t^{0.5}$ for the I-Ag, OSP, and ENIG finish respectively. The slope of each curve gave the value of the diffusion coefficient (D), whcih were 2.317×10^{-15} cm^{2}s^{-1}, 2.994×10^{-16} cm^{2}s^{-1} and 8.873×10^{-15} cm^{2}s^{-1} for the I-Ag, ENIG, and OSP finish respectively. These results had suggested that the diffusion growth rate in OSP was larger than I-Ag and ENIG. Results show that diffusion coefficients were smaller than those stated in the literature determined in the absence of a finish [13]. Results also reveal that the ENIG finish had a very low diffusion coefficient value.

The possible mechanism for the OSP finish having highest diffusion rate is that the organic preservative vaporized immediately upon being soldered in the molten state. Theoretically, the diffusion mechanism in the solid-state reaction should be very similar to pure Cu PCB metallizations without any diffusion barrier. It was reasonable for I-Ag to experience a lower diffusion rate as the immersion Ag may affect the diffusion and IMC formation, and for ENEG to have a lowest growth rate as Ni is a good barrier to block the interdiffusion of the Cu and Sn. Based on these results, it had suggested that the use of PCB finish could affect the diffusion process at the interface during the liquid and solid-state reaction.

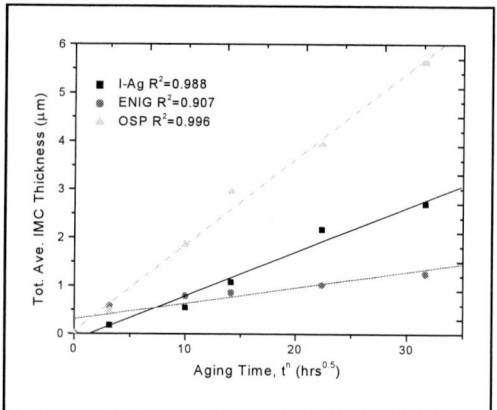

Fig.6 Ave. IMC thickness versus aging time

In order to determine the activation energies, the samples were aged at four different temperatures (100, 120, 150, 170 °C) for different times. Based on experiment data, the the diffusion coefficient (D) of the IMC formation in SAC solder/Cu interface on I-Ag and OSP finish were calculated in terms of equation 1 and summarized in Table 2. For both finishes, diffusion coefficient (D) of all the intermetallic layers increased with respect to the ageing temperatures. Most of the D values in OSP finish were slightly larger than the I-Ag finish.

Arrhenius relationship is used to determine the activation energy for intermetallic growth in this study:

$$D = D_o \exp(\frac{-Q}{kT}) \qquad (2)$$

where D is the interdiffusion coefficient, D_0 the pre-exponential constant of interdiffusion coefficient, Q the activation energy, k the Boltzmann constant, and T the absolute temperature. The activation energies (Q) for I-Ag and OSP finishes could be determined by calculation of the slope of the Arrhenius plot curves and found to be 30.83 kJ/mol and 32.67 kJ/mol respectively.

Table 2 Diffusion coefficient (D) at different temperature for I-Ag and OSP finishes

Ageing Temp (°C)	Immersion Ag (10^{-19} in m²/s)	OSP (10^{-19} in m²/s)
100	14.36	17.21
120	41.83	79.52
150	108.65	195.55
170	252.63	365.21

4. Conclusions

The effects of surface finishes on the microstructure evolution, morphology and growth of the intermetallic compound layer in Sn3.5Ag0.7Cu lead-free solders joints were investigated. The type of intermetallic layer formation such as Cu-Sn, Ni-Sn, and Cu-Ni-Sn compounds depends on the composition of the finish materials on the PCB Cu metallization. In comparison of intermetallic growth for different finishes after 1000 hours of ageing, generally, the OSP finish had the largest growth rate, and the ENIG finish served as a good diffusion barrier with the smallest intermetallic growth rate among all finishes. As for the I-Ag, OSP and I-Sn finish, the finish materials diffused into the solder or burn out upon reflow. Results show that the diffusion rate at the solder joint interface without a diffusion barrier such as with I-Ag, OSP and I-Sn was significantly higher than those with a diffusion barrier like nickel (ENIG). Interemtallic growth rate and activation energy of Sn3.5Ag0.7Cu solder on PCB Cu metallizations with I-Ag and OSP finish were determined and the growth of Cu-Sn intermetallic layer obeyed diffusion-controlled mechanism. The selection of PCB finishes played an important role in the morphology, microstructure, growth of the intermetallics formed in the solder joint interface.

Acknowledgments

The authors would like to acknowledge the financial support of Guangdong Nature Science Foundation of China (8151064101000014) for this work.

References

1. E. P. Wood and K. L. Nimmo, "In search of new lead-free electronic solders," *Journal of Electronic Materials*, vol. 23, No. 8 (1994), pp. 709-713.
2. Salam, N. N. Ekere, D. Rajkumar, "Study of the interface microstructure of Sn-Ag-Cu lead-free solders and the effect of solder volume on intermetallic layer formation," 51st Electronic Components and Technology Conference, Orlando, FL, USA, May, 2001, pp. 471-477.
3. M. R. Harrison, J. H. Vincent, H. A. H Steen, "Lead-free reflow soldering for electronics assembly," *Soldering & Surface Mount Technology*, Vol. 13, No. 3 (2001), pp. 21-38.
4. Shivap. Gadag and Susant Patra, "Numerical Prediction of Mechanical Properties of Pb-Sn Solder Alloys Containing Antimony, Bismuth and or Silver Ternary Trace Elements," *Journal of Electronic Materials*, Vol. 29, No. 12 (2000), pp. 1392-1397.
5. T. Y. Lee, W. J. Choi, K. N. Tu, J. W. Jang, S. M. Kuo, J. K. Lin, "Morphology, kinetics, and thermodynamics of solid-state aging of eutectic SnPb and Pb-free solders (Sn-3.5Ag, Sn-3.8Ag- 0.7Cu and Sn-0.7Cu) on Cu," *Journal of Materials Research*, Vol. 17, No. 2 (2002), pp. 291-301.
6. R. Frear, J. W. Jang, J. K. Lin, and C. Zhang, "Pb-Free Solders for Flip-Chip Interconnects," *Journal of Metals*, Vol. 53, No. 6 (2001), pp. 28-32, 38.
7. K. N. Tu and K. Zeng, "Tin-lead (SnPb) solder reaction in flip chip technology," *Materials Science and Engineering*, R34, No.1 (2001), pp.1-54.
8. M. Abtew and G. Selvaduray, "Lead free solders in microelectronics," *Mater. Science and Engrg*, Vol. 27 (2000), pp.95-141.
9. J.S. Hwang, "Lead-free Solders: Technology", *SMT magazine*, 1999, pp. 16-18.
10. B. Salam, N.N. Ekere, D, Rajkumar, "Study of the interface microstructure of Sn-Ag-Cu Lead-free Solders and the effect of Solder Volume on Intermetallic Layer Formation," Elect. Comp. And Tech. Conference, 2001.
11. J.S. Hwang, "Lead-free Soldering: the Sn/Ag/Cu system," *SMT magazine,* July 2000, pp. 18-21.
12. K. Suganuma, "Advances in lead-free electronics soldering," *Solid State and Mater. Sci.*, Vol. 5 (2001), pp. 55-64.
13. P. L. Liu and J.K. Shang, "Thermal stability of electroless-nickel/ solder interface: Part A. Interface chemistry and microstructure," *Metallurgical and Material Transactions A*, Vol. 31A (2000), pp. 2857-2866.

The Effect of Tape Casting Slurry System for Process Performance of Green Ceramic Tape Used for Electronic Packaging

Pengyuan Shi, Caihua Ren, Huajiang Jin, Ling Gao
Hebei Semiconductor Research Institute
P.O.Box 179-41, Shijiazhuang Hebei, P.R.China
E-mail:chinapackage@163.com, Phone: 0311-87091042

Abstract

Ceramic is widely used in electronic packages, the process of multilayer green ceramic tapes and co-firing is very popular now. Tape casting process is the main method of forming the green ceramic tape. The method involves mixing ceramic powder and binder, adding chemical agents to the components of the ceramic tape to manipulate cross linking of the binder, and casting the mixture of the ceramic powder. The binder and the binder manipulating chemical agent are casted then as a green tape. The chemical structure and molecular weight (Mw) of plasticizer on the physical properties of green tapes were investigated.

The Al_2O_3 green tapes prepared with phthalate-based plasticizers (DOP) were found to be more stable in Al_2O_3/PVB system.

1. Introduction

The technology of electronic packages is being consistently progressed to improve for high electrical properties and reliability. To achieve these objectives, it is important technical keys to manipulate the interparticle force between ultra fine particles suspended in liquid media and also impart sufficient strength and elongation to thin-layered green tape for subsequent processes in electronic packages fabrication, such as via forming, printing, laminating and firing.

Tape casting has been generally used for the fabrication of microelectronic components such as electronic packages and a powerful and economic process for producing thin and flat sheets of ceramics. For tape casting process, various organic additives have to be added to the suspension to control the stability and rheological behavior of the suspension as well as the strength and flexibility of green tape.

During firing, the organic additives must be completely removed from the tape before densification reaches an advanced stage, otherwise, residues may be retained within the tape and alter the desired properties of the thermoelectric materials. The removal of the organic additives is generally called the pyrolysis process or organic burnout, and consists of thermal decomposition and evaporation of the organic additives and the subsequent removal of the volatile compounds from the thermoelectric tape.

Various binders for tape casting in a nonaqueous media, such as vinyl binders and acrylic binders, have been studied. The commercially available PVB binder was chosen among various binders for this study since it is recently chosen as a binder for the commercial manufacture of tape cast Al_2O_3 due to the excellent green strength, solubility in many volatile solvents, and compatibility with other organic additives. However, most polymeric binders are prone to form a relatively strong, stiff, and brittle sheet without the addition of plasticizer.

2. Experimental procedure

The powder used in this study was Al_2O_3, provided by the manufacturer. To remove adsorbed water on the powder surface, the powder was kept in desiccator for 24 h prior to process. NP-10, the most common commercial dispersant for Al_2O_3 powder, was used as a dispersant. Poly vinyl butyral (PVB) served as a binder. The mixture solvent of toluene and ethanol was chosen based on our previous study.

Di-octyl phthalate (DOP), BBP, and two kinds of polyethylene glycol (PEG400, and PEG1530) were selected as plasticizers to impart the plasticity and flexibility to green tape. Poly ethylene glycols (PEG400 and PEG1530) have an aliphatic structure, and DOP and BBP have an aromatic unit. The structures of typical PVB binder and plasticizers used in this study are depicted in Fig. 1.

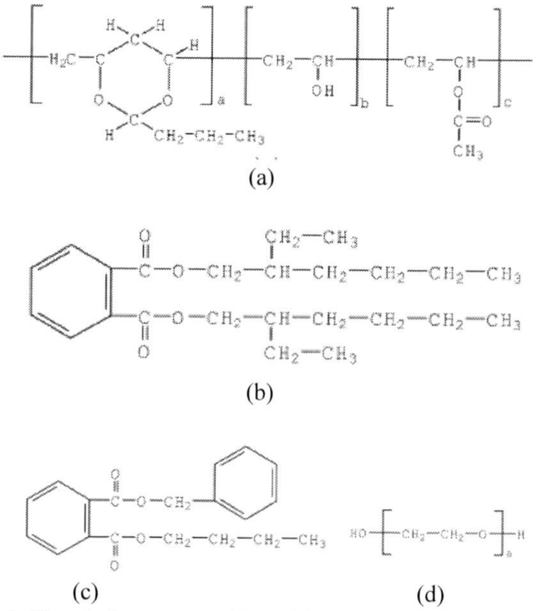

(a)

(b)

(c)   (d)

Fig.1 Chemical structure of PVB binder and plasticizers: (a) PVB, (b) DOP, (c) BBP, (d) PEG

In order to measuring the stability of green tape, The final dimensions of the green tape after cutting were $127 \times 127 \times 0.1 mm^3$, A square pattern of 4-mil diameter holes was punched through each sheet using a precision via-punching machine . Thus each sheet of tape had a hole near each corner, and the location of the holes was precisely controlled by the via punching machine. Distance between holes was therefore precisely known. The tape sheets were then "stabilized" by heating in baking box, which is intended to drive off residual solvent and promote tape shrinkage. After that, the distance between holes was precisely measured by holding the sheet of punched tape in a fixed position on a flat, and locating each hole on a high-precision X-Y translator equipped with high-magnification video system. X-Y coordinates for each hole on each sheet of tape were recorded. The difference in spacing between holes before and after stabilization was calculated as percent shrinkage. In this manner, the stability of each tape is determined by the shrinkage. In addition, for the purpose of observing cutting surface with various plasticizers, the green block without inner electrode was prepared and characterized by optical microscope.

3. Results and discussion

3.1 Rheological behavior of Al_2O_3 suspensions

The suspension viscosity was decreased with the addition of plasticizer into the Al_2O_3/PVB suspension. The degree of shear thinning is similar one another, the magnitude of the suspension prepared with short-chain plasticizers is slightly smaller than that of long-chain plasticizer. Lower Mw plasticizer (DOP, BBP and PEG 400) indicated lower viscosity than higher Mw PEG 1530. The plasticizer of low Mw is preferred to fabricate the flexible green tapes and it makes better and easier for plasticizer to access and penetrate to the cross-linked network structure of PVB chains. Role of plasticizer in the suspension was previously demonstrated as follows; the reduction of frictional force between PVB chains as a lubricant, the breakdown of network structure formed by PVB chains due to a decrease of the attractive interaction between PVB chains. It was found that all plasticizers in this study had a positive effect on the viscosity of Al_2O_3/PVB suspension. Especially, in the case of PEG containing OH functional group, a stronger plasticizer–binder interaction through hydrogen bonding is expected.

However, the influence of OH functional group on the flow behavior is not exactly differentiated by the viscosity measurement since suspension viscosity is a macroscopic property that measures the collective effects from many coincidental contributions to flow behavior. Rheological behavior of Al_2O_3 suspension is mainly dependent on the formation of particle agglomeration which is generally governed by the interparticle forces. Also it is well known that these forces are contributed by dispersant. From the viscosity measurement, it was apparent that the Mw of plasticizer strongly affected the viscosity. Therefore, the lower Mw plasticizer enables green tapes to be laminated, providing a sufficient flexibility for handling, screen-printing, and machining. Fig.2 shows the roughness status of the green tape using different unit of PVB and plasticizer.

3.2 Mechanical properties of Al_2O_3 green tapes

In general, the physical properties of polymer products are strongly dependent on the interaction between polymer chains. That is, their high interaction is contributed to variation of not only the tensile strength and tensile modulus

but also the elongation. In the case of PVB copolymer used as a binder, interaction between polymer chains can be formed during tape casting due to the strong hydrogen bonding of vinyl alcohol units.

roughness (μm)	1	2	3	4	5	6	7	8	AVER
10%PVB/ 5%BBP	0.48	0.38	0.34	0.42	0.47	0.50	0.48	0.62	0.44
10%PVB/ 5%DOP	0.21	0.20	0.22	0.21	0.19	0.21	0.20	0.20	0.21
10%PVB/ 5%PEG400	0.44	0.35	0.56	0.45	0.47	0.41	0.41	0.36	0.40
10%PVB/ 5%PEG1530	0.44	0.48	0.39	0.41	0.42	0.47	0.34	0.52	0.45
9%PVB/ 5%DOP	0.21	0.24	0.34	0.37	0.33	0.35	0.28	0.30	0.32

Fig.2 Roughness result of different slurry system

	Before baking		After baking		Stability(%)		Average stability (%)
	X	Y	X	Y	X	Y	
1#	90	89.98	89.91	89.93	0.10	0.06	0.074
	90.02	89.99	89.96	89.94	0.07	0.06	
	90.02	89.97	89.95	89.89	0.08	0.09	
2#	89.99	90.01	89.93	89.91	0.07	0.11	0.085
	90.01	89.99	89.91	89.96	0.11	0.03	
	89.99	90	89.9	89.92	0.10	0.09	
3#	90	90	89.95	89.93	0.06	0.08	0.055
	90.01	89.98	89.97	89.94	0.04	0.04	
	90	89.99	89.96	89.93	0.04	0.07	
4#	90.01	89.99	89.92	89.9	0.10	0.10	0.090
	90.01	89.98	89.93	89.9	0.09	0.09	
	90.01	89.97	89.93	89.89	0.09	0.09	
5#	89.97	89.96	89.9	89.89	0.08	0.08	0.075
	89.97	89.98	89.91	89.89	0.07	0.1	
	89.99	89.96	89.94	89.89	0.06	0.08	

Fig.3 Stability of different rate of PVB and different kind of plasticizer

1# 10%PVB/ 5%BBP
2# 10%PVB/ 5%PEG400
3# 10%PVB/ 5%DOP
4# 10%PVB/ 5%PEG1530
5# 11%PVB/ 5%DOP

Fig. 3 is the result of the stability of 4 different PVB/plasticizer unit. 3#(10% PVB and 5% DOP) shows the best stability after baking at 70℃, the stability data is only 0.055%.

The lower Mw of plasticizer decreases the concentration of hydrogen bonding between PVB chains as accessibility and penetration of plasticizer to PVB chains become easier during slurry process. Referring to the mechanism of plasticizer action, this phenomenon is also explained by free volume theory. The higher concentration of terminal groups of plasticizer resulting from lower Mw increases the breakage of many attachment points formed by hydrogen bonding between PVB chains. Therefore, the Mw of plasticizer was a dominant factor on the mechanical properties of green tape. Also, it was recognized that the green tapes with phthalate-based DOP and BBP plasticizers yielded higher elongation than that with PEG 400 although they have similar Mw, indicating that plasticizer containing an aromatic unit has more significant effect on plasticization of green tape using PVB binder than that with an aliphatic linear structure.

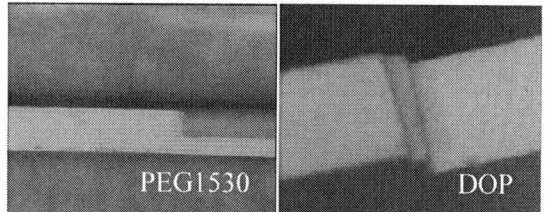

Fig.4 Optical micrographs in cutting surface of green blocks without inner electrode

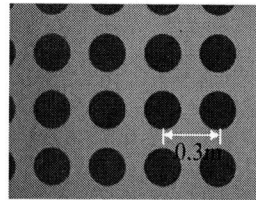

Fig.5 Optical micrographs in via-punching on the green tape using 10%PVB and 5% DOP

Fig. 4 shows the effect of plasticizer in the cutting surface of the laminated green block without inner electrode. The green block prepared with DOP exhibited well-laminated and flawless cutting surface. In the case of green block prepared with PEG 1530, however, the delamination was observed. This is due to lower adhesion during lamination and cutting processes. That is, less flexibility and less tacky property of green tape, which resulted from the low degree of plasticization of PVB binder, led to undesirable defects of green products before firing. Fig. 5 shows good process performance when punching the 0.2mm hole on the green tape of 10%PVB/5%DOP slurry system.

4. Conclusions

In the case of plasticizer with similar Mw, the plasticizer with an aromatic unit, such as DOP and BBP, has a dominant effect on the elongation properties of green tape, than that with an aliphatic structure such as PEG, showing high plasticity in DOP and BBP. The green tapes prepared with phthalate-based plasticizers (DOP and BBP) having an aromatic unit were more stable than those prepared with glycol-based plasticizer (PEG) having an aliphatic structure in Al_2O_3/PVB system, showing a decrease of the elongation of green tape by time in both green blocks prepared with PEG plasticizer. The physical properties of green tape, such as tensile strength and longation, were greatly dependent on structural characteristics as well as Mw of plasticizer.

References

1. Wei-Fang Su, Deborah P.Partlow, Ceramic tape formulations with green tape stability, 1998
2. Rao R. Tummala, Miccroelectronics packaging handbook, 2001
3. Moulson, A. J., Herbert, J. M., Electroceramics. Chapman & Hall, London, 1990.
4. Mizuno, Y., Okino, Y., Kohzu, N., Chazono, H., Kishi, H.,Jpn. J. Appl. Phys, 1998, 37, pp. 5227-5231.
5. Mistler, R. E., Shanefield, D. J. and Runk, B., Ceramic Processing Before Firing, In G. Y. Onoda and L. I. Hench. Wiley, NewYork, 1978, pp. 411-448.
6. Hellebrand, H., Processing of Ceramics Part I, Materials Science and Technology: a Comprehensive Treatment, In R. W. Cahn,P. Haasen and E. J. Kramer. VCH, New York, 1996, pp. 190-260.
7. Schwartz, B., "Multilayer Ceramic Devices." In J. B. Blum and W. R. Cannon. The Am. Ceram. Soc, Westerville, OH, 1986,pp. 13-14.
8. Moreno, R., Am. Ceram. Soc. Bull., 1992, 71, pp. 1521-1531.
9. Moreno, R., Am. Ceram. Soc. Bull., 1992, 71, pp. 1647-1657.
10. Paik, U., Hackley, V. A., Choi, S. C. and Jung, Y. G., Coll. Surf.A, 1998, 135, pp. 77-88.
11. Mistler, R. E. and Twiname, E. R., Tape Casting. The Am.Ceram. Soc. 735 Ceramic Place, Westerville, OH, 2000.
12. Song, J. K., Um, W. S., Lee, H. S., Kang, M. S., Chung, K. W.and Park, J. H., J. Euro. Ceram. Soc., 2000, 20, pp. 685-688.

Studies on Microstructure and Mechanical Properties of Sn-Zn-Bi-Cr Lead-free Solder

Tingbi Luo, Anmin Hu, Ming Li, Dali Mao

Lab of Microelectronic Materials & Technology, School of Materials Science and Engineering,
Shanghai Jiao Tong University, Shanghai 200240, China
Email: huanmin@sjtu.edu.cn, +86-21-34202748

Abstract

In this paper, Traces of Cr were added into Sn-8Zn-3Bi solder alloys to study the influences of Cr on mechanical properties and high temperature reliability of solder alloys by room temperature tensile test and high-temperature aging treatment. With observation of microstructure and fracture surface analysis, the strengthening mechanisms of Cr in solder alloys were discussed. Experiments show that: the microstructure was refined after traces of Cr added and two kinds of intermetallic compound (IMC) formed. Therefore the addition of Cr improved the plasticity of Sn-8Zn-3Bi alloy. The elongation reached up to 50.96% after 0.1% Cr added; and the fracture mechanism converted from quasi-cleavage fracture into ductile fracture. But after more than 0.3% Cr added, the brittleness of the alloy increased. During 0, 4, 9, 16 days aging, the mechanical properties of Sn-8Zn-3Bi-0.3Cr alloy were improving slightly with aging time. Through the microstructure analysis it was found that the Sn-Zn-Cr phase was transited after aging and Zn in alloy was consumed, so that primary Zn phase was reduced and microstructure was improved.

1 Introduction

With the density of electronic packaging improvement, solder joints subject to mechanical, electrical and thermal loads become higher and higher. This requires a new type of lead-free solder with higher mechanical properties, especially strength and plasticity to meet the needs of the electronics industry on reliability. Studies have shown that Bi alloying elements added to provide the wettability of Sn-Zn series solder. However, the Sn-Zn alloy added Bi not only expands the pasty range, but also reduces the plasticity of the alloy [1-4]. In this paper, mechanical properties and microstructure of Sn-Zn-Bi-Cr solder were researched. Influences of Cr addition in mechanical properties and microstructure of Sn-Zn series solder were discussed. Combined with high-temperature aging experiments, mechanical properties measurement and microstructure observation to research the high temperature reliability of solder alloy and the influences of Cr were discussed.

2 Experimental Materials and Methods

2.1 Preparation of solder alloys

The master alloy method was used in this experiment to prepare solder alloys. At first, Sn metal particle and Cr particle was heated to 1200-1300°C in a electromagnetic induction furnace and protected by Ar atmosphere to prepare the Sn-0.6Cr master alloy. Second, the Sn-0.6Cr alloy and Zn, Bi, Sn or other metal were placed in a corundum crucible at a certain ratio; the surface was covered with KCl-LiCl salt in order to prevent evaporation and oxidation of the elements. Then the crucible would be placed in a box-type resistance furnace at 500°C, insulated 30mins to promote melting and stir every 5mins during heat preservation to homogenize of the alloy.

2.2 Preparation of solder tensile specimen

The solder alloy was cast in to a stainless steel mold at 300°C approximately, and then the casting was prepared into the tensile test sample by the method of GB/228-2002 "metallic materials at room temperature tensile test method". The "Zwick" German materials testing machine was used in the tensile test, tensile rate was 2mm/min. The whole process of stretching was controlled by computer and calculated the corresponding performance data automatically. Each alloy was repeated the test three times, then taking the average of three data values for the experimental data.

2.3 Tensile test of aged sample

The Sn-8Zn-3Bi and Sn-8Zn-3Bi-0.3Cr solder alloys were used for aging at 150°C to study the influences on microstructure and mechanical properties by Cr addition at a high temperature. The tensile specimens were placed into an oven and took out respectively after 4, 9, 16 days aging, then observed the microstructure of aged sample and took the tensile test.

2.4 Microstructure observation and component analysis

Specimens were sampled from prepared solder alloy and cold set by epoxy resin. After grinding, polishing, etching (etching agent was 2 vol.% HNO_3 alcohol), the specimens was prepared to metallographic sample. The FEI SIRION 200 field emission scanning electron microscope, backscattered electron imaging (SEM-BSE) micro-analysis was used to microstructure observation, while the INCA OXFORD energy dispersive spectrometer (EDX) was used in component analysis.

2.5 Differential scanning thermal analysis

The differential scanning thermal analysis (DSC) was used to study the relationship between microstructure and solidification of solder alloys. Specimens were sampled from prepared solder alloy, each specimen is 20 mg approximately and analyzed by MDSC2910 analyzer. The heating rate was 10°C/min form room temperature to 150°Cand 2°C/min form 150°C to 250°C in Ar atmosphere.

3 Results and Discussion

3.1 The microstructure of Sn-Zn-Bi-Cr solder alloy

Fig. 1 is microstructure of Sn-8Zn-3Bi-Cr solder alloys with different Cr content. Coarse primary Zn phase, stripe-like Sn-Zn eutectic phase and Bi precipitated can be found in Sn matrix (Fig. 1(a)). When 0.1wt.% Cr added into solder alloy, the size of primary Zn phase decreased significantly and Sn-Zn eutectic phase was refined. Besides, there was granular Sn-Zn-Cr Intermetallic compound (IMC) formed, diffuse distribution in the alloy. The IMC particles are less than 10μm in diameter.

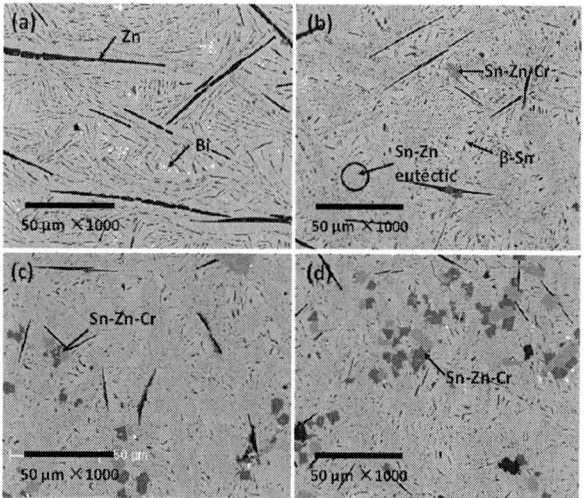

Fig. 1 SEM micrographs of Sn-8Zn-3Bi-Cr alloys
(a) Sn-8Zn-3Bi; (b) Sn-8Zn-3Bi-0.1Cr;
(c) Sn-8Zn-3Bi-0.3Cr; (d) Sn-8Zn-3Bi-0.5Cr

There are three possible reasons of microstructure refined after Cr addition:

1. It can be seen from Table 1 that the addition of Cr reduced the pasty range of solder alloys, so solidification rate increase and microstructure refined;

2. Zn_xCr[5] or Sn-Zn-Cr IMC might form at crystal surface and inhibit the growth of crystal.

3. The primary Zn phase, Sn-Zn eutectic phase and ¯Sn phase might heterogeneous nucleate at IMC surface, so that nucleation rate increased.

Table 1 DSC results of Sn-8Zn-3Bi-xCr alloys

Components	Solidus/°C	Melting point/°C	liquidius/°C	Pasty range/°C
Sn-9Zn-3Bi	193.33	196.71	202.96	9.64
Sn-9Zn-3Bi-0.1Cr	192.64	196.07	199.64	7.00
Sn-9Zn-3Bi-0.3Cr	192.61	197.2	200.23	7.62
Sn-9Zn-3Bi-0.5Cr	192.12	196.44	199.78	7.66

After 0.3 wt.% Cr added, microstructure of solder alloy further refined, addition of Cr up to 0.5 wt.%, the IMC phase increased and the primary Zn phase almost disappeared. From Fig. 1(d) it can be seen that there are two kinds of IMC phases in alloy, the dark-colored particles and the light-colored particles. The dark-colored particles have uniform particle size which is 5μm approximately. But the light-colored particle size is large than the dark-colored particle mostly, some particles are large than 10μm. Fig. 2 shows the spectrum analysis of IMC. Both IMC composed with Sn, Zn, Cr elements and the Cr content much higher than average level,

but the Zn content of the light-colored particles at average level and the dark-colored particles much higher than average level, so that the formation of dark-colored IMC consumed vast Zn, it makes the reduction of primary Zn phase when more than 0.3 wt.% Cr added. Furthermore, the formation of IMC would consume Sn, Zn and Cr, it means the Bi content in melted alloy will increase and the pasty range of alloy will enlarge after 0.3 wt.% Cr added [2], it also can be seen in Table 1 that the pasty range of solder alloys enlarged slightly after more that 0.3wt% Cr added. The crystal constants and physical properties of Sn-Zn-Cr IMC are not reported yet, which needs further research.

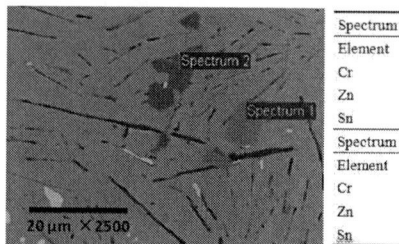

Fig.2 EDS analysis of IMCs

3.2 The tensile properties of Sn-Zn-Bi-Cr lead-free solder

The Sn-Zn-Bi-Cr solders different in Cr content was used in mechanical properties experiments. The stress-strain curves of Sn-8Zn-3Bi-xCr alloys are shown in Fig. 3 and the measured tensile strength and the elongation are listed in Table 2. Sn-8Zn-3Bi alloy tensile strength is 73.56MPa, higher than Sn-9Zn solder which tensile strength (55MPa); but only 12.47 percent elongation, which is lower than Sn-9Zn solder (25 ~ 30%) [3, 4]. It shows that the addition of Bi decreased the plasticity of solder alloys.

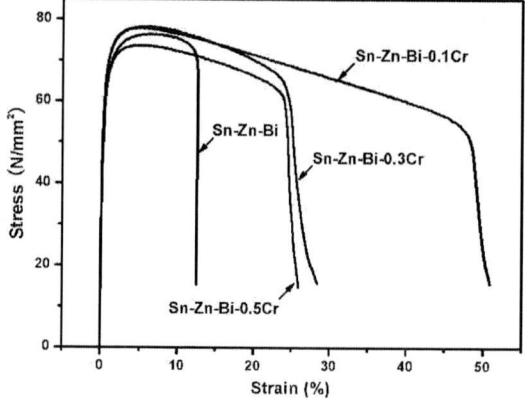

Fig. 3 Stress-strain curves of Sn-8Zn-3Bi-xCr alloys

On the other hand, Cr has little influence on the tensile strength of Sn-8Zn-3Bi alloy, the tensile strength of solder alloys with different Cr content is between 75 ~ 80MPa. But Cr has clear impact on plasticity of solder alloy, which can be seen from Fig. 3 and Table 2. The elongation of the solder alloy improved markedly after Cr was added. When 0.1wt.%

978-1-4244-4658-2/09 $25.00 © 2009 IEEE

Cr was added, the elongation of the solder alloy was the highest, reached up to 50.96%, which might be caused by the grain refinement. When more than 0.3wt.% Cr was added, the elongation of the alloy decreased slightly, which might be caused by the formation of brittle IMC. But the elongation was still higher than the Sn-8Zn-3Bi alloy, reached up to 25~30%, which was similar to elongation of Sn-9Zn alloy [4].

Table 2 Tensile Strength & Elongation of Sn-Zn-Bi-Cr alloys

Alloys	σ/MPa	δ/%
Sn-8Zn-3Bi	73.56	12.47
Sn-8Zn-3Bi-0.1Cr	75.97	50.96
Sn-8Zn-3Bi-0.3Cr	77.33	28.41
Sn-8Zn-3Bi-0.5Cr	74.54	25.92

3.3 Tensile fracture surface of Sn-8Zn-3Bi-xCr alloys

Fig. 4 is the tensile fracture surface of Sn-8Zn-3Bi-xCr alloys. Fig. 4(a) shows that the fracture surface of Sn-8Zn-3Bi alloy is cleavage plane mostly and plastic deformation take place partly, so that the fracture mechanism of Sn-8Zn-3Bi alloy is Quasi-cleavage fracture [6]. After 0.1 wt% Cr added, the small dimples and plastic deformation can be seen in Fig. 4(b), it's obvious that the fracture mechanism converts into ductile fracture. When 0.3 wt% and 0.5 wt% Cr added, two kinds of fracture can be seen in fracture surface, the fracture zone 1 is formed by dimples and plastic deformation, the fracture zone 2 is cleavage plan and its area smaller than fracture 1. So that the fracture mechanism remains ductile fracture mostly, but brittle fracture takes place partly (Fig. 4(c), (d)). Therefore, the plasticity of Sn-8Zn-3Bi-xCr alloys will decrease after more than 0.3 wt% Cr added.

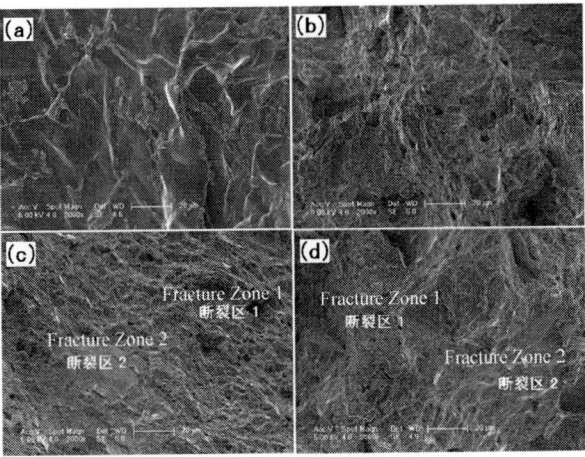

Fig. 4 SEM micrographs of tensile fracture surface of Sn-8Zn-3Bi-xCr alloys
(a) Sn-8Zn-3Bi; (b) Sn-8Zn-3Bi-0.1Cr; (c) Sn-8Zn-3Bi-0.3Cr; (d) Sn-8Zn-3Bi-0.5Cr

3.4 Microstructure and mechanical properties of Sn-Zn-Bi-xCr alloy after aging

Fig. 5 is the microstructure of Sn-Zn-Bi-xCr solders after aging. After comparing Fig. 5(a) with Fig. 1(a), it can be found that the precipitated Bi phase was solid dissolved into β–Sn matrix and the Sn-Zn eutectic phase converted into spheroidized structure.

From Fig. 5(b) and Fig. 1(c), it can be seen that except Bi dissolving and structure spheroidizing, the light-colored IMC was transformed into the dark-colored IMC, the profile of IMCs converted into petaloid and the particle size reduced. There may be intermetallic reactions taken place as follows:

IMC(light) + Sn-Zn eutectic → IMC(dark) + Sn
IMC(light) + Zn(primary) → IMC(dark) + Sn

The large light-colored IMC phase might be divided into small dark-colored IMC and Sn phase, otherwise the IMC might grow and outspread into primary Zn or eutectic phase, therefore the profile of IMC converted into petaloid.

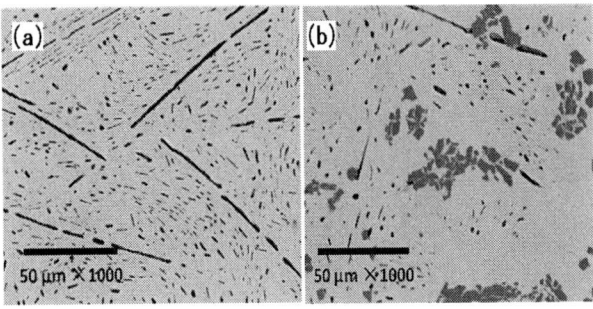

Fig. 5 SEM micrographs of solders alloy after aging for 9 days   (a) Sn-8Zn-3Bi; (b) Sn-8Zn-3Bi-0.3Cr

Table 3 is the mechanical properties of Sn-Zn-Bi-xCr alloys. The mechanical properties of Sn-8Zn-3Bi were not significant influenced by aging time, the elongation is still below 15%. But after 0.3% Cr is added, the Solid-state phase transition of IMC would consume Zn in the alloy and the large light-colored IMC crushed, therefore the mechanical properties would be improved with aging time. After 16 days aging, the tensile strength of Sn-8Zn-3Bi-0.3Cr had improved 8% and the elongation had improved 20% approximately, therefore the addition of Cr will improve the reliability of Sn-8Zn-3Bi alloy at high temperature.

Table 3 Tensile Strength & Elongation of Sn-Zn-Bi-Cr alloys after aging

Aging Time (day)	Sn-8Zn-3Bi		Sn-8Zn-3Bi-0.3Cr	
	σ/MPa	δ/%	σ/MPa	δ/%
0	73.56	12.47	77.33	28.41
4	75.62	14.04	81.35	31.87
9	74.69	13.83	83.02	32.67
16	74.22	13.26	83.13	33.06

Conclusions

1) The addition of traces Cr into Sn-8Zn-3Bi alloy will refine the microstructure of the alloy and form two kinds of IMCs.

2) After traces Cr added into Sn-8Zn-3Bi alloy, the elongation of the alloy was increased, but once more than 0.3 wt% Cr added, the elongation would decrease.

3) The fracture mechanism of Sn-8Zn-3Bi alloy is Quasi-cleavage fracture. After 0.1 wt% Cr added, the fracture mechanism converts into ductile fracture. But if more than 0.3 wt% Cr added, brittle fracture takes place partly.

4) The addition of traces Cr into Sn-8Zn-3Bi alloy will improve the mechanical properties of the alloy at high temperature.

Acknowledgments

This work is sponsored by International Science and Technology Cooperation of China (No. 2008DFA51680), National Natural Science foundation of China (60876071), Shanghai nano technology promotion center (No. 0852nm06300). We thank the Instrumental Analysis Center of Shanghai Jiao Tong University, for the use of the SEM equipment.

References

1. K. Suganuma, "Advances in lead-free electronics soldering," *Current Opinion in Solid State and Materials Science*, 5 (2001), pp. 55-64.

2. Y.S. Kim, K.S. Kim, C.W. Hwang, "Effect of composition and cooling rate on microstructure and tensile properties of Sn-Zn-Bi alloys." *Journal of Alloys and Compounds*, 352 (2003), pp. 237-245.

3. Y. Fukuda, M.G. Pecht, K. Fukuda, "Lead-free soldering in the Japanese electronics industry," *IEEE Transactions on Components, Packaging and Manufacturing Technology*, 3 (2003), pp. 616-624.

4. I. Shohji, T. Yoshida, T. Takahashi, "Comparison of low-melting lead-free solders in tensile properties with Sn-Pb eutectic solder," *Journal of Materials Science: Materials in Electronics*, 4 (2004), pp. 219-223.

5. Z. Moser1 and L. A. Heldt, "The Cr-Zn (chromium-zinc) system," *Journal of Phase Equilibria.*, 13 (1992), pp. 172-176.

6. Yuexian Chui, Changli Wang, Analysis of metal fracture, Harbin Institute of Technology, (Harbin, 1985).

Solder Joints Reliability with Different Cu Plating Current Density in Wafer Level Chip Scale Packaging (WLCSP)

Kenny Cao, KH Tan, CM Lai, Li Zhang
Jiangyin Changdian Advanced Package Company
No. 275, Binjiang middle Road, Jiangyin City
Kennycao@jcap.com.cn, 051086854189-2719

Abstract

Electroplated copper is becoming increasingly important in UBM (under bump metallurgy). The Cu is used as a wetting material for the solder (ball). After reflow the Cu and solder joints are the main contact method for mechanical and electrical interconnection to the pads of the chip. One of the critical factors affecting solder joint reliability is the formation of intermetallic compound (IMC); that is generated at the interface of Cu/solder alloy. The formation of IMC is the merging of the plated copper and solder during the reflow process. The IMC will continue to grow when the structure is subjected to temperature. In this study, different current densitys (0.5ASD, 1.0ASD, 1.5ASD, 2.0 ASD, 2.5ASD, 3.0 ASD) are used on the Cu plating process, which produces different microstructure in the plated Cu layer. It will also impact IMC formation and evolution. The investigation shows that only Cu_6Sn_5 is generated in as-reflow condition; Cu_3Sn is brought forth after being stored at 150 degree C. The early failure was observed at 48hrs, the fracture is located at the interface of Cu and Cu_3Sn in ball shear test. The analysis shows that Kirkendall void formation during aging treatment is the key factor for failure.

1. Introduction

WLCSP (Wafer level chip scale package) technology is being rapidly adopted in portable devices such as battery packs, PDAs, cellular phones, MP3 player, and notebook computers [1-2]. WLCSP technology has many advantages: small die size package (Die size=Package size), high speed and high performance, bumping processing and backend processing at the wafer level prior to single, bumping and flip Chip Technology Compatible with SMT assembly, Shorter manufacturing cycle time, low cost, IC to PCB inductance is minimized, and thermal conduction characteristics are enhanced. In addition, by using wafer-level CSP, peripheral arrays can be redistributed to full area with more convenience bump pitch without using an intermediate substrate. The micro package is directly mounted to the printed wiring board, thus completely eliminating the substrate or interposer, resulting in minimal signal delay [2-4].

WLCSP process involves passivation layer process, sputter metal seed layers, photoresist processing, electroplating process, grinding process, ball drop process, and electronic test process. Developing and understanding the correct process parameters is critical to achieving bumps with good reliability.

The reliability of WLCSP package is always a concern of thin interfacial Cu-Sn intermetallic compound (IMC) layer, because the IMC layer provides good connection for the substrate and chip. However, too thick an interfacial IMC may weaken the joint. Excessive growth of the IMC may promote brittle failure, weaken the solder joint strength, and affect the long term reliability [5-8]. Therefore, it is essential to investigate the effect of IMC growth (IMC thickness and structure at the solder-pad interface) on the integrity of solder joints (usually in term of solder ball attachment strength); IMC growth varies significantly with under bump metallurgy (UBM) microstructure.

The microstructure of electroplated copper UBM is determined by many parameters including current density, bath temperature, electrolyte composition, additive amounts, all of which determine the performance of the bump [8,9]. One of the most critical factors is the plating current density; generally a large current density causes Cu to become extremely rough. However, low plating current density results in a fine Cu grain, a smooth surface, a lengthening in the plating process and a decrease in production throughput. The understanding of the relationship between plating current density and solder bump reliability is critical to the optimization of reliability, yield and production throughput.

2. Experiment

2.1 Sample Preparation

8-inch wafers were coated with 5µm (after cured) polyimide, then sputter 1000Å Ti and 4000Å Cu adhesive and barrier seed layers. A positive photoresist opening pattern was used as mask for plating 8.0µm Cu UBM. Then photoresist and exposed sputter metal seed layers were removed. After removing, flux was coated and balls were placed on the Cu UBM for the whole wafer, the reflow process was conducted under a typical reflow profile with a peak temperature of 255°C. Fig.1 shows the bump structure after the reflowing process. Six wafers proceeded in the same condition except plating current density, different plating Cu UBM microstructure were prepared by using 0.5ASD, 1.0ASD, 1.5ASD, 2.0ASD, 2.5ASD, 3.0ASD plating current density. Then these samples (Table1) were stored at 150°C for 0h, 48h, 96h and 200h.

2.2 Reliability Tests and Sample Inspection

These samples were high temperature aged at 150°C for 0h, 48h, 96h and 200h. The ball shear test was performed on the lead-free solder balls after thermal aging using Dage 4000 microtester. The samples for different thermal aging time conditions were cross-sectioned using epoxy molding, grinding and polishing to characterize IMC microstructure of WLCSP samples. IMC morphology was observed using FIB (Focus Ion Beam) and electron microscopy (SEM). The mean thickness of IMC layers was measured by image processing software system based on digital technique (Fig.2); the thickness of IMC can give by the following equation:

$$THK = \frac{N_{total}}{N_{width}} \cdot \frac{l_{gauge}}{N_{gauge}}$$

Where l_{gauge} denotes gauge length, N_{gauge} is the pixel value for gauge length, N_{total} is the total pixel value for whole IMC area, and N_{width} is the pixel value in width direction for IMC.

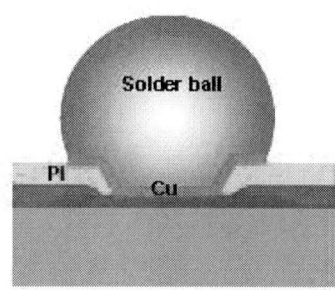

Fig.1. Bump structure after reflowing process

Table1 Samples at different conditions

Time Current Density	After Reflow	48h	96h	200h
0.5ASD	A1	A2	A3	A4
1.0ASD	B1	B2	B3	B4
1.5ASD	C1	C2	C3	C4
2.0ASD	D1	D2	D3	D4
2.5ASD	E1	E2	E3	E4
3.0ASD	F1	F2	F3	F4

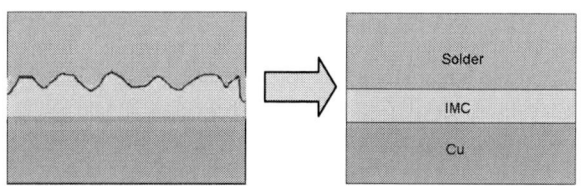

Fig 2 Equivalent diagram for IMC thickness measurement

3. Result and Discussion
3.1 Cu UBM microstructure and plating process

During the copper electroplating process, six samples (A1, B1, C1, D1, E1 and F1) were prepared using the same plating parameters except plating current density; it was found that Cu surface roughness is obviously different and increased rapidly as plating current density increased (Fig.3). From the FIB picture (Fig.4), it also can be found that the grain become larger with plating current increasing, the Cu grain size of F1 is about 1-2μm, but the Cu grain size of A1 is about 200-300nm.

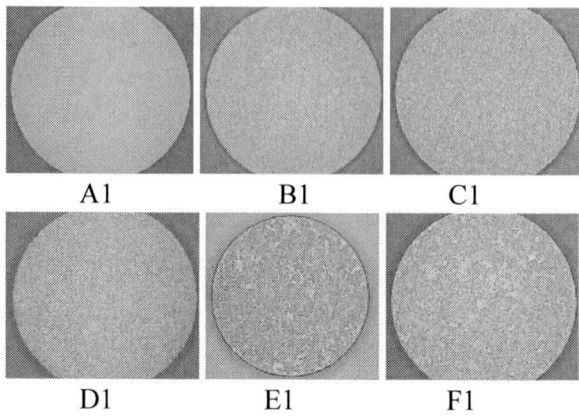

Fig.3. Plating Cu UBM surface picture for different plating current density

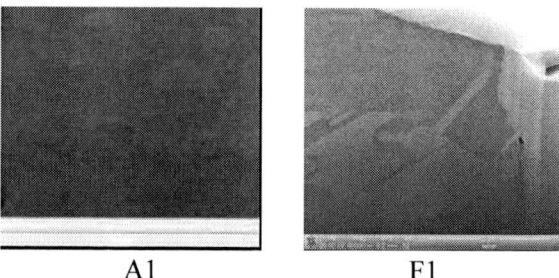

Fig.4. Cu grain picture for different plating current density

3.2 Shear test

Shear strength tests of these samples (A1-F1) were performed. Twenty bumps were sheared for each sample, the results are shown in Fig.5, and shear mode for these samples is shown in Fig.6.

The initial shear mode for these samples occurred between solder to solder (Fig.6. a); the shear strength tests for theses samples was not very different (Fig.5) and was determined by the solder properties. After 48h high temperature storage, the shear mode of A2 almost occurred at interface between IMC and plating copper (Fig.6. b), the shear mode for B2 to F2 still remained solder to solder. The shear strength of A2 obviously decreased, other samples did not decrease much as the aging time increased, and the shear strength of Cu-Sn IMC layer became lower than solder ball because there were voids in the IMC layer [5]. Large voids were observed on the fracture surface for sample A2. This result showed that the Cu UBM structure determined by the plating process affects the adhesion of plating copper and solder ball.

3.3 IMC growth and Cu microstructure

Cross section SEM photographs of these samples (C1, C2, C3 and C4) are shown in (Fig.7). A distinct reaction region of Cu-Sn IMC layer between the solder and the copper for aging time was found. A very thin Cu_6Sn_5 layer at the Sn/Cu interface after reflow for these samples were observed [8]. After a 48 hour aging time, a second intermetallic layer (Cu_3Sn) formed between the original Cu_6Sn_5 layer and Cu layer, and Kirkendall voids also occurred inside the Cu_3Sn

layer. The IMC layer thickness (including Cu_6Sn_5 and Cu_3Sn), the number, and size of voids increased with time.

interconnection, and result in the failure of shear test along the interface.

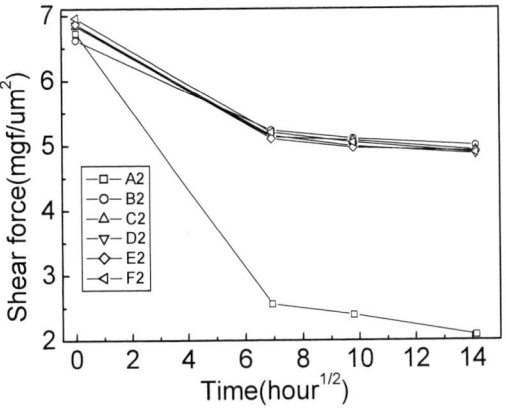

Fig.5. Shear strength results after thermal aging test for different Cu UBM structure

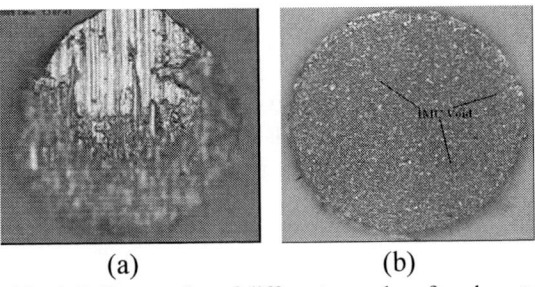

(a)   (b)

Fig.6. Failure modes of different samples after shear test

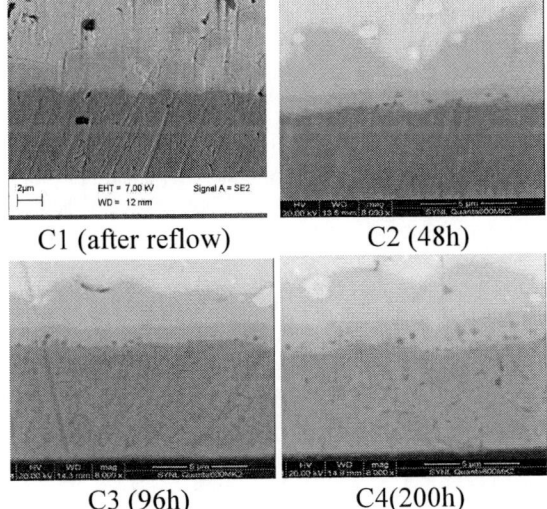

C1 (after reflow)   C2 (48h)

C3 (96h)   C4(200h)

Fig.7. SEM pictures of IMC layers at different aging test

After aging 48h at 150°C, the morphology of IMC layer for all samples is different (Fig.8). The thickness of IMC layers decreased with plating current increased; the voids of sample with lower plating current density more serious than others, the voids of sample A2 have been linked together and form microcrack (Fig.8.A2), which weakens the IMC and Cu

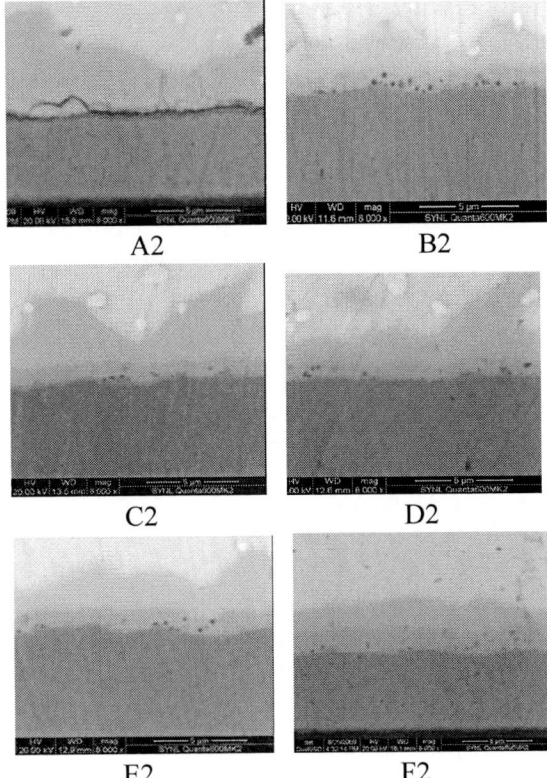

A2   B2

C2   D2

E2   F2

Fig.8. Cross section SEM picture of samples aged 48h

The mean thickness of IMC layers was measured by image processing software system. The thickness of total IMC and Cu_3Sn layers versus the square root of time at 150°C was shown in Fig.9, a linear relationship between the thickness of the IMC layer and the square root of time can be established according to the parabolic growth rate.

$$d=(Dt)^{1/2}$$

Where d is the thickness of IMC layer, D is the diffusion coefficient, and t is the time.

Table2 Diffusion coefficient for different Cu structure sample

D	A	B	C	D	E	F
IMC	0.189	0.181	0.179	0.173	0.168	0.161
Cu_3Sn	0.170	0.151	0.145	0.129	0.131	0.114

It was found from Table2 that the diffusion coefficient value of IMC and Cu_3Sn increased while the plating current density decreased [8]. The thickness of IMC layers depended on the plating Cu structure, the plating Cu grain is smaller, and the thickness of IMC and Cu_3Sn is thicker. For the same aging time, the IMC thickness and Cu_3Sn layer thickness of A was largest whereas that of F is thinnest.

978-1-4244-4658-2/09 $25.00 © 2009 IEEE   741

According to the report [10], the polycrystal Cu (8.4×10^{-17} cm/s) has greater diffusion coefficient than single Cu (2.9×10^{-17}) whatever the crystal orientation is. The diffusion coefficient deference is due to the grain boundary, in polycrystal Cu, the grain boundary would be as the direct path for atom diffusion, which have much greater diffusion rate than that in gain itself. The Cu grain size for larger plating current density is much greater than that lower plating current density, grain boundary supplies the easier diffusion path for fine grain structure, when the exchange between Cu and Sn is not in balance, the Kirkendall voids will be easily formed at the interface of Cu/Cu_3Sn. The diffusion of polycrystal Cu has two types: one is in grain, another is along grain boundary, and the diffusion is schematically shown as Fig 8.

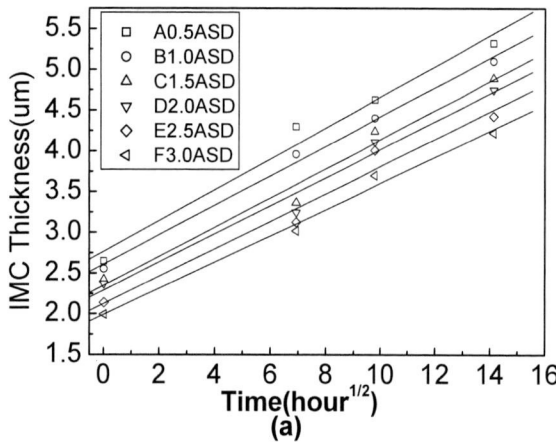

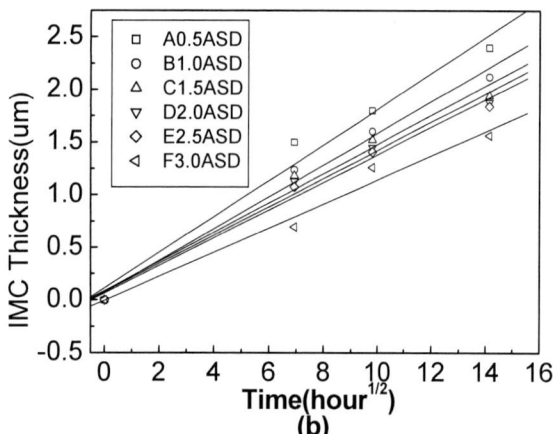

Fig.9. IMC layer thickness and Cu_3Sn versus time

Based on the analysis, in fine grain microstructure vacancy should start forming at the site of grain boundary and Cu_3Sn interface, then the vacancies were accumulated together and formed the void, when the grain size is very small, voids will be linked and the initial crack generate.

In addition, fine grain Cu (nanometer grain size) have higher internal stress and higher energy, so atom is in unstable status, and easy to migrate. While the big grain has low internal stress and atom energy, the atom is stable.

In general, the growth of IMC and void showed a different behavior depending on the initial morphology. The plating Cu grain is finer; the growth rate of Cu-Sn IMC layer and voids inside IMC layer is faster. After long aging time, the Cu will be consumed to form IMC layer. The Kirkendall voids will grow and form crack for exchange between Cu and Sn is not in balance, so there will be delamination between IMC and UBM layer. It was clear that the Cu UBM microstructure determined by the Cu plating process affected the failure modes and reliability of solder joints through the increasing rate of IMC growth.

4. Conclusion

The relation between plating Cu UBM microstructure, IMC layer growth and Cu plating process were investigated.

1) The Cu UBM surface roughness and Cu grain is affected by Cu plating current density. Larger current density will generate rougher surface and larger Cu grain, current size, surface roughness and grain size.

2) The growth kinetics of IMC was affected by the Cu UBM structure. The growth rate of total IMC layer and Kirkendall voids was related to Cu UBM microstructure and thermal aging time, the growth increased with Cu grain decreased and thermal aging time increased., after long aging time, there will be delamination between IMC and UBM layer.

3) The shear strength and failure mode after thermal aging test is affected by the Cu UBM microstructures. After thermal aging, the shear strength of solder bumps with fine Cu surface was lower and was determined by the IMC layers properties. Cu UBM microstructure determined by the Cu plating process affected adhesion of plating copper and solder ball and reliability of solder joints.

Acknowledgments

Thanks for the support from Institute of Metal Research Chinese Academy of Science, especially Dr Zhang Lei for his contribution on providing cross section and SEM and FIB analysis.

References

1. P. Garrou, "Wafer Level Packaging Has Arrived," *Semiconductor International*, October 2000, pp. 119-128.

2. Li Zhang, "Numerical and experimental analysis of large passivation opening for solder joint reliability improvement of micro SMD packages," *Microelectronics Reliability*, No. 44 (2004), pp. 533-541.

3. V. Patwardhan, "Lead-Free Wafer Level-Chip Scale Package: Assembly and Reliability," Electronic Components and Technology Conference, 2002, pp. 4-8.

4. Xiaowu Zhang, "Board level solder joint reliability analysis of a fine pitch Cu post type wafer level package (WLP)," *Microelectronics Reliability*, No. 48 (2008), pp. 602-610.

5. Y.C. Chan, "Growth kinetic studies of Cu-Sn intermetallic compound and its effect on shear strength of LCCC SMT

solder joints," *Materials Science and Engineering*, No. B55 (1998), pp. 5-13.

6. L. Xu, "Intermetallic Growth and Failure Study for Sn-Ag-Cu/ENIG PBGA Solder Joints Subject to Thermal Cycling," Proc. 55th Electronic Components & Technology Conference, Orlando, FL, June (2005), pp. 682-686.

7. Won Kyoung Choi, "Effect of Soldering and Aging Time on Interfacial Microstructure and Growth of Intermetallic Compounds between Sn-3.5Ag Solder Alloy and Cu Substrate," *Journal of ELECTRONIC MATERIALS*, Vol. 29, No. 10 (2000), pp. 1207-1213.

8. Guo-Wei Xiao, "Effect of Cu Stud Microstructure and Electroplating Process on Intermetallic Compounds Growth and Reliability of Flip-Chip Solder Bump," *IEEE TRANSACTIONS ON COMPONENTS AND PACKAGING TECHNOLOGIES*, Vol. 24, No. 4 (2001), pp. 682-690.

9. Y. Liu, "Influence of plating parameters and solution chemistry on the voiding propensity at electroplated copper-solder interface: Plating in acidic copper solution with and without polyethylene glycol," *J Appl Electrochem*, No. 38 (2008), pp. 1695-1705.

10. H.F. Zou, "Morphologies, orientation relationships and evolution of Cu_6Sn_5 grains formed between molten Sn and Cu single crystals," *Acta Materialia*, No. 56 (2008), pp. 2649-2662.

A Novel Research on Micro-ball Placement Machine Used for Wafer Level Package

Jinsong Liu
Shanghai Athele Automation Equipment Co.
1-2F, No.90 Bldg, 1122 Qin Zhou Bei Road, Shanghai, China
021-64956756 / 021-64956755, liu.jinsong@athlete-fa.co.jp.

Abstract

Wafer Level Package (WLP) is a good way to resolve the fine pitch high density IC production. In formal way, it is done by producing bumps on wafers. Several ways to produce bumps are available, such as stencil printing, imprint & photolithography. But the bumps height can not be controlled well in the above methods. Since the micro balls are mass produced in the same size (general less than 3% error on balls diameter), if we can place the micro balls on the wafer in the a proper way, it is a good solution for WLP. A new method and a machine presented here are introduced. The minimum ball size is 60μm in diameter. Using this kind of machine, the bumps producing process can be shorted and cost down and avoid producing mistakes. In this paper, the basic apparatus introduction and ball placing mechanical pictures are given. The machine has been used in the IC research and fab production since 2008.

1 why use micro ball placement

Semiconductors are very cost-sensitive and can only remain competitive and low-cost technologies. Screen printing appears to be a good choice for bumping, however, since stencil printing is limited and cannot transfer enough solder volume to allow for under fill free process flows during assembly. Ball placement of 200μm to 500μm solder spheres is well established to the substrates and has proven to be the most suitable technology for those devices. But to wafers, less than 150μm solder spheres are always needed, even to 60μm diameter solder balls.

Chip-scale package is one of the technologies demanding rapid increases in capacity for placing solder balls onto wafers, strips and substrates. Identifying the critical success factors for suitable processes helps to establish a set of best practices to ensure throughput, yield and cost targets.

Rapid increases in the performance and density of processors, memories, digital signal processors and a host of application-specific devices are being driven by market demands placed on handheld, personal computing and Internet devices. Advanced packaging technologies, such as wafer-level chip-scale package (WL-CSP) and package-on-package(POP) as shown in Fig. 1, are keys to meeting these performance and density targets.

Fig.1 POP structure graph

2 Introduction to the wafer ball placing and process

Wafer Level Chip Scale Packages are available in both area array and peripheral array formats in accordance with the specified guidelines. In an area array, the solder balls are arranged in a matrix of m columns by n rows. The rows and columns may be arranged in an aligned or staggered fashion.

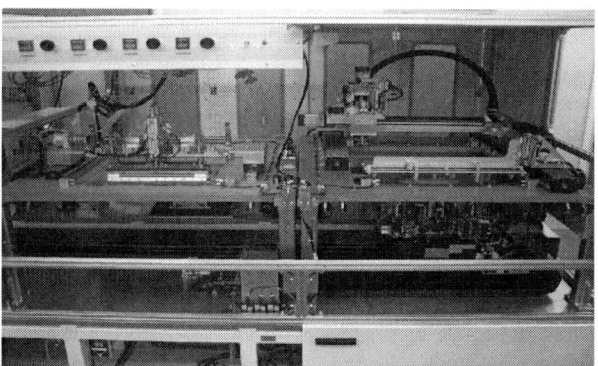

Fig. 2 The picture of micro ball placement machine

Here, athlete Co. develops a new kind of machine which holds flux printing function and ball placement function shown as Fig. 2. The left side of the picture is flux printing process, the right side is ball placement process.

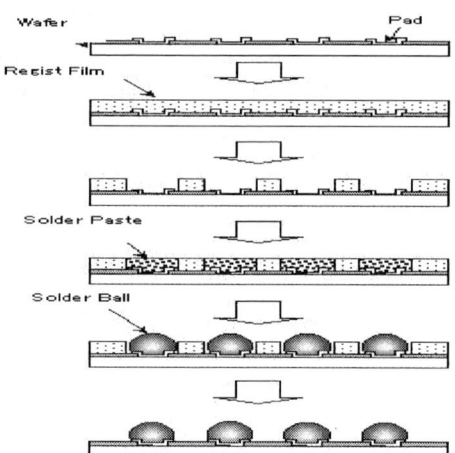

Fig. 3 Traditional bump producing method

In traditional bump producing methods, the most popular method is as Fig. 3 in short brief introduction. It is difficult to handle and costs valuable. Previous compliant WLP technologies used electroplated copper traces for first level electrical connection and routing and used solder balls for second level electrical connection.

WLP with one metal routing layer typically involves the following process steps:

1) Spin coating the dielectric layer;

2) Photo lithographically defining the pad openings for plating;

3) Vacuum depositing an adhesion layer/ plating seed layer prior to electroplating;

4) Photo resist and lithographic processing to define routing traces and solder ball pads;

5) Electroplating metal traces and solder pads;

6) Photo resist removal;

7) Wet etching and plasma etching to remove the adhesion/ plating seed layer;

8) Spin coating of a second dielectric layer to protect the electroplated copper traces ;

9) Photo lithographically defining the openings in the solder pad areas.

The process steps for vacuum deposited UBM and electroplated solder are as follows:

10a) Seed layer formation by vacuum deposition;

11a) Photo resist and lithographic processes to define areas for solder plating;

12) Solder plating;

13) Photo resist removal;

14) UBM etch;

15) Solder reflow.

The process with UBM and solder ball attachment involves the following steps:

10b) Electroless Ni/Au UBM by wet chemistry;

11b) Solder ball attachment by ball placement & reflow.

The full process has at least three costly lithographic steps and at least one vacuum deposition step. With the ball placing technology and etched compliant WLP process includes the different steps with described above. The new process reduces cost through:

1) Elimination of the vacuum deposition step for seed metal deposition;

2) Reduction of the number of photo lithographic steps from three to two;

3) Substitution of high cost photosensitive compliant materials such as silicone, polyimide or BCB with low cost epoxy resins.

Added process steps such as wire bonding, ball placing and bond window encapsulation are low cost, mature technologies that fit into existing infrastructure.

WLP on reconstituted wafers solves die shrinkage concerns. With a re-constituted wafer, the package size can be kept constant regardless of die size. An added benefit of reconstituted wafers is enhanced packaging yield since only known good dies are packaged. Because wafer yield also has a significant impact on the unit package cost, reconstituted wafers also enable cost reduction. Another benefit is die edge protection if the die is smaller than the package. Combination of reconstituted wafers and laminated and etched WLP technology dramatically reduces cost.

It is feasible to take a batch-processing approach using this equipment, and flux one wafer before performing ball placement on each fluxed unit. Alternatively, units in one cassette could be fluxed simultaneously with ball placement on a previously fluxed cassette. However, since the equipment used in this experiment was not configured for high volume manufacture, all three operations (flux print, ball placement, and inspection) were performed sequentially for each wafer, before the next wafer was processed.

In the industry generally, ball placement inspection is sometimes performed post-reflow. However, in this case it was carried out post placement so that no reflow-generated defects could influence the data. Therefore the placement process alone was measured. Inspection was performed manually and with both stereo zoom and video measuring microscopes.

Fig. 4 shows the 8 & 12 inch wafers after ball placement. The ball diameter is 100μm. It is hard to see the ball clearly without microscopes. The total ball amount on 12 inch wafer is over 2.2 millions.

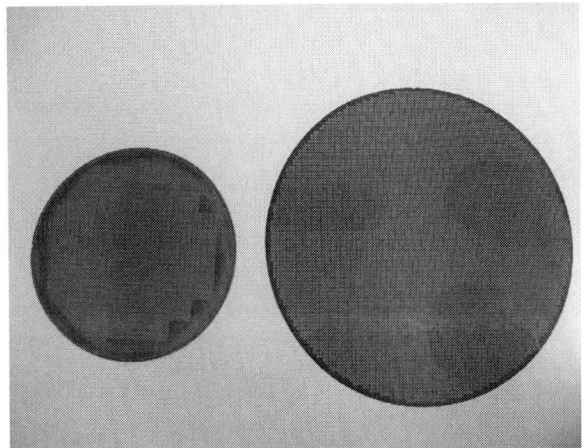

Fig.4   8 & 12 inches wafer after ball placement

3 The contrast between substrate ball mounting & Wafer ball placement

Technology roadmaps for electronic packaging and 3D integration show the continuing trend of increasing input/ output connection density between the semiconductor chip and the package or between two different ICs. Bump pitch requirements for 3D applications such as the integration of memory and logic are even tighter. These fine pitch applications exceed the capabilities of traditional wafer bumping processes such as solder screening or ball placement. For BGA or CSP packaging, it also has been used widely and low cost.

Here, a brief introduction about the difference between the substrate ball mounting process and the wafer ball placing process. Both of the two types of machines, we have presented to the mass industry. Shown as Fig. 5, the above side introduces the theory of wafer ball placing, the below side introduces the method of substrate ball mounting. Briefly, the wafer ball placing uses a mounting squeegee to clip the balls above the mounting stencil, every hole leaves a ball, take the mounting stencil up, balls are leaving on the wafer. The method of substrate ball mounting uses ball vacuum jig, that means, at first, absorbs balls from ball hopper; then by CCD alignment, mount balls to the substrate.

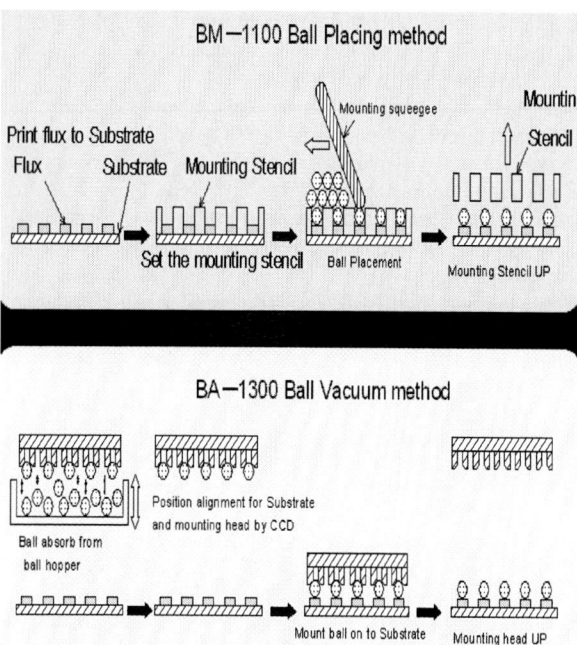

Fig. 5   two ball placing methods

4 Application with wafer ball mounting technology

Wafer back grinding to thicknesses 100μm is now achievable. This demands that package assemblers take special care to verify and optimize the characteristics of the wafer pallet. Distortion in ultra thin wafers is a major challenge to all companies currently developing WLP. For solder ball placement using mass imaging, custom wafer pallets are a selective solution. Another alternative is to use a porous aluminum wafer pallet, which can be machined to minimize distortion of the wafer while ensuring flatness and co-planarity to achieve successful solder ball placement.

WLP technology simplifies processing and reduces cost by eliminating substrate and under-fill. Using preformed balls results in a large overall collapse height, which meets solder joint reliability requirements for commercial applications.

To 2010, the market is expected to increase 20 times dominated primarily by flash EEPROM and high speed DRAM compared to 10 years ago. In 1999, WL-CSPs were primarily used in cell phones, but in 2009 other end-use applications include memory modules, PDA's, laptops, digital camcorders, and digital cameras. By 2010, over 600 million units of WL-CSPs will be used in cell phones, compared to 16.6 million in 1999. The number of WL-CSPs used in memory modules will increase from 2 million in 1999 to 320 million in 2010. The growth of WL-CSP uses in digital cameras, digital camcorders, PDAs and notebook computers will account for an additional 3300 million units.

Wafer-level packages based on solder ball area arrays have been thought of as glorified flip-chips, with a polymer or other material providing mostly moisture resistance. They have typically used a large solder ball to take up stresses from CTE (coefficient of thermal expansion) mismatch. Though good moisture resistance has been demonstrated for these devices, thermal cycling tests have not always been adequate. Now the packaging done at the wafer level is helping the solder ball take up some of the CTE mismatch stress. Though the added material does not necessarily take up the stress, the solder ball is supported to help prevent solder joint failure on the die.

Flip Chip Technology has developed what it calls polymer collar technology as an addition to its Ultra CSP package. A polymer material is added that wets up to the solder balls, forming a supporting collar.

Another developing package trend on next generation CPUs is wafer on wafer type. One of the key technologies is to producing 50-60μm bumps on the thin wafer. For sub-50μm technology nodes, interconnect scaling presents numerous challenges including nano-scale fabrication and control, increased RC time delay, and decreased reliability. To ensure and address these challenges, wafer ball mounting technology is still needed to be developed deeply.

Conclusions

This paper presents a new developed micro ball placement machine and the experiment has proven that it is possible to successfully automate the placement of 60μm to 0.3-mm balls for WL-CSPs with ball yield greater than 99.99%. It has also proven that the process is relatively stable and controllable.

With the development of compact end-use electrical applications, the high density IC packaging techniques are needed to research deeply. Wafer level package is the IC packaging trend in the future. Micro ball placement on wafers is one of the keys technologies. Since our developed this kind of machine and the process, we hope it can be used in the mass production and will help the fab to reduce cost and make many kinds of great functions ICs for the world.

References

1. John Baliga, "Wafer-Level Packages to Include Solder Ball Support," *Semiconductor International*, (2000).
2. David Foggie and Jens Katschke, "Factors for Successful Wafer-Level Solder Ball Placement," *Semiconductor International*, (2007).
3. M. Whitmore, M. Staddon, etc, "The Development Of Balling Technologies For Wafer Level Devices With Pitches Down To 0.4mm," *International Wafer-Level Packaging Conference*, 2005.
4. J. Baliga, "Wafer-Level Packages to Include Solder Ball Support," *Semiconductor International*, 23, 13 (2000), p. 58.
5. D.H. Kim, P. Elenius, S. Barrett, "A Polymer Reinforced WLP/Why It Has Superior Solder Joint Reliability," *Proceedings of IMAPS 2001*, Oct. 2001.

Developing on IC Flip Chip Bonder Machine & Process

JIAN GUO

Shanghai Micson Semiconductor Equipment Co.

A104, No.189 Xin Jun Huan Road, Shanghai, CHINA

Suiyan1@hotmail.com, 021-34637022/021-34637056

Abstract

Flip chip bonding process is a new solution for nowadays IC production. Since of small amount & much kinds of IC types, evenly to much package process, it also needs a flexible machine who can deal with many requirements in lower cost and quickly response. The MFC108 is new designed Flip Chip Bonder for above demands. In this paper, the designing methods are also introduced. By the 3D software of mechanical designing, it spent fewer time to finish the function design with simulation that can be fount out almost all designing mistakes. On the electrical soft design, it is controlled the three kinds of parameters in the operators' selection. The parameters are temperature, pressure and position. By the mixed process designed by users, MFC108 can finish the given bonding process. The machine bonding accuracy is ±5.

1 What is Flip Chip Package and why use it.

Flip chip technology has been practiced by AT & T since the 1950s, IBM developed the controlled-collapse chip connection (C4 called) in the early 1960s for its mainframe computer applications. However, due to the high cost of the ceramic substrate, it was not popular until the late 1980s, when IBM Japan researcher developed the successful implementation of flip chip low cost materials and bonding process. Even to nowadays IC fab, the infrastructure of flip chip is not well established; flip chip expertise is not commonly available; wafer bumping is still too costly; bare die/wafer is not commonly available; bare die/wafer handing is not so easy, and so on. The demands to use flip chip technology are also ingressive. The designed machine MFC108 is developed for the low cost needs and easy operation. We hope it can be used in the research institute at first in China, then it will also be potential used in the mass production fab.

Flip chip microelectronic assembly is the direct electrical connection of face-down (hence, "flipped") electronic components onto substrates, circuit boards, or carriers, by means of conductive bumps on the chip bond pads. In contrast, wire bonding, the older technology which flip chip is replacing, uses face-up chips with a wire connection to each pad.

Flip chip components are predominantly semiconductor devices; however, components such as passive filters, detector arrays, and MEMS devices are also beginning to be used in flip chip package form. Flip chip is also called Direct Chip Attach (DCA), a more descriptive term, since the chip is directly attached to the substrate, board, or carrier by the conductive bumps.

C4 is a die-to-package interconnection technology used in place of interconnection technologies such as wire bonding or TAB(Tape Automated Bonding). Unlike wire bonding and

TAB, C4 bond pads are not limited to the outer perimeter, but can be placed anywhere on the die surface. Shown as Fig.1, this results in higher interconnect density, and higher performance devices. The mostly used flip chip type package is FCBGA. The FCBGA is lower cost than FCPGA. It is belonging to the high level IC package technology.

Fig.1 C4 Flip Chip Bonding IC

2 The design for MFC-108

The IC package is finished by bonder machine. We here in MICSON Co. developed this new kind of semi-auto simple mechanical design apparatus named MFC-108. Used with 3-D solidworks CAD software, we finish it with short time and less mistakes. The picture of the machine is as Fig.2.

Fig.2 Flip Chip Bonder MFC-108 Machine

MFC108 is consisted of six parts. Part1, IC bonding head driven by AC servo motor with bonding tool and heating system max to 450C; Part 2, Flux stage with IC pick up stage; Part 3, Substrate stage with heating system; Part 4, 2-sides CCD vision system; Part 5, A monitor shows the picture of IC side and the picture of substrate side; Part 6, the controlling system with PLC.

By the 3D mechanical designing, it spent fewer time to finish the function design with simulation that can be fount out almost all designing mistakes. On the soft design, it is

978-1-4244-4658-2/09 $25.00 © 2009 IEEE

controlled the three kinds of parameters in the operators' selection. The parameters are temperature, pressure and position.

Seen from Fig.3, the CCD system can be adjusted by manual in X, Y, Z axis to be suitable for the above side IC and below side substrate. The bonding stage can also be adjusted by manual in X, Y and θ axis. The heating system max to 120°C is optional for the bonding stage. Sometimes, to heat the substrate is a good choice for Flip Chip Bonding.

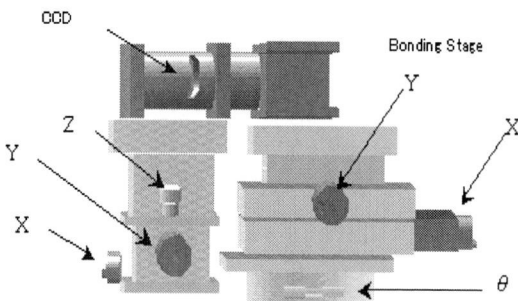

Fig.3 The structure of bonding stage

3 Bonding Process used MFC108

By the combined process given by users, MFC108 can finish the given bonding process in high response and easy setup, also can store the process parameters in the controlling system. The machine bonding accuracy is ±5 shown as Fig.4. It is token after bonding by enlarge under camera. The two lines denotes to the line of IC side and the other line of substrate side. The difference error between two lines is the bonding accuracy.

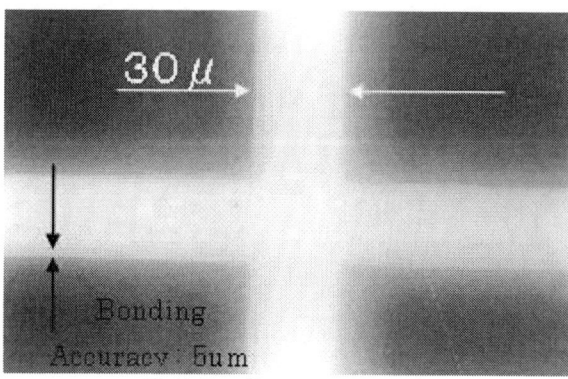

Fig.4 Bonding Accuracy: 5μm

The bonding process is as follows. The Fig.5 shows the bonding system mechanical structure. At first, the bonding head fixed with absorption tool moves to the pick up point taught before. Secondly, the Flux stage has another function called IC pick up stage. The flux stage moves in right direction to the pickup teaching point, the bonding head moves down and pickup the IC, then the bonding head goes back to the up point also taught before. Thirdly, the flux stage is transferred with flux on the given stage which is different

from the IC pickup stage. Then, the bonding stage moves down again to the flux transfer teaching point. And then touches the flux stage, it means the flux will be transferred to the IC. The bonding head goes back again to the up point also taught before. The flux stage does also move back at the left point. Fourthly, the bonding head moves down to the adjusting point taught before. Move the 2-sides CCD system to the place between the bonding head and the substrate stage. By adjusting X, Y and θ axis manually, the lines at the IC side and the substrate showed in the monitor are coincide. Then move back the 2-sides CCD system. Fifthly, the bonding head moves down directly to the bonding point taught before. Since the IC is also toughing the substrate, during the boding head moves down lower than the bonding point, the pressure is also slowly up. The bonding head stops when the pressure is reaching the given value.

The above process is the basic bonding theory, as introduced before, during the bonding process, the temperature and the pressure and the height value can be control mixed together in order to get the good bonding result.

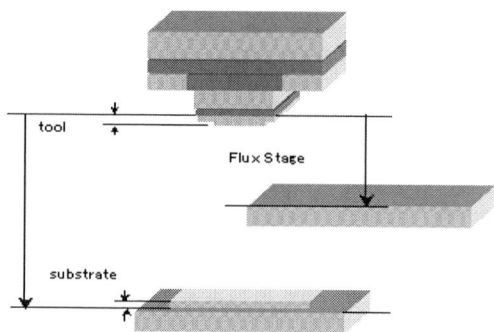

Fig.5 Bonding system mechanical structure

4 Bonding Process used MFC108

The flip chip bonding technology is widely used in the CPU and other high-density package IC. The next logical evolution of the flip-chip technology is 3D integration of chips on interposers or stacked chips. Chip and wafer thinning with through-silicon vias (TSVs) and new Cu/Cu and copper column/pillar interconnects at ultra fine pitches (50 μm) are now in development. In addition to the need to address the ongoing increases in power densities for advanced microprocessors, we believe the advent of stacked chips is creating an unusually challenging thermal management roadmap.

Conventionally, CMOS transistor scaling in accordance with Moore's law has enabled the exponential performance improvement in silicon integrated circuits. CMOS transistor scaling requires local interconnect pitch scaling. For sub-50nm technology nodes, interconnect scaling presents numerous challenges including nano-scale fabrication and control. This kind of developing IC package trend needs wafer level Flip Chip technology.

Packaging continues to push the envelope of materials science and technology. Although we view carbon nanotube (CNT) interconnects as still years away, the use of nanomaterials as a filler for underfills, thermal materials and laminate composites is likely in the near future.

There are also new trends in the deposition of materials, in addition to C4NP, there appears to be interest in the use of a traditional BGA attach method known as 'ball drop' for the formation of the flip-chip interconnect. Fig.6 is the typical application IC package with flip chip technology.

Fig.6 IC after flip chip bonding

Conclusions

Although IC flip chip assembly is not a new technology, it is gaining increasing popularity as a packaging strategy due to the benefits it can provide. By using the entire surface of the die for establishing interconnect, the need for wire bond interconnect is eliminated and package size can be reduced. More importantly for high-performance applications, directly connecting the die to a substrate or board greatly shortens the signal path while reducing the interconnect inductance and capacitance, all of which serves to greatly improve electrical performance.

MICSON Co. in china has been developed the MFC108 flip chip bonder. It can finished many kinds of IC assembly by flip chip package technology. Although MFC108 is a semi-auto machine, it is high accuracy, high performance, compact structure and cheap cost. We hope it can be used widely in the world.

References

1. JOHN H. LAU, Low Cost Flip Chip Technologies for DCA, WLCSP and PBGA Assemblies, McGraw-Hill Published, ISBN 0-07-135131-8.
2. LIU Jinsong *et al*, "BGA/CSP Packaging Techniques", *Journal of Harbin Institute of Technology*, Vol.35, No.5(May, 2003).
3. Sally Cole Johnson, "Flip chip packaging becomes competitive", *Semiconductor International*, Jan. 2009.
4. Daniel Blass, "Flip Chip on Flex. Area Array," *Consortium 2003*, Surface Mount Technology Laboratory, Universal Instruments Corporation, Binghamton, New York 13902.
5. Li Yi, "Flip Chip Assembly Process and its Requests for SMT Equipment", *Equipment for Electronic Products Manufacturing*, Vol.12(2007).
6. David R. Halk Joachim Pajonk, "Variable Flip Chip Assembly for High-volume Production", *Advanced Packaging*, June. 2004.

Effect of Ni Addition on the Sn-0.3Ag-0.7Cu Solder Joints

Lingling Wang, Fenglian Sun, Yang Liu, Lifeng Wang
College of Material Science & Engineering, Harbin University of Science and Technology
Harbin, China, 150080, E-mail: sunfengl@hrbust.edu.cn

Abstract

It is gradually recognized that Sn-Ag-Cu is better alloy system for application prospect in the lead-free solders. In order to cut down the cost and improve properties, researchers have always been searching for proper compositions and adding elements to improve properties actively. This paper has investigated the effect of Ni addition on Sn-0.3Ag-0.7Cu (SAC0307) low-Ag solder joint.

The results of EDS analysis showed that the major IMC formed between the SAC0307 solders and Cu substrate was Cu_6Sn_5, while for solders with Ni addition the main IMC was $(Ni_xCu_{1-x})_6Sn_5$. After aging, in the case of 0.05% Ni addition, the thickness of IMC layer was the thinnest regardless of aging time. SAC0307-0.05Ni solder is effective to reduce the formation of IMC at the interface during the reflow process and to inhibit the growth of IMC after aging. Additionally, the atom ratio of Ni to Cu in the $(Ni_xCu_{1-x})_6Sn_5$ phase after reflow and after aging were also studied. Some findings of this study can be rationalized by the Cu-Ni-Sn isotherm.

Introduction

In general, the Sn-Ag-Cu system solders are recognized as the most promising lead-free solders, but they are more expensive due to the use of silver [1]. The production cost is one of the principal concerns of the manufacturing system solders.

Recently, to improve the properties of solders and interfacial reactions, microelements, such as Ni, have been added into solders [2]. Previous studies showed adding Ni in Sn-Ag-Cu solder could improve its microstructure and its properties. It has been reported that the Ni concentration in $(Ni_xCu_{1-x})_6Sn_5$ phase may vary under different experimental conditions.

This study investigated the effect of a small addition of Ni on the behavior of the interfacial reaction between SAC0307 and Cu substrate. Numerous investigations on the solid-state growth of the Cu-Sn IMC have been reported [3-4]. The relationship between the thickness of the IMC layer and aging time was generally described; the growth rate of IMC during aging was also studied. The relationship between the thickness of IMC layer and aging time is generally described as:

$$X - X_0 = \sqrt{Dt} \tag{1}$$

Where $X-X_0$ is the thickness of the IMC layer, D is the growth rate constant and t is the aging time. The growth rate of the IMC layer depends on the aging temperature.

Experimental Procedure

Four different composition low-Ag solders were SAC0307-XNi (X=0, 0.05, 0.10, 0.15) (wt% and Sn balanced) respectively. The substrate was Cu layer of 99.99% purity (thickness: 0.2mm). The reflow temperature was controlled at 250°C and the reflow time was 10 minutes. During the aging process, the as-soldered Cu substrate specimens were aged at 180°C for 0, 24h, 96h, 216h and 384h respectively. The cross sections of the specimens were polished with 0.1μm diamond grains and then were etched by 5%HCl +95% alcohol. Scanning electron microscopy (SEM) was used to observe the interface between the solders and substrate. Energy dispersive spectroscopy (EDS) and electron probe microanalysis (EPMA) were used to investigate the thickness and the composition of the IMC, the growth rate of the IMC can be then explored. For every measured data, at least three measurements were made, and the average value was reported.

Results and Discussion

Figure 1 shows SEM micrographs of the interfacial reaction between Cu substrate and solders with different Ni addition which reflowed at 250°C for 10 min. The Cu_6Sn_5 phase is visible in Fig. 1.a. Cu_6Sn_5 IMC layer is continuous and void-free. In the literature, morphology of the Cu_6Sn_5 was reported to be almost round or rod type. For the 0.05Ni-doped solder as in Fig. 1b, the IMC morphology tended to be worm-shaped.

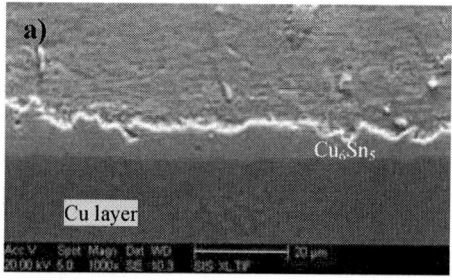

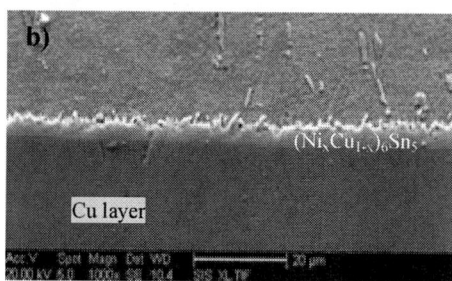

Fig.1. Effect of Ni on the IMC of SAC0307 solder joints
a) SAC0307 b) SAC0307-0.05Ni

The EDS analysis results indicated that the IMC layer was $(Ni_xCu_{1-x})_6Sn_5$ In contrast to Fig.1a, it can be found that the thickness of $(Ni_xCu_{1-x})_6Sn_5$ layer is thinner than that of Cu_6Sn_5 layer. They were of the same crystal structure though the $(Ni_xCu_{1-x})_6Sn_5$ crystal was reported to be cylinder with a hexagonal cross section. The reason why it is to be expressed as $(Ni_xCu_{1-x})_6Sn_5$ is that the compositional ratio of Ni to Cu (6:5) between (Cu+Sn) and Ni is so similar. The atom diameter of Ni is so similar to which of Cu, Ni and Cu atoms are replaceable in the compound easily. This may explain the forming of $(Ni_xCu_{1-x})_6Sn_5$ phase well.

Figure 2 shows the evolution of interfacial IMC formed between solders and Cu substrate after aging at 180°C for 384 hours. In contrast to Figure 1, thickness of IMC layer for two kind of solders increases with the increase of aging time respectively. However, the IMC thickness in the Fig.2b is thinner than that in Fig.2a obviously. Cu_3Sn layer can be detected for both cases. It was reported that the more the IMC grains are coarsened, the thicker the IMC layers becomes. It is well known that excessive IMC growth has a negative effect and a thick IMC layer at the interface usually degrades the reliability of the joint [5].

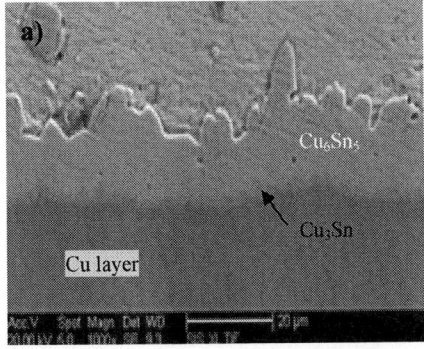

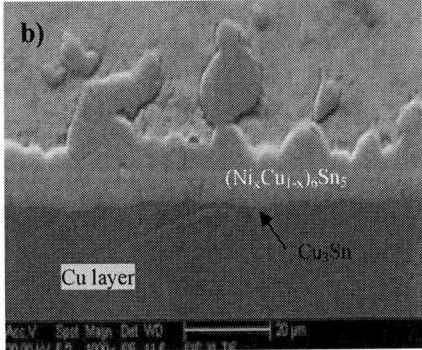

Fig.2.Evolution of IMC after aging at 180°C for 384hs
a) SAC0307 b) SAC0307-0.05Ni

Figure 3 shows the effect of aging time on the IMC layer thickness formed between different solders and Cu substrate. It can be found that it is a nearly linear relationship between the thickness of IMC and the square root of aging time. This confirms that the growth of the IMC layer is a diffusion-controlled mechanism.

With increasing aging time, the thickness of IMC all increase for them. Furthermore, that is smaller for Ni-doped solders, especially for SAC0307-0.05Ni solder.

It has been made clear that the addition of 0.05wt% Ni in the SAC0307 solder is certainly effective for reducing the

formation of IMC at the interface during the reflow process and for inhibiting the growth rate of IMC during the aging process. The growth rate of IMC for SAC0307 and SAC0307-0.05Ni are $5.37\times10^{-5}\mu m^2/s$ and $3.02\times10^{-5}\mu m^2/s$ respectively. The reason why the growth rate of IMC for Ni added solder is slower than that for SAC0307 solder is that Cu atom that has dissolved from Cu substrate seemed to be consumed in advance to change solders in the IMC layer into IMC in the case of Ni added solder. Lower growth rate of $(Ni_xCu_{1-x})_6Sn_5$ layer means higher growth activation energy for it.

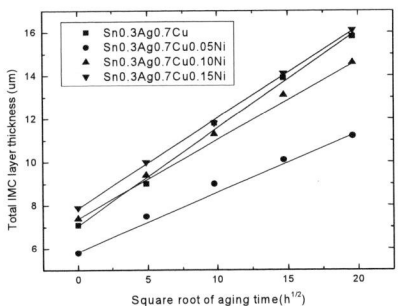

Fig.3. Effect of aging time on IMC layer thickness

Figure 4 shows the effect of Ni content in SAC0307-XNi solder on Ni concentration in $(Ni_xCu_{1-x})_6Sn_5$ phase formed between solders and Cu. From the figure, it can be found that the value of Ni:Cu in $(Ni_xCu_{1-x})_6Sn_5$ gradually increases with the Ni content increasing in the solder, the maximum atom ratio is about 0.77. That is, the Ni concentration is about 4.5Ni at% when it is doped with 0.15Niwt%. There is some relevant information available in the literature points out some reasons as follow: the atomic diameter of Ni is so similar to Cu that they are replaceable in the compound easily [6]. The tie lines connect these two phases, and a slight increase of Ni content in Sn could push the tie-line toward the Cu_6Sn_5 phase that has a much higher Ni content. Furthermore, the affinity of Ni for Sn is different from that of Cu for Sn, which may affect the substitution of Ni into Cu_6Sn_5 phase.

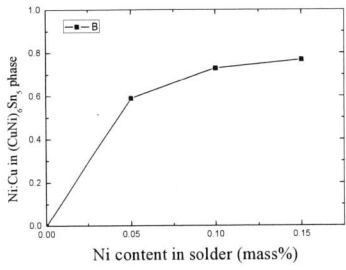

Fig.4. Atom ratio of Ni to Cu in the $(Ni_xCu_{1-x})_6Sn_5$

Figure5 illustrates the variation about the Chemical Composition of $(Ni_xCu_{1-x})_6Sn_5$ for SAC0307-0.05Ni solder after aging. Since the inter-diffusion between Ni and Cu atoms at aging temperature occurs inevitably, which could lead to the varied $(Ni_xCu_{1-x})_6Sn_5$ composition. As aging time increasing, the Ni concentration in the $(Ni_xCu_{1-x})_6Sn_5$ slightly decreased for the SAC0307-0.05Ni solder. The ratio value is about from 0.59 to 0.44, composition varies from $(Cu_{0.62}Ni_{0.38})_6Sn_5$ to $(Cu_{0.69}Ni_{0.31})_6Sn_5$ and the average Ni concentration in $(Ni_xCu_{1-x})_6Sn_5$ is about 19-21at% after aging for 384h. It was reported that 0–25at%Ni in $(Ni_xCu_{1-x})_6Sn_5$ phase formed at 240°C [6-8]. In contrast to the initial composition of $(Ni_xCu_{1-x})_6Sn_5$, the result is 21-23 at% Ni 34-36 at% Cu 45-47at% Sn. Because a new Cu_3Sn layer between Cu and $(Ni_xCu_{1-x})_6Sn_5$ can be found and grew during the aging process, which may include a very small concentration of Ni.

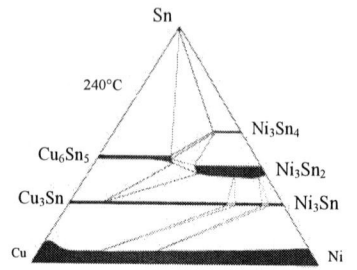

Fig.6. The Cu-Ni-Sn ternary isotherm at 240°C. This isotherm was adapted from Lin.

Conclusions

In the case of 0.05% Ni addition, the thickness and the growth rate of IMC layer was the smallest regardless of aging time. SAC0307-0.05Ni solder is effective on inhibiting the growth of IMC after aging. After aging at 180°C for 384h, atom ratio of Ni to Cu in $(Ni_xCu_{1-x})_6Sn_5$ phase gradually increases with the increase of the Ni content in the solder, the maximum atom ratio is about 0.77 when it is doped with 0.15 wt% Ni. With increasing aging time, the Ni concentration in $(Ni_xCu_{1-x})_6Sn_5$ slightly decreased.

Acknowledgments

The authors acknowledge the financial support of this work from the Science Innovation Foundation of Harbin Science and Technology Bureau. 2008RFXXG010

References

1. Bradley Edwin, "Lead-free solder assembly: Impact and opportunity." *Proceedings-Electronic Components and Technology Conference*, 2003, pp. 41-46.
2. Wang, Gu, "Effect of Ni on the Microstructures and Mechanical Properties of Sn-0.7Cu Lead-free Solder," *Electronics Technology*, Vol. 28, No. 1 (2007), pp. 17-19.
3. H. T. Lee, M. H. Chen, "Influence of intermetallic compounds on the adhesive strength of solder joints." *Materials Science and Engineering A*, Vol 333, Issues 1-2 (2002), pp.24.
4. C. M. Chuang, K. L. Lin, "Effect of Microelements Addition on the Interfacial Reaction between Sn-Ag-Cu Solders and the Cu Substrate." *Journal of Electronic Materials*, Vol. 32, No. 12 (2003), pp. 1426.
5. J. W. Kim, D. G. Kim, W. S. Hong, "Evaluation of Solder Joint Reliability in Flip-Chip Packages during Accelerated Testing." *Journal of Electronic Materials*, Vol. 34, No. 12, (2005), pp. 1552.
6. S. W. Chen, S. H. Wu, S. W. Lee, "Interfacial reactions in the Sn-(Cu)/Ni, Sn-(Ni)/Cu, and Sn/(Cu,Ni) Systems." Journal of Electronic Materials. Vol. 32, (2003), pp. 1188.
7. C. H. Lin, S. W. Chen, C. H. Wang, "Phase Equilibria and Solidification Properties of Sn-Cu-Ni Alloys." *Journal of Electronic Materials*, Vol. 31, No.9, (2002), pp. 907.
8. M. L. Huang, T. Loeher et al, "Morphology and Growth Kinetics of Intermetallic Compounds in Solid-State Interfacial Reaction of Electroless Ni-P with Sn-Based

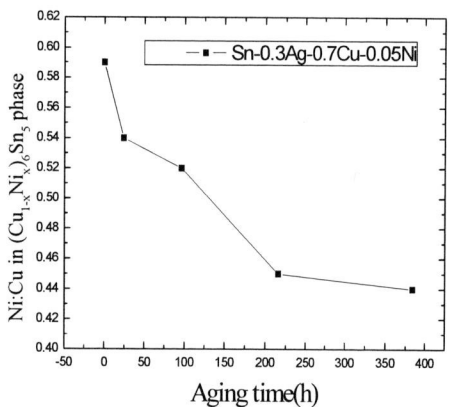

Fig.5. Effect of Aging time on atom ratio in $(Cu_{1-x}Ni_x)_6Sn_5$

The Cu-Ni-Sn ternary isotherm at 240°Cshown in Fig.6 is drawn based on the previously published isotherms.[9]The temperature used in this study (250°C) is not far from the temperature of the isotherm (240°C), so it can be used to explain the current system safely. Although there is another element Ag in our system, Ag was fixed as Ag_3Sn and did not participate in the reaction. So Cu-Ni-Sn ternary isotherm can be used to explain the behavior and relevant results in the current system. According to this isotherm, the phase fields that are in equilibrium with the Sn phase include one three-phase field, (Sn) + $(Cu_{0.54}Ni_{0.46})_6Sn_5$ + $(Ni_{0.77}Cu_{0.23})_3Sn_4$, and two two-phase fields, (Sn) + $(Ni_xCu_{1-x})_3Sn_4$ and (Sn) + $(Ni_xCu_{1-x})_6Sn_5$. Several diffusion paths are possible. As we know when the Cu concentration is over 0.6%, the diffusion path may pass through (Sn) + $(Ni_xCu_{1-x})_6Sn_5$ two-phase fields. In other words, the $(Ni_xCu_{1-x})_6Sn_5$ forms next to the (Sn) phase. When Cu concentration is low, the diffusion path may pass through the (Sn) + $(Ni_xCu_{1-x})_3Sn_4$ two-phase field. In this study, with the Sn-0.3Ag-0.7Cu-XNi solder, it is found that the diffusion path always prefers to pass through the (Sn) + $(Ni_xCu_{1-x})_6Sn_5$ two-phase field. The concentration of Ni does not change the type of IMC.

Lead-Free SoldersElectron." *Journal of Electronic Materials*, Vol. 35, No. 1 (2006), pp. 182.

9. C. E. HO, R. Y. TSAI et al, "Effect of Cu Concentration on the Reactions between Sn-Ag-Cu Solders and Ni." *Journal of Electronic Materials*, Vol. 31, No. 6 (2002), pp. 585.

Electrically Conductive Adhesives with Sintered Silver Nanowires

Zhongxian Zhang, Xiangyan Chen, Haowei Yang, Huiying Fu, Fei Xiao[*]

Department of Materials Science, Fudan University,
220 Handan Road, Shanghai, 200433, China.
Phone: + 86 21 6564 3267, E-mail: feixiao@fudan.edu.cn

Abstract

Concerning the hazard to human and environment by lead, the lead-free solders and the electrically conductive adhesives (ECAs) have been considered as the most promising alternatives of the tin-lead solder. Electrically conductive adhesives offer numerous advantages compared to conventional solder technology such as environmental friendliness, mild processing conditions, low stress on the substrates, and fine pitch interconnect capability. It has received considerable attention in microelectronic packaging that the nano-size metal particles fuse at much lower temperature than the melting point of the bulk metal. The sintering of nano-size silver in ECAs can reduce the number of contact points between fillers and increase the conductivity of ECAs. The silver nanowires with a slenderness ratio of 50~60 were successfully synthesized through a polyol process and characterized by field emission scanning electron microscopy (SEM) in this paper. The silver nanowires began to sinter at 200°C and conductive adhesives filled with silver nanowires were studied.

1. Introduction

Microelectronic packaging technology has developed rapidly in the last few years to meet the requirements of information, communications and other high-tech industries. Isotropic conductive adhesives (ICAs) used as a kind of interconnect bonding materials in electronic packaging offer numerous advantages compared to conventional solder technology such as environmental friendliness, mild processing conditions, low stress on the substrates, and fine pitch interconnect capability. However, conductive adhesive technology is still not mature yet, and concerns and limitations do exist. Main limitations of current commercial ICAs include lower conductivity, unstable contact resistance with non-noble metal finished components, and poor mechanical impact performance. ICAs usually consist of metallic silver in a polymer binder with typical filler loadings of 70-80 wt %. At these loadings, the materials have achieved the percolation threshold and are electrically conductive in all directions after the materials are cured. The total contact resistance of an ICA joint includes the bulk resistance of silver, the contact resistance between silver fillers, and the interfacial resistance between the ICA and the substrate or component (refer to Fig. 1). The bulk resistance of the metals does not change during aging. Therefore, changes of the contact and interfacial resistance will cause total resistance shifts [1-3].

The fact that nano-size metal particles melt at much lower temperature than the melting point of the bulk metal has received considerable attention in microelectronic packaging [4-5]. The sintering of nano-size silver, which fuse each other and form metallurgical contacts, can reduce the number of contact points between fillers and increase the conductivity of

ICAs. Jiang et al reported a ultrahigh conductive polymer composite by incorporating silver flakes and surface functionalized silver nanoparticles into epoxy [6]. At a filler loading of 80 wt %, the resistivity of the composite was dramatically reduced to as low as 5×10^{-6} $\Omega \cdot cm$ owning to low-temperature sintering of silver nanoparticles.

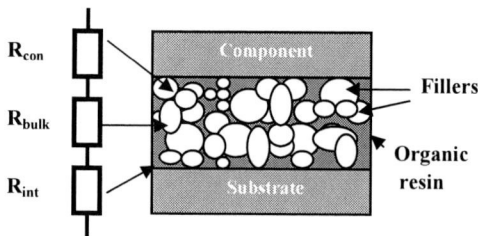

Fig. 1 Contact resistance model of an ICA joint

One-dimensional nanostructures such as nanowires have been the focus of many recent studies due to their potential applications in nanoelectronics, nanophotonics, chemical sensing, and biological imaging [7-10]. A great variety of effective chemical methods have been used to synthesize silver nanowires [11-12]. In this paper, the silver nanowires with a slenderness ratio of 50~60 were successfully synthesized through a polyol process and characterized by SEM. The sintering of silver nanowires at different temperatures was analyzed by SEM. Then, conductive adhesives filled with silver flake and low-temperature sintered silver nanowires were formulated and studied.

2. Experiments

Silver nitrate, ethylene glycol, aliphatic acids, polyvinyl pyrrolidone, sodium chloride and solvents were purchased from Sinopharm Chemical Reagent Co., Ltd. Diglycidyl ether of bisphenol A (Dow 332) was purchased from Dow Chemical Company. 4-methylhexahydrophthalic anhydride (MHHPA) was supplied by Shanghai Zhengrui Chemical Co., Ltd. 2-Ethyl-4-methylimidazole (2E4MZ) was supplied by Shikoku Chemicals Crop. The silver flakes (6-9 μm) was purchased from Hendera Company. Cobalt （Ⅱ） acetylacetonate ($Co(acac)_2$) was synthesized in lab.

The curing behavior of epoxy and the weight loss of the surface modified silver nanowires were measured using a Shimadzu DTG-60H simultaneous TG/DTA system under nitrogen. Sample (10-20 mg) was loaded in an aluminum pan and heated from room temperature to 300°C at a ramping rate of 10°C/min. The N_2 flow rate was 40 mL/min. The morphology of silver nanowires was observed by JEOL JSM-6701F field emission scanning electron microscopy.

Resistivity of the ICA was measured by a Shanghai Qianfeng SB100A/2 four-point technique as shown in   Fig.

978-1-4244-4658-2/09 $25.00 © 2009 IEEE

2 [13]. The bulk resistivity, ρ, were calculated using following equations:

$$\rho = \frac{\pi}{\ln 2} \times \frac{V}{I} \times t$$

where *I*, *V*, and *t* are the current , voltage , and thickness of sample, respectively.

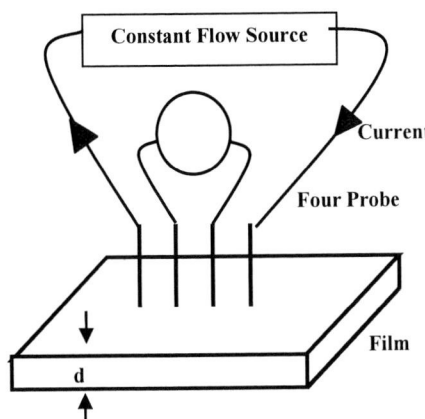

Fig. 2 Four-point technique

2.1 Synthesis of Silver Nanowires

The silver nanowires was synthesized through a polyol process in a similar methord to the literature.[11-12] Ethylene glycol solution (0.1M) of silver nitrate and ethylene glycol solution (0.1M) of polyvinyl pyrrolidone with sodium chloride (1 mM) were simultaneously added dropwise into ethylene glycol in a three-necked flask refluxed at 170°C for 60 min (the solution volume ratio is 5:5:2). Magnetic stirring was continuously applied throughout the entire process of reduction and wires growth. The reaction mixture was diluted with acetone (3 times by volume) and centrifuged to give PVP coated silver nanowires. The PVP on the silver nanowire surface can be removed by sonication in ethanol. The active silver nanowire surface was then modified by sonication in ethanol solution of aliphatic acids.

2.2 Sample Preparation

For better performance of anhydride cured epoxy, 1:0.85 epoxy/anhydride molar ratio was used and 1 wt % of 2E4MZ or Co(acac)₂ was added before curing as accelerator. The surface modified silver fillers were incorporated into the mixture of Dow 332 and MHHPA. The polymer composite was sonicated for 15 min and then mixed with the catalyst. Two strips of polyimide tape were applied on to a pre-cleaned glass slide with a gap width of 1 cm. The formulated composite was bladed into the space between the two strips. The polyimide tapes were removed after curing at desired temperature for 90 min. The thickness of the cured film was controlled by polyimide tapes.

3. Results and Discussion

Synthesis and surface treatment of silver nanowires

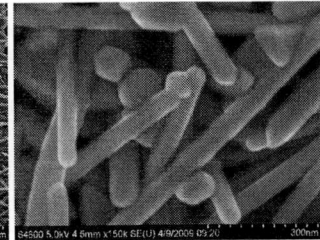

Fig. 3 SEM images of silver nanowires

Fig. 3 shows the morphologies and aspect ratios of the synthesized silver nanowires after purification by acetone centrifugation process. The diameters of most of the silver nanowires are in the range of 30-50 nm, and lengths of   1-3 μm. Silver nanoparticles with diameters of 50-100 nm were also observed in the Fig. 3. The protecting coating on the silver nanowires surface were examined by TGA measurement as shown in Fig. 4. After sonication in acetone and ethanol, most of the PVP on the surface was removed and the residue on the silver nanowires surface was less than 2 wt % which decomposed between 300-500°C. The fresh silver surface was further treated by sonication in ethanol solutions of an aliphatic acid. The precipitated silver nanowires after centrifugation were rinsed with ethanol and drying in vacuum for 2 h at room temperature. The TGA shows that about 1.4 wt % of the acid remained on the silver surface which prevent the nanowires from oxidization and help the dispersion in the epoxy [14].

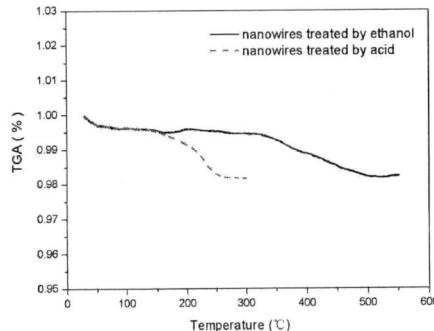

Fig. 4 TGA curves of nanowires

3.1 Sintering of Silver Nanowires

In order to study the sintering behavior, the acid modified silver nanowires were annealed on a glass slide for 30 min at 100°C, 150°C, 200°C and 250°C, respectively. Fig. 5 shows the morphology of the annealed silver nanowires. The obvious sintering occurred after annealing at 200°C. The nanowires sintered with each other or with nanoparticles mainly through the ends of wires because the surface to volume ratios of the end is higher than that of the body of wires. After heating at 250°C for 30 min, silver nanowires fused to large chunks as shown in the Fig. 5(d).

978-1-4244-4658-2/09 $25.00 © 2009 IEEE

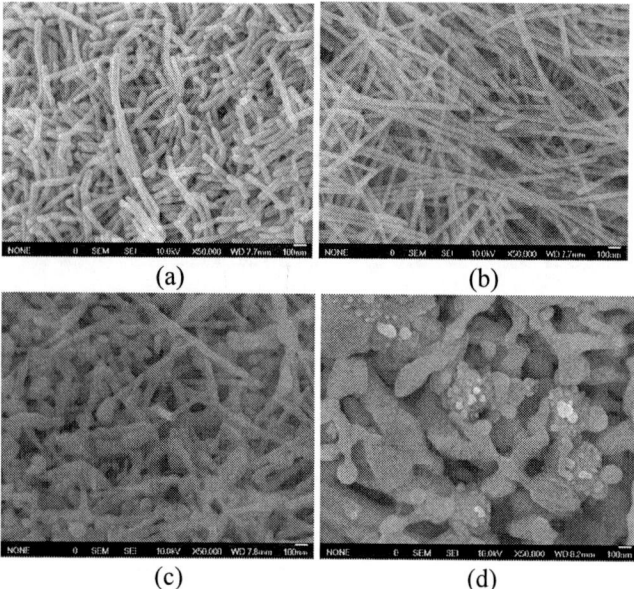

(a)   (b)

     (c)   (d)

Fig. 5 SEM images of silver nanowires annealed at
(a) 100°C; (b) 150°C; (c) 200°C; (d) 250°C

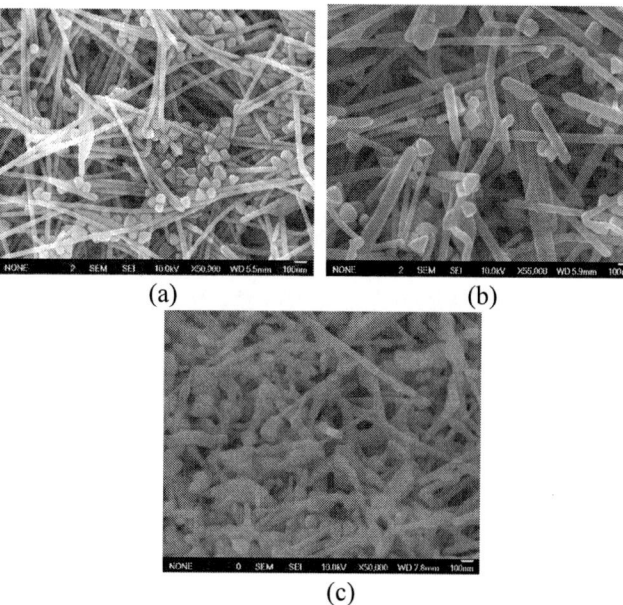

(a)   (b)

(c)

Fig. 6 SEM images of silver nanowires with sonication
in different solution: (a) acetone; (b) ethanol and (c)
aliphatic acid (annealed at 200°C for 30 min)

The surface condition plays an important role in the silver
nanowires sintering process. Fig. 6 shows the SEM
photographs of silver nanowires annealed at at 200°C for 30
min. The freshly prepared nanowires were treated by
sonication in acetone, ethanol, and aliphatic acid solution,
respectively. The silver nanowires do not sinter whether the
surface is coated with PVP or clean (Fig. 6(a) and 6(b)).
When the surface is modified by the aliphatic acid, the silver
nanowires are activated to sinter each other. The behavior is
in agreement with the literature result of sintering of
nanoparticles [6]. Unlike PVP on the surface which is difficult
to be removed, the acids on the surface of silver nanowires
prevent them from oxidization at room temperature and then

easily get rid of the surface at high temperature to leave a
highly activated surface for sintering.

3.2 Properties of ECA with silver nanowires

The electrically conductive adhesive was formulated by
incorporating 76 wt % of the acid modified silver nanowires
into epoxy. Fig. 7 shows the SEM photographs of the ECA
before and after curing at 200°C for 90 minutes. Although the
surface of silver nanowires was covered with epoxy in the
composite which may prohibit sintering, silver nanowires did
sinter to clumps with reducing in the slenderness ratios after
curing at 200°C. There still exist many separated nanowires
and particles in the epoxy resulting in large contact resistance.
There is no conductive network throughout the epoxy matrix
by fused silver nanowires, thus the resistivity of the ECA is
high (7.1×10^{-4} Ω·cm).

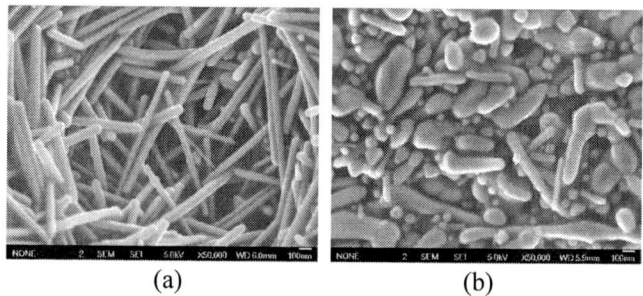

(a)   (b)

Fig. 7 SEM of ECA with 76 wt % of the silver nanowires:
(a) before and (b) after curing at 200°C for 90 minutes

To reduce the resistivity of ECA, a mixture of silver flakes
and nanowires was used as conductive fillers. Fig. 8 shows the
resistivity of ECA with 75 wt % of surface modified silver
flakes and nanowires (molar ratio 3:1) at different curing
conditions. It can be seen that the resistivity of ECA without
catalyst was considerably lower than that with 2E4MZ as
catalyst. The resistivity of the ECA catalyzed with 2E4MZ
increases with the increase in curing temperature, while the
resistivity of ECA without catalyst decreases slightly with the
increase in curing temperature. The result implies that the
cured epoxy will effectively prohibit the sintering of silver
nanowires.

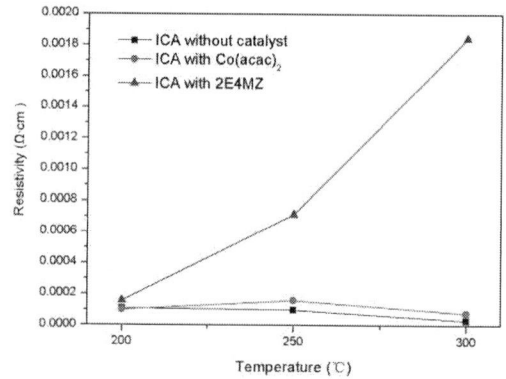

Fig. 8 Resistivity of the ICA filled with 75 wt. % of the
surface functionalized silver flakes and nanowires (3:1)

978-1-4244-4658-2/09 $25.00 © 2009 IEEE

The curing profiles of epoxy catalyzed with 2E4MZ and Co(acac)$_2$ are displayed in Fig. 9. The exothermic peak of curing with 2E4MZ and Co(acac)$_2$ are 156 and 200°C, respectively. As previously discussed, the surface modified silver nanowires begin to sinter at 200°C. With 2E4MZ as catalyst, the epoxy cured very fast above 200°C and hindered the contact of nanowires to sinter to large chunks. When 2E4MZ is substituted with Co(acac)$_2$, the curing of epoxy at 200°C becomes much slower and the silver nanowires can sinter before the epoxy cured. As shown in Fig. 8, the resistivity of ICA with Co(acac)$_2$ is close to that without catalyst, lower than that with 2E4MZ.

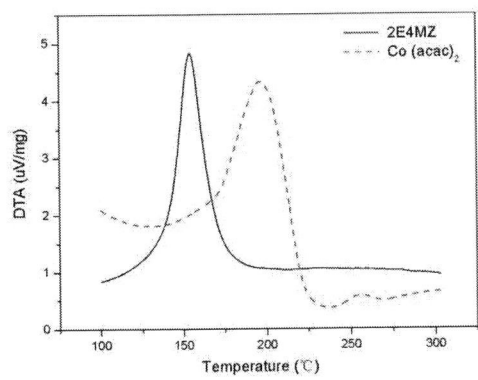

Fig. 9   Curing behavior of epoxy catalyzed with 2E4MZ and Co(acac)$_2$

4. Conclusions

The silver nanowires with a slenderness ratio of 50~60 were successfully synthesized through a polyol process and modified on the surface by aliphatic acids. The modified silver nanowires began to sinter at 200°C and became shorter and thicker gradually, and eventually formed large chunks at higher temperature. The electrically conductive adhesives filled with silver nanowires were formulated and characterized. The sintering of silver nanowires in epoxy can be facilitated by high-temperature catalyst such as Co(acac)$_2$ and cured at temperature above 200°C.

Acknowledgments

This work was supported by Shanghai-Applied Materials Research and Development Fund (No. 07SA09) and Science and Technology Commission of Shanghai (No. 08JC1411600).

References

1. Yi Li, K. Moon, and C.P. Wong, "Electronics without lead," *Science*, Vol. 308, No.5727 (2005), pp. 1419-1420.
2. Yi Li, C. P. Wong, "Recent Advances of Conductive Adhesives as a Lead-free Alternative in Electronic Packaging: Materials, Processing, Reliability and Applications," *Mater. Sci. Eng.*, Vol. R 51 (2006), pp. 1-35.
3. Daoqiang Daniel Lu, C. P. Wong, "Overview of Recent Advances on Isotropic Conductive Adhesives," *HDP'06. Conf*, Shanghai, CN, Jun. 2006, pp. 218-227.
4. G. L. Allen, R. A. Bayles et al, " Small Particle Melting of Pure Metals," *Thin Solid Films*, Vol.144, No. 2 (1986), pp.297-308.
5. James E. Morris, "Nanopackaging: Nanotechnologies and Electronics Packaging," *HDP'06. Conf*, Shanghai, CN, Jun. 2006, pp. 218-227.
6. Hongjin Jiang, C. P. Wong et al, "Surface Functionalized Silver Nanoparticles for Ultrahigh Conductive Polymer Composites," *Chem. Mater*, Vol.18, No. 13 (2006), pp. 2969-2973.
7. Yu Huang, Xiangfeng et al, "Directed Assembly of One-Dimensional Nanostructures into Functional Networks," *Science*, Vol. 291, No. 5504 (2001), pp. 630-633.
8. Rongchao Jin, YunWei Cao et al, "Photoinduced Conversion of Silver Nanospheres to Nanoprisms," *Science*, Vol. 294, No. 5548 (2001), pp. 1901-1903.
9. Matt Law, Donald J. Sirbuly et al, "Nanoribbon Waveguides for Subwavelength Photonics Integration," *Science*, Vol. 305, No. 5688 (2004), pp. 1269-1273.
10. Chang Chen, Li Wang et al, "Effect of Silver Nanowires on Electrical Conductance of System Composed of Silver Particles," *J Mater Sci*, Vol. 42 (2007), pp. 3172-3176.
11. Y Xia, P Yang et al, "One-Dimensional Nanostructures: Synthesis, Characterization, and Application," *Adv. Mater.*, Vol. 15, No. 5 (2003), pp. 353-389.
12. Chang Chen, Li Wang et al, "The Influence of Seeding Conditions and Shielding Gas Atmosphere on the Synthesis of Silver Nanowires through Polyol Process," *Nanotechnology*, Vol.17 (2006), pp. 466-474.
13. Wagendristel A, Wang Y, <u>An Introduction to Physics and Technology of Thin Films</u>, World Scientific Publishing (London, 1994).
14. Yi Li, Kyoung-Sik Moon, and C.P. Wong, "Electrical Property Improvement of Electrically Conductive Adhesives Through In-Situ Replacement by Short-Chain Difunctional Acids," *IEEE Trans-CPMT-A*, Vol. 29, No. 1 (2006), pp. 173-178.

Al/Ni Multilayer Used as a Local Heat Source for Mounting Microelectronic Components

Jun Zhang[1,2], Feng-shun Wu*[1,2], Jian Zou[1,2], Bing An[1,2], Hui Liu[1]

[1]Wuhan National Laboratory for Optoelectronics, Huazhong University of Science and Technology,
Wuhan, P. R. China, 430074

[2]State Key Laboratory of Materials Processing and Die & Mould Technology, Huazhong University of Science and Technology, Wuhan, P.R.China, 430074

*Corresponding author: fengshunwu@mail.hust.edu.cn Tel.: +86-27-87558275

Abstract

This paper describes a novel local heating technique for microelectronic mounting. Aluminum/nickel (Al/Ni) multilayer foils show self-propagating exothermic reactions, driven by a reduction in atomic bond energy which can provide rapid bursts of heat and can act as a local heat source to melt solder layers and join materials. In this work, we demonstrate the validity of the Al/Ni reactive foils as a heat source in microelectronic mounting by joining a tiny sheet coppers to a printed circuit board. The Al/Ni multilayer was made by cooled rolling. Thermal analysis of the exothermic reactions in Al/Ni multilayer films was made to make sure that the bonding process is feasible theoretically. The interface of the bonding and the reaction product of the reactive Al/Ni foils was characterized by using Scanning Electron Microscope (SEM) and Energy Dispersive X-ray Detector (EDX).

1 Introduction

The traditional methods of mounting components onto a Printed Circuit Board (PCB) include adhesives, mechanical fastening, and conventional solder reflow [1-5]. Adhesive joints are seldom used because of its poor electrical and thermal conductivity, low strength, and degrade over time with exposure to air. Mechanical fastening requires complex assembly processes that add to cost and design restrictions. The most commonly used alternative is conventional reflow soldering. Conventional reflow processes require that all components be exposed to a temperature higher than the melting temperature of the solder, which could potentially damage temperature sensitive components. Reflow processes also require the application of flux to the surfaces of the components to be joined, necessitating additional cleaning steps. Often, multiple components must be reflowed onto a board using solder alloys with different melting temperatures, thus creating complicated thermal profiles. To reduce cost, expensive components are usually attached later in the process. This process decreases flexibility in the order that components are joined with special solders that are used for a particular component.

Reactive multilayer foil is a new joining process [6] that enables flux less, lead-free soldering of similar and dissimilar materials at room temperature with no thermal damage to surrounding components. The joining process is based on the use of active multilayer foil as a local heat source. The foils are a new kind of material, which consist thousands of alternating layers comprised of elements with large negative heats of mixing. With a small thermal or electrical stimulus, a controlled, self-propagating reaction can be initiated in these foils at room temperature as illustrated in Fig. 1. By inserting a multilayer foil between two solder layers and two

components, heat generated by the reaction melts the solder and consequently bonds the components. Since the heat generated is localized to the bonding interface, components are not exposed to high temperature and hence thermal damage is avoided. Materials with dissimilar coefficients of thermal expansion can also be joined, due to the localized heating of the components.

This paper focuses on an application where tiny sheet coppers are joined to printed circuit boards using a eutectic Sn-Pb solder alloy. The Al/Ni multilayer was made by cooled rolling. Thermal analysis of the exothermic reactions in Al/Ni multilayer films were made to make sure that the bonding process is feasible theoretically. EDX was employed to investigate film structures of the reacted Al/Ni multilayer films. And the interface of the bonding was characterized by using SEM.

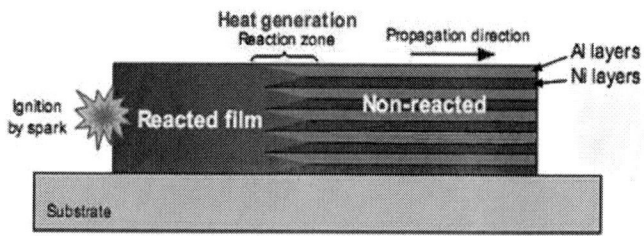

Fig. 1 Schematic diagram of Al/Ni films exothermic reaction

2 Experiment

2.1 Fabricate of the Al/Ni multilayer

A laboratory rolling mill with rollers diameter of 10cm was introduced to fabricate the Al/Ni multilayer. Initial material was thin sheets (5cm square) of pure elements of Ni and Al with initial thickness of 100μm (SCRC company, both Al and Ni with purity of 99wt %). To produce the films with a 1:1 atomic ratio of Al to Ni, the relative Al and Ni layer thicknesses were rolling to 60μm(Al) and 40μm(Ni). Fig. 2 shows the schematic of the cold rolling procedure. After that they were stacked together and then rolled to form a tube. A vise was used to flatten the tube. After that the flattened tube was put into the rolling mill and rolled a few times until to it's half original thickness. Afterwards, double up the multilayer to recover the original thickness. Then the stacked foils were cold rolled without changing the distance between the rollers. The thickness of the foil was reduced to half in one rolling pass. Subsequently, the resulting foils were cut into halves, stacked together, and the rolling procedure mentioned above was repeated for several times until a uniform multilayer foil

978-1-4244-4658-2/09 $25.00 © 2009 IEEE   758

was achieved. The total thickness of the foil was around 250 μm as show in Fig. 3.

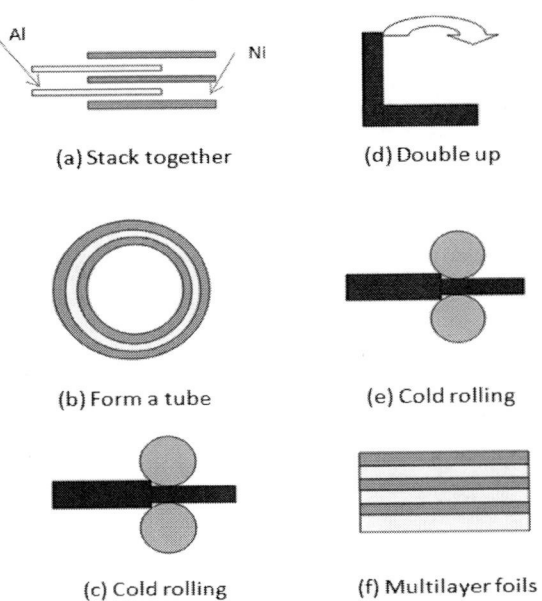

(a) Stack together   (d) Double up

(b) Form a tube   (e) Cold rolling

(c) Cold rolling   (f) Multilayer foils

Fig. 2 Schematic of the cold rolling procedure

Fig. 3 Image of unreacted Al/Ni foil

2.2 Al/Ni reactive film soldering

To demonstrate the validity of exothermic reactions in cold rolled Al/Ni multilayer films as a local heating source in microelectronic mounting, we addressed reactive film soldering between copper sheet and PCB. How the foils and the solder were stack together and the relative position of them was shown in Fig. 4.

By inserting an Al/Ni multilayer foil between two solder layers and two components, heat generated by the reaction melts the solder and consequently bonds the components. The thickness of the solder was 100μm and cut into small round with a diameter of 0.5mm. The copper sheet was cut into small pieces with dimension of 5mm×5mm×0.25mm, and the PCB is bigger which used as substrate in the soldering. A small ceramic block were put upon the Cu sheet to provide the pressure which allows the solder to flow and wet all surfaces,

A YAG optical laser were used to ignite the foil, one end of the reactive foil was extended out of the bonding package.

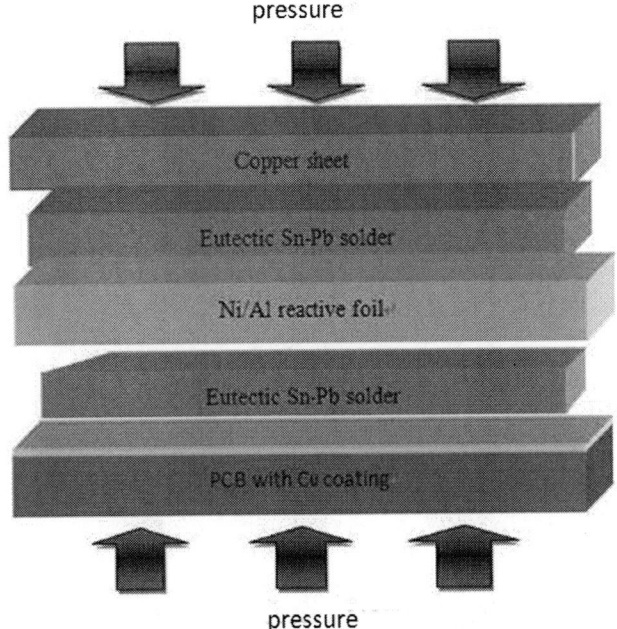

Fig. 4 Schematic illustration of soldering process

That used as the ignition point, and some ZnO powders were put on the surface to make sure that more laser power was absorbed.

3 Result and Discussions

3.1 Thermal analysis of the exothermic reactions in Al/Ni multilayer films

Wang et al [7] measured the heat of reaction for Al/Ni multilayer foils with different bilayer thickness. They found that the heat of reaction decreased as the bilayer thickness decreased. This indicated that the volume percentage of intermixing between layers, which occurred during the deposition, was significant and led to heat losses, and consequently a reduction in the measured heats of reaction. They suggested a formula to calculate the heat of reaction for the Al/Ni multilayer foils with different bilayer thickness.

$$\Delta H = \Delta H_o \left(1 - \frac{2\omega}{\lambda} \right) \tag{1}$$

where ΔH is the enthalpy of formation for the produced compound, ω is the intermixing thickness, and λ is the bilayer thickness. Eq. (1) is based on the assumptions that the intermixing thickness at the Al/Ni interface is unchanged with bilayer thickness and heat loss is proportional to ω/λ. With this, T.Namazu [8] et al conducted the Eq. (2) specifically for the Al/Ni multilayer films with different thickness and bilayer thickness which will make us get the reactive temperature more easily.

$$T = (t_o + 295) + \ln \lambda^{3.54 t_o + 6.8} \tag{2}$$

In the equation above, T is the reactive temperature. t_o is the thickness of the foil and λ is the bilayer thickness. With the former experiment parameter, we can get the reactive temperature is over 1000K, which is much higher than the eutectic temperature(approximate 466K) of Sn-Pb solder , The exothermic reactions of Al/Ni multilayer films provide enough heat quantity to finish soldering.

Reaction velocities for Al/Ni multilayer films was reported to be very rapid, according to Gavens, it's over 5 m/s [9],which means the bonding process is very quick and the temperature will cold down in a short time without any thermal damage to the temperature-sensitive component.

3.2 Ingredient and structural analysis by EDX and SEM measurement

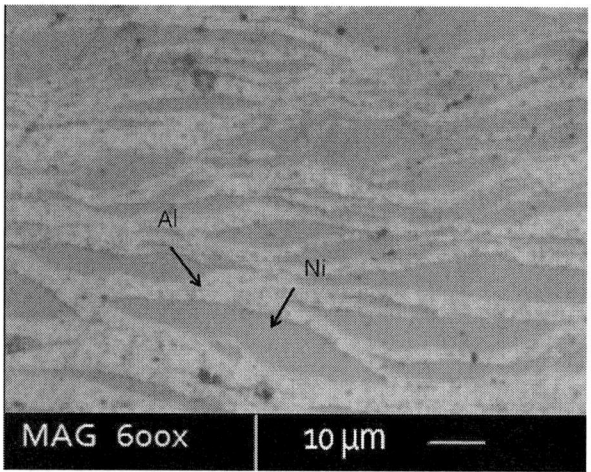

Fig. 5 Optical microscope images of Al /Ni multilayer foil

Fig. 5 shows optical microscope images of a cold rolled Al/Ni multilayer foil. No signal of reaction can be seen in this cold rolled Al/Ni multilayer foil. Necked Ni particles are embedded in the Al matrix and aligned along the rolling direction because of the better ductibility of aluminum. Most of the Ni particles possess a wavy surface. No initial intermixing layers can be identified at the Al/Ni interfaces.

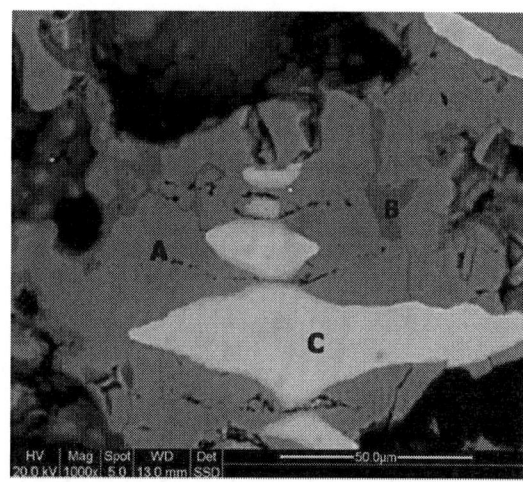

Fig. 6 SEM micrograph of reacted Al/Ni multilayer foil
Three different phases can be seen in the picture.

Fig. 6 shows the SEM images of a quenched Al/Ni multilayer foil. We can see three different phase (section A, B and C). A is the main product of the reaction, it was a whole integer and connect together with different phases into it. And B is a kind of phase that distributed at isolated sites, and there

are very little of them. The bright slab-sided phase C is bigger than B and distributed irregularly inside the phase A just like phase B. In addition, there are many cracks inside the quenched Al/Ni multilayer foil. This is attributed to the fact that when the foils react they contract due to densification; they also contract due to cooling from the high reaction temperatures. Both sources of contraction can be constrained by the surrounding material, thereby leading to those cracking .EDX has been taken to indentify the component of the different phase. The EDX result of section A can be seen in Fig. 7, and the Al/Ni atomic ratio is approximately 1/1. Which means Al/Ni is the main product of the reaction.

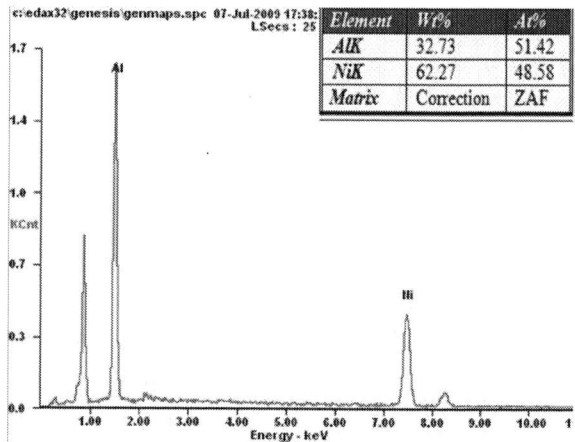

Fig. 7 Typical EDX curve and the results of Al/Ni multilayer foils in section A

In the same way phase B can be seen in Fig. 8, the Al Ni atomic ratio is 1/3 which means phase B is an Al rich phase Al_3Ni, an intermediate product of the reaction. Phase C is the unreacted Ni.

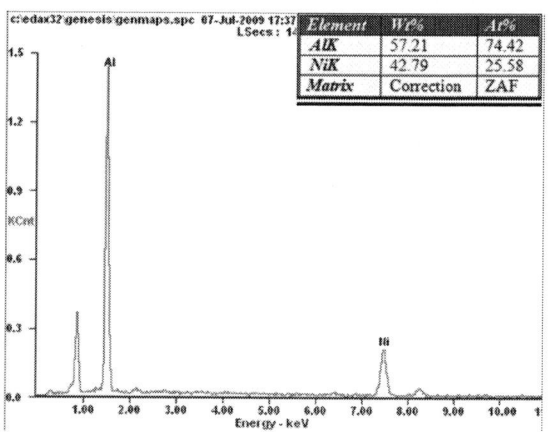

Fig. 8 Typical EDX curve and the results of Al/Ni multilayer foils in section B

In the Coffey's[10] model for the two-stage formation process of Al_3Ni. At the end of the reaction, there are three phases in the foil: the reaction product Al/Ni, the intermediate produce Al_3Ni and the remaining unreacted Ni. That's exactly in line with the result of our experiment.

3.3 SEM of the bounding interface

Fig. 9 is a image under the optical microscope which shows the overall interface situation at one side of the bonding, because the exthernic situation and material between the two solder bilayer are familer,we just discuss one of them. From the image we can see that the interface between the Al/Ni reactive foil and the Sn-Pb solder is uneven compare with the interface between solder and the Cu. And there are many thick dendrites inside the Sn-solder which is formed in the rapid cooling process.

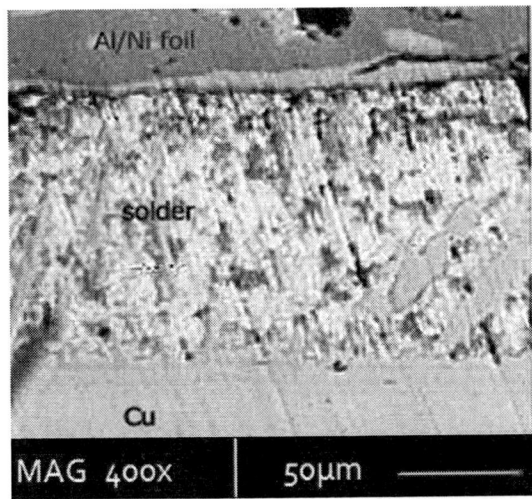

Fig. 9 Optical microscope images of overall situation at one side of the bonding

Fig. 10 shows the interface between Al/Ni foil and the Sn-Pb solder, the surface of the reactive foil is uneven with many cracking which were filled by solder. This is because the pressure in the welding process which enables the molten solder flow into these cracks .It can be observed that there is intermixing between the solder and the reactive foil at the interface.

Fig. 10 SEM micrograph of interface between Al/Ni foil and solder

Fig. 11 shows the interface between Sn-Pb solder and Cu sheet. The two parts were very closely connected and IMC(Inter-Metallic Compound) can been seen from the image which means the joining is successful.

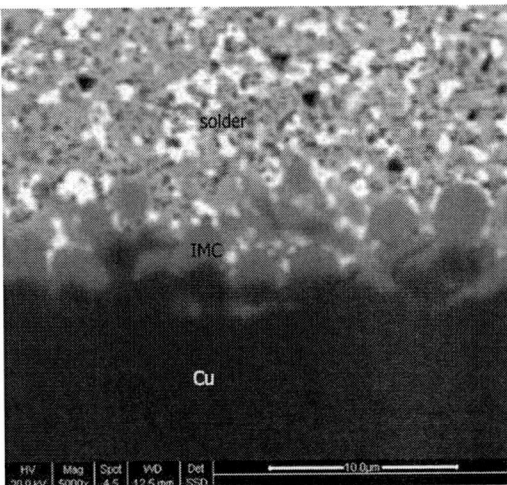

Fig. 11 SEM micrograph of interface between solder and Cu

4 Conclusions

The bonding process use cold rolled Al/Ni reactive multilayer foils were investigated by EDX and SEM. The dominant product after exothermic reaction was ordered B2 Al/Ni compound. The Al/Ni reactive multilayer foils were successfully used as local heat sources to melt Sn-Pb solder layers and bond Cu sheet to a PCB. The validity of exothermic reactions in cold rolled Al/Ni multilayer films as a local heat source in microelectronic mounting has been demonstrated in this paper. Using cold rolled multilayer Al/Ni foil to join components to PCB's add flexibility to the process by allowing joining at any time during fabrication without subjecting the components to high temperatures. At the same time, there are some shortages were found in this new mounting method, a bad wetting property with the traditional solder; poor mechanical behavior because of carves formed in the reacted foils and so on. Those are the main problems to be researched in the coming future. And the Al/Ni multilayer foil is expected to have the potential as a local heat source for microelectronic mounting.

References

1. Lu Daoqiang, Wong C P, "Overview of recent advances on isotropic conductive adhesives," *Conference on high density microsystem design and packaging and component failure analysis.* Shanghai, 2006, pp. 218-227.

2. Ramkumar, SManian, Srihari, et al., "A novel anisotropic conductive adhesive for lead free surface mount electronics packaging," *Journal of Electronic Packaging*, Vol.129, No.2, pp. 149-156.

3. Glazer J. "Metallurgy of low temperature Pb-free solders for electronic assembly," *IntMater Rev.,* Vol.40, No.7 (1995), pp. 65-93.

4. Chen C M, Wang K J, Chen K C, "Isothermal solid-state aging of Pb-5Sn solder bump on Ni/Cu/Ti under bump metallization," *Alloy Comp,* Vol. 432 (2007), pp. 122-128.

5. Takaku Y, Felicia L, Ohnuma I, et al., "Interfacial reaction between Cu substrates and Zn-Al base high-temperature Pb-free solders," *Electron Mater,* Vol. 37, No. 3 (2008), pp. 314-323.

6. Xiaotun Qiu, Jiaping Wang, "Reactive Multilayer Foils for Silicon Wafer Bonding," *Mater. Res. Soc. Symp. Proc.* Vol. 968 (2007), pp. 02-06.

7. J. Wang, E Besnoin, A. Duckham, S. J. Spey, M. E. Reiss, O. M. Knio and T. P. Weihs, "Jointing of stainless-steel specimens with nanostructured Al/Ni foils," *Appl. Phy.,* vol. 95, No. 1 (2004), pp. 248-256.

8. T. Namazu, H. Takemoto, H. Fujita, Y. Najai, S. Inoue, "SELF-PROPAGATING EXPLOSIVE REACTIONS IN NANOSTRUCTURED AL/NI MULTILAYER FILMS AS A LOCALIZED HEAT PROCESS TECHNIQUE FOR MEMS," *IEEE MEMS 2006,* Istanbul, Turkey, 2006, pp. 286-289.

9. A. J. Gavens, D. Van Heerden, A. B. Mann, M. E. Reiss, T. P. Weihs, "Effect of intermixing on self-propagating exothermic reactions in Al/Ni nanolaminate foils," *Appl. Phys.,* Vol. 87, No. 3 (2000), pp. 1255-1263.

10. K. R. Coffey, L. A. Clevenger, K. Barmak, D. A. Rudman, C. V. Thompson, "Experimental evidence for nucleation during thin-film reactions," *Appl. Phys. Lett.,* Vol. 55, No. 9 (1989), pp. 852-854.

Dramatic Morphological Change of Interfacial Prism-type Cu_6Sn_5 in the Sn3.5Ag/Cu Joints Reflowed by Induction Heating

Ling Wang[1], Hongbo Xu[2], Ming Yang[2], Mingyu Li[2, *], Yonggao Fu[1]

[1]China National Electric Apparatus Research

[2]State Key Laboratory of Advanced Welding Production Technology

Harbin Institute of Technology Shenzhen Graduate School

Abstract

Ball grid array (BGA) solder interconnecting is one of the key technologies in electronic packaging and assembly. Legislation of lead-free process has made the application of lead-free solder become wider in electronic products. Compared to the lead-tin solder, the relatively higher melting points of most lead-free solders call for a higher reflow temperature. Thus conventional integral-heating process at an elevated temperature would induce severe warpage of components and substrates and reduce the in-service reliability of solder joint. In this dissertation, a new concept of selective heating was proposed to address the reliability issues of integral heating process in lead-free BGA interconnection. By using this concept a novel BGA interconnection technology, named as induction heating reflow, is developed. The significance of this research lies in the potentials of acceleration of the lead-free process in electronic productions, promotion in the applicability of lead-free BGA packaging, and the enhancement of the reliability of BGA components. Compared with conventional hot-air reflow, the thermal characteristics of liquid-solid interfacial reaction and solid-solid interfacial reaction, the formation of interfacial intermetallic compound (IMC) and their growth kinetics were all studied. The result shows that with the increasing of temperature, Cu_6Sn_5 grains change from scallop-type into prism-type. In the solid-state aging process the evolution of interfacial IMC is merely controlled by the diffusion of Cu atoms from pad, and its thickness grows linearly with the square root of aging time.

Introduction

In the past decade, area array soldering has become the main interconnection technology for the packaging and assembly of IC components with a large number of I/Os [1]. The typical area array soldering processes include forced convective hot-air reflow and infrared reflow. Although these conventional heating methods have the merit of high throughput in mass industrial production, however, sometimes they may cause the warpage of components or printed circuit boards (PCBs) due to excessive heat supply and/or prolonged heating time. This issue may affect the reliability of electronic products to some extent [2,3]. Furthermore, lead-free soldering has been widely implemented in the electronics manufacturing industry in recent years [4,5]. Currently the melting points of most lead-free solder alloys are higher than that of the conventional SnPb eutectic solder [6]. Therefore, it is foreseeable that the aforementioned drawback of conventional universal heating methods will be aggravated by the substantially increased reflow temperature in lead-free soldering.

Localized heating may have promising effects on reflow soldering to solve these problems. Current localized heating methods that have been developed and applied for electronic packaging include laser reflow, solder ball bonding (SBB), and solder jetting technologies. Tian et al. developed a laser reflow method for plastic ball grid array (PBGA) solder balling with rapid localized heating and cooling rates [7,8]. This method may resolve the reliability issue of thermal warpage and improve the microstructure of solder balls. Hayes et al. invented the solder jetting that solder droplets can solidify to form bumps after being jetted onto the pads at a certain speed [9]. This technology may be automated for high throughput mass production and has been adopted by researchers such as Oppert et al. for laser-based solder jetting reflow [10]. However, in both cases of SBB and solder jetting, the equipment cost is tremendously high. In addition, it is rather difficult (if not impossible) to use these methods for the board level soldering of area array packages because the solder joints are hiding between the components and the PCB.

An innovative reflow method using induction heating was proposed in 2003 [11]. Several previous publications reported the studies on interfacial reaction, aging characteristics and shear strength of eutectic SnPb and lead-free solder balls on solder pads with Ni/Au surface finish [12,13]. In 2005, Yang et al. implemented a SnPb solder bonding process using a high frequency electromagnetic field for the hermetic seal of micro-electro-mechanical system (MEMS) devices [14]. Their experimental results demonstrated that reflow soldering by induction heating could provide adequate bonding strength and hermeticity for MEMS devices applications.

The present study is a continuing effort to discuss the metallurgical reaction during the induction heating reflow. In this paper, experimental investigations on the IMC morphology transformation during reflow and solid-state aging will be presented. A theoretical analysis was also performed to estimate the kinetic behavior of interfacial IMC growth.

Experimental Procedure

Lead free solder Sn-3.5 wt.% Ag balls with a diameter of 762.0μm produced by Cookson Electronics Corporation were chosen in this study and they were simply called Sn3.5Ag hereafter. The pads are chosen the structure of electroplated Cu over substrate. The thickness of Cu layer is about 18μm and the diameter is 0.8mm. The component of the substrate is FR4 with a thickness of 0.4mm.

Before the experiment the pads should be dipped in 3.7% HCl solution for 5 seconds and then in 50% C_2H_5OH solution for 30 seconds in order to remove the oxide and organic contamination on the Cu pad. The substrate was cleaned in an ultrasonic bath for 5 minutes before the solder balls were manually placed on the pads. The solder balls and the substrate were heated at the center of the coil and then cooled down to the room temperature. In this study the bumps needed were prepared by induction heating and hot air, respectively. Before induction heating the treated pads should be laid on some Hasaconi FL2002 (T) flux, and the solder should be adhered to the centre of the pad. Then put the pad and solder in the centre of the loop and heat. The heating current is 29A and the holding is 17A. The heating parameters are heating 2s and heating 2s plus holding 60s, respectively. The bumps were cooled at room temperate. The temperature of aging test was 150°C.

In order to observe the morphologies of interfacial IMC, we treat the specimens prepared by induction or hot air in different parameter metallographically and observer in cross-section and top view respectively. For cross-section direction observing the specimens were mounted in room temperature with E-44 epoxy, and then abraded by 180#, 400#, 600#, 800#, 1200# and 2000# sand paper sequentially, finally polished by diamond spray with a diameter of 0.5μm. To enhance the contrast under the SEM, the polished samples were lightly etched with 0.5%HCl-3%HNO$_3$-12%H$_2$O-84.5%C$_2$H$_5$OH solution for 3~5s. For top view direction observing the bump should be abrade close to the interface and etched by 10w% HNO$_3$ solution for several minutes and then dipped in 100% C$_2$H$_5$OH with ultrasonic vibration. After repeat several times, the IMC grains could be observed. Then the morphologies of interfacial IMC were observed under the scanning electron microscope (SEM) from cross-section and top view and analyzing IMCs' component under the energy dispersive spectrometry (EDS).

Discussion

Fig. 1 shows the morphologies of Sn3.5Ag/Cu interfacial IMC by induction heating and hot air reflow in different parameters. As shown in Fig. 1(a) and (b), the morphology of IMC obtained by induction heating for 2s is scallop-type. The diameters of scallops are about 1μm. Their sizes are uniformed. As shown in Fig. 1(c) and (d), it can be seen that with the increase of the holding time the IMC grains change from scallop-type into prism-type, and the prism-type grains become longer and bigger. The diameter is almost 5μm and the length is around 20μm. It could be concluded that with the increase of the holding time the scallops become prism-type gradually. The EDS result confirms that the composition of prism-type IMC grains is Cu$_6$Sn$_5$, as shown in Fig. 2. That means the two types of IMC have the same composition. The prism-type grains are evolved by scallop-type grains.

When liquid solder reacts with solid-state Cu pad the morphology of the IMC grains and the grain growth is mainly determined by the following factors: the diffusion of the Cu atomic, solder/pad interfacial reaction, the physical characteristics and crystallography of the IMC and so on. The ripening flux J^R and the interfacial reaction flux J^I coulde be expressed as below, respectively.

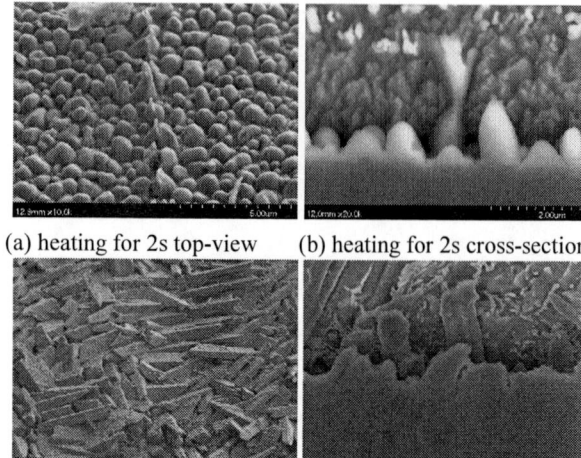

(a) heating for 2s top-view   (b) heating for 2s cross-section

(c) heating for 60s top-view   (d) heating for 60s cross-section

Fig. 1 Morphology of Sn3.5Ag/Cu interfacial IMC by induction heating

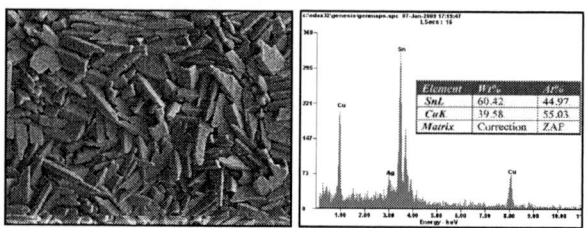

Fig. 2 EDS of Sn3.5Ag/Cu interfacial IMC of different morphology

$$J^R = \frac{2DM\gamma C_0}{LRT\rho}\frac{1}{r^2} \tag{1}$$

$$J^I = \frac{\rho N_A A v(t)}{2\pi M N_P(t)}\frac{1}{r^2} \tag{2}$$

where M is the molar volume of Cu$_6$Sn$_5$, γ is the interfacial energy per unit area between Cu$_6$Sn$_5$ and molten Sn, R is a gas constant, T is the absolute temperature, r is the radii of IMC semisphere, ρ is the density of Cu, $v(t)$ is the consumption rate of Cu foil, $N_p(t)$ is the total number of Cu$_6$Sn$_5$ grains at the interface.

For scallop-type grains, with the increase of the scallops' size, the channel of Cu atomic direct to molten solder will decrease. So the growth of the scallops is mainly caused by the ripening flux, i.e. J^R. This growth mechnism is quite similar tos the Ostwald ripening.

As for prism-type grains, since r is infinite, from Equation (1), $J^R = 0$. So the growth of the prism-type Cu$_6$Sn$_5$ grains was only affected by the interfacial reaction flux, i.e. J^I. That is to say when the morphology of the IMC grains becomes prism-type, the growth of the grains is mainly along the length direction not the wide direction. So with the increase of the reaction time the prism –type grains become longer and longer. Besides, since the main driving force behind the growth of prism-type grains is J^I, the consumption of Cu pad is comparative more than that of scallop-type grains.

978-1-4244-4658-2/09 $25.00 © 2009 IEEE

Fig. 3 shows the top-view and cross-section observation of interfacial IMC transformation of solder bumps reflowed by induction heating for 2s. It can be seen that with the increase of aging time, the interfacial IMCs grows and changes into layer-type. The IMC layer becomes thicker and flatter with time. From these pictures, we can see in solid-state aging it grows a thicker layer of Cu_3Sn. The formation of Cu_3Sn is accompanied with a large number of Kirkendall void in the layer and especially in the interface between Cu_3Sn and Cu pad, which is obviously showed in the SEM pictures of aging for 9 and 16 days.

Fig. 4 shows the top-view and cross-section observation of interfacial IMC transformation of solder bumps reflowed by induction heating for 30s. From these figures we can see that at the beginning the morphology of the two specimens is different. The morphology of the IMC grains showed in Fig. 4(a) and (b) is scallop-type and showed in Fig. 4(a) and (b) is prism-type. While with the increase of the aging time the two different morphologies become more and more similar, especially after aging for 4 days. So we can conclude that in the solid-state aging test, the growth mechanism of the IMC with different morphologies is similar.

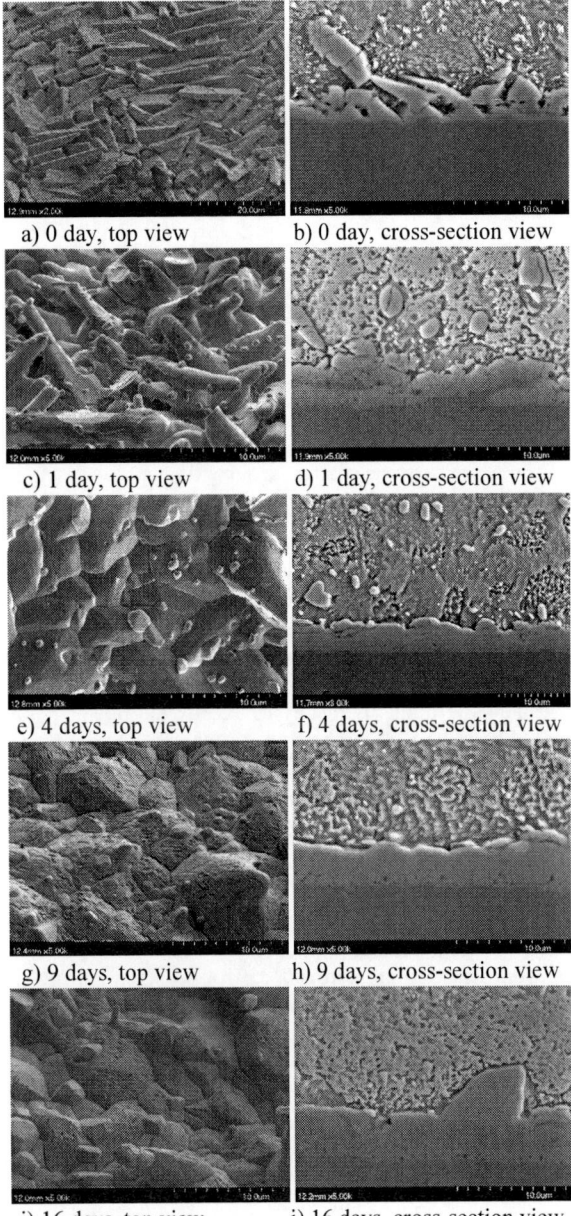

(a) 0 day, top view   (b) 0 day, cross-section

(c) 1 day, top view   (d) 1 day, cross-section

(e) 4 days, top view   (f) 4 days, cross-section

(g) 9 days, top view   (h) 9 days, cross-section

(i) 16 days, top view   (j) 16 days, cross-section

Fig. 3 Top-view and cross-section of interfacial IMC of solder bumps reflowed by induction heating for 2s in a high parameter at different aging time

a) 0 day, top view   b) 0 day, cross-section view

c) 1 day, top view   d) 1 day, cross-section view

e) 4 days, top view   f) 4 days, cross-section view

g) 9 days, top view   h) 9 days, cross-section view

i) 16 days, top view   j) 16 days, cross-section view

Fig. 4 Top-view and cross-section of interfacial IMC of solder bumps reflowed by induction heating for 2s in a high parameter plus 30s in a low parameter at different aging time

The general expression of the relation between diffusion coefficient and thickness of the IMC layer could be expressed by

$$X(t,T) = X_0 + k_0 \cdot e^{-\frac{Q}{RT}} \cdot t^n \tag{3}$$

where X_0 is the thickness of the IMC layer without aging.

The height of the bump in this study is about 600μm which is much bigger than the IMC layer's thickness, so we consider approximately that is semi-infinite Cu-Sn diffusion couples problem, which n is 0.5. That is to say the thickness of the IMC layer is direct proportion to the square root of aging time. Fig. 5 is the linear relationship between the thickness of interfacial IMC layer and the solid-state aging time. The dots are obtained by experiment and the lines are the fitting to these dots. As Fig. 5 shown the experiment data are consistent to the academic calculate.

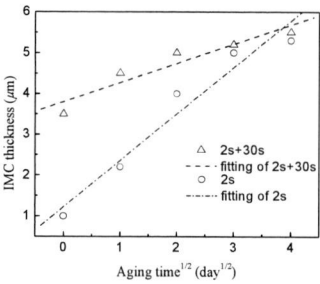

Fig. 5 Linear relationship of the thickness of interfacial IMC and the solid-state aging time

Conclusions

In this paper, the morphologies of interfacial IMC Cu_6Sn_5 obtained by induction heating in different parameters from cross-section and top view direction were observed. With the increase of reaction time, the morphology of the IMC grains becomes prism-type from scallop-type gradually. The growth of scallop-type grains is mainly caused by the ripening flux while the growth of prism-type grains is caused by the interfacial reaction flux. In solid-state aging the IMC layer becomes thicker and flatter with the aging time, and the average thickness of the IMC layer is direct proportion to the square root of aging time.

Acknowledgments

The authors would like to express their gratitude to the National Natural Science Foundation of China for supporting this work under grant No. 50405010 and No. 50875063. Additionally, this work was also supported by the Guangdong Province Lead Free Roadmap Project under grant No. 2007B080402003 of China and the NURI project from Republic of Korea.

References

1. Wojciechowski, Chan, D., M., and Martone, F., "Lead-free plastic area array BGAs and polymer stud grid arraysTM package reliability," *Microelectronics Reliability*, Vol. 41, No. 11 (2001), pp. 1829-1839.

2. Sawada, Y., Harada, K., and Fujioka, H., "Study of package warp behavior for high-performance flip-chip BGA," *Microelectronics Reliability*, Vol. 43, No. 3 (2003), pp. 465-471.

3. Ni, C. Y., Liu, D. S., and Chen, C. Y., "Procedure for design optimization of a T-cap flip chip package," *Microelectronics Reliability*, Vol. 42, No. 12 (2002), pp. 1903-1911.

4. Abtew, M. and Selvaduray, G., "Lead-free solders in microelectronics," *Mater. Sci. and Eng. R*, Vol. 27, No. 5-6 (2000)), pp. 95-141.

5. Suganuma, K., "Advances in lead-free electronics soldering," *Curr. Opin. Solid St. M.*, Vol. 5, No. 1 (2001), pp. 55-64.

6. Wu, C. M. L., Yu, D. Q., Wang, L., and Law, C. M. T., "Properties of lead-free solder alloys with rare earth element additions," *Mater. Sci. and Eng. R*, Vol. 44, No. 1 (2004), pp. 1-44.

7. Tian, Y. H., Wang, C. Q., Ge, X. S., Liu, P., and Liu, D.M., "Intermetallic compounds formation at interface between PBGA solder ball and Au/Ni/Cu/BT PCB board after laser reflow processes," *Mater. Sci. and Eng. B*, Vol. 95, No. 3 (2002), pp. 254-262.

8. Tian, Y. H., Wang, C.Q., and Liu, D.M., "Thermomechanical bechavior of PBGA package during laser and hot air reflow soldering," *Modelling Simul. Mater. Sci. Eng.*, Vol. 12, No. 2 (2004), pp. 235-243.

9. Hayes, D. J., Wallace, D. B., Boldman, M. T., and Marusak, R. E., "Picoliter solder droplet dispensing," *Int. J. Microcircuits Electron. Packag.*, Vol. 16, No. 3 (1993), pp. 173-180.

10. Oppert, T., Titerle, L., Zakel, E., Azdasht, G., and Teutsch, T., "Placement and reflow of solder balls for FC, BGA, wafer-level-CSP, optoelectronic components and MEMS by using a new solder jetting method," *Proc. of International Symposium on Microelectronics*, Denver, USA, 2002, pp. 145-150.

11. Li, M. Y., An, R., and Wang, C. Q., "Self excite reflowing and interfacial reaction of SnPb eutectic solder on BGA pad under alternate electromagnetic radiation," *Acta Metallurgica Sinica*, Vol. 40, No. 10 (2004), pp. 1093-1098.

12. An, R., Li, M. Y., and Wang, C. Q., "A Novel reflow method for electronic area array packaging," *Proc. of 1st International EcoDesign Electronics Symposium*, Shanghai, China, 2004, pp. 117-121.

13. Li, M., Xu, H., KIM, J., and KIM, H., "Failure Modes of Lead Free Solder Bumps Formed by Induction Spontaneous Heating Reflow," *J. Mater. Sci. Technol.*, Vol. 23, No. 1 (2007), pp. 61-67.

14. Yang, H., Wu, M., and Fang, W., "Localized Induction Heating Solder Bonding for Wafer Level MEMS Packaging," *J. Micromech. Microeng.*, Vol. 15, No. 2 (2005), pp. 394-399.

Wetting Behavior of Electrolyte in Fine Pitch Cu/Sn Bumping Process by Electroplating

Jin Jiang[1], Jinglin Bi[1], Zhuo Chen[1], Ming Li[1], Dali Mao[1], Tadatomo Suga[2]

[1]Lab of Microelectronic Materials & Technology, State Key Laboratory of Metal Matrix Composites
School of Materials Science and Engineering, Shanghai Jiao Tong University, Shanghai 200240, China
[2]Department of Precision Engineering, School of Engineering, the University of Tokyo
Email: mingli90@sjtu.edu.cn, Tel: +86-21-34202542

Abstract

This work reports on the wetting behavior of electrolyte in fine pitch Cu/Sn bumping process by electroplating. Three methods containing adding complex wetting agent to electrolyte, plasma treatment to photo-resist and ultrasonic vibration were taken to improve the wettability between electrolyte and related materials. Contact angles of electrolyte containing different amount of complex wetting agent with sputtered copper and photo-resist, before and after plasma treatment, were measured respectively. Certain amount of wetting agent can decrease the contact angles of photo-resist with electrolyte by nearly 20°, and wettability of photo-resist was further enhanced by plasma treatment that makes the contact angles decrease by about 40°. The contact angle of sputtered copper and electrolyte can be decreased by nearly 10°. SEM observations of cross sections and top view of bumps revealed that complex wetting agent help electrolyte get to the bottom of patterned figure quickly. In tin bumping process, this is the last step of wetting behavior for the bubbles in patterned figure would be discharged in electroplating process. While in copper bumping process, ultrasonic vibration was used to help remove the bubbles for the bubbles cannot be small enough by surface tension to overcome the adhesion effect of sidewall. Copper and tin bumps with a diameter of 60μm, height of 60μm and pitch of 180μm were fabricated by rational use of methods above.

Instruction

Concerning the ability to define sizes and pitch, electroplating is one of the best technologies to fabricate bumps [1]. With bump sizes and pitch decreased, there's nearly no other method alternative to achieve them economically. In fine pitch bumping process, copper and tin were paid the most attention on. Copper has excellent electrical and mechanical properties, while tin is the main component of lead-free solders. These pure metal bumps have great potential of application in a variety of electronic interconnecting structure [2-6].

The key issue of fabricating bumps by electroplating is to make electrolyte soak patterned figure. There are three main reasons for this:

(1) Photo-resists used to make patterned figure show poor wettability with electrolyte.
(2) Macromolecular wetting agent was usually used to improve the wettability of electrolyte, but they cannot get to the patterned figure deeply because of the huge molecule. So the excellent wetting characteristic of electrolyte will be disappeared gradually in the wetting process.

(3) Bubbles in patterned figures cannot be removed easily because of the adhesion effect of sidewall.

To overcome these, we took three methods as follows:

(1) Plasma treatment to photo-resist.
(2) Adding complex wetting agent containing small-molecule substance to electrolyte.
(3) Use ultrasonic vibration as pretreatment.

In this paper, effects of the three methods on wetting behavior of electrolyte were investigated by contact angles measurement and SEM observations. It was showed that all the three methods were beneficial to the wetting process. Whether ultrasonic vibration was used depends on the capability of bubbles to overcome the adhesion effect of sidewall. Through rational use of the three methods, Copper and tin bumps with a diameter of 60μm, height of 60μm and pitch of 180μm were fabricated. The wetting behavior of electrolyte in bumping process by electroplating was inferred according to the results above. It is expected to help smaller sized bumping process.

Experiment

The substrates used in contact angles test were copper layer sputtered on glass with a thickness of 100nm and positive photo-resist coated on silicon with a thickness of 60μm. Solutions with a difference of the complex wetting agent contents were prepared as shown in Table.1 and Table.2. The additive agents of copper and tin solution were provided by Sinyang Electronics Chemicals Co. Ltd. Contact angle measurements were performed with Contact Angle System OCA using sessile drop method at room temperature.

Table.1 Copper plating solution compositions

Chemical	Parameter	
	Solution A	Solution B
Cu^{2+}	40g/L	
H_2SO_4	60g/L	
Cl⁻	50ppm	
310A (accelerator)	4ml/L	
310S (suppressor)	10ml/L	
310L (leveler)	5ml/L	
Complex wetting agent	No	Yes

Table.2 Tin plating solution compositions

Chemical	Parameter	
	Solution C	Solution D
Methanesulfonic acid	170g/L	
Tin methanesulfonate	45g/L	
Additive agent	40ml/L	
Complex wetting agent	little	Suitable

Dies with patterned figure made by positive photo-resist was prepared as shown in Fig.1. The diameter of eyelets is 60μm. The depth is 60μm, and the pitch is 180μm.

The copper bumps were fabricated at room temperature with a current density of $3.0A/dm^2$. The tin bumps were fabricated at room temperature with a current density of $5.0A/dm^2$.

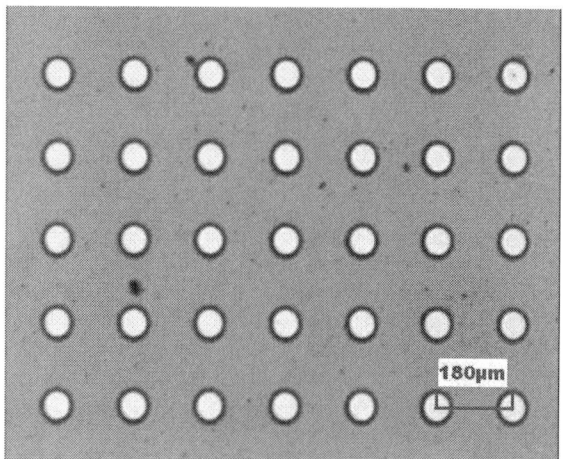

Fig. 1 Patterned figure made by positive photo-resist

The observations of cross sections and top view of bumps were performed with Scanning Electron Microscope (FEI SIRION 200).

Results and Discussion

During the wetting process of electrolyte into patterned figure, electrolyte will contact with photo-resist and sputtered copper layer as UBM at the bottom. Surface modification of photo-resist and complex wetting agent adding to electrolyte was used to affect the wetting behavior in the process. As shown in Fig. 2, plasma treatment to positive photo-resist decreased the contact angle from 89°to 48.9°. This remarkable change was because of the polar groups leaded by plasma treatment. The polar groups were combined with photo-resist surface, thus improving the surface tension and making contact angle smaller [7].

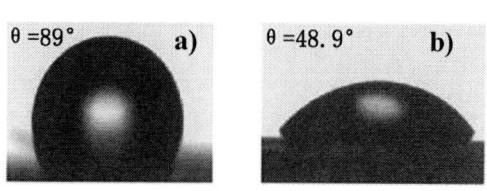

Fig. 2 The effect of plasma treatment to positive photo-resist on contact angles: contact angle between a) solution B and photo-resist before plasma treatment, b) solution B and photo-resist after plasma treatment.

Complex wetting agent also decreased the contact angle between electrolyte and photo-resist or sputtered copper layer. As shown in Fig. 3, the contact angle between electrolyte and sputtered copper was decreased from 78.9°to 65.6°, and the contact angle between electrolyte and photo-resist was decreased from 67.4°to 48.9°.

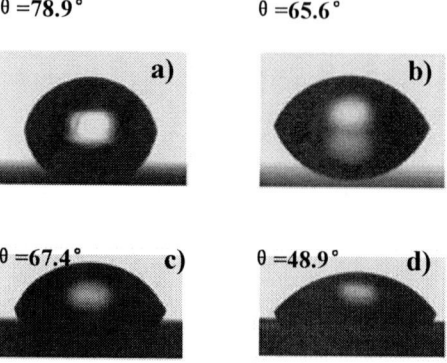

Fig.3 The effect of complex agent on contact angles: contact angle between a) solution A and sputtered copper layer, b) solution B and sputtered copper layer, c) solution A and photo-resist after plasma treatment, d) solution B and photo-resist after plasma treatment.

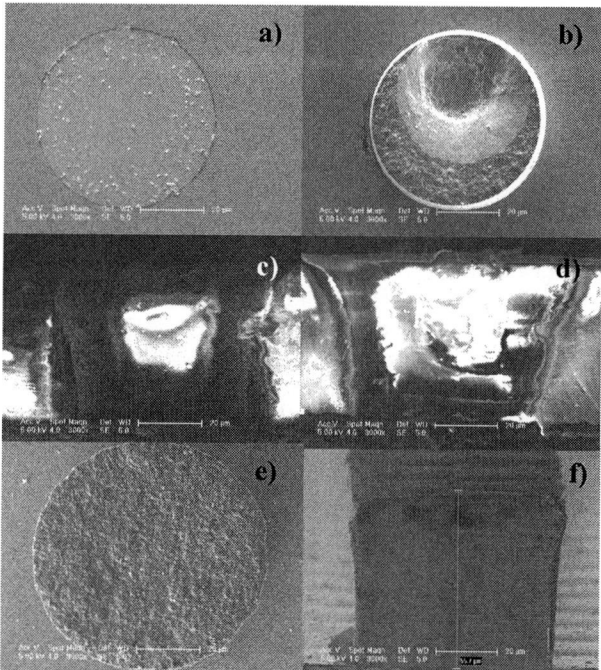

Fig. 4 The effect of complex wetting agent and ultrasonic vibration on copper bumping process: SEM images of top view and cross section of bumps a), c) made in copper solution A; b), d) made in copper solution B; e), f) with ultrasonic vibration in solution B for 1 minute as pretreatment.

In the complex wetting agent, macromolecule substance has great effect on contact angle, while small-molecular substance was used to keep the contact angle small in the whole wetting route by its high mobility. To confirm this, two bumps samples were made in the condition of soaking in copper solutions A or B for 5 minutes as pretreatment and electroplating for 90 minutes. As shown in Fig. 4(a) ~ (d), electrolyte cannot reach to the bottom of patterned figure without complex wetting agent. In copper bumping process, perfect bump cannot be fabricated only by adding wetting agent. Ultrasonic vibration was used to further improve the wetting effect then. After soaking in solution B for 2 minutes and vibration in it for 1 minute as pretreatment, perfect bumps were made at last. Fig. 4(e), (f) indicates that ultrasonic vibration help remove the bubbles at the bottom.

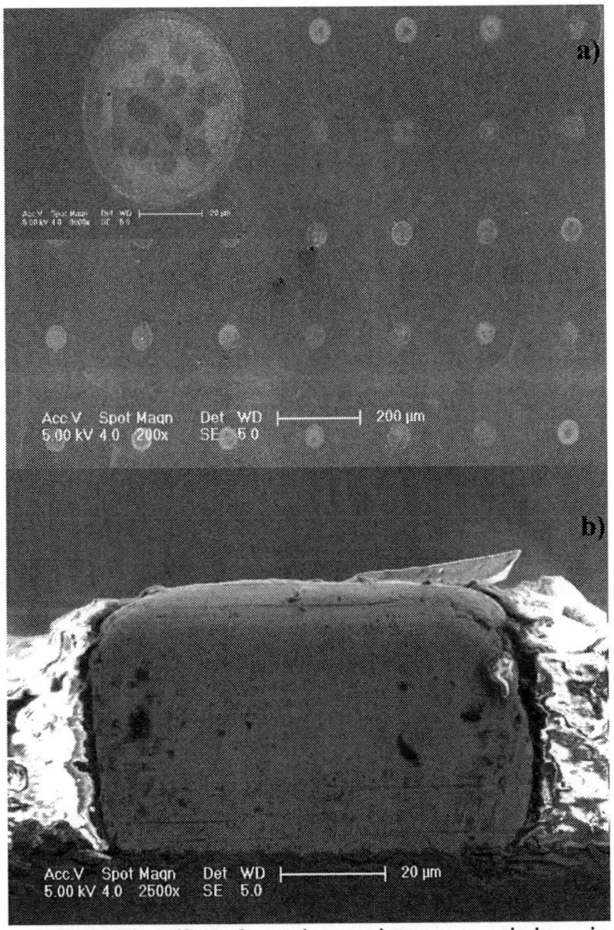

tension so that they can overcome the adhesion effect of sidewall at last. In this bumping process, dies was soaking in solution D for 2 minutes as pretreatment, and bumps was plating for 30 minutes.

Through rational use of the three methods referred to, Copper and tin bumps with a diameter of 60μm, height of about 60μm and pitch of 180μm were fabricated, as shown in Fig. 6. Patterned figures of dies used in the bumping process were all treated by plasma. The copper bumps were fabricated in the condition of plating in solution B for 90 minutes with the pretreatment of soaking in solution for 2 minutes and then ultrasonic vibration in solution for 1 minute. The tin bumps were fabricated in the condition of plating in solution D for 40 minutes with the pretreatment of soaking in solution for 2 minutes.

Fig. 6 SEM images of copper and tin bumps: a) copper bumps; b) tin bumps.

Fig. 5 The effect of complex wetting agent on tin bumping process: a) SEM images of 60-degree tilt view of bumps made in tin solution C; b) SEM images of cross section of bumps made in tin solution D.

However, ultrasonic vibration was not needed in tin bumping process. Fig. 5 illustrates the complex wetting agent was also effective in tin electrolyte and ideal bumps could be fabricated without the help of ultrasonic vibration. This is because bubbles will shrink to small enough under the surface

Conclusions

The wetting behavior of electrolyte in bumping process by electroplating was inferred according to the results above, as shown in Fig. 7.

The effect of plasma treatment to positive photo-resist is to decrease θ_1 (contact angle between photo-resist and electrolyte). Macromolecular substance in the complex wetting agent has the same effect. The complex wetting agent has the ability to keep θ_1 small in the whole wetting routine under the effect of high mobility of small-molecular substance. And it can decrease θ_2 (contact angle between sputtered copper layer and electrolyte), which makes the moistening of sputtered copper layer at the bottom easier.

After the sputtered copper at the bottom is moistened, only liquid-gas interface remains. The bubble will shrink under the surface tension. If the surface tension is great enough, the bubble will become small enough. So it can overcome the adhesion effect of sidewall, and leave the patterned figure at last. However, ultrasonic vibration should be used if the bubble cannot overcome the adhesion effect. The ultrasonic vibration would break the bubble and provide energy for bubble's leaving.

In summary, all the three methods mentioned were beneficial to the wetting process. Whether ultrasonic vibration was used depends on the capability of bubbles to overcome the adhesion effect of sidewall.

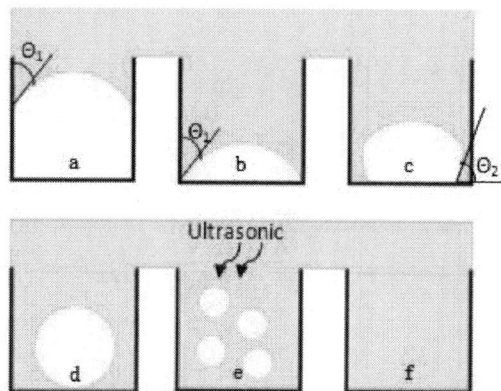

Fig. 7 The schematic diagram of the wetting process of electrolyte into patterned figure.

Acknowledgments

This work is sponsored by International Science and Technology Cooperation of China (No. 2008DFA51680), National Natural Science foundation of China (60876071), Shanghai nano technology promotion center (No. 0852nm06300). We thank the Instrumental Analysis Center of Shanghai Jiao Tong University, for the use of the SEM equipment.

References

1. Enric Cabruja, Marc Bigas, Miguel Ullan, Giulio Pellegrini, Manuel Lozano, "Special bump bonding technique for silicon pixel detectors," *Nuclear Instruments and Methods in Physics Research Section A: Accelerators, Spectrometers, Detectors and Associated Equipment*, Vol. 576, No. 1 (2007), pp. 150-153.

2. Daniel Lu, C.P. Wong, Materials for Advanced Packaging, Springer US (New York, 2009), pp. 547-600.

3. Lannon J., Gregory C., Lueck M., Huffman A., Temple D., "High density Cu-Cu interconnect bonding for 3-D integration," *Proc 59th Electronic Components and Technology Conference*, San Diego, CA, May. 2009, pp. 355-359.

4. Harr Kyoung-Moo, Kim Young-Min, Lim Dae Hwan, Kim Young-Ho, Kim Jin-Gu, Yi Sung, "A new COF bonding technique using Sn bumps and a non-conductive adhesive (NCA) for image sensor packaging," *Proc 59th Electronic Components and Technology Conference*, San Diego, CA, May. 2009, pp. 1475-1478.

5. Ju-Heon Yang, Young-Ho Kim, Jae-Seung Moon, Won-Jong Lee, "Chip-To-Chip Interconnection by Mechanical Caulking Using Reflowed Sn Bumps," *Proc 8th International Conference on Electronic Packaging Technology*, Shanghai, China, Aug. 2007, pp. 1-3.

6. Byeung-Gee Kim, Sang-Mok Lee, Young-Ho Kim, "The Reliability of 30 μm pitch COG joints fabricated using Sn/Cu bumps and non-conductive adhesive," *International Conference on Electronic Materials and Packaging*, Washington, DC, Nov. 2007, pp. 1-3.

7. Zengfu Hu, Surface and Interface of Material, East China University of Science and Technology (Shanghai, 2007), p.125.

Effects of the Matrix Shrinkage and Filler Hardness on the Thermal Conductivity of TCA

Cong Yue[1], Yan Zhang[1*], Zhili Hu[1], Johan Liu[1, 2], Zhaonian Cheng[1]

[1]Key Laboratory of Advanced Display and System Applications, Ministry of Education, SMIT Center
School of Mechatronics Engineering and Automation, Shanghai University, P. O. B. 282, Shanghai 200072, China
[2]SMIT Center & Bionano Systems Laboratory, Department of Microtechnology and Nanoscience
Chalmers University of Technology, SE-412 96 Gothenburg, Sweden
*Email: yzhang@shu.edu.cn

Abstract

Thermal conductive adhesives (TCA) have been widely used as the thermal interface material (TIM). The TCA are usually epoxy or silicone based mixtures containing fillers. The epoxy in TCA also applies a pressure on the filler particles as it shrinks during the curing process. This pressure leads to an increase in the contact area between the filler particles and thus improves the thermal conductivity. In our attempt to discover the effect of filler modulus on the contact areas between two particles we constructed a cell model assay. In order to study the different epoxy characteristics, 28 cases were simulated with different mixtures of epoxy and filler. The thermal conductivity of each case was calculated by FEM. When the shrinkage was higher than 1%, TCA with a filler modulus of 83Gpa (silver) has the largest contact area, which suggests silver has the best TCA performance among the studied materials. Furthermore the viscoelasticity of the epoxy was simulated to evaluate the fully relax time of epoxy and the effect on the thermal conductivity of TCA.

Introduction

With the increase in microprocessor power, the role of thermal management for electronic systems has become more crucial to overall system performance. Applications of power electronics are also significantly limited by the inability to transfer heat across interfaces into heat-sinks [1]. Thermal Interface Material (TIM) used for heat transfer includes thermal conductive adhesives (TCA), grease, phase change materials (PCMs) and thermal pads [2]. Here the TCAs, usually epoxy or silicone based formulations containing conductive fillers, offer a superior mechanical bond that can reduce the size and weight of a system, thus becoming one of the best potential options for thermal interface material. Much experimental work has been done [3-5] in order to improve its performance.

Unfortunately, there has been no prominent increase in the conductivity of TCA in recent years. Samples of Isotropic Conductive Adhesives (ICA) show thermal conductivity properties at k ≈ 7W/mK (vertical direction) [6], which is at present one of the highest k-values obtained for TCA.

Problem Description

There are various kinds of TCAs with respect to filler properties, including single model, bi-model and tri-model. In order to simplify the analysis, single-model TCA was studied first, composed of silver flake filled epoxy. The question is posed whether electrons or phonons are the main contributors to thermal conductivity. In accordance with the Wiedemann-Franz law, which states that the ratio between the electronic contribution to the thermal conductivity (k), and the electrical conductivity (σ) of a metal is proportional to the temperature T [7]:

$$\sigma = \frac{1}{\rho} = \frac{K}{LT} \qquad (1)$$

The Lorenz number of silver is 2.31 at the ambient temperature 273K. It has previously been shown that the in-plane electrical resistivity ρ of a TCA sample is about 4.8×10^{-5} Ω/cm [8]. Using these values, thermal conductivity by electron conductance is $K = 13.14$ W/mK, which approaches its in-plane K of TCA (i.e. 27W/mK). This implies that electrical conductivity is the main contributor to thermal conductivity. Moreover, based on tests on ICA made on silver flake filled epoxy, Inoue et. al. [6] made the same conclusion. Electron conductance is therefore premised to be the thermal conductor in single-model TCA.

Apparently, electrical conductance relies on the interconnection between the metal fillers' contact areas. Thus it is important to find out why the filler particles are in contact with each other. One possibility is that the pressure between the particles deforms the particles and flattens their tangential contact into a larger and possibly circular area. For anisotropic conductive adhesives the pressure consists mainly of external mechanical forces when the chip is attached onto the board. Likewise, TCA could also obtain conductivity by applying pressure to its filler particles. However, instead of external pressure, the pressure comes mainly from shrinkage of epoxy in TCA, which happens possibly during the curing process. Similar views were shown in previous studies, which indicate that the thermal conductivity of curing depends on the shrinkage of the epoxy and its filler particles' contact areas [9]. The ref. [10] also revealed this by simulating the TCA with silver particles filled low temperature glass transition (Tg) epoxy after curing.

This paper investigates the effects of matrix shrinkage on the thermal conductivity of a single-model TCA. Furthermore, the method is extended to other kinds of TCA fillers, e.g. BN or SiC, to study the effects of filler hardness since all aforementioned TCAs have similar internal structure.

Contact Area Modeling

As mentioned above, contact between particles is vital to TCA's thermal conductivity. To investigate this contact, in the present paper, a 3D model describing two filler particles in the epoxy cell was built in the ABAQUS environment, with the geometric model in the simulation shown in Fig. 1. The model is a quarter of a unit in the TCA at micro scale. Symmetric boundary conditions were adopted. As a dimensionless analysis, the radius of particle (R) was set to 1.

978-1-4244-4658-2/09 $25.00 © 2009 IEEE

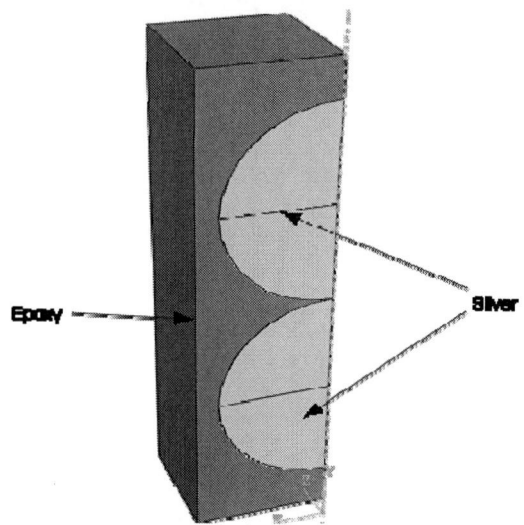

Fig.1 The model built in the ABAQUS environment

In the first step, the different shrinkages of epoxy and different elastic moduli of filler particles were simulated to obtain the circular contact areas between two particles in each case. The shrinkage was controlled by setting different coefficients of thermal expansion (CTE) of the filler and epoxy, and thereafter the model temperature was set to different values in the later steps. The elasticity of epoxy was set to constant 3.4Gpa and the CTE varied in each case.

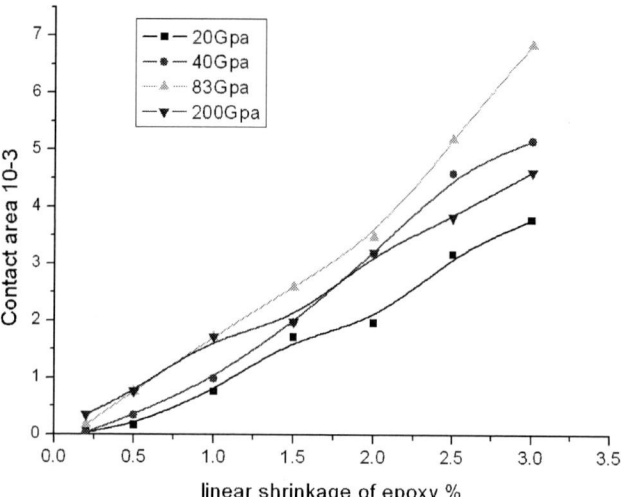

Fig.2 The contact area in each modulus

In order to simplify the simulation in this paper, all results were non-dimensional so as to compare with each other in the same conditions. Fig. 2 shows the sequence of contact areas between the four kinds of modulus depending on the shrinkage percentage. In contrast to the widely held assumption that contact area increases if particles became softer, in this paper, results indicate that higher elastic modulus (< 83Gpa) were more likely to cause larger contact areas. When the shrinkage was smaller than 1%, the larger the modulus were, the larger the contact area appears. With the increase of shrinkage, the contact areas with smaller modulus increased more rapidly than that with the larger modulus.

When the areas for each case had been determined, the ratio of r/R could easily be deduced in order to calculate the thermal conductivity. Here r stands for the radius of the contact area as shown in Fig. 3.

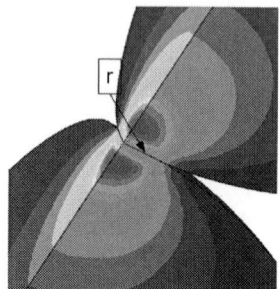

Fig. 3 Sketch of the contact area between two fillers

An analysis was then developed. In Fig. 4, L is the length of the two particles. After shrinkage, L decreased due to the epoxy's compression. The initial value of L was set to 4, the total deform displacement was appointed to be ΔL. The computed filler modulus ranging from 20Gpa to 200Gpa are shown in Table 1, indicating that the modulus has great effects. And the 20Gpa case compared with 200Gpa had significantly higher ΔL but a smaller contact area.

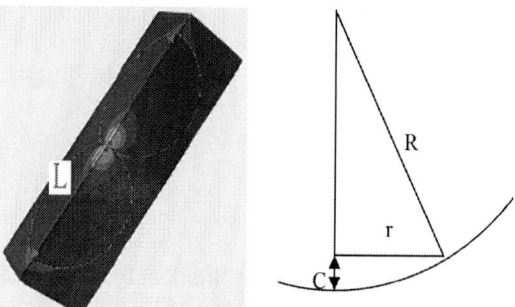

Fig. 4 Geometric parameters

This demonstrates that since softer fillers are more compressible, the compression caused by epoxy shrinkage will possibly be buffered into shrinkage of the filler itself, thus reducing compression pressure as well as contact area. Here the expression is as follows:

$$C = R - \sqrt{R^2 - \frac{4A}{\pi}} \qquad (2)$$

where C is the effective length of deformation that creates the contact area. Harder fillers result in higher C/ΔL values. ΔL consists of 2 parts, the length of deformation C and the deformation of filler bulk. At high shrinkage (3%), the area of 83Gpa has better properties compared with 200Gpa. This demonstrates that while the 200Gpa pass the compression pressure more efficiently, the filler particles are less compressed and thus create smaller contact areas. Harder fillers have higher C/ΔL but yet lower ΔL. As a result, with this modulus the highest C cannot be obtained. Therefore, it is crucial to find the suitable filler Young's modulus in different shrinkage of epoxy. Our study indicates that when the shrinkage is high (above 1%) 83Gpa will possess the highest thermal conductivity.

Table 1 The computed values with various filler modulus and matrix shrinkage

Case	1/4 Area	ΔL	C	C/ΔL (%)
20Gpa 0.2%	0.0411e-3	2.76e-3	2.6159e-5	0.95%
20Gpa 3%	3.777e-3	4.778e-2	2.4086e-3	5.04%
83Gpa 0.2%	0.1735e-3	1.7344e-3	1.105e-4	6.37%
83Gpa 3%	6.8519e-3	2.5573e-2	4.3738e-3	17.1%
200Gpa 0.2%	0.345e-3	1.336e-3	2.1977e-4	16.45%
200Gpa 3%	4.6136e-3	1.582e-2	2.943e-3	18.6%

Relationship between the thermal conductivity and contact area

In order to test the thermal conductivity of each case, an ABAQUS efficient model was built. The thermal conductivity of each filler was set to 430W/mK (i.e. that of Ag), in order to focus the comparison on the effect of contact area. Simulation results are shown in Fig. 5, which indicates that the thermal conductivity depends on the epoxy shrinkage, which is linked to the contact area.

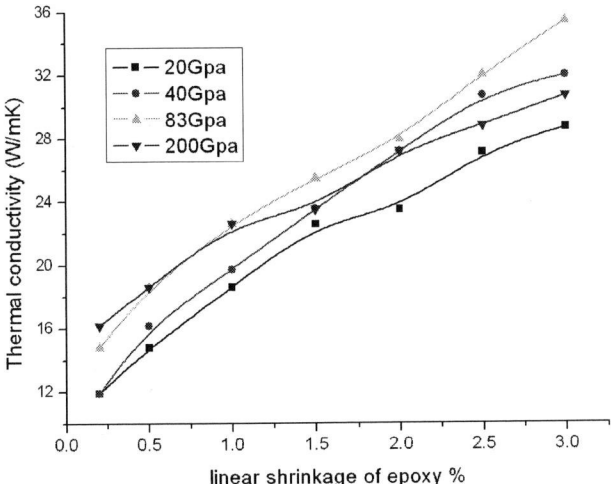

Fig. 5 Thermal conductivity in each modulus

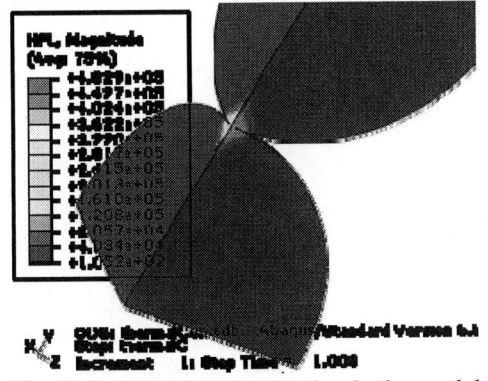

Fig. 6 The heat flux distribution in the model

Simulation also showed that the heat flux was concentrated on the contact area, as shown in Fig. 6, confirming that the contact area between two particles is the key point for conductivities.

Viscoelastic analysis

The epoxy in TCA is a viscoelastic material [10]. The modulus for viscoelastic materials is not constant but continuously changes with a temperature dependent speed. To accurately describe this process, it is assumed in our simulation that the viscoelastic material is defined by a Prony series expansion of the dimensionless relaxation modulus as:

$$\tau(t) = G_0(\theta)(\gamma - \int_0^t \dot{g}_R(\xi(s))\gamma(t-s)ds) \qquad (3)$$

where N, $\overline{g}_i^P$ and τ_i^G, $i=1,2,\ldots,N$ are material constants. And, the real time modulus is

$$G(t) = G(0)g_R(t) \qquad (4)$$

The constants are listed in Table 2. Accordingly, instantaneous modulus G(0) is 3.4e9Pa/2/(1+v), and the long term $G(\infty)$ is 51.4MPa.

Table 2 Constants for Prony series of the epoxy

$\overline{g}_i^P$	τ_i^G
0.002997	2.12E-10
0.028907	4.57E-09
0.078985	9.85E-08
0.111952	2.12E-06
0.218877	4.57E-05
0.233282	0.0009854
0.208339	0.02123
0.065354	0.45738
0.014018	9.8541
0.00406	212.3

On the other hand, the effect of temperature, θ on the material behavior is introduced through the dependence of the instantaneous stress, τ_0 on temperature and through a reduced time concept. The expression for the linear-elastic shear stress is rewritten as

$$\tau(t) = G_0(\theta)(\gamma - \int_0^t \dot{g}_R(\xi(s))\gamma(t-s)ds) \qquad (5)$$

where the instantaneous shear modulus G_0 is temperature dependent, $\dot{g}_R(\xi) = dg_R / d\xi$ and $\xi(t)$ is the reduced time, defined by

$$\xi(t) = \int_0^t \frac{ds}{A(\theta(s))} \quad (6)$$

where $A(\theta(t))$ is a shift function at time t. This reduced time concept is usually referred to as thermo-rheologically simple (TRS) temperature dependence. Often the shift function is approximated by the Williams-Landell-Ferry (WLF) form:

$$\log(A) = -\frac{C_1(\theta - \theta_0)}{C_2 + (\theta - \theta_0)} \quad (7)$$

where $\theta_0 = 150°C$ is the reference temperature at which the relaxation data are given and $C_1 = 17.66$, $C_2 = 170.58$ are calibration constants obtained at this temperature. If $\theta \leq \theta_0 - C_2$, deformation changes will be elastic, based on the instantaneous module. And the θ_0 here is in fact Tg.

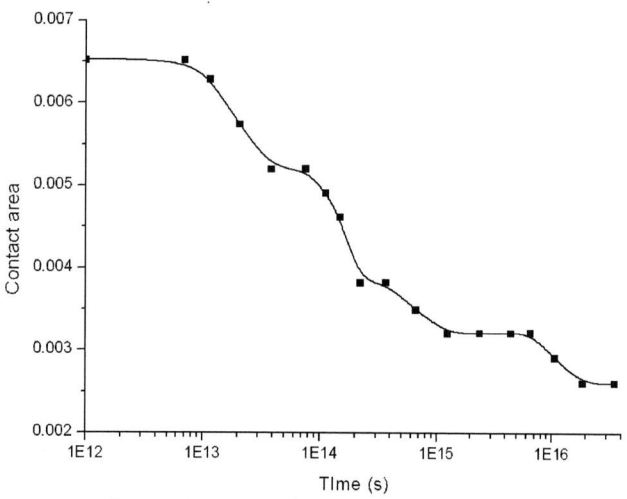

Fig. 7 The contact area with time increase

The simulated results of the contact area, with Young's modulus 83Gpa and 3% epoxy shrinkage, are shown in Fig. 7. As the epoxy's TG was very high (150°C), the calculations indicate that it will take too long time for the epoxy to become fully relaxed. However, when the epoxy is fully relaxed, the contact area and thermal conductivity will be 40% and 60% respectively of its initial values.

Conclusions

In the present work, a cell model was built to study the effect of filler modulus to the contact areas between two particles. Here four kinds of fillers with Young's modulus as 20Gpa, 40Gpa, 83Gpa and 200Gpa were set to compare with each other. Seven different epoxy shrinkage (values ranging from 0.2% to 3%) needed to be studied; therefore 28 cases were simulated in order to study the two factors. From the FEA results, in the low shrinkage cases, harder fillers produced larger contact areas, and as a result higher thermal conductivity. When the shrinkage was higher than 1%, a filler modulus of 83Gpa (silver) was optimal. This can be explained as resulting from the deformation of its components. The thermal conductivity of each case was calculated with FEM. Lastly, we simulated the viscoelasticity of the epoxy to find out its fully relax time and how it effects the thermal conductivity of TCA. The studies show that the present TCA

with high TG needs 2×10^{16}s to be fully relaxed. When the epoxy was fully relaxed, the contact area and thermal conductivity were 40% and 60% respectively of its initial values.

Acknowledgments

Support by NSFC (10702037) and Shanghai Pujiang Program (08PJ14054) is also greatly appreciated. This paper is also supported by Innovation Program of Shanghai Municipal Education Commission (09YZ01), Shanghai University Graduate Innovation Fund, and "SEC E-Institute: Shanghai High Institutions Grid" project.

The authors acknowledge the financial support from the 863 program (No.2008AA04Z301), from the Shanghai Science and Technology Commission with the contract no: 075007004 and the NSFC project (50876057).

This work is also supported by the FP7 IP Nanopack program, from the Swedish National Science Foundation under the project "Nanointerconnect" (621-2007-4660), from the Swedish Foundation for Strategic Research under the ProViking Program (PV08.08).

References

1. Mahajan R., Nair R., Wakharkar V., Swan J., Tang J., Vandentop, "Emerging Directions For Packaging Technolgies", *Intel Technolgy Journal*, Vol. 6, No. 2 (2002), pp.62-75.

2. Johan Liu, Björn Carlberg, Teng Wang, Masahiro Inoue, "Overwiew of Recent Progress of Thermal Interface Materials", *ESTC 2008*, London, UK, Sep. 2008, pp.351-358.

3. Fan, L., et al., "Electrical and thermal conductivities of polymer composites containing nano-sized particles", *Proceeding of the 54th Electronic Components and Technology Conference*, Las Vegas, NV, USA, Jun. 2004, pp. 148-154.

4. Lee, G.W., et al., "Enhanced thermal conductivity of polymer composites filled with hybrid filler", *Composites Part A: Applied Science and Manufacturing*, Vol. 37, No. 5, (2006), pp. 727-734.

5. Lee, T.M., et al., "High thermal efficiency carbon nanotube-resin matrix for thermal interface materials", *Proceedings of 55th Electronic Components & Technology Conference*, Lake Vuena Vista, FL, USA, May 31-June 3, 2005, pp. 55-59.

6. Masahiro Inoue, et al, "Physical Factors Determining Thermal Conductivities of Isotropic Conductive Adhesives", *Journal of Electronic Materials*, Vol. 38, No. 3 (2009), pp. 430-437.

7. Jones, William; March, Norman H. (1985). <u>Theoretical Solid State Physics</u>, Courier Dover Publications, 1985.

8. Cong Yue, et al, "Influences of Filler Geometry and Content on Effective Thermal Conductivity of Thermal Conductive Adhesive", *Proceedings of 59th Electronic Components and Technology Conference*, San Diego, CA, USA, May. 2009, pp. 2055-2059.

9. Bin Su, Jianmin Qu, "A Micromechanics Model for Electrical Conduction in Isotropically Conductive Adhesives during Curing", *Proceedings of 9th Int'l*

Symposium on Advanced Packaging Materials, Singapore, Dec. 2004, pp. 145-151.

10. Tomasz Falat, et.al., "Influence of matrix viscoelastic properties on thermal conductivity of TCA – Numerical approach", *Journal of Microelectronics Reliability*, Vol. 47 (2007), pp.1989–1996.

Synthesis and Characterization of Nano BaTiO₃/Epoxy Composites for Embedded Capacitors

Suibin Luo[1], Rong Sun[1], Jingwei Zhang[2], Shuhui Yu[1], Ruxu Du[1], Zhijun Zhang[2]

[1]Shenzhen Institute of Advanced Technology, Chinese Academy of Sciences, Shenzhen, China, 518067

[2]The Laboratory for Special Functional Materials, Ministry of Education Henan University, Kaifeng, China, 475001

rong.sun@siat.ac.cn

Abstract

The miniaturization trend of integrated circuits (ICs) calls for replacing discrete passive components with embedded passives. Among the passive components, the embedded decoupling capacitors which are used for simultaneous switching noise suppression have drawn great attention. It is because decoupling capacitors should be placed as near as possible to a chip to reduce parasitic inductance. Important requirements for embedded capacitor materials are high dielectric constant, low capacitor tolerance, good processibility and low cost. BaTiO₃-filled epoxy composite is a promising material to meet the above requirements. It utilizes the high dielectric constant of ceramic powders and processibility of polymers. The dielectric behavior of the composite is influenced by the crystal phase type, grain size and dispersability of the BaTiO₃ particles distributing in it. In this study, nano-sized BaTiO₃ powders have been synthesized with a modified hydrothermal reaction method. The powers possess tetragonal crystal phase with a narrow size range around 50nm. The dielectric constant of the epoxy filled with BaTiO₃ is 19.4 at the frequency of 10 kHz when the loading of BaTiO₃ was 50 vol% and the dielectric loss factor tanδ is about 0.02. It is believed that the high dielectric constant and low loss are attributed to the pure tetragonal phase and good dispersing of the nano particles.

1 Introduction

A continuing trend in the electronic industry is the miniaturization of electronic circuits, and the drive toward higher and higher circuit element density. On the conventional printed wiring boards today, a large fraction of the surface area is occupied by surface-mounted capacitors and other passive devices. The industry has recognized that one way to further increase circuit element density is to eliminate surface-mounted passives and embed or integrate the passive structures in the circuit boards themselves. There are some added advantages to place the capacitors much closer to the active components, such as reducing electrical lead length and lead inductance, thereby improving circuit speed and reducing signal noise [1~3].

The general requirements for embedded capacitor dielectrics include high dielectric constant, low dissipation factor, low processing temperature, low leakage current and high breakdown voltage [4~6]. Polymer-ceramic composites have been extensively studied as a candidate dielectric for embedded capacitors, because the combination of polymer and filler brings in the advantages from both sides such as the low-temperature ($<$ 200°C) processability of the organic polymer matrix and the ultra-high dielectric properties of the filler [7~9]. In general, the effective permittivity of the composites is a function of the frequency ν, the temperature

T, the permittivities of the polymer matrix and ceramic particles, ε_p and ε_c, the volume filling factor, $f = V_{particles}/V_{total}$, as well as the microstructure of the particles [10~12].

In this study, ferroelectric ceramic barium titanate, BaTiO₃, was used as the ceramic filler to prepare the epoxy composites. For BaTiO₃, tetragonal phase is desired because the dielectric constant of tetragonal phase is higher than cubic phase due to the spontaneous polarization. For BaTiO₃ nanoparticles, the literatures indicate a possible conflict between the nanoparticle size ($<$100 nm) and the tetragonal phase. The critical size below which the crystal structure changes from tetragonal to cubic phase has been reported to be from 10 nm to 110 nm. Uchino et al. [13] combined various means of crystal size determination and XRD data to place the critical size at 0.12 μm. Xu and Gao [14] reported that hydrothermal reaction method lead to the tetragonal-BaTiO₃ particles with an average particle size of 70 nm. In addition, a large lattice elongation was observed in the ferroelectric phase near a critical size of 10 nm [15]. Therefore, to produce BaTiO₃ nanoparticles of 30 to 110 nm is a critical step to understand the size-dependent properties of BaTiO₃ [16].

In this paper, we investigate the microstructure and dielectric behavior of the epoxy composites containing BaTiO₃ nano particles with tetragonal crystal phase which was synthesized with a modified hydrothermal reaction method.

2 Experiments

2.1 Fabrication of embedded capacitor paste

The BaTiO3 nano pariticles were synthesized with a modified hydrothermal reaction method. The BaTiO3 powders were heated at 350°C for 15 hours in order to remove the impurities containing carbons or water on the surface of BaTiO3 and hydroxyls in crystal lattice. The epoxy resin was also heated before using at 110°C in order to remove the water.

BaTiO₃ powders were mixed with epoxy resin, methyl ethyl ketone (MEK) and phosphate ester. The volume ratio of MEK to Epoxy resin was about 5:1. The phosphate ester act as the dispersant and its weight was about 1.5wt% relative to BaTiO₃ powders. Then the mixture was stirred at 2000 revolution per minute (rpm) for 40 minutes to achieve a uniform paste. The tetraethylenepentamine (TEPA) was chosen as the curing agent. It was dissolved in MEK and then mixed with the BaTiO₃-epoxy paste. The paste was ultrasonicly treated in order to remove the air bubble before preparing the embedded capacitor film.

2.2 Fabrication of embedded capacitor and measurement

The embedded capacitor film was fabricated with bar coating method. The thickness of the film was determined by the viscosity of the paste and the size of coating bar. After

978-1-4244-4658-2/09 $25.00 © 2009 IEEE

coating, two films were laminated at an optimized condition. In order to measure the thickness of the capacitor, three parts of the capacitor were chosen to make slices to achieve the exact value. Fig. 1 shows one of the pictures of slice.

The crystal structure of the BaTiO$_3$ was investigated at room temperature using a powder X-ray diffractometer (XRD) (Philips X' Pert Pro MPD, Cu-Kα, λ=0.15418 nm). The microstructure was examined using a JEOL JEM-100CX transmission electron microscope (TEM) at an accelerating voltage of 100 kV. The curing process and the glass transition process of the epoxy resin were measured with a differential scanning calorimeter (DSC, TA Instruments) under N$_2$ with the ramping speed of 5°C/min. Different loading levels of fillers were mixed with epoxy matrix. The dielectric properties of the composites were measured with Agilent4294 impedance analyzer at room temperature.

3 Results and discussion

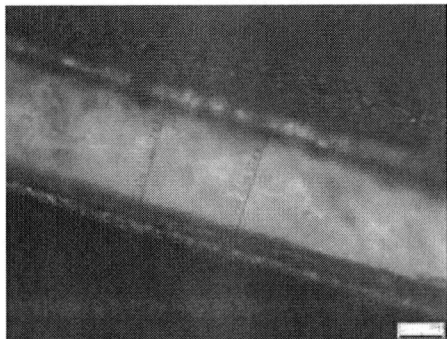

Fig. 1 Image of slice

3.1 Characteristics of BaTiO$_3$

Fig. 2 shows the XRD patterns of BaTiO$_3$ powders synthesized with a modified hydrothermal method. Both the BaTiO$_3$ powders calcined at 700°C and 900°C, respectively, have the standard XRD pattern of the perovskite phase, as shown in Fig. 2(a).

Fig. 2(b) shows the XRD pattern achieved with a slower scanning speed and enlarged the signal around 45°. The expect position of the tetragonal (002) and (200) peaks were obviously detected. The splitting of the reflection was a result of the distortion of the unit cell and the characteristic of tetragonal-BaTiO$_3$. The results of Fig.2(b) indicate that the both BaTiO$_3$ powders calcined at 700°C and 900°C were tetragonal. The tetragonal phase is more evident in the 900°C calcined BaTiO$_3$ powders, indicating an improved crystallization.

Fig. 3 shows the TEM micrograph of the hydrothermally synthesized BaTiO$_3$ powders calcined at 700°C. The BaTiO$_3$ particles were homogeneously monodispersed. And the shape of the BaTiO$_3$ particles was tetragonal with a particle size around 50 nm. The ultra fine BaTiO$_3$ particles with tetragonal phase added into the epoxy matrix was an effective way to achieve ultra thin film with high capacitance density.

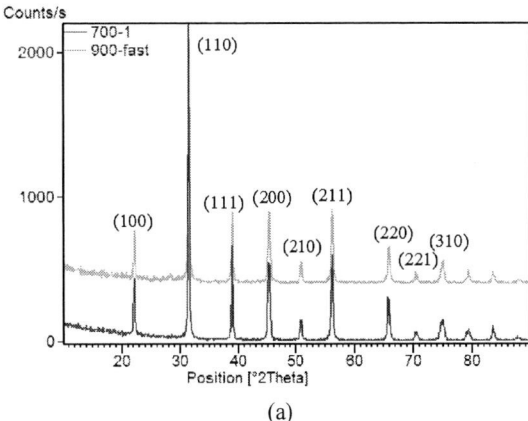

(a)

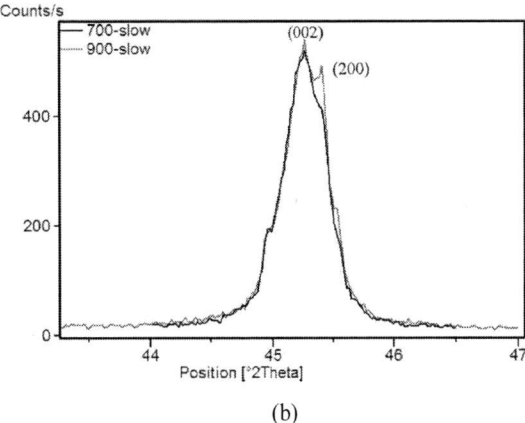

(b)

Fig.2 XRD patterns of hydrothermally synthesized BaTiO$_3$ powders (a) A comparison of BaTiO$_3$ calcined at 700°C and 900°C with a quick scanning speed; (b) Achieved with a slower scanning speed and enlarged XRD signal at 45°

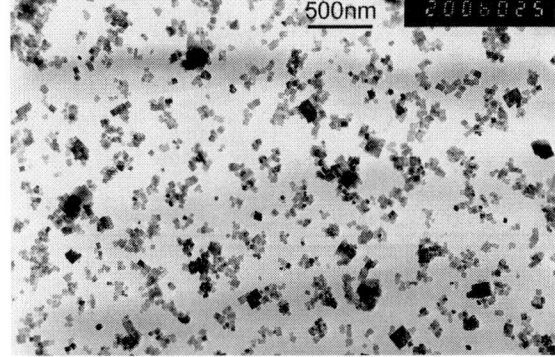

Fig. 3 TEM micrograph of hydrothermally synthesized BaTiO$_3$ powders calcined at 700°C

3.2 Curing property

The chemical structure of epoxy resin and curing agent are as follows:

Bisphenol A epoxy resin:

Tetraethylenepentamine (TEPA):
$NH_2(CH_2CH_2NH)_3 CH_2CH_2NH_2$

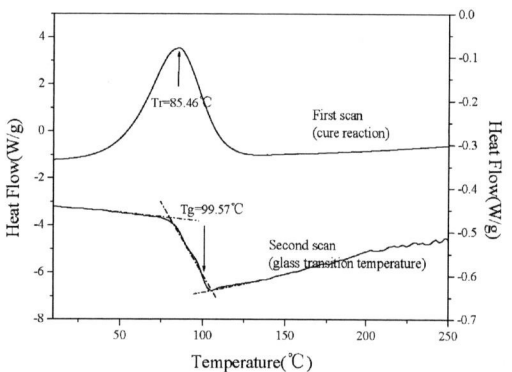

Fig. 4 Reaction mechanism of TEPA as the curing agent with a Bisphenol A epoxy resin

Fig. 4 shows the reaction mechanism of TEPT as the curing agent. It has two epoxy groups for each epoxy resin molecular. TEPA curing agent has a reactive –N-H group, which forms a chemically resistant C-N bond after reaction with the epoxy resin. Based on the reaction mechanism of TEPA as the curing agent with a Bisphenol A epoxy resin, the mass ratio of epoxy resin to TEPA was set to 1:0.186.

Fig. 5 shows the DSC curve of the curing process and the glass transition process. The epoxy resin can react with the curing agent at room temperature. The reaction temperature (Tr) was 85.46°C and the glass transition temperature (Tg) was 99.57°C.

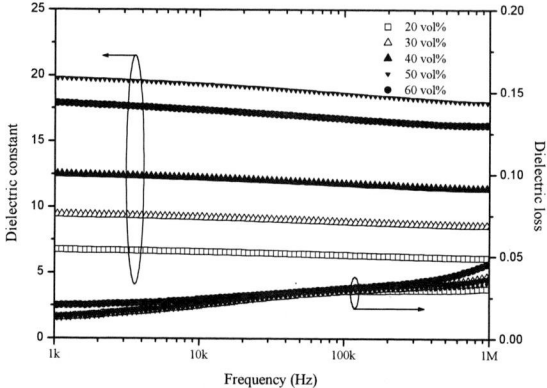

Fig. 5 Differential scanning calorimeter (DSC) of epoxy resin with 18.6 wt% TEPA

3.3 Dielectric property

Fig.6 shows the dielectric properties of the $BaTiO_3$-Epoxy composites. The dielectric constant of the composites is increased with the content of $BaTiO_3$ before the loading of $BaTiO_3$ reaches 50 vol%. The maximum dielectric constant of the composites is 19.4 (10 kHz) with the loading of $BaTiO_3$ at 50 vol%. However, when the content of $BaTiO_3$ is further increased, the dielectric constant of the composites would be reduced. When the fillers exceed the biggest stack density, some of the fillers can not be enwrapped by the polymer matrix which would result in air hole or interspaces and other defects. Due to the low dielectric constant of the defects, the dielectric constant of the composites is reduced. The dielectric

loss of the composites is about 0.02 (10 kHz) and does not change much with increasing the content of $BaTiO_3$. When the loading of $BaTiO_3$ is fixed, the dielectric constant and loss of the composites are decreased and increased with frequency, respectively. The reason is that the main mechanism of polarization is interfacial polarization in the frequency range of 1 kHz~1 MHz [17].

Fig. 6 Dielectric properties of $BaTiO_3$-Epoxy composites with different loading of $BaTiO_3$ as a function of frequency

Conclusions

Tetragonal $BaTiO_3$ nano particles were synthesized with a modified hydrothermal reaction method. The particles were homogeneously dispersed with a size about 50 nm. The epoxy matrix has a glass transition temperature (Tg) above 90°C which achieves the basic requirement of the embedded capacitor material. The synthesized $BaTiO_3$ added into the epoxy matrix obtained the maximum dielectric constant of 19.4 with the dielectric loss around 0.02 at 10 kHz. There is no apparent decrease for the dielectric consant in the frequency range of 1k~1MHz.

References

1. Smith, N, Fan, J, Andresakis, J, *et al*, "Embedded capacitor technology: A real world example," *Proc 58th Electronic Components and Technology Conf*, San Jose, CA, October. 2008, pp. 1919-1925.
2. Lee, M, Chan, C.Y, Tang, C.S, "Embedding capacitors and resistors into printed circuit boards using a sequential lamination technique," *Journal of Materials Processing Technology*, Vol.207 (1-3), (2008), pp. 72-88.
3. Jillek, W, Yung, W.K.C, "Embedded components in printed circuit boards: a processing technology review," *International Journal of Advanced Manufacturing Technology*, Vol.25 (3-4), (2005), pp.350-360.
4. Bai, Y, Cheng, Z.Y, Bharti, V, *et al,* "High-dielectric-constant ceramic-powder polymer composites," *Applied Physics Letters,*Vol.76 (2000), pp. 3804-3806.
5. Rao, Y, Ogitani, S, Kohl, P, *et al*, "Novel polymer-ceramic nanocomposite based on high dielectric constant epoxy formula for embedded capacitor application," *Journal of Applied Polymer Science*, Vol.83, (2002), pp. 1084-1090.

6. Rao, Y, Wong, C.P, "Ultra high dielectric constant epoxy silver composite for embedded capacitor application," *Proc 54th Electronic Components and Technology Conf,* San Diego, CA, Jun. 2002, pp.920-923.

7. Dasgupta, D.K, Doughty, K, "Polymer Ceramic Composite-Materials with High Dielectric-Constant," *Thin Solid Film,* Vol.158, (1988), pp.93-105.

8. Liang, S.R, Chong, S.R, Giannelis, E.P, "Barium titanate epoxy composite dielectric materials for integrated thin film capacitors," *Proc 48th Electronic Components and Technology Conf,* Settle, WA, May. 1998, pp. 171-175.

9. Windlass, H, Raj, P.M, Balaraman, D, *et al*, "Processing of polymer-ceramic nanocomposites for system-on-package applications," *Proc 51th Electronic Components and Technology Conf,* Orlando, FL, May. 2001, pp. 1201-1206.

10. Cho, S.D, Lee, J.Y, Paik, K.W, "Effects of particle size on dielectric constant and leakage current of epoxy/barium titanate (BaTiO$_3$) composite films for embedded capacitors," *Advances in Electronic Materials and Packaging,* Jejuisl, Nov. 2001, pp, 63-68.

11. Yang, X, Yang, Z, Mao, C, *et al*, "Dependence of dielectric properties on BT particle size in EP/BT composites," *Rare Metals,* Vol.25, (2006), pp. 250-254.

12. Dang, Z.M, Yu, Y.F, Xu, H.P, *et al*, "Study on microstructure and dielectric property of the BaTiO$_3$/epoxy resin composites," *Composites Science and Technology,* Vol.68, (2008), pp.171-177.

13. Uchino, K, Sadanagae, E, Hirose, T, "Dependence of the crystal-structure on particle-size in Barium-Titanate, " *Journal of the American Ceramic Society,* Vol.72, (1989), pp.1555-1558.

14. Xu, H.R, Gao, L, "Tetragonal nanocrystalline barium titanate powder: Preparation, characterization, and dielectric properties," *Journal of the American Ceramic Society,* Vol.86, (2003), pp.203-205.

15. Tanaka, M, Makino, Y, "Finite size effects in submicron barium titanate particles," *Ferroelectrics Letters Section,* Vol.24, (1998), pp.13-23.

16. Tsunekawa, S, Ito, S, Mori, T, Ishikawa, K, Li, Z.Q, Kawazoe, Y, "Critical size and anomalous lattice expansion in nanocrystalline BaTiO$_3$ particles," *Physical Review B,* Vol.62, (2000), pp.3065-3070.

17. Xu, J.W, Wong, C.P, "Dielectric behavior of ultrahigh-k carbon black composites for embedded capacitor applications," *Proc 55th Electronic Components and Technology Conf,* Lake Buena Vista, FL, May. 2005, pp. 1864-1869.

Influence of Leveler Concentration on Copper Electrodeposition for Through Silicon Via Filling

Huiqin Ling[1], Haiyong Cao[1], Yuliang Guo[1], Han Yu[2], Ming Li[1], Dali Mao[1]

[1]School of Materials Science and Engineering, Shanghai Jiao Tong University, 800 Dongchuan Road, Shanghai, China
[2]Shanghai Sinyang Semiconductor Materials Co., Ltd, 1268 Wenhe Road, Shanghai, China
hqling@sjtu.edu.cn, 86-21-34202741

Abstract

Through silicon via technology is one of the critical and enabling technologies for 3D packaging. 300μm deep vias with a diameter of 50μm were filled by copper electroplating with $CuSO_4$ and H_2SO_4 as base electrolyte. Chloride ions, accelerator and leveler were added. The effect of leveler concentration on filling performance was studied. Electrochemical measurements were used to investigate the cathode process and the action of additives. It was found that mass transportation of copper in via became the slowest step in deep vias, small current is necessary to obtain void free deposit and only conformal growth was obtained. And with increasing of leveler concentration filling performance became better.

Introduction

It is well known that 3D integration will be the next generation of packaging. Among all the 3D integration technologies, electrical interconnection between chips or chip and interposer by Through Silicon Via (TSV) owns many benefits compared with wire bonding and flipchip bonding, such as shorter connection, higher density, reduced RC delay [1]. One of the key technologies in TSV process is filling deep vias, which is also the most expensive step.

The common method to filling vias is copper electroplating, since it has been widely used in Damascene process and via filling in printed circuit board. To obtain void free filling, it is essential that copper should be plated with bottom up mode. That is via bottom is plated at a higher speed than orifice and sidewall. Three organic additives named suppressor, accelerator and leveler and Cl⁻ should be added to the electrolyte to gain bottom up filling. And the mechanisms of these additives have been widely studied [2,3]. Curvature enhanced adsorbate coverage model [3,4] developed by T. P. Mofffat are widely accepted to explain bottom up filling.

However, most researches focused on shallow and small via filling, in which mass transportation was overlooked. For TSV filling, since the depth of via is over 50μm, mass transportation must be considered. Dixit considered the cupric concentration distribution and adopted a stepped current to fill a 500μm deep through hole with an aspect ratio as high as 15 [5]. As to common TSV filling, it is essential to adopt very small plating current, while additives' performances at small current are seldom studied. Apart from plating current, leveler concentration plays a great role in via filling. Since the leveler deactivates the adsorbed accelerator, interface motion during filling is largely dominated by accelerator coverage and the impact of leveler on this coverage [4].

In this paper, effect of leveler concentration on 300μm deep via filling was studied. Several electrochemical methods such as linear sweep voltammetry (LSV) and chronopotentiometry were employed to characterize the effect of leveler concentration on the absorption behavior and inhibition ability of leveler at low plating current for the electrolyte with different leveler concentration.

Experimental Details

The basic electrolyte was made up of 40g/L Cu^{2+}, 60g/L H_2SO_4, 50mg/L Cl⁻. The suppressor concentration is 300mg/L, and accelerator 5mg/L. The concentration of leveler ranged from 2 to 30mg/L.

In electrochemical measurement, gold flake with an area of 1cm² was used as work electrode and platinum flake as counter electrode. And saturated calomel was used as reference electrode (SCE). In LSV measurement, the potential swept from 0.2V to -0.4V vs. SCE with a scan rate of 5mV/s. In chronopotentiometry, potential as a function of time (V-t) was measured at different current densities ranging from 0.5mA/cm² to 4mA/cm².

The vias used in our work were 300μm in depth and 50μm in diameter. The vias were filled at a small current to investigate the effect of leveler concentration on filling performance. After filling, the vias were incised and cross section images were observed by optical microscope.

Results and discussion

The effect of leveler concentration on the voltammetric behavior of the electrolyte is shown in Fig. 1. At low potential (0.03V~-0.13V vs. SCE), the leveler had accelerated effect on copper plating, which weakened with concentration increasing. At higher overpotential, it performed obvious suppressant action, and with concentration increasing, suppression became larger. These results are consistent with Wang's work [6], which corresponded to potential related adsorption, desorption and decomposition of the leveler. The suppression behavior of the leveler derived mostly from its decomposition product.

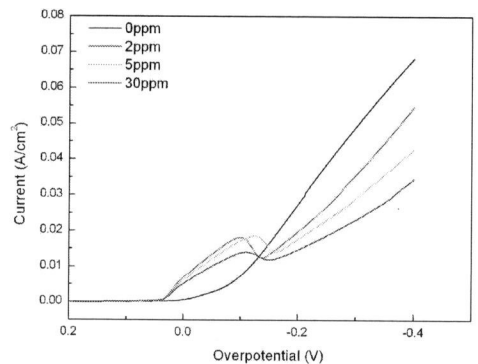

Fig. 1 LSV cures of electrolytes with different leveler concentration

Since the plating speed at the bottom of the deep via, the vias were filled with very small current. So chronopotentiometric performance of the electrolyte was measured at small current, such as 0.5mA/cm², 2mA/cm² and 4mA/cm². The results were shown in Fig. 2.

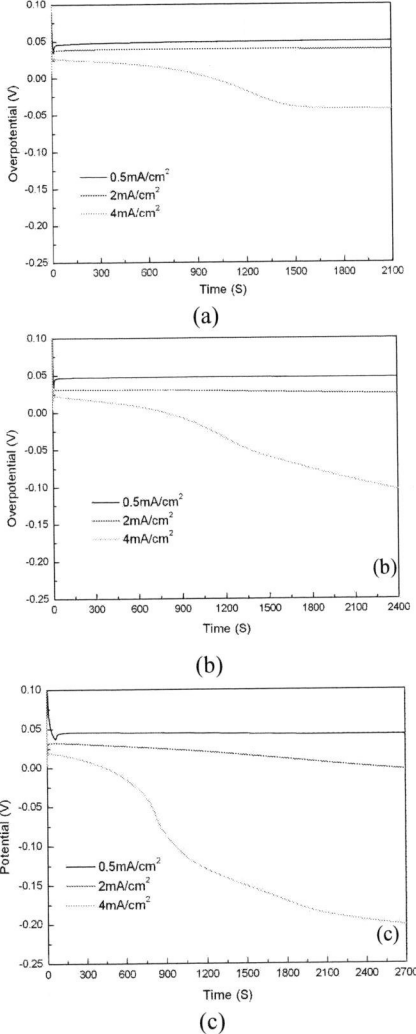

(a)

(b)

(c)

Fig. 2 Chronopotentiometric performance of the electrolyte with different leveler concentrations at various current density: (a) 2mg/L, (b) 5mg/L, (c) 30mg/L

At 0.5mA/cm² current (Fig. 2(a)), for all the leveler concentration, the overpotential remained constant about 0.05V vs. SCE during test time up to 2100s. Trace negative shift of overpotential with increasing leveler concentration was consistent with LSV results. At 2mA/cm² current (Fig. 2(b)), three electrolytes had various chronopotentiometric performance. For the electrolyte with 2mg/L leveler, the overpotential still remained constant until 2100s. When leveler was 5mg/L, the overpoteantial shifted gradually toward negative. While when the leveler was 30mg/L, the overpotential shifted faster and greatly toward negative. When the current increased to 4 mA/cm², overpotential of three electrolytes decreased with plating time increasing, and the decrease extent and the start point became larger and earlier with the increase of leveler concentration. For clearly seeing,

the chronopotentiometric performance of the electrolyte with different leveler concentrations at a current density of 0.5mA/cm² and 4mA/cm² were redrawn in Fig. 3.

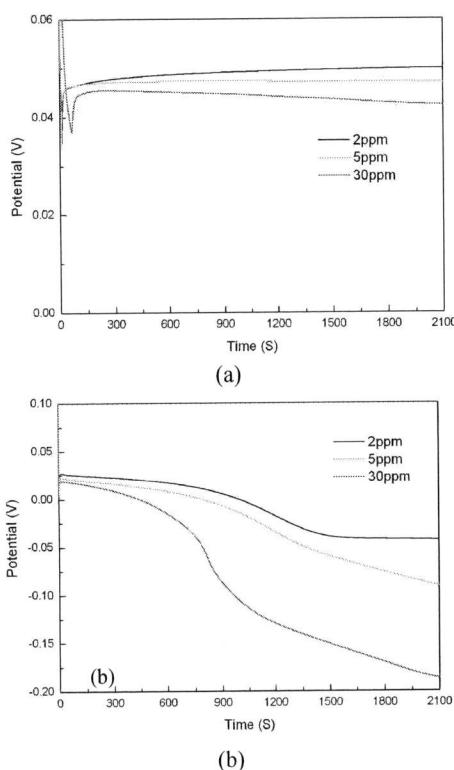

(a)

(b)

Fig. 3 Chronopotentiometric performance of the electrolyte with different leveler concentrations at a current density of (a)0.5mA/cm² and (b) 4mA/cm².

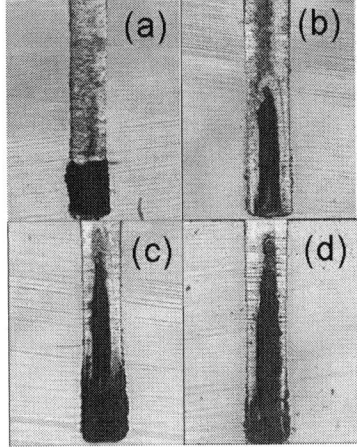

Fig. 4 Cross section images of TSV vias plated at different current with leveler concentration of 30mg/L. (a) 1mA/cm², (b) 2mA/cm², (c) 3mA/cm², (d) 5mA/cm²

From Fig. 2 and Fig. 3, we can see that the electric quantity passing through the electrolyte and leveler concentration decide the start point and extent of inhibit effect of leveler. And it is not concentration but electric quantity that

978-1-4244-4658-2/09 $25.00 © 2009 IEEE

promotes the action of the leveler. So, there are two ways to obtain the required action of leveler. The fist one is increasing plating current, the second one is increasing leveler concentration. Since the vias are so deep that it is difficult for enough cupric irons to diffuse to the bottom of via, void may be formed at the bottom of via.

The cross section images of the vias plated with different current is shown in Fig. 4. The concentration of the leveler was 30mg/L for all the deposition. Voids were formed in all the vias plated at the currents ranging from 1mA/cm^2 to 5mA/cm^2. And with the increase of plating current void became larger. So the vias could only be plated with very small current, and high leveler concentration is necessary to obtain sufficient leveling effect quickly.

The cross section images of TSV vias plated at a current smaller than 1mA/cm^2 with different leveler concentration in the electrolytes were shown in Fig. 5.

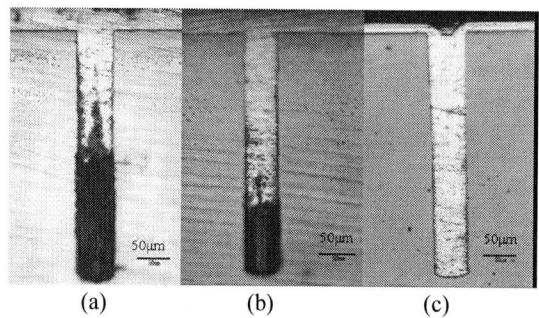

Fig. 5 Cross section images of TSV vias plated in the electrolytes with different leveler concentration. (a) 2mg/L, (b) 5mg/L, 30mg/L

There is obvious different filling performance of three electrolytes. When leveler concentration is 2mg/L, the void length was about a half of via depth. And the void length decreased with leveler increase. It was below 1/3 of via depth when the leveler concentration was 5mg/L. And void was disappeared when the leveler was 30mg/L. And the copper thickness at the surface of the dies decreased with the increase of leveler concentration. This phenomenon resulted from two reasons. The first is that the copper deposited at the surface was those which should be deposited in via. The second, and the most likely was that with the plating going on, the suppress effect of the leveler became stronger and higher leveler concentration corresponding to stronger suppression.

In LSV curve (Fig .1), when the current is 0.5mA/cm^2, the leveler had a accelerating effect for the concentration ranging from 2mg/L to 30mg/L. In chronopotentiometric measurement, when the current is 0.5mA/cm^2 (Fig. 3a), during test period, the leveler didn't worked as a weak suppressor, especially for the electrolytes with leveler concentration of 2mg/L and 5mg/L. While for 30mg/L leveler concentration, acceleration effect was weakened gradually with test time. When and how the effect of the leveler changes to inhibition should be studied detailedly.

Conclusions

300μm deep TSV vias were filled through electroplating. The effect of the leveler concentration on filling performance was studied. Electrochemistry measurement indicated that the leveler adopted in this work had weak accelerating action at low overpotential, and suppression action at high ovepotential, and the accelerating and suppression effect were weakened and enhanced respectively with the increase of leveler concentration. Chronopotentiometry measurement indicated that the transition point between acceleration and suppression was related to electric quantity passing through the electrolyte, as well as the concentration of the leveler. To obtain sufficient leveling action quickly, high concentration is required. In this work, the electrolyte with 30mg/L leveler had best filling performance.

Acknowledgments

This work is sponsored by International Science and Technology Cooperation of China (No. 2008DFA51680), National Natural Science foundation of China (60876071), Shanghai nano technology promotion center (No. 0852nm06300).

References

1. http://www.itrs.net/Links/2007ITRS/Home2007.htm
2. Ko, S.L., Lin, J. Y., Wang, Y. Y., Wan, C. C., "Effect of the Molecular Weight of Polyethylene Glycol as Single additive in Copper Deposition for Interconnect Metallization," *Thin Solid Films*, Vol. 516 (2008), pp. 5046-5051.
3. Moffat, T. P., Wheeler, D., Edelstein, M. D., Josell, D., "Superformal Film Growth: Mechanism and Quantification", *IBM J. Res.& Dev.,* Vol. 49, No. 1 (2005), pp. 19-36.
4. Moffat, T. P., Wheeler, D., Kim, S.-K., Josell, D., "Curvature Enhanced Adsorbate Coverage Model for Electrodeposition," *Journal of The electrochemical Society,* Vol. 152, No. 2 (2006), pp.C127-C132.
5. Dixit, P., Miao, J. M., "Aspect-Ratio-Dependent Copper Electrodeposition Technique for Very High Aspect-Ratio Through-Hole Plating," *Journal of The electrochemical Society,* Vol. 153, No. 6 (2006), pp. G552-G559.
6. Li, Y. B., Wang, W., Li, Y. L., "Adsorptaion Behavior and Related Mechanism of Janus Green B during Copper Via-Filling Process," *Journal of The electrochemical Society,* Vol. 156, No. 4 (2009), pp. D119-D124.
7. Shu, William K., "PBGA Wire Bonding Development," *Proc 46th Electronic Components and Technology Conf,* Orlando, FL, May. 1996, pp. 219-225.

Characterization of Ag Nanofilm Metallization on Copper Chip Interconnect and Its Ultrasonic Bondability

Yanhong Tian*, Shaowei Zhao, Chunqing Wang

State Key Lab of Advanced Welding Production Technology, Harbin Institute of Technology, Harbin, P. R. China, 150001

E-mail: tianyh@hit.edu.cn

Abstract

To improve the bondability of copper chip interconnect, a Si/Ti-Cu-Ag structure was designed and fabricated using evaporation method. Some analytical methods such as SEM, AFM, XRD and XPS were used to characterize the Ag nanofilm metallization and its bondability. It was found that the evaporated Ag metallization has good surface state and samll particle size, which could diffuses readily in the wire bonding process and thus improve the bondablilty. XPS showed the Ag metallization was pure silver without oxidation and sulfuration when stored at the room temperature. And XRD showed the crystal face index of the top surface polycrystalline Ag layer was (111), which agree with the principle of minimum surface energy. The bondability and shear strength of the Au ball bump on the Si/Ti-Cu-Ag structure were excellent at room temperature and 150°C, better than the oxygen free copper, and meet the JEDEC Standard 22-B116. After high temperature storage at 200°C for 16 days and temperature cycling for 1000 cycles, the interfaces of the Si/Ti-Cu-Ag structure were still very tight and clear, showing high reliability.

1. Introduction

As the IC feature sizes continue to shrink, the RC delay of the interconnect wire has become the main delay of the system, instead of the intrinsic gate delay. To get the higher speed response, the resistance of the interconnect wire should be reduced. The traditional aluminum is being replaced with copper for its lower resistivity (1.7 vs.2.7μΩ·cm) and higher electro-migration resistance [1-3]. However, Wire bonding on a bare copper pad has always been a challenge to semiconductor packaging [4]. Copper oxidizes easily in the air, however, there is no stable self-passivation layer formed during oxidation of copper, which will inhibit wire bonding and decrease the yield. Several methods have been emerged to protect the Cu pad, such as the inert gas shielding, the plasma cleaning and the top surface protective coatings, which is applied most widely. A review on the various form of coating to enhance the bondability of copper pad surface by Harman and Johnson indicated three types of surface coatings, namely metallic top surface, inorganic film and organic layer [5].

Jong-Ning AOH [6] applied argon-shielding atmosphere to prevent the copper pad from oxidizing, and got 100% gold wire attached on a copper pad at 180°C and above. Also Plasma cleaning could ensure adequate copper bond pad cleanliness for the first bond. The process time window is, however, readily short and then oxidation will reoccur making it difficult to bond [7]. Thin inorganic films represent a different way of surface protection, and which will be broken during the wire bonding process and be pushed aside as the wire/ball and the bond pad deform [8-9]. However, this process might damage the chips. The most popular used method is the metallization deposition on the top surface, such as gold, platinum, silver and palladium, [10-12] on the copper pad to improve the bondability. J.N. Aoh applied silver surface metallization to achieve 100% bondability and sufficient bonding strength [13]. The silver bonding layer does not degrade the electrical performance of chips with copper interconnects as silver has an even lower electrical resistance than copper.

In this paper, 100nm Ti film and 1μm Cu film were evaporated on the p type silicon to simulate the copper interconnect, and then 400nm Ag film was fabricated by the same method as the top surface protective film, which was described as Si/Ti-Cu-Ag. The morphology and chemical composition of the Ag metallization were characterized and the ultrasonic bondability of Si/Ti-Cu-Ag was investigated.

2. Experiments

PRO-500 was used to evaporate the Si/Ti-Cu-Ag structure and the K&S 4522 wire bonder was used to make the Au ball bump on the substrate. The bonding process parameters were listed in Table 1. The diameter of the Au wire was 1mil, which was produced by K&S. The cross section morphology was investigated by HITACHI-S-4700 scanning electron microscopy (SEM), and the surface morphology was tested by Nanoscope IIIa AFM. D/max-rB rotating anode XRD. X-ray Photoelectron Spectroscopy (XPS) was used to measure to the chemical composition of the Ag metallization.

To investigate the bondability of the Ag metallization, the Dage 4000 bond tester was used to get the shear strength of the Au ball bump on the Ag metallization. The test was in accordance with the standards of JEDEC Standard 22-B116. The high temperature storage and temperature cycling of the Au wire bond on the Ag metallization were tested in Espec PHH201 aging oven at 200°C, and the temperature cycling test was performed in Espec TSG-71L temperature cycling oven, which set high temperature: 150°C, keeping 10minutes, low temperature -55°C, keeping 10 minutes.

Table 1 Parameters of ultrasonic wire bonding of Au wire

Modes	Power	Time	Force	Search height	Loop	Tail	Ball Size
Ball Bump	2.0	5	1.5	3.0	8	4	2.0

3. Results and Discussion

3.1 The characterization of the Ag metallization

The cross section of the Si/Ti-Cu-Ag structure was shown in Fig.1. From the SEM photograph, the thickness of the Ag, Cu, Ti and Si was measured as 370nm, 800nm and 90nm respectively, which was well coincident with the designed

978-1-4244-4658-2/09 $25.00 © 2009 IEEE

values. Furthermore, the interface of the films were very neat and the thickness of the films was uniform, which indicates the evaporation was available to get the film with the desired thickness.

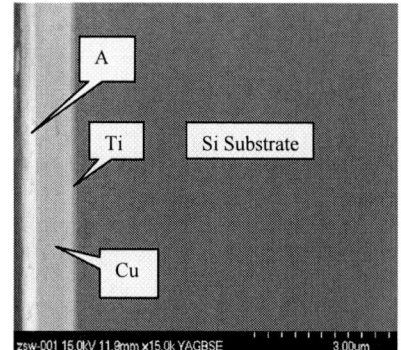

Fig.1 The cross morphology of the Si/Ti-Cu-Ag

Surface morphology and roughness for the top Ag layer was tested by AFM as shown in Fig.2. The mean roughness (Ra) of the surface Ag layer is 3.3 nm, and the average particle size (mean diameter) of Ag layer is 6.8 nm. Since the surface energy increases with the decrease of particle size, the nano particle has great surface energy, and which would diffuses readily in the wire bonding process. In this way, the Ag metallization may improve the bondablilty.

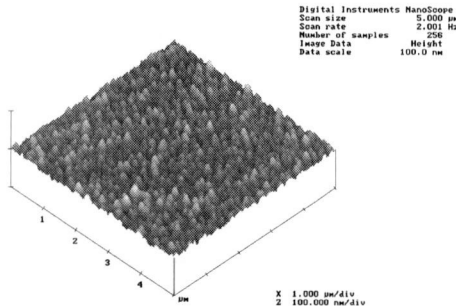

Fig. 2 AFM test for the surface of Ag layer

Fig.3 shows the XRD spectrum of the Si/Ti-Cu-Ag structure, and it is confirmed that the crystal face index of the top surface polycrystalline Ag layer was (111) orientation, which is the strong plane of the FCC structure. In the thin polycrystalline film, the anisotropy of surface and/or interface energies can supply an additional driving force for preferred orientation growth or texture [14]. This phenomenon can be explained with the principle of minimum surface energy.

After one months' storage at the room temperature in the air, XPS test was preformed. The surface of the film was pure silver without oxidation and sulfuration, and this can be evidenced by Fig.4. The binding energy of standard pure Ag $3d_{5/2}$ peak is 368.3 eV, and the binding energy of the surface Ag $3d_{5/2}$ in the test is 368.33 eV. It illustrates that Ag was hardly oxidized at the room temperature and the sulfuration procession of Ag require a long time.

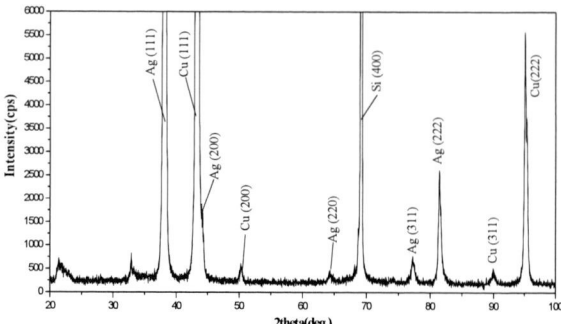

Fig.3 XRD spectrum of the Si/Ti-Cu-Ag structure

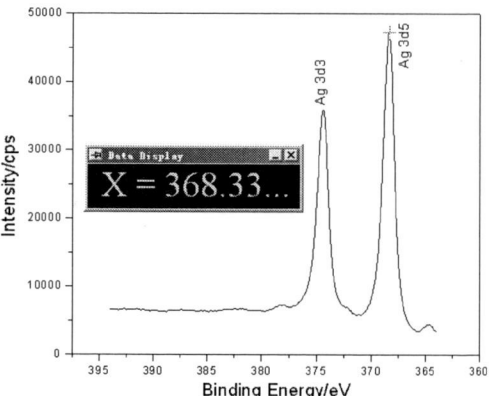

Fig.4 XPS spectrum of surface Ag layer

3.2 The bondability of the Ag metallization

In this study, the definition of J. N. Aoh about bondability was applied [13], which was described as the percentage of successful and gold ball bonds onto a copper pad divided by the number of total bonding actions. The number of the bonding actions in this study was 50 and the oxygen free copper sheet was compared in the test. The bond temperature was at the room temperature and 150°C, and the results were showed in Fig.5. It clearly shows that the Si/Ti-Cu-Ag structure has the better bondability than the oxygen free copper sheet both at room temperature and 150°C, which achieves 98% at these temperatures.

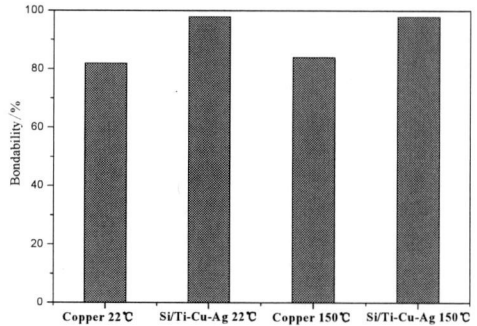

Fig.5 The bondabilty test of the Si/Ti-Cu-Ag and the copper

The shear strength of Au ball bump on the Si/Ti-Cu-Ag structure and the oxygen free copper sheet were performed by Dage4000 shear tester at room temperature and 150°C. The shear strength was achieved from the average of 20 bonds, and Fig.6 also shows the distribution of the data. The Au ball bump on the Si/Ti-Cu-Ag structure could get the better shear strength, as shown in Fig.6. Under 150°C bonding temperature, Au wire could not be bonded on the bare copper sheet easily due to the oxidation, while on the Ag metallization, excellent shear strength (about 46.19g) can be achieved. Furthermore, the data of shear strength on the Si/Ti-Cu-Ag structure(Standard deviation 2.81 at 22°C and 3.23 at 150°C) was more concentrated than on the copper(Standard deviation). According to the JEDEC Standard 22-B116, the minimum shear average of the 3.3mil ball bond diameter was 38.1g, the bump on the copper was not meet the standard, while the bump on the Si/Ti-Cu-Ag was available both at the room temperature and 150°C.

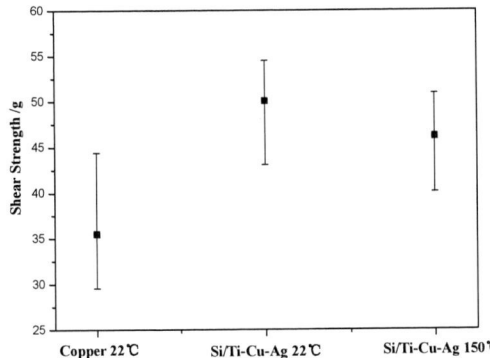

Fig.6 Shear strength of Au ball bump on the Si/Ti-Cu-Ag and the copper

Fig.7 shows the trace after shearing test on Si/Ti-Cu-Ag structure, and we can see the shear failure was at the interface between Ag layer and Cu layer from the EDX result. From the result, we can deduce that improving the adhesive strength between the Ag layer and the Cu layer could get the bigger strength.

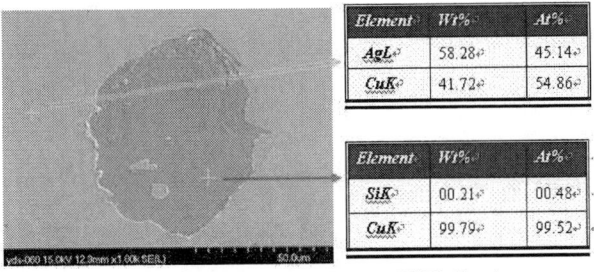

Fig.7 The trace after shearing on Si/Ti-Cu-Ag

The high temperature storage tests of the oxygen free copper sheet and the Si/Ti-Cu-Ag structure were performed at 200°C for 16 days. From Fig.8, we can see that there are two different diffusion layers between the Au layer and the Cu layer, marked as A and B. The thickness of A and B are both less than 0.5μm, and the content ratio of Au and Cu is 1:1 at

A, 1:4 at B. Since the EDX is not so accurate at this small region, we can deduce the phase of A is AuCu compound and $AuCu_3$ for B from the reference [15]. These compounds are easy to form Kirkendall void at the high temperature, which would decrease the strength and reliability [16]. Fig.9 shows the interface of the Au ball bump on the Si/Ti-Cu-Ag, there was no compound layer between the layers, and the interfaces were still very tight and clear. From the test, we can know that the Ag metallization has the better reliability than the copper.

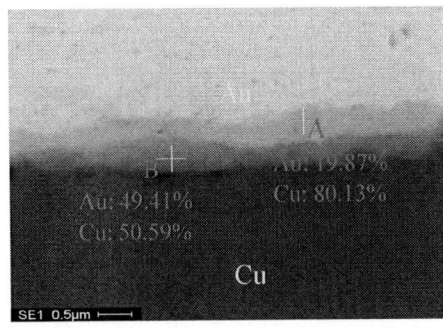

Fig.8 The interface of Au ball bump on the copper substrate after aging 16 days at 200°C

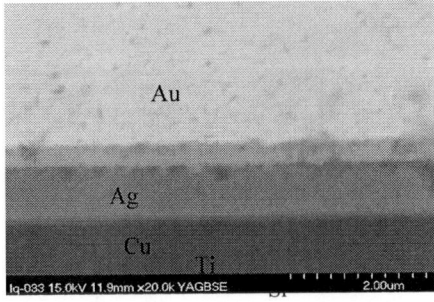

Fig.9 The interface of Au ball bump on the Si/Ti-Cu-Ag substrate after aging 16 days at 200°C

The interface morphology of Au ball bump on the Si/Ti-Cu-Ag after 1000 temperature cycles is shown as Fig.10. We can see that the stress produced by the changes of temperature didn't make obvious damage to the bonding interface, the interfaces of different layers were still clear, the reliability was good.

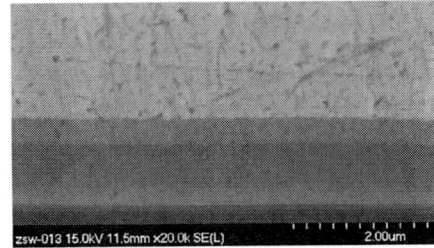

Fig.10 The interface of Au ball bump on the Si/Ti-Cu-Ag after 1000 temperature cycles

978-1-4244-4658-2/09 $25.00 © 2009 IEEE

4. Conclusions

In this paper, a Si/Ti-Cu-Ag structure was designed and fabricated, different characterizations for the Ag metallization were tested and the bondability of the structure was evaluated. All the results prove that the Ag metallization was available as the surface protective layer to improve the bondability on the copper chip interconnects.

(1) The evaporated Ag metallization has good surface state and samll particle size, which could diffuses readily in the wire bonding process and improve the bondablilty.

(2) At room temperature, the Ag metallization is hardly oxidized and sulfureted, and (111) is preferred orientation.

(3) The bondability and shear strength of the Au ball bump on the Si/Ti-Cu-Ag structure were excellent at room temperature and 150°C, better than the oxygen free copper , and meet the JEDEC Standard 22-B116.

(4) In the high temperature storage and temperature cycling test, the Si/Ti-Cu-Ag structure showed high reliability.

Acknowledgments

The authors gratefully acknowledge the financial support from the National Science Foundation of China (Grant No. 50705021).

References

1. Marc A Nieolet, "Temarya Morphous Metallie Thin Films as Diffusion Barriers for Cu Metallization", *Applied Surface Seience.* Vol. 91, No. 1-4 (1995), pp. 269-270.

2. P Haneda, K Aramaki, "Protection of Copper Corrosion by an Ultra Thin two dimensional Polymer Film of Alkanethiol Monolayer", *Electrochem Soc.* Vol. 145, No. 6 (1998), pp. 1856-1861.

3. T.Kodas, M.Hampten-Smith, The Chemistry of Metal CVD ,VCH(Weinheim,1994).

4. International Roadmap Committee. The International Technology Roadmap for Semiconductors 2003 Edition Executive Summary[R]. Apr. 2003.

5. Harman GG, Johnson CE, "Wire bonding to advanced copper-low-k integrated circuits, the metal/dielectric stacks,and material considerations", *Proc. 34th International Symposium on Microelectronics, IMAPS* 2001,Baltimore, USA, 9–11 Oct. 2001.

6. Jong-Ning Aoh, Cheng-Li Chuang, "Development of a Thermosonic Wire Bonding Process for Gold Wire Bonding to Copper Pads Using Argon Shielding", *Journal of Electronic Materials*, Vol.33, No.4(2004), pp.300-309.

7. Hong Meng Ho et al, "Direct gold and copper wires bondings on copper", *Microelectronics Reliability*, Vol.43,No.6(2003), pp.913-923.

8. G George. Harm, E Christian and Johnson, "Wire Bonding to Advanced Copper, Low-K Integrated Circuits, the Metal/Dielectric Stacks, and Materials Considerations", *IEEE Transactions on Components and Packaging Technologies*, Vol.125, No.4(2002), pp.667-683.

9. R.W.Cheung, M.-R.Lin,"Advanced Copper Interconnect System that is Compatible with Existing IC Wire Bonding Technology", U.S.Patent 5785236, Jul. 28. 1998.

10. J F Rohan, G O'Riordan, J Boardman,"Selective electroless nickel deposition on copper as a final barrier/bonding layer material for microelectronics applications," *Applied Surface Science*,Vol.185,No.3-4(2002), pp.289-297.

11. J.G.Strandjord,S.Popelar, and C.Jauernig, "Interconnecting to aluminum and copper-based semiconductors," *Microelectronics Reliability*, Vol.42, No.2(2002), pp.265-283.

12. Cheng-Li Chuang, "Increasing bondability and bonding strength of gold stud bumps onto copper pads with a deposited titanium barrier layer," *Microelectronic Engineering*, Vol.84,No.4(2007), pp.551-559.

13. J. N. Aoh, C. L. Chuang, "Thermosonic Bonding of Gold Wire onto Copper pad with Titanium Thin Film Deposition," *Journal of Electronic Materials*, Vol.33, No.4(2004), pp.290-299.

14. Jianmin Zhang, Fei Ma, Kewei Xu,"Caleulation of Surface Energy of FCC Metals with the Modified Embedded-Atom Method," *Chinese Physies*, Vol.36, No.4(2004). pp.355-359.

15. Vernon A. PITT, Christopher R.S. Needes," Thermosonic Gold Wire Bonding to Copper Conductors," *IEEE Transactionson Components, Hybrids and Manufacturing Technology*, Vol.5, No.4 (1982), pp.435-440.

16. Feinstein, L.G,"The Failure of Aged CuAu Thin Film by Kirkendall Porosity," *Journal of Vacuum Science and Technology*, Vol.16, No.2 (1979), pp.155-158.

Electroless Plating of Copper Nano-coned Array for High Reliability Packaging

Zhongwen Pan, Anmin Hu, Tao Hang, Yingying Duan, Ming Li, Dali Mao

Lab of Microelectronic Materials & Technology, School of Materials Science and Engineering,
Shanghai Jiao Tong University, Shanghai 200240, China
mingli90@sjtu.edu.cn, 021-34202542

Abstract

Nano-coned array of Cu were prepared by electroless plating with special crystallization conditioning agent. The morphologies of nano-coned array were observed by FE-SEM. The influences of active time in $PdCl_2$, concentration of $NiSO_4 \cdot 6H_2O$, crystallization conditioning agent, pH and temperature of plating solution were discussed. Then, the possible reasons of the influence on the growth of nano-coned array were investigated. $NiSO_4 \cdot 6H_2O$ was proved as catalytic active agent. The optimum preparation conditions were fixed. The crystal crystallographic orientation and elements of the nano-coned array were measured using TEM and EDX. The nano-cones consist of 95.57% copper and 4.43% nickel. The crystal crystallographic orientation of them is typical copper FCC single crystal.

1 Introduction

In the development of integrated circuit and high density industry package, the reliability of various package materials' interfaces becomes a critical problem [1-3]. For example, the failures are caused by the cracks between plastic resin and copper connection in lead frame, BGA substrate, and PCB. Thus, to solve these problems is a hot research topic at present [4].

In order to enhance the adhesive strength between the interfaces, various methods have been attempted, including physical and chemical methods [5]. For instance, resin of excellent properties and brown oxides treatment has been applied to enhance the cohesive force between the BGA substrates and copper wirings. Roughening the interfaces is also an effective method [6]. We have reported that Ni nano-coned array structure is an effective solution to improve the adhesion strength between the resin and the Ni/Pd/Au PPF (Pre-Plated Frame) [7]. Enlarged surface area, improved interface chemical activity and the effect of physical inlay may result in the enhancement of adhesion strength 2 times. As copper and its alloy have excellent electrical properties, they are widely adopted in lead frame, BGA substrates, and PCB. [8] Consequently, copper nano-coned array can be more widely applied than nickel nano-coned array. Only a few examples of metallic conical structures have been reported and the synthesis routes are mainly based on template methods [9]. These approaches are limited to the micrometer scale, and the preparation processes are quite complicated and expensive.

In this paper, copper nano-coned array was prepared by electroless plating with special crystallization conditioning agent. This method is rather simple and economical, and is expected to be appropriate for commercial application. The influence of bath composition, temperature, pH, and deposition velocity of electroless plating were discussed. The

crystal crystallographic orientation and composition of the nano-cones were measured. The possible mechanisms of influence factors on electroless plating were also investigated.

2 Experimental Materials and Methods

The nano-coned array of copper was electroless plated onto 2.5cm×2cm Cu plates. These Cu plates were first anodized at about $0.8A/cm^2$ in an oil removal solution for 20 seconds. After anodizing, the plates were rinsed with pure water, and vibrated in a solution of 20% H_2SO_4 for 30 seconds, thus, rinsed with pure water again, and vibrated in a solution of 0.02g/L active solution for 90 seconds. Finally, they were taken into plating solution for electroless plating. The plating solution contains $CuSO_4 \cdot 5H_2O$, $NiSO_4 \cdot 6H_2O$, $NaH_2PO_2 \cdot H_2O$, $Na_3C_6H_5O_7 \cdot 2H_2O$ and H_3BO_3 in a special ratio at pH=9.0 (adjusted with pure NaOH). $PdCl_2$ is added as the active solution. Electroless plating parameters were changed to obtain the best copper nano-coned array. The parameters are shown in the Table 1. In this study, if not special illumination, the conditions of electroless plating are active time: 90seconds, concentration of active solution: 0.02g/L, crystallization conditioning agent 1.28g/L, pH=9.0, temperature 65°C, electroless plating time 10minutes. The morphologies of copper nano-coned array were measured by means of field emitting scan electronic microscope (FE-SEM, FEI SIRION 200/INCA OXFORD, FEI, U.S.A). TEM images, electron diffraction (ED) patterns and EDX spectra were acquired using high-resolution transmission electron microscopy (JEM-2010/INCA OXFORD, Japan/U.S.A).

Table 1 The parameters of electroless plating

Influent factor	Parameters
Active time in $PdCl_2$(s)	0-120
Concentration of $NiSO_4 \cdot 6H_2O$ (g/L)	0-8.0
Crystallization conditioning agent (g/L)	1.28
pH	7.5-9.5
Temperature (°C)	45-75

3 Results and Discussion

3.1 The influence of the active time in $PdCl_2$

Fig. 1 is SEM images of Cu electroless plating with different active times. Activation is necessary for the growth of nano-cones. When the Cu plates were not activated, there was no nano-cone. When the active time was 60 seconds, the nano-cones grew up. And with the increase of active time, the nano-cones grew larger. Among the SEM images, the morphologies of nano-cones for active time 90 seconds were the best. As the active time was 120 seconds, the dendritic crystal grew up on the surface of nano-cones. So the optimum

978-1-4244-4658-2/09 $25.00 © 2009 IEEE

active time was 90 seconds. According to the results, the morphologies of the Cu nano-coned array were greatly influenced by the active time. No conical structures could be formed with no activation; it was believed that activation in $PdCl_2$ contributed to the formation of nano-cones. Based on the proposed mechanism for the growth of one-dimensional nanostructure under the confinement of an active agent [10], we here proposed that one possible function of activation was to kinetically control the growth rates of different crystalline faces of Cu nano-cones by interacting with these faces through adsorption and desorption [11].

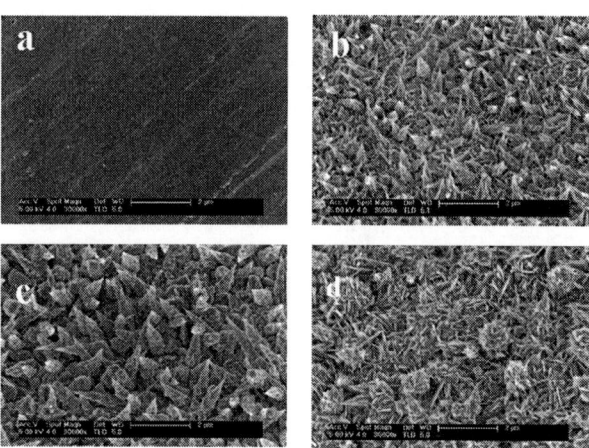

Fig.1 The FE-SEM images of Cu electroless plating with different active times: (a) 0s; (b) 60s; (c) 90s; (d) 120s

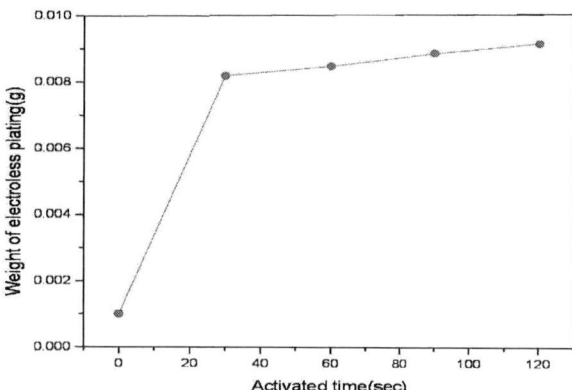

Fig.2 The weight of electroless deposit with different active times

Fig. 2 is the weight of electroless deposit with different active times. When the Cu plates were not activated, the weight of electroless deposit was nearly 0g. It illustrated the electroless plating did not carry out without activation. It was identical to the results of SEM. This proved that activation was necessary for the electroless plating; it could relatively reduce the energy of the reactions. After being activated with different times, the weight of electroless deposit onto the plates was almost the same to each other. The possible reason was that the catalytic reactions were saturated when it was above 30 seconds.

3.2 Influence of the concentration of $NiSO_4·6H_2O$ as the active agent

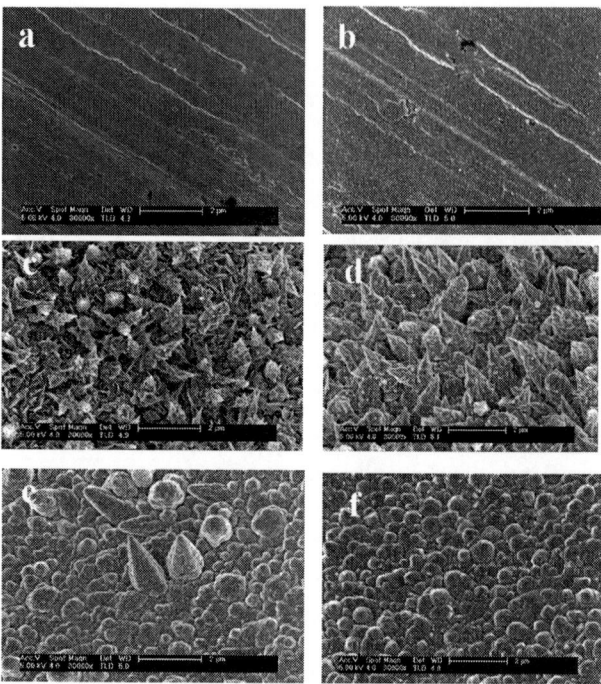

Fig. 3 The FE-SEM images of Cu electroless plating with the different concentrations of $NiSO_4·6H_2O$: (a) 0g/L; (b) 1.5g/L; (c) 2g/L; (d) 4g/L; (e) 6g/L; (f) 8g/L

Fig. 3 is SEM images of Cu depositing at different concentrations of $NiSO_4·6H_2O$. $NiSO_4·6H_2O$ as a catalytic agent was essential for electroless plating. The reactions can carry on persistently with it. According to the Fig. 3, the concentrations of $NiSO_4·6H_2O$ in the electroless plating solution greatly influence the morphologies of nano-coned array. When the concentrations were lower (0g/L, 1.5g/L), there was no nano-cones on the surface of plates. The possible reason was that Ni ions were not enough to activate the electroless plating reaction persistently. The nano-cones may only grow on some active points. The nano-cones grew up as the concentrations were high (2g/L, 4g/L). Above 2g/L, the reaction could persistently carry on with enough Ni ions. When the concentrations were even higher (6g/L, 8g/L), the nano-cones almost disappeared. Especially the concentration was 8g/L; the surface was smooth comparing to the ones at low concentrations. The possible reason was that too many Ni ions stayed on the surface of former nano-cones, causing the secondary nano-cones to grow on it, finally filling the gaps between the nano-cones, and reducing the roughness of the surfaces of plates. So the nano-cones seem perfect as the concentration of $NiSO_4·6H_2O$ is 2-4g/L.

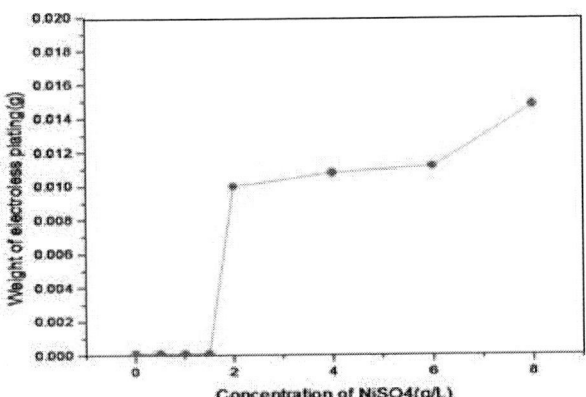

Fig.4 The weight of electroless deposit with different concentrations of NiSO$_4$·6H$_2$O

Fig. 4 shows the weight of electroless deposit at different concentrations of NiSO$_4$·6H$_2$O. The weights of electroless deposit on the plates were nearly zero g as the concentrations were below 2g/L. This proved the results of SEM of low concentrations (0g/L, 1.5g/L). When the concentration was 2g/L, the weight of electroless deposit raised remarkably. Then, with the increase of the concentration, the weight increment of electroless deposit was getting smaller. The curve illustrates that electroless plating carried on as the concentration of NiSO$_4$·6H$_2$O was above 2g/L. Above the concentration of NiSO$_4$·6H$_2$O at 2g/L, the reaction velocities of electroless plating almost kept the same. It was identical to the results of SEM. As the concentration was above 2g/L, there were enough Ni ions to activate the reaction persistently, the weight of eletroless deposit increased slowly, and the morphologies didn't change any more (Fig. 4(c), 4(d)). While the concentrations were higher, especially at 8g/L, the weights raised remarkably, because the secondary nano-cones filled the gaps between the nano-cones.

3.3 Influence of crystallization conditioning agent

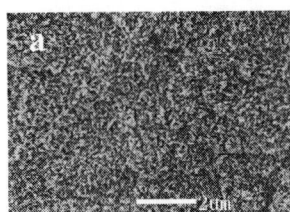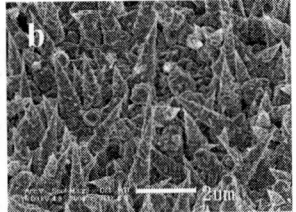

Fig. 5 FE-SEM images of Cu electroless plating (a) without crystallization conditioning agent (b) with crystallization conditioning agent

Fig. 5 shows the difference between the electroless plating without and with specific crystallization conditioning agent. The crystallization conditioning agent was an important factor for electroless deposit. Various conditioning agents were added into solution to control the morphologies of the electroless deposit [12]. The nano-coned array could not be obtained without them. Fig. 5(a) shows the morphology of electroless plating without crystallization conditioning agent,

there was only few nano-cones on the surface of clad layer. In Fig. 5(b), with the help of crystallization conditioning agent, the copper nano-cones appeared obviously, the diameter range of the nano-cones were about 0.2~0.5μm, and the heights were range from1.0μm to 2.5μm. It reveals that electroless plating can be stimulated by adding some agents into the solution, such as crystallization conditioning agent. The possible mechanism was that the crystallization conditioning agent can accelerate the absorption of Cu ions, then promoting the growth of nano-cones. At the early stage, the Cu atoms were deposited onto the Cu plates through the reduction process. The majority of the Cu was precipitated as nano-particles. In the following step, the larger nanoparticles, in the shape of pyramid, served as the seeds for the growth of nano-cones which developed into long nano-cones, while the smaller ones were buried among them [13].

3.4 Influence of pH

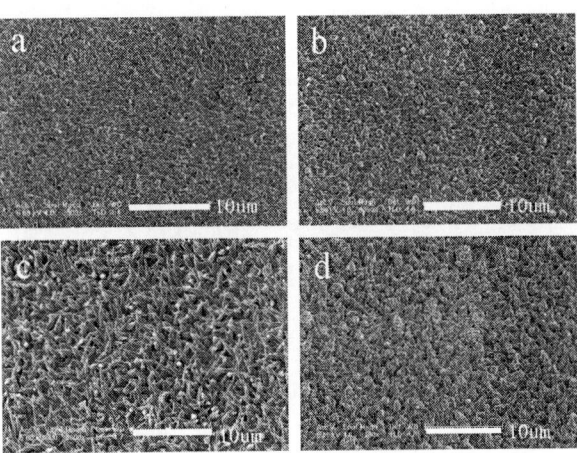

Fig. 6 The FE-SEM images of Cu electroless plating at different pH: (a) pH=7.5; (b) pH=8.5; (c) pH=9.0; (d) pH=9.5

Fig. 6 shows the SEM images of Cu electroless plating at different pH value. The pH was one of the main factors in electroless plating [14]. The morphologies of Cu nanocones deposited at three different pH were investigated in this paper. From the SEM images, when the pH was relatively low (pH=7.5 and pH=8.5), there was almost no nano-cones. With the increasing of pH to 9, the nano-cones had come to being. The nano-cones appeared cladodification at pH=9.5. So the proper pH for the sythesis of nano-cones was about 9.0. The possible reason was that high pH stimulated the reaction of anode reaction. It was reported that the velocity of electroless plating depended on the anode reaction [15-17]. The anode reaction in this eletroless plating is $H_2PO_2^- + H_2O \rightarrow H_2PO_3^- + 2H^+ + 2e^-$. For the reaction, higher pH was corresponding to the lower concentration of H^+, anode reaction was then accelerated. Thus, higher pH value would improve the growth of nano-cones.

3.5 Influence of temperature

Fig. 7 shows SEM images of Cu electroless plating at different temperatures. Temperature was a very important influence factor for chemical reaction [18,19]. High temperature can accelerated the proceeding of reactions. From

978-1-4244-4658-2/09 $25.00 © 2009 IEEE

the SEM images, the morphologies of nano-cones varied remarkably. When the temperatures were relatively low (45°C, 55°C), there was no nano-cones but the strip-shaped crystal. With the increase of temperature, the morphologies of nano-cones become obvious. The possible reason was that high temperature accelerated the reaction of electroless plating, then, nucleation had enough energy to carry on at the beginning. When the temperature was above 65°C, the activition of ions was intense. However, the temperature can not be too high, because H_3BO_3 would be dehydrated to be barium metaborate. So the proper temperature is about 65°C.

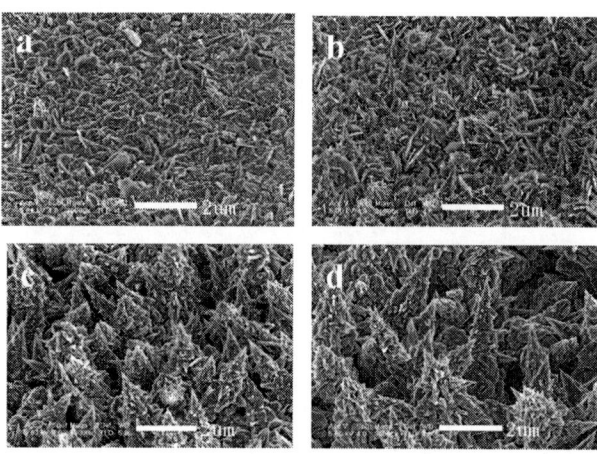

Fig. 7 The FE-SEM images of Cu electroless plating at different temperatures (a)45°C; (b)55°C; (c)65°C; (d)75°C

3.6 The TEM and EDX of the nano-cones

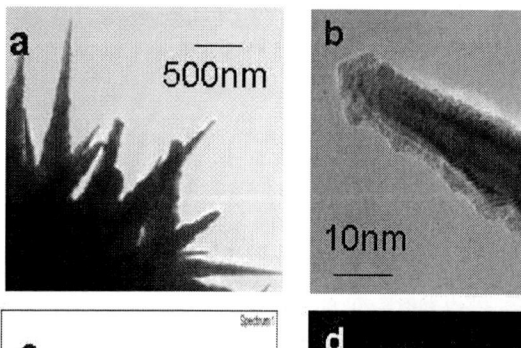

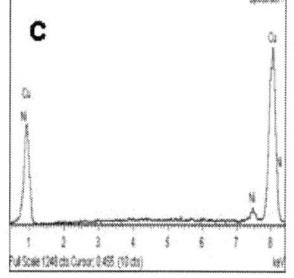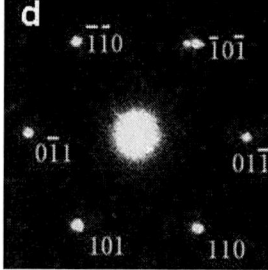

Fig. 8 (a) Bright-field image recorded of Cu nano-cones; (b) High magnification (TEM) image of single nanocone;(c) EDX spectrum taken from the nano-cones shown in (a); (d) Micro-diffraction pattern for that single nano-cone

The TEM samples were prepared by directly plating onto copper grids. Fig. 8(a) and (b) showed that nano-cones had very sharp tips with relatively smooth surface. The average tip angle of the nano-cones is about 15°. From Fig. 8(c), besides the existence of Cu as the main element of the deposit, energy-dispersive X-ray (EDX) analysis on several nano-cones also showed the presence of Ni. The elements of nano-cones consist of 95.57% copper and 4.43% nickel. No nickel oxide was detected, so the amorphous coating around the surface of the nano-cone may not be copper oxide. The Fig. 8(d) diffraction pattern was in agreement with the typical image of FCC copper along the zone axis of [111]. The diffraction spots formed a normal hexagon. It indicated that the nano-cone was a perfect FCC single crystal.

Conclusions

In conclusion, several important factors for electroless plating of Cu nanocones were investigated. The proper active time in $PdCl_2$ is about 90 seconds; the proper concentration of nickel sulfate as the catalytical agent was 2-4g/L; the crystallization conditioning agent was necessary for obtaining the nano-coned array; high pH was easier to get the nano-cones array, the proper pH was about 9.0; the optimum temperature was about 65°C. And the diameter range of the nano-cones was about 0.2~0.5μm, and the height range was about 1.0~2.5μm. From the EDX image, the elements of nano-coned array were mainly 95.57% copper and 4.43% nickel. It proved that Ni ion was the catalytical agent. From the results of TEM, the nano-cone was a perfect FCC single crystal. The detailed study of performance of nano-coned array was in progress in the lab.

Acknowledgments

This work is sponsored by International Science and Technology Cooperation of China (No. 2008DFA51680), National Natural Science foundation of China (60876071), Shanghai nano technology promotion center (No. 0852nm06300). We thank the Instrumental Analysis Center of Shanghai Jiao Tong University, for the use of the SEM equipment.

References

1. S.C. Park, C. Cho, and S.K. Paek, "Electronics Manufacturing Technology Symposium," *IEMT 2003, IEEE/CPMT/ SEMI 28th International IEEE Publisher*, 2003, pp. 31-37.

2. A. Fujii, M. Andoh, I. Yamamoto, H. Ibuki, K. Uchida, A. Yoshizumi, "Electronics Manufacturing Technology Symposium," *23rd IEEE/CPMT, IEEE Publisher*, 1998, pp. 478-491.

3. J. Park, Y. Kim, S.W. Wang, S.W. Lee, and H. Jeon, "Electromagnetic Scattering From Multiple Slant Strips," *IEEE Trans. Dev. Mater. Reliab.*, Vol. 6, No.1 (2006), 33

4. S.M. S. I. Dulal, HyeongJin Yun, CheeBurmShin,Chang-KooKim. "Electrodeposition of CoWP film III. Effect of pH and temperature," *Electrochimica Acta.*, Vol. 53 (2007), pp. 934-943.

5. Chang L.M., Guo H.F., An M.Z., "Electrodeposition of Ni-Co/Al$_2$O$_3$ Composite Coating by Pulse Reverse

Method Under Ultrasonic Condition," *Materials Letters*, Vol. 62, Issue 19, 15 July 2008.

6. Walter, EC, Favior, F, Penner, "Electronic devices from electrodeposited metal nanowires," *M. Anal Chem*, Vol. 27 (2002).

7. Heydon G P, Hoon S R, Farley A N, Tomlinson S L,Valera M S, Attenborough K and SchwarzacherW, J. Phys. D, "Magnetic force microscopy of soft magnetic materials," *Appl. Phys.*, 30, 1083, (1997).

8. K. Cho and E.C. Cho, "Phosphorus-doped silicon quantum dots for all-silicon quantum dot tandem solar cells Adhes," *Sci. Technol.*, Vol. 14, No. 11 (2000), pp. 1333

9. Abyaneh, M Y, Fleischmann M., "The Electrocrystallisation of Nickel," *Electroanal Chem*, 1981.

10. Tao Hang, Ming Li, Qin Fei and Dali Mao, "Characterization of nickel nanocones routed by electrodeposition without any template," *Nanotechnology*, 19, (2008) 035201 (5pp).

11. Hangtao, Huiqin Ling, Zhengji Xiu, MingLi, Dali Mao, "Study on the Adhesion Between Epoxy Molding Compound and Nanocone-Arrayed Pd Preplated Leadframes," *Journal of Electronic Materials*, Vol. 36, No. 12 (2007).

12. Shen G, Bando Y, Zhi C and Golberg D, "ELECTRON-ATOM-PHOTON INTERFACTIONS IN A LASER FIELD," *J. Phys.Chem.*, B 110 10714-9, (2006).

13. Paunovic, "Thermal degradation kinetics of isomeric poly (dipropyl itaconates)," *M.Plating*, (1968), 55: 1161.

14. Wernsdofer W, Doudin B, Mailly D, Hasselbach K, Benoit A, Meier J, Ansermet J-Ph and Barbara B, "Effect of Disorder on the Quantum Coherence in Mesoscopic Wires," *Phys. Rev. Lett.*, (1996) 77 1873.

15. L.D. Karpov, A.P. Genelev, Y.V. Mirgorodski, and A.N.Tikhonski, "Patterning of vertical thin film emitters in field emission arrays and their emission characteristics," The 9th International Vacuum Microelectronics C*onference*, St. Petersburg IEEE Publisher, 1996, pp. 542.

16. S.P. Davis, M.R. Prausnitz, and M.G. Allen, "Minimally invasive protein delivery with rapidly dissolving polymer microneedles," The 12th International Conference on Solid State Sensors Boston: Actuators and Microsystems, Bontham Science Publisher, 2003, pp. 1435.

17. Liu Z, Li S, Yang Y, Peng S, Hu Z and Qian Y, "Indentation of an oriented transparent polyamide," *Adv. Mater.*, 15 1946, 2003.

18. Fangzu Yang, Bin Yang, Jianting Huang, Liqiong Wu, Ling Huang, Shaoming Zhou., "Sodium-nickel alloy of copper prepared by sodium hypophosphite electroless plating," *Electroplating & Finishing*, Vol. 25, No. 7.

19. Dandan Shen, Fangzu Yang, Huihuang Wu., "The influence of 2,2'2 bipyridine and ferrous potassium cyanide on the glyoxylate electroless plationg," *Electrochemistry*, Vol. 13, No. 1 (2007).

Research on Self-constrained Sintering Low-temperature Cofired Ceramic

Yongda Hu, Taohua Liang, Yuanxun Li, Bangchao Yang, Yun Lu

State Key Laboratory of Electronic Thin Films and Integrated Devices, Chengdu, China, 610054

No. 4, Section 2, North Jianshe Road, Chengdu, P. R. China

Email: yongnet169@sina.com

Abstract

The effects of an inner constraint layer and alumina particles on the microstructure, strength, and shrinkage of the laminated low-temperature cofired ceramic (LTCC) green sheet were investigated. A crystalline glass mixed with alumina particles was used as inner constraint layer. Sintering shrinkage in the x-y direction of the LTCC is related to the bending strength of the alumina particle and glass-ceramic layer used for an inner-constraint layer. The crystallization temperature about crystalline glass is important. In the self-constrained sintering LTCC, the soft point of crystalline glass should be above 870°C. Smaller alumina particles in the inner-constraint layer produced a substrate with a high bending strength.

Introduction

Glass-based ceramic materials offer significant advantages, including low sintering temperature, and low dielectric constant, relative to high-temperature co-fired ceramics for electronic packaging applications. These materials, known as low temperature co-sintered ceramics (LTCCs), enable the integration of electronic, mechanical, photonic, fluidic, and bio-chemical functions in compact 3-D architectures.

At present the unconstrained sintering (UCS) is one of the most used LTCC processes. Despite many advantages the application field of freely sintered LTCC substrates is limited. One reason for this is the low dimensional accuracy. The resulting tolerances in shrinkage during sintering might be responsible for component assembly problems. Constrained sintering essentially eliminates almost completely the x/y shrinkage and reduces tolerances.

Different constrained sintering processes are known such as the self constrained sintering (SCS), the pressureless constrained sintering (PLAS) and pressure assisted constrained sintering (PAS).

PLAS and particularly the PAS process use a laminated release tape on the external faces of the module to constrain the LTCC during sintering. The release tapes are based on ceramic particles which sinter at higher temperatures than those used during the LTCC firing. So they do not shrink and mostly prevent the rest of the substrate from shrinking in x/y direction.

Self Constrained Sintering is quite a new technique and works without external constraints [1]. It is handled using standard free-sintered LTCC processes. The mechanism of this process is based on the glass ceramic composition of the tape itself. At present the choice of compatible pastes is limited.

Our research works on Self Constrained Sintering LTCC.

Experiment

1. Materials

A commercially available α-Al_2O_3 powder was used for this study. MgO, CaO, SrO and SiO_2 were chosen to made RO-Al_2O_3-SiO_2 glass for the inner-constraint layer about Self Constrained Sintering LTCC. P_2O_5 was chosen to be additives of the glass-ceramics [2, 3]. Cylindrical specimen were first pressed using a pressure of 100 MPa.

2. Microstructure characterization

Crystallization of the self-constrained LTCC material during sintering was studied using DTA equipment (STA 409C, NETZSCH). The DTA samples were prepared by debinding of the LTCC material to remove the organic binders. The testing parameters were 5°C/min from 30 to 900°C in air environment. X-ray diffraction (XRD, Philips X'Pert ProMPD DY1291 with CuKa) was used to follow phase evolution. Microstructure development from green compacts was observed by SEM (JSM-6490LV) on fracture.

Results and Discussions

1. Mechanism

The mechanism of Self Constrained Sintering LTCC was showed in Fig. 1. The inner-constraint layer is porous ceramic with glass-ceramic and alumina composite. It contains zero shrinkage at high temperature. The glass melts and infiltrates into porous ceramic and LTCC substrate completed.

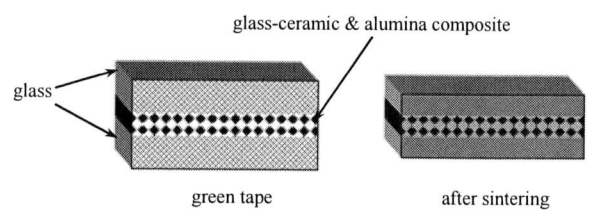

Fig. 1 The mechanism of Self Constrained Sintering LTCC

2. XRD analysis

RO-Al_2O_3-SiO_2 glass and Al_2O_3 powder was involved in the inner-constraint layer. Al_2O_3 powder with an average particle size of about 5μm was used for this study. It reacts with glass and $MgAl_2O_4$, the spinel structure was detected by XRD as shown in Fig. 2 when LTCC has been sintered at 865°C. Thus glass-ceramic was formed at inner-constraint layer. The melting point of $MgAl_2O_4$ spinel is about 2135°C, which is higher than the LTCC sintering temperature(865°C).

978-1-4244-4658-2/09 $25.00 © 2009 IEEE

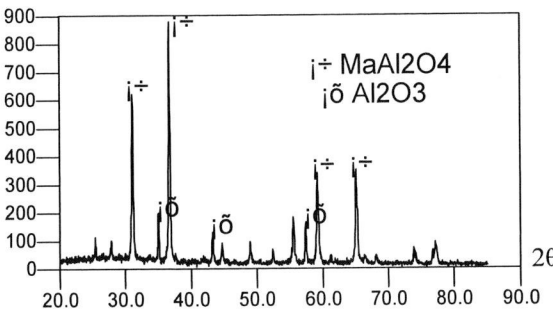

Fig. 2 XRD analysis on inner-constraint layer

3. DTA

MgO can prompt crystallization behavior of glass-ceramic and decrease crystallization temperature. P_2O_5 is crystal activator in glass-ceramic. With increasing P_2O_5 content, glass transition temperature and crystallization temperature all increased until P_2O_5 was contained about 4%wt in the glass-ceramic composite.

If 17%wt~23%wt MgO and 2.7%wt P_2O_5 were added in the glass-ceramic composite, the crystallization temperature can decrease to 710°C that was shown in Fig. 3. The crystallization temperature of glass-ceramic that was used to inner-constraint layer is lower than the soft point of 730°C (Fig. 4) about glass that was used on both side of inner-constraint layer.

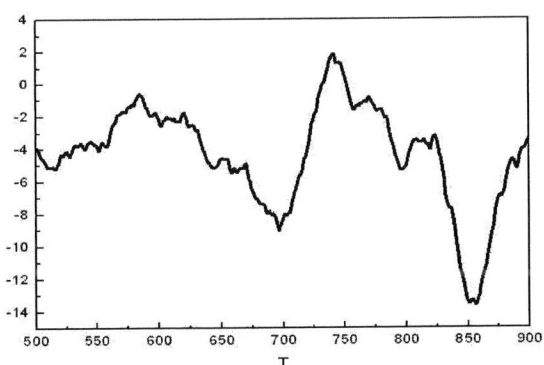

Fig. 3 Crystallization temperature on inner-constraint layer by DTA

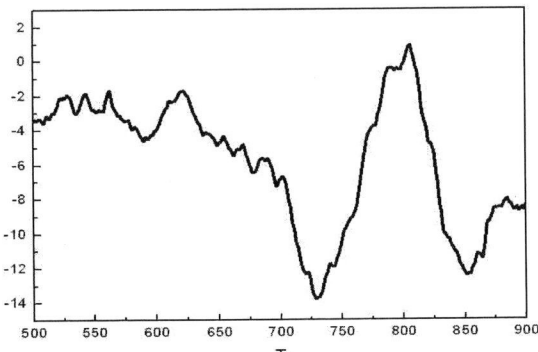

Fig. 4 soft point about glass that was used on both side of inner-constraint layer by DTA

4. SEM

If the quality proportion between alumina and glass-ceramic is 10:1 to 7:1, zero shrinkage porous ceramic structure was fabricated. The glass-ceramic crystallization temperature is lower than that glass softening-point where it is on both sides of inner-constraint layer. If the glass-ceramic portion were below 9%wt, it should not interlink alumina powder together and the inner-constraint layer should be broken. If the glass-ceramic portion were over 12%wt, the inner-constraint layer should be shrinkage in the LTCC manufacturing process and the substrate should be warp. The Fig. 5 shows the micro-structure about Self Constrained Sintering LTCC.

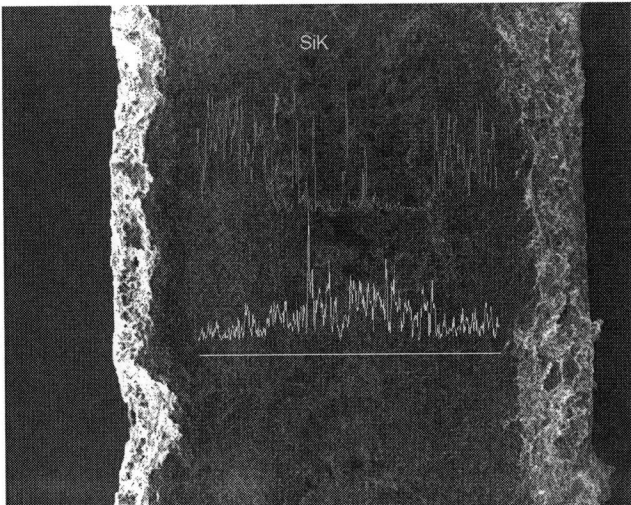

Fig. 5 The micro-structure about Self Constrained Sintering LTCC

Conclusions

A novel method, based on Self Constrained Sintering, was presented. The inner-constraint layer was made up of alumina particles and glass-ceramic. If the quality proportion between alumina and glass-ceramic is 10:1 to 7:1, zero shrinkage porous ceramic structure was fabricated. Spinel structure $MgAl_2O_4$ forms and the inner-constraint layer became solid in sintering stage. Glass, which is on both side of inner-constraint layer, melts and infiltrates into porous ceramic. A compact LTCC substrate was completed.

References

1. Barnwell, P., Amaya, E., Lautzen, "HeraLock Self-constrained LTCC Tape," *IMAPS Nordic*, Stockholm 2002.
2. J. Fu, "Effects of M^{3+} ions on the conductivity of glasses and glass-ceramics in the system $Li_2O-M_2O_3-GeO_2-P_2O_5$ (M=Al, Ga, Y, Dy, Gd, and La)," *J. Am. Ceram. Soc.*, Vol. 83, No. 4 (2000), pp. 1004-1006.
3. B. A. Vazquez, A. Caballero, P. Pena. "Quaternary system $Al_2O_3-CaO-MgO-SiO_2$: II Study of the crystallization volume of $MgAl_2O_4$," *J. Am. Ceram. Soc.*, Vol. 88, No. 7 (2005), pp, 1949-1957.

Fine Pitch and High Density Sn Bump Fabrication

Jinglin Bi[1], Jin Jiang[1], Anmin Hu[1], Ming Li[1], Dali Mao[1], Tadatomo Suga[2]

[1]Lab of Microelectronic Materials & Technology, State Key Laboratory of Metal Matrix Composites
School of Materials Science and Engineering, Shanghai Jiao Tong University, Shanghai 200240, China
[2]Department of Precision Engineering, School of Engineering, the University of Tokyo
Email: mingli90@sjtu.edu.cn, Tel: +86-21-34202542

Abstract

Three-dimensional packaging technology, which requires fine pitch and high density of solder bumps, has been developed recently for system-in-package applications. There are several methods being used for solder bumping process for now. As the sphere pitch decreases to below 100μm, electrodepositing has an advantage over robotic ball placement and screen printing in the cost per ball, according to the study done by ROHM HAAS electronic materials. The current environmentally conscious manufacturing moves toward Pb-free schemes for electronic devices and components. Sn–Ag and Sn–Cu alloy system is an acknowledged candidate among Pb-free solders. However, it is difficult to fabricate alloy-plated bumps. In this paper, the fabrication of lead-free bumps made of pure-tin, as a basic study, is described. Area-array tin solder bumps each of size 60μm diameter on an 180μm pitch with very tight height variation were obtained. In addition, tin bumps can also be applied in a new chip-to-chip interconnection method, in which bonding between the chips is achieved by deformation-injection of tin stud bump on a chip into the through via hole in the other chip. SEM was used to observe the microstructure of tin bumps. At last, the reflow was carried out in a glycerol bath in order to get a high quality of the solder bumps.

Introduction

At present, there are several technologies mainly used in bump fabrication, screen printing, ball placement and electroplating. As the distance between bumps gets small and the number of bumps becomes high, electroplating is the best concerning the ability to define bump sizes [1] and pitch and total cost [2].

On the other hand, the solder material used should be lead-free complying with the ban on lead-containing materials, and the Sn-Ag and Sn-Cu systems are acknowledged candidates among Pb-free solder materials, with tin constituting a major fraction of the solder on the basis of [3]:

(1) worldwide resources and availability of tin

(2) the comparable cost of the lead-free alloys to lead-tin solder

(3) the fact that these bumps can be fabricated using conventional low-cost electroplating techniques

(4) the compatibility of tin-based solders with current reflow processes, materials, and surface mount equipment

(5) the familiarity of the electronics manufacturing and assembly industry in handling tin alloys

However, the composition of alloy-plated bumps can not be easily obtained because metal contents of an alloy-plating solution are varied with increasing alloy-plating depositions [4]. To eliminate this concern, the two-step electroplating process using separate reactors for each element has been developed, in which Ag is first electroplated and then tin is deposited on the Ag layer.

In this paper, the fabrication of lead-free bumps made of pure-tin, as a basic study, is described.

Experiment

Electrodepositing is performed in the mesylate electrolyte as shown in Table 1 while an external power supply injects direct current between the anode (tin) and the cathode (wafer or sheet copper). Anode and cathode are placed face-to-face quite close.

Table 1 Tin plating solution compositions

Chemical	Parameter
Methanesulfonic acid	170g/L
Tin methanesulfonate	45g/L
Additive agent	40ml/L
Complex wetting agent	suitable

Dies with patterned figure made by positive photo-resist was prepared as shown in Fig. 1. The diameter of eyelets is 60μm. The depth is 60μm, and the pitch is 180μm.

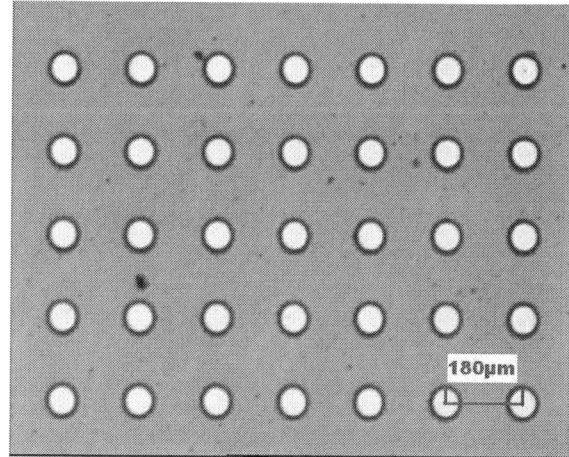

Fig. 1 Patterned figure made by positive photo-resist

The tin bumps were fabricated at room temperature with a current density of 5.0-30.0A/dm² with or without stirring.

Positive photo-resist and UBM under photo-resist are chemically removed before reflow. The reflow is performed in a glycerol bath because glycerol provides a uniform

978-1-4244-4658-2/09 $25.00 © 2009 IEEE

temperature along the wafer surface and performs similar oxidation prevention than the standard solder fluxes [1]. First, the wafer with the electrodeposited bumps was immersed into a glycerol bath at 265°C and kept for 90 seconds. And then, the cooling step is a sharp temperature decrease obtained by simply immersing the wafer into a 25°C glycerol bath.

The observations of longitudinal sections and top view of bumps were performed with Scanning Electron Microscope (FEI SIRION 200).

Results and Discussion

From a mass production point of view, production efficiency is the very important value that should be concerned. Thus, we study the effect of current density on current yield to calculate deposition rate that we can control the production efficiency. As shown in Fig. 2, current yield decreased steadily , as current density increased , from 98.5% with a current density of 5 A/dm^2 to 52.7% with a current density of 30 A/dm^2.

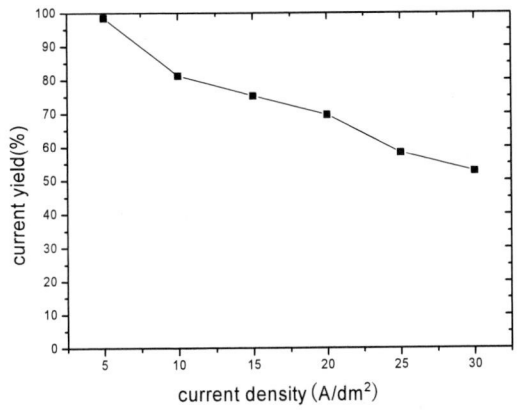

Fig. 2 Effect of current density on current yield

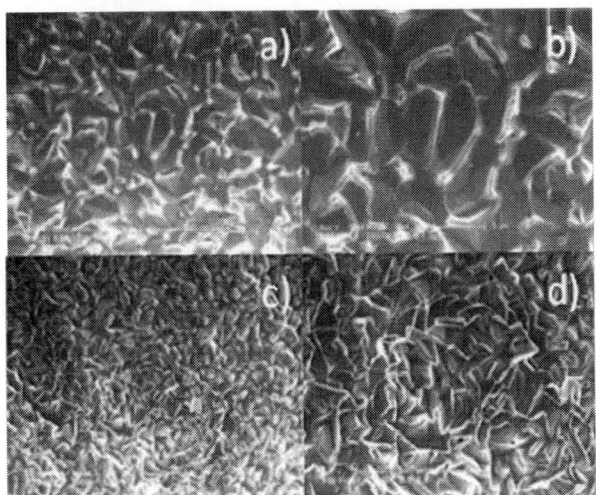

Fig. 3 SEM images of top view of electrodeposited tin layer under different current density in different enlargement ratio: a) 10A/dm^2, 5000×, b) 10A/dm^2, 10000×, c) 30A/dm^2, 5000×, d) 30 A/dm^2, 10000×

Besides, current density had a great influence on the microstructure of electrodeposited layer. As shown in Fig. 3, the crystal grains became much finer as the current density increased from 10A/dm^2 to 30A/dm^2.High current density means high energy that more crystal nucleus formed in the surface of cathode. These great numbers of nucleus grew until they touch each other, as a result, crystal grains got refined.

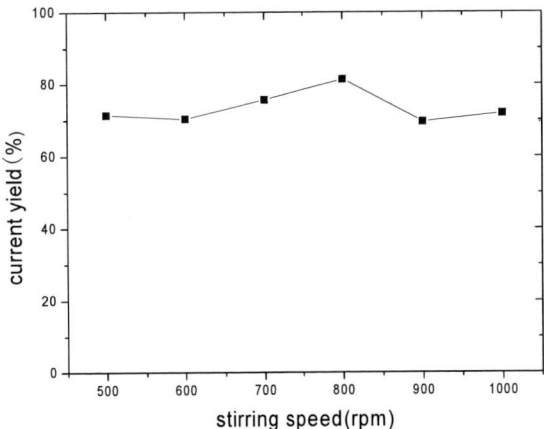

Fig. 4 Effect of stir speed on current yield

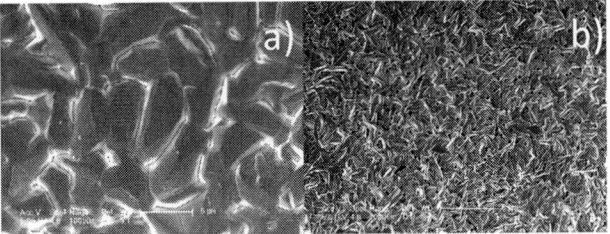

Fig. 5 SEM images of top view of electrodeposited layer a) with stirring and b) without stirring

Stirring rate is another factor that influences current yield. Current yield increased from 60% without stirring to 71.5% with a stirring rate of 500 rpm. However, higher rate didn't contribute to higher current yield, from figure 4. As stirring rate changed from 500 rpm to 1000 rpm, current density didn't get promoted remarkably but fluctuated rather narrowly. This is caused by the following three aspects: (1) concentration polarization reduced in a proper stirring intensity, and deposition rate got promoted. (2) the solution foamed with a too high stirring rate. (3) more air got into the solution with a higher rate, leading to the oxidation of tin ion from bivalence to quadrivalence. In face, stirring rate and current density are coupled, in other words, high current density asks for high stirring rate, but it is not wise to increase the stirring rate as high as possible with a certain current density. Besides, microstructure of the electrodeposited layer changed after stirring was adopted, because stirring lowered cathodic polarization which resulted in fine grains.

Although crystal grains got finer without stirring, electrodeposited layer turned dim. Moreover, little cords appeared on the surface as shown in Fig. 6, because they tend to discharge at points when there are little metallic ions nearby the cathode. Thus, suitable stirring rate is necessary to obtain bumps of high quality.

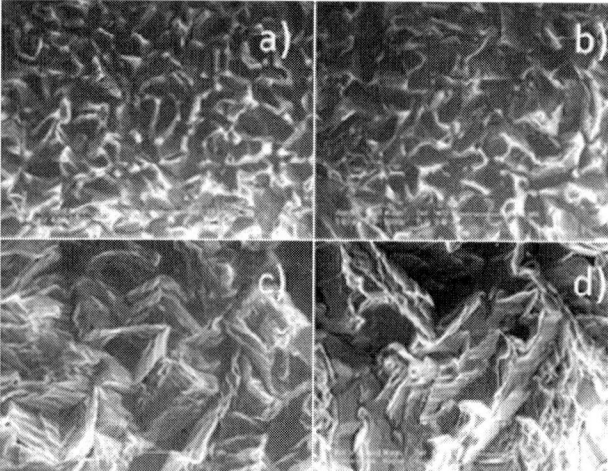

Fig. 6 Microstructure of surface cord on the electrodeposited tin layer without stirring, 500×.

Fig. 7 SEM images of top view of electrodeposited tin layer of a) 12.3μm, b) 18.1μm, c) 63.9μm, d) 152.3μm thick

The effect of thickness (electrodepositing time) to the microstructure of electrodeposited layer is shown in Fig. 7.

As the time of electrodepositing expanded, the microstructure of the layer changed obviously. When the layer was very thin, crystal grains are fine and neat as the growth of grains were almost determined by the lattice structure of substrate. However, as the thickness increased to a certain extend, the grains turned disordered caused by anisotropic growth. Fig. 7 (c) (d) shows river-shaped surroundings around the grain. This is probably due to the liberation of hydrogen which leads to an increase of the pH value of the cathode area. As a result, subsalt or hydroxide was formed in the layer. As what we can see in the Fig. 7(d), there are a lot of mill pores in the grains owing to severe liberation of hydrogen.

For future application in production, we finally prepare bumps. As we studied previous, we chose a lower current density and stirring rate can't be too high. Tin bumps were fabricated at room temperature with a current density of 5.0A/dm^2 with a stirring rate of 500 rpm. Fig. 8 shows bumps in different height from top view, cross section at 60° angle. The height of bumps can be controlled accurately by the time

of electrodepositing time, and tin bumps of 60μm can be exactly fabricated.

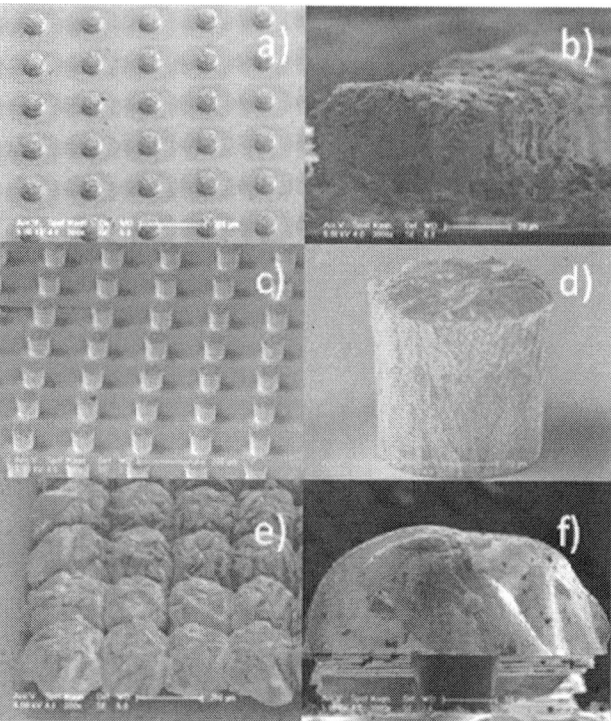

Fig. 8 SEM images of tin bumps in different height: a) b) 20μm; c) d) 60μm; e) f) above 100μm

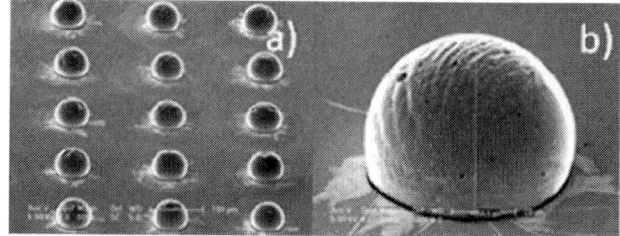

Fig. 9 SEM images of tin bumps after reflow

Finally, bumps were reflowed in a glycerol bath in order to get a high quality of the solder bumps and SEM images of tin bumps after reflow were shown in Fig.9. From figure 9, all the bumps achieved a perfect spherical shape and the surface of bumps were smooth.

Conclusions

1. The influence of current density and stirring rate to current yield and the microstructure of the electrodeposited layer was studied in the mesylate electrolyte. The result shows that current yield decreased as current density increased. The higher current density was, the finer the crystal grain was. Current yield increased as stirring rate was speeded up properly. Without stirring, the grain was much more refined, but little cords appeared on the surface. Besides, we also made a study on how the microstructure changed as electrodeposited layer thickened. It proved that when the layer

was very thin, crystal grains are fine and neat, but as the thickness increased to certain extend, the grains turned disordered.

2. Area-array tin solder bumps each of size 60μm diameter on an 180μm pitch in different height were obtained at room temperature with a current density of 5.0A/dm^2.

3. The bumps were reflowed in a glycerol bath, and perfect spherical bump shape with smooth surface was achieved.

Acknowledgments

This work is sponsored by International Science and Technology Cooperation of China (No. 2008DFA51680), National Natural Science foundation of China (60876071), Shanghai nano technology promotion center (No. 0852nm06300). We thank the Instrumental Analysis Center of Shanghai Jiao Tong University, for the use of the SEM equipment.

References

1. Enric Cabruja, Marc Bigas, Miguel Ullan, Giulio Pellegrini, Manuel Lozano, "Special bump bonding technique for silicon pixel detectors," *Nuclear Instruments and Methods in Physics Research Section A: Accelerators, Spectrometers, Detectors and Associated Equipment,* Vol. 576, No.1 (2007), pp. 150-153.

2. CL Wong and James How, "Low Cost Flip Chip Bumping Technologies," *1stIEEVCPMT Electronic Packaging Technology Conference(1997),* Singapore, pp. 244-250.

3. Zaheed S. Karim and Rob Schetty, "Lead-Free Bump Interconnections for Flip-Chip Applications," *26th IEEE/CPMT Int'l Electronics Manufacturing Technology Symposium, (2000),* Boulder, Colorado, pp. 274-278.

4. Hirokazu Ezawa, Masahiro Miyata, Soichi Honma, Hiroaki Inoue, Tsuyoshi Tokuoka, Junichiro Yoshioka, and Manabu Tsujimura, "Eutectic Sn–Ag Solder Bump Process for ULSI Flip Chip Technology," *IEEE transactions on electronics packaging manufacturing,* Vol. 24, No. 4 (2001), pp. 275-281.

Effect of [Au]/[Na₂SO₃] Molar Ratio on Co-electroplating Au-Sn Alloys in Sulfite-based Solution

Xiangyong Qing, Mingliang Huang*, Jianlin Pan

Electronic Packaging Materials Laboratory, School of Materials Science & Engineering, Dalian University of Technology, Dalian, 116024, China

*Corresponding Author, E-mail: huang@dlut.edu.cn, Tel: 86-411-84706595

Abstract

LEDs (Light Emitting Diodes) that assembled using flip-chip technology are today used as long-life, energy efficient, environmentally friendly light sources. However, the flip chip solder joints have to meet high requirements. Therefore, their performance and quality are crucial for the integrity of the assembly, which in turn is vital to the overall function of the LED. Au-30at.%Sn eutectic alloy is the generally used solder in electro-optical assemblies due to its excellent thermal and mechanical properties. Au-Sn solder bumps can be obtained by sequential electroplating of Au and Sn layers or by co-electroplating Au-Sn alloys from a single solution. In the present work, Au-Sn alloys have been co-electroplated from a non-cyanide, sulfite-based stable solution which contains $Na_3Au (SO_3)_2$ (gold sodium sulfite) as the source of gold and $SnSO_4$ (stannous sulfate) as the source of Sn. Na_2SO_3 (sodium sulfite) is added as the complexing agent for gold and an additional commercial complexing agent for Sn. The effect of the [Au]/[Na₂SO₃] molar ratio in the plating solution on the composition of the deposits, surface morphology and plating rate has been investigated. It was shown that the [Au]/[Na₂SO₃] molar ratio of 1/24 proved to be the best one with respect to plating rate and surface morphology in the present experiment. When the Sn^{2+} concentration is 0.03 mol/L, the optimum concentration for co-electroplating Au-Sn alloys is Au(I) concentration of 0.02 mol/L and Na_2SO_3 concentration of 0.48 mol/L, corresponding to the [Au]/[Na₂SO₃] molar ratio of 1/24.

Introduction

LEDs, fabricated with III-V compound semiconductors, are capable of transforming electric energy into light energy with high efficiency and a life-span up to thousands of hours. In order to decrease the thermal resistance and improve heat transport away from the LED, flip chip technology has been developed to fabricate high-performance of LEDs. Thus bumping becomes a critical step in FC technology. The performance and quality of the solder bumps are crucial for the integrity interconnection, which in turn is vital to the overall function of the LEDs.

As a joining material, solder provides electrical, thermal and mechanical continuity in electronics assemblies. Pb-Sn solders are well-known as the bumping material. However, with the increase of the environmental concerns, the use of lead free solders for electronic products has become a major driving force for technology developments [1, 2]. The Au-30at.%Sn alloy appears to be an attractive alternative, with a melting temperature of 280°C. Moreover, the Au-30at.%Sn alloy owns superior mechanical and thermal properties to conventional Pb-Sn solders, and have no negative effect on the environment.

Electroplating Au-30at%Sn alloy has been studied extensively because of its low cost and simplicity of operation. Au-30at.%Sn alloy is usually deposited by sequential electroplating of pure Au and Sn layers from separate Au and Sn solutions [3~6]. However, the available information concerning the co-electrodeposition of Au-Sn alloys from a single solution is virtually confined to the patent literature [7, 8]. Moreover, the plating rate, which is important to the competitiveness with available technologies used in industry, is very low.

In the present work, the effect of the [Au]/[Na₂SO₃] molar ratio in the plating solution on the composition of the deposits, surface morphology and plating rate was investigated. The final goal is to develop an Au-Sn co-electroplating process which can provide a fast plating rate.

Experiment

The chemical solution used for electroplating of the Au-Sn alloys was sulfite-based. It contained $Na_3Au (SO_3)_2$ (gold sodium sulfite) as the source of gold and $SnSO_4$ (stannous sulfate) as the source of Sn. Na_2SO_3 (sodium sulfite) is added as the complexing agent for gold and an additional commercial complexing agent for Sn. All solutions for this study were prepared with de-ionized water and pH was adjusted to 8~9.

In the experiments, the gold sodium sulfite concentration was varied from 0.02 mol/L to 0.04 mol/L and the dosage of sodium sulfite was varied according to the [Au]/[Na₂SO₃] molar ratios in each Au(I) concentration. For example, when the concentration of $Na_3Au(SO_3)_2$ was 0.02 mol/L, the dosage of Na_2SO_3 was 0.16, 0.24, 0.32, 0.4, 0.48 mol/L, corresponding to [Au]/[Na₂SO₃] (Sodium sulfite) molar ratio of 1/8, 1/12, 1/16, 1/20, 1/24.

Si wafers diced into pieces which had been sputtered with Ti (0.2μm) /Au (1μm) blanket metallization were used as cathodes. Ti acted as an adhesion layer and Au as a seed layer. A stainless steel plate sputtered with Ti (0.2μm) /Au (1μm) /Pt (0.2μm) was used as anode.

Co-electroplating of Au-Sn alloys were carried out under periodic reversal pulse current with a cycle period of 10 ms and $t_{on1}:t_{off1}:t_{on2}:t_{off2}=2:4:1:3$, as shown in Fig. 1. Electroplating was performed for 30 min and the electroplating temperature was 45°C.

The scheme of electroplating equipment is shown in Fig. 2. The plating solution was heated in the water bath and agitation was introduced to increase the diffusion of the ions in the solution in order to homogenize the plating bath.

All electroplated samples were examined using a scanning electron microscope (SEM). The chemical composition of electroplated Au-Sn film was identified by energy dispersive x-ray (EDX) analysis. An accelerating voltage of 15 KV was used for both imaging and composition analysis.

978-1-4244-4658-2/09 $25.00 © 2009 IEEE

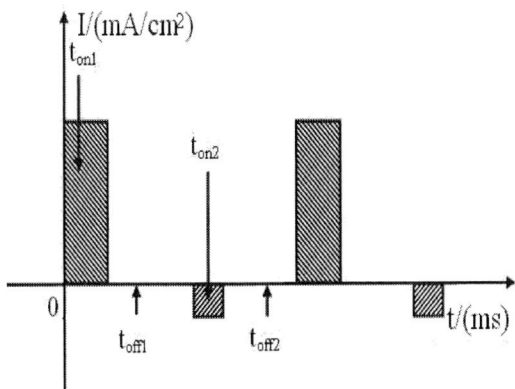

Fig. 1 The schematic of the electroplating current

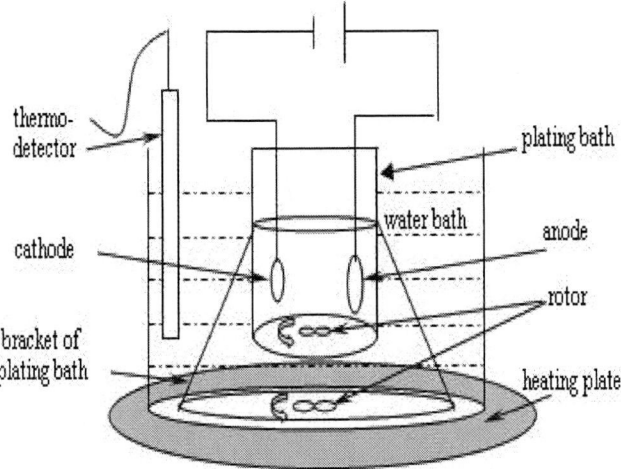

Fig. 2 The scheme of electroplating equipment

Results and Discussion

In the plating solution, $Na_3Au(SO_3)_2$ acted as the source of the gold and Na_2SO_3 mainly functioned as Au complexing agent, as well as buffer and conducting salt. Both of them played important roles in the electrodeposition and it is very likely that the $[Au]/[Na_2SO_3]$ molar ratio in the plating solution strongly influences the reaction   kinetics in the electrodeposition process.

As shown in Fig. 3, when the Au(I) concentration was 0.02 mol/L, the increase of Na_2SO_3 concentration had no significant effect on the deposits composition. However at higher Au(I) concentrations, e.g. 0.04 mol/L Au(I), the Sn content in deposits decreased with the increase of Na_2SO_3 concentration.

This phenomenon is related to the $[Au]/[Sn]$ molar ratio. In the Sn rich solution, two plateaus for 15at.%Sn (Au_5Sn phase) and 50at.%Sn (AuSn phase) deposits were plated out at high current density , as shown in Fig. 4 [9, 10]. But in the Au rich solution, the plateau for 50at.%Sn deposits could not be plated out at high current density, where the Sn content in Au-Sn deposits increased initially with increasing current density and then decreased steadily, as shown in Fig. 5 [11].

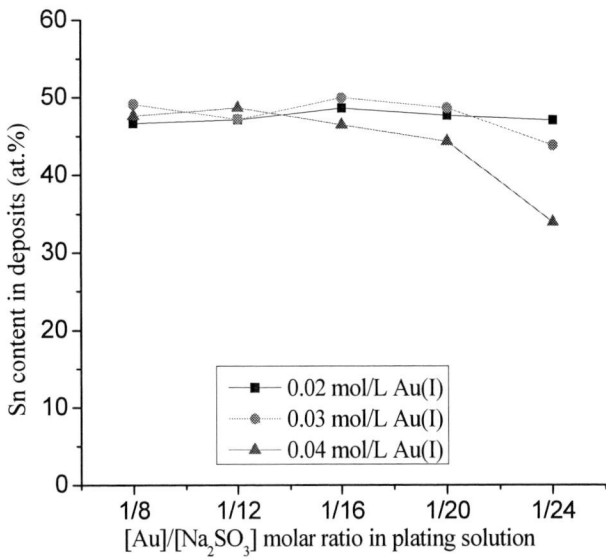

Fig. 3 Sn content in deposits versus the $[Au]/[Sn]$ molar ratios in the sulfite-based solution.

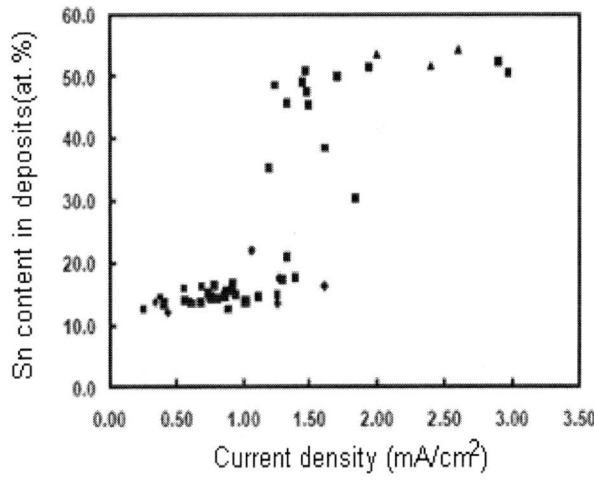

Fig. 4 Deposits tin content versus current density. Electroplating was performed at room temperature from the solution containing $KAuCl_4$ concentration of 5 g/L, corresponding to $[Au]/[Sn]$ molar ratio of 0.6 [9, 10].

When the Au(I) concentration was 0.02 mol/L, corresponding to $[Au]/[Sn]$ salt ratio of 2/3, the solution was confirmed to be Sn rich. It is very likely that a plateau for 50at.%Sn deposits was plated out at high current density. It was considered that the higher Na_2SO_3 concentration resulted higher complexation of Au ions, which would act as a barrier for the deposition of Au and shifted the plateau to lower current density, thus the plateau for 50at.%Sn deposition was still plated out when increasing the Na_2SO_3 concentration, Sn content in deposits have no significant change at high current density.

978-1-4244-4658-2/09 $25.00 © 2009 IEEE   799

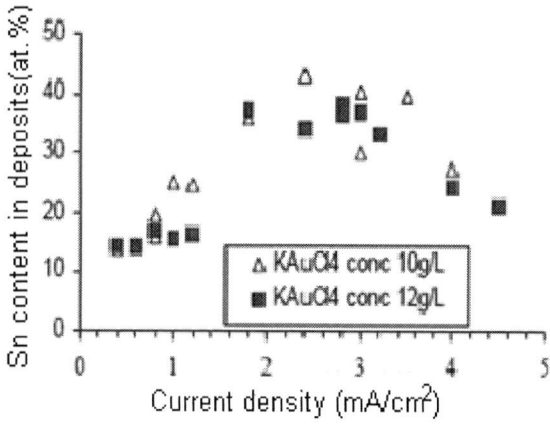

Fig. 5 Sn content in deposits obtained at various current densities from baths obtaining $KAuCl_4$ concentration of 10 g/L and 12 g/L, corresponding to [Au]/[Sn] molar ratio of 1.2 and 1.4 [11].

When increasing the concentration of the Au salt, the solution became Au rich. The plateau for 50at.%Sn deposits could not be plated out at high current density. The increase of the Na_2SO_3 concentration moved the curve of Sn content in deposits versus current density to the left. So in the low current density region, increasing the Na_2SO_3 concentration led to an increase of Sn content in deposits, while in the high current density region, it would have a decrease of Sn content.

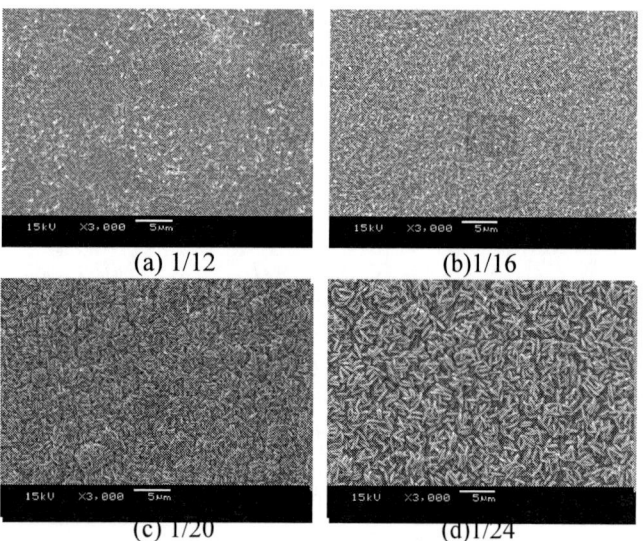

Fig. 6 SEM top view of Au-Sn deposits showing the effect of the different [Au]/[Na_2SO_3] molar ratio (a) 1/12, (b) 1/16, (c) 1/20 and (d) 1/24 in the plating solution on surface morphology when Au(I) concentration was 0.02 mol/L.

The morphologies of Au-Sn film obtained at different [Au]/[Na_2SO_3] molar ratios in the plating solution with the Au(I) concentration fixed at 0.02 mol/L were shown in Fig. 6. Crystal grains became finer when Na_2SO_3 concentration was increased, which could be related to the increasing of the over potential.

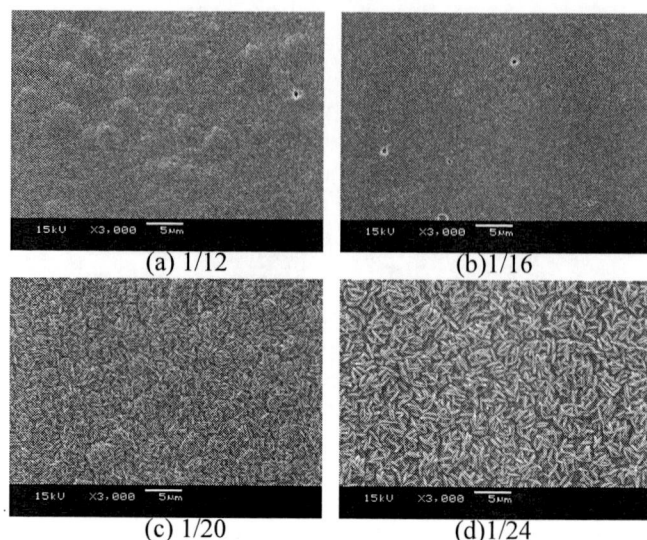

Fig. 7 SEM top view of Au-Sn deposits showing the effect of the different [Au]/[Na_2SO_3] molar ratio (a) 1/12, (b) 1/16, (c) 1/20 and (d) 1/24 in the plating solution on surface morphology when Au(I) concentration was 0.04 mol/L.

Fig. 7 showed the top view of Au-Sn deposits obtained at different [Au]/[Na_2SO_3] molar ratios when the Au (μ) concentration was 0.04 mol/L. The same phenomenon had been observed at the [Au]/[Na_2SO_3] molar ratio from 1/12 to 1/16. However, at the higher Na_2SO_3 concentration, the surface became coarser. This phenomenon was contributed to the high viscosity when a large amount of Na_2SO_3 added in the plating solution. The Na_2SO_3 concentration corresponding to the Fig. 7 (c) and (d) were 0.8 mol/L and 0.96 mol/L, which increased markedly the viscosity of the plating solution and retard the diffusion of metal ions to the cathode surface during plating, which led to a decreased activity of plating metal ions. Thus hydrogen evolution became significant, result in a coarser surface.

Fig. 8 showed the relationship between the plating rate and the [Au]/[Na_2SO_3] molar ratio in the plating solution. When the [Au]/[Na_2SO_3] molar ratio was varied from 1/8 to 1/20, increasing the Na_2SO_3 concentration had no significant effect on the plating rate. When the [Au]/[Na_2SO_3] molar ratio was enhanced to 1/24, plating rate had a sharply increase, although the deposits electroplated from the solution contained 0.04 mol/L Au(I) presented a coarser surface and lower Sn content in deposits compared to the others. So in terms of the plating rate, the [Au]/[Na_2SO_3] molar ratio of 1/24 was the best one for electroplating.

For the [Au]/[Na_2SO_3] molar ratio of 1/24, when the Au(I) concentration was 0.02 mol/L, plating rate was about 9 μm/30min. While increasing the Au(I) concentration, plating rate decreased. So increasing the Au(I) concentration was not surely increasing the plating rate. Reducing the Au(I) concentration had the added benefit of reducing the cost of plating bath.

Since the large amount of Na_2SO_3 markedly increased the viscosity of the plating solution, the Na_2SO_3 concentration could not be too high. The [Au]/[Na_2SO_3] molar ratio of 1/24 and Au(I) concentration of 0.02 mol/L had been proved to be the best parameters with respect to plating rate and surface morphology in the present work.

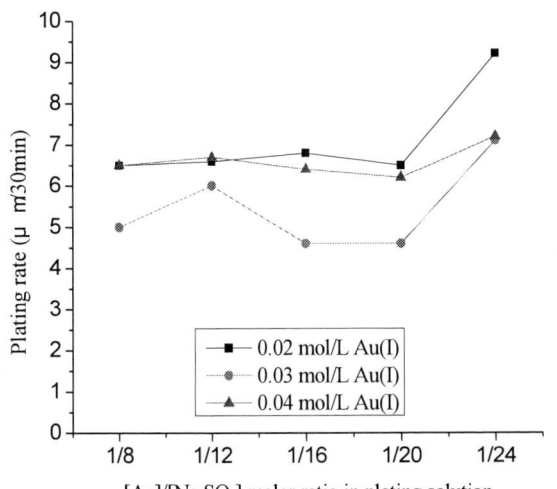

Fig. 8 Plating rate versus the [Au]/[Na$_2$SO$_3$] molar ratio in plating solution

Conclusions

In general, when the [Au]/[Na$_2$SO$_3$] molar ratio varied from 1/8 to 1/24, there was no significant effect on the Sn content in the Au-Sn deposits, when electroplated from Sn rich solutions at high current density. However, in the Au rich solutions, the Sn content of Au-Sn deposits decreased with the increasing Na$_2$SO$_3$ concentration at high current density.

A smoother, denser deposit was obtained from plating solutions containing a higher Na$_2$SO$_3$ concentration. However, if the Na$_2$SO$_3$ concentration was too high, hydrogen evolution became significant, resulting in a coarser surface of Au-Sn deposits.

The plating rate had no significant change at the [Au]/[Na$_2$SO$_3$] molar ratio from 1/8 to 1/20. When the ratio was increased to 1/24, the plating rate increased sharply.

When the Sn^{2+} concentration was 0.03 mol/L, the optimum concentration for electroplating Au-Sn alloys were Au(I) concentration of 0.02 mol/L and Na$_2$SO$_3$ concentration of 0.48 mol/L, corresponding to the [Au]/[Na$_2$SO$_3$] molar ratio of 1/24.

Acknowledgments

The authors wish to thank Dalian Lumei Optoelectronics Corporation for providing the Si substrates for electroplating and other technical help. This work is supported by National Key Technologies R&D Program (2006BAE03B02-2), NSFC key program (U0734006), Key Laboratory Program in Liaoning Province (20060133) and Key Program in Dalian (2006A11GX005).

References

1. K. N. Tu, A. M. Gusak, M. Li, "Physics and materials challenges for lead-free solders", *J. Appl. Phys.*, 93 (2003), p. 1335.
2. D. Suraski, K. Seelig, "The current status of lead-free solder alloys", *IEEE Trans Electron., Pack, Manufact.*, 24, (4) (2001), 244.
3. Anqiang He, Qi Liu, Douglas G. Ivey, "Electrodeposition of tin: a simple approach", *J Mater Sci: Mater Electron*, (2008), 19:553-562.
4. Kai Wang, Rozalia Beica and Neil Brown et al, "Soft Gold Electroplating from a Non-cyanide Bath for Electronic Applications", *Proceedings of the IEEE/CPMT International Electronics Manufacturing Technology (IEMT) Symposium*, 2004, pp. 242-246.
5. T. A. Green and S. Roy, "Speciation Analysis of Au(I) Electroplating Baths Containing Sulfite and Thiosulfate", *Journal of The Electrochemical Society*, 153 (3) (2006), pp. C157-C163.
6. Todd A. Green, "Gold Electrodeposition for Microelectronic, Optoelectronic and Microsystem Applications", *Gold Bulletin*, 40 (12), (2007), pp. 105-114.
7. Uchida et al, "Non-cyanide-type Gold-tin Alloy Plating Bath", US patent 6544398 B2.
8. Ivey et al, "Codepositing of Gold-tin Alloys", US patent 6245208.
9. W. Sun, D.G. Ivey, "Development of an electroplating solution for codepositing Au-Sn alloys", *Materials Science and Engineering B*, 65 (1999), pp. 111-122.
10. W. SUN, D. G. IVEY, "Microstructural study of co-electroplated Au/Sn alloys", *Journal of Materials Science*, 36 (2001), pp.757-766.
11. Yahui Zhang and Douglas G. Ivey, "Phase Formation in Gold-Tin Alloys Electroplated from a Non-cyanide Bath", *2003 International Conference on Compound Semiconductor Manufacturing Technology*, CS MANTECH, 2003.

Sequential Non-cyanide Electroplating Au/Sn/Au Films for Flip Chip-LED Bumps

Yang Liu, Mingliang Huang*

Electronic Packaging Materials Laboratory, School of Materials Science & Engineering, Dalian University of Technology
Dalian, 116024, China
*E-mail: huang@dlut.edu.cn, Tel: 86-411-84706595

Abstract

LEDs (Light Emitting Diode) have a high potential to replace the conventional light bulb as the long-life, energy efficient, environmentally friendly and multi-use light source in the future. Using flip-chip (FC) technology, the thermal dissipation and luminescence efficiency of high-power LEDs can be improved. Therefore FC packaging also attracts great research interests for high brightness LEDs (HB-LED). However, solder bumping is a critical step in FC technology. In the present article an Au-30Sn (at.%) eutectic alloy bumping process developed for high-power LED flip-chip technology has been described. Au-Sn solder bumps can be manufactured by sequential non-cyanide electroplating of Au and Sn layers. This paper focuses on the formation of Au bumps and the optimization of electroplating parameters for pure Au. The quality of the Au layers and the deposition rates were studied in terms of electroplating temperature and sodium sulfite concentration in baths. A series of electroplated tests at different sodium sulfite concentrations ranging from 0.135 mol/L to 0.675 mol/L were performed in order to study the effect of sodium sulfite concentration on the quality and depositing rate of Au layers. After the optimization of the Au plating parameters, Au/Sn/Au triple-layer films for FC-LED bumps were fabricated.

1. Introduction

Solder bumps used in the electronic/photoelectron flip chip packaging serve three major functions: mechanical support, heat dissipation and electrical connection. Of the gold eutectic solders, Au-30Sn (at.%) eutectic solder is the preferred alloy because of its relatively low melting point (280°C), low elastic modulus, high thermal conductivity, and high strength compared with the other solders commonly used in electronic packaging. Moreover, the properties of high gold content results in low tendency to oxidize during the welding process, so this alloy is suitable for soldering without flux [1~2]. Therefore, the Au-30Sn eutectic solder has a high potential for use in high-power LED chip packaging.

Au-30Sn bumps are usually deposited by sequential electroplating of pure Au and Sn layers from separate Au and Sn solutions. The alternative co-deposition technique has also been studied extensively in the past. In the present study bumps were manufactured by sequential electroplating of Au and Sn using a periodic reversal square wave pulse technique.

Two types of Au-plated solutions are available: one, containing cyanide has been widely used and the technology is quite mature; the other based on a non-cyanic solution which is still under extensive development. There are unparalleled advantages of the cyanic solution over the alternative so far. However, due to its toxicity, it raises profound concerns in process safety and waste management. In addition, from a the materials point of view the CN⁻ ions

produced during the electroplating process are not compatible with photosensitive anti-corrosion agent widely used in the electronic industry. These ions are detrimental to the interface between the photoresist and substrate, causing part of the photo resist to dissolve [3~4].

In the non-cyanic solutions, sulfite (ammonium salt and sodium) and thiosulfate are mainly used as the complexing agent for the electroplating. Nowadays, sulfite based gold-plating solutions are most widely used, but such solutions are still not stable when sulfite only is used as complexing agent. $Au^+ + 2SO_3^{2-} = [Au(SO_3)_2]^{3-}$ $\beta = 10^{10}$, β is the stability coefficient for the complex, which is much smaller than that of $[Au(CN)_2]^-$ of 2×10^{38}. Therefore the addition of a second or a third complexing or organic acid chelating agent is necessary to keep the solution stable.

The non-cyanic electroplating solution for gold provides more choices for the electronics industry However, for the majority of non-cyanide electroplating solutions problems have been reported due to the poor stability of the baths. Therefore, development of a stable non-cyanide electroplating solution for gold and the electroplating process is of high value for the potential application.

In the present paper, pulsed current electrodeposition were formed using a Sn plating solution based on stannous sulfate ($SnSO_4$), Potassium pyrophosphate ($K_4P_2O_7 \cdot 3H_2O$), L-ascorbic acid, pyrocatechol and nickelous chloride ($NiCl_2$). Sequential electroplating method was used to deposit three-layer Au/Sn/Au structures onto Si substrates. The study focuses on the optimization of electroplating parameters for Au and Sn. Microstructure characterization and deposition rates served as critical parameters for the quantification of results.

2. Experiments

The electroplating solutions used in this paper were prepared using de-ionized water. The separate Au and Sn electroplating solutions were used to sequentially depositing Au and Sn layers on Si test chips (4mm×9mm). The Si chips, as cathodes, were metalized with 0.2μm Cr and 1μm Au. Cr acts as an adhesion and barrier layer, while Au as a seed layer for electroplating. The backsides of the wafers were coated with stop-off lacquer to prevent backside plating. The plating area ranged between 0.08 and 0.28 cm², and the samples were cleaned in de-ionized water and dried in air.

The present paper focused on the formation of Au and Sn layers. In some paper on the subject it is claimed, that Cl⁻-ions in the solution has a detrimental effect on the formation of Au layers. Therefore, a stable non-cyanide solution, based on $Na_3Au(SO_3)_2$ was used instead of $NaAuCl_4$ as Au source, which has especially been developed for Au electroplating. The solution consisted of $Na_3Au(SO_3)_2$, sodium sulfite (Na_2SO_3) and EDTA , which respectively acted as main salt ,

978-1-4244-4658-2/09 $25.00 © 2009 IEEE

main complexing agent for Au, and the second complexing agent to Au and additive, respectively. The composition of solutions, electroplating temperature and peak current density were the major plating parameters on Au electroplating [5]. The electroplating process parameters in the present work were as follows: T=50~80°C, pH=7.0~9.0, J=0.1~2.5A/dm². The quality of the Au layers and the deposition rate were studied as a function of sodium sulfite concentration in baths and electroplating temperature.

In the Sn solutions, stannous sulfate ($SnSO_4$) salts were the sources of the initial Sn (II) ions. Potassium pyrophosphate ($K_4P_2O_7 \cdot 3H_2O$) acted as a complexing agent for Sn, and increased the conductivity of the solution. In order to prevent Sn(II) ions to spontaneously translate into Sn (IV), ascorbic acid was added as a chelating agent to prevent hydrolysis of Sn in solution. Pyrocatechol and nickelous chloride ($NiCl_2$) acted as an antioxidant and a grain refiner, respectively.

A CS300 CorrTest system was used for depositing Au and Sn layers with a periodic reversal current. Based on the previous work, the periods were set at 10ms, with a forward on-time of 2ms and a forward off-time of 4ms, moreover, a reverse on-time of 1ms and a reverse off-time of 3ms. An 88-1-type Magnetism Msier was used to keep the temperature of plating solutions constant. All specimens were characterized by optical microscope or JSM-5600LV scanning electron microscope (SEM). The composition of the electroplated layers was measured using energy dispersive spectroscopy (EDS).

3. Results and Discussions

3.1 Optimizing Au Electroplating Temperature

In order to find the appropriate temperature, we electroplated for a duration of 30 min, at 50°C, 60°C, 70°C and 80°C respectively. Three test specimens were made for each temperature setting, the results of which were averaged in the end.

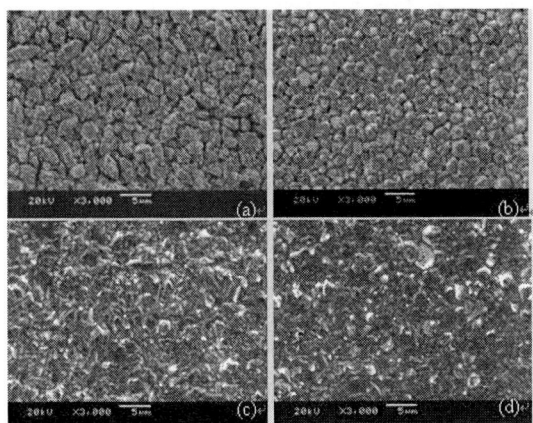

Fig. 1 The SEM top view of Au layers at different temperatures (×3000 , t=10min)
(a) T = 50°C (b) T = 60°C (c) 70°C (d) 80°C

Fig. 1 shows the top view of Au layers electroplated at different temperatures. It is clear that the deposition

temperature has a significant effect on surface topography of the layers. When the plating temperature was 60°C or lower, the arrangement of Au grains (or clusters) was very sparse, with pores between them. It was worth to note minor particles were found on the surface of the layer. The coronary minor particles can be attributed to a gathering of small grains [6]. It was found that with higher the deposition temperature, the minor particles became larger in size. The reason may be that the diffusion rate of the solution continuously increases at a higher temperature and the capacity for Au ion supply at the surface of cathode is also improved. The diffusion distance of the atomic absorption was also increased. The size of the minor particle increased with the diffusion distance. This consistent with the theory that the higher the temperature is, the coarser the grains grow.

The SEM cross sections of Au layers are shown in Fig. 2. From these micrographs the smoothness of the deposited layers was quantified and the thickness was measured.

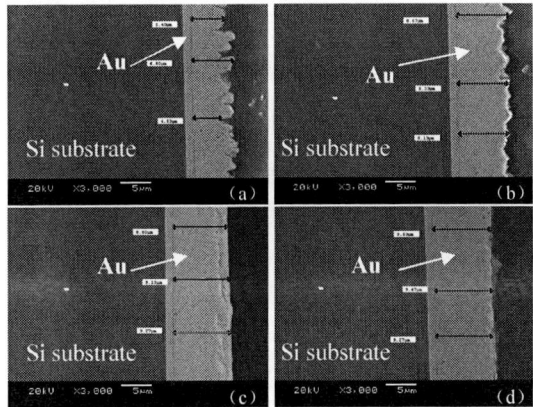

Fig. 2 The SEM cross section micrographs of Au layers at different temperatures (×3000, t=10min)
(a) T = 50°C (b) T = 60°C (c) 70°C (d) 80°C

Fig. 2 shows that the Au layers obtained at temperatures of 70°C and 80°C were denser and have smoother surfaces than those obtained at 50°C and 60°C. Although we could observe a few effects of temperature on the size of minor particles (grains) at the surface, they did obviously not affect the planarity of the surfaces. Using the same electroplating times, the growth rate of the layers increased with the increasing temperature. A possible reason is that ion supply rate increased along with the temperature. The effect was equivalent to raising the Au ion concentration near the cathode. From the cross section micrographs of Au layers we know that it is in principle better to use a higher plating temperature for this kind of electroplating solution. However, Au precipitated when the solution was heated up to 80°C. So the optimal temperature for electroplating was decided to be 70°C in the present work.

3.2 The Effect of Concentration of Complexing Agent

Sodium sulfite acts as a main complexing agent in the Au plating solution. Under the aforementioned experimental conditions, with the optimal plating temperature at 70°C, the effects of the sodium sulfite concentration on the growth rate of Au layers and their comprehensive quality were investigated. A series of electroplated tests at different sodium sulfite concentrations ranging from 0.135mol/L to 0.675mol/L, i.e. the ratios of Au ions concentration to sodium sulfite concentration of 1:3, 1:6, 1:9, 1:12 and 1:15 were carried out.

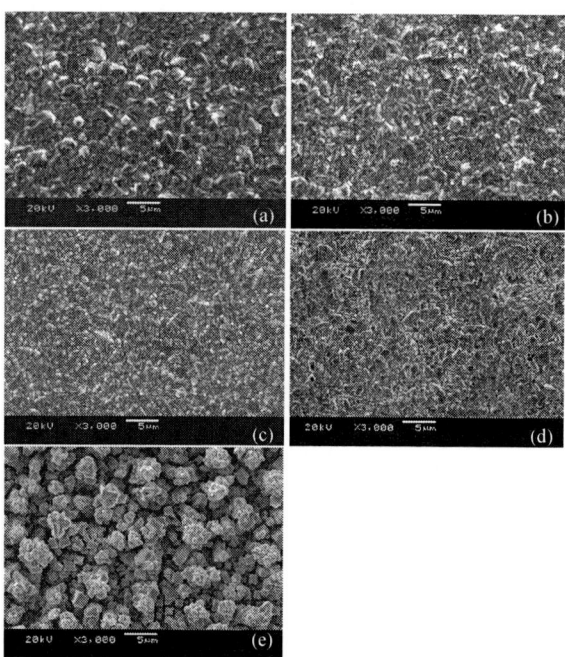

Fig. 3 The SEM top view of Au layers at different ratios of Au ions to Na_2SO_3 concentration (×3000 , t=10min)
(a)1:3   (b)1:6   (c)1:9   (d)1:12   (e)1:15

Fig. 3 shows that the top view images of the Au layers, the grain (cluster) size of Au layers became finer with the increase of Na_2SO_3 concentration ratio. However, dendrites formed when the ratio reached 1:15. The Au grains grow with a single orientation and a corallite-like structure, resulting in the dendrite morphology. This finding can be tentatively attributed to the fact that the cathode current efficiency declines along with the increase of free complexing agent contents. Once the free complexing agent content is excessive, plating in areas with lower current density is impeded. Therefore a rough surface topography arises, which finally leads to the formation of the dendrite structures [7].

Fig. 4 shows the secondary electron micrographs of cross sections from samples, where Au layers had been deposited under different ratios of Au ions to Na_2SO_3 concentration. (a) ~ (e), after plating for 10min, the Au layers, 7.78μm, 9.10μm, 8.58μm, 8.5μm and ~5.0μm in thickness, respectively. In (a) ~ (d) images, the Au layers were uniformly and continuously distributed on the Si substrate. The Au layer in (b) has a better planarity and a faster deposition rate. It is probable that the Au ions in that bath cannot be completely chelated by sodium sulfite since the concentration of sodium sulfite is low. Therefore the stability of the plating bath is relatively poor when the ratio is 1:3. As the concentration of sodium sulfite increased, the thickness of

Au layers slightly declined. The reason is that higher Na_2SO_3 concentration leads to the higher viscosity of bath, so that the diffusion of Au ions is affected. Therefore, the optimal ratio of Au ions to Na_2SO_3 concentration is 1:6, as derived from our experimental findings.

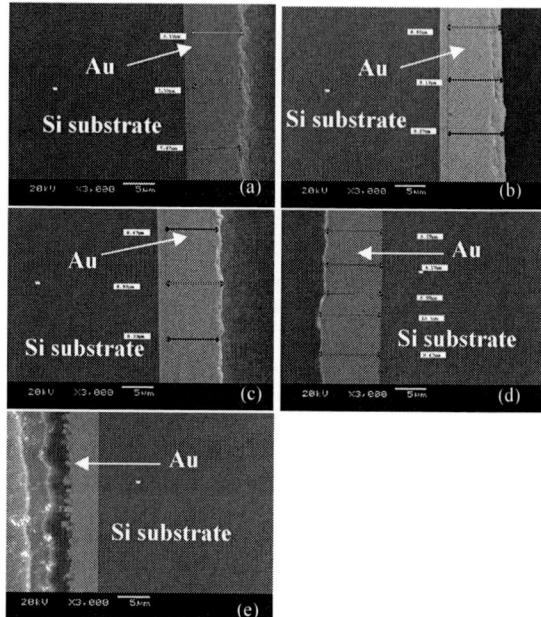

Fig. 4 The SEM cross section micrographs of Au layers at different ratios of Au ions to Na_2SO_3 concentration
(×3000 , t=10min)
(a)1:3   (b)1:6   (c)1:9   (d)1:12   (e)1:15

To sum up the above arguments, the optimal process was 1:6 ratio of Au ions to sodium sulfite concentration, T=70°C.

3.3 Sequential Electroplating of Au/Sn/Au Triple-layer Films

After optimizing the plating parameters for Au, triple-layer Au/Sn/Au films were fabricated. The thickness of the Au layer and the deposition time has a linear relationship so that we can control the thickness of Au and Sn layers by deposition time. For Au-30 Sn (at.%) eutectic solders, the ratio of Au layer to Sn layer is approximately 3:2, which is also ascertained by calculation. Fig. 5 shows a secondary electron image of a cross section sample for electroplating Au/Sn/Au triple-layer films. Au (I) and Au (III) layers were electroplated with a current density of 1.0 A/dm² for 5 min at 70°C and Sn (II) layer in the middle was electroplated with a current density of 1.0 A/dm² for 10min at 45°C. An obvious phenomenon was the presence of a reaction zone between the Au side and Sn side of the Au/Sn/Au triple-layer films. The reaction between Au and Sn quite noticeably occurred during the time of preparing the sample. Nakahara et al. [8] have reported similar phenomena for an Au/Sn thin foil couple at room temperature.

978-1-4244-4658-2/09 $25.00 © 2009 IEEE

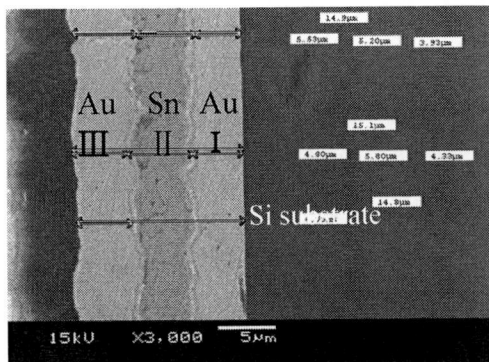

Fig. 5 Secondary electron image of a cross section sample of electroplated Au/Sn/Au triple-layer films on Si substrate

4. Conclusions

1. The grain size and roughness of the Au layer surface increased with the increasing temperature, so did the deposition rate of the Au layers. Au layers obtained at the temperature of 70°C and 80°C were relatively denser and the surfaces were smoother than those obtained at 50°C and 60°C. The deposition rates were about 0.91 and 0.93μm/min at 70°C and 80°C, respectively, which were higher than those at 50°C and 60°C. Au precipitates were observed when the solution was heated up to 80°C. It was concluded that the optimal Au electroplating temperature was 70°C in the present work.

2. A series of electroplated tests were performed at different sodium sulfite concentrations ranging from 0.135mol/L to 0.675mol/L, i.e. the ratios of Au ions concentration to sodium sulfite concentration of 1:3, 1:6, 1:9, 1:12 and 1:15, and all after plating for 10min, the thickness of the Au layers were 7.78μm, 9.10μm, 8.58μm, 8.5μm and ~5.0μm, respectively. The stability of the plating bath was relatively poor when the concentration of sodium sulfite was as low as 1:3. As the concentration of sodium sulfite increased, the thickness of Au layers slightly declined. Therefore, the optimal ratio of Au ions to Na_2SO_3 concentration is 1:6 in the present work.

3. The electroplating Au/Sn/Au triple-layer films were obtained. Both Au (I) and Au (III) layers had 4.5μm in thickness after plating for 5min, and the thickness of Sn (II) layer in the middle was about 5.8μm after plating for 10min. (total Au/Sn ~ 9μm /5.8μm).

4. The solid/solid interfacial reaction occurred at the Au/Sn interfaces. As a result, continuous Au-Sn intermetallic compound layers were observed at the Au and Sn interfaces of the Au/Sn/Au triple-layer films on Si substrate, even laid at room temperature for 10 h.

Acknowledgements

The authors wish to thank Dalian Lumei Optoelectronics Corporation for providing the Si substrates for electroplating and other technical help. This work is supported by National Key Technologies R&D Program (2006BAE03B02-2), NSFC key program (No.U0734006), Key Laboratory program in Liaoning Province (20060133) and Key program in Dalian (2006A11GX005).

References

1. Jeong-Won Yoon, Hyun-Suk Chun, Seung-Boo Jung, "Reliability evaluation of Au-20Sn flip chip solder bump fabricated by sequential electroplating method with Sn and Au, " *Materials science and engineering A, Structural materials : properties, microstructure and processing.* Vol. 473, No. 1-2 (2008), pp. 119-125.

2 Tao Zhou, Tom Bob, *et al*, "An Introduction to Eutectic Au/Sn Solder Alloy and Its Preforms in Microelectronics/Optoelectronic Packaging Applications," *Electronics and Packaging*, 1681-1070 (2005), pp. 5-8.

3. T. Osaka, A. Kodera, T. Misato, T. Homma, Y. Okinaka, and O. Yoshioka, "Electrodeposition of Soft Gold from a Thiosulfate-Sulfite Bath for Electronics Applications," *J. Electrochem. Soc.*, 144 (1997).

4. X. Wang, N. Issaev, and J. G. Osteryoung, *Journal of the Electrochemical Society*, Vol. 144, No.10 (1997), pp. 3462-3469.

5. Qingxia Mi, "The Study of Sequential Electroplating Au/Sn Bumps for FC-LEDs," Master's thesis , supervisor: Mingliang Huang, Dalian University of Technology(2008).

6. Tohru Watanabe, Nano-Plating: Microstructure Control Theory of Plated Bump and Data Base of Plated Bump Microstructure [M], Zhuping Chen, Guang Yang. Beijing, Chemical Industry Press, 2006, pp. 19-20.

7. Chengdian Wang, *et al*, "Study on Tin-nickel-copper Alloy Electrodeposition," *Contemporary Chemical Industry*, 1671- 0460 (2006)06-0392-04.

8. S. Nakahara, R.J. McCoy, L. Buene, J.M. Vandenberg, "Room temperature interdiffusion studies of gold/tin thin film couples," *Thin Solid Films*, 84 (2), (2006), pp. 185-96.

Study of Interfacial Reactions between Sn3.5Ag0.5Cu Alloys and Cu Substrate

L. C. Tsao[1], S. Y. Chang[2], W. H. Sun[3], S. F. Yen[4]

[1, 3]Department of Materials Engineering, Pingtung University of Science & Technology, Taiwan, China
[2]Department of Mechanical Engineering, Yunlin University of Science & Technology, Taiwan, China
[4]Microsystems Technology Center, Industrial Technology Research Institute, Taiwan, China
1, Hseuhfu Road, Neipu, Pingtung 91201, Taiwan, China
E-mail: tlclung@mail.npust.edu.tw Tel: 886-8-7703202 ext.7560

Abstract

Interfacial reactions between Sn3.5Ag0.5Cu lead-free solders (SAC) and Cu were investigated during soldering reactions between liquid SAC and Cu substrate at 250, 260, 275, 300 and 325°C for various reaction times. Experimental results show that a scallop-shaped layer of Cu_6Sn_5 intermetallic compounds formed during the soldering. Kinetics analysis shows that the growth of such interfacial Cu_6Sn_5 intermetallic compounds is diffusion controlled with an activation energy of 65.69 kJ/mol. This interesting behavior of IMC dissolution is attributed to the relatively high solubility of Cu in liquid solders.

Introduction

Sn-Pb eutectic solders have been widely used in electronics industries due to their low cost, good solderability, low melting temperature (Tm=183°C), and satisfactory mechanical properties. Recently, Pb is being removed from electronic products for environmental and health considerations. Many lead free solders have been studied as replacements for Sn-Pb solders. Among the various alloy systems, Sn-Ag-Cu solder (SAC) is one of the earliest commercially available lead-free solders and is also the most attractive candidate for electronic assembling, as it provides better mechanical properties than those of eutectic Sn37Pb solder [1-3]. During the soldering process, metallurgical reaction between liquid solder and copper pads or pads with Ni/Au metallization forms a layer of intermetallic compound (IMC) at the solder/metallization interface. The interfacial reactions of Sn thin-film, Sn-Pb, Sn-Ag, and Sn-Ag-Cu bulk solders with Cu pads have been investigated by a number of researchers [4-9]. They all showed that the intermetallic compounds formed during these interfacial reactions were of the Cu_3Sn (ε) and Cu_6Sn_5 (η) phase. However, since information on the growth kinetics of such intermetallic compounds at the SAC/Cu interface is scarce, this study focuses on that area. For this purpose, the intermetallic compounds appearing during the soldering reactions between liquid Sn3.5Ag0.5Cu(SAC) and Cu substrates in a wide temperature range of 250°C to 325°C were identified. The morphologies of such Cu_6Sn_5 intermetallic compounds formed at various temperatures were observed and their thicknesses measured as a function of reaction time. Through analysis of the kinetics, the activation energy for the growth of these intermetallic compounds can be obtained.

Experiment

Pure tin (99.99 pct), silver (99.99 pct), and copper (99.99 pct) were melted in a vacuum furnace at 650°C for 2.5 hours to produce solder alloys of eutectic Sn3.5Ag0.5Cu (wt.%). Cu substrates with a dimension of 8 mm×12 mm were cut from a 1 mm-thick Cu plate (99.95 pct), ground with SiC paper, and polished with 1 μm and 0.3μm Al_2O_3 powders. For the study of interfacial reactions, the SAC solder foil was placed on the Cu substrate and heated in a furnace under a vacuum of 10^{-3} torr. Through a water cooling system installed within the furnace, the specimens were cooled to room temperature in two minutes. Soldering reactions were conducted at temperatures between 250°C and 325°C for various heating time.

For the observations of the morphology of intermetallic compounds formed at the SAC/Cu interface after soldering reactions, scanning electron microscopy (SEM) was used. The specimens were cross-sectioned, ground with SiC paper, polished with 1μm and 0.3μm Al_2O_3 powders, and etched with 5% HCl and 95% H_2O solution. The chemical compositions of intermetallic compounds were analyzed by energy-dispersion spectroscopy (EDS). For growth kinetics analysis, the thickness of intermetallic compounds formed with various temperatures and time periods were calculated through dividing the total area of intermetallic cells spread out on the micrograph by the width of these intermetallics.

Results and discussion

Fig.1 shows the cross-sectional view SEM micrographs of the interface between SAC solder and Cu substrate soldering reactions at lower temperatures such as 250°C for different reaction times. A continuous layer of scallop-shaped intermetallic compounds can be seen at the solder/Cu interface. EDS analysis shows that the intermetallic compounds comprise only one layer of Cu_6Sn_5 IMC. In addition, accompanying the formation of Ag_3Sn intermetallic compounds identified by EDS analysis, a large number of needle-shaped precipitates are observed around the SAC/Cu interface (Fig. 1a). It can be seen that with increased reaction temperature and time, the long precipitate needles broke into particles and short needles, which can be attributed to the decrease of interfacial energy ($\gamma_{Ag_3Sn/SAC}$). Furthermore, the Ag_3Sn intermetallic compounds floated out from around the SAC/Cu interface. This implies that the quick diffusion of Cu atoms benefits the formation and growth of IMCs around the interface.

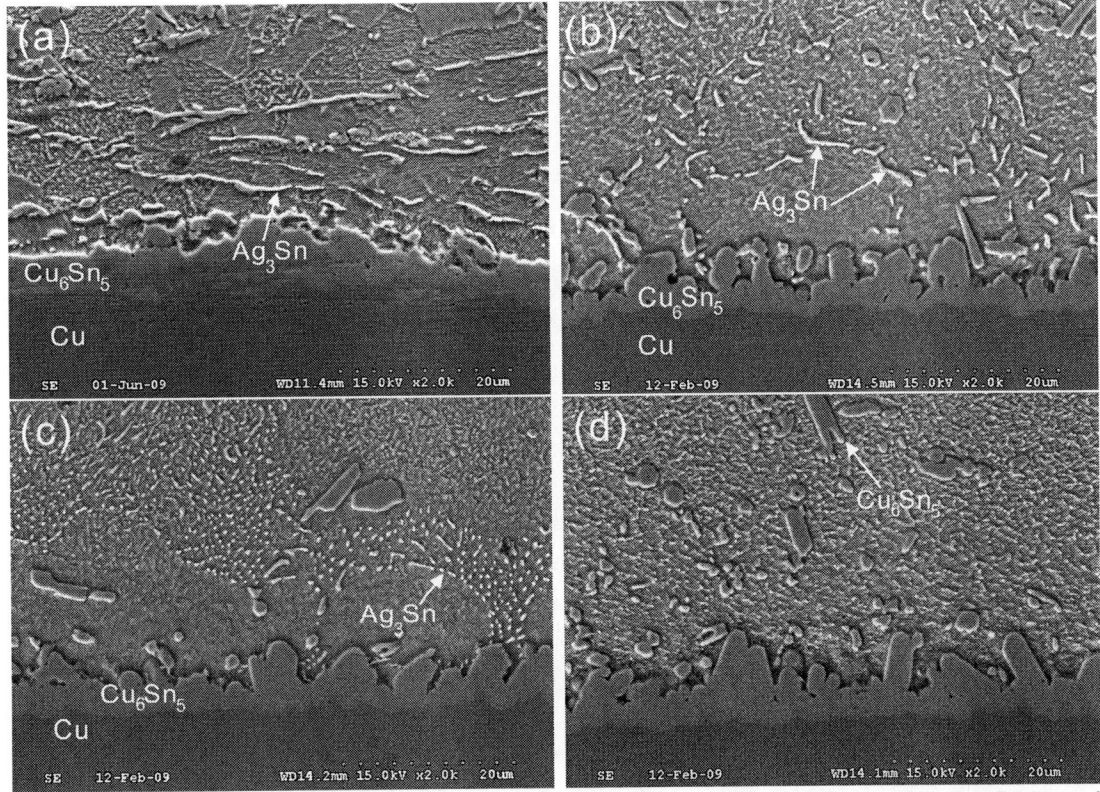

Fig.1 Typical morphology of intermetallic compounds formed at the SAC/Cu interface and in the SAC matrix after soldering reactions between liquid SAC and Cu substrates at 250°C for various time periods.(a: 15min, b: 30min, c: 45min, d: 60min).

Fig.2 shows that the solder/Cu interface exhibits a duplex structure of Cu_6Sn_5 (η) and Cu_3Sn (ε) at a long reaction time of 45 mins for different reaction temperatures. It was found that at higher temperatures, the scallop-shaped Cu_6Sn_5 intermetallic compounds broke away from the SAC/Cu interface and floated into the SAC matrix, as can be seen in Figs.2c and 2d (Marked A). The scallop-shaped intermetallic compounds at the SAC/Cu interface grew with the increase of reaction temperatures and time periods.

This result can be attributed to the relatively high solubility of Cu in molten SAC solder. The severe dissolution of Cu into the liquid SAC solders also causes the precipitation of island-shaped Cu_6Sn_5 phase in the SAC matrix after solidification of SAC solder. Adding an adequate amount of Cu element into the SAC solder can prevent such consumption of the Cu substrate.

Numerous investigations on the solid-state growth of the Cu-Sn IMC have been reported [10-11]. The IMC growth behavior is generally described as [12]:

$$X = X_0 + (Dt)^{1/2} \qquad (1)$$

where D is the interdiffusion coefficient Arrhenius expression, X is the thickness of the IMC layer after reflowing process, and t is the interface reaction time. For kinetics analysis, the thickness (x^2) of the Cu_6Sn_5

intermetallic compounds at the SAC/Cu interface versus the square root of the reaction time (t) are plotted in Fig.3 for various soldering temperatures. It can be seen that the plots are linear. The result indicates that the growth of Cu_6Sn_5 intermetallic compounds during the interfacial reactions between liquid SAC and Cu substrates is diffusion-controlled.

In order to have a further understanding of the growth rate of the IMC layer is given by the Arrhenius equation in terms of the interdiffusion coefficient [12]:

$$D = D_0 \exp-(Q/RT) \qquad (2)$$

where D_0 is the interdiffusion constant, Q is activation energy for growth of the interfacial IMC layer, R is the gas constant, and T is the absolute temperature. The growth rate constants ($D=x^2/t$) for various reaction temperatures are calculated and plotted in an Arrhenius relation as shown in Fig.4. From the slope of the Arrhenius plot, the activation energy (Q) for the growth of Cu_6Sn_5 intermetallic compounds during the SAC/Cu soldering reactions is estimated to be 65.69 kJ/mol. An early report on the growth of interfacial layers between liquid tin and solid copper estimated the activation energy for growth of the Cu_6Sn_5 (η) to be $Q_\eta = 33.4$ kJ/mol[13]. Coarsening of the ternary Sn–Ag–Cu eutectic is governed and the activation energy for growth of the Cu_6Sn_5 rods is 69 ± 5 kJ/mol[14]. Table 1 shows the activation energies for diffusion of Cu [15] and also the heats of solution for Cu in Sn [14]. The effective activation energy Q is the sum $(Q_S + Q_D)$. Q_S is the heat of solution for the rate-controlling species in the matrix, and Q_D is the activation energy for diffusion of the rate-controlling species in the matrix. From these values, the effective activation energy (Q) would be 69 kJ/mol for Cu diffusion

control. This implies that the growth of Cu_6Sn_5 intermetallic compounds during the SAC/Cu soldering reactions was predominantly controlled by the diffusion of Cu into the Sn solders, consistent with the morphology observation in Fig. 1.

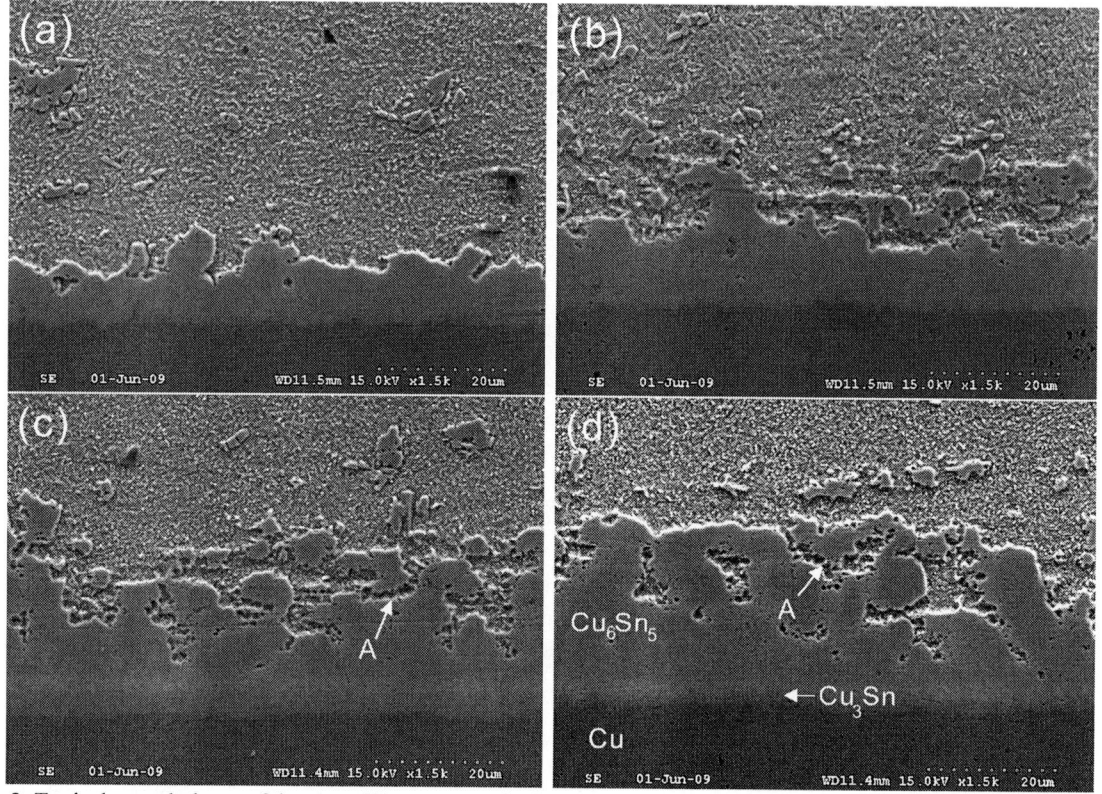

Fig.2 Typical morphology of intermetallic compounds formed at the SAC/Cu interface and in the SAC matrix after soldering reactions between liquid SAC and Cu substrates at various temperatures for 45 min.(a: 260°C, b: 275°C, c: 300°C, d: 325°C).

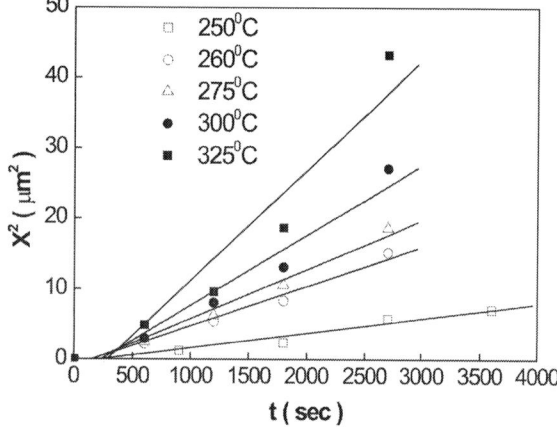

Fig.3 Thickness(x^2)of Cu intermetallic compounds as a function if the square root of reaction time(t)after soldering reactions between liquid Sn and Ag substrate at various temperatures.

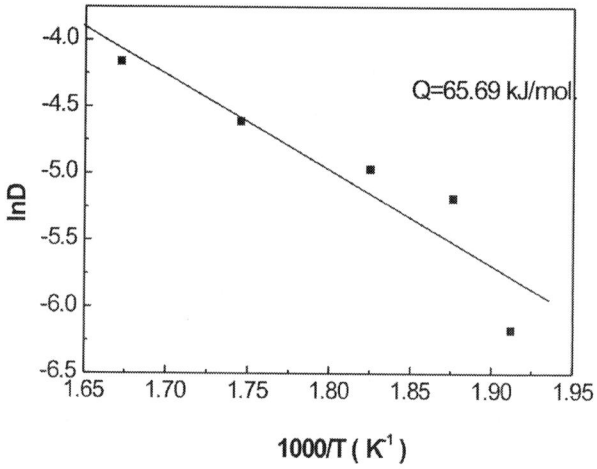

Fig.4 Arrhenius plot of the growth rate constants(D)of Cu_6Sn_5 intermetallic compounds formed after the SAC/Cu interfacial reactions

Table 1 Activation energies for diffusion (D) of copper in tin and the heats of solution (S) of copper in tin.

name	Solute	Solvent	Q (kJ/mol)	Reference
Q_D	Cu	Sn (a axis)	33 (D)	13
Q_S	Cu	Sn	36 (S)	14
Q			69	

$Q = Q_S + Q_D$

Conclusions

A continuous layer of scallop-shaped Cu_6Sn_5 intermetallic compounds appears at the SAC/Cu interface after the soldering reactions between liquid SAC and Cu substrate. Accompanying such a soldering reaction, the Cu substrate is found to dissolve severely into the molten SAC, which results in the formation of a large number of island-shaped Cu_6Sn_5 precipitates in the SAC matrix after solidification. The growth kinetics appears to be diffusion-controlled with an activation energy of 65.69 kJ/mol, which is near that of the diffusion control of Cu into Sn (69 kJ/mol). This result implies that the growth of Cu_6Sn_5 intermetallic compounds during the SAC/Cu soldering reaction is dominated by the diffusion of Cu into Sn solders.

Acknowledgments

The authors acknowledge the financial support of this work from the Science Council of Taiwan under Project No. NSC97-2218-E-020-004.

References

1. Anderson I., "Development of Sn-Ag-Cu and Sn-Ag-Cu-X alloys for Pb-free electronic solder applications," *Journal of Materials Science: Materials in Electronics*, Vol. 18 (2007), pp.55-76.

2. Glazer, J. "Microstructure and Mechanical Properties of Pb-Free Solder Alloys for Low-Cost Electronic Assembly: A Review," *J. Electron. Mater., Journal of Electronic Materials*, Vol. 23 (1994), pp.693- 700.

3. Glazer, J. "Metallurgy of low temperature P-Free solders for electronic assembly," *International Materials Reviews*, Vol.40, No.2 (1995), pp. 65-92.

4. Richard, L. H. Shih, Danny Y.K. Lau, Raymund W.M Kwok, "Metallurgy And Stability Of The Sn/Cu Interface For Lead-Free Flip," *5th International Conference on Electronic Packaging Technology*, Shanghai, China, 2003.

5. Romig, A. D. Jr. Chang, Y. A. Stephens, J. J. Frear, D. R. Marcotte V. Lea, C. Solder Mechanics: A State of the Art Assessment, Frear, D. R., Jones, W. B., Kinsman, K. R., Eds., TMS (Warrendale, PA, 1991), pp. 29-104.

6. Lee, J. H. Park, J. H. Kim,Y. S. and Shin, D. H. "Stability of channels at a scallop-like Cu6Sn5 layer in the solder interconnections," *J. Materials Research*, Vol. 16, No. 5 (2001), pp. 1227-1230.

7. Qi, L. Huang, J. Li, H. Zhao, X. Zhang, H. "Growth Behavior of IMCs and Fracture Forming Mechanism at Sn-Ag-Cu/Cu Interfaces under Thermal-Shearing Cycling Condition," *7th International Conference on Electronics Packaging Technology*, Shanghai, China, 2006.

8. Zribi, A. Clark, A. Zavalij, L. Borgesen, P. and Cotts, E. J., "The Growth of Intermetallic Compounds at Sn-Ag-Cu Solder/Cu and Sn-Ag-Cu Solder/Ni Interface and the Associated Evolution of the Solder Microstructure," *Journal of Electronic Materials*, Vol.30, No.6 (2001), pp. 1157-1164.

9. Kim, K. S. Huh, S. H. and Suganuma, K., "Effects of fourth alloying additive on microstructures and tensile properties of Sn-Ag-Cu alloy and joints with Cu," *Microelectronic Reliability*, Vol. 43 (2003), pp. 259–267.

10. So, A. C. K. and Chan, Y. C., "Aging Studies of Cu-Sn Intermetallic Compounds in Annealed Surface Mount Solder Joints," *Proc 46th Electronic Components and Technology Conf*, Orlando, FL, May. 1996, pp.1164-1171.

11. Pang, John H L, Xu, Luhua, Shi, X Q, Zhou, W, Ngoh, S L, "Intermetallic Growth Studies on Sn-Ag-Cu Lead-Free, Solder Joints," *Journal of Electronic Materials*, Vol. 33, No. 10 (2004), pp. 1219-1226.

12. Wassink, R. J. K. Soldering in Electronics, Electrochemical Publications Ltd., (Ayr, Scotland, 1989), pp. 149–159.

13. K. Kumar, A. Moscaritolo, and M. Brownawell, "Intermetallic Growth Dependence on Solder Composition in the System Cu-(Pb-Sn Solder)," *J. Electrochem. Soc. : Electrochemical Science and Technology*, Vol. 128, (1981), pp. 2165-2166.

14. Allen, S. L.; Notis, M. R. Chromik, R. R. Vinci, R. P. "Microstructural evolution in lead-free solder alloys: Part I. Cast Sn–Ag–Cu eutectic," *Journal of Materials Research*, Vol. 19, No. 5,(2004), pp. 1417-1424.

15. Dyson, B. F. Thomas, A. R. and Turnbull, D. "Interstitial diffusion of copper in tin," *Journal of Applied Physics*, Vol. 38 (1967), pp. 3408-3049.

Study on the Microstructure and the Shear Strength of Sn-0.7Cu-xZn

Yan-jun Gao, Zhong-bing Luo, Jie Zhao, Lai Wang*

School of Materials Science and Engineering, Dalian University of Technology, Dalian, 116024, China.
E-mail: wangl@dlut.edu.cn; Tel: 86-411-84707636; Fax: 86-411-84709284

Abstract

The effects of 0.3 wt.% Zn and 1 wt.% Zn addition on the melting point, the microstructure and the shear strength of Sn-0.7Cu were investigated. With small amount of Zn addition, the melting point decreased a little, and refined microstructure was observed especially for Sn-0.7Cu-1Zn and the fraction of eutectic region increased. Cu-Zn intermetallic compounds (IMCs) were detected in Sn-0.7Cu-1Zn at the phase boundary between β-Sn and eutectic region. Shear strength of solders increased with strain rate lineally in log-log plot. The strength of Sn-0.7Cu-1Zn was much higher than that of Sn-0.7Cu, which was similar to that of Sn-0.7Cu-0.3Zn. Ductile fracture was observed in all solders.

Introduction

The development of environmental-friendly solder, namely, lead-free solder, has been widely focused in recent years with the promulgating of legislations to limit the use of lead [1]. Among many solders, Sn-0.7Cu eutectic alloy is one of the most promising candidates to replace Sn-Pb alloy in the dip and wave soldering applications due to its versatile properties, such as low-cost and no limitation of patent. Unfortunately, the interfacial Cu-Sn IMCs would grow continually in the solid state under thermal loading. What's worse, Cu6Sn5 would coarsen when kept at high temperature for dozens of hours. This plays a degrading effect on the long-term reliability of solder joints [2].

Recently, several researchers reported that the properties of alloys could be improved effectively when adding trace elements. Peng Sun and coworkers reported that the stability of microstructure of Sn-Cu alloy increased with a small addition of element Co [3]. C. Bailey and coworkers revealed that during wetting and aging, the growing rate of IMC layer of Sn-Cu was suppressed after adding the element Ni [4]. Wang and coworkers reported that with the addition of Zn, the microstructure of bulk alloy was refined and the growth of IMC layer at the interface during thermal aging was depressed. Furthermore, the interfacial IMC composition completely changed from Cu-Sn to Cu-Zn after adding 1.0 wt.% Zn [5].

With the increasing facility of portable electronic products, the bump size tends to be smaller and smaller. Consequently, the reliability of interconnects, especially mechanical behaviors of solder and joint become more important. Due to the strain-sensitive feature of Sn matrix and the mismatch of thermal expansion coefficient between the package components and substrate, the study on shear strength at different strain rates is valuable. The results could not only afford reliability data, but also offer the corresponding basis for the construction of constitutive model.

In this work, the melting point, microstructure and shear strength of Sn-0.7Cu were investigated by adding 0.3 and 1 wt.%Zn, respectively. Room temperature shear test of the solder was conducted at the rate of 0.1, 1 and 10 rad/s.

Experimental procedure

Three Pb-free solders, Sn-0.7Cu, Sn-0.7Cu-0.3Zn and Sn-0.7Cu-1Zn (wt.%, hereafter designated as SC, SC-0.3Zn, SC-1Zn, respectively) were selected. The as-cast samples of bulk alloy were prepared by vacuum melting at 500°C for 5 hours and then cooled to 100°C with furnace. The ingots were re-melted at 270°C for 20 minutes timed from melting and then cast into a graphite mould which was pre-heated at 190°C about half an hour. The cooling rates were determined with numeral thermocouple, inserting into the center of the liquid alloy. The as-cast 20mm×20mm×30mm ingot (about 90g) was shown in Fig. 1(a). The test sample (dotted line) was also pointed out. All samples were wire-cut into a number of small strips along the height direction. Ultimate specimens for shear test about 2mm×2mm×20mm were obtained after they were ground and polished. The examination of microstructure and melting point was conducted at the plane shown in Fig. 1(a). Samples were ground, polished and etched with a solution of 5vol.% HNO₃, 2vol.% HCl and 93vol.% CH₃OH for several seconds. The microstructure and fractography were examined by JEOL JSM-5600LV scanning electron microscopy (SEM).

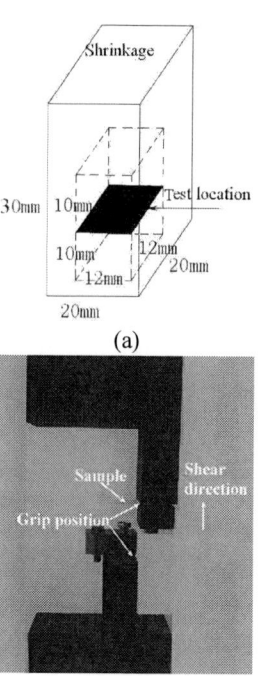

(a)

(b)

Fig. 1 Schematic diagram of as-cast solder ingot and tool for shear testing: (a) as-cast solder ingot; (b) tool for shear testing

The phases were identified using energy dispersive x-ray spectroscopy (EDX). Melting point of three alloys was measured by differential scanning calorimetry (DSC) at a constant heating and cooling rate of 5°C/min between 180°C and 260°C in Ar atmosphere.

Shear test was performed at room temperature on SHIMADZU servo-hydraulic test system (1KN Load cell). Tool for shear testing is shown in Fig. 1(b). Shear strain rate was 0.1, 1 and 10 rad/s and the shear gap between grips was 0.6mm which was designed to compare with the solder gap in undergoing joints. Four samples were tested in each condition.

Results and Discussion

Thermal Properties

Fig. 2 presents the DSC curves of three solders at the temperature of 180-260°C, showing that the melting point of SC was slightly decreased with the Zn addition, but it did not change significantly. Three melting peaks corresponded to the eutectic transformation as follows:

$$\beta \text{-} Sn + Cu_6Sn_5 \rightarrow L \tag{1}$$

M. McCormack et al. has proved that Zn atoms preferentially react with Cu atoms when 1 wt.% Zn was added [6]. By the way, there was no Cu-Zn peak appeared in the DSC results. The possible reason might be due to the small amount of Cu-Zn phase and the poor sensitivity of instrument.

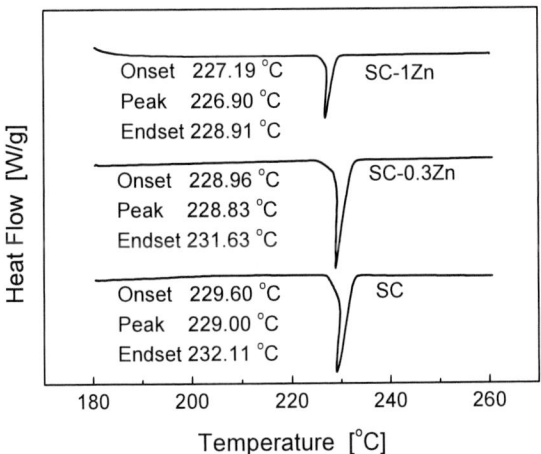

Fig. 2 DSC thermograms of SC, SC-0.3Zn and SC-1Zn

Microstructural Feature

The microstructure of three as-cast samples are shown in Fig. 3. It can be seen that SC and SC-0.3Zn were composed of β-Sn phase (black) and the eutectic region (gray), and no distinct difference was observed. Compared with Wang's result [5], which revealed that 0.2 wt.% Zn addition also refined the microstructure, this difference may result from different cooling rate. In this study, the determined cooling rate was 1.01K/s. Nevertheless, the microstructure of SC-1Zn alloy was refined effectively and approximate to equiaxed grain. New phase, in the size of 2-3 μm, was detected in SC-

1Zn at the phase boundary between β-Sn and eutectic region. It was identified as Cu-Zn IMCs by EDX.

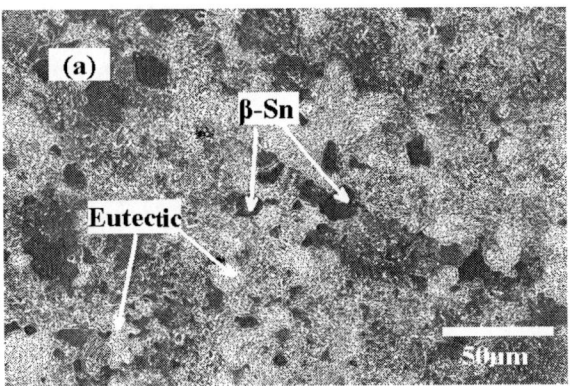

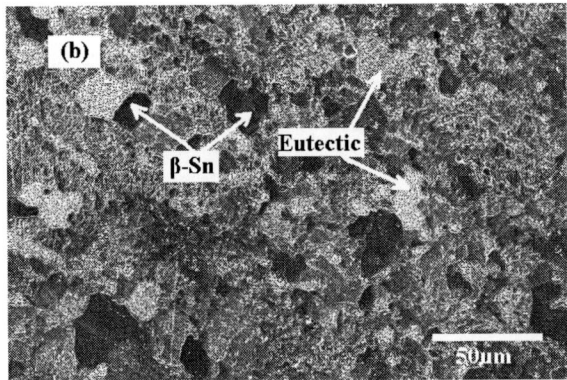

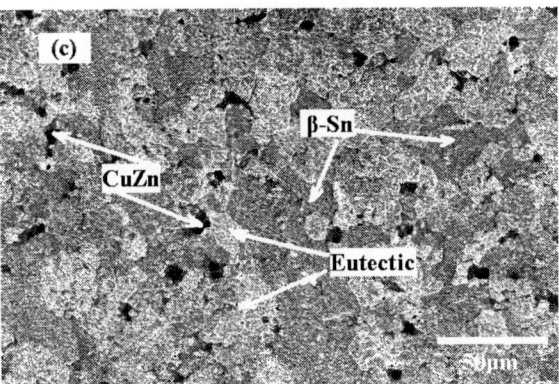

Fig. 3 SEM images of three bulk solders (a) SC; (b) SC-0.3Zn; (c) SC-1Zn

From the Sn-Zn and Cu-Zn phase diagram, small part of Zn would solute in the Sn matrix, and with further addition, the rest of Zn would precipitate as Cu-Zn IMCs, shown as in Fig. 4. These IMCs would serve as the primary nucleations and inhabit the supercooling of Sn [7].

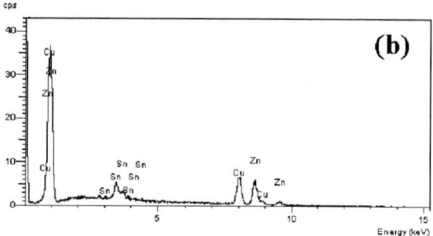

Fig.4. SEM micrograph and EDX result of Cu-Zn IMC in SC-1Zn: (a) BEI; (b) EDX spectra at point A

Shear Strength

Curves of shear strength versus strain rate about three solders in log-log plot are shown in Fig. 5. Error bars are the standard deviation of the data. The shear strength increased linearly with the increasing of shear strain rate. The strength of SC-1Zn was much higher than that of SC, which was similar to that of SC-0.3Zn.

The relationship between maximum engineering stress and strain rate could be approximated by the following equation:

$$\sigma = C(\varepsilon)^m \qquad (2)$$

By applying natural logarithms to both sides of Eq. 2, a new equation can be written as follow:

$$\lg(\sigma) = m \lg(\varepsilon) + \lg C \qquad (3)$$

Where σ is the maximum engineering stress, C is constant for a certain state, ε is stress for the strain rate, m is strain sensitivity index.

Table 1 Fitting Parameters of relationship between the maximum engineering stress and engineering strain rate at room temperature.

sample	m	C	r^2
SC	0.090	28.549	0.999
SC-0.3Zn	0.113	29.228	0.999
SC-1Zn	0.108	34.419	0.999

Table 1 is the fitting parameters of relationship between the maximum engineering stress and engineering strain rate, which is presented in Fig. 5. r^2 is the correlation coefficient. It can be seen that the experimental data shows good agreement

with the linear equation, with the correlation coefficient values of ~0.999 at different content of Zn. The scope of m is in the range of 0.09-0.11, which indicates that the strain sensitivities of the solders are very close.

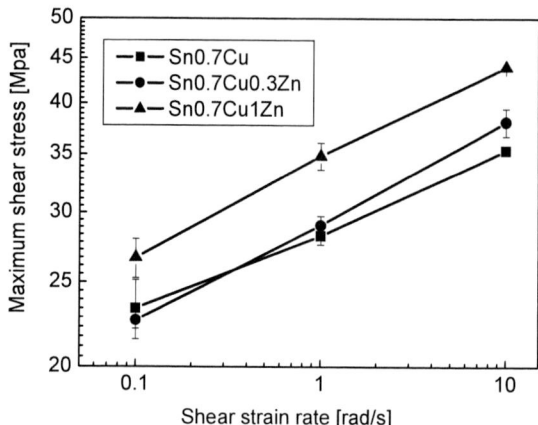

Fig. 5 Relationship between shear strength and shear strain rate of three kinds of solder at room temperature in log-log plot

From the above analysis of microstructure, Zn atoms served the role of solid-solution strengthening in SC-0.3Zn. However, owing to small solubility of Zn in Sn, solid-solution strengthening may be negligible. At the same time, the microstructure varied a little compared with SC, so the effect of fine-grain strengthening may be also not dominant.

While for SC-1Zn, the microstructure of bulk alloy was refined with 1 wt.% addition and it was more effective to strengthen the solder. Therein, parts of Zn will solute to the Sn matrix, which will act as solid-solution strengthening atoms. The rest will react with Cu and form Cu-Zn particles which will act as dispersion-strengthening. According to Ochoa [8] and Mei [9], the grain boundaries of equiaxed grains were easy to slide. As a result, particles at grain boundary would inhabit the movement and increase the strength of alloy. In this study, the Cu-Zn phase dispersed at the boundary between the β-Sn phase and the eutectic region. At room temperature, the strengthening effect of second-phase particles is more evident. Unfortunately, the formation of Cu-Zn phase will consume Cu_6Sn_5 which acts as ideal strengthening phase and the loss of Cu_6Sn_5 will reduce the strength. All in all, the remarkably higher strength of SC-1Zn resulted from the compositive effect mentioned above and Cu-Zn phase would play a more important role.

Shear fractography

Fig. 6 lists typical morphology of shear fracture under different content of Zn and shear rates. Comparing Fig. 6(a) with (b), shear fractography of SC and SC-1Zn under the same shear rate of 10rad/s is similar, so the content of Zn has tiny influence on the shear fractography. Fig. 6(c) and (d) present shear fractography of SC-0.3Zn under shear rates of 0.1 and 10rad/s , respectively.

Dimples in all fracture surface indicated that ductile fracture happened. Moreover, the size of dimples of the fracture varied with strain rate. The higher the strain rate, the smaller the dimples.

At low strain rate, larger dimples were observed which means severe plastic deformation happened. Thus, the corresponding shear strength was low. At high strain rate, such as 10rad/s, dimples became to be much smaller, indicative the less ductile behavior and higher shear strength. This is in accordance with previous analysis of mechanical properties.

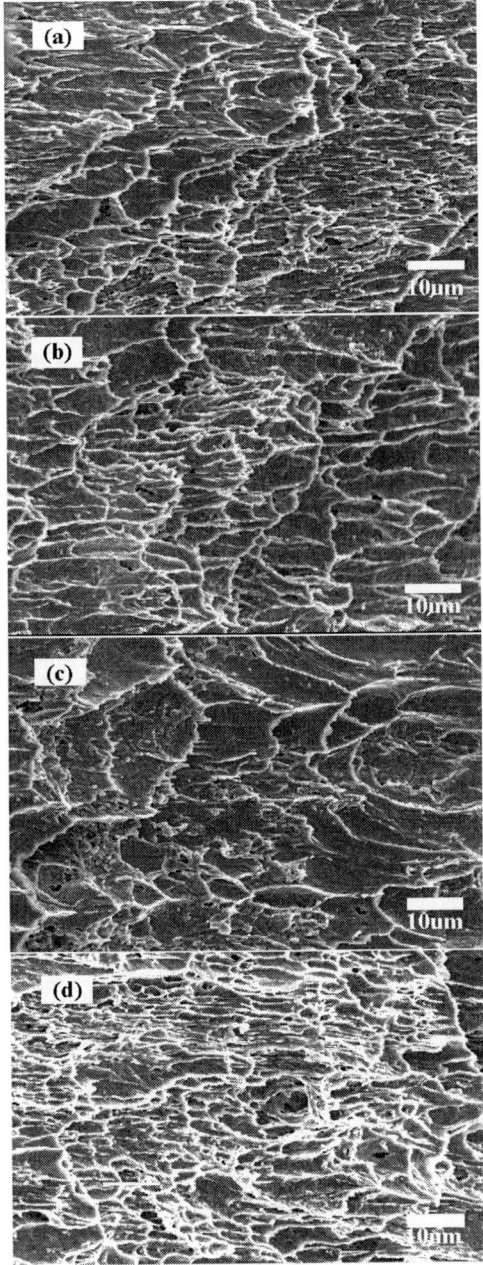

Fig. 6 Typical shear fractography under different content of Zn and shear strain rates: (a) and (b) are fractography of SC and SC-1Zn under 10rad/s; (c) and (d) are fractography of SC-0.3Zn under 0.1 and 10rad/s, respectively.

Conclusions

The effects of 0.3 and 1 wt.% Zn on the melting point, the microstructure and the shear strength of Sn-0.7Cu were investigated. The conclusions were obtained as follows:

(1) The melting point of SC decreased a little with Zn addition.

(2) Refined microstructure was observed especially for Sn-0.7Cu-1Zn and the fraction of eutectic region increased. Cu-Zn IMCs were detected in Sn-0.7Cu-1Zn at the boundary between β-Sn and eutectic region.

(3) Shear strength of solders increased lineally with strain rate in log-log plot. Because Cu-Zn particles inhabited the sliding of grain boundary, the strength of Sn-0.7Cu-1Zn was much higher than that of Sn-0.7Cu, which was similar to that of Sn-0.7Cu-0.3Zn. Ductile fracture was observed in all solders.

Acknowledgments

This work has been supported by the National Key Technologies R&D Program in the 11th Five-year Plan (2006BAE03B02-2) and the Joint Funding of NSFC (National Nature Science Foundation of China) and Guangdong Province (U0734008).

References

1. Abtew M., Selvaduray G, "Lead-free Solders in Microelectronics," *Materials Science and Engineering R*, Vol. 27, No. 5-6 (2000), pp. 95-141.

2. Frear D. R., "the Mechanical Behavior of Interconnect Materials for Electronic Packaging," *JOM*, Vol. 48, No. 5 (1996), pp. 49-53.

3. Andersson C, Sun P, Liu J, "Tensile properties and microstructural characterization of Sn–0.7Cu–0.4Co bulk solder alloy for electronics applications," *J. Alloys Comp.*, Vol. 475, No. 1-2 (2008), pp. 97-105.

4. Bailey C., Rizvi M.J., Chan, Y.C., M.N. Islam, H. Lu, "Effect of adding 0.3 wt% Ni into the Sn–0.7 wt.% Cu solder," *J. Alloys Comp.*, 438 (2007), pp.122–128.

5. Fengjiang Wang, Xin Ma, Yiyu Qian, "Improvement of microstructure and interface structure of eutectic Sn–0.7Cu solder with small amount of Zn addition," *Scrip. Mater.*, Vol. 53, No. 6 (2005) , pp. 699-702.

6. McCormack, N., Kammlott, G.W., Chen, H.S., et al, "Newlead-free, Sn-Ag-Zn-Cu solder alloy with improved mechanical properties," *Appl. Phys. Lett.*, Vol. 65, No.10 (1994), pp. 1233-1235.

7. S.K. Kang, D.Y. Shih, D. Leonard, D.W. Henderson, T. Gosselin, S. Cho, J. Yu, and W.K. Choi. "Controlling Ag3Sn plate formation in near-ternary-eutectic Sn-Ag-Cu solder by minor Zn alloying," *JOM*, Vol.56, No.6 (2004), pp. 34-38.

8. Ochoa F, Williams J J, Chala N. "Effects of Cooling Rate on the Microstructure and Tensile Behavior of A Sn-3.5wt.%Ag solder," *J Electron. Mater.*, Vol. 32 ,No. 12 (2003), pp. 1414-1420.

9. Z. Mei, J. W. Morris, M. C. Shine and T. S. E. Summers., "Effects of Cooling Rate on Mechanical Properties of Near-eutectic Tin-lead Solder Joints," *J Electron. Mater.*, Vol. 20, No. 10 (1991), pp. 599-608.

Investigation of the Fundamental Interactions among the Ingredients of Flux by the Group Contribution Method

Yunxia Jin[1], Jun Hu[1,*], Daniel Lu[2]
[1]East China University of Science and Technology
[2]Henkel Corporation Electronic Materials
[1]130 Rd. Meilong, Shanghai, 200237, [2]90 Zhujiang Road, Yantai, China
*corresponding author: Tel.: 86-21-64252922, Fax: 86-21-64252921, E-mail: junhu@ecust.edu.cn

Abstract

A clear understanding of the complicated interactions among various key ingredients of flux will be very beneficial for developing future high performance solder pastes. In this work, the group contribution method was introduced to estimate and predict the properties of compound and their interactions, especially, the chemical interactions. Based on the assumption that the effects of the individual groups are additive, the properties of a compound are calculated as a function of structurally dependent parameters, which can be determined by summing the number frequency of each group multiplied by its contribution. To be accurate, some corrections such as interaction between the neighbor groups, vaporization enthalpy contribution, and symmetrical effect are considered in the estimation. The thermodynamic properties of the formation enthalpy ($\Delta_f H$), the molar entropy (ΔS), the boiling temperature (T_b), the vaporization enthalpy ($\Delta_V H$) and the heat capacities (C_p) were calculated individually. Then, the chemical interaction properties such as the reaction enthalpy ($\Delta_r H$), the reaction entropy ($\Delta_r S$) can be calculated based on the chemical reaction which might happen among the various ingredients. And hence, the reaction Gibbs energy ($\Delta_r G$) and reaction constant (K) also can be estimated by the equations of $\Delta G = \Delta H - T\Delta S$ and $\Delta_r G = -RT\ln K$, respectively. When the parameters a, b, c in the relations of the heat capacity and the temperature of $C_p = a + bT + cT^2$ were determined, the dependence of $\Delta_r G$ and K with the temperature also can be determined through the formation enthalpy.

The flux media generally contains several chemicals: acids, bases, and additional compounds which can help to improve solder wettability. Usually, the main interactions between acids and bases are the neutralization when the temperature is low. For the cases in which acids have –COOH groups and bases have –OH groups, the calculation results suggested that $\Delta_r G$ of the esterification between these acids and bases were all smaller than zero in the temperature range, which meant the esterification were spontaneous reaction. Meanwhile, because $\Delta_r G$ decreased and K increased with the increasing temperature, it predicted that the higher the temperature was, the higher degree the esterification was. So the interaction of the neutralization changed into the esterification when the temperature is high enough. Containing multi-carboxylic groups in acids or multi-alkaline groups in bases, not only monoester but also diester, triester ect., multi-ester compounds or even polyester may be possibly formed at higher temperature due to the negative values of $\Delta_r G$ and the negative slopes of $\Delta_r G$ with temperature of multi-esterification.

Most of the calculation results were coincide with the experimental phenomens, this investigation may develop a method to evaluate the effects of various functional groups to the solder paste stability and wettability at different processing temperature ranges.

1 Introduction

Solder flux effective ingredients includes activator, solvent, thixotropic agent, thickener, surfactant, inhibitor and so on, which creates a extremely complicated system. In reality, the chemistry of flux interactions at solder surfaces is complicated and involves acid-base, oxidation-reduction, coordination and adsorption reactions [1]. Most fluxes react as Bronsted-Lowry acids with metallic oxides to form their respective salts and water. To clearly understand the complicated interactions among various key ingredients and the interactions between the ingredients and solder alloy surfaces will be very beneficial for developing future high performance solder pastes. Anyway, little work has appeared in the literature concerning the interactions among solder paste. Snyder et al. [2] studied the interactions between rosin flux and metal by diffuse reflectance Fourier transform-infrared (DRIFT), and determined which of the possible solder joint corrosion could be removed by the rosin flux by examining the evidence of carboxylate salt formation after heating. So far, no exact mechanism of flux ingredient interactions and flux influence on solder alloy are clear out. In this work, the group contribution method was introduced to aid in solving this problem. Moreover, experiments were also given to verify it.

2 Experimental and Theoretical Methods

2.1 Experimental method

Succinic acid ($HOOCCH_2CH_2COOH$, AC1), triethanolamine ($N(CH_2CH_2OH)_3$, AK1) and polyethylene glycol dibutyl ether ($C_4H_9(OCH_2CH_2)_nOC_4H_9$, where n=3-5, S1) were used without any pretreatment, mixing above three substances, ingredients of flux, and then heating with vigorous stirring and reflucne. The chemical bond properties of the pure chemicals at room temperature and that of the mixtures at different temperature stages of 298 K, 353 K (usual temperature of solder paste mixing) and 493 K (temperature of solder paste application) respectively, were determined by FTIR 380 (Nicolet).

2.2 Theoretical methods

The thermodynamic properties, such as reaction enthalpy $\Delta_r H$, reaction Gibbs energy $\Delta_r G$ and thermodynamic equilibrium constant K are chosen to analyze the reaction process which are estimated by Group Contribution method. The individual groups are defined as multiply bonded atoms with all their nearest neighbours. So it is obvious that all the atoms (except hydrogen) are considered twice (as central atom and a neighbor, respectively). For example, according to the group contribution method proposed by Benson et al. triethanolamine ($N(CH_2CH_2OH)_3$) can be divided into the following 4 kinds of groups and corresponding amounts: N-(3C), 1; CH_2-(C,N), 3; CH_2-(C,O), 3; OH-(C), 3. However, the group additivity rules reported differed a little from method to method. The whole scheme for the calculations is shown in Fig. 1. Where N_i represents the number frequency of individual group i and $S^0_{fi}(298\,K,g)$, $\Delta H^0_{fi}(298\,K,g)$ are contributions of i for calculation. And to be accurate, corrections must be considered. These will be introduced in detail in the following part.

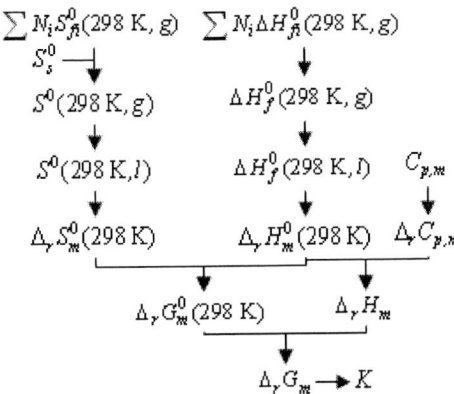

Fig. 1 The scheme for the calculations flow

2.2.1 The thermodynamic properties of substance

i) Enthalpy and entropy at 298 K in the ideal gaseous state

The enthalpy can be calculated directly as the sum of each group contribution, as shown in Eq. (1). For the entropy, besides the contribution of each group, the structure symmetry and isomers were also considered, as shown in Eq. (2-4), where S^0_s is the entropy correction, N_{oi} is the isomer number, N_{ts} is the product of external symmetry number N_{es} and inner symmetry number N_{is}.

$$\Delta H^0_f(298\,K,g) = \sum N_i \Delta H^0_{fi}(298\,K,g) \qquad (1)$$

$$S^0(298\,K,g) = \sum N_i S^0_{fi}(298\,K,g) + S^0_s \qquad (2)$$

$$S^0_s = R\,ln(N_{oi}) - R\,ln(N_{ts}) \qquad (3)$$

$$N_{ts} = N_{es} * N_{is} \qquad (4)$$

ii) Enthalpy and entropy at 298K in the liquid state

For the enthalpy and entropy at 298 K in the liquid state, the vaporization enthalpy and entropy should be subtracted from the above gaseous data as Eq. (5-6), respectively. The vaporization enthalpy can be obtained by using Eq. (7), where the contribution of the group interactions should be considered. Meanwhile, the vaporization entropy can be calculated by the boiling temperature as Eq. (8), where the boiling temperature is estimated by Eq. (9) [3], the group interactions also should be considered and N is atom number excluding hydrogen.

$$\Delta H^0_f(298\,K,l) = \Delta H^0_f(298\,K,g) - \Delta_v H(298\,K) \qquad (5)$$

$$S^0(298\,K,l) = S^0(298\,K,g) - \Delta_v S(298\,K) \qquad (6)$$

$$\Delta_V H(298\,K) = \sum N_i \Delta_{Vi} H(298\,K) \qquad (7)$$

$$\Delta_v S(298\,K) = A\,log\,T_b + B \qquad (8)$$

$$T_b = (\sum N_i * K_i) / (N^{0.6583} + 1.6868) + 84.3395 \qquad (9)$$

iii) Heat capacity

It can be expressed as the function of the temperature as Eq. (10), where A, B, D are three parameters which can be obtained by the group contribution as Eq. (11) [4].

$$C_{p,m} = R[A + B\frac{T}{100} + D(\frac{T}{100})^2] \qquad (10)$$

$$A = \sum N_i a_i \quad B = \sum N_i b_i \quad D = \sum N_i d_i \qquad (11)$$

2.2.2 The thermodynamic properties of the reaction

i) Reaction enthalpy, entropy and Gibbs energy below boiling point

Based on the reaction equations, the reaction enthalpy, entropy and Gibbs energy can be calculated by Eq. (12-14) where v has the positive sign for products and the negative sign for reactants.

$$\Delta_r H^0_m = \sum v \Delta H^0_f \qquad (12)$$

$$\Delta_r S^0_m = \sum v S^0_m \qquad (13)$$

$$\Delta_r G^0_m = \Delta_r H^0_m - T\Delta_r S^0_m \qquad (14)$$

By Kirchhoff equation, the reaction enthalpy at different temperature is given by Eq. (15), where $\Delta_r C_{p,m}$ is the reaction capacity change, which can be calculated from Eq. (16). Supposing all the heat capacity of reactants and products are the function of temperature expressed as Eq. (10), the reaction capacity change can be simplified as Eq. (17). Substituting into Eq. (15), the reaction enthalpy has an expression as Eq. (18).

$$\Delta_r H_m = \Delta_r H_m(298\,K,l) + \int_{298}^{T} \Delta_r C_{p,m} dT \qquad (15)$$

$$\Delta_r C_{p,m} = \sum v C_{p,m} \qquad (16)$$

$$\Delta_r C_{p,m} = a + bT + cT^2 \qquad (17)$$

$$\Delta_r H_m = \Delta_r H^0_m(298\,K,l) + aT + \frac{b}{2}T^2 + \frac{c}{3}T^3 - a\times298$$
$$-\frac{b}{2}\times298^2 - \frac{c}{3}\times298^3 = y + AT + BT^2 + CT^3 \qquad (18)$$

By Gibbs-Helmholtz Eq. (19), the reaction Gibbs energy can be calculated when the reaction enthalpy is obtained, or by integral Eq. (20).

$$[\partial(\Delta_r G_m / T) / \partial T]_p = -\Delta_r H_m / T^2 \qquad (19)$$

$$\Delta_r G_m = y + (\Delta_r G^0_m(298\,K) - y + A\times298\times ln\,298 +$$
$$B\times298^2 + \frac{C}{2}298^3) / 298 \times T - AT\,lnT - BT^2 - \frac{C}{2}T^3 \qquad (20)$$

All the parameters in above Eqs. (17), (18) and (20) are constants for a specified reaction.

ii) Reaction enthalpy, entropy and Gibbs energy above boiling point

As we know, if the reaction temperature is above the boiling point of components, the contribution of phase change to the reaction thermodynamic properties should be considered. Except water, the boiling points of all other components are all higher than 493K. For the reaction of neutralization, one of the products is water. So the above equations can only be used when temperature is lower than 373K. For temperature higher than 373 K Reaction enthalpy, Gibbs energy can be derived from Eq. (21) and (22), where $\Delta_r H_m^0(373\ \text{K}, g)$ and $\Delta_r G_m(373\ \text{K}, g)$ can be solved by Eq. (23) and (24), respectively.

$$\Delta_r H_m = \Delta_r H_m^0(373\ \text{K}, g) + a_1 T + \frac{b_1}{2}T^2 + \frac{c_1}{3}T^3 - (a_1 \times 373$$
$$+\frac{b_1}{2}\times 373^2 + \frac{c_1}{3}\times 373^3)\ = y_1 + A_1 T + B_1 T^2 + C_1 T \tag{21}$$

$$\Delta_r G_m = y_1 + (\Delta_r G_m^0(373\ \text{K}, g) - y_1 + A_1 \times 373 \times ln\,373 +$$
$$B_1 \times 373^2 + \frac{C_1}{2} 373^3)\ /\ 373 \times T - A_1 T\,lnT - B_1 T^2 - \frac{C_1}{2}T^3 \tag{22}$$

$$\Delta_r H_m^0(373\ \text{K}, g) = \Delta_r H_m^0(373\ \text{K}, l) + n \times \Delta_v H_m(373\ \text{K}) \tag{23}$$

$$\Delta_r G_m(373\ \text{K}, g) = \Delta_r G_m(373\ \text{K}, l) + n \times \Delta_v H_m(373\ \text{K})$$
$$-373 \times n \times \Delta_v S(373\ \text{K}) \tag{24}$$

iii) Thermodynamic equilibrium constant

The important thermodynamic relation of Eq. (25) enables us to predict the thermodynamic equilibrium constant K of any reaction from tables of thermodynamic data, and hence to predict the equilibrium composition of reaction mixtures.

$$RT\,ln\,K = -\Delta_r G_m^0 \tag{25}$$

3 Discussion and Results

3.1 Experimental

At low temperature, yellow liquid was isolated from solvent S1. In Fig. 3(a), bands in the range of 2900-3400 cm^{-1} and 1408 cm^{-1} are the typical hydroxyl group vibration peaks and 1716 cm^{-1} (peak II) is attributed to carboxyl acid group, which suggest there are lots of hydroxyl groups existing in the mixture and some –COOH groups remained neutral form. Theoretically, if tertiary ammonium salt produced, bands around 2250 cm^{-1} can be observed. In our work, frequency of the bands reduced to near 2000 cm^{-1}(peak I). It may be attributed to the strong hydrogen bonding between acid and base and the steric effect of base. So it can be considered that interaction between acid AC1 and base AK1 was as Eq. (26)

$$HOOCC_2H_4COOH + (HOCH_2CH_2)_3 N \rightarrow$$
$$(HOCH_2CH_2)_3 N:HOOCC_2H_4COOH \tag{26}$$

With temperature increasing from 353K to 493K, tertiary ammonium salt characteristics IR absorption band disappeared (Fig. 3(c), peak VIII) which suggests it was decomposed. In the same time, C=O stretching vibrations frequency changed from 1716 to 1730 cm^{-1}(Fig. 3, peak III, V) and the intensity of C-O-C stretching vibrations around 1160 cm^{-1} (Fig. 3(b), peak IV) obviously enhance. These mean ester groups are generated. Meanwhile, an enormous reduction of intensity of -OH stretching vibrations between 2900 and 3400 cm^{-1} (Fig.3c, peak VII) was observed, which indicates more –OH groups take part in esterification reaction

with temperature increasing. Since multi-functional groups containing in AC1 and AK1, esterification products may include monoester, diester, triester and cyclic ester, even polyester if temperature is high enough.

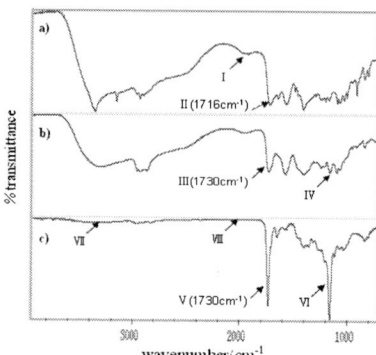

Fig. 3 Tthe IR spectrum of the mixture at (a) 353K, (b)453K and (c) 493 K.

3.2 Thermodynamic calculation

According to the calculation flow scheme in Fig. 1 and combining the equations from (1) to (11), the thermodynamic properties of all reactants and possible products can be calculated. Results are show in table 3.

Based on calculation methods defined as Eq. (12) to (25), it is possible to obtain the thermodynamic properties of reactions that predicted by experimental results. Concerning the contribution of water phase transition, during calculation of reaction thermodynamic properties with temperature increasing, functions were divided into two stages, one was within 298~373K and another 373~493K. Calculation results show in Table 4. In Fig. 4 it shows how the reaction Gibbs energy $\Delta_r G_m^0$ and the thermodynamic equilibrium constant K change with temperature increasing.

The reaction Gibbs energy for monoester, diester and triester reaction are all smaller than zero, so these reactions are possible to happen spontaneously under experimental condition in this work. Meanwhile, it can be also observed that all the reaction Gibbs energy of different types of esterification decrease with increasing of temperature, which suggest esterifications are apt to happen at high temperature. Furthermore, within 298~373K the thermodynamic equilibrium constants K show a little reduction with temperature increasing because the reaction enthalpy change is smaller than zero before 373 K. When the temperature is higher than 373 K, the reaction enthalpy is bigger than zero, all the thermodynamic equilibrium constants K increase with temperature increasing, which means higher temperature is favorable to shift reaction to positive direction for these endothermic reactions. Consequently, we can make conclusion that the higher temperature, the higher degree of esterification, which is coincided with the experimental observations and the IR spectrum analysis. The reaction Gibbs energy for cyclic ester reaction is bigger than zero which means the reaction can not happen spontaneously. Moreover, the thermodynamic equilibrium constant K of cyclic ester reaction is small which further indicates the low degree of the reaction. This is also coincident with IR test, no absorption   peak for carbonyl in circle.

Table 3 Te thermodynamic properties of pure component

Substance	ΔH_f^0(298 K, g) / kJ·mol^{-1}	S^0(298 K, g) / J·mol^{-1}·K^{-1}	$\Delta_v H$(298 K) / kJ·mol^{-1}	$\Delta_v S$(298 K) / J·mol^{-1}·K^{-1}	ΔH_f^0(298 K, l) / kJ·mol^{-1}	S^0(298 K, l) / J·mol^{-1}·K^{-1}
AK1	-558.41	545.69	123.92	147.25	-682.33	398.43
AC1	-823.78	403.78	101.34	199.12	-925.12	204.66
H2O(l)	-285.83	69.91	0.00	0.00	-285.83	69.91
H2O(g)	-241.82	188.83	0.00	0.00	-241.82	188.83
Monoester	-1160.76	769.40	165.05	133.48	-1325.81	635.92
Diester(2AC1)	-1763.11	989.74	206.18	139.01	-1969.29	850.72
Triester	-2365.46	1206.71	247.31	143.33	-2612.77	1063.38
Cyclic ester	-939.33	580.20	95.93	137.21	-1035.26	442.98
Diester(2AK1)	-1497.74	1135.02	228.76	141.15	-1726.50	993.87

Remark: 1.Where diester (2AC1) represents the product of the reaction of 2 AC1 molecules with 1 AK1 molecule. Similarly, diester (2AK1) represents the product of the reaction of 2 AK1 molecules with 1 AC1 molecule. 2. For calculations of ΔH_f^0(298 K, g), S^0(298 K, g) origin group contribution value from literature 3, for $\Delta_v H$(298 K) from the book writted by V.Majer et al.[5], and for $\Delta_v S$(298 K) from literature 3.

Table 4 Thermodynamics of the esterification of monoester, diester, triester and cyclic ester reactions at different temperature

T / K	$\Delta_r G_m$ /kJ·mol^{-1}					K				
	Monoester	Diester (2AC1)	Diester (2AK1)	Triester	Cyclic ester	Monoester	Diester (2AC1)	Diester (2AK1)	Triester	Cyclic ester
298	-34.81	-62.85	-47.77	-90.26	6.58	1.26E+06	1.04E+11	2.36E+08	6.63E+15	0.07
308	-35.83	-64.68	-49.09	-92.87	6.78	1.19E+06	9.33E+10	2.11E+08	5.63E+15	0.07
318	-36.86	-66.51	-50.41	-95.47	6.98	1.13E+06	8.41E+10	1.91E+08	4.82E+15	0.07
328	-37.88	-68.33	-51.72	-98.07	7.19	1.08E+06	7.62E+10	1.73E+08	4.15E+15	0.07
338	-38.90	-70.14	-53.02	-100.65	7.41	1.03E+06	6.91E+10	1.57E+08	3.59E+15	0.07
353	-40.41	-72.82	-54.95	-104.47	7.78	9.56E+05	5.97E+10	1.35E+08	2.88E+15	0.07
363	-41.41	-74.59	-56.21	-106.98	8.04	9.10E+05	5.41E+10	1.23E+08	2.48E+15	0.07
373	-39.00	-69.55	-50.68	-99.32	11.71	2.90E+05	5.49E+09	1.25E+07	8.12E+13	0.02
383	-41.06	-73.43	-54.06	-105.00	10.95	3.98E+05	1.03E+10	2.36E+07	2.09E+14	0.03
393	-43.08	-77.25	-57.37	-110.60	10.25	5.32E+05	1.85E+10	4.22E+07	5.02E+14	0.04
403	-45.08	-81.01	-60.63	-116.12	9.61	6.96E+05	3.17E+10	7.23E+07	1.12E+15	0.06
413	-47.04	-84.71	-63.83	-121.53	9.02	8.91E+05	5.19E+10	1.18E+08	2.35E+15	0.07
423	-48.97	-88.35	-66.96	-126.86	8.51	1.12E+06	8.14E+10	1.86E+08	4.63E+15	0.09
433	-50.87	-91.92	-70.02	-132.07	8.06	1.37E+06	1.23E+11	2.80E+08	8.58E+15	0.11
443	-52.73	-95.42	-73.01	-137.18	7.68	1.65E+06	1.78E+11	4.07E+08	1.50E+16	0.12
453	-54.55	-98.83	-75.92	-142.18	7.38	1.95E+06	2.49E+11	5.69E+08	2.48E+16	0.14
463	-56.34	-102.17	-78.76	-147.05	7.16	2.27E+06	3.37E+11	7.68E+08	3.90E+16	0.16
473	-58.08	-105.43	-81.51	-151.80	7.03	2.59E+06	4.40E+11	1.00E+09	5.82E+16	0.17
483	-59.78	-108.60	-84.17	-156.42	6.98	2.92E+06	5.56E+11	1.27E+09	8.27E+16	0.18
493	-61.43	-111.68	-86.75	-160.91	7.02	3.23E+06	6.81E+11	1.55E+09	1.12E+17	0.18

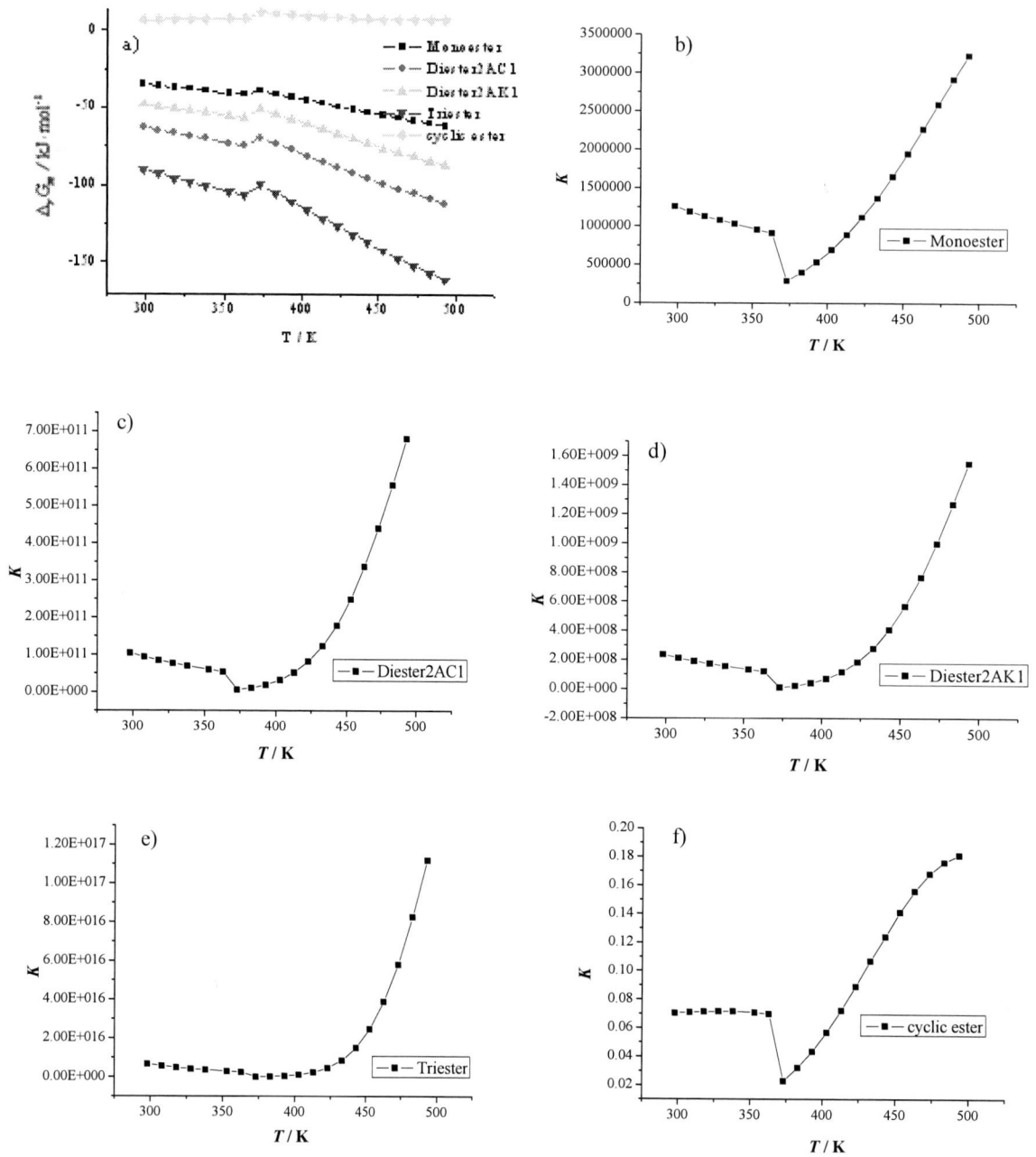

Fig. 4 (a) the reaction Gibbs energy and the thermodynamic equilibrium constant K of (b) monoester, (c) diester(2AC1), (d) diester(2Ak1), (e) trimester and (f) cyclic ester change with temperature

Conclusions

We have successfully established the experimental and theoretical methods for understanding the complicated interactions among various key ingredients of flux under various temperature conditions. Initially, the interactions between succinic acid and triethanolamine in the solvent of polyethylene glycol dibutyl ether at three different temperatures have been investigated by using IR spectrum as main analysis method. Meanwhile, the theoretical calculation was conducted by Group Contribution methods, the results of experiment and calculation were coincided well with each other. Details are as follows:

1. Reactions between succinic acid and triethanolamine are initially neutralization. With the increase of temperature, the esterification reactions are dominant.

2. The degree of esterifications increase with temperature increasing. When the temperature is at 493 K, the triamine salt decomposes, and only esters with different structures exist.

3. The Cyclic esterification is hard to generate.

978-1-4244-4658-2/09 $25.00 © 2009 IEEE   818

Acknowledgments

We appreciate the financial support from Henkel Corporation Electronic Materials.

References

1. Mark A. Barteau, "Organic Reactions at Well-Defined Oxide Surfaces," *Chem. Rev.*, Vol. 96 (1996), pp.1413-1430.

2. R. W. Snyder, "Diffuse Reflectance FT-IR Analysis of Rosin Flux-Metal Oxide Interactions," *Applied Spectroscopy*, Vol. 41, No. 3 (1987), pp. 460-463.

3. Y. Nannoolal et al., "Estimation of pure component properties, Part 1. Estimation of the normal boiling point of non-electrolyte organic compounds via group contributions and group interactions" *Fluid Phase Equilibria*, Vol. 226 (2004), pp. 45-63.

4. Bruce E Poling, John M. Prausnitz, John P O'Connell, The Properties of Gases and Liquids.5th ed., McGraw-Hill (New York, 2001), pp.120.

5. V.Majer,V.Svoboda,J.Pick, Heats of vaporization of fluids. Elsevier (Czechoslovakia, 1989), pp. 161-162.

Study of Stencil Printing Technology for Fine Pitch Flip Chip Bumping

Jin Yang[1,2,*], Jian Cai[1,2], Shuidi Wang[1,2], Songliang Jia[1,2]
[1]Institute of Microelectronics, Tsinghua University
[2]Tsinghua National Laboratory for Information Science and Technology (TNList)
Institute of Microelectronics, Tsinghua University, Beijing, P. R. China
* Corresponding author: Email: vickyyj@gmail.com, phone: +8610 62781852

Abstract

As miniaturization is the permanent pursuit of microelectronic industry, stencil printing technology for flip chip bumping has been contributing to this trend for almost half a century. Nowadays, it's still one of the lowest cost solutions to massive manufacture of IC packaging industry.

To meet the requirement of further miniaturization, this paper investigated the realization of fine pitch (about 100μm and sub 100μm) printing bumps on silicon wafers in-house. Electroformed stencil was fabricated and commercial printer was employed for bumping printing. Type-6 solder pastes (both leaded and lead-free), self-designed pallets, dummy wafers, etc., were applied in this report.

This paper closely investigated the practicable industry application of the fine pitch printing technology, and showed an integrated process to acquire industry-feasible fine pitch bumps including stencil design, dummy wafer design, materials and equipment preparation, etc. The essential parameters for printing process are presented as well.

Finally, the printing results showed that area arrays at pitches larger than 130μm and parallel arrays at pitches larger than 110μm were well achieved. Meanwhile, the problems on finer pitches' realization, aperture shapes, and solder wettability were brought on.

1. Introduction

Started from IBM's Controlled Collapse Chip Connection, also named C4 Technology [1], flip chip has been playing its important role in modern chip packaging for almost half a century. Unlike wire bonding and TAB, flip chip adopts bumping technology for the interconnection between chip I/O and peripheral circuit. As a critical factor of flip chip, bumping technology essentially contributes to the development in packaging density, volume, weight, electrical performance, etc.

Stencil printing has been proved to be a relatively simple way in bumping application. Moreover, due to its economical costs, stencil printing is widely adopted in the massive production. Three key factors for stencil printing will be discussed in this paper, i.e. printing materials (solder paste mostly), printing equipment, and stencil.

Due to the pursuit of miniaturization, fine pitch bumping is supposed to be one effective solution. For "fine", it denotes the pitch between two adjacent solder balls is around 100μm. Previous work has been carried out recent years. R. W. Kay, et al, [2] have reported printing at sub 100μm, and Fig. 1 shows the image of the reflowed bumps at 100μm pitch in their work. Fig. 2 shows the corresponding height frequency of the reflowed solder balls. Most of the solder ball heights are around 19-23μm (the pad size is 40μm).

Critical factors for stencil printing have also been studied closely. Jianbiao Pan, et al [3] reported that aperture size and stencil thickness are the two most critical variables for printing results using analysis of variance (ANOVA). However, the study was carried at pitches ranging from 760μm to 300μm which overflow our "fine" definition. Similar conclusions were made by R. Durairaj et al [4]. They applied hydrodynamics to analyze the whole printing process. And the results show that stencil parameters, which are called aspect ratios and area ratios, would be essential for the solder paste release. And area ratios seem to be dominant.

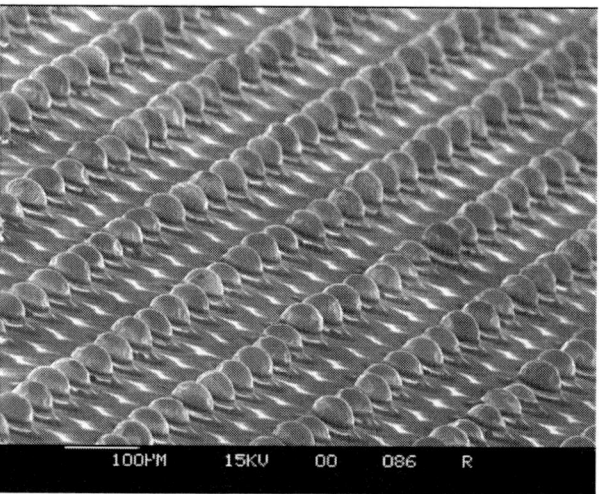

Fig. 1 100μm pitch side profile of an area array of reflowed solder bumps by R. W. Kay, et al.

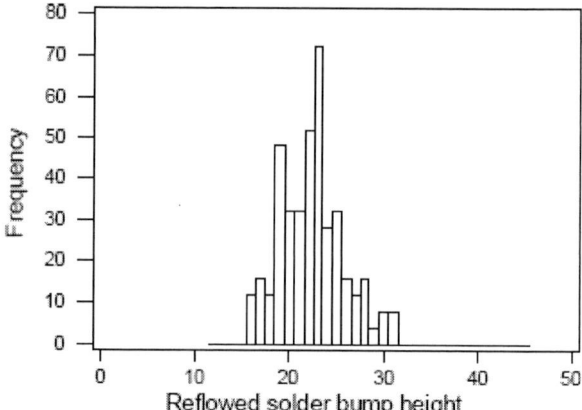

Fig. 2 100μm pitch, 70μm aperture diameter print height frequency by R. W. Kay, et al.

Based on the previous research outcomes, this paper will focus on the practicable industry application of the fine pitch printing technology, and show an integrated process to acquire industry feasible fine pitch bumps on silicon wafers including materials, equipment, design principles, technologic parameters, etc.

2. Experiment Preparation

2.1 Stencil design

Previous work has indicated that stencil was the most critical factor for bump printing. Important variables of a stencil include its fabrication method, thickness, and aperture size.

2.1.1 Stencil fabrication method

Three main stencil fabrication technologies have been discussed by Robert W. Kay, et al [5], i.e. chemical etching, laser cutting, and electroforming. However, both chemical etching and laser cutting can not produce smooth inner sidewalls. Furthermore, as the number of apertures increases, the production efficiency descends greatly. Electroformed stencils seem to be the best choice for sub 100μm pitch printing.

Electroforming is a bottom-up technology instead of a top-down one. It can create very smooth inner sidewalls. The common material of electroformed stencil is nickel, which has a higher rigidity (400-600HV) than stainless steel 304 (about 370HV). The thickness of the stencil is controllable. Moreover, electroformed stencils are easier to be cleaned and have a higher positional accuracy. However, the fabrication of electroformed stencils is much more complicated and then more expensive.

The stencils for experiment in this paper are electroformed and provided by PowerStencil Corp, which is the first Chinese corporation supplying electroformed stencils.

2.1.2 Stencil parameters

The stencil thickness and aperture size are the predominant parameters for stencil design. However, we can describe them in terms of area ratio and aspect ratio [6].

$$\text{Area ratio} = W / t$$
$$\text{Aspect ratio} = A_{ap} / A_{sw}$$

where W means aperture width, t means stencil thickness, A_{ap} means area of aperture, and A_{sw} means area of aperture sidewall.

Previous work [7] indicated that a critical value of area/aspect ratio should be reached so as to acquire an ideal solder paste transfer. These critical values for stencils by different fabrication technologies are shown in Table 1.

Table 1 The minimum values of area/aspect ratio required for different stencil technologies

	Chemical etching	Laser cutting	Electroforming
Area ratio	>0.66	>0.6	>0.5
Aspect ratio	>1.5	>1.2	>1.1

However, the solder transfer is a very complicated process. A number of factors can affect the final result. So this value not only depends on stencil fabrication method, but other integrated factual experiment factors. Hereby the figures provided in Table 1 can be referred to, they may not always work though.

In this paper, five pitches were chosen, i.e. 90μm, 100μm, 110μm, 130μm, 150μm. And there were two array categories – area array and parallel array for each pitch. Therefore, ten kinds of dummy chips were designed.

Srinivasa Aravamudhan, et al [8] reported that circular apertures have better release efficiency than the square apertures for a same nominal volume of paste to be deposited. Thus circular apertures were chosen for area arrays instead of square ones. While for parallel arrays, the aperture shapes do not have to be symmetrical. The direction not restricted can be prolonged so as to enlarge the nominal volume of paste. Thus rectangular apertures with fillets (increasing the release of solder) were used for parallel arrays.

In case slump happens after printing, the minimum gap between adjacent apertures was set at 30μm, which means, for example, apertures of 90A shouldn't have diameters larger than 60μm. Meanwhile, according to the area/aspect ratio principle, the thickness of stencil was therefore restricted. In this work, stencil thickness was set at 30μm. The designed stencil parameters are shown in table 2 & 3.

Table 2 Stencil parameters for area arrays

Chip category	Aperture diameter/μm	Area ratio	Aspect ratio
150A*	110	0.92	3.67
130A	90	0.75	3.00
110A	80	0.67	2.67
100A	70	0.58	2.33
90A	60	0.50	2.00

*150A stands for 150μm pitch area array, the rest can be deduced by analogy.

Table 3 Stencil parameters for parallel arrays

Chip category	Aperture width/μm	Aperture Length/μm	Area ratio	Aspect ratio
150P*	100	170	1.10	3.33
130P	90	150	0.99	3.00
110P	80	135	0.88	2.67
100P	70	120	0.78	2.33
90P	60	100	0.66	2.00

*150P stands for 150μm pitch parallel array, the rest can be deduced by analogy.

2.2 Dummy wafer design

Concerning the wafer design, future testing circuit and corresponding stencil productive feasibility should be taken into consideration.

Two types of testing circuit structures were used: daisy chain and Kelvin test structure. As the resistance of a single solder ball is extremely small (typically 10mΩ), the daisy chain structure (shown in Fig. 3) connects a number of solder balls and makes the measure for resistance possible.

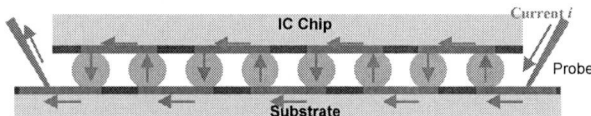

Fig. 3 Sketch of daisy chain testing structure

Kelvin testing structure [9] is a measure to avoid the contact resistance between the probes and pads. As shown in the Fig. 4, solder ball 1, 2, & 4 are connected on the IC chip. While on the substrate, solder ball 2 & 3 are connected.

Fig. 5 shows that if we applied a constant current i to pad 3 & 4, i.e. the current passing through solder ball 2, then the corresponding voltage applied to solder ball 2 can be measured precisely by the voltmeter between pad 1 & 2. Thus we can gain the accurate resistance of solder ball 2: $R_c=V_{12}/I_{34}$. Likewise, we can apply a constant voltage and measure the corresponding current to calculate the resistance as well.

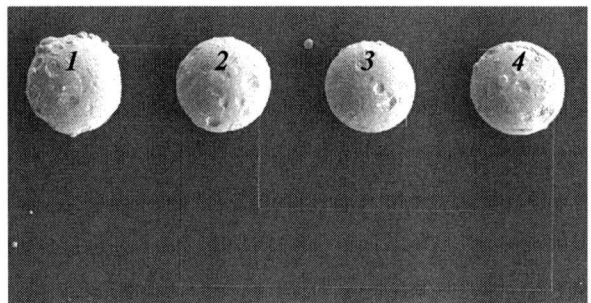

Fig. 4 Kelvin testing structure on the IC chip

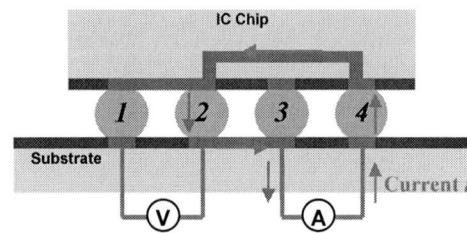

Fig. 5 Sketch of Kelvin testing structure

4-inch wafers were used to fabricate the testing circuit. Cu pads were electroplated onto the seed layer (TiW/Cu). Pad size for pitch 90μm, 100μm, & 110μm is 40μm×40μm, and 50μm×50μm for pitch 130μm & 150μm, which are smaller than the corresponding apertures on the stencil.

Co-design between patterns on wafer and stencil apertures should be set up during the bumping. As too many apertures cause the deformation of the stencil during the mesh stretching, they will lead to the degraded precision and aperture missing, and resultantly increase the difficulty of stencil fabrication. Within a 4-inch area, 8 defects occurred among 160,000 apertures, while only 3 defects among 100,000 according to our practical experience. Thus it's necessary to control the quantity of the apertures within a certain area during this co-design of both wafer and stencil.

2.3 Materials & equipment

In order to fulfill the sub-100μm pitch printing, the PSD (Particle Size Distribution) of solder paste smaller than 15μm is required, which means only type 6 and 7 solder pastes are fit at this fine pitch. In this paper, both leaded and non-lead type 6 solder pastes (PSD 5-15μm) were provided by Indium Corporation of America. Previous work [6] reported that type 6 solder can be used for minimum 110μm pitch, and type 7 for minimum 60μm pitch.

The essential equipment for fine pitch printing is a high-accuracy printer. Generally, the nominal translation accuracy of the printer should be within 20μm. The necessary rotation accuracy depends on the size of printing zone, e.g. for a 4-inch area, the rotation accuracy should be roughly 0.1°. In this work, a commercial Hitachi NP-O4XP printer was used.

With a conventional printer, a pallet was needed to hold and fixate the wafer during the printing process. For the printer doesn't own a vacuum system, a vacuum pump was used to cooperate with the pallet to tackle the wafer during the stencil-lifting process. This pallet with vacuum pump is shown in Fig. 6.

Fig. 6 Pallet with a vacuum pump for fixation

3. Experiment & Results

3.1 Experiment process

Printing parameters will affect the printing results to a large extent. The predominant printing parameters include printing speed, the pressure of squeegee, printing direction, separation speed, printing gap/snap off between the stencil and wafer, etc. [2-4, 10]

Printing speed impacts the printing results complexly. Normally, a printing speed between 10-30mm/s would be adoptable. In this work, 20mm/s printing speed was applied.

The pressure of squeegee should be high enough. Although a too high pressure will cause the deformation of both stencil and squeegee and even shorten the life of the extremely thin stencil. In this work, 0.1MPa pressure was applied.

Separation speed was set to 0.6mm/s and there was a very tiny snap off between the stencil and wafer. The experiment showed that if the snap off was too large (about larger than 0.5mm), it would lead to highly irregular printing bumps.

978-1-4244-4658-2/09 $25.00 © 2009 IEEE

In order to align the patterns on the wafer and stencil, a set of clear marks were needed. The Hitachi printer used in this experiment can identify circular marks with diameters of around 1mm. However, marks at this dimension would not be enough for the alignment accuracy. With a portable inspection microscope, the cupreous pads can be seen through the apertures for they're smaller than the apertures. Therefore, the position of stencil can be adjusted elaborately, and finally the precise alignment is able to be acquired.

To shorten the alignment process, more attention should be paid to the circumferential direction. As shown in Fig. 7, the alignment in the center seems to be good, while the sideward alignment is still a mess. At this circumstance, we should rotate the stencil clockwise/anticlockwise. Thus, as long as we ensure the alignment at both left & right side of the wafer, the adjustment is done. However, sometimes we still can't acquire a perfect alignment by doing this, because the alignment is also restricted by the precision of the apertures' locations & sizes, i.e. stencil technology.

Fig. 7 Sketch of the circumferential direction alignment

3.2 Experiment results

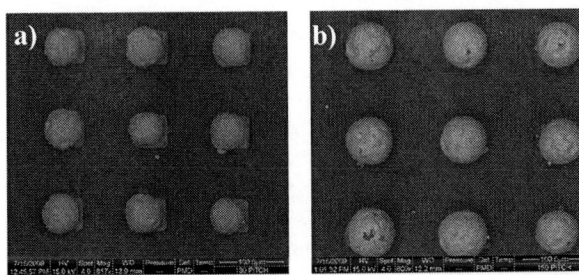

Fig. 8 SEM images of the reflowed solder balls of area arrays : a) 130A, b) 150A

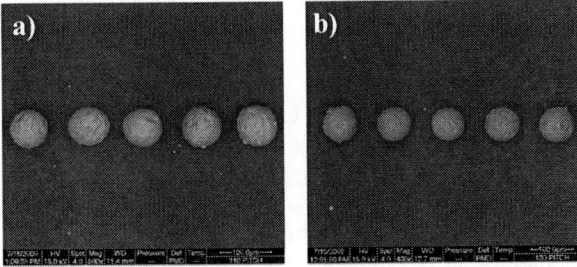

Fig. 9 SEM images of the reflowed solder balls of parallel arrays : a) 110P, b) 130P

In this experiment, we investigated 10 types of dummy chips as mentioned above, i.e. 90A, 90P, 100A, 100P, 110A, 110P, 130A, 130P, 150A, 150P. For area array, bumping at the pitches larger than 130μm were well achieved (i.e. 130A, 150A, as shown in Fig. 8). Here "well achieved" stands for about more than 95% of all the reflowed solder balls have an

ideal height and shape, and do not involve any bridging or skipping. And for parallel array, pitches larger than 110μm were well achieved (i.e. 110P, 130P, 150P, as shown in Fig. 9).

3.3 Problems encountered

A few problems might arise during the fine pitch bumping process. This paper brought some of them on.

The first problem is that high quality bump arrays are difficult to achieve when pitches get close to 100μm. Solder bridging and skipping occurred at this fine pitches. As we can see from Fig. 10, some 110A's bumps (unreflowed) are poor at solder transfer rate and uniformity.

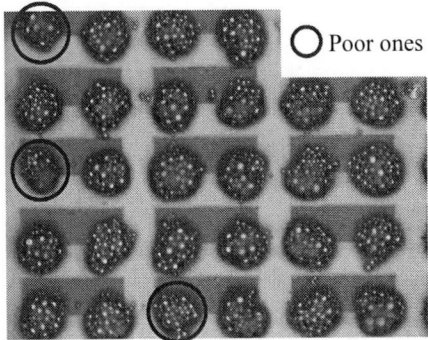

Fig. 10 Optical microscope image of 110A's bumps before reflow

Fig. 11 shows the SEM image of 110A after reflow. Apparently, some of the solder balls are smaller compared to other ones because of the poor transfer ratio.

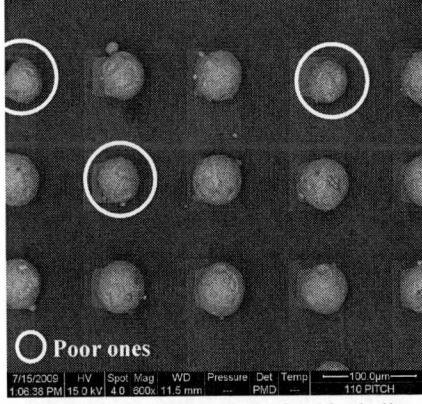

Fig. 11 SEM image of the reflowed solder balls of 110A

Second, the rectangular apertures obviously have a worse release ability than circular ones at the same pitch level. What's more, their transfer process depends on printing direction to a great extent. If longer side of the aperture is parallel to the squeegee moving direction, the printing result will be much worse than the opposite circumstance. However, parallel ones will still result in bigger bumps due to their bigger nominal volumes.

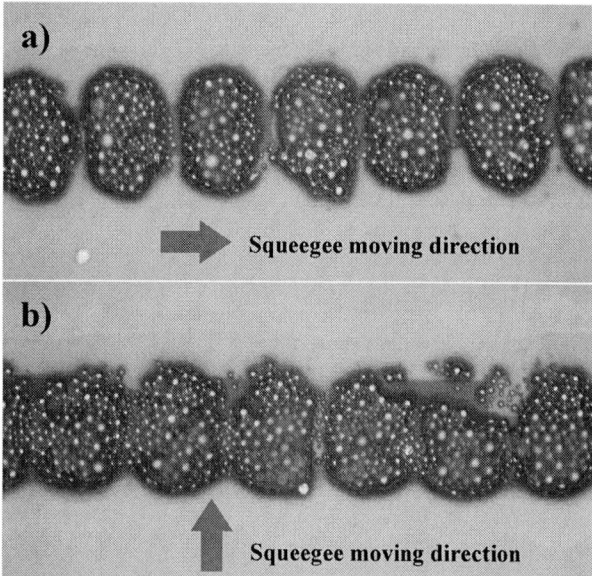

Fig. 12 Optical microscope image of 110P's bumps before reflow, longer side of the aperture is : a) parallel, b) perpendicular, to the squeegee moving direction

From Fig. 10 & 12, it can be concluded that the rectangular bumps are more likely to have the tendency to slump. Because rectangle isn't center-symmetrical, the slump along the longer side would be more serious. Fig. 13 shows a cross section image of the bridging bump caused by solder paste slump. Therefore, the minimum gap between two adjacent rectangular apertures (or other center-unsymmetrical shapes) should be sufficiently larger than circular ones. This might cause trouble in stencil design because large length/width ratio won't do any good to transfer ratio. In a word, circular apertures would be the best choice for this fine pitch printing.

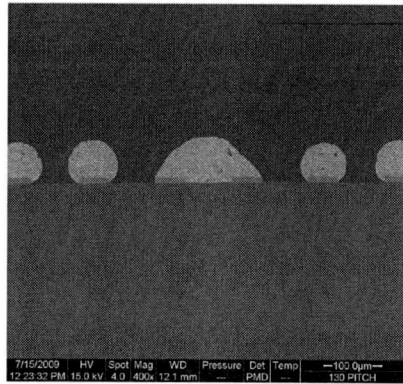

Fig. 13 SEM image of a bridging solder ball caused by slump

The third problem is the wettability at such fine pitches. The experiment encountered a serious wettability problem. As shown in Fig. 14, if the amount isn't enough, the solder paste isn't able to wet the Cu pad effectively. As a result, the reflowed solder balls merely stand on partial pads instead of the whole.

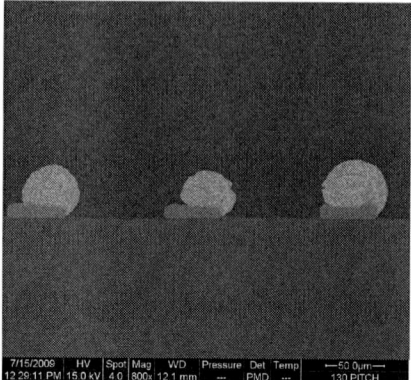

Fig. 14 SEM image of wettability problem at fine pitch

This problem may be caused by one or more of the following reasons:

i) The PSD of type 6 solder paste is so small that the metal particles are quite easy to be oxidated. The more oxide composition, the worse wettability it will be.

ii) The bare Cu pad surface is oxidated when exposed in the air or polluted during the printing process. The oxide (Cu_2O) on top of the Cu pad can degrade the wettability significantly. A plated Ni/Au/OSP layer can protect the pad from these negative factors effectively.

iii) The temperature profile of reflow is not appropriate for fine pitch bumps. Too long preheating period may cause the over-volatilization of flux for it seemed the flux didn't play their role at some circumstances.

Conclusions

The key factors of stencil printing include printing materials, printing equipment, and stencil.

Electroformed stencils are most suitable for fine pitch printing for its high quality inner sidewalls. The design principle of stencil is based on area/aspect ratio. Appropriate aperture shape, minimum gap, aperture size, and stencil thickness to meet the area/aspect ratio requirement are all essential parameters to be designed.

Dummy wafer should contain testing circuit if future electrical test was needed. Normally daisy chain and Kelvin testing structure are recommended. Meanwhile, the layout of the wafer should take corresponding stencil productive feasibility into account.

The solder pastes with PSD smaller than type 6 are necessary for fine pitch printing. The commercial printer should meet both the translation and rotation accuracy requirements. A tailor-made pallet system will be needed to fixate the wafer during the printing process.

Printing speed, the pressure of squeegee, printing direction, separation speed, printing gap / snap off between the stencil and wafer are the predominant printing parameters for printing. In order to acquire a perfect alignment, circumferential adjustment would be more important.

As a result, area arrays at pitches larger than 130μm and parallel arrays at pitches larger than 110μm were well achieved, which means the reflowed solder balls bear the desirable planarity and defect rate required by practicable industry use. However, to achieve the finer pitches is still to be investigated. Moreover, rectangular shape apertures and wettability problem are needed to be improved in the future.

Acknowledgments

The authors would like to appreciate PowerStencil Corp. for the great support on electroformed stencils. The authors would also thank Indium Corp. for supplying solder paste and Beijing Diantong Wintronic Electronic Co. for supplying printing & reflowing processes. This work also is supported by the National Engineering Laboratory on High Density IC Packaging Technology, China.

References

1. L.F. Miller, "Controlled collapse reflow chip joining," *IBM J. Res. Develop*, Vol. 13, No. 3 (1969), pp. 239-250.

2. R. W. Kay, E. de Gourcuff, M. P. Y. Desmulliez, G. J. Jackson, H. A. H. Steen, C. Liu, P. P. Conway, "Stencil Printing Technology for Wafer Level Bumping at Sub-100 Micron Pitch using Pb-Free Alloys," *2005 Electronic Components and Technology Conference*, 2005, pp. 848-854.

3. Jianbiao Pan, Gregory L. Tonkay, Robert H. Storer, Ronald M. Sallade, David J. Leandri, "Critical Variables of Solder Paste Stencil Printing for Micro-BGA and Fine-Pitch QFP," *IEEE Transactions on Electronics Packaging Manufacturing*, Vol. 27, No. 2 (2004), pp. 125-132.

4. R. Durairaj, T.A. Nguty, N.N. Ekere, "Critical factors affecting paste flow during the stencil printing of solder paste," *Soldering & Surface Mount Technology*, Vol. 13, No. 2 (2001), pp. 30-34.

5. Robert W. Kay, Stephen Stoyanov, Greg P. Glinski, Chris Bailey, Marc P.Y. Desmulliez, "Ultra-Fine Pitch Stencil Printing for a Low Cost and Low Temperature Flip-Chip Assembly Process," *IEEE Transactions on Components and Packaging Technologies*, Vol. 30, No. 1 (2007), pp. 129-136.

6. Dionysios Manessis, Rainer Patzelt, Andreas Ostmann, "Stencil Printing Technology for 100μm Flip Chip Bumping," *Global SMT & Packaging*, Vol. 4, No. 2 (2004), pp. 10-14.

7. H.W. Markstein, "Controlling the variables in stencil printing," *Electronic Packaging and Prod*, Vol. 37, No. 2 (1997), pp. 48-56.

8. Srinivasa Aravamudhan, Daryl Santos, Gerald Pham-Van-Diep, Frank Andres, "A Study of Solder Paste Release from Small Stencil Apertures of Different Geometries with Constant Volumes," *27th Annual IEEE/SEMI International Electronics Manufacturing Technology Symposium*, 2002, pp. 159-165.

9. Zhao Ying-wei, Pang Ke-jian, "Application of Making Resistance Measurement Technology Using Kelvin 4-Wire Method," *Package & Test Technology*, Vol. 30, No. 11 (2005), pp. 43-45.

10 T. Krebs, A. Brand, Richard Lathrop, R. W. Kay, T. Wang, I. Roney, "Statistical Evaluation for Type-6 Lead-free Solder Paste down to 200-Microns Pitch with Type 6 Pb-free Solder Paste," China SMT Forum, Shanghai, China, 2007, pp. 134-162.

Processing and Properties of Cu-base and Co-base Amorphous Wires

W.B. Liao, Y. Zhang

State Key Laboratory for Advanced Metals and Materials

University of Science and Technology Beijing, Beijing, 100083, China

drzhangy@skl.ustb.edu.cn; 0086-010-62334927

Abstract

Amorphous wires of CuZrAl and CoFeSiB alloys were fabricated by in-rotating-water melt quenching (IMQ) and glass coated Taylor (GCT) techniques, and the diameters of the amorphous wires prepared by the IMQ are in the range of about 50 to 200 μm, while by the GCT are in the range f about 5 to 18 μm. The stress-strain curves of the $Cu_{50}Zr_{46}Al_4$ amorphous wire show a nonlinear deformation after the elastic deformation, and the tensile strength is about 1800 MPa. The resistivity of the CoFeSiB amorphous wires exhibit no obvious size effect in the diameter range of 5 to 18 μm, the resistivity decreases after the annealing at 600°C.

1 Introduction

Amorphous wires have received increasing interest in recent years because their advantages of metallic glasses, such as high strength, wear resistance, and strong corrosion resistance [1-2]. The amorphous wires have the potential applications in the micro-sensor, bonding wire, micro-springs. The micro-forming and processing technology of the amorphous alloy have developed greatly in recent years [3-4]. Many research work showed that the amorphous alloy can be made into different shapes and can show different properties when the size of amorphous alloy reduced to micron or nano meter [5-6]. In this paper, in-rotating-water quenching method and glass coated taylor processing are used to produce the amorphous wires, and the CuZrAl and CoFeSiB alloys wereselected.

2 Experiment

The master alloys were prepared by arc melting the constituents, Cu, Zr and Al; and Co, Fe, Si, and B, respectively-in a Ti-gettered high-purity argon atmosphere. The ingots were cut into several parts, then degreased by ultrasonic cleaning in a bath of acetone followed by the same procedure in a bath of ethanol. The alloys were placed in a quartz tube with a nozzle at the bottom (Fig.1). Amorphous wires of different sizes were produced by the in-rotating-water quenching technique using the alloys as changing processing parameters such as the velocity of the drum, the diameter of the nozzle, the distance from the liquid level to nozzle, etc.

Differential scanning calorimetry (DSC) was performed with a Netzsch DSC 404C differential scanning calorimeter under argon atmosphere. A constant heating rate of 20k/min was employed. A Zeiss Supra 55 scanning electron microscope was used to examine the surface morphologies.

Tension experiments were conducted on an Instron 5848 microtester with a gauge length of ~10 mm and a strain rate of 0.1 mm/min at room temperature. A small gauge was used to calibrate and measure the strain during loading. The fracture surfaces were studied by scanning electron microscopy (SEM).

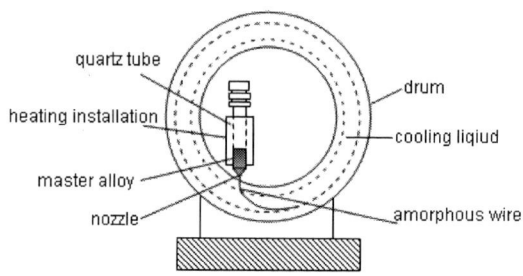

Fig.1 Abridge general view of in-rotating-water quenching method

3 Results and discussion

Fig.2 shows the shapes of micro-wires that made by the in-rotating-water quenching method. It can bee seen that with the increase content of Cu, the color of the micro-wires changes.

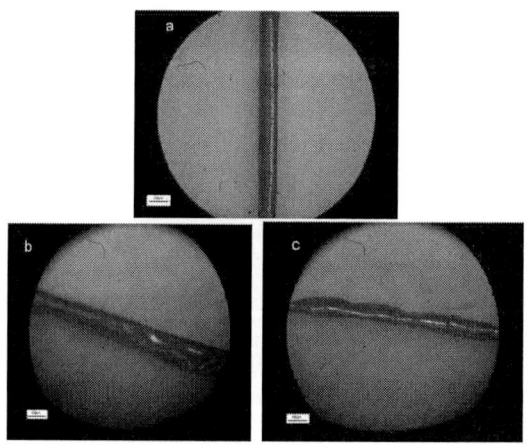

Fig.2. (a) $Cu_{38.5}Zr_{46.5}Al_{15}$ (b) $Cu_{50}Zr_{46}Al_4$ (c) $Cu_{80}Zr_{10}Al_{10}$

The diameters of the micro-wires were between 50μm and 200μm. The diameter of the micro-wire is mainly dependent on the diameter of the nozzle. If the arc gas flow from the above of the quartz tube and the water flow were not stable, it won't get the micro-wires but amount particles.

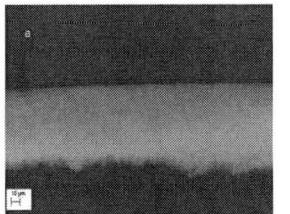 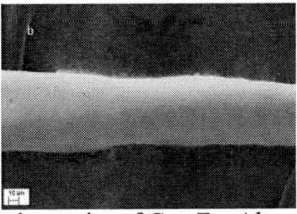

Fig.3. SEM images of the micro-wire of $Cu_{50}Zr_{46}Al_4$

Fig.3 shows that there are some defects on the micro-wire of $Cu_{50}Zr_{46}Al_4$ by the in-rotating-water quenching way. The side of the micro-wire that contacts with the water is smooth, and the other side is very rough (Fig. 3(a)). Fig. 3(b) shows that the size of the micro-wire in portrait is uneven. It's mainly influenced by the undulation of the water surface.

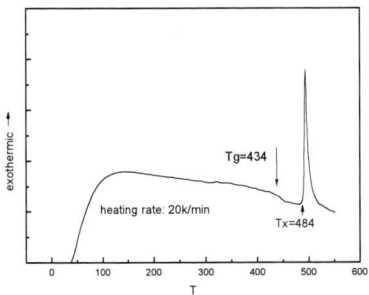

Fig.4 DSC trace of the micro-wire of $Cu_{50}Zr_{46}Al_4$

The DSC curve of the $Cu_{50}Zr_{46}Al_4$ micro-wire at heating rate of 20K/min under argon is shown in Fig.4. The thermogram of the alloy exhibits a glass transition endothermic (onset at T_g) and a crystallization exotherm (onset at T_x) appearing at 434 and 484°C.

The tensile engineering stress-strain curve of the micro-wire of $Cu_{50}Zr_{46}Al_4$ is shown in Fig.5.

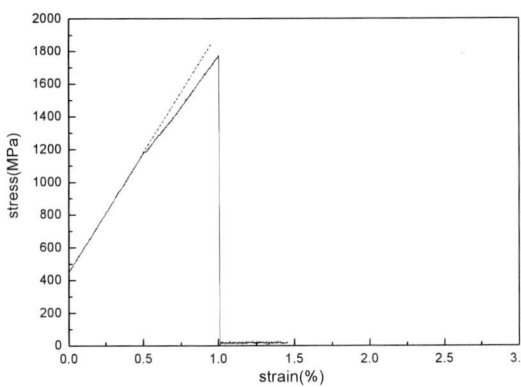

Figure 5 the stress-strain curve of $Cu_{50}Zr_{46}Al_4$ micro-wire

The tensile stress is about 1800 MPa which is almost the same as the compress stress [1]. The micro-wire of $Cu_{50}Zr_{46}Al_4$ initially deforms elastically, but at a certain stress the stress-strain curve deviates from the original straight line and shows a sign of nonlinear, inelastic deformation prior to failure, which is consistent with early experimental observation for other amorphous wires [7]. The flow defects play an important role during the plastic deformation [8].

Fig.6 shows the tensile fracture surface of the $Cu_{50}Zr_{46}Al_4$ micro-wire. Fig.6(a) depicts the shear bands forming and (b) shows the features of melting zone.

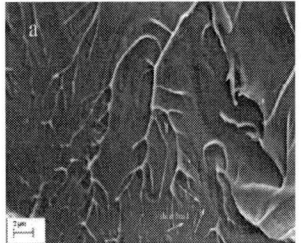 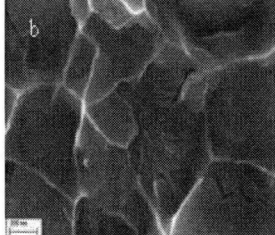

Fig.6 the tensile fracture surface of the $Cu_{50}Zr_{46}Al_4$ micro-wire

Figure 7 shows the morphologies of the CoFeSiB wires of 3 μm (a) and 12 μm (b). Figure 8 shows the resistivity changes of the wires before and after 600°C annealing, it was found that the resistivity decreases a bit after the annealing, and the size dependence of the resistivity seems not very sensitive.

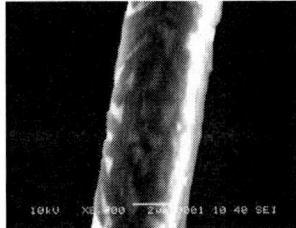 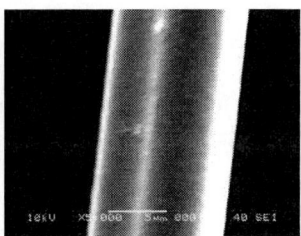

Figure 7 Amorphous wires of CoFeSiB alloy with 3 μm (a the left) and 12 μm (b the right) diameter.

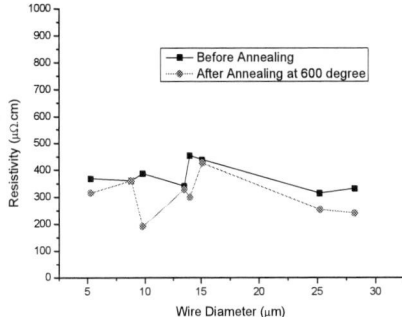

Figure 8 The relationship between the resistivity and the wire diameter, before and after annealing in forming gas environment (5% H_2/95% N_2) at 600°C for 20 mins. Multi segments of the microwire were measured and the data here is the average data.

4 Conclusions

In summary, the amorphous wires can be fabricated by in-rotating-water quenching technique or glass coated Taylor method, and the diameter is about 3~200μm. The original

978-1-4244-4658-2/09 $25.00 © 2009 IEEE

micro-wire still has some defects, but the tensile stress is almost equal to the compress stress. The curve of the stress-strain shows a nonlinear deformation, the mechanism still needs further study.

Acknowledgments

The work is supported by the National Natural Science Foundation of China. The resistivity measurements and annealing were performed by Cai H, Gan CL in the School of Materials Science and Engineering, Nanyang Technological University, Singapore 639798, Singapore.

References

1. J. Chen, et al. "Metallographic analysis of Cu-Zr-Al bulk amorphous alloys with yttrium addition," *Scripta Material,* Vol. 54 (2006), pp. 1351-1355.

2. A. Inoue, "Bulk non-equilibrium alloys and porous glassy alloys with unique mechanical characteristics," *Materials Science and Engineering: A,* Vol. 442 (2006), pp. 401-406.

3. X.H. Chen, et al. "A porous bulk metallic glass with unidirectional opening pores," *Electrochemical and Solid-State Letter,* Vol. 10, No. 12 (2007), pp. 21-23.

4. A.H.Brothers, D.C.Dunand, "Plasticity and damage in cellular amorphous metals," *Acta Materialia,* Vol. 53 (2005), pp. 4427-4440.

5. F.F. Wu, et al. "Size-dependent shear fracture and global tensile plasticity of metallic glasses," *Acta Materialia,* Vol. 57 (2009), pp. 257-266.

6. Z. Han, et al. "An instability index of shear band for plasticity in metallic glasses," *Acta Materialia,* Vol. 57 (2009), pp.1367-1372.

7. S. Takayama, "Serrated plastic flow in metallic glasses," *Scripta Metallurgica*, Vol. 13 (1979), pp. 463-467.

8. Y. Wu, et al. "Nonlinear tensile deformation behavior of samll-sized metallic glasses," *Scripta Materialia,* Vol. 61 (2009), pp. 564-567.

Absorption of Ag_3Sn on Cu_6Sn_5 Intermetallic Compounds at Sn-3.5Ag-xCu/Cu Interfaces

Ning Zhao[1,2], Lai Wang[1], Lixi Wan[2] and Liqiang Cao[2]

[1]School of Materials Science and Engineering, Dalian University of Technology, 2 Linggong Road, Dalian 116024, China
[2]Institute of Microelectronics of Chinese Academy of Sciences, 3 Beitucheng West Road, Beijing 100029, China
Email: zhaoning@ime.ac.cn; Tel: +86-10-82995720; Fax: +86-10-82995591

Abstract

The absorption behavior of Ag_3Sn particles on the surface of intermetallic compounds (IMCs) formed at Sn-3.5Ag-xCu/Cu (x=0, 0.7, 1.5 wt. %) interfaces was studied. The Sn-3.5Ag-xCu/Cu solder joints were prepared by reflow soldering at 260°C for different time. The X-ray diffraction results showed that the dominant IMCs formed at Sn-3.5Ag-xCu/Cu interfaces were Cu_6Sn_5. For Sn-3.5Ag/Cu and Sn-3.5Ag-0.7Cu/Cu joints, when the reflow time reached 120s and 60s respectively, besides Cu_6Sn_5, Ag_3Sn particles were also found at the interface. While for all the reflow time in this study, the Ag_3Sn particles were always detected at Sn-3.5Ag-1.5Cu/Cu interfaces. To observe the morphology of the interfacial IMCs, scanning electron microscope were performed. It was found that with the increase of the grain size of Cu_6Sn_5 the amount of absorbed Ag_3Sn particles increased. During the solidification, the Ag_3Sn particles could play the role of surface energy reducer for the interfacial Cu_6Sn_5 grains. There should be a critical grain size at which the absorption behavior of Ag_3Sn particles on Cu_6Sn_5 grains happened. According to this study, the critical grain size was determined to be about 2 μm.

1. Introduction

Due to the excellent combination property, Sn-Pb solders have been widely used in the electronic packaging industry, and many studies have been carried out to investigate their interfacial reactions [1-5]. However, the serviceable range of Pb is restricted by legislation all over the world because of its toxicity [6,7]. Then developing new lead-free solder turns to be necessary. Among the current developed lead-free solder alloys, Sn-Ag and Sn-Ag-Cu systems have become promising alternatives for Pb-bearing solders [8-10].

It is well known that one of the key issues that affect the reliability of solder joint is the growth of intermetallic compounds (IMCs) at the interface during soldering and aging/service [11,12]. In the process of soldering, Cu is widely used as the substrate. It has been found that when Ag-contained solder reacts with Cu substrate, Ag_3Sn particles can be found on the surface of interfacial IMCs [12,13]. The existence of Ag_3Sn particles would decrease the interfacial energy and hamper the growth of the IMC layer. Then the interfacial structure of the solder joint is affected consequently.

In this paper, the absorption behavior of tiny Ag_3Sn particles on the surface of Cu_6Sn_5 IMCs formed at Sn-3.5Ag-xCu/Cu (x=0, 0.7, 1.5 wt. %) interfaces was studied. The formation mechanism of these particles was discussed combining with the absorption theory and the liquid structure of alloy.

2. Experimental Procedures

The Sn-3.5Ag-xCu/Cu (x=0, 0.7, 1.5 wt. %) solder alloys were prepared from pure metals. The purity of element Sn, Ag and Cu is 99.95%, 99.9% and 99.999%, respectively. They were melted in a vacuum furnace at 500°C for 4 h. Cylindrical specimens were then cast using a steel mold. For the substrates, pure square Cu (99.9%) coupons of 0.1 mm thick and 10 mm wide were used. After polishing, the coupons were degreased in 50% water solution of HCl, followed by cleaning in ethanol. Small solder sheets with a dimension of Φ5×1.5 mm were prepared to perform soldering. The solder sheets covered by RMA flux were set on the Cu substrates, and then were put into the reflow soldering oven at 260°C for 10 s, 30 s, 60 s, 120 s and 300 s, respectively. After reflow the bulk solders of the joints were etched away by 5% nitric acid. Then the remainder parts were ultrasonic cleaned in water and in ethanol in turn for 2 min. XRD (X-ray diffraction analysis, XRD-6000) and EDX (energy dispersive X-ray analysis, Lin KIS6587) were performed to identify the composition of the IMCs formed at the interface. To facilitate the observation of the morphology of the interfacial IMCs, the SEM (scanning electron microscopes, JSM-5600LV) was also conducted. The average grain size of Cu_6Sn_5 IMCs was measured by using an image analysis software (Q500IW).

3. Results and Discussion

Fig. 1 shows the XRD patterns of IMCs formed at Sn-3.5Ag-xCu/Cu interfaces reflowed at 260°C for 300 s. The solders on top of the joints have been etched away. It's clear that Cu_6Sn_5 compounds were formed at the interface as expected. However, Ag_3Sn compounds were also found for each joint.

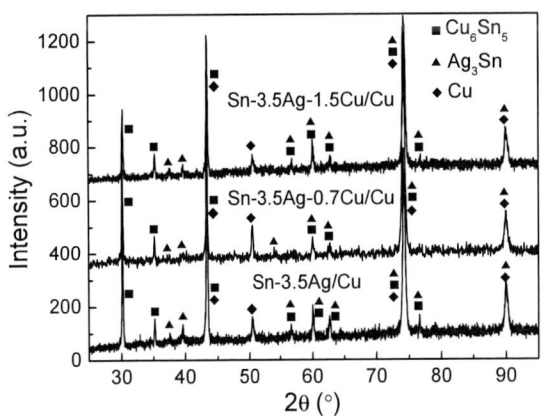

Fig. 1 XRD patterns of IMCs formed at Sn-3.5Ag-xCu/Cu interfaces reflowed at 260°C for 300 s

To reveal the morphology of the IMCs at Sn-3.5Ag-xCu/Cu interfaces SEM was carried out. Fig. 2 gives the top view of interfacial IMCs reflowed for 10 s. According to XRD and EDX analysis, the spherical grains were Cu_6Sn_5

compounds. Comparing the three joints, the grain size of Cu_6Sn_5 increased slightly with increasing Cu content in the solder. It indicated that the grain grew faster for higher Cu concentration solder. This is mainly driven by the different liquid structure of the solders. Solder with higher Cu content has larger and more Cu_6Sn_5 clusters during reflow [14,15]. These clusters served as the nucleation sites for the formation of Cu_6Sn_5 compounds, which accelerated the growth of its grains. It's interesting that a few tiny Ag_3Sn particles were detected on the surface of Cu_6Sn_5 grains just at Sn-3.5Ag-1.5Cu/Cu interface, while not on the other two.

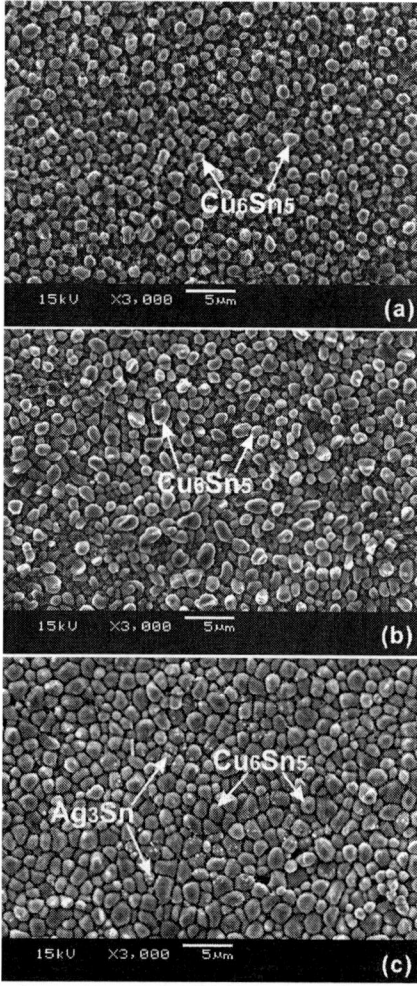

Fig. 2 Top view of interfacial IMCs reflowed for 10 s: (a) Sn-3.5Ag/Cu; (b) Sn-3.5Ag-0.7Cu/Cu; (c) Sn-3.5Ag-1.5Cu/Cu

The top view of the IMCs reflowed for 60 s is given in Fig. 3. All the Cu_6Sn_5 grains grew larger. While Sn-3.5Ag-1.5Cu/Cu still had the largest Cu_6Sn_5 grain size and there were even more absorbed Ag_3Sn particles, as shown in Fig. 3(c). Fig. 3 (d) is a high magnification view of Fig. 3 (c). The grain marked by "A" is the same one in Fig. 3(c) and Fig. 3(d). Moreover, larger plane of the grain tended to absorb more Ag_3Sn particles. Besides Sn-3.5Ag-1.5Cu/Cu, the absorption of Ag_3Sn particles also happened for Sn-3.5Ag-0.7Cu/Cu. However, no absorption happened for Sn-3.5Ag/Cu.

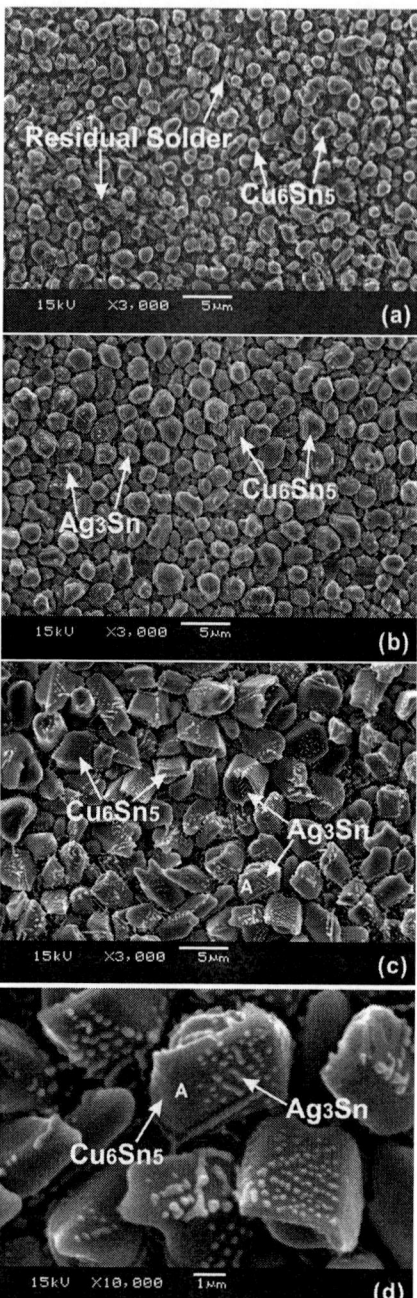

Fig. 3 Top view of interfacial IMCs reflowed for 60 s: (a) Sn-3.5Ag/Cu; (b) Sn-3.5Ag-0.7Cu/Cu; (c) and (d) low and high magnification of Sn-3.5Ag-1.5Cu/Cu

Fig. 4 shows the top view of interfacial IMCs reflowed for 300 s. The growth trend of Cu_6Sn_5 grain is consistent with that in Fig. 2 and Fig. 3. Meanwhile, the absorption of Ag_3Sn particles occurred at all the three interfaces. The Ag_3Sn particles even combined together to form a hat covering the top of Cu_6Sn_5 grains at Sn-3.5Ag-1.5Cu/Cu interface. Actually, based on the SEM observation the Ag_3Sn particles have already been found on Cu_6Sn_5 grains at Sn-3.5Ag/Cu interface when the reflow time reached 120 s.

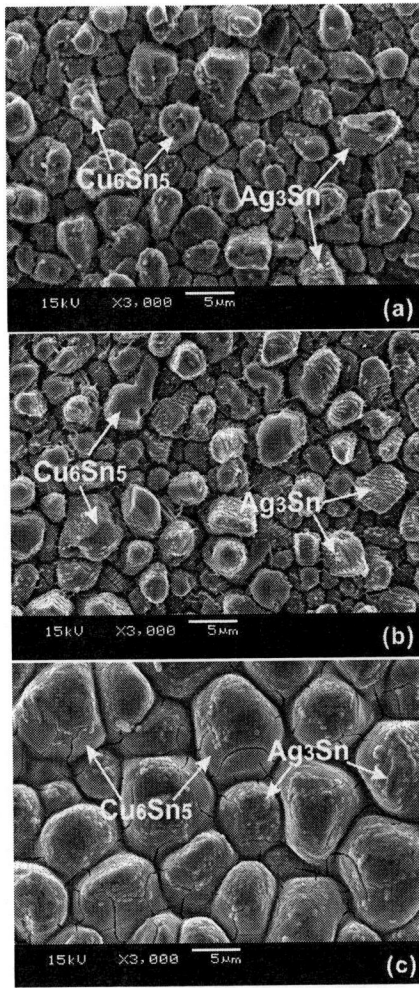

Fig. 4 Top view of interfacial IMCs reflowed for 300s: (a) Sn-3.5Ag/Cu; (b) Sn-3.5Ag-0.7Cu/Cu; (c) Sn-3.5Ag-1.5Cu/Cu

Huang et al have reported the absorption of Ag$_3$Sn particles on top of Cu$_6$Sn$_5$ grains [12]. They considered that Ag$_3$Sn particles nucleated on top of Cu$_6$Sn$_5$ grains during solidification or deposited on them while etching the solder away (Ag$_3$Sn has a higher corrosion resistance than Sn). The latter isn't reasonable. First, if Ag$_3$Sn particles deposited during etching, they should be found at all the interfaces. Secondly, the depositing Ag$_3$Sn particles can also be washed away by ultrasonic cleaning after etching. As bigger Cu$_6$Sn$_5$ grains had more Ag$_3$Sn particles attached, it was considered that Ag$_3$Sn nucleated and then grew on the surface of Cu$_6$Sn$_5$ grains during solidification, which could reduce the surface energy between Cu$_6$Sn$_5$ grains and liquid solder.

The interface free energy of crystal G can be given by [16]

$$G = \sigma A \qquad (1)$$

where σ is the surface energy and equals to the surface tension in numerical value; A is the surface area of the grain or the plane. Due to the surface tension between Cu$_6$Sn$_5$ grains and liquid solder is the same for these joints, based on Eq. (1) the larger Cu$_6$Sn$_5$ grains have larger interface free energy.

The Gibbs absorption equation is expressed as [17]

$$\Gamma = -\frac{x}{RT}\left(\frac{d\sigma}{dx}\right)_T \qquad (2)$$

where Γ is the differential concentration at unit interface between the solute at surface and that inside the whole solution; x is the gram-atom fraction; R is the general gas constant; T is the absolute temperature; and $(d\sigma/dx)_T$ gives the change of the surface tension with the concentration at certain temperature. If the solute can reduce the surface energy of the interface, its concentration at the interface will much higher than that inside the solution. In other words, the absorption phenomena at the interface will occur at this condition. So in this study, the Ag$_3$Sn particles played the role of surface energy reducer for the interfacial Cu$_6$Sn$_5$ grains during the solidification.

From Fig. 2, Fig.3 and Fig. 4, it is obvious that the absorption of Ag$_3$Sn particles didn't occur immediately when Cu$_6$Sn$_5$ compounds was formed. The Cu$_6$Sn$_5$ grains at Sn-3.5Ag-1.5Cu/Cu interface grew fastest and had the largest size, so the absorption phenomena happened initially. While Sn-3.5Ag/Cu took the longest reflow time, 120 s, to make it through. It seems that for Sn-3.5Ag-xCu/Cu joints there is a critical grain size of Cu$_6$Sn$_5$ beyond which the absorbed Ag$_3$Sn particles can be observed. That is to say the absorption can only occur when the surface energy of Cu$_6$Sn$_5$ grain reaches a certain value.

The mean grain size r_m of Cu$_6$Sn$_5$ at Sn-3.5Ag-xCu/Cu interfaces reflowed for different time was measured using an image analysis software. The results were plotted in Fig. 5. Fig. 5(a) shows the relationship between r_m and reflow time. For each joint, the Cu$_6$Sn$_5$ grains grew gradually with reflow time. On the other hand, for a given reflow time r_m increased with the increasing of Cu content in the solder. In this test, the

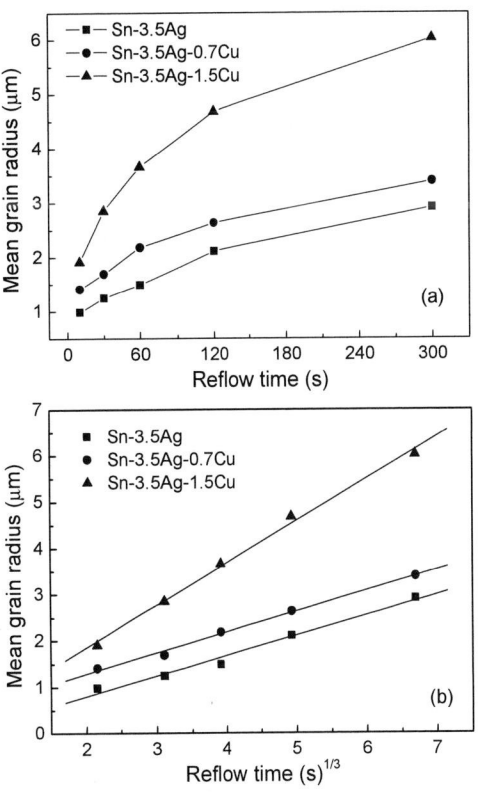

Fig. 5 Mean grain size of Cu$_6$Sn$_5$ at Sn-3.5Ag-xCu/Cu interfaces vs: (a) reflow time; (b) (reflow time)$^{1/3}$

Sn-3.5Ag/Cu, Sn-3.5Ag-0.7Cu/Cu and Sn-3.5Ag-1.5Cu/Cu is 120 s, 60 s and 10 s, respectively. And the corresponding r_m at these three points is 2.11 μm, 218 μm and 1.91 μm. According to the previous discussion, the critical grain size at which the absorption behavior of Ag_3Sn particles on Cu_6Sn_5 surface happened is about 2 μm.

Fig. 5(b) shows the plots of r_m vs the cube root of reflow time. A linear relationship was obtained for each joint. It can draw the conclusion that the growth of Cu_6Sn_5 grain at Sn-3.5Ag-xCu/Cu interfaces belongs to Ostwald ripening process [18,19]. The slope of the fitting line for Sn-3.5Ag/Cu, Sn-3.5Ag-0.7Cu/Cu and Sn-3.5Ag-1.5Cu/Cu is 0.44, 0.45 and 0.91, which indicates that the Cu_6Sn_5 grain at Sn-3.5Ag-1.5Cu/Cu interface grows much faster.

Conclusions

The absorption behavior of Ag_3Sn particles on the surface of Cu_6Sn_5 grains formed at Sn-3.5Ag-xCu/Cu interfaces was studied combining with the absorption theory and the liquid structure of alloy. Under the same soldering condition the Cu_6Sn_5 compounds grew faster with higher Cu content in the solder. A conception of critical grain size of Cu_6Sn_5 was proposed at which the absorption phenomena of Ag_3Sn particles would occur. This critical grain size was confirmed to be about 2 μm. When the absorption happened, larger Cu_6Sn_5 grains had more tiny Ag_3Sn particles on them. The Ag_3Sn particles on top of Cu_6Sn_5 grains could reduce the surface energy of the grains during solidification.

Acknowledgments

This work was supported by Hi-tech Research and Development Program of China (863 Program) No. 2007AA01Z2a6.

References

1. Kim, H. K. *et al*, "Three-dimensional Morphology of a Very Rough Interface Formed in the Soldering Reaction between Eutectic SnPb and Cu", *Applied Physics Letters*, Vol. 66, No. 18 (1995), pp. 2337-2339.

2. Lin, A. A. *et al*, "Spalling of Cu_6Sn_5 Spheroids in the Soldering Reaction of Eutectic SnPb on Cr/Cu/Au Thin Films", *Journal of Applied Physics*, Vol. 80, No. 5 (1996), pp. 2774-2779.

3. Kim, H. K. *et al*, "Kinetic Analysis of the Soldering Reaction Between Eutecitc SnPb Alloy and Cu Accompanied by Rippening", *Physical Review B*, Vol. 53, No. 23 (1996), pp. 16027-16034.

4. Prakash, K. H. *et al*, "Interface Reaction between Copper and Molten Tin-lead Solders", Acta Materialia, Vol. 49, No. 13 (2001), pp. 2481-2489.

5. Sharif, A. *et al*, "Dissolution Kinetics of BGA Sn–Pb and Sn–Ag Solders with Cu Substrates during Reflow", *Materials Science and Engineering B*, Vol. 106, No. 2 (2004), pp. 126-131.

6. Lee, N. C., "Getting Ready for Lead-free Solders", *Soldering & Surface Mount Technology*, Vol. 9, No. 2 (1997), pp. 65-69.

7. Vianco, P. T. *et al*, "Issue in the Replacement of Lead-Bearing Solders", *Journal of the Minerals Metals & Materials Society*, Vol. 45, No. 7 (1993), pp. 14-19.

8. Trumble, B., "Get the Lead Out!", *IEEE Spectrum*, Vol.35, No. 5 (1998), pp. 55-60.

9. Brodley, E. *et al*, "Lead-free Project Focuses on Electronics Assemblies", *Advanced Packaging*, Vol. 9, No. 2 (2000), pp. 34-42.

10. Kanchanomai, C. *et al*, "Effect of Temperature on Isothermal Low Cycle Fatigue Properties of Sn-Ag Eutectic Solder", *Materials Science and Engineering A*, Vol. 381, No. 1-2 (2004), pp. 113-120.

11. Yu, D. Q. *et al*, "Effects of Cu Contents in Sn-Cu Solder on the Composition and Morphology of Intermetallic Compounds at a Solder/Ni Interface", *Journal of Materials Research*, Vol. 20, No. 8 (2005), pp. 2205-2212.

12. Huang, M. L. *et al*, "Role of Cu in Dissolution Kinetics of Cu Metallization in Molten Sn-based Solders", *Applied Physics Letters*, Vol. 86, No. 18 (2005), p.181908

13. Yu, D. Q *et al*, "The Formation of Nano-Ag_3Sn Particles on the Intermetallic Compounds During Wetting Reaction", *Journal of Alloys and Compounds*, Vol. 389, No. 1-2 (2005), pp. 153-158.

14. Zhao, N. *et al*, "The Liquid Structure of Sn-based Lead-free Solders and the Correlative Effect on Liquid-solid Interfacial Reaction", *Journal of Physics: Conferences Series*, Vol.98 (2008), pp. U141-U144.

15. Zhao, N. *et al*, "Viscosity and Surface Tension of Liquid Sn-Cu Lead-Free Solders", *Journal of Electronic Materials*, Vol.38, No.6 (2009), pp. 828-833.

16. Xu, Z. Y *et al*, Thermodynamics of Materials, Science Press (Beijing, 2005), p. 39.

17. Xu, Z. Y *et al*, Thermodynamics of Materials, Science Press (Beijing, 2005), p. 78.

18. Yong, Q. L, "Ostwald Ripening of Second-phase Particles in Dilute Solution–I. Uuniversal Differential Equation", *Journal of Iron and Steel Research*, Vol. 3, No. 4 (1991), pp. 51-60.

19. Yong, Q. L. *et al*, "Ostwald Ripening of Second-phase Particle in Dilute Solution–II. Analytic Solution", *Journal of Iron and Steel Research*, Vol. 4, No. 1 (1992), pp. 59-66.

Prediction Model for Wire Bonding Process through Adaptive Neuro-Fuzzy Inference System

Jian Gao, Changhong Liu, Xin Chen, Detao Zheng, Ketian Li

Faculty of Electromechanical Engineering, Guangdong University of Technology
729 East Dongfeng Road, Guangzhou, 510090, P.R. China
Email: gaojian@gdut.edu.cn

Abstract

In the wire bonding process, different combinations of parameter values will directly affect wire bonding quality. The optimal combination of these parameter values is very important to ensure the overall process quality response. Therefore, it is necessary to investigate the effects and interactive relationship of the bonding parameters on the bonding quality. This paper chooses the response factors of shear strength and Squashed Ball Diameter (SBD) for the bonding quality evaluation. Through the design of experiments (DOE) method, 60 sets of experimental samples of the ball bond, varying 9 process parameters are manufactured and tested. The effect of the variation of these parameters on the shear force and SBD are analyzed and 6 out of 9 parameters are determined to be controlling factors in the prediction model. Considering the difficulty in analyzing the nonlinear and interactive relationships between the parameters and bonding quality, this paper proposes a process modeling approach based on an adaptive Neuro-Fuzzy Inference System. For the construction of the prediction model 50 sets of samples, where the 6 process control parameters are varied are prepared. These are used for the training of the process prediction model and a further 15 sets of samples are used for model validation. An error analysis is then performed to evaluate the model created. Based on the process prediction model, the characteristics of the bonding parameters affecting bonding quality are obtained, that can then be used for the optimization of wire bonding process.

Introduction

In the IC wire bonding process, there are many process parameters that have to be set properly to achieve a good bonding quality. It is evident that these wire bonding parameters are the main influencing factors not only on ball bond strength, but also on void formation [1]. Optimal combination of these parameter values will achieve a high bonding quality. Therefore, it is necessary to investigate the bonding parameters and find out the effects and relationship of these process parameters on the bonding quality [2].

Optimizing a wire bond process begins with a clear understanding of the machine set-up parameters, the response variables involved and their relationship to one another. Experimenting with these parameters is an important step towards developing a robust wire bond process. For this reason, a lot of effort has been spent on optimizing the bonding parameters and to reduce wire bond process variation [3-5]. Tan [3] presents an overview of thermionic bonding and the machine set-up parameters to interpret the results of wire bond design of experiment (DOE) data. It concludes that ultrasonic power has the strongest influence on bond quality and visual appearance, because it controls the level of deformation of the bonding wire. Insufficient power can result in narrow, under-deformed bonds and tail lifts. Excessive power results in wire bonds with a 'squashed' appearance, heel cracks, catering damage to the semiconductor die, undesirable build-up of residual bond pad material on the bonding tool and poor mechanical integrity of the wire bonds. Excessive bond deformation can occur if either the device being bonded is not properly secured or if the contact force applied to the bonding tool is too light. These can result in either mechanical chattering of the device or insufficient mechanical contact between the bonding wire, bonding tool and bond pad surfaces. Rooney [4] describes several different analytical techniques and the reliability testing used to evaluate the wire bonds. In the paper, an improved understanding of the wire bonding process was achieved by showing the dependence of the visual appearance of the wire bonds on wire bond process parameters. Liang [5] investigated the influence of wire bonding parameters on the bondability and ball bond reliability. By introducing the concept of a reduced bonding parameter, a combination of all bonding parameters, the authors are able to relate the bonding parameters to the bondability and ball bond reliability. Liu [6] has utilized the three-dimensional finite element method to simulate the wire looping process and studied the bonding loop parameter effect and residual stress distribution in the wire loop. Ding [7] studied the effects of bonding force on contact pressure across the bonded area and frictional energy in wire bonding. Shu [8] evaluated the important process criteria and bonding parameter interactions on loop profile characterization using a statistical method-"response surface methodology (RSM)". Tay [9] and Srikanth [10] employed the finite element code ABAQUS to simulate the wire bonding process and obtain a wire bond profile.

Therefore, the critical issue with increasingly finer pitch wire bonding required by industry at present is defining a prescribed combination of bonding parameters, so that each parameter can meet the packaging configuration constraints whilst maintaining the final bond quality. At present the packaging industry currently relies heavily on test data to determine the bonding parameters settings needed to meet the design requirements. This trial-and-error process is time consuming and cost ineffective [6]. Therefore, it is important to fully understand the effects of the bonding process parameters and their interactions to obtain the optimum parameter combination quickly. Considering the difficulty in determining and representing the nonlinear and interactive relationships between the bonding parameters and bonding quality, this paper proposes a process modeling approach using an Adaptive Neuro-Fuzzy Inference System (ANFIS). For the wire bonding modeling process, sets of experimental data are measured and used for the training ANFIS prediction model. The error of the prediction model is then analyzed with separate data and the effects and characteristics of varying the process control parameters on the bonding quality can be obtained and used to optimize the combination of bonding process parameter settings.

978-1-4244-4658-2/09 $25.00 © 2009 IEEE

In this paper, Section 2 describes the experimental platform and analyses the DOE experimental data. In Section 3, the prediction models for the response factors of shear force and Squashed Ball Diameter (SBD) are created through the ANFIS and the effects and characteristics of the process parameters to the quality response are discussed. The conclusions made during this work are summarized in Section 4.

DOE of Bonding Process Parameters

Design of Experiment

The analysis of the wire bond process optimisation is initially conducted using a DOE approach. One objective of the DOE study is to vary the bonder settings in a systematic way to maximise the test values of response variables, such as bond shear strength and pull strength. In this paper, due to the uncertainties of the bonding process, 9 parameters of the bonding process are investigated for the ball bond to determine which parameters influence the strength and quality of the bond. These process parameters are: ultrasonic power, binding force, time, temperature, contact velocity, ball size ratio, tail length, EFO (the Electronic Flame Off) gap and contact threshold. As Kumar [11] observed, the most important measured responses of bonding quality are ball diameter, height of the deformed ball and shear strength. The shear strength and SBD are selected in this paper as the quality response factors of the DOE. After designing the experiment, the actual bonding trials are carried out on a K&S1488Plus machine. Table 1 shows the levels and ranges of the process parameters designed.

Table 1 Values and ranges of process parameters selected

Parameters	Ranges	levels	Unit
A-Temperature	200-240	200, 210, 220, 230, 240	°C
B-Contact velocity	40-110	40, 60, 80,100, 110	mils/msecs
C-Time	8-20	8, 11, 14, 17, 20	msecs
D-Power	40-80	40, 50, 60, 70, 80	mWatts
E-Force	32-52	32, 37, 42, 47, 52	grams
F-Size ratio	2.0-2.8	2.0,2.2,2.4, 2.6, 2.8	no units
G-Tail length	100-160	100, 115, 130, 145, 160	tenth-mils
H-EFO gap	8-16	8, 10, 12, 14, 16	mils
I-Contact threshold	30-70	30, 40, 50, 60, 70	no units

Through these DOE trials, the combined effects of the 9 process parameters on the response variables can be measured and the process parameters that have the most influence on the bond quality can then be identified and used as the input parameters of process prediction model.

Analysis of Variance

The experimental data obtained from DOE trials is analysed by the method of ANOVA (Analysis of Variance) method. The ANOVA results for the shear force model and SBD model are shown in Tables 2 and 3. Through the comparison of the F-values for the shear force model and SBD model, we can determine that the ultrasonic power and then the temperature have the most effect on the bond shear force variable. The SBD is most influenced by the size ratio of the bond ball (EFO ball/wire) and the bonding temperature. From the analysis of the variance of DOE measurements and the F-value comparison, 6 main process parameters are chosen for the further consideration. These 6 parameters and their ranges are detailed in Table 4 and the other parameters are set to a fixed value in the next series of tests. These values are: contact velocity = 60(mils/msecs), the tail length = 130(tenth-mils) and the contact threshold = 50.

Table 2 ANOVA analysis results for the shear force model

Source	Sum of	dof	Mean	F-Value	p-value
A-TEMP	0.091	4	0.023	2.066	0.144
B-Velocity	0.018	4	0.004	0.408	0.799
C-TIME	0.027	4	0.007	0.621	0.656
D-POWER	0.141	4	0.035	3.201	0.049
E-FORCE	0.017	4	0.004	0.395	0.809
F-SIZE	0.04	4	0.01	0.916	0.484
G-TAIL	0.013	4	0.003	0.305	0.87
H-EFO GAP	0.023	4	0.006	0.527	0.718
I-Contact	0.028	4	0.007	0.628	0.651
Residual	0.143	13	0.011		
Cor Total	0.553	49			

Table 3 ANOVA analysis results for the SBD model

Source	Sum of Squares	dof	Mean Square	F-value	p-value
A-TEMP	363.314	4	90.8	2.108	0.138
B-Velocity	87.5606	4	21.9	0.508	0.7309
C-TIME	130.826	4	32.7	0.759	0.57
D-POWER	1038.73	4	260	6.027	0.0057
E-FORCE	101.059	4	25.3	0.586	0.6782
F-SIZE RATIO	1313.64	4	328	7.622	0.0022
G-TAIL LENGTH	34.0098	4	8.5	0.197	0.9354
H-EFO GAP	124.026	4	31	0.72	0.5936
I-CONTACT THRSHLD	87.6476	4	21.9	0.509	0.7306
Residual	560.139	13	43.1		
Cor Total	3523.97	49			

Table 4 Main process control parameters for the prediction model

Parameters	Ranges	Unit
A-temperature	200-240	°C
B-Time	8-16	msecs
C-Power	40-80	mWatts
D-Force	32-52	grams
E-Size ratio	2.0-2.6	no unit
F-EFO Gap	8-16	mils

Process Prediction Model and Effect Analysis

Prediction Model Establishment

The process model for the wire bonding can be described as follows:

$$[y_1(t), y_2(t)] = f(x_1(t), x_2(t), \cdots, x_n(t)) \qquad (1)$$

Her, $x_1(t), x_2(t), \cdots, x_n(t)$ represents process parameters, such as the control factors of ultrasonic power, bonding temperature, bonding time and contact force, etc. $y_1(t)$ and $y_2(t)$ are the response variables of shear force and SBD, respectively, where $f(\ )$ is a multi-variable nonlinear function.

The structure of the prediction model of wire bond process is illustrated in the Fig. 1. Based on the input of the process parameters, the model produces a prediction y'', the difference between y'' and the measured value y' will be fed back to and correct the prediction model, this procedure continues until the error is sufficiently small. Through the model created, the corresponding relationship between the process parameters and their response variables can be studied, and the optimisation of parameter combination can be achieved.

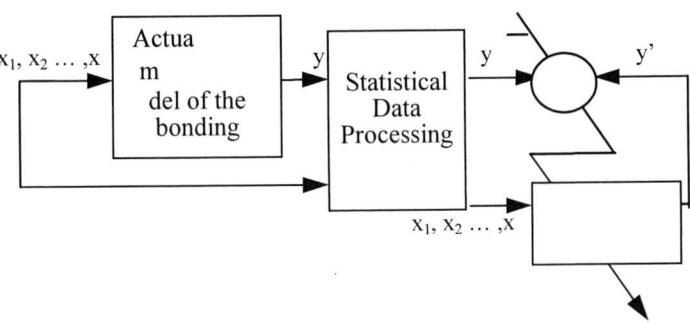

Fig. 1 Structure of prediction model of wire bonding process

There are many methods to establish a prediction model, such as the traditional methods of time sequence and regression equations, whilst the modern intelligent methods use neuro networks, fuzzy reasoning systems and genetic algorithms. Combining the fuzzy reasoning system with a neuro network, the adaptive Neuro-Fuzzy Inference System (ANFIS) has been widely applied in the prediction domain. The ANFIS adjusts the control parameters and response factors through the algorithm of back-propagation and least square methods.

The process prediction model in this paper is constructed by using the ANFIS method. For the construction and training of the model, 50 sets of trial samples are designated as training data. According to the ANFIS system, the numbers of input subjection functions for the shear force model and SBD model are set to [3 2 3 2 2 2] and [2 2 2 2 2 2], respectively. The input subjection functions are all set to be a Gauss function and the output subjection functions are set to be a constant value. Based on the basic settings, the ANFIS-based prediction model is implemented through the ANFIS Editor GUI module of the Matlab software. In the software, the hybrid-study algorithm is selected as the training method of the ANFIS. As a result, the shear force model is created and has an error of 2.5028 N after 25 iterative epochs, and the SBD model is also created and reaches an error of 0.35μm after 100 iterative epochs. Fig. 2 shows the training error curve of the two models. The shear force model produces 144 (= 3×2×3×2×2×2) rules, the SBD model produces 64 (= 2×2×2×2×2×2) rules. These rules cover the whole range of the parameters selected in Table 4.

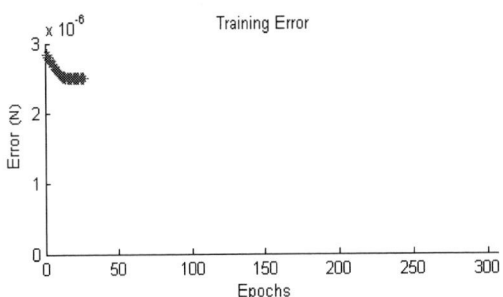

a) Predicted shear force model

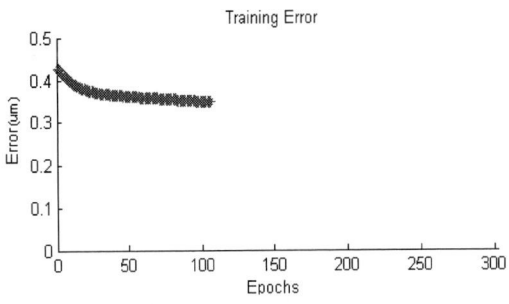

b) Predicted SBD model

Fig. 2 Training error curves for the prediction models

In order to verify and evaluate the prediction model created, 15 separate sets of test data from the trials are entered into the predicted model and the output values of the response factors of shear force and SBD from the prediction model are compared with the actual shear force and SBD measured. The prediction error can therefore be obtained. From the error analysis, it is found that the mean error and variance for the predicted shear force are 3.16% and 0.012 and for the predicted SBD are 1.24% and 1.37, respectively. From these results, it is concluded that the prediction models created reflect a true representation of the bonding process. The prediction model could be further improved by increasing sample sizes.

Analysis of Process Parameters

Based on the predicted process model, it is possible to identify how these control parameters affect bonding performance and quality. Fig. 3 presents the relationship of bonding temperature and power to the model response variables of shear force and SBD. This surface model is obtained under the parameter settings of bond time = 10ms, contact force = 45grams, ball size ratio = 2.4 and EFO gap = 14mils. In this figure, shear force is at a minimum when the temperature and power values are low, but increasing the temperature and power values to 220°C and 80mw respectively, the shear force increases quickly to a maximum (shown in Fig. 3a). The SBD variable has the similar relationship as the shear force, except that the power has more of an influence on the SBD than the temperature does, as shown in Fig. 3b). Therefore a considerable reduction in the SBD for minimal loss of bond shear strength can be achieved by reducing the power setting during the bonding process.

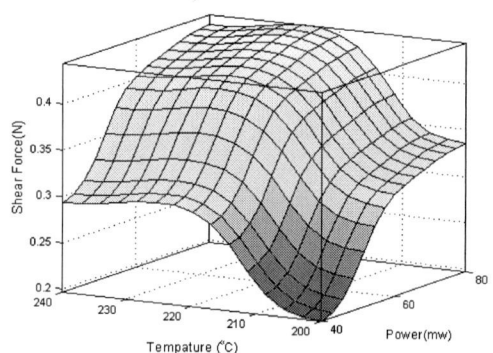

a) Shear force to temperature and power

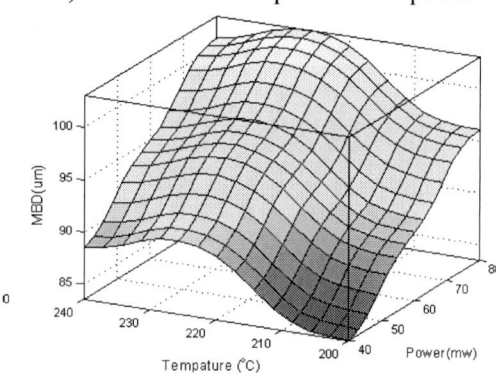

b) SBD to temperature and power

Fig. 3 Relationship of the bonding temperature and power to the response factors of shear force and SBD in the process prediction model

Conclusions

Based on the analysis of variance of DOEs, 6 process parameters are identified in this paper as control parameters for bonding quality control. The 6 parameters are bonding temperature, time, ultrasonic power, bonding force, ball size ratio and EFO gap. The shear force and SBD are chosen as response factors to determine bond quality. Through the ANFIS system, this paper constructed a prediction model for wire bonding process, which can be used to solve the difficulty of a mathematical representation of the bonding quality and process control parameters due to their nonlinear and interactive relationship. From the error analysis of the response variables for shear force and SBD from the prediction model created, it is found that the mean error and variance for the shear force and the SBD are 3.16% and 0.012, and 1.24% and 1.37, respectively. Based on the prediction model, the effects and relationships between the control parameters and the response factors are obtained, which can then be used to produce an optimal combination of process parameters for the wire bonding process. Further work is to be carried out on the development of a software tool to enable the optimisation of the wire bonding parameter combinations based on the work conducted in this paper.

Acknowledgments

This work is sponsored by Science and Technology Department of Guangdong Province, P. R. China, under the Research Grant of No. 2008A010300009 and No. 2006A10401003.

References

1. Harman G G, "The Ultrasonic Welding Mechanism as Applied to Aluminium and Gold Wire-Bonding in Microelectronics," *IEEE Transactions on Parts, Hybrids and Packaging*, v PHP-13, n 4, Dec. 1977, pp. 406-412.
2. Wang FL, *et al*, "Effect of bonding time on thick aluminium wire wedge bonding strength," *Transactions of the China Welding Institution*, 2006, 27(5), pp. 47-52.
3. Tan YC, *et al*, "Free Air Ball Modelling for Gold Wire Bonding for Different Wire Diameters," [online]. K&S website: Kulicke & Soffa. Available at: <http://www.kns.com/SendFile.asp?TID=58&FID =14384> Accessed 20 May 2008.
4. Rooney D.T., Nager D, *et al*, "Evaluation of wire bonding performance, process conditions, and metallurgical integrity of chip on board wire bonds," *Microelectronics Reliability*, 2005, 45, pp. 379-390.
5. Liang Z. N, Kuper,F. G. and Chen M. S., "A concept to relate wire bonding parameters to bondability and ball bond reliability," *Microelectronics Reliability*, 1998, 38, pp. 1287-1291.
6. Liu, D.S., *et al*, "Study of wire bonding looping formation in the electronic packaging process using the three-dimensional finite element method," *Finite Elements in Analysis and Design*, v 40, n 3, January, 2004, pp. 263-286.

7. Ding Y, Kim J.K, Tong P. "Effects of bonding force on contact pressure and frictional energy in wire bonding," *Microelectronics Reliability*, 2006, 46 (7), pp. 1101-1112.

8. Shu B., Lee S.S., Groover R., "Wire bond loop profile development for fine pitch-long wire assembly," *IEEE Trans. Semicond. Manuf. Technol. Part A* 18 (1) (1995) pp. 230-234.

9. Tay A. A. O., Seah B. C., Ong S. H., "Finite element simulation of wire looping during wirebonding," *Proceedings of the Advances in Electronic Packaging*, ASME, New York, 1997.

10. Srikanth N., Wen Y. Y., *et al*, "FEM based studies of ultrafine pitch wire bond process," *Proceedings of the APACK 2001 Conference on Advances in Packaging*, Singapore, 2001.

11. Kumar S., Florendo M., and Dittmer K., "A Wire Bond Process Optimisation Strategy for Very Fine Pitch Development," *SEMICON Singapore Assembly Seminar*, May 1999.

From Thin Cores to Outer Layers: Filling through Holes and Blind Micro Vias with Copper by Reverse Pulse Plating

Stephen Kenny and Bernd Roelfs
Atotech Germany
Erasmusstr. 20, 10553 Berlin
stephen.kenny@atotech.com; bernd.roelfs@atotech.com

Abstract

This paper presents new advancements in copper electroplating technology for both Blind Micro Vias and Through Holes. Processes and manufacturing technology are described as well as current limitations and requirements. As a highlight the complete filling of through holes with electroplated copper by Reversed Pulse Plating, RPP, is described. Both Blind Micro Via and Through Hole filling using this technology are already targeted for production at HDI and also at the packaging level.

The basis of the latest advancements has been the development of the so called SuperFilling[TM] process. SuperFilling[TM] combines the electrolytic reverse pulse deposition of copper into specific structures together with the simultaneous etching of copper from the substrate bulk surface in a continuous panel plating step. This combination reduces the amount of overplated Cu by up to 50% compared to conventional copper deposition processes for HDI fine line applications. The SuperFilling[TM] mechanism as well as manufacturing requirements and production supervision tools such as online analysis are described in detail.

The latest development, through hole filling by RPP, offers a viable alternative to the standard paste plugging for core processing in substrate manufacturing. Current core manufacturing requires a paste plugging process for through holes so that subsequent build up layers can be produced by sequential lamination, the flat core surface is essential for stacked via and also via in pad technology.

This paste plugging process requires additional process steps, each of which has its own limitations and contributes to the overall cost. Filling the core through vias by electroplating can eliminate the plugging process and significantly reduces the number of overall process steps which will also reduce costs. Moreover, it offers certain advantages such as potentially higher reliability in accelerated aging tests and an improved thermal management as the thermal conductivity of a completely copper filled through via is significantly higher than a paste plugged through via.

First manufacturing experiences with the Through Hole filling technology are presented along with a discussion of its manufacturing prerequisites and limitations.

Introduction

The manufacturing process for substrate cores requires production of a planar surface which is subsequently used as the starting point for the high-density build-up layers. The core is normally mechanically drilled to produce the required through-via connections which are then metalized and plugged with a thermally cured resin material. There is a current tendency for reduced core thickness and this has implications for yield, quality and ultimately for cost for

product from the complete process. The plugging process itself is relatively labor-intensive and requires, as part of the sequence, a mechanical abrading or brushing process after resin cure, which can cause problems of dimensional stability, particularly for substrate cores less than 100μm thick. The plugging resin itself has disadvantages in that it is a high-solid content material, which has a different coefficient of thermal expansion (CTE) to that of the surrounding material, including the copper metal in the via and also to that of the dielectric of the core itself, typically a glass-reinforced resin material.

The disadvantages of the existing plugging process can be eliminated by using pure copper to produce the planar surface. For the next-generation process, copper is deposited into the through vias as an integral part of the metallization process. The drilled dielectric is made conductive and a thin layer of copper metal is deposited to give a seed layer for the copper deposition process. The through-vias are then completely filled by a modified electrolytic copper deposition process, which can be accomplished in a single, fully automatic continuous processing line.

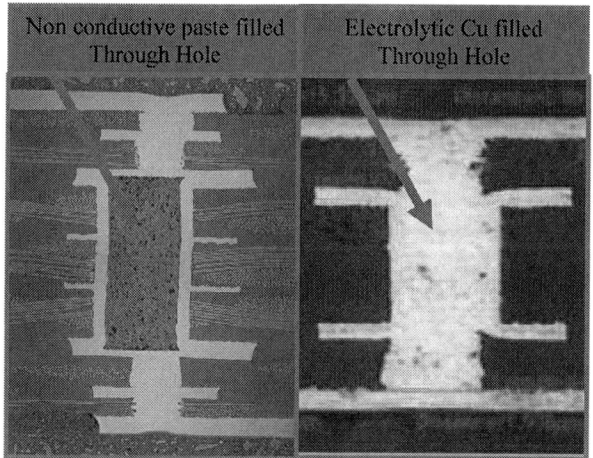

Fig. 1 X-sections showing a paste plugged through via in a core (left) and an electrolytically Cu filled through via (right), both with blind micro via layers on top (Via in pad).

The use of pure copper has obvious advantages in that its thermal characteristics are significantly better than any type of resin material available. This fact can give more design options to utilize the improved thermal transfer capability of vias in a substrate than are currently available. The CTE of the copper-filled core is dependant only on the copper metal and the glass-supported resin of the drilled dielectric. The copper

978-1-4244-4658-2/09 $25.00 © 2009 IEEE

structures in the subsequent layers may be positioned directly above the copper-filled through-vias with no reliability implications. In fact, the conductive path within the substrate may be designed to utilize the more direct and parallel connection from one side of the substrate to the other.

Table 1 Process flow for paste plugging and Cu plugging for through holes

	Conventional Process (Paste)	Electrolytic Process
1	Through hole drilling	Through hole drilling
2	PTH (Hole metallization)	PTH (Hole metallization)
3	Panel plate 18-25μm	Panel plate 3μm
4	Resin plug	Electrolytic through hole fill (12-20μm Cu)
5	Cure	Dry film
6	Grind / Brush	-
7	Resin metallization	-
8	Panel plate 18-25μm	-
9	Dry film	-

Via hole plugging [1] as a process is used to produce a planar surface which enables subsequent sequential lamination for production of build up structures, the planar surface is particularly important to ensure uniform dielectric spacing between circuit layers critical for controlled impedance in high frequency applications.

The standard process is to use a plugging paste which is applied by methods such as stencil printing or roller coating. On top of the plugging paste there will either be the next resin layer which may be glass reinforced, or a plated copper layer. This copper surface may be used to allow production of staggered or stacked blind micro vias which are conformably copper plated or copper filled depending on the application requirements. An example for an HDI board with a paste plugged core is given in Fig. 1 (left) as well as an example for a copper filled core. Both examples show also filled BMVs directly on top of the through hole (Via in pad technology). This example illustrates well the technique of blind micro via filling to ensure void free sequential lamination. The direct positioning of the BMV on the core through hole will save space in the substrate design.

Blind Micro Via Filling and SUPERFILLING™

The need for finer lines and higher reliability has driven the development for filled BMVs. The main drivers of the BMV filling technology were IC substrate manufacturers who are very sensitive on reducing real estate on their product and apply the stacked via technology which is based on perfectly filled Blind Micro vias of small dimensions (60μm×40μm). HDI manufacturers have followed this approach and tried to use the filling technology for their application. Unfortunately, HDI manufacturing requires filling of much larger BMVs (e.g. 125μm×75μm) at a significantly higher production throughput. Where IC substrates are produced at about 1-1.5

ASD, HDI boards are often produced with 5-10 ASD. This requires a significantly different production scenario. The key for success at such high current densities is a perfect distribution of copper, since the BMV filling is much more sensitive to copper thickness variations than normal conformal plating applications. Not only the distribution within the panel but also the panel to panel and batch to batch distribution needs to be optimized. For this very specific application we designed a completely horizontal conveyorized plating line comprising of a desmear process, initial metallization of the surface and hole by electroless copper and an electrolytic copper plating step. A conveyorized approach is extremely beneficial as the panel to panel variation in such a system is minimized. For perfect distribution within a panel both the e'less and the electrolytic copper deposition was optimized.

The first step was to improve the distribution the e'less deposition of copper. It was necessary to improve the amount of copper in a BMV since drilling through a glass fiber reinforced material for HDI often results in a non regular BMV shape. Fig. 2 depicts such a non optimal BMV shape but the newly developed e'less process Printoganth U Plus covers every part of the hole. To make this visible the board was run 10 times through a uniplate line otherwise the e'less layer would be too thin for a microscopic evaluation. This development required a variety of changes in the electrolyte and system set up.

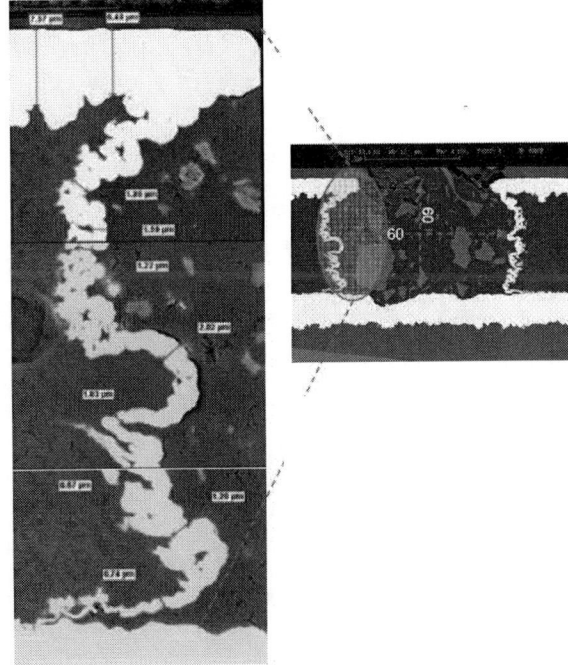

Fig. 2 BMV after exposure to the latest eless copper version Printoganth U Plus. Coverage is also achieved under wedges.

The biggest advancement in the subsequent electrolytic copper deposition was the development of the so called SUPERFILLING™ process for our Inpulse system. The Inpulse system is based on a combination of inert anodes and reverse pulse plating. The process is in fact a combination of a simultaneous electrolytic copper plating and copper etching in one electrolyte. This is realised by using an electrolyte which contains copper and $Fe^{2+/3+}$ at the same time.

Originally, the Fe redox system was introduced as a replenishment aid [2] for copper ions and as reaction partner for the inert anodes instead of additives and water. It reduces the consumption of the additives significantly. Moreover, it increased the life time of the expensive Iridium oxide coverage for such anodes since anodes are operating at the lower oxidative potential of the Fe redox couple. But the Fe^{3+} can be further used to minimize the amount of surface copper which is the basic idea behind the SUPERFILLING™ approach.

The principle of this SUPERFILLING™ process is shown in Fig. 3.

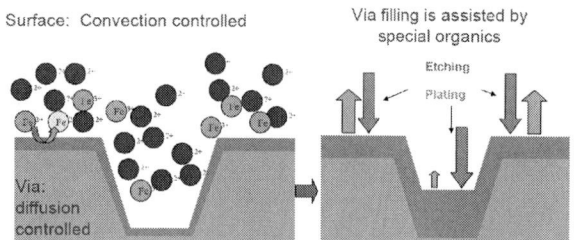

Fig. 3 Showing the principle of simultaneous plating and etching with preferred surface etching

This illustrates well the difference between a surface and the bottom of the BMV. Transport of ions and additives is for the surface mainly convection controlled due to the extremely high flow rate whereas a BMV behaves completely differently. The transport is diffusion controlled and as such is orders of magnitude slower. This leads to a concentration difference of Fe^{3+} ions between surface and BMV bottom. The Fe^{3+} ions are able to etch by Equation 1 preferentially from the surface.

$$2Fe^{3+} + Cu \rightarrow 2Fe^{2+} + Cu^{2+} \qquad (1)$$

By running this system with extremely high Fe^{3+} concentrations of 4-6g/l it is possible to reduce the amount of surface copper by up to 50% compared to a conventional non redox system.

Through Hole Filling with Electrolytic Copper-Mechanism

In the development work for the blind micro via filling SUPERFILLING™ process it was seen in early experiments that through vias were excessively plated at the centre of the hole. This result is normal when reverse pulse plating is used however the extent of the plating thickness in the hole centre was a surprise, as can be seen in Fig. 4.

How is such an exaggerated plating possible? The answer to the questions lies in the mechanism of the reverse pulse plating and its influence on the electrolyte additives. A copper electrolyte usually consists (not only) of a well balanced mixture of suppressing and accelerating additives. The suppressing additives (carriers and levellers) slow down the

copper deposition whereas the accelerating additives (brighteners) act as a catalyst for the copper deposit.

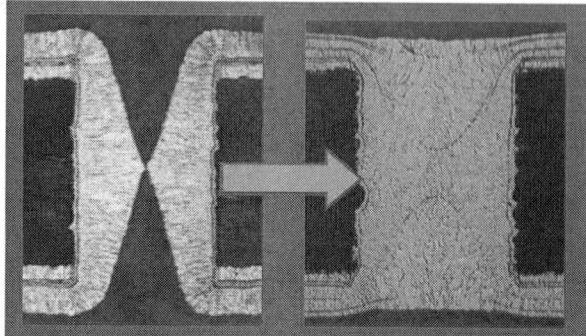

Fig. 4 Two step process for through hole filling. 1st X plating with extreme reverse pulse and 2nd BMV filling

The application of the reversed pulse now interferes with the distribution of these additives in a hole and on the surface. The reverse pulse is able to modify the concentration equilibrium in a hole and on the surface in such a way that the accelerating catalyst concentrates in the middle of a hole and is depleted from the surface. This is due to the fact that under reverse (=anodic) pulse conditions the catalyst is desorbing together with a small amount of copper. Since the concentration of the brightener is relatively small it takes a while to re-adsorb again and this catalyst is now partly missing for the subsequent cathodic pulse. The hole is compared to the surface electrically shielded so that the reverse pulse does not reach its maximum strength especially in the middle of the hole. Here, the desorption of the catalyst is incomplete and it can react immediately after switching the pulse to cathodic values. As a consequence the deposition speed in the middle of the hole is increasing compared to the surface. This mechanism is depicted in Fig. 5.

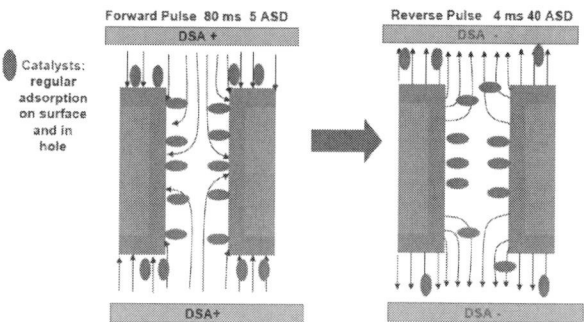

Fig. 5 Schematic model for the exaggerated X- plating. Left: Cathodic pulse with regular distribution of copper deposition catalyst. On right: Depletion of catalyst from surface during reverse anodic pulse (DSA= dimensionally stable anodes)

Through Hole Filling-History and Outlook

When we started with our development in 2005 we did not see an immediate benefit of this technology due to the fact that a considerable amount of surface copper was necessary to fill through vias. In fact, more than 50μm panel plated copper was necessary to fill a 100μm thin panel with 100μm diameter holes (Fig. 6). Together with the copper laminate of 17μm it was not possible to apply this technology for HDI boards or IC substrates. HDI boards were in general based on sub 75μm and now on < 50μm structures. These boards undergo an etching step for the final circuitization. The accessible fine line resolution is, by rule of thumb, about twice that of the overall copper thickness. To enter the 50μm lines and space arena it was necessary to reduce the amount of plated copper significantly down to about 20μm and work with thinner laminated copper (e.g. 3μm copper).

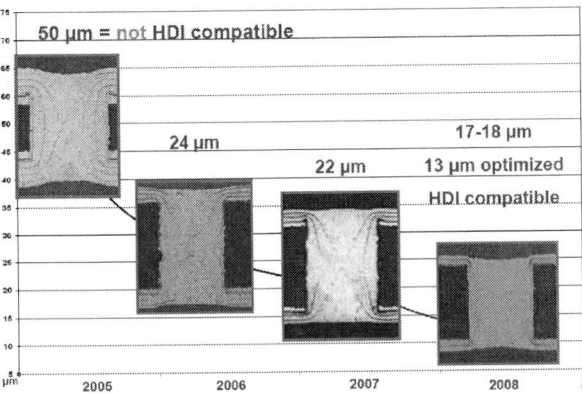

Fig. 6 Development of the amount of necessary surface copper to fill a 100μm hole in a 100μm thin panel for TH filling capability

There is also the possibility to reduce plated copper thickness by subsequent etching or even mechanically abrading to reach the thickness tolerances for HDI however this option was discarded during development as any copper removal process adds processing steps, wastes raw material and increases cost. Thus the only option was to optimize both equipment and the plating process.

We constantly improved the performance in the last years and as a result of our development work we decreased the amount of plated copper down to the imaginary 20μm border line for which HDI with this technology becomes feasible.

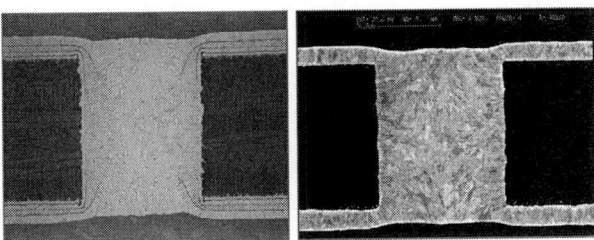

Fig 7 Showing an etched X-section and an SEM micrograph of a copper filled through hole plated with the latest technology

We not only decreased the amount copper but also had to improve the crystal structure and physical properties of the deposit to meet the IPC norms. In Fig. 4 especially on the left picture one can see rather orientated copper (columnar growth), which may arouse concerns about the reliability.

Thus, a specially developed electrolyte system has been developed and optimized for this special application.

Fig. 7 shows the copper deposit of the present electrolyte. Both the etched and the SEM micrograph show the normal polygonal copper structure as we know it from numerous investigations. A variety of measurements on ductility, tensile strength and differential scanning calorimetry (DSC) showed that this deposit is comparable to the normal HDI copper quality and withstands the usual requirements.

Two significant approaches made it possible to reduce the amount of copper so significantly: 1st the optimization of aforementioned SuperFilling™ process and 2nd the development of new rectification system. This new system allows improvement in especially the X- plating performance and reduces the amount of plated copper drastically to form a bridge between the hole wall sides.

We are constantly working to further reduce the amount of plated copper so that this panel plate process will keep pace with the requirements of future HDI technology.

Process Advantages for copper through hole filling

Numerous advantages of copper filled through holes can be described.

Not the least should be the expected increase in reliability. Reliability issues in accelerated aging test occur usually when the CTEs of adherent materials are very different. This is in fact the case for every PCB. The CTE of copper is approx. 17ppm/K and the CTE of typical FR4 base material is approx. 110ppm/K. The CTE of plugging pastes for the standard through hole filling process is at 30-40ppm/K in between that of the other materials. This means that a copper barrel in a through hole comes under stress from outside and inside the barrel. This often results in corner and barrel cracks

Such a stress for a completely copper filled through hole can only be induced from outside but a thicker copper barrel is a mechanically strengthened system due to the superior mechanical properties of copper layer

Another benefit of copper filled through holes could be the application for thermal vias. Copper has obvious advantages for thermal vias due to its high thermal conductivity; this is at 360 W/m K much higher than that of any plugging material. Normal plugging materials have thermal conductivity values of about 1W/m K, specially developed high thermal conductivity paste up to 8W/m K. Compared to these pastes the thermal conductivity of copper is about 40-50 times better. Simple calculations show that more than 90-99% of the heat transfer is still carried by the copper barrel depending on which paste is used for plugging. This effectively reduces the application of such paste plugged thermal vias. A comparison of the capability can be easily done by the introduction of a thermal resistance concept [3]. The through hole can be seen as thermally conductive wire for which the thermal resistance "R_{th}" can be described by,

$$R_{th} = d[\mu m] / (\lambda\ [W/mK].A[m^2]$$

In the formula "d" is the length of the heat conductor, in this case the thickness of the substrate and λ is the thermal conductivity of in this case either copper or the plugging material. A is the cross sectional area of the heat conductor, the plated copper in the barrel of the hole or the plugging material. The theoretical calculation for a via diameter of

978-1-4244-4658-2/09 $25.00 © 2009 IEEE

120μm and panel thickness 150μm shows that the thermal characteristics of a plated through hole are determined mostly by the plated copper thickness in the barrel of the hole. To improve the thermal performance it is more effective to increase the plated copper thickness in a barrel than to use a plugging material. The best result from the theoretical calculation was found, as could be expected by simulating a fully copper filled barrel.

Table 2 Comparison of heat transfer capabilities

Through via 20μm copper	Through via 30μm copper	Through via 20μm copper plugged	Through via copper filled
84 K/W	62 K/W	80 K/W	47 K/W

Also due to the increased thermal and electrical characteristics of a solid copper core the number of thermal vias required may be reduced whilst maintaining an equivalent thermal capability. This again can free up surface which could be required for circuitry so allowing further miniaturization. Build up layers may be positioned directly above the filled core with no loss in reliability and this design gives the shortest and most direct circuit path from one side of a substrate to the other.

Limitations of the Through Hole filling technology:

So far this technology can be applied to thin cores only ranging from 50μm to max. 400μm. For thicker cores the aspect ratio becomes too difficult so that small voids in the holes may be left. It is still unclear to which extent these voids are critical for the accelerated aging test such as TCT and IST but this will be the topic for the future.

As the next step in our development we are trying to apply this technology to a pattern plating sequence. Pattern plating is currently process of record for IC substrates with lines and spaces dimensions of below 20μm. The IC substrate manufacturer are widely using the via in pad technology and would benefit significantly from the copper plugging process. First tests with the existing system showed promising results. The adoption of the copper through hole filling technology for a pattern plating process requires both changes in the plating equipment and electrolyte system. The fine line structure of the pattern layout is very sensitive to mechanical damages and requires a touchless transport system. Moreover, modifications in the electrolyte system are necessary to generate a rectangular line shape.

Production experience

SUPERFILLING™ technology for filling BMVs has been recently adapted and can be seen as state of the art for HDI manufacturing. It is now part of the reliable production sequence for several mobile phone suppliers.

Qualification of the Inpulse 2THF through hole filling process has been successfully passed at a Taiwanese IC substrate manufacturer for 0.1mm thin cores with up 500 k holes per board. These boards with areas of clustered and isolated through holes have been successfully plated with 20μm copper on 3μm copper laminate.

Summary

This paper describes the progress in filling both BMV and through holes by electrolytic copper. Whereas BMV filling technology has been widely adapted by the HDI manufacturer, TH filling by copper is a new technology and is still waiting for its introduction into mass manufacturing. We have shown that through hole filling can be an interesting alternative for thin cores to the standard paste plugging process as it simplifies the production sequence and reduces significantly the number of production steps. This in turn can minimize costs for the manufacturing process. Several other benefits such as increased thermal conductivity and higher stress resistance are described. Qualification for IC substrates has been successfully passed and the application for the HDI segment with its thicker cores will be the next task. So far this technology is limited to thin cores up to 400mm only and the process is a pure panel plating process. The amount of surface copper can be limited to below 20μm which makes this process feasible for fine line application below 50μm.

The development of TH filling in a pattern plating process is under way to apply this technology for ultra fine line application.

References

1. M. Carano, "Via Hole Plugging Technology," *Circuitree*, May, 2007.
2. Jürgen Barthelmes "Acid CuPlating with insoluble Anodes-A novel Technology in PCB manufacturing," *ECWC*, Sept., 1999.
3. Christoph Lehnberger, "Strategien zur Elektronik-Kühlung," *Leiterplatten-Design*, October 2004, pp. 24.

Investigation of Thin Small Outline Package (TSOP) Solder Joint Crack after Accelerated Thermal Cycling Testing

L. N. Lu[1], H. Z. Huang[1]*, B. Y. Wu[1], Q. Zhou[1], X.X. Su[3], M. Cai[2]

[1]School of Mechatronics Engineering, University of Electronic Science and Technology of China
Chengdu, Sichuan, 610054, P. R. China
[2]Guilin University of Electronic Technology, P. R. China
[3]Flextronics Mobile Consumer, Zhuhai, P. R. China
hzhuang@uestc.edu.cn, Tel: 028-61830248, Fax: 028-61830229

Abstract

Solder joint crack is a common failure mode of printed circuit board assembly (PCBA) for electronic products. In order to investigate the crack behavior of fine-pitch SMT solder joints, accelerated thermal cycling (ATC) up to 1500 cycles was performed on advanced PCBAs with low-profile thin small outline package (TSOP). The functional examination result shows that the failure rate of TSOP solder joint open after ATC is 2.5%, 7.5% and 27.5% after 1000, 1250 and 1500 cycles, respectively. Optical inspection by microscopy demonstrates obvious solder joint cracks at many locations. Cross sectional study has confirmed serious crack in the bulk solder, resulting in electrical failures. Also, microstructural information has been obtained by metallurgical analysis with the aid of Scanning electron microscopy (SEM). In addition, finite element analysis (FEA) was performed to simulate the thermal-mechanical stress in solder joints under ATC. Based on the modelling, the fatigue life of solder layer was calculated. Both experimental study and modelling confirmed that solder fatigue crack upon concentrated stress during ATC is a major contributor to the open failure of TSOP solder joints.

Introduction

Lead-free solders are rapidly replacing lead-based solders as the EU legislations on RoHS and WEEE took effective in July 2006. The lead-free alloy recommended by the National Electronics Manufacturing Initiative (NEMI), Sn3.9Ag0.6Cu is gaining wide spread application with a narrow composition range of Sn (3-4) Ag (0.5-1) Cu. With the development of electronic assembly towards low cost, high density and high performance, the long-term reliability of fine-pitch solder joints is of a great concern for current for lead-free application. Che et al compared the solder joint reliability of PBGA, PQFP and TSSOP and found that PBGA solder joint is more sensitive to thermal fatigue than the other two [1]. Ma et al conducted a comparative study of lead-free and Sn37Pb solder joints and found that lead-free solder joints creep at higher rates than Sn37Pb one for the same stress level at elevated temperature aging. However, such effect was not observed at room temperature [2]. Emerick et al evaluated the reliability of TSOP assembly and found low compliance in the lead result in high strains in the solder which contributed to the reduced reliability of the TSOP [3]. Yoon et al found that lead-free solder systems were compatible with the conventional Sn37Pb solder with respect to board-level and mechanical solder joint reliability [4].

In this study, accelerated thermal cycling (ATC) from 0°C to 100°C and modeling simulation for TSOP assembly

of Sn-4.0Ag-0.5Cu solder was investigated to check the assembly reliability. Experimental failure analysis of TSOP assembly was performed to find the failure mode and mechanism. Fatigue life prediction was conducted using fatigue life prediction model and FEA analysis.

Experimental

In this study, FR4 substrates with OSP/Cu pads were used to assemble various components by Sn4Ag0.5Cu solder paste through standard SMT process. TSOP fine-pitch solder joints were one of the weakest interconnections, which were set as our focus of interest. This component body was 17mm in length and 12mm in width with a thickness of 1mm. There were 48 leads with a pitch of 0.5mm. The base material of leadframe was FeNi 42 alloy and the surface coating was Sn_4Bi with a thickness of 0.02mm. All assemblies were tested at normal production functional test stations and 40 passed PCBAs were selected for investigation.

ATC test was conducted in a one-zone temperature cycling chamber according to the JEDEC Standard, JESD22-A104C, with 0°C to 100°C up to 1500 cycles [5]. Inside the chamber, there were 9 thermocouples placed on samples to record the actual thermal profile during testing. Fig. 1 showed an example of real plot of actual thermal profile measured by one thermocouple. Each cycle lasted for 60 minutes, including 15min dwell at 0°C and 100°C, and 15 min for ramp up and cooling down. Function tests were performed every 250 cycles to detect the failure. For any sample failed with TSOP, microscopy study and cross sectional examination was performed on the solder joints to check the crack.

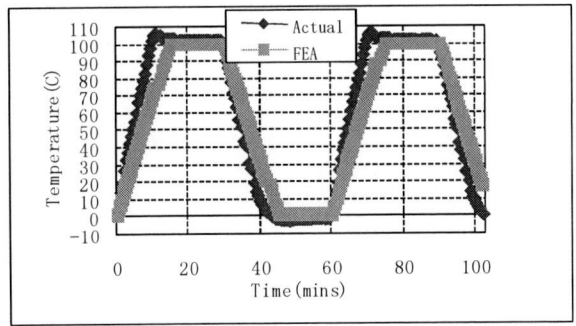

Fig. 1 ATC thermal profiles for actual measurement and FEA

At the same time, FE modelling was carried out to simulate the stress and strain distribution in the solder joint during ATC. Based on which, the fatigue life was calculated.

978-1-4244-4658-2/09 $25.00 © 2009 IEEE

Table I and II listed the physical properties of different materials used in the modelling [1, 6].

Table I Material properties for TSOP in modelling

Materials	Modulus (Gpa)	Possion's ratio	CTE (ppm/°C)
Solder	Table II	0.35	24.5
FR4 PCB	x,z:20 y:9.8	x,z:0.28 y:0.11	z:50 x,y:14
Mold	17	0.25	x:9 y:5
Coating Sn4Bi	60	0.29	17
FeNi 42 alloy	180	0.31	5.6
Copper	155.17	0.34	16

Table II Temperature dependent material properties

Temperature	0°C	25°C	50°C	100°C
Solder modulus(GPa)	46	42	37	27
FR4 PCB modulus GXY(GPa)	17	15.6	13.7	10

Results and Discussion

The functional examination result at every 250 cycles shows that until 750 cycles, no failure occurs at the TSOP location for all 40 PCBAs. Although we could not tell whether there is a partial crack or not, it is believed that at least the TSOP solder joint still can carry electrical continuity at this moment. However, units start to fail due to solder joint open after 750 cycles. As shown in Table III, the failure rate of TSOP solder joint open after 1000, 1250 and 1500 cycles is 2.5%, 7.5% and 27.5%, respectively. It is obvious that after 1250 cycles, the failure rate increases substantially.

Table III Functional test result related to TSOP open

ATC cycles	TSOP failure rate
750	0
1000	2.5%
1250	7.5%
1500	27.5%

According to microscopic view of the solder joints of failed units, the open location is identified, as shown by the arrows in Fig. 2. Serious deattachment of the leads from PCB is observed at both sides of the component. The impedances of these cracked solder joints are measured to be very much higher than normal ones. Based on which, it is suspected that crack propagate across the entire interconnect to cause electrical open failure. This is further confirmed by the cross sectional metallographic image as shown in Fig. 3. It is clear that crack occurs in the bulk solder right below the component lead, leaving a thin layer of solder on both the component lead and pad surface. This is a classic solder fatigue-induced bulk solder crack after ATC. As shown in Fig. 4, the microstructures under polarized light demonstrate a significant recrystallization of tin, which suggests that the solder joints experience a substantial thermal stress during ATC.

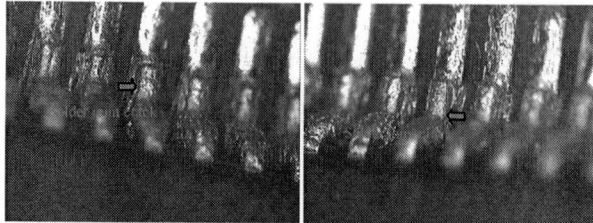

Fig. 2 Optical inspection image

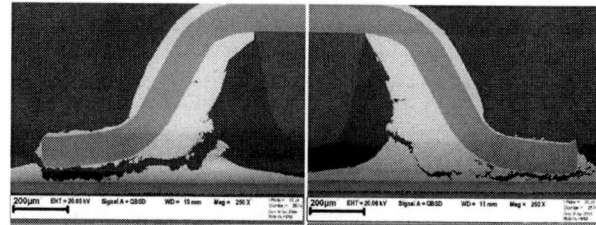

Fig. 3 Cross sectional SEM image

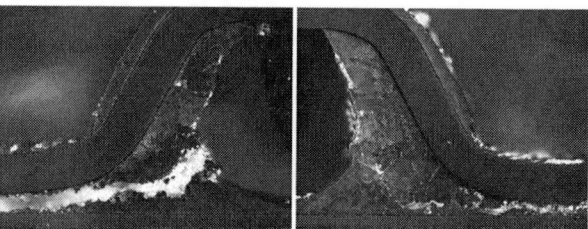

Fig. 4 Metallographic image under polarized light

ANSYS is an effective finite element modelling software. The TSOP device is modelled by ANSYS using submodel of the device. Sub-modeling is also known as the cut-boundary displacement method. It is necessary to verify first of all that the cut boundaries are far enough away from the stress-strain concentration region of interest area at the solder joint interfaces [7]. Submodel of one TSOP solder joint was selected for analysis. During the simulation, the fixed restriction and symmetrical restriction were added in the bottom face and the left face respectively. The initial temperature was set at 25°C. One cycle of thermal cycling was simulated for the sample and the temperature profile applied is described in Fig. 1.

Fig. 5 and 6 show the simulated distribution of stress in the overall solder joint and only solder layer respectively. It clear that stress is concentrated in the lead and solder joint. Due to coefficient of thermal expansion (CTE) mismatch, significant stress occurs in the system upon temperature variation in ATC. Particularly, the solder layer is in high tension and a maximum stress of 33.9MPa is identified. Meanwhile, it is understood that among the metal constitutes, solder is the weakest part. Therefore, the solder is prone to crack up overstress. This modelling result matches well with the experimental findings as shown in Fig. 3.

978-1-4244-4658-2/09 $25.00 © 2009 IEEE

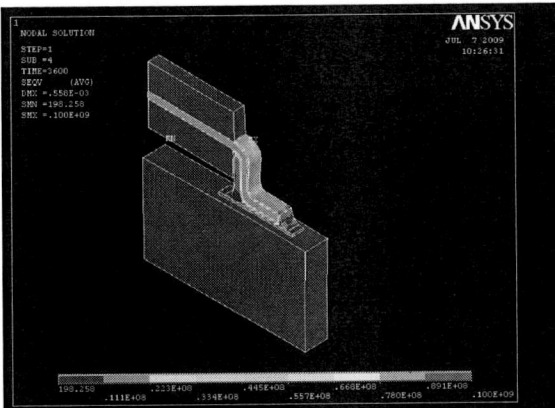

Fig. 5 Stress distribution in the submodel of solder joint

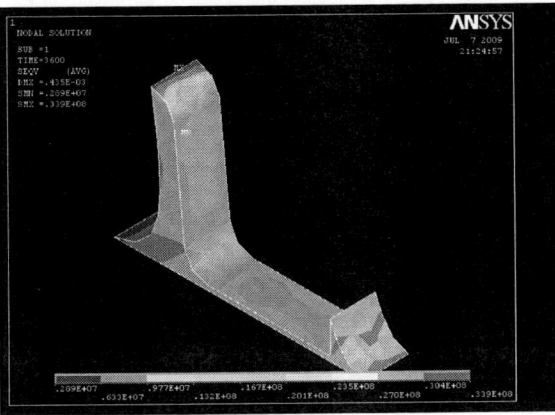

Fig. 6 Stress distribution in the solder

According to the simulation result, plastic work density or plastic strain energy density accumulated per cycle is extracted from numerical result as a failure parameter for fatigue life prediction. The volume-averaged method is used for plastic work density calculation as shown in Equation (1) and (2) [8, 9]. ΔW_{ave} is the total volume weighted average inelastic (visco-plastic) strain energy density (sometime referred as plastic work) accumulated per thermal cycle. It can be derived from Equation (1).

$$\Delta W_{ave} = \frac{\sum_{i=1}^{n} \Delta W_i * V_i}{\sum_{i=1}^{n} V_i} \tag{1}$$

$$W = \int \sigma_{ij} d\varepsilon_{ij} \tag{2}$$

Where ΔW_i is the plastic work density in the ith element which has volume of V_i, "n" is the number of elements. In the ANSYS calculation, an element table manipulation is used to generate the energy density in the post processing stage [10-12]. Since the solder under lead is very thin, the whole solder layer is selected to predict fatigue life. ΔW_i, V_i of the solder joint are obtained from the ANSYS

simulated result. There are 307 elements in all and ΔW_{ave} is calculated with 0.0298MPa.

The energy based Morrow fatigue model is fitted to the fatigue test result for Sn4.0Ag0.5Cu solder low cycle fatigue test over a wide range temperatures. The life prediction equation uses a power law fit by Equation (3).

$$N = C_1 (\Delta W_{ave})^{C_2} \tag{3}$$

Where N is the cycles to failure being calculated depending on the life prediction coefficients for the 1st failure, media life (N50-50% population failure) and characteristic life (N63.2-63.2% population failure). In this paper, characteristic failure life is calculated. C_1 and C_2 are given as constants by 2083.9 and -0.1204, respectively [6]. Then, the fatigue life of solder layer is calculated to be 3181 cycles. This is much longer than the actual failure life of 1500 cycles as recorded in the ATC test, as shown in Fig. 7. As mentioned before, there are 37.5% failures related to TSOP solder joint open in all after 1500 cycles.

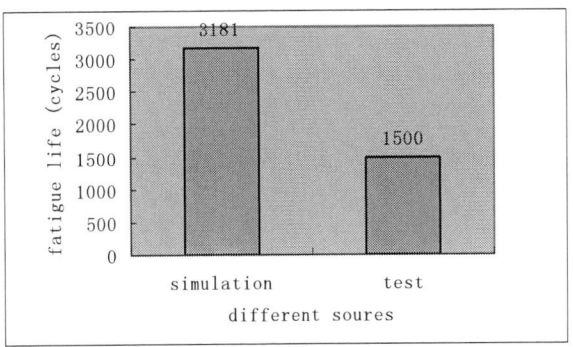

Fig. 7 Comparison of fatigue life of simulation with test

The difference of fatigue life in simulation and test is mainly due to the element type and element selection. Che *et al.* selected the upper and outmost ring for the fatigue life prediction and stated that different element selection will lead to different fatigue life in simulation [1]. However, in the model of this study, the solder joint between lead and copper pad is very thin. The whole solder layer is considered in the calculation of fatigue life which may lead to an inaccuracy. The following research will be based on element selection for the TSOP assembly and add quarter model for comparison.

Conclusions

In this study, ATC test from 0-100°C up to 1500 cycles and FE simulation were conducted for lead-free electronic assemblies of TSOP solder joints. After ATC cycle of 1000, 1250, 1500, failures were detected by functional tests. The failure analysis experiments were performed with optical inspection, cross section and SEM. It is found that due to CTE mismatch serious crack in the bulk solder occurred to cause electrical open. FEA has confirmed substantial stress concentration in the solder layer, which matches well with the crack location in the experimental study. The fatigue life of TSOP solder joint is calculated to be 3181 cycles, which is much higher than that measured in the ATC experiment.

Acknowledgements

This research was partially supported by the National Natural Science Foundation of China under the contract number 60806029.

References

1. Che, F. X., Pang, J. H. L., Xiong, B. S., Luhua Xu and Low, T. H., "Lead Free Solder Joint Reliability Characterization for PBGA, PQFP and TSSOP Assemblies," *Proc 55th Electronic Components and Technology Conference*, Vol. 1, May. 2004, pp. 916-921.

2. Hongtao Ma, Suhling, J. C., Yifei Zhang, Lall, P. and Bozack, M. J., "The Influence of Elevated Temperature Aging on Reliability of Lead Free Solder Joints," *Proc 57th Electronic Components and Technology Conference*, May. 2007, pp. 653-668.

3. Emerick, A., Ellerson, J., McCreary, J., Noreika, R., Woychik, C. and Viswanadham, P., "Enhancement of TSOP solder joint reliability using encapsulation," *Proc 43rd Electronic Components and Technology Conference*, Orlando, FL, USA, Jun. 1993, pp. 187-192.

4. Seung Wook Yoon, Jun Ki Hong, Hwa Jung Kim and Kwang Yoo Byun, "Board-level reliability of Pb-free solder joints of TSOP and various CSPs," *IEEE Trans-Electronics Packaging Manufacturing*, Vol. 28, No. 2, (2005), pp. 168-175.

5. JEDEC Standard, JESD22-A104C, Temperature Cycling, May, 2005.

6. Zahn, B.A, "solder joint fatigue life model methodology for 63Sn37Pb and 95.5Sn4Ag0.5Cu materials," *Proc 53rd Electronic Components and Technology Conf*, Pan Pacific Hotel, Singapore, May. 2003, pp. 83-94.

7. Pang, J.H.L., Low, P.T.H. and Xiong, B.S., "Lead-free 95.5Sn-3.8Ag-0.7Cu solder joint reliability analysis for micro-BGA assembly," *Proc 9th Intersociety Conference of Thermal and Thermomechanical Phenomena in Electronic Systems*, Vol.2, Jun. 2004, pp. 131-136.

8. Che, F.X., and Pang, H.L.J., "Thermal Fatigue Reliability Analysis for PBGA with Sn-3.8Ag-0.7Cu Solder Joints", *Proc 6th Electronics Packaging Technology Conference*, Dec. 2004, pp. 787-792.

9. Pang, J. H. L., Low, P. T. H. and Xiong, B. S., "Lead-free 95.5Sn-3.8Ag-0.7Cu solder joint reliability analysis for micro-BGA assembly," *Proc 9th Intersociety Conference of Thermal and Thermomechanical Phenomena in Electronic Systems*, Vol. 2, Jun. 2004, pp. 131-136.

10. Song Chen,Taekoo Lee, Jaisun Lee and Nufeng Feng, "Solder Joint Thermal Fatigue Analysis of 48-FBGA," *Proc 7th International Conference of Electronic Packaging Technology*, Aug. 2006, pp. 1-4.

11. J. H. L. Pang, Xiong, B. S. and Low, T. H., "Creep and fatigue characterization of lead free 95.5Sn-3.8Ag-0.7Cu solder", *Proc 54th Electronic Components and Technology Conference*, Vol. 2, Jun. 2004, pp. 1333-1337.

12. Pang, J. H. L., Chong, D. Y. R. and Low, T. H., "Thermal cycling analysis of flip-chip solder joint reliability," *IEEE Trans- Components and Packaging Technologies*, Vol. 24, No. 4 (2001), pp. 705-712.

On Variable Frequency Microwave Processing of Heterogeneous Chip-on-Board Assemblies

T. Tilford[1], S. Pavuluri[2], C. Bailey[1] and M. P. Y. Desmulliez[2]

[1] School of Computing and Mathematical Sciences, University Of Greenwich, Park Row,
Greenwich, London, SE10 9LS, United Kingdom

[2] MicroSystems Engineering Centre (MISEC), School of Engineering & Physical Sciences,
Earl Mountbatten Building, Heriot-Watt University, Edinburgh, EH14 4AS, United Kingdom

E-Mail: T. Tilford@gre.ac.uk

Abstract

Variable Frequency Microwave (VFM) processing of heterogeneous chip-on-board assemblies is assessed using a multiphysics modelling approach. The Frequency Agile Microwave Oven Bonding System (FAMOBS) is capable of rapidly processing individual packages on a Chip-On-Board (COB) assembly. This enables each package to be processed in an optimal manner, with temperature ramp rate, maximum temperature and process duration tailored to the specific package, a significant benefit in assemblies containing disparate package types. Such heterogeneous assemblies may contain components such as large power modules alongside smaller modules containing low thermal budget materials with highly disparate processing requirements.

The analysis of two disparate packages has been assessed numerically to determine the applicability of the dual section microwave system to curing heterogeneous devices and to determine the influence of differing processing requirements of optimal process parameters.

Introduction

Semiconductor dies utilised in microelectronics applications generally require encapsulation to protect them from environmental factors such as moisture, ingress of dust or residual humidity. Thermosetting polymer materials, which are often used for this purpose, require heat to initiate or expedite the cure process.

The most common heating system for this process is the convection oven, although approaches such as ultra-violet (UV) and Infrared (IR) heating are also used [reference]. These processes take a relatively long time, often several hours, to bring the encapsulant material up to temperatures which result in a significant cure rate. Convection heating raises the temperature of the gas in contact with the surface of the polymer material and heat energy is transferred into the bulk through thermal conduction. The rate of energy absorption is proportional to the difference in temperature between the package being heated and the ambient temperature in the oven. The temperature profile of the package is therefore asymptotic toward the ambient oven temperature. Electromagnetic energy at infrared or ultraviolet wavelengths will penetrate a small distance (approx. 200-400 nm for UV and 1-1000 µm for IR) into the load but conduction still limits heat transfer into the bulk material.

An alternative method for polymer curing is the use of Variable Frequency Microwave (VFM) systems, which have been shown to be capable of curing encapsulant materials in substantially shorter times [1]. Electromagnetic energy at microwave frequencies (1-30GHz.) will generally penetrate several millimetres into the material, leading to much more rapid volumetric heating. The heating rate in this type of system is governed by the operating power and the dielectric properties of the material. Although the dielectric properties of many materials are temperature dependant, the field intensity can be controlled, enabling the heating rate to be varied independently of the temperature of the material. This results in the capability to accurately control the temperature profile, significantly increase heating rates, control how the material will cure, and ultimately increase productivity rates of the packaging of electronic components and their reliability.

A recent innovation in VFM technology is the dual-section VFM system, proposed by Sinclair et al. [2]. In general, VFM systems require the entire board assembly to be placed in an oven and irradiated with electromagnetic energy. The dual-section system is open-ended and is designed to be placed over an individual component. This enables the curing of single components, eliminating thereby unnecessary heating of the remaining components and board assembly, resulting in reduction of thermomechanical stresses and associated reliability improvements. Furthermore, it is possible for the system to be linked to a temperature monitor, providing the capability to accurately control the heating pattern and rate, allowing the process parameters to be optimised for the specific component.

The composition of a dual-section microwave system is depicted schematically in Figure 1. The oven primarily consists of a ceramic (or polymer) block partially filling an open-ended lightweight metal box, with a coaxial cable feed penetrating the ceramic. The system is designed in a way in that electromagnetic fields are induced within the ceramic block and within the section of the oven which is not filled with ceramic but prevents radiation of fields from the open end of the oven.

The system comprises of the oven unit attached to a gantry system placed over a manufacturing line. The oven is attached to the gantry using a tool head which also links the oven to a microwave power source and is able to provide additional tool interchange and temperature sensing functionality. The gantry system enables the oven to be placed over an individual component on a printed circuit board. Once in position, the microwave source is activated to heat the encapsulant and initiate the cure process. Temperature data is fed back to the microwave source enabling accurate control of the induced temperature profile. Once the process is complete, the oven can be placed over the second component and where the process can be repeated. The system is able to use a different temperature profile for each component processed.

978-1-4244-4658-2/09 $25.00 © 2009 IEEE

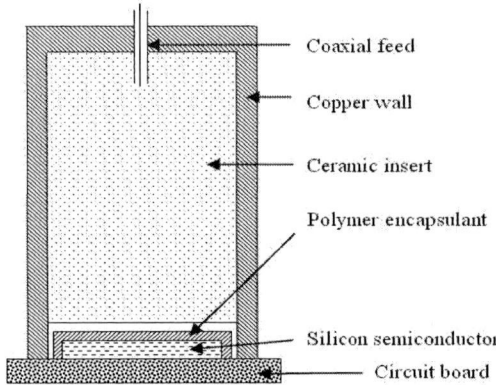

Figure 1 Diagrammatic representation of a dual-section microwave system

This paper aims to numerically assess the curing of encapsulant material in two simplified microelectronics packages with disparate processing requirements. Primary process parameters such as temperature variations, cure rate, degree of cure, and thermomechanical stresses need to be monitored throughout the process, with comparisons indicating benefits and weaknesses of the different heating approaches.

Curing of thermosetting polymer materials is a complex process, with interactions between temperature, curing and stress development. In addition, the analysis of the curing process must also take into account the complex interactions between the component and the electromagnetic fields present due to the microwave energy needed to cure the materials. In order to accurately model the curing process, a holistic approach has been taken, in which the process is not considered to be a sequence of discrete steps, but as a complex coupled system.

Experimental assessment of the key process parameters is extremely complicated, especially if real time/in-situ data are desired. A multiphysics approach to numerical analysis has therefore been adopted to analyse the key process parameters.

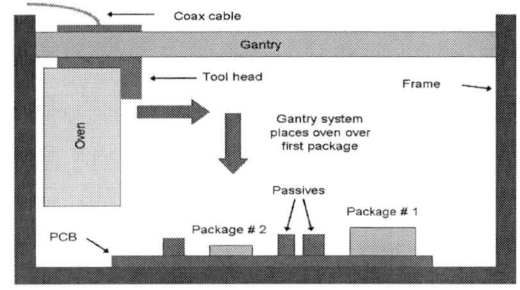

Figure 2 (a) Oven system initially 'parked' to the side of the line

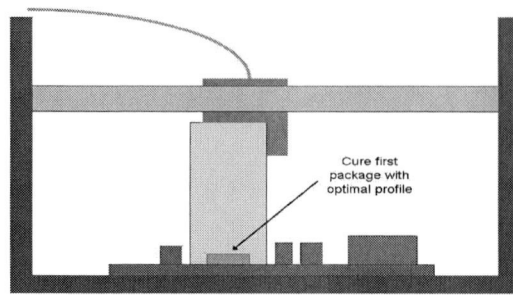

Figure 2 (b) Gantry system moves the oven head over an individual component and the oven is activated to process the component

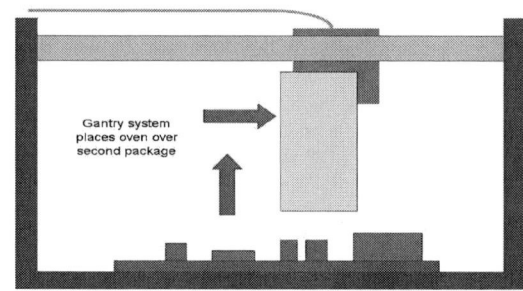

Figure 2 (c) After processing is complete the oven is moved to the next component

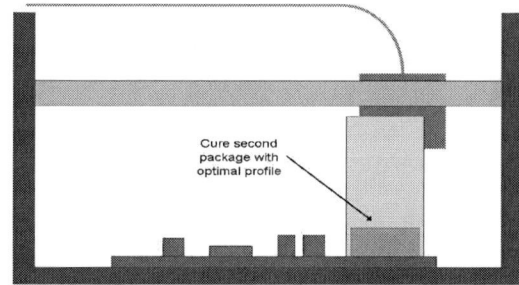

Figure 2 (d) After processing of the second component the oven returns to the starting position

Numerical Model

A numerical model capable of modelling variable frequency microwave heating has been developed. The model is highly complex, reflecting the intricacy of the process it is required to analyse. The major components of the model are algorithms to solve for the electromagnetic (EM) fields, temperature distribution, cure kinetics and induced thermomechanical stresses. The processes are coupled in a complex manner, as outlined in Figure 3, requiring a holistic approach to analyse the problem

The EM fields induce heating in the package, which expedites the cure process. Thermal expansion and cure shrinkage induce stresses within the package. The temperature rise and variation of degree of cure within the material significantly alter the dielectric properties, affecting the EM fields and therefore the heating rate and pattern. Furthermore, the cure reaction is predominantly exothermic, further complicating the relationship between cure rate and temperature.

The numerical model developed adopts a multi-domain FDTD-UFVM approach. The model comprises a Yee scheme [3] Finite Difference Time Domain (FDTD) electromagnetic solver coupled with an unstructured finite volume method (UFVM) multi-physics package [4]. Electromagnetic and thermophysical solutions are obtained within independent numerical domains, with coupling implemented through an inter-domain cross mapping process. The thermophysical analysis mesh is confined to the load material while the electromagnetic mesh occupies the entire oven domain (encompassing the thermophysical domain). This approach enables the FDTD mesh to be varied without requiring modification to the thermophysical mesh. This is of great benefit in cases in which the dielectric properties of the load material (and therefore local propagation wavelength) vary during the heating process. The cross mapping algorithm transfers relevant data between the two domains based on spatial coordinate sampling approach. The model has been shown to provide accurate analysis of complex microwave processing problems [5, 6].

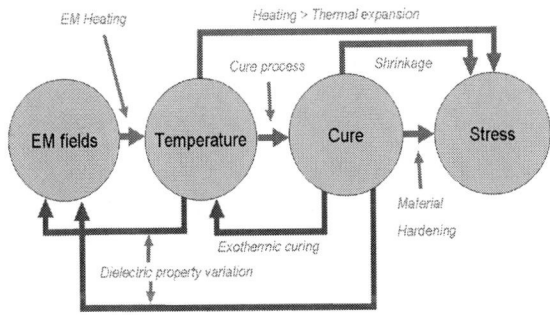

Figure 3 Complex interaction in microwave cure process

Test Strategy

Each of the two packages to be analysed have been considered in a simplified form, consisting of a 'silicon die' covered in a layer of encapsulant material, as illustrated in Figure 4. The dimensions of the packages are outlined in table I. A commercially available thermosetting epoxy material has been used for this assessment. Differential Scanning Calorimetry (DSC) data has been analysed to determine the cure behaviour and exothermal energy released by the material. The coefficients used in the implemented Arrhenius cure model have been fitted to the DSC data using a separate numerical integration scheme. This scheme considered a very large number of permutations of the coefficients, assessing fit between experimental and model data for each coefficient set. Optimal coefficients resulted in a fair agreement between experimental and model cure-time-temperature curves. A more detailed DSC analysis is required to more accurately assess cure behaviour and to enable the cure model

coefficients to be determined in a more intelligent manner. The thermomechanical load material properties are detailed in Table II. Experimental assessment of dielectric properties has been performed [7] using an Agilent Technologies 85070E dielectric probe kit. Obtained results indicated that both the real and imaginary components of the complex permittivity decrease with degree of cure. In the implemented numerical model, these components have been approximated by the functions given in Equations 1 and 2.

$$\varepsilon_r = 4.06 + \left(0.0062\left(T - 298\right)\right) - \left(0.35\ \alpha^*\right) \quad (1)$$

$$\varepsilon'' = 0.41 + \left(0.0040\left(T - 298\right)\right) - \left(0.50\ \alpha^*\right) \quad (2)$$

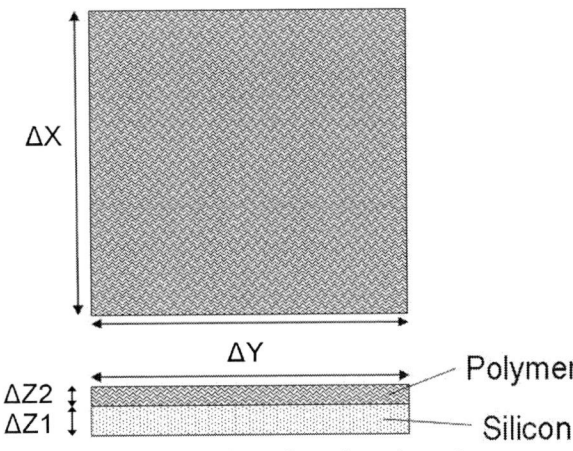

Figure 4 Package dimension schematic

Table I Package dimensions

Symbol	Variable	Package #1	Package #2
ΔX	Package width	7.5 mm	22.5 mm
ΔY	Package length	7.5 mm	22.5 mm
ΔZ1	Silicon thickness	2.0 mm	4.0 mm
ΔZ2	Polymer thickness	1.0 mm	2.0 mm

Table II Load material properties

Property	Silicon	Polymer	Unit
Density	2230	1100	kg m^{-3}
Thermal conductivity	f(T)	0.33	W M^{-1} K^{-1}
Specific heat	970	700	J kg^{-1} K^{-1}
Thermal expansion	2.6e-6	1.2e-4	K^{-1}
Young's modulus	120	f(α)	GPa.
Poisson's ratio	0.30	0.29	-
Dielectric constant	12.0	f(T)	F m^{-1}
Loss factor	0.108	f(T)	H m^{-1}

Results

Packages #1 and #2 were considered to be subjected to the temperature profiles illustrated in Figure 5. Numerical analysis of the process was performed, with cure kinetics, temperature distribution and thermomechanical stress magnitude monitored continuously. The maximum and minimum degree of cure for each package is plotted in Figure 6. From these plots it is evident that the cure process for package #1 is relatively rapid and results in a high maximal degree of cure with limited variation in degree of cure over the package. In contrast, package # is not fully cured with a maximum degree of cure of approximately 0.8 and a wider variability in the cure over the package. Investigation of the numerical data suggests that the corners of the package are being insufficiently heated. This is likely to be due to the relative size of the package and the oven cross section. The fields vary sinusoidally across the cross section, with zero magnitude at the oven walls. There would therefore appear to be a requirement for the oven cross section to be greater than the package cross section. The relatively low degree of cure results in a lower Young's modulus within the polymer material. This is likely to be the cause of the disparity in the effective (von Mises) stress induced in the silicon die during the process as shown in Figure 7.

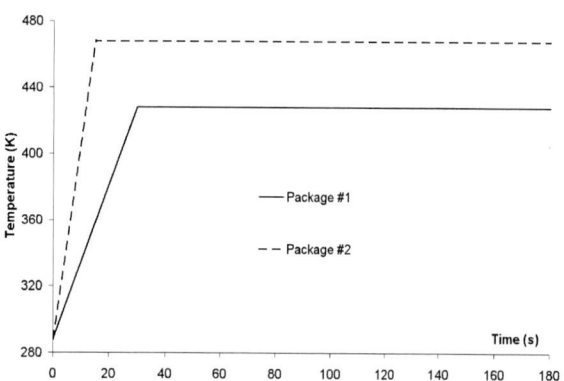

Figure 5 Temperature profiles for packages #1 and #2

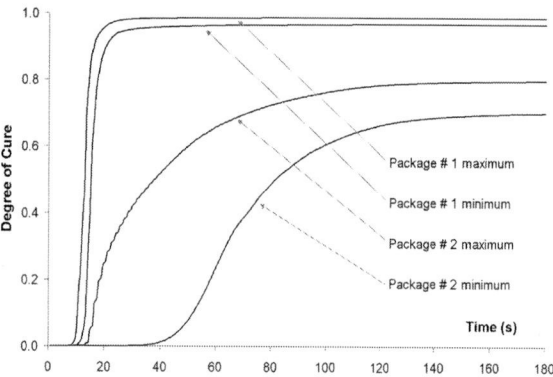

Figure 6 Evolution of degree of cure in packages #1 and #2 during a process time of 180s.

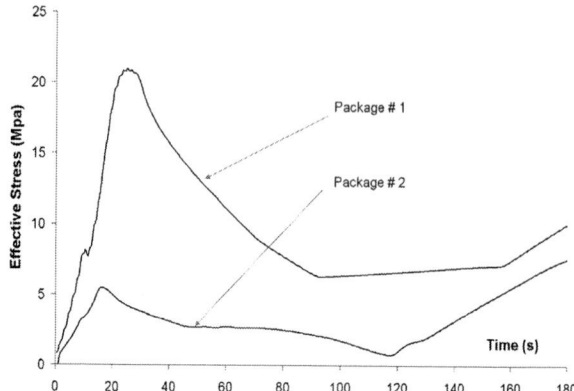

Figure 7 Evolution of effective stress in packages #1 and #2 during 180 second process time

Conclusions

Results obtained show that the VFM technology can be utilised to rapidly cure encapsulants in microelectronics applications. In cases in which the package is significantly smaller than the oven cross section, the degree of cure within the encapsulant is extremely high with little variation. However, issues are apparent when using a package with lateral dimensions comparable to those of the oven. This issue can be eliminated through adoption of a multi-oven tool head, which is capable of automatically swapping between oven units if differing dimensions.

Development of the FAMOBS system is continuing as part of a three-year project, supported through European Union 'Framework 7' funding. The project will focus on improvement of the system design to enhance performance and efficiency whilst decreasing cost. Materials used in the process will be characterised to determine their interaction with the microwave fields and the resulting cure process. New VFM-specific material may additionally be developed. Reliability of the VFM processed packages will be compared with conventionally cured packages through the use of a variety of accelerated lifetime testing approaches

Further information on this EU project can be found at the projects website: www.famobs.eu

Acknowledgments

The authors wish to acknowledge funding and support from the European Union Framework 7 programme (FP7-SME-2007-2), contract number 218350. This project also benefits from the financial support of our partners, Eesti Innovatsiooni Instituut, Fraunhofer Gesellschaft (IPA), Camero di Commercio Industria, Artigianato e Agricoltura di Milano, Mikrosystemtechnik Baden-Württemberg e.V., the National Microelectronics Institute, the European Society for Precision Engineering and Nanotechnology, ACI-ecotec GmbH & Co. KG, Kepar Electronica S.A., Satel S.A., Industrial Microwave Systems Ltd. and Ribler GmbH.

References

1. T. Tilford, K. I. Sinclair, G. Goussetis, C. Bailey, M. P. Y. Desmulliez, A. K. Parrott and A. J. Sangster, "Comparison of Encapsulant Curing with Convection and Microwave Systems," *Proceedings 33rd International Electronics*

Manufacturing Technology Conference, Penang, Malaysia, 2008.

2. Sinclair, K. I., Desmulliez, M. P. Y. and Sangster, A. J., 2006, "A novel RF-curing technology for microelectronics and optoelectronics packaging," *Proc. IEEE Electronics Systemintegration Technology Conference*. 2006, Vol. 2, pp. 1149-1157.

3. Yee, K., 1966, "Numerical solution of initial boundary value problems involving Maxwell's equations in isotropic media,". *IEEE Transactions on Antennas and Propagation*, 14: PP. 302-307.

4. PHYSICA (1996-2007). Physica Ltd, 3 Rowan Drive, Witney, Oxon, United Kingdom, http://www.physica.co.uk.

5. T. Tilford, K. I. Sinclair, C. Bailey, M. P. Y. Desmulliez, G. Goussettis, A. K. Parrott and A. J. Sangster, "Multiphysics Simulation of Microwave Curing in Micro-Electronics Packaging Applications," *Journal of Soldering and Surface Mount Technology*, Volume 19, Issue 3, 2007, pp. 26-33.

6. T. Tilford, E. Baginski, J. Kelder, A. K. Parrott and K. A. Pericleous, (2007) "Microwave Modelling and Validation in Food Thawing Applications," *Journal of Microwave Power and Electromagnetic Energy*, 41(4), pp. 30-45.

7. K. I. Sinclair, T. Tilford, G. Goussetis, C. Bailey, M. P. Y. Desmulliez, A. K. Parrott and A. J. Sangster, "Advanced Microwave Oven for Rapid Curing of Encapsulant," *proceedings 2nd Electronics Systemintegration Technology Conference* 2008, Greenwich, UK, pp. 551-556, ISBN 978-1-4244-2813-7.

Advanced Chip to Wafer Bonding: A Flip Chip to Wafer Bonding Technology for High Volume 3DIC Production Providing Lowest Cost of Ownership

A. Sigl[1], S. Pargfrieder[1], C. Pichler[2], C. Scheiring[2] and P. Kettner[1]
[1]EV Group, DI Erich Thallner Strasse 1, St. Florian/Inn, 4782, Austria
[2]Datacon Technology GmbH, Innstraße 16, Radfeld, 6240, Austria

Abstract

The shrinkage and the integration of various functionalities into electrical devices, like computers or mobile phones, lead to an ongoing need for shrinkage of the integrated semiconductor units. One possibility for manufacturing of highly integrated electrical devices is the System in Package (SiP) approach where various semiconductor chips with different functionalities are stacked and electrically connected to each other. The shrinkage affects all levels of the SiP, e.g. the transistor size, the die thickness, the height of the die stack and also the dimension and shape of interconnects between the dies.

The shrinkage of the die interconnects can cause difficulties of the existing widely used joint technologies, e.g. solder bumping, because of low amount of involved solder, so that the assembly yields drops and the reliability of the interconnects lowers.

The Advanced Chip to Wafer (AC2W) bonding is a two step process for stacking and bonding dies on wafers. First all dies are aligned and tacked on the wafer and in the second step all dies are bonded simultaneously permanently to the wafer. This process allows having force while bonding the dies on the wafer. In that way low solder volume interconnects can be formed on a wafer level with high assembly yield and throughput. The Cost of Ownership (CoO) connected with the throughput of the AC2W process can be an order of magnitude smaller then for comparable chip to wafer bonding processes and therefore the AC2W offers a low cost chip to wafer bonding process for high volume production.

This paper will show the AC2W bonding process in detail, some issues at die joint shrinkage and a comprehensive throughput and CoO comparison between the AC2W and comparable process flows.

Introduction

Collapsible solder bump interconnects are widely used for flip chip connections at various applications. However, as structure size on semiconductor devices decreases to follow the well known "Moore's Law", also the dimension of the interconnects have to shrink. Thereby of course the solder volume at the individual interconnects decreases. The solder has among others also the function to compensate height variations caused by the substrates and functional layers itself or by their warp and bow. With less amount of solder the interconnect losses this ability and therefore only with application of force while bonding a sufficient high manufacturing yield can be guaranteed.

While forming solder interconnects the used metals diffuse into each other and usually convert to intermetallic compounds (IMC) at the interface. At solder connects used up

to now this is an undesirable effect that occurs with a certain growth rate that depends on temperature and the used metals. Therefore the thickness of this layer depends on physics and is nearly of the same thickness at low solder volume interconnects (Figure 1). So with the shrinkage of the connection the portion of the IMC rises (Figure 2). Also the growth of the IMC doesn't stop after the heat cycle for forming the connection completely. At room temperature the conversion goes on but with a very long time constant. This means solder interconnects are not thermodynamically stable. Therefore an approach was to bring the connection into a thermodynamically stable state, which means to convert the metals completely to IMC and also use the IMC directly as material for the interconnect. In that case the volumes of the metal components are adjusted to each other in a way that all material can convert to IMC while bonding. This approach is similar to the well known eutectic bonding. The density of the IMC material usually is higher then the density of the metal components and therefore the volume shrinks while forming the connection. To guarantee a good contact and allow a sufficient high metal diffusion application of force is necessary for bonding.

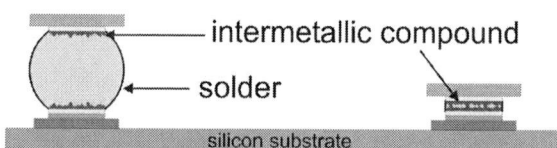

Figure 1 Comparison between C4 solder interconnect and low volume solder interconnect

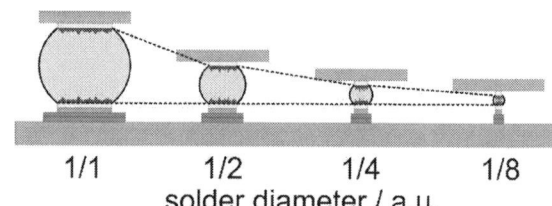

Figure 2 Portion of the intermetallic compound rises with shrinking solder volume because the thickness of it is constant

A more sophisticated approach for making small metal interconnects is the thermocompression bonding, which is well known from wafer to wafer bonding. It is well suited for making very small interconnects with dimensions in the single digit micrometer range. At this approach force is needed during bonding to overcome the micro roughness and bring the two metal surfaces near too each other so that the diffusion of the metal atoms can occur to form a metal bond.

978-1-4244-4658-2/09 $25.00 © 2009 IEEE

Several major semiconductor manufacturers evaluated these types of forming interconnects or use it actually for volume production [1-7]. As shown at all approaches for small interconnects application of force is necessary to form a proper bond. For the application of force for chip to wafer bonding a sophisticated process, the Advanced Chip to Wafer (AC2W) bonding process, is available which was designed especially for high volume production.

Advanced Chip to Wafer Bonding Process

The AC2W bonding process is a process flow for chip to wafer bonding especially designed for application of force while forming the bond at a throughput appropriate for volume production [11]. The concept of separation of aligning substrates and then bonding the substrates to each other is well known and widely used for wafer to wafer bonding. At the AC2W bonding process the same concept is adapted to chip to wafer bonding. The AC2W process is a two step process (Figure 3). First all chips are aligned and tacked to the wafer, second all chips are bonded in parallel simultaneously permanently to the wafer.

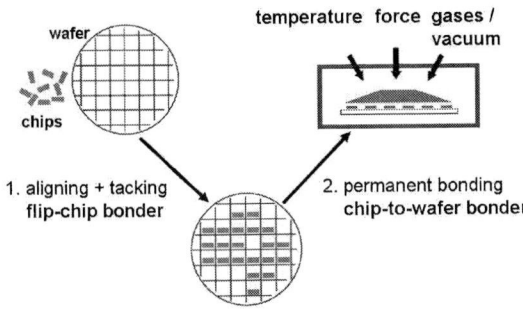

Figure 3 The AC2W process flow

In the first step of the AC2W bonding process the dies are aligned to the wafer and then the alignment is fixed by tacking the chips to the wafer. For the tacking are three different methods available. Each of them has its distinct application.

1) Tacking by a temporary adhesive [11]: As temporary adhesive for example bibenzyl can be used. This material has the properties that it is solid below 48°C, liquid above that temperature and it evaporates without residues in the temperature range of 250-280°C. The die and the adhesive is heated up over 53°C for aligning and applying the adhesive. Then it is cooled down and the adhesive gets solid. The adhesive completely evaporates while the permanent bonding step.

2) Tacking by pre applied underfill [12]: There are various manners to apply underfill before bonding either to the target wafer or to the dies. At that time the underfill has to have by its own or by special treatment the property to be sticky enough to fix the die to the wafer and secure the alignment. The temperature where the final curing for the underfill has to be done has to be above the bonding temperature of the metallic interconnection, so that first the metallic bond is formed and afterwards the mechanical connection of die and

wafer is made by the underfill. The underfill can either be fully cured in the permanent bonder by tempering for a longer time or raising the temperature or it can also be cured in a different oven. The usage of pre applied underfill has the big advantage that underfill has usually to be applied at stacked substrates anyway and therefore it covers two functions and saves costs.

3) Tacking by ultrasonic [13]: Ultrasonic tacking can be used at metals with a low melting point. In this case a fragile connection between the surfaces of the metal pads is formed by ultrasonic energy. This method is preferable for MEMS fabrication because no organic materials are involved. Furthermore no consumables are used.

All of the three methods fix the dies and thereby the alignment of the dies while the populated wafer is transported from the flip chip bonder (e.g. Datacon 8800 CHAMEO) to the permanent bonder (e.g. EVG540C2W). All of them can be performed at high speed, which means that the flip chip bonder can operate in a production mode in the range of the highest specified throughput for the equipment. After tacking the populated wafer is brought to the permanent bonder.

The permanent bonder has unique features for the chip to wafer bonding, as 1) a closed chamber for vacuum encapsulation or process gasses, 2) a pressure plate and force application system, 3) a point of application of force movement system and 4) a compliant layer. Therefore it allows to make the permanent bond process at high temperature (up to 550°C) with application of force and with vacuum (down to 1×10^{-5} mbar) or process gases.

1) The closed chamber for vacuum encapsulation or process gasses allows having a specific controlled atmosphere while the bond is formed. In that way gasses or vacuum can be encapsulated in semiconductor devices, e.g. pressure sensors, or bonding improving gasses can by used.

2) The pressure plate allows the application of force simultaneously on all dies on the wafer during the whole process of forming the bond as this is necessary to have a high process yield for small interconnects.

3) The point of application of force movement allows the usage of Known Good Dies (KGD) only, which is a big cost saving. At KGD the wafer is only partially populated, only tested chips on dies on the wafers which are also tested and working. To guarantee the same force on each die the point of application of force has to be at the center of gravity of the die population, as shown in Figure 4, which is achieved by the point of force application movement system.

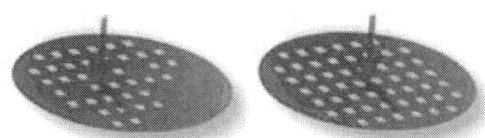

Figure 4 At Known Good Dies (KGD) usage the wafer is only partially populated and the point of application of force has to be moved to the new center of gravity.

4) The compliant layer is a soft material that compensates for thickness variations of the dies which occur when the dies

originate from different source wafers. The variations are usually in the range of tens of micrometers or below. With the compliant layer yield issues caused by small interconnects can be overcome as shown in Figure 5.

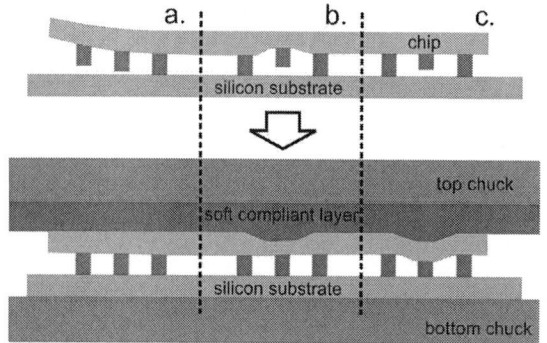

Figure 5 Open contacts can occur e.g. by a. bow or warpage, b. unevenness of the substrates, c. height variation of the interconnects (top). These issues can be overcome by application of force and usage of a soft compliant layer while bonding (bottom).

While the permanent bonding process force is applied on the dies on the wafer, then the stack is heated up to the bonding temperature under vacuum or certain process gasses and the bond is formed of all dies simultaneously. At this time also the temporary adhesive is either evaporated or cured depending on the tacking method that is used. Then the bonded stack is cooled down and can be unloaded from the chamber.

The AC2W bonding concept is a new concept compared to the well known Controlled-Collapse Chip Connection (C4) for flip chip, the wafer to wafer bonding and the chip to wafer bonding by only using a flip chip bonder. All of these concepts have their advantages and there is a field of application where the Advanced Chip to Wafer bonding approach is the best and in terms of cost of ownership the cheapest process.

Comparison of C4 to AC2W Process

The most popular process for making collapsible interconnections is the Controlled-Collapse Chip Connection (C4) process [8]. It mainly consists out of four steps, which you can see in Figure 6 on top: 1. applying of under bump metallization (UBM), 2. applying of solder with ball forming reflow, 3. fluxing and placing the chip on the substrate and 4. assembly reflow

The process flow of AC2W is quite similar to C4 as shown in Figure 6 bottom: 1. applying of under bump metallization, 2. applying of solder without ball forming reflow, 3. aligning and tacking on substrate by temporary adhesive, 4. assembly reflow with force.

In principle the concept of the two process flows are the same. In both cases the chips are fixed by an adhesive for the last step (flux at C4, temporary adhesive at AC2W). The difference is when and how the alignment is done. At the C4 process the chips undergo a self alignment when the solder

melts at the final assembly reflow. At that point the surface tension of the solder pulls the chips to the correct position. This concept only works when the amount of solder is big enough so that there is a droplet of solder that has a high enough surface tension to move the chip. At small solder volume interconnects the floating resistance is too high and this concept fails. Also as already shown the process yield drops because of open contacts.

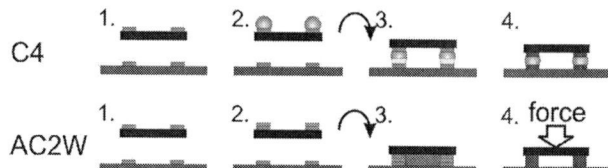

Figure 6 C4 and AC2W (for Cu-Sn-Cu eutectic bond) process flow comparison

At the AC2W process the alignment is already completed at the placement of the chip on the wafer and secured by the temporary adhesive and by the application of force while the final bonding step. The alignment is completely done by the flip chip bonder and not depending on material properties of the interconnection material. Therefore higher accuracies with smaller interconnects can be achieved compared to the C4 process and open contacts are avoided by the application of force while bonding.

Therefore the C4 process is well suited and offers high throughput for high volume solder interconnects but for low solder volume interconnects the AC2W process is favored.

Comparison of AC2W to Wafer to Wafer Bonding Process

Wafer to wafer (W2W) bonding is a widely used process approach for connecting and stacking devices on wafers on each other. The main advantage of the W2W bonding is the very high achievable alignment accuracy in the sub-micrometer range with high throughput.

At flip chip bonders the throughput depends strongly on the targeted alignment accuracy. Therefore the throughput at the first step of the AC2W drops when accuracies below 2 micrometer are targeted while at W2W it stays nearly constant for all. At the second step the high throughput of W2W bonding is also achieved by the AC2W bonding process since it is very comparable to W2W bonding as there are also multiple dies bonded to the target wafer at once. Also the process time and process recipe is quite the same.

In contrast to W2W at AC2W bonding Known Good Dies can be used. This offers a high cost saving. Especially when stacking multiple dies at W2W the yield can decrease dramatically as it can be seen in Table 1 where the results of a simplified yield calculation is shown. Here a yield of 80 % of the devices on the processed wafers is taken into account. It is assumed that all stacked wafers have the same yield. The yields for all other yield decreasing issues, like bonding, testing and chip yield, are assumed to be a 100%, since they are usually very high. At W2W the yield decreases exponentially while at AC2W it stays constant at the starting level of 80%.

When dies with different sizes have to be stacked to each other additional losses occur at the W2W approach because the patterns on the different wafer have to match regardless of the chip size on the wafer. An example should be shown here. Assuming two dies are stacked on each other and one device has a quarter of the size of the other. The two wafers (wafer 1, wafer 2) cost the same at processing. When stacking the devices with W2W approach three quarter of the wafer 1 have to be without devices since the patterns on the two wafers have to match each other and the devices on wafer 1 have to be at a certain position to land at the right place on the wafer 2 while bonding. Therefore on wafer 1 only one quarter of the wafer area is used with chips. On the other area of the wafer 1 chips could have been processed too, but it is not utilized for making devices. When looking on the costs, the wafers cost the same, so each of them 50% of total costs. At W2W 75% of the wafer area from wafer 1 is unused and possible dies are lost which would be used at the AC2W. Altogether for this example the costs for using W2W compared to AC2W are 37% higher caused by the chosen bonding process.

The W2W approach is the method of choice for applications where alignment accuracy in the sub micrometer range is necessary. The AC2W bonding offers an equal throughput simultaneously with higher yield and is therefore cheaper for stacking multiple dies.

Table 1 Yield advantage of AC2W through KGD depending on the number of stacked dies only taking a process yield of 80% of the devices on the wafers into account. All other yields (bonding, testing, chip) are assumed to be 100%.

# of stacked dies	W2W yield / %	AC2W yield / %
0	80	80
1	64	80
2	51	80
3	41	80
4	33	80
5	26	80
6	21	80
7	17	80
8	13	80

Comparison of Flip Chip Bonder Only to AC2W Process

The flip chip bonder only process flow is to do both steps, the alignment and bonding, in serial directly after each other for each individual die in the flip chip bonder. So all dies are bonded in serial to the wafer. As shown actually used die interconnects (eutectic bond, thermocompression bond) form by diffusion processes that need a certain time. This time is usually in the range of tens of seconds. At the flip chip bonder only process this time is needed to bond each individual die and therefore lowers the throughput dramatically. At the first step of the AC2W process the flip chip bonder can work with very high throughput. The time needed for diffusion contributes to the second step of the AC2W process, but since all dies are bonded in parallel the overall time contribution is

only again in the range of tens of seconds and therefore negligible.

For better understanding here an example will be given. It is assumed that 2641 chips with a size of $5{\times}5$ mm^2 are bonded to a 300 mm wafer. The type of interconnection is Cu-Sn which usually needs 20 s to form. The alignment time on a flip chip is about 0.5s. This means that at the flip chip bonder only process each individual die needs 20.5s to be bonded to the wafer which results in a throughput of about 175 dies per hour or 0.066 wafer per hour.

At the AC2W process the throughput of the two steps has to be evaluated separately. At the first AC2W step the throughput is determined by the flip chip bonder alignment time (0.5s) and therefore about 7200 dies per hour or 2.73 wafers per hour. The throughput of the second step is mainly determined by equipment restrictions and in the range of about 20 minutes per wafer or 3 wafers per hour. The 20 s diffusion time to form the interconnection which would contribute at the second bonding step has therefore no negative throughput effect at the AC2W process.

Besides the throughput there are several variables that affect the cost of ownership of the AC2W process, as the die size, wafer size, bond process and alignment accuracy. For the given example the cost of ownership (CoO) was calculated and can be seen in Figure 8. The CoO advantage of the AC2W process can easily be seen.

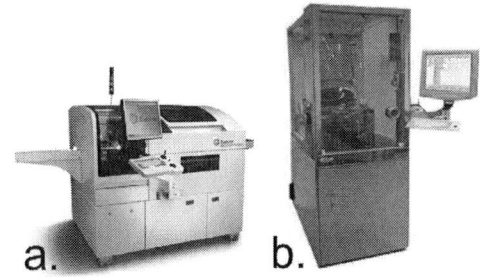

Figure 7 AC2W equipment: Step 1: a) flip chip bonder (e.g. Datacon 8800 CHAMEO), Step 2: b) permanent bonder (e.g. EVG540C2W) [9-10]

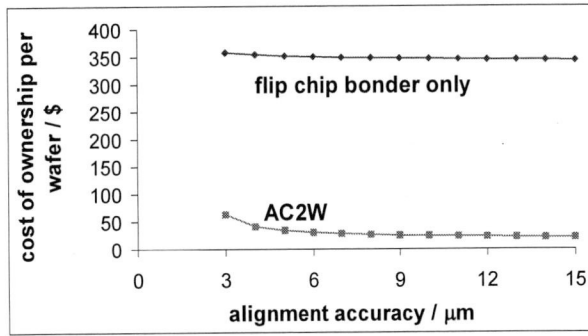

Figure 8 Cost of Ownership comparison of Flip chip bonder only and AC2W bonding process (CuSn bond, 300mm wafer, 2641 dies)

Conclusions

For actual die to wafer interconnects the application of force is necessary while forming the connection. The Advanced Chip to Wafer bonding process is a sophisticated approach for chip to wafer bonding with application of force while bonding. Compared to other approaches like flip chip bonder only or wafer to wafer bonding it offers a much higher yield and throughput and therefore a lower cost of ownership. These are the reasons why AC2W is best suited for high volume chip to wafer production.

References

1. Ph. Garrou *et al*, Handbook of 3D Integration, Wiley-VCH, Volume 1, pp. 40-43, 2008.

2. H. Huebner, O. Ehrmann, M. Eigner, W. Gruber, A. Klumpp, R. Merkel, P. Ramm, M. Roth, J. Weber and R. Wieland, "Face-to-Face Chip Integration with Full Metal Interface," *Proc. Advanced Metallization Conference, Materials Research Society*, pp. 53, San Diego, 2002.

3. A. Longford and D. James, "Copper pillar bumping in Intel microprocessors-one approac to lead free," *Presentation in Advanced Packaging Conference*, Semicon Europe, April, 2006.

4. T. Mitsuhashi, Y. Egawa, O. Kato, Y. Saeki, H. Kikuchi, S. Uchiyama, K. Shibata, J. Yamada, M. Ishino, H. Ikeda, N. Takakashi, Y. Kurita, M. Komuro, S. Matsui and M. Kawano, "Development of 3D-Packaging Process Technology for Stacked Memory Chips," *Mater. Res. Soc. Symp. Proc.*, Vol. 970, Materials Research Society, 2007.

5. "Sony Leads the Industry in Mass Production Devices that Achieve the Same Performance as embedded DRAM LSIs in the Low-Cost SiP Process," *CX-News*, Vol. 50, Nov. 2007, http://www.sony.net/Products/SC-HP/cx_news.

6. K. Sakuma, P. S. Andry, C. K. Tsang, S. L. Wright, B. Dang, C. S. Patel, B. C. Webb, J. Maria, E. J. Sprogis, S. K. Kang, R. J. Polastre, R. R. Horton, J. U. Knickerbocker, "3D chip-stacking technology with through-silicon vias and low-volume lead-free interconnections," *IBM J. Res. & Dev.*, Vol. 52, No. 6, pp. 611-622, Nov. 2008.

7. S. Pozder, A. Jain, R. Chatterjee, Z. Huang, R. E. Jones, E. Acosta, B. Marlin, G. Hillmann, M. Sobczak, G. Kreindl, S. Kanagavel, H. Kostner and S. Pargfrieder, "3D Die-to-wafer Cu/Sn Microconnects Formed Simultaneously with an Adhesive Dielectric Bond Using Thermal Compression Bonding," *Interconnect Technology Conference*, IITC International, Burlingame, CA, USA, pp. 46, 2008.

8. L. F. Miller, "Controlled Collapse Reflow Chip Joining", IBM J. Res. & Dev., Vol. 13, p. 239, 1969.

9. Homepage of EVGroup, http://www.evgroup.com.

10. Homepage of Datacon, http://www.datacon.at.

11. S. Pargfrieder, T. Matthias, C. Schaefer, M. Wimplinger, C. Scheiring, H. Kostner, "3D Packaging via Advanced-Chip-to-Wafer (AC2W) bonding enables Hybrid System-in-Package (SiP) Integration," *IWLP – Wafer-Level Packaging Congress*, November 2-4, San Jose, California, 2005.

12. S. Pozder, A. Jain, R. Chatterjee, Z. Huang, R. E. Jones, E. Acosta, B. Marlin, G. Hillmann, M. Sobczak, G. Kreindl, S. Kanagavel, H. Kostner and S. Pargfrieder, *Interconnect Technology Conference*, IITC International, Burlingame, CA, USA, pp. 46, 2008.

13. N. Marenco, H. Kostner, W. Reinert, G. Hillmann, "Hybrid Chip-Scale Integration of Inertial MEMS by Chip-to-Wafer Vacuum Bonding," *2nd European Conference & Exhibition on Integration Issues of Miniaturized Systems-MOMS, MOEMS, ICs and Electronic Components*, Smart System Integration, Barcelona, 2008.

The Effect of Plasma Etching Process on Rigid Flex Substrate for Electronic Packaging Application

K. C. Yung[1], H. M. Liem[1], H. S. Choy[1], H. F. Zheng[1], Tao Feng[2], T. M. Yue[1]
[1]Department of Industrial and Systems Engineering, the Hong Kong Polytechnic University
Hung Hom, Kowloon, Hong Kong, P. R. China
Email: mfkcyung@inet.polyu.edu.hk, Telephone: +852-27666599
[2]BMP Application Development, the Linde Group, Shanghai, P. R. China
Email: edward.feng@linde.com, Telephone: +86 (21) 58993535

Abstract

The demand for high-speed signal transmission circuits triggers the use of polymer substrates having the superiorly mechanical and electrical reliability properties. Foremost among them, rigid-flex polymer substrates are being used in the printed circuit board (PCB) of electronic packaging technologies by utilizing their property compatible with device miniaturization and packaging integration. Modifying the polymer interfacial property further before subsequent processing, plasma treatment is employed to adjust the physical and chemical changes of the surface and subsurface. One special issue of plasma-induced change is the surface roughness (hydrophobic or hydrophilic) depending on every detail of process cycle conditions. Thus, it is difficult to be precisely controlled so as to deliver the desirable outcome.

This study employs Taguchi experimental method to find the parameter critical to plasma etching particularly on rigid flex substrate. In contrast to the traditional use of atomic force microscopy (AFM) and/or scanning electron microscopy (SEM) to characterize the surface roughness morphology and thus deduce the optimal process conditions. It adopts a set of standard orthogonal arrays to determine parameters configuration and analysis results. These kinds of arrays use a small number of experimental runs but obtain maximum information and have high reproducibility and reliability. Taguchi method provides the experimenter with a systematic and efficient approach for conducting experiments to determine near optimum parameter settings for performance and cost.

Introduction

Advancement in PCB material science and the use of both isolated dense geometries has risen the standard for plasma processing at multiple steps in the manufacturing cycle. The conventional chemical method uses permanganate chemistry is limited by the fluid's ability to penetrate the fine vias used in multilayer PCBs. Chemical desmearing consists of three main steps, swelling, oxidization and neutralization. In fact, permanganate cannot clean the tiny via holes effectively, located at the inner layer level of laser-drilled via holes, nor high aspect ratio vias. In contrast, the inherent penetrating nature of plasma as the agent in plasma processing overcomes this limitation, and has become the preferred method for desmearing multi-layer printed circuits

The material surfaces in contact with the plasma experience interactions which may give rise to chemical and physical modifications at the surface, e.g. producing more reactive sites, or changes in cross-linking or molecular weight. In this way, without affecting the bulk properties, materials with desired properties can be obtained, such as an improved wettability and better adhesion [1]. Wettability is a vital parameter [2] to judge the quality of surface plasma treatment since it can be expressed as surface energy using Young's equation [3]. The water contact angle used in wettability measurement is an indicator of the adhesion trend [4]. Contact angle actually indicates the wettability of a surface and thereby its hydrophilicity or hydrophobicity. Both of these two types of surfaces exhibit self-cleaning property through the nature of the interaction of water on the surface, the former by rolling droplets and the latter by sheeting water that carries away dirt. A decrease in contact angle corresponding to the hydrophilicity increase will enhance the adhesion strength of PCB materials for subsequent electroplating process [5]. Therefore, studies on optimization of plasma treatment parameter design to achieve adequate wettability is of considerable importance.

Parameter Design Methodology

Taguchi's parameter design method serves as a fine search tool for optimizing the performance characteristic of a product/process, identifying and designing the settings of the process parameters that optimize the chosen quality characteristic and are least sensitive to noise factors. A set of standard orthogonal arrays are selected to determine parameters configuration and analysis results. This method can reduce research and development costs by simultaneously studying a large number of parameters [6]. In order to analyze the results, the Taguchi method uses a statistical measure of performance called the signal-to-noise (S/N) ratio. The S/N ratio takes both the mean and the variability into account. The S/N equation depends on the criterion for the quality characteristic to be optimized.

The cause-and-effect diagram illustrating the possible effects of the process parameters on the plasma treatment performance is shown in Fig. 1.

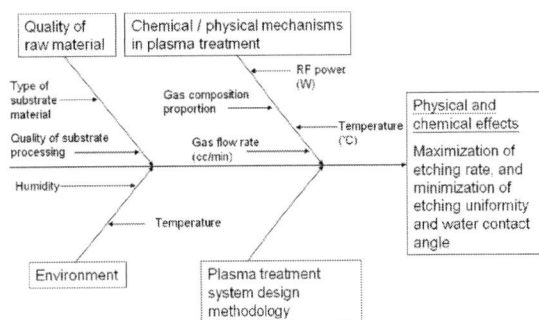

Fig. 1 Cause-and-effect diagram for surface plasma treatment on materials.

Selection of control factors and their levels with the ranges of input and fixed parameters are made on the basis of some previous work [7], review of literature, industrial testing experience, and some preliminary investigations to effectively encompass the wide variety of etching recipes. Four control factors such as gas composition ratio, flow rate, RF power and Temperature are selected for the study. Each of the four control factors is treated at three levels, as shown in Table 1. The choice of three levels has been made because the effect of these factors on the performance characteristic may vary nonlinearly.

Table 1 Operating parameters and their levels for FR4, PI and PTFE materials

Symbol	Operating parameter	Level	Input for FR4 (gas to be used: CF₄ to O₂)	Input for PI (gas to be used: CF4 to O2)	Input for PTFE (gas to be used: H₂ to N₂)	Unit
A	Gas composition proportion (by volume)	1	0.25			NA
		2	1			
		3	4			
B	Gas flow rate (total)	1	750	700	750	cc/min
		2	850	1000	850	
		3	950	1300	950	
C	RF power	1	1500	1700	1500	Watts
		2	1800	2000	1800	
		3	2100	2300	2100	
D	Temperature	1	150			°F
		2	200			
		3	250			

Design of Experiment

The experiments were designed based on the orthogonal array technique. An orthogonal array is a fractional factorial design with pair wise balancing property.

Table 2 Experimental layout using an L9 (34) orthogonal array

Experiment number	Operating parameter level			
	A	B	C	D
	Gas composition proportion	Gas flow rate	RF power	Temperature
1	1	1	1	1
2	1	2	2	2
3	1	3	3	3
4	2	1	2	3
5	2	2	3	1
6	2	3	1	2
7	3	1	3	2
8	3	2	1	3
9	3	3	2	1

Using orthogonal array design the effects of multiple process variables on the performance characteristic can be estimated simultaneously while minimizing the number of test runs. An L9 (34) standard orthogonal array [8] as shown in Table 2 was employed for the present investigation. This array

is most suitable to provide the minimum degrees of freedom as 9 [=1+4 × (3-1)] required for the experimental exploration.

Table 2 represents the layout of the experimental design, which has been obtained by assigning the selected factors and their levels to appropriate columns of L9 orthogonal array. This array has 9 rows and 4 columns and each row represents a trial condition while each column accommodates a specific process parameter. Moreover, the notation 3^4 implies that at most 4 factors, each at 3 levels can be investigated using this orthogonal array and their main effects can be estimated provided all other interactions are negligible which is assumed in the present case. The numbers in each column indicate the levels of specific factors (A, B, C, and D). However, while conducting the experiments the test runs have been randomly made to avoid the unidentified noise sources, which are not considered but could have an adverse impact on the response characteristic.

Materials, Procedure and Equipment

The materials used in the study are the major component materials building up rigid-flex PCBs, namely PI (polyimide from Dupont), PTFE (polytetrafluoroethylene from Rogers, 3003) and FR-4 (epoxy with glass fiber from Nelco, 4000-13EP). The specimen coupon dimension is 3" x 3" with a thickness of 30, 3, and 30 mils for FR4, PI, and PTFE respectively. The 5 coupons are evenly located with a configuration in the plasma chamber for one batch uniformity testing. The weight of each coupon is measured before and after plasma etching to calculate the change in weight percentage. A couple of dummy boards are also loaded to fill up the rest of the chamber space to simulate the real condition in industrial process. Each of the nine parameter settings in the L9 orthogonal array was performed five times, so a total of 45 samples were tested for etching. The quality characteristic is an average data that come from measuring five locations on the three material film samples with the same assigned parameters.

Etching took place in an industrial plasma etching system MK-II-1 from Plasma Etching Inc. Subsequent analysis was completed using the MiniTab v14.20 statistical package for conducting the statistical analysis [Mathews, 2005].

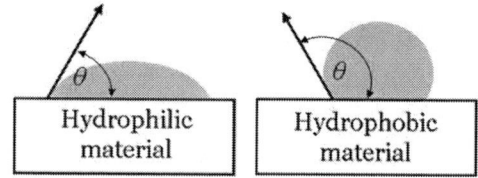

Fig. 2 Contact angle for different material types, (left) hydrophilic, and (right) hydrophobic.

The value of contact angle is the surface data of observation of polymer substrate. Some properties like hydrophilic or hydrophobic on the polymer surface can be obtained as shown in Fig. 2. Cutting the samples to standard size of 1cm×1cm after plasma treatment with different conditions, than a deionized water droplet is dropped on the sample and the value of contact angle is observed after putting the sample into contact angle measurement equipment

(Goniometer model 100). To lessen the effect of gravity, the volume of each drop was regulated to about 0.2mL by a micro syringe. The measurement was carried out at a 20°C and humidity of 45%RH. The averaged value of the angles of the both sides of each drop was counted as one measurement. Each contact angle was determined from an average of 10 measurements with a standard deviation of 1°.

Results and Discussion

The experimental results for etching rate, etching nonuniformity and contact angle are tabulated in Table 3.

Table 3 Experimental results for etching rate, etching nonuniformity and contact angle.

Experiment number	Etching rate			Uniformity			Contact angle (in degree)		
	FR4	PI	PTFE	FR4	PI	PTFE	FR4	PI	PTFE
1	0.0015967	0.01232	0.000226	14.34	14	15.72	77.43	36.62	91.23
2	0.0023476	0.01316	9.09E-05	15.56	14.5	16.54	79.64	28.77	87.33
3	0.0018129	0.02332	8.09E-05	18.36	16.2	17.36	92.68	32.14	85.20
4	0.0015376	0.02555	9.34E-05	16.72	15.7	16.89	81.33	40.66	79.63
5	0.0011724	0.04181	0.000143	17.81	16.5	18.67	84.32	48.63	76.86
6	0.0104449	0.05212	7.86E-05	12.52	9.49	13.56	75.16	46.3	75.34
7	0.0008174	0.01026	0.000191	18.67	16.5	18.61	87.54	42.12	84.72
8	0.0010146	0.00864	0.000109	13.33	11.1	14.63	76.33	36.24	82.13
9	0.0006002	0.00591	0.000082	17.49	14.9	16.21	82.16	40.52	83.64

For case 1, the importance order of the performance characteristics is etching rate (w1 = 0.5), then etching nonuniformity (w2 = 0.2), and then water contact angle (w3 = 0.3). However, the importance order of the performance characteristics for case 2 is changed to etching rate (w1 = 0.5), then etching nonuniformity (w2 = 0.3), and then water contact angle (w3 = 0.2).

Since the experimental design (Table 2) is orthogonal, it is then possible to separate out the effect of each operating parameter at different levels. For example, the mean of the multi-response S/N ratio for the gas composition proportion at level 1, 2, and 3 can be calculated by averaging the multi-response S/N ratios for the experiments 1±3, 4±6, and 7±9, respectively (Table 4).

Table 4 Multi-response signal-to-noise table for Case 1 (w1 = 0.5, w2 = 0.2, w3 = 0.3)

Symbol	Operating parameter		Mean multi-response S/N ratio (dB)				Priority of influence
			Level 1	Level 2	Level 3	Max-min	
A	Gas composition proportion	FR4	3.56	3.73	1.64	2.10	1
		PI	4.21	3.79	1.99	2.20	1
		PTFE	1.35	1.56	1.78	0.43	4
B	Gas flow rate	FR4	2.88	3.25	2.80	0.45	4
		PI	3.26	3.49	3.24	0.24	4
		PTFE	2.31	1.78	0.60	1.71	1
C	RF Power	FR4	4.19	2.59	2.15	2.05	2
		PI	3.80	2.95	3.24	0.85	3
		PTFE	1.98	0.91	1.80	1.07	2
D	Temperature	FR4	2.30	3.59	3.03	1.29	3
		PI	2.41	3.79	3.80	1.39	2
		PTFE	2.04	1.46	1.18	0.86	3

Total mean multi-response S/N ratios for FR4, PI and PTFE are 2.98, 3.33 and 1.56 respectively.

Table 5 Multi-response signal-to-noise table for Case 2 (w1 = 0.5, w2 = 0.3, w3 = 0.2)

Symbol	Operating parameter		Mean multi-response S/N ratio (dB)				Priority of influence
			Level 1	Level 2	Level 3	Max-min	
A	Gas composition proportion	FR4	3.62	3.81	1.65	2.16	2
		PI	3.80	3.99	1.95	2.04	1
		PTFE	1.49	1.50	1.81	0.31	4
B	Gas flow rate	FR4	2.88	3.30	2.90	0.41	4
		PI	3.09	3.36	3.29	0.27	4
		PTFE	2.35	1.79	0.66	1.68	1
C	RF Power	FR4	4.39	2.58	2.11	2.28	1
		PI	4.05	2.71	2.98	1.34	3
		PTFE	2.14	0.95	1.71	1.19	2
D	Temperature	FR4	2.31	3.69	3.08	1.39	3
		PI	2.34	3.84	3.56	1.50	2
		PTFE	2.07	1.49	1.24	0.83	3

Total mean multi-response S/N ratios for FR4, PI and PTFE are 3.03, 3.25 and 1.60 respectively.

The mean of the multi-response S/N ratio for each level of the other operating parameters can be computed in a similar manner. The mean of the multi-response S/N ratio for each level of the operating parameters is summarized and it is called the multi-response S/N table (Table 4 and 5). In addition, the total mean of the multi-response S/N ratio for the nine experiments is also calculated and listed in Tables 4 and 5. For case 1, the optimal operating level for FR4, PI and PTFE is A2B2C1D2, A1B2C1D3 and A3B1C2D1 respectively. As for case 2, the optimal operating level for FR4, PI and PTFE is A2B2C1D2, A2B2C1D2 and A3B1C1D1 respectively.

Apart from the above, some further testing on PTFE is performed using Scanning electron microscope (SEM) and atomic force microscope (AFM). The results substantially demonstrated that cleaning effect and surface activation had taken place during the plasma treatment as shown in Fig. 3 and Fig. 4.

Chemical method   Plasma method

Fig. 3 Hole demearing of PTFE PCB using chemical method and plasma method (SEM)

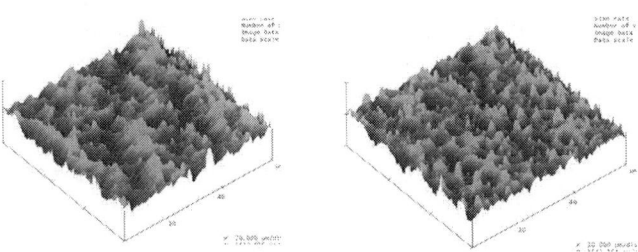

Control, no treatment   Plasma treatment

Fig. 4 Surface morphology for PTFE before and after plasma treatment (AFM)

Conclusions

We characterized various polymer films used for rigid-flex PCBs etched in different combinations of $H_2/N_2/O_2/CF_4$ plasma. The etch rate, uniformity and wettability were the etch responses to be addressed. A four factors 3-level orthogonal array experiment was designed and conducted to characterize the etch rate, uniformity, wettability to PI, FR4 and PFTE using a comprehensive set of controlling parameters. Using Design of Experiment, important factors were collected and related to etch responses. Etch behaviors were examined using analysis of S/N ratio. Competing effects of gases were illustrated with respect to etch response. Potential effects of accompanying physical phenomena were also separated and explained as a function of factors. From the qualitative analysis, several relationships of practical importance were estimated between etch responses and controlling parameters.

The empirical formulation presented herein are of great interest for a wide range of technological areas where the surface roughness is crucial such as a) ultraviolet lithography in nano-patterning, with aim to minimize surface roughness, b) formation of anti-reflection and super-hydrophobic coatings, with aim to maximize surface roughness, and c) plasma-directed spontaneously formed periodic topography fabrication, where a well-organized polymer array is expected.

Among the four process parameters, the gas composition proportion was having significant effect on the quality characteristic for FR4 and PI. On contrary, gas composition proportion has the significant effect on PTFE.

• The optimal operating level for FR4, PI and PTFE Case 1 is A2B2C1D2, A1B2C1D3 and A3B1C2D1 respectively.

• The optimal operating level for FR4, PI and PTFE Case 2 is A2B2C1D2, A2B2C1D2 and A3B1C1D1 respectively.

• Thus the predictions made by Taguchi's parameter design technique were in good agreement with the confirmation results.

• The results of the present investigation were valid within the specified range of the process parameters along with their chosen levels and for the specific combination of etching recipes.

• Additional testing results on PTFE were provided to demonstrate the cleaning effect and surface activation by plasma through SEM and AFM analysis.

Acknowledgments

The authors express their gratitude to the Linde Group for cooperation on this study. The authors would also thank the Department of Industrial and Systems Engineering of the Hong Kong Polytechnic University for providing all the testing facilities as required in this research.

References

1. Egitto, F.D., "Plasma etching and modification of organic polymers," *Pure Appl. Chem.*, 62, 1990, pp. 1699-1708.

2. Ruiz, A., Valsesia, A., Ceccone, G., Gilliland, D., Colpo, P. and Rossi, F., "Fabrication and characterization of plasma processed surfaces with tuned wettability," *Langmuir*, 23, 1990, pp. 12984-12989.

3. Paproth, A., Wolter, K.J., Herzog, T. and Zerna, T., "Influence of plasma treatment on the improvement of surface energy," *Proc. 24th International Spring Seminar Electronics Technology*, Romania, 2001, pp. 37-41.

4. Shin, D.K., Han, H.K., Lee, D.H, Song, Y.H. and Im, J., "Effect of surface conditions on interfacial adhesion between PCB and EMC," *International Conference on Electronic Materials and Packaging*, 2007, pp. 1-5.

5. Hu, Y. C., Lin, B. K., Du, Y. J., Sheen, I. H., Ding, P. W. and Tao, W. H., "Proceedings of the 4th International Symposium on electronic materials and packaging," 2002, pp. 145-149.

6. Grove D. M. and Davis T. P., Engineering quality and experimental design, (Longman Scientific and Technical, Harlow, 1997), UK.

7. Yung K. C., Wang J., Huang, S. Q., Lee. C. P. and Yue T. M., "Modeling the etching rate and uniformity of plasma-aided manufacturing using statistical experimental design," *Mater Manuf Process*, 21, 2006, pp. 899-906.

8. Montgomery DC., Design and analysis of experiments, 4th Edition, (New York, 1997): Wiley.

A Study on the Characterization of Quasi-Three-Dimensional PN Junction Capacitor

Huijuan Wang, Yao Lv, Wei Gao, Lixi Wan

Institute of Microelectronics of Chinese Academy of Sciences

B503, 3#, BEITUCHENG West, CHAOYANG District, Beijing, 100029, China

Telephone: 86-010-82995675-8006   Fax: 86-010-82995591

Email: wanghuijuansink@126.com

4681608@163.com   galaxyvenus@126.com   lixi.wan@gmail.com

Abstract

This paper discusses a novel quasi-three-dimensional PN Junction capacitor, which has a high density of capacitance ($200nF/cm^2$) with a relatively high working frequency (3GHz). These properties are because of its unique trench structure. To identify the performance of this kind of capacitor, some tests were carried out at both low frequency and high frequency. Moreover, a typical power supply filter network was designed to compare the embedded capacitor with a commercial 2nF SMD 0402 capacitor.

1. Introduction

In modern electronics packaging area, System-in-Package (SiP) is a remarkable technology with huge potential to increase packaging density [1]. As for the integration of capacitors, it is the key point of SiP because of its relatively low parasitic resistance and inductance, compared with the surface mounted device (SMD). With these advantages, the embedded capacitors could be applied in electronic systems working at much higher frequency and higher speed. Silicon is usually used as the substrate of passive devices to avoid degradation of the solder joint reliability. As a result, the capacitors fabricated on the silicon substrate are getting much more attentions than ever before. However, the structure of these kinds of capacitors still remains in Metal-Dielectric-Metal (MIM) type which has higher requirements of the dielectric layer and relative complexity process [2-4].

When PN junctions work at the reverse voltage, the leak current is puny for a large range. Meanwhile, the PN junction has a unique property, as the applied voltage change, the number of electrons and holes in the diffusion region will change too. As a result, the number of space charges in this region will change [5]. It is equal to the electric quantity charged and discharged in a capacitor. There will be only Barrier Capacitance when the voltage reverses. Therefore, making good use of Barrier Capacitance as a capacitor at reversed voltage is a promising technique for capacitor. In order to increase the capacitance of the capacitors, Micro-Electron-Mechanical System (MEMS) technology is used to form "Quasi-Three-Dimension" structure to increase the effective area.

To accurately characterize the properties of the PN junction capacitor, this paper gives the test result at the DC with the Agilent4284A LC test equipment, and at RF with the Agilent HP 8510C vector network analyzer (VNA). The S parameter from the test results will be applied to analyze the RF properties of this capacitor. Furthermore, this paper will introduce a set of low pass filters connected to the PN junction capacitor by wire-bonding to form a passive circuit. For comparison, an SMD capacitor with similar capacitance

was studied to replace the PN junction capacitor in the passive circuit. This comparison gives a method to verify the use condition at high frequency.

In order to illustrate and investigate the theory, structure and use condition of the PN junction capacitor in detail, this paper is formed of three parts. The first part gives an introduction of the basic scheme and the unique physical structure of the capacitor. The second part presents the testing results under the DC and high frequency. The last part supplies an application to compare the semiconductor capacitor to the commercial SMD component.

2. Physical structure

For a traditional PN junction capacitor, the reverse capacitor is the same as the barrier capacitor, the capacitor density can expressed as below [5].

$$C' = \left\{ \frac{q\varepsilon_r\varepsilon_0 N_a N_d}{2(V_{bi}+V_R)(N_a+N_d)} \right\}^{1/2} \quad (1)$$

In the equation above, q is electronic charge, ε_r is the relative dielectric constant, ε_r=11.7. Er is the vacuum dielectric constant. N_a and N_d are the density of accepter doping and donor doping respectively. Uniformly doped p-type semiconductor substrate is used to study the solid solubility in silicon. The resistivity of doped substrate is between 0.03 and 0.04. The doping concentration is 2.5×10^{18} atoms/cm^3. For the n-type substrate, p-n junction's capacitance density can reach $150nF/cm^2$ when doping concentration is high. It's equivalent to or higher than ordinary materials and embedded capacitor made by current manufactory process. When the doping concentration of N area reach the magnitude of $\times10^{19}$, the capacitance density can be higher than $250nF/cm^2$. However, as the doping concentration rises further, the change rate of capacitance density will slow down. This phenomenon suggests that the depletion layer capacitance is a function of the doping concentration of the low-doped zone. In other words, raising the doping concentration of the low-doped zone can effectively increase the density of the PN junction capacitance [5].

$$C = \frac{\varepsilon_r\varepsilon_0 A}{X_D} \quad (2)$$

Equation (2) shows the dependents of the parallel-plate capacitor. Comparing the Equation (2) to the Equation (1), it can be easily find that the PN junction capacitor formula has similar expression to the parallel-plate capacitor. Based on the formula, capacitance density can be improved in three ways: 1) finding a material with larger ε_r, 2) reducing the distance between the plates (In p-n junction, the distance is the width

978-1-4244-4658-2/09 $25.00 © 2009 IEEE

of space charge region), 3) increasing the effective area of parallel-plate. In this paper, the P-N junction is made on silicon. The width of depletion region is depend on the doping concentration when ε_r is a constant. To enhance capacitance, increasing the effective area of PN junction is one of the most effective ways [2].

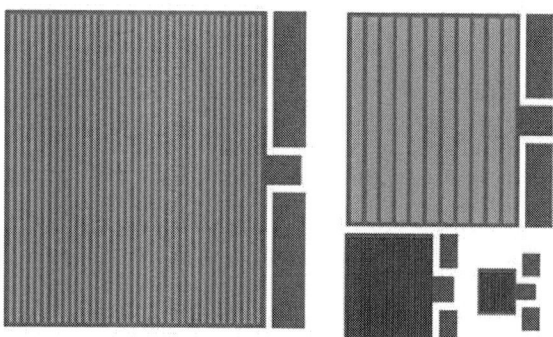

Fig. 1 The layout of the trench capacitors

Fig. 1 shows the layout for fabricating the capacitors. To compare the test results, four capacitors with different surfaces and channel width were designed. The capacitors have different effective areas. As capacitance is closely related to the effective surface area, the capacitance is different. Fig. 2 shows the cross-section of the physical structure of the capacitor, which also illustrates the basic processes to fabricate the capacitor.

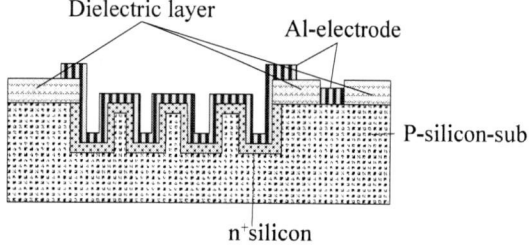

Fig. 2 Cross-section of the capacitor

As shown in Figures 1-2, three-dimensional shape of silicon etching can be carried out by using micro-processing technology. Based on that, the effective area of the two flat-panels can be increased greatly in the two-dimensional surface. The trench-like and porous surface area increase most significantly. High-density capacitor makes a significant improvement for meeting the requirements of current high density packaging which strictly restricts area and volume.

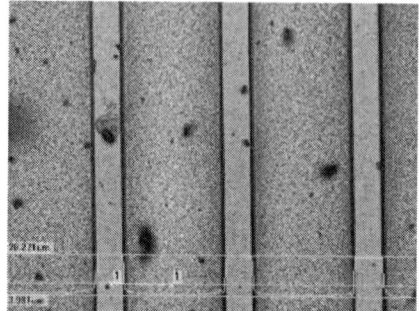

Fig. 3(a) Top view under optical microscope

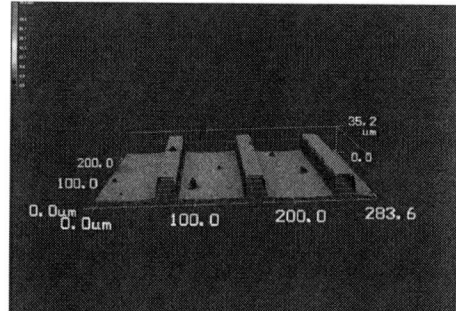

Fig. 3(b) Capacitor under optical microscope

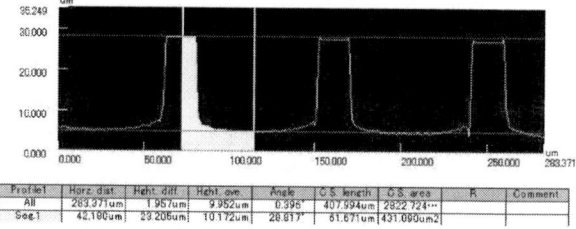

Fig. 3(c) Basic length under optical microscope

In order to obtain a high quality vertical channel, inductively coupled etching (ICP) is used. In this investigation, the luminance power and etching gas composition are strictly controlled. The channel depth is about 25μm. Fig. 3 (a)-(c) show the graphics of the capacitor in optical microscope. The capacitor observed have an effective surface area of 1500×1500μm², a bottom channel width of 65μm and a height of 25μm. For the finished capacitor as shown above, the trench depth is 23μm and the channel width is 67μm. The data measured above is consistent with expected length, which accounts for the accuracy of the process.

3. Measurement and Analysis

The trench PN junction capacitor based on silicon has both the characteristic of the PN junction semiconductor and the SMD capacitor. DC and AC characteristics are the two aspects investigated during the test. For DC testing, reverse breakdown voltage and leakage current are tested by Agilent HP 4155A Semiconductor Parametric Tester. Low-frequency capacitance measurement is performed by using 4284A LCR testing device. For RF testing, Agilent HP 8510c Vector Network Analyzer is used to test the device's S parameter, which is used for modeling and simulation.

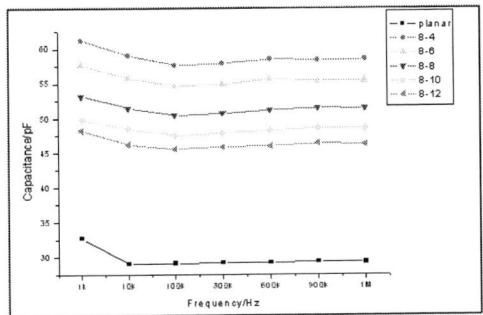

Fig. 4 Capacitance measured under changed frequency

Considering the capacitance of the capacitor under low frequency, this work measured the curve of capacitance by 4284A LCR. The test condition is a reverse voltage of 1V. The capacitor was tested with a surface area of $200\times200\mu m^2$. As shown in Fig. 4, the capacitance increases as the density of the groove becomes higher which enlarged surface area. Therefore, the capacitance is the smallest when there is no groove on the surface of the capacitor, as indicated by the black curve in Fig. 4. It is also indicated that the capacitance will remain nearly unchanged with the frequency changing from 1kHz to 1MHz.

Fig. 5 shows the high-frequency capacitance equivalent model of plate-capacitor which considered the parasitic parameters [6].

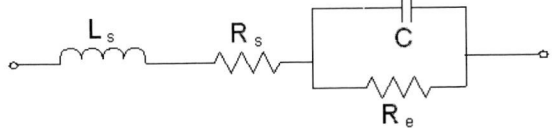

Fig. 5 Equivalent model of plate-capacitor under high-frequency

In the equivalent circuit, L_s is the parasitic inductor on the wire. R_s is the series resistor of the loss on the contact wire. R_e is the dielectric loss resistor. In this investigation, the capacitor is a PN diode. The parallel pattern of capacitor and resistor can also be used in this high-frequency equivalent model in Fig. 5. L is the lead inductance and R_s is the parasitic series resistance. The L_s is decreased for the design of the electrode capacitance, while the large area of metal is deposited on the surface of PN junction. This inducts to a distribution inductor L_d. Fig. 6 show a high-frequency equivalent circuit after L_d is inducted. Therefore, we can get the equivalent model of PN junction capacitor under high-frequency.

Using the equivalent model in Fig. 6, we can get the S parameter by ADS software. Meanwhile, to investigate the characteristic of the capacitor under high frequency, a two-port VNA and microprobes were used. We can also get the S parameter from the testing. Simulation results and measurement results were compared to draw the corresponding parameters. Fig. 7 shows the establishment of the model.

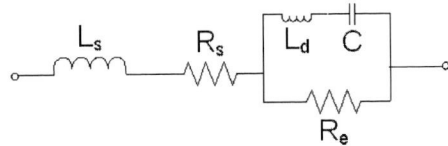

Fig. 6 Equivalent model of PN junction capacitor under high-frequency

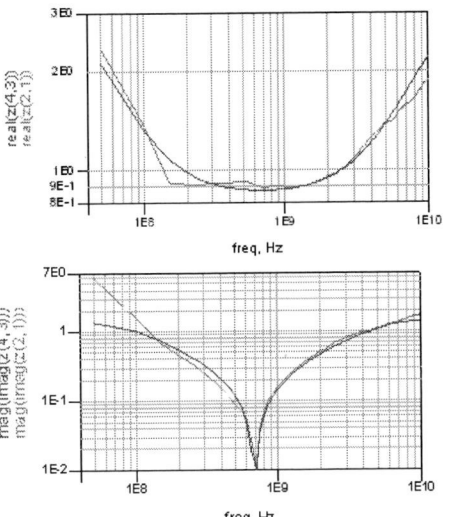

Fig. 7 Comparison between simulation (red curve) and testing (blue curve)

The red curve is the S21 came from the test result, and the blue curve is the simulation result came from the equivalent model as show in the Fig. 7. It can be seen that the two curves matches well.

4. Application example

In order to investigate the use condition of the capacitor, this study inducts a passive circuit to validate the usage of the PN junction capacitor. In this experiment, an embedded capacitor with two or more connections on an electrode plane is used as a two or multi-ports low pass filter network. As an example, an embedded capacitor with two connections, which could be a filter with two ports, one serves as an input, the other as an output, the same as the conventional low pass filter network composed of passive components. As seen in Fig. 8, the low pass filter can be called as Embedded Capacitor Filter (ECF) [7-8].

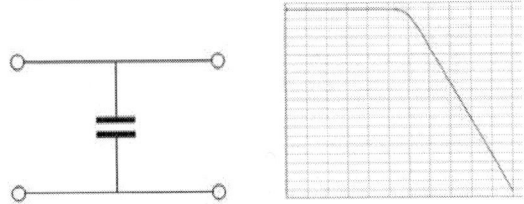

Fig. 8 ECF (Embedded Capacitor Filter)

978-1-4244-4658-2/09 $25.00 © 2009 IEEE

The ECF is used as a substrate and the incised capacitors are glued to both sides of the filter. The PN junction is connected to the lower and upper electrode respectively. As show in Fig. 9, an ECF is welded with ceramic capacitor as shown in the left part in the picture, and a PN junction capacitor is connected to another identical ECF as shown in the right part of the picture.

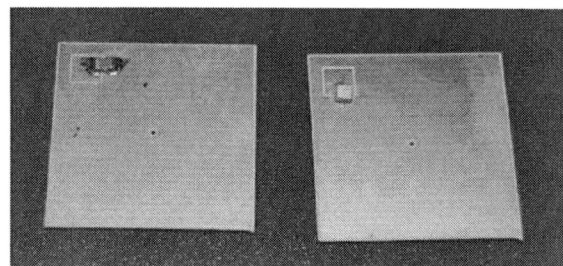

Fig. 9 ECF with the two kinds of capacitor

For the actual ECF without any connection with other components, we can get the Z parameter as show in Fig. 10(a). Fig. 10(b) shows the results when the ECF connected with the PN junction capacitor and the SMD.

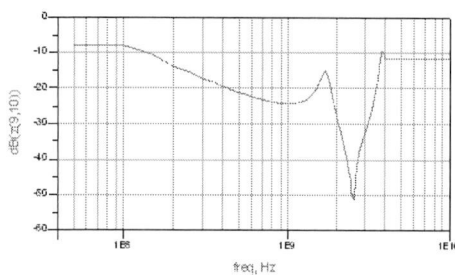

Fig. 10(a) the Z parameter about the used ECF

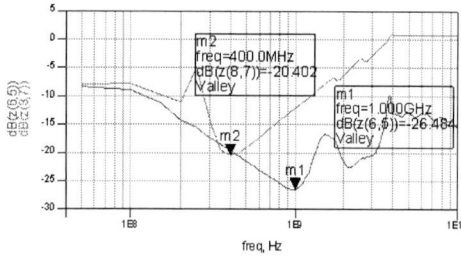

Fig. 10 (b) comparison between SMD capacitor (red curve) and the PN junction capacitor (blue curve)

The red curve is the Z21 came from the SMD connected to the ECF, and the blue curve is the Z21 came from the PN junction capacitor connected to the ECF. The comparison between SMD capacitor and the PN junction capacitor indicates that the PN junction capacitor have a better performance in the high frequency.

Conclusions

In this paper, a unique quasi-three-dimensional PN Junction capacitor is studied. An overview about its physical structure and design scheme is given. The parameters about a finished capacitor are illustrated. An equivalent passive circuit

is applied to replace the PN junction under high frequency. An embedded capacitor filter is inducted to give a comparison between SMD capacitor and the PN junction capacitor. It can be conclude that the PN junction capacitor has a good performance under high frequency.

Acknowledgement

This work was supported by Hi-tech Research and Development Program of China (863 Program) No. 2007AA01Z2a6.

References

1. Roozeboom, F. *et al.*, "System-in-Package Integration of Passives Using 3D Through-Silicon Vias," *Solid State Technology*, [EB/OL] http://www.solid-state.com.05/01/2008.
2. Yao, L. *et al.*, "The Capacitor Applied the PN Junction," *Electronic component and matreial*, 2009.
3. Philippe, P. and A. Oruk, "A Highly Miniaturized 2.4GHz Bluetooth Radio Utilizing an Advanced System-in-Package Technology," *European Microwave Week.*, Amsterdam, October 2004.
4. Leonard, W. Schaper and Chris Thomason, "High Density Tantalum Pent oxide Decoupling Capacitors," *Electronic Components and Technology Conference*, 2006, pp. 510-514.
5. Donald A. neamen, <u>Semiconductor Physics and Devices Basic Principles</u>, Electronic Industry Press 2005.
6. F. Roozeboom, W. Dekkers, K. Jinesh *et al.*, "Ultrahigh-Density (>0.4μF/mm^2) Trench Capacitors in Silicon," *First Int. Workshop on Power Supply On Chip (PowerSoC08)*, Sept. pp. 22-24, 2008, Cork, Ireland.
7. Wei Gao, *ect.*, "A Novel Lowpass Using Embedded Capacitor Technology for High Density Packaging with a Large Bandwidth in Multi-GHz PCBs, " *ECTC*, 2007.
8. Lixi Wan, *etc.*, "Design, Simulation and Measurement Techniques for Embedded Decoupling Capacitors in Multi-GHz Packages/PCBs," *ECTC*, 2005.

Effect of PIII on the Adhesion Behavior of Epoxy Molding Compound-Nickel Interface

Lilong Liu, Qian Lu, Yipeng Wang, Weifeng Dai, Xin Zhang, Yuesheng Li, Xiaojing Wu*

Department of Materials Science, Fudan University

wuxj@fudan.edu.cn

Abstract

The Cu leadframes pre-plated with nickel-coating were implanted with oxygen ion or nitrogen ion by Plasma Immersion Ion Implantation (PIII) technique. The contact angle measurement showed that ion-implantation lead to an obvious change of wettability. The chemical situation of the leadframe surfaces was analyzed by X-ray Photoelectron Spectroscopy (XPS), while the morphology of the leadframes was observed by scanning electron microscope (SEM). A pull-test was employed to examine the influences of ion-implantation on adhesion between the leadframe surface and molding compound.

1. Introduction

As electronic devices were further developed, interfacial stability and adhesion had become very important factors for these devices in determining their durability. Delaminations at various interfaces are one of the most critical reliability issues in plastic packages. Delaminations, especially those occurring between leadframe and molding compound in plastic packages, often lead to popcorning during/after the solder reflow process due to the presence of absorbed moisture within the plastic encapsulant, resulting in the cracking of the whole package [1]. Copper alloys have been the most widely used as leadframe material because of their high electrical and thermal conductivity and relatively low cost. However, legislation to restrict the use of lead in consumer electronics has brought unanticipated negative side effects despite best effects in predictive testing, molding, and planning [2-3]. Consequently, in recent years significant research efforts have been directed toward improved performance of interface adhesion. Thermal treatment and chemical treatment process may result in micro-roughening of the copper surfaces. At the same time, such treatments may form robust film on the leadframe surface that enhance mechanical interlocking and chemical bond [2, 4-7]. It is widely known that thin oxide films on copper leadframe surface can increase the adhesion significantly. Another promising approach is by altering the molding compound formulations for low moisture solubility, and high strength and low stress at high temperatures.

On the other hand, plating techniques have been also widely employed to improve the interface adhesion between molding compound and copper alloy leadframe [8]. The leadframe plating material should adhere well to the base material, as well as also minimize corrosion and enhance solderability. So far, several plating materials have been widely used in the industry, including Ni, Ag, Au, Pd coatings. However, in many cases such coating films may result in a weak bonding between the molding compound and the plating materials on leadframe [8]. Therefore, how could we find a new technique, which can enhance the adhesion remarkably, becomes an important issue?

In this paper, the surface treatment of Ni/Cu leadframes by PIII technique with oxygen ion or nitrogen ion was studied. The surface situation was examined by contact angle measurements, XPS, and SEM observation. The influence of such ion-implantation treatment on adhesion was measured by a pull test.

2. Experiments

The Ni/Cu leadframes are commercial goods. On the Cu-based leadframe, Ni thin film with a thickness of about 1 □m was coated by chemical method. The PIII technique was employed to implant oxygen ions or nitrogen ions into the leadframes at a 5kV with different dose. The implantation conditions for the samples were listed in Table1.

Table 1 S_0 was the one without treatment. SO: the samples with oxygen ion implantation. SN: the samples with nitrogen ion implantation. The implantation dose was calculated by Child- law [9].

Sample	Time (min.)	Voltage (kV)	Dose (cm^{-2})
S_0	0	0	0
SO_1	15	5	4.71E+15
SO_2	30	5	9.43E+15
SO_3	60	5	1.89E+16
SO_4	120	5	3.77E+16
SN_1	15	5	4.27E+15
SN_2	30	5	8.53E+15
SN_3	60	5	1.71E+16
SN_4	120	5	3.41E+16

Contact angles were measured by contact angle tester (OCA15, Dataphysics) at room temperature. All droplets were pure water with a volume of 2μL.

XPS measurements were carried out on a RBD upgraded PHI-5000C ESCA system with Mg Kα radiation (hν=1253.6 eV). The pass energy was fixed at 23.5eV to ensure sufficient resolution and sensitivity. Binding energies were calibrated by using the containment carbon (C1s =284.6eV). XPSPEAK was used to separate peaks.

The morphology of the samples was observed by a field emission SEM under an accelerating voltage of 5kV.

The leadframes were encapsulated with a molding compound using a transfer molding machine at 175°C. The specimens were post mould cured (PMC) at 175°C for 6 hours in an air circulated oven. The specimens were subsequently soaking in 60°C/60% relative humidity (RH) for 40 hours. The pull test was carried out under both of PMC and moisture

978-1-4244-4658-2/09 $25.00 © 2009 IEEE

sensitive level (MSL) conditions at the room temperature with a cross-head speed of 2.54mm/min.

3 Results and Discussions

The influence of implanting time on surface contract angle was shown in figure1. It is known that the lower water contact angle means better surface wettability to polar liquid [10]. As shown in Figure 1, the water contract angle dropped off sharply at an early stage of ion implantation and then rose gradually with the increase of implanting time.

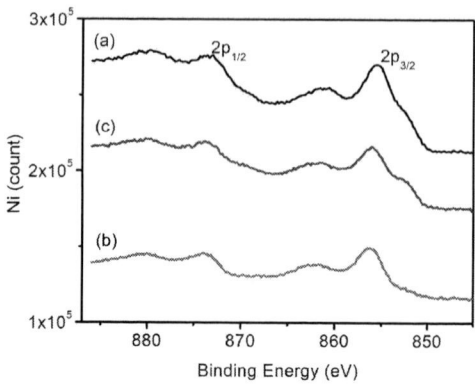

Figure 2 Ni-2p spectra of the samples (a) S_0, (b) SO_3 and (c) SN_3.

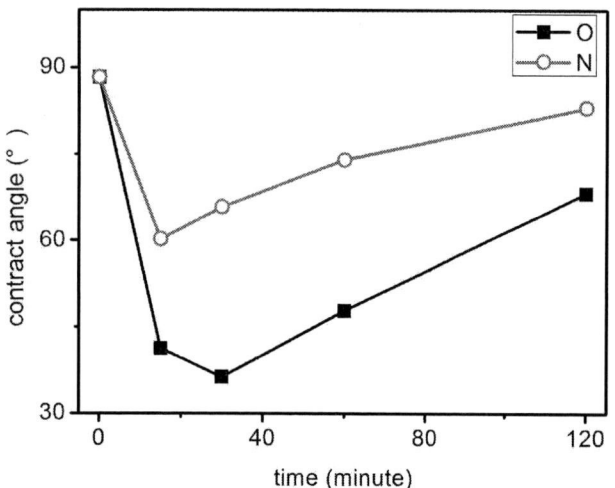

Figure 1 contract angles of water as a function of the plasma ion implantation time.

XPS analysis was conducted to obtain chemical state information on the leadframe surfaces. It was found from Figure 2 that the Ni 2p$_{1/2}$ peaks became lower in samples SO_3 and SN_3 due to much less hydroxide radical(-OH) combined with Nickel (II), indicating that the treated leadfreams absorbed less moisture.

The peaks of Ni-2p spectra from the samples S_0, SO_3 and SN_3 could be separated by XPSPEAK, and the results were shown in Figure 3. The nickel metal peak at 852.5eV completely disappeared in Figure 3 (b), indicating that the surface of SO_3 was totally coated by nickel oxide. On the other hand, on the surfaces of S_0 and SN_3, both Ni metal and nickel oxide peaks could be found [13]. These results implied that the implanted oxygen ions were easily bonded with nickel.

The peak at 856.1eV in Figure 3 (b) was come from Ni (III). Oxygen ions bonding with nickel had changed the chemical state from Ni (II) to Ni (III). The peak appeared at the position of 855.4eV in Figure 3 (a) and (c) was due to NiO [13].

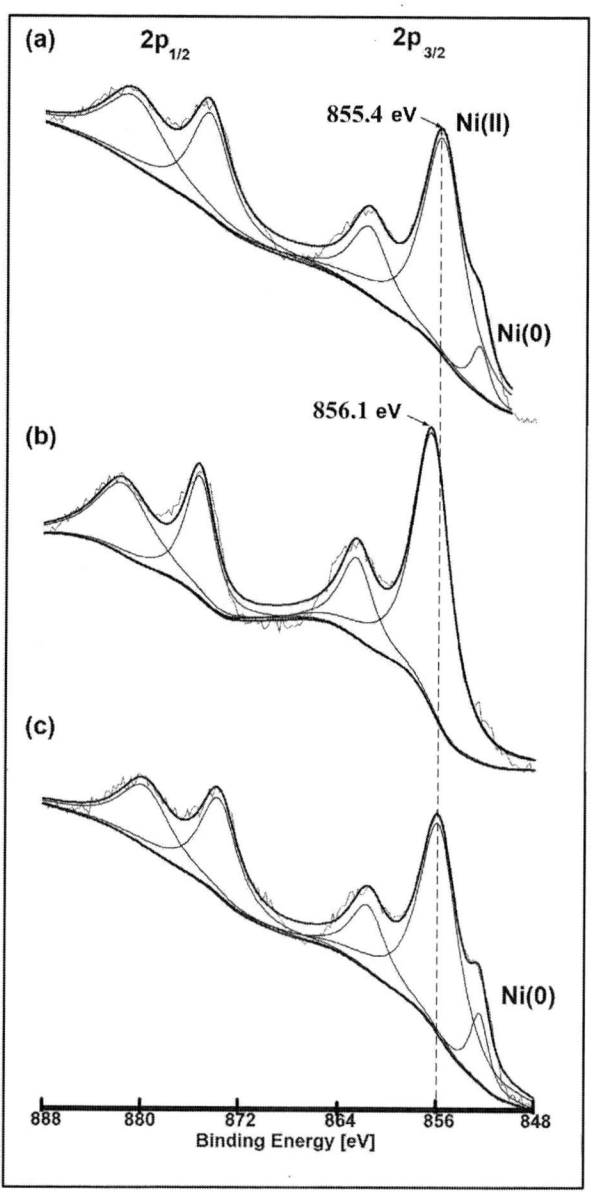

Figure 3 Ni-2p spectra of the sample (a) S_0, (b) SO_3 and (c) SN_3. A Shirley background under the entire Ni 2p spectrum was used for all spectra.

Figure 4 (a) and 4 (b) show the morphology of S_0 and SO_3 observed by SEM. Comparing Figure 4 (b) with 4 (a), it was found that the surface roughness decreased obviously. From Figure 4(a), it was seen that the surface of S_0 consisted of small particles with a size of about several hundred nanometers, while in Figure 4 (b), the surface of SO_3 was rather flat. Such morphology variation could be owed to the etching effect during ion-implantation.

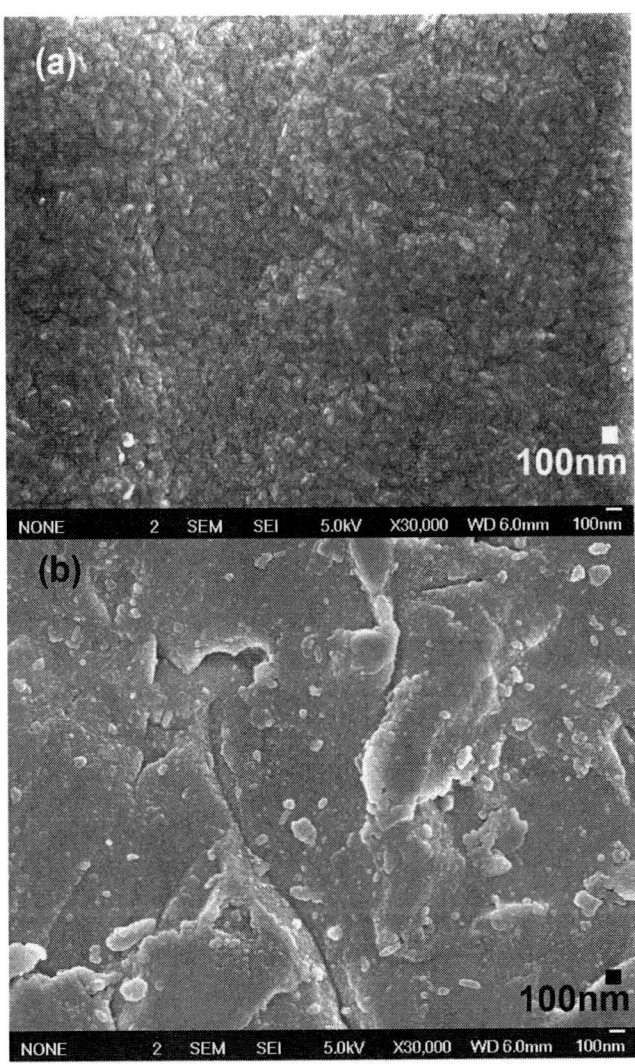

Figure 4 the morphology of leadframes: (a) S0, (b) SO_3.

The pull test results for the $S_0 \sim SO_x$ and $S_0 \sim SN_x$ (x=1~4) series were plotted in the Figures 5 and 6, respectively. It was found from figure 5 that the adhesion increased significantly in the MSL pull test after 1 hour oxygen ion implantation, but no improvement at all for other oxygen ion implanted samples. For the samples with nitrogen ion implantation, the adhesion almost kept as a constant for either the PMC or MSL3 test.

Three factors were involved in changing the contact angle. First of all, plasma cleaning may remove the surface contaminants [11-12], resulting in a change of surface tension. At the early stage of PIII, plasma cleaning effect was dominated, and then the contact angle dropped rapidly as shown in Figure 1. Secondly, the surface roughness is a factor

to determine contact angle, too. It was well-known that as the surface roughness decreasing, the apparent contact angle should increase when the intrinsic contract angle is less than 90°. As the ion implantation time increasing, the surface roughness decreased obviously as shown in Figure 4, resulting in a gradually increase of the contact angle as observed in Figure 1. Finally, the chemical composition change may also influence on the wettability. From XPS analysis, Ni(III) appeared in the samples with oxygen ion implantation, being remarkably different from the samples with nitrogen ion implantation. From figure 1, it was found that the contact angle of SO_x was lower than SN_x (x = 1~4) as much as about 20°.

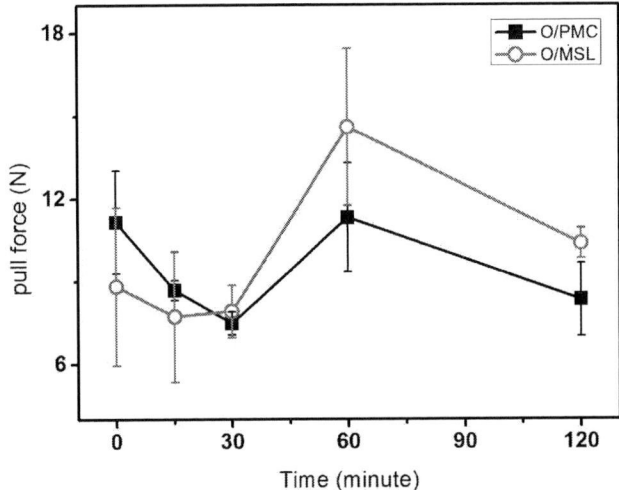

Figure 5 the pull force between molding compound and oxygen ions implanted leadframes at different times in PMC and MSL styles.

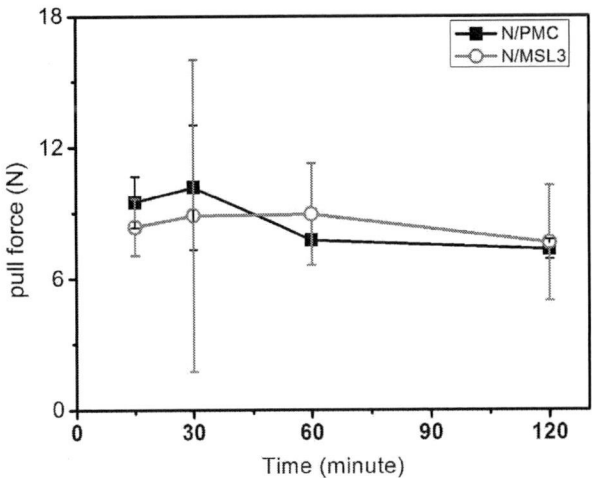

Figure 6 the pull force between molding compound and nitrogen ions implanted leadframes at different time in PMC and MSL styles.

It is well-known that there are two major factors which may give big influences on adhesion when the moulding compound was selected. The first one is the surface chemical situation of the leadframe, and the second one is the surface roughness of the leadframe. Generally, a rough surface will be in favor for enhancing adhesion due to a larger surface area.

In the present samples, the contract angle and XPS measurement showed an evident variation on the surface chemical situation after ion implantation, while the SEM observation indicated that the surface became much smooth by ion-etching effect. The adhesion of the leadframes will be determined by total contribution from both surface morphology and chemical situation. Due to a smooth surface in SO_3 as shown in Figure 4 (b) acted as a negative factor for enhancing adhesion, the increase of pull force as shown in figure 5 for SO_3 must be mainly from the contribution of surface chemical situation improvement. In Figure 3 (b), Ni (III) was detected in the sample SO_3 and it may have good effect on adhesion. The improvement of adhesion in MSL tests indicated that oxygen ion implantation not only enhanced the bond between the molding compound and leadframes, but also prevented moisture ingress along interfacial areas. In other words, oxygen ion implantation seems a good method to improve the adhesion between the leadframe and moulding compound. Meanwhile, the nitrogen ion implantation did not give any contribution in increasing the adhesion strength. It means that comparing with nickel nitride the nickel oxide has much more strong bonding with molding compound.

4. Conclusions

1) The increment in the surface wettability at the early stage of ion implantation may mainly come from the plasma cleaning effect, while as the ion implantation time increasing, the wettability decreased, ie., the contact angle increased, with the roughness decreased.

2) Ni (III) was detected at the leadframe surface in the sample SO_3. It may play a key factor to enhance the bonding strength between the molding compound and leadframe.

3) Oxygen ion implantation has a notable influence on the adhesion between the leadframe and the epoxy molding compound, while the influence from nitrogen ion implantation is tiny.

References

1. Nguyen, L. T., "Reliability of postmolded IC packages," *ASME J. Electronic Packaging*, 115 (1993), pp. 346-355.
2. Dan Hart, Bruce Lee, "Increasing IC Leadframe Package Reliability," *10th Electronics packaging Technology Conference*, (2008), pp. 1209-1213.
3. Gene Kim, James Hurley, "Improving Mold Compound Adhesion to Ni/Pd/Au Pre-plated Lead Frames," *Electronic Components and Technology Conference*, (2006), pp. 1436-1441.
4. HO-YOUNG LEE, JIANMIN QU, "Microstructure, adhesion strength and failure path at a polymer /roughened metal interface," *J. Adhesion Sci. Technol.*, (2003), Vol. 17, No. 2, pp. 195-215.
5. H. Y. Lee, Jin Yu, "Effects of oxidation treatments on the fracture toughness of leadframe:epoxy interfaces," *Materials Science and Engineering*, A277, (2000), pp. 154-160.
6. Narasimalu Srikanth, Lewis Chan, Charles J. Vath III, "Adhesion improvement of EMC–leadframe interface

using brown oxide promoters," *J. Thin Solid Films*, 504 (2006), pp. 397-400.
7. Paular W.K.CHUNG, "Effect of Copper Oxide on the Adhesion Behavior of Epoxy Molding Compound-Copper Interface," *Electronic Components and Technology Conference*, (2002), pp. 1665-1670.
8. Jang-Kyo KIM, Mohamed LEBBAI, "Interface Adhesion Between Copper Lead frame and Epoxy Moulding Compound: Effects of Surface Finish, Oxidation and Dimples" *Electronic Components and Technology Conference*, (2000), pp. 601-608.
9. R. A. Stewart, M. A. Lieberman, "Model of plasma immersion ilon implantation for voltage pulses with finite rise and fall times," *J. Appl. Phys.*, 70 (7), 1 October 1991, pp. 3481-3487.
10. Soon-Jin Cho, "The Effect of the Oxidation of Cu–Base Leadframe on the Interface Adhesion Between Cu Metal and Epoxy Molding Compound," *IEEE Transactions On Components, Packaging And Manufacturing Technology-Part B*, Vol. 20, NO. 2, May, 1997, pp. 167-175.
11. Sung Yi, "Bonding strengths at plastic encapsulant-gold-plated copper leadframe interface," *J. Microelectronics Reliability*, 40 (2000), pp. 1207-1214.
12. SUNG YI, "Effect of oxidation and plasma cleaning on adhesion strength," *J. Adhesion Sci. Technol.*, 13 (1999), pp. 789-804.
13. Mark C. Biesinger, "X-ray photoelectron spectroscopic chemical state quantification of mixed nickelmetal, oxide and hydroxide systems," *J. Surf. Interface Anal.*, 2009, 41, pp. 324-332.

Improving Board Assembly Yield through PBGA Warpage Reduction

Li Li, Ken Hubbard and Jie Xue
Cisco Systems Inc.
San Jose, CA 94135, USA
lili2@cisco.com

Abstract

Since its introduction in the early 1990s, plastic ball grid array (PBGA) package had become the "package of choice" due to its good electrical performance, lower cost, high assembly yield and self-alignment during board assembly process. Thermo-mechanical behavior of PBGA is highly dependent on the properties of the constituent components. The relative mechanical compliances and thermal expansion mismatch between the silicon chip, the mold compound material and the organic laminate substrate are particularly important to the design and performance the package. Strong coupling between the chip and the packaging materials can cause thermal deformation and deviation from an ideal state of uniform planar flatness, i.e., package warpage. If it is not well controlled, the temperature dependent package warpage can result in open or bridge BGA solder connections when mounting the device to a printed circuit board (PCB) using the surface mount (SMT) solder reflow process. The problem can be more severe as we migrate to lead free SMT soldering process.

In this study attention has been focused on improving PBGA SMT process yield through package warpage reduction. Combined experimental and modeling methods were used to investigate the thermo-mechanical behavior and the mechanisms controlling PBGA package warpages through reflow temperatures. Materials effect of mold compound and die encapsulation was first studied for minimizing the chip-package thermo-mechanical coupling over temperatures. Packaging process factors such as encapsulation curing time and temperatures were also investigated. Fully assembled PBGA packages with two different mold compound materials were evaluated. Thermo-mechanical response of the package was measured and analyzed using thermal shadow moiré and numerical modeling technique. The experiments and modeling were correlated with a well controlled manufacturing build with over 10,000 boards built. The combined experimental and numerical analysis confirmed our selection of the packaging materials and demonstrated that significantly improved board assembly yield can be achieved by controlling the PBGA warpage during board mount assembly process. It is also concluded the importance of a package warpage and the shape of the warpage at not only room temperature but also throughout reflow temperatures.

Introduction

PBGA is an extension of ceramic BGA (CBGA) that is built with organic laminate substrates as chip carriers. The distinct difference between CBGA and PBGA is that organic chip carrier has higher coefficient of thermal expansion (CTE) and more flexible than the ceramic substrate in CBGA.

Thermo-mechanical behavior of PBGA is highly dependent on the properties of the constituent components.

The relative mechanical compliances and thermal expansion mismatch between the silicon chip, the mold compound material and the organic laminate substrate are particularly important to the design and performance the package. Strong coupling between the chip and the packaging materials can cause thermal deformation and deviation from an ideal state of uniform planar flatness, i.e., package warpage. If it is not well controlled, the temperature dependent package warpage can result in open or bridge BGA solder connections when mounting the device to a PCB using the SMT solder reflow process. The problem can be more severe as we migrate to lead free SMT soldering process.

One of the failure modes for the BGA joints formed is the so called "pillow joint" as shown in Figure 1 & Figure 2. It is very difficult to catch the "pillow joint" fully during routine inspection and tests on the factory floor. It usually requires more advanced X-ray inspection tools such as the Agilent 5DX 3D Automated X-ray Inspection which is very time consuming and costly.

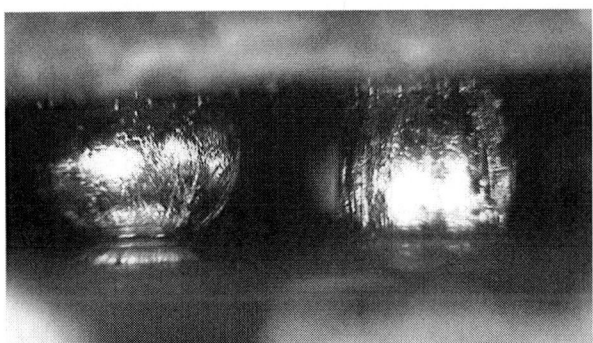

Figure 1 Optical picture of a "pillow joint"

Figure 2 Cross-section view of a "pillow joint"

For most JEDEC standard PBGA packages, the coplanarity is specified in the package outline drawing per JEDEC Publication 95, Design Guide 4.14 for Ball Grid Array Package [1]. During PBGA manufacturing, the package coplanarity is measured as one of the inspection steps at room temperature following JEDEC standard, JESD22-B108A [2].

However, the PBGA warpage is temperature dependent. It is important to characterize the warpages of PBGA packages and printed circuit boards under the simulated reflow conditions. In 2005, another JEDEC standard, JESD22-B112, was introduced [3]. A test method and procedure based on the shadow moiré method was outlined in JESD22-B112 for characterizing the package warpage at elevated temperatures.

Shadow moiré method is an optical technique of measuring topography and out-of-plane displacement on a specimen surface [4]. By using reference gratings with different frequencies, a range of measurement sensitivities can be achieved. Shadow moiré technique has been used successfully to measure warpages of various electronic components. The optical setup of shadow moiré is shown schematically in Figure 3. Commercial shadow moiré systems are now made available with both automatic data acquisition and temperature controls.

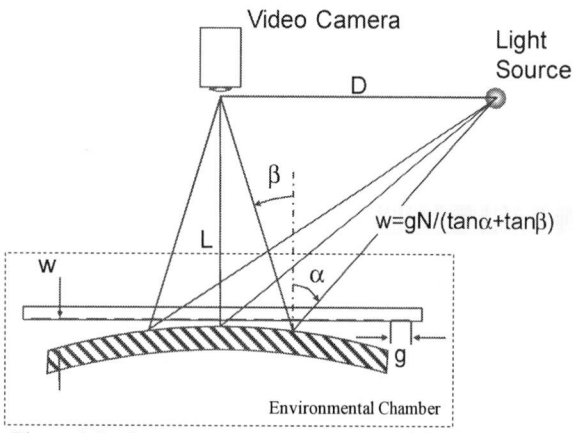

Figure 3 Package warpage measurement with Shadow Moiré method

For simplicity, sign convention for the package warped directions with respect to the BGA seating plane is given in Figure 4 [2-3]. Since most of the PBGA packages are symmetric, they follow this convention well.

Convex: "+" warpage   Concave: "-" warpage

Figure 4 Package warpage convention

Recent international standardization activities have resulted in the establishment of a maximum permissible package warpage at elevated temperatures as part of package qualification for BGA packages with different ball pitch and diameter [5]. Also, JEDEC recently included SPP-024, Issue A as part of JEP 95 [6] and provided the procedures for using component land side flatness during simulated reflow (warpage at elevated temperatures) when:

- The component does not meet the corresponding, registration, standard, or Design Guide coplanarity requirements.

- The component meets the corresponding Design Guide coplanarity requirements, but SMT solder reflow losses exceed customer expectations, attributable to package warpage (solder ball bridging, or non-wet opens).

There are many factors that can affect the formation of good solder joints. These include package warpage, PCB warpage, and SMT process parameters. The rest of the paper will focus on reducing package warpage in order to improve PBGA board assembly yield.

Package Warpage Reduction

The package included in this study is a 35mm×35 mm, 1.0mm pitch, 748 pin over-molded PBGA package. A top view of the package is shown in Figure 5.

Figure 5 Top view of the 35mm×35mm PBGA.

Shadow moiré technique was used to characterize the package warpage during a simulated solder reflow process. The contour maps of the package warpage are shown in Figure 6 to 9 for various temperatures.

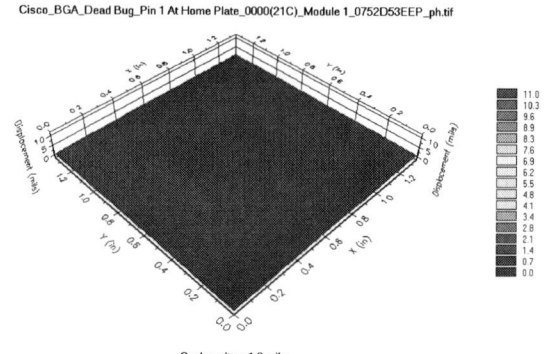

Figure 6 Package warpage measured at 21°C

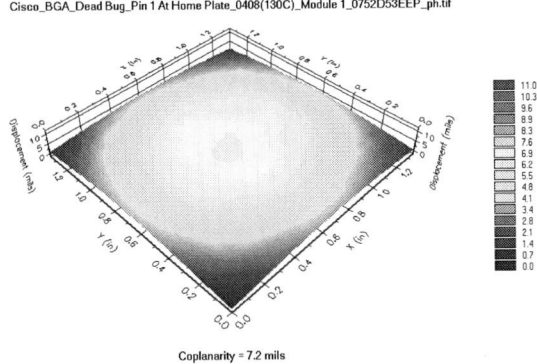

Coplanarity = 7.2 mils

Figure 7 Package warpage measured at 130°C during heating up

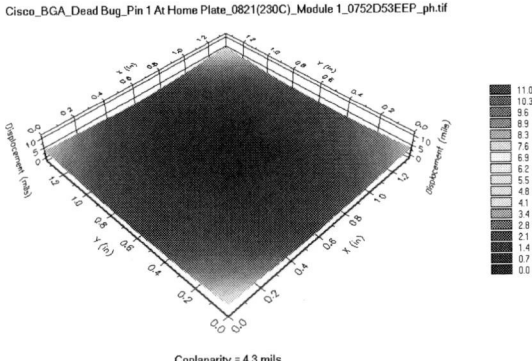

Coplanarity = 4.3 mils

Figure 8 Package warpage measured at 230°C

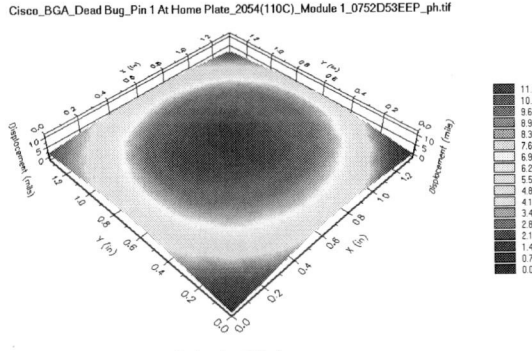

Coplanarity = 11.0 mils

Figure 9 Package warpage measured at 110°C during cool down

The maximum package warpages measured during the simulated reflow process are shown in Figure 11 for a total of six samples with mixed production date codes and lot codes. As we can see from the charts in Figure 11, the maximum package warpage measured at elevated temperature is not only way above the package coplanarity limit specified at 0.2 mm but also exceeds the limits specified in JEITA ED-7306 and in JEDEC JEP 15, SPP-024, Issue A.

The large package warpage measured during reflow makes this PBGA component very sensitive to the board assembly process and can introduce defects such as "pillow" solder joints causing electrical opens.

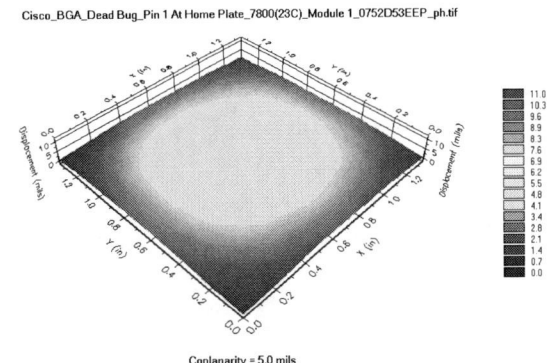

Coplanarity = 5.0 mils

Figure 10 Package warpage measured at 23°C

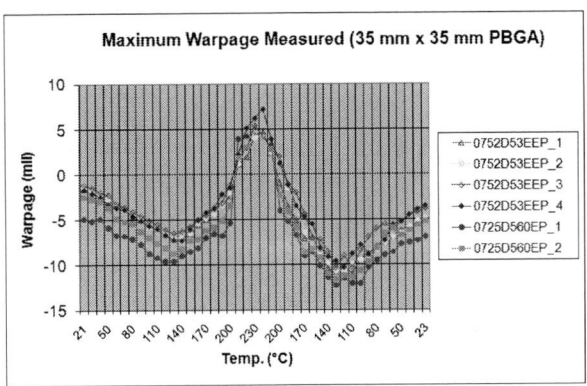

Figure 11 Maximum package warpage measured during the simulated reflow process

Finite Element Modeling

Packaging material change has been considered as the solution for controlling package warpage which is due to the thermal expansion mismatch among the packaging materials and the silicon die of this 35mm PBGA.

A finite element model was used to aid in the package material and process parameter selection. Figure 12 shows the finite element model developed for the 35mm PBGA package. The predicted package out-of-plane displacement due to a temperature excursion between 25°C and 140°C is shown in Figure 13.

Materials effect of mold compound and die encapsulation was studied for minimizing the package warpage over temperatures. The results are shown in Figure 14 for the four cases which involved two mold compound materials and two die encapsulation choices. As we can see from the modeling results, packages with the mold compound and die encapsulation choice included in Case 1 and Case 2 showed the best results in reducing maximum package warpage during the temperature excursion. Please note it is quite difficult to model the package shape changes during the reflow process,

978-1-4244-4658-2/09 $25.00 © 2009 IEEE

the temperature dependent material properties and volume shrinkage of mold compound during curing process have to be included in the model.

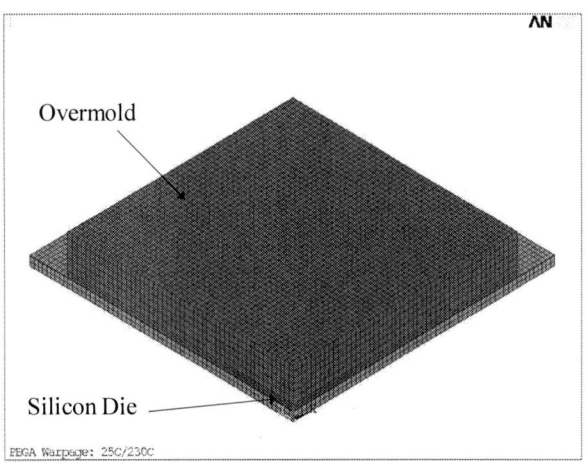

Figure 12 A finite element model for the 35mm PBGA package

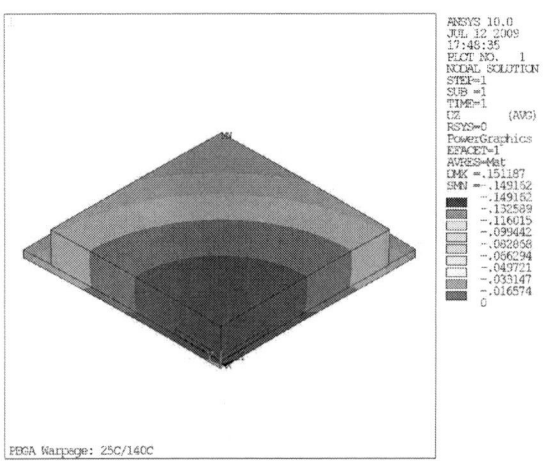

Figure 13 Predicted package out-of-plane displacement due to a temperature excursion between 25°C and 140°C

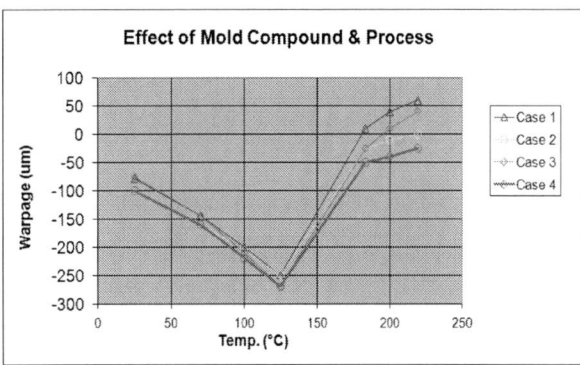

Figure 14 Effect of mold compound materials on PBGA warpage

Experimental Verification

In addition to the modeling effort, a 16-leg design of experiment (DOE) was conducted, which involved the same two mold compound materials used in modeling study and various molding and post mold curing process parameters. The DOE data have led to an optimized mold compound material and process that helped to reduce the package warpage significantly as shown in Figure 15. Warpage measurement results for four package samples made with the "new" mold compound and processes are shown Figure 15 together with the results from two packages made with the "old" material and process.

In order to quantify package warpage reduction with more data points and statistical analysis, we also run shadow moiré measurement on additional 30 parts from the production build to reinforce the DOE results.

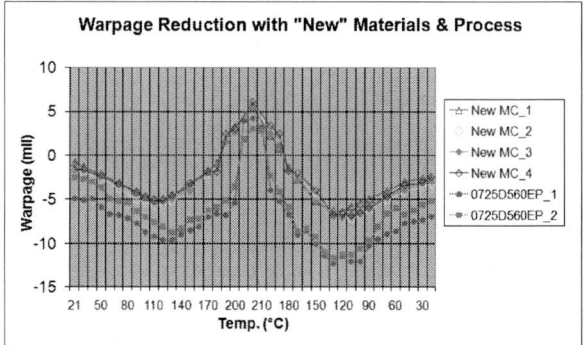

Figure 15 Package warpage reduction through optimized packaging process

After the optimal package material set and process were obtained, the standard component level reliability testing was used to qualify the "new" package. The qualification results are shown in Table 1.

Table 1 Qualification of the optimized package materials and processes

Reliability Test	Condition & Duration	Sample Size	Results
Preconditioning	JEDEC MSL-3, with 225°C peak reflow temp	All samples	Pass
C-SAM	Before and after reliability testing	All samples	Pass
Temp. Cycling	-55/125°C, 1000 cycles	25×3 (lots)	Pass
uHAST	130C/85%RH, 96 hrs	15×3 (lots)	Pass
HTS	150°C, 1000 hrs	15×3 (lots)	Pass

For production support, we also sampled four parts per lot using shadow moiré measurement with simulated reflow conditions for the initial ten production lots until the process stabilized. After that, a well controlled Cisco manufacturing build which included 10,000 product boards was carried out to demonstrate that significantly improved board assembly yield can be achieved by controlling the PBGA warpage during board mount assembly process. The controlled manufacturing build showed the defect level of "pillow" solder joint was reduced to zero parts per million (ppm) with the improved process and PBGA warpage reduction.

Conclusions

Strong coupling between the chip and the packaging materials can cause package deviation from an ideal state of uniform planar flatness, i.e., package warpage. If it is not well controlled, the temperature dependent PBGA package warpage can result in open, bridge or "pillow" BGA solder connections when mounting the device to a printed circuit board (PCB) using the surface mount solder reflow process. The problem can be more severe as we migrate to lead free SMT soldering process.

Standard testing methods and practicing procedures published by JEDEC and JEITA make it easier to characterize the package warpage and to establish a maximum permissible package warpage during the solder reflow process for the BGA components.

Through packaging material and process optimization, significantly improved board assembly yield can be achieved by controlling the PBGA warpage during board mount assembly processes.

Acknowledgments

The support from Cisco Manufacturing Ops team, its component suppliers and manufacturing partners is greatly appreciated.

References

1. JEDEC, "Design Requirments for Outlines of Solid State and Related Products," *JEDEC Publication 95*, *Design Guide 4.14, Ball Grid Array Packages, Issue D*, JEDEC Solid State Technology Association, 2002.
2. JEDEC, "Coplanarity Test for Surface-Mount Semiconductor Devices," *JESD22-B108A*, JEDEC Solid State Technology Association, 2003.
3. JEDEC, "High Temperature Package Warpage Measurement Methodology," *JESD22-B112*, JEDEC Solid State Technology Association, 2005.
4. V. J. Parks, "Geometric Moiré," in *Handbook on Experimental Mechanics*, A. S. Kobayashi, Ed. Englewood Cliffs, NJ: Prentice-Hall, pp. 282-313.
5. JEITA, "Measurements Methods of Package Warpage at Elevated Temperature and the Maximum Permissible Warpage," *JEITA ED-7326*, Japan Electronics and Information Technology Industries Association, 2007.
6. JEDEC, "Reflow Flatness Requirements for Ball Grid Array Packages," *JEDEC Publication 95, SPP-024 Issue A*, JEDEC Solid State Technology Association, 2009.

A Jetting System for Chip on Glass Package

Haili Jia[1], Zikai Hua[1], Maoyu Li[1], Jinsong Zhang[1], Jianhua Zhang[2]*

[1.] School Engineering and Engineering and Automation, Shanghai University, Shanghai City, 200072

[2.] Key Laboratory of Advanced Display and System Applications of Ministry of Education, Shanghai University, Shanghai City, 200072

Email: jhzhang@staff.shu.edu.cn, Tel: +86-21-56331976, Fax: +86-21-56331977

Abstract

Jetting is regarded as the next generation dispensing technology due to its features of non-contact and high precision dispensing. In this paper, we developed a jetting system for chip-on-glass package consisting of a jetting dispenser and a 3-axis movement system. The jetting dispenser applied a concept of modular design to construct four modules, which were nozzle, actuator, feeder and temperature controller. To analyze the factors influencing the fluid jetting, the proper fluid and structure models were set up. In the jetting dispenser, we used a solenoid valve (f >100Hz) to provide a high-speed movement for a needle. In a dispensing process, a piston drove a needle to move reciprocate, and its stroke was measured and controlled by a micrometer. A 3-axis system was a platform for the installation of jetting dispenser, and supported a reciprocating motion to follow the programmed route.

1. Introduction

Fluid dispensing systems have been widely employed in electronics industry, in which fluid materials are delivered somewhere by a controllable manner [1-3]. Currently there are six different technologies used for fluid dispensing systems: pin transfer, screen-printing, time pressure, rotary auger screw, piston displacement pump, and adhesive jetting.

Jetting is often regarded as the next generation technology due to its features of non-contact and high precision dispensing. For the jetting technology, a kind of fluid directly adheres to a substrate jetted by a nozzle, independent on the gravity or surface tension of the fluid [4]. During a jetting process, there is no motion for the Z axes of the dispenser, and this can improve production efficiency because of no fluid backhaul [1-3]. The jetting technology has another merit that is the nozzle dispenses the fluid in a slot in high quality comparing with other traditional technologies [1, 3].

This work developed a jetting system driven by compressed air, which was controlled by a solenoid valve. The structural parameters of design had been obtained based on a fluid and structure analysis.

2. Design layout and modeling

To achieve a jetting dispenser, we analyzed and elucidated its working principle at first. Figure 1 shows a design principle scheme of a jetting dispenser. For a concept of modular design, a jetting dispenser is divided into three modules, which are nozzle, feeder and actuator.

The nozzle module has a nozzle and seat, in which a chamber can contain adhesive for jetting. The feeder module includes an adhesive syringe, pneumatic pump and air regulator and it stores and provides the adhesive for jetting.

The actuator module consists of a needle, cylinder, piston, spring, solenoid valve, compressed air valve and pulse exciter and it controls the valve to drive the needle reciprocating motion.

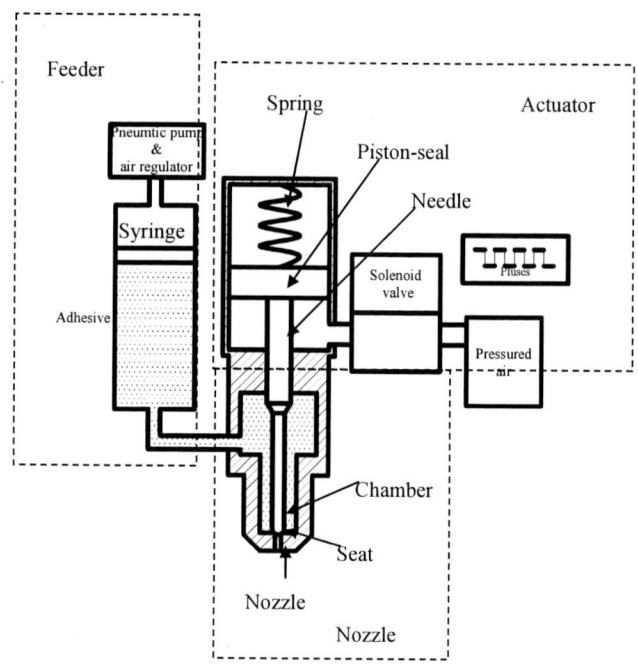

Figure 1 A scheme of design principle for a jetting dispenser

As a controlling signal excites the actuator to start work, the solenoid valve is on when the voltage is at the high level. Then, the compressed air passes the solenoid valve and flows into the cylinder to lift the piston and needle moving upside in the nozzle chamber. At the same time, the compressed air in the syringe pushes the adhesive to flow and fill the nozzle chamber. A pressure governor valve maintains a constant pressure to the adhesive; hereby, the adhesive volume in each jetting will be consistent and equivalent. This step is the first stage of the adhesive dispensing process.

When the driving voltage turns to the low level, the solenoid valve is off and the compressed air existing in the cylinder is exhausted through the hole on the wall. So that, the compressed spring begins to restore the original state, and this forces the piston and needle moving downside. As a result, the needle impacts the adhesive to spray on the substrate. A total dispensing process completes. Figure 2 shows a flow chat of one adhesive dispensing process.

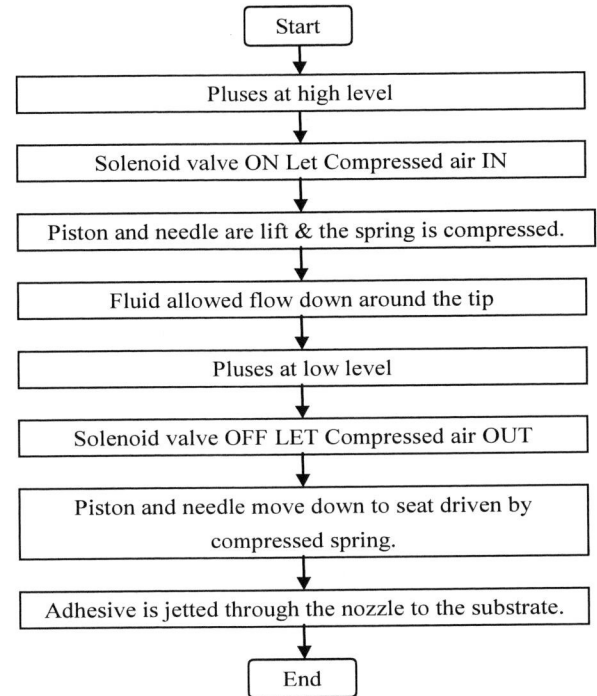

Figure 2 A flow chart of a reciprocating motion for the needle

2.1 Fluid modeling

For a jetting dispensing system, several adhesives can be used in different applications. Considering different adhesives have dissimilar fluidic properties, the fluid modeling becomes more important for the design of the nozzle module.

The adhesive volume of each adhesive dispensing is a key parameter for design, and it also determines a jetting system performance. The classical power-law model in hydrokinetics states the mathematic relationships between the adhesive volume and the flow rate and jetting velocity.

The flow rate is a medium parameter, which is dependent on the jetting velocity and dominates the adhesive volume. Hereby, we propose some assumptions before the fluid modeling: (1) no gravitational effect occurs since the adhesive is a high viscosity fluid; (2) the flow is laminar in a adhesive dispensing process; (3) the fluid flow in the radial direction is negligible; (4) the non-linear convective acceleration terms are negligible; (5) the material properties of the adhesive are constants [5, 6].

Table 1 lists the fluid properties of the adhesive and equations 1-4 give the formulas to calculate the adhesive volume by the flow rate and jetting velocity.

Table 1 the fluid properties of the adhesive

	parameters	Values
Fluid properties (epoxy)	Density (ρ)	980kg/m^3
	Consistency Index (K)	1
	Flow Behavior Index (n)	0.80
	Bulk Modulus (B)	1

As the adhesive is pushed by the compressed air to flow in the nozzle, its unsteady behaviors can be expressed as following [4, 5].

$$v(t, \varepsilon) = v_{max}(t)(1 - \varepsilon^{1+1/n}) \tag{1}$$

where

$v(t, \varepsilon)$ --the time –dependent velocity of the flow

ε --the dimensionless radius, equal to r / R_{noz}

n --the flow behavior index of the power fluid

$v_{max}(t)$ --the velocity of the flow in the center of the nozzle.

$v_{max}(t)$ is derived from the following equation [4]:

$$\frac{P_S - P_a}{L_{noz}} - \frac{2K}{R_{noz}^{n+1}}(1+1/n)^n v_{max}(t)^n = \rho \frac{dv_{max}(t)}{dt} \tag{2}$$

P_a --the atmospheric pressure

P_s --the fluid pressure in the region between the needle and seat

L_{noz} --the length of the nozzle

R_{noz} --the radius of the nozzle

ρ --the density of the adhesive

K—the consistency index

The dispensing flow rate via the nozzle is:

$$Q_{noz} = \int_0^1 2\pi R_{noz}^2 \varepsilon v(t, \varepsilon) d\varepsilon = \frac{n+1}{3n+1}\pi R_{noz}^2 v_{max}(t) \tag{3}$$

As mentioned above, if the flow rate of the fluid is identified, the adhesive volume V_{total}, be calculated by the integration of the flow rate with time.

$$V_{total} = \int_0^T Q_{noz} dt = \frac{n+1}{3n+1}\pi R_{noz}^2 \int_0^T v_{max}(t) dt \tag{4}$$

where

V_{total} -- the adhesive volume per dispensing

T--the period time per dispensing, determined by the duration of one cycle for the needle down loading

In equation 2, P_s can be calculated by the following:

$$P_s = P - \Delta P \tag{5}$$

where

P --the fluid pressure in the chamber, determined by the pressure in the syringe

ΔP --the pressure difference with respect to the needle movement

When the needle moves down to jet the adhesive, ΔP is:

$$\Delta P = -\frac{1}{2C_d^2}\rho\left[\frac{1}{2}\left(1-\frac{A_{con}}{A_1}\right)^{\frac{3}{4}}\right]\left(\frac{Q^2}{A_{con}^2}\right)$$

(6)

and in which,

$$Q = \pi R_{n1} v_n$$

(7)

$$A_{con} = 2\pi R_{con}(gap - x_n)$$

(8)

$$A_1 = \pi(R_{cham}^2 - R_b^2)$$

(9)

where

v_n -- the needle velocity

R_{n1} -- the radius of the needle

R_b -- the radius of the tip of the needle

C_d -- the coefficient of discharge

gap -- the distance between needle and the seat

R_{cham} -- the radius of the hole of the chamber

R_{con} -- the radius of the contact area

When the adhesive volume is obtained, the mass of it M_d is:

$$M_d = \rho V_{total}$$

(10)

From the fluid modeling, it is clear that the fluid parameters, such as the adhesive volume V_{total}, and the adhesive mass M_d, are the dominant factors for the jetting dispensing. And they are controlled by the structure parameters, such as the distance between the needle and the seat gap, the radius of nozzle R_{noz}, the radius of the tip of the needle R_b, and the length of the nozzle L_{noz}.

2.2 Structure analysis

The above statements reveal the relationship between the fluid modeling and the structure parameters. In a jetting system, the movement of the needle should be discussed detailed to ensure the liability and reliability of the structure design [7].

A scheme of motion principle for the needle is shown in Figure 3.

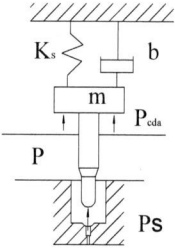

Figure 3 A scheme of motion principle for the needle

From Figure 3, the forces of the needle motion in a jetting dispensing process are deduced as following:

$$ma + F_f + K_s x + F_o = F + P_s A_{eq} + P\pi(R_{n2}^2 - R_{n1}^2)$$

(11)

m -- the mass of the movement components, including the masses of the needle and the piston

F_f -- the friction of the component

x -- the displacement of the piston (is equal to that of the needle)

K_s -- the spring's elastic coefficient

P_s -- the fluid pressure in the region between the needle and seat

A_{eq} -- the equivalent area, $A_{eq} = \pi R_{n1}^2$

R_{n2} -- the radius of the needle stem

P -- the fluid pressure in the chamber (is equal to the pressure in syringe)

F_o -- the pre-pressure of the spring

Herein, the pre-pressure of the spring is a product of K_s by x_0.

$$F_o = K_s x_0$$

x_0 -- the spring's decrement

In equation (11), we found the order of magnitude of the item of $P\pi(R_{n2}^2 - R_{n1}^2)$ is one or two smaller than that of the items of F and $P_s A_{eq}$. So that, the item of $P\pi(R_{n2}^2 - R_{n1}^2)$ can be ignored and the simplified equation is:

$$ma + F_f + K_s x + F_o = F + P_s A_{eq}$$

(12)

In equation (12), F is defined as the force acting on the piston from the compressed air.

$$F = P_{cda}\pi(R_C^2 - R_{n2}^2)$$

(13)

P_{cda} -- the pressure of clean dry air

R_C -- the cylinder radius

In the first stage of the reciprocating motion, the compressed air pushes the piston and needle to move upside, and the fluid pressure P_s in the region between the needle and seat is smaller than the fluid pressure P in the chamber from Equation 5. Therefore, the item of $P_s A_{eq}$ in Equation 12 is also negligible, and the equation is expressed as:

$$ma + F_f + K_s x = F - F_0$$

(14)

Taking Equation 13 to Equation 14, P_{cda} is obtained:

$$P_{cda} = \frac{ma + K_s x_0 + F_f}{\pi(R_C^2 - R_{n2}^2)}$$

(15)

According to the value of P_{cda}, the working pressure in the pump of compressed air is determined.

In the second stage of the reciprocating motion, the compressed air exhausts out of the chamber, the spring obliges the piston and needle to move downside. When the needle contacts the seat of the nozzle, the force acting on the piston F is zero. Equation 12 is expressed by:

$$ma + F_f + K_s x + F_o = P_s A_{eq} \qquad (16)$$

Thus, the spring's elastic coefficient is:

$$K_s = \frac{ma + P_s A_{eq} + F_f}{x + x_0} \qquad (17)$$

From the above structure analysis, we deduced the expressions of the two crucial parameters for structure design, which were P_{cda} and K_s. As all other parameters are known, the values of P_{cda} and K_s will be carried out to determine the design parameters of air pressure and the spring.

3. Design of a jetting dispenser

Combining the analysis of the fluid modeling and structure calculation, several important factors affecting the design of a jetting dispenser have been proposed, which are listed in Table 2.

Table 2 Key factors for the design of a jetting dispenser

Structure parameters	Nozzle radius R_{noz} & length L_{noz}
	Seat radius R_s
	Radius of chamber R_{cham}
	Needle radius R_{n1}, R_{n2}, R_b
Actuator parameters	Stroke x
	Acceleration
	Frequency f
Fluid Properties	Viscosity
	Density ρ
External conditions	Feed pressure P

As we state above, a jetting dispenser contains three modules: nozzle, feeder and actuator. In the following, the design parameters of them will be discussed in detail.

Nozzle module:

The fluid modeling and structure analysis indicates that some geometric sizes of different components are key parameters for design [8]. We calculated their values and listed them in Table 3.

Feeder module:

This module applies a time-pressure pneumatic pump. The air pressure corresponding to different adhesives can be set up by an air regulator (Range: 0.02-0.8MPa), which attaches to the pneumatic pump (Allowable stress: <0.3MPa).

Actuator module:

This module is consisted of a cylinder, spring, needle, solenoid valve and compressed air valve. Because the design

target requires the frequency of dispensing is very high, we select a solenoid valve with a high frequency to 100Hz. All other design parameters listed in Table 4 are obtained from the equations in the part two of this paper.

Furthermore, the real acceleration of the needle moving downside dominates the dispensing volume and quality. A high acceleration will cause atomizing effects in the adhesive dispensing process. A low acceleration decreases the pressure drop between the top and bottom of the nozzle, when the needle is moving to the seat. This induces the nozzle can not jet the adhesive continuously and efficiently.

Table 3 Design parameters of the nozzle module

	Parameters	Values
Nozzle	Radius(R_{noz})	0.05mm
	Lengthen(L_{noz})	3mm
Seat	Radius of seat(R_s)	2.75mm
	Radius of Hole(R_{SH})	0.5mm
	Length of Hole(L_{SH})	1mm
Chamber	Radius of cross(R_{cham})	3mm

For a reliable solution, we set up the pre-pressure of the spring by adjusting the original displacement using a micrometer, which is installed at the top of the cylinder. And then, a real acceleration can be achieved by the Equation 16.

Table 4 Design parameters of the actuator module

	parameters	values
Needle	Ball radius(R_b)	2.5mm
	Needle Radius(R_{n1})	2mm
	Needle Radius(R_{n2})	3mm
	Length	70mm
Piston	Radius(R_p)	10mm
Spring	Constant(K_s)	1000N/m
Micrometer	Range	15mm
Cylinder	Radius (R_C)	21mm
	Stroke(S)	0.5-2mm
Compressed air	Pressure(P_{CDA})	0.5-0.7MPa

In sum, all the design parameters have been obtained, and a structure figure is shown in Figure 4. For protect the needle avoiding friction damage in the movement, a guide sleeve is added surrounding the needle. In addition, a buffer board, located at the bottom of the cylinder, is used to reduce noise.

978-1-4244-4658-2/09 $25.00 © 2009 IEEE

4. Design of a jetting system

A jetting system not only includes a dispenser, but also needs a controlled motion platform. For a non-contact jetting system, a 3-axes platform can satisfy the movement requirement since the dispenser only does a reciprocating motion in Z-axis after the platform dwells at the expected position. The selected motion platform has one servo motor in the drive-ball-screw of each axis. When a servo motor runs, a drive-ball-screw transfers a rotation into a linear motion. Because the servo motor can control the displacement with a high precision, an accurate position will be reached for the platform. The detail parameters of the motion platform are shown in Table 5.

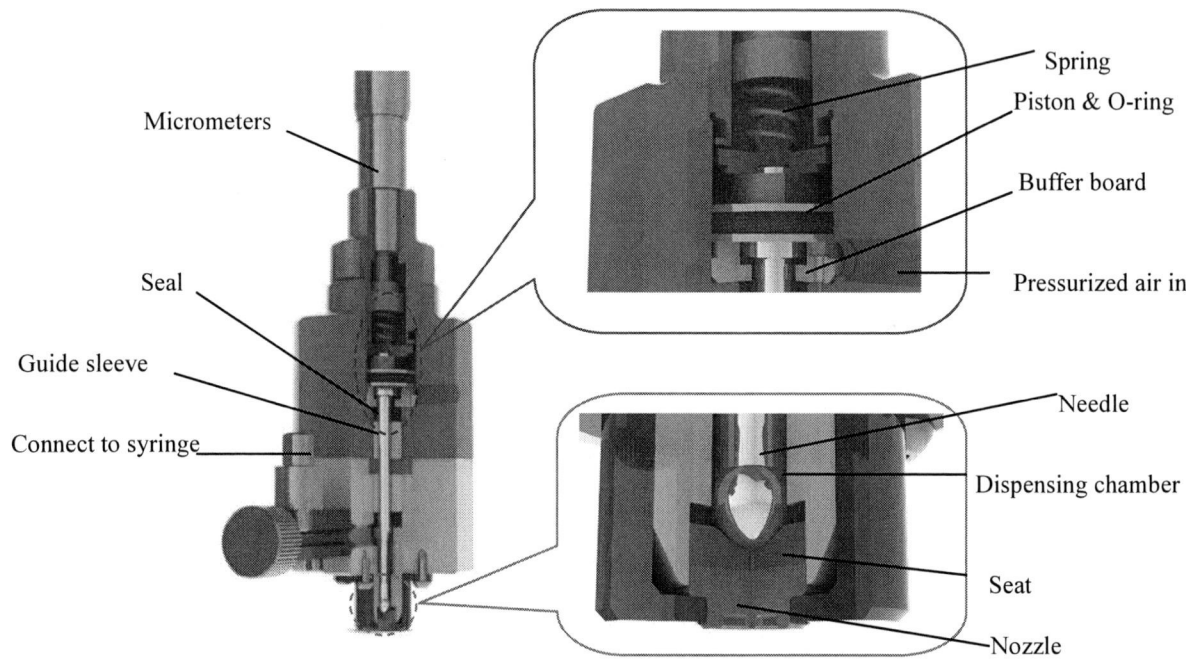

Figure 4 A designed jetting dispenser

Table 5 Main parameters of the motion platform

	Parameters	Values
X Axis (Y Axis)	Stroke	500mm
	Velocity	500mm/s
	Acceleration	4.9m/s^2
	Positional accuracy	0.05mm
	Repetition accuracy	0.03mm
Z Axis	Stroke	100mm
	Velocity	200mm/s
	Acceleration	2.5m/s^2
	Positional accuracy	0.03mm
	Repetition accuracy	0.02mm

Mounting the jetting dispenser to the 3-axes platform, an integrated jetting system has been fabricated. Figure 5 shows a prototype of the jetting dispensing system.

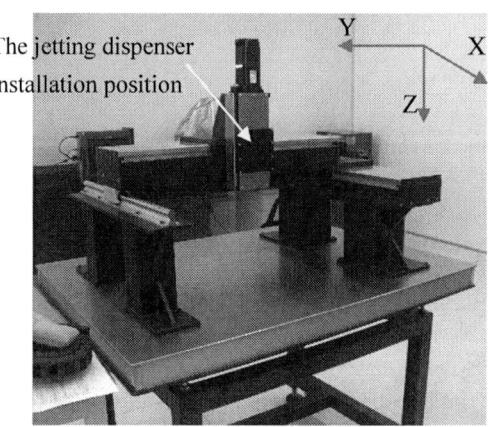

Figure 5 A jetting dispensing system

Figure 6 is a general framework of the jetting dispensing system, and it elucidates the different components and their relationships. This figure includes three units, which are mechanism unit, controlling unit, pneumatic unit.

The mechanism unit has a jetting dispenser, 3-axes motion platform. The controlling unit contains a personal computer with special software, two A/D & D/A conversion cards, an amplifier circuit, a solenoid valve and a temperature controller. The pneumatic unit consists of a gas pump, pressure governor valve, adhesive syringe and pipes.

The former parts of this paper state the working principle and structure of the mechanism unit particularly.

In a controlling unit, a personal computer provides a human-machine interface for operation, which is the core of the control system. Using the special software, we can realize a control of the motion trace by programming. Two standard PCI interface cards plugged in the main board of the personal computer, one is used to actualize A/D and D/A conversion,

and another is employed to control the 3-motion platform and the solenoid valve.

The pneumatic unit executes the performance by driving the piston with a needle to jet adhesive in a reciprocating motion. Through a pressure governor valve, a gas pump supplies a constant pressure air to the adhesive in the syringe, so that the jetting dispenser can spray the adhesive continuously and uniformly. Moreover, a temperature controller has been used to improve a low viscosity and a high flow property of the adhesive.

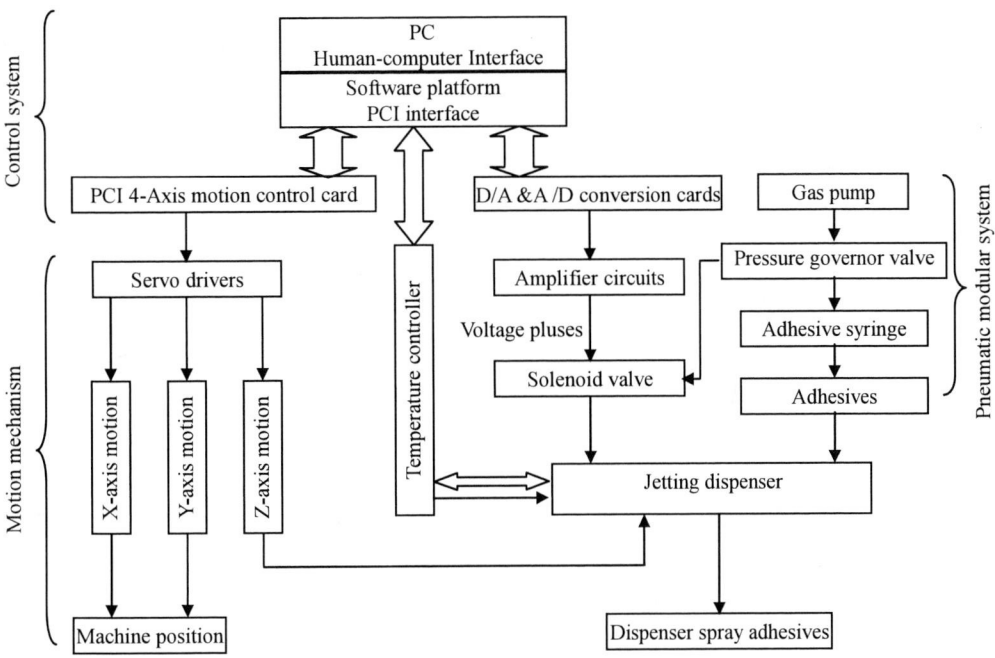

Figure 6 A general framework of the jetting system

5. Conclusions

In this paper, a jetting system for chip on glass had been developed. First we set up a fluid model to analyze the relationship between the fluid and structure parameters. It was found that the adhesive volume was determined by some structure parameters, such as the distance between the needle and seat gap, the radius of the nozzle R_{noz}, the radius of the tip of the needle R_b, and the length of the nozzle L_{noz}. And then, we adopted these structure parameters to design geometric sizes for different components in a jetting dispenser by the method of modular design. As a result, four modules including nozzle, actuator, feeder and temperature controller had been achieved to comply with the design targets. Finally, we fabricated a 3-axes movement platform to provide a high precision motion controlled by a servo motor. Thus, a jetting system had been developed completely.

In the future work, the jetting system will be installed the special software and detectors for automation.

Acknowledgments

This work was supported by Program for New Century Excellent Talents in University under the grant number NCET-07-0535, and Shanghai Science and Technology Committee under the grant number 075007004 and 08QH14007.

References

1. X. B. Chen, J. Kai, M. Hashemi, "Evaluation of Fluid Dispesnsing Systems Using Axiomatic Design Principles," *Transactions of the ASME, 129*, 2007, pp. 640-648.

2. B. Nguon , M. Jouaneh, "Design and characterization of a precision fluid dispensing valve," *Int J Adv Manuf Technol*, 2004, 24, pp. 251-260.

3. Liu Yanwei, Deng Guiling, "The Influence of Fluid Viscosity of Fluid Jetting Dispensing," http://ieeexplore.ieee.org/servlet/opac?punumber=4283545

4. Horatio Quinones, Alec Babiarz, Christian Deck, "Fluid Jetting for Next Generation Packages,"

http://www.asymtek.com/news/articles/2002_04_fluid_jetting_for_next_generation_packages.pdf.

5. Yi-Xiang Zhao, Han-Xiong Li, Han Ding, You-Lun Xiong, "Integrated modelling of a time-pressure fluid dispensing system for electronics Manufacturing," *Int J Adv Manuf Technol*, 2005, pp. 1-9.

6. Quoc Hung Nguyen, Young-Min Han, Seung-Bok Choi, and Seung-Min Hong, "Dynamic Characteristics of a New Jetting Dispenser Driven by Piezostack Actuator," *IEEE Transactions on Electronics Packaging Manufacturing*, 31(3), (2008), pp. 248-259.

7. Chen Kuiyu, Deng Guiling, "The influence Regularity of Structural Parameters of Fluid Jetting Dispensing," http://ieeexplore.ieee.org/servlet/opac?punumber=4283545.

8. Meng Aihua, "Research on Theory of Pulse Jet On-Off Valve and Application in BCP," Doctoral thesis of Zhejiang University, (2006), pp. 20-46.

A Multilayer Low Pass Filter Fabricated by Ferrite and Ceramic Cofiring System Based on LTCC Technology

Yuanxun Li[1], Yingli Liu[1], Huaiwu Zhang[1], Likun Han[2], Zongbao Yang[3]
[1]State Key Laboratory of Electronic Thin Film and Integrated Devices
University of Electronic Science and Technology of China Chengdu, 610054, China
[2]Institute of Astronautics and Aeronautics
University of Electronic Science and Technology of China Chengdu, 610054, China
[3]Integrated microcircuit company of Anhui Province, Hefei, 230088, Anhui, China

Abstract

LTCC (Low Temperature Co-fired Ceramics) has been become the key technology of packaging for the integrated of passive components due to its higher performance of thermal sink, reliability and plays an important role in increasing higher frequency, decreasing the loss, minimize the volume, etc. This paper mainly focuses on the design and fabrication of low pass filter based on ferrite and ceramic cofiring system by LTCC technology. Firstly, the match conditions of co-firing characteristics for ferrite and ceramic cofiring system were carefully studied by TMA measurement and the excellent shrinkage controlling was obtained. The low-pass filters with cut-off frequency at 120MHz were fabricated using Co/Ti doped barium ferrite and ULF140 material as the dielectrics to further validate the circuit model and the samples were tested with high consistence with the simulated data.

I. Introduction

With the current explosive growth of communication technologies and a wide variety of applications being found for high density packaging of electronic components, LTCC (Low Temperature Cofired Ceramics) with materials that possess different characteristics have been rapidly developed, due to its greater multifunctionality, sub-miniaturization and higher performances [1-3]. Especially, the multilayer ceramics based on ferrites and ceramics are absorbing more and more interest for their excellent magnetic and electric properties [3-7]. For example, as Figure 1 illustrates the ferrite/ceramic co-firing system for the LTCC applications. In Fig. 1 (a), an embedded LTCC MCM (multichip module) can be constructed. The ferrite tape layers could be used for embedded inductors and transformers and the high dielectric constant tape layer for embedded capacitors. Fig. 1 (b) describes the integration of passive components for high density packaging.

Whereas the considerable attentions for their applications have been paid to, the sintering problems in these multilayer materials have not yet been thoroughly resolved [8-9]. And the co-fired mismatch of these ferrites and ceramics is often serious. This has greatly affected the reliability and electronic properties of the multilayer devices.

In this paper, we proposed a manufacturing method of low pass filters (LPFs) based on the ferrite and ceramic cofiring system by LTCC technology. By employing the proposed design method, the multilayer chip LPFs configuration can be made more compact and flexible. The designed chip-type LPFs were realized by implementing the multilayer chip inductors and capacitors. The inductors and capacitors were implemented using Co/Ti doped M type ferrites and ceramic material respectively. The dielectric constant and loss tangent of the ceramics were chosen to be 14 and 0.0015. The permeability and cut-off frequency of the ferrites were 13.5 and beyond 1GHz. The designed LPFs were fabricated by 20um thickness films and constructed by LTCC technology with excellent consistency between simulation and fabrication with the merits of simple structures, sub-miniaturization and low cost for wide applications.

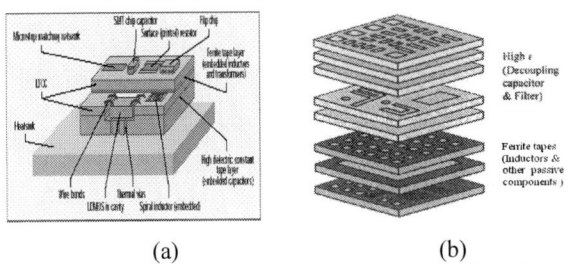

(a)   (b)

Fig. 1 The MCM module (a) and integration of passive components (b) constructed by ferrite/ceramic co-firing system

II. Circuit and Stimulation

Circuit Model

The LPFs were constructed by traditional three-order Butterworth filter configuration based on the lumped-element L-C filter circuit. The typical circuit for LPFs was shown in Fig. 2 which is composed of inductors and capacitors by circuit stimulation.

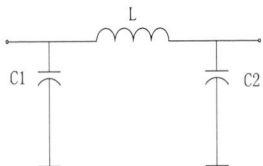

Fig. 2 The circuit model for low pass filters

Component Implementation

The physical prototype for LPFs was shown in Fig. 3. After optimized, the characteristics of the filters simulated can be achieved and presented in Fig. 4.

978-1-4244-4658-2/09 $25.00 © 2009 IEEE

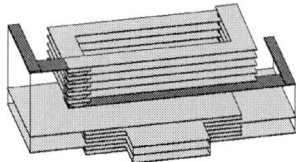

Fig. 3 The physical prototype for band-pass filters

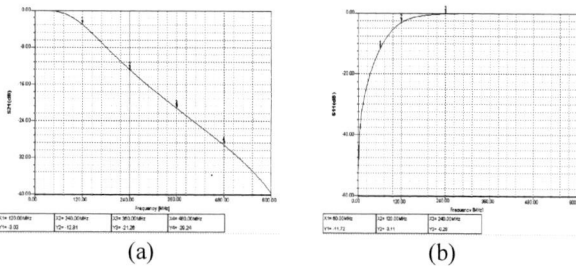

(a)   (b)

Fig. 4 The stimulated results of low-pass filters by HFSS software with the cut-off frequency at 120MHz (a) for the parameter S_{21} center frequency at 1.8GHz, (b) for the parameter S_{11}

From the response of three dimension configurations in Fig. 4., a cut off frequency at 120MHz,more than 20dB of the attenuation above 360MHz frequency band can be achieved, which will satisfy the performances of LPFs.

Fig. 5 shows the graphics of silk screen and relative sizes for one unit which will guarantee the good connection for each layer and formation of capacitors and inductors with the packaging size of 0805.

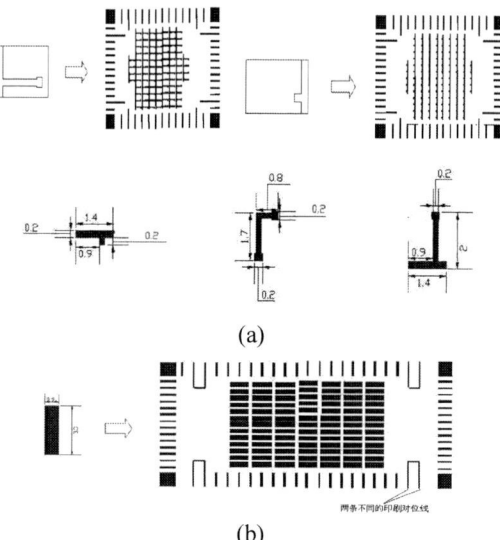

(a)

(b)

Fig. 5 Silk screen (a), (b) for the fabrication of LPFs by LTCC technology with the packaging size of 0805

III. Experimental

For LTCC process technology, the compatibility of different materials with respect to shrinkage, thermal expansion coefficient and chemical compatibility must be considered. And the different shrinkage rate of the fired specimens during co-firing make the materials distort, which cause deviation of the designed component.

A. Casting

After careful casting process, the diaphragm with good quality of ceramic and ferrite were shown in Fig. 6.

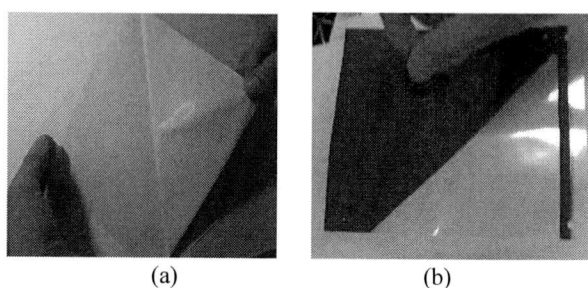

(a)   (b)

Fig. 6 The diaphragms made by ULF 140 ceramics (a) and M type barium ferrites (b)

B. Ferrite and Ceramic Cofiring Behavior

Fig. 7 shows that the sintering behaviors of the barium ferrites and ceramics. The shrinkage profiles of ferrites gives an outstanding resemblance with that of the ceramics, which demonstrates a good co-firing matching condition between these two materials.

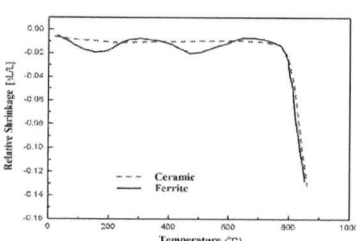

Fig. 7 The densification characteristics of the ferrite and ceramic

Fig. 8 gives back-scatter morphology near the interface between ferrites and ceramics in a multilayer sample prepared by tape casting and sintered at 920°C for 6h. Alternate heterogeneous layers are in good connection, with no evidence of cracks or delimitations. Both ferrite and ceramic layer show a dense microstructure and the corresponding grains grow normally.

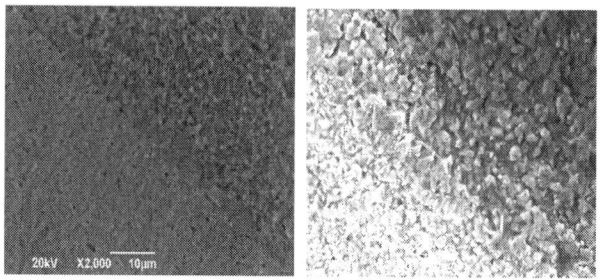

Fig. 8 The cross-sectional microstructure of a bi-layer sample

C. The Manufacture of LPFs by LTCC Technology

Based on the ferrite and ceramic cofiring system, the laminated LPFs samples were fabricated by the conventional LTCC technology process with the outline dimension of 2.0mm×1.2mm×0.9mm shown in Fig. 9.

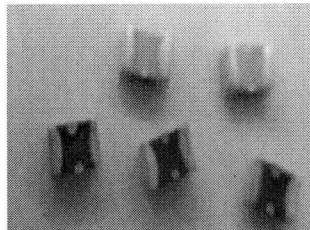

Fig. 9 Experimental prototype for the samples

D. The Measurements of the Performances for LPFs Samples

The measurements were carried on by Agilent 8722ES and the collected data was then calibrated to the desired reference plane by the thru-reflect line (TRL) technique through carefully designed calibration standards embedded in the same LTCC tie. The measured responses of the filters are shown Fig. 10. From Fig. 10, it can be clearly seen that the measured results agree with the simulated data basically. The cut-off frequency is about 130MHz (3dB) and the attenuation beyond 30dB (360MHz~1GHz).Thus, a good coincidence between simulated and measured data is observed. The difference between the measurement and the stimulation is that the cut-off frequency is a little larger than the designed. It is believed that the deviation comes from the additional inductive and capacitive parasitic effects. On the other hand, the process of LTCC technology also brings many errors which will introduce the manufacturing inaccuracy. Firstly, during the co-firing, the organic solvent in the silver conductor cannot dispel all air bubbles, which cause the value of inductance and capacitance smaller and make the self-resonance frequency larger. Secondly, the effective value of the component cannot be controlled accurately due to the difficulties of tiny manipulation problems including the printing and laminating.

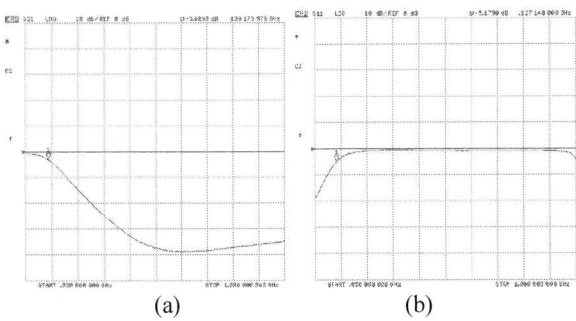

(a)   (b)

Fig. 10 The testing results of low pass filter with cut-off frequency at 120MHz (a) for S_{21} parameter varied with frequency (b) for S_{11} parameter varied with frequency

Conclusions

The structure analysis of LTCC-based passive components is reported for the design of a small multilayer chip LPFs based on the ferrite and ceramic cofiring system. The LPFs fabricated by LTCC process has small size (2.0mm×1.2mm×0.9mm).The cut-off frequency for LPFs samples is about 130MHz and the attenuation beyond 30dB (360MHz~1GHz) respectively. The testing results are in a good agreement with the simulated data which will be helpful for the manufacture of high integrated components.

Acknowledgments

This work was supported by the Youth Fund of Sichuan Province under Grant No 08ZQ026-013, the Foundation for Innovative Research Groups of the NSFC under Grant No. 60721001, the Youth Fund of University of Electronic Science and Technology of China under Grant No. L08010301JX0725.

References

1. Peng Te-Ming Hsu, Rung-Tsung, Jean Jau-Ho, "Low-fire processing and properties of ferrite/dielectric ceramic composite," *Journal of the American Ceramic Society*, Vol. 89, No. 9 (2006), pp. 2822-2827.

2. Sung-Hun Sim, Chong-Yun Kang, Ji-Won Choi, *et al*, "A compact lumped-element lowpass filter using low temperature cofired ceramic technology," *Journal of the European Ceramic Society,* Vol. 23, No. 5 (2003), pp. 2717-2720.

3. Higuchi Yukio, Sugimoto Yasutaka, Harada Jun, *et al*, "LTCC system with new high-ε_r and high-Q material co-fired with conventional low-εr base material for wireless communications," *Journal of the European Ceramic Society*, Vol. 27, No. 8-9 (2007), pp. 2785-2788.

4. Peng Te-Ming Hsu, Rung-Tsung, Jean Jau-Ho, "Low-fire processing and properties of ferrite/dielectric ceramic composite," *Journal of the American Ceramic Society*, Vol. 89, No. 9 (2006), pp. 2822-2827.

5. Cui Xue-Min, Zhou Ji, Li Bo, *et al*, "Co-firing behavior and interfacial structure of BaO-TiO_2-B_2O_3-SiO_2 glass-ceramics/NiCuZn ferrite composites," *Materials and Manufacturing Processes*, Vol. 22, No. 2 (2007), pp. 251-255.

6. Matters-Kammerer M., MacKens U., Reimann K., *et al*, "Material properties and RF applications of high k and ferrite LTCC ceramics," *Microelectronics Reliability*, Vol. 46, No. 1, (2006), pp, 134-143.

7. Lee W H, Su C Y, "Improvement in the temperature stability of a $BaTiO_3$-Based multilayer ceramic capacitor by constrained sintering," *Journal of the American Ceramic Society*, Vol. 90, No. 10 (2007), pp. 3345-3348.

8. Hagymasi Marcel, Roosen Andreas, Karmazin Roman, *et al*, "Constrained sintering of dielectric and ferrite LTCC tape composites," *Journal of the European Ceramic Society*, Vol. 25, No. 12 (2005), pp. 2061-2064.

9. Zhang, H. W., Zhong, H., Liu, B. Y., *et al*, "Electromagnetic properties of a new ferrite-ceramic low-temperature cocalcined (LTCC) composite materials,"

IEEE Transactions on Magnetics, Vol 41, No 10 (2005),
pp. 3454-3456.

Research of SMT Product Assembly Quality Management System Based on J2EE

Huihuang Zhao[1], Dejian Zhou[2, 3], Zhaohua Wu[3], Xiaoyong Cheng[3]

[1]School of Mechano-Electronic Engineering, Xidian University, Xi'an, 710071, China
[2]Department of Mechanical Engineering, Guangxi University of Technology, Liuzhou, 545006, China
[3]School of Mechanical and Electronical Engineering Guilin University of Electronic Technology, Guilin, 541004, China

Abstract

SMT product assembly quality management system structure and content is designed, according to its characteristics and requirements. A developing method of SMT product assembly quality management system is proposed based on J2EE, in order to improve its reusability, portability and expansibility. An integrated architecture which consists of Struts,Spring and Ibatis is designed based on lightweight framework. Then the integrated architecture is used to develop the SMT product assembly quality management system. During the integrated architecture, Struts is used in presentation layer, Spring is used in business layer and Ibatis is used in persistence layer. The process and implementation of this management system of a function module are introduced. The analysis results have shown that the SMT product assembly quality management system based on J2EE has some advantages in reusability, portability and expansibility.

Discussion 1 Introduction

The surface mount technology (SMT) of circuit modules product has characters not only keeping electrical performance, but also ensuring the reliability of mechanical connection. Its assembly quality is the life of SMT product and directly determines the performance and reliability of SMT product. SMT product assembly quality management system is the core part in quality management system of SMT product enterprise. Arming to the characteristics of product production and high requirements in quality and efficiency, the SMT assembly quality management system can solve the assembly quality problem and management problem, which are universal existence in the process of product production, involving SMT product assembling process design, assembling quality verification, material quality management, process quality verification, assembling quality detection and control, product user service and other links. It reflects the advancement, reliability, efficiency, high quality economy of product assembly quality system outstandingly. And it has some obvious characteristics compared with other product assembly quality management systems [1].

Java is becoming the language of choice for developing and deploying enterprise-class server-side applications. Java2 Platform, Enterprise Edition (J2EE) technology standardizes the operating environment for server-side Java applications. The J2EE platform provides solid baseline standards on various functional components or containers for presentation and business logic with communication links to client side presentation, as well as back-end database and legacy systems. It also offers communication links to other remote J2EE systems [2-3].J2EE framework technology, which has some advantage in improving development speed and saving development cost, etc, can adapt to the enterprise applications [4-5].

Discussion 2 SMT Product Assembly Quality Management System Main Function and Performance

The design of SMT product assembly quality management system follows up ISO-9000 family standard and eight principles of modern quality management [6]. It is managed and controlled in quality by five factors as man, materials, machine, construction method and environment. Pursuing best product assembly quality is its goal. Prevention first and total quality management is its guiding principle. According to the requirements in quality control and management in the process of the assembly and produce, a reliable quality assurance system can be established. And computers, network database, expert system ect, are used to acquire, statistic, analysis and feedback quality data. Then the product quality detection and control in the process of SMT product assembly and produce can be realized. These ensure the product with qualified quality.

SMT product assembly quality management system main function includes as follows:

(1)Product designing quality verification;
(2)Material quality detection and control;
(3)Process quality, production quality control;
(4)Product quality statistical, analysis and evaluation;
(5)Inspection and test, including procedure detection and online test;
(6)Quality cost management.

SMT product assembly quality management system main performance includes as follows:

(1) Every technical in quality detection and management is in accordance with relevant national standards and stipulations; (2) Every technical has good maneuverability and feasibility of implementation; (3) The quality management system, software and interface design are in accordance with relevant national standards and stipulations; (4) The network database and its operation software have excellent openness and scalability; (5) The quality management system has a complete operation mechanism and a reliable implementation guarantee condition; (6) The software terminal has characteristics of simple operation and friendly user interface.

Discussion 3 SMT Product Assembly Quality Management System's Function Modules Design

SMT product assembly quality management system can be divided into five parts: Quality plan; Quality detection and control; Quality evaluation and improvement; Comprehensive quality information management, Background management. Then the five parts can be divided into 19 sub parts which can be showed as Figure 1.

978-1-4244-4658-2/09 $25.00 © 2009 IEEE

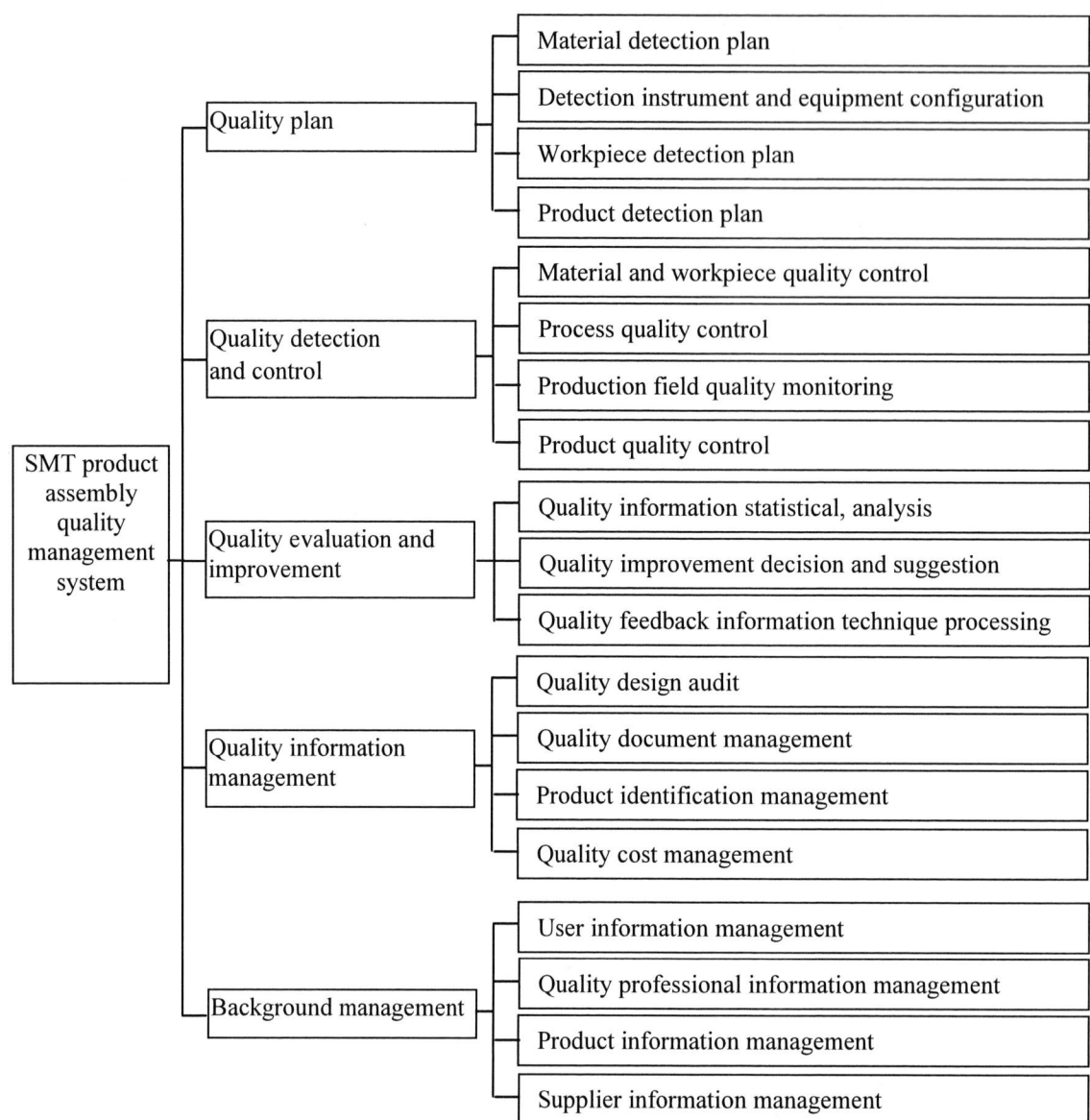

Figure 1 SMT product assembly quality management system function modules

(1) Quality plan

It mainly function is decomposing task document which is commanded by production management system, as a result, material detection plan, detection instrument and equipment configuration rules, and workpiece and product detection rules ,etc, are formed.

(2) Quality detection and control

It is an important embodiment of quality activity acted by quality management system. According to quality plan, acquisition, some quality performance parameters and quality information are acquired, processed, and analyzed. Then the relevant controls are done according to the quality statistical and analysis result.

(3) Quality evaluation and improvement

It is mainly used in quality statistical, analysis, processing, evaluation, feedback of quality dates, including quality analysis, decision and suggestion of quality improvement, quality feedback information technique processing.

(4) Quality information management

It is mainly used in managing the system internal information. Quality design audit mainly aims to audit the designed product. Quality document management includes procedure document of quality system, standards, tasks, reports and files of detection record. Quality cost management includes calculating the quality cost, makes the quality cost controllable.

(5) System background management

It mainly includes user information management, quality professional information management, supplier information management, ect., which are acted by administrator. The operation includes records adding, deleting, updating, querying, ect.

Discussion 4 Struts, Spring, Ibatis Introduction

When a mature framework is used during system development, there are lots of advantages in reducing the repeat work, decreasing work time, reducing cost, improving reusability, maintainability and expansibility. So it is very important for system development to choose a right framework [7].

Struts framework is an open-source framework for developing the web applications in J2EE, based on MVC architecture [8]. It uses and extends the Java Servlet API. Struts is the robust architecture and can be used for the development of application of any size. Struts framework makes it much easier to design scalable, reliable Web applications with Java.

Spring is an open source framework created to address the complexity of enterprise application development. One of the main advantages of the Spring framework is its layered architecture, which allows to be selective about which of its components you use while also providing a consistent framework for J2EE application development. Spring provides a consistent way of managing business objects and encourages good practices such as programming to interfaces, rather than classes. The architectural basis of Spring is an dependency injection (known as Inversion of Control) container designed to configure any POJO. And Spring provides a unique data access abstraction, including a simple and productive JDBC framework that greatly improves productivity and reduces the likelihood of errors. Spring's data access architecture also integrates with Ibatis, Hibernate, JDO, JPA and other O/R mapping solutions.

IBatis is a very lightweight persistence solution that gives us most of the semantics of an O/R mapping toolkit, without all the drama. In other words , iBatis strives to ease the development of data-driven applications by abstracting the low-level details involved in database communication (loading a database driver, obtaining and managing connections, managing transaction semantics, etc.), as well as providing higher-level ORM capabilities (automated and configurable mapping of objects to SQL calls, data type conversion management, support for static queries as well as dynamic queries based upon an object's state, mapping of complex joins to complex object graphs, etc.). iBatis simply maps JavaBeans to SQL statements using a very simple XML descriptor. Simplicity it is the key advantage of iBatis over other frameworks and object relational mapping tools. These can meet the development requirements of SMT product assembly quality management system.

Discussion 5 The Integrated Architecture Designing and Analysis

An integrated architecture is designed for meeting the requirement of SMT product assembly quality management system better. At first, the program development is divided into three layers. Then choose a right framework for every layer according to their characteristics. According to the above introduction of Struts, Spring and iBatis, during the integrated architecture, the presentation layer is Struts. Business layer is Spring, and persistence layer is iBatis. The integrate architecture model can be showed as Figure 2:

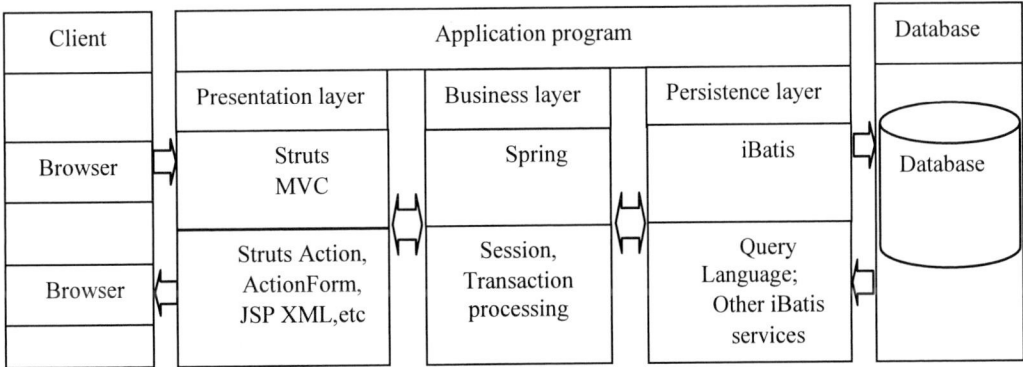

Figure 2 The integrate architecture model based on Struts, Spring, iBatis

(1) Presentation layer

Its mainly functions are data display and data submission. It is accomplished by Struts.

(2) Business layer

It takes charge of business logic, business check, transaction management of application program, and allowing business layer to depend on other layer's interface which can acts each other, etc. This layer is finished by Struts.

(3) Persistence layer

During the system development, there are many interactions with database. In order to make these interactions more effective and fast, the persistence layer is established between application program and database. It takes charge of saving the data in database and compiling data manipulation sentences and data control sentences, such as index, update,

delete, and so on, as a result , accomplish O/R mapping. iBatis is used in this layer.

The integrate architecture flow is follows: At first, Http request, which a user typed in a HTML form on a web page in client, is sent to controller (ActionServlet, a kind of Action) in Struts, and wait for be treated. ActionServlet reads and processed the request, and it generates results with the existing data. Then ActionServlet calls the business logic directly in the servlet or another class which come from business layer, which contain the logic or executes a database query.

The preprocessing of the business logic is taken charge by Spring, including defining how beans are created, configured, and managed. When a class or table is worked with, the Spring configuration file is the start place. The configuration file is an XML file (applicationContext.xml). Within it some properties and iBATIS SimpleDataSource are configured. Within the Spring configuration xml, the sqlMap property is wired with a reference to an iBATIS. Spring gets a SqlMapClient is through SqlMapClientFactoryBean. When the business logic model is called, the interaction with database can be finished by Ibatis in background.

The data is dealt with if it is necessary in business layer (Spring), then sent it to struts. The results will be formatted. Browser expects an answer in the HTML format. The results must be formatted in accordance with the standard. It is possible to return different formats of data with a servlet. (gif, jpeg, doc, etc).

The 19 sub function modules are in correspondence with 19 Actions. Each Action includes some operations such add, delete, update, query, etc, besides communicating with other Actions.

The advantages of the integrate architecture are:

(1) Reusability

There are a few codes in the web page. It has some merits for using and maintaining. Business logic is separates from view thoroughly. It is clear concepts, and reasonable structure. Each business logic result is conserved in a core javabean. If want to update the code, it just need to modify the javabean, and not to modify the web interface code.

(2) Expansibility

Application program logic separates from code. The application program logic is mainly defined by a configuration file while not by a code file. If the requirement changed, it just needs to modify the program logic and not to modify the program code greatly. At the same time, the integrate architecture can finish adding a new function module easily.

(3) Portability

The management system based on the integrate architecture has a good portability. If modify the corresponding configuration file, it can be run at different application servers and database. The integrate architecture has the features as clear division and levels. It can simplify development process, quicken development speed, and save time.

6 SMT product assembly quality management system realization based on integrate architecture

There are 19 sub function modules in SMT product assembly quality management system, and correspond with 19 actions, and product quality information management is a very important model in the management system. So product quality information management is taken as example to test the feasibility of the integrate architecture and introduce development flow.

Persistence layer stays between application program and database, includes compiling data manipulation sentences and data control sentences, accomplishing O/R mapping. SMT_CPXX is a table which is used to save the product information; its mapping class is SMT_cpxxVO.java. SMT_cpxxVO.xml includes all SQL sentences corresponded with every operation of SMT_CPXX table, such as add, delete, update, query, and so on.

There are corresponding codes in sql-map-config.xml ,as follow:

```
<typeAlias alias=
    "smtcpxx" type="com.company.vo.SMT_cpxxVO" />
<sqlMap resource=
    "com/company/vo/SMT_cpxxVO.xml"/>
```

This also accomplishes that the field names in query result are one to one correspondence with variables in SMT_cpxxVO.java. Some operations which aim at the table are defined in an interface class SMT_cpxxdao.java.

Business logic is finished by Spring. Get and Set function are defined in an interface of business object. And its implementation class can be accomplished easily, then inject into interface SMT_cpxxservice.java. SMT_cpxxserviceimpl.java is its implementation class, which includes the all implementation of business operations.

The dependency injection can be finished in applicationContext.xml

```
<bean
id=" smt _cpxxservice" parent= "baseTransactionProxy">
    <property name="target">
      <bean class=
         "com.company.service.SMT_cpxxserviceimpl">
      <property name=" smt _cpxxdao">
       <ref bean=" smt _cpxxdao" /></property>
      </bean>
    </property>
</bean>
```

Presentation operation is finished by Struts based on MVC architecture. The operation of product information saving into database can be divided into three parts. First is SMT_cpxxAction.java which is the sub class of Action; second is SMT_cpxxForm.java which is the sub class of ActionForm, last is corresponding JSP files. SMT_cpxxForm.java deals with the data which is input by a form. SMT_cpxxAction.java deals with some business operation, such add, delete, update, query, and so on, according to the different token value.

The web page of publishing product information can be showed as Figure 3.

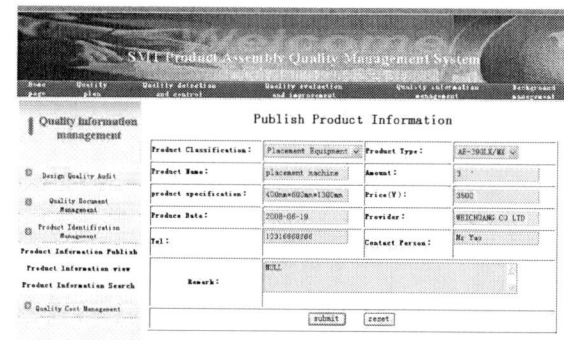

Figure 3 Publishing product information

When the operate is successful, we can access another web page. The result can be showed as Figure 4.

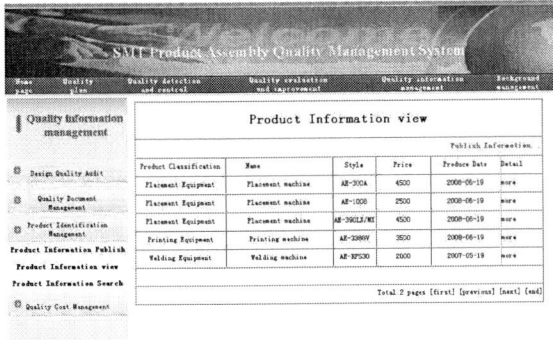

Figure 4 Product information browse

Lack of space forbids further treatment of content and process of other function modules here.

Conclusions

The principle and realization method of SMT product assembly quality management system based on J2EE are introduced in detail. The SMT product assembly quality management system structure is designed. An integrated architecture which consists of Struts, Spring and Ibatis is proposed too. The integrated architecture has some advantages in reusability, portability and expansibility. And it has the all characters of Struts, Spring and Ibatis. Then the integrated architecture is used to develop the SMT product assembly quality management system. A test shows that the method is feasible. The research has theoretical importance and practical engineering application value.

References

1. Dejian Zhou, *et al.* "The design of SMT product assembly quality management system," *Manufacturing Automation*, Vol. 3 (2008), pp. 52-56 (in Chinese).

2. Gutkowski, M., *et al.*, "Thesis Management Supporting System based on J2EE Platform," *CADSM '07,* Polyana, Ukraine, February, 2007, pp. 395-398.

3. Liu, Li, *et al.*, "Research of JFSSH application framework based on J2EE," *Control and Decision Conference 2008,* Shandong Yantai, July, 2008, pp. 688-692.

4. Bak, T., *et al.* "Development of advanced J2EE solutions based on lightweight containers on the example of e-department application," *Proceedings of the International Conference MIXDES 2006*, Gdynia, POLAND, June. 2006, pp. 779-782.

5. Johnson R., "J2EE development frameworks," *Computer.* Vol. 38, No. 1 (2005), pp. 107-110.

6. Su qing. <u>Modern Quality Management</u> (BeiJing 2005).

7. http://www.developersbook.com.

8. Wojciechowski J., *et al.*, "MVC model, struts framework and file upload issues in web applications based on J2EE platform," *TCSET'2004,* Lviv-Slavsko, Ukraine, February. 2004, pp. 342-345.

Test Scheme of SOC Test with Multi-constrained to Reduce Test Time

Chuanpei Xu, Jing Zhang, Min Zhang

School of Electronic Engineering, Guilin University of Electronic Technology, Guilin, 541004, China
Mailing address: Guilin University of Electronic Science and Electronic, Engineering College.
Zip code: 541004. Telephone: 0773-5601344. E-mail: xcp@guet.edu.cn.

Abstract

SOC integrates an intact system on one single chip, so that the size of chip is dwindled. However, the difficulty and complexity of system circuit testing is increased. Kinds of constraint conditions should be considered in testing cores, in order to meet the high-performance requirements of circuit system test, including the realization of parallel module test with test power and priority constrains. SOC test structural optimization is NP-hard problem, and it is hard to be solved using the common traditional arithmetic because of its complexity. While quantum search algorithm may reach to N magnitude acceleration, it is applicable to solve NP problems.

In this paper, by combining quantum algorithm with encapsulation standards based on test bus and IEEE P1500 test wrapper, the policy of TAM based on test bus is analyzed. Firstly a mathematical model of SOC test scheme with test power and priority constraints is presented based on the quantum algorithm, by distributing TAM width, choosing an appropriate parameter, and using the superiority of quantum bit in solving NP problems. Next correlative test scheme algorithm is designed. Then, in the paper partial SOC circuits in ITC'02 test benchmarks are taken as experimentation objects. Compared with other similar algorithms, experimental results showed that QA has a better performance and it gets a comparatively shorter testing time.

1. Introduction

Nowadays, integration levels and scales spread constantly, and intellectual property reusing to construct SOC becomes the mainstream of chip design. Much attention has been paid on SOC test technology based on IP cores, such as power and test priority, which come to be important matters in testing SOC cores [1].

In module testing, the states of inner circuit nodes change frequently, and the frequency would engender the test power. What's more, the power may accumulate up to a deviant scalar which is much higher than that in regular mode. The concentrated test power would destroy systems finally. Consequently, the quantity of IP cores must be limited and test power should be circumscribed under the permitted maximum test power; On the other hand, test scheme would be more and more complex in multi-testing, so an optimum test arbitration scheme is needed to keep test order in an appropriate way. That is to say, in SOC test scheme power and test priority constrains are indispensable. Consequently, a test scheme of SOC test with power and test priority constrains is designed to reduce test time, and the correlated algorithm design and verification are accomplished as well.

Many experts and scholars at home and abroad have made a very thorough study on SOC test scheduling algorithm design and test technology. According to different types of cores, design different testing methods. Present studies have shown the problems of TAM optimization and IP test scheduling are NP-hard problems. Consequently, in this topic quantum algorithm (QA) is used to solve scheduling problems of SOC testing with multi-constraints, which has a better performance compared to canonical algorithms.

2. SOC Test Scheme and Quantum Algorithm

SOC test scheduling of bus-based TAM architecture means [2]: Let the SOC consists of N cores, and the test width W is divided into B TAM partitions of widths {w_1, w_2, ..., w_B} to get a minimal test time. So it is necessary to distribute SOC test buses in reason, to unify TAM design with CTW design under constrains of test power and priority, and to design IP core optimizing testing schemes coordinated with effective scheduling algorithms.

Compared to classical algorithms, QA has the most essential feature in terms of its superposition and coherence of quantum states, the entanglement between quantum bits as well. The most important difference compared with classical algorithm is quantum parallelism. The greater the number of quantum bit, the more fully use of QA parallelism, so the superiority of quantum algorithm is more prominent and it is more suitable for solving NP problems [3].

3. Quantum Algorithm Mathematical Model Design

Considering all the cores are embedded deeply on chips, it is incapable to test integrated cores in the single-core test approach. So logic modules are needed to be connected to peripheral circuit, in order to generate and transmit test patterns. In that way core under test (CUT) is connected with external resources. IEEE P1500 [4] is the exactly thread to be proposed as embedded test structure.

3.1 Power-Constrained SOC Architecture Optimization

Firstly, assuming set C means n test core {$C_1, C_2, ..., C_n$}, and the test time of each core is represented by set T {$t_1, t_2, ..., t_n$}, also assuming power consumption of each core consumes P{$P_1, P_2, ..., P_n$}, a test sequence comes to a partial order consisted by test cores. Especially, P_i is defined to be the peak power of C_i in t_i, in order to calculate test power.

Secondly, L_{ij}, $(i, j) \in \{Test_1, Test_2, \cdots, Test_n\} \bigcap i \neq j$, is designed to estimate whether there is an overlap in test time when testing $Test_i$ and $Test_j$.

$$L_{ij} \begin{cases} 1 & \text{If Testi and Testj overlapped} \\ 0 & \text{If Testi and Testj not overlapped} \end{cases}$$

Then, the calculation must meet the following formula:

$$L_{ij} \times (P_i + P_j) \leq P_{max}$$

(P_{max} is defined as power limit.)

So, the discriminate function was defined:

$$power(L_{ij}, P_i, P_j) = \begin{cases} 1, L_{ij} \times (P_i + P_j) \leq P_{max} \\ 0, others \end{cases}$$

The power consumption of a test sequence is defined as the maximum power consumption of the test patterns in the test sequence. The problem is illustrated to a configuration using a 2-D bin-packing formulation shown in Figure 1.

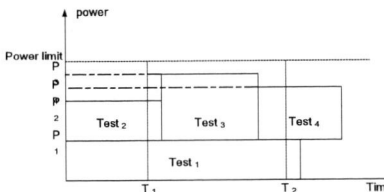

Figure 1 Power-constrained test schedule Mathematical Model Design

In Figure 1, an example of a power-constrained test schedule for four deterministic test sequences is shown. $Test_i$ (i=1,2,3,4) refers to test time for each core, P_i (i=1,2,3,4) refers to the power of $Test_i$ consumed, and power limit refers to the peak-power consumption. If tests overlap in a test time, add up overlapped test power. Put L_{ij}, P_i, P_j (i,j=1,2,3,4) into the above function, if the calculation result comes to be 1, it means the current test set meets the power constraint; Otherwise, it means the current test doesn't meet the power constraint.

3.2 Priority-Constrained SOC Architecture Optimization

First of all, assuming priority rating is fixed, $T=\{t_{11}, t_{12}, ..., t_{1M}\}$ comes to be the starting time set of core i (i=1,...,I), m is presented test resources (m=1,...,M), and the length of test set is $L=\{L_{11}, L_{12}, ..., L_{im}, ..., L_{IM}\}$. Then the final target of SOC architecture optimization is to get the shortest total test time.

Core i and core j will not come into conflict in $t_{im} \geq t_{jm} + l_{jm}$ or $t_{jm} \geq t_{ik} + l_{ik}$ in the same test resource, which means test times do not overlap when testing core i and core j in resource m. Similarity, test resources m and k will not come into conflict when core i testing different test resources and test time do not overlap, that is $t_{ik} \geq t_{jm} + l_{jm}$ or $t_{jm} \geq t_{ik} + l_{ik}$. In a word, if core q tests prior to core i, there is no conflicting priority happened when $t_{qr} \geq t_{im} + l_{im}$. Constraints condition may be expressed as follows:

Aim: minimize T, $T = \max\{t_{im} + l_{im}\}$;
Precedence constraints: $t_{qr} \geq t_{im} + l_{im}$;
t_{im}---starting time of core i in test resource m;
l_{im}---process time core i in test resource m;
i,q---core i, q=1,...,I;
m, k, r--- test cell, m, k, r=1, ...,M.
Priority ranking of ITC'02 SOC test benchmarks:

BIST > Larger cores > external vendors > in-house cores.

Defining test cores priority bases on test priority:

$$priority_{t_i} = \sum_{i=1}^{c} \delta_i \omega_i$$

C--- Priority category;
w_i--- Rank ordering of every sort of priority;
δ_i---Weighted value.
The larger δ_i goes, the higher priority is. If δ_i =1, the priority is BIST, priority ordering is the first; If δ_i =2, the priority is larger cores, priority ordering is the second; If δ_i =3, the priority is external vendors cores, priority ordering is the third; If δ_i =4, the priority is in-house cores, priority ordering is the last.

4. Algorithm Implementation

The SOC test scheduling problem is solved after choosing wrapper solutions from the sets of pre-optimized wrapper designs. Evolutionary algorithm and sequence-pair representation are used in this paper.

The flow of SOC test schedule is as follows: First of all, rank the priority of test cores according to the information of circuits and priority requirements of ITC'02 SOC test benchmarks. Second, apply QA to solve the problem P_{aw}. That is to say, fix TAM width and distribute SOC models to every TAM in reason to get the minimum test time. Specifically, the observation value of every unit is the sequence number of TAM which IP cores are distributed on, and the corresponding test time is needed to be calculated. In addition, the total power also needs to be distinguished whether it is satisfy the power constraint. If it doesn't fit the constraint, change genetic factor and wash it out. Next save the minimum test time as the current optimal value. Then modulate the corresponding probability angle and optimize units in colony. Finally find the one which satisfies power constraint in the colony.

5. Experimental Results

The algorithm is working in the environment of the Pentium 4 CPU 1.60GHz, 512M of memory, and VC + + 6.0. Using standard C programming, simulation results are based on partial hierarchical SOCs in ITC'02 Test Benchmark [7]: p22810, p93791 and d695 (Table 1 (a)(b)(c), Table 2).

Testing item including: W---TAM partition; Power limit---power constrained value; Time---test time; Ip--- test time improvement; Precedence---precedence constraint.

The results reveal QA a great superiority in raveling out SOC test scheduling with multiple constraints.

The highest value is 55% for d695, 46% for p22810, 50% for p93791 with power constraint; the best improved value of p93791 achieves 34% with precedence constraint.

6. Conclusions

QA has a great characteristic of quantum superposition, entanglement and interference. And it may carry out calculation in a multi-path parallel processing instead in a single line, which greatly enhances the ability of computing [3]. In this paper QA is adopted to design mathematical model of the SOC test structure on multi-constrained test planning. By means of dividing a number of cores into different sizes, the task group is tested, at the same time, tests scheduling is carried out to meet the power and test priority constraints.

The experimental results show that the parallel test algorithm meets the priority sequence of tests, which not only greatly reduces the test time, but also greatly reduces the test power. It provides a better method of SOC scheduling.

Acknowledgments

This work is supported by National Natural Science Fund under Grant No. 60766001.

References

1. V.Iyengar, K. Chakrabarty, "System-on-a-Chip Test Scheduling With Precedence Relationships Preemption, and Power Constraints," *IEEE transactions on computer-aided design of integrated circuits and systems*, VOL. 21, NO. 9, S, 2002, pp. 1088-1094.
2. Hey Tony, "Quantum computing: An introduction," *Computing & Control Engineering Journal*, 1996, 10 (3), pp. 105-112.
3. J.Pouget, E.Larsson *et al*, "An Efficient Approach to SoC Wrapper Design, TAM Configuration and Test Scheduling," *Proceedings of the Eighth IEEE European Test Workshop (ETW'03)*, 2003, pp. 51-56.
4. S. Chattopadhyay, K. S. Reddy, "Genetic algorithm based test scheduling and test access mechanism design for system-on-chips," In: *Proc .of Int'1 Conf. on VLSI Design. New Delhi: IEEE Pres*. 2003, pp. 341-346.
5. K. Chakrabarty, "Test scheduling for core-based systems using mixed integer linear programming," *IEEE Trans. CAD*, Vol. 19, October 2000, pp. 1163-1174.
6. Y. Huang, W. T. Cheng, C. C. Tsai, "Resource allocation and test scheduling for concurrent test of core-based SOC design," In: *Proc. of Asian Test Symposium. Kyoto:IEEE Press*, 2001, pp. 265-270.
7. S. K. Goel and E. J. Marinissen, "Cluster-based Test Architecture Design for System-on-chip," *Proceeding 20th IEEE VLSI Test Symposium(VTS 2002)*, 2002, pp. 259-264.

Table 1 Results (a)d695, compared with [5][6]

W	Power limit	Time	Time[6]	Ip [6]	Time[5]	Ip[5]	W	Time	Time[6]	Ip [6]	Time[5]	Ip[5]
32	P_{max} =1500	26677	45560	41%	43541	39%	80	18370	20941	12%	24369	25%
	P_{max} =1800	25829	44341	42%	42450	39%		17645	20467	14%	18774	6%
	P_{max} =2000	25828	43221	40%	42450	39%		17636	19206	8%	18691	6%
	P_{max} =2500	25828	43221	40%	41847	38%		15698	19206	18%	18691	16%
48	P_{max} =1500	23050	31028	26%	42450	46%	96	17927	20914	14%	23425	23%
	P_{max} =1800	22980	29919	23%	32054	28%		14297	18077	21%	18774	24%
	P_{max} =2000	22804	29419	22%	29106	22%		13906	17825	22%	17467	20%
	P_{max} =2500	22804	29023	21%	29106	22%		12663	15847	20%	17257	8%
64	P_{max} =1500	19080	27573	31%	42450	55%	128	17260	16841	-2%	19402	13%
	P_{max} =1800	18869	24454	23%	23864	21%		13573	14899	9%	16804	19%
	P_{max} =2000	18869	24171	22%	21942	14%		13573	14128	4%	14469	6%
	P_{max} =2500	18869	23721	20%	21931	14%		12376	12993	5%	13394	8%

(b) p22810, compared with [5]

W	Power limit	Time	Time[5]	Ip[5]	W	Time	Time[5]	Ip[5]
32	P_{max} =10000	260140	473418	45%	80	194989	195733	0%
	P_{max} =8000	262945	473418	44%		194152	195733	1%
	P_{max} =6000	261604	475951	46%		194305	209559	7%
	P_{max} =5000	262648	472026	44%		195076	364038	46%
	P_{max} =4000	262969	480223	45%		195505	285307	31%
	P_{max} =3000	266022	482963	45%		196250	356215	45%
48	P_{max} =10000	221426	352834	37%	96	154049	159994	4%
	P_{max} =8000	221279	352834	37%		154262	159994	4%
	P_{max} =6000	221362	346461	36%		152895	174928	13%
	P_{max} =5000	221215	382507	42%		181207	266166	32%
	P_{max} =4000	221374	389243	43%		184068	285814	36%
	P_{max} =3000	223549	392525	43%		183603	311632	41%
64	P_{max} =10000	208132	236186	12%	128	124462	128332	3%
	P_{max} =8000	207529	236186	12%		137632	142056	3%
	P_{max} =6000	209334	250487	16%		139282	157568	12%
	P_{max} =5000	208631	321930	35%		178525	246110	27%
	P_{max} =4000	207236	324478	36%		178236	268856	34%
	P_{max} =3000	214690	309255	31%		180959	293021	38%

(c) p93791, compared with [5]

W	Power limit	Time	Time[5]	Ip[5]	W	Time	Time[5]	Ip[5]
32	$P_{max}=30000$	922073	1827819	50%	80	560105	787588	29%
	$P_{max}=25000$	922070	1827819	50%		560108	821475	32%
	$P_{max}=20000$	922070	1827819	50%		560104	821575	32%
	$P_{max}=15000$	961195	1827819	47%		574305	848050	32%
	$P_{max}=10000$	1668308	1827819	19%		1026965	1091210	6%
48	$P_{max}=30000$	629449	1220469	48%	96	332315	639217	48%
	$P_{max}=25000$	629448	1220469	48%		332313	639217	48%
	$P_{max}=20000$	629447	1220469	48%		332313	658132	50%
	$P_{max}=15000$	657341	1220469	46%		341909	631214	46%
	$P_{max}=10000$	1125154	1220469	8%		578500	691866	16%
64	$P_{max}=30000$	605605	945425	36%	128	326328	457862	29%
	$P_{max}=25000$	605602	965383	37%		326327	493599	34%
	$P_{max}=20000$	605602	957921	37%		326328	472653	31%
	$P_{max}=15000$	633202	1014616	38%		338387	486469	30%
	$P_{max}=10000$	1099833	1117385	2%		574829	568734	-1%

Table 2 Results p93791, compared with [7]

W	Time [6]	Time QEA	Ip
12	2995271	2724496	9%
18	2126800	2213416	- 4%
24	1616442	1394149	14%
30	1266674	1176123	7%
36	1088861	938932	14%
42	989252	931911	6%
48	1020215	674809	34%
54	827545	690792	17%
60	752312	590659	21%

Design and Modeling of Jet Dispenser Based on Giant Magnetostrictive Material

Zhiqi Ge, Guiling Deng

Key Laboratory of Modern Complex Equipment Design and Extreme manufacturing, Central
South University, Ministry of Education
College of Mechanical and Electronic Engineering; Central South University; Changsha Hunan 410083; China
gzq821224@163.com
+8613786124572

Abstract

The rapid development of microelectronics packaging technology drives the development of packaging equipment manufacturing industry. As an important equipment of packaging technology, fluid dispensing system is developed rapidly. Non-contact jet dispensing is the latest technology of dispensing technology. Compared to the conventional contact dispensing methods, jet dispensing technology is a non-contact method and has the following advantages: (1) it has high flow rate, high dispensing frequency; (2) This method needn't z-axis motion; (3) it has high dispensing accuracy and smaller wet area; (4) Needle bending and Chip surface damage situation does not arise.

In this paper, we design a non-contact jet dispensing device based on giant magnetostrictive material. The giant magnetostrictive material (GMM) is a new type of function material with giant strain, high response speed, high power density and great output force. For this jet dispenser, the GMA (Giant Magnetostrictive Actuator) provides the driving force of dispensing. Because the magnetostrictive coefficient of GMM is small, we design a displacement amplifying mechanism with flexure hinges to amplify the output displacement of giant magnetostrictive actuator. The structure and principle of jet dispenser with a displacement amplifying mechanism are presented. In order to obtain the dynamic characteristics of the jet dispenser, its dynamic model and the transfer function block diagram of dispenser between excitation current and output displacement is established. The jet dispenser's dynamic characteristics are simulated though the use of matlab. The amplitude-frequency response characteristics and dynamic characteristics is obtained. The amplifying ratio N ($N = l_2 / l_1$) of displacement amplifying mechanism is verified, in this paper, the amplifying ratio N is 5.

Discussion 1 Non-contact fluid jetting technology

The non-contact fluid jetting technology has its own prominent merits: higher efficiency, little sensitivity to variations in board height, accurate-jetting with high repeatable and smaller dot, especially jetting a wide range of adhesive, such as surface mount adhesive, under-fill adhesive, silver epoxy. It is considered as the next generation fluid dispenser technology [1-2]. But, at present, only two companies have patents about non-contact fluid jetting and the driving force of needle is air pressure, see Fig.1. In order to break the patent protection, we design jet dispenser based on GMM.

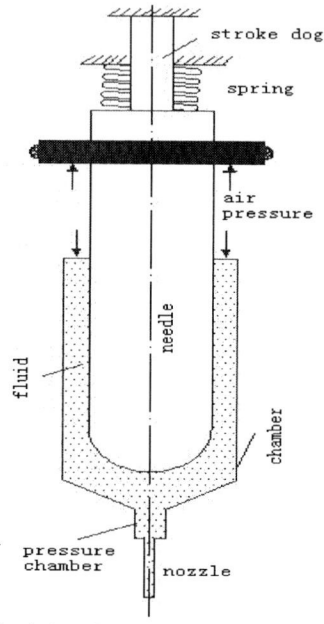

Fig. 1 the principle of air-spring driven jetting valve

Discussion 2 Design and modeling of jet dispenser

2.1 Design of jet dispenser based on giant magnetostrictive material

For this jet dispenser, the GMA (Giant Magnetostrictive Actuator) drives the needle, see Fig. 2. The jet dispenser with a displacement amplifying mechanism with flexure hinges can get big displacement of the needle, see Fig. 3. Accessing cycle square-wave current to the rod, cycle magnetic field can be generated. Then the giant magnetostrictive rod length changes periodically and drives the needle to achieve high-precision movement. When the needle contacts with the nozzle, the liquid dynamic pressure effect is generated between the needle's spheric surface and the nozzle's taper surface, see Fig. 4. Then the glue can be jetted out from the nozzle forming droplets.

978-1-4244-4658-2/09 $25.00 © 2009 IEEE   894

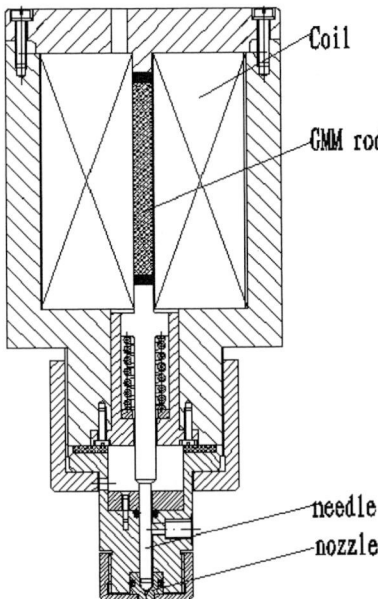

Fig. 2 Structure of jet dispenser

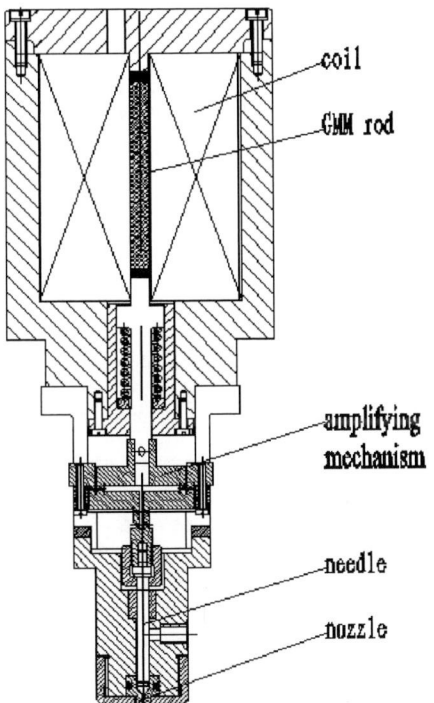

Fig. 3 Structure of jet dispenser with displacement amplifying
mechanism

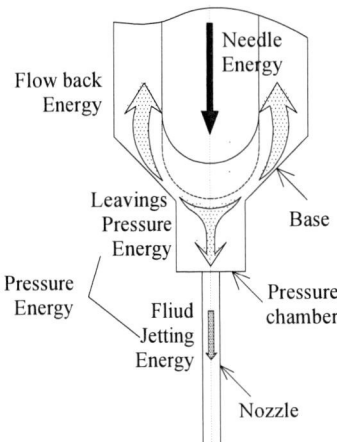

Fig. 4 Principle of fluid jetting

2.2 Displacement amplifying mechanism with flexure hinges

The displacement amplifying mechanism with flexure hinges is a new mechanism that uses the elastic deformation energy of the flexure hinges to achieve the transmission and conversion of movement and force. Flexure hinge is transmission component that is a small size, no mechanical friction, no gaps and high sensitivity. It is widely used in the requirements of small angular displacement and high-precision rotation. Pseudo-rigid body model is a means that the amplifying mechanism's flexure hinges is simplified rigid rod model [3-4]. For this paper's amplifying mechanism, the flexure hinges can be equivalent to a torsion spring and the other parts of amplifying mechanism can be equivalent to leverage. The amplifying ratio N ($N = l_2/l_1$) of amplifying mechanism we designed is 5, see Fig.5. The simplified model can be achieved by means of pseudo-rigid body mode [5-6], see Fig.6.

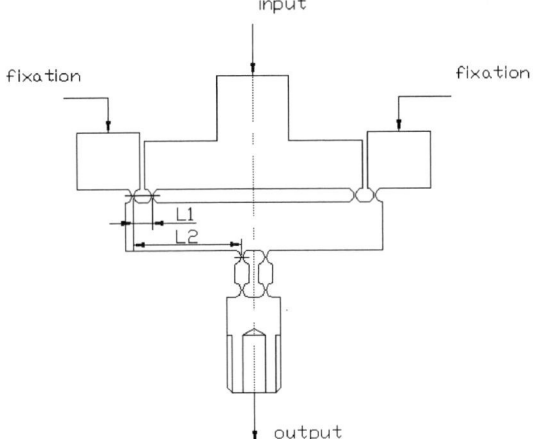

Fig. 5 Amplifying mechanism with flexure hinges

978-1-4244-4658-2/09 $25.00 © 2009 IEEE

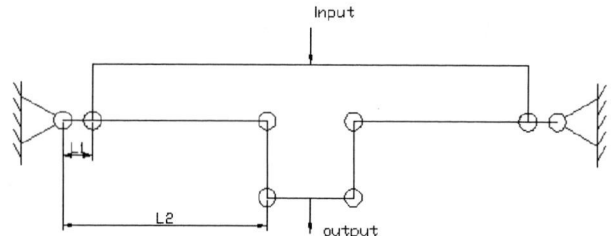

Fig. 6 Model of amplifying mechanism

The rotational stiffness of flexure hinge can be obtained by Bernoulli-Euler's equation, see Fig. 7.

$$\frac{\mathrm{d}\theta}{\mathrm{d}s} = \frac{M(\text{x})}{EJ(\text{x})} = \frac{1}{\rho} \qquad (1)$$

where E is Young's modulus of the material, M(x)is the bending moment applied to dx, J(x) is the moment of inertia of dx cross-section around z axis, ρ is radius of curvature.

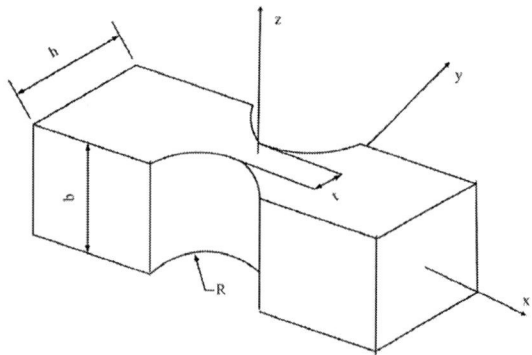

Fig. 7 Single-axis flexure hinge

From the knowledge of Higher Mathematics, we can obtain

$$\theta = \int_0^\pi \frac{12Mr\sin\alpha}{Eb(2R+t)^3}d\alpha \qquad (2)$$

Then the rotational stiffness formula can be obtained

$$\mathrm{k}_\theta = 1 \Big/ \int_0^\pi \frac{12R\sin\alpha}{Eb(2R+t)^3}\mathrm{d}\alpha \qquad (3)$$

In this paper, the material of amplifying mechanism is 65Mn, E=210GPa, the Poisson's ratio μ is 0.3, b=10mm, R×t=2mm×0.3mm.Then the rotational stiffness can be calculated $\mathrm{k}_\theta = 5.2791$.

2.3 Dynamic model of jet dispenser

When the input excitation is cycle current, the dynamic model of jet dispenser can be divided into three subsystems: flux equation, magnetostrictive force equation and force balance equation [7-9]. The dynamic model can be equivalent to a spring-mass-damper system, see Fig.8.

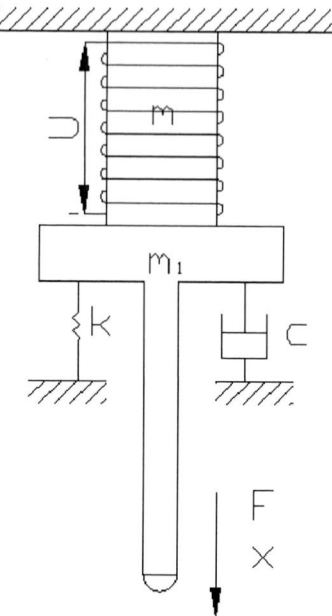

Fig. 8 Dynamic model of jet dispenser without amplifying mechanism

The magnetic momentum of the GMA magnetic circuit can be divided into two parts: one generated by the excitation current and the other generated by the stain of GMM rod. So the flux equation can be expressed as

$$\phi = \frac{NI(t) + \dfrac{x}{d'_{33}}}{P} \qquad (4)$$

Where d'_{33} is vertical piezomagnetic coefficient of the GMM rod, $d'_{33} \approx d_{33}$, P is the total magnetic reluctance, N is the coil turns.

The magnetostrictive force equation of GMM rod can be expressed as

$$F = \frac{\Phi}{d_{33}}K \qquad (5)$$

where d_{33} is magnetic-field coupling coefficient, K is eddy current coefficient. $\mathrm{K} = \dfrac{2\delta}{\mathrm{r}}$, r is the GMM rod radius.

$\delta = \sqrt{2/\omega\sigma\mu_r\mu_0}$, where ω, σ, μ_0 and μ_r are excitation frequency, GMM conductivity, vacuum permeability and GMM relative permeability.

2.4 The force balance equation of jet dispenser

2.4.1 The force balance equation of jet dispenser without amplifying mechanism

The force balance equation of jet dispenser without amplifying mechanism can be obtained by Newton's second law, see Fig.8, where m is the quality of GMM rod, m_1 is the mass of needle, k is the effective stiffness of jet dispenser, c is the effective damping coefficient.

$$m_e \overset{..}{x} + c_e \overset{.}{x} + k_e x = F \qquad (6)$$

Where $m_e = \dfrac{1}{3}m + m_1$.

2.4.2 The force balance equation of jet dispenser with amplifying mechanism

When the model of amplifying mechanism was simplified, the mechanical model of jet dispenser with amplifying mechanism can be obtained. It's was shown in Fig9. x is the output displacement of the needle, x_1 is the output displacement of the GMM rod, θ_1, θ_2, θ_3 and θ_4 is the rotation angle of flexure hinges, F is the output force of GMM rod that applied to the amplifying mechanism, F_1, F_2 is the force applied to the flexure hinges by rigid rod, m_1, m_2, m_3, m_4 and m_5 is the different part mass of amplifying mechanism. From the leverage theory, we can obtain

$$\theta_1 = \theta_2, \theta_3 = \theta_{14} = \theta_1 - \theta_4, l_2\cos\theta_1 + l_4\sin\theta_4 = l_2.$$

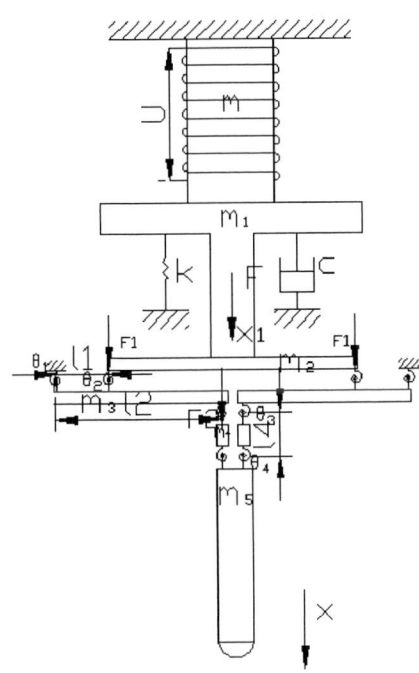

Fig. 9 Dynamic model of jet dispenser with amplifying mechanism

Because

$$\theta_1 = \frac{x}{Nl_1}, \quad \theta_2 = \theta_1 = \frac{x}{Nl_1}, \quad \sin\theta_4 = \frac{l_2}{l_4}(1 - \cos\theta_1)$$

,we can obtain by power series formula of sine and cosine

$$\theta_4 = \frac{l_2 x^2}{2N^2 l_4 l_1^{\,2}},$$

$$\theta_3 = \theta_{14} = \theta_1 - \theta_4 = \frac{10 l_1 l_4 x - l_2 x^2}{2N^2 l_1^{\,2} l_4}$$

then

$$\theta_4 = \frac{l_2 x^2}{2N^2 l_4 l_1^{\,2}}$$

Because x is tens of microns, $\dfrac{l_2 x^2}{2N^2 l_4 l_1^{\,2}}$ is very

small compared to $\theta_1 = \dfrac{x}{Nl_1}$, then $\theta_4 \approx 0$.

$$\theta_3 \approx \theta_1 = \theta_2 = \frac{x}{Nl_1}$$

Therefor

Newton's differential equations of m, m_1 and m_2 can be obtained by Newton's second law

$$\left(\frac{m}{3} + m_1 + m_2\right)\overset{..}{x}_1 + c_1 \overset{.}{x}_1 + kx_1 = F - 2F_1 \qquad (7)$$

where c_1 is the damping coefficient of GMM rod.

Rotation differential equations of leverage can be expressed as

$$J \overset{..}{\theta}_1 + k_\theta \theta_1 + k_\theta \theta_2 + k_\theta \theta_3 = F_1 l_1 - F_2 l_2 \qquad (8)$$

where J is the moment of inertia of leverage m_3, $J = \dfrac{1}{3}m_3 l_2^{\,2}$, $\overset{..}{\theta}_1 = \dfrac{\overset{..}{x}}{l_2}$, $\theta_1 = \dfrac{x}{l_2}$.

Newton's differential equation of the needle can be obtained by Newton's second law

$$(2m_4 + m_5)\overset{..}{x} + c_2 \overset{.}{x} = 2F_2 \qquad (9)$$

where c_2 is the transient fluid damping coefficient of glue.

From the geometric relationship of amplifying mechanism,

$$\frac{\overset{..}{x}_1}{\overset{..}{x}} = \frac{\overset{.}{x}_1}{\overset{.}{x}} = \frac{x_1}{x} = \frac{l_1}{l_2} = \frac{1}{N}$$

From Eqs. (7), (8) and (9), we can deduce differential equation of jet dispenser with amplifying mechanism

$$\left(\frac{m}{3N} + \frac{m_1}{N} + \frac{m_2}{N} + \frac{2m_3 N}{3} + 2Nm_4 + Nm_5\right)\overset{..}{x}$$

$$+ (c_1 + c_2)\overset{.}{x} + \left(\frac{k}{N} + \frac{6k_\theta}{l_1 l_2}\right)x = F \qquad (10)$$

Then Eqs. (10) can be written as

$$m_e \overset{..}{x} + c_e \overset{.}{x} + k_e x = F \qquad (11)$$

where m_e, c_e and k_e is the effective mass, effective damping coefficient and effective stiffness,

$c_e = \frac{2k_e\xi}{\omega_n} = 2\xi\sqrt{k_e m_e}$, ξ is damping ratio, in this paper, $\xi = 0.25$.

We can obtain the transfer function between the output displacement of the needle and the output force of the GMM rod from Eqs.(6) or (14) by the Laplace transform

$$G_1(s) = \frac{X(s)}{F(s)} = \frac{1}{m_e s^2 + c_e s + k_e} \quad (12)$$

From Eqs.(4), (5) and (12), we can obtain transfer function block diagram of the jet dispenser transfer function block diagram, see Fig.10.

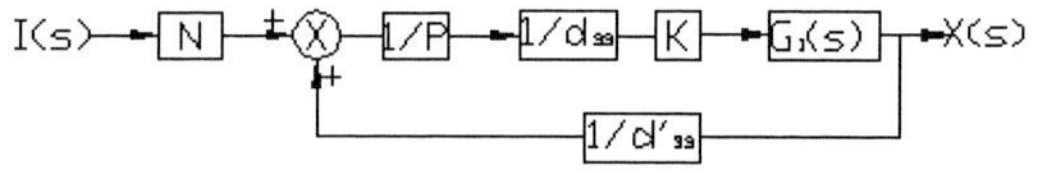

Fig. 10 Jet dispenser transfer function block diagram

Then the transfer function can be obtained

$$G(s) = \frac{X(s)}{I(s)} = \frac{NKd_{33}}{d_{33}^2 P(m_e s^2 + cs + k_e) - K} \quad (13)$$

Above equation parameters are in Table 1, 2, 3.

Table 1 Jet dispenser parameters

Name	Symbol	Units	Value
Vacuum permeability	μ_0	H/m	$4\pi\times10^{-7}$
GMM relative permeability	μ_r		10
GMM conductivity	σ	S/m	1.67×10^6
GMM rod stiffness	k	N/m	6.6×10^7
Coil turns	N		1300
GMM rod radius	r	mm	5
Piezomagnetic coefficient	d_{33}		2.9×10^{-6}
Coil resistance	R	Ω	3.09
Magnetic circuit reluctance	P	H	1.0×10^8
Damping ratio	ξ		0.25

Table 2 Parameters without amplifying mechanism

Name	Symbol	Units	Value
Effective mass	m_e	kg	0.19×10^{-3}
Effective stiffness	k_e	N/m	6.61×10^7

Table 3 Parameters with amplifying mechanism

Name	Symbol	Units	Value
Effective mass	m_e	kg	0.36×10^{-3}
Effective stiffness	k_e	N/m	1.32×10^7

2.5 Simulation results

Matlab is a powerful software tool for simulation. System simulation model can be obtained by transfer function block diagram. Imposing some input signal to the system simulation model, the output response and the dynamic characteristics of jet dispenser can be obtained, see Fig. 11, 12, 13.

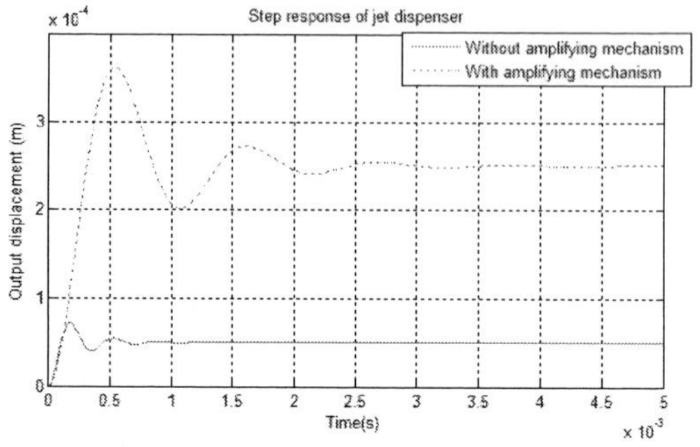

Fig. 11 Step response of jet dispenser

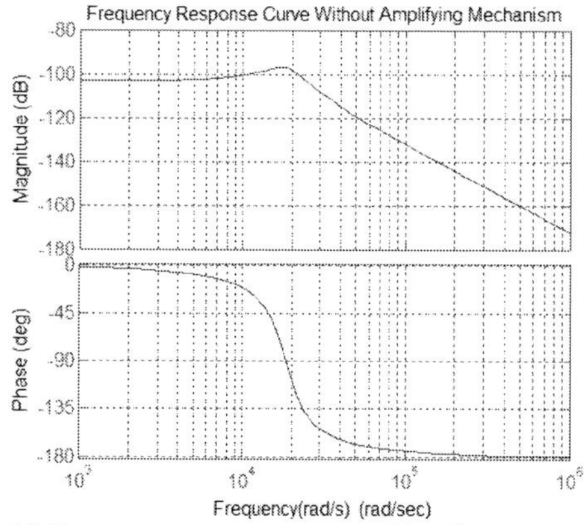

Fig. 12 Frequency response curve of jet dispenser without amplifying mechanism

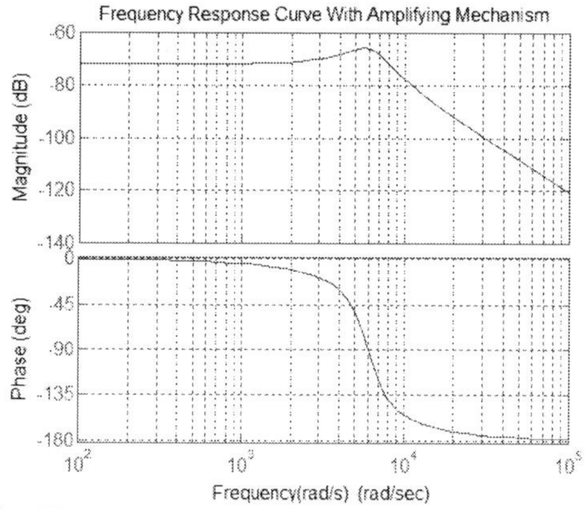

Fig. 13 Frequency response curve of jet dispenser with amplifying mechanism

Discussion 3 Outlook

The output response and the dynamic characteristics of jet dispenser were obtained in this paper. The next step is manufacturing the jet dispenser and debugging the experimental results.

Conclusions

1 Through the simulation of matlab, the feasibility of jet dispenser with amplifying mechanism was verified. The amplifying ratio is about 5 from the Fig11. It is in accordance with the design amplifying ratio. The jet dispenser has high frequency response characteristics from Fig.12, 13.

2 The parameters affected the dynamic characteristics can be obtained from Fig11. With increasing of the jet dispenser's effective mass, the rise time increases and the response speed slows down. The dynamic characteristics can be improved by reducing the mass of jet dispenser, such as the nozzle and amplifying mechanism.

3 The Rise time of jet dispenser with amplifying mechanism is less then 0.5 ms and the output displacement of it is about 0.25mm from Fig11. These two parameters meet design requirements.

Acknowledgments

This paper was sponsored by 863 Project (2007AA04Z350) and National Natural Science Project (50475138).

References

1. Steven J. Adamson, David Wang, "A Change in Dispensing Technology--Jetting Takes Off," *Semiconductor*, Vol. 29, No. 3, Mar. 2004, pp. 67-70.

2. Chen Kui-yu, Deng Gui-ling, "The Influence of Actuator Parameters on Fluid Jetting Dispensing," Computer Simulation, Vol. 24, No. 11, Nov.2007, pp. 289-292.

3. Howell L L, "The development of force deflection relationships for compliant mechanisms," *ASME, Machine Elements and Machine Dynamics*, Vol. 71, No. 4, Mar. 1994, pp. 501-508.

4. Ma Hao-quan, Hu De-jin, Zhang Kai, "Micro-displacement amplifying mechanism driven by piezoelectric actuator," *Journal of Southeast University*, Vol. 20, No. 1, Mar. 2004, pp. 75-79.

5. Chen Gui-min, Jia Jian-yuan, "On Right-circular Elliptical Flexure Hinges," *Machine Design and Research*, Vol. 21, No. 4, Mar. 2005, pp. 37-38.

6. Fan Ye-sen, Jia Wang San-min, "Equivalent Dynamic Analysis Method Lumped Compliant Mechanisns," *Mechanical Science and Technology for Aerospace Engineering*, Vol. 26, No. 7,July. 2007, pp. 880-884.

7. Xia Chun-lin, Ding Fan, Lu Yong-xiang, "Modeling of Giant Magnetostr ictive Actuator," *CHINA MECHANICAL ENGINEERING*, Vol. 11, No. 11, Nov. 2003, pp. 1288-1291.

8. Wang Wen-jing, Yu Yue-qing, Lu, "Dynamic Modeling and Analysis of Compliant Mechanism With Elliptical Flexure Hinges," *Machine Design and Research*, Vol. 23, No. 6, Dec. 2007, pp. 54-57.

9. Wng Chuan-li, Ding Fan, Zhang Kai-jun, "The Study on Simulation of Dynamic Characteristic of Rare-Earth GMA," *JOURNAL OF SYSTEM SIMULATION*, Vol. 15, No. 3, Mar. 2003, pp. 379-381.

Simulation and Experimental Study on Temperature Field of Fluid Jet-dispenser

Zhengrong Ding, Hao Hu, Guiling Deng
Key Laboratory of Modern Complex Equipment Design and Extreme manufacturing
Central South University, Ministry of Education Changsha 410083, China
Email:dingzhengrong@163.com, huhaodlu@163.com

Abstract

Fluid jet-dispensing technology has been widely used in micro-electronic encapsulation industry. The fluid viscosity varies as temperature changes, while it affects the process remarkably.Besides basic method on temperature field solution, some thermal behaviors, including heat conduction and heat exchange, boundary conditions and some physical parameters are overall analyzed in this paper. Based on the simulation software platform ANSYS, a two-dimensional transient temperature field model of jet-dispenser was built, and simulation programs on temperature field for two sorts of fluid with obviously different thermal conductivity were performed. Then heating experiments were finished to validate the feasibility of FEA, and results show that temperature distribution is homogeneous along fluid area and it meets the conditions of jetting.

1. Introduction

Fluid jet-dispensing technology is a high-speed, high-quality and low-cost dispensing possess, which is widely used in the microelectronic packaging industry and regarded as the next-generation fluid dispensing technology [1]. The mechanism of fluid-jetting is very complex, and one of the factors that influence dispensing effect and bonding quality is fluid viscosity. Each sort of fluid has its own viscosity, which varies with temperature remarkably. Generally speaking, with temperature increasing, thermal motion of fluid molecular becomes fiercer, so when the interaction force decreases, the fluid viscosity decreases, accordingly.

In practice, for the purpose of getting a stable-viscosity fluid, we placed a heater near the nozzle to keep fluid temperature under control. Different sorts of fluid have different sensitivities to temperature, what's more, in different temperature range, the effects that temperature imposes on fluid are different. Therefore, to get a good-consistency adhesive dot, it's necessary to keep temperature gradient in a small range, namely homogenize temperature field. Based on the relationship between temperature and viscosity, the author did some heating research on different fluid with obviously different thermal conductivity. Modal heat flux and temperature field in fluid area was confirmed by simulation, and temperature characteristic under certain heating conditions of the dispenser was obtained.

2. Basic theory of temperature field

Where there are temperature grads in a body or among objects, the heat transfers from high-temperature parts to lower ones. Here, the heat generated by the heater was transferred through the chamber, fluid filed and then to ball-needle. This kind of heat transfer mode is called conduction of heat. The temperature distribution of a certain point at any time in the dispenser could be described with three-dimension transient thermal conductivity differential equation [2]:

$$\rho c \frac{\partial T}{\partial t} = \frac{\partial}{\partial x}(k\frac{\partial T}{\partial x}) + \frac{\partial}{\partial y}(k\frac{\partial T}{\partial y}) + \frac{\partial}{\partial z}(k\frac{\partial T}{\partial z}) + \dot{q} \tag{1}$$

Where ρ is density, C is specific heat,k is coefficient of thermal conductivity, $\dot{q}$ is inner origin of heat.

The surface of the chamber was surrounded by air and heat transferred only because of the temperature differences. As the air flows when being heated, this heat transfer mode is called natural convection which is expressed by Newton cooling equation

$$q=h \times \Delta t \tag{2}$$

Where q is heat flux, h is convection coefficient, Δt is temperature difference between the chamber and air. The heating process of the fluid jet-dispenser belonged to transient heat transfers, and the temperature, the heat flux, the boundary condition and the system internal energy changed obviously in different time. According to the principle of conservation of energy, transient heat balance can be written as (matrix)

$$[C]\{\dot{T}\} + [K]\{T\} = \{Q\} \tag{3}$$

[K] is transfer matrix including thermal conductivity, convection, radiance and shape coefficient;

[C] is specific heat matrix, considering the increase of system internal energy;

{T} is vector of nodal temperature;

$\{\dot{T}\}$ is equal to $\frac{\partial T}{\partial t}$;

{Q} is vector of nodal heat current, including heat generation.

3. Thermal analysis simulation

The working principle of the fluid jetting dispensing is that the nozzle needle repeatedly acts, driven by driving and restoring force, and moves up and down in the fluid chamber. When the ball-needle stroke down, the fluid around the ball-needle tip is extruded, and the fluid near the nozzle ejects out and forms a dot. It is clear that controlling the viscosity of the fluid, especially the fluid near the nozzle, which influences the flow of the fluid in the chamber, by controlling the temperature, is a key factor to the quality of the fluid jetting dispensing.

3.1 Physical model

The structure of the fluid jet-dispenser has been shown in the fig.1; the heater wires the surface of the cylindrical nozzle and heat flows into it and conduct. It's a three dimensional heat conduction process, so that it's difficult to describe the situation of the heat conduction exactly. Therefore, according to practical working situation, we should take some appropriate assumptions and simplifying steps as well as abandon some minor factors to solute and analyze [3-4].

1. Considering the axial symmetry of geometry structure, loaded heat flux and boundary condition of the fluid jetting dispenser, it was sensible to choose a half of the dispenser along longitudinal profiles and build the finite element model.

2. Because the dispenser was manufactured very precisely, so thermal contact resistance can be neglected, and the dispenser can be considered as a whole body, seamless and no thermal resistance.

3. Neglecting the adhesive pipeline that was small size and it was favorable of meshing finite model.

4. Supposing the uniform heat flow generated by heater flowed into the jetting dispenser totally. Neglecting the energy wasted by heating the heater itself.

5. Neglecting the thermal radiation because of the low temperature.

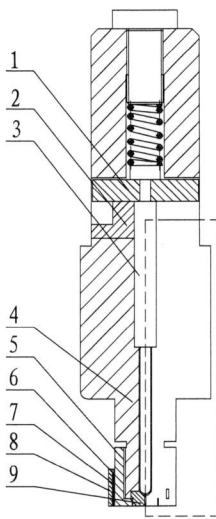

Fig. 1 Structural schematic of jet-dispenser 1.piston 2. stroke stop 3. ball-needle 4.valve body 5. press shell 6. electric heating film 7. insulating layer 8. sleeve 9. nozzle (In view of technical privacy protection, this figure has been simplified and changed from original drawing.)

Heat is generated from the electric heating film (part 6), transmitting in radial direction and to other parts. According to the assumption 2, the fluid jet-dispenser can be considered as a close-coordinated and seamless body without any thermal resistance. Considering the vacancy between the stop (part 2) and the piston (part 1), and compared with the injector (45 steel), the thermal conductivity of air is very small, so it can be regarded as a convective heat transfer with the air when the heat passed by this part. Although the heat can still be transmitted up along the shell (part 4), but compared with fluid area, it is very thin and can only transmit a little heat, so it has little influence. Therefore we set up a physical model based on the area under the stop (as shown in the dashed line pane in Figure 1), as shown in the Figure 2.

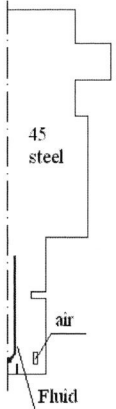

Fig. 2 Geometrical model of jet-dispenser for thermal analysis

3.2 The Physical Parameters of the Material

Two sorts of fluid with obviously different thermal conductivity are studied in this issue: conductive silver colloid and silicone resin. Their heat conduction coefficients change with temperature, but just a little bit under a low temperature, so the changes with temperature are ignored here. Their physical parameters are shown in the Table 1 as follows:

Table 1 Physical Parameters of the Material

	heat conduction coefficient W/(m·K)	density kg/m^3	specific heat capacity J·(kg·K)$^{-1}$
45 steel	50.2	7800	448
silicone resin	0.35	972	1550
conductive silver colloid	25.8	3500	1050
air	2.59×10^{-2}	1.205	1005

3.3 Determination of the thermal loading

An electric heating film is used as the heater to offer heat for fluid jet-dispenser and the heat is assumed to flow directly into the dispenser. It can be expressed as a formula as following:

$$-k \frac{\partial T}{\partial n}\Big|_{\Gamma} = q \qquad (4)$$

Where q is heat flux generated by the heater; T is wall temperature; k is coefficient of thermal conductivity; Γ is the contacting boundary.

According to the formula, Q=Aq, thermal power could be transferred to heat flux. A is the surface area of the chamber. As the height of the chamber bottom is 10mm, diameter is 17mm, the surface area A= 533.8×10^{-6} m^2. Besides, the practical resistance of electric heating film is 37Ω, and the practical voltage is 24.2V, so the actual power turns out to be 15.83W. Therefore, the heat flux q=Q/A=29652W/m^2, which also means the heat flux flown into the dispenser. It was loaded to the simulation model, as shown in Fig.1.

3.4 Boundary condition

When the chamber contacts with the air, thermal transmission occurs as a form of convection heat transfer, which can be categorize into the third boundary condition that can be described by equation (5):

$$-k\frac{\partial T}{\partial n}\Big|_{\Gamma} = h\left(T - T_f\right)\Big|_{\Gamma} \tag{5}$$

Where α is the coefficient of convection heat transfer; T_f is the media temperature; other variables are idem.

The convection boundary condition was set up on the boundary of the chamber and air. The convection coefficient was set up to 2 W/(m² · ℃), and air temperature was 293K (20°C).

Besides, on the top of the model, a large part of the dispenser is exposed in the air and far away from the heating center and temperature area which has little connection with the temperature field. Therefore, it was also set up to an adiabatic boundary condition.

The center line of the physical model satisfies the adiabatic boundary condition and it can be explained as the formula following:

$$-k\frac{\partial T}{\partial n}\Big|_{\Gamma} = 0 \tag{6}$$

3.5 Initial condition

Before heated, the temperature of the chamber was considered to be evenly distributed which can also be considered as the environmental temperature:

$$T=T_0 \tag{7}$$

T_0 was the environmental temperature, 20°C.

3.6 Calculation of the heat transferring simulation

In this issue, two sorts of fluid with dramatically different heat conduction coefficient have been analyzed by simulation separately, while simulation environment and parameters were set as the same. And analysis unit chosen for thermal analysis was plane thermal element with 4-node, plane55. Total element number of the elements was 6550 while node number 6803. Temperature distribution of fluid area near the nozzle head was the focus of the analysis. As amplified in Fig.3, there were 224 elements and 279 nodes.

Heat flux was loaded to the finite element modal and the boundary condition was fixed as shown in Fig.3. The simulation termination time was set up to 240 seconds and automatic time stepping was turned off and time step was set up to 5 seconds with 48 load steps in total. The initial temperature of the model was considered the same as the room temperature, 293K (20°C). And related options were also set for the transient solution.

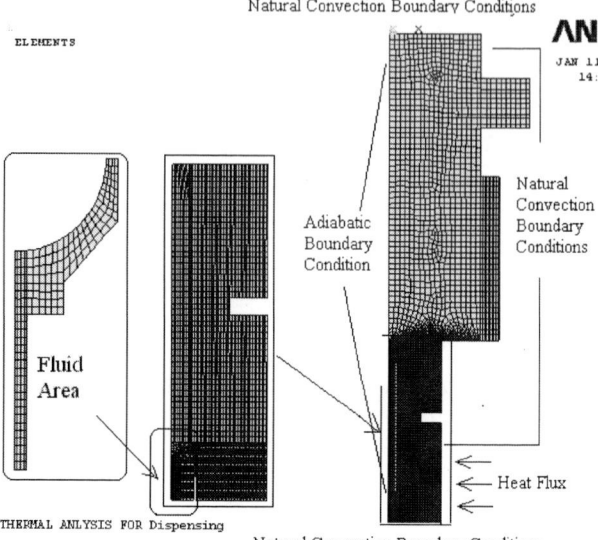

Fig. 3 Mesh model, load and boundary conditions of jet-dispenser for thermal analysis

3.7 simulation results of heat transmission

After solution, the results read from POST1 show that the temperature of both fluid under the ball-needle achieved the scheduled temperature (30~60°C) in 10~70 seconds [5]. General temperature and partial silicone resin fluid temperature profile in jet-dispenser are shown as Fig. 4 and Fig. 5.

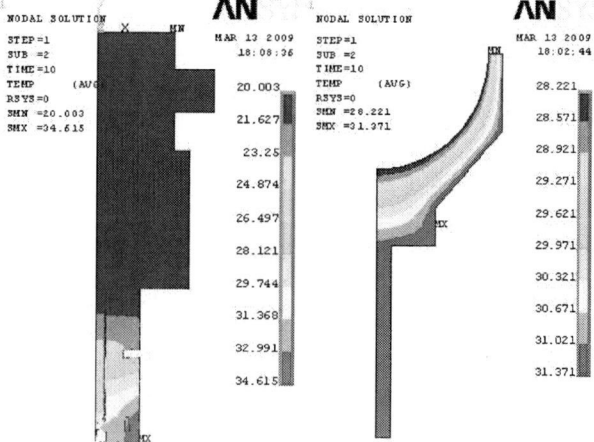

Fig. 4 General and fluid area temperature profile in jet-dispenser at 10s

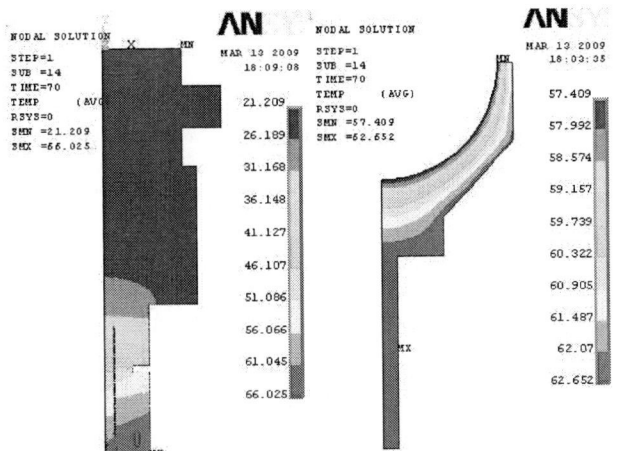

Fig. 5 General and fluid area temperature profile in jet-dispenser at 70s

As shown in Fig. 4 and Fig. 5, general temperature gradient is large, and temperature nearby the heater is high. And in the area far away from fluid area, temperature increased just a little bit. So it is easy to conclude that heat affected zone is small, and fluid temperature distribution is relatively uniform.

Besides, in different load steps, with ANSYS internal function ×VGET, node temperature can be extracted from database to obtain element temperature of fluid under the ball-needle. With function series ×VSFUN, average temperature (MEAN), maximum temperature (MAX), minimum temperature (MIN) and temperature standard deviation (STDV) can all be obtained. Temperature of silicone resin varied as time, as is shown in fig.6. The figure shows that there are no large difference among MAX, MIN and MEAN, and it gives a further illustration of the uniform temperature distribution in fluid area. After 70s, average temperature of fluid under the ball-needle raised from 20°C to 60°C and average temperature rised by 0.43°C/s, with increasing rate of temperature fast followed slow. It satisfied the scheduled requirement and the increasing rate of temperature in jet-dispensing process.

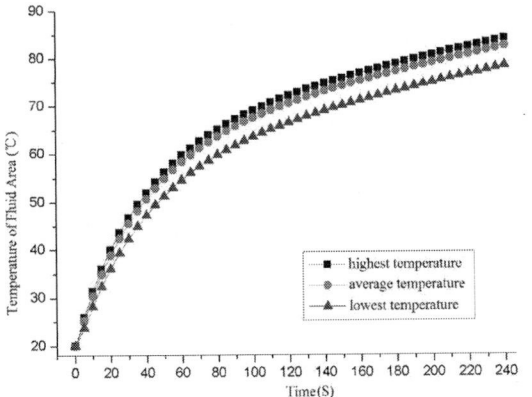

Fig. 6 Highest temperature, average temperature and lowest temperature of silicone resin at different times

The simulation analysis results of conductive silver colloid, with the same parameters, show that general temperature distribution in the dispenser is almost the same as silicone resin one, what's more, temperature raising curves are almost the same, as shown in Fig.10. The differences is that temperature standard deviation of former is smaller, that's to say, the temperature field is more uniform, as shown in Fig. 7 and Fig. 8.

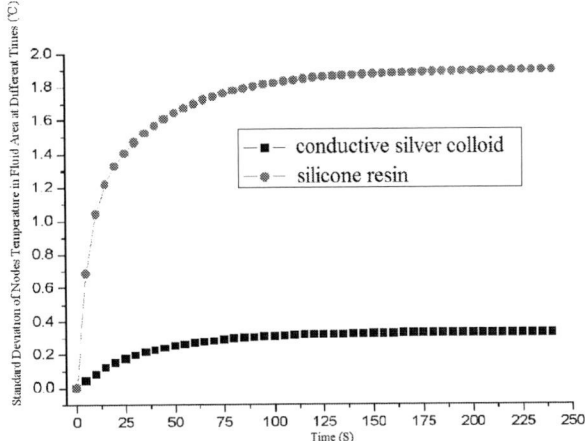

Fig. 7 Temperature standard deviation of different fluid at different times

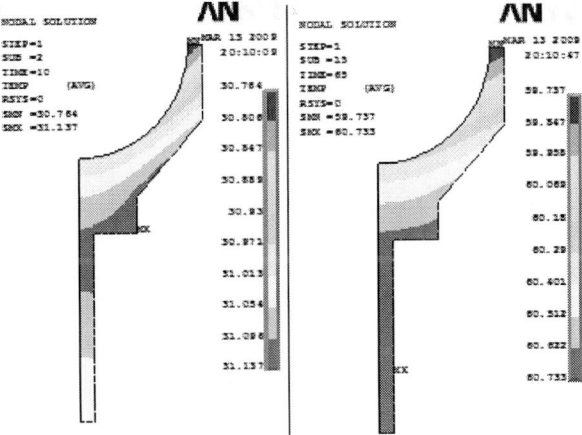

Fig. 8 Temperature distribution of conductive silver colloid at 10s and 65s

As shown in Fig. 7, temperature standard deviation increases as time, that is, fluid temperature distribution has an uneven tendency. And the tendency is first fast then followed slow, finally to balance, namely, inconsistency of temperature distribution would saturate finally. At needed temperature 60°C, temperature standard deviation of silicone resin and conductive silver colloid are 1.742 and 0.282 respectively. Compared with Fig. 5, Fig. 6 and Fig. 8, we know that temperature distribution of conductive silver colloid is more uniform, that is because its thermal conductivity is more than 80 times larger than that of silicone resin, and it is similar to dispenser material, 45 steel. That means it is easier for the heat to spread in conductive silver colloid, so temperature distributes more uniformly. Generally speaking, fluid temperature field meets initial demands, and it provides beneficial conditions for fluid-structure interaction analysis and possibility for fluid consistency, at the same time, it proves that the heater is feasible.

By the way, if heat flux in simulation is loaded smaller, fluid temperature distribution would be more uniform, but the heating time should be prolonged. If heat flux is decreased by half, heating time to get a scheduled temperature would increase several times. So in industrial production, to get the best dispensing quality and highest productivity, both heating power and time should be taken into consideration.

4. Experimental verification

In the experiment, polyimide electric heating film was sticked on the lower part of dispenser nozzle, and platinum resistance PT100 was the temperature sensor. Their installation position is shown in Fig. 9. Intellectual accommodate instrument, LU-906K, was used here to collect temperature data and display in real time.

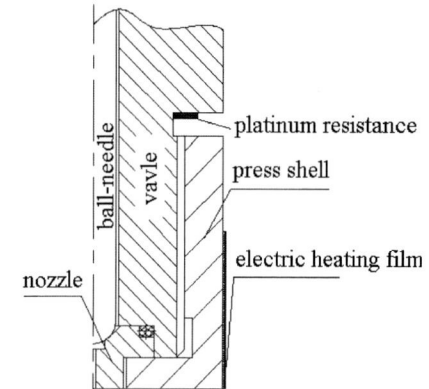

Fig. 9 Installation schematic of platinum resistance and electric heating film

Electric heating film, practical resistance 37Ω, was collected with 24V DC power supply. Temperature data changing with time were recorded, as shown in Fig.10. Temperature was measured twice, and the values were almost the same. Temperature data of all elements in the area where PT100 was installed should be extracted, then to be averaged. The results show that when heating began, simulation value was higher than experimental ones, but both of them increased with the time, and then tend to coincide. All of these prove that simulation analysis is reliable, and the heater designed is feasible. What is more, compared real-time fluid temperature with temperature in measured area, it's easy to get the difference, through which we can control the practical fluid temperature by measuring and modifying a certain point in outer surface.

In addition, the following problems may cause simulation and experimental errors: first of all, fluid physical parameter is not very definite, and the ones used in simulation come from reference while in experiment, they cannot be precisely measured for lacking of certain experimental conditions, so their practical values cannot be obtained. Fortunately, fluid volume is very small compared with dispenser, which means fluid has little effect on heat diffusion and temperature measurement. Secondly, external dispenser surface where electric heating film was installed was machined as prismatic for easily dismounting, while prismatic-shape causes some troubles for firm attachment and heat conduction. Thirdly, heat transmission begins after electric heating film itself reaches a relatively high temperature, and the temperature difference with heated objects is usually 100°C or more. But in simulation, all the heat generated by the film is assumed to flow into dispenser, without any consideration of self-heating and heat exchange with the air. That is why temperature in simulation is higher than that in experiment. Last but not least, positive temperature coefficient of electric heating film, stability of DC power, installation error of platinum resistance, system error and reading error, and external temperature variation are all the factors that may affect experiment results

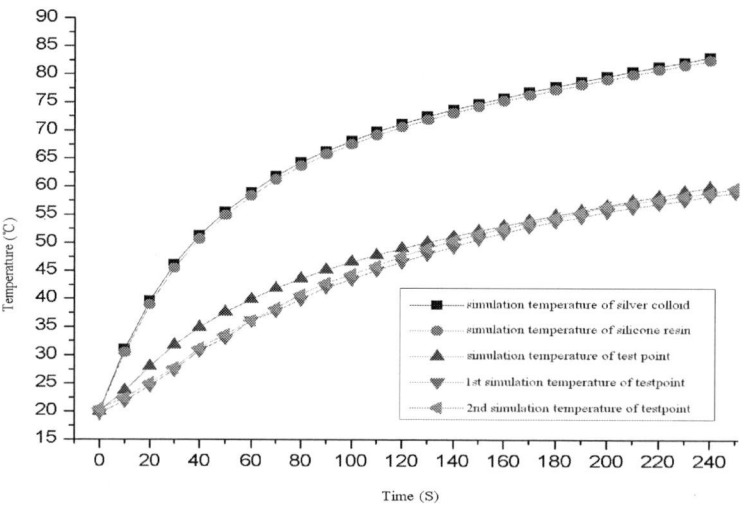

Fig. 10 Simulation temperature and experimental temperature of test point and real-time temperature of fluid

978-1-4244-4658-2/09 $25.00 © 2009 IEEE   904

5. Conclusions

(1)From the point of the basic theory of the temperature, a full picture of the physical and boundary parameters, and some thermal behavior such as heat conduction and heat exchange have been given in this issue.

(2)Two-dimensional transient temperature field model of fluid jet-dispenser were built, based on which two sorts of fluid with dramatically different heat conduction coefficient have been analyzed by FEA.

(3)Heat experiments were done on the basis of simulation and temperature characteristic of fluid jet-dispenser under certain heating condition is gained. The results of the simulation and experiment show that the temperature distribution is very uniform in fluid area, which also proves the heater designed is feasible.

Acknowledgments

The work was supported by a grant from National Natural Science Foundation of China (No. 50475138) and a grant from Hi-Tech Research and Development Program of China (No. 2007AA04Z350).

References

1. H. Quinones, A. Babiarz, C. Deck. Fluid jetting for next generation packages. Pac tech, Berlin, April, 2002.

2. Yang Shi-ming, Tao Wen-quan, Heat Transfer(3nd Edition), Beijing:Higher Education Press,1998.

3. Hu Hao, Deng Gui-ling. "The Influence Discipline of Temperature of High Viscosity Fluid Jetting," *2008 International Conference on Electronic Packaging Technology & High Density Packaging (ICEPT-HDP).* July 2008, pp. 231.

4. Li Shi-hui. "3-D Numerical Simulation of Adhesive flow in Jet-dispenser and the Design of Structure Parameters of Jet-dispenser," *Changsha: College of Mechanical and Electrical Engineering,* Central South University,2006.

5. H. Quinones, A. Babiarz, L. Fang. Jetting Technology for Microelectronics. IMAPS Nordic, tockholm, Sweden, September, 2002.

Automatic Plating Technology for Ceramic Packaging

Congge Lu, Shengqian Liu, Lei Zhang
Hebei Semiconductor Research Institute
P. O. Box 179-41, Shijiazhuang Hebei, P. R. China
E-mail: chinapackage@163.com
Phone: 0311-87091654

Abstract

The plating technology meets many difficulties due to the shape variety, frame complexity and material diversity of the ceramic packages. Nowadays much of the plating still relies on handwork, which limits both the batch production and quality stability of ceramic packages. One method to solve the problem is the automatization of plating by using advanced equipments, stable plating process and control methods. Therefore, in the paper the automatic plating for ceramic package was reported in three aspects, i.e., plating equipments, plating process and control methods. The technique of this work has been successfully applied in practical production, and supported batch production of ceramic packages.

The increasing demand on the production of ceramic packages requires more advanced techniques, among which plating is one indispensable process. The automatization plating technique can benefit the batch production of ceramic packages, though it is difficult to be realized due to the variety or even non-standard requirement of ceramic packages. Especially in the batch production of mixed types of packages, the different configurations, surface materials and plating procedures have high requirements on the automatization process. Only the production line that combines with multi-processes in a range of operational parameters can take full advantage of the automatization method. Thus, in this paper, the development of automatic production lines for plating ceramic packages was reported according to the plating equipment, plating process and control method.

Discussion 1 Automatic plating equipment

Automatic plating equipment is necessary in the plating automatization production. In our work, PLC [1] method was used to control the running of production lines, circulating filter pump, temperature control circuit, spray clean device. The running of production lines was programmed with pre-set process parameters, and the plating procedures can be realized automatically. The maintenance and replenishment of plating solution can be achieved by using online solution analyzer. This automatic analytic instrument is independent of the operation system of production line, and can set different working cycles according to the whole production cycle. Such performance largely improved the stability of the plating solution and ensured the smooth running of the automatic production. Fig. 1 shows the automatic production line on site.

1.1 Automatic analytic and supplemental equipment for plating solution

The automatic analytic and supplemental instrument for plating solutions can analyze the Ni^{2+}, reductant concentration

and pH value in a short time. Different from another supplemental way, i.e., only analyzing Ni^{2+} and adding reductant proportionally, the method used in our work can more reasonably control the solution concentration and make the plating solution more stable.

Fig. 1 Automatic production line for plating ceramic packages

The analysis operation of this equipment combines chemical titration, absorbency, weighing methods. Thus, it is an accurate, rapid and easily operational method. Especially, it is automatically sampling based on the presetting cycles, and replenishing according to the analysis result, so that the automatization of plating nickel production is stable and reliable. The equipment configuration is shown in Fig. 2.

Fig. 2 Automatic supplemental equipment for plating solution

1.2 PLC program for automatic process of plating nickel

PLC function can realize the programming operation of different plating process, which can largely reduce the veracity of artificial operation and improve the accurate control of process parameters. 10 sets of process parameters can be stored in PLC, which can satisfy the usage for most kinds of plating nickel process. This can be a shortcut for the operation and also improve the efficiency of mix production for various plating parts. The equipment is shown in Fig. 3.

Fig. 3 PLC and electric control device

Discussion 2 Plating process

With the advanced equipment, the improvement of the quality, reliability and integrality of the plating is also important to the high quality of the process.

The ceramic package is welding from multi-layer ceramic metallic basement and the metallic components (leading wire frame, sealing circle and cooling fin etc.) use silver-copper solder. It is not easy to get high quality of gold and nickel plating on the varieties of materials.

Processing

Preceding cleaning → chemical degreasing → electro-degreasing → etch cleaning → HCl activation → nickel impinging → nickel plating → gold impinging → gold plating → rear treatment → drying → heat treatment

The five processes before nickel impinging are called preceding treatment, which is determined the finishing quality of the ceramic package. From the productive practice, the reasons for most plating problems are influenced by the preceding treatment, especially on the properties of plating cohesion, corrosion resistant and smoothness. Therefore it is very important to have appropriate preceding treatment which is the determinant of the finishing quality, to get high quality of plating layer.

The preceding Cleaning is focus on the clean of the dust and graphite on the surface [2]. The "heel" or "sentus" will be found when there is dust or metallic powder on the surface of the package, particularly on the bonding area which is influenced strongly on the bonding properties. In addition, the short circuit and electric resistance decreasing will happen when there is graphite during the welding. Accordingly, in order to avoid waster after gold plating, it is important to get rid of dust in the preceding cleaning.

As it is cheap and controllable, alkaline degreasing is widely used in chemical degreasing. The alkaline materials which can emulsify and disperse the grease, including sodium hydrate (NaOH), sodium carbonate, sodium bicarbonate, sodium silicate, sodium phosphate and so on. But the silicate film will be formed when the work during high temperature treatment, which will determine to bad cohesion and pitting, so it is very important to wash properly after treatment.

The electro-degreasing is including cathode electrolysis, anode electrolysis and periodic reversing electrolysis. The mechanism of cathode electrolysis is use the hydrogen generated from the cathode to get rid of the dirt and oxide on the surface. The bulk material will not dissolve in this method, so it is appropriate to nonferrous metal. But the impurity of metallic ion in the solution will cause electro-deposition on the surface which will decrease the cohesion between the plating layers. Anode electrolysis can not only remove the oxide film and metallic chemicals but also can activation the surface, which is more wildly used than cathode electrolysis for the ferrous bulk material, while for the nonferrous material, the color of the surface will change and lead to inferior finishing quality. Periodic reversing electrolysis is the effective method through using cathode and anode electrolysis alternately to electro-degreasing. But it should be noticed that the anode electrolysis is used during exit tank.

After the electro-degreasing, the package need etch cleaning to get rid of the oxidation, and acid activation before the plating. Nickel impinging is used to increase the cohesion between the nickel layer and bottom layer. The nickel impinging solution including dark-nickel, low stress nickel and nickel sulfamic acid. With the purity, high temperature behavior and high dispersion property, the dark-nickel and low stress nickel are mostly used for the normal plating as the basic layer for gold plating. Because of the lowest stress and inferior of dispersion, the nickel sulfamic acid is used for the thick nickel plating. It is useful to improve the corrosion resistant property of the plating after optimizing the plating layers.

Gold impinging before gold plating is used to increase the cohesion with the nickel layer. Litmusless gold plating is usually used to ceramic package because it prevents the color change at high temperature and is easy to pressure bonding. The purity of the gold plating is required to be above 99.9%. The rigidity of the gold plating is asked to be below 90.

After the gold plating the package needs to be cleaned seriously in order to ensure the surface without any smear or salt.

Discussion 3 technology control

On the basis of a cleaned and well surface treated shell before electroplating, the controlling of the electroplating process becomes critical in terms of obtaining a coherent electroplating layer, which has well coherence with the shell matrix as well as satisfied thickness and purity. It is necessary for the test of various properties and also the demand of manufacture, storage and application.

The electroplating process controlling includes three aspects:

Chemical aspect---including the composition of the contents, the PH value et al.

A well chemical controlling can be achieved by increasing the refill times and the decrease the amount of every refill.

978-1-4244-4658-2/09 $25.00 © 2009 IEEE   907

During the electroplating process, it is important to minimize the variation of composition and PH value of the electrolyte and keeps the composition constant all through the electroplating process.

Physical aspect---including temperature, the stirring speed, the liquid surface, the area ratio of anode and cathode, the distance between anode and cathode, the voltage and current, *et al.*.

On the basis of well controlled composition and pH value, the variation of the physical parameters mentioned have great influence on the characteristic of the electroplating layer. A well electroplating process need to be carried out at a best working condition. Therefore, the optimizing of these physical parameters is also very important. All these parameters are monitored and adjusted by meters and sensors. These parameters should be maintained in a demanded range during the electroplating process.

Contaminations---including indiscerptible impurity, organic impurity and metallic impurity.

The contaminations in the electrolyte can also change the characteristic of the electroplating layer. The contaminations mainly come from the reaction of the anode and cathode, the corrosion of the electroplating instruments and the improper operations. The contaminations have great influence on the purity, the joint force, thickness (electroplating rate) of the electroplated layer. Though it is impossible to eliminate all these contamination thoroughly, the contaminations can still be controlled at a certain level through regular analysis and instrument maintenance. For example, filtrating the indiscerptible impurity by recyclable filtration, reducing the organic and complex metallic impurities by regular active carbon treatment.

Conclusions

In a word, we should combine the plating equipment, plating process with control method well. If we can do it successfully, the ceramic package can be produced automatically and achieved batch production.

References

1. Chunquan Nie, "Automatic Control System of Plating Line," *Industrial Control Technique*, 2008, pp. 8-12.
2. Liming Feng, Electroplating Techonology and Equipment, Beijing , Chemical Industry Press,2005

Analysis and Compensation of Perpendicularity Error for High Speed and High Precise IC Assembly Equipment Base on Coordinate Transformation

Haichen Qin, Jianzhou Quan, Bo Peng, Zhouping Yin
School of Mechanical Science & Engineering, Huazhong University of Science & Technology
Wuhan, 430074, China
Qinhaichen@163.com, quan-jzh@163.com

Abstract

The present paper relates to a new scheme used in high speed and high precise IC assembly equipment. Firstly, the author presents a new Hand-Eye Separation motion platform. Secondly, according to the analysis of assembly and manufacturing errors influence on perpendicularity in this platform, this paper presents a new method for error compensate base on coordinate transformation. An application at RFID Packaging and Assembly Equipment is used to provide realism. The result shows that this scheme can help ensure the stability of equipment operation and improve positioning accuracy.

Discussion 1 Introduction

In modern IC Packaging area, pick and place machines are provided with pick up heads having a vacuum nozzle movable in the vertical or Z direction. The head or heads are mounted on an X-Y gantry which permits any component picked form a supply and placed on a substrate or PC board. In order to assure more accurate pick-and-place of components, usually use machine vision guided positioning operation. For example SMT (Surface Mounted Technology) surface packaging equipment [9]. The machine vision system is used as the sensor to measure X-Y coordinates of placement position and then the motion system control terminal actuator to accomplish put the components placed corresponding position. In this X-Y gantry structure, machine vision system and terminal actuator usually installed in together mostly and the machine vision system move together with terminal actuator, two parts unable concurrent working, thus affects equipments efficiency.

Recently, with the chips is getting smaller and request of placement accuracy is getting more accurate, in order to meet the need of large-scale production, further improve production efficiency, need to higher efficient coordination between machine system and servo system.

For the above reasons, this paper presents a new Hand-Eye Separation motion platform [11-12]. The "Hand" is terminal actuator in IC Packaging equipment is placement head, the "Eye" is machine vision in IC Packaging equipment is machine vision system, there are assembled on different motion platform. "Hand" and "Eye" are working simultaneity, the results of treated vision signal transfer to the servo controller real time, and leading the placement head to accomplish the work of pick and place. This structure has many advantages. First of all, machine vision system and terminal actuator are parallel workflow, greatly improve the production efficiency. Second, has reduce the load weight of each single motion platform, it is provided favorable

conditions for platform moving with higher speed and higher acceleration [7].

Although this scheme has potential advantages, it does not come without additional concerns. Because "Hand" and "Eye" are assembled on different motion platform, after an X-Y coordinate of placement location have been caught by machine vision system need to coordinate transformation to the terminal actuator coordinate system can be able to lead the placement head to accomplish the work of pick and place [1]. So, it is the key that the accuracy of coordinate transformation from machine vision motion platform to terminal actuator motion platform. There are multiple factors affect the accuracy of coordinate transformation such as assembly accuracy and machining accuracy of mechanical elements. This can bring about the coordinate system of motion platform is not perfect Cartesian coordinate system. And then, after a long period work the equipment has lost working accuracy lead to coordinate transformation inaccuracy.

In this paper, according to the analysis of assembly and manufacturing errors influence on perpendicularity in pick and place machine, presents a new method for error compensate base on coordinate transformation. An application in RFID Packaging and Assembly Equipment is used to provide realism. The result showed that this method can help ensure the stability of equipment operation and improve positioning accuracy.

Discussion 2 Hand-Eye Separation Motion Platform

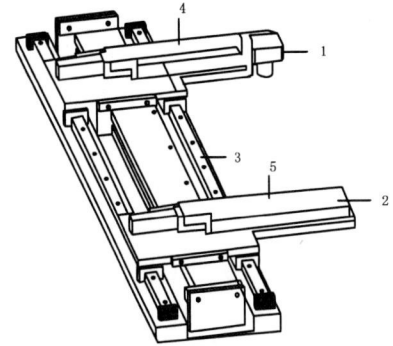

Fig. 1 Hand-Eye Separation Motion Platform

1——Visual CCD

2——Installation position of placement head

3——The Linear guides of Y-axis

4——X-axis module of machine vision system

5——X-axis module of placement head

As shown in Fig. 1 a new Hand-Eye Separation motion platform has been present. The "Hand" is terminal actuator in IC Packaging equipment is placement head, the "Eye" is machine vision in IC Packaging equipment is machine vision system, there are assembled on different motion platform. "Hand" and "Eye" are working simultaneity, the results of treated vision signal transfer to the servo controller real time, and leading the placement head to accomplish the work of pick and place.

Discussion 3 Analysis and Compensation of Perpendicularity Error

3. 1 Coordinate system

As shown in Fig. 2, the coordinates system in IC Packaging equipment, including $O_C X_C Y_C$ is the coordinate system of machine vision and $O_V X_V Y_V$ is coordinate system of placement head. Grid origin of $O_V X_V Y_V$ Y-shift b and X-shift a, and the Y-axis has been reversed will be $O_C X_C Y_C$.

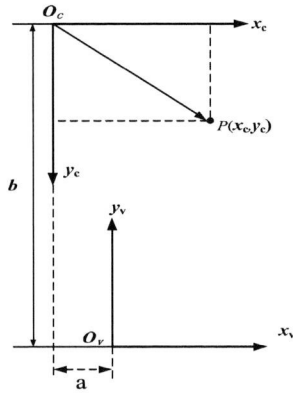

Fig. 2 Coordinate system of machine vision platform and placement head platform

Assumption (X_C, Y_C) is the coordinate of point P in $O_C X_C Y_C$, as following homogeneous coordinate transformation:

$$\begin{bmatrix} x_v \\ y_v \\ 1 \end{bmatrix} = \begin{bmatrix} x_c - a \\ -y_c + b \\ 1 \end{bmatrix} \quad (1)$$

Put (1) into matrix form:

$$\begin{bmatrix} x_v \\ y_v \\ 1 \end{bmatrix} = T \begin{bmatrix} x_c \\ y_c \\ 1 \end{bmatrix} = \begin{bmatrix} 1 & 0 & -a \\ 0 & -1 & b \\ 0 & 0 & 1 \end{bmatrix} \begin{bmatrix} x_c \\ y_c \\ 1 \end{bmatrix} \quad (2)$$

The matrix T is transformation matrix.

$$T = \begin{bmatrix} 1 & 0 & -a \\ 0 & -1 & b \\ 0 & 0 & 1 \end{bmatrix}$$

$$\begin{bmatrix} x_c \\ y_c \\ 1 \end{bmatrix} = T^{-1} \begin{bmatrix} x_v \\ y_v \\ 1 \end{bmatrix} = \begin{bmatrix} 1 & 0 & a \\ 0 & -1 & b \\ 0 & 0 & 1 \end{bmatrix} \begin{bmatrix} x_v \\ y_v \\ 1 \end{bmatrix} \quad (3)$$

Following (3) we can to find (x_c, y_c).

In the actual system, the two actual x-y coordinate systems is not completely parallel, even in the respective x-y coordinate system, x-axis and Y-axis is perpendicular to each other, because of assembly and mismachining tolerance.

As shown in Fig. 3, there are several different situations.

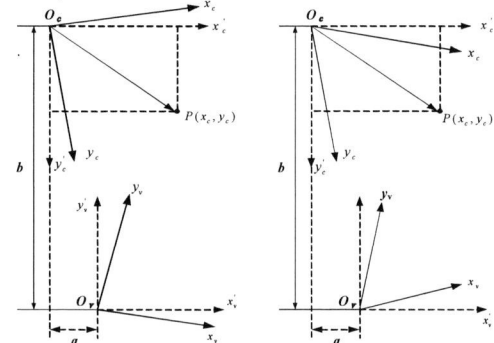

Fig. 3 Different situations of two actual coordinate systems

3.2 Analysis of non-perpendicularity

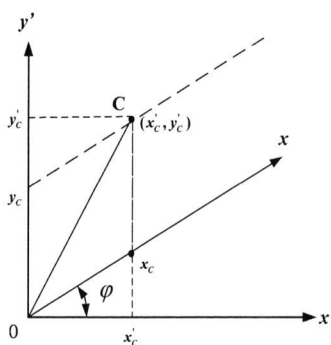

Fig. 4 The situation of non-perpendicularity coordinate system

As shown in Fig. 4, it is a situation of non-perpendicularity x-y coordinate system. A discretional point C, (X_C, Y_C) is a coordinate of ideal coordinate system $OX'Y'$, (X_C', Y_C') is a coordinate of actual coordinate system $O_C X_C Y_C$, φ is an angular separation between the actual x-axis and ideal x-axis.

According to Fig. 4 we can get the following conclusions.

$$\begin{cases} x_c' = x_c \cos \varphi \\ y_c' = y_c + x_c \sin \varphi \end{cases} \quad (4)$$

Put (4) into matrix form:

$$\begin{bmatrix} x_c^{'} \\ y_c^{'} \\ 1 \end{bmatrix} = T_c^{c'} \begin{bmatrix} x_c \\ y_c \\ 1 \end{bmatrix} = \begin{bmatrix} \cos\varphi & 0 & 0 \\ \sin\varphi & 1 & 0 \\ 0 & 0 & 1 \end{bmatrix} \begin{bmatrix} x_c \\ y_c \\ 1 \end{bmatrix}$$

(5)

$$T_c^{c'} = \begin{bmatrix} \cos\varphi & 0 & 0 \\ \sin\varphi & 1 & 0 \\ 0 & 0 & 1 \end{bmatrix}$$

Inverse transformation is:

$$\begin{bmatrix} x_c \\ y_c \\ 1 \end{bmatrix} = T_{c'}^{c} \begin{bmatrix} x_c^{'} \\ y_c^{'} \\ 1 \end{bmatrix} = \begin{bmatrix} \cos^{-1}\varphi & 0 & 0 \\ -tg\varphi & 1 & 0 \\ 0 & 0 & 1 \end{bmatrix} \begin{bmatrix} x_c^{'} \\ y_c^{'} \\ 1 \end{bmatrix}$$

(6)

$$T_{c'}^{c} = \begin{bmatrix} \cos^{-1}\varphi & 0 & 0 \\ -tg\varphi & 1 & 0 \\ 0 & 0 & 1 \end{bmatrix}$$

So:

$$\begin{cases} x_c = \dfrac{x_c^{'}}{\cos\varphi} \\[2mm] y_c = y_c^{'} - x_c tg\varphi \end{cases}$$

(7)

3.3 Homogeneous transformation of non-standard coordinate

Based on the above analysis we can set up the mapping relationship for machine vision coordinate to placement head coordinate.

Fig. 4 depicts the actual coordinate relation of the machine vision coordinate system and placement head coordinate system after actual measurement. Including, $O_C X_C Y_C$ is the coordinate system of machine vision, $O_V X_V Y_V$ is coordinate system of placement head. Grid origin of $O_V X_V Y_V$ x-shift and y-shift were a and b, and the y-axis has been reversed will be $O_C X_C Y_C$.

Shown in Fig. 5, the machine vision coordinate system and placement head coordinate system is not vertical. In the y-direction, as a result of the marble as a base substrate, the two linear motion actuator are shared the same linear guide, so we can think of no deflection.

In order to set up the mapping relationship between actual machine vision coordinate system and actual placement head coordinate system, we has redefined ideal machine vision coordinate system $OX_C'Y_C'$ and ideal placement head coordinate system $O_V'X_V'Y_V'$, ideal origin O_C' and ideal y_c'-axis respectively coincide with the actual coordinate system, x_c' and y_c' perpendicular to each other. In like manner, ideal origin O_V' and ideal y_V'-axis respectively coincide with the actual coordinate system, x_V' and y_V' perpendicular to each other. The α is an angular separation between x_c-axis of the actual machine vision coordinate system and ideal x_c'-axis, β is a angular separation between x_V-axis of the actual placement head coordinate system and ideal x_V'-axis.

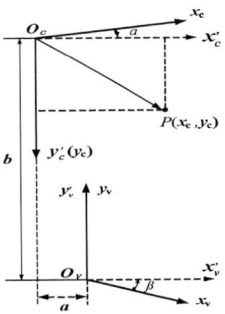

Fig. 5 The actual coordinate relation of the machine vision and placement head coordinate system

Assumption (X_C, Y_C) is the coordinate of point P in the actual machine vision coordinate system $O_C X_C Y_C$, according to formula (5) could be able to reach the following results:

$$\begin{bmatrix} x_c^{'} \\ y_c^{'} \\ 1 \end{bmatrix} = T_c^{c'} \begin{bmatrix} x_c \\ y_c \\ 1 \end{bmatrix} = \begin{bmatrix} \cos a & 0 & 0 \\ -\sin a & 1 & 0 \\ 0 & 0 & 1 \end{bmatrix} \begin{bmatrix} x_c \\ y_c \\ 1 \end{bmatrix}$$

(8)

$$T_c^{c'} = \begin{bmatrix} \cos a & 0 & 0 \\ -\sin a & 1 & 0 \\ 0 & 0 & 1 \end{bmatrix}$$

In like manner could be able to get the mapping relations of ideal and real placement head coordinate system.

$$\begin{bmatrix} x_v \\ y_v \\ 1 \end{bmatrix} = T_{v'}^{v} \begin{bmatrix} x_v^{'} \\ y_v^{'} \\ 1 \end{bmatrix} = \begin{bmatrix} \cos^{-1}\beta & 0 & 0 \\ tg\beta & 1 & 0 \\ 0 & 0 & 1 \end{bmatrix} \begin{bmatrix} x_v^{'} \\ y_v^{'} \\ 1 \end{bmatrix}$$

(9)

$$T_{v'}^{v} = \begin{bmatrix} \cos^{-1}\beta & 0 & 0 \\ tg\beta & 1 & 0 \\ 0 & 0 & 1 \end{bmatrix}$$

According to above-mentioned analysis, it is homogeneous translate transformation between ideal machine vision coordinate system to ideal placement head coordinate system. Transformational relations are as follows:

$$\begin{bmatrix} x_v \\ y_v \\ 1 \end{bmatrix} = T_{c'}^{v'} \begin{bmatrix} x_c^{'} \\ y_c^{'} \\ 1 \end{bmatrix} = \begin{bmatrix} 1 & 0 & -a \\ 0 & -1 & b \\ 0 & 0 & 1 \end{bmatrix} \begin{bmatrix} x_c^{'} \\ y_c^{'} \\ 1 \end{bmatrix}$$

(10)

The matrix $T_{c'}^{v'}$ is translational transformation matrix.

$$T_{c'}^{v'} = \begin{bmatrix} 1 & 0 & -a \\ 0 & -1 & b \\ 0 & 0 & 1 \end{bmatrix}$$

According to formula (8) and (10) can get transformational relation from machine vision coordinate system to placement head coordinate, as follows:

$$T_c^v = T_{c'}^c T_{c'}^{v'} T_{v'}^v$$

$$= \begin{bmatrix} \cos a & 0 & 0 \\ -\sin a & 1 & 0 \\ 0 & 0 & 1 \end{bmatrix} \begin{bmatrix} 1 & 0 & -a \\ 0 & -1 & b \\ 0 & 0 & 1 \end{bmatrix} \begin{bmatrix} \cos^{-1}\beta & 0 & 0 \\ tg\beta & 1 & 0 \\ 0 & 0 & 1 \end{bmatrix}$$

$$= \begin{bmatrix} \dfrac{\cos a}{\cos \beta} & 0 & -a\cos a \\ -\dfrac{\sin a}{\cos \beta} - tg\beta & -1 & -1 + a\sin a + b \\ 0 & 0 & 1 \end{bmatrix} \quad (11)$$

So:

$$T_c^v = \begin{bmatrix} \dfrac{\cos a}{\cos \beta} & 0 & -a\cos a \\ -\dfrac{\sin a}{\cos \beta} - tg\beta & -1 & -1 + a\sin a + b \\ 0 & 0 & 1 \end{bmatrix} \quad (12)$$

Thus deduce the transformation equation between real machine vision coordinate system and real placement head coordinate system.

$$\begin{cases} x_v = x_c \dfrac{\cos a}{\cos \beta} - a\cos a \\ \\ y_v = y_c + (-\dfrac{\sin a}{\cos \beta} - x_c tg\beta) + a\sin a + b - 1 \end{cases} \quad (13)$$

3.4 Application of Coordinate transformation at RFID Picking and assembly equipment

Fig. 6 The part of packing and assembly module of our RFID equipment

Fig. 6 depicts the packing and assembly module of our RFID equipment. Machine vision system composed by Visual CCD and LED Lighting drove by LPMSM (permanent magnet linear synchronous motor) and servo motor. Terminal actuator composed by placement head and servo modules.

First of all, we will be the Visual CCD and placement head removed replacement for marker, let Visual CCD module and placement head module back to respective origin

(electrical origin) of x-axis and y-axis. To take Visual CCD module for an example, to maintain $y=0$ unchanged, let the module forward from the beginning $x=0$, this can map out a trajectory x_c at the work region. And then let the module return to origin, to maintain $x=0$ unchanged, make the module forward from the beginning $y=0$, as well as to map out a straight line y_c at the work region. Draw a straight line x_c^r perpendicular y_c through point (0, 0), this will construct an ideal coordinate system as has been stated, and the trajectory x_c is migrated x-axis.

Repeated measurements has found, x_c is not a straight line, this shows that offset angle α and β.

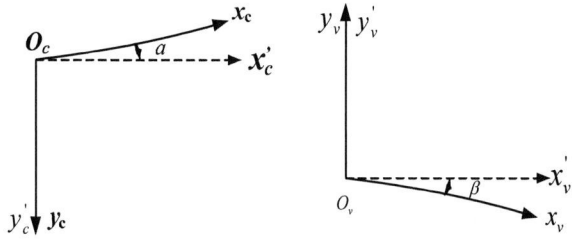

Fig. 7 Actual metrical X-Y Trajectories in RFID equipment

Used measurement method can be measured value b and a, but the two angular deviations α and β are difficulty to actual measure. If we make use of curve fitting method can get an approximate curve of deviation angle, but this approximate method will be bring into calculation error base on the measurement error. So we take the equivalent parameters δ_c and δ_v to replace α and β to reduce global error.

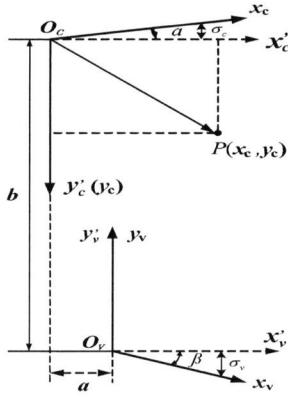

Fig. 8 Modules vertical displacement deviation

Combination of equipment installation method and actual measuring method, used δ_c and δ_v replace to trigonometric function in above-mentioned formula (13) could be able to reach the following results:

$$\begin{cases} \sqrt{x_c^2 - \delta_c^2} - a = \sqrt{x_v^2 - \delta_v^2} \\ (y_c - \delta_c) + (y_v - \delta_v) = b \end{cases}$$

$$\Downarrow$$

$$\begin{cases} x_v = \left[(x_c^2 - \delta_c^2) - 2a\sqrt{x_c^2 - \delta_c^2} + a^2 + \delta_v^2 \right] \\ y_v = -y_c + b + \delta_c + \delta_v \end{cases}$$

δ_c and δ_v —modules vertical error, Can be obtained through actual measurement, the parameter b is a distance between two electrical origin (null pick-up of optical Scale) of linear motion actuator active cell in the RFID equipment, easily measured. And parameter a is deviation between electrical origin (null pick-up of optical Scale) for x-direction of placement head relative electrical origin (null pick-up of optical Scale) for x-direction of Visual CCD, it has more difficult to measure.

In order to measure δ_c, δ_v and a, the largest work area of pick and place has been divided into 192 parts, a total of 221 measurement points. As shown in Fig.8. Row R-01 to R-17 is parallel to y_c and y_v, viz. parallel to the linear motor guide.

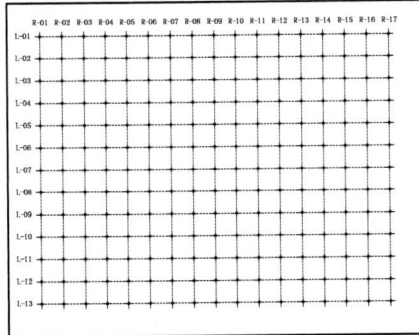

Fig. 9 Metrical zoning of working platform

First of all, the Visual CCD starting from the location of mechanical origin and get each measurement point coordinates of row R-01.These measurement points x-axis coordinate almost equal so recorded as $x_{c\text{-R01}}$. With the same measurement method we can be obtained $x_{c\text{-R02}} \sim x_{c\text{-R1}}$ and $x_{V\text{-R01}} \sim x_{V\text{-R17}}$ (obtained the x-axis coordinates of placement head module). However, it is different that the installation position of two modules' electrical origin relative to respective mechanical origin, and so lead to $x_{c\text{-R12}} \sim x_{c\text{-R17}}$ are different from $x_{V\text{-R01}} \sim x_{V\text{-R17}}$. Comparison of coordinates $x_{c\text{-R01}}$ and $x_{V\text{-R01}} \ldots \ldots x_{c\text{-R17}}$ and $x_{V\text{-R17}}$ could be obtained 17 deviation value $a_1 \sim a_{17}$. In order to ensure the accuracy of the coordinate transformation, the work place has been divided into 17 parts take R-01 to R-17 for boundary line.

From electrical origin to row R-17 total of 104746.5cts.

According to the x-axis coordinate extent of Visual CCD module used to table lookup can achieve value a when the RFID equipment run-time, this will not only guarantee the accuracy of coordinate transformation but also be able to avoid calculating time lag by higher degree equation of curve.

Table.1 X-axis coordinate extent and value a

X-axis coordinate extent	a (cts)
$0 \sim x_{c\text{-R01}}$	a_1
$x_{c\text{-R01}} \sim x_{c\text{-R02}}$	a_2
......	......
$x_{c\text{-R16}} \sim x_{c\text{-R17}}$	a_{17}

And then we need to measure δ_c and δ_v. To take the Visual CCD module for example, first of all, the Visual CCD starting from the location of mechanical origin and get each measurement point coordinates of line L-01. These measurement points y-axis coordinates should be equally in the context of no-perpendicularity error. But in fact, these y-axis coordinates is gradually increasing, just to be identical with the trajectory in Fig. 7. So we can get y-axis deviation of every measurement point compare with the point (L-01,R-01), these deviations a defined as $\delta_{C\text{-L01R01}} \sim \delta_{C\text{-L01R17}}$. By the same way we can get all Y-axis deviations from line L-02 to L-13. Comparison each deviation from row R-01 to R-17 $\delta_{C\text{-L01R01}} \sim \delta_{C\text{-L17R01}} \ldots \ldots \delta_{C\text{-L01R17}} \sim \delta_{C\text{-L17R17}}$, in theory should be equal to each row the deviation of all measurement points, but in fact there are subtle difference caused by Machining accuracy of Mechanical Elements, assembly accuracy and equipment physical depreciation. Averaging to y-axis deviation in several rows R-01 to R-17. Now we get 17 y-axis deviations $\delta_{C\text{-1}} \sim \delta_{C\text{-17}}$, as shown in following Table 2:

Table.2 Measurement points and deviations of Visual CCD module

Points	Deviations (cts)
1	0.5cts
2	1.0cts
3	1.5cts
4	2.0cts
5	5.0cts
6	5.8cts
7	7.8cts
8	8.9cts
9	10.5cts
10	12.0cts
11	13.0cts
12	15.0cts
13	15.5cts
14	16.0cts
15	16.5cts
16	17.0cts
17	17.0cts

Draw deviations distribution and sequences of measurement points, as shown in following Fig. 10:

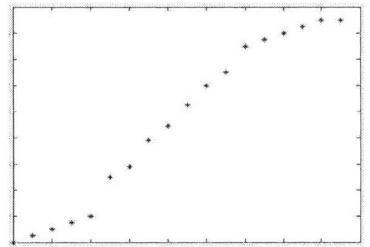

Fig. 10 Deviations distribution and sequences of measurement points

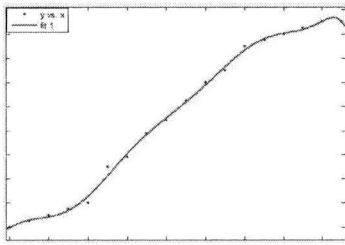

Fig. 11 Higher curve fitting

As shown in Fig. 11, is a 9th degree polynomial curve, if we want to get a better fitting result, this degree will continue to increase, will lessen the real-time of system.

According to the Fig. 10, we can segment all points in Fig. 10 into 10 parts as shown in Fig. 11. Therefore, ensure that all points in Fig. 10 can be expressed by a liner equation such as $f(x)=kx+b$, this will greatly improve the computation speed of control program under the premise of accuracy.

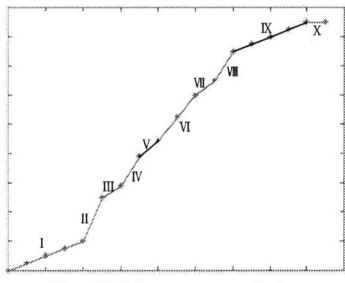

Fig. 12 Linear curve fitting

1. These deviations from origin to R-04 and R-12 to R-16 are equidifferent increasing, so part I and part IX can be expressed by a liner equation such as $\delta_c=kx+b$. We can easily obtain k and b.
2. Deviations at part VI are noncollinear points from R-08 to R-10, but be a very close approximation to a straight line. Fig.13 is the result of curve fitting.
3. As shown in Fig. 13 the result of linear curve fit very well.
4. Remaining parts are constituted by two points in Fig. 11.
5. X-axis coordinates extent of CCD module as shown in following table. We can easily obtain all k and b.

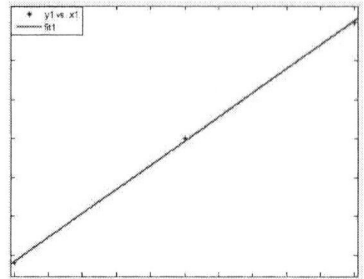

Fig. 13 Linear curve fit of part VI

Table.3 X-axis coordinates extent of CCD module and Deviation equation

X-axis coordinates extent	Deviation equation (cts)
$0 \sim x_{c\text{-}R04}$	$\delta_C=8.115 \times 10^{-5}x+0$
$x_{c\text{-}R04} \sim x_{c\text{-}R05}$	$\delta_C=4.868 \times 10^{-4}x-10$
$x_{c\text{-}R05} \sim x_{c\text{-}R06}$	$\delta_C=1.298 \times 10^{-4}x-1$
$x_{c\text{-}R06} \sim x_{c\text{-}R07}$	$\delta_C=3.246 \times 10^{-4}x-6.2$
$x_{c\text{-}R07} \sim x_{c\text{-}R08}$	$\delta_C=1.785 \times 10^{-4}x+0.1$
$x_{c\text{-}R08} \sim x_{c\text{-}R09}$	$\delta_C=2.516 \times 10^{-4}x-3.483$
$x_{c\text{-}R10} \sim x_{c\text{-}R11}$	$\delta_C=1.623 \times 10^{-4}x+2$
$x_{c\text{-}R11} \sim x_{c\text{-}R12}$	$\delta_C=3.246 \times 10^{-4}x-9$
$x_{c\text{-}R12} \sim x_{c\text{-}R16}$	$\delta_C=8.115 \times 10^{-5}x+9$
$x_{c\text{-}R16} \sim x_{c\text{-}R17}$	$\delta_C=17.0$

Make use of the same method to measure deviations δ_v of placement head module, and used X-axis coordinates of CCD module to partition, because the CCD module and the placement head module are worked in the same work area, when the CCD module obtained coordinates of pick and place position can be reached a deviation equation immediately. This can greatly reduce the computing time of motor control program.

Now we have obtained all parameter for coordinate transformation.

The above-mentioned coordinate transformation effectively improved the packaging error in RFID equipment, and the packing accuracy has reached $\pm20\mu m$ and fulfils the technological requirements.

Conclusions

In the actual measurement, the perpendicularity error will be influenced by many factors; even including the linear motor rail can't think theoretical straight line. It is necessary that put the liner motor rail divided in many parts for error measurement, and then make use of method of averaging to eliminate the influence on liner error to perpendicularity error compensation.

In this paper, the method for perpendicularity error compensation is able to eliminate the influence on perpendicularity error to packing accuracy. Not only can be applied in RFID packing to fix position at high speed and high accuracy but also can be applied in numerical control machine to eliminate the influence on perpendicularity error to machining accuracy in the way of precompilation compensation and so on.

Acknowledgments

The work is supported by the National Fundamental Research Program (973) (Project No. 2003CB716207) and the 863 program of China under Grant No. 2006AA04A110.

References

1. Dunin-Barkowski, Seung-Han Yang, Young-Suk Kim, Sang-Ryong Lee, "Error compensation method for a gantry robot and a laser-vision sensor-based chassis module measurement system," *Manuf Technol*, (2005) Vol. 27, pp. 329-333.

2. Wang Zhengping, "Laser accuracy measurement and calibration technique for machine tools," *Journal of China Jiliang University*, 2006, 17 (4).

3. CHEN G Q, YUAN J X, NI J, "A displacement measurement approach for machine geometric error assessment," *International Journal of Machine Tools & Manufacture*, 2001 ,41 (1), pp. 149-161.

4. PAN Shuwei, CAO Yongjie, FU Jianzhong, "Research on the Errors Measurement for CNC Machine Tools," *MACHINE TOOL & HYDRAULICS*, 2008, 36 (5).

5. LIAO Qiang, ZHOU Yi,MI Lin,XU Zhongjun, "The Applications of Machine Vision to Precision Measurement," *Journal of Chongqing University(Natural Science Edition)*, 2006, 25 (6).

6. Craig, J. J, Introduction to Robotics: Mechanics and Control, Third Edition, China Machine Press.

7. Wang Ying, "High-Speed/High Accuracy Motion Control of Servo System Driven by Liner Motor for IC Packing," *Doctoral dissertation of Shanghai Jiaotong University*, 2006.

8. Xiong Youlun, Robot Technology, Huazhong University of Science and Technology Press, Wuhan.

9. Zugen Yan, Lining Sun, Bo Huang, "Research of a Novel XY-table Base on error compensation," Harbin Institute of Technology. *Proceedings of the IEEE. International Conference on Mechatronics & Automation*, Niagara Falls, Canada, July, 2005.

10. Chen Guoliang, H. X, Wang Ming, "Micro-visual Servoing for Micro-assembly," *CHINESE JOURNAL OF MECHANICAL ENGINEERING*, 44 (2), Feb.2008.

11. Xiong Chunshan, Huang Xinhan, Wang Min, "Algorithm For Hand-Eye Stereo Vision And Implementation," *ROBOT*, 23 (2), March, 2001.

12. Wang Xueying, Liu Shugui, Zhang Hongtao, Du Jianjun, "Research on the Calibration of Robot Eye in Hand Vision Based on Parameters Separation," *ACTA METROLOGICA SINICA*,. April, 2007.

13. Wei Zhenzhong, Gao Ming, Zhou Fuqiang, Zhang Guangjun, "Robot Extended Eye-in-hand Calibration Method Based on an Assistant Camera," *Opto-Electronic Engineering*, Sep., 2008.

14. Cipolla R, Hollinghurst N, "Visually Guided Grasping in Unstructured Environments," *Robotics and Autonomous Systems*, 1997 ,19, pp. 337-346.

15. Motai Y, Kosaka A, "SmartView: Hand-Eye Robotic Calibration For Active Viewpoint Generation And Object Grasping," *IEEE International Conf on Robotics and Automation*, 2001, 3, pp. 2183-2190.

16. Shen T-S, Huang J, Meng C-H (1999), "Multiple-sensor integration for rapid and high-precision coordinate metrology," *Proceedings of the 1999 IEEE/ASME International Conference on Advanced Intelligent Mechatronics*, Atlanta, pp. 908-915.

Inspection of Miniaturised Interconnections in IC Packages with Nanofocus X-Ray Tubes and NanoCT

Zhenhui He, Quan Wen, Xiaojie Huang
GE Sensing & Inspection GmbH
Niels-Bohr-Str.7, 31515 Wunstorf, Germany
Zhenhui.he2@ge.com, quan.wen@ge.com, jedi.huang@ge.com

Abstract

Nanofocus tube technology and high resolution CT are the future inspection tools for IC packages. The elementary principles and techniques of latest 2D and 3D nanofocus techniques are briefly outlined and typical results of highest resolution failure analysis including bond wire defects, copper bond wire inspection, Flip Chip solder interconnection and microvia inspection are presented.

Introduction

In qualification and spot checks of IC packaging, X-ray systems are customarily used for the inspection of classical features such bond wires, die bonds, moldings and seals. New challenges to X-ray inspection equipment arise from three trends in package technology: miniaturisation, the use of novel materials and increasing device complexity. Package miniaturisation leads to higher density and smaller size of internal structures such as microvias and Flip Chip interconnections, demanding for resolution in the sub-micron range at highest magnifications as provided by novel nanofocus tube technology. The absorption of some materials like non-conductive die adhesives or copper bond wires either is to low to yield sufficient contrast in customary X-ray images or is strong enough to conceal other package features, as observed for copper-tungsten alloy caps. Highly sensitive a-Si 16-bit digital flat panel detectors and image processing techniques now have remarkably improved the detectability of such low contrast features. The complexity of devices with 3D set-ups such as stacked or multiple die including corresponding wire connections leads to confusing overlaps in the two-dimensional X-ray images, in other words, they must be inspected slice by slice or in 3D visualisations as provided by computed tomography (CT). However, up to now, many laboratory CT systems are using X-ray tubes of some microns focal spot size and maximum tube voltages around 100 kV. Since electronic devices contain very fine structures and strongly absorbing materials like gold or copper, this results in unsatisfying image resolution and strong image artefacts, respectively. In view of this situation, a nanoCT system was designed for highest resolution CT of electronic components with a 180 kV nanofocus tube providing excellent penetration at sub-micron detail detectability. Adjacently some typical failure and quality analysis results are presented.

Nanofocus X-ray Method

The following 2D investigations were performed by means of an automatic X-ray shadow microscope with 2 MPixel digital image chain for 24.000x magnification and a detail detectability of 200 – 300 nm (nanome|x 180).

In order to cover the widest possible range of samples the system is equipped with the first commercially available 180kV high power nanofocus (HPNF) tube. This source can be operated in four different modes. On the one hand in the so called nanofocus mode it provides an X-ray spot size of down to approximately $< 0.9\mu m$ which can be used for highest resolution X-ray images and CT scans with sub-micrometer voxelsize. Due to the penumbra effect, see fig. 1, the spot size predominates the images sharpness for extreme magnifications (for details cf. e.g. [1]). In the high power mode (up to 15 Watts at the target) on the other hand it has enough penetration power to examine high-absorption samples like copper, steel or tin alloys and thus allowing e.g. the analysis of new connection systems for electronic devices.

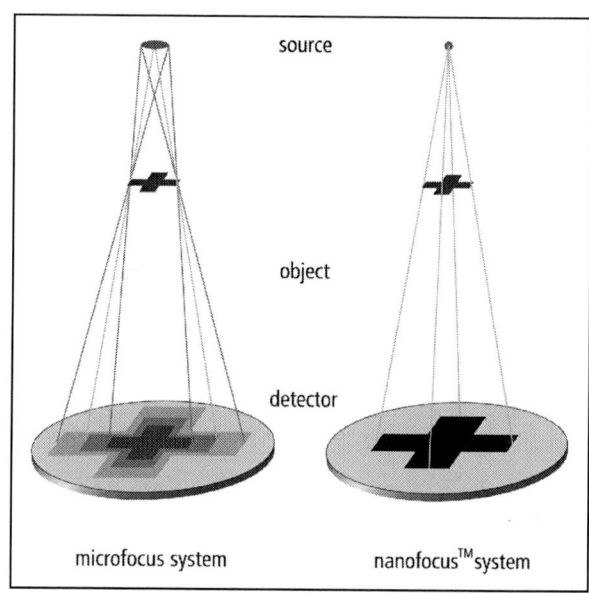

Fig. 1 Influence of focal spot size on image sharpness: The practically punctiform X-ray source of the nanofocus tube increases resolution, so that even *periodic* structures of 0.6µm are resolved.

In Fig. 2 the resolution capabilities of the high power nanofocus source are demonstrated. It shows that the 0.4µm structure (line width) of the JIMA test pattern (designed by the Japan Inspection Instruments Manufacturers' Association for testing high resolution X-ray equipment [2]) can clearly be resolved with more than approximately 20% of the CTF. Fig. 2 shows that for isolated structures of high absorbing material on a low absorbing substrate it is even possible to detect details of 0.5 µm size and below. The limit for this so called detail detectability with the HPNF tube is about 200 – 300 nm.

978-1-4244-4658-2/09 $25.00 © 2009 IEEE

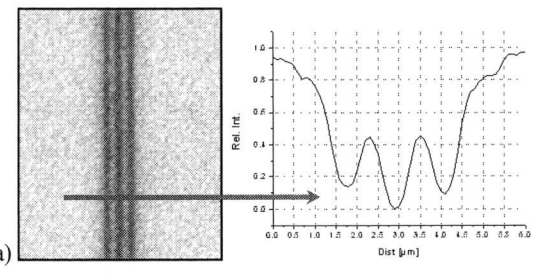

a)

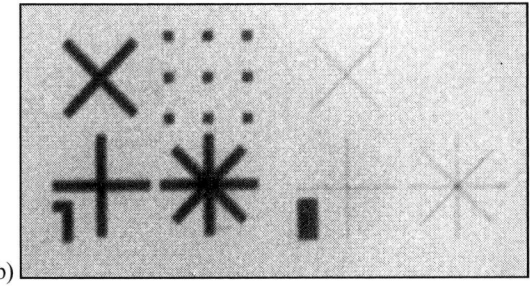

b)

Fig. 2 X-ray images of test patterns proving the resolution and detail detectability of phoenix|x-ray's high power nanofocus tube: In (a) the 0.6μm line pair structure of the JIMA test pattern is clearly resolved. (b) shows structures at the right, which are as small as 0.5μm and below [3].

The system can provide highly magnified images also at viewing angles up to 70°(ovhm) by tilting the detector as shown in Fig. 3.

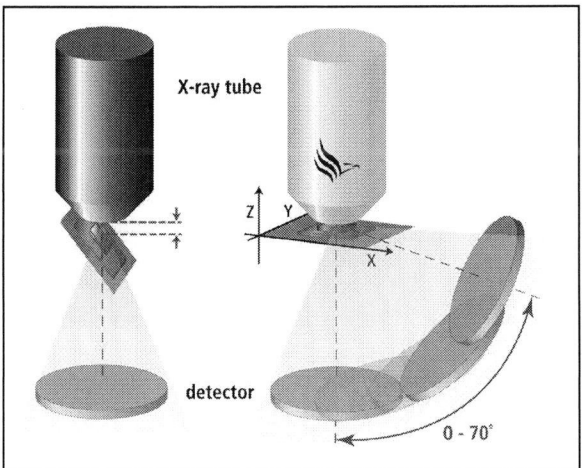

Fig. 3 Oblique views at highest magnification: Tilting the detector instead of tilting the sample keeps the source-object distance small and avoids the loss of magnification.

Nanofocus X-ray - Results

One of the most challenging tasks in 2D X-ray inspection is the proper imaging of the fine dendrites as created by electromigration: sub-micron spatial resolution at high magnification is required as well as outstanding contrast resolution. Fig. 4 sharply displays dendrites growing out of a copper conductor and causing a short circuit. The dendrite width is measured 10 μm. Another frequent inspection task is the detection of cracked and lifted up wire bonds. If the resulting gap is not only to be found but also measured and clearly displayed for further investigation of its cause, the image resolution must be much smaller than its size. In Fig. 5 a ball bond is shown by nanofocus X-ray technology to be lifted up about 2μm. Note that at slightly less resolution the edges of ball bond and bond pad could not be differentiated any more and the gap would be not visible. Due to nanofocus resolution the ripped of wedge bond displayed in Fig. 6 is not only detected, but also proved to be cracked at the rim of the weld area, since the edge of the bond capillaries footprint is still visible. Note that the images in Fig. 5 and 6 require precise adjustments of the viewing angle without loss of magnification (ovhm).

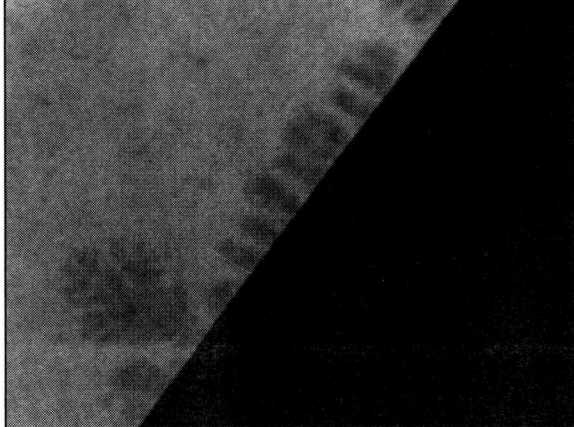

Fig. 4 Nanofocus X-ray image of dendrites growing out of a conductor. The measured dendrite diameter is 10μm.

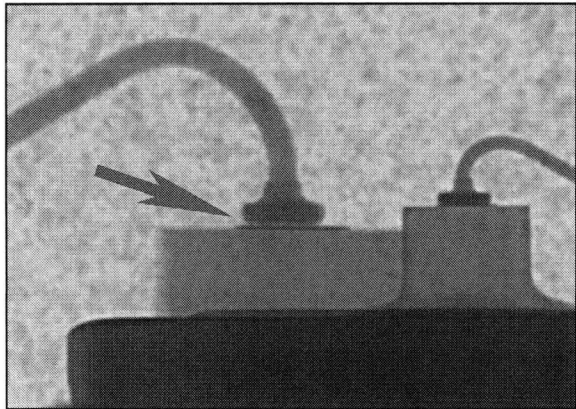

Fig. 5 Nanofocus X-ray image of a lifted ball bond at highest magnification.

In IC processor packages the internal bonds are Flip Chip solder joints which show features and defects similar to area array solder joints as known from electronic assemblies [4]. However, the Flip Chip interconnection may be more than 10 times smaller (25 to 100μm) and require for an appropriate image resolution to detect defects like voids and shape deviations in the micrometer range, see Fig.7. In particular,

978-1-4244-4658-2/09 $25.00 © 2009 IEEE

oblique views at sufficient magnification and resolution are absolutely necessary to detect open joints from deviation in their lateral outline and proportions, see Fig. 8.

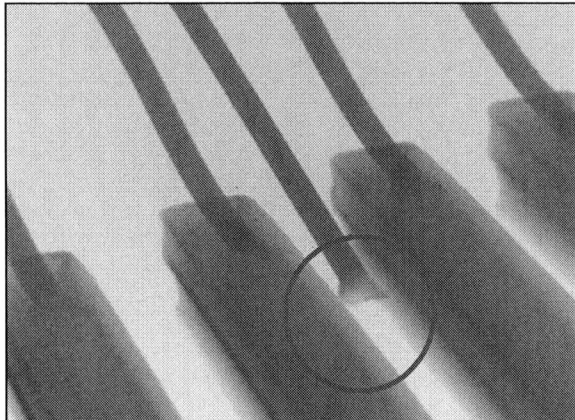

Fig. 6.Nanofocus X-ray image lifted wedge bond at 55° ovhm.

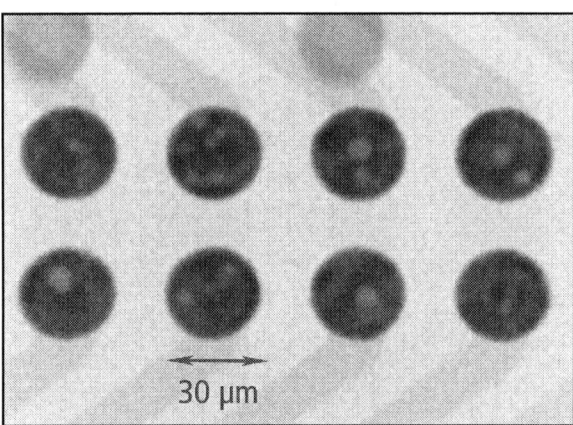

Fig. 7 Nanofocus X-ray image of Flip Chip solder joint in an electronic package revealing micrometer sized voids

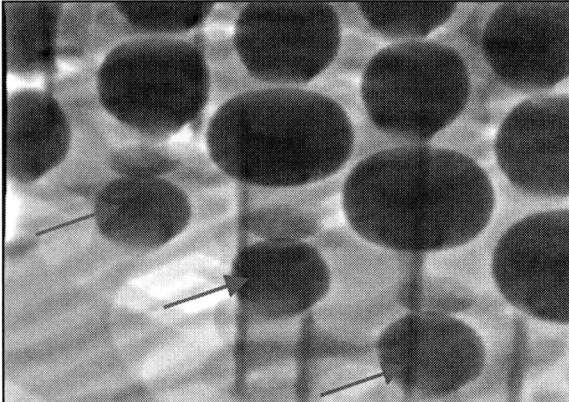

Fig. 8 Nanofocus X-ray images of Flip Chip solder joints in an electronic processor package. Three open solder joints are detected. The solder balls (bottom) are not connected to the lands (top). Note that the neighboured balls are much bigger than average possibly indicating solder shift after bridging.

In comparison to the highly absorbing gold wires as shown above copper bond wires yield a much poorer X-ray contrast that can be enhanced not only by using digital imaging but also by high magnification and sharp imaging. In Fig. 9 the 0.7mil copper wires of a chip scaled package including wedge and ball bond are displayed plainly and may be inspected in the usual way.

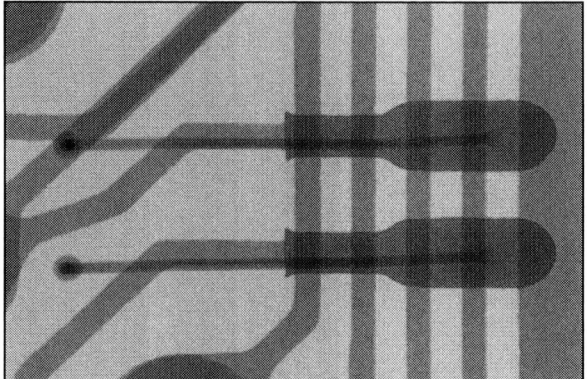

Fig. 9 Nanofocus X-ray image of 0.7mil (18μm) copper bond wires in a CSP.

NanoCT - Method

The nanoCT scans were carried out with a recently improved very compact laboratory CT system (nanotom® 180) dedicated to the analysis of small samples at sub-micron voxel resolution. The system comprises a 180kV / 15W high power nanofocus tube as described above and a 5 MPixel flat panel CMOS detector with a GOS scintillator deposited on a fibre optic plate. The set-up is shown in fig. 10. The pixel size of 50μm and a 3-position virtual detector (i.e. 360mm virtual detector width) give rise to a wide variety of experimental possibilities. The geometric set-up as shown in Fig. 11 enables a minimum voxel size < 500nm.

To avoid any influence of vibrations or thermal expansion, tube, detector and manipulator are mounted on a granite structure. Furthermore, special materials and construction details for e.g. the tube mounting are used to minimise the variation of the distance between focal spot and detector during scanning. In addition, minimal vibrations of the system are suppressed by air bearings of the rotation unit.

For reconstruction of the volume data phoenix|x-ray uses a proprietary implementation based on Feldkamp's cone beam reconstruction algorithm [5]. The reconstruction software contains several different modules for artefact reduction (such as beam hardening compensation, ring artefact reduction and residual drift compensation) to optimise the results.

NanoCT - Results

While 2D images of a memory cube with stacked wires and stacked dies (Fig. 12a) are not suitable for analysis due to overlaying features, 3D nanoCT virtual slices or sections allow to examine each individual die-attach for voids (Fig. 12b) as well as the flow of the bond wires. In stacked die and similar devices the wires are arranged in layers so that the wires overlap in the X-ray images and for example short circuits cannot be clearly told from crossings. In a 3D visualisations as provided by nanoCT the spatial arrangement of the wires may be examined slice by slice and their distances may be measured at any position.

A further example for advanced failure analysis with nanoCT is through silicon vias (TSV). In prior to the tomographic scan we took highly resolving 2D X-ray images for comparison and orientation, see Fig. 13. In oblique view at high magnification the vias and some internal void in the copper filling are plainly displayed, but some vias are concealed by the solder bumps. Adjacently the area shown in Fig. 13 was scanned by nanoCT system at voxel size of 1.25µm. The result is visualised in Figs. 14 and 15. The shape of the vias and bumps may be inspected under arbitrary angles view and virtual sections can be made in all positions and directions. In this way, the voids cannot only be seen and inspected but also automatically be detected in order to statistically evaluate their size and volume as shown in Figure 16.

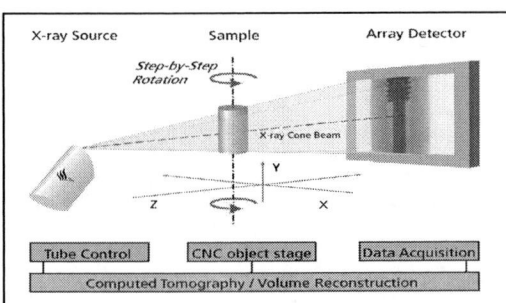

Fig. 10 Function principle of cone beam computed tomography: The system acquires a series of two-dimensional X-ray images while progressively rotating the sample step by step through a full 360° rotation at increments of less than 1° per step. These projections contain information on the position and density of absorbing object features within the sample. This accumulation of data is then used for the numerical reconstruction of the volumetric data.

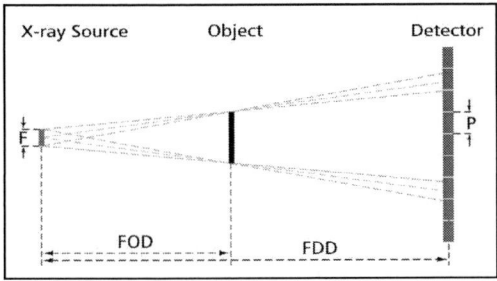

Fig. 11 Resolution of nanoCT: The geometric voxel resolution V is given by the pixel size P divided by the geometric magnification M=FDD/FDO, i.e.: V=P/M. With a pixel size P=50µm and a FDD= 500mm a voxel size of 0.5µm can be easily achieved at M=100. The final limitation of resolution is the focal spot size F of the X-ray tube, which causes an additional unsharpness on the detector (green lines). Hence, for sub-micron computed tomography an X-ray tube with a focal spot size below 1µm is required (nanofocus tube).

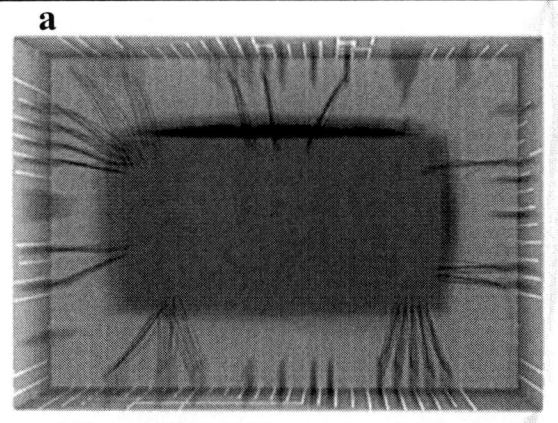

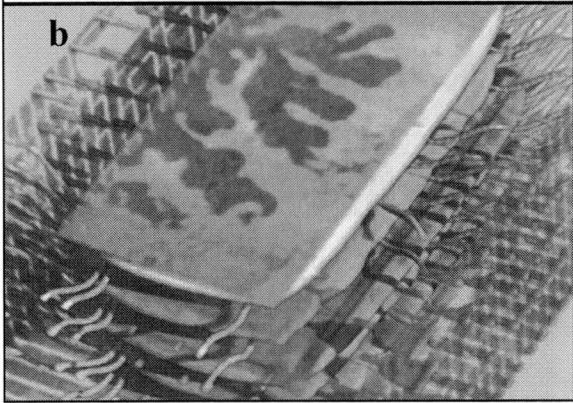

Fig. 12 a) Frontal 2D X-ray image of a memory cube with stacked dies and b) tomographic section visualising the die attach (by courtesy of 3D-Plus). Size of the sample is about 15mm x 10mm x 10mm.

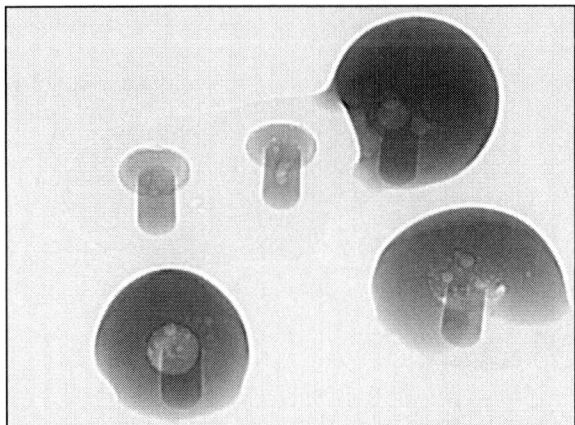

Fig. 13 Digital nanofocus X-ray image at 70kV/220 µA of a five TSVs. Voids in the copper via filling are detected, but some vias are concealed by the solder bumps which also contain voids.

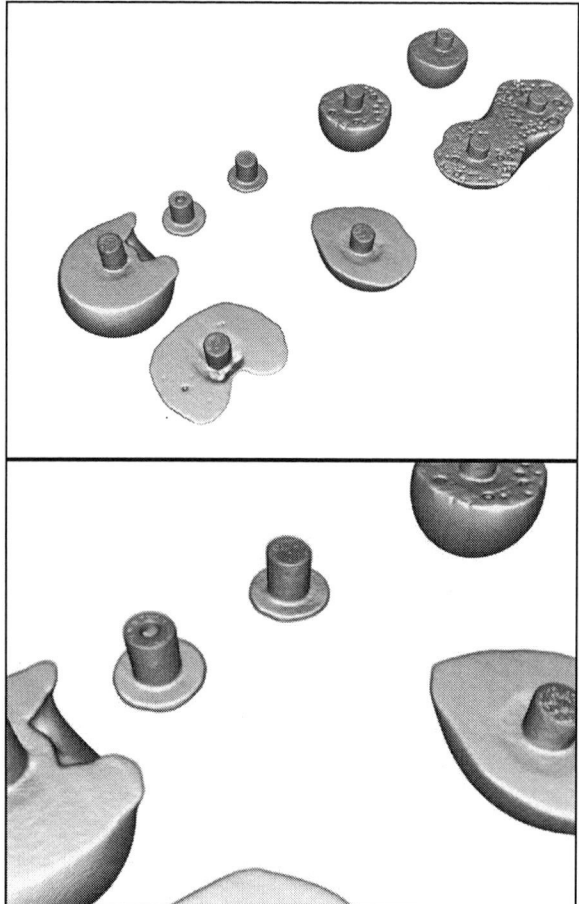

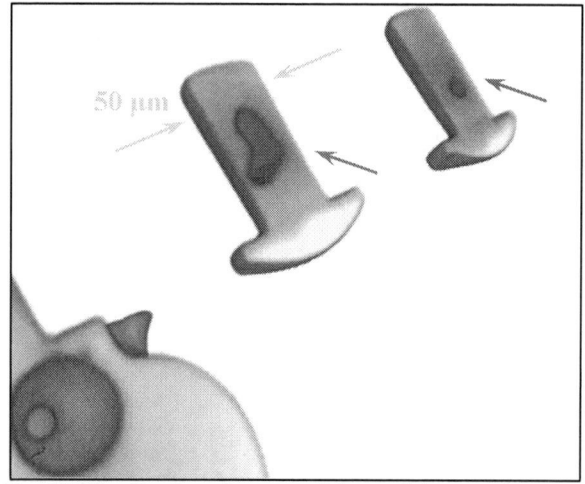

Fig.14 Visualisation of tomographic voxel data, virtually sectioned perpendicular to the via axis. Voids in the copper via filling are detected. The voxel size is 1.25μm, the via diameter is about 50μm.

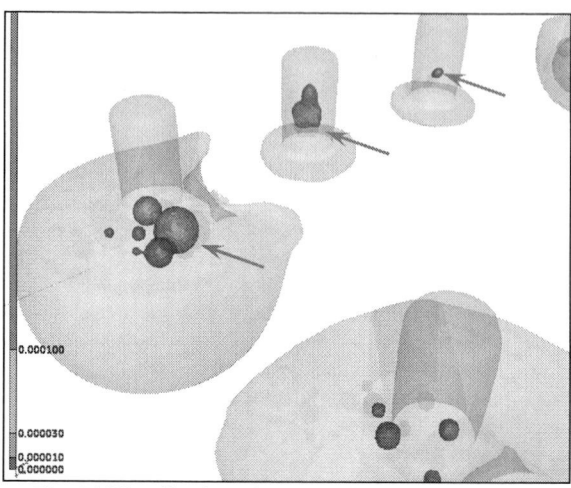

Fig. 16 Visualisation of tomographic voxel data, after automatic void detection. The surface of the metal is displayed transparently so that the voids are visible.

Fig. 17 Tomographic section visualisation of a ceramic IC package, showing vias and wire loop

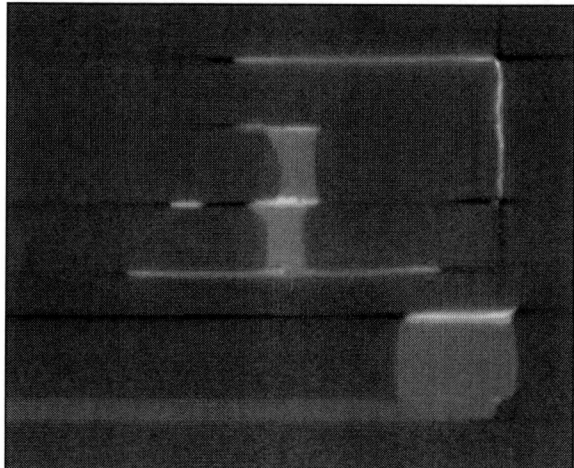

Fig. 15 Visualisation of tomographic voxel data, section along the via axis. An oblong void in the copper via filling is displayed. The voxel size is 1.25μm, the via diameter is about 50μm.

Fig. 18 Tomographic slice showing mis-registrated vias in the ceramics redistribution plane

978-1-4244-4658-2/09 $25.00 © 2009 IEEE

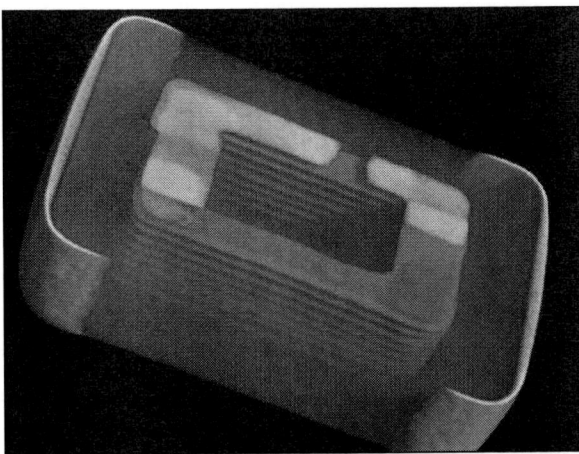

Fig. 19 Tomographic section visualisation of a 0805 chip inductor, clearly revealing a crack in the coil

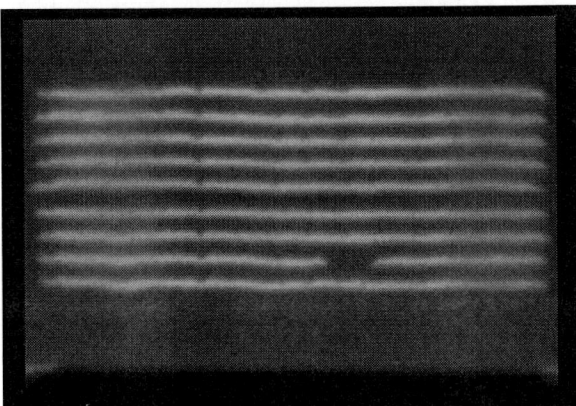

Fig. 20 Tomographic slice across the crack in the coil of the 0805 chip inductor (cf. Fig. 19)

Furthermore defects in concealed conductors in package redistribution planes, bond wires or the spatial localisation of package voids or defects in chip components are presented; see Fig. 17, 18, 19 and 20.

Conclusion

Digital nanofocus X-ray inspections systems with the capability of oblique views at highest magnification are a fast and most effective tool for the analysis of electronic packages and detection of defects of miniaturised interconnection down to the sub-micron range.

The resolution of such systems can be verified by determining the CTF or by a periodic test pattern like the JIMA test.

Typical applications are bond wire defects, copper bond wire inspection, Flip Chip solder interconnection and microvia inspection.

Based on the nanofocus tube technology nanoCT widely expands the spectrum of detectable microstructures in complex electronic devices and packages by 3D visualisation and slice-by-slice analysis. Thus nanofocus tube technology pushes computed tomography systems into application fields that very recently were exclusive to expensive synchrotron techniques [3].

Due to its microscopic resolution nanoCT will partially replace, support or supplement destructive methods saving costs and time per sample inspected.

References

1. Brockdorf, K. *et al*, "Sub-micron CT: visualization of internal structures," in Developments in X-ray Tomography VI, edited by Stuart Stock, *Proceedings of SPIE* Vol. 7078, (2008).
2. Japan Inspection Instruments Manufacturers' Association, http://www.jima.jp.
3. Brunke, O. *et al*, "Comparison between X-ray tube based and synchrotron radiation based µCT," *Proc. Developments in X-ray Tomography, SPIE Iint. Symposion of Opical Engineering & Applications* , San Diego, CA, Aug. 2008, to be published.
4. Roth, H., "Hochauflösende Röntgenanalyse und Computertomographie von Lötstellen," *GMM-Fachbericht* 50, (2006), pp. 285-294.
5. Feldkamp, L. A., Davis, L. C., Kress, J. W. "Practical cone beam algorithm," *Journal of the Optical Society of America A*, 1(6), 612-619 (1984).

Advanced Moisture Diffusion Model and Hygro-Thermo-Mechanical Design for Flip Chip BGA Package

Ming-Han Tsai*, Feng-Jui Hsu, Meng-Chieh Weng, Hsiang-Chen Hsu

Department of Mechanical and Automation Engineering, I-Shou University
No. 1, Sec. 1, Syuecheng Rd., Dashu Township, Kaohsiung County, Taiwan 84008
*E-mail: dan06234@hotmail.com

Abstract

In the present paper, a comprehensive moisture diffusion model and characterization for encapsulated plastic Flip Chip (FC) Ball Grid Array (BGA) package are investigated. The transient moisture diffusion analysis described by Fick's second law is performed to evaluate the overall moisture distribution. Diffusivities in the moisture desorption model are determined under Arrhenius behaviors. Hygroscopic swelling properties of polymeric materials are characterized by using an existing TMA/TGA extraction method. With the so-called "thermal-wetness" analogous technique, finite element analysis (FEA) is developed to evaluate the entire moisture distribution on FC BGA package. The analytical expression for total expansion strain due to hygro-thermo-mechanical coupled effect is implemented using finite element software ANSYS. Finite element predictions reveal the significance of contribution of hygroswelling induced strain. Reliability analysis for FC BGA is performed in accordance with JEDEC standard JESD22-A120. A series of comprehensive experimental works and parametric studies are conducted in this research.

1. Introduction

Polymer-based materials such as ultraviolet (uv) glue, die attach (DA), Epoxy Molding Compound (EMC), solder mask (SM) and Bismaleimide Triazine (BT) substrate have been widely applied to non-hermetic IC packages for the past two decades. Previous studies [1-11] reported that advanced electronic/optical packages are particularly sensitive to uncontrolled moisture absorption and temperature loading environments. The water is condensed into micropores and becomes moisture, which is absorbed through the polymer bulk exterior and the interface of polymer and metal. Polymers used on electronic packages expand upon absorbing moisture. Differential swelling occurs along the interfacial materials and consequent hygroscopic strains arise due to mismatch of coefficient of moisture expansion (CME). Reliability becomes an issue when the residual non-uniform moisture distribution induces hygroscopic swelling stress. Humidity and temperature gradient induced stresses are the critical factors for the useful life of electronic packages. The particular strain pattern at stressed region represents the potential failure site. Preliminarily results showed that the hygro-thermo coupling induced strain is responsible for the localized delamination for FC BGA packages.

Polymeric materials transport moisture primary by diffusion, which can be analogically modeled in Fickian transient diffusion equation. In moisture diffusion model, field variable (local moisture concentration) is found to be discontinuous across different material boundaries when it

exposed to the same temperature-humidity conditions [4]. This can be resolved by using the proposed temperature-wetness technique for modeling multi-material interfacial moisture diffusion. The same technique is also applied to model moisture desorption during dry baking and reflow process. In order to conduct the integrated thermo-hygro-mechanical analysis the hygroscopic swelling strain is treated as an additional term adding to the thermal strain. The developed moisture diffusion model is designated to evaluate the overall moisture distribution and the local moisture concentration at the critical interfaces.

A two-dimension FC BGA model based on finite element software ANSYS were developed to illustrate the moisture diffusion and the thermo-hygro induced strain distributions. However, the precise FEA prediction only depends on the accurate experimental material data. Therefore, moisture and hygroscopic swelling properties including the saturated moisture concentration, moisture diffusivity and CME were carefully measured and evaluated.

In accordance with JEDEC preconditioning test standard - Moisture Sensitivity Level (MSL), moisture diffusion analyses are performed at 30°C60%RH, 60°C60%RH, 85°C60%RH 85°C85%RH and 95°C 60% RH. The reflow process was conducted for moisture desorption from 25°C to 260°C within 5 minutes. Reliability analysis in this paper includes (1) material development (2) structure analysis (3) manufacturing process improvement.

2. Constitutive Models

The integrated modeling in this paper involves moisture diffusion (absorption/desorption), heat transfer, hygro-mechanical, thermo-mechanical, and hygro-thermo-mechanical models.

2.1 Moisture Diffusion Model

Polymers transport moisture primarily by diffusion which can be analogically modeled in standard Fickian transient diffusion equation. Equation (1) shows that field variable is C, the local moisture concentration which is defined as weight of water per volume in bulk material

$$\frac{\partial C}{\partial t} = D\left(\frac{\partial^2 C}{\partial x^2} + \frac{\partial^2 C}{\partial y^2} + \frac{\partial^2 C}{\partial z^2} \right) \tag{1}$$

where x, y and z are the spatial coordinates, t is the time and D is the moisture diffusivity which represents the rate of diffusion. However, $0 \le C \le C_{sat}$ where C_{sat} is the saturated moisture concentration and depends on temperature, relative humidity and material. Thus, C is discontinuous along the interfacial of material. Therefore, a new field variable, the "moisture wetness" w, is introduced to avoid differential difficulty and it is continuous across multi-material interface,

$$w = \frac{C}{C_{sat}} \qquad (2)$$

The new field variable w obeys the same Fickian equation [8-10]. The moisture diffusion model can then be re-written as

$$\frac{\partial w}{\partial t} = D\left(\frac{\partial^2 w}{\partial x^2} + \frac{\partial^2 w}{\partial y^2} + \frac{\partial^2 w}{\partial z^2}\right) \qquad (3)$$

where $0 \le w \le 1$, w=0 is the lower limit which implied complete dry condition and w=1 is the upper limit which implied fully saturated with moisture at relative humidity condition. It should be noted that materials such as metal and die are assumed to be impermeable to moisture, i.e. do not absorb moisture and have zero D and C_{sat}.

The mechanism of absorption is that the water vapour condenses in micropores and is internally transported. In desorption, the liquid water evaporates and is transported to the ambient [11]. Therefore, different materials constants are required to evaluate moisture absorption and moisture desorption.

2.1.1 Moisture Absorption

In moisture soaking test under given temperature and humidity, the moisture properties, i.e. D and C_{sat} can be determined by linearized curve-fitting the moisture data measured by weight gain method.

For moisture absorption, the initial condition is w=0 on the entire package and the boundary condition is w=1 along the external surface all the time at ambient moisture.

2.1.2 Moisture Desorption

Diffusivity in moisture desorption depends on temperature change during reflow process, i.e. the Arrhenius equation is used to model D,

$$D = D_o \exp(\frac{-E_d}{KT}) \qquad (4)$$

where D_o is the diffusivity coefficient, E_d is the activation energy, K is the Boltzmann constant and T is the absolute temperature.

Moisture varies as temperature change in reflow process. There will be a challenge in equation (2) since the saturation concentration could be zero during reflow process. By Henry's law, C_{sat} is linearly proportional to the relative humidity of the given condition,

$$C_{sat} = DS \qquad (5)$$

where D is equation (4) and S is solubility which is defined as

$$S = S_o \exp(\frac{-E_S}{KT}) \qquad (6)$$

where S_o is solubility coefficient, E_S is the activation energy. Equation (5) is generally true except at high relative humidity condition. Thus, it is most convenient to use the equation (5) to model moisture diffusion in reflow process.

For moisture desorption, the initial condition is the residual moisture distribution in the overall package and the boundary condition is w=0 all the time at the external surface.

2.2 Heat Transfer Model

The transient thermal diffusion (conduction) equation given in (7) can be solved to obtain the overall temperature distributions.

$$\frac{\partial T}{\partial t} = \alpha_T\left(\frac{\partial^2 T}{\partial x^2} + \frac{\partial^2 T}{\partial y^2} + \frac{\partial^2 T}{\partial z^2}\right) \qquad (7)$$

where T is the absolute temperature, x, y and z are the spatial coordinates, t is the time, $\alpha_T = k/(\rho C_P)$ is the thermal diffusivity, k is the thermal conductivity, C_P is the specific heat and ρ is the density.

The boundary condition used in the thermal model is fixed external surface temperature according to the reflow temperature profile within the JEDEC standards. The effects of convection and ambient temperature in different zones in oven are included in the reflow profile.

2.3 Hygro-Mechanical Model

The change in dimension and weight can be related to CME which is defined as the change of strain with moisture concentration. Moisture-induced hygroscopic strain can be obtained as

$$\varepsilon_H = \beta C \qquad (8)$$

where β is CME or the coefficient of hygroscopic swelling and C is local moisture concentration. Due to CME mismatch among various materials, the hygro-mechanical or hygroswelling stress is induced. Materials such as metal and die are assumed to be zero CME.

2.3 Thermo-Mechanical Model

Thermal strain due to change in temperature (ΔT) can be calculated by the expression

$$\varepsilon_T = \alpha\Delta T \qquad (9)$$

where α is coefficient of thermal expansion (CTE). Due to CTE mismatch among different materials, thermo-mechanical strain will be induced as the temperature increases during IR-reflow process.

2.4 Hygro-Thermo-Mechanical Model

The thermo-hygro-mechanical combined expansion strain can be determined by simply summing equations (8) and (9)

$$\varepsilon = \varepsilon_H + \varepsilon_T = \beta C + \alpha\Delta T = (\frac{\beta C}{\Delta T} + \alpha)\Delta T \qquad (10)$$

As can be seen, hygroscopic strain can be treated as an additional term in the thermal strain,

$$\alpha^* = \frac{\beta C}{\Delta T} + \alpha \qquad (11)$$

where α^* is the equivalent CTE. In any finite element commercial software, α^* can be easily substituted for α to conduct the hygro-thermo-mechanical analysis.

3. Experiment Works

Moisture saturated concentration (C_{sat}), diffusivity (D) and hygroscopic swelling CME (β) were carefully measured and evaluated in this paper.

3.1 Moisture Concentration Characterization

All the material properties were characterized as possible to reliability test conditions for new material development. An analytical expression for the absolute weight gain as a function of time is given as

$$\frac{M_t}{M_\infty} = 1 - \sum_{n=0}^{\infty} \frac{8}{(2n+1)^2 \pi^2} \exp\left(\frac{-D(2n+1)^2 \pi^2 t}{l_x^2}\right) \qquad (12)$$

$$C_{sat} = \frac{M_{sat}}{abc\left(100 - \text{vol\% of moisture}\right)} \quad (13)$$

where M_t is the instantaneous weight gain, M_∞ is the saturated weight gain, M_{sat} is the saturated mass of the sample, l_x is an equivalent length and a, b, c are the length, width and thickness of the sample, respectively. Previous study reported that equation (12) tends to be converged as the subscript n=16. D can then be obtained by employing non-linear regression technique extracted from TGA instrument.

For the new material EMC, samples were prepared in special size at $3*1.5*1.5$ mm^3 to fit the instrument. All the samples were baked at 125°C for 24 hours to remove any inside moisture. Fig. 1 illustrates the moisture weight gain (mg) as function of time (hrs) at 30°C60%RH, 85°C60%RH and 85°C85%RH, respectively. From equations (12) and (13), moisture absorption diffusivity D and moisture saturated concentration C_{sat} can then be determined and tabulated in Table 1.

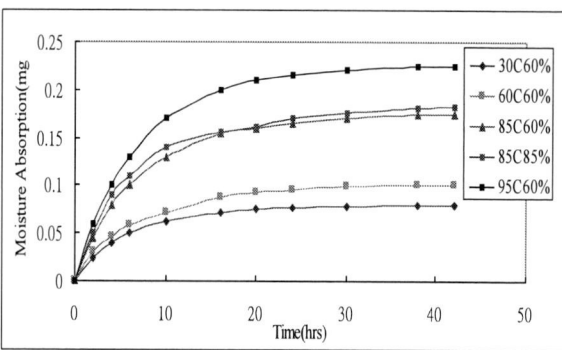

Fig. 1 Moisture weight gain (mg) vs. time (hr).

Table 1 Moisture absorption properties.

	D(cm^2/s)	C_{sat}(mg/cm^3)
30°C60%RH	2.31e-9	8.7
60°C60%RH	2.12e-8	11.2
85°C60%RH	5.96e-8	19.4
85°C85%RH	6.14e-8	20.9
95°C60%RH	---	24.9

3.2 Moisture Desorption Characterization

Moisture desorption was conducted at various temperatures and the diffusivity was observed to comply with the Arrhenius behavior in equation (4). A variable transform scheme is employed by taking natural logarithm on both sides in equation (4) and then the equation becomes

$$\ln(D) = \ln(D_O) - \frac{E_d}{KT} \quad (14)$$

Plot ln(D) versus (1/KT) in Fig. 2. By applying curve fitting scheme, the slope of fitting line should represent the activation energy E_d and the interception of fitting line and y-axes would be ln(D_O). For the case of 30°C60%RH,

$$y = -0.1419x + 7.1257$$

where the activation energy E_d is 0.1419 eV and

$$D_O = \exp(7.1257)*e-6 = 0.001244 \text{ cm}^2/\text{s}$$

From Fig. 2, E_d and D_O are also reported as 0.161 eV, 0.167eV and 0.0022 cm^2/s, 0.0027 cm^2/s for 85°C60%RH and 85°C85%RH, respectively.

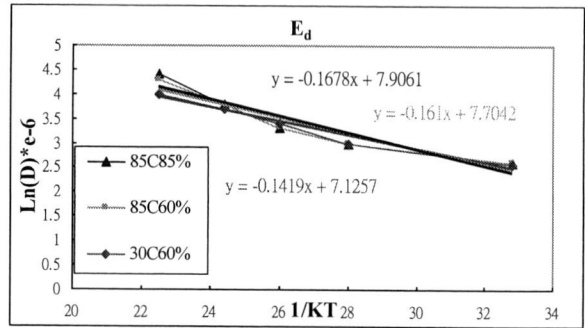

Fig. 2 Moisture desorption diffusivity Arrhenius behavior.

Table 2 lists the measured moisture desorption diffusivity at different temperatures. In moisture desorption model, an increase in temperature results in an increase in diffusivity. This implies that moisture desorption diffusion during reflow can only be illustrated by Arrhenius equation. From Tables 1 and 2, moisture diffusion in desorption is much faster than in absorption. Thermal and humidity are the major factors which impact moisture diffusion.

Table 2 Moisture desorption diffusivity (D).

	80°C	140°C	170°C	200°C	240°C
30°C60%RH	1.3e-7	1.9e-7	2.8e-7	4.2e-7	5.5e-7
85°C60%RH	1.4e-7	2.1e-7	2.9e-7	4.4e-7	7.5e-7
85°C85%RH	1.4e-7	2.1e-7	2.9e-7	4.5e-7	7.5e-7

Empirical constants used to model the solubility at given relative humidity (60%) are determined the C_{sat}. Fig. 3 demonstrates the solubility dependency on temperature.

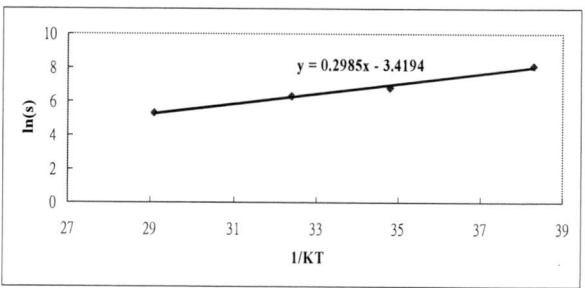

Fig. 3 Moisture desorption solubility Arrhenius behavior.

Taking natural logarithm on both sides in equation (6) and then the equation becomes

$$\ln(S) = \ln(S_O) - \frac{E_S}{KT} \quad (15)$$

Plot ln(S) versus (1/KT) and apply curve fitting scheme in Fig. 3. For the case of 60% relative humidity,

$$y = 0.2985x - 3.4194$$

where E_S =0.2985 eV and S_O=0.032 mg/cm^3.

3.3 Hygroscopic Swelling Characterization

Hygroscopic swelling characterization technique is previous developed by using TMA/TGA instruments [8-10]. The procedures consists of moistening and saturating two identical samples at the same temperature and humidity for equal duration, and then desorbs moisture from each sample isothermally in TMA/TGA. Graph dimensionless strain versus moisture concentration. The slope of the strain-moisture concentration curve defines the linear relationships between hygroscopic swelling and moisture content, which enabling the CME to be mathematically defined.

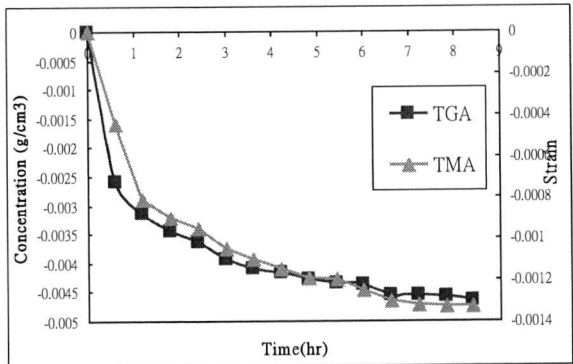

Fig.4 TGA/TMA results at 80°C (30°C60%RH).

Fig. 4 illustrates extract moisture dimensional change (ΔL) from TMA and moisture weight loss (ΔM) from TGA at 30°C60%RH. Plot dimensional change ($\varepsilon_h = \Delta L / L$) in y-axis versus moisture concentration (C=ΔM/V) in x-axis. The linearity slope should be CME (β). Fig. 5 presents the strain versus moisture concentration at 80°C where β = 0.1614 mm^3/mg. Due to temperature variation in the reflow process, CME is measured at 80, 140, 160, 200 and 240°C, respectively. The corresponding CMEs for 30°C60%RH, 85°C60%RH and 85°C85%RH are illustrated in Fig.6 to 8.

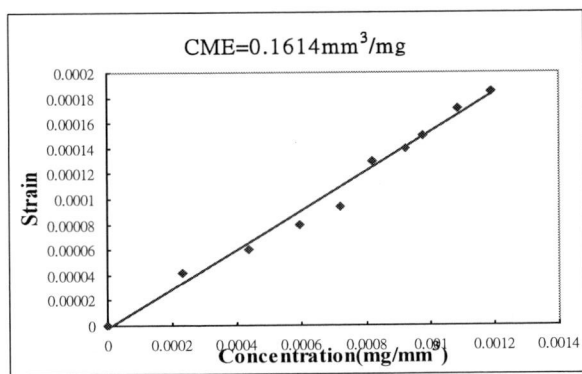

Fig. 5 Linearity of CME at 80°C (30°C60%RH).

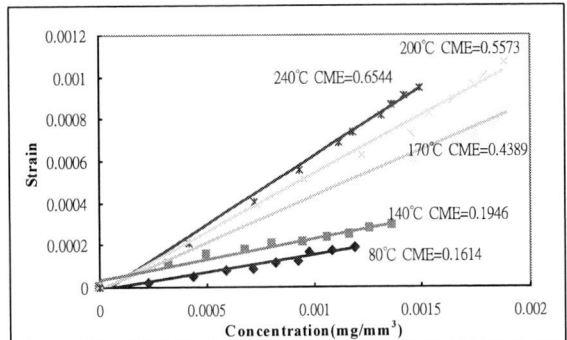

Fig. 6 CME over temperature range (30°C60%RH).

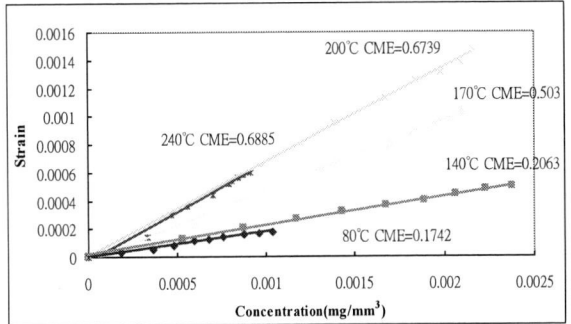

Fig.7 CME over temperature range (85°C60%RH).

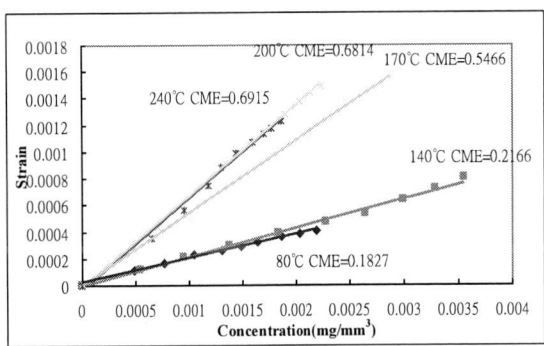

Fig8. CME over temperature range (85°C85%RH).

4. Finite Element Predictions

As shown in Fig. 9, an integrated finite element system-in-package (SiP) of FC BGA solid model has been developed to predict moisture diffusion, temperature distribution and hygro-thermo-mechanical coupled effects. Both 2-D and 3-D model are developed to examine the transportation of moisture diffusion through the exterior bulk material and the critical bi-material interface.

978-1-4244-4658-2/09 $25.00 © 2009 IEEE

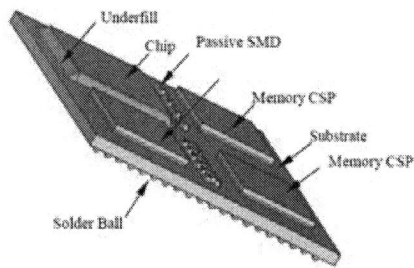

Fig. 9 Three dimensional FC BGA finite element model.

A 2-D finite element model is used to validate thermal induced in-plane deformation on FC BGA package. Fig. 10 to 13 shows the results for Moiré experiment and FEA prediction and the error are listed in Table 3.

Fig. 10 Moiré experiment along horizontal direction (ux).

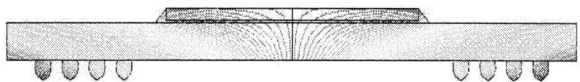

Fig.11 FEM prediction along horizontal direction (ux).

Fig. 12 Moiré experiment along vertical direction (uy).

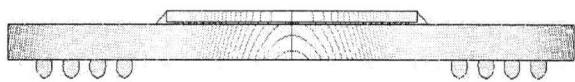

Fig.13 FEM prediction along vertical direction (uy).

Table 3 Thermal induced in-plane deformation.

	Moiré	Garofalo-Arrhenius FEM	Error
ux	2.085 um	2.035 um	-2.4%
uy	10.842um	10.873 um	0.3%

It should be noted that material for lead-free solder bump in FC BGA is Sn4.0Ag0.5Cu (SAC 405). Reliability analysis involves in both creep and fatigue, which implies Garofalo-Arrhenius equation should be applied to FEA.

Most of commercial FE software is not available to perform the moisture diffusive analysis but can be extended as a thermal-moisture model. With appropriate parameters changed, the moisture transient diffusion can be analogically modeled by thermal diffusion [5-7]. Table 4 lists the parameters used for thermal analysis and moisture analysis in ANSYS. Mechanical and hygroswelling properties for all the materials used in this paper can be found in [1-2, 8-11].

Table 4 Parameters for thermal-moisture analyses in ANSYS.

	Thermal Analysis	Moisture Analysis
Field variable	Temperature, T	Wetness, w
Density	ρ	1
Conductivity	K	DC_{sat}
Specific Heat	Cp	C_{sat}
CTE	α	βC_{sat}

5. Results and Discussions

Based on JEDEC MSL 3A standard - 30°C 60%RH, the moisture absorption duration is 192 hours and the reflow process is performed from 25°C to 250°C in 5 minutes. Moisture distributions during soaking are shown in Fig. 14. Fig. 15 shows moisture distributions during dry baking.

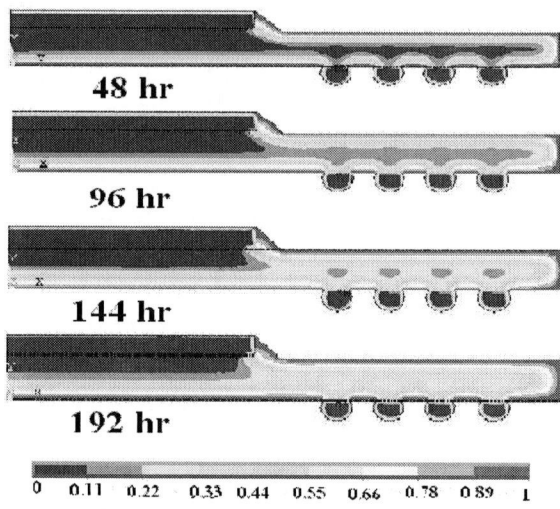

Fig.14 Wetness distribution during soaking test.

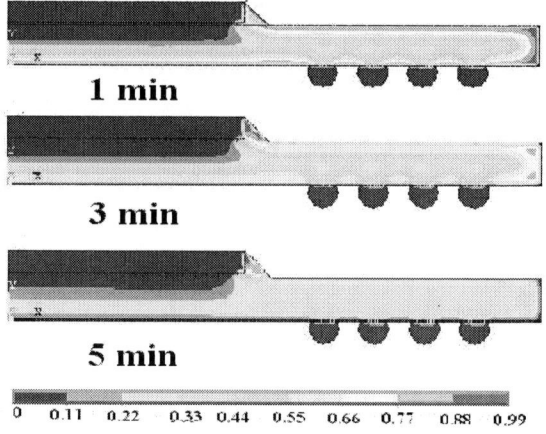

Fig.15 Wetness residual distribution during reflow process.

As can be seen, the moisture is transported through the exterior of underfill and BT substrate. The residual moisture in the underfill and BT substrate will induce hygroscopic swelling and hygro-thermo stress during high temperature reflow process.

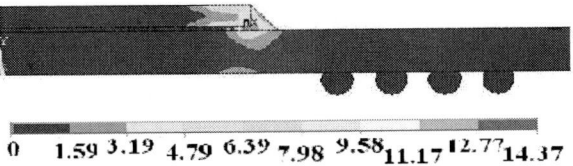

0   1.59 3.19 4.79 6.39 7.98 9.58 11.17 12.77 14.37

Fig. 16 Hygro-mechanical effective stress (MPa).

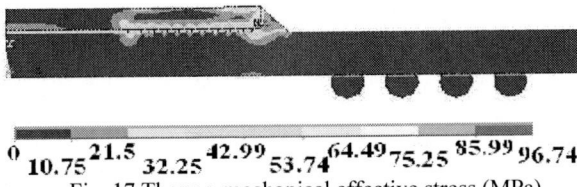

0   10.75 21.5 32.25 42.99 53.74 64.49 75.25 85.99 96.74

Fig. 17 Thermo-mechanical effective stress (MPa).

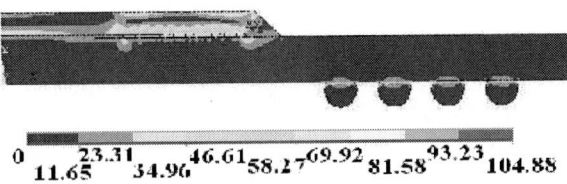

0   11.65 23.31 34.96 46.61 58.27 69.92 81.58 93.23 104.88

Fig. 18 Hygro-thermo-mechanical effective stress (MPa).

Fig.16 to Fig. 18 characterizes the hygro-mechanical, thermo-mechanical and hygro-thermo-mechanical effective stress distributions, respectively. The peak hygro-mechanical effective stress on the localize stressed area (in the adjacent of die/undefill/solder bump/BT substrate) is 14.37MPa. Fig. 17 shows the maximum thermo-mechanical effective stress is 96.74 Mpa. The ratio of hygroswelling stress to thermal stress is around 1:7. The combination of thermo-hygro-mechanical analysis is performed and the maximum effective stress is 104.88 Mpa in Fig. 18. However, the stressed areas have been slightly changed, which demonstrates the hygroswelling stress can alter the thermal stressed area. The residual stressed area is found along the interfacial of die/underfill/BT substrate, which illustrates the potential delamination in these regions after thermal cycling test, which is shown in Fig. 19..

Fig. 19 SEM picture for delamination and crack along the interfacial of die/underfill/BT substrate (after TCT).

6. Conclusions

Thermal induced in-plane micro deformations have been measured by Moiré technique and predicted by Garofalo-Arrhenius equation in FEA. It has been shown that the FEA

spredicted results have excellent agreement with Moiré experimental results. The FEA model can then be applied to predict thermal induced stress. The constitutive of hygro-thermo-mechanical model has been developed in this research. This paper also demonstrates an experimental procedure to determine moisture and hygroscopic swelling properties for polymeric materials. An integrated finite element model is developed to investigate the moisture diffusion and thermo-hygro-mechanical coupled effects. The residual moisture content in the interfacial actually degrades the adhesion, which results in an opening failure between two materials. Finite element predictions reveal the significance of contribution of hygroswelling induced stress. This localize stressed area imposes a potential failure region. The results can be directly applied to develop a new formula for new material and improve the manufacturing process.

Acknowledgments

This work was supported by I-Shou University, Taiwan, Grant No. ISU-97-In-01 and Science Council, Taiwan, Grant No. NSC-97-2218-E-214-003.

References

1. Hsu, H.C. and Hsu, Y.T., "Characterization of hygroscopic swelling and thermo-hygro-mechanical design on electronic package," *Journal of Mechanics,* vol. 25 (2009), pp. 225-232.

2. Hsu, H.C., Lee, H.Y., Hsu, Y.C. and Fu, S.L, "Thermal-Hygro-Mechanical Design and Reliability Analysis for CMOS Image Sensor," *Journal of Thermal Stress,* Vol.31, (2008), pp.914-931.

3. Shieh, W.L., Thermo-hygro-structure design and reliability analysis on plastic packaging MEMS pressure sensor, *Master Thesis*, Department of Mechanical and Automation Engineering, I-Shou University, May 2008.

3. Ma, X., Jansen, K.M.B., Zang, G.Q. and Ernst, L.J., "Hygroscopic effects on swelling and viscoelasticity of electronic packaging Epoxy," *IEEE ICEPT*, Shanghai, China, August 2006, pp. 262-266.

4. Karad, S.K., "Mechanisms of moisture absorption by Cyanate Ester modified Epoxy resin matrices:the Clustering of water molecules," *Journal of Polymer,* vol. 45 (2005), pp.2732-2738, August 2005.

5 The, L.K., Teo, M., Anto, E., Wong, C.C., Mhaisalkar, S.G., Teo, P.S. and Wang, E.H., "Moisture-induced failures of adhesive flip chip interconnects," *IEEE Transactions on Components and Packaging Technologies,* vol. 28, No. 3 (2005), pp. 506-516.

6. Zhang, X., Tee, T.Y., Ng, H.S., Teysseyre, J., Loo, S. and Mhaisalkar, S.,"Comprehensive hygro-thermo-mechanical modeling and testing of stacked die BGA module with molded underfill," *IEEE Electronic Components and Technology Conference (ECTC)* , 2005, pp.196-200.

7 Lahoti, P.S., Kallolimath, S.C. and Zhou, J., "Finite element analysis of thermo-hygro-mechanical failure of a flip chip package," *IEEE 5th Electronic Packaging Technology Conference (EPTC)*, 2004, pp. 180-183.

8 Wong, E.H., Rajoo, R., Koh, S.W. and Lim, T.B., "The mechanics and impact of hygroscopic swelling of

polymeric materials in electronic packaging," *J. Electronic Packaging (Trans. of ASME)*, vol. 124, No. 2 (2002), pp. 122-126.

9 Tee, T.Y. and Ng, H.S., "Whole field vapor pressure modeling of QFN during reflow with coupled hygro-mechanical and thermo-mechanical stresses," *IEEE Electronic Components and Technology Conference (ECTC)*, 2002, pp. 1552-1559.

10 Wong, E.H., Koh, S.W., Rajoo, R. and Lim, T.B., "Underfill swelling and temperature-humidity performance of flip chip PBGA package," *IEEE 1st Electronic Packaging Technology Conference (EPTC)*, 2000, pp. 258-262.

11 Galloway, J.E., and Miles, B.M., "Moisture Absorption and Desorption Predictions for Plastic Ball Grid Array Packages," *IEEE Transactions on Components, Packaging, and Manufacturing Technology*—Part A, Vol. 20, No. 3 (1997), pp.274-278.

Analysis the Performance of the Micro-Channels Cooler with Different Inlet Position

Xiaojing Wang, Wen Zhang, Hongjun Liu, Ling Chen, Zongshuo Li
Shanghai University
Shanghai University, 224mail box, 149 Yan Chang RD. Shanghai, 20072, China
xjwang@mail.shu.edu.cn, 86-21-66136117

Abstract

Electronic chips are now working at higher temperatures than they were before. It is impossible for electronic products to achieve 100% efficiency, the problems about heat manage is becoming more and more. Many experiments were performed to explore the benefits of micro-channel cooling. In the published experiments, the location of the fluid entrance are different. In order to analyze the effect of the inlet position on the performance of the micro cooler, in this paper, different inlet (outlet) position MCHS models, different distance between the inlet surface and the near ends of fins MCHS models and different inlet shape MCHS models are compared. The simulation model is established to analyze the heat distributing and highest temperature of the MCHS. The total heat flux amount is $250W/cm^2$. Water is chosen as the coolant and the velocity ranges from 0.1m/s to 5m/s. The results show that the highest temperature in z-axis direction inlet is much lower than other two positions, so that it is the best inlet and outlet position for temperature drop, where the entrance position is perpendicular to the direction of flow in the channels. The distance between the inlet and the end of fins has effect on the performance of the micro-channel cooler. The rectangular and circular inlet has nearly the same effect on the heat dissipation, but the inlet area is much important. The simulation results play an important role for the design of micro-channels cooler.

1 Introduction

Advanced very large-scale integration (VLSI) technology has resulted in significant improvements in the performance of electronic systems in the past decades. With the trend toward higher circuit density and faster operation speed, there is a steady increase in the dissipative heat flux at the component, module, and system levels. This leads to an increasing demand for highly efficient electronic cooling technologies. To meet this demand, various electronic cooling schemes have been developed. The parallel-plain fin (PPF) array structure is widely applied in convective heat sinks in order to create extended surface for the enhancement of heat transfer [1]. For investigating the influences of designing parameters of PPF heat sink with an axial-flow cooling fan on the thermal performance, a systematic experimental design based on the response surface methodology is used. The thermal resistance and pressure drop are adopted as the thermal performance characteristics. A typical microchannel cooler configuration is a finned structure, which is cooled by forced convection (Fig. 1). The power is dissipated on the circuit side and the heat is conducted through the substrate to the fins where it is transferred to the coolant [2]. The inlet and outlet position of this experimental device is set perpendicular to the direction of flow at the top of the cooler, and the shape is rectangular. Weilin Qu, Issam Mudawar [3] discussed some new experimental results which provide new physical insight into the unique nature of flow boiling in narrow rectangular micro-channels. The micro-channel heat sink contained 21 parallel channels having a $231\times713\mu m$ cross-section. Tests were performed with deionized water over a mass velocity range from $135kg/m^2s$ to $402kg/m^2s$. Fig. 2 illustrated the construction of the test module with the inlet location which parallels the direction of flow. And the shape of inlet is circular. Charlotte Gillot et al [4] performed some experiments to assess the feasibility of single and two-phase micro heat exchangers applied to the cooling of insulated gate bipolar transistor (IGBT) power components. The evaporator was made of circular channels. Its sizes were chosen according to criteria given by Bower and Mudawar [5] to ensure a good distribution of the heat flux all over the periphery of the channels. Six channels were machined in a piece of copper (see Fig. 3). The position of inlet and outlet, whose shape is circular, is set perpendicular to the direction of flow at the same side of the device.

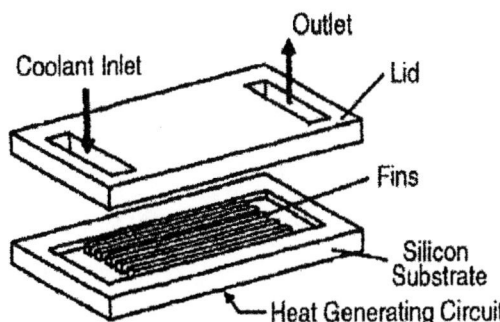

Fig. 1 Configuration of the microchannel cooler

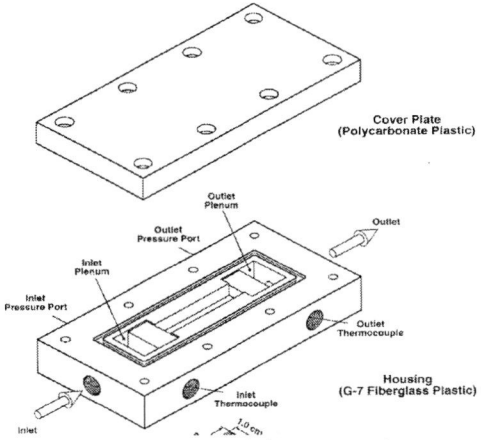

Fig. 2 Test module construction

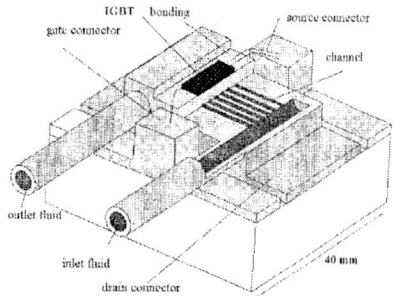

Fig.3. Global view of the two-phase heat exchanger

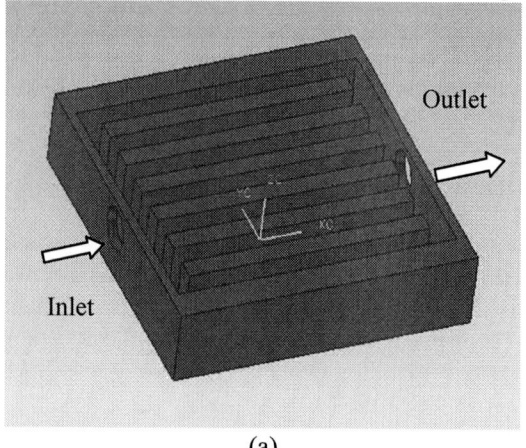

(a)

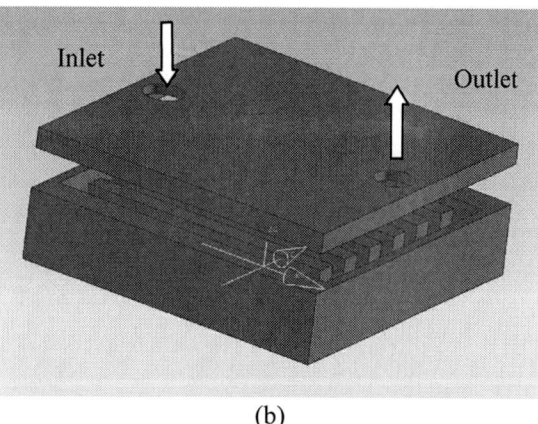

(b)

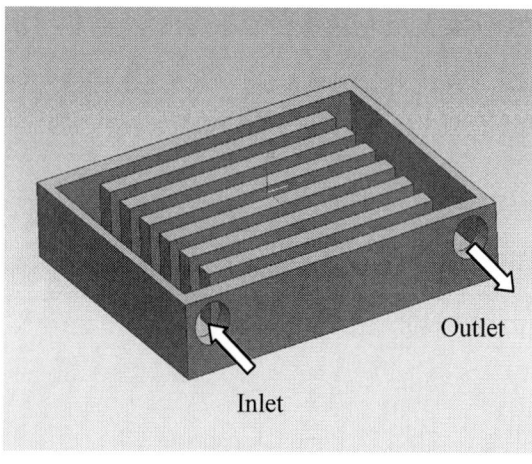

(c)

Fig.4 Schematic of vertical and horizontal direction between inlet (outlet) velocity and flow

As the test constructions are shown above, the performances of the cooler with inlet and outlet positions are different. The experimental results are unable to be compared with the different structure and parameters. In this paper, with the simulation of the computation fluid dynamics software FLUENT, several types of micro-channel heat sink cooler models are established to compare the temperature-rise of the MCHS with different inlet and outlet positions, different shapes and different distances between the inlet surface and the near ends of fins. The directions of inlet velocity and flow are vertical or horizontal (Fig. 4). The shapes of the installation inlet are circular and rectangular. And the distances between inlet and fins are 2mm, 3mm and 4mm. The temperature distribution and highest temperature-rise of these different models are studied with the inlet velocity from 0.1m/s to 5m/s.

2 Theoretical analyze

2.1 Governing Equations

Computational fluid dynamics (CFD) is used to tackle this problem. The fluid is assumed to be incompressible and the flow field is continuous (Knudsen number is small enough in the present study). Based on the computational domain, the conductions and convection in the micro-channels are simulated in the same time to capture the heat transfer. Steady state continuity, momentum and energy equations are solved.

Continuity equation:

$$\frac{\partial U}{\partial x} + \frac{\partial V}{\partial y} + \frac{\partial W}{\partial z} = 0 \tag{1}$$

Momentum equation:

$$\begin{cases} U\frac{\partial U}{\partial X} + V\frac{\partial U}{\partial Y} + W\frac{\partial U}{\partial Z} = \frac{1}{Re}(\frac{\partial^2 U}{\partial X^2} + \frac{\partial^2 U}{\partial Y^2} + \frac{\partial^2 U}{\partial Z^2}) - \frac{\partial P}{\partial X} \\ U\frac{\partial V}{\partial X} + V\frac{\partial V}{\partial Y} + W\frac{\partial V}{\partial Z} = \frac{1}{Re}(\frac{\partial^2 V}{\partial X^2} + \frac{\partial^2 V}{\partial Y^2} + \frac{\partial^2 V}{\partial Z^2}) - \frac{\partial P}{\partial Y} \\ U\frac{\partial W}{\partial X} + V\frac{\partial W}{\partial Y} + W\frac{\partial W}{\partial Z} = \frac{1}{Re}(\frac{\partial^2 W}{\partial X^2} + \frac{\partial^2 W}{\partial Y^2} + \frac{\partial^2 W}{\partial Z^2}) - \frac{\partial P}{\partial Z} \end{cases} \tag{2}$$

Energy equation:

$$\frac{\partial}{\partial t}(\rho E) + \nabla \bullet (\vec{v}(\rho E + p)) = \nabla \bullet (k_{eff}\nabla T - \sum h_j \overline{J_j} + (\overline{\overline{\tau}}_{eff} \bullet \vec{v})) + S_h \tag{3}$$

Where k_{eff} is the effective conductivity ($k_{eff} = k + k_t$, where k_t is the turbulent thermal conductivity, defined according to the turbulence model being used), and J_j is the diffusion flux of species j. The first three terms on the right-hand side of Equation (3) represent energy transfer due to conduction, species diffusion, and viscous dissipation, respectively. S_h includes the heat of chemical reaction, and any other volumetric heat sources user have defined.

2.2 Model specification and boundary conditions

The description of the MCHS models in this simulation are as follows, the substrate copper plate of 30mm×36mm×2mm, eight fins with 26mm length and 2mm width and 5mm height distribute on the plate uniformly. The inlet and outlet position are designed at three different places. Other two distances between the inlet (outlet) and the end of fins and two types of inlet shape are designed with the hydraulic diameter (de) is calculated by Equation (4) and (5).

$$h = \frac{Nu \cdot \lambda_f}{de} \qquad (4)$$

$$de = A / L \qquad (5)$$

Where h is heat transfer coefficient of water, A is the effective sectional area of the total flow, and L is the contact length between fluid and solid wall.

The uniform velocity and temperature are applied in the inlet of the micro-channel. The total flow rate can be determined by the total number of micro-channels used in a micro-channel heat sink module and the velocity of individual micro-channel. The constant pressure of 1atm is applied in the exit of the micro-channel. The heat flux of $250W/cm^2$ is applied at the bottom of the heat sink. Heat is assumed to be removed by working fluid only. For the uniform heat source simulation the heat flux is applied at the bottom (30mm×36mm) of the heat sink average.

3 Simulation Results

The models in the simulation are described as follows: eight fins with 26mm length and 2mm width and 5mm height distribute on the plate uniformly. Seven fluid channels are 2mm×26mm×5mm.

3.1 Comparison with different inlet positions

Three different inlet (outlet) position models are established. Circular inlet and outlet are located at the same line parallels y-axis, z-axis or x-axis. Type 1 is shown in Fig. 4(a), whose inlet and outlet position parallels the direction of flow (y-axis). Fig. 4(b) shows the type 2, whose inlet and outlet position is perpendicular to the direction of flow at the top of the MCHS (z-axis). And type 3 is shown in Fig. 4(c), whose inlet and outlet position is also perpendicular to the direction of flow, but at one side of the cooler (x-axis). The highest temperatures of three models' heating surface are shown in Table 1.

Table 1 Highest temperature under different inlet(outlet) positions

Inlet velocity (m/s)	Type 1 (k)	Type 2 (k)	Type 3 (k)
0.1	394	388	398
0.2	373	364	372
0.3	361	355	361
0.4	358	352	356
0.5	356	350	353
0.6	351	349	351
1	346	348	346
2	344	343	340
5	340	339	338

The effect of inlet velocity with different inlet positions is list in Table 1. It can be seen that the temperature in type 2 is much lower than other two positions. But with the velocity increasing, the temperature drops not as quickly as x-axis direction inlet. Then with the huge inlet velocity, the temperature varies small. Fig. 5 shows the temperature distributions of three models in 0.5m/s inlet velocity. Fig. 5(a) is the temperature distribution graph of type 1 0.5m/s inlet

velocity. Fig. 5(b) and Fig. 5(c) are the temperature distribution graphs of type 2 and type 3 with 1mm height and 2mm radius circular fluid inlet and outlet area.

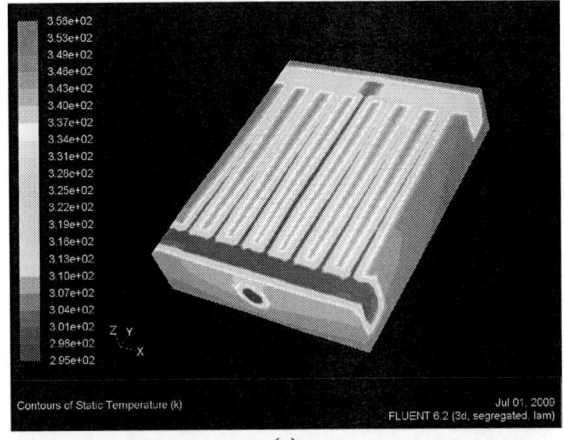

(a)

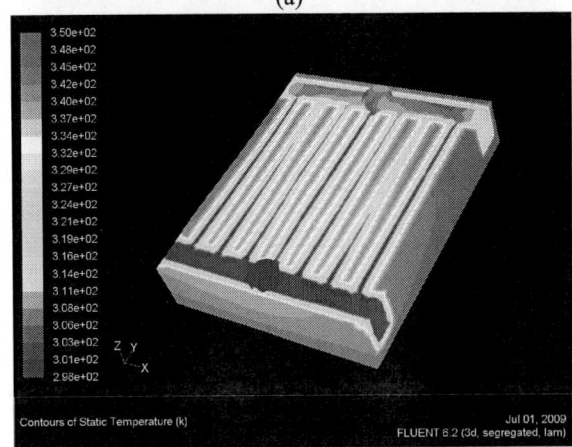

(b)

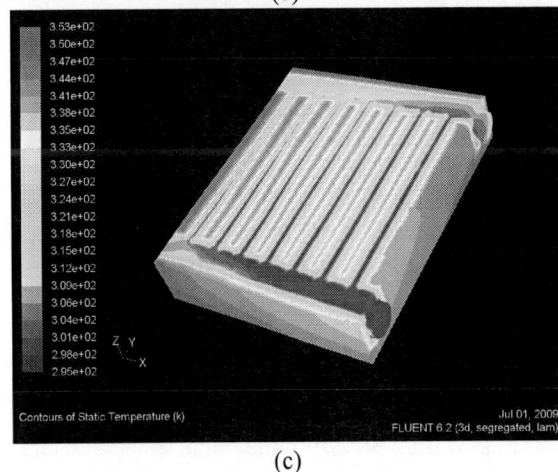

(c)

Fig. 5 Temperature distribution of three different inlet position models. (a) Inlet position parallels to y-axis (type 1). (b) Inlet position parallels to z-axis (type 2). (c) Inlet position parallels to x-axis (type 3).

3.2 Comparison with different distances of fluid channels between the inlet surface and the near ends of fins

With the comparison of different inlet and outlet position, type 2 (Fig. 4(b)) is chosen to research the temperature-rise of different distances between the center of circular inlet and one ends of fins models. Three models have different distances between the center of circular inlet and one ends of fins of the fluid channel, which are 2mm, 3mm and 4mm, which are case1, case2 and case3. Table 2 shows the highest temperatures of heating surface with these three cases of distances between inlet and one ends of fins.

Table 2 Highest temperature under different distances between inlet position and fins

Inlet velocity (m/s)	Case1 (k)	Case2 (k)	Case3 (k)
0.1	388.5	388.7	389.3
0.2	363.6	363.5	364.4
0.3	355.5	355.3	356.0
0.4	352.0	351.5	351.8
0.5	350.1	349.2	349.4
0.6	349.2	347.9	347.7
1	347.7	345.4	344.6
2	342.5	342.1	342.4
5	339.1	339.4	341.1

It can be seen from Table 2 that the highest temperatures of these three models are very close. With the increasing of inlet velocity, the temperatures of 2mm distance model (case 1) are the highest among three models. Small distance of the fluid entrance channel may lead to the fluid flow inadequate, and reduce the heat dissipation. Compare with the results of case 2 and case 3, 3mm distance is the ideal one, except in 0.6m/s inlet velocity there is a slight increase of the highest temperature. Fig. 6 shows the temperature distributions of 3mm distance between the center inlet and one ends of fins in 0.5m/s inlet velocity model (case 2) and 4mm distance model (case 3).

3.3 Comparison with different inlet and outlet shapes of MCHS models

The different inlet and outlet shapes of experimental micro-flow apparatus are pointed out from the Fig. 1-Fig. 3. It is necessary to simulate different shape models to compare the temperature rise, in order to effectively improve the cooling performance. In the principle of the area equivalent, 4mm×3mm rectangular inlet MCHS model is established to compare the highest temperature with the 2mm radius circular inlet MCHS model. And then, the radius is decreased to 1.5mm. All of three models use the inlet position of type1 (Fig. 4(a)) and 4mm distance between inlet surface and near ends of fins. Table 3 shows the highest temperatures of heating surface with different inlet shapes, which are 1.5mm, 2mm radius circular and 4mm×3mm rectangular.

The results of Table 3 show that the temperature of rectangular inlet MCHS is nearly the same as the circular one which has the same inlet area in different inlet velocity, but the rectangular inlet should pay attention on sealing. The smaller circular inlet is not more effective on cooling. The temperature of 1.5mm radius circular inlet model is much higher than 2mm radius inlet model, because of the inadequate flow in entrance. Fig. 7 is the temperature

distributions plot of rectangular inlet MCHS in 0.5m/s inlet velocity.

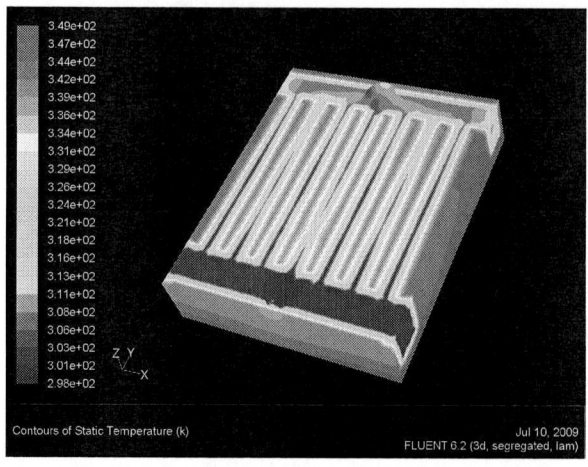

(a)

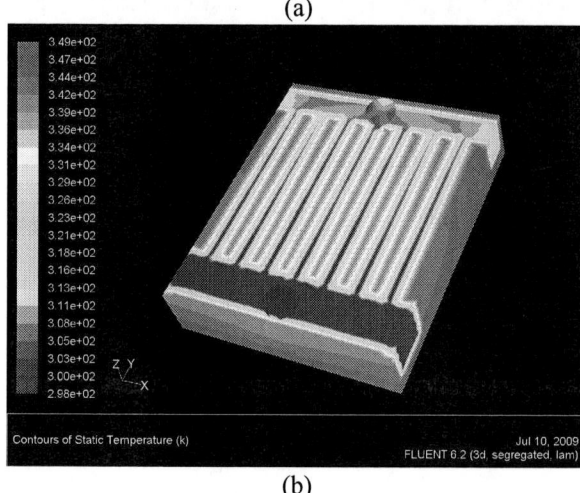

(b)

Fig. 6 Temperature distribution of different distances fluid channels MCHS. (a) Temperature distribution of 3mm distance between the center inlet and one ends of fins in 0.5m/s inlet velocity. (b) Temperature distribution of 4mm distance between the center inlet and one ends of fins in 0.5m/s inlet velocity.

Table 3 Highest temperature under different inlet shape

Inlet velocity (m/s)	Temperature of Rectangular inlet (k)	Temperature of circular inlet (k) r=2mm	Temperature of circular inlet (k) r=1.5mm
0.1	395	394	445
0.2	371	373	406
0.3	361	361	396
0.4	358	358	389
0.5	353	356	385
0.6	351	351	381
1	348	346	372
2	344	344	371
5	340	340	347

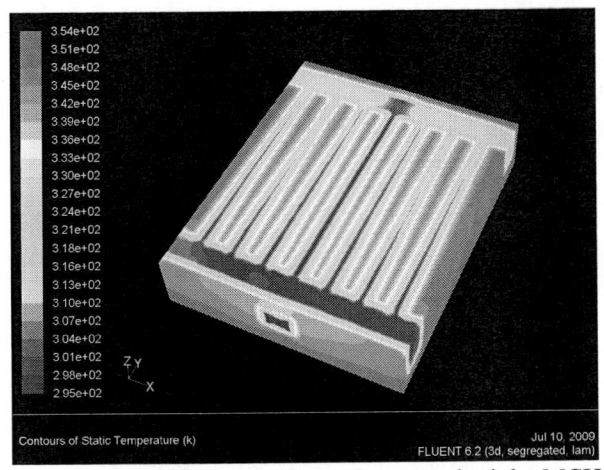

Fig. 7 Temperature distributions of rectangular inlet MCHS model (0.5m/s inlet velocity)

Conclusions

The comparison has three parts, different inlet and outlet positions, different distances between inlet location and near ends of fins, and different inlet and outlet shapes. It gives a simulation basis for the efficiency of experiment. The results are as follows,

1. In the micro-channel cooler which inlet position is perpendicular to the direction of flow in the channels. The high temperature is much lower than other two inlet positions, so that it is the best inlet and outlet position for heat dissipation.

2. The distance between the inlet and the end of fins has effect on the performance of the micro-channel cooler.

3. The rectangular and circular shape of the inlet have nearly the same effect on the heat dissipation. The inlet area plays a important factor on the performance.

Acknowledgments

The authors acknowledge the financial support from 863 program (No. 2008AA04Z301), NSFC project (No. 50876057) and Shanghai Municipal Education Commission (No. 08YZ15).

References

1. Ko-Ta Chiang, "Modeling and Optimization of Designing Parameters for A Parallel-plain Fin Heat Sink with Confined Impinging Jet Using the Response Surface Methodology," *Applied Thermal Engineering*, Vol. 27, (2007), pp. 2473-2482.

2. D. B. Tuckerman, R. F. W. Pease, "High-performance Heat Sinking for VLSI," *IEEE Electron Device Lett.*, Vol. EDL-2, (1981), pp. 126-129.

3. Q. Weilin, I. Mudawar, "Flow Boiling Heat Transfer in Two-phase Micro-channel Heat Sinks--I. Experimental Investigation and Assessment of Correlation Methods," *International Journal of Heat and Mass Transfer*, Vol. 46, (2003), pp. 2755-2771.

4. C. Gillot, L. Meysenc, C. Schaeffer, "Integrated Single and Two-Phase Micro Heat Sinks Under IGBT Chips," *IEEE Transactions on Components and Packaging Technology*, Vol. 22, No. 3 (1999), pp. 234-236.

5. M. B. Bower, I. Mudawar, "Two-phase Electronic Cooling Using Mini-channel and Micro-channel Heat Sink--Part 1: Design Criteria and Heat Diffusion Constraints," *Electron. Packag.*, Vol. 117, (1994), pp. 290-297.

6. A. Shah, B. G. Sammakia, H. Srihari, K. Ramakrishna, "A Numerical Study of the Thermal Performance of an Impingement Heat Sink--fin Shape Optimization," *IEEE Transations on Components and Packaging Technologies*, Vol. 27, No. 4 (2004), pp. 710-717.

7. Y. Mishan, A. Mosyak, E. Pogrebnyak, G. Hetsroni, "Effect of Developing Flow and Thermal Regime on Momentum and Heat Transfer in Micro-scale Heat Sink," *International Journal of Heat and Mass Transfer*, Vol. 150, Issues 15-16 (2007), pp. 3100-3114.

Sn Whisker Concern in IC Packaging for High Reliability Application

Jeffrey Chang, Bing Lee
IST-Integrated Service Technology
19, Pu-Ding Rd., Hsin-chu 30072, Taiwan, China
E-mail:Jeffrey_lee@istgroup.com

Abstract

In the study, the Ni underlayer plus matt Sn plated IC packaging of PLCC, PDIP and LQFP were subject to lead free (SAC) and SnPb surface mounting and wave soldering, respectively, then followed by TCT (-55°C to 85°C) 1000 cycle, THT (60°C /90%RH) 3000hrs to investigate whisker growth propensity. The practical whisker performance confirmation on the PCB beyond reflow simulation and component level was concluded to understand the Sn whisker potential in term of Ni underlayer efficiency. Additionally, various Sn thickness over Ni layer in packaging level was also applied to explore the Sn thickness effect on the whisker growth over TCT. The Ni underlayer as Cu migration barrier to mitigate Sn whisker growth efficiency was concluded for high reliability application.

Introduction

EU and China RoHS identified $Pb/Hg/Cd/Cr^{+6}$, PBBs and PBDEs were banned since July/2006 and Mar/2007 respectively, which was inevitable requirement for the whole electronic supply chain to satisfy it. Pb elimination in 6 banned substances was most complicated to find potential alternative due to material, process and equipment involved. In terms of lead free package development, optimum lead free solder ball composition for laminate package, lead free solder plating for lead frame package and lead free solder bump composition for flip chip package especially for low k devices, have not been fully aligned in the industry, especially in Japan and EU region. The solder ball composition of SAC305 and SAC405 even SAC105 with dopant or not for laminate package were highly expected to be standardized for IC packaging application, as well as the lead frame package using SnBi, SnCu and matt Sn plating in Japan and US/EU.

Matt Sn plating as SnPb alternative for lead frame package has been consensus in Europe and U.S based on cost and compatibility in backward and forward conversion even though its whisker concern from high reliability application. The IC packaging in high reliability application, such as telecom network infrastructure product, is still Pb exemption allowable in EU RoHS, which is RoHS5 compliant. But it is true that most suppliers will not be able to support SnPb component due to limited market share in the high end field.

The current whisker acceptance criteria and its test condition defined by IPC/JEDEC have been standardized in the industry [1-3] since 2005. Various design concepts to approach so-called whisker free requirement from chemical supplier were proposed for user selection, as well the IPC/JEDEC JP002 [2] about whisker guidelines was proposed for practice, such as post baking, Ni or Ag as Cu barrier, thicker Sn deposit, conformal coating and so on. No matter what kind of best practice, the practical service life could not be predicted exactly due to more extra factors involved from surface mounting.

The lead frame package after mounting on the PCB, the characteristic of solder deposit will vary with the reflow or wave soldering even hand soldering condition. The internal stress during solder deposit might be re-distributed to lead to different whisker potential. Currently, most study on Sn whisker was the test results from package level. But actually, the Sn whisker observation should include entire solder joint and non-wetting area across the lead frame to simulate the practical service life, as shown in Fig. 1.

In the paper, we will take Ni underlayer skill as example to explore its efficiency on whisker mitigation for high reliability application.

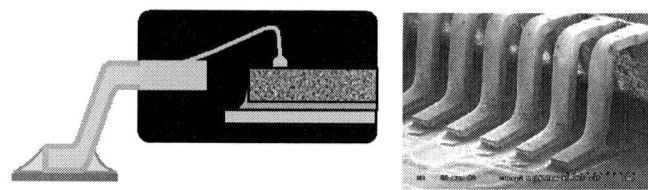

Fig. 1 The outline of solder joint of lead frame package on board for whisker observation

Experiment

3 IC package vehicles (LQFP/PLCC/PDIP) with Ni (0.5~0.7um) underlayer plus matt Sn (10~12um) finish was electrically plated after molding, followed by the rest of back end assembly process, as shown the package profile in Table 1.

The 3 studies were conducted to investigate Ni underlayer efficiency.

1. Ni layer consumption with the reflow simulation:

The above 3 packages were subject to 220°C and 260°C reflow, respectively, to explore the Ni layer consumption and IMC evolution with Sn and Cu by way of X-section. Furthermore, FIB and element mapping were applied to check the Ni layer consumption in detail after multi-reflow at 260°C.

2. Whisker on board performance with SnPb and SnAgCu mounting

The above 3 packages were subject to board mounting with optimum lead free SAC305 surface mounting at 245°C hat type profile and wave soldering at 260°C peak temperature sequentially. In parallel, backward conversion was checked with SnPb surface mounting at 220°C angle type profile and wave soldering at 220°C peak temperature sequentially. The packages on board sample, as shown in Fig. 2, were subject to whisker testing with TCT (-55°C /85°C) 1000cycle, THT (60°C /90%RH) 3000hrs, respectively. The longest whisker length was recorded following JESD22A121 [1] by SEM observation across solder joint and non-wetting area at 2 sample size to show whisker intensity.

3. Sn thickness effect on the whisker performance of Ni+Sn layer in LQFP package

The TCT (500/1000/1500cycles) were applied to check the Sn thickness effect in terms of most critical LQFP208 package in order to understand the Sn thickness efficiency against large CTE mismatch effect between Ni and Sn to induce significant thermal stress. The surface morphology over various Sn thickness was also observed by SEM analysis. Furthermore, the solderability test was conducted with dip and look method, as well as the practical surface mounting to check X-section of solder joint and lead pull strength over the temperature cycle.

Table 1 The package profile for on board whisker DOE

Packages	LQFP	PLCC	PDIP
Leadframe	Olin 7025	Olin 151	Olin 194
Base metal	Cu/Ni/Si/Mg	Cu/Zr	Cu/Fe/P/Zn
Sn Thickness	8~12um		
Ni thickness	0.5um~0.7um		
C content	<300ppm		
Lead shape	Gull wing	J-bend	Dual in line

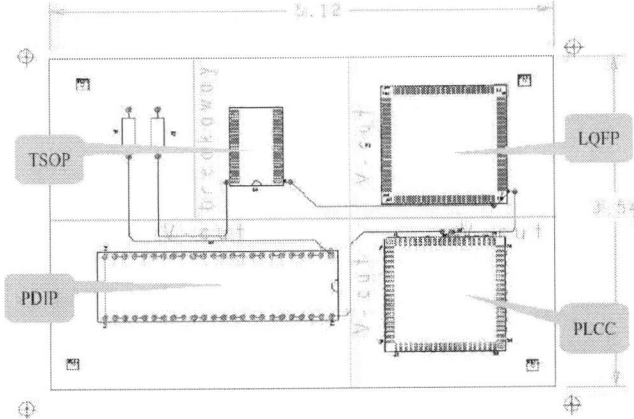

Fig. 2 The outline of PCB with 3 package vehicles (TSOP was removed for Ni+Sn plated package test)

Result and discussion

1. Ni layer consumption with the reflow simulation:

Fig. 3 showed the X-section of LQFP, PLCC and PDIP package over 220°C and 260°C reflow at one time. It was obvious that Ni layer between Cu and Sn was around 0.6~0.8um prior to reflow, as show the evidence in Fig. 4 by way of FIB with element mapping in terms of LQFP package. After 220°C and 260°C reflow, the Ni started consumption to react with Cu and Sn to form (NiCu) Sn IMC, and the total thickness of Ni+(NiCu)Sn in the interface increased over reflow due to thermal excursion involved. It was not very clear to judge the relative thickness between Ni and (NiCu) Sn from the X-section view by mechanical polishing.

Fig. 5 showed Ni consumed fully after 5 times multi-reflow at 260°C, which illustrated the Ni layer efficiency will be possibly reduced after long term used condition or over

reflow process. This will be concerned by high reliability product when in long term service life.

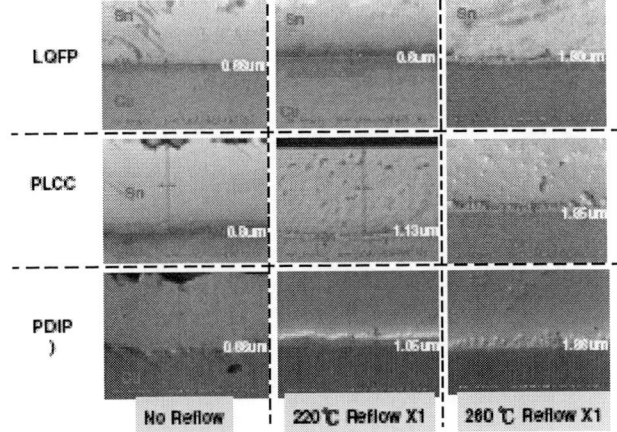

Fig. 3 The X-section of package over reflow

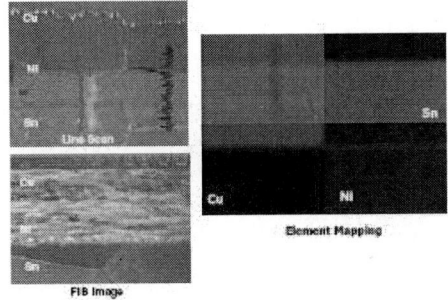

Fig. 4 The X-section of LQFP package by way of FIB/element mapping

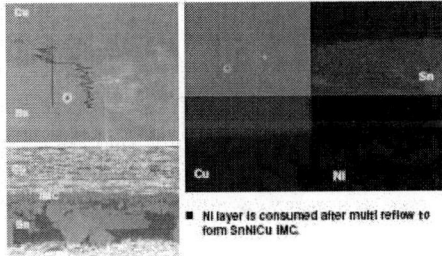

Fig. 5 The X-section of LQFP package after 5 times multi-reflow at 260°C by way of FIB/element mapping

2. Whisker on board performance with SnPb and SnAgCu mounting

Fig. 6 showed the significant whisker growth after THT 3000hrs exposure in terms of 3 IC packaging. Apparently, the corrosion happened seriously to drive long whisker out no matter what package type in SnPb or SnAgCu mounting, which illustrated Ni underlayer did not show any efficiency to mitigate corrosion generation. In other words, whisker mechanism induced by corrosion was verified to be related to the compressive stress to the Sn layer from the SnO_2 or SnO, less effect from significant interfacial Cu_6Sn_5 IMC growth, so that the Ni layer function for the mitigation of Cu migration would not show any efficiency to prevent it. The temperature effect from SnAgCu and SnPb mounting did not play critical role to mitigate corrosion happening in the case. Higher reflow temperature was expected to melt Sn grain completely might be more efficient to reduce moisture penetration from outside environment.

Sn+Ni packaged on board whisker at THT3000hrs
Up : SnPb mounting   Bottom : SAC mounting

Fig. 6 Whisker growth due to corrosion after THT 3000hrs

Temperature cycling test was always applied to test whisker performance on Sn deposit over Cu lead frame to verify thermal mechanical stress effect. Fig. 7 showed the whisker result after TCT 1000cycle in terms of LQFP package on SnPb and SnAgCu mounting. In Fig. 7-1 with SnPb mounting, the large CTE mismatch between Sn deposit and Ni layer to induce great thermal mechanical stress drive high density whisker growth, especially on non-wetting area of lead bending location with higher stress and top of toe without solder paste coverage. There was no whisker found on the wetting area due to solder coverage to depress whisker initiation. From previous study published by author in ECTC [4] about whisker growth without Ni layer in LQFP package after TCT 1000cycle, the contrast was very remarkable.

In Fig. 7-2 with SnAgCu mounting, whisker density was reduced apparently compared to SnPb mounting. It was assumed the partial Sn deposit melting to mitigate Sn atom migration along the interface boundary, so only a few shorter whiskers were observed.

Sn+Ni LQFP packaged on board whisker at TCT1000 cycle
SnPb mounting

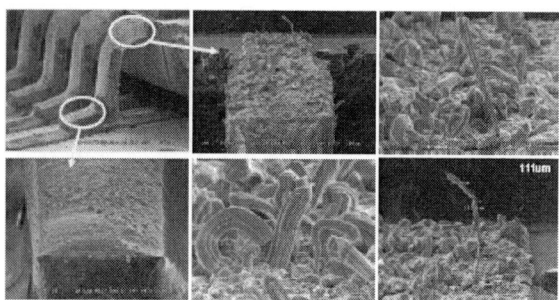

Fig. 7-1 Whisker in LQFP with SnPb mounting

Sn+Ni LQFP packaged on board whisker at TCT 1000 cycle
SnAgCu mounting

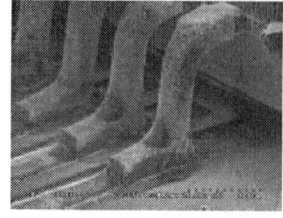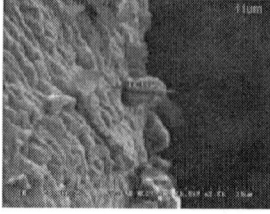

■ Whisker growth at SAC mounting is less and shorter than that at SnPb mounting.

Fig. 7-2 Whisker in LQFP with SnAgCu mounting

Fig. 8 showed the observation in PLCC package after SnPb surface mounting. The longer whisker happened in larger non-wetting area due to no solder coverage and Sn deposit melting. The shorter whisker in SnAgCu mounting was observed at same location due to partial Sn deposit melting to mitigate Sn atom migration. Lower whisker density in PLCC was observed than in LQFP because of different lead shape effect. It can be ignored from the Cu base effect between Olin 7025 Olin 151 due to Ni layer to be barrier for Cu migration.

Sn+Ni PLCC package on board whisker at TCT 1000 cycle

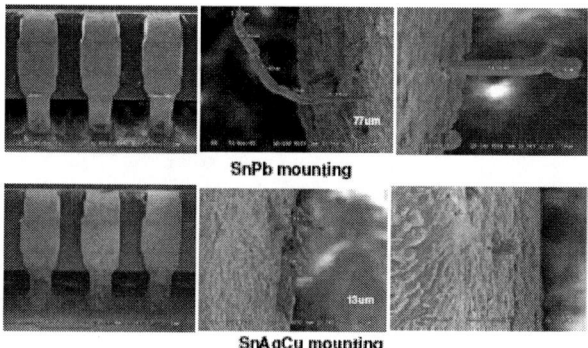

Fig. 8 The whisker in PLCC after TCT 1000cycle

Sn+Ni PDIP packaged on board whisker at TCT 1000 cycle

Fig. 9 The whisker in PDIP after TCT 1000cycle

Fig. 9 showed the whisker observation in PDIP subject to wave soldering. The large non-wetting area on the positive side of PCB whatever SnPb or SnAgCu soldering to show equal whisker potential due to slow thermal transportation from back of PCB. Likewise, the whisker density was far less than that in LQFP due to straight lead shape effect.

In practical, whisker will be concerned more in LQFP package due to its fine pitch structure, less concern in PLCC and PDIP due to wider lead pitch.

3. Sn thickness effect on the whisker performance of Ni+Sn layer in LQFP package

In terms of most concerned LQFP package, the further study was done to explore various Sn thickness effect over Ni by way of thermal cycling condition with 500, 1000 and 1500 cycle. Table 2 showed Sn thickness with 2.5mm, 5mm,10mm and 15mm were plated over 0.6~0.8mm Ni layer, and 10um Sn deposit without Ni layer was taken as control. Actual Sn thickness after plating on the flat frame was measured in different location by XRF to verify well controlled Sn thickness distribution across the flat frame.

The surface morphology of various Sn thicknesses on the flat frame was shown in Fig. 10. There was no significant difference on the grain size among Sn thickness above 5μm, but finer grain size was observed on the 2.5μm thickness.

Table 2 Sn thickness measurement by XRF

location	Sn thickness (μm)									
	Top side					Back side				
leg	1	2	3	4	5	6	7	8	9	10
Ni+Sn/2.5	2.6	2.1	2.4	2.0	1.5	2.3	2.3	2.4	2.2	1.6
Ni+Sn/5	6.0	5.6	6.3	3.4	4.0	6.9	6.1	5.6	4.5	4.0
Ni+Sn/10	12.2	11.5	10.8	11.0	11.2	11.7	12.2	11.5	10.5	10.4
Ni+Sn/15	15.5	15.5	17.2	13.8	15.0	18.0	15.3	17.2	14.7	12.5
Sn/10	12.4	12.2	12.1	12.1	9.4	10.2	12.5	12.6	11.7	11.5

The flat frame with various Sn thickness were subject to trim-forming by standard packaging assembly process to generate final LQFP package, then subject to TCT for whisker testing.

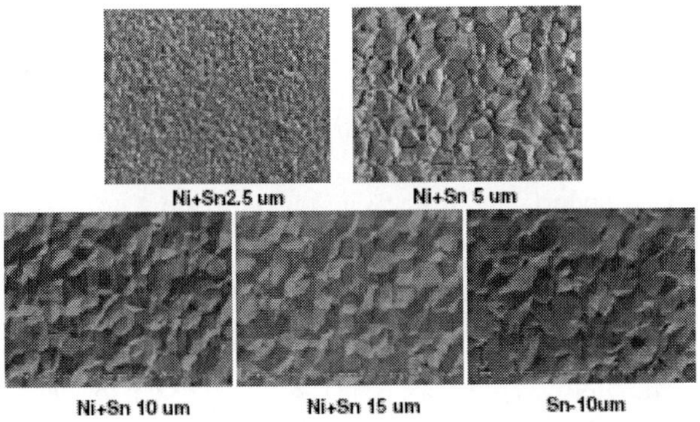

From the discussion above, lower Sn thickness +Ni underlayer could be a solution to achieve whisker free. The further study to mount package on the PCB will be implemented later.

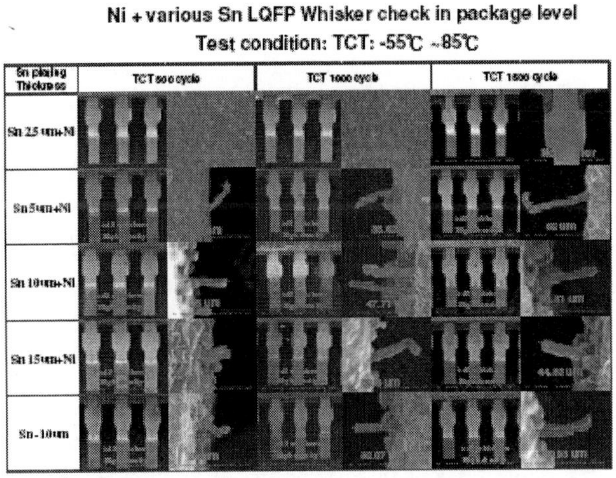

Fig. 10 The surface morphology of various Sn thickness

Fig. 11 Whisker result over various Sn thickness in TCT

Fig. 11 showed the whisker result over various Sn thicknesses. It was very interesting there was no whisker observed on 2.5mm Sn thickness over 500/1000/1500cycle.

When Sn thickness increased to 5μm, whisker was found and grew over the thermal cycling number. With the Sn thickness increased to 10 and 15μm, the whisker growing behavior was similar as in 5μm.

This finding illustrated Sn thickness has to be controlled under certain lower thickness to reduce compressive stress inside Sn deposit during thermal cycling period when combining with Ni underlayer. When stress level became tensile, whisker can be mitigated. Additionally, the whisker was also found in pure Sn deposit with 10μm thickness and presented similar behavior as Ni+Sn with 5um above.

When finding lower Sn thickness plus Ni layer can achieve whisker free performance during thermal cycling test, the related performance has to be considered not to maintain. Fig. 12 showed X-section and appearance of solder joint when SnAgCu surface mounting at 245°C. Based on IPC610D criteria, the criteria can be satisfied from lower to higher Sn thickness.

978-1-4244-4658-2/09 $25.00 © 2009 IEEE

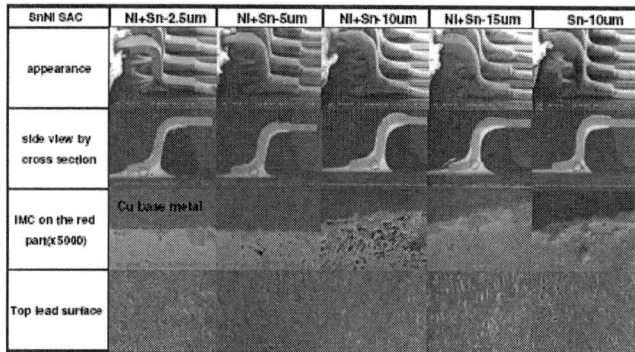

Fig. 12 Solderability observation over Sn thickness

Fig. 13 showed solder joint strength over Sn thickness and TCT by lead pull testing at 245°C SnAgCu surface mounting. There was no great solder joint difference over Sn thickness and the strength degradation over TCT can be maintained above 50% of initial strength.

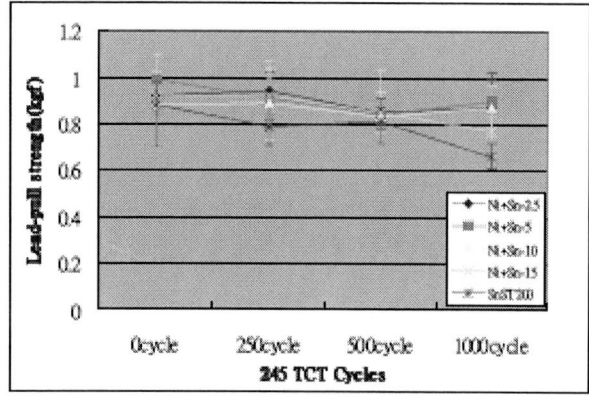

Fig. 13. Solder joint strength over Sn thickness and TCT by lead pull testing at 245°C SnAgCu reflow

Conclusions

Matt Sn plating in lead frame IC packages has become main stream in lead free conversion due to well backward and forward compatibility. Whisker was a major concern to limit its application comprehensively, especially for high reliability product operating in harsh environments application like telecom, server, storage and automotive electronics, even aerospace and defense products. High-reliability products usually require very long field lifetimes, typically ranging from 10 to 20 years with very low failure rates. It was necessary to find a good solution to reduce potential risk of Sn whisker if applied matt Sn in high reliability products. In the study, 3 typical IC packages with Ni underlayer plus Sn deposit were discussed about whisker growth potential in terms of on-board condition to simulate practical service life. It can be concluded zero whisker level will be impossible to achieve under long term thermal humidity testing due to corrosion issue. Furthermore, large CTE mismatch between Ni and Sn will generate significant compressive stress inside Sn deposit during thermal cycling period, so that the highly whisker potential will be drive.

Sn thickness reducing to 2.5μm over Ni layer seemed to be a direction for whisker mitigation in the TCT condition, but tight plating process control will be another major issue to limit its application.

The lead free SnAgCu surface mounting condition in practical process was not efficient to suppress whisker growth if soldering condition was not enough to melt the Sn deposit completely, especially in large PCB assembly for telcom and server product. Using higher reflow temperature or longer wetting time to deliver enough heat to the various thermal mass components will be able to satisfy the situation, but on the contrary, the overheating will influence adversely component and PCB reliability. Further development from the material and process perspective in component, PCB and SMT will be suggested for high end product application.

References

1. Test Method for Measuring Whisker Growth on Tin and Tin Alloy Surface Finishes, *JESD 22-A121A*, July, 2008.
2. Current Tin Whiskers Theory and Mitigation Practices Guideline, *IPC/JEDEC JP002*, Mar 2006.
3. Environmental Acceptance Requirements for Tin Whisker Susceptibility of Tin and Tin Alloy Surface Finishes, *JESD201A*, Sep 2008.
4. Jeffrey C. B Lee, "The IC package whisker growth on the PC board assembly," *ECTC*, Reno, May 2007.
5. INEMI, Recommendations on Lead-Free Finishes for Components Used in High-Reliability Products, v3, May 2005.
6. J. W. Osenbach, "Lead free package and Sn whisker," *ECTC*, Las Vegas, 2004, pp. 1314-1324.
7. Marc Dittes, *et al*, "Two ste approach for the release of lead-free component finish with respect to whisker risk," *International IPC/JEDEC conference on Lead-Free Electronic Components and Assemblies*, Frankfurt Oct, 2004.
8. P.Obernodorff, *et al.*, "Whisker formation on matt Sn influence of high humidity," *ECTC 2005* pp 429 to 433.
9. Jie-Hua Zhao, *et al.*, "Microstructure-based stress modeling of Sn whisker growth," *ECTC 2005* pp. 137-144.
10. Mars Dittes, "Humidity effect on Sn whisker formation," *Whisker workshop in ECTC*, Orlando. 2005.
11. J.W.Osenbach, "Sn corrosion and its influence on whisker ," *Whisker workshop in ECTC*, Orlando. 2005.
12. Peng Su, *et al.*, "A statistical study of Sn whisker population and growth during elevated temperature and humidity storage test," *Whisker workshop in ECTC*, Orlando. 2005.
13. Joe, Smetana, *et al.*, "Theory of Sn whisker growth," *Whisker workshop in ECTC*, Orlando. 2005.

Investigation of Mechanism for Spontaneous Zinc Whisker Growth from an Electroplated Zinc Coating

Alongheng Baated[1, a], Keun-Soo Kim[2], Katsuaki Suganuma[2]

[1]Graduate School of Engineering, Osaka University, Osaka, Japan
[2]Institute of Scientific and Industrial Research, Osaka University, Osaka, Japan
Address: Mihogaoka 8-1, Ibaraki, Osaka 567-0047, Japan
[a] Corresponding author. E-mail address: aroohan@eco.sanken.osaka-u.ac.jp, Tel.: +81-6-6879-8521; Fax: +81-6-6879-8522

Abstract

Zinc (Zn) whiskers are tiny hair-like electrically conductive filaments of Zn that sometimes grow from Zn coated surfaces (e.g., electroplated, hot dip). Zn coatings are commonly used as anti-corrosion coatings for iron (Fe) based structures. One of many common applications for Zn coated Fe is found on raised-floor tiles and support structures utilized in computer data centers. The formation of Zn whiskers threatens the reliable operation of electronic equipments due to the electrical shorting hazard they present. As with tin whiskers (much more broadly researched than Zn whiskers), the mechanism of formation is still not clear. This work investigated the Zn whisker growth mechanism for an electroplated Zn coating above carbon steel substrate from a raised floor tile using recent technology methods. Iron-zinc (Fe-Zn) Intermetallics and Zn oxides were identified by X-ray diffraction analysis (XRD). EDS (energy disperse spectroscopy) and EPMA (electron probe micro analysis) identified Fe-Zn intermetallic compounds on the surface of the Zn layer in addition to the interface between Zn coating and carbon steel substrate. Zn oxides formed primarily on the surface of Zn coating. Consequently, we speculate that Fe-Zn intermetallic compounds and Zn oxide formation can be the source of compressive stress effect on Zn whiskers growth on electroplated Zn coating above carbon steel substrate.

Introduction

The formation of metallic whiskers is a critical issue for the reliability of electronics because whiskers can cause short circuiting. Metallic whiskers grow spontaneously from zinc, tin and cadmium coatings at ambient temperatures but their formation mechanisms are still unknown. Researchers have focused all their attention on the study of tin whiskers and have investigated the mechanism of tin whisker growth [1-2]. Although Zn whiskers were discovered in the late 1940's on Zn plated wall brackets, few articles describe Zn whiskers [3-4]. It is well-known that Zn coatings are commonly used for corrosion resistance on component structures such as computer room floors, automotive sensors, computer equipment racks, chassis and air distribution systems. In other words, there is a greater need to address problems related to Zn whiskers than tin whiskers because Zn coatings are more broadly used. Furthermore, Zn whiskers may be circulated through an air distribution system resulting in potential health problems. In this work, we studied the mechanism of Zn whisker formation on Zn electroplated carbon steel that was taken from a raised floor tile in a data center. We attempted to compare the formation mechanism of Zn whiskers and tin whiskers.

Experimental procedures

For this work, we used a carbon steel substrate coated with electroplated Zn that was taken from a raised floor tile in a data center. It is gathered that the sample was present on the underside of a raised floor tile in an ambient temperature room for 20 years. No further detailed information about this sample exists. From the examined cross-section of the sample, it is determined that the Zn coating was approximately 20μm thick and was electroplated onto a 960μm thick carbon steel substrate.

The surface morphology of the electroplated Zn coated specimen is investigated by scanning electron microscopy (SEM, JEOL JSM-5510). X-ray diffraction (XRD, RIGAKU RINT2500) analysis was also used to identify the crystal structure of the electroplated Zn coating and the iron substrate. Cross-sectional samples of the electroplated Zn coating were prepared by cross section polisher (CP, JEOL SM09010) which uses an argon ion beam. Cross-sectional Zn whisker samples were also fabricated using focused ion beam microscopy (FIB, HITACHI FB-2100). The elemental composition was identified using electron probe micro-analysis (EPMA, JEOL JXA-8800R).

Results and discussions

Figures 1 and 2 show typical surface morphologies of the electroplated Zn coating taken from a raised floor tile that was obtained from a data center. Plenty of Zn whiskers were observed on the surface of the specimen. These Zn whiskers were characterized by lengths between 0.5mm to 2mm and diameters between 1μm to 20μm. It is clear that the surface of a Zn whisker has a striated trace which seems to be extruded from the Zn electroplated coating. Nodule-like microstructures had formed on the roots of some Zn whiskers. Comparing with the spontaneous growth of tin whiskers under ambient conditions, Zn whiskers are far larger in diameter and many Zn whiskers grow from the same root. It is clear that Zn whiskers are not single crystals.

To investigate the compositions of the Zn coating specimen, X-ray diffraction analysis was carried out on the Zn electroplated coating and the carbon steel substrate (Fig. 3). Besides Zn, Zn oxides and Fe-Zn intermetallic compounds were identified on the surface of the Zn electroplated coating. Fe was present on the substrate.

To further investigate Zn oxide and Fe-Zn intermetallic formation on the Zn electroplated coating, an EPMA analysis were carried out. Figure 4 shows the elemental analysis of the electroplated Zn coating's surface, which was above the carbon steel substrate. From the surface of the Zn coating, it is determined that the Zn coating mainly consists of Zn, Fe, O, C, S and Cl. A qualitative analysis of the specimen surface is given in Fig. 4b. It is obvious that S and Cl are surface

978-1-4244-4658-2/09 $25.00 © 2009 IEEE

residues from electrolytic solutions. Figure 5a and b shows the cross-sectional microstructure of the Zn electroplated coating specimen and was obtained using CP. From this measurement, we determine that the thickness of the Zn electroplated coating was approximately 20 μm. Elemental analysis of this cross-section of the Zn coating clearly showed that the Zn whiskers consist of Zn and that the Zn coating mainly consists of Zn, Fe and O (Fig. 5c). Combined to the result of X-ray diffraction analysis, it is clear that Fe-Zn intermetallic compounds had formed whole of Zn coating layer and that Zn oxides had formed near the surface of the Zn layer. From the substrate, Fe atoms must have diffused into the Zn coating and formed Fe-Zn intermetallic compounds and would have reached the surface of the Zn coating. On the contrary, the Zn coating would have been oxidized from the surface. O atoms can diffuse into the Zn coating and form Zn oxides on the surface. Zn oxides formation and Fe-Zn intermetallic formation is a major source of compressive stress and this stress acts on Zn atoms, which results in Zn whiskers.

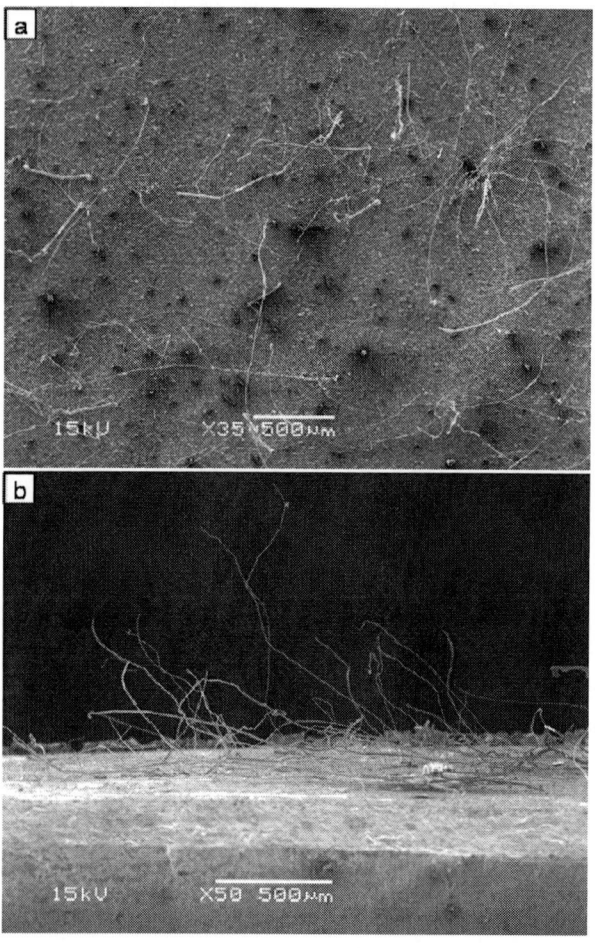

Fig. 1 Surface morphologies of an electroplated Zn coating on floor tile in a data center. (a) a full-frontal image of Zn coating, (b) a side ward image of Zn coating.

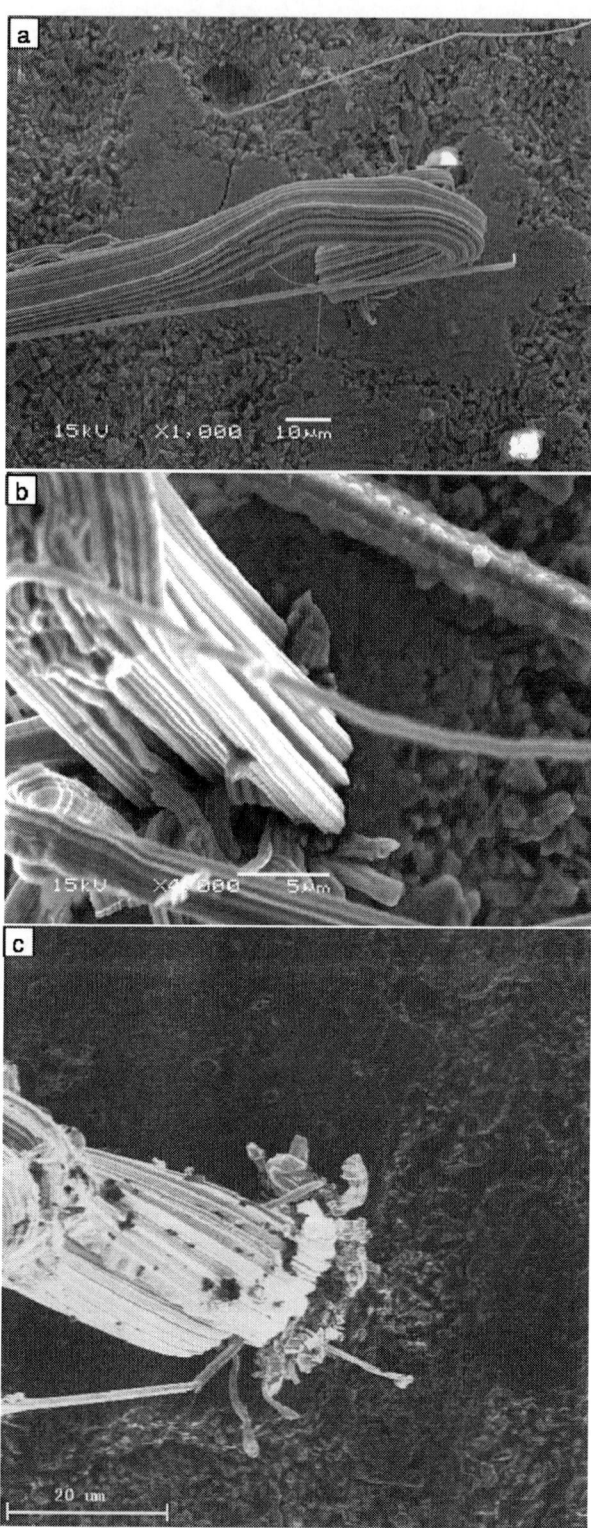

Fig. 2 Magnified images of Zn whiskers root growth from the electroplated Zn coating on carbon steel substrate. (a), (b) are SEM images, and (c) is SIM (scanning ion microscopy) image obtained by FIB.

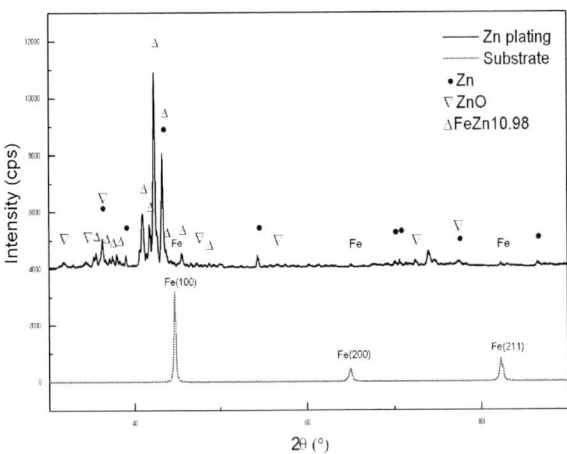

Fig. 3 X-ray diffraction analyses of the Zn coating and the substrate.

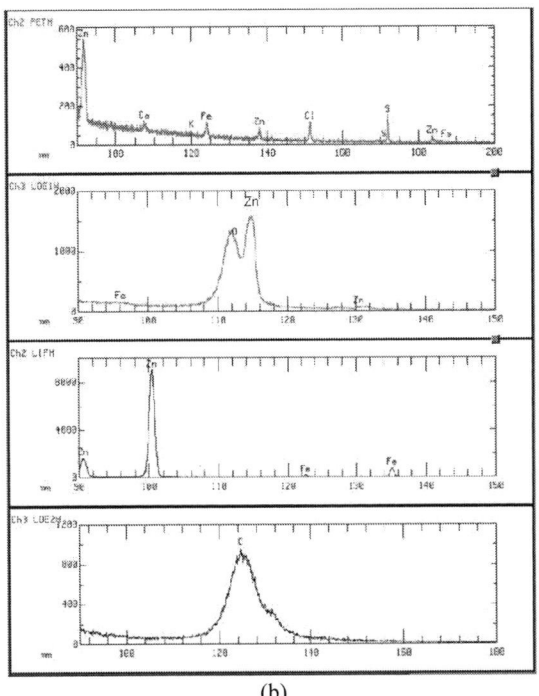

(b)

Fig. 4 Elemental analysis obtained by EPMA on the surface of Zn coating with growth Zn whisker on carbon steel substrate: (a) Elemental mapping, (b) a qualitative analysis.

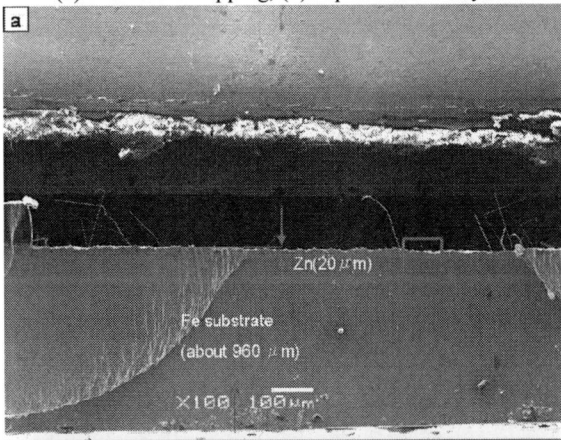

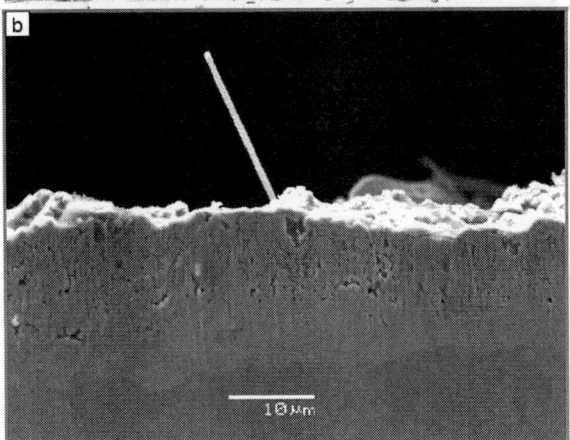

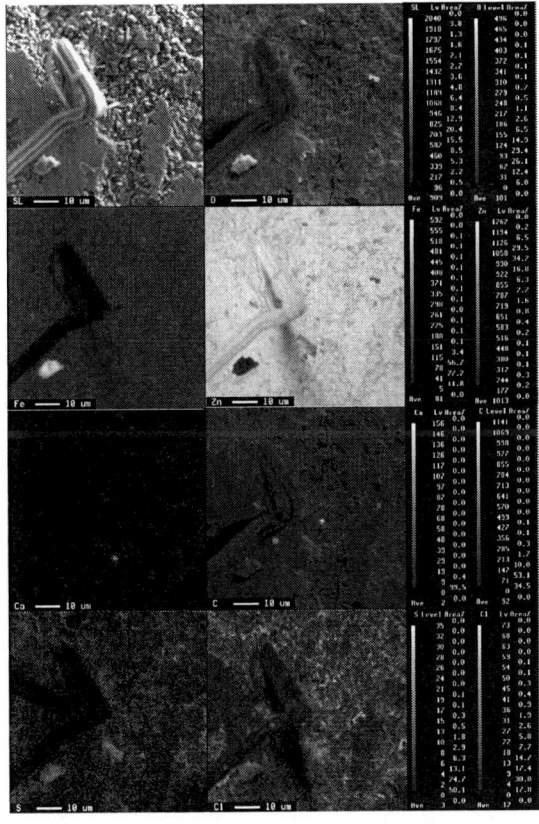

(a)

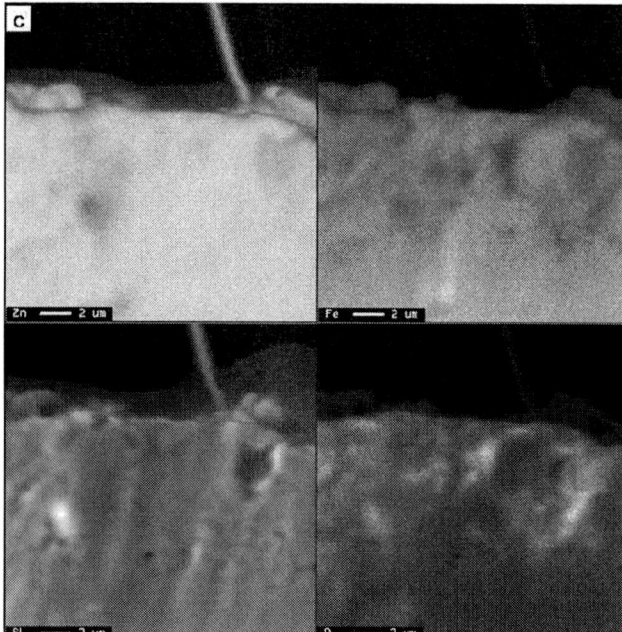

Fig. 5 Elemental analysis on cross sectional microstructure of electroplated Zn coating on carbon steel substrate using EPMA. (a) A cross sectional microstructures of Zn coating sample obtained CP milling, (b) magnified image of the red squared area, (c) elemental mapping around a Zn whisker.

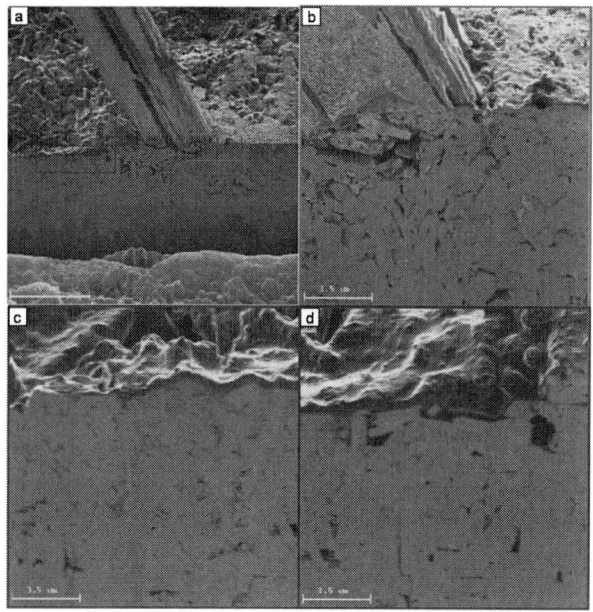

Fig. 6 Cross sectional microstructures of a Zn whisker on the electroplated Zn coating was obtained by FIB.

To investigate the effect of formation of Zn oxides formation and Fe-Zn intermetallic on Zn whisker growth, cross-sectional microstructures of a Zn whisker was examined (Fig. 6). It was observed clearly that a thin layer of Zn oxide formed on the surface of the Zn coating and lump-like microstructures had formed in whole of Zn coating even under the Zn whisker. Accord with the elemental analysis on cross-section (Fig. 5 c), it is determined that the Fe-Zn intermetallics formed with lump-like shape. Further

investigation is required to identify the cross-sectional microstructures of Zn whisker.

Therefore, it is obtained from those results that in addition to the formation of Fe-Zn intermetallics and Zn oxide, extra Zn atoms diffused to the free surface because of compressive stress. Thus, the Zn whiskers grow out from the Zn coating surface.

Consequently, it was recommended that a much stable substrate material against Zn should be selected or some barrier layer should be employed onto carbon steel to mitigate the Zn whisker formation. Further study is required on this issue.

Conclusions

In this work, the mechanism of Zn whisker growth from an electroplated Zn coating on a carbon steel substrate was investigated. The formation of Fe-Zn intermetallics and Zn oxides were observed from the Zn coating specimen. Those formations in the Zn coating result in the generation of compressive stress inside the coating which can promote Zn atoms diffusion to the free surface forming Zn whiskers. Hence, it was proposed that the employ of a different substrate or an underlayer below the Zn coating should be recommended to inhibit the formation of Zn whiskers.

Acknowledgments

This work was supported by a Grant-in-Aid for Science Research (A). The authors would like to thank Dr. Jay Brusse (NASA GSFC Tin Whisker Investigation Team) for supplying zinc whisker specimen. And the authors would also like to thank Mr. T. Tanaka and Mr. T. Ishibashi for their assistance in EPMA measurements and TEM observations.

References

1. George T. Galyon, "A history of Tin Whisker Theory: 1946 to 2004," *SMTAI International conference*, Chicago (2004).
   (http://www.nemi.org/projects/ese/tin_whikser_activities.html).
2. Joe Smetana, "Theory of Tin Whisker Growth: The End game," *IEEE Trans-EPM*, Vol. 30, No. 1 (2007), pp. 11-22.
3. U. Lingborg, "Observations of the Growth of Whisker Crystals from Zinc Electroplate," *Metallurgical Tansactions A*, Vol. 6A, August (1975), pp. 1581-1586.
4. Jay Brusse and Mickael Sampson, "Zinc Whiskers: Hidden Cause of Equipment Failure," *IEEE Computer Sciety*, Nov. (2004), pp. 43-47.

Effects of Thermal Aging on the Electrical Resistance of Sn-3.5Ag Micro SOH Solder Joints

Jin Peng[1], Fengshun Wu[1, 2], Hui Liu[1], Longzao Zhou[1], Qilin Pan[1]

[1]Wuhan National Laboratory for Optoelectronics, Huazhong University of Science and Technology, Wuhan, P. R. China, 430074

[2]State Key Laboratory of Materials Processing and Die & Mould Technology, Huazhong University of Science and Technology, Wuhan, P. R. China, 430074

Abstract

The relationship among the electrical resistance, the thickness of the IMC layer and the aging timer of the micro SOH solder joints is investigated in present paper. It's found that the composition of compositions of the intermetallic compounds are Cu_3Sn near the Cu side and Cu_6Sn_5 near the solder side; With the increasing of aging time, the thickness of the Cu_3Sn increased while the thickness of the Cu_6Sn_5 decreased. The total thickness of IMCs (Cu_6Sn_5 and Cu_3Sn) increased as the aging time increased, while the resistance of the micro SOH solder joints increased at first and then decreased. The resistance of solder joints increased at first and then decreased as the total thickness of IMCs (Cu_6Sn_5 and Cu_3Sn) increased.

Introduction

The lead and lead contained compounds' toxicity to human body and the environment and the restriction of legislation, therefore, lead-free solders have became the tendency of the electronic packing materials. However, there is a typical problem during the application of lead-free solders: thick Cu-Sn intermetallic compounds (IMCs) are formed between the solder and the Cu substrate. The formation and distribution of intermetallic compounds in the microstructure of lead-free solders which acts as the thermal, electronic and mechanical connections, directly affect the soldering performance during the microelectronic packaging process [1].

Demands are being placed on the production of low power consumption, low weight, and compact packaging technologies for very large scale integration (VLSI) integrated circuits, aerospace, and military applications, The trend in flip-chip and ball grid array (BGA) packaging to increase I/O count drives the interconnection solder joins to be smaller in size and lower in the stand-off height (SOH). The thickness of the IMC and the electrical resistance of micro SOH solder joints, which have a serious influence on the electrical performance and reliability of the joints, are variable to the thermal aging, therefore, it is urgent to establish the relationship among the electrical resistance, the thickness of the IMC and the aging time of the micro SOH solder joints [2-4].

In the present paper, near eutectic composition Sn-3.5Ag alloys are selected as the research material, which was widely used in the industrial production. Four probes methods were used to measure the electrical resistance of the micro SOH sold joints. In order to establish the relationship among the electrical resistance, the thickness of the IMC and the aging time of the micro SOH solder joints, the thickness of the Cu_6Sn_5 and the Cu_3Sn IMC layers was detected respectively.

Experimental procedures

In order to investigate the relationship among the electrical resistance, the thickness of the IMC layer and the aging timer of the micro SOH solder joints, the Cu/Sn-3.5Ag/Cu with 100μm, 50μm, 20μm and 10μm SOH were prepared. Fig. 1 shows the structure of the Cu/Sn-3.5Ag/Cu joint, which is composed by two Cu bars on the both ends and a Sn-3.5Ag bump in the middle. The diameter of the Cu bar is 0.9mm and its purity is 99.99%, which were cut from a commercial grade Cu wire.

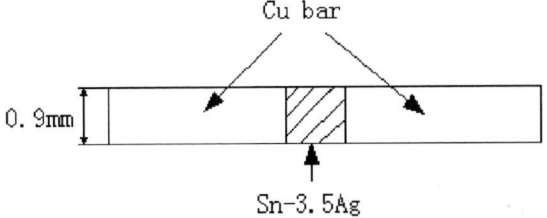

Fig. 1 The structure of the Cu/Sn-3.5Ag/Cu joint

The device for controlling the height of the solder joints was designed to accurately control the height of the solder joints as 100μm, 50μm, 20μm and 10μm. It is constituted by the base the parallels, the holding devices, the spring and the adjusting knob. The part of holding devices is used to fixate the Cu bars on both sides, while, the function of the springs are making the Cu bars closely contact with the solder. We can use the adjusting knob, whose precision is 1μm, to control the space between the two Cu bars. With the parallels, the moving direction of Cu bar be constrained to the horizontal axis, the dislocation of the Cu bars in Vertical direction can be averted.

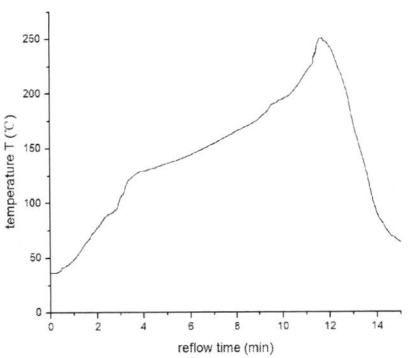

Fig. 2 Reflow temperature graph

978-1-4244-4658-2/09 $25.00 © 2009 IEEE

The melting point of the Sn-3.5Ag is about 243°C, and its reflow temperature in practical production will reach to 250°C~260°C. Fig. 2 shows the reflow temperature graph applied to our experiments. The including four temperature areas are 185°C-225°C-250°C-185°C, respectively. The prepared micro SOH solder joints of Cu/Sn-3.5Ag/Cu with 100μm, 50μm, 20μm and 10μm was placed in an aging box, which can keep a constant temperature of 150°C.

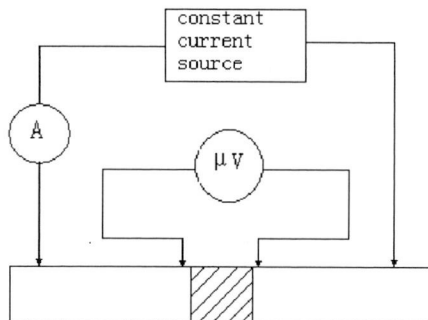

Fig. 3 Schematic diagram of four probes method

Four probes method is used for measuring the electrical resistance of the different SOH joints after aging for 100h, 200h, 300h and 500h. Fig. 5 shows the schematic diagram of four probes method. Four metal probes contact with the surface of the sample when measuring. Current generate by the constant current source passed through the offside two probes, while, the voltage of the test sample was measured by the accurate potentiometer.

The microstructures the IMC layers were analyzed and measured with a scanning electron microscope (SEM). The energy-dispersive X-ray spectrometer (EDX) was used for analyzing the compositions of the IMC layers.

Results and discussion

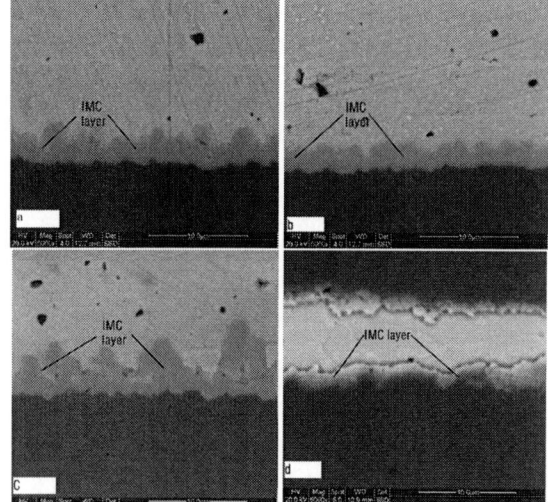

Fig. 4 SEM images of the Cu/Sn-3.5Ag/Cu solder joint with (a) 100μm SOH (b) 50μm SOH (c) 20μm SOH and (d) 10μm SOH.

Fig. 4(a)- (d) show the SEM images of the Cu/Sn-3.5ag/Cu solder joints with 100μm, 50μm, 20μm and 10μm SOH without aging, respectively. Fig. 5(a)- (d) show the SEM images of the Cu/Sn-3.5ag/Cu solder joints with 100μm, 50μm, 20μm and 10μm SOH without aging 100 hours, respectively.

SEM and EDX shows that there are two IMC layers between the Sn-3.5Ag solder and the Cu substrate, the one near to the Cu substrate is thin Cu_3Sn, while the other one near to the Sn-3.5Ag solder is thicker Cu_6Sn_5. There is also clumpy Ag_3Sn in the solder joints.

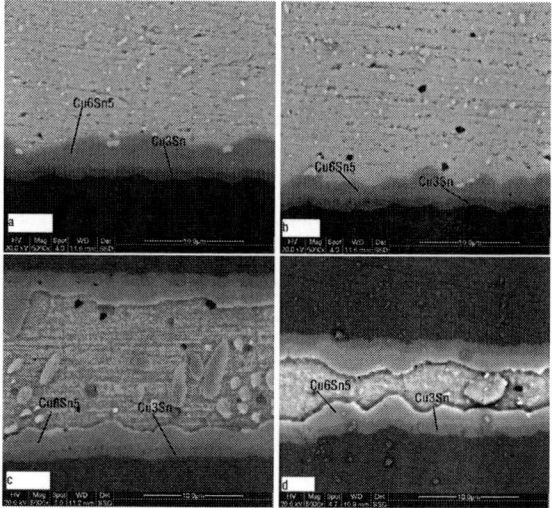

Fig. 5 Images of the Cu/Sn-3.5Ag/Cu solder joint with (a) 100μm SOH (b) 50μm SOH (c) 20μm SOH and (d) 10μm SOH after 100h aging

It's obvious that the total thickness of IMCs (Cu_6Sn_5 and Cu_3Sn) increased after aging for 100 hours, but after precise calculate, we found that the thickness of the Cu_6Sn_5 and the Cu_3Sn IMC layers showing the different trend.

The IMCs at the interface of the Sn-3.5Ag and the Cu bar have a complex appearance with a scalloped edge, which are not uniform at the different positions. But after aging for a long time, the thickness of the IMC layers become more well-distributed and smooth. In present paper we measured the thickness of the IMC layers by software called Image tool. We import the photos of the IMCs from the optical microscope into the Image tool, and then calculate the areas of the irregular IMC layers. We can get the average thickness of the IMC lays if we got the area and the length of the irregular IMC layers.

Fig. 6 shows the thickness of the Cu_6Sn_5 IMC layers of 100μm, 50μm, 20μm and 10μm SOH vary with the aging time. Fig. 7 shows the thickness of the Cu_3Sn IMC layers of 100μm, 50μm, 20μm and 10μm SOH vary with the aging time. Fig. 8 shows the total thickness of IMC layers of 100μm, 50μm, 20μm and 10μm SOH vary with the aging time.

978-1-4244-4658-2/09 $25.00 © 2009 IEEE

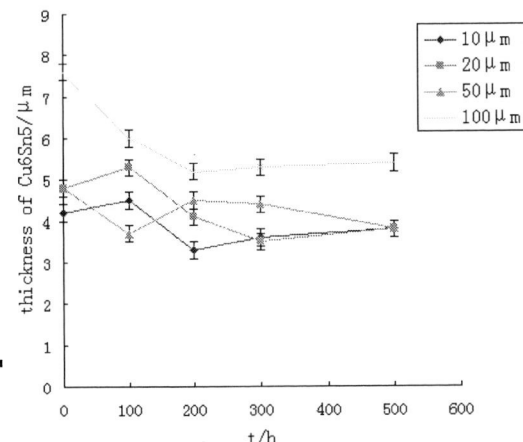

Fig. 6 Relationship between thickness of Cu_6Sn_5 and aging time

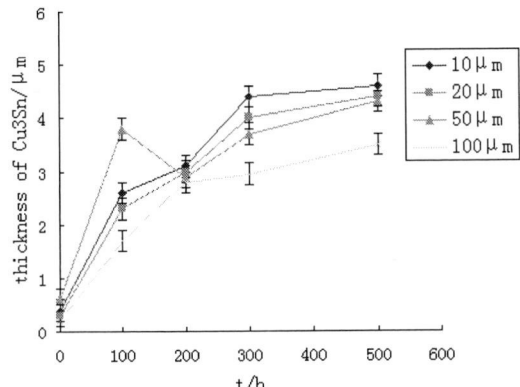

Fig. 7 Relationship between thickness of Cu_3Sn and aging time

It is found that, on the one hand, thickness of the Cu_6Sn_5 IMC layers decreased with the aging time increasing, on the other hand, the thickness of the Cu_3Sn IMC layers increased with the aging time increasing.

There were only Cu_6Sn_5 IMC lays formed at the beginning of the formation of the solder joints. From the Sn-Cu binary phase diagram we can find that the Cu_6Sn_5 and the Cu are not at an equilibrium state. With the increasing of the aging time, more and more Cu diffused into the Cu_6Sn_5 and formed Cu_3Sn, while, there is an equilibrium state between the Cu_6Sn_5 and the Sn, so a little Cu_6Sn_5 formed in the aging process, therefore, the thickness of Cu_6Sn_5 will decrease and the thickness of Cu_3Sn will increase during the aging process [5].

Sum up the above two graphs we can get the relationship between the total thickness of IMC layers (Cu_6Sn_5 and Cu_3Sn) and the aging time. We can found that the total thickness of IMCs increased with the aging time. It is because that the further diffusion of Cu will form the Cu_3Sn phase, even react with the Sn phase to produce grain of Cu_6Sn_5 when cut across the IMC layers.

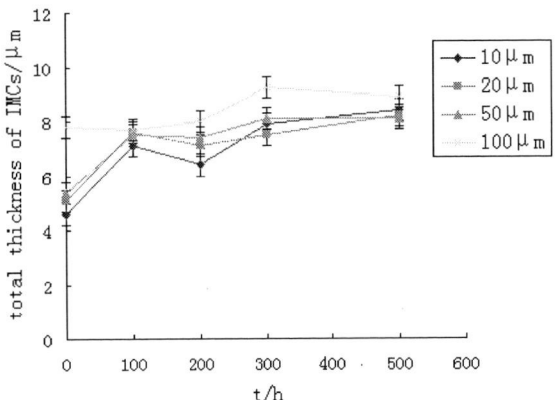

Fig. 8 Relationship between total thickness of IMCs and aging time

The electrical resistance of micro SOH solder joint with 100μm, 50μm, 20μm and 10μm after aging was measured. Fig. 9 shows the relationship between the electrical resistance of the solder joints and the aging time. It's found that, with the increasing of the aging time, the electrical resistance of the solder joints raised at the beginning and then designed after aging for 100h. This phenomenon can be explained by the change of the IMC layers. In the first 100 hours the increase of the electrical resistance can be attributed to the rapid increase of the IMC lays.

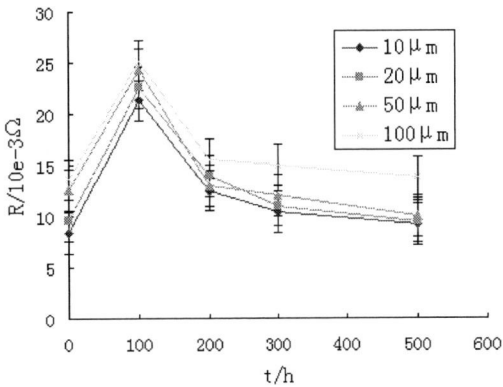

Fig. 9 Relationship between resistance of Sn-Ag solder joint and aging time

We can know from the above analysis that with the increasing of the aging time, the total thickness of the IMC lays increased with a great change on its composition. According to researchers, the electric resistivity of Cu_6Sn_5 is 17.5μΩcm, 8.9μΩcm for Cu_3Sn, and 10.1μΩcm for Sn. As Cu diffused into the Cu_6Sn_5 and formed the Cu_3Sn after aging for 100 hours, the total thickness of the IMC layers keep almost constant, so, the electrical resistance of the solder joints decreased, and then leveled off.

We can also interpret this phenomenon as the interface stability. Because that the Sn surface would be microscopically rough, the interfaces between solder bumps and Cu bar were not completely contacted during reflow process, During the aging process, because of the interdiffusion , good contact formed on the interface between the solder and the Cu bars. It is believed that descending of the contact resistance was due to the interdiffusion and phase transformation during aging at high temperature. With the aging time increasing, the interdiffusion occurred more extensively, therefore, after aging for 100 hours, the electrical resistance of the solder joints decreased and then leveled off.

As mentioned above, on one hand, the Cu_6Sn_5 phase gradually transformed into Cu_3Sn during aging treatment, and the Cu_6Sn_5 phase faded in the end, it was indicated that the contact resistance of Cu_3Sn should also be lower than that of Cu_6Sn_5, on the other hand, the contact resistance of the Cu/Sn-3.5Ag decreased during aging process because of interdiffusion. So, it was suggested to be the reason why the contact resistance found a descending trend and then leveled off after aging for 100 hours [6].

Conclusions

SEM and EDX shows that there are two IMC layers between the Sn-3.5Ag solder and the Cu substrate, the one near to the Cu substrate is thin Cu_3Sn, while the other one near to the Sn-3.5Ag solder is thicker Cu_6Sn_5. There is also clumpy Ag_3Sn in the solder joints. Thickness of the Cu_6Sn_5 IMC layers decreased while thickness of the Cu_3Sn IMC layers increased with the aging time increasing. Because that Cu diffused into the Cu_6Sn_5 phase and formed the Cu_3Sn phase.

The electrical resistance of the solder joints raised at the beginning increased because of the rapid increase of the IMC layers after aging for 100 hours, the electrical resistance of the solder joints decreased gradually then leveled off. There are two reasons can interpret this phenomenon, on one hand, the Cu_6Sn_5 phase gradually transformed into Cu_3Sn during aging treatment, and the Cu_6Sn_5 phase faded in the end, on the other hand, the contact resistance of the Cu/Sn-3.5Ag decreased during aging process because of interdiffusion.

References

1. V.Sivasubramaniam, N.S.Bosco, J.JanczakRush, J.Cugnoni, J.Botsis, "Interfacial Intermetallic Growth and Strength of Composite Lead-Free Solder Alloy Through Isothermal Aging," *Journal of ELECTRONIC MATERIALS*, Vol. 37, No. 10, 2008, pp. 1598-1604.

2. Anupam, Choubey, Haoyu, Michael, Osterman, "Intermetallics Characterization of Lead-Free Solder Joints under Isothermal Aging," *Journal of ELECTRONIC MATERIALS*, Vol. 37, No. 8, 2008, pp. 1131-1137.

3. Albert T. Wu, Ming-Hsun Chen, Ciou-Nan Siao "The Effects of Solid-State Aging on the Intermetallic Compounds of Sn-Ag-Bi-In Solders on Cu Substrates," *Journal of ELECTRONIC MATERIALS*, Vol. 38, No. 2, 2009, pp. 252-256.

4. J. Liang, N. Dariavach, D. Shangguan, "Solidification Condition Effects on Microstructures and Creep Resistance of Sn-3.8Ag-0.7Cu Lead-Free Solder," The Minerals, *Metals & Materials Society and ASM International 2007* ,Vol 38A, pp. 1530-1538.

5. Sun-Kyoung Seo, Sung K. Kang , Da-Yuan Shih, Hyuck Mo Lee, "The evolution of microstructure and microhardness of Sn-Ag and Sn-Cu solders during high temperature aging," *Microelectronics Reliability.* NO.49 (2009), pp. 288-295.

6. Seung-Hyun Lee, Hee-Ra Roh, Zhi GangChen, Young-Ho Kim, "Contact Resistance and Shear Strength of the Solder Joints Formed Using Cu Bumps Capped with Sn or Ag/Sn Layer," *Journal of ELECTRONIC MATERIALS*, Vol. 34, No. 11, 2005, pp. 1446-1454.

Instability and Failure Analysis of Film-substrate Structure under Electrical Loading

Qinghua Wang, Huimin Xie*, Jia Liu, Xue Feng, Fulong Dai
AML, Department of mechanics and engineering, Tsinghua University
Tsinghua University, Beijing 100084, China
*xiehm@mail.tsinghua.edu.cn, +86-10-62792286

Abstract

Delamination between the film line and the substrate is a main influence factor for instability and failure of the film-substrate structure, which exists widely in the large scale integrated circuit. The Buckle-driven delamination of constantan film line on polymer substrate under electrical loading was investigated in this paper. The post-buckling theory for beam was introduced to quantitatively analyze the residual strain and the residual stress. The maximal tensile stress and the maximal compressive stress of the buckled film line was found to be 1.69GPa and 2.03GPa respectively, after bearing 3.14×10^8A/m^2 dc for 370.7 hours. The maximal compressive axial stress was 0.17GPa, showing the bending stress contributes to the majority of the residual stress of the film line. The instability and the failure behavior of the film-substrate structure were studied.

1 Introduction

Film-substrate structure is widely employed in micro-electromechanical system (MEMS) and semiconductor integrated circuit, whose reliability and service life are directly related to the instability and the failure behavior of the film-substrate structure. Structure of thin metal film line on polymer substrate is the common service form, where the film line always bears the coupling effect of force, electric and heating in working circumstance. Delamination and electromigration are the main influencing factors for the instability and the failure of the film-substrate structure [1-3]. A substantial amount of studies mainly concentrated on delamination under force, laser or thermal loading [4-6], and electromigration under electrical loading [7-8]. Another phenomenon of delamination of the film line under electrical loading has been reported [9] and arouses the authors' attention. As the constantan film line is extensively used as the main component of resistance strain gauge and thermocouple in electrocircuit, the buckle-driven delamination of the constantan film line on the polymer substrate under electrical loading are studied in this paper, through electrifying the film line to simulate its functional mode.

2 Sample preparation and experiment

The sample was prepared by photographic lithographic technique, through depositing constantan alloy onto the surface of a polymer substrate with a photoresist mask, without any anneal after the deposition. The constantan alloy comprises 55% copper, 44~45% nickel and a little manganese, while the polymer is composed of 90% polyvinyl formal-acetal and 10% epoxy novolac. The constantan film line distributes in a shape of narrow 'S' and so all the parallel segments are series-wound, as drawn in Fig. 1. The width, the thickness, the spacing between two adjacent parallel segments, and the length of one parallel segment of the film line are b=28μm, h_f=6μm, d=108μm, l=3100μm respectively.

The thickness of the substrate is h_s=100μm, while the length and the width of the substrate are large enough to hold the film line. All these dimension parameters were measured by a large-field-depth microscope and a common microscope. In order to electrify and observe conveniently, the polymer substrate was affixed to a piece of organic glass whose thickness is one order of magnitude lager than that of the polymer.

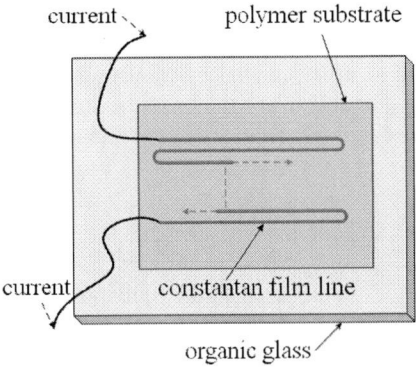

Fig. 1 Schematic diagram of the sample, with the constantan film line on the polymer substrate

Before electrification, the glassy temperature of the polymer was measured by differential scanning calorimetry (DSC). Electrical loading experiments were conducted in a simple closed circuit at room temperature environment. The exerted direct current (dc) was provided by a stabilized voltage supply. During the process of electrification, the temperature of the film line was measured by an infrared thermal imager-ThermaCAM P60, and the resistance of the film line was obtained from the voltage divided by the current. A scanning electric microscopy (SEM) was used to observe the morphology of the sample in low vacuum, avoiding the influence of gold spraying in high vacuum.

3 Results and discussions

The glassy temperature of the polymer substrate is 393.24K by DSC. When electrifying, the temperature of the film line rises rapidly from 300K, and the temperature almost keeps constant with little change after around 20s. The temperature rises monotonously as the current density increases. The maximum temperature of the film line is less than 385K under dc, at the maximal current density the film line can bear. The resistance of the film line is 350Ω before electrification, and it has a small change of less than 3Ω in the electrifying process.

Buckle-driven delamination of the thin film line on the polymer substrate was observed when bearing appropriate electrical loading for a period of time, as the configuration

978-1-4244-4658-2/09 $25.00 © 2009 IEEE   947

shown in Fig. 2. Since the constantan film line in this study can be treated as a beam, the post-buckling analysis of beam model is used to determine the residual strain and the residual stress of the buckled film line, based on energy minimization and governing equations of beam [10].

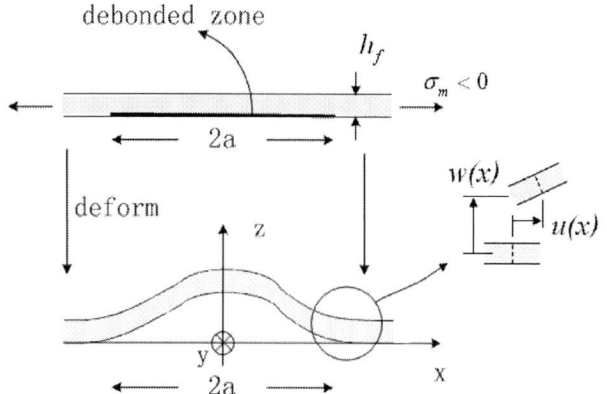

Fig. 2 Sketch of the bending beam for post-buckling analysis of a thin film line on an infinitely thick substrate, where $2a$ is the delamination length, h_f is the thickness of the film line, σ_m is the axial stress, $w(x)$ is the deflection and $u(x)$ is the displacement.

For simplicity, the compressive residual stress in the film line is treated to be homogeneous and biaxial. Compared with the deformation of the film line, the polymer substrate is considered to be rigid, and the clamped boundary condition is used as granted in planar film [11].

The equilibrium equations of beam are:

$$EIw''' + Nw'' = 0, \quad dN/dx = 0 \quad (1)$$

where E, I, w, N are the elastic modulus, the bending stiffness, the deflection and the axial force of the buckled beam respectively. As to the symmetric domains of buckles, the bending axis is assumed to be a sinusoidal curve.

The total energy of the buckled film line is constituted of the elastic energy of axial compression and the energy of bending. Minimizing the total energy, the bending moment of the film line can be obtained:

$$M(x) = -\frac{\pi^2}{a^2}\frac{Eh_f^3 w_0}{24}\cos\left(\frac{\pi x}{a}\right) \quad (2)$$

where a is half of the delamination length and w_0 is the maximal deflection of the film line.

Then, the residual strain distribution can be expressed by the sum of the axial strain and the bending strain:

$$\varepsilon = -\frac{\pi^2 h_f^2}{12a^2} + \frac{\pi^2 w_0}{2a^2}\cos\left(\frac{\pi x}{a}\right)z' \quad (3)$$

where z' is the distance between the calculation point and the neutral layer of the film line. In elastic state, the constitutive relation between stress and strain is represented by:

$$\sigma = E\varepsilon \quad (4)$$

As seen from Eq. (3) and Eq. (4), the residual strain depends on the thickness of the film line, half of the delamination length, and the maximal deflection.

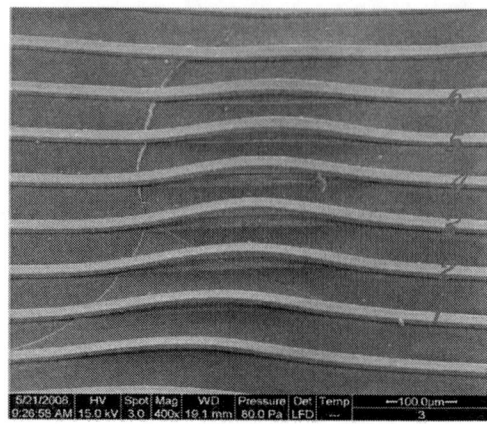

Fig. 3 SEM photograph of the buckled film line on the polymer substrate after bearing 3.14×10^8A/m^2 dc for 370.7 hours, with six buckled parallel segments, when the SEM stage tilting 45 degrees.

Take a typical experiment as the example to analyze. Fig. 3 shows a SEM photograph of the buckled film line on the polymer substrate after bearing 3.14×10^8A/m^2 dc for 370.7 hours, with six buckled parallel segments, when the SEM stage tilting 45 degrees. There are three buckled regions on the surface of the sample under this electrical loading, and it is one region in Fig. 3, similar to the other two regions. The buckled region consists of six numbered parallel segments of the film line, with the delamination length and the maximal bending deflection different everywhere. Delamination is mainly contributed to the difference between the thermal expansion coefficients of the film line and the substrate. Since the thermal expansion coefficient of constantan is smaller than that of the polymer, the film line acquires a biaxial compression upon cooling after electrification which can generate Joule heat. The compressive residual stress and residual strain in the film line under electric loading are the main parameters for the stability assessment of the film-substrate structure. Besides, there is a crack in the polymer substrate in Fig. 3, caused by the tensile stress in the polymer after cooling.

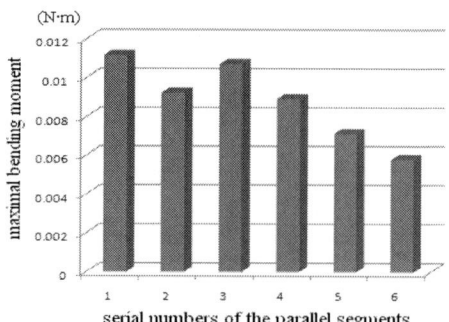

Fig. 4 The maximal bending moment of the buckled film line in Fig. 3. The abscissa values 1-6 represent the numbers of the parallel segments.

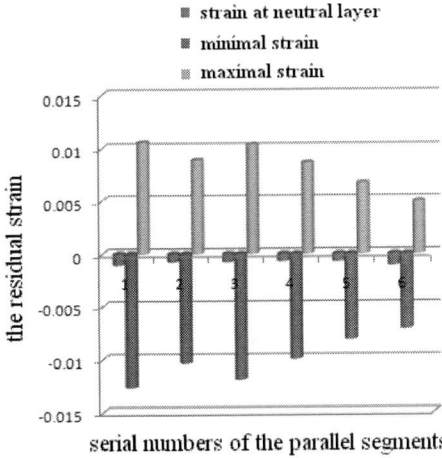

- strain at neutral layer
- minimal strain
- maximal strain

Fig. 5 The residual strain distribution of the buckled film line in Fig. 3. The abscissa values 1-6 represent the numbers of the parallel segments.

The maximal bending moment of the six parallel segments of the buckled film line calculated from Eq. (2) was described in Fig. 4. The residual strain at neutral layer, the maximal residual strain at the top edge and the minimal residual strain at the bottom edge of the film line when the deflection is maximal are obtained through Eq. (3), as illustrated in Fig. 5. The residual stress can be calculated from Eq. (4), with the elastic modulus of constantan E=160GPa. The residual stress has the same distribution as the residual strain except the value due to the linear relationship. The maximal tensile strain at the top edge of the film line is 1.05%, and the maximal compressive strain at the bottom edge is 1.27% as seen from Fig. 5. Correspondingly, the maximal tensile stress and the maximal compressive stress are 1.69GPa and 2.03GPa respectively. But the maximal compressive axial stress, i.e., the stress at the neutral layer of the film line is only 0.17GPa from Fig. 5, suggesting the stress in the buckled film line mainly comes from the buckling stress. This can also be deduced from the similarity of the shapes of the maximal bending moment and the maximal strain as well as the minimal strain. When the maximal stress reaches or exceeds the critical yield stress of the film line, the film-substrate structure will be liable to fail to work.

Conclusions

In this paper, the instability and the failure behavior of the film-substrate structure under dc electrical loading were investigated. Buckle-driven delamination of the thin film line on the polymer substrate was studied. The residual strain and the residual stress of the buckled film line were obtained through the post-buckling analysis of beam model. The residual stress in the buckled film line mainly comes from the buckling stress. This work provides a valuable reference for the stability assessment of the film-substrate structure under electrical loading.

Acknowledgments

The work was supported by the National Basic Research Program of China (Grant 2004CB619304), National Natural Science Foundation of China (Grants 10625209, 10732080, 10472050), Beijing Natural Sciences Foundation (Grant 3072007), Program for New Century Excellent Talents (NCET) in Universities, and Chinese Ministry of Education (Grant NCET-05-0059).

References

1. Basile Audoly, "Stability of Straight Delamination Blisters," *Physical Review Letters*, Vol. 83, No. 20 (1999), pp. 4124-4127.
2. Gioia G., Ortiz M., "Delamination of Compressed Thin Films," *Advances in Applied Mechanics*, Vol. 33 (1997), pp. 120-92.
3. Suo Z., Wang W., Yang M., "Electromigration Instability: Transgranular slits in interconnects," *Applied Physics Letters*. Vol. 64, No. 15 (1994), pp. 1944-1946.
4. Cotterell B., Chen Z., "Buckling and crackling of thin films on compliant substrate under compression," *International Journal of Fracture*, Vol. 104 (2000), pp. 169-79.
5. McDonald J. P., Mistry V. R., Ray K. E., Yalisove S. M., Nees J. A., and Moody N. R., "Femtosecond pulsed laser direct write production of nano- and microfluidic channels," *Applied Physics Letters*, Vol. 88, No. 18 (2006), pp. 183113-1-3.
6. Liu B. C., Tong R. C., "Buckle-driven delamination of metal film/ceramic substrate systems under thermal loading," *Journal of Tsinghua University*, Vol. 34, No. 2 (1994), pp. 75-80.
7. Kraft O., Arzt E., "Electromigration mechanisms in conductor lines: Void shape changes and slit-like failure," *Acta Materialia*, Vol. 45, No. 4 (1997) pp. 1599-1611.
8. Ju Y. S., and Goodson K. E., "Thermal mapping of interconnects subjected to brief electrical stresses," *IEEE Electron Device Letters*, Vol. 18, No. 11 (1997), pp. 512-514.
9. Wang Q. H., Xie H. M., Feng X., Chen Z. J., Dai F. L., "Delamination and Electromigration of Film Lines on Polymer Substrate Under Electrical Loading," *IEEE Electron Device Letters*, Vol. 30, No. 1 (2009), pp. 11-13.
10. Hutchinson J. W., and Suo Z., "Mixed mode cracking in layered materials," *Advances in Applied Mechanics*, Vol. 29 (1992), pp. 63-191.
11. Freund L. B. and Suresh S., Thin Film Materials: Stress, Defect Formation and Surface Evolution, Cambridge University Press (Cambridge, 2004).

Reliability Study of RFID Flip Chip Assembly by Isotropic Conductive Adhesive through Computer Simulation

Edward K L Chan, Bo Gao, Matthew M F Yuen

Department of Mechanical Engineering, The Hong Kong University of Science and Technology

Clear Water Bay, Kowloon, Hong Kong

edwardc@ust.hk, megb@ust.hk, meymf@ust.hk, Tel: (852) 23588814

Abstract

Radio Frequency Identification (RFID) is quickly gaining a foothold in the identification and security industry. However, one major roadblock that prevents companies in adopting RFID technologies is its high manufacturing cost, particularly assembly cost. One approach to reduce the assembly cost is using surface mount technology. Currently anisotropic conductive adhesives (ACA) tape, isotropic conductive adhesives (ICA) and non-conductive adhesives (NCA) are being used for RFID flip chip assembly. Typically the ACA tape and NCA provide the necessary height control in the assembly. However, few literatures discuss about the possibility of using ICA as flip chip connection for different RFID product. In this work, we have developed techniques based on finite-element method. We use the full wave analysis approach that combines full wave simulator HFSS (High Frequency Structure Simulator) with circuit simulator ADS (Advanced Design System) to simulate the different packaging approach and the electromagnetic effects on the transponder strap. From the results, design guideline by using low cost ICA as RFID flip chip interconnect can be obtained. The result is helpful in height control for the batch printing process for different RFID products.

Introduction

Radio Frequency Identification (RFID) is quickly gaining a foothold in the identification and security industry. It is used in areas of healthcare, cashless ticketing system, logistics and security identification. Transponders are attached to various kinds of objects which differ in shape and material. Data transfer between a transponder and a reader is carried out by a backscattering of electromagnetic waves. Transponder partially rectifies electromagnetic waves from a reader and uses it as a power source. Then, the transponder sends a signal back to the reader. Many transponder antennas have been developed depending on materials to which they are attached.

An RFID transponder consists of a transponder chip combined with an antenna in a compact package; the packaging is structured to allow the RFID transponder chip to be attached to an antenna. A major contributor to low cost transponder is throughput in packaging. For this reason flip chip bonding is preferred to wire bonding. However, one major roadblock that prevents companies in adopting RFID technologies is its high manufacturing cost, particularly assembly cost. One approach to reduce the assembly cost is using surface mount technology [1]. Currently anisotropic conductive adhesives (ACA) tape and non-conductive adhesives (NCA) are being used for RFID flip chip assembly. ACAs come in either a paste or film form contain conductive particles made of either a polymer bead coated with nickel and gold or just simply solid gold particles which are

deposited in the adhesive resin and only conduct in the Z direction. Similar to that of NCA, the ACA is first partially cured during the initial individual chip thermo-compression followed by post cure to fully cure it which also requires precise alignment [2-3].

ICAs, made up of a composition of polymer resin and conductive silver fillers, provide electrical conductivity in X, Y and Z directions when the silver flakes pack together due to the shrinkage of the ICA after being cured [3]. As no thermo-compression is required for this approach, the overall assembly time will be greatly reduced as compared to that of NCA and ACA. Moreover ICAs, due to its composition, can be batch screen printed which further promotes higher throughput as compared to ACAs and NCAs. Isotropic conductive adhesives (ICA) joints however have a more sensitive affect from moisture and mechanical stress. Typically the ACA tape and NCA provide the necessary height control in the assembly [1]. However, few literatures discuss about the possibility of using ICA as flip chip connection for different RFID product.

HFSS Model setup

In this work, we have developed techniques based on finite-element method. We use the full wave analysis approach that combines full wave simulator HFSS (High Frequency Structure Simulator) to simulate the different packaging approach and the electromagnetic effects on the transponder strap. For RFID purposes, 13.56MHz are considered high frequencies, 860-960MHz and 2.4GHz-2.5GHz are considered ultra high frequency (UHF) and microwave frequency respectively. Different frequencies require suitable bump height for impedance matching and loss calculation. In this study, the finite element based simulation is conducted in Ansoft HFSS for the RFID mounted on PET substrate with ICA as flip chip interconnect.

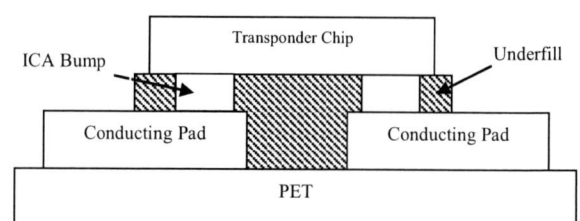

Figure 1 Structure of RFID Transponder with ICA interconnect

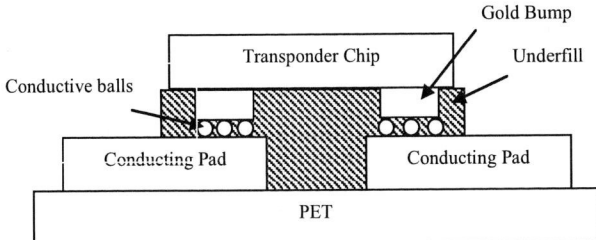

Figure 2 Structure of RFID Transponder with ACF interconnect

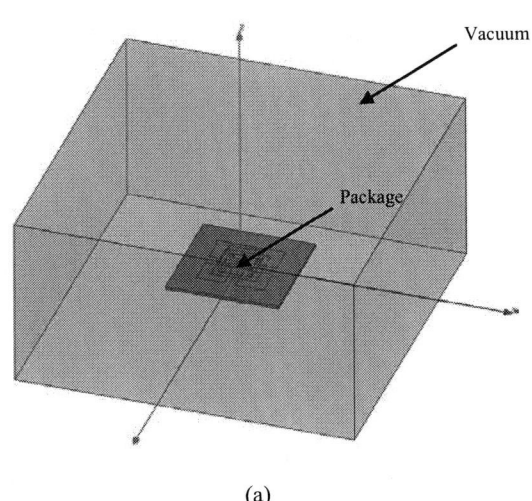

(a)

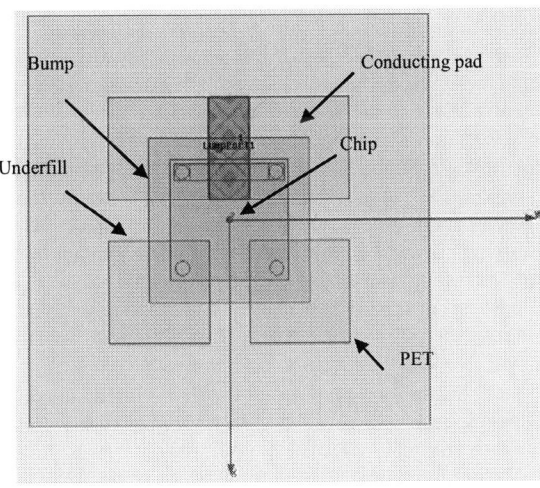

(b)

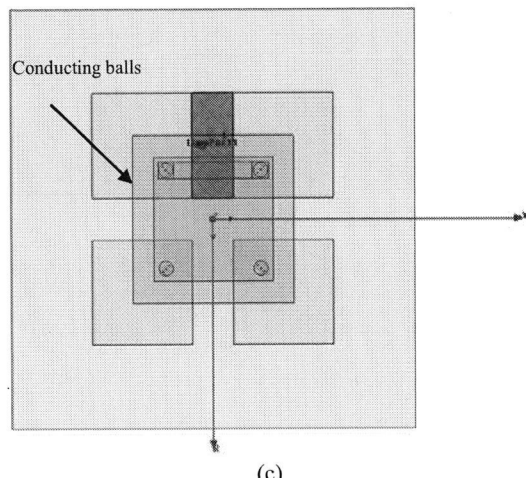

(c)

Figure 3 Simulation model for transponder strap. (a) 3D view (b) top view for model with ICA interconnect (c) top view for model with ACF interconnect

In our simulation model, the transponder chip is expressed as pure silicon. The ICA and gold bumps are presented by circular cylinder with the equal height. The conductive traces are constructed as two-dimensional surface with finite conductivity. Substrate is represented as 50μm PET that has a relative dielectric constant 3 and underfill around the chip is modified as pure epoxy resin which have a relative dielectric constant 10. The conductive balls beneath the gold bumps are presented by circular cylinder with three balls under each bump and the height is 2μm which is the real diameter of conductive ball. A lumped port is located at one side of the chip. Sweeping of frequency from 860MHz to 2.5GHz was conducted and impedance variation due to bump height variation from 10μm to100μm is recorded.

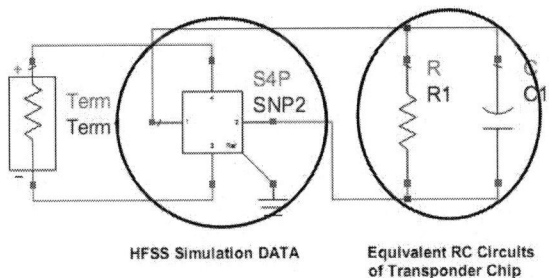

Figure 4 HFSS simulation data is combined with the equivalent RC circuits of transponder [5].

The full wave HFSS simulation is combined with transponder equivalent RC circuits. The impedance of transponder chip is converted to equivalent RC circuits to consider the impedance variation due to the change of frequency. Simulations with bump height is varied from 10μm to100 stepped by 5μm is being conducted but the figures (Figure 5 to Figure 8) showed three bump height and frequency from 1.6GHz to 2.5GHz for simplicity.

978-1-4244-4658-2/09 $25.00 © 2009 IEEE

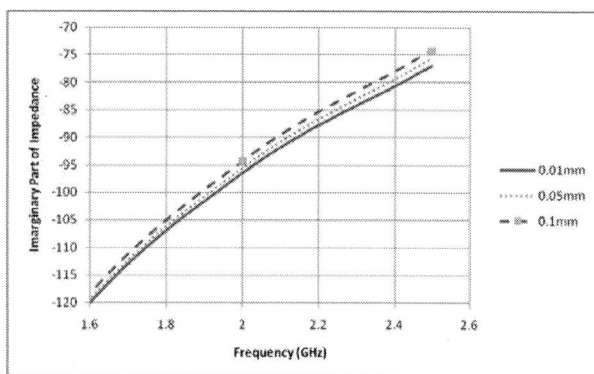

Figure 5 HFSS Simulated impedance data for imaginary part for package with ICA as interconnect as a function of bump height and frequency. Bump height is varied from 10μm to 100μm stepped by 5μm.

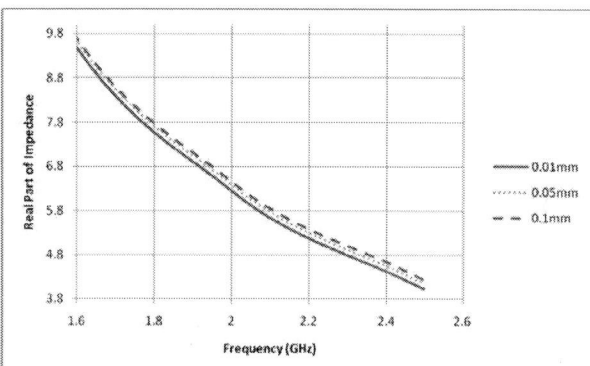

Figure 6 HFSS Simulated impedance data for real part for package with ICA as interconnect as a function of bump height and frequency. Bump height is varied from 10μm to100 stepped by 5μm.

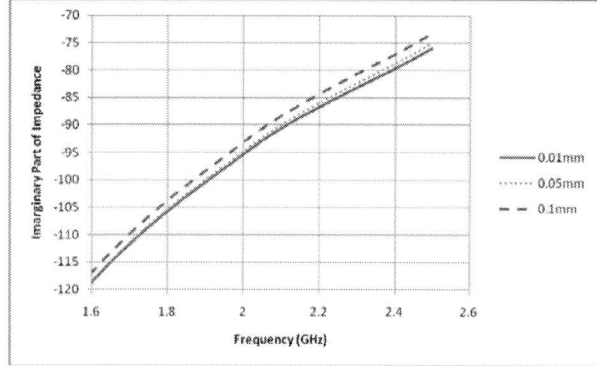

Figure 7 HFSS Simulated impedance data for imaginary part as a function of bump height and frequency. Bump height is varied from 10μm to100 stepped by 5μm.

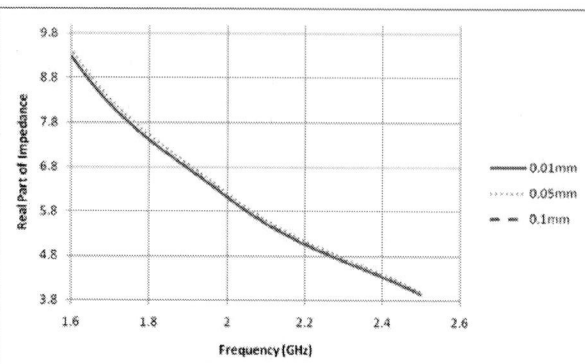

Figure 8 HFSS Simulated impedance data for imaginary part as a function of bump height and frequency. Bump height is varied from 10μm to100 stepped by 5μm.

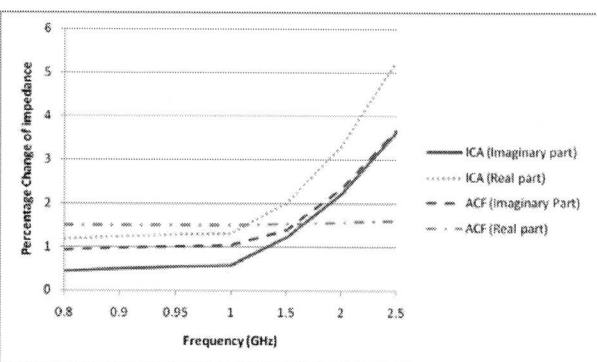

Figure 9 Percentage error for impedance data as a function of frequency.

From the Figure 5 and Figure 6, the impedance change for the RFID package with ICA as interconnect has around 1-5% variation when bump height changes. The package with gold bumps and silver particles as interconnect on the other hand has less variation for bump height changes, especially the real part as shown in Figure 9. On higher frequencies, typically the UHF region, the percentage change due to the bump height variation is more significant for the ICA packages. Larger variation may come from the majority of using the conductive adhesive having lower conductivity compared to the gold bumping used in the ACF packages. This leads to different response to the RF signal. Attenuation frequency and read range will then change for different bump height.

Typical ICA screen printing for the RFID interconnect has a good resolution of xy- plane but since the ICA may slump or bleed when flip-chip is placed onto the substrate. Height control is needed when compared to the ACF tape. However the cost of controlling the bump height is still less than the material cost of using ACF tape.

Conclusion

In this paper, we present the full wave simulation model for flip-chip bonding onto a PET substrate. We have considered the chip, gold bumps, electrically conductive adhesive, and substrate. These five are interdependent, and therefore it is important to see the full picture in order to

978-1-4244-4658-2/09 $25.00 © 2009 IEEE   952

design a low cost transponder and efficiently produce the transponder. The S-parameter of simulated model and equivalent RC-circuits for chip impedance are used to study the influence of bump height in flip-chip bonding. Our results indicate that the bump height under the chip can significantly affect on the impedance of strap for using ICA as interconnect. This means that parameter must be treated more carefully in RFID Transponder production in order to move to lower cost RFID by using ICA.

References

1. Lim, S.Y.L., Ser Choong Chong, Lihui Guo, Wai Yin Hnin, "Surface mountable low cost packaging for RFID device," *Electronics Packaging Technology Conference, 2006. EPTC '06. 8th*, Vol., No., pp. 255-259, 6-8 Dec. 2006.

2. Miessner, R., Nieland, C., Aschenbrenner, R., Reichl, H., "Evaluation of flip chip bonding using ACA on polyester substrates ," *Adhesive Joining and Coating Technology in Electronics Manufacturing, 2000. Proceedings. 4th International Conference on* , Vol., No., pp. 46-51, 2000.

3. Huabin Chu, Bing An, Fengshun Wu, Yiping Wu, "RFID Tag Packaging with Anisotropically Conductive Adhesive," *Electronic Packaging Technology, 2006. ICEPT '06. 7th International Conference on* , Vol., No., pp. 1-4, 26-29 Aug. 2006.

4. Morris, J.E., Jeahuck Lee, Liu, J., "Isotropic Conductive Adhesive Interconnect Technology in Electronics Packaging Applications," *Polymers and Adhesives in Microelectronics and Photonics, Polytronic, 2005. Polytronic 2005. 5th International Conference on* , Vol., No., pp. 45-52, 23-26 Oct. 2005.

5. JeongKi Ryoo, JaeYul Choo, Hwan Park, JinKook Hong, Jeongjoon Lee, "Full Wave Simulation of Flip-Chip Packaging Effects on RFID Transponder," *RFID, 2007. IEEE International Conference on*, Vol., No., pp. 37-40, 26-28 March 2007.

Effect of Thermomigration in Eutectic SnPb Solder Layer

Yuan Tao[1,2], Lan Ding[1], Yuhui Yao[1], Bing An[1,2], Fengshun Wu[1,2], Yiping Wu[1,2]

[1]Wuhan National Laboratory for Optoelectronics,
Huazhong University of Science and Technology, Wuhan, P. R. China, 430074
[2]State Key Laboratory of Materials Processing and Die & Mould Technology,
Huazhong University of Science and Technology, Wuhan, P. R. China, 430074

Abstract

Due to the miniaturization trend and functional demand in high-density microelectronic packaging, the thermomigration in flip chip solder joints owing to the joule heating becomes a serious reliability issue. In this study, a novel apparatus which can provide a sufficient temperature gradient cross the solder joint in the specimen was used to carry on the thermomigration experiment separated from electromigration behavior. A constant temperature gradient above 2000°C/cm was applied on the eutectic SnPb solder layer of specimen and the load duration were 20h, 40h, and 80h respectively. The result reveals that in eutectic SnPb solder layer of specimen, an obvious mass diffusion from hot side to cold side occurs and Pb is estimated as the primary diffusion element under a high enough temperature gradient. EDX result also proves that the percentage of Pb element in the cold side is higher than that in the hot side. With the temperature gradient load time goes up from 20h to 80h, the phenomenon of Pb migration to cold side turns to be much severer. Due to the redistribution of both Pb and Sn element, the morphology and thickness of IMC in both hot side and cold side changes obviously comparing that in as-reflowed solder layer. Meanwhile, the tensile test also shows that the mechanical strength of eutectic SnPb solder layer becomes weaker with the thermomigration load duration increases gradually, and this may be owing to the defects existing in the solder caused by mass diffusion.

1. Introduction

With the miniaturization trend and functional demand in high-density microelectronic packaging, the feature sizes of interconnection structure, such as metal line and solder joint, have decreased rapidly in recent years. Meanwhile, in order to satisfy the demand of industrial application, it has been reported that the average current density crossing the solder joint in flip chip exceeds 10^4A/cm^2 [1]. Consequently, the interconnection traces in the silicon chip act as the primary heat source due to the joule heating in a typical flip chip module, which brings on an obvious temperature gradient across the tiny solder joint. As we know that when the temperature gradient is up to 1000°C/cm, the thermomigration can be triggered in the SnPb solder joints [2-3]. If the phenomenon of thermomigration turns to be seriously, the defects such as voids and phase accumulation, which are similar with that in electromigration, will lead to an open circuit or short circuit in the devises [4]. Other research work also confirms that the thermomigration significantly dominates the failure process of the solder joints in flip chip under high electric current density [5]. Hence, thermomigration in the solder joint should become one of the major concerns in the development of the electronic packages nowadays.

In recent years, a majority of researches have been devoted to understanding the failure mechanism of the thermomigration accompanied by the electromigration behavior; however, only a few works are focused on the thermomigration phenomenon separated from the electromigration and other damages induced by the high current density, and there is scarcely any experimental data about the relationship between thermomigration and mechanical properties of solder joint. So in this study, a novel apparatus which has been described in our previous study [6] was used to create a sufficient temperature gradient across the solder layer without the effect of electrical current. An constant temperature gradient above 2000°C/cm was applied on the eutectic SnPb solder layer and the load duration were 20h, 40h and 80h respectively. The microstructure of cross-section in eutectic SnPb solder layer was inspected by SEM and EDX analysis. Furthermore, the tensile test was used to estimate the effect of thermomigration with different grades on the mechanical property of eutectic SnPb solder layer.

2. Apparatus and specimen structure

The thermomigration apparatus which is designed by our team is shown in Figure 1. The solder iron heater and water circulation system can make the temperature gradient accumulate in the solder layer of the specimen. And the temperature gradient can be controlled by setting the power of the solder iron and the water velocity in the water circulation system. As the recycled water is at the upper portion of this apparatus, a layer of fluid sealant is need at the connection point between the specimen and the water, ensuring the upper water cannot seep through the tiny hole. Moreover, in order to force the most part of the heat flow goes through the solder, an insulation layer is also needed between the specimen and the container.

Figure 2 displays the structure of the specimen which is fit for the apparatus in this study. Cu plate/SnPb solder layer/Cu bar is chosen to simulate the sandwich structure in flip chip. The exterior surface of Cu plate is connected with the solder iron heater, and the Cu bar is immersed into the recycled water to eliminate heat by convection. Since the coefficient of thermal conductivity of Cu is quite higher than that of SnPb solder, most of the temperature gradient can accumulate in the solder layer. So this type of apparatus and specimen can easily resolve the problem of creating a high temperature gradient across such a tiny solder layer.

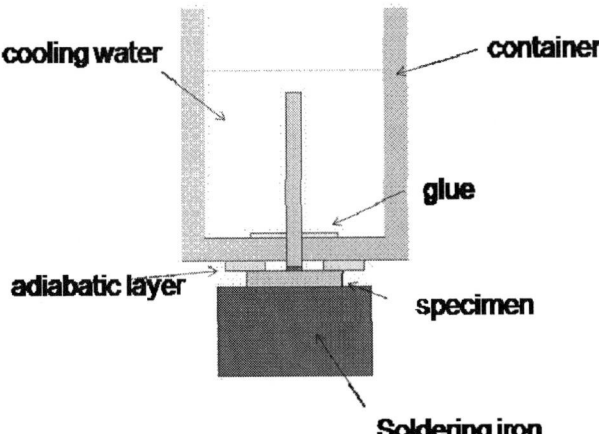

Figure 1 the schematic graph of the thermomigration apparatus

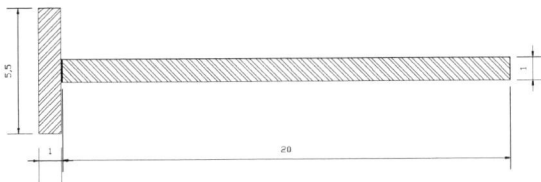

Figure 2 the schematic graph of the specimen structure

3. Experimental

To prepare this kind of specimen, a Cu plate with 10mm in thickness, a Cu bar with 10mm in diameter and a layer of SnPb solder were firstly assembled in an especial clamping fixture designed by our group. This fixture can accurately obtain the solder layer with the thickness in range of 20μm to 400μm. And then the whole specimen was put into a forced convection reflow oven to carry out soldering process. The peak temperature of the reflow profile was 215°C, and the duration above the melting point was about 60s.

As the specimens have been prepared, they were divided into two parts for the subsequent experiment. One part was loaded a constant temperature gradient above 2000°C on the solder layer for 20h, 40h and 80h. The result of three-dimensional thermal finite element simulation proved that the temperature of the hot side and the cold side in the solder layer was about 172°C and 165°C, respectively. The other part was kept in an oven at 170°C for ageing test, and the time span was the same as that in thermomigration experiment. Subsequently, some of the specimens were mounted in epoxy and metallurgically polished in preparation for characterization. The reaction zone for each specimen was examined using a scanning electron microscope (SEM) and energy dispersive x-ray spectroscopy (EDX). Furthermore, the mechanical properties of the specimen were also examined by the tensile test, and the load velocity is 0.1mm/min.

4. Result and discussion

Microstructure

The original cross-section microstructure of the SnPb solder layer just after one time reflow is shown in Figure 3(a), and Figure 3(b) and (c) displays the enlarged view of the

connection interface in the both sides of this specimen. The Pb-rich phrases are dispersed in Sn matrix randomly, and the sizes of these phases are nearly uniform. It is obvious that there is no appearances of phase accumulation after one time reflow. At the interfacial area between SnPb solder and Cu, the IMC layer is scallop shape and the thickness is a little small. The composition of it is Cu_6Sn_5, which is examined by EDX analysis. It is worth to mention that there is no evident differences in morphology and thickness between the interfaces of the both sides because of the same heat process in one time reflow.

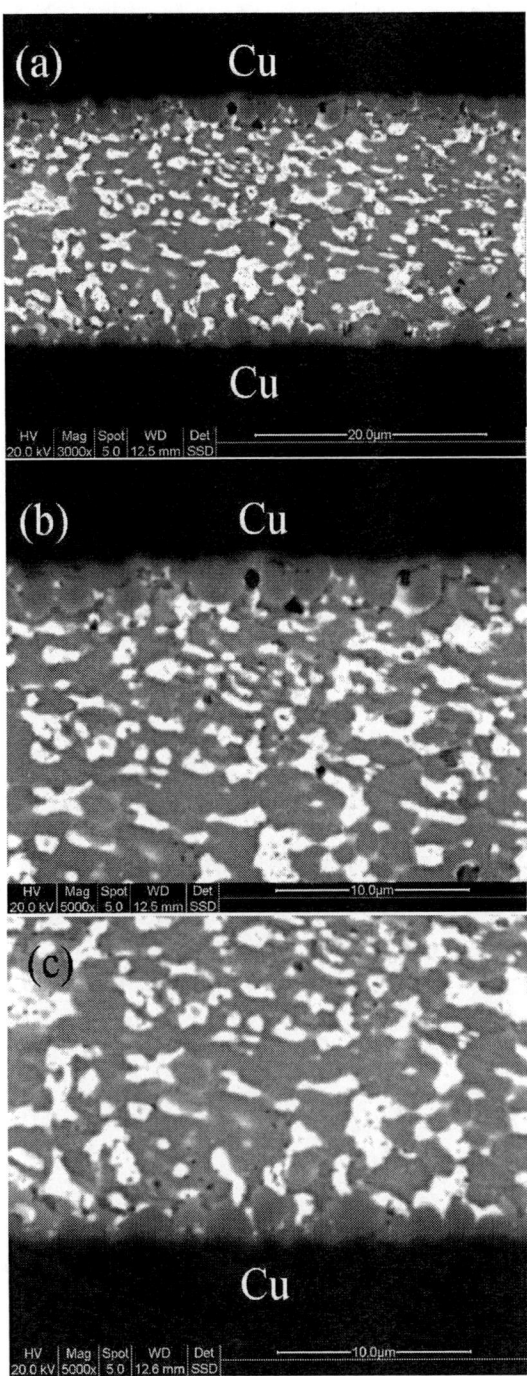

Figure 3 the microstructure of the original SnPb solder layer

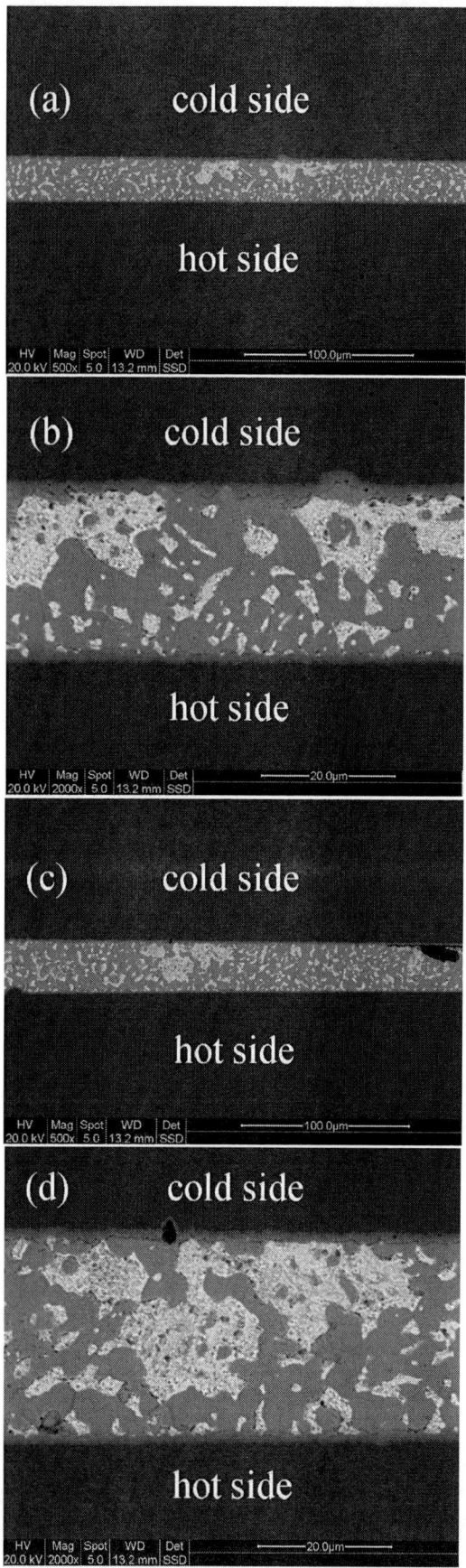

Figure 4 the microstructure of the SnPb solder layer after thermomigration 20h

Figure 4(a) and (c) show the cross-section microstructures of the SnPb solder layer after thermomigration 20h, and (b), (d) are the partial enlarged photographs. The upper part is the cold side and the lower part is the hot side, which corresponds to the apparatus we described above accurately. Comparing to the Figure 3, the morphology and composition distribution in SnPb solder layer have been changed evidently. There are some Pb-rich phrases gathering at the interface near the cold side, and the shapes are mostly anomalous, but the area of phase's accumulation is not so large. The Pb-rich phases in other region are in homogeneous distribution, which are coarser after thermomigration 20h. The morphology of the IMC layer becomes smoother than before. But there is no obvious change in the thickness of IMC layer in both sides, which proves that the thermomigration for a short period cannot affect the IMC layer adequately.

The sectional microstructure of SnPb solder layer after thermomigration for 40h is shown in Figure 5(a), and the partial enlarged image is Figure 5(b). It is clear that there are more regions where the Pb-rich phases gather at the cold side of the solder layer. The total area of phase's accumulation does not increase too much, but all the location of it is much closer to the cold side of the solder layer than that in thermomigration for 20h, which means the phenomenon of composition redistribution turns to be much severer. Meanwhile, the IMC layer does not change a lot comparing to that in 20h thermomigration.

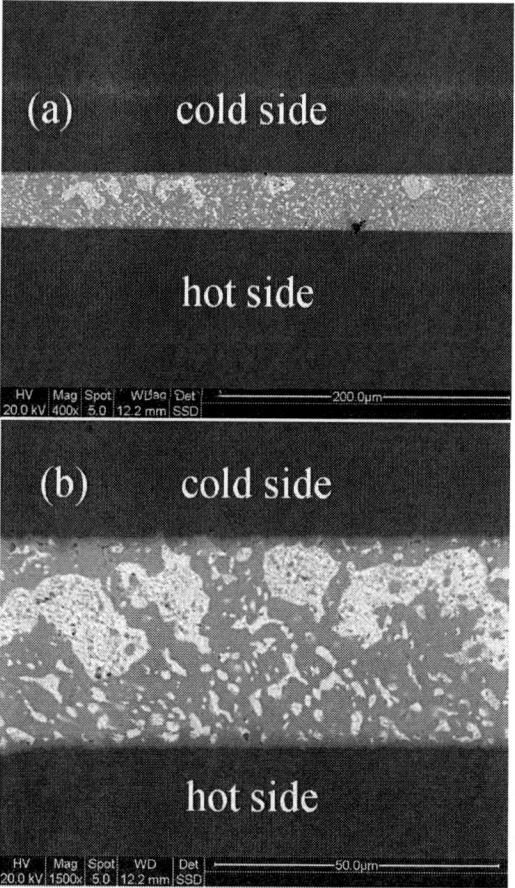

Figure 5 the microstructure of the SnPb solder layer after thermomigration 40h

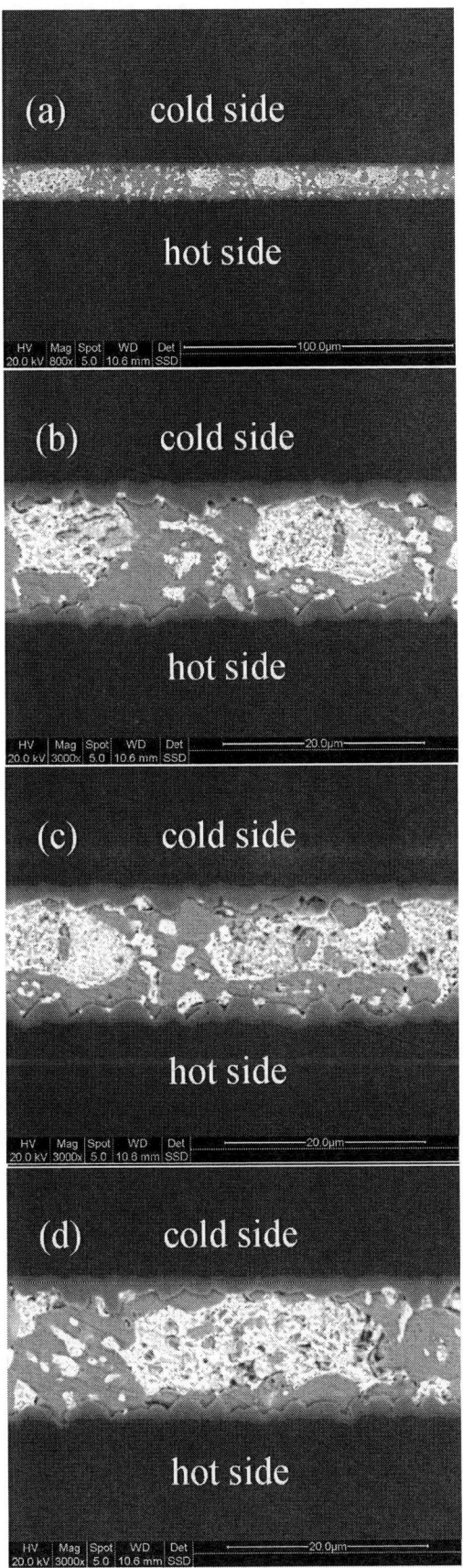

Figure 6 the microstructure of the SnPb solder layer after thermomigration 80h

Figure 6(a) displays the cross-section microstructures of the SnPb solder layer after thermomigration 80h, and (b), (c), (d) are the partial enlarged photographs. Most of the Pb-rich phases are forced to accumulate at the cold side of SnPb solder layer, and the total area of that has increased a lot. The other Pb-rich phases which are dispersed in Sn matrix are also coarse severely. These phenomena can be explained as follow: Firstly, as the sufficient temperature gradient exists at the vertical direction across the solder layer, the Pb atoms are driven to migrate towards the cold side. The IMC at the cold side can serve as a barrier layer, so a mass of Pb atoms have to gather at the interface between the solder and the IMC. In order to reduce the surface free energy, the Pb-rich phases trend to appear as the block shape. Besides, the coarse phenomenon of the rest Pb-rich phases is similar with that in ageing experiment, so it can be considered as the normal result for ageing 80h at 170°C.

Moreover, the thickness of the IMC layer in both sides increases a lot after thermomigration for 80h. It is obvious that there is a thin IMC layer growing at the interface between Cu_6Sn_5 and Cu, which is Cu_3Sn IMC examined by EDX analysis. Meanwhile, it is interesting to mention that the thickness of the IMC layer at the hot side is 4.61μm, which is bigger than that at the cold side 3.29μm. This phenomenon may due to the following two reasons: On the one hand, the higher average temperature at the hot side can accelerate the growth and thickening of IMC layer; on the other hand, most of the Pb atoms migrate towards to the cold side, and the Sn atoms are forced to diffuse to the area near the hot side, so it can supply a higher concentration of Sn atoms for the Cu_6Sn_5 IMC grows more rapidly.

Tensile test

In order to make a round analysis for the effect of thermomigration to the mechanical property of SnPb solder layer, the original sample and the test specimens which underwent thermomigration or ageing experiment for 20h, 40h and 80h were all taken into account for the tensile test. Figure 7 shows the measured tensile strength versus the duration of both thermomigration and ageing. It reveals that no matter the thermomigration or ageing for 20h and 40h, the tensile test of the solder layer are both higher than that of the original sample. As we know that the IMC layer is the key part determining the mechanical behavior of the solder joint. It will be the weak spot of the connection structure in the active device if the thickness is not appropriate. The IMC layer in the specimen for thermomigration or ageing 20h and 40h reaches the right scale of the thickness, and there is no micro voids forming in this layer after such a short period of thermal history, thus the tensile strength are higher than before.

Meanwhile, the tensile strength of the specimen ageing 80h decreases a little, but it is still higher than that of original sample. It is important to mention that the reduction of mechanical property of specimen for 80h thermomigration is fully obvious, which is nearly one third of the original value. It can be inferred that due to a mass of Pb atoms migrate to the cold side of the solder layer, and the Sn atoms cannot be duly pushed back to the hot side for filling up the vacancy left by Pb atoms, so the micro voids are able to grow up gradually under high temperature. Since the sizes of the micro voids are too tiny, we cannot see them clearly by SEM. But they may serve as the source of the crack because of the stress concentration during the tensile test, which will make a negative effect on the mechanical property of solder layer directly.

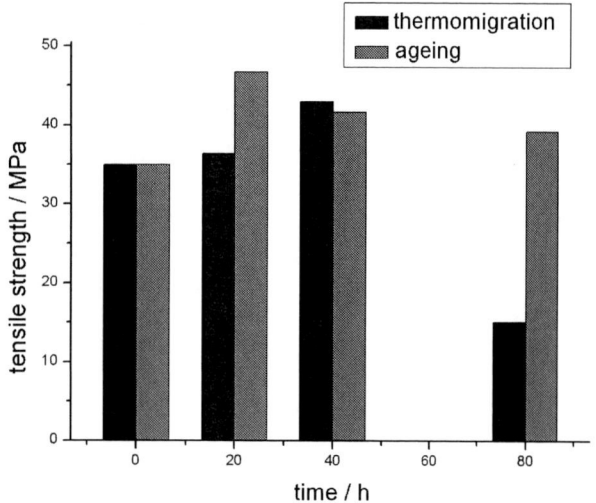

Figure 7 the tensile strength versus time of thermomigration and ageing

5. Conclusion

A novel apparatus which can provide the constant temperature gradient was used for thermomigration experiment in this study. Solder layers with composition of eutectic 63Sn37Pb was tested for thermomigration 20h, 40h and 80h with the temperature gradient above 2000°C/cm. Microstructure analysis indicates that Pb is the primary diffusion element from hot side to cold side under a high enough temperature gradient. The thickness and morphology of IMC at the interface between SnPb solder layer and Cu change evidently comparing that in the original specimen. And the tensile test also reveals that the mechanical strength of eutectic SnPb solder layer decreases obviously after a long period of thermomigration, and this may be owing to the micro voids growing in the IMC layer caused by Pb mass diffusion. All of these results illuminate that the thermomigration phenomenon under large temperature gradient in high-density microelectronic packaging cannot be ignored nowadays.

Acknowledgment

The authors would like to acknowledge the financial support by National Natural Science Foundation of China (No. 60876070 and No. 60776033).

References

1. K.N. Tu, "Recent advances on electromigration in very large scale integration of interconnects," *Journal of Applied Physics*, Vol.94(2003), pp. 5451-5473.
2. Annie T. Huang, A. M. Gusak, K. N. Tu, "Thermomigration In SnPb composite flip chip solder joints," *Applied Physics Letters*, Vol.88, (2006), pp. 141911.
3. Fan-Yi Ouyang, Annie T. Huang, K. N. Tu, "Thermomigration in SnPb composite solder joints and wires," *Electronic Components and Technology Conference*, 2006, pp. 1974-1978.
4. Annie T. Huang, K. N. Tu, "Effect of the combination of electromigration and thermomigration on phase migration and partial melting in flip chip composite SnPb solder joints," *Journal of Applied Physics*, Vol.100, (2006), pp. 33512.
5. C. Basaran, H. Ye, D. C. Hopkins. "Failure modes of flip chip solder joint under high electric current density," *Journal of Electronic Packaging*, Vol.127, (June 2005), pp. 157-163.
6. Yuan Tao, Gan-ran Tang, Bo Wang.et al, "Experimental design and simulation for thermomigration in the solder joint," *International Conference on Electronics Packaging*, 2009.

Adhesion Behavior between Epoxy Molding Compound and Different Leadframes in Plastic Packaging

Li Xu[1], Xiuzhen Lu[1*], Johan Liu[1,3], Xinyu Du[2], Yan Zhang[1], Zhaonian Cheng[3]

[1]Key Laboratory of Advanced Display and System Applications, Ministry of Education
and SMIT Center, School of Mechatronics Engineering and Automation,
Shanghai University, Shanghai 200072, China
[2]Henkel Corporation, 15350 Barranca Pkwy, Irvine CA 92618, USA
[3]SMIT Center & Bionano Systems Laboratory, Department of Microtechnology and Nanoscience,
Chalmers University of Technology, SE-412 96 Gothenburg, Sweden
*Corresponding Author's e-mail: xzlu@staff.shu.ed u.cn

Abstract

Adhesion has been identified as one of the key elements in solving failure problems in electronic packaging. Understanding and improving the adhesion between epoxy molding compound (EMC) and leadframes is thought to be a key step towards improving package performance. The objective of this work was to study the effect on adhesion behavior and delamination of plastic packages using various metal coated leadframes that had been subjected to preconditions. In this work, the effects of moisture absorption testing and reflow process on the interface adhesion strength of the EMC/metal leadframes were studied. Three kinds of metal coated leadframes were used: copper, silver coated copper and nickel/palladium/gold coated copper. Adhesion strength was measured using the tab pull test. These data were correlated to some extent with the findings of package delamination observed by the C-mode Scanning Acoustic Microscope (C-SAM). Scanning Electron Microscopy (SEM) and Energy Dispersive Spectroscopy (EDS) were employed to characterize the surface of EMC and metal leadframes after separation along the vertical plane.

Introduction

Interfacial delamination between epoxy molding compound (EMC) and leadframes is one of the common failure modes in electronic plastic packages [1]. Delamination, especially those occurring between EMC and leadframes, often lead to popcorning during or after the reflow process due to the presence of absorbed moisture within the plastic encapsulation. In order to improve reliability, it is essential to understand the characterization of EMC and leadframe interface adhesion in plastic packages.

A number of approaches have been developed to investigate interfacial adhesion behavior. The epoxy surface was treated with either an Ar or O_2 plasma etch to study the chemical reaction between metal and polymer substrates before depositing the thin film metals [2]. It has previously been suggested [3] that a thiol compound which bonded readily and formed a self-assembly monolayer with copper would improve interfacial adhesion between copper and EMC. Extensive research has been done to improve the adhesion between copper leadframe and epoxy molding compound. Strong adhesion of metal to polymer has been achieved mainly with different surface finishes [4-5].

In this paper, the effects of moisture absorption and reflow process on the interface adhesion strength of the EMC/metal leadframe were studied. Interface delamination and surface morphology of the samples was studied using C-SAM and SEM. The correlations of adhesion force amongst three kinds of samples after moisture absorption testing and reflow progress was discussed.

Experimental Procedure

1. Design of Experiments

The epoxy molding compound used in this study was procured commercially. Three kinds of metal coated leadframes consisting of copper, silver coated copper and nickel/palladium/gold coated copper were used. All the samples were molded and kept in a desiccators. To allow the absorbed moisture from the molding process to evaporate, the samples were dried at 80°C for four days. Three case studies were carried out as seen Table 1. In the first case, three kinds of EMC/metal leadframes were subjected to the precondition MSL 1 (85°C /85%RH for 168 hours). In the second case, the three kinds of samples underwent humidity and temperature testing as above, followed by 3 time reflow with 260°C peak temperature for 2 minutes after humidity and temperature testing. In the third case, the three kinds of EMC/metal leadframes were treated with reflow processing only.

Table 1 Design of Experiments

Compound	Condition	Leadframe-coated	ID
EMC	Humidity & Temperature Testing	Cu/Cu	1
		Ag/Cu	2
		Ni/Pd/Au/Cu	3
	Humidity & Temperature Testing and Reflow	Cu/Cu	4
		Ag/Cu	5
		Ni/Pd/Au/Cu	6
	Reflow only	Cu/Cu	7
		Ag/Cu	8
		Ni/Pd/Au/Cu	9

2. Mechanical Tests

The tab pull method [6] was used to measure the interfacial strength between leadframe and EMC as show in Figure 1. The measured adhesion was defined as the maximum amount of pull force needed to pull the tab away from the molding compound. The total surface area of the tab in contact with the molding compound was 0.784 in 2.

The self-designed pull-apart method was performed as described in Figure 2. SEM and EDS was used for surface characterization of the EMC and the leadframe after separation along the vertical plane (pull-apart method).

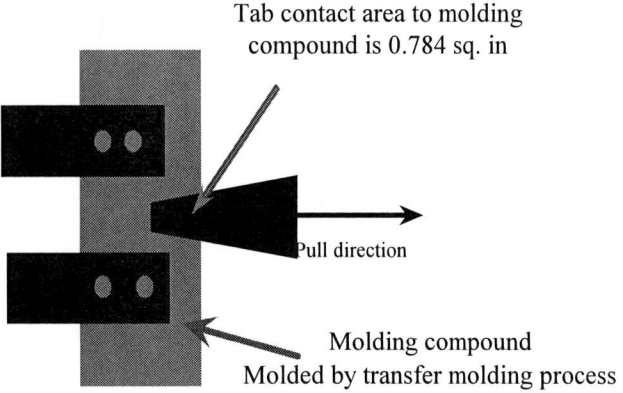

Tab contact area to molding compound is 0.784 sq. in

Pull direction

Molding compound
Molded by transfer molding process

Figure 1 Tab pull test

Results and Discussion

Delamination in the different treatments

Mode C of Scanning Acoustic Microscopy (C-SAM) was used to locate the adhesion and delamination area of the EMC/metal. The samples were chosen for the non-destructive C-SAM test after the different treatments. The C-SAM results of the EMC/Cu, EMC/Ag and EMC/Ni/Pd/Au specimens at different treatment are shown in Figure 3.

The interface for the EMC/Cu specimens subjected to humidity and temperature testing show little interface delamination, while partial interface delamination occurred after humidity & temperature testing and reflow. No delamination area can be seen in the specimens subjected only to reflow. Interface delamination can be seen in the EMC/Ag specimens after moisture absorption testing and after reflow, while the result of the specimens subject to the other two treatments was contrary to that of EMC/Cu specimens. Half delamination happened in EMC/ Ni/Pd/Au specimens treated with humidity and temperature testing and almost total interface delamination was found under the other two different conditions.

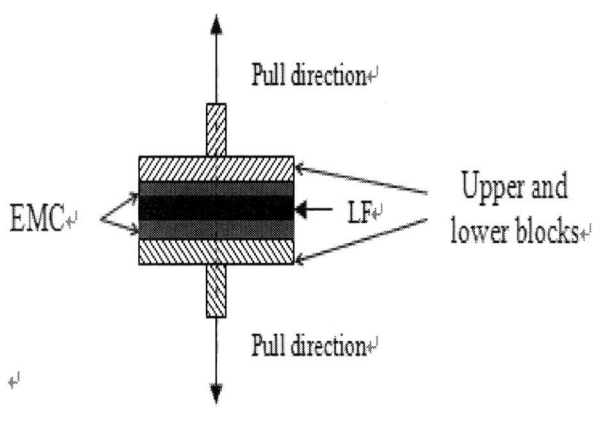

Pull direction

EMC

LF

Upper and lower blocks

Pull direction

Figure 2 Pull-apart method

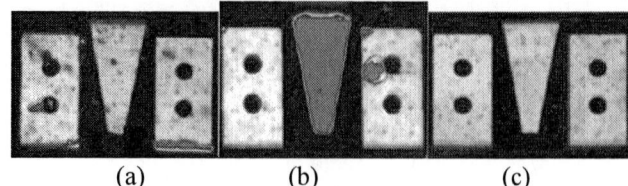

(a)   (b)   (c)

Figure 3(a) The C-SAM images taken with the EMC/Cu samples at different stages after (a) after Humidity & Temperature Testing; (b) after Humidity & Temperature Testing and Reflow; (c) after Reflow only

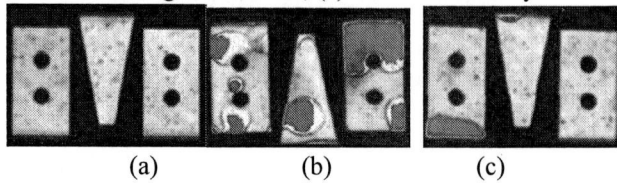

(a)   (b)   (c)

Figure 3(b) The C-SAM images taken with the EMC/Ag samples at different stages after (a) after Humidity & Temperature Testing; (b) after Humidity & Temperature Testing and Reflow; (c) after Reflow only

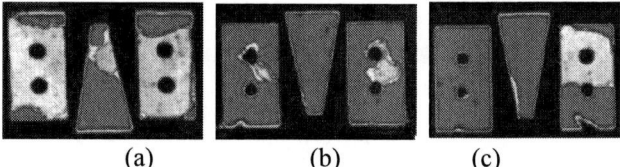

(a)   (b)   (c)

Figure 3(c) The C-SAM images taken with the EMC/ Ni/Pd/Au samples at different stages after (a) after Humidity & Temperature Testing; (b) after Humidity & Temperature Testing and Reflow; (c) after Reflow only

Results of the tab pull test

Adhesion strength to three kinds of metal coated leadframes was also tested by the tab pull method and the results are shown in Figure 4. The results showed that the interface adhesion strength of the EMC/Cu leadframe was the best amongst the three kinds of specimens whilst the EMC/Ni/Pd/Au leadframe showed poor adhesion strength. It was also found that the EMC/Cu specimens were more susceptible to moisture absorption in a negative fashion than the EMC/Ag specimens. On the contrary, the reflow process seems to decrease the adhesion strength of the EMC/Ag specimens more than the EMC/Cu specimens. However, the reflow process generally had a greater influence on the interface adhesion strength of the EMC/metal specimens after moisture absorption testing, which resulted in great degradation of the interface adhesion strength; especially for the EMC/Ni/Pd/Au samples, for which the adhesion strength was nearly zero.

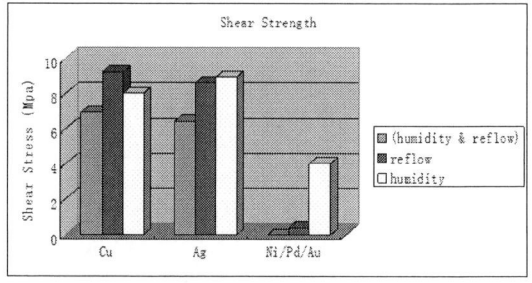

Figure 4 Schematic of Tab pull test results

The results indicate a strong relationship between the C-SAM and the tab pull test. The adhesion strengths correlated with the delamination observations: the smaller the surface area that was adhered to, the weaker the measurable pull strength was.

Analysis of SEM and EDS

The surface characteristics of the EMC and leadframe were observed after the EMC and leadframe were separated along the vertical plane using the SEM. The morphology results of EMC/Ag sample are shown in Figure 5. The EMC/Ag specimens exhibited wave form and valley microstructure on the surface. The microstructure matches well with the EMC surface hole when the samples were cured, which provided a better mechanical force. In contrast, the surface of Ni/Pd/Au coated leadframe appeared smooth with no residue adhered to the surface (Figure 6), which explained the weakest adhesion strength. The surface morphology and elements for EMC/Cu samples with the treatment of moisture absorption are shown in Figure 7(a) and Figure 7(b), respectively. It was found that a large amount of residue was adhered to the Cu leadframe surface. Using Energy Dispersive Spectroscopy (EDS), the elements C, O, Cu, Si and Fe were detected. Residues on the Cu leadframe surface were shown to consist of epoxy, which corresponds to the best adhesion strength amongst three kinds of samples. Teck-Gyu. Kang *et al.* [7] studied the correlation between the interface segregation of copper alloys and the failure path after adhesion test. The interface adhesion strength can be decreased by segregation because the segregated atoms can cause lattice distortion at interface and thus create failure paths. In this case, it was deduced that the adhesion failure occurred at the EMC/oxide copper interface due to large amounts of copper and oxygen on the surface.

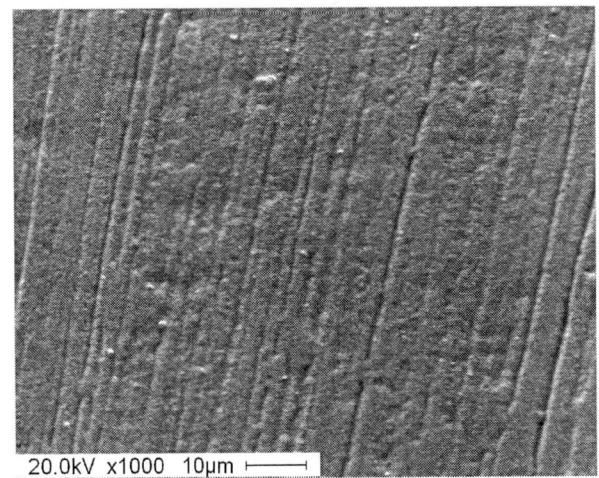

Figure 6 SEM picture of the surface of Ni/Pd/Au specimen after reflow

(a)

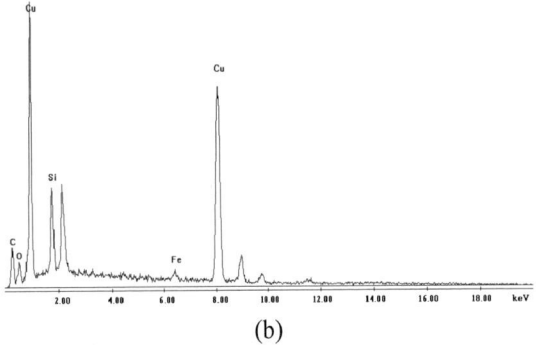

(b)

Figure 7 (a) SEM picture of the surface of Cu specimen after humidity & temperature testing (b) EDS schematic of the surface of Cu specimen after humidity & temperature testing

Figure 5 SEM picture of the surface of Ag specimen after humidity & temperature testing

Conclusion

The adhesion strength between different metal coated leadframes and EMC was measured after MSL 1 condition and 260°C reflow. Amongst the three kinds of samples, the adhesion strength of EMC/Cu specimens was the best while the EMC/Ni/Pd/Au specimens showed the poorest adhesion strength. In addition, the reflow process has great influence on the interface adhesion strength of EMC/metal specimens after the samples underwent humidity and temperature testing.

From SAM, it was found that the delamination area of different EMC/metal specimens correlated well with the adhesion strength, which also identified the relationship: the smaller the surface area that was adhered to, the weaker the measurable pull strength. SEM analysis indicated that the rough surface of EMC/Ag specimens could promote improved adhesion due to its wave form and valley microstructure. It would provide a better mechanical means for molding compound to adhere to the leadframe surface. A large quantity of epoxy residue adhered to the surface of EMC/Cu samples, which could explain the stronger adhesion strength. The adhesion behavior between EMC and leadframes will be investigated in our future work.

Acknowledgment

This work was supported by the Henkel Company's Shanghai Region-Henkel Joint Electronics Materials Research and Failure Analysis Center project. We are grateful to Mr. Wei Tan from the Henkel Company for his help in preparing the samples.

References

1. H. B. Fan, *et al.*, "A new method to Predict delamination in electronic packages," *Electronic Components and Technology Conference*, 2005, pp. 145-150.
2. Lara J. Martin, *et al.*, "Chemical and mechanical adhesion mechanisms of sputter-deposited metal on epoxy dielectric for high density interconnect printed circuit boards," *IEEE*, 2000, pp.416-424.
3. Cell K. Y. Wong, Hongwei Gu, Bing Xu, and Matthew M.F. Yuen, "A New Approach in Measuring Cu-EMC Adhesion Strength by AFM," *Components and Packaging Technologies*, VOL.29, No.3, September 2006, pp. 543-550.
4. M. Lebbai, *et al.*, "Effect of black oxide on interface adhesion between copper substrate & glob-top resin," *IEEE*, 2000, pp. 61-69.
5. Jang-Kyo KIM, *et al.*, "Interface adhesion between copper leadframe and epoxy moulding compound: effect of surface finish, oxidation and dimples," *Electronic Components and Technology Conference*, 2000, pp. 601-608.
6. Xinyu Du, Guangchao Xie, Wei Tan, Suqiong Qin and Xingming Cheng, "Stress Reduction of Epoxy Molding Compound and Its Effect on Delamination", *Proceedings of HDP'07*, 2007, pp. 50-54.
7. Teck-Gyu. Kang, Ik-Seong Park *et al.*, "Characterization of Oxidized Copper Leadframes and Copper/Epoxy Molding Compound Interface Adhesion in Plastic Package," *IEEE*, 1998, pp. 106-111.

Effect of Zn Addition on Microstructure of Sn-Bi Joint

Q. S. Zhu[1], H. Y. Song[1], H. Y. Liu[1], Z. G. Wang[1], and J. K. Shang[1, 2]

[1]Shenyang National Laboratory for Materials Science

Institute of Metal Research, Chinese Academy of Sciences, Shenyang 110016, China

[2]Department of Materials Science and Engineering

University of Illinois at Urbana-Champaign, Urbana, IL 61801, USA

*Corresponding authors: qszhu@imr.ac.cn, jkshang@imr.ac.cn

Abstract

Zn was added into Sn-Bi solder to investigate the effect of Zn on the microstructure of the alloy. Zn addition was found to refine and redistribute the Bi-rich phase. With Zn addition, a different intermetallic compound, Cu_5Zn_8 was formed at the Sn-Bi-Zn/Cu interface upon reflow. After aging, the Bi segregation was not observed at the Sn-Bi-Zn/Cu interface while a continuous Bi segregation layer appeared in the Sn-Bi/Cu interface. The Zn addition into the solder freed Sn near the interface from interfacial reaction, which suppressed the occurrence of Bi segregation layer.

1. Introduction

With the development of high-density packaging technologies, the smaller sizes and more functions in electronic components demand a high quality and reliability of solder interconnection. During recent years, in view of the toxicity of Pb, Pb-based alloys can severely cause human body and environment problems, and have been comprehensively prohibited. Accordingly, the search for Pb-free solders that have good integrated properties as the substitutes for Sn-Pb alloy has intensified worldwide in microelectronic packaging industry [1-3]. Nowadays, several Pb-free solders systems such as Sn-Zn, Sn-Ag, Sn-Cu, Sn-Bi and Sn-Ag-Cu have been developed and widely used. Among them, the eutectic Sn-Bi alloy is a possible Pb-free solder for low temperature soldering due to its low melting temperature. When devices to be soldered are sensitive to thermal damage, low temperature soldering is necessary. Low temperature soldering can also reduce the damage of thermal cycling caused by thermal expansion mismatch among various materials in an electronic package. However, the 42Sn-58Bi solder alloy also exhibits noticeable microstructural coarsening during thermal aging [4-5]. The microstructure instability at the elevated temperature can be suppressed by incorporating fine dispersoid particles into the eutectic Sn-Bi solder alloys. Moreover, since the Bi and Cu can not form intermetallic compound, the consumption of Sn will create a serious Bi segregation layer along the interface, which can greatly decrease the reliability of the solder joint.

Zn is one of very useful and cheap alloy elements, and has been paid much attention in the research of the Pb-free alloys. In addition, being very active, Zn is usually the first to react with the substrate metals to form the intermetallic compounds [6-7]. It is shown that minor Zn addition into Pb-free solders can apparently depress the growth of intermetallic compounds (IMCs), especially restraining the growth of Cu_3Sn and the formation of Kirkendall voids during isothermal aging [8-9]. Recently, Chang et al. reported that 0.5wt.% Zn doped into

the Sn-3Ag-0.5Cu-0.5Ce solder can inhibit the growth of whisker on the surface [10]. In this work, the 3wt.% Zn was added into the eutectic Sn-Bi alloy. The results have shown that the microstructure was apparently refined and the Bi segregation layer was effectively inhibited through the addition of Zn element into the alloy.

2. Experimental procedure

Ingots of the 41Sn-56Bi-3Zn alloy were prepared by melting high purity Sn, Bi and Zn (purity 99.99%) in vacuum furnace under 800°C for 60 min. The ingots were remelted and stirred at 200°C for 15min and then cast into a plate in a steel mold. The alloys were cooled down to room temperature, and then were aged for two weeks to obtain a stable microstructure. Small button samples for the microstructural observations were cut from the plates. The samples were coarsely grounded with 1μm Al_2O_3 abrasive paste and finely polished with 0.05μm colloidal silica suspension and then etched in a solution of 3% nitric acid and 97% methanol for several seconds. The observation of microstructure of the alloys were carried out under a scanning electron microscope (SEM) equipped with an energy dispersive X-ray spectrometer (EDS).

The two copper cubes were aligned and fixed before the assembly was heated in an oven where the solder was reflowed at 200°C. The samples for interfacial observation were cut from the soldered cubes. Some of them were thermal aging at 120°C for 7 days. At last, the samples were carefully polished and the interfacial microstructures were observed under SEM.

3. Results and discussion

Fig. 1 (a) shows the typical microstructure of the eutectic Sn-Bi alloy. The Sn-Bi alloy had a dual coexistence structure, consisting of a Sn-rich phase (the gray area) of several microns in size and a Bi-rich phase (the white area). As shown in Fig. 1 (b), a small addition of Zn into the Sn-Bi alloy produced a relatively oriented microstructure. The dark needle-like Zn-rich phase was uniformly dispersed within the eutectic network. From the magnified image in Fig. 1 (c), it is seen that the individual Bi-rich phase (the white area) had a rod or needle shape, and these phases formed a fishbone-like formation. The mean intercept length of Bi-rich phase in Sn-Bi-Zn alloy was notably smaller than that in the eutectic Sn-Bi alloy. That is, the Sn-Bi-Zn exhibited a finer structure compared to the eutectic Sn-Bi alloy. According to the Sn-Bi-Zn phase diagram, the Sn-Bi eutectics form after the primary crystallized Zn-rich phase. It appears that the microstructure of eutectic Sn-Bi alloy may be strongly dependent on the primary Zn-rich phase.

978-1-4244-4658-2/09 $25.00 © 2009 IEEE

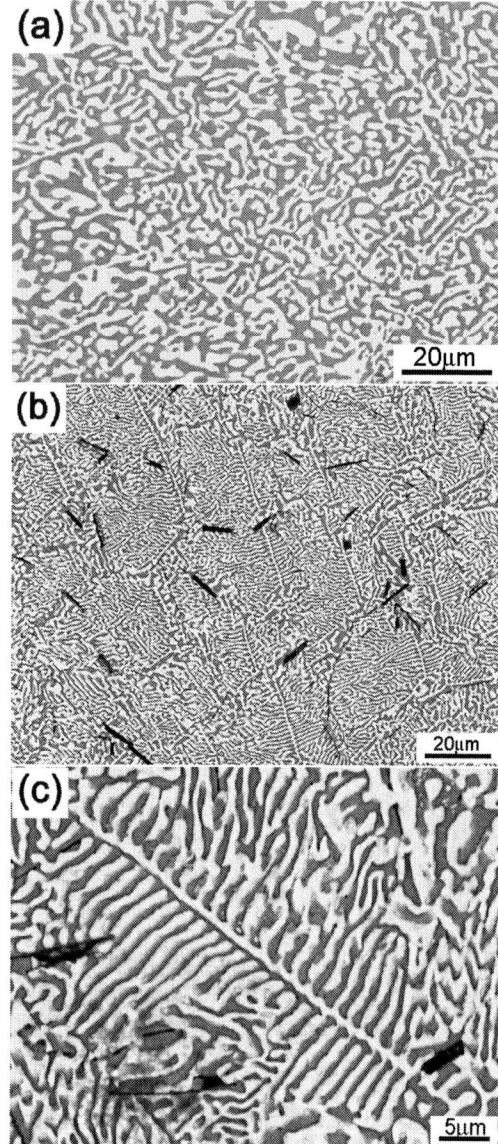

Figure 1 Microstructure of bulk materials: (a) eutectic Sn-Bi alloy (b) Sn-Bi-Zn alloy (c) magnified image

Fig. 2 (a) shows the interfacial microstructure between the eutectic Sn-Bi solder and Cu substrate in the as-reflowed condition. During the reflowing process, Cu reacted selectively with the liquid Sn to form a thin layer of Cu_6Sn_5 intermetallic compound phase. The thickness of the IMC layer is about 1~2μm. After aging for 7 days, the microstructure of the solder was obviously coarsened and the interfacial IMC phase grew to above 7μm, as shown in Fig. 2 (b). A thinner Cu_3Sn IMC was also detected between the Cu and Cu_6Sn_5 IMC. These observations were consistent with previous studies on SnBi solder on Cu substrate [11-12]. Since the Bi element does not form IMC phase with Cu, the residual Bi from eutectic Sn-Bi solder had to accumulate along the IMC/solder interface, when Sn was consumed by the interfacial reactions. As shown in Fig. 2 (b), the Bi phase had grown to a continuous layer. Since the Bi phase has a

very low toughness, such a Bi segregation layer will greatly deteriorate the reliability of the interconnection.

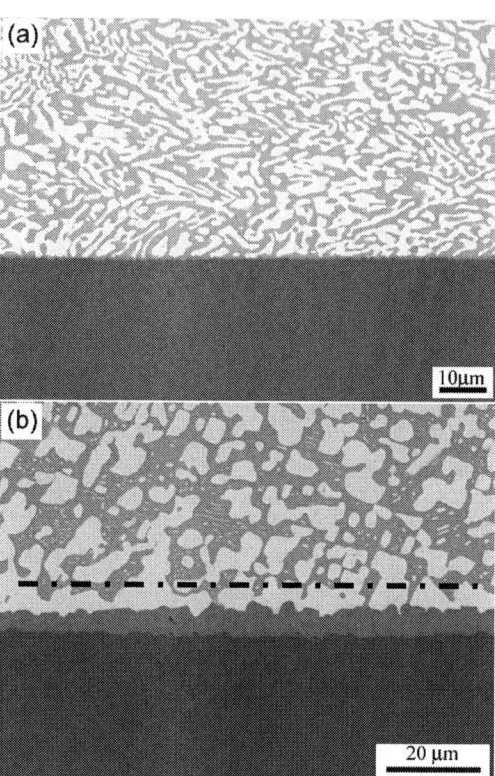

Figure 2 Interfacial microstructure between Sn-Bi and Cu substrate: (a) after reflowing for 2 minutes (b) aging at 120°C for 7 days.

Fig. 3 shows the cross-sectional SEM images of the interfaces between Sn-Bi-Zn solder and Cu substrate. It is shown that a very thin and planar morphology IMC layer formed along the interface after reflowing for 2 minutes, as shown in Fig. 2(a). The average thickness of this planar IMC layer was less than 1μm. EDS analysis indicated that the IMC layer contained Zn and Cu elements. The Zn element existed in the form of long whiskers, some of which were originated at the interface and extended far into the solder. These Zn whiskers may supply Zn atoms to form the Cu-Zn IMC at the interface.

After isothermal solid-state aging for 7 days, the average thickness of the IMC layer in the Sn-Bi-Zn/Cu interface had a significant increase, as shown in Fig. 2 (b). The average thickness was about 10μm. The result of EDS analysis demonstrated that the ratio of Zn and Cu was close to be 8:5, and this IMC was further determined to be Cu_5Zn_8 phase. In addition, there was not Sn element to be detected in the thickness IMC layer. This result is consistent with the previous study, where the formation of Cu-Zn IMC proceeded the formation of the Cu-Sn IMC when the Zn element content was above a certain value for the Sn-Ag-Zn alloy [13]. With the consumption of Zn element near the interface, a wide Zn-free zone was left, as shown in Fig. 2(b). The width of the Zn-free zone continuously increased when the aging time was further prolonged.

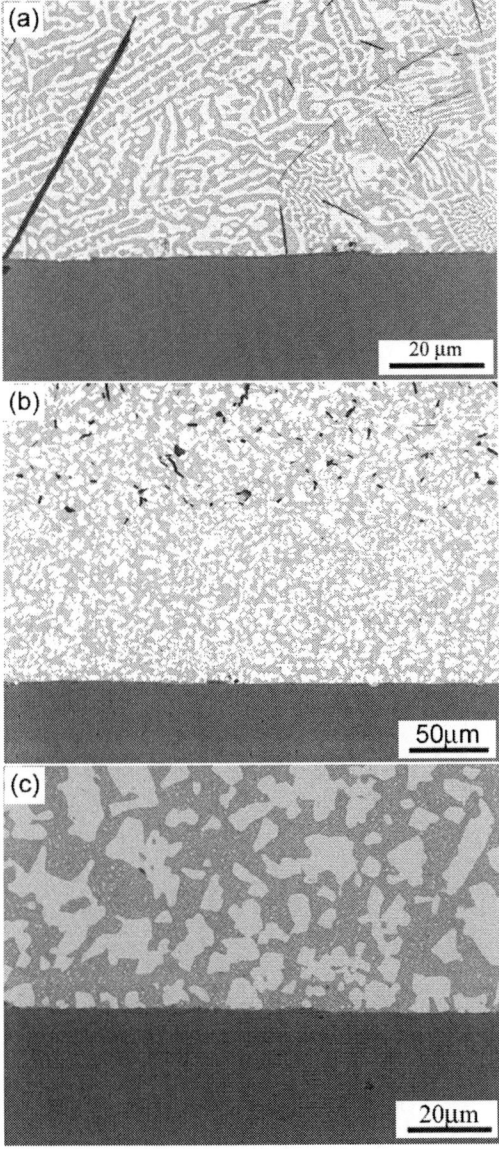

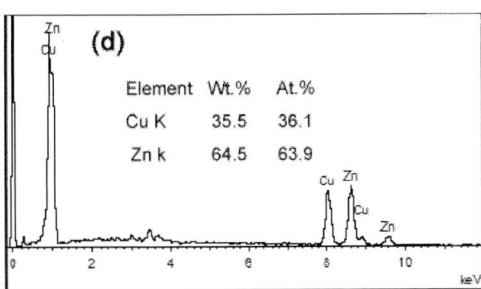

Figure 3 Interfacial microstructure between Sn-Bi-Zn and Cu substrate: (a) after reflowing for 2 minutes (b) aging at 120°C for 7 days (c) magnified image (d) EDAX spectrum of the IMC.

The Zn-rich phase had not been observed in the magnified image in Fig. 3 (c). Besides, there was not a trace of Bi phase segregation at the interface. The microstructure near the interface was similar to that of the interior, which indicated that the Sn had not reacted with the Cu substrate. Of course, the Sn may begin to be consumed when the Zn is consumed away or the distance between the Zn phase and interface is long enough. Undoubtedly, the addition of Zn into the Sn-Bi alloy effectively inhibited the interfacial Bi phase segregation that resulted from the Sn consumption during aging.

Conclusions

By adding Zn into the Sn-Bi alloy, the alloy exhibited a refined microstructure. After reflowing, the Cu_5Zn_8 IMC instead of Cu_6Sn_5 occurred at the interface for the Sn-Bi-Zn/Cu connection. While the aged Sn-Bi/Cu interface exhibited a notable Bi segregation layer, a beneficial effect of the Zn addition was to suppress the Bi segregation by freeing Sn from the interfacial reactions. The present research may offer a solution to present Bi segregation in the application of Sn-Bi solder system.

Acknowledgments

This work was financially supported by National Basic Research Program of China under grant No. 2004CB619306

References

1. K. Zeng, K. N. Tu, "Six cases of reliability study of Pb-free solder joints in electronic packaging technology," *Materials Science and Engineering: R: Reports*, 2002 (38), pp. 55-105.

2. M. Abtew, G. Selvaduray, "Lead-free solders in microelectronics," *Materials Science & Engineering R-Reports*, 2000 (27), pp. 95-141.

3. E. P. Wood, K.L. Nimmo, "In search of new lead-free electronic solders," *Journal of Electronic Materials*, 1994 (23), pp. 709-713.

4. M. McCormack, S. Jin, and G. W. Kammlott, "Suppression of microstructural coarsening and creep deformation in a lead-free solder," *Applied Physics Letters*, 1994 (64), pp. 580-582.

5. C. H. Raeder, L. E. Felton, V. A. Tanzi, and D. B. Knorr, "The effect of aging on microstructure, room temperature deformation, and fracture of SnBi/Cu solder joints," *Journal of Electronic Materials*, 1994 (23), pp. 611-617.

6. K. Suganuma, K. Niihara, T. Shoutoku, and Y. Nakamura, "Wetting and interface microstructure between Sn-Zn binary alloys and Cu," *Journal of Materials Research*, 1998 (13), pp. 2859-2865.

7. I. Shohji, T. Nakamura, F. Mori, and S. Fujiuchi, "Interface reaction and mechanical properties of lead-free Sn-Zn alloy/Cu joints," *Materials Transactions*, 2002 (43), pp.1797-1801.

8. F. J. Wang, F. Gao, X. Ma, Y. Y. Qian, "Depressing effect of 0.2wt.%Zn addition into Sn-3.0Ag-0.5Cu solder alloy on the intermetallic growth with Cu substrate during isothermal aging," *Journal of Electronic Materials*. 2006 (35), pp.1818-24.

9. M. G. Cho, S. K. Kang, D. Y. Shih, H. M. Lee. "Effects of Minor Additions of Zn on Interfacial Reactions of Sn-Ag-

Cu and Sn-Cu Solders with Various Cu Substrates during Thermal Aging," *Journal of Electronic Materials*. 2007 (36), pp.1501-9.

10. T. H. Chuang, H. J. Lin, "Inhibition of Whisker Growth on the Surface of Sn-3Ag-0.5Cu-0.5Ce Solder Alloyed with Zn," *Journal of Electronic Materials*. 2009(38), pp.420-4.

11. P. L. Liu, J. K, "Shang, Interfacial segregation of bismuth in copper/tin-bismuth solder interconnect," *Scripta Materialia*, 2001 (44), pp. 1019-1023.

12. Q. S. Zhu, Z. F. Zhang, Z. G. Wang, and J. K. Shang, "Inhibition of interfacial embrittlement at SnBi/Cu single crystal by electrodeposited Ag film," *Journal of Materials Research*, 2008, 3(1), pp. 78-82.

13. Y. K. Jee, Y. H. Ko, J. Yu, "Effect of Zn on the intermetallics formation and reliability of Sn-3.5Ag solder on a Cu pad," *Journal of Materials Research*, 2007 (22), pp. 1879-1887.

Deformation Characteristics of Sn-3Ag-0.5Cu/Cu/Ni-xCu/Ti Joints after Mechanical Test

Cung-Nan Peng[1], Jenq-Gong Duh[1,*]

[1]Department of Materials Science and Engineering, Tsing Hua University, Hsinchu, Taiwan, China

*corresponding author: E-mail: jgd@mx.nthu.edu.tw Fax: 886-3-5712686

Abstract

In BGA packages and flip-chip packages, a Ni layer is often used in the under bump metallurgy (UBM) to serve as a diffusion barrier. However, unstable phase of Ni_3Sn_4 compound formed between Ni UBM and Sn-rich solder, leading to a critical spalling issue. To avoid the formation of a $(Ni, Cu)_3Sn_4$ interlayer between Ni versus the solder, it is known that to increase Cu concentration in the solder was an effective way for the reactive interface [1-8].

In this study, a systematic design of UBM with Cu/Ni-xCu/Ti (x = 0-20wt.%) were adopted to react with Sn-3Ag-0.5Cu ball (φ = 300μm). Multi-reflow altered the morphology of the IMCs formed at the interface. FE-EPMA was used to quantitatively analyze intermetallic compounds and to observe fracture surface. Furthermore, the pull and shear tests were employed to measure the mechanical properties of Sn-3Ag-0.5Cu/Cu/Ni-xCu/Ti joints. It was revealed that the shear strength increased with Cu contents in Ni layer and reflow times. The thickness of $(Cu,Ni)_6Sn_5$ IMC affected the shear strength. The shear mode changed from ductile (failure inside the bulk solder) to partially brittle (failure at the solder/IMC or IMC/UBM interfaces) in the Sn-3Ag-0.5Cu/Cu/Ni-xCu/Ti (x = 0-20wt.%) joints with the reflow times up to 5. The result of pull test was different from that of shear test, and force direction of tool was critical. The pull strength increased with the number of reflow times. Although the thickness of $(Cu, Ni)_6Sn_5$ increased with reflow times, it was too thin to affect pull strength. The Cu concentration of solder did play a critical role in pull strength. It was argued that the Cu concentration of solder could decrease Sn grain size, leading to the higher pull strength. The similar tread of shear and pull results was also found in Chen's [9] and Lehma's [10] studies. The results of this study suggest that Cu-rich Ni-xCu UBM can be used to suppress interfacial spalling and to improve both shear and pull strength in the solder joint.

Results and discussion

Figure 1 was the morphological cross section of Sn-3Ag-0.5Cu/Ni-xCu/Ni after multiple reflows (x = 0-20wt.%). $(Cu, Ni)_6Sn_5$ IMC were formed up to 5 reflow. The thickness of $(Cu, Ni)_6Sn_5$ between the solder and Cu/Ni-xCu/Ti joints increased with the number of reflow times.

The Cu concentration in Sn near the $(Cu, Ni)_6Sn_5$ IMC was obtained through detailed quantitative analysis, as shown in Table 1. The lowest Cu concentration was 0.62wt.%, and the highest was 0.77wt.%.

According to the phase diagram by Lin et al. [11] and Zeng et al. [12-13], an enlarged Sn-corner of Sn-Cu-Ni ternary isotherm was proposed to illustrate the formation of $(Cu, Ni)_6Sn_5$ IMC. When the concentration of Cu in Sn was greater than 0.6wt.%, $(Cu,Ni)_6Sn_5$ IMC might form. The

thickness of $(Cu, Ni)_6Sn_5$ was increased with the number of reflow times, as shown in Fig. 2.

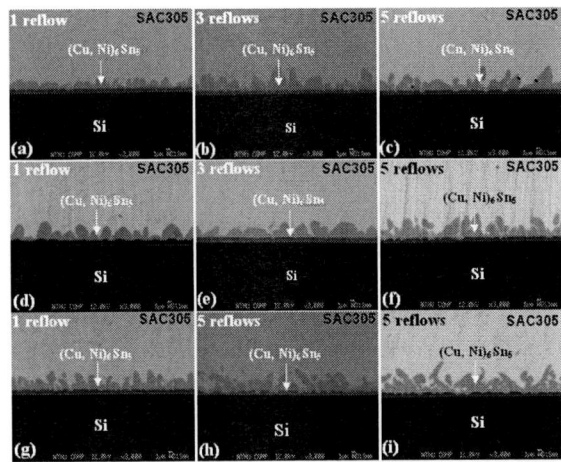

Figure 1 Interfacial morphology of the Sn-3Ag-0.5Cu solder joints sheared at 240°C after various reflows. (a) 1 reflow for Cu/Ni/Ti, (b) 3 reflows for Cu/Ni/Ti, (c) 5 reflows for Cu/Ni/Ti, (d) 1 reflow for Cu/Ni-5Cu/Ti, (e) 3 reflows for Cu/Ni-5Cu/Ti, (f) 5 reflows for Cu/Ni-5Cu/Ti, (g) 1 reflow for Cu/Ni-20Cu/Ti, (h) 3 reflows for Cu/Ni-20Cu/Ti, and (i) 5 reflows for Cu/Ni-20Cu/Ti.

Table 1 Quantitative analysis results in the Sn-3Ag-0.5Cu/ Cu/Ni-xCu/Ti after various reflows. (x = 0-20wt.%)

Reflow times	Cu concentration in Sn (wt.%)		
	SAC305/Cu/Ni/Ti	SAC305/Cu/Ni-5Cu/Ti	SAC305/Cu/Ni-20Cu/Ti
1	0.68	0.72	0.77
3	0.63	0.67	0.68
5	0.62	0.64	0.65

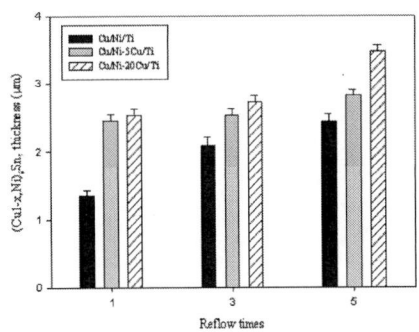

Figure 2 Thickness of the $(Cu_z,Ni_{1-z})_6Sn_5$ IMC formed between the Sn-3Ag-0.5Cu solder and Cu/Ni-xCu/Ti joints (x = 0, 5, 20wt.%) after various reflows.

Practically, the solder joints often suffer shearing stress in working circumstance because of the thermal expansion coefficient between PCB and integrated circuit. It will cause microcrack in the Cu_6Sn_5 compound at the interface because of the highest hardness among the solder, Ag_3Sn particles and Cu_6Sn_5 compound [14]. A summary of the shear strength in Sn-3Ag-0.5Cu/Cu/Ni-Cu/Ti joints after multi-reflow during shear testing is provided in Fig. 3.

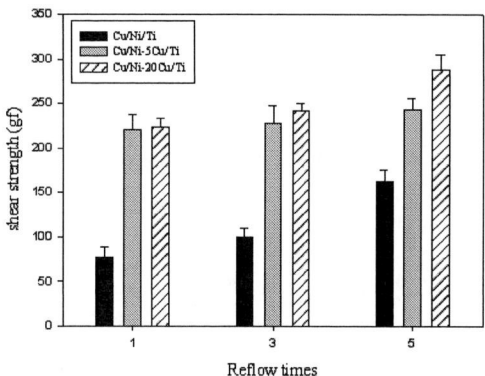

Figure 3 Shear strength of the Sn-3Ag-0.5Cu/Cu/Ni-xCu/Ti joints (x = 0, 5, 20wt.%) after various reflows.

The shear strength increased with Cu concentration and the number of reflow times, and the shear strength of Sn-3Ag-0.5Cu/Cu/Ni-20Cu/Ti joint increased 8.7% and 30% after 3 and 5 reflows, respectively, as compared to 1 reflow. With the maximum thickness of $(Cu, Ni)_6Sn_5$ IMC around 5μm, the shear height was close to $(Cu, Ni)6Sn5$ IMC after ball shear test, as shown in Fig. 2. It was argued that the resistance was provided by $(Cu, Ni)_6Sn_5$ IMC in ball shear test, and the creep resistance increased with the thickness of $(Cu, Ni)_6Sn_5$ IMC.

The solder joints exhibiting undesirable deformation measured from pull and shear tests was due to test software limit. Solder ball shear samples typically have a flattened surface at the contact region between the shear tool and solder ball, with a related bulge at the opposite non-contact surface. The extent of this deformation varies not only with solder composition, and shear tool placement, but also with test speed, due to the highly strain-rate dependent properties of most solder alloys. A typical concern is raised to discuss the effect of this non-shear deformation on the fracture mechanics and resultant interface. This study was aimed to provide some answers to this query.

To quantify the structural integrity of a given interconnect in relation to its ability to a low speed shear impact, a scaled rank-based on visual inspection of the sheared interconnect surface was developed for low speed shear test. In this ranking system, sheared interconnect surfaces were assigned by a number percentage of brittle fracture based on the amount of the surface that shows brittle fracture behavior. Fig. 4 (a) and (b) illustrate this grading methodology.

A brief graphical summary of solder ball shear test result is shown in Fig. 5. The deformation behavior after shear test was explained by shear fracture surface. Not only pure ductility but also brittleness was formed in the joint after

shear test. The most fracture mode failures were mixed type. For shear testing, the percentage of brittle fracture mode failures increased with the thickness of $(Cu, Ni)_6Sn_5$ IMC.

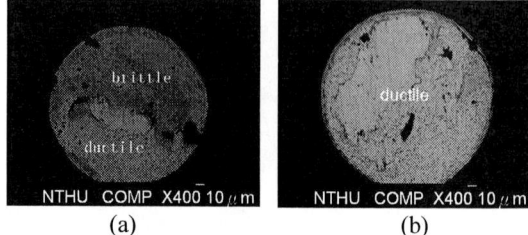

(a)   (b)

Figure 4 Shear failure modes under low speed: (a) 50% ductile, and (b) 100% ductile.

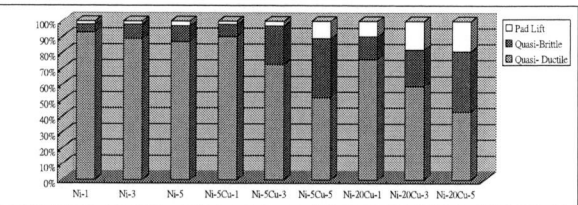

Figure 5 Failure mode distribution of the Sn-3Ag-0.5Cu/Cu/Ni-xCu/Ti joints in ball shear tests after various reflows. (Ni-5Cu-3 means the case of Ni-5Cu UBM after 3 reflows.)

In addition, ball pull method is widely acknowledged by the industry in the recent years to fully determine the weak interface and the joint strength. Ball pull is found to be more stringent than ball shear because of the pulling mechanicism that minimizes ball deformation above the bond site, renders the bond not to be supported by the solder pad cavity wall, and thus exposes the true bond strength. To evaluate the effect of the Cu concentration near the solder/$(Cu,Ni)_6Sn_5$ interface and the fracture mode of solder bumps, a bump pull test was carried out. The result of pull test was different from that of shear test. The pull test is a commonly employed method for estimating the strength of the joint interface. Fig. 6 represents the mechanical behavior of Sn-3Ag-0.5Cu/UBM joint during pull test after multiple reflows.

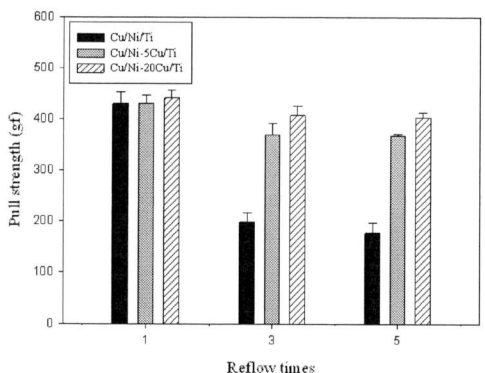

Figure 6 Pull strength of Sn-3Ag-0.5Cu/Cu/Ni-xCu/Ti joints (x = 0, 5, 20wt.%) after various reflows.

It was observed that pull strength increased with the number of reflow times. The normal force is given to solder balls during ball pull test process, which makes the solder detach from interface. Although the thickness of $(Cu,Ni)_6Sn_5$ was increased with the number of reflow times (Fig. 2), the thickness of $(Cu,Ni)_6Sn_5$ IMC was too thin to affect pull strength. The Cu concentration of solder is crucial in the pull strength. In this study, the Cu concentration of solder in the Sn-3Ag-0.5Cu/Cu/Ni-20Cu/Ti joint was 0.77wt.%, 0.68wt.%, and 0.65wt.% after 1, 3, and 5 reflows, respectively. It was reported that the Cu concentration of solder could decrease Sn grain size, which led to the higher pull strength [9, 14].

The results above indicate that the Cu content in Ni layer affects the low speed pull strength of joint after multiple reflows. However, due to the vertical pull with low speed cold bond pull as opposed to horizontal shear with low speed shear, the fracture modes appeared different for low speed bond pull. Fig. 7 (a) and (b) illustrate fracture modes from the low speed bond pull testing.

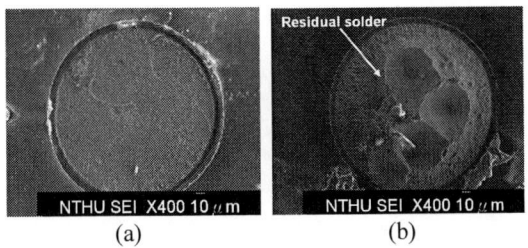

(a)   (b)

Figure 7 Pull failure modes under low speed: (a) 0% Residual solder (b) 100% Residual solder

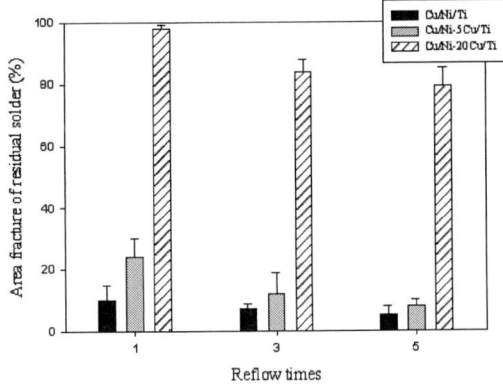

Figure 8 Estimated areas of percentage of residual solder by the pull test of the Sn-3Ag-0.5Cu/Cu/Ni-xCu/Ti joints after various reflows.

In the mechanical testing, although the extent of deformation is adjustable through machine settings and clamping jaw geometries, some non-tensile deformations of the pull samples are unavoidable, even in an optimized set-up. In addition, it was argued that the annular deformation due to the jaw clamping would significantly alter the fracture mechanics, uniformity, and repeatability. The nonideal solder ball deformation of the pull test samples might be of less

significance than that for the shear test samples [15-16]. Depending on the test conditions, the low speed bond pull test results could range from 0-100% brittle as with low speed shear. The residual solder increased with increasing pull strength and decreased with increasing reflow times, as shown in Fig. 8.

The degradation of the joining ductility, as demonstrated by the increase of the brittle failure, is primarily a result of the morphological change and the intermetallics growth in the solder, as shown in Fig. 9.

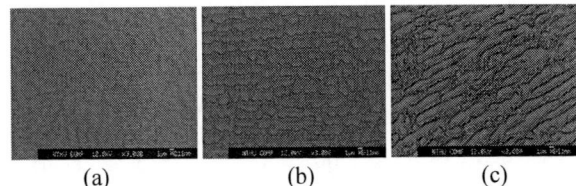

(a)   (b)   (c)

Fig. 9 SEM micrographs of the Sn-3Ag-0.5Cu/Cu/Ni-xCu/Ti joints at 240°C after 1 reflow (a) Sn-3Ag-0.5Cu/Cu/Ni/Ti joint, (b) Sn-3Ag-0.5Cu/Cu/Ni-5Cu/Ti joint, and (c) Sn-3Ag-0.5Cu/Cu/Ni-20Cu/Ti joint.

With increasing concentrations of Cu in the Sn-3Ag-0.5Cu/Cu/Ni-xCu/Ti joints, the ratio of the eutectic microconstituent to the primary β phase increases. The β phase provided the observed resistance in the ball pull test. Therefore, most residual solders were observed in the Sn-3Ag-0.5Cu/Cu/Ni-20Cu/Ti joint. In contrast, a slight change in the residual solder in Sn-3Ag-0.5Cu/Cu/Ni/Ti joint and Sn-3Ag-0.5Cu/Cu/Ni-5Cu/Ti joint were revealed.

Conclusions

This study investigated the failure mechanism during low-speed ball shear/pull testing in the Sn-3Ag-0.5Cu/Cu/Ni-xCu/Ti (x = 0-20wt.%) assembly. The Sn-3Ag-0.5Cu/Cu/Ni-20Cu/Ti joint exhibited the maximum strength. The shear strength increased with the number of reflows, while the pull strength decreased with an increase in the number of reflows. It was argued that the $(Cu,Ni)_6Sn_5$ IMC provided the observed resistance in the ball shear test, and the resistance increased with the thickness of the $(Cu, Ni)_6Sn_5$ IMC. The percentage of brittle fracture mode failures increased with the thickness of the $(Cu,Ni)_6Sn_5$ IMC. However, the amount of residual solder decreased with an increase in the number of reflows. The results of both shear test and pull test after multiple reflows demonstrated that Sn-3Ag-0.5Cu/Cu/Ni-20Cu/Ti joints exhibited better optimal mechanical properties among all the joints investigated.

Acknowledgments

The financial support from Science Council of Taiwan, under the contract NSC-97-2811-E-007-022 is acknowledged.

References

1. F. Q. Li, C. Q. Wang, Y. H. Tian, "Investigation of interfacial reaction between eutectic Sn–Pb solder droplet and Au/Ni/Cu pad," *Materials Science and Technology*, Vol. 24 (2008) pp. 744.

2. H. T. Chen, C. Q. Wang, M. Y. Li, D. W. Tian, "Effect of Cu diffusion through Ni on the interfacial reactions of Sn3.5Ag0.75Cu and SnPb solders with Au/Ni/Cu substrate during aging," *Mater Letters*, Vol. 60 (2006) pp. 13.

3. H. J. Lin, T. H. Chuang, "Intermetallic Reactions in Reflowed and Aged Sn-9Zn Solder Ball Grid Array Packages with Au/Ni/Cu and Ag/Cu Pads," *Journal of Electronic Materials*, Vol. 35 (2006) pp. 154.

4. Y. S. Lai, C. W. Lee, Y. T. Chiu, Y. H. Shao, "Electromigration of 96.5Sn-3Ag-0.5Cu Flip-chip Solder Bumps Bonded on Substrate Pads of Au/Ni/Cu or Cu Metallization," *2006 Electronic Components and Technology Conference*, pp. 641.

5. Sharif, Y. C. Chan, "Effect of Reaction Time on Mechanical Strength of the Interface Formed between the Sn-Zn(-Bi) Solder and the Au/Ni/Cu Bond Pad," *Journal of Electronic Materials*, Vol. 35 (2006) pp. 1812.

6. H. Oppermann, R. Kalicki, S. Anhoeck, C. Kallmayer, M. Klein, R. Aschenbrenner, H. Reichl, "Reliability Investigations of Hard Core Solder Bumps Using Mechanical Palladium Bumps and SnPb Solder," *IEEE Transactions on Electronics Packaging Manufacturing*, Vol. 25 (2002) pp. 210.

7. T. Takenaka, M. Kajihara, N. Kurokawa, K. Sakamoto, "Reactive diffusion between Pd and Sn at solid-state temperatures," *Materials Science and Engineering: A*, 406 (2005) pp. 134.

8. Q. Zhang, A. Dasgupta, D. Nelson, H. Pallavicini, "Systematic Study on Thermo-Mechanical Durability of Pb-Free Assemblies: Experiments and FE Analysis," *Journal of Electronic Packaging*, Vol. 127 (2005) pp. 415.

9. T. Chen, I. Dutta, "Effect of Ag and Cu Concentrations on the Creep Behavior of Sn-Based Solders," *Journal of Electronic Materials*, Vol. 37 (2008) pp. 347.

10. L. P. Lehman, S. N. Athavale, T. Z. Fullem, A. C. Giamis, R. K. Kinyanjui, M. Lowenstein, K. Mather, R. Patel, D. Rae, J. Wang, Y. Xing, L. Zavalij, P. Borgesen, E. J. Cotts, "Growth of Sn and intermetallic compounds in Sn-Ag-Cu solder," *Journal of Electronic Materials*, Vol. 33 (2004) pp. 1429.

11. C. Y. Li, J. G. Duh, "Phase equilibria in the Sn-rich corner of the Sn–Cu–Ni ternary alloy system at 240°C," *Journal of Material Research*, Vol. 20 (2005) pp. 3118.

12. M. Li, F. Zhang, W. T. Chen, K. Zeng, K. N. Tu, H. Balkan, P. Elenius, "Interfacial microstructure evolution between eutectic SnAgCu solder and Al/Ni(V)/Cu thin films," *Journal of Material Research*, Vol. 17 (2002) pp. 1612.

13. K. Zeng, V. Vuorinen, J. K. Kivilahti, "Intermetallic Reactions between Lead-free SnAgCu Solder and Ni(P)/Au Surface Finish on PWBs," *2001 Electronic Components and Technology Conference*, pp. 1384.

14. Ahmed Sharif, Y. C. Chan, "Effect of indium addition in Sn-rich solder on the dissolution of Cu metallization," *Journal of Alloys and Compounds*, 390: (2005) pp. 67.

15. J. W. Kim, J. Joo, D. J. Quesnel, S. B. Jung, "Correlation between displacement rate and shear force in shear test of Sn-Pb and lead free solder joints," *Materials Science and Technology*, Vol. 21 (2005) pp. 373.

16. J. Liang, N. Dariavach, P. Callahan, "Effects of thermal history on the intermetallic growth and mechanical strength of Pb-free and Sn-Pb BGA solder balls," *Soldering & Surface Mount Technology*, Vol. 19 (2007) pp. 4.

Studies on Microstructure of Epoxy Molding Compound (EMC)-Leadframe Interface after Environmental Aging

Xiuzhen Lu[1], Li Xu[1], Huaxiang Lai[1], Xinyu Du[2], Johan Liu[1, 3, *] and Zhaonian Cheng[1]

[1]Key Laboratory of Advanced Display and System Applications, Ministry of Education
and SMIT Center, School of Mechatronics Engineering and Automation,
Shanghai University, Shanghai 200072, China
[2]Henkel Corporation, 15350 Barranca Pkwy, Irvine CA 92618, USA
[3]SMIT Center & Bionano Systems Laboratory, Department of Microtechnology and Nanoscience,
Chalmers University of Technology, SE-412 96 Gothenburg, Sweden
*E-mail: jliu@chalmers.se

Abstract

The Leadframe-Epoxy molding compound (EMC) interface is known to be one of the weakest interfaces in an electronic packaging exhibiting delamination during reliability test. Interfaces of EMC and leadframes with different metal coatings exhibit different failure mode behavior after environmental aging because of the different adhesion strengths. In this paper, the interface microstructure of EMC-leadframes with different metal coatings was studied using Scanning Electron Microscopy (SEM). The leadframes used in this study were copper, copper with Ag coating, Ni coating and Ni/Pd/Au coating. The results of tab pull testing showed the adhesion strength of the EMC/copper leadframe interface was strongest while that of the EMC/leadframe with Ni/Pd/Au coating interface was weakest. Little delamination appeared on the EMC/Cu sample, including the samples after the moisture and reflow treatment. Delamination appeared on the interface of EMC/leadframe with Ag coating after treatment, and serious delamination was found on the interface of EMC/leadframe with Ni coating. The SEM micrographs showed that there were some microcracks between the filler of EMC and the leadframe.

Introduction

Epoxy molding compound (EMC) used in semiconductor and IC package has become the leading packaging material in today's microelectronic packaging industry. The reliability of electronic package is strongly influenced by the properties of the interface between EMC and the leadframe. Due to the inherent differences in thermo-mechanical properties and thermal mismatches between the EMC and leadframe material, the adhesion between them is generally poor. Lack of adhesion between leadframe and epoxy molding compound (EMC) is a prime reason for the failure of EMC-leadframe interfaces. Especially after moisture absorption, cracking may occur during the subsequent reflow process because of the vapor pressure generated in the space between the EMC and leadframe. With the increase of reflow temperature resulting from the application of lead-free solder, the reliability of the interface between the EMC and leadframe is more susceptible to moisture [1-2]. Adhesion between EMC and Cu leadframe were though to be related to the thickness and type of oxide of Cu [3-4]. Most of the leadframes need to be coated with different metal films before molding due to the electronic and mechanical performance of the devices. The reliability of the bond between EMC and metal coated Cu leadframes needs further study after environmental loading. In this paper, the microstructure of the interface between EMC and Cu, Ag coated copper and Ni/Pd/Au coated is studied by SEM.

Experimental Procedure

1. Specimen preparation

The leadframes used in this study were Cu, Ag coated Cu, Ni coated Cu and Ni/Pd/Au coated Cu. All the samples were prepared by Henkel Company and dried at 80 for 96 hours before treatment. They were divided into 4 groups for the next treatment, which is shown in Table 1. Every group consisted of samples with 4 types of leadframes. MSL1 (85°C, 85%, 168hours) was selected as the condition of moisture absorption testing. The peak temperature of reflow was 260°C according to the application of lead-free solder used in the electronic devices.

Table 1 Treatments for different samples

No.	Treatment
A	No treatment
B	Moisture absorption
C	Reflow
D	Moisture absorption & Reflow

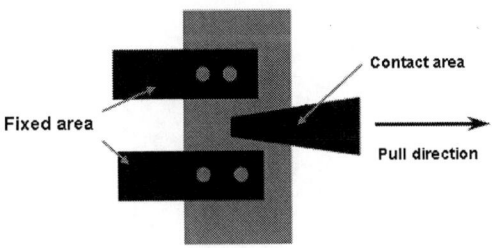

Fig. 1 Tab pull adhesion testing

2. Mechanical Adhesion Testing

Adhesion strength was measured using tab pull testing (Fig. 1). The gray area indicates EMC, and the black area indicates leadframe. The overlap area is the adhesion area between the EMC and leadframe. The contact area was pulled open during the tab pull testing.

3. SEM Experiment

The microstructure of the interface between the EMC and leadframe was observed using a JSM-6700F SEM instrument after the preparation of cross section samples. The interface between the filler and leadframe was observed in detail.

Result and Discussion

1. Effect of coating material

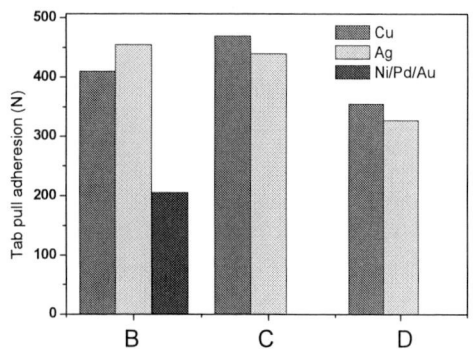

Fig. 2 Adhesion to Cu, Ag and Ni/Pd/Au of EMC

The results of tab pull testing shown in Fig. 2 indicate that the adhesion to Cu leadframes without coating is the strongest, and the adhesion to Ni/Pd/Au coated Cu leadframes is the weakest. The EMC/Ni/Pd/Au samples subjected to condition C and D broke during the preload of pull testing, so no data was obtained for these EMC/Ni/Pd/Au specimens. The poor adhesion strength of the interface between EMC and Ni/Pd/Au coated leadframe is related to the Ni coating. Therefore, we use EMC/Ni coated leadframe specimens for the microstructure study.

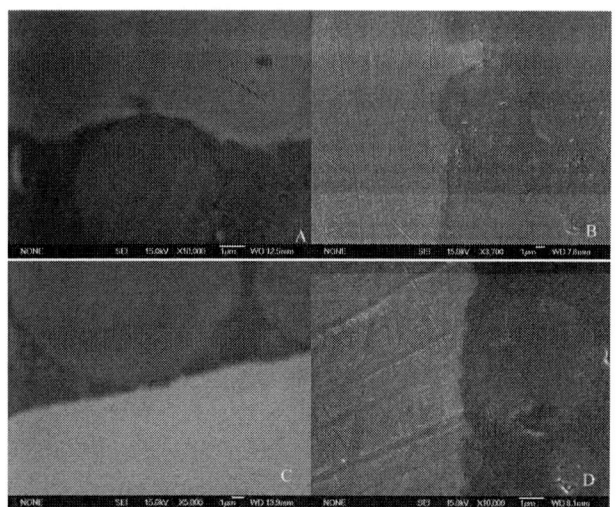

Fig. 3 SEM images of interface of EMC/Cu specimens subjected to A) no treatment, B) moisture absorption testing, C) reflow, D) moisture absorption & reflow

Fig. 3 shows the microstructure of interface of EMC/Cu specimens subjected to no treatment, moisture absorption testing, reflow and moisture absorption & reflow respectively. No delamination appeared in EMC/Cu samples subjected to treatment A, B, C and D. The interface between EMC and Cu leadframes is very smooth, which indicates the adhesion between EMC and Cu is very tight.

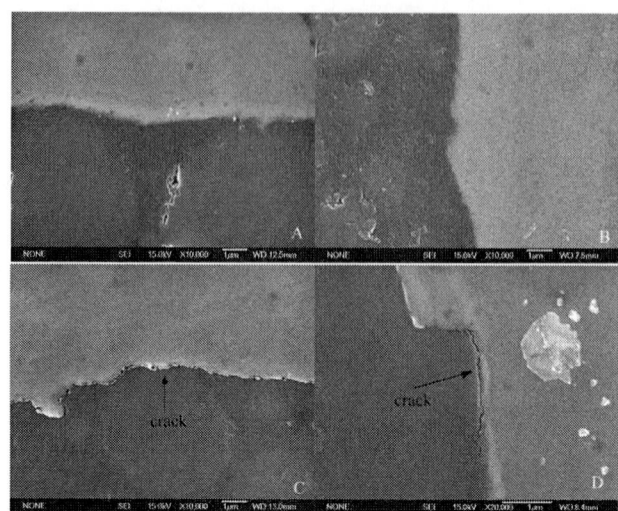

Fig. 4 SEM images of interface of EMC/Ag specimens subjected to A) no treatment, B) moisture absorption testing, C) reflow, D) moisture absorption & reflow

No delamination was found on the interface of the EMC/Ag specimens, which is shown in fig. 4A and 4B. The perfect interface demonstrates moisture absorption itself does not have much effect on adhesion of EMC/Ag if there is no subsequent reflow process. Obvious crack appeared in the SEM micrograph of interface of EMC/Ag subjected to reflow or reflow after moisture absorption testing, which is shown in Fig. 4C and 4D. It is well known that thermal mismatch exists between EMC and leadframes, which results in thermal stress on the interface when the environmental temperature changes [5-6]. Cracks, delamination and other failures appear when the thermal stress reaches a certain degree. The main failure cause for the EMC/Ag specimen subjected to reflow is thought to be the thermal mismatch between the EMC and leadframe with Ag coating. There are two ways for ingress of water into semiconductor devices. One is moisture absorption by EMC; the other is moisture absorption through the interface of the EMC and leadframe [7]. The poorer the adhesion of the interface is, the more serious the moisture absorption is. Moisture transforms into vapor at high temperatures during the reflow process. With the combined effect of thermal mechanical stress and high vapor pressure, delamination appears in the EMC and leadframe interface, which is one of the main causes of cracks in EMC/Ag specimens subjected to reflow after moisture absorption.

Delamination was found on the EMC/Ni interface without treatment (shown in Fig. 5A). Serious delamination appeared on the interface of EMC/Ni samples after moisture absorption testing (Fig. 5B) or reflow processing (Fig. 5C). Fracture zones formed in the EMC side, which was evidence of the very poor adhesion strength between the EMC and Ni coating. The sample of EMC/Ni subjected to reflow after moisture absorption testing broke during the preparation of cross section specimens because of its serious delamination.

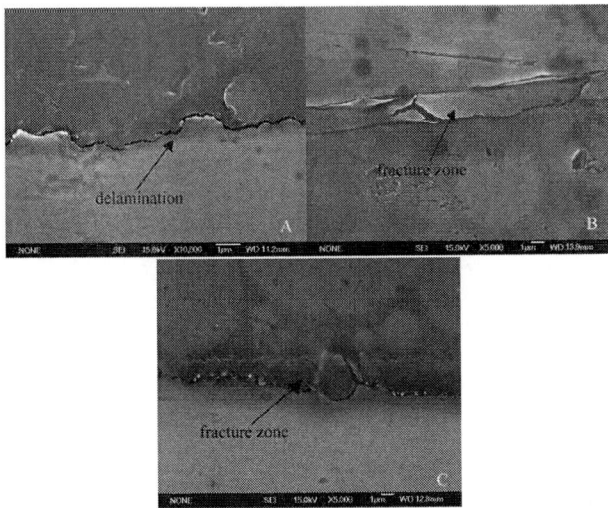

Fig. 5 SEM images of interface of EMC/Ni specimens subjected to A) no treatment, B) reflow, C) moisture absorption & reflow

2. Effect of filler of EMC

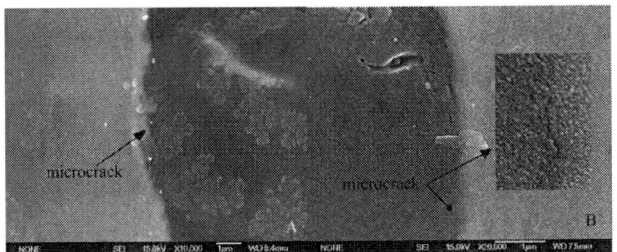

Fig. 6 SEM images of interface between filler of EMC and leadframe. Inset: Enlarged image of a microcrack

SEM images of the interface between the filler of EMC and leadframe is shown in Fig. 6A Microcrack was found on the interface between the filler and leadframe. When thermal stress or vapor pressure appears during the reflow process, these microcracks will expand and transform into cracks and delamination. These microcracks will also become containers for water during the moisture absorption process.

Conclusions

Interface between EMC and leadframes with different coatings subjected to 4 different treatments was studied using SEM and Tab pull testing.

The adhesion of interfaces of EMC/Cu samples is the strongest among the 3 kinds of specimens, and the EMC/Ni/Pd/Au interface is the poorest.

There was no obvious crack founded on the interface of EMC/Cu samples subject to treatment A, B, C and D. Delamination appeared on the interface of EMC/Ag specimens subjected to reflow with and without prior moisture absorption testing. Obvious delamination existed before any treatment for the EMC/Ni specimens. Serious delamination and fracture zones were found in them subjected to moisture absorption testing or reflow. The result of SEM agreed well with the result of tab pull testing.

There was a microcrack on the interface of the filler and leadframe, which would act as an extended source during the penetration of moisture or generation of temperature difference and would be harmful to the reliability of electronic devices.

Acknowledgments

This work was supported by Shanghai Region-Henkel Joint Electronics Materials Research and Failure Analysis Center project from the Henkel Company. We are grateful to Mr. Wei Tan from Henkel Company for his help in preparing the samples.

References

1. Joseph Fauty, Leonorina G. Cada and Michal Stana, "Effect of 269 reflow on the ability of mold compounds to meet moisture sensitivity level one," *IEEE Trans. Compon. Packag. Technol.*, Vol. 28, No. 4 (2005), pp. 841-851.

2. Xinyu Du, Guangchao Xie, Wei Tan, Suqiong Qin and Xingming Cheng. "Stress Reduction of Epoxy Molding Compound and Its Effect on Delamination," *Proceedings of HDP'07*, 2007, pp. 50-54.

3. Haleh Ardebili, Ee Hua Wong and Michael Pecht, "Hygroscopic swelling and Sorption characteristics of epoxy molding compounds used in electronic packaging," *IEEE Trans. Compon. Packag. Technol.*, Vol. 26, No. 1 (2003), pp. 206-214.

4. E. H. Wong, S. W. Koh. K. H. Lee and R. Rajoo, "Comprehensive treatment of moisture induced failure-recent advances," *IEEE Trans. Electron. Packag. Manufact.*, Vol. 25, No. 3 (2002), pp. 223-230.

5. Jongwoo Park, Hyun-Joon Cha etal., "inteacial degration mechanism of Au/Al and Alloy/Al bonda under high temperature storage test: contamination, epoxy molding compound, wire and bonding strength," *IEEE Trans. Compon. Packag. Technol.*, Vol. 30, No. 4 (2007), pp. 731-744.

6. H. shirangi, *et al.*, "Characterization of dual-stage moisture diffusion, residual moisture content and hygroscopic swelling of epoxy molding compounds," *9th Int. Conf. on Thermal Mechanical and Multiphysics Simulation and Experiments in Micro-Electronics and Micro-System*, EuroSimE 2008, Freiburg, Germany, April, 2008, pp. 1-8.

7. Rainer Dudek, *et al.*, "Numercal Analysis for Thermo-mechanical reliability of polymers in electronic packaging," *IEEE Polytronic 2007 Conference*, Tokyo, Japan, January, 2007, pp. 220-227.

The Effects of Bonding Parameters on the Reliability Performance of Flexible RFID Tag Inlays Packaged by Anisotropic Conductive Adhesive

Xiong-hui Cai[1], Xian-cai Chen[2], Bing An[1], Feng-shun Wu[1], Yi-ping Wu[1, 3]

[1]College of Materials Science and Engineering,
[2]State Key Lab of Manufacturing Equipment & Technology,
[3]Wuhan National Laboratory for Optoelectronics
Huazhong University of Science and Technology, Wuhan, 430074, China
Corresponding author: Yi-ping Wu. E-mail: ypwu@mail.hust.edu.cn Tel.: +86-27-87792402

Abstract

In this work, ACA was prepared by mixing micro-sized spherical Ag particles into thermo-set epoxy resin, and RFID flip chips were assembled on the Al/PET antennae through hot-press process. The effect of bonding parameters, such as the curing degree, the curing rate and the temperature combination of down and up hot-press heads on the contact resistance and shear strength of ACA bonding joints for flexible RFID application were studied. And the reliability test (high-temperature and humidity test, 85°C, 85% relative humidity, 288 hrs) was also used to investigate the reliability of RFID tag inlays. It was found that these bonding parameters had great effect on the mechanical and electrical performance of bonding joints. For this ACA prepared here, the optimum bonding parameters for flexible RFID tag inlays was that: the curing degree is 85%, the curing rate is 15s / 170°C, and the temperature of down and up hot-press heads are 180°C/160°C.

1. Introduction

Radio frequency identification (RFID) is a small tag containing an integrated circuit chip and an antenna, and has the ability to respond to radio waves transmitted from the RFID reader. For its advantages, it is recognized one of the information technologies with the most future potential. And assembly the radio Frequency Integrated Circuit (RFIC) on the flexible antennae using anisotropic conductive adhesive (ACA) through flip-chip technology is the simple way to accomplish the low cost and large-scale manufacture the flexible RFID tag inlays. However, the reliability of flexible RFID tag inlays packaged by ACA is still a critical problem despite its wide application. Although it is recognized that the nature of polymer adhesive is the main cause, the effects of bonding parameters on the reliability performance of RFID tag inlays cannot be neglected. The different bonding parameters can lead to the shift of contact resistance of ACA joints and the decrease of adhesive strengths during the various environmental test. They are two critical cause decrease the reliability of ACA joints. Therefore, the electrical performance and adhesive strengths must both be considered to determine the optimum bonding parameters for reliability of interconnection of flexible RFID tag inlays packaged by ACA.

ACA can be filled with polymer coating metal layers conductive particles and metallic conductive particles. Apparently, the cost of ACA filled with the former is much more expensive than that of the ACA filled with the latter. Importantly, the ACA filled with polymer coating metal layers conductive particles is not suitable for the packaging of the some flexible RFID tag inlays, especially for the etched Al/PET substrates which is mostly used. The ACA filled with the metallic conductive particles is more appropriate to the assembly of flexible RFID tag inlays. The effects of bonding parameters on the reliability of chip on flexible substrates (COF) packaged by ACA filled with polymer coating metal layers conductive particles have been studied [1-4]. The reliability of joints packaged by ACA filled with metallic conductive particles is rarely studied, especially for the flexible RFID packaging application.

In this paper, an ACA was prepared by uniformly mixing the uniform micro-sized spherical metallic conductive particles, latent curing agent and other additives in the thermo-set epoxy resin. And RFID flip chips were assembled on the Al/PET substrate through hot-press process. The effect of bonding parameters on the reliability performance of RFID tag inlays packaged by anisotropic conductive adhesive was studied.

2. Experiment

An ACA was prepared by uniformly mixing the uniform micro-sized spherical metallic conductive particles (the mean diameter was 3μm), latent curing agent and other additives into thermo-set epoxy. The flexible substrate was the etched Al/PET film. And the thickness of PET film and aluminum line were 30 and 36μm respectively. And the test chips were the modified normal radio frequency integrated circuits (RFICs) [5].

Flexible RFID assemblies were prepared by assembling the RFICs on the pads of the antenna with the ACA through hot-press process [6]. And different samples were carried out according to different bonding parameters.

The change of contact resistance and shear strength of bonding joints during the reliability test (high-temperature and humidity test, 85°C, 85% relative humidity, 288 hrs) was used to evaluate the reliability performance of flexible RFID tag inlays assembled by ACA. And the measurement method of them had been illustrated in our previous work [5]. The contact resistance of bonding joints was detected every 72 hours. And the shear strength of them was determined after the whole aging test. And three pieces of samples for one bonding parameters were tested. The mean value was used as the final results. After the chips were sheared off from the substrate, the split interfaces were also studied by the Scanning electron microscopy (SEM).

The curing degree was determined according to the kinetics equation of the ACA, which was regressed by the data of non-isothermal differential scanning calorimetry (DSC) technique at different heating rates. And the

978-1-4244-4658-2/09 $25.00 © 2009 IEEE

relationship between the bonding temperature, the bonding time and the curing degrees could be concluded from the different integration formulas of the kinetics equation. When the bonding temperature was fixed, the curing degree increased with the increase of bonding time. At the same time, there were different combinations of bonding temperature and bonding time to attain the same curing degree. The higher was the bonding temperature, the shorter was the bonding time, and the higher was the curing rate of adhesive. And the bonding temperatures were controlled by the temperatures of the down and up hot-press heads during the hot-press process. So there were different combinations of them to reach the same bonding temperature.

3. Results and discussion

3.1 The effect of curing degrees

As mentioned above, the curing degree was determined by controlling the bonding time at the bonding temperature of 170°C and the bonding pressure of 4.0MPa, when the temperatures of the down and up press head were 180°C and 160°C respectively. The results of the effect of curing degree on shear strength and contact resistance of bonding joints of RFID tag inlays packaged by ACA were shown in Figure 1 and Figure 2.

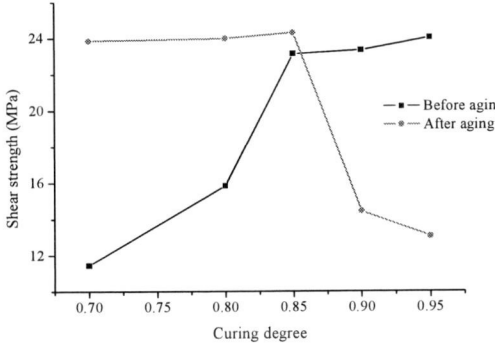

Fig. 1 The evolution of shear strength vs. the curing degree

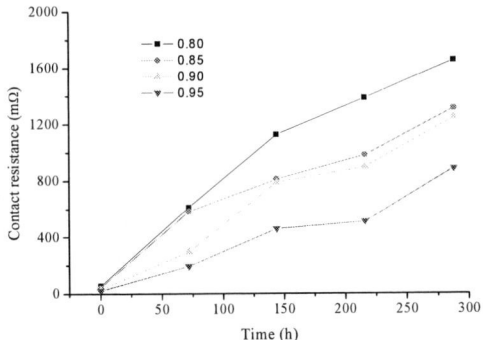

Fig. 2 The evolution of contact resistance vs. time in hot-humidity test for the bonding joints with different curing degrees

It was clear that the shear strength of nonageing samples increased firstly with the increase of curing degree. And it changed a little after the curing degree reached 0.85. But it was reverse to the ageing samples. The shear strength did not change firstly and then decreased rapidly when curing degree reached 0.85. It was concluded that the effect of the curing degree on shear strength of bonding joints was evident. As the adhesive cured further, the microstructure of polymer matrix became complicated and the shear strength increased. But when the curing degree reached a value (0.85), its effect on the adhesive strength was little. This could be approved by the microstructure of split interface of bonding joints. The results were shown in Figure.3. It was clear the there was no residual cured adhesive on the substrates when the curing degree was 0.8. But it was different when the curing degrees were 0.85 and 0.9. It means that there was a rupture of cured adhesive when the chips were sheared off from the substrate when the curing degrees were 0.85 and 0.9. But the adhesive of bonding joints could cure further during the aging test and fully cured in 48 hours hot-humidity test [5]. The thermal expansion of materials also could lead the release of thermal stress during the aging test. It would cause the decrease of shear strength. The higher was the curing degree, the bigger was the thermal stress. And when the curing degree exceeded one value (0.85), the release of thermal stress could not be neglected. It lead the rapidly decrease of shear strength.

The contact resistance increased with the increase of aging time for all samples with different curing degrees. But the curing degree was higher; the change trend was smoother. As mentioned in our previous work, the shift of contact resistance for flexible RFID tag inlays using Al/PET as substrates was mainly caused by the oxidation of Al [5]. The lower was the curing degree, the easier could the water vapor penetrate into the polymer matrix. It lead did the oxidation reaction occur more easily. And the contact resistance increased more rapidly.

As a whole, the curing degree had much effect on the electrical and mechanical performance on the flexible RFID tag inlays using Al/PET as the substrate. The curing degree of 0.85 maybe was suitable for this kind flexible RFID tag inlays.

3.2 The effect of the curing rate of adhesive

As mentioned above, different combinations of bonding temperature and bonding time could be designed to attain the same curing degrees. But the different bonding time to achieve the same curing degree would imply the different curing rates of adhesive. In this work, the samples were prepared at the bonding parameters: 175°C/10s, 162°C/17s and 152°C/27s, the bonding pressure of 4.0MPa. The curing degrees of adhesive all were 0.85. The results of the effect of curing rate on shear strength and contact resistance of bonding joints of RFID tag inlays packaged by ACA were shown in Figure 4 and Figure 5. There, the different bonding time were used to represent the different curing rates of the ACA.

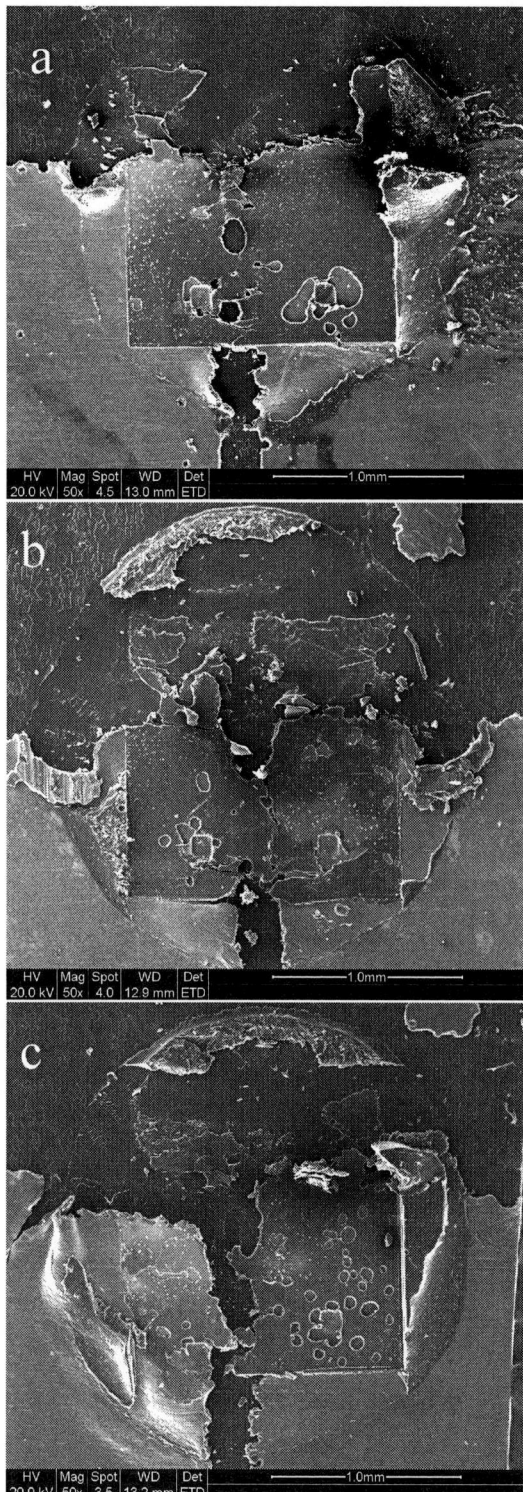

Fig. 3 The microstructure of split interface on the sides of substrate for the curing degrees of 0.80 (a), 0.85(b) and 0.90 (c)

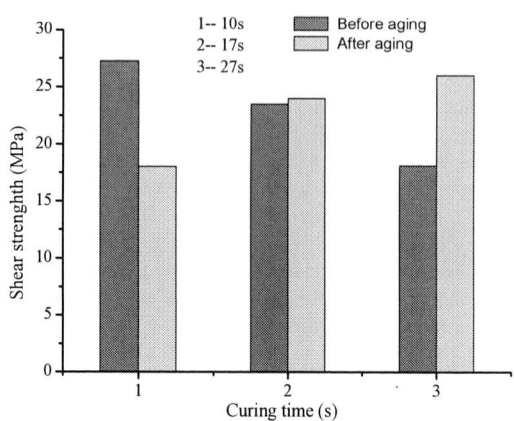

Fig. 4 The shear strength of bonding joints with the different curing rates

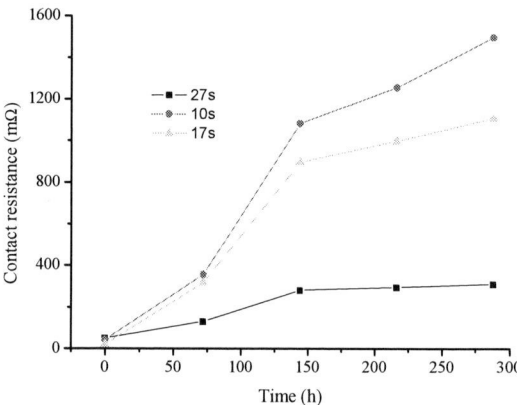

Fig. 5 The evolution of contact resistance vs. time in hot-humidity test for the bonding joints with different curing rates

It was clear that with the decrease of curing rate, the shear strength of nonageing samples decreased, and that of ageing samples increased. There were three different materials: chips, cured adhesive and substrate in all flexible RFID tag inlays. They had different coefficients of thermal expansion (CTE). And they also experienced the thermal expansion during the curing process of adhesive. If the effect of thermal expansion could not be neglected, it would lead the decrease of adhesive strength between the adhesive interfaces. Especially when the temperature of heat process approached to or exceeded the T_g (glass transition temperature) of polymer materials, this effect was very evident. In this paper, all the bonding temperatures had exceeded the T_g of cured adhesive (about 135°C) and PET film (about 80°C). So, the longer was the bonding time, and the bigger were the effect of thermal expansion. It caused the decrease of shear strength of bonding joints of RFID tag inlays. And during aging test, the thermal stress could be released with slow rate for its low temperature. But the higher was the curing rate during the curing process, the higher were the accumulation of the thermal stress. And more rapidly did it release during the aging test, poorer was the shear strength.

The contact resistance increased with the increase of aging time for all different curing rates. But the smaller curing rate leads the smoother change trend. It maybe was connected with the release of thermal stress as motioned above.

On all accounts, the curing rate also had much effect on the reliability of bonding joints packaged by ACA. There exists a suitablbe curing rate of adhesive. Too high lead the anti-aging performance decrease. Too low, it would lead the initial performance poor. In this work, the suitable curing rate was 17s to achieve the curing degree of 0.85.

3.3 The effect of temperature combination of down and up hot-press heads

The bonding temperature was controlled by the temperatures of the down and up hot-press heads. There were different combinations of them to reach the same bonding temperature. Taking the T_g of polymer substrates into account, the temperature of up hot-press head was usually lower than that of the down hot-press head. To attain the bonding temperature of 170°C, the temperature of the down and up hot-press heads could be 245°C/125°C, 200°C/140°C and 170°C /170°C. The test samples were assembled at the theses bonding temperatures, pressure of 4.0MPa and bonding time of 17s. The results were shown in Figure 6 and Figure 7.

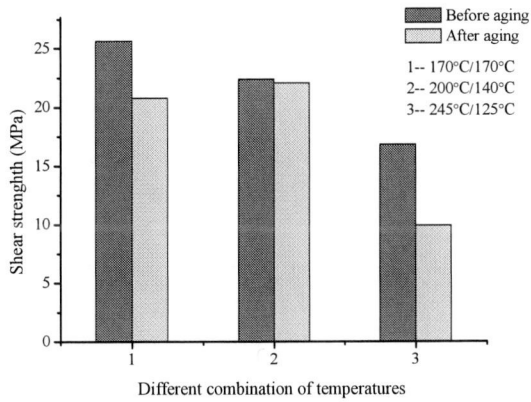

Fig. 6 The shear strength of bonding joints for different temperature combinations of the hot-press heads

It was clear that with the increase of the temperature difference between the down and up hot-press heads, the shear strengthen of nonageing samples decreased. And that of the ageing samples firstly changed little, but when the temperature difference exceeds one value, it decreased rapidly. On one hand, it was maybe because that the higher was the temperature of down hot-press head, the larger was the mismatch of the cured adhesive and the chip, which lead the adhesive strength between the chip and substrate decrease. And on the other hand, the temperature of up and down hot-press heads also affected the theology of the adhesive during the curing process. The result of it was the air bubbles existing in the cured adhesive. It was approved by the microstructure of the cured adhesive on the sides of chip. The results were shown in Figure 8. It was clear that there were more air bubbles in the cured adhesive when the temperature of up hot-press head was 125°C. It greatly affected the shear strength of bonding joints. This maybe because that when the temperature of up hot-press head was higher, the faster did the viscosity decrease, and the easier could the air bubbles release from the flowing adhesive.

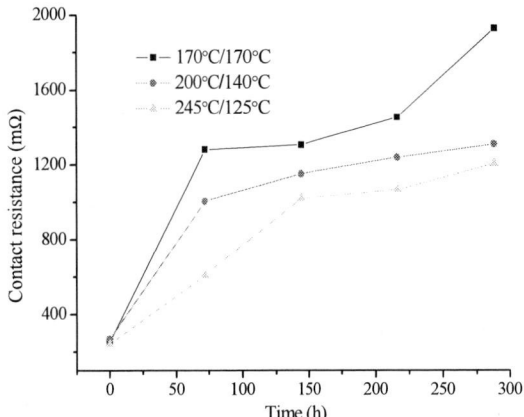

Fig. 7 The evolution of contact resistance vs. time in hot-humidity test for different temperatures combinations of the hot-press heads

The contact resistance increased with the increase of aging time. But the change trend decreased with the increase of the difference between the up and down hot-press heads. It may be because that the low temperature of up hot-press head could lead smaller mismatch between adhesive and substrate. And it implied that it was benefit for the adhesion between the cured adhesive and substrate.

In a words, the temperatures of down and up hot-press heads had great influence on the reliability of flexible RFID tag inlays packaged by ACA through did affect the materials' thermal expansion and the theology of the adhesive in curing process. Lower of them were preferred to attaining high reliability performance. For this ACA, the combination of 180°C/160°C for them maybe was suitable.

Conclusions

It was found that the curing degree, the curing rate and the temperature of down and up hot-press heads had much effect on the reliability of flexible RFID tag inlays using Al/PET substrate. The optimal bonding parameters are that: the curing degree of 0.85, and there was a suitable curing rate. For this ACA, the curing rate was 17s to attain the curing degree of 0.85. The temperature of up and down hot-press heads are 180°C /160°C.

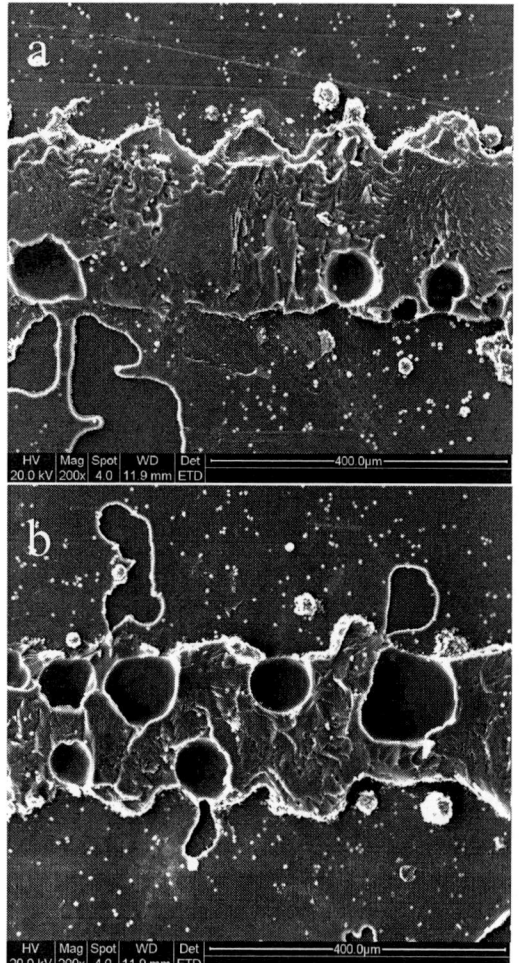

Fig. 8 The microstructure of cured adhesive on the sides of chip for temperatures combinations of the hot-press heads: (a) 200°C/140°C and (b) 245°C/125°C.

Acknowledgments

The authors acknowledge the financial support by National High Technology Research and Development Program of China (863 Program) (No. 2006AA04A110), and National Natural Science Foundation of China (No. 60776033).

References

1. Y C Chan, D YLuk, "Effects of bonding Parameters on the reliability Performance of anisotropic conductive adhesive intereonnects flip-chip-on-flex Packages assembly I. Different bonding temperature," *Microelectronics Reliability*, 42, 2002. pp. 1185-1194.

2. X Chen, J Zhang, C L Jiao, *et al.*, "Effects of different bonding Parameters on the electrical Performance and Peeling strengths of ACF interconnection," *Microelectronics Reliability*, 46 (5-6) 2006. pp. 774-785.

3. L Q Cao, Z H Lai, J Liu, "Effect of curing condition of adhesion strength and ACA flip-chip contact resistance," *Proceeding of the Sixth IEEE CPMT Conference on High Density Microsystem Design and Packaging and Component Failure Analysis(HDP,04)*, Shanghai,china, June, 2004, pp. 254-258.

4. Y C Chan,D Y Luk, "Effects of bonding Parameters on the reliability performance of anisotropic conductive adhesive interconnects flip-chip-on-flex Packages assembly II.Different bonding pressure. Microelectronics Reliability," 42, 2002. pp. 1195-1204.

5. Xiong-hui Cai, Bing An, Yi-ping Wu, Feng-shun Wu, Xiao-wei Lai, "Research on the contact resistance and reliability of flexible RFID tag inlays packaged by anisotropic conductive paste," *9th International Conference on Electronic Packaging Technology & High Density Packaging(ICEPT-HDP'08)*, Shanghai, July, 28-31 2008, pp. 129-133.

6. Xiong-hui Cai, Bing An, Xiao-wei Lai, Yi-ping Wu, Feng-shun Wu, "Reliability Evaluation on Flexible RFID Tag Inlay Packaged by Anisotropic Conductive Adhesive," *8th International Conference on Electronics Packaging Technology(ICEPT'07)*,Shanghai, China, August 14-17, 2007, pp. 716-719.

Effects of Design, Structure and Material on Thermal-Mechanical Reliability of Large Array Wafer Level Packages

Bhavesh Varia[1], Xuejun Fan[1,2], Qiang Han[2]
[1]Department of Mechanical Engineering
Lamar University
PO Box 10028, Beaumont, Texas 77710, USA
[2]College of Civil Engineering and Transportation
South China University of Technology, Guangzhou, China
bhaveshvaria.mech@gmail.com; xuejun.fan@lamar.edu

Abstract

In this paper, thermo-mechanical reliability of a variety of state-of-art wafer level packaging (WLP) technologies is studied from a structural design point of view. Various WLP technologies, such as Ball on I/O with and without redistribution layer (RDL), Ball on Polymer with and without under bump metallurgy (UBM) process, and encapsulated Copper Post WLPs, are investigated for their structural characteristics and reliability performance. Ball on I/O WLP, in which solder balls are attached directly to the metal pads on silicon wafer, is used as a benchmark for the analysis. 3-D finite element modeling is performed to investigate the effects of WLP structures, UBM process, polymer film material properties (in Ball on Polymer), and encapsulated epoxy material properties (in Copper Post WLP). Fundamentals underlying thermomechanical reliability mechanisms are uncovered through detailed parametric studies. Experimental tests with various parameters were conducted to validate simulation results. Both Ball on Polymer and Copper Post WLPs have shown great reliability improvement in thermal cycling. Encapsulated copper post WLP showed the best performance.

1. Introduction

Wafer level packaging (WLP) is one of the fastest growing segments in semiconductor packaging industry due to the rapid advances in integrated circuit (IC) fabrication and the demands of a growing market for faster, lighter, smaller, yet less expensive electronic products. Higher performance, low cost compared to die level packaging, and small form factor are three primary advantages of WLP. So it becomes a pioneer in the recently growing market of handheld and mobile electronic systems (Fan et al., 2008, 2009) [1-2].

Thermo-mechanical reliability of wafer level packages is still a major concern because solder joint thermal cycling reliability is the weakest point of technology. In order to withstand against thermal cycling, WLP packages must have relatively smaller die size and I/Os. However, there has been a demand for larger number of I/Os and larger size because of more integrated functionality.

A WLP is basically a chip scale package because the final package is same as die. A typical WLP on board has three major elements: silicon die, solder balls and the printed circuit board. The large difference between the coefficient of thermal expansion (CTE) of silicon (~2.6ppm/°C) and PCB (~17ppm/°C) limits the solder ball thermal cycling fatigue performance. In order to improve the reliability, a variety of

WLP technologies have been developed (Fan et al., 2008,Reche et al., Kim et al., 2002)[1][3][4]. There are several materials involved with different WLP structures. Dimensions, structural design, and material properties have great influence on reliability.

In this paper, various state-of-art WLP technologies and the corresponding thermo-mechanical reliability are analyzed from a structural design point of view. These WLP technologies include standard WLP (ball on I/O), WLP with redistribution layer (RDL) and under bump metallurgy (UBM) process (ball on polymer with UBM), WLP with redistribution layer without UBM process (ball on polymer without UBM), as well as encapsulated copper post WLP [6-10]. Detailed structures of each WLP are described. 3-D finite element models are created for various WLP structures. The effects of various WLP structures, UBM process, polymer film material properties (in Ball on Polymer), and encapsulated epoxy material properties (in Copper post WLP) are studied. The following parameters are considered in the study,

1) Different WLP structures
2) Ball pitch effect
3) Redistribution layer geometry effect
4) Redistribution layer material properties
5) UBM effect
6) Array size
7) Solder ball shape
8) PCB design

The simulation results are compared to the experimental test data and failure analysis. The mechanisms in enhancing thermo-mechanical reliability of WLP are discussed.

2. WLP Descriptions

Four different WLP packages, as shown in Table 1, are studied.

Table 1 WLP descriptions

WLP Structures	Descriptions
WLP Structure A	Standard WLP (Ball on I/O)
WLP Structure B	Ball on polymer without UBM
WLP Structure C	Ball on polymer with UBM
WLP Structure D	Encapsulated copper post

2.1 WLP Structure A: Standard WLP

Ball on I/O WLP is a standard wafer level packaging technology and the process is very similar to a typical flip chip technology. As shown in Figure 1, the ball is attached to the aluminum pad directly through under bump metallurgy (UBM). The bumps are directly attached to the final I/O metal pad. Passivation opening, overlapped by UBM, provides a seal to the under laying I/O aluminum pad. The solder ball in this structure is connected to silicon base directly. Figure 2 is a schematic view of details of WLP Structure A.

RDL pad on a stack of polymer dielectric materials. Two-layers of dielectric materials (usually polyimide) are processed, named Polymer 1 and Polymer 2, respectively, to serve as passivation and redistribution layers. Redistribution copper traces connect final metal pads to solder balls with UBM incorporated. Polymer layer serves as stress buffer when thermal-mechanical stress is subjected due to thermal mismatch between PCB and silicon during temperature change. Detailed ball structure is shown in Figure 6.

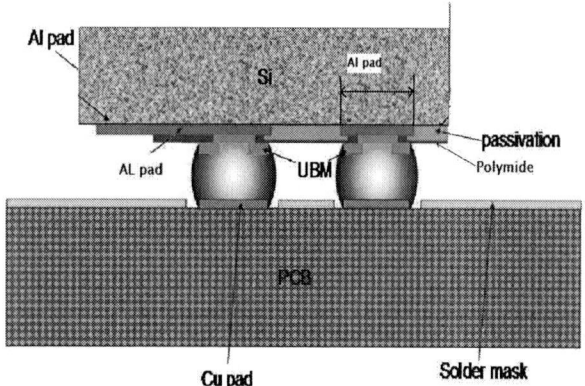

Figure 1 WLP structure A-Standard WLP

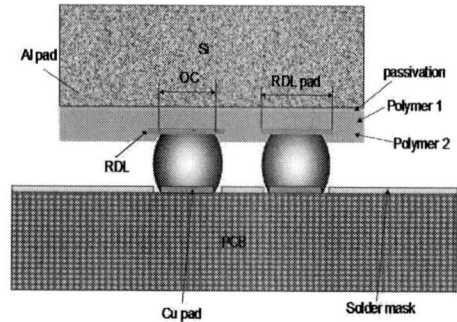

Figure 3 A schematic diagram of WLP Structure B-Ball on Polymer without UBM

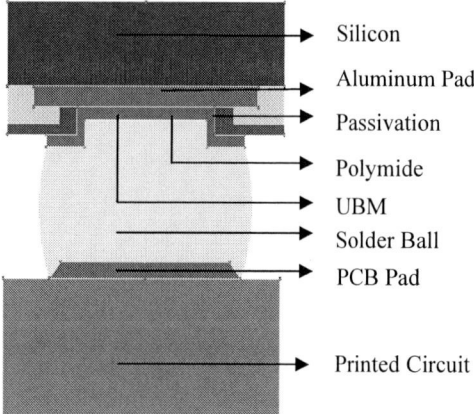

Figure 2 A schematic cross section view of bump structure for WLP Structure A

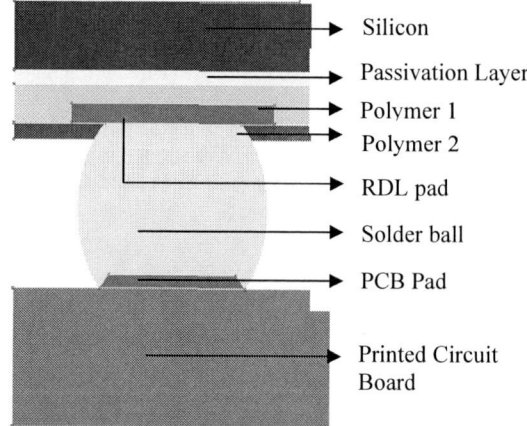

Figure 4 A schematic cross section view of bump structure for WLP Structure B

2.2 WLP Structure B: Ball on Polymer without UBM

Figure 3 shows a schematic diagram of ball on polymer without UBM. Redistribution traces and pads are usually processed with electroplating using copper, which makes it possible to attach solder balls directly on RDL pads without UBM. Detailed structure is shown in Figure 4. Solder balls in this structure sit on a dielectric polymer film layer to avoid a direct connection with silicon base.

2.3 WLP Structure C: Ball on Polymer with UBM

Figure 5 is a schematic of ball on polymer WLP structure with UBM. In this structure the solder ball is placed over

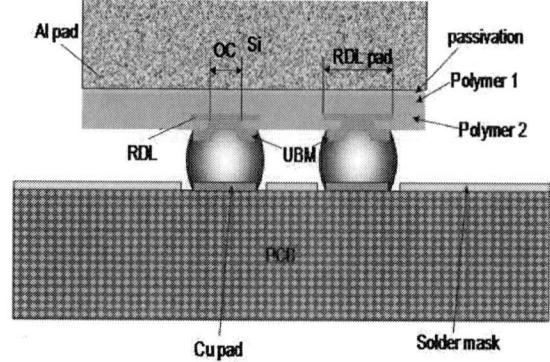

Figure 5 A schematic diagram of ball on polymer with UBM WLP

978-1-4244-4658-2/09 $25.00 © 2009 IEEE   980

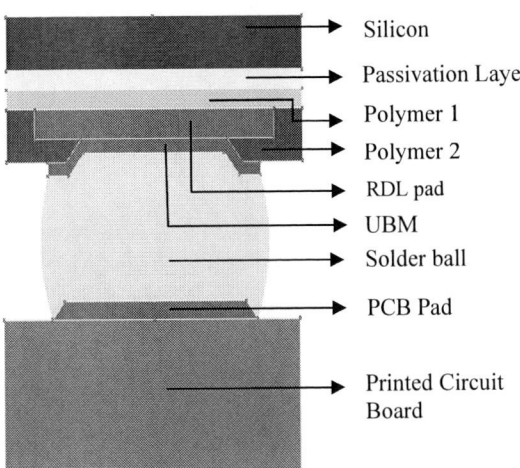

Figure 6 A schematic cross-section view of bump structure for WLP Structure C

2.4 WLP Structure D: Encapsulated Copper Post WLP

In encapsulated copper post WLP technology, bond pads are redirected into an array of interconnects. They are in the form of electroplated copper posts instead of pads to provide enough confrontation for the active wafer surface to be encapsulated in low stress epoxy by transfer molding, exposing only the top portions of the posts where the solder balls will be attached, as shown in Figure 7. Detailed ball structure is shown in Figure 8.

In the following study, the ball pitch used for all structures is 0.5mm unless otherwise stated, and solder ball opening diameter on silicon side is fixed as 0.25mm. The PCB side is assumed non-solder mask defined. Other important geometrical dimensions of all four structures are shown in Table 2.

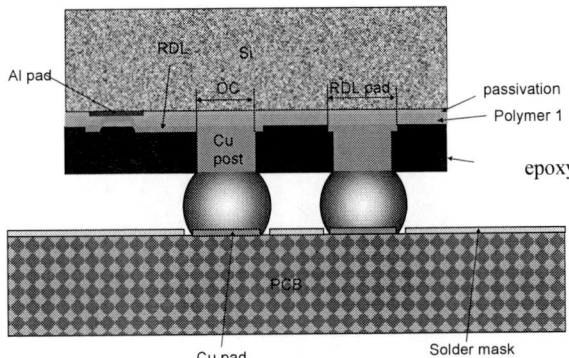

Figure 7 A schematic diagram of Encapsulated copper post WLP

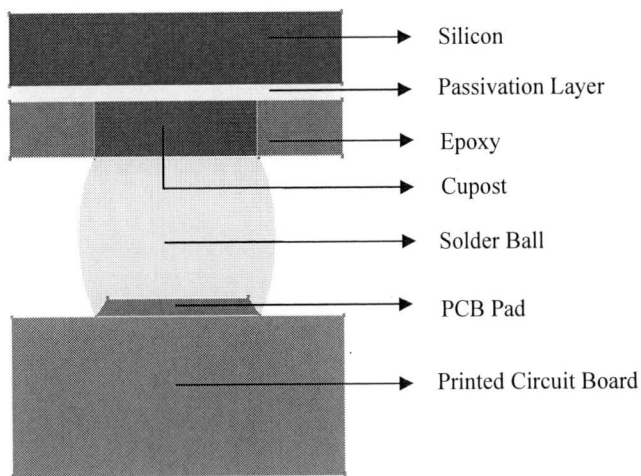

Figure 8 A schematic cross section view of bump structure for WLP Structure D

Table 2 Geometrical dimensions of WLP packages

WLP Structure	Dimensions (μm)			
	A	B	C	D
Silicon thickness	400	400	400	400
Solder ball diameter	310	310	310	310
Solder ball standoff height	240	240	240	240
Solder ball opening diameter	250	250	250	250
PCB pad diameter	250	250	250	250
PCB thickness	1000	1000	1000	1000
Wafer Passivation thickness	4	4	4	4
Pitch Size	500	500	500	500
UBM combined thickness	2.5	-	2.5	-
Epoxy/Copper post thickness	-	-	-	70
Polymer film 1 thickness	-	5	5	-
Polymer film 2 thickness	-	5	5	-

3. Finite Element Model

In the present study we consider only the equal array size of WLP packages in both directions. Thus one eighth model is created using symmetry condition. The top view of a WLP package is shown in Figure 9. Figure 10 shows finite element model of one eighth part of package. ANSYS 11.0 is used for all the finite element analysis. VISCO 107 linear element is used to mesh solder joint which allows using viscoplastic material properties. The Solid 45 element is used to mesh all other materials. In order to reduce the possible edge effect of PCB board on the outermost solder ball stress analysis, the PCB size in the model is extended at least 2.5 times of the package size, as shown in Figure 10. The boundary condition for the one eighth model involves applying symmetry boundary conditions on symmetric planes, the node at the origin is constrained so as to have zero displacement in all the three direction.

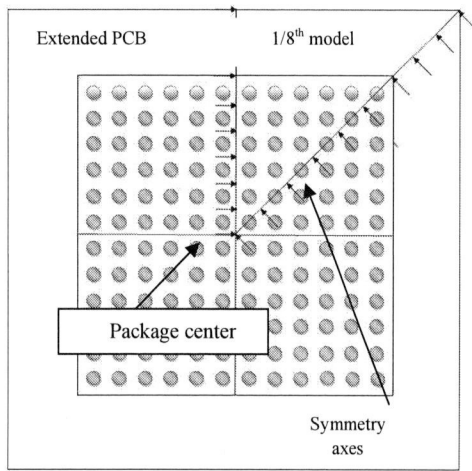

Figure 9. Top view of an equal array WLP package

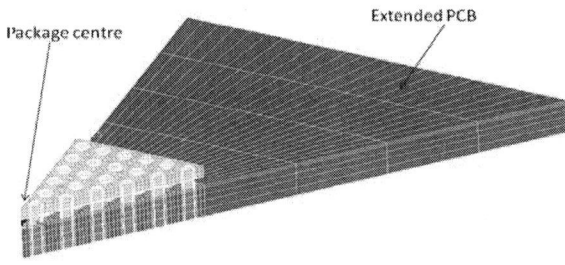

Figure 10 One-eighth finite element model for WLP due to symmetry

Four WLP Structures A, B, C, and D are molded in detail and are shown in Figure 11. Figure 12 shows the experimental observations of failure mode at solder bulk due to fatigue on package side. In finite element models, a fixed-thickness layer (10μm) is created for each structure to extract damage parameter.

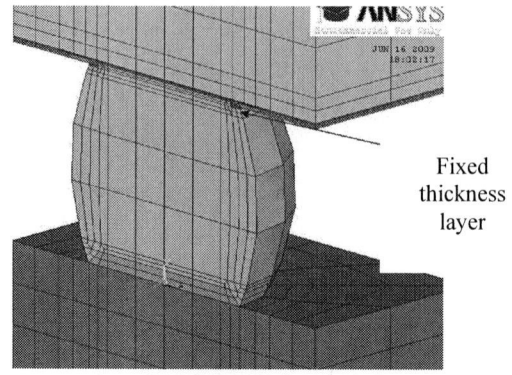

a) WLP structure A

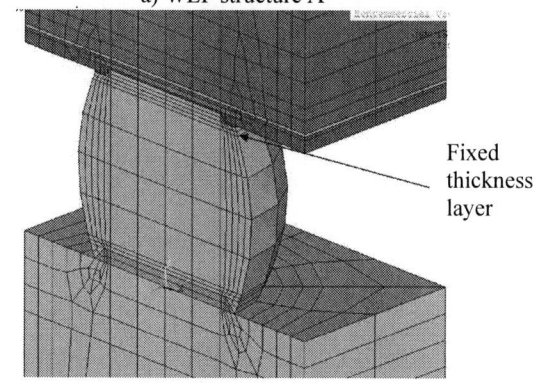

b)WLP structure B

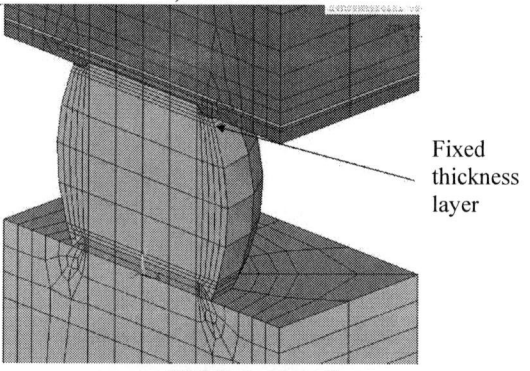

c)WLP structure C

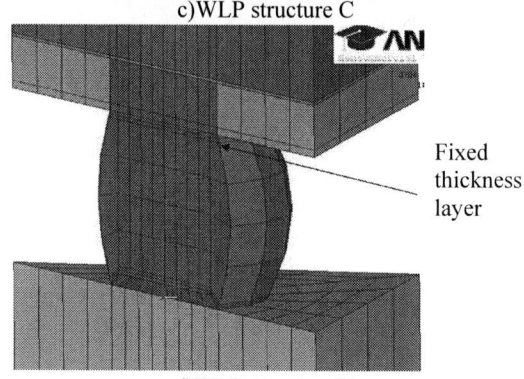

d)WLP structure D

Figure 11 Finite element models of various WLP structures a). WLP Structure A; b). WLP Structure B; c). WLP Structure C; d). WLP Structure D

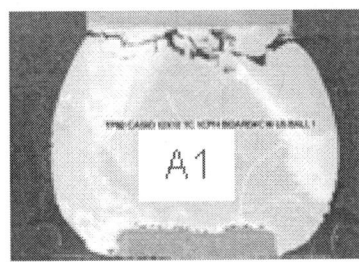

Figure 12 Solder bulk crack on package side due to fatigue

3.1 Material Properties

WLP packages are made of different materials. A summary of the materials used for analysis is shown in Table 3. All materials used in the analysis are modeled as linear elastic and the temperature dependency is taken into consideration whenever the glass transition temperature T_g is within the thermal cycling range of -45°C to 125°C. The PCB is fiber reinforced epoxies which makes the properties differ in out of plane direction. Orthotropic properties are therefore used for these materials.

Table 3 Material properties

Materials	Modulas of Elasticity (GPa)	Coefficient of Thermal Expansion (ppm/°C)	Poisson's ratio
Silicon	130	2.6	0.278
Passivation	105	11	0.24
UBM	50	16	0.35
Aluminum Pad	69	24	0.32
RDL pad	130	16.8	0.34
Epoxy	14	20	0.24
Polymide	1.2	52	0.34
Polymer 1	1.2	52	0.34
Polymer 2	1.2	52	0.34
Cu Post	130	16.8	0.34
Solder Ball	50	24.5	0.35
Pcb pad	130	16.8	0.34
PCB	25	16	0.39

The solder joint used for the analysis is SAC305 alloy. This alloy is modeled as rate-dependant viscoplastic material property using ANAND model (Anand 1985) [5]. The Anand model in commercial software ANSYS has been used here to characterize the rate-dependent creep behavior of solder alloys at varying temperatures. In Anand model, the flow equation is,

$$d^p = Ae^{-\frac{Q}{R\theta}}\left[\sinh\left(\zeta\frac{\sigma}{s}\right)\right]^{\frac{-1}{m}}$$

$$s = \left\{h_0\left(|B|\right)^a \mathrm{sgn}\left(B\right)\right\}d^p$$

$$B = 1 - \frac{s}{s^*}$$

$$s^* = s\left[\frac{d^p}{A}e^{\frac{Q}{R\theta}}\right]^n \qquad (1)$$

where, d = effective inelastic deformation rate,
σ = the effective Cauchy stress,
s = the deformation resistance,
s^* = the saturation value of deformation resistance,
$\acute{s}$ = the time derivative of deformation resistance,
θ = the absolute temperature.

Darveaux (1995, 2000) [7] [8] gave the nine material constants in Equation (1) for eutectic solder alloys. This has led the Anand model very popular in solder joint reliability modeling. For SnAgCu alloys, this paper uses the material constants given by Reinkainen et al. (2005) [11]. He has fitted 9 constants for SAC305 alloy. Table 4 lists the Anand's model constants used in SAC 305 alloy.

Table 4 Material parameters of viscoplastic Anand model (Reinkainen et al., 2005) [11]

Constant	Constant Value
s_0, MPa	1.3
Q/R, K	9000
A, sec⁻¹	500
ξ	7.1
m	0.3
h_0, MPa	5900
$\hat{s}$, MPa	39.4
n	0.03
α	1.5

3.2 Loading Condition

Stress free initial temperature is important consideration before subjecting any package to loading profile (Fan et al., 2006) [9]. It is temperature of a material corresponds to the temperature at which the material has either been cured or assembled. There are three commonly used initial stress-free temperature conditions. One is the solidus temperature of solder material (e.g., for SAC305, this temperature is 217°C). This condition considers that the solder joints start to provide mechanical support as soon as the solder material is solidified during the reflow process. The second one is the room temperature as initial stress-free (e.g. 25°C). This assumes that the shipping and storage time is sufficient to relax all the residual stresses in solder joints from the assembly process. The last one uses the high dwell temperature of thermal cycle or operating conditions (denoted as Tmax, e.g. =125°C for thermal cycling from -45°C to 125°C). This assumes that after several thermal cycles, the package reaches a stabilized cyclic pattern where the lowest stresses are seen at the end of the high temperature dwell period.

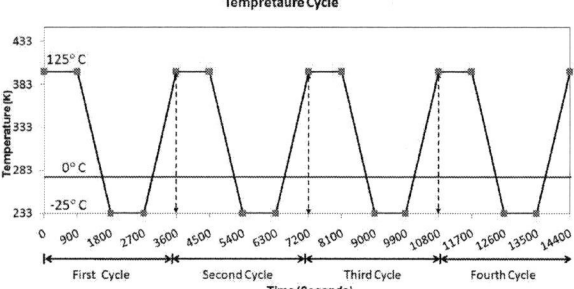

Figure13 Temperature cycle used for simulation

In this analysis, we have used thermal cycle condition -40°C to 125°C to ensure a reliable package performance. Figure shows a loading profile for simulation which includes four cycles, each cycle is 60 minutes with 15 minutes ramp up and down time and 15 minutes of dwell time at -40°C and 125°C. According to previous studies, viscous material, used for solder ball, readjust the stress state and reach 'a near stress free' at high dwell temperature after few cycles. It also indicates that the stabilized values of strain or strain energy density per cycle are independent of the initial stress free temperature setting and results showed that the stabilization happened within the first cycle (Fan *et al.*, 2006) [9].

3.3 Analysis approach

There are many methods to evaluate solder joint reliability, e.g., stress based, plastic/creep strain based, energy based, and damage accumulation based. In this paper, the analysis is based on damage accumulation method. Usually per-cycle inelastic strain (or creep strain) or inelastic strain energy density is used as damage metrics to evaluate solder joint reliability. To prevent any mesh dependency and stress singularity effect at geometry edge, Darveaux (1995, 2000) [7-8] has used a solid thin layer of elements near package/solder interface for volume averaging. All four WLP structures are modeled with a 10μm thin disk with two layers of element as shown in Figure 14.

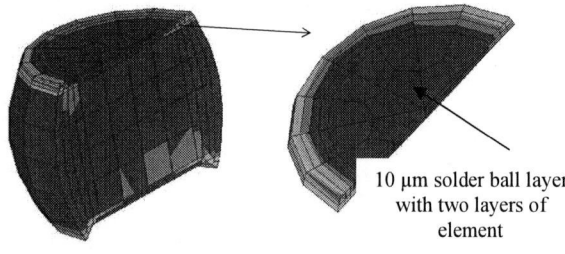

10 μm solder ball layer with two layers of element

Figure 14 Volume averaging on a fixed thickness layer

The volume averaged inelastic energy density is defined as over the thin disk of two layered element as follows,

$$\Delta W_{ave} = \frac{\sum \Delta W V}{\sum V} \qquad (2)$$

where,

ΔW_{ave} = Average viscoplastic strain energy density accumulated per cycle for interface element

ΔW = Average viscoplastic strain energy density accumulated

V = Volume of each element

This equation gives accumulated inelastic energy density per cycle and accumulation comes from all four time periods during each cycle, i.e., ramp up and down and dwell at extreme high temperature and extreme low temperature.

3.4 Important Observations
1) Results show that the maximum damage occurs in the diagonally outer most solder ball, which is also known as the critical solder ball.
2) We have used four complete cycles of thermo mechanical loadings. It has been found that the averaged inelastic strain energy accumulation per cycle is approximately same for all cycles.

Therefore, for analysis purpose, the accumulated damage, e.g. inelastic energy density, is calculated only for critical solder ball in the following from the first cycle result.

4. Modeling Results-Parametric Study
4.1 Effect of Pitch Size

Figure 15 shows the inelastic strain energy density for two pitches, 0.4mm and 0.5mm, respectively. When the pitch decreases, the solder joint fatigue life is improved.

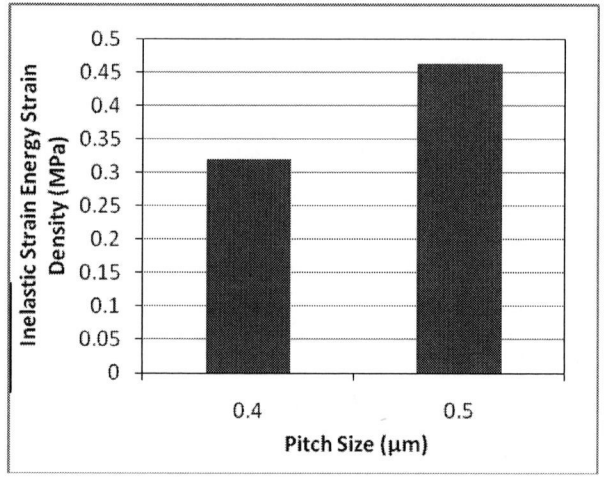

Figure 15 Effect of pitch size

4.2 Effect of Array Size

With increasing array size, inelastic strain energy density continues to increase, therefore fatigue life continues to decreases. Figure 16 shows the results from 6×6 array size to 12×12 array.

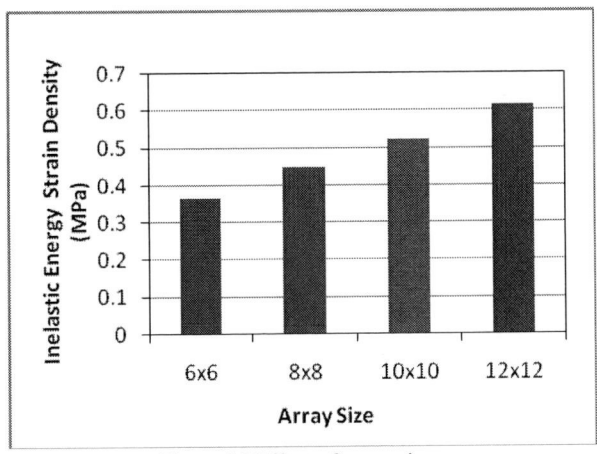

Figure 16 Effect of array size

4.3 Effect of Solder Ball Diameter

By keeping other parameters same, increasing solder ball diameter will increase inelastic strain energy density, and therefore decreases fatigue life under thermal cycling. Figure 17 shows inelastic energy density results for 220μm, 300μm, and 350μm diameters, respectively.

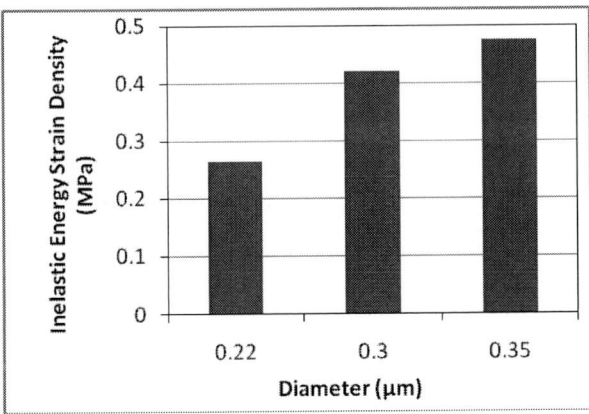

Figure 17 Effect of solder ball diameter

4.4 Effect of Solder Ball Opening Diameter

A larger solder ball opening diameter increases the total contact/interface area, and therefore it takes a longer time for solder ball crack propagations throughout contact interface. Figure 18 shows increasing die size opening diameter from 0.250μm to 0.280μm increases solder joint reliability. It becomes obvious that increasing the contact area-solder ball opening diameter is the most direct way to improve the solder joint reliability because failures often take place at solder bulk near solder ball/package interface.

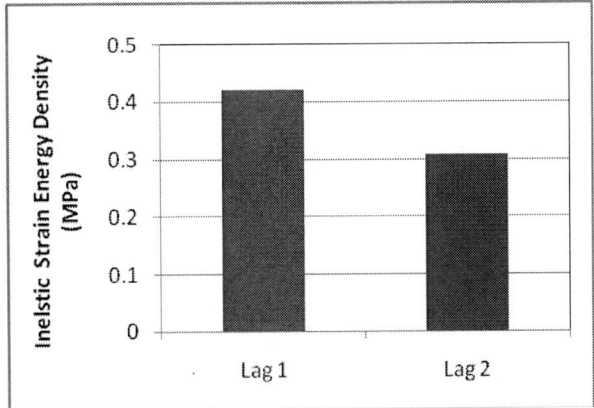

Figure 18 Effect of solder ball opening diameter

4.5 Effect of UBM

Figure 19 shows comparison of inelastic strain energy density of two structures B and C. The difference between two structures is that structure B does not have UBM layer. Results indicate that the UBM layer has slightly beneficial effect on thermo-mechanical performance of solder joint reliability.

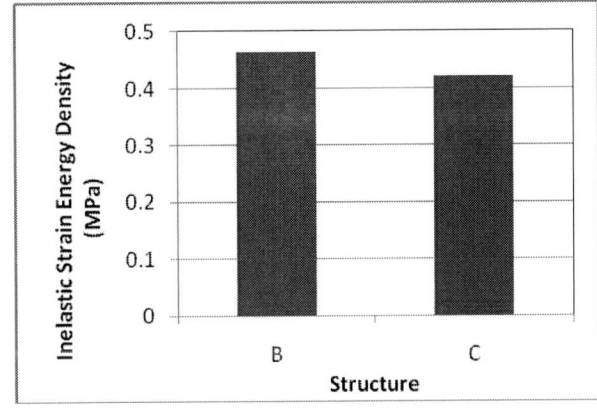

Figure 19 Effect of UBM

4.6 Effect of Passivation Layer

Thermal cycling finite element modeling results show insignificant effect of passivation layer in both structures B and C. Figure 20 shows that eliminating passivation layer in finite element model present good results for both Structure B and C.

978-1-4244-4658-2/09 $25.00 © 2009 IEEE

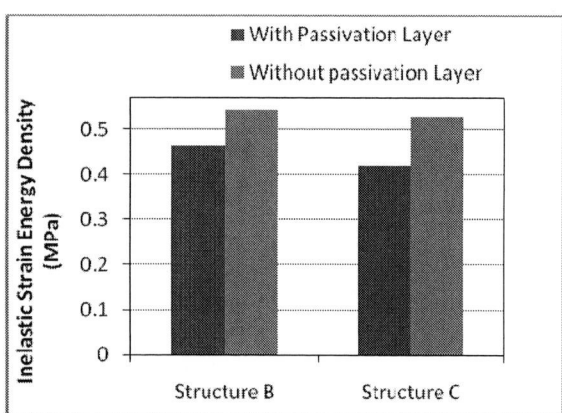

Figure 20 Effect of Passivation Layer

4.7 Effect of Material Properties of Polymer Film 1

Polyimide film is usually used for polymer 1 and 2 in WLP Structures B and C. From Table 3, it can be seen that polyimide is very compliant with a Young's modulus of 1.2GPa, and a coefficient of thermal expansion of $52\times10^{-6}/°C$ respectively. The extreme compliance of polyimide film is often attributed to be the reason for thermal-mechanical performance improvement in solder joints.

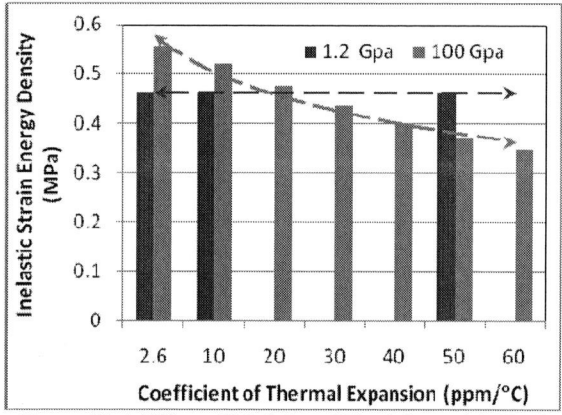

Figure 21 a) Effect of Modulas of Elasticity and CTE of Polymer film in Structure B

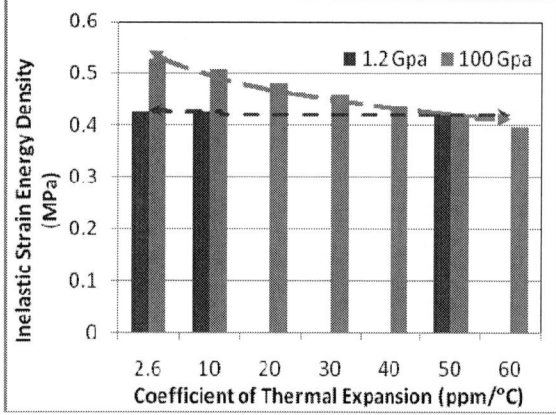

Figure 22 b) Effect of Modulas of Elasticity and CTE of Polymer film in Structure C

A parametric matrix study is performed to understand the effects of Young's modulus and CTE of the film, as shown in Figures 21 (a) and (b), respectively. When the modulus is 1.2GPa, which means that film is extremely compliant, the CTE of the film has no effect on solder joint behavior. However, when film modulus is 100GPa, solder joint stress decreases significantly with the increasing of film CTE. When the CTE is above $50\times10^{-6}/°C$, solder joint stress is even lower than that the case with the film Young's modulus of 1.2GPa. Such results indicate that the stress buffer effect can be realized either with extreme compliant material or 'hard' material with relatively large CTE. For a very soft film, solder joint stresses are relieved due to large deformation of film. For a hard film with larger CTE, the overall CTE of the combined silicon/film structure increases, therefore, the thermal mismatch with PCB is reduced. Fig. 21 b) show the effects of material properties of polymer film 1 for Structure C, and same conclusion can be reached. Figure 21 c) shows comparison of solder joint reliability of both structures at higher modulas of elasticity of polymer film. It indicates solder joint reliability decreases for UBM structure at higher CTE. It means Structure B has better reliability at higher modulas of elasticity and CTE.

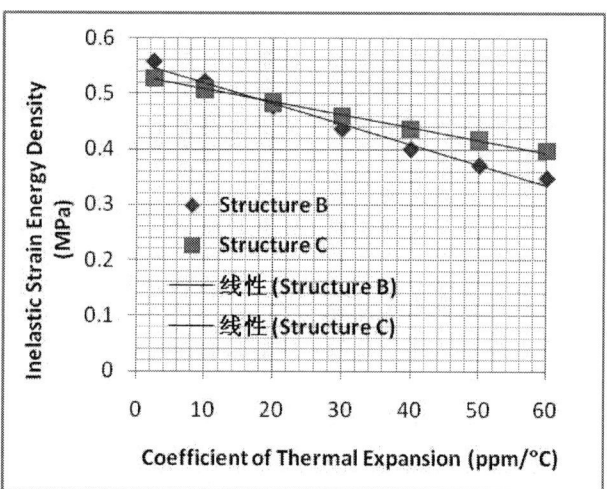

Figure 21 c) Comparison of solder joint reliability of both structures at higher Modulas of Elasticity of Polymer film

These results show that there might be an optimal point for both CTE and modulus to achieve the maximum benefit for solder joint reliability improvement. By optimizing polymer material properties, Structure B and C WLP reliability can be further enhanced.

4.8 Effect of Thickness of Polymer Thickness

Increasing polymer thickness in both structures B and C improves solders joint reliability, as shown in Figure 22. (Leg 1 film thickness < Leg 2 film thickness).

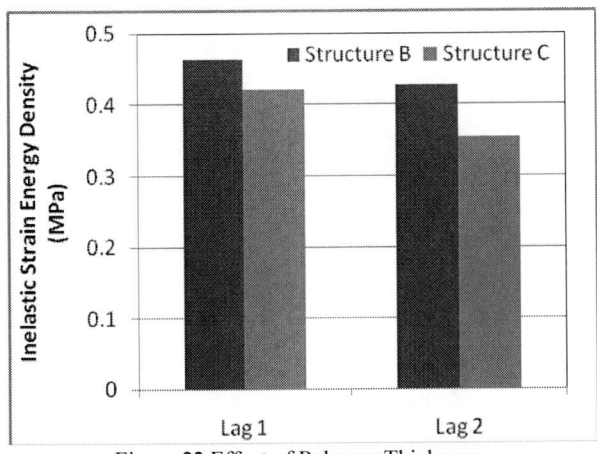

Figure 22 Effect of Polymer Thickness

4.9 Effect of Epoxy

The CTEs of the encapsulated epoxy and copper post in WLP Structure D are $20\times10^{-6}/°C$ and $17\times10^{-6}/°C$, respectively, which are much greater than silicon's. Therefore, the effective CTE of the encapsulated silicon increases effectively, which reduces the thermal mismatch with the PCB. As a result, solder joint stresses are reduced. In order to understand the effect of material properties of epoxy, a parametric study is performed, as shown in Figs. 23 and 24, respectively. When the CTE of the epoxy is kept at $20\times10^{-6}/°C$, the modulus of epoxy has nonlinear relationship with ΔW. It seems an optimal value is around 70GPa for the lowest solder joint stress. On the other hands, in Fig. 24, it can be seen that further increasing epoxy CTE from $20\times10^{-6}/°C$ to $40\times10^{-6}/°C$ will reduce stresses in solder joint, but stress will not go down further from $40\times10^{-6}/°C$ to $60\times10^{-6}/°C$. These results show that there might be an optimal point for both CTE and modulus to achieve the maximum benefit for solder joint reliability improvement. By optimizing epoxy material properties, Copper Post WLP reliability can be further enhanced.

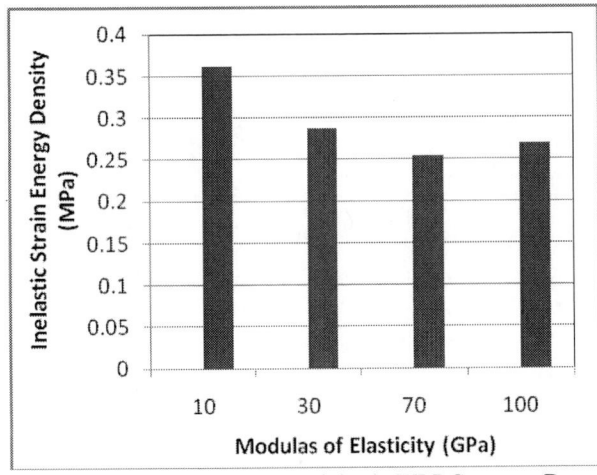

Figure 23 Effect of epoxy modulus in WLP Structure D
(CTE=20ppm/°C)

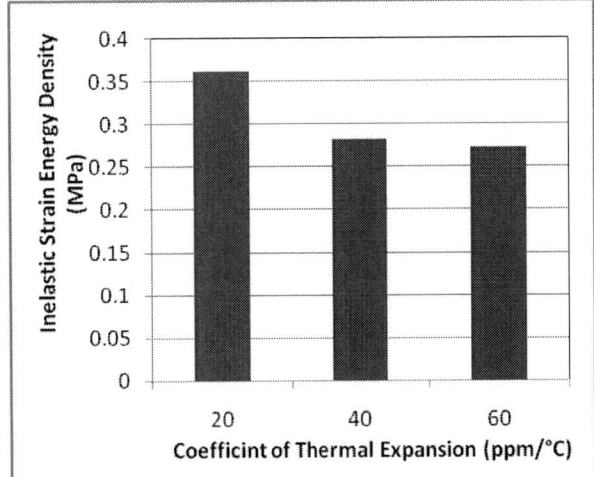

Figure 24 Effect of epoxy CTE in WLP Structure D
(E=14GPa)

4.10 Effect of Material Properties of PCB

Thermal-mechanical stresses developed in solder joints are induced by the thermal mismatch between PCB and silicon. Lowering the CTE of PCB can also reduce the stresses in solder joints. Figure 25 shows the results of inelastic strain energy density for three set of PCB CTEs. When low CTE PCB core material is used, the fatigue life can be increased greatly. It has been demonstrated that a polymer film layer between solder balls and silicon with a larger CTE can increase the overall effective CTE of silicon, and thus reduce the thermal mismatch with PCB. Similar concept can be developed at PCB side to include a layer of material between PCB and solder balls. The detailed studies will be reported separately.

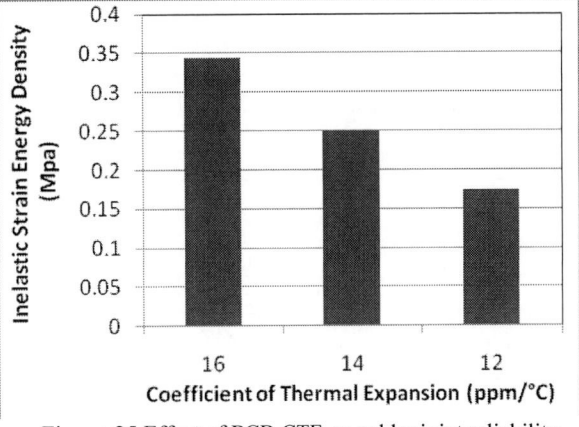

Fig.ure 25 Effect of PCB CTE on solder joint reliability

4.11 Effect of WLP Structure

Fig. 26 shows the per-cycle inelastic strain energy densities for four WLP Structures A, B, C, and D, respectively, for a 12×12 array packages with 0.5mm pitch.

978-1-4244-4658-2/09 $25.00 © 2009 IEEE

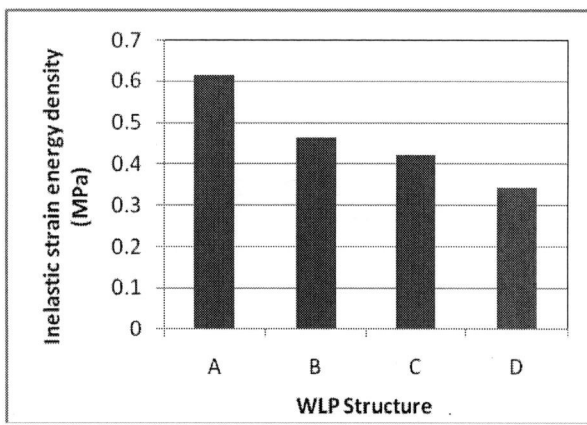

Figure 26 Effect of WLP structure

Compared to the Structure A, all other three Structures B, C and D showed more than 30% reduction in the accumulated inelastic strain energy density per cycle. This means that, with the incorporation of a dielectric polymer film between solder ball and silicon, or an encapsulated copper post layer, the stresses in solder joints can be reduced significantly compared to a 'rigid' connection in Structure A. Structure D with encapsulated copper post showed the best performance. Experimental data have shown that Structure A WLPs can survive only up to 6x6 array size while all other three structures can pass thermal cycling reliability requirement up to 12×12 array sizes (Fan *et al.*, 2009) [2]. The finite element modeling results are consistent with experimental observations.

4.12 Solder Ball Damage Map

The inelastic energy dissipation plot is given by Figure 27. It is seen that the corner solder ball shows the highest energy dissipation and therefore the largest damage accumulation during temperature cycling. The energy dissipation decays rapidly both along die edge and diagonal direction towards package center. Results show if the corner solder balls are not electrically connected the WLP reliability would be greatly enhanced. This agrees well with the test results.

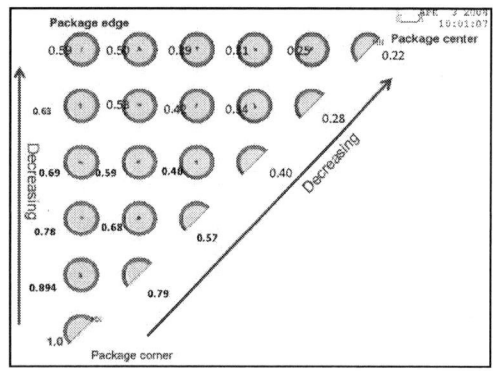

Figure 27 Inelastic Energy Density Contour for Corner Solder Joint

In order to understand the locations of solder joint crack initiation, von Mises stress plot of the solder joints are presented (Figure 28). It is seen that there is stress concentration at both sides for each solder ball. This explains the observation of cracks initiation from both sides of the solder joint. Furthermore, it is seen that the stress is higher at the inner side. This suggests that crack initiates first from inner side. This observation agrees with the findings from the failure analysis.

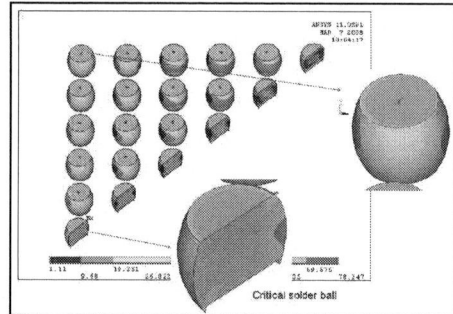

Figure 28 von Mises Stress in Solder Balls for identifying the Critical Ball

5. Experimental Validations [12]

To assess the wafer level package capability and technology limit, the following parameters are considered in experimental study to correlate with simulation results (Rahim *et al.* 2009) [12].
1. Different WLP structures
2. Ball pitches of 0.5 and 0.4 mm.
3. Array sizes of 12×12 and 10×10

5.1 Array Size

A comparison of failure data for 10×10 and 12×12 array sizes for WLP-C is shown in Table 5. It is observed that WLP-C, 10×10 array has 20% longer fatigue life than 12×12 array. In this case the ratio of WLP size between 12×12 and 10×10 array is 1.2. It appears that the fatigue life is inversely proportional to the package size.

Table 5 Normalized Characteristic Life for 12×12 and 10×10 Array WLP

Array Size	12x12	10x10
Characteristic life	1	1.2

5.2. WLP Ball Pitch

12×12 array WLPs with 0.5 and 0.4mm pitches are tested. WLP fatigue lives are compared against each other (Table 6). It is observed that 0.4mm pitch has 30% longer fatigue life than 0.5mm.

Table 6 Normalized Characteristic Life for 0.5mm and 0.4mm Pitch WLP

Pitch	0.5 mm	0.4 mm
Characteristic Life	1	1.3

Conclusions

Four different WLP structures are studied to investigate the effect of WLP structures on solder joint reliability with a combined modeling, test and failure analysis approach. These four WLP structures are Bump on I/O with and without RDL, Bump on Polymer with UBM, Bump on Polymer without UBM, and Encapsulated Copper Post. Although Ball on I/O WLP structure is limited for the application in small array packages, it has been used as a benchmark here for the analysis. Finite element models for these WLP structures have been created and analyzed. Results showed that Ball on Polymer with and without UBM, as well as Copper Post WLPs had a great improvement in thermo-mechanical reliability performance over the Bump on I/O WLPs.

Thermo-mechanical reliability of solder joints of WLP packages can also be improved by optimizing ball geometry and array size and pitch. The more compliant solder ball is, the greater thermo-mechanical reliability is achieved. Therefore, reducing individual solder ball volume or using more compliant materials such as polymer cored solder balls will improve reliability performance. As solder balls become the weakest link during thermal cycling, increasing the contact area-solder ball opening diameter is the most effective way to improve the solder joint reliability. Also, small array size and lower pitch size has better reliability compare to higher array size and pitch size.

Materials also play important roles in enhancing solder joint reliability. In Ball on Polymer WLP structure, polymer film between silicon and solder balls creates a 'cushion' effect to reduce the stresses in solder joints. Such cushion effect can be achieved either by an extremely compliant film or a 'hard' film with large coefficient of thermal expansion. In the later case, the reduction of solder joint stresses is due to the overall increase of the combined film/wafer effective CTE. It has been found that a 'hard' layer with a large CTE can reduce solder joint stress beyond a compliant film. This has been validated by encapsulated Copper Post WLP structure, which showed the best performance on all four structures in terms of solder joint reliability.

The crack is in bulk solder at package side. The cracks initiate from both sides of the solder joint. The cracks propagate from edge toward the center of the solder ball. The corner balls are most susceptible to solder joint failures. Based on test data, making corner balls electrically not connected improves the WLP reliability by 20%. It is concluded that for a given ball array size, smaller pitch gives better solder joint life.

References

1. Fan, X.J., Han, Q., 2008. "Design and reliability in wafer level packaging," *Proc of IEEE 10th Electronics Packaging Technology Conference (EPTC)*, pp. 834-841, 2008

2. Fan, X.J., Liu, Y. 2009. "Design, Reliability and Electromigration in Chip Scale Wafer Level Packaging," *ECTC Professional Development Short Course Notes*.

3. Reche, J.J.H., and Kim, D.H. 2003. "Wafer level packaging having bump-on-polymer structure, Microelectronics Reliability," 43, pp. 879-894.

4. Kim, D.H., Elenius, P., Johnson, M., and Barraett, S. 2002. "Solder joint reliability of a polymer reinforced wafer level package," *ECTC*.

5. Anand, L., 1985. "Constitutive equations for hot working of metals," *J. Plasticity*, 1, pp. 213-231.

6. Bumping Design Guide, [online], Available: http://www.flipchip.com/

7. Darveaux, R., Banerji, K., Mawer, and Dody, G. 1995. Reliability of plastic ball grid array assembly, Ball Grid Array Technology, Lau, J. ed, McGraw-Hill, New York

8. Darveaux, R. 2000. "Effect of simulation methodology on solder joint crack growth correlation," 2000, *Proc. ECTC*.

9. Fan, X. J., Pei, M., and Bhatti, P.K. 2006. "Effect of finite element modeling techniques on solder joint fatigue life prediction of flip-chip BGA packages," *Proc. Of IEEE Electronic Components and Technology Conference (ECTC)*, May 30-June 2, San Diego, CA.

10. Kawahara, T. 2002. SuperCSPs, IEEE Transactions on Advanced Packaging, v.23, No. 2.

11. Reinikainen, T.O, Marjamäki, P., Kivilahti, J.K. 2005. "Deformation characteristics and microstructural evolution of SnAgCu solder joints," *EuroSimE*.

12. Rahim, M.S.K., Zhou, T., Fan, X.J., Rupp, G. 2009. "Board level temperature cycling study of large array wafer level packages," *Proc of Electronic Components and Technology Conference (59th ECTC)*, pp. 898-902.

13. Syed, A. 2001. "Predicting solder joint reliability for thermal, power, & bend cycle within 25% accuracy," *51st ECTC*, pp. 255-263.

Development of High Speed Cold Ball Pull as a Quick Turn Monitor for Solder Joint Reliability

Yu Wang, Liqiang Cao
Institute of Microelectronics of Chinese Academy of Science
NO.3 Beitucheng west Road, Chaoyang District, Beijing, 100029, China
Tel: +86-10-82995675-8007
Email: yuwang9@gmail.com

Abstract

This paper investigates the cold ball pull methodology as a quick turn monitor for characterizing the solder joint reliability of electronic packages. Since cold ball pull is a relatively new metrology, there is still no standard to regulate this testing method. In this paper, the dependence of cold ball pull test result on pulling speed is investigated. It is found that the strength of solder joints is sensitive to strain rate. For lead-free solders, high pulling speed should be employed to better expose the failures at brittle IMC interfaces. Then the high speed cold ball pull is compared with JEDEC standard board level drop test at 1500g/0.5ms to evaluate its accuracy. It is found that the high speed cold ball pull test can qualitatively match the JEDEC standard board level drop test. It can be employed as an effective quick turn monitor for evaluating solder joint reliability performance in electronic package development and high volume manufacturing.

Introduction

The reliability of IC (integrated circuit) assemblies in handheld and portable electronic devices such as mobile phones and PDAs has become a great concern, because they are prone to experience drop damage resulted from mishandling during transportation or customer usage environment. The most commonly seen failure caused by the dropping events is the cracking of the solder interconnections between ICs and printed circuit boards (PCBs). The standardized method to evaluate and compare drop performance of surface mount components for handheld electronic products is the board level drop test (following JESD22-B111). However, the drop test has some inherent disadvantages within which the most problematic ones may be the long test building and executing time. Therefore, due to increasing demand for short time-to-market, it is quite necessary to conduct in-depth studies to develop reliable and robust quick turn monitors for characterizing solder joint integrity both in product development and during high volume manufacturing.

After comparing numerous options, the cold ball pull test is investigated in this work due to its inherent superiority over other alternatives. Compared to ball shear test, cold ball pull exposes a symmetrical stress distribution on the test bond, so it is more prone to reveal the weak interfaces in the solder joint [1]. The same as drop test, cold ball pull is performed on the solder joints at room, or "cold", temperature, so it does not take any risk of changing the solder morphology and intermetallic structure at the solder/package interface as observed in hot ball pull test [2].

Former works have shown that the results of cold ball pull test are dependent on many test condition parameters [3]. This paper focuses on studying the dependence of cold ball pull test on strain rate and the correlation of cold ball pull test with JEDEC standard board level drop test. The goal is to develop this metrology into a reliable and robust quick turn monitor for characterizing solder joint integrity.

Experimental setups

In this work, the cold ball pull test is performed with a Dage 4000 microtester, which is shown in Figure 1. For ball pull application, the Dage 4000 microtester is equipped with a set of pull jaws with different sizes. As shown in Figure 2, the pull jaw is composed of two cylindrical cavity shaped tips which can hold the solder balls in a firm manner. To test specific sized solder balls, the proper pull jaw has to be installed on the machine and the corresponding clamping pressure has to be set. Improper pull jaw or clamping pressure may extrude the solder balls and induce noise into the test outputs.

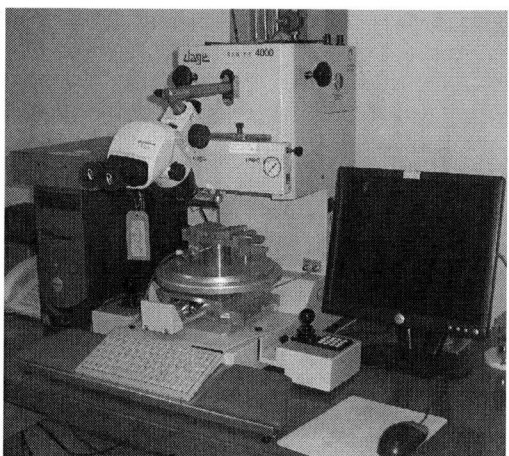

Figure 1 Dage 4000 microtester

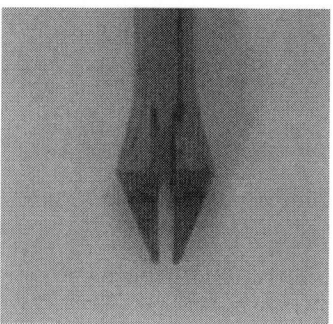

Figure 2 Pull jaw

The cold ball pull test starts with setting up the test parameters in the software and installing the sample on the clamp. The process flow of performing the pull is schematically shown in Figure 3. The pull jaw is first aligned over the target solder ball. Then it descends to substrate surface, clamps the target ball and pulls it up with pre-defined speed. By inspecting the solder balls after pulling, four typical failure modes can to be identified and recorded according to the schematics shown in Figure 4. Of the four failure modes, Mode 1 reflects substrate strength issues. Mode 2 is ductile failure through the bulk solder and indicates for a good bond. Mode 3 reflects improperly selected test conditions or contamination of the pull jaw, so it should be avoided in tests. Mode 4 is brittle failure caused by the fracture of the brittle intermetallic compound (IMC) between the solder and pad surface finish.

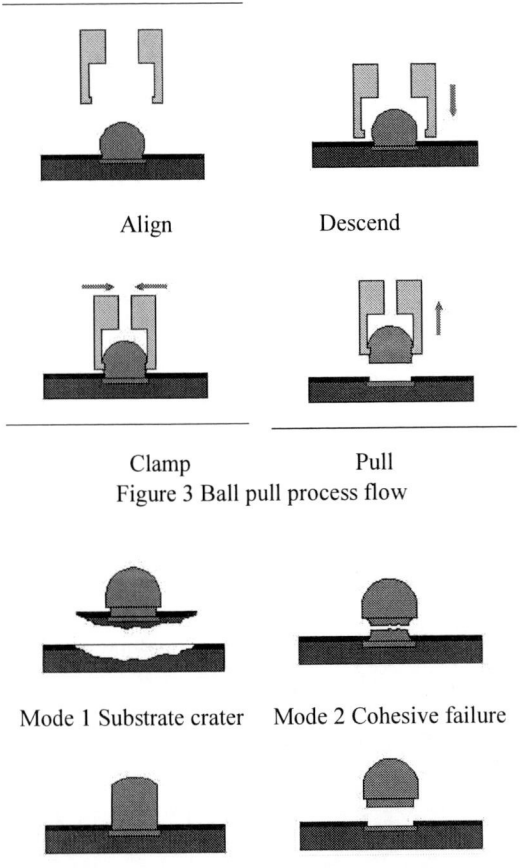

Figure 3 Ball pull process flow

Mode 1 Substrate crater   Mode 2 Cohesive failure

Mode 3 Ball extrusion   Mode 4 Brittle IMC failure
Figure 4 Schematic failure modes

Experimental procedures

The test samples under evaluation in this work are ball grid array (BGA) packages with a sphere diameter of D. The chemical composition of lead-free solder alloy was x%Sn-y%Ag-z%Cu. The solder bond pads were solder mask defined with an opening of 0.85D. The surface finish of the bond pads was Ni/Au. The substrates for solder ball attachment were BT laminates with a thickness of 0.6D. To evaluate the robustness of the cold ball pull metrology, the test samples were divided into four batches with different solder joint reliability performances resulted from the variation in substrate process and supplier. All the solder balls were attached to the bond pads using the same reflow profile. After reflow, they experienced the same aging time under room temperature. The pull jaw size selected for these tests were 0.85D.

The cold ball pull tests were performed with three pulling speeds of 1mm/s, 5mm/s and 100mm/s. 5 components from each batch were tested at each pulling speed. 12 solder balls were selected on each package. The failure mode results after ball pulling were collected and recorded for analysis.

Standard board level drop test following JEDEC standard shock level of 1500g/0.5ms was also performed using samples from the same four batches. All the packages were assembled on the same type of printed circuit board (PCB) with the same surface mount process parameters. Failure analysis was performed on the failed solder balls to confirm the failure mode. The solder ball life distribution results were collected to evaluate the performance of the cold ball pull metrology.

Results and discussion

Figure 5 provides the cold ball pull test results as a function of pull speed. It is shown that with the pull speed of 1mm/s and 5mm/s, no differences in the solder ball failure mode were detected among the four tested sample batches. All the tested samples showed cohesive failures through the bulk solder as shown in Figure 6. However, with the elevated pull speed of 100 mm/s, a significantly increased rate of brittle failures through the IMC layer between the solder and pad was observed as shown in Figure 7. This transition of failure mode can be illustrated by the effect of high strain rate on the strength behavior of solder alloys. As shown in Figure 8, the yield strength of the solder increases at high strain rates [4]. On the other hand, the brittle fracture strength of the IMC layer is not sensitive to strain rate and falls within a narrow band of values. At low strain rates where the UTS of the solder is lower than the brittle fracture strength of the IMC, the cohesive failures through the bulk solder occur before the brittle IMC fractures. Therefore, only mode 2 failures were observed in the low speed cold ball pull experimental results. However, at high strain rates as the solders experience in drop test, the UTS can be much higher than the brittle fracture strength. In this case, the IMC can be much more readily exposed as the weakest link in the force chain.

As shown in Figure 5, high speed cold ball pull test results also indicated quite different rates of brittle IMC failures among the four sample batches. Batch 1 and 4 showed much higher rate of brittle IMC failures than batch 2 and 3. It can be concluded that the solder interconnections in batch 1 and 4 have much high risk of breaking under drop impact.

978-1-4244-4658-2/09 $25.00 © 2009 IEEE

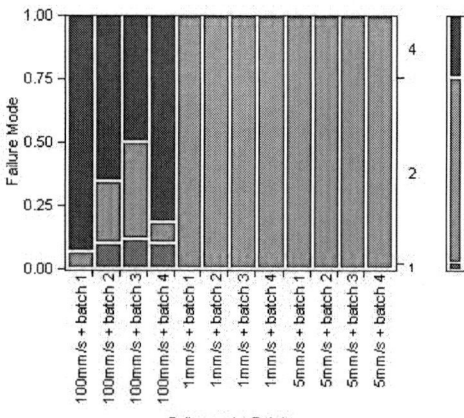

Figure 5 cold ball pull test failure mode results

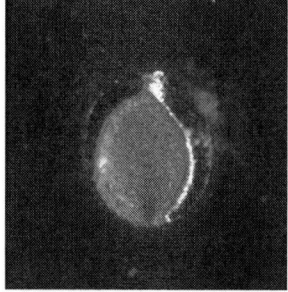

Figure 6 cold ball pull induced cohesive failure through the bulk solder

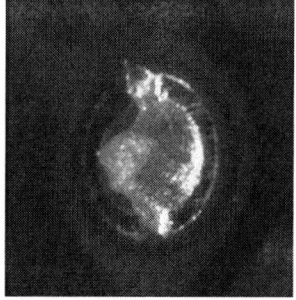

Figure 7 High speed cold ball pull induced brittle failures through the IMC

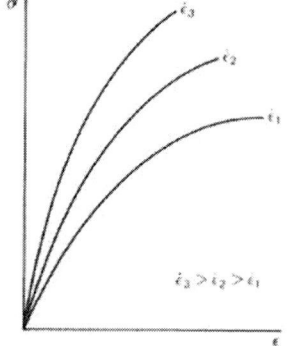

Figure 8 Yield strength change as a function of strain rate

Figure 9 provides the life distribution results from JEDEC standard board level drop test at 1500g/0.5ms using sample packages from the same four batches. The plots showed the failure rates of the solder balls as a function of drop cycles. From the well distributed Weibull plots, it can be concluded that every batch of samples fails in only one single failure mode. To understand the failure mode, dye and pry was conducted on all the tested samples and cross section was done partially. It was confirmed that most of the package side failures were caused by fractures through the brittle IMC layer. Figure 10 shows a failed solder ball after dye and pry. It can be observed that the whole bond pad surface has been covered by red color which means that the dye has penetrated completely through the solder to pad interface. In Figure 11, cross section of the interconnection indicated that cracking occurred thoroughly through the IMC layer. From all the observations above, it can be concluded that high speed cold ball pull test can produce the same mode of failures as JEDEC standard board level drop test.

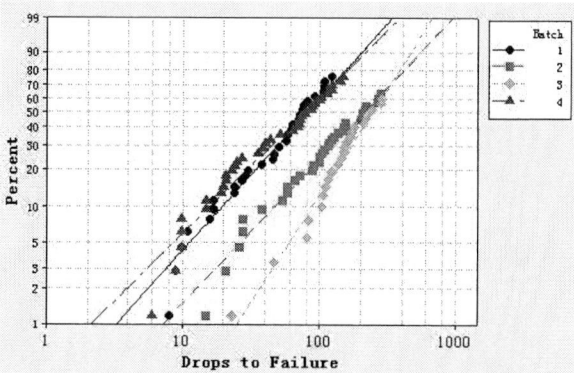

Figure 9 Solder joint life distribution results under drop test

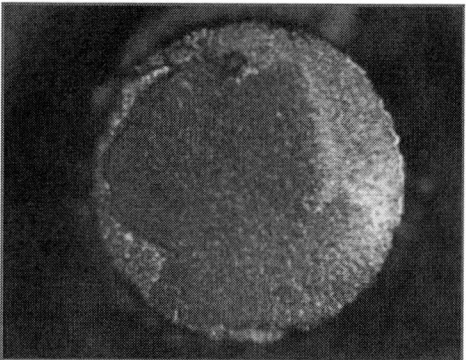

Figure 10 IMC fracture after dye and pry

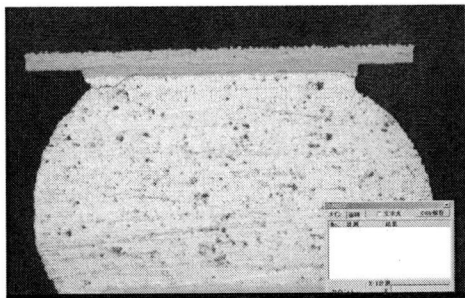

Figure 11 IMC fracture after cross section

As shown in Figure 9, life distribution results from drop test also indicated that the solder interconnections in sample batch 1 and 4 failed much earlier than the ones in batch 2 and 3. These results correlated well with the high speed cold ball pull test results. It can be concluded that the high speed cold ball pull test can qualitatively predict the solder joint reliability performance under drop impact.

Conclusion and recommendation

In the present work, experimental investigation was performed to develop the cold ball pull metrology for characterizing solder joint integrity. The effect of ball pull speed was studied and the robustness of this metrology was evaluated by comparing with JEDEC standard board level drop test. The conclusions come that the strength of the solder joints is sensitive to strain rate and relatively high pull speed should be employed to better expose the brittle IMC interface. The high speed cold ball pull test can qualitatively predict the solder joint reliability performance under drop impact and it can be employed as an effective quick turn monitor for evaluating solder joint reliability performance in product development.

Reference

1. Coyle, R.J. Serafino, A.J. Solan, P.P. (2002), "Ball shear versus ball pull test methods for evaluating interfacial failures in area array packages," *27th Annual IEEE/SEMI International Electronics Manufacturing Technology Symposium*, pp. 200-205.

2. Newman, K. (2005), "BGA brittle fracture-alternative solder joint integrity test methods," *55th Electronic Components and Technology Conference Proceedings*, Vol. 2, pp. 1194-1201.

3. Song, Fubin. Lee, S.W.R., "Effects of testing conditions and multiple reflows on cold bump pull test of Pb-free solder balls," *Electronic Packaging Technology, 2005 6th International Conference on*, pp. 474-480.

4. Hertzberg, Richard W. (1996), Deformation and fracture mechanics of engineering materials, J. Wiley & Sons.

Influencing Factors and Solutions for Ball Short during Wire Bonding

Zhong Meng, Yusheng Feng, Sunggug Lee
Samsung Electronics (Suzhou) Semiconductor Co., Ltd. (SESS)
Address: No. 15, Jin Ji Hu Road, Suzhou-industrial Park, Suzhou, China
Email: zhong.meng@samsung.com

Abstract

During cold season (from November to April), ball short failure caused by ball deformation often occurred and pad corrosion during PCT is also detected simultaneously, so it is urgent to find root cause and assure wire bond quality. The paper reveals that ball deformation and pad corrosion are mainly caused by S and Cl in atmosphere composition. The solution of adding chemical filter is presented and a mini-environment is setup to confirm efficiency, finally ball deformation and PCT failures are eliminated.

1. Introduction

Ball short failure is one of the typical issues during wire bonding process, and it is often caused by ball deformation due to initial golf ball as shown from Fig. 1.

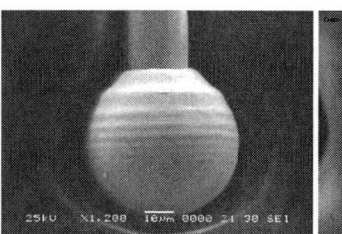
Fig. 1.1 Initial good ball

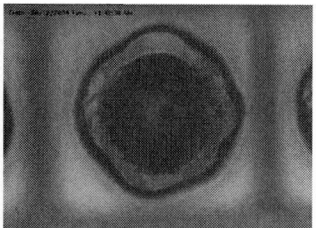
Fig. 1.2 Good ball on pad

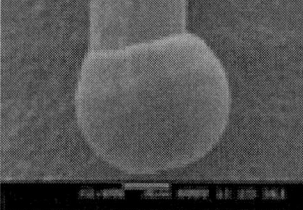
Fig. 1.3 Initial golf ball

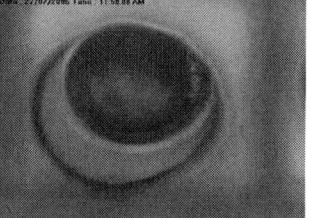
Fig. 1.4 Ball deformation on Pad

Fig.1 Initial good ball and initial golf ball

For a long time in SESS mass production line, it is detected that there is a very high PPM ball deformation ratio in wire bonding process, but the initial ball is very good, so this time ball deformation is not caused by initial golf ball.

And a very interesting phenomenon was found for ball deformation issue: according to Monthly deformation QCN (Quality Control Notification) data from 2006 to 2008 which is shown in Fig. 2, it is obviously that annual deformation ratio (in ppm level) is in the same trend and it has a very strong relationship with seasons, the ratio is very high during cold seasons especially in winter compared with other seasons in the year.

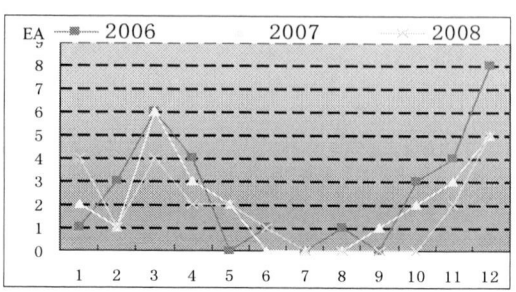
Fig. 2 Monthly deformation QCN data from 2006 to 2008

Purpose of the paper is to reveal the root cause of ball deformation.

2. Experiment & results

Many factors such as wire bond parameters, wire bonder status as listed in Tab. 1 have been taken into consideration first but ball the situation can't be improved.

Tab.1 Check item to deal with ball deformation issue

Check Item		Measurements	Result
Material	Wire	Vendor contrast	NO
	L/F	L/F SEM & EDX analysis	NO
	Cap'y	New cap'y test	NO
Operation		New finger cot test	NO
Machine		Wire path cleaning	NO
		Parameter optimized	NO
Line humidity		Line humidity up to50%	NO

Finally the suspicion is focus on capillary contaminations. Fig. 3 explains formation of ball deformation: Free Air Ball (FAB) is formed by sparking, and normally FAB should be in the center of capillary hole. If the capillary is contaminated by particles, the FAB could not be centered in capillary, finally causes ball deformation.

Fig. 3.1 Centered FAB

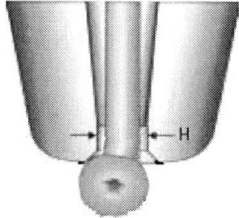
Fig. 3.2 Off-center

Fig. 3 Formation of ball deformation

Further SEM observation confirmed that capillary hole is contaminated by particles.

Fig. 4 Capillary contamination (particles)

Also serious pad corrosion was found simultaneously during Pressure Cooker Test (PCT) test.

Fig. 5 Pad corrosion

2.1 Analysis

Particles inside capillary hole are examined using Scanning Electron Microscope (SEM) & Energy Dispersive X-Ray Analysis (EDX). The result is showed in Fig. 6 and Sulfur with 27.03% atom ratio is detected. In contrast, Sulfur couldn't be found in the capillary where there is no ball deformation (Fig. 7).

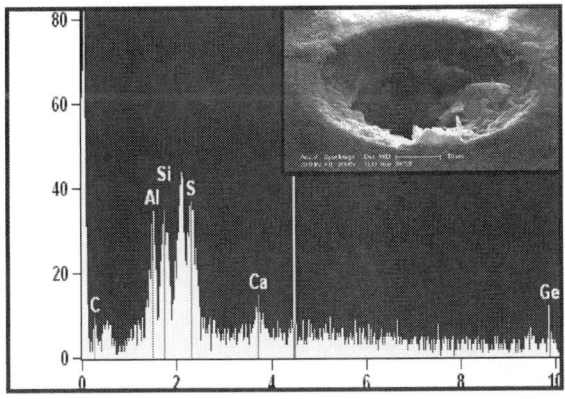

Fig. 6 EDX result of contaminated capillary

And for corrosion pad, Cl with atom ratio of 7.1% is detected. The result is showed in Fig. 8.

For comparison, no Cl could be detected in normal pad after PCT test.

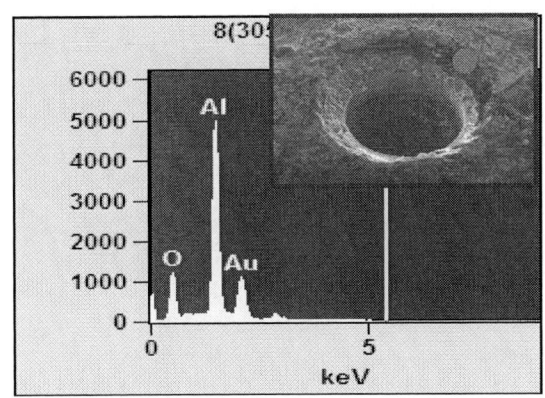

Fig.7 EDX result of normal capillary

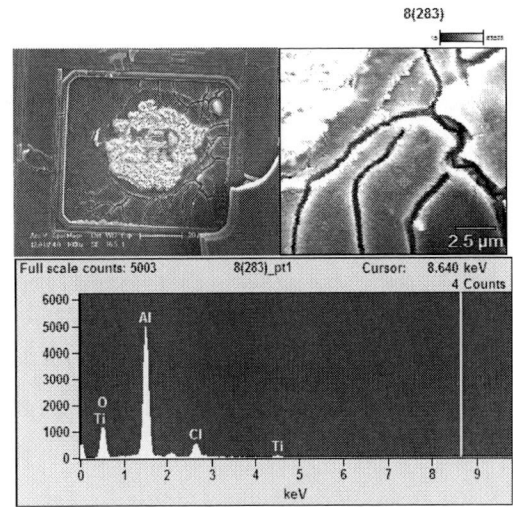

Fig. 8 EDX of pad corrosion

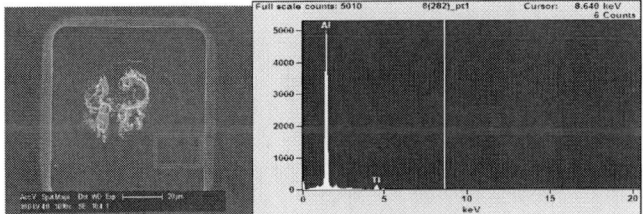

Fig. 9 EDX of normal pad

And used capillaries in different lifetime are also observed by EDX. Cl had been detected when capillary lifetime is 300K whereas S had been detected when capillary lifetime is 2000K. And S & Cl acid ion can't be detected for capillary in normal process. Results are shown in Fig. 10 and Fig. 11.

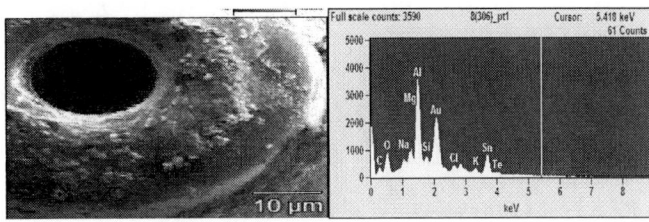

Fig. 10 EDX for capillary with 300K lifetime

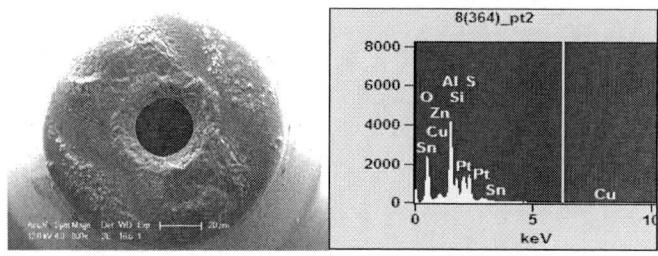

Fig. 11 EDX for capillary with 2000K lifetime

Based on analysis above, we got the following conclusions:

First, S ion will react with metal on capillary surface during sparking fire process; the reactants will accumulate on capillary tip. If the capillary hole is not smooth, FAB can't be centered, and then ball deformation will occur.

Second, acid compound in capillary tip may also contaminates bond pad, corrupts Al pad during PCT test, and causes pad corrosion.

Contamination of capillary was confirmed, but the source of contamination is still a great concern for the issue.

2.2 Investigation for source of S & Cl

With the development of modern industry, atmosphere quality becomes worse, especially in cold seasons like winter. And the source of contaminations are suspected coming from atmosphere inside clean room.

To find the atmosphere composition difference, we've checked air circulation system (shown in Fig. 12, OAC is fresh air supplier for AHU, and AHU is filter. Air was circulated between AHU and cleaning room) of the line at 5th Dec. 2008. Totally 4 different points are checked and Tab. 2 gives the result.

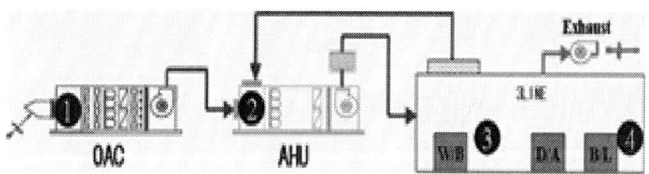

Fig. 12 Air circulation system

Tab.2 Clean room atmosphere composition (Unit: $\mu g/m^3$)

Analysis Item	Detect system	Check point			
		1	2	3	4
HCl(Cl⁻)	≤ 0.2	0.3	0.3	0.8	0.9
$NO_2(NO_3^-)$	≤ 0.2	0.3	0.4	0.7	0.8
$SO_2(SO_3^{2-})$	≤ 0.2	8.3	8.7	16.2	16.7

It is concluded that S & Cl existed in all the points and their concentrations in cleaning room (point 3, 4) was higher than fresh air (point 1, 2). Although we didn't analyze atmosphere composition for other seasons but it is believed air quality in winter is the worst and S & Cl was lower in other seasons. This explains the reason why deformation ratio is very high during winter.

2.3 Chemical filter estimation

Since ball deformation and pad corrosion is caused by S & Cl, a chemical filter was suggested to improve cleaning room airborne molecular contaminants.

The principle of chemical filter was shown in Fig. 13. Chemical substance in filter would react with S & Cl, and then S & Cl in air could be filtered.

To confirm efficiency of chemical filter, we built a mini-environment: Two wire bonders were put in a close room and a chemical filter was on top of the close room to filter the air. Mini-environment model is shown as Fig. 14, and FFU was a fan to supply power for air circulation.

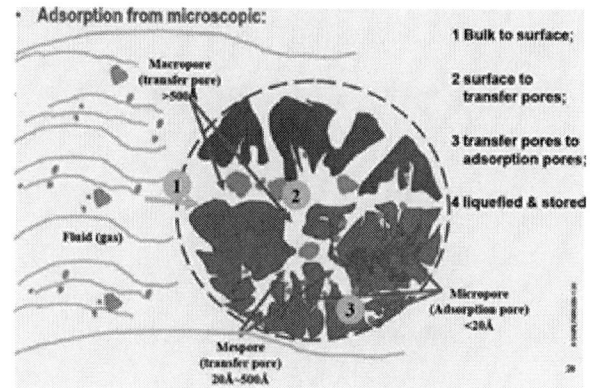

Fig. 13 Chemical principle

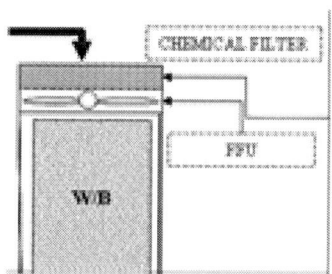

Fig. 14 Mini-environment

Then molecular contaminants inside mini-environment were checked, and the data is listed in Tab. 3.

Tab.3 Mini-environment atmosphere composition (Unit: $\mu g/m^3$)

Analysis Item	Detect System	Mini-Environment
HCl(Cl⁻)	≤ 0.2	0.2
$NO_2(NO_3^-)$	≤ 0.2	<0.2
$SO_2(SO_3^{2-})$	≤ 0.2	0.5

Compared with polluted air with SO_2 concentration more than 16.0 and Cl concentration up to 0.8 in Tab. 2, air condition in mini-environment is greatly improved.

Also bonding quality in mini-environment is checked, capillary lifetime, wire bond process yield were in compared with other machines outside mini-environment in Fig. 15.

Bonding quality and yield inside mini-environment is dramatically raised, and ball deformation failure eliminated, S

& Cl etc. acid ion is no longer detected on capillary tip, also capillary lifetime reached to up limit.

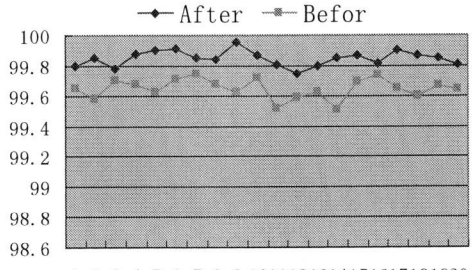

(15.1) Wire bond yield

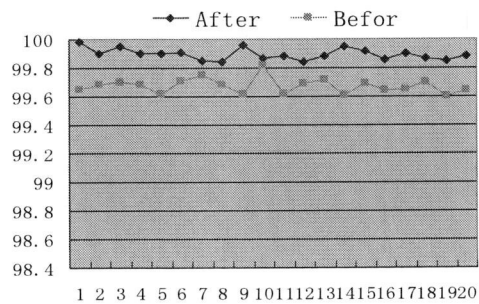

(15.2)Test yield

Fig. 15 Bonding quality comparison

Inside & outside of mini-environment

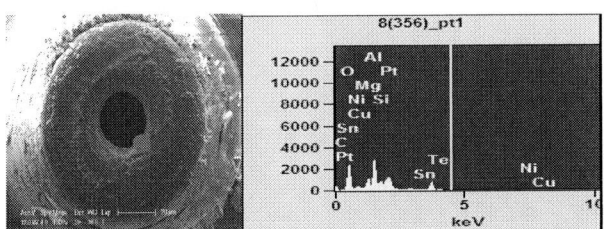

Fig. 16 EDX of capillary tip inside mini-environment

Setup of mini-environment has proved efficiency of chemical filter, so in January in 2009, the chemical filtering system was installed for the whole mass production line.

Conclusions

In this paper, we've confirmed that ball deformation and pad corrosion was caused by airborne acidity molecule especially S and Cl.

The solution with chemical filter was presented, a mini-environment with chemical filter was setup to confirm its efficiency, Ball deformation failure and pad corrosion during PCT can be eliminated.

Acknowledgments

The authors would like to thank Cmafil engineers and Utility department for their support and assistance in experiments and measurement.

References

1. Wang yunhua, Chemical material theory, pp. 112-115.
2. Xiaohua Yu, Air circulation system, pp. 23-28.
3. Fu haiming, Filter theory and development, pp. 8-10.
4. Tawakol, "Wire bond FAB technology," *Tokyo semiconductor*, pp. 72-75.

Modeling Electrochemical Migration through Plastic Microelectronics Encapsulations

M. van Soestbergen[1, 2,*] A. Mavinkurve[3], R.T.H Rongen[3], L. J. Ernst[2], G.Q. Zhang[2]
[1]Materials innovation institute
[2]Delft University of Technology, Fundamentals of Microsystems Engineering
[3]NXP Semiconductors
*Mekelweg 2, 2628 CD, Delft, the Netherlands, m.vansoestbergen@tudelft.nl

Abstract

Plastic encapsulations will absorb moisture in humid environments due to their hydrophilic nature, this in combination with the inherent ionic contamination of the plastic will result in an electrolyte. This electrolyte might pose several reliability issues for the package and the encapsulated microelectronic circuit, such as electrochemical migration and bond pad corrosion. The corresponding failure mechanisms are associated with ionic currents through the package and electrochemical processes at e.g. the leads or bond wires.

In this paper we present a generic mathematical framework for modeling ionic currents and electrochemical processes. We will apply this framework to electrochemical migration of metal between the leads of a plastic encapsulation, and show results for the transient formation of migration fluxes through the plastic encapsulation.

Introduction

Electrochemical migration (i.e. dendrite growth) is a mechanism in which metal is electrochemically removed from one location and deposited at another location [1-4]. Although this mechanism is more common for connections at the exterior of the encapsulation [5-6], it can also occur within the encapsulation itself [7-8]. The mechanisms will initially result in a leakage current between the metal sites [5-6]. For prolonged time the deposited metal can bridge the metal sites, such that a short-circuit occurs (Fig. 1) [8]. The deposited metal can have a dendritic structure, hence the name of the failure mechanisms. Note that the initial leakage current is reversible and might thus lead to intermittent failures whereas the deposition of metal is irreversible leading to permanent failures. Though electrochemical migration generally occurs under moisturized condition, failures under relatively dry conditions are reported as well [7]. Furthermore, the composition of the metal sites is a major factor determining the migration susceptibility, it has been reported that copper, lead and silver are most prone to show migration [1-4].

In the present work we will focus on modeling the initial leakage current between two leads of a package, rather than the development of the deposit which leads to a short-circuit. To do this we present a generic mathematical framework for modeling an electrochemical cell, where a cation (metal ion) is formed at the anode and deposits at the cathode. We assume that the electrolyte phase between the metal leads contains both cat- and anions, which represent the low level of contamination present in the package [9]. We will use the generalized Frumkin-Butler-Volmer equation and the Poisson-Nernst-Planck theory to model the electrochemical processes and the transport of ions, respectively, as we will explain in the next section. (Note that this model is based on earlier work by the authors, cf. refs. [10-11] and refs. herein)

Fig. 1 Cross-section of a package with a metal deposit between two leads

Theory

In this section we will discuss the model as depicted in Fig. 2. The model has four variables we solve for, namely, (1) reactive metal ions, C+, (2) inert cations, Cc, (3) inert anions, Ca, (4) and the electrical potential, V.

We assume that the ions cannot reside infinitesimally close to the metal, such that there is a plane of closest approach for the ions adjacent to the electrodes. The spacing between this plane and the electrode is called the Stern layer. Further, we assume that at the plane of closest approach we have the electrochemical reaction, $C^+ + e^- \leftrightarrow Me$, where Me denotes metal. Consequently, the plane of closest approach coincides with the reaction plane. The electrochemical reaction leads to a flux of metal ions at this reaction plane described by the generalized Frumkin-Butler-Volmer equation,

$$J = K_R C^+ \exp(-\alpha_R \cdot f \cdot \Delta V_s) - J_O \exp(\alpha_O \cdot f \cdot \Delta V_s) , \quad (1)$$

where KR is the reduction rate constant, JO the oxidation constant, □R and □O are the transfer coefficient and f= e/kBT, which is the reciprocal of the thermal voltage (~25.6 mV at room temperature). Further, in Eq. (1) we have the potential drop across the Stern layer, ΔV_s, which is the potential difference between the potential of the metal and the reaction plane, equal to $\Delta V_s = V_{elec} - V_{rp}$. Where V_{rp} is the potential of the reaction plane and V_{elec} is the potential of the electrode, i.e. the potential we can apply or measure. We can compute the potential drop across the Stern layer using the electrical field at the reaction planes, as we will discuss later.

The ion flux at the reaction plane can be coupled to the ion flux through the electrolyte phase, which is given by the Nernst-Planck equation,

$$J_i = -D_i \cdot (\nabla C_i + z_i \cdot f \cdot C_i \cdot \nabla V) \quad (2)$$

where D_i is the diffusion coefficient of species i, C_i their concentration, z_i their valence and V is the local electrostatic potential. The relation between electrostatic potential and space charge density (and thus the ion concentration) is given by Poisson's law,

$$\varepsilon_r \varepsilon_0 \nabla^2 V = -e\big(C^+ + C_c - C_a\big) \qquad (3)$$

where ε_r is the relative dielectric permittivity of the polymer and ε_0 the permittivity of vacuum. If we combine Eq. (2) and (3) with the conservation law, $\dot{C}_i = -\nabla \cdot J_i$, we obtain the time-dependent Poisson-Nernst-Planck transport theory.

To generalize the equations presented above we will normalize the model parameters. First, we introduce the dimensionless electrostatic potential, $\phi = f \cdot V$, which is the ratio of electrical to thermal energy (Note that we implicitly assume that all ionic species we consider have a valence equal to ±1). Next, we scale the ion concentrations to the ionic strength, $c_i = C_i / C_\infty$. Where the ionic strength equals the summation of ion concentration at $t=0$, i.e., $C_\infty = \frac{1}{2}\sum_i C_{i,t=0}$. Further, we scale the spatial coordinate system to the lead spacing, $x_i = X_i / L$, where i indicates the x and y direction. Adjacent to the reaction plane we have a local region containing a non-zero space charge density, the so-called diffuse layer, (Fig. 2) with a thickness in the order of the Debye length, $\lambda_D = \sqrt{\varepsilon k_B T / 2e^2 C_\infty}$. We introduce $\in = \lambda_D / L$ as the Debye length on the dimensionless length scale. Next, we define the dimensionless Stern layer thickness as, $\delta = \lambda_s / \lambda_D$, where λ_s is the dimensional Stern layer thickness. Further, we use the diffusion-limited current to scale all ionic currents in the system to $j = J/J_{DL} = J \cdot L / 4DC_\infty$ (Note that we assume equal diffusion coefficients for all ionic species we consider). Consequently, we have $k_R = K_R C_\infty / J_{DL}$ and $j_O = J_O / J_{DL}$ for the reduction and oxidation rate constant. Finally, we have the dimensionless time, $\tau = t \cdot D / L^2$.

We now solve the dimensionless Nernst-Planck equation,

$$\frac{dc_i}{d\tau} = \nabla \cdot \big(\nabla c_i \pm c_i \cdot \nabla \phi\big) \qquad (4)$$

and the dimensionless Poisson equation,

$$\in^2 \cdot \nabla^2 \phi = -\tfrac{1}{2}\big(c^+ + c_c - c_a\big) \qquad (5)$$

at the interior of the model, where the ±-sign refers to the positive (c^+, c_c) and negative ions (c_a), respectively.

The boundary conditions that are applicable to the model are as follows. Near the electrodes we have the reaction plane as the boundary for transport, where we have

$$\big(\nabla c^+ + c^+ \cdot \nabla \phi\big) \cdot \boldsymbol{n} = 4 \cdot j \qquad (6)$$

for the reactive metal ions and

$$\big(\nabla c_c + c_c \cdot \nabla \phi\big) \cdot \boldsymbol{n} = \big(\nabla c_a - c_a \cdot \nabla \phi\big) \cdot \boldsymbol{n} = 0 \qquad (7)$$

for the inert ions of the background electrolyte, where

$$j = k_R c^+ \exp\big(-\tfrac{1}{2}\Delta \phi_s\big) - j_O \exp\big(\tfrac{1}{2}\Delta \phi_s\big) \qquad (8)$$

and $\boldsymbol{n}$ is the outward normal vector. We can also determine the potential drop across the charge free Stern layer from the electrical field at the reaction plane according to,

$$\Delta \phi_s = \varepsilon \cdot \delta \cdot \nabla \phi \cdot \boldsymbol{n} \qquad (9)$$

Consequently, we can apply the following boundary condition for the potential to the reaction plane,

$$\phi_{rp} = \phi_{elec} - \Delta \phi_s \qquad (10)$$

As a result we do not require to model the Stern layer (since we implement it via the boundary conditions), such that the reaction plane is the outer boundary of the model.

The speed of electrons through the leads is not infinitesimal fast. Consequently, the electrode potential will not rise instantaneously. Therefore we use,

$$\phi_{elec} = \phi^0 \big[1 - \exp(-\beta \cdot \tau)\big] \qquad (11)$$

where $1/\beta$ is the characteristic rise-time of the potential.

The far field condition in Fig. 2 represents the position in the model where we assume both the electrical field and the transport of ions to vanish since the distance from the electrodes to this position is large. The remaining symmetry boundary conditions, as shown in Fig. 2, are trivial.

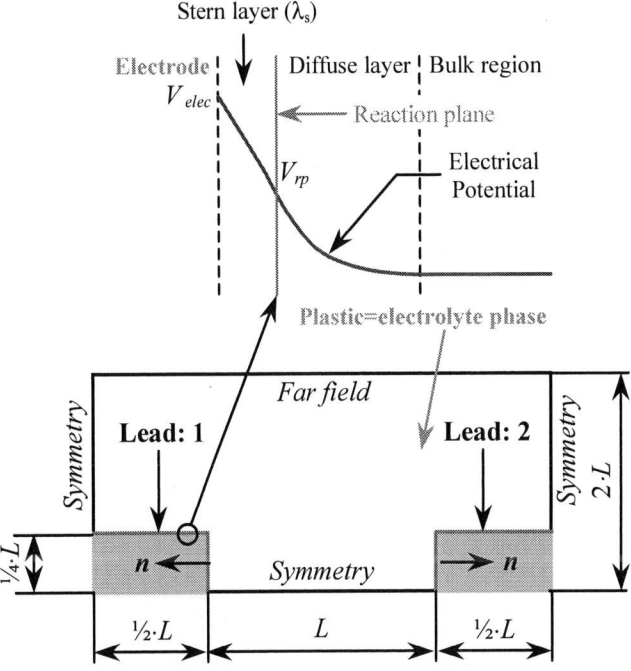

Fig. 2 Schematics of the geometry of the model

Results and Discussion

In this section we show results (i) for the reactive ion flux at the leads as function of time and (ii) for the flux throughout the model at different times. Besides, we will show the final distributions of reactive ions throughout the model, i.e. at the steady-state.

First, we discuss the input parameters that are applicable to the model. The Debye length is typically in the order of several nm for commercially available plastic encapsulants, while the lead spacing is in the order of tenths of mm, such that we take $\in = 10^{-4}$. The Stern layer thickness is in the order of an atomic layer, but can increase further due to e.g. an oxide layer, therefore we set δ equal to 1. Next, we require initial condition for the ion concentration. We use $\chi_{a=0}=1$, $c_{c,\tau=0}=1\cdot10^{-6}$ and $c^+_{\tau=0}=10^{-6}$. Note that $c^+_{\tau=0}$ equal to zero would make the model computational much more stringent. Further, we consider leads of similar metal composition, such that there is zero potential difference between them at open-circuit conditions. Consequently, the kinetic rate constants for both leads are identical. Besides, we assume that these constants are sufficiently high, such that the current is limited by the diffusion of reactive ions. Consequently, we set the kinetic parameters for both electrodes equal to $k_R=10^6$ and $j_O=1$. Finally, we apply $\phi°=0$ to lead 1 (left side) and $\phi°=100$ to lead 2 (right side), while we take $\beta=100$.

The applied electrode potential as function of time is presented in Fig. 3. The corresponding flux of reactive particles, c^+, (integrated over the length of the leads) as function of time is presented in Fig. 3 as well. Fig. 3 shows that the flux at lead 2 initially increases with the applied electrode potential until it reaches a maximum, after which it decays to the final steady-state flux. The flux at lead 1 decays (after a very initial increase) with increasing electrode potential until the electrode potential reaches its maximum. After reaching this maximum the flux at lead 1 increases to the steady-state flux. The initial increase of the flux at lead 1 is the results of the removal of reactive ions in the near vicinity of the electrodes. Due to the deficit of ions near lead 1 for more prolonged time the flux will decrease again until the supply of reactive ions from lead 2 has reached lead 1. The propagation of this supply is presented in Fig. 4. Here it is shown that initially the current is mainly at lead 2, whereas it distributes along the whole model at the steady state. The initial increase of the reactive particle flux at lead 2 is due to the increasing electrode potential, and thus the increasing potential difference across the Stern layer. However, as the fluxes at both leads are not equal, the concentration of reactive species throughout the whole model will increase. Consequently, the back-reaction rate in Eq. (8) will increase, which will decrease the particle flux at lead 2.

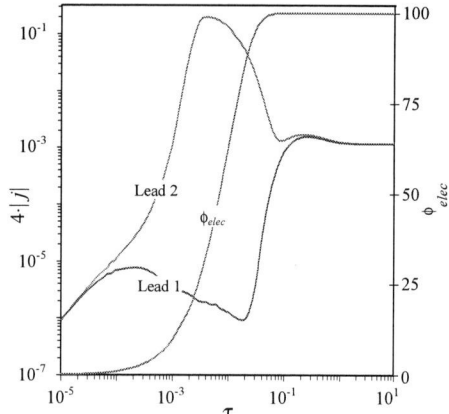

Fig. 3 Reactive particle flux at both leads and their potential difference as function of time; kR=106, jO=1, c+τ=0=10-6, δ=1, ∈=10-4, φ°=100 and β=100 (lead 2).

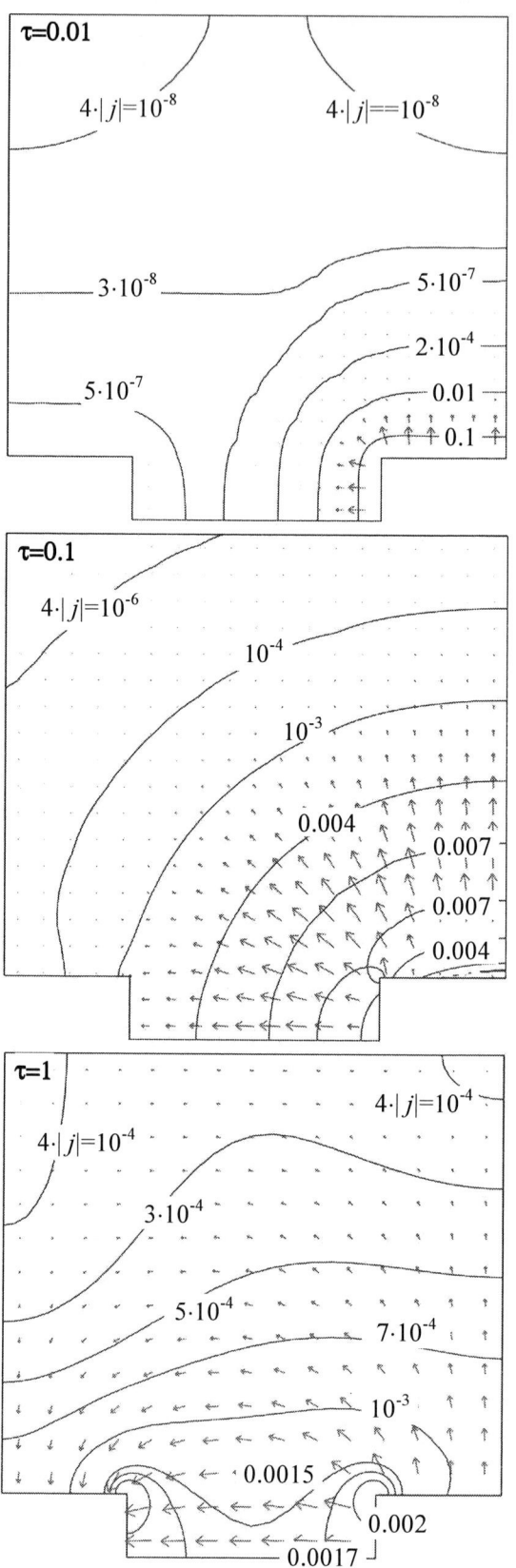

Fig. 4 Reactive particle flux, j, throughout the model for τ=0.01, 0.1 and 1; lines represent the norm of j, arrows represent a vector plot indicating the direction of the reactive ion flux; model parameters identical to Fig. 3.

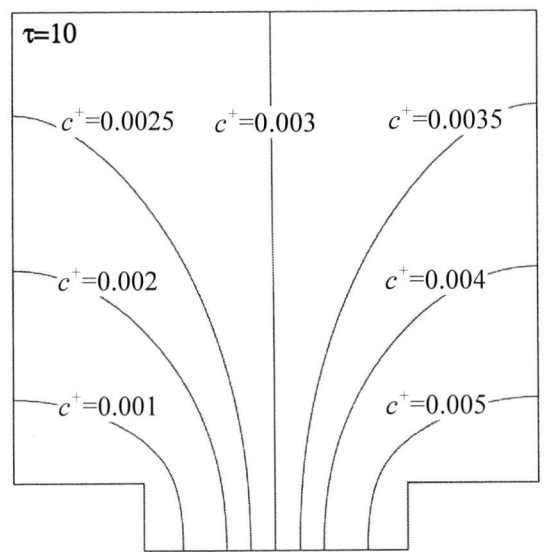

Fig. 5 Reactive particle concentration distribution at $\tau=10$; model parameters identical to Fig. 3.

The steady-state distribution of reactive particles is presented in Fig. 5. Obviously, the plastic encapsulation can store some charge, which is reflected by the increased reactive ion concentration. However, the concentration distribution does not show a leakage path between the leads as might be expected (Fig. 1). Such a path originates from an imperfection at the lead surface (i.e. a nucleation site), which is not included in our model.

Conclusions

We have presented a generic framework to model the ionic currents throughout plastic encapsulations. We have applied this framework to model the initial electrochemical migration of metal between two electrodes (i.e. leads) in a plastic encapsulation. The results show that the electrochemical processes significantly increase the overall reactive (metal) ion concentration in the system. Further, the total flux of ions at the electrodes is not a monotonic function of time as it has a distinct maximum (or minimum) until it reaches the steady-state. Next, the steady-state is reached at the characteristic diffusion-time of ions between the electrodes. Finally, the formation of the dendritic structure and thus the leakage path are *not* modeled as they originate from imperfections at the electrodes, which are not included in this model. Nevertheless, the framework presented here can be a very powerful tool in predicting ionic currents in plastic encapsulation and their corresponding failures.

Acknowledgments

This research was carried out under project number MC3.05236 in the framework of the Research Program of the Materials innovation institute M2i (www.m2i.nl).

References

1. Steppan, J. J. *et al.*, "A Review of Corrosion Failure Mechanisms during Accelerated Tests," *Journal of the electrochemical society* Vol. 134, (1987), pp. 175.
2. Harsanyi, G., Inzelt, G. "Comparing Migratory Resistive Short Formation Abilities of Conductor Systems Applied in Advanced Interconnection Systems," *Microelectronics Reliability* Vol. 41, (2001), pp. 229.
3. Harsanyi, G., "Electrochemical Processes Resulting in Migrated Short Failures in Microcircuits," *IEEE CPMT* Vol. 18, (1995), pp. 602.
4. Manepalli, R., "Silver Metallization for Advanced Interconnects," *IEEE Trans. Adv. Packaging* Vol. 22, (1999), pp. 4.
5. Yang, S., Wu, J., and Pecht, M., "Electrochemical migration of land grid array sockets under highly accelerated stress conditions," *IEEE Proceedings of the 51st Holm Conference on Electrical Contacts*, Chicago, IL, Sep. 26-29, 2005, pp. 238.
6. Yang, S., Wu, J., and Chistou, A., "Initial stage of silver electrochemical migration degradation," *Microelectronics Reliability* Vol. 46, (2006), pp. 1915.
7. Tan, S.H., Ong, S.H., "A Dry Migration? Copper Dendrite Growth in Adhesive Tape During Burn-in," *Proc. Int. Symp. Physical and Failure Analysis of Ics*, Singapore, 2001, pp. 178.
8. Pecht, M., Deng, Y., "Electronic device encapsulation using red phosphorus flame retardants," *Microelectronics Reliability* Vol. 46, (2006), pp. 53.
9. Lantz, L., Hwang, S., Pecht, M., "Characterization of plastic encapsulant materials as a basline for quality assessment and reliability testing," *Microelectronics Reliability* Vol. 42, (2002), pp. 1163.
10. Van Soestbergen, M. *et al.*, "On Electrochemical Cell Modelling as Basis for Predicting Corrosion Failures in Plastic Encapsulated Microelectronics," *10th EurosimE Conference*, Delft, the Netherlands, 2009, pp. 176.
11. Van Soestbergen, M., *et al.*, "Modelling Ion Transport through Molding Compounds and its Relation to Product Reliability," *9th ICEPT-HDP Conference*, Shanghai, China, 2008.

Microstructure Changes and Compound Growth Dynamic at Lead-free/Cu Interface under Different Conditions

Lihua Qi[1,2], Jihua Huang[1], Jing Niu[2], Long Yang[2], Yaorong Feng[2], Xingke Zhao[1], Hua Zhang[1]

[1]School of Material Science and Technology, University of Science and Technology Beijing

[2]Tubular Goods Research Center of CNPC

E-mail address: qlh1973@163.com

Abstract

The atom diffusion and growth behavior of intermetallic compound (IMC) at Sn3.5Ag0.5Cu/Cu interfaces under isothermal aging and thermal-shearing cycling conditions were investigated. The results show that the morphology of Cu_6Sn_5 IMCs formed at Sn3.5Ag0.5Cu/Cu interface changed gradually from scallop-like to chunk-like, and IMCs thickness developed with the isothermal aging and thermal-shearing cycling times increasing. Furthermore, Cu_6Sn_5 IMC growth rate under the thermal-shearing cycling condition was higher than that of under isothermal aging. Compared to isothermal aging condition, only one Cu_6Sn_5 layer was formed and developed at the interface between SnAgCu solder and Cu substrate after 720 cycles. IMC growth dynamic equation is $Y = K_\varepsilon \sqrt{D_{Cu}^\varepsilon t} + K_\eta \sqrt{D_{Cu}^\eta t}$ with K_η is about $10^{-5} \mu m^2/s$ under isothermal aging, and that is $Y = y_0 + K_\eta \sqrt{D_{Cu}^\eta t}$ under the thermal-shearing cycling condition with dynamic parameter $K_\eta = \dfrac{\sqrt{2}\left(C_\eta^{\eta/\varepsilon} - C_\eta^{\eta/\beta}\right)^{\frac{1}{2}}}{\left(C_\eta^{\eta/\beta} - C_\beta^{\beta/\eta}\right)^{\frac{1}{2}}}$ is about 10^{-5} to $10^{-4} \mu m^2/s$.

1. Introduction

SnAgCu solders system was considered as one of prominent lead-free alloys replaced the Sn-Pb solder system because its microstructure is denser and IMC growth speed is lower than that of Sn-Zn, Sn-Cu and Sn-Ag systems, and lots of papers were mainly focus on the growth dynamics of IMC at the interface under the isothermal aging condition [1-4]. Interfacial microstructure has significant impacts on the reliability of solder joints in microelectronic assemblies. At present, IMCs microstructure transformation and growth behavior at SnAgCu/Cu interface have been researched during different reflowing temperature and isothermal aging

time [1, 5-6]. Practically, Solder joints often suffer shear cycle to lose their function because of periodical stress-strain generated by thermal expansion coefficient difference between print circuit board (PCB) and integrated circuit (IC) under the thermal cycling condition. Furthermore, because of lower-melting point and lower-recrystallization temperature of Sn-base solder, the value of stress-strain has an important function on atom diffusion and IMC growth behavior at the interfaces. However, there is little research on the atoms diffusion, growth behavior and growth dynamic of IMC under the thermal-shearing cycling condition. Specifically, the present paper has investigated the atoms diffusing dynamic and IMC growth behavior at Sn3.5Ag0.5Cu/Cu interface under the thermal-shearing cycling at 25~125°C, -55~125°C and isothermal aging at 125°C conditions.

2. Experimental

A special specimen was designed to study the interfacial microstructure and the IMC growth dynamic in the solder joints during thermal-shearing cycling process. On the one hand, the chip was taken place of tungsten bar because of the similar thermal expansion coefficient ($4.6 \times 10^{-6}/°C$), and copper bar replaced the PCB. Due to the poor wettability of SnAgCu solder on tungsten, nickel foils were selected to join the both ends of tungsten through BNi₂ brazing alloy, and has good weldability with SnAgCu solder. On the other hand copper and tungsten bars are rigid bodies, and low melting solder of SnAgCu is viscoelastic material [7]. Deformation caused by copper and tungsten bars was mainly absorbed by the SnAgCu solder during experimental process. In the sample, a tungsten bar was firstly brazed with the nickel foils at both ends with BNi₂ brazing alloy at 1120°C for 20 mins under a vacuum of 1.6×10^{-4} Pa and then soldered with a copper bar using Sn3.5Ag0.5Cu eutectic alloy, as shown in Fig. 1.

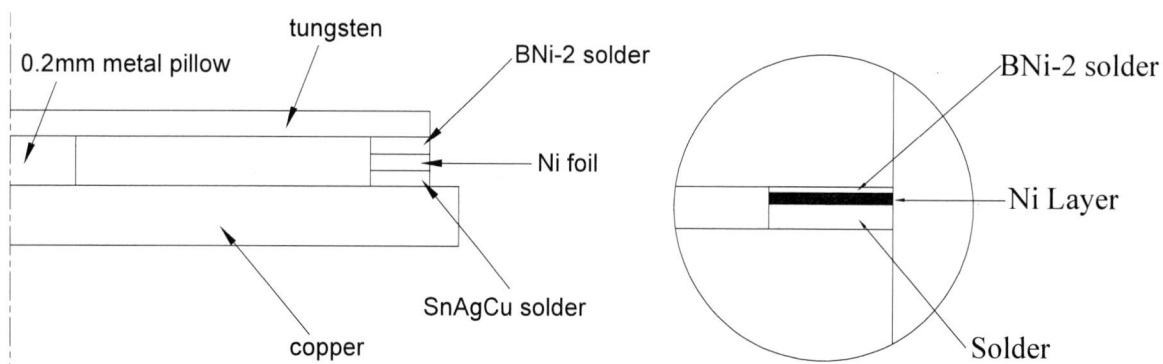

Fig. 1 Schematic of thermal-shearing cycling test (a) half-sample and (b) amplificatory picture of joint

978-1-4244-4658-2/09 $25.00 © 2009 IEEE

The experimental process is as follow: the samples were put into the TSA-71S-A cold & hot blow test box, heated up from 25°C to 125°C in 5 minutes and preserved for 25 minutes at extreme temperature of 125°C, and then cooled down from 125°C to 25°C in 5 minutes and preserved for 25 minutes at extreme temperature of 25°C. It is a cycle period. The samples were handled 24, 200, 400 and 720 cycles, respectively. Another thermal cycles from -55~125°C for 12, 100, 200 and 360 cycles, as remained for 27.5 minutes at the minimum temperature of -55°C and the maximum temperature of 125°C, respectively. Three types of samples were made to explore the effects of different shear-stress on atom transmission and IMC growth tropism at the interface. The specimens are 180×10×10mm (big-sized sample), 90×10×10mm (medium-sized sample) and 45×10×10mm (small-sized sample). The samples for isothermal aging test were put into a constant temperature box at 125°C for 12, 100, 200, 360 and 720 hours, respectively. Specimens were prepared by linear cutting into small pieces (5×5×2mm). The specimens were first cold mounted with epoxy, and then polished and etched with 5% HCl+95% H_2O for better observation of their microstructure and IMC growth behavior by Scanning Electron Microscopy (SEM) and Energy Dispersion Spectroscopy (EDS) and X-ray Diffraction (XRD).

3. Results and discussions

3.1 Intermetallic morphology

Fig. 2 is SEM pictures of IMCs at the Sn-3.5Ag-0.5Cu/Cu interface after isothermal aging for 12, 360, 720 hours. There is an intermetallic compound layer, Cu 58.14%, Sn 41.86% at Sn3.5Ag0.5Cu/Cu interface, formed after reflowing. Increasing isothermal aging, the morphology of compound varies gradually from scallop-like after reflowing to plate-like after aging. After 360 hours, there is another compound layer, Cu_3Sn, formed between Cu_6Sn_5 and Cu substrate, and whose thickness increases remarkably. As we all know that diffusing rate of Cu atom from substrate to interface is higher in liquid state than that in solid state, it leads growth rate of compound at the interface very slow in solid state. Therefore, the same IMC thickness formed for 3 seconds in liquid state will take about 720 hours forming at the interface in solid state. As a result, atom diffusion and compound growth are a slow dynamic process in solid state [8]. With regard to binary diagram of Cu-Sn, Cu_3Sn can be formed under the enough supplying of Cu atom circumstance. Because of low diffusing rate of Cu atom in solid state, some part of Cu atoms has difficult to get through the thick layer of Cu_6Sn_5 and compel to react with Cu_6Sn_5 forming Cu_3Sn at Cu_6Sn_5/Cu interface, and others diffuses through Cu_6Sn_5 lay to react with Sn in the solder. Due to higher activation energy of Cu_3Sn slightly than that of Cu_6Sn_5 at normal temperature [9], the growth rate of Cu3Sn is lower than that of Cu_6Sn_5. All these phenomena prove that diffusion of Cu atom control the development of compound at the interface.

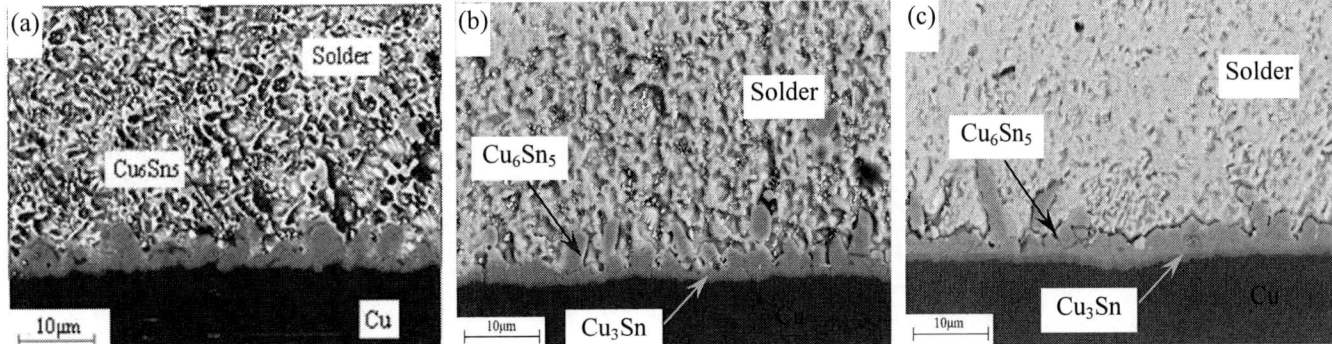

Fig.2 SEM images of Sn-3.5Ag-0.5Cu/Cu interface after isothermal aging at 125°C for (a) 12, (b) 360, (c) 720h

Fig. 3 is SME pictures of inter-metallic at Sn3.5Ag0.5Cu/Cu interface under thermal-shearing cycle condition between -55~125°C for 24, 200 and 720 cycles. There is a scallop-like IMC layer formed between the solder and Cu base, whose thickness is about 2.3μm after reflowing. With the number of thermal-shearing cycling increasing, the IMC thickness in the valley of every adjacent two scallop-like increases significantly, while IMC thickness on the top of IMC grows slightly. After 200 cycles, the morphology of compound transforms plate-like from scallop-like at the interface. Till thermal-shearing cycling 720 cycles, there is only one layer of Cu_6Sn_5 compound, which component is Cu 53.43%, Sn 46.57%. As know, the quick diffusion of Cu atom benefits formation and growth of compound at the interface. Moreover, interface energy between liquid solder and scallop-like compound is lower than that between solid solder and scallop-like compound, so scallop-like compound exists stably in liquid state. However, in solid state, the diffusing rate of Cu atom reduces remarkably and formation of compound drops accordingly, it makes plate-like compound that has the minimum surface energy and interface energy exist stably. Therefore, the morphology of Cu_6Sn_5 trends gradually plate-like. According to the transformation of morphology of Cu_6Sn_5, Cu atom diffuses firstly from the valley of every adjacent two scallop-likes to Sn3.5Ag0.5Cu/Cu interface and forms Cu_6Sn_5 with Sn at the interface layer. Furthermore, the shear stress makes the crystal lattice disfigurement adding to promote Cu atom diffuse, there is only one Cu_6Sn_5 layer after 720 cycles under the thermal-shearing cycling condition.

978-1-4244-4658-2/09 $25.00 © 2009 IEEE

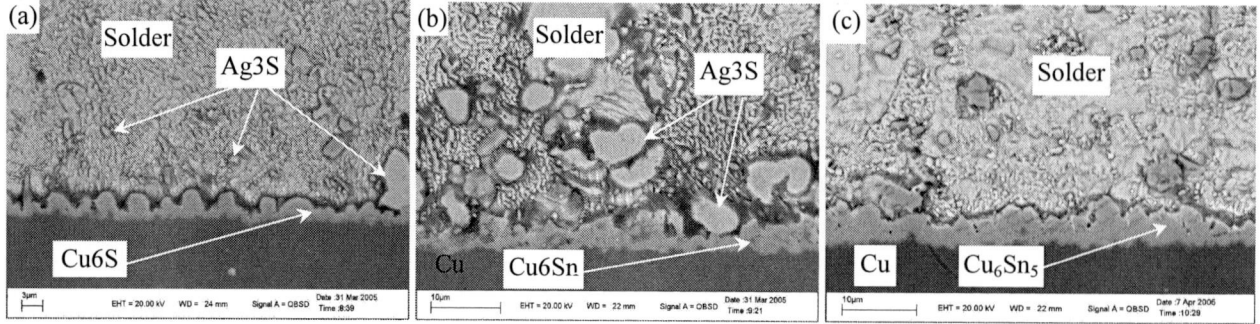

Fig. 3 SEM pictures of IMCs under thermal-shearing cycling condition for (a) 24, (b) 200 and (c) 720 cycles

3.2 Intermetallic growth dynamic

Figs. 4 and 5 are the pictures of the IMC growth behavior at the SnAgCu/Cu interface under the isothermal aging and thermal-shearing cycling conditions, which can be described by one-dimensional growth parameter. It suggested that the atom diffusion control IMC growth mechanism that is related to the square root of the cyclic time [10]:

$$Y = Y_0 + (Dt)^{1/2} \qquad 1$$

$$D = D_0 \exp(-\frac{Q}{RT}) \qquad 2$$

Where D is the diffusion coefficient given by an Arrhenius expression; Y_0 is the IMC thickness after reflow process; D_0 is the diffusion constant; Q, the activation energy; R, the Boltzmann constant and T, the absolute temperature.

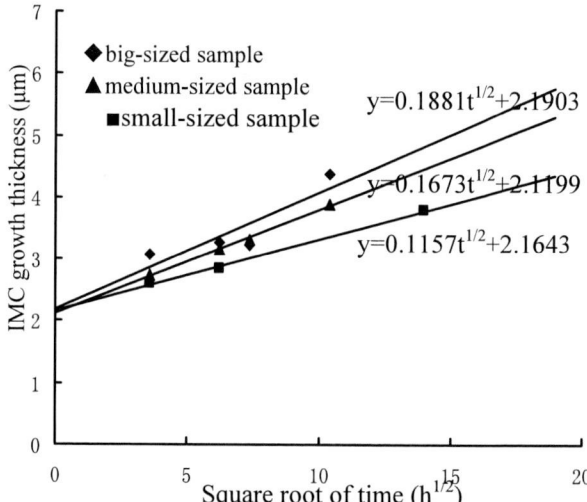

Fig. 5 the relationship between IMC thickness and square root of time at Sn3.5Ag0.5Cu/Cu interface

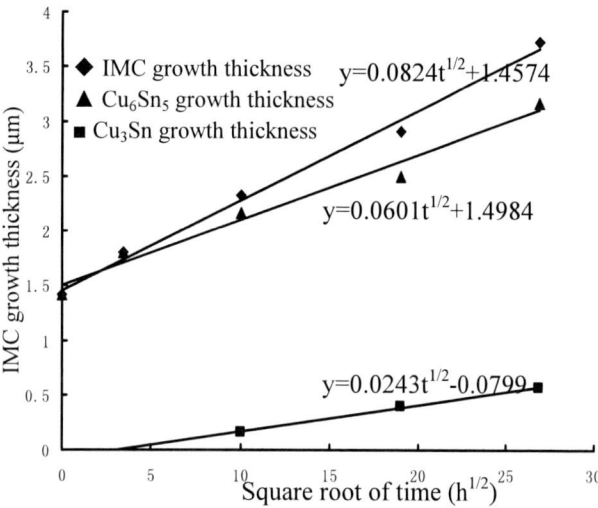

Fig. 4 the relationship between IMC thickness and square root of time after isothermal aging

Therefore, IMC growth model including Cu_6Sn_5 (η) and Cu_3Sn (ε) were founded, and X_ε, X_η are IMC thickness of Cu_3Sn and Cu_6Sn_5 of the T time, respectively, as shown in Fig. 6. According to the quality balance principle, the IMC growth behavior dynamic, Cu_6Sn_5 and Cu3Sn, were as below:

$$\left(C_\varepsilon^{\varepsilon/\eta} - C_\eta^{\eta/\varepsilon} \right)\frac{dX_\varepsilon}{dt} = -\left(D_{Cu}^\varepsilon \frac{\partial C_\varepsilon}{\partial x}\bigg|_{x=X_\varepsilon^{-0}} - D_{Cu}^\eta \frac{\partial C_\eta}{\partial x}\bigg|_{x=X_\varepsilon^{+0}} \right) \qquad 3$$

$$\left(C_\eta^{\eta/\beta} - C_\beta^{\beta/\eta} \right)\frac{dX_\eta}{dt} = -\left(D_{Cu}^\eta \frac{\partial C_\eta}{\partial x}\bigg|_{x=(X_\varepsilon+X_\eta)^{-0}} - D_{Cu}^\beta \frac{\partial C_\beta}{\partial x}\bigg|_{x=(X_\varepsilon+X_\eta)^{+0}} \right) \qquad 4$$

Where D_{Cu}^ε, D_{Cu}^η and D_{Cu}^β are diffusion coefficient of Cu atom in the Cu_3Sn, Cu_6Sn_5 and Sn phase, respectively; $\frac{dX_\varepsilon}{dt}$ and $\frac{dX_\eta}{dt}$ are IMC growth velocity of Cu_6Sn_5 and Cu_3Sn of the T time.

Finally, IMCs growth dynamic equation, Cu_6Sn_5 (η) and Cu_3Sn (ε), were gained under the isothermal aging condition, which is $Y = K_\varepsilon \sqrt{D_{Cu}^\varepsilon t} + K_\eta \sqrt{D_{Cu}^\eta t}$. While under the thermal-shearing cycling condition, only one Cu_6Sn_5 (η) compound was formed which growth kinetic equation is $Y = y_0 + K_\eta \sqrt{D_{Cu}^\eta t}$ with dynamic

$$K_\eta = \frac{\sqrt{2}\left(C_\eta^{\eta/\varepsilon} - C_\eta^{\eta/\beta}\right)^{\frac{1}{2}}}{\left(C_\eta^{\eta/\beta} - C_\beta^{\beta/\eta}\right)^{\frac{1}{2}}}$$

parameter . In accord with Figs. 4 and 5, atom average diffusion coefficients at the SnAgCu/Cu interface under the two conditions was gained.

The diffusion coefficient D_{Cu}^η of Cu atoms in the Cu_6Sn_5 phase under thermal-shearing cycling condition is from 10^{-5} to $10^{-4} \mu m^2/s$, which is one-digit growth higher than that after the isothermal aging, as shown in Table1. The results indicate that shear stress-strain is useful to atomic diffusion and IMC growth.

Table1 atom average diffusion coefficient at the SnAgCu/Cu interface under the two conditions ($\mu m^2/s$)

conditions	small-size sample D_{Cu}^η	medium-size sample D_{Cu}^η	big-size sample D_{Cu}^η
thermal-shearing cycling (25~125°C)	9.887×10^{-5}	1.408×10^{-4}	——
thermal-shearing cycling (-55~125°C)	1.200×10^{-4}	2.510×10^{-4}	3.173×10^{-4}
isothermal aging	D_{Cu}^η 5.236×10^{-5}		

The compound growth rate is controlled by the diffusing rate of Cu atom in the solder and substrate. Therefore, together-effects of thermal-effect and shearing tension make the crystal lattice disfigurement and the access of atom diffusion adding. It makes compound growth rate higher under the thermal-shearing cycling condition than that after isothermal aging. The thickness of compound increases with thermal-shearing cycling increasing and isothermal aging that can be described parabola growth kinetics.

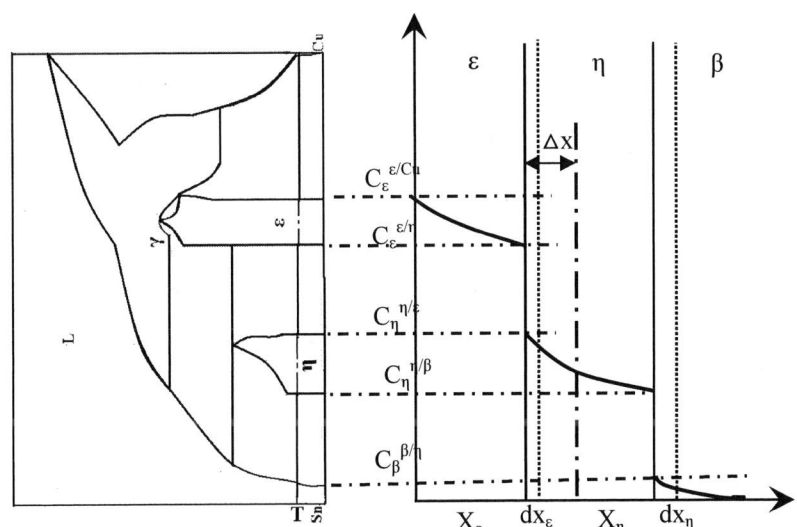

Fig. 6 Cu_3Sn and Cu_6Sn_5 phase growth model in solid state

Conclusions

There is only one Cu_6Sn_5 IMCs layer formed at Sn3.5Ag0.5Cu/Cu interface after 720 cycles under the thermal-shearing cycling condition. The compounds morphology changes gradually from scallop-like to planar-like, and planar-like IMC that has lower surface energy to the solid solder is stable in solid state.

The IMCs formation and growth at Sn3.5Ag0.5Cu/Cu interface are controlled by the dissolution and diffusion of Cu atoms after the isothermal aging and thermal-shearing cycling. The co-effects of thermal-effect and shearing-stress resulting in the crystal lattice distortion increasing promote Cu atoms diffusion and growth of IMC. The equation of compounds growth was gained, and the average diffusion coefficient of Cu atoms in IMCs is from 10^{-5} to 10^{-4} $\mu m^2/s$ at the interface under thermal-shearing cycling condition, which is higher than that under the isothermal aging condition at 125°C.

Acknowledgement

This work was supported by the National Natural Science Foundation of China (No. 50371010).

References

1. J.Y.Tsai, C.R.Kao, "The effect of Ni on the interfacial reaction between Sn-Ag solder and Cu metallization," *Proceedings of the 4th International Symposium on Electronic Materials and Packaging Conference*, Taiwan, China: IEEE Press, 2002, pp. 271-276.
2. Y. S. Kim, K.S. Kim, C.W. Hwang, *et.al.*, "Effect of compositipm and cooling rate on microstructure and

978-1-4244-4658-2/09 $25.00 © 2009 IEEE

tensile properties of Sn-Zn-Bi alloys," *J. Alloys and Compd.*, 2003, 352(3), pp. 237-245.

3. X. Ma, F.J. Wang, Y.Y. Qian, *et.al.*, "Development of Cu-Sn intermetallic compound at Pb-free solder/Cu joint interface," *Materials Letters*, 2003, 57(22-23), pp. 3361-3365.

4. K.S. Kim, J.M. Yang, C.H. Yu, I.O. Jung, *et.al.*, "Analysis on interfacial reactions between Sn-Zn solders and the Au/Ni electrolytic-plated Cu pad," *J. Alloys and Compd.*, 2004, 379(1/2), pp. 314-318.

5. C. L. Wei, C. R. Kao, "Liquid/solid and solid/solid reactions between SnAgCu lead-free solders and Ni surface finish," *Proceedings of the 4th International Symposium on Electronic Materials and Packaging Conference*, Taiwan, China: IEEE Press, 2002, pp.330-334.

6. C.M.Chang, P.C.Shi, K.L.Lin, "Interfacial reaction between Sn-Ag-Cu, Sn-Ag-Cu-Ni-Ge lead-free solders and metallic substrates," *Proceedings of the 4th International Symposium on Electronic Materials and Packaging Conference*, Taiwan, China: IEEE Press, 2002, pp. 360-366.

7. S. Wiese, F. Feustel, E. Meusel, "Characterization of Constitutive Behavior of SnAg, SnAgCu, and SnPb Solder in Flip Chip Joints," *Sensors and Actuators A*, 99 (2002), pp. 188-193.

8. K.N.Tu, K.Zeng, "Tin-lead (Sn-Pb) solder reaction in flip chip technology," *Materials Science and Engineering*, R34, (2001), pp. 1-58.

9. H.L.J.Pang, K.H.Tan, X.Q.Shi *et al.*, "Microstructure and intermetallic growth effects on shear and fatigue strength of solder joints subjected to thermal cycling aging," *Materials Science and Engineering A*, 307 (2001), pp. 42-50.

10. L. Qin, J. Zhao, L. Wang. *et.al.*, "Microstructure evolution in lead-free solder joints after wave soldering and reflow soldering," *Electronic Process Technology*, 2004, 25(2), pp. 64-67.

11. H.L.J. Pang, T.H. Low, B.S. Xiong, *et.al.*, "Thermal cycling aging effects on Sn-Ag-Cu solder joint microstucture IMC and strength," *Thin Solid Films*, 462-463 (2004), pp. 370-375.

12. I.Dutta, D.Pan. R.A.Marks, *et.al.*, "Effect of thermo-mechanically induced microstructural coarsening on the evolution of creep response of SnAg-based microelectronic solders," *Materials Science and Engineering A*, 2005, 410-411(25), pp. 48-52.

Research of Structural Factors Effects on Drop Reliability

Jianhui Wang, Xingming Fu, Xiaoqiang Xie, Jianwei Zhou, Qian Wang, Zaisung Lee
Samsung Semiconductor (China) R&D CO., LTD
No.15, Jin Ji Hu Road, Suzhou Industrial Park, Suzhou, China
Phone: (86512) No. 62888288-8823 Fax: (86512) No. 62888388 Email: jianhui.wang@samsung.com

Abstract

This paper discusses the effects of chip size, PCB thickness, fixture tightness, via information on the board level drop impact performance of common IC package for hand held electronic product applications namely FBGA, when subjected to the JESD22-B111 test methodology. The experiment results reveal significant difference in the reliability performance between different via information and different PCB thickness. In addition, simulation work using input G method was performed and verified experimentally and the results are coincided with experiment results.

Introduction

At a time when handheld electronics becomes more and more popular, capability for products to sustain accidental drops has become a major issue. Drop impact reliability is a great concern to handheld or portable electronic products such as mobile phones and PDAs. The JEDEC organization came out a standard [1] about drop test but not took all the effecting factors into consideration. Many important factors have been considered in the recent experiment and numerical studies, such as PCB pad structure and PWB build-up layer [2]. With the development of high density interconnect; via-in-pad has emerged as one of the key enabling technologies for increasing the I/O density, but also it is a great threat to drop reliability. The drop performance of via in pad structure may be quite different from normal pad due to its special structure, as is shown in Fig. 1. Because of the trend and the risk, we compared the drop performance of via in pad structure with no via in pad structure.

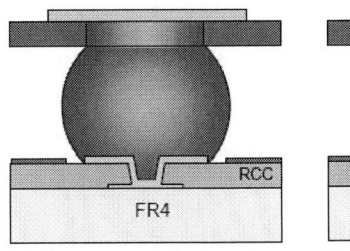 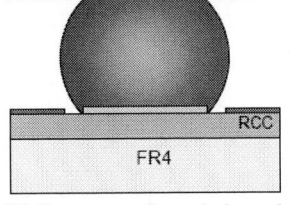

(a) Structure of via in pad   (b) Structure of no via in pad

Fig. 1 Sketch map of pad structure

Since the portable products is smaller and thinner, printed circuit board (PCB) needs to be thinner. The effect of package thickness has been studied by Zhang Jing already [3], but the effect of PCB thickness has not been studied by researchers. According to JEDEC standard, PCB is 8 layers and the thickness is 1.0mm. Considered the vibration mode of thin PCB and thick PCB maybe different, 8 layers (1.0mm), 6

layers (0.73mm) and 4 layers (0.44mm) PCB are designed to check the effect of PCB thickness.

The component technologies used in the portable products is variable, and the chip scale inside the package is different due to different functions. However there is a lack of documentation in the chip size influences on drop impact reliability. In this paper, we will also discuss the effect of chip size on drop impact reliability.

Screw is also a very important factor in drop reliability. The drop impacts for 4-screw and 6 screw configurations and different board sizes are investigated and compared under different test conditions by Fong Kuan Ng [4]. There is no description about screw tightness in JEDEC standard, but during the assembly process, the screw tightness maybe different and causes different drop results consequently. As is shown in Fig. 2, (1) is the sketch map of normal screw tightness, and the yellow arrowhead represents the transmission direction of stress wave. When a drop impact happens, the stress wave will transmit along that direction. (2) is the sketch map of loose screw tightness, and after the drop impacting, the PCB vibrates along horizontal direction besides vertical direction. What's more, in the case of over tight, the stress may be concentrated in the vicinity of screws. As considered above, the tightness of screw is really worth of research. In this project, the screw tightness is defined by a torque, and the data is 150cNm (Tight), 95cNm (Normal), and 40cNm (Loose).

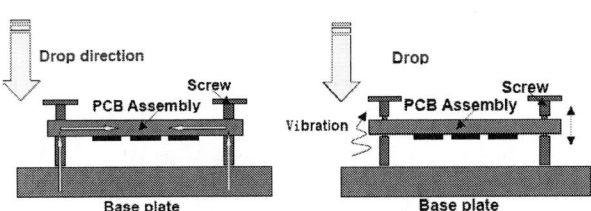

(a) Normal screw tightness   (b)Loose screw tightness

Fig. 2 Sketch map of drop assembly

In this paper, based on finite element modeling and design of experiment (DOE), a novel approach was discussed in detail to investigate the effects of via information, chip size, PCB thickness and screw tightness as well as their weights and interactions on the drop reliability of FBGA packages.

In-put G method has been widely used in drop impact simulation since introduced by Tee in 2004 [5]. Its timesaving and high accuracy advantages have been proved by many researchers as well as our previous project. Tee [6] put forward that the peeling stress or normal stress was the dominant stress component affecting the solder joint reliability. Tee and his team also found that the peeling stress

978-1-4244-4658-2/09 $25.00 © 2009 IEEE

was induced mainly by PCB bending and the bending was the major failure mechanism. P. Towashiraporn [7], Don-Son Jiang *et al.* [8], Pekka Marjamaki [9] and Kailin Pan *et al.* [10] also found the effect of peeling as a major reason on package failures from their simulation results. In our experiment, the simulation work was done by in-put G method and peeling stress was calculated out to verify the failure position.

Experiment setup

A. Device Information

56FBGA is chosen as the investigation vehicles with ball diameter 0.4mm and ball pitch 0.8mm. The designed chip size information is listed in Table 1.

Table 1 Package information

PKG	Chip size	PKG size
56FBGA	6mm×9mm 4.52mm×6.78mm 2.2mm×3.3mm	8.00mm×11.60mm×1.20mm

B. Test board design

Test board is daisy chain designed according to JEDEC standard, NSMD, OSP finished, 3×5 components. The packages are mounted at the following locations, seen in Fig. 3.

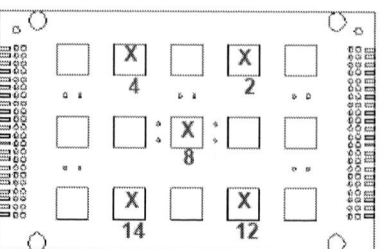

Fig. 3 Package location at test board

C .DOE Matrix

Based on statistic theories, DOE is a structured, organized method for determining the relationship between factors and the output of a process. In our study, via information, chip size, PCB thickness and screw tightness are chosen as the four input factors with 2 levels for each factor separately, and the drop failure life (number of drops to fail) is the output factor. It is known that different DOE approaches lead to different DOE setup and different number of runs to accomplish the analysis. The current study used the 1/2 factorial design, and the DOE legs setup is shown in Table 2. The input acceleration of this experiment is the 1500 G and the duration time is 0.5ms.

Table 2 DOE Matrix

Std Order	Run Order	Via	chip size (mm^2)	PCB Thk (layer)	Screw tightness (cNm)
9	1	no via	30.65	6	95
2	2	Via in pad	7.26	4	150
4	3	via in pad	54	4	40
6	4	via in pad	7.26	8	40
8	5	via in pad	54	8	150
5	6	no via	7.26	8	150
3	7	no via	54	4	150
1	8	no via	7.26	4	40
7	9	no via	54	8	40
10	10	via in pad	30.65	6	95

Result and discussion

A. Main factors and interaction

In this study, all the data was analyzed using Minitab's Reliability/Survival Analysis functions. According to JEDEC spec, the board shall be dropped until 80% of all the devices failed. There were totally 5 devices in PCB board, so we took the fourth life data as the failure life of the board. For each condition nine to thirteen PCBs were tested to satisfy the quantity requirement of JEDEC. Since the number of drops-to-failure follows the Weibull distribution, cumulative failure plots were generated, as summarized in Fig. 4.

978-1-4244-4658-2/09 $25.00 © 2009 IEEE

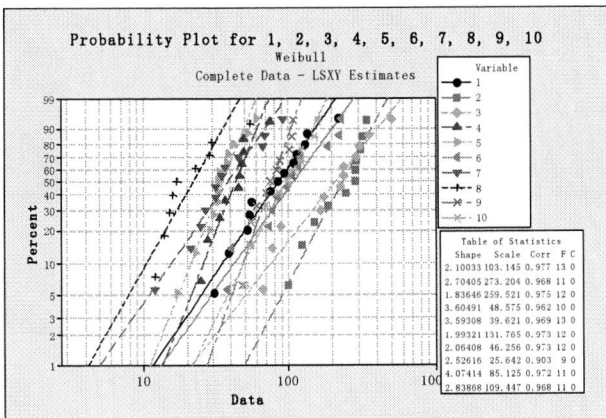

Fig. 4 Probability plot for samples

Based on this plot and the calculated life time, Minitab, the DOE software, can generate an explicit solution and calculate the weight of each factor. The main effect of life and the interactions among factors are shown in Fig. 5 and Fig. 6 separately.

In Fig. 5, slope rate relates to the effect on the drop reliability. The bigger the slope rate, the bigger the effect. From Fig. 5, we can conclude that: the effects of via and PCB thicknesses are significant and the effects of chip size and screw tightness are comparatively slight. The smaller the chip size, the tighter the screw tightness, the longer the life time. Generally speaking, the life time of via on pad is longer than no via on pad samples. The life time of 4 layers PCB is also longer than the 8 layers PCB. But this is not always necessarily the case, the interaction of PCB thickness and via information is huge enough to topple over this rule in some situation, and this will be explained in the following in detail.

In Fig. 6, nonparallel lines represent higher interactions. It is easy to observe that: the interaction of via information and PCB thickness has shown the strongest influence to drop reliability. When there is via on the pad, and PCB thickness is small, the sample performs best. On the other hand, it's also reasonable to confirm that the interaction between PCB thickness and screw tightness is small for the drop reliability. This information quantified some common senses and revealed some unknown fact, therefore, is very valued.

Take the interactions and main effect together, we can draw a Pareto chart of parameters to get a weight factor, as is shown in Fig. 7. The interaction between via information and PCB thickness is the main effect among all the parameters. Following the interactions, via information and PCB thickness are also un-ignored factors. Screw tightness and interactions of via information and screw tightness have an effect on the drop reliability in an extent.

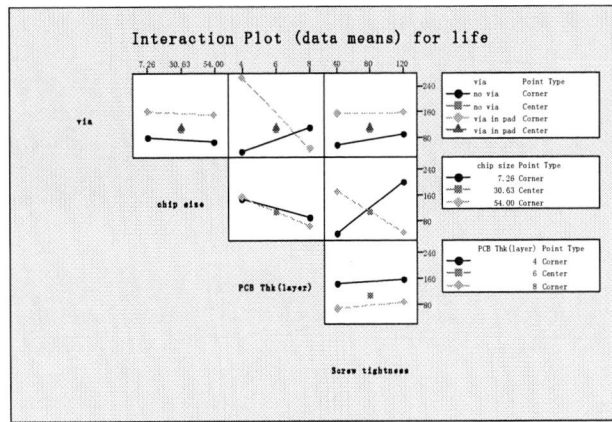

Fig. 6 Interaction plot for life

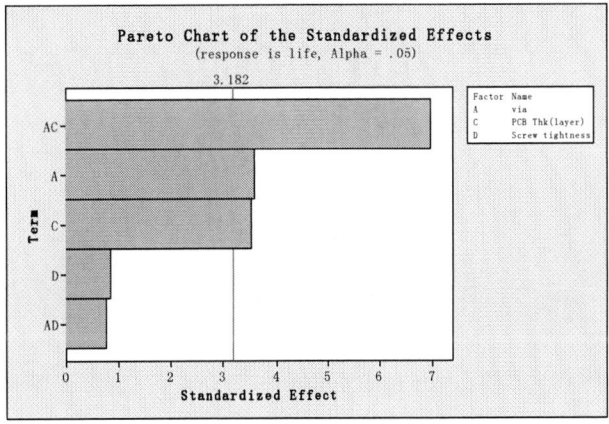

Fig. 7 Pareto chart of parameters

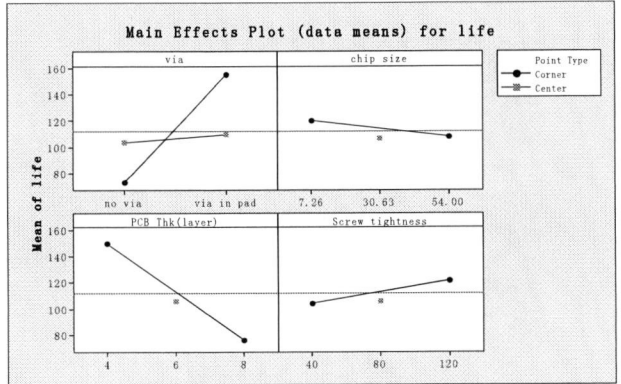

Fig. 5 The main effects plot for life

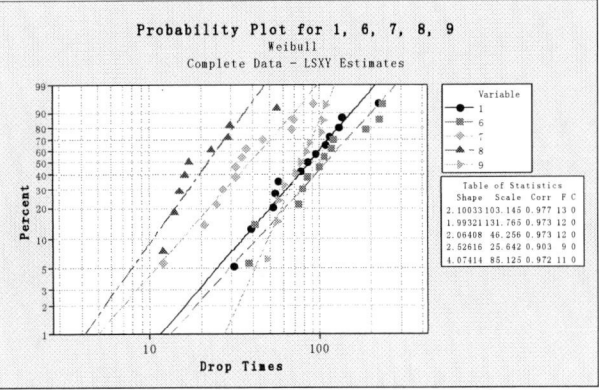

Fig. 8 Probability plot for samples without via in pad

B. Effect of via information

(a) no via in pad

Since via information and PCB thicknesses are the most significant factors to decide drop life, we choose the sample 1, 6, 7, 8, 9 out to compare their life time. The common character of sample 1, 6, 7, 8, 9 is that they don't have via in pad. The probability plot of sample 1, 6, 7, 8, 9 is shown in Fig. 8, from which we can conclude clearly that: the order of drop lives is that: 6# > 1#> 9#> 7#> 8#.

In the case of no via in pad, the order of drop life is: 8 layers PCB≈ 6 layers PCB > 4 layers PCB. Simulation work was done to research the PCB distortion during impact. Fig. 9, Fig. 10 and Fig. 11 show the result comparisons of 3#, 5#, 7# and 9#. It can be seen that 5# and 9# have close displacement, strain and maximum peeling stress. This also happens between 3# and 7#. Fig. 23 shows that the displacements and intervals of 4 layers are twice of 8 layers. Although there are no significant differences between each model on strains and stresses, it can still be seen that 8 layer samples have larger stresses and smaller strains than 4 layers because of their higher stiffness.

The simulation result reveals that the displacement of 4 layers PCB is nearly twice of the 8 layers PCB. In other words, the solder ball of 4 layers PCB will endure a longer distance to achieve its utmost position than the 8 layers PCB, like the sketch map Fig. 12 shows. Obviously this over movement makes the 4 layers PCB samples fail faster than 8 layers PCB. We believe that the 4 layers PCB has probably gone with a mixed mode vibration during drop impact, and this needs more research.

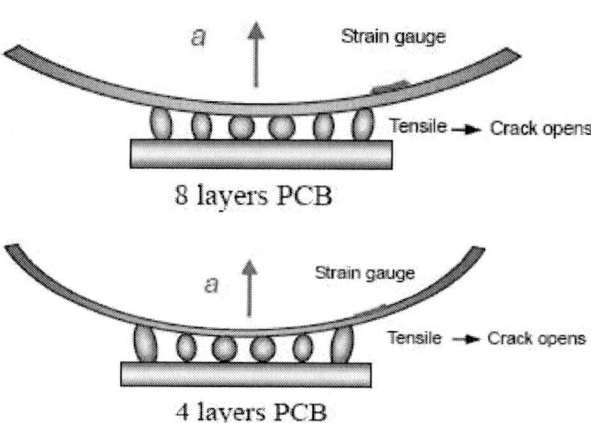

Fig. 12 The configuration of PCB at the utmost position

The mainly failure mode of drop test is pad lift, which has been proved in former research. According to the results of section polish both the 4 layers PCB and the 8 layers PCB have the same failure position, just under the pad, which is shown in Fig. 13.

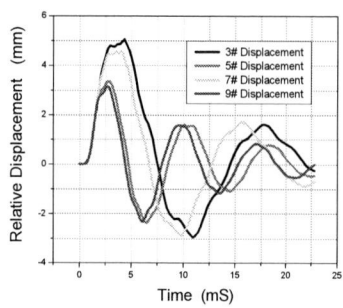

Fig. 9 PCB deflection

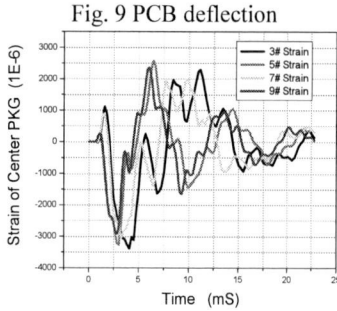

Fig. 10 Strain of center package

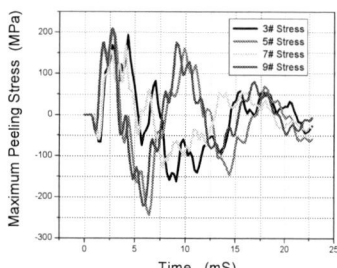

Fig. 11 Max. Peeling stress

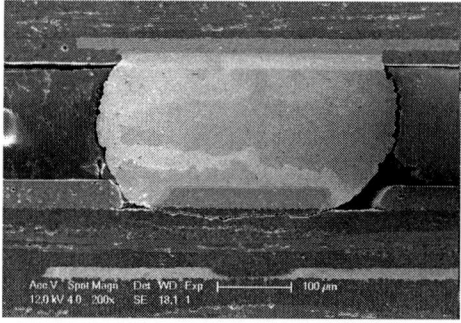

(a) 4 layers (no via in pad)

(b) 8 layers (no via in pad)

Fig. 13. SEM images of failed no via in pad samples' solder joints after drop impact

(b) via in pad

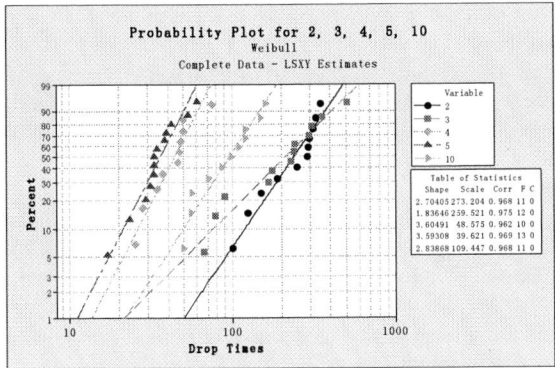

Fig. 14 Probability plot for samples with via in pad

Fig. 14 shows the probability plot of samples with via in pad. The result obviously shows that: 2# ≈ 3# > 10# > 4# ≈ 5# , from which we see a phenomenon that in the case of via in pad, drop life of 4 layers PCB is longest, and the drop life of 6 layers PCB is middle, and the drop life of 8 layers PCB is the shortest.

This result is interesting, since the results are totally different in the case of via in pad and no via in pad, and it is different from the common sense. To have a better understand, further factors such as structure effect are taken into consideration. The pad structure of via in pad seeing in Fig. 1.According to section polish results, shown in Fig. 15, the failure positions of different thickness PCBs are different. Generally speaking, the failure position of 8 layers PCB sample is just under the pad, and it is relatively in the upper of the via ,seen in Fig. 15 (a). The failure position of 6 layers PCB is mainly in the middle of via, and it is deeper compared to samples of 8 layers PCB, seen in Fig. 15 (b). When it comes to 4 layers PCB samples, the failure position is in the bottom of via, quite deep of via, and the whole via is lifted at the root position, seen in Fig. 15(c). We concluded this different failure position determines the different drop times. For 4 layers PCB samples, they has to overcome the extra resistance force came from the first dielectric layer, so they are not as easily to fail as 8 layers PCB samples. The failure position of 8 layers PCB samples is superficial thus it is much easier to be lifted.

The simulation result is corresponding to the experiment result. From the section view of Table 3, it can be seen that the maximum peeling stress located close to the PCB side. Additionally, the maximum peeling stress of 5, 9# is higher than that of 3, 7#. This might because that 3, 7# is a 4 layers PCB structure, which is a much more flexible material than 8 layers PCB (5, 9#), and it can absorb more impact energy compared with 8 layers PCB. It can be concluded that the thinner PCB thickness is, the higher failure probability of PCB; the thicker PCB thickness is, the higher failure probability of solder joint. That's say, for 8 layers samples, the stress is concentrated on the solder ball, and for 4 layers samples, the PCB undergoes more serious impact. This simulation result can also explain why the failure position of

8 layers sample is much more superficial, which is near the solder ball.

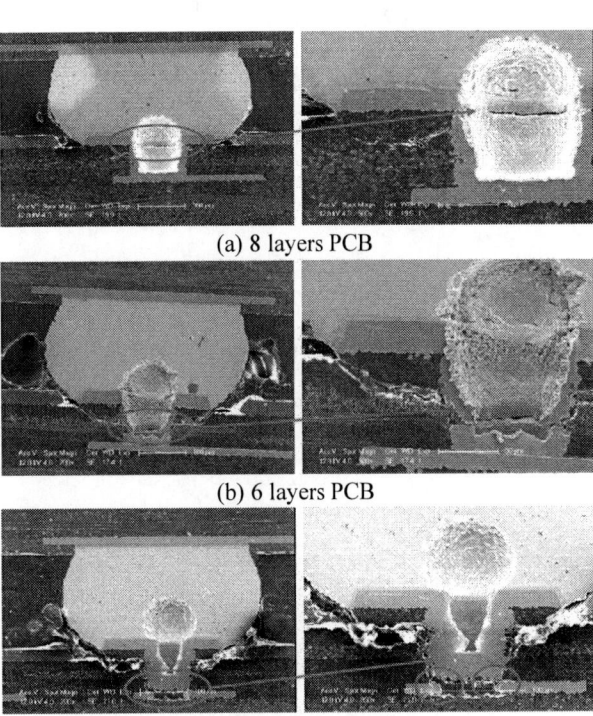

(a) 8 layers PCB

(b) 6 layers PCB

(c) 4 layers PCB

Fig. 15 SEM images of failed via in pad samples' solder joints after drop impact

Table 3 Maximum peeling stress distributions

No.	Critical ball section	
3#		
5#		
7#		
9#		

C. Effect of chip size

Simulation work is done to confirm the effect of chip size. Fig. 16, Fig. 17, and Fig. 18 plot the three factors of 6# and 9#, they have different chip sizes. Because EMC and chip have close densities, and their differences are so tiny that can be ignored compared with the large PCB, 6# and 9# have almost the same dynamic responses. Although 9# has little larger values in displacement, strain and stress than those of 6#, it can be concluded that the chip size has very little influence on dynamic response.

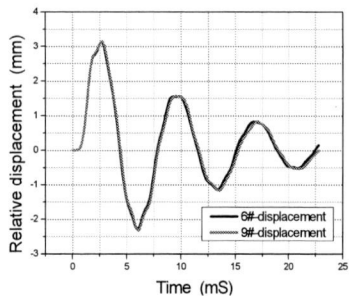

Fig. 16 PCB Deflection

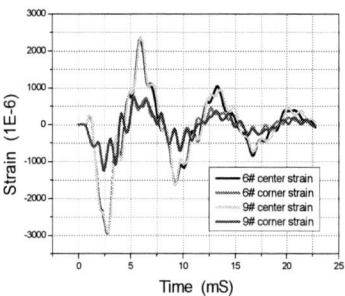

Fig. 17 Strain of center package

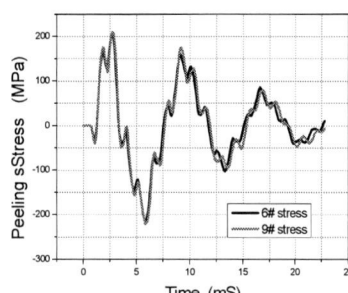

Fig. 18 Max Peeling stress

Conclusions

The drop performance of FBGA device has been studied in our research, and the following conclusion can be drawn:

(1) The effects of via and PCB thicknesses are significant and the effects of chip size and screw tightness are comparatively slight. The smaller the chip size, the tighter the screw tightness, the longer the life time. The interaction of via information and PCB thickness has shown the strongest influence to drop reliability.

(2) In the case of no via in pad, the order of drop life is: 8 layers PCB≈ 6 layers PCB > 4 layers PCB. In terms of via in pad, the order of drop life is: 4 layers PCB > 6 layers PCB > 8 layers PCB.

(3) The failure position of via in pad samples with different thickness is different. For 8 layers sample, the failure position is just under the pad, for 6 layers sample, the failure position is in the middle of the pad, and 4 layers sample fails at the root of via.

(4) Samples with via fails faster than samples without via for 8 layers PCB; Samples without via fails sooner than samples with via for 4 layers PCB; for 6 layers PCB, life time doesn't have difference.

(5) Simulation results using input G method are coincided with experiment results.

References

1. JEDEC Standard JESD22-Blll, Board Level Drop Test Method of Components for Handheld Electronic Products, 2003.
2. Jong Gi Lee, Hyun Jong Woo, Ji Seok Hong, Pyoung Wan Kim and Young Hee Song, "Effect of PCB pad structure and PWB build-up layer on solder joint life under thermal cycling and drop condition," *9th Electronics Packaging Technology Conference*, 2007
3. Jing Zhang, Maohua Du, Nufeng Feng, "Board Lever Drop Test Reliability for MCP Package," *7th International Conference on Electronics Packaging Technology*, 2006.
4. Fong Kuan Ng, Chwee Teck Lim, Eric Pek, "Study of PCB strain and component position under board lever drop test," *Electronics Packaging Technology Conference*, 2006, pp. 248-254.
5. Tee TY, Ng HS, Luan JE, "Comparison of Free-fall vs. Input-G Drop Impact Models," *6th EMAP Conference Proc*, Malaysia, 2004, pp. 48-55.
6. Tee TY, Luan JE, *et al.*, "Advanced Experimental and Simulation Techniques for Analysis of Dynamic Responses during Drop Impact," *Electronic Components and Technology Conference*, 2004, pp. 1088-1094.
7. P. Towashiraporn, "Cohesive Modeling of Solder Interconnect Failure in Board Level Drop Test," *IEEE*, 2006, pp, 817-825.
8. Don-Son Jiang, Yuan Lin Tzeng, yu-Po Wang *et al.*, "Board Level Drop Test and Simulation of CSP for Handheld Application," *7th International Conference on Electronics Packaging Technology*, 2006.
9. Pekka Marjamaki, Toni Mattila, Jorma Kivilahti, "Finite Element Analysis of Lead-Free Drop Test Boards," *Electronic Components and Technology Conference*, 2005, pp. 422-426.
10. Kailin Pan, Bin Zhou, Yilin Yan, "Simulation Analysis of Dynamic Responses for CSP under Board Level Drop Test," *7th International Conference on Electronics Packaging Technology*, 2006.

Effect of Stand-off Height on the Reliability of Cu/Sn-4.8Bi-2Ag/Cu Solder Joint

Hui Liu[2], Longzao Zhou[2], Jun Li[1,2], Fengshun Wu*[1,2], Yiping Wu[1,2]

[1]Wuhan National Laboratory for Optoelectronics, Wuhan, China, 430074

[2]Department of Materials Science and Engineering, Huazhong University of Science and Technology, Wuhan, China, 430074

Corresponding author: Fengshun Wu, E-mail: fengshunwu@mail.hust.edu.cn, Tel: +86-27-87558275

Abstract

This study investigates intermetallic compound (IMC) growing behavior at different stand-off heights (SOH) in miniature Cu/Sn-4.8Bi-2Ag/Cu solder joints. The solder joints with SOH of 10, 20, 50 and 100µm will be studied, and the microstructures and compositions will be discussed. Meanwhile the tensile strength and the fracture mode of Sn-4.8Bi-2Ag solder joints will be also studied. The results show that Bi appears as particles spreading over the bulk of the solder joint, which will increase the strength of the solder joint. It is also found that the content of the IMC layer increases as the SOH reduces. The SOH of the solder joint plays an important role in the tensile strength. Tensile strength of the solder joint decreases as the SOH reduces, which correlates with the change of microstructures and compositions in the solder joints. The fractured path of the solder joint transfers from the bulk of the solder joint to the interface between IMC and solder, and the fracture mode tends to be brittle fracture.

Introduction

The trends of multi-function, slighter, and portable consumer electronics requires minimize chip sizes, which leads to the subsequent reduction of the stand-off height (SOH) to solder joint [1]. As SOH dramatically decreases to minimize the chip, the solder joints will face the challenge of reliability. The thickness of interfacial IMC (intermetallic compounds) layer will change. The reliability of the solder joints will be affected by the intrinsic brittleness of the IMC. Now, many studies are taken on the reliability of the solder joints. Ahmed et al. studied the effects of the changing of the volume of solder joints on the interfacial reaction. The results show that the thickness of the IMCs increases as the solder volume decreases. As the SOH decreases, the reliability of the solder joints decreases [2-5].

Lead-free is another trend of electronic manufacturing. Recently, Lead-free solder alloy has been widely used in electronic products because of the environmental concerns and government legislation. Sn-Bi-Ag solder paste is widely studied due to its lower melting temperature and superior properties [6-7].

The reliability of the Sn-4.8Bi-2Ag alloy in small size (under 100µm) used in lead-free solder joints was less studied. In this paper, the solder joints prepared by Sn-4.8Bi-2Ag solder with SOH of 10, 20, 50 and 100µm will be studied, and the microstructures and compositions changing will also be discussed.

Experimental procedures

To study the effect of the SOH on the reliability of solder joints, the samples with different SOH are prepared. A sandwich structure of Cu/Sn-4.8Bi-2Ag/Cu is designed to study the interfacial reactions during the reflow process as shown in Fig. 1. The sample has two components, which are the Cu bars with the length of 30mm and the solder with the diameter of 1mm.

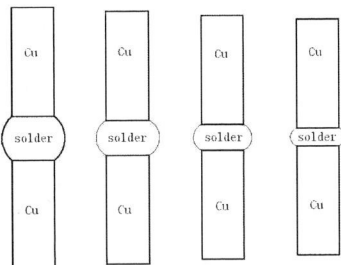

Fig. 1 Samples with different SOH

Samples with the SOH of 10, 20, 50 and 100µm are prepared by reflow process, respectively. The temperature profile of reflow process is shown in Fig. 2. The temperature above the liquidus is maintained for about 200s during which the highest temperature reaches 250°C. After the experiments, the specimens are mounted, sectioned and polished perpendicularly to the Cu/solder/Cu interfaces for observation.

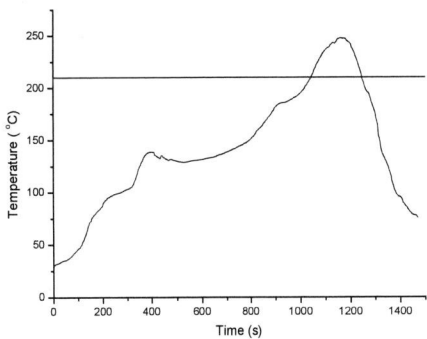

Fig. 2 The reflow profile

Five samples of each kind of solder joints are selected for the tensile test. The tensile specimens are clamped on a RGD0.1 testing machine. Tensile tests are carried out at a crosshead speed of 0.1mm/min at room temperature to obtain the maximum tensile load. Force and displacement are recorded.

Scanning electron microscope (SEM) with energy dispersive x-ray (EDX) is executed to characterize the interfacial microstructures and phase compositions. The mean

978-1-4244-4658-2/09 $25.00 © 2009 IEEE

thickness of 5 different IMC was calculated in the same image with SEM.

Results and discussion

Effect of SOH on the microstructures of solder joints

Fig. 3 shows the cross-sectional BEI (back-scattered electron image) micrographs of the Cu/Sn-4.8Bi-2Ag/Cu solder joint with different SOH.

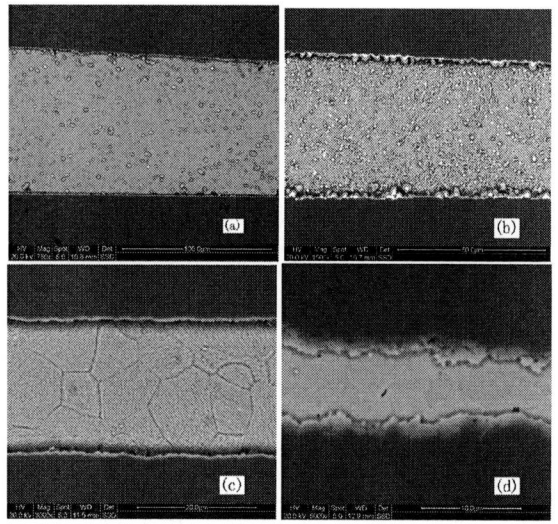

Fig. 3 BEI micrographs of the solder joint with different SOH: (a) 100μm; (b) 50μm; (c) 20μm; (d) 10μm.

The solder joint contains the bilateral IMC layers and the intermediate solder bulk layer. As shown in Fig. 3, it is found that the SOH has a significant effect on the solder joint bulk. The interfacial microstructure of each side forms nearly the same morphology in solder joints with different SOH, like scallop-type. The SOH has a significant effect on the microstructures of the solder joints. When the SOH decreases, the IMCs become sharper.

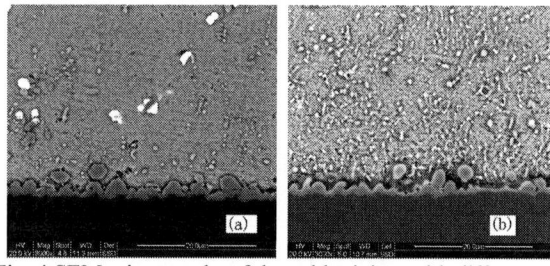

Fig. 4 SEM micrographs of the solder joints with different SOH: (a) 100μm; (b) 50μm.

As we know, the solder joint consists of IMC layer and solder bulk layer. As shown in Fig. 4, the microstructures of solder bulk in Cu/Sn-4.8Bi-2Ag/Cu solder joint consists of Sn-rich phase, Sn-Ag eutectic zone, Bi phase and some dissociative intermetallic compounds. From the observation on the as-soldered interface microstructures, the IMC layers have round scallop-type morphology; the compounds are identified to be Cu_6Sn_5 by EDX analysis. Some IMC layers have separated. In the four Cu/Sn-4.8Bi-2Ag/Cu solder joints, the Sn-Ag eutectic zone gives rise to small particles and it is identified to be Ag_3Sn by EDX analysis. At the same time, the Bi phase gives rise to white particles spreading in the solder bulk; the dissociative intermetallic compounds are identified to be Cu_6Sn_5 by EDX analysis.

The mean thickness of the IMC at both interfaces is calculated, and the relationship between the IMC thickness and the SOH is shown in Fig. 5.

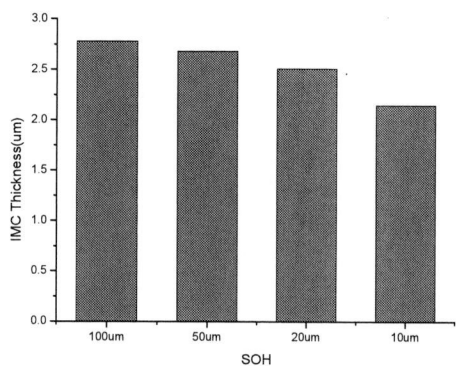

Fig. 5 The relationship between IMC thickness and SOH

As shown in Fig. 5, when the SOH decreases, the IMC thickness of the solder joint also decreases. However, the proportion of IMC layer in the solder joint is very different, and the IMC layer proportion is the ratio of interfacial IMC thickness to SOH. The IMC layer proportion of Cu/Sn-4.8Bi-2Ag/Cu solder joints with the SOH of 100, 50, 20 and 10μm are 5.56%, 10.72%, 25.1% and 43%. So the IMC layer proportion of solder joint is from 5.56% to almost half of the SOH. In the four solder joints, some ball-type dissociative IMCs closed to the IMC layer at Cu/solder interface can also be observed in the solder bulk. Just as IMCs in the solder joint with 10μm SOH, those IMCs are almost half of the solder bulk. The IMC layer portion increases as the SOH decreases. Due to IMCs intrinsic brittleness, the reliability of the solder joint will be weakened.

Effect of SOH on tensile strength of Cu/Sn-4.8Bi-2Ag/Cu solder joint

Fig. 6 shows the SEM secondary electron images of tensile fracture surface of Cu/Sn-4.8Bi-2Ag/Cu solder joints with different SOH.

For the Cu/Sn-4.8Bi-2Ag/Cu solder joints with the SOH of 100μm and 50μm, as shown in Fig. 6(a) and Fig. 6(b), there is an obvious honeycombed passages on the fracture surfaces. It indicates that the failure occurs at the interface between the solder bulk and the Cu_6Sn_5 IMC. The fracture surface contains honeycombed passages and round scallop-type morphology particles. The compositions of these particles are identified to be Cu_6Sn_5 by EDX analysis. So the failure occurs at the IMC layer when the SOH are 20μm and 10μm.

The ultimate tensile strengths (UTS) of Cu/Sn-4.8Bi-2Ag/Cu solder joints with different SOH are shown in Fig. 7.

From Fig. 7, it can be found that when the SOH decreases, the UTS of the solder joint also decreases. The UTS of the solder joints are 90.37, 71.56, 56.4 and 34.45MPa as the SOH are 100, 50, 20 and 10μm, respectively. As shown in Fig. 5, as the SOH decreases, the IMC layer proportion of the solder joints increases. The reliability of the solder joint is affected by the IMC layer proportion. Therefore, the fracture firstly occurs at the interface between the solder bulk and the Cu_6Sn_5 IMC when the SOH is 100μm. When the SOH is 10μm the fracture occurs at the IMC layer. The tensile strength decreases with the SOH decreases.

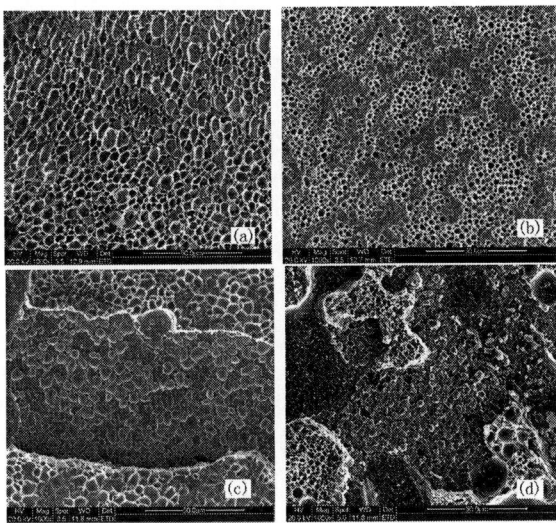

Fig. 6 SEM images of tensile fracture surface of Cu/Sn-4.8Bi-2Ag/Cu solder joint with different SOH: (a) 100μm; (b) 50μm; (c) 20μm; (d) 10μm.

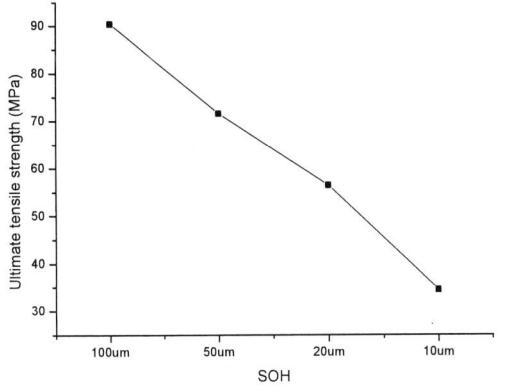

Fig. 7 The ultimate tensile strength of Cu/Sn-4.8Bi-2Ag/Cu solder joint with different SOH

Conclusions

As the SOH decreases, the microstructures and tensile strength of Cu/Sn-4.8Bi-2Ag/Cu solder joint are affected significantly. The following conclusions can be made from this study:

(1) In the interface of Cu/Sn-4.8Bi-2Ag/Cu solder joints, a single Cu_6Sn_5 IMC layer with round scallop-type morphology is formed. The thickness of the IMC decreases with the SOH decreases. The proportion of IMC layer in the solder joint increases as the SOH decreases.

(2) As the SOH decreases, the ultimate tensile strength of Cu/Sn-4.8Bi-2Ag/Cu joint decreases. And the failure occurs at the interface between the solder bulk and the Cu_6Sn_5 IMC when the SOH is 100μm. When the SOH is 10μm the fracture occurs at the IMC layer.

Acknowledgments

The authors acknowledge the financial support from National Natural Science Foundation of China (No. 60776033 and No. 60876070).

References

1. R.Plieninger, M.Dittes, K.Pressel, "Modern IC Packaging Trends and their Reliability Implications," *Microelectronics Reliability*, Vol.46, (2006), pp. 1868-1873.
2. Ahmed Sharif, Y.C. Chan, Rashed Adnan Islam, "Effect of volume in interfacial reaction between eutectic Sn–Pb solder and Cu metallization in microelectronic packaging," *Materials Science and Engineering*, B106, 2004, pp. 120-125.
3. C.K. Wong, John H.L. Pang, Y.F. Sun, F.L. Ng, J.W. Tew and W. Fan, "Influence of Solder Volume on Interfacial Reaction between Sn-Ag-Cu Solder and TiW/Cu/Ni UBM," *2005 Electronics Packaging Technology Conference*.
4. E. B. Liao, Andrew A. O. Tay, Simon, "Fatigue and Bridging Study of High-Aspect-Ratio Multicopper-Column Flip-Chip Interconnects Through Solder Joint Shape Modeling," *IEEE Transactions on Components and Packaging Technologies*, 2006, 29(3), pp. 560-569.
5. Chih-Tang Peng, Kuo-Ning Chiang, Terry Ku, "Design, Fabrication and Comparison of Lead-Free/Eutectic Solder Joint Reliability of Flip Chip Package. Thermal and Mechanical Simulation and Experiments in Microelectronics and Microsystems," *2004. EuroSimE 2004. Proceedings of the 5th International Conference*, 2004, pp. 149-156.
6. Hoh Huey Jian, Eu Poh Leng, Min Ding, "A Study On Lead Free Sn-Ag-Cu Solder System," *International Electronic Manufacturing Technology*, IEMT 2006, Putrajaya, Malaysia, pp. 450-455.
7. Zhi-tian Hu,Qian-jin He,Dao-rong Xu, "The Developments and Trends of the Lead-Free Solder on the Domestic and International," *Welding Technology*, Vol. 34, No. 3 (2005), pp. 4-6.

Enhancement of TBGA Substrate in Packing Drop Test

Kelvin Pun, C Q Cui
Compass Technology Co., Ltd.
Suite 10, 5/F, Chiaphua Centre, 12 Siu Lek Yuen Road, Shatin, NT, Hong Kong
Email: Kelvin_pun@cgth.com

Abstract

The drawback of Sn/Ag/Cu solder alloy on selective Ni/Au plating (SMOBC) TBGA substrate is weakened solder joint, which promote the embrittlement and affects long-terms reliability. The ability of electronic packages and assemblies to resist ball drop failure is becoming a growing concern recently. In this paper, test vehicles of Sn-Ag-Cu (SAC) solder alloys with different Ni thickness of 3, 5, 8 and 10μm are prepared. Firstly, the effect of Ni thickness on the mechanical strength of solder joint was investigated. Secondly, packing drop test was performed to correlate the ball drop level with different Ni thickness metallization. In experimental, the ball shear strength on the SMOBC TBGA with thin Ni was found to be lower as-reflow and in the early stage of high temperature storage, compared to that with thick Ni. This is identified due to the solder mask (SM) tail in thin Ni, where Cu diffusion could get the simple channel through SM and also provides a fracture location for the weakened strength of solders joints and mixed brittle fracture failure in the shear test. With the extend high temperature storage and multiple reflows, it is found that the brittle TIMC (Ternary-intermetallic-compound) growth is governed by the amount of copper diffusion and the dissolution rate of Ni layer. With thin Ni, more Ni was consumed and the resistance of copper diffusion was dropped, resulting to the acceleration of interfacial TIMC growth and to reduce the shear strength in shear test with the brittle fracture failure. The weakened resistance of copper diffusion by thin Ni could be directly attributed to the missing ball issue on the assembled package in the dropping test. In contrast, a good diffusion barrier is demonstrated by thick Ni solder system, exhibiting a slower IMC growth and lower consumption rate of Ni layer, which evidences a higher shear loads and good performance in the dropping test. In conclusions, a minimum 5μm thick Ni outperforms 3um thin Ni in shear test and passes all of packing drop test cycles. Thus, this gives the evidence that the increase in Ni thickness for Ni/Au finish plays a critical role for improving the robustness of Sn-Ag-Cu solder joint in the SMOBC TBGA.

Introduction

In the Cu/Ni/Au UBM of BGA package, Ni acts as a barrier layer to prevent Cu diffusion and also as an adhesion promoter. A certain thickness of Ni is considered to maintain for preventing Cu diffusion into solder during reflow.

Compass normally used a minimum 2.5μm Ni in the UBM for lead-free application. Many researchers have suggested typical Ni and Au thickness of 3-5μm and 0.15-0.25μm, respectively [1-2]. In fact, on lead-free TBGA device, the selective Ni/Au plating structure with solder mask over bare copper (SMOBC) (Fig. 1) has more preferences because of the amount of Au reduction and better interfacial bond strength between solder mask and copper to prevent delamination issue for reliability concerns. However, a certain level of Cu recession is usually formed on using this structure, due to the essential step of copper pre-etching step prior to selective Ni/Au plating. The current industrial trend shows that Sn-Ag-Cu (SAC) solder alloys display promising joint strength and ductility and are rapidly gaining acceptance as a preferred lead-free solder for electronic assembly [3-4]. However, the combination of Sn-Ag-Cu solders joints and SMOBC TBGA shows a high failure rate of missing ball with extremely brittle behavior during reflow, handling and shipping. The brittle fracture is normally occurred within the interfacial $(Ni,Cu)_3Sn_4$ IMC/Ni layer, which causes an unpredictable yield loss in assembles [4-5]. The failure location is normally occurred on the outer row and corner portion of the unit. It is known that the percentage of IMC brittle fracture increases with the increase in Cu content within the Sn-Ag-Cu (SAC) solder alloy [4]. From this view of the point, the Sn-Ag-Cu lead-free solder was soldered with the different selective Ni/Au plated TBGA substrates. The effect of Ni thickness on the interfacial reactions with Pb-free solder, the purities of Ni, the formations of IMCs, the shear load and the consumption rate of UBM are investigated thoroughly, subjected to the thermal aging test. In addition, the effects of Ni thickness in relationship to the missing ball issue during packing drop test are also explored.

Experimental

Compass, 740 I/O 1-metal layer tape ball grid array (1-ML TBGA) substrate with selective Ni/Au plating and internal wiring Cu covered with solder mask was used as a test vehicle. In the design, solder mask opening was 0.45mm for 1.00mm ball pitch. Lead-free Sn3.5Ag0.5Cu (wt.%) (SAC) solder balls with a 0.76mm (30mil) diameter and water soluble flux WS609 made by Alpha Metal were used for solder ball attachment, under the same reflow profile at an actual peak temperature of 250°C in a 5-zone reflow oven. Solder ball shear test of 3μm, 6μm and 8μm test vehicles was performed at time zero and after different isothermal storage at 250°C with a Dage series 4000 bond tester. The ball shear height was kept at 0.04mm over solder mask surface with a shear speed of 0.3mm/s. The sample size for ball shear test was 5 sample/each test vehicles. 80 outer balls were tested with 20 balls at each corner of the sample. Scanning electronic microscope (SEM) was used to analyze the fracture surface of sheared ball pads and the IMC structure in the cross-sectional samples. Ni and Au thickness were measured with X-ray thickness tester. The purities of Ni and Au layer on soldering pad for each test vehicles were measured by Auger spectroscopy analysis. As shown in Fig. 2 (a) and (b), only Au and Ni were found as atomic composition without

978-1-4244-4658-2/09 $25.00 © 2009 IEEE

any impurities after ion sputtering of 5000A and 6000 for each Ni test vehicles. For high temperature storage, the samples were aged at 250°C for 5, 10 and 20 min. The schematic structure of test vehicle was shown in Fig. 1. Packing drop test was carried out using center pedestal support trays to assess the solder joints integrity at a manufacturing level as shown in Fig. 3. Thin Ni at 3μm and thick Ni at 6μm, 8μm & 10μm respectively were chosen for this test. All units were gone through simulated burn-in process prior to this drop test. Drop height from floor for all test vehicles were set at 39 inch. Drop point on the box was 10 points with 10 angles per requirement. Sample size was 60 good assembly units per each Ni thickness lots. 6 units/tray with 1 unit at 4 corners and 2 units at the center and rest are dummy unit were designed. Total 5 full trays / bundle plus one empty tray on top and 2 bundles / box were packed. Samples were dropped up to 11 cycles. After each cycle, the samples were inspected for any missing ball failure. Samples with different electrolytically plated Ni and Au thickness are tabulated in Table 1 in the range of Ni at 2.5μm-12.04μm, and Au at 0.50-0.61μm.

Table 1 XRF result of Ni & Au thickness data

	Ni Thickness (um)				Au Thickness (um)			
	Min	Max	Mean	Std-Dev.	Min	Max	Mean	Std-Dev.
1	2.51	3.69	3.14	0.434	0.50	0.56	0.54	0.03
2	5.53	8.19	6.01	0.610	0.52	0.61	0.56	0.04
3	7.54	10.69	8.18	0.691	0.54	0.58	0.56	0.02
4	9.62	12.04	10.01	0.865	0.52	0.58	0.56	0.03

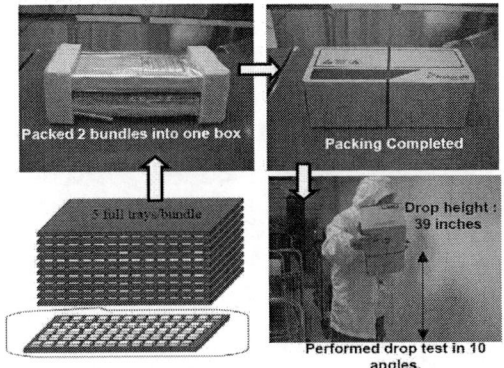

Fig. 3 Procedure of packing drop test with drop height from floor set at 39 inch and 10 drop point on the box

Results and Discussion

Ball shear Test results

The average shear strength of thick Ni/solder system is found to be outperformed thin Ni/solder system at time zero and all of high temp storage conditions. The changes in shear strength during extended thermal aging for all Ni test vehicles are shown in Fig. 4. Related failure modes distribution for 3μm thin, and 6μm/ 8μm thick Ni test vehicles are given in Figs. 5 and 6 respectively. Generally, thin Ni at 3μm shows lower shear strength and also degraded at a faster pace, compared to thick Ni.

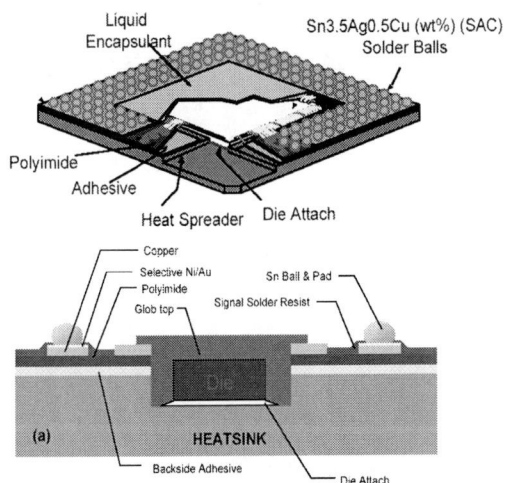

Fig. 1 Schematic diagram of 740 1-ML TBGA package

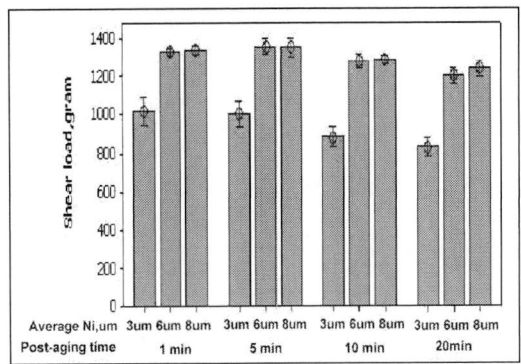

Fig. 4 Average shear load of each Ni thickness solder joints as a function of high temp storage time.

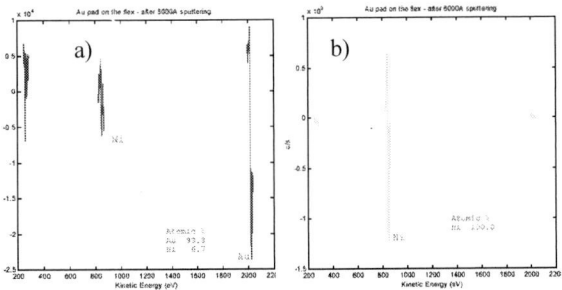

Fig. 2 Auger Electron Spectroscopy Analysis of Au and Ni layer on solder ball pad for both thin and thick Ni test vehicles. (a) 93.3% Au and 6.7% Ni in atomic % are found after 5000A ion sputtering. (b) 100% Ni in atomic % is found after 6000A ion sputtering

Fig. 5 (a)-(d) show typical pictures of the four types of fracture mode for thin Ni/solder system, including ductile mode, brittle mode and mixed mode. The mixed mode includes two modes such as <50% brittle fracture and >50%

978-1-4244-4658-2/09 $25.00 © 2009 IEEE

brittle fracture. During high temperature storage, the amount of IMC brittle fracture increases with extended aging time. After 20 minutes high temp storage, mainly 100% brittle fracture is mainly observed on thin Ni solder joints with an average shear load drop down to 831g/f (Figs. 4, 5(e) and 6(a)).

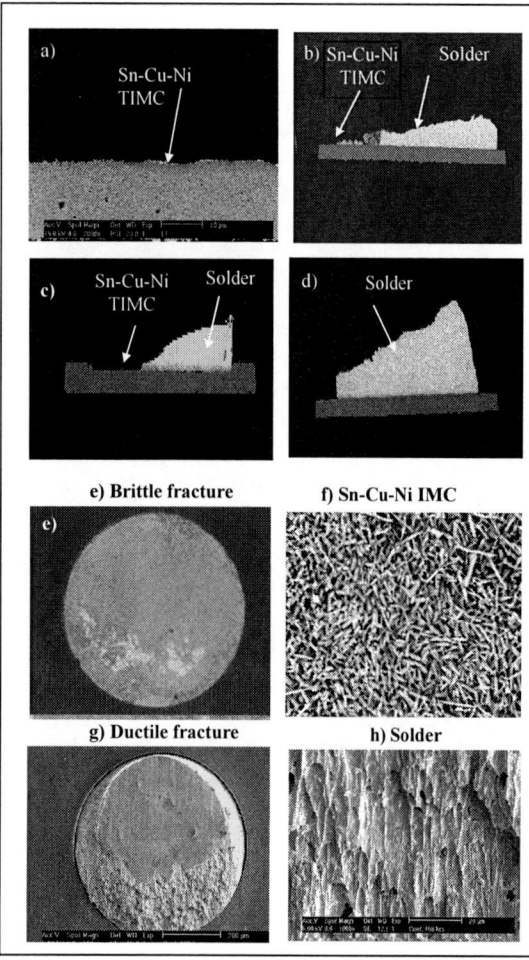

Fig. 5 Cross-sectional view of fracture mode (a) 100% brittle fracture occurs within the IMC, (b) <50% brittle fracture occurs within the IMC and solder (c) >50% brittle fracture occurs within the solder and IMC, (d) 100% ductile fracture within the solder, top view (e and f) 100% brittle fracture and (g and h) 100% ductile fracture surface of solder.

Analysis on the fracture surface of solder pad exhibits a flat brittle fracture with a smooth cleavage surface, mainly happens along the interface between Ni and Sn-Cu-Ni IMC layer (Figs. 5(a), 5(e), 5(f)). Shear load highly depends on the fracture site. This result showed that the brittleness of solder joint increases with thin Ni and extended thermal aging. For the thick Ni/ solder joint, shear strength up to 5 minutes aging times maintains with ductile fracture (Fig. 4, 5(g)), with average shear strength of above 1200 gf. No significant

differences in shear strength and fracture mode between 6um and 8um Ni is observed.

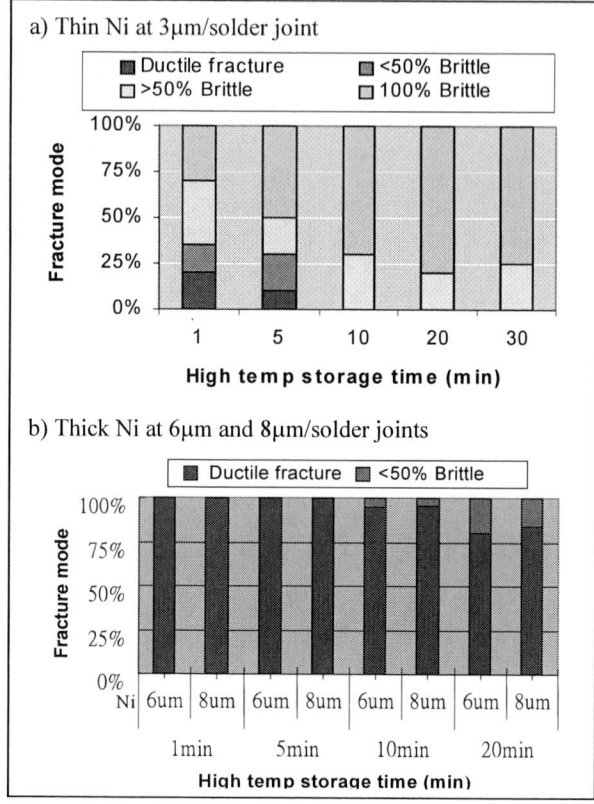

Fig. 6 Fracture failure mode percentage subjected to high temperature storage time (a) Thin 3μm Ni, (b) Thick 6μm Ni, 8μm Ni/solder (shear speed 300um/s).

The enlarged SEM micrograph showed that the fracture occurs within the solder (Fig. 5(h)). The fracture site exhibited a dentate surface, which is a character of ductile deformation. After 20 minutes thermal storage, shear strength has slightly decreased due to mixed mode (<50% brittle) occurs in some solder ball site for 6μm, 8μm Ni (Figs. 4, 6(b)) respectively. These results demonstrate that thick Ni/solder systems have greater solder joint integrity against high temp storage. The reasons for such different trends in mechanical load and fracture mode could be explained by in-depth study of the interface.

Investigation of solder ball pad profile

In perspective view of thin Ni test vehicle, it was found that the solder pads profile was containing a gap between Ni/Au plating and surrounding solder mask (Fig. 7(a)). EDX analysis on the surface of interfacial gap reveals a thin films layer of solder mask. In porosity test, the samples were first cleaned with ethanol and then placed in nitric acid vapor filled desiccators for 1 hour, it is found that the reagent attacks the expose base material through the porosity of solder mask and generates corrosion products around the solder pad of the thin Ni substrate (Fig. 7(b)).

978-1-4244-4658-2/09 $25.00 © 2009 IEEE

Subsequence EDX analysis evidenced that there is an exposed Cu product occurred in the corrosion region, evidencing significant amount of porosity occurred in the SM tail. In contrast, for the solder pad plated with thick Ni, no interfacial gap or black corrosion product is found before and after porosity test as shown in Fig. 8.

A gap between Ni/Au plating and surrounding SM

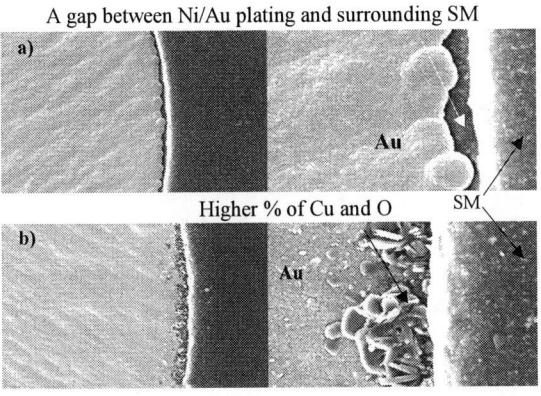

Fig. 7 Thin Ni, (a) Before Nitric acid porosity test, and (b) After Nitric acid porosity test.

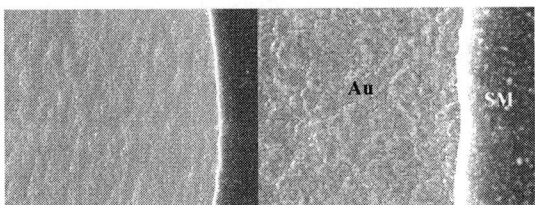

Fig. 8 Thick Ni (6μm) ball pad after Nitric acid vapor porosity test.

From x-sectional view of thin Ni solder pad, it is found that the solder mask profile exhibits a SM tail (Fig. 9(a)), leading to form an interfacial gap as shown from perspective view. Increasing Ni thickness is demonstrated effectively where the Ni is plated over the SM tails. In worse case, the interfacial gap could be increased with more Cu recession and or reducing the Ni plating. As shown in Figure 9(b) and thicker Ni at 6μm demonstrates a good Ni coverage with uniform metal deposits to the SM tail and the surrounding solder mask.

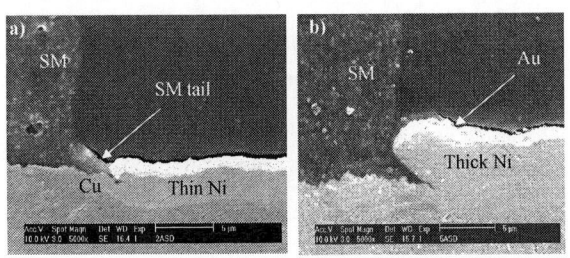

Fig. 9 (a) Poor Ni coverage of SM tail for thin Ni, b) Good Ni coverage of SM tail for thick Ni. Cu.

SEM/EDAX analysis on intermetallic reactions with as-reflow and extended thermal aging test

Microstructure and failure mode are closely related for Pb-free solder joints. EDAX analysis on UBM structure reveals a continuous $Sn_{0.58}Cu_{0.21}Ni_{0.21}$ TIMC layer, containing medium Cu (copper 15-35at.%), formed on top of Ni layer. (Figure 10) Initially, the growth rate of interfacial TIMC for all the Ni thickness solder joints is almost same after as-reflow. However, for thin Ni, more TIMCs containing higher Cu, and low Ni ($Sn_{0.58}Cu_{0.32}Ni_{0.10}$ at%) were also found at the edge of solder joint and surrounding solder mask (Fig. 10(a)), it is due to the copper atom diffused to the solder in a shorter path for forming the TIMC where it is quickly resettled to the interface during extended thermal aging. It is seen that the IMC forms over the SM tail not only provide an extra diffusion channels, but also forming a pre-existing crack around the edge of solder joint. Such a pre-crack zone could be served as a wedge, leading to have high stress concentration with promoting fracture failure along the brittle IMC interface eventually (Figure 11(a)). Therefore, the SM tail with thin Ni is responsible for the low shear loads with mixed and brittle fracture failure occurred in time zero and the early stage of high temperature storage. In contrast, thick Ni is proven to provide a robust solder joint as-reflow. With thicker Ni, the distance of TIMC layer is far away from the solder mask tail, assuring solder fracture failure occurred (Fig. 11(b)).

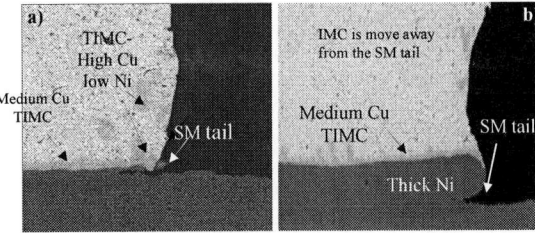

Fig. 10 (a) Pre-existing crack on SM tail of thin Ni/solder joint, (b) Normal solder joint of thick Ni/solder system.

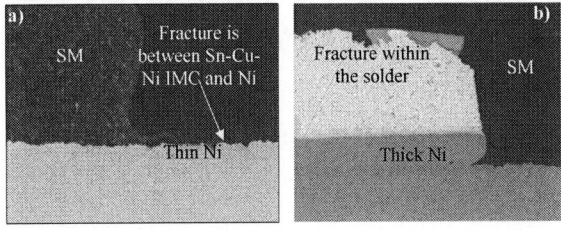

Fig. 11 (a) Brittle fracture with propagation of pre-cracking of solder Joint for thin Ni (b) Ductile fracture failure within solder for thick Ni.

Fig. 12(a) shows cross-sectional images of interfacial reactions of solder joints with 10-minutes high temp storage. For thin Ni, the growth rate of interfacial TIMC containing medium Cu was increased significantly. The average IMC thickness is measured to be 6.06μm, which is 1.34 times higher than that with thick Ni. Such a rapid growth of interfacial TIMC layer is probably responsible for weakened

978-1-4244-4658-2/09 $25.00 © 2009 IEEE

strength of solder joint, and increase in brittle fracture percentages in the shear test (Fig. 6(a)). Moreover, large amounts of Cu-Sn and Cu-Sn-Au compounds found in the bulk solder (Fig. 12(b)) could be directly attributed to the rapid diffusion of copper atoms into the molten solder.

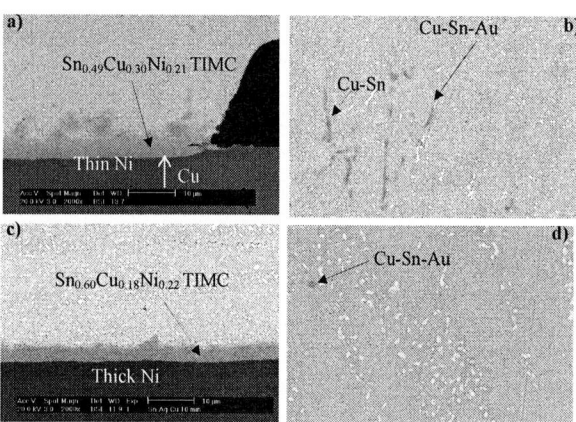

Fig. 12 Electrolytic Ni/solder joints after 10 minutes high temp storage-(a and b) Thin Ni/ solder system, and (c and d) Thick Ni/solder system.

It is suggested that Cu atoms could get a simple channel for diffusing into the molten solder through the solder mask tail continuously, leading to form more IMCs containing high Cu within the bulk solder and corresponded resettled to the interface, which accelerates the interfacial TIMC growth at this stage. For the thick Ni /solder joint (Fig. 12(c)), the interfacial TIMC thickness was still sluggish after 10 min high temp storage. As shown in Fig. 12(d), the amount of Cu-Sn compound particles within the bulk solder was insignificant, implying no extra Cu diffusion could gets into the molten solder at this stage. In both cases, Au does not resettle to the interfaces, due to the formation Cu-Sn-Au IMC within the solder, which prevent this resettlement.

The interfacial TIMCs growth on thin Ni solder joint has increased dramatically after reaction with 20-minutes high temperature storage (Fig. 13(a)). The average interfacial TIMC thickness of thin Ni is measured to be 9.18μm, which is almost 1.35 times higher than it is for thick Ni. The particles of Cu_6Sn_5 IMC with a specific morphology and more Cu-Sn-Au IMC have also found in the bulk solder at this stage (Figure 13(b)). During long time thermal aging, it is suggested that large amount of Cu atoms have been diffused into molten solder for thin Ni. For this reason, most of Cu-Sn and Cu-Sn-Au IMC phases that originally participated in the upside solder have been resettled to the interface of UBM, providing a driving force to have a rapid interfacial TIMC growth. More importantly, spalling of TIMC is occurred for thin Ni solder join, evidencing the resistance of Ni diffusion barrier layer is dropped, due to most of the thin Ni has been consumed. The formation of spalling could be attributed to the micro-pores of remaining thin Ni layer, whereas the Cu could be directly diffused though the crystal boundary of thin Ni, resulting in a rapid interfacial reaction with the molten solder. With more Ni consumed, a new Cu-Sn IMC phase

were also formed between the medium-Cu TIMCs and the Cu layer (Fig. 13(a)), which poses an important factor for weakening the solder joints strength in the shear test.

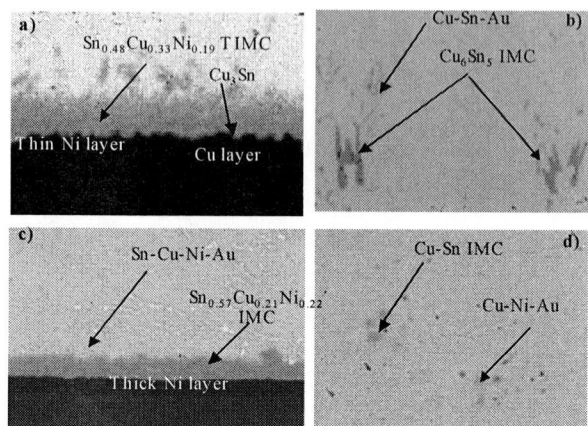

Fig. 13 Electrolytic Ni/solder joints after 20 minutes high temp storage-(a, b) Thin Ni, and (c, d) Thick Ni/ solder joint.

On the contrary, the growth of the interfacial TIMC layer for thick Ni was still very sluggish after 20 min high temp storage, the chemistry and morphology of the interfacial IMC is totally different with flat structure, compared to thin Ni (Figs. 13(a) & (c)). In this stage, small amount of Au has resettled to the interface for forming a new Sn-Cu-Ni-Au IMC, which slightly reduce the shear load and causes mixed brittles fracture failure as shown in Figs. 13(c) and 5(b). This shows thick Ni maintains a good diffusion barrier to withstand long time thermal aging.

Fig. 14 shows the thickness of the TIMC layer as a function of high temperature storage time. The TIMC compound growth showed almost a linear relationship with high temperature storage time. With rapid increase in TIMC thickness for thin Ni solder joint, it is implied that the continuous copper diffusion process with higher Ni consumption rate might control the growth rate of interfacial TIMC.

Fig. 15 shows the amount of the Ni layer consumed as a function of high temperature storage time. It is striking that the amount of Ni layer consumed in the cases of thinner Ni was much higher than that of thicker Ni during high temp storage. For thinner Ni, the driving force of accelerating the Ni consumption is suggested related to faster diffusion rate of copper with thin Ni, and higher growth rate of IMC at the thin Ni interface. During 20 min high temp storage, around 2.56μm thin Ni layer has been consumed by molten solder. The average consumption rate of the Ni layer is 0.128μm/minute. For thick Ni, about 1.74μm of the Ni layer has been consumed after 20 minutes high temp storage. The average consumption rate of this Ni layer is 0.087μm/minute, which is approximately 1.5 times lower than the thin Ni/solder system. With high Ni consumption rate, more Cu could be diffused into the solder and the percentage of IMC brittle fracture increases with the increase in Cu contents within the Sn-Ag-Cu solder alloy. These results is able to explain why the thicker Ni (above 6μm) outperforms 3μm

978-1-4244-4658-2/09 $25.00 © 2009 IEEE

thinner Ni on reducing brittle fracture percentages during extended high temp storage.

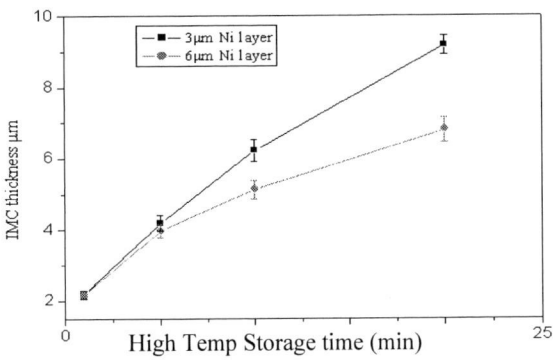

Fig. 14 IMCs thickness as a function of high temp storage times at 250°C

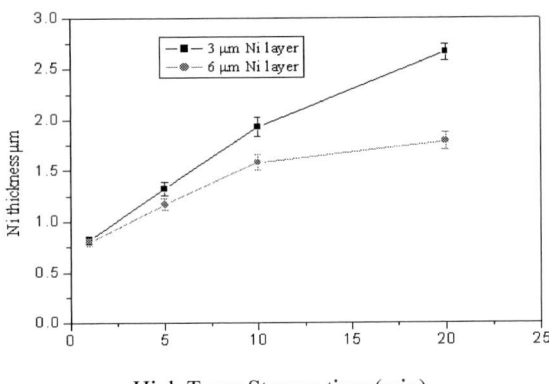

Fig. 15 Consumed Ni layer thickness as a function of high temp storage at 250°C

Fig. 16 shows the mechanism of higher and lower IMC formations for thin and thick Ni. With thin Ni, a rapid growth rate of interfacial IMC after 10 min and 20 min thermal storage could be attributed to extra Cu diffusion channels. In this case, the diffusion paths of copper is suggested to be passed through from two ways 1) The Cu transports directly to the molten solder through the SM tail, 2) The Cu transports from the remaining thin Ni and TIMC interface to the molten solder, which accelerates the interfacial TIMC growth and corresponded increases the consumption rate of Ni layer. As the Cu pad underneath the interface is much thicker than the Ni layer, it acts as a substantial source of Cu into the molten solder. With thicker Ni, longer the diffusion path as is, which act as a good diffusion barrier for preventing copper comes in contact with molten solder. This can be explained why thicker Ni layer/solder system has lower TIMC thickness and corresponded lower brittle fracture failure rate.

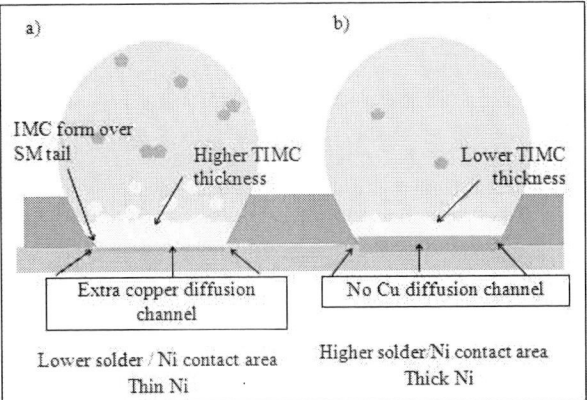

Fig. 16 Schematic diagram of solder joint after 20 min high temp storage, a) thin Ni/solder system and b) thick Ni/solder joint.

Packing drop test results

All samples were packed in box according to test method (Fig. 3) and then dropped test were performed to evaluate the performance of different Ni/solder joint under mechanical impact. The drop test results are analyzed in Table 2.

Table 2 Packing drops test results.

No. of Drop Cycles	Missing ball			
	3μm Ni	6μm Ni	8μm Ni	10μm Ni
3	Yes	No	No	No
7	Yes	No	No	No
11	Yes	Yes	No	No

In the results, it is evidenced that 3μm thinner Ni fails to withstand 3 cycles of packing drop test (Table 2, Fig. 17). For 6μm Ni, missing balls start to occur after 11 cycles of packing drop test. No missing ball is found in 8μm and 10μm Ni during 11 cycles drop test. Thicker Ni (6μm, 8μm and 10μm Ni) outperforms 3μm Ni largely to resist drop impact, which shown more than 3 times increase in the drop test cycles. Therefore, the above results indicate that higher Ni thickness has a higher mechanical strength to overcome the situation of missing balls against vibration and shock impact during handling or transportation. This also shows that drop test result correlates to the previous findings on ball shear strength, fracture mode and UBM structure.

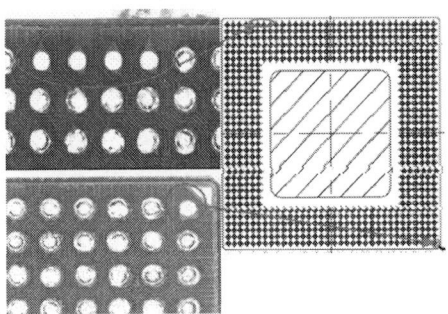

Fig. 17 Missing solder ball after drop test.

Conclusions

During the reaction of molten Sn3.5Ag0.5Cu (wt.%) (SAC) solder ball on SMOBC TBGA with different Ni metallization, the mechanical strength of solder joint, IMC structure, and dissolution rate of UBM were investigated to correlate with the ball drop level. During time zero or early stage of high temperature storage, the poor shear strength and higher brittle fracture failure observed in thin Ni are identified due to SM tails, where the Cu diffusion into molten solder could gets in a simple channel through the solder mask and provide a brittle fracture location, which are responsible for mixed and brittle fracture in the shear test. During 10 min high temp storage, the TIMC thickness of thin Ni has almost 1.35 times higher than it is for thick Ni, more Cu-Sn IMC phase found in the bulk solder could be directly attributed to the continuous Cu diffusion, leading to rapid increase in interfacial TIMC thickness. During 20 min high temp storage, the average consumption rate of thin Ni is approximately 1.5 times higher than it is for thick Ni, evidencing the resistance of copper diffusion was dropped due to more Ni was consumed with extended aging time, which accelerates the dissolution rate of Ni layer and corresponded increase the growth rate of interfacial TIMC layer. Therefore, the higher Ni consumption rate and thicker interfacial TIMC layer are identified to be responsible for increasing brittle fracture percentages during thermal aging test for the thin Ni solder joint. On the contrary, a higher shear loads, lower IMC thickness, lower consumption rate of Ni layer is demonstrated by thick Ni solder system at all of high temperature conditions. In final, a minimum 5µm thick Ni outperforms 3µm thin Ni in solder ball shear test and pass all cycles of packing drop test. This shows the drop test result correlates with the findings on mechanical ball shear test and UBM structure.

Acknowledgements

The authors would like to acknowledge the support provided by Compass Technology Co. Ltd. and all the extensive supports from Freescale Semiconductor Malaysia on assisting the assembly packaging and packing drop test.

References

1. Anton Zoran Miric, Angela Grusd, "Lead Free Alloys," *Soldering and Surface Mount Technology*, 10/1, 1998, pp. 19-25.

2. Ibrahim Ahmad, B. Y. Majlis, A. Jalar, Eu Poh Leng, "Optimization of Nickel Thickness On Substrate For TBGA Using SAC387 Solder Material," *Int'l Electronics Manufacturing Technology Symposium*, 2007 IEEE, pp. 192.

3. J. H Lau, C. P. Wong, N. C. Lee and S. W. R. Lee, Electronic Manufacturing with lead free, Halogen-free and conductive-adhesive materials, McGraw-Hill, New York, 2003.

4. Ashok Anand, YC Mui, Jaime weidler, Nelson Diaz, "Impact of substrate finish on Sn/Ag/Cu alloy solder joint," *ECTC*, 2004, pp. 335-338.

5. Jang, J. W. Lin J. Frear D. R., "Failure Morphology after drop Impact Test of the Ball Grid Array Package with Sn 3.8Ag 0.7Cu solder with Cu and Ni metalization," *Journal of Electronic Materials*, Vol. 36, No. 3, March 2007, pp. 207-213.

Quality and Reliability Challenges for Ultra Mobile Computing and Communication Application Processor Packaging

Dongming He and Wonjae Kang
Quality and Reliability Engineering
Intel Corporation
Prairie City Road, Folsom, CA 95630, USA
Phone: (916)-356-5449; Email: dongming.he@intel.com

Abstract

Market drivers, technology scaling and integration trends of application processor packaging are first presented. Component level and board level quality and reliability challenges are then discussed in the areas of thin die and thin core package warpage, lead free flip chip die to package interconnect mechanical integrity, and lead free package to motherboard solder joint reliability. Challenges from package qualification methodology and use condition are also addressed.

Introduction

In addition to continuous evolution of notebook computers and cellular phone devices in their own markets, silicon and package technology innovations are now enabling the convergence of ultra mobile computing and communication around the internet. The creation and phenomenal growth of new categories of devices such as smart phone [1], netbook [2] and mobile internet device (MID) [3] are good examples. Form factors of these devices vary from 2 to 3 inch screen sizes for typical smart phones, 3 to 5 inch screen sizes for typical MIDs, and 8 to 10 inch screen sizes for typical netbooks. These devices may support single or multiple wireless wide area network (WWAN) and wireless local area network (WLAN) protocols. In addition, some of these devices can support global positioning system (GPS) and have one or more integrated cameras.

Microprocessor, chipset, and application specific processor are the powerhouses inside these ultra mobile devices to support operation system, memory, graphics and audio processing. Due to form factor size and battery life constraints, smart phones desire to have the largest integration scale, such as monolithic silicon approach called system on chip (SoC) to realize these functions. Operation systems for smart phones are currently much less complicated than a personal computer offering limited internet performance. While larger form factor devices such as MID and netbook currently have a microprocessor and a separate chipset to provide improved internet user experience. In this paper, application processor is used as a general term for these components.

While integration at monolithic silicon level will continue to include additional functions such as radio, package integration such as package on package (PoP), package in package (PiP) [4], system in package (SiP), and system on package (SoP) are attractive cost effective alternatives or additions to provide fast improving internet, graphics and audio experience "on the go". Figure 1 illustrates three types of packaging options namely single chip, PoP and SiP for an application processor. Both silicon and package integration

trends drive smaller package sizes with decreasing die to package interconnect and package to mother board interconnect pitches to meet increasing IO density requirement. System z-height constraint, die stacking and package stacking drive thinner die and package thickness. Silicon backend scaling leads to fragile low k and lower k silicon interlayer dielectrics (ILD). Environmental hazardous material regulation introduces new materials such as lead free die to package and package to board interconnects, and halogen free package and assembly materials. Silicon and package level integration, geometry scaling and new materials introduce new and exacerbate existing packaging quality and reliability challenges.

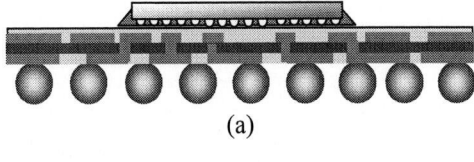

(a)

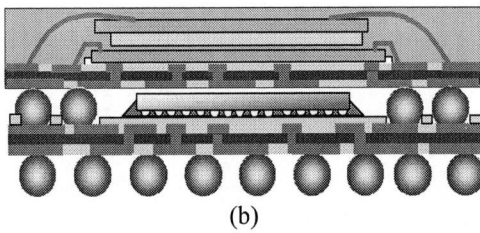

(b)

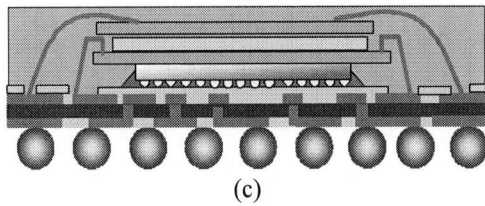

(c)

Figure 1 Flip chip application processor packaging options: (a) single chip package; (b) PoP with memory package stacked on top of bottom application processor package; (c) SiP with memory die stacked on top of application process die inside a single package.

Technology trends

Application processors are typically the largest, most complicated and most expensive component surface mounted on mother board inside ultra mobile devices. Due to silicon integration scale, power and signal integrity requirements, and geometry constraints from the final system, there is no unified packaging solution for application processors. As a matter of fact, even the same silicon or same family of silicon designs can use more than one package type to address specific system geometry, backwards compatibility, and cost requirements. In addition, time to market, product volume, business model such as internal development versus subcontracting, and memory device supply chain management play important roles in defining package architecture. Nonetheless, the trends are continuously increasing integration scale and shrinking key feature sizes for ultra mobile device packaging. According to International Technology Roadmap for Semiconductors (ITRS) [5], the die size of single chip package for hand held device is smaller than 100 mm^2 with power less than 3 W. The minimum single chip package z height usually from a smaller component than the application processor is 0.3mm in 2009 and 0.2mm around 2015. Table 1 lists some key silicon and package features for a couple of representative products currently on the market. The trend is clearly near chip scale packages with shrinking z height, and solder ball pitch for small form factor devices especially smart phones.

Table 1 Key feature size of application processors

Product	Intel Atom Z5xx CPU [6]	Intel Atom SCH* [7]	TI OMAP 3530/25 [8, 9]
Silicon	45nm node	130nm node	65nm node
Die size	~ 26 mm^2	~ 100mm^2	~ 60mm^2
Pacakge size: (mm)	13×14×1.6	22×22×1.9	CBC: 14×14×1.0 PoP** CBB: 12×12×0.9 PoP** CUS: 16×16×1.4
BGA pitch (mm)	0.6	0.6	CBC: 0.5 bottom, 0.65 top CBB: 0.4 bottom, 0.5 top CUS: 0.65
Ball count	441	1249	CBC: 515 bottom, 152 top CBB: 515 bottom, 168 top CUS: 423

*SCH: system control hub that includes graphics, audio, memory interface and general purpose IO interfaces.
**PoP z height is only for bottom application processor package.

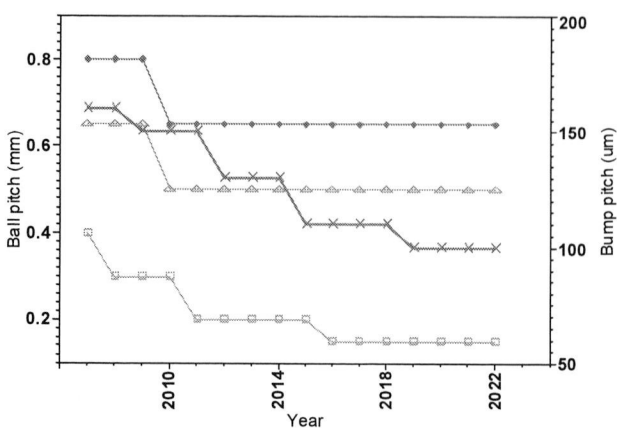

Left Scale: □—Ball pitch - FBGA   ◇—Ball pitch - PBGA   △—Ball pitch - TBGA
Right Scale: ×—Bump pitch

Figure 2 Bump pitch and ball pitch scaling trends for organic flip chip packaging. FBGA: fine pitch BGA, PGBA: plastic BGA, TBGA: tape BGA.

Both flip chip (controlled collapse chip connect-C4) and wire bond technologies can be employed for die to package interconnect. Due to large integration scale needed for application processor silicon, flip chip has many advantages over wire bond to achieve higher IO density, better power delivery for various power rails, and better signal integrity than wire bond for high speed signals. Organic flip chip package substrates are the primary choice for consumer electronics primary due to much lower cost than ceramic package. Figure 2 shows the organic packaging flip chip bump pitch and BGA ball pitch scaling trends according to ITRS. As an example, minimum bump pitch of 175µm was reported starting volume production by Intel using copper die bump with eutectic SnPb substrate solder for 65nm node in 2005 [10]. SnAgCu substrate solder was selected to achieve Pb free interconnect for 45nm silicon with volume production starting in 2007 [11]. Bump pitch of 146µm was reported in 2008 for 32nm Intel silicon with copper bump and lead free solder [12]. Figure 3 shows the package thickness road map [5]. At the same time, motherboard thickness and feature size are shrinking. For mobile phones, board thickness currently ranges from 0.3 to 1.2mm for four-layer to ten-layer boards with eight-layer board thickness as thin as 0.55mm in 2007 and 0.45mm in 2009 [13].

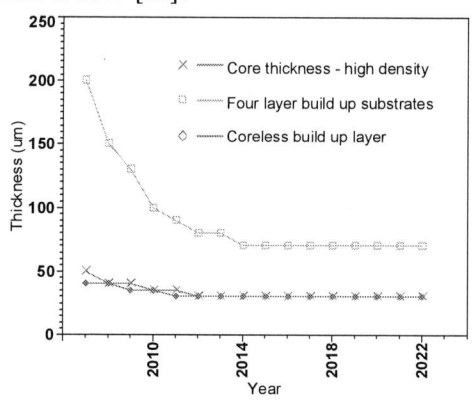

Figure 3 Tend of package substrate and core thickness.

New materials are continuously being introduced to support technology scaling and environmental regulation. Copper silicon back end interconnects and mechanically weak low k and lower k ILD plus lower k etch stop layer are now in high volume production with further k reduction expected by introducing molecular level doping during CVD film deposition or microscale porosity such as air gap. Final silicon passivation layer can be a silicon nitride layer or polymer on top of silicon nitride layer as additional dielectrics for redistribution layer or as a stress buffer layer before bumping.

Several flip chip die bump materials are used for flip chip die to package interconnect. High Pb with 3 to 5% Sn or eutectic SnPb are still being widely used with expectation to be replaced with Pb free interconnect in next few years. Pb free bumping examples include SnAg or SnCu solder bumps, gold stud bump and copper bump. Figure 4 is a comparison between Cu die bump and solder die bump. Different bump material requires different under bump metallurgy. Copper bump uses simple Ti/Cu seed layers while gold stud bump uses Al pad or Al cap on top of Cu pad. Solder bump requires a complex metal stack with an adhesion layer to the Cu metal pad such as Ti or Cr, and a barrier layer that can include Al, Ni and Cu to prevent Sn from diffusing and then reacting with the Cu metal pad [14-15]. During reflows, Sn from solder reacts with Cu/Ni in UBM stack forming intermetallic compound (IMC) between bulk solder and metal pad.

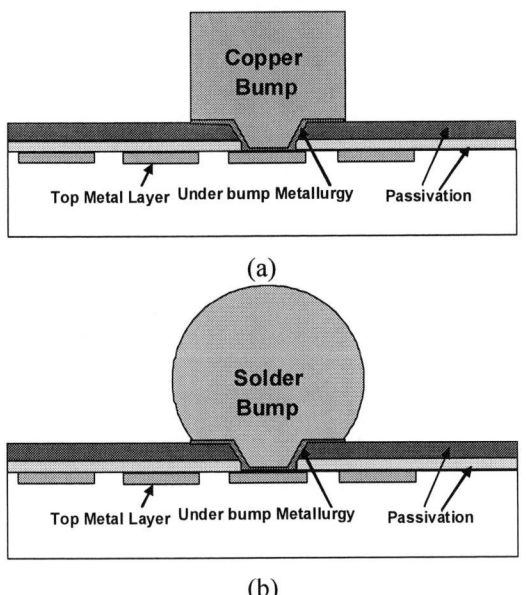

(a)

(b)

Figure 4 Copper (a) and Solder C4 bumps (b).

For package to board connections, Pb free solders have become main stream. In applications where long term reliability under thermal cycling load is required, SnAgCu solders with a composition of 3 to 4% of Ag and approximately 0.5% of Cu (SAC305, SAC405) have become the primary choices. In handheld devices, SAC105 with 1% Ag content is used to improve drop reliability however SAC105 is still not as good as SnPb solder joints. There are numerous reports in recent years on SACX solders by adding a small percentage of nano particles such as 0.05Ni% [16], and many other type of the fourth element such as Co, Pt, Al,

P, Zn, Ge, Ag, In, Sb, Au, Ce, Fe [17-19] to improve Pb free solder joint drop and temperature cycle reliability. In addition to solder, package surface finish plays a key role in board level reliability. There are several types of surface finish for BGA lands and metal pads on mother board including organic solder preserve (OSP), immersion Sn, electroplated NiAu, electroless Ni immersion Au (ENIG), electroless Ni electroless Pd immersion Au (ENEPIG).

New substrate materials and fabrication methods are being introduced for production or under development to support increasing interconnect density, reducing z height, halogen free (HF) and better compatibility with Pb free reflow temperature. Reinforcement with fillers is introduced to both core and build up layers to lower material coefficient of thermal expansion (CTE) and dielectric constant. New dielectric materials such as liquid crystal polymers (LCP) and coreless substrate are also emerging.

Quality and reliability challenges

A flip chip BGA is a multi-layered heterogeneous structure build with materials of different CTE and modulus, etc. Organic materials involved can have different glass transition temperatures. Array interconnects between die and package go through multiple high temperature processing steps. Bulk silicon CTE is 3ppm/°C. Cu CTE is 17ppm/°C. Equivalent CTE of a typical organic flip chip package today is about 16ppm/°C. Managing deformation and stress caused by CTE mismatch and high temperature process steps are critical to yield and reliability. Underfill is used to reduce stress on the C4 solder joint caused by CTE mismatch between silicon and package at the expense of increasing package warpage at room temperature. In addition, stress in the silicon and solder joint during or after chip attach reflow can be high enough to cause silicon ILD, passivation and C4 interconnect cracking. Figure 5 shows in plane and out of plane CTE trend to address CTE mismatch with bulk silicon. The modulus of rigid core, build up with and without reinforcement material are 30, 26, and 5GPa respectively in addition to CTE differences. Organic packages with CTE down to 3ppm/°C using negative CTE fiber for core and ~0ppm CTE build up layers has been reported for single chip package and SoP [20].

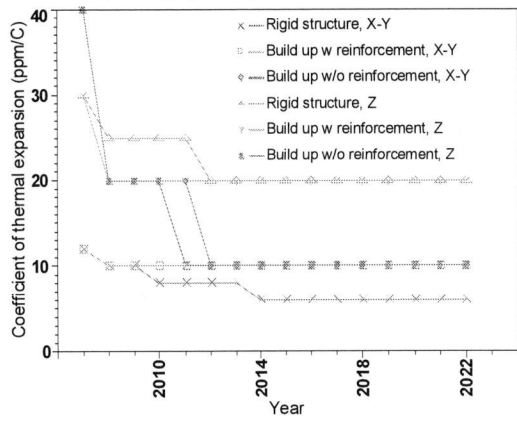

Figure 5 Trends of substrate material in plane and out of plane CTE.

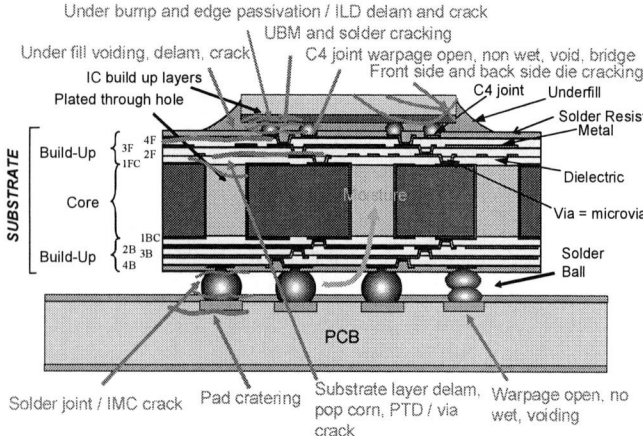

Figure 6 Typical failure modes for bare die flip chip BGA.

Figure 6 summarizes some of the key single chip flip chip packaging failure modes during manufacturing, shipping, storage and field use caused by CTE mismatch and moisture exposure. Geometry scaling, increasing interconnect density, and new silicon and package materials, bring in additional challenges. Complicated PoP and SiP architectures also introduce new interfaces and interconnect hierarchy such as bottom PoP and top PoP package connections or die to die stacking interfaces. Some of key challenges are discussed below.

1. Warpage and its impact to yield and reliability

One of the key challenges is to manage silicon and package warpage or mating solder array coplanarity to form reliable solder joints for thin die and thin core packaging. Thin die and thin package increase both silicon and package warpage. Smaller C4 interconnect sizes and BGA solder balls associated with shrinking pitch to prevent solder bridging reduce process tolerance to warpage. Figure 7 is an example of silicon warpage vs. die thickness.

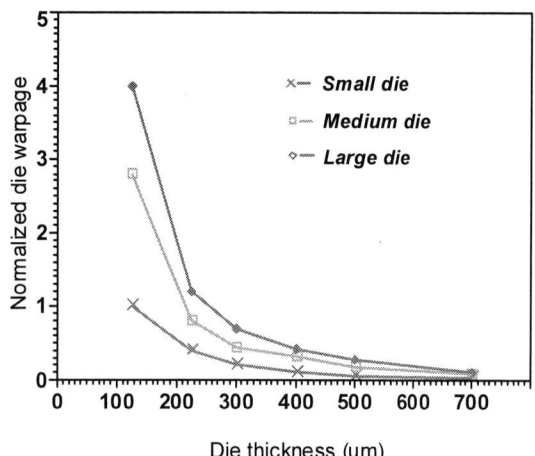

Figure 7 Die warpage at chip attach reflow temperature-220°C. Small die 8×8mm, medium die 13.7×13.7mm, large die 17.4×17.4mm.

Figure 8 shows package warpage vs. core and die thicknesses. It can be seen that warpage of thin die and thin core can be several times higher than that of thick die and thick core packaging. When the solder array coplanarity tolerance of top and bottom parts exceeds the solder collapse height at reflow temperature, open or defective joint form as result. Vice versa, solder bridging can occur if solder is being squeezed out due to warpage.

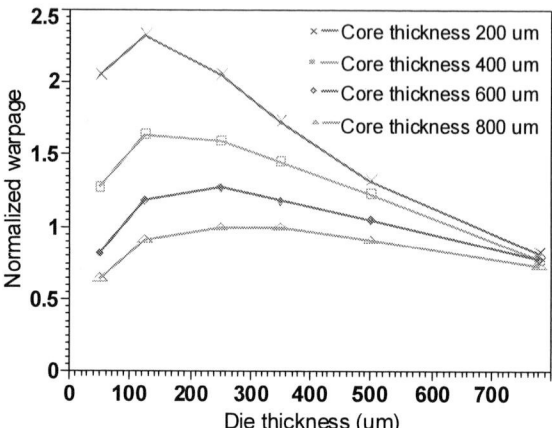

Figure 8 Package warpage as function of core and die thickness.

Both the shape and the magnitude of warpage are important. In addition, warpage is a function of temperature so the warpage shape and magnitude may not match room temperature measurement as shown in an example illustrated in Figure 9. Characterizing warpage vs. temperature is important to ensure good interconnect yield and reliability in additional to reduce room temperature warpage and collect SMT yield data [21-23]. A reflow temperature flatness requirements spec is released by JEDEC as an alternative to room temperature warpage requirements [24]. Figure 10 shows examples of good versus bad joints caused by warpage. In addition to causing interconnect test yield loss, defective C4 joints caused by warpage can fail in downstream process such as C4 open during BGA SMT or during reliability temperature cycling stress. Using BGA solder joint reliability (SJR) as an example, about a 30% temperature cycle life reduction is reported for 27mm and 35mm BGA when initial room temperature board warpage is increased from 1700μm to 3000μm for BGA mounted on a 203×140×0.63mm PWB [25].

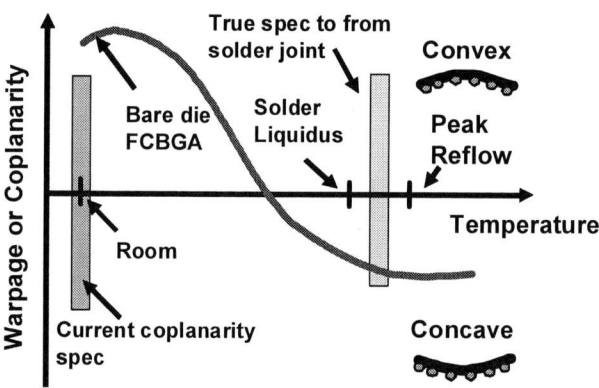

Figure 9 A BGA package warpage vs. temperature

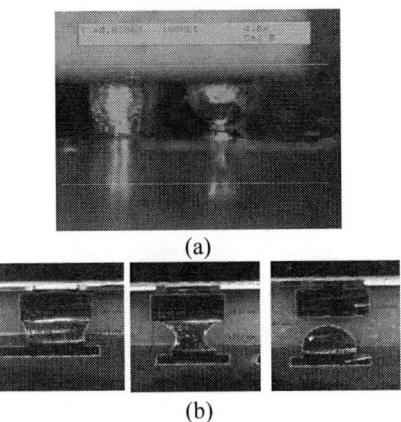

(a)

(b)

Figure 10 Warpage impact to BGA ball joint and C4 joints.
(a) Good joint vs. a head on pillow BGA joint. (b) Good vs.
Stretched and open C4 joints with copper die bump.

Material properties such as glass transition temperature, CTE and modulus affect shape and magnitude of warpage in addition to die and package thicknesses and sizes [26]. Design features such as metal density and balancing have been known to modulate PWB and package warpage. These effects have also been reported for PoP packages [27]. Modeling has been effective to study the effect of material properties, geometry and layout for warpage reduction. To solve the warpage challenge of further package z height reduction, molding is introduced for new PoP structures such as fan-in PoP [28] and through mold via PoP [29].

2. Lead free impact to low k ILD and solder joint reliability

Stiffer Pb free solder, smaller solder joint size, lower standoff height, and higher processing temperature exacerbate stress to the interconnects, silicon ILD and package. For Cu / low k silicon ILD, there are two major failure mechanisms. The first one is ILD cohesive cracking or interfacial delamination under C4 bump caused by peeling stress transferred by C4 joints due to silicon and package CTE mismatch after chip attach reflow [10, 30]. The Second one is area delamination propagated from die corner or edge due to defects caused by die singulation and sometimes exacerbated by additional defects such as die to underfill delamination during temperature cycling stress [31]. From one stress modeling example, ILD peeling stress of PbSn die bump / substrate solder and SnAg die bump / substrate solder is 70% and 40% less than Cu bump/SnAg solder respectively. Copper bump improves C4 electro-migration to allow smaller die passivation opening size to improve on die power line layout. It also simplifies bumping process by removing bump reflow and using thin Ti/Cu UBM stack other than complex UBM stack for Sn containing solder die bump. However, silicon passivation and ILD stress management is much more challenging than Sn based Pb free solder bump. For solder bump, stress to under bump ILD and passivation is reduced, however UBM cracking due to brittle Sn based IMC formed with Ni / Cu from UBM has to be carefully characterized [32]. In addition to material properties and passivation structure, design features such as UBM size and passivation opening sizes will impact stress to UBM and ILD. Not only stress reduction is important, reducing and controlling material defects during process are critical to prevent ILD crack and UBM crack initiation.

For Pb free BGA interconnect reliability, both good drop performance and temperature cycling performance are important for hand held devices. One of the major challenges is to identify one primary or a couple of generic surface finishes and solder materials as PbSn solder replacement. Higher Ag containing SAC solder such as SAC405 and NiAu surface finish are popular in notebook and desktop computers, which is primarily concerned with solder fatigue during temperature cycle. However multiple Pb free solders and different surface finishes are widely used in ultra mobile devices. Since the two key solder joint failure locations are either intermetallic interface between bulk solder and metal pad or inside bulk solder adjacent to IMC, small amounts of Cu and additional alloy elements can reduce IMC thickness, improve IMC uniformity, reduce voiding at IMC or change bulk solder grain structure resulting in improved reliability for drop and temperature cycle. It was reported that drop performance of OSP surface finish is two times that of a 5µm NiAu surface finish with LF35 (Sn-1.2%Ag-0.5%Cu-0.05%Ni) solder while SAC305 is only 25% of the drop performance of LF35 / OSP for both surface finishes [16].

In addition to optimizing solder materials and surface finishes, board level underfill or corner glue can be applied during SMT process to improve drop reliability however sometimes at the expense of degrading temperature cycle reliability performance with low T_g underfill to facilitate rework. Low CTE and higher T_g underfill with filler is found to be preferred choice to optimize drop and temperature cycle reliability. Corner only dispense pattern can mitigate impact to temperature cycle performance [33]. Board level IO routing method such as via off pad (VOP) and via in pad (VIP) can also impact the failure mechanisms, as well as drop and temperature cycle performance. At the system level, solder joint reliability is impacted by system PCB design, layout, chassis, mounting screws and EMI shielding etc., which influence board bending and flexing.

3. Qualification methodology and use condition

There are two basic methods to qualify IC packaging, i.e. standard based stress test driven qualification [34] and application specific knowledge based qualification [35]. Table 2 lists stress conditions and criteria for standard based package component level qualification. The duration is based on historical failure acceleration models and generic temperature and humidity conditions. For example, temperature cycle requirement is scaled based on Coffin-Manson power law $(\Delta T_{stress}/\Delta T_{use})^n$ with factor n = 2. For board level reliability, results have a strong dependency on board design, SMT process and test methods. JEDEC and IPC have established specs on test board design and test method for temperature cycle, shock and bend for commercial applications [36-38]. Standard based package qualification is still widely used in the industry due to its simplicity. However, it can set up unnecessarily higher requirements for certain market segments or specific failure mechanisms with a possibility of underestimating the requirement for a certain mechanism or market segment. Application specific knowledge based qualification allows for qualification stress requirements based on specific failure mechanism acceleration and failure rate expectation. This usually involves different stress conditions to establish acceleration models for new package architecture and material set. Understanding both failure acceleration and field use condition is required to apply knowledge based qualification. Published reliability models and acceleration parameters are documented and updated regularly in JEDEC JEP122 for reference [39]. However, models for new package architecture and material sets are still lacking.

Table 2 JESD47 nonhermetic package qualification stress requirement: sample size 3 lots×25 units with zero failure [34].

Stress	Condition	Duration
High temperature storage	150°C	1000 hours
Temperature cycling (additional condition in [34])	-55-125°C or -65-150°C or -40-125°C	700 cycles 500 cycles 850 cycles
Temperature humidity: V_{cc} max or unbiased if cannot run biased test	85°C/85% RH or 110°C/85% RH or 130°C/85% RH	1000 hours 264 hours 96 hours

Application specific processors for ultra mobile devices have low power and long standby time so that high temperature storage or bake hours can be reduced. Assume the device operates every day running 8 hours at T_j of 90°C for 5 years. With a lower bound Ea of 1.0eV for IMC formation and Kirkendall voiding which are the packaging specific bake failure mechanisms documented [39], package bake duration of only about 160 hours is needed at 150°C. However, bake can cause memory device charge loss or gain with lower bound Ea of 0.75eV. For a memory product qualification, this translates to 500 bake hours at 150°C if data is to be collected for packaged devices. Longer stress duration on the other hand does compensate the drawback of small sample sizes to reduce the chance of a wear out failure mechanism being missed. But it introduces the possibility that an application specific product that meets use condition expectation cannot be qualified if the product starts to wear out at longer stress durations or additional failure mechanisms occur at later durations outside the use condition requirement. If stress duration is reduced based on use condition then a large sample size such as 230 units is sometimes recommended for zero defect sampling plan to meet less than 1% failure rate at 90% confidence or with an acceptable failure rate predicted by a reliability model.

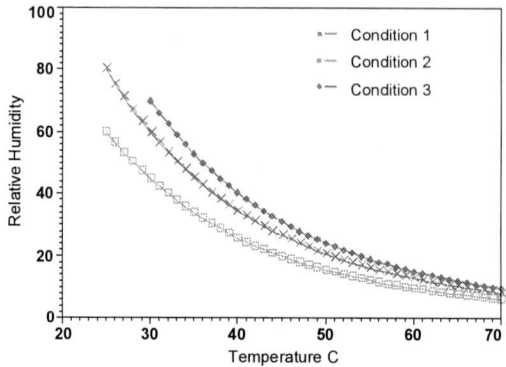

Figure 11 Relative humidity vs. external ambient and location temperature

Table 3 Biased HAST and THB stress hours vs. Ea (n = 2)

E_a (eV)	130°C / 85 % RH bHAST	85°C / 85% RH THB
0.4	400 hours	1700 hours
0.5	155 hours	943 hours
0.6	60 hours	524 hours
0.7	24 hours	291 hours
0.8	9 hours	162 hours

Condition: 3 years standby at 30°C / 70% RH as example [35]

On the other hand, low power ultra mobile devices have a more severe field moisture usage requirement than notebook and desktop due to long stand by hours with always on always connected signals and some power planes still active during standby / sleep in order to wake up when receiving calls or messages. Peck's model is often used for temperature humidity failure mechanism acceleration with E_a and power law for relative humidity. Typical lower bound for E_a is 0.7eV and lower bound for power law humidity acceleration is about 2 [39]. T_j increase reduces location relative humidity as shown in Figure 11 as local saturation vapor pressure increases with local temperature increase. Since MID/phone device's standby/sleep state power is very low to save battery life, die and package temperature and relative humidity are essentially the same as ambient so that there is little humidity acceleration compared with stress condition of 85% RH. Table 3 shows that bHAST and THB stress duration requirement is very sensitive to E_a. If standby state temperature can be 10°C higher than ambient at failure location such as notebook or desktop computer standby temperature, relative humidity is decreased from 70% to 40%. Stress duration can be reduced by a factor of 3 with a power law exponent of 2. In order to develop a new acceleration model on Ea, at least three different stress temperatures are preferred with consideration of sample size and process variations. THB is closer to actual use temperature but requires long stress durations while 130°C bHAST runs into risk of over accelerating or causing unrelated failure modes since T_g of many underfill materials and package solder resist are close to or lower than 130°C.

For temperature cycle stress, there are two major challenges. The first one is that there can be multiple failure mechanisms with wide ranges of $(\Delta T)^n$ power law parameter n for different kinds of materials in the packaging structure. For a ductile material such as solder, n can range from 1 to 3 with hard materials such as IMC, Si and ILD having higher n values. For BGA solder joint reliability, the three-parameter Coffin-Mansion or so called Norris-Landzberg model is typically used to include solder creep related to dwell time or frequency and maximum solder temperature T_{max}. Different n values are currently reported in the literature for BGA Pb free solder joint along with different T_{max} and frequency acceleration parameters, such as 1.8 [40] and 2.3 [41]. A Model where n and T_{max} acceleration parameters vary with temperature cycling duration has been recently proposed with n = 0.92 from 0 to 500x, n = 0.87 from 500 to 1000x, n = 2.1 from 1000x to 1500x and n = 1.6 from 1500x to 2000x of 0 to 100°C temperature cycling stress respectively [42].

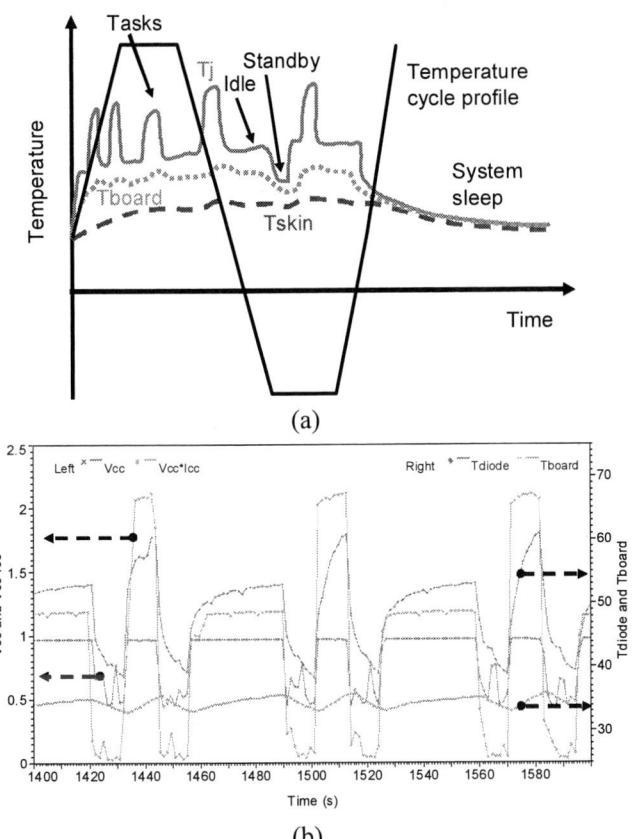

(a)

(b)

Figure 12 (a) Example of usage temperature profile vs. temperature cycle stress for a system with enclosure. (b) An example with an application processor subject to period work load pattern. Application processor is SMT to a board without enclosure. Board reference is measured by thermocouples 1 mm away from package.

Another key challenge in defining the temperature cycle stress requirement is to translate from use condition temperature profile to stress condition temperature profile even after an acceleration model is picked. Since ultra mobile devices are power conscious, system power and device power can be managed based on the temporary workloads or tasks. Different system and device power states can be introduced by software and hardware designs. The system and device can switch to high power for short period time to download

internet data or decoding video. After that, both system and application processor can go to low power state or idle state when switching to low activity task such as web browsing or working on emails. For ultra mobile devices like smart phones, there is limited or no thermal solution used such as without passive heat spreader attached to die. With small thermal mass, silicon temperature can react quickly to work load with a time constant on the order of 10 seconds. Thermal time constant of package and board reference near package is on the order of one minute while system skin time constant is on the order of 10 minutes. Two examples are shown in Figure 12. There can be multiple temperature cycles with ΔT as function of workload power and duration depending on application and user behavior for a single continuous use. The dwell time can be on the order of minutes for these additional ΔT cycles. On the other hand, there can be a large transient temperature gradient among silicon, package, board reference and device skin reference. The magnitude of temperature response decreases moving away from silicon to device skin. Fortunately, silicon CTE is the smallest compared to packaging materials though it has the highest temperature perturbation. However, package surface material is close to die temperature causing stress value changes due to CTE mismatch during workload change. Understanding failure mechanism and acceleration model also enable proper assessment for this kind of system and human behavior.

4. Other challenges

For thin die and thin core packages with small BGA ball sizes, handling during assembly, test and SMT process is another challenge. For wire bond stacked die packaging, die thickness can be thinner than 100μm nowadays while flip chip die thickness can be as thin as 100 to 200μm. Die and package damage such as die crack, mold crack, or missing solder ball can occur if handling and test equipment are not well designed and characterized for high volume production. Manual handling during development and reliability stress can also cause visible or latent die and package damage.

New package architecture and materials also bring in challenges to detect failures and defects as well as fault isolation and failure analysis. Some of the failure modes especially precursors are not sensitive to electrical test. For example, a die to package or package to board interconnect barely touching as small as a few percent of normal contact area has a large chance to pass DC open test electrically especially under test socketing pressure since solder joint resistance is on the order of milli-Ohm. Offline RF test is more sensitive to detect marginal joints [43] but cannot be easily implemented for volume data collection during manufacturing without significant amount of test cost increase. Nondestructive detection methods such as CSAM are being increasingly used as development monitors for low k ILD cracking, underfill voiding/delam, UBM cracking, mold cracking and delamination, package layer delamination etc. with added overhead during assembly process and reliability characterization. 2D X-ray is being used to examine solder joint alignment and voiding. 3-D X-ray is emerging to examine solder joint shape but with long throughput time. In situ electrical monitoring, physical cross section, and dye and pry are being used during stress and post stress for solder joint reliability. Focused ion beam (FIB) is used for ILD crack,

UBM crack failure analysis with TEM being used to examine IMC layers at UBM interfaces.

Designing an electrical sensitive test vehicle with excellent fault isolation capability for various failure mechanisms such as die to package C4 interconnect, under bump low k ILD crack and die edge low K ILD crack is critical to reduce data turnaround time during technology development to meet short time to market challenge [44]. Improving product IO test coverage and test sensitivity to interconnect resistance increase or even implementing some key thermomechanical test structures are critical for product qualification and manufacturing excursion management. Designing product with design rules or guidelines to reduce packaging quality and reliability risk has becoming increasingly important. From manufacturing side, process window characterization with skewed experiments for process parameters also help to evaluate yield and reliability risk using electrical test failure detection.

Conclusion

New packaging materials, technology integration and scaling, thin die and thin core, and new market segment usage model introduce significant quality and reliability challenges for ultra mobile device application processor packaging. Thin die and thin core packages increase the risk of flip chip die to package and BGA to mother board interconnect opens or short failures during manufacturing and reliability stress. Warpage changes with temperature and is modulated by material properties, design and assembly process. UBM integrity of Sn based solder bump and Cu bump stress to low k ILD are two key Pb free die to package interconnect risks. For Pb free solder board level reliability, key challenges are to select the right solder and surface furnish material or develop board level underfill/corner glue solution to meet both drop and temperature cycle requirements.

For product qualification, lacking of acceleration models for new material sets is one of the key challenges. Ultra mobile devices have higher humidity bias accelerated life test requirement than notebook and desktop computers due to low power, long stand by or sleep time and always on always connected usage model assuming the same the similar life time requirement. Workload based power management introduce challenges to translate usage temperature profile to equivalent stress requirements. Handling of thin die and thin core package and failure/defect detection are some other challenges for ultra mobile device application processor packaging.

Acknowledgments

The authors would like to thank Intel Assembly Technology Development Director and Vice President Nasser Grayeli, Mobility Group Quality and Reliability Manager Thomas Marieb, Quality and Reliability Engineers Tom Moss, Sudarshan Rangaraj and Yeen San Yip for reviewing this paper and providing valuable inputs. We also want to thank to Intel Assembly Technology Development for modeling and cross section referred in this paper.

References

1. http://en.wikipedia.org/wiki/Smartphone
2. http://en.wikipedia.org/wiki/Netbook
3. http://en.wikipedia.org/wiki/Mobile_Internet_Device
4. Pendse, R., "Flip Chip Package-in-Package (fcPiP): A New 3D Packaging Solution for Mobile Platforms", 2007 ECTC.
5. http://www.itrs.net/reports.html
6. http://download.intel.com/design/processor/datashts/319535.pdf
7. http://download.intel.com/design/chipsets/embedded/datashts/319537.pdf
8. http://focus.ti.com/lit/ds/symlink/omap3530.pdf
9. http://www.embedded.com/underthehood/210101486
10. Yeoh, A. et al., "Copper Die Bumps (First Level Interconnect) and Low-K Dielectrics in 65nm High Volume Manufacturing," 2006 ECTC, pp. 1611-1615.
11. Mistry, K. et al., IEDM Tech. Dig., pp. 247, 2007.
12. Natarajan, S. et al., "A 32nm Logic Technology Featuring 2nd-Generation High-k+Metal-Gate Transistors, Enhanced Channel Strain and $0.171\mu m^2$ SRAM Cell Size in a 291Mb Array," IEDM, 2008.
13. http://www.kcg.co.kr/pr_lounge/brochure/kcc_brochure.pdf.
14. Ebersberger B. et al., "Qualification of SnAg solder bumps for lead free flip chip applications," 2004 ECTC, pp. 683-691.
15. Lin X. et al., "The growth and influencing factors if voids in SnAg solder bump and their imapct on interfacical bond strengtheth," IEEE Xplore, 2007.
16. Tanaka M. et al., "Improvement in Drop Shock Reliability of Sn-1.2Ag-0.5Cu BGA Interconnects by Ni Addition," 2006 ECTC, pp. 78-84.
17. Amagai M. et al., "A Study of Nano Particles in SnAg-Based Lead Free Solders for Intermetallic Compounds and Drop Test Performance," 2006 ECTC, pp. 1170-1190.
18. Lau J. et al., "Reliability of Sn 3%Ag 0.5%Cu 0.019%Ce (SACC) Solder Joints," 2009 ECTC, pp. 418-422.
19. Hutter, M. et al., "Effects of Additional Elements (Fe, Co, Al) on SnAgCu Solder Joints," 2009 ECTC, pp. 54-60.
20. Yamanaka, K. et al., "Advanced Surface Laminar Circuit Packagin with Low Coefficient of Thermal Expansion and High Wiring Density," 2009 ECTC, pp. 329-332.
21. Xie D. et al., "Head in Pillow (HIP) and Yield Study on SIP and PoP Assembly," 2009 ECTC, pp. 752-758.
22. Cho S. et al., "Validation of Warpage Limit for Successful Component Surface Mount (SMT)," 2008 ECTC, pp. 899-906.
23. Garner L. et al., "High Temperature Alternatives to Ball Grid Coplanarity Measurements to Improve Correlation to Surface Mount Quality," 2008 ECTC, pp. 406-411.
24. SPP-024, JEP No. 95, http://www.jedec.org/
25. Tan W., et al., "Effects of Warpage on Fatigue Reliability of Solder Bumps: Experimental and Analytical Studies," 2008 ECTC, pp. 131-138.
26. Vijayaragavan N. et al., "Package on Package Warpage - Impact on Surface Mount Yields and Board Level Reliability," 2008 ECTC, pp. 389-396.
27. Zhao J., et al., "Effects of Package Design on Top PoP Package Warpage," 2008 ECTC, pp. 1082-1088.

28. Carson F. *et al.*, "The Development of the Fan-in Package-on-Package," *2008 ECTC*, pp. 956-963.

29. Kim J. *et al.*, "Application of Through Mold Via (TMV) as PoP Base Package," *2008 ECTC*, pp. 1089-1092.

30. Pendse R. *et al.*, "Innovative Approaches in Flip Chip Packaging for Mobile Applications," *2009 ECTC*, pp. 285-292.

31. Bansal A. *et al.*, "Reliability of High-end Flip-Chip Package with Large 45nm Ultra Low-k Die," *2008 ECTC*, pp. 1357-1361.

32. Gu Y. *et al.*, "Interfacial Delamination Near Solder Bumps and UBM in Flip-Chip Packages," *ASME Journal of Electronic Packaging*, SEPTEMBER 2001, Vol. 123, pp. 295-301.

33. Lee J. *et al.*, "Study on the Board Level Reliability Test of Package on Package (PoP) with 2nd Level Underfill," *2008 ECTC*, pp. 1905-1910.

34. JEDEC JESD47, http://www.jedec.org/

35. JEDEC JESD94, http://www.jedec.org/

36. JEDEC JESD22-B111, http://www.jedec.org/

37. JEDEC JESD22B113, http://www.jedec.org/

38. IPC-9701, Performance Test Methods and Qualification Requirements for Surface Mount Solder Attachments.

39. JEDEC JEP122, http://www.jedec.org/

40. Vasudevan V. *et al.*, "An Acceleration Model for Lead-Free (SAC) Solder Joint Reliability under Thermal Cycling," *2008 ECTC*, pp. 139-145.

41. Lall P. *et al.*, "Principal Component Analysis Based Development of Norris-Landzberg Acceleration Factors and Goldmann Constants for Leadfree Electronics," *2009 ECTC*, pp. 251-261.

42. Ahmad M. *et al.*, "Parametric Acceleration Transforms for Lead-Free Solder Joint Reliability under Thermal Cycling Conditions," *2009 ECTC*, pp. 682-691.

43. Kwon *et al.*, "Detection of Solder Joint Failure Precursors on Tin-Lead and Lead-Free Assemblies using RF Impedance Analysis," *2009 ECTC*, pp. 663-667.

44. He D. *et al.*, "Test Vehicle to Characterize Silicon to Organic Flip Chip Package Thermomechanical Interactions," *2004 ECTC*, pp. 712-717.

Electromigration Analysis and Electro-Thermo-Mechanical Design for Semiconductor Package

Hsiang-Chen Hsu[*], Shen-Wen Ju, Jie-Rong Lu, Hong-Shen Chang, Hong-Hau Wu
Department of Mechanical and Automation Engineering, I-Shou University
No. 1, Sec. 1, Syuecheng Rd., Dashu Township, Kaohsiung County, Taiwan 84008
*E-mail: hchsu@isu.edu.tw

Abstract

In this paper, an advanced electro-thermo coupling model is developed to investigate the electromigration and electro-thermo-mechanical effects on electronic packaging, especially on Package-on-Package (POP). POP packaging involves in ultra thin gold wire ($\phi = 1 mil$) on wirebonding and Sn4.0Ag0.6Cu (SAC405) solder ball on package. The current density arising in the aluminum pad (wirebonding) and in the Copper trace above SAC405 solder ball imply the hot spot where results in an electromigration along the current direction. Finite element predictions reveal the maximum electro-thermo-mechanical effective stress is located at the regions where electromigration potentially occurred. Reliability on electro-thermo-mechanical for wirebonding and SAC405 solder ball is evaluated. Current crowding, temperature distribution and electro-thermo induced effective stress distribution are predicted. A series of comprehensive parametric studies were conducted in this research.

1. Introduction

For the past two decades, electromigration on Integrated Circuit (IC) packaging has been a serious reliability issue, which drives many researchers and engineers concentrated on this study. Recently, a new packaging technology, flip chip on Ball Grid Array (BGA) has been developed for high density interconnection. Solder bump on flip chip and solder ball on BGA serves as an electronic path and a thermal channel as well as the supporting structure. Reliability becomes an issue when the integrity of interconnection has been failed. Previous studies experimentally demonstrated [1-3] that electromigration on the solder bump/solder ball is the most tenacious cause affected reliability. Copper ionic corrosion always occurs under high current density at leveled temperature. Working temperature and current density are the main factors which directly impact electromigration behaviors. Higher thermal resistance and lower resistivity implies higher anti-eletromigration capability. Current crowding is reported on the corner where geometry dramatically changes. Joule heat is then accumulated as current density is increased. Thermal-induced stress results in larger inner displacement and damage in the structure. Structure deformed due to thermal expansion while current density varies as structure deformed. Electro-thermo-mechanical coupled analysis is required to predict electrically thermal-induced strain in the structure. An increase in temperature results in a decrease in the useful life time.

Aside experiment works, several numerical predictions [4-7] were performed to gain more fundamental understanding of the mechanism of electromigration on solder bump. In these papers, the magnitudes and distributions of electric current densities around the current crowding region have been displayed. Current crowding occurred in the vicinity of the locations where copper traces connect the solder bump. However, the consequent electro-thermo induced strains were excluded.

In this research, electric analysis is first performed to evaluate the current density and current crowding and follow-by electro-thermo coupled design on the POP package. Because the current is crowded due to complex geometry of structure, a single solder ball and an entire bond wire are modeled to simply the problem. As electricity analysis is continued, the selected solder ball is electrified to observe the current crowding. Black's equation is employed to determine the predicted equivalent life time for SAC405 solder ball. Material properties, such as electrical resistivity and temperature coefficient of resistance are crucial and vary from working temperature. Mesh density is required to evaluate the convergence for structure integrity. A submodel scheme is applied for evaluation of equivalent life time of solder ball.

2. Constitutive Models

The integrated modeling in this research involves electrical analysis, heat transfer, electro-thermo-mechanical, and Black's MTTF models [8].

2.1 Electrical Model

The basic equation for electrical analysis is Ohm's law

$$V = IR \qquad (1)$$

where V is the voltage, I is the current and R is the resistance. Resistivity is defined as

$$\beta = \frac{RA}{l} = \frac{VA}{Il} = \frac{1}{\sigma} \qquad (2)$$

where l is length and σ is electrical conductivity. Electrical field strength E is dined as E=σJ, where J is current density

$$J = \frac{I}{A} = E\sigma \qquad (3)$$

Joule heat, H is defined as

$$H = I^2 Rt = ms\Delta T \qquad (4)$$

where t is time, m is mass, s is specific heat and ΔT is temperature change.

2.2 Heat Transfer Model

The electrical-induced transient thermal diffusion (conduction) equation given in (5) can be solved to obtain the overall temperature distributions.

$$\frac{\partial T}{\partial t} = \alpha_T \left(\frac{\partial^2 T}{\partial x^2} + \frac{\partial^2 T}{\partial y^2} + \frac{\partial^2 T}{\partial z^2} \right) \qquad (5)$$

where T is the absolute temperature, x, y and z are the spatial coordinates, t is the time, $\alpha_T = k/(\rho C_P)$ is the thermal diffusivity, k is the thermal conductivity, C_P is the specific heat and ρ is the density.

2.3 Electro-Thermo Model

The heat generation rate per volume, Q due to electricity on package is given as

$$Q = \nabla \cdot ([k][\Delta T]) + \{J\}^T [\sigma]^{-1} \{J\} - \Delta T([\alpha_E][J]) \quad (6)$$

where α_E is the thermoelectric power. In (6), the first term is heat conduction, the second term is Joule heat and the third term is thermoelectric.

2.4 Black's MTTF Model

Mean time to failure (MTTF) is first developed by Black [8] to evaluate the equivalent life time for electromigration and is defined as

$$(MTTF)^{-1} = AJ^2 \exp(-\frac{E_q}{RT}) \quad (7)$$

where A is a constant, E_q is the activation energy, R is the Boltzmann constant and T is the absolute temperature.

3. Finite Element Model

The current flow direction is from PWB to solder ball and then through second bond at BT substrate to Au bond wire to first bond at Al pad in chip. As mentioned before, current crowding is always located in the vicinity of geometry sharply changed, where the high current density is found. Fig. 1 presents the solder ball electrical model and Fig. 2 shows bond wire electrical model, respectively.

For electro-thermo-mechanical coupling analysis, Fig. 3 demonstrates the entire strip model for POP package and a submodel of the far-most solder ball for further more accurate analysis. Reliability test is performed after electro-thermo-mechanical coupling analysis. MTTF can only be determined by using the single solder ball model in Fig. 1(b). A non solder mask defined (NSMD) model is used in this paper.

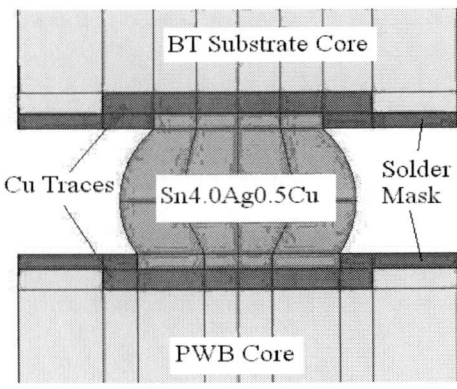

Fig. 1(a) Electrical model including BT substrate and PWD

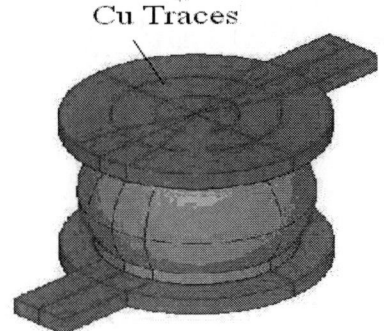

Fig. 1(b) 3D single SAC 405 solder ball with Cu traces

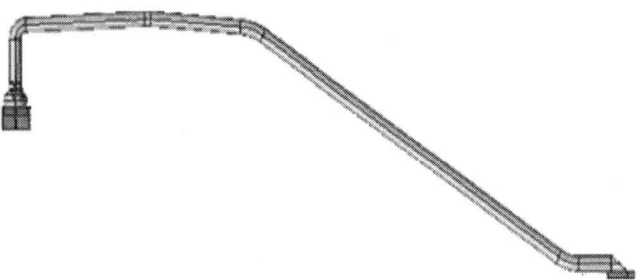

Fig. 2(a) Wirebonding electrical model

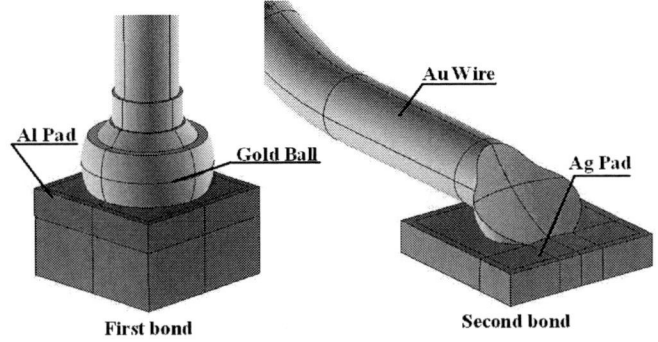

Fig. 2(b) First bond and second bond regions

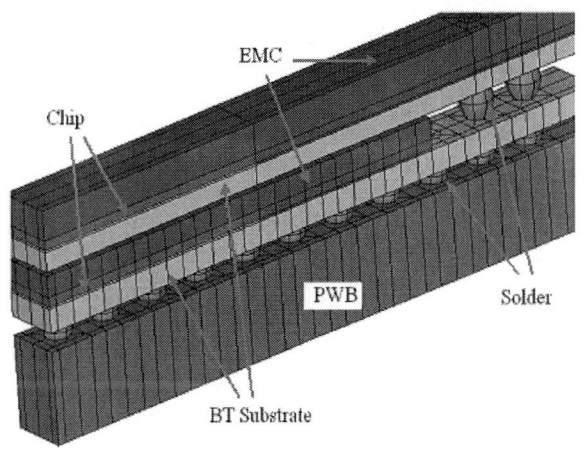

Fig. 3(a) Strip POP model for electro-thermo-mechanical coupling analysis

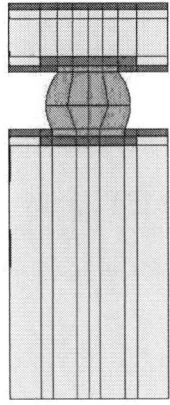

Fig. 3(b) Submodel of the far-most solder ball

978-1-4244-4658-2/09 $25.00 © 2009 IEEE

4. Results and Discussions

4.1 Electrical Analysis and Electromigration Analysis

It is required to perform mesh density convergence test before all the analyses are started. Fig. 4 illustrates current density tends to converged when the total number of element becomes 466, 848 of 3D single ball with Cu traces model in Fig. 1(b). The relationships between current density and number of element can be derived as

$$Current\_density = 220.76(No.\_of\_element)^{0.0501} \qquad (8)$$

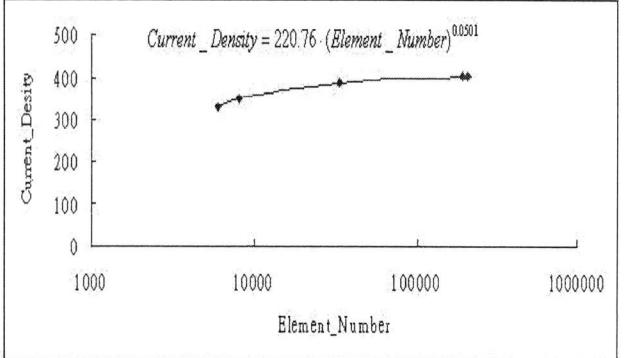

Fig. 4 Convergent for current density and number of element

The boundary conditions are given as zero voltage at bottom Cu trace and 1 ampere at top Cu trace. Temperature is initially set to be room temperature (25°C) and material electrical properties are resistance for SAC405 is 13e-5 Ω-mm and for Copper is 0.158 Ω-mm, while temperature coefficient of resistance for Copper is 4.3e-3. As electricity is continuously implemented, the voltage difference between top and bottom of solder ball is increased up to 10.868 V. Current crowding is also reported at the region where geometry and flow direction abruptly change. These can be seen in Fig. 5 and Fig. 6, respectively.

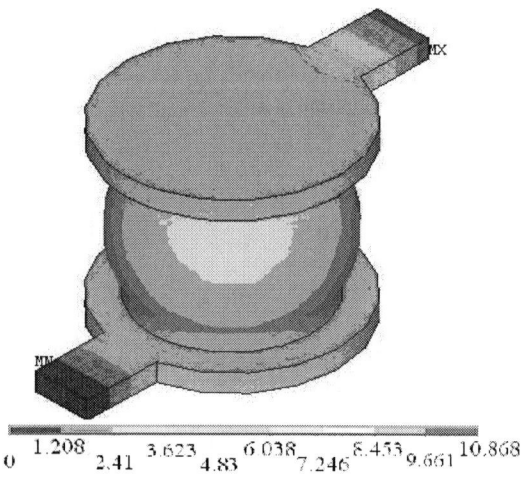

Fig. 5 Electrical voltage distribution (V)

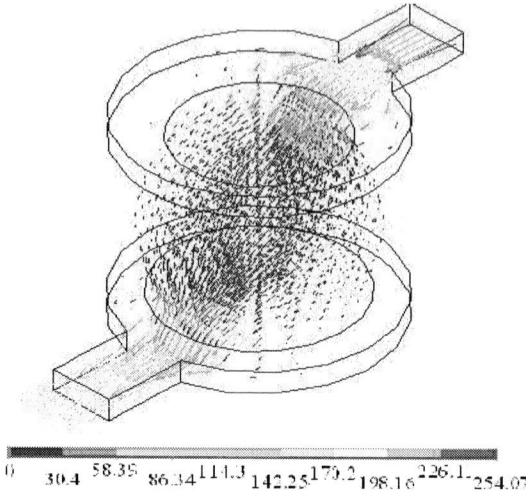

Fig. 6 Current flow direction

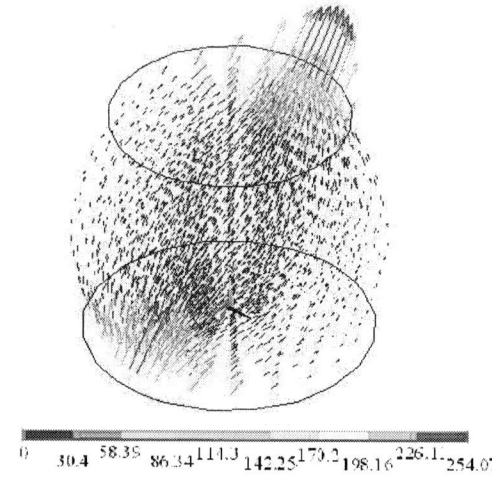

Fig. 7 Electron flow direction (reversed from Fig.6)

From Fig. 6 and Fig. 7, current crowding always located at the top and the bottom Cu traces, where imply eletromigration would occur at these regions. Fig.8 illustrates the electrical-induced Joule heat distribution. It can be seen that the predicted hot spots are located at the inlet and of outlet solder ball, where are close to the vicinity of current crowding.

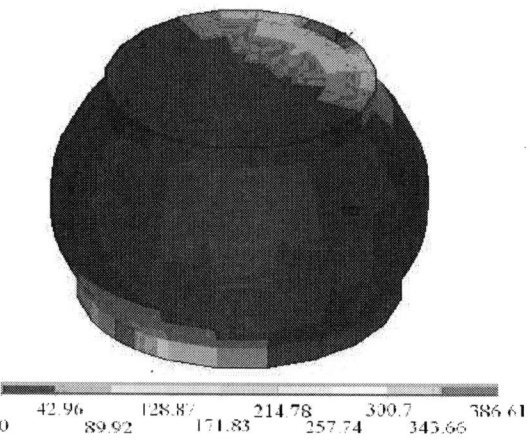

Fig. 8 predicted Joule heat distribution (W)

Due to flexible geometry of solder ball and input current, parametric studies are performed to evaluate various Joule heat from solder ball height and current. The results are shown in Fig. 9 and Fig. 10, respectively.

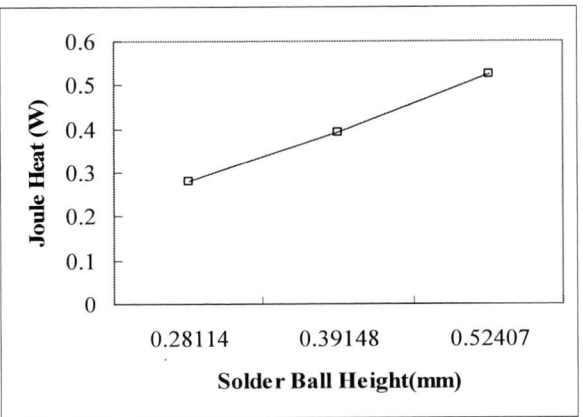

Fig. 9 Joule heat vs. solder ball height.

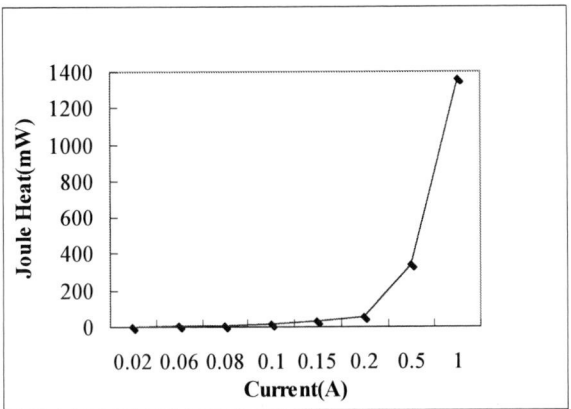

Fig. 10 Joule heat vs. input current.

The 99.99% gold bonding wire did not report any electromigration due to its low activation energy. However, the inter-metallic compound (IMC) exists beneath first bond and Al pad. Higher working temperature induced from Joule heat will result an increase in the depth of IMC. Electrical analysis is also conducted at this tiny region. The current is 0.48-0.6A from the second bond and the voltage is given 3.3V. Fig. 11 and Fig. 12 depict the current density and Joule heat at first bond, respectively. The predicted maximum current density is $0.483653e-3\mu A/(\mu m)^2$ and is located at the interfacial of gold ball bottom and top of Al pad, where the maximum Joule heat (0.021406 W) is occurred. Definitely, IMC will continuous grow as the electricity is input.

4.2 Electro-Thermo-Mechanical Analysis

For the material of SAC405, creep is essentially occurring at room temperature. In the consequent electro-thermo analysis, working temperature is always higher than room temperature. Viscoplastic material behavior needed to be included in electro-thermo-mechanical analysis for SAC405. Element type should be very carefully selected in ANSYS software. The coupling procedure is: 3D Solid 231 element is first used for electro-thermo-mechanical analysis and 3D Solid 186 element is applied to interpolation scheme of

submodel to evaluate the mesh convergence. In the final analysis, Solid 185 element is selected for SAC405 and Solid 45 element is for Cu traces. Fig. 13 depicts the element type used in this analysis.

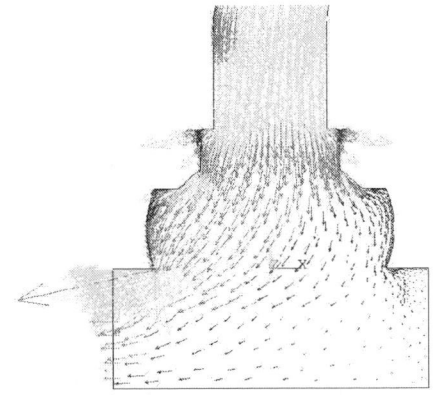

Fig. 11 Predicted current density at first bond (Au wire)

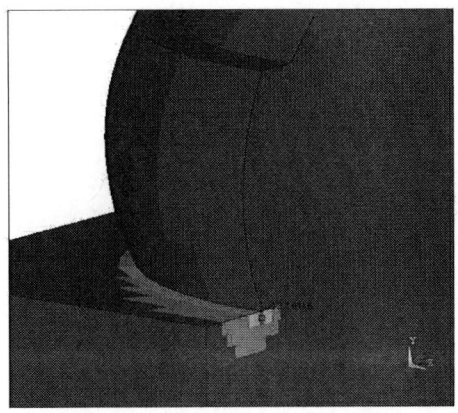

Fig. 12 Predicted Joule heat at at first bond (Au wire)

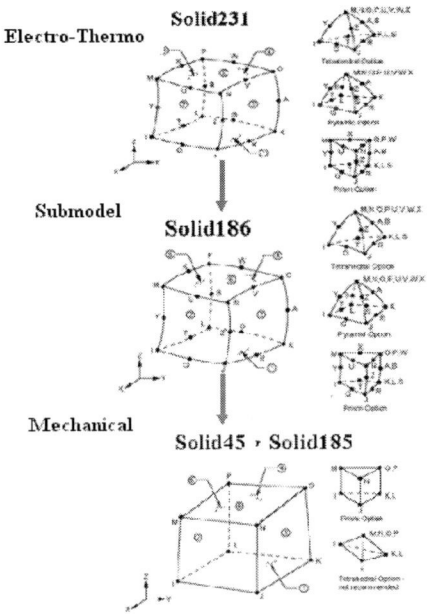

Fig. 13 Element types used in electro-thermo-mechanical coupling analysis

Fig. 14 and Fig. 15 demonstrate the electro-thermo induced total equivalent strain distribution in the single ball and with/without traces, respectively. The peak stain is 0.01351 at bottom copper trace in Fig. 14 and 0.10189 at solder ball adjacent to copper trace in Fig. 15.

Fig. 14 Total equivalent strain distribution (ball and traces)

Fig. 15 Total equivalent strain distribution (ball only)

The corresponding equivalent stress distributions are shown in Fig. 16 and Fig. 17, respectively. The predicted peak equivalent stress is 34.7MPa in Fig. 16 and 28.7MPa in Fig. 17. The stressed areas are very close the inlet and outlet of solder ball.

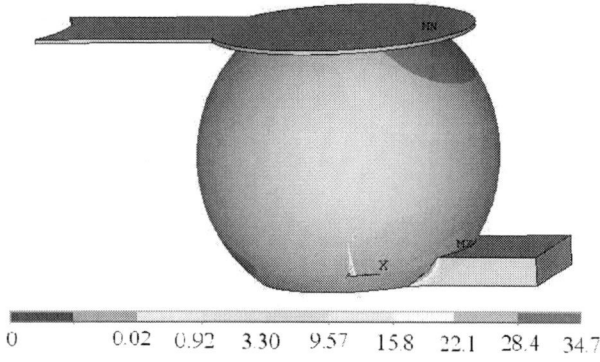

Fig. 16 Equivalent stress distribution (MPa)

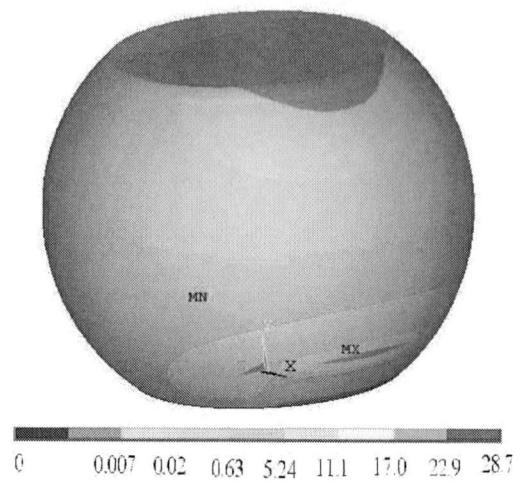

Fig. 17 Equivalent stress distribution in solder ball (MPa)

Electro-thermo-mechanical analysis is also performed at Au wire first bond. The maximum equivalent stress is 18.19MPa which located in the bottom surface of ball bond. The interfacial stressed area implies high working temperature at this region which results in an increase in the depth of IMC. As electrical current is continuously input, the depth of IMC will also be enlarged. The interfacial shear stress is then dramatically reduced.

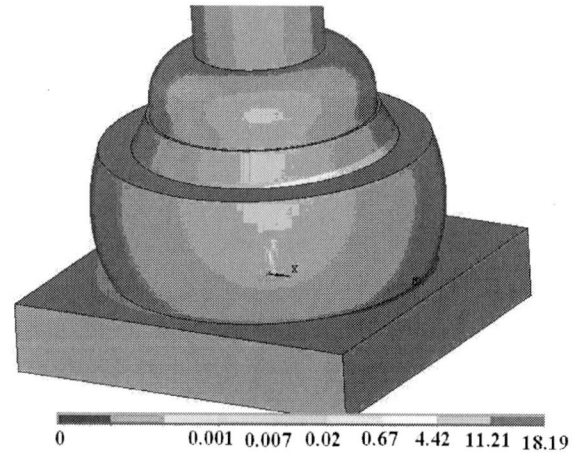

Fig. 18 Equivalent stress distribution in first bond (MPa)

4.3 Reliability Analysis

Based on Black's equation in (7), current density can be obtained from the above numerical results and A is found to be 391.9322 and E_q is 0.98eV for SAC406 solder. Mean time to failure at room temperature (25°C) is calculated as 0.27887e14 hours under input current of 1A. However, the predicted useful life would dramatically drop as working temperature is increasing (current density is increasing). Fig. 19 illustrates the deceasing of useful life as the temperature is creased. MTTF=0.84366e9 hrs at 150°C (423K).

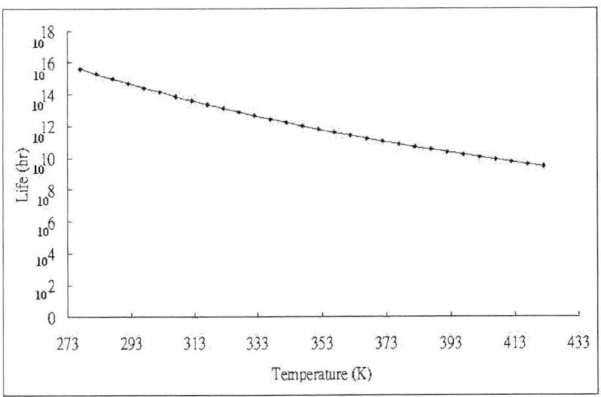

Fig. 19 Useful life time vs. working temperature (hrs)

Conclusions

The constitutive of electromigration and electro-thermo-mechanical models have been derived in this paper. An integrated finite element model based on ANSYS is developed to evaluate the current density and Joule heat in the area current is crowding. FEA predicted results show that current would be crowded at the region geometry is abruptly changed. As electrical current is continuously input, electromigration occurs at a tiny region where current is crowding. Current and geometry are the major factors for Joule heat. Thermal strain is then induced by the consequent working temperature. Electro-thermo-mechanical coupled analysis is performed to predict electrically thermal-induced strain in the structure. This localize stressed area enlarges the depth of IMC and imposes a potential failure region in the package. Reliability analysis shows that the useful life of SAC405 solder ball decreases as the working temperature is increased. In this paper, FEA predictions reveal the insight of electromigration mechanism. The predicted results show a very good agreement with previous experimental works.

Acknowledgments

This work was supported by I-Shou University, Taiwan, Grant No. ISU-97-In-01.

References

1. Tao, J. and Liew, B. K., "Electromigration under time-varying current stress," *Microelectron. Reliab.*, Vol. 38, No. 3 (1998), pp. 295-308.
2. Hu, Y. C., Kin, Y. H., Kao, C. R. and Tu, K. N., "Electromigration failure in flip chip solder joints due to rapid dissolution of copper," *J. Mater. Res.*, Vol. 18, No. 11 (2003), pp. 2544-2548.
3. Wang, L., F. Wu, Y. and Zhang, J., "The effect of current crowding on electromigration in lead-free flip chip bump interconnect," *High Density Packaging (HDP)*, 2004.
4. Gee, S. and Kelkar, N., "Lead free and PbSn bump electromigration testing," *(IPACK)*, 2005, pp. 263-268.
5. Lee, T. Y. and Tu, K. N., "A study of electromigration in 3D flip chip solder joint using numerical simulation of heat flux and current density," *IEEE Electronic Component and Technology Conference (ECTC)*, 2001, pp. 558-563.
6. Lai, Y. S. and Kao, C. L., "Calibration for electromigration reliability of flip chip packages," *IMAPS*

Taiwan International Technical Symposium, 2006, pp. 128-135.

7. Chiu, S. H., Chen, C., Lin, S. S., Chou, C. M., Liu, Y. C. and Chen, K. H., "Joule Heating Effect in Flip-Chip Solder Joints," *IMAPS Taiwan International Technical Symposium*, 2006, pp. 161-166.
8. Black, J. R., "Electromigration on failure modes in Aluminum metallization for semiconductor devices," *IEEE, Trans. CPMT*, Vol. 57, No. 9 (1969), pp. 1145-1150.

Comparison of Thermal Fatigue Reliability of SnPb and SAC Solders under Various Stress Range Conditions

Chaoran Yang[1], Yuen Sing Chan[1], S. W. Ricky Lee[1]*, Yuming Ye[2], Sang Liu[2]

[1]Department of Mechanical Engineering, Hong Kong University of Science and Technology
Clear Water Bay, Kowloon, Hong Kong SAR
*Tel: +852-2358-7203, Fax: +852-2358-1543, E-mail: rickylee@ust.hk
[2]Process Technology Section of Manufacture Technologies Center, Huawei Technologies Co., Ltd.
Huawei Industrial Base, Bantian Longgang, Shenzhen 518129, P. R .China

Abstract

Most accelerated temperature cycling (ATC) tests reported the observations that SnAgCu (SAC) solders have better thermal fatigue reliability than SnPb solders due to their lower creep strain rate. But material studies revealed that this is true only when stress level is below a certain level. When the stress level in the solder increases, eventually the creep strain rate of SAC solder may outgrow that of the SnPb solder. Therefore, lead-free soldering under thermal cyclic loading may lead to more reliability concern than people expected. Further investigations on ATC tests with various stress ranges are still in demand.

In the present study, by making use of distance-from-the-neutral-point (DNP) effect, a simple but effective custom-designed dummy package with various solder joint spacing is designed and manufactured to evaluate the thermal fatigue reliability of SnPb and SAC solders under different stress levels. It was found that SnPb solder could outperform SAC solder under a higher stress level condition. For the lower stress level condition, although the ATC test is still in progress, based on the present preliminary result, it is reasonable to predict that SAC solder would be better than SnPb solder in this case. The finite element analysis using the same package configuration was also conducted. The computational creep strains in the solders agreed with the ATC test results. Further discussion on the thermal fatigue life prediction is also presented in this paper.

1. Introduction

For the implementation of lead-free soldering process, the board level solder joint reliability is one of the major concerns [1-3]. Most of the results from different researchers tend to support that the SAC alloy family solders can improve the ATC test performance compared with the SnPb solder [4-6]. Nevertheless, different conclusions have also been drawn [7-9]. It is found that the competition between SAC and SnPb is closely related to the constraint condition of the solder joint, which may be induced either by the package type or by the ATC test condition. Schubert [7] used the flip chip on board as the test vehicle with different ATC temperature profiles. He found that, under the temperature range of -50°C to 20°C, SAC performs better than SnPb. But if the temperature range is from 50°C to 120°C, both SAC and SnPb perform worse than the results under -50°C to 20°Cbut SnPb becomes better than SAC. A comparison study between SAC and SnPb solder joint reliability with different types of package was conducted by B. Vandevelde et al. using finite element modeling [10]. Their conclusion is: for the CSP and flip chip

packages, SAC scores worse in more extreme loading conditions, while for QFN, SAC is always better.

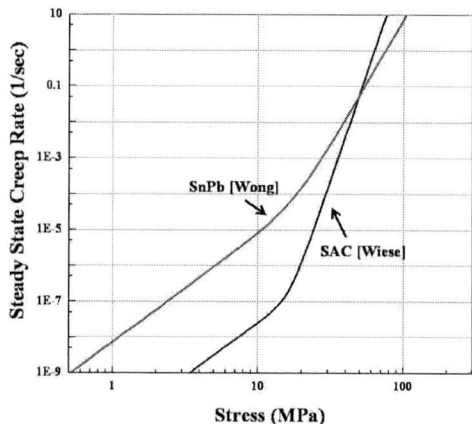

Figure 1 Constitutive Creep Models of SnPb and SAC Solders [11-12]

These two contradictory conclusions can be explained from the material response point of view. The creep behaviors of SnPb and SAC solders have different degrees of stress-dependence; i.e., when the stress level increases, the SAC solder will have a larger creep strain increment than the SnPb solder (Figure 1). At the low stress level, which is also the stress condition for most of ATC tests, the SAC solder has a smaller creep strain rate. This results in a better thermal fatigue reliability score for the SAC solders. When the stress exceeds a certain level, the situation will be reversed and the SnPb solder becomes having a lower creep strain rate. But there also exist several arguments about the application of this theory on the thermal fatigue analysis of solder joints. One is that most of the results of this material behavior study were conducted using bulky specimen under isothermal loading conditions. But according to S. Wiese et al. [13], material behaviors are different between solder joints and bulk materials. Another argument is about the ATC test. The type of the tested packages is usually fixed and the condition with a more severe constraint at the solder joint can only be applied by changing the ATC temperature profile. In this case, another factor, namely, temperature, will also be induced.

In this study, a special dummy package is designed and fabricated. With different solder joint spacings, two stress levels in the solder joints are created. The ATC test is

978-1-4244-4658-2/09 $25.00 © 2009 IEEE

conducted to compare the thermal fatigue reliability of SnPb and SAC solder joints. The finite element analysis is also performed and will be discussed later in this paper.

2. Procedures of Experiment

2.1. Sample Preparation

The design concept of the present package configuration is based on the DNP effect. The definition of DNP is the distance of a solder joint to the center point of a package. The loading applied on the solder joints in the ATC test is due to the CTE mismatch between the package and PCB. The induced stress/strain is in general proportional to the DNP of each solder joint. By adjusting the DNP, the stress level in the solder joints may be tuned to various ranges.

In the present study, simple dummy samples of BGA-PCB assembly were introduced as shown in Figure 2. Although this configuration is one-dimensional in nature, four solder joints are required in order to maintain the balance of the samples and to complete the daisy chains for real-time monitoring. By definition, the spacing between solder joints in the longitudinal direction represent two times of the DNP. For the purpose of enhancing the CTE mismatch effect, the substrate of a commercial daisy-chained CBGA package was cut into strips to represent the package part (see Figure 3) of the dummy sample. In this study, two solder materials with two different solder joint spacings were planned as given in Table 1. Note that the solder pad pitch of the original CBGA substrate is 1.0 mm. Therefore, the solder joint spacing can be easily defined by counting the number of pads between solder joints. 0.635 mm diameter solder balls are used to make the solder joints.

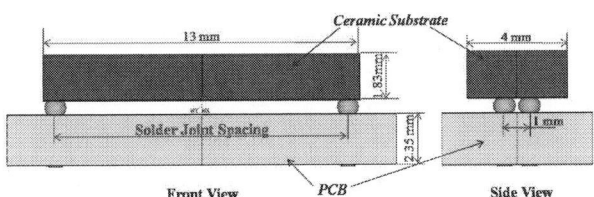

Figure 2 Schematic Diagram of Dummy Sample of BGA-PCB Assembly

Table 1 Sample Preparation Matrix

Solder Alloy	Solder Joint Spacing (mm)	Number of Dummy Samples
Sn37wt.%Pb	12	16
	4	16
Sn4.0wt.%Ag0.5wt.%Cu	12	16
	4	16

The PCBs used in the present study were regular FR4 boards with organic solderability preservative (OSP) surface finish. The solder pad opening was non solder mask defined (NSMD). The assembly of dummy samples followed the conventional SMT process. Each PCB carried four ceramic strips as shown in Figure 4. Note that the current PCB design

served as a common test vehicle for several different kinds of reliability tests. Therefore, in Figure 5, redundant pads exist and jump wires are required in order to complete the necessary daisy chains.

Figure 3 Strips of CBGA Ceramic Substrate as the Package Part of the Dummy Sample (Top and Bottom Views)

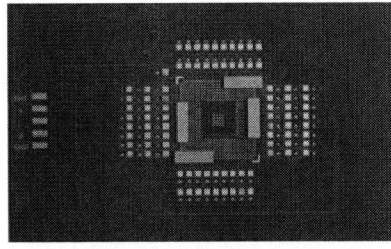

Figure 4 Four Ceramic Substrate Strips Assembled on one PCB by the SMT Process

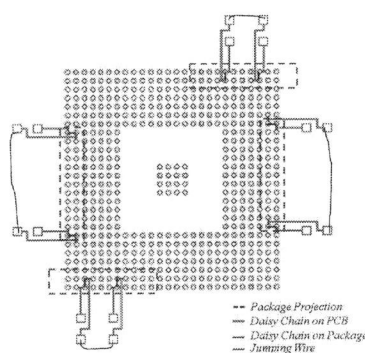

Figure 5 Daisy Chains with Jump Wires

After the preparation of samples, initial inspections were carried out. Figures 6 and 7 show typical X-ray and cross-section images, respectively. All samples looked normal and acceptable for the subsequent ATC test.

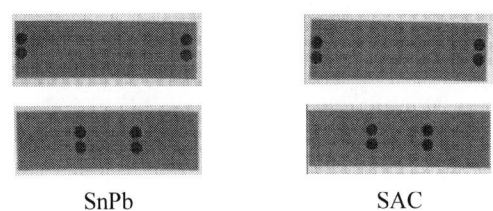

SnPb   SAC

Figure 6 X-ray Inspection of Solder Joints

978-1-4244-4658-2/09 $25.00 © 2009 IEEE   1039

Figure 7 Cross-section of a Typical Solder Joint

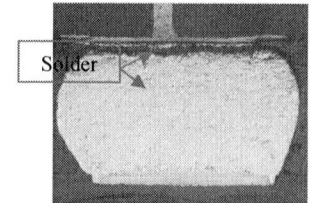

Figure 10 Cross-section of a Typical Failed Solder Joint

2.2. ATC Test

The current ATC test followed the Condition G of JESD22-A104-C (-40~125°C, one hour cycles). The temperature cycling profile is given in Figure 8. The daisy chains were monitored by data loggers. The failure criterion was set at 100 ohm. The collected data were analyzed by the standard Weibull analysis.

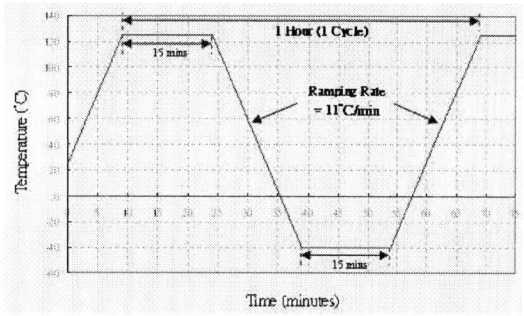

Figure 8 Temperature Profile of the ATC Test

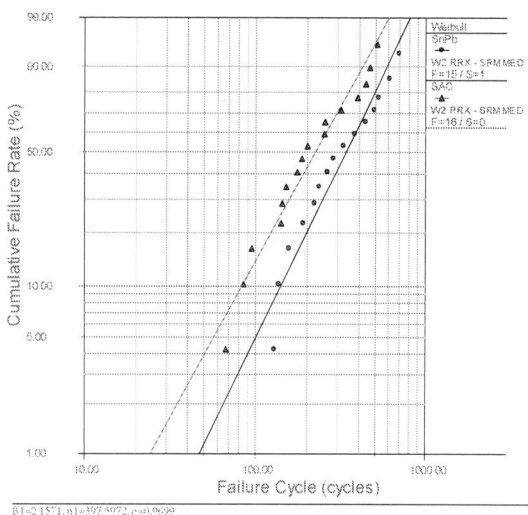

Figure 9 Weibull Analysis of ATC Test Results of 12mm Solder Joint Spacing Samples

3. Results and Discussion

The present study is an ongoing research. By the time of writing this paper, 1000 cycles have been finished. Almost all samples with 12mm solder joint spacing have failed, including both SnPb and SAC samples. However, for samples with 4 mm solder joint spacing configuration, less than 50% of samples failed so far. Therefore, Figure 9 only reports the Weibull results of 12mm solder joint spacing samples. It can be seen that for this configuration, the SnPb solder has a better performance than the SAC solder. The cross-section of a typical failed solder joint is shown in Figure 10. It can be seen that the crack is at the ceramic substrate side and the propagation path is within the bulk solder near the interface.

For 4mm solder joint spacing samples, the most updated test results are listed in Table 2. Based on the current data, it is reasonable to project that, in this lower stress level configuration, the SAC solder should have a longer thermal fatigue life than the SnPb solder.

Table 2 ATC Test Data of 4 mm Solder Joint Spacing Samples

Solder Alloy	Cycle Number of the First Failure	Number of Failed Samples	Number of Total Samples
SnPb	479	6	16
SAC	726	3	16

4. Finite Element Analysis

4.1. Model Description

Using this package structure, the finite element analysis is performed using ANSYS®. A quarter model was built by taking advantage of the symmetry as illustrated in Figure 11. Mesh details of the solder joint is presented in Figure 12. The 3-D 8-node structural element SOLID185, which contains three degrees of freedom at each node, is adopted for the model development. Totally 5 models are developed with solder joint spacing of 1mm, 4mm, 8mm, 12mm, and 16mm, respectively. Each model has two versions: with SnPb solder joint and with SAC solder. The temperature cycling profile is the same as that used in the ATC test. All models ran for three cycles to get a stabilized creep response. Volume-averaging method was used for the post-processing and to determine the thermal fatigue life in the life prediction equations.

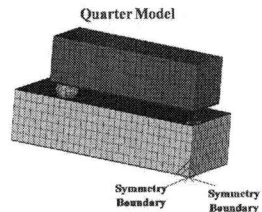

Figure 11 Finite Element Model

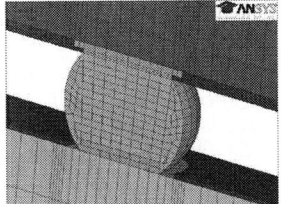

Figure 12 Mesh Details of Solder Joint

For the creep model, the constitutive relations of Wong *et al.* [11] and Wiese *et al.* [12] were used for SnPb and SAC solders, respectively. Both models take the form of the double power law, which can be implemented using the creep model 11 in ANSYS®. Table 3 summarizes the parameters required for the finite element analysis input. These parameters defined the equivalent stress σ in [MPa], temperature T in [K], and time t in [s]. E is the elastic modulus with unit in [MPa] and ε_{cr} is the equivalent creep strain. Linear elastic material properties of the other materials are summarized in Table 4.

Table 3 Creep Models for Simulation

Creep Model 11:		
$$\varepsilon_{cr} = \dfrac{C_1 \sigma^{C_2} t^{C_3+1} \exp\left(\dfrac{-C_4}{T}\right)}{C_3+1} + C_5 \sigma^{C_6} t \exp\left(\dfrac{-C_7}{T}\right)$$		
	Wong *et al* [11] (SnPb)	Wiese *et al* [12] (SAC)
E	56024-88T	59533-66.7T
C_1	$1.7 \times 10^{12}/E^3$	4×10^{-7}
C_2	3	3
C_3	0	0
C_4	5433.5	3223
C_5	$8.9 \times 10^{24}/E^7$	1×10^{-12}
C_6	7	12
C_7	5433.5	7348

Table 4 Other Material Properties Used for Simulation

Material	E (MPa)	v	α (ppm/K)
SnPb	-88T+56024	0.35	24.5
SAC	-59T+61251	0.36	20
Cu	117000	0.35	17.7
Ceramic	380000	0.22	7.4
FR4	21600	0.39	16
Solder Mask	5000	0.35	95

4.2. Simulation Results and Discussion

The accumulated equivalent creep strain range (ε_{cr}) after three cycles is plotted in Figure 13. It can be seen that, with the increase of the solder joint spacing, the ε_{cr} increases for both SnPb and SAC solders, but with different rates. For the samples with less than 10mm solder joint spacing, the SAC

solder joint has smaller creep strain. For the samples having 12mm and 16mm solder joint spacing, SnPb has smaller creep strain. This trend coincides with the results observed in the ATC test.

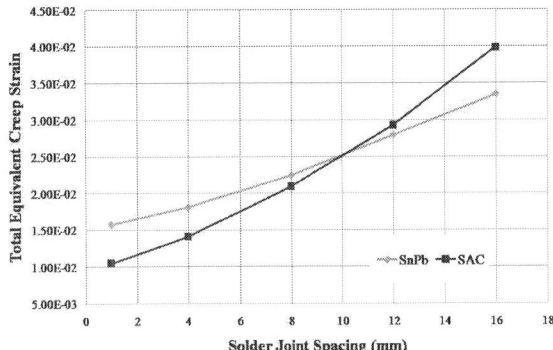

Figure 13 Accumulated Equivalent Creep Strain Range (ε_{cr}) vs. Solder Joint Spacing

After the creep strain calculation, Syed's approach [14] is adopted for the life prediction. The equations for SnPb and SAC solders are listed as follows:

$$N_{SnPb} = \left(0.061\Delta\varepsilon_{cr1} + 0.193\Delta\varepsilon_{cr2}\right)^{-1} \quad (1)$$

$$N_{SAC} = \left(0.1968\Delta\varepsilon_{cr}\right)^{-1} \quad (2)$$

where N_{SnPb} and N_{SAC} are the predicted thermal fatigue life of the SnPb and the SAC solder joints, respectively. For the SnPb solder, $\Delta\varepsilon_{cr1}$ is the cyclic accumulated equivalent creep strain due to grain boundary sliding and $\Delta\varepsilon_{cr2}$ is the cyclic accumulated equivalent creep strain due to dislocation. For the SAC solder, the total cyclic accumulated equivalent creep strain $\Delta\varepsilon_{cr}$ is used.

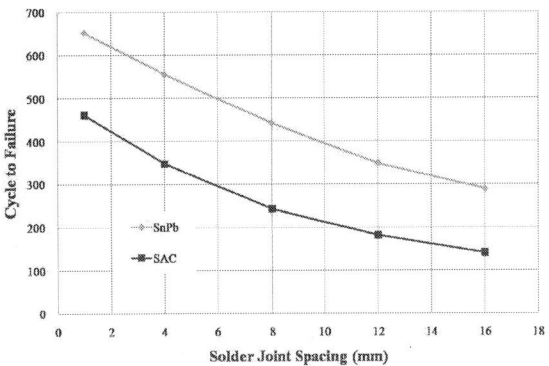

Figure 14 Predicted Thermal Fatigue Life vs. Solder Joint Spacing

After substituting the creep strains in the SnPb and SAC solder joints into their respective life prediction equations, the predicted thermal fatigue life can be obtained, as shown in Figure 14. It is found that, even with the 1mm solder joint spacing configuration, the SnPb samples still have a longer failure life than the SAC package. The discrepancy between the comparison of creep strain and the comparison of thermal fatigue life should be due to the parameters in the life

978-1-4244-4658-2/09 $25.00 © 2009 IEEE

prediction equations. For SnPb and SAC solders, the equations have different parameters based on the curve fitting of different experiments, which have uncertainties and inconsistencies in test configurations. Therefore, the universal applicability of those life prediction equations may still require further investigations and validations.

5. Conclusions

This paper presented a study to compare the thermal fatigue performance of SnPb and SAC solders under different stress levels. A special dummy sample configuration was proposed. The feature of the proposed sample configuration is to use different solder joint spacings to create various stress levels in the solder joints during the temperature cycling.

Both the ATC test and the finite element analysis were conducted. For the ATC test, 1000 cycles have been finished. The 12mm solder joint spacing samples have almost all failed while the test of the 4mm solder joint spacing samples is still in process. At the higher stress level, it is observed that SnPb scored better than the SAC solder. For the 4 mm solder joint spacing samples, based on the current result, it is reasonable to project that the SAC solder should perform better. Therefore, a general statement saying that the SAC solder is better than the SnPb solder under temperature cycling should not be valid. The comparison of solder joint thermal fatigue performance should also depend on the stress level during the ATC test.

For the simulation study, the trend of cyclic creep strains of the SnPb solder and the SAC solder coincides with the ATC test results. However, after converting the cyclic creep strain into thermal fatigue life using empirical life prediction equations, the SnPb solder always outperforms the SAC solder, even at the lowest stress level. This prediction is different from our experimental observation. This discrepancy should be due to the uncertainties in the curve fitting parameters used in the life prediction equations. The universal applicability of those life prediction equations may still require further investigations and validations.

Acknowledgments

The authors would like to thank Mr. Tong Jiang and Mr. Minshu Zhang of EPACK Lab at HKUST for their effective assistance in sample preparation and inspection. Special thanks are due to SUN Microsystems Inc. for their providing of CBGA ceramic substrates.

References

1. E. Bradley, *et al.*, Lead-free Electronics: iNEMI Projects Lead to Successful Manufacturing, John Wiley & Sons, Hoboken, New Jersey, 2007.
2. J. H. Lau, C. P. Wong, N. C. Lee and S. W. R. Lee, Electronics Manufacturing with Lead-free, Halogen-free and Conductive-adhesive Materials, McGraw-Hill, New York, NY, 2003.
3. S. Ganesan, M. Pecht, Lead-free Electronics, John Wiley & Sons, Hoboken, New Jersey, 2006.
4. Syed, "Reliability and Au Embrittlement of Lead Free Solders for BGA Applications," *Proc. International Symposium on Advanced Packaging Materials*, 2001, pp. 143-147.

5. W. Tan, I. Ume, "Effects of Warpage on Fatigue Relibility of Solder Bumps: Experimental and Analytical Studies," *Proc. 58th ECTC*, Orlando, FL, May 2008, pp. 131-138.
6. M. Farooq *et al.*, "Thermo-Mechanical Fatigue Reliability of Pb-Free Ceramic Ball Grid Arrays: Experimental Data and Life Prediction Modeling," *Proc. 53rd ECTC*, New Orleans, LA, May 2003, pp. 827-833.
7. Schubert *et al.*, "Fatigue Life Models for SnAgCu and SnPb Solder Joints Evaluated by Experiments and Simulations," *Proc. 53rd ECTC*, New Orleans, LA, May 2003, pp. 603-610.
8. T. Woodrow, "Reliability and Leachate Testing of Lead-Free Solder Joints," *IPC-JEDEC 2002*, USA, pp. 116-125.
9. M. Osterman, A. Dasgupta, "Life Expectancies of Pb-Free SAC Solder Interconnects in Electronic Hardware," *Journal of Material Science: Materials in Electronics*, 2007, Vol. 18, pp. 229-236.
10. B. Vandevelde *et al.*, "Thermal Cycling Reliability of SnAgCu and SnPb Solder Joints: A Comparison for Several IC-Packages," *Microelectronics Reliability*, 2007, Vol. 47, pp. 259-265.
11. B. Wong, D. E. Helling and R. W. Clark, "A Creep-Rupture Model for Two-Phase Eutectic Solders," *IEEE Trans. on CHMT*, Vol. 11, No. 3, 1998, pp. 284-290.
12. S. Wiese, E. Meusel and K-J. Wolter, "Microstructural Dependence of Constitutive Properties of Eutectic SnAg and SnAgCu Solders," *Proc. 53rd ECTC*, New Orleans, LA, May 2003, pp. 197-206.
13. S. Wiese et al., "Constitutive Behavior of Lead-free Solders vs. Lead-containing Solders-Experiments on Bulk Specimens and Flip-Chip Joints," *Proc. 51st ECTC*, Orlando, FL, May 2001, pp.197-206.
14. Syed, "Accumulated Creep Strain and Energy Density Based Thermal Fatigue Life Prediction Models for SnAgCu Solder Joints," *Proc. 54th ECTC*, Las Vegas, NV, June 2004, pp. 737-746.

Board Level Reliability Assessments of Thru-Mold Via Package on Package (TMV™ PoP)

Tae-Kyung Hwang*, Dong-Joo Park, Jin-Seong Kim, Jin-Young Kim, Jae-Dong Kim, Choon-Heung Lee
Amkor Technology Korea, Inc.
280-8, 2-ga, Seongsu-dong, Seongdong-gu, Seoul, 133-706, Korea
*Phone: 82-2-460-5191, FAX: 82-505-460-5581, E-mail: tkhwang@amkor.co.kr

Abstract

In recent years, Package-on-package (PoP) has been adopted as major application package platform in 3D integration of logic and memory devices. However, as electronic technology developed, higher technology requirements are requested in packaging. Amkor Technology, Inc introduced the next generation PoP solution to meet the next generation technology requirements in 2008 by the using of TMV™ technology which incorporates a laser ablation process that is conducive to current matrix-molded semiconductor assembly techniques. The next generation PoP platform named as TMV™ PoP has been qualified in all package level qualification tests. Also in board level reliability tests, BLR TC & drop performances are similar or better than those of the conventional PoP.

Introduction

3D stacked packages (Package on package, PoP) have been developed for products requiring efficient memory architectures including multiple buses, high density memory and improved performance, while reducing mounted area. After the introduction of PoP platform to electronic markets, it has been rapidly and widely adopted for 3D integration of logic and memory module for many kinds of portable devices and then PoP becomes the one of the main stream package platforms. However, existing methods of making the PoP base package may not be applicable for next generation applications that will require increased-integration, miniaturization and performance without requiring development of a new SMT stacking infrastructure or adding cost. These are challenging requirements to meet given the increased interconnect densities associated with new memory and signal processing architectures; which include: [1]

Signal processing

- Multi-core SOC or dual chip designs that integrated a baseband modem and applications processor in a single chip or stacked die configurations. Either configuration can greatly increases first level interconnect density challenges.
- Processor core clock speeds have tended to increase significantly with each new CMOS node (currently at 1GHz for the 45nm node).
- The above and additional performance factors have driven a strong transition from wire bond to flip chip designs, which is tending to accelerate below the 90nm node.
- These integration and signal speed trends are driving increased BGA densities (higher ball counts at finer pitches to 0.4mm).
- With higher speed, density and higher I/O processors, we see requirements for integrating decoupling capacitors within the package for signal integrity and form factor requirements.

Memory Architecture and Interfaces

- Higher speed memory interfaces trending from SDRAM –> DDR –> LP DDR2
- Wider DDR memory bus architectures 16 to 32 bit paired with non-volatile memory devices
- From a shared data bus to split bus to new 2 channel (wide bus) architectures
- The above and additional performance factors driving an increase in memory interface densities to 0.5 and 0.4mm pitch.

PoP Form Factor

- Smaller package body sizes which increased die to package ratios, interconnect and substrate wiring densities.
- Thinner PoP stacks which coupled with BGA densities and fine pitch requirements demand tight warpage control for bottom and top package technologies.

Amkor Technology, Inc introduced the next generation PoP solution with new technologies to create interconnect vias through the mold cap, naming this technology as thru-mold mold via (TMV™) in 2008 to meet the requirements of next generation PoP solution. [2-3] Figure 1 illustrate the key elements of the bottom TMV™ PoP that was developed by Amkor Technology, Inc.

(a) Cross section view

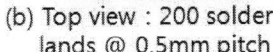

(b) Top view : 200 solder lands @ 0.5mm pitch

(c) Bottom view : 620 BGAs @ 0.4mm pitch

Figure 1 Cross section, top and bottom views of new bottom PoP package technology using TMV™ technology

The bottom TMV™ PoP is based on a standard matrix molded FBGA style package with wire bond, flip-chip or stacked devices. After molding, a blind via through the mold compound is created to expose the solder on pad stacking interface of the substrate. The resulting structure is a fully

978-1-4244-4658-2/09 $25.00 © 2009 IEEE

molded FBGA package with a conductive blind via to form the PoP stacking interface.

Packages incorporating TMV technology experience many of the benefits required for next generation PoP applications, which include:

- TMV technology removes the pitch vs. package clearance bottlenecks to support future memory interface density requirements. Figure 2 shows the PoP size reduction benefits, as TMV enables the memory interface to scale with CSP pitch reduction trends.
- TMV improves warpage control and bottom package thickness reduction requirements, by utilizing a balanced fully-molded structure.
- TMV provides an increased die to package size ratio (also represented in Figure 5), studies have shown TMV can provide over a 30% increase in maximum die size. For example TMV rules indicate up to a $97mm^2$ die should mechanically fit in the $12\times12mm$ 216 pad interface PoP required for the new JEDEC 4 proposal.
- TMV supports wire bond, FC, stacked die and passive integration requirements.
- The TMV structure leverages strong technology roadmaps and high volume scale, from FBGA, stacked die, flip chip CSP, and system in a package (SiP) platforms. Integrates proven laser ablation technology available from a host of laser process equipment suppliers.
- The TMV structure can improve board level reliability of the stacked memory interface through design rules that Amkor has developed.

• Size reduction through memory interface pitch reduction
• Baseline design: 7x7mm die, 200 I/O top package IF, 2 row perimeter

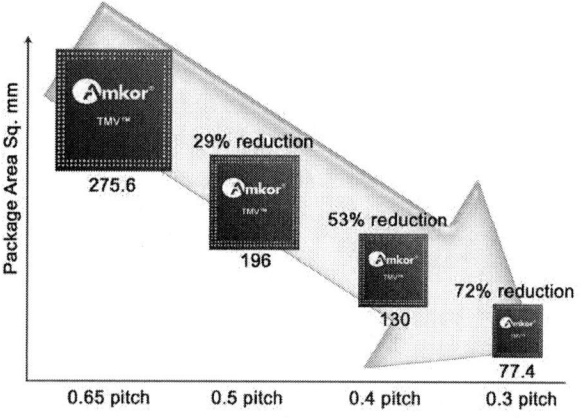

Figure 2 Package scaling benefits of TMV™ technology

Package Description

Test vehicles which were used for internal qualification are shown in figure 1.14mm×14mm matrix molded 2 sided PoP base package has the bottom 33×33 BGA array which contains 620 balls at 0.4mm pitch. The top 27×27 PoP BGA array contains 200 PoP bond pads in a JEDEC compliant 2 row, 0.5mm pitch perimeter array. The base package contains 1 flip-chip daisy chain die. The die size is 7.10mm×6.97mm and die thickness is 0.127mm. Minimum flip-chip pad pitch is

0.225mm and the die is bumped with plated Pb-free bumps. Total substrate thickness is 0.21mm and the mold cap thickness is 0.40mm measured from the top of the substrate to the top of the package. This mold cap thickness was chosen based on availability but can be thinner for future TMV enabled packages. The top package is a 14mm FBGA that is representative of a standard PoP memory package with a 200 BGA balls arranged in a 2 row, 0.5mm pitch perimeter array to match the PoP bond pads of the base package. BGA solder alloy is Sn/1.0Ag/0.5Cu. The top package has a 0.13mm thick 2 layer substrate and the mold cap thickness is 0.40mm measured from the top of the substrate to the top of the package. Total PoP stacked structure (top+bottom) thickness is 1.32mm.

Assembly Process

TMV™ PoPs follow a conventional flip-chip or FBGA assembly process flow. But there are two different process steps which did not apply in the conventional process flows-pre solder ball attach and via generation.

Before the EMC mold process, solder balls are pre attached on top PoP interface lands. And then the flip-chip die is underfilled during the transfer mold process. Void free underfill was achieved using a proven molded underfill transfer mold process. Amkor uses in production for other FC based package technologies. After molding, solder balls on top PoP interface lands were exposed by generating vias through the mold compound. Pre solder ball attach method enable to reduce the laser ablation time by reducing the EMC volume to be ablated.

The TMV™ structure provides design and process flexibility for engineer to optimize stack height and top solder joint mechanical reliability.

Pre solder ball attach and via generation

To generate via, 0.3mm diameter solder ball was attached on PoP interface land before molding then laser drilling was performed to expose the solder ball which is interconnected to top package solder ball. Figure 3 shows the process flow for via generation.

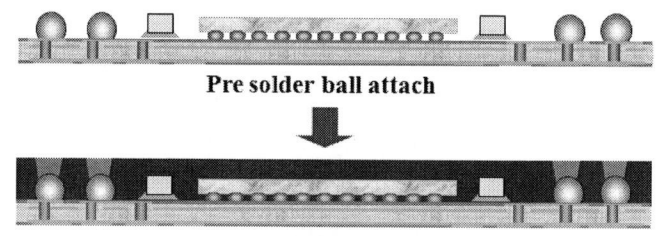

Pre solder ball attach

Mold and laser drill

Figure 3 Process flow for via generation

Figure 4 shows the via cone shape base on 0.45mm top diameter and 0.3mm bottom diameter and no visible mold compound residue or contamination on the solder ball surface. This shape helps to achieve stable top package loading during SMT stacking. A residue free solder ball surface of the PoP interface is critical for solder wetting and PoP stacking electrical continuity.

978-1-4244-4658-2/09 $25.00 © 2009 IEEE

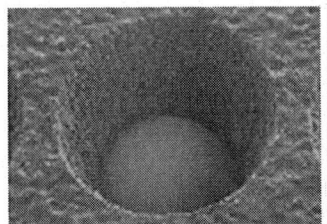

Figure 4 Through Mold Via shape

After laser drilling, we measured via diameter to check the laser drilling accuracy and measurement data are shown in Table 1.

Table 1 Through Mold Via diameter data

Position	AVE (µm)	min (µm)	Max (µm)	STD
Top	449	440	462	5.909
Bottom	313	309	321	3.465

Pre solder attach also enable to establish the uniform bump height and stable process flow. Solder bump height in the vias was measured using Hisomet and Table 2 shows measurement data.

Table 2 Solder bump height after reflow

Average (µm)	min (µm)	Max (µm)	STD
249	245	258	4.690

Package flatness

Package flatness TMV™ PoP packages were measured over a lead free reflow temperature profile using shadow moiré. Figure 5 shows the warpage profile for the molded TMV™ PoP as well as the bare die base package along with the top daisy chain (memory like) package. At room temperature, both packages show (+) "crying" face shaped convex warpage. At the elevated reflow profile temperature measurement points the TMV™ PoP shows maximum (-) 55µm "smiling" face concave warpage while the bare die base package had higher peak temperature warpage showing a maximum warpage of (-)137µm.

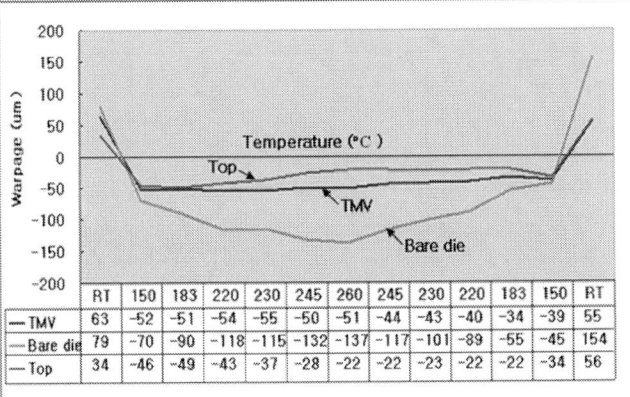

Figure 5 Warpage of TMV™ PoP, bare die PoP package and top package over simulated reflow profile based on shadow moiré measurements.

In PoP stacking, the top and bottom packages should have stable warpage profiles over the reflow profile to ensure good stacking SMT yield both for the package to package solder joints as well as the bottom package to PWB solder joints. The warpage profile of the TMV™ PoP and the bare die package can be compared to top memory package. Table 3 shows the difference in package flatness between the top memory package and the respective base packages. During heating condition (RT~260°C) the TMV™ PoP shows excellent compatibility with the top package with 28.8µm maximum warpage gap. The bare die flip-chip base package had a maximum 179.8µm gap. After cooling to room temperature (260°C~RT), the TMV™ PoP shows 11µm warpage gap compared to 170.4µm for the bare die package.

Table 3 Warpage gap between top and bottom packages

Temperature (°C)	RT	150	220	260	220	150	RT
Top & TMV™ PoP (µm)	3.6	0.4	9.8	28.8	11	8.8	9.4
Top & bare die (µm)	80.4	55.4	132	180	115	13.2	170

SMT Process

Package stacking was performed to evaluate stacking yield, solder joint geometry and gap height between the top and bottom packages. An automated IC placement machine equipped with a flux or solder paste dipping module was used to place the top package. The top packages were dipped in either flux or solder paste to coat the BGA balls of top package before being placed on the top surface of bottom package. After placement, the BGA balls of the top package rest inside the vias of the bottom package and make contact with the solder bumps on the bottom package. Figure 6 shows the stacking process flow.

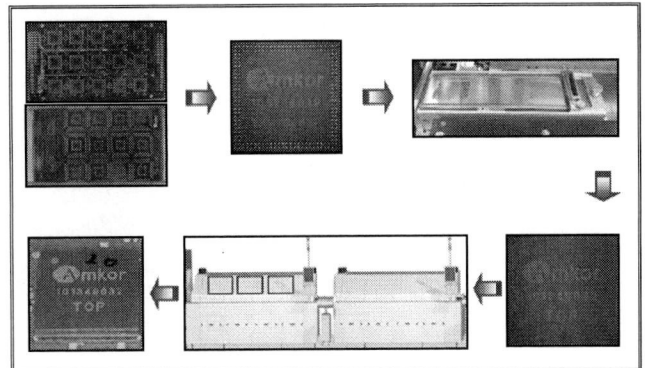

Figure 6 Stacking process flow

After solder reflow, x-ray and cross section analysis was performed. 100% stacking yield was observed for both flux dip and paste dip stacking processes.

Figure 7 Gap between top & bottom package after stacking process

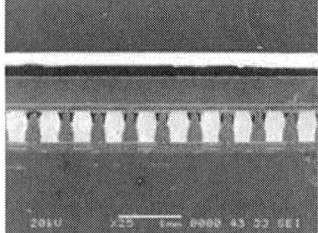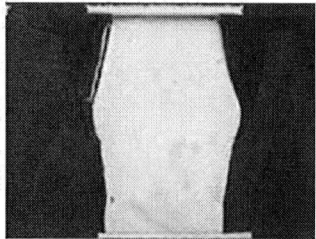

Figure 8 Cross section of TMV™ solder joints after package stacking

Gap height between the top package and bottom package was measured. In the case of the flux dip stacking method, the gap height between the packages was 10~15μm. In the case of solder paste dip stacking method, the gap was 20~25μm. These values are in line with typical gap height observed in standard straddle mount PoP stacks. Figure 7 shows the gap between top package and bottom package after pre stacking. A cross section of the resulting solder joint is shown in Figure 8. The solder joint conforms to, but does not wet, the via walls up to the approximate midpoint of the via. This solder joint geometry allows for a tall solder joint at fine pitch with no risk of solder shorting between PoP solder joints.

Reliability Tests

To apply the next generation 3D package platform, TMV™ PoP has been qualified per JEDEC MSL L3 260°C requirements including 4 reflow cycles indicated to support PoP pre-stacking flows. The results for other package level qualification requirements are shown as below.

Internal qualification results

- MRT (L3 260°C, 4× reflow)   : Passed
- Temp cycle (-55/125°C, 1000x)   : Passed
- HAST (130°C 85%RH / 96hrs)   : Passed
- HTS (150°C / 1000hrs)   : Passed

Board level reliability assessment is one of the most important factors in the applications of 3D package platform. Because major fields of 3D package applications are mobile and handheld products, harsh environments tests such as mechanical drop and accelerated temperature cycle will be evaluated.

Board level temperature cycling test [4]

In board level temperature cycling test (BLR TC test), JEDEC JESD22-A104C, condition G (-40~125°C) was applied. In BLR TC test, 14×14mm TMV™ PoP DC samples which were shown in figure 1 were used. In DC samples, there are 5 separated monitoring nets which consisted of top package corner, top package inner, bottom package, flipchip bump and passive components nets. Those 5 separated nets except passive components net were in situ monitored during BLR TC test to detect the exact failure cycles for each nets. Schematic diagrams for each daisy chain nets are shown in Figure 9.

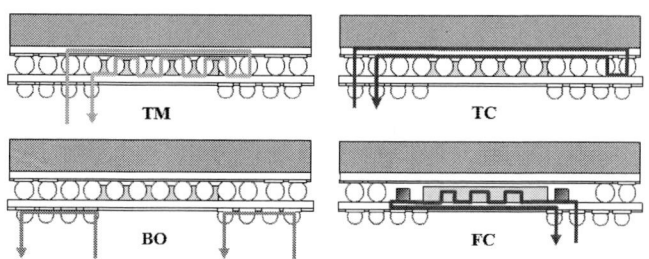

> Orange line : TM net (Input (V8&W8), Output (T8&U8))
> Blue line : TC net (Input (H19&H20), Output (H17&H18))
> Brown line : FC net (Input (R26&T26), Output (U26&V26))
> Pink line : BO net (Input (AN6), Output (AF18))

Figure 9 Schematic diagrams for daisy chain nets

Table 4 Weibull analysis results for BLR TC test

DOE	Net Name	min	TM	TC	BO	FC
NiAu with SAC105	NBR of failure	36 / 36	33 / 33	35 / 35	27 / 34	9 / 36
	1st failure	1082	1082	1306	2571	2387
	Mean life	1702	1708	2014	3518	4676
	Char. Life	1827	1835	2140	3728	5098
	Slope	6.46	6.38	7.86	8.33	4.92
OSP with SAC125Ni	NBR of failure	35 / 35	32 / 32	35 / 35	33 / 33	30 / 34
	1st failure	384	803	384	584	1183
	Mean life	937	1785	933	2204	2448
	Char. Life	1035	1939	1035	2485	2698
	Slope	3.96	5.27	3.64	2.46	4.07

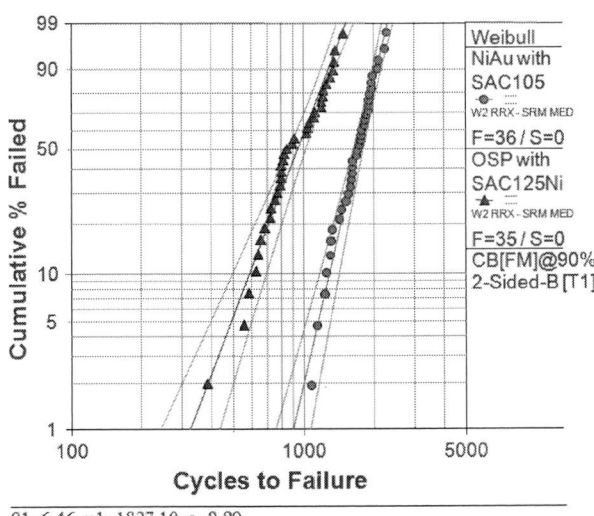

β1=6.46, η1=1827.10, ρ=0.99
β2=3.96, η2=1034.74, ρ=0.98

Figure 10 Weibull distributions of BLR TC test

Standard TMV™ PoP material sets were applied except bottom ball pad finish type & solder alloy. In bottom ball pad finish & solder alloy of TMV™ PoP, NiAu with SAC105 (Sn/1.0Ag/0.5Cu) and OSP with SAC125Ni (Sn/1.2Ag/0.5Cu/0.05Ni) were applied. In via pad, NiAu with SAC105 pre attached solder ball was applied.

As of 4000 cycles completed, Weibull analysis results for BLR TC test were shown in Table 4 & Figure 10. In minimum failure data set, NiAu with SAC105 has better performance than OSP with SAC125Ni. Mean life of NiAu with SAC105 is longer than that of OSP with SAC125Ni by about 2 times. Among the daisy chain nets, bottom package has better performance than top package nets (top corner & top inner nets). Mean life of bottom package is longer than those of top package at least by 2 times and these differences are larger than the cases of conventional PoP. Normally, top package net fail faster than bottom package but the difference is not so much large in the conventional PoP. The main reason can be guessed from the failure analysis results.

Figure 11 shows the solder joint crack patterns which were induced from BLR TC test in TMV™ PoP. In top solder joint failures, crack location is different with the conventional PoP case. Because solder joints are surrounded by mold cap in bottom region, mold cap can be affected such as 2nd level

underfill. Due to those effects, solder joint cracks were initiated from the similar point which was defined by solder joint and mold cap top surface of TMV.

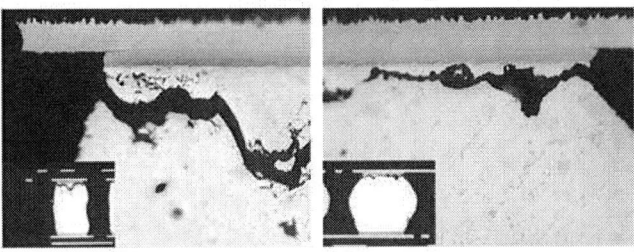

(a) Solder joints between top SCSP & TMV™ PoP

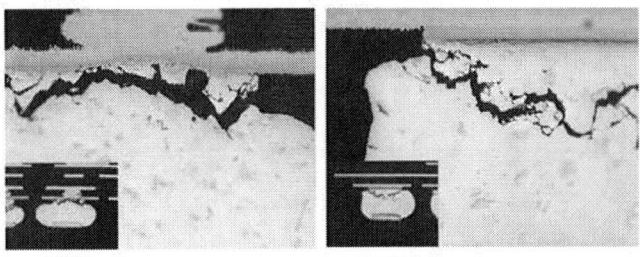

(b) Solder joints between TMV™ PoP & test board
Figure 11 Cross section views of solder joint cracks in TMV™ PoP under BLR TC test

Board level drop test [5]

In board level drop test (BLR drop test), JEDEC JESD22-B111 was applied. In BLR drop test, 1500G peak acceleration during 0.5msec half sine pulse was applied and same DC samples with BLR TC test were used. As of 1200 drops completed, Weibull analysis results for BLR drop test were shown in Table 5 & Figure 12. In minimum failure data set, NiAu with SAC105 has better performance than OSP with SAC125Ni. During the monitoring nets, most of failures were occurred at bottom package net which was consisted of bottom solder joints of TMV™ PoP. Because top solder joints of TMV™ PoP were surrounded by mold cap, only small numbers of top solder joints were failed and less than the conventional PoP case. Therefore in BLR drop test, mold cap of TMV™ PoP affected BLR drop performances such as 2nd level underfill.

Table 5 Weibull analysis results for BLR drop test (critical analysis group only)

DOE	Net Name	min	TM	TC	BO	FC
NiAu with SAC105	NBR of failure	24 / 24	12 / 22	0 / 22	23 / 23	0 / 24
	1st failure	17	401	N/A	17	N/A
	Mean life	275	847	N/A	252	N/A
	Char. Life	285	944	N/A	264	N/A
	Slope	1.10	3.27	N/A	1.13	N/A
OSP with SAC125Ni	NBR of failure	23 / 24	13 / 15	8 / 17	23 / 23	8 / 22
	1st failure	14	251	14	33	693
	Mean life	168	593	1039	161	1065
	Char. Life	185	670	1131	181	1141
	Slope	1.43	2.33	5.06	1.82	6.76

Figure 13 shows the cross section views of failed solder joints. Solder joint crack patterns are normal and have similar trends with thick mold cap thickness FBGA.

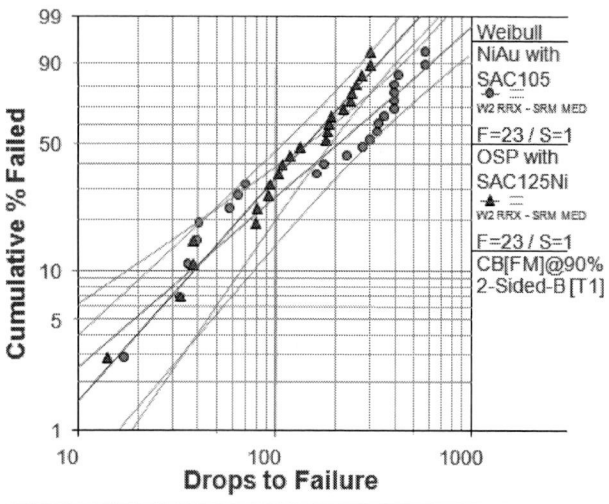

β1=1.09, η1=288.32, ρ=0.96
β2=1.43, η2=185.27, ρ=0.99

Figure 12 Weibull distributions of BLR drop test (critical analysis group only)

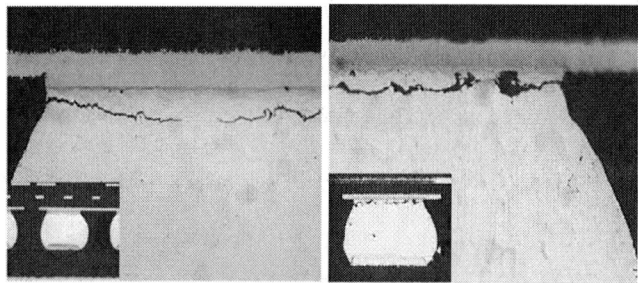

Figure 13 Cross section views of solder joint cracks between TMV[TM] PoP & test board under BLR drop test

Conclusions

A unique interconnection technology has been developed to address the requirements for next-generation PoP applications. This innovative technology, known as Amkor's TMV™, incorporates a laser ablation process that is conducive to current matrix-molded semiconductor assembly techniques. Dramatic improvements in interconnect density; package warpage and package size reduction are exhibited by this technology.

To meet the requirements for next-generation PoP packages, Amkor has completed to qualify the TMV[TM] PoP in package level qualification and evaluated board level reliability performances in TC & drop tests.

Future work will focus on more detailed DOE evaluations (material and process effects) in board level reliability to improve performances. Also, we anticipate that finer pitch PoP interfaces including 0.4 and even 0.3mm will be enabled with TMV technology, which will extend the application space for PoP device combinations.

References

1. Zwenger, C. *et al.*, "Next Generation Package-on-Package (PoP) Platform with Through Mold Via (TMV[TM]) Interconnection Technology," *Proc IMAPS Device Packaging Conference*, Scottsdale, AZ, March 2009.
2. Kim, J. S. *et al.*, "Application of Through Mold Via (TMV) as PoP base package," *Proc 58th Electroni Components & Technology Conference*, Orlando, FL, May 2008.
3. Zwenger, C. *et al.*, "Surface mount assembly and board level reliability for high density PoP utilizing through mold via interconnection technology," *Proc SMTA International Conference,* Orlando, FL, August 2008.
4. JEDEC JESD22-A104C, "Temperature Cycling".
5. JEDEC JESD22-B111, "Board Level Drop Test Method of Components for Handheld Electronic Products".

Prediction of IMC Formation during Interfacial Reactions: Application of CALPHAD Approach to Electronic Package

Huashan Liu*, WenJun Zhu, ZhanPeng Jin

School of Materials Science and Engineering, Central South University

ChangSha, HuNan 410083, People's Republic of China

* Corresponding author: hsliu@mail.csu.edu.cn

Abstract

Intermetallic compound (IMC) formed in the interfacial reaction may exert much influence on the reliability of soldering joint. So, effective prediction of IMC formation at soldering joint may do help to optimize the packaging process. Recently, a semi-empirical criterion has been proposed in our group to predict the formation of intermetallic compounds (IMC) at the interface between different pure elements (J. Mater. Res., 2007, pp. 1502). The advantage of this criterion over other models is that it involves both thermodynamic and kinetic factors thus give more reliable prediction. Even so, only thermodynamic calculation, i.e. CALPHAD method, is involved. Applications of the present criterion to the Cu/In, Cu/Sn, Ni/Sn and Co/Sn systems binary couples and Sn-Zn/Ni ternary couple confirm the reasonality of the new model. Application of this model to other ternary systems is in progress.

1 Introduction

Due to interdiffusion of the elemental atoms across the interface, interfacial reaction may happen between Under Bump Metallurgy (UBM) layers, or between the UBM and soldering alloy in electronic packaging. Such reaction may result in the formation of other intermetallic compound (IMC). If the interfacial reaction is predictable, one should be able to understand better the physical metallurgy of the joint and make reliable electrical connection either by controlling the microstructure of the joint or providing clues for novel solder material design.

It is generally accepted that the IMC forming firstly may not be the most stable one during interfacial reaction. So, how to theoretically predict the interfacial reactions has attracted much research interest. Until now, a few theoretical studies dealing with the prediction of the formation of IMC at the interface have ever been reported. These theories can be roughly classified into 2 catalogues. The one is based on kinetics (shortened as KT here) and the other mainly on thermodynamics (TT, ibid).

The TT only compares the possibility of phase formation. Several other methods based on thermodynamics have also been provided. Van Loo [1] proposed the chemical potential criterion, which has been successfully employed to explain the formation of compounds at the Ni/Sn-Cu [2] and Ni/Sn-Ag [3] interfaces. The method does not take into account of kinetic factors and therefore its reliability is not adequate, especially at low temperature where diffusion plays a key role in phase formation.

Differing from the TT model, the KT treats both nucleation and growth of the possible phases. There are 2 methods provided for KT: one is called competitive nucleation theory (CNT), the other is competitive growth theory (CGT).

CGT treats that the phase with higher growth rate forming and growing faster. Under such condition, Nishizawa and Chiba [4] and GÖsele and Tu [5] proposed different formula to analyze the phase formed at interface reaction. However, some important parameters such as the interfacial reaction constant etc. are unknown in CGT which makes the CGT theories unpractical for qualitative analysis.

For CNT, there are three methodologies proposed by Johnson [6], d'Heurle [7], Clevenger and Thompson [8] and Thompson [9] in order to qualitatively explain the formation of IMC at the interface between 2 metals. Later, in order to quantitatively analyze IMC formation at interface, a new method has been proposed by Lee *et al* [10]. According to [10], it is the phase with the largest driving force of nucleation appearing first at the interface. This is regarded as the largest driving force criterion. More recently, taking account of the effect of interface energy, Choi *et al.* [11] have modified the Lee's model [10] and proposed that it is the phase with the lowest nucleation activation energy forming first. This method is then called the lowest nucleation-work model.

Both CNT and CGT theories have their own limitations. The CNT cannot tell why the metastable phases appear prior to those stable ones [12-13], while the CGT cannot explain the formation of metastable phases [8].

As will be discussed later, the above theories or methods can only explain partial experimental phenomena, and are applicable only to special systems on special situation at a certain temperature range. It is easy to understand that phase formation at the interface reaction is the common result of thermodynamics and kinetics, which should involve atomic interdiffusion, nucleation and growth. Whenever a phase forms, it must satisfy both the thermodynamic and kinetic conditions at the same time. All the above-mentioned models pay attention to only part of these interactions. Considering the shortcomings of the formerly proposed models, a new model [14] was recently developed in our group to predict the interface reaction between 2 metals with different crystal structures as introduced in the next section.

2 The criterion to judge the phase formation

As shown in Fig. 1, a diffusion couple of A/B is welded and annealed at a chosen temperature to study the formation sequence of IMCs at the interface between 2 metals of different structures. Assuming there are 2 IMCs, γ and δ co-existing in this A-B system, as illustrated in Fig. 2. We would like to know which phase could form and appear first. In light of the well-known J-M equation (J-M equation) [15] and take into account of the deduction by Thompson [9] that the interface is the initial site for phase formation, and only after the local equilibrium is reached can the IMC form, a semi-empirical criterion [14] has been proposed in our group to predict the interface reaction as

$$y_1 = -\frac{b}{64m^2}\left[\ln\frac{x_B^\alpha}{x_B^X x_B^\beta}\right]^3 - 3|\Delta x| \qquad (1)$$

Here m is the dimensionless factor relating to driving force, equal to $\Delta G_d / RT$. According to the method of Lee *et al* [10] and Christian [16], the driving force of nucleation and composition of X can be calculated. As an example shown in Fig. 2, we can get the driving force of nucleation and composition of γ and δ phases at the interface between A and B. For the calculations, it is worth emphasizing here that the phases α and β are corresponding to the stable states of A and B respectively. Here the diffusivity of A in β is assumed to be far larger than that of B in α. If an IMC takes larger y_1, it will form faster. As seen, all parameters occurring in the model are available through CALPHAD approach. As long as the Gibbs energy of respective phase is available, the criterion for the corresponding phase can be obtained.

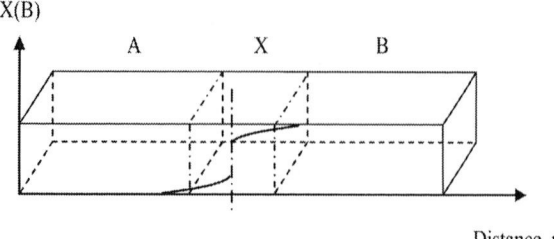

Fig. 1 Schematic diagram of a diffusion couple and the location of the initially formed phase

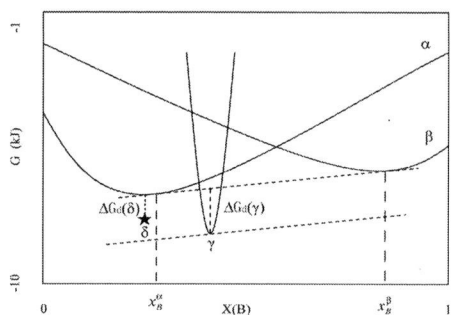

Fig. 2 Driving force calculation for the phases in the studied system

Table 1. Comparison of experiments and predicitions by different models on IMC formation sequence in the diffusion couples of Cu/Sn[a]. Thermodynamic data are referred to Ref. [22]

System	Temperature	Criterion	IMC forming first	
			Predicted	Observed
Fcc(Cu)/Bct(Sn)	50°C	$\Delta G_d / RT$ (Cu$_6$Sn$_5$)=1.1962	Cu$_6$Sn$_5$ [10] $\Delta G_d / RT$ max	Cu$_6$Sn$_5$[19]
		$\Delta G_d / RT$ (Cu$_3$Sn)=1.1357		
		y_1(Cu$_6$Sn$_5$)= -0.8035	Cu$_6$Sn$_5$ Present, y_1 max	
		y_1(Cu$_3$Sn)= -1.4403		
Fcc(Cu)/Liq(Sn)	400°C	$\Delta G_d / RT$ (Cu$_6$Sn$_5$)=0.035	Cu$_6$Sn$_5$ [10] $\Delta G_d / RT$ max	Cu$_6$Sn$_5$[21]
		$\Delta G_d / RT$ (Cu$_3$Sn)=0.028		
		$\Delta H_m^3 / \Delta G_d^2$ (Cu$_6$Sn$_5$)=14256	Cu$_6$Sn$_5$ [11] $\Delta H_m^3 / \Delta G_d^2$ min	
		$\Delta H_m^3 / \Delta G_d^2$ (Cu$_3$Sn)=26887		
		y_1 (Cu$_6$Sn$_5$)=0.04768	Cu$_6$Sn$_5$ Present, y_1 max	
		y_1 (Cu$_3$Sn)=2.438e-54		

[a]: Here only the phase with positive driving force is considered. Because the model proposed by Choi *et al* [11] is only applicable to the case with liquid reactant, we didn't calculate the nucleation work at low temperature, such as that at 50°C.

Table 2 First formed IMC in the diffusion couples of Ni/Sn[b]: predicitions and experiments.
Thermodynamic data are cited from Ref. [26]

System	Temperature	Criterion	IMC forming first	
			Predicted	Observed
Fcc(Ni)/Bct(Sn)	200°C	$\Delta G_d / RT$ (Ni$_3$Sn$_4$)=1.197	Ni$_3$Sn$_4$ [10] $\Delta G_d / RT$ max	Ni$_3$Sn$_4$[17, 20]
		$\Delta G_d / RT$ (Ni$_3$Sn$_2$)=0.384		
		y_1 (Ni$_3$Sn$_4$)=-1.051	Ni$_3$Sn$_4$ Present, y_1 max	
		y_1 (Ni$_3$Sn$_2$)=-1.9908		
Fcc(Ni)/Liq(Sn)	520°C	$\Delta G_d / RT$ (Ni$_3$Sn$_4$)=0.7729	Ni$_3$Sn$_4$ [10] $\Delta G_d / RT$ max	Ni$_3$Sn$_4$[24]
		$\Delta G_d / RT$ (Ni$_3$Sn$_2$)= 0.6842		
		$\Delta H_m^3 / \Delta G_d^2$ (Ni$_3$Sn$_4$)=45	Ni$_3$Sn$_4$ [11] $\Delta H_m^3 / \Delta G_d^2$ min	
		$\Delta H_m^3 / \Delta G_d^2$ (Ni$_3$Sn$_2$)=71		
		y_1 (Ni$_3$Sn$_4$)=-0.3606	Ni$_3$Sn$_4$ Present, y_1 max	
		y_1 (Ni$_3$Sn$_2$)=-1.0632		
Fcc(Ni)/Liq(Sn)	650°C	$\Delta G_d / RT$ (Ni$_3$Sn$_4$)=0.5503	Ni$_3$Sn$_2$ [10] $\Delta G_d / RT$ max	Ni$_3$Sn$_4$[25]
		$\Delta G_d / RT$ (Ni$_3$Sn$_2$)=0.8991		
		$\Delta G_d / RT$ (Ni$_3$Sn)=0.4394		
		$\Delta H_m^3 / \Delta G_d^2$ (Ni$_3$Sn$_4$)=56.2	Ni$_3$Sn$_2$ [11] $\Delta H_m^3 / \Delta G_d^2$ min	
		$\Delta H_m^3 / \Delta G_d^2$ (Ni$_3$Sn$_2$)=25.9		
		$\Delta H_m^3 / \Delta G_d^2$ (Ni$_3$Sn)=105.3		
		y_1 (Ni$_3$Sn$_4$)=-0.3048	Ni$_3$Sn$_4$ Present, y_1 max	
		y_1 (Ni$_3$Sn$_2$)=-0.4066		
		y_1 (Ni$_3$Sn)=-2.6769		

[b]: The driving forces of Ni$_3$Sn at 200 and 550°C are negative, not considered in the Table.

3 Application
3.1 To Binary systems

Many researchers have investigated the interfacial reactions between the Sn-base soldering alloys and their corresponding Cu or Ni substrates. It is widely accepted that Cu$_6$Sn$_5$ is the first formed IMC in the studied temperature range below 500°C. For example, for bulk couples, Su [17] and Gagliano and Fine [18] found that Cu$_6$Sn$_5$ formed first when heated at 300°C and 250-325°C, respectively. For thin film couples, same results were reported for the Cu/Sn couple at temperature near 100°C [19, 20] and at 240-300°C [21]. Comparison of the results from prediction and experiment for Cu/Sn is listed in Table 1. It is clear that all models can reproduce similar results in agreement with experiments and it is hard to say which one is better.

Bader et al [21] have found that Ni$_3$Sn$_4$ is the first formed IMC at 240-300°C in the Ni/Sn couple and similar results at 200°C have been reported by Chen et al. [23]. According to

Kang and Ramachandran [24], Ni_3Sn_4 is still the first formed IMC at the Ni/Sn couple at 520°C and at even higher temperature 650°C [25]. It is then believed that Ni_3Sn_4 is the first formed IMC in the Ni/Sn couple as long as the temperature is lower than the decomposition temperature of Ni_3Sn_4, no matter whether Sn exists as solid or liquid in the interfacial reaction. Along with the experimental observation, the results from predication are listed in Table 2. Although all models reproduce similar results for Ni/Sn couple at 200°C and 520°C, only our model works fine at 650°C.

Moreover, typical results of experiments and predictions of the reaction in the Cu/In couples annealed at different temperatures are listed in Table 3. It is $Cu_{11}In_9$ rather than $Cu_{16}In_9$ that always forms first regardless of the temperature

change (whether Indium is appeared as tetragonal or liquid) in both thin film and bulk couples [17, 27-28]. It is worth noting that agreement between different model is okey at 300°C, but diagreement occurs in the case at 100°C where Indium takes solid form. For the Cu/In couple at 100°C, only the result from current model agrees well with observation.

Looking Tables 1 to 3, it is clear that the previous models [10-11] can correctly predict the phase sequence in the Cu/Sn system but failed to do so for Ni/Sn and Cu/In systems. Apparently, only our model works fine for all Cu/Sn, Ni/Sn and Cu/In systems. Up to now we found our model wins advantages on former 2 models.

Table 3. Comparison of IMC between observations and predictions for the interface reaction in the Cu/In diffusion couples. Thermodynamic parameters are refered to H.S.Liu *et al* [29]

System	Temperature	Criterion	IMC forming first	
			Predicted	Observation
Fcc(Cu)/Liq(In)	300°C	$\Delta G_d / RT$ (Cu_3In)=0.4978	$Cu_{16}In_9$ [10] $\Delta G_d / RT$ max	$Cu_{11}In_9$[17]
		$\Delta G_d / RT$ ($Cu_{16}In_9$)=0.5776		
		$\Delta G_d / RT$ ($Cu_{11}In_9$)=0.4906		
		$\Delta H_m^3 / \Delta G_d^2$ (Cu_3In)=31.0	Cu_3In [11] $\Delta H_m^3 / \Delta G_d^2$ min	
		$\Delta H_m^3 / \Delta G_d^2$ ($Cu_{16}In_9$)=50.3		
		$\Delta H_m^3 / \Delta G_d^2$ ($Cu_{11}In_9$)= 53.2		
		y_1 (Cu_3In)=-2.18	$Cu_{11}In_9$ Present, y_1 max	
		y_1 ($Cu_{16}In_9$)=-1.313		
		y_1 ($Cu_{11}In_9$)=-0.964		
Cu/In	100°C	$\Delta G_d / RT$ (Cu_3In)=0.4676	$Cu_{11}In_9$ [10] $\Delta G_d / RT$ max	$Cu_{11}In_9$[27, 28]
		$\Delta G_d / RT$ ($Cu_{16}In_9$)=0.6901		
		$\Delta G_d / RT$ ($Cu_{11}In_9$)=0.6924		
		y_1 (Cu_3In)=-4.058	$Cu_{11}In_9$ Present, y_1 max	
		y_1 ($Cu_{16}In_9$)=-2.227		
		y_1 ($Cu_{11}In_9$)=-1.7801		

Table 4 IMC formed first in the Co/Sn system: Comparison between experiments and predictions from different model. Thermodynamic data are cited from Jiang *et al* [30]

System	Temperature	Criterion	IMC forming first	
			Predicted	Observed
Hcp(Co)/Bct(Sn)	50°C	$\Delta G_d / RT$ (CoSn)=5.910	CoSn [10] $\Delta G_d / RT$ max	Amorphous[31]
		$\Delta G_d / RT$ (Co$_3$Sn$_2$)=4.772		
		$\Delta G_d / RT$ (CoSn$_2$)=4.367		
		$\Delta G_d / RT$ (CoSn$_3$)=3.327		
		$\Delta G_d / RT$ (liquid)=1.938		
		y_1 (CoSn)=0.6334	"Liquid"(amorphous) Present, y_1 max	
		y_1 (Co$_3$Sn$_2$)=0.6178		
		y_1 (CoSn$_2$)=0.8754		
		y_1 (CoSn$_3$)=1.772		
		y_1 (liquid)=6.004		
Hcp(Co)/Bct(Sn)	100°C	$\Delta G_d / RT$ (CoSn)=5.077	CoSn [10] $\Delta G_d / RT$ max	Amorphous[31]
		$\Delta G_d / RT$ (Co$_3$Sn$_2$)=4.099		
		$\Delta G_d / RT$ (CoSn$_2$)=3.768		
		$\Delta G_d / RT$ (CoSn$_3$)=2.885		
		$\Delta G_d / RT$ (liquid)=1.614		
		y_1 (CoSn)=0.4945	"Liquid"(amorphous) Present, y_1 max	
		y_1 (Co$_3$Sn$_2$)=0.4348		
		y_1 (CoSn$_2$)=0.6178		
		y_1 (CoSn$_3$)=1.297		
		y_1 (liquid)=5.166		

The important and interesting predictions are further carried out to Co/Sn couples at variant temperatures. With the thermodynamic parameters of the Co-Sn binary system [30], interfacial reaction in the Co-Sn binary couples is predicted as listed in Table 4. Although the new criterion predicts that formation of liquid will precedes other phases at 50 and 100°C. Guilmin *et al* [31] detected the amorphous layer at the interface between Co and Sn at room temperature. It should be pointed out that supercooled liquid at such low temperature should take its solid status amorphous. In fact, the amorphous and supercooled liquid in our model were given the same Gibbs energy. So, it is easy to see what our model predicted should really be the amorphous phase. This means that the prediction from the present criterion is consistent with the experiments, but the other models can not predict the amorphous formation in the interfacial reaction.

3.2 To Ternary system

Effort has further been made to apply the new criterion to predict IMC formation in simple ternary system. Based on the constituent binary systems [26, 32-33] and the information about the possible ternary phase [34], metastable phase equilibria together with isothermal sections of the Ni-Sn-Zn ternary system at 463 and 673K were preminarily and simply extrapolated asshown in Fig. 3. As seen, the solubility of Sn or Zn in Fcc (Ni) is much larger than that of Ni in Bct (Sn) or Hcp (Zn) at the temperatures studied. This indicates that Ni diffuse in Bct (Sn) or Hcp (Zn) faster than Sn or Zn in Fcc (Ni). Similar case also happens to the diffusion of Ni in liquid Sn-Zn. Because Ni-Sn compounds or τ_1 are compositionally covered by Fcc(Ni) and Ni-Zn compounds are located between Fcc(Ni) and Sn-Zn alloy (Fig. 4), it is easy to concluded that the diffusion of Sn or Zn determine formation of Ni-Sn compounds or τ_1 whereas diffusion of

Ni controls formation of γ-Ni$_5$Zn$_{21}$ or δ-NiZn$_8$. Considering the difference in composition and diffusivity of various compounds discussed above, formation of the Ni-Zn compounds γ-Ni$_5$Zn$_{21}$ or δ-NiZn$_8$ is much easier than the Ni-Sn binary or the ternary compounds τ_1. That is to say, γ-Ni$_5$Zn$_{21}$ or δ-NiZn$_8$ may form preferably.

In order to further illuminate the fact that γ-Ni$_5$Zn$_{21}$ rather than δ-NiZn$_8$ forms first in Sn-Zn/Ni couple, prediction by the new model was carried out. Because of the negligible solubility of Sn in γ-Ni$_5$Zn$_{21}$ or δ-NiZn$_8$, the effect of Sn on the formation sequence of these 2 compounds are excluded. Thus the model for a binary couple is adopted to predict the formation of IMC at the Sn-Zn/Ni (simplified to Zn/Ni). The predictions are listed in Table 5. Clearly, prediction by the model indicates that formation of γ-Ni$_5$Zn$_{21}$ has the priority over other phases when interfacial reaction happens at 463K or 673K. Such prediction was confirmed by our recent experiment (Fig. 4).

Table 5 Calculated model criterion value for various phases in Sn-Zn/Ni couples

Couple	Temperature	Phase	Status	Driving Force(J/mol)	Criterion, y_1	First Phase
Zn/Ni	463K	Fcc(Ni)	Entered	0	0	γ-Ni$_5$Zn$_{21}$ (y_1 max)
		Hcp(Zn)	Entered	0	0	
		γ-Ni$_5$Zn$_{21}$	Dormant	9373.95	- 0.11	
		δ-NiZn$_8$	Dormant	5829.85	- 0.17	
	673K	Fcc(Ni)	Entered	0	0	γ-Ni$_5$Zn$_{21}$ (y_1 max)
		Liquid	Entered	0	0	
		γ-Ni$_5$Zn$_{21}$	Dormant	6505.16	- 0.21	
		δ-NiZn$_8$	Dormant	1693.33	- 0.69	

4 Discussion

The present model covers both thermodynamics and kinetics, and includes estimation of the effect of interfacial energy change. It seems more complicated in quantitative prediction than those by Lee [10] and Choi [11]. In fact, the new model involves only thermodynamic calculation, i.e. all parameters needed in this model can be obtained through thermodynamic calculation. As long as the Gibbs energies of all the phases in the system of interest are available, the product of the interface reaction can be predicted.

Now let's analyze the influence of thermodynamics and kinetics. According to our experience, for the IMC, which may be possible to form in the thermodynamic meaning, the corresponding driving force satisfies the condition $\Delta G > 0.1RT$. It may be helpful to check the magnitudes of the first and second terms in Eq. (1). By assuming $b = 20$, $m = 1$, $x_B^\alpha = 0.2$, $x_B^X = 0.5$ $x_B^\beta = 0.8$ and plug them into Eq. (1), we can undertake the following estimation:

$$-\frac{b}{64m^2}\left[\ln\frac{x_B^\alpha}{x_B^X x_B^\beta}\right]^3 = 0.1 \qquad (2)$$

Here, $\left|\dfrac{b}{64m^2}\left[\ln\dfrac{x_B^\alpha}{x_B^X x_B^\beta}\right]\right|$ is likely varying between 0.01 and 10 in general and the composition change Δx_B^X between 0 and 1. Obviously, the magnitudes of these two terms do not differ greatly. This indicates that we cannot simply compare the driving force, as Lee [10] did, to predict which phase will form first in the interfacial reaction. The driving force may be crucial in some ranges, and the compositions and/or the composition change (close related to diffusion) may play a key role in other ranges.

The better applicability is attributed to the consideration of the effect of both nucleation and growth on IMCs forming at the interface. However, during the deduction of the present model, some approximations such as the treatment of interfacial energy are not refined, so further confirmation of the validity of the present model in other binary or multicomponent systems is needed.

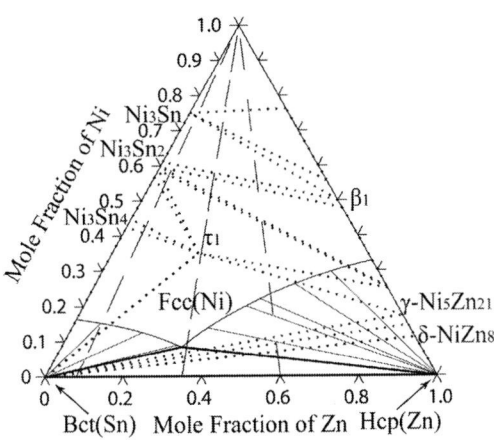

Fig. 3 Metastable phase equilibria imposed with the isothermal section of the Ni-Sn-Zn ternary system at 463K — metastable equilibria, …stable equilibria, --- starting couples.

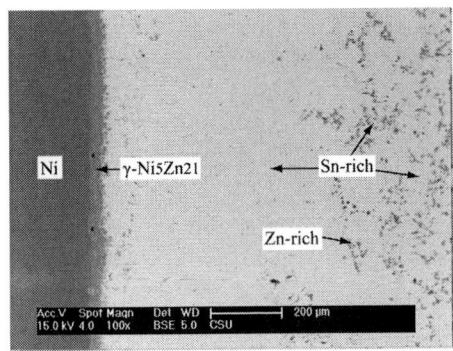

Fig. 4 Sn-8at.%Zn/Ni couples annealed at 463K for 1200 hours.

5 Conclusion

Based on thermodynamic calculations a novel methodology capable of predicting the products in the interfacial reaction between 2 dissimilar materials has been provided in the present work. This new model considers almost all thermodynamic factors (composition, driving force etc.) exerting influence on the formation of IMCs at the interface. The application to the interface of Ni/Sn, Cu/Sn, Cu/In and Co/Sn binary pairs and Ni/Sn-Zn ternary couple is successful and the prediction agrees very well with experimental observations. Further application of this model is in progress.

Acknowledgement

This work was financially supported by National Science Foundation of China (Grant No. 50671122).

References

1. F.J.J. van Loo et al., Diffusion in Solids, ed. M.A. Dayananda and G.E. Murch. TMS, Warrendale, PA, 1985, pp. 231-259.
2. K. Zeng et al., "Use of Multicomponent Phase Diagrams for Predicting Phase Evolution in Solder/Conductor Systems," *J. Electr. Mater.*, 30 (2001), pp. 35-44.
3. C.L. Liu et al., "Application of CALPHAD in Soldering of Electronic Materials," *Chin. J. Nonferr. Metals*, 13 (2003), pp. 1343-1349.
4. T. Nishizawa et al., "Phenomenological Consideration on Inter-phase Equilibrium in Diffusion Couple," *J. Jpn. Inst. Met.*, 34 (1970), pp. 629-637.
5. U. Gösele et al., "Growth Kinetics of Planar Binary Diffusion Couples: Thin-film Case versus Bulk Cases," *J. Appl. Phys.*, 53 (1982), pp. 3252-3260.
6. W.L. Johnson, "Thermodynamic and kinetic aspects of the crystal to glass transformation in metallic materials," *Prog. in Mater. Sci.*, 30 (1986), pp. 81-134.
7. F.M. d'Heurle, "Nucleation of a New Phase from the Interaction of Two Adjacent Phases: Some Silicides," *J. Mater. Res.*, 3 (1988), pp. 167-195.
8. L.A. Clevenger et al., "Nucleation-limited Phase Selection during Reactions in Nickel/ Amorphous-silicon Multilayer Thin Films," *J. Appl. Phys.*, 67 (1990), pp. 1325-1333.
9. C.V. Thompson, "On the Role of Diffusion in Phase Selection during Reactions at Interfaces," *J. Mater. Res.*, 7 (1992), pp. 367-373.
10. B.J. Lee et al., "Prediction of Interface Reaction Products between Cu and Various Solder Alloys by Thermodynamic Calculation," *Acta Mater.*, 45 (1997), pp. 1867-1874.
11. W.K. Choi et al., "Prediction of Primary Intermetallic Compound Formation during Interfacial Reaction between Sn-based Solder and Ni Substrate," *Scripta Mater.*, 46 (2002), pp. 777-781.
12. C.V. Thompson et al., Thin film Structures and Phase Stability, Ed. Clemens BM and Johnson WL, Mater. Res. Soc. Symp. Proc., Pittsburgh, PA, 1990.
13. R.J. Highmore et al., "Transient Nucleation Model for Solid State Amorphisation," *Mater. Lett.*, 6 (1988), pp. 401-405.
14. H.S. Liu et al., "Prediction of Formation of Intermetallic Compounds in Diffusion Couples," *J. Mater. Res.*, 22 (2007), pp. 1502-1511.
15. W.A. Johnson et al., "Reaction Kinetics in Processes of Nucleation and Growth," *Trans. AIME.*, 135 (1939), pp. 416-458.
16. J.W. Christian, The Theory of Transformations in Metals and Alloys, 1st ed., Pergamon Press, Oxford, 1965, pp. 537.
17. L.H. Su et al., "Interfacial Reactions in Molten Sn/Cu and Molten In/Cu Couples," *Metall. Mater. Trans.*, 28B (1997), pp. 927-934.
18. R.A. Gagliano et al., "Thickening Kinetics of Interfacial Cu6Sn5 and Cu3Sn Layers during Reaction of Liquid Tin with Solid Copper," *J. Electr. Mater.*, 32 (2003), pp. 1441-1447.
19. K.N. Tu, "Interdiffusion and Reaction in Bimetallic Cu-Sn Thin Films," *Acta Metall.*, 21 (1973), pp. 347-354.
20. R. Chopra et al., "Low Temperature Compound Formation in Cu/Sn Thin Film Couples," *Thin Solid Films*, 94 (1982), pp. 279-288.
21. S. Bader et al., "Rapid Formation of Intermetallic Compounds by Interdiffusion in the Cu-Sn and Ni-Sn

Systems," *Acta Metall. Mater.*, 43 (1995), pp. 329-337.

22. J.H. Shim *et al.*, "Thermodynamic Assessment of the Cu-Sn System," *Z. Metallkd.*, 87 (1996), 205-212.

23. S.W. Chen, *et al.*, "Mechanical Properties and Intermetallic Compounds Formation at the Sn/Ni and Sn-0.7wt%Cu/Ni Joints," *J. Electr. Mater.*, 32 (2003), pp. 1284-1289.

24. J.H. Kim *et al.*, "Morphological Transition of Interfacial Ni3Sn4 Grains at the Sn-3.5Ag/Ni Joint," *J. Electr. Mater.*, 32 (2003), pp. 1228-1234.

25. S.K. Kang *et al.*, "Growth Kinetics of Intermetallic Phases at the Liquid Sn and Solid Ni Interface," *Scripta Metall.*, 14 (1980), pp. 421-424.

26. H.S. Liu *et al.*, "Thermodynamic Optimization of the Ni-Sn Binary System," *CALPHAD*, 28 (2004), pp. 363-370.

27. I. Manna *et al.*, "Interdiffusion between In layer and Bulk Cu or Cu-In Alloy," *Phys. Status Solidi*, A119 (1990), K9-K13.

28. R. Roy *et al.*, "Formation of Intermetallics in Cu/In Thin Films," *J. Mater. Res.*, 7 (1992), pp. 1377-1386.

29. H.S. Liu *et al.*, "Thermodynamic Assessment of the Cu-In Binary System," *J. Phase Equil.*, 23 (2002), pp. 409-415.

30. M. Jiang *et al.*, "A Thermodynamic Assessment of the Co-Sn System," *CALPHAD*, 28 (2004), pp. 213-220.

31. P. Guilmin *et al.*, "Amorphization of crystalline Co and Sn Multilayers by Solid State Reaction," *Physics Letters A*, 109 (1985), pp. 174-178.

32. J. Miettinen, "Thermodynamic Description of the Cu-Ni-Zn System above 600°C," *CALPHAD*, 27 (2003) pp. 263-274.

33. B.J. Lee, "Thermodynamic assessment of the Sn-Zn and In-Zn binary systems," *CALPHAD*, 20 (1996), pp. 471-480.

34. M.Y. Chiu *et al.*, "Intermetallic compounds formed during interfacial reactions between liquid Sn-8Zn-3Bi solders and Ni substrates," *J. Electr. Mater.*, 31 (2002) pp. 494-499.

Establishing Mixed Mode Fracture Properties of EMC-Copper (-oxide) Interfaces at Various Temperatures

A. Xiao[1], G. Schlottig[2], H. Pape[2], B. Wunderle[3], O. van der Sluis[1], K. M. B. Jansen[1], L. J. Ernst[1]

[1]Delft University of Technology, Mekelweg 2, 2628 CD Delft, the Netherlands
[2]Infineon Technologies AG, 81726 Munich, Germany
[3]Fraunhofer IZM, 13355 Berlin, Germany
Email: A.Xiao@tudelft.nl, Phone: +31 (0)15 27 83726

Abstract

Interfacial delamination is known as one of the root causes of failure in microelectronic industry. In order to explore the risk of interface damage, FE simulations for the fabrication steps as well as for the testing conditions are generally made in the design stage. In order to be able to judge the risk for interface fracture, the critical fracture properties of the interfaces being applied should be available, for the occurring combinations of temperature and moisture preconditioning. As a consequence there is an urgent need to establish these critical interface fracture parameters. For brittle interfaces such as between epoxy molding compound (EMC) and metal (-oxide) substrates the critical energy release rate (or delamination toughness, Gc) can be considered as the suitable material parameter. This material parameter is strongly dependent on the temperature, the moisture content of the materials involved and on the so-called mode mixity of the stress state near the crack tip. The present study deals with experimental investigation of the delamination toughness of EMC-Copper lead-frame interfaces as can directly be obtained from the production line. The experimental set-up as designed for this purpose was previously reported [1], together with some measurement results and toughness evaluations for room temperature fracture tests. This study deals with the experimental and simulation procedures to establish the interfacial fracture toughness from fracture test results at different temperatures, especially in the glass transition temperature region of epoxy molding compound. In order to calculate accurate fracture toughness, the viscoelastic material properties of molding compound are measured and considered. A special test procedure used to investigate the fracture properties in the glass transition temperature region of EMC will be introduced. The FE model used to simulate the viscoelastic material behavior will be discussed. The delamination toughness as a function of mode mixity at different temperatures will be given in the result section.

1 Introduction

It is well known that semiconductor packages are fabricated from highly dissimilar materials. Usually, the fabrication processes take place at high temperature and then the products are cured to room temperature. Subsequent thermal cycling occurs in qualification tests, during reflow soldering as well as during use in applications. Due to different thermo-mechanical properties of the dissimilar materials, interfacial delamination related failures appear to be one of the dominant reliability issues. Among other interface problems, the EMC to copper interfaces seem to be very critical. To predict delamination at high temperature,

especially in the glass transition region of the EMC, the material properties need to first to be characterized as a function of temperature and time. Detailed material characterization procedure and results are shown in [2]. In the paper, the material model of the EMC is also validated by experimentally measuring sample warpage caused by curing and thermal stresses and calculating from both 2D and 3D simulations. Further, for the fracture simulation, this warpage calculation is taken into account.

Characteristic for the procedure of establishing the fracture toughness from fracture test results is that adequate FE simulations of the fracture test are required. A necessary 1st FE analysis step is performed to establish the residual stress state due to chemical shrinkage (due to the cure of the EMC) and the subsequent cooling down to room temperature. Here it occurred to be quite important that suitable visco-elastic constitutive models are used for the EMC. After the residual stress state is established, the actually interface delamination experiment is simulated in a 2nd step. Since the EMC behaves visco-elastic, in principle, the stress state in the sample is time-dependent, and so the stress singularity at the crack tip is. When the position of the crack tip is changing during the experiment, a second time dependency, related to the crack propagation rate is interacting with the visco-elastic time dependencies. This should be well accounted for in the FE simulations. In the 2nd, the critical energy release rate is calculated by Griffith's energy balance approach. This approach is shown to be suitable to calculate time-dependant fracture process in a bi-material cracked along the interface [3-4].

The goal of this study is to establish the temperature and mode mixity dependant delamination toughness of an EMC - Cu lead frame interface that is directly obtained from production line packaging. The test procedure and results as well as the simulation approach and results will be discussed.

2 Test condition and result

In order to guarantee that the established delamination properties will be identical to those of real products, in the present study on EMC-Cu leadframe interfaces, the test specimens are created using the runners on a matrix leadframe from the production line. Here Cu lead frames were applied, that did not possess degating holes beneath the runner positions of the lead frames and care was taken that the Cu at the runner positions has identical surface treatments as at the product positions. The singulation of the bi-material sample from the molded leadframe appeared to be a quite delicate process because of the desired small sample width of about

1mm. After singulation the sides of the samples were polished to reduce possible flaws.

Final Test samples are 35mm long, 1mm wide and 1.2mm thick. They are sawn along both sides of the runner on the copper lead pad of a matrix lead frame [1]. Before delamination test starts, all the samples are heated to the molding temperature and subsequently cooled down to room temperature under controlled heating and cooling rate.

Due to the fact that after production, the sample is stored in the lab environment for some time period, the internal stress in the structure is considerably influenced by physical ageing of the molding compound and by moisture absorption. Often, the storage time and moisture content in unknown. Therefore, this additional step is necessary and used to re-establish a well defined stress state. Measuring sample without this step will result in lower critical force value and instable delamination path (result shown in Figure 1, force is normalized to the maximum value in the pre-conditioned curve). As a result, the critical energy release calculated from such a curve will not be accurate unless the sample storage history and moisture content is calculated accurately.

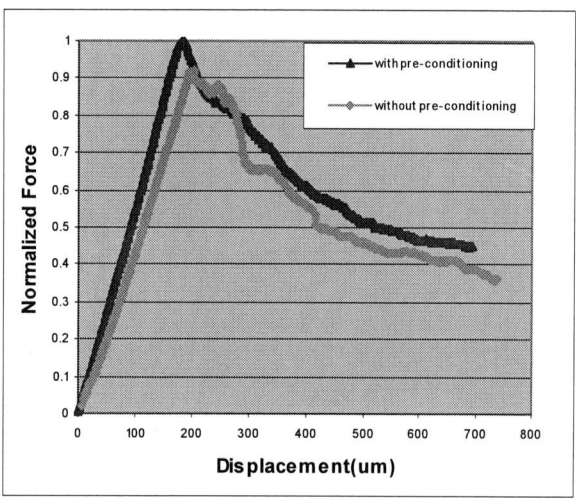

Figure 1 Experiment results with and without pre-conditioned samples

After pre-conditioning, the sample was clamped in our mixed mode bending (MMB) tool, surrounded by an oven. The oven was then heated to the ideal test temperature. It took about 15 minutes to stabilize the temperature in the oven. The procedure was repeated for each measurement.

The experiments were done at 85°C, 120°C and 175°C. Four loading mixities were executed at each temperature. At least two experiments were performed at each loading mixity. Some results are shown in Figs. 2-5. Figs. 2 and 3 show the force/displacement curves at 85°C for mixed mode bending (MMB) and load angle $\psi=0°$, which means a double cantilever beam (DCB) loading, respectively. Fig. 4 and 5 show similar results at 175°C. Force values are normalized to the maximum force value of the MMB at 85°C.

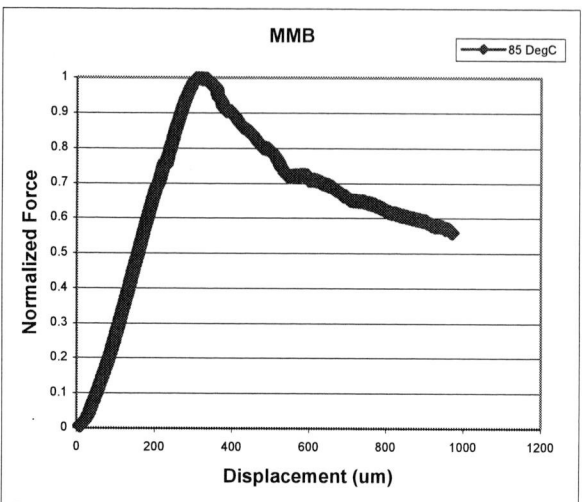

Figure 2 MMB test at 85°C

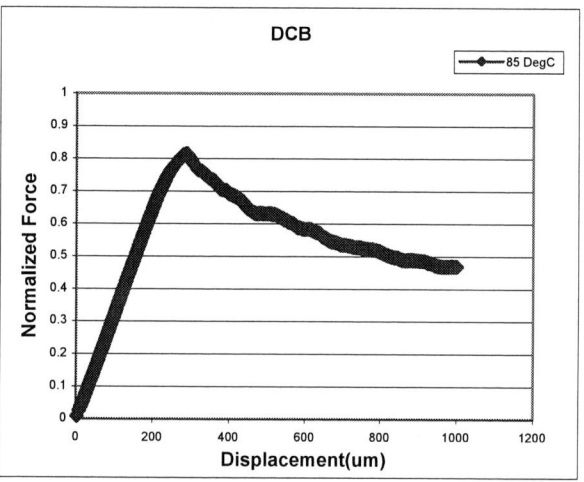

Figure 3 DCB test at 85°C

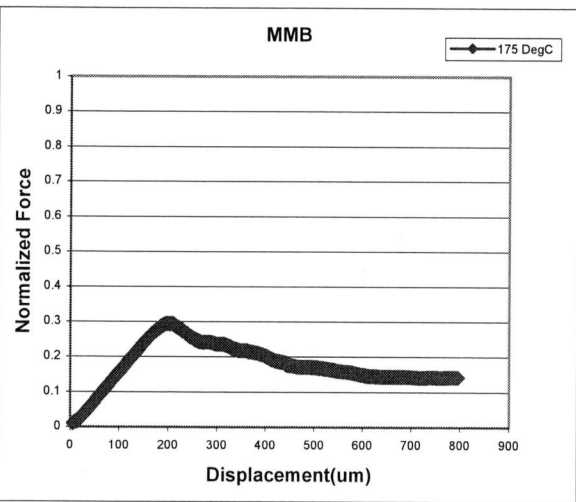

Figure 4 MMB test at 175°C

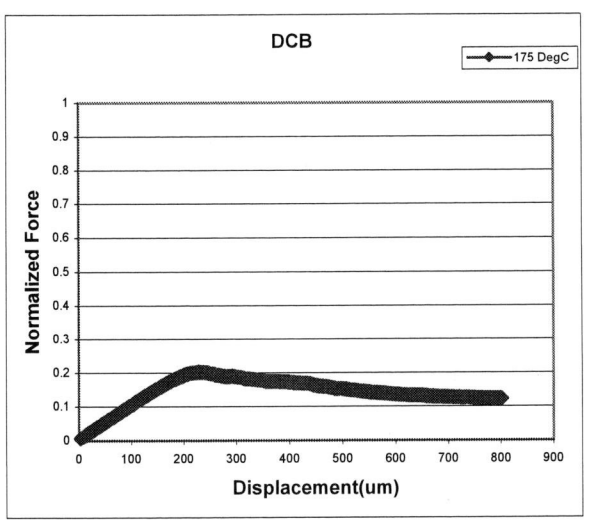

Figure 5 DCB test at 175°C

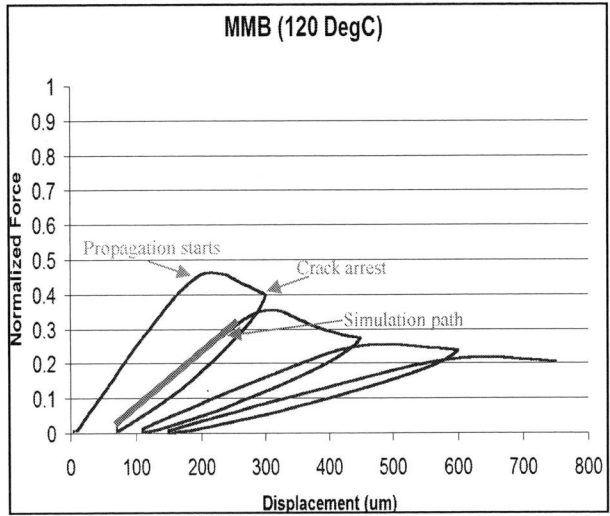

Figure 6 MMB test at 120°C

The measurement procedure was modified for test at 120°C, which is about the glass transition temperature of the EMC being considered. The following steps are defined:

1) The test sample is first loaded with a constant loading rate of 20μm/min.

2) At a certain moment, after the initial crack starts to propagate, the sample is unloaded (till the force is about zero).

3) Then, the sample displacement was kept steady for 30 minutes (according to the relaxation curves for the EMC being under consideration, the relaxation time at 120°C is about 1000s). This step is necessary for stress relaxation and stress redistribution.

4) Next the subsequent loading cycle is performed.

Figure 6 and 7 show the results from DCB and MMB tests at 120°C. Force values are normalized to the maximum force value of the MMB at 85°C.

As expected, it is observed that here the unloading and loading paths are almost identical and crack propagation restarts at forces slightly lower than at crack arrest level of a previous cycle. The FE simulations for 120°C experiments are performed on FE models where the crack length corresponds to experimentally established crack tip positions at each cycle.

It is known that during crack propagation, the stress state in the sample is changed due to time dependent behavior of EMC. It is found quite time consuming in FE analysis to simulate the real crack growth. Therefore, in the model, the crack growth is not simulated. Instead the force/displacement and the crack length of each loading cycle at the moment where the crack continues propagating after the arrest period are used. E. g. in the first loading cycle, the moment that the pre-crack starts to propagate is taken into account. For the second cycle, the moment that the extended (pre-) crack starts to grow is considered, etc. In between the two cycles, stress relaxation is considered in the FE analysis.

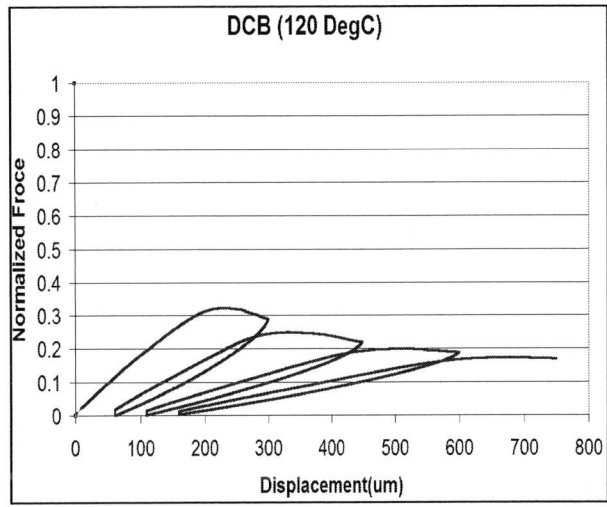

Figure 7 DCB test at 120°C

The same experiment procedure has also been performed at room temperature (The result is shown in Figure 8); where the EMC is expected to be elastic. It can also be seen from this result that the unloading curve coincides with the re-loading line. This proves that for delamination measurements at room temperature and for the EMC under consideration, therefore simple continuous crack testing can be applied. This is because the material can be considered as "glassy elastic" during the crack propagation.

In this case a simplification of the simulation procedure is allowed. In this simplified procedure the actual propagation of the crack during loading is not simulated. Instead, for a chosen crack tip position, the previously discussed 1st step is performed using the complete viscoelastic constitutive model to (approximately) establish the crack tip singularity, due to cure, cooling down and preconditioning. Next in a 2nd elastic step the external loading is added to the already cracked sample. For verification the critical energy release rate established with this simplified procedure was compared to the result based on simulation of the complete cyclic test procedure.

This simplified procedure for establishing the critical energy release rate and mode mixity for a chosen crack tip position can also be used for temperatures far above the glass transition temperature. For example, at 175°C the relaxation of stress occurs within a fraction of a second. We thus can perform an elastic simulation in the 2nd step, this time while using the "rubbery modulus" values as quasi-elastic modulus.

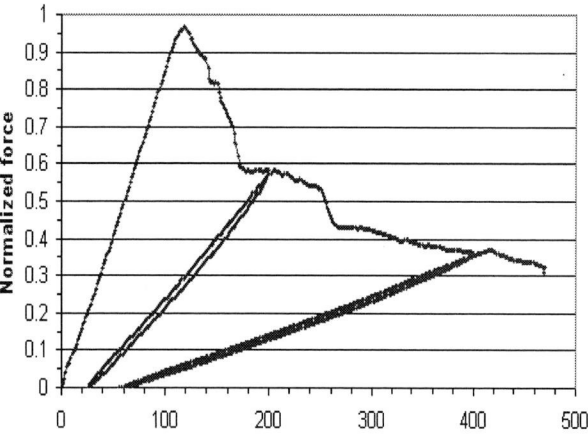

Figure 8 The modified test procedure for delamination testing applied at room temperature (Force-displacement data at 23°C)

3 Finite element modeling and results

It is found that for the present measurement setup with the bending sample, the lever system and "razor blade" load transfer, FEM-fracture mechanics simulations must be geometrically non-linear, while all the parts must be included in the model. This is because during loading and crack propagation, several parts of the system undergo a non-negligible rotation. As described before, a necessary 1st FE analysis step is performed to establish the residual stress state due to chemical shrinkage and the subsequent cooling down to room temperature. Here it occurred to be quite important that suitable viscoelastic constitutive models are used for the EMC.

The actual loading procedure (for testing in the glass transition temperature region, until crack re-propagation) is simulated in the 2nd step. Since the EMC behaves viscoelastic, in principle, the stress state in the sample is time-dependent, and so the stress singularity at the crack tip is. Therefore, it will be clear that the FE interpretation of the measurement results requires a dedicated viscoelastic constitutive model for the EMC, for which the temperature dependent relaxation data are established beforehand [2]. When the position of the crack tip is changing during the experiment due to crack propagation, a second time dependency, related to the propagation rate is interacting with the viscoelastic time dependencies. This should be well accounted for in the FE simulations.

We found that in many standard FEM packages used the calculation of the energy release rate through the implemented J-integral gave erroneous results in case of viscoelastic material behavior. Therefore, we used the following Griffith's energy balance approach for post processing calculation. The energy balance for a crack of length a is given by the relation

$$W - U - S = 0 \qquad (1)$$

Where W is the total input energy at the boundary, U is the strain energy density and S is the surface energy per unit area.

When the material is considered to be elastic, Equation (1) can be written as:

$$G = S \qquad (2)$$

Where G is the energy release rate and is given by the well known relation

$$G = \left| \frac{\partial U}{\partial a} \right| \qquad (3)$$

In the viscoelastic analysis, the process of crack propagation is assumed to be time dependent. Meaning, at any instant time t, the crack is assumed to have length a, and it is then assumed to further propagate by a small amount of Δa in a small time interval Δt. The Δt is assumed to be small enough such that the stress variations in the time interval can be considered to be negligible, so that the dissipation effects are neglected over this time step. During crack propagation, the temperature is assumed to be constant, so the other energy contributions like kinetic energy and heat energy are neglected. Therefore the elastic strain energy release rate relationship (3) can be used in this time step.

The energy release rate G is evaluated by calculating the strain energy change in the small time step. Consider a crack length a at time t which is further loaded in the next time interval (t+Δt), then strain energy at the end of this time is U1 (a, t+Δt). Now consider the same crack length at time t and let it grow by an amount Δa (a to a+Δa) in the time interval Δt (t to t+Δt), the strain energy at the end of this time is U2 (a+Δa, t+Δt).

The energy difference $\Delta U = U1-U2$ is the energy released in creating the new surface of length Δa and therefore the strain energy release rate G in this small time step is

$$G(t + \Delta t) = \left| \frac{\partial U}{\partial a} \right| \approx \left| \frac{\Delta U}{\Delta a} \right| \qquad (4)$$

It can be rewritten as:

$$G(t + \Delta t) \approx \left| \frac{U_2 \left(a + \Delta a, t + \Delta t \right) - U_1 \left(a, t + \Delta t \right)}{da} \right| \qquad (5)$$

In order to verify this approach, a benchmark study is done to compare Gc value from the energy approach with J-integral at room temperature and compare Gc from the energy approach with an analytical solution at high temperatures. It is found from the benchmark study that the energy approach is a suitable and accurate approach for Gc calculations when viscoelastic material has to be considered.

In our FE model, we chose Δa as 1/10 of the size of the adjacent crack tip element size. We calculate the energy change at the end of the loading in the same time interval.

Based on the experimental results, the critical energy release rate of copper-epoxy molding compound was calculated using 2D non linear simulation and the mode mixity was calculated using the crack surface displacement extrapolation method [9].

The result of fracture toughness and mode mixity at different temperatures is shown in Figure 9. Generally speaking, at high temperatures in and above the glass transition temperature, the critical energy release rate and mode mixity have a similar relation to those at room temperature. When shifting the loading closer to mode II (corresponding from lower mode mixity to higher mode mixity in Figure 9) the energy release rate and mode mixity increased. When looking at room temperature, the Gc is about 3 times higher at mode II loading than mode I loading. While, at high temperatures, the difference in Gc due to change in loading is getting smaller and smaller. At 175°C, Gc values of mode II loading are smaller than the mode I values.

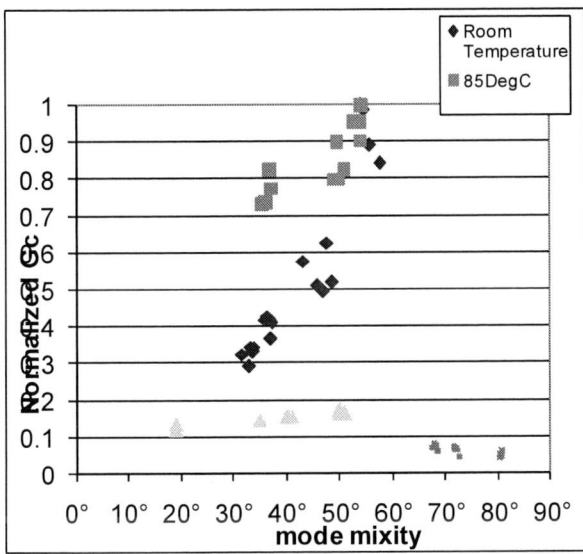

Figure 9 Normalized Gc as function of mode mixity at different temperature

Compared with the previous room temperature result, Gc at 85°C is almost higher than at room temperature at all mode mixities tested here. It is found that at and above glass transition temperature, the fracture toughness is significantly reduced. At the highest mode mixity, let's consider the fracture toughness at room temperature to be at 100%, relative Gc reduces up to 70% at 120°C and further up to 90% at 175°C. We think the main reason for this phenomenon is that at 120°C or above, compared to the state at room temperature, the material strength is very much reduced and so is the interface fracture toughness,

Apart from Gc, looking at the temperature effect on the calculated load angle in a fixed load geometry of the setup, it seems that below 120°C, mode mixity is not much influenced by temperature. This can be seen in table1. While at 175°C, mode mixity much shifted to very high load angles, i.e. shear

loading is achieved, below 120°C the maximum is just 56° at room temperature.

Table 1 Comparison of Gc and maximum load angle achieved in the setup at different temperatures

MMB2	Normalized G_c [-]	Mode mixity
Room temperature	1	56°
85°	0.98	54°
120°	0.34	51°
175°	0.08	72°

Conclusions

In this study, the mixed mode bending test method was used to characterize the fracture toughness of EMC/copper as a function of mode mixity at different temperature. The EMC material model considered cure shrink and viscoelasticity, which were measured before. The characterization procedure and results were previously reported. The sample geometry and sample preparation procedure were explained. It is worthwhile to mention that the fabricated test samples have identical surfaces as in the real product. The test procedure was built and the interfacial fracture toughness and mode mixity were calculated. A special test procedure was introduced for glass transition temperature range of the EMC. Before delamination experiment, the test samples were pre-conditioned to re-establish residual stresses, which are calculated in the FE simulation. It can be seen that at high temperature, the interface fracture toughness and mode mixity showed a similar relation as it was found at room temperature. Further, the fracture toughness values at the glass transition temperature (and above) reduced significantly compared to the room temperature values.

In the near future, our focus will be on the development of a full range of temperature and also moisture dependent interface fracture data, for various EMC-copper interfaces.

Acknowledgment

We thank the European Commission for partial funding of this work in FP7 under project NanoInterface (NMP-2008-214371).

References

1. A. Xiao et. al., "Mixed Mode Interface Characterization Considering Thermal Residual Stress," International Conference on Electronic Packaging Technology and High Density Packaging, JUL 28-31, 2008 Shanghai P R CHINA, pp. 937-943, 2008.

2. A. Xiao et al., "Establishing Fracture Properties of EMC-Copper Interfaces in the Visco-Elastic Temperature Region," 59th Electronic Components and Technology Conference, Page: 239-245, 2009.

3. M. V .Srinivas and G. Ravichandran, "Interfacial crack propagation in a thin viscoelastic file bonded to an elastic

substrate," *International Journal of Fracture* 65, pp. 31-47, 1994.

4. H. K. Mueller, W. G. Knauss, "Crack propagation in a linearly viscoelastic strip," *Journal of Applied Mechanics*, 38 series E (1971) pp. 483-488.

5. W. D. van Driel *et al.*, "Characterization of Interface Strength as function of Temperature and Moisture Conditions," *6th International conference on electronic package technology*, pp. 1-6, ISBN: 0-7803-9449-6.

6. J. de Vreugd *et al.*, "Advanced viscoelastic material model for predicting warpage of a QFN panel," *Proceeding Electronic Components and Technology conference*, IEEE 2008, pp. 1635-1640.

7. G. Q. Song *et al.*, "Effect of Loading Mode, Temperature and Moisture on Interface Fracture Toughness of Silicon/Underfill/Silicon Sandwiched System," *Intersociety Conference on Thermal and Thermo mechanical Phenomena in Electronics Systems*, pp. 1147-1152, May 2006.

8. A. A. O. Tay, Y Ma, T Nakamura, and S H Ong, "A Numerical and Experimental Study of Delamination of Polymer-Metal Interfaces in Plastic Packages at Solder Reflow Temperatures," *The 9th Intersociety Conference on Thermal and Thermomechanical Phenomena in Electronic Systems*, 2004. ITHERM 2004., 1-4 June 2004, pp. 245-252.

9. Andrew Tay, "Application of fracture mechanics in microelectronics," *Lecture note of EuroSime 2006 short course*.

10. Weidong Xie, Suresh K. Sitaraman, "Investigation of Interfacial Delamination of a Copper-Epoxy Interface Under Monotonic and Cyclic Loading: Experimental Characterization," *IEEE Transactions on Advanced Packaging*, Vol. 26, NO. 4, Nov 2003.

978-1-4244-4658-2/09 $25.00 © 2009 IEEE

Effect of Thermal Cycling on Interfacial IMCs Growth and Fracture Behavior of SnAgCu/Cu Joints

Xiaoyan Li, Fenghui Li

School of Materials Science and Engineering, Beijing University of Technology, Beijing, 100022, P. R. China

Abstract

The morphological features and growth kinetics of interfacial IMCs have shown significant effect on the mechanical properties, and hence, on the reliability of lead free solder joints. The growth behavior of the interfacial IMCs of SnAgCu/Cu solder joint, in reflow and thermal cycling, was investigated with the focus on the influence of reflow dwell time and the cyclic parameters on the growth kinetics. The three-dimensional IMCs feature was explored by etch the solder matrix out of the SnAgCu/Cu interface. The phases of IMCs were identified by energy dispersive X-ray (EDX). The thickness of the IMCs was measured by element mapping and phase constitution analysis. The thermal cycling of SnAgCu/Cu soldered joints was performed within the temperature region of -25°C to 125°C and -40°C to 125°C respectively. The corresponding IMCs growth rates were formulated according to the data from various thermal cycles. The growth kinetic of the IMCs was analyzed in the framework of diffusion principles. The shear strength of the joint was evaluated and the fracture mechanism was analyzed in accordingly. It was found that Cu_6Sn_5 was formed and followed by rapid coarsening at the solder and Cu interface during reflowing. During thermal cycling, however, IMCs grain coarsening and breaking were noted. The thickness of IMCs was found increases with the thermal cycles; however, the growth rate is less than that of thermal aging. The dwell time in high temperature portion of a thermal cycle was found has significant influence on the growth rate of the IMCs. The growth of the IMCs, both for reflowing and thermal cycling, was found follows diffusion model. The shear strength of the joints was found decreases both with the increase of the thermal cycles as well as with the decrease of the cooling dwell temperature in the thermal cycles.

1 Introduction

In most electronic assembles, solder joints provide both electric connection and mechanical support to the components. It has been demonstrated that the reliability of electric devices depends strongly on the reliability of the soldered joints while the later one was controlled, mainly, by the formation and growth of the interfacial intermetallic compounds (IMCs) between the solder matrix and the substrates. In the past decades, research works were made to explore the formation mechanisms and growth kinetics of the interfacial IMCs with the focus on the morphological features and growth rate under isothermal and thermal cyclic conditions. Various results were reported by different researchers on this topic. Nevertheless, the so far proposed mechanisms on the formation and growth of IMCs were not widely accepted and need to be further investigated.

Actually, after years of research and development, the transition of the microelectronics industry to lead free soldering has become into practice. Many efforts were made by researchers, in the past decades, either to develop lead free solders, or to modify the soldering profiles in assemblies or to study the reliability of the packages and to estimate the lifetime of the devices. There are indications that soldered structures may be weakened in ways specific to the combination of component, substrate and process [1-3]. Such degradation is often believed to be originated from the formation and evolution of the interfacial intermetallic compounds. On the other hand, the simply implementation of the findings from SnPb soldered joints to Pb-free soldered joints are frequently found misleading [4]. The formation and growth of the IMCs are proven to be controlled by chemical reaction mechanism during reflow process and by diffusion mechanism during thermal aging and thermal cycling, however, some researches reported that the kinetics often dose not follow Fick's diffusion law well [5-7]. Current proprietary research is showing the above outstanding problems to be associated with the unique characteristics of the interfacial microstructure evolution of Pb-free soldered joints [8, 9]. With the aims focused on the exploration of the influences of reflow time and thermal cycling on the formation and growth kinetics of IMCs as well as the joint mechanical properties, the present study was performed and the findings on the SnAgCu/Cu interface were analyzed.

2 Experimental Procedures

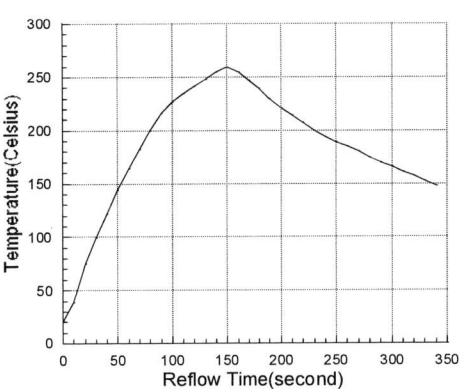

Figure 1 Reflow Temperature Profile

Commercial Cu bar with diameter of 3mm was used as the substrates in the present study. A Sn-3.8Ag-0.7Cu Pb-free solder ball was then soldered, by using an infrared reflow oven, in between the two pieces of Cu pads to form a sandwich-like soldered joint specimen. In order to study the effects of the reflow time on the formation and growth of the interfacial IMCs, the specimens were reflowed at peak temperature of 260°C for 20s and 200s respectively. The preheat temperature and the time, however, were kept the same for all the specimens. The typical temperature profile employed in the reflowing is shown in Figure 1.

978-1-4244-4658-2/09 $25.00 © 2009 IEEE

During thermal cycling, the temperature changes were fluctuating within the region of -25°C to 125°C and -40°C to 125°C respectively. For metallographic observation of the interfacial microstructure, the specimens were first cross-sectioned perpendicular to the solder-Cu interface of the joints and polished, then etched slightly by using 5% HCl-95%C_2H_5OH solutions. For the observation of three-dimensional morphology of the IMCs, the solders on the interface were etched deeply by using 10% HCl-90% C2H5OH solution for 8h followed by etching with 10% HNO_3-90%C_2H_5OH solution for 3h. Scanning electron microscopy (SEM) was used to study the microstructural morphology of the IMCs. X-ray diffraction (XRD) was used to observe the microstructure evolution of the intermetallic grains and to identify the IMC phase constitutions. Energy dispersive x-ray spectrometry (EDS) and XRD were used to characterize the composition of the IMCs and to analysis the elementary distribution. The mean thickness of the IMCs was measured according the elementary mapping and image analysis. Tensile tests the microhardness impressions were performed at different aging time. The fractography of the solder joints, after the shear tests, were also examined by SEM.

3 Results and Discussion

Intermetallic compounds (IMCs), actually, comprise an important and integral part of the solder-joint structure. IMCs in SnAgCu lead free system are frequently found to be the secondary phases of the microstructure of the solder joint and the size and dispersion affect the joint properties. During reactive wetting, IMCs usually form layered structures between the molten solder and wetted substrate.

The morphologies of the interfacial Cu-Sn intermetallic layers for reflow time 20s and 200s were shown in Fig. 2. The light gray scalloped interfacial regions on the optical micrographs are the IMCs. The composition of the IMCs was verified by EDX microprobe analysis. It was found that only the η-phase Cu_6Sn_5 was formed in as-soldered samples. The intermetallic layer is very rough and irregular. The interface between the η phase and solder displays a scalloped-like morphology. The formation of η-phase Cu_6Sn_5 intermetallic layers in soldered joint during the reflow process is believed to be originated from the interfacial reactions between its constituting species, Sn from the solder and Cu from the copper pad. In order to observe the three-dimensional morphology of the interfacial IMC, a deep and selective etching was utilized to remove the solder and expose the intermetallic layer. In this way, the interfacial Cu-Sn intermetallic layer could be viewed in the SEM looking down from the solder side of the interface, as show in Fig. 2. The scalloped morphology of the intermetallic layer was clearly visible while the rounded grains and deep channels between the grains could be easily distinguished. It was noted that, the intermetallic layer thickness increases with increasing reflow time. In addition, the size of the scallops increases with reflow time. The increasing size of the scallops could be due to some combination of particle agglomeration and ripening or a competitive grain growth phenomenon according to the literature.

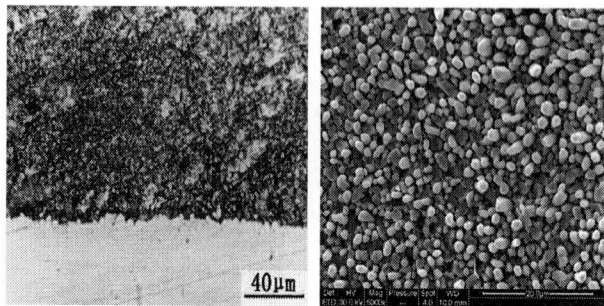

Cross-sectional view   Top view

(a) Reflow for 20s

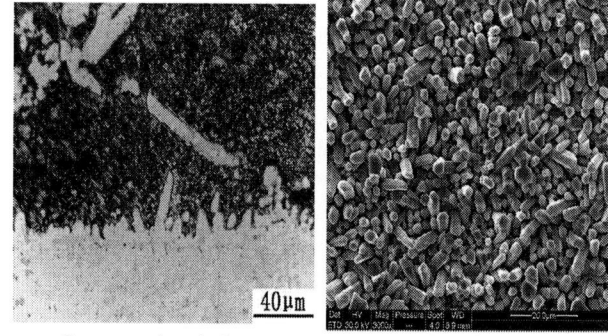

Cross-sectional view   Top view

(b) Reflow for 200s

Figure 2 Micrographs of cross-sectional view and top view of interfacial IMCs for different reflow dwell time

Thermal cycling of solder joints is very important as far as the reliability of the solder joint is concerned. Nevertheless, the impact of microstructure evolution and IMCs growth kinetics on the reliability of the joints was not well understood. It was reported that the IMCs layer thickness is smaller in thermal cycled samples compared to isothermally annealed samples [9]; however, the effect of thermal cycling temperature profile was not clear presented. In the present paper, the thermal cycling of SnAgCu/Cu soldered joints was performed within the temperature region of -25°C to 125°C and -40°C to 125°C respectively. The dwell time at the up portion of the temperature cycles was 30 minutes while the dwell time at the low portion of the temperature cycles was 10 minutes for both above temperature regions. The thermal cycling profiles were shown in Figure 3.

It was found that, the intermetallic layer was very rough and irregular. The interface between the η-phase and solder displays a scalloped morphology. With the increase of thermal cycles, the flatten of the scallop-like IMCs was noted. The IMCs growth rate for the thermal cycling within -40°C to 125°C is found greater than that of the thermal cycling within -25°C to 125°C, as shown in Figure 4 and Figure 5.

978-1-4244-4658-2/09 $25.00 © 2009 IEEE

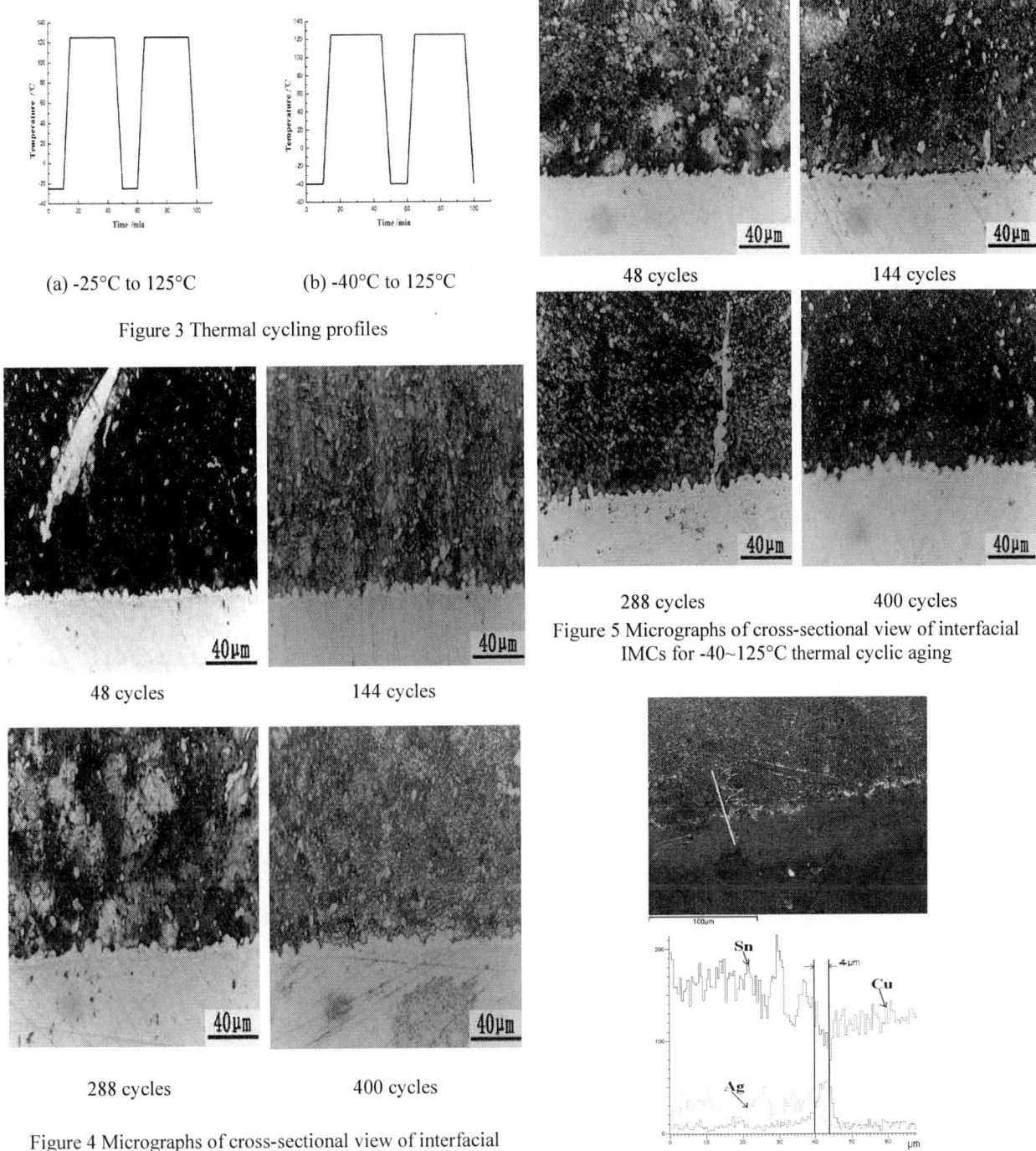

(a) -25°C to 125°C   (b) -40°C to 125°C

Figure 3 Thermal cycling profiles

48 cycles   144 cycles

288 cycles   400 cycles

Figure 5 Micrographs of cross-sectional view of interfacial
IMCs for -40~125°C thermal cyclic aging

48 cycles   144 cycles

288 cycles   400 cycles

Figure 4 Micrographs of cross-sectional view of interfacial
IMCs for -25~125°C thermal cycling

Figure 6 SEM micrographs and EDX spectrums
across the interfaces of Sn3.8Ag0.7Cu/Cu joints
(as soldered)

In order to analysis the element diffusion during reflow soldering and thermal cycling, the element mapping, across the IMC lays, was performed at different time, as shown in Figure 6. Consequently, the thickness of the Cu_6Sn_5 η-phase layer and the Cu_3Sn ε-phase layer were measured based on the composition and phase constitution analysis.

The changes of the thickness of IMCs layers, during soldering and thermal cycling, was measured in accordance with the above methods, the results were shown in Figure 7 and Figure 8 respectively. It was found that, the IMC layer thickness increases rapidly in the short soldering time region, the region when the soldering time less than 60s in the

978-1-4244-4658-2/09 $25.00 © 2009 IEEE

present study, while the layer thickness increases slowly when the aging time was further elongated. The variation of the layer thickness with the soldering time may be due to the changes of the mechanisms of IMC formation and growth. Once the solder melts during reflow process, the tin in molten solder immediately reacts with the Cu pad and develops very minute centers of crystalline or nuclei which quickly grow up and form scallop-like islands of the IMC at the interface. This was mainly controlled by the reaction mechanism and the so formed IMC's islands become large enough and connect each other to form a continuous layer of intermetallic. Following the reaction dominated IMC formation process, the grain boundary diffusion and the volume diffusion processes dominated the formation and growth of the IMCs afterward. However, the diffusion of tin or Cu, during the grain boundary diffusion stage and volume diffusion stage, was prohibited by the IMC layer which was formed during reaction stage and finally resulted in the slow down of the growth rate of the IMCs.

The growth kinetic analysis of IMCs in the thermal cycling was rare in the literature due to that the IMC layer growth kinetic analysis is not as simple as for isothermal annealing because the diffusivity of elements through the IMC is temperature and time dependant during the thermal cycling profile. Nevertheless, the IMCs layer thickness, demonstrated by the present experimental data, is smaller in thermal cycled samples compared to isothermally annealed samples. Note that the dwell time at peak temperature in thermal cycling is rather shorter. The smaller IMCs layer thickness could be expected because of shorter time of exposure at peak temperature of a thermal cycle. With the enlarging of the temperature fluctuating region, the IMCs layer thickness was found increased. This phenomenon, however, could be attributed to the influence of large difference in the coefficient of thermal expansion (CTE) of the solder matrix and the substrate. The CTE mismatching is believed might cause the breaking of the previous formed IMCs which may resulted in the increase of the grain boundary. The increase of the grain boundary might accelerate the grain boundary diffusion process and resulted in the increase of the IMCs layer thickness in the thermal cycling with a larger fluctuating temperature region.

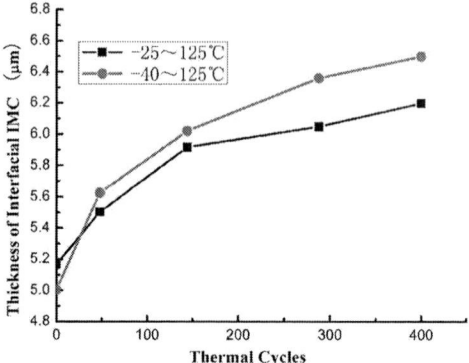

Figure 8 IMCs growth during thermal cycling

Shear tests were performed for the sandwich-like soldered specimens on different thermal cycles. Figure 9 shows the changes of the shear strength of the joints which were thermal cycling aged within the temperature region of -25°C to 125°C and -40°C to 125°C respectively. It was noted that the shear strength of the joint decreases with the further thermal cycling aging. This could be due to the influence of the growth of the interfacial IMCs layer. With the further growth of the interfacial IMCs, the embitterment of the IMCs may plays the predominate role in the deformation of the joint which resulted in the decrease of the shear strength of the joint. In addition, it was noted that the decrease of the shear strength for -40°C to 125°C thermal cycling is more than that for -25°C to 125°C thermal cycling. This may be due to that the larger temperature fluctuating region of a thermal cycling usually produce a thicker IMCs layer which may results in a lower shear strength of a joint.

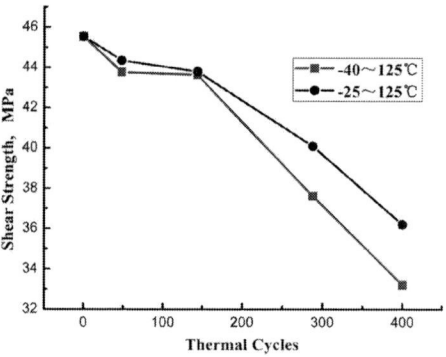

Figure 9 Shear strength of the joints after various thermal cycles

In order to have a further understanding on the fracture behavior of the soldered joints, shear fracture tests were performed for the joints which were thermal cycling aged for various cycles. Fractographic examination was also performed with SEM. The results were shown in Figure 10. It was noted that plastic deformation trace was left on the fracture surface which indicated that the shear fracture of the joint was deformation predominated. Further examination on the fracture of the joints indicated that the fracture was initiated and propagated along the interface of IMCs layer

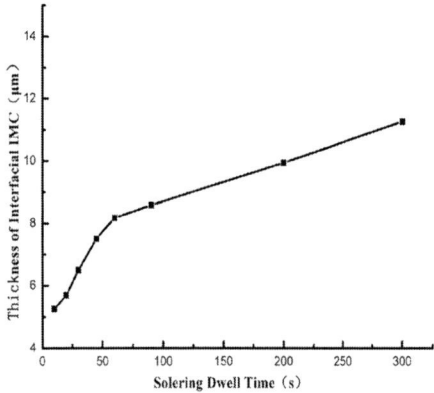

Figure 7 IMCs growth during soldering

and solder matrix. However, the fracture was found at solder side.

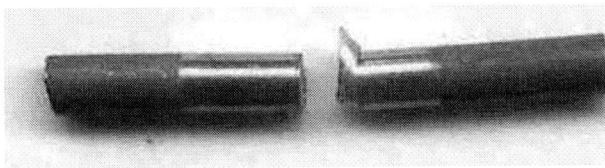

(a) Shear fracture

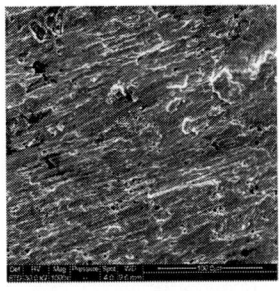 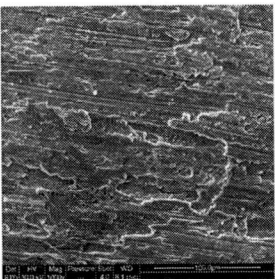

(b) Fractograph at Cu substrate side

(c) Fractograph at solder side

Figure 10 Shear Fractographs of the joint after 300 cycles (-40~125°C)

Conclusions

Only the η-phase Cu_6Sn_5 was formed in the reflow process, the formation of the IMC was dominated by reaction mechanism. The growth rate of the IMC was faster at shorter reflow time stage compare with that of longer reflow time stage.

The growth rate of the IMCs layer in thermal cycling is less than that of reflowing aging, however, the growth rate of the IMCs is bigger in thermal cycling with larger temperature fluctuating region compare with that in smaller temperature fluctuating region.

The shear strength of the joints decreases rapidly with the increase of thermal cycles. The joints which exposed to a larger temperature fluctuating shown the lower shear strength.

Acknowledgements

This study was partially supported by National Natural Science Foundation of China under the grant No. 50871004, the Nature Science Foundation of Beijing under the grant No. 2082003 and Guangdong-Hong Kong Joint Research Project under the grant number 2008A092000007, which was acknowledged.

Reference

1. Y. G. Lee abd J. G. Duh, "Interfacial Morphology and Concentration Profile in the Unleaded Solder/Cu jont Assambly," *J. Materi. Sci.: Mater. E.ectron.*, Vol.11, 2000, pp. 33-43.
2. Xiaoyan Li, Zhisheng Wang, "Thermo-fatigue Life Evaluation of SnAgCu/Cu Solder Joints in Flip Chip Assemblies," *J. Mater. Process. Tech.*, Vol. 183 , 2007, pp. 6-12.
3. Luhua Xu, John H. L. Pang, Kithva H. Prakash and T. H. Low, "Isothermal and Thermal Cycling Aging on IMC Growth Rate in Lead-Free and Lead-Based Solder Interface," *IEEE Transaction on Components and Packing Technologies*, Vol. 28, 2005, pp. 408-414.
4. L. P. Lehman, R. K. Kinyanjui, J. Wang, Y. Xing, L. Zavalij, P. Borgesen, E. J. Cotts, "Microstructure and Damage Evolution in Sn-Ag-Cu Solder Joints," *2005 Electronic Components and Technology Conference*, pp. 674-681, May 31-June 3, 2005, Lake Buena Vista, FL, USA.
5. J. H. Lee, J. H. Park, Y. H. Lee, Y. S. Kim and D. H. Shin, "Stability of Changes at a Scalloplicke Cu_6Sn_5 Layer in Solder Interconnections," *J. Mater. Res.*, Vol.16, 2001, pp. 1227-1230.
6. K. Zeng and K. N. Tu, "Six Cases of Reliability Study of Pb-free Solder Joints in Electronic Packaging Technology," *Mater. Sci. Eng.:R:Reports*, Vol. 38, 2002, pp. 55-105.
7. J. K. Chen, J. E. Beraun and D. Y. Tzou, "A DualpPhase Lag Diffusion Model for Predicting Intermetallic Compound Layer Growth in Solder Joints," *J. Electron. Packing*, Vol. 123, 2001, pp. 53-57.
8. Area Array Consortium Research, Universal Instruments Corporation, Binghamton, NY. USA.
9. Xiaoyan Li, Xiaohua Yang and Fenghui Li, "Effect of Isothermal Aging on Interfacial IMC Growth and Fracture Behavior of SnAgCu/Cu Soldered Joints", *2008 International Conference on Electronic Packaging Technology & High Density Packaging (ICEPT-HDP 2008).*

Effect of Miniaturization on the Microstructure and Mechanical Property of Solder Joints

Bo Wang[1,2], Fengshun Wu*[1,2], Jin Peng[2], Hui Liu[2], Yiping Wu[1,2], Yuebo Fang[3]

[1] Wuhan National Laboratory for Optoelectronics
Huazhong University of Science and Technology, Wuhan, P. R. China, 430074
[2] State Key Laboratory of Materials Processing and Die & Mould Technology
Huazhong University of Science and Technology, Wuhan, P. R. China, 430074
[3] Ningbo Kangqiang Electronics Co., LTD, Ningbo, P.R.China, 315105
* Corresponding author: fengshunwu@mail.hust.edu.cn

Abstract

In present paper, the effect of miniaturization on the microstructure and mechanical property of solder joints is investigated. With the miniaturization of solder joints, the thickness of IMC decreases, while, the IMC proportion to the solder joint increases; meanwhile, the concentrations of base materials in the bulk also increase with the reducing stand-off height (SOH). Due to the interaction of interfacial reactions in the Cu/Sn/Ni solder joints, the IMC layers with incompact connection are formed at Ni side, and tensile test results show that the incompact IMC are the weakest part in the solder joints. When the SOH reduces to 10μm, the solder bulk contains only one grain in height in the Cu/Sn/Cu and Cu/Sn/Ni solder joints, leading to dramatic changes in fracture mode and the ultimate tensile strength (UTS). When the SOH is lowered to 10μm, the UTS of Cu/Sn/Ni solder joints decreases dramatically to be lower than that of Cu/Sn/Cu solder joints.

1. Introduction

With the demands of miniaturization and higher electrical performance in the electronic products, the area array density of interconnections increases rapidly, leading to a concomitant decrease in the pitch/diameter of the solder bumps. As a result, the size of solder joints continues to shrink, and this leads to the miniaturization in the stand-off height (SOH) of the solder joint. SOH is the total height of the bilateral IMC layers and the solder bulk layer. For example, the SOH reduces from 80~100μm to less than 20μm when the bumps pitch decrease from 200μm to 50μm in the micro-C4 joining technology [1-3]. And we can see in the future, the SOH of solder joints will decrease further. Due to the reduction in SOH, the microstructure and mechanical property of the solder joints would be changed. This has a serious influence on the interconnection reliability, which needs to be highlighted.

Solder joints with Cu/solder/Ni structure are commonly used in most solder interconnection [4-6]. With the miniaturization, the interaction between the Cu/solder and solder/Ni interfacial reactions would become stronger, which is because the inter-diffusion path is narrowed in the sandwich structure solder joint. The interaction will affect the compositions and microstructures of the IMC layers on the both interfaces of the solder joints, and it will have a serious influence on the device reliability. Therefore, for the further understanding of solder joints in high density interconnection, it is very important to investigate the effect of the miniaturized SOH lower than 100μm on the microstructure and mechanical property of the Cu/solder/Ni solder joints.

However, very few works have been focused on that. Meanwhile, in order to understand the influence of SOH on the solder joints without the interaction of the two different interfacial reactions, Cu/Sn/Cu solder joints are also studied to compare with the Cu/Sn/Ni solder joints.

Tensile test was carried out to investigate the mechanical properties of solder joints. The tensile test can offer a uniform and equivalent stress across every section layer, including the soft solder bulk layer, the intermetallic compounds layer and the substrate layer. This method can be reliably used to find out the weak bonding layer in solder joints with micro SOH. In addition, tensile fracture surfaces will be better preserved and could provide better evidence to help identify the failure mechanism [7].

2. Experimental procedures

In order to investigate the microstructure and mechanical behavior of the miniaturized solder joint, the Cu/Sn/Cu and Cu/Sn/Ni solder joints with 100μm, 50μm, 20μm and 10μm SOH were prepared respectively. Pure Sn foil was used as the solder to join the terminal faces of the two bars (Cu/Cu or Cu/Ni bars). The bars of Cu and Ni have a length of 30mm, which were cut by spark cutting from a commercial grade Cu (99.99%) wire and a commercial grade Ni wire with diameter of 0.9mm. The commercial grade Cu wire is supplied by Ningbo Kangqiang Electronics Co., LTD. Then the cutting surfaces of Cu bars and Ni bars were polished using 1.0μm Al_2O_3 powder pastes to ensure the constant soldering surfaces. A clamping apparatus was used to control and regulate the distance of Cu and Ni terminal faces. When the two bars with parallel surfaces were held in alignment with the distance of expected space in clamp apparatus, the flux was coated on the side surfaces. Subsequently, a pure Sn (99.99%) foil was put in between the two aligned bars. Then, the clamping apparatus with the bars was inserted into a hot-air convection oven for reflow. After that, the Cu/Sn/Cu and Cu/Sn/Ni solder joints with specific SOH were prepared. Fig.1 illustrates the schematic diagram of the solder joint. The samples were mounted in epoxy and polished to reveal their cross-sectional microstructure. The polished samples were slightly etched with a 99%CH_3OH-1%HCl (in vol.%) solution to delineate the morphology of the microstructure of solder joints and the IMC layer. The reaction zones in the samples were analyzed with a scanning electron microscope (SEM). Using the image analyzing software "UTHSCSA ImageTool", the thicknesses of the IMC layers were measured through dividing the areas of the layers by the linear interface length. The compositions of the IMC layers and the elemental distribution across the reaction zone were analyzed using an energy-dispersive X-ray spectrometer (EDS).

The tensile test was carried out to study the mechanical behavior of solder joints at room temperature with a constant crosshead speed of 0.1mm/min (1.6×10^{-3}mm/s). Force and displacement parameters were recorded. In order to obtain the average tensile result, 10 specimens with the same SOH was prepared as a group for the tensile test. And, the fractograph of solder joint was observed using SEM.

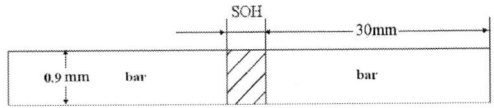

Figure 1 Schematic diagram of the solder joint with the specific SOH

3. Results and discussion

3.1 Effect of SOH on the microstructure of the Cu/Sn/Cu solder joints

Figure 2a-d show the cross-sectional back-scattered electron (BSE) mode SEM images of the Cu/Sn/Cu solder joint with 100μm, 50μm, 20μm and 10μm SOH, respectively. As we know, the solder joint consists of two Cu_6Sn_5 IMC layers and a solder bulk layer. According to Figure 2a-d, it is also found that SOH has an obvious effect on the Cu_6Sn_5 IMC thickness and proportion.

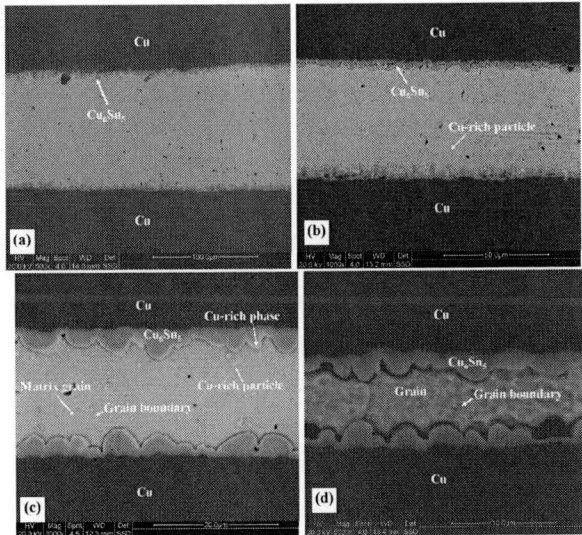

Figure 2a-d Cross-sectional backscattered electron (BSE) images of the Cu/Sn/Cu solder joints with SOH of (a) 100μm (b) 50μm (c) 20μm and (d) 10μm.

In present study, the IMC thickness means the totaling thickness of IMC at the both sides of the solder joint, the mean value of IMC thickness is obtained by dividing the measuring area of IMC by the length of interfacial IMC in the solder joint. A software program "UTHSCSA ImageTool" is used to measure the area and length of the IMC in the SEM cross-section images. The IMC proportion is the ratio of mean IMC thickness to the SOH. And Figure 3 shows the changes of IMC thickness and proportion in the solder joints with the

reducing SOH. Referring to Figure 3, the total IMC thickness decreases with the reducing SOH, and it is about 9μm, 8μm, 5.4μm and 3.6μm in the solder joints with 100μm, 50μm, 20μm and 10μm SOH respectively; however, their corresponding IMC proportion increases from 9%, 16%, 27% to 36%. It also can be found that SOH has a remarkable effect on the number of grains contained in the bulk. According to Figure 2c-d, the grain number decreases with the reducing SOH. For example, there are about 3-5 Sn grains in height in the solder joint with 20μm SOH, while, there is only one grain in height in the bulk of the solder joint with 10μm SOH, comparatively. The changes of IMC proportion and grain number in height have important influences on the mechanical property of solder joints.

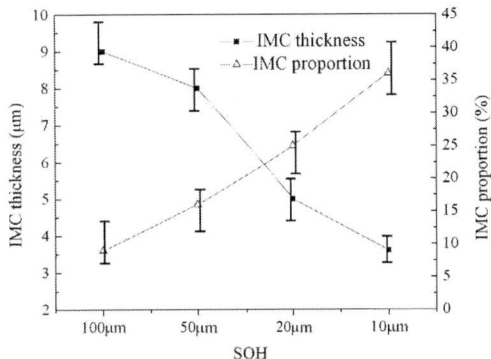

Figure 3 IMC thickness and proportion changes in the Cu/Sn/Cu solder joints with 100μm, 50μm, 20μm and 10μm SOH

3.2 Effect of SOH on the microstructure of the Cu/Sn/Ni solder joints

Fig. 4a-d show the cross-sectional back-scattered electron (BSE) images of the Cu/Sn/Ni solder joint with 100μm, 50μm, 20μm and 10μm SOH, respectively. It is found that the microstructure of the Cu/Sn/Ni solder joints differs much from the microstructure of the Cu/Sn/Cu solder joints due to the interaction of different base materials.

In the Cu/Sn/Ni solder joint, the formed IMC at the Cu side is $(Cu,Ni)_6Sn_5$ with a solubility of 3.4-3.8at.% Ni. The morphology of the $(Cu,Ni)_6Sn_5$ IMC shows a layer-type with irregular protrudent parts, which is different from the scallop-type of the uniphase Cu_6Sn_5 after reflow. This indicates that the dissolved Ni atoms in the $(Cu,Ni)_6Sn_5$ obviously change the morphology of Cu_6Sn_5. Ni atoms participate in the reaction of Cu and Sn to form the ternary $(Cu,Ni)_6Sn_5$ by diffusing across the molten solder during the reflow process. And Fig. 5 displays the thickness of Cu_6Sn_5 at the single Cu side in the Cu/Sn/Cu solder joints and $(Cu,Ni)_6Sn_5$ at Cu side in the Cu/Sn/Ni solder joints. From Fig.5, the thickness of IMC decreases with the reducing SOH in both the Cu/Sn/Cu and Cu/Sn/Ni solder joints. Compared with the Cu/Sn/Cu solder joint, the growth of IMC at Cu side in the Cu/Sn/Ni is markedly suppressed, which is because that part of the dissolved Cu diffuses into the Ni side to form the $(Cu,Ni)_6Sn_5$ and other Cu-Ni-Sn ternary IMC in the Cu/Sn/Ni solder joint, therefore, the formed IMC is lower in the Cu/Sn/Ni than that

978-1-4244-4658-2/09 $25.00 © 2009 IEEE

in Cu/Sn/Cu. For example, when the SOH is 100µm, the average Cu_6Sn_5 thickness is about 4.5µm at the single Cu side in the Cu/Sn/Cu solder joints, while, the average $(Cu,Ni)_6Sn_5$ thickness is about 2.58µm at the Cu side in the Cu/Sn/Ni solder joints with 100µm SOH; however, the difference in the IMC thickness between the Cu/Sn/Cu and Cu/Sn/Ni solder joints is narrowed with the reducing SOH, and when the SOH is lowered to 10µm, the thickness difference can be ignored. The reason would be that the mass amount of dissolved Cu is reduced with the lower SOH, leading to form less IMC in the solder joints and narrowing the IMC thickness difference.

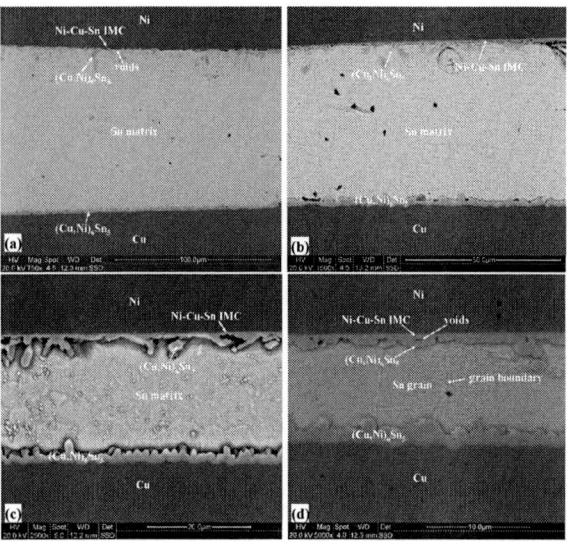

Figure 4a-d Cross-sectional backscattered electron (BSE) images of the Cu/Sn/Ni solder joint with (a) 100µm SOH (b) 50µm SOH (c) 20µm SOH and (d) 10µm SOH.

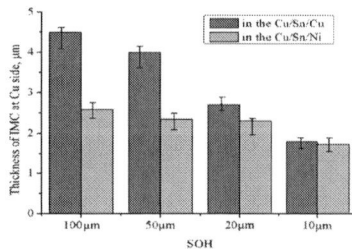

Figure 5 the average IMC thickness on the one Cu side in the Cu/Sn/Cu and Cu/Sn/Ni solder joints with the reducing SOH

By contrast, the IMC on the Ni/bulk interface shows a complex and incompact structure. It could be composed by two IMCs. One layer is a very thin layer adjoining the Ni with the thickness of about 0.4µm and this layer is identified to be Ni (25-35at.%)-Cu (25-30at.%)-Sn (35-50at.%) ternary intermetallics by EDS analysis. The compositions of this ternary IMC are variational in the solder joints, which is difficult to be definitely determined within the resolution of EDS. Since the reaction rate of Ni and molten Sn is about two orders of magnitude slower than that of Cu and molten Sn [8], the formed Ni-Cu-Sn IMC layer is much thinner than the (Cu, $Ni)_6Sn_5$ at Cu side. It has been reported that Ni_3Sn_4 compound

can be formed when pure Ni reacts with Sn [5], and the Ni_3Sn_4 compound can dissolve Cu as much as 8.5 at.%. Once the Cu concentration in the Ni_3Sn_4 compound is greater than threshold value, the Ni_3Sn_4 compound phase would be inclined to transforms into the Cu_6Sn_5 compound [4]. Therefore, it can be deduced that the formed Ni-Cu-Sn compound would be the transient phase from Ni_3Sn_4 to Cu_6Sn_5.

The other IMC is made up of the faceted $(Cu,Ni)_6Sn_5$ intermetallics with a solubility of 4-6at.% Ni. When the concentration of Cu arrival at the Ni/bulk interface was greater than 0.6wt.% which is the maximal solubility of Cu in the molten Sn-Cu-Ni solution [4], the excessive dissolved Cu atoms will precipitate as $(Cu,Ni)_6Sn_5$ IMC on the formed thin ternary Ni-Cu-Sn IMC layer. Similarly, the dissolved Ni in the IMC changes the morphology of Cu_6Sn_5 from the scallop-type into the needle-type. More Ni atoms are involved in the $(Cu,Ni)_6Sn_5$ at the Ni side compared with the $(Cu,Ni)_6Sn_5$ at the Cu side, which is because the concentration of Ni in the molten solder at the Ni side is higher than that at the Cu side. Similar to the change of IMC thickness at the Cu side with the reducing SOH, the maxlength of the formed faceted $(Cu,Ni)_6Sn_5$ is measured statistically, and it is about 8.5µm, 7.0µm, 4.5µm and 2µm in the solder joints with 100µm, 50µm, 20µm and 10µm SOH, respectively. The reason for the decreasing trend is also due to the decreasing amount of the dissolved Cu in the bulk with the reducing SOH.

The point is that the Ni-Cu-Sn IMC and $(Cu,Ni)_6Sn_5$ IMC are connected incompactly, and voids can be observed between them. Fig. 6 shows the incompact interconnection between the two IMC layers in the as-soldered Cu/Sn/Ni solder joint with 100µm SOH. This incompact interconnection is very detrimental to the mechanical property of solder joints due to the brittle nature of IMC, which is propitious for crack propagation. This will be proved by the following tensile test.

Additionally, similar to the Cu/Sn/Cu solder joint with 10µm SOH, when SOH is reduced to 10µm, the bulk of Cu/Sn/Ni solder joint also contains only one grain in height, like shown in Fig. 4d. This microstructural change would have significant influence on the mechanical property of solder joint.

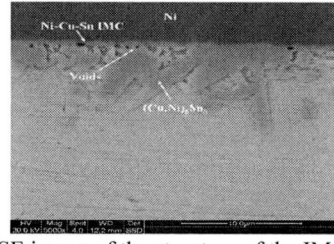

Figure 6 BSE image of the structure of the IMC layer at Ni side in the Cu/Sn/Ni solder joint with 100µm SOH

3.3 Effect of SOH on the compositions of the bulk in the Cu/Sn/Cu and Cu/Sn/Ni solder joints

Figure 7 shows the change of mean concentration of base materials in the solder bulk with the reducing SOH in the Cu/Sn/Ni and Cu/Sn/Cu solder joints. According to results of EDS analysis on the whole solder bulk area, in the Cu/Sn/Cu solder joint, the mean Cu concentration in the solder bulk increases with the reducing SOH, and it is about 1.6wt.% (2.9at%), 2.8wt.% (5.1at%), 4.0wt.% (7.3at%), 4.6wt.% (8.3at%) in the bulk of the solder joints with 100μm, 50μm, 20μm and 10μm SOH respectively. The dissolved Cu in the bulk exists in the form of solid solution and Cu-rich particles. For example, a thin layer of Cu-rich phase can be observed along the scalloped Cu_6Sn_5 IMC, like shown in Figure 2c. Meanwhile, Cu-rich phase particles also can be observed in the solder bulk, as shown in Figure 2b-c, and they mainly distribute at the grain boundaries. EDS results show that the composition of the Cu-rich phase is about 92wt.% Sn-8wt.%Cu in the solder joint with 20μm SOH, compared with the composition of the matrix grain 97wt.%Sn-3wt.%Cu. The Cu-rich phase is the result of rapid cooling during reflow process and it is in an unstable state. It will finally transform into Cu_6Sn_5 IMC after thermal aging treatment.

Similarly, in the Cu/Sn/Ni solder joints, the mean concentrations of both Cu and Ni in the solder bulk also increase with the reducing SOH. The Cu concentration is about 1.4wt.% (2.53at.%), 1.65wt.% (3.01at.%), 2.32wt.% (4.19at.%) and 2.84wt.% (5.09at.%) in the solder joint with 100μm, 50μm, 20μm and 10μm SOH respectively, which is lower than the Cu concentration in the Cu/Sn/Cu solder joints with the corresponding SOH. Ni concentration is about 1.08wt.% (2.05at.%), 1.15wt.% (2.12at.%), 1.42wt.% (2.78at.%) and 1.75wt.% (3.39at.%) in the solder joint with 100μm, 50μm, 20μm and 10μm SOH respectively. The mean Ni concentration in the solder bulk is much lower than the Cu concentration in the solder bulk since the slower dissolving rate of Ni in the wetting reaction [9].

From Fig. 7, it can be concluded that the base materials concentrations increase with the reducing SOH. The reasons for the concentration increase are presented as follows. The solder volume will become smaller when the SOH reduces from 100μm to 50μm, 20μm and 10μm. Given the constant dissolving rate of base materials, the molten Sn solder can be saturated earlier in the solder joints with lower SOH than the solder joints with higher SOH, leading to the higher base materials concentrations in the solder joints with lower SOH; however, the gross amount of the dissolved base materials is smaller. In the cooling period of reflow process, the dissolved base materials will react with Sn to form IMC, so, the formed IMC is thinner in the solder joints with lower SOH.

3.4 Effect of SOH on the fracture mode of the solder joints
3.4.1 Effect of SOH on the fracture mode of the Cu/Sn/Cu solder joints

Figure 8a-d shows the SEM tensile fracture surface images of solder joints with 100μm, 50μm, 20μm and 10μm SOH respectively. From Figure 8a-c, equiaxed dimples can be observed on the fracture surface of the solder joints with 100μm, 50μm and 20μm SOH. And with the reducing SOH, the dimples become shallower and smaller. Moreover, the hillock-like Cu_6Sn_5 IMC can be observed at the bottom of the dimples on the fracture surface of the solder joint with 20μm SOH. According to Fig. 8a-c, the fractures occur within the solder bulk, showing a typical ductile characteristic for the solder joints with 100μm, 50μm and 20μm SOH. However, when the SOH reduces to10μm, fracture displays a remarkable difference. From Figure 8d, the cleavage fracture

surface of the scallop-type IMC indicates a brittle failure in the IMC layer. In Figure 8d, the fractured IMC delaminating from the opposite Cu bar can be observed clearly, and it adheres to the solder. While, the remaining cleaved IMC adhering to the copper presents a lotus shape at the bottom of dimples. Accordingly, Figure 8(a-d) indicate that a transition from a fracture in bulk to the fracture in IMC occurs when the SOH reduces from 100μm, 50μm and 20μm to 10μm, meanwhile, the fracture mode also transfers from ductile to brittle.

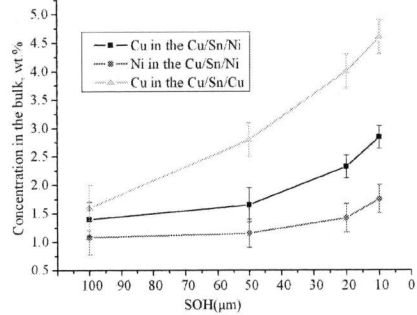

Figure 7 The mean concentration of base materials in the solder bulk with the reducing SOH

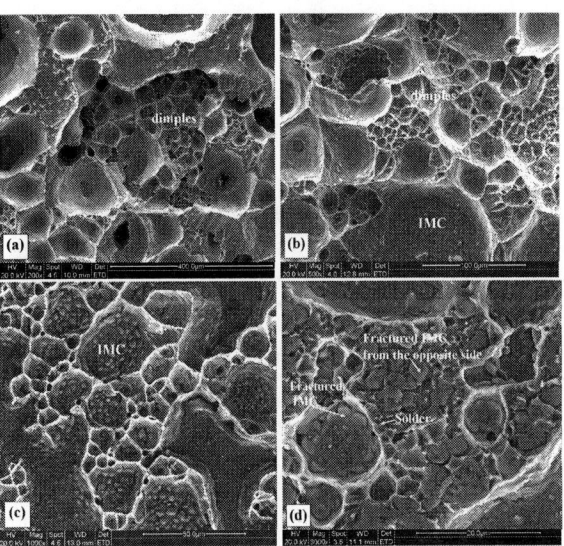

Figure 8a-d SEM tensile fracture surface images of solder joints with SOH of (a) 100μm, (b) 50μm, (c) 20μm and (d) 10μm

3.4.2 Effect of SOH on the fracture mode of the Cu/Sn/Ni solder joints

Figure 9a-d show the SEM tensile fracture surface images on the Cu bar of the Cu/Sn/Ni solder joints with 100μm, 50μm, 20μm and 10μm SOH. From Figure 9a-d, IMC layer delaminating from Ni side can be observed, adhering to the solder bulk. According to the EDX analysis, the IMC is identified to be $(Cu,Ni)_6Sn_5$. Accordingly, fractures mainly occur in the IMC layer at Ni side. Interestingly, this is very different from the fracture in the Cu/Sn/Cu solder joints. Due to the incompact connection between $(Cu,Ni)_6Sn_5$ and Ni-Cu-Sn IMC, cracks are inclined to propagate along the incompact connection interface. Therefore, the interface between $(Cu,Ni)_6Sn_5$ and Ni-Cu-Sn IMC becomes the weakest part in the solder joints. And the fractures show a brittle characteristic under tensile stress due to the brittle nature of IMC.

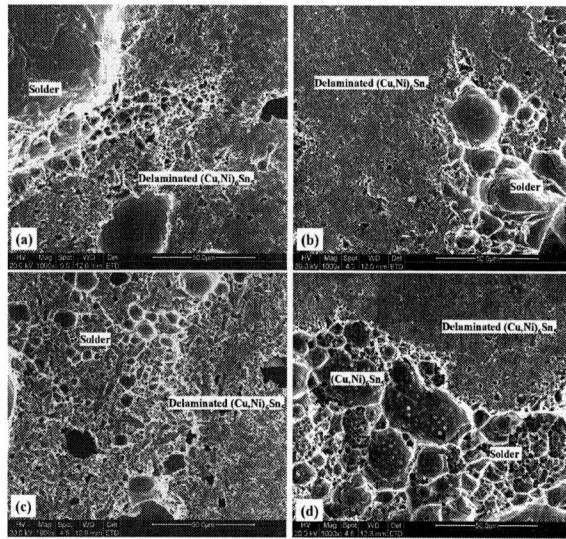

Figure 9a-d SEM tensile fracture surface images on the Cu bar of the Cu/Sn/Ni solder joints with SOH of (a) 100μm, (b) 50μm, (c) 20μm and (d) 10μm

3.5 Effect of SOH on the ultimate tensile strength of the Cu/Sn/Cu and Cu/Sn/Ni solder joints

Figure 10 shows the ultimate tensile strength of the Cu/Sn/Cu and Cu/Sn/Ni solder joints with 100μm, 50μm, 20μm and 10μm SOH. According to the Figure 10, the UTS of the Cu/Sn/Cu solder joints increases with the reducing SOH, especially, when the SOH is 10μm, the UTS has a dramatic increase due to the fracture in the IMC layer. This is different from the Cu/Sn/Ni solder joints. The UTS of Cu/Sn/Ni also increases with the SOH reducing from 100μm to 20μm; however, it experiences a dramatic decrease when the SOH is 10μm. Therefore, it can be concluded that when the SOH is lowered to 10μm, both the UTS of the Cu/Sn/Cu solder joint and the UTS of the Cu/Sn/Ni solder joints experience a dramatic change. And these changes correlate closely with the microstructural changes that one grain in height when SOH is 10μm. Moreover, as the SOH reduces from 100μm to 20μm, the UTS of Cu/Sn/Ni solder joints are higher than that of Cu/Sn/Cu solder joints. By contrast, when the SOH is 10μm, the UTS of Cu/Sn/Ni solder joint decreases dramatically to be lower than that of the Cu/Sn/Cu solder joint with 10μm SOH.

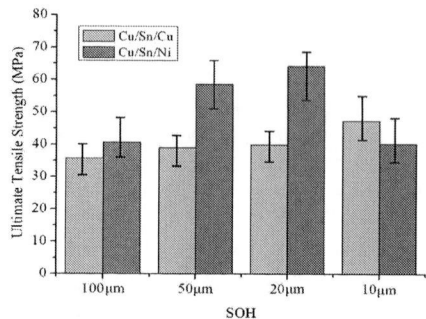

Figure 10 the ultimate tensile strength of the Cu/Sn/Cu and Cu/Sn/Ni solder joints with 100μm, 50μm, 20μm and 10μm SOH

4. Conclusion

With the miniaturization of the solder joints, both the microstructure and the mechanical property of solder joints experience significant changes.

In the Cu/Sn/Cu solder joints, the thickness of the formed IMC decreases with the miniaturization of solder joint, while, the IMC proportion to the solder joint increases. Meanwhile, the concentration of dissolved Cu in the bulk increases with the reducing SOH. When the SOH is reduced to 10μm, the bulk contains only one grain in height, which would correlate closely with the dramatic UTS increase and fracture mode transition from ductile to brittle under tensile test.

In the Cu/Sn/Ni solder joints, due to the diffusion of Ni towards the Cu in the reflow process, Ni atoms participate in the reaction of Cu and Sn to form the $(Cu,Ni)_6Sn_5$ intermetallic, which is different from the Cu_6Sn_5 IMC in the Cu/Sn/Cu solder joint. Similarly, because of the Cu diffusion towards Ni, Cu also influences the interfacial reaction at Ni side, and the IMC layers with incompact connection are formed. And this incompact connection interface between the two IMC layers is the weakest part in the solder joints. Under tensile stress, brittle fracture occurs between the two IMC layers. With the SOH lowered to 10μm, UTS experiences a dramatic decrease, which would be related to the one grain height of the solder bulk.

Compared with Cu/Sn/Cu solder joint, Cu/Sn/Ni solder joints have higher UTS when the SOH reduces from 100μm to 50μm and 20μm; however, a dramatic change occurs when the SOH is lowered to 10μm, and the UTS of Cu/Sn/Ni solder joint decreases to be lower than that of Cu/Sn/Cu solder joint with 10μm SOH.

Acknowledgments

The authors acknowledge the financial support from National Natural Science Foundation of China (No. 60776033) and the State Key Laboratory of Advanced Welding Production Technology (welding 09013). The authors also acknowledge the Analytical and Testing Center of HUST for their analytical work.

References

1. Dang B., Wright S. L., Andry P. S., *et al.*, "3D chip stacking with C4 technology", *IBM J. Res. Dev.* Vol. 52, (2008), pp. 599-609.

2. Plieninger R., Dittes M., Pressel K., "Modern IC Packaging Trends and their Reliability Implications", *Microelectronics Reliability* Vol. 46 (2006), pp. 1868-1873.

3. H. Huebner, *et al.*, "Microcontacts with sub-30 μm pitch for 3D chip-on-chip integration", *Microelectronic Engineering* Vol. 83 (2006), pp. 2155-2162.

4. S. J. Wang, C. Y. Liu, "Study of interaction between Cu-Sn and Ni-Sn interfacial reactions by Ni-Sn3.5Ag-Cu sandwich structure", *J.Electron. Mater.* Vol. 32 (2003), pp. 1303-1309.

5. T. Laurila, V. Vuorinen, J. K. Kivilahti, "Interfacial reactions between lead-free solders and common base materials", *Mater. Sci. Eng. R.* Vol. 49 (2005), pp. 1-60.

6. S. J. Wang, C. Y. Liu, "Asymmetrical solder microstructure in Ni/Sn/Cu solder joint", *Scripta Mater.* Vol. 55 (2006), pp. 347-350.

7. K. H. Prakash, T. Sritharan, "Tensile fracture of tin-lead solder joints in copper", *Mater. Sci. Eng. A* Vol. 379 (2004), pp. 277-285.

8. K. N. Tu, K. Zeng, "Tin-lead (SnPb) solder reaction in flip chip technology," *Mater. Sci. Eng. R* Vol. 34 (2001), pp. 1-58.

9. K. Zeng, K. N. Tu, "Six cases of reliability study of Pb-free solder joints in electronic packaging technology," *Mater. Sci. Eng. R.* Vol. 38 (2002), pp. 55-105.

Numerical Simulation on Variable Width Multi-Channels Heat Sinks with Non-uniform Heat Source

Xiaojing Wang, Wen Zhang, Hongjun Liu, Ling Chen, Zongshuo Li
Shanghai University
Shanghai University, 224mail box, 149 Yan Chang RD. Shanghai, 20072, China
xjwang@mail.shu.edu.cn, 86-21-66136117

Abstract

The micro-channel heat sink (MCHS) is almost using a separate production of silicon or copper MCHS, which is indirectly on package dimensions. Using this package structure of the heat sink, the temperature of its central region is much higher than the surrounding region. It makes the surface of the hot load non-uniformly. At present, most of the studies have adopted the uniform thermal load, regardless of the way through trial or through the means of simulation. Traditional micro-channel has effects on treating the uniform thermal load, however, its structure needs to be further improved in order to reduce the hot concentrated caused by the non-uniformity. Few papers have done the research about the non-uniform heat source distributions. In this paper, the non-uniform heat source distributions are studied in the micro-channel heat sink (MCHS) cooler. The simulation model is established to analyze the temperature and pressure distributing of the MCHS with different channel width-dimensions. Water is chosen as the coolant for its superior hot properties and the velocity range is from 0.01m/s to 10m/s. With the simulation of the computation fluid dynamics software FLUENT, results show that the non-equal displacement of fins can effectively decrease the temperature rise under the same conditions cooling a non-uniform heat source.

1 Introduction

As the integrated level of electronic device increases continuously, the heat management of the high density power devices such as diode laser array CPU has become a key technology. Technology of micro channel heat sinks has been concerned increasingly by the experts in the field at home and abroad. The MCHS is almost using a separate production of silicon or copper MCHS, that is, the current cooler not directly effect on the chip, but indirectly on package dimensions, for the restrictions on the processing technology or the sealing problem that cannot be solved. The thermal concentration caused by the package structure makes the surface of the hot load non-uniformly. Because of the effect of the non-uniform heat source, the heat transfer capability drops. Therefore, the development of the performance of the micro channel heat sinks under the non-uniform heat source is very important.

Tuckerman et al [1] first demonstrated the configuration of micro cooler with many micro channels inside can remove 790W/cm^2 with forced convective water at a substrate-to-coolant temperature difference of 71K. Fig. 1 shows the schematic configuration of the MCHS. Following the work of Tuckerman and Pease, many research works have been conducted for microchannel heat sinks, as reviewed by Phillips [2]. In most of these studies, the fin approach is employed. As well known, the fin approach is an effective tool to analyze the transport of heat in a number of practical applications, and has been used recently to study the heat transfer properties of open-celled metal foam and honeycomb heat sinks [3-5].

Mudawar and Bowers [6] employed a stainless steel micro-tube in their heat transfer experiments. The microtube had an inner diameter of 902μm, wall thickness of 89μm and length of 5.8mm. They demonstrated heat fluxes as high as 3000W/cm^2 for single-phase water flow. Comprehensive reviews of the different heat transfer techniques employed in electronic cooling were provided by Yeh [7] and Mudawar [8].

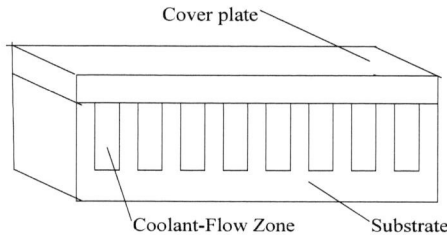

Fig. 1 Schematic configuration of micro-channel heat sink system

In this paper, new types of micro-channel heat sinks are established, and loaded with non-uniform heat source. The study mainly aims at the parameter of hydraulic diameter, with the premise of the same total flow, changes the width of micro-channel heat sink hoping to increase as the region of the coolant flow velocity, reduce heat unevenly and thermal stress. With the simulation of the computational fluid dynamics software FLUENT, the models of the multi-lines heat sinks with variable width channels are established. The temperature distribution and pressure distribution are studied under the non-uniform heat source with different inlet velocity.

2 Theoretical analyze

2.1 Model specification

Nguyen and Mochizuki [9] have tested the micro-channel on the hydraulic diameter of 540μm and 2.54mm, and conclusions showed that these dimensions of coolers had higher cooling capacity than the limit cooling capacity of air forced, and can meet the demands of heat for the next generation of chips. At the same time, the pressure dropped smaller at the both ends of the micro-channel with 2.54mm hydraulic diameter. According to these conclusions, macro-channel was not very different from the micro-channel in the working principle, except that on cooling capability.

Therefore, the macro-channel will be taken to study the performance of micro-channel under the non-uniform thermal load. Table 1 shows the parameter of the model.

Table 1 Dimensions of micro-channel model

Structure Parameters	Dimensions
Model dimension (mm)	30×30×7
District heating region (mm)	10×10(Center)
Micro-channel (mm)	26×2×5
Fins (mm)	26×2×5
Channel Number	7

There are rare reports about the theory of non-equivalent width micro-channel. Therefore, it cannot design the dimensions based on a mature theory, but only suppose the width changing of the micro-channel and verify the feasibility of this approach. Table 2 shows the size of three types of micro-channel heat sink. The equivalent width of micro-channel heat sink 1 is 2mm, and the width of heat sink 2 is 1mm. Variable width channel model consists of 1mm and 2mm width channels, and each dimension has four channels. All of these models (Fig. 2 & Fig. 3) will be used to simulate a concentrated thermal load on central region.

Table 2 Dimensions of three MCHS models

Models	Total width (mm)	Width (mm)×Number	Number of chips
Equivalent width channel 1	30	2×7	8
Equivalent width channel 2	30	1×10	11
Variable width channel	30	1×6 (narrow) 2×4 (wide)	11

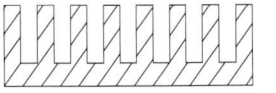 

(a)   (b)

Fig. 2 Section view of equivalent width micro-channel heat sink. (a) Heat sink 1. (b) Heat sink 2.

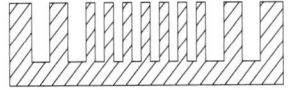

Fig. 3 Section view of variable width micro-channel heat sink

2.2 Boundary conditions

The uniform velocity and temperature are applied in the inlet of the micro-channel. The total volume flow rate can be determined by the total number of micro-channels used in a micro-channel heat sink module and the area and velocity of individual micro-channel. The constant pressure of 1atm is applied in the exit of the micro-channel. The heat flux of $270W/cm^2$ is applied at the bottom of the heat sink. Heat is assumed to be removed only by working fluid. For the uniform heat source simulation the heat flux is applied at the bottom (3cm×3cm) of the heat sink average, and for the non-uniform one the heat flux concentrate on the center zone (1cm×1cm) of the bottom (Fig. 4). The models of the multi-channel heat sink under the non-uniform heat source are established.

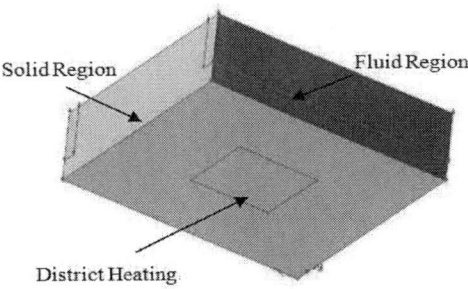

Fig. 4 Model of non-uniform heat source

3 Simulation Results

3.1 Temperature comparison with non-uniform heat source

The maximum temperatures of the surfaces of three kinds coolers are compared under different inlet velocity, and results are shown in Table 3.

Table 3 Maximum surface temperature-rise of three different micro-channel structures (N/A: the temperature is higher than the boiling point)

Inlet velocity (m/s)	Maximum surface temperature (K)		
	Heat sink 1	Variable width channel	Heat sink 2
0.02	N/A	N/A	N/A
0.03	N/A	370	369
0.04	N/A	365	359
0.05	367	355	352
0.1	361	348	349
0.5	356	344	346
1	355	343	345
5	355	340	345
10	355	331	345

From Table 3, the results show that the variable width channels heat sink is better than two equivalent width channel heat sink on the capability of the maximum temperature control. When the flow velocity is below a certain value, the temperature of surface center of the equivalent width channel heat sink has been above the boiling point of water, and the coolant may boil away. So the numerical computation result cannot reflect its true value. The critical flow velocity of boiling occurrence is 0.02m/s for the variable width channel heat sink and equivalent width channel heat sink 2.

Fig. 5 is the surface temperature distribution of three different models with 1m/s inlet velocity, and Fig. 6 is the total temperature distribution of three models. After the increase of flow velocity, the edge temperature of equivalent width channel heat sink 1 drops from 318K to 314K, which equals to the central temperature, so does the equivalent width channel heat sink 2. However, the edge temperature of variable width channel heat sink remains at about 317K, while the central temperature drops significantly from 348K to 343K. When the velocity is up to 10m/s, the surface temperatures of heat sink 1and heat sink 2 are no longer remained at 355K and 345K. While, the highest central temperature of variable width channel heat sink further drops to 331K. Above all, using variable width channel heat sink can delay the limit point of heat elimination, and also reduce the thermal stress coursed by temperature difference of cooling surface.

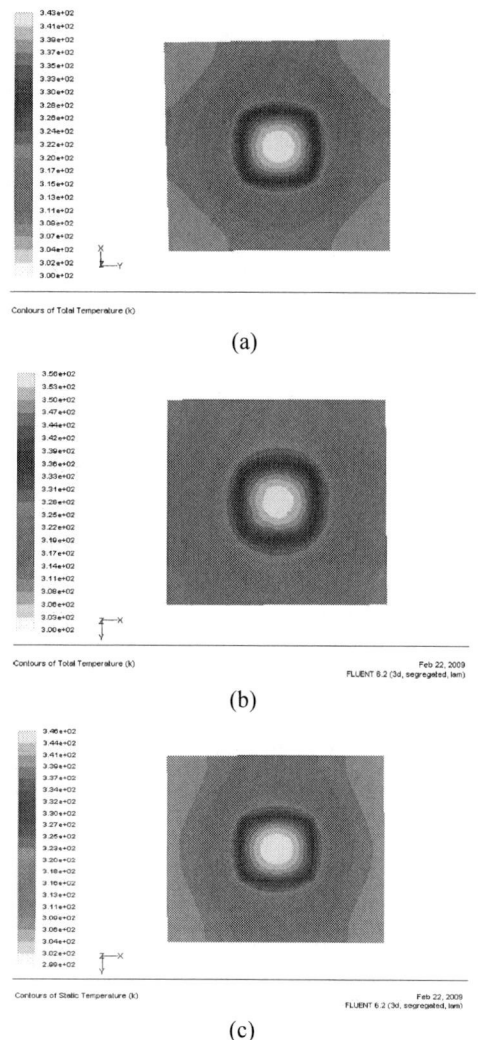

(a)

(b)

(c)

Fig. 5 Surface temperature distribution of three different models with 1m/s inlet velocity. (a) Variable width channel heat sink. (b) Equivalent width channel heat sink 1. (c) Equivalent width channel heat sink 2.

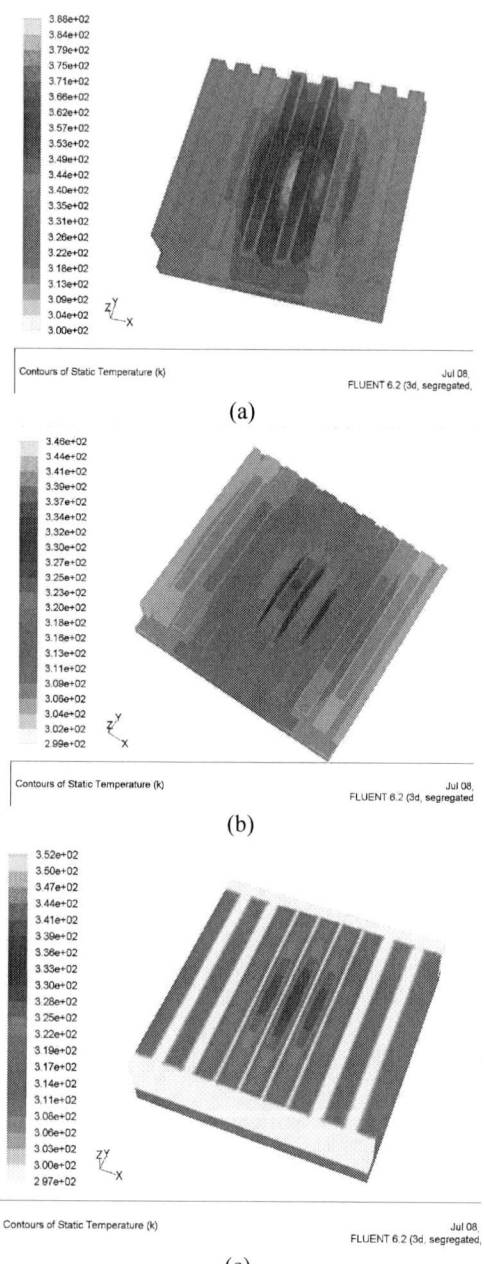

(a)

(b)

(c)

Fig. 6 Total temperature distribution of three models with 1m/s inlet velocity. (a) Variable width channel heat sink. (b) Equivalent width channel heat sink 1. (c) Equivalent width channel heat sink 2.

When inlet velocity is from 0.1m/s to 1m/s, the temperatures of heating surface are falling slowly both the equivalent width channel heat sink 1 and 2, but the pressures begin to rise. However, the maximum surface temperatures of variable width channel heat sink reduce with the increasing of inlet velocity. When the inlet velocity rises to 10m/s, the highest temperature is 331K.

3.2 Pressure drop comparison with non-uniform heat source

The original intention of variable width channel design is to strengthen the flow of coolant of central region. Fig. 7 is the inlet pressure distribution graph of variable width channel heat sink with 0.5m/s inlet velocity. It can be seen from the figure that the pressure of the central region is higher 0.3bar than the edge. The pressure at the edge of four channels is less than the central region, though it changes with different flow velocity. Compare with the pressure distribution of traditional micro-channel, which is nearly uniform distribution. Simulation results show that reducing the width of the channel is conductive to strengthen the flow in these channels.

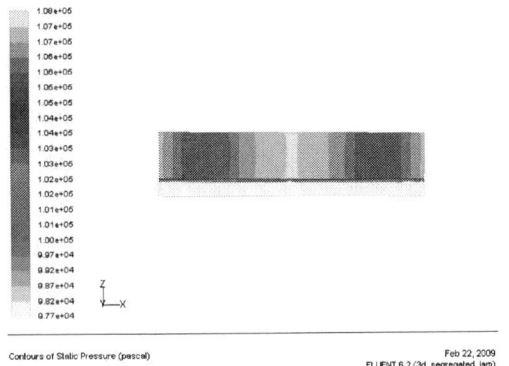

Fig. 7 Pressure drop distribution graph of variable width channel

Comparing with the inlet pressure of three different models with different flow velocity, the relevant results are shown in Table 4. The inlet pressures of three models are 1.01bar with 0.1m/s inlet velocity and below 0.1m/s. The pressure distribution of whole flow field is basically no changing with inlet velocity below 0.1m/s, which is the same as the outlet pressure. So that, in this process the development of the flow field is not fully. However, with the increasing velocity, the highest pressures of three models grow fast. The increasing of inlet pressure of variable width channel heat sink is in the middle of two equivalent width channel heat sinks.

Table 4 Inlet pressures drop of three models with different inlet velocity

Inlet velocity (m/s)	Pressure drop(bar)		
	Heat sink 1	Variable width channel	Heat sink 2
0.1	1.01	1.01	1.01
0.5	1.03	1.03	1.04
1	1.07	1.08	1.10
5	2.23	2.55	2.90
10	6.22	7.48	8.36

By contrast, it is not difficult to find that changing the width of micro-channel can effectively delay the emergence of cooling limit, meanwhile, the pressure of both ends of the channels will not increase obviously.

Conclusions

Traditional equivalent width micro-channel has weakness on treating the non-uniform thermal load, because of the appearance of heat point. When decreasing the width of fluid channel, the highest surface temperature of heat sink falls, however, the pressure drop rises.

In the premise of the same length, the variable width micro-channel heat sink model is designed to reduce the hot concentrated caused by the non-uniformity. The special structure of intensive intermediate channels has effects on reducing the temperature-rise. Meanwhile the pressure drop has insignificant difference with the traditional structure. The research shows that variable width micro-channel heat sink is the effective means of resolving thermal concentration problem.

Acknowledgments

The authors acknowledge the financial support from 863 program (No.2008AA04Z301), NSFC project (No.50876057) and Shanghai Municipal Education Commission (No.08YZ15).

References

1. D. B. Tuckerman, R. F. W. Peaser, "High performance heat sinking for VLSI," *IEEE Electron Device Letters*, Vol. Edl-2, No. 5 (1981), pp. 126-129.
2. R. J. Philips, "Micro-channel Heat Sinks, in: A. Bar-Cohen, A.D. Kraus (Eds.), Advances in Thermal Modelling of Electronic Components and Systems," *ASME*, Vol. 2, (1990), pp. 346-351.
3. T. J. Lu, H. A. Stone, M. F. Ashby, "Heat Transfer in Open-cell Cetal Foams," *Acta Mater*, Vol. 46, (1998), pp. 3619-3635.
4. S. Gu, T. J. Lu, A. G. Evans, "On the Design of Two-dimensional Cellular Metals for Combined Heat Dissipation and Structural Load Capacity," *Heat Mass Transfer*, Vol. 44, (2001), pp. 2136-2175.
5. T. J. Lu, "Heat Transfer Efficiency of Metal Honeycombs," *Heat Mass Transfer*, Vol. 42(1999), pp. 2031-2040.
6. I. Mudawar, M. B. Bowers, "Ultra-high Critical Heat Flux (CHF) for Subcooled Water Flow Boiling--I: CHF Data and Parametric Effects for Small Diameter Tubes," *Heat Mass Transfer*, Vol. 42, (1999), pp. 1405-1428.
7. L. T. Yeh, "Review of Heat Transfer Technologies in Electronic Equipment," *ASME J. Electron. Package*, Vol. 11, No. 7 (1995), pp. 333-339.
8. Mudawar, "Assessment of High-heat-flux Thermal Management Schemes," Proc 7[th]Intersociety Conference on Thermal and Thermomechanical Phenomena in Electronic Systems, 2000, pp. 1-20.
9. T. Nguyen, M. Mochizuki, "Advanced Cooling System Using Miniature Heat Pipes in Mobile PC," *Proceedings of 6[th]Intersociety Conference on Thermal and Thermomechanical Phenomena in Electronic Systems*, March 1998, pp. 507-511.

Characteristic Analysis of Transducer Drive Current in Ultrasonic Wire Bonding Process

Shaohua Liu, Fuliang Wang
Key Laboratory of Modern Complex Equipment Design and Extreme Manufacturing, Ministry of Education,
Central South University, Changsha 410083, China
liushaohua291@163.com, Tel: 013787105893

Abstract

In the ultrasonic wire bonding process, the impedance change of transducer system was reflected by the current amplitude, the change of the system's resonance frequency was represented by the change of the current's frequency. The current signals contained much information about the bonding process and the resulted bond quality. The characteristics of current were depending on the tip states. Free, fixed, or damped tips led to different characteristics of current in some way. The bond formation process and the related physical process can be reflected by the current characteristics during bonding process. In this paper the current monitoring hardware system was set up for ultrasonic wire bonding, and the relevant software was designed. The reliable current signals had been acquired by data acquisition system and the bonding quality data was gathered by measuring the bonding shearing force. The current signals were analyzed by using the wavelet time-frequency method. Several characteristics were selected. The analysis results indicated that the frequency, the amplitude value and the change tendency of current signals are related to the bonding quality in certain extent. The fundamental resonance frequency deviation, very small current, and exceptional current changing tendency always mean failed bond.

1 Introduction

It is well known that the overall lifetime and reliability of electronic systems strongly depends on the quality of electrical interconnections which are so far provided widely by the ultrasonic wire bonding in the microelectronics industry. Recently, the bonding quality evaluation is mainly realized by measuring destructively pulling and shearing force offline. This method not only increases the workload largely, but also is inefficient. Developing on-line methods for monitoring bonding process and evaluating bond quality has been a continual effort in the past decades. Published and patented methods are based on various working principles. However, no on-line method or technique is universal and applied as yet. Characteristic analysis of transducer drive current was achieved in this paper which could be used to develop the real-time quality monitoring method of ultrasonic wire bonding process.

In the past decades some of these studies are as following. In 1998, S.W.Or et.al. measured the mechanical vibrations of the transducer which is embedded a piezoelectric sensor, and the bond quality was evaluated based on these signals [1-2]. Similarly, Michael et.al. [3]. indicated that bond quality monitoring based on the signals of an additional sensor integrated into the transducer allows a much better discrimination between "good" and "bad" bonds, especially for the case of bonds on contaminated surfaces. These methods could detect the bond failure, but the embedded

sensor influenced the properties of the transducer badly, so these methods were not applicable.

And some other methods such as measurement of wire-deformation [4-6], measurement of the wedge vibrations [7-10] and so on, need increase costly measuring equipments whose values are several times of the bond machine, so are not possible to apply into the industry.

Developing on-line methods for bond quality monitoring by measuring the drive signals of transducer has received the favor of whole field because of not influencing the transducer's properties and not needing more investments for hardware. This method was based on the cognition to the bond mechanism. According to the understanding of bond mechanism currently, the ultrasonic bond process was composed of three stages: Firstly, the wire moves to contact the bond pad under the bond pressure and the contamination and oxide layer covering the interface is removed. Secondly, as the contamination and oxide layer is broken down or removed, the underlying fresh metal surfaces are brought into contact under the bond force and diffuse. Finally interatomic forces will come into operation and atomic bonding will occur and the final bond strength forms [1, 11-12]. In this process, the impedance and natural frequency of transducer system varies as the bond surface conditions change, which causes the variety of the drive current and voltage of transducer and the tool vibrations, and the variety is different if the bond strength varies. Therefore, bond quality on-line monitoring would be realized as long as the characteristics which represent the strength forming are found.

In this paper, a data acquisition system was designed and lots of data was gathered for the research. The current signals were analyzed by using the wavelet time-frequency method, and the relations between current and bond strength were revealed, which is hoped to be applied in the on-line monitoring system.

2 Data acquisition and experiments

The ultrasonic wire bonding system mainly consists of the ultrasonic generator, the piezoelectric transducer, the coupler, as well as the wedge tool and so on.

The experiments were carried out on the large aluminum wire bonder U3000 produced by the WESEL Company in Shenzhen. Because of the similarity of ultrasonic bond, the same methods can be applied in thermosonic wire bonding and thermosonic flip-chip bonding.

The signals, including driver voltage, current, tool vibration and bonding force during bonding process, were acquired, and saved synchronously with data acquisition card PCI6110, which programmed by LabView8.2 and the bonding quality data was gathered by measuring the bonding shearing force using Dage-Series-4000. The pressure sensor M200B01 produced by the PCB Company was adopted to measure the bonding pressure P. The data of capillary vibration was recorded by Doppler Laser Vibrometer PSV-400-M2 and transferred into PCI6110.

Fig. 1 is the signals sensing circuit, where U1, U2, U3, U4 are the measured signals of current, voltage, bond pressure and capillary vibration of the first bond respectively. The relations between the measured signals and the real signals are as following:

Current: I=U1/9 (A)

Voltage: U=U2 (V)

Bond Pressure: P=U3×8.4 (N)

Capillary Vibration: V=U4 (m/s)

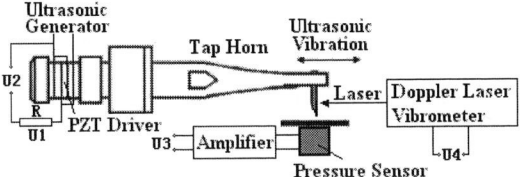

Fig. 1 Signal acquisition connection sketch map

The front panel of data acquisition soft likes Fig. 2. The soft is composed of front panel and block diagram. The channel parameters, timing parameters and trigger parameters are set in the front panel and can be adjusted manually according to the specific bond conditions. Besides the data acquisition, the acquired data can be saved in the specified files and shown in the front panel, the user's operations can be responded, exceptional errors can be processed and the frequency results can be shown in the soft front panel.

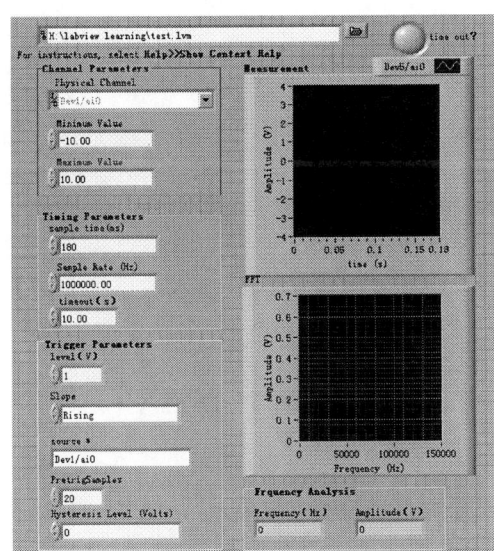

Fig. 2 Front panel of LabView data acquisition

The experiments were carried out on the large aluminum wire bonder U3000 with parameters as following: the operating frequency is 57 kHz, bonding time is 10-500ms, the available aluminum wire's diameter is 75-500μm, and the attainable bonding pressure is 30-1200g. The aluminum wire

diameter in the experiment is 300μm, the capillary model is LW300, work holder uses the aluminum with a nickel surface (the size is 40×50×1mm). The bond parameters were set as follows: bond force is 4 scale (about 14.1N), bond time is 3 scale, bond power is 4 scale(about 0.8W), sample rate is 1MHz, sample time 240ms, trigger source is voltage channel, rising slope trigger, pretrigger samples is 5000, trigger level is 1V. Running the program before the bond starting, the bond signals would be acquired. Run the program for each bond process. The typical current signal likes Fig. 3.

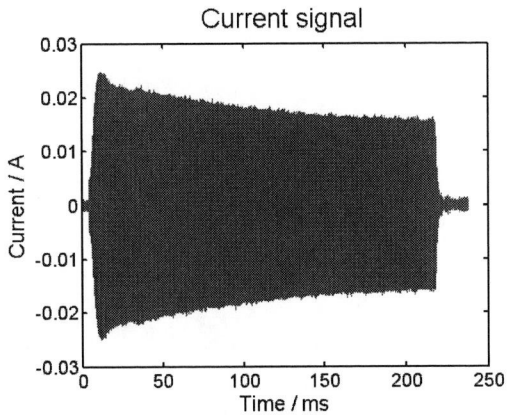

Fig. 3 A typical current signal

3 Wavelet analysis and characteristic extraction

Lots of data had been acquired using the above data acquisition system. These signals would contain the characteristics which represented the bond quality. An on-line monitoring method could be found if these characteristics could be picked up.

The current signals contained much frequency information, also contained the signals related to the resonance frequency which played a significant role in the bond process. The resonance frequency would be analyzed because it contained most energy of the original signal. The current signals were analyzed first using Fast Fourier Transformation, shown as Fig. 4.

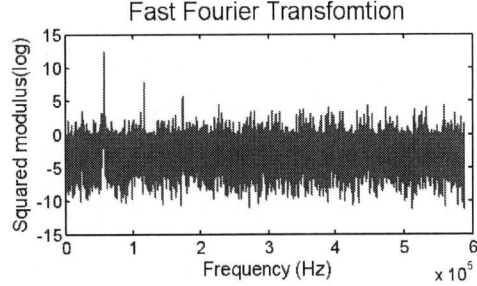

Fig. 4 Energy spectrum of the current signal in Fig. 3

The natural frequency of the transducer was 57kHz, so the fundamental resonance frequency of current also was about 57kHz and other harmonic frequencies were approximately integral multiple of 57kHz. In order to acquire the related characteristics of harmonics, the current signals were analyzed and processed utilizing orthogonal wavelet packet transformation adopting wavelet *db40*. The frequency bands containing the harmonics were detected by picking up the 8th layer signals after transformation, shown in Fig.5. Then the effective values of all harmonic bands were calculated using formula (1).

$$D(i) = \sqrt{\frac{1}{200} \sum_{n=i*200}^{(i+1)*200} S(n)^2} \qquad i = 0,1,\cdots\cdots length(S)/200 \tag{1}$$

Where S is the signal, D is the effective value.

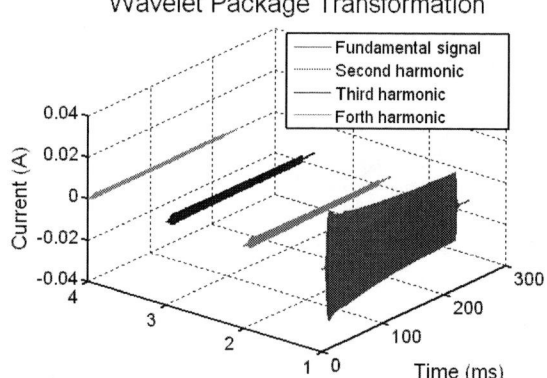

Fig. 5 The current harmonics by wavelet package transformation

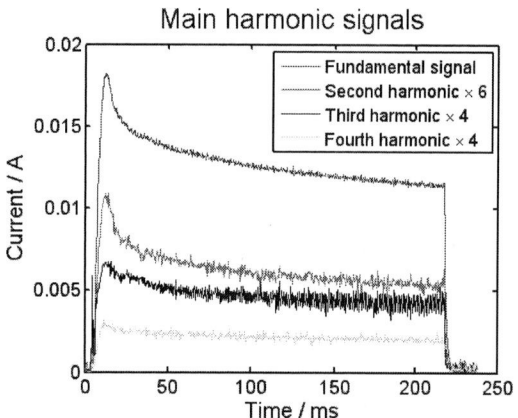

Fig. 6 Effective value curves of main harmonic signals

The effective value curves are shown in Fig. 6. Lots of data indicated that the amplitudes and the peak values of different signals were different. The signals of round 20ms of resonance frequency varied and were all different. Based on the above observation, we chose characteristics as following:

1) Fundamental, second, third, forth harmonic band signals' variance and average of time 8.5-34ms.
2) Fundamental resonance frequency band signal's average, peak value, peak value's time, average of time 85-102ms and time 187.1-204ms and their difference.
3) Fundamental resonance frequency f.

After analysis, some relations between the above characteristics and the bond quality are as following:

I Fundamental resonance frequency band signal's average *avereng*

This feature was the average of the effective value of fundamental resonance frequency band signal, which was calculated by formula (2).

$$avereng = \frac{1}{1150} \sum_{i=51}^{1200} D_1(i) \tag{2}$$

Where D1 is the fundamental resonance frequency band signals' effective value, shown in Fig. 7. If *avereng* is very small, the bond would fail. The experiments indicated that the threshold would be different if the bond conditions changed.

The bond process was a consuming energy process. If the current was small, the power which the bond consumed was not enough which led to the contamination and oxide layer removing failure or the atom diffusing failure, then led to bond failure.

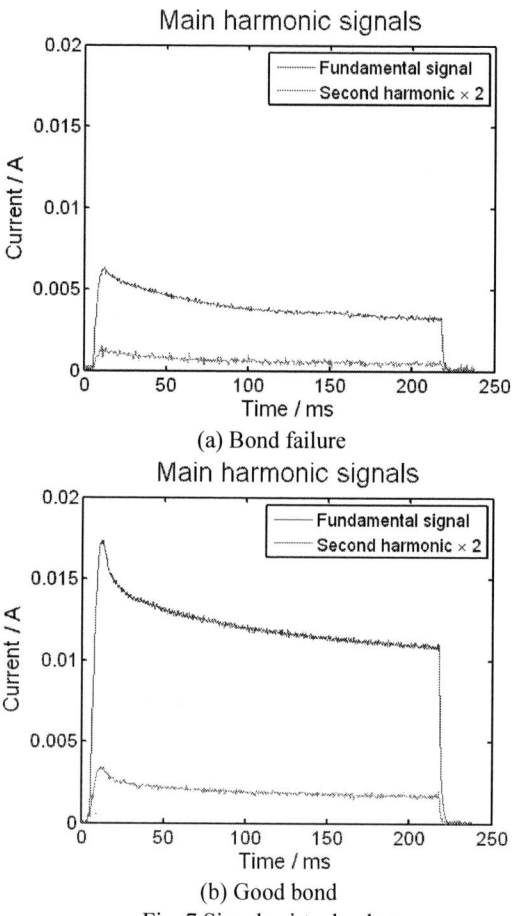

(a) Bond failure

(b) Good bond

Fig. 7 Signals virtual value

II Signals change tendency feature *diff*

This feature is the difference of the current fundamental resonance frequency band signals' average of time 85-102ms and time 187.1-204ms, which represented the rate of current change. We obtained *diff*:

978-1-4244-4658-2/09 $25.00 © 2009 IEEE

$$diff = \frac{1}{100}\sum_{i=501}^{600}D_1(i) - \frac{1}{100}\sum_{i=1101}^{1200}D_1(i) \tag{3}$$

Shown in Fig. 8, the bond would fail if *diff* is very small, viz. the amplitude of current's second half doesn't decrease or decrease a little or increase instead, and the corresponding first half surely decease fast.

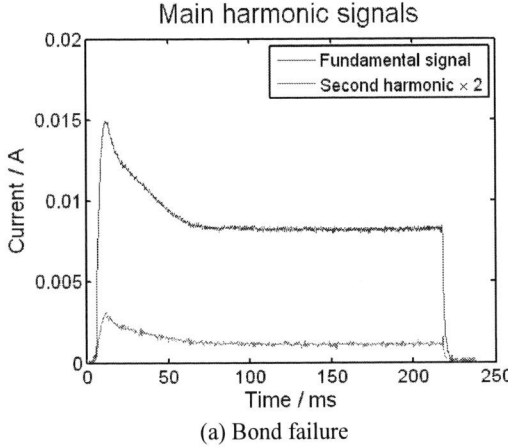

(a) Bond failure

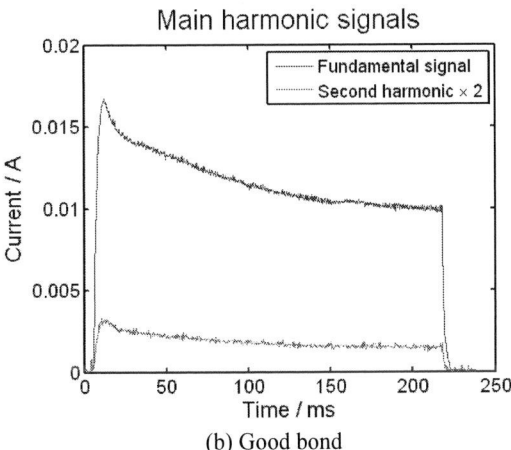

(b) Good bond

Fig. 8 Signals changing tendency

In the bond process, the impedance of the transducer system was changing as the contamination and oxide layer removing and atom diffusing. If the bond quality was formed in a short time, for example, in 60ms, the impedance and the bond strength tended to stabilization soon, but the bond was still in progress which might influence the formed bond even destroy it, so the bond failed.

III The fundamental band signal's average of time 85-102ms: *averfor*

This feature is the fundamental resonance frequency band signal's average of time 85-102ms, calculated by formula (4).

$$averfor = \frac{1}{100}\sum_{i=501}^{600}D_1(i) \tag{4}$$

averfor is more sensitive than *avereng* because *averfor* is the very bonding signals. In the conditions of bond force 4, power 3, bond time 4, the feature *averfor*'s distribution is shown in Fig. 9. If *averfor* is very small, the bond would fail, though the bond quality might not be good while the *averfor* is normal, same as *avereng* and *diff*. The experiments indicated that the threshold would be different if the bond conditions changed.

IV Fundamental resonance frequency *f*.

We can obtain the fundamental resonance frequency *f* of current in the data acquisition system, viz. the resonance frequency of transducer system. Compared with transducer's natural frequency 57kHz, if *f* is far from 57kHz, provisionally estimated above 70kHz or below 50kHz, the phase locked of transducer system would fail and bond failed.

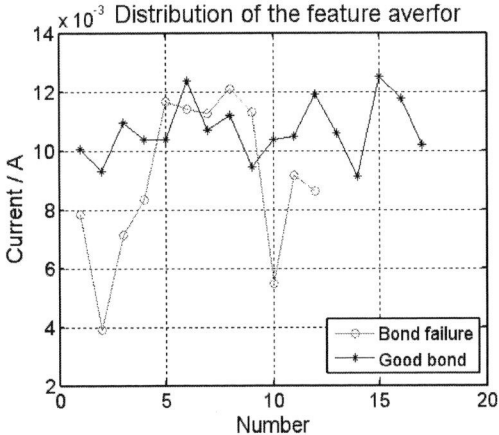

Fig. 9 *averfor* distribution

Associating with the aforementioned features, most bond failure could be found out. These features would be applied and play important roles in the on-line monitoring system. With more features of voltage, vibration and pressure which would be studied further, a perfect quality on-line monitoring system would be set up.

4 Conclusions

A data acquisition sensing circuit was designed and the data acquisition soft was programmed with PCI6110 and Labview8.2. Lots of data had been acquired using the data acquisition system. The current signals were analyzed using wavelet packet transformation in Matlab. The analysis results indicated that the bonding always failed if the mean of the fundamental signature virtual value was very small, also if the changing tendency of harmonic was exceptional. In addition, the frequency deviation of fundamental signature caused bond failure.

Acknowledgments

This work has been supported by the China Department of Science & Technology Program 973 (No. 2009CB 724203).

References

1. S. W. Or, H. L.W. Chan, V. C. Lo, and C. W. Yuen, "Ultrasonic wire bond quality monitoring using piezoelectric sensor," *Sens. Actuators A*, vol. 65 (1998), pp.69-75.

2. Uthe, P. Michael, and Bamburak, "Ultrasonic wire bonding quality monitor and method", *USA patent*, 4815001, 1989.

3. Michael Brökelmann, Jörg Wallaschek, "Bond Process Monitoring via Self-Sensing Piezoelectric Transducers", *Frequency Control Symposium and Exposition Proceedings of the 2004 IEEE International*, Montreal, Canada, 2004, pp.125-129.

4. B. Goebel and A. Ziemann, "Quality control for wire bonding", *United States patent*, 4984730, 1991.

5. Farassat and F. Farhad, "Wire bonding ultrasonic control system responsive to wire deformation", *USA patent*, 5314105, 1994.

6. M. J. Hight, R. V. Winkle, and J. R. Dale, "Ultrasonic bonding apparatus", *USA Patent* ,4040885, 1977.

7. R. Pufall, "Automatic Process Control of wire bonding", *Proceedings of 43rd Electronic Components and Technology Conference*, Orlando, FL, USA, 1993, pp.159-162.

8. K. U. Raben, "Überwachung von Bondparametern während des Bondvorganges", *European patent*, 4854494, 1989.

9. G. Zschimmer, "Ultraschall-Sonotrode", *German patent*, 2946154A1, 1979.

10. O. E. Gibson, W. J. Gleeson, L. D. Burkholder and B. K. Benton, "Bond Signature Analyzer", *USA patent*, 4998664, 1989.

11. WU Fu-liang, LI Jun-hui, HAN Lei *et al*, "Vibration Transfer Process at Thermosonic Flip Chip Bond Interface", *Chinese Journal of Mechanical Engineering*, Vol.44, No.2 (2008), pp.68-73.

12. WU Fu-liang, HAN Lei, ZHONG Jue, Effect of ultrasonic power on the wire bonding strength[J], *Chinese Journal of Mechanical Engineering*, Vol.43 No.3, (2007) pp.107-111.

Corrosion Characterization of Sn37Pb Solders and With Cu Substrate Soldering Reaction in 3.5wt.% NaCl Solution

L. C. Tsao

Department of Materials Engineering, Pingtung University of Science & Technology, Taiwan, China
1, Hseuhfu Road, Neipu, Pingtung 91201, Taiwan, China
E-mail: tlclung@mail.npust.edu.tw   Tel: 886-8-7703202 ext.7560

Abstract

The electrochemical corrosion behavior of Sn37Pb solder, Cu_6Sn_5, Cu_3Sn and Cu in 3.5% NaCl solution was investigated by using Galvanic corrosion testing, potentiodynamic polarization methods. The galvanic current densities for Sn37Pb solder in 3.5wt.% NaCl solution are 38, 16 and 5μA/cm^2 for Cu_3Sn, Cu, and Cu_6Sn_5, respectively. Polarization studies revealed that an increase in Cu content and formation of IMCs shifted the corrosion potential ($E_{corr.}$) towards more noble values and increased the corrosion current density ($I_{corr.}$).

1 Introduction

Solder is extensively used for the electronics industry for packaging applications. Especially, lead-tin solder is widely used for chip-to-package connection. It has very excellent properties, with a unique combination of electrical, chemical, physical, thermal and mechanical behavior [1-2]. According to the International Technology Roadmap for Semiconductors (ITRS), it has been projected that the pad pitch may fall below 20μm by the year 2016 [3]. In some flip chip packages, solder balls of 20μm in size are used to connect the pads on the chip and the print circuit board (Fig.1). In connected metals, all the common base materials, coatings, and metallizations such as Cu, Ni, Ag, Ag-Pd, and Au, form intermetallic compounds (IMCs) with Sn, which is the major element in Sn solders. Cu is the material most frequently used for leads and pads on flip chip substrates and printed wiring boards. It is now known that in the Cu/solder interfacial reaction, Sn reacts rapidly with Cu to form Cu_3Sn (ε) and Cu_6Sn_5 (η) [4-5]. These intermetallic compounds are generally more brittle than the base metal, which can have an adverse impact on the solder joint reliability [6-7]. In addition to air, the solder is exposed to moisture and other corrosives such as chlorine and sulfur compounds. Therefore, many researchers have studied the corrosion behavior of Pb-free solders [8-14], but few have studied the galvanic corrosion properties of Sn-Ag solder, Sn-Al-Zn solder, and Sn37Pb solder with respect to electroless Ni-P, Ni-Cu-P deposits and Cu substrate [15]. No literature is available on the effect of Cu_6Sn_5 and Cu_3Sn on the corrosion resistance of Sn37Pb/Cu solder. In the present study, 3.5wt.% NaCl solution was used to simulate sea water, and the corrosion properties of Sn37Pb solder, Cu_6Sn_5, Cu_3Sn, and Cu substrate in this solution were studied through potentiodynamic polarization and galvanic corrosion tests.

2 Experimental

In the present study, alloys containing Sn37Pb, Cu_6Sn_5, and Cu_3Sn were made by weighing and melting commercially purity metals (nominally 3N) on the vacuum quartz tube by 10mm in diameter, and then quenched to water at room temperature. The intermetallic compounds ingot was homogenized temperature at 670°C (Cu_3Sn) and 415°C (Cu_6Sn_5) in 40 days.

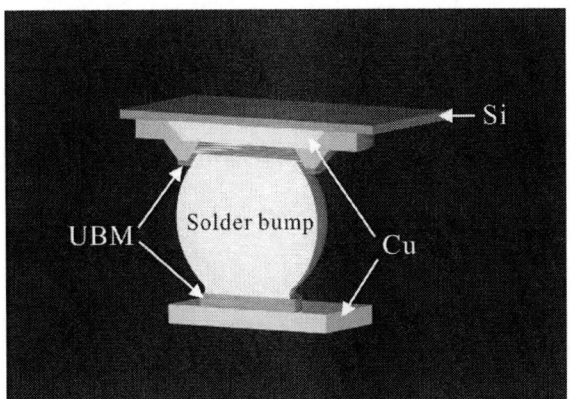

Figure 1 Schematic structure of solder joint in a flip chip package.

The electrochemical measurement experiments was measured with a typical three-electrode cell, no stirring, and degassing of the solution at room temperature by an EG & G M273A potentioset. The reference potential was a saturated calomel electrode (SCE) and platinum (Pt) counter electrode (Φ 1.5mm by 20cm). All electrolytes were prepared by dissolving high-grade chemicals in high purity deionized water (Millipore Milli-Q SP, 18MΩ.cm). The area of the specimens in the experiment was 0.2829cm^2. For dynamic polarization testing, the specimen was immersed in the electrolyte for 1 hour, and then the potential began at -800mV$_{SCE}$ and scanned in the noble direction to an anodic 1V$_{SCE}$ at a scanning rate of 1mV/s. Galvanic corrosion current densities were measured with a zero resistance meter and A/D recorder for Cu, Cu_6Sn_5 and Cu_3Sn. The electrode area was 2.83cm^2. The specimens were cut from rod-shaped ingots rod and wet polished with 240-grit to 2000-grit silicon carbide (SiC), rinsed with acetone and deionized water, and cleaned with ultrasonic cleaning.

The phases of the alloy as observed in the microstructure were identified with energy-dispersive spectroscopy (EDS) and scanning electron microscopy.

3 Results and discussion

Many studies have shown that Sn solders and Cu substrate interfacial reactions result in Cu_6Sn_5 and Cu_3Sn intermetallic compounds. Figure 2 is an optical micrograph showing the microstructure of Sn37Pb solder and copper substrate at the interfacial reactions formed at 420°C for 10 min. Multiple layers can be clearly seen in the micrograph.

978-1-4244-4658-2/09 $25.00 © 2009 IEEE

The layers are in the order of solder, Cu_6Sn, Cu_3Sn, and Cu, respectively.

The joining of materials with Sn37Pb solders generally results in a multi-layer structure in which intermetallic compounds are formed between substrate and solders (Fig.2). Such a structure in a flip chip packaging is a galvanic couple. The galvanic corrosion behavior of the solder bump structures has a great effect upon reliability. The galvanic current densities of the Sn37Pb solder with respect to the intermetallic compounds Cu_6Sn_5 and Cu_3Sn, and base Cu were investigated (Fig. 3). It appears that Sn37Pb solder has a greater galvanic current density, and thus is very subject to corrosion, especially more so in coupling with the formation of Cu_3Sn layers than with Cu_6Sn layers. The galvanic current densities for the Sn37Pb solders with respect to Cu_3Sn, Cu, and Cu_6Sn_5 are about 38, 16, and 5 ($\mu A/cm^2$), respectively. It can be seen that the galvanic corrosion behavior of Cu_3Sn is generally greater than that of Cu_6Sn for the flip chip packaging in a 3.5wt.% NaCl solution environment. This indicates that the formation of Cu_3Sn and Cu_6Sn_5 layers causes many problems with corrosion behavior and reliability.

of the Sn37Pb solder, Cu_6Sn_5, Cu_3Sn, and Cu substrate in 3.5% NaCl solution. These IMCs exhibit corrosion potentials shifting toward noble as the amount of Cu increases. This indicates that the corrosion resistance of Sn solder increases with increasing Cu content.

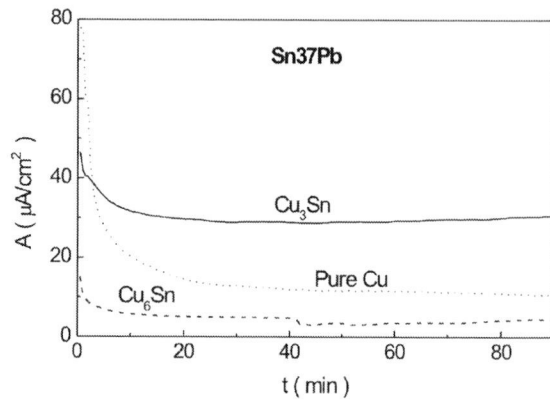

Figure 3 The galvanic current densities of the Sn37Pb solder with respect to intermetallic compound Cu_6Sn_5, Cu_3Sn and Cu substrate in a 3.5wt.% solution.

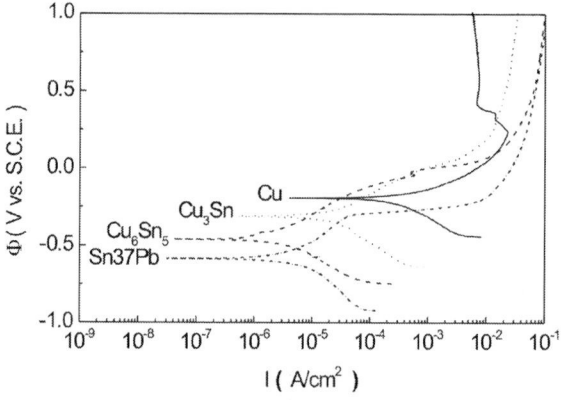

Figure 4 The potentiodynamic polarization curves of Sn37Pb solder, Cu_6Sn_5 IMC, Cu_3Sn IMC and pure Cu samples in a 3.5% NaCl solution.

Figure 2 Typical optical micrograph images of Sn37Cu solder joints with Cu substrate at 420°C for 10 min.

Figure 4 shows the polarization curves of the Sn37Pb solders, Cu_3Sn, Cu_6Sn_5, and base Cu in a 3.5wt.% NaCl solution. The corrosion current densities of all specimens were calculated through the Tafel method and are summarized with other corrosion parameters in Table 1. The corrosion potential (Φ_{corr}) and corrosion passivation current density (I_{corr}) of the Sn37Pb solder were $-584.4mV_{SCE}$ and $6.48\mu A/cm^2$, respectively. In the solder/Cu interface, Sn reacts rapidly with Cu to form Cu-Sn intermetallic compounds (IMCs), thus leading to a signification shifting of the equilibrium corrosion potential towards noble values, -457.7 and $-309.0mV_{SCE}$ for the IMC Cu_6Sn_5 and Cu_3Sn, respectively. The Cu substrate exhibits the highest equilibrium corrosion potential, $-192.1m\ V_{SCE}$. Also, the maximum corrosion current density of the base Cu is $391.6\mu A/cm^2$, which is higher than that of 48.17, and $2.61\mu A/cm^2$ for the IMC Cu_3Sn and Cu_6Sn_5, respectively. The corrosion current densities of the Sn37Pb solder and IMC Cu_6Sn_5 are very similar. Figure 5 presents the effects of Cu content on both Φ_{corr} and I_{corr} values during polarization

On the other hand, the breakdown potential (Φ_b) of IMC becomes much more noble after the Cu/solder interface reaction forms Cu-Sn IMCs, and the larger value of $\Delta\Phi$ ($=\Phi_b-\Phi_{corr}$; Φ_{corr}=corrosion potential) for the Cu_6Sn_5 specimens reveals a characteristic of more stable passivation. This indicates that the pitting corrosion tendency of Cu-Sn IMC at the solder/Cu interface can be alleviated after solder packaging. The Cu substrate exhibits a passivation behavior at above $239mV_{SCE}$ with current densities of $22.3mA/cm^2$. However, the passive behavior of other specimens does not present a prominent peak. The passivation current densities of all specimens at above $460mV_{SCE}$ are around $10^{-1}A/cm^2$ with the declining sequence of $Sn37Pb \geq Cu_6Sn_5 > Cu_3Sn > Cu$.

978-1-4244-4658-2/09 $25.00 © 2009 IEEE

Table 1 Corrosion properties in a 3.5wt.% NaCl solution for the Sn37Pb solder with different heat treatments.

Specimens	Φ_{corr} (mV$_{SCE}$)	Φ_b (mV$_{SEC}$)	$\Delta\Phi$ (mV)	I_{corr} (μA/cm^2)	I_p (mA/cm^2)
Sn37Pb	-584.4	-303.0	281	6.48	67.7
Cu$_6$Sn$_5$	-457.7	-45	412	2.61	56.9
Cu$_3$Sn	-309.0	-8.9	300	48.17	18.3
Cu	-192.1	236	428	391.6	6.5

Φ_{corr} : corrosion potential; I_{corr} : corrosion current density; Φ_b: breakdown potential;

$\Delta\Phi = \Phi_{corr} - \Phi_b$; Φ_p: passivation range of solder alloy;

I_p: passivation current density at above 460mV$_{SCE}$.

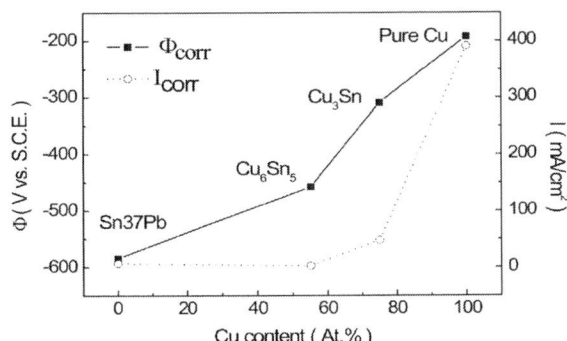

Figure 5 Effect of Cu content on the both Φ_{corr} and I_{corr} value during polarization of the Sn37Pb solder, Cu$_6$Sn$_5$, Cu$_3$Sn and Cu substrate in 3.5% NaCl solution.

Conclusions

The corrosion behavior of Sn37Pb solder, Cu$_3$Sn, Cu$_6$Sn$_5$, and base Cu were studied by means of galvanic corrosion and potentiodynamic polarization in 3.5 wt.% NaCl solution. The galvanic current densities of Sn37Pb solder in 3.5wt.% NaCl solution are 38, 16, and 5μA/cm^2 for Cu$_3$Sn, Cu, and Cu$_6$Sn$_5$, respectively. Increasing the copper content, which reacts with Sn to form IMCs, led to a significant improvement in the corrosion resistance of solders and increased the corrosion current density (I_{corr}). The passivation current densities of all specimens at above 460 mV$_{SCE}$ were around 10^{-1}A/cm^2, with the declining sequence of Sn37Pb $\geq$ Cu$_6$Sn$_5$ > Cu$_3$Sn > Cu.

Acknowledgments

The authors acknowledge the financial support of this work from the Science Council of Taiwan under Project No. NSC97-2218-E-020-004.

References

1. Glazer, J. "Metallurgy of low temperature Pb-free solders for electronic assembly," *International materials Reviews*, Vol. 40, No.2 (1995), pp. 65-93.
2. R. J. Klein Wassink, Soldering in Electrons, 2nd ed. Electrochemical Publeications, (British Isles,1989).
3. Tummala, R. R, Semiconductor International, June (2003).
4. Kim, H. K. Liou H. K. and Tu, K. N, "Three-dimensional morphology of a very rough interface formed in the soldering reaction between eutectic SnPb and Cu," *Applied Physics Letters*,Vol. 66 (1995), pp. 2337-2339.
5. Kim, H. K. and Tu, K. N, "Kinetic Analysis of the Sodering Reaction between Eutectic SnPb Alloy and Cu Accompanied by Ripening," *Physical Review B*, Vol. 53, (1996), pp. 16027-16034.
6. Hedges, E. S, Tin and its Alloys, Edward Awnold, (London, 1960).
7. Harris, P. G. Chaggar, K. S, "The role of intermetallic compounds in lead-free soldering," *Soldering & Surface Mount Technology*, Vol.10, No.3 (1998), pp.38-52.
8. Oulfajrite, H. *et al*, "Electrochemical behavior of a new solder material (Sn-In-Ag)," *Materials Letters*, Vol.57, (2003) , pp.4368-4371.
9. Rosalbino, F. Angelini, E. Zanicchi, G. and Marazza. R, "Corrosion behaviour assessment of lead-free Sn-Ag-M (M = In, Bi, Cu) solder alloys," *Materials Chemistry and Physics*, Vol. 109, No. 2-3 (2008), pp. 386-391.
10. Lin, K. L. and Liu, T. P, "The electrochemical corrosion behaviour of Pbfree Al-Zn-Sn solders in NaCl solution," *Materials Chemistry and Physics*, Vol.56, (1998), pp. 171-176.
11. Lin, K. L. Chung, F. C. Liu, T. P, "The potentiodynamic polarization behavior of Pb-free XIn-9(5Al-Zn)-YSn solders," *Materials Chemistry and Physics*, Vol.53, (1998), pp. 55-59.
12. Mohanty, U. S. Lin, K. L, "The effect of alloying element gallium on the polarisation characteristics of Pb-free Sn-Zn-Ag-Al-XGa solders in NaCl solution," *Corrosion Science*, Vol. 48 (2006), pp. 662-678.
13. Mori. M, Miura. K, Sasaki. T, Ohtsuka. T, "Corrosion of tin alloys in sulfuric and nitric acids," *Corrosion Science*, Vol. 44(2002), pp. 887-898.
14. Chang, T. C. Wang, J. W. Wang, M. C. Hon, M. H, "Solderability of Sn-9Zn-0.5Ag-1In lead-free solder on Cu substrate Part 1. Thermal properties, microstructure, corrosion and oxidation resistance," *Journal of Alloys and Compounds*, Vol. 422 (2006), pp. 239-243.
15. Lin, K. L. and Liu, T. P, "The electrochemical corrosion behaviour of Pb-free Al-Zn-Sn solders in NaCl solution," *Materials Chemistry and Physics*, Vol. 56 (1998), pp. 171-176.

Shape and Fatigue Life Prediction of Chip Resistor Solder Joints

Guanqun Zheng[1], Chunqing Wang[2]

[1]Shenzhen Polytechnic, Shenzhen Guangdong 518055, Tel: 86-755-26731244, E-mail: zhengguanqun@sina.com
[2]Harbin Institute of Technology, Harbin Heilongjiang 150001, Tel: 86-451-86418725, E-mail: wangcq@hit.edu.cn

Abstract

In order to increase the fatigue life of chip resistor, it is necessary to optimize the shape of solder joints. Shape and fatigue life of chip resistor solder joint were predicted by using finite element analysis methods. Through changing the solder volume, four typical solder joint shape prediction were conducted, and three-dimensional mechanical model of fatigue life analysis was set up. The distribution characteristics of the stress and strain in solder joints under thermal cycle load were analyzed. Based on this, fatigue life of solder joints with different solder volumes was predicted. Analysis results show that under thermal cycling conditions solder joints of tiny concave acquired the best mechanical properties, and the its life is the longest.

Discussion 1 Preface

Most circuit modules are assembled in the SMT (Surface Mount Technology) production line, solder joint failure accounts for a great part of electronics failure, therefore, how to improve the fatigue life of solder joints is an important issue. The main ways to improve the solder joint life expectancy are to improve the mechanical properties of solder, to develop CTE matching materials, to optimal design solder joint shape, and to improve the component layout. Because surface mount solder joints are of the small geometric size, it makes the use of experimental method very difficult, and causes great waste in manpower and material resources, so the finite element numerical simulation has been widely used.

The shape and fatigue life of surface mount resistor solder joint are predicted. Surface Evolver software was used to predict resistor solder joint shapes with different solder paste volume. ANSYS software was used to establish the stress-strain analysis model of resistor solder joint under thermal cycling, and to predict the fatigue life of solder joint. By changing the solder paste volume and using finite element analysis the solder joint shape corresponding to maximum thermal cycle fatigue life was found out. The result is of important practical value in the engineering practice.

Discussion 2 Establishment of 3D mathematical model for solder joint shape prediction

In the process of SMT reflow soldering, the effect of intermetallic compound between the molten solder and the metal pad on the solder joint shape is too little to be mentioned. Therefore the forming process of SMT solder joints can be describe as the forming process integrated gravity and surface tension acting on the molten solder metal[1]. According to the principle of energy minimization, when the solder wetted on the pad to achieve balance, the solder joint system consisting of the pad, the molten solder and the lead is in the steady state energy minimization. Here

the total energy E equals to surface potential energy E_s plus gravitational potential energy E_g :

$$E = E_s + E_g = \iiint_V \rho g z dV + \left(\iint_{A_o} T dA + \iint_{A_i} -T \cos\theta dA \right)$$

Where g is the acceleration due to gravity, ρ is the solder density, V is the solder volume, z is vertical coordinate, T is the surface tension of molten solder, A_o is the total area of free surface, A_i is the total area of the solid-liquid interface, and θ is wetting angle.

If solder volume is V_0, then the volume limitation is:

$$\iiint 1 dV - V_0 = 0$$

Using functional:

$$I = \iint_{A_o} T dA + \iint_{A_i} -T \cos\theta dA + \iiint_V \rho g z dV + \lambda \iiint_V 1 dV - V_0$$

Where: λ is the Lagrange coefficient. The stationary point can be acquired when the functional meet the Euler- Lagrange equation, thereby the solder joints shape can be predicted. Fig 1 is the structure and the three-dimensional shape of RC chip solder joint after simulation.

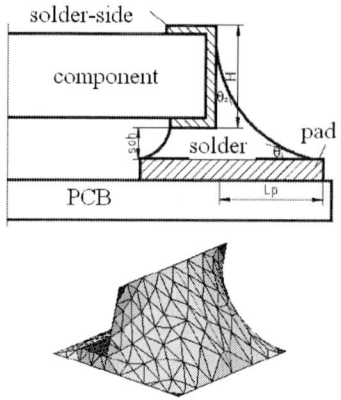

Fig 1. The structure and the three-dimensional shape of RC chip solder joint after simulation

Discussion 3 The finite element model for mechanical properties analysis of solder joints

In order to study the reliability difference of solder joint with different shapes, R1206 is used as the typical component for solder joint shape and fatigue life analysis, the gap between the component and the PCB(printed circuit board) 0.1mm, solder volume 0.2, 0.35, 0.5, 0.65mm³(C1, C2, C3, C4), respectively.

The component is chip resistor, FR4 substrate select linear elastic material isotropy, assume the material property have

978-1-4244-4658-2/09 $25.00 © 2009 IEEE   1086

no change with temperature. The Eutectic solder Sn63Pb37 is temperature dependence elastic-plastic materials. Material parameters were show in Table 1.

Table 1 Material parameters of Sn63Pb37, Al₂O₃ and FR4 [2-3]

Material	Elastic modulus E(MPa)	Poisson's ratio ν	Coefficient of thermal expansion CTE (ppm/°C)
Sn63Pb37	35366-161×T	0.35	24.5
Al₂O₃	3.79×10^5	0.21	5.3
FR4	1.11×10^4	0.28	20

Under thermal cycling load, the creep behavior of Sn63Pb37 solder material is described by Darveaux creep constitutive equation [2]:

$$\varepsilon_{cr} = A_1 \left(\frac{E}{T} \right) \left[\sinh \left(\beta \frac{\sigma}{E} \right) \right]^n \exp \left(-\frac{Q}{kT} \right)$$

Where ε_{cr} is equivalent creep strain rate, A_1=6.14K/MPa·S, n=3.3, β =2027, Q/k =6352K, σ : VonMises Stress, E : Elastic modulus, T : temperature K.

Mechanical analysis of solder joints is carried out by using ANSYS / Multiphysics software. Finite element model is shown in Figure 2.

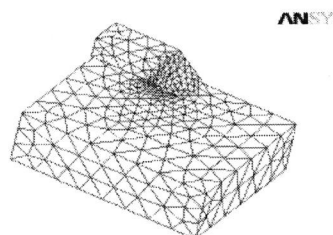

Fig. 2 The finite element model for analyzing mechanical behavior of solder joints

In finite element analysis of solder joint mechanical properties, the load conditions are: temperature range is T= -55°C ~ +125°C, holding time at peak and valley temperature is 10min, time rising rate is 36°C/min, the thermal cycle frequency is f = 2cycle/hr, the reference temperature at the state of zero stress-strain is Tref =25°C (study shows that Tref value variety have no effect with the result during the stress and strain analysis under cycling thermal load). It generally need go through a four-cycle calculation in finite element analysis [4].

Discussion 4 Stress-strain variation under thermal cycling load

The stress-strain of solder joint varies periodically during the thermal cycling, and generally tends to be stable in the fourth cycle. So the eigenvalues of fourth cycle are used in analyzing the mechanical behavior of solder joints during thermal cycling, there are effective stress and effective non-elastic strain of center section of solder joint (x=0).

Figure 3 shows the distribution of effective stress in solder joint at four points of time profile during the fourth thermal cycle: the beginning of low-temperature, the end of low-temperature, the beginning of high temperature and end of high temperature. The gap between the component and the substrate is the location of high stress concentration during the whole thermal cycle. For the solder joint with the shape of tiny concave, at the beginning of -55°C during the thermal cycle, the high stress concentration area distribute in solder along the solder-side of component, the highest stress is close to 60MPa, significantly more than the yield limit of Sn63Pb37, after a certain period of time of constant temperature, stress level decrease due to stress relaxation. When the temperature reaches +125°C, the overall level of stress decline rapidly, the high stress area mainly locate in the solder under the component. During the holding time of high temperature, the stress level of solder joint decrease to the lowest point of the thermal cycle due to stress relaxation.

Figure 4 shows the distribution of effective strain in solder joint at four points of time profile during the fourth thermal cycle: the beginning of low-temperature, the end of low-temperature, the beginning of high temperature and end of high temperature. The gap between the component and the substrate is the location of high non-elastic strain concentration during the whole thermal cycle. During the time of constant temperature, the strain level in the solder joint increased under equivalent creep load condition. The strain level at low temperature is significantly more than that at high temperature.

From the above analysis, results can be drawn: the weakest part of the solder joint locate at the solder in the gap between the component and the substrate, and the solder near the upright solder-side of component, cracks are most likely to initiate and propagate in this region, as shown in Figure 5 (center section of cracks in failure solder joints after thermal cycle loading).

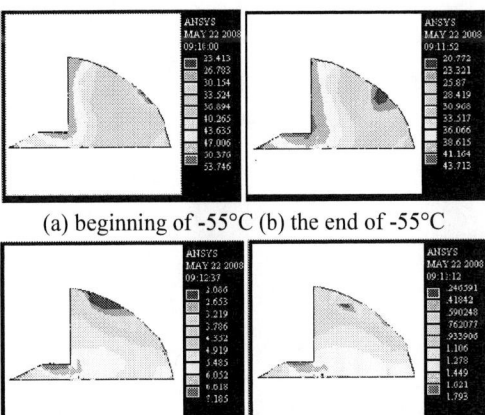

(a) beginning of -55°C (b) the end of -55°C

(c) beginning of 125°C (d) the end of 125°C

Fig. 3 Distribution of effective stress in solder joint during thermal cycle (σ, MPa)

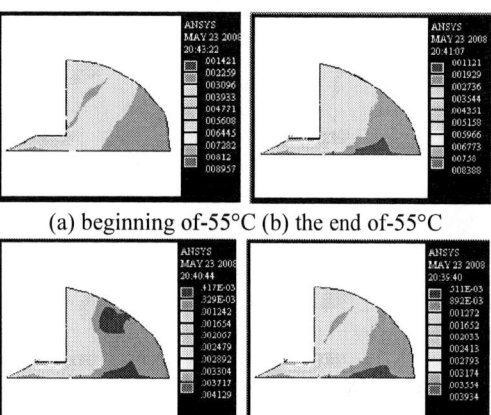

(a) beginning of-55°C (b) the end of-55°C

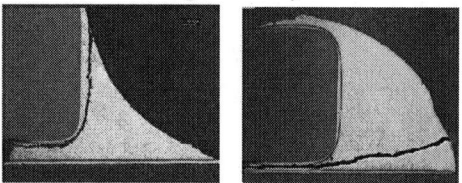

(c) beginning of 125°C (d) the end of 125°C

Fig. 4 Distribution of equivalent plastic strain in solder joint during thermal cycle

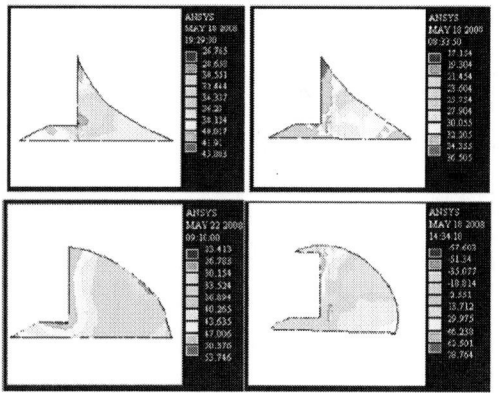

Fig. 5 Cracks in failure solder joints (section)

Discussion 5 Solder joint shape vs. stress-strain

Figure 6 shows the distribution of effective stress in solder with different joint shape. For the joint shape of concave and tiny concave, the high stress area concentrates at the interface between solder and component and between solder and substrate. For the joint shape of convex and tiny convex, the high stress area locate at the interface between solder and component. At the same time, the stress level of solder joint is obviously different from each other: the stress level is the lowest for joint shape of tiny concave (17.15~37.5MPa), for the shape of concave (26.77~43.8MPa)and tiny convex (23.4~53.7MPa), the stress level is comparatively higher, the convex joint shape have the highest level of stress (0.7~78.8MPa), and a large distribution of high stress.

Fig. 6 Distribution of effective stress in solder with different joint shape at start point of-55°C

Figure 7 shows the distribution of equivalent plastic strain at center cross section in solder with different joint shape at start point of -55°C. For concave and tiny concave solder joint, the high strain area locate in the solder near the component upright solder-side, for the tiny convex solder joint, the high strain area locate in the solder under the component, as for the convex solder joint, the whole solder joint have high strain level.

To sum up, during the thermal cycling, tiny concave solder joints have the minimum stress and strain level and the best mechanical properties.

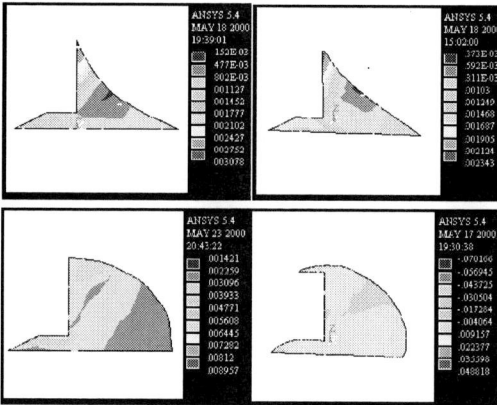

Fig. 7 Distribution of equivalent plastic strain in solder with different joint shape at start point of-55°C

Discussion 6 Solder joint fatigue life prediction

The volume averaging method established by Bilgic *et al* is used in solder joint fatigue life prediction [5]. This approach assumed the mechanical properties of solder joints as a whole rather than a point lead to fatigue failure of solder joints. The failure criterion of solder joint is the accumulated strain energy which generate when irreversible deformation occurred in the solder joints.

The accumulated strain energy of solder joint during one thermal cycle: $\overline{W}_{total} = \overline{W}_{av} \cdot \sum V^e$

The fatigue life of solder joint is: $N_f = \left(\dfrac{\Delta \overline{W}_{total}}{W_0} \right)^{1/k}$

Where $\Delta \overline{W}_{total}$ is the accumulated strain energy of solder joint during one thermal cycle, W_0 and k are material constant, for Sn60Pb40 $k = -0.8165$, $W_0 = 0.358 N \cdot mm$.

Figure 8 shows the hysteresis curve of stress-strain in solder joint during thermal cycle. Despite the different solder joint shape, stress-strain curves are similar, and stabilized in the first few cycles during thermal cycling. However, for different solder joint shape, stress-strain variation range is different. To area surrounded by the hysteresis curve in one cycle for criterion, the area surrounded by the hysteresis curve of concave joint shape is the biggest, that of concave and tiny convex take second place, and that of tiny concave is the smallest. The fatigue life of the joint is inverse ratio to the area. Results are showed in Table 2.

978-1-4244-4658-2/09 $25.00 © 2009 IEEE   1088

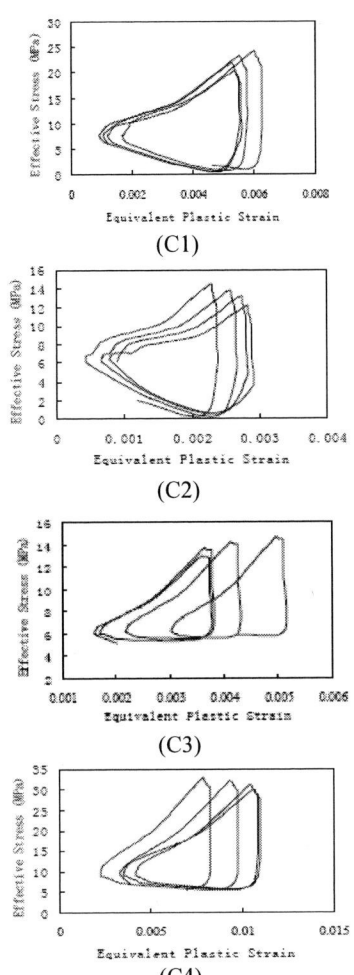

(C1)

(C2)

(C3)

(C4)

Fig. 8 Hysteresis of stress-strain in solder joint during thermal cycle

Table 2 Effects of shape on the fatigue life of solder joints

Serial number	Solder joint shape	Accumulated strain energy (N.mm)	Range of shear strain $\gamma\Delta(mm)$	Thermal cycle life Nf(cycles)
C1	Concave	0.00030	2.45×10^{-3}	5786
C2	tiny concave	0.00021	1.84×10^{-3}	9120
C3	tiny Convex	0.00028	2.22×10^{-3}	6166
C4	Convex	0.00033	2.87×10^{-3}	5214

Conclusions

1) Surface Evolver, software based on the minimum energy principle and finite element analysis, effectively modeled and predicted the joint shape of surface mount resistor.

2) Solder joint stress-strain analysis shows that: solder joints of tiny concave have the best mechanical properties, in the solder joint circumference of the solder along the component solder-side interface is a high stress-strain zone, which is the critical location of fracture failure of resistor solder joint.

3) The fatigue life analysis of resistor solder joints shows that: joint shape of surface mount resistor has a impact on its fatigue life, the plump solder joints have relatively shorter fatigue life, the joint of tiny concave has the longest fatigue life, and nearly 1.5 ~ 1.8 times longer than other three types of shape.

References

1. Zhao Xiujuan, "Optimal Design of Solder Interconnection Joint Shape in Microelectronics Packaging and Assembly," Harbin Institute of Technology. Chapter 2, 2000, 6, pp. 19-55.
2. Vandevelde, Beyne B, Zhang E *et al*, "Solder parameter sensitivity for CSP life - time prediction using simulation - based optimization method," *Electronic components and technology conference*, Florida, USA, 2001, pp. 281-287.
3. John H Lau, SW Ricky Lee, "Effects of build - up printed circuit board thickness on the solder joint reliability of a wafer level chip scale package (WLCSP)," *IEEE transactions on components and packaging technology*, 2002, 25 (1), pp. 4-6.
4. N. H. Paydar, Y. Tong and H. U.Akay, "A Finite Element Study of Factors Affecting Fatigue Life of Solder Joints," *ASME Journal of Electronic Packaging*, 1996, pp. 67-82.
5. C. Basaran, C. S. Desai and T. Kundu, "Thermomechancial Fine Element Analysis of Problems in Electronic Packaging Using the Distributed State Concept: Part 2-Verification and Application," *ASME Journal of Electronic Packaging*, 1998, 120 (1), pp. 54-60.

Effect of Bonding Temperature and Power Setting on Transducer Velocity Using Principal Components Analysis in Thermosonic Bonding

Ya'nan Zhang, Lei Han

Key Laboratory of Modern Complex Equipment Design and Extreme manufacturing, Central South University
Ministry of Education, Changsha 410083, China
E-mail: zhang_yanan007@163.com

Abstract

In thermosonic wire bonding process, the bonding power setting and bonding temperature are important factors to the vibration of transducer system, which plays a crucial role in bonding formation. In a number of bonding experiments which were carried out at different power setting and temperature, vibration velocity at the tip of transducer was monitored. Vibration Signals were analyzed by wavelet decomposition and principal component analysis (PCA). The results show that principal component analysis is an effective method for vibration velocity analysis. The first principal component contains the most variability. And the component loadings on the major indexes are much larger than those on others, which can be neglected. These experimentally obtained data and results could conduce to further studying about the interaction of bonding parameters.

1. Introduction

The continual increasing performance of microelectronics products places a high demand on packaging technologies. Common wire bonding technique interconnects the I/O pad and lead with peripheral circuit using Al or Au wire. It will remain one of, if not the, major chip and packaging interconnection technology in the foreseeable future [1]. Many process parameters (such as power input, bonding pressure, bonding time, stage temperature, transducer configuration) affect the final performance of bonding [2, 4]. Noticeably, the vibration of the transducer system is one of the factors in the overall wire bonding. It is a characterization for the kinetic energy of transducer.

The influence of only one parameter on axial vibration velocity of transduce was studied in the past investigations, but there are few publications to address the case about two parameters changing simultaneously. The combining influence of bonding temperature and ultrasonic power setting on the velocity is discussed in this paper. And principal components analysis (PCA) is applied in this domain for the first time.

2. Experiment

2.1 Experiment principle

Bonding temperature and ultrasonic power were changed simultaneously and regularity in the condition that other bonding parameters were fixed. The select bonding temperature was 50, 140, 230 and 320°C. At the same time, the ultrasonic power setting varied from 2 to 7 grade (respectively 0.036, 0.106, 0.368, 0.633, 0.953 and 1.408W). 100 times bonding were done at every combination of the above two parameters,while the transducer axial vibration velocity was acquired as shown in Fig.1. In thermosonic wire bonding, compared with the 2nd bonding process, the 1st bonding is more representational, thus we only studied the 1st bonding process.

2.2 Test Equipment and parameters setting

1) Bonding equipment

Thermosonic gold ball bonding was performed with a ball bonder (WT2310, WETEL) with the working frequency of approximately 62.5kHz. Parameters of the 1st bonding are shown in Table I, while those of the 2nd bonding were fixed at 0. The gold wire was 23μm in diameter and had an elongation of 2-7%. The substrate (silver coated aluminum) was with size of 50mm×20mm×0.5mm.

Table I Bonding parameters setting

1st Bonding		Balling		Others
Time (Grade)	Pressure (Grade)	Time (Grade)	Current (Grade)	(Grade)
1.5	4	4	4.5	4

2) Test equipment of transducer velocity

The vibration signal of the transducer system of the bonder was measured by using a Polytec® PSV-400-M2 Laser Doppler Vibrometer (LDV) and software Labview. The sensing beam of LDV was aligned to the testing point of transducer system. Triggered by the ultrasonic loading signal, the velocity versus time profiles then were recorded at a sampling rate of 1MHz. Test point A is the axial direction of the transducer terminal. The experimental apparatus are shown as Fig. 1. Typical velocity signal measured is shown in Fig. 2.

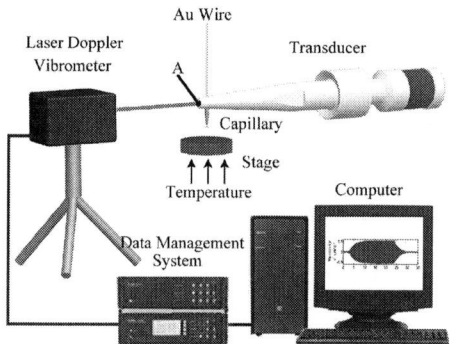

Fig. 1 Experimental schematic diagram

2.3 Experimental procedure

1) Bonding temperature was adjusted to one selected value, while other parameters were kept.

2) We repeated 100 times bonding at every selected ultrasonic power, monitoring the velocity signal.

3) Repeated 1) until all bondings at selected power and temperature were carried out.2400 times bondings have been done totally (24 groups of 100 repeats each). Actually, the effective ultrasonic power may change when the temperature is different, even the ultrasonic power setting is the same. But its changes resulting from temperature alteration are much smaller, comparing with the changes caused by different power setting. We will discuss the phenomenon above in another paper.

3. Data Process

Frequency division and period division must be done in order to find out the variation of each frequency component and period of vibration velocity, while the temperature alters.

3.1 Period of time Division

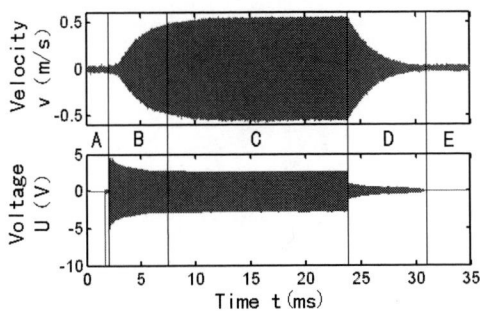

Fig. 2 Period Division of Velocity

Typical velocity and voltage for a thermosonic wire bonding process is shown as in Fig. 2. Five periods (A, B, C, D and E) can be divided as shown in Fig. 2 and Table II.

Table II Period division

Symbol	Meaning	Period of time(ms)
A	Pre-triggering Period	0-2
B	Loading Period	2-7.5
C	Stable Period	7.5-23.6
D	Unloading Period	23.6-31
E	Completing Period	31-35

3.2 Frequency division by wavelet decomposition

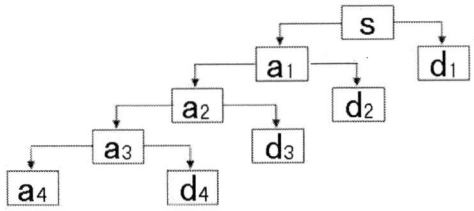

Fig. 3 Wavelet decomposition tree

Describing a nonlinear process, the time-frequency characteristics of the signal features is proven to be very helpful. Wavelet analysis is a promising tool due to its capability of time-frequency localization. Unlike conventional techniques, wavelet decomposition produces a family of hierarchically organized decompositions. The selection of a suitable level for the hierarchy will depend on the signal and experience. There are more than a dozen wavelet families included in the Wavelet Toolbox built on the MATLAB numeric computing environment ranging from sym6 to cof3. However unsymmetrical Daubechies wavelet db30 is sufficiently sensitive to bonding dynamics [5-6].The discrete signals were transferred to Matlab for off-line processing. The db30 wavelet was used for this analysis due to its capability. The velocity signal is broken down into five lower resolution components [7]. This is called the wavelet decomposition tree as shown in Fig. 4.

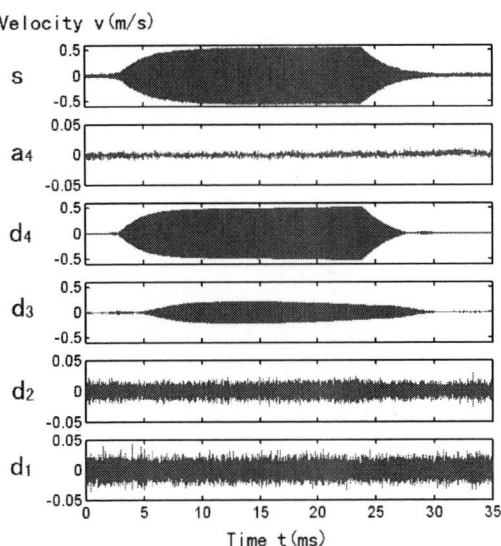

Fig. 4 Frequency division of velocity by wavelet decomposition

Five components could be available from the original velocity signal as follows (Fig. 5) by Daubechies wavelet db30 decomposing:

d1—256-512 kHz;
d2—128-256 kHz;
d3—64-128 kHz;
d4—32-64 kHz;
a4—0-32 kHz

3.3 Data Process

Each velocity signal can be divided into five periods, and each period can be broken down into five frequency components by Daubechies wavelet db30 decomposing. Each velocity signal can be divided into 25 parts with the methods mentioned above. Each part would be transferred its RMS value, which can be regard as the average power for each part. The energy of each part can be obtained by multiplying the average power by its duration. Then we calculated the average over 100 runs. The bonding uncertainty and inherent variability, if any, can be reduced by repeated measurements while systematic dependent components cannot. Fill the energy values in Table III as a 24-by-25 data matrix V comprising a set of observations of 25 variables. In order to reduce the data, find the most important index and uncover unknown trends of velocity signal, principal component analysis was carried out by as follows Table III.

Table III Period and frequency division

Period division			A					B	C	D	E
			a4	d4	d3	d2	d1	a4-d1	a4-d1	a4-d1	a4-d1
Serial No.	Frequency component Power setting (x_j) (grade)	Index Temperature (°C)	X1	X2	X3	X4	X5	X6-X10	X11-X15	X16-X20	X21-X25
1	2	50						...	...	...	...
2		140									
3		230									
4		320									
5-8	3	50-320			...			...	...	...	...
9-12	4	50-320			...			...	...	...	...
13-16	5	50-320			...			...	...	...	...
17-20	6	50-320			...			...	...	...	...
21-24	7	50-320			...			...	...	...	...

3.4 Principal Components Analysis

Principal component analysis (PCA) involves a mathematical procedure that transforms a number of possibly correlated variables into a smaller number of uncorrelated variables called principal components. The first principal component accounts for as much of the variability in the data as possible, and each succeeding component accounts for as much of the remaining variability as possible. So low-order components often contain the "most important" aspects of the data. PCA is theoretically the optimum transform for given data in least square terms. PCA is the simplest of the true eigenvector-based multivariate analyses, and is recommended as an exploratory tool to uncover unknown trends in the data. Often, its operation can be thought of as revealing the internal structure of the data in a way which best explains the variance in the data. The results of a PCA are usually discussed in terms of component scores and loadings [8-11].

PCA is mathematically defined as an orthogonal linear transformation that transforms the data to a new coordinate system such that the greatest variance by any projection of the data comes to lie on the first coordinate (called the first principal component), the second greatest variance on the second coordinate, and so on [10].

Computing PCA on the energy matrix V using follow steps:

1) Organize the data set

Standardize matrix V using the covariance method, for the reason that observations have the same unit and value range.Find the 25-by-25 covariance matrix N from matrix V.

2) Find the eigenvectors and eigenvalues of the covariance matrix N

Compute the eigenvectors and eigenvalues of the matrix N and Rearrange the eigenvectors and eigenvalues. Sort the eigenvectors and eigenvalues in the order of decreasing eigenvalue.

The eigenvalues of the matrix N is

$$\lambda_i (i = 1, 2, \cdots, 25)$$

where $\lambda_1 \geq \lambda_2 \geq \cdots \geq \lambda_{25}$, their corresponding eigenvectors is

$$e_i (i = 1, 2, \cdots, 25)$$

e_{ij} denotes the jth component of e_i.

3) Compute the cumulative energy content for each eigenvector

The eigenvalues represent the distribution of the source data's energy among each of the eigenvectors, where the eigenvectors form a basis for the data.

The energy content α_i for the ith eigenvector is

$$\alpha_i = \frac{\lambda_i}{\sum_{k=1}^{25} \lambda_k} (i = 1, 2, \cdots, 25)$$

The cumulative energy content β_i for the ith eigenvector is the sum of the energy content across all of the eigenvalues from 1 through 25:

$$\beta_i = \frac{\sum_{k=1}^{i} \lambda_k}{\sum_{k=1}^{25} \lambda_k} \quad (i = 1, 2, \cdots, 25)$$

Select a subset comprising first 1 eigenvectors as basis vectors:

$$\beta_l \geq g$$

Often $g = 75 \sim 90\%$.

The eigenvectors, energy content and energy content of first two principal components are shown as Table IV. The cumulative energy content of the first principal component accounts for 98.55%, which means that the first principal component includes the most variability in the data so that the following principal component could be ignored.

4) Compute the component loadings

The ith component loading on the jth index (x_j) is

$$\eta_{ij} = \sqrt{\lambda_i} e_{ij} (i, j = 1, 2, \cdots, 25)$$

The first principal component loadings are shown in Fig. 5.

Table IV The eigenvectors, energy content and energy content

	First Principal Component	Second Principal Component
eigenvectors	5.500	0.038
energy content	98.55%	0.70%
cumulative energy content	98.55%	99.25%

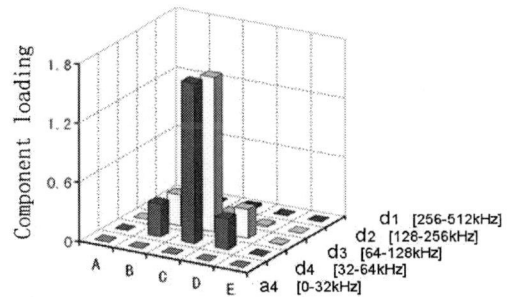

Fig. 5 The component loading

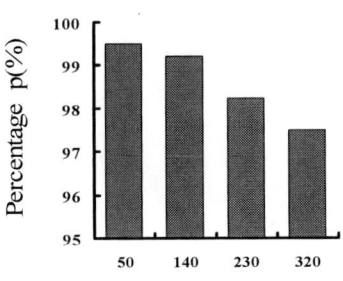

Fig. 6 The energy content

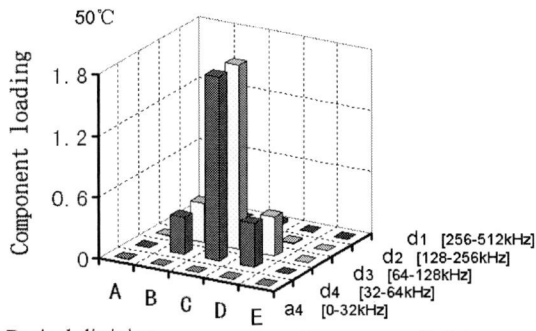

(a) 50°C

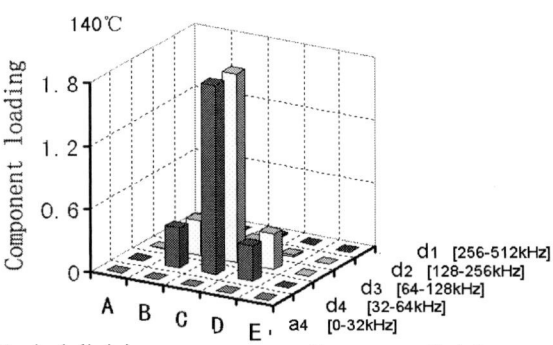

(b) 140°C

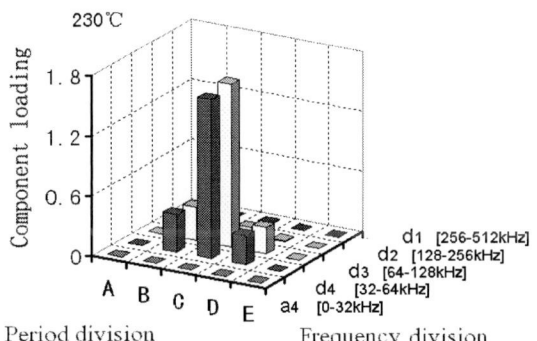

(c) 230°C

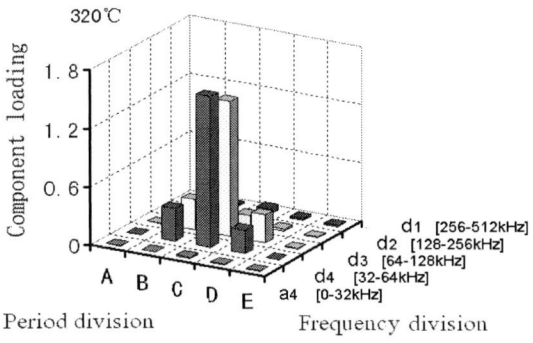

(d) 320°C

Fig. 7 The first principal component loading at different temperature

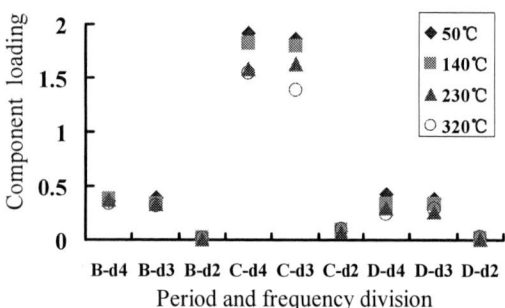

Fig. 8 The first principal component loadings on major indexes

The first principal component loadings denote the influence level on first principal component by the index (x_j). The larger loading is, the greater influence on the first principal component the index has. Fig. 5 shows the component loading on 32-128 kHz of Period C are the largest, while the Component loading on 32-128 kHz of B and D take the second place. The component loadings on other indexes are very small, which can be neglected. That means above indexes affect the first principal component profoundly.

In order to find out how temperature affects velocity, we analyzed the matrix V_T (T=50, 140, 230, 320), which consists by the vibration energy values at the same temperature using above methods. For instance, a 6-by-25 matrix V_{50} includes the 1st, 5th, 9th, 13th, 17th and 21st row of V. The results are shown in Fig. 6 and Fig. 7. The energy contents for the first principal components in different temperature are all greater than 95% and they decrease with the increase of temperature. We can see how the temperature influences above 9 indexes from Fig. 8. The first principal component loadings on the major indexes decrease with the increase of temperature. It shows that the elevation of temperature has a negative influence on the kinetic energy of transducer in above main indexes.

4. Conclusions

1) Principal component analysis is proved to be an effective method for vibration velocity analysis.

2) The first principal component includes the most information (more than 95%), and therefore it can be regarded as a evaluation indicator for vibration analysis .

3) It is verified by PCA that most energy is concentrated on 32-128 kHz of period B, C and D of transducer velocity.

4) The elevation of temperature plays a negative role in the kinetic energy of transducer on above main indexes.

Acknowledgments

This work has been supported by the China Department of Science & Technology Program 973 (No.: 2009CB 724203).

References

1. A. W. Topol, *et al.*, "Three-dimensional Integrated Circuits," *IBM Journal of Research and development*, 50 (2006), pp. 491-506.

2. Han Lei, Zhong Jue, "Effect of Tightening Torque on Transducer Vibration and Bond Strength," *2007 International Symposium on High Density Packaging and Microsystem Integration (HDP'07)*. June 26-28, 2007, pp. 172-177.

3. Medding J. Mayer M., "In situ ball bond shear measurement using wire bonder head," *IEEE/CPMT/SEMI. 28th International Electronics Manufacturing Technology Symposium*, 2003, pp. 59-63.

4. Long Zhili, Han Lei, Wu Yunxin, Zhong Jue, "Experiment Study of Temperature Parameter Effect on Bonding Process and Quality in Thermosonic Wire Bonding," *6th International Conference on Electronics Packaging Technology*, 2005, Shenzhen, pp. 186-193.

5. S. Aijun, L. Han, H. X. Li, "Experimental identification of parasitic vibrations on ultrasonic bonding transducer," *The 8th IEEE CPMT Conference on High Density Microsystem Design and Packaging and Component Failure Analysis (HDP'06)*, Shanghai, June 27-30, 2006, pp. 205-208.

6. Lei Han, Rongzhi Gao, Jue Zhong, Hanxiong Li, "Wire bonding dynamics monitoring by wavelet analysis," *Sensors and Actuators*, Vol. A 137 (2007), pp. 41-50.

7. ASTM Standard Test Method, Test Methods for Destructive Shear Testing of Ball Bonds, in Annual Book of ASTM Standards, ASTM, West Conshohocken, (PA, USA, 1995).

8. "Principal Components Analysis," Agilent Technologies, Inc. (2005), http://www.chem.agilent.com/cag/bsp/products/gsgx/Downloads/pdf/pca.pdf.

9. "A tutorial on Principal Component Analysis," Lindsey Smith (2002), http://www.cs.otago.ac.nz/cosc453/student_tutorials/principal_components.pdf.

10. "Principal component analysis," Wikimedia Foundation, Inc.http://en.wikipedia.org/wiki/Principal_components_analysis.

11. Diana D. Suhr, Ph.D, "Principal Component Analysis vs. Exploratory Factor Analysis," *SUGI 30 Proceedings*, April 10-13, 2005, pp. 1-10.

Study of Thermal Fatigue Lifetime of Fan-in Package on Package (FiPoP) by Finite Element Analysis

Xueyou Yan* and Guoyuan Li

School of Electronic and Information Engineering, South China University of Technology, Guangzhou 510641, China
*Email: Xueyou.YAN@mail.scut.edu.cn

Abstract

Thermal fatigue lifetime of the Fan-in Package on Package (FiPoP) was analyzed based on the plastic strain model by finite element analysis (FEA). Both the stress and strain of FBGA and PBGA solder joints were studied under thermal cycling. It was found that the outmost solder joint on the PCB was the most dangerous position. The fatigue lifetime for the key solder joint on the PCB was estimated using Engelmaier modified Coffin-Manson model. Results showed that the fatigue lifetime of the solder joint of the FiPoP was up to 2073 temperature cycles. The effects of different Ag contents of the lead-free solder paste used for board level assembly on the fatigue lifetime of solder joints of the FiPoP were studied. Results revealed that Sn-4.0Ag-0.5Cu solder had a longest fatigue lifetime of 2441 cycles, the Sn-1.0Ag-0.5Cu had a shortest fatigue lifetime of 1830cycles, and the Sn-3.0Ag-0.5Cu which was used in previous work showed a middle lifetime. The effects of thermal cycling temperature profile on the fatigue lifetime were also investigated. Results showed that the faster temperature ramp rate decreased fatigue lifetime significantly. When the ramp rate increased 16.5°C/min to 33°C/min, the fatigue lifetime dropped to 1686 cycles. On the other hand, the shorter the temperature dwell time, the longer the lifetime. When the dwell time reduced to 10mins, the fatigue lifetime was improved to 2572cycles. It meant that the creep suffered by solder joints had a contribution to the lifetime of the solder joints as well.

1. Introduction

3D packaging technology grows rapidly because today's consumers expect the electronics industry to offer the products that are smaller, with more functionality, better performance and lower cost [1]. The package-on-package (PoP) is one of the 3D packaging solutions, which typically integrates a high-density digital logic processor at the bottom package with high capacity memory dies on the top package [2]. The two components which are sourced from different IC suppliers allows for burn-in and testing prior to assembly. Because of the flexibility and fast time-to-market, PoP has been rapidly used in portable products in recent years [3-4]. However, the traditional PoP approach has some limitations. One limitation is that the top package must be the same size as the bottom package. Another is the overall height of the PoP solution is thicker than that of an equivalent stack-die package. The Fan-in Package on Package (FiPoP) addresses these limitations [5]. Fig. 1 shows the typical configuration of the FiPoP. The connection between top and bottom package is made by means of a center ball array, thereby decoupling the size of the top package from that of the bottom package. Although the warpage during reflowing to the motherboard is reduced due to the FiPoP configuration enabling the use of thinner (and smaller) substrate and allowing overall reduction in mounted height, the reliability of the FiPoP still needs to be inspected and verified from various kinds of tests. Some researchers reported that the FiPoP has the excellent surface mount technology (SMT) yield and drop impact reliability [5-6]. Recently, the FiPoP is intended to implement in a mobile electronic device that work in the higher temperature environment. Hence in addition to the drop impact reliability, the reliability of solder interconnects under thermal cycling loads is also a crucial concern in application and needs to be investigated.

This work aims to investigate the thermal fatigue lifetime of board level FiPoP based on the Coffin-Manson model. The total stress, total strain and plastic strain suffered in BGA of the FiPoP under temperature cycling loads were studied. Based on the plastic strain determined by FEA, the location of the critical solder joint was determined. The effects of Ag contents in Sn-Ag-Cu solder and thermal cycling profile on the fatigue lifetime of the solder joints were also explored.

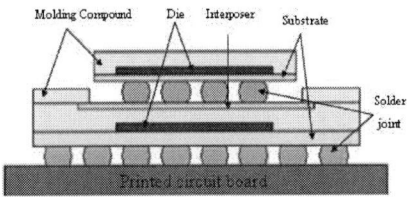

Fig. 1 Fan-in Package on Package (FiPoP) structure

2. Finite Element Model

As shown in Fig. 1, the FiPoP structure consists of a top package and a bottom package. The top package is a fine-pitch ball grid array (FBGA) assembly and the bottom package is a plastic ball grid array (PBGA) assembly. There is an exposed interconnect land array on the top center surface of the bottom package. The FBGA and PBGA packages are joined together by the reflow process, and then mounted on the PCB. Considering the symmetry property, three-dimensional quarter finite element models for FiPoP are made as shown in Fig. 2. The detailed geometries of FiPoP model are exhibited in Table 1.

978-1-4244-4658-2/09 $25.00 © 2009 IEEE

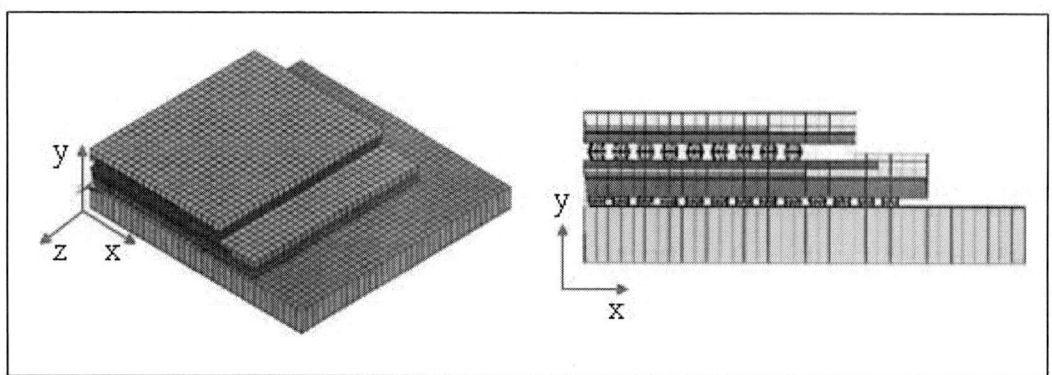

Fig. 2 Three-dimensional quarter finite element model for FiPoP

Table 1 The detailed geometries of FiPoP module

		Materials	Length (mm)	Width (mm)	Thickness (mm)
FBGA		Substrate	11	14	0.13
		Die-attach Film	7.5	9	0.02
		Die	7.5	9	0.075
		Mold Cap	11	14	0.3
PBGA		Substrate	14	14	0.26
		Die-attach Film	9	9	0.02
		Die	9	9	0.075
		Mold Cap	14	14	0.36
Interposer		BT	12	14	0.13
PCB		RF-4	18	18	1
Solder joints		Materials	Pitch (mm)	Radius (mm)	Height (mm)
		SAC305 (FBGA)	0.50	0.175	0.25
		SAC305 (PBGA)	0.50	0.175	0.15

There are also some assumptions made in this study as below: (a) All materials are homogeneous and isotropic; (b) All material interfaces have perfect adhesion [7]. Linear elastic behavior is assumed for the package materials unless otherwise stated. The material properties used in simulation are referred from references [2, 9-13] and listed in Table 2. The constants required for Anand's viscoplastic constitutive model for lead-free solder alloys Sn-1.0Ag-0.5Cu (SAC105), Sn-3.0Ag-0.5Cu (SAC305), and Sn-4.0Ag-0.5Cu (SAC405) are referred from references [8-10] and shown in Table 3.

Table 2 Material properties used for the FiPoP module

		E(GPa)	Poisson Ratio	CTE (ppm/°C)
Substrate		22.6	0.25	16
Die-attach Film		4.8	0.40	40
Die		130	0.30	2.81
Molding Compound		28.5	0.25	20
PCB		22.0	0.28	20
Interposer		15	0.20	19.23
Solder Alloys	SAC105	33	0.30	25
	SAC305	38.7	0.35	21
	SAC405	40	0.38	20

Table 3 Anand constants for lead-free SAC105, SAC305 and SAC405

Description	Symbol	Solder Alloys		
		SAC105	SAC305	SAC405
Initial value of deformation resistance	s_0(Mpa)	2.3479	45.9	20
Activation energy	Q/R(K)	8076	7460	10561
Pre-exponential factor	A(1/s)	3.773	5870	325
Stress multiplier	£	0.9951	2	10
Strain rate sensitivity of stress	m	0.4454	0.0945	0.32
Hardening coefficient	h_0(Mpa)	4507.5	9350	800000
Coefficient of deformation resistance saturation value	S(Mpa)	3.5833	58.3	42
Strain rate sensitivity of saturation value	n	0.012	0.015	0.02
Strain rate sensitivity of hardening coefficient	a	2.1669	1.5	2.57

The temperature loading profile based on MIL-STD-883 is shown in Fig. 3. The temperature cycles between -40°C and 125°C, maintains for 20 mins at the highest and lowest temperature respectively, and the ramp rate is 16.5°C/min. The thermal stress is supposed to be zero at the temperature of 125°C. The stress and strain of solder joints vary periodically under the thermal cycling, and become stable in the third cycle [14]. So, in this work, three cycling loads are applied.

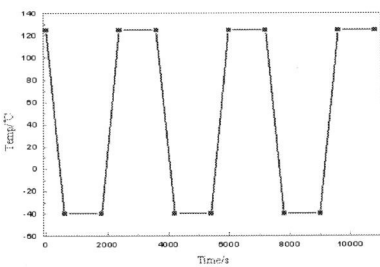

Fig. 3 Temperature cycling profile used in modeling

3. Results and Discussions

3.1 Analysis of the position of the key solder joint

In order to identify the site of the most dangerous solder joint, the total stress, total elastic strain and plastic strain of the FBGA and PBGA are studied. Fig. 4-6 show the distributions of the equivalent stress, equivalent elastic strain and equivalent strain in FBGA and PBGA solder joints respectively.

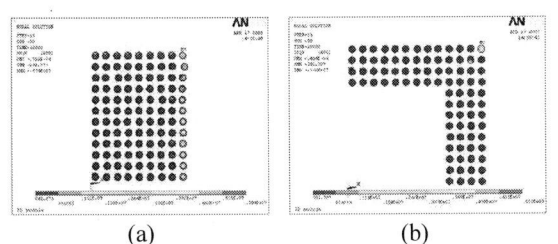

(a)   (b)

Fig. 4 Von Mises stress distribution: (a) FBGA solder joints; (b) PBGA solder joints

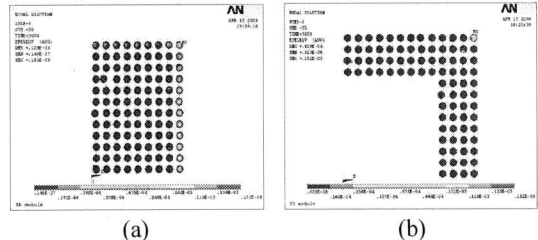

(a)   (b)

Fig. 5 Von Mises elastic strain distribution: (a) FBGA solder joints; (b) PBGA solder joints

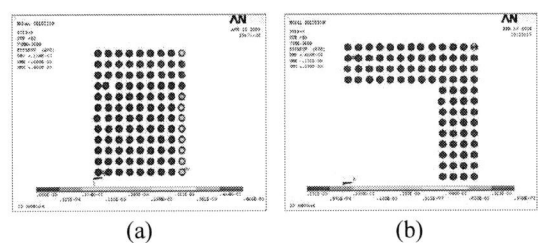

(a)   (b)

Fig. 6 Von Mises plastic strain distribution: (a) FBGA solder joints; (b) PBGA solder joints

From the Fig. 4 and Fig. 5, it could be found that for the FBGA, the maximum stress of 5.94Mpa and the maximum elastic strain of 1.53×10^{-4} are located in the corner solder joint. This may be due to the thermal expansion coefficient (TEC) mismatch of FBGA substrate and interposer in the process of thermal cycling loading. The deformation in the edge of FBGA substrate is larger than that in the center, which causes the largest stress and strain in the outermost solder joint. For the PBGA, owing to the different TEC of PCB and PBGA substrate, the outmost solder joint has the largest stress of 5.52Mpa and strain of 1.52×10^{-4}. Exploring the Von Mises plastic strain shown in Fig. 6, the maximum plastic strain of 5.79×10^{-4} in PBGA is bigger than the maximum plastic strain of 4.66×10^{-4} in FBGA, so the outmost solder joint in the PBGA is the most dangerous.

In view of the results above, the outmost solder joint in the PBGA will be chosen to be the critical solder joint in the following study.

3.2 Analysis of the stress and strain for the critical solder joint

After identification of key solder joint and its maximum plastic strain point, the time history of the stress and elastic strain at this point were drawn in Fig. 7. The time history of plastic strain energy density at the maximum plastic strain point is shown in the Fig. 8.

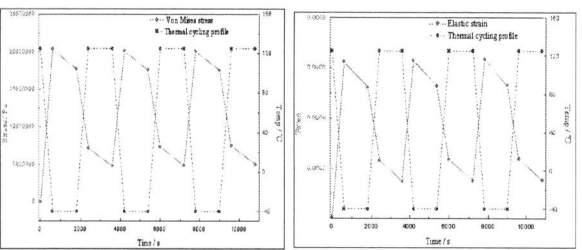

(a)Time history of von Mise stress   (b) Time history of elastic strain

Fig. 7 Time history of stress and elastic strain at the critical point

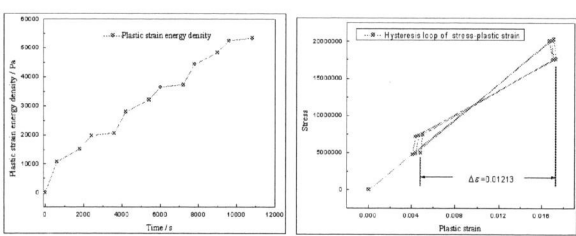

Fig. 8 Time history of plastic strain energy density   Fig. 9 Hysteresis loop of stress-plastic strain

From Fig. 7, it was found that the high stress and strain happen in the low temperature phase, while low stress and strain correspond to high temperature phase. There exists stress relaxation in the high and low temperature dwell. From Fig. 8, it is obviously revealed that under thermal cycling loading, plastic strain of the dangerous point is a cumulative process, which will ultimately cause the failure of the solder joint.

3.3 Thermal fatigue lifetime prediction of the FiPoP solder joint

Solder joint thermal fatigue lifetime prediction is carried out by the Werner Engelmaier modified the Coffin-Manson model as shown below [15].

$$N_f = \frac{1}{2}\left(\frac{\Delta\gamma}{2\varepsilon_f'}\right)^{\frac{1}{c}} \quad (1)$$

Where N_f is the number of cycles to failure, $\Delta\gamma$ is shear strain range, which can be calculated by $\Delta\gamma = \sqrt{3}\Delta\varepsilon$, $\Delta\varepsilon$ is strain range that can be obtained from hysteresis loop shown in Fig. 9, ε_f' =0.325, which is fatigue ductility coefficient, and c is fatigue ductility index that can be expressed as

$$c = -0.442 - 6\times10^{-4}T_M + 1.74\times10^{-2}In(1+f) \quad (2)$$

where T_M is the mean temperature in the thermal cycling process, $T_M = (T_{Max}+T_{Min})/2$ =42.5°C, T_{Max} is the maximal cycle temperature, T_{Min} is the minimal cycle temperature, f is thermal fatigue frequency, f = 24(cycles/day). From these

data, c is calculated as -0.412. From modeling, the fatigue lifetime (cycles to failure) of the FiPoP solder joint was obtained and shown in Table 4.

Table 4 Thermal fatigue lifetime of solder joint

Solder Alloy	$\Delta\varepsilon$	N_f(cycles)
SAC105	0.01283	1830
SAC305	0.01213	2073
SAC405	0.01134	2441

3.4 Effects of Ag contents in Sn-Ag-Cu solder alloys

In order to study the effects of Ag contents in Sn-Ag-Cu solder alloys on the thermal fatigue lifetime, the contrast of the three lead-free solder alloys on the fatigue lifetime of the PBGA is plotted in Fig. 10.

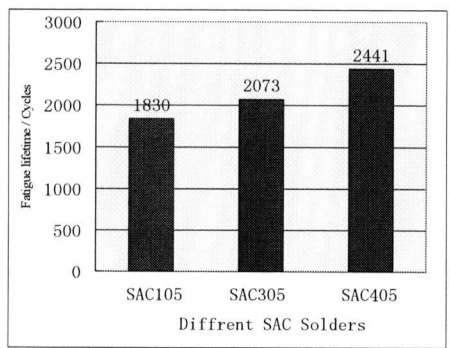

Fig. 10 Fatigue lifetime contrasts of the three Lead-free solder alloys

It can be observed that the Ag contents in SAC solder have significant impact on the FiPoP solder joint fatigue lifetime. Among three solder alloys, SAC405 has longest fatigue lifetime and SAC105 has the shortest fatigue lifetime. This result is attributed to strengthening by an increase in volume fraction of Ag_3Sn dispersions in the higher silver-content alloy. The higher the Ag contents in solder alloy, the higher the young's modulus and the lower TEC with it. Seeing from Table 2, SAC405 has the higher young's modulus and the lower TEC than other two solder alloys. The thermal mismatch among the PBGA substrate, SAC405 solder ball and PCB is smallest. So the SAC405 shows the smallest plastic deformation and longest fatigue lifetime during the thermal cycling. The results are in accordance with the results in the literature [16].

3.5 Effects of the thermal cycling profile

Table 5 shows relationship between ramp rate and fatigue lifetime. It can be seen that the increase in ramp rate decreases fatigue lifetime significantly. The mechanism may be that the higher the temperature change rate, the larger the strain rate suffered by solder joints, which leads to earlier to fail for the solder joints. This is the reason that thermal shock is more stringent than thermal cycling test (slower ramp rate). On the other hand, the shorter the temperature dwell time, the longer the lifetime. As the creep and stress relaxation of solder joint occur during the dwell time, shortening the dwell time reduces the creep and stress relaxation, which leads to increase in fatigue lifetime.

Table 5 Effects of thermal cycling profile

Effects	Descriptions	Fatigue lifetime (cycles)
Ramp/Dwell time (mins)	10/20	2073
	5/20	1686
	10/10	2572

Conclusions

In this study, the thermal analysis of the Fan-in Package on package (FiPoP) has been investigated using finite element modeling. It is found that the critical solder joint locates at the outmost solder joint of the PBGA and the fatigue lifetime of the solder joint is predicted as 2073 cycles. The Ag contents in SAC solder have significant impact on the FiPoP solder joint fatigue lifetime. Among three solder alloys, SAC405 has longest fatigue lifetime of 2441 cycles and SAC105 has the shortest fatigue lifetime of 1830 cycles. It is also found that the faster ramp rate decreases fatigue lifetime significantly. On the other hand, the shorter the temperature dwell time, the longer the lifetime. The mechanism to influence the lifetime has been suggested.

Acknowledgment

The authors would like to acknowledge the financial support of Guangdong Nature Science Foundation of China (8151064101000014) for this work.

References

1. Dreiza, M. *et al*, "High Density PoP (Package-on-Package) and Package Stacking Development," *Electronic Components and Technology Conference*, Reno, Nevada, May.2007, pp. 1397-1402.
2. Xie, B. et al, "Design Advisor for Package-on-Package (PoP) Manufacturing," International Conference on Electronic Packaging Technology & High Density Packaging (ICEPT-HDP 2008), Shanghai, China, Jul.2008, pp. 1-7.
3. Vijayaragavan, N, "Package on Package Warpage - Impact on Surface Mount Yields and Board Level Reliability," Electronic Components and Technology Conference, Orlando, FL, May.2008, pp. 389-396.
4. Lee, J. Y. et al, "Study on the Board Level Reliability Test of Package on Package (PoP) with 2nd Level Underfill," Electronic Components and Technology Conference, Reno, Nevada, May.2007, pp. 1905-1910.
5. Carson, F, "The Development of the Fan-in Package-on-Package," Electronic Components and Technology Conference, Orlando, FL, May.2008, pp. 956-963.
6. Lai, Y. S. et al, "Examination of board-level drop reliability of package-on-package stacking assemblies of different structural configurations," Microelectronic Engineering, Vol.84, No. 1 (2007), pp. 87-94.
7. Tzeng, Y.L. et al, "Warpage and Stress Characteristic Analyses on Package-on-Package (PoP) Structure," 9th Electronics Packaging Technology Conference, Singapore, Singapore, Jan.2007,pp.482-487.
8. Bhate, Dhruv. et al, "Constitutive Behavior of Sn3.8Ag0.7Cu and Sn1.0Ag0.5Cu Alloys at Creep and Low Strain Rate Regimes," IEEE Transactions on Components and Packaging Technologies, Vol. 31, No.3 (2008), pp. 622-633.
9. ZHANG, L. et al, "Fatigue life prediction of SnAgCu soldered joints of FCBGA device," TRANSACTIONS OF THE CHINAWELDING INSTITUTION, Vol. 29, No. 7 (2008), pp. 85-88.
10. Wang ,Q. et al, "Experimental Determination and Modification of Anand Model Constants for Pb-Free Material 95.5Sn4.0Ag0.5Cu," EurosimE2007, London, England, Apr.2007, pp. 1-9.
11. Li, C. G. et al, "Thermal reliability analysis of package in PBGA," Electronic components and materials, Vol. 27, No. 1 (2008), pp. 65-68.
12. Dudek, R. et al, "Fatigue Life Prediction and Analysis of Wafer Level Packages with SnAgCu Solder Balls," Electronics System integration Technology Conference, Dresden, Germany ,Sep.2006, pp. 903-911.
13. Liu, K.C. et al, "Manufacturability and Reliability of a High-Speed CSP SRAM on an Interposer Package," Electronic Components and Technology Conference, Orlando, FL, May.2008, pp. 374-381.
14. Ma, X. et al, "Finite element analysis of stress-strain distribution characteristics in SMT solder joints (1)-Dynamic feature of stress-strain field distribution," The Chinese Journal of Nonferrous Metals, Vol. 10, No. 3 (2000), pp. 404-410.
15. Engelmaier, W, "Fatigue life of leadless chip carrier solder joint during power cycling," IEEE CHMT, Vol. 6, No. 3 (1983), pp. 232-237.
16. Kittidacha, W. et al, "Effect of SAC Alloy Composition on Drop and Temp cycle Reliability of BGA with NiAu Pad Finish," Electronics Packaging Technology Conference, Singapore, Singapore, Dec.2008, pp. 1074-1079.

Thermal-Mechanical Fatigue Reliability of PbSnAg Solder Layer of Die Attachment for Power Electronic Devices

Xinpeng XIE[1*], Xiangdong BI[2], Guoyuan LI[1],

[1]School of Electronic and Information Engineering, South China University of Technology, Guangzhou 510641, China
[2]Guangdong Yuejing High Technology Co. Ltd., Guangzhou 510663, China
*Email: xinpeng.xie@mail.scut.edu.cn

Abstract

PbSnAg solder was widely used in die attachment for high power chip packaging, and the thermal-mechanical reliability of PbSnAg solder layer is a key factor to evaluate the quality of high power devices packaging. Viscoplastic finite-element simulation methodologies were utilized to predict Pb92.5Sn5Ag2.5 solder joint reliability for die attachment under accelerated temperature cycling conditions (-55°C to +125°C, 10min ramps/20min dwells). The behavior of solder under accelerated temperature cycling was described by Anand's viscoplastic constitutive equations and the fatigue life prediction was investigated in volume-average-energy method by R.Darveaux. Three finite-element models (2D plane model, 3D slice model and 3D quarter model) were established to validate the results, and the response of different chip size for die attach was discussed to optimize the fatigue life. The results show that, the maximum plastic stress, strain, and plastic energy were found at the corner or on the edge of the solder layer, which is the week point leading to initial crack damage in the leadframe-solder-die interfaces. With increasing in temperature cycles, the stress-strain hysteretic loops of the dangerous position have a steady trend. The plastic energy accumulated within temperature cycle becomes larger while the increment of plastic energy per cycle trends to be stable. Also, chip size has some influences on the thermal reliability. With increasing in chip size, the maximum plastic stress and strain at the dangerous position increase, which could lead to poor reliability of the die attachment.

1 Introduction

The coefficient of thermal expansion (CTE) mismatch between substrate, heat sink and die for high power device package requires the die bond to be compliant to prevent large stress in the silicon die. Soft solder such as Pb92.5Sn5Ag2.5 is more suitable to be used here, since it has low modulus of elasticity, low yield strength, and relatively high ductility. Due to the high thermal mismatch result from high power, solder degradation may occur frequently and causes different types of failure [1]. In recent years, more and more researchers focus on the reliability of solder joint in die attachment for high power devices. From literature review [2-4], two types of simulation, rate-independent plastic and creep of the solder layer were performed. Results show that solder creep is the dominant factor affecting long term reliability of power semiconductor die attachment solder layer. Metallographic experimentations show that cracks initiate at the border of the solder joint and grow through the die bond. The loss of die bonds will increase in die temperature and effectively reduce the safe operating area of the device, which will lead to the fatal failure to the thermal and electrical performance. Hence the thermal-mechanical reliability and precise fatigue life prognosis of solder layer for power die attachment under thermal cycling test is significant to estimate the quality of power device packages. Finite-element analysis based thermo-mechanical reliability prediction and performance optimization for solder or adhesive joints are widely used in packaging industry. Various investigations use a semi-empirical approach to predict fatigue life. A number of researchers have shown interest in the energy based methods for fatigue life prediction since energy is a scalar quantity and can be assumed to have a cumulative effect. Results show that the volumetric average energy approach was employed successfully to estimate the solder fatigue life for several types of packages, such as the ball grid array, chip resistor, and quad flat package (QFP) [5-9]. However, no research was found on the die attachment solder fatigue reliability for the power device in this method. This paper aims to predict the die attachment solder fatigue life subjected to thermal cycling by volumetric average energy approach.

2 Modeling methodology

2.1. Viscoplastic Anand's model for PbSnAg solders

Eutectic SnPb solder used in electronic devices usually works at high homologous temperatures which is usually in excess of the homologous temperature of 0.5 T_m (melting temperature). High homologous temperatures experienced by the solder and the thermal strains imposed on the solder due to the CTE mismatch among the materials in a package always cause a complex deformation behavior. This deformation behavior is associated with the irreversible, temperature and rate (or time) dependent inelastic characteristics, which is known to be viscoplastic behavior.

Anand L. developed a constitutive equation for the rate-dependent deformation of metals at high temperatures (i.e. in excess of the homologous temperature of $0.5T_m$). There are two basic features in the Anand's model. Firstly, this model needs no explicit yield condition and no loading/unloading criterion. Secondly, this model employs a single scalar as an internal variable, the deformation resistance (s), to represent the averaged isotropic resistance to plastic flow. Anand's model unifies the creep and rate-independent plastic behavior of the solder by making use of a flow equation and an evolution equation, which are listed as following [10]:

$$\varepsilon'_p = A \exp(-\frac{Q}{RT})[\sinh(\xi \frac{\delta}{s})]^{1/m} \quad (1)$$

where ε'_p is the inelastic strain rate, A is the pre-exponential factor, Q is the activation energy, R is gas constant, T is absolute temperature, ξ is the multiplier of stress, δ is the equivalent stress, and m is strain rate sensitivity.

The evolution Equation can be expressed as:

$$s' = \{h_0\left(\left|1-\frac{s}{s^*}\right|\right)^a \times sign(1-\frac{s}{s^*})\} \times \varepsilon_p'; \ a > 1 \qquad (2)$$

in which, $s^* = s(hat)[\frac{\varepsilon_p'}{A}\exp(\frac{Q}{RT})]^n$ $\qquad (3)$

where h_0 is the hardening/softening constant, a is the strain rate sensitivity of hardening/softening. The quantity s' represents a saturation value of deformation resistance, $s(hat)$ is a coefficient, and n is the strain rate sensitivity for the saturation value of deformation resistance.

There are nine material parameters of the unified rate-dependent Anand's model for Pb92.5Sn5Ag2.5 solder, which are listed in Table 1 [4].

Table 1 Anand's model parameters of Pb92.5Sn5Ag2.5 solder

A s^{-1}	Q/R K	ξ	m	h_0 MPa
1.03E+7	11010	7	0.241	1432
$s(hat)$ MPa	n	a	s_0 MPa	
33.07	0.002	1.3	23.07	

2.2. Fatigue life prediction methodology

By measuring the crack growth rate of actual solder joints, Robert Darveaux [5-6] established four crack growth correlation constants (K_1 through K_4) along with two equations, shown as Equation (5) and (6), to predict thermal fatigue life of the solder alloy. Finite element simulation results could be used to calculate thermal cycles to initial crack and crack propagation rate per thermal cycle. Because of sensitivity to the finite element modeling procedure, the element thickness at the interface of solder should be controlled and element volumetric average plastic work technology was used to predict the solder fatigue life.

An accumulated viscoplastic strain energy density in the volumetrically averaged form, shown as Equation (4), is employed to reduce solution sensitivity in regard to the mesh density.

$$\Delta W_{avg} = \frac{\sum_{element} \Delta W_{elem} \times V_{elem}}{\sum_{element} V_{elem}} \qquad (4)$$

In Equation (4), ΔW_{elem} is the viscoplastic strain energy density accumulated per test cycle in an element within the volumetric region, and V_{elem} is the volume of the corresponding element. The prediction methodology equations by Robert Darveaux for thermal cycle fatigue life of the test solder is listed as below:

Crack initiation: $\qquad N_0 = K_1(\Delta W_{avg})^{K_2}$ $\qquad (5)$

Crack growth rate: $\qquad \frac{dl}{dN} = K_3(\Delta W_{avg})^{K_4}$ $\qquad (6)$

Characteristic life: $\qquad N = N_0 + \frac{l}{dl/dN}$ $\qquad (7)$

Failure free life: $\qquad N_{ff} = N/2$ $\qquad (8)$

In the equations above, l is the smallest characteristic dimension of the selected volumetric region of the solder

joint, which is obtained when the crack propagation has lead to 5% decrease of solder joint surface area [11]. For this paper, the value of l is selected to be 200um.

The parameters (K_1 through K_4) in Equation (5) and (6) for the Pb92.5Sn5Ag2.5 solder alloy are listed in Table 2.

Table 2 Life prediction model parameters of Pb92.5Sn5Ag2.5 solder

K_1/ cycles/psi	K_2	K_3/ in/cycles/psi	K_4
22400	-1.52	5.86×10^{-7}	0.98

3 Finite element analysis model

Three FEA models, as shown in Fig. 1, were established using FEA software. There are three materials in the MOSFET power device package modeling: copper lead frame, solder layer, silicon chip. The bottom part is lead frame material, and the top is silicon chip. Solder material is located between substrate and chip. The size of substrate is 10.0mm×9.0mm×1.5mm, and the size of silicon chip is 4.0mm×4.0mm×0.29mm. The dimension of the solder material generally is somewhat larger than that of the silicon chip but is assumed to be the same size with the silicon chip, in this case except the thickness is 0.05mm. Table 1 shows the Anand's model parameters for solder used for simulation. Table 3 shows the physical and mechanical properties of the materials used in the simulation. The accelerated temperature cycling profile is shown in Fig. 2.

The viscoplastic behavior of the solder is modeled by using element VISCO107 and the Cu lead frame and chip exhibiting elastic behavior are modeled by using SOLID45 element for linear material behavior. Based on the thermal cycling profile in Fig. 2, four complete thermal cycles are simulated in order to establish a stable stress-strain hysteresis loop.

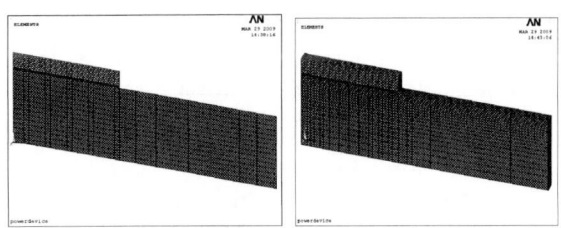

(a) 2D plane model;   (b) 3D slice model;

(c) 3D quarter model.

Fig. 1 FEA models for die attachment

Table 3 Material physical parameters

Materials	Thermal Conductivity (W/m*K)	CTE ($10^{-6}K^{-1}$)	Young Modulus (MPa)	Poisson's Ratio
Silicon Die	136	2.5	148×10^3	0.25
Solder joints	40	29	$24.7(10^3)$-$37.3T(°C)$	0.35
Cu Substrate	398	17	117×10^3	0.3

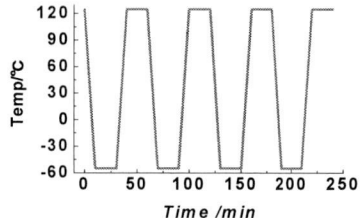

Fig. 2 Accelerated temperature cycling profile (-55°C to +125°C, 10min ramps/20min dwells)

4 Simulation results and discussion

4.1 Stress, strain and plastic energy analysis for model (c)

Fig. 3 shows the calculated result maps of von Mises stress, von Mises plastic strain and plastic energy distribution within the solder joint at the end of both temperature dwells (after stress relaxations) in the last thermal cycle. It can be seen that, the maximum values of stress, plastic strain and plastic energy were at the corner or on the edge of the solder joints generally, which is the most dangerous points in die attachment solder layers and the first failure position within the accelerated thermal cycling test according to the failure analysis criterion. Furthermore, the maximum values were found on the interface of solder layer with the chip side or Cu lead frame side due to the max thermal expansion (CTE) mismatch within the solder layer. However, because of the intermetallic compounds (IMC) within the solder layer interfaces formed in the die attachment process at nearly 360°C, which have good mechanical strength properties than solder itself , the failure (such as cracks) positions were found in the solder layer inside along the IMC line usually. The maximum values of stress and plastic strain at the end of high temperature dwell, which is 1.72MPa and 0.0025 respectively, are smaller than that of the low temperature dwell, which is 23.2MPa and 0.0136 respectively, since the reference temperature (T_{ref}) is 125°C. However, the maximum value of accumulate plastic energy of 1.19MPa at the end of high temperature dwell is larger than 1.06MPa at the low temperature dwell due to the scalar quantity of the energy and its cumulative effect. Also, it can be concluded that the maximum value or the dangerous position located in the side, where is near the Cu lead frame boundary, of solder layer, because of the Cu lead frame is a rectangle shape (10mm×9mm).

The von Mises stress and plastic strain evolution histories along with time at the dangerous points within solder layer were shown in Fig. 4 and Fig. 5. The values trended to have a reverse change discipline with the time history profile. In the stress evolution history, a small fluctuation occurred at the end of low temperature dwell. It is probably because the solder layers couldn't release the too big stress quickly due to its property and the fluctuation can be considered to a buffer. The stress-strain hysteresis loop in the die attachment solder joints was shown in Fig. 6, and the loop has a trend of moving toward right and up direction due to the simulation results of von Mises stress and plastic strain increasing with the thermal cycles. Generally the loop trended to be a stable state after four thermal cycles. Also there is a small fluctuation in the loop due to the stress fluctuation at the end of low temperature dwell.

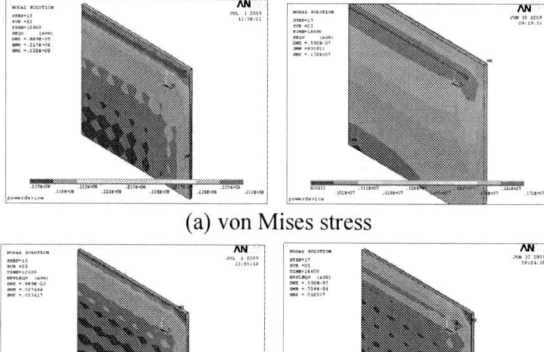

(a) von Mises stress

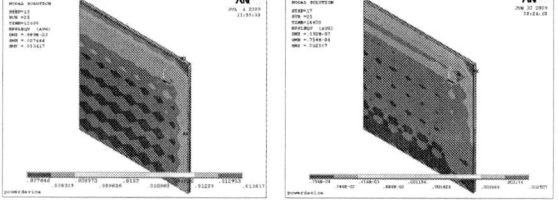

(b) von Mises plastic strain

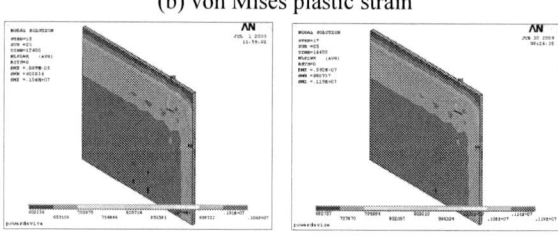

(c) Plastic energy distribution maps

Fig. 3 Simulation results at the end of low (-55°C) and high (+125°C) temperature dwells for the last cycle for model (c): (Left map: low temp dwell; right map: high temp dwell)

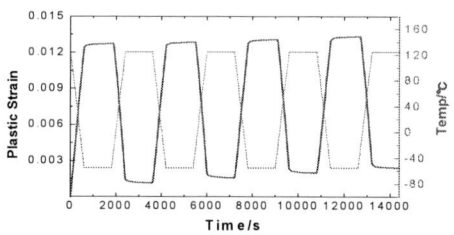

Fig. 4 Plastic strain evolution in the dangerous point of solder joints

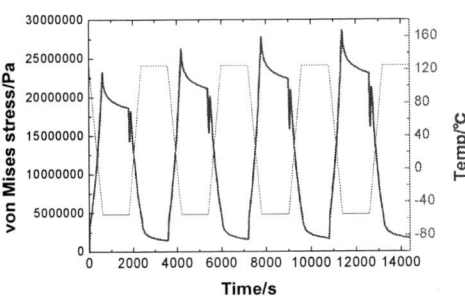

Fig. 5 Von Mises stress evolution in the dangerous point of solder joints

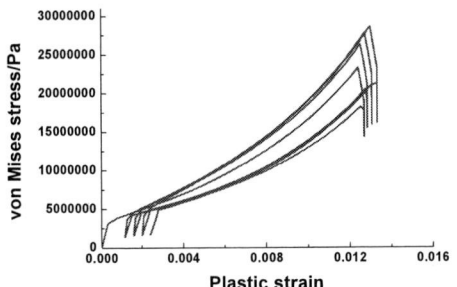

Fig. 6 Stress-Strain hysteresis loop of the solder joints

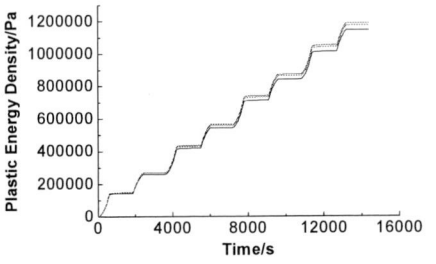

Fig .7 Accumulated plastic energy in the dangerous points of solder joints

The evolution history of plastic energy density within solder layer was shown as Fig. 7, which increased with time slowly and stably, and the increasing quantity per cycle was approximately the same value.

4.2 Fatigue life prediction based on volumetric average plastic energy approach

The fatigue life for the solder joints was calculated based on volume-averaged viscoplastic strain energy density accumulated per cycle (ΔW_{avg}) by Robert Darveaux method, which is described in Equation (4) through (8). Considering the thickness of solder joints in die attachment is very small, nearly 0.05mm, entire solder layer area was employed to calculate the results of viscoplastic strain energy density for more precise fatigue life cycles.

The obtained results are listed in Table 4, where the crack initiation, crack growth rate and characteristic life cycles of three models were compared.

Table 4 Fatigue life prediction for die attachment models

Life prediction items	Model (c)	Model (b)	Model (a)
Delta plastic energy per cycle (MPa)	0.209099 E+00	0.188369 E+00	0.245123 E+00
Delta plastic energy per cycle (psi)	30.3272	27.3206	35.5521
Crack initiation (cycles)	125	147	98
Crack growth rate (mm/cycle)	0.4216 E-3	0.3806 E-3	0.4927 E-3
Solder joints failure criterion (mm)	0.200	0.200	0.200
Crack propagation (cycles)	474	526	406
Characteristic life (cycles)	599	673	504

(1pis = 6.894757×10^{-3}MPa; 1inch=25.4mm)

As expected, results show that, the crack initiation, crack growth rate and characteristic life cycles have approximate results, which can verify the validity of thermal fatigue life prognosis methodology and the established finite element models. Model (b) has the highest characteristic life of 673 cycles, while model (a) has the lowest characteristic life of 504 cycles. These two models cannot describe all of property of die attachment solder layers, so in some extent, it can be concluded that model (c) has a more precise and appropriate prediction life of 599 cycles and this fatigue prognosis life has more convince to understand due to the more detail description in model (c).

Volumetric average energy approach was employed to predict thermal fatigue life and performance optimization of solder joints widely, and it has some advantageous points than the plastic stain based solder fatigue life prediction methodology which is also very popular, such as Coffin-Mason relationship and Engelmaeier's model. Strain based solder fatigue description is not adequate as the fatigue life of the solder may also be function of stress, which may be a very important parameter during the fatigue of solders. Furthermore, energy is a scalar quantity and can be assumed to have a cumulative effect. Especially, the volumetric average plastic energy approach proposed, which has the advantages of easy to carry out and less sensitive to the mesh size of the finite element model by linking the volumetric average cyclic accumulated equivalent plastic work density to the mean cycles to initial crack and a crack propagation rate, is adequate to estimate the solder fatigue reliability.

4.3 Effects of chip size on the thermal fatigue reliability

The thermal fatigue reliability of the solder layer was evaluated based on stress, plastic strain, plastic energy density and fatigue reliability life. And the effects of die size on the thermal fatigue reliability were also listed in Table 5. In power MOSFET chip packaging process, the chip size varies from 3mm to 5mm approximately according to the requirements of microelectronic design and manufacturing process. In this part, we will focus on the influences of chip size on power chip packaging reliability.

Table 5 Effects of chip size on the thermal reliability of solder joints

Thermal reliability items	3mm×3mm	4mm×4mm	5mm×5mm
Maximum von Mises stress(Mpa)	1.52	1.72	1.92
Maximum plastic strain	0.001224	0.002531	0.004667
Plastic energy density(MPa)	1.05	1.19	1.48
Delta plastic energy per cycle (MPa)	0.203185 E+00	0.209099 E+00	0.221413 E+00
Characteristic life (cycles)	618	599	563

The values of thermal reliability items listed in Table 5 of the solder layer under accelerated thermal cycling in model (c) were the maximum results at the end of the last cycle respectively. The results show that the chip size has some obviously influences on the thermal reliability. With increasing in chip size, the maximum von Mises stress, von Mises plastic strain and the plastic energy density of solder layer at the dangerous position increased with discipline. And the characteristic prediction life based volumetric average energy approach decreased with the increase in chip size. It means that smaller chip size has higher reliability level and vice versa. So in the process of power MOSFET chip packaging, the chip size should be selected and controlled suitably in order to make the electrical, mechanical property and the thermal reliability items of overall packaging to be the best tradeoff. Beside that, the lead frame of MOSFET chip packaging should also be considered thoughtfully. Because the distance of chip boundary with Cu lead frame boundary has some effects on the thermal reliability, and the nearer they are, the worse reliable thermal quality they have, which was discussed in Part 4.1.

Conclusions

The thermal-mechanical fatigue reliability of PbSnAg die attachment solder layer was studied in detail by finite element analysis in the way of von Mises stress, von Mises plastic strain, and accumulated plastic energy density under accelerated thermal cycling loading. The results show that the maximum plastic stress, strain, and plastic energy were found at the corner or on the edge of the solder layer, which could lead to initial crack damage in the solder layer according to the failure analysis criterion. With increasing in temperature cycles, the stress-strain hysteretic loops of the dangerous position have a steady trend. The plastic energy accumulated within temperature cycle becomes larger while the increment of plastic energy per cycle trends to be stable. Furthermore, chip size has some influences on the thermal reliability. With increasing in chip size, the maximum plastic stress and strain

at the dangerous position increase, which could lead to poor reliability of the die attachment.

Acknowledgment

The authors would like to acknowledge the financial support of Guangdong Nature Science Foundation of China (8151064101000014) and Guangdong Yuejing High Technology Co. Ltd. for this work.

References

1. Hua Ye, Minghui Lin, Cemal Basaran, "Failure modes and FEM analysis of power electronic packaging," *Finite Elements in Analysis and Design*, Vol. 38, No. 3 (2002), pp. 601-612.
2. Zhuang, W. D. *et al*, "Effect of solder creep on the reliability of large area die attachment," *Microelectronics Reliability*, Vol. 41, No. 12 (2001), pp. 2011-2021.
3. Rodriguez M *et al*, "Static and dynamic finite element modeling of thermal fatigue effects in IGBT modules," *Microelectronics Reliability*, Vol. 40, No. 3 (2000), pp. 455-463.
4. Zhang S. H. *et al*, "Reliability of PbSnAg solder layer of power modules under thermal cycling in electronic packaging," *The Chinese Journal of Nonferrous Metals*, Vol. 11, No. 1 (2001), pp. 120-124.
5. Robert Darveaux, "Effect of Simulation Methodology on Solder Joint Crack Growth Correlation and Fatigue Life Prediction," *Transactions ASME Journal of Electronic Packaging*, Vol. 124, No. 3 (2002), pp. 147-155.
6. Robert Darveaux, "Effect of Simulation Methodology on Solder Joint Crack Growth Correlation," *Proc 50th Electronic Components and Technology Conf*, Las Vegas, NV, May. 2000, pp. 1048-1058.
7. Vijay Sarihan, "Energy Based Methodology for Damage and Life Prediction of Solder Joints Under Thermal Cycling," *IEEE Transactions on Components, Packaging, and Manufacturing Technology, Part B*, Vol. 17, No. 4, (1994), pp. 626-631.
8. Akay, H. U. *et al*, "Fatigue Life Predictions for Thermally Loaded Solder Joints Using a Volume-Weighted Averaging Technique," *Transactions ASME Journal of Electronic Packaging*, Vol. 119, No. 4 (1997), pp. 228-236.
9. Changwoon Han, Byeongsuk Song, "Development of Life Prediction Model for Lead-free Solder at Chip Resistor," *Proc 56th Electronic Components and Technology Conf*, Singapore, Dec.2006, pp. 781-786.
10. Anand L, "Constitutive Equations for the Tate-Dependent Deformation of Metals at Elevated Temperature," *Transactions ASME Journal of Engineering Materials and Technology*, Vol. 104, No. 1, (1982), pp. 12-17.
11. Bouarroudj M. *et al*, "Thermo-mechanical investigations on the effects of the solder meniscus design in solder joint lifetime for power electronic devices," *Thermal, Mechanical and Multi-Physics Simulation Experiments in Microelectronics and Micro-Systems*, 2007, EuroSime 2007, London, April.2007, pp. 1-7.

Effects of Strain Rate and Temperature on Mechanical Behavior of SACB Solder Alloy

Guozheng Yuan, Xuexia Yang, Xuefeng Shu

Institute of Applied Mechanics & biomedical engineering, Taiyuan University of Technology,
Taiyuan, Shanxi 030024, China,
yuanguozheng@tyut.edu.cn, shuxf@tyut.edu.cn, 0351-6014455

Abstract

With the deep research of the reliability of electronic package, more and more investigator realizes that the study on the high strain rate behavior of materials is necessary. Drop impact reliability has become an important criterion when assessing the reliability of portable electronics. The quality of solder joints directly determines the drop-impact reliability of the product. Therefore, to study the dynamic stress-strain constitutive relationship and the fracture/failure behavior of the solder material due to an external impact rate loading become important. Solder balls, as a structural member of the electronic product, are used to connect the microchip and the associated printed circuit board by a proper joining technique such as the surface mount technology (SMT).

The static compressive properties of Sn3.0Ag0.7Cu3.0Bi lead-free solder are studied through quasi-static compression testing. In addition, The mechanical behavior of the two lead-free alloy, at strain rates ranging from 3×10^2 to $1.2 \times 10^3 s^{-1}$ and temperatures between 25°C and 140°C, was investigated with the compressive split-Hopkinson pressure bar (SHPB). The dynamic property of Sn-3.0Ag-0.7Cu-3.0Bi alloy was assessed by means of split Hopkinson pressure bar test technique. The strain rate sensitivity of the lead-free solder is characterized using a modified split Hopkinson pressure bar system. The experimental results indicate that, the strain rate sensitivity parameter increases along with increasing strain and strain rate, but decreases with increasing temperature. The addition of Bi delays the sensitivity of strain rate of the lead-free solder. The influence of the addition of Bi on the dynamic stress–strain constitutive relationship and the fracture/failure behavior of the solder material is investigated. The activation energy varies inversely with the flow stress, and has a low value at high deformation strain rates or low temperatures.

In this paper, Quasi-static and high strain rate properties of the lead-free solder with temperature are given. Basing on these experimental data, the constitutive relation of Johnson-cook model for this kind of material was built up.

1 Introduction

In recent years, many lead-free solder alloys begin to replace the lead-containing solder alloys in microelectronic industry and come into use in some packaging processes and interconnects of certain microelectronic components. Many lead-free solder mechanical properties have not been clarified and understood very well due to a short term use in microelectronics assemblies [1]. And compared to other lead-free solder, Sn-Ag-Cu lead-free solder-joint reliability is equal or better than SnPb solder-joint reliability, and has better creep performance than SnPb solder at low stresses and is equal to SnPb solder at high-creep stresses. So this solder alloy has been widely used in microelectronic packaging devices [2]. But Sn-Ag-Cu solder's melting temperature is higher than SnPb solder, is not conducive to welding. The addition of Bi reduced effectively the melting temperature and improved wetting ability compared with Sn-Ag-Cu solder alloy. Its dynamic mechanical properties have not been clarified very well.

Therefore, the purpose of this paper to study the dynamic mechanical behavior of Sn-3Ag-0.5Cu-Bi lead-free solder tested at strain rates from 3×10^2 to $1.2 \times 10^3 s^{-1}$.The effect of temperature and strain rate on dynamic conventional properties of lead-free solder alloy Sn-3.0Ag-0.7Cu-3.0Bi was obtained. In this paper, Quasi-static and high strain rate properties of the lead-free solder with temperature are given. Basing on these experimental data, the constitutive relation of Johnson-cook model for this kind of material was built up.

2 Experimental

2.1 Specimen preparation

The geometry of specimen should meet the condition to minimize the effect of inertia. So the diameter and the length of the cylindrical rod type solder samples are both set to be 5mm, and the samples were fabricated using the casting technique. These samples were used to investigate the high strain rate compression behavior of unleaded SACB solder materials. The quasi-static compression tests for the SACB solder were also conducted by a MTS material tester. Prior to the SHPB high strain rate test, the center line along the specimen, the input bar and the output bar was carefully aligned. In the present study, the solder sample was examined and the corresponding chemical composition is Sn-3.0Ag-0.7Cu-3.0Bi.

2.2 Experimental procedure

In order to achieve this level of high strain rate loading condition, the SHPB apparatus with its associated dynamic strain measuring and recording instruments was used to conduct experiments. Test setup of the SHPB experiments, containing an incident bar, a transmission bar, and a striker bar, is shown in Fig. 1.

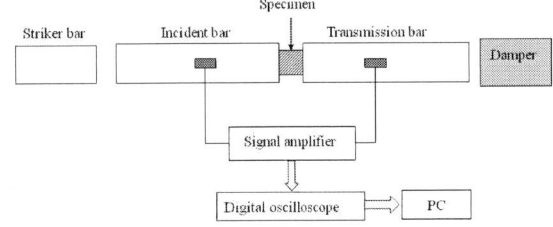

Fig. 1 Test setup for the SHPB apparatus

978-1-4244-4658-2/09 $25.00 © 2009 IEEE

Table 1 the mechanical properties for incident and transmission bars

Material	high strength alloying steel
Density	8000kg/m³
Length	800mm
Diameter	8.0mm
Bar wave speed	4900m/s
Yield strength	1.8GPa

In order to analyze the dynamic elastic wave propagation signals in the circular rods of the split Hopkinson pressure bars, high strength alloying steel was usually adopted experimentally. The mechanical properties of incident and transmission bars are summarized in Table 1. In conventional SHPB technique, the specimen is located in between incident and transmitted bars. When the striker bar impacts the incident bar, rectangular stress pulse is generated and travels along the incident bar until it hits the specimen. Part of the incident stress pulse reflects from the bar/specimen interface because of the material impedance mismatch, and part of it transmits through the specimen. The transmitted pulse emitted from the specimen travels along the transmitted bar until it hits the end of the bar [3]. In conventional SHPB analysis, the deformation history of the specimen is extracted from the signals in the strain gages mounted on the incident and transmission bars. Analysis is accomplished by referencing the three signals. ε_I, ε_R and ε_T stand for the strain due to incident wave, reflection wave and transmitted wave, respectively. The stress, strain and strain rate in the specimen can be obtained in terms of the recorded strains of the two bars as follows:

$$\sigma(t) = \frac{E A_0}{A} \varepsilon_T(t) \qquad (1)$$

$$\dot{\varepsilon}(t) = -\frac{2c}{l} \varepsilon_R(t) \qquad (2)$$

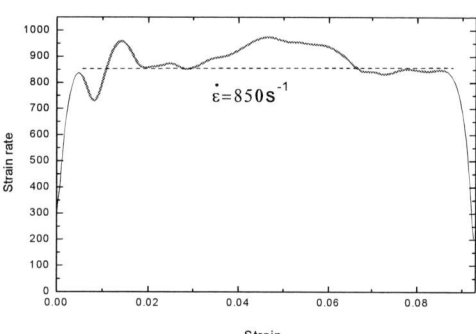

Fig. 2 Relationship between strain rate and strain for specimen deformed at average strain rate of $850\,s^{-1}$

$$\varepsilon(t) = -\frac{2c}{l} \int_0^t \varepsilon_R(t) d\tau \qquad (3)$$

Fig. 2 shows the relationship between time and the derived strain rate for a specific SACB SHPB test. An averaged strain rate of 850s⁻¹ for this test is calculated using the data extracted from the heavy red curve shown in Fig. 2.

3 Result and discussion

The static compressive properties of Sn3.0Ag0.7Cu3.0Bi lead-free solder are studied through quasi-static compression testing. Fig. 3 shows the stress-strain curve of Sn-3.0Ag-0.7Cu -3.0Bi under compression test of quasi-static.

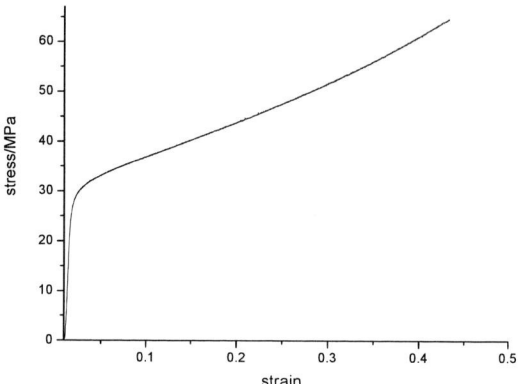

Fig. 3 The stress-strain curve for Sn-3.0Ag-0.7Cu-3.0Bi solder under compression of quasi-static

3.1 Temperature effect

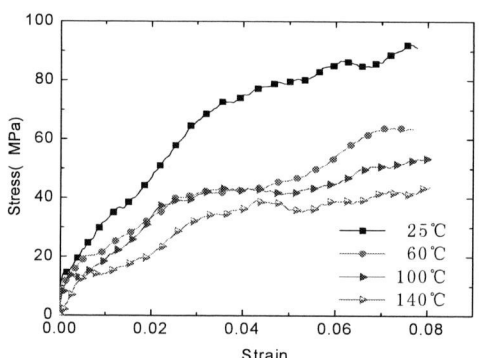

Fig. 4 The stress-strain curves of Sn-Ag-Cu-Bi solder at different temperature and constant strain rates of $700\,s^{-1}$

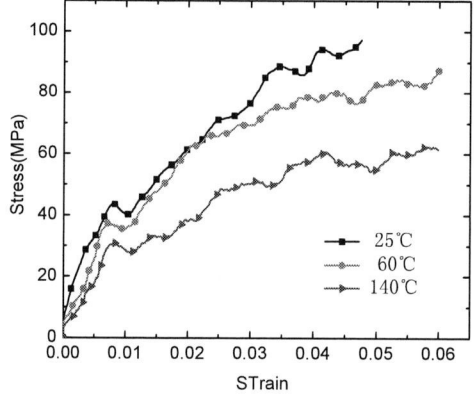

Fig. 5 The stress-strain curves of Sn-Ag-Cu-Bi solder at different temperature and constant strain rates of $850\,s^{-1}$

Fig. 4 show the stress-strain curves of Sn-3.0Ag-0.7Cu-3.0Bi solder at different temperature and constant strain rates of $700\,s^{-1}$, Fig. 5 show the stress-strain curves of Sn-3.0Ag-0.7Cu-3.0Bi solder at different temperature and constant strain rates of $850\,s^{-1}$.

The stress-strain curves presented in Fig. 4 and Fig. 5 shows that the temperature has a significant effect on the flow stress of Sn-Ag-Cu-Bi alloy. For a constant strain rate, the fracture strain and work hardening rate reduce with increasing temperature at temperatures between 25°C and 140°C.

3.2 Strain rate effect

Fig. 6 show the stress-strain curves of Sn-Ag-Cu-Bi solder at different temperature and the room temperature (25°C) Fig.7 show the stress-strain curves of Sn-Ag-Cu-Bi solder at different strain rates and the same temperature of 140°C.

The stress-strain curves presented in Fig.6 and Fig. 7 shows that the plastic flow is affected not only by the temperature, but also by the strain rate. For a given temperature, the fracture strain and work hardening rate increase with increasing strain rate.

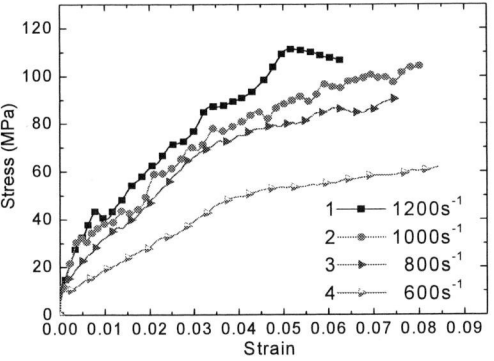

Fig. 6 The stress-strain curves of Sn-Ag-Cu-Bi solder at different temperature and the room temperature (25°C)

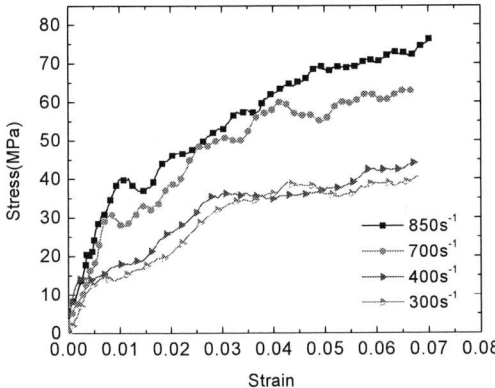

Fig. 7 The stress-strain curves of Sn-Ag-Cu-Bi solder at different strain rates and the same temperature of 140°C

4 the constitutive relation of Johnson-cook model

In order to better describe the temperature and strain rate on the effects of constitutive, Johnson-cook model was used. In Johnson-Cook model [4], the flow stress σ_e is given by

$$\sigma_e = \left[A + B \left(\varepsilon_e^p \right)^n \right] \left(1 + C \ln \dot{\varepsilon}^* \right) \left[1 - \left(T^* \right)^m \right] \quad (4)$$

Where A, B, C, n and m are constitutive parameters, ε_e^p is the average plastic strain, $\dot{\varepsilon}^*$ is effective plastic strain rate, $T^* = (T - T_0)/(T_m - T_0)$ is an integration variable of the circumstance temperature, the melting temperature and the room temperature. T_m is the melting temperature of the material, T_0 is the room temperature. In Equation (4), $\left[A + B \left(\varepsilon_e^p \right)^n \right]$, $\left(1 + C \ln \dot{\varepsilon}^* \right)$, $\left[1 - \left(T^* \right)^m \right]$ decried hardening effect of plastic , strain rate effect and Softening effect of temperature, respectively. The material constants in Johnson-Cook model were determined from experiments.

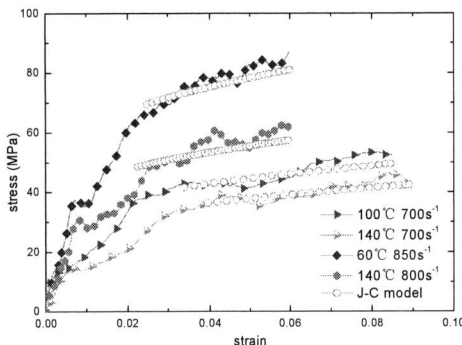

Fig. 8 Comparison of the true flow stress-true strain relation ship of Sn3.0Ag0.7Cu3.0Bi for the experimental data and the Johnson-Cook constitutive equation

Using Johnson-cook model, base on the date obtained by the experiment, through fitting with the result of experiment, we get the A, B, n, C and m of material's parameters, then the dynamic constitutive model that can describe the relationship that plastic flow stress with equivalent plastic strain, equivalent plastic strain rate and temperature effect was gotten. In different temperatures and different strain rate coupling, the model can predict the material's plastic flow stress effectively under the actual condition, in order to provide the basis for further researching the reliability of solder joints. Fig. 8 shows the comparison result of the true flow stress-true strain relation ship of Sn3.0Ag0.7Cu3.0Bi for the experimental data and the Johnson-Cook constitutive equation.

Basing on the experimental data, the five material's parameters were objected, and the Johnson-Cook constitutive equation is

$$\sigma_e = \left[27.3 + 60 \left(\varepsilon_e^p \right)^{0.7} \right] \left(1 + 0.159 \ln \dot{\varepsilon}^* \right) \left[1 - \left(T^* \right)^{1.36} \right] \quad (5)$$

This equation can describe the relationship that plastic flow stress with equivalent plastic strain, equivalent plastic strain rate and temperature effect.

Conclusions

This study has investigated the effects of strain rate and temperature on the dynamic mechanical response of Sn-3.0Ag-0.7Cu-3.0Bi solder alloy. The results have shown that the flow stress increases with increasing strain rate, but decreases with increasing temperature. For a given temperature, the fracture strain and work hardening rate increase with increasing strain rate. However, for a constant strain rate, the fracture strain and work hardening rate reduce with increasing temperature at temperatures between 25°C and 140°C.

Basing on the date obtained by the experiment, the dynamic constitutive relation of Johnson-cook model was built up. The dynamic constitutive relation including strain-rate effect and temperature effects of this solder joints provide a research foundation for further study the reliability of electronic.

Acknowledgements

The authors gratefully acknowledge the financial support provided to this study by the National Natural Science Foundation (The experimental and numerical research of dynamic failure mode and some relative mechanics problem of lead-free solder material electronic package chip, No. 10672113).

References

1. S. T. Jenq, Hsuan-Hu Chang, Yi-Shao Lai, Tsung-Yueh Tsai, "High strain rate compression behavior for Sn-37Pb eutectic alloy, lead-free Sn-1Ag-0.5Cu and Sn-3Ag-0.5Cu alloys," *Microelectronics Reliability*, Vol. 49, No. 3, (2009), pp. 310-317.
2. Fulong Zhu, Honghai Zhang, Rongfeng Guan, Sheng Liu, "Effects of temperature and strain rate on mechanical property of Sn96.5Ag3Cu0.5," *Journal of Alloys and Compounds*, 438 (2007) pp. 100-105.
3. O. S. Lee, M. S. Kim, "Dynamic material property characterization by using split Hopkinson pressure bar (SHPB) technique," *Nuclear Engineer and Desigh*, 226 (2003), pp. 119-125.
4. John O. Hallquist, "LS-DYNA Theoretical manual," Livermore software technology corporation, 1998, pp. 16-26.

Thermal Stress Analysis and Structural Optimization of Ultra-thin Chip Stacked Package Device

Li Li, Xiao-song MA, Xi Zhou
Dept of Mechanical & Electronic Engineering
Guilin University of Electronic Technology, Guilin 541004, China
E-mail: lilissfe@163.com

Abstract

Electronic package devices often endure a substantial number of thermal loading during working. Due to the mismatch of the materials' CTE, the thermal stress can accumulate in the device's interior, which would cause device failure such as die crack, warpage and so on. Especially for the popular multi-chip stacked package, single-chip requires thinner, the tensile strength of chip becomes very small correspondingly. Under the thermal stress the reliability of ultra-thin chip appears more important. At present, little is seen on the study of ultra-thin chip stacked package's reliability at home. In this paper, a 2D parametric finite element model of six ultra-thin die stacked package by QFN is built through element analysis software. Thermal stress distribution and warpage deformation of the device after reflow loading were analyzed. Based on this, by adopting uniform design method and regression analysis, select part of sensitivity structural parameters to optimize the device. The purpose is minimizing thermal stress of the ultra-thin die. The results show that the maximum stress appears at the bottom passive chip; the structure of the largest warpage is at the upper corner of the EMC. From center to out of the package, the deformation appears more and more serious; the thickness of copper pad, die and EMC is the key factors. Thermal stress can be effectively lowered down by choosing optimal structural parameters. The results provide a theoretical basis for the structural size design and could improve the reliability of the package.

1 Introduction

In recent years, with the development of network technology, electronic equipment requires multi-function, high reliability, small size, easy to carry, the demands of device dimension is also getting smaller and smaller [1]. The development and progress in many respects such as requirements of device miniaturization, improvement of package structure, as well as to reduce thermal resistance and improve heat dissipation capability of the chip, require the chip become increasingly thin and high quality correspondingly [2]. For the current prevalence multi-chip stacked package, as try to maintain the total thickness of package invariability, the single-chip requires thinner, so the tensile strength of chip becomes smaller correspondingly. The reliability problem of ultra-thin chip appears all the more important under thermal stress, such as ease to appear warpage, crack and other failure phenomena [3-4]. These problems are mainly caused by the mismatch of various material properties and geometric dimensions, as well as non-uniform temperature distribution in the packaging process and so on. They will seriously affect the reliability of electronic devices, welding performance and yield, which would become an obstacle to further development of electronic packaging technology. Therefore, it is necessary to study the reliability issues of ultra-thin chip stacked package.

In this paper, adopt finite element numerical simulation method, select multi-layer ultra-thin chip stacked package device as study objective. Analyzed warpage and thermal stress distribution of the device after reflow loading. For the crack problems of stacked chip, investigated the impact of sensitivity material thickness on thermal stress by using uniform experimental design method combined uniform design software. Then optimized the regression equation through Matlab software. Finally a more optimal QFN stacked package structure was obtained.

2. Finite element model

2.1 Geometry model

The geometry model of ultra-thin chip stacked package is shown in Fig. 1. The six ultra-thin chips have the same size. Between the six chips micro-solder joints interconnection method is used and the filler is underfill. The passive chips and the active chip of the device connect by flip-chip structure. The exchange of information with outside through lead frame is by virtue of solders. The structural size is as follows: the whole size: $6 \times 6 \times 0.95 \text{mm}^3$; passive die: $5.1 \times 5.1 \times 0.05 \text{mm}^3$; active die: $2.1 \times 2.1 \times 0.2 \text{mm}^3$; copper pad: $3.5 \times 3.5 \times 0.07 \text{mm}^3$; the height of glue: 0.06mm; the height of underfill: 0.015mm; the height of lead frame: 0.2mm; the diameter of solder: 0.3mm; the diameter of bump: 0.02mm. Ultra-thin chips from down to up is named die1, die2, die3, die4, die5, die6 in order.

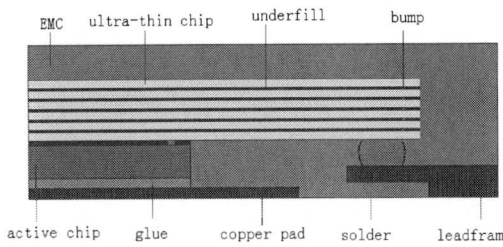

Fig. 1 Diagram of 1/2 geometry model structure

2.2 The built of finite element model and load

A 2-D plane strain parametric FEM model is established to simulate the thermal-mechanical of the device, which is assumed to undergo reflow loading. Because of the symmetry, only one half of the structure is established. All the solder bumps in device are structured to rectangular approximately in order to facilitating the calculation. The material property parameters are listed in Table1. Temperature-dependent Young's modulus of EMC and underfill are shown in Tab. 2 and Tab. 3, respectively. Reflow temperature profile is shown in Fig. 2. The entire reflow soldering process lasts 300 seconds, the peak temperature is 250°C. The initial temperature is 20°C at which the device is assumed stress-free.

Tab. 1 Material property parameters

Materials	E (GPa)	γ	CTE (ppm/°C)
EMC	See Tab. 2	0.35	T<Tg,α=8; T> Tg,
underfill	See tab. 3	0.3	T<Tg,α=23;T>Tg,
Glue	0.1	0.3	152
Die	169	0.23	3
Bump	Sx=32,	0.4	25
Solder	Sx=35,	0.4	21
Copper	120,	0.33	17.5

Tab. 2 The Young's modulus of EMC

t/°C	-65	25	50	75	100
E/MPa	20250	20241	19985	19345	12577
t/°C	125	175	200	225	250
E/MPa	2731	812	717	692	701

Tab. 3 The Young's modulus of underfill

t/°C	-65	25	50	75	100	125
E/Mpa	9000	8436	7220	6034	5822	4674
t/°C	140	175	200	225	250	140
E/Mpa	1075	790	347	205	145	1075

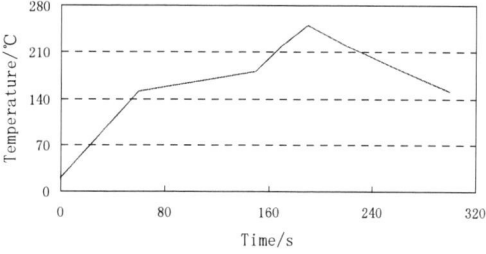

Fig. 2 Temperature profile of reflow

3 Analysis of thermal stress and warpage

Fig. 3 shows the Equivalent Von Mises Stress of package after the whole reflow process is over. Fig. 4 shows the relationships between stress and distance. Fig. 3 and Fig. 4 show that the maximum stress value locates in the bottom passive ultra-thin die; the suspension area of this die appears high stress concentration. And as the mismatch of the materials' CTE, the maximum stress value (116.2Mpa) lies in the junction of the bottom die, EMC and underfill materials. So it's easy to occur delaminate and crack along the die and underfill, and could lead to the solder failure. Besides, the stress reduces gradually along the die center and edge.

Fig. 5 shows the Equivalent Von Mises Stress of six ultra-thin dies changing with time, in which the six

maximum stress nodes of each ultra-thin die are selected respectively. The maximum stress value of each die appears in the high temperature reflow phase; the stress of the bottom die is largest all along, which is much larger than that of other five. The peak value of the bottom die reach up to 160.2MPa, and the more the die lies up, the smaller the stress is. The distribution of tensile stress is similar to the Equivalent Von Mises Stress. And the maximum value is as high as 95.6MPa. Tensile stress can be used as indicator to determine the vertical crack of die [5-6]. When the stress value approach or exceed the scope of tensile stress of 100-200MPa, or the compressive stress is more than -500~-600MPa, it is likely to occur vertical cracks in corresponding position of die so as to lead to failure. Especially when considering the defects which are formed in manufacturing process of ultra-thin die, it is more likely to crack [7].

Reflow temperature up to peak takes a very shorter period of time, and die would endure considerable thermal stress in instantaneous peak temperature, especially for ultra-thin chip. Therefore, it's worth while to design a reasonable reflow profile.

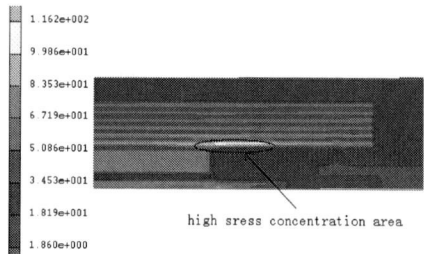

Fig. 3 Equivalent Von Mises Stress of package (unit: MPa)

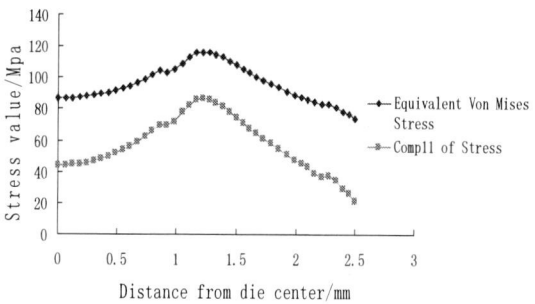

Fig. 4 The relationships between stress and distance

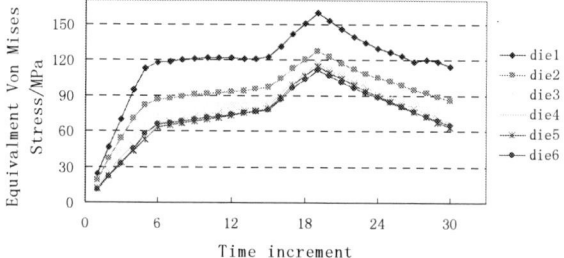

Fig. 5 The stress value of each ultra-thin die

978-1-4244-4658-2/09 $25.00 © 2009 IEEE

From the analysis above, for the typical ultra-thin chip stacked package product, the high-stress area usually appears in the bottom ultra-thin chip under reflow process. Compared with other dies, the bottom die is most likely to occur vertical cracking because of thermal stress, which is also the very dangerous area in the package body.

Fig. 6 shows warpage of the package body and ultra-thin chips (for ease of observation, the warpage model is enlarged 10 times). The structure of the largest warpage is at the upper corner of the EMC. From center to out of the package body, the more near the corner, the more severe is deformation. That is to say, four corners of the package body appear upward warpage trend and present concave. The largest warpage (0.02826mm) appears in the peak temperature. The ultra-thin dies have the similar warpage trend. The warpage maximum value (0.02209mm) appears at the outside edge of the top die. In addition, the package body keeps upward warpage trend all the time in the whole reflow professes.

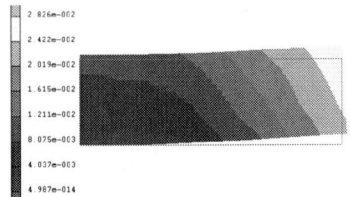

a. Warpage of package body (unit: mm)

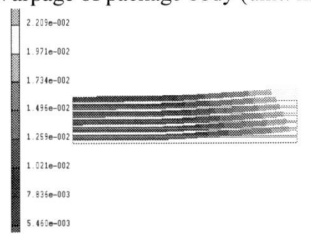

b. Warpage of ultra-thin dies (unit: mm)

Fig. 6 Cloud map of warpage in high temperature phase

4 Structural optimization

4.1 Experimental Design

Uniform design is a mature experimental design method, the basic idea of which is to make reasonable arrangements for the experimental program. It can reflect mathematical model characteristics with less number of experimental data [8]. It also can ensure the experimental points have the uniform distribution statistical properties. At present, uniform design has been used in many areas [9].

In this paper, uniform design is adopted to optimize structural parameters of the package. Based on the simulation results above, select the maximum stress of the bottom ultra-thin chip as optimization objective, take copper pad (X1), glue(X2), die1(X3), underfills between ultra-thin dies(X4), EMC(X5) as the optimization factors, and set five levels for each factor. The uniform design table U^*_{20} (20^7) is chosen to arrange experiments, compile corresponding parametric MARC process to simulate the

results. Uniform experimental designs and results are shown in Tab. 4.

4.2 Regression analysis of data and optimization

Uniform design only considers the nature of experimental point's uniform dispersion in experimental range, but not considers the nature of ordered and comparability. So it's necessary to use regression analysis method to analyze the experimental data so as to infer the optimal experimental condition [10]. In this paper, uniform design software (Version3.00) is introduced to analyze data with the whole linear regression analysis method.

Regression equation obtained is as follows:

Y=186.3+220.1*X1-14.68*X2-524.1*X3-52.65*X4-10.40*X5

The multiple correlation coefficient of model is R=0.9534, the amended coefficient of determination is Ra=0.9847, the analysis of variance is shown in Tab. 5. Test value is F_t=27.96. Critical value is F (0.05, 5, 14) =2.958. Therefore the equation appears significant level in a=0.05; residual standard deviation is s=1.834, the model has a high prediction accuracy. After the relevant statistical test, the effect of the regression equation is significant and highly relevant. So it can be used as the objective function of optimization.

Tab. 6 shows the regression contribution of each equation. U(i) is partial regression sum of square. It's evident that the impact of each factor on the objective: X(3)>X(1)>X(5)>x(2)>X(4).That is, the thickness of die1, copper pad and EMC is the main impact factor, as well as the thickness of glue and under fill is followed by.

Tab. 4 Uniform experimental designs and results

num	X1	X2	X3	X4	X5	result
1	0.03	0.08	0.1	0.015	0.85	136.9
2	0.05	0.06	0.1	0.025	0.85	138
3	0.07	0.04	0.08	0.015	0.85	148.3
4	0.09	0.02	0.08	0.025	0.85	148.2
5	0.11	0.1	0.06	0.01	0.95	158.4
6	0.03	0.06	0.06	0.025	0.95	144.5
7	0.05	0.04	0.04	0.01	0.95	161.8
8	0.07	0.02	0.04	0.025	0.95	166.7
9	0.09	0.1	0.02	0.01	1.05	188.6
10	0.11	0.08	0.02	0.02	1.05	196.1
11	0.03	0.04	0.1	0.01	1.05	134.7
12	0.05	0.02	0.1	0.02	1.05	137.2
13	0.07	0.1	0.08	0.005	1.15	149.5
14	0.09	0.08	0.08	0.02	1.15	145.3
15	0.11	0.06	0.06	0.005	1.15	160.4
16	0.03	0.02	0.06	0.02	1.15	143.6
17	0.05	0.1	0.04	0.005	1.25	155.5
18	0.07	0.08	0.04	0.015	1.25	160.5
19	0.09	0.06	0.02	0.005	1.25	183.3
20	0.11	0.04	0.02	0.015	1.25	195.6

978-1-4244-4658-2/09 $25.00 © 2009 IEEE

Fig. 7 shows the impact of each factor on the optimization objective when keeping other factors unchanged, which is done with uniform design software. It's very evident that thermal Stress increases with the thickness of copper pad increases, but reduces with the thickness of die1 increases obviously. It also reduces with the thickness of EMC, glue and underfill increased but not evident.

As limited number of experiments, much more interactions (for example: X1*X3, X2*X5, X3*X3) weren't been considered when calculated with software, which led to linear relationship between the factors (X) and the indicators (Y) of the regression equation. Therefore all the factors appear peak value when the indicator is supreme.

Then select the regression equation analyzed above as the objective function, using the optimization toolbox of MATLAB software optimize the equation. Boundary constraints are the range of the design variables. The optimized combination of size variables is as follows:

X1=0.03; X2=0.1; X3=0.1; X4=0.025; X5=1.25

The optimal solution is 124.7MPa. The result after experimental simulation verification is 127.2MPa, which is close to the optimization result. Compared to the original result, 160.2MPa, the value reduced to 20.6% or so. Moreover, the maximum warpage of the ultra-thin dies is also slightly declined correspondingly after this optimization (from the initial value, 0.02209mm, reduced to 0.01862mm). The result has achieved the aim of the experimental optimization.

Tab.5 Variance analysis of the regression equation

Variation source	Regression	Residuals	Summation
Sum of square	u=6528	Q=653.8	L=7182
Degree of freedom	k=5	N-1-K=14	N-1=19
Mean Sq	u/k=1306	Q/(N-1-K)=46.7	—
Sum Sq	Ft=27.96	α=0.05	F(0.05,5,14)=2.958

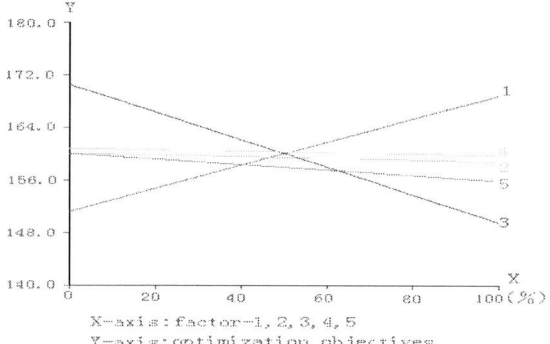

X-axis: factor-1, 2, 3, 4, 5
Y-axis: optimization objectives

Fig. 7 The impact of each factor on the optimization objective

Tab. 6 The contribution of each equation to Regression

	U(1)	U(2)	U(3)	U(4)	U(5)
U(i)	562.4	2.502	2630	1.659	25.85
u(i)/u	8.62%	0.038%	40.3%	0.025%	0.396%

Conclusions

(1) After reflow loading, the maximum stress value locates in the bottom passive ultra-thin die; the more the ultra-thin die is upward, the smaller the thermal stress is. The bottom ultra-thin die is most likely to occur vertical crack, which is also the very dangerous area in package body.

(2) For the whole reflow process, the structure of the largest warpage appears at the upper corner of EMC. Four corners of the package appear upward warpage trend and present concave. The ultra-thin dies have the similar warpage trend. The warpage maximum appears at the top die.

(3) Impact of the thickness of copper pad, bottom die and EMC on the objective is very obvious, yet the thickness of glue and underfill is not distinct relatively. After optimization, compared to the original result, the stress value reduced 20.5% or so.

In this paper, only from the perspective of optimize the material size, select part of sensitivity structural parameters to optimize the device. As lots of reliability issues such as die crack and warpage exist simultaneously, multi-objective optimization from various aspects such as material properties and process condition should be considered.

References

1. K. Hara, "The trend of packaging technology in Japan," *Proc. ISMP*, Seoul, Korea, Sep. 24-25, 2003, pp. 173-190.
2. CHONG C M, CHEUNG Y M, "Finite element stress analysis of thin die detachment profess," *pro5th Int. Confon. Electronic Packaging Technology*, CHINA: Shang hai, 2003. 44-51.
3. Van Driel, W. D., Zhang, G. Q., *et al.*, "Prediction and verification of process-induced thermal deformation of electronic packages using non-linear FEM and 3D interferometry," *Proc. EuroSimE*, 2002: 362-367.
4. Wu TY, Tsukad Y, Chen WT. "Materials and mechanics issues in flip-chip organic packaging," *Proceedings of 46th electronics and computer technology conference*, 1996, pp. 524-33.
5. Kruis RF, "Numerical Analysis of Thermal Stress Related Failures in IC Packages," *Philips Internal Report*, 1990.
6. Zhang G. Q, H P Stehouwer, "Simulation-base Optimization in Virtual Thermo-Mechanical Prototyping of Electronic Packages," *Benefiting from thermal and mechanical simulation in micro-electronics*, edited by G.Q. Zhang, L.J. Ernst, Kluwer Academic Publishers, 2000, pp. 151-164.
7. McLellan N, Fan N, Liu S, Lau K, Wu J., "Effect of wafer thinning condition on the roughness,

morphology and fracture strength of silicon die," *ASME J Elect Package* 2004, 126: 110-4.

8. Fang, K. T.and Ma Changxing, Orthogonal and uniform experiment design, Science press (Beijing, 2001).

9. Liang Yizeng, Fang Kaitai, Xu Qingsong, "Uniform design and its application sinchemistry and chemical engineering," *Chemometrics and Intelligent Laboratory Systems*, 2001, 58 (1), pp. 43-57.

10. Zhang Lin, Liang Yizeng, JiangJianhui, *et al.* "Uniform design applied to nonlinear multivariateca libration by ANN," *Analytica Chimica Acta*, 1998, 370 (1), pp. 65-77.

Prediction of Bending Reliability of BGA Solder Joints on Flexible Printed Circuit (FPC)

Jianlin Huang[1], Quayle Chen[2], Leon Xu[2], G. Q. Zhang[3]
[1]Guilin University of Electronic Technology,
No. 1 Jinji Road, Qixing Area, Guilin, Guangxi, China, 541004
ext-jianlin.huang@nokia.com
[2]Nokia Research Center (Beijing)
No. 5 Dong Huan Zhong Lu, BDA, Beijing, China 100176
Quayle.Chen@nokia.com
[3]Strategy and Business Development of NXP Semiconductors, Eindhoven, the Netherlands

Abstract

In this paper, the mechanical performance of solder joints of BGA mounted on flexible printed circuit (FPC) was studied using finite element method (FEM). To optimize the electric components layout during the design, it is necessary to predict the bending reliability of components on FPC. Traditionally, 3-point or 4-point bending are common used to study if the substrate is rigid. In order to simulate the free bending condition with soft substrate, a new bending method is taken. That's to fix the FPC at one side and move the other side to bend the FPC. Based on this novel model, the length of the FPC, the height and diameter of the solder joints are respectively taken as the optimized variables. With the variable change, there are many different cases. For one general case, the bending process is divided into two steps. At first, to rotate the FPC about 180° at one side and make the two fixtures parallel; then move the fixture to control the bending. It is found that the stress of the solder joints in first step linearly proportioned increase with the bending. While in the second step, the stress reaches the peak and declines gradually with the two ends of the FPC get closer. Simulation results indicate that by using the longer FPC can improve the reliability significantly. Normally, reliability can be also improved by increasing the height and diameter of the solder joints. While in the translating-induced bending, increasing the diameter will worsen the solder reliability unless the diameter is larger than 0.3mm.

1 Introduction

For the miniaturization of the consumer electronic products and the demands of novel products in the market, many researchers and manufacturers investigated on the development of bendable and stretchable electronics, such as flexible display and flexible printed circuit (FPC). Due to the excellent bendability and twistability, the FPC has been used not only as the connection between two rigid boards, but also as the circuit board instead of rigid PCB in some applications. The performance of FPC was studied by some researchers in recent years [1-2].

To guarantee the reliability of flexible products during services, mechanical reliability tests should be conducted for flexible components and devices, such as bending, twisting, drop etc. During these experiments, the failure phenomenon and statistic data can be collected. While the failure root cause can not be found. Generally, solder joint reliability is critical for the device reliability. Because stress is used for failure criteria of material, it is fine to get the values of solder joints for failure analysis and design optimization. Obviously, the

experiment method is expensive and time consuming. Finite Element Method is proven to be a very efficient tool in IC packaging for design analysis and optimization in the past few decades [3]. This paper uses the FEM to have the reliability prediction of FPC in bending to optimize the electric design.

The 3-point and 4-point bending test are the most often used for the rigid board in the past. The bending tests were conducted by resting the printed wiring board assembly on two support anvils while deflecting the board in the downward direction by displacing the load anvils, which was standardized in JESD22B13. With components mounted, the local strain in the component region will be different from the global PWB strain [3]. Based on the JESD22B13, many papers have been reported on the reliability of electronic devices [4-6]. 3-point bending and 4-point bending are available for traditional PCBs, because these PCBs are rigid enough to support the weight of it and all the components mounted on it. But for FPC, they are not suitable because the FPC will collapse due to rigid components mounted on it.

Some researchers have developed some different methods. For example, Drozd Zdzislaw *et al.* [7] introduced an apparatus which produce the curvature by loading a certain moment on the two fixtures simultaneously. The strain of the PCB could be induced by bending the ends of the PCB to a certain angle. Because of the rigid property of rigid PCBs, the range of the bending angles is very limited. Alexander Ptchelintsev [8] has also presented a methodology to study the FPCs under extreme loading conditions. Various FPCs were bent to different radii by moving one end of the FPC while the other end of the FPC was fixed to a rigid plane. In fact, these two methods were not standardized by the industry until now. It is a little bit troublesome and high-cost to specially design a device to perform such experiments, while it can be achieved easily through finite element method (FEM).

In this paper, based on the methodologies in publication [7] and [8], the simulation models are built to simulation a BGA mounted on the FPC. The length of the FPC, the height of the solder joints, and the diameter of the solder joints will be taken as driven parameters. By analyzing and comparing the Mises stress of the solder joint from the different simulation cases, the design suggestions were given.

2 Finite element model

In this simplified simulation model, there are three parts: the FPC, the fixtures, and the BGA. The dimension of FPC is l×30mm×0.1mm, where l is the length of the FPC. Because the thickness of the FPC is great smaller than the other two dimensions, shell elements are used on the FPC. The BGA

978-1-4244-4658-2/09 $25.00 © 2009 IEEE

dimension is 8mm×8mm×1.1mm. It includes two parts, the mold and the solder array. Normally, the height of the solder joints is varied in practical products. To keep the whole height of the BGA as constant, the height of mold should be changed to a certain value, responding to the changes of the solder joints. For example, if the height of the solder joints was rose from 0.15mm to 0.17mm, in order to keep the whole thickness of the BGA as constant, the height of the mold should be decreased from 0.95mm down to 0.93mm, but the whole thickness was always 1.1mm. For length of FPC and sizes of solder joints are varied from different cases, the detailed size of the FPC and the solder joints are sequentially given in next section. A detailed BGA should be composed of mold, silicon, filling material, die, solder mask. For simplifying the model, only mold and solder joints are modeled. The whole model is shown in Figure 1. It is assumed that the BGA was mounted in the central area of the FPC. Because the model is symmetry, only half of the model is taken to computation.

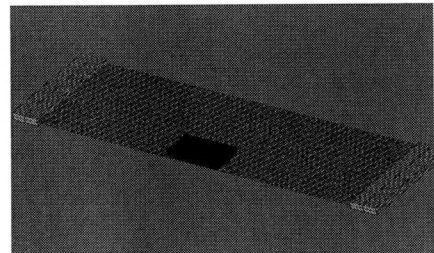

Figure 1 The FEA model

The materials are shown in Table 1 below. The two clamps were assumed to be rigid body due to their high stiffness. FPC and mold were modeled as elastic, while solder joints as plastic.

Table 1 model material data

	Elastic Modulus (MPa)	Poisson Ratio	Yield Stress (MPa)
FPC	2700	0.37	-
Mold	25500	0.3	-
Solder	54000	0.4	38

3 Bending process

The loadings could be divided roughly in two steps. During the two steps, one of the fixtures is fixed called the fixed end. The other fixture is moving and is called moved end. Firstly, the moved end will rotate along the y-axis about 180 degrees. This process is called rotating-induced bending. At the end of this step, the two ends of the FPC faced towards each other and paralleled. Then, translate the moved end of the FPC along the z-axis to control the bending degree. This process is called translating-induced bending. With the distance of two ends of the FPC becoming smaller and smaller, the FPC is bent more and more serious. In the translation-induced bending process, set h as the distance between the moved end and the fixed end of the FPC; and h/l as the bending severity which will be discussed in the following sections. The higher the value, the more seriously

the FPC was bent. The smallest bending severity modeled is 0.1 (h/l) in all cases. The bending process is shown in Figure 2.

To understand the impaction of the FPC length to the reliability of BGA mounted on FPC, six different cases with FPC length of 50mm, 60mm, 70mm, 80mm, 90mm and 100mm were considered. The width of FPCs in these models is the same. The constant height and diameter of the BGA solder joints is assumed in this model. The diameter and height of the solder joints are 0.3mm and 0.22mm respectively. The bending process has two directions according to the FPC bent down or up. When the FPC is bent down, the BGA mounted on it would be outside of the FPC after the FPC bent as an arc; whereas, the BGA is inside the arc. It shows in Figure 3.

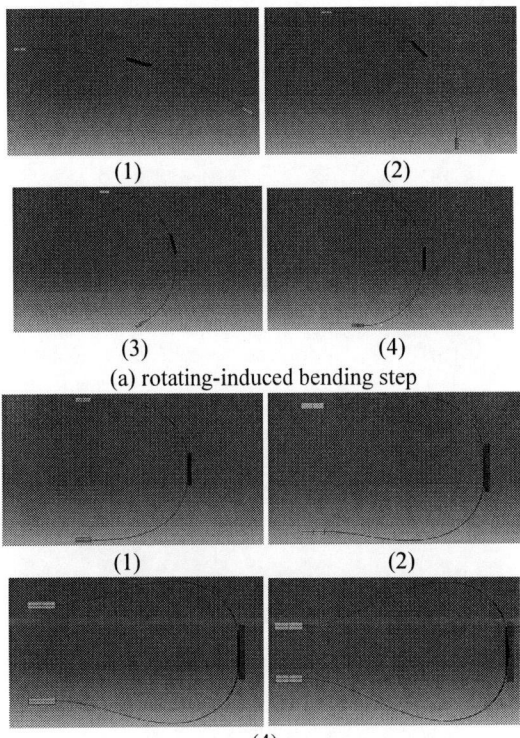

(a) rotating-induced bending step

(b) translating-induced bending step
Figure 2 Bending process, only bend down is shown

(a) component inside the arc   (b) component outside the arc
Figure 3 Bend directions with component inside/outside the arc of FPC

To understand the impaction of the solder joints height, according to the IPC-7095, five kinds of solder ball with different heights are constructed in the model. As Figure 4(a) shown, the diameters of all the five solder balls are 0.3mm,

978-1-4244-4658-2/09 $25.00 © 2009 IEEE

the heights are 0.15mm, 0.17mm, 0.19mm, 0.21mm and 0.23mm respectively, and the length of the FPC is 70mm. When it turns to the simulation of impaction of the solder joints' size on the reliability, only bending down was conducted, as shown in Figure 2.

Effects of the solder balls' diameters to the reliability of the BGA were also studied in this simulation. Solder balls with different diameters were modeled in this section. Because the diameters of the solder joints are different, the pad sizes also differ from each other. According to the IPC-7095, the pad size should be designed as 0.7~0.8 times of the diameter of the solder balls. It is recommended that the pad size for solder joints whose diameters are no more than 0.5mm should be defined as 0.7 times of the diameter. The diameters of the solder joints used in this section are 0.25mm, 0.275mm, 0.3mm, 0.325mm, and 0.35mm, the corresponding pad sizes is designed as 0.2mm, 0.22mm, 0.24mm, 0.26mm and 0.28mm respectively. The height of all these solder joints are 0.21mm, as Figure 4(b) shown. The length of the FPC is 70mm.

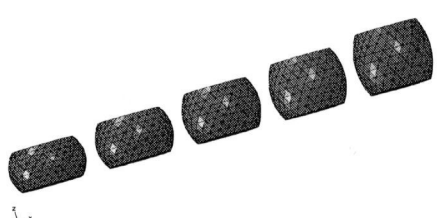

(a) Different heights of solder balls

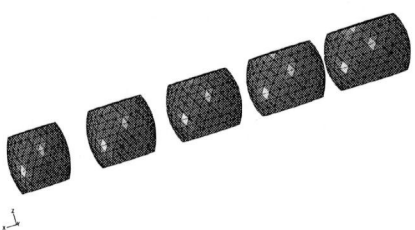

(b) Different diameters of solder balls
Figure 4 solder joints with different sizes

4 Results

1. Length of FPC

From the simulation results, it is found that distribution of the Von Mises stress on the solder joints are symmetry, and the max value occurred at the corner solder joint of the outermost row. The max stress during the rotation-induced bending process occurred at the same location in all the six simulation cases. In the central area of the BGA, the stresses of the solder joints are very small at the final state of the bending. And the outermost solder row which paralleled with the fixtures sustained most of the load during the bending, as shown in Figure 5.

As shown in Figure 6, whenever the FPC was bent down or bent up, the maximum Von Mises stress on solder joints are consistently increased with the bend angle increase. The stress in case of bend up is larger than that of bend down.

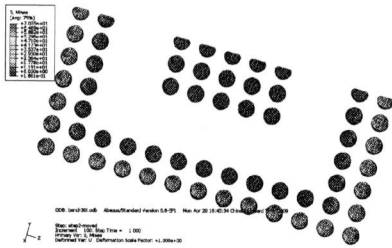

Figure 5 Stress distribution

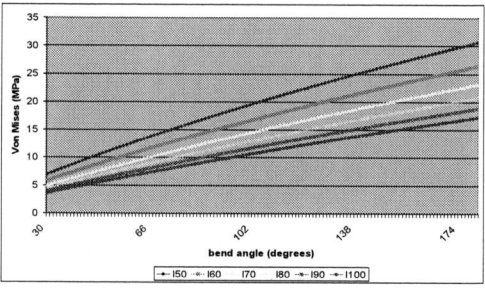

(a) FPC bent down with BGA outside the FPC arc

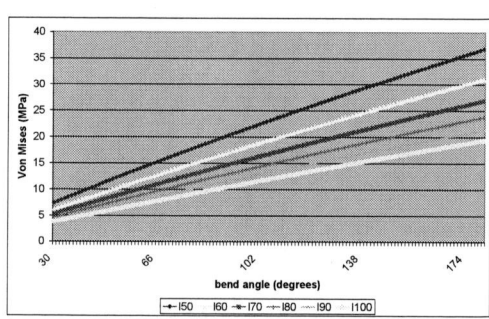

(b) FPC bent up with BGA inside the FPC arc
Figure 6 the stress comparison with different bend angle

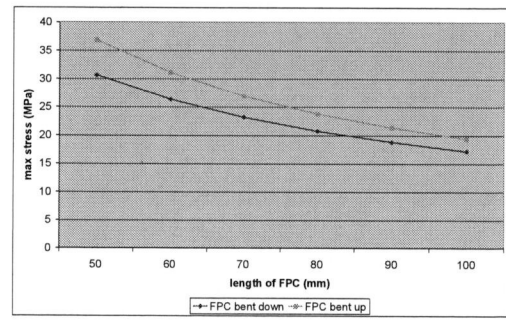

Figure 7 the max stress in case of different FPC length

The Figure 7 shows the max stress during the bending decrease gradually with the increasing of the FPC length. In the cases that the FPC is bent down, the max stress declined by 43.9% when the length of the FPC varied from 50 mm to 100mm; and in the cases that the FPC is bent up, the max stress declined by 47.7%, which is a little higher than stress declining in the cases that FPC is bent down. Compared with the stresses in the same condition, the max stresses in the case that FPC is bent up are higher about 12.8%~20.4% than that of the FPC is bent down. It indicates that it is a little bit safety when layout the BGA in the outer of the FPC bent curve than the inner of the FPC bent curve. Increasing the length of PFC can improve dramatically the reliability of solder joints.

The state of the stress during the translating-induced bending process is more complicated than that in rotating-induced bending process. As Figure 8 shown, are the stress distributions of the FPC with 70mm in length. Both cases, namely FPC bending down and bending up from the bending severity of 0.5 (h/l) to 0.1 (h/l) are plotted in the figure. The stress is increasing at the beginning. After it reaches the peak value, it will decline gradually. It also shows that it is faster that the stress reaches the peak value in case of the FPC bent up than the FPC bent down. When the FPC is bent up, the maximum stress is not proportional with the bending severity, the peak value happened at the case that h/l equals about to 0.23. While in case of FPC bent down, the stress is proportional with the bending severity. With the bending severity drop down, the stress is increasing.

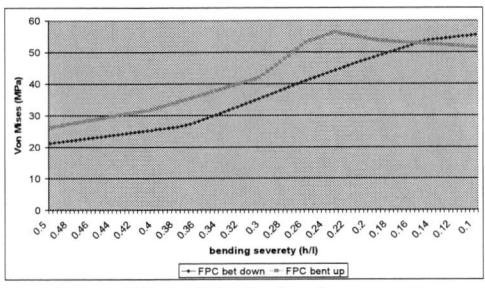

Figure 8 Stress VS. the bending severity

Where h represents the distance between two ends of the FPC, and l is the length of the FPC, h/l is defined as the degree of bending severity. It is the same in the following sections if not specially noted.

The max stresses for the two bending directions during the translating-induced bending process are different. In the case that the FPC is bent down, the max stress declined linearly with the FPC length increasing. The difference of the stress between the case to using 50mm length of FPCs and 100mm length of FPCs is about 34%. While for the case that the FPC is bent up, the max stresses for 6 kinds of FPCs does not change a lot. The values were about 54.8~56.2MPa, and the biggest difference among them is 2.5%. It shows in Figure 9. When the length of the FPC is less than 70mm, one can confirm that BGA mounted inside the FPC bending arc is safer than that outside the FPC bending arc. However, when the length of the FPC is larger than 70mm, BGA mounted outside the FPC bending arc is much safer than that inside the FPC bending arc.

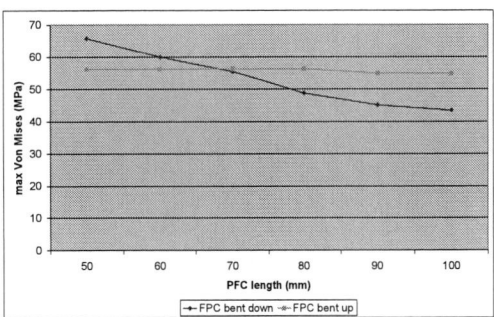

Figure 9 max stress VS. different length of FPCs

2. The height of the solder joints

Figure 10 shows the relationship of stress VS. bend angle in the rotating-induced bending step. The stress increases linearly when the FPC bent from 0° to 180°. Further more, the results from the simulation in this section indicated that the lower the height of the solder balls, the higher stress in the bending.

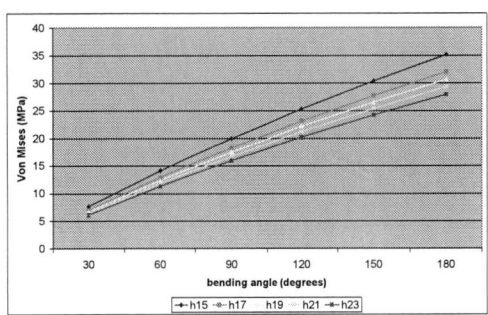

Figure 10 Stress VS. bend angle

Figure 10 also demonstrates that the max stress declines slightly when the height of the solder balls increases from 0.15mm to 0.23mm. The reduction of the stress is more obvious when the height of the solder joint is smaller relatively. The max stress declines by 8.7% if the height of the solder ball is changed from 0.15mm to 0.17mm. While the reduction of the max stress declines by 4.7% if the height of the solder ball is changed from 0.21mm to 0.23mm. It indicates that the reliability of the solder joints can be improved by increasing the height of the solder balls during the rotating-induced bending.

It shows that the height of the solder ball strongly impacts the stress of solders during the translating-induced bending process. Figure 11 shows that the stress of the solder joint whose height is 0.15mm reached its peak value with higher bending severity value. It is about 58.86MPa, and occurred when the bending severity was 0.416 (h/l). Note that stress of the solder joints whose height is 0.23mm does not reach its peak value even the FPC is bent to the bending severity equal to 0.1 (h/l). It indicates that shorter solder joints could experience much less bending severity, compared to the higher solder joints.

978-1-4244-4658-2/09 $25.00 © 2009 IEEE   1117

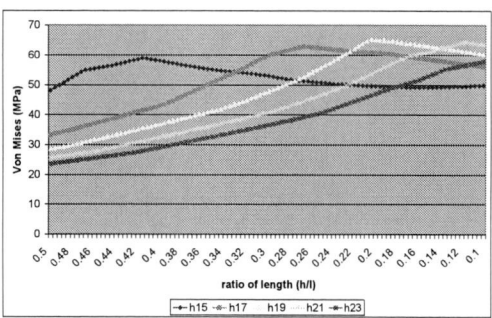

Figure 11 stress VS. bending severity

The Table 2 shows the reliability impaction of the height of solder ball. The max stress in case of using different height of solder ball does not change so much. And the maximum variation between them is about 11.5%.

Table 2 peak stress occurred with the bending severity

Height of solder joint (mm)	0.15	0.17	0.19	0.21	0.23
Peak stress (MPa)	58.86	62.68	65.29	64.34	-
Bending severity (h/l)	0.416	0.268	0.204	0.116	-

3. The diameter of solder balls

Figure 12 shows the results in the rotating-induced bending step. With the diameter of the solder joints increased from 0.25mm to 0.35mm, the maximum stress is decreasing. In this rotating-induced bending step, reliability can be improved by increasing the diameter of the solder joints. The effect is slightly more significant if the diameter is in a relatively small range.

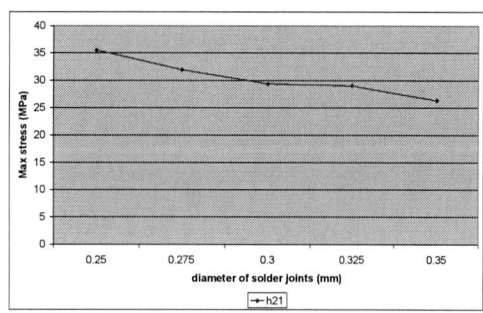

Figure 12 Stress VS. solder joint diameter

As shown in Figure 13, in the translation-induced bending step, when the moved end of the PFC move close to the fixed end, the max stress on solder joints increase firstly, and then decline gradually. There is one difference with the impaction of solder ball height that the peak stresses in different solder ball diameter are more concentrated in bending severity between 0.2~0.1(h/l). As shown in Figure 14, the max peak stress is 64.34MPa and occurred in case of the solder ball diameter is 0.3mm. The peak stress increases when the diameter of the solder joint increased from 0.25mm to 0.3mm. It is decreased with the diameter augmentation. The peak

stress increased by 10.6% when the diameter changed from 0.25mm to 0.3mm.

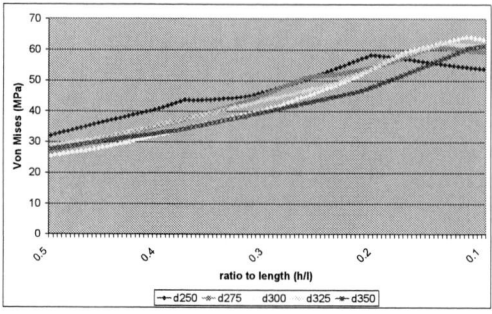

Figure 13 the peak stress concentrated in a small range

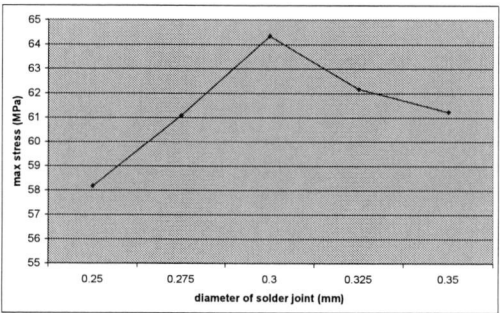

Figure 14 peak stress VS. diameter

Conclusions

The bending reliability of a BGA mounted on the FPC was investigated in this paper. The length of the FPC, the height and diameter of the solder joint were considered as the affecting factors. From the simulations based on the new bending method, it is found that increasing the length of the FPC can dramatically improve the solder joints reliability whenever the BGA is mounted inside or outside of the bended FPC arc. When the length of the FPC is less than 70mm, BGA mounted inside of the bended FPC arc is better than that mounted outside of the arc. The stress in case of using the higher solder joints is lower than that using the shorter ones with the same bending severity (h/l). The stresses are more sensitive with the solder joints height changing within relatively low value. In the rotating-induced bending step, the reliability can be improved by increasing the diameters of the solder joint. In the translating-induced bending step, the reliability can be also improved if the diameter of the solder joints is varied from 0.3mm to 0.35mm. However, if the diameter of the solder joints is changed from 0.25mm to 0.3mm, the reliability would become worse.

References

1. Yin, C.Y *et al.*, "Thermal, Mechanical and Multiphysics Simulation and Experiments in Micro-Electronics and Micro-systems," *EuroSime2006. 7th International Conference*, April.2006, PP. 1-6.
2. Wang, L. *et al.*, "Mechanical characterization of microelectronics embedded in flexible and stretchable

substrate," *Electronics Packaging Technology conference*, December, 2006, pp. 766-772.

3. Jing-En Luan *et al.*, "Advanced Numerical and Experimental Techniques for Analysis of Dynamic Responses and Solder Joint Reliability During Drop Impact," *Components and Packaging Technologies, IEEE Transactions*, Volume 29, No. 3 (2006), pp. 449-456.

4. Che, F.X *et al.*, "Cyclic bend fatigue reliability investigation for Sn-Ag-Cu solder joints," *Electronics Packaging Technology Conference*, December, 2006, pp. 313-317.

5. Graver Chang J. *et al.*, "Cyclic Bending Testing condition effect on the SnAgCu Solder Interconnects in TFBGA Package on Board," *Microsystems, Packaging, Assembly & Circuits Technology Conference*, October, 2008, pp. 181-184.

6. Fa Xing Che *et al.*, "Design for Cyclic Bending Reliability of Large PBGA Assembly Using Experimental and Numerical Methods," *Electronics Packaging Technology Conference*, December, 2008, pp. 1171-1177.

7. Zdzislaw, D. *et al.*, "Failure Modes and Fatigue Testing Characteristics of SMT Solder Joints," *Electronics Systemintegration Technology Conference*, September, 2006, pp. 1187-1193.

8. Ptchelintsev, A., "Automated Modeling and Fatigue Analysis of Flexible Printed Circuits Thermal," EuroSime2006. 7th *International Conference*, April, 2006, pp. 1-6.

Failure Evaluation of Flexible-Rigid PCBs by Thermo-Mechanical Simulation

Luciano Arruda[1,3], Quayle Chen[2], Jairo Quintero[1]
[1]Instituto Nokia de Tecnologia
[2]Nokia Research Center (NRC)
[3]Universidade Estadual do Amazonas
Rodovia Torquato Tapajós, 7200, Colônia Terra Nova, CEP: 69054-415, Manaus, Brazil
ext-luciano.arruda@nokia.com

Abstract

In this present work the finite element method has been used for the simulation models in order to develop tools for the early stages of product design. The objective is to develop simulation models for Flexible Printed Circuits Boards (PCBs) in a flex-rigid concept in the shape of a wrist device to evaluate its critical stress and strain when this device is submitted to thermal loading considering FR4 and Polyimide substrates as the constitutive materials of the board with BGA attached components. The critical points are the mismatch of the coefficient of thermal expansion of different materials as well as the thermoset viscoelastic nature of the polyimide. In the end of this present study it will be shown that additional research should be done to the final product. In this paper we will show the preliminary results of strain and stress distribution induced by thermo loading using a commercial finite element package.

Introduction

Flexible Printed Circuit Boards are being used more extensively in portable electronic devices due to their reduced thickness and ability to bend and adapt to various shapes. These characteristics have a significant role in miniaturization and therefore PCBs are going to prevail in the future of electronic devices.

One of the main causes of failure in electronic devices is related to heat generated during field use. When analyzing reliability, it is necessary to study induced loads in different fields. In the electronics industry, thermo-mechanical simulation, which analyzes the mechanical loads induced by the temperature field, is gaining ground. The temperature causes the structure to deform, which in turn causes mechanical failure in the electronic device [1].

The PCBs are going to replace rigid boards in numerous electronic devices. Durability and long-term performance are among the primary concerns for the use of these devices in consumer products [2]. For a PCB subjected to long-term exposure at elevated temperatures, the viscoelastic nature of the polymer matrix contained in the flexible part will contribute to macroscopic changes in composite stiffness, strength, and fatigue life [3]. Thermal shock tests are widely used by electronics industry and are being a common method to simulate fatigue. This type of experimental test has extreme importance in the electronics industry since they are usually reliable to accelerate the long term behavior of electronic products [1].

Over time, changes in the polymer due to physical aging will have profound effects on the viscoelastic compliance of the material, and hence will affect its long-term durability [3].

In this study the polyimide material is used with visco-elastic material [4-5]. The linear model is the simpler one, but the polyimide as substrate material used in the flexible board of a product is a thermosetting visco-elastic material, which increases the complexity of thermal analysis. It is the glass transition temperature of the polymer matrix that limits the maximum useful service temperature of the polymer composite system [3].

This work is a preliminary step to the main goal which is to obtain the strain level for the product design employing numerical modeling. The finite element method here will be applied in order to improve the reliability analysis techniques to evaluate the failure of electronic boards made with polyimide and FR4 substrates.

Finite Element Modeling

The prototype was proposed in order to evaluate the reliability of flex-rigid interconnection. It consisted of a rigid PWB segmented into three parts which were connected by flexible parts (PCBs). The PCB layout is shown in Figure 1.

The simulation model is shown in Figure 4 in a 3D solid element mesh. The rigid-flex board was modeled in ANSYS [6] using the solid element type of SOLID45 for each layer. It is important to observe that this model is a full 3D mesh where each layer was meshed with this 3D solid element. Table 1 below summarizes the solution options adopted.

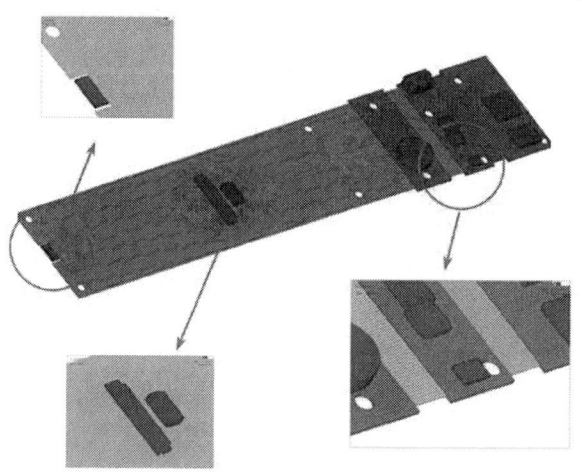

Fig. 1 Global and Detail's mesh used in a flex-PCB model.

Table 1 FEM Analysis Options Used

Options	Employed
Problem Dimensionality	3-D
Degrees of Freedom	Ux, Uy and Uz
Analysis Type	Static (Steady-State)
Offset Temperature from Absolute Zero	273
Equation Solver Option	SPARSE
Newton-Raphson Option	Program Chosen
Globally Assembled Matrix	symmetric

Table 2 The Elements of Mesh in 3D Geometry [6]

Element Type	Name	Description
20 Node Quadratic Hexahedron	Solid186	20 Node Structural Solid
20 Node Quadratic Wedge	Solid186	20 Node Structural Solid
Quadratic Quadrilateral Contact	Conta174	3D 8 Node Surface to Surface Contact
Quadratic Quadrilateral Target	Target170	3D Target Segment
Quadratic Triangular Contact	Conta174	3D 8 Node Surface to Surface Contact
8 Node Quadratic Quadrilateral Shell	Surf154	3D Structural Surface Effect

Figure 2 shows the BGA components attached to the board by solder joints considered in a full 3-D geometry model.

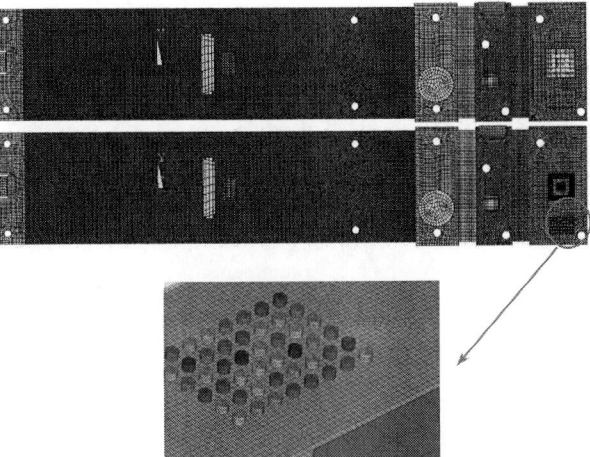

Fig. 2 The solder joints in a flex-PCB and in a 3D mesh detail

The mesh has some complexity to approximate the FEM model to the product design purpose. Different 3D element configuration was used like 20 node quadric hexahedron and wedge 20 node element; quadratic quadrilateral contact and target elements; quadratic triangular contact elements and 8 node quadratic quadrilateral shell element [6]. Table 2 below summarizes these elements related to the software employed in all analyses.

The electronic components were glued to the flexible substrate by contact element. The 3D solder joints geometry was used only in the BGA components. This is the approximation that was chosen in this preliminary stage of the product design. In this way it was possible to study the strains and stresses between the flexible parts and attached components in order to choose the best attachment. Additionally the contact elements were employed to glue the flexible parts to the rigid parts. The position of the contact areas can be seen in Figure 3.

The FE model summary is shown in Table 4 below. The geometry model was built automatically considering the CAD geometry of the prototype [6].

Material Properties

Table 5 shows the linear material properties considered for the computational modeling. The FR4 was considered to be orthotropic with its Elastic Modulus [E], Poison Ratio and Coefficient of Thermal Expansion (CTE). The copper was considered linear isotropic due the stress that was found in the linear region of the strain-stress curve.

Table 3 Typical Contact Properties

Description	Value
Contact Algorithm: Penalt Method	-
Contact detection at: Gauss integration point	-
Contact stiffness factor FKN	10.000
The resulting contact stiffness	0.11916E07
Default penetration tolerance factor FTOLN	0.10000
The resulting penetration tolerance	0.35000E-01
Default opening contact stiffness OPSF	-
Default tangent stiffness FKT	1.000
Default Max. friction stress TAUMAX	0.1000E21
Average contact surface length	0.37615
Average contact pair depth	0.35000
Default pinball region factor PINB	0.25000
The resulting pinball region	0.87500E-01

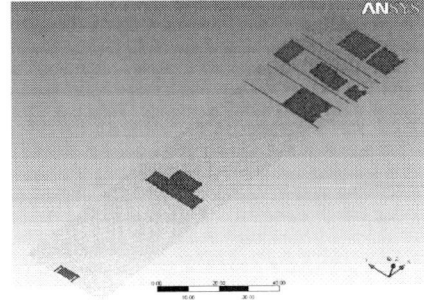

Fig. 3 Contact areas considered in the FE Model

Table 4 FE Model Summary [6]

Description	Quantity
Total Nodes	200662
Total Elements	42601
Total Body Elements	29619
Total Contact Elements	12982
Material	16
Contacts Region	30

Table 5 Linear Material Properties

Material	Elastic Modulus (MPA)	Poisson Ratio	Coefficient of Thermal Expansion ppm/°C
FR4	$E_x = 25,000$	$\nu_{xy} = 0.11$	$\alpha_x = 17E\text{-}6$
	$E_y = 25,000$	$\nu_{xz} = 0.39$	$\alpha_y = 17E\text{-}6$
	$E_z = 11,000$	$\nu_{yz} = 0.39$	$\alpha_z = 60E\text{-}6$
Copper	$E_x = 117,000$	$\nu = 0.30$	$\alpha_x = 17.7E\text{-}6$

The nonlinear properties of polyimide (PI-C) considering the Young modulus and CTE can be seen in Figures 4 and 5 respectively [4]. The solder joint material was represented by Anand Model [6] considering the Darveaux values for SnPbAg [7] in Table 6.

Table 6 Anand Model for ANSYS 62Sn36Pb2Ag Solder [7]

ANSYS	Parameter	Value	Definition
C1	So (psi)	1800	initial value of deformation resistance
C2	Q/k (1/k)	9400	activation energy/ Boltzmann's constant
C3	A (1/sec)	4.0E6	pre-exponential factor
C4	ξ	1.5	multiplier of stress
C5	m	0.303	strain rate sensitivity of stress
C6	ho (psi)	2.0E5	hardening constant
C7	s^(psi)	2.0E3	coefficient for deformation resistance saturation value
C8	n	0.07	strain rate sensitivity of saturation (deformation resistance) value
C9	a	1.3	strain rate sensitivity of hardening

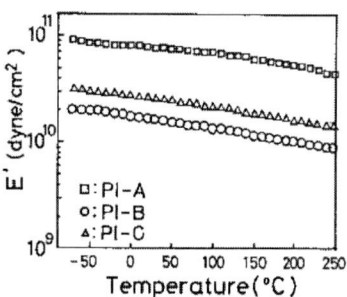

Fig. 4 Temperature dependence of modulus of PI-C [4]

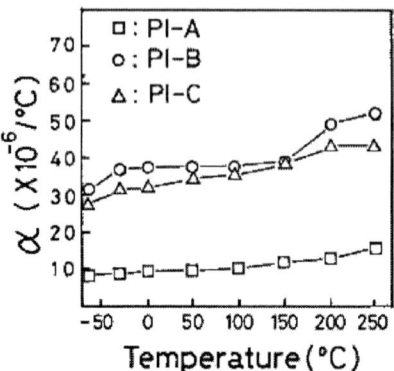

Fig. 5 Temperature dependence of thermal expansion of PI-C [4]

Load Profile

The temperature profile was applied as shown in Figure 6. This type of profile is used to check the mechanical reliability of components and substrates of circuit boards in a thermo-mechanical load conditions. In the Finite Element Method the distribution of temperature is interpreted by contour plot. The critical point of this analysis is the different CTEs (coefficient of thermal expansion) of polyimide, copper, solder joint and ceramics material of populated components.

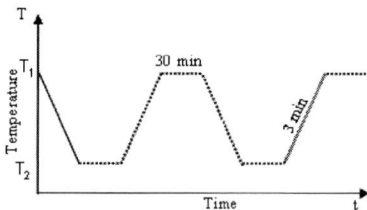

Fig. 6 Applied Thermal Loading. T_1=-40°C; T_2=+125°C

Results

The most critical strain points in the system level analysis have been found to be located near holes, corners of connections in flex-rigid, and near electronic components. The results obtained from simulations (see Figures 7, 8, 9, 10, 11 and 12) provides useful information for the electric layout design and manufacture of electronic products in order to improve the product's quality.

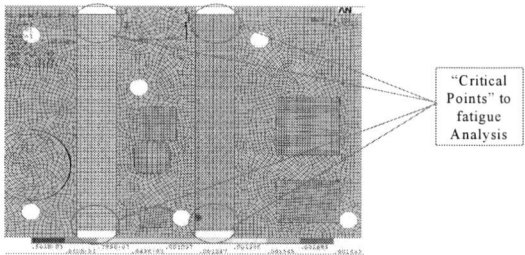

Fig. 7 Details of Von Mises strain measure on Flexible segments between PWB parts

The critical points of strain can be seen in Figure 7. This high concentration of stress is caused by the difference in thickness of the polyimide substrate in relation to the FR4 part, as well as difference in materials.

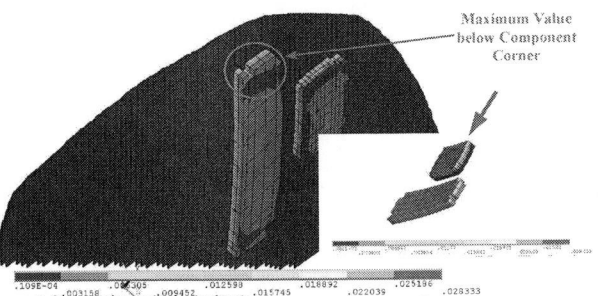

Fig. 8 Von Mises strain between components and flexible substrate part

Figure 8 shows another von Mises strain concentration due to the mismatch of coefficient of thermal expansion of different materials: ceramic (component) and polyimide (substrate). The high makes the glue material fail mainly in shear direction (see Figure 9). This interaction of forces between component-glue-substrate is of major interest to final product design and this simulation shows the necessity to make a new analysis with this focus.

Figure 10 shows the contour of full flex-PCB board and the stress concentration. The maximum stress concentration found in solder joints is shown in Figure 11.

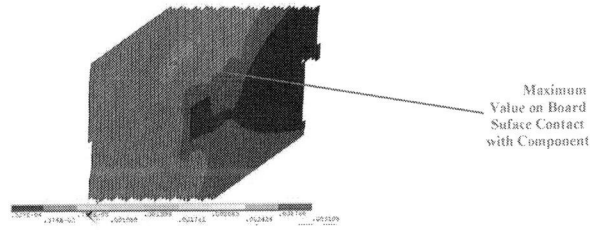

Fig. 9 Von Mises strain on below the components shown in Figure 8

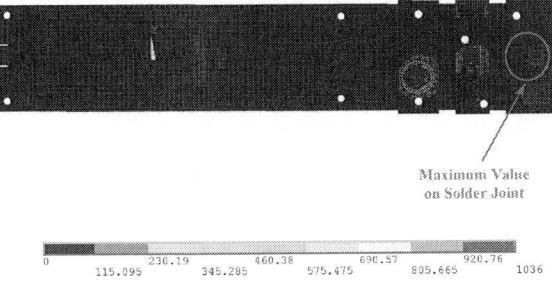

Fig. 10 Von Mises stress contour of full flex-PCB board.

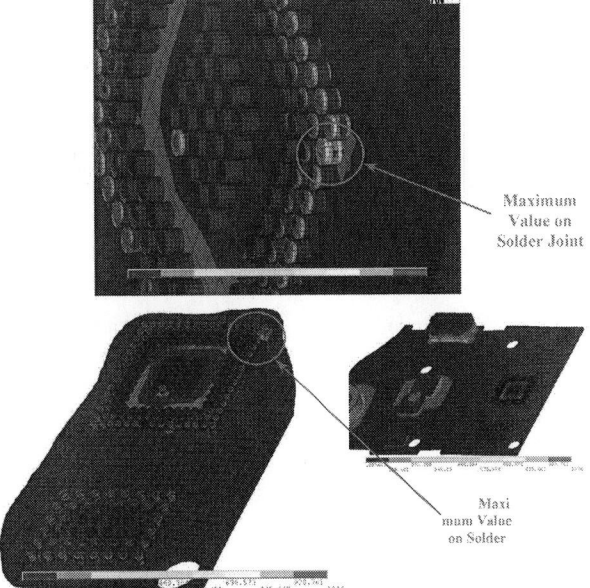

Fig. 11 Von Mises stress contour in solder joints.

The strain distribution in solder joints is critical to the product design quality. Its contour plot is shown in Figure 12. It is important to note that the position on of solder joint with the highest strain is different from the position of the solder joint with the highest stress. The Anand's model is a nonlinear analysis that permits this type of observation as we can find it in the real situation.

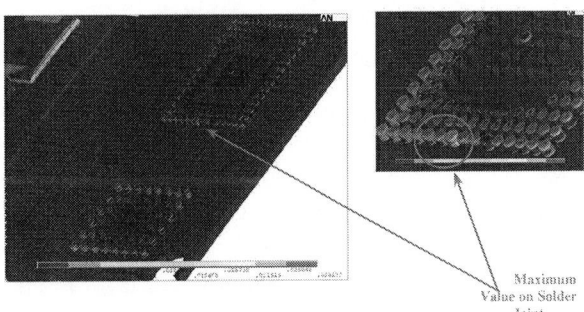

Fig. 12 Von Mises strain contour in solder joints

Conclusions

The failure analysis shows that the probability of microcracks would be in the connection between the rigid and flexible boards and components and flexible parts. The large number of failures is probably related to the design of the board and the differences between the thermal expansion values of materials. Thus, both factors (design and thermal expansion) could contribute to the occurrence of high stresses in the connections.

The proposed simulation model is able to identify the critical regions. Thus, this FEM approach may be considered to be successful when applied in early stages of design. Enhanced FEM simulation models considering non-linear material properties and the fracture model would be the next steps to provide a better model for the evaluation of product design.

Acknowledgments

The authors would like to thank Germano Freitas for supervising the program. Thanks and appreciation also extended to Antti Salo from NRC Beijing as well as all the INdT and NRC members that provided support for this work directly and indirectly

References

1. Arruda L., Bonadiman R., Costa J., Reinikainen T., "Cracking Phenomena on Flexible-Rigid Interfaces in PWBs under Thermo Cycling Loading," *Circuit World*, Vol. 35; No. 2 pp. 18-22 (2009).

2. Coombs' Jr, Clyde F., Coombs' Printed Circuits Handbook, McGraw Hill (2001), pp. 56.3-56.6.

3. Nicholson, L. M. *et al.*, "The Combined Influence of Molecular Weight and Temperature on the Aging and Viscoelastic Response of a Glassy Thermoplastic Polyimide," *NASA/TM -2000-210312. Technical Memorandum*, September, 2002. Hampton, Virginia, USA. pp. 1-2.

4. Takeda, T., Tokoh, A., "Low-Stress Polyimide Resin for IC," *Proceedings of 38th Electronics Components Conference*, Los Angeles, CA, 1988, pp. 420-424.

5. Ward, I. M., Sweeney, J., An Introduction to the Mechanical Properties of Solid Polymers, 2nd Edition, John Wiley & Sons, (2004), England.

6. ANSYS, Software Electronics Manual. Version 11.0 2008, USA.

7. Darveaux, R., "Effect of simulation methodology on solder joint crack growth correlation," *Proceedings 50th Electronic Components and Technology Conf*, Orlando, FL, May. 2000.

8. Zhang, G. Q.; van Driel, W. D.; and X. J. Fan, Mechanics of Microelectronics Solid Mechanics and Its Applications. Springer, 2006, Netherlands.

9. Antonakakes, J. N., Bhargava, P., Chuang, K. C., Zehnder, A. T., "Linear Viscoelastic Properties of HFPE-II Polyimide," *Journal of Applied Polymer Science*, Vol. 100, pp. 3255-3263(2006).

10. Bhargava, P., Zehnder, A. T., "High Temperature Shear Strength of T650-35 HFPE-II-52 polyimde matrix unidirectional composite," *Journal of Experimental Mechanics*, Vol. 46, pp. 245-255 (2006).

11. Banevicius, P., Gydas, J., "Analysis of Stress Relaxation Process in Polymers and High and Low Temperatures," *Materials Science.*, Vol. 10. No. 2, pp. 237-243 (2004).

12. Arruda L., "On the Simulation of Flexible Circuits Boards," *EMPC2007 6th European Microelectronics and Packaging Conference*, Oulu, Finland, June, 2007, pp. 984-989.

13. Arruda, L., Bonadiman, R., Costa, J. & Freitas, G "Experimental and Numerical Analyses of Flexible PCBs under Various Loading Conditions," *Thermal, Mechanical and Multiphysics Simulation and Experiments in Micro-*

Electronics and Micro-Systems, EUROSIME 2009, Delft, NL, April, 2009, pp. 618-627.

14. Wang, J. Z., Parvatareddy, H., Chang, T., Iyengar, N., Dillard, D. A., and Reifsnider, K. L., "Physical Aging Behavior of High-Perfomance Composites," *Composites Science and Technology.* Vol. 54, 1995, pp. 405-415.

Fast Qualification Using Thermal Shock Combined with Moisture Absorption

Xiaosong Ma[1,2], G. Q. Zhang[1,2], K. M. B. Jansen[1], W. D van Driel[3], O. van der Sluis[1], L. J. Ernst[1]
Charles Regard[4,5,6], Christian Gautier[4,5], Hélène Frémont[6]
[1] Delft University of Technology, Mekelweg 2, 2628 CD Delft, the Netherlands
[2] Guilin University of Electronic Technology, Guilin 541004, China
[3] NXP, Gerstweg 2, 6534 AE Nijmegen, the Netherlands
[4] NXP Semiconductors,
[5] LaMIPS, Université de Caen, 2, rue de la Girafe, 14000 Caen, France
[6] IMS Bordeaux Université de Bordeaux, 351 cours de la libération, 33405 Talence, France
Phone: +31-(0)15-2782859, Fax: 31-(0)15-2782150 e-mail:X.Ma@tudelft.nl

Abstract

Time to market is becoming one of the most important factors because of the fierce market competition. However, traditional reliability and interface toughness characterization tests take very long time. For example, moisture sensitivity level assessment (MSL1) will take 168 hours pre conditioning at 85°C/85%RH and tradition thermal cycling takes even longer time. The long preconditioning times are chosen to ensure that also the thicker sections of a package are completely saturated. Thinner package, however, are already saturated after one to two days. In this study, we therefore investigated whether it would be possible to speed up the qualification process by shortening the preconditioning time. We focus in particular on the interface toughness. From our four point bending test and analysis, it is found that temperature has great effects on the interface toughness and moisture also has small effects on the interface toughness. In order to do the fast qualification test, thermal shock cycling tests combined with moisture absorption are performed. Experiments show that moisture can speed up the delamination.

1 Introduction

All packages must pass the qualification tests before they are put into mass production. By tradition qualification, such as MSL and thermal cycling test [1-2], it takes a long time. How could the qualifications and properties characterization be accelerated? The standard tests are stress test driven qualification of integrated circuits [3-4]. Knowledge based qualification are studied by some researchers [5-6]. Knowledge-based reliability qualification differs from standards methods in that it comprehends the end-user environment and failure mechanism based-methods. With this approach, the challenge and tradeoffs of qualification can be more flexibly addressed than with standards qualification methods. In this paper, a moisture absorption model [7-8] is used to determine the saturation time and the interface toughness at different temperatures is measured.

2 Experimental samples and equipment

2.1 Tri-materials sample

In order to characterize the interface toughness, samples are made according to the real package process. The test samples are tri material strips sample with copper (0.2mm in thickness), molding compound (0.6mm in thickness) and a die attach adhesive (60μm) layer in between. The sample dimension is $60\times10\times0.85$ mm^3.

In order to make a uniform glue layer, a foil stencil is made. A thin layer of adhesive glue is dispensed on the surface of the copper lead frame by using the flexible foil stencil, which is fixed in a frame. The thickness of the adhesive glue on the surface of the lead frame is controlled by the thickness of the stencil foil.

After adhesive glue dispensing, samples are placed into the pre heated oven for 15 minutes at 180°C for curing. Molding is finished in 60 seconds in a pre heated mold at 180°C and post cured in the mold for 90 seconds. The final map mold is shown in Fig. 1. Before cutting the sample, the map mold is post cured at 175°C for 4 hours to ensure that epoxy molding compound and glue are fully cured. Then the map molds are cut into strips with the size of 60×9mm^2.

Fig. 1 Map mold   Fig. 2 Cut samples

To trigger the interface delamination, a pre-defined notch (0.5mm wide and 85% deep of EMC) is created in the molding compound epoxy materials. The geometry and dimensions of the sample is shown in Fig. 3.

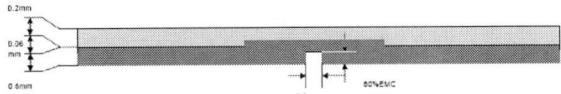

Fig. 3 Geometry and dimensions of the sample

2.2 Setup of Four Point Bending

A special four point bending tool is designed and manufactured to investigate the interface toughness. The schematic test setup is shown in Fig. 4. The four point bending tool, see Fig. 5, consists three parts. The first part is the four point bending frame which is used to support the two rollers. The second parts are the two rollers which are used to support the test sample. For decreasing friction

978-1-4244-4658-2/09 $25.00 © 2009 IEEE   1125

between the rollers and molding compound layer of the test sample, two rollers are allowed to rotate in bearings. The third part is the loading head which apply displacement or load to the sample.

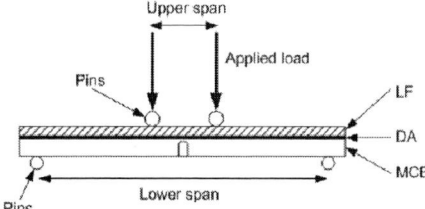

Fig. 4 Schematic overview of four point bending test setup

Fig. 5 Four bending tool

2.3 Loading system and optical system

See Fig. 6, a universal tester Zwick/Roell Z005 is used to apply displacement and measures the reaction force of the loading head. A Keyence optical camera system is placed at the back of the four bending tool. The optical camera focuses on the notch and monitors the deformation and delamination of glue between copper and epoxy molding compound (EMC).

3 Four Point Bending Test Results

3.1 Four point test crack and delamination processes

In order to get more reliable interface toughness date, four point bending tests are performed at different temperatures. Fig. 7 to Fig. 10 typically visualizes the processes of four point bending tests. Fig. 10 shows the relation between displacement and load. Load speed is 0.1mm/min and load speed has no effects on "allowable load" according to our tests at room temperature.

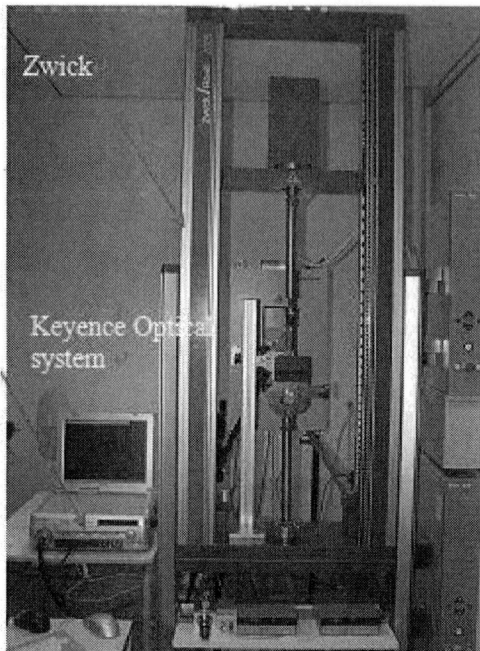

Fig. 6 Loading and Optical system

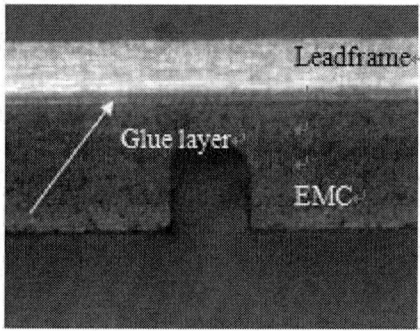

Fig. 7 Initial state of the notch

From Fig. 7 to Fig. 10, local three layers are shown. They are leadframe, glue and EMC from top to bottom layers.

Fig. 7 shows the initial state of the sample before load head touches the sample. The thickness of glue layer is uniformly distributed between copper and molding compound layer due to the controlled stencil foil and is approximately 58μm.

When the load head touches the test sample, the test sample deforms and the response is elastic, see Fig. 11. The applied load increases with the displacement and the load gradually reaches the highest point. At some critical displacement, the load suddenly drops. The EMC notch cracks and penetrates through the molding compound towards the interface, see Fig. 8.

After that, the crack penetrates through glue layer very quickly and reaches the interface between glue and copper, see Fig. 9. The load still increases until interface delamination starts at both sides see Fig. 10. Finally the load begins to decrease, see Fig. 11.

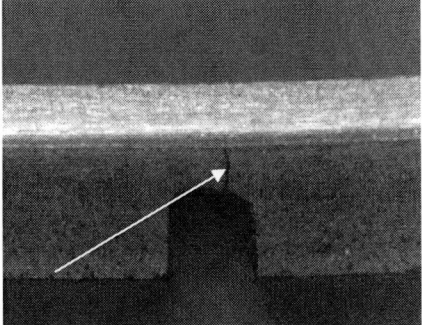

Fig. 8 Initial molding compound crack

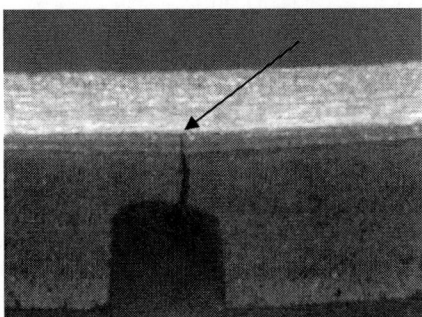

Fig. 9 Initial delamination

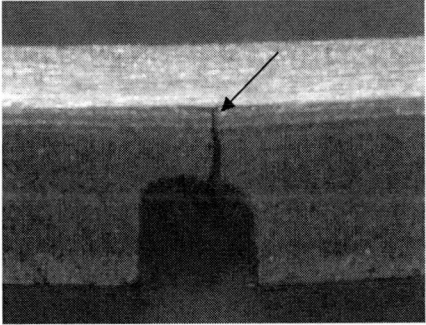

Fig. 10 Delamination propagation

When this load starts decreasing the delamination starts. After both side delamination start then the crack propagation load "stabilizes" around a constant value. From this constant allowable load, the interface fracture toughness value is derived combined with simulation.

3.2 Test result at different temperatures of dry samples

Fig. 12 shows response for four repetitive tests. The stable crack propagation force is reproducible within about 0.1N. The test was repeated for 40, 60, 85 and 150°C. Only the detail results for the 150°C tests are shown here (Fig. 13). Fig. 14 shows the response of four points bending at different temperatures and the results shows that temperature has great effects on the critical crack propagation load. This means that the interface toughness decreases with the increasing of temperature.

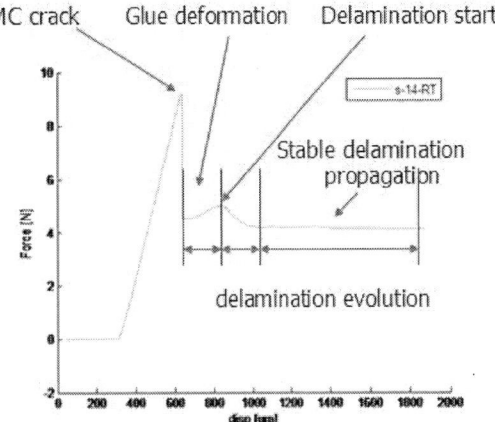

Fig. 11 Relation between displacement and load

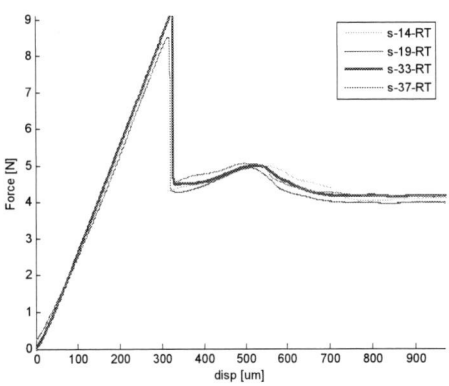

Fig. 12 Response of four point bending at room temperature

3.3 Four Point Bending Test results of PRECON Samples

In order to do fast interface toughness characterization, test samples are put in the humidity oven at 85°C/85%RH for at least 48 hours in order to reach moisture saturation. Then moisture absorption time (tradition 168 hours) is greatly shortened. Four point bending tests are performed at room temperature and 85°C respectively after this PRECON.

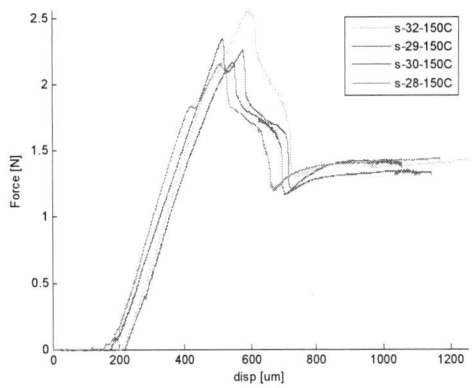

Fig. 13 Response of four point bending at 150°C

978-1-4244-4658-2/09 $25.00 © 2009 IEEE

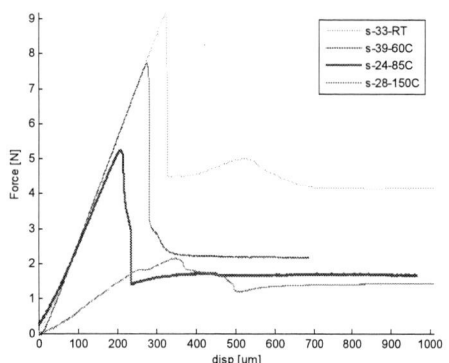

Fig. 14 Response of four point bending at different temperatures

Test results show that at room temperature the average crack propagate load of the PRECON sample decreases a little compared to that of the dry sample. This is the same as other researchers' results because experiments are all done at room temperature [9]. However, at 85°C the average crack propagate load of the PRECON sample increases a little compared to that of the dry sample, see Table 2. This maybe is because moisture could make the glue even stickier such that the interface toughness increases. Some other glue experiments also show that the interface toughness at high temperature is higher than that of low temperature. The experiment is finished in about 15 minutes therefore only a very small amount of moisture evaporates.

3.4 Calculation of the interface toughness

3.4.1 Analytical model

The critical interface fracture toughness, G_c can be deduced analytically by recognizing that it is simply the difference in the strain energy in the uncracked and cracked beam. Since there is negligible strain energy in the beam above the crack, G_c can be deduced from consideration of the energies in the uncracked section, and in a section of the lower part below the crack. Using the Euler-benoulli theory and plane strain conditions, these energies can be expressed in terms of the applied moment M as [10]

$$U = \left(1 + v^2\right) M^2 / 2EI \qquad (1)$$

where U is the strain energy per unit cross-section and I is the second moment of area per unit width. Resulting in

$$G_C = \frac{M^2 \left(1 - v_2^2\right)}{2E_2} \left(\frac{1}{I_2} - \frac{\lambda}{I_C} \right) \qquad (2)$$

where I_2 and I_c are second moment of inertia per unit cross-sectional area for the bottom layer and the composite beam, respectively, and

$$\lambda = E_2 \left(1 - v_1^2\right) / E_1 \left(1 - v_2^2\right) \qquad (3)$$

$$I_2 = \frac{1}{12} h_2^3 \qquad (4)$$

$$I_C = \frac{1}{12} h_1^3 + \frac{\lambda}{12} h_2^3 + \frac{\lambda h_1 h_2 \left(h_1 + h_2\right)^2}{4\left(h_1 + \lambda h_2\right)} \qquad (5)$$

The subscript 1 indicates quantities relevant to the top layer, whereas the subscript 2 denotes the corresponding quantities for the bottom layer. Subscript C refers to the composite beam. Note that the moment per unit width M= PI/2B, with P being the constant load and I the spacing between inner and outer span. According to Eq. (2) analytical Gc can be obtained. This Gc value can be used initially estimate the crack propagation load in the simulation model. Change the Gc value until it equal to crack propagation load, then Gc is obtained.

3.4.2 Numerical model

In addition to the analytical model, the interface toughness was also derived using numerical simulation.

The four point bending method geometry is reconstructed in the finite element package Marc with its graphical user interface Mentat. Due to the symmetry of the model, only half of the specimen is modeled, see Fig. 15. Furthermore, a roller support fixes the vertical displacement at the place of the lower supports, the half of the critical load is applied as a nodal load at the place of the upper supports and a initial crack length is made in the model.

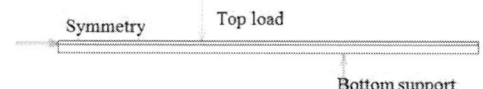

Fig. 15 2D four point bending model

Marc has a library of so-called interface elements, which can be used to simulate the onset and progress of delamination. The constitutive behavior of these elements is expressed in the terms of tractions versus relative displacements between the top and bottom edge/surface of the elements, see one element in the Fig. 16.

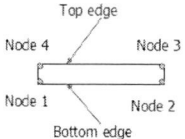

Fig. 16 2D Interface element

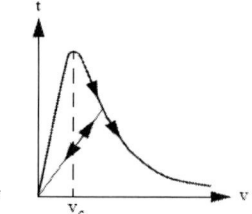

Fig. 17 exponential model

The effective traction t is introduced as function of effective opening displacement and is characterized by an initial reversible response followed by an irreversible response as soon as a critical effective opening displacement v_c has been reached. The irreversible part is characterized by increasing damage range from 0 (onset of delamination) to 1 (full delamination). An exponential function is used in the simulation, see Fig. 17.

978-1-4244-4658-2/09 $25.00 © 2009 IEEE

$$t = G_c \frac{v}{v_c^2} e^{\frac{-v}{v_c}} \qquad (6)$$

in which G_c is the energy release rate (cohesive energy)[11].

3.4.3 Cohesive material mode

It can easily be verified that the maximum effective traction t_c, corresponding to the critical effective opening displacement v_c is given by [11]:

$$t_c = \frac{G_c}{ev_c} \qquad (7)$$

where e=2.718. So if the maximum effective traction is known, the critical or effective opening displacement can be determined by:

$$v_c = \frac{G_c}{et_c} \qquad (8)$$

Damage is defined as:

$$D = \frac{\int_{v_c}^{v_i} t dv}{\int_{v_c}^{v_m} t dv} \qquad (9)$$

$v_c < v_s < v_m$

If D =1 the element is fully damaged.

v_m is at right of v_c but t almost equal to 0.

G_c is obtained from 4 point bending tests combined with simulation of the fitted crack propagation load.

A temperature profile is prescribed as function of time. The temperature profiles consists of different parts. The first part is cooling down from 175°C (stress free temperature) to room temperature in 100 seconds. The following step is heat up to test temperature and to do the four point bending test at assumed temperature. Fig. 18 shows one typical 4 point bending simulation results.

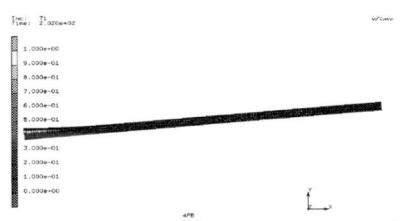

Fig. 18 4 point bending delamination simulation

Table 1 G_c value of Equation and Simulation

Temp	Load (N)		G_c (e^{-3}J/mm^2)		
(°C)	Dry	PRECON	Eq.	Dry	PRECON
23	4.1	4.0	27.2	26.5	25
40	2.75		12.2	13	
60	2.1		7.1	8	
85	1.7	1.76	4.7	5.6	5.8
150	0.71		3.0	1.9	

Table 1 shows the crack propagation load and G_c values of Equation (2) and simulation. From Fig. 19, it can be seen that these curves fit well. Temperature has great effects on the G_c value. It can be seen that the G_c value decreases greatly with the increasing of temperature. Moisture effects on G_c value is small.

3.5 Measurement of the interface critical deformation

Interface toughness and critical opening are two important parameters for simulation. In order to extract critical opening from glue deformation, a special optical camera system is installed combined with universal tester Zwick. The Keyence optical camera system is placed at the back of the four point bending tool frame and focuses on the notch, especially the interface glue location. A series of particular pictures was taken, especially when the load curve starts to decrease after molding compound crack, see Fig. 7 to Fig. 10.

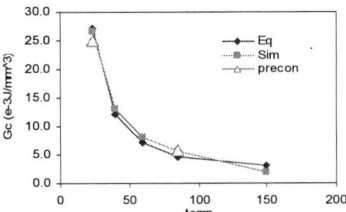

Fig. 19 Temperature and moisture effects on Gc value

According to the pictures taken during the period of the delamination start and propagation, analysis shows that if the deformation is around or less than 3μm the delamination starts. After the delamination start the delamination propagation is very stable and the crack propagation load is almost constant. After fitting the deformation of the glue between the copper and molding compound with simulations, the critical opening parameter can be estimated.

4 Fast Qualifications Design of Experiment (DOE)

In order to know the diference between classical thermal shock and combined moisture/thermal shock (PRECON 15 hours at 85°C /85%RH. A series DOEs are performed. Theses parts of DOEs are performed at Caen France.

4.1 DOE 1

In this DOE, the impact of the PRECON (85°C/85%RH 168 hours) on failure is investigated. Components are not assembled to the printed circuit board. Each batch is composed of 25 parts and combined/thermal cycles are performed up to 200 cycles. Acoustic analyses are performed every 10 cycles and X-ray analysis is performed at 200 cycles. Temperatures at highest and lowest are 150 and -65°C respectively and duration is 15 min respectively.

Table 2 Matrix and results

Non-assembled part			
Number	PRECON	Moisture before shocks	Glue crack rate
batch 1	No	No	24%
batch 1	Yes	No	61%
batch 1	No	Yes	59%
batch 1	Yes	Yes	59%

These results show that the no delamination is found by acoustic analysis. X-ray show that without PRECON the quantity of cracks is much lower. In combined tests,

PRECON has no impact. This means that the accelerated PRECON is equal to the combined moisture/thermal shock.

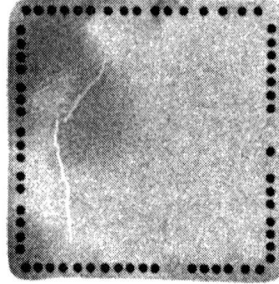 

Fig. 20 crack of glue   Fig. 21 thermal test locations

4.2 DOE 2

Components are assembled on the mini-board. Two batches are divided and each batch is composed of 30 parts. Thermal cycles are performed up to 100 cycles. Following analyses are performed. Thermal measurements are performed every 10 cycles, acoustic analyses are performed at 0, 10, 50 and 100 cycles. X-ray analyses are performed at 0, 10, 50, 100 cycles.

Table 3 DOE Matrix

Assembled part		
Number	PRECON	Moisture before shock
Batch 1	Yes	No
Batch 2	Yes	Yes

No delamination was found by acoustic analysis. X-ray did not show glue crack. Thermal measurements show that the evolution occurs only at the corner pin, which means delamination. Further more, the evolution do not depend on the tests.

5 Results and discussion

From a series four points bending tests, it is found that temperature has a great effect on the interface toughness. G_c value decreases greatly with increasing temperature. Moisture has only a small effect on the G_c value. At room temperature, moisture decreases the interface toughness compared to that of dry condition. However, at 85°C moisture increases the interface toughness compared to that of dry condition. Another commercial glue had been tested. It was found that this kind of glue the interface toughness at 85°C is higher than that at room temperature. The reason maybe is that the glue is stickier at high temperature. Moisture can soften the glue [12] and make glue stickier.

For the DOE 1 tests moisture accelerated the crack failure and the results of combined moisture/thermal shock equals to that of acceleration PRECON. For DOE 2, thermal measurements can show the electrical evolution at corner pin, which means delamination. Combining DOE 1 and DOE 2, it can be seen that the glue crack can be between 100 to 200 thermal shock cycles.

Acknowledgement

Thanks to W. D van Driel, D. G. Yang and Jeroen to help me with the sample making. And thanks to Rob, Jos, Harry and Patrick for the construction of the test setup.

References

1. Moisture/Reflow sensitivity classification for no hermetic solid surface mound device, *IPC/JEDEC J-STD 020D*, June 2007.
2. Temperature Cycling, *JEDEC Standard*, JESD22-A104, 2005.
3. Stress test driven qualification of IC, *JEDEC Standard*, JESD47, 2007.
4. Tress test driven qualification of and failure mechanism associated with assembled solid surface mount components, *JEDEC*, JEP150, May 2005.
5. Understanding and developing knowledge based qualification of silicon devices, *international SEMATECH*, 2003.
6. Knowledge based reliability qualification tests of silicon devices, *international SEMATECH*, 2000.
7. Xiaosong Ma etc, "Charactrization and modeling of moisture absorption of underfill for IC pacakging," *ICEPT Proceedings*, 2007; pp. 380-384.
8. Xiaosong Ma, etc, "Charactrization of moisture properties of polymers for IC pacakging," *Microelectronics Reliability*, Vol. (2007), pp. 1685-1689.
9. W. D van Driel, "Characterization of interface stength as function of temperature and moisture," *IEEE CEPT* 2005.
10. Charalambides P. G., "A Test Specimen for Determining the Fracture Resistance of Bimaterial Interface," *J. Appl. Mech.*, Vol. 56, pp. 77-82, 1989.
11. Marc.mentat regerence, volue A, 2008 r1
12. Xiaosong Ma, etc, "Moisture effects on the creep of thermosetting IC packaging polymers," *Eurosime* 2006, pp. 262-266.

Controlling the Morphology and Orientation of Cu_6Sn_5 through Designing the Orientations of Cu Single Crystals

H. F. Zou, H. J. Yang, Z. F. Zhang[*]

Shenyang National Laboratory for Materials Science, Institute of Metal Research
Chinese Academy of Sciences, Shenyang, 110016, P. R. China
Tel: 0086-24-23971043, Email: zhfzhang@imr.ac.cn

Abstract

It is found that the morphologies and orientations of Cu_6Sn_5 can be well controlled through designing orientations of Cu single crystal substrates. On (001) and (111) Cu single crystals, Cu_6Sn_5 grains display regular prism-type morphology and align either along two perpendicular directions or along three directions having an angle of 60° between each other. By electron backscatter diffraction (EBSD) method, the preferred orientations of the Cu_6Sn_5 grains are determined to strongly depend on the orientations of the Cu single crystals.

1 Introduction

Sn-Pb alloys have widely been used in industry for many years [1-2]. However, much effort has gone into developing replaceable solders because Sn-Pb alloys could cause serious environmental pollution. Therefore, some lead-free solders (Sn-Ag, Sn-Ag-Cu and Sn-Cu solders) have been developed [1-7]. Among these lead-free alloys, eutectic Sn-Ag-Cu alloy is one of the most promising solders to replace Sn-Pb alloys because of its low melting point; good solder ability and physical properties [4]. In the past, many researchers paid much attention to the reactions and physical properties of the joints between Sn-Ag-Cu solders and Cu [1, 2, 8-10]. But these researchers mainly used polycrystalline Cu substrate, then the scallop-type Cu_6Sn_5 grains with random orientation are often formed on polycrystalline Cu surface, as displayed in Fig. 1.

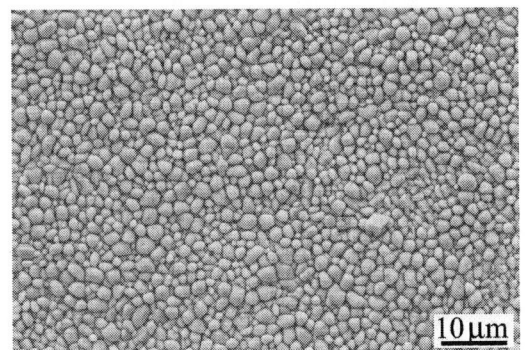

Figure 1 Top-view scanning electron microscope (SEM) image of Cu_6Sn_5 scallop-type grains on polycrystalline Cu

However, in Tu's recent study [11], they found that the rooftop-type instead of scallop-type Cu_6Sn_5 grains were formed on (001) Cu single crystal. In particular, these rooftop-type Cu_6Sn_5 grains have strong preferred orientation. In addition, Prakash et al. [12] observed that Cu_6Sn_5 layer displayed a strong texture feature. Therefore, the nucleation,

growth, and ripening behaviors of Cu_6Sn_5 on Cu single crystal substrate may be quite different from the conventional case of wetting on polycrystalline Cu surface. This gives rise to an interesting question, whether we can control the morphology and orientation of the interfacial intermetallic compounds (IMCs) during soldering or not through designing the orientations of the Cu single crystal substrates. In this article, we employ Sn-3.8wt%Ag-0.7wt%Cu/Cu, eutectic SnBi/Cu, eutectic SnPb/Cu and Sn/Cu couple as examples to demonstrate the effect of crystallographic orientations of Cu single crystals on the morphologies and orientation relationships of the formed Cu_6Sn_5 grains during liquid-state reaction.

2 Experimental procedures

In this study, Cu single crystals were used as substrate and Sn-3.8wt%Ag-0.7wt%Cu, eutectic SnBi, eutectic SnPb alloy or Sn was employed as solder. Firstly, bulk Cu single crystal plate with a dimension of $40\times150\times10mm^3$ was grown from Cu bar with a purity of 99.999% by the Bridgman method in a horizontal furnace [13]. By the EBSD method, the orientation of the Cu single crystal plate was determined. Secondly, some pieces of Cu thin plates with a dimension of $10\times10\times1mm^3$ were spark-cut from the grown Cu single crystal plate, ensuring that the surfaces are parallel to (001) or (111) planes, respectively. These Cu single crystals samples were ground with 800#, 1000#, and 2000# SiC papers and then carefully polished with the 2.5, 1.5 and 0.5μm polishing pastes. Wetting samples were prepared by reacting small Sn-Ag-Cu or Sn solder balls (~0.7mm in diameter) with (001), (111) Cu single crystals at 260°C for 60s. Finally, these samples were quenched in air to room temperature. After that, samples were deeply etched with the 5%HCl+3%HNO$_3$+CH$_3$OH (wt.%) etchant solution to remove the excess Sn phase so that the reactive phases can be well exposed. All the samples were observed with LEO super35 scanning electron microscope (SEM) to detect the IMC morphology.

3 Experimental results and discussion

Figure 2 shows the morphologies of Cu_6Sn_5 grains formed on the (001) and (111) Cu single crystals, respectively. The regular prism-type Cu_6Sn_5 grains displayed on both (001) and (111) Cu single crystal surfaces, which is quite different from that in Fig. 1. And also the morphologies of Cu_6Sn_5 grains on (001) and (111) Cu single crystals are different. The prism-type Cu_6Sn_5 grains on (001) Cu single crystal align along two perpendicular directions, as shown in Fig. 2(a), which is well consistent with Tu's observations [11]. However, the prism-type Cu_6Sn_5 grains align along three different directions on (111) Cu single crystal, in particular, their interaction angles are all equal to 60°, forming many equilateral triangles, as

978-1-4244-4658-2/09 $25.00 © 2009 IEEE

shown in Fig. 2 (b). These phenomena were also detected between molten Sn and (001), (111) Cu single crystals, as displayed in Fig. 2 (c) and (d).

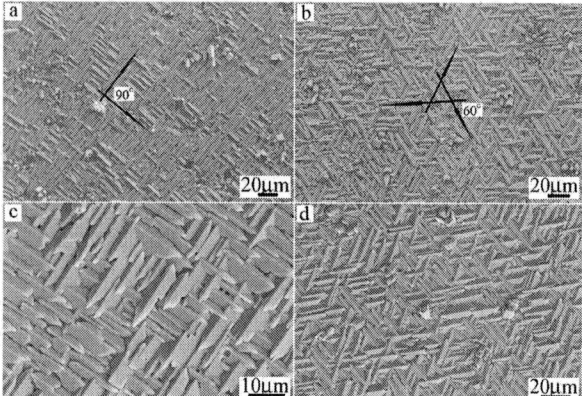

Figure 2 Morphologies of prism-type Cu_6Sn_5 grains formed on (a) (001) and (b) (111) Cu single crystals with Sn-Ag-Cu solder; and (c) (001) and (d) (111) Cu single crystals with Sn solder

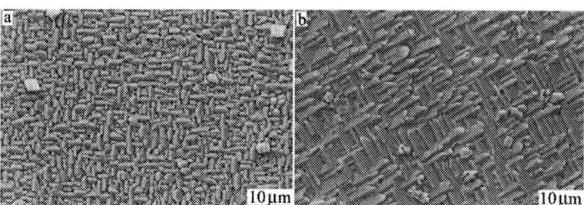

Figure 3 Morphologies of prism-type Cu_6Sn_5 grains formed on (001) Cu with (a) eutectic SnPb solder; (b) eutectic SnBi solder

Besides, the SnPb/(001) Cu and SnBi/(001) Cu were also investigated in this study. It can be found that the regular prism-type Cu6Sn5 grains were also detected, as displayed in Fig. 3. Based on the Fig. 3, it can be found that the size of IMC for the SnBi solder is larger than that for the SnPb solder. It should be attributed to the melting point of the solder. In this study, the reactive temperature is 190°C for the SnPb/Cu, and it is 200°C for the SnBi/Cu. The different of reactive temperature is only 10°C. But the melting points are 183°C, 139°C for SnPb and SnBi solder, respectively. It indicated that the Sn diffusion is more quickly for the SnBi solder than that for the SnPb solder. As a result, the grain size of IMC for the SnBi solder is larger than that for the SnPb solder. According to these phenomena, it can be concluded that the morphologies of the Cu6Sn5 grains can be purposefully controlled to have preferred elongation directions through designing the orientations of Cu single crystals. And also the average length of prism-type Cu6Sn5 grains was measured from Fig. 2. The measured values of the lengths (L) are 9.09μm and 11.38μm for (001) and (111) Cu single crystals, respectively. Figure 4 shows the length distribution of the prism-type Cu6Sn5 grains on the two Cu single crystals; where the ratios $L/\overline{L}$ of the measured length

to the average length were calculated. It is found that the distributions of the Cu6Sn5 grains on the two (001) and (111) Cu single crystals are approximately similar.

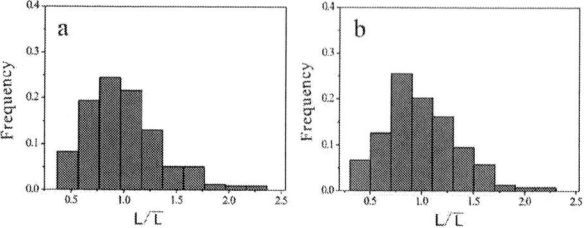

Figure 4 Normalized particle length distributions of the prism-type Cu_6Sn_5 grains formed on (a) (001) and (b) (111) Cu single crystals with Sn solder for 60s

As mentioned above, there should be preferential orientation relationship between the elongated Cu_6Sn_5 grains and the Cu single crystals. And the electron backscattered diffraction (EBSD) technique is a good method to establish the orientation relationships between Cu_6Sn_5 grains and the Cu single crystals. However, the surfaces of these etched samples are so rough that EBSD mapping can not be performed on the whole surface because a smooth surface is required. In addition, another problem for EBSD mapping is the phase structure of Cu_6Sn_5. Larsson et al. have reported that the η-Cu_6Sn_5 phase is stable only at high temperature (above 350°C) and the η'-Cu_6Sn_5 phase is stable at low temperature [14-15]. Thus, the η'-Cu_6Sn_5 phase with a monoclinic structure (C_2/c a=11.022Å, b=7.282Å, c=9.827Å, β=98.84°) was used in our EBSD experiment because the current experimental temperature was 260°C [14]. It is easy to detect the orientation of each Cu_6Sn_5 grain because each small prism-type Cu_6Sn_5 grain in Fig. 2 can be considered as a small Cu_6Sn_5 single crystal. Therefore, the orientation of each Cu_6Sn_5 grain can be obtained by using the Kikuchi band because performing Kikuchi band only requires a rather smooth spot to determine the crystallographic orientation for a small Cu_6Sn_5 grain.

Figure 5 shows the Kikuchi bands of Cu single crystals and the formed Cu_6Sn_5 grains on them [16]. In Fig. 5(a) and (b), the Kikuchi bands were obtained from the (001), (111) Cu single crystals before wetting experiment. Figure 5(c) and (d) are the Kikuchi bands of Cu_6Sn_5 grains with or without indexing elongated along one direction formed on (001) Cu single crystal. In fact, many similar Kikuchi bands can be gotten when we performed EBSD mapping on the prism-type IMC along two perpendicular elongated directions. Further analysis illustrates that the (102) plane of Cu_6Sn_5 grain is parallel to the (001) plane of Cu. Figure 5(e) and (f) are the Kikuchi bands of the elongated Cu_6Sn_5 grains with or without indexing formed on (111) Cu single crystal Through careful analysis, the crystallographic planes of the Cu_6Sn_5 grains are approximately indicated as (010). Therefore, the (010) plane of Cu_6Sn_5 grain is parallel to the (111) plane of Cu for the Cu_6Sn_5 grain formed on (111) Cu. After analyzing all the Kikuchi bands, four orientation relationships between Cu_6Sn_5

grains and (001), (111) Cu single crystals were obtained as follows:

$$(010)_{Cu6Sn5} \parallel (001)_{Cu} ; (102)_{Cu6Sn5} \parallel (001)_{Cu} ,$$

$$(010)_{Cu6Sn5} \parallel (111)_{Cu} ; (102)_{Cu6Sn5} \parallel (111)_{Cu}$$

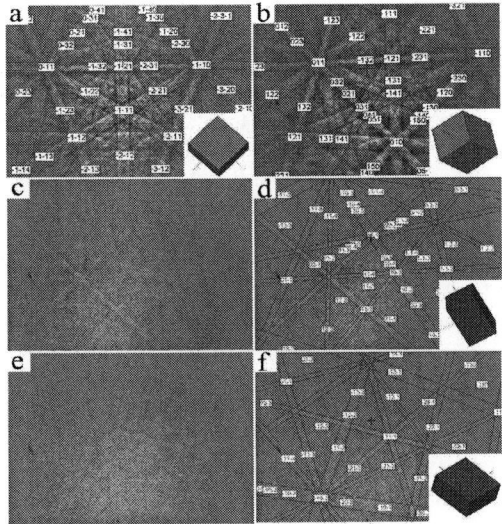

Figure 5 Kikuchi bands of (a) (001) and (b) (111) Cu single crystals; (c) and (d) show Kikuchi bands of Cu_6Sn_5 grains along on direction with and without on (001) Cu single crystal; (e)-(f) Kikuchi bands of Cu_6Sn_5 grains with and without indexing along one direction on (111) Cu single crystal.

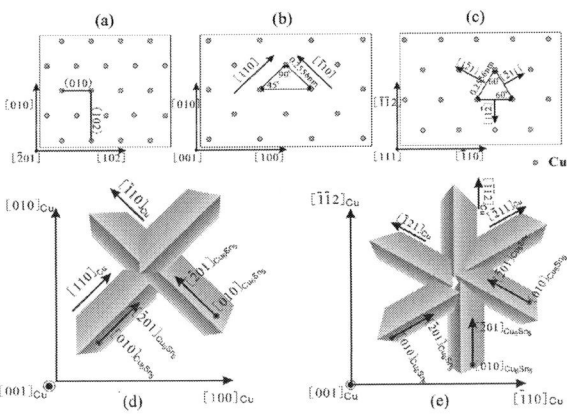

Figure 6 (a) Structure of Cu_6Sn_5 cell projected from $[\bar{2}01]$ direction; (b) structure of Cu cell projected from [001] direction; (c) structure of Cu cell projected from [111] direction; (d) and (e) schematic diagrams of the morphologies and orientations of Cu6Sn5 grains formed on (001) and (111) Cu single crystals.

In our EBSD experiments, the analysis illustrates that the crystallographic planes of Cu_6Sn_5 grains are not exact (102) or (010) plane. The main reason is that the surfaces of the Cu

single crystal samples are not exactly parallel to the (001) or (111) planes due to the error (within 5°) caused by cutting and the subsequent polishing.

As mentioned above, the strong texture of Cu_6Sn_5 grains was only formed on Cu single crystals instead of polycrystalline Cu [11]. It is indicated that the regular array of atoms plays an important role in the formation of those strong textures in Cu_6Sn_5 grains. Figure 6(a) shows the arrays of Cu atoms in Cu_6Sn_5 along $[\bar{2}01]$ direction. Figure 6(b) and (c) show the arrays of Cu atoms in Cu single crystal along [010] and [111] directions, respectively. According to the structure of Cu_6Sn_5, the space between two Cu atoms plane in Cu_6Sn_5 is 0.25478 nm along $[\bar{2}01]$ direction. However, the distance of Cu atom is 0.25560 nm along [110] or $[\bar{1}10]$ direction on (001) plane of Cu, which is the same as that along $[\bar{1}21]$, $[\bar{2}11]$ or $[1\bar{1}2]$ directions on (111) plane of Cu, as illustrated in Figs. 6(b) and (c). Therefore, based on the difference in the atom spaces above, the misfit (α) of Cu atoms would trend to decrease to the minimum value when $[\bar{2}01]$ direction of Cu_6Sn_5 elongated along [110], $[\bar{1}10]$ directions on (001) Cu single crystal surface, and elongated along $[\bar{1}21]$, $[\bar{2}11]$ and $[1\bar{1}2]$ directions of Cu:

$$\alpha = (0.25560-0.25474)/0.25474 = 0.34\%.$$

The result indicated that the misfit of Cu atoms between Cu and Cu_6Sn_5 is extremely low (only 0.34%). In order to minimize the interfacial energy, Cu_6Sn_5 should preferentially nucleate along these low misfit directions after the liquid-state reaction, leading to the formation of the strong texture in Cu_6Sn_5 grains. According to the crystallographic relations illustrated in Fig. 6 (b), it can be found that [110] and $[\bar{1}10]$ directions in Cu are perpendicular to each other. Therefore, the morphology of Cu_6Sn_5 grains formed on (001) Cu single crystal should be preferentially elongated along two perpendicular directions, which is the same as Tu's results [11], as shown in Fig. 2(a). However, the elongated directions of Cu_6Sn_5 grains formed on (111) Cu single crystal should become $[\bar{1}21]$, $[\bar{2}11]$ and $[1\bar{1}2]$. It should be pointed out that the angle between each two $[\bar{1}21]$, $[\bar{2}11]$ and $[1\bar{1}2]$ directions in Cu_6Sn_5 is equal to 60°, as illustrated in Fig. 6(c). This is why the angle between each two elongated directions of Cu_6Sn_5 grains formed on (111) Cu single crystal is 60°, as shown in Figs. 2(b) and (d), which is a new finding from Suh and Tu's result [11]. Figure 6(d) and (e) clearly illustrate the orientation relationships between Cu_6Sn_5 grains and Cu single crystals. For Cu_6Sn_5 grains on (001) Cu single crystal, we have

$$[\bar{2}01]_{Cu6Sn5} \parallel [\bar{1}10]_{(001)Cu} \text{ and } [\bar{2}01]_{Cu6Sn5} \parallel [110]_{(001)Cu}$$

For Cu_6Sn_5 grains on (111) Cu single crystal, the orientation relationships follow

978-1-4244-4658-2/09 $25.00 © 2009 IEEE

$$[\bar{2}01]_{Cu6Sn5} \parallel [1\bar{2}1]_{(111)Cu}, \ [\bar{2}01]_{Cu6Sn5} \parallel [\bar{1}\bar{1}2]_{(111)Cu}$$

$$[\bar{2}01]_{Cu6Sn5} \parallel [\bar{2}1\bar{1}]_{(111)Cu}$$

It is suggested that the regular array and the minimum misfit of Cu atoms should lead to the great differences in the morphologies and orientation relationships of Cu_6Sn_5 grains formed on (001) and (111) Cu single crystals. Morphologies, orientation relationships and evolution of the Cu_6Sn_5 grains formed between molten solder and Cu, Ag single crystals with different orientations can be well discussed elsewhere [16-17].

Conclusions

In summary, the current study demonstrates that the morphologies orientation relationships of Cu_6Sn_5 grains can be well controlled through designing the orientations of (001) and (111) Cu single crystals after liquid reactions with solders. On (001) Cu single crystal surface, the regular prism-type Cu_6Sn_5 grains align along the two perpendicular preferred directions. However, on (111) Cu single crystal surface, the prism-type Cu_6Sn_5 grains are elongated along three preferred directions with the equal angle of 60° between each two directions. In terms of EBSD method, it is confirmed that the preferred orientations of Cu_6Sn_5 grains formed on those (001) and (111) Cu single crystals with low index should take the minimum misfit principle into account.

Acknowledgments

The authors would like to thank W. Gao, P. Li, H. H. Su, X. H. Sang, G. L. Gong, J. T. Liu, F. Yang, J. T. Fan and Q. S. Zhu for sample preparation, SEM observations, crystallographic analysis and stimulating discussion. This work was financially supported by National Basic Research Program of China under Grant No. 2004CB619306, the "Hundred of Talents Project" by Chinese Academy of Science and the National Outstanding Young Scientist Foundation under Grant No. 50625103.

References

1. M. Abtew, and G. Selvaduray, "lead-free solders in microelectronics," *Mater. Sci. Eng.* R27, 95 (2000).
2. K. Zeng and K. N. Tu, "six cases of reliability study of Pb-free solder joints in electronic packaging technology," *Mater. Sci. Eng.*, R38, 55 (2002).
3. S. Park, R. Dhakal, L. Lehman and E. Cotts, "Measurement of deformations in SnAgCu solder interconnects under in situ thermal loading," *Acta Mater.*, 55, 3253 (2007).
4. F. Ren, J. W. Nah, K. N. Tu, B. S.Xiong, L. H. Xu and J. H. L. Pang, "Electromigration induced ductile-to-brittle transition in lead-free solder joints," *Appl. Phys. Lett.*, 89, 141914 (2006).
5. M. Kerr and N. Chawla, "Creep deformation behavior of Sn–3.5Ag solder/Cu couple at small length scales," *Acta Mater.*, 52, 4527 (2004).
6. K. S. Bae and S. J. Kim, "Microstructure and adhesion properties of Sn–0.7Cu/Cu solder joints," *J. Mater. Res.*, 17, 743 (2002).

7. F. J. Wang, X. Ma and Y. Y. Qian, "Improvement of microstructure and interface structure of eutectic Sn–0.7Cu solder with small amount of Zn addition," *Scripta Mater.*, 53, 699 (2005).
8. T. Y. Lee *et al.*, "Morphology, kinetics, and thermodynamics of solid-state aging of eutectic SnPb and Pb-free solders (Sn-3.5Ag, Sn-3.8Ag-0.7Cu and Sn-0.7Cu) on Cu," *J. Mater. Res.*, 17, 291 (2002).
9. B. Y. Wu, H. W. Zhong, Y. C. Chan and M. O. Alam, "Shearing tests of solder joints on tape ball grid array substrates," *J. Mater. Res.*, 21, 2224 (2006).
10. Fouassier, J. M. Heintz, J. Chazelas, P. M. Geffroy and J. F. Silvain, "Microstructural evolution and mechanical properties of SnAgCu alloys," *J. Appl. Phys.*, 100, 043519 (2006).
11. J. O. Suh, K. N. Tu and N. Tamura, "Dramatic morphological change of scallop-type Cu_6Sn_5 formed on (001) single crystal copper in reaction between molten SnPb solder and Cu," *Appl. Phys. Lett.*, 91, 051907 (2007).
12. K. H. Prakash and T. Sritharan, "texture growth of Cu/Sn intermetallic compounds," *J. Electron. Mater.*, 31, 1250 (2002).
13. Z. F. Zhang, Z. G. Wang and Z. M. Sun, "Evolution and microstructural characteristics of deformation bands in fatigued copper single crystals," *Acta Mater.*, 49, 2875 (2001).
14. K. Larsson, L. Stenberg and Lidin, "The superstructure of Domain-twinned η´–S," *Acta Crys.*, B50, 636 (1994).
15. K. Larsson, L. Stenberg and S. Lidin, "Crystal structure modulations in η´- Cu_5Sn_4 Z," *Kristallogr.*, 210, 832 (1995).
16. H. F. Zou, H. J. Yang and Z. F. Zhang, "Morphologies, orientation relationships and evolution of Cu_6Sn_5 grains formed between molten Sn and Cu single crystals," *Acta Mater.*, 56, 2649 (2008).
17. H. F. Zou, H. J. Yang, J. Tan and Z. F. Zhang, "Prefertial growth and orientation relationship of Ag_3Sn grains formed between molten Sn and (001) Ag single crystal," *J. Mater. Res.*, 24, 2141 (2009).

Thermal-Mechanical Failure and Life Analysis on CBGA Package used for Great Scale FPGA Chip

Wenchang Li[1], Xiaojun Zhang[2]
[1]State Key Laboratory of Electronic Thin Films and Integrated Devices, UESTC,
No. 4 Section 2, North Jianshen Road, Chengdu, P. R. China,
Telephone: 028-85177737-270 Email: wenchang@csmsc.com
[2]Hebei Semiconductor Research Institute,
No. 113 Hezuo, Road, Shijiazhuang, P. R. China
Telephone: 0311-87091547-5002 Email: chinapackage@163.com

Abstract

CBGA (Ceramic Ball Grid Array) is one kind of advanced package for FPGA, which can fulfill the demands of high interconnect density, high thermal and electronic performance, high chip-assembling yields, high reliability. But in multiple loading environments, CBGA has the invalidation mode of solder joint thermal-mechanical failure, which will influence the long term reliability of FPGA circuits. But there is few research on the China native made CBGA package. In this paper research of finite element analysis and solder joint fatigue predictive equation have been made on some of CBGA packages, and some failure prediction and ATC test results have been given.

Foreword

Ceramic Ball Grid Array Package (CBGA) is widely used in Large Scale Integrated circuit package, such as high reliable FPGA. But second level assemble reliability is always a problem should be solved by manufacturers and users. The main mode of invalidation of second grade assemblage is the invalidation of solder ball thermal fatigue. This text analyzed the phenomenon and mechanism of thermal fatigue at first, and has provided the corresponding prediction method, at last it provided the test result.

Thermal-Mechanical Failure mechanism

Ball of CBGA package should not only has good electrical connections, but also has enough intensity used for guarantee device connection at being in transport and vibration require. Linear expansion coefficient of ceramic is $6\sim8\times10^{-6}/°C$, But linear expansion coefficient used for making epoxy resin of PCB board is $12\sim16\times10^{-6}/°C$, Under the actual using condition, because their linear expansion coefficients have relatively great differences, there has very big shear stress in ball, it is very easy to get invalid.

Figure 1 shows a sketch map of fatigue of CBGA assembled on PCB, and a photo of cracked joint [1].

When the temperature change, will produce big thermal stress in ball and solder joint of two kinds of ceramic to lead to the fact ball ruptures in solder joint. And according to the difference in parameters such as PCB board thickness, ceramic thickness, solder plate size, the position that stress concentrate is different to some extent.

At the same packages, the ball one relative deformation values and shear stress with as package in the center from increase and increasing, as shown in Figure 2 [2].

Figure 1 a photo of cracked joint

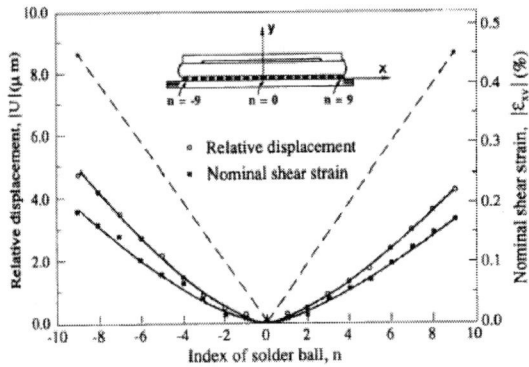

Figure 2 Deformation with stress up to ball increase from the increase of in the center

Take CBGA208 package as an example, adopt finite element analysis result of 1/4 of the models as follows.

The temperature field in one Cooling cycle is distributed as shown in Fig. 3:

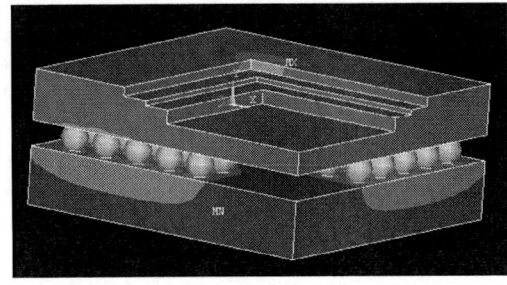

Figure 3 The temperature field in one Cooling cycle

After a circulation, produce greater deformation in ball, as shown in Fig. 4:

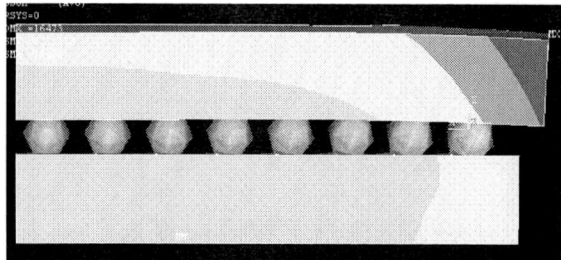

Figure 4 Greater deformation in ball

We can find out is it out of shape size in ball since picture with from package in the center from increase and increasing. Because PCB board deformation is often limited in using actually, so, it is the largest that the place of ceramic package corner is out of shape.

The stress that CBGA208 package produces because of hot mismatch is distributed the cloud chart as shown in Figure 5:

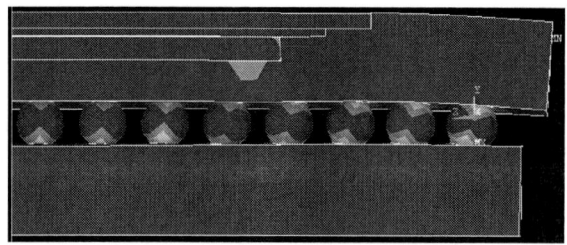

Figure 5 The stress that CBGA208 package

Fig. 5 shows that, the stress caused by deformation increased along with the increased distance to the center of package. The farthest ball from the center (the ball in the diagonal) under the biggest stress, and the ceramic lateral stress is bigger than the PCB board lateral stress, this point is the easiest to get cracking failure. This is coincide with the real experiment result of CBGA208 package.

In the course of circulating in temperature, solder joint will produce plastic deformation and creep deformation at the same time, creep deformation will make grain boundary produce cavity and crack, in creep deformation course of solder, the production of the defect impels fatigue crack to take shape and expand, cause solder joint to rupture and lose efficiency finally. Among them, creep strain exerts a great influence on solder joint fatigue and life.

At the same time, relevant experimental study show too, equivalent plastic strain of solder joint increases with increase of cycle frequency and temperature change pace, drop with increase of time, however, creep strain is contrary to this. Viscoplastic stain increases with increase of the number of times of circulation totally, among them creep strain occupies the main position.

Thermal-Mechanical Failure and Life prediction

At first, some men think Thermal-Mechanical Failure and Life of CBGA package solder joint accords with Manson-Coffin equation.

In this equation, m and C are constants, $\Delta \varepsilon P$ is plasticity strain range. With the deepening studying, people found that this equation ignored the effects of cycle temperature and cycle frequency. After considering these two factors, the low cycle fatigue life of CBGA soldered spot can denoted by Manson-Coffin equation:

$$N_f = \frac{1}{2}\left(\frac{\Delta \gamma}{2\,\varepsilon'_f}\right)^{1/C}$$

In this equation, Nf is the average cycle number; ε'_f is Failure toughness coefficient (≈ 0.325); $\Box \gamma$ is shear stress range of solder joint, $\Box \gamma = \sqrt{3}\,\Box \varepsilon$; $\Box \varepsilon$ is total strain range;

$$C = 0.442 - \left(6 \times 10^{-4}\right)T_m + \left(1.74 \times 10^{-2}\right)\ln(1+f)$$

In this equation, T_m is the average temperature of fatigue test, f is frequency of fatigue load ($1 \leq f \leq 1000$ Circle/day).

This equation is the most commonly used approximation model of solder joint's thermal-mechanical fatigue. In this equation, as we get the equivalent strain range $\Box \varepsilon$, we can predict the cycle times when the fatigue invalidation occurred.

Besides software simulation and equation calculation, the more practical prediction method is experiment. The most precise prediction is experiment which strictly according to the practice using. But the situations of practice are defer in thousands ways. It's impossible to do experiments according every possible situation. And because of the low stress standard, it needs a very long time to get the result. Generally we do the experiment through the method of accelerating in the laboratory, and using some equation to calculate the experiment result into reality.

The relation between solder joint's thermal recycling life in the experiment situation and in the reality can be denoted by the factor which deducted by "Norris and Landzberg", the equation is as follow [3]:

$$AF = \left(\Delta T_l / \Delta T_f\right)^{1.9}\left(f_f / f_l\right)^{1/3} \times e^{\left[1414\left(1/T_{\max f} / T_{\max l}\right)\right]}$$

In this equation, AF is accelerating factor, $T_{\max}$ is the highest temperature of the thermal cycle. ΔT is the difference between the highest and lowest temperature of thermal cycle. f is frequency of thermal cycle, the subscript l is designated situation of experiment, f represent situation of reality. Through the accelerating factor, we can get the soldered spot's fatigue life in the practical using situation from the experiment.

Because the chipsets which use CBGA package generally have multi power plane and filed plane, and have plentiful balls connected with the field or power, individual ball's invalidation will not affect much about the whole circuit. People concern more about the cycle times when the invalid balls reached a certain percentage.

The cycle times of 50% invalidation of CBGA soldered spot is an important index, the Manson-Coffin equation of sample N50 is as follow [4]:

978-1-4244-4658-2/09 $25.00 © 2009 IEEE

In this equation, C and n are constant, $\Delta\xi$ in is the every cycle of non-elastic strain's scope, usually we got this from software finite element analysis. C and n are defined by the experiment and software stimulation curve.

A research indicated that, the equation of CBGA package N50 which used 63Sn37Pb as solder material can denote below [4]:

$$N50 = 82.4\left(\Delta\varepsilon_{in}\right)^{-.863}$$

So, as long as different temperature cycle parameters, stimulated the Non-elastic strain range $\Delta\xi$ in by software, then we can do the prediction of thermal fatigue life.

Even the equations above have been widely used in the prediction of fatigue life, there's still some scarcity, in order to define $\Delta\xi$ in, we must do finite element analysis in softwares. So Andy Perkins and Suresh researched a new prediction equation without finite element, as follow [5]:

$$N50 = g(A, B, C, D, E, \Delta T, T_{peak}, f)$$

In this equation, A is package size, B is the difference of PCB board's expanded coefficient and ceramic expanded coefficient, C is the thickness of package, D is thickness of PCB board, E is the pitch of balls, ΔT is the changing range of Kelvin temperature, T_{peak} is the crest value denoted by Kelvin temperature, f is cycle times in each one hour.

In above equation, it considered the possible factors which affect the fatigue life, but if we want to get an explicit equation, we need to go on analyzing. The process is using mathematics model, and do regression analysis. Andy Perkins and Suresh analyzed that, and got the equation below: (0~100°C, 2 cycle per hour) [5]:

$$CM-life = 12349 - 70.1A - 434B - 1031C - 930D - 272E + 302CD$$

The significances ranks of the factors which affect life are B, C, A, D, CD. From this equation we can see that, the product of package thickness and PCB board thickness has a reinforcing effect on the fatigue life.

From this equation, we can also find the changes of solder plate, ball, solder amount while neglecting pitch is changing. But pitch has influence for life span, when pitch reduces, the ball diameter and Assemblage height reduce correspondingly, this is unfavorable for the thermal-mechanical failure. But, when the out diameter is unchanged, the lesser pitch has more outlet terminal, and then the load factor on each ball is reduced correspondingly, so it has some compensate effect on the effect of reducing assemblage height. Moreover, when the number of outlet terminal is unchanged, the out diameter of lesser pitch is relatively small, it can also has compensate effect on the shorter life span which resulted by the reducing assemble height.

According to above the equation, joint with the accelerate factor of norris and, we can get N50's life span as follows:

$$N50 = \left(\left(\frac{100}{\Delta T}\right)^{1.9}\left(\frac{f}{2}\right)^{1/3}e^{1414\left(\frac{1}{T_{peak}}-\frac{1}{373}\right)}\left[12439 - 70.1A - 434B - 1301C - 930D - 272E + 302CD\right]\right)$$

We don't need the finite element analysis of using this equation, it's easy to use, can get the Thermal-Mechanical Failure's life span easily at the beginning of design.

For some applications, 50% solder joint with failure often fail to describe the mechanical failure life. We can use the equation below to get the cycle index at any percentage of solder joint failure [6].

$$N_f(x\%) = N_f(50\%)\left[\frac{\ln(1-0.01x)}{\ln(0.5)}\right]^{\frac{1}{\beta}}$$

In this equation, β is Weibull shape factor, It determines failure frequency around the average invalid time, for CBGA package, β is 12.3.

However, due to neglecting the effect of ball diameter changes, solder plate size, solder amount when the pitch is changed. This equation has big error, usually the error is $\pm$ 20%. In use, we can combine this equation with Manson-Coffin equation (got from the finite element analysis) together.

Using 10 packages to calculate N50 (50%), N1 (1%), N01 (0.1%), N001 (0.01%) life cycle of failure in Table 1. Temperature from -55 to 125 °C, keep 4 minutes, Conversion time of 1 minute, the ceramic Coefficient of Thermal expansion 7.3ppm/°C, PCB Coefficient of Thermal expansion 16.5ppm/°C, thickness is 1.5mm.

Table 1 Life cycle of failure of 10 packages

	N50	N1	N01	N001
CBGA152	1389	984	816	677
CBGA48	1617	1146	950	789
CBGA100	1635	1159	961	797
CBGA132	1541	1092	906	752
CBGA256	1234	874	725	602
CBGA72	1641	1163	964	800
CBGA144	1603	1136	942	782
CBGA208	1508	1069	886	735
CBGA324	1366	968	803	666
CBGA252	1349	956	793	658
CBGA208 (-65~175°C)	578	409	339	282

Regarding above tables in CBGA152, CBGA256, and CBGA208 have carried on the ball hot endurance test, measured that the test plate uses the FR4 material, thickness 1.5mm, the experimental situation distinction is as follows:

CBGA152 solder plate superficial is nickel plating, thickness is 5~7μm. The limiting temperature maintains 4 minutes, the switching time is 1 minute. Assembles separately on the experimental board uses 5 ball of diameter 0.89mm CBGA152, uses 5 ball of diameter 0.76mm CBGA152. Because we did not know when the expiration will appear, Therefore first has carried on 100 temperature cycles according to the above condition, and then carries on the electric interlock test and the range estimate immediately. The result showed that the electricity connection test does not have an expiration, the range estimate outward appearance situation is good. Then continues according to every 100 circulations to carry on the plan which a time tests to carry on the experiment, after 500 circulations still had not the expiration situation. Considered that 500 circulations can definitely satisfy the product's operation requirements, therefore stop experiment.

978-1-4244-4658-2/09 $25.00 © 2009 IEEE

CBGA256 solder plate's surface is nickel plated, thickness of nickel is 5~7μm, the situation of experiment is same as CBGA152. We assembled 5 CBGA256 packages of 0.89mm balls on the PCB board. Because we have the result of CBGA152, we directly made temperature cycling test for 500 times, and immediately put the electricity connection test and visual test. All the electricity connection tests has been break, there's no invalidation of visual test on the ball, but on the PCB board, the copper solder plate has the evidence of turning up, so stop the experiment.

CBGA208 solder plate's surface is nickel plated, thickness of nickel is 5~7μm, the thicknesses of gold are 0.1~0.5μm, 0.6~1.0μm, 1.1~1.8μm separately, there are only 5 pieces of each gold thickness, experiment situation is from -65 to 175°C. The limiting temperature maintains 4 minutes, the switching time is 1 minute. We operate an electricity test and a visual test at each 25 times, until 250 times, there's still hadn't any invalidation phenomenon. At the 275 time's test, all of the three different coating thicknesses have a circuit break, The average invalid welded points are 2.67, it's coincide with the N001 (0.01%) test. Observe under the microscope, the ball which at the top corner, near the part of ceramic, appeared circuit break, other parts hadn't appeared circuit break, this is relatively coincide with the finite element simulation on the side stress.

Conclusions

The Thermal-Mechanical Failure of CBGA package is a very important factor during the research and usage of high reliability FPGA circuits. This paper studied the invalidation phenomenon and theory, and used interrelated equation on the life prediction. The practice experiment results showed that, the prediction results are relatively coincide with the experiment results.

References

1. ATMEL Crop. <u>PowerPC Products Packaging Offer</u>, Dec. 2001, pp. 6.
2. Michael Howieson. "CBGA to FR4 Printed Circuit Board with No Underfill Thermal Mismatch Study," *2001 ECTC* pp. 369-375.
3. Zhang Cheng-jing,Wang Chun-qing. "Overview on Two Types of Cer amic Ar r ay Package and Their Solder Joint Reliability," *Equipment for Electronic Products Manufacturing*, 2006, Vol. 139, pp. 10-17.
4. Amaneh Tasooji, Reza Ghaffarian, "DESIGN PARAMETERS INFLUENCING RELIABILITY OF CCGA ASSEMBLY: A SENSITIVITY ANALYSIS," *2001 IEEE*, pp. 1056-1063.
5. Andy Perkins, Suresh K. Sitaraman, "Predictive Fatigue Life Equation for CBGA Electronic Packages based on Design Parameters," *2004 IEEE*, pp. 253-256.
6. R. Master, *et. al.*, "Solder Column Interposer for Single Chip Ceramic Packaging," *1999 Electronic Components and Technology Conference*, pp. 118-127.

Finite Element Analysis of Sn-Ag-Cu Solder Joint Failure under Impact Test

Ganran Tang[1], Bing An[1], Yiping Wu[1,2], Fengshun Wu[1]
[1]Huazhong Univercity of Science & Technology
[2]Wuhan National Laboratory for Optoelectronics
Wuhan, China, 430074
Email: ypwu@gmail.hust.edu.cn Tel.: +86-27-87792402

Abstract

BGAs packaging offer high pin counts and lower interconnecting space, and are suitable for high density packaging. However, it is difficult to inspect individual solder joints on BGA assembly by conventional visual methods and need a complicated practice on rework. Ball impact test is a useful method to estimate the reliability of BGA solder joint.

In this study, the three-dimensional explicit finite element analysis is employed to carry out dynamic responses of solder joints under ball impact test. Through a three-dimensional explicit element analysis incorporated with contact, fracturing and fragmentation mechanisms, transient fracturing of the solder joint subjected to high speed impact test is investigated.

Different IMC strengths are specified and corresponding structural responses and failure modes are examined in this paper. The impact force histories with respect to different IMC strengths are clearly show by the method of finite element simulation. From the results, three kinds of failure modes can be found which tally with the actual failure modes of ball impact test. The IMC strength plays an important role in determining the failure mode in ball impact test. Model 1 (totally brittle break in IMC layer) occurs when the IMC strength is relatively low (below 300MPa) with very short impact duration (within 30μs); Model 2 (interfacial break with some solder remaining on the fracture surface) occurs when the IMC strengths are between 400MPa and 700Mpa, the impact duration increase to 40~100μs; and when IMC strength exceeds 700MPa; Model 3 (totally ductile break in solder bulk) happens with impact duration above 100μs. The impact curves of Model 1 and Model 2 are similar to half-sine profile while Mode 3 is followed by a prolonged vibration pattern due to interaction between the shear tool and solder.

1 Introduction

Lead-free solder bump behaves brittleness under dynamic impact loading. In conventional quasi-static strength test, fracture always takes place in the solder bulk; while under high speed impact load conditions, it is frequently observed that fracturing occurs around the interface between the solder joint and the bonding pads, where intermetallic compounds (IMC) are formed [1], Fig. 1. Many testing methods could be adopted to evaluate brittleness of solder joint, such as board-level drop test, high speed shear test, bending test and peel test, etc.

In previously investigation [2], differential flexing has been identified as the dominant failure driver for components mounted near the centre of the PCB in board-level drop test. The z-direction stress from axial stretching is the most critical stress component, Fig. 2. The outer most interconnection of the package is most vulnerable under this failure driver.

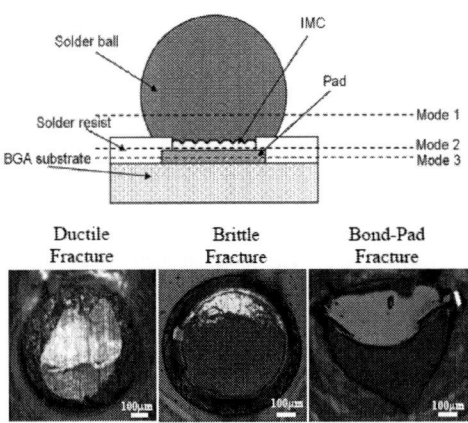

Fig. 1 Failure models of solder joint under impact loading

Fig. 2 Model of drop failure

Geometrical stress concentration will be utmost critical in the drop impact due to the suppression of plastic yielding. Endeavour shall be make to minimize geometrical stress concentration in the interconnection design. The reliability of interconnections at drop impact hinges on the impact strength of brittle materials such as intermetallic.

In the impact test, characteristics of the impact force profile in typical package-level impact test are defined according to Fig. 3.

The impact force increases after the shear tool hits the solder joint while decreases after fracturing within the solder joint initiates. The impact force profile oscillates subsequently as a result of fixture vibrations, which is determined by the natural frequency of impact test system and the rigidity of shear tool. A second peak, however, is sometimes observed which refers to the contact between the solder joint and the solder mask. The first half-sine part of the impact force profile is considered, which represents the structural behavior of the solder joint from the initiation of the impact load till fracturing completes.

978-1-4244-4658-2/09 $25.00 © 2009 IEEE   1139

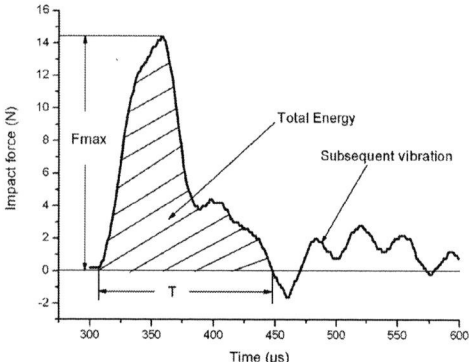

Fig. 3 Characteristics of impact force profile

Those derived from Fig. 3 are described in the following:

F_{max}: The maximum impact force, which represents the IMC strength.

T_r: The duration of the first half-sine part of the impact force profile, which stands for the ductility of the solder joint.

Total energy absorbed: The area below the first half-sine part of the impact force profile, which represents the toughness of the solder joint. In this investigation, total energy absorbed is considered as a key characteristic of solder joint anti-shock ability.

Investigation [3] has revealed the proportional relation of force and energy in experimental result between board-level drop test and package-level ball impact test, Fig. 4.

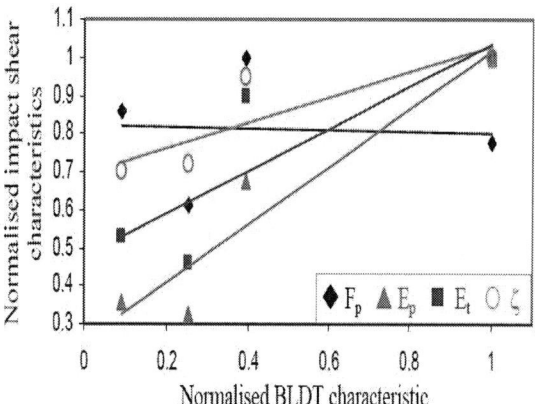

Fig.4 Correlation between shear test & drop test

E. Shuir has proved theoretically and drawn the following conclusion [4]: the drop test conditions could be adequately mimicked by shock test, if the lower (fundamental) frequency of the vulnerable structural element is known and the duration of loading in shock tests is chosen sufficiently short, as close to an instantaneous impulse conditions as possible.

Since dominate failure model in board-level drop test and package-level ball impact test are both brittle fracture in the interconnections between solder bump and bonding pad, the IMC strength is the key factor to relationship of the two testing methods. We deem that board-level drop reliability of electronic products is determined by the material properties of IMC, which exist on both package side and PCB side. In the

dropping moment, the ball arrays undergo a fiercely instantaneous impact loading, which has a great acceleration and short duration. We expect to adopt the simple impact test to substitute the complex, expansive and time consuming drop test. In this investigation, ANSYS LS-DYNA is employed for the transient analysis of structural responses of solder joint subjected the impact load.

2 Finite element analysis of ball impact test

Through a three-dimensional explicit element analysis incorporated with contact, fracturing and fragmentation mechanisms, transient fracturing of the solder joint subjected to high speed impact test is investigated. Different IMC strengths are specified and corresponding structural responses and failure modes are examined in the following text.

2.1 Finite element modeling

In order to save computing time, only a half 3D model is built according to the actual size of ball impact test samples, Fig. 5. After modeling, linear hexahedral solid elements are applied except the exterior region of solder joint where force concentration and large deformation are expected. Instead, linear tetrahedral elements are applied for a more accurate simulation, see Fig. 6.

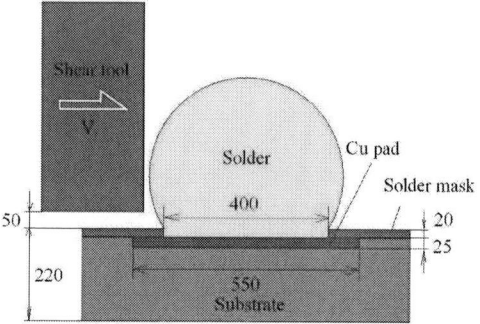

Fig. 5 physical model of ball impact test (units: μm)

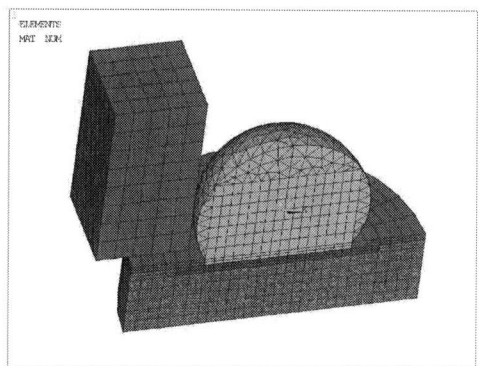

Fig. 6 Half-symmetry of finite element model

2.2 Materials Parameters and boundary conditions

Elastic properties for the constituent components are listed in Table 1. In the table, E is the Young's modulus, υ the Poisson's ratio, and ρ the mass density. The shear tool is assumed to be rigid.

Table 1 Elastic properties of constituents

Material	$\rho(kg/m^3)$	E (GPa)	υ
SAC305*	7.44×10^3	39.5	0.4
Cu pad	8.91×10^3	121	0.38
Solder mask	1.91×10^3	2.41	0.4
Substrate	1.91×10^3	22.5	0.38
Shear tool	7.9×10^3	Rigid	-

In consideration of plastic deformation during impact process, plastic properties of solder and Cu pad are defined. Depicted in Fig. 7 and summarized in Table 2, the pad is presumed a bilinear elastic-plastic material while the solder alloy a trilinearly elastic-plastic material [5].

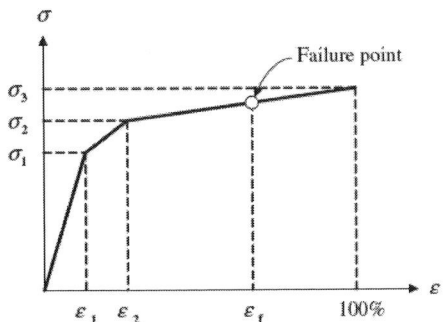

Fig. 7 Trilinear material constitutive model

Table 2 plastic properties

Material	σ_1(Mpa)	σ_2(Mpa)	σ_3(Mpa)	ε_1(%)	ε_2(%)	ε_3(%)
SAC305	55.3	76	2200	0.14	0.40	38
Cu pad	330	3355	-	0.27	50	50

Symmetry boundary condition is applied on the XY=0 planes and all degree of freedom of nodes on the bottom surface of substrate are fixed. The speed of shear tool is specified with a constant speed of 2m/s.

The automatic nodes-to-surface contact is applied between the shear tool and solder ball, while tiebreak nodes-to surface contact between the solder and Cu pad. Tiebreak contact links adjacent meshes and confines the movements of nodes until the bond force is exceeded; the bond failure is characterized by [6]

$$\left(\frac{|f_n|}{S_n}\right)^2 + \left(\frac{|f_s|}{S_s}\right)^2 \geq 1 \tag{1}$$

where the subscripts n and s denote normal and shear, respectively, and f and S the calculated nodal force and the given ultimate nodal force at which the bond breaks, respectively. Failure envelope of formula (1) is show in Fig. 8.

For simplicity, normal and shear stress are set equal and denoted the IMC strength, different IMC strength are then specified in order to evaluate its effect on failure modes.

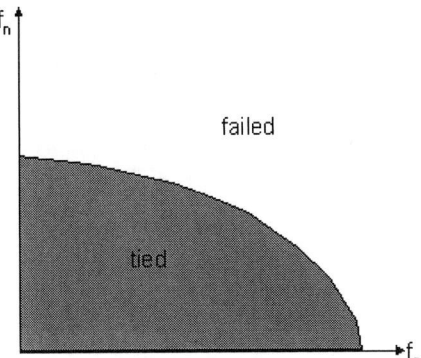

Fig. 8 Failure envelope of tiebreak contact

3 Results

Fig. 9 clearly shows the impact force histories with respect to different IMC strengths. From the results, three kinds of failure modes can be found which tally with the actual failure modes of ball impact test. Model (totally brittle break in IMC layer) occurs when the IMC strength is relatively low (below 300MPa) with very short impact duration (within 30µs); Model 2 (interfacial break with some solder remaining on the fracture surface) occurs when the IMC strengths are between 400MPa and 700Mpa, the impact duration increase to 40~100µs; and when IMC strength exceeds 700MPa, Model 3 (totally ductile break in solder bulk) happens with impact duration above 100µs. The impact curves of Model 1 and Model 2 are similar to half-sine profile while Mode 3 is follow by a prolonged vibration pattern due to interaction between the shear tool and solder. The maximum impact force is also proportional to IMC strength, from 1.3N to 13N.

Figs. 10~18 show the animation of solder bump during fracture process and stress distribution in the entire models, solder bumps and Cu pads, respectively. For Model 1, interfacial fracture takes place within 45µs and the solder bump is sheared off by the shear tool, Fig. 10. Soon as the break completed, stress on the solder is immediately released and the ball remains intact with little deformation. Stress distribution on the Cu pad indicates that the left side of pad endures a tensile stress while the right side a compressive one, see Figs. 11~12.

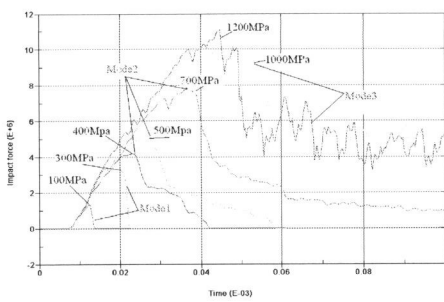

Fig. 9 Impact force histories with respect to IMC strength

978-1-4244-4658-2/09 $25.00 © 2009 IEEE

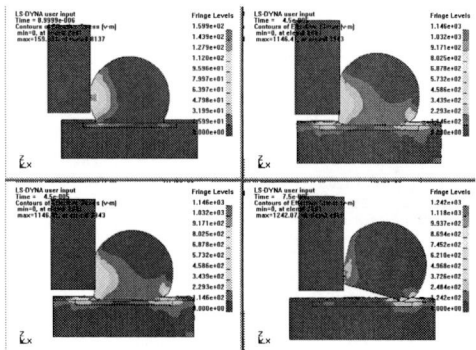

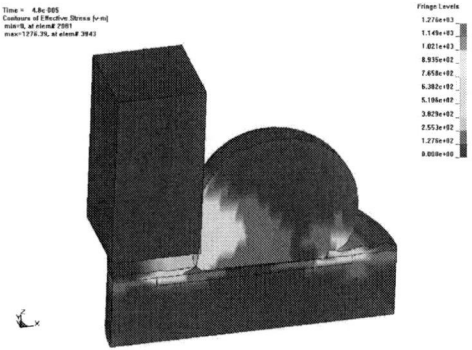

Fig. 10 Animations of failure Model 1

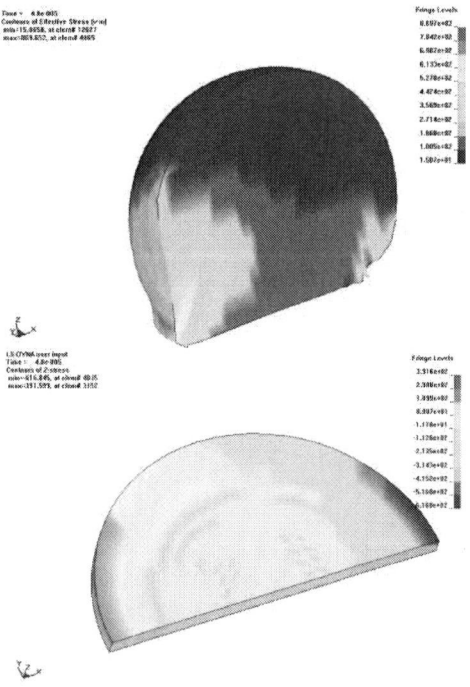

Fig. 11 Stress distribution in the entire model

Fig. 12 Stress distribution in the solder and Cu pad

It is shown in Fig. 13 that with a much more IMC strength (500MPa or more), solder surfers a higher stress during the shear impact process. Fracture initiates at 33μs on the left edge of interface till 60μs when solder bulk begins to fail due to overload of shear force. As a result, fracture surface appears to be some solder remaining, and the amount of remained solder has relation to the strength of IMC. Figs. 14~15 show that the solder bump has a larger deformation than that in Model 1.

If IMC strength exceeds 700MPa, as has discussed above, totally ductile break in the bulk happens. Fig. 16 shows that the IMC strength is too high to break that failure only takes place in the solder. The interacting time between shear tool and solder is relatively longer. Even after 132μs, half of the ball is still remaining on the pad. Large deformation of solder appeared during the impact process. See Figs. 16~18.

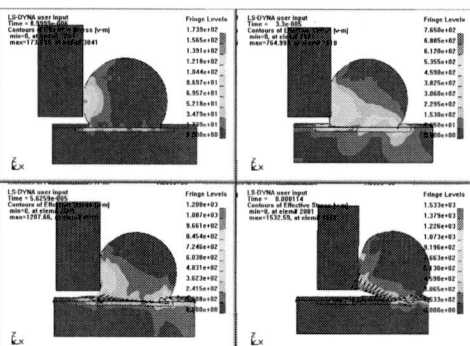

Fig. 13 Animations of failure Model 2

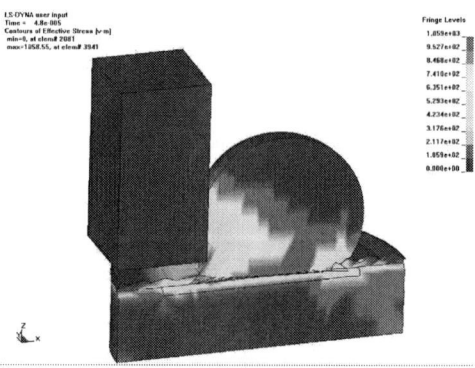

Fig. 14 Stress distribution in the entire model

4 Conclusions and Discussions

In this investigation, ANSYS LS-DYNA is employed for structural response of solder joints under high speed impact test. Numerical simulation results reveal that, three types of failure modes in ball impact test are simulated based on tiebreak contact defined model. Fracture time increases with the growth of IMC strength. Failure time is 30μs, 40μs~100μs, and above 100μs for brittle break, medial break and ductile break, respectively. And maximum shear force is between 1.3N to 13N, which is much lower than actual impact force, owing to neglecting of hardening mechanism of material subjected to high strain ratio loading.

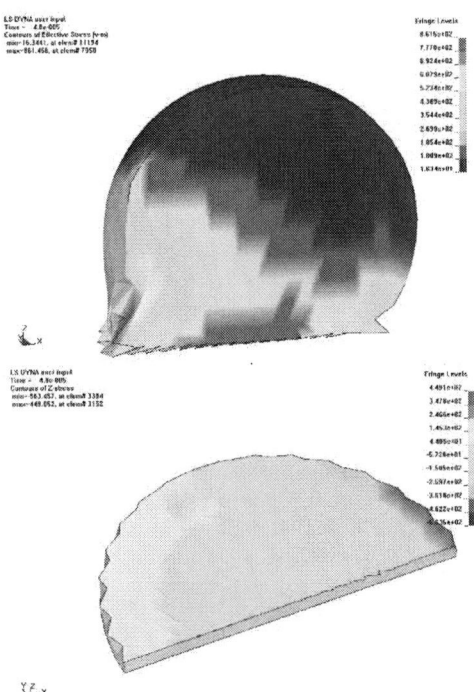

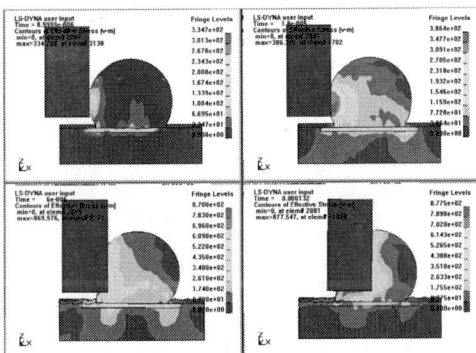

Fig. 15 Stress distribution in the solder and Cu pad

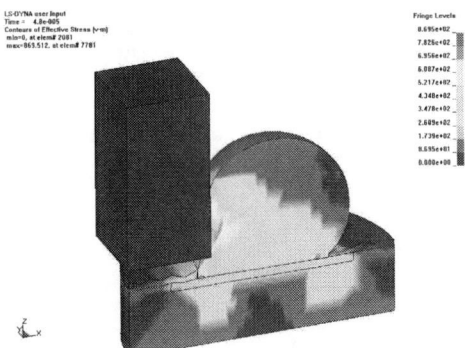

Fig. 16 Animations of failure Model 3

Fig. 17 Stress distribution in the entire model

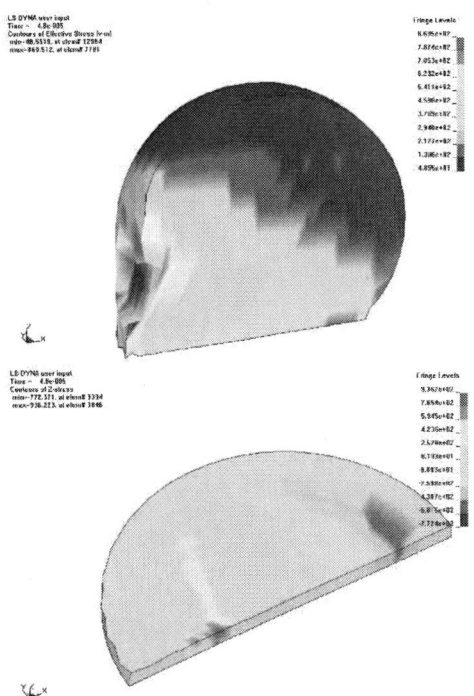

Fig. 18 Stress distribution in the solder and Cu pad

References

1. Shengquan Ou, Yuhuan Xu and K. N. Tu, "Micro-Impact Test on Lead-Free BGA Balls on Au/Electrolytic Ni/Cu Bond Pad," *Proc.55th Electronic Components and Technology Conf.*, 2005, pp. 467-471.

2. E. H. Wong, K. M. Lim, Norman Lee, *et al.*, "Drop Impact Test-Mechanics & Physics of Failure," *Proc. Electronics Packaging Technology Conf.*, 2002, pp. 327-333.

3. E. H. Wong, Y-W Mai, R. Rajoo, K. T. Tsai, F. Liu, S. K. W. Seah, C-L Yeh, "Micro Impact Characterization of Solder Joint for Drop Impact Application," *Proc. Electronic Components and Technology Conf.*, 2006, pp. 64-71.

4. E. Suhir. "Could Shock Tests Adequately Replace Drop Tests?," *Electronic Components and Technology Conference*, 2002, pp. 563-573.

5. Chang-Lin Yeh, Yi-Shao Lai, "Transient fracturing of solder joints subjected to displacement-controlled impact loads," *Microelectronics Reliability*, 46 (2006) pp. 885-895.

6. ANSYS Inc., ANSYS v.10.0 Manuals, 2006.

Study on Moisture Behavior in Flip Chip BGA Packages and Bake Process Optimization

W.Q. Dai[1], Z.K. Hua[2], J.H. Dai[1], E.W.Pang[1], L.Jiang[1], C.Y. Li[2], P. Liao[2], J.H. Zhang[2]*

[1]Intel Products (Shanghai) Co., Ltd.

[2]Key Laboratory of Advanced Display and System Applications (Shanghai University), Ministry of Education,
School of Mechanical & Electronic Engineering and Automation
Shanghai University, Shanghai, 200072, China
*E-mail address: jhzhang@staff.shu.edu.cn

Abstract

The Lead-Free (LF) and Halogen-Free (HF) "green" initiatives are driving the advanced packaging manufacturers to develop new generation materials and assembly technologies. However, the moisture related reliability issues become the significant technical challenge to meet stringent reliability and quality standards comparing to the previous Lead and Halogen technology. In this study, the moisture absorption and desorption performance of Halogen Free and Lead Free material were investigated. The experimental data revealed that after fully baked process there was still some resident moisture which was mainly contained at substrate level. The modeling data of moisture absorption and desorption behavior with 125°C bake comparison was discussed and was fully aligned with experimental data. Moreover, a 'soft bake' method via Nitrogen and dry air was introduced. In low RH effect test, under 25°C / 60%RH precondition process, all the moisture absorbed for 1100 hours can be removed by 'soft' bake. This indicates that not only no moisture absorption in low RH environment, but also an additional 'soft bake' process occurring during the storage time. The results demonstrated that the storage of moisture-sensitive material in the optimized environment is an attractive facility solution which can reduce the risk of popcorn and cracking problems.

1. Introduction

Moisture impact is considered as one of the most serious electronic package reliability issues. The delamination and crack caused by moisture make the device sensitive to external environment and easily weak. And other failures, such as lower bump reliability, the displacement and deformation of the metallization layer, leakage current, and open circuit caused by moisture corrosion [1]. Moisture can cause the oxidation of metal as interconnecting materials and the oxide formation are shown to be one mechanism explaining the rate of resistance change in a humid environment [2-3]. Also, "popcorn" phenomenon is directly linked to the moisture [4-5].

Currently, the Lead-Free (LF) [6] and Halogen-Free (HF) [7] "green" initiatives are driving the advanced packaging manufacturers to develop new generation materials and assembly technologies. The moisture related reliability issues become the significant technical challenge to meet stringent reliability and quality standards comparing to the previous Lead and Halogen technology. This is due to the HF flame retardant containing more Metal hydrate which could release more decomposition / water during higher Pb-free reflow temperatures (~260°C). This brings a challenge specifically for ball grid array (BGA) component reliability, as a

significant amount of moisture release during repeated Pb-free reflows (for ball attach, board mount, etc.) can facilitate delamination in the substrate [8].

As a solution to this moisture issue, a "Bake Out" process is required to remove absorbed moisture from packages. The key factor of this process is to effectively reduce the moisture content of the package to the level required for shipping and storing and to prevent 'popcorn' at the board assembly reflow process [9]. Therefore, it is quite necessary to understand the moisture behavior in the different form factor flip chip ball grid array (FCBGA) packages, substrates and materials.

In this study, the moisture absorption and desorption performance of HF and LF material were investigated. Results showed that after fully baked process there was still some resident moisture which was mainly contained at substrate level. Moreover, a 'soft bake' method via Nitrogen and dry air was introduced. The moisture absorption and desorption behavior with soft bake and normal bake comparison was discussed. The results demonstrated that the storage of moisture-sensitive material in the optimized environment was an attractive facility solution which can reduce the risk of popcorn and cracking problem.

2. Method and material

2.1 Experiment

4 HF and LF FCBGA units, together with 4 epoxy samples and 6 substrates (two main components in the unit) were tested in the experiment. Moisture diffusivity and solubility data were gathered by periodically weighing samples soaked in a humidity chamber, controlled to an accuracy of 1°C and 1% relative humidity (RH). A Sartorius analytical balance, having an accuracy of ±0.01mg was used for weighing. By using an $50 \times 50 \times 5$ mm aluminum plate for heat sink capability, no measurable loss in moisture uptake occurred in the test specimens from the moment they were removed from the environmental chamber, allowed to cool to room temperature, and experimentally tested.

All the samples were pre-baked in an oven at 125°C for 48 hours for completely drying and base line establishment. Then, the samples were preconditioned at 85°C/85%RH levels according to the J-STD-20 MSL standard [10]. The sample was considered dry when successive measurements resulted in a deviation less than 0.002%.

In low-RH-effect experiment ('soft' bake), 6 FCBGA units were tested in the dry air condition. After completely pre-baked, samples were preconditioned at 25°C/60%RH, and placed into an oven then. The oven condition was strictly controlled at 1% to 5% RH by nitrogen gas flow, while the temperature is at 25°C. Weight gain method was also applied for evaluation.

978-1-4244-4658-2/09 $25.00 © 2009 IEEE

2.2 Simulation

Finite Element Method was used to study the moisture distribution in the FCBGA package device (Fig. 1). Fick's second law was applied to describe the moisture diffusion process (Eq. (1)). Wetness was used to provide continuous bimaterial interface for moisture diffusion [11].

$$\frac{1}{D}\frac{\partial w}{\partial t} = \frac{\partial^2 w}{\partial x^2} + \frac{\partial^2 w}{\partial y^2} + \frac{\partial^2 w}{\partial z^2} \qquad (1)$$

Where D is the tensor of diffusion coefficient, W=W (t, x) is moisture concentration, t is the time and x is the Cartesian coordinate.

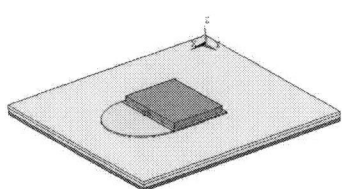

Fig. 1 Schematic of the finite element model of FCBGA package

3. Result and discussion

3.1 Resident moisture

In order to determine the moisture absorption and desorption performance, 4 HF and LF FCBGA units, together with 4 epoxy samples and 6 substrates (two main components in the unit) were investigated. The result of FCBGA unit, epoxy and bare substrate's moisture absorption at 85°C/85%RH and desorption at 125°C baking were shown respectively in Fig. 2.

In Fig.2 (a), it is found that after 85°C/85% absorption then 125°C bake, there is still about 0.02% moisture resident in the units. This means that not all the moisture absorbed can be reversed through the bake process. In epoxy and substrate test, results show that all the moisture absorbed in epoxy is baked out (Fig. 2 (b)) while resident moisture is found in substrate (Fig. 2 (c)). Moreover, the absorbed moisture content in substrate is similar to that in the unit. T. Ferguson *et.al.* [12-13] evaluated the recovery of epoxies upon drying after moisture absorption; it was found that the tested material properties, e.g. elastic modulus and interfacial fracture toughness, could not be fully recovered due to the moisture impact. This experimental data revealed that after fully baked process there was still some resident moisture which was mainly contained at substrate level.

The resident moisture found in this study is one of the proper answers to the material property loss phenomenon. Generally, moisture can affect package through two primary mechanisms [14]. The first mechanism is free water, the direct moisture presence in materials and interfaces, which is free from the action of the intermolecular attraction and is only held by capillary force. Such moisture can be removed through bake process. However, the resident moisture found in the experiment is attributed to the second mechanism,

bound water. Moisture has polar molecules, which consist of a negative hydroxyl (OH) fraction. The hydroxyl groups are negatively charged electrically. Therefore, the free hydroxyl groups in the units can attract and hold moisture by hydrogen bonding.

The hydrogen bond is a strong fixed dipole-dipole van der Waals-Keesom force, but weaker than covalent, or ionic bonds. The energy of hydrogen bond is typically 5 to 30kJ/mol [15]. In bake process, due to the temperature variation, the acquired thermal energy of the resident moisture (Q) can be evaluated as below.

$$Q = C_{mio} m_{re} (T_1 - T_2) \qquad (2)$$
$$m_{re} = 1.234 \times 10^{-3} g$$
$$C_{moi} = 1.901 kJ/(kg \cdot K)$$

Where m_{re} is the weight of the resident moisture, C_{moi} is the heat capacity ratio of moisture, T_1 and T_2 represent the environment temperature (25°C) and bake temperature (125°C) respectively.

Assuming all the resident moisture found in this study is hydrogen bond, from Eq.2, energy acquired through the bake process (Q) is 2.34×10^{-4} kJ/mol, which is much lower than O-H hydrogen bond energy [16]. This indicates that the acquired energy can not break the hydrogen bond and moisture is locked in the units.

As shown in Fig. 2, we also find that the resident moisture only exists in substrates and all the moisture absorbed in epoxy is reversed. In substrates, different layers are used for interconnection, mechanical buildup and electrical insulation [17]. More polymers, especially the cross-linked, are applied in these structures. C.Soles and A.Yee [18-19] have shown that the equilibrium moisture content increases by increasing the cross-link density of the material. The localized moisture in these areas is relatively difficult to bake out comparing with those in epoxy. Hence, the moisture is more likely to be bounded in substrates. However, it should be pointed out that too many factors during the test have posed experimental difficulties and so far precluded a detailed theoretical understanding of the often intricate processes involved.

FEA of moisture diffusion was performed for 400 hours absorption under 5°C /85%RH to validate moisture absorption and desorption behavior. Finite element analysis is performed based on the material properties acquired from the humidity experiment. Fig. 3 shows the moisture distribution after 400 hours absorption under 85°C/85%RH. It is found that substrate absorbs more moisture than the epoxy, which dominates the hygroscopic performance of the whole units. And this agrees well with the experiment. On the other hand, because moisture barrier effect of the chip above, the epoxy is somehow prevented from the direct exposure to moisture, which causes less moisture concentration. Vice versa, it is believed that in bake process, moisture localized in this area will be difficult to be desorbed, which is a really critical part in manufacture. Some other important factors, such as vapor pressure [20], thermal stress [21] and hygroscopic stress [22] should be taken into consideration for the future work. This FEA results indicate the substrate determine hygroscopic performance of the units, which aligned well those finding in the above experiment.

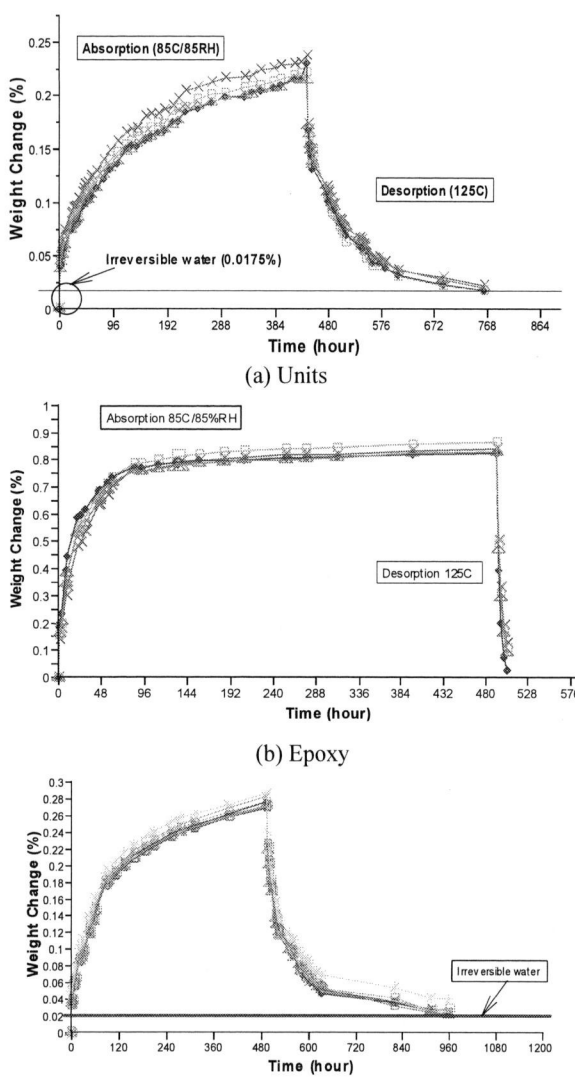

(a) Units

(b) Epoxy

(c) Substrate

Fig. 2 Moisture absorption & desorption profile for FCBGA units, epoxy and substrates at 85°C/85%RH & 125°C

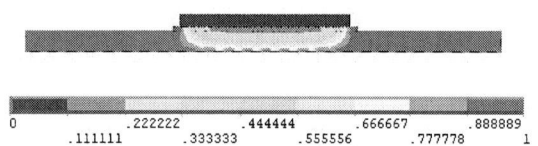

Fig. 3 Moisture distribution in FCBGA package

3.2 Low RH storage & 'soft' bake effect

Fig. 4 shows the result of low-RH-effect experiment. It is found that all the moisture absorbed in precondition process is reversed. The residual moisture in the package can be removed by this new introduced 'soft bake'. This indicates not only no moisture absorption in low RH environment, but also an additional 'soft' bake process [23] occurring during the storage time. Due to the low absolute humidity, that is less moisture in the surrounding air, it is easy to understand why no additional moisture has been absorbed in the units during the storage time. As to the 'soft' bake effect, the absolute humidity is also an important fact, which will cause the moisture content difference, and finally desorbs the moisture.

Moisture absorption or desorption depends on the nature of the driving force, such as moisture gradient and partial vapor pressure. In external condition, lower RH result in lower partial water vapor pressure, hence the external boundary condition for absorption or desorption is changed, and pressure difference guides the moisture movement direction. Within the package, moisture has to redistribute due to the change of the external boundary condition. It is regarded that the chemical potential of absorbed moisture is a function of the moisture content in the materials. Hence, a moisture gradient content is accompanied by a gradient of chemical potential. If there exists difference in chemical potential, absorption or desorption will happen. Finally, moisture will redistribute itself throughout the units until the chemical potential is uniform. This may give a proper answer why the 'soft' bake happens. Moreover, in Fig. 3, it is also found that the diffusion rate of 'soft' bake is higher than that of the absorption. In diffusion, higher temperature results in higher diffusion rate value. In this study, both absorption and desorption were operated under 25°C, but higher diffusion rate was found in low RH condition. This means that diffusion rate is not only determined by temperature. In Arrhenius equation (Eq. 3), besides temperature, activation energy is also an important factor. As the Gibbs free energy per mole is usually expressed as the chemical potential, the difference in chemical potential may also cause the change of activation energy.

$$D = D_0 \exp(\frac{-E_d}{RT}) \qquad (3)$$

Furthermore, it is quite inspiring to find that no resident moisture is found in the low RH effect test, comparing with the standard humidity test results. It indicates that 'soft' bake may provide a more safe process towards units. It should be pointed out that whether the resident moisture is generated during high temperature absorption or 125°C bake process is still unknown. However, high temperature bake process does have some negative impact on electronic packages, such as intermetallic layer growth and additional chemical reaction. 'Soft' bake process, which may provide a more safe condition, is an attractive solution towards these problems.

4. Conclusion

In this study, the moisture behavior in FCBGA packages was studied. After 85°C/85%RH soaking and fully 125°C bake, the resident moisture contained at substrate level was found in the units. FEA results indicate the substrate determine hygroscopic performance of the units, which agrees well those found in the experiment. In calculation, it is found the acquired energy in bake process can not break the hydrogen bond and moisture is locked in the units. In low RH effect test, under 25°C/ 60%RH precondition process, all the moisture absorbed for 1100 hours can be removed by 'soft' bake. This indicates that not only no moisture absorption in low RH environment, but also an additional 'soft bake' process occurring during the storage time. This may provide a more safe solution towards those problems in high temperature bake. Moreover, if the unit is saturated, moisture in the area below the chip needs more bake time, which will be more critical for moisture localization. In future work, research will be carried on to investigate the cause of resident moisture generation and reliability study of the 'soft' bake process.

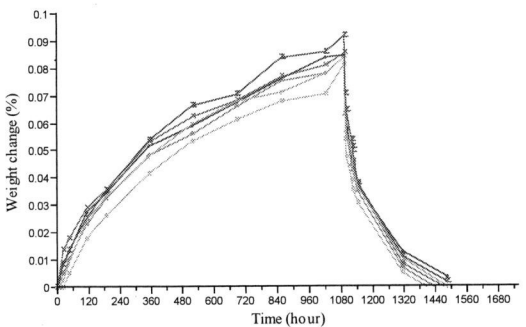

Fig. 4 Moisture absorption & desorption profile for FCBGA at 25°C/60%RH & 25°C/5%RH

Acknowledgments

The study is financially supported by the Intel Higher Education Program, PO Number: 4507445982.

References

1. Lin, R., Blackshear, E., and Serisky, P., "Moisture induced package cracking in plastic encapsulated surface mount components during solder reflow process," *Proc 26th Rel Phys Symp*, 1988, pp. 83-89.
2. Yim, M. J., Hwang, J. S., Kwon, W. S., Jang, K. W., and Paik, K. W., "High reliable nonconductive adhesives for flip chip CSP applications," *Proc. 52nd Electronic Components and Technology Conf*, 2002, pp. 1385-1389.
3. Aschenbrenner, R., Gwiasda, J., Eldring, J., Zakel, E., and Reichl, H., "Flip chip attachment using nonconductive adhesives and gold ball bumps," *Int J Microcircuits Electron Packag*, Vol. 18 (1995), pp. 154-161.
4. Hua, Z. K., Li, C. Y., Zhang, J. H., Cao, L. Q., and Luo Y. X., "Hygro-thermal Finite Element Analysis of Green Stacked Die Package," *32nd Proc Int'l Electronics Manufacturing Tech Conf*, San Jose, USA, 2007, pp. 78-85.
5. Li, C. Y., Hua, Z. K., Luo, Y. X., Cao, L. Q., Zhang, J. H., "Investigation of the moisture impact on the stacked die package," *2nd Electronics System-Integration Technology Conf*, Greenwich, UK, 2008, pp. 1175-1178.
6. Suraski, D., Seelig, K., "The current status of lead-free solder alloys," *IEEE Trans Packag Manufact*, Vol. 24, No. 4 (2001), pp. 244-248.
7. Bergendahl, C.G., "Electronics goes halogen-free: international driving forces and the availability and potential of halogen-free alternatives," *Electronics and the Environment*, 2000, pp. 54-58.

8. Mukul P. Renavikar, etc. "Materials Technology for Environmentally Green Packaging", *Intel Technology Journal*, Vol. 12, Issue 1, 2008
9. Nguyen, T. G., "PQFP Moisture Bake Out Process Optimization." *Proc Electro 98 Professional Program*, 1998, pp. 21-35.
10. IPC/JEDEC J-STD-20 MSL Classifications.
11. Wong, E. H., Koh, S. W., Lee, K. H., and Rajoo, R., "Comprehesive treatment of moisture induced failure-recent advances," *IEEE Trans Packag Manufact*, Vol. 25, No. 3 (2002), pp. 223-230.
12. Ferguson, T., and Qu, J., "Moisture absorption analysis of interfacial fracture test specimens composed of noflow underfill materials," *J Electron Packaging*, Vol. 125 (2003), pp. 24-30.
13. Ferguson, T., and Qu, J., "Effect of moisture on the interfacial adhesion of the underfill/soldermask interface," *J Electron Packaging*, Vol. 124 (2002), pp. 106-110.
14. Strumillo, C., and Kudra, T., <u>Drying: Principles, Applications and Design</u> (New York, 1986), pp. 448.
15. http://en.wikipedia.org/wiki/Hydrogen_bond, 2009.
16. Jeffrey, G. A., <u>An introduction to hydrogen bonding</u>, (Oxford University Press, 1997).
17. Michael, G. P., "Moisture sensitivity characteriazation of build-up ball grid array substrates," *IEEE Trans Ad. Packag*, Vol. 22, No. 3 (1999), pp. 515-523.
18. Soles, C., Chang, F., Gidley, D., and Yee, A., "Contributions of the nano void structure to the kinetics of moisture transport in epoxy resins," *J Polym Sci Polym Phys*, Vol. 38 (2000), pp. 776-791.
19. Soles, C., and Yee, A., "A discussion of the molecular mechanisms of moisture transport in epoxy resins," *J Polym Sci Polym Phys*, Vol. 38 (2000), pp.792-802.
20. Liu, P., Cheng, L., and Zhang, Y. W., "Interface delamination in plastic IC packages induced by thermal loading and vapor pressure -A micromechanics model," *IEEE Tran Adv Packag*, Vol. 26, No. 1 (2003), pp. 1-9.
21. Pecht, M. G., and Govind, A., "In-situ measurements of surface mount IC package deformations during reflow soldering," *IEEE Trans Comp Packag Manufact Technol, C*, Vol. 20, No. 3 (1997), pp. 207-212.
22. Stellrecht, E., Han, B., and Pecht, M., "Characterization of hygroscopic swelling behavior of mold compounds and plastic packages," *IEEE Trans Comp Packag Manufact Technol*, Vol. 27, No. 3 (2004), pp. 499-506.
23. Theriault, M., Carsac, C., and Blostein, P., "Evaluating nitrogen storage as an alternative to baking moisture/reflow sensitive components," *Proc of NEPCON WEST 2000*, CA, USA, 2000, pp. 1-12.

Study on Thermo-mechanical Reliability of Embedded Chip during Thermal Cycle Loading

Ligang Niu, D. Yang, Mingjun Zhao
School of Mechanical and Electrical Engineering, Guilin University of Electronic Technology,
Guilin, Guangxi 541004, China
gstnlg@yahoo.com.cn

Abstract

With the development trend of microelectronic system with small size, high speed, high frequency and high density, passive and active components are directly embedded into a core or high-density-interconnect layers. This System-in-Package (SiP) technology could shorten interconnection between the die and substrate and reduce the inductance and noise interference. However, there are many electrical and mechanical reliability issues including the reliability issue for embedded structure. An embedded structure was chosen in this study. The embedded chip was surrounded by epoxy. An epoxy was selected as the adhesive to embed the chip. The active surface of the chip was face up, to form a planar surface with the substrate. Benzocylobutene (BCB) was chosen as the dielectric polymer for embedding technology because of its low curing temperature. One quarter 3D model of embedded structure was loaded on six thermal cycles according to the temperature cycling standards JESD22-A104C. The thermo-mechanical reliability was investigated and the modified Coffin-Manson equation was employed to predict the fatigue life of copper films. FEA simulation results revealed that the fatigue life of copper film is 35.7 cycles. Stresses in the die always lead to various failures in manufacturing and using process, so equivalent von Mises stress and peel stress were also analyzed in this study.

Introduction

The coming generation of portable products will require significant improvements of integration and packaging technologies, mainly due to increasing signal frequencies and demand for higher density of functions at acceptable cost [1]. System-in-Package (SiP) can meet the demand of small size, lightweight, multi-function and system integration for consuming electronic devices. The obvious benefits of SiP structure include simplified and shortened wiring, better electrical performance by reduction of parasitic especially in GHz high frequency application, saving of surface-space and improvement of high frequency characteristics. Moreover, due to the elimination of solder joints between flip chips and copper, high thermo-mechanical reliability can be achieved and heat dissipation can be substantially improved due to shortened thermal paths between chip hot spots and copper planes [2-3].

The new concept, called "Chip in Polymer" technology, was first introduced by Fraunhofer IZM and TU Berlin [4]. It is based on the embedding of thin chips into build-up layers by using printed circuit board (PCB) technology. GE molds plastic around the chips and then builds up a multilayer interconnect over the top of the chip and the substrate using polyimide films and laser via formation [5]. Intel, Helsinki university and others are developing an embedded active technology that utilizes organic core such as BT (bismaleimide triazine) laminate and FR4 instead of the plastic molded substrate of GE and Lockheed Martin [6]. In NXP Semiconductors, under the technology framework of Lampack, a MOSFET power package based on the embedded die technology was developed and the demonstrators were built [7].

However, there are many electrical and mechanical reliability issues in embedded structure. A mismatch of coefficient of thermal expansion (CTE) among packaging materials and devices can lead to delamination, crack or copper trace break in the package. And thermo-mechanical problems may usually arise at interfaces between several materials due to their geometry, different CTE and stiffness, or at interconnects provoked by their small thickness [4]. Thermal deformations and thermal stresses may occur due to mismatches of the coefficient of thermal expansion among the packaging materials during packaging process as well as service conditions, which lead to serious quality problems and failure of the products [8-9].

In this study, a System-in-Package structure [10], as shown in Figure 1, was conducted to analyze the thermo-mechanical reliability during thermal cycle loading. At the beginning, a 3D FE model was created; FE calculations were carried out in order to study the thermo-mechanical reliability, especially at interconnects between the hidden die and the epoxy surrounded the die as well as the interface between the hidden die and epoxy, and the modified Coffin-Manson equation was employed to predict the fatigue life of copper films. Then, the interfacial stresses between silicon die and epoxy were analyzed, and pointed out that the stresses may lead to delamination or interfacial failure.

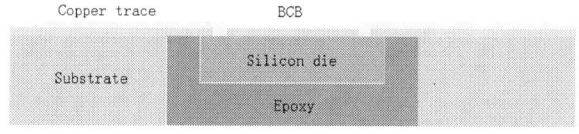

Fig. 1 Schematic of SiP structure (not to scale)

Finite Element Modeling

Geometrical model

FEA modeling on SiP structure was conducted in three dimension analysis using standard FE program to investigate the thermo-mechanical reliability. Due to the symmetry of packaging structure, only a quarter of the SiP structure was modeled, as shown in Figure 2. It consists of 21484 nodes and 17404 elements. And the entire model was meshed with hexahedral elements. Finite element modeling was used to simulate the stresses and strains of the package during thermal loading.

The chip was surrounded by epoxy and embedded in a substrate. An epoxy was selected as the adhesive to embed the chip. The active surface of the chip was face up, to form a planar surface with the substrate. Benzocylobutene (BCB) was chosen as the dielectric polymer for embedding technology because of its low curing temperature. The dimensions of the substrate were 18×18mm square and 400μm thick. The chip was a 6×6mm square and 200μm thick. The BCB was 20μm thick. And the copper trace was 50μm thick and 50μm wide. The connections between the chip and copper line were rectangular pads 100μm on one side, and 50μm on the other [10].

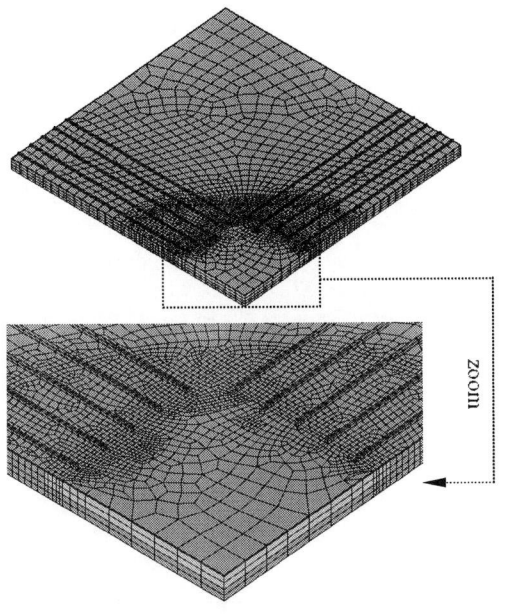

Fig. 2 3D FEM model for SiP

Material properties

DMA Q800 (TA Instrument) was employed to obtain the temperature-dependent properties of epoxy under 1Hz. TMA Q400 (TA Instrument) was conducted to obtain the coefficient of thermal expansion (CTE). The glass transition temperature is about 150°C, and the coefficients of thermal expansion (CTE) below and above the glass transition temperature is 27.16ppm/°C and 68.02ppm/°C, respectively.

For other materials, the silicon die, BCB and copper were assumed to be isotropic, except for the FR4 substrate which was considered to be orthotropic. The elastic material properties for the silicon die, BCB and FR4 in thermal cycle loading, but the copper was considered to be elastic-plastic material and the yield stress was 65MPa [11]. The material properties used in the FEM simulation are shown in Table 1.

Table 1 Material properties

Materials		E(MPa)	V	G(MPa)	CTE(ppm/°C)
Silicon die		131000	0.3	--	2.7
Epoxy		Temperature-dependent	0.35	--	27.16 (T<Tg)
BCB		2900	0.34	--	52
Copper		102000	0.36	--	16
FR4	In plane	197000	0.18	3700	17.6
	Out of plane	9000	0.39	2900	54.2

Thermal loading

The thermal loading was conducted according to the temperature cycling standards JESD22-A104C [12], thermal cycling test condition G and soak mode condition 4 were chosen as thermal loading condition, which is assumed to be within the state-steady condition. The thermal loading started from room temperature with -40°C~+125°C temperature range, and then ramp up to maximum temperature of 125°C at which dwell time of 10 minutes was used. Each cycle spent 60 minutes and six cycles were used. In JEDEC standard, typical ramp rate for this situation was 15°C/minute or less for any portion of each cycle, with a preferred rate of 10°C to 14°C/minute, and the ramp rate of 11°C/minute was used in this study. The device was assumed stress-free at 25°C. Thermal cycling in this study is shown in Figure 3.

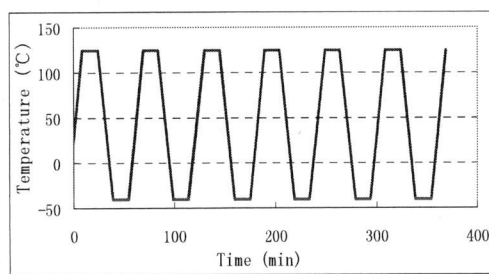

Fig. 3 Thermal cycling profile

Results and Discussion

Fatigue life of copper film

The fatigue behavior of metallic thin films, especially copper films, has already been shown to obey a Manson-Coffin type relation [13] for the critical cycle number N_f:

$$N_f = \left(\frac{\Delta\sigma}{\sigma_0}\right)^{-b} + \left(\frac{\Delta\varepsilon_{pl}}{\varepsilon_f}\right)^{-c} \qquad (1)$$

Where the first term describes high cycle fatigue, i.e. fatigue caused by cyclic straining in the elastic deformation range, measured by the cyclic stress amplitude $\Delta\sigma$ in relation to a characteristic stress σ_0, and the second term describes low cycle fatigue, where the damage stored in each cycle is related to the cyclic plastic strain amplitude $\Delta\varepsilon_{pl}$ divided by a constant ε_f. Sometimes this constant can be related to the static fracture strain. Because this cycle number is above the usually required limit in electronics, our investigations are more related to the second term, which relates the cycles-to-failure to the cyclic plastic strain [14]. Then the Equation (1) could be simplified as below:

$$N_f = \left(\frac{\Delta\varepsilon_{pl}}{\varepsilon_f} \right)^{-c} \qquad (2)$$

In which c is a constant, which is within the range of 1.4 to 1.7 for ductile metals. Using 1.5 for the constant c, the fatigue life can be estimated based on the model (2) [16].

2D and 3D parametric simulations have been performed. Though 3D model can give more accurate results than 2D model, much more time for computer calculation and for result analysis are needed. Therefore, in this study 3D modeling is mainly used for identifying the hotspots and studying the trends.

Figure 4 shows the plastic strain in the copper line at the thermal loading. Critical locations are at the copper line above the connections between silicon die and epoxy and the connections between epoxy and FR4 substrate.

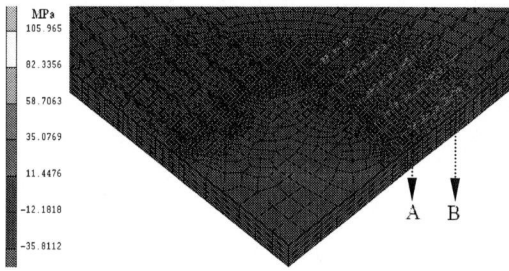

Fig. 4 Contour plot of normal stress in X direction

Figure 5 depicts the equivalent plastic strain accumulated during the thermal loading, in which the critical points A and B are simulated. According to the Figure 5, the $\Delta\varepsilon_{pl}$ at points A and B are 0.12% and 0.1%, respectively. Equation (2) is used to estimate the fatigue of copper film, and the fatigue life at point A and B are 35.7 cycles and 46.9 cycles respectively. So, the fatigue life of copper film is no more than 35.7 cycles under the thermal loading conditions.

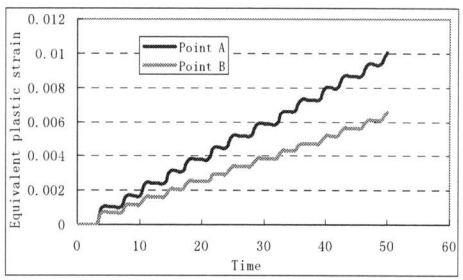

Fig. 5 Total equivalent plastic strain of Cu at points A and B

Stresses in the silicon die

The equivalent von Mises stress mainly occurs at the interface between chip and epoxy, as shown in the Figure 6. The maximum equivalent von Mises stress is located very closely to the chip corner because of the epoxy outside the chip corner is thin.

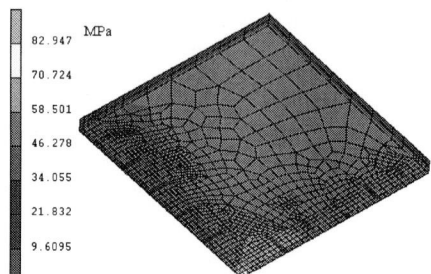

Fig. 6 Contour plot of equivalent von Mises stress in silicon die

Figure 7 presents the maximum peel stress, σ_{zz}, during the thermal loading in the silicon die. The positive peel stress will lift the die, while the negative peel stress will push the die, they will lead to fracture failure. It can be seen from Figure 7 that the critical locations are the areas between silicon die and copper trace. If the stress higher than the joint strength between die and copper trace, delamination or crack will happen at the interface.

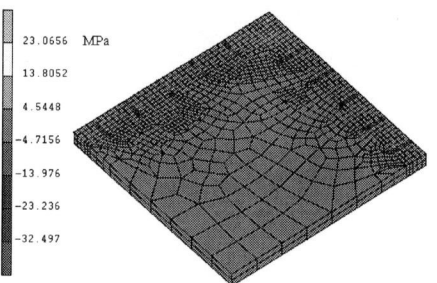

Fig. 7 The maximum peel stress, σ_{zz}, in the silicon die

Conclusions

One quarter of 3D FE models were established to investigate the thermo-mechanical reliability performance under thermal cycle loading.

The modified Coffin-Manson equation was employed to predict the fatigue life of copper films. FEA simulation results revealed that the plastic strain of copper film in each cycle is 0.12%, and the fatigue life is 35.7 cycles.

The equivalent von Mises stress mainly occurs between the interface of chip and epoxy, and the maximum equivalent von Mises stress, 82.947MPa, located at the corner of the silicon die.

The critical locations of the peel stress are the areas between silicon die and copper trace. If the peel stress is higher than the joint strength between die and copper trace, delamination or crack will happen at the interface.

Acknowledgments

The research work in this paper is financially supported by the National Natural Science Foundation of China (NSFC) (Grant No. 60666002).

References

1. Chien-Wei Chien, Li-Cheng Shen, et al., "Wafer Level Chip Stacked Module by Embeded IC Packaging Technology," IMPACT, 2007, pp. 136-140.

2. Gregor Langer, Markus Wuchse, et al., "Integrated Smart System Solution," Electronics Systemintegration Technology Conference, Vol. 2, (2006), pp. 1137-1142.

3. Dionysios Manessis, Shiu-Fang Yen, et al., "Technical Understanding of Resin-Coated-Copper (RCC) Lamination Processes for Realization of Reliable Chip Embedding Technologies", 57th Electronic Components and Technology Conference, May 29-June 1, 2007, pp. 278-285.

4. Sommer, J.P., Michel, B., Ostmann, A., "Electronic Assemblies with Hidden Dies-Design Support by Means of FE Analysis," Electronics Systemintegration Technology Conference, 2006, pp. 1088-1095.

5. US patent 3,903,590, "Multiple Chip Fabricating Integrated Circuit Module," 1994.

6. Mahajan, R. et al., "Emerging Directions for Packaging Technology," Intel Technoloty Journal, Vol. 6, No. 2 (2002) pp. 62-75.

7. Van der Lugt, A., Peels, W., "Embedded Actives Technology, from Functional Densification to Fanout Redistribution," Proceedings of EMPC2007, 2007.

8. Low, B. J., Kao, C. H., et al., "On the Study of Piezoresistive stress sensors for Microelectronic Packaging," J. Electron. Pack., Vol. 124, (2002) pp. 22-26.

9. van Driel, W. D., Zhang, G. Q., et al., "Prediction and verification of process-induced thermal deformation of electronic packages using non-linear FEM and 3D interferometry," Proc. EuroSim2002, pp. 362-367.

10. Liu Chen, et al., "Characterization of Substrate Materials for System-in-Package Applications," JUNE 2004, Vol. 126, Journal of Electronic Packaging, pp. 195-201.

11. Xiuzhen Lu, et al., "Reliability Analysis of Embedded Chip Technique with Design of Experiment Methods," International Symposium on Electronics Materials and Packaging, 2005, pp. 43-49.

12. JEDEC Solid State Technology Association. JESD22-A104C, "Temperature Cycling", May, 2005.

13. W. Engelmaier, "Manufacturing and Reliability Issues of Small-Diameter/High-Aspect-Ratio Plated Through-Holes Vias," Short Course Proceeding, IZM-ZVE, Oberpfaffenhofen, 1994.

14. Dudek, R., Walter, H., Zapf, J., and Michel, B., "Investigations on Low Cycle Fatigue of Electrodeposited Thin Copper and Nickel Films," Proceedings of Eurosime2003, France, 2003.

15. Daoguo Yang, Martien Kengen, W. G. M. Peels, et al., "Reliability Modeling on a MOSFET Power Package Based on Embedded Die Technology," EuroSimE 2009 in Delft, The Netherlands.

The Influence of Plastic-package on the Voltage Shift of Voltage Reference in Analog Circuit

Yanfeng Jiang, Jiaxin Ju

Microelectronic Center, College of Information Engineering, North China University of Technology, Beijing, 100144, China

E-mail address: yfjiang@ncut.edu.cn

Abstract

Bandgap references, packaged in plastic, have been known to shift in voltage, a pre-package to post-package voltage variation. The package shift has been analytically discussed and experimentally investigated in this paper. Once the reference is encapsulated, a package-induced stresses present, which lead to a systematic voltage shifts ranging from 3 to 7mV. The variation is closely related to package type and processing. Two kinds of shifting have been discussed, one is systematic package shift, which can be trimmed and its temperature coefficient compensated. The other is random package shift. The method on how to decrease them has been discussed, too. Structure and method of minimizing package-shift effects in integrated circuits is implemented by using a thick metallic overcoat applied after the deposition and patterning of the conventional insulating protective overcoat.

1. Introduction

The demand for precision is relentless, from 0.1% voltage references and 0.5μV offset operational amplifiers to 190MS/second 14-bit analog-to-digital (A/D converter) converters, yet fundamental error sources like transistor mismatch errors over process and temperature and RMS switching noise effects continue to plague performance. Shrinking supply voltages, budget-constrained test times, and rising bandwidth requirements only exacerbate the problem. State-of-the-art analog circuit solutions therefore succumb to trimming or switching networks to mitigate these process-induced errors. Unfortunately, neither of these techniques is especially attractive to the designer because trimming is costly and switching is noisy, which is why alternate circuit solutions are desired.

Bandgap references are used in a wide variety of integrated systems where accurate and precise voltage references with excellent line regulation and temperature-drift performance are required [1]. Since bandgap references play an important role in determining the accuracy of integrated systems, designers employ different types of trimming techniques and algorithms to compensate for process variations, temperature, and complex second-order and third-order effects [2]. However, bandgap references encapsulated in plastic packages exhibit a characteristic shift in voltage. Once it is packaged in plastic, the bandgap reference's output voltage differs from its original, nonpackaged value. This package shift, unfortunately, is not completely consistent from unit to unit, even if the same encapsulant and packaging technique is used. This randomness is detrimental since designers cannot easily account for this variant in the design phase. Structure and method of minimizing package-shift effects in integrated circuits is implemented by using a thick metallic overcoat applied after the deposition and patterning of the conventional insulating protective overcoat. The metallic overcoat most preferably comprises a layer of

electrolytically deposited copper approximately 15μm thick that is patterned to provide for electrically independent regions; but an unbroken area of the metallic overcoat is left over any sensitive analog circuitry, such as a bandgap reference circuit. The thick metallic coating, in addition to minimizing package-shift effects, is also useful as a low-resistance routing layer. The metallic overcoat is sufficiently thin to allow low-profile packaging. The method employs a conductive overcoat that is significantly thin compared to conventional insulating conformal overcoats. Two kinds of shifting have been discussed, one is systematic package shift, which can be trimmed and its temperature coefficient compensated. The other is random package shift. The method on how to decrease them has been discussed, too.

2. Reduction of Systematic Shift in Voltage Reference

The systematic shift is mainly caused by the mismatch of layout and process variation, such as the error in different directions, etc. An efficient way now during technology is using Trimming method.

Trimming is a post-fabrication circuit adjustment aimed at correcting the process-induced offsets of various components. The temperature-drift dependence of this adjustment should track that of the offset. Typically, one or more strategically placed resistors are tuned to offset the mismatch errors of two or more devices.

The resistance is varied by:

(1) Fabricating a number of binarily weighted resistors and open- and/or short-circuiting them with onchip fuses or

(2) Reshaping and therefore resizing a resistor with a laser [1].

The accuracy of the former is limited by the reach and resolution of the trim resistors, that is to say, the initial mismatch accuracy performance that sets the full-scale trim-range resistance and the silicon area and test-time boundaries that limit the total number of bits that can be afforded. Laser trimming, on the other hand, is more accurate and area efficient and therefore often used in high performance data converter applications, but its inherent cost in test time and equipment is many times prohibitive.

The reason why trimming in general is so attractive is that many process-induced errors have an almost linear temperature dependence, like several of the mismatch and offset errors in a bandgap reference [2] and a bipolar differential pair, and consequently trimming at one temperature, for instance, room temperature is sufficient to cancel the temperature drift of the offset [1]. Its cost in manufacturing time, however, can account for 25% of the total cost of a power management IC [3], and this is only to correct first-order (linear) errors. The temperature dependence of higher order errors present in bandgap circuits and MOS and BiCMOS amplifiers are not compensated, only their absolute offsets at the trimming temperature (for example, room temperature) are reduced. Compounded to this are package stress induced offset errors because most trimming procedures are performed at wafer level to circumvent the increased costs of post-package EEPROM trimming procedures. Package shift offset effects can be reduced by adding post-fabrication low-stress mechanically compliant layers to the IC before encapsulating it with plastic [1], but again, adding these compounds is costly.

3. Reduction of Random Package Shift

Due to the wide variation expected threshold voltage of MOS devices from die-to-die and within-die during the life time of a process, present leakage current estimation techniques provide lower and upper bounds on the leakage current. The upper and lower bounds are at least an order of magnitude apart and leakage power of most chips lies between the two bounds as shown in [1]. In older technology generations, basing system design on the two leakage current bounds was acceptable since leakage power was a negligible component of the total power. In most systems, the worst case bound is assumed for the design. In future technology generations where as much as half of the system power during active mode can be due to leakage, depending the worse case bound will lead to extremely pessimistic and expensive design solutions. One cannot base the system design on the lower bound since it will lead to overly optimistic and unreliable design solutions. Therefore, it will be crucial to estimate leakage current as accurately as possible. The upper and lower bound estimate equations and measurements are provided in the next part of this section. The lower bound leakage current estimation of a chip is given as follows, To include the impact of within-die threshold voltage or channel length variation it is necessary to consider the entire range of leakage currents, not just the mean leakage or the worst-case leakage. Let us assume that the within-die threshold voltage or channel length variation follows a normal distribution with respect to transistor width, with being the mean and being the sigma of the distribution. Let Io be the leakage of the device with the mean threshold voltage or channel length. Then by performing the weighted sum of devices of different leakage, we can estimate the total leakage of the chip. This is achieved by integrating the threshold voltage or channel length distribution multiplied by the leakage power measurements on several samples of a 0.18μm 32-bit microprocessor were carried out. Using these individual device measurements, with w_p and w_n obtained from the design the leakage power was calculated using the I_{leak}-l, I_{leak}-u, and I_{leak}-w formulae. In addition, we assumed that on an average half of the devices will be in off state, that is, $m_p = m_n = 2$. The three calculated leakages are then compared with the measured leakage.

One solution to the problem of ever-increasing leakage is to force a non-stack device to a stack of two devices without affecting the input load, as shown in Figure 1. By ensuring iso-input load, the previous gate's delay and the switching power will remain unchanged. Logic gates after stack forcing will reduce leakage power, but incur a delay penalty, similar to replacing a low-V_t device with a high-V_t device in a dual-V_t design [4]. In a dual-V_t design, the low-V_t devices are used in performance critical paths and the high-V_t devices in the rest [6]. Usually a significant fraction of the devices can be

high-V_t or forced-stack since a large number of the paths are non-critical. This will reduce the overall leakage power of the chip without impacting operating clock frequency. This stack forcing technique either can be used in conjunction with dual-V_t or can be used to reduce the leakage in a single-V_t design.

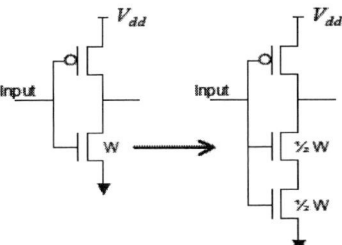

Fig. 1 Trade-off between standby leakage and performance by forcing a two-stack under iso-input load. An NMOS two-stack will reduce leakage when input stays at logic "0"

It is possible to facilitate delay-leakage trade-off by increasing the channel length of devices [1] that are in non-critical paths. To maintain iso-input load the channel width will have to be reduced along with increase in the channel length. Figure 2 shows the mean leakage reduction achievable by increasing the channel length. In Figure 2 the channel length of interest is given by x=0.18μm and stack leakage is for a stack of two devices with $w_u = w_l = \frac{1}{2}w$. As it is clear from Figure 1, the channel length has to be increased 3 times as that of the nominal channel length to match the mean leakage of a two-stack of 0.18μm devices. The reason for such a large increase is attributed to the reverse short channel effect that is present due to halo doping [4] where V_t reduces with increase in channel length. It is important to note that stacking two devices of nominal channel length is different from doubling the channel length due to the two dimensional nature of barrier-lowering and drain induced barrier lowering.

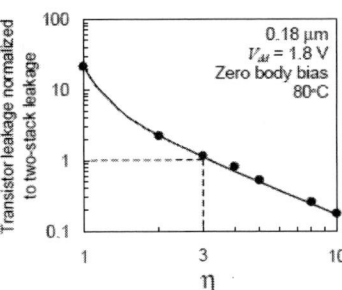

Figure 2 Comparing device leakage reduction due to channel length increase with two-stack leakage. The channel length is given by x=0.18μm. Stack leakage is a two stack of devices with $w_u = w_l = \frac{1}{2}w$. Leakage numbers are obtained from simulation under iso-input load.

Two-stack assignment of low-V_t transistors was applied in 0.13μm technology. Bandgap references, packaged in plastic, have been known to shift in voltage, a pre-package to post-package voltage variation. The package shift has been analytically discussed and experimentally investigated in this paper. Once the reference is encapsulated, a package-induced stresses present, which lead to a systematic voltage shifts ranging from 3 to 7mV. The variation is closely related to package type and processing. Two kinds of shifting have been discussed, one is systematic package shift, which can be trimmed and its temperature coefficient compensated. The other is random package shift. The method on how to decrease them has been discussed, too. Structure and method of minimizing package-shift effects in integrated circuits is implemented by using a thick metallic overcoat applied after the deposition and patterning of the conventional insulating protective overcoat.

4. Conclusion

Package-induced offsets, unfortunately, vary from unit to unit while roughly conforming to a Gaussian distribution. From a designer's perspective, the problem is addressed in two ways: 1) compensating the mean offset as well as the mean TC and 2) minimizing the effects of the mechanisms that cause random variation. The best approach to compensate for the mean package shift offset and its TC, which turns out to be mostly linearly positive in the temperature range of interest, is to include it in the design of the circuit itself. As a result, characterization of the bandgap circuit within its particular package is required.

Additionally, if the process technology permits, adding a moderately thick, yet thin relative to dropper-applied and spin-on overcoats, layer of elastic material between the die and the plastic mold yields significant improvements, a marginal cost that, depending on the application, may be worthwhile.

Acknowledgement

This work was Supported by Program for New Century Excellent Talents in University (NCET) by Ministry of Education.

References

1. Hamada A., "A new aspect of mechanical stress effects in scaled MOS devices," *IEEE Trans Electron Dev.*, Vol.38, No.4 (1991), pp.895-899.
2. Le Cam, Guyader F., de Buttet, *et.al.*, "A low cost drive current enhancement technique using shallow trench isolation induced stress for 45nm node," *Proc 2006 VLSI Technology*, Honolulu, HI, May.2006, pp. 82-83.
3. Alavi M., Bohr H., "A PROM element based on salicide agglomeration of poly fuses in a CMOS logic process," *Proc of 1997 IEDM Tech Dig*, Minesota, MN, Jan.1997, pp.197-198.
4. Bianchi R.A., Bouche G., "Accurate Modeling of Trench Isolation Induced Mechanical Stress effects on MOSFET Electrical Performance," *Proceeding of 2002 IEDM Tech Dig*, Newyork, NW, Jan.2002, pp. 117-118.
5. Schenkel M., Mettler S., Reiner W. *et al.*, "Measurements and 3D simulations of full-chip potential distribution at parasitic substrate current injection," *Proceeding of European Solid-state Device Research Conference*, Cork,Ireland, May, 2000,pp.600-603.
6. Gallon C., Reimbold G., Ghibaudo G., *et al.*, "Electrical analysis of external mechanical stress effects in short channel MOSFETs on (001) silicon," *Solid-state Electronics*, Vol. 48, No.3 (2004), pp. 561-566.

7. Bradley AT, Jaeger RC, Suhling JC, *et al.*, "Piezoresistive characteristics of short-channel MOSFETs on (100) silicon," *IEEE Trans Electron Dev*, Vol 48, No.9 (2001), pp. 2009-2015.
8. Taur .Y., Zicherman DS, Lombardi DR, *et al.*, "A new shift and ratio method for MOSFET channel length extraction," *IEEE Electrons Dev Lett*, Vol. 13, No.5 (1992), pp. 267-269.
9. En WG, Ju DH, Chan D, *et al.*, "Reduction of STI-active stress on 0.18um SOI devices through modification of STI process," *Proceeding of IEEE Int SOI Conference*, Trie, Italy, July, 2001, pp. 85-86.

Electrical Analysis of Mechanical Stress Induced by Shallow Trench Isolation

Yanfeng Jiang, Jiaxin Ju

Microelectronic Center, College of Information Engineering, North China University of Technology, Beijing, 100144, China
E-mail address: yfjiang@ncut.edu.cn

Abstract

In many modern technologies, shallow trench isolation (STI) exhibits a potential application, especially for power devices or SOI ones. During its application, technologies have found the mechanical stress which originated from STI technology. This paper describes the usage of STI on power devices, which fulfills 700V technology on 100V BCD technology. Main results are the mobility variations with stress, the strong effect of Rsd on transistors. Then using the same approach on short devices with different distances gate edge to STI, we show how to evaluate stress distribution induced by STI as well as its mean value under the gate of the devices. These results help to understand, minimize or optimize stress effects.

1. Introduction

Smart Power integrated circuits (SPICs) with blocking voltage capability of about 700V are of growing interest for off-line applications in industrial and consumer products. If terminals of integrated bipolar power devices like IGBTs (insulated gate bipolar transistor) and freewheeling diodes are on increased potential with respect to substrate (high side switch of a half bridge configuration) a dielectric insulation (DI) process must be used for the realization of such high-voltage (HV) SPICs [1]. The different approaches towards DI HV devices can be classified in which way the voltage is distributed along the vertical direction of the device.

In a first class of technologies the blocking voltage develops only in the (n-doped) silicon of the DI island the power device is embedded. An n^+ layer at the bottom of the DI island prevents the penetration of the electrical field into the bottom oxide. In the case for 500V blocking voltage island depth of about 50μm is needed. In the "classical" method for the fabrication of such a DI island a silicon wafer is structured by an anisotropic etch, thermally oxidized and replenished with very thick poly-silicon layer. Upside down the poly-silicon layer carries the DI islands, which get bare after the initial single crystal wafer is thinned out [2]. With direct wafer bonded silicon (BSOI) wafer[3] and a trench etch and refill process for the lateral insulation the drawbacks of the deposited poly-silicon wafers may be eliminated, but the high expenses for the deep trench etch and refill generally abandon the introduction as a production process [4].

A second class of processes supports the blocking voltage of the insulated device both in the silicon island and the insulation oxide [5-6]. For low voltage (LV) demands (<150V) with few μm thick active silicon films such processes are well established for the production of SPICs with focus on automotive applications. Up to now the resulting expenses for trench etch and refill of a 20μm thick active silicon on top of 2μm buried oxide-necessary to support about 700V-impede a broad access to production scale for off-line applications.

One way to overcome the problem of thick DI islands is to use instead of the common highly doped substrate wafer a low p-doped one, so that portions of vertical voltage drop of the blocking device can also be supported by this wafer side (partial DI) [7]. To prevent the formation of an inversion layer an additional n-doped region must be provided at the substrate side of the buried oxide which has to be connected to the high terminal of the device. The realization of such a contact not only results in a significant increase of process complexity and process cost, the DI islands are no more completely dielectric insulated and high leakage and displacement currents from the substrate space charge region will appear.

If instead of a low p-doped silicon an insulator is used as carrier wafer for the active devices, support of most of the blocking voltage in vertical direction by the substrate is possible without the drawbacks of partial DI. For these reasons SOS (silicon on sapphire) technology has evolved. But because of the restricted quality of the active silicon layer and the high cost of the wafers it is not suited for the production of cost-critical HV SPICs [8]. Mechanical stress may be generated in MOS transistors at many technology process steps, as each of them generally implies different process temperature as well as material with different mechanical properties, thermal coefficient mismatch and so on [1]. As CMOS devices continue to be scaled down, these effects become more important. In the case of hetero-structures as SiGe devices, stress is used to improve performances. Otherwise it is mainly detrimental. For back-end part of the process, stress voiding and mechanical weakness of low k materials are of prime importance. For front end part, Shallow Trench Isolation (STI) is today the dominant source of stress in MOSFET channels [2-3]. Stress induced by STI can in worst case affect yield through dislocations and in consequence the leakage current increases. For lower stress, MOSFETs drivability depends on design layout [4], and each design must be carefully electrically characterized.

While piezoresistivity has been deeply studied in bulk silicon from both theoretical and practical points of view (for sensors), few works have been performed on advanced deep submicron CMOS technologies. Stress is difficult to measure locally and to simulate, since there is a critical lack of data for many thin film materials used in process.

The aim of this paper is to propose a methodology to evaluate the STI induced internal stress in the MOSFETs channel by using a four-point bending method.

2. Design and Fulfillment of Power MOSFET under Trench Technology

The layout of the mesh array transistor was built using a technique similar to lego construction where elementary parts or bricks can be put together to construct different size and shape transistors. This mesh array topology can also be implemented automatically resorting to adequate software in order to speed up error free design layout of large analogue blocks [9]. A sketch of one of these elementary parts is depicted in Fig. 1. Fig. 1 a) represents a mesh array implementation of a CMOS standard cell. An high voltage cell, based on the extended drain high voltage technique [5-6], is represented in Fig. 1 b), where L_D represents the width of the drain diffusion, L_{GD} the drift length between the induce channel and the drain, W_2 is half the length of the poly gate over thin oxide, L_P poly gate over thick oxide and W_1 the width of the elementary block. The n-well mask is aligned with the inner active area border.

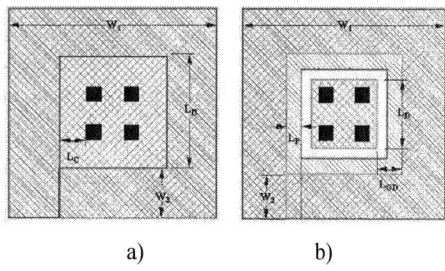

a)   b)

Figure 1 Elementary transistor of the mesh array: a) low voltage cell; b) high voltage cell.

The high voltage elementary cell was optimized with the aid of a 2D simulator, regarding minimum channel length to avoid punch through and minimum L_{GD} length to assure no breakdown degradation. Fig. 2 shows simulated and experimental results at off state for the high voltage structure with the minimum channel length allowed by design rules for this particularly high voltage cell ($3\mu m$). As can be seen, the transistor evidences an avalanche rather than a punch through breakdown above 21V. The same optimization methodology provided a secure margin for L_{GD} length above $1\mu m$. According to the simulation study the high voltage cell pitch with optimized channel length and drift path was set to $1\mu m$.

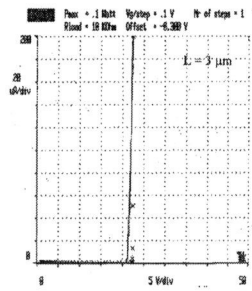

Figure 2 Simulated and experimental I_D-V_{DS} curves at off state ($V_{GS} = 0V$) for the high voltage cell.

Fig. 3 shows a layout of the proposed mesh array of transistors, where the drains are completely surrounded by poly-silicon gate. Transistor source surrounds the arrangement at the edge of the layout and is connected by the lower level metal layer to the inner source points. The interconnection of common drains and sources is also made by means of 45° parallel strips of first metallization.

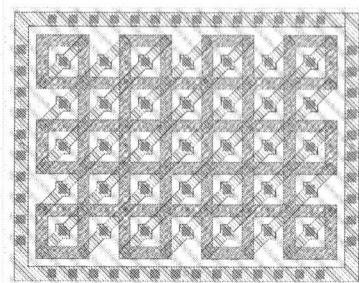

Figure 3 General aspect of the mesh array of transistors NMOS transistor in a 0.35µm technology process.

The connection to upper metal layers is made using stacked vias over the metal/diffusion contacts. In order to reduce the interconnection resistance with the pads a staircase like shape is used, to enlarge the interconnecting stripes. In this case, routing to pads must be made according to the arrows directions. This interconnection solution is important as it contributes to reduce debiasing effect due to parallel running of metal interconnects of both source and drain [10] and [11].

Fig. 4 shows the cross-sectional SEM pictures taken at the situ-locations. Part (a) is showing an overview of the full trench, while part (b) only concentrates on the top part of the trench. From picture (b) it is clear that we realized to have the filler polysilicon etch back process developed in such a way that the original dimple in the poly (immediately after the filler deposition) is taken away during the etch back process.

During the development phase of the deep trench processing sequence, we have been facing severe silicon stress situations leading to defects within the silicon material. An extensive process design of experiments was required in order to get finally the complete structure free of defects.

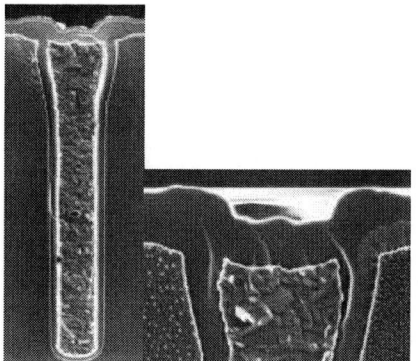

(a) full DTI   (b) only top part of DTI
Figure 4 Cross-sectional SEM pictures

3. Measurement and Discussion

We tested n and p MOS transistors fabricated on (100) silicon substrates for bulk and SOI 0.13µm technologies. Similar results are obtained. We focus on SOI results in this paper. Studied devices have a gate oxide thickness

T_{ox}=2nm, a channel width W=10µm and a long (L=10µm) or short (L=0.13µm) channel length. The I_d-V_d and I_d-V_g

characteristics of the device are plotted below. The characteristics are very close to the actual 90nm characteristics that can be found in the literature.

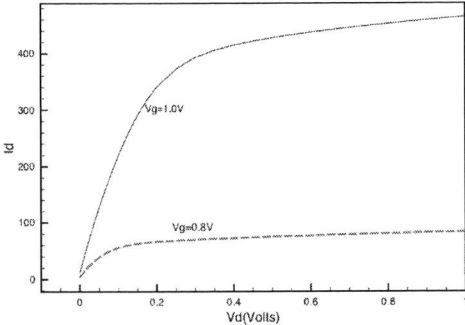

Fig. 5 I_d-V_d characteristics of the device

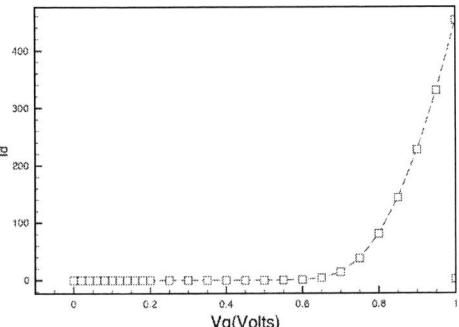

Fig. 6 I_d-V_g characteristics of the device

Mechanical stress on semiconductor induces a change in the band-structure this in turn affects the carrier mobility. This effect is well known [3] and can be explained by the historically significant theory known as deformation potential theory. Mathematically the stress tensor σ_{ij} and the generalized Hook's law for anisotropic materials are used to calculate the strain tensor. Depending on the crystal one can use symmetry arguments and one can reduce the elasticity modulus to 3 components corresponding to the perpendicular, parallel and shear components. Once the strain is calculated the next step is to calculate the change in carrier statistics. In the deformation potential theory the strains are considered relatively small. The main impact of strain on the carrier statistics is through a change in the band-structure itself. Strain changes the number of carrier sub-valleys and eventually a change in the actual band-gap in the material [4].

Finally, we discuss the impact of strain on mobility in these devices. The carrier redistribution that takes place between the various sub-valleys causes the change in mobility. The strain can have either a detrimental or enhancing effect. This depends on the materials involved. As an example strained silicon grown relaxed SiGe can actually enhance the electron mobility. The mobility enhancement is attributed increase in the occupancy of the conduction band valleys [5]; there is a reduction in inter-valley scattering. It has been shown that the perpendicular effective mass is much

lower than the longitudinal mass. Consequently we get an enhancement in mobility.

The aim of this study was to explore the process needed to model stress induced effects in submicron. Modeling stress effects in submicron semiconductor devices is a necessity for the semiconductor industry. With rapidly reducing device sizes these effects have become all the more relevant. What were thought to be second order effects just a few years back is now affecting device performance in s big way. We have undertaken the task of modeling this complex problem.

Modeling stress-induced effects is complicated because not much data exists on the impact of stress on the several thin films that are used in these processes. We use data from work done by researchers previously. As a first stab we use mobility change calculations using strain induced mobility models to estimate these changes. Although the trends are as expected but the results show us that there is a discord between model and actual experimental data [6]. The reasons for this can be many, our model neglects both visco-elastic and piezo effects. We might also be underestimating the stress, but that alone cannot account for the difference. These effects are very process dependent and can make a big difference in the accuracy of the model.

4. Conclusion

We have shown that we can get many valuable information on piezoresistance and on stress effects on advanced technologies. We have assessed the stress response on short transistors. We have proved a similar behavior for small and long transistors once Rsd effects have been taken into account. To this end, the 2D stress profiles (shape and quantitative values) induced by STI have been extracted and permit to understand and optimize the initial process. The visco-elastic model and other time dependent phenomena can be modeled in multiphysics. The visco-elastic model is a time dependent problem and is efficiently modeled in system. The inclusion of piezo effects has not yet been assessed. Another approach could be using experimental data that assesses the impact of piezo effects on mobility. These two effects will make the model more complete and can closely mirror actual device data.

Acknowledgement

This work was supported by Program for New Century Excellent Talents in University (NCET) by Ministry of Education.

References

1. Hamada A., "A new aspect of mechanical stress effects in scaled MOS devices," *IEEE Trans Electron Dev.,* Vol.38, No.4 (1991), pp. 895-899.
2. Le Cam, Guyader F., de Buttet, *et.al.,* "A low cost drive current enhancement technique using shallow trench isolation induced stress for 45nm node," *Proc 2006 VLSI Technology,* Honolulu, HI, May.2006, pp. 82-83.
3. Alavi M., Bohr H., "A PROM element based on salicide agglomeration of poly fuses in a CMOS logic process," *Proc of 1997 IEDM Tech Dig,* Minesota, MN, Jan.1997, pp. 197-198.
4. Bianchi R.A., Bouche G., "Accurate Modeling of Trench Isolation Induced Mechanical Stress effects on MOSFET

Electrical Performance," *Proceeding of 2002 IEDM Tech Dig*, Newyork, NW, Jan.2002, pp. 117-118.

5. Schenkel M., Mettler S., Reiner W. *et al.*, "Measurements and 3D simulations of full-chip potential distribution at parasitic substrate current injection," *Proceeding of European Solid-state Device Research Conference*, Cork, Ireland, May, 2000, pp. 600-603.

6. Gallon C., Reimbold G., Ghibaudo G., *et al.*, "Electrical analysis of external mechanical stress effects in short channel MOSFETs on (001) silicon," *Solid-state Electronics*, Vol 48, No.3 (2004), pp.561-566.

7. Bradley AT, Jaeger RC, Suhling JC, *et al.*, "Piezoresistive characteristics of short-channel MOSFETs on (100) silicon," *IEEE Trans Electron Dev*, Vol 48, No.9 (2001), pp. 2009-2015.

8. Taur Y., Zicherman DS, Lombardi DR, *et al.*, "A new shift and ratio method for MOSFET channel length extraction," *IEEE Electrons Dev Lett*, Vol 13, No.5 (1992), pp. 267-269.

9. En WG, Ju DH, Chan D, *et al.*, "Reduction of STI-active stress on 0.18um SOI devices through modification of STI process," *Proceeding of IEEE Int SOI Conference*, Trie, Italy, July, 2001, pp. 85-86.

978-1-4244-4658-2/09 $25.00 © 2009 IEEE   1158

The Effect of Thermal Cycling on Nanoparticle Reinforced Composite Lead-free Solder

Si Chen[1], Zhaonian Cheng[2], Johan Liu[1, 2], Yulai Gao[3] and Qijie Zhai[3]

[1]Key Laboratory of Advanced Display and System Applications
& SMIT Center, School of Automation and Mechanical Engineering,
Shanghai University, No. 149 Yanchang Road, Shanghai 200072, P.R. China

[2]SMIT Center & Bionano Systems Laboratory, Department of Microtechnology and Nanoscience,
Chalmers University of Technology, SE-412 96 Goteborg, Sweden

[3]School of Materials Science and Engineering, Shanghai University, No. 149 Yanchang Road, Shanghai 200072, P.R. China

Corresponding author Email: jliu@chalmers.se

Abstract

The effects of thermal cycling on shear strength and fracture mode of the nanosized Sn-3.0Ag-0.5Cu particulates reinforced Sn-58Bi composite solder were investigated in this paper. By using a self developed top-down method named Consumable-electrode Direct Current Arc technique, the Sn-3.0Ag-0.5Cu nanoparticles were successfully manufactured. The primary particle size of Sn-3.0Ag-0.5Cu nanoparticles ranged from 20nm to 80nm. Sn-3.0Ag-0.5Cu nanoparticles with different weight percentages were mixed into commercial Sn-58Bi solder paste in order to develop a composite solder paste which is lead-free and possess high strength and low melting point. Following the conventional surface-mount technology process, the 1206 chip resistor and ENIG/Cu pad were joined by the composite solder. Scanning electron microscope, transmission electron microscope and optical microscope were employed to observe the morphology of nanoparticles, microstructure of solder matrix, fracture mode after shear test and crack after thermal cycling. The experimental results indicated that before thermal cycling all composite solders' shear strength increased greatly compared to Sn-58Bi solder making them comparable to Sn-3.0Ag-0.5Cu solder. The fracture surfaces of all composite solder joints occurred at the interface between the solder matrix and the resistor termination. After thermal cycling, the shear strength of the composite solders was at a constant value. However, when the weight percentages of Sn-3.0Ag-0.5Cu nanoparticles exceeded a certain value, the shear strength of composite solder joints decreased rapidly and the case of solder brittle fracture increased as the nanoparticles content increased.

Introduction

Lead-free Sn-based solders have been applied in electronic packaging industry for several years to eliminate lead's negative effects on human health and the environment. Unfortunately, higher melting temperature is still a critical issue for lead-free solders. It is well known that higher melting point result in higher process temperature which in turn results in higher defect rates. Though Sn-52In and Sn-58Bi solder alloys have low melting points and have been used for step soldering and non-heatproof device for a long time, the high price of indium and the brittleness of Sn-58Bi limit their development. Therefore the development of new solders that possess good mechanical and electrical performance as well as low melting point is essential. An effective attempt is to add certain nanoparticles into

conventional solder paste to form a kind of composite solder. Liu et al. [1] added Ag nanoparticles into Sn-Pb solder and proved that the nanosized Ag particle-reinforced composite solder improved the creep resistance of solder joints. Hsiao et al. [2] mechanically mixed Cu_6Sn_5 nanopowders into Sn-3.5Ag solder paste to form a kind of nano-sized Cu_6Sn_5 doped Sn-Ag-Cu solder paste. Their results indicated the ball shear strength of Cu_6Sn_5-contained joint was higher than that of the bare Sn-Ag-Cu one due to the nano Cu_6Sn_5 reinforcement. Shi et al. [3] produced a series of composite solders by blending Sn37Pb and Sn0.7Cu solder micro-sized powders with different volume percentages of nano-sized Cu, Ag, Al_2O_3 and TiO_2 reinforcement particles. According to their results, the creep resistance of each composite solder was improved. Gain et al. [4] studied what effects on the shear strength the nano Ni additions would cause in Sn-9Zn and Sn-8Zn-3Bi solders with Au/Ni/Cu pad metallization in ball grid array applications. Nano Ni powder was added to Sn-based solders; thereafter the shear loads were increased. The studies concluded that adding nano-sized particulates into a conventional solder alloy to form a kind of composite solder was a successful method of enhancing the mechanical performance of solders joints. Conventional lead-free solders with high melting points were widely used as the base material. Thus the process temperature was still high. Shin et al. [5] added SiC nanoparticles into Sn-58Bi low melting point solder in an attempt to form a new low melting point composite solder. From their results, the shear strength increased 13% compared to Sn-58Bi. However while SiC possesses great thermal conductivity, its electrical conductivity is significant low. So, the purpose of this paper is to present a lead-free composite solder with low melting point, high mechanical and electrical performance and good thermo-mechanical fatigue resistance by adding Sn-3.0Ag-0.5Cu nanoparticles into Sn-58Bi solder paste.

Experimental procedure

Sn-3.0Ag-0.5Cu nanoparticles were manufactured by using a self developed top-down method called the Consumable-electrode Direct Current Arc (CDCA) technique. [7-8] The nanoparticles exhibited a near spherical morphology with particle sizes ranging from 20nm to 80nm. Figure 1 shows the scanning electron microscope (SEM) image of Sn-Ag-Cu nanoparticles which were manufactured by the CDCA method [6].

The composite solders were prepared by stirring Sn-58Bi solder paste with different weight percentages of Sn-3.0Ag-0.5Cu nanoparticles (1%, 2%, 3% and 4%) for 30 min to form a homogeneous distribution of the reinforcement particles in the solder matrix.

Figure 1 SEM image of nanometer-sized SAC particles

The FR-4 PCBs with ENIG/Cu pad were designed and manufactured for this study. Every PCB consists of fifty 1206 chip resistors, as shown in Figure 2. Following the conventional surface mount technology, the chip resistors were mounted on pads. The reflow soldering was conducted in a reflow oven with ten different temperature zones. For composite solder, a temperature profile with a peak temperature of 180°C and a total duration time of 6 min was used. The dwell time above the melting temperature was approximately 60 seconds. The samples were air-cooled until they reached room temperature. The Sn-58Bi and Sn-3.0Ag-0.5Cu solder without nanoparticles were also used for comparing with the composite solder.

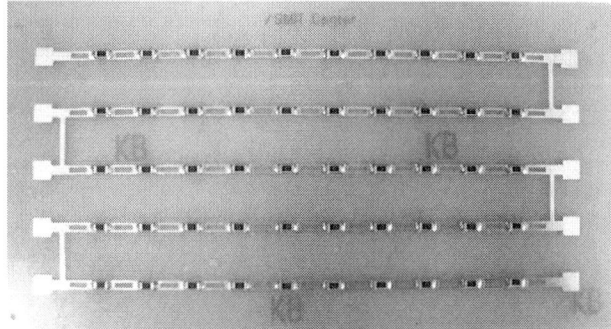

Figure 2 FR-4 PCB with 1206 chip resistors

For evaluating the effect of thermal cycling (TC) on shear strength and fracture mode of the composite solder, the test samples were divided into two groups. One group was shear tested directly after soldering, and another group endured 500 cycles of thermal cycling before the shear test. In all 12 cases (see Table 1) were studied in the present work.

Table 1 Design of Experiment

	Solder matrix	Nanoparticles (wt %)	TC (cycles)
Case 1	Sn-58Bi	0	0
Case 2	Sn-58Bi	0	500
Case 3	Sn-3.0Ag-0.5Cu	0	0
Case 4	Sn-3.0Ag-0.5Cu	0	500
Case 5	Sn-58Bi	1	0
Case 6	Sn-58Bi	1	500
Case 7	Sn-58Bi	2	0
Case 8	Sn-58Bi	2	500
Case 9	Sn-58Bi	3	0
Case 10	Sn-58Bi	3	500
Case 11	Sn-58Bi	4	0
Case 12	Sn-58Bi	4	500

The shear test was performed on samples by using a kind of Shear Tester. The shear height and the test speed of the shear test were 20μm and 700μm/s respectively. The thermal cycling was carried out in a single chamber. The temperature profile was in a range of -40°C and 125°C, and the dwell time of 19 min was set for the high and low temperature zones. The heating/cooling rate was fixed at 15°C/min, and the total duration of the thermal cycling profile was about 60 min.

Results and Discussion
Microstructure observation

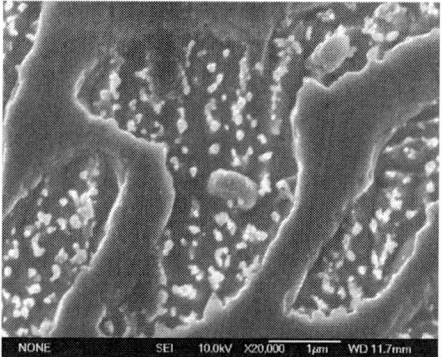

(a) Morphology of the eutectic Sn-Bi solder joint

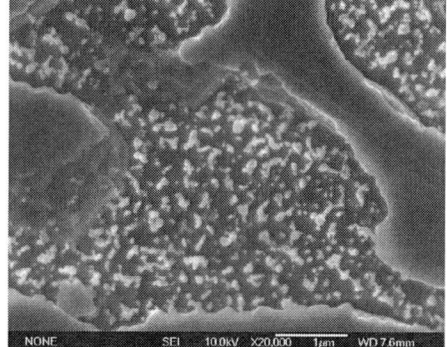

(b) Morphology of the composite solder joint
Figure 3 SEM micrographs of the eutectic Sn-Bi solder and the composite solder joint

Figure 3 shows the microstructures of the Sn-58Bi solder joint and the composite solder joint. The features observed are quite typical of a eutectic Sn-Bi solder and consist of fine alternating lamellae of the tow constituent phases. The grains of composite solder are obviously finer than pure eutectic solder. A possible explanation is that the nanoparticles dispersed throughout the composite microstructure obstruct the grains' movement of dislocation and prevent their growth [9-10].

The morphology of composite solder was also observed by transmission electron microscopy (TEM) as shown in Figure 4. From Figure 4(a), it can be seen that the nanoparticles were distributed uniformly and only a small amount of the nanoparticles were agglomerated. Figure 4(b) shows the nanoparticles tended to gather into crystal grain boundaries or near the grain boundaries. More detailed analysis about microstructure of composite solder joint was conducted in our previous work [11].

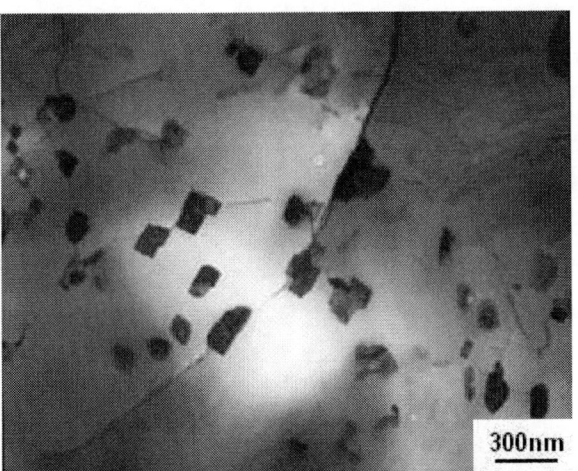

(a) Nanoparticles distribution in composite solder

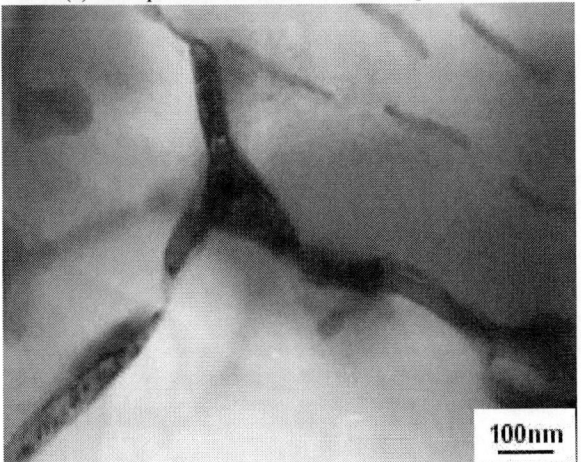

(b) Nanoparticles in crystal grain boundaries

Figure 4 TEM images of composite solder

Shear strength

Figure 5 shows the measured shear strength for composite solder joints with two kinds of pure eutectic solders. The measurement data for reinforced composite solder include four different weight percentages (1%, 2%, 3% and 4%) of nanoparticles. The shear strength before and after thermal cycling are plotted in figure 5.The top and bottom sides of the rectangles represent the 75 percentage and 25 percentage values respectively. From Figure 5, the shear strength of all composite solders was around 30MPa. That was far higher than the shear strength of Sn-58Bi and near the shear strength of Sn-3.0Ag-0.5Cu. The enhancement of shear strength of composite solder could attribute to the addition of nanoparticles in solder matrix which resulted in finer grains. A possible reason is that the fine grains reduced dislocation movements near the grain boundaries, since with the nanoparticles; much more energy was needed for the dislocation movement to travel across the boundaries [5].

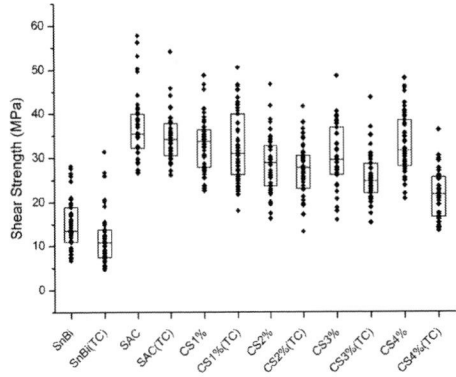

Figure 5 Shear strength of composite solder with two conventional eutectic solders before and after thermal cycling

From Figure 5, the shear strength of composite solder, after 500 cycles of thermal cycling test, maintained a stable value when the weight percentage of nanoparticles was lower than 2wt.%. However, when the weight percentage of nanoparticles exceeded 2wt.%, the shear strength of composite solder after thermal cycling decreased significantly. In addition, the shear strength of pure eutectic Sn-58Bi also decreased after thermal cycling. The shear strength of Sn-3.0Ag-0.5Cu remained constant.

The results of the shear strength tests after thermal cycling indicated that the thermomechancial fatigue (TMF) resistance of Sn-58Bi was enhanced by adding the Sn-3.0Ag-0.5Cu nanoparticles to the solder matrix. It is known that grain boundary sliding is the predominant mode of TMF damage that occurs during the high-temperature dwell in a TMF cycle [12]. In the present test, the grain boundary slide could be constrained effectively during thermal cycling due to the nanoparticles gathering in grain boundaries as above mention. However, too much nanoparticles resulted in degradation of shear strength. When the weight percentage of nanoparticles was too high, the micro-interfaces between nanoparticles and solder matrix might increase, and would easily result in crack initiation and propagation during thermal cycling.

(a) Crack in Sn-58Bi solder joint after thermal cycling

(b) Crack in composite solder joint with 3wt% nanoparticles

(c) Crack in composite solder joint with 4wt% nanoparticles
Figure 6 Cross-section observations after thermal cycling

A cross-section inspection was made to detect any crack initiation and propagation in solder matrix after thermal cycling. The most obvious fatigue crack in Sn-58Bi and composite solders with 3wt.% and 4wt.% nanoparticles are shown in Figure 6. For Sn-3.0Ag-0.5Cu and composite

solders with 1wt.% and 2wt.% nanoparticles, no crack could be found after 500 cycles of thermal cycling. The results of cross-section corresponded well with the data of shear strength, indicating that the initiation and propagation of crack in solder joint resulted in the decrease of shear strength.

Fracture mode

All samples after shear testing were inspected under optical microscope (OM) to identify the fracture mode. From the inspection results, two kinds of fracture modes were found. The first kind was when the fracture surface is within the solder itself, as shown in Figure 7(a). The second kind is when the fracture surface was located at the interface between resistor terminations and solder joint, as shown in Figure 7(b). In order to clearly show the fracture location, the chip resistors are also included in the pictures.

(a) Fracture in solder

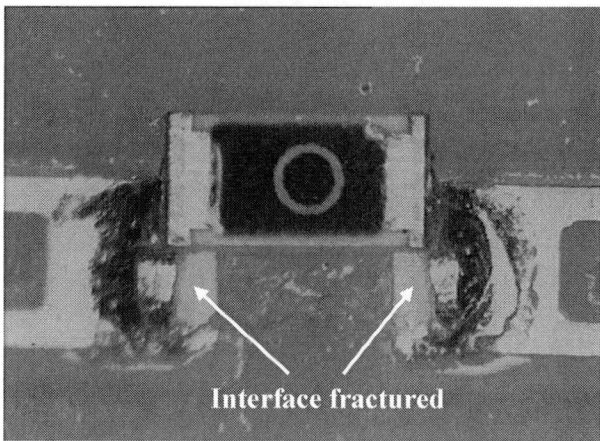

(b) Fracture at interface
Figure 7 Fracture surface observations after shear testing

Figure 8 shows the percentages of the two fracture modes in all 12 kinds of samples. There were 40 shear samples for every testing of 12 cases. Only two fracture modes, fracture in solder and fracture at interface, were observed in this work and the samples with fracture in solder and with fracture at interface counted respectively. The percentage of fracture mode in the Y axle of Figure 8 indicates the ratio of samples

978-1-4244-4658-2/09 $25.00 © 2009 IEEE   1162

of each fracture mode to the total number of samples in every case. Before thermal cycling, the fracture within solder could be observed in the Sn-58Bi samples because the Sn-58Bi solder was often brittle [13]. However for reinforced composite solders, dispersed nanoparticles could provide a classical dispersion strengthening of the material and therefore could enhance the strength and microhardness [6]. In other words, the solder itself became stronger when nanoparticles were mixed into the solder matrix. So, during the process of shear testing, the interface became relatively weaker than the solder itself, which caused interface fracture. After thermal cycling, in addition to Sn-58Bi samples, fracture in solder could also be detected in composite solders with 3wt.% and 4wt.% nanoparticles. This might be also caused by the increase of micro-interface between nanoparticles and solder matrix.

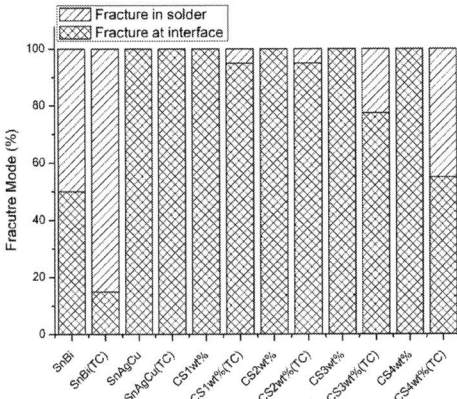

Figure 8 Proportion of two fracture modes after shear testing

Conclusions

A reinforced composite solder with low melting point was developed and investigated in this work. The composite solder was produced by mixing Sn-3.0Ag-0.5Cu nanoparticles into conventional Sn-58Bi solder paste. The nanoparticles were manufactured with a self developed top-down method called CDCA. Shear strength of composite solders with different weight percentages of nanoparticles was studied and compared to pure eutectic Sn-58Bi and Sn-3.0Ag-0.5Cu solder. Observations made by SEM and TEM indicated that nanoparticles contained composite solder provided finer grains than eutectic Sn-58Bi solder as the nanoparticles were distributed uniformly in solder matrix. The finer grains in composite solder could effectively constrain the dislocation movement and the shear strength of composite solder was far higher than Sn-58Bi solder, while too many nanoparticles decreased the resistance of TMF, moderate concentration of uniformly dispersed nanoparticles greatly decreased the effects of TMF. The inspection of fracture mode after the shear test showed lots of fractures within the solder in Sn-58Bi solder joints, and all composite solders had fractures at the interface between component termination and solder before thermal cycling. The solder itself was enhanced by nanoparticles. After thermal cycling, too much nanoparticles

resulted in solder joint damage and in turn, increased the probability of fracture in solder during shear test.

Acknowledgements

We acknowledge the financial support from the Chinese Ministry of Science and Technology (2008AA04Z301), from 863 program on nanosolder paste (2006AA03Z339) and from the Shanghai Science and Technology Commission with the contract no: 075007004. This work is also supported by the FP7 IP Nanopack program, from the Swedish National Science Foundation under the project "Nanointerconnect" (621-2007-4660), from the Swedish Foundation for Strategic Research under the ProViking Program (PV08.08).

References

1. J.P. Liu, F. Guo, Y.F. Yan, W.B. Wang, and Y.W. Shi, "Development of Creep-Resistant, Nanosized Ag Particle-Reinforced Sn-Pb Composite Solders," *Electronic Materials*, Vol. 33, No. 9 (2004), pp. 958-963.

2. Li-Yin Hsiao, Guo-Jyun Chiou, Jeng-Gong Duh, and Su-Yueh Tsai, "Synthesis and Application of Novel Lead-Free Solders Derived from Sn-based Nanopowders," *Proceedings of the 7th IEEE CPMT International Conference on Electronics Packaging Technology*, 2006, pp. 358-363.

3. Yaowu Shi, Jianping Liu, Zhidong Xia, Yongping Lei, Fu Guo, and Xiaoyan Li, "Creep property of composite solders reinforced by nano-sized particles," *J Mater Sci: Mater Electron*, No. 19 (2008), pp. 349-356.

4. Asit Kumar Gain, Y.C. Chan, K.C. Yung, Ahmed Sharif, and Lafir Ali, "Effect of Nano Ni Additons on the Structure and Properties of Sn-9Zn and Sn-8Sn-3Bi Solder in Ball Grid Array Packages," *Proceedings of the 2nd IEEE CPMT Electronics Systemintegration Technology Conf*, Greenwich, UK, 2008, pp. 1291-1294.

5. Yue-Seon Shin, Se-hyung Lee, Chang-Woo Lee, Seung-Boo Jung, and Jeong-Han Kim, "Effect of Dispersed SiC Nano-particles in Eutectic Sn58Bi Solder Micro-Bumps of Wafer Level Package By Electroplating," *Proceedings of the 10th IEEE CPMT 10th Electronics Packaging Technology Conf*, Dec 2008, Singapore, pp. 279-284.

6. Johan Liu, Cristina Andersson, Yulai Gao, and Qijie Zhai, "Recent Development of Nano-solder Paste for Electronics Interconnect Applications," *Proceedings of the 10th IEEE CPMT Electronics Packaging Technology Conf*, Dec 2008, Singapore, pp. 84-93.

7. Xinzhi Xia, Changdong Zou, Yulai Gao, Johan Liu, and Qijie Zhai, "Preparation Techniques and Characterization for Sn-3.0Ag-0.5Cu Nanopowders," *Proceedings of the 2007 IEEE CPMT International Symposium on High Density Packaging and Microsystem Integration*, Shanghai, P.R.China, 2007, pp. 302-304.

8. Wanbing Guan, Suresh Ghand Verma, Yulai Gao, Cristina Andersson, Qijie Zhai, and Johan Liu, "Characterization of Nanoparticles of Lead Free Solder Alloys," *Proceedings of the 1st IEEE CPMT Electronics Systemintegration Technology Conf*, Dresden, Germany, 2006, pp. 7-12.

9. D.C. Lin, S. Liu, T.M. Guo, G.X. Wang, T.S. Srivatsan, and M. Petraoli, "An investigation of nanoparticles addition on solidification kinetics and microstructure development of tin-lead solder," *Material Science and Engineering A*, Vol. 360 (2003), pp. 285-292.

10. D.C. Lin, G.X. Wang, T.S. Srivatsan, Meslet Al-hajri, and M. Petraroli, "Influence of titanium dioxide nanopowder addtion on microstructural development and hardness of tin-lead solder," *Materials Letters*, Vol. 57 (2003), pp. 3193-3198.

11. Lili Zhang, Wenkai Tao, Johan Liu, Yan Zhang, Zhaonian Cheng, Cristina Andersson, Yulai Gao, and Qijie Zhai, "Manufacture, Microstructure and Microhardness Analysis of Sn-Bi Lead-Free Solder Reinforced with Sn-Ag-Cu Nano-particles," *Proceedings of the 2008 IEEE CPMT International Conference on Electronics Packaging Technology & High Density Packaging*, 2008, pp. 167.

12. Andre Lee and K.N. Subrammanian, "Development of Nano-Composile Lead-Free Electronic Solders," *Electronic Materials*, Vol. 34, No. 11 (2005), pp. 1399-1407.

13. Kikuchi S., Nishimura M., Suetsugu K., and Ikari T., "Strength of bonding interface in lead-free Sn alloy solders," *Materials Science and Engineering A*319-321 (2001), pp. 475-479.

Study of Isothermal Bending Fatigue Test

Minyi Lou, Long Wen, Zhengrong Chen, Jianwei Zhou, Qian Wang, Jaisung Lee
Samsung Semiconductor (China) R&D Co., Ltd
No. 15, Jin Ji Hu Road, Suzhou Industrial Park, Suzhou, China
Minyi.lou@samsung.com, 86-512-62888288-8813

Abstract

TC test is very important for solder joint reliability, but the test time for TC test is very long, normally it cost about 2 months. In order to shorten TC test time, many researchers study the new test method to replace traditional TC test, such as bending fatigue test. However, there's still some limit for this field study. Considering the actual user condition, the final product not only experience mechanical loading, but also thermal loading. Therefore, isothermal bending test (-25°C, 25°C, 75°C, 125°C) were studied to research the relationship of lifetime and failure mechanism between TC test and bending fatigue test. ANSYS was used to simulate isothermal bending fatigue test and also compare with traditional TC test.

1 Introduction

For real user environment of portable products with key press function, several environment condition changes at the same time, such as temperature, humidity etc. So the electronic components in portable products usually need to be reliable against both thermal and mechanical load. Much study of thermal and mechanical loading on electronic packages has been finished respectively, the thermal and bending test method has been developed, and the influence of experiment condition on the package reliability properties has also been studied.

In order to research the mechanical loading under very high temperature, John H. L. Pang and Fa-Xing Che [1] researched isothermal bending fatigue test method with SAC solder joints at room temperature (25°C) and high temperature (125°C). Torres-Montoya, Duffek, *et al.* [2] studied mechanical loading applied in addition to thermal loading.

In our paper, isothermal bending fatigue was studied systematically to find out the relationship of lifetime and failure mechanism between TC test and bending fatigue test.

Also, bending fatigue test under high temperature (125°C) is proved to be substitute of TC test and shorten the test time.

2 Experiment

2.1 Sample Preparation

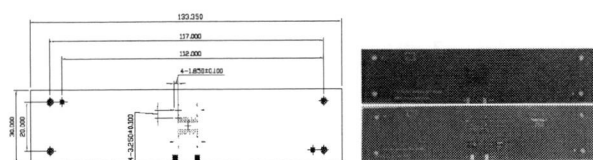

Fig. 1 60BOC Board Level Sample

2.2 Bending Fatigue Test

The bending fatigue system is depicted in Fig. 2. MTS 858 system with specific 4-point bending fatigue test fixture was used in this test. This fixture consists of the moving inner roller and the fixed outer roller, which supports the test board and prevents its movement in the test procedure. The test board is placed between these rollers.

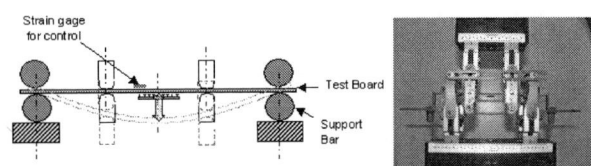

Fig. 2 Bending Fatigue Test System

2.3 Bending Fatigue Test Condition

The test condition was listed in tables below.

Table 1 Test Condition I

Load Span	Support Span	S/B Condition	Strain Rate	Failure Criterion	Sampling Rate
55mm	105mm	Φ0.45 (Sn2.5Ag0.5Cu)	16000με/s	1000με	1024Hz

Table 2 Test Condition II

Temperature Condition	Strain Range [με]						
	1000	1500	2000	2500	3000	4000	6000
-25°C	×	×	√	√	√	√	√
25°C	√	√	√	√	√	√	√
75°C	×	×	√	√	√	√	√
125°C	×	√	√	√	√	√	√

1. 5 specimens per test condition (temperature, strain range) were prepared and uniaxial strain gage was attached to every specimen.

2. The failure was all regarded to happen as the monitored resistance of daisy-chain circuit beyond 1000με.

2.4 TC Test Condition

Temperature cycling -25-125°C is chosen as TC condition.

2.5 Failure Analysis

2.5.1 Dye and Pry

Failure specimens are immerged into fine particle red ink in a vacuum chamber for 1 hour and baked at 100°C for 30 minutes. Then the packages are sheared from the test board by DAGE 5000 and reveal the failure position marked by red particle absorbed.

2.5.2 OM & SEM

Cross-section OM and SEM photos are used to observe the failure modes of the failure specimens.

3 Results Analysis and Discussion

3.1 TC Results and Analysis

3.1.1 TC Lifetime Analysis

50 specimens were tested for TC condition. According to these results, Weibull distribution probability plot can be drawn as Fig. 3.

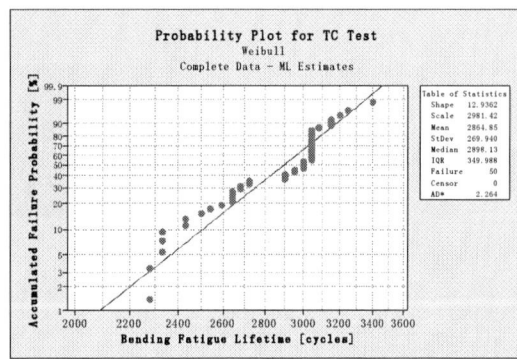

Fig. 3 Probability plot for bending fatigue test depend on strain range condition

3.1.2 Failure Mode Analysis

3.1.2.1 Dye-Pry Analysis

Dye-Pry method is used to observe the failure position of the failure specimen. The dye-pry photos showed three main failure positions, component side, board side and pad lift. The results of TC failure specimens are almost concentrated on component side failure. 4 specimens were observed and the results showed that the failure position of TC test is concentrated in component side.

3.1.2.2 OM & SEM Analysis

1 specimen cross-section was used for OM failure mode investigation. The cross-section location is showed as Fig. 4, which located at outmost column.

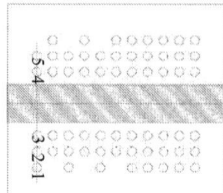

Fig. 4 Cross-section Location

4 failure modes are investigated, component side bulk crack, component side IMC interface crack, board side bulk crack and pad lift. Fig. 5 depicted the SEM photos for the typical failure mode. For the crack in solder bulk, crack initiates from the corner of the solder ball and propagates along the copper pad direct in the solder inner. For IMC interface crack, crack initiates from the corner of the solder ball and propagates along the interface between IMC layer and solder bulk. For the pad lift, the crack initiates from the corner of the solder ball and propagates along the board PCB materials.

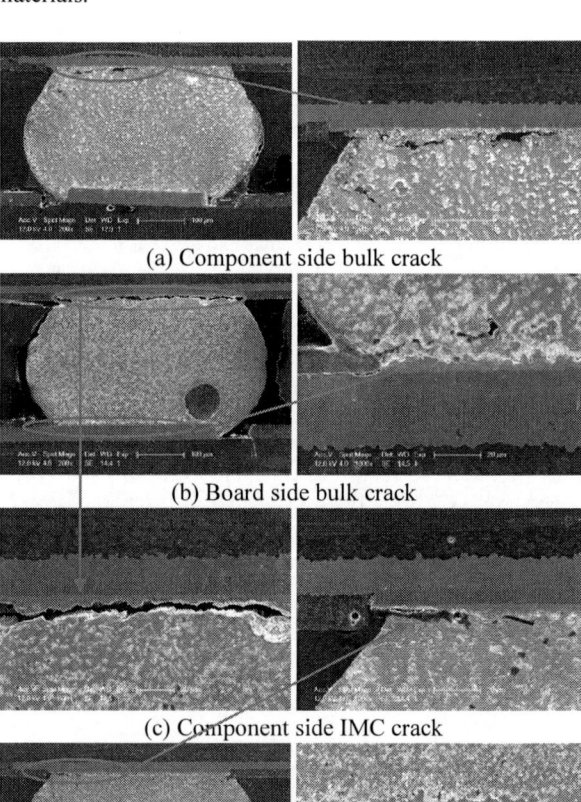

(a) Component side bulk crack

(b) Board side bulk crack

(c) Component side IMC crack

(d) Pad lift

Fig. 5 Typical TC Failure Modes

The OM results are listed in Table 3 and the results are slimier as SEM analysis.

Table 3 TC Failure Mode OM Analysis

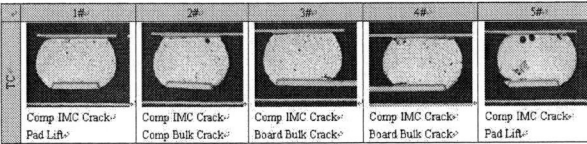

3.2 Isothermal Bending Fatigue Test Results and Analysis

3.2.1 Lifetime Analysis

According to test results, Weibull distribution probability plot can be drawn as Fig. 6 (depend on test environment temperature condition) and Fig. 7 (depend on test strain range). Under same test temperature, along with the increase of strain range, the life time decreases exponentially. However, under same strain range, when temperature increases, the character life decreases, but the difference is not significant.

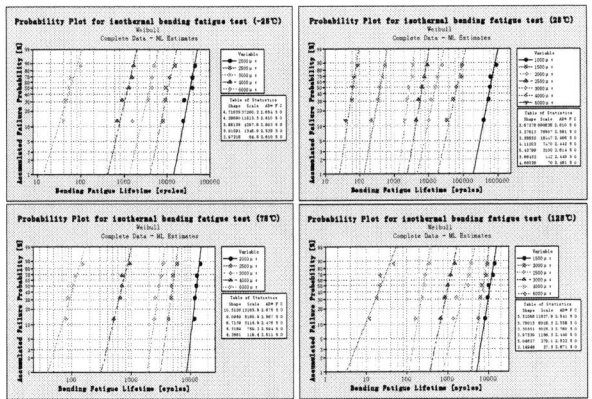

Fig. 6 Probability Plot for Bending Fatigue Test Depend on Environment Temperature Condition

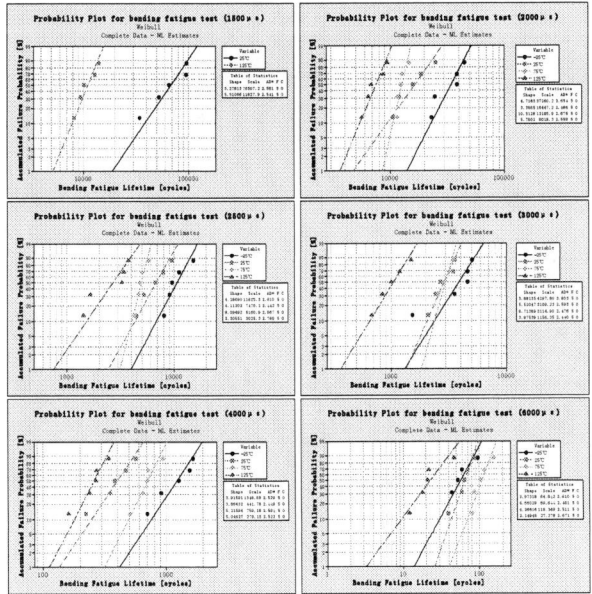

Fig. 7 Probability plot for bending fatigue test depend on strain range condition

For the same strain range loading, the plastic strain of solder ball will increase with the increase of temperature it is endured, and the lifetime decreases. And under high temperature, the crystal defect is active and the large defect is easy to form, which also leads to the lifetime decreases. On the opposite aspect, under high temperature, the material becomes soft and hard to crack. Another effect factor is the strain we controlled in our test is the board strain not the solder ball strain, and when the temperature changes, the two strains cannot be changes identically. Many factors affect the lifetime at the same time, so the lifetime is not linear with temperature.

The character lifetime of the bending fatigue test is the lifetime when 63.2% specimens are failed, and in another word, is the scale parameter of Weibull distribution probability plot. The character lifetime of our test groups is listed in the table below.

Table 4 Character Lifetime under Different Test Condition

	-25°C	25°C	75°C	125°C
±500με	×	690838	×	×
±750με	×	76507.2	×	11827.9
±1000με	37260.2	18447.2	13185.9	8018.3
±1250με	11612.3	7470.2	5160.9	3025.3
±1500με	4297.8	3100.4	3114.9	1156.3
±2000με	1348.9	441.8	759.2	279.1
±3000με	64.8	69.6	118.4	27.3

3.2.2 Failure Mode Analysis

3.2.2.1 Dye-Pry Analysis

Dye-Pry method is also used to observe the failure position of isothermal bending fatigue test. Three main failure positions, component side, board side and pad lift are investigated as the photos shown in below.

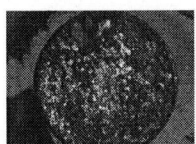 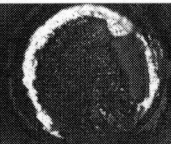 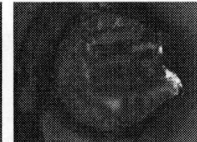

(a) Component Side Crack (b) Board Side Crack (c) Pad Lift
Fig. 8 Dye-Pry OM Photo

3 specimens under every test temperature are used for failure position statistic and the results are in Table 5. The color meanings of the ball map are listed in Table 6.

The ball map depicted that:

From the small strain range to high strain range, the failure position changed from component side to board side gradually. As for the failure mode of solder ball, it seems that the crack often occurs at component side due to the maximum plastic strain when the solder ball experiences low strain range loading. However, with the increase of strain range, the maximum localized plastic strain will occur at board side because the NSMD pad structure will result in strain concentration. This phenomenon is also proved by simulation results.

Table 5 Failure Position Map

	-25°C	25°C	75°C	125°C
1000με	×	(map)	×	×
1500με	×	(map)	×	(map)
2000με	(map)	(map)	(map)	(map)
2500με	(map)	(map)	(map)	(map)
3000με	(map)	(map)	(map)	(map)
4000με	(map)	(map)	(map)	(map)
6000με	(map)	(map)	(map)	(map)

Table 6 Color Tag List

○	Comp Side (1 Sample)	◉	Board Side (1 Sample)	◉	Pad Lift (1 Sample)
○	Comp. Side (2 Samples)	●	Board Side (2 Samples)	◉	Pad Lift (2 Samples)
◉	Comp. Side (3 Samples)	●	Board Side (3 Samples)	◉	Pad Lift (3 Samples)

From low temperature to high temperature, the component side failure decreased, however the board side failure and pad lift increased. When the test temperature increases, under same test strain range, the displacement increases. Maybe because, the solder ball strain increases larger than board strain, when the temperature increases, then the actual strain range is larger than the test control strain range, which results in the lifetime decrease and failure modes change to board side crack and pad lift.

The failure position distribution also proved the law that from the small strain range to high strain range, the failure position changed from component side to board side

gradually. Under low temperature isothermal bending fatigue test, the law is more obvious than under high temperature.

3.2.2.2 OM & SEM Analysis

The OM & SEM analysis cross-section location is similar as TC test, which is located at outmost column where the solder balls concentrate higher stress and plastic deformation than others and regarded as the most easy to fail location.

5 failure modes are investigated, component side bulk crack, component side IMC interface crack, board side bulk crack, board side IMC interface crack and pad lift. Fig. 9 depicted the SEM photos for the typical failure mode.

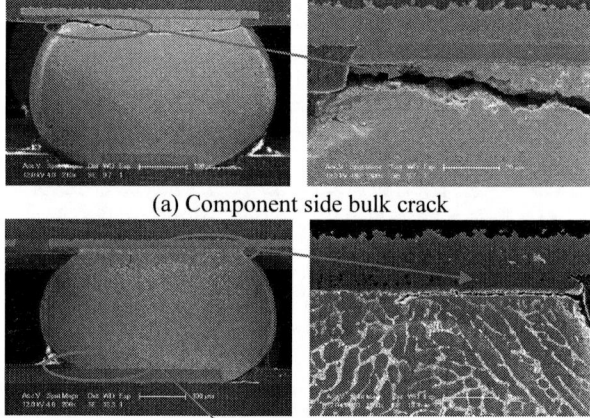

(a) Component side bulk crack

(b) Component side IMC crack

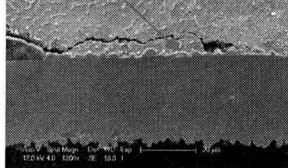

(c) Board side bulk crack

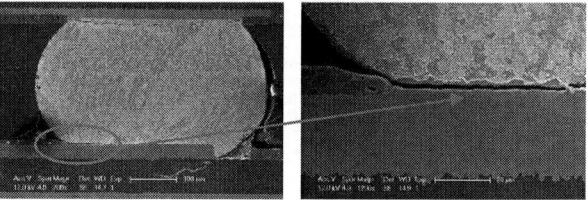

(d) Board side IMC crack

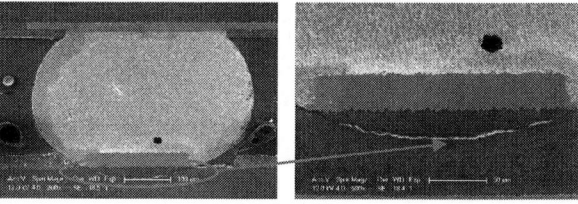

(e) Pad lift

Fig. 9 Typical Failure Mode in Bending Fatigue Test

According to the statistic data, 3-dimension bar is drawn in below. It's depicted that:

(1) Component IMC crack is concentrated in low temperature (-25°C) @ all strain range except the very

large strain range (6000με) or room temperature (25°C) @ low strain range (1000με, 1500με and 2000με).

(2) Component bulk crack is concentrated in not very high temperature (125°C) @ middle strain range (2000με, 2500με and 3000με).

(3) For board IMC crack, except for the very small strain range (1000με), almost at every test condition this failure mode happens.

(4) Failure mode of board bulk crack is infrequently appeared, only happened at temperature 25°C @ middle strain range 3000με and temperature 75°C @ middle strain range (2500με, 3000με and 4000με).

(5) Pad lift is the most familiar failure mode, happens at all test condition except for very low strain range (1000με and 1500με)

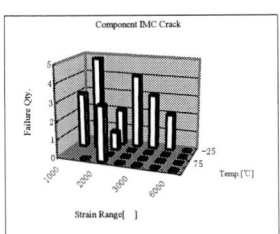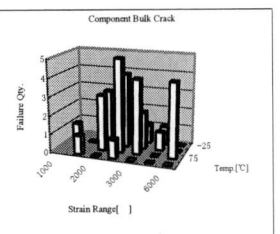
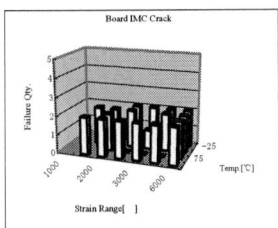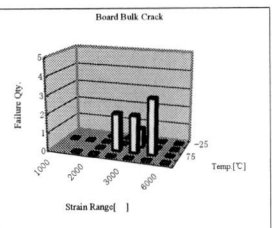
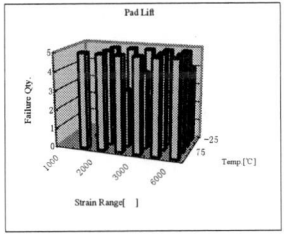

Fig. 10 Failure Mode 3-dimension statistic bar

3.3 Comparison of TC Test & Isothermal Bending Fatigue Test

3.3.1 Failure Mode Comparison

3.3.1.1 Dye-Pry Analysis

Though above dye-pry analysis, board level sample undergo TC test, the failure position is mainly concentrated in component side. Among isothermal bending test, under following test condition the failure position is similar as TC test, -25°C@2000με and 2500με; 25°C@1000με, 1500με, 2000με and 2500με; 75°C@2000με and 2500με; 125°C@1500με and 2000με, totally 10 test condition, which also highlight and mark "√" in Table 7.

Table 7 Similar Failure Position Test Condition Matrix

	-25°C	25°C	75°C	125°C
1000με		√		
1500με		√		√
2000με	√	√	√	√
2500με	√	√	√	

3.3.1.2 OM Analysis

Failure modes of TC test are mainly component IMC crack with little component bulk crack, board bulk crack and pad lift. Then compare above mentioned 10 similar test groups by OM analysis, the ones whose failure modes are similar as TC test are listed in Table 8, highlighted and marled "√√", totally 6 test condition.

Table 8 Reference Test Condition Matrix

	-25°C	25°C	75°C	125°C
1000με		√√		
1500με		√√		√
2000με	√√	√√	√	√√
2500με	√√	√	√	

3.3.2 Test Time Comparison

Lifetime of reference test groups is listed in Table 9. Calculate the lifetime not only in cycles but also in minutes.

Table 9 Lifetime of Reference Test Group

	-25°C		25°C		125°C	
	Cycles	Time [min]	Cycles	Time [min]	Cycles	Time [min]
1000με			690838	5325.21		
1500με			76507.2	629.59		
2000με	37260.2	326.03	18447.2	161.41	8018.3	70.16
2500με	11612.3	118.94				

3.3.2 Integrated Comparison

Compare above 6 reference test groups, whose failure mode is similar to TC test mainly by test time and failure mode similarity degree, and three test group was chosen to simulate general TC test.

(1) -25°C @ 2500με:
  - ➢ Test time is short
  - ➢ Failure mode is most close to TC test
  - ➢ Test condition is simple, but needs liquid N₂.

(2) 125°C @ 2000με:
  - ➢ Test time is shortest
  - ➢ Failure mode is also similar to TC test
  - ➢ Test condition is also easy

Table 10 Test Condition Comparison

Environment Condition	Strain Range	Test Time	Test Simpleness	Remark
-25℃	2000με	④	☆	Should consume LN2
	2500με	②		LN₂ Price: RMB 590/ 200L
				LN₂ consumption: about 12hr / 200L
25℃	1000με	⑥	☆☆☆	
	1500με	⑤		
	2000με	③		
125℃	2000με	①	☆☆	

Note:
Cost Time: "①" means cost time is shortest.
Test Simpleness: "☆☆☆"means test is simplest.

4. Simulation

4.1 Simulation for Isothermal Bending Test

4.1.1 Structural Response Analysis

Finite element analysis software ANSYS is used to simulate the structural response and internal stress/strain history of package during cyclic loading. To save running time, some simplifications are made and only a quarter of model is built. The finite element model is shown in Fig. 11.

Table 11 Bending Fatigue Test Conditions

<1> Simulation for Isothermal Bending					
Temperature (°C)	Strain Range (με)				
	1500	2000	2500	3000	4000
-25	×	√	√	√	√
25	√	√	√	√	√
75	×	√	√	√	√
125	√	√	√	√	√
<2> Simulation for TCT: -25°C~125°C					

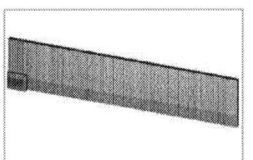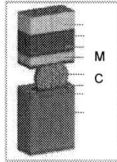

Fig. 11 Finite Element Model (1/4)

All constitutive materials except metal material (solder ball, copper trace) are assumed to be elastic. The temperature-dependent elastic modulus and CTE are used in finite element model. The brief material properties are list in Table 12.

Table 12 Material Properties

Material	Elastic modulus (GPa)	Shear modulus (GPa)	Poisson's Ratio	CTE (ppm)	Density (g/cm³)
EMC	29.8 (RT)	--	0.35	Temp dependent	1.9
Chip	168.9 (X,Y), 130.2 (Z)	50.9(XY), 79.4(YZ,XZ)	0.065(XY), 0.36(YZ,XZ)	2.6	2.4
Adhesive	0.26 (RT)	--	0.35	Temp dependent	1.2
PSR	3.66 (RT)	--	0.29	Temp dependent	1.3
Copper	112	--	0.343	17.3	8.9
BT core	26 (X,Y), 11 (Z)	11.7 (XY), 70.4 (YZ,XZ)	0.11(XY), 0.39 (YZ,XZ)	11 (X,Y), 58(Z)	1.8
S/B	36	--	0.3	25	7.5
FR-4	22 (X,Y), 10 (Z)	9.91 (XY), 6.52 (YZ,XZ)	0.11(XY), 0.28 (YZ,XZ)	16 (X,Y), 84(Z)	2.3

Anand developed constitutive equations for the rate-dependent deformation of metals at high temperature. Although aimed at hot working of steels and other structural metals, it has been adopted successfully to represent the viscoplastic behavior of solders.

A simple form of evolution equation was given by Anand as follows:

$$\dot{s} = \left\{ h_0 \left| 1 - \frac{s}{s^*} \right|^a sign\left(1 - \frac{s}{s^*} \right) \right\} \cdot \dot{\varepsilon}_p \quad (1)$$

With

$$s^* = \hat{s} \left[\frac{\dot{\varepsilon}_p}{A} \exp\left(\frac{Q}{RT} \right) \right]^n \quad (2)$$

Where, A is the pre-exponential factor, Q is the activation energy, ξ is the multiplier of stress, m is the strain rate sensitivity, h_0 is the strain hardening/softening constant, $\hat{s}$ is a coefficient, and n is the strain rate sensitivity for the saturation value of deformation resistance, respectively, a is the strain rate sensitivity of hardening/softening and s_0 is the initial value of the deformation resistance, which is needed to determine the evolution of deformation resistance.

From the above viscoplastic ANAND model, there are nine material parameters: A, Q, ξ, m, h_0, $\hat{s}$, n, a, and s_0.

It was found that reference [3] and [4] published all nine constants for Sn2.0Ag0.5Cu, which is close in composition to Sn2.5Ag0.5Cu, the alloy under investigation in this report. According to this, the strain-stress relationship at different temperature and strain rate are inversely estimated using Origin software, see Fig. 12.

From the strain-stress curves we can see that the ANAND model can commendably represent the rate and temperature-behavior of solders. With the increase of temperature, the solder seems to be "softer", and has lower yield strength and lower stress level. While the strain rate increases, the material becomes "harder", and has higher yield strength.

978-1-4244-4658-2/09 $25.00 © 2009 IEEE

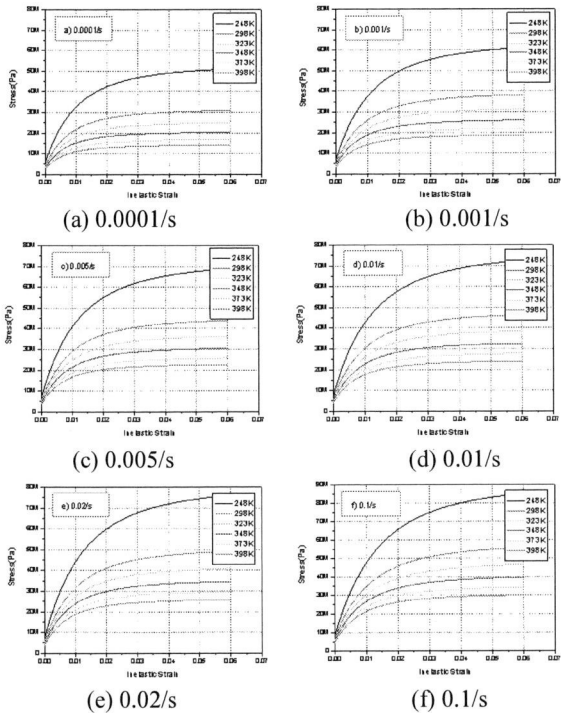

(a) 0.0001/s   (b) 0.001/s

(c) 0.005/s   (d) 0.01/s

(e) 0.02/s   (f) 0.1/s

Fig. 12 Temperature dependent material properties at different strain rate

The boundary conditions are shown in Fig. 13 The nodes on test board at support anvils position are constrained in Y&Z direction, while the nodes at load anvils position are applied a cyclic displacement load which is the displacement history of loading anvils recorded by MTS. The temperature is applied to all nodes using ANSYS BF command. The first two loading cycles are simulated using transient analysis.

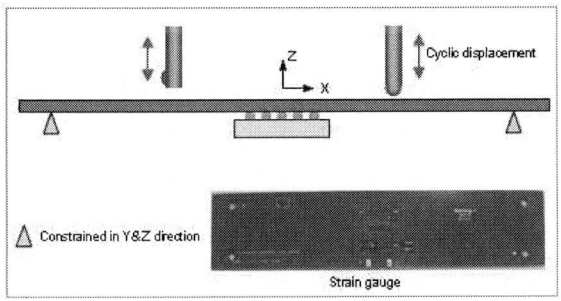

Fig. 13 Boundary conditions

4.1.2 Isothermal Condition (-25°C, 125°C)

Considering the total length of this paper, we just put the most important results of -25°C and 125°C bending fatigue simulation results here.

The deflections of test board center as response to input displacement and the simulation results of PCB strain in length direction compared with experimental results are both shown in Figs. 14-15. It was observed that the simulation results are coincidental with experimental very well.

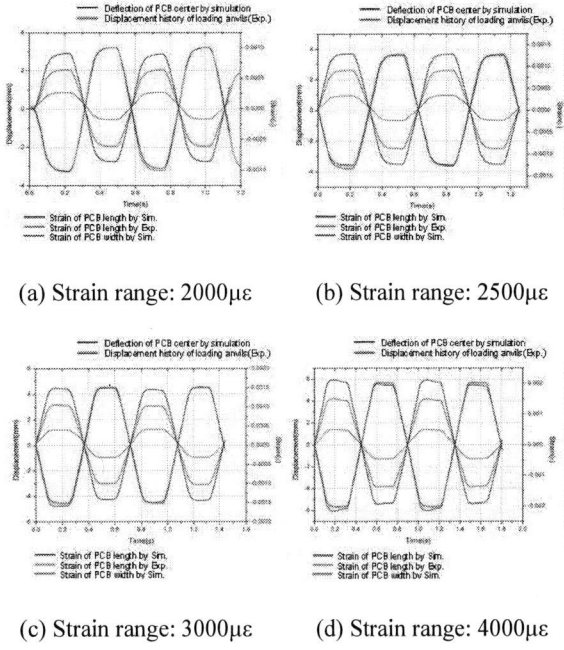

(a) Strain range: 2000με   (b) Strain range: 2500με

(c) Strain range: 3000με   (d) Strain range: 4000με

Fig. 14 Structural response under different conditions (@-25°C)

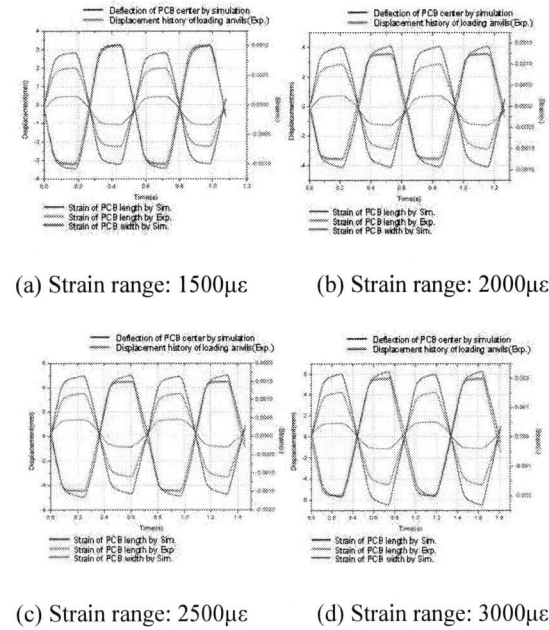

(a) Strain range: 1500με   (b) Strain range: 2000με

(c) Strain range: 2500με   (d) Strain range: 3000με

Fig. 15 Structural response under different conditions (@125°C)

The stress/strain states in solder ball during cyclic bending for different loading conditions, as well as the deflection contour of whole structure, are shown in Tables 12-13. To verify the simulation results, the simulation results are also compared with the experimental results.

Table 12 Stress/strain state for solder ball (Bending down@-25°C)

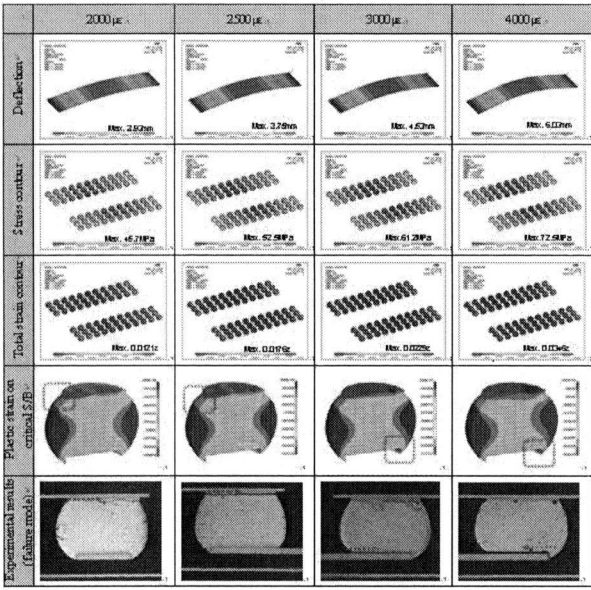

Table 13 Stress/strain state for solder ball (Bending down@125°C)

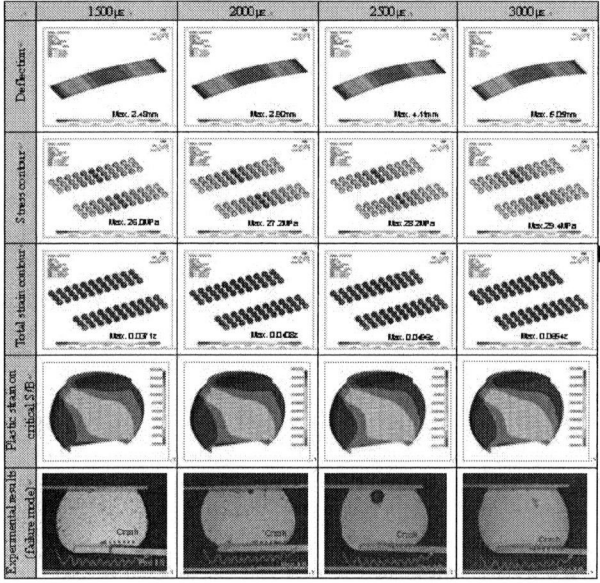

4.1.3 Summary

During cyclic loading, localized plastic deformation may occur at highest stress site. This plastic deformation induced permanent damage to the component and a crack develops. Also, it is investigated that the Von Mises stress on solder ball has no big difference between different strain ranges for each isothermal condition. However, for each isothermal condition, the total strain on solder ball increases remarkably with the increase of strain range. Among these total strains, the plastic strain is the primary component, account for more than 90%.

It was also observed that the solder balls locating at outermost columns have much more possibility to fail due to occurrence of higher stress and plastic deformation than other solder balls.

For the same strain range loading, the plastic strain of solder ball will increase with the increase of temperature it is endured, and for each isothermal condition, the plastic strain will increase with the increase of strain range.

As for the failure mode of solder ball, it seems that the crack often occurs at component side due to the maximum plastic strain when the S/B experiences low strain range loading. However, for each isothermal condition, with the increase of strain range, the maximum localized plastic strain will occur at board side because the NSMD pad structure will result in strain concentration. This phenomenon is proved by experimental results.

4.1.5 Fatigue Life Simulation

When predicting solder joint fatigue failure, a metric of the damage occurring, called a damage parameter, must be chosen. The damage parameter ψ is a scalar quantity representing the damage that causes the fatigue failure. In this report, ΔW_{avg} (accumulated plastic work density per cycle) for the worst-case solder joint was used to evaluate the fatigue life. The time history plastic work under different strain range during first two bending cycles is shown in Fig. 16.

A volume-weighted average is used to resolve the inherent problem of singularity in FEM when choosing which nodes/elements to take the damage parameter from. There are two layers of elements with total thickness of 1mil at solder joint interface as shown in Fig. 17 in which the damage parameter per cycle ψ is volume-averaged over.

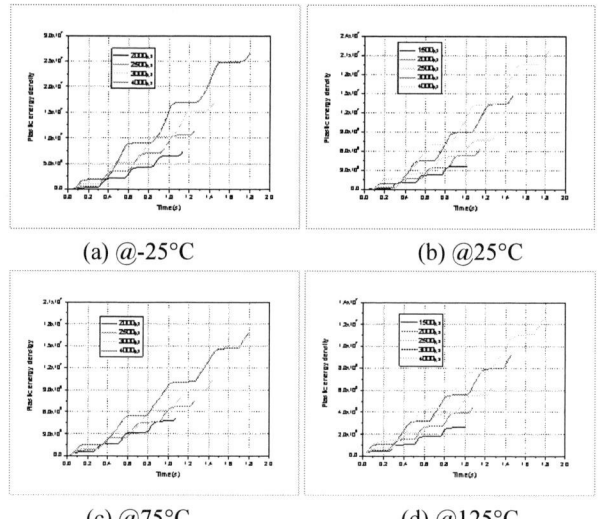

(a) @-25°C   (b) @25°C

(c) @75°C   (d) @125°C

Fig. 16 Variation of plastic work density for first two cycles

ΔW_{avg} is element volumetric average of the stabilized change in plastic work between the first cycle and second cycle. The calculated ΔW_{avg} under different strain range are shown in Table 14. The power law regression is conducted to get the Coffin-Manson fatigue model, shown in Fig. 18. It was found that for this BOC package under cyclic bending conditions, the fatigue life N_f was related to ΔW_{avg} in the form such as:

$$N_f = 8253.6(\Delta W_{avg})^{-1.84047} \qquad (3)$$

Table 14 ΔW_{avg} for different testing conditions

Strain range / Temp		-25°C	25°C	75°C	125°C
150	ΔW_{avg}	N/A	0.4048	N/A	0.6103
0	Life	N/A	58890	N/A	11828
200	ΔW_{avg}	0.4406	0.6387	0.7917	0.9315
0	Life	37260	18447	13186	8018
250	ΔW_{avg}	0.8394	1.1015	1.1740	1.3725
0	Life	11612	7470	5161	3025
300	ΔW_{avg}	1.3832	1.9399	1.7350	2.0718
0	Life	4298	3100	3115	1156
400	ΔW_{avg}	2.5934	3.2159	3.0856	3.0278
0	Life	1349	442	759	279

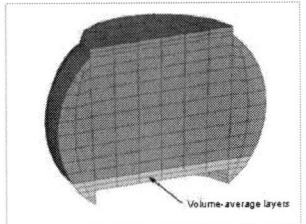

Fig. 17 Volume-average layers

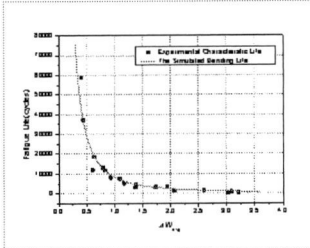

Fig 18 Calibration for in S/B Coffin-Manson fatigue model

This is a unified finite element modeling methodology for cyclic bending test, and the Coffin-Manson fatigue model can be used in future bending reliability evaluation once the accumulated plastic work density per cycle ΔW_{avg} is calculated using FEM.

4.2 Simulation for TC Test

The same three-dimensional FE model for 60BOC assembly as bending simulation is used here. The solder balls were also built using VISCO107 with consideration of the viscoplasticity properties through ANAND model. The boundary conditions were applied to the quarter model and subjected to a temperature cycling condition of -25°C to 125°C as shown in Fig. 19 and Fig. 20 respectively. Two consecutive thermal cycles were simulated. The cycling

begins by ramping the model from ambient temperature (25°C) to the high temperature extreme. Thus, for the cycling phase of the analysis, the zero strain reference temperature is set as the ambient temperature. The underlying assumption behind the zero strain reference temperature setting is that enough time has passed before the onset of cycling for all residual stress to creep out of the solder material. The relationship between time and temperature is defined as a 2D table, and the temperature is ramped from their previous temperature over the time step by invoking the KBC, 0 command. In all cases, the cyclic temperature loads are applied to all nodes using the BF command.

Fig. 19 FE model

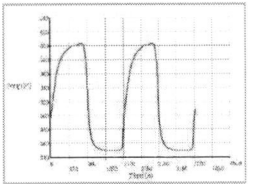

Fig.20 Temperature profile Boundary condition

The simulation results are shown in Table 15. It is observed that the package will warp with the change of temperature during TC, due to the CTE mismatch between constitutive materials. At high temperature extreme, it behaves as concave warpage of 26.3μm, while at low temperature extreme, convex warpage of 21.2μm is observed. Obviously, the warpage will make the maximum stress and strain occur at the outermost solder joint.

The maximum stress at low temperature extreme (about 29MPa) is much higher than that of high temperature (about 9.3MPa). It is because the solder material seems to be very "soft" at high temperature. The maximum plastic work which indicates the accumulated plastic deformation is located at the component side. So it implies that the failure mode of IMC crack at package side will occur during temperature cycle test. This agrees well with the failure mode of solder ball resulting from experiments.

Table 15 Stress/strain state for solder ball (TC)

	Warpage distribution	Stress distribution	Plastic strain contour	Plastic work on critical solder ball
@125°C	Max: 26.3μm	Max: 9.3MPa	Max: 0.02	
@-25°C	Max: 21.2μm	Max: 29.9MPa	Max: 0.0145	

4.4 Failure Mechanism Analysis for TC vs. Bending

Fig. 21 shows the deformation shape of package mounting on PWB during TC and ambient bending test. During TC test, with the cyclic change of temperature, the package will also cyclically warp up and down due to the CTE mismatch between the constitutive materials. But for the test board, there is very small warpage due to the symmetric construction. Therefore, the stress or strain concentration of solder ball will always occur at the component side. And failure mode of IMC crack is the main failure mode for TC testing. Of course, if the interface strength of component side is much higher than that of board side interface, the IMC crack at board side is possible.

And another main failure cause for TC is the creep of solders. At high temperature extreme of TC, the creep behavior of solders is very obvious and significant. The long dwell time at high temperature will enlarge creep deformation at strain concentration site. Therefore, the creep of solder accelerates the failure of package.

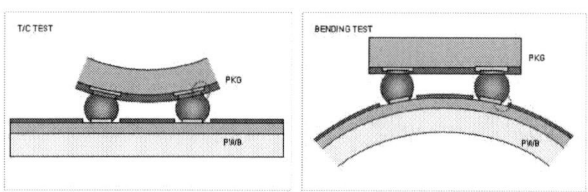

Fig. 21 Schematic of deformation during TC and bending

For the ambient bending test, the package may have warpage due to assembly process (e.g. molding), but the followed reflow process will release the residual stress in SB. So the bending of PCB will primarily induce stress/strain concentration at the board side interface, and the stress/strain concentration at component side interface is secondary. If the tested assembly is bended under higher or lower isothermal conditions, the solder ball will deform under the loading combining package warpage and PCB bending.

If the curvature of PCB is very small (i.e. low strain range condition), the stress/strain concentration at component interface and board interface is near similar, because the NSMD structure of board side has higher mechanical strength than SMD structure of package side, the failure always occur at component interface. However, with the increase of strain range, the curvature of PCB will also increase, ant it makes the stress/strain concentration at board is much higher than that of component interface. So the failure will occur at the solder board side.

5. Conclusion

The TC character lifetime of 60BOC board level sample is 2981.42cycles (89442.6min or 2months). The failure position of TC test is concentrated in component side crack and pad lift. The failure modes are mainly component IMC crack.

Under same test temperature, along with the increase of strain range, the life time decreases exponentially. However, under same strain range, when temperature increases, the character life decreases, but the difference is not significant.

From the small strain range to high strain range, the failure position changed from component side to board side gradually. From low temperature to high temperature, the component side failure decreased, however the board side failure and pad lift increased.

In simulation results, when experiencing same movement status (e.g. bending up or bending down), there is no big difference in the stress of solder ball for different loading conditions, but the plastic strain will remarkably increase with the increase of board deflection.

The simulation results coincide with experimental results well.

There are 2 commended test conditions: 125°C @ 2000με (lifetime: 70.16min) and -25°C @ 2500με(lifetime: 118.94min), whose failure mode is similar to TC test and the test time is much shorter than TC test.

References

1. Shi X. Q., Zhou W., Pang H. L. J., *et al.*, "Effect of Temperature and Strain Rate on Mechanical Properties of 63Sn/37Pb Solder Alloy," *Journal of Electronic Packaging*, 1995, 121, pp. 179-185.
2. Darveaux, R., Syed, A., "Reliability of Area Array Solder Joints in Bending," *SMTA International*, 2000.
3. T.O. Reinikainen, P. Marjamaki, "Deformation characteristics and microstructural evolution of SnAgCu solder joints," *6th. Int. Conf. on Thermal, Mechanical and Multiphysics Simulation and Experiments in Micro-Electronics and Micro-Systems*, EuroSimeE, 2005, pp. 91-98.
4. Hun Shen Ng, Tong Yan Tee, Kim Yong Goh, "Absolute and relative fatigue life prediction methodology for virtual qualification and design enhancement of lead-free BGA," *Electronic Components and Technology Conference*, 2005, pp. 1282-1291.

Testing Failure of Solder-Joints by ESPI on Board-Level Surface Mount Devices

Yunxia Gao, Jun Wang
Department of Materials Science, Fudan University
No. 220, Handan Road, Shanghai 200433, China
jun_wang@fudan.edu.cn

Abstract

Fast and nondestructive welding quality inspection on board-level BGA and PLCC devices is relatively difficult due to small size and special location of solder joints. In this study, a novel testing method combing Electronic Speckle Pattern Interferometry (ESPI) and Finite Element Analysis (FEA) was proposed to detect joint failures in surface-mount devices. ESPI technology was utilized to compare the difference of assembly deformation between damage solder joints and the good counterparts under simple mechanical loadings. Using finite element analysis, the location and failure modes of damage joints were identified. The testing results from pseudo BGA sample and real PLCC device illustrated the validity of the method; especially for testing corner solder joints.

Introduction

Ball Grid Array (BGA) and PLCC devices are mounted to PCB by solder reflow process. The solder joints are served as interconnections between device and PCB. As the solder joints are small and usually hidden below the device body, fast and nondestructive quality inspection on solder joints is relatively difficult.

Electronic Speckle Pattern Interferometry (ESPI) is a kind of nondestructive optical testing technology. The sensitivity and precision of ESPI are very high and can be in nanometer range [1]. There have been reported that ESPI was used to characterize thermal deformation of electronic packages under temperature cycling [2-5]. Taking the advantages of ESPI, it is possible to compare the difference of assembly deformation due to damage solder joints and the good counterparts under simple mechanical loadings. Hence the failure joints can be then identified and located with further analysis.

In this study, combing ESPI and Finite Element Analysis (FEA) is proposed to inspect quality of solder joints in surface-mount devices. Under controlled mechanical loading, the surface deformation of a pseudo six-ball BGA sample mounted on PCB was tested by ESPI technology when the joints were good and damage. Comparing with finite element simulation results, the location and failure modes of damage joints were analyzed. Testing of actual PLCC device mounted on PCB was also performed to verify the method.

The Optical System and Principle of ESPI

The out-of-plane displacement sensitive ESPI system used in the experiments is sketched in Fig. 1.

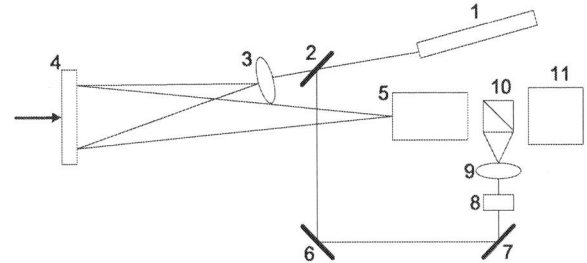

1. helium neon laser; 2. beam splitter; 3, 9. beam expander; 4. sample and loadings; 5. zoom lens; 6. mirror; 7. PZT and its power; 8. double polarizer; 10. prism; 11. CCD.

Fig. 1 ESPI out-of-plane-displacement-sensitive setup

In the above setup, the laser beam emitted from the helium neon laser is first incident on the beam splitter and is divided into a transmitted object beam and a reflected reference beam. The object beam is expanded by beam expander and projected to the surface of testing sample. The reference beam is reflected by the mirror and PZT to change its direction. After the intensity modulation and light expansion, the reference beam finally irradiates the prism. CCD device receives both beams and transmits the light intensity information to computer for further calculation.

In the initial state, the intensity of the two coherent laser beams are

$$\begin{cases} I_0(x,y) = A_0 \exp[i\varphi_0(x,y)] \\ I_r(x,y) = A_r \exp[i\varphi_r(x,y)] \end{cases} \quad (1)$$

where A_0 and A_r are the amplitude of the object beam and the reference beam respectively. φ_0 and φ_r are their phases. The intensity on the CCD optical target is

$$I(x,y) = A_0^2 + A_r^2 + 2A_0A_r \cos(\varphi_0 - \varphi_r) \quad (2)$$

When the sample deformed, the amplitude of the object beam speckle field remains approximately the same, whereas its phase changes from φ_0 to $\varphi_0 - \Delta\varphi$

$$I_0'(x,y) = A_0 \exp[\varphi_0(x,y) - \Delta\varphi] \quad (3)$$

As the reference beam is constant during the deformation, the intensity on the CCD optical target becomes

$$\begin{aligned} I'(x,y) = &A_0^2 + A_r^2 \\ &+ 2A_0A_r \cos[\varphi_0(x,y) - \varphi_r(x,y) - \Delta\varphi] \end{aligned} \quad (4)$$

The fringe pattern can be obtained by subtracting the two intensity diagrams,

$$\begin{aligned} \Delta I(x,y) &= |I'(x,y) - I(x,y)| \\ &= 4A_0A_r \left| \sin\left(\varphi_0 - \varphi_r - \frac{\Delta\varphi}{2}\right) \sin\frac{\Delta\varphi}{2} \right| \end{aligned} \quad (5)$$

where, $\sin\left(\varphi_0 - \varphi_r - \dfrac{\Delta\varphi}{2}\right)$ is the high frequency carrier noise, and $\sin\dfrac{\Delta\varphi}{2}$ is the low frequency speckle fringe.

According to the wave equation and the geometry of the ESPI optical setup, we have

$$\Delta\varphi(x,y) = \frac{2\pi}{\lambda}[w(1+\cos\theta) + u\sin\theta] \qquad (6)$$

Where, λ is the wave length of the laser beam, w is the out-of-plane deformation of the sample, u is in-plane deformation of the sample; θ is the incident angle of the object beam. At relatively small incident angle, the relationship between out-of-plane deformation and phase difference satisfies

$$\Delta\varphi = \frac{4\pi}{\lambda}w \qquad (7)$$

According to (7), the displacement distribution can be obtained from the intensity deviation (phase difference).

Tests on Pseudo BGA Samples

A pseudo six-ball BGA sample was prepared and mounted on PCB by solder reflow. The assembly was tested by the ESPI optical system shown in Fig. 1. Fig. 2 illustrates the perspective view of setup of the BGA sample, and the clamping and loading method in the ESPI tests.

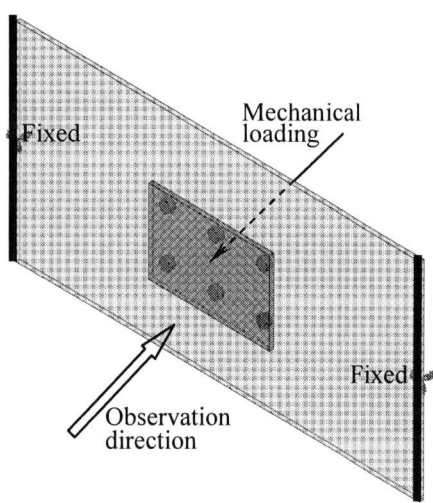

Fig. 2 Sketch of pseudo BGA sample testing

The ESPI fringe patterns are shown in Fig.3. Fig. 3(a), (b) and (c) represent the cases of the down-right corner solder joint were good, delaminated and missed, respectively.

The difference in Fig. 3 is obvious. The ESPI fringe pattern of normal BGA exhibits good symmetry, whereas abrupt changes in fringe occur near the damage solder joint in Fig. 3(b) and (c). Accordingly, it is possible to judge the occurrence of joint failure and approximately locate the failed ball from ESPI tests.

To further analyze the failure modes, finite element analysis was carried out to simulate the deformation of sample in the same setting when the down-right corner solder joint was good, delaminated or missed. The FEM model is shown

in Fig. 4 and material parameters used in the simulation are listed in Tab. 1.

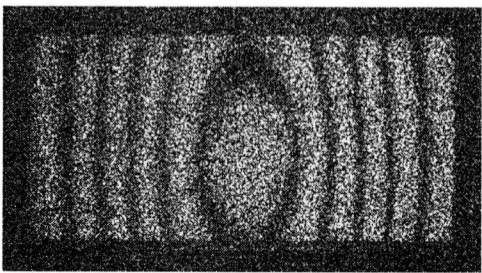

(a) normal BGA

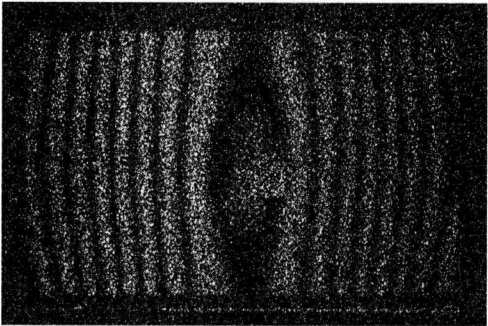

(b) delaminated BGA

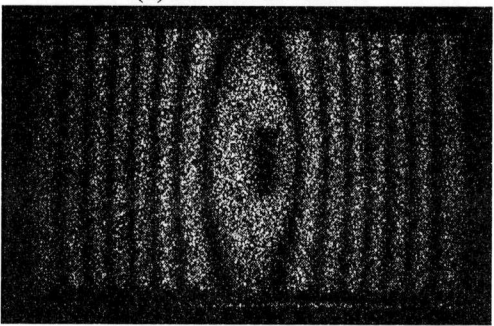

(c) ball-missing BGA

Fig. 3 Out-of-plane ESPI fringe patterns of pseudo BGA samples

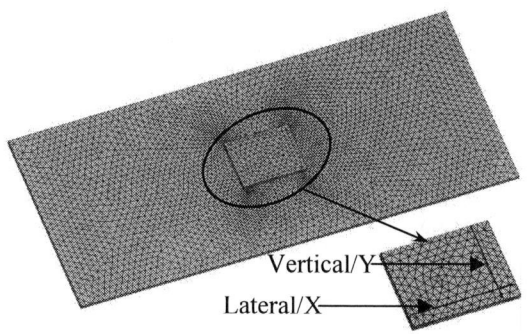

Fig. 4 FEM model of the pseudo BGA sample

Tab. 1 Material parameters used in FEM

Mat.	E/GPa			PR			G/GPa		
	X	Y	Z	XY	YZ	XZ	XY	YZ	XZ
FR-4 PCB	17	17	7.5	0.13	0.42	0.42	3	2.4	2.4
Sn63Pb37	30			0.35			-		

Relative displacement distributions along the lateral/X path shown in Fig. 4 were compared between FEM results and experimental data in Fig. 5.

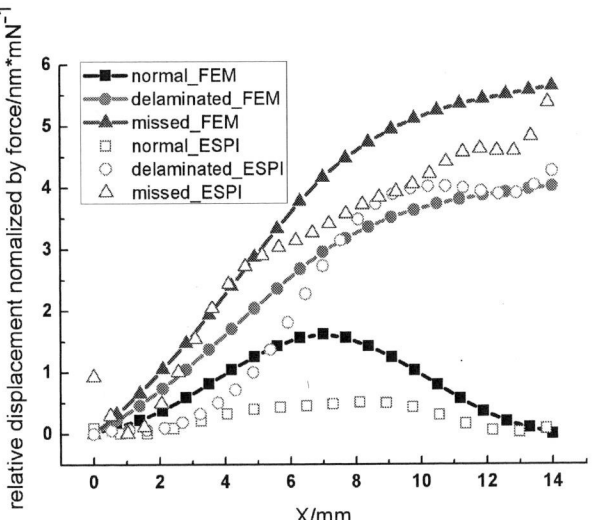

Fig. 5 Normalized relative surface displacement along Lateral/X path in Fig. 4

In Fig. 5, the hollow scattered triangle, circle and square points represent the ESPI test results of good BGA joint, delaminated BGA joint and ball-missing samples respectively; the curves with corresponding solid points are the FEM data. The trend of data revealed, out-of-plane displacement tended to increase at the damage solder joints. Furthermore, the severer the damage was, the greater the slope of the displacement curve became. Among the simulation displacement curves, the case of the ball-missing BGA sample had the biggest gradient, the gradient of delaminated BGA sample followed, and the curve of normal BGA sample was relatively smooth, which well conformed with the ESPI experiments. Therefore, failure modes of solder joints can be identified by comparing ESPI tests data to finite element analysis results.

Tests of Real PLCC Devices

In order to further verify the above method, a commercial PLCC device mounted on PCB was prepared and tested. The testing setup was identical to that of pseudo BGA sample.

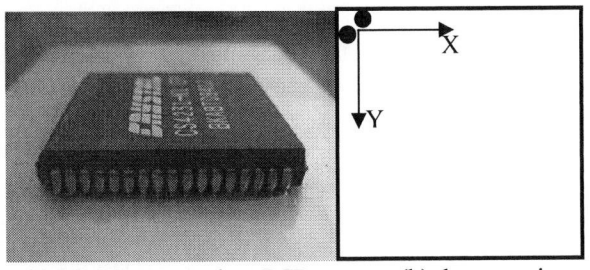

(a) PLCC mounted on PCB   (b) damage pins
Fig. 6 Photo of PLCC sample and damage pin location

Fig. 6(a) is the photo of mounted the PLCC device. The device has 17 I/O pins along each side. When the top-left corner joints, illustrated in Fig. 6(b), were good or

delaminated, the out-of-plane fringe patterns from ESPI are shown in Fig. 7.

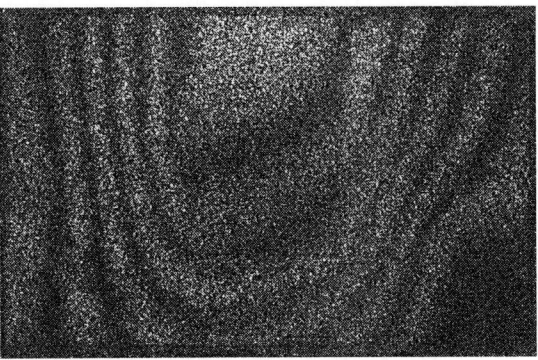

(a) normal PLCC

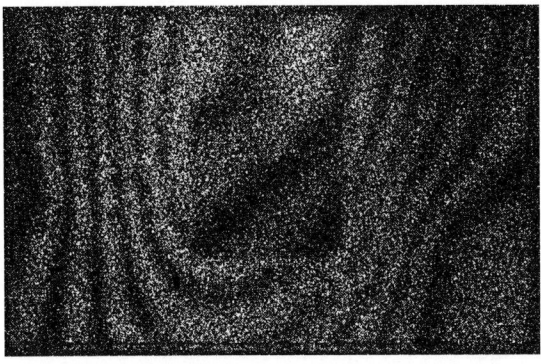

(b) delaminated PLCC

Fig. 7 Out-of-plane fringe patterns of PLCC samples

It was hard to mount the PLCC device exactly in the center of the PCB, for the solder reflow process was finished manually. The ESPI fringe patterns were hence asymmetric. Remarkable fringe difference was still observed: a new fringe appeared at the top-left corner when the corner joints are delaminated. Therefore, it is possible to detect solder joint failure by ESPI in real surface-mount devices.

Furthermore, finite element analysis was conducted. Fig. 8(a) and Tab. 2 shows the FEM model and the material parameters. The simulation result was compared to experimental data in Fig. 8(b).

Tab. 2 Material parameters for PLCC model

Mat.	E/GPa			PR			G/GPa		
FR-4	X	Y	Z	XY	YZ	XZ	XY	YZ	XZ
PCB	17	17	7.5	0.13	0.42	0.42	3	2.4	2.4
Sn63Pb37	30			0.35			-		
Cu	117			0.3			-		
EMC	19.6			0.4			-		
Si	16.1			0.2			-		

978-1-4244-4658-2/09 $25.00 © 2009 IEEE

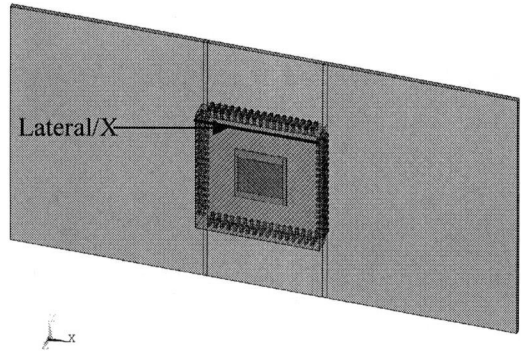

(a) FEM model and Lateral/X path

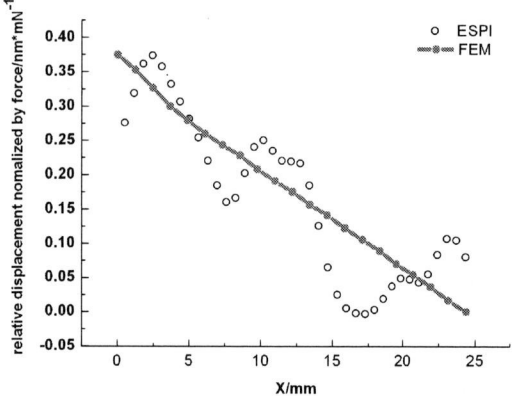

(b) Normalized relative displacement along Lateral/X

Fig. 8 Comparison of ESPI and FEM results for PLCC device mounted on PCB

The difference of relative displacement between PLCC with delaminated corner joints and PLCC with good joints was calculated, which value along the lateral/ X path in Fig. 8(a) is illustrated in Fig. 8(b). The hollow scattered circle points are from ESPI data, and the curve with solid circle points is FEM simulation results. Similar to the pseudo BGA samples, out-of-plane displacement increased at the damage I/O pins. Failure locations can be thus identified from the trend of the displacement curve.

Conclusions

In this study, a novel testing method combing ESPI and FEM is proposed and verified by surface-mounted pseudo BGA samples and real PLCC device mounted on PCB. The results indicate that the method can be used to inspect quality of solder joints in surface-mount devices quickly and nondestructively. The results from pseudo BGA sample and real PLCC device illustrate the validity of the method, especially for testing corner joints.

Acknowledgments

This work was supported by Shanghai-Applied Materials Research and Development Fund (No. 07SA09) and Science and Technology Commission of Shanghai (No. 08JC1411600).

References

1. LE K. D., ZHOU X., TANG J. Y., *et al.*, "A Study on Laser whole-field Modal Measurement Technology," *ACTA Photonica Sinica*, 2003, 32(5), pp. 608-611. (In Chinese)
2. LEE B. W., JANG W., KIM D. W., *et al.*, "Application of electronic speckle-pattern interferometry to measure in-plane thermal displacement in flip-chip packages," *Materials Science and Engineering A*. 2004, 380, pp. 231-236.
3. DILHAIRE S., JOREZ S., CORNET A., *et al.*, "Optical method for the measurement of the thermomechanical behavior of electronic devices," *Microelectronics Reliability*. 1999, 39, pp. 981-985.
4. CLAEYS W., DILHAIRE S., JOREZ S., *et al.*, "Laser probes for the thermal and thermomechanical characterization of microelectronic devices," *Microelectronics Journal*. 2001, 32, pp. 891-898.
5. XIONG X. M., HU F. R., LI G., "The Application of ESPI in Study Thermal Reliability of IC Packaging," *Chinese Journal of Lasers*. 2006, 33, pp. 395-397. (In Chinese)

Study on Shear Strength and J_c of EMC/Cu Interface with Cu Oxidation and Moisture Absorption

Xing FANG[1], Qiang FANG[1], Jun WANG[1], Hongkun YU[1], Xuefeng SHAO[2]

[1]Department of Materials Science, Fudan University
No. 220, Handan Road, Shanghai 200433, China
[2]Fairchild Semiconductor (Suzhou) Co., Ltd, China
Email: 072030014@fudan.edu.cn Mobile Phone: 13917092873

Abstract

In plastic power devices, the interfaces of Cu/EMC are most likely to delaminate under thermal loading, especially when moisture diffuses into the interface through EMC. In this work, the bare die samples were fabricated in standard commercial process. Shear test of Cu/EMC interface was designed to measure the strength of the interface with different lead-frame oxidation and moisture absorption time. The samples with pre-cracks on Cu/EMC interfaces were studied by the same shear test method and the critical value of J-integral was investigated by FEA (Finite Element Analysis). The results indicated that the shear strength of Cu/EMC interfaces and the value of J-integral declined dramatically with moisture absorption time. However, the effect of lead-frame oxidation is complex. The 2-D analysis of power device showed that the small initial crack, as small as 0.13 mm, will propagate along Cu/EMC interface when temperature is decreased to -55°C.

Introduction

In power devices with plastic packaging, copper alloys are usually served as lead-frames. The unit is encapsulated by molding compound and the interfaces between Cu and EMC are extensive in the package. As the CTE mismatch is large between Cu and EMC, the value of stresses on the Cu/EMC interfaces is high. The concentration of stresses on the Cu/EMC interface is critical for the reliability of the package. The delamination of Cu/EMC often takes place when the package is subjected to hygrothermal conditions which are applied in package during fabrication or testing processes. For example, the Fig. 1 illustrates "popcorn" delamination due to hygrothermal loadings in the package. The delamination is along interface of Cu/EMC and extend to out surface of EMC [1-2].

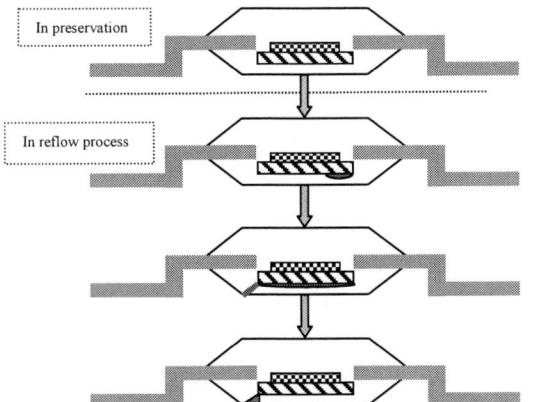

Fig. 1 The "popcorn" delamination in plastic packages

The delamination along interface of Cu/EMC is attributed to the competition of crack driving force and resistance of the interface. The delamination driving forces are usually hygrothermal or mechanical loadings. The resistance is in terms of characteristics of interface, e.g. the moisture absorption and Cu oxidation etc. Epoxy based EMC is hydrophilic material and the moisture will diffuse into the package through EMC surface in humidity environment. The interface of Cu/EMC can be weakened when moisture concentration is high at the interface. Additionally, the hygro-stress due to swelling of moisture in the package may not be ignored [3].

On the other hand, Cu oxidation layer can react with humidity in the atmosphere and become hydrated. Thus, the metal surface is covered with hydroxyl groups. The surface hydroxyl groups readily form hydrogen bonds with water in atmosphere and so the metal surface at the air interface is covered eventually with several layers of water molecules. The tight bonding between the polymer adhesive and the metal surface requires that the polar or active center of the polymer must reach through the layers of water molecules to the hydroxyl group terminated metal surface to form a tight bond of covalent or ionic character [4].

Soon-Jin Cho studied the influence of oxidation at a low temperature on Cu/EMC adhesiveness. They found oxidation can enhance the interface strength in 150-300°C. When the thickness of the oxidation layer is 20-30μm, the strength is in its highest value [5]. In mode I cracking, Kuan H. Lu et al. used the DCB (Double Cantilever Beam) method to measure the fracture toughness with moisture absorbing [6]. Tay et al. studied the critical stress intensity factor at crack tip on the interface by the same method [7].

In this study, the bare die sample was fabricated in standard commercial process. The lead-frame oxidation and moisture absorption time were varied. With mechanical polishing the Cu/EMC interface was prepared. Then the samples with/without pre-crack were subject to shear tests to measure interface strength. The FEA based on the experiments were performed to obtain Jc at interface cracks. Furthermore, the package with small defect under low temperature was analysis by finite element method.

Experiments and Results

In order to obtain interfaces of Cu/EMC in actual devices, the bare die package samples were fabricated by standard commercial process. The lead-frames in samples were oxide at 165°C for 0, 5, 30 and 150 minutes respectively before molding. The photo of final bare die samples are shown in Fig. 2 and a single sample's length, width and thickness are 8mm, 5mm and 4.62mm, respectively. After lead-frame trimming and mechanical polishing, the sample contains interface of EMC/Cu was prepared. The photo of polished sample is shown in Fig. 3. The black is EMC and the golden part is Cu pad.

Fig. 2 Photo bare die Samples

Fig. 3 Sample preparation with Cu/EMC interface

The moisture can affect the strength of the interface. To study the moisture effect on the interface of Cu/EMC, the samples were first desorption at 60°C for 24 hours. The samples were then subjected to moisture absorption process under 85°C/85%RH. The moisture absorption time is 24h, 72h, 168h, respectively.

Fig. 4 Image of sample holder with sample mounted

Fig. 5 Photo of shear test setup

As the samples were small, the sample holder for shear testing was designed, which is shown in Fig. 4. The sample can be mounted in the holder and fixed tightly. The sample holder was mounted on the electrical universal tensile machine (in Fig. 5) and the shear test of Cu/EMC interface was performed by pushing the Cu pad. The sketch of shear test for sample with/without pre-crack is shown in Fig. 6 and Fig. 7, respectively.

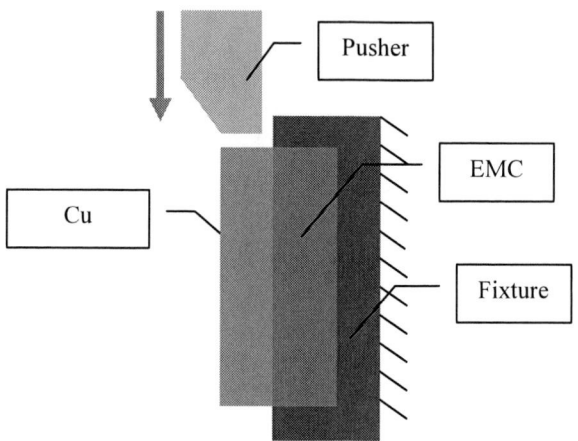

Fig. 6 Shear test of sample without the pre-crack

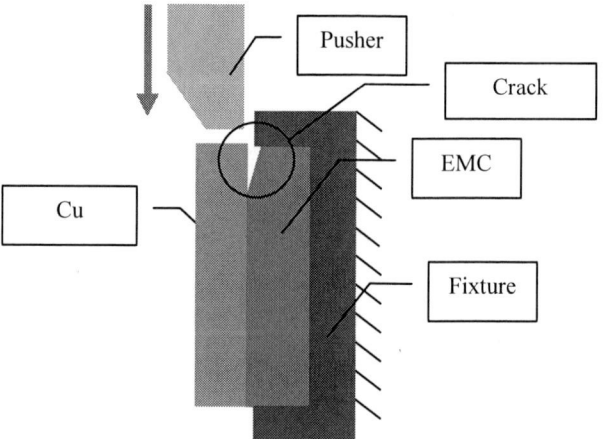

Fig. 7 Shear test of sample with the pre-crack

The maximum shear forces were recorded and the shear strength was calculated when the interface area was measured. For the samples that had no pre-crack and their lead-frame was 5 minutes oxidation, the shear strength varied with moisture absorption time is plotted in Fig. 8. The curve demonstrated that the average shear strength of the dry samples (moisture absorption time is zero) was 10.79MPa. The strength decreased to 9.22Mpa after 24h moisture absorption and to 8.17MPa after 72h water absorbing. Finally, the strength is reduced to 7.48MPa when moisture absorption time is 168h.

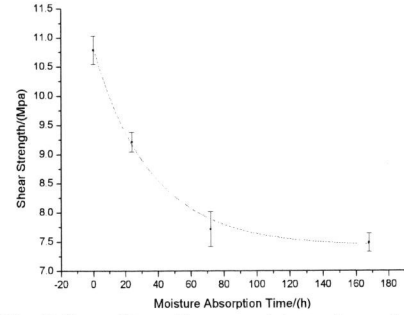

Fig. 8 Shear Strength vs. moisture absorption

When the oxidation time of lead-frame of samples is different, the shear strength of dry and wet sample (moisture absorption time is 168h) is compared in Fig. 9. The shear strength decreased when oxidation time is 30 minutes but increased when the oxidation is 150 minutes. For the moisture absorption cases, the trend of data is in the same but the value of strength is decrease more than 20% compare to their dry counterparts.

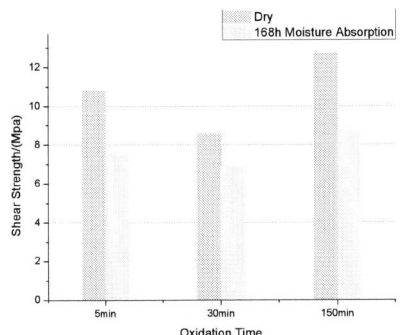

Fig. 9 Shear strength varied with oxidation and moisture absorption

The SEM image of delaminated surface on Cu is illustrated in Fig. 10 when interface is dry or wet. The dry interface is relatively flat. However, some small oxide particle can be observed on wet surface.

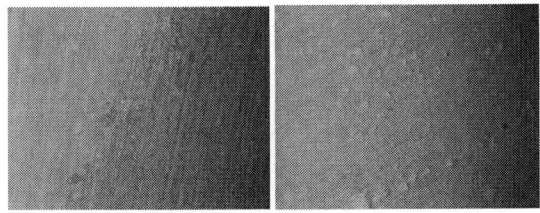

(a) dry surface   (b) wet surface
Fig. 10 SEM image of delaminated Cu surface

Value of J_c and Finite Elements Analysis

For pre-crack samples, the crack propagation will take place at critical shear force. The J-integral around crack tip can be calculated when the stress and strain filed in the package is know. The 2-D finite element models were built according to the size of the samples. The typical mesh near

the crack is shown in Fig. 11. The meshes around crack tip were refined to reduce the effect of stress singularity. The material properties are listed in Table 1.

Table 1 Material properties used in FEA

	Temp. (°C)	Young's Modulus (GPa)	Possion's Ratio	CTE (ppm/°C)
Cu	-73	132.79	0.34	15.2
	25	117	0.3	17.3
	127	124.11	0.35	17.5
	327	114.45	0.35	18.8
EMC	25	19.61	0.3	12
	149			
	151	0.93		45
	300			
Si	-73	164.65	0.22	1.4
	27	161.06	0.21	2.6
	327	157.27	0.21	3.8
92.5Pb/5Sn/2.5Ag		13.79		28.7

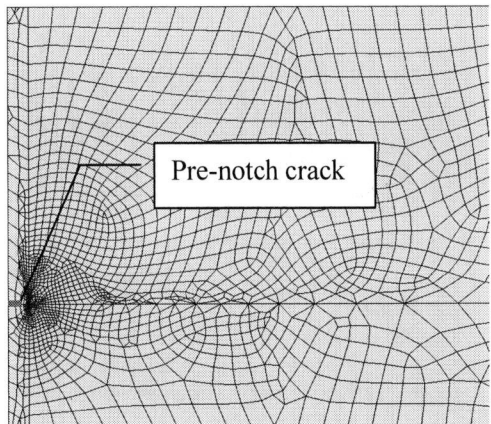

Fig. 11 Typical finite element meshes near the crack

When the lead-frame was not oxide, the J_c is plotted vs. moisture absorption time is illustrated in Fig. 13. The value of J_c falls about 38% when moisture absorption time is 168h, which is from 31.614 J/m^2 to 19.661J/m^2.

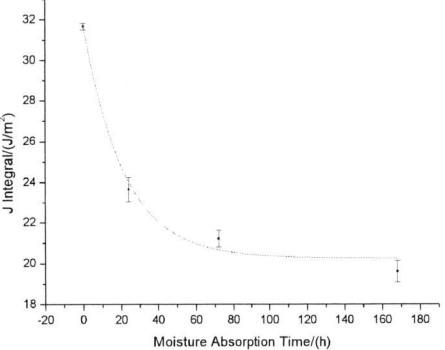

Fig. 12 J integral vs. moisture absorption time(85°C/85H)

For the real package, the initial crack will propagate when applied J-integral exceeds the J_c. The 2-D finite element

978-1-4244-4658-2/09 $25.00 © 2009 IEEE   1181

analysis was carried out to study the delamination possibility when the package is under -55°C environment. The finite element model is shown in Fig.14. The initial crack was assumed 0.13 mm near the edge of Cu/EMC interface, which is demonstrated in Fig. 15. The reference temperature was set to 175°C and the material properties are listed in the Table 1. The applied J-integral value is found to be 39.7J/m^2 which exceeds the J$_c$ of Cu/EMC interface even when the interface is dry. Hence, the delamination will most probably take place, especially when the interface is wet and the temperature is dropped quickly.

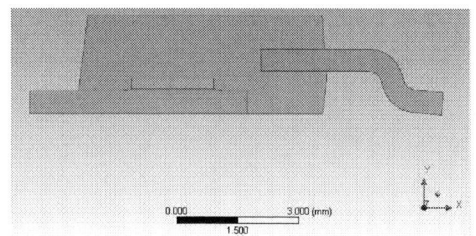

Fig. 13 2-D finite element model for real package

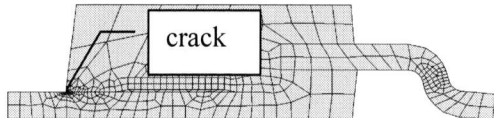

Fig. 14 Finite element meshes for real package

Concluding Remarks

The bare die samples were fabricated following standard commercial processes to obtain the Cu/EMC interface. The shear tests were designed and performed when Cu lead-frame oxidation time and moisture absorption time were varied. The shear strength and J$_c$ of the Cu/EMC interface are both declined when moisture absorption time is increased. The decreasing rate is slower after 72h moisture absorption. According to the FEA for the real package, the delamination will take place at -55°C when the initial crack size is about 0.13mm near the edge of Cu/EMC. Effect of Lead-frame on shear strength of Cu/EMC interface is ambiguous that the strength of the interface is not always decrease with increasing of oxidation time. The characterization of micro-region on interface and explanation will be present in the followed paper.

Reference

1. Eiji TAKANO, Toshikazu MINO, Kenji TAKAHASHI, Kanako SAWADA, "The Oxidation Control of Copper Leadframe Package for Prevention of Popcorn Cracking," *1997 Electronic Components and Technology Conference*, pp. 78-83.

2. Xiran Su, Haiping Luo, D. G. Yang, "Investigation of Thermal-Mechanical and Moisture Driven Delamination in Lead-free QFN Packages during SMT Reflow Soldering Process," *IEEE. 2006 7th International Conference on Electronics Packaging Technology*, pp. 1-6.

3. Andrew A. O. Tay, and T. Y. Lin, "Influence of Temperature, Humidity, and Defect Location on Delamination in Plastic IC Packages," *IEEE Transactions on Components and Packaging Technology*, Vol, 22, No. 4, pp. 512-518.

4. Hidetoshi Yamabe, "Stabilization of the polymer-metal interface," *Progress in Organic Coatings*,Volume 28, Issue 128 (1996), pp. 9-15.

5. Socin-Jin Cho, Kyung-Wook Paik, Young-Gil Kim, "The effect of the oxidation of Cu-base leadframe on the interface adhesion between Cu metal and epoxy molding compound," *Components, Packaging, and Manufacturing Technology. Part B: Advanced Packaging, IEEE Transactions*, Volume 20, Issue 2 (1997), pp. 167-175.

6. Kuan H. Lu, Brook Chao, Zhiquan Luo, Lijuan Zhang, Hualiang Shi, Jay Im, and Paul S. Ho, "Moisture Transport and its Effects on Fracture Strength and Dielectric Constant of Underfill Materials," *2007 Electronic Components and Technology Conference, ECTC 07. Proceedings. 57th*, pp. 1040-1044.

7. Tay, A.A.O.;Tan, G.L. and Lim, T.B, "Predicting Delamination in Plastic IC Packages and Determining Suitable Mold Compound Properties," *IEEE Transactions on Components, Packaging, and Manufacturing Technology, Part B, Advanced Packging*, Vol.17, No.2 (1994), pp. 201-208.

XPS Study on Epoxy/Ni Interface

Li Liu*, Wenting Xv

Dept. of Polymer Science, School of Material Science and Engineering Technology, Shanghai Univ.
Chengzhong Road, 20, Jiading District, Shanghai, 201800, China
Email: liuli@staff.shu.edu.cn; Tel : (+86)-021-69982311; MP: (+86)-021-13916119130

Abstract

The interface between epoxy molding compound (EMC) and nickel has been investigated by using X-ray photoelectron spectroscopy (XPS). The change of binding energy of main element indicated that there was interaction between metallic and non-metallic element and new bond might exist in the interface.

Introduction

Although some experimental investigations on metal /epoxy interface have already been reported [1-3] in the past decades, the problem of adhesion between metal and epoxy still remains an open issue. When the epoxy is applied onto metallic substrates and cured, an interface, having chemical, physical and mechanical properties quite different from that of epoxy molding plastics, is created between the substrate and the epoxy. The interfacial bonding and consequently adhesion are influenced directly by the way in which the interface is formed. Therefore, investigating the interface is very important for understanding the adhesion behavior.

The chemical bonding at a metal/epoxy interface is believed to play an important role in adhesion [4]. XPS is an effective method to characterize the interface if chemical reaction occurs. In this paper, we attempt to use XPS to characterize the chemical reaction existing between Ni/epoxy.

Experiment

Sample description

During our whole experiments, same resin system was used in all samples and its main component is listed in Table 1. The main raw materials of EMC using in this article included epoxy resin, phenolic resin, catalyst, silica, coupling agent, pigment, release agent and some promoter. The epoxy resin used in this research was a type of biphenyl while the phenolic resin was used as harder.

Table 1 Major component of EMC

EMC	Component	Description
Filler	SiO$_2$	Sphericity type
Polymer	Epoxy resin	Diphenyl type
	Bakelite	Xylok
	Coupling agent	With mercapto group
Additive	Parting agent	Polythene
	Stress absorbent	CTBN

Notes: CTBN is carboxylic acrylonitrile butadiene rubber

Fig. 1 shows the schematic diagram of the final product. Part that drawn in solid line referred to the metal substrate, and the resin was cured outside the substrate in the dotted line region which forming the interface between metal and epoxy.

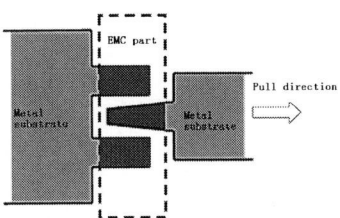

Fig. 1 Scheme of EMC sample

XPS study of epoxy/ Ni interface

XPS is an effective method to characterize the interface between two different phases. The model of XPS using in this article is Thermo ESCALAB 250. Operating conditions were: X-ray source = Al K$_\alpha$ (hv =1486.6eV); power = 150W; beam spot = 500μm; fixed transmission energy of energy analyzer = 20eV; depth = 10nm.

We used XPS method to test Ni substrate (bulk), epoxy (bulk), Ni and epoxy interface. We called them Ni$_0$, EP$_0$, Ni$_s$ and EP$_s$ respectively, and detail account of these is shown in Table 2. As shown in Fig. 1, along the direction of the arrow, we got the interfacial samples of Ni$_s$ and EP$_s$.

Table 2 Description of XPS samples

No.	Sample
Ni$_0$	Metal substrate (after H.T.)
Ni$_s$	Interfacial (Metal)
EP$_s$	Interfacial (Molding)
EP$_0$	Molding body

Notes: "H.T." means heat treatment (under 175°C for 3min), and the pure metal samples were preheated before XPS tests in order to simulate the actual manufacturing process.

Results and discussions

In the current system, only nickel is coated on the cooper sheet. Data from XPS tests are all listed in Table 3.

In Ni$_0$ sample, the peak of 852.58eV referred to pure Ni and Ni/Cu, which meant that there was no oxide on metal sheet before curing. In Ni$_s$ sample, we observed peaks of both the oxide of Nickel (856.08eV) and pure Ni (66.79eV) [5]. In EP$_s$ sample, we failed to find the peak of pure nickel and the peak of 858.29eV appeared. Compared with the Ni$_s$ samples, the binding energy increased from 856.08eV to 858.29eV which illustrated that nickel provided electrons to other group existing in our molding system. If the binding energy of some element existing in our system decreased, we may presume that nickel had formed new bonds with that group. In our system, non-metallic elements like nitrogen and silicon decreased from epoxy to interfacial part, binding energy of nitrogen is weaker in the interface than in the bulk (decreased

978-1-4244-4658-2/09 $25.00 © 2009 IEEE

from 399.80eV to 399.22eV), which shows the chemical environment of group -CN (cyanogen radical) existing in stress absorbent (CTBN) has been changed; for Silicon, 103.60eV referred to the SiO_2 and 101.92eV referred to the SiO_x in the EP_0 sample. In the interfacial part, the binding energy of Silicon decreased to 101.71eV and 101.22eV. The binding energy of nickel increased while the other elements of nitrogen and silicon all decreased, from which we initially speculated that it did exist some chemical or physical bonds in the interfacial part between Ni and non-metallic elements.

As for other elements, no sulfur peak was found in molding body but found in the interface, so we presumed that -SH (mercapto group) existing in coupling agent might transfer to the boundary face to form new bonds. And for cooper, we only found it in Ni_0 and failed to find that in the other parts because of the detection depth of XPS instrument, only 10nm.

Table 3 Binding energy of main elements (eV)

Element	Ni_0	Ni_s	EP_s	EP_0
Ni	852.58	856.08 66.79	858.29	--
Cu	932.14	N/A	N/A	--
S	--	162.17 168.32	168.80	N/A
N	--	400.06	399.22	399.80
Si	--	101.22	101.71	101.92 103.60
O	--	532.12	532.11	532.64

Notes: "N/A" means "not any"

Conclusions

In this work, we studied the interface of epoxy/nickel using XPS. We investigated the binding energy of typical elements on different site by using XPS. Data showed the binding energy change of main elements. We presumed that new bonds may form in the interface between nickel and non-metallic elements like nitrogen and silicon which existed in the form of cyanogen radical and silicon dioxide. And our further research on Ni/Epoxy interface is still in progress.

Acknowledgments

We greatly thank for the help from Henkel Huawei, especially Dr. Xingyu Du and Wei Tan who gave us a great deal of support during our work.

References

1. Roche A.A. *et al.*, "Formation of epoxy-diamine/metal interphases," *Internatinal Journal of Adhesion & Adhesives*, Vol. 22 (2002), pp. 431-441.
2. Ramos Marta M.D. *et al.*, "Atomistic modeling interfacial bonding at metal/polymer interface," *Journal of Materials Processing Technology*, Vol. 92-93 (1999), pp. 147-150.
3. Zhang J. *et al.*, "Theoretical study on polyimide-Cu(100)/ Ni(100) Adhesion," *Chemical Mater*, Vol. 18 (2006), pp. 5312-5316.
4. Gao S. C., "The XPS analysis of the interface between steel and adhesion," *Journal of Solid Rocket Technology*, Vol. 27, No. 2 (2004), pp. 154-156.
5. Website, http://srdata.nist.gov/xps/

Shock Performance Study of Solder Joints in Wafer Level Packages

Amarinder Singh Ranouta[1], Xuejun Fan[1,2], Qiang Han[2]
[1]Department of Mechanical Engineering
Lamar University
PO Box 10028, Beaumont, TX 77710, USA
[2]College of Civil Engineering and Transportation
South China University of Technology, Guangzhou, China
xuejun.fan@lamar.edu

Abstract

In this paper, an integrated testing, finite element modeling and failure analysis approach for drop test reliability of wafer level packages is developed to examine the shock performance of large array wafer level packages. For standard JEDEC drop test, it has been found that corner component group (group A) failed first for 12×12 array packages. This is different from previously reported failure test data of BGA packages. Careful analysis concluded that the high failure rate of group A is mainly due to the effect of mounting screws rather than the intrinsic strength of the package. For a given WLP, corner balls always fail first during drop test. The crack initiates at inner side of the solder joint and propagate towards the opposite side. The primary failure is always on the intermetallic compound (IMC) at WLP side. It has been found that drop reliability significantly decreases with array size increasing. Novel finite element modeling approach has been developed to correlate with experimental data. The finite element model was validated with experimental board strain data, and frequency analysis. In-plane principal strain at corner locations and maximum peeling stress in IMC at critical solder joints are used to correlate with experimental data. Excellent agreement was reached to predict the failure rate of components in each group. Two new findings have been observed and validated. One is that existing JEDEC board design will lead group A components fail first for certain array size of wafer level packages. Another finding is that PCB board strain does not always correlate with maximum peeling stress in solder joints when array size changes.

1. Introduction

Reliability of handheld electronic devices such as cellular phones due to drop and impact event is a major concern in electronics industry. During a drop/impact event, printed circuit board (PCB) assembly inside casing vibrates causing a flexural/bending motion of the board [1]. The PCB bending results in transient dynamic stresses or strains on solder joints of electronic components. It ultimately leads to the failure in solder joints. The failure can occur at package side or PCB side. Other failure modes such as pad-crater and broken board traces are also observed [2].

The dynamics and reliability of electronic components under board level drop test have been well studied. A board level drop test method has been standardized through Joint Electronic Device Engineering Council (JEDEC), JESD22-B111 (2003) [3], to evaluate the performance of IC packages under drop conditions. Multi-channel real-time monitoring system has been applied to record electrical connections of daisy-chained components, accelerations, and in-plane strains at various locations of PCB using strain gages and accelerometers [4-7]. High speed cameras have been applied to capture the images of board assembly during impact to extract displacement and deformation [7-11]. Digital Image Correlation (DIC) system integrated with the cameras has been developed to analyze the acquired images to give dynamic deformation, shape and strain over the entire surface of board [8-16].

Previously, various shock/impact modeling techniques have been developed to predict board dynamic strains and transient solder joint stresses. Explicit dynamics has been applied in both product and board levels [7-11 and 17-19]. Several special treatments such as equivalent layer models for solder interconnects [11], shell element in global models [20], solid-to-solid sub-modeling technique using half PCB board [21-23], shell-to-solid sub-modeling using beam-shell-based quarter symmetry models [8-11 and 17], shell-to-solid sub-modeling without any assumption of symmetry [9-11], have been developed to reduce the computational time required for simulation. The board level model can be analyzed by using the drop table acceleration as input loading. This so-called Input-G method decouples the board finite element model from the system model [24]. There are several approaches in implementing the input-G loading method. Tee used explicit dynamics analysis by directly applying acceleration impulse using DYNA-3D [7]. Syed introduced the large mass method to convert acceleration input into force input by multiplying the acceleration with a large mass with implicit dynamics [20]. Irving proposed the input-D method, in which the acceleration input is integrated twice to obtain the displacement boundary condition over time [25]. Loh used mode superposition method for a linear system under impact loading [5].

In this paper, a comprehensive study is carried out to examine the shock performance of large array wafer level packages (WLPs). Copper post wafer level packages are used with different array sizes to investigate the failure characteristics under JEDEC setting. Experimental work for controlled JEDEC drop test is conducted. It is found that the primary failure mechanism of WLP drop test failures is fracture of intermetallic compound (IMC) at WLP side. Transient board strains and accelerations at various locations are measured during impact to correlate with failure life of each component. The fundamental frequencies of test board are extracted through FFT transformation. Statistical analysis is performed to analyze the drop life for each group. Finite

978-1-4244-4658-2/09 $25.00 © 2009 IEEE

element modeling using newly developed direct acceleration input method (DAI) is applied. Global/local modeling is adopted to capture both board strains and solder joint stresses accurately. Experimental results are compared to the simulation data. The effects of array size and failure locations are studied in detail. The correlation between board strain and solder joint stress is described. Several new findings through both test and simulation are discussed.

2. Experimental Setup [6]

In this study, a JEDEC test board has been used with dimensions 132mm×77mm×1mm. The test board has 15 copper post wafer level packages with different array sizes. The packages are populated on one side in a three-row, five-column format, as shown in Figure 1.

Figure 1 JEDEC test board and strain gauge rosette attachments

Figure 2 is a schematic view of solder bump structure for a copper post wafer level package. A thick copper post, which is encapsulated by epoxy, is formed on wafer level before ball attachment. The geometric dimensions of the WLP are given in Table 1. The ball pitch is 0.5mm. The test assemblies have been subjected to a 1500g, 0.5ms pulse consistent with the JESD22-B111. The drop height and the pulse shape have been adjusted using pulse shapers between the impacting surfaces. A half-sine pulse has been achieved. Figure 3 shows the schematic of shock test platform, acceleration profile of shock table, and the arrangement of components (face-down) and numbering.

Table 1 Geometrical dimensions of copper post WLP

	Dimensions (μm)
Silicon thickness	400
Solder ball diameter	310
Solder ball standoff height	240
Solder ball opening diameter	250
PCB pad diameter	250
PCB thickness	1000
Wafer Passivation thickness	4
Epoxy/Copper post thickness	70

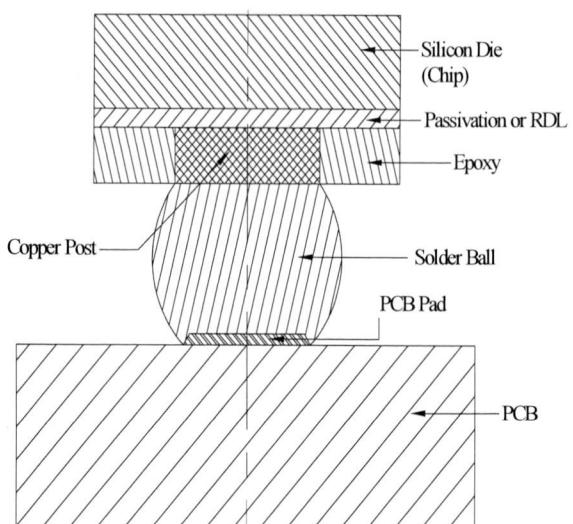

Figure 2 Cross-section view of solder bump structure for a copper post wafer level package

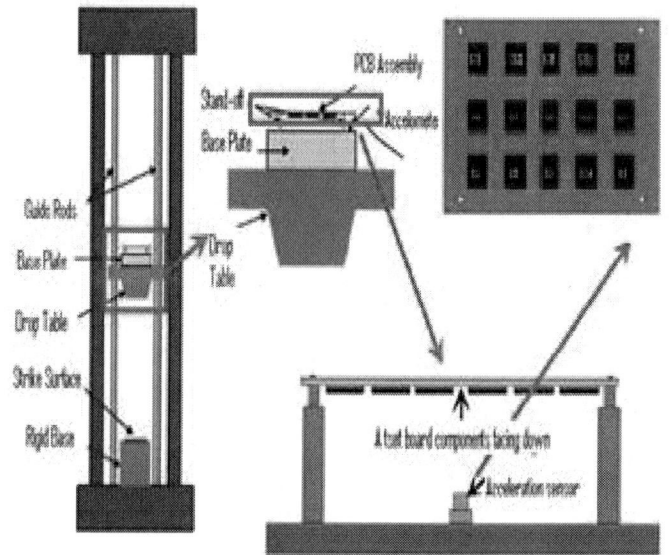

Figure 3 Schematic of experimental setup for controlled JEDEC drop test

45° strain gauge rosettes, which are attached with 1mm offset from component corner in both horizontal and vertical directions as per IPC9704 at U8, U10 and U15 locations, respectively (as shown in Figure 1), are used to measure board strain transient responses. A typical strain data measurement is given in Figure 4 for the central component U8 at 0°, 45°, and 90°, respectively [6]. Frequency spectrum of board vibration is generated a by strain data through fast Fourier transform (FFT). Figure 5 shows the frequency spectrum of PCB strains. As is seen the first resonant frequency is registered at 230Hz, and second one is found at ~ 650Hz.

978-1-4244-4658-2/09 $25.00 © 2009 IEEE   1186

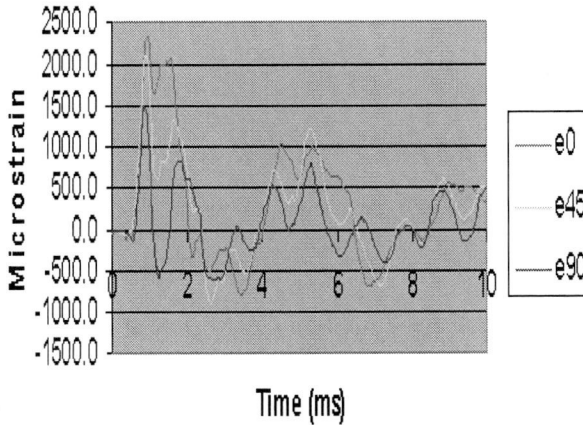

Figure 4 Strain components as function of time at U8 location

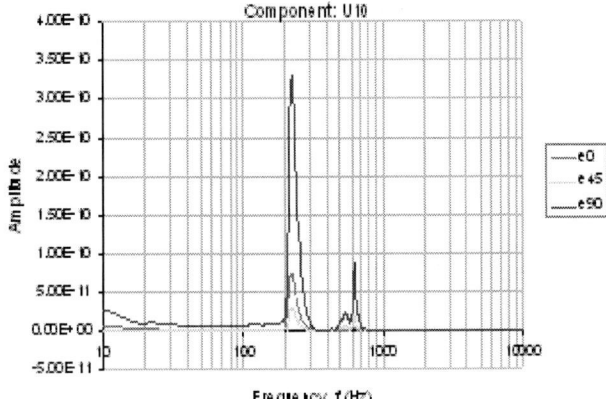

Figure 5 Frequency spectra through FFT

3. Finite Element Modeling

There are 15 components on a JEDEC board and each component has hundreds of solder balls. In order to handle this very large model without sacrifice of accuracy, special considerations are implemented in both global and local model levels. In global model, a quarter model is created due to symmetry. Solder balls are simplified as rectangular blocks with one 3-D solid element for each ball. Copper post, epoxy, passivation layer, and PCB pad are neglected in the global model, as shown in Figure 6. 3-D elements are used for entire structure including PCB board and silicon chips. Direct acceleration input (DAI) method is used to apply impulse loading [21-23 and 26]. The damping coefficient for PCB is determined by correlating with experimental strain data (to be discussed in next section).

A local model is constructed next. The script is developed to build a local model at any desired location of components. Figure 7(a)-(c) show an example of a local model for component U1. In the local model, the PCB is extended to 2mm away from component corner in both x and y directions, respectively to create cut boundary and DOF constraints taken from global model.

Solder bump arrays

Figure 6 A quarter global finite element model

In order to further reduce the model size, all solder balls in the local model are modeled as rectangular blocks, except critical solder balls with refined meshes and detailed structures. It has been shown that such a local model can produce almost same results compared to a local model with all refined solder balls [21-23]. Figure 7(d) and 7(e) describe the details of solder ball structures in the local model with rectangular blocks and refined structures, respectively. Since the primary failure is at the intermetallic layer on WLP side [6], a 10μm layer with two layers of elements is created at solder/copper post interface.

Table 2 defines the material properties used for both global and local finite element models. All the materials are considered as elastic ones.

Table 2 Material Properties

Material	Mechanical Properties		
	Modulus (GPa)	Poisson Ratio	Density (gm/cm³)
PCB	22	0.25	2.1
Solder (SAC)	51	0.36	7.2
Silicon Die	130	0.278	2.5
Underfill	10	0.3	2.0
Copper Post	128.8	0.34	8.3
Epoxy	4.7	0.38	2.2
Passivation	2.89	0.34	2.2
PCB Pad	128.8	0.34	8.3

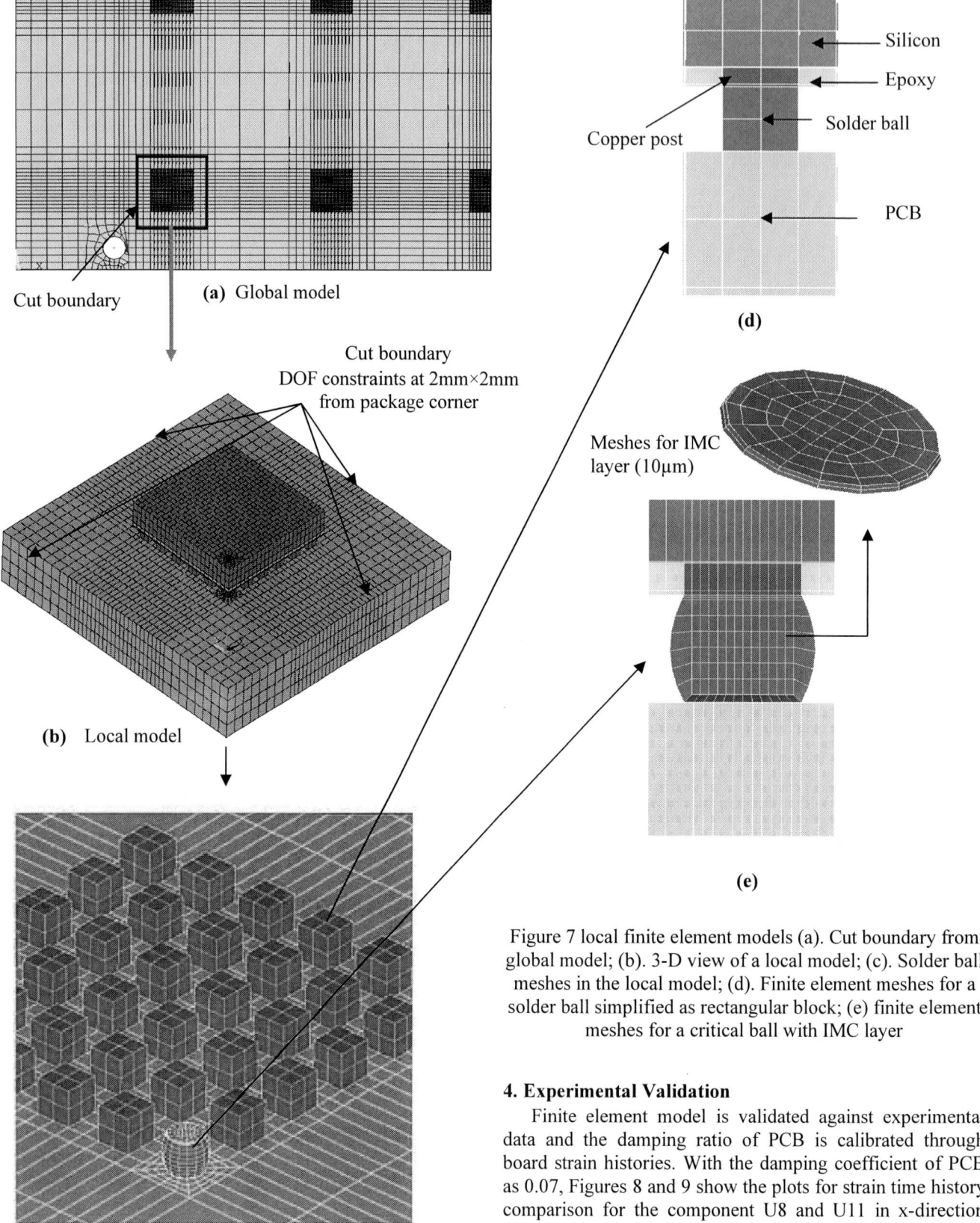

Figure 7 local finite element models (a). Cut boundary from global model; (b). 3-D view of a local model; (c). Solder ball meshes in the local model; (d). Finite element meshes for a solder ball simplified as rectangular block; (e) finite element meshes for a critical ball with IMC layer

4. Experimental Validation

Finite element model is validated against experimental data and the damping ratio of PCB is calibrated through board strain histories. With the damping coefficient of PCB as 0.07, Figures 8 and 9 show the plots for strain time history comparison for the component U8 and U11 in x-direction respectively. Similarly, Figures 10 and 11 show the plots for strain time history comparison for the component U8 and U11 in y-direction respectively. Overall, the FEA predicts the board strain dynamic responses very well.

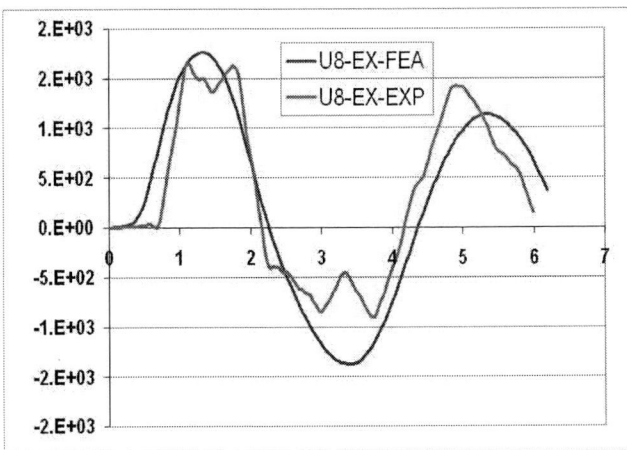

Figure 8 Strain time history comparison for experimental and FEA prediction for U8 in x-direction

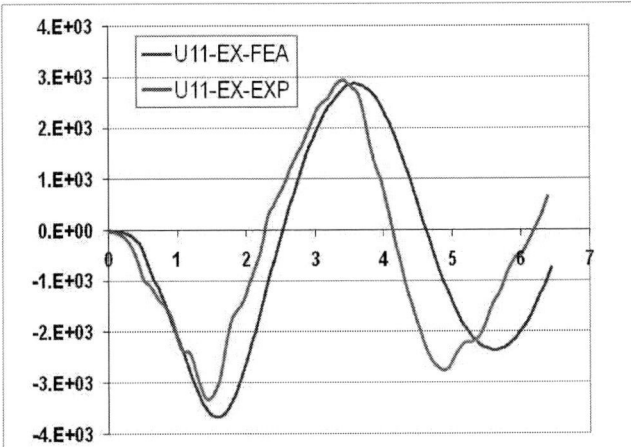

Figure 9 Strain time history comparison for experimental and FEA prediction for U11 in x-direction

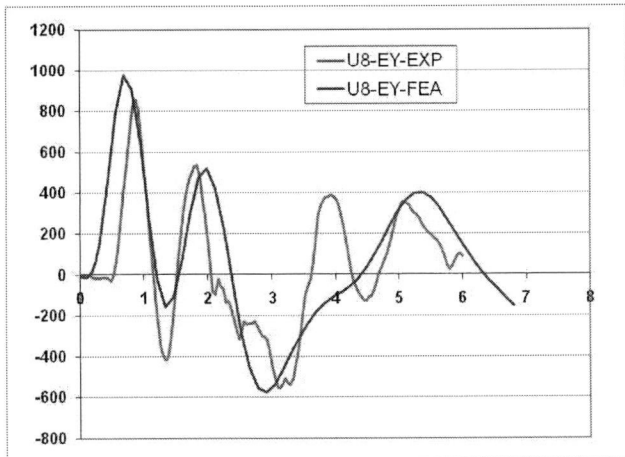

Figure 10 Strain time history comparison for experimental and FEA prediction for U8 in y -direction

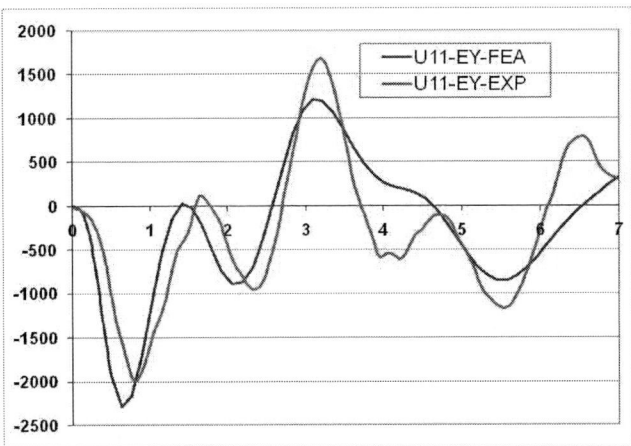

Figure 11 Strain time history comparison for experimental and FEA prediction for U11 in y -direction

Modal analysis is also performed with the global finite element model. The first two symmetrical modes and the corresponding natural frequencies are calculated as 220Hz and 654Hz, respectively from modeling. It is seen that at fundamental frequency, the mode shape is ε_x dominant. While at 654Hz the mode shape is εy dominant. Modeling results correlate very well with measured data in Figure 5 (230Hz and 650Hz).

5 Board Strains Analysis

5.1 Corner Strain Analysis

Previous studies have shown that board strains at package corner locations can correlate well with solder joint failures [5]. In the following study, the exact corner locations on component side are picked for strain data evaluation. In other words, corner strains in the following are defined as strains on PCB at component side at the exact left corner location for each component. ε_x, ε_y, and ε_{xy} can be extracted from finite element results. In-plane principal strain ε_1 and ε_2 can be calculated as follows,

$$\varepsilon_{1,2} = \frac{1}{2}\left[\left(\varepsilon_x + \varepsilon_y\right) \pm \sqrt{\left(\varepsilon_x + \varepsilon_y\right)^2 + \gamma_{xy}^2}\right] \quad (1)$$

where $\varepsilon_{1,2}$= Principal strains

ε_x = Strain in x-direction (PCB board long-side)

ε_y = Strain in y-direction (PCB board short-side)

γ_{xy}= Shear strain in x-y plane (board plane)

Since only positive board strain (component side) generates tensile stress in solder balls, the first principal strain ε_1 will be analyzed only. Figure 12 and Figure 13 plot the ε_x and ε_y time history for all components (U1, U2, U3, U6, U7, and U8) respectively. From these figures it is quite clear that ε_x is dominant for most components except U6, in which ε_y is dominant. In Figure 14, it is shown that the maximum principal strain at U6 occurs when the maximum ε_y is reached. At the same time, the ε_x is negative.

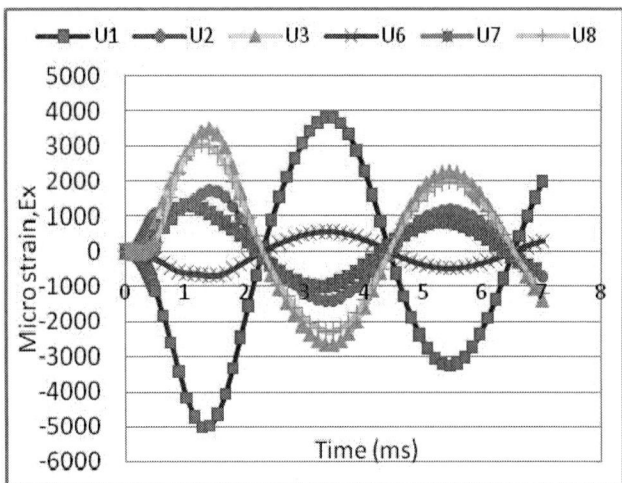

Figure 12 Strain time history plot for globl model of array 12×12 in x-direction

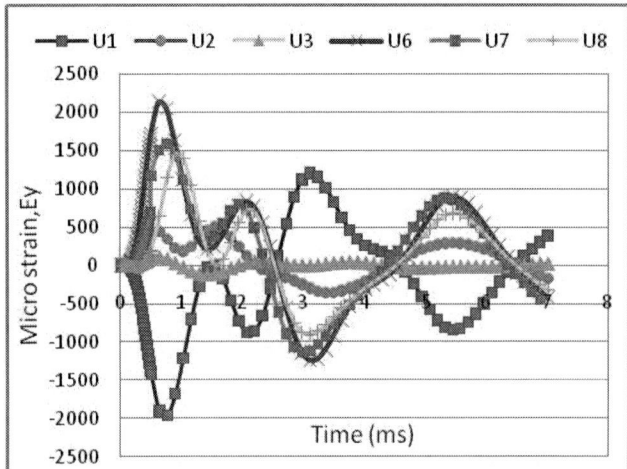

Figure 13 Strain time history plot for global model of array 12×12 in y-direction

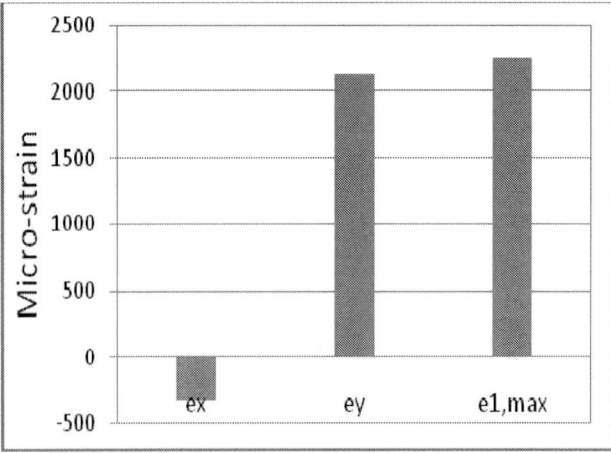

Figure 14 Maximum principal strain at U6 and corresponding ε_x and ε_y (12×12 array)

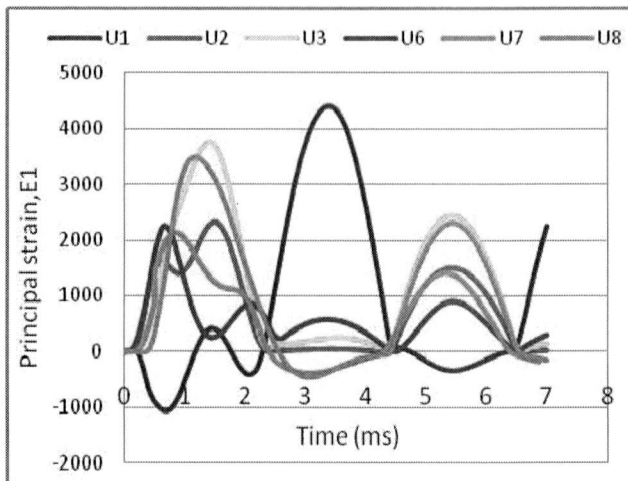

Figure 15 Principal strain history plot (12×12 array)

Figure 15 plots the principal strain ε_1 history. It is clear that U1 has the maximum board strain among all components, followed by U3 and U8.

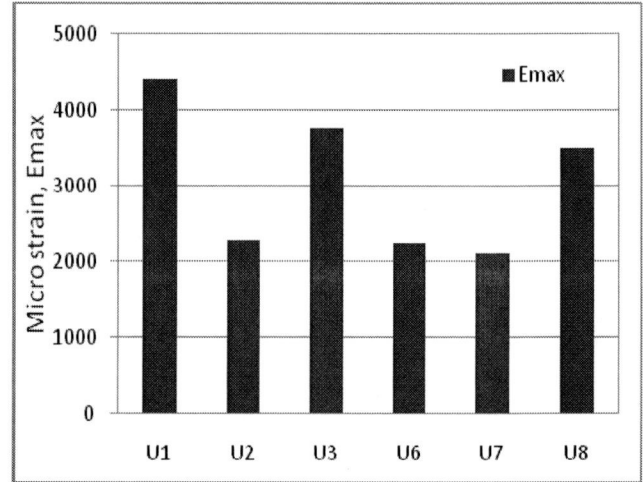

Figure 16 Maximum principal strains induced in different components of array 12×12

Figure 16 depicts the maximum values of principal corner strain induced at each component of JEDEC board of array size 12×12. The pattern shown in the figure clearly indicates that the maximum values of principal strain in components U1, U3 and U8 are much higher than components U2, U6 and U7. The difference between two groups is almost by 50%. Overall, the strains induced in PCB board can be ranked as U1>U3>U8>U2>U6>U7. It is fairly clear that U1, U3 and U8 are going to fail first than U2, U6 and U7. Therefore U1, U3 and U8 are more important components in JEDEC board. In the subsequent analysis only results from U1, U3 and U8 are presented in this paper.

5.2 Effect of Array Size

In previous section, it is discovered that U1 has maximum board corner strain. Here its behavior has been tested for different array sizes. Figure 17 plots the maximum principal strain and maximum x-strain at U1 in array sizes from 6×6 to

28×28. As array size increases beyond 20×20 (package size 10mmx10mm), strain decreases in PCB board. This nature is found not only with maximum principal strain but the same as with strain in x-direction.

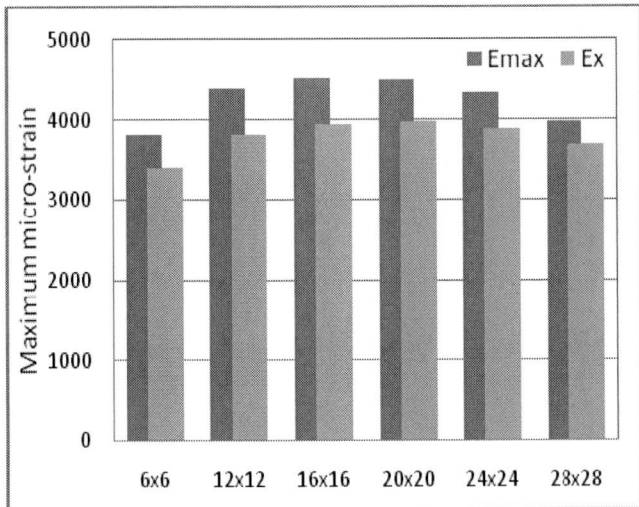

Figure 17 Plot for maximum principal strains and strains in X-direction at U1 of different arrays

Now let us look at behavior of strains induced at U3 and U8 components with different array sizes in JEDEC board, as shown in Figure 18 and 19. It clearly shows the fact that with increase in array size, both principal strains and strains induced in x-direction at component U3 and U8 increase.

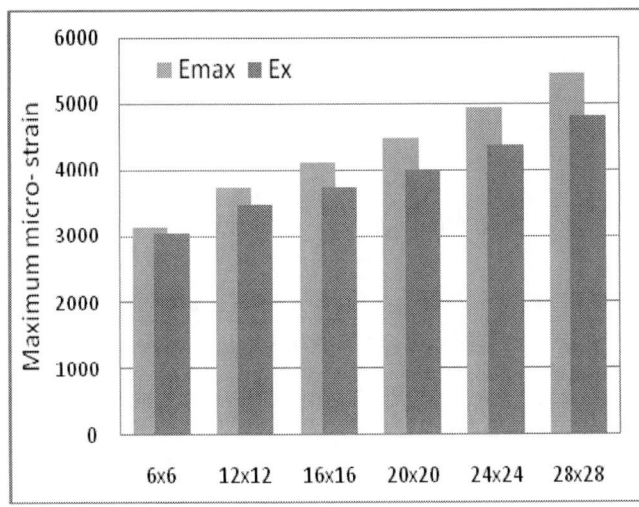

Figure 18 Plot for maximum principal strains and strains in x-direction at U3 of different arrays

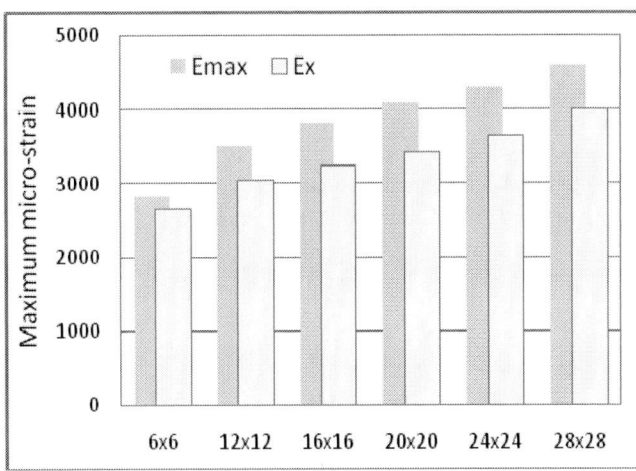

Figure 19 Plot for maximum principal strains and strains in x-direction at U8 of different arrays

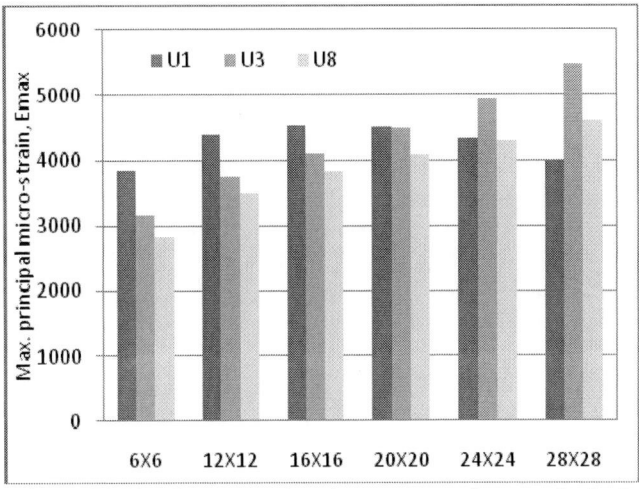

Figure 20 Plot for comparison between maximum principal strains induced at U1, U3 & U8 of different arrays

Figure 20 plots the compiled strain data for components U1, U3 and U8 for different array sizes. From this figure U1, U3 and U8 are ranked for various array sizes as shown in Table 3. It can be seen that the rank changes with array size. This implies that with large array size, the first failure may shift from the component U1 to U3 and U8.

Table 3 Ranking of U1, U3 and U8 based on maximum principal strain with different sizes

Array Size	Rank
6x6	U1 > U3 > U8
12x12	U1 > U3 > U8
16x16	U1 > U3 > U8
20x20	U1 > U3 > U8
24x24	U3 > U1 > U8
28x28	U3 > U8 > U1

6. Strain Comparison between Global and Local Models

To check whether global/local model built is accurate or not, corner strains from global model and local model with array size of 12 are compared in Figure 21. Figure 22 is a time strain history plot for U1, U3 and U8 components for 12×12 global and local models. It validates that corner strains in local and global model are same.

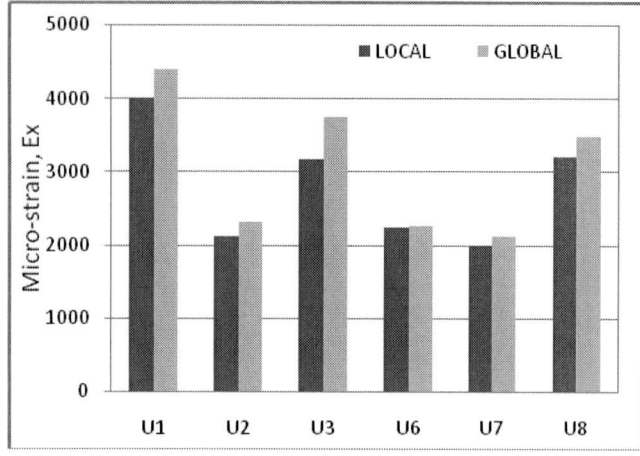

Figure 21 Plot for comparison between corner strains induced in local and global model of array 12×12 in x-direction

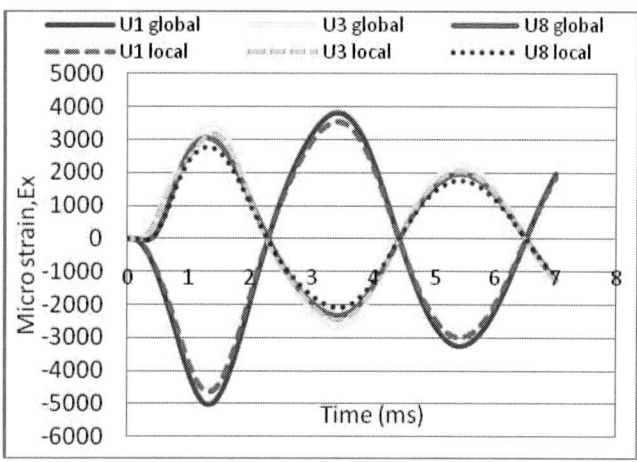

Figure 22 Comparison between time strain history plot for global and local model

7. Solder Joint Stress and Experimental Validation

Since the stresses and strains cannot be found in solder balls, strain measurement at PCB at corner locations are considered as a tool to calculate the stress level in solder balls. It is recognized that board strain at location near the package corner would determine the limit of PCB loading, regardless of package types and loading conditions. Figure 23 shows the correlation between maximum principal corner strain and maximum peeling stress for different components in 12×12 array WLP. It is noticed that the relation between board strain and peeling strees in critical layer holds very well.

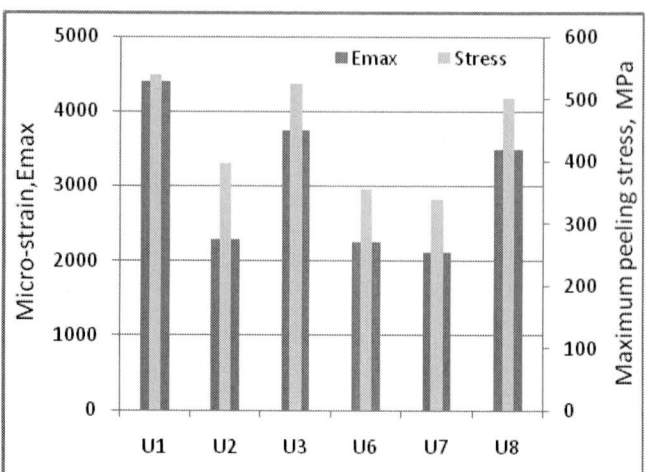

Figure 23 Plot for correlation between maximum principal strains and maximum peeling stresses in different components of array 12×12

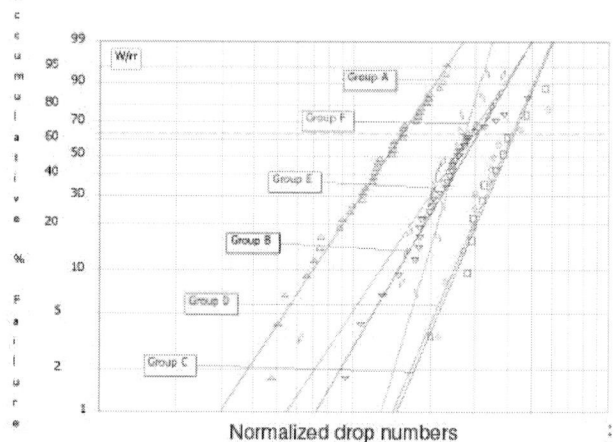

Figure 24 Weibull Plots for a 12x12 array 0.5mm Pitch WLP Components [?]

Figure 24 shows the experimental data (Weibull plots) of a 0.5mm pitch WLP for JEDEC JESD22-B111 standard test board during the drop test [6]. It is noted from this figure that component in group A (U1, U5, U11 and U15) failed first during the drop test. The failure of component by this experimental data is as U1>U3 or U8>U2>U6>U7, which is well matched with finite element analysis result data given in Figure 23.

8. Effect of Array Size on Solder Joint Stress

Further investigation is done to find out the effect of array size on maximum peeling stress generated in critical layer which causes the failure of solder ball. At first, individual behavior of each component is checked for different array sizes and then these are ranked based on maximum peeling stress. Figure 25 shows the maximum peeling stress in critical layer of solder ball at component U1 in different array sized WLPs. The trend is very clear in the figure that up to 20×20 array maximum peeling stress increases but it decreases afterwards.

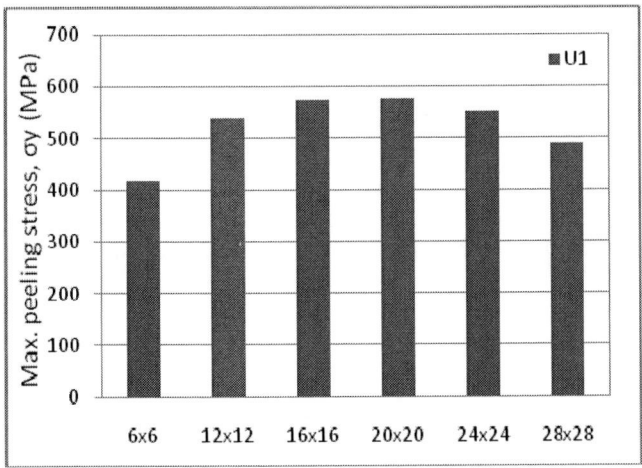

Figure 25 Plot for maximum peeling stresses at U1 of different array sized WLPs

Figure 26 and Figure 27 are the patterns of maximum peeling stresses induced in critical layers of solder balls of WLP models of various array sizes. There is continuous increase in peeling stress at both U3 and U8 components when chip size increases.

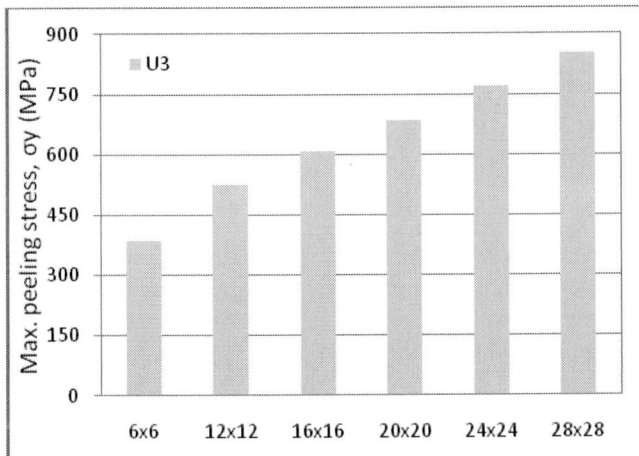

Figure 26 Plot for maximum peeling stresses at U3 of different array sized WLPs

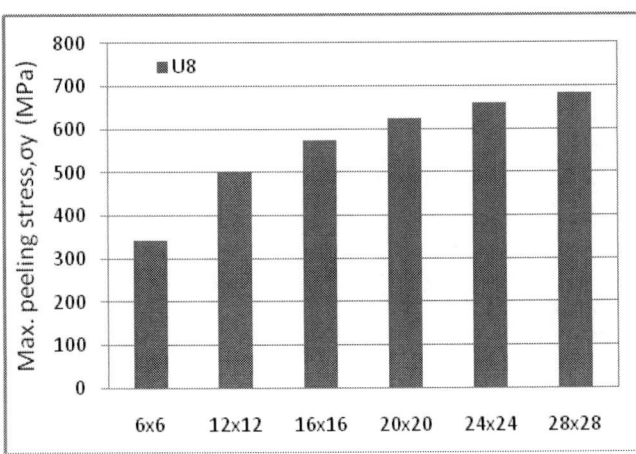

Figure 27 Plot for maximum peeling stresses at U8 of different array sized WLPs

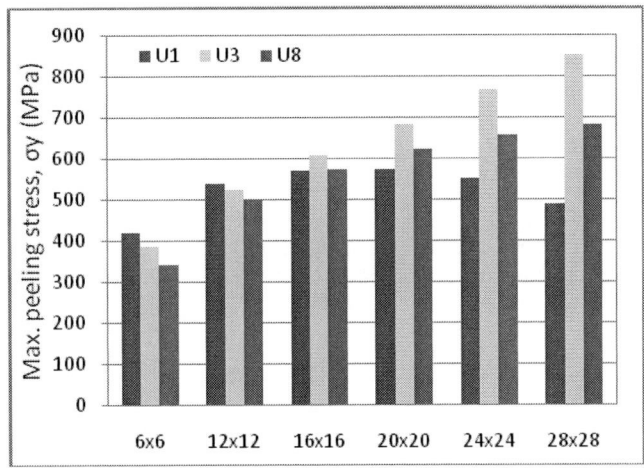

Figure 28 Comparison between maximum peeling stresses at U1, U3 & U8 of different array sized WLPs

Figure 28 shows the comparison between maximum peeling stresses produced at U1, U3 and U8 in various WLPs of array size from 6×6 to 28×28. Using this figure, Table 4 is generated which depicts the trend of failure of components in different array sized WLPs. It can be seen that the first failure location will shift from U1 location to U3 location when array size increases. Compared to the Table 3, it is found that the correlation does not hold based on the ranking from strain and the ranking from stress. This means that although corner strain is a good indicator for solder joint failures, but exact correlation with solder joint stress is more complicated.

Table 4 Failure trend in components for different array sizes

Array size	Order in which components fail		
	1st	2nd	3rd
6×6	U1	U3	U8
12×12	U1	U3	U8
16×16	U3	U8	U1
20×20	U3	U8	U1
24×24	U3	U8	U1
28×28	U3	U8	U1

9. Conclusions

In this study, large array WLP drop test reliability has been studied with an integrated testing, finite element modeling and failure analysis approach. For standard JEDEC drop test, it has been found that corner component group (group A) failed first for 12×12 array packages. This unexpected result is different from previously reported failure data of BGA packages. The high failure rate of group A is mainly due to the effect of mounting screws rather than the intrinsic strength of the package. For a given WLP, corner balls always fail first during drop test. The crack initiates at inner side of the solder joint and propagate towards the opposite side. The primary failure is always on the intermetallic compound (IMC) at WLP side. It has been found that drop reliability significantly decreases with array size increases.

Novel finite element modeling approach has been developed in this paper to correlate with experimental data. The direct acceleration input method has been applied to apply impulse loading effectively. IMC layer has been created in the local model to capture dynamics solder joint stresses accurately. The finite element model was validated with experimental board strain data, and frequency analysis. In-plane principal strain at corner locations and maximum peeling stress in IMC at critical solder joints are used to correlate with experimental data. Excellent agreement was reached to predict the failure rate of components in each group.

For different array sizes board strains have been studied at different component locations. At U1, board strain increases when array size increases from 6×6 to 20×20. However, board strain starts to decrease beyond 20×20 arrays. U3 and U8 board strain keeps increasing as array size increases. Similar trends have been found for peeling stress at IMC in critical solder joints. This implies that for very large array size package, U3 and U8 will first earlier than U1, which are consistent with experimental data of BGA packages.

The orders to fail for different components are different based on board strain and peeling stress, respectively. The correlation between solder ball stresses and board strains is investigated by results from local model and global model of 12×12 arrays respectively. It is found that board strain is able to capture change of solder ball stress when chip size is changed. But some caution must be taken while using board strain alone as parameter to analyze solder joint performance under drop impact.

References

1. Wong, E. H., Lim, C. T., Field, J. E., Tan, V. B. C., Shim, V. P. M., Lim, K. T., Seah, S. K. W. 2003. "Tackling the Drop Impact Reliability of Electronic Packaging," *ASME InterPAK*, July 6-11, Maui, pp. 1-9.
2. Tee, T.Y., Tan, T.B., Anderson, R., Ng, H.S., Low, J.H., Khoo, C.P., Moody, R., and Rogers, B., "Advanced analysis of WLCSP copper interconnect reliability under board level drop test," *Proceedings of 2008 Electronics Packaging Technology Conference*, pp 1086-1095.
3. JEDEC Standard JESD22-B111. 2003. Board Level Drop Test Method of Components for Handheld Electronic Products.
4. Fan, X.J., Han, Q. 2008. "Design and reliability in wafer-level packaging," *Proc of IEEE 10th Electronics Packaging Technology Conference (EPTC)*, pp. 1-8.
5. Loh W. K.; Hsiang L.Y.; Munigayah, A. 2005. "Nonlinear dynamic behavior of thin PCB board for solder joint reliability study under shock loading," *International Symposium on Electronics Materials and Packaging*, pp. 268-274.
6. Zhou, T., Derk, R., Rahim, K., Fan, X.J. 2009. "Larger array fine pitch wafer level package drop test reliability," Interpack. IPACK2009-89018
7. Tee, T.Y., Luan, J.E., Pek, E., Lim, C.T., and Zhong, Z.W. 2004. "Advanced Experimental and Simulation

Techniques for Analysis of Dynamic Responses During Drop Impact," pp. 1089-1094.
8. Lall, P., Gupte, S., Choudhary, P., Suhling, J. 2006. "Solder-Joint Reliability in Electronics under Shock and Vibration using Explicit Finite-Element Sub-modeling," *Proceedings of the 56th ECTC*, pp. 428-435.
9. Lall, P., Choudhary, P., Gupte, S., Suhling, J., Hofmeister, J. 2007a. "Statistical Pattern Recognition and Built-In Reliability Test for Feature Extraction and Health Monitoring of Electronics under Shock Loads," *57th Electronics Components and Technology Conference*, Reno, Nevada, pp. 1161-1178.
10. Lall, P., Panchagade, D., Iyengar, D., Shantaram, S., Suhling, J., Schrier, H. 2007b. "High Speed Digital Image Correlation for Transient-Shock Reliability of Electronics," *Proceedings of the 57th ECTC*, Reno, Nevada, pp. 924-939.
11. Lall, P. Panchagade, D., Liu, Y., Johnson, W., Suhling, J. 2007c. "Smeared Property Models for Shock-Impact Reliability of Area-Array Packages," *ASME Journal of Electronic Packaging*, Volume 129, pp. 373-381.
12. Park, S., Shah, C., Kwak, J., Jang, C., Pitarresi, J. 2007. "Transient dynamic simulation and full-field test validation for a slim-PCB of mobile phone under drop impact," *Proceedings of the 57th ECTC*, Reno, Nevada, pp. 914-923.
13. Scheijgrond, P.L.W., Shi, D.X.Q., Driel, W.D.V., Zhang, G.Q., Nijmeijer, H. 2005. "Digital Image Correlation for Analyzing Portable Electronic Products during Drop Impact Tests," *6th International Conference on Electronic Packaging Technology*, pp. 121-126.
14. Song, G., Shi, X., Qin, F., He, C., 2006. "Effect Of Loading Mode, Temperature And Moisture On Interface Fracture Toughness Of Silicon/Underfill/Silicon Sandwiched System," *Proceedings of ITherm Conference*, pp. 1147-1152.
15. Yogel, D., Grosser, V., Schubert, A., Michel, B. 2001. "MicroDAC Strain Measurement for Electronics Packaging Structures," *Optics and Lasers in Engineering*, Vol. 36, pp. 195-211.
16. Zhang, F., Li, M., Xiong, C., Fang, F., Yi, S., 2005, "Thermal Deformation Analysis of BGA Package by Digital Image Correlation Technique," *Microelectronics International*, Vol. 22, No. 1, pp. 34-42.
17. Ren, W., Wang, J. 2003. "Shell-based simplified electronic package model development and its application for reliability analysis," *Proceeding of Electronic Packaging Technology Conference*, pp. 217-222.
18. Wu, J., Song, G., Yeh, C., Wyatt, K., 1998. "Drop/impact simulation and test validation of telecommunication products," *InterSociety Conference on Thermal Phenomena*, pp. 330-336.
19. Zhu, L. 2003. "Modeling Technique for Reliability Assessment of Portable Electronic Product Subjected to Drop Impact Loads," *Proceedings of the 53rd ECTC*, pp. 100-104.
20. Syed Ahmer, Kim Mo Seung, Lin Wei, Khim Young Jin, Song, Sook Eun, Shin, Hyeon Jae, Panczak Tony. 2005.

"A Methodology for Droop Performance Prediction and Application for Design Optimization of Chip Scale Packages," *2005 Electronic Components and Technology Conference.*

21. Dhiman, H.S., Fan, X.J., Zhou, T., 2008a. "Modeling techniques for board level drop test for a wafer-level package," *Proc. of International Conference on Electronic Packaging Technology and High Density Packaging (ICEPT-HDP).*

22 Dhiman, H.S. 2008b. "Study on finite element modeling of dynamic behaviors of wafer level packages under impact loading," M.S. Thesis, Lamar University.

23. Dhiman, H.S., Fan, X.J., Zhou, T., 2009. "JEDEC board drop test simulation for wafer level packages (WLPs)," *2009 Electronic Components and Technology Conference*, pp. 556-564.

24. Jing-en Luan and Tong Yan Tee. 2004. "Novel board level drop test simulation using implicit transient analysis with Input-G method," *6th EPTC Conference*, Singapore.

25. Irving, S., Liu, Y. 2004. "Free drop test simulation for portable IC package by implicit transient dynamics FEM," *Proceedings of the 54th ECTC*, pp. 1062-1066.

26. Lianxi Shen. 2008. "Simulation of drop test board with 15 components using explicit and implicit solvers," *2008 International ANSYS Conference*, August 26 to 28 in Pittsburgh, Pennsylvania, U.S.A.

27. Fan, X.J., Liu, Y. 2009. "Design, Reliability and Electro migration in Wafer Level Packaging," *ECTC Professional Development Short Course Notes.*

28. Ren, W., Wang, J., Reinikainen, T. 2004. "Application of ABAQUS/Explicit submodeling technique in drop simulation of system assembly," *Proceeding of Electronic Packaging Technology Conference*, pp. 541-546.

29. Rahim, M.S.K., Zhou, T., Fan, X.J., Rupp, G., 2009. "Board level temperature cycling study of large array wafer level packages," *Proc of Electronic Components and Technology Conference (59th ECTC).*

30. Sun, Y., Pang, J., Shi, X., Tew, J. 2006. "Thermal Deformation Measurement by Digital Image Correlation Method," *Proceedings of ITherm Conference*, pp. 921-927, May 2006.

31. Wu J. 2000. "Global and local coupling analysis for small components in drop simulation," *6th International LSDYNA Users Conference*, pp. 11:17-11:26.

32. Xie, D., Minna Arra, Dongkai Shangkai, Hoang Phan, David Geiger and Sammy Yi, 2002. "Life prediction of lead free solders joints for handheld products, Telecom Hardware Solutions Conference," Plano, Texas, USA, and May 15-16.

33. Xie, D., Minna Arra,, Yi, S., Rooney, D., 2003. "Solder joint behavior of area array packages in board-level drop for handheld devices," *Proceedings of the 53rd ECTC*, pp. 130-135.

Fatigue Evaluation of Power Devices

Kazunori Shinohara[1], Qiang YU[2]
[1]Kanagawa Academy of Science and Technology
KSP, 3-2-1 Sakado, Takatsu-ku, Kawasaki City, Kanagawa, 213-0012, Japan
sinohara@swan.me.yun.ac.jp
[2]Yokohama National University
79-5, Tokiwadai, Hodogaya-ku Yokohama, 240-8501, Japan
qiang@swan.me.yun.ac.jp

Abstract

Semiconductors are solids whose electrical conductivity is intermediate between that of a conductor and an insulator. Semiconductor devices are active devices that consume, accumulate, and discharge the supplied electric power. They are used as electronic or power devices. The former are employed in the control systems of electrical products such as mobile phones and computers, which are controlled by electrical currents. Power devices are also controlled by currents, but these currents are the main energy sources of the electrical products. Therefore, the Joule heating generated in power devices is much larger than that generated in electronic devices. This heating is concentrated at a single point on the body of the device, where deterioration due to strain is localized. There is an increasing need for power devices that are lightweight and compact. However, reducing the weight of power devices would result in reduced stiffness, and greater compactness would results in increased current density, which, in turn, would increase the temperature threshold of the device. Thus, to realize lightweight, compact power devices with sufficient reliability, multiphysics fatigue analysis techniques that simultaneously consider electricity, heat, and stress should be developed. In the literature [1-16], a FEM-based lifetime model for Al wire has been derived. Methods for thermal analysis of power devices considering the wire connection layout have been analyzed using 3D FEM simulation, and a FEM-based evaluation method has been presented. The thermal fatigue reliability of power devices has also been described. In one such study, coupled thermal-electrical and thermal-structural analyses were performed to evaluate the fatigue behavior of these devices. Then, crack propagation in the device was evaluated. In the present study, to accurately predict the fatigue properties of power devices, evaluation techniques based on multiphysics analysis are proposed.

Introduction

Power devices consist of parts made of various materials connected by solder. These parts, with different coefficients of thermal expansion, deform by different amounts in response to temperature change, which then results in mechanical stresses. The yield stress of the solder material is generally smaller than that of the other materials in the electronic package. Therefore, the solder joints most often suffer fatigue and fracture. Such damage to the solder is the cumulative effect of high stress loading of the structure over a long period of time, and cyclical loading with small stresses. The initial cracks subsequently progress over time, finally leading to fracture. Material damage caused by deformation is classified into fatigue damage and creep damage. At temperatures much lower than the melting point, fatigue damage is dominant; at temperatures comparable with the melting point, creep damage is dominant. And since power devices generally operate between -55°C and 150°C, the fatigue failure of solder joints occurs via an overlap of the two damage modes.

In this study, to calculate the fatigue properties of power devices quantitatively, an evaluation method that is based on the cumulative damage rule and the Coffin-Manson law is proposed. The fatigue life cycle is calculated in terms of crack length in the solder.

Power device structure

The typical power device structure is shown in Fig. 1.

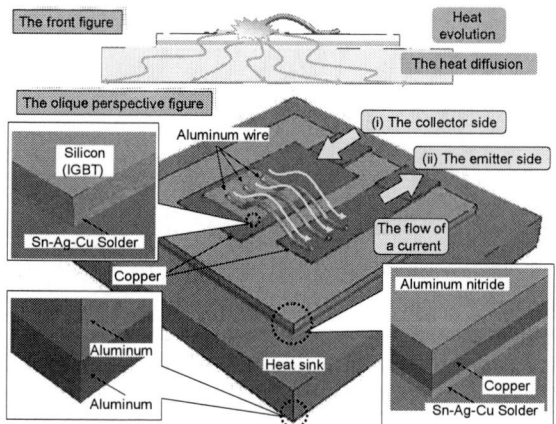

Fig. 1 Power device structure

This structure consists of aluminum wire, copper plate, aluminum nitride plate, silicon (IGBT), and aluminum plate (heat sink). The silicon is attached to the copper plate by the solder, which is an alloyed metal consisting of tin, silver, and copper. The aluminum nitride plate is attached to the aluminum plate through the copper plate. It is necessary to have a structure to absorb the internal stress which is caused by the heat concentration in the structure because of the expansion differences between the materials making up the structure. To prevent this expansion, it is necessary for the structure to release heat from the IGBT. Therefore, the contact area of aluminum nitride is larger than that of the laminated structure between the IGBT and the heat sink plate. In Fig. 1, the boundaries (i) and (ii) are on the collector side

978-1-4244-4658-2/09 $25.00 © 2009 IEEE

and emitter side, respectively. The current flows from the collector side (i) to the emitter side (ii) through the aluminum wires. The current on boundary (i) is the collector current. The voltage difference between (i) and (ii) is the collector-emitter voltage.

Power cycle test

The purpose of this study is to develop a system that can estimate the fatigue life of the power device. In estimating fatigue life, a stress cycle test is repeated until cracks appear; it consists of a test in which stress on the structure is loaded and relieved sequentially. Similarly, a thermal cycle test is repeated until cracks appear; it uses a test in which the temperature rises to a specified temperature, and then is reduced to room temperature. In this study, the power cycle test is employed [9]. By switching the electrical power supply on and off, stress between parts is generated by thermal expansion.

The time cycle of the switching operation (the power cycle) depends on the user, but generally it has time spans from several minutes to several weeks. These electrical power supply time cycles (on and off) can be modeled using the regularity time cycle. These cycles make up the power cycle. By using them, the fatigue life cycle in the power device is estimated by the power cycle test. Tests alternately turning the electricity on and off within these cycles are repeated. The switching spans correspond to a specific cycle (the on and off periods). As shown in Fig. 7, the electricity is turned on until a certain high temperature is reached. After that, the electricity is turned off until the temperature returns to room temperature. These processes together are defined as one cycle. In this study, by using one cycle (the on and off periods), the fatigue life cycle of a power device is estimated, as shown in Fig. 7. The fatigue life is the number of cycles that occur before the solder connection in the power device develops a crack.

Coffin-Manson Law

The fatigue life N is related to the strain range $\Delta\varepsilon$. In the double logarithmic graph, an approximate line correlating N and $\Delta\varepsilon$ is presented by Coffin and Manson. This relation is the Coffin-Manson Law, expressed as follows:

$$\Delta\varepsilon \cdot N^{\alpha} = C \qquad (1)$$

C and α are the constant of proportion and the fatigue coefficient, respectively. In this study, $\alpha = 1.24$ and $C = 52.5$. N is defined as follows:

$$N = 1000\left(\frac{\Delta\varepsilon}{0.01}\right)^{-1.24} \qquad (2)$$

Plastic strain range and creep strain range affect the fatigue life cycle. In this study, $\Delta\varepsilon$ is defined as follows:

$$\Delta\varepsilon = \frac{\Delta\varepsilon_p + \Delta\varepsilon_c}{2} \qquad (3)$$

$\Delta\varepsilon_p$ and $\Delta\varepsilon_c$ represent the plastic strain range and the creep strain range, respectively.

Cumulative damage rule (Miner rule)

In the real environment, steady repeating loads are rare and usually there are intricate changes in the load duration and amplitude. Therefore, it is difficult to quantitatively estimate the material damage. The cumulative damage rule is generally applied in order to approximately estimate the damage. This rule is the evaluation method based on damage superposition. The damage ratio for the range of strain experienced in one cycle is assumed to be $1/N$. The fatigue damage is derived from Eq. (1) as follows:

$$\frac{1}{N} = \left(\frac{\Delta\varepsilon}{C}\right)^{\frac{1}{\alpha}} \qquad (4)$$

Damage D caused by loading n times is assumed to be:

$$D = \frac{n}{N} \qquad (5)$$

These loads are under the same conditions. Cumulative damage caused by different loads is assumed to be as follows:

$$D = \sum_{i}\left\{\frac{n_i}{N_i}\right\} \qquad (6)$$

These loads are under different conditions. This relation is also called Miner's rule. The index i represents the phase of the crack progress as shown in Fig. 4. The phase indicates the time point which is defined by arbitrarily dividing the time span. If the value of damage reaches $D = 1$, a crack occurs in the material. In the case $i = 0$ (the initial condition), the material does not have any damage. The fatigue life cycle is defined as $N_1 = N$ (Eq. (1)).

After the phase index $i = 0$, additional damage in the phase $i = 1$ is caused. Therefore, the equation is as follows:

$$n_i = N_i(1 - D) \qquad (7)$$

The quantity in the parenthesis, $1-D$, indicates the ratio of the damage not yet caused. The variable N_i is obtained by the Coffin-Manson Law (Eq. (1)).

Algorithm

The algorithm is shown in Fig. 5. The analysis algorithm consists of five steps. In the first step, the boundary conditions and the material constants are set in the calculation model. In the second step, the electrical-thermal coupled analysis is performed iteratively. The algorithm determines the temperature data at each time step for all nodes. The power cycle from the start time to the end time is obtained, as shown in Fig. 7. In the third step, the thermal-stress coupled analysis is executed in order to obtain the strains (plastic and creep strains). The algorithm in the third step requires the temperature distribution data. In the fourth step, the number of cycles required to determine the fatigue life is calculated. The fourth step consists of five stages described as follows.

The first stage: In this study, it is assumed that the initial crack occurs at the point of maximum strain. After that, the solder domain replaces the crack domain by a threshold at every phase i. The threshold is decided by Manson-Coffin Low and cumulative damage rule. In order to calculate the crack length, one line is decided by how the crack progress. In the computational domain, crack domains show 3D domains. To decide a crack length (1D), crack domains should be projected to an axis (one line). Therefore, an axis (one line) is set along the crack progress direction after checking the 3D crack domain as shown in Fig. 14.

978-1-4244-4658-2/09 $25.00 © 2009 IEEE

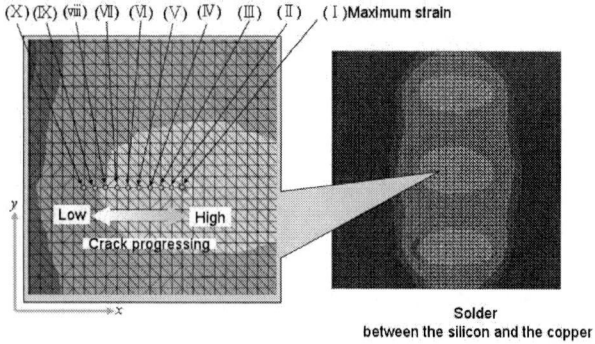

Fig. 2 Strain contours on the solder (see Fig. 1)

The second stage: The fatigue life cycle is calculated by the Coffin-Manson Law. Strain ranges (Eq. (3)) in the Coffin-Manson law are calculated by the finite element method.

The third stage: The damage at the node point (I) (the maximum strain) is defined as $D = 1$. Therefore, the initial crack occurs at the maximum strain point. In each phase i, the fatigue life cycle N at the maximum strain is defined as the variable in Eq. (6). By using N, the fatigue life cycle at the other node point is obtained.

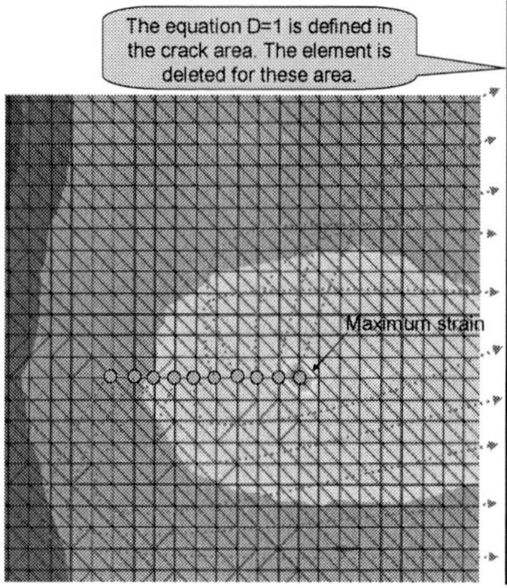

Coordinate	Fatigue Life cycle (Eq (2))	Damage Eq (5)
(I)	2162553	(=2162553/ 2162553)
(II)	3671684	0.59 (=2162553/ 3671684)
(III)	3481152	0.62 (=2162553/ 3481152)
(IV)	3609851	0.60 (=2162553/ 3609851)
(V)	3768876	0.57 (=2162553/ 3768876)
(VI)	3932370	0.55 (=2162553/ 3932370)
(VII)	4538109	0.48 (=2162553/ 4538109)
(VIII)	5920268	0.37 (=2162553/ 5920268)
(IX)	6708234	0.32 (=2162553/ 6708234)
(X)	7119610	0.30 (=2162553/ 7119610)

The equation D=1 is defined in the crack area. The element is deleted for these area.

Maximum strain

Fig. 3 Strain contours in the initial fatigue life cycle ($i=0$) (The third stage)

The fourth stage: At each node, the damage at each phase is added (Eq. (6)).

Tables 1 to 3 show the relation between the fatigue life cycle and the damage. For example, the accumulated damage of the coordinate (II) in Table 2 is calculated as follows:

$$D_{2,(II)} = D_{1,(II)} + \frac{n_{2,(II)}}{N_{2,(II)}} = 0.59 + \frac{9.4 \times 10^5}{22.9 \times 10^5} = 1.00 \quad (8)$$

The parameter $n_{2,(II)}$ is calculated as follows:

$$n_{2,(II)} = N_{2,(II)}\left(1 - D_{1,(II)}\right) = 22.9 \times 10^5 \times (1 - 0.59) = 9.4 \times 10^5 \quad (9)$$

The accumulated damage of the coordinate (III) in Table 3 is calculated as follows

$$D_{3,(III)} = D_{1,(III)} + D_{2,(III)} + \frac{n_{3,(III)}}{N_{3,(III)}} = 0.87 + \frac{3.2 \times 10^5}{24.4 \times 10^5} = 1.00 \quad (10)$$

The parameter $n_{3,(III)}$ is calculated as follows:

$$n_{3,(III)} = N_{3,(III)}\left(1 - D_{1,(III)} - D_{2,(III)}\right) = 24.4 \times 10^5 \times (1 - 0.87) = 3.2 \times 10^5 \quad (11)$$

The fifth stage: The relation between the fatigue life cycle and the crack length is obtained. In this study, the projection of the crack length on the x axis is shown in Fig. 2.

In fifth step, the analysis data (stress, thermal, and electrical distribution) is visualized.

Crack Length	Fatigue Life Cycle	Crack
$0\mu m$	22×10^5	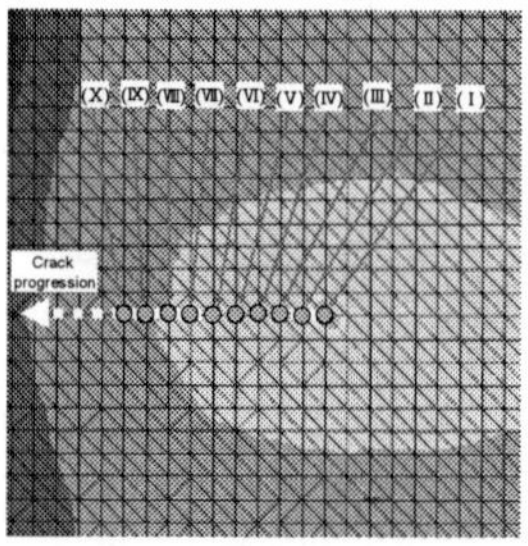 phase $i=1$ (Table 1)
$50\mu m$	23×10^5	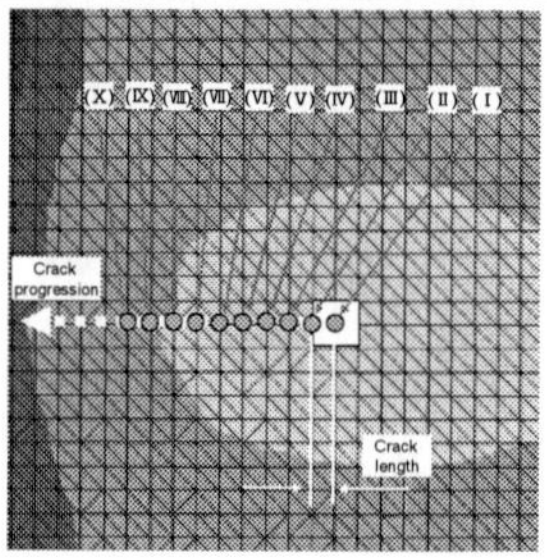 phase $i=2$ (Table 2)
$100\mu m$	77×10^5	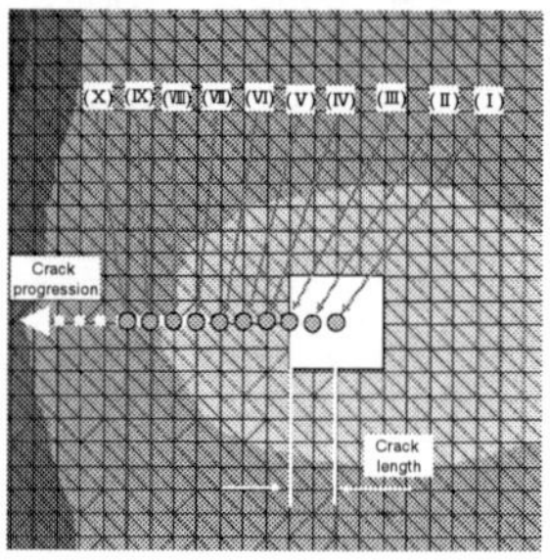 phase $i=3$ (Table 3)

Fig. 4 Crack progress

Table 1 Fatigue life cycle in the phase $i=1$

Coordinate	Fatigue Life Cycle (Eq.(2))	Accumulated damage (Eq.(5))
(I)	2162553	$1.00(=2.1 \times 10^5/2.1 \times 10^5)$
(II)	3671684	$0.59(=2.1 \times 10^5/3.6 \times 10^5)$
(III)	3481152	$0.62(=2.1 \times 10^5/3.4 \times 10^5)$
(IV)	3609851	$0.60(=2.1 \times 10^5/3.6 \times 10^5)$
(V)	3768876	$0.57(=2.1 \times 10^5/3.7 \times 10^5)$
(VI)	3932370	$0.55(=2.1 \times 10^5/3.9 \times 10^5)$
(VII)	4538109	$0.48(=2.1 \times 10^5/4.5 \times 10^5)$
(VIII)	5920268	$0.37(=2.1 \times 10^5/5.9 \times 10^5)$
(IX)	6708234	$0.32(=2.1 \times 10^5/6.7 \times 10^5)$
(X)	7119810	$0.30(=2.1 \times 10^5/7.1 \times 10^5)$

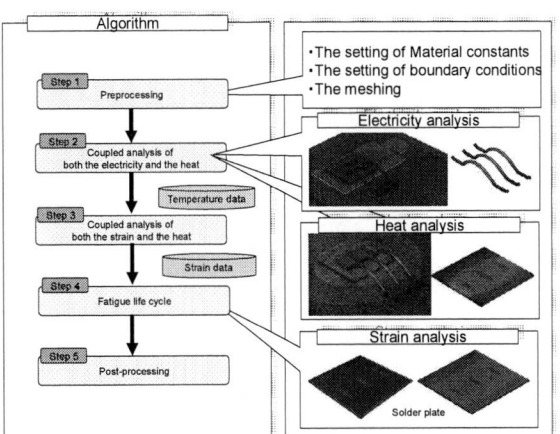

Fig. 5 Algorithm

Table 2 Fatigue life cycle in the phase $i=2$

Coordinate	Fatigue Life Cycle (Eq.(2))	Accumulated damage (Eq.(6))
(I)	Crack domain	Crack domain
(II)	2293763	1.00
(III)	3745064	0.87
(IV)	3600837	0.86
(V)	3756262	0.82
(VI)	3950283	0.79
(VII)	4615109	0.68
(VIII)	5911460	0.52
(IX)	6651597	0.46
(X)	7077491	0.44

Table 3 Fatigue life cycle in the phase $i=3$

Coordinate	Fatigue Life Cycle (Eq.(2))	Accumulated Damage (Eq.(6))
(I)	Crack domain	Crack domain
(II)	Crack domain	Crack domain
(III)	2440093	1.00
(IV)	3821137	0.94
(V)	3728167	0.91
(VI)	3913761	0.87
(VII)	4694502	0.70
(VIII)	5902676	0.58
(IX)	6595819	0.51
(X)	7035625	0.48

Table 4 Material constant

Material (Unit)	Heat Conductivity (W / (mmK))	Electrical conductivity (S/mm)	Density (kg/mm³)	Specific heat (J/(kg·K))	Young's modulus (GPa)	Poisson ratio (Nondimension)	Coefficient of thermal expansion (1/K)
Silicon (IGBT)	0.15	0.5	2.33×10^{-6}	700.0	131.0	0.28	4.2×10^{-6}
Aluminum	0.24	37.7×10^3	2.70×10^{-6}	900.0	70.0	0.30	2.15×10^{-5}
Copper	0.40	59.6×10^3	8.96×10^{-6}	380.0	120.0	0.30	1.70×10^{-5}
Solder (Sn-Ag-Cu)	0.05	9.1×10^3	7.40×10^{-6}	234.0	40.0	0.30	2.30×10^{-5}
Aluminum nitride	0.15	1.0×10^{-15} ($\fallingdotseq 0.0$)	3.40×10^{-6}	710.0	320.0	0.24	4.60×10^{-6}
Aluminum (The bottom)	0.24	37.7×10^3	2.70×10^{-6}	900.0	70.0	0.30	2.15×10^{-5}

(note)1.00 S =1.00 Ω-1 = 1.00 A/V

Calculation conditions

The mesh is shown in Fig. 3. The number of nodes and the number of elements are 36053 and 201219 respectively. To verify the effect of the thickness of the solder on the fatigue cycle, for each calculation, models with solder thicknesses of 12.5, 25, and 50 μm are created. In the electrical analysis, the collector side current on boundary (i) is 10.0 A. The voltage on the emitter side is set to 0.0 V. In the thermal analysis, the temperature on the bottom of the heat sink is set to 25°C (the room temperature) as shown in Fig. 6. In the structure analysis, the displacement with respect to the *x, y,* and *z* axes is fixed at the coordinate ○ in Fig. 6. The material properties used in the calculated model are summarized in Table 1. The time step is set to 0.1s. However, the program automatically subdivides a large time step into several smaller time steps if the algorithm has some problems.

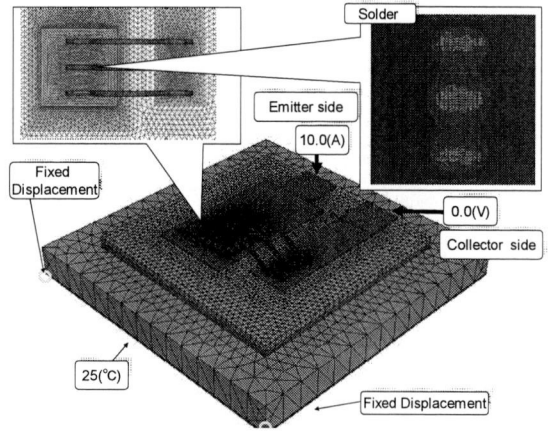

Fig. 6 Power device structure

Power cycle

The power cycle is shown in Fig. 7. The powers on time and off time are set to 2.0s and 18.0s, respectively. We select the long off time so that the temperature can return to room temperature.

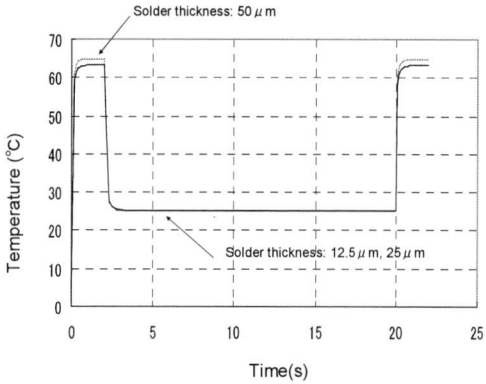

Fig. 7 Temperature cycle for one power cycle

Thermal-electrical analysis

Fig. 8 shows the temperature contour map of the surface of the power device. The contour map of temperature on the solder is shown in Fig. 9. Compared to the corners on the solder, the high temperatures occur below the bottom of the wire bonding (around the center on the solder). The temperature decreases with distance from the source of Joule heating. Based on the temperature distribution, inelastic strain range in the solder joint is calculated by thermal-mechanical analysis.

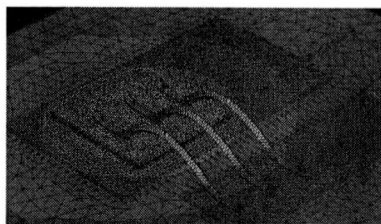

Fig. 8 Temperature distribution

Thermal-structural analysis

Fig. 9 shows the contour map of equivalent creep strain on the solder joint. The strain concentrates on the solder layer below the aluminum bonding wire. The distribution of the equivalent creep strain appears similar to that of temperature. The equivalent creep strain depends on the temperature distribution induced by Joule heating.

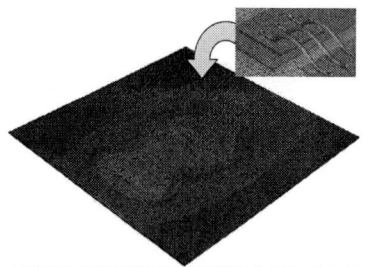

Fig. 9 Equivalent creep strain distribution

Plastic stain and creep strain

The plastic strain and the creep strain at the maximum temperature point are shown in Figs. 10 and 11, respectively. Each of the strains changes drastically when the power cycle (shown in Fig. 7) alternates between the ON and OFF positions.

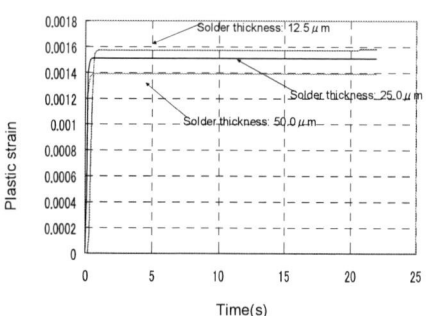

Fig. 10 Plastic strain

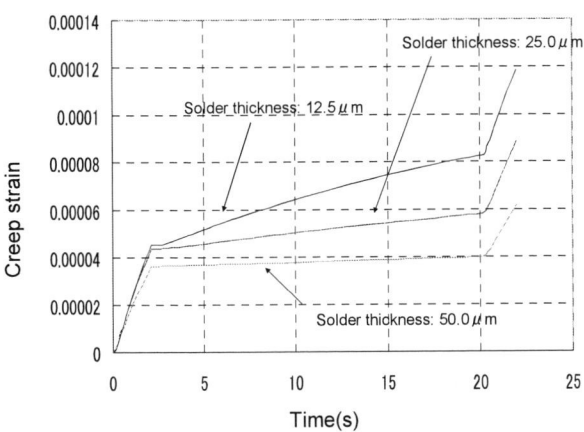

Fig. 11 Creep strain

Crack progress

The behavior of the crack propagation is shown in Figs. 12-14. The ellipsoidal crack initiates on the chip below the aluminum wire bonding and propagates concentrically. On the other hand, no cracks initiate at the corners. The size of crack increases according to the time in the power cycle.

The relation between fatigue life cycle and crack length is shown in Fig. 15. Below crack lengths of 100μm, the thinner the solder is, the greater the progress of the cracks is. When we comparing the model with solder thickness of 25μm with the model of solder thickness 50μm, we find that there is no difference in the fatigue life cycles for cracks of lengths greater than 100μm.

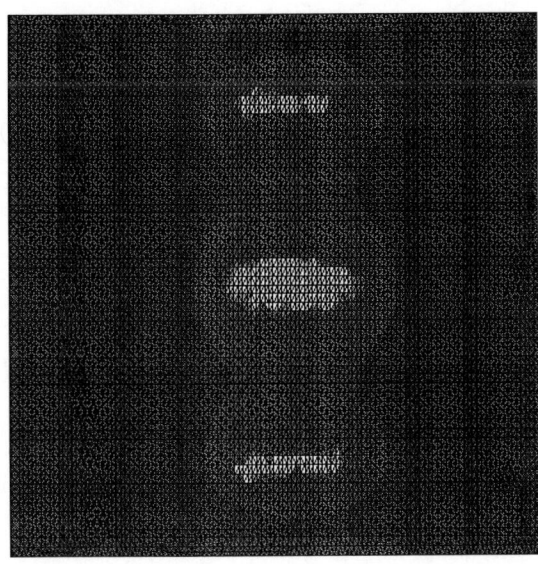

Fig. 12 Crack on the solder (Phase i=1) (the solder thickness is 25μm)

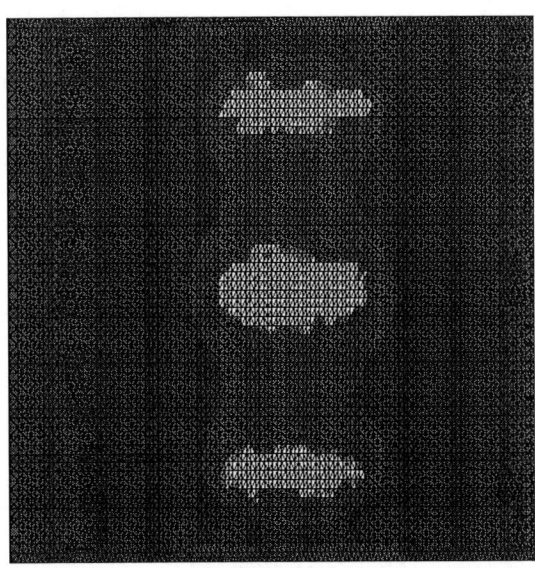

Fig. 13 Crack on the solder (Phase i=3) (the solder thickness is 25μm)

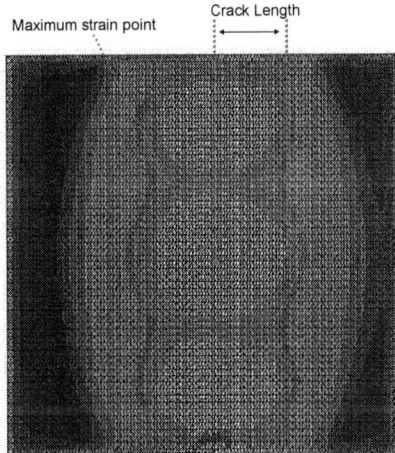

Fig. 14 Crack on the solder (Phase i=9) (the solder thickness is 25μm)

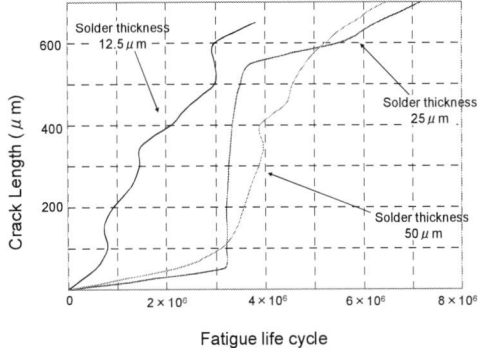

Fig. 15 Fatigue life cycle with respect to crack length

Conclusions

To quantitatively calculate the fatigue life cycle in power devices, an evaluation method is presented that uses the Coffin-Manson Law and the cumulative damage rule. Using this method, coupled thermal-electrical and thermal-structural analyses were performed to evaluate the failure behavior. Then, the crack propagation with respect to the fatigue life cycle was evaluated. The temperature distribution depends on Joule heating. Subsequently, the solder joint below the wire bonded area is subjected to thermal load. Crack path simulation was performed to investigate the destruction mechanism. The crack in the solder joint initiates below the bonded wire due to the concentration of inelastic strain. The crack occurs at the edge of the solder joint due to the mismatch of the coefficients of thermal expansion of the constituent parts. Destruction mechanisms and strain speed affect the fatigue life cycle of the power device. Finally, three power devices with solder thicknesses of 12.5, 25, and 50μm were modeled. It was found that with increasing solder thickness the fatigue life cycle increased.

This method is an important first step in improving the reliability of power devices because it helps in the design of power devices and their environments that limit temperature increases and prevent failure mechanisms such as delamination.

References

1. Baliga B.J., <u>Fundamentals of Power Semiconductor Devices</u>, Springer-Verlag, (2008).

2. Ramminger S., Seliger N. and Wachutka G., "Reliability Model for Al Wire Bonds subjected to Heel Crack Failures," *Microelectronics Reliability*, Vol. 40, (2000) pp. 1521-1525.

3. Ishiko M., Usui M., Ohuchi, T. and Shirai M., "Design concept for wire-bonding reliability improvement by optimizing position in power devices," *Microelectronics Reliability* , Vol. 37, (2006), pp. 262-268.

4. Shinohara K., Yu Q., Anzawa T., Ishii H., "High-Accuracy Fatigue Evaluation of Power Devices by Multi-Coupled Analysis," *Proceedings of IPACK2009*, 89043, (2009).

5. Yi-Ming J., Ying-Lung W. and Chih-Kai F., "Impact of the number of chips on the reliability of the solder balls for wire-bonded stacked-chip ball grid array packages," *Microelectronics and Reliability*, Vol. 46, (2006), pp. 386-399.

6. Huang C.Y., Wu Z.H. and Zhou, D.J., "Solder joint formation simulation and reliability study of quad flat no lead package," *Science and Technology of Welding & Joining*, Vol. 12, (2007), pp. 100-105.

7. Usui O., Muto H. and Kikunaga T., "Evaluation of Temperature Distribution of a Power Semiconductor Chip Using Electrothermal Simulation," *IEEJ Transactions on Industry Applications*, Vol. 124, (2003), pp. 108-115.

8. Takahashi H., Takashi K., and Mukai M., "Thermal Fatigue Life Simulation for Sn-Ag-Cu Lead-Free Solder Joints," *Japan Institute of Electronics Packaging*, Vol. 7,(2004), pp. 308-313.

9. Morozumi A., Yamada K., Miyasaka T., Sumi S. and Seki Y., "Reliability of power cycling for IGBT power semiconductor modules," *Industry Applications IEEE Transactions*, Vol. 39, (2003), pp. 665-671.

10. Ladani, L.J., "Reliability Estimation for Large-Area Solder Joints Using Explicit Modeling of Damage," *Device and Materials Reliability, IEEE Transactions*, Vol. 8, (2008), pp. 375-386.

11. Matsunaga T., Sudo S., "Evaluation of Fatigue Life Reliability and New Lead Bonding Technology for Power Modules," *Mitsubishi Electr Adv*, Vol. 113, (2006), pp. 13-16.

12. Yoshihara K., Ikeda Y., Iizuka Y. and Yamashita M., "A Study of Radiation Effect by Lead Frame Connection for Power Devices," *Proceedings of JIEP(Japan Institute of Electronics Packaging) Annual Meeting*, (2004).

13. Zhuang W.D., Chang P.C., Chou F.Y. and Shiue R.K., "Effect of solder creep on the reliability of large area die attachment," *Microelectronics Reliability*, Vol. 41, (2001), pp. 2011-2021.

14. Ye H., Lin M. and Basaran C., "Failure modes and FEM analysis of power electronic packaging," *Finite Elements in Analysis and Design*, Vol. 38, (2002), pp. 601-612.

15. Nagatomo Yoshiyuki, Nagase Toshiyuki and Shimamura Shoichi, "FEM Analysis of Thermal Cycle Properties of the Substrates for Power Modules," *Journal of Japan Institute of Electronics Packaging*, Vol. 3, (2000), pp. 330-334.

16. Usui M., Tanaka H., Hotta K., Kuwano S., Ishiko M., "Mechanical Stress Dependence of Power Device Electrical Characteristics," *IEEJ(The transactions of the Institute of Electrical Engineers of Japan) Transactions on Industry Applications*, Vol. 128, (2008), pp. 577-583.

Advanced High Density Interconnect Materials and Techniques

J. Wei, S. M. L. Nai, X. F. Ang, K. P. Yung
Singapore Institute of Manufacturing Technology,
71 Nanyang Drive, Singapore, 638075
Email: jwei@SIMTech.a-star.edu.sg, Tel. no.:+65 6793 8575

Abstract

The trend in micro/nanosystems is to be lighter, smaller and cheaper. At the same time there is a prodigious push for increasing functionalities. Such demands can only be fulfilled by progressively higher density integrated devices and circuits. With 2D IC reaching its physical limitation soon, 3D IC has attracted tremendous attentions and interests worldwide. For either 2D or 3D IC, the interconnection and packaging will be one of the major challenges for the development and commercialization of micro/nanosystems. Flip chip at chip level and wafer-level have the advantages of having the lowest possible inductance per lead, highest frequency response speed as well as the lowest cross talk and simultaneous switching noise. Challenges arise when interconnection method of flip chip is gradually growing into the mainstream in the integration and packaging industry where the issue of size becomes increasingly critical for interconnection and pitches. Therefore, the development of new interconnection materials and techniques is necessary to meet the ever-stringent requirements of mechanical, thermal and electrical properties of interconnection when the interconnection dimension and pitch size are reduced to very fine scales. Furthermore, such advanced interconnection materials and techniques can also be used to stack 3D IC. In this paper, the development of novel lead-free solder nanocomposites, room to low temperature Cu-Cu and Au-Au bonding, and carbon nanotube interconnection techniques will be reported. The developed micro/nanointerconnection techniques can be easily adopted by the industry to realize high density and multifunctional integration and packaging.

Introduction

Through the years, as micro/nanosystems technology advanced, the size of electrical components shrank and the number of input/output terminations increased. The trend in micro/nanosystems is to be smaller and lighter, such demands can only be fulfilled by the use of higher density integrated devices and circuits [1-5]. This will lead to the ever-stricter requirements for interconnection materials.

In conventional integrated circuits (ICs), signals transmit through interconnects patterned on a 2D planar design space. In the genesis of integrated circuits, designing on a plane was favored because manufacturing and miniaturizing the planar circuits was easier. All these were made possible by photolithography which eventually became the industry standard production method in the 1960s [4]. Even when more companies were driven by Moore's Law to make bigger chips at high yield to meet the demand for higher circuit density on the 2D space, putting more metal lines at finer dimensions for interconnection on the planar substrate can still be achieved by the progress from conventional lithography process to optical projection. Moore's Law

predicts a doubling of transistors integrated per microchip every two years. While researchers are still actively developing technologies to squeeze more circuits on the planar substrate, the continued demand for higher circuit density and further miniaturization forces conventional interconnection methods to their limits. As further scaling in feature size takes place to improve service and functional performance, the closely packed interconnects with smaller cross-sections and pitch, degrades signal integrity by increasing unwanted resistance, capacitance and inductance that inhibit propagation of signals [5].

Advances in micro/nanosystems technology can no longer guarantee the reliability of conventional interconnection materials. The development of electrical components becomes interconnection-constrained. Therefore to alleviate the existing interconnection-related issues, one of the possible solutions would be by three-dimensional integration which connects different functional layers in the vertical dimension.

Moreover, in order to fulfill the ever-stricter service and functional requirements, new interconnection materials equipped with improved material's properties need to be developed. As the trend progresses towards 3D integration, it is also essential to devise new innovative bonding techniques to form reliable interconnects between multiple stacked layers.

Accordingly in this study, (i) the development of novel lead-free solder nanocomposites, (ii) room to low temperature direct metal (Cu-Cu and Au-Au) bonding, and (iii) carbon nanotube interconnection techniques will be reported.

Lead-free Solder Nanocomposites

In recent years, with increasing functional requirement and with miniaturization of electronic components, conventional solder technology can no longer guarantee device reliability. Devices are shrinking in size and the spacing between the solder interconnect joints is getting closer. This will lead to the possibility that even small positional shifts in the joints would result in mechanical and electrical failures. Furthermore, solders in general operate at high homologous temperatures. This result in the solder interconnects being loaded under conditions where they can be susceptible to creep. In addition, as the electronic device is turned on and off, it undergoes thermal cycling and this result in thermo-mechanical fatigue due to thermally induced stresses that develop as a consequence of the coefficient of thermal expansion (CTE) mismatch between the solder and the various components in the device. Hence, to meet the ever-demanding challenges, a new generation of solders such as composite solders has been developed and it has received considerable attention in recent years. Composite solders are solders with intentionally incorporated reinforcements. Studies by several researchers have shown that by introducing second phases to a conventional solder alloy, the overall performance in the end applications can be convincingly improved [6-16]. The presence of these second phases has been proposed as a potential mechanism to control the reliability of the solder joints. The second phase fillers chosen in this study is carbon nanotube (CNT). CNTs have been the subject of increasing scientific and research interest ever since its discovery in 1991 [17], due to its unique physical, electrical, mechanical and thermal properties [17-19]. CNTs possess almost five times elastic modulus (~ 1 TPa) and 100 times tensile strength (~ 150 GPa) than those of high-strength steel alloys [20-22]. In view of this, in recent years several investigators have introduced CNTs as reinforcement in polymer, metal and ceramic matrices, in an attempt to overcome the performance limits of conventional materials [23-26].

In this study, varying amount of multi-walled carbon nanotubes were incorporated into commercially 95.8 Sn – 3.5 Ag – 0.7 Cu lead-free solder, using the powder metallurgy technique. Characterization studies were conducted to determine the physical, thermal, microstructural and mechanical properties of these nanocomposite solders.

The results of the melting point test (using the Differential Scanning Calorimeter) showed that there was no significant change in the melting point of the nanocomposite solder, despite the addition of CNTs in the solder. The melting point (T_m) of the materials ranged from 220.9 to 221.9 °C (see Table 1). This implied that when using such nanocomposite solders as interconnects, there is no need to make any modification to the existing soldering processes. Furthermore, with the incorporation of increasing amount of CNTs, the density of the solder samples exhibited a decreasing trend. This could be due to the much lower density value of CNTs (upperbound density = 2.6 g/cm^3), as compared to that of SnAgCu solder (ρ_{SnAgCu} = 7.44 g/cm^3).

Table 1. Melting point and density results of unreinforced and nanocomposite solders.

Material	MWCNTs (wt.%)	T_m (°C)	Density (g/cm^3)
SAC	-	221.9	7.440 ± 0.070
SAC-1CNT	0.01	221.2	7.371 ± 0.010
SAC-4CNT	0.04	220.9	7.364 ± 0.004
SAC-7CNT	0.07	221.4	7.337 ± 0.009

Wettability between the solder and substrate is an important issue in reliability of electronic packaging [1, 27]. It is generally described by the contact angle (θ) to the substrate and it has been accepted that the smaller the contact angle, the better the wettability. In this study, the wettability results revealed that addition of CNTs in solder improved the wetting behavior (see Figure 1). With the addition of 0.04 wt. % of CNT and 0.07 wt. % of CNT, the contact angles were decreased by 15.7% and 19.8% respectively.

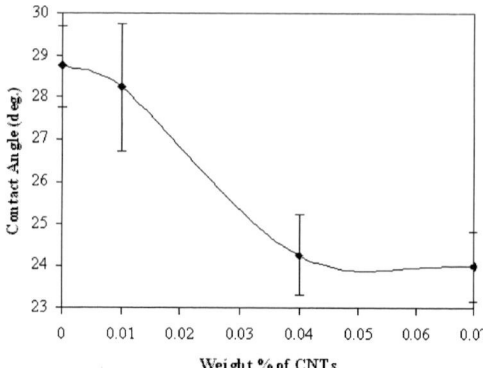

Fig. 1 Graphical relationship between amount of CNTs in the nanocomposite solders and their corresponding contact angles.

The CTE results of the composite solders showed reduced CTE values as compared to that of the unreinforced solder (see Table 2). This could be attributed to: (i) the judicious selection of reinforcement (CNTs), which has a much lower CTE value (CTE_{CNTs} = -5.86 x 10^{-9}/°C) than that of SnAgCu solder (CTE_{SnAgCu} = 22.9 x 10^{-6}/°C) and (ii) the ability of well bonded CNTs to effectively constrain the expansion of SnAgCu matrix. It may be noted that a reduction in CTE values for the composite solders, assists in reducing the thermal mismatch between such interconnection materials and device materials, or between interconnection materials and substrate material, due to less induced thermal stresses [28].

Table 2. CTE and tensile results of unreinforced and nanocomposite solders.

Material	CTE (x 10^{-6}/°C)	0.2% YS (MPa)	UTS (MPa)	Ductility (%)
SAC	22.9	31 ± 2	35 ± 1	41 ± 8
SAC-1CNT	19.8	36 ± 2	47 ± 1	36 ± 2
SAC-4CNT	19.3	36 ± 4	46 ± 6	37 ± 5
SAC-7CNT	19.3	33 ± 3	43 ± 5	35 ± 4

Microstructural analysis revealed a fair degree of uniformly distributed CNT clusters in the solder matrix. The length of the CNTs used in the present study is < 100 μm. Their high aspect ratio and the strong van der Waals forces cause the CNTs to attract one another [26,29,30]. Thus CNTs have a tendency to entangle together, resulting in CNT clustering rather than homogenous dispersion of individual CNTs. However, despite the clustering of CNTs, tensile tests results of the nanocomposite solders showed an improvement in 0.2% yield strength (0.2% YS) and ultimate tensile strength (UTS) over that of their unreinforced counterpart (refer to Table 2). 0.2% YS improved upto ~16% while UTS improved upto ~34%. However, beyond 0.04 wt.% of CNTs addition, a decreasing trend in mechanical properties was observed.

The improvement in strength of the nanocomposite solders can be attributed to: (a) the progressive increase in dislocation density due to coefficient of thermal expansion (CTE) mismatch ($\Delta\sigma_{CTE}$) [31] between SnAgCu and CNT, and (b) the elastic modulus (EM) mismatch ($\Delta\sigma_{EM}$) between solder matrix and CNT. The strength of a reinforced matrix can be defined by [31]:

$$\sigma_{my} = \sigma_{mo} + \Delta\sigma \qquad (1)$$

where σ_{my} and σ_{mo} are the yield strength of the reinforced and unreinforced matrix, respectively. $\Delta\sigma$ represents the total increment in yield stress of the reinforced SnAgCu matrix and is estimated by [32]:

$$\Delta\sigma = \sqrt{\left(\Delta\sigma_{CTE}\right)^2 + \left(\Delta\sigma_{EM}\right)^2} \qquad (2)$$

The presence of CNTs as reinforcements (Figure 2) acts as obstacles to restrict the initiation of dislocation motion in the solder matrix. Hence, a higher initial stress is necessary.

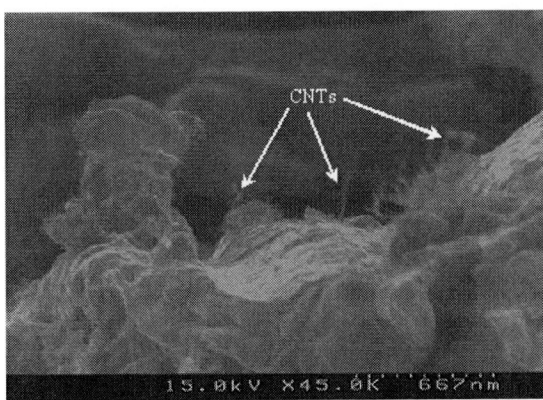

Fig. 2 Typical micrograph showing presence of CNTs on the fracture surface of nanocomposite solders.

It was also observed that the improvements in tensile properties were less significant at 0.04 wt.% and 0.07 wt.% of CNTs. This could be associated to the higher level of microporosity observed at higher weight percentage of CNTs. For the nanocomposite solders, bonding occurs only between: (i) the Sn-Ag-Cu solder and Sn-Ag-Cu solder particles and (ii) Sn-Ag-Cu solder particles and CNTs. Thus, with the addition of increasing amount of CNTs in the solder matrix, there will be some areas present in the solder matrix where CNTs come into contact with each other rather than with the solder particles. This is due to strong van der Waals forces between the CNTs, which resulted in mutual attraction of the nanotubes [26,29,30]. Small clusters of CNTs were formed and this consequently hindered effective bonding between the CNTs and the solder particles, resulting in cluster related porosity. Porosity acted as potential stress concentration sites that were favored for the formation of micro-cracking, which intensify failure. Existing literature [33] also reported that porosity at microscopic level could be detrimental to material's strength. Hence, the superior mechanical properties of CNTs cannot be fully realized in the synthesized nanocomposite solders due to the weak interfacial bonding

between the CNT and the solder material. This thus further limits the strength improvement of the nanocomposite solders with increasing addition of CNTs (more than 0.04 wt. % CNTs, see Table 2).

Ductility of the nanocomposite solders was observed to decrease slightly (taking into consideration the standard deviation values) with increasing amount of CNTs in the solder matrix (see Table 2). This could be attributed to the presence of CNTs serving as crack nucleation sites, leading to decreasing ductility under tensile loading conditions. This observation is also reported by several investigators working on other composite systems [34, 35].

For a low melting point solder material such as SnAgCu, its creep behavior is important as it is used at high homologous temperature. The creep results showed that the nanocomposite solders exhibited significantly improved creep resistance and the creep time to failure also increased. As shown in Figure 3, the steady-state creep rates of SAC-1CNT solder and SAC-4CNT solder were reduced by approximately 1 order of magnitude and 2 orders of magnitude respectively, as compared to that of the unreinforced SAC solder for the three test conditions. The improvement in creep resistance can be attributed to the presence of CNTs which aid to resist the motion of the dislocations [36,37]. For all test load conditions, a similar trend of decreasing creep rate with increasing weight percentage of CNTs addition was observed.

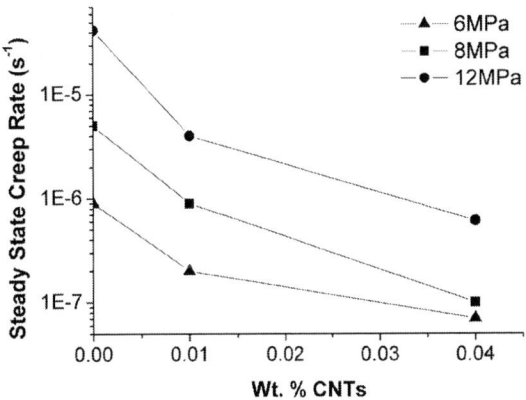

Fig. 3 Graph showing steady-state creep rates for unreinforced and nanocomposite solder joints, subjected to 6 MPa, 8 MPa and 12 MPa loads at 25 °C.

In essence, the results convincingly established that composite technology coupled with nanotechnology in electronic solders can lead to improvement in the solder material's properties. These advanced nanocomposite solders will thus benefit the microelectronics assembly and packaging industries.

Room to Low Temperature Direct Metal Bonding

In the earlier studies, gold has been the preferred material for thermocompression as it is known to be oxidation-free. It offers the conductive bonds using relatively low temperatures and pressures but it is limited by the organic contaminants present on the surface during bonding that will affect the mechanical strength of the bonds formed [38]. Due to the presence of these adventitious contaminants, bonding temperature required to form strong bonds ranges from 150°C to 620°C for gold ball bonding to gold-plated leads and gold metallization [39,40]. Similarly, forming good quality Cu bonds is not as easy as compared to other interconnection materials such as gold. The ease of oxidation of copper often entails a need for high bonding temperature (> 300°C) and/or ultrahigh vacuum conditions. A post-annealing process at temperature equivalent to or higher to the bonding temperature under inert atmosphere over an adequate amount of time is also reported to help in enhancing bond integrity [41,42].

Bond quality of the copper joints is governed by bonding parameters such as temperature, pressure and time [43]. As mentioned above, the highly reactive copper surface thus requires extreme caution in the process of both wet and dry cleaning to remove copper oxide prior to bonding. Environmental control during preparation and sample transfer between cleaning and actual bonding becomes important too, since often, a time lag between sample preparation and bonding leads to an inevitable formation of surface oxide, which necessitates more stressful bonding conditions for successful bonding.

In our previous study [44,45], feasibility studies on nanostructured organic coatings [NSOCs]-assisted direct gold thermocompression bonding were evaluated. The versatility of NSOCs due to the ease of modifying its chemical structure such as thickness, anchoring group to attach or adhere onto the substrate as well as end group which governs the surface properties, necessitates further work in exploring how thickness of the NSOC would influence the bondability of the gold joints. It was found that increasing thickness (NSOC 3) did not result in a significant improvement in bond integrity over a range of bonding temperature, or equivalently, higher amount of bonding temperature is required to achieve superior joint strength as compared to the uncoated gold joints, shown in Figure 4.

The results obtained provide interesting insights on the importance of surface passivation prior to bonding and influence of mechanical displacement at the bonding interface during bonding in creating a strong bond. While NSOCs showed superior surface passivating characteristic, it takes more effort in displacing them in order to expose bare gold region for bonding to take place, shown in Figure 5. The compression action was simulated at room temperature and surface chemical analysis using XPS revealed thicker NSOC i.e. NSOC 3 are able to retain more of its chemical constituents (main structure and surface anchoring group) as compared to the shorter NSOCs.

Successful bonding is defined when freshly exposed gold surfaces come in close contact and bond under the influence of temperature and/or pressure applied. Hence the displacement of any interlay present at the bonding interface is essentially important. For NSOCs-assisted direct gold bonding, it is apparent that the ease of mechanical displacement of the NSOCs layer dominates the bonding behaviour at the gold joints.

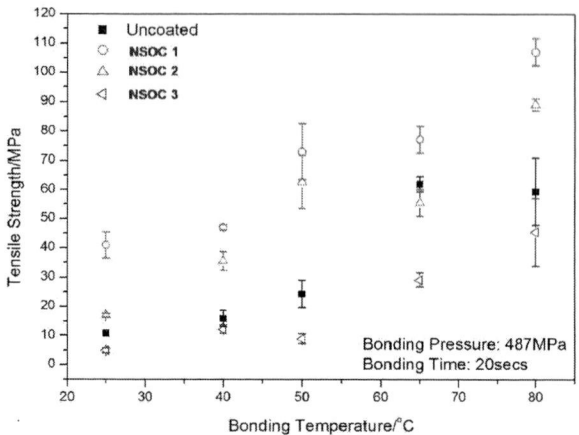

Fig. 4 Influence of NSOCs at various thicknesses (1, 2 and 3) on the tensile strength of the gold joints. NSOC 1 showed greatest strength enhancement even at room temperature.

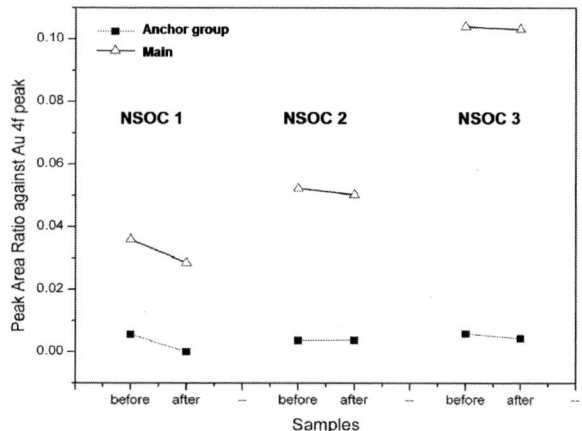

Fig. 5 Simulated bonding action on coated gold samples was carried out. The amount of anchoring group and main constituent left after normal compression was evaluated using X-ray Photoelectron Spectroscopy (XPS).

With many studies examining the influence of surface morphologies and bonding conditions on direct copper bonding over a range of sufficiently high bonding temperature and pressure, it is commonly understood that higher temperature aids strength enhancement. The role of bonding temperature in influencing the bonding from room temperature to 300°C was explored at a sufficient amount of bonding pressure. It was found that an unexpected behavior where an increase in bonding temperature beyond 80°C till 140°C does not improve bond strength (refer to Figure 6). An attempt was made to explain this drop in bond strength with temperature by a self-limiting role of temperature. Similarly, it was demonstrated that such degradation in strength can be avoided by applying NSOC onto the copper surfaces prior to bonding. By simply coating the copper surface with NSOC prior to bonding, the Cu joints can be successfully formed at close to ambient temperature without a vacuum, yielding joint shear strengths on the order of about 70MPa (see Figure 7). It is believed that the densely packed organic coating serves to passivate the copper surface against oxidation under ambient conditions. The ultrathin organic monolayer structure, as compared to a bulk oxide layer, could be easily displaced during the mechanical deformation at the bonding interface which accompanies thermocompression.

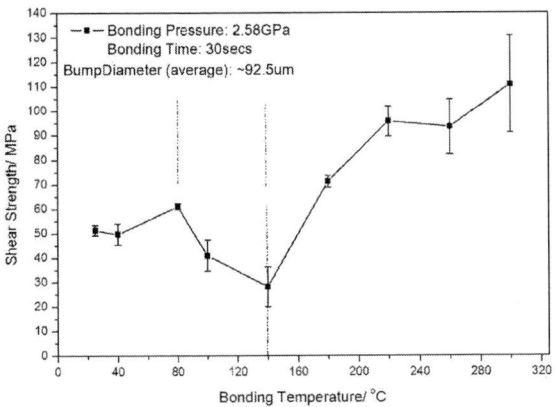

Fig. 6 Shear Strength of uncoated copper from 25°C to 300 °C at 2.58GPa for 30s.

Recently, our study demonstrated successful bonding between two copper surfaces under ambient condition at a bonding temperature below 180°C, yielding typical joint shear strength of 70MPa. Further studies were conducted to correlate the ability of NSOCs as a barrier layer on copper surfaces against surface oxidation to its assistance in enhancing bondability between copper surfaces at temperatures up to 180°C. In Figure 7, NSOCs–coated copper surfaces were found to improve bonding strength under equivalent bonding conditions (at a bonding temperature range of 60°C – 140°C) as compared to freshly clean copper, suggesting that much lower bonding temperature and pressure is possible for copper-copper bonding using NSOCs under ambient laboratory conditions. As discussed, the key consideration in enabling room temperature copper bonding is the need to resolve the issue of surface oxidation – either by passivating copper surfaces or providing means of oxide removal prior to bonding. It was speculated that the primary contribution by NSOCs in lowering bonding temperature and pressure is its ability to passivate copper surfaces very well prior to bonding. While air exposure is unavoidable since bonding is carried out in ambient, bond strength of coated copper samples remains high from 80°C to 140°C. This deviates from what was observed in uncoated copper samples, suggesting that closely packed NSOCs layer could protect copper surfaces even as bonding temperature increases to 140°C.

Our previous result revealed the capability of nanostructured organic coatings-NSOCs in reducing the bonding temperature needed to bond copper surfaces from 300°C to 60°C. Our recent work demonstrates room temperature copper bonding successfully with the help of the organic layers. Further investigation is made to evaluate the influence of thickness of NSOCs to alleviate bonding

temperature. It was found that all NSOCs (1, 2, 3) showed superior bond strength (>25MPa) as compared to that of the uncoated copper (<23MPa) at bonding temperatures from 25°C to 80°C. Since it is imperative that surface oxide has to be removed for bonding to take place, the enhancement exhibited is attributed to an effective surface passivation by the organic layer. It is postulated that this ultrathin layer, which behaves as a milder layer as compared to the bulk oxide layer, can be easily displaced for bond formation.

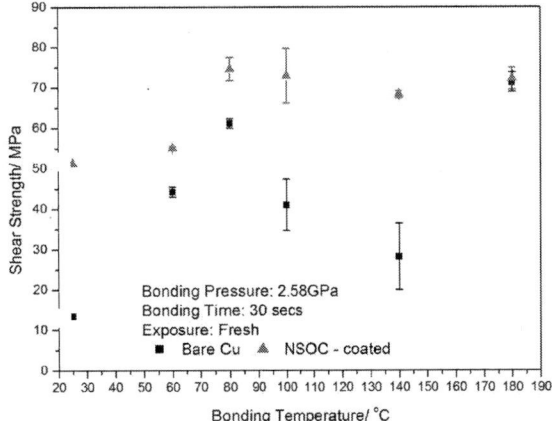

Fig. 7 Comparison of shear strength in coated copper against uncoated copper from 60°C to 180°C.

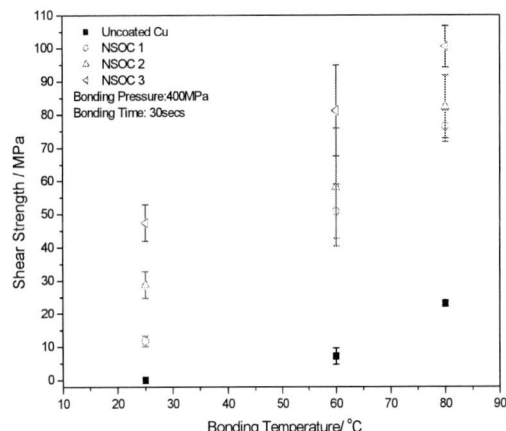

Fig. 8 Significant improvement in shear strength of Cu joints against uncoated joints.

Nano Interconnection Material and Technique

CNT is a novel material [17] which exhibits exceptional electrical, thermal and mechanical properties [46-48]. CNTs have been compared with equivalent gold wires on their electrical performance. For example, the nanotubes with smaller diameter (8.6 nm) have lower resistance than Au wire whereas nanotubes with larger diameter (15.3 nm) have higher contact resistance than Au wire. Nanotubes have no degradation after a period of 350 hours at current density exceeding 10^{10} A/cm^2. The great potential of this material prompted researchers from leading electronics manufacturers, such as NEC and Intel [49,50], to develop new on-chip interconnect technology using the CNTs. Despite the large research effort to use CNTs as the future interconnect material [51,52], there still exist fundamental issues which need to be resolved before CNT interconnects can be implemented. When using CNTs as interconnect, CNTs have to bridge at least two terminals with low electric contact resistance. However, carbon nanostructure assembly has proven to be difficult as its size in nano-scale hinders direct bonding. The available developed techniques to assemble such nanostructures involved high cost manual micromechanical methods and chemical processes which damage the CNT structures [53-57]. Although hypothetically possible, large scale assembly of CNT-based interconnect network using these techniques is not ideal. Therefore, innovations in fabrication methodology and new architectures of CNT assembly are required for the widespread technological application of CNT interconnect technology.

In this study, a new CNT joining method for CNT bump was developed. Silicon substrates were used to grow the CNT bump. A layer of thermal oxide which acts as the diffusion barrier layer, was grown on the Si substrate. A Ti/Cu/Cr metallization layer was then sputtered on the Si substrate for the electrode circuitries. Following this, the metallization layer was patterned with circuit mapping using the lithography technique, and etched accordingly. The Ni catalyst was subsequently sputtered and patterned by lift-off with openings on the bond pads. CNTs were then grown on the bonding site using the PECVD system. Figure 9 shows the morphology of the as-grown CNT.

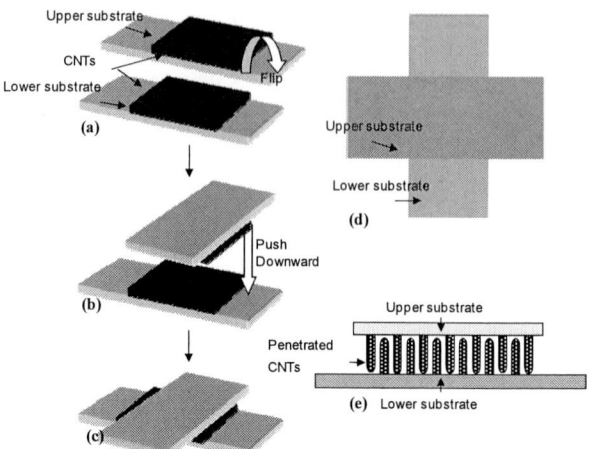

Fig. 10 Illustration of CNT insertion process sequence (a-c), top view of the assembled sample (d), and cross-sectional view of the assembled sample (e).

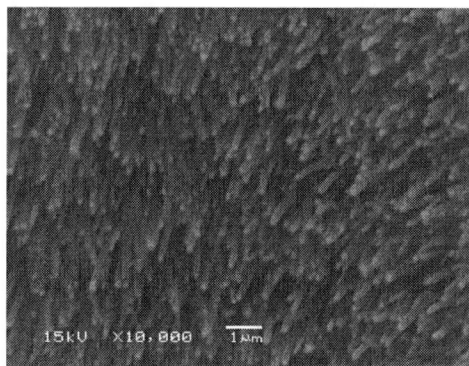

Fig. 9 High density and aligned CNTs.

After the CNT growth process, the nanostructure was assembled using an easy and practical method which involved the "insertion" of the carbon nanotube bumps grown directly on one surface of the substrate into another similar CNT bump surface (see Figure 10). Bonding was done at a controlled bonding temperature, with applied pressure on the backside of the top substrate for a fixed duration. The pressure applied allowed the 'insertion' of CNT bumps into each other. This thus creates conductive pathways between the two components as shown in Figure 10. Figure 11 shows the cross-sectional view of the CNT structures after the insertion assembly, under different applied pressures.

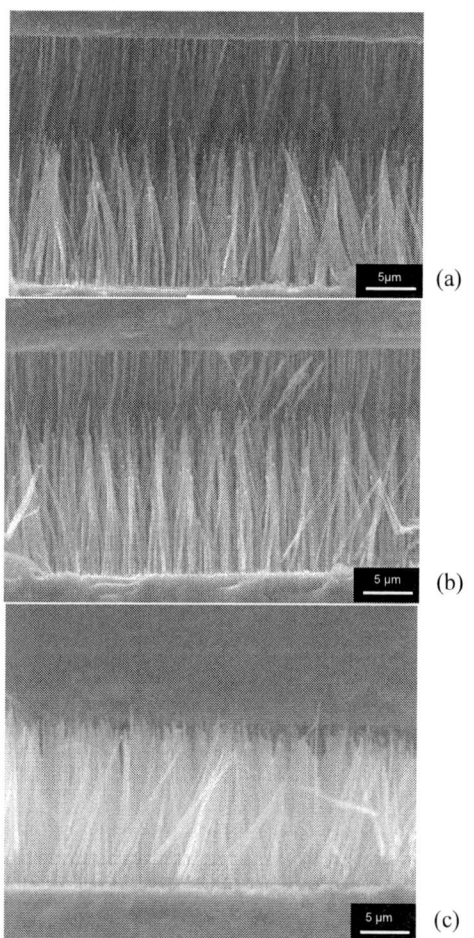

Fig. 11 SEM images showing the cross-sectional view of the CNT structures after the insertion assembly under applied pressure of: (a) 1 kgf/cm^2, (b) 2 kgf/cm^2 and (c) 4 kgf/cm^2.

Using the probe station, the I-V measurement tests were conducted on the assembled structure. The conductivity of the

CNT bump was found to be several magnitudes higher than the conductivity of silver paste as shown in Figure 12. These measurements revealed that CNT bumps have superior electrical performance over the widely used silver paste material. This developed CNT interconnection technique has great potential to be used for flip chip interconnect as shown in Figure 13.

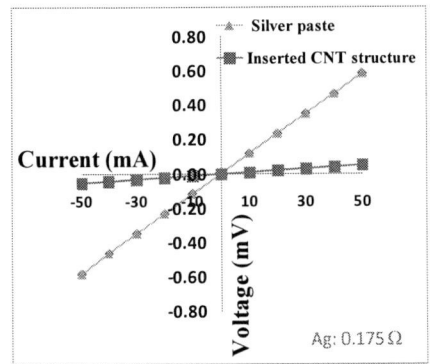

Fig. 12 I-V measurement of the fully inserted CNT structure and silver paste.

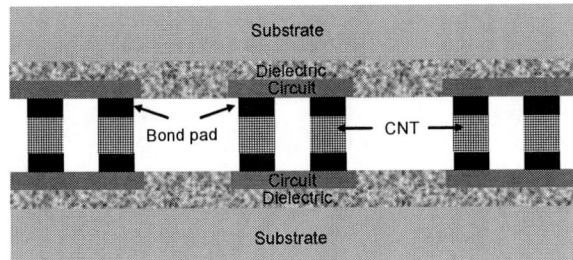

Fig. 13 Schematic diagram showing the CNT interconnection.

Conclusion

In this paper, the development of novel lead-free solder nanocomposites, room to low temperature Cu-Cu and Au-Au bonding, and carbon nanotube interconnection techniques has been reported. The developed new interconnection materials and techniques are necessary to meet the ever-stringent requirements of mechanical, thermal and electrical properties of interconnection when the interconnection dimension and pitch size are reduced to very fine scales. Furthermore, such advanced interconnection materials and techniques will not only be used for high density 2D IC but also for 3D IC stacking.

The developed micro/nanointerconnection techniques can also be easily adopted by the industry to realize high density and multifunctional integration and packaging, and push forward the advancement of various micro/nanosystems.

References

1. M. Abtew and G. Selvaduray, *Mater. Sci. Eng.: R: Reports,* vol. 27, (2000), p. 95.

2. S. F. Al-sarawi, D. Abbott and P. D. Franzon, *IEEE Trans. Comp., Packag., Manufact. Technol.,* B21, (1998), p. 2.

3. *International Technology Roadmap for Semiconductor* (2008).

4. R. R. Schaller, *Spectrum, IEEE,* vol. 34, (1997), pp. 52-59.

5. K. Banerjee, S. J. Souri, P. Kapur and K. C. Saraswat, *Proceedings of the IEEE,* vol. 89, (2001), pp. 602-633.

6. D. C. Lin, G. X. Wang, T. S. Srivatsan, Meslet Al-Hajri and M. Petraroli, *Mater. Lett.,* vol. 57, no. 21, (2003), p. 3193.

7. J. L. Marshall, J. Calderon, J. Sees, G. Lucey and J. S. Hwang, *IEEE Trans. Components, Hybrids and Manufacturing Technology,* vol. 14, no. 4, (1991), p. 698.

8. H. Mavoori and S. Jin, *JOM – Journal of the Minerals Metals & Materials Society,* vol. 52, no. 6, (2000), p. 30.

9. L. Wang, D. Q. Yu, S. Q. Han, H. T. Ma and H. P. Xie, *International Conference on the Business of Electronic Product Reliability and Liability,* vol. 50, (2004).

10. S. M. L. Nai, J. Wei and M. Gupta, *J. Electron. Mater.,* vol. 35, no. 7, (2006), pp. 1518-1522.

11. S. M. L. Nai, J. Wei and M. Gupta, Proceedings of ASME IMECE 2006, *International Mechanical Engineering Congress and Exposition,* Chicago, Illinois, USA, Nov 2006.

12. S. M. L. Nai, J. Wei and M. Gupta, *Thin Solid Films,* vol. 504, no. 1 – 2, (2006), pp. 401-404.

13. S. M. L. Nai, J. Wei and M. Gupta, *Solid State Phenomena,* vol. 111, (2006), pp. 59-62.

14. P. Babaghorbani, S. M. L. Nai and M. Gupta, *J. Mater. Sci.: Mater. in Electronics,* vol. 20, (2009), pp. 571-576.

15. M. E. Alam, S. M. L. Nai and M. Gupta, *J. Alloys and Compounds,* vol. 476, no. 1 – 2, (2009), pp. 199-206.

16. S. M. L. Nai, J. Wei and M. Gupta, *Journal of Alloys and Compounds,* vol. 473, no. 1 – 2, (2009), pp. 100-106.

17. S. Iijima, *Nature,* vol. 354, (1991), p. 56.

18. M. F. Yu, B. S. Files, S. Arepalli and R. S. Ruoff, *Phys. Rev. Lett.,* vol. 84, (2000), p. 5552.

19. H. Dai, *Surf. Sci.,* vol. 500, (2002), p. 218.

20. K. T. Lau and D. Hui, *Compos. Part B: Eng.,* vol. 33, (2002), p. 263.

21. R. B. Pipes and P. Hubert, *Compos. Sci. Technol.,* vol. 62, (2002), p. 419.

22. J. S. Delmotee and A. Rubio, *Carbon,* vol. 40, (2002), p. 1729.

23. L. Valentini, J. Biagiotti, J. M. Kenny and S. Santucci, *Compos. Sci. Technol.,* vol. 63, (2003), p. 1149.

24. C. Balazsi, Z. Shen, Z. Konya, Z. Kasztovszky, F. Weber, Z. Vertesy, L. P. Biro, I. Kiricsi and P. Arato, *Compos. Sci. Technol.,* vol. 65, no. 5, (2005), p. 727.

25. A. Mamedov, N. A. Kotov, M. Prato, D. M. Guldi, J. P. Wicksted and A. Hirsch, *Nat. Mater.,* vol. 1, (2002), p. 190.

26. S. I. Cha, K. T. Kim, S. N. Arshad, C. B. Mo and S. H. Hong, *Adv. Mater.,* vol. 17, (2005) p. 1377.

27. K. N. Tu and K. Zeng, *Mater. Sci and Eng. R: Reports,* vol. 34, no. 1, (2002), p. 1.

28. Dutta, A. Gopinath and C. Marshall, *J. Electron. Mater.,* vol. 31, no. 4, (2002), pp. 253-264.

29. Szleifer and R. Yerushalmi-Rozen, *Polymer,* vol. 46, no. 19, (2005), pp. 7803-7818.

30. D. W. Coffin, L. A. Carlsson and R. B. Pipes, *Compos. Sci. Tech.,* vol. 66, no. 9, (2006) pp. 1132 -1140.

31. L. H. Dai, Z. Ling and Y. L. Bai, *Compos. Sci. Technol.,* vol. 61, (2002), p. 1057.

32. T. W. Clyne and P. J. Withers, *An Introduction to Metal Matrix Composites,* Cambridge University Press, 1993.

33. H. Mavoori and S. Jin, *J. Electron. Mater.,* vol. 27, no. 11, (1998), p. 1216.

34. X. L. Zhong and M. Gupta, *Adv. Eng. Mater.,* vol. 7, no. 11, (2005), p. 1049.

35. M. Gupta and T. S. Srivatsan, *Mater. Letters,* vol. 51, (2001), p. 255.

36. F. G. Yost, F. M. Hosking and D. R. Frear, *The Mechanics of Solder Alloy Wetting and Spreading,* 1993, New York, Van Nostrand Reinhold.

37. F. Guo F, J. Lee, J. P. Lucas, K. N. Subramanian and T. R. Bieler, J. Electron. *Mater.,* vol. 30, no. 9, (2001), pp. 1222-1227 .

38. G. E. McGuire, J. V. Jones and H. J. Dowell, *Thin Solid Films,* vol. 45, (1977), pp. 59-68.

39. A. T. English and J. L. Hokanson, *Reliability Physics Symposium,* 1971. 9th Annual, 1971, pp. 178-186.

40. H. Ramsey, *Solid State Technology,* (1973), pp. 43-47.

41. K. N. Chen, C. S. Tan, *A. Fan and R. Reif, Electrochem. Solid-State Lett.,* vol. 7, (2004), p. G14.

42. K. N. Chen, A. Fan, C. S. Tan and R. Reif, *J. Electron Mat,* vol. 32, (2003), pp. 1371 -1374.

43. K. N. Chen, S. M. Chang, L. C. Shen and R. Reif, *J. Electron Mat,* vol. 35, (20065), pp. 1082-1086.

44. L. C. Chin, X. F. Ang, J. Wei, Z. Chen and C. C. Wong, *Thin Solid Films,* vol. 504, (2006), pp. 367-370.

45. X. F. Ang, F. Y. Li, W. L. Tan, Z. Chen, C. C. Wong and J. Wei, *Appl. Phys. Lett.,* vol. 91, (2007), p. 061913.

46. Q. Wei, R. Vajtai and P. M. Ajayan, *Appl. Phys. Lett.,* vol. 79, (2001), p. 1172.

47. M. A. Osman and D. Srivastava, *Nanotech.,* vol. 12, (2001), p. 21.

48. R. L. Jacobsen, T.M. Tritt, J.R. Guth, A.C. Ehrlich and D.J. Gillespie, *Carbon,* vol. 33, (1995), p. 1217.

49. M. Ishida, T. Ichihashi, J. Fujita, F. Nihey and Y. Ochiai, *Oyo Buturi,* vol. 75, no. 3, (2006), p. 322.

50. A. Keshavarzi., A. Raychowdhury, J. Kurtin, K. Roy and V. De, *IEEE Transactions on Electron Devices,* vol. 53, no. 11, (2006), p. 2718.

51. F. Kreupl, A. P. Graham, G. S. Duesberg, W. Steinhogl, A. Liebau, E. Unger and W. Honlein, *Microelectronic Engineering,* vol. 64, no. 1-4, (2002), p. 399.

52. A. Nieuwoudt and Y. Massoud, *IEEE Trans. on Ele. Devices,* vol. 53, no. 10, (2006), p. 2460.

53. K. Keren, R. Berman, E. Buchstab, U. Sivan and E. Braun, *Sci.,* vol. 302, (2003), p. 1380.

54. A Javey, J. Guo, Q. Wang, M. Lundstrom and H. Dai, *Nature,* vol. 424, (2003), p. 654.

55. T. Rueckes, K. Kim, E. Joselevich, G. Tseng, C. Cheung and C. Lieber, *Sci.,* vol. 289, (2000), p. 94.

56. N. Jonge, Y. Lamy, K. Schoots and T. Oosterkamp, *Nature,* vol. 420, (2002), p. 393.

57. B. Hinds, N. Chopra, T. Rantell, R. Andrews, V. Gavalas and L. Bachas, *Sci.,* vol. 303, (2004), p. 62.

The High Balance Symmetric Balun for WLAN and WiMAX Application Using the Integrated Passive Device (IPD) Technology

Sung-Mao Wu, Wang-Yu Lin, Kao-Yi Wang, Chien-Hsiang Huang, Wen-Kuan Yeh
Department of Electrical Engineering, University of Kaohsiung, Kaohsiung, Taiwan, China
No.700, Kaohsiung University Rd., Nan-Tzu District 811, Kaohsiung, Taiwan
Tel: 886-7-5919436, Fax: 886-7-5919374, E-mail: sungmao@nuk.edu.tw

Abstract

In this study, the design of planar symmetric balance circuit and perform the high-balance differential output frequency response for system-in-package (SiP) application using the integrated passive device (IPD) technology are proposed. Passive balun with single-end input and differential output is an important device in RF front-end communication system. The merits of passive component are no power dispersion and have wide frequency bands.

The design of this planar symmetric balun is broadband and covers the frequency band of 2 to 4.2 GHz. It can apply to the 2.4 GHz WLAN and 3.4 to 3.7 GHz WiMAX applications. The measured return loss is better than -10dB within the pass band. The measurement results of magnitude and phase imbalance are lower 0.8 dB and 0.03 degrees, respectively.

Introduction

In recent years, relationship researches and development of advance packaging, like System-in-Package (SiP) or TSV on 3D die stacked, increase the demand of micro- electrical packaging correlative technique investigation violently. The design of balanced circuit is a key and can convert the single-end into differential-end. In RF wireless communication system, the balun can connect some circuit like mixer, push-pull amplifiers and differential low-noise amplifier (DLNA) and provide the signal conversion of single-end to differential pair. Differential signals with good amplitude and phase balance are immune to common-mode noise and even-order distortion. The IPD technologies have the highly integrated capability of passive component for SiP application. The thick-Cu metal layer, high dielectric constant embedded metal-isolation-metal (MIM) capacitor layers and high-resistivity glass substrate are the main feature of IPD process that can reduce the loss and improve the circuit's electrical performance. The integrity and capability of IPD is important and necessary for the design of RF wireless front-end system.

Design and Implementation

The schematic circuit view of the standard balun is shown in Figure 1. It is a 3-port device and the center-tap of balanced output. The primary turns provide a single-end input and the second turns on differential output consists of port2 and port3 terminals. The midpoint of the second turns as center-tap is referenced to the ground of the circuit. This design of planar balun referred to the Rabjohn balun [1-2] that have an extremely symmetrical structure in horizontal direction but not in vertical direction. Both terminals (port2

and port3) of the balanced side are symmetrical and can perfectly provide the common reference to ground. However, it's within two disadvantages in this case. First, the ratios of length of primary and second turns are not actual. The inaccurate ratios of length cause the inaccurate ratio of voltage between primary and second turns. Second, the single-end input and differential output terminals are not coincident in the same metal layer of outside coils. In order to improve both disadvantages, we modify the Rabjohn balun and present the novel design of planar symmetric balun that shown in Figure 2.

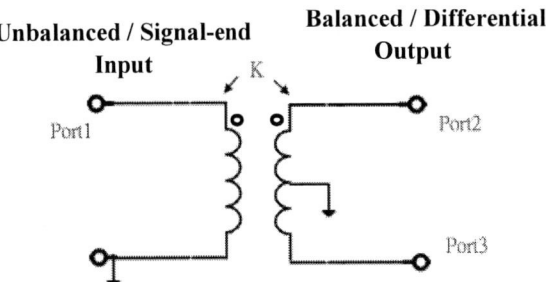

Figure 1. The schematic circuit of standard balun

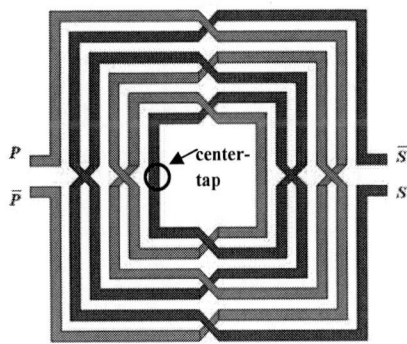

Figure 2. The planar symmetric balun

In this study, a 3:3 turns ratio for this design of planar symmetric balun is presented. For improving the performance of this balun, we will concentrate on technological improvements. First, reducing the return loss, we change the 90 degree bend corner of this balun to 45 degree [3]. Second, decreasing the total chip dimension and shifting the forbidden frequency band of transmission, we use the internal ground system to connect the all ground terminals of this balun. The design of internal ground system can create the shortest path to ground and decrease the capacitance effect between the signal path and ground path.

978-1-4244-4658-2/09 $25.00 © 2009 IEEE

Third, to promote the electrical performance in a selected bandwidth, we need to do match by adding the embedded capacitor in the circuit. Those capacitors can be placed by a way of series or shunt at the unbalanced or balanced side. In this design, we individually add the shunt capacitor at port1, port2 and port3 terminal. Figure 3 show the schematic circuit of standard balun adding the matching capacitor network. The 3D structure of symmetric planar balun is shown in Figure 4.

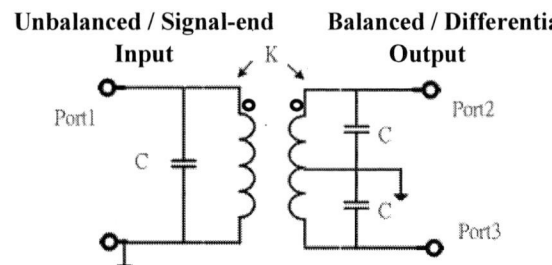

Figure 3. The schematic circuit of standard balun adding the matching capacitor network

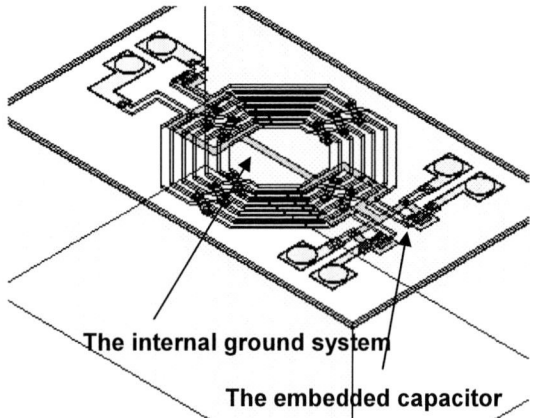

Figure 4. 3D structure of symmetric planar balun

Results and Discussion

The simulation and measurement results of this symmetric balun are shown in Figure 5. The implemented balun is simulated using Ansoft HFSS (High frequency structure simulation) full-wave 3D EM solver within standard 50Ω matching load at each port. The on wafer measurement data extracting was by Agilent Vector Network Analyzer 8753ES up to 6GHz and using short-open-load-through (SOLT) calibration. It is obviously that the results of comparison between simulation and measurement have good agreement. In Figure 5, the measured return loss is better than -10 dB from 2 to 4.2 GHz and better than -13dB from 2.2 to 3.9 GHz. At the frequency of 2.4 GHz and 3.4 to 3.7 GHz, the measured return loss are better than -17dB and -14 to -23 dB, respectively. The measurement data of difference of differential both output ports in phase are approximately 180 degree.

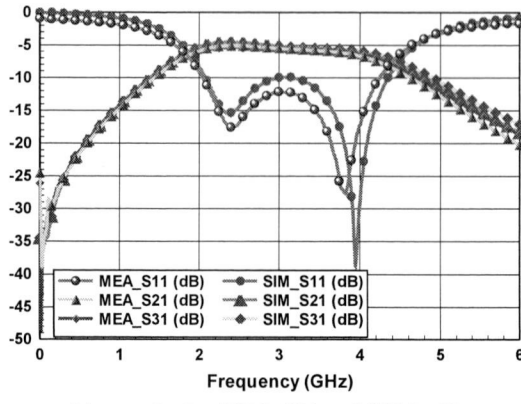

(a) magnitude of S11, S21 and S31 in dB

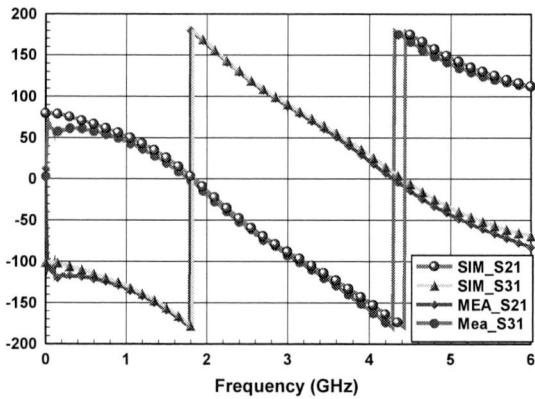

(b) phase of S21 and S31

Figure 5. Comparison of simulation and measurement results

To characterize the differential output performances of this balun with some parameters including magnitude imbalance (δ), phase imbalance (θ), insertion loss (IL) and common mode rejection ratio (CMRR), the above parameters are calculated by measured S-parameters as the following formulas (1) through (4), respectively. [4][5]

$$\delta = 20\log\left|\frac{S_{21}}{S_{31}}\right| \tag{1}$$

$$\theta = 180 - \left|\tan^{-1}(\frac{\mathrm{Im}(S_{21})}{\mathrm{Re}(S_{21})}) - \tan^{-1}(\frac{\mathrm{Im}(S_{31})}{\mathrm{Re}(S_{31})})\right| \tag{2}$$

$$\mathrm{IL} = -10\log(\left|S_{21}\right|^2 + \left|S_{31}\right|^2)$$

$$-10\log(\frac{\left|S_{21}\right|^2 + \left|S_{31}\right|^2 + 2\left|S_{21}\right|\left|S_{31}\right|\cos\theta}{\left|S_{21}\right|^2 + \left|S_{31}\right|^2 + 2\left|S_{21}\right|\left|S_{31}\right|}) \tag{3}$$

$$\mathrm{CMRR} = 20\log\left|\frac{S_{31} - S_{21}}{S_{31} + S_{21}}\right| \tag{4}$$

For the formulas (3), the first segment accounts for loss amplitude attenuation, and the second segment accounts for phase imbalance. When the phase imbalance is small, the second segment can be neglected and the insertion loss can be approximated as formulas (5). CMRR is defined as the ratio of the differential mode signal versus the common mode signal. The advantage of performance of balance circuit can be characterized in CMRR.

$$IL \approx -10\log(|S_{21}|^2 + |S_{31}|^2)$$ (5)

According to the formula (1) and (2), it can be seen that the measured magnitude and phase imbalance of this balun in Figure 6 and Figure 7. It exhibit excellent electrical performance as low as 0.8 dB and 0.03 degrees from the frequency band of 2 to 4.2 GHz.

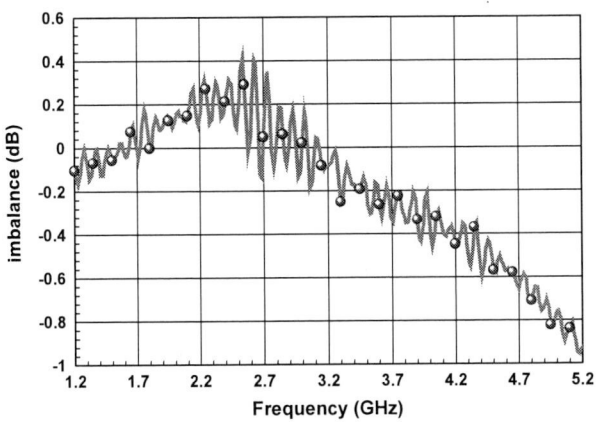

Figure 6. The measurement result of magnitude imbalance

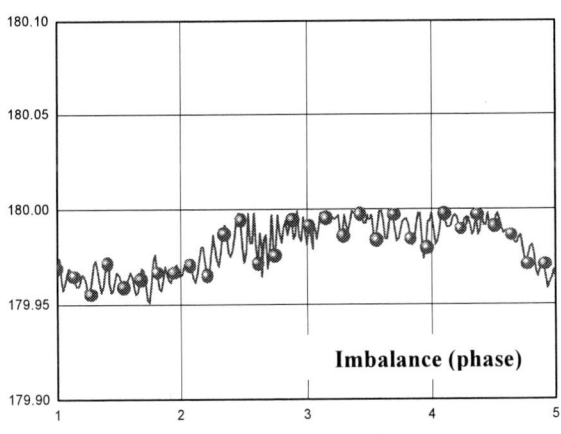

Figure 7. The measurement result of phase imbalance

Figure 8 show the measured insertion loss of this balun from the formula (3), and the insertion loss is less than 4.2 dB from 2 ~ 4.2 GHz. At the frequency of 2.4 GHz and 3.4 ~ 3.7GHz, the insertion loss is about 2.2dB and 2.9dB, respectively. The measured CMRR of this balun is evaluated by formula (4) and is shown in Figure 9. It is better than 30 dB in the measured CMRR and perform a high performance in differential output terminals.

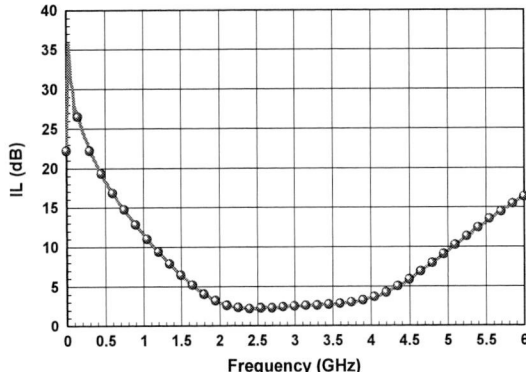

Figure 8. The measurement result of insertion loss

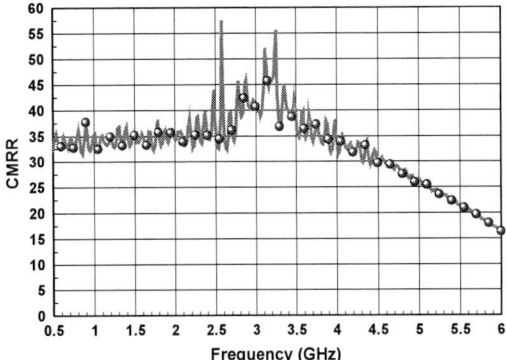

Figure 9. The measurement result of CMRR

A 2 - 4.2 GHz broadband symmetrical balun fabricated with IPD process for WLAN and WiMAX application is presented. The symmetry performance of this balun is excellent and the balun dimension is obviously reduced. The photograph of symmetrical balun fabrication is shown in Figure 10 and the overall chip dimension is 0.82mm×1.45mm. The comparisons among other reports of balun are summarized in Table 1.

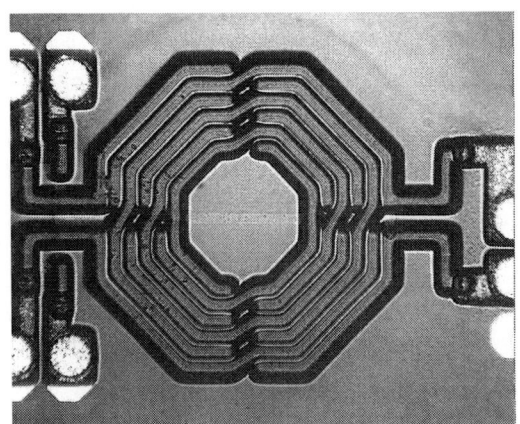

Figure 10. The photograph of symmetrical balun

978-1-4244-4658-2/09 $25.00 © 2009 IEEE

Table 1 Summary and comparisons among other reports of balun

Ref	Tech.	Frequency	Chip Dimension	Magnitude Imbalance	Phase Imbalance	Insertion Loss	CMRR
This work (measurement)	IPD	2-4.2 GHz	0.82mm*1.45 mm	<0.8 dB	<0.03 degree	< 4.2 dB	> 30 dB
[6] (simulation)	0.13μm CMOS	3-11 GHz		<0.02 dB	<0.2 degree	< 1.4 dB	
[7] (measurement)	FCBGA	2.3-2.6 GHz	1.6mm*1.7mm	<1.45 dB	<3 degree		
[7] (measurement)	FCBGA	4.9-5.9 GHz	1.35mm*1.43mm	<0.3 dB	<2 degree		
[8] (measurement)		2-3 GHz	1.37mm*1.68mm	<0.5dB	<5 degree		
[8] (measurement)	·	4-7 GHz	1.14mm*1.54mm	<0.3ddB	<6 degree		

Conclusions

In this paper, we design the symmetric balun for WLAN and WiMAX application using the IPD thick-films process. The symmetric balun implemented show a good electrical performance and present a compact on chip planar balun design with tuning capacitor. The measured magnitude imbalance, phase imbalance, insertion loss and CMRR are excellent and wide band over frequency from 2 to 4.2 GHz. The overall chip dimension is 0.82mm×1.45mm. This symmetric balun not only reduces the size and cost of the circuit, but also reduces the return loss and improve the insertion loss.

Acknowledgments

The authors would like to thank the Chip Implementation Center (CIC), Hsinchu, Taiwan and Advanced Semiconductor Engineering (ASE), Kaohsiung, Taiwan for chip fabrication.

References

[1] J. R. Long, "Monolithic transformers for silicon RF IC design," *Solid-State Circuits, IEEE Journal of,* vol. 35, pp. 1368-1382, 2000.

[2] G. G. Rabjohn, "Monolithic microwave transformers," *M.Eng. thesis*, Carleton University, Ottawa, ON, Canada, Apr. 1991.

[3] S. Kapur, D. E. Long, R. C. Frye, C. Yu-Chia, C. Ming-Hsiang, C. Huai-Wen, O. Jun-Hong, and H. Bigchoug, "Synthesis of Optimal On-Chip Baluns," *in Custom Integrated Circuits Conference, 2007. CICC '07. IEEE,* pp. 507 - 510 , 16-19 Sept. 2007 2007, pp. 507-510.

[4] Y. J. Yoon, L. Yicheng, R. C. Frye, M. Y. Lau, P. R. Smith, L. Ahlquist, and D. P. Kossives, "Design and characterization of multilayer spiral transmission-line baluns," *Microwave Theory and Techniques, IEEE Transactions on,* vol. 47, pp. 1841-1847, 1999.

[5] W. R. Eisenstadt, B. Stengel, and B. M. Thompson, " Microwave Differential Circuit Design Using Mixed-Mode S-parameters," *Norwood: Artech house,* 2006.

[6] X. Jun, W. Zhigong, and G. Xuefeng, "3-11 GHz Monolithic Balun in CMOS Technology for UWB Applications," *in Microwave and Millimeter Wave Technology,* 2007. ICMMT '07. International Conference on, 2007, pp. 1-3.

[7] E. Davies-Venn and T. Kamgaing, "LC-based WiFi and WiMAX Baluns embedded in a multilayer organic flip-chip ball grid array (FCBGA) package substrate," *in Electronic Components and Technology Conference,* 2008. ECTC 2008. 58th, 2008, pp. 169-174.

[8] E. Davies-Venn and T. Kamgaing, "Miniaturized rf transformer-based baluns for 802.11a/b/g WLAN modules embedded in organic package substrate," *in Radio and Wireless Symposium, 2008 IEEE,* pp. 359 - 362, 22-24 Jan. 2008 2008, pp. 359-362.

[9] Agilent Technologies, "Tuning, optimization and statistical design," *Agilent Inc.,* Sep. 2006.

New Packaging Technology Enabling Integration of Magnetics and Semiconductors in One Component

Abel Pot[1], Horst Roehm[2], Rinus v.d. Berg[1], Tamim P. Sidiki[3]
[1]DSM Engineering Plastics, Geleen, The Netherlands
[2]NXP Semiconductors, Hamburg, Germany
[3]DSM Engineering Plastics, Sittard, The Netherlands
DSM Engineering Plastics, Global Research & Technology
Urmonderbaan 22, 6167 RD Geleen, P.O. Box 604, 6160 AP Geleen, The Netherlands

Abstract

The continuous trend towards convergence and miniaturization is recently generating significant interest in new technologies for Electronics. This requires the integration of Semiconductors and Magnetics, two entirely different industries with different players in the value chain. In this paper, we demonstrate a packaging technology which allows three dimensional stacking of Magnets and Semiconductors. We realized the integration of a Semiconductor chip - which provides protection against electro static discharge (ESD)-and a common mode filter (CMF) into one thermoplastic package. For the first time ever, this filter is integrated directly into the thermoplastic part which is used as the substrate, filter and housing at the same time.

Laser direct structuring in combination with Stanyl® ForTii™ as an ultra high performance, entirely halogen-free high temperature thermoplastic omits any wires for the realized coil, and also facilitates high flexibility in design and manufacturing, allowing ultra small footprints and the realization of components suitable for surface mount technology.

As an example of this new technology, we demonstrate a component which can provide full ESD protection and common mode filtering for a high speed USB2.0 interface.

ESD Protection

Impact of Moore's Law on ESD protection of advanced CMOS ICs

The continuous trend of feature-size miniaturization has enabled semiconductor manufacturers over decades to improve chip performance, reduce power consumption, and drive cost down by squeezing billions of transistors into a single IC. Despite all obvious advantages, there is one major disadvantage in miniaturization of sub- circuits: integration of sufficiently robust ESD protection.

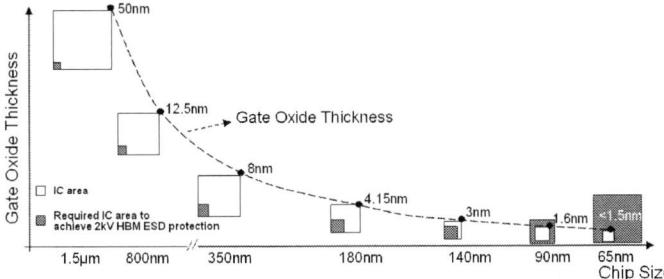

Figure 1: ESD considerations for advanced CMOS ICs

Figure 1 shows the reduction of the total IC area for various technology nodes. The red boxes within each of these ICs indicate schematically the required area to implement a minimum 2-kV ESD protection into the IC [1].

With each technology node the relative area required for ESD protection increases. The reason is that ESD protection scales with the area of the diodes and these diodes can not be shrunk at the same scale as transistors required for logic functions. It is obvious that for very advanced technology nodes there is a physical and economical limitation to integrate robust enough ESD protection. Advanced ICs are optimized for power consumption and speed, not for ESD protection. An optimization for ESD protection would blow up the chip above any acceptable limit.

Smaller feature sizes (channel length) related with thinner and smaller gate oxides drive down the maximum gate (e.g. for CMOS90 below 1.5V) and drain-source voltages (e.g. for CMOS90 <1.6V). Such ICs are very sensitive to over voltage and therefore especially sensitive to ESD discharges, which destroy sub- circuits already at very low ESD levels. As such, external board-level ESD protection becomes a must if developers of consumer/computer appliances want to build "CE"-compliant devices and furthermore want to prevent high field return rates due to ESD and other discharge issues. In general, one can say that today's ESD issue will become tomorrow's nightmare when even smaller feature sizes are applied.

External interfaces to other appliances are subject to ESD damage. In specific, higher speed, hot-plug interfaces such as HDMI, USB or Display port are most critical. Users can connect any sink or source equipment while at least one of the applications is still running, i.e. there is a supply voltage at the port. Needless to say, such a powered port will be affected by serious ESD issues. The question is not if, but only when the related transceivers (standalone or integrated) will be seriously damaged.

The high interface speed in conjunction with "hot plug" characteristics implies stringent requirements for an ESD protection solution, including:

- very high diode switching speed (nsec) and ultra low line capacitance (<1pF) can ensure signal integrity
- robust ESD protection without degradation after several ESD strikes
- low leakage even after several hundred ESD discharges

Based on main stream monolithic silicon technology, NXP Semiconductors provides ESD protection ICs fulfilling highest performance and meeting today's and tomorrow's requirements of OEMs in Electronics like:

978-1-4244-4658-2/09 $25.00 © 2009 IEEE

- the required low-cost solution for the mass consumer and computer market
- ultra-low total line capacitance of below 0.5 pF (Silicon chip incl. bonding wires, package and any existing parasitic)
- no degradation even after thousands of high-level ESD strikes (IEC61000-4-2)
- a fast diode reaction time (nsec range) to ESD pulses
- highest integration
- full compliance with high speed interfaces such as HDMI 1.3, Display Port or USB3.0

EMI Filtering

High speed digital interfaces like HDMI, USB or Display Port are widely used in the mobile as well as in the computer and TV area [2]. As base for data exchange, all these interfaces use differential signals to exchange data which means, that two complementary signals were sent on two separate wires. As long as ideal differential signals are transmitted, no electromagnetic interference (EMI) will occur. Unfortunately, in a real electronic system phase lag between differential signals, potential differences between differential signals and rise (fall) time lag between differential signals leads to common mode signals and therefore to EMI. This affects other electrical circuits by electromagnetic conduction or electromagnetic radiation from external antennas. It is obvious that EMI has to be suppressed in electronic systems.

In systems with differential signals common mode filter (CMF) are widely used to suppress the unwanted EMI generated by common mode signals. Especially, if unshielded twisted pair (UTP) cables – which acts as an antenna for common mode signals - are used as interconnects between devices, the use of CMF is a must.

High Speed Interface Protection

Today, state of the art solutions are using separated devices for ESD protection and EMI filtering to protect the highly integrated silicon chips and to suppress unwanted EMI (Figure 2).

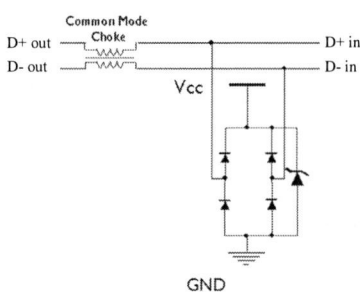

Figure 2: Schematic of a differential signal interface ESD & EMI protection

To overcome disadvantages of the discrete solution like space requirement, performance mismatch, inventory costs, etc. the integration of ESD protection and EMI filtering in one device is the next step, a straight forward approach.

Because CMFs are, in principle, built with wires winded around a ferrite core but on the other hand the ESD protection devices consists of diodes diffused in a block of silicon connected to a lead frame and covered by plastic the main challenge was to combine two different technologies in one package.

Global, cross-industry collaboration

Since this project involves entirely new technology, four companies have been working closely together to make it happen. The design of the package concept was proposed by DSM Engineering Plastics in The Netherlands, where a package was crafted enabling the integration of Semiconductors and Magnetics into one thermoplastic package based on injection molding. The injection molding of the package was done at NTM (NanoTechnology Mfg. Pte. Ltd.) in Singapore, the transfer of the EMI filter was achieved by Laser Direct Structuring (LDS) at Laser Micronics in Germany. In order to meet the high requirements of this technology to the thermoplastic, a new high temperature polyamide called Stanyl ForTii has been selected.

Market need for new package concept

OEMs in Electronics industry with their strong drive of application conversion seek for increased functionality integration and reduction of form factor to focus on PCB space and component count reduction. Since all external interfaces, in specific those operating at higher speeds such as HDMI, USB or Display Port, do require EMI filtering as well as ESD protection, two different components populate such interfaces and eat up valuable real estate on the PCB: ESD protection devices and EMI filters. From various discussions with leading OEMs, it is clear that a component which can integrate both these functionalities and at same time offers a space and component count reduction is highly appreciated and can solve some of the existing issues of OEMs realizing easier and denser application designs.

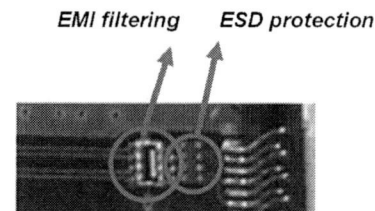

Figure 3: Optical microscopy of a typical external interface in Consumer Electronics using two individual components for ESD protection (first part after the interface pins) and EMI filtering (second component) in the electrical path on its way from the interface to the transceiver IC

We have designed and realized a package which integrates both these components into one package. This is a breakthrough technology which involves many industry first actions.

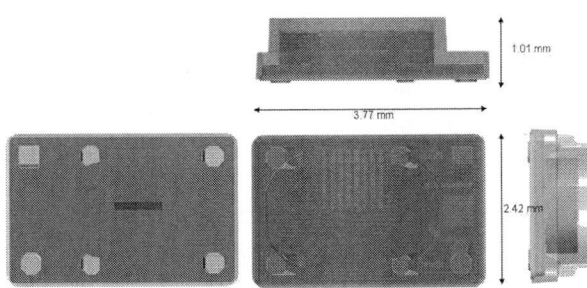

Figure 4: View of the package from the top, bottom and side showing total package dimensions

Figure 4 shows the sizes of the new package. With a total footprint of 3.77mm x 2,42mm x 1,01mm this is currently the world's smallest package integrating a fully EMI filters and an ESD protection for a high speed interface such as e.g. USB. The equivalent space reduction on the PCB is >75% by integration of these two components into one package. The small footprint in combination with low height enables PCB space reduction for OEMs.

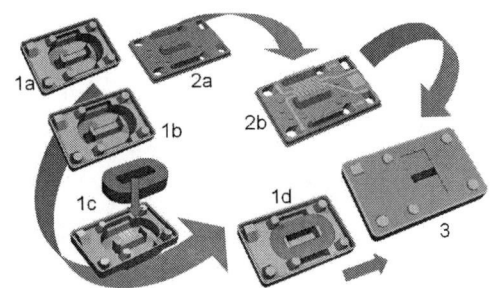

Figure 5: Schematics of manufacturing steps to realize a package integrating IC and magnet

Figure 5 shows the various manufacturing steps. In step 1a and 2a the top and cap layer of the package are molded in Stanyl ForTii, a high temperature polyamide suitable for lead free surface mount assembly due to high melting temperature Tm=320°C, a high glass transition temperature of Tg=135°C and a stiffness across a broad temperature range. From similar work in air cavity packages which are used for e.g. MEMS sensors it is well known that co-planarity is a key issue since it can lead to delamination of ICs mounted on lead frames. Due to the high stiffness and a comparable CTE (Coefficient of Thermal Expansion) values in the parallel and vertical flow direction, Stanyl ForTii was selected for this application. In specific the high stiffness at lead free reflow temperature range between 260-288°C makes Stanyl ForTii an unbeatable solution for such applications. Although reflow temperature is typically 260°C, we have also looked into a higher range up to 288°C in order to account for potential hot spots during assembly. Due to its high toughness before and after reflow (flexural strength), Stanyl polyamide family is one of few materials enabling such designs as applied in this concept. Any other halogen free material which would fit the temperature requirements of reflow soldering such as Liquid Chrystal Polymers (LCPs) are commonly known to be very brittle and would hence fail during later stages of package assembly.

After the top and bottom part are molded, the parts will be exposed to laser Direct Structuring (LDS) to transfer all electrical tracks (Figure 5: steps 1b and 2b). Later in this paper the process is described in more detail. At a next stage, chip and magnet are inserted (Figure 5: steps 1c and 1d) and finally the two parts of package are put together (Figure 5: step 3).

Package Molding

The specific design expertise at NTM allows design and building of tools and concepts with the help of mold simulation to seek application approval before starting with tool fabrication. 2D & 3D drawings are drafted to tooling specialist's to start with micro tool fabrication. The micro tools are fabricated using state-of-art machines like Kern Pyramid Nano, Charmilles Robform Die Sinker & Hauser Jig Grinder for highly accuracy and finishing. Every process is precisely machined and quality controlled before the tool are being assembled.

The micro tool is set on the Battenfeld microsystem injection molding machine in the 10k clean room facility for better control and cleanness. The raw material is dried according to DSM material recommendation before injection molding in the micro tool starts. Micro molding process Engineer carefully process control the micro molding machine to ensure the parts are produced with high precision and stringent quality. Final inspection is supported by state of art measurement equipment and methodology to meet customer satisfaction.

Making use of the good rheological properties of the high temperature polyamide Stanyl ForTii, the cover and base parts could easily be molded in an injection molding machine fit for micro molding [3,4]. Even smallest details of the mold were easily transferred into the plastic.

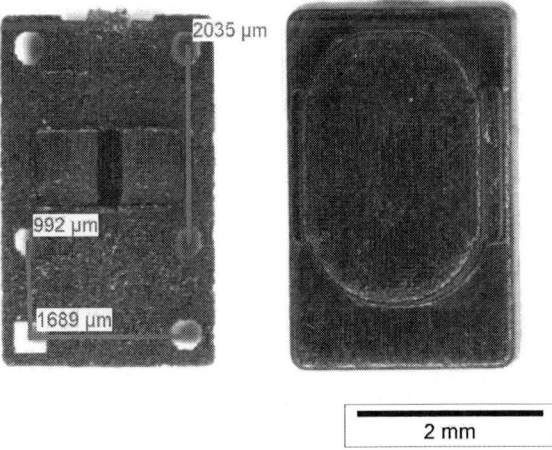

Figure 6: Top and bottom cap of the package after being micro molded at NTM

Figure 6 shows the outcome of the micro molding. Despite the ultra small feature sizes, all features have been transferred perfectly into Stanyl ForTii. Process setup is easy and fast and no flashing occurs with Stanyl ForTii.

Stanyl ForTii

The thermoplastic materials for micro molding also should be carefully selected to fit the process and required part design. Here the brand new high temperature polyamide Stanyl ForTii was selected for its good combination of thermal, mechanical and rheological properties. Parts of this material can withstand lead-free soldering conditions without degradation. Therefore the material is suitable for manufacturing miniaturized, small footprint SMT components. Stanyl ForTii is currently the only available high temperature polyamide enabling such a high flexibility in design and full compliance to lead free reflow SMT assembly. Stanyl ForTii furthermore is entirely halogen free, fully meeting OEM specifications.

With Stanyl (polyamide 46), Stanyl ForTii (polyamide 4T) and Xantar (PC/ABS) DSM Engineering Plastics is offering the broadest available portfolio of LDS grades and covers the entire temperature range in Electronics applications. A global presence of DSM Engineering Plastics, a presence since more than 20 years in the Electronics industry as well as a high application and design support level makes DSM Engineering Plastics a strong partner to OEMs and connector and component manufacturer.

LDS to selectively create tracks on thermoplastic parts

Molded parts with integrated tracks are called MID's (Molded Interconnected Devices). There are several technologies available to create MID's. Laser Direct Structuring (LDS) is the technology with the most design freedom to selectively plate plastic parts. The process to create a MID with LDS technology consists of three basic steps: molding of the thermoplastic part, selective laser ablation of the track-layout on the part and plating of the track layout.

Plastic parts of all shapes and materials fulfilling all kinds of functions are daily practice. Examples are housing parts of household appliances, mobile phones, interior parts of cars and even lunch boxes. A common and widely available process to manufacture plastics parts is injection molding. This process allows a lot of design freedom to come to 3D design solutions. This can also be translated in miniaturization. In order to do so, tooling and molding machines need to be adapted to the situation. Additional design rules should be considered. Not every molder has the capability for micro molding. The parts used in this study are molded in Nano Technology Manufacturing; they are specialized in ultra-precision- manufacturing.

After molding, the next step is laser ablation. During the selective laser ablation, the track pattern is written in the surface top layer of the plastic parts. The LDS additive in modified polymer is transformed into micro metal cores they appear on the surface of the tracks and are fixed in the polymer matrix. This allows electroless Cu plating of the tracks.

Design solutions

The design in our study demonstrates several design solutions. Between two parts a spool around a ferrite core is created, contact or soldering pads are created with only one sided lasering and mechanical fixation between the cover and base part is created.

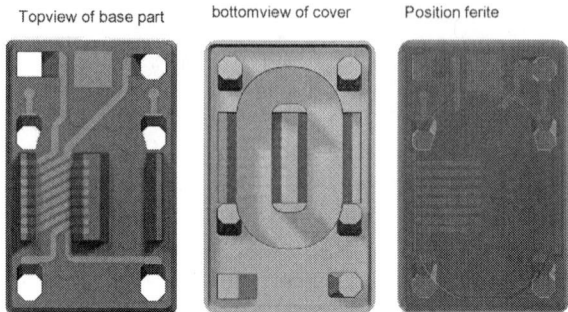

Figure 7: Top and bottom view schematics of the package as well as position of the ferrite

Compared to stitching, the windings of the spool can be more close to the ferrite core. There is no need to compensate the design for the use of stitching the wires. Hindering of the stitching head does not occur is simply no hindering. Apart form the used spool design in this study other design solutions are possible.

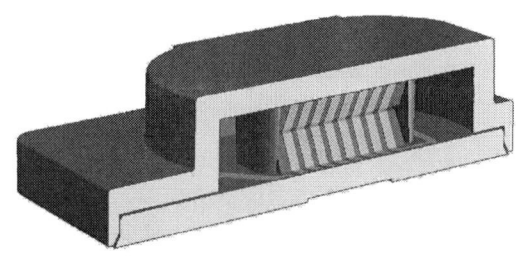

Figure 8: Assembly of top and bottom part closes the Cu tracks around the magnet and puts together both parts of the EMI filter

For ease of manufacturing it is chosen to use only one sided laser structuring to create the 3D track. Of coarse multi sided laser structuring is also possible. This allows even greater design freedom to create additional functions.

The solder pads for mounting on the PCB are part of the cover design. Due to through contacting of the legs of the cover and the tracks in the receiving holes interconnection with the tracks on the base is established. It demonstrates interconnection between tracks on 3D surfaces of different parts. In this way stacks of layers and via's are easily created. Again, only with one sided laser structuring.

Apart from fixation of the assembly on the PCB the cover and base should also be mechanically fixated. In micro-molded parts the standard snap fit solutions in plastics is not possible. There is no design space available leading to any mechanical strength. The demonstrated solution is a track, isolated from the electronic circuit, all around the inside of the cover and outside of the base. The soldering seals the parts into one assembly.

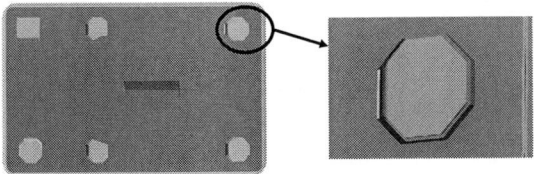

Figure 9: Realization of the solder pads by LDS

The last step in LDS process is the plating of the tracks. Due to the laser exposure the LDS additive in the polymer is changed into Cu-cores. Those metal cores are sensitive for electroless plating. For this plating, 15 µm standard commercial chemical processes are available, like MacDermid bath and Rohm &Haas. Once the tracks are covered with Cu additional layers can be added by galvanic plating.

Figure 10: Optical microscopy of the two assembled LDS parts including the Semiconductor chip and magnetic coil

Stanyl ForTii

Stanyl ForTii is the very first entirely new high temperature thermoplastic introduced by any Chemical company in this millennium. Stanyl ForTii is a polyamide 4T, which DSM is marketing under the Stanyl brand family and which enlarges the DSM portfolio into the ultra high performance polymers best suitable for lead free reflow soldering applications. Stanyl ForTii is entirely halogen free meeting latest industry requirements.

Stanyl ForTii is an ultra high performance polymer meeting the highest requirements of lead free reflow soldering with a unique balance of thermal, mechanical and electrical properties. Stanyl ForTii is most suitable for demanding applications in the electronics industry/lighting/automotive/aerospace. In this project, we have selected this material due to its best fit to the required properties. In addition to the regular high performance requirements demanded by lead-free reflow soldering, the integration of components inside the package do require an excellent co-planarity of the package in order to avoid possible delamination of the chip from the polymer. Stanyl ForTii with its very high stiffness at reflow temperatures does enable this.

Conclusions

In summary, we have successfully shown the integration of Semiconductor ICs and Magnets in one thermoplastic package. LDS and micro molding technologies have enabled aggressive space reduction which can be used by OEMs to add additional functionality onto their PCBs or to simply reduce PCB size. Furthermore, OEMs can omit standalone components and hence also reduce component count which directly reduced their assembly cost.

The availability of an ultrahigh performance thermoplastic such as Stanyl ForTii from DSM Engineering Plastics has opened the door to realize LDS concepts on 3D designs in air cavity packages with high toughness, high co-planarity and excellent fit to lead free reflow soldering temperatures fulfilling JEDEC MSL 1 standard. The ease of processing of Stanyl ForTii enables an excellent material fit to the stringent requirements of micro molding.

References

1. Dr. Tamim P. Sidiki, D. F. *et al*, "ESD protection for HDMI 1.3 interfaces", Network systems Designline, september 06, 2006, http://www.networksystemsdesignline.com/showArticle.jhtml?printableArticle=true&articleId=192503738
2. NXP, "Vollintegrierte VGA-Interfacechips unterstützen UXGA für PCs und 1080p für Fernsehgeräte" 07.12.2006. http://www.electronic-data.com/een/hro/11_1776.asp
3. Heckle M. *et al*, "Review on micro molding of thermoplastic polymers", *Journal of Micromechanics and Microengineering*, Volume 14, Issue 3, pp. R1-R14 (2004).
4. Zhao J. *et al*, "Polymer micromould design and micromoulding process", *Plastics, Rubber and composites*, Volume 32, Number 6, August 2003 , pp. 240-247(8)

Parametric Study of Electroplating-based Via-filling Process for TSV Applications

K. Y. K. TSUI*, S. K. YAU, V. C. K. LEUNG, P. SUN, D. X. Q. SHI

Hong Kong Applied Science & Technology Research Institute (ASTRI)
2 Science Park East Avenue, Hong Kong Science Park, Shatin, New Territories, Hong Kong, China
*Email: kolotsui@astri.org

Abstract

In this study, the effects of different influence factors on electroplating-based via-filling process were studied in a systematic manner. A through-silicon-via (TSV) chip was firstly designed, the chip size was 16x10mm^2 with the TSV interconnects. The via was in diameter of 50µm and depth of 75µm. The deep reactive ion etching (DRIE) technique was employed to fabricate the vias onto the silicon wafer. In order to fill up the vias by using the electroplating process, the SiO$_2$ isolation layer was firstly prepared onto the sidewall and the defined surface of the wafer using the plasma enhanced chemical vapor deposition (PECVD) technique, the TiW barrier layer was prepared onto the SiO$_2$ layer with the sputtering process, the Cu seed layer was then deposited onto the barrier layer by the sputtering process to provide a foundation for copper growth during electroplated deposition. To achieve the "bottom-up" via-filling, the addictives were added into the copper plating solution. The results showed that the void free Cu filling can be achieved with an optimal additives ratio. The variation in both via opening and depth were found to be key factors influencing the via-filling quality. The process parameters such as current density, power waveform and so on were found to affect the vial filling quality. A comparison on the effects of key process parameters was made, showing that the current density, voltage waveform, and pulse reserve are three key parameters that affect the filling quality more than other parameters.

1. Introduction

The demand of electronics devices has soared especially for the market of superior performance portable electronic device in terms of multifunctional with lighter in weight and slim in outlook. Therefore, it is a driving force to push both industries and academic institutions working together for developing advanced packaging technologies that are feasible to apply on electronics device to fulfill the market requirements. One of advanced packaging technologies is the three dimensional packaging (3D packaging). General speaking, 3D packaging is describing the structure of modules in which two or even more chips are assembled vertically by interconnects such as wire bonds or solder bumps associated with TSV. The benefits from 3D packaging are higher silicon density package, shorter interconnection and lower profile system-in-package (SIP) [2-8].

TSV structure plays a key role in 3D packaging because it can accompany with bumps to form the shortness interconnection between two chips. In general, the TSV structure is used as a path to provide the way for transferring electrical signal from one chip to another chip vertically such like an elevator to provide a way for people (electrical signal) to reach different floors (different chips in the 3D package) in a building (the 3D package). In the area of 3D packaging with TSV structure, a number of researchers were focused on the methods to overcome the problems in fabricating high aspect ratio TSV. In practice, most of problems come from TSV filling which are influenced by several factors. In case of studying well in TSV filling, researchers classified and expressed the factors in a systematic way such as Fig. 1. And a group of researchers are studying feasible ways to apply TSV structure in the market existing functional chips without any change of the inner structure of those functional chips. One of the ways to realize this purpose is to use the dog-bone design to redistribute the bonding pads to the TSV locations. In practice, this approach is good but not suitable for all kinds of existing functional chips because it may not have enough spacing where is available for TSV structure pass through due to the condensed circuitry inside the silicon body. Therefore, via-in-pad (VIP) approach has been developed which is describing a TSV structure built underneath / inside the existing bond pad.

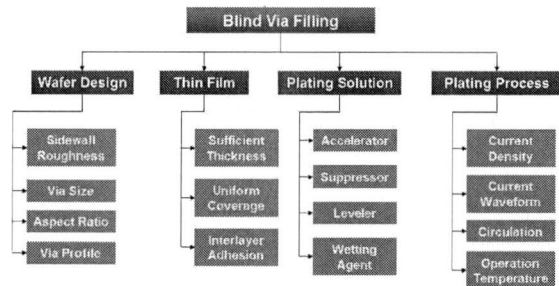

Fig. 1 Summary of influence factors of electroplating-based via-filling.

Memory chip is one of the most common areas to apply this advance technology especially of flash memory due to simple structure, small number of I/O and widely use in portable electronic device [1, 10-11]. In addition, the bond pad diameter of flash memory chip is around 100µm so the diameter of VIP should not be greater than that, otherwise the bond pad will be totally rebuilt. Moreover, the thickness of a flash memory chip after thinning process is under 100µm which is much thinner than a piece of bare silicon wafer with 400µm in thickness. Under these conditions, there is a potential application in developing low aspect ratio TSV which is under 2. In this study, it is focusing on the factors affecting the copper filling with via aspect ratio under 2; in the meanwhile, the diameter is not larger than 100µm [10-11].

2. Experimental Methods

In this study, bare silicon wafers with 4 inch in diameter and 400μm in thickness were used. The micro-vias were fabricated into the bare silicon wafers first, the copper filling process was then employed to fill up the micro-vias. In details, a layer of silicon oxide with 3μm in thickness was prepared onto the surface of the silicon wafer as the etching mask for via drilling. Before via drilling, an etching mask was fabricated by photolithograph as showed in Fig. 2. With the via layout defined by the silicon oxide etching mask, deep reactive ion etching (DRIE) was conducted to drill via on the wafer. When the appropriate via diameter and depth were achieved, the etching mask was removed as showed in Fig. 3.

Fig. 2 Silicon wafer with DRIE etching mask.

Fig. 3 TSV formed in the silicon wafer after DRIE.

Before the copper filling process, few thin film layers were deposited. First, a thin layer of silicon oxide with 1μm in thickness was deposited onto the surface of the silicon wafer and the sidewall of the via as the insulation layer by using the plasma enhanced chemical vapor deposition (PECVD) method. In order to fill the via using electroplating method, a thin layer of copper with sufficient thickness is needed to be prepared as the seed layer for the copper electroplating process. A thick copper seed layer around 2μm in thickness was deposited by using the sputtering process. Since the copper seed layer has poor adhesion with the SiO2 layer and the SiO2 layer can not prevent the Cu diffusion, an intermediate adhesion and barrier layer is required. In this study, TiW was used as the intermediate layer. The sputtering method was employed to prepare the TiW layer with the thickness of about 0.1um. Table 1 summarizes the information of thin film layers and the schematic diagram of corresponding structure showed in Fig. 4.

Table 1 Summary of the specifications of thin film layers

Layer	Materials	Thickness
Insulation Layer	SiO$_2$	1μm
Barrier & Adhesion Layer	TiW	0.1um
Seed Layer	Cu	2μm

Fig. 4 TSV wafer after thin film deposition.

Copper electroplating can be performed using an electroplating mask to define the copper plating area over the wafer. Both organic photosensitive film and conventional spinning coat thick photoresist can be used as the plating mask as showed in Fig. 5.

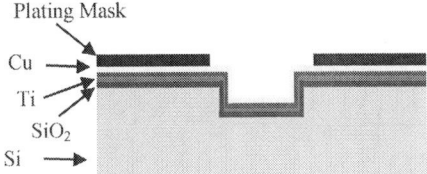

Fig. 5 TSV wafer after thin film deposition.

In this study, the copper plating solution was supplied by two different suppliers named as Supplier A and Supplier B. The solution from Supplier A is sulfuric acid based copper electrolytic solution plus three additives namely accelerator, suppressor and leveler. The solution from Supplier B is methanesulfonic based copper electrolytic solution plus two additives namely accelerator and suppressor.

Table 2 Summary of copper electrolytic solutions

Supplier	Electrolytic Solution	Additives
A	Sulfuric based	Accelerator
		Suppressor
		Leveler
B	Methanesulfonic based	Accelerator
		Suppressor

In this study, several key parameters were studied to understand their effects on the copper filling process. The parameters include: a) the via depth and the via opening; b) the composition ratio of additives added in the electroplating solution; c) the current density and d) the waveform of the power source.

3. Experimental Results and Discussion
3.1 Effect of Via Geometry

In the copper filling process, the via geometry was found to be one of key parameters that affected the via-filling quality. In general, there are two main measures to describe the via. They are the shape of the via opening and the depth of the via. On the one hand, the shape of the via could be in circular, square, rectangular or any other possible shape. On the other hand, the depth of the via also plays a critical role in the copper filling process. In case of maintaining the opening size & shape constant, the more deep of the via is, the higher probability of the void inside the via forms. A ratio of the via depth over the via opening is usually used to describe the via geometry. In this study, the circular shape via was selected with two different diameters and depths. One was 100μm in opening diameter with 150μm in depth, another one was 50μm in opening diameter with 75μm in depth, the aspect ratio for both was 1.5.

978-1-4244-4658-2/09 $25.00 © 2009 IEEE

Figs. 6 and 7 show the results using the electrolytic solution from Supplier B. In Fig. 6, the via was in circular shape with the 100μm in diameter and 150μm in depth, while in Fig. 7, the via was also in circular shape with the 50μm in diameter and 75μm in depth. For the two results, all other parameters such as the electroplating duration, the addictives concentration, the current density and so on were remained unchanged By comparing Figs. 6 and 7, it was found that although the aspect ratio was same for both cases, the small via had sufficient via-filling and the big via has insufficient via-filling. For both cases, there was no any void observed. Table 3 summarizes the effect of via geometry on the via-filling quality.

Table 3 Summary of effect of via geometry on the via-filling quality.

Sample	Via Diameter (μm)	Via Depth (μm)	Via-filling Quality
1	100	150	50% via-filling
2	50	75	100% via-filling

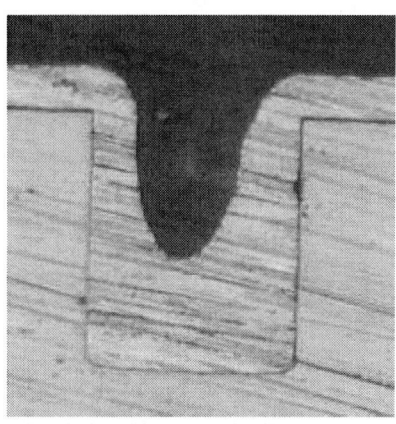

Fig. 6 Cu filled via with 100μm in diameter and 150μm in depth.

Fig. 7 Cu filled via with 50μm in diameter and 75μm in depth.

3.2 Effect of Additive Concentration

Accelerator concentration in the electroplating solution was found to be an important factor that affect the via-filling quality. Both Supplier A and Supplier B use it as an essential additive. The function of accelerator was to enhance the copper deposition at the bottom of the via than that at the top of the via. As the copper deposition rate at the bottom of the via is accelerated, the bottom-up deposition and the void-free via-filling can be achieved.

Figs. 8 and 9 show the results of copper filled vias using the electrolytic solution from Supplier B but with different accelerator concentration. The via was 100μm in diameter and 150μm in depth. By comparing Figs. 8 and 9, it was noted that higher accelerator concentration could enhance the copper deposition rate at the bottom of the via, in other words, the bottom-up copper deposition could be achieved if the higher accelerator concentration was used. Also higher accelerator concentration could help reduce the void formation and improve the copper pad uniformity. It was also found that there was no big variation in the copper deposition rate for adjacent vias, in other words, the Cu volume deposited in each via would be quite similar. Therefore, if a void was formed inside a via, the surplus volume would be deposited onto the surface of the wafer, causing the bridging between two vias as shown in Fig. 8. Table 4 summarizes the effect of additive concentration on the via-filling quality.

Fig. 8 Filled vias using Supplier B's solution with the accelerator concentration of X ml/L and maintaining other parameters unchanged.

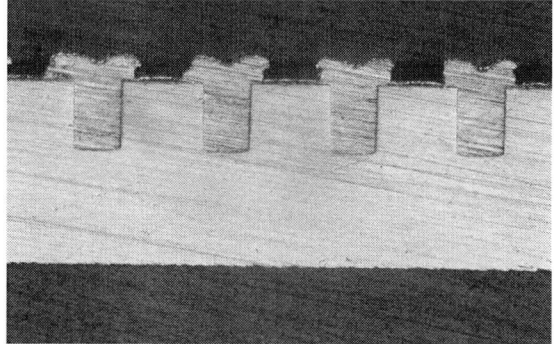

Fig. 9 Filled vias using Supplier B's solution with the accelerator concentration of 1.2X ml/L and maintaining other parameters unchanged.

978-1-4244-4658-2/09 $25.00 © 2009 IEEE

Table 4 Summary of the effect of accelerator concentration on the via-filling quality.

Sample	Concentration of Accelerator (ml/L)	Via-filling Quality
1	X	Void inside the via; Bridging between two vias.
2	1.2X	100% filled via

3.3 Effect of Current Density

The current density of the power source was found to be one key parameters that could be used to improve the via-filling quality. In general, higher current density can accelerate the electroplating process, faster Cu deposition rate can be obtained. In other words, less plating duration can be achieved by maintaining other parameters unchanged. However, in practice, the increasing of Cu deposition rate was limited by increasing the current density, because high current density would create other problems such as the over plating, the void formation inside the via and so on.

In the study, two samples were electroplated by using the electrolytic solution from Supplier B. The via was 50μm in diameter and 75μm in depth. All the parameters were remained same, only difference between the two samples were the current density. The results are summarized in Table 5. As seen, when the current density was increased from 1.0I ASD to 1.5I ASD, the vias could be well-filled (100% filling) from the partial filling (50% filling). Also, higher current density could over-plate the vias, the volume of the over-plated copper was found to be large enough to form a copper stud bump on the top of wafer surface. It gave the potential opportunity for developing the interconnection such as stud bumps together with adhesive between two chips for TSV structure 3D integration applications.

Table 5 Summary of the effect of the current density on the via-filling quality.

Sample	Current Density	Via-filling Quality
1	1.0I ASD	50% via-filling
2	1.5I ASD	100% via-filling

3.4 Effect of Power Source Waveform

The waveform of the power sources was found to affect the via-filling quality as well. More than changing the current density level in a direct current (DC) power source, it is possible to interpret another waveform of power source. One of the most common waveforms used in copper filling application is the pulse reverse current (PREC) power source. It can help to minimize the probability of void formation inside the TSV due to slow down the copper deposition rate along the opening of TSVs by reducing excess charging.

In the study, two samples were electroplated by using the electrolytic solution from Supplier A. The via was 100μm in diameter and 150μm in depth. Different with the copper filling using the direct current, where the high charging level can only be accumulated around the opening of the via, by using the PREC power source, the high charging level can be generated around the via opening, which could increase not only the Cu deposition rate at the top of the via but also the Cu deposition rate at the bottom of the via. As a result, the voids were observed in the vias electroplated by using the direct current method, while the void-free filled vias were obtained by using the pulse reverse current method. Table 6 summarizes the effect of the power source waveform on the via-filling quality.

Table 6 Summary of the effect of power source waveform on the via-filling quality.

Sample	Power Source Waveform	Via-filling Quality
1	Direct current	100% via-filling with voids
2	Pulse reverse current	100% via-filling without voids

4. Conclusions

Via-filling is a key process for developing TSV-based 3D packages. The electroplating filled Cu TSV interconnect not only provides the shortest way for signal transmission between silicon chips but also reduces the power consumption in the package as compared to other interconnection technologies. However, TSV interconnect manufacturability and reliability is influenced by many factors when the electroplating technique is used. In this study, several key parameters were studied in a systematic manner to understand their effects on the via-filling quality: a) the via depth and the via opening; b) the composition ratio of the additives in the electroplating solution; c) the current density level and d) the waveform of the power source.

By maintaining other parameters unchanged, although the aspect ratio was same, the vias with 50μm in diameter and 75μm in depth could be filled fully (100% filling), while the vias with 100μm in diameter and 150μm in depth could only be filled partially (50% filling). In order to achieve the void-free copper deposition, the accelerator concentration in the solution was increased by 50%, the results showed that the method could achieve better "bottom-up" deposition.

Short electroplating duration is one of major target for industry. When the current density was increased by 50%, the results demonstrated the method could speed-up the deposition rate by 100%. To minimize the void formation inside the via, the pulse reverse current method was applied and the results showed that the method could achieve the void-free via-filling.

Other parameters such as sidewall roughness, high aspect ratio, via profile, seed layer thickness, additives ratio, operation temperature and so on were also investigated in a systematic manner, as a result, a full set of database was established in ASTRI. With it, optimal process recipes were developed for filling vias with different profiles, aspect ratios and densities.

References

1. M. Sunohara, A. Shiraishi, Y. Taguchi, K. Murayama, M. Higashi, M. Shimizu, "Development of Silicon Module with TSVs and Global Wiring (L/S=0.8/0.8μm)", *Proceeding of 59th Electronic Components and Technology Conf*, San Diego, California, USA, May. 2009, pp. 25-31.

2. M. Nagai, Y. Tamari, N. Saito, F. Kuriyama, A. Fukunaga, A. Owatari, Masashi, Shimoyama, C. Moore, "Electroplating Copper Filling for 3D Packaging", *Proceeding of 59th Electronic Components and Technology Conf*, San Diego, California, USA, May. 2009, pp. 648-653.

3. H. H. Chang, Y. C. Shih, Z. C. Hsiao, C. W. Chiang, Y. H. Chen and K. N. Chiang, "3D Stack Chip Technology Using Bottom-up Electroplated TSVs", *Proceeding of 59th Electronic Components and Technology Conf*, San Diego, California, USA, May. 2009, pp. 1177-1184.

4. K. Kumagai, Y. Yoneda, H. Izumino, H. Shimojo, M. Sunohara, T. Kurihara, M. Higashi, and Y. Mabuchi, "A Silicon Interposer BGA Package with Cu-Filled TSV and Multi-Layer Cu-Plating Interconnect", *Proceeding of 58th Electronic Components and Technology Conf*, Lake Buena Vista, Florida, USA, May. 2008, pp. 571-576.

5. R. Beica, C. Sharbono, T. Ritzdorf, "Through Silicon Via Copper Electrodeposition for 3D Integration", *Proceeding of 58th Electronic Components and Technology Conf*, Lake Buena Vista, Florida, USA, May. 2008, pp. 577-583.

6. R. Beica, P. Siblerud, C. Sharbono, M. Bernt, "Advanced Metallization for 3D Integration", *Proceeding of 10th Electronics Packaging Technology Conf*, Singapore, Dec. 2008, pp. 212-218.

7. W. Worwag, T. Dory, " Copper Via Plating in Three Dimensional Interconnects", *Proceeding of 57th Electronic Components and Technology Conf*, Reno, Nevada, USA, May. 2007, pp. 842-846.

8. D. M. Jang, C. Ryul, K.Y. Lee, B.H. Cho, J. Kiml, T.S. Oh, W.J. Lee and J. Yu , "Development and Evaluation of 3-D SiP with Vertically Interconnected Through Silicon Vias (TSV)", *Proceeding of 57th Electronic Components and Technology Conf*, Reno, Nevada, USA, May. 2007, pp. 847-852.

9. B. Kim, C. Sharbono, T. Ritzdorf, and D. Schmauch, "Factors Affecting Copper Filling Process Within High Aspect Ratio Deep Vias for 3D Chip Stacking", *Proceeding of 56th Electronic Components and Technology Conf*, San Diego, California, USA, May. 2006, pp. 838-843.

10. R. Rieske, R. Landgraf, K.J. Wolter, "Novel Method for Crystal Defect Analysis of Laser Drilled TSVs", *Proceeding of 59th Electronic Components and Technology Conf*, San Diego, California, USA, May. 2009, pp. 1139-1146.

11. M. Dellutri, P. Pulici, D. Guamaccia, P. Stoppino, G. Vanalli, T. Lessio, F. Vassallo, R.D. Stefano, G. Labriola, A. Tenerello, F.L. Iacono, G. Campardo, "1 Gb Stacked Solution of Mutlilevel NOR Flash Memory Packaged in a LFBGA 8mm by 10mm by 1.4,, of thickness", *Proceeding of 7th International Conference on Thermal, Mechanical and Multiphysics Simulation and Experiments in Micro-Electronics and Micro-Systems*, EuroSime, Como, Italy, Apirl. 2006, pp. 1-5.

Packaging and Assembly of 12-Channel Parallel Optical Transceiver Module

Zhihua Li, Wei Gao, Jian Song, Baoxia Li, Lixi Wan
Institute of Microelectronics, Chinese Academy of Sciences
3# BEITUCHENG West, CHAOYANG District, Beijing 100029, China
Email: lizhihua@ime.ac.cn, Phone: 86+10-82995720

Abstract

The fabrication process of a 12-channel parallel optical transceiver module developed in our group is presented in this paper. The module is composed of a VCSEL array, a PIN PD array, a VCSEL driver chip, a TIA/LA chip and supporting PCB and connector. A SiOB and its vertical assembly are emphasized as the highlights of the structure of this module, which is promising to effectively reduce the package cost and improve the optical coupling efficiency.

Introduction

The demand for higher bandwidth interconnects continues to increase for all levels of interconnection within high performance computing and switch/router systems, whether these are system level rack-to-rack links or board-to-board links or even module-to-module on a single board[1]. Advances in processor speeds, multicore processors, parallel systems with a large number of processors, and wider data buses present steadily increasing aggregate bandwidth, channel data rate, and density requirements that are progressively more difficult to meet using electrical interconnect technology[2]. Optical interconnects have been increasingly used to replace copper interconnects since they provide significant advantages for applications with longer links as well as supporting higher data rates, increased bandwidth density, and more compact cables and connectors. In supercomputers, core routers, switches, high-end servers and demonstration of high-speed DSP, optical interconnection becomes a practical necessity and some commercial parallel optical interconnect modules (POIMs) have been used in these applications[3-8].

Vertical cavity surface emitting lasers (VCSELs) and planar PIN photodetector are dominating optoelectronic devices in parallel optical modules because of the mature fabrication process, high performance and low cost. However, the integration of VCSEL and PIN PD in optical interconnects requires some means of turning a light beam by 90 degree, as seen in Figure 1. In most published work, 45 degree mirrors and relay lenses are used in order to couple light from the waveguide or fiber to the optoelectronic devices[8-12]. The fabrication of micro mirrors and micro lenses and the complicated alignment processes are not only major cost issues in high density parallel optical interconnects, but after more than 10 years of development optical alignment is still a major issue. T. Ishikawa et al[13] suggested a bent fiber connector for the light beam turning, but it would increase the cost of the optical interconnection and make it more complicated.

In this paper, we suggested a new structure of optical transceiver module which has vertical silicon optical bench (SiOB) assembly and eliminates the using of micro mirror and lenses. The packaging and assembly process will be described.

Optical subassembly

Here, optical subassembly is the part including optoelectronic devices and chips and the optical coupling structure. Specially, it composed of a VCSEL array or a PIN PD array, a driver chip or a TIA/LA chip, a SiOB with holes array, a copper block and a MPO optical connector with pigtail of 12-channel fiber ribbon

SiOB is the most critical part of the optical subassembly and provides mechanical support and electrical connection for chips attached on it. For its excellent thermal conductivity, high mechanical strength, and suitable coefficient of thermal expansion (CTE), silicon is usually used a device carrier. In this work, we designed and fabricated the SiOB based on a high resistivity p-type (111) Si substrate, shown as figure 1. The fabrication procedure is as follow. First, an oxide layer about 1μm was formed by thermal oxidation. Next, the 200/5000 Å Ti/Au layer was deposited by e-beam evaporation for the transmission lines and wire bonding metal. The transmission lines are arrayed with 125 μm pitch, including 65μm width of line and 60 μm gap. Wafer electroplating was adopted to form the PbSn solder bumps with 70μm diameter and 30μm height. Last, the 12 through holes arrayed with 250 μm were formed by RIE-etching. All the positions of the bumps, transmission lines and holes were designed according the structure of optoelectronic devices and chips.

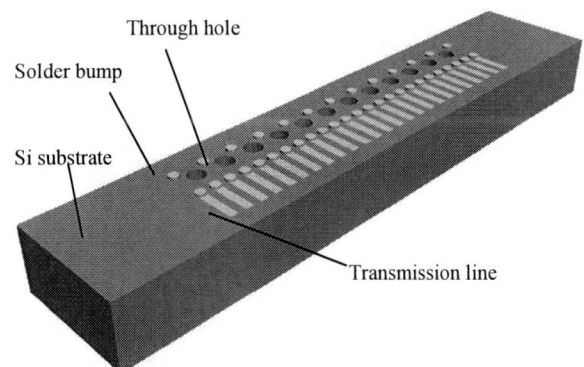

Figure 1. Schematic structure of SiOB

After SiOB fabrication, the optical subassembly was assembled as following processes. First, a 12-channel VCSEL array was flip-chiped on the SiOB. The pads and active areas of the VCSEL are in one-to-one correspondence with the bumps and through holes in the SiOB. Then, the VCSEL driver chip was wire bonded with the SiOB, and a fine-tuning appliance was used to adjust the SiOB to be perpendicular to the driver chip. Next, the SiOB and driver chip were mounted on a copper block and fixed with heat conduction glue. Last, optical alignment and coupling is implemented. All the optical assembly is shown in figure2.

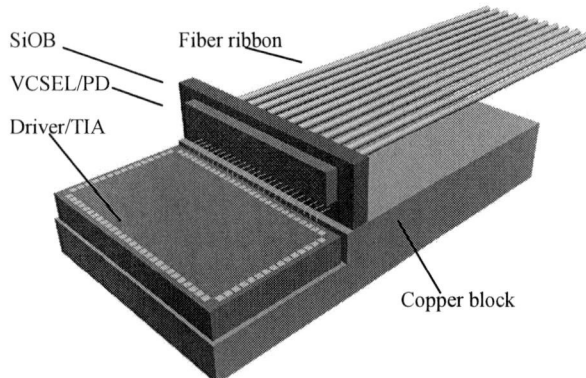

Figure 2. Schematic structure of the optical subassembly

Optical alignment and coupling

As well known, the requirement of high precise alignment and coupling of light to components is always a major problem of the parallel optical interconnects. The alignment contains 6-degree of freedom motion, namely, lateral (x-axis and y-axis), longitude (z-axis) and angular (rotation around x/y/z-axis). Simple optical coupling and alignment processing is one of the most efficient approaches to cut cost of an optical module.

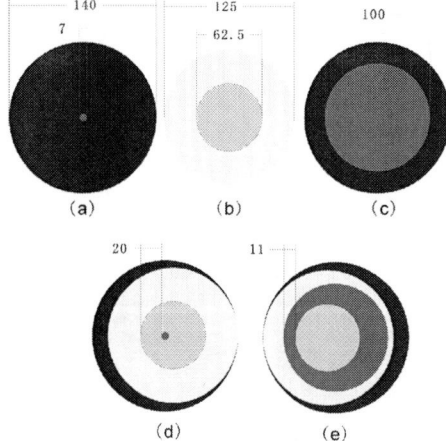

Figure 3. Schematic of alignment between fiber and VCSEL/PD.(a) the hole and active area of VCSEL, (b) the fiber core and cladding, (c) the hole and active area of PD, (d) alignment of fiber and VCSEL, (e) alignment of fiber and PD

In this work, we used the most direct passive alignment technique method for optical alignment and coupling. A 5-degree Newport Precision Stage was applied to put the fiber array into the holes array and to contact the surface of VCSEL or PD slightly. Figure 3 shows the alignment structure between the fiber and VCSEL or PD. According to the data sheets, the fiber core and cladding are 62.5 μm and 125 μm in diameter respectively, and the active areas of the VCSEL and PD are 7 μm and 100 um respectively. The hole diameter in the SiOB is about 140 μm which is decided by design and

RIE-etching process. As figure 3 (d) and (e) show, the VCSEL active area is completely in the fiber core area, and fiber core area is also in the PD active area. That is to say, if the reflection and transmission loss are not considered, the light beam from the VCSEL will completely couple into the fiber and then into the PD. In fact, there is position deviation between the center of the hole and the center of active area of VCSEL or PD due to the error of flip-chip process. Figure 3 (d) and (e) show that the permissible position errors for of VCSEL and PD are 20 μm and 11 μm, respectively. That is very relaxed for flip-chip process. After alignment, the MMF ribbon was fixed on the SiOB and the copper block by UV curing adhesive.

Module assembly and measurement

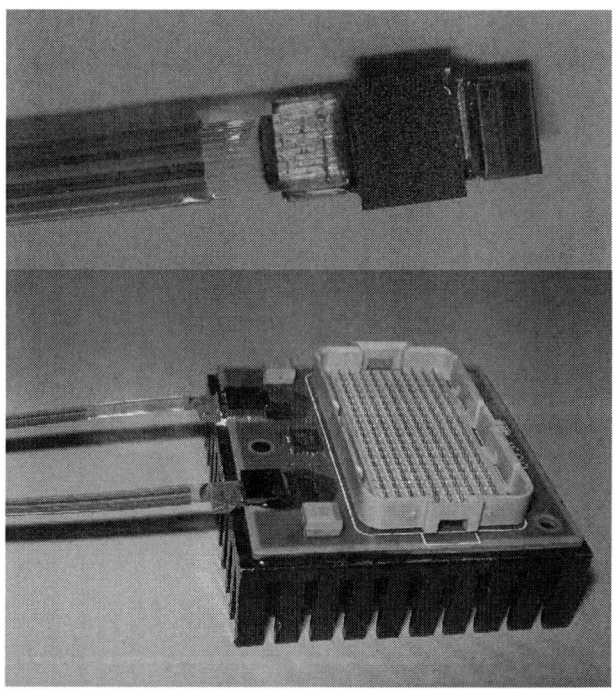

Figure 4. Photography of optical subassembly (top) and optical module (bottom).

A 3cm×3cm 6-layer PCB was designed and fabricated to house the SiOB by wire bonding. A .microcontroller chip, a 200-pin FCI plug connector and two filter capacitors for power supply were mounted on the surface of PCB by using the conventional SMT package process. 100-Ω differential transmission lines were applied on board for routing high-speed signals from I/O of high-speed chips to the pads of electrical connector.

A test board with receiver connector for the optical module was designed and fabricated. The back-to-back eye-diagram was measured as shown in Figure 5. In this case, NRZ 2^{23}-1 pseudo random bit stream (PRBS) generator drove the VCSELs to produce optical signals. Through the optical coupling structure and the multimode fiber and optical coupling structure again, optical signals was converted back to electrical signals at PDs within the same transceiver module. Limited by the speed of generator output, the data rate was 2.7Gb/s in the measurement.

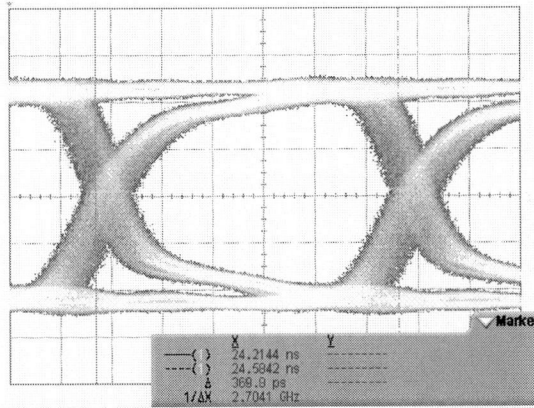

Figure 5. 2.7Gb/s eye-diagram of the links between 1 transmitter channel and 1 receiver channel in a single transceiver

Conclusion

We demonstrated a 12-channel optical transceiver module with SiOB vertically packaged in the optical subassembly. This structure eliminated the using of micro mirrors and lenses. Direct passive alignment and coupling are performed by putting the fiber into the SiOB holes and accessing to the surface of the optoelectronic devices which are flip-chiped on the SiOB. The analysis indicated that high coupling efficiency with relax permissible error of assembly can be expected.

Acknowledgments

This work was supported by Hi-tech Research and Development Program of China (863 Program) No. 2007AA01Z2a6.

The authors would like to thank Shennan Circuits Corporation for their supporting of high speed PCB fabrication. And we gratefully acknowledge Hi-tech Research and Development Program of China.

References

1. Doany, F. E. et al, "160 Gb/s Bidirectional Polymer-Waveguide Board-Level Optical Interconnects Using CMOS-Based Transceivers," *Advanced Packaging, IEEE Transactions,* vol. 32, (2009), pp. 345-359.
2. Benner, A. et al, "Exploitation of optical interconnects in future server architectures," *IBM J. Res. Develop.* vol. 49,, (2005), pp. 755-775.
3. Cook, C. et a, "A 36-channel parallel optical interconnect module based on optoelectronics-on-VLSI technology," *Selected Topics in Quantum Electronics, IEEE Journal of,* vol. 9, (2003) pp. 387-399.
4. Windover, L. A. B. et al, "Parallel optical interconnects > 100 Gb/s," *J. Lightw. Technol.,* vol. 22, (2004), pp. 2055-2063.
5. Kuchta, D. M. et al, "120 Gb/s VCSEL based parallel optical interconnect and custom 120 Gb/s testing station," *J. Lightw. Technol.* vol. 22, (2004), pp. 2200-2212.

6. Han, S. P. et a, "A high density two dimensional parallel optical interconnection module," *IEEE Photon. Technol. Lett.,* vol. 17, (2005), pp. 2448-2450.
7. Sung Hwan, H. et al, "Parallel optical transmitter module using angled fibers and a V-grooved silicon optical bench for VCSEL array," *Advanced Packaging, IEEE Transactions on,* vol. 29, (2006). pp. 457-462.
8. Watts, P. et al, "Experimental demonstration of real-time DSP with FPGA-based optical transmitter," *Transparent Optical Networks, 2008. ICTON 2008. 10th Anniversary International Conference on (2008),* pp. 202-205.
9. McCarthy, A. et al, "Fabrication and characterisation of direct laser-written multimode polymer waveguides with out-of-plane turning mirrors," *Conference on Lasers and Electro-Optics(Europe, 2005),* pp. 477.
10. Young, I. "Intel Introduces Chip-to-Chip Optical I/O Interconnect Prototype," *Technology@Intel Magazine April,* (2004), pp1-7
11. Wang,F. F. et al, "45 Degree Polymer Micromirror Integration for Board-Level Three-Dimensional Optical Interconnects," *Optics Express,* vol. 17, (2009), pp. 10514-10521.
12. Suzuki, A. et al, "High optical coupling efficiency using 45°-ended fibre for low-height and low-cost optical interconnect modules," *Electronics Letters,* vol. 44, (2008) pp. 724-725.
13. Ishikawa, T. et al, "High-density and Low-cost 10-Gbps x 12ch Optical Modules for High-end Optical Interconnect Applications," *Optical Fiber communication/National Fiber Optic Engineers Conference,* (2008), pp. 1-3.

Recent Advances in Laser Assisted Polymer Intermediate Layer Bonding for MEMS Packaging

Changhai Wang*, Jun Zeng, Yufei Liu
School of Engineering & Physical Sciences,
Heriot Watt University, Edinburgh EH14 4AS, UK
E-mail: c.wang@hw.ac.uk, Tel: +44 131 451 3903

Abstract

This paper presents the recent advances in the development of a laser assisted fast polymer bonding method for electronic packaging applications. In this method a high power diode laser is used to cure a polymer adhesive material to bond substrates together. A unique beam forming method using custom designed optical phase plate elements was developed to transform a fiber delivered laser beam into top-hat and frame shaped beam profiles for energy efficient polymer bonding for electronics manufacturing. In addition it has been found that the frame shaped beam profile can produce a desirable temperature distribution for MEMS packaging in which the surface temperature at the center of the substrate is lower by ~50°C than the packaging temperature (~300°C) when bonding a glass cap to a silicon substrate. An accurate temperature monitoring method using an embedded thin film microsensor array has been developed and has been used successfully for process monitoring. Defect-free fast bonding (~10 seconds) of transparent cap (glass) and non-transparent cap (silicon) to a silicon chip has been demonstrated. This work illustrates the potential of the diode laser based photonic technology for advanced MEMS packaging and electronics manufacturing.

1. Introduction

Polymer is a material that has been widely used in microelectronic manufacture for bonding and protection of IC chips and components. In recent years there has been a growing interest in polymers as bonding materials for MEMS and microsystem packaging. Polymer bonding offers several advantages over the other bonding methods such as low cost, low temperature, high bond strength, low thermal stress and high tolerance to non-uniform substrate surfaces [1]. The low temperature bonding characteristic is particularly attractive for MEMS packaging since excessive processing temperature in the packaging process can cause thermal mechanical failure of the micromechanical structures in the devices [2]. One polymer material that has been studied widely for MEMS packaging is the benzocyclobutene (BCB). It is a thermosetting material with a curing temperature between 200°C-350°C. The BCB polymer exhibits minimal outgassing, low moisture uptake and possesses excellent electrical properties. The BCB based polymer bonding approach has been investigated at chip level [3, 4] and wafer level [5-7]. The BCB polymer layer can be patterned easily to create sealing rings for housing MEMS devices by using dry etching or more conveniently by UV photolithography for the photosensitive BCB [1]. Strong, chemically and thermally stable BCB bonds have been obtained for various substrates. It has been shown that by cladding the BCB bond with an additional diffusion barrier [8] or by using an embedded barrier [9], it is possible to improve the hermeticity of the polymer seal. A unique property of the BCB polymer is that it can be cured rapidly with a curing time of seconds at temperatures above 300°C. This makes it ideal for laser curing of the BCB polymer for fast substrate bonding applications. In addition selective bonding can be obtained utilizing the localized nature of the laser heating effect.

The development of diode laser technology for industrial applications has resulted in high power, robust and highly efficient systems for direct application in materials processing. The wall plug efficiency of diode lasers is as high as 50% and further improvement is being made. These energy efficient lasers have already found successful applications in laser welding of plastic materials [10]. In our previous demonstration of laser based polymer bonding, a poor quality, free space beam was used resulting in an inefficient process [11]. In this paper we present our recent work on laser assisted polymer bonding using a fiber coupled laser and custom designed beam forming optical elements. By combining the polymer bonding technique with the laser assisted curing approach, the processing time can be reduced significantly and the bonding efficiency is improved greatly. We'll show that it is possible to realize reliable bonding within seconds using the laser based method. A new method for accurate temperature monitoring using an embedded sensor array for process control is also presented in the paper.

2. Principle of laser assisted intermediate polymer layer bonding

In the laser based intermediate polymer layer substrate bonding method, a laser is used to cure a polymer adhesive layer to join two substrates together to form a substrate assembly as illustrated in Fig. 1. The laser radiation can be absorbed by one of the substrates or the polymer layer itself to produce the necessary temperature change to cure the polymer layer to form a strong bond between the substrates. Fig. 1(a) shows the schematic of bonding a transparent substrate to a non-transparent substrate while Fig. 1(b) illustrates bonding of two non-transparent substrates. The substrate configuration in Fig. 1(a) can be reversed to have the non-transparent substrate facing the laser beam in the same approach as shown in Fig. 1(b). In MEMS packaging both transparent and non-transparent caps are used for encapsulation. Glass cap is commonly used for optical MEMS such as the micromirrors to provide a window for the input and output light beams. Silicon to silicon bonding provides a perfect CTE (coefficient of thermal expansion) match between the substrates improving the thermal mechanical reliability of the resultant package. Silicon can be considered as a non-transparent material for laser wavelengths below 1.1 μm.

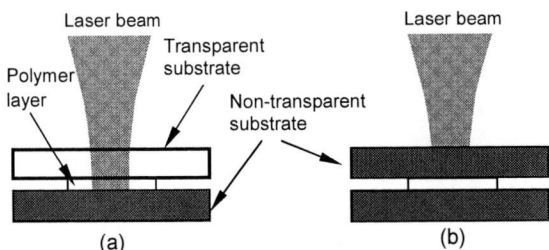

Fig. 1. Illustration of laser assisted polymer intermediate layer bonding of substrates, (a) absorption of laser radiation by the bottom substrate and (b) absorption of laser radiation by the top substrate.

3. Diode laser bonding of glass-silicon and silicon-silicon substrate assemblies using BCB polymer

In the laser bonding method a laser beam is used to produce a temperature increase to cure an intermediate polymer layer between two substrates to bond them together to form a cavity for MEMS packaging [11]. A high-power fiber-coupled diode laser at the wavelength of 970 nm was used in the work described in this paper. The diameter of the optical fiber is 200 µm. A beam delivery module and a beam forming element is used to produce a suitable beam profile for energy efficient polymer bonding. Fig. 2 shows the beam profiles produced using two different beam forming elements. Fig. 2 (a) and (b) show the 3D plots of the measured intensity distribution for a top-hat beam and a frame shaped beam. The outer dimensions of the beam for laser bonding are about 6mmx6mm for both beam profiles. The ring width of the frame shaped beam was designed to be 1 mm. Fig. 1 (c) and (d) show examples of the 2D plots of intensity distribution for each beam. It can be seen that most of the laser power can be directed to the area of bonding.

The BCB polymer bonding rings were produced on caps using the method described previously [4]. A BCB film was deposited on a glass or silicon wafer using spin-coating. The film was then baked and patterned using UV photolithography. The thickness of the resultant BCB film was about 10 µm. The dimensions of the BCB rings are 5mmx5mm. The track width is 400 µm. For chip scale investigation of the laser bonding method, the wafer was diced to produce individual capping substrates each with a BCB ring.

A bonding setup assembled in-house was used to investigate the laser assisted polymer bonding method for MEMS packaging. The substrates to be bonded, one with a BCB ring, are aligned and placed on a stainless steel supporting platform. In order to improve thermal efficiency a 0.9 mm thick ceramic plate was inserted between the platform and the substrate. A computer controlled force applicator (bonding arm) was used to press the substrates together to apply a suitable bonding force/pressure. A window on the bonding arm provides the optical path for the laser beam to reach the substrate assembly to raise the required temperature through the absorption of laser light by the silicon substrate.

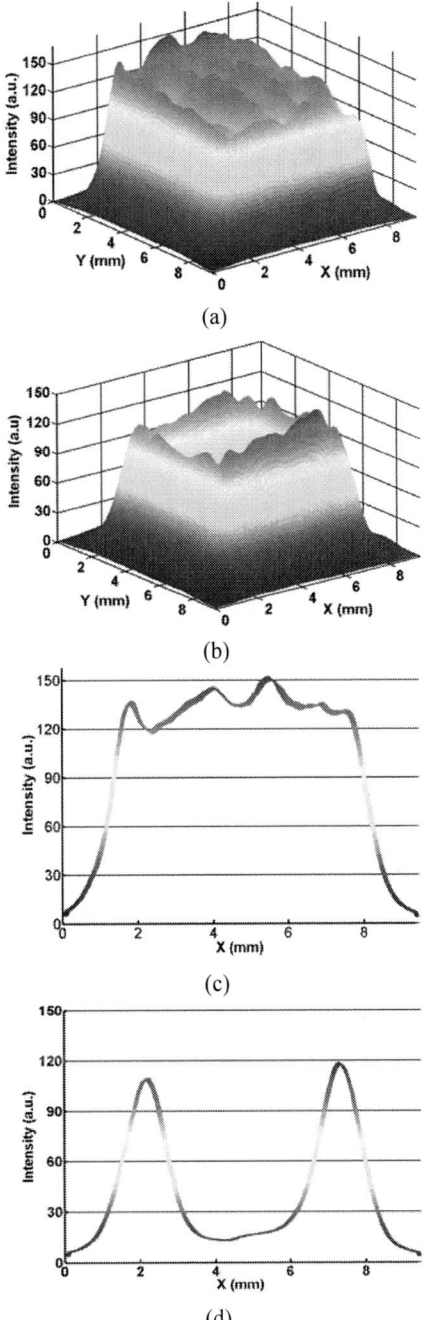

Fig. 2. Characteristics of the laser beam profiles produced using beam shaping optical elements. (a) top-hat beam, (b) frame shaped beam, (c) illustration of intensity distribution along the x-axis in (a) for the top hat beam and (d) illustration of intensity distribution for the frame shaped beam.

Fig. 3 shows optical pictures of laser bonded glass-silicon substrate assemblies using the top-hat and frame shaped beam profiles respectively. The laser power was 50 W for the top-hat beam and 60 W for the frame shaped beam. The bonding time was 10 seconds in both cases. It can be seen that defect free bonding has been achieved. The bonding process was carried out using the substrate configuration shown in Fig. 1(a). The light beam was transmitted through the glass substrate and absorbed by the silicon substrate. Fig. 4 shows an optical picture of a glass substrate bonded to a silicon substrate but the bonding process was carried out using the reverse arrangement of the substrate configuration shown in Fig. 1(a). The silicon substrate was arranged to face the laser beam. The thickness of the silicon substrates for this work was ~0.5 mm. All of the laser radiation entering the silicon substrate was absorbed and converted to thermal energy. In this configuration it only requires 30 W of laser power to raise the temperature to ~300°C for curing the BCB polymer. The bonding force and time were the same as 1 kgf and 10 seconds as before. The significant reduction of the laser power is due to the improvement of the thermal efficiency since the thermal conductivity of the glass is two orders of magnitude lower than that of silicon.

Precise temperature control is important in the laser assisted polymer bonding process. The laser induced temperature change depends not only on the laser power and beam profile but also on the dimensions and properties of the substrates as well as the thermal arrangement of the bonding setup. Secondly the curing time of the BCB polymer is highly dependent on the curing temperature [10], for example the curing time is of order 15 minutes at 250°C but seconds at ~300°C. More importantly excessive bonding temperature can cause thermal mechanical failure of MEMS devices. In the previous work, the conventional methods based on infrared detection [11] and thermo-sensitive paints [12] were used to monitor the temperature of the laser heating effect. But these methods could not provide the precise information about the temperature change within the polymer bonding track and the thermal distribution within the packaging cavity.

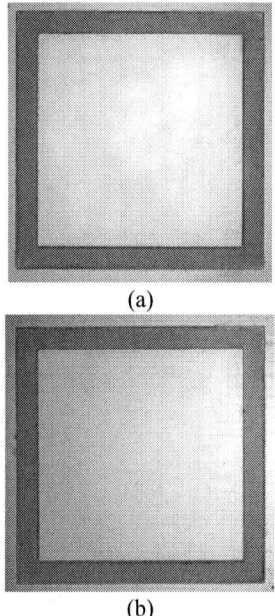

(a)

(b)

Fig. 3. Examples of laser bonded defect free glass-silicon substrate assemblies using the substrate configuration shown in Fig. 1(a) and the top-hat beam (a) and frame shaped beam (b) respectively.

Fig. 4 Optical picture of a laser bonded defect free glass-silicon substrate assembly using the reverse arrangement of the configuration shown in Fig. 1(a), i.e. the silicon substrate.

4. Temperature monitoring using a microsensor array

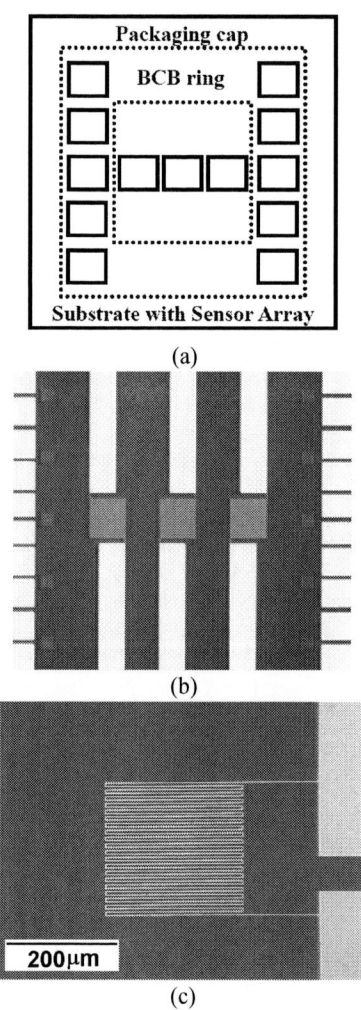

Fig. 5. (a) Plan view of a substrate assembly showing the alignment of the sensor array with respect to the BCB bonding ring for temperature monitoring in laser bonding, (b) Optical picture of a sensor array on glass substrate and (c) the detail of a peripheral sensor in the sensor array.

In order to monitor the packaging temperature accurately and to determine the temperature distribution at the surface of the device substrate, we have developed thin film resistive microsensor arrays for accurate temperature monitoring. Platinum based sensor arrays were designed and fabricated by plasma etching. Fig. 5(a) shows a schematic of the overlapping of the sensor array on a substrate and the BCB bonding ring on the cap for temperature monitoring. The peripheral sensors are embedded under the polymer track for monitoring the in situ temperature during the bonding process. Fig. 5 (b) and (c) show a fabricated sensor array on a glass substrate and the detail of a peripheral sensor. The footprint of the peripheral sensors is only 240μm×250μm. Meander design was used for the sensors for small footprint essential for this work.

In order to embed the peripheral sensors of a sensor array under the BCB bonding ring, a pre-bonding step was used to attach a cap with a BCB bonding ring to the sensor array on another substrate. The pre-bonding process was carried out on a flip chip bonder at 100°C at which the BCB material is soft and bondable. Fig. 6 shows an optical image of a sensor array on glass bonded to a silicon substrate. It can be seen that all of the peripheral sensors in the array are well placed under the BCB sealing track. Electrical leads were attached to the contact pads of the sensor array for measurement during the laser bonding process.

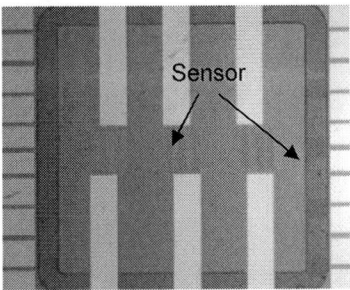

Fig. 6. Optcal picture of a sensor array on a glass substrate bonded to a silicon substrate after the pre-bonding step showing good alignment of the peripheral sensors to the BCB bonding ring.

In the temperature monitoring experiments, a pre-bonded cap and substrate assembly with the embedded sensors was placed on the bonding platform with an intermediate 0.9 mm thick ceramic plate between the sensor substrate and the stainless steel platform. The bonding area was aligned to the beam pattern. The top-hat and frame shaped laser beams were used to investigate the dependence of temperature change on the laser power and the beam profile. Fig. 7 shows the measured results of temperature change from the sensor array for the frame shaped beam profile. The cap is silicon and the sensor substrate is glass. At the onset of the laser radiation, the temperature increases rapidly and then slows down before reaching the steady-state temperature following an exponential characteristic. It can be seen that a temperature difference of about 50°C between the temperatures from the peripheral sensor and the center sensor can be obtained using

the frame shape beam profile. Fig. 8 shows the results of temperature change in bonding of a silicon cap to a silicon sensor substrate at different laser powers for both of the beam profiles. The results show that there is no significant difference between the top-hat and the frame shaped beam profiles. This is due to that the thermal conductivity of silicon is much higher than that of glass. Another effect of silicon as the bottom substrate is that a laser power of 50 W is required to produce a temperature of ~ 275°C, much higher than the 20 W for silicon-glass assembly. More details of the temperature monitoring work will be published elsewhere.

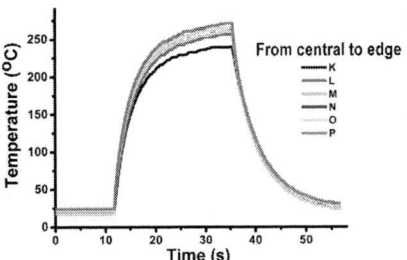

Fig. 7. Temperature monitored by the sensors in the array at a laser power of 20 W using the frame shaped beam.

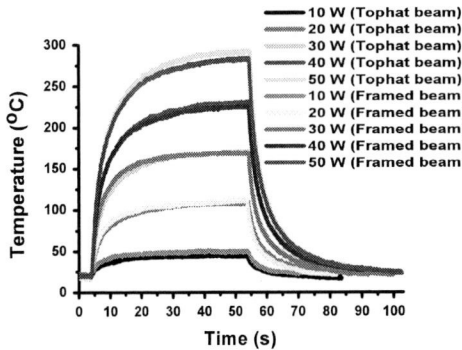

Fig. 8. Comparison of temperature change monitored by a peripheral sensor in a silicon-silicon assembly at different laser powers for both profiles.

5. Reliability studies

Shear test and leak test were conducted to assess the quality of the laser bonded substrate assemblies. For fine leak test the through-hole based approach was used [11]. The shear strength of the lased produced polymer bond is of order 20 kgf. This is comparable to that obtained for the similar BCB ring dimensions using the conventional bonding method on a device bonder [3] but having a much shorter bonding time. The fine leak rate was measured to be better than the sensitivity of the helium detector (~10^{-9} mbar·l/s).

6. Conclusions

A laser assisted fast polymer bonding process for MEMS packaging has been developed. Energy efficient polymer bonding has been demonstrated using a fiber coupled diode laser and custom designed beam forming optical elements.

978-1-4244-4658-2/09 $25.00 © 2009 IEEE

Defect free bonding of glass to silicon was achieved in 10 seconds using 50 W and 60 W of laser power for the top-hat and frame shaped beam profiles respectively when the laser beam was transmitted through the glass substrate before reaching the silicon substrate. For bonding the same assembly but having the silicon as the top substrate a laser power of only 30 W was necessary due to the improvement of thermal efficiency since the rate of heat dissipation through the glass substrate is lower than silicon. The quality of the laser bonded samples was assessed through optical inspection, shear and leak test. It was found that the quality of the laser cured polymer seal is as good as that of the samples bonded using the global heating approach, but having a short bonding time of seconds as compared to the duration of minutes to hours for the latter method. Diode lasers are the most efficient laser sources. Demonstration of direct application of a diode laser for MEMS packaging shows the potential of low cost photonic technology for electronics manufacturing.

An accurate temperature monitoring method for process control was developed using a microsensor array. The thin film resistive sensors have a small footprint and can be embedded underneath the polymer bonding track to provide in situ monitoring of the bonding temperature. It has been shown that the frame shaped beam profile can produce a lower temperature at the center of device substrate than the bonding temperature. This is a highly desirable effect for packaging of temperature sensitive MEMS and sensors. The high power diode laser technology has the potential to be incorporated in production equipment for energy and process efficient manufacture of MEMS and sensor products.

Acknowledgments

This work was supported by Scottish Enterprise and the European Fund through the Proof of Concept Programme (PoCP), and the Innovative Electronics Manufacturing Research Centre (IeMRC) of the UK Engineering and Physical Sciences Research Council (EPSRC). The authors would like to thank Dr Roy McBride of PowerPhotonic Ltd (UK) for assistance in measurement of the beam profiles.

References

1. Oberhammer, J. et al, "Selective Wafer-level Adhesive Bonding with Benzocyclobutene for Fabrication of Cavities," *Sensors and Actuators,* Vol. 105 No. 3 (2003), pp. 297–304.
2. Lin, L., "MEMS Post-Packaging by Localized Heating and Bonding," *IEEE Transactions on Advanced Packaging,* Vol. 23, No. 4 (2000), pp. 608-616.
3. Jourdain, A. et al, "Mechanical and Electrical Characterization of BCB as A Bond and Seal Material for Cavities Housing (RF-) MEMS Devices," *Journal of Micromechanics and Microengineering,* Vol. 15, No. 7 (2005), pp. S89–S96.
4. Wang, C. H. et al, "Chip Scale Studies of BCB Based Polymer Bonding for MEMS Packaging," *Proc 58th Electronic Components and Technology Conf,* Lake Buena Vista, FL, May 2008, pp. 1869-1873.
5. Niklaus, F. et al, "Low-temperature Full Wafer Adhesive Bonding," *Journal of Micromechanics and Microengineering,* Vol. 11, No. 2 (2001), pp. 100–107.

6. Niklaus, F. et al, "Low-Temperature Wafer-Level Transfer Bonding," *Journal of Microelectromechanical Systems,* Vol. 10, No. 4 (2001), pp. 525-531.
7. Oberhammer, J. et al, "Selective Wafer-level Adhesive Bonding with Benzocyclobutene for Fabrication of Cavities," *Sensors and Actuators,* Vol. 105, No. 3 (2003), pp. 297–304.
8. Oberhammer, J. et al, "Sealing of Adhesive Bonded Devices on Wafer Level," *Sensors and Actuators,* Vol. 110, No. 1-3 (2004), pp. 407–412.
9. Wang, C. H. et al, "A Novel Encapsulation Method for Wafer Level MEMS Packaging,", *Proc of IMAPS-UK Conf on Biosensors and MEMS Packaging,* Edinburgh, 2-3 March 2009.
10. Bachmann F. G. et al, "Laser welding of polymers using high power diode lasers," *Proc SPIE,* Vol. 51, 2003, pp. 385-398.
11. Bardin F. et al "Laser Bonding of Glass to Silicon Using Polymer for Microsystems Packaging," *IEEE/ASME Journal of Microelectromechanical Systems,* Vol.6, No.3, (2007), pp. 571-580.
12. Wu, Q. et al, "Localised Laser Joining of Glass to Silicon with BCB Intermediate Layer," *Proc of 3rd Pacific International Conference on Application of Lasers and Optics,* Beijing, Apr. 2008, pp. 202.

Wafer-Scale Hermetically Packaged MEMS Switches with Liquid Gallium Contacts

Qingquan Liu

Nanjing University of Information Science and Technology

College of Electronic and Information Engineering, Micro/Nano Electronics Research Team, Nanjing, China 210044

Email: q.liu@ieee.org, Phone: +86-25-58731271

Abstract

A wafer-scale hermetically packaged self-healing RF MEMS switch, which utilizes liquid gallium (Ga) contacts to take the place of the traditional solid metal contacts, is proposed in this paper. Liquid Ga contact switch features significantly improved power handling capability compared to conventional MEMS switches. The switch is driven by an electrostatic actuator, which includes upper electrodes on a silicon nitride bridge and lower electrodes on the substrate. Small liquid Ga droplets serve as a self-healing interface between the upper and lower contact electrodes. To minimize RF loss, a quartz wafer is used as the substrate material. To prevent oxidation of the liquid gallium, a hermetical packaging process with frit glass seal is used. A pyrex glass cap wafer is connected to the quartz device wafer by the frit glass seal. Finite element analysis is used to simulate the electro-thermal heating of the package. The measured package leakage rate is below 1 Torr/year.

Introduction

Microelectromechanical systems (MEMS) contact switch, also named as MEMS relay, can allow electric current or radio frequency (RF) signal to pass through two metal contact electrodes, controlled by an external actuation voltage. They generally feature low contact resistance, good isolation and linearity [1]. Since 1990's, a significant amount of effort has been focused on MEMS contact switches, which have shown great potential to replaced transistor and diode switches used in traditional RF systems. Nevertheless, MEMS contact switches suffer from contact welding, pitting, hardening, oxidation, corrosion, and mechanical bouncing. Heat generated on the microscopic contact areas limits the power handling capability. Although some switches have survived multi-billion-cycle contact tests with 10 mA current, they cannot survive current on the order of 1 A [2]. Liquid metal switches, e.g. gallium (Ga) or mercury switch, may offer high power handling capability [3]. The application of mercury switch is limited by environmental concerns. Hence, it is desirable to have a MEMS switch with integrated liquid Ga contacts. Furthermore, a MEMS-based liquid metal switch needs a hermetic package, which not only provides long-term sealing to prevent oxidation of Ga but also features high power handling capability.

Design and Fabrication of the Liquid Gallium Switches

This paper proposes a wafer-scale packaged RF MEMS switch with liquid Ga contacts. For harsh environment applications (e.g. in aerospace), the package of the switch is required to be reliable in a wide range of temperature. It has to be noticed that, since the melting point of the Ga is close to room temperature, external heating is needed for low temperature applications. The switch illustrated in Fig. 1 is designed to handle 1 A DC current. Electrostatic actuation is

used to drive a silicon nitride bridge with upper electrodes. When the silicon nitride bridge is pulled down, small liquid Ga droplets shown in Fig. 2 work as a contact medium between the upper and lower electrodes. To minimize substrate loss of the RF signal, a 400 μm thick quartz wafer is used. A 2-μm-thick electroplated gold layer is used to construct a coplanar waveguide (CPW), which carries electric signal from one side of the package to the other. The gold CPW also serves as a path for heat transfer, because the thermal conductivity of the quartz substrate is smaller than that of gold. A 10×10 array of 4-μm-diameter Ga droplets is deposited on the lower contact electrodes at the center of the CPW. Fig. 3 is a scan electron microscopy (SEM) image of the Ga droplet array.

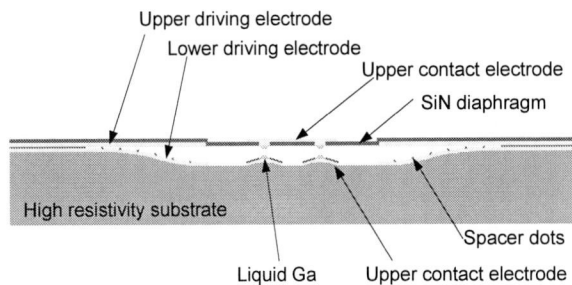

Fig. 1. Cross-section of a liquid Ga contact switch with an s-shaped electrostatic actuator. The switch is on its off-state.

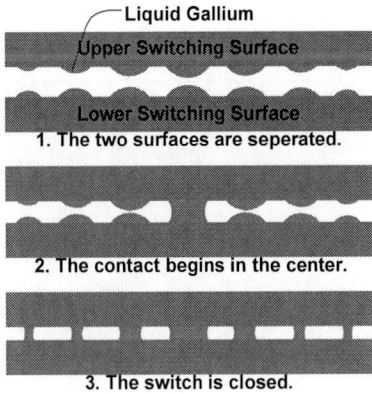

Fig. 2. Conceptual cross-section of the upper and lower contact electrodes covered by liquid Ga droplets.

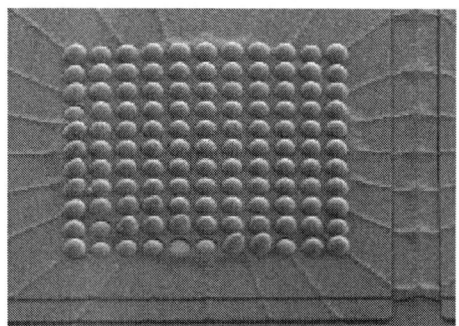

Fig. 3. SEM image of an array of Ga droplets deposited on the electrodes located on the signal line of the CPW.

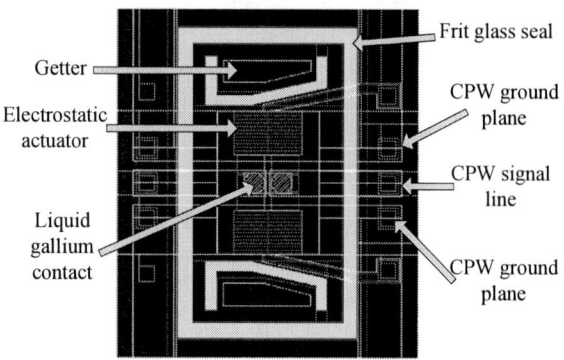

Fig. 4. Layout design of a liquid Ga contact RF MEMS switch with a hermetic package. The rectangular frit glass seal pattern encloses the liquid Ga contacts, electrostatic actuators, and titanium getter.

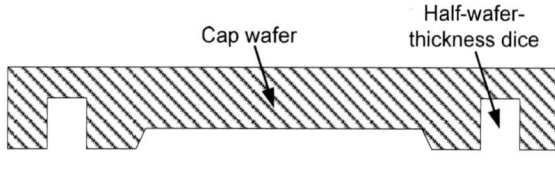

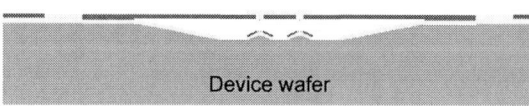

Fig. 5. Cross-sectional schematics of the cap wafer with half-way cut trenches.

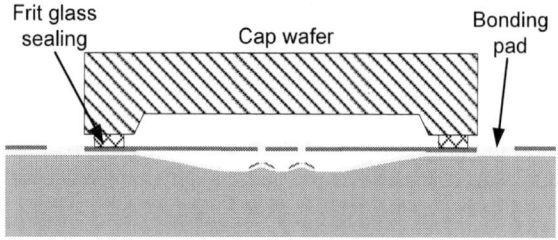

Fig. 6. Cross-sectional schematics of the completely cut cap wafer bonded to the device wafer.

Wafer-Scale Hermetic Packaging

To reduce cost and improve yield, a wafer-scale packaging process is developed. The layout design is shown in Fig. 4. The package consists of a quartz wafer that hosts the switch device, a pyrex glass cap wafer sitting on the top of the device wafer, and a frit glass sealing pattern, which provides hermetic sealing between the device wafer and the cap wafer. The packaging process is as follows. First, a deep reactive ion etching (DRIE) or a half-way saw dicing process step produces trenches approximately 200 μm deep on the cap wafer, as shown in Fig. 5. The rectangular frit glass pattern is screen printed on the device wafer. The cap wafer is then bonded to the device wafer. This step is illustrated in Fig. 6. In order to minimize oxidation and contamination, titanium getter is deposited on the cap wafer before bonding. After bonding, a dicing saw can be employed to dice the cap wafer along the half-way trenches to cut the cap wafer completely. Then the device wafer is diced. Gold wires are used as bonding wires.

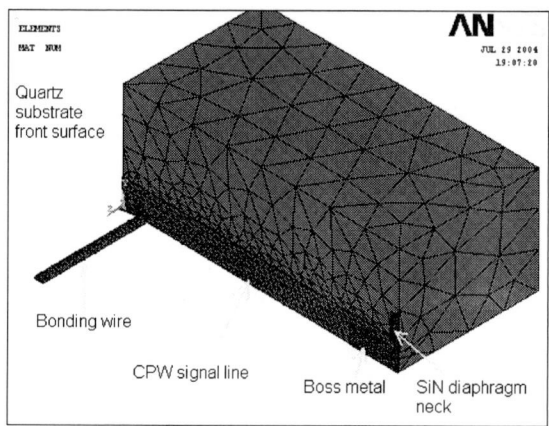

Fig. 7. Finite element model of the device, including the quartz substrate and the bonding wire.

Design and Fabrication of the Liquid Gallium Switches

When a 10 mA current is applied, heating of the package can be negligible. However, a 1 A DC current or high power microwave signal induces significantly more heat than small currents.

The liquid Ga contact temperature and the temperature distribution of the switch package are importance characteristics of the switch. Excessive resistive heating of the liquid Ga droplet can cause evaporation and loss of liquid Ga, thus compromising the reliability of the device. Excessive heating may also increase the stress and strain of the frit glass seal. The wire bonding reliability might also be threatened by the temperature increase of the bonding wire and the package. Since the switch is expected to operate in harsh environment of a broad temperature range, it is necessary to investigate the thermal characteristic of the entire switch including bonding wires.

Since the electrothermal simulation of this liquid Ga contact switch is complicated, the modeling work starts with a simplified model that includes a part of quartz wafer with half of the CPW signal line and a liquid Ga contact. A finite element analysis (FEA) model shown in Fig. 7 is used to simulate electrical and thermal characteristics of the package. Since the device is symmetric, only a quarter of the model is shown. ANSYS is used again here as the FEA software. Fig. 8 is a 3-dimensional view of the center of the switch.

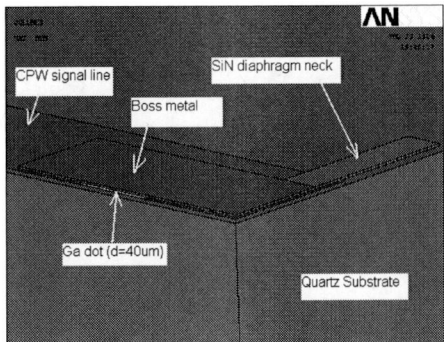

Figure 8. 3-dimensional image of the center of the switch. Only 1/4 of the device is shown.

For simplicity purposes, it can be assumed that a liquid Ga contact is a cylindrical droplet. The resistive heating power P of a volume can be given by

$$P = I^2 R = I^2 \rho \frac{t}{S} \quad (1)$$

where R is resistance; I is current; ρ is resistivity; t is thickness of the volume (the travel distance of the current in the volume); and S is cross-sectional area of the droplet. The volume is $t \cdot S$. Then the resistive heating power per unit volume is

$$\frac{P}{tS} = \frac{I^2 \rho}{S^2} \quad (2)$$

In the FEA model, 1 A DC current is applied on the signal line of CPW. The heat generation rate in the model is calculated using Eq. (2). The geometry parameters in this FEA model are given in Table 1.

Table 1. Parameters used in the simulation.

Parameters	Values
Quartz wafer thickness	400 μm
Gold bonding wire diameter	25 μm or 50 μm
Gold bonding wire length	300 μm or 1000 μm
CPW signal line width	75 μm
CPW signal line length	800 μm
CPW signal line thickness	2 μm
Ga dot diameter	40 μm
Ga dot thickness	1 μm
Gold boss metal thickness	1 μm
Si3N4 diaphragm thickness	1 μm
Si3N4 diaphragm neck width	20 μm

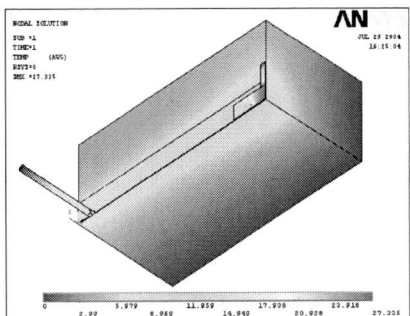

Figure 9. Contour plot of temperature distribution.

Since the two pieces of CPW ground lines have smaller resistance than the signal line, the resistive heat generated in the ground lines is less than that of the signal line. To simplify the model, the ground lines are not included in the analysis. To compensate the reduced resistive power, the width of quartz substrate model is reduced to 300 μm. The bonding wires have been included in the model. It is assumed that the back surface of the quartz substrate, the outer end of the silicon nitride diaphragm, and the outer surface of the bonding wire have a boundary condition of constant reference temperature. Air convection and radiation are neglected. The results have been obtained by electrothermal analysis of a die with a bonding wire of 300 μm length and 25 μm diameter. Due to relatively low thermal conductivity of the quartz substrate, the CPW and bonding wires play an important role in thermal conduction. Since the heat conduction to the cap wafer is negligible, the cap wafer is not included in the thermal conduction model. The modeling results of the temperature distribution are illustrated in Fig. 9. The maximum temperature increase of approximately 29 K occurs at the center of the switch, when a 1 A DC current is applied. Modeling results also indicate that this design can survive 1 A DC current in aerospace applications. The temperature increase of the liquid Ga droplet ranges from 13 to 16 K. The maximum thermal flux and thermal gradient appear near the interface between the liquid Ga dots and a solid metal electrode on the silicon nitride bridge. The result of the heat flux out of the package is listed in Table 2.

Table 2. Result summary of heat flow.

Thermal flux	Power (mW)
Thermal flux flowing through the bonding wire	5.29
Thermal flux flowing through the Si3N4 diaphragm	0.19
Thermal flux flowing through the substrate to the package	30

It can be seen from the thermal flux modeling results that the primary heat sinks are the bonding wires and the bottom of the substrate. The substrate thickness and the bonding wire dimensions have significant impacts on the heat conduction pattern. The resistive heating power of the CPW signal line is more than 4 times as much as that of the bonding wire. The resistive heating power of the liquid Ga dot is negligibly small compared to that of the CPW signal line. In the simulation, a 0.15 V voltage is applied on the RF signal path of the switch. Therefore the voltage applied on the half die model is 0.075 V.

Simulations of different bonding wire dimensions are also performed. In another model, the bonding wire features a length of 1000 μm and a diameter of 25 μm. ANSYS gives the maximum temperature increase of 44 K at the bonding wire. This is caused by the resistive heating of the relatively long bonding wire with a small diameter. The Ga dot and substrate temperature distributions are close to the previous result of a 300 μm long bonding wire. The applied voltage on the half die is 0.1063 V. The temperature distribution is given in Fig. 10.

Another switch model with a bonding wire length of 1000 μm and diameter of 50 μm is investigated. To obtain a DC current of approximately 1 A, the applied voltage on the half die model is 0.728 V. The distribution of temperature increase is shown in Fig. 11. The maximum temperature increase is approximately 28 K.

An SEM image the package is shown in Fig. 12. An accelerated leakage rate test has been performed. The measured package leakage rate is below 1 Torr/year.

Figure 12. SEM image of the hermetic package. The bottom die is from the device wafer. On the top is the die from the cap wafer. To provide wire bonding spaces, the length of the cap die is shorter than that of the device die.

Conclusions

This paper demonstrated a wafer scale hermetic packaging technology customized for a liquid Ga contact RF MEMS switch. Quartz substrate is used to reduce substrate RF loss. Frit glass sealing is used to provide long term hermiticity. The modeling results of electrothermal heating and thermal conduction of the package suggest that the maximum temperature increase is in the range of 29 to 44 K when a 1 A DC current is applied. Measurement results of less than 1 Torr/year leak up rate has been obtained.

Acknowledgments

The research is supported by the "HERMIT" program of the Defense Advanced Research Program Administration (DARPA) of the United States, and by Honeywell. The publication of this article is also sponsored by the "Six Great Talents" Project of Jiangsu Province, China, by the Research Funding of Nanjing University of Information Science and Technology (NUIST), China, and by the Micro/Nano Electronics Research Team of NUIST. The author appreciates Professor Norman Tien, Case Western Reserve University, Dr. Daniel McCormick, Advanced MEMS Inc., and Dr. Daniel Youngner, Honeywell, for their valuable help.

References

1. Wang Y. *et al.*, "A low-voltage lateral MEMS switch with high RF performance," *J. Microelectromechanical Syst.*, vol. 13, no. 6 (2004), pp. 902-911.
2. Chan R. *et al.*, "Low-actuation voltage RF MEMS shunt switch with cold switch lifetime of seven billion cycles," *J. Microelectromechanical Syst.*, vol. 12, no. 5 (2003), pp. 713-720.
3. Shen W. *et al.*, "Controlling the adhesion force for electrostatic actuation of microscale mercury drop by physical surface modification," *Proc. MEMS '02*, 2002. p. 52-55.

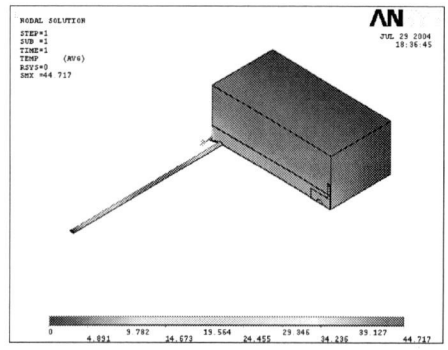

Figure 10. Temperature distribution (bonding wire length 1000 μm, diameter 25 μm)

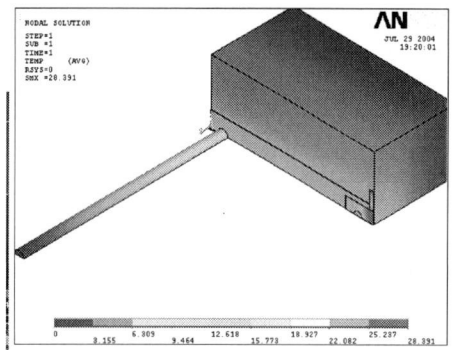

Figure 11. Temperature distribution of the package driven by 1 A DC current. (bonding wire length 1000μm, diameter 50μm)

Underfill Study for Large Dice Flip Chip Packages

Antony Lin[1], CY Li[1], Meng-Kai Shih[1], Yi-Shao Lai[1], Bernd Appelt[2], Andy Tseng[2]

[1]Advanced Semiconductor Engineering, Inc.

26 Chin 3rd Rd., Nantze Export Processing Zone, Nantze, Kaohsiung, Taiwan, China

[2]ASE (US) INC. 3590 Peterson Way, Santa Clara, California, 95054, USA,

Email: antony_lin@aseglobal.com, cy_li@aseglobal.com, ericmk_shih@aseglobal.com, yishao_lai@aseglobal.com.

bernd.appelt@aseus.com, andy.tseng@aseus.com, Tel: +1-408-986-6502

Abstract

Flip chip packages are becoming more popular due to many factors such as electrical performance, functionality and high I/O interconnections. To fulfill such needs in different applications, chip sizes are gradually becoming larger. Due to the large die sizes with high pin count, small bump pitch and low-K inter-metal-dielectric material, reliability concerns are arising at the interfaces of die, solder bumps and substrate. Of concern are package warpage issues, bump cracks, underfill void/delamination/cracks and die cracks, etc. The reliability issues can be solved by selecting more appropriate underfill materials to relief mechanical stress from CTE mismatch. Many commercial brands of underfill materials are available in the market and the underfill properties such as Tg, modulus, CTE, viscosity, flow characteristics, and adhesion need to be characterized before implementation. In this project, the underfill properties are studied and discussed and stress modeling for large dice in large packages is performed. Package data such as warpage, bump crack and delamination are measured for verification. The optimum underfill material for large die flip chip packages has been implemented in mass production. [1-2]

Introduction

Due to high electrical performance demands and more functionality with high I/O needs, the flip chip packages are becoming more popular and common. To fulfill such needs in different applications, the chip sizes are gradually becoming larger. Due to large die sizes with high pin count, small bump pitch and low-K inter-metal-dielectric material, reliability concerns are arising at the interfaces of die, solder bumps and substrate. Of concern are packaging warpage, bump cracks and underfill void/ delamination, etc. The reliability issues can be solved by the selecting more appropriate underfill materials to relief mechanical stress, and therefore, material properties are becoming critical for package reliability. Many commercial brands of underfill materials are available in the market and the underfill properties such as T_g, modulus, CTE, viscosity, flow characteristics, and adhesion need to be characterized and are discussed in this paper. Stress modeling of underfill materials for large dice in large packages is performed and package data such as warpage, bump crack and delamination are measured for verification.

In this study, twelve commercial underfills have been implemented and studied in a 37.5×37.5mm HFCBGA package with 17×18mm die. Two candidates out of twelve underfills with best performance will be chosen as the underfills for 21×18mm die in 42.5×42.5mm HFCBGA (device A) and 22×22 mm die in 45×45 mm HFCBGA packages (device B). Simulation modeling of stresses for large die size and package is performed and package data such as warpage, package stress and bump crack/delamination are measured for verification.

The underfill selection criteria are based on a number of performance indicators as listed in the following: Tg (temperature of glass transition), viscosity, CTE (coefficient of temperature expansion), modulus, toughness and curing conditions. These values provide only an indicator, as the actual optimized values of the parameters will depend on the type of package and specific process conditions. [3-5]

- Viscosity: -- Higher viscosity materials most likely have higher filler loadings which cause higher moduli and lower CTEs and may be more reliable material properties to protect solder bumps. But they may cause die cracking and lower throughput to impact mass production.

- Tg (Temperature of glass transition): -- Higher Tg materials can prevent phase changes during temperature cycling and provide better protection for solder bumps due to higher moduli. But a higher Tg induces higher warpage and higher residual stress on components and a decreasing trend on TC performance for low-K wafers. High Tg materials get better performance in HAST test due to more robust material structures and can resist moisture penetration.

- CTE (Coefficient of Thermal Expansion): -- Low CTE is always positive but most likely it comes with higher viscosity and higher modulus which increase die cracking, high stress and warpage trends on the components. Modulus: -- High modulus can provide good protection on solder bumps but will cause high warpage and higher initial stress which increase the impact on wafers with low K materials.

- Fracture toughness: -- Higher fracture toughness is better prevention for underfill cracking and moisture penetration.

- Alpha Ray Emission: -- Low alpha ray emission is required. Since almost all fillers are synthetic, underfill materials are low alpha materials.

- Curing condition: -- Underfills need to be cured under specified curing profiles. The material properties need to be certified again for any changes in curing profiles.

Warpage, Stress and Strain Simulation

Table 1 The properties of undefill UA, UB, UC and UD for ANSYS 10.0 simulations.

	Type	E (GPa)	α (ppm / °C)	Tg	v
Die	Silicon	131	2.8	-	0.3
Underfill	UA	7	32 / 110	70	0.28
	UB	9.2	30 / 110	85	0.28
	UC	10	30 / 100	74	0.28
	UD	8.5	28 / 90	100	0.28
Adhesive	AD-1	8	48 / 99	75	0.35
Solder mask	SM-1	2.7	50 / 140	100	0.3
BT	BT-1	4	XY: 46 / 120	156	0.3
			Z: 47 / 155		
Copper	-	121	16.9	-	0.3
Substrate (Core)	Core-1	27	XY: 14	160	0.3
			Z: 25 / 110		
Eutectic solder	63Sn/37Pb	C0= 75.84	25.4	-	0.35
		C1=-0.15			
TIM	TIM-1	20.2	320	-	0.35

Simulation for four underfill, UA, UB, UC and UD, used in large dice Heat spreader Flip Chip BGA (HFCBGA) has been done. The package is 45×45mm body size and 22×22mm die size. The table 1 shows the properties of four undefills. And the full solid model is shown in Figure 1.1, 1.2 and 2.

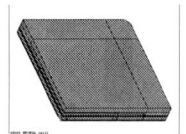

Figure 1.1 Figure 1.2

Figure 1.1 and 1.2 full solid model

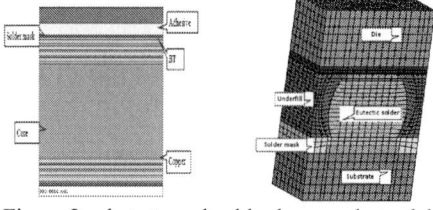

Figure 2 substrate and solder bump sub-model.

The package warpage data is shown in table 2. Package warpage with underfill UC is the smallest. The modulus of underfill UC is higher than others. I.e. the more constrain between die and underfill and its CTE is closed to soldermask's CTE, hence, the CTE mismatch is smaller than others.

Table 2 Package warpage simulation data.

Undefill	UA	UB	UC	UD
Warpage (um)	58.1	57.0	56.4	58.7

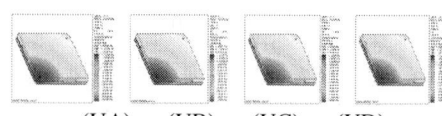

(UA)   (UB)   (UC)   (UD)

Figure 3 1/4 Package warpage contours of using underfill of UA, UB, UC and UD.

The potential of die crack through investigation of 1st principal stress has been simulated and is shown in table 3. The stress over of package also shows in figure 4.

Table 3 Package stress simulation data.

undefill	UA	UB	UC	UD
(MPa)	229	254	271	213

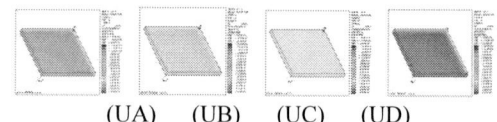

(UA)   (UB)   (UC)   (UD)

Figure 4 1/4 Package stress contours of using underfill of UA, UB, UC and UD.

The die is immune of crack if the stress is in compressive. Figure 4 shows the σ_1 principal stress distribution on the die at 25°C. The maximum stresses are located on the corner of top surface. The stress (σ_1) of underfill UD is smallest. The high modulus of underfill is high constrained between underfill and die. Ie, the thermal stress can't be released by shear deformation of the underfill. Beside, the maximum 1st principle stress in die occurs as interface between die and underfill. The data of the potential of solder bump failure through investigating of the von Mises total strain is shown in table 4. To examine the failure of the solder bump, a sub-

978-1-4244-4658-2/09 $25.00 © 2009 IEEE   1238

model for the outermost solder bump on die edge is set. Figure 5 shows contours of the strain on the critical solder bump.

Table 4 Package strain simulation data.

undefill	UA	UB	UC	UD
Von. Mises strain	1.58%	1.66%	1.67%	1.91%

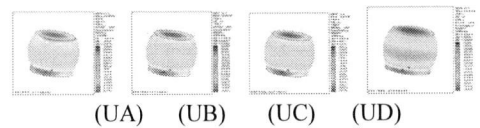

(UA)   (UB)   (UC)   (UD)

Figure 5 Package strain contours of using underfill of UA, UB, UC and UD.

The lower region close to substrate is much strain than upper region on die side. The CTE of solder bump with underfill UA, UB, UC and UD is 53.6, 48.5, 48.98 and 39.1 ppm/°C, respectively. The solder mask's CTE at 25°C is 86 ppm/°C. The solder bump would have larger deformation in Underfill UD due to large CTE mismatch.

Overall, the T_g, modulus and CTE of underfill are major factors that cause package bigger warpage, higher stress and bump cracking issues occur.

Design of Experiments

Viscosity impacts underfill flow. A fast underfill flow is no doubt a significant improvement since the underfill process is usually the bottleneck process in flip chip assembly due to long underfill flow time especially for large die size FCBGA. It means the throughput of Units-Per-Hour (UPH) of underfill process can be significantly increased when fine filler underfill is applied in standard underfill. The elimination of flow striations is a preferred result as it means more homogenous underfill and less risk of voids, and hence better package reliability as well. For effective under filling process via capillary flow the underfill material is expected to have low viscosity at dispense temperature, low contact angle for excellent wetting to the interfaces, and high surface tension for quick capillary underfill process. Further the underfill is expected to have uniform flow front, so that no flow voids are created. There are two capillary force components in the underfill process, one is caused by the surface tension between chip and substrate that is the function of gap height (horizontally), and the other is caused by the space between solder bumps which is the function of the pitch (vertically). By considering both capillary forces, it was found that on a full array solder bump pattern, the capillary flow was actually faster for finer pitch case in some arrangement. [5]

A DOE (design of experiment) has been composed and twelve underfills from four suppliers are used for this study. The material properties are shown in Table 1 and 2.

Table 5-1 Underfill properties

Vendor	Underfill	R1	R2	R3	R4	T1	T2
Filler Size (Avg)	Um	2	2	2	2	0.7	1.5
Filler (by weight)	%	55%	60%	60%	55%	60%	65%
Viscosity @25°C	Pa.s	60	30	50	20	14	60
Tg	'C	70	85	100	135	125	120
CTE α_1	ppm/C	32	30	30	40	29	26
CTE α_2	ppm/C	110	110	110	120	98	92
Modulus (<Tg)	G Pa	7	9.2	10	7.9	7.6	10
Curing Condition	C/min.	165/90	150/120	150/120	165/90	165/90	150/120

Table 5-2 Underfill properties

Vendor	underfill	S1	S2	S3	V1	V2	V3
Filler Size (Avg)	um	0.8	0.8	1	1	1	1
Filler (by weight)	%	60%	60%	40%	65%	60%	65%
Viscosity @25°C	Pa.s	40	28	13.0	20	11	30
Tg	'C	118	88	140	110	115	110
CTE α_1	ppm/C	32	32	50	31	32	34
CTE α_2	ppm/C	105	100	125	101	110	98
Modulus (<Tg)	G Pa	6.9	6.7	4.8	9.4	7.4	8.2
Curing Condition	C/min.	165/120	165/120	165/120	165/120	165/120	165/120

HFCBGA Package with 37.5×37.5mm body size is used and the size of 65nm low-K die with polyimide passivation is 17×18 mm. The bump composition is high lead with bump pitch 200 um. Substrate is 3-2-3 buildup substrate with 800um core thickness. Total package thickness is 2.55mm. The flip chip assembly process is shown in Figure 6. The

paxkage is shown in Figure 7. The wafer is high lead (Sn/Pb 95/5 %) bumped. In parallel with wafer mount and wafer saw, the buildup substrate is baked at 150°C/2hours and then processed through plasma clean, chip-cap pre-solder and chip-cap mount on build-up substrate. This is followed by flip chip bonding, underfill dispensing and underfill curing, heat spreader attach and curing, ink marking, solder ball attach and solder flux cleaning, and then final visual inspection.

After flip chip assembly, the parts go through warpage measurement and multiple reflows with readout points after 3 times, 5 times and 10 times, and then going through pre-condition, and TCB test with readout points at 200, 400, 600, 800, and 1000 cycles. The test results are shown in Tables 3.

Figure 7 HFC BGA package before Heat Spreader Attach. Die size 17×18 mm is bonded on 37.5×37.5mm HFCBGA.

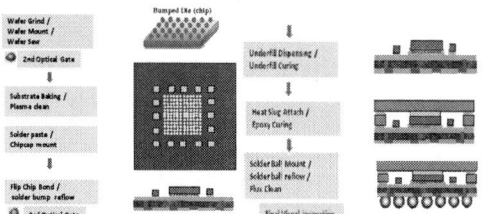

Figure 6 Flip Chip assembly process flow.

Table 6 Reliability test result of twelve underfills

UF	Tg	Modulus	Warpage	Multi-Re-flow	Pre-con	TCT 200	TCT 400	TCT 600	TCT 800	TCT 1000
R1	70	7	O	O	O	O	O	O	O	O
R2	85	9.2	O	O	O	O	O	O	O	O
R3	100	10	X	O	O	O	O	X		
R4	135	7.9	X	O	X					
T1	125	7.6	X	O	O	X				
T2	120	10	X	O	O	X				
V1	110	9.4	O	X	O	O	X			
V2	115	7.4	X	X	O	O	X			
V3	110	8.2	X	O	O	O	X			
S1	118	6.9	X	O	O	O	X			
S2	88	6.7	O	O	O	O	O	O	X	
S3	140	4.8	X	O	O	X				

Parts encountered underfill void / delamination or die crack or substrate popcorn and delamination as indicated by 'x', except for underfill materials R1 and R2. Both R1 and R2 passed TCT 1000cycles. The failure modes are shown in figure 8~15 – as defined in [6-8].

Figure 8 R1/R2 passed TCB 200, 400, 600, 800 and 1000 cycles.

Figure 9 R3 Underfill crack and tiny delam after TCB 600.

Figure 10 R4 Lower left corner delam after precondition.

Figure 11 T1/T2 Underfill Tiny delam after TCB 200

Figure 12 V1/V2/V3 Underfill tiny delam after TCB 400.

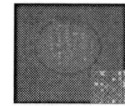

Figure 13 S1 Underfill tiny delam after TCB 200.

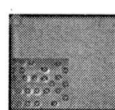

Figure 14 S2 Underfill Tiny delam after TCB 800

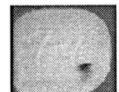

Figure 15 S3 Underfill large area delam after precond.

Underfill voids are one of the most common assembly defects, which depended on the die standoff, substrate and die surface conditions, process temperature, and material properties. The solder bump is usually in the way of the material flow and it is typical to see a series of underfill voids on one side of the bumps. It is also possible that some air adjacent to a bump is trapped by the material flows from two sides of a bump. After the underfill is cured, the air bubble is left in the epoxy. For big area voids, so called underfill delamination, results from a mismatch of material properties or the surface condition of the substrate: adhesion of underfill to substrate, adhesion to die passivation, and adhesion to solder bump. Underfill cracking is another major defect mode and is a combination of high stress and die defects, such as handling damage, edge cracking and stepping from dicing. Also, underfill cracking is generally related to underfill and passivation material properties. Strong adhesion between underfill and passivation, high underfill modulus, and brittle underfill and passivation material could result in cracking or delamination. Underfill properties contribute much to the die stress. A compliant layer such as polyimide could reduce the risk of cracking failure mode.

The small voids or delamination around bumps in R3/T1/T2/V1/V2/V3/S1/S3 were possibly the result of poor adhesion of underfill with passviation during reliability test. On S3/R4 large delamination on the die were found, the possible root cause is high Tg of underfill materials. The appropriate underfills will be selected from lower Tg for next DOE trial run.

Package warpage (CPL) data was collected for understanding the relationship of underfill properties and package warpage. Figure 16 shows that R1, R2, V2 and S2 have better coplanarity results than those underfills with higher T_g values. If the reliability data is combined with all the assembly data, underfills R1 and R2 will be considered two suitable candidates for bigger die sizes and larger package sizes. Underfill R1 and R2 were selected for further data collection, i.e. reliability and more coplanarity data.

Figure 17 shows the coplanarity prediction by using JMP optimization analysis. All the properties of different underfills were input and the result show that higher T_g underfill material will cause higher warpage and it fits a linear model.

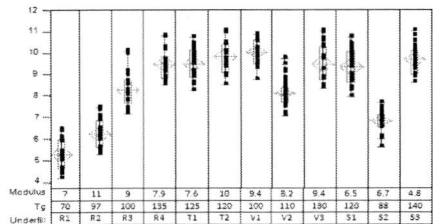

Figure 16 12 underfill coplanarity comparison chart

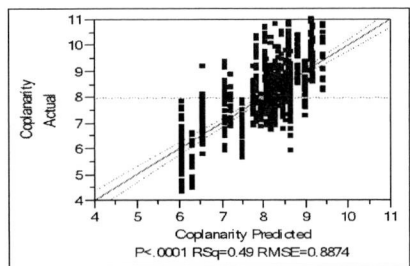

Figure 17 Coplanarity prediction chart

The prediction results in Figure 18 show that if T_g is lower (about 61°C) and the modulus is lower (about 4.8Gpa), the coplanarity could have reach the lowest value.

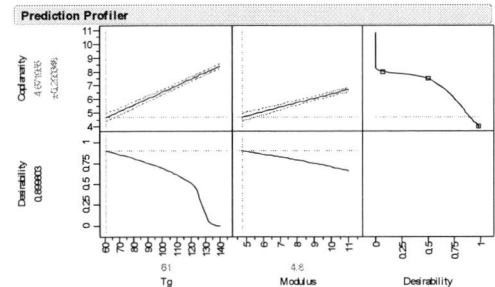

Figure 18 Warpage Prediction

Design of Experiment for Larger Dice Packages

Two candidates are used for larger die and larger package body size for the next study. The detail package information is shown in Table 7. The assembly process was the same as in DOE 1.

Table 7 Package information: Ring = stiffener, 1 pc HS = one piece heat spreader, 2pcs HS-two pieces heat spreader.

Device	Wafer Tech.	Passivation	Bump Pitch	Bump Comp	Die Size mm	PKG type	Heat Spreader	Substrate	Substrate Core
A	45nm	PI	170	EU	21×18	HFC BGA	ring, 1HS, 2HS	4-4-4	800μm
B	55nm	PI	200	EU	22×22	HFC BGA	ring, 1HS, 2HS	4-2-4	800μm

Results and Discussion

Table 8 Large die reliability test result.

Device	underfill	Ring only		1pcs HS		2pcs HS	
		Assembly Coplanarity	Reliability TCT 1000 cycles	Assembly Coplanarity	Reliability TCT 1000 cycles	Assembly Coplanarity	Reliability TCT 1000 cycles
A	R2	X	O	O	X		O
	R1	O	O	O	O	O	O
B	R2	X	O	O	X	X	X
	R1	X	O	O	O		

(A) Ring only (stiffener) FCBGA structure:

All underfill fail at Package coplanarity (CPL) test except for underfill R1 which passed in device A. Both underfill R1 and R2 passed reliability test in device A and device B. No reliability concerns are found for R1 and R2 underfill materials. Package warpage (CPL) improvement is required.

(B) One pieces heat spreader FCBGA structure:

Underfill R1 and R2 are pass package coplanarity test in both devices. Underfill R1 pass reliability test and R2 has delamination issue after TCB test in both devices.

(C) Two pieces heat spreader FCBGA structure:

Underfill R1 passed final coplanarity test but R2 fail. Both of underfill R1 and R2 passed reliability test on device A. R2 has delamination in device B. only R1 underfill can meet the reliability for package type A.

Conclusion

An appropriate underfill can reduce overall package coplanarity/warpage and reliability issues. To find a set of suitable underfill properties for large die size and big body side package, the fundamental understanding and the modeling data has been studied, the underfill properties, Tg, modulus and CTE, are the major indexes for underfill selection. High Tg material can prevent phase changes during temperature cycling and provide better protection for solder bump due to high modulus. But high Tg induces high warpage and high residual stress on components and has a negative effect on TC performance for low-K wafers. High Tg materials get better performance in HAST tests due to more robust material structures and can resist moisture penetration. High moduli can provide good protection on solder bumps but cause high warpage and higher initial stress which increase the impact on wafers with low-K layers. Low CTE is always positive but most likely it comes with higher viscosity and higher moduli which cause die cracking, high stress and warpage on components. Based on the material properties data, the underfill R1 was selected as the underfill material to solve coplanarity and reliability issues of large die sizes and large body size packages.

Acknowledgments

The authors would like to thank ASE R&D Lab for their assistance in the underfill modeling. Special thanks also to the ASE A3 Flip Chip team for flip chip assembly process. Special thanks to underfill suppliers for their kind support.

Reference

1. Fine P., Cobb B., Nguyen L., "Flip Chip Underfill Flow Characteristics and Prediction", *Proceedings of the 49th Electronic Components and Technology Conference*, 1999.
2. N. Iwamoto, M. Nakagawa, G.G.W. Mustoe, "Simulating Underfill Flow for Microelectronics packaging," *Proceedings of the 49th Electronic Components and Technology Conference*, 1999. pp294-301
3. Qing Tan, Rebecca Cole, Addi Mistry, and Craig Beddingfield, "Failure Mechanisms of Flip Chip DCA Assembly Using Eutectic Solder," *Proceedings of the 50th Electronic Components andTechnology Conference*, 2000.
4. Pei-Haw Tsao, Chender Huang, Mirng-Ji Lii, Bob Su and Nun-Sian Tsai, "Underfill Characterization for Low-k Dielectric/Cu Interconnect IC Flip-chip Package Reliability," *Proceedings of the 53th Electronic Components and Technology Conference*, 2004. Pp767-770.
5. Stone, Bill et al., "High Performance Flip Chip PBGA Development," *Proc 51th Electronic Components and Technology Conference*, May 2001.
6. Ahn, Eun-Chul et al., "Reliability of Flip Chip BGA Package on Organic Substrate," *Proc 50th Electronic Components and Technology Conference*, May 2000, pp. 1215-1220. [conference]
7. Shim, Jong-Bo et al., "Mechanism of Die and Underfill Cracking in Flip Chip PBGA Package," *2000 International Symposium on Advanced Packaging Materials*, pp. 201-205.
8. J-STD-030, "Guideline for Selection and Application of Underfill Material for Flip Chip and other Micropackages", *JEDEC,* Draft 7, Dec 2000, pp. 9-10.

Development of a Novel Cost-Effective Package-on-Package (PoP) Solution

P. SUN*, V. C. K. LEUNG, D. YANG, D. X. Q. SHI

Hong Kong Applied Science & Technology Research Institute (ASTRI)
2 Science Park East Avenue, Hong Kong Science Park, Shatin, New Territories, Hong Kong, China
*Email: psun@astri.org

Abstract

Package-on-Package (PoP) is one of the major 3D packaging approaches. It vertically combines discrete memory and logic Ball Grid Array (BGA) packages, where one package rests on the top of another. Recently, PoP technologies have attracted more interests, especially for portable electronics related products and applications.

For the existing PoP solutions, they have the following disadvantages:

- As the next-generation PoP module incorporates more features, top package requires more memory inside and finer pitch is needed. Finer pitch of top package will translate into lower stand-off height between top and bottom packages. It means the mold cap of the bottom package should be kept as thin as possible.
- The logic function of the bottom package will be expanded with two stacked dies. The mold cap of the bottom package should be thicker than that of the original design which has only one die inside. It causes the conflict between the stand-off height issue and expanded logic functionality.
- The bottom package is a Plastic Ball Grid Array (PBGA) package format normally. Due to its partial mold-cap design, the package's warpage could be large especially non-uniform on a panel since the structure is not well balanced.
- The bottom package with customized mold chase with the top pin gate molding method has the high cost issue.

To solve the above disadvantages, ASTRI has developed a new PoP structure, which employs a new bottom package that is over molded Fine-pitch Ball Gird Array (FBGA) format with mechanically balanced package structure. For the top package, it's a commercial FBGA format with two die stacking inside.

Since the new bottom package structure is based on an existing FBGA format, the mold chase is independent of the mold cap and size of the bottom PoP package. The final mold layer for bottom package has a thickness of 0.5mm. The top interface of the bottom PoP package has 136 pads with 0.65mm pitch and a two-row peripheral format, while the bottom interface has 272 pads with 0.65mm pitch and a four-row peripheral format. The size of both top and bottom packages is $14 \times 14 mm^2$.

The configuration of the package is evaluated by scanning electron microscopy (SEM), which shows a favorable interconnection structure. The warpage of the bottom PoP package after assembly process is characterized by shadow moiré system. The warpage is well controlled by adopting the fully molded structure. The average value is around 30μm that is well below the warpage target of 80μm widely used in the industry.

1. Introduction

Package-on-Package (PoP) is an integrated circuit packaging technology to allow vertically combining discrete logic and memory Ball Grid Array (BGA) packages into a single module. The obvious benefits of PoP design are motherboard space saving, a mix and match logic with multiple memories, flexible combination and assembly [1, 2]. PoP has become an attractive solution for higher density package in the market of mobile, PDA and digital camera [3, 4]. Besides die stacking within a single FBGA solution, PoP will be the next most popular form. It is expected to grow from 316M units in 2007 to 1733M units in 2012, as shown in Fig. 1 [5].

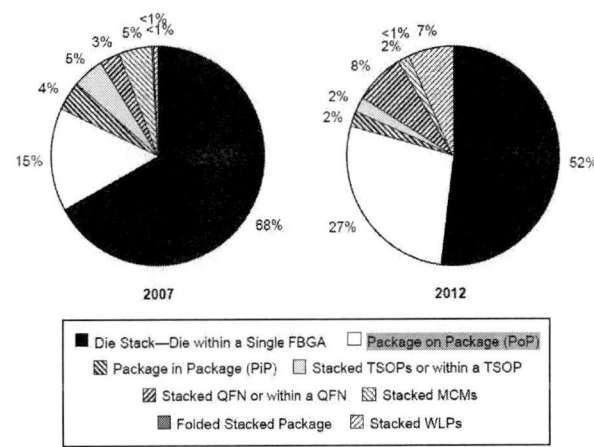

Fig. 1 Market forecast of stacked package units: 2007 vs. 2012.

In traditional PoP design, the top package is a stacked die memory product and the bottom package contains a logic processor. The perimeter area of the top surface of the bottom package arranges BGA land pads. The solder ball on the bottom surface of the top package substrate is soldered on the BGA land pads located on the top surface of the bottom package substrate. The solder joints provide Z direction space for the logic device mounted on the top surface of the bottom package. Since future applications require PoP memory unit with increased pin count new memory architecture, the ball pitch of the top memory package will be reduced in order to increase I/Os, the smaller solder joints will translate into lower stand off height between the top and bottom packages [4, 6, and 7].

Furthermore, increasing pin count with decreasing Z direction height will be a great challenge to PoP warpage

978-1-4244-4658-2/09 $25.00 © 2009 IEEE

control because the tolerance between the warpage of the top and bottom modules is decreased.

In traditional PoP solution, the top and bottom packages are in over molded FBGA format and partial molded PBGA format respectively, they exhibit different warpage behavior in the assembly process flow. Consequently, existing PoP design will become more difficult to achieve the higher memory pin count target and warpage control requirement.

To solve the above-mentioned issues in the PoP solutions, several new PoP package solutions are proposed and developed, as illustrated in Fig. 2 [3, 7-15]. As seen, major products and developments of the major players such as Amkor, ASE and Stats Chippac are compared.

Many companies participate into PoP development nowadays, some active players are:

OSAT: Amkor, ASE, SPIL and Stats ChipPAC;

Device supplier: Hynix, Intel, NEC, Numonyx, Samsung, Sony, Spansion, STMicroelectronics, Tessera and TI;

All the emerging PoP solutions can be classified into three categories:

PoP with opening interposer [12];

PoP base package with through mold via (TMV) technology [7];

Stacked fan in PoP (FiPoP) [13].

Table 1 summarizes the advantages and disadvantages of the three emerging PoP solutions as compared with traditional PoP design.

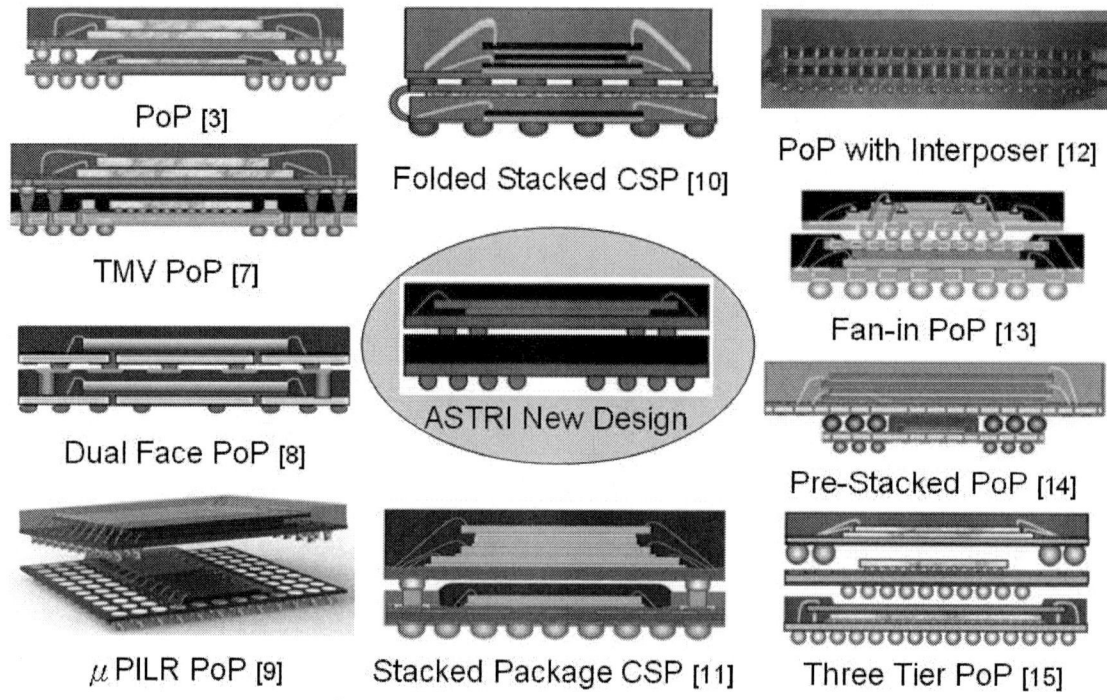

Fig. 2 Comparison of different PoP solutions.

Table 1 Comparison of three emerging PoP solutions and traditional PoP design.

	Traditional	Opening Interposer	Through Mold Via	Fan-in PoP
Typical Structure				
Top Package	Commercial package	Commercial package	Commercial package	Custom package
Mold Chase for Bottom Package	Custom mold chase	Custom mold chase	Commercial mold structure	Custom mold chase
PoP Thickness	Thin	Thick	Thin	Thick
Warpage Behavior	Poor	Poor	Good	Good
Circuit Length	Short	Short	Short	Long
Cost Factor	Custom mold chase	Custom mold chase	Drilling & filling process for solder in the via	Custom mold chase

2. Newly Designed PoP Module

To address the issues in traditional PoP structure, ASTRI develops a new PoP solution based on over molded FBGA package stacking structure. It remains the advantages of full testability, flexible combination, compatible with existing processes & equipments and low lost, but most importantly it solves the challenges such as stand-off height and warpage control in traditional PoP structure.

The new PoP stacked solution comprises:
1) A first package for memory units in FBGA format;
2) A second package for controller units with embedded substrate molded in FBGA format;
3) The embedded substrate comprises the conductive elements for interconnection purpose.

This new emerging solution demonstrates three notable advantages as compared with traditional PoP design:
1) FBGA format for both top and bottom packages, the mold chase is independent of the mold cap layout and package size;
2) Large stand-off height for die stacking in the bottom package;
3) Low warpage behavior of bottom package with mechanically balanced package structure.

The top package is one over-molded FBGA with two stacked dies inside and its body size is 14mm×14mm×0.7mm (length×width×height). It represents one memory package with 136 solder balls arranged in two rows, 0.65mm pitch perimeter array to match the soldering pads on the top surface of the bottom package. Ball alloy of the top package is Sn-3.0Ag-0.5Cu and surface finish is the electroless Ni/Pd/Au. The top package has a 0.20mm thick two layer BT substrate, which consists of two Cu layers and one layer of BT/prepreg core material, Mitsubishi HL-0832. The mold cap thickness is 0.5mm when measured from the top of the substrate to the top of the package.

The bottom package is also one over-molded FBGA with one die inside and the body size is 14mm×14mm×0.8mm (length×width×height). It represents one controller unit with 272 solder balls arranged in four rows, 0.65mm pitch perimeter array because logic package requires more connections to the motherboard. Solder alloy of the bottom package is Sn-3.0Ag-0.5Cu and surface finish on both sides is the electroless Ni/Pd/Au. The bottom package has 0.30mm thick four layer BT substrate, which consists of four Cu layers, one layer of BT core material, Mitsubishi HL-0832 and two layers of glass fibers prepreg material. For this design, the mold cap thickness is 0.5mm. The mold cap thickness of the bottom package is changeable, and is dependent on the total thickness of the embedded structure molded inside. As a merit of the design, the interposer allows for more or less die stacking by adjusting its standing height, which makes it possible to achieve the balance between the stand-off and extend function issues. The geometry information of both top and bottom package is summarized in Table 2.

Table 2 Package geometry of the top and bottom packages.

	Top Package	Bottom Package
Die Size (mm)	8×7×0.1	6×6×0.1
Number of Die	2	1
Mold Body Size (mm)	14×14×0.5	14×14×0.5
BT Core Substrate Size (mm)	14×14×0.2	14×14×0.3
Die Attach Material	put thickness here	put thickness here
Ball Pitch (mm)	0.65	0.65
Ball Diameter (mm)	0.45	0.45
Alloy of Ball Material	SAC305	SAC305
Number of Ball	136	272

3. Manufacturing of the Novel PoP Module

Comparing to the top gate molding method used in traditional PoP bottom package manufacturing process, the new bottom package design is proposed based on over mold FBGA format. The mold window size of the bottom package is 46mm×46mm as shown in Fig. 3. It should be noted that the Cu conductive element could be covered by epoxy molding compound after transfer molding process. To expose the pads of Cu for soldering connection, a precise depth-controlled polishing is developed. Fig. 4 illustrates daisy-chain patterns in the top package and bottom package, which are specifically designed to monitor the electrical interconnection continuity during PoP stacking process and board level reliability testing.

Following the top and bottom package assembly flow, the PoP stacking process is also evaluated on the basis of the traditional flux dipping method. Firstly, solder paste is screen-printed on the testing PCB with a steel stencil and the bottom package is mounted on the board. Then the top package is dipped into a tacky flux tray. After flux dipping process, the top package is placed on the top of the bottom package. When the mounting process is fulfilled, the top and bottom packages on the testing board are reflowed at the same time.

Fig. 3 Bottom PoP panel after transfer molding.

978-1-4244-4658-2/09 $25.00 © 2009 IEEE

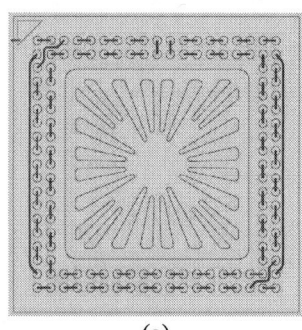

(a)

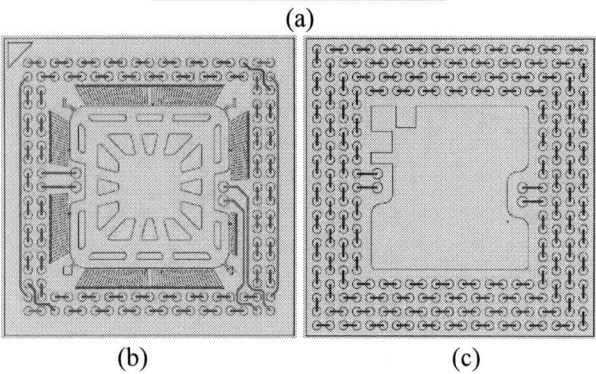

(b)   (c)

Fig. 4 The top and bottom package daisy chain nets: (a) top package netlist, (b) top side netlist of the bottom package and (c) bottom side netlist of the bottom package.

4. Performance Assessment of the Novel PoP Module

Fig. 5 PoP module stacked on the testing board.

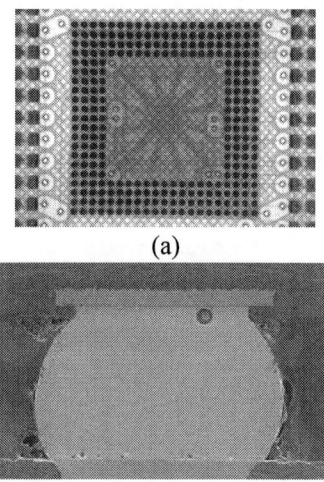

(a)

(b)

Fig. 6 Stacked PoP module: (a) X-ray inspection image of the stacked PoP module and (b) solder ball interconnection between top and bottom packages.

Fig. 5 shows the image of the PoP module stacked on the testing board. The electrical tests demonstrate the SMT yield is 100%. After PoP stacking reflow, the visual observation, X-ray inspection, SEM examination are successively carried out. The results are shown in Fig. 6.

X-ray inspection shows a well-aligned solder interconnection. There is no ball bridging or missing. From the cross-sectional analysis, it is speculated that the solder joints connecting the top and bottom packages was ideally formed during reflow process. As compared with the same size solder ball (0.45mm in diameter) used in traditional PoP interconnection between the top and bottom packages, the solder joints in this case exhibit a normal configuration (naturally collapse down) without tensile and compressive stresses introduced. Instead, as illustrated by Fig. 7 in our previous study [16], the shape of the balls has been vertically extended due to the warpage issue. If the warpage direction is opposite or the warpage magnitude of both packages is significantly different even on the same direction, the deformation of the connecting solder balls will be inevitably generated during the stacking process.

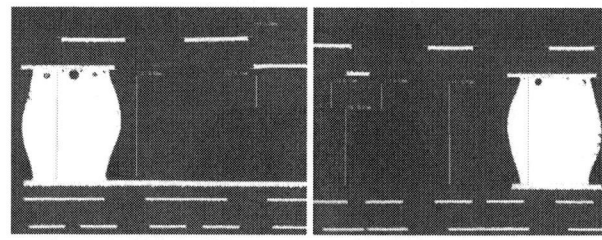

Fig. 7 Solder ball interconnection between the top & bottom packages in traditional PoP structure.

All the warpage are measured by LineMoiré PS-16M system with the measurement accuracy of ±2.5μm. The warpage direction definition is shown in Fig. 8. The concave shape is defined as smiling warpage and the convex shape is defined as crying warpage. The data of concave shape warpage is negative value and the data of convex shape warpage is positive one.

(a)   (b)

Fig. 8 Definition of package warpage: (a) concave and (b) convex.

The typical warpage results of the top and bottom molded blocks, after post mold curing (PMC) and reflow process, are illustrated in Fig. 9. According to our previous research [16], 4 hours of PMC are suitable to achieve a fully-curing and optimal warpage performance. As shown in Figs. 9 (a) and (b), the warpage of final molded top and bottom blocks are around 223μm and 189μm, respectively. Then the reflow process is simulated without attaching solder balls, and the warpage are measured to be 345μm and 270μm for the top and bottom blocks, respectively (see Figs. 9 (c) and (d)), which is a little higher than that after PMC. This could be understood because high-temperature reflow process enhances the thermal mismatch of different materials and the warpage is correspondingly increased.

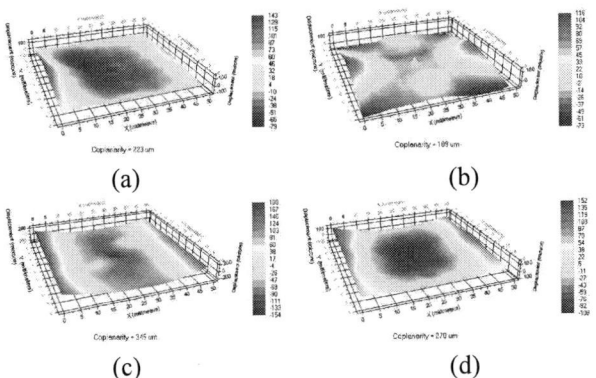

(a) (b)

(c) (d)

Fig. 9 Warpage patterns: (a) top block after PMC, (b) bottom block after PMC, (c) top block after reflow and (d) bottom block after reflow.

Also, the unit warpage is measured after block singulation, as shown in Fig. 10. The top package exhibits a concave pattern at room temperature. Likewise, the bottom package shows a similar one. It is different from other research work on the warpage study [17, 18]. The difference is believed to be caused by the variation in the specific configuration of the bottom packages. The warpage direction depends on the package design, material selection and process modification. Both concave and convex warpage shape have been found and reported. It is suggested that more attention should be paid to control the magnitude and the direction of the package warpage [19].

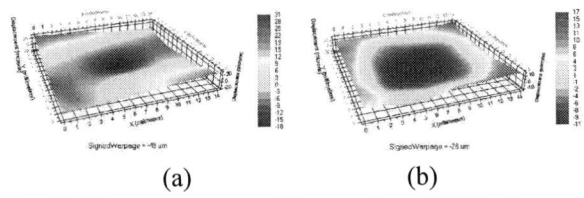

(a) (b)

Fig. 10 Warpage patterns: (a) top unit and (b) bottom unit.

The data of ten units (each for top and bottom packages) is collected. The average for top package and bottom package is about 46μm and 34μm, respectively. It is well within the warpage target of 80μm widely used in the industry. This low warpage would be compatible with even a finer pitch PoP (e.g., 0.4mm). Significantly, the warpage performance of bottom module is much improved by allowing for a fully molded structure. Also, to ensure good stacking yields of the solder joints between the top and the bottom packages and between the bottom package and the printed circuit board, the warpage difference between the top and the bottom packages after PoP stacking is evaluated. The result is found to be about 26μm in the average. It is also supported by the above micro-structural examination that the warpage difference is relatively mitigated and thus the stacking configuration demonstrates well through the new design. It is therefore concluded that the newly designed PoP structure and developed manufacturing processes can solve well both the warpage and stand-off issues in traditional PoP module.

5. Conclusions

In this work, a new PoP structure has been developed to address the stand-off and warpage issues observed in the traditional PoP module. Full set of PoP solution including the package design and manufacturing processes is established and numbers of prototyping samples were built. The electrical test results demonstrate an interconnect yield of 100%. The failure analyses including X-ray and micro-structural analysis further proves a favorable interconnection performance. The warpage behaviors of both top and bottom packages are characterized by shadow moiré technique. The warpage of the bottom package is well controlled and the average warpage value of different packages measured is around 30μm that is far below the typical warpage requirement of 80μm widely used in the industry. In summary, the newly developed PoP solution exhibits three notable advantages as compared with traditional PoP designs.

a) FBGA format for both top and bottom packages, the mold chase is independent of the mold-cap layout and package size;

b) Large stand-off height for die stacking in the bottom package;

c) Low warpage behavior of the bottom package with mechanically balanced package structure.

References

1. http://en.wikipedia.org/wiki/Package_on_package
2. David Geiger, Dongkai Shangguan, Samuel Tam, Dan Rooney, "Package Stacking in SMT for 3D PCB Assembly", *28th International Symposium on Electronics Manufacturing Technology*, 2003, pp. 261-264
3. Moody Dreiza, Akito Yoshida, Jonathan Micksch and Lee Smith, "Stacked Package-on-Package Design Guidelines", *Chip Scale Review*, July 2005.
4. Joanna Kristine Wildhart, Moody Dreiza, "Challenges for high density PoP (package on package) utilizing SoP (solder on pad)", *Global SMT & Packaging*, April 2008.
5. Electronic Trend Publications, Inc., "*Advanced IC Packaging Markets and Trends*", 2008 Edition
6. Moody Dreiza, Akito Yoshida, Kazuo Ishibashi, Tadashi Maeda, "High Density PoP (Package-on-Package) and Package Stacking Development, *Proceeding of 57th ECTC*, 2007, pp.1397-1402.
7. Jinseong Kim, Kiwook Lee, Dongjoo Park, Taekyung Hwang, Kwangho Kim, Daebyoung Kang, Jaedong Kim, Choonheung Lee, Christopher Scanlan, CJ Berry, Curtis Zwenger, Lee Smith, Moody Dreiza, Robert Darveaux, "Application of Through Mold Via (TMV) as PoP Base Package", *Proceeding of 58th ECTC*, 2008, pp.1089-1092.
8. Masamichi Ishihara, Yasuyuki Takehara, Takumi Yano, Yoshihiko Ino, Takashi Kurogi, Kenji Hashimoto, Hirotada Kawano, "A Dual Face Package Using a Post with Wire Component: Novel Structure for PoP, Wafer Level CSP and Compact Image Sensor Packages", *Proceeding of 58th ECTC*, 2008, pp.1093-1098.

978-1-4244-4658-2/09 $25.00 © 2009 IEEE

9. http://www.tessera.com/technologies/microelectronics/Pages/µpilr.aspx

10. Vijay Wakharkar, Chris Matayabas, Ed Lehman, Rahul Manepalli, Mukul Renavikar, Saikumar Jayaraman, Vassou LeBonheur, "Materials Technologies for Thermomechanical Management of Organic Packages", *Intel Technical Journal*, Vol.9, No.4 (2005), pp.309-323

11. Telesphor Kamgaing, Kinya Ichikawa, Xiang Yin Zeng, Kyu Pyung Hwang, Yongki Mi, Jiro Kubota, "Future Package Technologies for Wireless Communication Systems", *Intel Technical Journal*, Vol.9, No.4 (2005), pp.353-364.

12. Wei Chung Wang, Fred Lee, GL Weng, Willie Tai, Michael Ju, Ron Chuang, Weileun Fang, "Platform of 3D Package Integration" *Proceeding of 57th ECTC*, 2007, pp.743-747.

13. http://www.statschippac.com/services/packagingservices/3dsdsp/prestackedpop.aspx

14. http://www.statschippac.com/services/packagingservices/3dsdsp/fipop.aspx

15. Flynn Carson, Kazuo Ishibashi, Yeong Cheol Kim, "Three-Tier PoP Configuration Utilizing Flip Chip Fan-in PoP Bottom Package", *Proceeding of 59th ECTC*, 2009, pp.313-318.

16. Peng Sun, Vincent Chi-Kuen LEUNG, Bin XIE, Vivian Wei MA and Daniel Xun-Qing SHI, "Warpage Reduction of Package-on-Package (PoP) Module by Material Selection & Process Optimization", *Proceeding of ICEPT-HDP 2008*, 2008.

17. Flynn Carson, Seong Min Lee and In Sang Yoon, "The Development of the Fan-in Package-on-Package", *Proceeding of 58th ECTC*, 2008, pp.956-963.

18. Curtis Zwenger, Lee Smith and Jin Seong Kim, "Next generation PoP platform with TMV interconnection technology", *Proceedings of the IMAPS Device Packaging Conference*, 2009.

19. Niranjan Vijayaragavan, Flynn Carson and Addi Mistry, "Package-on-package warpage – impact on surface mount yields and board level reliability", *Proceeding of 58th ECTC*, 2008, pp.389-396.

EcoDesign Technical Committee of JIEP (Japan Institute of Electronics Packaging) and Its Activity

Hidetaka Hayashi
NPO Ecodesign Promotion Network
Eco-Design & Jisso Technical Committee of JIEP
Email: hhayashi@cba.att.ne.jp

Abstract

The discussions on environmental issue in electronics industry in the past are mostly on the regulation oriented technical issues to follow the needs of rapid increase on environmental regulations. We have reported in JIEP (Japan Institute of Electronics Packaging) session in EcoDesign2001 as EcoDesign JIEP Tomorrow [1]. Even if there are still many disputes, we could show the electronics industry in Japan is about to full fill most of all requirements stated in EcoDesign for tomorrow that was based on the global regulative scheme until 2010. Then our focus should shift to the future task that is more radical and challenging but difficult one.

This paper introduces activity of JIEP and EcoDesign & Jisso Technical Committee in the past and direction of study forward to the future. We pay special attention on saving, reliability on RoHS alternative material, reliability on easy decomposition structure, lifetime extension and emerging new technologies for the future. Those subjects are sometimes in trade-off relations of performance and environmental aspect.

Introduction

JIEP was founded in 1998 merging Japan Institute for Interconnection and Packaging Electronic Circuits (JIPC) and Society for Hybrid Microelectronics (SHM) collecting researchers and engineers in Electronic Jisso (Interconnection and Packaging) Industry in Japan.

There are 3066 members, 209 affiliates and 11 Technical Committees as of 31 Mar. 2009 and coves from material to Ecodesign. (Table 1)

Table 1 JIEP membership and Technical Committee

		As of Apr. 1 1998	As of Mar 31 2009
No of members	Regular	2,416	2,907
	Student	39	159
	Total	2,455	3,066
	Affiliate	329	209
Technical Committee(TC)	Material TC		
	Circuit & Jisso Design TC		
	Electromagnetic TC		
	Printed circuit manufacturing TC		
	Reliability TC		
	Electronic Component & Jisso TC		
	Inspection & Measurement TC		
	Optical Circuit Jisso TC		
	EcoDesign & Jisso TC		
	Micro-mechatronics Jisso TC		
	IC Packaging TC		

EcoDesign and Jisso TC started in 1998 succeeding the activity of Environment Protection and Recycling TC that had

been started in 1994 as a TC of JIPC (Japan Institute for Interconnection and Packaging Electronic Circuits).

The subject of the EPR TC committee in JIPC had started its activity mostly to prevent environment from chemicals used in the manufacturing process such as VOC free and waste water treatment and then shifted to recycling of WEEE and lead free soldering that is getting concerns of environmental regulations in Japan and EU directives. JIPC first co-organized Environmentally Conscious Engineering in Electronics '96 cooperating with LCA forum of Japan and University of Tokyo / Inverse Manufacturing Laboratory in 1996.

Activity of EcoDesign and Jisso TC

Soon after the start of activity the WEEE issue and RoHS issue were the major concerns of world electronics industry. At Ecodesign2001 Symposium we carried a survey on Ecodesign consciousness of JIEP affiliate companies and confirmed that lead free and halide free are well understood in JIEP. This survey also covered related Ecodesign consciousness such as energy saving, material saving and recyclability material. In Fig.1 more than 70% has plan, evaluating and already applied these issues.

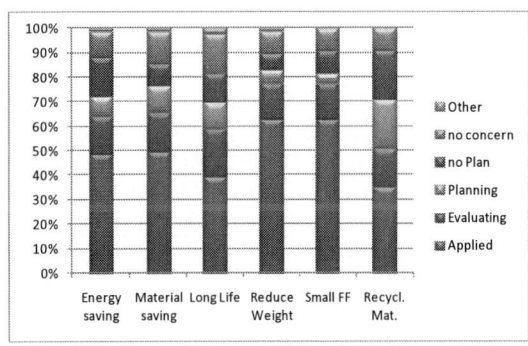

Fig.1 Consciousness of Jisso EcoDesign in 2001
(Affiliate of JIEP) additional to Regulation Issues

Since then protection environment, WEEE & RoHS and these saving issues are the subjects of EcoDesign and Jisso TC of JIEP. In 2006 we added reliability issue and summarized our subject as in Table 2 and Table 3 [2],[3].

Around 2006 global legislative requirement on WEEE (Waste of Electric and Electronic Equipment) is enhanced reflecting EPR (Extended Producer's Responsibility [4]) concept intending to clear global environmental issue caused by WEEE. RoHS alternative material technology, Recycling and Legislative and social framework in Table 2 related to EPR. Saving technology also related to EPR in a broader sense. But it has more rational meaning. The focal point of Ecodesign & Jisso TC is gradually shifting from simple EPR compliant to development of rational solutions.

978-1-4244-4658-2/09 $25.00 © 2009 IEEE

Table 2 Revised Subjects of EcoDesign and Jisso TC (2006)

Category	Material	Production process	Design and Software
RoHS Alternative Material Technology	Lead free solder Conductive adhesive Halide-free plastics (Flame retardant), Substrate for PWB wire and cable Pb eliminated dielectrics	Decomposition: Wet/ Dry process Cleaning :Wet/ Dry process	Modular Design for Hazardous Components Chemicals DB Analytical Equipment and Method for Hazardous substance
2R Technology	Bio plastics for Cabinet /Case Biodegradable Package Shape-memory material	Detachment, Removal, Recovery Room Tempe-rature repair	Design for easy de-composition CAD and DB for Recyclable Design
Saving Technology (Energy) (Material) (CO2)	2nd Fuel cell Catalyst Film based inter-connection Functionally purified water	Cu/Cu direct bonding, Wiring by plating Ink Jet wiring Nano-particle Wiring	Low power consumption design for equipment Miniature Low Power supply Low power cooling
Legislative and Social Framework		VOC free painting, RFID, LCA, LCC, Environment accounting, Environment Efficiency, CSR, Regulation, Technology Road map, Social System	

Reliability issue is basic issue in industry and commercial product. Historically, selection of material in a product is chosen affirming reliability spending very long time through variety of application. There are some acceleration methods to prove reliability of material but it must be confirmed by certain field test or commercial test. This means long time aging is mandate to establish reliability. Therefore it should be also mandate to assure reliability of RoHS alternative material and also assure reliability of easy disjoining and recyclable structure. Whisker in lead free solder and migration issue of Br alternative flame retardant in printed circuit board are typical issues for RoHS alternative material. Subjects listed in Table 3 are long time reliability issues related to establish material systems of our concern. There are special issues for each component categories.

Table 3 Reliability Issue for Green Components and Material

Components	Part of alteration (Replaced material)	Example of alternatives	Issue for reliability
PWB	Flame retardant additives in plastics (Br/Sb)	Organic phosphorous, Metal hydroxide, No additive	Reliability test
Encapsulation Resin	Flame retardant additives in plastics (Br/Sb)	Organic phosphorous, Metal hydroxide, No additive	Contents of P, Sb, Rigidity,
Wire & Cable	Frame retardant additives in PE (PBB,PBDE), Stabilizer (PVC)(Pb), Pigments (Cd; Red, Yellow:Pb,Cr6+)	Compound(Hydrotalcite +Zinc Stearate), Metallic soap, Organic pigment, Composite oxide yellow	Mechanical strength vs. flammability, Electric characteristics, long time insulation resistance
Connector, Fuse, Inductor	Pin plating, Terminal Plating (SnPb Solder)	Sn, Sn-Cu, Sn-Bi	Tin Whisker
Resistor, Fuse, Relay, Switch	Resistor(Cd), Fuse alloy (SnPbBi), Contacts (Ag-CdO)	Cd Free, Pb Free(Bi, In Alloy), Ag-SnO2,Ag-Ni	Poor lifetime, Temperature deviation(Fuse), Poor abrasion resistance
Screw	Anti corrosion (Cr6+ Chromate)	Cr3+ Chromate	Less resistive

Our focal point in Ecodesign

Together with the subjects listed in Table 2 and Table 3, there are very important following focal points for our activities. They are recycling, rational material saving and Silicon alternative active components (best mix of technologies). These subjects are chosen as the best mix of solutions to confront increasing EEE (Electric and Electronic Equipment) demand that is very challenging to construct global sustainable system globally.

Recycling

Recycling is an activity to collect WEEE to keep environment clean and recover material used in WEEE. It is a

978-1-4244-4658-2/09 $25.00 © 2009 IEEE

part of EPR activity. It is also effective system to resist material shortage. We also admit above thinking but we have another viewpoint additionally. That is value and effective supply of EEE to the market. These viewpoints are not purely engineering point of view. But as industry or engineering must be the tools to make society comfortable, these points of view should be qualified as first priority importance. Introducing these we found that merely recycling is not enough for sustainability. Prolonging lifetime of EEE in the market has the first priority. Fig. 2 and Fig. 3 explain above importance [4].

Relation of loosing mass and value increase in manufacturing

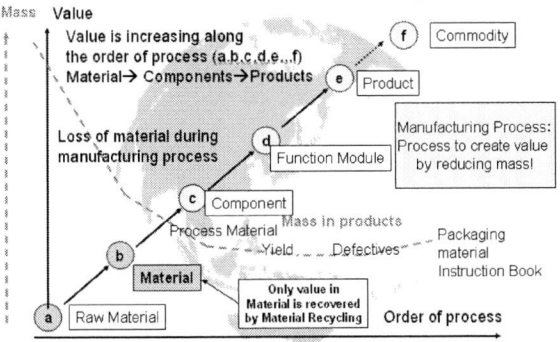

Fig.2 By recycling the value of "b" is recovered

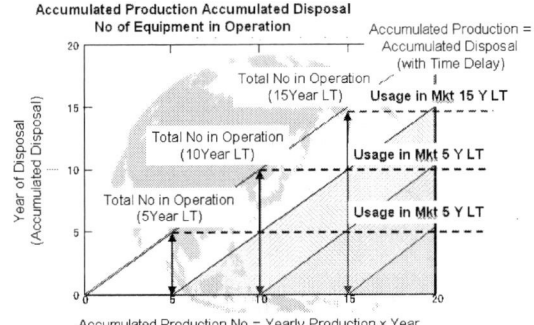

Fig. 3 Relation of Lifetime (LT) , Number of Production, Usage, Disposal of Equipment

Fig. 2 shows procedures of value increase during production process. Therefore expected value recovered by material recycling will be very small. In Japanese home appliance the recovered portion is less than 5% [5]. For high value product such as mobile phone the portion is less than home appliances. This relation is not merely shows the relation of value but shows the relation of loss in human activity by recycling. Fig 3 shows the relation of production and waste in terms of EEE lifetime. Shorter lifetime needs bigger production number and produce bigger waste.

Therefore our subject to improve recycling system is clear. That is, modular based equipment design, indicator of modular interface and power consumption recorder to improve EEE performance after distribution. One model is shown in Fig 4 (Integrated inverse distribution and production

system). Flexible circuit and disjoining technology have been focused.

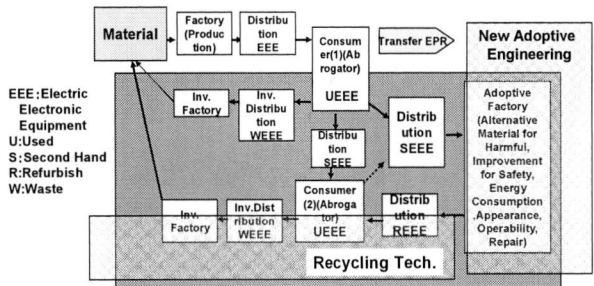

Fig. 4 Integrated inverse distribution and production system

Rational material saving

Our next focal point is very rational. That is elimination of insulating material in printed circuit. Of course it is not realistic to eliminate totally but it is feasible to replace most of material by very small supporting mean. Skeleton Circuit Structure (SCS) is one solution [6]. Basic idea of SCS is free-standing structure of conductors. As without supporting a structure cannot keep standing in the gravitational field, SCS employs little amount but negligible amount of material for supporting. (1) Casing, (2) Components, (3) Partial support are the feasible supporting structures.

Casing, Components are the structures with no excess material for supporting because both of them are composed with packaging material of insulation to be used for supporting. But it has less freedom to make printed circuit. Production procedure shown Fig 5 is one of them. Components are mounted on the casing or supporting tool before making SCS. It is difficult to assemble printed circuit prior to make equipment.

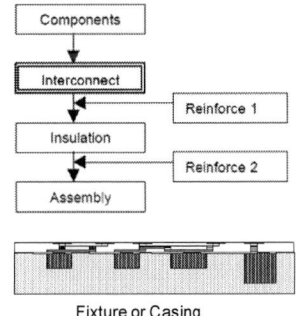

Fig. 5 Assembly procedure of SCS

The structure is shown in Fig 6. This structure has a power and ground sandwich structure on the base composed with thin dielectrics and conductor layers (SCS on base layer). By this structure we can be free from the restriction on process sequence to make printed circuit. The additional material used to make the thin structure of dielectrics and conductors is not considered as excessive material but considered as a part of power source component in which low characteristic impedance is favorable and essential. It is preferable to use high permittivity material in dielectric thin layer for making low characteristic impedance sandwich structure. SCS structure has little meaning on this power ground layers. Partial support is original idea of SCS and universally applicable to all parts of SCS.

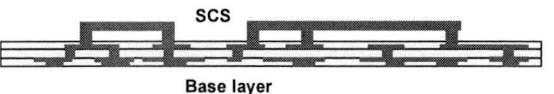

Fig. 6 SCS on base layer (Basic Structure)

The production process is now in progress based on the idea shown in Fig. 7. A supporting layer of temporary material is formed on base structure. Via holes are made by laser drilling or mechanical drilling at the location of via holes. The supporting structure is formed filling thermo set plastics in drilled via. A seed copper layer is made on the temporary layer by either sputtering or electro-less plating method. The copper seed layer is increased thickness by electric plating.

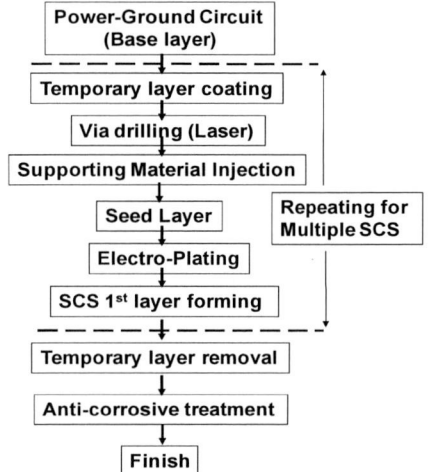

Fig.7 Production Process of SCS on Base layer

Best mix of technologies

As Silicon based technology has excellent characteristics of performance and process capability, miniaturization of device and integration was accomplished on silicon wafer. As a result ultra high density IC made by tens of nanometer technology has billions of transistors in a processor with dimensional disjunction between IC and interconnection on IC. Therefore to fulfill the interconnection gap between IC and Printed circuit board is very urgent and fascinating subject for the engineers of Jisso industry. But shrinkage in device increases the gap between interconnection in IC and outside. As outside IC is connected to human scale device best mix of technology will be combination of at least two different technologies. One is very high speed, very high density and miniature the other is low speed, human scale density and human size. Our next focal point is the latter part technology. In this part, size of device does not require silicon size. Therefore the material saving and energy saving has

greater meaning. Organic semiconductor technology has a capability of our concern (Table 4).

Table 4. Silicon vs. Organic Semiconductors

Item	Si	Organic
Major player of Characteristics	Dopant Impurities	πelectron, Impurities
Concentration of Impurities	≈10ppm	≈1%
Process Temperature	≈1000°C (Diffusion)	≈150°C (Curing)
Flexibility	No	Possible
Printable Ink	Impossible	Possible
Attachment of Device on FPC	Adhesives	Integrated in Printing

Conclusions

Environment conscious design in Electronics is getting more attention to make sustainable electronic industry. RoHS alternative material and recycling have partial importance. But more important focal points of research have been become clear by extending the framework of research to value or economical aspect of industry. They are prolonging lifetime of equipment, rational saving of material and more radical solution. JIEP EcoDesign & Jisso TC provides one of the best play ground for the researchers who are thinking to contribute "Green Jisso Technology" for sustainable global industrial system.

References

1. Hayashi, H., "JIEP EcoDesign Tommorrow," *Proc EcoDesign2001 International Symposium*, Tokyo, Dec. 1999, pp. 1018-1019.
2. Environmental Conscious Technical Committee. "Electronics Jisso Roadmap-Green Jisso Technology for Sustainable Era" *JJIEP* Vol9 No1 pp35-39 2006 (in Japanese).
3. Hayashi, H. Yoneda, Y., Wasserman, Y, "Green Jisso Technology for Sustainable Evolution," *Proc 1st EcoDesign2006 Asia Pacific Symposium*, Tokyo, Dec. 2006, pp. JI-1-4.
4. Hayashi, H., "Suggestions on China's Secondhand E-product Market Control System," *Proc. 2007 International Forum on Environmental Legislation and Sustainable Development*, Dec. 2007 Beijing, Dec. 1-2 2007.
5. Hayashi, H. Suga, T., "Investigation on value increase in inverse distribution process Part ll," *Proc EcoDesign2007 International Symposium*, Tokyo, Dec. 2007, pp.OS-2-01.
6. Hidetaka, H., "Skeleton Circuit Structure (SCS) for advanced electronic applications," *Proc ICEP2009 International Conference on Electronics Packaging*, Kyoto, Apr. 2009, pp. 130-134.

Artificial Neural Network Application in Vertical Interconnection Modeling

Yuanjun Liang[1,2], Lei Li[1]

[1]Shenzhen Institute of Advanced Integration Technology
Chinese Academy of Sciences/The Chinese University of Hong Kong
1068 Xueyuan Avenue, Shenzhen University Town, Nanshan District, China
[2]Institute of Microelectronics, Chinese Academy of Sciences
Email: yj.liang@siat.ac.cn

Abstract

This paper proposes a neural network-based method for modeling vertical interconnect balls. The π equivalent circuit with lumped element was used to characterize the electrical performance of balls. The values of lumped elements were extracted from the full-wave simulate result, and the extracted data was used to train back-propagation (BP) neural networks to obtain the relationship between the lumped element values and the physical size of the balls. Then this model was used to estimate the electrical character of ball with other layout parameters.

Introduction

In recent years, microsystem continues to move forward to higher speed and smaller. It brings electronic packaging into three-dimensional (3-D) structure and system integration called system in package (SIP). SIP requires high–density interconnection, which needs much shorter and finer connector [1]. Vertical interconnections are applied in 3-D structure package, and they allow designers much more option, but also lead to difficulties in signal routing. Solder-plated polymer balls are used as vertical connector between conductors. [2]

As the physical environment surrounding the interconnections affects frequency response, 3-D integration and high-density packages increase the need for electromagnetic (EM) modeling. [3] A good model can characterize electromagnetic property of the package before manufacture and provide prototype for electronic simulation. EM models are built up by 3-D full wave simulation tools, such as High Frequency Structure Simulator (HFSS). Based on the EM simulation, an equivalent lump SPICE model is created for vertical interconnection and then polynomial approximation is used to approximate the relationship between EM model and physical size of polymer balls. [4] However, there is no certain obvious function between physical size and EM model sometimes, and polynomial approximation is not suitable for the mapping. Then some other methods are used. This paper introduces Artificial Neural Network (ANN) into the EM modeling, the EM simulation data is used to train back-propagation (BP) neural networks to get an accurate model as a function of layout parameter. The prediction error can be reduced to 5%, even 1% by increasing the layers of the network or the numbers of nerve cells. This article presents a 3 layers BP neural network to model different sizes of Solder-plated polymer balls in vertical interconnection, and builds up an accuracy and fast relationship between the physical size and electromagnetic character of the ball.

Modeling of Vertical Interconnection

The analyzed vertical interconnect structure is shown in Fig.1. EM simulation was done with HFSS to obtain S-parameter and Y-parameter of frequency response. Since the structure symmetrical, $S_{11}=S_{22}$ and $S_{12}=S_{21}$. The equivalent circuit model of vertical interconnection was developed on the basis of physical geometry of the structure, [3] shown in Fig.2, where C1 and C2 denote the capacitance between conductor and ground (C1=C2), L1 denotes the inductance of the ball, R1, R2, L2 denote the frequency-dependent losses due to finite conductivity metals loss and substrate and dielectric loss, [5] Fig.3 presents a circuit and HFSS simulated S_{11} result, they match closely in the frequency range 1Hz-10GHz.

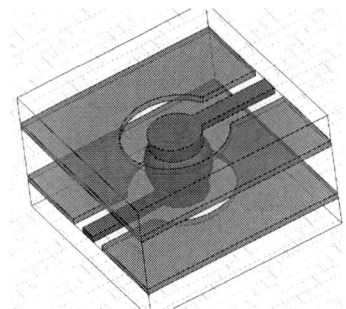

Fig.1. Structure of the vertical interconnection.

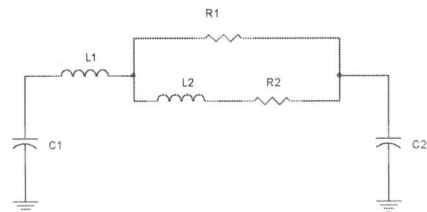

Fig.2.The equivalent circuit of vertical interconnection

The Y-parameters of the equivalent circuit are given as follows [3]

$$Y_{11} = Y_{22} = sC_1 + 1/(R(s)+1/sL_1) \qquad (1)$$
$$Y_{21} = Y_{12} = -1/(1/R(s)+1/sL_1) \qquad (2)$$

R(s) is frequency-dependent and denotes the equal resistor of loss, it can be seen as a constant when physical size changes and be calculated by the expression in [3],[6]

By solving (1), (2), L_1 and C_1 can be expressed as

$$C_1 = imag(Y_{11}+Y_{21})/s \qquad (3)$$
$$L_1 = imag(-Y_{21})/s \qquad (4)$$

978-1-4244-4658-2/09 $25.00 © 2009 IEEE   1253

Then the mean values of the capacitance and inductance over the entire frequency range are used to approximately represent the frequency independence.

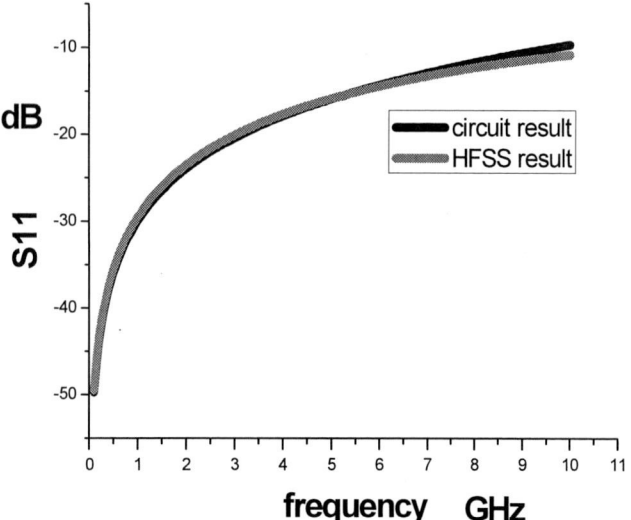

Fig.3 A circuit and HFSS simulated S_{11} result

In this paper, the diameter of the vertical interconnection ball was variable from 100um to 300um, 32 balls with different diameters were performed by HFSS and equated with lumped element. The numerical values of lumped elements were extracted by solving (1) ~ (4), and used to train the BP neural network

Neural Network Modeling

Artificial neural networks have been used widely in modeling complex, nonlinear and uncertain relationships. It is a mathematical model or computational model that tries to simulate the structure and/or functional aspects of biological neural networks. It consists of an interconnected group of artificial neurons and processes information using a connectionist approach to computation. In most cases an ANN is an adaptive system that changes its structure based on external or internal information that flows through the network during the learning phase, a simple ANN structure is shown in Fig.4

BP neural network is the most commonly used artificial neural network. It is a supervised learning method in which an input vector is fed into the network, and output is calculated by summing the weighted input connections of each layer and filtering this sum with a hyperbolic tangent activation function. BP neural network uses a gradient decent technique to train the network fast and accurately.

hyperbolic tangent activation function. BP neural network uses a gradient decent technique to train the network fast and accurately.

The performance of the network is evaluated by the root-mean-squared error (RMSE), given by

$$RMSE = \sqrt{\frac{1}{n-1}\sum_{i=1}^{n}(y_i - \bar{y}_i)^2} \qquad (5)$$

where n is the number of the data and y_i is the goal, and $\bar{y}_i$ is the neural network model output, the error is then back-propagation to change the weight between two nodes. The change in weight is given by

$$\Delta w_{ijk} = -\eta \frac{\partial error}{\partial w_{ijk}} \qquad (6)$$

where i denotes a node in layer k, j is a node in preceding layer ($k-1$), constant η is called learning rata, which is in the range of 0-1, a new error will be obtained by (5) and compared to the target error. This process would iterate unless the error matches the target error.

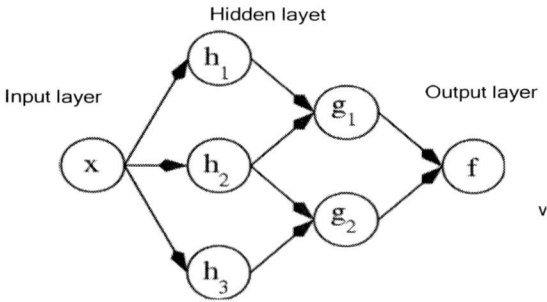

Fig.4 A simple neural network

A 3-layer neural network was used to model the relationship between layout sizes and lumped element values in this paper. The input layer had one neuron, corresponding to the layout diameter. The minimum number of neurons in the hidden layer was selected as 8 to get greater accuracy and fewer training iteration. The output layer consisted of 2 neurons representing the lumped capacitance and inductance. Network training was done by MATLAB using its ANN tools. The training performance is shown in Fig.5, after 854 iterations, the goal was reached and the RMSE of this neural model was 0.0005. Fig.6 and Fig.7 shows the extracted and modeled values of inductance and capacitance, they fit well at the whole diameter range.

Verify the Model

To verify this model, 4 different size balls with diameters of 105um, 200um, 250um and 295um were modeled by the proposed method, their modeled capacitance and inductance were listed in Tab. 1

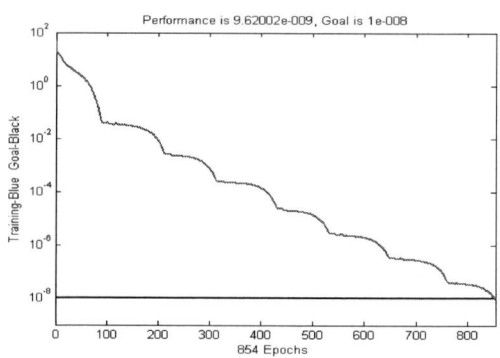

Fig.5 Training performance of network

978-1-4244-4658-2/09 $25.00 © 2009 IEEE

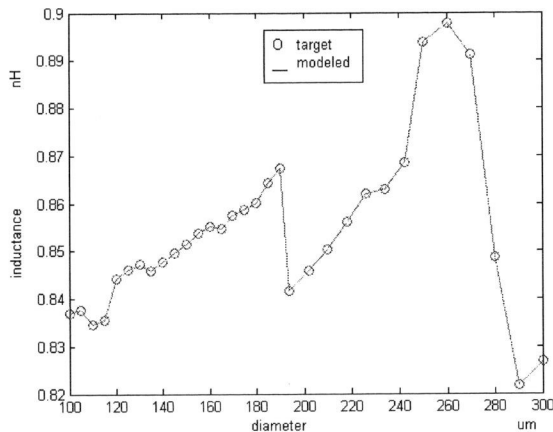

Fig.6 Extracted and modeled inductance

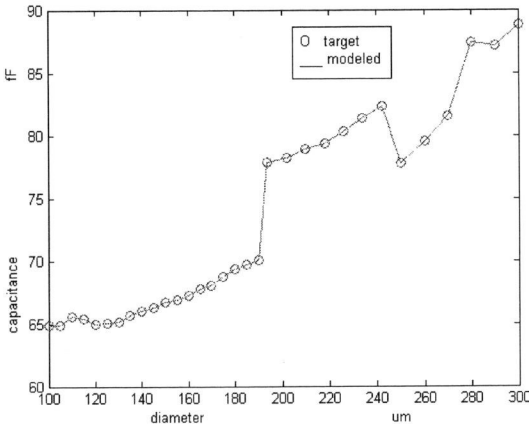

Fig.7 Extracted and modeled capacitance

Tab.1 The modeled L and C

Diameter	ANN modeled	
	C (fF)	L (nH)
105um	64.864	0.83697
200um	78.189	0.84573
250um	77.772	0.89391
295um	87.122	0.82166

These values of lumped elements were used to simulate in a circuit simulator such as Hspice using the model in Fig.2 and the results were compared with the results simulated by HFSS. Fig.8 (a) ~ (d) were the simulated S-parameter results, they were closely matched in the frequency range.

Conclusions

This paper has presented a neural network to model the relationship between the values of lumped elements and layout diameter, it was trained by the data extracted from the full-wave simulation tool HFSS and used to predicted 4 diameters in the range of 1Hz-10GHz, the results showed a good accuracy of this model.

In this paper, only one variable was used in the input of the network, future work could include other variables such as the relative permittivity of the dielectric, the height between conductor and the ground plane. These parameters will make the model more accurate.

Acknowledgements

This work was supported by National High-tech Project (863) (Optical Interconnect Technology for High Speed Chips and Verification Platform NO. 2007AA01Z2a6.)

References

1. Sundaram,V.;Tummala,R.R. etc "Next-generation microvia and global wiring technologies for SOP," *Advanced Packaging, IEEE Transactions*, Vol. 27, Issue 2 (May 2004), pp. 315-325.
2. Umemoto, M.; Tanida, K etc "High-performance vertical interconnection for high-density 3D chip stacking package," *Electronic Components and Technology Conference, 2004. Proceedings. 54th*, Vol. 1, 1-4, (June 2004), pp. 616-623.
3. Ghouz, H.H.M. El-Sharawy, E.-B. "An accurate equivalent circuit model of flip chip and via interconnects," *Microwave Theory and Techniques, IEEE Transactions*, Vol. 44, Issue 12, Part 2 (Dec. 1996), pp. 2543-2554.
4. Staiculescu, D. Sutono, A. Laskar, J. "Wideband scalable electrical model for microwave/millimeter wave flip chip interconnects" *Advanced Packaging, IEEE Transactions*, Vol. 24, Issue 3 (Aug. 2001), pp. 255-259.
5. Kang, M.; Gil, J.; Hyungcheol Shin; "A simple parameter extraction method of spiral on-chip inductors," *Electron Devices, IEEE Transactionson* Vol. 52, Issue 9, Sept. (2005), pp. 1976-1981.
6. Mantysalo, M.; Ristolainen, E.O.; "modeling and analyzing vertical interconnection," *Advanced Packaging, IEEE Transactions*, Vol. 29, Issue 2 (2006), pp. 335-342.

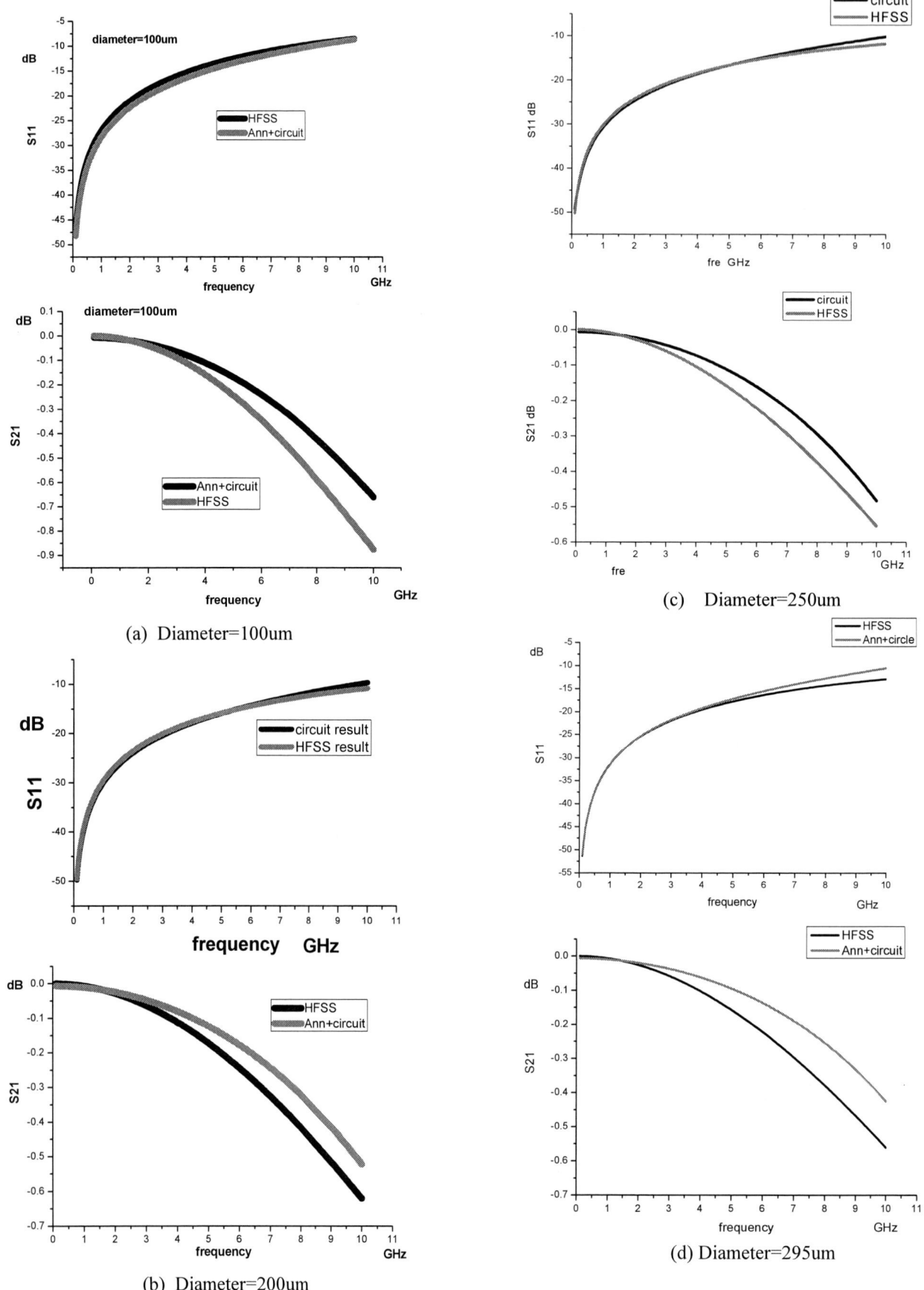

Fig.8 (a) ~ (d) the results of simulation with different diameter

A New Method to Fabricate Sidewall Insulation of TSV Using a Parylene Protection Layer

Ming Ji[1,2], Yunhui Zhu[2], Shenglin Ma[2], Xin Sun[2], Min Miao[2,3], Yufeng Jin[1,2*]

[1]Peking University Shenzhen Graduate school, Shenzhen, 518055, P. R. China

[2]National Key Laboratory on Micro/Nano Fabrication, Peking University, Beijing, 100871, P. R. China

[3]Inst. of Information Microsystem, Beijing Information Science & Technology University, Beijing, 100085, P. R. China

*Telephone: (86-10) 62752536-22, Email: jinyf@ime.pku.edu.cn

Abstract

This paper focused on the process of forming sidewall insulation of through silicon via (TSV) which was a challenging bottleneck in 3D integration technologies. In traditional way, etching silicon oxide on via bottom would reduce the thickness of sidewall insulation layer inevitably, which might lead to the failure of TSV sidewall insulation and electrical interconnection characteristic.

In this paper, the parylene-C (called parylene herein) film was used to cover the silicon oxide of via in the anisotropic etching process, which prevented the silicon oxide at the sidewall from being etched. Using this method, a well sidewall insulation layer was fabricated successfully. The sidewall thickness of oxide layer gained 0.93μm, 0.49μm, and 0.40μm at the top part, middle part of the via and the via-base respectively after etching process.

Introduction

The development of three-dimensional system in packaging (3D-SiP) technology is driven by the strong demand for high speed, high density, small size, and multifunctional electronic devices. It is believed that through silicon via (TSV) interconnection is the ultimate way for 3D integration due to its shortest interconnection distance and fastest speed. [1] There are several key steps involved in TSV processes which could successfully address the limitations of today's packaging technologies, including via forming, sidewall insulating, via filling, wafer thinning and wafer or die stacking. [2] Sidewall insulating is one of the challenging bottlenecks. TSV sidewall needs to be insulated to prevent shorting between metal and silicon which is rather important for device reliability. To ensure the expected insulation properties with higher breakdown voltage, no leakage, and no cracking, the insulation layer needs good coverage and uniformity, lower stress, and process temperature compatibility. [3]

Based on such considerations, a series of sidewall insulation experiment of TSV with high aspect ratio was proposed. We chose silicon oxide by plasma enhanced chemical vapor deposition (PECVD) as insulation material. Silicon oxide is used for insulation material frequently in semiconductor process. PECVD is a preferred method due to its advantages of lower deposition temperature (350℃), well-established good adhesion and close CTE to silicon, and so on. [1].

Sidewall Insulation Process

There were three steps in this experiment. First, via was formed by inductively coupled plasma (ICP) etching. Second, silicon oxide was deposited by PECVD. Third, re-etching was operated to remove the bottom silicon oxide and form sidewall insulation layer.

By traditional ICP etching on silicon, via sidewall shows scallop-like shapes consisting of many micro-concaves, [4] which causes the conformal coverage characters of PECVD silicon oxide in the via with high aspect ratio rather poor. Fig. 1 shows the whole profile and points out every part of the via after PECVD process. The result was disappointing because the thickness of sidewall insulation layer decreased from via-top to via-base significantly. Sidewall insulation performance depends on the minimum thickness of silicon oxide, so we should pay more attention on via-base where's the weak point of PECVD process.

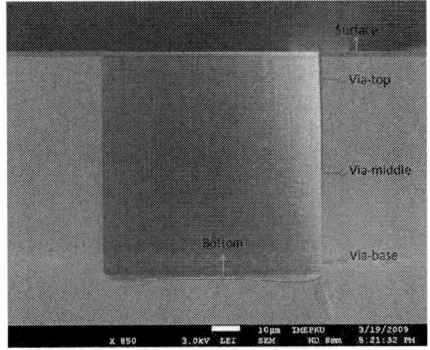

Fig. 1 the profile of the via after PECVD process (2 microns SiO2 as an example)

To demonstrate conformal coverage character of PECVD insulation layers, silicon oxide with three different thicknesses from 1μm to 2μm were studied for the via with the diameter of 50 μm and depth of 75μm. The results could be imagined. The thickness of silicon oxide at every part of the via was not uniform. Thickness from via-top to via-base was decreasing significantly. Fig. 2, Fig. 3 and Fig. 4 show the experimental results respectively. The numerical results were shown in Table 1.

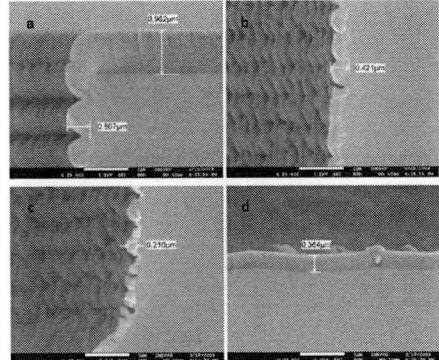

Fig. 2 PECVD one micron: (a)via-top, (b)via-middle, (c)via-base, (d)bottom

978-1-4244-4658-2/09 $25.00 © 2009 IEEE

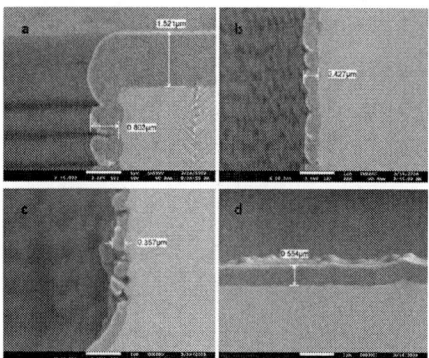

Fig. 3 PECVD one and a half microns: (a)via-top, (b)via-middle, (c)via-base, (d)bottom

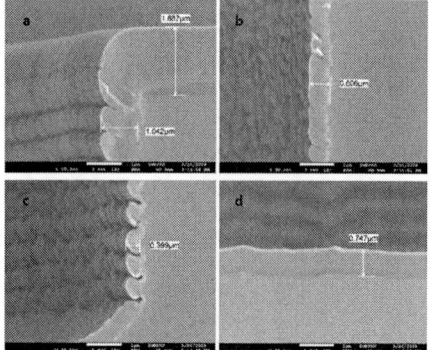

Fig. 4 PECVD two microns: (a)via-top, (b)via-middle, (c)via-base, (d)bottom

Table 1 Thickness of SiO_2 at every part of the via after PECVD process (Unit: μm)

	Surface	Via-top	Via-middle	Via-base	Bottom
1μm	0.962	0.507	0.421	0.210	0.364
1.5μm	1.521	0.803	0.427	0.357	0.554
2μm	1.887	1.042	0.606	0.399	0.747

Silicon oxide on the bottom of via is expected to be removed by ICP etching in this process. Although ICP is a typical anisotropic etching method, it would reduce the thickness of sidewall film and deteriorate the sidewall insulation character inevitably. After ICP etching, the results were shown in Fig. 5, Fig. 6 and Fig. 7. Table 2 shows the numerical results.

Table 2 Thickness of SiO_2 at every part of the via after ICP process (Unit: μm)

	Surface	Via-top	Via-middle	Via-base	Bottom
1μm	0	0.466	0.210	~0	0
1.5μm	0	0.681	0.404	0.343	0
2μm	0	0.880	0.560	0.340	0

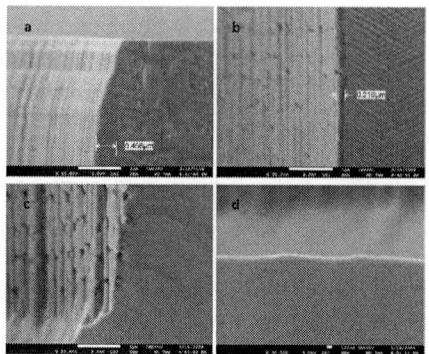

Fig. 5 one micron after etching: (a) via-top, (b) via-middle, (c) via-base, (d) bottom

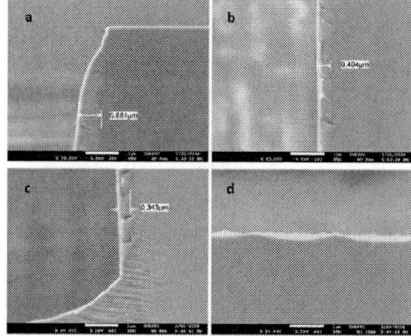

Fig. 6 one and a half microns after etching: (a) via-top, (b) via-middle, (c) via-base, (d) bottom

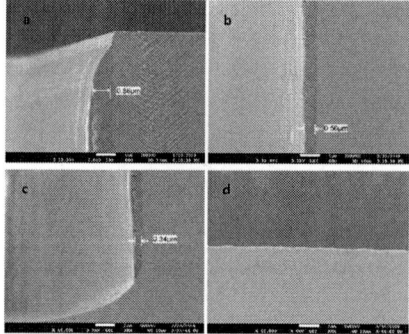

Fig. 7 two microns after etching: (a)via-top, (b)via-middle, (c)via-base, (d)bottom

We could see that silicon oxide with the thickness of 2μm achieved the best result. Meanwhile, in this process, silicon oxide layer was almost etched up on the via-base without protection as shown in Fig. 5(c), which would result in electrical interconnection problem and lead to the failure of TSV sidewall insulation.

So, two points were proved critical to achieve the ideal insulation layer: one is keeping conformal coverage character of PECVD insulation layer, the other is protecting insulation layer from being etched in the course of ICP etching silicon oxide.

Process Optimization

It is difficult to improve conformal coverage characteristics of PECVD due to equipments' limitation, but etching problem can be solved skillfully. We could find a special material to form a sidewall protection layer in etching process.

Parylene owned its unique features which is competent to be the protection layer in this new method. First, parylene has good conformal coverage character so it can protect the entire sidewall. Second, etching parylene is an anisotropic course. Third, the condition of etching parylene is different from etching silicon oxide and neither would be affected. Based on these three factors, a new process was exploited in which parylene was utilized masterly and fully.

Using this method, an intact sidewall insulation layer was fabricated successfully. The thickness of oxide layer gained 0.93μm, 0.49μm, and 0.40μm at the top part, middle part of the via and the via-base respectively, as shown in Fig. 8.

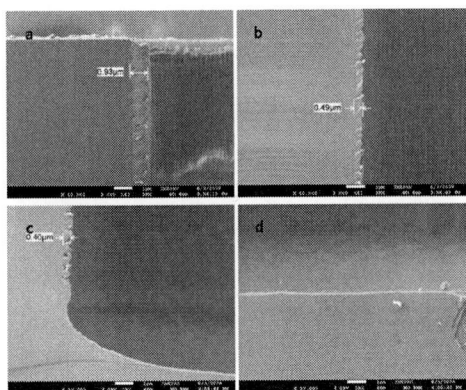

Fig. 8 the thickness of sidewall silicon oxide under protection of parylene: (a)via-top, (b)via-middle, (c)via-base, (d)bottom

New Process Using Parylene

Parylene is widely used in MEMS recently owing to its promising properties such as excellent conformality, chemical inertness, and mechanical performance. [5] Besides, an outstanding merit of the parylene film is its unique vapor deposition polymerization (VDP) coating process at room temperature. [6] In this new process, parylene was deposited first, and then parylene on the bottom was removed, the remaining parylene at the sidewall formed protection layer which could prevent sidewall silicon oxide from being etched in ICP process.

In this paper, the PDS 2010 LABCOTER™2 parylene deposition system was employed to deposit the parylene films, the deposition pressure was set at 22 mTorr, and the thickness of the film was about one micron. As shown in Fig. 9, parylene had good conformal coverage character in deposition process.

Parylene could be etched in oxygen plasma atmosphere. The direction showed a little anisotropic. Etching time must be controlled appropriately. Because parylene on the bottom of via could not be etched up if time was less, while more time may lead to the disappearance of the sidewall parylene.

After etching, parylene on the surface and via bottom was etched off, while sidewall parylene still keeps considerable thickness, as shown in Fig. 10.

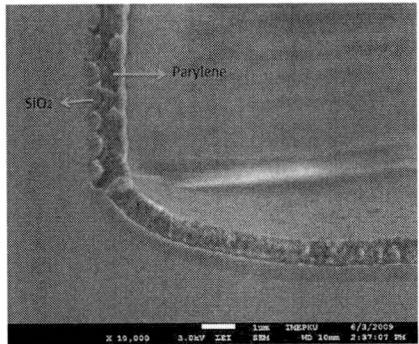

Fig. 9 Parylene with the thickness of 1 micron was deposited (2 microns SiO_2 as an example)

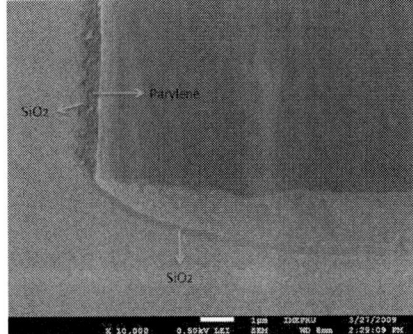

Fig. 10 Parylene was etched anisotropically (1 micron SiO_2 as an example)

With the sidewall silicon oxide under protection, etching silicon oxide on the bottom of via could be achieved easily. Etching gas was $C_4F_8/H_2/He$ which didn't impact on parylene. So etching time could be set more than expected to ensure silicon oxide was etched up. Fig. 11 shows the silicon oxide at the bottom was removed while sidewall was under protection. Then, parylene on the sidewall was cleared away by oxygen plasma, which had no effect to silicon oxide in the same way. An intact sidewall insulation layer was done finally.

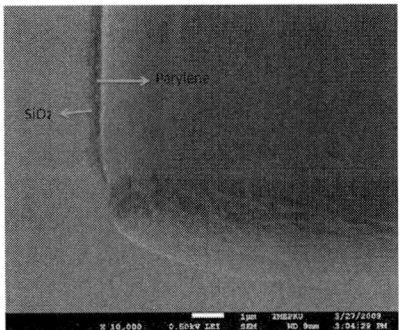

Fig. 11 Silicon oxide at the bottom was removed while sidewall was under protection (1 micron SiO_2 as an example)

Conclusions

Three different thicknesses of silicon oxide were compared in PECVD and ICP process to demonstrate the sidewall insulation character in the via with high aspect ratio, 2 microns achieved the best result.

Using parylene, the thickness of sidewall silicon oxide ultimately gained 0.93μm, 0.49μm, and 0.40μm at the top part, middle part of the via and the via-base respectively. The insulation character was proved better which brought about high reliability of TSV.

Next, we will focus on testing the insulation characters of sidewall silicon oxide, including the breakdown voltage, the leakage current, and so on.

Acknowledgments

The authors would like to appreciate Xianju Huang, Xinyi Zhang and Yinhua Lei for sample preparation. The authors would like to thank Prof. Zhihong Li and Dr. Wei Wang for discussion.

References

1. Jiang, T. et al, "3D Integration - Present and Future," *Proc 10th Electronics Packaging Technology Conf,* Singapore, Dec. 2008, pp. 373-378.

2. Miao Min, et al, "Research on Deep RIE-based Through-Si-Via Micromachining for 3-D System-in-package Integration," *Proc of the 4th Annual IEEE International Conf on Nano/Micro Engineered and Molecular Systems,* Shenzhen, China, Jan. 2009, pp. 90-93.

3. Beica R. et al, "Advanced Metallization for 3D Integration," *Proc 10th Electronics Packaging Technology Conf,* Singapore, Dec. 2008, pp. 212-218.

4. Cho Byeong-Hoon, et al, "Filling of Very Fine Via Holes for Three Dimensional Packaging by Using Ionized Metal Plasma Sputtering and Electroplating," *International Conf on Electronic Materials and Packaging,* Daejeon, Korea, Nov. 2007, pp. 1-3.

5. Sun C. M., et al, "A novel double-side CMOS-MEMS post processing for monolithic sensor integration," *Proc of 21st IEEE International Conf on Micro Electro Mechanical Systems,* Tucson, AZ, Jan. 2008, pp. 90-93.

6. Ji Xu, et al, "Parylene film for sidewall passivation in SCREAM process," *Science in China Series E: Technological Sciences,* Vol. 52, No. 2 (2009), pp. 357-362.

The Electrical, Mechanical Properties of Through-Silicon-Via Insulation Layer for 3D ICs

Sang-Woon Seo[1], Jae-Hyun Park[2], Min-Seok Seo[3], and Gu-Sung Kim[1,*]

[1]Kangnam University, Yongin 446-702, Korea
[2]EPWorks Co.,Ltd., ESIP Lab. Gyeonggi 464-070, Korea
[3]Hynix Semiconductor Co., Ltd., PKG development team, Icheon 467-866, Korea
E-mail : sangwoon.s@gmail.com, *gkim@kangnam.ac.kr

Abstract

This paper descibes variety of methods to examine the electrical and physical characteristics of the isolation layer deposited on TSV(Through-Si-Via). A sample was manufactured for the experiment with a diameter of 10μm and a depth of 50μm using Deep-RIE(Reactive Ion Etching). SiO_2 thin-film was deposited on the TSV sample by two separate procedures: PECVD (Plasma Enhanced Chemical Vapor Deposition) and PETEOS (PE Tetra-Ethyl-Ortho-Silicate). The insulating layer of TSV is supposed to decrease inter-diffusion between materials that fill the wall and its interior, improve adhesion and prevent electrical leakage. Hence, physical deposition characteristics, such as the surface step coverage, deposition rate, and film's density were observed and analyzed in order to determine if the deposited layer met the above criteria. The results confirmed that the thin layer deposited by PETEOS deposition was superior to that formed by PECVD in every category considered. Moreover, in order to assess the electrical characteristics, the interior of via hole was filled with copper (Cu) using the damascene process to create a sample. I-V was measured using the Time Dependent Dielectric Breakdown (TDDB) method for the sample, The measurement values were used to check the voltage level where the leakage current appeared. These experiment results indicate that the failure rate of the insulating layer depends upon the film's thickness and the deposition process. This assertion provides clues for conjecturing the main causes of insulation destruction. In this study, we determined the best deposition process for insulating the interior of TSV and the optimal insulating layer thickness in relation to the usage voltage.

1. Introduction

Study on 3-D integration technology, which is a method implementing a multi chip that requires diverse and complicated functions and high integration, are in active progress.[1] Similarly, studies regarding System-in-Package (SiP) technology are also currently being performed. Among the 3-D integration technologies, TSV forms via holes on the semiconductor wafer, filling them with conductor for the purpose of enabling interconnection among the chips.[2] 3-D lamination of chips using TSV is appropriate for high-end devices that operate at high frequencies because of direct signal connection between the chips without the need for a substrate. As a result, TSV is a technology that overcomes spatial limitations and maximizes the effectiveness of electric power distribution.

In general, TSV follows a common process where via holes are created on a wafer. After the holes are made, via isolation, via filling, wafer thinning, dicing, and chip stacking are then performed.[1][3] Via holes are created using Deep

RIE and the insulating layer is subsequently formed on via hole walls that secures insulation against through electrode. When the insulating layer has a defect, current leakage results, which can significantly affect the performance of the device. Current leakage leads to malfunction, which eventually makes the device less reliable. Hence, the insulating layer to be deposited in the interior of the TSV must have superior step coverage and conformal deposition characteristics.[4][5] Moreover, since the packaging process is one of the BEOL (Back End of the Line) processes performed after the IC is already mounted, high temperatures can lead to residual stress in the wafer, significantly influencing the stability of the device. Meanwhile, a seed layer is required when filling the TSV so that the metal, which is the material used to fill the TSV, can grow it. However, seed layer easily delamination due to the use of a low-silicon substrate and the process of adhesion, which calls for the insulating layer's role as a middle layer that can improve adhesion. Finally, the most important characteristic of the insulating layer is the dielectric constant. A very low dielectric constant value is required in order to minimize the signal delay caused by parasitic capacitance due to the coupling between the metal used to fill the via hole and the adjacent metal. On the other hand, it is a well-known fact that the lower the dielectric constant, the higher the breakdown field where the destruction of film occurs. Among the various materials that may be used as an insulating layer, SiO_2 has a comparative advantage in that it shows excellent deposition characteristics, even at low temperatures. Furthermore, SiO_2 has a relatively low dielectric constant (3.8-3.9) and the electrical insulation characteristics of SiO_2 do not change up to 10 eV because of the broad energy band gap in case of crystallization or activation, while showing superior interface characteristics with Si.[6][7] We created a TSV sample by lithography and etching after designing via with a variety of diameters and trench with patterns. This was done for the purpose of evaluating the deposition characteristics of SiO_2 during each process. The deposition process, which occurred at low temperatures, was analyzed to be compressed using PECVD and PETEOS deposition. The SiO_2 thin layer was subsequently deposited. Next, the deposition characteristics and the film's properties were compared to determine the best deposition process for insulating the interior of TSV and the optimal insulating layer thickness in relation to the usage voltage.

2. Experimental Method

In this experiment, the insulating layer was deposited on via and trench that have high aspect ratios with diameters of 10μm and depths 50μm. The characteristics of the insulating layer were then examined. Specifically, via holes and trenches with depths of 50μm were formed on a silicon wafer by Deep

978-1-4244-4658-2/09 $25.00 © 2009 IEEE

RIE. SiO_2 was deposited as an insulating layer and its deposition characteristics were evaluated afterwards. Moreover, the sample was manufactured by copper electroplating after depositing the diffusion barrier and seed layer in order to examine the electrical characteristics.[8]

2.1 Deposition of the SiO_2 insulation layer

The purpose of this study is to form an insulating layer on the interiors of via holes using the via-last method, which is a method that creates TSV after a chip is mounted on a silicon substrate. We also evaluated various deposition characteristics of the layer. For this purpose, a temperature range that does not hamper the performance of the mounted chip was selected when depositing the SiO_2 layer. For the ICs that are currently created, the process should be performed below 400°C when the memory chip is mounted.

At the first, the SiO_2 insulation layer was deposited by PECVD on TSV. a (100) silicon wafer substrate was inserted before depositing an SiO_2 layer that was 1.5μm thickness. To compare the characteristics of the thin layer at different temperatures, three separate substrate temperatures of 250°C, 200°C, and 150°C were used. The thin layer was deposited after the substrate's temperature was stabilized. An optimized flow rate for the deposition of the SiO_2 layer (i.e., SiH_4 10 sccm, O_2 18 sccm, and Ar 20 sccm) was maintained constantly. The chamber pressure was maintained at 3 mTorr in vacuum conditions and the RF power was applied at 25 W to generate plasma. In the second place, it deposited by PETEOS on TSV. The thin layer was grown while maintaining the substrate's temperature at 350°C so that the PETEOS could be decomposed and form a thin layer at a stable manner. The quantity of vaporized TEOS solvent, which is a liquid at room temperature, was 800 sccm, while 50 sccm of O_2 and 560 sccm of He were used as reaction and carrier gases, respectively. A pressure of 5.5 torr and an RF power of 580 W were applied.

2.2 Structure formation for electrical analysis

To determine whether the SiO_2 layer deposited on the interior of the TSV sample satisfies the requirement for being an insulator between silicon and metal, the conductor was filled in the sample where the insulating layer was deposited by the deposition process. It was manufactured using the Cu damascene process and described below.

Before proceeding with the plating process, Ti/Cu was deposied as seed layer by sputter process. Ti, which is widely used as an adhesion and diffusion barrier, was formed at a thickness of 2000Å by DC sputtering in order to increase the adhesion between SiO_2 and Cu, as well as to prevent Cu ions from diffusing to the silicon. The Cu layer was deposited at a thickness of 8000Å over the front of the wafer using DC sputtering in order to create a seed layer that flows electricity and allows the Cu ions to grow. In order to completely fill the interior of TSV with Cu, the plating process was performed by applying electricity in a rack type electroplaing bath using the pulse/reverse-pulse method. The reaction bath was filled with a plating solution consisting of Cu ions, an accelerator, and a moderator, and the seed layer was formed on the wafer. The wafer was attached to a cathode and inserted in the reaction bath, which was spun at a constant rpm to create an environment where the Cu ions could be deoxidized at an identical rate over the entire portion of the wafer. However, it is very difficult to circulate the plating solution deep into the interior of electrode while maintaining a constant density. Since more deoxidization occurs at the corner of the sample than in the interior, the entrance is often blocked before the interior of electrode is filled. Hence, we applied a reverse electric field during plating in order to remove excess Cu at the entrances of via holes. At the last, the Cu formed on the upper part of the wafer should have been removed by CMP(Chemical Mechanical Polishing) process. However, when the insulating layer that is formed between the Cu and the silicon wafer is also removed, then an electric field may also be formed along the silicon interface. Hence, a sample was created using CMP so that only the Cu that formed on the front of the wafer was removed, leaving at least 0.5μm of the insulating layer. Fig 2.1 shows TSV sample that CMP was completed.

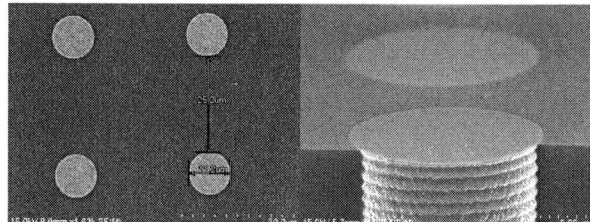

Fig. 2.1 FESEM image of TSV smaple that was manufactured by electroplating and CMP

3. Experiment Results

3.1. Physical deposition characteristics of the thin layer

3.1.1. Analysis by FESEM

In order to examine the deposition characteristics of the insulating layer created inside the TSV, the insulating layer was deposited using two deposition processes. Images were taken of the upper and lower portions of the wafer by FESEM(Field Emission Scaning Electron Microscope), as is shown in Fig. 3.1. The image in Fig.3.1 (d) is the lower portion of the insulating layer created by PETEOS deposition. As can be seen, the film is very smooth. Moreover, no void was observed in any part, confirming the excellent status of deposited film. The image in Fig. 3.1 (a) is of the thin layer deposited by PECVD at 150°C, and it can be seen that the film is not very dense. Indeed, the film density is the most important point in film evaluation. Furthermore, the film is not uniformly distributed. Fig. 3.1(b) shows an image of an insulating layer deposited inside via holes at 200°C by PECVD. which was measured for comparison with trench. The film density and smoothness of the film's upper corner is slightly improved compared to that shown in Fig. 3.1(a). However, it also was not visually good condition; its step coverage was lower than that of the trench by about 10%, as is shown on the wall of via hole. At the bottom, the thin layer was observed to be less than several dozen Å, implying that the quality of the insulating layer is significantly decreased in via holes when compared to trenches. Fig. 3.1(c) shows a layer that was deposited by PECVD at 250°C. Indeed, the condition of the film is quite improved compared to the films

978-1-4244-4658-2/09 $25.00 © 2009 IEEE

deposited at 150°C and 200°C. However, the condition of the film in via hole walls showed little improvement. All in all, a number analyses were conducted in order to determine the characteristics of the thin layer formed by PECVD at different temperatures. However, this type of deposition showed poor results with regard to important characteristics such as step coverage and uniformity. Hence, in order to apply PECVD, future studies are required to determine process parameters such as gas flow, pressure, and plasma density for TSV structures in improved and stabilized temperature conditions.

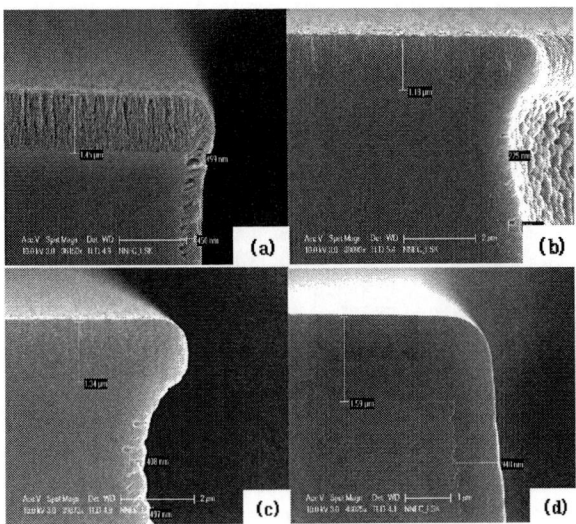

Fig. 3.1 A cross-sectional FESEM image of the SiO_2 layers formed by PETEOS deposition and PECVD

3.1.2 Measurement of deposited film

Deposited isulation layer was evaluated through analysis of step coveage and deposition rate. At frist, step coverage was computed based on the statistics results relating to the data collected and classified according to each process variable. Step-coverage is important estimation element among deposition properties because generation of leakage current by concentration of electric field, properties decline in post process by void creation can occur from part that thickness is thin.[9] The step coverage measured on the walls of via holes was 60% for PETEOS deposition and 33%, 22%, and 18% for PECVD at 250°C, 200°C, and 150°C, respectively. When it was measured in the bottom of the TSV, the overall values decreased from those measured on via walls. PETEOS deposition showed step coverage of 42%, which was about 20% less than the previous measurement, while the PECVD measurements were all less than 8%. This indicates that the uniformity of the insulating layer significantly decreases at the bottoms of via holes.

Second, the time it took to complete each deposition process was measured in order to compute the deposition rate. As a result of mesurement of deposition rate, PECVD showed 357 Å/min, 454 Å/min, and 510 Å/min at 150°C, 200°C, and 250°C, respectively, indicating a deposition rate increase of approximately 50 Å/min at each increase in temperature. However, the increment was not large. For PETEOS deposition, the deposition rate was determined to be

5000Å /min, which is more than ten times higher than that of PECVD.

3.2. Electrical characteristics of the thin layer

TDDB, which is generally determined to detect gate oxide, was used after modifying the method with a purpose of observing insulation strength and insulation destruction of the insulating layer deposited on via and confirming its reliability. This testing method increases electrical stress on the insulating layer as time goes by and allows us to observe the destruction of the insulating layer after a certain amount of time.[10] The electrical analysis was conducted by measuring the leakage current, measuring the I–V through Probe station. Since there are a number of factors that affect the destruction of an insulating layer and thereby causing current leakage, finding the direct cause of the destruction is not a straightforward matter. However, it is known that SiO_2 insulating layers may be destroyed in three ways.[11] First, destruction at the lowest voltage level may be caused by a defect in the film, such as pin hole on the oxide layer (sometimes called a ternary peak). Insulation destruction that occurs at the second level is called a secondary peak and is caused by electrical weakness of the oxide layer. Finally, intrinsic insulation destruction, known as a primary peak and occurs when the internal pressure limit of the oxide layer is exceeded. Both the first and the second types of destruction occur from local defects in the oxide layer. The main causes include particles on surface of the silicon wafer, segregation of carbon to lattice defect, metal contamination, and contamination inside the oxide layer. Hence, great efforts should be made to avoid contamination of the insulating layer. Indeed, it is also important to form compact layers by strictly controlling the process. Fig. 3.2 shows structures for electricaltest, I-V was measured over ten times in the structure between silicon and electrode for each process and thickness.

Fig. 3.2 Illustration of analysis structure for electrical test

These measurements were subsequently used to evaluate the leakage current level for each thickness to determine the optimal thickness for the insulation of the electrode. Fig. 3.3 presents the measurement results relating to the silicon and the via structure. In case of PECVD, the leakage current level increased by 2V for every 0.5μm increase in the thickness of the layer. For PETEOS deposition, the leakage current level increased by 4 V for every 0.5μm increase in the thickness of the layer. Based on these values, it was confirmed that the leakage current level increases in proportion to the thickness of the insulating layer on via wall. It should also be noted that

978-1-4244-4658-2/09 $25.00 © 2009 IEEE

the leakage current can differ according to the process used to form the insulating layer. Among the various causes for insulation layer destruction, it appears in this case that the Cu ion that was activated by the current application was diffused to the insulating layer, causing a defect that eventually destroyed the insulating layer.

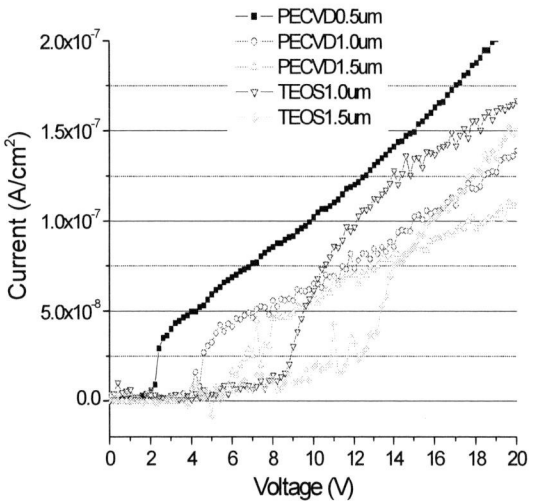

Fig. 3.3 Voltage ramping test results regarding the leakage of Si to via by the I-V graph.

4. Conclusions

In this study, chips that were interconnected by TSV were laminated in 3-D. A SiO_2 layer was deposited on the interior of electrode and electrode to insulate the TSV. Evaluations of the physical and electrical characteristics of two deposition processes were conducted. The processes were assessed according to their suitability for forming an insulating layer with the proper thickness. For this purpose, via and trench with high aspect ratios of 5:1 were formed and insulating layers were created with thicknesses of 1.5μm using PECVD and PETEOS deposition. Indeed, these processes are generally used to create SiO_2 layers at low temperatures. In the case of PECVD, deposition was performed at 150°C, 200°C, and 250°C, while deposition was performed at 350°C, which is an optimal temperature where thin layer can be deposited stably in case of PETEOS. In order to examine the physical characteristics of the SiO_2 layer formed inside the TSV, step coverage, and deposition rate were compared. According to the results of the comparison, film deposited by PECVD tends to improve its characteristics as the temperature increases, but the improvement was negligible. The SiO_2 film formed by PETEOS deposition was better than that formed using PECVD in every category that we examined. Moreover, I-V was measured to verify the electrical insulation characteristics of actual chips in an operational environment. Based on the results, the voltage of the leakage current was conjectured. This value was subsequently used to evaluate the optimal deposition process and film thickness. The evaluation

results indicated that PETEOS deposition produces a higher-quality film that PECVD. Considering that the leakage current was 4 V for every 0.5μm of thickness, an optimal insulating layer thickness was determined that is suitable for the usage of TSV.

References

1. K. Takahashi, M. Sekiguchi, "Through Silicon Via and 3-D Wafer/Chip Stacking Technology", VLSI Circuits 2006 Digest of Technical Papers, (2006), pp.89-92.
2. C.H.Yun, T.J.Brosnihan, W.A. Webster, J.Villarreal, "Wafer Level Packaging of MEMS Accelerometers with Through-Wafer Interconnects", ECTC, 55th (2005), pp.320-323.
3. Z. Wang, L. Wang, N.T. Nguyen, Wim A.H., H. Schellevis , P.M. Sarro, "Silicon micromachining of high aspect ratio, high- density through-wafer electrical interconnects for 3-D multi chip packaging", IEEE Trans. Advanced Packaging, Vol 29, No 3 (2006), pp.615-622.
4. A.M.Mahajan, L.S.Patil, J.P.Bange, D.K.Gautam, "Growth of SiO_2 films by TEOS-PECVD system for microelectronics applications", Surface & Coatings Tech, 183th (2004), pp.295-300.
5. T.Kondo, Y.Sawada, "Step coverage study of indium-tin-oxide thin films by spray CVD on non-flat substrates at different temperature". Thin solid Films (2007).
6. Puchkar Jain, Eugene J. Rymaszewski, "Thin-Film Capacitors for Packaged Electronics", Kluwer Academic Pub., Messachusetts (2004), pp.27-31.
7. C Zhang, Najafi, K, "Fabrication of thick silicon dioxide layers using DRIE, oxidation and trench refill", IEEE int. conf. MEMS, 15th (2002), pp.160-163.
8. "3-D TSV Interconnects", Equipment & materials, 2008 report, Yole developpement (2008).
9. Sorab K. Ghandhi, "VLSI Fabrication Principles", Silicon and Gallium Arsenide, 2nd ED., Wiley- Interscience, New York (2002), pp.522-527.
10. Ogawa, E. T. Kim, J. Haase, G. S. Mogul, H. C. McPherson, J. W. ,"Leakage, Breakdown, and TDDB Characteristics of Porous Low-k Silica-Based Interconnect Dielectrics", IEEE International Reliability physics proceeding, Vol.41 (2003) , pp.166-172.
11. Lloyd, J. R. Murray, C. E. Ponoth, S. Cohen, S. Liniger, E. "The effect of Cu diffusion on the TDDB behavior in a low-k inter-level dielectrics", Microelectronics and reliability, vol.46, no.9-11 (2006), pp.1643-1647.

Through Silicon Via Filling by Copper Electroplating in Acidic Cupric Methanesulfonate Bath

Qi Li[1], Huiqin Ling[1], Haiyong Cao[1], Zuyang Bian[2], Ming Li[1*], Dali Mao[1]

[1]Lab of Microelectronic Materials & Technology, State Key Laboratory of Metal Matrix Composites, School of Materials Science and Engineering, Shanghai Jiao Tong University, Shanghai 200240, China

[2]Sinyang Semiconductor Material Co., Ltd., 1268 Wenhe Road, Shanghai, China

*Email: mingli90@sjtu.edu.cn, Tel: +86-21-34202542

Abstract

Copper electrodeposition in acidic cupric methanesulfonate bath with organic additives is discussed in this paper. The influence of poly(ethylene glycol) (PEG) and bis-(3-sodiumsulfopropyl disulfide) (SPS) on copper deposition were studied by means of linear sweep voltammetry, cyclic voltammetry and chronoamperometry. These electrochemical analysises revealed a competition of PEG and SPS on electrode surface site. The swiftness of SPS chemisorption and the subsequent displacement by the passivating film of PEG exerted an extra wave at small overpotential on the negative-going sweep. The following polarization curve indicated the firmness of the passivating film. All these features of additives in acidic cupric methanesulfonate bath suggested a novel method to achieve superconformal or bottom-up filling which was proved by actual TSV plating.

Introduction

3D packaging has been a hot topic in microelectronic industry in the last several years due to the continuous demand for higher density in microelectronic devices. Among the genius prototypes of 3D packaging, Through Silicon Via (TSV) promises small RC delay because of its short interconnection length, high density and the possibility of compiling hetero-functional devices in one packaging. Attribute to this, TSV is the most probable one to approach mass production. Nevertheless, there are still some longstanding difficulties in the way. One of them is to fill the vias with high efficiency and low cost.

Copper damascene process offers an effective approach of via filling. However, this process can not be applied in through silicon via filling directly due to the great difference of scale. Those micro features in damascene process are usually in submicron level while through silicon vias are commonly 100~300 microns in depth and 20~60 microns in diameter. This makes the diffusion of cupric ions into the vias a crucial problem. Furthermore, the diluteness of cupric ion will greatly restrain the deposition rate, especially at the bottom of vias where cupric ions can not diffuse into easily. This will make TSV plating time-consuming and low efficient and thereby costly. One of the solutions is to raise the concentration of cupric ion. Thus acidic cupric methanesulfonate was assumed because of its higher solubility.

Electrolyte used for micro feature filling usually contains organic additives as suppresser, accelerator and leveler in order to achieve superconformal or bottom-up filling. In cupric sulfate bath, PEG can form a passivating film synergistically with chloride ion and cuprous ion and inhibits copper deposition [1~5]. Whereas short chain disulfide (SPS) or thiol (MPS) molecules with a sulfonate-end group can absorb on the electrode surface subsequently, leading to the disruption of displacement of the inhibition film and accelerate copper deposition [5, 6]. The small molecule weight of SPS and MPS makes them easier to diffuse into the micro features than PEG. Furthermore, the chemisorption of SPS or MPS will be enhanced at concave curvature section while the surface area decreases along with copper deposition. This will lead to an enrichment of accelerators and thereby a superconformal filling. Janus Green B (JGB), a suppresser per se, is usually taken on as leveler due to its special absorption feature. Containing tertiary amino group, JGB is always positively charged in solution. So it tents to absorb on the cathode surface, especially on convex sections where electric fluxlines concentrate whereby takes the effect of leveling.

In this study, behaviors of the additives in acidic cupric methanesulfonate bath will be explored by means of electrochemical analysis as well as actual die depositions.

Experiments

Basic electrolyte is either $0.625mol/L$ $Cu(CH_3SO_3)_2$ with $0.625mol/L$ HCH_3SO_3 or $1.25mol/L$ $Cu(CH_3SO_3)_2$ with $0.625mol/L$ HCH_3SO_3 depending on each experiment. PEG (Mw=4000), SPS and JGB were prepared in a stock solution with concentration of 100g/L, 1g/L and 1g/L respectively. The stock solutions were stored in refrigerator at 0°C and pipette injected when used.

Electrochemical analysises were performed in a three-electrode cell. Working electrode is high-purity gold electrode which working area is 1.0 cm^2. Gold electrode was adopted so that copper could be deposited at cathode process and dissolve at anode process whereby electrode surface could be restored to initial statue. Counter electrode is platinum-coated titanium electrode with working area of 7.1 cm^2. Saturated calomel electrode was assumed as reference electrode. Solution used in each analysis is 250 ml and no forces convection was applied during data sampling.

Results and Discussion

The influence of each additive on copper electrodeposition in acidic cupric methanesulfonate bath was explored respectively. The basic electrolyte referred herein is 0.625 mol/L $Cu(CH_3SO_3)_2$ with 0.625 mol/L HCH_3SO_3.

PEG

Linear sweep voltammetry and chronoamperometry were performed to examine the effect of PEG. As shown in Fig. 1, the behavior of PEG strongly depended on its concentration. When only 1~3 mg/L PEG was added in basic electrolyte, noticeable acceleration of copper deposition was observed. As mentioned hereinbefore, PEG could absorb on electrode and form a passivating film whereby suppressing copper deposition. While its concentration is too low to build the inhibition layer efficiently, the incomplete film could

978-1-4244-4658-2/09 $25.00 © 2009 IEEE

somehow increase deposition rate. Very different phenomenons were reported in cupric sulfate electrolyte. Low concentration of PEG would hardly influence copper deposition rather than increase it [1~5]. This implies a different mechanism of PEG's interaction on electrode surface. Subsequent results shown in Figure 1, b) tell that the increase of PEG concentration produces no more acceleration but the onset of suppressing effect. The inhibition reached saturated when concentration is 100mg/L and higher.

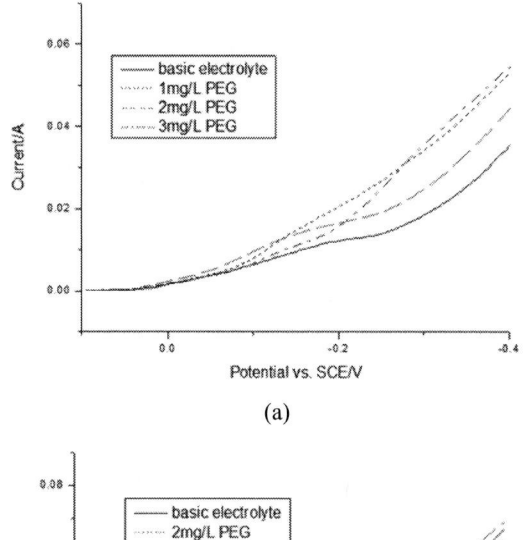

(a)

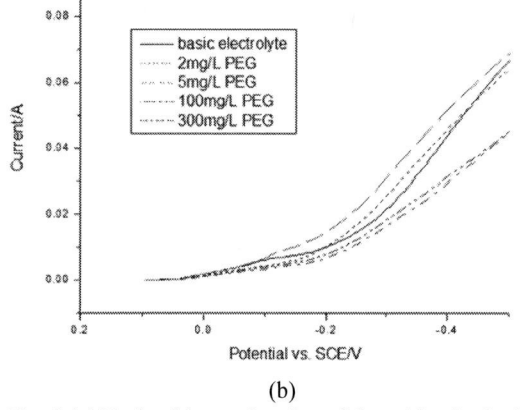

(b)

Fig. 1 (a) Noticeable acceleration of deposition is obtained when small amount of PEG was added in basic electrolyte; (b) Inhibition of copper deposition takes place when high concentration of PEG is reached, even without chloride ion

Please note that there is no chloride ion in basic electrolyte which means that even without the presence of chloride ion, PEG could form the passivating film as soon as a threshold concentration is reached. Taken as a whole, the voltammetry results indicate that in methanesulfonate bath, PEG has different interactions on electrode, probably with methanesulfonic radical. The initial intermediate may accelerate copper deposition whereas form an inhibition film when there are enough PEG molecules absorbed on electrode surface. To study the chemisorption of PEG, chronoamperometry was performed on a PEG-derivatized electrode. Gold electrode was derivatized in 300mg/L PEG

with 0.625mol/L HCH_3SO_3 for time ranging from 10s to 600s. The weakness of chemisorption of PEG on electrode when not charged can be indicated from the results shown in Fig. 2.

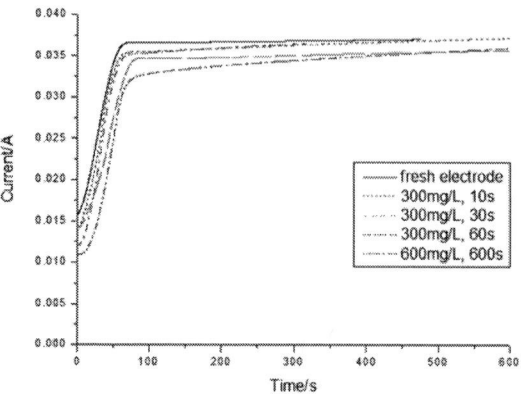

Fig. 2 PEG derivatized electrode shows barely any inhibition of copper deposition which suggests the weak chemisorption of PEG.

SPS

Interaction of SPS, chloride ion and cuprous ion was reported in cupric sulfate bath [7~9]. Analogous to PEG, SPS is not able to take effect when chloride ion is absent. In methanesulfonate bath, however, from results shown in Fig. 3, evident acceleration arose as soon as SPS is added in the basic electrolyte. This voltammetry result, together with that of PEG, suggest that methanesulfonic radical ion substitute or partially substitute the role of chloride ion in the intermediate forming interaction of PEG and SPS. A tendency of extra wave at low overpotential was observed then SPS concentration was greater than 3mg/L. This most possibly arose from interaction between SPS and cuprous ion [6]. The initial step of copper deposition, $Cu^{2+} + e^- \rightarrow Cu^+$ as well as

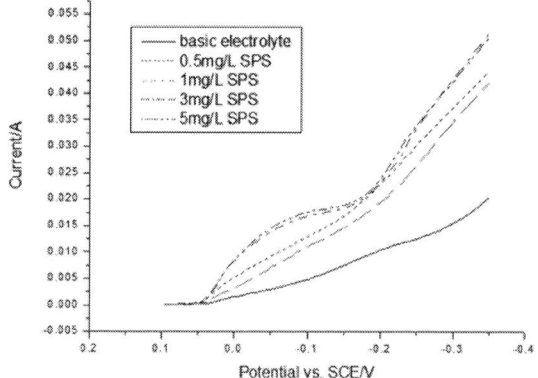

Fig. 3 Accelerating effect is obtained when SPS is added in basic electrolyte. An extra wave at low overpotential is obtained when SPS concentration is beyond 3mg/L

the oxidation of formerly deposited copper by cupric ion, $Cu^{2+} + Cu \rightarrow 2Cu^+$, would produce amount of cuprous ion which would act synergistically with SPS to accelerate copper deposition [10, 11]. A number of rotating ring-disc experiments revealed the homogeneous interaction between SPS and Cu^+, especially at low overpotential where a substantial Cu^+ activity is present [1, 12~14].

Chronoamperometry was also performed on SPS-derivatized electrode to explore its chemisorption. Gold electrode is derivatized in a 200mg/L SPS and 0.625mol/L HCH_3SO_3 solution for 60s. Evident raise of current was observed at both -0.05V and -0.25V.

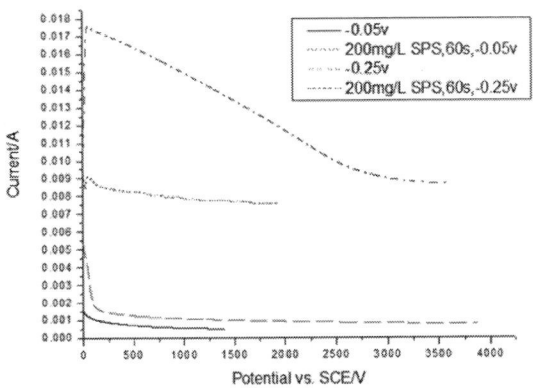

Fig. 4 Evident acceleration of copper deposition on SPS derivatized electrode reveals its strong chemisorption. The deactivation of SPS which leads to the decrease of current over time is way faster at lower overpotential.

Difference lay in the subsequent displacement by PEG. The deactivation of SPS was way faster at smaller overpotential which indicated that SPS was more competitive when overpotential was more negative.

Competition of Additives

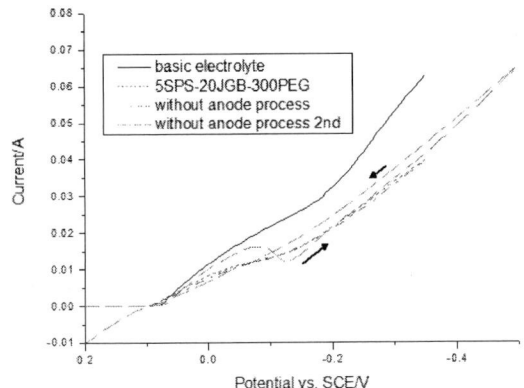

Fig. 5 An extra wave or peak at low overpotential was observed on the negative-going sweep and disappeared on the back sweep and linear sweep voltammetry which were performed immediately after copper deposition without anode process

When performing actual TSV plating, the electrolyte usually contains both suppresser and accelerator, as well as leveler (i.e. JGB) and wetting agent (i.e. small molecule

weight PEG), to achieve superconformal filling. In order to simulate this situation, 300mg/L PEG, 5mg/L SPS and 20mg/L JGB are added into basic electrolyte, 1.25mol/L $Cu(CH_3SO_3)_2$ and 0.625mol/L HCH_3SO_3, to study the synergistic activation or competition among the additives.

An apparent extra wave or peak was observed on the negative-going sweep of cyclic voltammetry result shown in Fig. 5. As a consequence of dynamical equilibrium, the peak on the polarization curve does not necessarily mean higher current at lower overpotential which was proved by the chronoamperometry transient curve shown in Fig. 6. The back sweep section of cyclic voltammetry also supported this conclusion. As discussed hereinbefore, the chemisorption of PEG was relatively weak whereas SPS would form steady absorbate on electrode surface. Furthermore, inasmuch as the interaction between SPS and cuprous ion, extra acceleration at low overpotential was also observed in Fig. 3. Therefore, when a fresh electrode dipped in the electrolyte, a quick chemisorption of SPS took place. When charging set off, acceleration took effect immediately. As the over potential went negatively, the interaction channel was cut off because of the deactivation of cuprous ion. Beside, the absorption of PEG is no longer negligible. Taken as a whole, these resulted in decrease of current along with the negative-going overpotential.

Usually, anode processes were performed to dissolve deposited copper and restore electrode surface after cathode process, like voltammetry or chronoamperometry. Two linear sweep voltammetry were sampled following chronoamperometry without anode process (Fig. 5). Analogous to back sweep section of cyclic voltammetry, no extra wave was observed on neither curve. This phenomenon suggested the firmness of PEG inhibition film once it was formed meaning SPS could not disrupt or displace it even at low overpotential where extra wave arose.

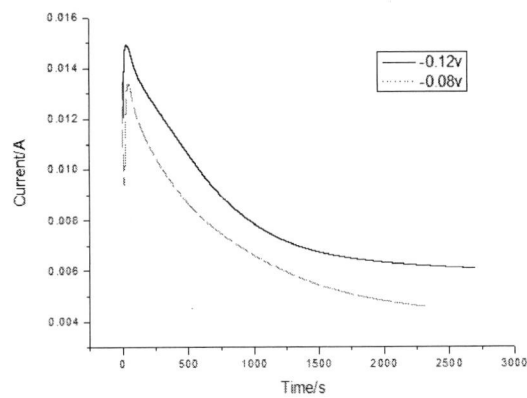

Fig. 6 Current at -0.08V, which was the overpotential of peak current on polarization curve, was actually lower than that of -0.12V, overpotential of trough current. "Rise-and-fall" was also observed on both transient curves.

Consider the chronoamperometry analysis hereinabove, after the initial relaxation stage, a quick current increase took place, followed by a drastic fall of deposition rate. Reason being analogous to that of voltammetry, this phenomenon supported the conclusion that swift acceleration by SPS and subsequent displacement by PEG.

To sum up, in acidic cupric methanesulfonate bath, with the absence of chloride ion, PEG and SPS behaved much differently form that in sulfate bath. Not that SPS-Cl⁻-Cu⁺ complex disrupted or displaced the PEG inhibition film which resulted in hysteresis in cyclic voltammetry, PEG supplanted the swiftly forming SPS absorbate. This produced the extra wave at low overpotential on the negative-going sweep section of cyclic voltammetry and the "rise-and-fall" chronoamperometry transient curve.

Filling Model

To take advantage of this competition feature, a special injection treatment was adopted to ensure highly concentrated SPS inside the vias. Then plating was carried out in electrolyte containing very dilute SPS and relatively high-concentrated PEG as well as JGB.

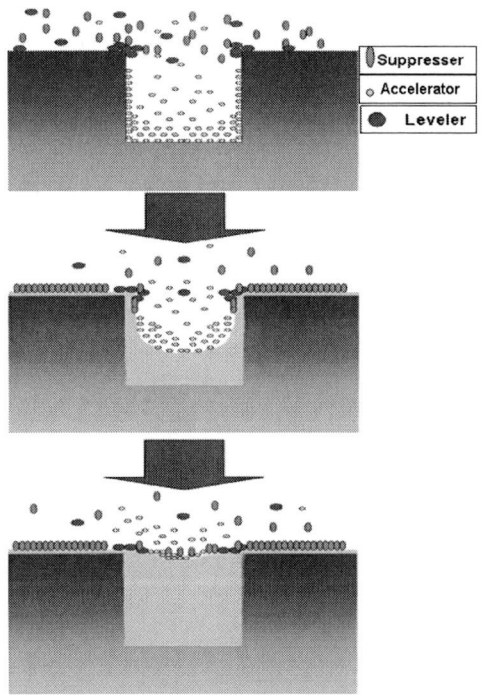

Fig. 7 Schematic sketch of via filling model in PEG-SPS-JGB system

As presented in the sketch hereinabove (Fig. 7), totally 3 steps would take while via filling.

First, high concentration of PEG as well as its swift chemisorption and acceleration feature ensures the initially high deposition rate inside the vias. But at the opening of via and the surface, due to the diluteness of SPS in body electrolyte, deposition rates are restrained.

When plating goes on, more and more SPS molecules will diffuse out of vias because of the concentration gradient and thereby SPS is not dilute any more. However, at this point, PEG has already dominated the surface site as the opening and form steady inhibition layer, therefore the diffusion of

SPS will not lead to unexpected fast deposition around opening which may end up with voids or seams. In addition, result from its large molecule weight, PEG is difficult to diffuse into the vias to displace SPS, so deposition inside the vias remains high.

This step develops and a superconformal is obtained.

Besides the high concentration of SPS inside the vias, another key is to plate at a properly high current density to enhance the accelerating effect, especially at the upper section of vias where diffused PEG may displace SPS and end up with a conformal filling. On top of that, inappropriately high concentration of SPS may lead to fast deposition not only inside vias but also around the opening. Due to geometric reasons, electric fluxlines concentrate at the opening whereby causes extra acceleration and finally voids inside. So SPS concentration and current density must be adjusted carefully to achieve a defect-free filling.

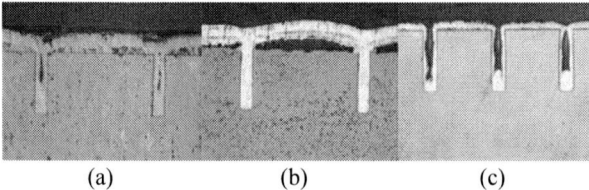

| (a) | (b) | (c) |

Fig. 8 TSV plating at different conditions: (a) low SPS concentration; (b) optimum condition; (c) high current density

Vias in Fig. 8 are 20 microns in diameter and 100 microns in depth. As discussed hereinbefore, too low concentration of SPS, as shown in Fig. 8(a), could not resist the displacement of diffused PEG at the upper section of vias. Therefore copper deposition was restrained there uniformly and led to conformal filling whereby seams were left. On the other hand, when plating at too high current density, due to the enhancement of SPS competitive power, acceleration was obtained not only inside the vias but also around the opening where diffused SPS located. Acting synergistically with the concentration of electric fluxlines, additional acceleration was obtained and ended up with huge voids inside (Fig. 8(c)). Only when appropriate SPS concentration, current density, as well as other practical condition were guaranteed, like that in Fig. 8(b), defect-free via fill could be achieved.

Conclusions

The influences of organic additives, PEG and SPS, on copper deposition in acidic cupric methanesulfonate bath were studied by means of linear sweep voltammetry and chronoamperometry respectively. Being different from that in sulfate bath, PEG and SPS could take effect without the presence of chloride ion which implied a different mechanism of inhibition/acceleration. Competition between PEG and SPS was revealed by the cyclic voltammetry and chronoamperometry results. The extra wave on negative-going sweep of cyclic voltammetry and the "rise-and-fall" chronoamperometric transient curve arose from the swift chemisorption and acceleration by SPS and the subsequent displacement by PEG per se.

Considering the special features of PEG and SPS, a novel filling model was brought in this paper. Highly concentrated

978-1-4244-4658-2/09 $25.00 © 2009 IEEE

SPS inside the vias and properly high current density guaranteed superconformal fillings. Following this filling model, actual TSV plating were carried out and defect-free filling was achieved.

Acknowledgments

This work is sponsored by International Science and Technology Cooperation of China (No. 2008DFA51680), National Natural Science foundation of China (60876071), Shanghai nano technology promotion center (No. 0852nm06300).

References

1. J. R. White, "Reverse Pulse Plating of Copper from Acid Electrolyte : A Rotating Ring Disc Electrode Study," *J. Appl. Electrochem.*, Vol. 17 (1987), pp. 977-982.

2. D.Stoychev, C. TSVetanov, "Behavior of Poly(Ethylene Glycol) During Electrodeposition of Bright Copper Coating in Sulfuric Acid Electrolytes," *J. Appl. Electrochem.*, Vol. 26 (1996), pp. 741-749.

3. J. J. Kelly, A. C. West, "Copper Deposition in the Presence of Poly(Ethlene Glycol) I," *J. Electrochem. Soc.*, Vol. 145 (1998), pp. 3472-3476.

4. J. J. Kelly, A. C. West, "Copper Deposition in the Presence of Poly(Ethlene Glycol) II," *J. Electrochem. Soc.*, Vol. 145 (1998), pp. 3477-3481.

5. T. P. Moffat, D. Wheeler, D. Josell, "Electrodeposition of Copper in the SPS-PEG-Cl additive System I," *J. Electrochem. Soc.*, Vol. 151, No. 4 (2004), pp. C262-C271.

6. T. P. Moffat, J. E. Bonevich, etc., "Superconformal Electrodeposition of Copper in 500-90 nm Features," *J. Electrochem. Soc.*, Vol. 147, No. 12 (2000), pp. 4524-4535.

7. J. P. Healy, D. Pletcher, and M. Goodenough, "The Cehemistry of the Additives in Acid Copper Electroplating Bath, Part II," *J. Electroanal. Chem.*, Vol. 338 (1992), pp. 167-177.

8. D. Josell, D. Wheeler, W. H. Huber, J. E. Bonevich, and T. P. Moffat, "A simple equation for Predicting Superconformal Electrodeposition in Submicrometer Trenches," *J. Electrochem.Soc.*, Vol. 148 (2001), pp. C767-C773.

9. D. Josell, B. Baker, C. Witt, D. Wheeler, and T. P. Moffat, "Via Filling by Electrodeposition," *J. Electrochem. Soc.*, Vol. 149 (2002), pp. C637-C641.

10. M. A. Pasqual, L. M. Gassa, A. J. Arvia, "Copper Electrodeposition from An Acidic Plating Bath Containing Accelerating and Ihibiting Organic Additives," *Electrochemica. Acta*, Vol. 53 (2008), pp. 5891-5904.

11. R. Winand, P. Van Ham, R. Colin, D. Milojevic, "An Attempt to Quantify Electrodeposit Metallographic Growth Structures," *J. Electrochem. Soc.*, 144 (1997), pp. 428-436.

12. U. Bertocci, "Application of Elctrochemical Theory to The Behavior of Copper in Cupric and Cuprous Solutions," *Electrochim. Acta*, Vol. 11 (1966), pp.1261-1277.

13. A. Molodov, G. N. Markosyan, and V. V. Losev, "Regularities of Low-Valency Intermidiate Accumulation During A Step-Wise Elecrtode Process," *Electrochim. Acta*, Vol. 17 (1972), pp. 701-721.

14. E. Gileadi and V. Tsionsky, "Study of Electroplating Using An EQCM. I. Copper and Silver on Gold," *J. Electrochem. Soc.*, Vol. 147 (2000), pp. 567-74.

Package Heat Dissipation with Integrated Carbon Nanotube Micro Heat Sink

Xiaojing Wang, Hongjun Liu, Jia Wang, Wen Zhang, Zongshuo Li
Shanghai University
Shanghai University, 224mail box, 149 Yan Chang RD. Shanghai, 20072, China
Email: xjwang@mail.shu.edu.cn, Tel: 86-21-66136117

Abstract

Micro-channel cooler is a very promising approach to meet the requirements of microelectronics package cooling. A lot of investigations about micro-channels have been undertaken in the past years. A silicon micro-channel can remove 790W/cm2 heat with a temperature rise of 71 degrees between the substrate and the coolant. The width and height of the silicon channel are 50um and 302um separately [1]. However, as the trends in the electronics industry moves towards higher packaging density, the high-pressure drop problem limits the performance of traditional silicon heat sink. Replacing the silicon fins with nanotube fins to enhance the thermal exchange rate between cooling liquid and substrate is one way to overcome this problem. Growing aligned nanotubes on the whole substrate is another one [2].

Carbon nanotubes (CNTs) are a new form of carbon which was discovered in 1991 by Iijima [3]. CNTs can be grown directly on the surface of silicon accurately according to pre-defined small-scale catalyst patterns normally transferred by standard photolithography processes. It was reported that nanotubes have an extremely high thermal conductivity over 3000W/ (km) [4], so in the present work, the thermo physical properties of CNTs are obtained via molecular dynamics (MD) method. Fig.1 shows one single-wall CNT simulation by LAMMPS. Several CNTs-fin heat sinks (1cm*1cm) with different fin width varied from 2um to 200um are tested (Fig.2). Then the aligned CNTs heat sink is also simulated to compare with the long CNTs-fin model.

The heat transfer coefficients and pressure drop of these structures are computed. Results indicate replacing the silicon-fin with CNTs-fin is a promising method and the long CNT-fin model shows better performance than the aligned CNTs for generating little pressure drop. Among the models, the best one is chosen for experiments in next work.

Introduction

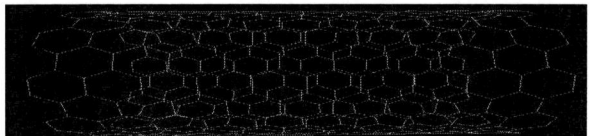

Fig.1 Single-wall CNT simulation

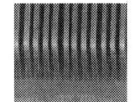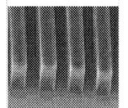

Fig.2 CNTs-fins with different fin width

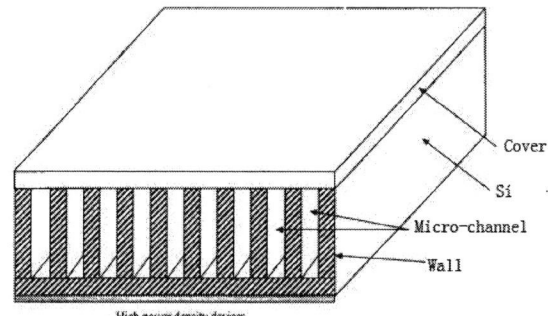

Fig.3 Micro-channel heat sink structure

The problem of micro system and the equipment cooling consumption, the micro-channel cooler as a new heat transfer device was generated. A typical micro-channel cooler includes cover, substrate and micro-channel, described in Fig.3. Channels are etched and covered by plexiglass on the top and ultimately formed a complete structure. The basic principle of micro-channel heat is that bottom is in touch with the heat and fluid flows through the entrance to the export to take away heat. In practice, with increasing heat, when the micro-cooler maximum temperature exceeds the fluid's temperature, convection heat is generated between wall and fluid until the heat balance is stabilized and micro-cooler works into the stable working condition.

Replacing the silicon fins with nanotube fins or growing aligned nanotubes on the whole substrate to enhance the thermal exchange rate between cooling liquid and substrate is one way to overcome this problem [2].

The channel width plays a very important part in the MCHS due to the high pressure drop problem and effect on heat transfer coefficient. Several CNTs-fin heat sinks (1cm*1cm*0.8mm) with different fin width varies from 0um to 200um are tested (Fig.2).Only 2D simulations are done by previous researchers, so in the paper 3D problems are simulated. The thermal resistance and pressure drop of different models are computed and compared, and among them the best one is chosen for experiments in next work.

Theoretical Analyses

A coolant flows through a micro channel heat sink described in Fig.3 takes away heat from heat component attached below. The top face which is made of glass is insulated and the bottom material is Silicon. Because of the large amounts of channels, it is difficult to mesh grids and compute. For simplicity, one typical channel in Fig.4 is chosen in our simulation due to symmetry. The heat transfer contains two parts: conduction in the solid and convection between the solid and coolants. By continuities of temperature and heat flux, the solid region and fluid region are coupled. Some simplifying assumptions are considered as follows:

978-1-4244-4658-2/09 $25.00 © 2009 IEEE

(1) Laminar;

(2) Incompressible;

(3) Hydro dynamically and thermally fully developed;

(4) No radiation of the wall;

(5) Negligible convection of air;

(6) The physical properties of Si and water are constant, but some properties of CNTs are not constant such as thermal conductivity.

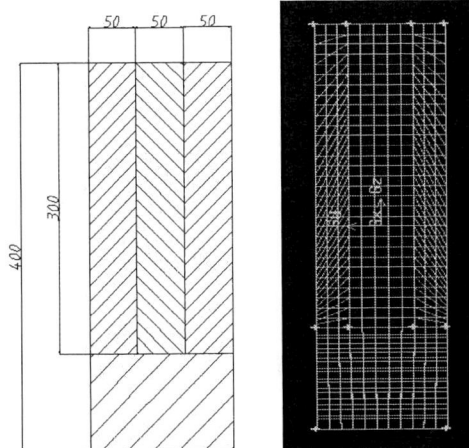

Fig.4 Single micro-channel heat sink model section and mesh diagram

According to the above assumptions, the continuity, momentum and energy equations are expressed, respectively, as

$$\nabla \cdot \vec{u} = 0 \qquad (1)$$

$$(\vec{u} \cdot \nabla)\vec{u} = -\frac{1}{\rho}\nabla p + \frac{u_{eff}}{\rho}\nabla^2 \vec{u} \qquad (2)$$

$$(\vec{u} \cdot \nabla)T = \frac{k_{eff}}{\rho C_p}\nabla^2 T + \frac{u_{eff}}{\rho C_p}(\frac{\partial u_i}{\partial x_j} + \frac{\partial u_j}{\partial x_i})\frac{\partial u_i}{\partial x_j} \qquad (3)$$

$$\nabla^2 T = 0 \quad \text{(Liquid region)} \qquad (4)$$

The airebo pair style [5] computes the Adaptive Intermolecular Reactive Empirical Bond Order (AIREBO) Potential for a system of carbon and/or hydrogen atoms. The potential consists of three terms: the REBO, LJ and torsion term.

$$E = \frac{1}{2}\sum_i \sum_{j \neq i}\left[E_{ij}^{REBO} + E_{ij}^{LJ} + \sum_{k \neq i, j}\sum_{l \neq i,j,k} E_{kijl}^{TORSION}\right] \qquad (5)$$

The REBO term and the LJ term are added to the nanotube potential to account for the short-ranged and long-ranged carbon-carbon interactions, and the TORSION term is turned off. The Lennard-Jones term is:

$$U_{(r_{ij})} = 4\varepsilon_{CC}\left[\left(\frac{\sigma_{CC}}{r_{ij}}\right)^{12} - \left(\frac{\sigma_{CC}}{r_{ij}}\right)^6\right] \qquad (6)$$

In the present study, the carbon nanotube represents an infinitely long tube, and end-effects are neglected by using axial boundary conditions. The thermal conductivity of SWCNT is calculated by reverse non-equilibrium method based on the Muller-Plathe algorithm [6]. The nanotube is divided into layers. The bottom layer and middle layer are cold and hot regions respectively. The thermal flux is maintained via energy exchange between the hot and cold regions. The rest atoms in the boundary layers interact with the hot and cold atoms, so the temperature gradient along the whole tube is maintained. The heat flux, J is computed from

$$J = \frac{\frac{1}{2}\sum_{i=1}^{N} m_i (v_i'^2 - v_i^2)}{A dt} \qquad (7)$$

Where A is the cross-sectional area of the SWCNT. The thermal conductivity λ is determined from

$$\lambda = \frac{J}{\nabla T} \qquad (8)$$

Results and discussion

The thermal conductivity of SWCNT is changed gradually as the rise of temperature. The block symbol represents thermal conductivity in different temperatures and the red line is the second-degree fitting curve. The fitting curve is adopted during the simulation procedure.

In order to compare CNT to silicon micro-channel, four different models with different fluid widths which are 0um, 50um, 100um, 200um are established. 0um fluid width group means that the model only has coolant and supposes fluid width of 50um.

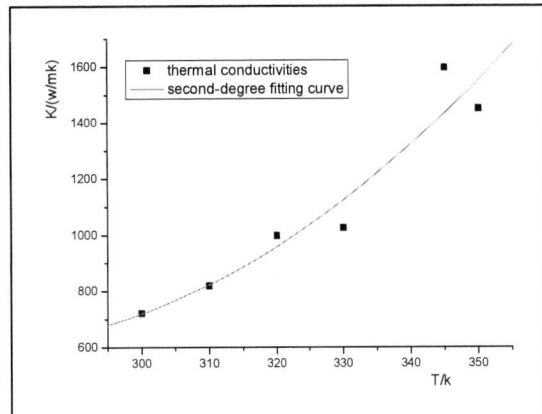

Fig 5 Thermal conductivities of the Carbon Nanotube

Heat sink is adopted rectangular micro-channel, and channel sizes are depicted in Fig. 6 and Table 1 and both carbon nanotubes and silicon micro coolers are set up at the same size. Boundary conditions are as follows: the inlet velocity is 5m/s, the inlet temperature is at room temperature and the pressure outlet is 0psi.

978-1-4244-4658-2/09 $25.00 © 2009 IEEE

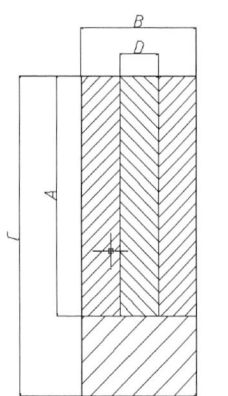

Fig. 6 Single micro-channel heat sink model section

Table 1 micro-channel size

	A μm	B μm	C μm	D μm
1	300	50	400	50
2	300	100	400	50
3	300	200	400	100
4	300	400	400	200

As seen in Fig.7, when the material is transformed from silicon to CNT, the pressure drop of the micro-channel cooler changes little because the pressure drop is not related with the material, only with the fluid flow rate, density hydraulic diameter, and resistance coefficient. Pressure drop of 4 groups is showed a downward trend and in the fourth group it reaches the minimum.

In Fig.8, when the fin material is changed, the block one (carbon nanotubes group) data is smaller than the rhombus one (silicon) and the temperature rise is dropped significantly; at the same time two groups are showed "V"-shaped; the minimum of temperature rise is the second group. The second group (fin width is 50um) shows the best cooling performance. So cooling performance of CNT and Si are compared under different inlet velocities and powers using the structure size as group 2.

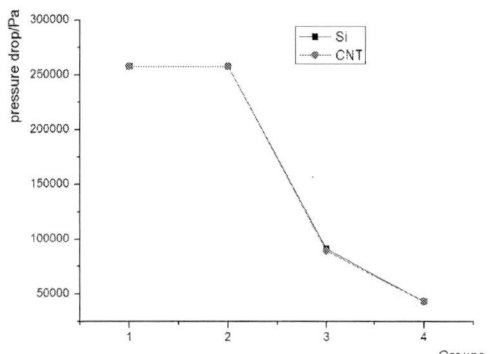

Fig.7 Pressure drop of silicon and carbon nanotubes

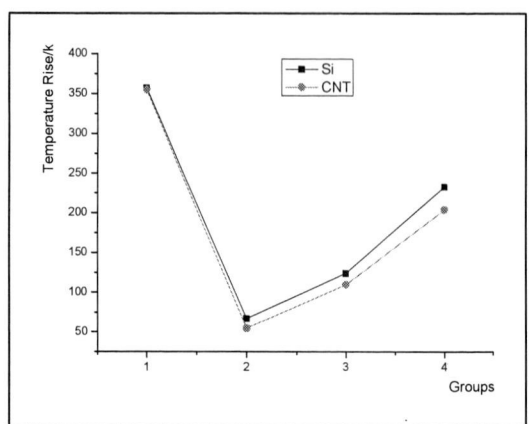

Fig. 8 Temperature rise of silicon and carbon nanotubes

When inlet velocity or wall heat flux is changed, models are simulated to observe the impact on the results. Structure size is adopted Fig.6 Group B; inlet velocity is set up 0.1, 0.2, 0.3…0.9, 1, 2, 3, 4…10m/s; wall heat flux is set up 100-1000 W/cm^2; other boundary condition is the same. The results are shown in Fig.9 and Fig.10.

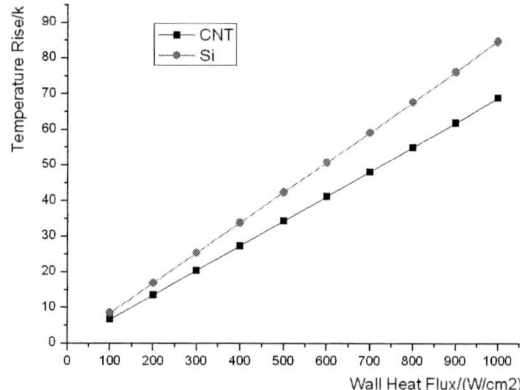

Fig.9 Temperature rise of different wall hear flux

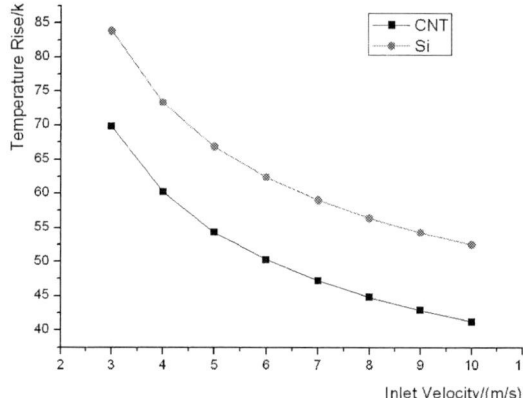

Fig.10 Temperature rise of different inlet velocity

978-1-4244-4658-2/09 $25.00 © 2009 IEEE

From Fig.9, with the increase of the wall heat flux, the temperature rise of Si and CNT is increasing linear but temperature rise of Si is more quickly than that of CNT. From Fig.10, the temperature rise is decreasing slowly with the increasing of the inlet velocity.

Conclusions

On the basis of simulation of Si and CNT micro heat sink, the performance of CNT one shows the better advantages. The results are as follows,

1. According to four groups of different structures, the results showed that the second group whose channel width is $50\,\mu m$ is the best structural design of heat dissipation.

2. The thermal conductivity of CNT is increasing with temperature. Silicon micro-channel replaced by carbon nanotubes micro-channel, pressure drop is the same but temperature rise of the carbon nanotubes is lower than the temperature rise of silicon.

3. For Si and CNT, the temperature rise of both substances is increasing with the increase of wall heat flux. However, because of the property of CNT, the temperature rise of Si is more quickly than that of CNT.

4. For Si and CNT, the temperature rise is decreasing with the increase of inlet velocity and both of them have the same decreasing speed.

Acknowledge

The authors acknowledge the financial support from 863 program (No.2008AA04Z301), NSFC project (No.50876057) and Shanghai Municipal Education Commission (No. 08YZ15).

References

1. D.B. Tuckerman, R. Pease. "High-performance heat sinking for VLSI". *IEEE ELECTRON DEVICELETT*, VOL.2 (1981) PP126-129

2. T. Wang, M. Jönsson, E. Nyström, Z.M. Mo, E.B. Eleanor, Campbell and J. Liu. *2006 Electronics System integration Technology Conference*, Dresden, Germany, pp.881-885

3. Iijima. "Single-shell carbon nanotubes of 1-nm diameter". *Nature*, 363 (1993), pp.603-605

4. P. Kim, M. L. Shi, McEuen, "Thermal Transport Measurements of Individual Multiwalled Nanotubes," *Phys. Rev. Lett.*, 87 (2001) 215502-215504.

5. Stuart, Tutein, and Harrison, *J. Chem. Phys.* 112(2000), pp.6472-6486.

6. F. Muller-Plathe, *Phys. Rev. E* 59 (1999) pp.4894-4898.

AUTHOR INDEX

An, B. 88, 758, 954, 974, 1139
An, R. 535
An, T. 133, 184, 294, 299
Ang, X. F. 1203
Aoh, J. 649
Appelt, B. 1237
Aria, P. 31
Arruda, L. 1120
Baated, A. 939
Bai, S. 57, 454, 525
Bai, S. L. 251
Bailey, C. 103, 129, 198, 847
Bao, S. 683
Berg, R. V. D. 1215
Bi, J. 767, 794
Bi, X. 1100
Bian, Z. 1265
Cai, J. 20, 23, 71, 76, 428, 557, 820
Cai, M. 52, 191, 843
Cai, X. 974
Calata, J. N. 385
Cao, B. 402
Cao, H. 27, 780, 1265
Cao, J. 317
Cao, K. 739
Cao, L. 829, 990
Cao, X. 172
Cao, Y. 5
Carlberg, B. 405
Chan, E. K. L. 950
Chan, Y. S. 1038
Chang, C. 454
Chang, G. 476
Chang, H. 1032
Chang, J. 934
Chang, S. 476, 806
Chen, B. 312, 673
Chen, C. 513, 594
Chen, F. 367
Chen, H. 88, 241, 486, 909
Chen, J. 57, 204, 251, 308
Chen, K. 444
Chen, L. 154, 590, 929, 1074
Chen, M. 149
Chen, Q. 1114, 1120
Chen, S. 1159
Chen, T. 289, 657
Chen, W. 191, 259, 649
Chen, X. 362, 385, 505, 754, 833, 974
Chen, Y. 541, 566, 570
Chen, Z. 111, 353, 402, 767, 1165
Cheng, C. 444
Cheng, J. 666
Cheng, X. 214, 646, 885
Cheng, Z. 179, 771, 959, 971, 1159

Chew, C. S. 715
Cho, M. 501
Choy, H. S. 857
Chuang, C. 649
Cong, Y. 586, 710
Conway, P. P. 619
Cui, C. Q. 1016
Cui, Y. 517
Dai, F. 947
Dai, J. H. 1144
Dai, W. 99, 865, 1144
Davignon, J. 551
Deng, G. 894, 900
Desmulliez, M. P. Y. 103, 847
Ding, D. 15, 454, 525, 646
Ding, L. 954
Ding, X. 9
Ding, Z. 900
Dong, M. 71, 76, 557
Dou, X. 23, 428
Du, R. 513, 776
Du, X. 547, 959, 971
Duan, Y. 787
Duh, J. 482, 967
England, L. 228
Ernst, L. J. 998, 1057, 1125
Fan, H. B. 158
Fan, J. 84, 223
Fan, X. 586, 710, 979, 1185
Fang, Q. 1179
Fang, X. 1179
Fang, Y. 1068
Feng, T. 71, 857
Feng, X. 947
Feng, Y. 994, 1002
Fremont, H. 1125
Fu, H. 551, 754
Fu, X. 1007
Fu, Y. 179, 763
Gai, J. 241
Galuschki, K. 525
Gan, Z. 353
Gang, F. 9
Gao, B. 436, 950
Gao, G. 162
Gao, J. 833
Gao, L. 732
Gao, W. 93, 861, 1225
Gao, Y. 810, 1159, 1175
Gautier, C. 1125
Ge, Z. 894
Geng, T. 223
Gong, A. 525
Goosen, J. F. L. 722
Guan, R. 706

AUTHOR INDEX

Gui, D. 312, 673
Guido, L. 505
Guo, F. 509, 677
Guo, J. 747
Guo, X. 405, 414
Guo, Y. 780
Hall, S. H. 551
Han, L. 303, 881, 1090
Han, M. 661
Han, Q. 979, 1185
Han, Y. D. 464
Hang, T. 608, 787
Hao, H. 509
Haseeb, A. S. M. A. 579, 715
Hayashi, H. 1249
He, D. 1023
He, Z. 916
Higurashi, E. 399
Hong, T. 541, 570
Hou, H. N. 414
Hsu, C. 541
Hsu, F. 922
Hsu, H. 922, 1032
Hsu, K. 444
Hu, A. 608, 625, 735, 787, 794
Hu, H. 900
Hu, J. 625, 719, 814
Hu, Y. 525, 792
Hu, Z. 405, 771
Hua, L. 414
Hua, Z. 874
Hua, Z. K. 1144
Huang, C. 277, 282, 1211
Huang, H. Z. 52, 843
Huang, J. 1002, 1114
Huang, M. 517, 590, 611, 798, 802
Huang, Q. 1, 574
Huang, T. 212
Huang, W. 268
Huang, X. 916
Hubbard, K. 869
Hutt, D. A. 380
Hwang, K. 395
Hwang, T. 1043
Jansen, K. M. B. 1057, 1125
Ji, M. 1257
Jia, H. 874
Jia, S. 20, 820
Jiang, J. 65, 767, 794
Jiang, L. 1144
Jiang, X. 702
Jiang, Y. 1152, 1155
Jin, H. 472, 732
Jin, P. 81
Jin, Y. 57, 251, 255, 391, 450, 814, 1257

Jin, Z. 1049
Jing, H. Y. 464
Johan, M. R. 579, 715
Ju, J. 1152, 1155
Ju, S. 1032
Kang, N. 468
Kang, R. 40
Kang, W. 1023
Kao, C. R. 458
Ke, L. 331
Kenny, S. 838
Kersaudy-Kerhoas, M. 103
Kettner, P. 852
Kim, G. 468, 501, 1261
Kim, J. 501
Kim, Jae-Dong 1043
Kim, Jin-Seong 1043
Kim, Jin-Young 1043
Kim, K. 939
Kivilahti, J. K. 486
Koo, Y. 468
Kugler, A. 117
Lai, C. M. 739
Lai, H. 971
Lai, Y. 1237
Lam, A. 241
Lee, B. 228, 934
Lee, C. 1043
Lee, D. W. 395
Lee, H. 541, 570
Lee, J. 444, 501, 598, 1165
Lee, S. 994
Lee, S. W. R. 1038
Lee, Y. B. 129
Lee, Z. 1007
Lei, G. 385, 505
Lei, J. 357
Leu, J. 444
Leung, V. C. K. 1220, 1243
Li, B. 93, 166, 1225
Li, C. Y. 1144, 1237
Li, D. 317
Li, D. X. 493, 521
Li, F. 1063
Li, G. 727, 1095, 1100
Li, J. 207, 299, 349, 486, 1013
Li, K. 833
Li, L. 139, 259, 331, 341, 395, 547, 869, 1109, 1253
Li, M. 15, 27, 454, 525, 608, 625, 735, 763,
767, 780, 787, 794, 874, 1265, 486
Li, Q. 27, 722, 1265
Li, S. 496, 706
Li, T. 282
Li, W. 1135
Li, X. 1063

AUTHOR INDEX

Li, Y.................... 36, 286, 418, 450, 454, 792, 865, 881
Li, Z.................... 154, 289, 345, 929, 1074, 1225, 1270
Liang, L...235
Liang, T...792
Liang, Y. 282, 1253
Liao, C...207
Liao, P...1144
Liao, W. B...826
Liao, X...125
Liem, H. M...857
Lin, A...1237
Lin, C...149
Lin, W...1211
Ling, H. 15, 27, 780, 1265
Liu, B...677
Liu, C. 214, 380, 421, 619, 696, 833
Liu, D...286
Liu, F...235
Liu, H. 154, 758, 929, 943, 963, 1013, 1049, 1068, 1074, 1270
Liu, J. 84, 179, 312, 322, 405, 619, 673, 744, 771, 947, 959, 971, 1159
Liu, L...317, 865, 1183
Liu, Q...1233
Liu, S. 111, 149, 218, 353, 367, 402, 496, 615, 906, 1038, 1078
Liu, W...149, 486, 566
Liu, X...341, 362, 375, 611
Liu, Y. ...36, 111, 228, 235, 286, 418, 750, 802, 881, 1228
Liu, Z...122, 367, 432, 493, 521
Lou, M...1165
Lu, C...906
Lu, D...814
Lu, G...172, 385, 505
Lu, H...198, 308
Lu, J...440, 1032
Lu, L. N...52, 843
Lu, M...65
Lu, Q...865
Lu, X...959, 971
Lu, Y...792
Luo, L...1, 5, 9, 574
Luo, S...385, 513, 776
Luo, T...735
Luo, X...218, 367
Luo, Z...810
Lv, D...683
Lv, Y...861
Ma, L...683
Ma, S...450, 1257
Ma, X...317, 1109, 1125
Ma, Y...65
Mao, D.................. 15, 27, 454, 525, 608, 625, 735, 767, 780, 787, 794, 1265
Mao, Z...218

Martin, S. 228
Mattila, T. 486
Mavinkurve, A. 998
Mei, H. 547
Meng, Z. 994
Miao, M. 57, 251, 1257
Miao, X. 312
Michel, B. 117
Min, Y. 395
Mu, F. 57, 61
Mu, W. 277
Nadia, A. 531
Nai, S. M. L. 464, 1203
Ngo, K. 505
Ngo, K. D. T. 172
Nie, L. 557
Ning, C. 454
Ning, W. 1
Niu, J. 1002
Niu, L. 191, 246, 687, 1148
Niu, X. 289, 657
Okuno, A. 372
Osterman, M. 40, 557
Pan, H. 642
Pan, J. 798
Pan, K. 47, 317, 322
Pan, Q. 943
Pan, Z. 787
Pang, E. W. 1144
Pape, H. 1057
Pargfrieder, S. 852
Park, D. 1043
Park, J. 501, 1261
Park, S. 468
Patel, M. K. 103
Pavuluri, S. 847
Pechr, M. 40
Pecht, M. 557
Peng, B. 909
Peng, C. 967
Peng, J. 943, 1068
Pfahl, R. C. 551
Pichler, C. 852
Pomerleau, R. 31
Pot, A. 1215
Priest, J. 31
Pu, Y. 71
Pun, K. 1016
Qi, L. 1002
Qian, K. 586, 710
Qin, F. 133, 184, 263, 294, 299
Qin, R. 454
Qin, Y. 696
Qing, X. 798
Qiu, W. 47, 322

AUTHOR INDEX

Qiu, Y. 391
Quan, J. 909
Quintero, J. 1120
Rank, H. 117
Ranouta, A. S. 1185
Rausch, M. 551
Regard, C. 1125
Ren, C. 263, 732
Roehm, H. 1215
Roelfs, B. 838
Rong, Y. 20
Rongen, R. T. H. 998
Ruan, Z. 1
Savic, J. 31
Sawada, R. 399
Scheiring, C. 852
Schlottig, G. 1057
Seo, M. 1261
Seo, S. 1261
Seo, W. 468
Shang, J. K. 493, 521, 594, 963
Shang, P. J. 493, 521
Shao, X. 1179
Shi, D. X. Q. 1220, 1243
Shi, P. 732
Shi, Y. 509, 677
Shih, K. 541, 570
Shih, M. 1237
Shinohara, K. 1196
Shrivastava, A. 40
Shu, X. 289, 657, 1105
Sidiki, T. P. 1215
Sigl, A. 852
Silberschmidt, V. V. 214
Sommer, J. P. 117
Song, B. 615
Song, H. Y. 963
Song, J. 93, 327, 1225
Song, X. 353
Song, Z. 496
Stevens, B. 380
Strusevich, N. 129
Su, F. 642
Su, X. X. 52, 843
Suga, T. 143, 399, 440, 767, 794
Suganuma, K. 939
Suhir, E. 362
Sun, F. 750
Sun, J. 509
Sun, L. 566
Sun, P. 1220, 1243
Sun, R. 513, 776
Sun, W. H. 806
Sun, X. 255, 450, 1257
Tai, F. 677

Takigawa, R. 399
Tan, C. M. 464
Tan, K. H. 739
Tan, W. 547, 646
Tanaka, O. 372
Tang, C. 727
Tang, G. 1139
Tang, J. 327
Tao, B. 335
Tao, Y. 954
Tay, S. L. 579
Tian, D. 312
Tian, H. 666
Tian, X. 692
Tian, Y. 212, 380, 535, 566, 783
Tilford, T. 847
Tisdale, S. 551
Topham, D. 103
Tsai, M. 922
Tsai, S. 482
Tsao, L. C. 806, 1083
Tseng, A. 1237
Tseng, H. C. 204
Tsui, K. Y. K. 1220
Tu, Y. 496
Van Beek, J. T. M. 722
Van Der Sluis, O. 1057, 1125
Van Driel, W. D. 603, 1125
Van Keulen, F. 722
Van Soestbergen, W. 998
Varia, B. 979
Verheijden, G. J. A. M. 603
Wan, L. 93, 139, 207, 341, 349, 829, 861, 1225
Wang, B. 1068
Wang, C. 143, 212, 357, 409, 535, 566, 619, 696, 783, 1086, 1228
Wang, F. 1078
Wang, H. 162, 861
Wang, J. 47, 362, 375, 428, 496, 586, 710, 1007, 1175, 1179, 1270
Wang, K. 367, 482, 1211
Wang, L. 598, 611, 750, 750, 763, 810, 829
Wang, N. 23, 357
Wang, Q. 598, 702, 947, 1007, 1165
Wang, R. 476
Wang, S. 20, 179, 223, 428, 615, 820, 395
Wang, T. 345, 505, 586, 710
Wang, X. 154, 179, 632, 929, 1074, 1270
Wang, Y. 71, 76, 139, 341, 440, 525, 865, 990, 458
Wang, Z. 57, 61, 122, 335, 432, 666, 963
Wassay, A. 696
Wei, J. 464, 1203
Wei, S. 327
Wen, L. 1165
Wen, Q. 418, 916

AUTHOR INDEX

Weng, L. 513
Weng, M. 922
Whalley, D. 421
Wilcox, G. D. 696
Wong, C. K. Y. 158
Wu, B. Y. 52, 843
Wu, C. M. L. 611
Wu, D. 375
Wu, F. 88, 758, 943, 954, 974, 1013, 1068, 1139
Wu, H. 1032
Wu, J. 241, 632, 646
Wu, N. 81
Wu, S. 1211
Wu, X. 865
Wu, Y. 88, 954, 974, 1013, 1068, 1139
Wu, Z. 277, 885
Wunderle, B. 1057
Xi, Y. 149
Xia, S. 57
Xia, Z. D. 677
Xiang, Y. 476
Xiao, A. 1057
Xiao, F. 754
Xie, H. 947
Xie, X. 719, 727, 1007, 1100
Xie, Y. 418
Xiong, Y. 335
Xu, C. 890
Xu, G. 1, 509, 574
Xu, H. 763
Xu, J. 353
Xu, L. 615, 959, 971, 1114
Xu, X. 702
Xue, J. 31, 869
Xue, K. 241
Xue, X. 103
Xv, W. 1183
Yan, B. 661
Yan, H. 353
Yan, X. 727, 1095
Yang, B. 792
Yang, C. 1038
Yang, D. 191, 246, 687, 1148, 1243
Yang, G. 162
Yang, H. 754, 1131
Yang, J. 414, 820
Yang, L. 255, 1002
Yang, M. 763
Yang, S. 212
Yang, W. 23
Yang, X. 15, 1105
Yang, Y. 166, 308
Yang, Z. 65, 692, 881
Yao, Y. 954
Yau, S. K. 1220

Ye, J. 683
Ye, Y. 496, 1038
Yeh, W. 1211
Yen, S. F. 806
Yin, C. 129
Yin, L. 166
Yin, Z. 335, 909
Yu, C. 308
Yu, H. 780, 1179
Yu, Q. 1196
Yu, R. 345
Yu, S. 513, 776
Yu, W. 125
Yu, X. 15, 27, 450
Yu, Z. 666
Yuan, C. 47, 322
Yuan, G. 1105
Yuan, Z. 228, 362, 375
Yue, C. 405, 771
Yue, T. M. 857
Yuen, M. M. F. 158, 436, 950
Yung, K. C. 857
Yung, K. P. 1203
Zaal, J. J. M. 603
Zeng, G. 312
Zeng, J. 619, 1228
Zhai, Q. 1159
Zhang, B. 586, 710
Zhang, E. 375
Zhang, G. Q. 317, 603, 687, 722, 998, 1114, 1125
Zhang, H. 36, 418, 637, 881, 1002
Zhang, J. 57, 251, 657, 758, 776, 890, 1144
Zhang, Jianhua 874
Zhang, Jinsong 874
Zhang, K. L. 464
Zhang, L. 594, 739, 906
Zhang, M. 890
Zhang, Q. 615
Zhang, S. 40
Zhang, W. 154, 432, 472, 929, 1074, 1270
Zhang, X. 93, 166, 865, 1135
Zhang, Y. 57, 61, 84, 179, 223, 251, 375, 421, 632, 771, 826, 959, 1090
Zhang, Z. 637, 702, 754, 776, 1131
Zhao, H. 885
Zhao, J. 496, 810
Zhao, K. 409, 696
Zhao, L. 255, 391
Zhao, M. 191, 246, 385, 509, 586, 710, 1148
Zhao, N. 829
Zhao, S. 783
Zhao, W. 706
Zhao, X. 1002
Zhao, Y. 611
Zhao, Z. 598

AUTHOR INDEX

Zheng, D...833
Zheng, G. 1086
Zheng, H. F. 857
Zheng, Q. 345
Zhong, Q. 81
Zhou, D. 885
Zhou, F. 646
Zhou, H. 303
Zhou, J. 1007, 1165
Zhou, L. 88, 943, 1013
Zhou, P. 331
Zhou, Q. 81, 843
Zhou, S. 402
Zhou, X. 1109
Zhou, Y. 349
Zhu, D. 357
Zhu, H. 162
Zhu, Q. S. 963
Zhu, W. 259, 702, 1049
Zhu, Y. 1257
Zou, H. F. 1131
Zou, J. 758
Zou, K. 27